FORMULAS/EQUATIONS

Distance Formula   The distance from (x_1, y_1) to (x_2, y_2) is $\sqrt{(x_2 - }$

Midpoint Formula   The midpoint of the line segment with endpoints (x_1, y_1) and (x_2, y_2) is $\left(\dfrac{x_1 + x_2}{2}, \dfrac{y_1 + y_2}{2} \right)$.

Standard Equation of a Circle   The standard equation of a circle of radius r with center at (h, k) is
$$(x - h)^2 + (y - k)^2 = r^2$$

Slope Formula   The slope m of the line containing the points (x_1, y_1) and (x_2, y_2) is
$$\text{slope } (m) = \frac{\text{change in } y}{\text{change in } x} = \frac{y_2 - y_1}{x_2 - x_1} \quad (x_1 \neq x_2)$$
m is undefined if $x_1 = x_2$

Slope-Intercept Equation of a Line   The equation of a line with slope m and y-intercept $(0, b)$ is $y = mx + b$

Point-Slope Equation of a Line   The equation of a line with slope m containing the point (x_1, y_1) is $y - y_1 = m(x - x_1)$

Quadratic Formula   The solutions of the equation $ax^2 + bx + c = 0$, $a \neq 0$, are $x = \dfrac{-b \pm \sqrt{b^2 - 4ac}}{2a}$

If $b^2 - 4ac > 0$, there are two distinct real solutions.

If $b^2 - 4ac = 0$, there is a repeated real solution.

If $b^2 - 4ac < 0$, there are two complex solutions (complex conjugates).

GEOMETRY FORMULAS

Circle $r =$ Radius, $A =$ Area, $C =$ Circumference
$A = \pi r^2 \quad C = 2\pi r$

Triangle $b =$ Base, $h =$ Height (Altitude), $A =$ area
$A = \frac{1}{2}bh$

Rectangle $l =$ Length, $w =$ Width, $A =$ area, $P =$ perimeter
$A = lw \quad P = 2l + 2w$

Rectangular Box $l =$ Length, $w =$ Width, $h =$ Height, $V =$ Volume, $S =$ Surface area
$V = lwh \quad S = 2lw + 2lh + 2wh$

Sphere $r =$ Radius, $V =$ Volume, $S =$ Surface area
$V = \frac{4}{3}\pi r^3 \quad S = 4\pi r^2$

Right Circular Cylinder $r =$ Radius, $h =$ Height, $V =$ Volume, $S =$ Surface area
$V = \pi r^2 h \quad S = 2\pi r^2 + 2\pi rh$

CONVERSION TABLE

1 centimeter $\approx$ 0.394 inch

1 meter $\approx$ 39.370 inches
$\approx$ 3.281 feet

1 kilometer $\approx$ 0.621 mile

1 liter $\approx$ 0.264 gallon

1 newton $\approx$ 0.225 pound

1 joule $\approx$ 0.738 foot-pound

1 gram $\approx$ 0.035 ounce

1 kilogram $\approx$ 2.205 pounds

1 inch $\approx$ 2.540 centimeters

1 foot $\approx$ 30.480 centimeters
$\approx$ 0.305 meter

1 mile $\approx$ 1.609 kilometers

1 gallon $\approx$ 3.785 liters

1 pound $\approx$ 4.448 newtons

1 foot-lb $\approx$ 1.356 Joules

1 ounce $\approx$ 28.350 grams

1 pound $\approx$ 0.454 kilogram

FUNCTIONS

Constant Function	$f(x) = b$
Linear Function	$f(x) = mx + b$, where m is the slope and b is the y-intercept
Quadratic Function	$f(x) = ax^2 + bx + c, a \neq 0$ or $f(x) = a(x - h)^2 + k$ parabola vertex (h, k)
Polynomial Function	$f(x) = a_n x^n + a_{n-1} x^{n-1} + \cdots + a_1 x + a_0$
Rational Function	$R(x) = \dfrac{n(x)}{d(x)} = \dfrac{a_n x^n + a_{n-1} x^{n-1} + \cdots + a_1 x + a_0}{b_m x^m + a_{m-1} x^{m-1} + \cdots + b_1 x + b_0}$
Exponential Function	$f(x) = b^x, b > 0, b \neq 1$
Logarithmic Function	$f(x) = \log_b x, b > 0, b \neq 1$

GRAPHS OF COMMON FUNCTIONS

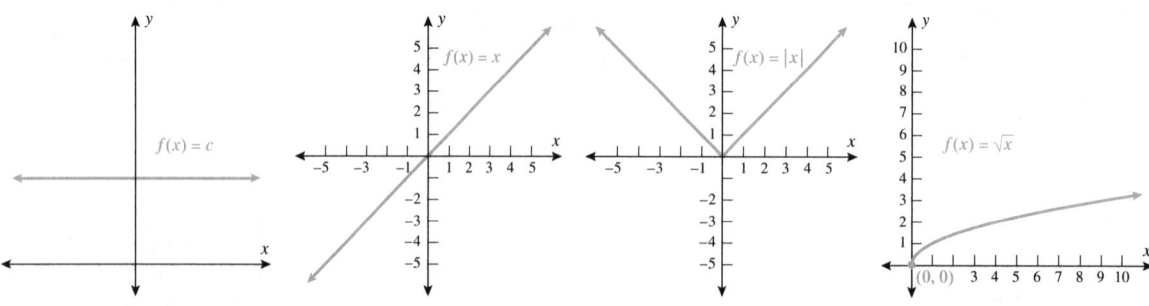

Constant Function	Identity Function	Absolute Value Function	Square Root Function

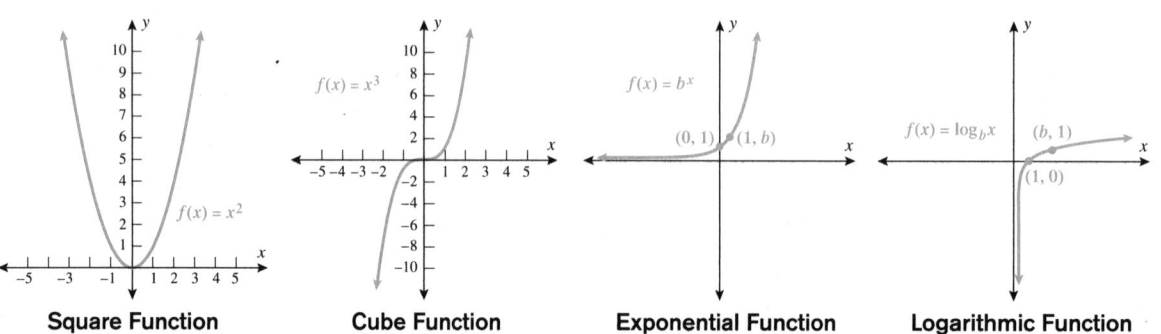

Square Function	Cube Function	Exponential Function	Logarithmic Function

TRANSFORMATIONS

In each case, c represents a positive real number.

Function		Draw the graph of f and:
Vertical translations	$\begin{cases} y = f(x) + c \\ y = f(x) - c \end{cases}$	Shift f upward c units. Shift f downward c units.
Horizontal translations	$\begin{cases} y = f(x - c) \\ y = f(x + c) \end{cases}$	Shift f to the right c units. Shift f to the left c units.
Reflections	$\begin{cases} y = -f(x) \\ y = f(-x) \end{cases}$	Reflect f about the x-axis. Reflect f about the y-axis.

HERON'S FORMULA FOR AREA

If the semiperimeter, s, of a triangle is

$$s = \frac{a + b + c}{2}$$

then the area of that triangle is

$$A = \sqrt{s(s - a)(s - b)(s - c)}$$

Precalculus

CYNTHIA Y. YOUNG | *Professor of Mathematics*

UNIVERSITY OF CENTRAL FLORIDA

WILEY

John Wiley & Sons, Inc.

This book is dedicated to teachers and students.

VICE PRESIDENT AND PUBLISHER	Laurie Rosatone
SENIOR DEVELOPMENT EDITOR	Ellen Ford
DEVELOPMENT ASSISTANT	Marcus Van Harpen
ASSISTANT EDITOR	Jeffrey Benson
SENIOR PRODUCTION EDITOR	Sujin Hong
V.P. DIRECTOR OF MARKETING	Susan Elbe
MARKETING MANAGER	Sarah Davis
SENIOR ILLUSTRATION EDITOR	Sigmund Malinowski
PHOTO EDITOR	Sarah Wilkin
MEDIA EDITOR	Melissa Edwards
DIRECTOR, CREATIVE SERVICES	Harry Nolan
COVER DESIGNER	Jeof Vita
COVER PHOTO	John Kelly/Getty Images Inc.
INTERIOR DESIGNER	Brian Salisbury

This book was set in 10/12 Times by Aptara and printed and bound by Quebecor. The cover was printed by Quebecor.

This book is printed on acid free paper. ∞

To order books or for customer service, please call 1-800-CALL WILEY (225-5945).

ISBN: 978-0-471-75684-2
ISBN: 978-0-470-55665-8 (BRV)

Printed in the United States of America

10 9 8 7 6 5 4 3 2 1

The Wiley Faculty Network

Where Faculty Connect

The Wiley Faculty Network is a faculty-to-faculty network promoting the effective use of technology to enrich the teaching experience. The Wiley Faculty Network facilitates the exchange of best practices, connects teachers with technology, and helps to enhance instructional efficiency and effectiveness. The network provides technology training and tutorials, including *WileyPLUS* training, online seminars, peer-to-peer exchanges of experiences and ideas, personalized consulting, and sharing of resources.

Connect with a Colleague

Wiley Faculty Network mentors are faculty like you, from educational institutions around the country, who are passionate about enhancing instructional efficiency and effectiveness through best practices. You can engage a faculty mentor in an online conversation at www.wherefacultyconnect.com

Participate in a Faculty-Led Online Seminar

The Wiley Faculty Network provides you with virtual seminars led by faculty using the latest teaching technologies. In these seminars, faculty share their knowledge and experiences on discipline-specific teaching and learning issues. All you need to participate in a virtual seminar is high-speed internet access and a phone line. To register for a seminar, go to www.wherefacultyconnect.com

Connect with the Wiley Faculty Network

Web: www.wherefacultyconnect.com
Phone: 1-866-4FACULTY

Cynthia Y. Young is a native of Tampa, Florida. She is currently a Professor of Mathematics at the University of Central Florida (UCF) and author of *College Algebra, Trigonometry, Algebra and Trigonometry,* and *Precalculus.* She holds a B.A. degree in Secondary Mathematics Education from the University of North Carolina, an M.S. degree in Mathematical Sciences from UCF, and was awarded a graduate fellowship sponsored by Kennedy Space Center (KSC) to pursue doctoral studies at the University of Washington (UW). While earning her M.S. in Electrical Engineering and Ph.D. in Applied Mathematics from UW, she worked in the failure analysis and optics labs at KSC as a graduate intern and taught developmental mathematics at Shoreline Community College in Seattle, Washington. The summer after completing graduate studies at UW, she worked in the optics branch at Boeing Space and Defense and simultaneously earned her scuba diving certification and private pilot's license. Dr. Young turned down several industrial offers as a senior scientist in laser propagation to accept a faculty position at the University of Central Florida in order to pursue her passion for teaching. In 2001, the Office of Naval Research selected her for the Young Investigator Award, and in 2006 she was named a Fellow of the International Society for Optical Engineers. Although she excels in her research field, she considers teaching her true calling. She is a recipient of the UCF Excellence In Undergraduate Teaching Award, the UCF Scholarship of Teaching and Learning Award, and a two-time recipient of the UCF Teaching Incentive Program. Dr. Young is committed to improving student learning and has shared her techniques and experiences with colleagues around the country through talks at colleges, universities, and conferences.

Dr. Young and her husband, Dr. Christopher Parkinson, enjoy spending time outdoors and competing in Field Trials with their Labrador retrievers Ellie and Wiley in both the United States and Canada.

Dr. Young is pictured with Wiley (left) and Ellie (right).

Preface

To the Instructor

"I understand you in class but when I get home I am lost."
"I can't read the book."

How many times have we heard these statements from our students? Students come together in Precalculus with varied backgrounds; many have some level of math anxiety, some may have poor study habits, insufficient prerequisites, or different learning styles. As a result, there is a gap that exists between when we have our students in our classrooms and when they are on their own. My goal in writing this book was to bridge that gap and to help them succeed. I recognize that it is difficult to get students to love mathematics, but you can get students to love *succeeding* at mathematics. In order to accomplish this goal, I wrote with two main objectives: that the voice of the book should be your voice (*seamless with how you teach*) so that your students hear you when they read this book, and to enable the students to become independent, successful learners by giving them a book *they can read* (without sacrificing the rigor needed for concept mastery).

Distinguishing Features

The following features from the Young series have been well received and support my goal and distinguish this book from others.

PARALLEL WORDS AND MATH

In most mathematics books, the mathematics is performed in the center of the page and if a student doesn't understand a step, they look to the right for some brief description or hint. But, that's not how we teach. In class we don't write the mathematics on the board and then say "Class, guess what I just did." Instead, we lead them, telling them what step is coming, and then performing that mathematical operation together. I have tried to parallel our classroom experience in this book. In examples, the student is told the step that is coming, then they see the step—reading naturally from left to right.

SKILLS AND CONCEPTUAL OBJECTIVES AND EXERCISES

In order to give students a true understanding of any college mathematics course, it is important to emphasize both *skills* and *concepts*. In my experience, skills help students at the micro level while concepts help students at the macro level of understanding. Therefore, I have separated section objectives into *skills objectives* and *conceptual objectives* and have labeled exercises as *skill* or *conceptual* questions.

CATCH THE MISTAKE

When a student comes to your office hours and brings you a problem they have unsuccessfully tried to solve, the first question they generally ask is "Where did I go wrong?" Instead of painstakingly going over their work to see where their solution went awry, we often find it far easier to simply start over with a fresh piece of paper. Catch the Mistake Exercises are found in every section, facilitating a higher level of mastery by putting students into the role of teacher.

vii

Features and Benefits at a Glance

Each of the features in this book has been designed to help instructors bridge the gap that exisits between class work and homework.

FEATURE	BENEFIT TO STUDENT
Chapter Opening Vignette	Peaks the students interest with a real-world application of material presented in that chapter. Later in the chapter, the same concept from the vignette is reinforced.
Chapter Overview/Flowchart	Students see the big picture of how topics relate to each other.
Skills and Conceptual Objectives	Emphasize conceptual understanding and push students to think at a more global level on the topics presented.
Clear, Concise, and Inviting Writing Style, Tone, and Layout	Students are able to **read** this book, which reduces math anxiety and encourages student success.
Parallel Words and Math	Increases students' ability to read and understand examples.
Common Mistake/ Correct vs. Incorrect Boxes	Addresses different learning styles (by counter example). Demonstrates common mistakes so students understand why a step is incorrect and reinforce the correct mathematics.
Color for Pedagogical Reasons	A student sees that a function is written in red in the text, then its graph is also red, whereas another function is written in blue and its graph is blue. Particularly helpful for visual learners in piecewise defined functions, graphing by transformations, and graphing inverse functions.
Study Tips	Serves to reinforce notes you would write on the board in class.
Technology Tips	Demonstrates use of technology to confirm analytic solutions. Kept in the margins so that faculty can choose to use at their discretion.
Your Turn	Engages students during class, build student confidence, and assist instructor in assessing class understanding.
Catch the Mistake Exercises	Encourages students to take the role of teacher—demonstrates higher mastery level.
Conceptual Exercises	Teaches students to think globally. For example, students have to answer questions about all types of functions, not a specific function.
Preview to Calculus	Demonstrate where the current precalculus topics will be used in calculus. This helps answer the question "when will I use this?".
Modeling Your World	Engages students through timely issues: global climate change. When students use mathematical models they develop as a result of this course, they value the course more.
Chapter Review (Key Ideas)	Key formulas and ideas are presented section by section in a chart format—improves study skills.
Chapter Review Exercises	Improves study skills.
Chapter Practice Test	Offers self-assessment and improves study skills.
Cumulative Test	Improves retention.

ACKNOWLEDGMENTS

I want to thank all of the instructors and students who have class tested this book and sent me valuable feedback. This is *our* book. I want to especially thank Mark McKibben who has written the solutions manuals for this series and always stayed in sync with me on a day to day basis throughout the process. It's so refreshing to have a colleague to whom I can send an email at 6 AM and get an immediate response! Special thanks to Pauline Chow for her Herculean effort on the Technology Tips, Technology Exercises, and accuracy checking.

I'd like to thank the entire Wiley *team*. Every year that I am with Wiley, I realize that I have the best partner in this market. Thanks to my developmental editor, Ellen Ford, who coordinated this book from manuscript through production and who somehow kept us all on schedule and could always be counted on for a laugh. To Jeff Benson, assistant editor, who coordinated the AIE, reviewers, accuracy checkers, and all of the supplements. Thanks to Sarah Davis for your novel marketing efforts and your ability to help our sales force have great conversations with faculty about student learning. A special thanks to Laurie Rosatone, publisher, who has shared my vision for this series since its inception. To my production editor, Sujin Hong, and the illustration and photo editors, Sigmund Malinowski and Sarah Wilkin, thank you for your hard work, dedication, and patience with schedules and revisions. And, finally, I want to thank the entire Wiley sales force for your unwavering commitment and constant enthusiasm for my series of books.

To the greatest man I know, my husband Christopher, thanks for all your support. To our daughter Caroline, you give everyone around you such overwhelming joy— especially mommy.

My sincere appreciation goes out to all the reviewers, class testers, and accuracy checkers on this book. Their input, comments, and detailed feedback have been invaluable in improving and fine tuning this edition.

Commitment to Accuracy

From the beginning, the editorial team and I have been committed to providing an accurate and error-free text. In this goal, we have benefited from the help of many people. In addition to the reviewer feedback on this issue and normal proofreading, we enlisted the help of several extra sets of eyes during the production stages and during video filming. I wish to thank them for their dedicated reading of various rounds of pages to help me insure the highest level of accuracy.

Marwan Abu-Sawwa, *Florida State College at Jacksonville*
Khadija Ahmed, *Monroe County Community College*
Raul Aparicio, *Blinn College*
Jan Archibald, *Ventura College*
Erika Asano, *University of South Florida*
Holly Ashton, *Pikes Peak Community College*
Donna Bailey, *Truman State University*
Raji Baradwaj, *University of Maryland, Baltimore*

Mickey Beloate, *Northwest Mississippi Community College*
Mary Benson, *Pensacola Junior College*
Dave Bregenzer, *Utah State University*
Zhao Chen, *Florida Gulf Coast University*
Diane Cook, *Northwest Florida State College*
Donna Densmore, *Bossier Parish Community College*
Johnny Duke, *Floyd Junior College*
Mike Ecker, *Penn State Wilkes-Barre*

Angela Everett, *Chattanooga State Community College*
Hamidullah Farhat, *Hampton University*
Peng Feng, *Florida Gulf Coast University*
Elaine Fitt, *Bucks County Community College*
Maggie Flint, *Northeast State Community College*
Joseph Fox, *Salem State College*
David French, *Tidewater Community College*
Alicia Frost, *Santiago Canyon College*
Ginger Harper, *Kaplan University*
Celeste Hernandez, *Richland College*
Aida Kadic-Galeb, *University of Tampa*
Mohammad Kazemi, *University of North Carolina—Charlotte*
Judy Lalani, *Central New Mexico Community College*
Carl Libis, *Cumberland University*
Nancy Matthews, *University of Oklahoma-Norman*
Teresa McConville, *Southeast Technical Institute*
Susan McLoughlin, *Union County College*
Jennifer McNeilly, *University of Illinois at Urbana-Champaign*
Mary Ann Moore, *Florida Gulf Coast University*

Scott Morrison, *Western Nevada Community College*
Luis Ortiz-Franco, *Chapman University*
C. Altay Ozgener, *Manatee Community College*
Edmon Perkins, *East Central University*
Charlotte Pisors, *Baylor University*
Richard Pugsley, *Tidewater Community College*
Alexander Retakh, *Stony Brook University*
Patricia Rhodes, *Treasure Valley Community College*
Dawn Sadir, *Pensacola Junior College*
Radha Sankaran, *Passaic County Community College*
Morteza Shafii-Mousavi, *Indiana University South Bend*
Delphy Shaulis, *University of Colorado at Boulder*
Walter Sizer, *Minnesota State University Moorhead*
Joseph Stein, *University of Texas at San Antonio*
Ba Su, *Saint Paul College*
Robert Talbert, *Franklin College*
Philip Veer, *Johnson County Community College*
Kurt Verderber, *State University of New York at Cobleskill*
David Vinson, *Pellissippi State Community College*
Derald Wentzien, *Wesley College*
Deborah Wolfson, *Suffolk County Community College*

Supplements

Instructor Supplements

INSTRUCTOR'S SOLUTIONS MANUAL (ISBN: 9780470532056)
- Contains worked out solutions to all exercises in the text.

INSTRUCTOR'S MANUAL (ISBN: 9780470583722)
Authored by Cynthia Young, the manual provides practical advice on teaching with the text, including:
- sample lesson plans and homework assignments
- suggestions for the effective utilization of additional resources and supplements
- sample syllabi
- Cynthia Young's Top 10 Teaching Tips & Tricks
- online component featuring the author presenting these Tips & Tricks

ANNOTATED INSTRUCTOR'S EDITION (ISBN: 9780470561676)
- Displays answers to all exercise questions, which can be found in the back of the book.
- Provides additional classroom examples within the standard difficulty range of the in-text exercises, as well as challenge problems to assess your students mastery of the material.

POWERPOINT SLIDES
- For each section of the book, a corresponding set of lecture notes and worked out examples are presented as PowerPoint slides, available on the Book Companion Site (www.wiley.com/college/young) and WileyPLUS.

TEST BANK (ISBN: 9780470532072)
Contains approximately 900 questions and answers from every section of the text.

COMPUTERIZED TEST BANK
Electonically enhanced version of the Test Bank that
- contains approximately 900 algorithmically-generated questions.
- allows instructors to freely edit, randomize, and create questions.
- allows instructors to create and print different versions of a quiz or exam.
- recognizes symbolic notation.
- allows for partial credit if used within WileyPLUS.

BOOK COMPANION WEBSITE (WWW.WILEY.COM/COLLEGE/YOUNG)
- Contains all instructor supplements listed plus a selection of personal response system questions.

Student Supplements

STUDENT SOLUTIONS MANUAL (ISBN: 9780470532034)
- Includes worked out solutions for all odd problems in the text.

BOOK COMPANION WEBSITE (WWW.WILEY.COM/COLLEGE/YOUNG)
- Provides additional resources for students, including web quizzes, video clips, and audio clips.

WileyPLUS

This online teaching and learning environment integrates the **entire digital textbook** with the most effective instructor and student resources to fit every learning style.

With *WileyPLUS*:

- Students achieve concept mastery in a rich, structured environment that's available 24/7.
- Instructors personalize and manage their course more effectively with assessment, assignments, grade tracking, and more.

WileyPLUS can complement your current textbook or replace the printed text altogether.

For Students

PERSONALIZE THE LEARNING EXPERIENCE
Different learning styles, different levels of proficiency, different levels of preparation—each of your student is unique. *WileyPLUS* empowers them to take advantage of their individual strengths:

- Students receive timely access to resources that address their demonstrated needs, and get immediate feedback and remediation when needed.
- Integrated, multi-media resources—including audio and visual exhibits, demonstration problems, and much more—provide multiple study-paths to fit each student's learning preferences and encourage more active learning.
- *WileyPLUS* includes many opportunities for self-assessment linked to the relevant portions of the text. Students can take control of their own learning and practice until they master the material.

For Instructors

PERSONALIZE THE TEACHING EXPERIENCE

WileyPLUS empowers you with the tools and resources you need to make your teaching even more effective:

- You can customize your classroom presentation with a wealth of resources and functionality from PowerPoint slides to a database of rich visuals. You can even add your own materials to your *WileyPLUS* course.
- With *WileyPLUS* you can identify those students who are falling behind and intervene accordingly, without having to wait for them to come to office hours.
- *WileyPLUS* simplifies and automates such tasks as student performance assessment, making assignments, scoring student work, keeping grades, and more.

The Wiley Faculty Network—Where Faculty Connect

The Wiley Faculty Network is a faculty-to-faculty network promoting the effective use of technology to enrich the teaching experience. The Wiley Faculty Network facilitates the exchange of best practices, connects teachers with technology, and helps to enhance instructional efficiency and effectiveness. The network provides technology training and tutorials, including *WileyPLUS* training, online seminars, peer-to-peer exchanges of experiences and ideas, personalized consulting, and sharing of resources.

Connect with a Colleague

Wiley Faculty Network mentors are faculty like you, from educational institutions around the country, who are passionate about enhancing instructional efficiency and effectiveness through best practices. You can engage a faculty mentor in an online conversation at www.wherefacultyconnect.com.

Participate in a Faculty-Led Online Seminar

The Wiley Faculty Network provides you with virtual seminars led by faculty using the latest teaching technologies. In these seminars, faculty share their knowledge and experiences on discipline-specific teaching and learning issues. All you need to participate in a virtual seminar is high-speed Internet access and a phone line. To register for a seminar, go to www.wherefacultyconnect.com.

Connect with the Wiley Faculty Network

Web: www.wherefacultyconnect.com
Phone: 1-866-4FACULTY

Table of Contents

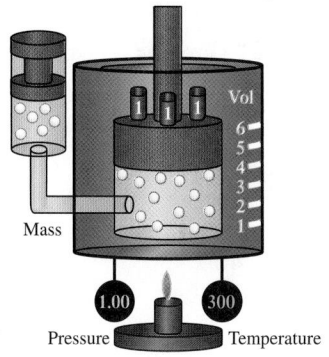

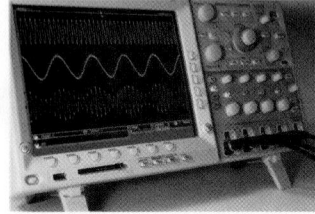

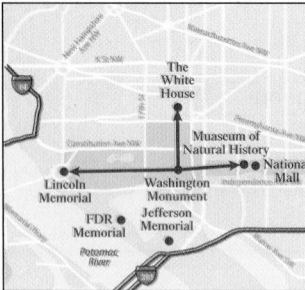

A Note from the Author to the Student

I wrote this text with careful attention to ways in which to make your learning experience more successful. If you take full advantage of the unique features and elements of this textbook, I believe your experience will be fulfilling and enjoyable. Let's walk through some of the special book features that will help you in your study of Precalculus.

Prerequisites and Review (Chapter 0)

A comprehensive review of prerequisite knowledge in Chapter 0 provides a brush up on knowledge and skills necessary for success in precalculus.

Clear, Concise, and Inviting Writing

Special attention has been made to present an engaging, clear, precise narrative in a layout that is easy to use and designed to reduce any math anxiety you may have.

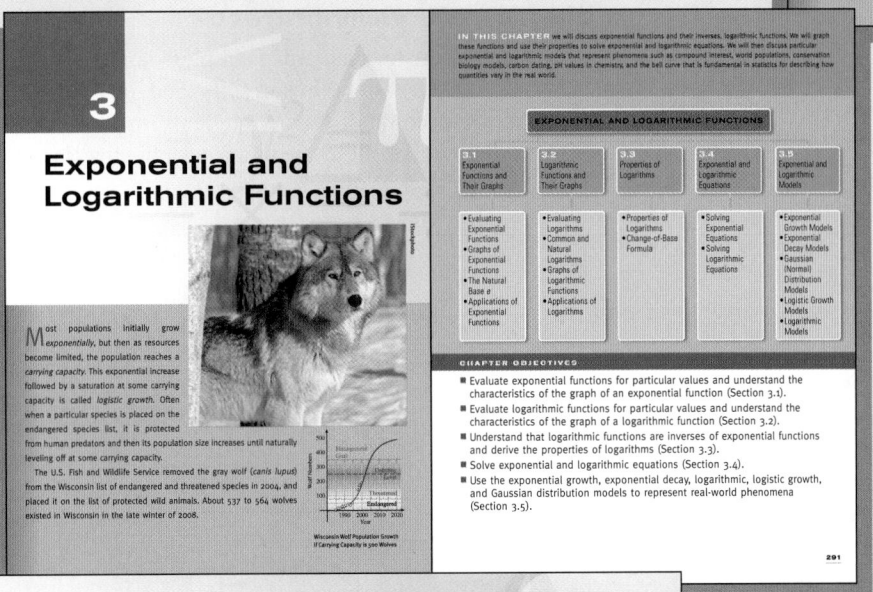

Chapter Introduction, Flow Chart, Section Headings, and Objectives

An opening vignette, flow chart, list of chapter sections, and targeted chapter objectives give you an overview of the chapter to help you see the big picture, the relationships between topics, and clear objectives for learning in the chapter.

Skills and Conceptual Objectives

For every section, objectives are further divided by skills *and* concepts so you can see the difference between solving problems and truly understanding concepts.

Examples

Examples pose a specific problem using concepts already presented and then work through the solution. These serve to enhance your understanding of the subject matter.

EXAMPLE 1 Operations on Functions: Determining Domains of New Functions

For the functions $f(x) = \sqrt{x-1}$ and $g(x) = \sqrt{4-x}$, determine the sum function, difference function, product function, and quotient function. State the domain of these four new functions.

Solution:

Sum function: $f(x) + g(x) = \sqrt{x-1} + \sqrt{4-x}$

Difference function: $f(x) - g(x) = \sqrt{x-1} - \sqrt{4-x}$

Product function:
$$f(x) \cdot g(x) = \sqrt{x-1} \cdot \sqrt{4-x}$$
$$= \sqrt{(x-1)(4-x)} = \sqrt{-x^2 + 5x - 4}$$

Quotient function: $\dfrac{f(x)}{g(x)} = \dfrac{\sqrt{x-1}}{\sqrt{4-x}} = \sqrt{\dfrac{x-1}{4-x}}$

The domain of the square root function is determined by setting the argument under the radical greater than or equal to zero.

Domain of $f(x)$: $[1, \infty)$

Domain of $g(x)$: $(-\infty, 4]$

The domain of the sum, difference, and product functions is
$$[1, \infty) \cap (-\infty, 4] = [1, 4]$$

The quotient function has the additional constraint that the denominator cannot be zero. This implies that $x \neq 4$, so the domain of the quotient function is $[1, 4)$.

--- --- ---

■ **YOUR TURN** Given the function $f(x) = \sqrt{x+3}$ and $g(x) = \sqrt{1-x}$, find $(f + g)(x)$ and state its domain.

Your Turn

Immediately following many examples, you are given a similar problem to reinforce and check your understanding. This helps build confidence as you progress in the chapter. These are ideal for inclass activity or for preparing for homework later. Answers are provided in the margin for a quick check of your work.

COMMON MISTAKE

The most common mistake in calculating slope is writing the coordinates in the wrong order, which results in the slope being opposite in sign.

Find the slope of the line containing the two points (1, 2) and (3, 4).

⊘ CORRECT	☒ INCORRECT
Label the points.	The **ERROR** is interchanging the coordinates of the first and second points.
$(x_1, y_1) = (1, 2)$	
$(x_2, y_2) = (3, 4)$	
Write the slope formula.	$m = \dfrac{4-2}{1-3}$
$m = \dfrac{y_2 - y_1}{x_2 - x_1}$	The calculated slope is **INCORRECT** by a negative sign.
Substitute the coordinates.	$m = \dfrac{2}{-2} = -1$
$m = \dfrac{4-2}{3-1}$	
Simplify. $m = \dfrac{2}{2} = \boxed{1}$	

Common Mistake/ Correct vs. Incorrect

In addition to standard examples, some problems are worked out both correctly and incorrectly to highlight common errors students make. Counter examples like these are often an effective learning approach for many students.

Parallel Words and Math

This text reverses the common textbook presentation of examples by placing the explanation in words *on the left* and the mathematics in parallel *on the right*. This makes it easier for students to read through examples as the material flows more naturally from left to right and as commonly presented in class.

WORDS	MATH
Step 1: In a table, list several pairs of coordinates that make the equation true.	

x	$y = x^2$	(x, y)
0	0	(0, 0)
−1	1	(−1, 1)
1	1	(1, 1)
−2	4	(−2, 4)
2	4	(2, 4)

Step 2: Plot these points on a graph and connect the points with a smooth curve. Use arrows to indicate that the graph continues.

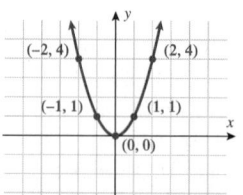

Study Tips

These marginal reminders call out important concepts.

Study Tip

The determinant of a 2×2 matrix is found by finding the product of the main diagonal entries and subtracting the product of the other diagonal entries.

Technology Tips

These marginal notes provide problem solving instructions and visual examples using graphing calculators.

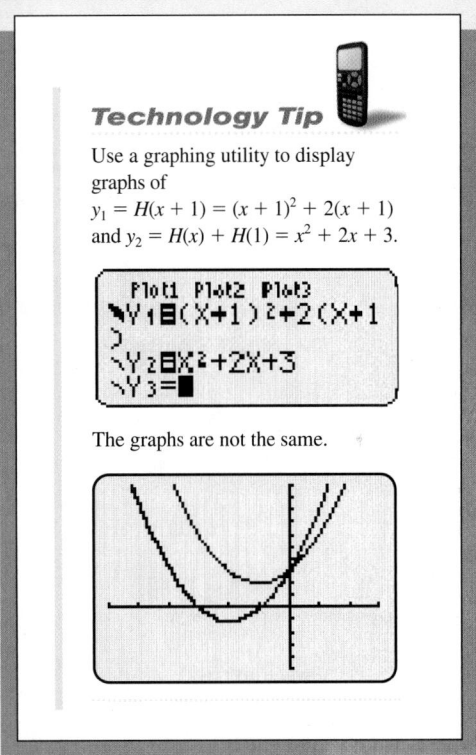

Technology Tip

Use a graphing utility to display graphs of
$y_1 = H(x + 1) = (x + 1)^2 + 2(x + 1)$
and $y_2 = H(x) + H(1) = x^2 + 2x + 3$.

The graphs are not the same.

Video icons

Video icons appear on selected examples throughout the chapter to indicate that the author has created a video segment for that element. These video clips help you work through the selected examples with the author as your "private tutor."

EXAMPLE 10   Determining the Domain of a Function

State the domain of the given functions.

a. $F(x) = \dfrac{3}{x^2 - 25}$ **b.** $H(x) = \sqrt[4]{9 - 2x}$ **c.** $G(x) = \sqrt[3]{x - 1}$

Solution (a):

Write the original equation. $F(x) = \dfrac{3}{x^2 - 25}$

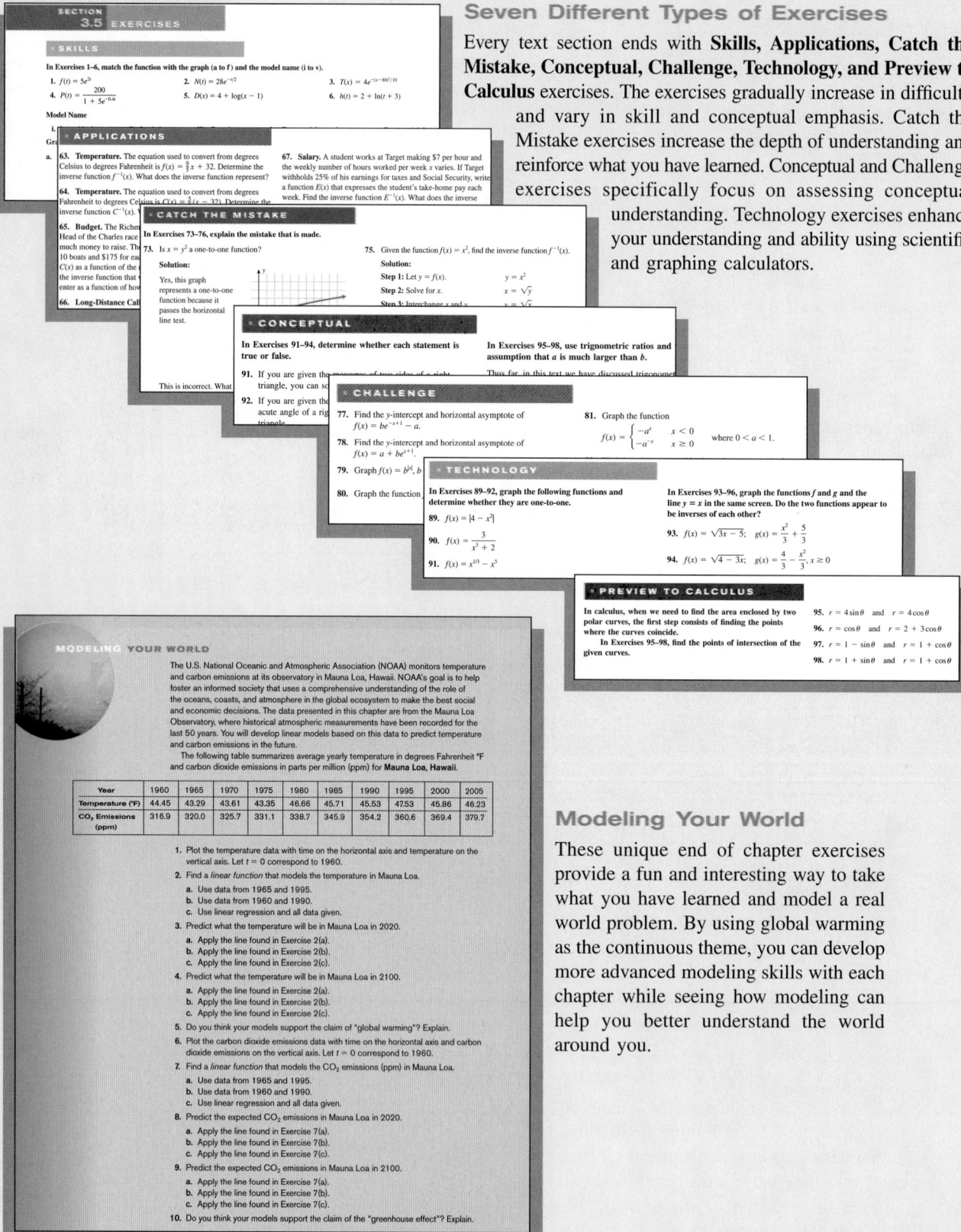

Seven Different Types of Exercises

Every text section ends with **Skills, Applications, Catch the Mistake, Conceptual, Challenge, Technology, and Preview to Calculus** exercises. The exercises gradually increase in difficulty and vary in skill and conceptual emphasis. Catch the Mistake exercises increase the depth of understanding and reinforce what you have learned. Conceptual and Challenge exercises specifically focus on assessing conceptual understanding. Technology exercises enhance your understanding and ability using scientific and graphing calculators.

SECTION 3.5 EXERCISES

SKILLS

In Exercises 1–6, match the function with the graph (a to f) and the model name (i to v).

1. $f(t) = 5e^{2t}$
2. $N(t) = 28e^{-t/2}$
3. $T(x) = 4e^{-(x-80)/10}$
4. $P(t) = \dfrac{200}{1 + 5e^{-0.4t}}$
5. $D(x) = 4 + \log(x - 1)$
6. $h(t) = 2 + \ln(t + 3)$

Model Name

APPLICATIONS

a.
63. **Temperature.** The equation used to convert from degrees Celsius to degrees Fahrenheit is $f(x) = \frac{9}{5}x + 32$. Determine the inverse function $f^{-1}(x)$. What does the inverse function represent?
64. **Temperature.** The equation used to convert from degrees Fahrenheit to degrees Celsius is $C(x) = \frac{5}{9}(x - 32)$. Determine the inverse function $C^{-1}(x)$.
65. **Budget.** The Richm... Head of the Charles race... much money to raise. Th... 10 boats and \$175 for ea... $C(x)$ as a function of the... the inverse function that... enter as a function of how...
66. **Long-Distance Cal...**

67. **Salary.** A student works at Target making \$7 per hour and the weekly number of hours worked per week x varies. If Target withholds 25% of his earnings for taxes and Social Security, write a function $E(x)$ that expresses the student's take-home pay each week. Find the inverse function $E^{-1}(x)$. What does the inverse

CATCH THE MISTAKE

In Exercises 73–76, explain the mistake that is made.

73. Is $x = y^2$ a one-to-one function?

Solution:

Yes, this graph represents a one-to-one function because it passes the horizontal line test.

This is incorrect. What

75. Given the function $f(x) = x^2$, find the inverse function $f^{-1}(x)$.

Solution:

Step 1: Let $y = f(x)$. $y = x^2$
Step 2: Solve for x. $x = \sqrt{y}$
Step 3: Interchange x and y. $y = \sqrt{x}$

CONCEPTUAL

In Exercises 91–94, determine whether each statement is true or false.

91. If you are given the measure of two sides of a right triangle, you can so...
92. If you are given the acute angle of a rig... triangle

In Exercises 95–98, use trigonometric ratios and assumption that a is much larger than b.

Thus far, in this text we have discussed trigonome...

CHALLENGE

77. Find the y-intercept and horizontal asymptote of $f(x) = be^{-x+1} - a$.
78. Find the y-intercept and horizontal asymptote of $f(x) = a + be^{x+1}$.
79. Graph $f(x) = b^{|x|}, b$...
80. Graph the function...

81. Graph the function
$$f(x) = \begin{cases} -a^x & x < 0 \\ -a^{-x} & x \geq 0 \end{cases} \quad \text{where } 0 < a < 1.$$

TECHNOLOGY

In Exercises 89–92, graph the following functions and determine whether they are one-to-one.

89. $f(x) = |4 - x^2|$
90. $f(x) = \dfrac{3}{x^3 + 2}$
91. $f(x) = x^{1/3} - x^3$

In Exercises 93–96, graph the functions f and g and the line $y = x$ in the same screen. Do the two functions appear to be inverses of each other?

93. $f(x) = \sqrt{3x - 5}; \quad g(x) = \dfrac{x^2}{3} + \dfrac{5}{3}$
94. $f(x) = \sqrt{4 - 3x}; \quad g(x) = \dfrac{4}{3} - \dfrac{x^2}{3}, x \geq 0$

PREVIEW TO CALCULUS

In calculus, when we need to find the area enclosed by two polar curves, the first step consists of finding the points where the curves coincide.

In Exercises 95–98, find the points of intersection of the given curves.

95. $r = 4\sin\theta$ and $r = 4\cos\theta$
96. $r = \cos\theta$ and $r = 2 + 3\cos\theta$
97. $r = 1 - \sin\theta$ and $r = 1 + \cos\theta$
98. $r = 1 + \sin\theta$ and $r = 1 + \cos\theta$

MODELING YOUR WORLD

The U.S. National Oceanic and Atmospheric Association (NOAA) monitors temperature and carbon emissions at its observatory in Mauna Loa, Hawaii. NOAA's goal is to help foster an informed society that uses a comprehensive understanding of the role of the oceans, coasts, and atmosphere in the global ecosystem to make the best social and economic decisions. The data presented in this chapter are from the Mauna Loa Observatory, where historical atmospheric measurements have been recorded for the last 50 years. You will develop linear models based on this data to predict temperature and carbon emissions in the future.

The following table summarizes average yearly temperature in degrees Fahrenheit °F and carbon dioxide emissions in parts per million (ppm) for **Mauna Loa, Hawaii.**

Year	1960	1965	1970	1975	1980	1985	1990	1995	2000	2005
Temperature (°F)	44.45	43.29	43.61	43.35	46.66	45.71	45.53	47.53	45.86	46.23
CO_2 Emissions (ppm)	316.9	320.0	325.7	331.1	338.7	345.9	354.2	360.6	369.4	379.7

1. Plot the temperature data with time on the horizontal axis and temperature on the vertical axis. Let $t = 0$ correspond to 1960.
2. Find a *linear function* that models the temperature in Mauna Loa.
   a. Use data from 1965 and 1995.
   b. Use data from 1960 and 1990.
   c. Use linear regression and all data given.
3. Predict what the temperature will be in Mauna Loa in 2020.
   a. Apply the line found in Exercise 2(a).
   b. Apply the line found in Exercise 2(b).
   c. Apply the line found in Exercise 2(c).
4. Predict what the temperature will be in Mauna Loa in 2100.
   a. Apply the line found in Exercise 2(a).
   b. Apply the line found in Exercise 2(b).
   c. Apply the line found in Exercise 2(c).
5. Do you think your models support the claim of "global warming"? Explain.
6. Plot the carbon dioxide emissions data with time on the horizontal axis and carbon dioxide emissions on the vertical axis. Let $t = 0$ correspond to 1960.
7. Find a *linear function* that models the CO_2 emissions (ppm) in Mauna Loa.
   a. Use data from 1965 and 1995.
   b. Use data from 1960 and 1990.
   c. Use linear regression and all data given.
8. Predict the expected CO_2 emissions in Mauna Loa in 2020.
   a. Apply the line found in Exercise 7(a).
   b. Apply the line found in Exercise 7(b).
   c. Apply the line found in Exercise 7(c).
9. Predict the expected CO_2 emissions in Mauna Loa in 2100.
   a. Apply the line found in Exercise 7(a).
   b. Apply the line found in Exercise 7(b).
   c. Apply the line found in Exercise 7(c).
10. Do you think your models support the claim of the "greenhouse effect"? Explain.

Modeling Your World

These unique end of chapter exercises provide a fun and interesting way to take what you have learned and model a real world problem. By using global warming as the continuous theme, you can develop more advanced modeling skills with each chapter while seeing how modeling can help you better understand the world around you.

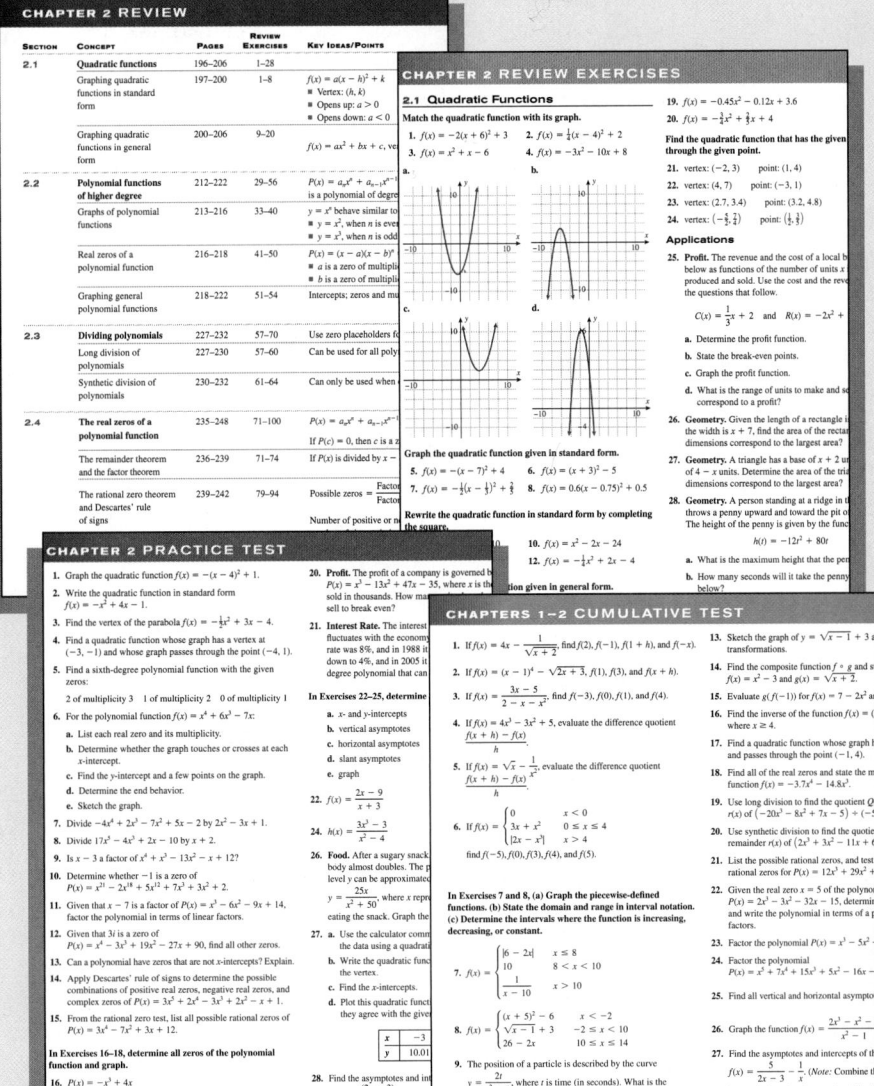

Chapter Review, Review Exercises, Practice Test, Cumulative Test

At the end of every chapter, a summary review chart organizes the key learning concepts in an easy to use one or two-page layout. This feature includes key ideas and formulas, as well as indicating relevant pages and review exercises so that you can quickly summarize a chapter and study smarter. Review Exercises, arranged by section heading, are provided for extra study and practice. A Practice Test, without section headings, offers even more self practice before moving on. A new Cumulative Test feature offers study questions based on all previous chapters' content, thus helping you build upon previously learned concepts.

0

Review: Equations and Inequalities

Boyle's Law

For a given mass, at constant temperature, the product of the pressure and the volume is constant.

$$pV = C$$

Have you ever noticed when you open the twist-off cap of a soda bottle that more pressure seems to be released if the soda is warm than if it has been refrigerated? The combined gas law in chemistry says that the pressure of a gas is directly proportional to the temperature of the gas and inversely proportional to the volume of that gas. For example, if the volume stays the same (container of soda), and the temperature of the soda increases, the pressure also increases. However, for a constant temperature, the pressure of a gas is inversely proportional to the volume containing the gas: As the volume decreases, the pressure increases. If the gas contained in a 4-ml container at a temperature of 300 kelvins has a pressure of 1 atmosphere, what is the resulting pressure if the volume of the container changes to 3 ml?*

*See Section 0.7 Exercise 52.

REVIEW: EQUATIONS AND INEQUALITIES

0.1 Linear Equations	0.2 Quadratic Equations	0.3 Other Types of Equations	0.4 Inequalities	0.5 Graphing Equations	0.6 Lines	0.7 Modeling Variation
• Solving Linear Equations in One Variable • Applications Involving Linear Equations	• Factoring • Square Root Method • Completing the Square • The Quadratic Formula	• Rational Equations • Radical Equations • Equations Quadratic in Form: u-Substitution • Factorable Equations • Absolute Value Equations	• Graphing Inequalities and Interval Notation • Linear Inequalities • Polynomial Inequalities • Rational Inequalities • Absolute Value Inequalities	• The Distance and Midpoint Formulas • Point-Plotting • Using Intercepts and Symmetry as Graphing Aids • Circles	• Graphing a Line • Slope • Slope–Intercept Form • Finding Equations of Lines • Parallel and Perpendicular Lines	• Direct Variation • Inverse Variation • Joint Variation and Combined Variation

CHAPTER OBJECTIVES

- Solve linear equations in one variable (Section 0.1).
- Solve quadratic equations in one variable (Section 0.2).
- Solve other types of equations that can be transformed into linear or quadratic equations (Section 0.3).
- Solve inequalities in one variable (Section 0.4).
- Graph equations in two variables in the Cartesian plane (Section 0.5).
- Find the equation of a line (Section 0.6).
- Use equations to model variation (Section 0.7).

SKILLS OBJECTIVES

- Solve linear equations in one variable.
- Solve application problems involving linear equations.

CONCEPTUAL OBJECTIVE

- Understand the mathematical modeling process.

An **algebraic expression** (see Appendix) consists of one or more terms that are combined through basic operations such as addition, subtraction, multiplication, or division; for example,

$$3x + 2 \qquad 5 - 2y \qquad x + y$$

An **equation** is a statement that says two expressions are equal. For example, the following are all equations in one variable, x:

$$x + 7 = 11 \qquad x^2 = 9 \qquad 7 - 3x = 2 - 3x \qquad 4x + 7 = x + 2 + 3x + 5$$

To **solve** an equation in one variable means to find all the values of that variable that make the equation true. These values are called **solutions**, or **roots**, of the equation. The first of these statements shown above, $x + 7 = 11$, is true when $x = 4$ and false for any other values of x. We say that $x = 4$ is the solution to the equation. Sometimes an equation can have more than one solution, as in $x^2 = 9$. In this case, there are actually two values of x that make this equation true, $x = -3$ and $x = 3$. We say the **solution set** of this equation is $\{-3, 3\}$. In the third equation, $7 - 3x = 2 - 3x$, no values of x make the statement true. Therefore, we say this equation has **no solution**. And the fourth equation, $4x + 7 = x + 2 + 3x + 5$, is true for any values of x. An equation that is true for any value of the variable x is called an **identity**. In this case, we say the solution set is the **set of all real numbers**.

Two equations that have the same solution set are called **equivalent equations**. For example,

$$3x + 7 = 13 \qquad 3x = 6 \qquad x = 2$$

are all equivalent equations because each of them has the solution set $\{2\}$. Note that $x^2 = 4$ is not equivalent to these three equations because it has the solution set $\{-2, 2\}$.

When solving equations, it helps to find a simpler equivalent equation in which the variable is isolated (alone). The following table summarizes the procedures for generating equivalent equations.

Generating Equivalent Equations

ORIGINAL EQUATION	DESCRIPTION	EQUIVALENT EQUATION
$3(x - 6) = 6x - x$	■ Eliminate the parentheses. ■ Combine like terms on one or both sides of the equation.	$3x - 18 = 5x$
$7x + 8 = 29$	Add (or subtract) the same quantity to (from) *both* sides of the equation. $7x + 8 - \mathbf{8} = 29 - \mathbf{8}$	$7x = 21$
$5x = 15$	Multiply (or divide) both sides of the equation by the same nonzero quantity: $\dfrac{5x}{5} = \dfrac{15}{5}$.	$x = 3$
$-7 = x$	Interchange the two sides of the equation.	$x = -7$

Solving Linear Equations in One Variable

You probably already know how to solve simple linear equations. Solving a linear equation in one variable is done by finding an equivalent equation. In generating an equivalent equation, remember that whatever operation is performed on one side of an equation must also be performed on the other side of the equation.

Technology Tip

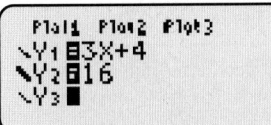

Use a graphing utility to display graphs of $y_1 = 3x + 4$ and $y_2 = 16$.

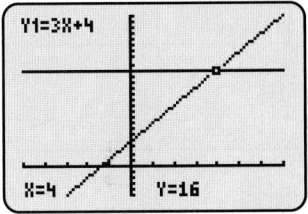

EXAMPLE 1 **Solving a Linear Equation**

Solve the equation $3x + 4 = 16$.

Solution:

Subtract 4 from both sides of the equation.

$$3x + 4 = 16$$
$$\underline{ -4 \quad -4}$$
$$3x = 12$$

Divide both sides by 3.

$$\frac{3x}{3} = \frac{12}{3}$$

The solution is $x = 4$.

$$\boxed{x = 4}$$

The solution set is $\{4\}$.

The x-coordinate of the point of intersection is the solution to the equation $3x + 4 = 16$.

■ YOUR TURN Solve the equation $2x + 3 = 9$.

■ **Answer:** The solution is $x = 3$. The solution set is $\{3\}$.

Example 1 illustrates solving linear equations in one variable. What is a linear equation in one variable?

DEFINITION **Linear Equation**

A **linear equation in one variable**, x, can be written in the form

$$ax + b = 0$$

where a and b are real numbers and $a \neq 0$.

What makes this equation linear is that x is raised to the first power. We can also classify a linear equation as a **first-degree** equation.

Equation	Degree	General Name
$x - 7 = 0$	First	Linear
$x^2 - 6x - 9 = 0$	Second	Quadratic
$x^3 + 3x^2 - 8 = 0$	Third	Cubic

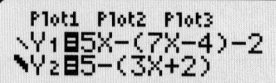
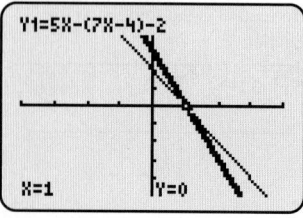
■ Answer: The solution is $x = 2$.
The solution set is $\{2\}$.

EXAMPLE 2   Solving a Linear Equation

Solve the equation $5x - (7x - 4) - 2 = 5 - (3x + 2)$.

Solution:

Eliminate the parentheses.

Don't forget to distribute the negative sign through *both* terms inside the parentheses.

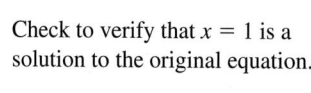

$$5x - 7x + 4 - 2 = 5 - 3x - 2$$

Combine like terms on each side.
$$-2x + 2 = 3 - 3x$$

Add $3x$ to both sides.
$$\frac{+3x \qquad\qquad +3x}{x + 2 = 3}$$

Subtract 2 from both sides.
$$\frac{-2 \quad -2}{x = 1}$$

Check to verify that $x = 1$ is a solution to the original equation.
$$5 \cdot 1 - (7 \cdot 1 - 4) - 2 = 5 - (3 \cdot 1 + 2)$$
$$5 - (7 - 4) - 2 = 5 - (3 + 2)$$
$$5 - (3) - 2 = 5 - (5)$$
$$0 = 0$$

Since the solution $x = 1$ makes the equation true, the solution set is $\{1\}$.

■ YOUR TURN Solve the equation $4(x - 1) - 2 = x - 3(x - 2)$.

To solve a linear equation involving fractions, find the least common denominator (LCD) of all terms and multiply both sides of the equation by the LCD. We will first review how to find the LCD.

To add the fractions $\frac{1}{2} + \frac{1}{6} + \frac{2}{5}$, we must first find a common denominator. Some people are taught to find the lowest number that 2, 6, and 5 all divide evenly into. Others prefer a more systematic approach in terms of prime factors.

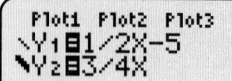
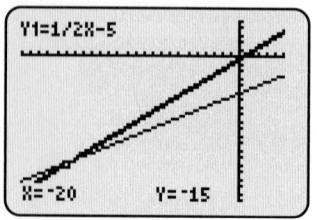
■ Answer: The solution is $m = -18$.
The solution set is $\{-18\}$.

EXAMPLE 3   Solving a Linear Equation Involving Fractions

Solve the equation $\frac{1}{2}p - 5 = \frac{3}{4}p$.

Solution:

Write the equation.
$$\frac{1}{2}p - 5 = \frac{3}{4}p$$

Multiply each term in the equation by the LCD, 4.
$$(4)\frac{1}{2}p - (4)5 = (4)\frac{3}{4}p$$

The result is a linear equation with no fractions.
$$2p - 20 = 3p$$

Subtract $2p$ from both sides.
$$\frac{-2p \qquad\qquad -2p}{-20 = p}$$

$$\boxed{p = -20}$$

Since $p = -20$ satisfies the original equation, the solution set is $\{-20\}$.

■ YOUR TURN Solve the equation $\frac{1}{4}m = \frac{1}{12}m - 3$.

Solving a Linear Equation in One Variable

STEP	DESCRIPTION	EXAMPLE
1	Simplify the algebraic expressions on both sides of the equation.	$-3(x-2)+5 = 7(x-4)-1$ $-3x+6+5 = 7x-28-1$ $-3x+11 = 7x-29$
2	Gather all variable terms on one side of the equation and all constant terms on the other side.	$-3x+11 = 7x-29$ $\underline{+3x \qquad\quad +3x}$ $11 = 10x-29$ $\underline{+29 \qquad\quad +29}$ $40 = 10x$
3	Isolate the variable.	$10x = 40$ $x = \dfrac{40}{10}$ $\boxed{x = 4}$

Applications Involving Linear Equations

We now use linear equations to solve problems that occur in our day-to-day lives. You typically will read the problem in words, develop a mathematical model (equation) for the problem, solve the equation, and write the answer in words.

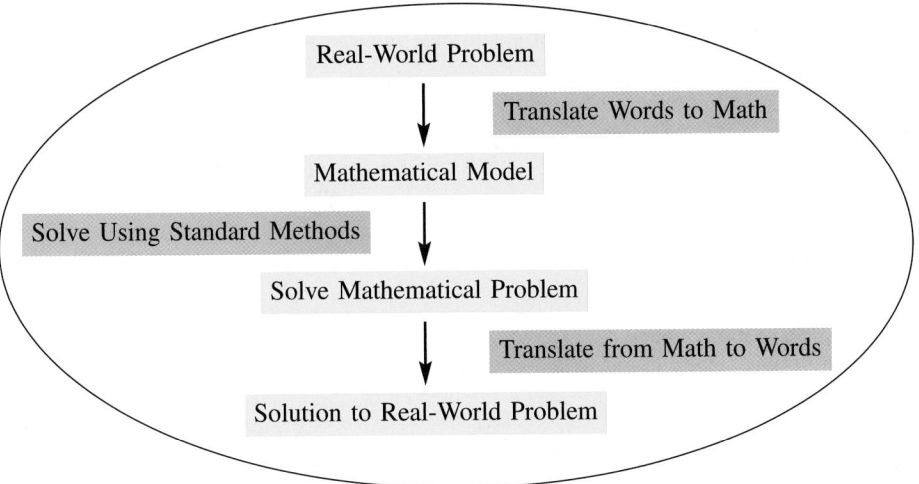

You will have to come up with a unique formula to solve each kind of word problem, but there is a universal *procedure* for approaching all word problems.

PROCEDURE FOR SOLVING WORD PROBLEMS

Step 1: Identify the question. Read the problem *one* time and note what you are asked to find.

Step 2: Make notes. Read until you can note something (an amount, a picture, anything). Continue reading and making notes until you have read the problem a second* time.

Step 3: Set up an equation. Assign a variable to represent what you are asked to find.

Step 4: Solve the equation.

Step 5: Check the solution. Substitute the solution for the variable in the equation, and also run the solution past the "common sense department" using estimation.

*Step 2 often requires multiple readings of the problem.

EXAMPLE 4 How Long Was the Trip?

During a camping trip in North Bay, Ontario, a couple went one-third of the way by boat, 10 miles by foot, and one-sixth of the way by horse. How long was the trip?

Solution:

STEP 1 **Identify the question.**

How many miles was the trip?

STEP 2 **Make notes.**

Read	**Write**
... one-third of the way by boat	BOAT: $\frac{1}{3}$ of the trip
... 10 miles by foot	FOOT: 10 miles
... one-sixth of the way by horse	HORSE: $\frac{1}{6}$ of the trip

STEP 3 **Set up an equation.**

The total distance of the trip is the sum of all the distances by boat, foot, and horse.

Distance by boat + Distance by foot + Distance by horse = Total distance of trip

Distance of total trip in miles = x

Distance by boat = $\frac{1}{3}x$

Distance by foot = 10 miles

Distance by horse = $\frac{1}{6}x$

$$\overbrace{\frac{1}{3}x}^{\text{boat}} + \overbrace{10}^{\text{foot}} + \overbrace{\frac{1}{6}x}^{\text{horse}} = \overbrace{x}^{\text{total}}$$

STEP 4 **Solve the equation.**

$$\frac{1}{3}x + 10 + \frac{1}{6}x = x$$

Multiply by the LCD, 6. $\qquad 2x + 60 + x = 6x$

Collect x terms on the right. $\qquad 60 = 3x$

Divide by 3. $\qquad 20 = x$

> The trip was 20 miles. $\qquad\qquad x = 20$

STEP 5 **Check the solution.**

Estimate: The boating distance, $\frac{1}{3}$ of 20 miles, is approximately 7 miles; the riding distance on horse, $\frac{1}{6}$ of 20 miles, is approximately 3 miles. Adding these two distances to the 10 miles by foot gives a trip distance of 20 miles.

■ **Answer:** The distance from their car to the gate is 1.5 miles.

■ **YOUR TURN** A family arrives at the Walt Disney World parking lot. To get from their car in the parking lot to the gate at the Magic Kingdom, they walk $\frac{1}{4}$ mile, take a tram for $\frac{1}{3}$ of their total distance, and take a monorail for $\frac{1}{2}$ of their total distance. How far is it from their car to the gate of the Magic Kingdom?

Geometry Problems

Some problems require geometric formulas in order to be solved.

EXAMPLE 5   Geometry

A rectangle 24 meters long has the same area as a square with 12-meter sides. What are the dimensions of the rectangle?

Solution:

STEP 1 **Identify the question.**
What are the dimensions (length and width) of the rectangle?

STEP 2 **Make notes.**

Read	**Write/Draw**
A rectangle 24 meters long	w [rectangle], $l = 24$
... has the same area	rectangle area $= l \cdot w = 24w$
... as a square that is 12 meters long	area of square $= 12 \cdot 12 = 144$

STEP 3 **Set up an equation.**

The area of the rectangle is equal to the area of the square.   rectangle area $=$ square area

Substitute in known quantities. $24w = 144$

STEP 4 **Solve the equation.**

Divide by 24. $w = \dfrac{144}{24} = 6$

The rectangle is 24 meters long and 6 meters wide.

STEP 5 **Check the solution.**
A 24 m by 6 m rectangle has an area of 144 m^2.

....

■ **YOUR TURN**   A rectangle 3 inches wide has the same area as a square with 9-inch sides. What are the dimensions of the rectangle?

■**Answer:** The rectangle is 27 inches long and 3 inches wide.

Interest Problems

In our personal or business financial planning, a particular concern we have is interest. **Interest** is money paid for the use of money; it is the cost of borrowing money. The total amount borrowed is called the **principal**. The principal can be the price of our new car; we pay the bank interest for loaning us money to buy the car. The principal can also be the amount we keep in a CD or money market account; the bank uses this money and pays us interest. Typically, interest rate, expressed as a percentage, is the amount charged for the use of the principal for a given time, usually in years.

 Simple interest is interest that is paid only on the principal during a period of time. Later we will discuss *compound interest*, which is interest paid on both the principal and the interest accrued over a period of time.

> **DEFINITION   Simple Interest**
>
> If a principal of P dollars is borrowed for a period of t years at an annual interest rate r (expressed in decimal form), the interest I charged is
>
> $$I = Prt$$
>
> This is the formula for **simple interest**.

EXAMPLE 6   Multiple Investments

Theresa earns a full athletic scholarship for college, and her parents have given her the $20,000 they had saved to pay for her college tuition. She decides to invest that money with an overall goal of earning 11% interest. She wants to put some of the money in a low-risk investment that has been earning 8% a year and the rest of the money in a medium-risk investment that typically earns 12% a year. How much money should she put in each investment to reach her goal?

Solution:

STEP 1   Identify the question.

How much money is invested in each (the 8% and the 12%) account?

STEP 2   Make notes.

Read	Write/Draw

Theresa has $20,000 to invest.

If part is invested at 8% and the rest at 12%, how much should be invested at each rate to yield 11% on the total amount invested?

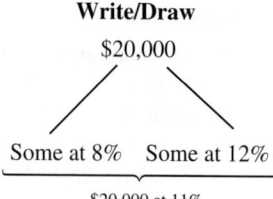

STEP 3   Set up an equation.

If we let x represent the amount Theresa puts into the 8% investment, how much of the $20,000 is left for her to put in the 12% investment?

Amount in the 8% investment: x

Amount in the 12% investment: $20{,}000 - x$

Simple interest formula: $I = Prt$

INVESTMENT	PRINCIPAL	RATE	TIME (YR)	INTEREST
8% Account	x	0.08	1	$0.08x$
12% Account	$20{,}000 - x$	0.12	1	$0.12(20{,}000 - x)$
Total	$20{,}000$	0.11	1	$0.11(20{,}000)$

Adding the interest earned in the 8% investment to the interest earned in the 12% investment should earn an average of 11% on the total investment.

$$0.08x + 0.12(20{,}000 - x) = 0.11(20{,}000)$$

STEP 4 **Solve the equation.**

Eliminate the parentheses. $0.08x + 2400 - 0.12x = 2200$

Collect x terms on the left,
constants on the right. $-0.04x = -200$

Divide by -0.04. $x = 5000$

Calculate the amount at 12%. $20{,}000 - 5000 = 15{,}000$

> Theresa should invest \$5,000 at 8% and \$15,000 at 12% to reach her goal.

STEP 5 **Check the solution.**

If money is invested at 8% and 12% with a goal of averaging 11%, our intuition tells us that more should be invested at 12% than 8%, which is what we found. The exact check is as follows:

$$0.08(5000) + 0.12(15{,}000) = 0.11(20{,}000)$$

$$400 + 1800 = 2200$$

$$2200 = 2200$$

■ **YOUR TURN**   You win \$24,000 and you decide to invest the money in two different investments: one paying 18% and the other paying 12%. A year later you have \$27,480 total. How much did you originally invest in each account?

■ **Answer:** \$10,000 is invested at 18% and \$14,000 is invested at 12%.

Mixture Problems

Mixtures are something we come across every day. Different candies that sell for different prices may make up a movie snack. New blends of coffees are developed by coffee connoisseurs. Chemists mix different concentrations of acids in their labs. Whenever two or more distinct ingredients are combined, the result is a **mixture**.

Our choice at a gas station is typically 87, 89, and 93 octane. The octane number is the number that represents the percentage of iso-octane in fuel. 89 octane is significantly overpriced. Therefore, if your car requires 89 octane, it would be more cost-effective to mix 87 and 93 octane.

EXAMPLE 7   Mixture Problem

The manual for your new car suggests using gasoline that is 89 octane. In order to save money, you decide to use some 87 octane and some 93 octane in combination with the 89 octane currently in your tank in order to have an approximate 89 octane mixture. Assuming you have 1 gallon of 89 octane remaining in your tank (your tank capacity is 16 gallons), how many gallons of 87 and 93 octane should be used to fill up your tank to achieve a mixture of 89 octane?

Solution:

STEP 1 **Identify the question.**

How many gallons of 87 octane and how many gallons of 93 octane should be used?

STEP 2 **Make notes.**

Read	**Write/Draw**
Assuming you have 1 gallon of 89 octane remaining in your tank (your tank capacity is 16 gallons), how many gallons of 87 and 93 octane should you add?	

89 octane + 87 octane + 93 octane = 89 octane
[1 gallon]   [? gallons]   [? gallons]   [16 gallons]

STEP 3 Set up an equation.

$$x = \text{gallons of 87 octane gasoline added at the pump}$$

$$15 - x = \text{gallons of 93 octane gasoline added at the pump}$$

$$1 = \text{gallons of 89 octane gasoline already in the tank}$$

$$0.89(1) + 0.87x + 0.93(15 - x) = 0.89(16)$$

STEP 4 Solve the equation.

$$0.89(1) + 0.87x + 0.93(15 - x) = 0.89(16)$$

Eliminate the parentheses. $\quad 0.89 + 0.87x + 13.95 - 0.93x = 14.24$

Collect x terms on the left side. $\quad -0.06x + 14.84 = 14.24$

Subtract 14.84 from both sides
of the equation. $\quad -0.06x = -0.6$

Divide both sides by -0.06. $\quad x = 10$

Calculate the amount of 93 octane. $\quad 15 - 10 = 5$

> Add 10 gallons of 87 octane and 5 gallons of 93 octane.

STEP 5 Check the solution.

Estimate: Our intuition tells us that if the desired mixture is 89 octane, then we should add approximately 1 part 93 octane and 2 parts 87 octane. The solution we found, 10 gallons of 87 octane and 5 gallons of 93 octane, agrees with this.

■**Answer:** 40 ml of 5% HCl and 60 ml of 15% HCl

■**YOUR TURN** For a certain experiment, a student requires 100 ml of a solution that is 11% HCl (hydrochloric acid). The storeroom has only solutions that are 5% HCl and 15% HCl. How many milliliters of each available solution should be mixed to get 100 ml of 11% HCl?

Distance–Rate–Time Problems

The next example deals with distance, rate, and time. On a road trip, you see a sign that says your destination is 90 miles away, and your speedometer reads 60 miles per hour. Dividing 90 miles by 60 miles per hour tells you that if you continue at this speed, your arrival will be in 1.5 hours. Here is how you know.

If the rate, or speed, is assumed to be constant, then the equation that relates distance (d), rate (r), and time (t) is given by $d = r \cdot t$. In the above driving example,

$$d = 90 \text{ miles} \qquad r = 60 \frac{\text{miles}}{\text{hour}}$$

Substituting these into
$d = r \cdot t$, we arrive at

$$90 \text{ miles} = \left[60 \frac{\text{miles}}{\text{hour}} \right] \cdot t$$

Solving for t, we get

$$t = \frac{90 \text{ miles}}{60 \dfrac{\text{miles}}{\text{hour}}} = 1.5 \text{ hours}$$

EXAMPLE 8   Distance–Rate–Time

It takes 8 hours to fly from Orlando to London and 9.5 hours to return. If an airplane averages 550 mph in still air, what is the average rate of the wind blowing in the direction from Orlando to London? Assume the wind speed is constant for both legs of the trip.

Solution:

STEP 1 **Identify the question.**

At what rate in miles per hour is the wind blowing?

STEP 2 **Make notes.**

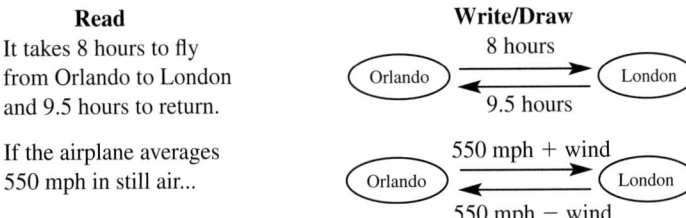

Read	Write/Draw
It takes 8 hours to fly from Orlando to London and 9.5 hours to return.	8 hours / 9.5 hours
If the airplane averages 550 mph in still air...	550 mph + wind / 550 mph − wind

STEP 3 **Set up an equation.**

The formula relating distance, rate, and time is $d = r \cdot t$. The distance d of each flight is the same. On the Orlando to London flight, the time is 8 hours due to an increased speed from a tailwind. On the London to Orlando flight, the time is 9.5 hours and the speed is decreased due to the headwind. Let w represent the wind speed.

Orlando to London: $d = (550 + w)8$

London to Orlando: $d = (550 - w)9.5$

These distances are the same, so set them equal to each other:

$$(550 + w)8 = (550 - w)9.5$$

STEP 4 **Solve the equation.**

Eliminate the parentheses. $4400 + 8w = 5225 - 9.5w$

Collect w terms on the left, constants on the right. $17.5w = 825$

Divide by 17.5. $w = 47.1429 \approx 47$

The wind is blowing approximately $\boxed{47 \text{ mph}}$ in the direction from Orlando to London.

STEP 5 **Check the solution.**

Estimate: Going from Orlando to London, the tailwind is approximately 50 mph, which when added to the plane's 550 mph speed yields a ground speed of 600 mph. The Orlando to London route took 8 hours. The distance of that flight is (600 mph) (8 hr), which is 4800 miles. The return trip experienced a headwind of approximately 50 mph, so subtracting the 50 from 550 gives an average speed of 500 mph. That route took 9.5 hours, so the distance of the London to Orlando flight was (500 mph)(9.5 hr), which is 4750 miles. Note that the estimates of 4800 and 4750 miles are close.

- -

■ **YOUR TURN** A Cessna 150 averages 150 mph in still air. With a tailwind it is able to make a trip in $2\frac{1}{3}$ hours. Because of the headwind, it is only able to make the return trip in $3\frac{1}{2}$ hours. What is the average wind speed?

■ **Answer:** The wind is blowing 30 mph.

EXAMPLE 9   Work

Connie can clean her house in 2 hours. If Alvaro helps her, together they can clean the house in 1 hour and 15 minutes. How long would it take Alvaro to clean the house by himself?

Solution:

STEP 1   **Identify the question.**

How long does it take Alvaro to clean the house?

STEP 2   **Make notes.**

- Connie can clean her house in 2 hours, so Connie can clean $\frac{1}{2}$ of the house per hour.
- Together Connie and Alvaro can clean the house in 1 hour and 15 minutes, or $\frac{5}{4}$ of an hour. Therefore together, they can clean $\frac{1}{5/4} = \frac{4}{5}$ of the house per hour.
- Let x = number of hours it takes Alvaro to clean the house by himself. So Alvaro can clean $\dfrac{1}{x}$ of the house per hour.

	AMOUNT OF TIME TO DO ONE JOB	AMOUNT OF JOB DONE PER UNIT OF TIME
Connie	2	$\dfrac{1}{2}$
Alvaro	x	$\dfrac{1}{x}$
Together	$\dfrac{5}{4}$	$\dfrac{4}{5}$

STEP 3   **Set up an equation.**

Amount of house Connie can clean per hour

Amount of house Alvaro can clean per hour

Amount of house they can clean per hour if they work together

$$\frac{1}{2} \quad + \quad \frac{1}{x} \quad = \quad \frac{4}{5}$$

STEP 4   **Solve the equation.**

Multiply by the LCD, $10x$. $5x + 10 = 8x$

Solve for x. $x = \dfrac{10}{3} = 3\dfrac{1}{3}$

It takes Alvaro ⎸3 hours and 20 minutes⎸ to clean the house by himself.

STEP 5   **Check the solution.**

Estimate: Since Connie can clean the house in 2 hours and together with Alvaro it takes 1.25 hours, we know it takes Alvaro longer than 2 hours to clean the house himself.

SECTION
0.1 SUMMARY

To solve a linear equation

1. Simplify the algebraic expressions on both sides of the equation.
2. Gather all variable terms on one side of the equation and all constant terms on the other side.
3. Isolate the variable.

In the real world, many kinds of application problems can be solved through modeling with linear equations. Some problems require the development of a mathematical model, while others rely on common formulas. The following procedure will guide you:

1. Identify the question.
2. Make notes.
3. Set up an equation.
4. Solve the equation.
5. Check the solution.

SECTION
0.1 EXERCISES

SKILLS

In Exercises 1–26, solve for the indicated variable.

1. $9m - 7 = 11$
2. $2x + 4 = 5$
3. $5t + 11 = 18$
4. $7x + 4 = 21 + 24x$
5. $3x - 5 = 25 + 6x$
6. $5x + 10 = 25 + 2x$
7. $20n - 30 = 20 - 5n$
8. $14c + 15 = 43 + 7c$
9. $4(x - 3) = 2(x + 6)$
10. $5(2y - 1) = 2(4y - 3)$
11. $-3(4t - 5) = 5(6 - 2t)$
12. $2(3n + 4) = -(n + 2)$
13. $2(x - 1) + 3 = x - 3(x + 1)$
14. $4(y + 6) - 8 = 2y - 4(y + 2)$
15. $5p + 6(p + 7) = 3(p + 2)$
16. $3(z + 5) - 5 = 4z + 7(z - 2)$
17. $7x - (2x + 3) = x - 2$
18. $3x - (4x + 2) = x - 5$
19. $2 - (4x + 1) = 3 - (2x - 1)$
20. $5 - (2x - 3) = 7 - (3x + 5)$
21. $2a - 9(a + 6) = 6(a + 3) - 4a$
22. $25 - [2 + 5y - 3(y + 2)] = -3(2y - 5) - [5(y - 1) - 3y + 3]$
23. $32 - [4 + 6x - 5(x + 4)] = 4(3x + 4) - [6(3x - 4) + 7 - 4x]$
24. $12 - [3 + 4m - 6(3m - 2)] = -7(2m - 8) - 3[(m - 2) + 3m - 5]$
25. $20 - 4[c - 3 - 6(2c + 3)] = 5(3c - 2) - [2(7c - 8) - 4c + 7]$
26. $46 - [7 - 8y + 9(6y - 2)] = -7(4y - 7) - 2[6(2y - 3) - 4 + 6y]$

Exercises 27–38 involve fractions. Clear the fractions by first multiplying by the least common denominator, and then solve the resulting linear equation.

27. $\dfrac{1}{5}m = \dfrac{1}{60}m + 1$
28. $\dfrac{1}{12}z = \dfrac{1}{24}z + 3$
29. $\dfrac{x}{7} = \dfrac{2x}{63} + 4$
30. $\dfrac{a}{11} = \dfrac{a}{22} + 9$
31. $\dfrac{1}{3}p = 3 - \dfrac{1}{24}p$
32. $\dfrac{3x}{5} - x = \dfrac{x}{10} - \dfrac{5}{2}$
33. $\dfrac{5y}{3} - 2y = \dfrac{2y}{84} + \dfrac{5}{7}$
34. $2m - \dfrac{5m}{8} = \dfrac{3m}{72} + \dfrac{4}{3}$
35. $p + \dfrac{p}{4} = \dfrac{5}{2}$
36. $\dfrac{c}{4} - 2c = \dfrac{5}{4} - \dfrac{c}{2}$
37. $\dfrac{x - 3}{3} - \dfrac{x - 4}{2} = 1 - \dfrac{x - 6}{6}$
38. $1 - \dfrac{x - 5}{3} = \dfrac{x + 2}{5} - \dfrac{6x - 1}{15}$

APPLICATIONS

39. Puzzle. Angela is on her way from home in Jersey City to New York City for dinner. She walks 1 mile to the train station, takes the train $\frac{3}{4}$ of the way, and takes a taxi $\frac{1}{6}$ of the way to the restaurant. How far does Angela live from the restaurant?

40. Puzzle. An employee at Kennedy Space Center (KSC) lives in Daytona Beach and works in the vehicle assembly building (VAB). She carpools to work with a colleague. On the days that her colleague drives the car pool, she drives 7 miles to the park-and-ride, continues with her colleague to the KSC headquarters building, and then takes the KSC shuttle from the headquarters building to the VAB. The drive from the park-and-ride to the headquarters building is $\frac{5}{6}$ of her total trip and the shuttle ride is $\frac{1}{20}$ of her total trip. How many miles does she travel from her house to the VAB on days when her colleague drives?

41. Budget. A company has a total of $20,000 allocated for monthly costs. Fixed costs are $15,000 per month and variable costs are $18.50 per unit. How many units can be manufactured in a month?

42. Budget. A woman decides to start a small business making monogrammed cocktail napkins. She can set aside $1,870 for monthly costs. Fixed costs are $1,329.50 per month and variable costs are $3.70 per set of napkins. How many sets of napkins can she afford to make per month?

43. Geometry. Consider two circles, a smaller one and a larger one. If the larger one has a radius that is 3 feet larger than that of the smaller circle and the ratio of the circumferences is 2:1, what are the radii of the two circles?

44. Geometry. The length of a rectangle is 2 more than 3 times the width, and the perimeter is 28 inches. What are the dimensions of the rectangle?

45. Biology: Alligators. It is common to see alligators in ponds, lakes, and rivers in Florida. The ratio of head size (back of the head to the end of the snout) to the full body length of an alligator is typically constant. If a $3\frac{1}{2}$-foot alligator has a head length of 6 inches, how long would you expect an alligator to be whose head length is 9 inches?

46. Biology: Snakes. In the African rainforest there is a snake called a Gaboon viper. The fang size of this snake is proportional to the length of the snake. A 3-foot snake typically has 2-inch fangs. If a herpetologist finds Gaboon viper fangs that are 2.6 inches long, how big a snake would she expect to find?

47. Investing. Ashley has $120,000 to invest and decides to put some in a CD that earns 4% interest per year and the rest in a low-risk stock that earns 7%. How much did she invest in each to earn $7,800 interest in the first year?

48. Investing. You inherit $13,000 and you decide to invest the money in two different investments: one paying 10% and the other paying 14%. A year later your investments are worth $14,580. How much did you originally invest in each account?

49. Investing. Wendy was awarded a volleyball scholarship to the University of Michigan, so on graduation her parents gave her the $14,000 they had saved for her college tuition. She opted to invest some money in a privately held company that pays 10% per year and evenly split the remaining money between a money market account yielding 2% and a high-risk stock that yielded 40%. At the end of the first year she had $16,610 total. How much did she invest in each of the three?

50. Interest. A high school student was able to save $5,000 by working a part-time job every summer. He invested half the money in a money market account and half the money in a stock that paid three times as much interest as the money market account. After a year he earned $150 in interest. What were the interest rates of the money market account and the stock?

51. Chemistry. For a certain experiment, a student requires 100 ml of a solution that is 8% HCl (hydrochloric acid). The storeroom has only solutions that are 5% HCl and 15% HCl. How many milliliters of each available solution should be mixed to get 100 ml of 8% HCl?

52. Chemistry. How many gallons of pure alcohol must be mixed with 5 gallons of a solution that is 20% alcohol to make a solution that is 50% alcohol?

53. Communications. The speed of light is approximately 3.0×10^8 meters per second (670,616,629 mph). The distance from Earth to Mars varies because their orbits around the Sun are independent. On average, Mars is 100 million miles from Earth. If we use laser communication systems, what will be the delay between Houston and NASA astronauts on Mars?

54. Speed of Sound. The speed of sound is approximately 760 mph in air. If a gun is fired $\frac{1}{2}$ mile away, how long will it take the sound to reach you?

55. Boating. A motorboat can maintain a constant speed of 16 miles per hour relative to the water. The boat makes a trip upstream to a marina in 20 minutes. The return trip takes 15 minutes. What is the speed of the current?

56. Aviation. A Cessna 175 can average 130 mph. If a trip takes 2 hours one way and the return takes 1 hour and 15 minutes, find the wind speed, assuming it is constant.

57. Distance–Rate–Time. A jogger and a walker cover the same distance. The jogger finishes in 40 minutes. The walker takes an hour. How fast is each exerciser moving if the jogger runs 2 mph faster than the walker?

58. Distance–Rate–Time. A high school student in Seattle, Washington, attended the University of Central Florida. On the way to UCF he took a southern route. After graduation he returned to Seattle via a northern trip. On both trips he had the same average speed. If the southern trek took 45 hours and the northern trek took 50 hours, and the northern trek was 300 miles longer, how long was each trip?

59. Distance–Rate–Time. College roommates leave for their first class in the same building. One walks at 2 mph and the other rides his bike at a slow 6 mph pace. How long will it take each to get to class if the walker takes 12 minutes longer to get to class and they travel on the same path?

60. Distance–Rate–Time. A long-distance delivery service sends out a truck with a package at 7 A.M. At 7:30 the manager realizes there was another package going to the same location. He sends out a car to catch the truck. If the truck travels at an average speed of 50 mph and the car travels at 70 mph, how long will it take the car to catch the truck?

61. Work. Christopher can paint the interior of his house in 15 hours. If he hires Cynthia to help him, together they can do the same job in 9 hours. If he lets Cynthia work alone, how long will it take her to paint the interior of his house?

62. Work. Jay and Morgan work in the summer for a landscaper. It takes Jay 3 hours to complete the company's largest yard alone. If Morgan helps him, it takes only 1 hour. How much time would it take Morgan alone?

63. Work. Tracey and Robin deliver Coke products to local convenience stores. Tracey can complete the deliveries in 4 hours alone. Robin can do it in 6 hours alone. If they decide to work together on a Saturday, how long will it take?

64. Work. Joshua can deliver his newspapers in 30 minutes. It takes Amber 20 minutes to do the same route. How long would it take them to deliver the newspapers if they worked together?

65. Sports. In Super Bowl XXXVII, the Tampa Bay Buccaneers scored a total of 48 points. All of their points came from field goals and touchdowns. Field goals are worth 3 points and each touchdown was worth 7 points (Martin Gramatica was successful in every extra point attempt). They scored a total of 8 times. How many field goals and touchdowns were scored?

66. Sports. A tight end can run the 100-yard dash in 12 seconds. A defensive back can do it in 10 seconds. The tight end catches a pass at his own 20 yard line with the defensive back at the 15 yard line. If no other players are nearby, at what yard line will the defensive back catch up to the tight end?

67. Recreation. How do two children of different weights balance on a seesaw? The heavier child sits closer to the center and the lighter child sits further away. When the product of the weight of the child and the distance from the center is equal on both sides, the seesaw should be horizontal to the ground. Suppose Max weighs 42 lb and Maria weighs 60 lb. If Max sits 5 feet from the center, how far should Maria sit from the center in order to balance the seesaw horizontal to the ground?

68. Recreation. Refer to Exercise 67. Suppose Martin, who weighs 33 lb, sits on the side of the seesaw with Max. If their average distance to the center is 4 feet, how far should Maria sit from the center in order to balance the seesaw horizontal to the ground?

69. Recreation. If a seesaw has an adjustable bench, then the board can be positioned over the fulcrum. Maria and Max in Exercise 67 decide to sit on the very edge of the board on each side. Where should the fulcrum be placed along the board in order to balance the seesaw horizontally to the ground? Give the answer in terms of the distance from each child's end.

70. Recreation. Add Martin (Exercise 68) to Max's side of the seesaw and recalculate Exercise 69.

■ CATCH THE MISTAKE

In Exercises 71–72, explain the mistake that is made.

71. Solve the equation $4x + 3 = 6x - 7$.

Solution:

Subtract $4x$ and add 7 to the equation.	$3 = 6x$
Divide by 3.	$x = 2$

This is incorrect. What mistake was made?

72. Solve the equation $3(x + 1) + 2 = x - 3(x - 1)$.

Solution:
$$3x + 3 + 2 = x - 3x - 3$$
$$3x + 5 = -2x - 3$$
$$5x = -8$$
$$x = -\frac{8}{5}$$

This is incorrect. What mistake was made?

■ CONCEPTUAL

73. Solve for x, given that a, b, and c are real numbers and $a \neq 0$:
$$ax + b = c$$

74. Find the number a for which $y = 2$ is a solution of the equation $y - a = y + 5 - 3ay$.

In Exercises 75–82, solve each formula for the specified variable.

75. $P = 2l + 2w$ for w

76. $P = 2l + 2w$ for l

77. $A = \frac{1}{2}bh$ for h

78. $C = 2\pi r$ for r

79. $A = lw$ for w

80. $d = rt$ for t

81. $V = lwh$ for h

82. $V = \pi r^2 h$ for h

■ **CHALLENGE**

83. Tricia and Janine are roommates and leave Houston on Interstate 10 at the same time to visit their families for a long weekend. Tricia travels west and Janine travels east. If Tricia's average speed is 12 mph faster than Janine's, find the speed of each if they are 320 miles apart in 2 hours and 30 minutes.

84. Rick and Mike are roommates and leave Gainesville on Interstate 75 at the same time to visit their girlfriends for a long weekend. Rick travels north and Mike travels south. If Mike's average speed is 8 mph faster than Rick's, find the speed of each if they are 210 miles apart in 1 hour and 30 minutes.

■ **TECHNOLOGY**

In Exercises 85–88, graph the function represented by each side of the equation in the same viewing rectangle and solve for x.

85. $3(x + 2) - 5x = 3x - 4$

86. $-5(x - 1) - 7 = 10 - 9x$

87. $2x + 6 = 4x - 2x + 8 - 2$

88. $10 - 20x = 10x - 30x + 20 - 10$

89. Suppose you bought a house for $132,500 and sold it 3 years later for $168,190. Plot these points using a graphing utility. Assuming a linear relationship, how much could you have sold the house for had you waited 2 additional years?

90. Suppose you bought a house for $132,500 and sold it 3 years later for $168,190. Plot these points using a graphing utility.

Assuming a linear relationship, how much could you have sold the house for had you sold it 1 year after buying it?

91. A golf club membership has two options. Option A is a $300 monthly fee plus $15 cart fee every time you play. Option B has a $150 monthly fee and a $42 fee every time you play. Write a mathematical model for monthly costs for each plan and graph both in the same viewing rectangle using a graphing utility. Explain when Option A is the better deal and when Option B is the better deal.

92. A phone provider offers two calling plans. Plan A has a $30 monthly charge and a $.10 per minute charge on every call. Plan B has a $50 monthly charge and a $.03 per minute charge on every call. Explain when Plan A is the better deal and when Plan B is the better deal.

SECTION

0.2 QUADRATIC EQUATIONS

SKILLS OBJECTIVES

■ Solve quadratic equations by factoring.
■ Solve quadratic equations by means of the square root method.
■ Complete the square to solve quadratic equations.
■ Solve quadratic equations by means of the quadratic formula.
■ Solve real-world problems that involve quadratic equations.

CONCEPTUAL OBJECTIVES

■ Choose appropriate methods for solving quadratic equations.
■ Interpret different types of solution sets (real, imaginary, complex conjugates, repeated roots).
■ Derive the quadratic formula.

In a linear equation, the variable is raised only to the first power in any term where it occurs. In a *quadratic equation*, the variable is raised to the second power in at least one term. Examples of *quadratic equations*, also called second-degree equations, are

$$x^2 + 3 = 7 \qquad 5x^2 + 4x - 7 = 0 \qquad x^2 - 3 = 0$$

A **quadratic equation** in x is an equation that can be written in the **standard form**

$$ax^2 + bx + c = 0$$

where a, b, and c are real numbers and $a \neq 0$.

Study Tip

In a quadratic equation the variable is raised to the power of 2, which is the highest power present in the equation.

There are several methods for solving quadratic equations: *factoring*, the *square root method*, *completing the square*, and the *quadratic formula*.

Factoring

FACTORING METHOD

The **factoring method** applies the **zero product property**:

WORDS	**MATH**
If a product is zero, then at least one of its factors has to be zero.	If $B \cdot C = 0$, then $B = 0$ or $C = 0$ or both.

Consider $(x - 3)(x + 2) = 0$. The zero product property says that $x - 3 = 0$ or $x + 2 = 0$, which leads to $x = -2$ or $x = 3$. The solution set is $\{-2, 3\}$.

When a quadratic equation is written in the standard form $ax^2 + bx + c = 0$, it may be possible to factor the left side of the equation as a product of two first-degree polynomials. We use the zero product property and set each linear factor equal to zero. We solve the resulting two linear equations to obtain the solutions of the quadratic equation.

EXAMPLE 1 **Solving a Quadratic Equation by Factoring**

Solve the equation $x^2 - 6x - 16 = 0$.

Solution:

The quadratic equation is already in standard form.	$x^2 - 6x - 16 = 0$
Factor the left side into a product of two linear factors.	$(x - 8)(x + 2) = 0$
If a product equals zero, one of its factors has to be equal to zero.	$x - 8 = 0$ or $x + 2 = 0$
Solve both linear equations.	$\boxed{x = 8 \quad \text{or} \quad x = -2}$

The solution set is $\boxed{\{-2, 8\}}$.

YOUR TURN Solve the quadratic equation $x^2 + x - 20 = 0$ by factoring.

■ **Answer:** The solution is $x = -5, 4$. The solution set is $\{-5, 4\}$.

▼ CAUTION

Do not divide by a variable (because the value of that variable may be zero). Bring all terms to one side first and then factor.

Technology Tip

Use a graphing utility to display graphs of $y_1 = 2x^2$ and $y_2 = 3x$.

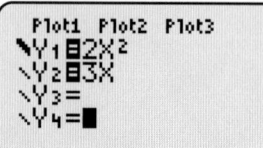

The x-coordinates of the points of intersection are the solutions to this equation.

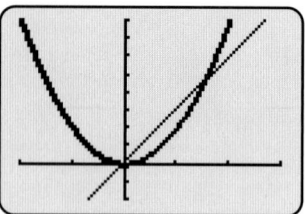

EXAMPLE 2   Solving a Quadratic Equation by Factoring

Solve the equation $2x^2 = 3x$.

COMMON MISTAKE

The common mistake here is dividing both sides by x, which is not allowed because x might be 0.

✪ CORRECT

Write the equation in standard form by subtracting $3x$.

$$2x^2 - 3x = 0$$

Factor the left side.

$$x(2x - 3) = 0$$

Use the zero product property and set each factor equal to zero.

$$x = 0 \quad \text{or} \quad 2x - 3 = 0$$

Solve each linear equation.

$$\boxed{x = 0 \quad \text{or} \quad x = \frac{3}{2}}$$

The solution set is $\boxed{\left\{0, \frac{3}{2}\right\}}$.

✖ INCORRECT

Write the original equation.

$$2x^2 = 3x$$

The **error** occurs here when both sides are divided by x.

$$2x = 3$$

In Example 2, the root $x = 0$ is lost when the original quadratic equation is divided by x. Remember to put the equation in standard form first and then factor.

Square Root Method

The square root of 16, $\sqrt{16}$, is 4, *not* ± 4. In the Appendix, the **principal square root** is discussed. The solutions to $x^2 = 16$, however, are $x = -4$ and $x = 4$. Let us now investigate quadratic equations that do not have a first-degree term. They have the form

$$ax^2 + c = 0 \qquad a \neq 0$$

The method we use to solve such equations uses the square root property.

SQUARE ROOT PROPERTY

WORDS	MATH
If an expression squared is equal to a constant, then that expression is equal to the positive or negative square root of the constant.	If $x^2 = P$, then $x = \pm\sqrt{P}$.

Note: The variable squared must be isolated first (coefficient equal to 1).

EXAMPLE 3   Using the Square Root Property

Solve the equation $3x^2 - 27 = 0$.

Solution:

Add 27 to both sides.	$3x^2 = 27$
Divide both sides by 3.	$x^2 = 9$
Apply the square root property.	$x = \pm\sqrt{9} = \pm 3$

The solution set is $\boxed{\{-3, 3\}}$.

If we alter Example 3 by changing subtraction to addition, we see in Example 4 that we get imaginary roots, as opposed to real roots which is reviewed in the Appendix.

EXAMPLE 4   Using the Square Root Property

Solve the equation $3x^2 + 27 = 0$.

Solution:

Subtract 27 from both sides.	$3x^2 = -27$
Divide by 3.	$x^2 = -9$
Apply the square root property.	$x = \pm\sqrt{-9}$
Simplify.	$x = \pm i\sqrt{9} = \pm 3i$

The solution set is $\boxed{\{-3i, 3i\}}$.

■ **YOUR TURN**   Solve the equations:

$\quad$ **a.** $y^2 - 147 = 0$ **b.** $v^2 + 64 = 0$

■ **Answer:**
a. The solution is $y = \pm 7\sqrt{3}$. The solution set is $\left\{-7\sqrt{3}, 7\sqrt{3}\right\}$.
b. The solution is $v = \pm 8i$. The solution set is $\{-8i, 8i\}$.

EXAMPLE 5   Using the Square Root Property

Solve the equation $(x - 2)^2 = 16$.

Solution:

If an expression squared is 16, then the expression equals $\pm\sqrt{16}$.

$$(x - 2) = \pm\sqrt{16}$$

Separate into two equations.

$$x - 2 = \sqrt{16} \quad \text{or} \quad x - 2 = -\sqrt{16}$$
$$x - 2 = 4 \qquad\qquad x - 2 = -4$$
$$x = 6 \qquad\qquad\quad x = -2$$

The solution set is $\boxed{\{-2, 6\}}$.

It is acceptable notation to keep the equations together.

$$(x - 2) = \pm\sqrt{16}$$
$$x - 2 = \pm 4$$
$$x = 2 \pm 4$$
$$\boxed{x = -2, 6}$$

Completing the Square

Factoring and the square root method are two efficient, quick procedures for solving many quadratic equations. However, some equations, such as $x^2 - 10x - 3 = 0$, cannot be solved directly by these methods. A more general procedure to solve this kind of equation is called **completing the square**. The idea behind completing the square is to transform any standard quadratic equation $ax^2 + bx + c = 0$ into the form $(x + A)^2 = B$, where A and B are constants and the left side, $(x + A)^2$, has the form of a **perfect square**. This last equation can then be solved by the square root method. How do we transform the first equation into the second equation?

Note that the above-mentioned example, $x^2 - 10x - 3 = 0$, cannot be factored into expressions in which all numbers are integers (or even rational numbers). We can, however, transform this quadratic equation into a form that contains a perfect square.

WORDS	MATH
Write the original equation.	$x^2 - 10x - 3 = 0$
Add 3 to both sides.	$x^2 - 10x = 3$
Add 25 to both sides.*	$x^2 - 10x + 25 = 3 + 25$
The left side can be written as a perfect square.	$(x - 5)^2 = 28$
Apply the square root method.	$x - 5 = \pm\sqrt{28}$
Add 5 to both sides.	$x = 5 \pm 2\sqrt{7}$

*Why did we add 25 to both sides? Recall that $(x - c)^2 = x^2 - 2xc + c^2$. In this case $c = 5$ in order for $-2xc = -10x$. Therefore, the desired perfect square $(x - 5)^2$ results in $x^2 - 10x + 25$. Applying this product, we see that $+25$ is needed. If the coefficient of x^2 is 1, a systematic approach is to take the coefficient of the first degree term of $x^2 - 10x - 3 = 0$, which is -10. Divide -10 by 2 to get -5; then square -5 to get 25.

SOLVING A QUADRATIC EQUATION BY COMPLETING THE SQUARE

WORDS	MATH
Express the quadratic equation in the following form.	$x^2 + bx = c$
Divide b by 2 and square the result, then add the square to both sides.	$x^2 + bx + \left(\dfrac{b}{2}\right)^2 = c + \left(\dfrac{b}{2}\right)^2$
Write the left side of the equation as a perfect square.	$\left(x + \dfrac{b}{2}\right)^2 = c + \left(\dfrac{b}{2}\right)^2$
Solve using the square root method.	

EXAMPLE 6   Completing the Square

Solve the quadratic equation $x^2 + 8x - 3 = 0$ by completing the square.

Solution:

Add 3 to both sides.

$$x^2 + 8x = 3$$

Add $\left(\frac{1}{2} \cdot 8\right)^2 = 4^2$ to both sides.

$$x^2 + 8x + 4^2 = 3 + 4^2$$

Write the left side as a perfect square and simplify the right side.

$$(x + 4)^2 = 19$$

Apply the square root method to solve.

$$x + 4 = \pm\sqrt{19}$$

Subtract 4 from both sides.

$$\boxed{x = -4 \pm \sqrt{19}}$$

The solution set is $\boxed{\left\{-4 - \sqrt{19}, -4 + \sqrt{19}\right\}}$.

Technology Tip

Graph $y_1 = x^2 + 8x - 3$.

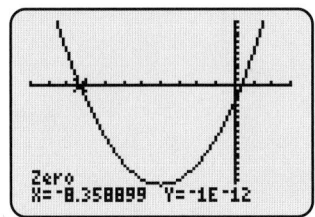

The x-intercepts are the solutions to this equation.

In Example 6, the leading coefficient (the coefficient of the x^2 term) is 1. When the leading coefficient is not 1, start by first dividing the equation by that leading coefficient.

Study Tip

When the leading coefficient is not 1, start by first dividing the equation by that leading coefficient.

EXAMPLE 7   Completing the Square When the Leading Coefficient Is Not Equal to 1

Solve the equation $3x^2 - 12x + 13 = 0$ by completing the square.

Solution:

Divide by the leading coefficient, 3.

$$x^2 - 4x + \frac{13}{3} = 0$$

Collect the variables to one side of the equation and constants to the other side.

$$x^2 - 4x = -\frac{13}{3}$$

Add $\left(-\frac{4}{2}\right)^2 = 4$ to both sides.

$$x^2 - 4x + 4 = -\frac{13}{3} + 4$$

Write the left side of the equation as a perfect square and simplify the right side.

$$(x - 2)^2 = -\frac{1}{3}$$

Solve using the square root method.

$$x - 2 = \pm\sqrt{-\frac{1}{3}}$$

Simplify.

$$x = 2 \pm i\sqrt{\frac{1}{3}}$$

Rationalize the denominator (Appendix).

$$\boxed{x = 2 - \frac{i\sqrt{3}}{3}, \ x = 2 + \frac{i\sqrt{3}}{3}}$$

The solution set is $\left\{2 - \frac{i\sqrt{3}}{3}, 2 + \frac{i\sqrt{3}}{3}\right\}$.

■ **YOUR TURN** Solve the equation $2x^2 - 4x + 3 = 0$ by completing the square.

Technology Tip

Graph $y_1 = 3x^2 - 12x + 13$.

The graph does not cross the x-axis, so there is no real solution to this equation.

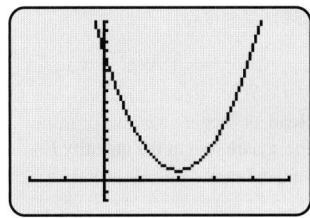

■ **Answer:** The solution is $x = 1 \pm \frac{i\sqrt{2}}{2}$. The solution set is $\left\{1 - \frac{i\sqrt{2}}{2}, 1 + \frac{i\sqrt{2}}{2}\right\}$.

The Quadratic Formula

Let us now consider the most general quadratic equation:

$$ax^2 + bx + c = 0 \qquad a \neq 0$$

We can solve this equation by completing the square.

WORDS	MATH
Divide the equation by the leading coefficient a.	$x^2 + \dfrac{b}{a}x + \dfrac{c}{a} = 0$
Subtract $\dfrac{c}{a}$ from both sides.	$x^2 + \dfrac{b}{a}x = -\dfrac{c}{a}$
Square half of $\dfrac{b}{a}$ and add the result $\left(\dfrac{b}{2a}\right)^2$ to both sides.	$x^2 + \dfrac{b}{a}x + \left(\dfrac{b}{2a}\right)^2 = \left(\dfrac{b}{2a}\right)^2 - \dfrac{c}{a}$
Write the left side of the equation as a perfect square and the right side as a single fraction.	$\left(x + \dfrac{b}{2a}\right)^2 = \dfrac{b^2 - 4ac}{4a^2}$
Solve using the square root method.	$x + \dfrac{b}{2a} = \pm\sqrt{\dfrac{b^2 - 4ac}{4a^2}}$
Subtract $\dfrac{b}{2a}$ from both sides and simplify the radical.	$x = -\dfrac{b}{2a} \pm \dfrac{\sqrt{b^2 - 4ac}}{2a}$
Write as a single fraction.	$x = \dfrac{-b \pm \sqrt{b^2 - 4ac}}{2a}$

We have derived the **quadratic formula**.

THE QUADRATIC FORMULA

If $ax^2 + bx + c = 0$, $a \neq 0$, then the solution is

$$x = \dfrac{-b \pm \sqrt{b^2 - 4ac}}{2a}$$

Note: The quadratic equation must be in standard form ($ax^2 + bx + c = 0$) in order to identify the parameters:

$$a\text{—coefficient of } x^2 \qquad b\text{—coefficient of } x \qquad c\text{—constant}$$

We read this formula as *negative b plus or minus the square root of the quantity b squared minus 4ac all over 2a*. It is important to note that negative b could be positive (if b is negative). For this reason, an alternate form is "opposite b. . . ." The quadratic formula should be memorized and used when simpler methods (factoring and the square root method) cannot be used. The quadratic formula works for *any* quadratic equation.

EXAMPLE 8   Using the Quadratic Formula and Finding Two Distinct Real Roots

Use the quadratic formula to solve the quadratic equation $x^2 - 4x - 1 = 0$.

Solution:

For this problem, $a = 1$, $b = -4$, and $c = -1$.

Write the quadratic formula.

$$x = \frac{-b \pm \sqrt{b^2 - 4ac}}{2a}$$

Use parentheses to avoid losing a minus sign.

$$x = \frac{-(\square) \pm \sqrt{(\square)^2 - 4(\square)(\square)}}{2(\square)}$$

Substitute values for a, b, and c into the parentheses.

$$x = \frac{-(-4) \pm \sqrt{(-4)^2 - 4(1)(-1)}}{2(1)}$$

Simplify. $x = \dfrac{4 \pm \sqrt{16 + 4}}{2} = \dfrac{4 \pm \sqrt{20}}{2} = \dfrac{4 \pm 2\sqrt{5}}{2} = \dfrac{4}{2} \pm \dfrac{2\sqrt{5}}{2} = \boxed{2 \pm \sqrt{5}}$

The solution set $\boxed{\{2 - \sqrt{5}, 2 + \sqrt{5}\}}$ contains two distinct real numbers.

▪ **YOUR TURN** Use the quadratic formula to solve the quadratic equation $x^2 + 6x - 2 = 0$.

Study Tip

Using parentheses as placeholders helps avoid ± errors.

$$x = \frac{-b \pm \sqrt{b^2 - 4ac}}{2a}$$

$$x = \frac{-(\square) \pm \sqrt{(\square)^2 - 4(\square)(\square)}}{2(\square)}$$

▪ **Answer:** The solution is $x = -3 \pm \sqrt{11}$. The solution set is $\{-3 - \sqrt{11}, -3 + \sqrt{11}\}$.

EXAMPLE 9   Using the Quadratic Formula and Finding Two Complex Roots

Use the quadratic formula to solve the quadratic equation $x^2 + 8 = 4x$.

Solution:

Write this equation in standard form $x^2 - 4x + 8 = 0$ in order to identify $a = 1$, $b = -4$, and $c = 8$.

Write the quadratic formula.

$$x = \frac{-b \pm \sqrt{b^2 - 4ac}}{2a}$$

Use parentheses to avoid overlooking a minus sign.

$$x = \frac{-(\square) \pm \sqrt{(\square)^2 - 4(\square)(\square)}}{2(\square)}$$

Substitute the values for a, b, and c into the parentheses.

$$x = \frac{-(-4) \pm \sqrt{(-4)^2 - 4(1)(8)}}{2(1)}$$

Simplify. $x = \dfrac{4 \pm \sqrt{16 - 32}}{2} = \dfrac{4 \pm \sqrt{-16}}{2} = \dfrac{4 \pm 4i}{2} = \dfrac{4}{2} \pm \dfrac{4i}{2} = \boxed{2 \pm 2i}$

The solution set $\boxed{\{2 - 2i, 2 + 2i\}}$ contains two complex numbers. Note that they are complex conjugates of each other.

▪ **YOUR TURN** Use the quadratic formula to solve the quadratic equation $x^2 + 2 = 2x$.

▪ **Answer:** The solution set is $\{1 - i, 1 + i\}$.

EXAMPLE 10   Using the Quadratic Formula and Finding One Repeated Real Root

Use the quadratic formula to solve the quadratic equation $4x^2 - 4x + 1 = 0$.

Solution:

Identify a, b, and c. $\qquad\qquad\qquad\qquad$ $a = 4, b = -4, c = 1$

Write the quadratic formula. $\qquad\qquad$ $x = \dfrac{-b \pm \sqrt{b^2 - 4ac}}{2a}$

Use parentheses to avoid losing a minus sign. $\qquad\qquad$ $x = \dfrac{-(\square) \pm \sqrt{(\square)^2 - 4(\square)(\square)}}{2(\square)}$

Substitute values $a = 4, b = -4, c = 1$. $\qquad$ $x = \dfrac{-(-4) \pm \sqrt{(-4)^2 - 4(4)(1)}}{2(4)}$

Simplify. $\qquad\qquad\qquad\qquad$ $x = \dfrac{4 \pm \sqrt{16 - 16}}{8} = \dfrac{4 \pm 0}{8} = \dfrac{1}{2}$

The solution set is a repeated real root $\left\{\dfrac{1}{2}\right\}$.

Note: This quadratic also could have been solved by factoring: $(2x - 1)^2 = 0$.

■ **Answer:** $\left\{\dfrac{1}{3}\right\}$

■ **YOUR TURN** Use the quadratic formula to solve the quadratic equation $9x^2 - 6x + 1 = 0$.

TYPES OF SOLUTIONS

The expression inside the radical, $b^2 - 4ac$, is called the **discriminant**. The discriminant gives important information about the corresponding solutions or roots of $ax^2 + bx + c = 0$, where a, b, and c are real numbers.

$b^2 - 4ac$	SOLUTIONS (ROOTS)
Positive	Two distinct real roots
0	One real root (a double or repeated root)
Negative	Two complex roots (complex conjugates)

In Example 8, the discriminant is positive and the solution has two distinct real roots. In Example 9, the discriminant is negative and the solution has two complex (conjugate) roots. In Example 10, the discriminant is zero and the solution has one repeated real root.

Applications Involving Quadratic Equations

In Section 0.1, we developed a procedure for solving word problems involving linear equations. The procedure is the same for applications involving quadratic equations. The only difference is that the mathematical equations will be quadratic, as opposed to linear.

EXAMPLE 11 Stock Value

From 1999 to 2001 the price of Abercrombie & Fitch's (ANF) stock was approximately given by $P = 0.2t^2 - 5.6t + 50.2$, where P is the price of stock in dollars, t is in months, and $t = 1$ corresponds to January 1999. When was the value of the stock worth $30?

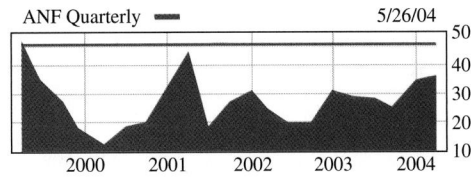

Solution:

STEP 1 Identify the question.
> When is the price of the stock equal to $30?

STEP 2 Make notes.
> Stock price:
$$P = 0.2t^2 - 5.6t + 50.2$$
$$P = 30$$

STEP 3 Set up an equation.
$$0.2t^2 - 5.6t + 50.2 = 30$$

STEP 4 Solve the equation.
> Subtract 30 from both sides.
$$0.2t^2 - 5.6t + 20.2 = 0$$

> Solve for t using the quadratic formula.
$$t = \frac{-(-5.6) \pm \sqrt{(-5.6)^2 - 4(0.2)(20.2)}}{2(0.2)}$$

> Simplify.
$$t \approx \frac{5.6 \pm 3.9}{0.4} \approx 4.25, 23.75$$

Rounding these two numbers, we find that $t \approx 4$ and $t \approx 24$. Since $t = 1$ corresponds to January 1999, these two solutions correspond to $\boxed{\text{April 1999 and December 2000}}$.

STEP 5 Check the solution.
> Look at the figure. The horizontal axis represents the year (2000 corresponds to January 2000), and the vertical axis represents the stock price. Estimating when the stock price is approximately $30, we find April 1999 and December 2000.

Technology Tip

The graphing utility screen for
$$\frac{-(-5.6) \pm \sqrt{(-5.6)^2 - 4(0.2)(20.2)}}{2(0.2)}$$

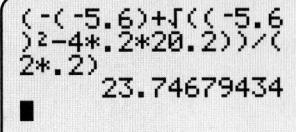

Study Tip

Dimensions such as length and width are distances, which are defined as positive quantities. Although the mathematics may yield both positive and negative values, the negative values are excluded.

SECTION
0.2 SUMMARY

The four methods for solving quadratic equations are factoring, the square root method, completing the square, and the quadratic formula. Factoring and the square root method are the quickest and easiest but cannot always be used. The quadratic formula and completing the square work for all quadratic equations. A quadratic equation can have three types of solutions: two distinct real roots, one real root (repeated), or two complex roots (conjugates of each other).

SECTION
0.2 EXERCISES

■ **SKILLS**

In Exercises 1–22, solve by factoring.

1. $x^2 - 5x + 6 = 0$ **2.** $v^2 + 7v + 6 = 0$ **3.** $p^2 - 8p + 15 = 0$ **4.** $u^2 - 2u - 24 = 0$

5. $x^2 = 12 - x$ **6.** $11x = 2x^2 + 12$ **7.** $16x^2 + 8x = -1$ **8.** $3x^2 + 10x - 8 = 0$

9. $9y^2 + 1 = 6y$ **10.** $4x = 4x^2 + 1$ **11.** $8y^2 = 16y$ **12.** $3A^2 = -12A$

13. $9p^2 = 12p - 4$ **14.** $4u^2 = 20u - 25$ **15.** $x^2 - 9 = 0$ **16.** $16v^2 - 25 = 0$

17. $x(x + 4) = 12$ **18.** $3t^2 - 48 = 0$ **19.** $2p^2 - 50 = 0$ **20.** $5y^2 - 45 = 0$

21. $3x^2 = 12$ **22.** $7v^2 = 28$

In Exercises 23–34, solve using the square root method.

23. $p^2 - 8 = 0$ **24.** $y^2 - 72 = 0$ **25.** $x^2 + 9 = 0$ **26.** $v^2 + 16 = 0$

27. $(x - 3)^2 = 36$ **28.** $(x - 1)^2 = 25$ **29.** $(2x + 3)^2 = -4$ **30.** $(4x - 1)^2 = -16$

31. $(5x - 2)^2 = 27$ **32.** $(3x + 8)^2 = 12$ **33.** $(1 - x)^2 = 9$ **34.** $(1 - x)^2 = -9$

In Exercises 35–46, solve by completing the square.

35. $x^2 + 2x = 3$ **36.** $y^2 + 8y - 2 = 0$ **37.** $t^2 - 6t = -5$ **38.** $x^2 + 10x = -21$

39. $y^2 - 4y + 3 = 0$ **40.** $x^2 - 7x + 12 = 0$ **41.** $2p^2 + 8p = -3$ **42.** $2x^2 - 4x + 3 = 0$

43. $2x^2 - 7x + 3 = 0$ **44.** $3x^2 - 5x - 10 = 0$ **45.** $\dfrac{x^2}{2} - 2x = \dfrac{1}{4}$ **46.** $\dfrac{t^2}{3} + \dfrac{2t}{3} + \dfrac{5}{6} = 0$

In Exercises 47–58, solve using the quadratic formula.

47. $t^2 + 3t - 1 = 0$ **48.** $t^2 + 2t = 1$ **49.** $s^2 + s + 1 = 0$ **50.** $2s^2 + 5s = -2$

51. $3x^2 - 3x - 4 = 0$ **52.** $4x^2 - 2x = 7$ **53.** $x^2 - 2x + 17 = 0$ **54.** $4m^2 + 7m + 8 = 0$

55. $5x^2 + 7x = 3$ **56.** $3x^2 + 5x = -11$ **57.** $\frac{1}{4}x^2 + \frac{2}{3}x - \frac{1}{2} = 0$ **58.** $\frac{1}{4}x^2 - \frac{2}{3}x - \frac{1}{3} = 0$

In Exercises 59–74, solve using any method.

59. $v^2 - 8v = 20$ **60.** $v^2 - 8v = -20$ **61.** $t^2 + 5t - 6 = 0$ **62.** $t^2 + 5t + 6 = 0$

63. $(x + 3)^2 = 16$ **64.** $(x + 3)^2 = -16$ **65.** $(p - 2)^2 = 4p$ **66.** $(u + 5)^2 = 16u$

67. $8w^2 + 2w + 21 = 0$ **68.** $8w^2 + 2w - 21 = 0$ **69.** $3p^2 - 9p + 1 = 0$ **70.** $3p^2 - 9p - 1 = 0$

71. $\frac{2}{3}t^2 - \frac{4}{3}t = \frac{1}{5}$ **72.** $\frac{1}{2}x^2 + \frac{2}{3}x = \frac{2}{5}$ **73.** $x^2 - 0.1x = 0.12$ **74.** $y^2 - 0.5y = -0.06$

■ **APPLICATIONS**

75. Stock Value. From June 2003 until April 2004 JetBlue airlines stock (JBLU) was approximately worth $P = -4t^2 + 80t - 360$, where P denotes the price of the stock in dollars and t corresponds to months, with $t = 1$ corresponding to January 2003. During what months was the stock equal to $24?

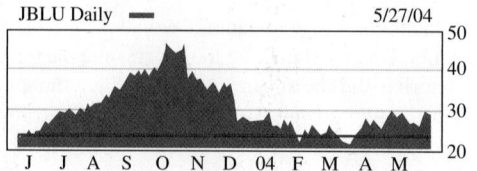

76. Stock Value. From November 2003 until March 2004 Wal-Mart Stock (WMT) was approximately worth $P = 2t^2 - 12t + 70$, where P denotes the price of the stock in dollars and t corresponds to months, with $t = 1$ corresponding to November 2003. During what months was the stock equal to $60?

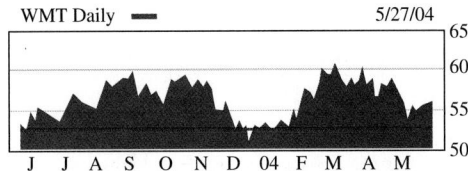

77. Television. A standard 32-in. television has a 32 in. diagonal and a 25 in. width. What is the height of the 32-in. television?

78. Television. A 42-in. LCD television has a 42 in. diagonal and a 20 in. height. What is the width of the 42-in. LCD television?

79. Numbers. Find two consecutive numbers such that their product is 306.

80. Numbers. Find two consecutive odd integers such that their product is 143.

81. Geometry. The area of a rectangle is 135 square feet. The width is 6 feet less than the length. Find the dimensions of the rectangle.

82. Geometry. A rectangle has an area of 31.5 square meters. If the length is 2 meters more than twice the width, find dimensions of the rectangle.

83. Geometry. A triangle has a height that is 2 more than 3 times the base and an area of 60 square units. Find the base and height.

84. Geometry. A square's side is increased by 3 yards, which corresponds to an increase in the area by 69 square yards. How many yards is the side of the initial square?

85. Falling Objects. If a person drops a water balloon off the rooftop of a 100-foot building, the height of the water balloon is given by the equation $h = -16t^2 + 100$, where t is in seconds. When will the water balloon hit the ground?

86. Falling Objects. If the person in Exercise 85 throws the water balloon downward with a speed of 5 feet per second, the height of the water balloon is given by the equation $h = -16t^2 - 5t + 100$, where t is in seconds. When will the water balloon hit the ground?

87. Gardening. A square garden has an area of 900 square feet. If a sprinkler (with a circular pattern) is placed in the center of the garden, what is the minimum radius of spray the sprinkler would need in order to water all of the garden?

88. Sports. A baseball diamond is a square. The distance from base to base is 90 feet. What is the distance from home plate to second base?

89. Volume. A flat square piece of cardboard is used to construct an open box. Cutting a 1 foot by 1 foot square off of each corner and folding up the edges will yield an open box (assuming these edges are taped together). If the desired volume of the box is 9 cubic feet, what are the dimensions of the original square piece of cardboard?

90. Volume. A rectangular piece of cardboard whose length is twice its width is used to construct an open box. Cutting a 1 foot by 1 foot square off of each corner and folding up the edges will yield an open box. If the desired volume is 12 cubic feet, what are the dimensions of the original rectangular piece of cardboard?

91. Gardening. A landscaper has planted a rectangular garden that measures 8 feet by 5 feet. He has ordered 1 cubic yard (27 cubic feet) of stones for a border along the outside of the garden. If the border needs to be 4 inches deep and he wants to use all of the stones, how wide should the border be?

92. Gardening. A gardener has planted a semicircular rose garden with a radius of 6 feet, and 2 cubic yards of mulch (1 cubic yard = 27 cubic feet) is being delivered. Assuming she uses all of the mulch, how deep will the layer of mulch be?

93. Work. Lindsay and Kimmie, working together, can balance the financials for the Kappa Kappa Gamma sorority in 6 days. Lindsay by herself can complete the job in 5 days less than Kimmie. How long will it take Lindsay to complete the job by herself?

94. Work. When Jack cleans the house, it takes him 4 hours. When Ryan cleans the house, it takes him 6 hours. How long would it take both of them if they worked together?

■ CATCH THE MISTAKE

In Exercises 95–98, explain the mistake that is made.

95. $t^2 - 5t - 6 = 0$
$(t - 3)(t - 2) = 0$
$t = 2, 3$

96. $(2y - 3)^2 = 25$
$2y - 3 = 5$
$2y = 8$
$y = 4$

97. $16a^2 + 9 = 0$
$16a^2 = -9$
$a^2 = -\dfrac{9}{16}$
$a = \pm\sqrt{\dfrac{9}{16}}$
$a = \pm\dfrac{3}{4}$

98. $2x^2 - 4x = 3$
$2(x^2 - 2x) = 3$
$2(x^2 - 2x + 1) = 3 + 1$
$2(x - 1)^2 = 4$
$(x - 1)^2 = 2$
$x - 1 = \pm\sqrt{2}$
$x = 1 \pm\sqrt{2}$

▪ CONCEPTUAL

In Exercises 99–102, determine whether the following statements are true or false.

99. The equation $(3x + 1)^2 = 16$ has the same solution set as the equation $3x + 1 = 4$.

100. The quadratic equation $ax^2 + bx + c = 0$ can be solved by the square root method only if $b = 0$.

101. All quadratic equations can be solved exactly.

102. The quadratic formula can be used to solve any quadratic equation.

103. Write a quadratic equation in standard form that has $x = a$ as a repeated real root. Alternate solutions are possible.

104. Write a quadratic equation in standard form that has $x = bi$ as a root. Alternate solutions are possible.

105. Write a quadratic equation in standard form that has the solution set $\{2, 5\}$. Alternate solutions are possible.

106. Write a quadratic equation in standard form that has the solution set $\{-3, 0\}$. Alternate solutions are possible.

In Exercises 107–110, solve for the indicated variable in terms of other variables.

107. Solve $s = \frac{1}{2}gt^2$ for t.

108. Solve $A = P(1 + r)^2$ for r.

109. Solve $a^2 + b^2 = c^2$ for c.

110. Solve $P = EI - RI^2$ for I.

111. Solve the equation by factoring: $x^4 - 4x^2 = 0$.

112. Solve the equation by factoring: $3x - 6x^2 = 0$.

113. Solve the equation using factoring by grouping: $x^3 + x^2 - 4x - 4 = 0$.

114. Solve the equation using factoring by grouping: $x^3 + 2x^2 - x - 2 = 0$.

▪ CHALLENGE

115. Show that the sum of the roots of a quadratic equation is equal to $-\dfrac{b}{a}$.

116. Show that the product of the roots of a quadratic equation is equal to $\dfrac{c}{a}$.

117. Write a quadratic equation in standard form whose solution set is $\{3 - \sqrt{5}, 3 + \sqrt{5}\}$. Alternate solutions are possible.

118. Write a quadratic equation in standard form whose solution set is $\{2 - i, 2 + i\}$. Alternate solutions are possible.

119. **Aviation.** An airplane takes 1 hour longer to go a distance of 600 miles flying against a headwind than on the return trip with a tailwind. If the speed of the wind is a constant 50 mph for both legs of the trip, find the speed of the plane in still air.

120. **Boating.** A speedboat takes 1 hour longer to go 24 miles up a river than to return. If the boat cruises at 10 mph in still water, what is the rate of the current?

121. Find a quadratic equation whose two distinct real roots are the negatives of the two distinct real roots of the equation $ax^2 + bx + c = 0$.

122. Find a quadratic equation whose two distinct real roots are the reciprocals of the two distinct real roots of the equation $ax^2 + bx + c = 0$.

123. A small jet and a 757 leave Atlanta at 1 P.M. The small jet is traveling due west. The 757 is traveling due south. The speed of the 757 is 100 mph faster than that of the small jet. At 3 P.M. the planes are 1000 miles apart. Find the average speed of each plane. (Assume there is no wind.)

124. Two boats leave Key West at noon. The smaller boat is traveling due west. The larger boat is traveling due south. The speed of the larger boat is 10 mph faster than that of the smaller boat. At 3 P.M. the boats are 150 miles apart. Find the average speed of each boat. (Assume there is no current.)

▪ TECHNOLOGY

125. Solve the equation $x^2 - x = 2$ by first writing in standard form and then factoring. Now plot both sides of the equation in the same viewing screen ($y_1 = x^2 - x$ and $y_2 = 2$). At what x-values do these two graphs intersect? Do those points agree with the solution set you found?

126. Solve the equation $x^2 - 2x = -2$ by first writing in standard form and then using the quadratic formula. Now plot both sides of the equation in the same viewing screen ($y_1 = x^2 - 2x$ and $y_2 = -2$). Do these graphs intersect? Does this agree with the solution set you found?

127. **a.** Solve the equation $x^2 - 2x = b$, $b = 8$ by first writing in standard form. Now plot both sides of the equation in the same viewing screen ($y_1 = x^2 - 2x$ and $y_2 = b$). At what x-values do these two graphs intersect? Do those points agree with the solution set you found?

   b. Repeat (a) for $b = -3, -1, 0$, and 5.

128. **a.** Solve the equation $x^2 + 2x = b$, $b = 8$ by first writing in standard form. Now plot both sides of the equation in the same viewing screen ($y_1 = x^2 + 2x$ and $y_2 = b$). At what x-values do these two graphs intersect? Do those points agree with the solution set you found?

   b. Repeat (a) for $b = -3, -1, 0$, and 5.

SKILLS OBJECTIVES

- Solve rational equations.
- Solve radical equations.
- Solve equations that are quadratic in form.
- Solve equations that are factorable.
- Solve absolute value equations.

CONCEPTUAL OBJECTIVES

- Transform a difficult equation into a simpler linear or quadratic equation.
- Recognize the need to check solutions when the transformation process may produce extraneous solutions.
- Realize that not all polynomial equations are factorable.

Rational Equations

A **rational equation** is an equation that contains one or more rational expressions (Appendix). Some rational equations can be transformed into linear or quadratic equations that you can then solve, but as you will see momentarily, you must be certain that the solution to the resulting linear or quadratic equation also satisfies the original rational equation.

EXAMPLE 1 **Solving a Rational Equation That Can Be Reduced to a Linear Equation**

Solve the equation $\dfrac{2}{3x} + \dfrac{1}{2} = \dfrac{4}{x} + \dfrac{4}{3}$.

Solution:

State the excluded values (those which make any denominator equal 0).

$$\frac{2}{3x} + \frac{1}{2} = \frac{4}{x} + \frac{4}{3} \qquad x \neq 0$$

Multiply *each term* by the LCD, $6x$.

$$6x\left(\frac{2}{3x}\right) + 6x\left(\frac{1}{2}\right) = 6x\left(\frac{4}{x}\right) + 6x\left(\frac{4}{3}\right)$$

Simplify both sides.

$$4 + 3x = 24 + 8x$$

Subtract 4.

$$\frac{-4 \qquad\qquad -4}{3x = 20 + 8x}$$

Subtract $8x$.

$$\frac{-8x \qquad\qquad -8x}{-5x = 20}$$

Divide by -5.

$$\boxed{x = -4}$$

Since $x = -4$ satisfies the original equation, the solution set is $\{-4\}$.

■ **YOUR TURN** Solve the equation $\dfrac{3}{y} + 2 = \dfrac{7}{2y}$.

Technology Tip

Use a graphing utility to display graphs of $y_1 = \dfrac{2}{3x} + \dfrac{1}{2}$ and $y_2 = \dfrac{4}{x} + \dfrac{4}{3}$.

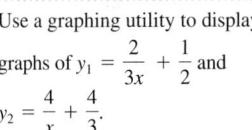

The x-coordinate of the point of intersection is the solution to the equation $\dfrac{2}{3x} + \dfrac{1}{2} = \dfrac{4}{x} + \dfrac{4}{3}$.

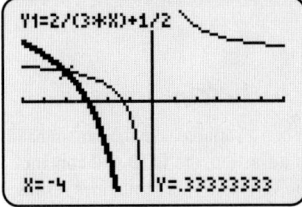

Study Tip

Since dividing by 0 is not defined, we exclude values of the variable that correspond to a denominator equaling 0.

■ **Answer:** The solution is $y = \frac{1}{4}$. The solution set is $\left\{\frac{1}{4}\right\}$.

Technology Tip

Use a graphing utility to display graphs of $y_1 = \dfrac{3x}{x-1} + 2$ and $y_2 = \dfrac{3}{x-1}$.

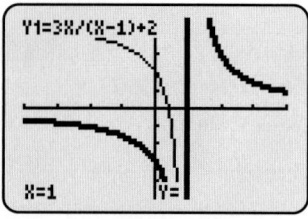

The x-coordinate of the point of intersection is the solution to the equation $\dfrac{3x}{x-1} + 2 = \dfrac{3}{x-1}$.

No intersection implies no solution.

■ **Answer:** no solution

EXAMPLE 2   Solving Rational Equations: Eliminating Extraneous Solutions

Solve the equation $\dfrac{3x}{x-1} + 2 = \dfrac{3}{x-1}$.

Solution:

State the excluded values (those which make any denominator equal 0).

$$\frac{3x}{x-1} + 2 = \frac{3}{x-1} \qquad x \neq 1$$

Eliminate the fractions by multiplying each term by the LCD, $x - 1$.

$$\frac{3x}{x-1} \cdot (x-1) + 2 \cdot (x-1) = \frac{3}{x-1} \cdot (x-1)$$

Simplify.

$$\frac{3x}{x-1} \cdot (x-1) + 2 \cdot (x-1) = \frac{3}{x-1} \cdot (x-1)$$

$$3x + 2(x-1) = 3$$

Distribute the 2. $3x + 2x - 2 = 3$

Combine x terms on the left. $5x - 2 = 3$

Add 2 to both sides. $5x = 5$

Divide both sides by 5. $x = 1$

It may seem that $x = 1$ is the solution. However, the original equation had the restriction $x \neq 1$. Therefore, $x = 1$ is an extraneous solution and must be eliminated as a possible solution.

Thus, the equation $\dfrac{3x}{x-1} + 2 = \dfrac{3}{x-1}$ has $\boxed{\text{no solution}}$.

■ **YOUR TURN**   Solve the equation $\dfrac{2x}{x-2} - 3 = \dfrac{4}{x-2}$.

Study Tip

When a variable is in the denominator of a fraction, the LCD will contain the variable. This sometimes results in an extraneous solution.

In order to find a *least* common denominator of more complicated expressions, it is useful to first factor the denominators to identify common multiples.

Rational equation: $\dfrac{1}{3x-3} + \dfrac{1}{2x-2} = \dfrac{1}{x^2 - x}$

Factor the denominators: $\dfrac{1}{3(x-1)} + \dfrac{1}{2(x-1)} = \dfrac{1}{x(x-1)}$

LCD: $6x(x-1)$

EXAMPLE 3 **Solving Rational Equations**

Solve the equation $\dfrac{1}{3x + 18} - \dfrac{1}{2x + 12} = \dfrac{1}{x^2 + 6x}$.

Solution:

Factor the denominators. $\qquad\qquad \dfrac{1}{3(x + 6)} - \dfrac{1}{2(x + 6)} = \dfrac{1}{x(x + 6)}$

State the excluded values. $\qquad\qquad x \neq 0, -6$

Multiply the equation by the LCD, $6x(x + 6)$.

$$6x(x + 6) \cdot \dfrac{1}{3(x + 6)} - 6x(x + 6) \cdot \dfrac{1}{2(x + 6)} = 6x(x + 6) \cdot \dfrac{1}{x(x + 6)}$$

Divide out the common factors.

$$6x\cancel{(x + 6)} \cdot \dfrac{1}{3\cancel{(x + 6)}} - 6x\cancel{(x + 6)} \cdot \dfrac{1}{2\cancel{(x + 6)}} = 6\cancel{x}\cancel{(x + 6)} \cdot \dfrac{1}{\cancel{x}\cancel{(x + 6)}}$$

Simplify. $\qquad\qquad\qquad\qquad\qquad 2x - 3x = 6$

Solve the linear equation. $\qquad\qquad\qquad x = -6$

Since one of the excluded values is $x \neq -6$, we say that $x = -6$ is an extraneous solution. Therefore, this rational equation has $\boxed{\text{no solution}}$.

■ YOUR TURN Solve the equation $\dfrac{2}{x} + \dfrac{1}{x + 1} = -\dfrac{1}{x(x + 1)}$.

■ Answer: no solution

EXAMPLE 4 **Solving a Rational Equation That Can Be Reduced to a Quadratic Equation**

Solve the equation $1 + \dfrac{3}{x^2 - 2x} = \dfrac{2}{x - 2}$.

Solution:

Factor the denominators. $\qquad\qquad\qquad 1 + \dfrac{3}{x(x - 2)} = \dfrac{2}{x - 2}$

State the excluded values
(those which make any
denominator equal to 0). $\qquad\qquad\qquad x \neq 0, 2$

Multiply each term by
the LCD, $x(x - 2)$. $\qquad 1 \cdot x(x - 2) + \dfrac{3}{x(x - 2)} \cdot x(x - 2) = \dfrac{2}{(x - 2)} \cdot x(x - 2)$

Divide out the common factors. $\quad x(x - 2) + \dfrac{3}{\cancel{x(x - 2)}}\cancel{x(x - 2)} = \dfrac{2}{\cancel{(x - 2)}}x\cancel{(x - 2)}$

Simplify. $\qquad\qquad\qquad\qquad\qquad x(x - 2) + 3 = 2x$

Eliminate the parentheses. $\qquad\qquad\qquad x^2 - 2x + 3 = 2x$

Write the quadratic equation
in standard form. $\qquad\qquad\qquad\qquad x^2 - 4x + 3 = 0$

Factor. $\qquad\qquad\qquad\qquad\qquad (x - 3)(x - 1) = 0$

Apply the zero product property. $\qquad \boxed{x = 3 \quad \text{or} \quad x = 1}$

Since $x = 3$ or $x = 1$ both satisfy the original equation, the solution set is $\{1, 3\}$.

Radical Equations

Radical equations are equations in which the variable is inside a radical (that is, under a square root, cube root, or higher root). Examples of radical equations follow:

$$\sqrt{x-3} = 2 \qquad \sqrt{2x+3} = x \qquad \sqrt{x+2} + \sqrt{7x+2} = 6$$

Often you can transform a radical equation into a simple linear or quadratic equation. Sometimes the transformation process yields **extraneous solutions**, or apparent solutions that may solve the transformed problem but are not solutions of the original radical equation. Therefore, it is very important to check your answers in the original equation.

EXAMPLE 5   Solving an Equation Involving a Radical

Solve the equation $\sqrt{x-3} = 2$.

Solution:

Square both sides of the equation. $\qquad\qquad (\sqrt{x-3})^2 = 2^2$

Simplify. $\qquad\qquad\qquad\qquad\qquad\qquad x - 3 = 4$

Solve the resulting linear equation. $\qquad\qquad \boxed{x = 7}$

The solution set is $\{7\}$.

Check: $\sqrt{7-3} = \sqrt{4} = 2$

■**Answer:** $p = 7$ or $\{7\}$

■ **YOUR TURN** Solve the equation $\sqrt{3p+4} = 5$.

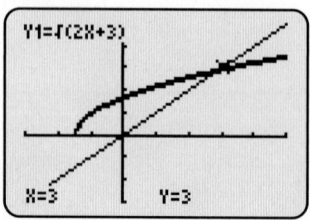
EXAMPLE 6   Solving an Equation Involving a Radical

Solve the equation $\sqrt{2x+3} = x$.

Solution:

Square both sides of the equation. $\qquad\qquad (\sqrt{2x+3})^2 = x^2$

Simplify. $\qquad\qquad\qquad\qquad\qquad\qquad 2x + 3 = x^2$

Write the quadratic equation in standard form. $\qquad x^2 - 2x - 3 = 0$

Factor. $\qquad\qquad\qquad\qquad\qquad (x-3)(x+1) = 0$

Use the zero product property. $\qquad\qquad x = 3 \quad \text{or} \quad x = -1$

Check these values to see whether they *both* make the original equation statement true.

$x = 3$: $\quad \sqrt{2(3)+3} = 3 \Rightarrow \sqrt{6+3} = 3 \Rightarrow \sqrt{9} = 3 \Rightarrow 3 = 3$ (True)

$x = -1$: $\quad \sqrt{2(-1)+3} = -1 \Rightarrow \sqrt{-2+3} = -1 \Rightarrow \sqrt{1} = -1 \Rightarrow 1 \neq -1$ (False)

The solution is $\boxed{x = 3}$. The solution set is $\{3\}$.

■**Answer:** $t = 4$ or $\{4\}$

■ **YOUR TURN** Solve the equation $\sqrt{12+t} = t$.

■**Answer:** $x = -1$ and $x = -3$ or $\{-3, -1\}$

■ **YOUR TURN** Solve the equation $\sqrt{2x+6} = x + 3$.

Examples 5 and 6 contained only one radical each. We transformed the radical equation into a linear (Example 5) or quadratic (Example 6) equation with one step. The next example contains two radicals. Our technique will be to isolate one radical on one side of the equation with the other radical on the other side of the equation.

★ CORRECT

Square the expression.

$$\left(3 + \sqrt{x + 2}\right)^2$$

Write the square as a product of two factors.

$$\left(3 + \sqrt{x + 2}\right)\left(3 + \sqrt{x + 2}\right)$$

Use the FOIL method.

$$9 + 6\sqrt{x + 2} + (x + 2)$$

✖ INCORRECT

Square the expression.

$$\left(3 + \sqrt{x + 2}\right)^2$$

The **error** occurs here when only individual terms are squared.

$$\neq 9 + (x + 2)$$

EXAMPLE 7 Solving an Equation with More Than One Radical

Solve the equation $\sqrt{x + 2} + \sqrt{7x + 2} = 6$.

Solution:

Subtract $\sqrt{x + 2}$ from both sides.

$$\sqrt{7x + 2} = 6 - \sqrt{x + 2}$$

Square both sides.

$$\left(\sqrt{7x + 2}\right)^2 = \left(6 - \sqrt{x + 2}\right)^2$$

Simplify.

$$7x + 2 = \left(6 - \sqrt{x + 2}\right)\left(6 - \sqrt{x + 2}\right)$$

Multiply the expressions on the right side of the equation.

$$7x + 2 = 36 - 12\sqrt{x + 2} + (x + 2)$$

Isolate the term with the radical on the left side.

$$12\sqrt{x + 2} = 36 + x + 2 - 7x - 2$$

Combine like terms on the right side.

$$12\sqrt{x + 2} = 36 - 6x$$

Divide by 6.

$$2\sqrt{x + 2} = 6 - x$$

Square both sides.

$$4(x + 2) = (6 - x)^2$$

Simplify.

$$4x + 8 = 36 - 12x + x^2$$

Rewrite the quadratic equation in standard form.

$$x^2 - 16x + 28 = 0$$

Factor.

$$(x - 14)(x - 2) = 0$$

Solve.

$$x = 14 \quad \text{and} \quad x = 2$$

The apparent solutions are 2 and 14. Note that $x = 14$ does not satisfy the original equation; therefore, it is extraneous. The solution is $\boxed{x = 2}$. The solution set is {2}.

■ **YOUR TURN** Solve the equation $\sqrt{x - 4} = 5 - \sqrt{x + 1}$.

Technology Tip

Use a graphing utility to display graphs of

$$y_1 = \sqrt{x + 2} + \sqrt{7x + 2}$$

and $y_2 = 6$.

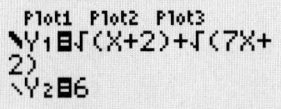

The x-coordinate of the point of intersection is the solution to the equation

$$\sqrt{x + 2} + \sqrt{7x + 2} = 6.$$

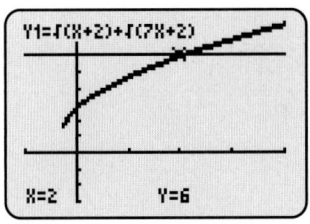

■ **Answer:** $x = 8$ or {8}

PROCEDURE FOR SOLVING RADICAL EQUATIONS

Step 1: Isolate the term with a radical on one side.
Step 2: Raise both (*entire*) sides of the equation to the power that will eliminate this radical, and simplify the equation.
Step 3: If a radical remains, repeat Steps 1 and 2.
Step 4: Solve the resulting linear or quadratic equation.
Step 5: Check the solutions and eliminate any extraneous solutions.

Note: If there is more than one radical in the equation, it does not matter which radical is isolated first.

Equations Quadratic in Form: *u*-Substitution

Equations that are higher order or that have fractional powers often can be transformed into a quadratic equation by introducing a *u*-substitution. When this is the case, we say that equations are **quadratic in form**. In the table below, the two original equations are quadratic in form because they can be transformed into a quadratic equation given the correct substitution.

ORIGINAL EQUATION	SUBSTITUTION	NEW EQUATION
$x^4 - 3x^2 - 4 = 0$	$u = x^2$	$u^2 - 3u - 4 = 0$
$t^{2/3} + 2t^{1/3} + 1 = 0$	$u = t^{1/3}$	$u^2 + 2u + 1 = 0$

For example, the equation $x^4 - 3x^2 - 4 = 0$ is a fourth-degree equation in x. How did we know that $u = x^2$ would transform the original equation into a quadratic equation? If we rewrite the original equation as $(x^2)^2 - 3(x^2) - 4 = 0$, the expression in parentheses is the *u*-substitution.

Let us introduce the substitution $u = x^2$. Note that squaring both sides implies $u^2 = x^4$. We then replace x^2 in the original equation with u, and x^4 in the original equation with u^2, which leads to a quadratic equation in u: $u^2 - 3u - 4 = 0$.

WORDS	**MATH**
Solve for x.	$x^4 - 3x^2 - 4 = 0$
Introduce *u*-substitution.	$u = x^2$ (Note that $u^2 = x^4$.)
Write the quadratic equation in u.	$u^2 - 3u - 4 = 0$
Factor.	$(u - 4)(u + 1) = 0$
Solve for u.	$u = 4$ or $u = -1$
Transform back to x, $u = x^2$.	$x^2 = 4$ or $x^2 = -1$
Solve for x.	$\boxed{x = \pm 2 \quad \text{or} \quad x = \pm i}$

The solution set is $\{\pm 2, \pm i\}$.

It is important to correctly determine the appropriate substitution in order to arrive at an equation quadratic in form. For example, $t^{2/3} + 2t^{1/3} + 1 = 0$ is an original equation given in the above table. If we rewrite this equation as $\left(t^{1/3}\right)^2 + 2\left(t^{1/3}\right) + 1 = 0$, then it becomes apparent that the correct substitution is $u = t^{1/3}$, which transforms the equation in t into a quadratic equation in u: $u^2 + 2u + 1 = 0$.

PROCEDURE FOR SOLVING EQUATIONS QUADRATIC IN FORM

Step 1: Identify the substitution.
Step 2: Transform the equation into a quadratic equation.
Step 3: Solve the quadratic equation.
Step 4: Apply the substitution to rewrite the solution in terms of the original variable.
Step 5: Solve the resulting equation.
Step 6: Check the solutions in the original equation.

EXAMPLE 8 Solving an Equation Quadratic in Form with Negative Exponents

Find the solutions to the equation $x^{-2} - x^{-1} - 12 = 0$.

Solution:

Rewrite the original equation. $\left(x^{-1}\right)^2 - \left(x^{-1}\right) - 12 = 0$

Determine the u-substitution. $u = x^{-1}$ (Note that $u^2 = x^{-2}$.)

The original equation in x corresponds to a quadratic equation in u. $u^2 - u - 12 = 0$

Factor. $(u - 4)(u + 3) = 0$

Solve for u. $u = 4$ or $u = -3$

The most common mistake is forgetting to transform back to x.

Transform back to x. Let $u = x^{-1}$. $x^{-1} = 4$ or $x^{-1} = -3$

Write x^{-1} as $\frac{1}{x}$. $\frac{1}{x} = 4$ or $\frac{1}{x} = -3$

Solve for x. $\boxed{x = \frac{1}{4}}$ or $\boxed{x = -\frac{1}{3}}$

The solution set is $\left\{-\frac{1}{3}, \frac{1}{4}\right\}$.

■ **YOUR TURN** Find the solutions to the equation $x^{-2} - x^{-1} - 6 = 0$.

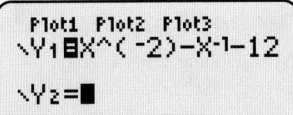
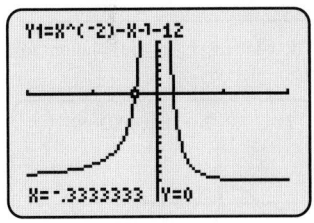
■ **Answer:** The solution is $x = -\frac{1}{2}$ or $x = \frac{1}{3}$. The solution set is $\left\{-\frac{1}{2}, \frac{1}{3}\right\}$.

EXAMPLE 9 **Solving an Equation Quadratic in Form with Fractional Exponents**

Find the solutions to the equation $x^{2/3} - 3x^{1/3} - 10 = 0$.

Solution:

Rewrite the original equation. $\qquad (x^{1/3})^2 - 3x^{1/3} - 10 = 0$

Identify the substitution as $u = x^{1/3}$. $\qquad u^2 - 3u - 10 = 0$

Factor. $\qquad (u - 5)(u + 2) = 0$

Solve for u. $\qquad u = 5 \quad$ or $\quad u = -2$

Let $u = x^{1/3}$ again. $\qquad x^{1/3} = 5 \qquad x^{1/3} = -2$

Cube both sides of the equations. $\qquad (x^{1/3})^3 = (5)^3 \qquad (x^{1/3})^3 = (-2)^3$

Simplify. $\qquad \boxed{x = 125} \qquad \boxed{x = -8}$

The solution set is $\boxed{\{-8, 125\}}$, which a check will confirm.

■ **Answer:** $t = 9$ or $\{9\}$

■ **YOUR TURN** Find the solution to the equation $2t - 5t^{1/2} - 3 = 0$.

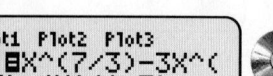

Factorable Equations

Technology Tip

Use a graphing utility to graph $y_1 = x^{7/3} - 3x^{4/3} - 4x^{1/3}$.

```
Plot1 Plot2 Plot3
\Y1❚X^(7/3)-3X^(
4/3)-4X^(1/3)
```

The x-intercepts are the solutions to this equation.

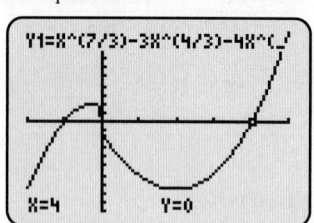

Some equations (both polynomial and with rational exponents) that are factorable can be solved using the zero product property.

EXAMPLE 10 **Solving an Equation with Rational Exponents by Factoring**

Solve the equation $x^{7/3} - 3x^{4/3} - 4x^{1/3} = 0$.

Solution:

Factor the left side of the equation. $\qquad x^{1/3}(x^2 - 3x - 4) = 0$

Factor the quadratic expression. $\qquad x^{1/3}(x - 4)(x + 1) = 0$

Apply the zero product property. $\qquad x^{1/3} = 0 \quad$ or $\quad x - 4 = 0 \quad$ or $\quad x + 1 = 0$

Solve for x. $\qquad \boxed{x = 0} \quad$ or $\quad \boxed{x = 4} \quad$ or $\quad \boxed{x = -1}$

The solution set is $\{-1, 0, 4\}$.

EXAMPLE 11 **Solving a Polynomial Equation Using Factoring by Grouping**

Solve the equation $x^3 + 2x^2 - x - 2 = 0$.

Solution:

Factor by grouping (Appendix). $\qquad (x^3 - x) + (2x^2 - 2) = 0$

Identify the common factors. $\qquad x(x^2 - 1) + 2(x^2 - 1) = 0$

Factor. $\qquad (x + 2)(x^2 - 1) = 0$

Factor the quadratic expression. $\qquad (x + 2)(x - 1)(x + 1) = 0$

Apply the zero product property. $x + 2 = 0$ or $x - 1 = 0$ or $x + 1 = 0$

Solve for x. $\boxed{x = -2}$ or $\boxed{x = 1}$ or $\boxed{x = -1}$

The solution set is $\{-2, -1, 1\}$.

■ **YOUR TURN** Solve the equation $x^3 + x^2 - 4x - 4 = 0$.

Absolute Value Equations

The **absolute value** of a real number can be interpreted algebraically and graphically.

DEFINITION   **Absolute Value**

The **absolute value** of a real number a, denoted by the symbol $|a|$, is defined by

$$|a| = \begin{cases} a, & \text{if } a \geq 0 \\ -a, & \text{if } a < 0 \end{cases}$$

Study Tip

Algebraically: $|x| = 5$ implies
$x = -5$ or $x = 5$.
Graphically: -5 and 5 are 5 units
from 0.

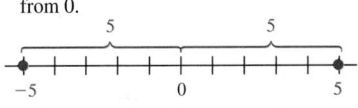

When absolute value is involved in algebraic equations, we interpret the definition of absolute value as follows.

DEFINITION   **Absolute Value Equation**

If $|x| = a$, then $x = -a$ or $x = a$, where $a \geq 0$.

In words, "If the absolute value of a number is a, then that number equals $-a$ or a." For example, the equation $|x| = 7$ is true if $x = -7$ or $x = 7$. We say the equation $|x| = 7$ has the solution set $\{-7, 7\}$. *Note:* $|x| = -3$ does not have a solution because there is no value of x such that its absolute value is -3.

EXAMPLE 12   **Solving an Absolute Value Equation**

Solve the equation $|x - 3| = 8$ algebraically and graphically.

Solution:

Using the absolute value equation definition, we see that if the absolute value of an expression is 8, then that expression is either -8 or 8. Rewrite as two equations:

$$x - 3 = -8 \qquad \text{or} \qquad x - 3 = 8$$
$$x = -5 \qquad\qquad\qquad x = 11$$

The solution set is $\boxed{\{-5, 11\}}$.

Graph: The absolute value equation $|x - 3| = 8$ is interpreted as "What numbers are 8 units away from 3 on the number line?" We find that 8 units to the right of 3 is 11 and 8 units to the left of 3 is -5.

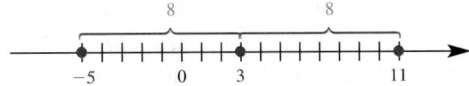

Technology Tip

Use a graphing utility to display
graphs of $y_1 = |x - 3|$ and $y_2 = 8$.

The x-coordinates of the points of
intersection are the solutions to
$|x - 3| = 8$.

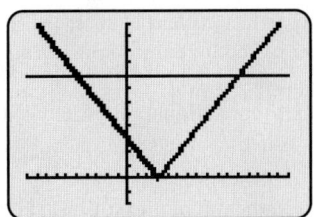

■ **YOUR TURN** Solve the equation $|x + 5| = 7$.

EXAMPLE 13 Solving an Absolute Value Equation

Solve the equation $2 - 3|x - 1| = -4|x - 1| + 7$.

Solution:

Isolate the absolute value expressions to one side.

Add $4|x - 1|$ to both sides. $2 + |x - 1| = 7$

Subtract 2 from both sides. $|x - 1| = 5$

If the absolute value of an expression
is equal to 5, then the expression is $x - 1 = -5$ or $x - 1 = 5$
equal to either -5 or 5. $x = -4$ $x = 6$

The solution set is $\boxed{\{-4, 6\}}$.

■ Answer: $x = -4$ or $x = 12$.
The solution set is $\{-4, 12\}$.

■ YOUR TURN Solve the equation $3 - 2|x - 4| = -3|x - 4| + 11$.

EXAMPLE 14 Solving a Quadratic Absolute Value Equation

Solve the equation $|5 - x^2| = 1$.

Solution:

If the absolute value of an expression $5 - x^2 = -1$ or $5 - x^2 = 1$
is 1, that expression is either -1 or 1, $-x^2 = -6$ $-x^2 = -4$
which leads to two equations. $x^2 = 6$ $x^2 = 4$

$x = \pm\sqrt{6}$ $x = \pm\sqrt{4} = \pm 2$

The solution set is $\boxed{\{\pm 2, \pm\sqrt{6}\}}$.

■ Answer: $x = \pm\sqrt{5}$ or $x = \pm 3$.
The solution set is $\{\pm\sqrt{5}, \pm 3\}$.

■ YOUR TURN Solve the equation $|7 - x^2| = 2$.

SECTION
0.3 SUMMARY

Rational equations, radical equations, equations quadratic in form, factorable equations, and absolute value equations can often be solved by transforming them into simpler linear or quadratic equations.

■ **Rational Equations:** Multiply the entire equation by the LCD. Solve the resulting equation (if it is linear or quadratic). Check for extraneous solutions.

■ **Radical Equations:** Isolate the term containing a radical and raise it to the appropriate power that will eliminate the radical. If there is more than one radical, it does not matter which

radical is isolated first. Raising radical equations to powers may cause extraneous solutions, so check each solution.

■ **Equations Quadratic in Form:** Identify the u-substitution that transforms the equation into a quadratic equation. Solve the quadratic equation and then remember to transform back to the original variable.

■ **Factorable Equations:** Look for a factor common to all terms or factor by grouping.

■ **Absolute Value Equations:** Transform the absolute value equation into two equations that do not involve absolute value.

SECTION
0.3 EXERCISES

■ SKILLS

In Exercises 1–20, specify any values that must be excluded from the solution set and then solve the rational equation.

1. $\dfrac{x}{x-2} + 5 = \dfrac{2}{x-2}$

2. $\dfrac{n}{n-5} + 2 = \dfrac{n}{n-5}$

3. $\dfrac{2p}{p-1} = 3 + \dfrac{2}{p-1}$

4. $\dfrac{4t}{t+2} = 3 - \dfrac{8}{t+2}$

5. $\dfrac{3x}{x+2} - 4 = \dfrac{2}{x+2}$

6. $\dfrac{5y}{2y-1} - 3 = \dfrac{12}{2y-1}$

7. $\dfrac{1}{n} + \dfrac{1}{n+1} = \dfrac{-1}{n(n+1)}$

8. $\dfrac{1}{x} + \dfrac{1}{x-1} = \dfrac{1}{x(x-1)}$

9. $\dfrac{3}{a} - \dfrac{2}{a+3} = \dfrac{9}{a(a+3)}$

10. $\dfrac{1}{c-2} + \dfrac{1}{c} = \dfrac{2}{c(c-2)}$

11. $\dfrac{n-5}{6n-6} = \dfrac{1}{9} - \dfrac{n-3}{4n-4}$

12. $\dfrac{5}{m} + \dfrac{3}{m-2} = \dfrac{6}{m(m-2)}$

13. $\dfrac{2}{5x+1} = \dfrac{1}{2x-1}$

14. $\dfrac{3}{4n-1} = \dfrac{2}{2n-5}$

15. $\dfrac{t-1}{1-t} = \dfrac{3}{2}$

16. $\dfrac{2-x}{x-2} = \dfrac{3}{4}$

17. $x + \dfrac{12}{x} = 7$

18. $x - \dfrac{10}{x} = -3$

19. $\dfrac{4(x-2)}{x-3} + \dfrac{3}{x} = \dfrac{-3}{x(x-3)}$

20. $\dfrac{5}{y+4} = 4 + \dfrac{3}{y-2}$

In Exercises 21–46, solve the radical equation for the given variable.

21. $\sqrt{u+1} = -4$

22. $-\sqrt{3-2u} = 9$

23. $\sqrt[3]{5x+2} = 3$

24. $\sqrt[3]{1-x} = -2$

25. $(4y+1)^{1/3} = -1$

26. $(5x-1)^{1/3} = 4$

27. $\sqrt{12+x} = x$

28. $x = \sqrt{56-x}$

29. $y = 5\sqrt{y}$

30. $\sqrt{y} = \dfrac{y}{4}$

31. $s = 3\sqrt{s-2}$

32. $-2s = \sqrt{3-s}$

33. $\sqrt{2x+6} = x+3$

34. $\sqrt{8-2x} = 2x-2$

35. $\sqrt{1-3x} = x+1$

36. $\sqrt{2-x} = x-2$

37. $\sqrt{x^2-4} = x-1$

38. $\sqrt{25-x^2} = x+1$

39. $\sqrt{x^2-2x-5} = x+1$

40. $\sqrt{2x^2-8x+1} = x-3$

41. $\sqrt{2x-1} - \sqrt{x-1} = 1$

42. $\sqrt{8-x} = 2 + \sqrt{2x+3}$

43. $\sqrt{3x-5} = 7 - \sqrt{x+2}$

44. $\sqrt{x+5} = 1 + \sqrt{x-2}$

45. $\sqrt{2+\sqrt{x}} = \sqrt{x}$

46. $\sqrt{2-\sqrt{x}} = \sqrt{x}$

In Exercises 47–66, solve the equations by introducing a substitution that transforms these equations to quadratic form.

47. $x^{2/3} + 2x^{1/3} = 0$

48. $x^{1/2} - 2x^{1/4} = 0$

49. $x^4 - 3x^2 + 2 = 0$

50. $x^4 - 8x^2 + 16 = 0$

51. $2x^4 + 7x^2 + 6 = 0$

52. $x^8 - 17x^4 + 16 = 0$

53. $4(t-1)^2 - 9(t-1) = -2$

54. $2(1-y)^2 + 5(1-y) - 12 = 0$

55. $x^{-8} - 17x^{-4} + 16 = 0$

56. $2u^{-2} + 5u^{-1} - 12 = 0$

57. $3y^{-2} + y^{-1} - 4 = 0$

58. $5a^{-2} + 11a^{-1} + 2 = 0$

59. $z^{2/5} - 2z^{1/5} + 1 = 0$

60. $2x^{1/2} + x^{1/4} - 1 = 0$

61. $6t^{-2/3} - t^{-1/3} - 1 = 0$

62. $t^{-2/3} - t^{-1/3} - 6 = 0$

63. $3 = \dfrac{1}{(x+1)^2} + \dfrac{2}{(x+1)}$

64. $\dfrac{1}{(x+1)^2} + \dfrac{4}{(x+1)} + 4 = 0$

65. $u^{4/3} - 5u^{2/3} = -4$

66. $u^{4/3} + 5u^{2/3} = -4$

In Exercises 67–82, solve by factoring.

67. $x^3 - x^2 - 12x = 0$

68. $2y^3 - 11y^2 + 12y = 0$

69. $4p^3 - 9p = 0$

70. $25x^3 = 4x$

71. $u^5 - 16u = 0$

72. $t^5 - 81t = 0$

73. $x^3 - 5x^2 - 9x + 45 = 0$

74. $2p^3 - 3p^2 - 8p + 12 = 0$

75. $y(y - 5)^3 - 14(y - 5)^2 = 0$

76. $v(v + 3)^3 - 40(v + 3)^2 = 0$

77. $x^{9/4} - 2x^{5/4} - 3x^{1/4} = 0$

78. $u^{7/3} + u^{4/3} - 20u^{1/3} = 0$

79. $t^{5/3} - 25t^{-1/3} = 0$

80. $4x^{9/5} - 9x^{-1/5} = 0$

81. $y^{3/2} - 5y^{1/2} + 6y^{-1/2} = 0$

82. $4p^{5/3} - 5p^{2/3} - 6p^{-1/3} = 0$

In Exercises 83–104, solve the absolute value equation.

83. $|p - 7| = 3$

84. $|p + 7| = 3$

85. $|4 - y| = 1$

86. $|2 - y| = 11$

87. $|3t - 9| = 3$

88. $|4t + 2| = 2$

89. $|7 - 2x| = 9$

90. $|6 - 3y| = 12$

91. $|1 - 3y| = 1$

92. $|5 - x| = 2$

93. $\left|\frac{2}{3}x - \frac{4}{7}\right| = \frac{5}{3}$

94. $\left|\frac{1}{2}x + \frac{3}{4}\right| = \frac{1}{16}$

95. $|x - 5| + 4 = 12$

96. $|x + 3| - 9 = 2$

97. $2|p + 3| - 15 = 5$

98. $8 - 3|p - 4| = 2$

99. $5|y - 2| - 10 = 4|y - 2| - 3$

100. $3 - |y + 9| = 11 - 3|y + 9|$

101. $|4 - x^2| = 1$

102. $|7 - x^2| = 3$

103. $|x^2 + 1| = 5$

104. $|x^2 - 1| = 5$

■ APPLICATIONS

For Exercises 105–108, refer to this lens law.

The position of the image is found using the thin lens equation

$$\frac{1}{f} = \frac{1}{d_o} + \frac{1}{d_i}$$

where d_o is the distance from the object to the lens, d_i is the distance from the lens to the image, and f is the focal length of the lens.

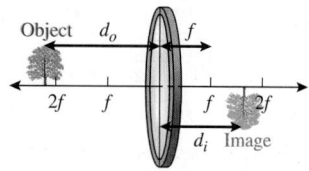

105. Optics. If the focal length of a lens is 3 cm and the image distance is 5 cm from the lens, what is the distance from the object to the lens?

106. Optics. If the focal length of the lens is 8 cm and the image distance is 2 cm from the lens, what is the distance from the object to the lens?

107. Optics. The focal length of a lens is 2 cm. If the image distance from the lens is half the distance from the object to the lens, find the object distance.

108. Optics. The focal length of a lens is 8 cm. If the image distance from the lens is half the distance from the object to the lens, find the object distance.

109. Speed of Sound. A man buys a house with an old well but does not know how deep the well is. To get an estimate, he decides to drop a rock at the opening of the well and count how long it

takes until he hears the splash. The total elapsed time T given by $T = t_1 + t_2$, is the sum of the time it takes for the rock to reach the water, t_1, and the time it takes for the sound of the splash to travel to the top of the well, t_2. The time (seconds) that it takes for the rock to reach the water is given by $t_1 = \dfrac{\sqrt{d}}{4}$, where d is the depth of the well in feet. Since the speed of sound is 1100 ft/s, the time (seconds) it takes for the sound to reach the top of the well is $t_2 = \dfrac{d}{1100}$. If the splash is heard after 3 seconds, how deep is the well?

110. Speed of Sound. If the owner of the house in Exercise 109 forgot to account for the speed of sound, what would he have calculated the depth of the well to be?

111. Physics: Pendulum. The period (T) of a pendulum is related to the length (L) of the pendulum and acceleration due to gravity (g) by the formula $T = 2\pi\sqrt{\dfrac{L}{g}}$. If gravity is 9.8 m/s^2 and the period is 1 second, find the approximate length of the pendulum. Round to the nearest centimeter. (*Note:* 100 cm = 1 m.)

112. Physics: Pendulum. The period (T) of a pendulum is related to the length (L) of the pendulum and acceleration due to gravity (g) by the formula $T = 2\pi\sqrt{\dfrac{L}{g}}$. If gravity is 32 ft/s^2 and the period is 1 second, find the approximate length of the pendulum. Round to the nearest inch. (*Note:* 12 in. = 1 ft.)

For Exercises 113 and 114, refer to the following:

Einstein's special theory of relativity states that time is relative: Time speeds up or slows down, depending on how fast one object is moving with respect to another. For example, a space probe traveling at a velocity v near the speed of light c will have "clocked" a time t hours, but for a stationary observer on Earth that corresponds to a time t_0. The formula governing this relativity is given by

$$t = t_0 \sqrt{1 - \frac{v^2}{c^2}}$$

113. Physics: Special Theory of Relativity. If the time elapsed on a space probe mission is 18 years but the time elapsed on Earth during that mission is 30 years, how fast is the space probe traveling? Give your answer relative to the speed of light.

114. Physics: Special Theory of Relativity. If the time elapsed on a space probe mission is 5 years but the time elapsed on Earth during that mission is 30 years, how fast is the space probe traveling? Give your answer relative to the speed of light.

▪ CATCH THE MISTAKE

In Exercises 115–118, explain the mistake that is made.

115. Solve the equation $\sqrt{3t + 1} = -4$.

    Solution:
$$3t + 1 = 16$$
$$3t = 15$$
$$t = 5$$

This is incorrect. What mistake was made?

116. Solve the equation $x = \sqrt{x + 2}$.

    Solution:
$$x^2 = x + 2$$
$$x^2 - x - 2 = 0$$
$$(x - 2)(x + 1) = 0$$
$$x = -1, x = 2$$

This is incorrect. What mistake was made?

117. Solve the equation $\dfrac{4}{p} - 3 = \dfrac{2}{5p}$.

    Solution:

Cross multiply.
$$(p - 3)2 = 4(5p)$$
$$2p - 6 = 20p$$
$$-6 = 18p$$
$$p = -\frac{6}{18}$$
$$p = -\frac{1}{3}$$

This is incorrect. What mistake was made?

118. Solve the equation $\dfrac{1}{x} + \dfrac{1}{x - 1} = \dfrac{1}{x(x - 1)}$.

    Solution:

Multiply by the LCD, $x(x - 1)$.
$$\frac{x(x - 1)}{x} + \frac{x(x - 1)}{x - 1} = \frac{x(x - 1)}{x(x - 1)}$$

Simplify.
$$(x - 1) + x = 1$$
$$x - 1 + x = 1$$
$$2x = 2$$
$$x = 1$$

This is incorrect. What mistake was made?

▪ CONCEPTUAL

In Exercises 119–120, determine whether each statement is true or false.

119. The solution to the equation $x = \dfrac{1}{1/x}$ is the set of all real numbers.

120. The solution to the equation $\dfrac{1}{(x - 1)(x + 2)} = \dfrac{1}{x^2 + x - 2}$ is the set of all real numbers.

121. Solve for x, given that a, b, and c are real numbers and $c \neq 0$.
$$\frac{a}{x} - \frac{b}{x} = c$$

122. Solve the equation for y: $\dfrac{1}{y - a} + \dfrac{1}{y + a} = \dfrac{2}{y - 1}$.

Does y have any restrictions?

▪ CHALLENGE

123. Solve the equation $\sqrt{x + 6} + \sqrt{11 + x} = 5\sqrt{3 + x}$.

124. Solve the equation $3x^{7/12} - x^{5/6} - 2x^{1/3} = 0$.

125. Solve the equation for x in terms of y: $y = \dfrac{a}{1 + \dfrac{b}{x} + c}$.

126. Solve for t: $\dfrac{t + \dfrac{1}{t}}{\dfrac{1}{t} - 1} = 1$.

127. Solve the equation $\sqrt{x-3} = 4 - \sqrt{x+2}$. Plot both sides of the equation in the same viewing screen, $y_1 = \sqrt{x-3}$ and $y_2 = 4 - \sqrt{x+2}$, and zoom in on the x-coordinate of the point of intersection. Does the graph agree with your solution?

128. Solve the equation $2\sqrt{x+1} = 1 + \sqrt{3-x}$. Plot both sides of the equation in the same viewing screen, $y_1 = 2\sqrt{x+1}$ and $y_2 = 1 + \sqrt{3-x}$, and zoom in on the x-coordinate of the points of intersection. Does the graph agree with your solution?

129. Solve the equation $x^{1/2} = -4x^{1/4} + 21$. Plot both sides of the equation in the same viewing screen, $y_1 = x^{1/2}$ and $y_2 = -4x^{1/4} + 21$. Does the point(s) of intersection agree with your solution?

130. Solve the equation $x^{-1} = 3x^{-2} - 10$. Plot both sides of the equation in the same viewing screen, $y_1 = x^{-1}$ and $y_2 = 3x^{-2} - 10$. Does the point(s) of intersection agree with your solution?

SECTION
0.4 INEQUALITIES

SKILLS OBJECTIVES

- Use interval notation.
- Solve linear inequalities.
- Solve polynomial inequalities.
- Solve rational inequalities.
- Solve absolute value inequalities.

CONCEPTUAL OBJECTIVES

- Apply intersection and union concepts.
- Understand that inequalities may have one solution, no solution, or an interval solution.
- Understand zeros and test intervals.
- Realize that a rational inequality has an implied domain restriction on the variable.

Graphing Inequalities and Interval Notation

We will express solutions to inequalities four ways: an inequality, a solution set, an interval, and a graph. The following are ways of expressing all real numbers greater than or equal to a and less than b:

Inequality Notation	Solution Set	Interval Notation	Graph/Number Line
$a \le x < b$	$\{x \mid a \le x < b\}$	$[a, b)$	

In this example, a is referred to as the **left endpoint** and b is referred to as the **right endpoint**. If an inequality is a strict inequality ($<$ or $>$), then the graph and interval notation use *parentheses*. If it includes an endpoint ($\ge$ or $\le$), then the graph and interval notation use *brackets*. Number lines are drawn with either closed/open circles or brackets/parentheses. In this text the brackets/parentheses notation will be used. Intervals are classified as follows:

Open (,)   Closed [,]   Half open (,] or [,)

LET x BE A REAL NUMBER.

x IS...	INEQUALITY	SET NOTATION	INTERVAL	GRAPH
greater than a and less than b	$a < x < b$	$\{x \mid a < x < b\}$	(a, b)	
greater than or equal to a and less than b	$a \leq x < b$	$\{x \mid a \leq x < b\}$	$[a, b)$	
greater than a and less than or equal to b	$a < x \leq b$	$\{x \mid a < x \leq b\}$	$(a, b]$	
greater than or equal to a and less than or equal to b	$a \leq x \leq b$	$\{x \mid a \leq x \leq b\}$	$[a, b]$	
less than a	$x < a$	$\{x \mid x < a\}$	$(-\infty, a)$	
less than or equal to a	$x \leq a$	$\{x \mid x \leq a\}$	$(-\infty, a]$	
greater than b	$x > b$	$\{x \mid x > b\}$	(b, ∞)	
greater than or equal to b	$x \geq b$	$\{x \mid x \geq b\}$	$[b, \infty)$	
all real numbers	$\mathbb{R}$	$\mathbb{R}$	$(-\infty, \infty)$	

1. *Infinity* (∞) is not a number. It is a symbol that means continuing indefinitely to the right on the number line. Similarly, *negative infinity* ($-\infty$) means continuing indefinitely to the left on the number line. Since both are unbounded, we use a parenthesis, never a bracket.
2. In interval notation, the lower number is always written to the left.

    Write the inequality in interval notation: $-1 \leq x < 3$.

    ⊕ **CORRECT** $[-1, 3)$ ☒ **INCORRECT** $(3, -1]$

EXAMPLE 1   Expressing Inequalities Using Interval Notation and a Graph

Express the following as an inequality, an interval, and a graph:

a. x is greater than -3.
b. x is less than or equal to 5.
c. x is greater than or equal to -1 and less than 4.
d. x is greater than or equal to 0 and less than or equal to 4.

Solution:

Inequality	Interval	Graph
a. $x > -3$	$(-3, \infty)$	
b. $x \leq 5$	$(-\infty, 5]$	
c. $-1 \leq x < 4$	$[-1, 4)$	
d. $0 \leq x \leq 4$	$[0, 4]$	

Since the solutions to inequalities are sets of real numbers, it is useful to discuss two operations on sets called **intersection** and **union**.

DEFINITION   **Union and Intersection**

The **union** of sets A and B, denoted $A \cup B$, is the set formed by combining all the elements in A with all the elements in B.

$$A \cup B = \{x \mid x \text{ is in } A \text{ \textbf{or} } B \text{ \textbf{or} both}\}$$

The **intersection** of sets A and B, denoted $A \cap B$, is the set formed by the elements that are in both A and B.

$$A \cap B = \{x \mid x \text{ is in } A \text{ \textbf{and} } B\}$$

The notation "$x \mid x$ is in" is read "all x such that x is in." The vertical line represents "such that."

EXAMPLE 2   Determining Unions and Intersections: Intervals and Graphs

If $A = [-3, 2]$ and $B = (1, 7)$, determine $A \cup B$ and $A \cap B$. Write these sets in interval notation, and graph.

Solution:

Set	Interval notation	Graph
A	$[-3, 2]$	
B	$(1, 7)$	
$A \cup B$	$[-3, 7)$	
$A \cap B$	$(1, 2]$	

Answer:

$$C \cup D = [-3, 5]$$

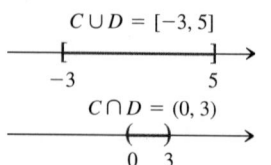

$$C \cap D = (0, 3)$$

■ **YOUR TURN** If $C = [-3, 3)$ and $D = (0, 5]$, find $C \cup D$ and $C \cap D$. Express the intersection and union in interval notation, and graph.

Linear Inequalities

Study Tip

If you multiply or divide an inequality by a negative number, remember to change the direction of the inequality sign.

If we were to solve the linear equation $3x - 2 = 7$, we would add 2 to both sides, divide by 3, and find that $x = 3$ is the solution, the *only* value that makes the equation true. If we were to solve the *linear inequality* $3x - 2 \leq 7$, we would follow the same procedure: Add 2 to both sides, divide by 3, and find that $x \leq 3$, which is an *interval* or *range* of numbers that make the inequality true.

In solving linear inequalities, we follow the same procedures that we used in solving linear equations with one general exception: *If you multiply or divide an inequality by a negative number, then you must change the direction of the inequality sign.*

INEQUALITY PROPERTIES

Procedures That Do Not Change the Inequality Sign

1. Simplifying by eliminating parentheses and collecting like terms.

$$3(x - 6) < 6x - x$$
$$3x - 18 < 5x$$

2. Adding or subtracting the same quantity on both sides.

$$7x + 8 \geq 29$$
$$7x \geq 21$$

3. Multiplying or dividing by the same *positive* real number.

$$5x \leq 15$$
$$x \leq 3$$

Procedures That Change (Reverse) the Inequality Sign

1. Interchanging the two sides of the inequality.

$x \leq 4$ is equivalent to $4 \geq x$

2. Multiplying or dividing by the same *negative* real number.

$-5x \leq 15$ is equivalent to $x \geq -3$

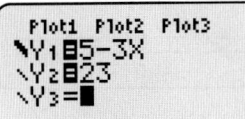
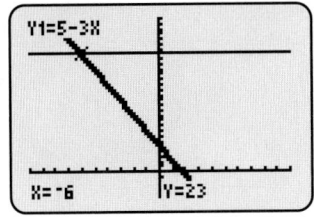

EXAMPLE 3   Solving a Linear Inequality

Solve and graph the inequality $5 - 3x < 23$.

Solution:

Write the original inequality.	$5 - 3x < 23$
Subtract 5 from both sides.	$-3x < 18$
Divide both sides by -3 and reverse the inequality sign.	$\dfrac{-3x}{-3} > \dfrac{18}{-3}$
Simplify.	$x > -6$

Solution set: $\boxed{\{x \mid x > -6\}}$ Interval notation: $\boxed{(-6, \infty)}$ Graph:

■ **YOUR TURN**   Solve the inequality $5 \leq 3 - 2x$. Express the solution in set and interval notation, and graph.

EXAMPLE 4   Solving a Double Linear Inequality

Solve the inequality $-2 < 3x + 4 \leq 16$.

Solution:

This double inequality can be written as two inequalities.	$\overbrace{-2 < 3x + 4 \leq 16}$
Both inequalities must be satisfied.	$-2 < 3x + 4$ and $3x + 4 \leq 16$
Subtract 4 from both sides of each inequality.	$-6 < 3x$ and $3x \leq 12$
Divide each inequality by 3.	$-2 < x$ and $x \leq 4$

Combining these two inequalities gives us $-2 < x \leq 4$ in inequality notation; in interval notation we have $(-2, \infty) \cap (-\infty, 4]$ or $(-2, 4]$.

Use a graphing utility to display graphs of $y_1 = -2$, $y_2 = 3x + 4$, and $y_3 = 16$.

The solutions are the x-values such that the graph of $y_2 = 3x + 4$ is between the graphs of $y_1 = -2$ and $y_3 = 16$ and overlaps that of $y_3 = 16$.

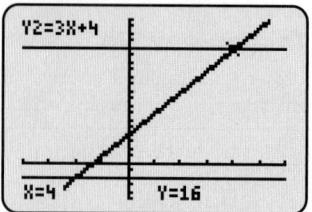

Notice that the steps we took in solving these inequalities individually were identical. This leads us to a **shortcut method** in which we solve them together:

Write the combined inequality.	$-2 < 3x + 4 \le 16$
Subtract 4 from each part.	$-6 < 3x \le 12$
Divide each part by 3.	$-2 < x \le 4$

Interval notation: $\boxed{(-2, 4]}$

EXAMPLE 5   Comparative Shopping

Two car rental companies have advertised weekly specials on full-size cars. Hertz is advertising an \$80 rental fee plus an additional \$.10 per mile. Thrifty is advertising \$60 and \$.20 per mile. How many miles must you drive for the rental car from Hertz to be the better deal?

Solution:

Let x = number of miles driven during the week.

Write the cost for the Hertz rental.	$80 + 0.1x$
Write the cost for the Thrifty rental.	$60 + 0.2x$
Write the inequality if Hertz is less than Thrifty.	$80 + 0.1x < 60 + 0.2x$
Subtract $0.1x$ from both sides.	$80 < 60 + 0.1x$
Subtract 60 from both sides.	$20 < 0.1x$
Divide both sides by 0.1.	$200 < x$

$\boxed{\text{You must drive more than 200 miles for Hertz to be the better deal.}}$

Polynomial Inequalities

A polynomial must pass through zero before its value changes from positive to negative or from negative to positive. **Zeros** of a polynomial are the values of x that make the polynomial equal to zero. These zeros divide the real number line into **test intervals** where the value of the polynomial is either positive or negative. For $x^2 + x - 2 < 0$, if we set the polynomial equal to zero and solve:

$$x^2 + x - 2 = 0$$
$$(x + 2)(x - 1) = 0$$
$$x = -2 \quad \text{or} \quad x = 1$$

we find that $x = -2$ and $x = 1$ are the zeros. These zeros divide the real number line into three test intervals: $(-\infty, -2)$, $(-2, 1)$, and $(1, \infty)$.

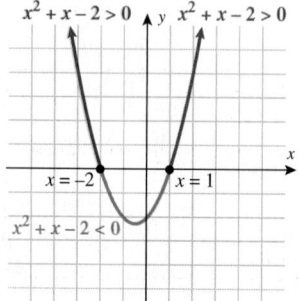

Since the polynomial is equal to zero at $x = -2$ and $x = 1$, we select one real number that lies in each of the three intervals and test to see whether the value of the polynomial at each point is either positive or negative. In this example, we select the real numbers $x = -3$, $x = 0$, and $x = 2$. At this point, there are two ways we can determine whether the value of the polynomial is positive or negative on the interval. One approach is to substitute each of the test points into the polynomial $x^2 + x - 2$.

$x = -3$ $(-3)^2 + (-3) - 2 = 9 - 3 - 2 = 4$ Positive

$x = 0$ $(0)^2 + (0) - 2 = 0 - 0 - 2 = -2$ Negative

$x = 2$ $(2)^2 + (2) - 2 = 4 + 2 - 2 = 4$ Positive

The second approach is to simply determine the sign of the result as opposed to actually calculating the exact number. This alternate approach is often used when the expressions or test points get more complicated to evaluate. The polynomial is written as the product $(x + 2)(x - 1)$; therefore, we simply look for the sign in each set of parentheses.

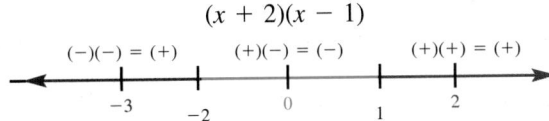

In this second approach we find the same result: $(-\infty, -2)$ and $(1, \infty)$ correspond to a positive value of the polynomial, and $(-2, 1)$ corresponds to a negative value of the polynomial.

In this example, the statement $x^2 + x - 2 < 0$ is true when the value of the polynomial (in factored form), $(x + 2)(x - 1)$, is negative. In the interval $(-2, 1)$, the value of the polynomial is negative. Thus, the solution to the inequality $x^2 + x - 2 < 0$ is $(-2, 1)$. To check the solution, select any number in the interval and substitute it into the original inequality to make sure it makes the statement true. The value $x = -1$ lies in the interval $(-2, 1)$. Upon substituting into the original inequality, we find that $x = -1$ satisfies the inequality $(-1)^2 + (-1) - 2 = -2 < 0$.

PROCEDURE FOR SOLVING POLYNOMIAL INEQUALITIES

Step 1: Write the inequality in *standard form*.
Step 2: Identify zeros of the polynomial.
Step 3: Draw the number line with zeros labeled.
Step 4: Determine the sign of the polynomial in each interval.
Step 5: Identify which interval(s) make the inequality true.
Step 6: Write the solution in interval notation.

Note: Be careful in Step 5. If the original polynomial is <0, then the interval(s) that correspond(s) to the value of the polynomial being negative should be selected. If the original polynomial is >0, then the interval(s) that correspond(s) to the value of the polynomial being positive should be selected.

Technology Tip

Use a graphing utility to display graphs of $y_1 = x^2 - x$ and $y_2 = 12$.

The solutions are the x-values such that the graph of y_1 lies above the graph of y_2.

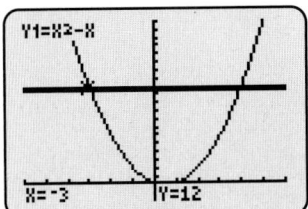

EXAMPLE 6   Solving a Quadratic Inequality

Solve the inequality $x^2 - x > 12$.

Solution:

Write the inequality in standard form. $x^2 - x - 12 > 0$

Factor the left side. $(x + 3)(x - 4) > 0$

Identify the zeros. $(x + 3)(x - 4) = 0$

                                   $x = -3$ or $x = 4$

Draw the number line with the zeros labeled.

Test each interval.

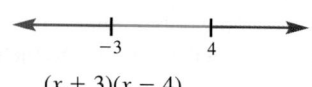

$(x + 3)(x - 4)$

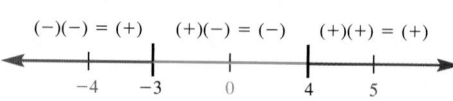

Intervals in which the value of the polynomial is *positive* make this inequality true.

$(-\infty, -3)$ or $(4, \infty)$

Write the solution in interval notation.

$(-\infty, -3) \cup (4, \infty)$

■ **Answer:** $[-1, 6]$

■ **YOUR TURN** Solve the inequality $x^2 - 5x \le 6$ and express the solution in interval notation.

The inequality in Example 6, $x^2 - x > 12$, is a strict inequality, so we use parentheses when we express the solution in interval notation $(-\infty, -3) \cup (4, \infty)$. It is important to note that if we change the inequality sign from $>$ to $\ge$, then the zeros $x = -3$ and $x = 4$ also make the inequality true. Therefore, the solution to $x^2 - x \ge 12$ is $(-\infty, -3] \cup [4, \infty)$.

EXAMPLE 7 Solving a Quadratic Inequality

Solve the inequality $x^2 > -5x$.

▼ **CAUTION**

Do not divide inequalities by a variable.

COMMON MISTAKE

A common mistake is to divide by x. Never divide by a variable, because the value of the variable might be zero. Always start by writing the inequality in standard form and then factor to determine the zeros.

⭐ **CORRECT**

Write the inequality in standard form.

$$x^2 + 5x > 0$$

Factor.

$$x(x + 5) > 0$$

Identify the zeros.

$$x = 0, x = -5$$

Draw the number line and test the intervals.

$x(x + 5)$

$(-)(-) = (+)$ $(-)(+) = (-)$ $(+)(+) = (+)$

$-6 \quad -5 \quad -1 \quad 0 \quad 1$

Intervals in which the value of the polynomial is *positive* satisfy the inequality.

$(-\infty, -5)$ and $(0, \infty)$

Express the solution in interval notation.

$(-\infty, -5) \cup (0, \infty)$

❌ **INCORRECT**

Write the original inequality.

$$x^2 > -5x$$

ERROR:

Divide both sides by x.

$$x > -5$$

Dividing by x is the mistake. If x is negative, the inequality sign must be reversed. What if x is zero?

EXAMPLE 8   Solving a Quadratic Inequality

Solve the inequality $x^2 + 2x < 1$.

Solution:

Write the inequality in standard form.	$x^2 + 2x - 1 < 0$
Identify the zeros.	$x^2 + 2x - 1 = 0$
Apply the quadratic formula.	$x = \dfrac{-2 \pm \sqrt{2^2 - 4(1)(-1)}}{2(1)}$
Simplify.	$x = \dfrac{-2 \pm \sqrt{8}}{2} = \dfrac{-2 \pm 2\sqrt{2}}{2} = -1 \pm \sqrt{2}$

Draw the number line with the intervals labeled.

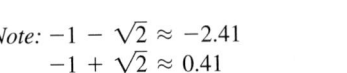

Note: $-1 - \sqrt{2} \approx -2.41$
$-1 + \sqrt{2} \approx 0.41$

Test each interval.

$\left(-\infty, -1 - \sqrt{2}\right)$	$x = -3$:	$(-3)^2 + 2(-3) - 1 = 2 > 0$
$\left(-1 - \sqrt{2}, -1 + \sqrt{2}\right)$	$x = 0$:	$(0)^2 + 2(0) - 1 = -1 < 0$
$\left(-1 + \sqrt{2}, \infty\right)$	$x = 1$:	$(1)^2 + 2(1) - 1 = 2 > 0$

Intervals in which the value of the polynomial is *negative* make this inequality true.

$$\boxed{\left(-1 - \sqrt{2}, -1 + \sqrt{2}\right)}$$

■ **YOUR TURN** Solve the inequality $x^2 - 2x \geq 1$.

EXAMPLE 9   Solving a Polynomial Inequality

Solve the inequality $x^3 - 3x^2 \geq 10x$.

Solution:

Write the inequality in standard form.	$x^3 - 3x^2 - 10x \geq 0$
Factor.	$x(x - 5)(x + 2) \geq 0$
Identify the zeros.	$x = 0, x = 5, x = -2$

Draw the number line with the zeros (intervals) labeled.

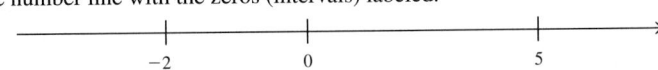

Test each interval.

Intervals in which the value of the polynomial is *positive* make this inequality true.

$$\boxed{[-2, 0] \cup [5, \infty)}$$

■ **YOUR TURN** Solve the inequality $x^3 - x^2 - 6x < 0$.

Technology Tip

Using a graphing utility, graph $y_1 = x^2 + 2x$ and $y_2 = 1$.

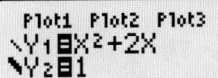

The solutions are the x-values such that the graph of y_1 lies below the graph of y_2.

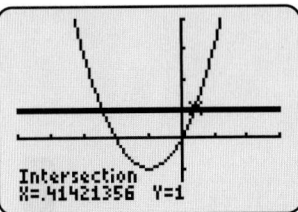

Note that

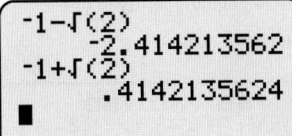

■ **Answer:**
$$\left(-\infty, 1 - \sqrt{2}\,\right] \cup \left[1 + \sqrt{2}, \infty\right)$$

■ **Answer:** $(-\infty, -2) \cup (0, 3)$

Rational Inequalities

A rational expression can change signs if either the numerator or denominator changes signs. In order to go from positive to negative or vice versa, you must pass through zero. To *solve* rational inequalities such as $\dfrac{x-3}{x^2-4} \geq 0$, we use a similar procedure to the one used for solving polynomial inequalities, with one exception. You must eliminate from the solution set values for x that make the denominator equal to zero. In this example, we must eliminate $x = -2$ and $x = 2$ because these values make the denominator equal to zero. Rational inequalities have implied domains. In this example, $x \neq \pm 2$ is a domain restriction and these values ($x = -2$ and $x = 2$) must be eliminated from a possible solution.

We will proceed with a similar procedure involving zeros and test intervals that was outlined for polynomial inequalities. However, in rational inequalities once expressions are combined into a single fraction, any values that make *either* the numerator *or* the denominator equal to zero divide the number line into intervals.

Study Tip

Values that make the denominator equal to zero are always excluded.

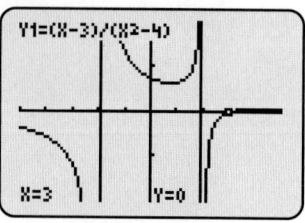

Technology Tip

Use a graphing utility to display the graph of $y_1 = \dfrac{x-3}{x^2-4}$.

```
Plot1 Plot2 Plot3
\Y1冒(X-3)/(X²-4)
```

The solutions are the x-values such that the graph of y_1 lies on top and above the x-axis, excluding $x = \pm 2$.

```
Y1=(X-3)/(X²-4)



X=3          Y=0
```

EXAMPLE 10 Solving a Rational Inequality

Solve the inequality $\dfrac{x-3}{x^2-4} \geq 0$.

Solution:

Factor the denominator. $\qquad\qquad \dfrac{(x-3)}{(x-2)(x+2)} \geq 0$

State the domain restrictions on the variable. $\qquad x \neq 2, x \neq -2$

Identify the zeros of numerator and denominator. $\qquad x = -2, x = 2, x = 3$

Draw the number line and divide into intervals.

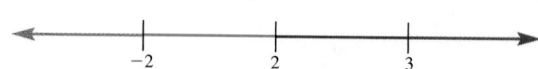

Test the intervals.

$$\dfrac{(x-3)}{(x-2)(x+2)}$$

$$\underset{-3}{\overset{\frac{(-)}{(-)(-)}=(-)}{}} \Big|_{-2} \underset{0}{\overset{\frac{(-)}{(-)(+)}=(+)}{}} \Big|_{2} \underset{2.5}{\overset{\frac{(-)}{(+)(+)}=(-)}{}} \Big|_{3} \underset{4}{\overset{\frac{(+)}{(+)(+)}=(+)}{}}$$

Intervals in which the value of the rational expression is *positive* satisfy this inequality. $\qquad (-2, 2)$ and $(3, \infty)$

Since this inequality is greater than or equal to, we include $x = 3$ in our solution because it satisfies the inequality. However, $x = -2$ and $x = 2$ are not included in the solution because they make the denominator equal to zero.

The solution is $\boxed{(-2, 2) \cup [3, \infty)}$.

■ **Answer:** $[-2, 1)$

■ **YOUR TURN** Solve the inequality $\dfrac{x+2}{x-1} \leq 0$.

EXAMPLE 11   Solving a Rational Inequality

Solve the inequality $\dfrac{x}{x+2} \le 3$.

COMMON MISTAKE

Do not cross multiply.

⊛ **CORRECT**

Subtract 3 from both sides.

$$\frac{x}{x+2} - 3 \le 0$$

Write as a single rational expression.

$$\frac{x - 3(x+2)}{x+2} \le 0$$

Eliminate the parentheses.

$$\frac{x - 3x - 6}{x+2} \le 0$$

Simplify the numerator.

$$\frac{-2x - 6}{x+2} \le 0$$

Factor the numerator.

$$\frac{-2(x+3)}{x+2} \le 0$$

Identify the zeros of the numerator and the denominator.

$$x = -3 \text{ and } x = -2$$

Draw the number line and test the intervals.

$$\frac{-2(x+3)}{x+2} \le 0$$

$$\frac{(-)(-)}{(-)} = (-) \quad \frac{(-)(+)}{(-)} = (+) \quad \frac{(-)(+)}{(+)} = (-)$$

Intervals in which the value of the rational expression is *negative* satisfy the inequality. $(-\infty, -3]$ and $(-2, \infty)$. Note that $x = -2$ is not included in the solution because it makes the denominator zero, and $x = -3$ is included because it satisfies the inequality.

The solution is

$$\boxed{(-\infty, -3] \cup (-2, \infty)}$$

☒ **INCORRECT**

ERROR:
Do not cross multiply.

$$x \le 3(x+2)$$

Absolute Value Inequalities

To solve the inequality $|x| < 3$, look for all real numbers that make this statement true. If we interpret this inequality as distance, we ask *what numbers are less than 3 units from the origin?* We can represent the solution in the following ways:

Inequality notation: $-3 < x < 3$

Interval notation: $(-3, 3)$

Graph:

Similarly, to solve the inequality $|x| \geq 3$, look for all real numbers that make the statement true. If we interpret this inequality as a distance, we ask *what numbers are at least 3 units from the origin?* We can represent the solution in the following three ways:

Inequality notation: $x \leq -3$ or $x \geq 3$

Interval notation: $(-\infty, -3] \cup [3, \infty)$

Graph:

This discussion leads us to the following equivalence relations.

PROPERTIES OF ABSOLUTE VALUE INEQUALITIES

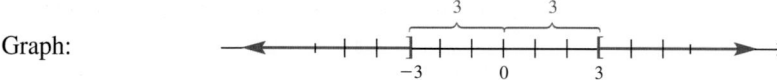

1. $|x| < a$ is equivalent to $-a < x < a$

2. $|x| \leq a$ is equivalent to $-a \leq x \leq a$

3. $|x| > a$ is equivalent to $x < -a$ or $x > a$

4. $|x| \geq a$ is equivalent to $x \leq -a$ or $x \geq a$

It is important to realize that in the above four properties, the variable x can be any algebraic expression.

EXAMPLE 12   Solving an Inequality Involving an Absolute Value

Solve the inequality $|3x - 2| \leq 7$.

Solution:

We apply property (2) and squeeze the absolute value expression between -7 and 7.

$$-7 \leq 3x - 2 \leq 7$$

Add 2 to all three parts.

$$-5 \leq 3x \leq 9$$

Divide all three parts by 3.

$$-\frac{5}{3} \leq x \leq 3$$

The solution in interval notation is $\boxed{\left[-\frac{5}{3}, 3\right]}$.

Graph:

■ **YOUR TURN** Solve the inequality $|2x + 1| < 11$.

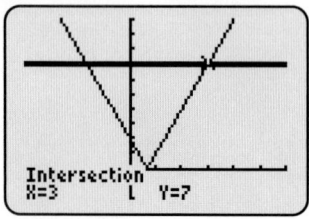

Technology Tip

Use a graphing utility to display graphs of $y_1 = |3x - 2|$ and $y_2 = 7$.

```
Plot1 Plot2 Plot3
\Y1■abs(3X-2)
\Y2■7
```

The values of x where the graph of y_1 lies on top and below the graph of y_2 are the solutions to this inequality.

```
Intersection
X=3          Y=7
```

■ **Answer:**
Inequality notation: $-6 < x < 5$
Interval notation: $(-6, 5)$

It is often helpful to note that for absolute value inequalities,

- *less than* inequalities can be written as a single statement (see Example 12).
- *greater than* inequalities must be written as two statements (see Example 13).

EXAMPLE 13   Solving an Inequality Involving an Absolute Value

Solve the inequality $|1 - 2x| > 5$.

Solution:

Apply property (3).	$1 - 2x < -5$ or $1 - 2x > 5$
Subtract 1 from all expressions.	$-2x < -6$ $-2x > 4$
Divide by -2 and reverse the inequality sign.	$x > 3$ $x < -2$
Express the solution in interval notation.	$\boxed{(-\infty, -2) \cup (3, \infty)}$

Graph: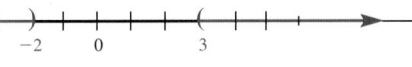

■ **YOUR TURN** Solve the inequality $|5 - 2x| \geq 1$.

■ **Answer:**
Inequality notation: $x \leq 2$ or $x \geq 3$
Interval notation: $(-\infty, 2] \cup [3, \infty)$

Notice that if we change the problem in Example 13 to $|1 - 2x| > -5$, the answer is all real numbers because the absolute value of any expression is greater than or equal to zero. Similarly, $|1 - 2x| < -5$ would have no solution because the absolute value of an expression can never be negative.

EXAMPLE 14   Solving an Inequality Involving an Absolute Value

Solve the inequality $2 - |3x| < 1$.

Solution:

Subtract 2 from both sides.	$-|3x| < -1$
Multiply by (-1) and reverse the inequality sign.	$|3x| > 1$
Apply property (3).	$3x < -1$ or $3x > 1$
Divide both inequalities by 3.	$x < -\dfrac{1}{3}$ or $x > \dfrac{1}{3}$
Express in interval notation.	$\boxed{\left(-\infty, -\dfrac{1}{3}\right) \cup \left(\dfrac{1}{3}, \infty\right)}$

Graph.

SECTION
0.4 SUMMARY

In this section, we used interval notation to represent the solution to inequalities.

- **Linear Inequalities:** Solve linear inequalities similarly to how we solve linear equations with one exception—when you multiply or divide by a negative number, you must reverse the inequality sign.

- **Polynomial Inequalities:** First write the inequality in standard form (zero on one side). Determine the zeros, draw the number line, test the intervals, select the intervals according to the sign of the inequality, and write the solution in interval notation.

- **Rational Inequalities:** Write as a single fraction and then proceed with a similar approach as used in ploynomial inequalities—only the test intervals are determined by finding the zeros of either the numerator or the denominator. Exclude any values from the solution that result in the denominator being equal to zero.

- **Absolute Value Inequalities:** Write an absolute value inequality in terms of two inequalities that do not involve absolute value:
  - $|x| < A$ is equivalent to $-A < x < A$.
  - $|x| > A$ is equivalent to $x < -A$ or $x > A$.

SECTION
0.4 EXERCISES

■ **SKILLS**

In Exercises 1–10, rewrite in interval notation and graph.

1. $-2 \leq x < 3$

2. $-4 \leq x \leq -1$

3. $-3 < x \leq 5$

4. $0 < x < 6$

5. $x \leq 6$ and $x \geq 4$

6. $x > -3$ and $x \leq 2$

7. $x \leq -6$ and $x \geq -8$

8. $x < 8$ and $x < 2$

9. $x > 4$ and $x \leq -2$

10. $x \geq -5$ and $x < -6$

In Exercises 11–20, graph the indicated set and write as a single interval, if possible.

11. $(-\infty, 4) \cap [1, \infty)$

12. $(-3, \infty) \cap [-5, \infty)$

13. $[-5, 2) \cap [-1, 3]$

14. $[-4, 5) \cap [-2, 7)$

15. $(-\infty, 4) \cup (4, \infty)$

16. $(-\infty, -3] \cup [-3, \infty)$

17. $(-\infty, -3] \cup [3, \infty)$

18. $(-2, 2) \cap [-3, 1]$

19. $(-\infty, \infty) \cap (-3, 2]$

20. $(-\infty, \infty) \cup (-4, 7)$

In Exercises 21–32, solve each linear inequality and express the solution set in interval notation.

21. $3(t + 1) > 2t$

22. $2(y + 5) \leq 3(y - 4)$

23. $7 - 2(1 - x) > 5 + 3(x - 2)$

24. $4 - 3(2 + x) < 5$

25. $\frac{2}{3}y - \frac{1}{2}(5 - y) < \frac{5y}{3} - (2 + y)$

26. $\frac{s}{2} - \frac{(s - 3)}{3} > \frac{s}{4} - \frac{1}{12}$

27. $-3 < 1 - x \leq 9$

28. $3 \leq -2 - 5x \leq 13$

29. $0 < 2 - \frac{1}{3}y < 4$

30. $3 < \frac{1}{2}A - 3 < 7$

31. $\frac{1}{2} \leq \frac{1 + y}{3} \leq \frac{3}{4}$

32. $-1 < \frac{2 - z}{4} \leq \frac{1}{5}$

In Exercises 33–50, solve each polynomial inequality and express the solution set in interval notation.

33. $2t^2 - 3 \leq t$

34. $3t^2 \geq -5t + 2$

35. $5v - 1 > 6v^2$

36. $12t^2 < 37t + 10$

37. $2s^2 - 5s \geq 3$

38. $8s + 12 \leq -s^2$

39. $y^2 + 2y \geq 4$

40. $y^2 + 3y \leq 1$

41. $x^2 - 4x < 6$

42. $x^2 - 2x > 5$

43. $u^2 \geq 3u$

44. $u^2 \leq -4u$

45. $x^2 > 9$

46. $t^2 \leq 49$

47. $x^3 + x^2 - 2x \leq 0$

48. $x^3 + 2x^2 - 3x > 0$

49. $x^3 + x > 2x^2$

50. $x^3 + 4x \leq 4x^2$

In Exercises 51–66, solve each rational inequality and express the solution set in interval notation.

51. $\dfrac{s+1}{4-s^2} \geq 0$

52. $\dfrac{s+5}{4-s^2} \leq 0$

53. $\dfrac{3t^2}{t+2} \geq 5t$

54. $\dfrac{-2t-t^2}{4-t} \geq t$

55. $\dfrac{3p-2p^2}{4-p^2} < \dfrac{3+p}{2-p}$

56. $-\dfrac{7p}{p^2-100} \leq \dfrac{p+2}{p+10}$

57. $\dfrac{x^2+10}{x^2+16} > 0$

58. $-\dfrac{x^2+2}{x^2+4} < 0$

59. $\dfrac{v^2-9}{v-3} \geq 0$

60. $\dfrac{v^2-1}{v+1} \leq 0$

61. $\dfrac{2}{t-3} + \dfrac{1}{t+3} \geq 0$

62. $\dfrac{1}{t-2} + \dfrac{1}{t+2} \leq 0$

63. $\dfrac{3}{x+4} - \dfrac{1}{x-2} \leq 0$

64. $\dfrac{2}{x-5} - \dfrac{1}{x-1} \geq 0$

65. $\dfrac{1}{p-2} - \dfrac{1}{p+2} \geq \dfrac{3}{p^2-4}$

66. $\dfrac{2}{2p-3} - \dfrac{1}{p+1} \leq \dfrac{1}{2p^2-p-3}$

In Exercises 67–82, solve the absolute value inequality and express the solution set in interval notation.

67. $|x-4| > 2$

68. $|x-1| < 3$

69. $|4-x| \leq 1$

70. $|1-y| < 3$

71. $|2x| > -3$

72. $|2x| < -3$

73. $|7-2y| \geq 3$

74. $|6-5y| \leq 1$

75. $|4-3x| \geq 0$

76. $|4-3x| \geq 1$

77. $2|4x| - 9 \geq 3$

78. $5|x-1| + 2 \leq 7$

79. $9 - |2x| < 3$

80. $4 - |x+1| > 1$

81. $|x^2-1| \leq 8$

82. $|x^2+4| \geq 29$

■ **APPLICATIONS**

83. Lasers. A circular laser beam with a radius r_T is transmitted from one tower to another tower. If the received beam radius r_R fluctuates 10% from the transmitted beam radius due to atmospheric turbulence, write an inequality representing the received beam radius.

84. Electronics: Communications. Communication systems are often evaluated based on their signal-to-noise ratio (SNR), which is the ratio of the average power of received signal, S, to average power of noise, N, in the system. If the SNR is required to be at least 2 at all times, write an inequality representing the received signal power if the noise can fluctuate 10%.

The following table is the 2007 Federal Tax Rate Schedule for people filing as *single:*

TAX BRACKET #	IF TAXABLE INCOME IS OVER–	BUT NOT OVER–	THE TAX IS:
I	$0	$7,825	10% of the amount over $0
II	$7,825	$31,850	$782.50 plus 15% of the amount over $7,825
III	$31,850	$77,100	$4,386.25 plus 25% of the amount over $31,850
IV	$77,100	$160,850	$15,698.75 plus 28% of the amount over $77,100
V	$160,850	$349,700	$39,148.75 plus 33% of the amount over $160,850
VI	$349,700	No limit	$101,469.25 plus 35% of the amount over $349,700

85. Federal Income Tax. What is the range of federal income taxes a person in tax bracket III will pay the IRS?

86. Federal Income Tax. What is the range of federal income taxes a person in tax bracket IV will pay the IRS?

87. Profit. A Web-based embroidery company makes monogrammed napkins. The profit associated with producing x orders of napkins is governed by the equation

$$P(x) = -x^2 + 130x - 3000$$

Determine the range of orders the company should accept in order to make a profit.

88. Profit. Repeat Exercise 87 using $P(x) = x^2 - 130x + 3600$.

89. Car Value. The term "upside down" on car payments refers to owing more than a car is worth. Assume you buy a new car and finance 100% over 5 years. The difference between the value of the car and what is owed on the car is governed by the expression $\dfrac{t}{t-3}$, where t is age (in years) of the car. Determine the time period when the car is worth more than you owe $\left(\dfrac{t}{t-3} > 0\right)$.

When do you owe more than it's worth $\left(\dfrac{t}{t-3} < 0\right)$?

90. Car Value. Repeat Exercise 89 using the expression $-\dfrac{2-t}{4-t}$.

91. Bullet Speed. A .22 caliber gun fires a bullet at a speed of 1200 feet per second. If a .22 caliber gun is fired straight upward into the sky, the height of the bullet in feet is given by the equation $h = -16t^2 + 1200t$, where t is the time in seconds with $t = 0$ corresponding to the instant the gun is fired. How long is the bullet in the air?

92. Bullet Speed. A .38 caliber gun fires a bullet at a speed of 600 feet per second. If a .38 caliber gun is fired straight upward into the sky, the height of the bullet in feet is given by the equation $h = -16t^2 + 600t$. How many seconds is the bullet in the air?

93. Sports. Two women tee off of a par-3 hole on a golf course. They are playing "closest to the pin." If the first woman tees off and lands exactly 4 feet from the hole, write an inequality that describes where the second woman lands to the hole in order to win the hole. What equation would suggest a tie? Let d = the distance from where the second woman lands to the tee.

94. Electronics. A band-pass filter in electronics allows certain frequencies within a range (or band) to pass through to the receiver and eliminates all other frequencies. Write an absolute value inequality that allows any frequency f within 15 Hz of the carrier frequency f_c to pass.

■ CATCH THE MISTAKE

In Exercises 95–98, explain the mistake that is made.

95. Solve the inequality $2 - 3p \le -4$ and express the solution in interval notation.

Solution:
$$2 - 3p \le -4$$
$$-3p \le -6$$
$$p \le 2$$
$$(-\infty, 2]$$

This is incorrect. What mistake was made?

96. Solve the inequality $u^2 < 25$.

Solution:

Take the square root of both sides. $u < -5$

Write the solution in interval notation. $(-\infty, -5)$

This is incorrect. What mistake was made?

97. Solve the inequality $3x < x^2$.

Solution:

Divide by x. $3 < x$

Write the solution in interval notation. $(3, \infty)$

This is incorrect. What mistake was made?

98. Solve the inequality $\dfrac{x+4}{x} < -\dfrac{1}{3}$.

Solution:

Cross multiply. $3(x+4) < -1(x)$

Eliminate the parentheses. $3x + 12 < -x$

Combine like terms. $4x < -12$

Divide both sides by 4. $x < -3$

This is incorrect. What mistake was made?

■ CONCEPTUAL

In Exercises 99–102, determine whether each statement is true or false. Assume that a is a positive real number.

99. If $x < a$, then $a > x$.

100. If $-x \ge a$, then $x \ge -a$.

101. If $x < a^2$, then the solution is $(-\infty, a)$.

102. If $x \ge a^2$, then the solution is $[a, \infty)$.

■ CHALLENGE

In Exercises 103 and 104, solve for x given that a and b are both positive real numbers.

103. $\dfrac{x^2 + a^2}{x^2 + b^2} \ge 0$

104. $\dfrac{x^2 - b^2}{x + b} < 0$

105. For what values of x does the absolute value equation $|x + 1| = 4 + |x - 2|$ hold?

106. Solve the inequality $|3x^2 - 7x + 2| > 8$.

■ TECHNOLOGY

107. a. Solve the inequality $x - 3 < 2x - 1 < x + 4$.
  b. Graph all three expressions of the inequality in the same viewing screen. Find the range of x-values when the graph of the middle expression lies above the graph of the left side and below the graph of the right side.
  c. Do (a) and (b) agree?

108. a. Solve the inequality $x - 2 < 3x + 4 \le 2x + 6$.
  b. Graph all three expressions of the inequality in the same viewing screen. Find the range of x-values when the graph of the middle expression lies above the graph of the left side and on top of and below the graph of the right side.
  c. Do (a) and (b) agree?

109. Solve the inequality $\left|\dfrac{x}{x+1}\right| < 1$ by graphing both sides of the inequality, and identify which x-values make this statement true.

110. Solve the inequality $\left|\dfrac{x}{x+1}\right| < 2$ by graphing both sides of the inequality, and identify which x-values make this statement true.

SECTION
0.5 GRAPHING EQUATIONS

SKILLS OBJECTIVES
■ Calculate the distance between two points and the midpoint of a line segment joining two points.
■ Graph equations in two variables by point-plotting.
■ Use intercepts and symmetry as graphing aids.
■ Graph circles.

CONCEPTUAL OBJECTIVES
■ Expand the concept of a one-dimensional number line to a two-dimensional plane.
■ Relate symmetry graphically and algebraically.

HIV infection rates, stock prices, and temperature conversions are all examples of relationships between two quantities that can be expressed in a two-dimensional graph. Because it is two-dimensional, such a graph lies in a **plane**.

Two perpendicular real number lines, known as the **axes** in the plane, intersect at a point we call the **origin**. Typically, the horizontal axis is called the **x-axis** and the vertical axis is denoted as the **y-axis**. The axes divide the plane into four **quadrants**, numbered by Roman numerals and ordered counterclockwise.

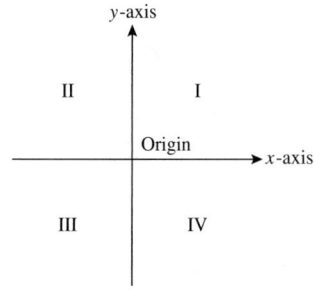

Points in the plane are represented by **ordered pairs**, denoted (x, y). The first number of the ordered pair indicates the position in the horizontal direction and is often called the x-coordinate or **abscissa**. The second number indicates the position in the vertical direction and is often called the y-coordinate or **ordinate**. The origin is denoted $(0, 0)$.

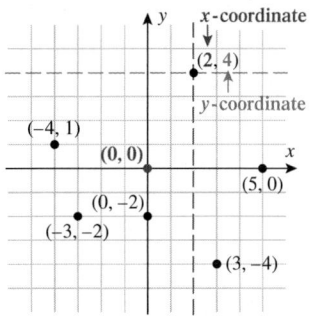

Examples of other coordinates are given on the graph to the left.

The point $(2, 4)$ lies in quadrant I. To **plot** this point, start at the origin $(0, 0)$ and move to the right two units and up four units.

All points in quadrant I have positive coordinates, and all points in quadrant III have negative coordinates. Quadrant II has negative x-coordinates and positive y-coordinates; quadrant IV has positive x-coordinates and negative y-coordinates.

This representation is called the **rectangular coordinate system** or **Cartesian coordinate system**, named after the French mathematician René Descartes.

The Distance and Midpoint Formulas

Study Tip

It does not matter which point is taken to be the first point or the second point.

DEFINITION   Distance Formula

The **distance** d between two points $P_1 = (x_1, y_1)$ and $P_2 = (x_2, y_2)$ is given by

$$d = \sqrt{(x_2 - x_1)^2 + (y_2 - y_1)^2}$$

The distance between two points is the square root of the sum of the square of the difference between the x-coordinates and the square of the difference between the y-coordinates.

EXAMPLE 1   **Using the Distance Formula to Find the Distance Between Two Points**

Find the distance between $(-3, 7)$ and $(5, -2)$.

Solution:

Write the distance formula. $d = \sqrt{[x_2 - x_1]^2 + [y_2 - y_1]^2}$

Substitute $(x_1, y_1) = (-3, 7)$
and $(x_2, y_2) = (5, -2)$. $d = \sqrt{[5 - (-3)]^2 + [-2 - 7]^2}$

Simplify. $d = \sqrt{[5 + 3]^2 + [-2 - 7]^2}$

$d = \sqrt{8^2 + (-9)^2} = \sqrt{64 + 81} = \sqrt{145}$

Solve for d. $\boxed{d = \sqrt{145}}$

■ **Answer:** $d = \sqrt{58}$

■ **YOUR TURN** Find the distance between $(4, -5)$ and $(-3, -2)$.

DEFINITION   Midpoint Formula

The **midpoint**, (x_m, y_m), of the line segment with endpoints (x_1, y_1) and (x_2, y_2) is given by

$$(x_m, y_m) = \left(\frac{x_1 + x_2}{2}, \frac{y_1 + y_2}{2} \right)$$

The midpoint can be found by averaging the x-coordinates and averaging the y-coordinates.

EXAMPLE 2 **Finding the Midpoint of a Line Segment**

Find the midpoint of the line segment joining the points $(2, 6)$ and $(-4, -2)$.

Solution:

Write the midpoint formula.

$$(x_m, y_m) = \left(\frac{x_1 + x_2}{2}, \frac{y_1 + y_2}{2} \right)$$

Substitute $(x_1, y_1) = (2, 6)$ and $(x_2, y_2) = (-4, -2)$.

$$(x_m, y_m) = \left(\frac{2 + (-4)}{2}, \frac{6 + (-2)}{2} \right)$$

Simplify.

$$(x_m, y_m) = (-1, 2)$$

One way to verify your answer is to plot the given points and the midpoint to make sure your answer looks reasonable.

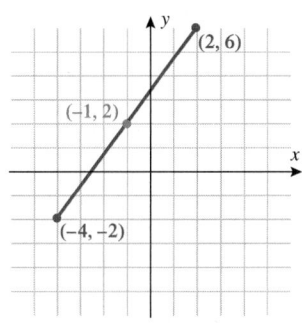

Technology Tip

Show a screen display of how to enter $\dfrac{2 + (-4)}{2}$ and $\dfrac{6 + (-2)}{2}$.

Scientific calculators:

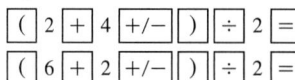

Or,

$(2+ {}^-4)/2$
                    -1
$(6+ {}^-2)/2$
                    2

■ **Answer:** midpoint $= (4, 2)$

■ **YOUR TURN** Find the midpoint of the line segment joining the points $(3, -4)$ and $(5, 8)$.

Point-Plotting

The **graph of an equation** in two variables, x and y, consists of all the points in the xy-plane whose coordinates (x, y) satisfy the equation. A procedure for plotting the graphs of equations is outlined below and is illustrated with the example $y = x^2$.

WORDS	MATH

Step 1: In a table, list several pairs of coordinates that make the equation true.

x	$y = x^2$	(x, y)
0	0	$(0, 0)$
-1	1	$(-1, 1)$
1	1	$(1, 1)$
-2	4	$(-2, 4)$
2	4	$(2, 4)$

Step 2: Plot these points on a graph and connect the points with a smooth curve. Use arrows to indicate that the graph continues.

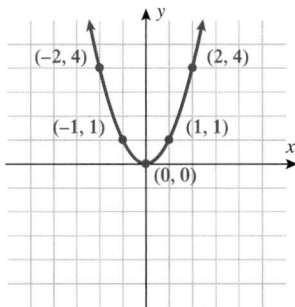

In graphing an equation, first select arbitrary values for x and then use the equation to find the corresponding value of y, or vice versa.

EXAMPLE 3   Graphing an Equation by Plotting Points

Graph the equation $y = x^3$.

Solution:

STEP 1 In a table, list several pairs of coordinates that satisfy the equation.

x	$y = x^3$	(x, y)
0	0	$(0, 0)$
-1	-1	$(-1, -1)$
1	1	$(1, 1)$
-2	-8	$(-2, -8)$
2	8	$(2, 8)$

STEP 2 Plot these points on a graph and connect the points with a smooth curve, indicating with arrows that the curve continues in both the positive and negative directions.

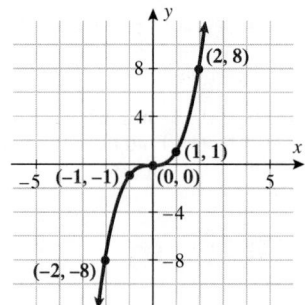

Using Intercepts and Symmetry as Graphing Aids

Intercepts

When point-plotting graphs of equations, which points should be selected? Points where a graph crosses (or touches) either the x-axis or y-axis are called **intercepts** and identifying these points helps define the graph unmistakably.

An **x-intercept** of a graph is a point where the graph intersects the x-axis. Specifically, an x-intercept is the x-coordinate of such a point. For example, if a graph intersects the x-axis at the point $(3, 0)$, then we say that 3 is the x-intercept. Since the value for y along the x-axis is zero, all points corresponding to x-intercepts have the form $(a, 0)$.

A **y-intercept** of a graph is a point where the graph intersects the y-axis. Specifically, a y-intercept is the y-coordinate of such a point. For example, if a graph intersects the y-axis at the point $(0, 2)$, then we say that 2 is the y-intercept. Since the value for x along the y-axis is zero, all points corresponding to y-intercepts have the form $(0, b)$.

It is important to note that graphs of equations do not have to have intercepts, and if they do have intercepts, they can have one or more of each type.

One x-intercept
Two y-intercepts

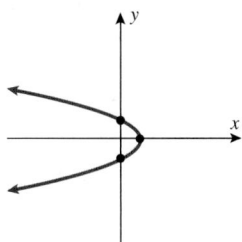

No x-intercepts
One y-intercept

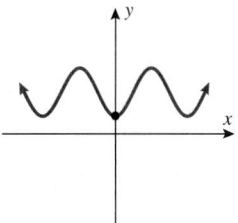

No x-intercepts
No y-intercepts

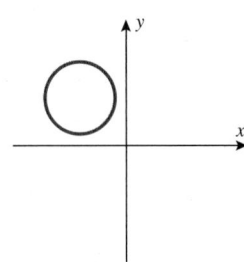

Three x-intercepts
One y-intercept

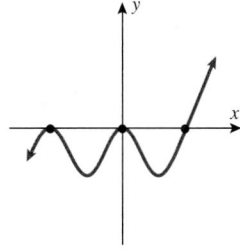

Note: The origin $(0, 0)$ corresponds to both an x-intercept and a y-intercept.

EXAMPLE 4 **Finding Intercepts from an Equation**

Given the equation $y = x^2 + 1$, find the indicated intercepts of its graph, if any.

a. x-intercept(s)   **b.** y-intercept(s)

Solution (a):

Let $y = 0$. $0 = x^2 + 1$

Solve for x. $x^2 = -1$ no real solution

> There are no x-intercepts.

Solution (b):

Let $x = 0$. $y = 0^2 + 1$

Solve for y. $y = 1$

The y-intercept is located at the point $\boxed{(0, 1)}$.

■ YOUR TURN For the equation $y = x^2 - 4$

a. find the x-intercept(s), if any.   **b.** find the y-intercept(s), if any.

■ **Answer:**
 a. x-intercepts: -2 and 2
 b. y-intercept: -4

Symmetry

The word **symmetry** conveys balance. Suppose you have two pictures to hang on a wall. If you space them equally apart on the wall, then you prefer a symmetric décor. This is an example of symmetry about a line. The word (water) written below is identical if you rotate the word 180 degrees (or turn the page upside down). This is an example of symmetry about a point. Symmetric graphs have the characteristic that their mirror image can be obtained about a reference, typically a line or a point.

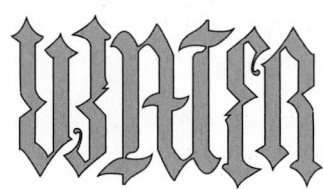

Symmetry aids in graphing by giving information "for free." For example, if a graph is symmetric about the y-axis, then once the graph to the right of the y-axis is found, the left side of the graph is the mirror image of that. If a graph is symmetric about the origin, then once the graph is known in quadrant I, the graph in quadrant III is found by rotating the known graph 180 degrees.

It would be beneficial to know whether a graph of an equation is symmetric about a line or point before the graph of the equation is sketched. Although a graph can be symmetric about any line or point, we will discuss only symmetry about the x-axis, y-axis, and origin. These types of symmetry and the algebraic procedures for testing for symmetry are outlined below.

Types and Tests for Symmetry

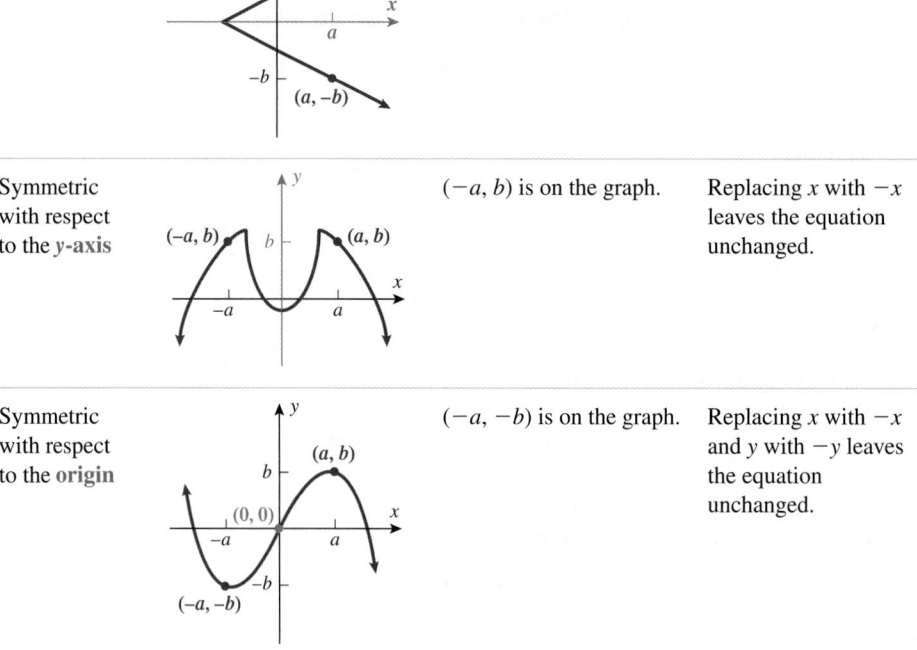

TYPE OF SYMMETRY	GRAPH	IF THE POINT (a, b) IS ON THE GRAPH, THEN THE POINT . . .	ALGEBRAIC TEST FOR SYMMETRY
Symmetric with respect to the x-axis		$(a, -b)$ is on the graph.	Replacing y with $-y$ leaves the equation unchanged.
Symmetric with respect to the y-axis		$(-a, b)$ is on the graph.	Replacing x with $-x$ leaves the equation unchanged.
Symmetric with respect to the **origin**		$(-a, -b)$ is on the graph.	Replacing x with $-x$ and y with $-y$ leaves the equation unchanged.

Study Tip

Symmetry gives us information about the graph "for free."

When testing for symmetry about the x-axis, y-axis, and origin, there are *five* possibilities:

- No symmetry
- Symmetry with respect to the x-axis
- Symmetry with respect to the y-axis
- Symmetry with respect to the origin
- Symmetry with respect to the x-axis, y-axis, and origin

EXAMPLE 5 Testing for Symmetry

Determine what type of symmetry (if any) the graphs of the equations exhibit.

a. $y = x^2 + 1$ **b.** $y = x^3 + 1$

Solution (a):

Replace x with $-x$. $y = (-x)^2 + 1$

Simplify. $y = x^2 + 1$

The resulting equation is equivalent to the original equation, so the graph of the equation $y = x^2 + 1$ is symmetric with respect to the y-axis.

Replace y with $-y$. $(-y) = x^2 + 1$

Simplify. $y = -x^2 - 1$

The resulting equation $y = -x^2 - 1$ is not equivalent to the original equation $y = x^2 + 1$, so the graph of the equation $y = x^2 + 1$ is not symmetric with respect to the x-axis.

Replace x with $-x$ and y with $-y$. $(-y) = (-x)^2 + 1$

Simplify. $-y = x^2 + 1$

                                                                     $y = -x^2 - 1$

The resulting equation $y = -x^2 - 1$ is not equivalent to the original equation $y = x^2 + 1$, so the graph of the equation $y = x^2 + 1$ is not symmetric with respect to the origin.

> The graph of the equation $y = x^2 + 1$ is **symmetric with respect to the y-axis**.

Solution (b):

Replace x with $-x$. $y = (-x)^3 + 1$

Simplify. $y = -x^3 + 1$

The resulting equation $y = -x^3 + 1$ is not equivalent to the original equation $y = x^3 + 1$. Therefore, the graph of the equation $y = x^3 + 1$ is not symmetric with respect to the y-axis.

Replace y with $-y$. $(-y) = x^3 + 1$

Simplify. $y = -x^3 - 1$

The resulting equation $y = -x^3 - 1$ is not equivalent to the original equation $y = x^3 + 1$. Therefore, the graph of the equation $y = x^3 + 1$ is not symmetric with respect to the x-axis.

Replace x with $-x$ and y with $-y$. $(-y) = (-x)^3 + 1$

Simplify. $-y = -x^3 + 1$

                                                                     $y = x^3 - 1$

The resulting equation $y = x^3 - 1$ is not equivalent to the original equation $y = x^3 + 1$. Therefore, the graph of the equation $y = x^3 + 1$ is not symmetric with respect to the origin.

> The graph of the equation $y = x^3 + 1$ exhibits **no symmetry**.

■ **YOUR TURN** Determine the symmetry (if any) for $x = y^2 - 1$.

Technology Tip

Graph of $y_1 = x^2 + 1$ is shown.

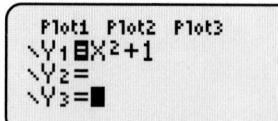

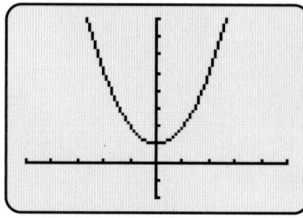

Graph of $y_1 = x^3 + 1$ is shown.

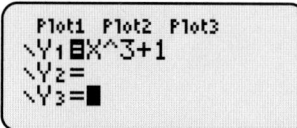

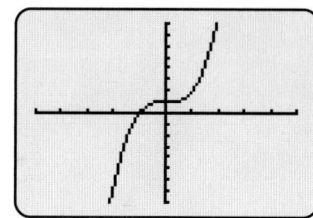

■ **Answer:** The graph of the equation is symmetric with respect to the x-axis.

EXAMPLE 6 Using Intercepts and Symmetry as Graphing Aids

For the equation $x^2 + y^2 = 25$, use intercepts and symmetry to help you graph the equation using the point-plotting technique.

Solution:

STEP 1 **Find the intercepts.**

For the x-intercepts, let $y = 0$. $\qquad\qquad x^2 + 0^2 = 25$

Solve for x. $\qquad\qquad\qquad\qquad\qquad x = \pm 5$

The two x-intercepts correspond to the points $(-5, 0)$ and $(5, 0)$.

For the y-intercepts, let $x = 0$. $\qquad\qquad 0^2 + y^2 = 25$

Solve for y. $\qquad\qquad\qquad\qquad\qquad y = \pm 5$

The two y-intercepts correspond to the points $(0, -5)$ and $(0, 5)$.

STEP 2 **Identify the points on the graph corresponding to the intercepts.**

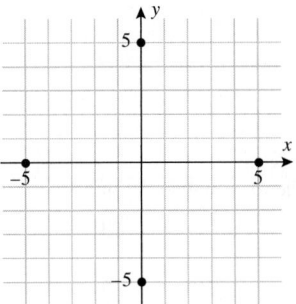

STEP 3 **Test for symmetry with respect to the y-axis, x-axis, and origin.**

Test for symmetry with respect to the y-axis.

Replace x with $-x$. $\qquad\qquad\qquad\qquad (-x)^2 + y^2 = 25$

Simplify. $\qquad\qquad\qquad\qquad\qquad\qquad x^2 + y^2 = 25$

The resulting equation is equivalent to the original, so the graph of $x^2 + y^2 = 25$ is symmetric with respect to the y-axis.

Test for symmetry with respect to the x-axis.

Replace y with $-y$. $\qquad\qquad\qquad\qquad x^2 + (-y)^2 = 25$

Simplify. $\qquad\qquad\qquad\qquad\qquad\qquad x^2 + y^2 = 25$

The resulting equation is equivalent to the original, so the graph of $x^2 + y^2 = 25$ is symmetric with respect to the x-axis.

Test for symmetry with respect to the origin.

Replace x with $-x$ and y with $-y$. $\qquad (-x)^2 + (-y)^2 = 25$

Simplify. $\qquad\qquad\qquad\qquad\qquad\qquad x^2 + y^2 = 25$

The resulting equation is equivalent to the original, so the graph of $x^2 + y^2 = 25$ is symmetric with respect to the origin.

We need to determine solutions to the equation on only the positive x- and y-axes and in quadrant I because of the following symmetries:

■ Symmetry with respect to the y-axis gives the solutions in quadrant II.
■ Symmetry with respect to the origin gives the solutions in quadrant III.
■ Symmetry with respect to the x-axis yields solutions in quadrant IV.

Technology Tip

To enter the graph of $x^2 + y^2 = 25$, solve for y first. The graphs of
$y_1 = \sqrt{25 - x^2}$ and
$y_2 = -\sqrt{25 - x^2}$ are shown.

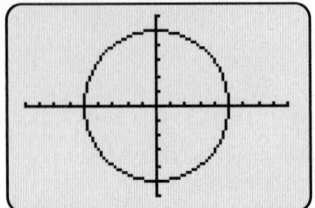

Solutions to $x^2 + y^2 = 25$.

Quadrant I: (3, 4), (4, 3)

Additional points due to symmetry:

Quadrant II: $(-3, 4), (-4, 3)$

Quadrant III: $(-3, -4), (-4, -3)$

Quadrant IV: $(3, -4), (4, -3)$

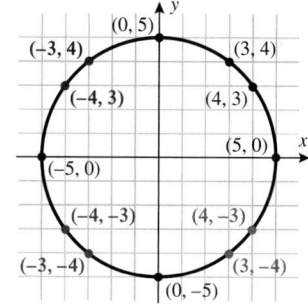

Circles

DEFINITION   Circle

A **circle** is the set of all points in a plane that are a fixed distance from a point, the **center**. The center, C, is typically denoted by (h, k), and the fixed distance, or **radius**, is denoted by r.

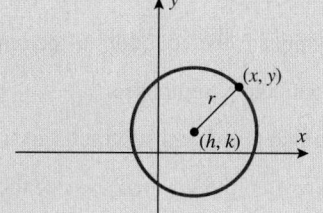

EQUATION OF A CIRCLE

The standard form of the equation of a **circle** with **radius** r and **center** (h, k) is

$$(x - h)^2 + (y - k)^2 = r^2$$

For the special case of a circle with center at the origin $(0, 0)$, the equation simplifies to $x^2 + y^2 = r^2$.

UNIT CIRCLE

A circle with radius 1 and center $(0, 0)$ is called the **unit circle**:

$$x^2 + y^2 = 1$$

The unit circle plays an important role in the study of trigonometry. Note that if $x^2 + y^2 = 0$, the radius is 0, so the "circle" is just a point.

 EXAMPLE 7   Finding the Center and Radius of a Circle

Identify the center and radius of the given circle and graph.

$$(x - 2)^2 + (y + 1)^2 = 4$$

Solution:

Rewrite this equation in standard form. $[x - 2]^2 + [y -(-1)]^2 = 2^2$

Identify h, k, and r by comparing this equation with the standard form of a circle: $(x - h)^2 + (y - k)^2 = r^2$. $h = 2, k = -1$, and $r = 2$

$$\boxed{\text{Center } (2, -1) \text{ and } r = 2}$$

Technology Tip

To enter the graph of $(x - 2)^2 + (y + 1)^2 = 4$, solve for y first. The graphs of

$y_1 = \sqrt{4 - (x - 2)^2} - 1$ and

$y_2 = -\sqrt{4 - (x - 2)^2} - 1$ are shown.

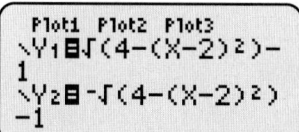

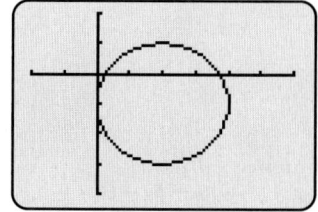

To draw the circle, label the center $(2, -1)$. Label four additional points 2 units (the radius) away from the center: $(4, -1)$, $(0, -1)$, $(2, 1)$, and $(2, -3)$.

Note that the easiest four points to get are those obtained by going out from the center both horizontally and vertically. Connect those four points with a smooth curve.

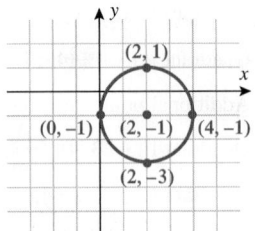

■ Answer: Center: $(-1, -2)$
Radius: 3

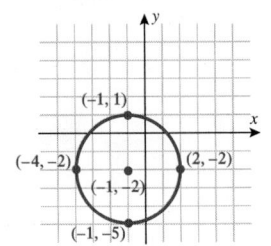

■ YOUR TURN Identify the center and radius of the given circle and graph.

$$(x + 1)^2 + (y + 2)^2 = 9$$

Let's change the look of the equation given in Example 7.

In Example 7, the equation of the circle was given as $(x - 2)^2 + (y + 1)^2 = 4$

Eliminate the parentheses. $x^2 - 4x + 4 + y^2 + 2y + 1 = 4$

Group like terms and subtract 4 from both sides. $x^2 + y^2 - 4x + 2y + 1 = 0$

We have written the *general form* of the equation of the circle in Example 7.

> The **general form** of the **equation of a circle** is
>
> $$x^2 + y^2 + ax + by + c = 0$$

Suppose you are given a point that lies on a circle and the center of the circle. Can you find the equation of the circle?

EXAMPLE 8 **Finding the Equation of a Circle Given Its Center and One Point**

The point $(10, -4)$ lies on a circle centered at $(7, -8)$. Find the equation of the circle in general form.

Solution:

This circle is centered at $(7, -8)$, so its standard equation is $(x - 7)^2 + (y + 8)^2 = r^2$.

Since the point $(10, -4)$ lies on the circle, it must satisfy the equation of the circle.

Substitute $(x, y) = (10, -4)$. $(10 - 7)^2 + (-4 + 8)^2 = r^2$

Simplify. $3^2 + 4^2 = r^2$

The distance from $(10, -4)$ to $(7, -8)$ is 5 units. $r = 5$

Substitute $r = 5$ into the standard equation. $(x - 7)^2 + (y + 8)^2 = 5^2$

Eliminate the parentheses and simplify. $x^2 - 14x + 49 + y^2 + 16y + 64 = 25$

Write in **general form**. $\boxed{x^2 + y^2 - 14x + 16y + 88 = 0}$

■ Answer:
$x^2 + y^2 + 10x - 6y - 66 = 0$

■ YOUR TURN The point $(1, 11)$ lies on a circle centered at $(-5, 3)$. Find the equation of the circle in general form.

If the equation of a circle is given in general form, it must be rewritten in standard form in order to identify its center and radius. To transform equations of circles from general to standard form, complete the square on both the x- and y-variables.

 EXAMPLE 9   Finding the Center and Radius of a Circle by Completing the Square

Find the center and radius of the circle with the equation:

$$x^2 - 8x + y^2 + 20y + 107 = 0$$

Solution:

Our goal is to transform this equation into standard form.

$$(x - h)^2 + (y - k)^2 = r^2$$

Group x and y terms, respectively, on the left side of the equation; move constants to the right side.

$$(x^2 - 8x) + (y^2 + 20y) = -107$$

Complete the square on both the x and y expressions.

$$(x^2 - 8x + \Box) + (y^2 + 20y + \Box) = -107$$

Add $\left(-\frac{8}{2}\right)^2 = 16$ and $\left(\frac{20}{2}\right)^2 = 100$ to both sides.

$$(x^2 - 8x + 16) + (y^2 + 20y + 100) = -107 + 16 + 100$$

Factor the perfect squares on the left side and simplify the right side.

$$(x - 4)^2 + (y + 10)^2 = 9$$

Write in standard form.

$$(x - 4)^2 + [y - (-10)]^2 = 3^2$$

> The center is $(4, -10)$ and the radius is 3.

■ **YOUR TURN**   Find the center and radius of the circle with the equation:

$$x^2 + y^2 + 4x - 6y - 12 = 0$$

Technology Tip

To graph $x^2 - 8x + y^2 + 20y + 107 = 0$ without transforming it into a standard form, solve for y using the quadratic formula:

$$y = \frac{-20 \pm \sqrt{20^2 - 4(1)(x^2 - 8x + 107)}}{2}$$

Next, set the window to $[-5, 30]$ by $[-30, 5]$ and use $\boxed{\text{ZSquare}}$ under $\boxed{\text{ZOOM}}$ to adjust the window variable to make the circle look circular. The graphs of

$$y_1 = \frac{-20 + \sqrt{20^2 - 4(x^2 - 8x + 107)}}{2}$$

and

$$y_2 = \frac{-20 - \sqrt{20^2 - 4(x^2 - 8x + 107)}}{2}$$

are shown.

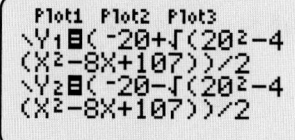

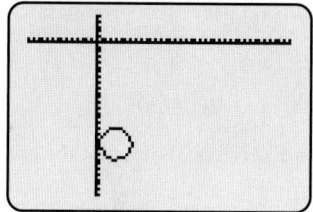

■ **Answer:** Center: $(-2, 3)$
Radius: 5

SECTION 0.5 SUMMARY

Distance between two points

$$d = \sqrt{(x_2 - x_1)^2 + (y_2 - y_1)^2}$$

Midpoint of segment joining two points

$$(x_m, y_m) = \left(\frac{x_1 + x_2}{2}, \frac{y_1 + y_2}{2}\right)$$

Intercepts

- x-intercept: Let $y = 0$ and solve for x.
- y-intercept: Let $x = 0$ and solve for y.

Symmetry

- About the x-axis: Replace y with $-y$ and the resulting equation is the same.
- About the y-axis: Replace x with $-x$ and the resulting equation is the same.
- About the origin: Replace x with $-x$ and y with $-y$ and the resulting equation is the same.

Circles

$$(x - h)^2 + (y - k)^2 = r^2 \quad \text{center } (h, k) \text{ and radius } r$$

▪ **SKILLS**

In Exercises 1–12, calculate the distance between the given points, and find the midpoint of the segment joining them.

1. $(1, 3)$ and $(5, 3)$

2. $(-2, 4)$ and $(-2, -4)$

3. $(-1, 4)$ and $(3, 0)$

4. $(-3, -1)$ and $(1, 3)$

5. $(-10, 8)$ and $(-7, -1)$

6. $(-2, 12)$ and $(7, 15)$

7. $(-3, -1)$ and $(-7, 2)$

8. $(-4, 5)$ and $(-9, -7)$

9. $(-6, -4)$ and $(-2, -8)$

10. $(0, -7)$ and $(-4, -5)$

11. $\left(-\frac{1}{2}, \frac{1}{3}\right)$ and $\left(\frac{7}{2}, \frac{10}{3}\right)$

12. $\left(\frac{1}{5}, \frac{7}{3}\right)$ and $\left(\frac{9}{5}, -\frac{2}{3}\right)$

In Exercises 13–18, graph the equation by plotting points.

13. $y = -3x + 2$

14. $y = 4 - x$

15. $y = x^2 - x - 2$

16. $y = x^2 - 2x + 1$

17. $x = y^2 - 1$

18. $x = |y + 1| + 2$

In Exercises 19–24, find the x-intercept(s) and y-intercepts(s) (if any) of the graphs of the given equations.

19. $2x - y = 6$

20. $y = 4x^2 - 1$

21. $y = \sqrt{x - 4}$

22. $y = \dfrac{x^2 - x - 12}{x}$

23. $4x^2 + y^2 = 16$

24. $x^2 - y^2 = 9$

In Exercises 25–30, test algebraically to determine whether the equation's graph is symmetric with respect to the x-axis, y-axis, or origin.

25. $x = y^2 + 4$

26. $y = x^5 + 1$

27. $x = |y|$

28. $x^2 + 2y^2 = 30$

29. $y = x^{2/3}$

30. $xy = 1$

In Exercises 31–36, plot the graph of the given equation.

31. $y = x^2 - 1$

32. $x = y^2 + 1$

33. $y = \dfrac{1}{x}$

34. $|x| = |y|$

35. $x^2 - y^2 = 16$

36. $\dfrac{x^2}{4} + \dfrac{y^2}{9} = 1$

In Exercises 37–44, write the equation of the circle in standard form.

37. Center $(5, 7)$
 $r = 9$

38. Center $(2, 8)$
 $r = 6$

39. Center $(-11, 12)$
 $r = 13$

40. Center $(6, -7)$
 $r = 8$

41. Center $(5, -3)$
 $r = 2\sqrt{3}$

42. Center $(-4, -1)$
 $r = 3\sqrt{5}$

43. Center $\left(\frac{2}{3}, -\frac{3}{5}\right)$
 $r = \frac{1}{4}$

44. Center $\left(-\frac{1}{3}, -\frac{2}{7}\right)$
 $r = \frac{2}{5}$

In Exercises 45–50, state the center and radius of the circle with the given equations.

45. $(x - 2)^2 + (y + 5)^2 = 49$

46. $(x + 3)^2 + (y - 7)^2 = 81$

47. $(x - 4)^2 + (y - 9)^2 = 20$

48. $(x + 1)^2 + (y + 2)^2 = 8$

49. $\left(x - \frac{2}{5}\right)^2 + \left(y - \frac{1}{7}\right)^2 = \frac{4}{9}$

50. $\left(x - \frac{1}{2}\right)^2 + \left(y - \frac{1}{3}\right)^2 = \frac{9}{25}$

In Exercises 51–60, find the center and radius of each circle.

51. $x^2 + y^2 - 10x - 14y - 7 = 0$

52. $x^2 + y^2 - 4x - 16y + 32 = 0$

53. $x^2 + y^2 - 2x - 6y + 1 = 0$

54. $x^2 + y^2 - 8x - 6y + 21 = 0$

55. $x^2 + y^2 - 10x + 6y + 22 = 0$

56. $x^2 + y^2 + 8x + 2y - 28 = 0$

57. $x^2 + y^2 - 6x - 4y + 1 = 0$

58. $x^2 + y^2 - 2x - 10y + 2 = 0$

59. $x^2 + y^2 - x + y + \dfrac{1}{4} = 0$

60. $x^2 + y^2 - \dfrac{x}{2} - \dfrac{3y}{2} + \dfrac{3}{8} = 0$

■ APPLICATIONS

61. Travel. A retired couple who live in Columbia, South Carolina, decide to take their motor home and visit their two children who live in Atlanta and in Savannah, Georgia. Savannah is 160 miles south of Columbia and Atlanta is 215 miles west of Columbia. How far apart do the children live from each other?

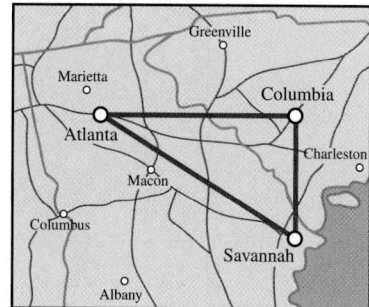

62. Sports. In the 1984 Orange Bowl, Doug Flutie, the 5 foot 9 inch quarterback for Boston College, shocked the world as he threw a "hail Mary" pass that was caught in the end zone with no time left on the clock, defeating the Miami Hurricanes 47–45. Although the record books have it listed as a 48 yard pass, what was the actual distance the ball was thrown? The following illustration depicts the path of the ball.

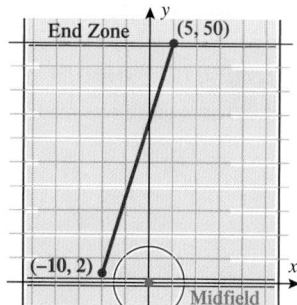

63. NASCAR Revenue. Action Performance, Inc., the leading seller of NASCAR merchandise, recorded $260 million in revenue in 2002 and $400 million in revenue in 2004. Calculate the midpoint to estimate the revenue Action Performance, Inc. recorded in 2003. Assume the horizontal axis represents the year and the vertical axis represents the revenue in millions.

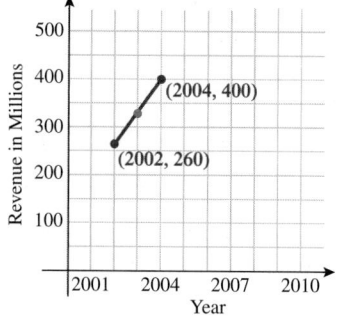

64. Ticket Price. In 1993 the average Miami Dolphins ticket price was $28 and in 2001 the average price was $56. Find the midpoint of the segment joining these two points to estimate the ticket price in 1997.

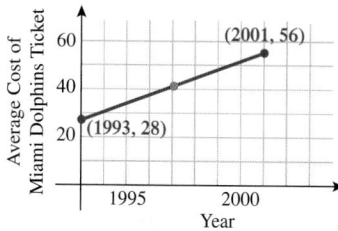

65. Design. A university designs its campus with a master plan of two concentric circles. All of the academic buildings are within the inner circle (so that students can get between classes in less than 10 minutes), and the outer circle contains all the dormitories, the Greek park, cafeterias, the gymnasium, and intramural fields. Assuming the center of campus is the origin, write an equation for the inner circle if the diameter is 3000 feet.

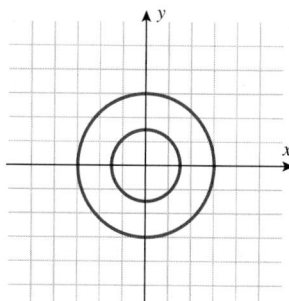

66. Design. Repeat Exercise 65 for the outer circle with a diameter of 6000 feet.

67. Cell Phones. A cellular phone tower has a reception radius of 200 miles. Assuming the tower is located at the origin, write the equation of the circle that represents the reception area.

68. Environment. In a state park, a fire has spread in the form of a circle. If the radius is 2 miles, write an equation for the circle.

▪ CATCH THE MISTAKE

In Exercises 69–72, explain the mistake that is made.

69. Graph the equation $y = x^2 + 1$.

Solution:

x	$y = x^2 + 1$	(x, y)
0	1	$(0, 1)$
1	2	$(1, 2)$

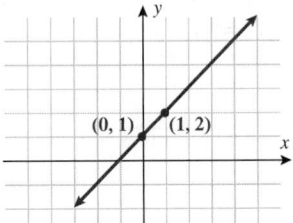

This is incorrect. What mistake was made?

70. Use symmetry to help you graph $x^2 = y - 1$.

Solution:

Replace x with $-x$. $(-x)^2 = y - 1$

Simplify. $x^2 = y - 1$

$x^2 = y - 1$ is symmetric with respect to the x-axis.

Determine points that lie on the graph in quadrant I.

y	$x^2 = y - 1$	(x, y)
1	0	$(0, 1)$
2	1	$(1, 2)$
5	2	$(2, 5)$

Symmetry with respect to the x-axis implies that $(0, -1)$, $(1, -2)$, and $(2, -5)$ are also points that lie on the graph.

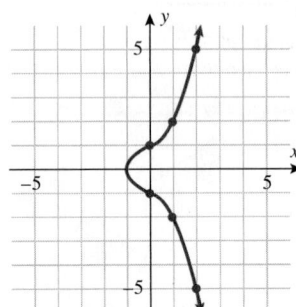

This is incorrect. What mistake was made?

71. Identify the center and radius of the circle with equation $(x - 4)^2 + (y + 3)^2 = 25$.

Solution: The center is (4, 3) and the radius is 5.

This is incorrect. What mistake was made?

72. Identify the center and radius of the circle with equation $(x - 2)^2 + (y + 3)^2 = 2$.

Solution: The center is $(2, -3)$ and the radius is 2.

This is incorrect. What mistake was made?

▪ CONCEPTUAL

In Exercises 73–76, determine whether each statement is true or false.

73. If the point (a, b) lies on a graph that is symmetric about the x-axis, then the point $(-a, b)$ also must lie on the graph.

74. If the point (a, b) lies on a graph that is symmetric about the y-axis, then the point $(-a, b)$ also must lie on the graph.

75. If the point $(a, -b)$ lies on a graph that is symmetric about the x-axis, y-axis, and origin, then the points (a, b), $(-a, -b)$, and $(-a, b)$ must also lie on the graph.

76. Two points are all that is needed to plot the graph of an equation.

77. Describe the graph (if it exists) of

$$x^2 + y^2 + 10x - 6y + 34 = 0$$

78. Describe the graph (if it exists) of

$$x^2 + y^2 - 4x + 6y + 49 = 0$$

▪ CHALLENGE

79. Determine whether the graph of $y = \dfrac{ax^2 + b}{cx^3}$ has any symmetry, where a, b, and c are real numbers.

80. Find the intercepts of $y = (x - a)^2 - b^2$, where a and b are real numbers.

81. Find the equation of a circle that has a diameter with endpoints (5, 2) and (1, −6).

82. Find the equation of a circle that has a diameter with endpoints (3, 0) and (−1, −4).

83. For the equation $x^2 + y^2 + ax + by + c = 0$, specify conditions on a, b, and c so that the graph is a single point.

84. For the equation $x^2 + y^2 + ax + by + c = 0$, specify conditions on a, b, and c so that there is no corresponding graph.

▪ TECHNOLOGY

In Exercises 85–86, graph the equation using a graphing utility and state whether there is any symmetry.

85. $y = 16.7x^4 - 3.3x^2 + 7.1$

86. $y = 0.4x^5 + 8.2x^3 - 1.3x$

In Exercises 87 and 88, (a) with the equation of the circle in standard form, state the center and radius, and graph; (b) use the quadratic formula to solve for y; and (c) use a graphing utility to graph each equation found in (b). Does the graph in (a) agree with the graphs in (c)?

87. $x^2 + y^2 - 11x + 3y - 7.19 = 0$

88. $x^2 + y^2 + 1.2x - 3.2y + 2.11 = 0$

SECTION 0.6 LINES

SKILLS OBJECTIVES

- Graph a line.
- Calculate the slope of a line.
- Find an equation of a line.
- Graph lines that are parallel or perpendicular to given lines.

CONCEPTUAL OBJECTIVES

- Classify lines as rising, falling, horizontal, and vertical.
- Understand slope as a rate of change.

Graphing a Line

First-degree equations such as

$$y = -2x + 4 \qquad 3x + y = 6 \qquad y = 2 \qquad x = -3$$

have graphs that are straight lines. The first two equations given represent inclined or "slant" lines, whereas $y = 2$ represents a horizontal line and $x = -3$ represents a vertical line. One way of writing an equation of a straight line is called *general form*.

EQUATION OF A STRAIGHT LINE: GENERAL* FORM

If A, B, and C are constants and x and y are variables, then the equation

$$Ax + By = C$$

is in **general form** and its graph is a straight line.

Note: A or B (but not both) can be zero.

*Some books refer to this as standard form.

The equation $2x - y = -2$ is a first-degree equation, so its graph is a straight line. To graph this line, list two solutions in a table, plot those points, and use a straight edge to draw the line.

x	y	(x, y)
-2	-2	$(-2, -2)$
1	4	$(1, 4)$

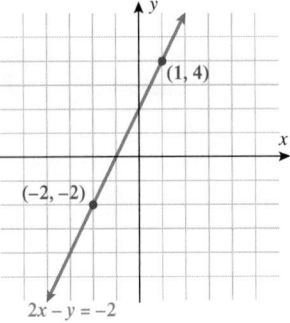

Slope

If the graph of $2x - y = -2$ represented an incline that you were about to walk on, would you classify that incline as steep? In the language of mathematics, we use the word **slope** as a measure of steepness. Slope is the ratio of the change in y over the change in x. An easy way to remember this is *rise over run.*

SLOPE OF A LINE

A nonvertical line passing through two points (x_1, y_1) and (x_2, y_2) has slope m given by the formula

$$m = \frac{y_2 - y_1}{x_2 - x_1}, \text{ where } x_1 \neq x_2 \text{ or}$$

$$m = \frac{\text{rise}}{\text{run}} = \frac{\text{vertical change}}{\text{horizontal change}}$$

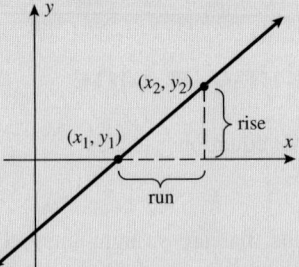

Note: Always start with the same point for both the x-coordinates and the y-coordinates.

Let's find the slope of our graph $2x - y = -2$. We'll let $(x_1, y_1) = (-2, -2)$ and $(x_2, y_2) = (1, 4)$ in the slope formula:

$$m = \frac{y_2 - y_1}{x_2 - x_1} = \frac{[4 - (-2)]}{[1 - (-2)]} = \frac{6}{3} = 2$$

Notice that if we had chosen the two intercepts $(x_1, y_1) = (0, 2)$ and $(x_2, y_2) = (-1, 0)$ instead, we still would have found the slope to be $m = 2$.

COMMON MISTAKE

The most common mistake in calculating slope is writing the coordinates in the wrong order, which results in the slope being opposite in sign.

Find the slope of the line containing the two points $(1, 2)$ and $(3, 4)$.

▼ CAUTION

Interchanging the coordinates will result in a sign error in a nonzero slope.

★ CORRECT

Label the points.

$$(x_1, y_1) = (1, 2)$$
$$(x_2, y_2) = (3, 4)$$

Write the slope formula.

$$m = \frac{y_2 - y_1}{x_2 - x_1}$$

Substitute the coordinates.

$$m = \frac{4 - 2}{3 - 1}$$

Simplify. $m = \dfrac{2}{2} = \boxed{1}$

✖ INCORRECT

The **ERROR** is interchanging the coordinates of the first and second points.

$$m = \frac{4 - 2}{1 - 3}$$

The calculated slope is **INCORRECT** by a negative sign.

$$m = \frac{2}{-2} = -1$$

When interpreting slope, always read the graph from *left to right*. Since we have determined the slope to be 2, or $\frac{2}{1}$, we can interpret this as rising 2 units and running (to the right) 1 unit. If we start at the point $(-2, -2)$ and move 2 units up and 1 unit to the right, we end up at the x-intercept, $(-1, 0)$. Again, moving 2 units up and 1 unit to the right put us at the y-intercept, $(0, 2)$. Another rise of 2 and run of 1 take us to the point $(1, 4)$. See the figure on the left.

Lines fall into one of four categories: rising, falling, horizontal, or vertical.

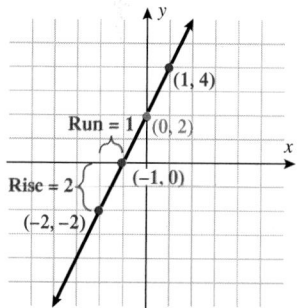

Line	Slope
Rising	Positive ($m > 0$)
Falling	Negative ($m < 0$)
Horizontal	Zero ($m = 0$), hence $y = b$
Vertical	Undefined, hence $x = a$

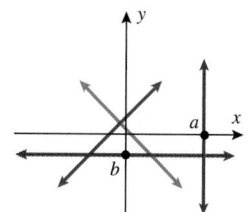

The slope of a horizontal line is 0 because the y-coordinates of any two points are the same. The change in y in the slope formula's numerator is 0, hence $m = 0$. The slope of a vertical line is undefined because the x-coordinates of any two points are the same. The change in x in the slope formula's denominator is zero; hence m is undefined.

EXAMPLE 1   Graph, Classify the Line, and Determine the Slope

Sketch a line through each pair of points, classify the line as rising, falling, vertical, or horizontal, and determine its slope.

a. $(-1, -3)$ and $(1, 1)$ **b.** $(-3, 3)$ and $(3, 1)$
c. $(-1, -2)$ and $(3, -2)$ **d.** $(1, -4)$ and $(1, 3)$

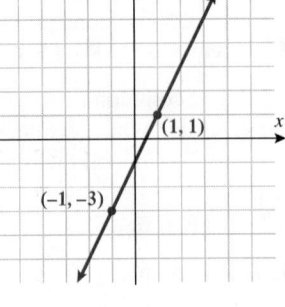

Solution (a): $(-1, -3)$ and $(1, 1)$

This line is rising, so its slope is positive.

$$m = \frac{1 - (-3)}{1 - (-1)} = \frac{4}{2} = \frac{2}{1} = 2.$$

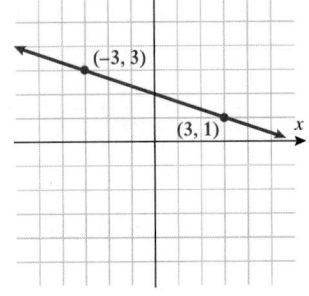

Solution (b): $(-3, 3)$ and $(3, 1)$

This line is falling, so its slope is negative.

$$m = \frac{3 - 1}{-3 - 3} = -\frac{2}{6} = -\frac{1}{3}.$$

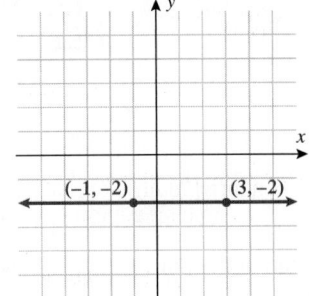

Solution (c): $(-1, -2)$ and $(3, -2)$

This is a horizontal line, so its slope is zero.

$$m = \frac{-2 - (-2)}{3 - (-1)} = \frac{0}{4} = 0.$$

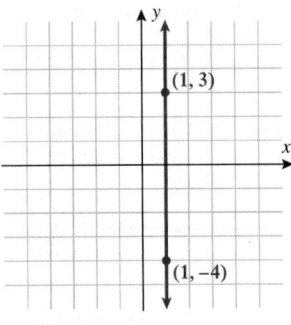

Solution (d): $(1, -4)$ and $(1, 3)$

This is a vertical line, so its slope is undefined.

$$m = \frac{3 - (-4)}{1 - 1} = \frac{7}{0}, \text{ which is undefined.}$$

■**Answer:**
  a. $m = -5$, falling
  b. $m = 2$, rising
  c. slope is undefined, vertical
  d. $m = 0$, horizontal

■ **YOUR TURN** For each pair of points, classify the line that passes through them as rising, falling, vertical, or horizontal, and determine its slope. Do not graph.

a. $(2, 0)$ and $(1, 5)$ **b.** $(-2, -3)$ and $(2, 5)$
c. $(-3, -1)$ and $(-3, 4)$ **d.** $(-1, 2)$ and $(3, 2)$

Slope–Intercept Form

As mentioned earlier, the general form for an equation of a line is $Ax + By = C$. A more standard way to write an equation of a line is in slope–intercept form, because it identifies the slope and the y-intercept.

> ### EQUATION OF A STRAIGHT LINE: SLOPE–INTERCEPT FORM
>
> The **slope–intercept form** for the equation of a nonvertical line is
>
> $$y = mx + b$$
>
> Its graph has slope m and y-intercept b.

For example, $2x - y = -3$ is in general form. To write this equation in **slope–intercept form**, we isolate the y variable:

$$y = 2x + 3$$

The **slope** of this line is **2** and the **y-intercept** is **3**.

EXAMPLE 2   Using Slope–Intercept Form to Graph an Equation of a Line

Write $2x - 3y = 15$ in slope–intercept form and graph it.

Solution:

STEP 1 *Write in slope–intercept form.*

Subtract $2x$ from both sides. $-3y = -2x + 15$

Divide both sides by -3. $y = \dfrac{2}{3}x - 5$

STEP 2 *Graph.*

Identify the slope and y-intercept.   Slope: $m = \dfrac{2}{3}$ y-intercept: $b = -5$

Plot the point corresponding to the y-intercept $(0, -5)$.

From the point $(0, -5)$, rise 2 units and run (to the right) 3 units, which corresponds to the point $(3, -3)$.

Draw the line passing through the two points.

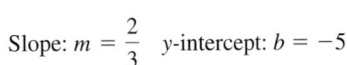

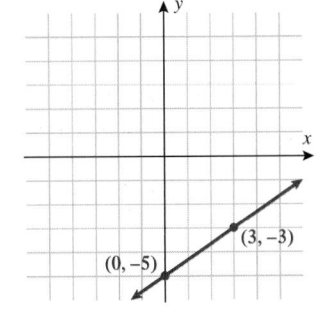

■ **YOUR TURN** Write $3x - 2y = 12$ in slope–intercept form and graph it.

Technology Tip

To graph the equation $2x - 3y = 15$, solve for y first. The graph of $y_1 = \frac{2}{3}x - 5$ is shown.

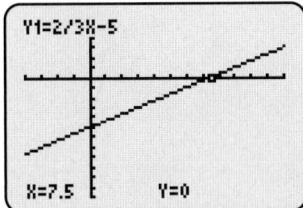

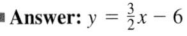

■ **Answer:** $y = \frac{3}{2}x - 6$

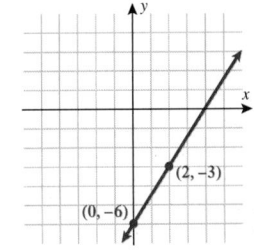

Finding Equations of Lines

Instead of starting with equations of lines and characterizing them, let us now start with particular features of a line and derive its governing equation. Suppose that you are given the y-intercept and the slope of a line. Using the slope–intercept form of an equation of a line, $y = mx + b$, you could find its equation.

 EXAMPLE 3 Using Slope–Intercept Form to Find the Equation of a Line

Find the equation of a line that has slope $\frac{2}{3}$ and y-intercept $(0, 1)$.

Solution:

Write the slope–intercept form of an equation of a line. $\qquad\qquad y = mx + b$

Label the slope. $\qquad\qquad m = \dfrac{2}{3}$

Label the y-intercept. $\qquad\qquad b = 1$

The equation of the line in slope–intercept form is $\boxed{y = \frac{2}{3}x + 1}$.

■ **Answer:** $y = -\frac{3}{2}x + 2$

■ **YOUR TURN** Find the equation of the line that has slope $-\frac{3}{2}$ and y-intercept $(0, 2)$.

Now, suppose that the two pieces of information you are given about an equation are its slope and one point that lies on its graph. You still have enough information to write an equation of the line. Recall the formula for slope:

$$m = \frac{y_2 - y_1}{x_2 - x_1}, \qquad \text{where } x_2 \neq x_1$$

We are given the slope m, and we know a particular point that lies on the line (x_1, y_1). We refer to all other points that lie on the line as (x, y). Substituting these values into the slope formula gives us

$$m = \frac{y - y_1}{x - x_1}$$

Cross multiplying yields

$$y - y_1 = m(x - x_1)$$

This is called the *point–slope form* of an equation of a line.

EQUATION OF A STRAIGHT LINE: POINT–SLOPE FORM

The **point–slope form** for the equation of a line is

$$y - y_1 = m(x - x_1)$$

Its graph passes through the point (x_1, y_1), and its slope is m.

Note: This formula does not hold for vertical lines since their slope is undefined.

EXAMPLE 4 Using Point–Slope Form to Find the Equation of a Line

Find the equation of the line that has slope $-\frac{1}{2}$ and passes through the point $(-1, 2)$.

Solution:

Write the point–slope form of an equation of a line.
$$y - y_1 = m(x - x_1)$$

Substitute the values $m = -\frac{1}{2}$ and $(x_1, y_1) = (-1, 2)$.
$$y - 2 = -\frac{1}{2}(x - (-1))$$

Distribute.
$$y - 2 = -\frac{1}{2}x - \frac{1}{2}$$

Isolate y.
$$y = -\frac{1}{2}x + \frac{3}{2}$$

We can also express the equation in general form $\boxed{x + 2y = 3}$.

■ **YOUR TURN** Find the equation of the line that has slope $\frac{1}{4}$ and passes through the point $\left(1, -\frac{1}{2}\right)$.

■ **Answer:** $y = \frac{1}{4}x - \frac{3}{4}$ or $-x + 4y = -3$

Suppose the slope of a line is not given at all. Instead, two points that lie on the line are given. If we know two points that lie on the line, then we can calculate the slope. Then, using the slope and *either* of the two points, we can derive the equation of the line.

EXAMPLE 5 Finding the Equation of a Line Given Two Points

Find the equation of the line that passes through the points $(-2, -1)$ and $(3, 2)$.

Solution:

Write the equation of a line.
$$y = mx + b$$

Calculate the slope.
$$m = \frac{y_2 - y_1}{x_2 - x_1}$$

Substitute $(x_1, y_1) = (-2, -1)$ and $(x_2, y_2) = (3, 2)$.
$$m = \frac{2 - (-1)}{3 - (-2)} = \frac{3}{5}$$

Substitute $\frac{3}{5}$ for the slope.
$$y = \frac{3}{5}x + b$$

Let $(x, y) = (3, 2)$. (Either point satisfies the equation.)
$$2 = \frac{3}{5}(3) + b$$

Solve for b.
$$b = \frac{1}{5}$$

Write the equation in slope–intercept form.
$$y = \frac{3}{5}x + \frac{1}{5}$$

Write the equation in general form.
$$\boxed{-3x + 5y = 1}$$

Study Tip

When two points that lie on a line are given, first calculate the slope of the line, then use either point and the slope–intercept form (shown in Example 6) or the point–slope form:
$$m = \frac{3}{5}, (3, 2)$$
$$y - y_1 = m(x - x_1)$$
$$y - 2 = \frac{3}{5}(x - 3)$$
$$5y - 10 = 3(x - 3)$$
$$5y - 10 = 3x - 9$$
$$\boxed{-3x + 5y = 1}$$

■ **YOUR TURN** Find the equation of the line that passes through the points $(-1, 3)$ and $(2, -4)$.

■ **Answer:** $y = -\frac{7}{3}x + \frac{2}{3}$ or $7x + 3y = 2$

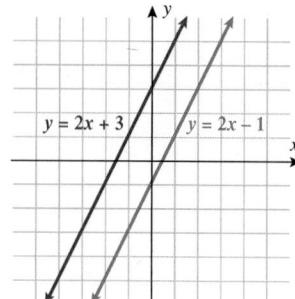

Parallel and Perpendicular Lines

Two distinct nonintersecting lines in a plane are *parallel*. How can we tell whether the two lines in the graph on the left are parallel? Parallel lines must have the same steepness. In other words, parallel lines must have the same slope. The two lines shown on the left are parallel because they have the same slope, 2.

DEFINITION	**Parallel Lines**

Two distinct lines in a plane are **parallel** if and only if their slopes are equal.

In other words, if two lines in a plane are parallel, then their slopes are equal, and if the slopes of two lines in a plane are equal, then the lines are parallel.

WORDS	**MATH**
Lines L_1 and L_2 are parallel.	$L_1 \| L_2$
Two parallel lines have the same slope.	$m_1 = m_2$

 EXAMPLE 6 **Finding an Equation of a Parallel Line**

Find the equation of the line that passes through the point $(1, 1)$ and is parallel to the line $y = 3x + 1$.

Solution:

Write the slope–intercept equation of a line.	$y = mx + b$
Parallel lines have equal slope.	$m = 3$
Substitute the slope into the equation of the line.	$y = 3x + b$
Since the line passes through $(1, 1)$, this point must satisfy the equation.	$1 = 3(1) + b$
Solve for b.	$b = -2$

The equation of the line is $\boxed{y = 3x - 2}$.

■ **Answer:** $y = 2x + 5$

■ **YOUR TURN** Find the equation of the line parallel to $y = 2x - 1$ that passes through the point $(-1, 3)$.

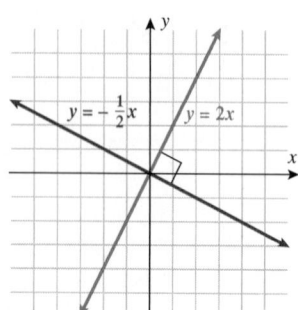

Two *perpendicular* lines form a right angle at their point of intersection. Notice the slopes of the two perpendicular lines in the figure to the right. They are $-\frac{1}{2}$ and 2, negative reciprocals of each other. It turns out that almost all perpendicular lines share this property. Horizontal ($m = 0$) and vertical (m undefined) lines do not share this property.

DEFINITION	**Perpendicular Lines**

Except for the special case of a vertical and a horizontal line, two lines in a plane are **perpendicular** if and only if their slopes are negative reciprocals of each other.

In other words, if two lines in a plane are perpendicular, their slopes are negative reciprocals, provided their slopes are defined. Similarly, if the slopes of two lines in a plane are negative reciprocals, then the lines are perpendicular.

WORDS	**MATH**	**Study Tip**
Lines L_1 and L_2 are perpendicular.	$L_1 \perp L_2$	If a line has slope equal to 3, then a line perpendicular to it has slope $-\frac{1}{3}$.
Two perpendicular lines have negative reciprocal slopes.	$m_1 = -\dfrac{1}{m_2}$ $m_1 \neq 0, m_2 \neq 0$	

EXAMPLE 7 Finding an Equation of a Line That Is Perpendicular to Another Line

Find the equation of the line that passes through the point $(3, 0)$ and is perpendicular to the line $y = 3x + 1$.

Solution:

Identify the slope of the given line $y = 3x + 1$.
$$m_1 = 3$$

The slope of a line perpendicular to the given line is the negative reciprocal of the slope of the given line.
$$m_2 = -\frac{1}{m_1} = -\frac{1}{3}$$

Write the equation of the line we are looking for in slope–intercept form.
$$y = m_2 x + b$$

Substitute $m_2 = -\frac{1}{3}$ into $y = m_2 x + b$.
$$y = -\frac{1}{3}x + b$$

Since the desired line passes through $(3, 0)$, this point must satisfy the equation.
$$0 = -\frac{1}{3}(3) + b$$
$$0 = -1 + b$$

Solve for b.
$$b = 1$$

The equation of the line is $\boxed{y = -\frac{1}{3}x + 1}$.

■ **YOUR TURN** Find the equation of the line that passes through the point $(1, -5)$ and is perpendicular to the line $y = -\frac{1}{2}x + 4$.

■ **Answer:** $y = 2x - 7$

SECTION 0.6 SUMMARY

Lines are often expressed in two forms:

■ General Form: $Ax + By = C$
■ Slope–Intercept Form: $y = mx + b$

All lines (except horizontal and vertical) have exactly one x-intercept and exactly one y-intercept. The slope of a line is a measure of steepness.

■ Slope of a line passing through (x_1, y_1) and (x_2, y_2):
$$m = \frac{y_2 - y_1}{x_2 - x_1} = \frac{\text{rise}}{\text{run}} \quad x_1 \neq x_2$$

■ Horizontal lines: $m = 0$
■ Vertical lines: m is undefined

An equation of a line can be found if either two points or the slope and a point are given. The point–slope form $y - y_1 = m(x - x_1)$ is useful when the slope and a point are given. Parallel lines have the same slope. Perpendicular lines have negative reciprocal (opposite) slopes, provided their slopes are defined.

SECTION
0.6 EXERCISES

▪ SKILLS

In Exercises 1–10, find the slope of the line that passes through the given points.

1. (1, 3) and (2, 6)
2. (2, 1) and (4, 9)
3. (−2, 5) and (2, −3)
4. (−1, −4) and (4, 6)

5. (−7, 9) and (3, −10)
6. (11, −3) and (−2, 6)
7. (0.2, −1.7) and (3.1, 5.2)
8. (−2.4, 1.7) and (−5.6, −2.3)

9. $\left(\frac{2}{3}, -\frac{1}{4}\right)$ and $\left(\frac{5}{6}, -\frac{3}{4}\right)$
10. $\left(\frac{1}{2}, \frac{3}{5}\right)$ and $\left(-\frac{3}{4}, \frac{7}{5}\right)$

For each graph in Exercises 11–16, identify (by inspection) the x- and y-intercepts and slope if they exist, and classify the line as rising, falling, horizontal, or vertical.

11.
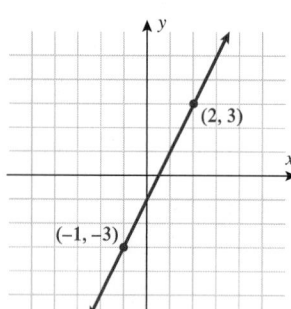

12.
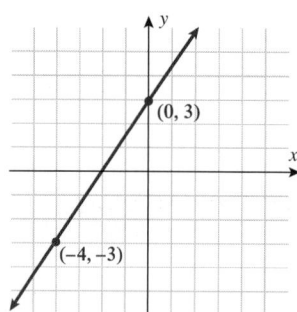

13.
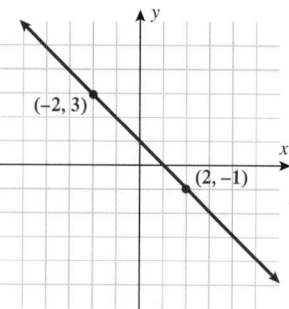

14.
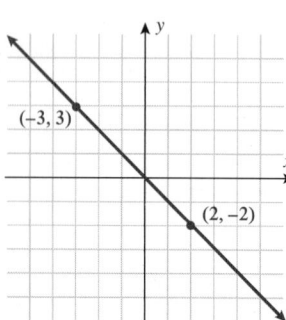

15.
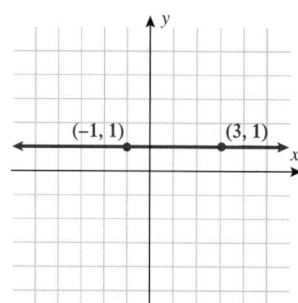

16.
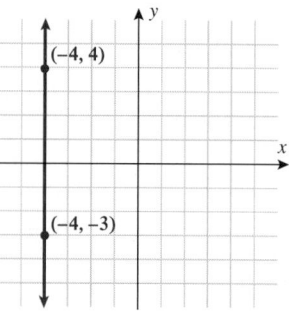

In Exercises 17–30, find the x- and y-intercepts if they exist and graph the corresponding line.

17. $y = 2x - 3$
18. $y = -3x + 2$
19. $y = -\frac{1}{2}x + 2$
20. $y = \frac{1}{3}x - 1$
21. $2x - 3y = 4$

22. $-x + y = -1$
23. $\frac{1}{2}x + \frac{1}{2}y = -1$
24. $\frac{1}{3}x - \frac{1}{4}y = \frac{1}{12}$
25. $x = -1$
26. $y = -3$

27. $y = 1.5$
28. $x = -7.5$
29. $x = -\frac{7}{2}$
30. $y = \frac{5}{3}$

In Exercises 31–42, write the equation in slope–intercept form. Identify the slope and the y-intercept.

31. $2x - 5y = 10$
32. $3x - 4y = 12$
33. $x + 3y = 6$
34. $x + 2y = 8$

35. $4x - y = 3$
36. $x - y = 5$
37. $12 = 6x + 3y$
38. $4 = 2x - 8y$

39. $0.2x - 0.3y = 0.6$
40. $0.4x + 0.1y = 0.3$
41. $\frac{1}{2}x + \frac{2}{3}y = 4$
42. $\frac{1}{4}x + \frac{2}{5}y = 2$

In Exercises 43–50, write the equation of the line, given the slope and intercept.

43. Slope: $m = 2$
y-intercept: (0, 3)

44. Slope: $m = -2$
y-intercept: (0, 1)

45. Slope: $m = -\frac{1}{3}$
y-intercept: (0, 0)

46. Slope: $m = \frac{1}{2}$
y-intercept: (0, −3)

47. Slope: $m = 0$
y-intercept: (0, 2)

48. Slope: $m = 0$
y-intercept: (0, −1.5)

49. Slope: undefined
x-intercept: $\left(\frac{3}{2}, 0\right)$

50. Slope: undefined
x-intercept: (−3.5, 0)

In Exercises 51–60, write an equation of the line in slope–intercept form, if possible, given the slope and a point that lies on the line.

51. Slope: $m = 5$
$(-1, -3)$

52. Slope: $m = 2$
$(1, -1)$

53. Slope: $m = -3$
$(-2, 2)$

54. Slope: $m = -1$
$(3, -4)$

55. Slope: $m = \frac{3}{4}$
$(1, -1)$

56. Slope: $m = -\frac{1}{7}$
$(-5, 3)$

57. Slope: $m = 0$
$(-2, 4)$

58. Slope: $m = 0$
$(3, -3)$

59. Slope: undefined
$(-1, 4)$

60. Slope: undefined
$(4, -1)$

In Exercises 61–80, write the equation of the line that passes through the given points. Express the equation in slope–intercept form or in the form $x = a$ or $y = b$.

61. $(-2, -1)$ and $(3, 2)$

62. $(-4, -3)$ and $(5, 1)$

63. $(-3, -1)$ and $(-2, -6)$

64. $(-5, -8)$ and $(7, -2)$

65. $(20, -37)$ and $(-10, -42)$

66. $(-8, 12)$ and $(-20, -12)$

67. $(-1, 4)$ and $(2, -5)$

68. $(-2, 3)$ and $(2, -3)$

69. $\left(\frac{1}{2}, \frac{3}{4}\right)$ and $\left(\frac{3}{2}, \frac{9}{4}\right)$

70. $\left(-\frac{2}{3}, -\frac{1}{2}\right)$ and $\left(\frac{7}{3}, \frac{1}{2}\right)$

71. $(3, 5)$ and $(3, -7)$

72. $(-5, -2)$ and $(-5, 4)$

73. $(3, 7)$ and $(9, 7)$

74. $(-2, -1)$ and $(3, -1)$

75. $(0, 6)$ and $(-5, 0)$

76. $(0, -3)$ and $(0, 2)$

77. $(-6, 8)$ and $(-6, -2)$

78. $(-9, 0)$ and $(-9, 2)$

79. $\left(\frac{2}{5}, -\frac{3}{4}\right)$ and $\left(\frac{2}{5}, \frac{1}{2}\right)$

80. $\left(\frac{1}{3}, \frac{2}{5}\right)$ and $\left(\frac{1}{3}, \frac{1}{2}\right)$

In Exercises 81–86, write the equation corresponding to each line. Express the equation in slope–intercept form.

81.
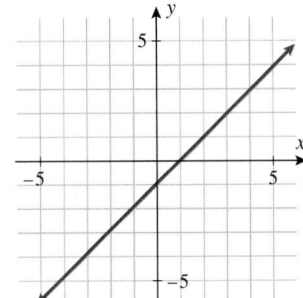

82.
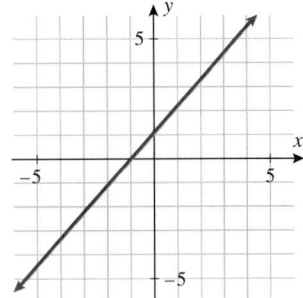

83.
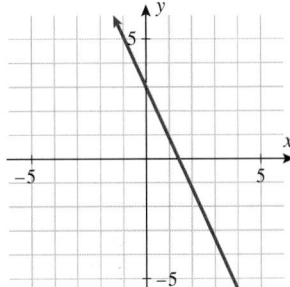

84.
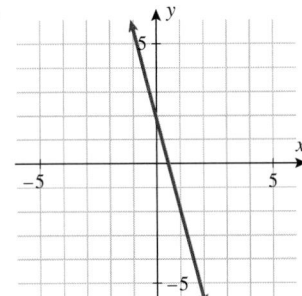

85.
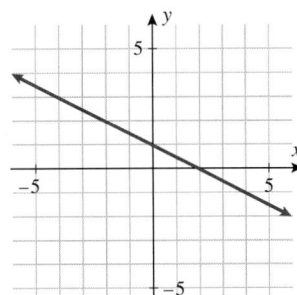

86.
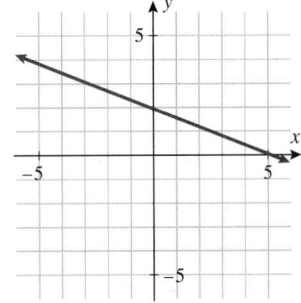

In Exercises 87–96, find the equation of the line that passes through the given point and also satisfies the additional piece of information. Express your answer in slope–intercept form, if possible.

87. $(-3, 1)$; parallel to the line $y = 2x - 1$

88. $(1, 3)$; parallel to the line $y = -x + 2$

89. $(0, 0)$; perpendicular to the line $2x + 3y = 12$

90. $(0, 6)$; perpendicular to the line $x - y = 7$

91. $(3, 5)$; parallel to the x-axis

92. $(3, 5)$; parallel to the y-axis

93. $(-1, 2)$; perpendicular to the y-axis

94. $(-1, 2)$; perpendicular to the x-axis

95. $(-2, -7)$; parallel to the line $\frac{1}{2}x - \frac{1}{3}y = 5$

96. $(1, 4)$; perpendicular to the line $-\frac{2}{3}x + \frac{3}{2}y = -2$

APPLICATIONS

97. Budget: Home Improvement. The cost of having your bathroom remodeled is the combination of material costs and labor costs. The materials (tile, grout, toilet, fixtures, etc.) cost is $1,200 and the labor cost is $25 per hour. Write an equation that models the total cost C of having your bathroom remodeled as a function of hours h. How much will the job cost if the worker estimates 32 hours?

98. Budget: Rental Car. The cost of a one-day car rental is the sum of the rental fee, $50, plus $.39 per mile. Write an equation that models the total cost associated with the car rental.

99. Budget: Monthly Driving Costs. The monthly costs associated with driving a new Honda Accord are the monthly loan payment plus $25 every time you fill up with gasoline. If you fill up 5 times in a month, your total monthly cost is $500. How much is your loan payment?

100. Budget: Monthly Driving Costs. The monthly costs associated with driving a Ford Explorer are the monthly loan payment plus the cost of filling up your tank with gasoline. If you fill up 3 times in a month, your total monthly cost is $520. If you fill up 5 times in a month, your total monthly cost is $600. How much is your monthly loan, and how much does it cost every time you fill up with gasoline?

101. Weather: Temperature. The National Oceanic and Atmospheric Administration (NOAA) has an online conversion chart that relates degrees Fahrenheit, °F, to degrees Celsius, °C. 77 °F is equivalent to 25 °C, and 68 °F is equivalent to 20 °C. Assuming the relationship is linear, write the equation relating degrees Celsius to degrees Fahrenheit. What temperature is the same in both degrees Celsius and degrees Fahrenheit?

102. Weather: Temperature. According to NOAA, a "standard day" is 15 °C at sea level, and every 500-feet elevation above sea level corresponds to a 1 °C temperature drop. Assuming the relationship between temperature and elevation is linear, write an equation that models this relationship. What is the expected temperature at 2500 feet on a "standard day"?

103. Life Sciences: Height. The average height of a man has increased over the last century. What is the rate of change in inches per year of the average height of men?

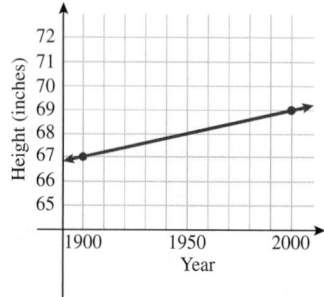

104. Life Sciences: Height. The average height of a woman has increased over the last century. What is the rate of change in inches per year of the average height of women?

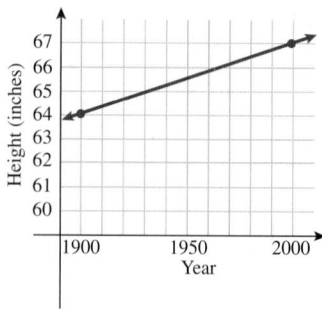

105. Life Sciences: Weight. The average weight of a baby born in 1900 was 6 pounds 4 ounces. In 2000 the average weight of a newborn was 6 pounds 10 ounces. What is the rate of change of birth weight in ounces per year? What do we expect babies to weigh at birth in 2040?

106. Sports. The fastest a man could run a mile in 1906 was 4 minutes and 30 seconds. In 1957 Don Bowden became the first American to break the 4-minute mile. Calculate the rate of change in mile speed per year.

107. Monthly Phone Costs. Mike's home phone plan charges a flat monthly fee plus a charge of $.05 per minute for long-distance calls. The total monthly charge is represented by $y = 0.05x + 35$, $x \geq 0$, where y is the total monthly charge and x is the number of long-distance minutes used. Interpret the meaning of the y-intercept.

108. Cost: Automobile. The value of a Daewoo car is given by $y = 11,100 - 1850x$, $x \geq 0$, where y is the value of the car and x is the age of the car in years. Find the x-intercept and y-intercept and interpret the meaning of each.

109. Weather: Rainfall. The average rainfall in Norfolk, Virginia, for July was 5.2 inches in 2003. The average July rainfall for Norfolk was 3.8 inches in 2007. What is the rate of change of rainfall in inches per year? If this trend continues, what is the expected average rainfall in 2010?

110. Weather: Temperature. The average temperature for Boston in January 2005 was 43 °F. In 2007 the average January temperature was 44.5 °F. What is the rate of change of the temperature per year? If this trend continues, what is the expected average temperature in January 2010?

111. Environment. In 2000 Americans used approximately 380 billion plastic bags. In 2005 approximately 392 billion were used. What is the rate of change of plastic bags used per year? How many plastic bags are expected to be used in 2010?

112. Finance: Debt. According to the Federal Reserve, Americans individually owed $744 in revolving credit in 2004. In 2006 they owed approximately $788. What is the rate of change of the amount of revolving credit owed per year? How much were Americans expected to owe in 2008?

■ CATCH THE MISTAKE

In Exercises 113–116, explain the mistake that is made.

113. Find the x- and y-intercepts of the line with equation $2x - 3y = 6$.

Solution:

x-intercept: set $x = 0$ and solve for y. $-3y = 6$
   $y = -2$

The x-intercept is $(0, -2)$.

y-intercept: set $y = 0$ and solve for x. $2x = 6$
   $x = 3$

The y-intercept is $(3, 0)$.

This is incorrect. What mistake was made?

114. Find the slope of the line that passes through the points $(-2, 3)$ and $(4, 1)$.

Solution:

Write the slope formula. $m = \dfrac{y_2 - y_1}{x_2 - x_1}$

Substitute $(-2, 3)$ and $(4, 1)$. $m = \dfrac{1 - 3}{-2 - 4} = \dfrac{-2}{-6} = \dfrac{1}{3}$

This is incorrect. What mistake was made?

115. Find the slope of the line that passes through the points $(-3, 4)$ and $(-3, 7)$.

Solution:

Write the slope formula. $m = \dfrac{y_2 - y_1}{x_2 - x_1}$

Substitute $(-3, 4)$ and $(-3, 7)$. $m = \dfrac{-3 - (-3)}{4 - 7} = 0$

This is incorrect. What mistake was made?

116. Given the slope, classify the line as rising, falling, horizontal, or vertical.
  a. $m = 0$ **b.** m undefined
  c. $m = 2$ **d.** $m = -1$

Solution:

  a. vertical line   **b.** horizontal line
  c. rising   **d.** falling

These are incorrect. What mistakes were made?

■ CONCEPTUAL

In Exercises 117–122, determine whether each statement is true or false.

117. A non-horizontal line can have at most one x-intercept.

118. A line must have at least one y-intercept.

119. If the slopes of two lines are $-\frac{1}{5}$ and 5, then the lines are parallel.

120. If the slopes of two lines are -1 and 1, then the lines are perpendicular.

121. If a line has slope equal to zero, describe a line that is perpendicular to it.

122. If a line has no slope (undefined slope), describe a line that is parallel to it.

■ CHALLENGE

123. Find an equation of a line that passes through the point $(-B, A + 1)$ and is parallel to the line $Ax + By = C$. Assume that B is not equal to zero.

124. Find an equation of a line that passes through the point $(B, A - 1)$ and is parallel to the line $Ax + By = C$. Assume that B is not equal to zero.

125. Find an equation of a line that passes through the point $(-A, B - 1)$ and is perpendicular to the line $Ax + By = C$. Assume that A and B are both nonzero.

126. Find an equation of a line that passes through the point $(A, B + 1)$ and is perpendicular to the line $Ax + By = C$.

127. Show that two lines with equal slopes and different y-intercepts have no point in common. *Hint:* Let $y_1 = mx + b_1$ and $y_2 = mx + b_2$ with $b_1 \neq b_2$. What equation must be true for there to be a point of intersection? Show that this leads to a contradiction.

128. Let $y_1 = m_1 x + b_1$ and $y_2 = m_2 x + b_2$ be two nonparallel lines ($m_1 \neq m_2$). What is the x-coordinate of the point where they intersect?

In Exercises 129–134, determine whether the lines are parallel, perpendicular, or neither, and then graph both lines in the same viewing screen using a graphing utility to confirm your answer.

129. $y_1 = 17x + 22$

$y_2 = -\frac{1}{17}x - 13$

130. $y_1 = 0.35x + 2.7$

$y_2 = 0.35x - 1.2$

131. $y_1 = 0.25x + 3.3$

$y_2 = -4x + 2$

132. $y_1 = \frac{1}{2}x + 5$

$y_2 = 2x - 3$

133. $y_1 = 0.16x + 2.7$

$y_2 = 6.25x - 1.4$

134. $y_1 = -3.75x + 8.2$

$y_2 = \frac{4}{15}x + \frac{5}{6}$

SECTION 0.7 MODELING VARIATION

SKILLS OBJECTIVES

- Develop mathematical models using direct variation.
- Develop mathematical models using inverse variation.
- Develop mathematical models using combined variation.
- Develop mathematical models using joint variation.

CONCEPTUAL OBJECTIVES

- Understand the difference between direct variation and inverse variation.
- Understand the difference between combined variation and joint variation.

In this section, we discuss mathematical models for different applications. Two quantities in the real world often *vary* with respect to one another. Sometimes, they vary *directly*. For example, the more money we make, the more total dollars of federal income tax we expect to pay. Sometimes, quantities vary *inversely*. For example, when interest rates on mortgages decrease, we expect the number of homes purchased to increase, because a buyer can afford "more house" with the same mortgage payment when rates are lower. In this section, we discuss quantities varying *directly*, *inversely*, or *jointly*.

Direct Variation

When one quantity is a constant multiple of another quantity, we say that the quantities are *directly proportional* to one another.

DIRECT VARIATION

Let x and y represent two quantities. The following are equivalent statements:

- $y = kx$, where k is a nonzero constant.
- y **varies directly** with x.
- y is **directly proportional** to x.

The constant k is called the **constant of variation** or the **constant of proportionality**.

In 2005 the national average cost of residential electricity was 9.53 ¢/kWh (cents per kilowatt-hour). For example, if a residence used 3400 kWh, then the bill would be $324, and if a residence used 2500 kWh, then the bill would be $238.25.

EXAMPLE 1   Finding the Constant of Variation

In the United States, the cost of electricity is directly proportional to the number of kilowatt · hours (kWh) used. If a household in Tennessee on average used 3098 kWh per month and had an average monthly electric bill of \$179.99, find a mathematical model that gives the cost of electricity in Tennessee in terms of the number of kilowatt · hours used.

Solution:

Write the direct variation model.

$$y = kx$$

Label the variables and constant.

x = number of kWh
y = cost (dollars)
k = cost per kWh

Substitute the given data $x = 3098$ kWh and $y = \$179.99$ into $y = kx$.

$$179.99 = 3098k$$

Solve for k.

$$k = \frac{179.99}{3098} \approx 0.05810$$

$$y = 0.0581x$$

In Tennessee the cost of electricity is 5.81 ¢/kWh .

■ **YOUR TURN** Find a mathematical model that describes the cost of electricity in California if the cost is directly proportional to the number of kWh used and a residence that consumes 4000 kWh is billed \$480.

■ **Answer:** $y = 0.12x$; the cost of electricity in California is 12 ¢/kWh.

Not all variation we see in nature is direct variation. Isometric growth, where the various parts of an organism grow in direct proportion to each other, is rare in living organisms. If organisms grew isometrically, young children would look just like adults, only smaller. In contrast, most organisms grow nonisometrically; the various parts of organisms do not increase in size in a one-to-one ratio. The relative proportions of a human body change dramatically as the human grows. Children have proportionately larger heads and shorter legs than adults. *Allometric growth* is the pattern of growth whereby different parts of the body grow at different rates with respect to each other. Some human body characteristics vary directly, and others can be mathematically modeled by *direct variation with powers*.

DIRECT VARIATION WITH POWERS

Let x and y represent two quantities. The following are equivalent statements:

■ $y = kx^n$, where k is a nonzero constant.
■ y **varies directly with the nth power** of x.
■ y **is directly proportional to the nth power** of x.

One example of direct variation with powers is height and weight of humans. Statistics show that weight (in pounds) is directly proportional to the cube of height (feet):

$$W = kH^3$$

EXAMPLE 2   Direct Variation with Powers

The following is a personal ad:

Single professional male (6 ft/194 lb) seeks single professional female for long-term relationship. Must be athletic, smart, like the movies and dogs, and have height and weight similarly proportioned to mine.

Find a mathematical equation that describes the height and weight of the male who wrote the ad. How much would a 5′6″ woman weigh who has the same proportionality as the male?

Solution:

Write the direct variation (cube) model for height versus weight.

$$W = kH^3$$

Substitute the given data $W = 194$ and $H = 6$ into $W = kH^3$.

$$194 = k(6)^3$$

Solve for k.

$$k = \frac{194}{216} = 0.898148 \approx 0.90$$

$$W = 0.9H^3$$

Let $H = 5.5$ ft.

$$W = 0.9(5.5)^3 \approx 149.73$$

A woman 5′6″ tall with the same height and weight proportionality as the male would weigh approximately $\boxed{150\ \text{lb}}$.

■ **Answer:** ≈ 200 pounds

■ **YOUR TURN** A brother and sister both have weight (pounds) that varies as the cube of height (feet) and they share the same proportionality constant. The sister is 6′ tall and weighs 170 pounds. Her brother is 6′4″. How much does he weigh?

Inverse Variation

Two fundamental topics covered in economics are supply and demand. Supply is the quantity that producers are willing to sell at a given price. For example, an artist may be willing to paint and sell 5 portraits if each sells for $50, but that same artist may be willing to sell 100 portraits if each sells for $10,000. Demand is the quantity of a good that consumers are not only willing to purchase but also have the capacity to buy at a given price. For example, consumers may purchase 1 billion Big Macs from McDonald's every year, but perhaps only 1 million filets mignons are sold at Outback. There may be 1 billion people who want to buy the filet mignon but don't have the financial means to do so. Economists study the equilibrium between supply and demand.

Demand can be modeled with an *inverse variation* of price: When the price increases, demand decreases, and vice versa.

INVERSE VARIATION

Let x and y represent two quantities. The following are equivalent statements:

- $y = \dfrac{k}{x}$, where k is a nonzero constant.
- y **varies inversely** with x.
- y is **inversely proportional** to x.

The constant k is called the **constant of variation** or the **constant of proportionality**.

EXAMPLE 3   **Inverse Variation**

The number of potential buyers of a house decreases as the price of the house increases (see graph on the right). If the number of potential buyers of a house in a particular city is inversely proportional to the price of the house, find a mathematical equation that describes the demand for houses as it relates to price. How many potential buyers will there be for a $2 million house?

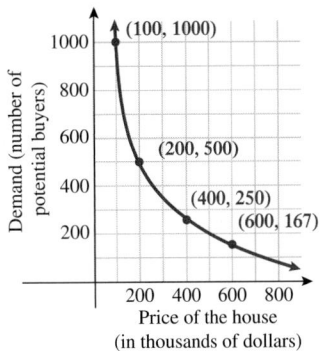

Solution:

Write the inverse variation model.

$$y = \frac{k}{x}$$

Label the variables and constant.

x = price of house in thousands of dollars
y = number of buyers

Select *any* point that lies on the curve.

(200, 500)

Substitute the given data $x = 200$ and $y = 500$ into $y = \frac{k}{x}$.

$$500 = \frac{k}{200}$$

Solve for k.

$$k = 200 \cdot 500 = 100,000$$

$$y = \frac{100,000}{x}$$

Let $x = 2000$.

$$y = \frac{100,000}{2000} = 50$$

There are only 50 potential buyers for a $2 million house in this city.

■ **YOUR TURN**   In New York City, the number of potential buyers in the housing market is inversely proportional to the price of a house. If there are 12,500 potential buyers for a $2 million condominium, how many potential buyers are there for a $5 million condominium?

■ **Answer:** 5000

Two quantities can vary inversely with the *n*th power of *x*.

If x and y are related by the equation $y = \frac{k}{x^n}$, then we say that y varies **inversely with the *n*th power of *x***, or y is inversely **proportional to the *n*th power of *x***.

Joint Variation and Combined Variation

We now discuss combinations of variations. When one quantity is proportional to the product of two or more other quantities, the variation is called **joint variation**. When direct variation and inverse variation occur at the same time, the variation is called **combined variation**.

An example of a **joint variation** is simple interest (Section 0.1), which is defined as

$$I = Prt$$

where

- I is the interest in dollars.
- P is the principal (initial) in dollars.
- r is the interest rate (expressed in decimal form).
- t is time in years.

The interest earned is proportional to the product of three quantities (principal, interest rate, and time). Note that if the interest rate increases, then the interest earned also increases. Similarly, if either the initial investment (principal) or the time the money is invested increases, then the interest earned also increases.

An example of **combined variation** is the combined gas law in chemistry:

$$P = k\frac{T}{V}$$

where

- P is pressure.
- T is temperature (kelvins).
- V is volume.
- k is a gas constant.

This relation states that the pressure of a gas is directly proportional to the temperature and inversely proportional to the volume containing the gas. For example, as the temperature increases, the pressure increases, but when the volume decreases, pressure increases.

As an example, the gas in the headspace of a soda bottle has a fixed volume. Therefore, as temperature increases, the pressure increases. Compare the different pressures of opening a twist-off cap on a bottle of soda that is cold versus one that is hot. The hot one feels as though it "releases more pressure."

 EXAMPLE 4   Combined Variation

The gas in the headspace of a soda bottle has a volume of 9.0 ml, pressure of 2 atm (atmospheres), and a temperature of 298 K (standard room temperature of 77 °F). If the soda bottle is stored in a refrigerator, the temperature drops to approximately 279 K (42 °F). What is the pressure of the gas in the headspace once the bottle is chilled?

Solution:

Write the combined gas law.

$$P = k\frac{T}{V}$$

Let $P = 2$ atm, $T = 298$ K, and $V = 9.0$ ml.

$$2 = k\frac{298}{9}$$

Solve for k.

$$k = \frac{18}{298}$$

Let $k = \frac{18}{298}$, $T = 279$, and $V = 9.0$ in $P = k\frac{T}{V}$.

$$P = \frac{18}{298} \cdot \frac{279}{9} \approx 1.87$$

Since we used the same physical units for both the chilled and room-temperature soda bottles, the pressure is in atmospheres.

$\boxed{P = 1.87 \text{ atm}}$

SECTION
0.7 SUMMARY

Direct, inverse, joint, and combined variation can be used to model the relationship between two quantities. For two quantities x and y, we say that

- y is directly proportional to x if $y = kx$.
- y is inversely proportional to x if $y = \dfrac{k}{x}$.

Joint variation occurs when one quantity is directly proportional to two or more quantities. Combined variation occurs when one quantity is directly proportional to one or more quantities and inversely proportional to one or more other quantities.

SECTION
0.7 EXERCISES

SKILLS

In Exercises 1–16, write an equation that describes each variation. Use k as the constant of variation.

1. y varies directly with x.
2. s varies directly with t.
3. V varies directly with x^3.
4. A varies directly with x^2.
5. z varies directly with m.
6. h varies directly with $\sqrt{t}$.
7. f varies inversely with λ.
8. P varies inversely with r^2.
9. F varies directly with w and inversely with L.
10. V varies directly with T and inversely with P.
11. v varies directly with both g and t.
12. S varies directly with both t and d.
13. R varies inversely with both P and T.
14. y varies inversely with both x and z.
15. y is directly proportional to the square root of x.
16. y is inversely proportional to the cube of t.

In Exercises 17–36, write an equation that describes each variation.

17. d is directly proportional to t; $d = r$ when $t = 1$.
18. F is directly proportional to m; $F = a$ when $m = 1$.
19. V is directly proportional to both l and w; $V = 6h$ when $w = 3$ and $l = 2$.
20. A is directly proportional to both b and h; $A = 10$ when $b = 5$ and $h = 4$.
21. A varies directly with the square of r; $A = 9\pi$ when $r = 3$.
22. V varies directly with the cube of r; $V = 36\pi$ when $r = 3$.
23. V varies directly with both h and r^2; $V = 1$ when $r = 2$ and $h = \dfrac{4}{\pi}$.
24. W is directly proportional to both R and the square of I; $W = 4$ when $R = 100$ and $I = 0.25$.
25. V varies inversely with P; $V = 1000$ when $P = 400$.
26. I varies inversely with the square of d; $I = 42$ when $d = 16$.
27. F varies inversely with both λ and L; $F = 20\pi$ when $\lambda = 1$ μm (micrometers or microns) and $L = 100$ km.
28. y varies inversely with both x and z; $y = 32$ when $x = 4$ and $z = 0.05$.
29. t varies inversely with s; $t = 2.4$ when $s = 8$.
30. W varies inversely with the square of d; $W = 180$ when $d = 0.2$.
31. R varies inversely with the square of I; $R = 0.4$ when $I = 3.5$.
32. y varies inversely with both x and the square root of z; $y = 12$ when $x = 0.2$ and $z = 4$.
33. R varies directly with L and inversely with A; $R = 0.5$ when $L = 20$ and $A = 0.4$.

34. F varies directly with m and inversely with d; $F = 32$ when $m = 20$ and $d = 8$.

35. F varies directly with both m_1 and m_2 and inversely with the square of d; $F = 20$ when $m_1 = 8$, $m_2 = 16$, and $d = 0.4$.

36. w varies directly with the square root of g and inversely with the square of t; $w = 20$ when $g = 16$ and $t = 0.5$.

■ APPLICATIONS

37. Wages. Jason and Valerie both work at Panera Bread and have the following paycheck information for a certain week. Find an equation that shows their wages W varying directly with the number of hours worked H.

EMPLOYEE	HOURS WORKED	WAGES
Jason	23	$172.50
Valerie	32	$240.00

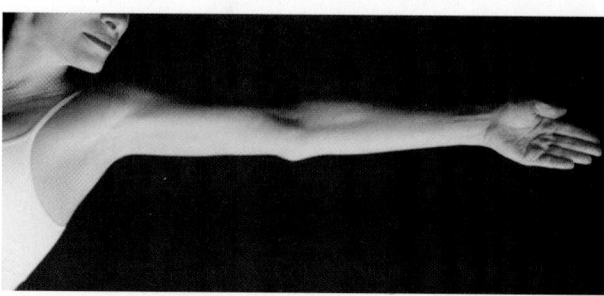

Kim Steele/Getty Images, Inc.

38. Sales Tax. The sales tax in Orange and Seminole Counties in Florida differs by only 0.5%. A new resident knows this but doesn't know which of the counties has the higher tax. The resident lives near the border of the counties and is in the market for a new plasma television and wants to purchase it in the county with the lower tax. If the tax on a pair of $40 sneakers is $2.60 in Orange County and the tax on a $12 T-shirt is $.84 in Seminole County, write two equations: one for each county that describes the tax T, which is directly proportional to the purchase price P.

For Exercises 39 and 40, refer to the following:

The ratio of the speed of an object to the speed of sound determines the Mach number. Aircraft traveling at a subsonic speed (less than the speed of sound) have a Mach number less than 1. In other words, the speed of an aircraft is directly proportional to its Mach number. Aircraft traveling at a supersonic speed (greater than the speed of sound) have a Mach number greater than 1. The speed of sound at sea level is approximately 760 mph.

39. Military. The U.S. Navy Blue Angels fly F-18 Hornets that are capable of Mach 1.7. How fast can F-18 Hornets fly at sea level?

40. Military. The U.S. Air Force's newest fighter aircraft is the F-22A Raptor, which is capable of Mach 1.5. How fast can a F-22A Raptor fly at sea level?

Exercises 41 and 42 are examples of the golden ratio, or phi, a proportionality constant that appears in nature. The numerical approximate value of phi is 1.618 (from www.goldenratio.net).

41. Human Anatomy. The length of your forearm F (wrist to elbow) is directly proportional to the length of your hand H (length from wrist to tip of middle finger). Write the equation that describes this relationship if the length of your forearm is 11 inches and the length of your hand is 6.8 inches.

42. Human Anatomy. Each section of your index finger, from the tip to the base of the wrist, is larger than the preceding one by about the golden (Fibonacci) ratio. Find an equation that represents the ratio of each section of your finger related to the previous one if one section is 8 units long and the next section is 5 units long.

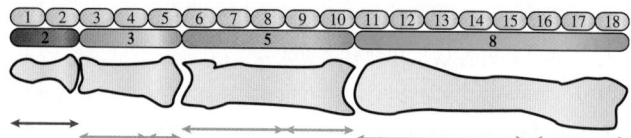

For Exercises 43 and 44, refer to the following:

Hooke's law in physics states that if a spring at rest (equilibrium position) has a weight attached to it, then the distance the spring stretches is directly proportional to the force (weight), according to the formula:

$$F = kx$$

where F is the force in Newtons (N), x is the distance stretched in meters (m), and k is the spring constant (N/m).

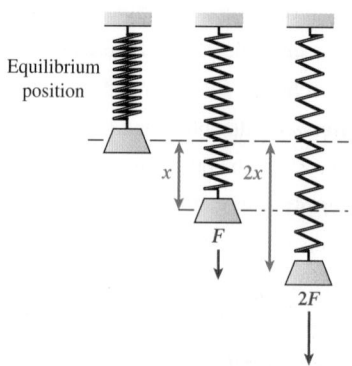

Equilibrium position

43. Physics. A force of 30 N will stretch the spring 10 cm. How far will a force of 72 N stretch the spring?

44. Physics. A force of 30 N will stretch the spring 10 cm. How much force is required to stretch the spring 18 cm?

45. Sales. Levi's makes jeans in a variety of price ranges for juniors. The Flare 519 jeans sell for about $20, whereas the 646 Vintage Flare jeans sell for $300. The demand for Levi's jeans is inversely proportional to the price. If 300,000 pairs of the 519 jeans were bought, approximately how many of the Vintage Flare jeans were bought?

46. Sales. Levi's makes jeans in a variety of price ranges for men. The Silver Tab Baggy jeans sell for about $30, whereas the Offender jeans sell for about $160. The demand for Levi's jeans is inversely proportional to the price. If 400,000 pairs of the Silver Tab Baggy jeans were bought, approximately how many of the Offender jeans were bought?

For Exercises 47 and 48, refer to the following:

In physics, the inverse square law states that any physical force or energy flow is inversely proportional to the square of the distance from the source of that physical quantity. In particular, the intensity of light radiating from a point source is inversely proportional to the square of the distance from the source. Below is a table of average distances from the Sun:

PLANET	DISTANCE TO THE SUN
Mercury	58,000 km
Earth	150,000 km
Mars	228,000 km

47. Solar Radiation. The solar radiation on Earth is approximately 1400 watts per square meter (W/m^2). How much solar radiation is there on Mars? Round to the nearest hundred watts per square meter.

48. Solar Radiation. The solar radiation on Earth is approximately 1400 watts per square meter. How much solar radiation is there on Mercury? Round to the nearest hundred watts per square meter.

49. Investments. Marilyn receives a $25,000 bonus from her company and decides to put the money toward a new car that she will need in 2 years. Simple interest is directly proportional to the principal and the time invested. She compares two different banks' rates on money market accounts. If she goes with Bank of America, she will earn $750 in interest, but if she goes with the Navy Federal Credit Union, she will earn $1,500. What is the interest rate on money market accounts at both banks?

50. Investments. Connie and Alvaro sell their house and buy a fixer-upper house. They made $130,000 on the sale of their previous home. They know it will take 6 months before the general contractor can start their renovation, and they want to take advantage of a 6-month CD that pays simple interest. What is the rate of the 6-month CD if they will make $3,250 in interest?

51. Chemistry. A gas contained in a 4-ml container at a temperature of 300 K has a pressure of 1 atm. If the temperature decreases to 275 K, what is the resulting pressure?

52. Chemistry. A gas contained in a 4-ml container at a temperature of 300 K has a pressure of 1 atm. If the container changes to a volume of 3 ml, what is the resulting pressure?

▪ CATCH THE MISTAKE

In Exercises 53 and 54, explain the mistake that is made.

53. y varies directly with t and inversely with x. When $x = 4$ and $t = 2$, then $y = 1$. Find an equation that describes this variation.

Solution:

Write the variation equation.	$y = ktx$
Let $x = 4$, $t = 2$, and $y = 1$.	$1 = k(2)(4)$
Solve for k.	$k = \dfrac{1}{8}$
Substitute $k = \frac{1}{8}$ into $y = ktx$.	$y = \dfrac{1}{8}tx$

This is incorrect. What mistake was made?

54. y varies directly with t and the square of x. When $x = 4$ and $t = 1$, then $y = 8$. Find an equation that describes this variation.

Solution:

Write the variation equation.	$y = kt\sqrt{x}$
Let $x = 4$, $t = 1$, and $y = 8$.	$8 = k(1)\sqrt{4}$
Solve for k.	$k = 4$
Substitute $k = 4$ into $y = kt\sqrt{x}$.	$y = 4t\sqrt{x}$

This is incorrect. What mistake was made?

▪ CONCEPTUAL

In Exercises 55 and 56, determine whether each statement is true or false.

55. The area of a triangle is directly proportional to both the base and the height of the triangle (joint variation).

56. Average speed is directly proportional to both distance and time (joint variation).

In Exercises 57 and 58, match the variation with the graph.

57. Inverse variation

58. Direct variation

a.

b.

▪ CHALLENGE

Exercises 59 and 60 involve the theory governing laser propagation through Earth's atmosphere.

The three parameters that help classify the strength of optical turbulence are the following:

▪ C_n^2, index of refraction structure parameter

▪ k, wave number of the laser, which is inversely proportional to the wavelength λ of the laser:

$$k = \frac{2\pi}{\lambda}$$

▪ L, propagation distance

The variance of the irradiance of a laser, σ^2, is directly proportional to C_n^2, $k^{7/6}$, and $L^{11/16}$.

59. When $C_n^2 = 1.0 \times 10^{-13} \,\text{m}^{-2/3}$, $L = 2$ km, and $\lambda = 1.55 \,\mu\text{m}$, the variance of irradiance for a plane wave σ_{pl}^2 is 7.1. Find the equation that describes this variation.

60. When $C_n^2 = 1.0 \times 10^{-13} \,\text{m}^{-2/3}$, $L = 2$ km, and $\lambda = 1.55 \,\mu\text{m}$, the variance of irradiance for a spherical wave σ_{sp}^2 is 2.3. Find the equation that describes this variation.

▪ TECHNOLOGY

For Exercises 61–64, refer to the following:

Data from 1995 to 2006 for oil prices in dollars per barrel, the U.S. Dow Jones Utilities Stock Index, New Privately Owned Housing, and 5-year Treasury Constant Maturity Rate are given in the table below. (These data are from Forecast Center's Historical Economic and Market Home Page at www.neatideas.com/djutil.htm.)

Use the calculator STAT EDIT commands to enter the table with L_1 as the oil price, L_2 as the utilities stock index, L_3 as number of housing units, and L_4 as the 5-year maturity rate.

JANUARY OF EACH YEAR	OIL PRICE, $ PER BARREL	U.S. DOW JONES UTILITIES STOCK INDEX	NEW, PRIVATELY OWNED HOUSING UNITS	5-YEAR TREASURY CONSTANT MATURITY RATE
1995	17.99	193.12	1407	7.76
1996	18.88	230.85	1467	5.36
1997	25.17	232.53	1355	6.33
1998	16.71	263.29	1525	5.42
1999	12.47	302.80	1748	4.60
2000	27.18	315.14	1636	6.58
2001	29.58	372.32	1600	4.86
2002	19.67	285.71	1698	4.34
2003	32.94	207.75	1853	3.05
2004	32.27	271.94	1911	3.12
2005	46.84	343.46	2137	3.71
2006	65.51	413.84	2265	4.35

61. An increase in oil price in dollars per barrel will drive the U.S. Dow Jones Utilities Stock Index to soar.

  a. Use the calculator commands $\boxed{\text{STAT}}$, $\boxed{\text{linReg}}$ $(ax + b)$, and $\boxed{\text{STATPLOT}}$ to model the data using the least squares regression. Find the equation of the least squares regression line using x as the oil price in dollars per barrel.

  b. If the U.S. Dow Jones Utilities Stock Index varies directly as the oil price in dollars per barrel, then use the calculator commands $\boxed{\text{STAT}}$, $\boxed{\text{PwrReg}}$, and $\boxed{\text{STATPLOT}}$ to model the data using the power function. Find the variation constant and equation of variation using x as the oil price in dollars per barrel.

  c. Use the equations you found in (a) and (b) to calculate the stock index when the oil price hit $72.70 per barrel in September 2006. Which answer is closer to the actual stock index of 417? Round all answers to the nearest whole number.

62. An increase in oil price in dollars per barrel will affect the interest rates across the board—in particular, the 5-year Treasury constant maturity rate.

  a. Use the calculator commands $\boxed{\text{STAT}}$, $\boxed{\text{linReg}}$ $(ax + b)$, and $\boxed{\text{STATPLOT}}$ to model the data using the least squares regression. Find the equation of the least squares regression line using x as the oil price in dollars per barrel.

  b. If the 5-year Treasury constant maturity rate varies inversely as the oil price in dollars per barrel, then use the calculator commands $\boxed{\text{STAT}}$, $\boxed{\text{PwrReg}}$, and $\boxed{\text{STATPLOT}}$ to model the data using the power function. Find the variation constant and equation of variation using x as the oil price in dollars per barrel.

  c. Use the equations you found in (a) and (b) to calculate the maturity rate when the oil price hit $72.70 per barrel in September 2006. Which answer is closer to the actual maturity rate at 5.02%? Round all answers to two decimal places.

63. An increase in interest rates—in particular, the 5-year Treasury constant maturity rate—will affect the number of new, privately owned housing units.

  a. Use the calculator commands $\boxed{\text{STAT}}$, $\boxed{\text{linReg}}$ $(ax + b)$, and $\boxed{\text{STATPLOT}}$ to model the data using the least squares regression. Find the equation of the least squares regression line using x as the 5-year rate.

  b. If the number of new privately owned housing units varies inversely as the 5-year Treasury constant maturity rate, then use the calculator commands $\boxed{\text{STAT}}$, $\boxed{\text{PwrReg}}$, and $\boxed{\text{STATPLOT}}$ to model the data using the power function. Find the variation constant and equation of variation using x as the 5-year rate.

  c. Use the equations you found in (a) and (b) to calculate the number of housing units when the maturity rate was 5.02% in September 2006. Which answer is closer to the actual number of new, privately owned housing units of 1861? Round all answers to the nearest unit.

64. An increase in the number of new, privately owned housing units will affect the U.S. Dow Jones Utilities Stock Index.

  a. Use the calculator commands $\boxed{\text{STAT}}$, $\boxed{\text{linReg}}$ $(ax + b)$, and $\boxed{\text{STATPLOT}}$ to model the data using the least squares regression. Find the equation of the least squares regression line using x as the number of housing units.

  b. If the U.S. Dow Jones Utilities Stock Index varies directly as the number of new, privately owned housing units, then use the calculator commands $\boxed{\text{STAT}}$, $\boxed{\text{PwrReg}}$, and $\boxed{\text{STATPLOT}}$ to model the data using the power function. Find the variation constant and equation of variation using x as the number of housing units.

  c. Use the equations you found in (a) and (b) to find the utilities stock index if there were 1861 new, privately owned housing units in September 2006. Which answer is closer to the actual stock index of 417? Round all answers to the nearest whole number.

For Exercises 65 and 66, refer to the following:

From March 2000 to March 2008, data for retail gasoline price in dollars per gallon are given in the table below. (These data are from Energy Information Administration, Official Energy Statistics from the U.S. Government at http://tonto.eia.doe.gov/oog/info/gdu/gaspump.html.) Use the calculator $\boxed{\text{STAT}}$ $\boxed{\text{EDIT}}$ command to enter the table below with L_1 as the year ($x = 1$ for year 2000) and L_2 as the gasoline price in dollars per gallon.

MARCH OF EACH YEAR	2000	2001	2002	2003	2004	2005	2006	2007	2008
RETAIL GASOLINE PRICE $ PER GALLON	1.517	1.409	1.249	1.693	1.736	2.079	2.425	2.563	3.244

65. a. Use the calculator commands [STAT] [LinReg] to model the data using the least squares regression. Find the equation of the least squares regression line using x as the year ($x = 1$ for year 2000) and y as the gasoline price in dollars per gallon. Round all answers to three decimal places.

b. Use the equation to determine the gasoline price in March 2006. Round all answers to three decimal places. Is the answer close to the actual price?

c. Use the equation to find the gasoline price in March 2009. Round all answers to three decimal places.

66. a. Use the calculator commands [STAT] [PwrReg] to model the data using the power function. Find the variation constant and equation of variation using x as the year ($x = 1$ for year 2000) and y as the gasoline price in dollars per gallon. Round all answers to three decimal places.

b. Use the equation to find the gasoline price in March 2006. Round all answers to three decimal places. Is the answer close to the actual price?

c. Use the equation to determine the gasoline price in March 2009. Round all answers to three decimal places.

MODELING YOUR WORLD

The Intergovernmental Panel on Climate Change (IPCC) claims that Carbon Dioxide (CO_2) production from industrial activity (such as fossil fuel burning and other human activities) has increased the CO_2 concentrations in the atmosphere. Because it is a greenhouse gas, elevated CO_2 levels will increase global mean (average) temperature. In this section, we will examine the increasing rate of carbon emissions on Earth.

In 1955 there were (globally) 2 billion tons of carbon emitted per year. In 2005 the carbon emissions more than tripled to reach approximately 7 billion tons of carbon emitted per year. Currently, we are on the path to doubling our current carbon emissions in the next 50 years.

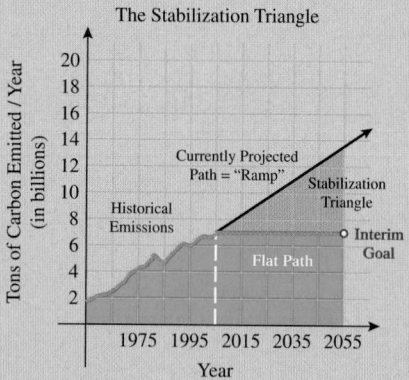

The Stabilization Triangle

Two Princeton professors* (Stephen Pacala and Rob Socolow) introduced the Climate Carbon Wedge concept. A "wedge" is a strategy to reduce carbon emissions over a 50-year time period from zero to 1.0 GtC/yr (gigatons of carbon per year).

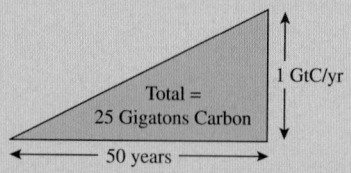

*S. Pacala and R. Socolow, "Stabilization Wedges: Solving the Climate Problem for the Next 50 Years with Current Technologies," *Science*, Vol. 305 (2004).

1. Draw the Cartesian plane. Label the vertical axis C, where C represents the number of gigatons (billions of tons) of carbon emitted, and label the horizontal axis t, where t is the number of years. Let $t = 0$ correspond to 2005.

2. Find the equations of the flat path and the seven lines corresponding to the seven wedges.

   a. Flat path (no increase) over 50 years (2005 to 2055)
   b. Increase of 1 GtC over 50 years (2005 to 2055)
   c. Increase of 2 GtC over 50 years (2005 to 2055)
   d. Increase of 3 GtC over 50 years (2005 to 2055)
   e. Increase of 4 GtC over 50 years (2005 to 2055)
   f. Increase of 5 GtC over 50 years (2005 to 2055)
   g. Increase of 6 GtC over 50 years (2005 to 2055)
   h. Increase of 7 GtC over 50 years (2005 to 2055) [projected path]

3. For each of the seven wedges and the flat path, determine how many **total** gigatons of carbon will be reduced over a 50-year period. In other words, how many gigatons of carbon would the world have to reduce in each of the eight cases?

   a. Flat path
   b. Increase of 1GtC over 50 years
   c. Increase of 2 GtC over 50 years
   d. Increase of 3 GtC over 50 years
   e. Increase of 4 GtC over 50 years
   f. Increase of 5 GtC over 50 years
   g. Increase of 6 GtC over 50 years
   h. Increase of 7 GtC over 50 years (projected path)

4. Research the "climate carbon wedge" concept and discuss the types of changes (transportation efficiency, transportation conservation, building efficiency, efficiency in electricity production, alternate energies, etc.) the world would have to make that would correspond to each of the seven wedges and the flat path.

   a. Flat path
   b. Wedge 1
   c. Wedge 2
   d. Wedge 3
   e. Wedge 4
   f. Wedge 5
   g. Wedge 6

SECTION	CONCEPT	PAGES	REVIEW EXERCISES	KEY IDEAS/POINTS
0.1	**Linear equations**	4–14	1–16	$ax + b = 0$
	Solving linear equations in one variable	5–7	1–12	Isolate variable on one side and constants on the other side.
	Applications involving linear equations	7–14	13–16	Five-step procedure: Step 1: Identify the question. Step 2: Make notes. Step 4: Solve the equation. Step 3: Set up an equation. Step 5: Check the solution. **Geometry problems:** Formulas for rectangles, triangles, and circles **Interest problems:** Simple interest: $I = Prt$ **Mixture problems:** Whenever two *distinct* quantities are mixed, the result is a mixture. **Distance–rate–time problems:** $d = r \cdot t$
0.2	**Quadratic equations**	18–27	17–40	$ax^2 + bx + c = 0$
	Factoring	19–20	17–20	If $(x - h)(x - k) = 0$, then $x = h$ or $x = k$.
	Square root method	20–21	21–24	If $x^2 = P$, then $x = \pm\sqrt{P}$.
	Completing the square	22–23	25–28	Find half of b; square that quantity; add the result to both sides.
	The quadratic formula	24–27	29–32	$x = \dfrac{-b \pm \sqrt{b^2 - 4ac}}{2a}$
0.3	**Other types of equations**	31–40	41–74	
	Rational equations	31–33	41–46	Eliminate any values that make the denominator equal to 0.
	Radical equations	34–36	47–54	Check solutions to avoid extraneous solutions.
	Equations quadratic in form: u-substitution	36–38	55–62	Use a u-substitution to write the equation in quadratic form.
	Factorable equations	38–39	63–70	Extract common factor or factor by grouping.
	Absolute value equations	39–40	71–74	If $|x| = a$, then $x = -a$ or $x = a$.
0.4	**Inequalities**	44–55	75–104	Solutions are a range of real numbers.
	Graphing inequalities and interval notation	44–46	75–78	■ $a < x < b$ is equivalent to (a, b). ■ $x \le a$ is equivalent to $(-\infty, a]$. ■ $x > a$ is equivalent to (a, ∞).
	Linear inequalities	46–48	79–86	If an inequality is multiplied or divided by a *negative* number, the inequality sign must be reversed.
	Polynomial inequalities	48–51	87–92	Zeros are values that make the polynomial equal to 0.
	Rational inequalities	52–53	93–98	The number line is divided into intervals. The endpoints of these intervals are values that make either the numerator or denominator equal to 0. Always exclude values that make the denominator equal to 0.
	Absolute value inequalities	54–55	99–104	■ $|x| \le a$ is equivalent to $-a \le x \le a$. ■ $|x| > a$ is equivalent to $x < -a$ or $x > a$.

SECTION	CONCEPT	PAGES	REVIEW EXERCISES	KEY IDEAS/POINTS
0.5	**Graphing equations**	59–69	105–130	
	The distance and midpoint formulas	60–61	105–112	$d = \sqrt{(x_2 - x_1)^2 + (y_2 - y_1)^2}$ $(x_m, y_m) = \left(\dfrac{x_1 + x_2}{2}, \dfrac{y_1 + y_2}{2}\right)$
	Point-plotting	61–62	121–126	List a table with several coordinates that are solutions to the equation; plot and connect.
	Using intercepts and symmetry as graphing aids	62–67	113–120	If (a, b) is on the graph of the equation, then $(-a, b)$ is on the graph if symmetric about the y-axis, $(a, -b)$ is on the graph if symmetric about the x-axis, and $(-a, -b)$ is on the graph if symmetric about the origin. **Intercepts:** x-intercept: let $y = 0$. y-intercept: let $x = 0$. **Symmetry:** The graph of an equation can be symmetric about the x-axis, y-axis, or origin.
	Circles	67–69	127–130	Standard equation of a circle with center (h, k) and radius r. $(x - h)^2 + (y - k)^2 = r^2$ General form: $x^2 + y^2 + ax + by + c = 0$ Transform equations of circles to the standard form by completing the square
0.6	**Lines**	73–81	131–140	General form: $Ax + By = C$
	Graphing a line	73–74		Vertical: $x = a$ Slant: $Ax + By = C$, Horizontal: $y = b$ where $A \neq 0$ and $B \neq 0$
	Slope	74–76		$m = \dfrac{y_2 - y_1}{x_2 - x_1}$, where $x_1 \neq x_2$ $\dfrac{\text{"rise"}}{\text{"run"}}$
	Slope–intercept form	77	131–138	$y = mx + b$ m is the slope and b is the y-intercept.
	Finding equations of lines	78–79	131–140	Point–slope form: $y - y_1 = m(x - x_1)$
	Parallel and perpendicular lines	80–81	139–140	$L_1 \parallel L_2$ if and only if $m_1 = m_2$ (slopes are equal). $L_1 \perp L_2$ if and only if $m_1 = -\dfrac{1}{m_2}\begin{cases} m_1 \neq 0 \\ m_2 \neq 0 \end{cases}$ (slopes are negative reciprocals).
0.7	**Modeling variation**	86–90		
	Direct variation	86–88		$y = kx$
	Inverse variation	88–89		$y = \dfrac{k}{x}$
	Joint variation and combined variation	89–90	141–144	Joint: One quantity is directly proportional to the product of two or more other quantities. Combined: Direct variation and inverse variation occur at the same time.

0.1 Linear Equations

Solve for the variable.

1. $7x - 4 = 12$

2. $13d + 12 = 7d + 6$

3. $20p + 14 = 6 - 5p$

4. $4(x - 7) - 4 = 4$

5. $3(x + 7) - 2 = 4(x - 2)$

6. $7c + 3(c - 5) = 2(c + 3) - 14$

7. $14 - [-3(y - 4) + 9] = [4(2y + 3) - 6] + 4$

8. $[6 - 4x + 2(x - 7)] - 52 = 3(2x - 4) + 6[3(2x - 3) + 6]$

9. $\dfrac{12}{b} - 3 = \dfrac{6}{b} + 4$

10. $\dfrac{g}{3} + g = \dfrac{7}{9}$

11. $\dfrac{13x}{7} - x = \dfrac{x}{4} - \dfrac{3}{14}$

12. $5b + \dfrac{b}{6} = \dfrac{b}{3} - \dfrac{29}{6}$

13. Investments. You win $25,000 and you decide to invest the money in two different investments: one paying 20% and the other paying 8%. A year later you have $27,600 total. How much did you originally invest in each account?

14. Investments. A college student on summer vacation was able to make $5,000 by working a full-time job every summer. He invested half the money in a mutual fund and half the money in a stock that yielded four times as much interest as the mutual fund. After a year he earned $250 in interest. What were the interest rates of the mutual fund and the stock?

15. Chemistry. For an experiment, a student requires 150 ml of a solution that is 8% NaCl (sodium chloride). The storeroom has only solutions that are 10% NaCl and 5% NaCl. How many milliliters of each available solution should be mixed to get 150 ml of 8% NaCl?

16. Chemistry. A mixture containing 8% salt is to be mixed with 4 ounces of a mixture that is 20% salt, in order to obtain a solution that is 12% salt. How much of the first solution must be used?

0.2 Quadratic Equations

Solve by factoring.

17. $b^2 = 4b + 21$

18. $x(x - 3) = 54$

19. $x^2 = 8x$

20. $6y^2 - 7y - 5 = 0$

Solve by the square root method.

21. $q^2 - 169 = 0$

22. $c^2 + 36 = 0$

23. $(2x - 4)^2 = -64$

24. $(d + 7)^2 - 4 = 0$

Solve by completing the square.

25. $x^2 - 4x - 12 = 0$

26. $2x^2 - 5x - 7 = 0$

27. $\dfrac{x^2}{2} = 4 + \dfrac{x}{2}$

28. $8m = m^2 + 15$

Solve by the quadratic formula.

29. $3t^2 - 4t = 7$

30. $4x^2 + 5x + 7 = 0$

31. $8f^2 - \frac{1}{3}f = \frac{7}{6}$

32. $x^2 = -6x + 6$

Solve by any method.

33. $5q^2 - 3q - 3 = 0$

34. $(x - 7)^2 = -12$

35. $2x^2 - 3x - 5 = 0$

36. $(g - 2)(g + 5) = -7$

37. $7x^2 = -19x + 6$

38. $7 = (2b^2 + 1)$

39. Geometry. Find the base and height of a triangle with an area of 2 square feet if its base is 3 feet longer than its height.

40. Falling Objects. A man is standing on top of a building 500 feet tall. If he drops a penny off the roof, the height of the penny is given by $h = -16t^2 + 500$, where t is in seconds. Determine how many seconds it takes until the penny hits the ground.

0.3 Other Types of Equations

Specify any values that must be excluded from the solution set and then solve the rational equation.

41. $\dfrac{1}{x} - 4 = 3(x - 7) + 5$

42. $\dfrac{4}{x + 1} - \dfrac{8}{x - 1} = 3$

43. $\dfrac{2}{t + 4} - \dfrac{7}{t} = \dfrac{6}{t(t + 4)}$

44. $\dfrac{3}{2x - 7} = \dfrac{-2}{3x + 1}$

45. $\dfrac{3}{2x} - \dfrac{6}{x} = 9$

46. $\dfrac{3 - 5/m}{2 + 5/m} = 1$

Solve the radical equation for the given variable.

47. $\sqrt[3]{2x - 4} = 2$

48. $\sqrt{x - 2} = -4$

49. $(2x - 7)^{1/5} = 3$

50. $x = \sqrt{7x - 10}$

51. $x - 4 = \sqrt{x^2 + 5x + 6}$

52. $\sqrt{2x - 7} = \sqrt{x + 3}$

53. $\sqrt{x + 3} = 2 - \sqrt{3x + 2}$

54. $4 + \sqrt{x - 3} = \sqrt{x - 5}$

Solve the equation by introducing a substitution that transforms the equation to quadratic form.

55. $y^{-2} - 5y^{-1} + 4 = 0$

56. $p^{-2} + 4p^{-1} = 12$

57. $3x^{1/3} + 2x^{2/3} = 5$

58. $2x^{2/3} - 3x^{1/3} - 5 = 0$

59. $x^{-2/3} + 3x^{-1/3} + 2 = 0$

60. $y^{-1/2} - 2y^{-1/4} + 1 = 0$

61. $x^4 + 5x^2 = 36$

62. $3 - 4x^{-1/2} + x^{-1} = 0$

Solve the equation by factoring.

63. $x^3 + 4x^2 - 32x = 0$

64. $9t^3 - 25t = 0$

65. $p^3 - 3p^2 - 4p + 12 = 0$

66. $4x^3 - 9x^2 + 4x - 9 = 0$

67. $p(2p - 5)^2 - 3(2p - 5) = 0$

68. $2(t^2 - 9)^3 - 20(t^2 - 9)^2 = 0$

69. $y - 81y^{-1} = 0$

70. $9x^{3/2} - 37x^{1/2} + 4x^{-1/2} = 0$

Solve the absolute value equation.

71. $|x - 3| = -4$

72. $|2 + x| = 5$

73. $|3x - 4| = 1.1$

74. $|x^2 - 6| = 3$

0.4 Inequalities

Graph the indicated set and write as a single interval, if possible.

75. $(4, 6] \cup [5, \infty)$

76. $(-\infty, -3) \cup [-7, 2]$

77. $(3, 12] \cap [8, \infty)$

78. $(-\infty, -2) \cap [-2, 9)$

Solve the linear inequality and express the solution set in interval notation.

79. $2x < 5 - x$

80. $6x + 4 \leq 2$

81. $4(x - 1) > 2x - 7$

82. $\dfrac{x + 3}{3} \geq 6$

83. $6 < 2 + x \leq 11$

84. $-6 \leq 1 - 4(x + 2) \leq 16$

85. $\dfrac{2}{3} \leq \dfrac{1 + x}{6} \leq \dfrac{3}{4}$

86. $\dfrac{x}{3} + \dfrac{x + 4}{9} > \dfrac{x}{6} - \dfrac{1}{3}$

Solve the polynomial inequality and express the solution set using interval notation.

87. $x^2 \leq 36$

88. $6x^2 - 7x < 20$

89. $4x \leq x^2$

90. $-x^2 \geq 9x + 14$

91. $4x^2 - 12 > 13x$

92. $3x \leq x^2 + 2$

Solve the rational inequality and express the solution set using interval notation.

93. $\dfrac{x}{x - 3} < 0$

94. $\dfrac{x - 1}{x - 4} > 0$

95. $\dfrac{x^2 - 3x}{3} \geq 18$

96. $\dfrac{x^2 - 49}{x - 7} \geq 0$

97. $\dfrac{3}{x - 2} - \dfrac{1}{x - 4} \leq 0$

98. $\dfrac{4}{x - 1} \leq \dfrac{2}{x + 3}$

Solve the absolute value inequality and express the solution set using interval notation.

99. $|x + 4| > 7$

100. $|-7 + y| \leq 4$

101. $|2x| > 6$

102. $\left| \dfrac{4 + 2x}{3} \right| \geq \dfrac{1}{7}$

103. $|2 + 5x| \geq 0$

104. $|1 - 2x| \leq 4$

0.5 Graphing Equations

Calculate the distance between the two points.

105. $(-2, 0)$ and $(4, 3)$

106. $(1, 4)$ and $(4, 4)$

107. $(-4, -6)$ and $(2, 7)$

108. $\left(\frac{1}{4}, \frac{1}{12}\right)$ and $\left(\frac{1}{3}, -\frac{7}{3}\right)$

Calculate the midpoint of the segment joining the two points.

109. $(2, 4)$ and $(3, 8)$

110. $(-2, 6)$ and $(5, 7)$

111. $(2.3, 3.4)$ and $(5.4, 7.2)$

112. $(-a, 2)$ and $(a, 4)$

Find the x-intercept(s) and y-intercept(s) if any.

113. $x^2 + 4y^2 = 4$

114. $y = x^2 - x + 2$

115. $y = \sqrt{x^2 - 9}$

116. $y = \dfrac{x^2 - x - 12}{x - 12}$

Use algebraic tests to determine symmetry with respect to the x-axis, y-axis, or origin.

117. $x^2 + y^3 = 4$

118. $y = x^2 - 2$

119. $xy = 4$

120. $y^2 = 5 + x$

Use symmetry as a graphing aid and point-plot the given equations.

121. $y = x^2 - 3$

122. $y = |x| - 4$

123. $y = \sqrt[3]{x}$

124. $x = y^2 - 2$

125. $y = x\sqrt{9 - x^2}$

126. $x^2 + y^2 = 36$

Find the center and the radius of the circle given by the equation.

127. $(x + 2)^2 + (y + 3)^2 = 81$

128. $(x - 4)^2 + (y + 2)^2 = 32$

129. $x^2 + y^2 + 2y - 4x + 11 = 0$

130. $3x^2 + 3y^2 - 6x - 7 = 0$

0.6 Lines

Write an equation of the line, given the slope and a point that lies on the line.

131. $m = -2$ $(-3, 4)$

132. $m = \frac{3}{4}$ $(2, 16)$

133. $m = 0$ $(-4, 6)$

134. m is undefined $(2, -5)$

Write the equation of the line that passes through the given points. Express the equation in slope–intercept form or in the form of $x = a$ or $y = b$.

135. $(-4, -2)$ and $(2, 3)$

136. $(-1, 4)$ and $(-2, 5)$

137. $\left(-\frac{3}{4}, \frac{1}{2}\right)$ and $\left(-\frac{7}{4}, \frac{5}{2}\right)$

138. $(3, -2)$ and $(-9, 2)$

Find the equation of the line that passes through the given point and also satisfies the additional piece of information.

139. $(-2, -1)$ parallel to the line $2x - 3y = 6$

140. $(5, 6)$ perpendicular to the line $5x - 3y = 0$

0.7 Modeling Variation

Write an equation that describes each variation.

141. C is directly proportional to r; $C = 2\pi$ when $r = 1$.

142. V is directly proportional to both l and w; $V = 12h$ when $w = 6$ and $l = 2$.

143. A varies directly with the square of r; $A = 25\pi$ when $r = 5$.

144. F varies inversely with both λ and L; $F = 20\pi$ when $\lambda = 10 \ \mu$m and $L = 10$ km.

Solve the equation.

1. $4p - 7 = 6p - 1$

2. $-2(z - 1) + 3 = -3z + 3(z - 1)$

3. $3t = t^2 - 28$

4. $8x^2 - 13x = 6$

5. $6x^2 - 13x = 8$

6. $\dfrac{3}{x - 1} = \dfrac{5}{x + 2}$

7. $\dfrac{5}{y - 3} + 1 = \dfrac{30}{y^2 - 9}$

8. $x^4 - 5x^2 - 36 = 0$

9. $\sqrt{2x + 1} + x = 7$

10. $2x^{2/3} + 3x^{1/3} - 2 = 0$

11. $\sqrt{3y - 2} = 3 - \sqrt{3y + 1}$

12. $x(3x - 5)^3 - 2(3x - 5)^2 = 0$

13. $x^{7/3} - 8x^{4/3} + 12x^{1/3} = 0$

14. Solve for x: $\left|\frac{1}{5}x + \frac{2}{3}\right| = \frac{7}{15}$.

Solve the inequality and express the solution in interval notation.

15. $3x + 19 \geq 5(x - 3)$

16. $-1 \leq 3x + 5 < 26$

17. $\dfrac{2}{5} < \dfrac{x + 8}{4} \leq \dfrac{1}{2}$

18. $3x \geq 2x^2$

19. $3p^2 \geq p + 4$

20. $|5 - 2x| > 1$

21. $\dfrac{x - 3}{2x + 1} \leq 0$

22. $\dfrac{x + 4}{x^2 - 9} \geq 0$

23. Find the distance between the points $(-7, -3)$ and $(2, -2)$.

24. Find the midpoint between $(-3, 5)$ and $(5, -1)$.

In Exercises 25 and 26, graph the equations.

25. $2x^2 + y^2 = 8$

26. $y = \dfrac{4}{x^2 + 1}$

27. Find the x-intercept and the y-intercept of the line $x - 3y = 6$.

28. Find the x-intercept(s) and the y-intercept(s), if any: $4x^2 - 9y^2 = 36$.

29. Express the line in slope–intercept form: $\frac{2}{3}x - \frac{1}{4}y = 2$.

30. Express the line in slope–intercept form: $4x - 6y = 12$.

Find the equation of the line that is characterized by the given information. Graph the line.

31. Passes through the points $(-3, 2)$ and $(4, 9)$

32. Parallel to the line $y = 4x + 3$ and passes through the point $(1, 7)$

33. Perpendicular to the line $2x - 4y = 5$ and passes through the point $(1, 1)$

34. Determine the center and radius of the circle $x^2 + y^2 - 10x + 6y + 22 = 0$.

In Exercises 35 and 36, use variation to find a model for the given problem.

35. F varies directly with m and inversely with p; $F = 20$ when $m = 2$ and $p = 3$.

36. y varies directly with the square of x; $y = 8$ when $x = 5$.

1

Functions and Their Graphs

You are buying a pair of running shoes. Their original price was $100, but they have been discounted 30% as part of a weekend sale. Because you arrived early, you can take advantage of door-buster savings: an additional 20% off the sale price. Naïve shoppers might be lured into thinking these shoes will cost $50 because they add the 20% and 30% to get 50% off, but they will end up paying more than that. Experienced shoppers know that the store will first take 30% off of $100, which results in a price of $70, and then it will take an additional 20% off of the sale price, $70, which results in a final discounted price of $56. Experienced shoppers have already learned *composition of functions*.

A composition of functions can be thought of as a function of a function. One function takes an input (original price, $100) and maps it to an output (sale price, $70), and then another function takes that output as its input (sale price, $70) and maps that to an output (checkout price, $56).

FUNCTIONS AND THEIR GRAPHS

1.1 Functions	1.2 Graphs of Functions	1.3 Graphing Techniques: Transformations	1.4 Combining Functions	1.5 One-to-One Functions and Inverse Functions
• Definition of a Function • Functions Defined by Equations • Function Notation • Domain of a Function	• Common Functions • Even and Odd Functions • Increasing and Decreasing Functions • Average Rate of Change • Piecewise-Defined Functions	• Horizontal and Vertical Shifts • Reflection About the Axes • Stretching and Compressing	• Adding, Subtracting, Multiplying, and Dividing Functions • Composition of Functions	• One-to-One Functions • Inverse Functions • Graphical Interpretation of Inverse Functions • Finding the Inverse Function

CHAPTER OBJECTIVES

- Evaluate a function for any argument using placeholder notation (Section 1.1).
- Determine characteristics of graphs: even or odd, increasing or decreasing, and the average rate of change (Section 1.2).
- Graph functions that are transformations of common functions (Section 1.3).
- Find composite functions and their domains (Section 1.4).
- Find inverse functions and their domains and ranges (Section 1.5).

SKILLS OBJECTIVES

- Determine whether a relation is a function.
- Determine whether an equation represents a function.
- Use function notation.
- Find the value of a function.
- Determine the domain and range of a function.

CONCEPTUAL OBJECTIVES

- Think of function notation as a placeholder or mapping.
- Understand that all functions are relations but not all relations are functions.

Definition of a Function

What do the following pairs have in common?

- Every person has a blood type.
- Temperature is some typical value at a particular time of day.
- Every working household phone in the United States has a 10-digit phone number.
- First-class postage rates correspond to the weight of a letter.
- Certain times of the day are start times for sporting events at a university.

They all describe a particular correspondence between two groups. A **relation** is a correspondence between two sets. The first set is called the **domain** and the corresponding second set is called the **range**. Members of these sets are called **elements**.

DEFINITION	**Relation**

A **relation** is a correspondence between two sets where each element in the first set, called the **domain**, corresponds to *at least* one element in the second set, called the **range**.

A relation is a set of ordered pairs. The domain is the set of all the first components of the ordered pairs, and the range is the set of all the second components of the ordered pairs.

PERSON	BLOOD TYPE	ORDERED PAIR
Michael	A	(Michael, A)
Tania	A	(Tania, A)
Dylan	AB	(Dylan, AB)
Trevor	O	(Trevor, O)
Megan	O	(Megan, O)

WORDS	MATH
The domain is the set of all the first components.	{Michael, Tania, Dylan, Trevor, Megan}
The range is the set of all the second components.	{A, AB, O}

A relation in which each element in the domain corresponds to exactly one element in the range is a **function**.

DEFINITION | **Function**

A **function** is a correspondence between two sets where each element in the first set, called the **domain**, corresponds to *exactly* one element in the second set, called the **range**.

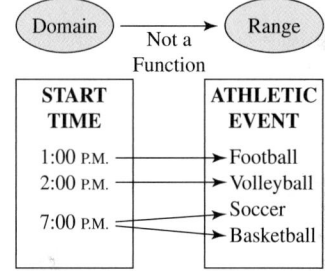

Note that the definition of a function is more restrictive than the definition of a relation. For a relation, each input corresponds to *at least* one output, whereas, for a function, each input corresponds to *exactly* one output. The blood-type example given is both a relation and a function.

Also note that the range (set of values to which the elements of the domain correspond) is a subset of the set of all blood types. Although all functions are relations, not all relations are functions.

For example, at a university, four primary sports typically overlap in the late fall: football, volleyball, soccer, and basketball. On a given Saturday, the table to the right indicates the start times for the competitions.

TIME OF DAY	COMPETITION
1:00 P.M.	Football
2:00 P.M.	Volleyball
7:00 P.M.	Soccer
7:00 P.M.	Basketball

WORDS	MATH
The 1:00 start time corresponds to exactly one event, Football.	(1:00 P.M., Football)
The 2:00 start time corresponds to exactly one event, Volleyball.	(2:00 P.M., Volleyball)
The 7:00 start time corresponds to two events, Soccer and Basketball.	(7:00 P.M., Soccer) (7:00 P.M., Basketball)

Because an element in the domain, 7:00 P.M., corresponds to more than one element in the range, Soccer and Basketball, this is not a function. It is, however, a relation.

Study Tip

All functions are relations but not all relations are functions.

EXAMPLE 1   Determining Whether a Relation Is a Function

Determine whether the following relations are functions:

a. $\{(-3, 4), (2, 4), (3, 5), (6, 4)\}$
b. $\{(-3, 4), (2, 4), (3, 5), (2, 2)\}$
c. Domain = Set of all items for sale in a grocery store; Range = Price

Solution:

a. No x-value is repeated. Therefore, each x-value corresponds to exactly one y-value.
  This relation is a function.

b. The value $x = 2$ corresponds to *both* $y = 2$ and $y = 4$. This relation is not a function.

c. Each item in the grocery store corresponds to exactly one price. This relation is a function.

▪ **YOUR TURN** Determine whether the following relations are functions:

  a. $\{(1, 2), (3, 2), (5, 6), (7, 6)\}$
  b. $\{(1, 2), (1, 3), (5, 6), (7, 8)\}$
  c. $\{(11:00 \text{ A.M.}, 83\,°F), (2:00 \text{ P.M.}, 89\,°F), (6:00 \text{ P.M.}, 85\,°F)\}$

▪ **Answer:** **a.** function
  b. not a function
  c. function

All of the examples we have discussed thus far are **discrete** sets in that they represent a countable set of distinct pairs of (x, y). A function can also be defined algebraically by an equation.

Functions Defined by Equations

Let's start with the equation $y = x^2 - 3x$, where x can be any real number. This equation assigns to each x-value exactly one corresponding y-value. For example,

x	$y = x^2 - 3x$	y
1	$y = (1)^2 - 3(1)$	-2
5	$y = (5)^2 - 3(5)$	10
$-\frac{2}{3}$	$y = \left(-\frac{2}{3}\right)^2 - 3\left(-\frac{2}{3}\right)$	$\frac{22}{9}$
1.2	$y = (1.2)^2 - 3(1.2)$	-2.16

Since the variable y *depends* on what value of x is selected, we denote y as the **dependent variable**. The variable x can be any number in the domain; therefore, we denote x as the **independent variable**.

Although functions are defined by equations, it is important to recognize that *not all equations define functions*. The requirement for an equation to define a function is that each element in the domain corresponds to exactly one element in the range. Throughout the ensuing discussion, we assume x to be the independent variable and y to be the dependent variable.

> **Equations that represent functions of x:** $\qquad y = x^2 \qquad y = |x| \qquad y = x^3$
>
> **Equations that do not represent functions of x:** $\qquad x = y^2 \qquad x^2 + y^2 = 1 \qquad x = |y|$

Study Tip

We say that $x = y^2$ is not a function of x. However, if we reverse the independent and dependent variables, then $x = y^2$ is a function of y.

In the "equations that represent functions of x," every x-value corresponds to exactly one y-value. Some ordered pairs that correspond to these functions are

$$y = x^2: \qquad (-1, 1) \, (0, 0) \, (1, 1)$$
$$y = |x|: \qquad (-1, 1) \, (0, 0) \, (1, 1)$$
$$y = x^3: \qquad (-1, -1) \, (0, 0) \, (1, 1)$$

The fact that $x = -1$ and $x = 1$ both correspond to $y = 1$ in the first two examples does not violate the definition of a function.

In the "equations that do not represent functions of x," some x-values correspond to *more than one* y-value. Some ordered pairs that correspond to these equations are

$$x = y^2: \quad (1, -1) \, (0, 0) \, (1, 1) \qquad x = 1 \text{ maps to } \textbf{both } y = -1 \text{ and } y = 1$$
$$x^2 + y^2 = 1: \quad (0, -1) \, (0, 1) \, (-1, 0) \, (1, 0) \qquad x = 0 \text{ maps to } \textbf{both } y = -1 \text{ and } y = 1$$
$$x = |y|: \quad (1, -1) \, (0, 0) \, (1, 1) \qquad x = 1 \text{ maps to } \textbf{both } y = -1 \text{ and } y = 1$$

Let's look at the graphs of the three **functions of x**:

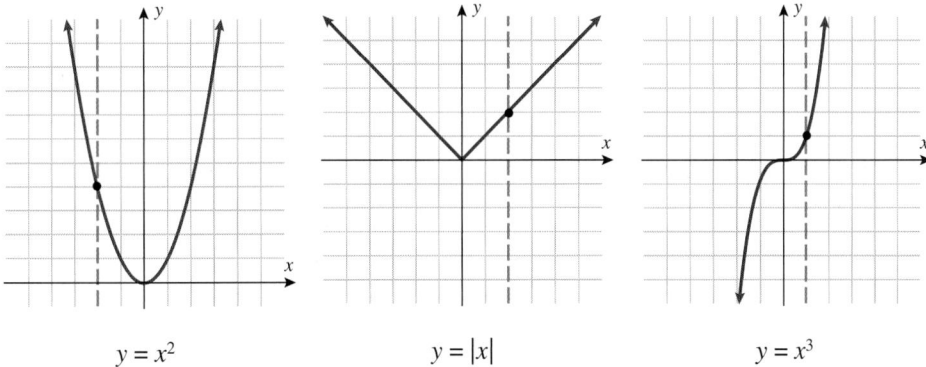

$y = x^2$ $y = |x|$ $y = x^3$

Let's take any value for x, say, $x = a$. The graph of $x = a$ corresponds to a vertical line. A function of x maps each x-value to exactly one y-value; therefore, there should be at most one point of intersection with any vertical line. We see in the three graphs of the functions above that if a vertical line is drawn at any value of x on any of the three graphs, the vertical line only intersects the graph in one place. Look at the graphs of the three equations that do **not** represent **functions of x**.

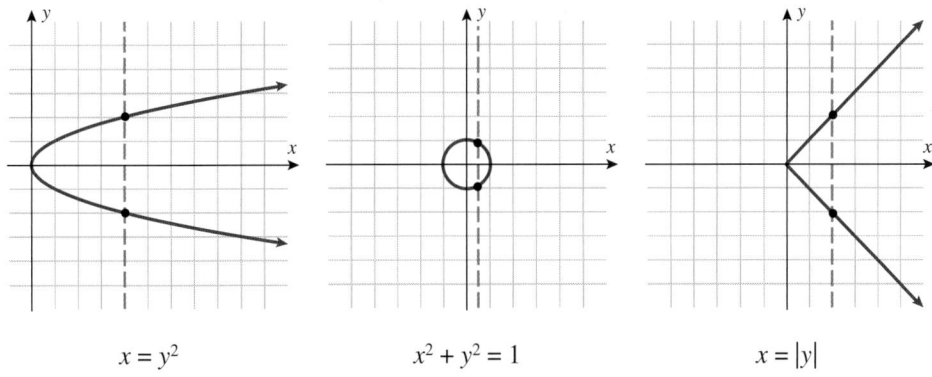

$x = y^2$ $x^2 + y^2 = 1$ $x = |y|$

A vertical line can be drawn on any of the three graphs such that the vertical line will intersect each of these graphs at two points. Thus, there is more than one y-value that corresponds to some x-value in the domain, which is why these equations do not define functions of x.

DEFINITION   **Vertical Line Test**

Given the graph of an equation, if any vertical line that can be drawn intersects the graph at no more than one point, the equation defines a function of x. This test is called the **vertical line test**.

Study Tip

If any x-value corresponds to more than one y-value, then y is **not** a function of x.

EXAMPLE 2   Using the Vertical Line Test

Use the vertical line test to determine whether the graphs of equations define functions of x.

a.   **b.**

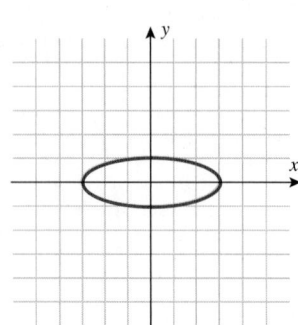   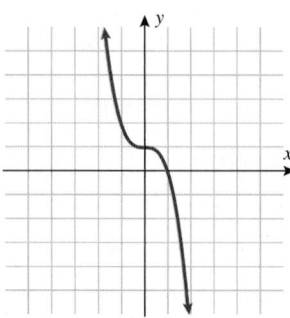

Solution:

Apply the vertical line test.

a.   **b.**

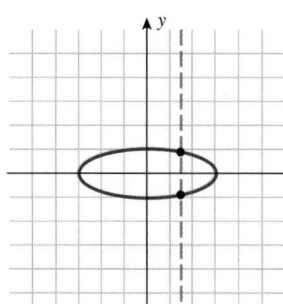   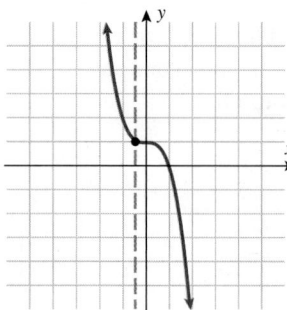

a. Because the vertical line intersects the graph of the equation at two points, this equation   does not represent a function .

b. Because any vertical line will intersect the graph of this equation at no more than one point, this equation   represents a function .

Answer: The graph of the equation is a circle, which does not pass the vertical line test. Therefore, the equation does not define a function.

■ **YOUR TURN** Determine whether the equation $(x - 3)^2 + (y + 2)^2 = 16$ is a function of x.

To recap, a function can be expressed one of four ways: verbally, numerically, algebraically, and graphically. This is sometimes called the Rule of 4.

Expressing a Function

VERBALLY	NUMERICALLY	ALGEBRAICALLY	GRAPHICALLY
Every real number has a corresponding absolute value.	$\{(-3, 3), (-1, 1), (0, 0), (1, 1), (5, 5)\}$	$y = |x|$	

Function Notation

We know that the equation $y = 2x + 5$ is a function because its graph is a nonvertical line and thus passes the vertical line test. We can select x-values (input) and determine unique corresponding y-values (output). The output is found by taking 2 times the input and then adding 5. If we give the function a name, say, "f", then we can use **function notation**:

$$f(x) = 2x + 5$$

The symbol $f(x)$ is read "f evaluated at x" or "f of x" and represents the y-value that corresponds to a particular x-value. In other words, $y = f(x)$.

INPUT	FUNCTION	OUTPUT	EQUATION
x	f	$f(x)$	$f(x) = 2x + 5$
Independent variable	Mapping	Dependent variable	Mathematical rule

It is important to note that f is the function name, whereas $f(x)$ is the value of the function. In other words, the function f maps some value x in the domain to some value $f(x)$ in the range.

x	$f(x) = 2x + 5$	$f(x)$
0	$2(0) + 5$	$f(0) = 5$
1	$2(1) + 5$	$f(1) = 7$
2	$2(2) + 5$	$f(2) = 9$

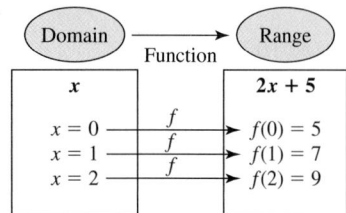

The independent variable is also referred to as the **argument** of a function. To evaluate functions, it is often useful to think of the independent variable or argument as a placeholder. For example, $f(x) = x^2 - 3x$ can be thought of as

$$f(\square) = (\square)^2 - 3(\square)$$

In other words, "f of the argument is equal to the argument squared minus 3 times the argument." Any expression can be substituted for the argument:

$$f(1) = (1)^2 - 3(1)$$
$$f(x + 1) = (x + 1)^2 - 3(x + 1)$$
$$f(-x) = (-x)^2 - 3(-x)$$

It is important to note:

- $f(x)$ does *not* mean f times x.
- The most common function names are f and F since the word function begins with an "f". Other common function names are g and G, but any letter can be used.
- The letter most commonly used for the independent variable is x. The letter t is also common because in real-world applications it represents time, but any letter can be used.
- Although we can think of y and $f(x)$ as interchangeable, the function notation is useful when we want to consider two or more functions of the same independent variable or when we want to evaluate a function at more than one argument.

Study Tip

It is important to note that $f(x)$ does not mean f times x.

EXAMPLE 3 **Evaluating Functions by Substitution**

Given the function $f(x) = 2x^3 - 3x^2 + 6$, find $f(-1)$.

Solution:

Consider the independent variable x to be a
placeholder.

$$f(\square) = 2(\square)^3 - 3(\square)^2 + 6$$

To find $f(-1)$, substitute $x = -1$ into
the function.

$$f(-1) = 2(-1)^3 - 3(-1)^2 + 6$$

Evaluate the right side.

$$f(-1) = -2 - 3 + 6$$

Simplify.

$$\boxed{f(-1) = 1}$$

EXAMPLE 4 **Finding Function Values from the
Graph of a Function**

The graph of f is given on the right.

a. Find $f(0)$.
b. Find $f(1)$.
c. Find $f(2)$.
d. Find $4f(3)$.
e. Find x such that $f(x) = 10$.
f. Find x such that $f(x) = 2$.

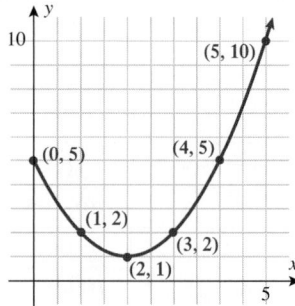

Solution (a): The value $x = 0$ corresponds to the value $y = 5$. $\boxed{f(0) = 5}$

Solution (b): The value $x = 1$ corresponds to the value $y = 2$. $\boxed{f(1) = 2}$

Solution (c): The value $x = 2$ corresponds to the value $y = 1$. $\boxed{f(2) = 1}$

Solution (d): The value $x = 3$ corresponds to the value $y = 2$. $4f(3) = 4 \cdot 2 = \boxed{8}$

Solution (e): The value $y = 10$ corresponds to the value $\boxed{x = 5}$.

Solution (f): The value $y = 2$ corresponds to the values $\boxed{x = 1}$ and $\boxed{x = 3}$.

■ **Answer: a.** $f(-1) = 2$
b. $f(0) = 1$
c. $3f(2) = -21$
d. $x = 1$

■ **YOUR TURN** For the following graph of a function, find

a. $f(-1)$ **b.** $f(0)$ **c.** $3f(2)$
d. the value of x that corresponds to $f(x) = 0$

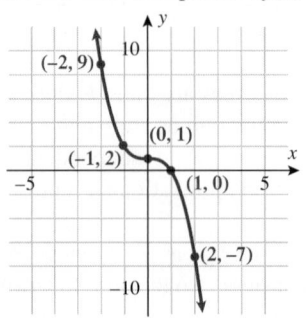

EXAMPLE 5   **Evaluating Functions with Variable Arguments (Inputs)**

For the given function $f(x) = x^2 - 3x$, evaluate $f(x + 1)$.

COMMON MISTAKE

A common misunderstanding is to interpret the notation $f(x + 1)$ as a sum: $f(x + 1) \neq f(x) + f(1)$.

⭐ **CORRECT**

Write the original function.

$$f(x) = x^2 - 3x$$

Replace the argument x with a placeholder.

$$f(\square) = (\square)^2 - 3(\square)$$

Substitute $x + 1$ for the argument.

$$f(x + 1) = (x + 1)^2 - 3(x + 1)$$

Eliminate the parentheses.

$$f(x + 1) = x^2 + 2x + 1 - 3x - 3$$

Combine like terms.

$$\boxed{f(x + 1) = x^2 - x - 2}$$

❌ **INCORRECT**

The **ERROR** is in interpreting the notation as a sum.

$$f(x + 1) \neq f(x) + f(1)$$

$$f(x + 1) \neq x^2 - 3x - 2$$

▼ **CAUTION**

$f(x + 1) \neq f(x) + f(1)$

■ **YOUR TURN**   For the given function $g(x) = x^2 - 2x + 3$, evaluate $g(x - 1)$.

■ **Answer:** $g(x - 1) = x^2 - 4x + 6$

EXAMPLE 6   **Evaluating Functions: Sums**

For the given function $H(x) = x^2 + 2x$, evaluate

a. $H(x + 1)$ **b.** $H(x) + H(1)$

Solution (a):

Write the function H in placeholder notation.

$$H(\square) = (\square)^2 + 2(\square)$$

Substitute $x + 1$ for the argument of H.

$$H(x + 1) = (x + 1)^2 + 2(x + 1)$$

Eliminate the parentheses on the right side.

$$H(x + 1) = x^2 + 2x + 1 + 2x + 2$$

Combine like terms on the right side.

$$\boxed{H(x + 1) = x^2 + 4x + 3}$$

Solution (b):

Write $H(x)$.

$$H(x) = x^2 + 2x$$

Evaluate H at $x = 1$.

$$H(1) = (1)^2 + 2(1) = 3$$

Evaluate the sum $H(x) + H(1)$.

$$H(x) + H(1) = x^2 + 2x + 3$$

$$\boxed{H(x) + H(1) = x^2 + 2x + 3}$$

Note: Comparing the results of part (a) and part (b), we see that

$$H(x + 1) \neq H(x) + H(1).$$

Technology Tip

Use a graphing utility to display graphs of
$y_1 = H(x + 1) = (x + 1)^2 + 2(x + 1)$
and $y_2 = H(x) + H(1) = x^2 + 2x + 3$.

The graphs are not the same.

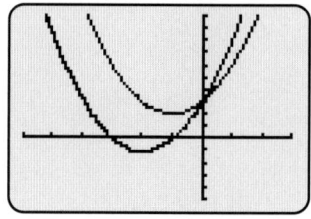

Technology Tip

Use a graphing utility to display graphs of $y_1 = G(-x) = (-x)^2 - (-x)$ and $y_2 = -G(x) = -(x^2 - x)$.

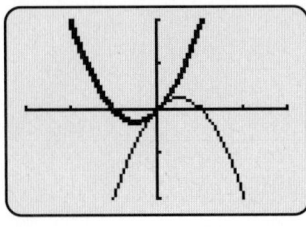

The graphs are not the same.

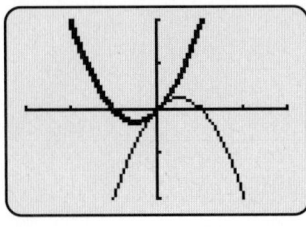

EXAMPLE 7   Evaluating Functions: Negatives

For the given function $G(t) = t^2 - t$, evaluate

a. $G(-t)$ **b.** $-G(t)$

Solution (a):

Write the function G in placeholder notation.	$G(\square) = (\square)^2 - (\square)$
Substitute $-t$ for the argument of G.	$G(-t) = (-t)^2 - (-t)$
Eliminate the parentheses on the right side.	$\boxed{G(-t) = t^2 + t}$

Solution (b):

Write $G(t)$.	$G(t) = t^2 - t$
Multiply by -1.	$-G(t) = -(t^2 - t)$
Eliminate the parentheses on the right side.	$\boxed{-G(t) = -t^2 + t}$

Note: Comparing the results of part (a) and part (b), we see that $G(-t) \neq -G(t)$. If $G(t)$ was an odd function, then $G(-t) = -G(t)$, but in general this is not true.

EXAMPLE 8   Evaluating Functions: Quotients

For the given function $F(x) = 3x + 5$, evaluate

a. $F\left(\dfrac{1}{2}\right)$ **b.** $\dfrac{F(1)}{F(2)}$

Solution (a):

Write F in placeholder notation.	$F(\square) = 3(\square) + 5$
Replace the argument with $\frac{1}{2}$.	$F\left(\dfrac{1}{2}\right) = 3\left(\dfrac{1}{2}\right) + 5$
Simplify the right side.	$\boxed{F\left(\dfrac{1}{2}\right) = \dfrac{13}{2}}$

Solution (b):

Evaluate $F(1)$.	$F(1) = 3(1) + 5 = 8$
Evaluate $F(2)$.	$F(2) = 3(2) + 5 = 11$
Divide $F(1)$ by $F(2)$.	$\boxed{\dfrac{F(1)}{F(2)} = \dfrac{8}{11}}$

▼ **CAUTION**

$$f\left(\frac{a}{b}\right) \neq \frac{f(a)}{f(b)}$$

Note: Comparing the results of part (a) and part (b), we see that $F\left(\dfrac{1}{2}\right) \neq \dfrac{F(1)}{F(2)}$.

■ **Answer: a.** $G(t - 2) = 3t - 10$
   b. $G(t) - G(2) = 3t - 6$
   c. $\dfrac{G(1)}{G(3)} = -\dfrac{1}{5}$
   d. $G\left(\dfrac{1}{3}\right) = -3$

■ **YOUR TURN**   Given the function $G(t) = 3t - 4$, evaluate

a. $G(t - 2)$ **b.** $G(t) - G(2)$ **c.** $\dfrac{G(1)}{G(3)}$ **d.** $G\left(\dfrac{1}{3}\right)$

Examples 6–8 illustrate the following in general:

$$f(a + b) \neq f(a) + f(b) \qquad f(-t) \neq -f(t) \qquad f\left(\frac{a}{b}\right) \neq \frac{f(a)}{f(b)}$$

Now that we have shown that $f(x + h) \neq f(x) + f(h)$, we turn our attention to one of the fundamental expressions in calculus: the **difference quotient**.

$$\frac{f(x + h) - f(x)}{h} \qquad h \neq 0$$

Example 9 illustrates the difference quotient, which will be discussed in detail in Section 1.2. For now, we will concentrate on the algebra involved when finding the difference quotient. In Section 1.2, the application of the difference quotient will be the emphasis.

 EXAMPLE 9   Evaluating the Difference Quotient

For the function $f(x) = x^2 - x$, find $\dfrac{f(x + h) - f(x)}{h}$.

Solution:

Use placeholder notation for the function $f(x) = x^2 - x$. $\qquad f(\square) = (\square)^2 - (\square)$

Calculate $f(x + h)$. $\qquad f(x + h) = (x + h)^2 - (x + h)$

Write the difference quotient. $\qquad \dfrac{f(x + h) - f(x)}{h}$

Let $f(x + h) = (x + h)^2 - (x + h)$ and $f(x) = x^2 - x$.

$$\frac{f(x + h) - f(x)}{h} = \frac{\overbrace{[(x + h)^2 - (x + h)]}^{f(x+h)} - \overbrace{[x^2 - x]}^{f(x)}}{h} \qquad h \neq 0$$

Eliminate the parentheses inside the first set of brackets.

$$= \frac{[x^2 + 2xh + h^2 - x - h] - [x^2 - x]}{h}$$

Eliminate the brackets in the numerator.

$$= \frac{x^2 + 2xh + h^2 - x - h - x^2 + x}{h}$$

Combine like terms.

$$= \frac{2xh + h^2 - h}{h}$$

Factor the numerator.

$$= \frac{h(2x + h - 1)}{h}$$

Divide out the common factor, h.

$$= \boxed{2x + h - 1} \qquad h \neq 0$$

■ YOUR TURN   Evaluate the difference quotient for $f(x) = x^2 - 1$.

■ Answer: $2x + h$

Domain of a Function

Sometimes the domain of a function is stated *explicitly*. For example,

$$f(x) = |x| \qquad \underset{\text{domain}}{\underline{x < 0}}$$

Here, the **explicit domain** is the set of all negative real numbers, $(-\infty, 0)$. Every negative real number in the domain is mapped to a positive real number in the range through the absolute value function.

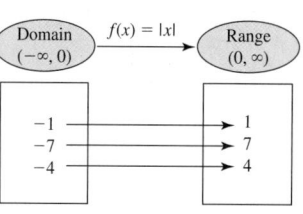

If the expression that defines the function is given but the domain is not stated explicitly, then the domain is implied. The **implicit domain** is the largest set of real numbers for which the function is defined and the output value $f(x)$ is a real number. For example,

$$f(x) = \sqrt{x}$$

does not have the domain explicitly stated. There is, however, an implicit domain. Note that if the argument is negative, that is, if $x < 0$, then the result is an imaginary number. In order for the output of the function, $f(x)$, to be a real number, we must restrict the domain to nonnegative numbers, that is, if $x \geq 0$.

FUNCTION	IMPLICIT DOMAIN
$f(x) = \sqrt{x}$	$[0, \infty)$

In general, we ask the question, "What can x be?" The implicit domain of a function excludes values that cause a function to be undefined or have outputs that are not real numbers.

EXPRESSION THAT DEFINES THE FUNCTION	EXCLUDED x-VALUES	EXAMPLE	IMPLICIT DOMAIN
Polynomial	None	$f(x) = x^3 - 4x^2$	All real numbers
Rational	x-values that make the denominator equal to 0	$g(x) = \dfrac{2}{x^2 - 9}$	$x \neq \pm 3$ or $(-\infty, -3) \cup (-3, 3) \cup (3, \infty)$
Radical	x-values that result in a square (even) root of a negative number	$h(x) = \sqrt{x - 5}$	$x \geq 5$ or $[5, \infty)$

Technology Tip

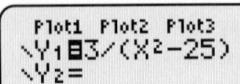

To visualize the domain of each function, ask the question: What are the excluded x-values in the graph?

Graph of $F(x) = \dfrac{3}{x^2 - 25}$ is shown.

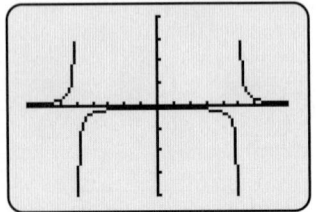

```
Plot1 Plot2 Plot3
\Y1■3/(X²-25)
\Y2=
```

The excluded x-values are -5 and 5.

EXAMPLE 10   Determining the Domain of a Function

State the domain of the given functions.

a. $F(x) = \dfrac{3}{x^2 - 25}$ **b.** $H(x) = \sqrt[4]{9 - 2x}$ **c.** $G(x) = \sqrt[3]{x - 1}$

Solution (a):

Write the original equation.

$$F(x) = \frac{3}{x^2 - 25}$$

Determine any restrictions on the values of x.

$$x^2 - 25 \neq 0$$

Solve the restriction equation.

$$x^2 \neq 25 \text{ or } x \neq \pm\sqrt{25} = \pm 5$$

State the domain restrictions.

$$x \neq \pm 5$$

Write the domain in interval notation.

$$\boxed{(-\infty, -5) \cup (-5, 5) \cup (5, \infty)}$$

Solution (b):

Write the original equation. $H(x) = \sqrt[4]{9 - 2x}$

Determine any restrictions on the values of x. $9 - 2x \geq 0$

Solve the restriction equation. $9 \geq 2x$

State the domain restrictions. $x \leq \dfrac{9}{2}$

Write the domain in interval notation. $\boxed{\left(-\infty, \dfrac{9}{2}\right]}$

Solution (c):

Write the original equation. $G(x) = \sqrt[3]{x - 1}$

Determine any restrictions on the
values of x.   no restrictions

State the domain. $\mathbb{R}$

Write the domain in interval notation. $\boxed{(-\infty, \infty)}$

■ **YOUR TURN**   State the domain of the given functions.

    a. $f(x) = \sqrt{x - 3}$ **b.** $g(x) = \dfrac{1}{x^2 - 4}$

■ **Answer:**
a. $x \geq 3$ or $[3, \infty)$
b. $x \neq \pm 2$ or
$(-\infty, -2) \cup (-2, 2) \cup (2, \infty)$

Applications

Functions that are used in applications often have restrictions on the domains due to physical constraints. For example, the volume of a cube is given by the function $V(x) = x^3$, where x is the length of a side. The function $f(x) = x^3$ has no restrictions on x, and therefore, the domain is the set of all real numbers. However, the volume of any cube has the restriction that the length of a side can never be negative or zero.

EXAMPLE 11   **The Dimensions of a Pool**

Express the volume of a 30 ft $\times$ 10 ft rectangular swimming pool as a function of its depth.

Solution:

The volume of any rectangular box is $V = lwh$, where V is the volume, l is the length, w is the width, and h is the height. In this example, the length is 30 ft, the width is 10 ft, and the height represents the depth d of the pool.

Write the volume as a function of depth d. $V(d) = (30)(10)d$

Simplify. $\boxed{V(d) = 300d}$

Determine any restrictions on the domain. $d > 0$

Relations and Functions (Let *x* represent the independent variable and *y* the dependent variable.)

TYPE	MAPPING/CORRESPONDENCE	EQUATION	GRAPH
Relation	Every *x*-value in the domain maps to **at least one** *y*-value in the range.	$x = y^2$	
Function	Every *x*-value in the domain maps to **exactly one** *y*-value in the range.	$y = x^2$	 Passes vertical line test

All functions are relations, but not all relations are functions. Functions can be represented by equations. In the following table, each column illustrates an alternative notation.

INPUT	CORRESPONDENCE	OUTPUT	EQUATION
x	Function	*y*	$y = 2x + 5$
Independent variable	Mapping	Dependent variable	Mathematical rule
Argument	*f*	*f(x)*	$f(x) = 2x + 5$

The **domain** is the set of all inputs (*x*-values) and the **range** is the set of all corresponding outputs (*y*-values). Placeholder notation is useful when evaluating functions.

$$f(x) = 3x^2 + 2x$$

$$f(\square) = 3(\square)^2 + 2(\square)$$

An explicit domain is stated, whereas an **implicit domain** is found by *excluding* *x*-values that

- make the function undefined (denominator = 0).
- result in a nonreal output (even roots of negative real numbers).

■ SKILLS

In Exercises 1–18, determine whether each relation is a function. Assume that the coordinate pair (*x, y*) represents the independent variable *x* and the dependent variable *y*.

1. {(0, −3), (0, 3), (−3, 0), (3, 0)}

2. {(2, −2), (2, 2), (5, −5), (5, 5)}

3. {(0, 0), (9, −3), (4, −2), (4, 2), (9, 3)}

4. {(0, 0), (−1, −1), (−2, −8), (1, 1), (2, 8)}

5. {(0, 1), (1, 0), (2, 1), (−2, 1), (5, 4), (−3, 4)}   **6.** {(0, 1), (1, 1), (2, 1), (3, 1)}

7. $x^2 + y^2 = 9$ **8.** $x = |y|$ **9.** $x = y^2$

10. $y = x^3$ **11.** $y = |x - 1|$ **12.** $y = 3$

13.
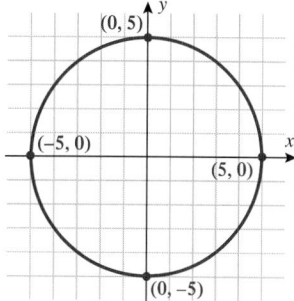

14.
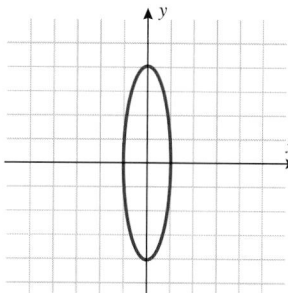

15.
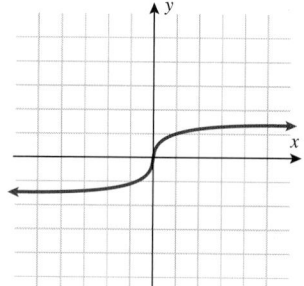

16.
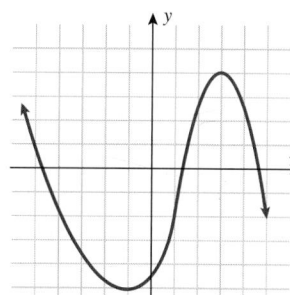

17.
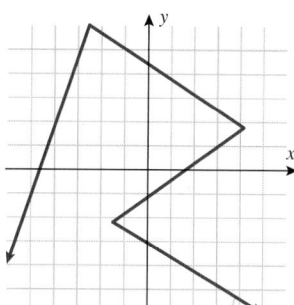

18.
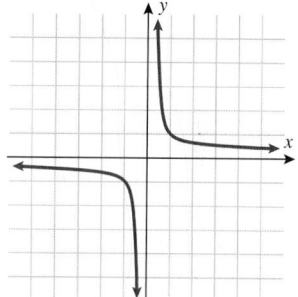

In Exercises 19–26, use the given graphs to evaluate the functions.

19. $y = f(x)$
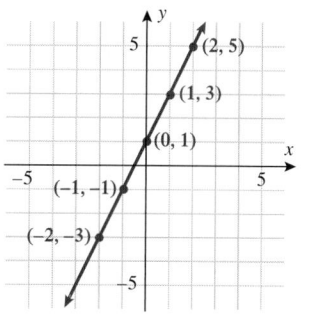
a. $f(2)$ **b.** $f(0)$ **c.** $f(-2)$

20. $y = g(x)$
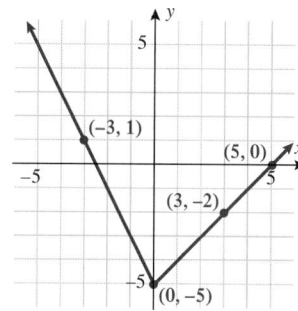
a. $g(-3)$ **b.** $g(0)$ **c.** $g(5)$

21. $y = p(x)$
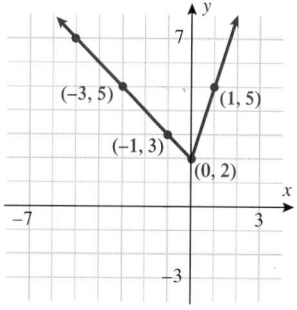
a. $p(-1)$ **b.** $p(0)$ **c.** $p(1)$

22. $y = r(x)$
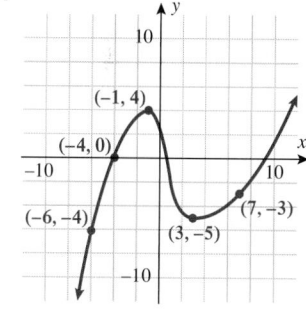
a. $r(-4)$ **b.** $r(-1)$ **c.** $r(3)$

23. $y = C(x)$
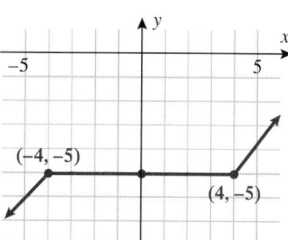
a. $C(2)$ **b.** $C(0)$ **c.** $C(-2)$

24. $y = q(x)$
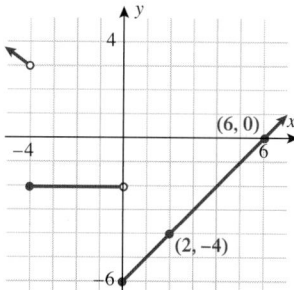
a. $q(-4)$ **b.** $q(0)$ **c.** $q(2)$

25. $y = S(x)$
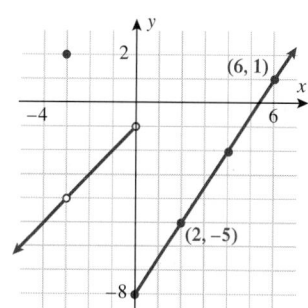
a. $S(-3)$ **b.** $S(0)$ **c.** $S(2)$

26. $y = T(x)$
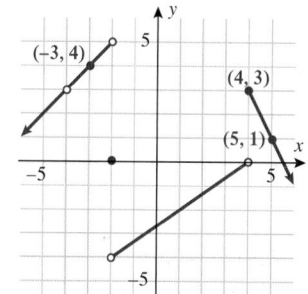
a. $T(-5)$ **b.** $T(-2)$ **c.** $T(4)$

27. Find x if $f(x) = 3$ in Exercise 19.   **28.** Find x if $g(x) = -2$ in Exercise 20.   **29.** Find x if $p(x) = 5$ in Exercise 21.

30. Find x if $C(x) = -7$ in Exercise 22.   **31.** Find x if $C(x) = -5$ in Exercise 23.   **32.** Find x if $q(x) = -2$ in Exercise 24.

33. Find x if $S(x) = 1$ in Exercise 25.   **34.** Find x if $T(x) = 4$ in Exercise 26.

In Exercises 35–50, evaluate the given quantities applying the following four functions:

$$f(x) = 2x - 3 \qquad F(t) = 4 - t^2 \qquad g(t) = 5 + t \qquad G(x) = x^2 + 2x - 7$$

35. $f(-2)$ **36.** $G(-3)$ **37.** $g(1)$ **38.** $F(-1)$

39. $f(-2) + g(1)$ **40.** $G(-3) - F(-1)$ **41.** $3f(-2) - 2g(1)$ **42.** $2F(-1) - 2G(-3)$

43. $\dfrac{f(-2)}{g(1)}$ **44.** $\dfrac{G(-3)}{F(-1)}$ **45.** $\dfrac{f(0) - f(-2)}{g(1)}$ **46.** $\dfrac{G(0) - G(-3)}{F(-1)}$

47. $f(x + 1) - f(x - 1)$ **48.** $F(t + 1) - F(t - 1)$ **49.** $g(x + a) - f(x + a)$ **50.** $G(x + b) + F(b)$

In Exercises 51–58, evaluate the difference quotients using the same f, F, G, and g given for Exercises 35–50.

51. $\dfrac{f(x + h) - f(x)}{h}$ **52.** $\dfrac{F(t + h) - F(t)}{h}$ **53.** $\dfrac{g(t + h) - g(t)}{h}$ **54.** $\dfrac{G(x + h) - G(x)}{h}$

55. $\dfrac{f(-2 + h) - f(-2)}{h}$ **56.** $\dfrac{F(-1 + h) - F(-1)}{h}$ **57.** $\dfrac{g(1 + h) - g(1)}{h}$ **58.** $\dfrac{G(-3 + h) - G(-3)}{h}$

In Exercises 59–90, find the domain of the given function. Express the domain in interval notation.

59. $f(x) = 2x - 5$ **60.** $f(x) = -2x - 5$ **61.** $g(t) = t^2 + 3t$ **62.** $h(x) = 3x^4 - 1$

63. $P(x) = \dfrac{x + 5}{x - 5}$ **64.** $Q(t) = \dfrac{2 - t^2}{t + 3}$ **65.** $T(x) = \dfrac{2}{x^2 - 4}$ **66.** $R(x) = \dfrac{1}{x^2 - 1}$

67. $F(x) = \dfrac{1}{x^2 + 1}$ **68.** $G(t) = \dfrac{2}{t^2 + 4}$ **69.** $q(x) = \sqrt{7 - x}$ **70.** $k(t) = \sqrt{t - 7}$

71. $f(x) = \sqrt{2x + 5}$ **72.** $g(x) = \sqrt{5 - 2x}$ **73.** $G(t) = \sqrt{t^2 - 4}$ **74.** $F(x) = \sqrt{x^2 - 25}$

75. $F(x) = \dfrac{1}{\sqrt{x - 3}}$ **76.** $G(x) = \dfrac{2}{\sqrt{5 - x}}$ **77.** $f(x) = \sqrt[3]{1 - 2x}$ **78.** $g(x) = \sqrt[5]{7 - 5x}$

79. $P(x) = \dfrac{1}{\sqrt[5]{x + 4}}$ **80.** $Q(x) = \dfrac{x}{\sqrt[3]{x^2 - 9}}$ **81.** $R(x) = \dfrac{x + 1}{\sqrt[4]{3 - 2x}}$ **82.** $p(x) = \dfrac{x^2}{\sqrt{25 - x^2}}$

83. $H(t) = \dfrac{t}{\sqrt{t^2 - t - 6}}$ **84.** $f(t) = \dfrac{t - 3}{\sqrt[4]{t^2 + 9}}$ **85.** $f(x) = (x^2 - 16)^{1/2}$ **86.** $g(x) = (2x - 5)^{1/3}$

87. $r(x) = x^2(3 - 2x)^{-1/2}$ **88.** $p(x) = (x - 1)^2(x^2 - 9)^{-3/5}$ **89.** $f(x) = \frac{2}{5}x - \frac{2}{4}$ **90.** $g(x) = \frac{2}{3}x^2 - \frac{1}{6}x - \frac{3}{4}$

91. Let $g(x) = x^2 - 2x - 5$ and find the values of x that correspond to $g(x) = 3$.

92. Let $g(x) = \frac{5}{6}x - \frac{3}{4}$ and find the value of x that corresponds to $g(x) = \frac{2}{3}$.

93. Let $f(x) = 2x(x - 5)^3 - 12(x - 5)^2$ and find the values of x that correspond to $f(x) = 0$.

94. Let $f(x) = 3x(x + 3)^2 - 6(x + 3)^3$ and find the values of x that correspond to $f(x) = 0$.

■ APPLICATIONS

95. Temperature. The average temperature in Tampa, Florida in the springtime is given by the function $T(x) = -0.7x^2 + 16.8x - 10.8$, where T is the temperature in degrees Fahrenheit and x is the time of day in military time and is restricted to $6 \leq x \leq 18$ (sunrise to sunset). What is the temperature at 6 A.M.? What is the temperature at noon?

96. Temperature. The average temperature in Orlando, Florida in the summertime is given by the function $T(x) = -0.5x^2 + 14.2x - 2.8$, where T is the temperature in degrees Fahrenheit and x is the time of the day in military time and is restricted to $7 \leq x \leq 20$ (sunrise to sunset). What is the temperature at 9 A.M.? What is the temperature at 3 P.M.?

97. Falling Objects: Baseballs. A baseball is hit and its height is a function of time, $h(t) = -16t^2 + 45t + 1$, where h is the height in feet and t is the time in seconds, with $t = 0$ corresponding to the instant the ball is hit. What is the height after 2 seconds? What is the domain of this function?

98. Falling Objects: Firecrackers. A firecracker is launched straight up, and its height is a function of time, $h(t) = -16t^2 + 128t$, where h is the height in feet and t is the time in seconds, with $t = 0$ corresponding to the instant it launches. What is the height 4 seconds after launch? What is the domain of this function?

99. Volume. An open box is constructed from a square 10-inch piece of cardboard by cutting squares of length x inches out of each corner and folding the sides up. Express the volume of the box as a function of x, and state the domain.

100. Volume. A cylindrical water basin will be built to harvest rainwater. The basin is limited in that the largest radius it can have is 10 feet. Write a function representing the volume of water V as a function of height h. How many additional gallons of water will be collected if you increase the height by 2 feet? (*Hint:* 1 cubic foot = 7.48 gallons.)

For Exercises 101 and 102, refer to the table below. It illustrates the average federal funds rate for the month of January (2000 to 2008).

101. Finance. Is the relation whose domain is the year and whose range is the average federal funds rate for the month of January a function? Explain.

102. Finance. Write five ordered pairs whose domain is the set of even years from 2000 to 2008 and whose range is the set of corresponding average federal funds rate for the month of January.

YEAR	FED. RATE
2000	5.45
2001	5.98
2002	1.73
2003	1.24
2004	1.00
2005	2.25
2006	4.50
2007	5.25
2008	3.50

For Exercises 103 and 104, use the following figure:

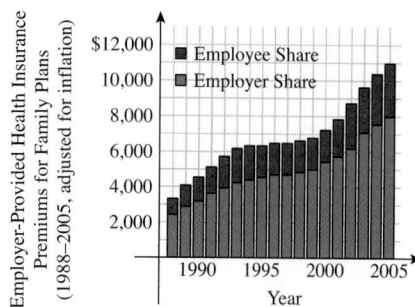

Source: Kaiser Family Foundation Health Research and Education Trust. *Note:* The following years were interpolated: 1989–1992; 1994–1995; 1997–1998.

103. Health-Care Costs: Fill in the following table. Round dollars to the nearest $1,000.

YEAR	TOTAL HEALTH-CARE COST FOR FAMILY PLANS
1989	
1993	
1997	
2001	
2005	

Write the five ordered pairs resulting from the table.

104. Health-Care Costs. Using the table found in Exercise 103, let the years correspond to the domain and the total costs correspond to the range. Is this relation a function? Explain.

For Exercises 105 and 106, use the following information:

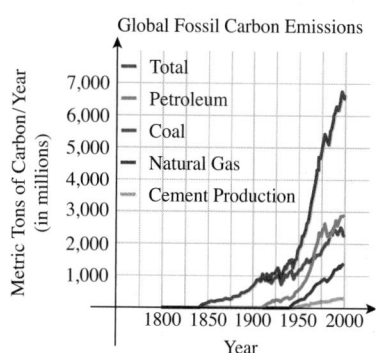

Source: http://www.naftc.wvu.edu

Let the functions f, F, g, G, and H represent the number of tons of carbon emitted per year as a function of year corresponding to cement production, natural gas, coal, petroleum, and the total amount, respectively. Let t represent the year, with $t = 0$ corresponding to 1900.

105. Environment: Global Climate Change. Estimate (to the nearest thousand) the value of

a. $F(50)$ **b.** $g(50)$ **c.** $H(50)$

106. Environment: Global Climate Change. Explain what the sum $F(100) + g(100) + G(100)$ represents.

▪ CATCH THE MISTAKE

In Exercises 107–112, explain the mistake that is made.

107. Determine whether the relationship is a function.

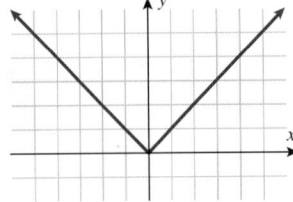

Solution:

Apply the horizontal line test.

Because the horizontal line intersects the graph in two places, this is not a function.

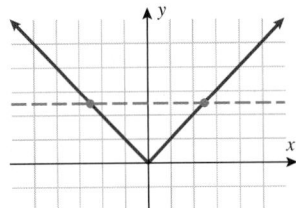

This is incorrect. What mistake was made?

108. Given the function $H(x) = 3x - 2$, evaluate the quantity $H(3) - H(-1)$.

Solution: $H(3) - H(-1) = H(3) + H(1) = 7 + 1 = 8$

This is incorrect. What mistake was made?

109. Given the function $f(x) = x^2 - x$, evaluate the quantity $f(x + 1)$.

Solution: $f(x + 1) = f(x) + f(1) = x^2 - x + 0$
$$f(x + 1) = x^2 - x$$

This is incorrect. What mistake was made?

110. Determine the domain of the function $g(t) = \sqrt{3 - t}$ and express it in interval notation.

Solution:

What can t be? Any nonnegative real number.
$$3 - t > 0$$
$$3 > t \quad \text{or} \quad t < 3$$
Domain: $(-\infty, 3)$

This is incorrect. What mistake was made?

111. Given the function $G(x) = x^2$, evaluate
$$\frac{G(-1 + h) - G(-1)}{h}.$$

Solution:

$$\frac{G(-1 + h) - G(-1)}{h} = \frac{G(-1) + G(h) - G(-1)}{h}$$
$$= \frac{G(h)}{h} = \frac{h^2}{h} = h$$

This is incorrect. What mistake was made?

112. Given the functions $f(x) = |x - A| - 1$ and $f(1) = -1$, find A.

Solution:

Since $f(1) = -1$, the point $(-1, 1)$ must satisfy the function. $-1 = |-1 - A| - 1$

Add 1 to both sides of the equation. $|-1 - A| = 0$

The absolute value of zero is zero, so there is no need for the absolute value signs: $-1 - A = 0 \Rightarrow A = -1$.

This is incorrect. What mistake was made?

▪ CONCEPTUAL

In Exercises 113–116, determine whether each statement is true or false.

113. If a vertical line does not intersect the graph of an equation, then that equation does not represent a function.

114. If a horizontal line intersects a graph of an equation more than once, the equation does not represent a function.

▪ CHALLENGE

119. If $F(x) = \dfrac{C - x}{D - x}$, $F(-2)$ is undefined, and $F(-1) = 4$, find C and D.

120. Construct a function that is undefined at $x = 5$ and whose graph passes through the point $(1, -1)$.

115. For $x = y^2$, x is a function of y.

116. For $y = x^2$, y is a function of x.

117. If $f(x) = Ax^2 - 3x$ and $f(1) = -1$, find A.

118. If $g(x) = \dfrac{1}{b - x}$ and $g(3)$ is undefined, find b.

In Exercises 121 and 122, find the domain of each function, where a is any positive real number.

121. $f(x) = \dfrac{-100}{x^2 - a^2}$

122. $f(x) = -5\sqrt{x^2 - a^2}$

123. Using a graphing utility, graph the temperature function in Exercise 95. What time of day is it the warmest? What is the temperature? Looking at this function, explain why this model for Tampa, Florida, is valid only from sunrise to sunset (6 to 18).

124. Using a graphing utility, graph the height of the firecracker in Exercise 98. How long after liftoff is the firecracker airborne? What is the maximum height that the firecracker attains? Explain why this height model is valid only for the first 8 seconds.

125. Let $f(x) = x^2 + 1$. Graph $y_1 = f(x)$ and $y_2 = f(x - 2)$ in the same viewing window. Describe how the graph of y_2 can be obtained from the graph of y_1.

126. Let $f(x) = 4 - x^2$. Graph $y_1 = f(x)$ and $y_2 = f(x + 2)$ in the same viewing window. Describe how the graph of y_2 can be obtained from the graph of y_1.

PREVIEW TO CALCULUS

For Exercises 127–130, refer to the following:

In calculus, the difference quotient $\dfrac{f(x + h) - f(x)}{h}$ of a function f is used to find a new function f', called the *derivative of f*. To find f', we let h approach 0, $h \to 0$, in the difference quotient. For example, if $f(x) = x^2$, $\dfrac{f(x + h) - f(x)}{h} = 2x + h$, and allowing $h = 0$, we have $f'(x) = 2x$.

127. Given $f(x) = x^3 + x$, find $f'(x)$.

128. Given $f(x) = 6x + \sqrt{x}$, find $f'(x)$.

129. Given $f(x) = \dfrac{x - 5}{x + 3}$, find $f'(x)$.

130. Given $f(x) = \sqrt{\dfrac{x + 7}{5 - x}}$, find $f'(x)$.

SECTION
1.2 GRAPHS OF FUNCTIONS

SKILLS OBJECTIVES

- Recognize and graph common functions.
- Classify functions as even, odd, or neither.
- Determine whether functions are increasing, decreasing, or constant.
- Calculate the average rate of change of a function.
- Evaluate the difference quotient for a function.
- Graph piecewise-defined functions.

CONCEPTUAL OBJECTIVES

- Identify common functions.
- Develop and graph piecewise-defined functions.
  - Identify and graph points of discontinuity.
  - State the domain and range.
- Understand that even functions have graphs that are symmetric about the y-axis.
- Understand that odd functions have graphs that are symmetric about the origin.

Common Functions

The nine main functions you will read about in this section will constitute a "library" of functions that you should commit to memory. We will draw on this library of functions in the next section when graphing transformations are discussed.

In Section 0.6, we discussed equations and graphs of lines. All lines (with the exception of vertical lines) pass the vertical line test, and hence are classified as functions. Instead of the traditional notation of a line, $y = mx + b$, we use function notation and classify a function whose graph is a *line* as a *linear* function.

LINEAR FUNCTION

$$f(x) = mx + b \qquad m \text{ and } b \text{ are real numbers.}$$

The domain of a linear function $f(x) = mx + b$ is the set of all real numbers $\mathbb{R}$. The graph of this function has slope m and y-intercept b.

LINEAR FUNCTION: $f(x) = mx + b$	SLOPE: m	y-INTERCEPT: b
$f(x) = 2x - 7$	$m = 2$	$b = -7$
$f(x) = -x + 3$	$m = -1$	$b = 3$
$f(x) = x$	$m = 1$	$b = 0$
$f(x) = 5$	$m = 0$	$b = 5$

One special case of the linear function is the *constant function* ($m = 0$).

CONSTANT FUNCTION

$$f(x) = b \qquad b \text{ is any real number.}$$

The graph of a constant function $f(x) = b$ is a horizontal line. The y-intercept corresponds to the point $(0, b)$. The domain of a constant function is the set of all real numbers $\mathbb{R}$. The range, however, is a single value b. In other words, all x-values correspond to a single y-value.

Points that lie on the graph of a constant function $f(x) = b$ are

$(-5, b)$

$(-1, b)$

$(0, b)$

$(2, b)$

$(4, b)$

$\ldots$

(x, b)

Domain: $(-\infty, \infty)$ Range: $[b, b]$ or $\{b\}$

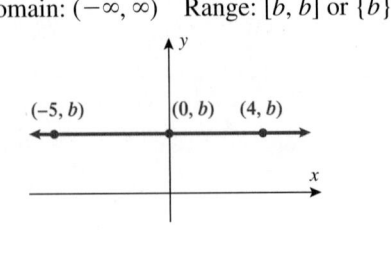

Another specific example of a linear function is the function having a slope of one ($m = 1$) and a y-intercept of zero ($b = 0$). This special case is called the *identity function*.

IDENTITY FUNCTION

$$f(x) = x$$

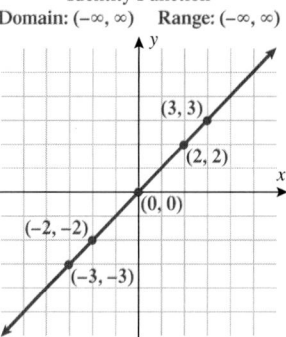

Identity Function
Domain: $(-\infty, \infty)$ Range: $(-\infty, \infty)$

The graph of the identity function has the following properties: it passes through the origin, and every point that lies on the line has equal x- and y-coordinates. Both the domain and the range of the identity function are the set of all real numbers $\mathbb{R}$.

A function that squares the input is called the *square function*.

SQUARE FUNCTION

$$f(x) = x^2$$

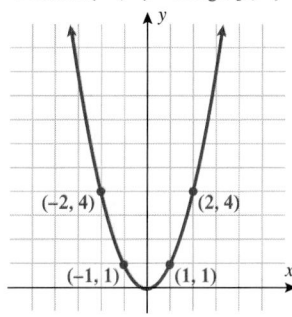

Square Function
Domain: $(-\infty, \infty)$ Range: $[0, \infty)$

The graph of the square function is called a parabola and will be discussed in further detail in Chapter 9. The domain of the square function is the set of all real numbers $\mathbb{R}$. Because squaring a real number always yields a positive number or zero, the range of the square function is the set of all nonnegative numbers. Note that the only intercept is the origin and the square function is symmetric about the y-axis. This graph is contained in quadrants I and II.

A function that cubes the input is called the *cube function*.

CUBE FUNCTION

$$f(x) = x^3$$

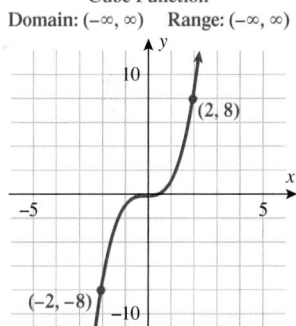

Cube Function
Domain: $(-\infty, \infty)$ Range: $(-\infty, \infty)$

The domain of the cube function is the set of all real numbers $\mathbb{R}$. Because cubing a negative number yields a negative number, cubing a positive number yields a positive number, and cubing 0 yields 0, the range of the cube function is also the set of all real numbers $\mathbb{R}$. Note that the only intercept is the origin and the cube function is symmetric about the origin. This graph extends only into quadrants I and III.

The next two functions are counterparts of the previous two functions: square root and cube root. When a function takes the square root of the input or the cube root of the input, the function is called the *square root function* or the *cube root function*, respectively.

SQUARE ROOT FUNCTION

$$f(x) = \sqrt{x} \quad \text{or} \quad f(x) = x^{1/2}$$

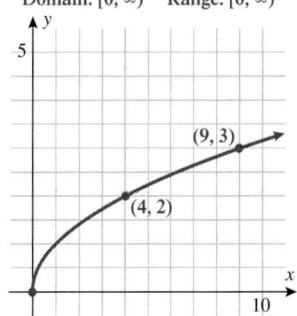

Square Root Function
Domain: $[0, \infty)$ Range: $[0, \infty)$

In Section 1.1, we found the domain to be $[0, \infty)$. The output of the function will be all real numbers greater than or equal to zero. Therefore, the range of the square root function is $[0, \infty)$. The graph of this function will be contained in quadrant I.

Cube Root Function
Domain: $(-\infty, \infty)$ Range: $(-\infty, \infty)$

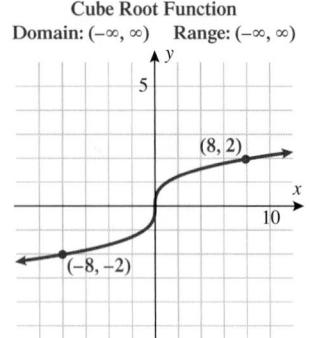

Absolute Value Function
Domain: $(-\infty, \infty)$ Range: $[0, \infty)$

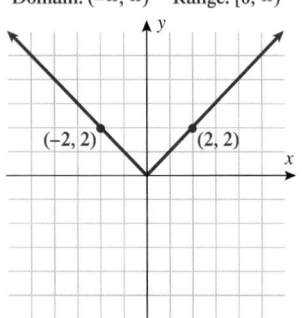

Reciprocal Function
Domain: $(-\infty, 0) \cup (0, \infty)$
Range: $(-\infty, 0) \cup (0, \infty)$

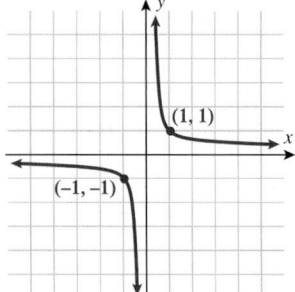

CUBE ROOT FUNCTION

$$f(x) = \sqrt[3]{x} \quad \text{or} \quad f(x) = x^{1/3}$$

In Section 1.1, we stated the domain of the cube root function to be $(-\infty, \infty)$. We see by the graph that the range is also $(-\infty, \infty)$. This graph is contained in quadrants I and III and passes through the origin. This function is symmetric about the origin.

In Sections 0.3 and 0.4, absolute value equations and inequalities were reviewed. Now we shift our focus to the graph of the *absolute value function*.

ABSOLUTE VALUE FUNCTION

$$f(x) = |x|$$

Some points that are on the graph of the absolute value function are $(-1, 1)$, $(0, 0)$, and $(1, 1)$. The domain of the absolute value function is the set of all real numbers $\mathbb{R}$, yet the range is the set of nonnegative real numbers. The graph of this function is symmetric with respect to the y-axis and is contained in quadrants I and II.

A function whose output is the reciprocal of its input is called the *reciprocal function*.

RECIPROCAL FUNCTION

$$f(x) = \frac{1}{x} \qquad x \neq 0$$

The only restriction on the domain of the reciprocal function is that $x \neq 0$. Therefore, we say the domain is the set of all real numbers excluding zero. The graph of the reciprocal function illustrates that its range is also the set of all real numbers except zero. Note that the reciprocal function is symmetric with respect to the origin and is contained in quadrants I and III.

Even and Odd Functions

Of the nine functions discussed above, several have similar properties of symmetry. The constant function, square function, and absolute value function are all symmetric with respect to the y-axis. The identity function, cube function, cube root function, and reciprocal function are all symmetric with respect to the origin. The term **even** is used to describe functions that are symmetric with respect to the y-axis, or vertical axis, and the term **odd** is used to describe functions that are symmetric with respect to the origin. Recall from Section 0.5 that symmetry can be determined both graphically and algebraically. The box below summarizes the graphic and algebraic characteristics of even and odd functions.

EVEN AND ODD FUNCTIONS

Function	Symmetric with Respect to	On Replacing x with $-x$
Even	y-axis or vertical axis	$f(-x) = f(x)$
Odd	origin	$f(-x) = -f(x)$

The algebraic method for determining symmetry with respect to the y-axis, or vertical axis, is to substitute in $-x$ for x. If the result is an equivalent equation, the function is symmetric with respect to the y-axis. Some examples of even functions are $f(x) = b$, $f(x) = x^2$, $f(x) = x^4$, and $f(x) = |x|$. In any of these equations, if $-x$ is substituted for x, the result is the same, that is, $f(-x) = f(x)$. Also note that, with the exception of the absolute value function, these examples are all even-degree polynomial equations. All constant functions are degree zero and are even functions.

The algebraic method for determining symmetry with respect to the origin is to substitute $-x$ for x. If the result is the negative of the original function, that is, if $f(-x) = -f(x)$, then the function is symmetric with respect to the origin and, hence, classified as an odd function. Examples of odd functions are $f(x) = x$, $f(x) = x^3$, $f(x) = x^5$, and $f(x) = x^{1/3}$. In any of these functions, if $-x$ is substituted for x, the result is the negative of the original function. Note that with the exception of the cube root function, these equations are odd-degree polynomials.

Be careful, though, because functions that are combinations of even- and odd-degree polynomials can turn out to be neither even nor odd, as we will see in Example 1.

EXAMPLE 1   **Determining Whether a Function Is Even, Odd, or Neither**

Determine whether the functions are even, odd, or neither.

a. $f(x) = x^2 - 3$ **b.** $g(x) = x^5 + x^3$ **c.** $h(x) = x^2 - x$

Solution (a):

Original function.	$f(x) = x^2 - 3$
Replace x with $-x$.	$f(-x) = (-x)^2 - 3$
Simplify.	$f(-x) = x^2 - 3 = f(x)$

Because $f(-x) = f(x)$, we say that $\boxed{f(x) \text{ is an } even \text{ function}}$.

Solution (b):

Original function.	$g(x) = x^5 + x^3$
Replace x with $-x$.	$g(-x) = (-x)^5 + (-x)^3$
Simplify.	$g(-x) = -x^5 - x^3 = -(x^5 + x^3) = -g(x)$

Because $g(-x) = -g(x)$, we say that $\boxed{g(x) \text{ is an } odd \text{ function}}$.

Solution (c):

Original function.	$h(x) = x^2 - x$
Replace x with $-x$.	$h(-x) = (-x)^2 - (-x)$
Simplify.	$h(-x) = x^2 + x$

$h(-x)$ is neither $-h(x)$ nor $h(x)$; therefore, the function $h(x)$ is $\boxed{\text{neither even nor odd}}$.

In parts (a), (b), and (c), we classified these functions as either even, odd, or neither, using the algebraic test. Look back at them now and reflect on whether these classifications agree with your intuition. In part (a), we combined two functions: the square function and the constant function. Both of these functions are even, and adding even functions yields another even function. In part (b), we combined two odd functions: the fifth-power function and the cube function. Both of these functions are odd, and adding two odd functions yields another odd function. In part (c), we combined two functions: the square function and the

Technology Tip

a. Graph $y_1 = f(x) = x^2 - 3$.

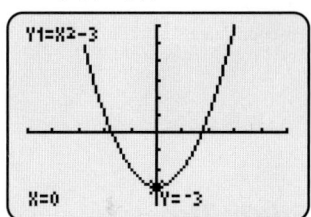

Even; symmetric with respect to the y-axis.

b. Graph $y_1 = g(x) = x^5 + x^3$.

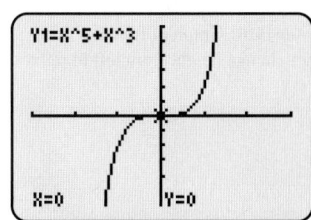

Odd; symmetric with respect to origin.

c. Graph $y_1 = h(x) = x^2 - x$.

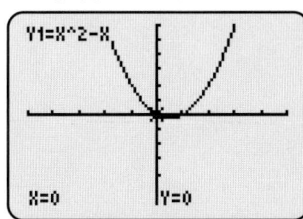

No symmetry with respect to y-axis or origin.

identity function. The square function is even, and the identity function is odd. In this part, combining an even function with an odd function yields a function that is neither even nor odd and, hence, has no symmetry with respect to the vertical axis or the origin.

■ **Answer: a.** even   **b.** neither

■ **YOUR TURN** Classify the functions as even, odd, or neither.

$$\textbf{a. } f(x) = |x| + 4 \qquad \textbf{b. } f(x) = x^3 - 1$$

Increasing and Decreasing Functions

Look at the figure on the left. Graphs are read from *left to right*. If we start at the left side of the graph and trace the red curve, we see that the function values (values in the vertical direction) are decreasing until arriving at the point $(-2, -2)$. Then, the function values increase until arriving at the point $(-1, 1)$. The values then remain constant ($y = 1$) between the points $(-1, 1)$ and $(0, 1)$. Proceeding beyond the point $(0, 1)$, the function values decrease again until the point $(2, -2)$. Beyond the point $(2, -2)$, the function values increase again until the point $(6, 4)$. Finally, the function values decrease and continue to do so.

When specifying a function as increasing, decreasing, or constant, the *intervals are classified according to the x-coordinate*. For instance, in this graph, we say the function is increasing when x is between $x = -2$ and $x = -1$ and again when x is between $x = 2$ and $x = 6$. The graph is classified as decreasing when x is less than -2 and again when x is between 0 and 2 and again when x is greater than 6. The graph is classified as constant when x is between -1 and 0. In interval notation, this is summarized as

Decreasing	**Increasing**	**Constant**
$(-\infty, -2) \cup (0, 2) \cup (6, \infty)$	$(-2, -1) \cup (2, 6)$	$(-1, 0)$

An algebraic test for determining whether a function is increasing, decreasing, or constant is to compare the value $f(x)$ of the function for particular points in the intervals.

Study Tip

- Graphs are read from left to right.
- Intervals correspond to the x-coordinates.

Study Tip

Increasing: Graph of function rises from left to right.
Decreasing: Graph of function falls from left to right.
Constant: Graph of function does not change height from left to right.

INCREASING, DECREASING, AND CONSTANT FUNCTIONS

1. A function f is **increasing** on an open interval I if for any x_1 and x_2 in I, where $x_1 < x_2$, then $f(x_1) < f(x_2)$.
2. A function f is **decreasing** on an open interval I if for any x_1 and x_2 in I, where $x_1 < x_2$, then $f(x_1) > f(x_2)$.
3. A function f is **constant** on an open interval I if for any x_1 and x_2 in I, then $f(x_1) = f(x_2)$.

In addition to classifying a function as increasing, decreasing, or constant, we can also determine the domain and range of a function by inspecting its graph from left to right:

- The domain is the set of all x-values where the function is defined.
- The range is the set of all y-values that the graph of the function corresponds to.
- A solid dot on the left or right end of a graph indicates that the graph terminates there and the point is included in the graph.
- An open dot indicates that the graph terminates there and the point is not included in the graph.
- Unless a dot is present, it is assumed that a graph continues indefinitely in the same direction. (An arrow is used in some books to indicate direction.)

EXAMPLE 2   **Finding Intervals When a Function Is Increasing or Decreasing**

Given the graph of a function:

a. State the domain and range of the function.

b. Find the intervals when the function is increasing, decreasing, or constant.

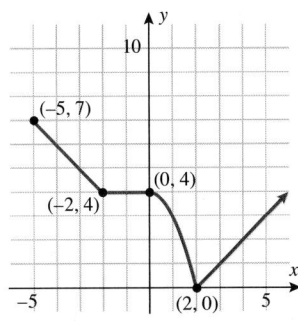

Solution:

Domain: $[-5, \infty)$

Range: $[0, \infty)$

Reading the graph from **left to right**, we see that the graph

- decreases from the point $(-5, 7)$ to the point $(-2, 4)$.

- is constant from the point $(-2, 4)$ to the point $(0, 4)$.

- decreases from the point $(0, 4)$ to the point $(2, 0)$.

- increases from the point $(2, 0)$ on.

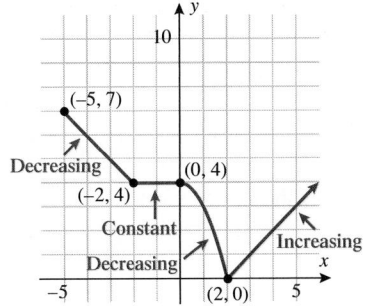

The intervals of increasing and decreasing correspond to the **x-coordinates**.

We say that this function is

- increasing on the interval $(2, \infty)$.

- decreasing on the interval $(-5, -2) \cup (0, 2)$.

- constant on the interval $(-2, 0)$.

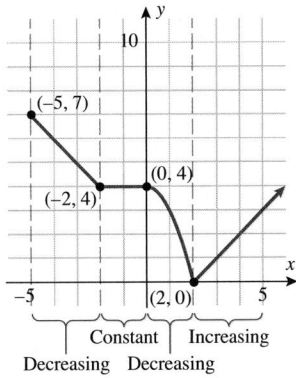

Note: The intervals of increasing or decreasing are defined on *open* intervals. This should not be confused with the domain. For example, the point $x = -5$ is included in the domain of the function but not in the interval where the function is classified as decreasing.

Average Rate of Change

How do we know *how much* a function is increasing or decreasing? For example, is the price of a stock slightly increasing or is it doubling every week? One way we determine how much a function is increasing or decreasing is by calculating its *average rate of change*.

Let (x_1, y_1) and (x_2, y_2) be two points that lie on the graph of a function f. Draw the line that passes through these two points (x_1, y_1) and (x_2, y_2). This line is called a **secant line**.

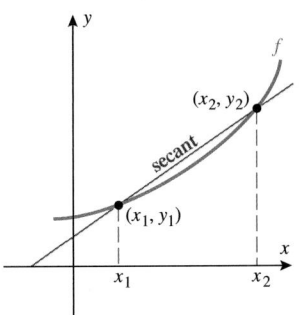

Note that the slope of the secant line is given by $m = \dfrac{y_2 - y_1}{x_2 - x_1}$, and recall that the slope of a line is the rate of change of that line. The **slope of the secant line** is used to represent the *average rate of change* of the function.

AVERAGE RATE OF CHANGE

Let $(x_1, f(x_1))$ and $(x_2, f(x_2))$ be two distinct points, $(x_1 \neq x_2)$, on the graph of the function f. The **average rate of change** of f between x_1 and x_2 is given by

$$\text{Average rate of change} = \dfrac{f(x_2) - f(x_1)}{x_2 - x_1}$$

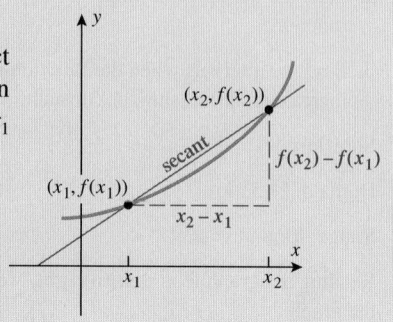

 EXAMPLE 3   Average Rate of Change

Find the average rate of change of $f(x) = x^4$ from

a. $x = -1$ to $x = 0$ **b.** $x = 0$ to $x = 1$ **c.** $x = 1$ to $x = 2$

Solution (a):

Write the average rate of change formula.

$$\dfrac{f(x_2) - f(x_1)}{x_2 - x_1}$$

Let $x_1 = -1$ and $x_2 = 0$.

$$= \dfrac{f(0) - f(-1)}{0 - (-1)}$$

Substitute $f(-1) = (-1)^4 = 1$ and $f(0) = 0^4 = 0$.

$$= \dfrac{0 - 1}{0 - (-1)}$$

Simplify.

$$= \boxed{-1}$$

Solution (b):

Write the average rate of change formula.

$$\dfrac{f(x_2) - f(x_1)}{x_2 - x_1}$$

Let $x_1 = 0$ and $x_2 = 1$.

$$= \dfrac{f(1) - f(0)}{1 - 0}$$

Substitute $f(0) = 0^4 = 0$ and $f(1) = (1)^4 = 1$.

$$= \dfrac{1 - 0}{1 - 0}$$

Simplify.

$$= \boxed{1}$$

Solution (c):

Write the average rate of change formula.

$$\dfrac{f(x_2) - f(x_1)}{x_2 - x_1}$$

Let $x_1 = 1$ and $x_2 = 2$.

$$= \dfrac{f(2) - f(1)}{2 - 1}$$

Substitute $f(1) = 1^4 = 1$ and $f(2) = (2)^4 = 16$.

$$= \dfrac{16 - 1}{2 - 1}$$

Simplify.

$$= \boxed{15}$$

Graphical Interpretation: Slope of the Secant Line

a. Between $(-1, 1)$ and $(0, 0)$, this function is decreasing at a rate of 1.

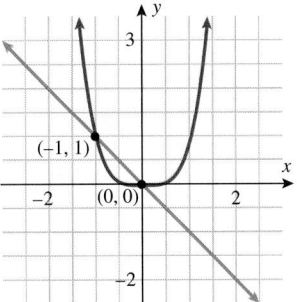

b. Between $(0, 0)$ and $(1, 1)$, this function is increasing at a rate of 1.

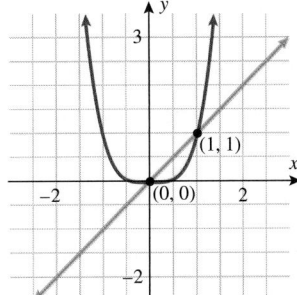

c. Between $(1, 1)$ and $(2, 16)$, this function is increasing at a rate of 15.

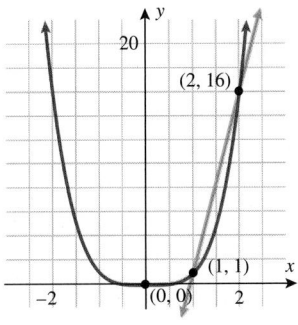

■ **YOUR TURN** Find the average rate of change of $f(x) = x^2$ from

       a. $x = -2$ to $x = 0$ **b.** $x = 0$ to $x = 2$

■ **Answer: a.** -2 **b.** 2

The average rate of change can also be written in terms of the difference quotient.

WORDS	MATH
Let the distance between x_1 and x_2 be h.	$x_2 - x_1 = h$
Solve for x_2.	$x_2 = x_1 + h$
Substitute $x_2 - x_1 = h$ into the denominator and $x_2 = x_1 + h$ into the numerator of the average rate of change.	$\text{Average rate of change} = \dfrac{f(x_2) - f(x_1)}{x_2 - x_1}$
	$= \dfrac{f(x_1 + h) - f(x_1)}{h}$
Let $x_1 = x$.	$\boxed{= \dfrac{f(x + h) - f(x)}{h}}$

When written in this form, the average rate of change is called the **difference quotient**.

DEFINITION   **Difference Quotient**

The expression $\dfrac{f(x + h) - f(x)}{h}$, where $h \neq 0$, is called the **difference quotient**.

The difference quotient is more meaningful when h is small. In calculus the difference quotient is used to define a *derivative*.

Study Tip

Use brackets or parentheses around $f(x)$ to avoid forgetting to distribute the negative sign:

$$\frac{f(x + h) - [f(x)]}{h}$$

EXAMPLE 4   **Calculating the Difference Quotient**

Calculate the difference quotient for the function $f(x) = 2x^2 + 1$.

Solution:

Find $f(x + h)$.

$$f(x + h) = 2(x + h)^2 + 1$$
$$= 2(x^2 + 2xh + h^2) + 1$$
$$= 2x^2 + 4xh + 2h^2 + 1$$

Find the difference quotient.

$$\frac{f(x + h) - f(x)}{h} = \frac{\overbrace{2x^2 + 4xh + 2h^2 + 1}^{f(x+h)} - \overbrace{(2x^2 + 1)}^{f(x)}}{h}$$

Simplify.

$$= \frac{2x^2 + 4xh + 2h^2 + 1 - 2x^2 - 1}{h}$$

$$= \frac{4xh + 2h^2}{h}$$

Factor the numerator.

$$= \frac{h(4x + 2h)}{h}$$

Cancel (divide out) the common h.

$$= \boxed{4x + 2h} \qquad h \neq 0$$

■ **Answer:**

$$\frac{f(x + h) - f(x)}{h} = -2x - h$$

■ **YOUR TURN** Calculate the difference quotient for the function $f(x) = -x^2 + 2$.

Piecewise-Defined Functions

Most of the functions that we have seen in this text are functions defined by polynomials. Sometimes the need arises to define functions in terms of *pieces*. For example, most plumbers charge a flat fee for a house call and then an additional hourly rate for the job. For instance, if a particular plumber charges $100 to drive out to your house and work for 1 hour and then $25 an hour for every additional hour he or she works on your job, we would define this function in pieces. If we let h be the number of hours worked, then the charge is defined as

$$\text{Plumbing charge} = \begin{cases} 100 & 0 < h \leq 1 \\ 100 + 25(h - 1) & h > 1 \end{cases}$$

We can see in the graph of this function that there is 1 hour that is constant and after that the function continually increases.

The next example is a piecewise-defined function given in terms of pieces of functions from our "library of functions." Because the function is defined in terms of pieces of other functions, we draw the graph of each individual function and, then, for each function darken the piece corresponding to its part of the domain.

EXAMPLE 5   Graphing Piecewise-Defined Functions

Graph the piecewise-defined function, and state the domain, range, and intervals when the function is increasing, decreasing, or constant.

$$G(x) = \begin{cases} x^2 & x < -1 \\ 1 & -1 \le x \le 1 \\ x & x > 1 \end{cases}$$

Solution:

Graph each of the functions on the same plane.

Square function:
$f(x) = x^2$

Constant function:
$f(x) = 1$

Identity function:
$f(x) = x$

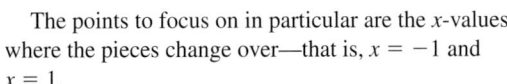
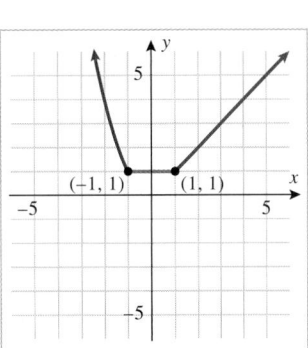

The points to focus on in particular are the x-values where the pieces change over—that is, $x = -1$ and $x = 1$.

Let's now investigate each piece. When $x < -1$, this function is defined by the square function, $f(x) = x^2$, so darken that particular function to the left of $x = -1$. When $-1 \le x \le 1$, the function is defined by the constant function, $f(x) = 1$, so darken that particular function between the x-values of -1 and 1. When $x > 1$, the function is defined by the identity function, $f(x) = x$, so darken that function to the right of $x = 1$. Erase everything that is not darkened, and the resulting graph of the piecewise-defined function is given on the right.

This function is defined for all real values of x, so the domain of this function is the set of all real numbers. The values that this function yields in the vertical direction are all real numbers greater than or equal to 1. Hence, the range of this function is $[1, \infty)$. The intervals of increasing, decreasing, and constant are as follows:

Decreasing: $(-\infty, -1)$

Constant: $(-1, 1)$

Increasing: $(1, \infty)$

Technology Tip

Plot a piecewise-defined function using the TEST menu operations to define the inequalities in the function. Press:

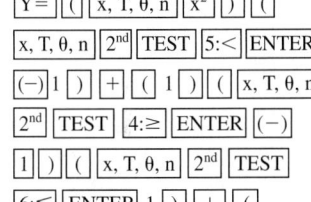

Set the viewing rectangle as $[-4, 4]$ by $[-2, 5]$; then press GRAPH.

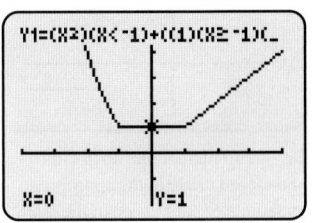

The term **continuous** implies that there are no holes or jumps and that the graph can be drawn without picking up your pencil. A function that does have holes or jumps and cannot be drawn in one motion without picking up your pencil is classified as **discontinuous**, and the points where the holes or jumps occur are called *points of discontinuity.*

The previous example illustrates a *continuous* piecewise-defined function. At the $x = -1$ junction, the square function and constant function both pass through the point $(-1, 1)$. At the $x = 1$ junction, the constant function and the identity function both pass through the point $(1, 1)$. Since the graph of this piecewise-defined function has no holes or jumps, we classify it as a continuous function.

The next example illustrates a *discontinuous* piecewise-defined function.

EXAMPLE 6 Graphing a Discontinuous Piecewise-Defined Function

Graph the piecewise-defined function, and state the intervals where the function is increasing, decreasing, or constant, along with the domain and range.

$$f(x) = \begin{cases} 1 - x & x < 0 \\ x & 0 \le x < 2 \\ -1 & x > 2 \end{cases}$$

Solution:

Graph these functions on the same plane.

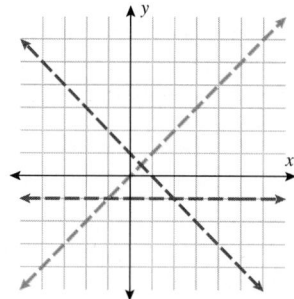

Linear function:
$f(x) = 1 - x$

Identity function:
$f(x) = x$

Constant function:
$f(x) = -1$

Darken the piecewise-defined function on the graph. For all values less than zero ($x < 0$), the function is defined by the linear function. Note the use of an open circle, indicating up to but not including $x = 0$. For values $0 \le x < 2$, the function is defined by the identity function.

The circle is filled in at the left endpoint, $x = 0$. An open circle is used at $x = 2$. For all values greater than 2, $x > 2$, the function is defined by the constant function. Because this interval does not include the point $x = 2$, an open circle is used.

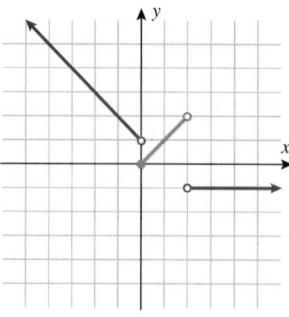

At what intervals is the function increasing, decreasing, or constant? Remember that the intervals correspond to the x-values.

Decreasing: $(-\infty, 0)$ Increasing: $(0, 2)$ Constant: $(2, \infty)$

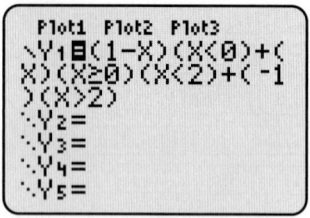
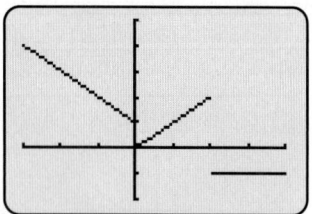
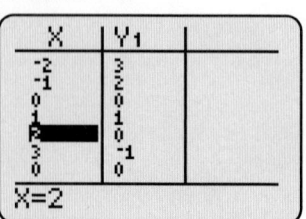

The function is defined for all values of x except $x = 2$.

$$\text{Domain:} \quad (-\infty, 2) \cup (2, \infty)$$

The output of this function (vertical direction) takes on the y-values $y \geq 0$ and the additional single value $y = -1$.

$$\text{Range:} \quad [-1, -1] \cup [0, \infty) \text{ or } \{-1\} \cup [0, \infty)$$

We mentioned earlier that a discontinuous function has a graph that exhibits holes or jumps. In Example 6, the point $x = 0$ corresponds to a jump, because you would have to pick up your pencil to continue drawing the graph. The point $x = 2$ corresponds to both a hole and a jump. The hole indicates that the function is not defined at that point, and there is still a jump because the identity function and the constant function do not meet at the same y-value at $x = 2$.

■ **YOUR TURN** Graph the piecewise-defined function, and state the intervals where the function is increasing, decreasing, or constant, along with the domain and range.

$$f(x) = \begin{cases} -x & x \leq -1 \\ 2 & -1 < x < 1 \\ x & x > 1 \end{cases}$$

■ **Answer:** Increasing: $(1, \infty)$
Decreasing: $(-\infty, -1)$
Constant: $(-1, 1)$
Domain:
$(-\infty, 1) \cup (1, \infty)$
Range: $[1, \infty)$

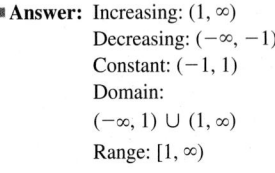

Piecewise-defined functions whose "pieces" are constants are called **step functions**. The reason for this name is that the graph of a step function looks like steps of a staircase. A common step function used in engineering is the **Heaviside step function** (also called the **unit step function**):

$$H(t) = \begin{cases} 0 & t < 0 \\ 1 & t \geq 0 \end{cases}$$

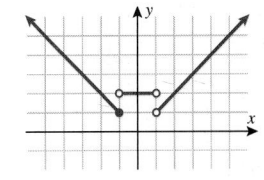

This function is used in signal processing to represent a signal that turns on at some time and stays on indefinitely.

A common step function used in business applications is the *greatest integer function*.

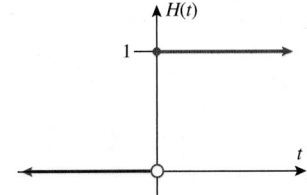

GREATEST INTEGER FUNCTION

$$f(x) = [[x]] = \text{greatest integer less than or equal to } x$$

x	1.0	1.3	1.5	1.7	1.9	2.0
$f(x) = [[x]]$	1	1	1	1	1	2

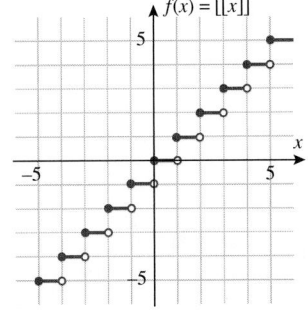

SECTION
1.2 SUMMARY

NAME	FUNCTION	DOMAIN	RANGE	GRAPH	EVEN/ODD
Linear	$f(x) = mx + b$, $m \neq 0$	$(-\infty, \infty)$	$(-\infty, \infty)$		Neither (unless $y = x$)
Constant	$f(x) = c$	$(-\infty, \infty)$	$[c, c]$ or $\{c\}$		Even
Identity	$f(x) = x$	$(-\infty, \infty)$	$(-\infty, \infty)$		Odd
Square	$f(x) = x^2$	$(-\infty, \infty)$	$[0, \infty)$		Even
Cube	$f(x) = x^3$	$(-\infty, \infty)$	$(-\infty, \infty)$		Odd
Square Root	$f(x) = \sqrt{x}$	$[0, \infty)$	$[0, \infty)$		Neither
Cube Root	$f(x) = \sqrt[3]{x}$	$(-\infty, \infty)$	$(-\infty, \infty)$		Odd
Absolute Value	$f(x) = |x|$	$(-\infty, \infty)$	$[0, \infty)$		Even
Reciprocal	$f(x) = \dfrac{1}{x}$	$(-\infty, 0) \cup (0, \infty)$	$(-\infty, 0) \cup (0, \infty)$		Odd

Domain and Range of a Function

- **Implied Domain:** Exclude any values that lead to the function being undefined (dividing by zero) or imaginary outputs (square root of a negative real number).

- Inspect the graph to determine the set of all inputs (domain) and the set of all outputs (range).

Finding Intervals Where a Function Is Increasing, Decreasing, or Constant

- **Increasing:** Graph of function rises from left to right.
- **Decreasing:** Graph of function falls from left to right.
- **Constant:** Graph of function does not change height from left to right.

Average Rate of Change $\dfrac{f(x_2) - f(x_1)}{x_2 - x_1}$ $x_1 \neq x_2$

Difference Quotient $\dfrac{f(x + h) - f(x)}{h}$ $h \neq 0$

Piecewise-Defined Functions

- **Continuous:** You can draw the graph of a function without picking up the pencil.
- **Discontinuous:** Graph has holes and/or jumps.

SECTION
1.2 EXERCISES

■ SKILLS

In Exercises 1–16, determine whether the function is even, odd, or neither.

1. $h(x) = x^2 + 2x$

2. $G(x) = 2x^4 + 3x^3$

3. $h(x) = x^{1/3} - x$

4. $g(x) = x^{-1} + x$

5. $f(x) = |x| + 5$

6. $f(x) = |x| + x^2$

7. $f(x) = |x|$

8. $f(x) = |x^3|$

9. $G(t) = |t - 3|$

10. $g(t) = |t + 2|$

11. $G(t) = \sqrt{t - 3}$

12. $f(x) = \sqrt{2 - x}$

13. $g(x) = \sqrt{x^2 + x}$

14. $f(x) = \sqrt{x^2 + 2}$

15. $h(x) = \dfrac{1}{x} + 3$

16. $h(x) = \dfrac{1}{x} - 2x$

In Exercises 17–28, state the (a) domain, (b) range, and (c) x-interval(s) where the function is increasing, decreasing, or constant. Find the values of (d) $f(0)$, (e) $f(-2)$, and (f) $f(2)$.

17.

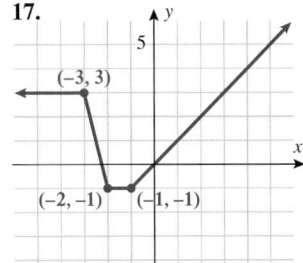

18.

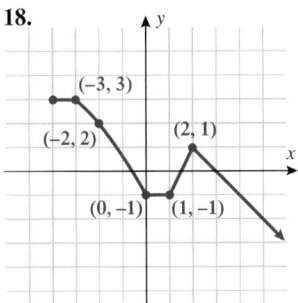

19.

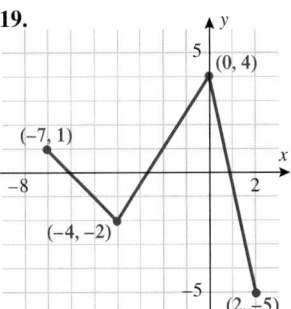

20.

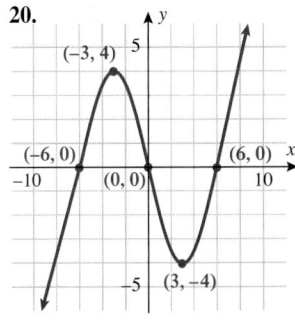

21.

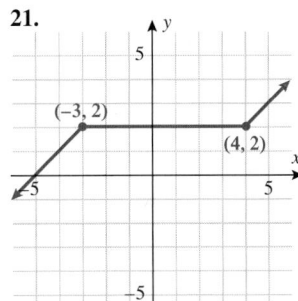

22.

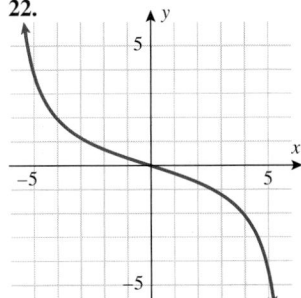

23.

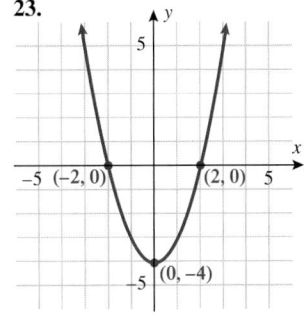

24.

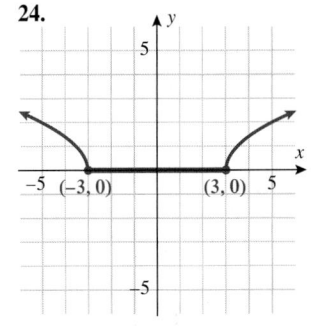

25.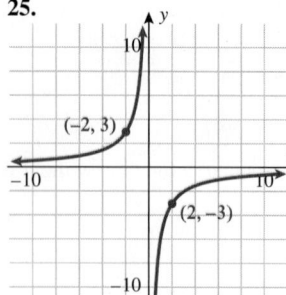

26.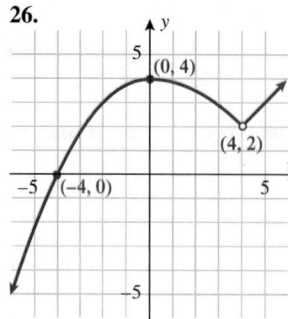

27.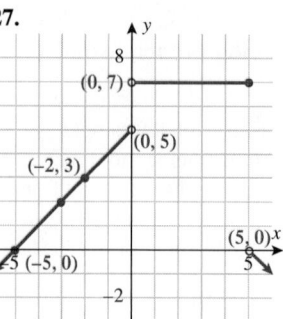

28.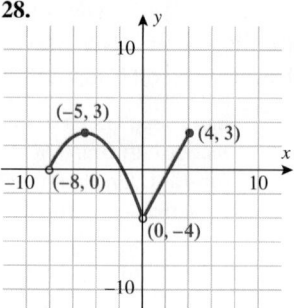

In Exercises 29–44, find the difference quotient $\dfrac{f(x + h) - f(x)}{h}$ for each function.

29. $f(x) = x^2 - x$

30. $f(x) = x^2 + 2x$

31. $f(x) = 3x + x^2$

32. $f(x) = 5x - x^2$

33. $f(x) = x^2 - 3x + 2$

34. $f(x) = x^2 - 2x + 5$

35. $f(x) = -3x^2 + 5x - 4$

36. $f(x) = -4x^2 + 2x - 3$

37. $f(x) = x^3 + x^2$

38. $f(x) = (x - 1)^4$

39. $f(x) = \dfrac{2}{x - 2}$

40. $f(x) = \dfrac{x + 5}{x - 7}$

41. $f(x) = \sqrt{1 - 2x}$

42. $f(x) = \sqrt{x^2 + x + 1}$

43. $f(x) = \dfrac{4}{\sqrt{x}}$

44. $f(x) = \sqrt{\dfrac{x}{x + 1}}$

In Exercises 45–52, find the average rate of change of the function from $x = 1$ to $x = 3$.

45. $f(x) = x^3$

46. $f(x) = \dfrac{1}{x}$

47. $f(x) = |x|$

48. $f(x) = 2x$

49. $f(x) = 1 - 2x$

50. $f(x) = 9 - x^2$

51. $f(x) = |5 - 2x|$

52. $f(x) = \sqrt{x^2 - 1}$

In Exercises 53–76, graph the piecewise-defined functions. State the domain and range in interval notation. Determine the intervals where the function is increasing, decreasing, or constant.

53. $f(x) = \begin{cases} x & x < 2 \\ 2 & x \geq 2 \end{cases}$

54. $f(x) = \begin{cases} -x & x < -1 \\ -1 & x \geq -1 \end{cases}$

55. $f(x) = \begin{cases} 1 & x < -1 \\ x^2 & x \geq -1 \end{cases}$

56. $f(x) = \begin{cases} x^2 & x < 2 \\ 4 & x \geq 2 \end{cases}$

57. $f(x) = \begin{cases} x & x < 0 \\ x^2 & x \geq 0 \end{cases}$

58. $f(x) = \begin{cases} -x & x \leq 0 \\ x^2 & x > 0 \end{cases}$

59. $f(x) = \begin{cases} -x + 2 & x < 1 \\ x^2 & x \geq 1 \end{cases}$

60. $f(x) = \begin{cases} 2 + x & x \leq -1 \\ x^2 & x > -1 \end{cases}$

61. $G(x) = \begin{cases} -1 & x < -1 \\ x & -1 \leq x \leq 3 \\ 3 & x > 3 \end{cases}$

62. $G(x) = \begin{cases} -1 & x < -1 \\ x & -1 < x < 3 \\ 3 & x > 3 \end{cases}$

63. $G(t) = \begin{cases} 1 & t < 1 \\ t^2 & 1 \leq t \leq 2 \\ 4 & t > 2 \end{cases}$

64. $G(t) = \begin{cases} 1 & t < 1 \\ t^2 & 1 < t < 2 \\ 4 & t > 2 \end{cases}$

65. $f(x) = \begin{cases} -x - 1 & x < -2 \\ x + 1 & -2 < x < 1 \\ -x + 1 & x \geq 1 \end{cases}$

66. $f(x) = \begin{cases} -x - 1 & x \leq -2 \\ x + 1 & -2 < x < 1 \\ -x + 1 & x > 1 \end{cases}$

67. $G(x) = \begin{cases} 0 & x < 0 \\ \sqrt{x} & x \geq 0 \end{cases}$

68. $G(x) = \begin{cases} 1 & x < 1 \\ \sqrt[3]{x} & x > 1 \end{cases}$

69. $G(x) = \begin{cases} 0 & x = 0 \\ \dfrac{1}{x} & x \neq 0 \end{cases}$

70. $G(x) = \begin{cases} 0 & x = 0 \\ -\dfrac{1}{x} & x \neq 0 \end{cases}$

71. $G(x) = \begin{cases} -\sqrt[3]{x} & x \leq -1 \\ x & -1 < x < 1 \\ -\sqrt{x} & x > 1 \end{cases}$

72. $G(x) = \begin{cases} -\sqrt[3]{x} & x < -1 \\ x & -1 \leq x < 1 \\ \sqrt{x} & x > 1 \end{cases}$

73. $f(x) = \begin{cases} x + 3 & x \leq -2 \\ |x| & -2 < x < 2 \\ x^2 & x \geq 2 \end{cases}$

74. $f(x) = \begin{cases} |x| & x < -1 \\ 1 & -1 < x < 1 \\ |x| & x > 1 \end{cases}$

75. $f(x) = \begin{cases} x & x \le -1 \\ x^3 & -1 < x < 1 \\ x^2 & x > 1 \end{cases}$

76. $f(x) = \begin{cases} x^2 & x \le -1 \\ x^3 & -1 < x < 1 \\ x & x \ge 1 \end{cases}$

■ APPLICATIONS

77. Budget: Costs. The Kappa Kappa Gamma sorority decides to order custom-made T-shirts for its *Kappa Krush* mixer with the Sigma Alpha Epsilon fraternity. If the sorority orders 50 or fewer T-shirts, the cost is $10 per shirt. If it orders more than 50 but fewer than 100, the cost is $9 per shirt. If it orders 100 or more the cost is $8 per shirt. Find the cost function $C(x)$ as a function of the number of T-shirts x ordered.

78. Budget: Costs. The marching band at a university is ordering some additional uniforms to replace existing uniforms that are worn out. If the band orders 50 or fewer, the cost is $176.12 per uniform. If it orders more than 50 but fewer than 100, the cost is $159.73 per uniform. Find the cost function $C(x)$ as a function of the number of new uniforms x ordered.

79. Budget: Costs. The Richmond rowing club is planning to enter the *Head of the Charles* race in Boston and is trying to figure out how much money to raise. The entry fee is $250 per boat for the first 10 boats and $175 for each additional boat. Find the cost function $C(x)$ as a function of the number of boats x the club enters.

80. Phone Cost: Long-Distance Calling. A phone company charges $.39 per minute for the first 10 minutes of an international long-distance phone call and $.12 per minute every minute after that. Find the cost function $C(x)$ as a function of the length of the phone call x in minutes.

81. Event Planning. A young couple are planning their wedding reception at a yacht club. The yacht club charges a flat rate of $1,000 to reserve the dining room for a private party. The cost of food is $35 per person for the first 100 people and $25 per person for every additional person beyond the first 100. Write the cost function $C(x)$ as a function of the number of people x attending the reception.

82. Home Improvement. An irrigation company gives you an estimate for an eight-zone sprinkler system. The parts are $1,400, and the labor is $25 per hour. Write a function $C(x)$ that determines the cost of a new sprinkler system if you choose this irrigation company.

83. Sales. A famous author negotiates with her publisher the monies she will receive for her next suspense novel. She will receive $50,000 up front and a 15% royalty rate on the first 100,000 books sold, and 20% on any books sold beyond that. If the book sells for $20 and royalties are based on the selling price, write a royalties function $R(x)$ as a function of total number x of books sold.

84. Sales. Rework Exercise 83 if the author receives $35,000 up front, 15% for the first 100,000 books sold, and 25% on any books sold beyond that.

85. Profit. A group of artists are trying to decide whether they will make a profit if they set up a Web-based business to market and sell stained glass that they make. The costs associated with this business are $100 per month for the Web site and $700 per month for the studio they rent. The materials cost $35 for each work in stained glass, and the artists charge $100 for each unit they sell. Write the monthly profit as a function of the number of stained-glass units they sell.

86. Profit. Philip decides to host a shrimp boil at his house as a fund-raiser for his daughter's AAU basketball team. He orders gulf shrimp to be flown in from New Orleans. The shrimp costs $5 per pound. The shipping costs $30. If he charges $10 per person, write a function $F(x)$ that represents either his loss or profit as a function of the number of people x that attend. Assume that each person will eat 1 pound of shrimp.

87. Postage Rates. The following table corresponds to first-class postage rates for the U.S. Postal Service. Write a piecewise-defined function in terms of the greatest integer function that models this cost of mailing flat envelopes first class.

WEIGHT LESS THAN (OUNCES)	FIRST-CLASS RATE (LARGE ENVELOPES)
1	$0.88
2	$1.05
3	$1.22
4	$1.39
5	$1.56
6	$1.73
7	$1.90
8	$2.07
9	$2.24
10	$2.41
11	$2.58
12	$2.75
13	$2.92

88. Postage Rates. The following table corresponds to first-class postage rates for the U.S. Postal Service. Write a piecewise-defined function in terms of the greatest integer function that models this cost of mailing parcels first class.

WEIGHT LESS THAN (OUNCES)	FIRST-CLASS RATE (PACKAGES)
1	$1.22
2	$1.39
3	$1.56
4	$1.73
5	$1.90
6	$2.07
7	$2.24
8	$2.41
9	$2.58
10	$2.75
11	$2.92
12	$3.09
13	$3.26

A square wave is a waveform used in electronic circuit testing and signal processing. A square wave alternates regularly and instantaneously between two levels.

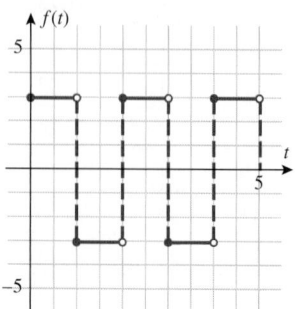

89. Electronics: Signals. Write a step function $f(t)$ that represents the following square wave:

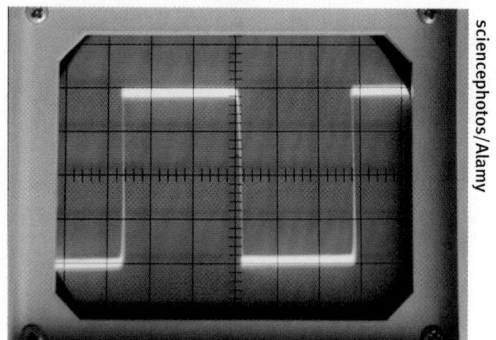

90. Electronics: Signals. Write a step function $f(x)$ that represents the following square wave, where x represents frequency in Hz:

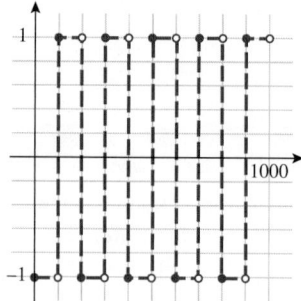

For Exercises 91 and 92, refer to the following table:

Global Carbon Emissions from Fossil Fuel Burning

YEAR	MILLIONS OF TONS OF CARBON
1900	500
1925	1000
1950	1500
1975	5000
2000	7000

91. Climate Change: Global Warming. What is the average rate of change in global carbon emissions from fossil fuel burning from

a. 1900 to 1950?   **b.** 1950 to 2000?

92. Climate Change: Global Warming. What is the average rate of change in global carbon emissions from fossil fuel burning from

a. 1950 to 1975?   **b.** 1975 to 2000?

For Exercises 93 and 94, use the following information:

The height (in feet) of a falling object with an initial velocity of 48 feet per second launched straight upward from the ground is given by $h(t) = -16t^2 + 48t$, where t is time (in seconds).

93. Falling Objects. What is the average rate of change of the height as a function of time from $t = 1$ to $t = 2$?

94. Falling Objects. What is the average rate of change of the height as a function of time from $t = 1$ to $t = 3$?

■ CATCH THE MISTAKE

In Exercises 95–98, explain the mistake that is made.

95. Graph the piecewise-defined function. State the domain and range.

$$f(x) = \begin{cases} -x & x < 0 \\ x & x > 0 \end{cases}$$

Solution:

Draw the graphs of
$f(x) = -x$ and
$f(x) = x.$

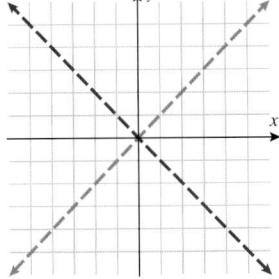

Darken the graph of
$f(x) = -x$ when $x < 0$
and the graph of $f(x) = x$
when $x > 0$. This gives
us the familiar absolute
value graph.

Domain: $(-\infty, \infty)$ or $\mathbb{R}$
Range: $[0, \infty)$

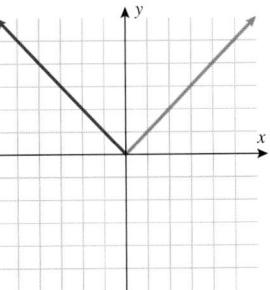

This is incorrect. What mistake was made?

96. Graph the piecewise-defined function. State the domain and range.

$$f(x) = \begin{cases} -x & x \le 1 \\ x & x > 1 \end{cases}$$

Solution:

Draw the graphs of
$f(x) = -x$ and
$f(x) = x.$

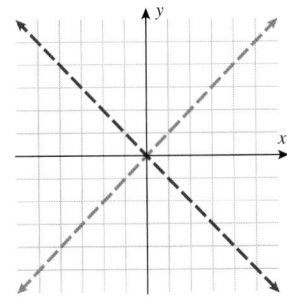

Darken the graph of
$f(x) = -x$ when $x < 1$
and the graph of $f(x) = x$
when $x > 1.$

The resulting graph is
as shown.

Domain: $(-\infty, \infty)$ or $\mathbb{R}$
Range: $(-1, \infty)$

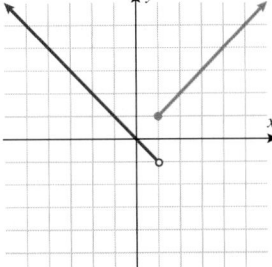

This is incorrect. What mistake was made?

97. The cost of airport Internet access is $15 for the first 30 minutes and $1 per minute for each additional minute. Write a function describing the cost of the service as a function of minutes used online.

Solution: $C(x) = \begin{cases} 15 & x \le 30 \\ 15 + x & x > 30 \end{cases}$

This is incorrect. What mistake was made?

98. Most money market accounts pay a higher interest with a higher principal. If the credit union is offering 2% on accounts with less than or equal to $10,000 and 4% on the additional money over $10,000, write the interest function $I(x)$ that represents the interest earned on an account as a function of dollars in the account.

Solution: $I(x) = \begin{cases} 0.02x & x \le 10{,}000 \\ 0.02(10{,}000) + 0.04x & x > 10{,}000 \end{cases}$

This is incorrect. What mistake was made?

▪ CONCEPTUAL

In Exercises 99 and 100, determine whether each statement is true or false.

99. If an odd function has an interval where the function is increasing, then it also has to have an interval where the function is decreasing.

100. If an even function has an interval where the function is increasing, then it also has to have an interval where the function is decreasing.

In Exercises 101 and 102, for a and b real numbers, can the function given ever be a continuous function? If so, specify the value for a and b that would make it so.

101. $f(x) = \begin{cases} ax & x \leq 2 \\ bx^2 & x > 2 \end{cases}$

102. $f(x) = \begin{cases} -\dfrac{1}{x} & x < a \\ \dfrac{1}{x} & x \geq a \end{cases}$

▪ CHALLENGE

In Exercises 103 and 104, find the values of a and b that make f continuous.

103. $f(x) = \begin{cases} -x^2 - 10x - 13 & x \leq -2 \\ ax + b & -2 < x < 1 \\ \sqrt{x - 1} - 9 & x \geq 1 \end{cases}$

104. $f(x) = \begin{cases} -2x - a + 2b & x \leq -2 \\ \sqrt{x + a} & -2 < x \leq 2 \\ x^2 - 4x + a + 4 & x > 2 \end{cases}$

▪ TECHNOLOGY

105. In trigonometry you will learn about the sine function, $\sin x$. Plot the function $f(x) = \sin x$, using a graphing utility. It should look like the graph on the right. Is the sine function even, odd, or neither?

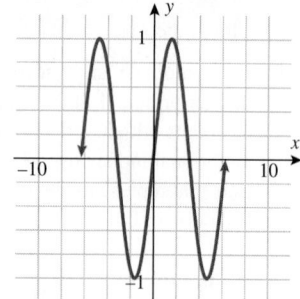

106. In trigonometry you will learn about the cosine function, $\cos x$. Plot the function $f(x) = \cos x$, using a graphing utility. It should look like the graph on the right. Is the cosine function even, odd, or neither?

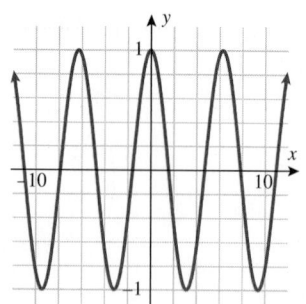

107. In trigonometry you will learn about the tangent function, $\tan x$. Plot the function $f(x) = \tan x$, using a graphing utility. If you restrict the values of x so that $-\dfrac{\pi}{2} < x < \dfrac{\pi}{2}$, the graph should resemble the graph below. Is the tangent function even, odd, or neither?

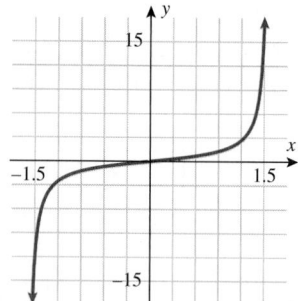

108. Plot the function $f(x) = \dfrac{\sin x}{\cos x}$. What function is this?

109. Graph the function $f(x) = [[3x]]$ using a graphing utility. State the domain and range.

110. Graph the function $f(x) = \left[\left[\frac{1}{3}x\right]\right]$ using a graphing utility. State the domain and range.

▪ PREVIEW TO CALCULUS

For Exercises 111–114, refer to the following:

In calculus, the difference quotient $\dfrac{f(x + h) - f(x)}{h}$ of a function f is used to find the derivative f' of f, by allowing h to approach zero, $h \to 0$. Find the derivative of the following functions.

111. $f(x) = k$, where k is a constant

112. $f(x) = mx + b$, where m and b are constants, $m \neq 0$

113. $f(x) = ax^2 + bx + c$, where a, b, and c are constants, $a \neq 0$

114. $f(x) = \begin{cases} 7 & x < 0 \\ 2 - 3x & 0 < x < 4 \\ x^2 + 4x - 6 & x > 4 \end{cases}$

SKILLS OBJECTIVES

- Sketch the graph of a function using horizontal and vertical shifting of common functions.
- Sketch the graph of a function by reflecting a common function about the x-axis or y-axis.
- Sketch the graph of a function by stretching or compressing a common function.
- Sketch the graph of a function using a sequence of transformations.

CONCEPTUAL OBJECTIVES

- Identify the common functions by their graphs.
- Apply multiple transformations of common functions to obtain graphs of functions.
- Understand that domain and range also are transformed.

The focus of the previous section was to learn the graphs that correspond to particular functions such as identity, square, cube, square root, cube root, absolute value, and reciprocal. Therefore, at this point, you should be able to recognize and generate the graphs of $y = x$, $y = x^2$, $y = x^3$, $y = \sqrt{x}$, $y = \sqrt[3]{x}$, $y = |x|$, and $y = \dfrac{1}{x}$. In this section, we will discuss how to sketch the graphs of functions that are very simple modifications of these functions. For instance, a common function may be shifted (horizontally or vertically), reflected, or stretched (or compressed). Collectively, these techniques are called **transformations**.

Horizontal and Vertical Shifts

Let's take the absolute value function as an example. The graphs of $f(x) = |x|$, $g(x) = |x| + 2$, and $h(x) = |x - 1|$ are shown below.

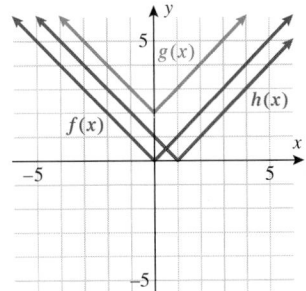

x	$f(x)$
-2	2
-1	1
0	0
1	1
2	2

x	$g(x)$
-2	4
-1	3
0	2
1	3
2	4

x	$h(x)$
-2	3
-1	2
0	1
1	0
2	1

Notice that the graph of $g(x) = |x| + 2$ is the graph of $f(x) = |x|$ shifted *up* 2 units. Similarly, the graph of $h(x) = |x - 1|$ is the graph of $f(x) = |x|$ shifted to the *right* 1 unit. In both cases, the base or starting function is $f(x) = |x|$.

Note that we could rewrite the functions $g(x)$ and $h(x)$ in terms of $f(x)$:

$$g(x) = |x| + 2 = f(x) + 2$$

$$h(x) = |x - 1| = f(x - 1)$$

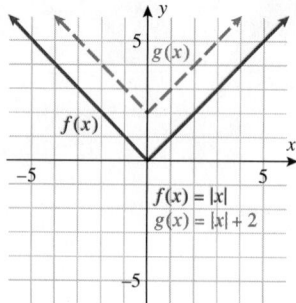
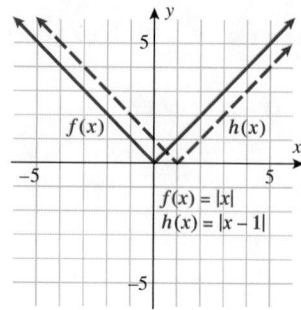

In the case of $g(x)$, the shift $(+2)$ occurs "outside" the function—that is, outside the parentheses showing the argument. Therefore, the output for $g(x)$ is 2 more than the typical output for $f(x)$. Because the output corresponds to the vertical axis, this results in a shift *upward* of 2 units. In general, shifts that occur *outside* the function correspond to a *vertical* shift corresponding to the sign of the shift. For instance, had the function been $G(x) = |x| - 2$, this graph would have started with the graph of the function $f(x)$ and shifted down 2 units.

In the case of $h(x)$, the shift occurs "inside" the function—that is, inside the parentheses showing the argument. Note that the point $(0, 0)$ that lies on the graph of $f(x)$ was shifted to the point $(1, 0)$ on the graph of the function $h(x)$. The y-value remained the same, but the x-value shifted to the right 1 unit. Similarly, the points $(-1, 1)$ and $(1, 1)$ were shifted to the points $(0, 1)$ and $(2, 1)$, respectively. In general, shifts that occur *inside* the function correspond to a *horizontal* shift opposite the sign. In this case, the graph of the function $h(x) = |x - 1|$ shifted the graph of the function $f(x)$ to the right 1 unit. If, instead, we had the function $H(x) = |x + 1|$, this graph would have started with the graph of the function $f(x)$ and shifted to the left 1 unit.

VERTICAL SHIFTS

Assuming that c is a positive constant,

To Graph	Shift the Graph of $f(x)$
$f(x) + c$	c units upward
$f(x) - c$	c units downward

Adding or subtracting a constant **outside** the function corresponds to a **vertical** shift that goes **with the sign**.

HORIZONTAL SHIFTS

Assuming that c is a positive constant,

To Graph	Shift the Graph of $f(x)$
$f(x + c)$	c units to the left
$f(x - c)$	c units to the right

Adding or subtracting a constant **inside** the function corresponds to a **horizontal** shift that goes **opposite the sign**.

EXAMPLE 1   Horizontal and Vertical Shifts

Sketch the graphs of the given functions using horizontal and vertical shifts.

a. $g(x) = x^2 - 1$
b. $H(x) = (x + 1)^2$

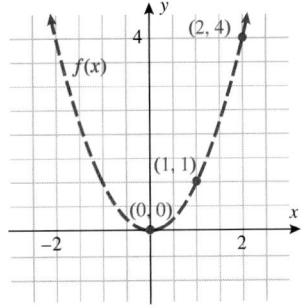

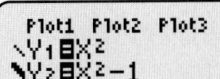

Technology Tip

a. Graphs of $y_1 = x^2$ and $y_2 = g(x) = x^2 - 1$ are shown.

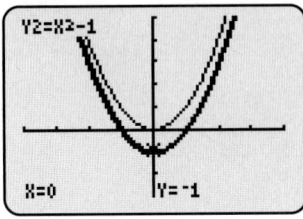

Solution:

In both cases, the function to start with is $f(x) = x^2$.

a. $g(x) = x^2 - 1$ can be rewritten as $g(x) = f(x) - 1$.

   1. The shift (1 unit) occurs *outside* of the function. Therefore, we expect a vertical shift that goes with the sign.

   2. Since the sign is *negative*, this corresponds to a *downward* shift.

   3. Shifting the graph of the function $f(x) = x^2$ down 1 unit yields the graph of $g(x) = x^2 - 1$.

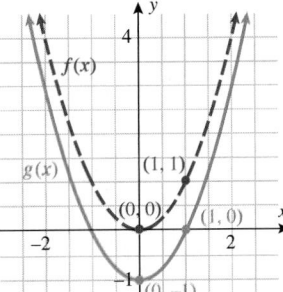

b. Graphs of $y_1 = x^2$ and $y_2 = H(x) = (x + 1)^2$ are shown.

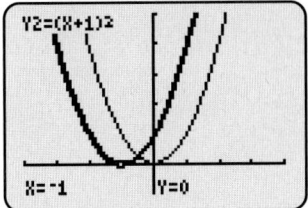

b. $H(x) = (x + 1)^2$ can be rewritten as $H(x) = f(x + 1)$.

   1. The shift (1 unit) occurs *inside* of the function. Therefore, we expect a horizontal shift that goes *opposite* the sign.

   2. Since the sign is *positive*, this corresponds to a shift to the *left*.

   3. Shifting the graph of the function $f(x) = x^2$ to the left 1 unit yields the graph of $H(x) = (x + 1)^2$.

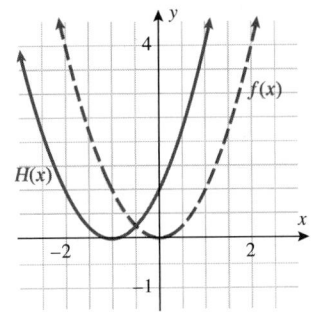

■ **Answer:**
a.

YOUR TURN Sketch the graphs of the given functions using horizontal and vertical shifts.

   a. $g(x) = x^2 + 1$ **b.** $H(x) = (x - 1)^2$

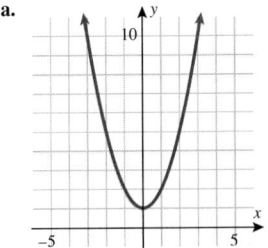

b.

It is important to note that the domain and range of the resulting function can be thought of as also being shifted. Shifts in the domain correspond to horizontal shifts, and shifts in the range correspond to vertical shifts.

EXAMPLE 2 **Horizontal and Vertical Shifts and Changes in the Domain and Range**

Graph the functions using translations and state the domain and range of each function.

a. $g(x) = \sqrt{x + 1}$

b. $G(x) = \sqrt{x} - 2$

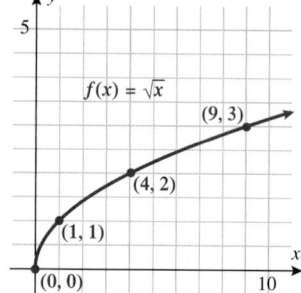

Solution:

In both cases the function to start with is $f(x) = \sqrt{x}$.

Domain: $[0, \infty)$

Range: $[0, \infty)$

a. $g(x) = \sqrt{x + 1}$ can be rewritten as $g(x) = f(x + 1)$.

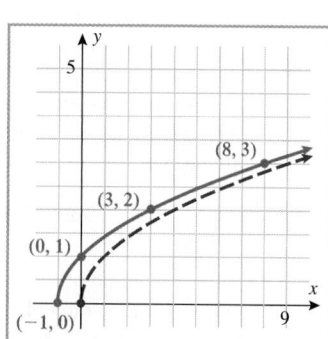

1. The shift (1 unit) is *inside* the function, which corresponds to a *horizontal* shift *opposite the sign.*

2. Shifting the graph of $f(x) = \sqrt{x}$ to the *left* 1 unit yields the graph of $g(x) = \sqrt{x + 1}$. Notice that the point $(0, 0)$, which lies on the graph of $f(x)$, gets shifted to the point $(-1, 0)$ on the graph of $g(x)$.

Although the original function $f(x) = \sqrt{x}$ had an implicit restriction on the domain $[0, \infty)$, the function $g(x) = \sqrt{x + 1}$ has the implicit restriction that $x \geq -1$. We see that the output or range of $g(x)$ is the same as the output of the original function $f(x)$.

Domain: $[-1, \infty)$ Range: $[0, \infty)$

b. $G(x) = \sqrt{x} - 2$ can be rewritten as $G(x) = f(x) - 2$.

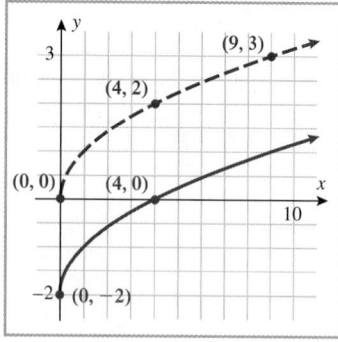

1. The shift (2 units) is *outside* the function, which corresponds to a *vertical* shift *with the sign.*

2. The graph of $G(x) = \sqrt{x} - 2$ is found by shifting $f(x) = \sqrt{x}$ down 2 units. Note that the point $(0, 0)$, which lies on the graph of $f(x)$, gets shifted to the point $(0, -2)$ on the graph of $G(x)$.

The original function $f(x) = \sqrt{x}$ has an implicit restriction on the domain: $[0, \infty)$. The function $G(x) = \sqrt{x} - 2$ also has the implicit restriction that $x \geq 0$. The output or range of $G(x)$ is always 2 units less than the output of the original function $f(x)$.

Domain: $[0, \infty)$ Range: $[-2, \infty)$

■ **YOUR TURN** Sketch the graph of the functions using shifts and state the domain and range.

a. $G(x) = \sqrt{x} - 2$ **b.** $h(x) = |x| + 1$

■ **Answer:**

a. $G(x) = \sqrt{x} - 2$

Domain: $[2, \infty)$ Range: $[0, \infty)$

b. $h(x) = |x| + 1$

Domain: $(-\infty, \infty)$ Range: $[1, \infty)$

The previous examples have involved graphing functions by shifting a known function either in the horizontal or vertical direction. Let us now look at combinations of horizontal and vertical shifts.

EXAMPLE 3 **Combining Horizontal and Vertical Shifts**

Sketch the graph of the function $F(x) = (x + 1)^2 - 2$. State the domain and range of F.

Solution:

The base function is $y = x^2$.

1. The shift (1 unit) is *inside* the function, so it represents a *horizontal* shift *opposite the sign*.
2. The -2 shift is *outside* the function, which represents a *vertical* shift *with the sign*.
3. Therefore, we shift the graph of $y = x^2$ to the left 1 unit and down 2 units. For instance, the point $(0, 0)$ on the graph of $y = x^2$ shifts to the point $(-1, -2)$ on the graph of $F(x) = (x + 1)^2 - 2$.

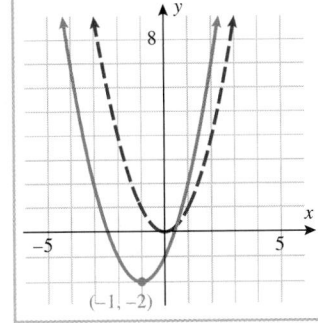

Domain: $(-\infty, \infty)$	Range: $[-2, \infty)$

■ YOUR TURN Sketch the graph of the function $f(x) = |x - 2| + 1$. State the domain and range of f.

All of the previous transformation examples involve starting with a common function and shifting the function in either the horizontal or the vertical direction (or a combination of both). Now, let's investigate *reflections* of functions about the x-axis or y-axis.

Reflection About the Axes

To sketch the graphs of $f(x) = x^2$ and $g(x) = -x^2$, start by first listing points that are on each of the graphs and then connecting the points with smooth curves.

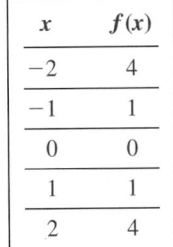

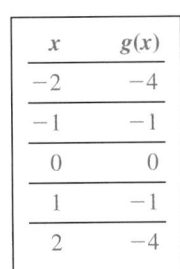

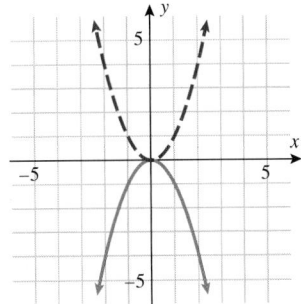

x	$f(x)$
-2	4
-1	1
0	0
1	1
2	4

x	$g(x)$
-2	-4
-1	-1
0	0
1	-1
2	-4

Note that if the graph of $f(x) = x^2$ is reflected about the x-axis, the result is the graph of $g(x) = -x^2$. Also note that the function $g(x)$ can be written as the negative of the function $f(x)$; that is $g(x) = -f(x)$. In general, **reflection about the x-axis** is produced by multiplying a function by -1.

Technology Tip

Graphs of $y_1 = x^2$, $y_2 = (x + 1)^2$, and $y_3 = F(x) = (x + 1)^2 - 2$ are shown.

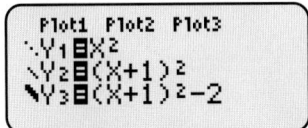

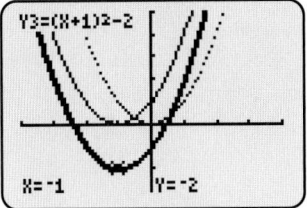

■ **Answer:**
$f(x) = |x - 2| + 1$
$f(x) = |x|$
Domain: $(-\infty, \infty)$
Range: $[1, \infty)$

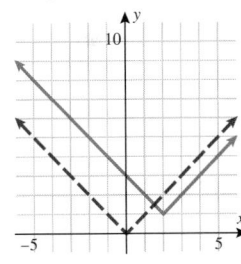

Let's now investigate reflection about the y-axis. To sketch the graphs of $f(x) = \sqrt{x}$ and $g(x) = \sqrt{-x}$, start by listing points that are on each of the graphs and then connecting the points with smooth curves.

x	$f(x)$
0	0
1	1
4	2
9	3

x	$g(x)$
-9	3
-4	2
-1	1
0	0

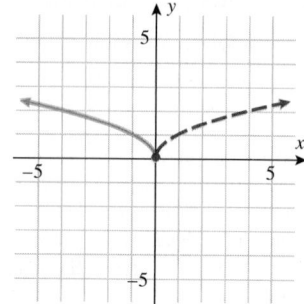

Note that if the graph of $f(x) = \sqrt{x}$ is reflected about the y-axis, the result is the graph of $g(x) = \sqrt{-x}$. Also note that the function $g(x)$ can be written as $g(x) = f(-x)$. In general, **reflection about the y-axis** is produced by replacing x with $-x$ in the function. Notice that the domain of f is $[0, \infty)$, whereas the domain of g is $(-\infty, 0]$.

REFLECTION ABOUT THE AXES

The graph of $-f(x)$ is obtained by reflecting the graph of $f(x)$ about the x-axis.
The graph of $f(-x)$ is obtained by reflecting the graph of $f(x)$ about the y-axis.

EXAMPLE 4   Sketching the Graph of a Function Using Both Shifts and Reflections

Sketch the graph of the function $G(x) = -\sqrt{x + 1}$.

Solution:

Start with the square root function.

$$f(x) = \sqrt{x}$$

Shift the graph of $f(x)$ to the left 1 unit to arrive at the graph of $f(x + 1)$.

$$f(x + 1) = \sqrt{x + 1}$$

Reflect the graph of $f(x + 1)$ about the x-axis to arrive at the graph of $-f(x + 1)$.

$$-f(x + 1) = -\sqrt{x + 1}$$

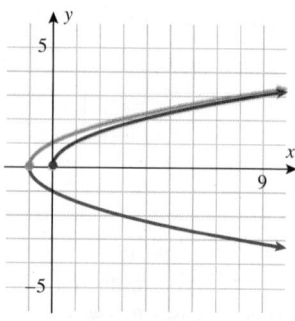

EXAMPLE 5 **Sketching the Graph of a Function Using Both Shifts and Reflections**

Sketch the graph of the function $f(x) = \sqrt{2 - x} + 1$.

Solution:

Start with the square root function. $\qquad$ $g(x) = \sqrt{x}$

Shift the graph of $g(x)$ to the left 2 units to arrive at the graph of $g(x + 2)$. $\qquad$ $g(x + 2) = \sqrt{x + 2}$

Reflect the graph of $g(x + 2)$ about the y-axis to arrive at the graph of $g(-x + 2)$. $\qquad$ $g(-x + 2) = \sqrt{-x + 2}$

Shift the graph $g(-x + 2)$ up 1 unit to arrive at the graph of $g(-x + 2) + 1$. $\qquad$ $g(-x + 2) + 1 = \sqrt{2 - x} + 1$

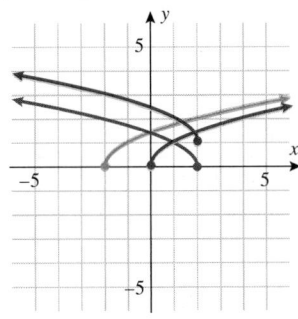

Technology Tip

Graphs of $y_1 = \sqrt{x}$, $y_2 = \sqrt{x + 2}$, $y_3 = \sqrt{-x + 2}$, and $y_4 = f(x) = \sqrt{2 - x} + 1$ are shown.

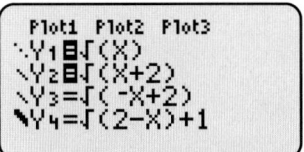

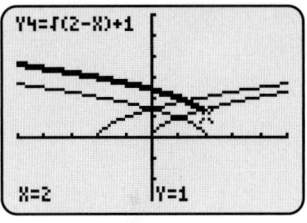

■ **Answer:**

Domain: $[1, \infty)$
Range: $(-\infty, 2]$

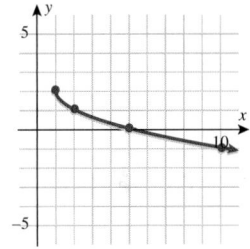

■ **YOUR TURN** Use shifts and reflections to sketch the graph of the function $f(x) = -\sqrt{x - 1} + 2$. State the domain and range of $f(x)$.

Look back at the order in which transformations were performed in Example 5: horizontal shift, reflection, and then vertical shift. Let us consider an alternate order of transformations.

WORDS	**MATH**
Start with the square root function.	$g(x) = \sqrt{x}$
Shift the graph of $g(x)$ up 1 unit to arrive at the graph of $g(x) + 1$.	$g(x) + 1 = \sqrt{x} + 1$
Reflect the graph of $g(x) + 1$ about the y-axis to arrive at the graph of $g(-x) + 1$.	$g(-x) + 1 = \sqrt{-x} + 1$
Replace x with $x - 2$, which corresponds to a shift of the graph of $g(-x) + 1$ to the right 2 units to arrive at the graph of $g[-(x - 2)] + 1$.	$g(-x + 2) + 1 = \sqrt{2 - x} + 1$

In the last step we replaced x with $x - 2$, which required us to think ahead, knowing the desired result was $2 - x$ inside the radical. To avoid any possible confusion, follow this order of transformations:

1. Horizontal shifts: $f(x \pm c)$

2. Reflection: $f(-x)$ and/or $-f(x)$

3. Vertical shifts: $f(x) \pm c$

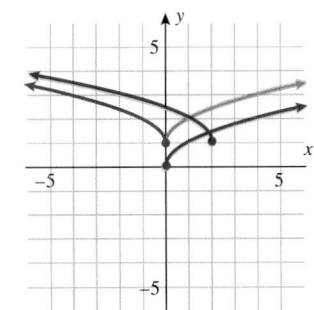

Stretching and Compressing

Horizontal shifts, vertical shifts, and reflections change only the position of the graph in the Cartesian plane, leaving the basic shape of the graph unchanged. These transformations (shifts and reflections) are called **rigid transformations** because they alter only the *position*. **Nonrigid transformations**, on the other hand, distort the *shape* of the original graph. We now consider *stretching* and *compressing* of graphs in both the vertical and the horizontal direction.

A vertical stretch or compression of a graph occurs when the function is multiplied by a positive constant. For example, the graphs of the functions $f(x) = x^2$, $g(x) = 2f(x) = 2x^2$, and $h(x) = \frac{1}{2}f(x) = \frac{1}{2}x^2$ are illustrated below. Depending on if the constant is larger than 1 or smaller than 1 will determine whether it corresponds to a stretch (expansion) or compression (contraction) in the vertical direction.

x	$f(x)$
-2	4
-1	1
0	0
1	1
2	4

x	$g(x)$
-2	8
-1	2
0	0
1	2
2	8

x	$h(x)$
-2	2
-1	$\frac{1}{2}$
0	0
1	$\frac{1}{2}$
2	2

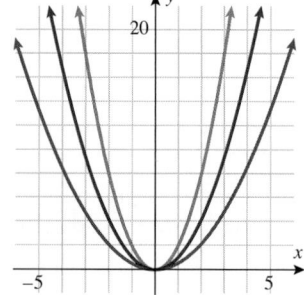

Note that when the function $f(x) = x^2$ is multiplied by 2, so that $g(x) = 2f(x) = 2x^2$, the result is a graph stretched in the vertical direction. When the function $f(x) = x^2$ is multiplied by $\frac{1}{2}$, so that $h(x) = \frac{1}{2}f(x) = \frac{1}{2}x^2$, the result is a graph that is compressed in the vertical direction.

VERTICAL STRETCHING AND VERTICAL COMPRESSING OF GRAPHS

The graph of $cf(x)$ is found by:

- **Vertically stretching** the graph of $f(x)$ if $c > 1$
- **Vertically compressing** the graph of $f(x)$ if $0 < c < 1$

Note: c is any positive real number.

EXAMPLE 6   **Vertically Stretching and Compressing Graphs**

Graph the function $h(x) = \frac{1}{4}x^3$.

Solution:

1. Start with the cube function.

$$f(x) = x^3$$

2. Vertical compression is expected because $\frac{1}{4}$ is less than 1.

$$h(x) = \frac{1}{4}x^3$$

3. Determine a few points that lie on the graph of h.

$$(0, 0) \quad (2, 2) \quad (-2, -2)$$

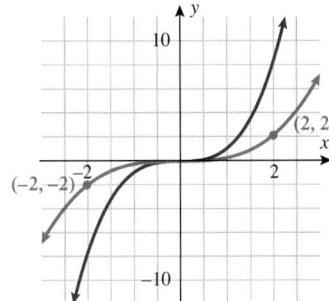

Conversely, if the argument x of a function f is multiplied by a positive real number c, then the result is a *horizontal* stretch of the graph of f if $0 < c < 1$. If $c > 1$, then the result is a *horizontal* compression of the graph of f.

HORIZONTAL STRETCHING AND HORIZONTAL COMPRESSING OF GRAPHS

The graph of $f(cx)$ is found by:

- **Horizontally stretching** the graph of $f(x)$ if $0 < c < 1$
- **Horizontally compressing** the graph of $f(x)$ if $c > 1$

Note: c is any positive real number.

EXAMPLE 7   **Vertically Stretching and Horizontally Compressing Graphs**

Given the graph of $f(x)$, graph

a. $2f(x)$ **b.** $f(2x)$

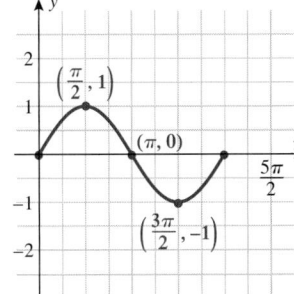

Solution (a):

Since the function is multiplied (on the outside) by 2, the result is that each **y-value** of $f(x)$ is ***multiplied*** **by 2**, which corresponds to vertical stretching.

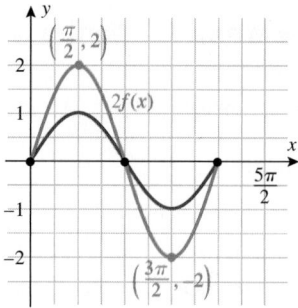

Solution (b):

Since the argument of the function is multiplied (on the inside) by 2, the result is that each **x-value** of $f(x)$ is **divided by 2**, which corresponds to horizontal compression.

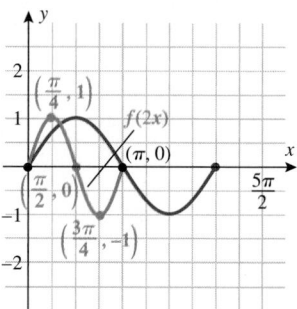

■ **Answer:** Vertical stretch of the graph $f(x) = x^3$.

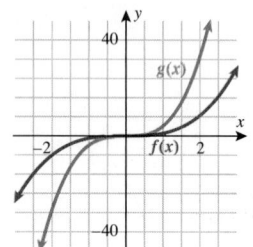

■ **YOUR TURN** Graph the function $g(x) = 4x^3$.

EXAMPLE 8 **Sketching the Graph of a Function Using Multiple Transformations**

Sketch the graph of the function $H(x) = -2(x - 3)^2$.

Solution:

Start with the square function. $f(x) = x^2$

Shift the graph of $f(x)$ to the right 3 units to arrive at the graph of $f(x - 3)$. $f(x - 3) = (x - 3)^2$

Vertically stretch the graph of $f(x - 3)$ by a factor of 2 to arrive at the graph of $2f(x - 3)$. $2f(x - 3) = 2(x - 3)^2$

Reflect the graph $2f(x - 3)$ about the x-axis to arrive at the graph of $-2f(x - 3)$. $-2f(x - 3) = -2(x - 3)^2$

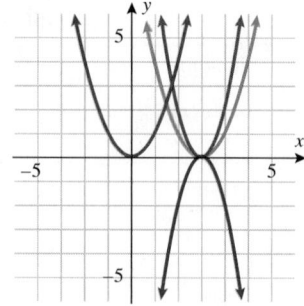

Technology Tip

Graphs of $y_1 = x^2$, $y_2 = (x - 3)^2$, $y_3 = 2(x - 3)^2$, and $y_4 = H(x) = -2(x - 3)^2$ are shown.

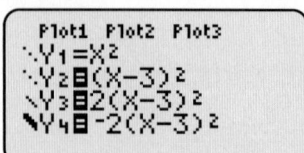

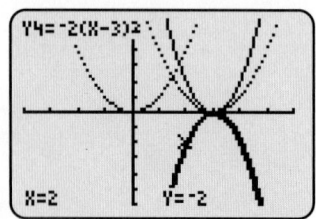

In Example 8 we followed the same "inside out" approach with the functions to determine the order for the transformations: horizontal shift, vertical stretch, and reflection.

SECTION
1.3 SUMMARY

TRANSFORMATION	TO GRAPH THE FUNCTION...	DRAW THE GRAPH OF f AND THEN...	DESCRIPTION
Horizontal shifts ($c > 0$)	$f(x + c)$ $f(x - c)$	Shift the graph of f to the left c units. Shift the graph of f to the right c units.	Replace x by $x + c$. Replace x by $x - c$.
Vertical shifts ($c > 0$)	$f(x) + c$ $f(x) - c$	Shift the graph of f up c units. Shift the graph of f down c units.	Add c to $f(x)$. Subtract c from $f(x)$.
Reflection about the x-axis	$-f(x)$	Reflect the graph of f about the x-axis.	Multiply $f(x)$ by -1.
Reflection about the y-axis	$f(-x)$	Reflect the graph of f about the y-axis.	Replace x by $-x$.
Vertical stretch	$cf(x)$, where $c > 1$	Vertically stretch the graph of f.	Multiply $f(x)$ by c.
Vertical compression	$cf(x)$, where $0 < c < 1$	Vertically compress the graph of f.	Multiply $f(x)$ by c.
Horizontal stretch	$f(cx)$, where $0 < c < 1$	Horizontally stretch the graph of f.	Replace x by cx.
Horizontal compression	$f(cx)$, where $c > 1$	Horizontally compress the graph of f.	Replace x by cx.

SECTION
1.3 EXERCISES

SKILLS

In Exercises 1–6, write the function whose graph is the graph of $y = |x|$, but is transformed accordingly.

1. Shifted up 3 units

2. Shifted to the left 4 units

3. Reflected about the y-axis

4. Reflected about the x-axis

5. Vertically stretched by a factor of 3

6. Vertically compressed by a factor of 3

In Exercises 7–12, write the function whose graph is the graph of $y = x^3$, but is transformed accordingly.

7. Shifted down 4 units

8. Shifted to the right 3 units

9. Shifted up 3 units and to the left 1 unit

10. Reflected about the x-axis

11. Reflected about the y-axis

12. Reflected about both the x-axis and the y-axis

In Exercises 13–36, use the given graph to sketch the graph of the indicated functions.

13.

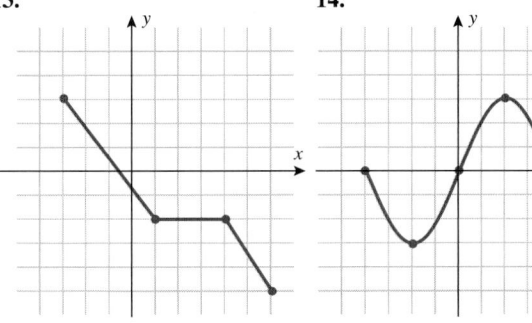

a. $y = f(x - 2)$
b. $y = f(x) - 2$

14.

a. $y = f(x + 2)$
b. $y = f(x) + 2$

15.

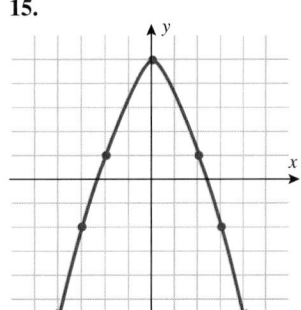

a. $y = f(x) - 3$
b. $y = f(x - 3)$

16.

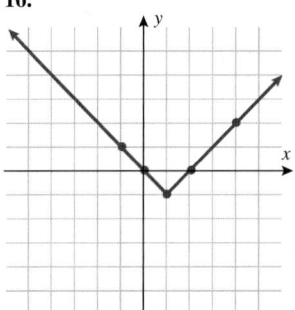

a. $y = f(x) + 3$
b. $y = f(x + 3)$

17.

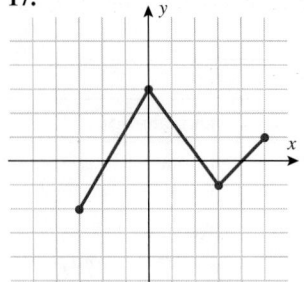

a. $y = -f(x)$
b. $y = f(-x)$

18.

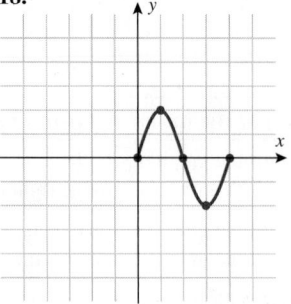

a. $y = -f(x)$
b. $y = f(-x)$

19.

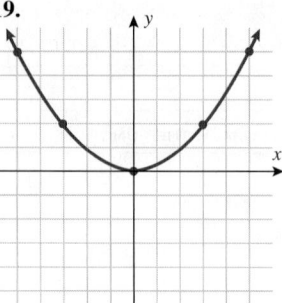

a. $y = 2f(x)$
b. $y = f(2x)$

20.

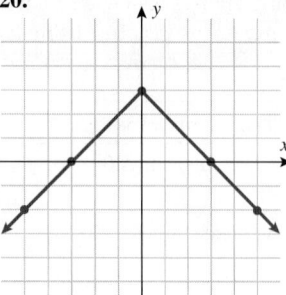

a. $y = 2f(x)$
b. $y = f(2x)$

21. $y = f(x - 2) - 3$

22. $y = f(x + 1) - 2$

23. $y = -f(x - 1) + 2$

24. $y = -2f(x) + 1$

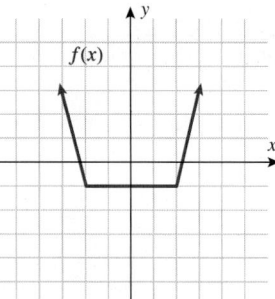

25. $y = -\frac{1}{2}g(x)$

26. $y = \frac{1}{4}g(-x)$

27. $y = -g(2x)$

28. $y = g\left(\frac{1}{2}x\right)$

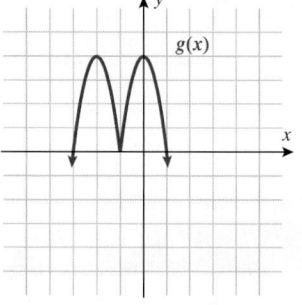

29. $y = \frac{1}{2}F(x - 1) + 2$

30. $y = \frac{1}{2}F(-x)$

31. $y = -F(1 - x)$

32. $y = -F(x - 2) - 1$

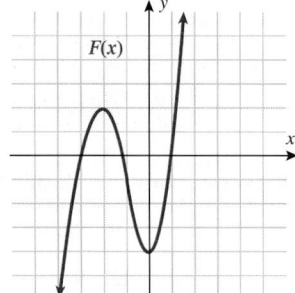

33. $y = 2G(x + 1) - 4$

34. $y = 2G(-x) + 1$

35. $y = -2G(x - 1) + 3$

36. $y = -G(x - 2) - 1$

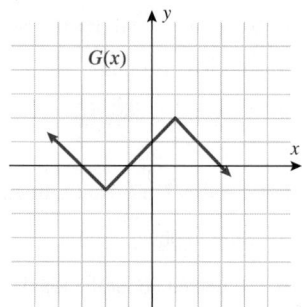

In Exercises 37–62, graph the function using transformations.

37. $y = x^2 - 2$

38. $y = x^2 + 3$

39. $y = (x + 1)^2$

40. $y = (x - 2)^2$

41. $y = (x - 3)^2 + 2$

42. $y = (x + 2)^2 + 1$

43. $y = -(1 - x)^2$

44. $y = -(x + 2)^2$

45. $y = |-x|$

46. $y = -|x|$

47. $y = -|x + 2| - 1$

48. $y = |1 - x| + 2$

49. $y = 2x^2 + 1$

50. $y = 2|x| + 1$

51. $y = -\sqrt{x - 2}$

52. $y = \sqrt{2 - x}$

53. $y = -\sqrt{2 + x} - 1$

54. $y = \sqrt{2 - x} + 3$

55. $y = \sqrt[3]{x - 1} + 2$

56. $y = \sqrt[3]{x + 2} - 1$

57. $y = \dfrac{1}{x + 3} + 2$

58. $y = \dfrac{1}{3 - x}$

59. $y = 2 - \dfrac{1}{x + 2}$

60. $y = 2 - \dfrac{1}{1 - x}$

61. $y = 5\sqrt{-x}$

62. $y = -\frac{1}{5}\sqrt{x}$

In Exercises 63–68, transform the function into the form $f(x) = c(x - h)^2 + k$, where c, k, and h are constants, by completing the square. Use graph-shifting techniques to graph the function.

63. $y = x^2 - 6x + 11$

64. $f(x) = x^2 + 2x - 2$

65. $f(x) = -x^2 - 2x$

66. $f(x) = -x^2 + 6x - 7$

67. $f(x) = 2x^2 - 8x + 3$

68. $f(x) = 3x^2 - 6x + 5$

▪ APPLICATIONS

69. Salary. A manager hires an employee at a rate of $10 per hour. Write the function that describes the current salary of the employee as a function of the number of hours worked per week, x. After a year, the manager decides to award the employee a raise equivalent to paying him for an additional 5 hours per week. Write a function that describes the salary of the employee after the raise.

70. Profit. The profit associated with St. Augustine sod in Florida is typically $P(x) = -x^2 + 14,000x - 48,700,000$, where x is the number of pallets sold per year in a normal year. In rainy years Sod King gives away 10 free pallets per year. Write the function that describes the profit of x pallets of sod in rainy years.

71. Taxes. Every year in the United States each working American typically pays in taxes a percentage of his or her earnings (minus the standard deduction). Karen's 2005 taxes were calculated based on the formula $T(x) = 0.22(x - 6500)$. That year the standard deduction was $6,500 and her tax bracket paid 22% in taxes. Write the function that will determine her 2006 taxes, assuming she receives a raise that places her in the 33% bracket.

72. Medication. The amount of medication that an infant requires is typically a function of the baby's weight. The number of milliliters of an antiseizure medication A is given by $A(x) = \sqrt{x} + 2$, where x is the weight of the infant in ounces. In emergencies there is often not enough time to weigh the infant, so nurses have to estimate the baby's weight. What is the function that represents the actual amount of medication the infant is given if his weight is overestimated by 3 ounces?

73. Taxi Rates. Victoria lives in a condo on Peachtree Street in downtown Atlanta and works at the Federal Reserve Bank of Atlanta, which is 1 mile north of her condo on Peachtree Street. She often eats lunch at Nava Restaurant in Buckhead that is x miles north of the Federal Reserve Bank on Peachtree Street. A taxi in downtown Atlanta costs $7.00 for the first mile and $.30 for every mile after that. Write a function that shows the cost of traveling from Victoria's office to Nava for lunch. Then rewrite the same function to show the cost of the taxi on days when Victoria walks home first to let her dog out and then takes the taxi from her condo to the Nava Restaurant.

74. Taxi Rates. Victoria (in Exercise 73) also likes to eat lunch at a sushi bar x miles south of the Federal Reserve Bank on Peachtree Street. Write a function that shows the cost of traveling from Victoria's office to the sushi bar for lunch. Then rewrite the same function to show the cost of the taxi on days when Victoria walks home first to let her dog out and then takes the taxi from her condo to the sushi bar.

75. Profit. A company that started in 1900 has made a profit corresponding to $P(t) = t^3 - t^2 + t - 1$, where P is the profit in dollars and t is the year (with $t = 0$ corresponding to 1950). Write the profit function with $t = 0$ corresponding to the year 2000.

76. Profit. For the company in Exercise 75, write the profit function with $t = 0$ corresponding to the year 2010.

▪ CATCH THE MISTAKE

In Exercises 77–80, explain the mistake that is made.

77. Describe a procedure for graphing the function
$f(x) = \sqrt{x - 3} + 2$.

Solution:
a. Start with the function $f(x) = \sqrt{x}$.
b. Shift the function to the left 3 units.
c. Shift the function up 2 units.

This is incorrect. What mistake was made?

78. Describe a procedure for graphing the function
$f(x) = -\sqrt{x + 2} - 3$.

Solution:
a. Start with the function $f(x) = \sqrt{x}$.
b. Shift the function to the left 2 units.
c. Reflect the function about the y-axis.
d. Shift the function down 3 units.

This is incorrect. What mistake was made?

79. Describe a procedure for graphing the function
$f(x) = |3 - x| + 1$.

   Solution:
   a. Start with the function $f(x) = |x|$.
   b. Reflect the function about the y-axis.
   c. Shift the function to the left 3 units.
   d. Shift the function up 1 unit.

   This is incorrect. What mistake was made?

80. Describe a procedure for graphing the function
$f(x) = -2x^2 + 1$.

   Solution:
   a. Start with the function $f(x) = x^2$.
   b. Reflect the function about the y-axis.
   c. Shift the function up 1 unit.
   d. Expand in the vertical direction by a factor of 2.

   This is incorrect. What mistake was made?

■ **CONCEPTUAL**

In Exercises 81–86, determine whether each statement is true or false.

81. The graph of $y = |-x|$ is the same as the graph of $y = |x|$.

82. The graph of $y = \sqrt{-x}$ is the same as the graph of $y = \sqrt{x}$.

83. If the graph of an odd function is reflected around the x-axis and then the y-axis, the result is the graph of the original odd function.

84. If the graph of $y = \dfrac{1}{x}$ is reflected around the x-axis, it produces the same graph as if it had been reflected about the y-axis.

85. If f is a function and $c > 1$ is a constant, then the graph of $-cf$ is a reflection about the x-axis of a vertical stretch of the graph of f.

86. If a and b are positive constants and f is a function, then the graph of $f(x + a) + b$ is obtained by shifting the graph of f to the right a units and then shifting this graph up b units.

■ **CHALLENGE**

87. The point (a, b) lies on the graph of the function $y = f(x)$. What point is guaranteed to lie on the graph of $f(x - 3) + 2$?

88. The point (a, b) lies on the graph of the function $y = f(x)$. What point is guaranteed to lie on the graph of $-f(-x) + 1$?

89. The point (a, b) lies on the graph of the function $y = f(x)$. What point is guaranteed to lie on the graph of $2f(x + 1) - 1$?

90. The point (a, b) lies on the graph of the function $y = f(x)$. What point is guaranteed to lie on the graph of $-2f(x - 3) + 4$?

■ **TECHNOLOGY**

91. Use a graphing utility to graph
   a. $y = x^2 - 2$ and $y = |x^2 - 2|$
   b. $y = x^3 - 1$ and $y = |x^3 - 1|$

   What is the relationship between $f(x)$ and $|f(x)|$?

92. Use a graphing utility to graph
   a. $y = x^2 - 2$ and $y = |x|^2 - 2$
   b. $y = x^3 + 1$ and $y = |x|^3 + 1$

   What is the relationship between $f(x)$ and $f(|x|)$?

93. Use a graphing utility to graph
   a. $y = \sqrt{x}$ and $y = \sqrt{0.1x}$
   b. $y = \sqrt{x}$ and $y = \sqrt{10x}$

   What is the relationship between $f(x)$ and $f(ax)$, assuming that a is positive?

94. Use a graphing utility to graph
   a. $y = \sqrt{x}$ and $y = 0.1\sqrt{x}$
   b. $y = \sqrt{x}$ and $y = 10\sqrt{x}$

   What is the relationship between $f(x)$ and $af(x)$, assuming that a is positive?

95. Use a graphing utility to graph $y = f(x) = [[0.5x]] + 1$. Use transformations to describe the relationship between $f(x)$ and $y = [[x]]$.

96. Use a graphing utility to graph $y = g(x) = 0.5\,[[x]] + 1$. Use transformations to describe the relationship between $g(x)$ and $y = [[x]]$.

▪ PREVIEW TO CALCULUS

For Exercises 97–100, refer to the following:

In calculus, the difference quotient $\dfrac{f(x + h) - f(x)}{h}$ of a function f is used to find the derivative f' of f, by letting h approach 0, $h \to 0$. Find the derivatives of f and g.

97. Horizontal Shift. $f(x) = x^2$, $g(x) = (x - 1)^2$. How are the graphs of g' and f' related?

98. Horizontal Shift. $f(x) = \sqrt{x}$, $g(x) = \sqrt{x + 5}$. How are the graphs of g' and f' related?

99. Vertical Shift. $f(x) = 2x$, $g(x) = 2x + 7$. How are the graphs of g' and f' related?

100. Vertical Shift. $f(x) = x^3$, $g(x) = x^3 - 4$. How are the graphs of g' and f' related?

SECTION
1.4 COMBINING FUNCTIONS

SKILLS OBJECTIVES

- Add, subtract, multiply, and divide functions.
- Evaluate composite functions.
- Determine domain of functions resulting from operations on and composition of functions.

CONCEPTUAL OBJECTIVES

- Understand domain restrictions when dividing functions.
- Realize that the domain of a composition of functions excludes values that are not in the domain of the inside function.

Adding, Subtracting, Multiplying, and Dividing Functions

Two functions can be added, subtracted, and multiplied. The domain of the resulting function is the intersection of the domains of the two functions. However, for division, any value of x (input) that makes the denominator equal to zero must be eliminated from the domain.

Function	Notation	Domain
Sum	$(f + g)(x) = f(x) + g(x)$	{domain of f} ∩ {domain of g}
Difference	$(f - g)(x) = f(x) - g(x)$	{domain of f} ∩ {domain of g}
Product	$(f \cdot g)(x) = f(x) \cdot g(x)$	{domain of f} ∩ {domain of g}
Quotient	$\left(\dfrac{f}{g}\right)(x) = \dfrac{f(x)}{g(x)}$	{domain of f} ∩ {domain of g} ∩ {$g(x) \neq 0$}

We can think of this in the following way: Any number that is in the domain of *both* the functions is in the domain of the combined function. The exception to this is the quotient function, which also eliminates values that make the denominator equal to zero.

EXAMPLE 1 Operations on Functions: Determining Domains of New Functions

For the functions $f(x) = \sqrt{x - 1}$ and $g(x) = \sqrt{4 - x}$, determine the sum function, difference function, product function, and quotient function. State the domain of these four new functions.

Solution:

Sum function: $f(x) + g(x) = \sqrt{x - 1} + \sqrt{4 - x}$

Difference function: $f(x) - g(x) = \sqrt{x - 1} - \sqrt{4 - x}$

Product function: $f(x) \cdot g(x) = \sqrt{x - 1} \cdot \sqrt{4 - x}$

$$= \sqrt{(x - 1)(4 - x)} = \sqrt{-x^2 + 5x - 4}$$

Quotient function: $\dfrac{f(x)}{g(x)} = \dfrac{\sqrt{x - 1}}{\sqrt{4 - x}} = \sqrt{\dfrac{x - 1}{4 - x}}$

The domain of the square root function is determined by setting the argument under the radical greater than or equal to zero.

Domain of $f(x)$: $[1, \infty)$

Domain of $g(x)$: $(-\infty, 4]$

The domain of the sum, difference, and product functions is

$$[1, \infty) \cap (-\infty, 4] = [1, 4]$$

The quotient function has the additional constraint that the denominator cannot be zero. This implies that $x \neq 4$, so the domain of the quotient function is $[1, 4)$.

■ **Answer:**

$(f + g)(x) = \sqrt{x + 3} + \sqrt{1 - x}$

Domain: $[-3, 1]$

■ **YOUR TURN** Given the function $f(x) = \sqrt{x + 3}$ and $g(x) = \sqrt{1 - x}$, find $(f + g)(x)$ and state its domain.

Technology Tip

The graphs of $y_1 = F(x) = \sqrt{x}$, $y_2 = G(x) = |x - 3|$, and

$y_3 = \dfrac{F(x)}{G(x)} = \dfrac{\sqrt{x}}{|x - 3|}$ are shown.

```
Plot1 Plot2 Plot3
\Y1■√(X)
·.Y2■abs(X-3)
\Y3■√(X)/abs(X-3
)
```

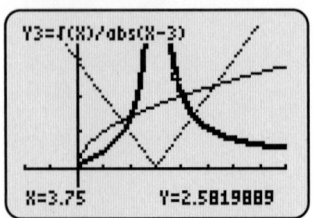

Y3=√(X)/abs(X-3)

X=3.75   Y=2.5819889

■ **Answer:**

$\left(\dfrac{G}{F}\right)(x) = \dfrac{G(x)}{F(x)} = \dfrac{|x - 3|}{\sqrt{x}}$

Domain: $(0, \infty)$

EXAMPLE 2 Quotient Function and Domain Restrictions

Given the functions $F(x) = \sqrt{x}$ and $G(x) = |x - 3|$, find the quotient function, $\left(\dfrac{F}{G}\right)(x)$, and state its domain.

Solution:

The quotient function is written as

$$\left(\frac{F}{G}\right)(x) = \frac{F(x)}{G(x)} = \frac{\sqrt{x}}{|x - 3|}$$

Domain of $F(x)$: $[0, \infty)$ Domain of $G(x)$: $(-\infty, \infty)$

The real numbers that are in both the domain for $F(x)$ and the domain for $G(x)$ are represented by the intersection $[0, \infty) \cap (-\infty, \infty) = [0, \infty)$. Also, the denominator of the quotient function is equal to zero when $x = 3$, so we must eliminate this value from the domain.

$$\text{Domain of } \left(\frac{F}{G}\right)(x): [0, 3) \cup (3, \infty)$$

■ **YOUR TURN** For the functions given in Example 2, determine the quotient function $\left(\dfrac{G}{F}\right)(x)$, and state its domain.

Composition of Functions

Recall that a function maps every element in the domain to exactly one corresponding element in the range as shown in the figure on the right.

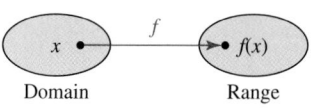

Suppose there is a sales rack of clothes in a department store. Let x correspond to the original price of each item on the rack. These clothes have recently been marked down 20%. Therefore, the function $g(x) = 0.80x$ represents the current sale price of each item. You have been invited to a special sale that lets you take 10% off the current sale price and an additional $5 off every item at checkout. The function $f(g(x)) = 0.90g(x) - 5$ determines the checkout price. Note that the output of the function g is the input of the function f as shown in the figure below.

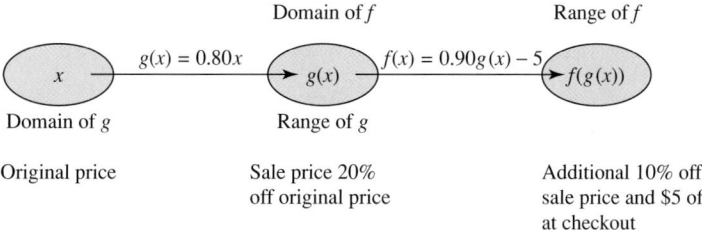

This is an example of a **composition of functions**, when the output of one function is the input of another function. It is commonly referred to as a function of a function.

An algebraic example of this is the function $y = \sqrt{x^2 - 2}$. Suppose we let $g(x) = x^2 - 2$ and $f(x) = \sqrt{x}$. Recall that the independent variable in function notation is a placeholder. Since $f(\square) = \sqrt{\square}$, then $f(g(x)) = \sqrt{g(x)}$. Substituting the expression for $g(x)$, we find $f(g(x)) = \sqrt{x^2 - 2}$. The function $y = \sqrt{x^2 - 2}$ is said to be a composite function, $y = f(g(x))$.

Note that the domain of $g(x)$ is the set of all real numbers, and the domain of $f(x)$ is the set of all nonnegative numbers. The domain of a composite function is the set of all x such that $g(x)$ is in the domain of f. For instance, in the composite function $y = f(g(x))$, we know that the allowable inputs into f are all numbers greater than or equal to zero. Therefore, we restrict the outputs of $g(x) \geq 0$ and find the corresponding x-values. Those x-values are the only allowable inputs and constitute the domain of the composite function $y = f(g(x))$.

The symbol that represents composition of functions is a small open circle; thus $(f \circ g)(x) = f(g(x))$ and is read aloud as "f of g." It is important not to confuse this with the multiplication sign: $(f \cdot g)(x) = f(x)g(x)$.

▼ **CAUTION**

$f \circ g \neq f \cdot g$

COMPOSITION OF FUNCTIONS

Given two functions f and g, there are two **composite functions** that can be formed.

NOTATION	WORDS	DEFINITION	DOMAIN
$f \circ g$	f composed with g	$f(g(x))$	The set of all real numbers x in the domain of g such that $g(x)$ is also in the domain of f.
$g \circ f$	g composed with f	$g(f(x))$	The set of all real numbers x in the domain of f such that $f(x)$ is also in the domain of g.

Study Tip

Order is important:

$$(f \circ g)(x) = f(g(x))$$
$$(g \circ f)(x) = g(f(x))$$

Study Tip

The domain of $f \circ g$ is always a subset of the domain of g, and the range of $f \circ g$ is always a subset of the range of f.

It is important to realize that there are two "filters" that allow certain values of x into the domain. The first filter is $g(x)$. If x is not in the domain of $g(x)$, it cannot be in the domain of $(f \circ g)(x) = f(g(x))$. Of those values for x that are in the domain of $g(x)$, only some pass through, because we restrict the output of $g(x)$ to values that are allowable as input into f. This adds an additional filter.

The domain of $f \circ g$ is always a subset of the domain of g, and the range of $f \circ g$ is always a subset of the range of f.

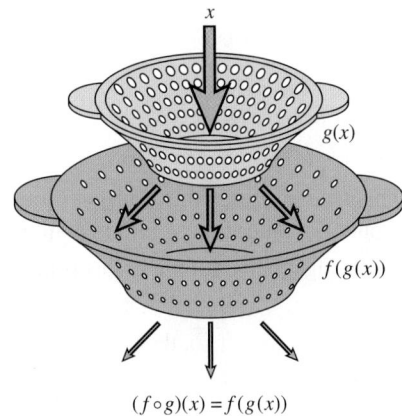

$$(f \circ g)(x) = f(g(x))$$

EXAMPLE 3 Finding a Composite Function

Given the functions $f(x) = x^2 + 1$ and $g(x) = x - 3$, find $(f \circ g)(x)$.

Solution:

Write $f(x)$ using placeholder notation.	$f(\square) = (\square)^2 + 1$
Express the composite function $f \circ g$.	$f(g(x)) = (g(x))^2 + 1$
Substitute $g(x) = x - 3$ into f.	$f(g(x)) = (x - 3)^2 + 1$
Eliminate the parentheses on the right side.	$f(g(x)) = x^2 - 6x + 10$

$$\boxed{(f \circ g)(x) = f(g(x)) = x^2 - 6x + 10}$$

■ **Answer:** $g \circ f = g(f(x)) = x^2 - 2$

■ **YOUR TURN** Given the functions in Example 3, find $(g \circ f)(x)$.

EXAMPLE 4   **Determining the Domain of a Composite Function**

Given the functions $f(x) = \dfrac{1}{x-1}$ and $g(x) = \dfrac{1}{x}$, determine $f \circ g$, and state its domain.

Solution:

Write $f(x)$ using placeholder notation.

$$f(\square) = \frac{1}{(\square) - 1}$$

Express the composite function $f \circ g$.

$$f(g(x)) = \frac{1}{g(x) - 1}$$

Substitute $g(x) = \dfrac{1}{x}$ into f.

$$f(g(x)) = \frac{1}{\dfrac{1}{x} - 1}$$

Multiply the right side by $\dfrac{x}{x}$.

$$f(g(x)) = \frac{1}{\dfrac{1}{x} - 1} \cdot \frac{x}{x} = \frac{x}{1 - x}$$

$$\boxed{(f \circ g) = f(g(x)) = \frac{x}{1 - x}}$$

What is the domain of $(f \circ g)(x) = f(g(x))$? By inspecting the final result of $f(g(x))$, we see that the denominator is zero when $x = 1$. Therefore, $x \neq 1$. Are there any other values for x that are not allowed? The function $g(x)$ has the domain $x \neq 0$; therefore, we must also exclude zero. The domain of $(f \circ g)(x) = f(g(x))$ excludes $x = 0$ and $x = 1$ or, in interval notation,

$$\boxed{(-\infty, 0) \cup (0, 1) \cup (1, \infty)}$$

■ **YOUR TURN**   For the functions f and g given in Example 4, determine the composite function $g \circ f$ and state its domain.

The domain of the composite function cannot always be determined by examining the final form of $f \circ g$, as illustrated in Example 4.

EXAMPLE 5   **Determining the Domain of a Composite Function (Without Finding the Composite Function)**

Let $f(x) = \dfrac{1}{x - 2}$ and $g(x) = \sqrt{x + 3}$. Find the domain of $f(g(x))$. Do not find the composite function.

Solution:

Find the domain of g. $[-3, \infty)$

Find the range of g. $[0, \infty)$

In $f(g(x))$, the output of g becomes the input for f. Since the domain of f is the set of all real numbers except 2, we eliminate any values of x in the domain of g that correspond to $g(x) = 2$.

Let $g(x) = 2$. $\sqrt{x + 3} = 2$

Square both sides. $x + 3 = 4$

Solve for x. $x = 1$

Eliminate $x = 1$ from the domain of g, $[-3, \infty)$.

State the domain of $f(g(x))$. $\boxed{[-3, 1) \cup (1, \infty)}$

Technology Tip

The graphs of $y_1 = f(x) = \dfrac{1}{x-1}$, $y_2 = g(x) = \dfrac{1}{x}$, and $y_3 = (f \circ g)(x)$

$= \dfrac{1}{\dfrac{1}{x} - 1} = \dfrac{x}{1 - x}$ are shown.

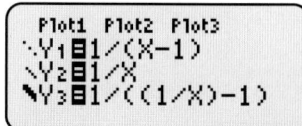

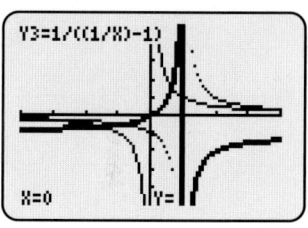

■ **Answer:** $g(f(x)) = x - 1$. Domain of $g \circ f$ is $x \neq 1$, or in interval notation, $(-\infty, 1) \cup (1, \infty)$.

▼ **CAUTION**

The domain of the composite function cannot always be determined by examining the final form of $f \circ g$.

EXAMPLE 6   Evaluating a Composite Function

Given the functions $f(x) = x^2 - 7$ and $g(x) = 5 - x^2$, evaluate

a. $f(g(1))$ **b.** $f(g(-2))$ **c.** $g(f(3))$ **d.** $g(f(-4))$

Solution:

One way of evaluating these composite functions is to calculate the two individual composites in terms of x: $f(g(x))$ and $g(f(x))$. Once those functions are known, the values can be substituted for x and evaluated.

Another way of proceeding is as follows:

a. Write the desired quantity. $f(g(1))$
   Find the value of the inner function g. $g(1) = 5 - 1^2 = 4$
   Substitute $g(1) = 4$ into f. $f(g(1)) = f(4)$
   Evaluate $f(4)$. $f(4) = 4^2 - 7 = 9$

$$\boxed{f(g(1)) = 9}$$

b. Write the desired quantity. $f(g(-2))$
   Find the value of the inner function g. $g(-2) = 5 - (-2)^2 = 1$
   Substitute $g(-2) = 1$ into f. $f(g(-2)) = f(1)$
   Evaluate $f(1)$. $f(1) = 1^2 - 7 = -6$

$$\boxed{f(g(-2)) = -6}$$

c. Write the desired quantity. $g(f(3))$
   Find the value of the inner function f. $f(3) = 3^2 - 7 = 2$
   Substitute $f(3) = 2$ into g. $g(f(3)) = g(2)$
   Evaluate $g(2)$. $g(2) = 5 - 2^2 = 1$

$$\boxed{g(f(3)) = 1}$$

d. Write the desired quantity. $g(f(-4))$
   Find the value of the inner function f. $f(-4) = (-4)^2 - 7 = 9$
   Substitute $f(-4) = 9$ into g. $g(f(-4)) = g(9)$
   Evaluate $g(9)$. $g(9) = 5 - 9^2 = -76$

$$\boxed{g(f(-4)) = -76}$$

■ **Answer:** $f(g(1)) = 5$
$g(f(1)) = -7$

■ **YOUR TURN** Given the functions $f(x) = x^3 - 3$ and $g(x) = 1 + x^3$, evaluate $f(g(1))$ and $g(f(1))$.

Application Problems

Recall the example at the beginning of this section regarding the clothes that are on sale. Often, real-world applications are modeled with composite functions. In the clothes example, x is the original price of each item. The first function maps its input (original price) to an output (sale price). The second function maps its input (sale price) to an output (checkout price). Example 7 is another real-world application of composite functions.

Three temperature scales are commonly used:

■ The degree Celsius (°C) scale
   • This scale was devised by dividing the range between the freezing (0 °C) and boiling (100 °C) points of pure water at sea level into 100 equal parts. This scale is used in science and is one of the standards of the "metric" (SI) system of measurements.

- The Kelvin (K) temperature scale
  - This scale shifts the Celsius scale down so that the zero point is equal to absolute zero (about $-273.15\,°C$), a hypothetical temperature at which there is a complete absence of heat energy.
  - Temperatures on this scale are called **kelvins**, *not* degrees kelvin, and kelvin is not capitalized. The symbol for the kelvin is K.
- The degree Fahrenheit (°F) scale
  - This scale evolved over time and is still widely used mainly in the United States, although Celsius is the preferred "metric" scale.
  - With respect to pure water at sea level, the **degrees Fahrenheit** are gauged by the spread from 32 °F (freezing) to 212 °F (boiling).

The equations that relate these temperature scales are

$$F = \frac{9}{5}C + 32 \qquad C = K - 273.15$$

EXAMPLE 7   Applications Involving Composite Functions

Determine degrees Fahrenheit as a function of kelvins.

Solution:

Degrees Fahrenheit is a function of degrees Celsius.

$$F = \frac{9}{5}C + 32$$

Now substitute $C = K - 273.15$ into the equation for F.

$$F = \frac{9}{5}(K - 273.15) + 32$$

Simplify.

$$F = \frac{9}{5}K - 491.67 + 32$$

$$\boxed{F = \frac{9}{5}K - 459.67}$$

SECTION 1.4 SUMMARY

Operations on Functions

Function	Notation
Sum	$(f + g)(x) = f(x) + g(x)$
Difference	$(f - g)(x) = f(x) - g(x)$
Product	$(f \cdot g)(x) = f(x) \cdot g(x)$
Quotient	$\left(\dfrac{f}{g}\right)(x) = \dfrac{f(x)}{g(x)} \qquad g(x) \neq 0$

The domain of the sum, difference, and product functions is the intersection of the domains, or common domain shared by both f and g. The domain of the quotient function is also the intersection of the domain shared by both f and g with an additional restriction that $g(x) \neq 0$.

Composition of Functions

$$(f \circ g)(x) = f(g(x))$$

The domain restrictions cannot always be determined simply by inspecting the final form of $f(g(x))$. Rather, the domain of the composite function is a subset of the domain of $g(x)$. Values of x must be eliminated if their corresponding values of $g(x)$ are not in the domain of f.

■ **SKILLS**

In Exercises 1–10, given the functions f and g, find $f + g$, $f - g$, $f \cdot g$, and $\dfrac{f}{g}$, and state the domain of each.

1. $f(x) = 2x + 1$

$g(x) = 1 - x$

2. $f(x) = 3x + 2$

$g(x) = 2x - 4$

3. $f(x) = 2x^2 - x$

$g(x) = x^2 - 4$

4. $f(x) = 3x + 2$

$g(x) = x^2 - 25$

5. $f(x) = \dfrac{1}{x}$

$g(x) = x$

6. $f(x) = \dfrac{2x + 3}{x - 4}$

$g(x) = \dfrac{x - 4}{3x + 2}$

7. $f(x) = \sqrt{x}$

$g(x) = 2\sqrt{x}$

8. $f(x) = \sqrt{x - 1}$

$g(x) = 2x^2$

9. $f(x) = \sqrt{4 - x}$

$g(x) = \sqrt{x + 3}$

10. $f(x) = \sqrt{1 - 2x}$

$g(x) = \dfrac{1}{x}$

In Exercises 11–20, for the given functions f and g, find the composite functions $f \circ g$ and $g \circ f$, and state their domains.

11. $f(x) = 2x + 1$

$g(x) = x^2 - 3$

12. $f(x) = x^2 - 1$

$g(x) = 2 - x$

13. $f(x) = \dfrac{1}{x - 1}$

$g(x) = x + 2$

14. $f(x) = \dfrac{2}{x - 3}$

$g(x) = 2 + x$

15. $f(x) = |x|$

$g(x) = \dfrac{1}{x - 1}$

16. $f(x) = |x - 1|$

$g(x) = \dfrac{1}{x}$

17. $f(x) = \sqrt{x - 1}$

$g(x) = x + 5$

18. $f(x) = \sqrt{2 - x}$

$g(x) = x^2 + 2$

19. $f(x) = x^3 + 4$

$g(x) = (x - 4)^{1/3}$

20. $f(x) = \sqrt[3]{x^2 - 1}$

$g(x) = x^{2/3} + 1$

In Exercises 21–38, evaluate the functions for the specified values, if possible.

$$f(x) = x^2 + 10 \qquad g(x) = \sqrt{x - 1}$$

21. $(f + g)(2)$

22. $(f + g)(10)$

23. $(f - g)(2)$

24. $(f - g)(5)$

25. $(f \cdot g)(4)$

26. $(f \cdot g)(5)$

27. $\left(\dfrac{f}{g}\right)(10)$

28. $\left(\dfrac{f}{g}\right)(2)$

29. $f(g(2))$

30. $f(g(1))$

31. $g(f(-3))$

32. $g(f(4))$

33. $f(g(0))$

34. $g(f(0))$

35. $f(g(-3))$

36. $g(f(\sqrt{7}))$

37. $(f \circ g)(4)$

38. $(g \circ f)(-3)$

In Exercises 39–50, evaluate $f(g(1))$ and $g(f(2))$, if possible.

39. $f(x) = \dfrac{1}{x}$, $g(x) = 2x + 1$

40. $f(x) = x^2 + 1$, $g(x) = \dfrac{1}{2 - x}$

41. $f(x) = \sqrt{1 - x}$, $g(x) = x^2 + 2$

42. $f(x) = \sqrt{3 - x}$, $g(x) = x^2 + 1$

43. $f(x) = \dfrac{1}{|x - 1|}$, $g(x) = x + 3$

44. $f(x) = \dfrac{1}{x}$, $g(x) = |2x - 3|$

45. $f(x) = \sqrt{x - 1}$, $g(x) = x^2 + 5$

46. $f(x) = \sqrt[3]{x - 3}$, $g(x) = \dfrac{1}{x - 3}$

47. $f(x) = \dfrac{1}{x^2 - 3}$, $g(x) = \sqrt{x - 3}$

48. $f(x) = \dfrac{x}{2 - x}$, $g(x) = 4 - x^2$

49. $f(x) = (x - 1)^{1/3}$, $g(x) = x^2 + 2x + 1$

50. $f(x) = (1 - x^2)^{1/2}$, $g(x) = (x - 3)^{1/3}$

In Exercises 51–60, show that $f(g(x)) = x$ and $g(f(x)) = x$.

51. $f(x) = 2x + 1$, $g(x) = \dfrac{x - 1}{2}$

52. $f(x) = \dfrac{x - 2}{3}$, $g(x) = 3x + 2$

53. $f(x) = \sqrt{x - 1}$, $g(x) = x^2 + 1$ for $x \geq 1$

54. $f(x) = 2 - x^2$, $g(x) = \sqrt{2 - x}$ for $x \leq 2$

55. $f(x) = \dfrac{1}{x}, \quad g(x) = \dfrac{1}{x}$ for $x \neq 0$

56. $f(x) = (5 - x)^{1/3}, \quad g(x) = 5 - x^3$

57. $f(x) = 4x^2 - 9, \quad g(x) = \dfrac{\sqrt{x + 9}}{2}$ for $x \geq 0$

58. $f(x) = \sqrt[3]{8x - 1}, \quad g(x) = \dfrac{x^3 + 1}{8}$

59. $f(x) = \dfrac{1}{x - 1}, \quad g(x) = \dfrac{x + 1}{x}$ for $x \neq 0, \ x \neq 1$

60. $f(x) = \sqrt{25 - x^2}, \quad g(x) = \sqrt{25 - x^2}$ for $0 \leq x \leq 5$

In Exercises 61–66, write the function as a composite of two functions f and g. (More than one answer is correct.)

61. $f(g(x)) = 2(3x - 1)^2 + 5(3x - 1)$

62. $f(g(x)) = \dfrac{1}{1 + x^2}$

63. $f(g(x)) = \dfrac{2}{|x - 3|}$

64. $f(g(x)) = \sqrt{1 - x^2}$

65. $f(g(x)) = \dfrac{3}{\sqrt{x + 1} - 2}$

66. $f(g(x)) = \dfrac{\sqrt{x}}{3\sqrt{x} + 2}$

■ APPLICATIONS

Exercises 67 and 68 depend on the relationship between degrees Fahrenheit, degrees Celsius, and kelvins:

$$F = \frac{9}{5}C + 32 \qquad C = K - 273.15$$

67. Temperature. Write a composite function that converts kelvins into degrees Fahrenheit.

68. Temperature. Convert the following degrees Fahrenheit to kelvins: 32 °F and 212 °F.

69. Dog Run. Suppose that you want to build a *square* fenced-in area for your dog. Fencing is purchased in linear feet.
a. Write a composite function that determines the area of your dog pen as a function of how many linear feet are purchased.
b. If you purchase 100 linear feet, what is the area of your dog pen?
c. If you purchase 200 linear feet, what is the area of your dog pen?

70. Dog Run. Suppose that you want to build a *circular* fenced-in area for your dog. Fencing is purchased in linear feet.
a. Write a composite function that determines the area of your dog pen as a function of how many linear feet are purchased.
b. If you purchase 100 linear feet, what is the area of your dog pen?
c. If you purchase 200 linear feet, what is the area of your dog pen?

71. Market Price. Typical supply and demand relationships state that as the number of units for sale increases, the market price decreases. Assume that the market price p and the number of units for sale x are related by the demand equation:

$$p = 3000 - \frac{1}{2}x$$

Assume that the cost $C(x)$ of producing x items is governed by the equation

$$C(x) = 2000 + 10x$$

and the revenue $R(x)$ generated by selling x units is governed by

$$R(x) = 100x$$

a. Write the cost as a function of price p.
b. Write the revenue as a function of price p.
c. Write the profit as a function of price p.

72. Market Price. Typical supply and demand relationships state that as the number of units for sale increases, the market price decreases. Assume that the market price p and the number of units for sale x are related by the demand equation:

$$p = 10{,}000 - \frac{1}{4}x$$

Assume that the cost $C(x)$ of producing x items is governed by the equation

$$C(x) = 30{,}000 + 5x$$

and the revenue $R(x)$ generated by selling x units is governed by

$$R(x) = 1000x$$

a. Write the cost as a function of price p.
b. Write the revenue as a function of price p.
c. Write the profit as a function of price p.

73. Environment: Oil Spill. An oil spill makes a circular pattern around a ship such that the radius in feet grows as a function of time in hours $r(t) = 150\sqrt{t}$. Find the area of the spill as a function of time.

74. Pool Volume. A 20 ft × 10 ft rectangular pool has been built. If 50 cubic feet of water is pumped into the pool per hour, write the water-level height (feet) as a function of time (hours).

75. Fireworks. A family is watching a fireworks display. If the family is 2 miles from where the fireworks are being launched and the fireworks travel vertically, what is the distance between the family and the fireworks as a function of height above ground?

76. Real Estate. A couple are about to put their house up for sale. They bought the house for $172,000 a few years ago; if they list it with a realtor, they will pay a 6% commission. Write a function that represents the amount of money they will make on their home as a function of the asking price p.

■ **CATCH THE MISTAKE**

In Exercises 77–81, for the functions $f(x) = x + 2$ and $g(x) = x^2 - 4$, find the indicated function and state its domain. Explain the mistake that is made in each problem.

77. $\dfrac{g}{f}$

Solution: $\dfrac{g(x)}{f(x)} = \dfrac{x^2 - 4}{x + 2}$

$$= \dfrac{(x - 2)(x + 2)}{x + 2}$$

$$= x - 2$$

Domain: $(-\infty, \infty)$

This is incorrect. What mistake was made?

78. $\dfrac{f}{g}$

Solution: $\dfrac{f(x)}{g(x)} = \dfrac{x + 2}{x^2 - 4}$

$$= \dfrac{x + 2}{(x - 2)(x + 2)} = \dfrac{1}{x - 2}$$

$$= \dfrac{1}{x - 2}$$

Domain: $(-\infty, 2) \cup (2, \infty)$

This is incorrect. What mistake was made?

79. $f \circ g$

Solution: $f \circ g = f(x)g(x)$

$$= (x + 2)(x^2 - 4)$$

$$= x^3 + 2x^2 - 4x - 8$$

Domain: $(-\infty, \infty)$

This is incorrect. What mistake was made?

80. $f(x) - g(x) = x + 2 - x^2 - 4$

$$= -x^2 + x - 2$$

Domain: $(-\infty, \infty)$

This is incorrect. What mistake was made?

81. $(f + g)(2) = (x + 2 + x^2 - 4)(2)$

$$= (x^2 + x - 2)(2)$$

$$= 2x^2 + 2x - 4$$

Domain: $(-\infty, \infty)$

This is incorrect. What mistake was made?

82. Given the function $f(x) = x^2 + 7$ and $g(x) = \sqrt{x - 3}$, find $f \circ g$, and state the domain.

Solution: $f \circ g = f(g(x)) = \left(\sqrt{x - 3}\right)^2 + 7$

$$= f(g(x)) = x - 3 + 7$$

$$= x - 4$$

Domain: $(-\infty, \infty)$

This is incorrect. What mistake was made?

■ **CONCEPTUAL**

In Exercises 83–86, determine whether each statement is true or false.

83. When adding, subtracting, multiplying, or dividing two functions, the domain of the resulting function is the union of the domains of the individual functions.

84. For any functions f and g, $f(g(x)) = g(f(x))$ for all values of x that are in the domain of both f and g.

85. For any functions f and g, $(f \circ g)(x)$ exists for all values of x that are in the domain of $g(x)$, provided the range of g is a subset of the domain of f.

86. The domain of a composite function can be found by inspection, without knowledge of the domain of the individual functions.

■ **CHALLENGE**

87. For the functions $f(x) = x + a$ and $g(x) = \dfrac{1}{x - a}$, find $g \circ f$ and state its domain.

88. For the functions $f(x) = ax^2 + bx + c$ and $g(x) = \dfrac{1}{x - c}$, find $g \circ f$ and state its domain.

89. For the functions $f(x) = \sqrt{x + a}$ and $g(x) = x^2 - a$ find $g \circ f$ and state its domain.

90. For the functions $f(x) = \dfrac{1}{x^a}$ and $g(x) = \dfrac{1}{x^b}$, find $g \circ f$ and state its domain. Assume $a > 1$ and $b > 1$.

▪ TECHNOLOGY

91. Using a graphing utility, plot $y_1 = \sqrt{x + 7}$ and $y_2 = \sqrt{9 - x}$. Plot $y_3 = y_1 + y_2$. What is the domain of y_3?

92. Using a graphing utility, plot $y_1 = \sqrt[3]{x + 5}$, $y_2 = \dfrac{1}{\sqrt{3 - x}}$, and $y_3 = \dfrac{y_1}{y_2}$. What is the domain of y_3?

93. Using a graphing utility, plot $y_1 = \sqrt{x^2 - 3x - 4}$, $y_2 = \dfrac{1}{x^2 - 14}$, and $y_3 = \dfrac{1}{y_1^2 - 14}$. If y_1 represents a function

f and y_2 represents a function g, then y_3 represents the composite function $g \circ f$. The graph of y_3 is only defined for the domain of $g \circ f$. State the domain of $g \circ f$.

94. Using a graphing utility, plot $y_1 = \sqrt{1 - x}$, $y_2 = x^2 + 2$, and $y_3 = y_1^2 + 2$. If y_1 represents a function f and y_2 represents a function g, then y_3 represents the composite function $g \circ f$. The graph of y_3 is only defined for the domain of $g \circ f$. State the domain of $g \circ f$.

▪ PREVIEW TO CALCULUS

For Exercises 95–98, refer to the following:

In calculus, the difference quotient $\dfrac{f(x + h) - f(x)}{h}$ of a function f is used to find the derivative f' of f by letting h approach 0, $h \to 0$.

95. Addition. Find the derivatives of $F(x) = x$, $G(x) = x^2$, and $H(x) = (F + G)(x) = x + x^2$. What do you observe?

96. Subtraction. Find the derivatives of $F(x) = \sqrt{x}$, $G(x) = x^3 + 1$, and $H(x) = (F - G)(x) = \sqrt{x} - x^3 - 1$. What do you observe?

97. Multiplication. Find the derivatives of $F(x) = 5$, $G(x) = \sqrt{x - 1}$, and $H(x) = (FG)(x) = 5\sqrt{x - 1}$. What do you observe?

98. Division. Find the derivatives of $F(x) = x$, $G(x) = \sqrt{x + 1}$, and $H(x) = \left(\dfrac{F}{G}\right)(x) = \dfrac{x}{\sqrt{x + 1}}$. What do you observe?

SECTION 1.5 ONE-TO-ONE FUNCTIONS AND INVERSE FUNCTIONS

SKILLS OBJECTIVES

- Determine algebraically and graphically whether a function is a one-to-one function.
- Find the inverse of a function.
- Graph the inverse function given the graph of the function.

CONCEPTUAL OBJECTIVES

- Visualize the relationships between domain and range of a function and the domain and range of its inverse.
- Understand why functions and their inverses are symmetric about $y = x$.

Every human being has a blood type, and every human being has a DNA sequence. These are examples of functions, where a person is the input and the output is blood type or DNA sequence. These relationships are classified as functions because each person can have one and only one blood type or DNA strand. The difference between these functions is that many people have the same blood type, but DNA is unique to each individual. Can we map backwards? For instance, if you know the blood type, do you know specifically which person it came from? No, but, if you know the DNA sequence, you know that the sequence belongs to only one person. When a function has a one-to-one correspondence, like the DNA example, then mapping backwards is possible. The map back is called the *inverse function.*

One-to-One Functions

In Section 1.1, we defined a function as a relationship that maps an input (contained in the domain) to exactly one output (found in the range). Algebraically, each value for x can correspond to only a single value for y. Recall the square, identity, absolute value, and reciprocal functions from our library of functions in Section 1.3.

All of the graphs of these functions satisfy the vertical line test. Although the square function and the absolute value function map each value of x to exactly one value for y, these two functions map two values of x to the same value for y. For example, $(-1, 1)$ and $(1, 1)$ lie on both graphs. The identity and reciprocal functions, on the other hand, map each x to a single value for y, and no two x-values map to the same y-value. These two functions are examples of *one-to-one functions.*

DEFINITION   **One-to-One Function**

A function $f(x)$ is **one-to-one** if no two elements in the domain correspond to the same element in the range; that is,

$$\text{if } x_1 \neq x_2, \text{ then } f(x_1) \neq f(x_2).$$

In other words, it is one-to-one if no two inputs map to the same output.

EXAMPLE 1   **Determining Whether a Function Defined as a Set of Points Is a One-to-One Function**

For each of the three relations, determine whether the relation is a function. If it is a function, determine whether it is a one-to-one function.

$$f = \{(0, 0), (1, 1), (1, -1)\}$$
$$g = \{(-1, 1), (0, 0), (1, 1)\}$$
$$h = \{(-1, -1), (0, 0), (1, 1)\}$$

Solution:

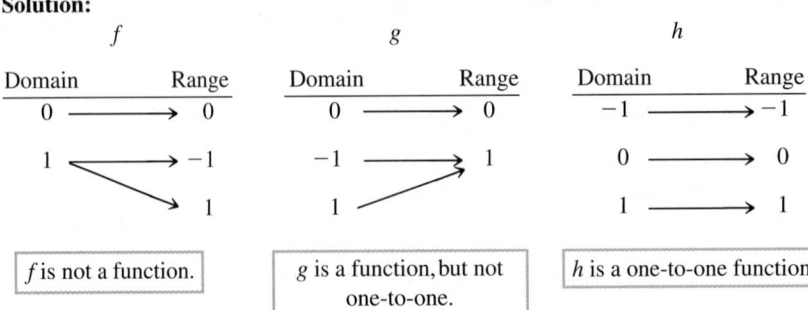

Just as there is a graphical test for functions, the vertical line test, there is a graphical test for one-to-one functions, the *horizontal line test*. Note that a horizontal line can be drawn on the square and absolute value functions so that it intersects the graph of each function at two points. The identity and reciprocal functions, however, will intersect a horizontal line in at most only one point. This leads us to the horizontal line test for one-to-one functions.

DEFINITION   Horizontal Line Test

If every horizontal line intersects the graph of a function in at most one point, then the function is classified as a one-to-one function.

 EXAMPLE 2   Using the Horizontal Line Test to Determine Whether a Function Is One-to-One

For each of the three relations, determine whether the relation is a function. If it is a function, determine whether it is a one-to-one function. Assume that x is the independent variable and y is the dependent variable.

$$x = y^2 \qquad y = x^2 \qquad y = x^3$$

Solution:

$$x = y^2 \qquad\qquad\qquad y = x^2 \qquad\qquad\qquad y = x^3$$

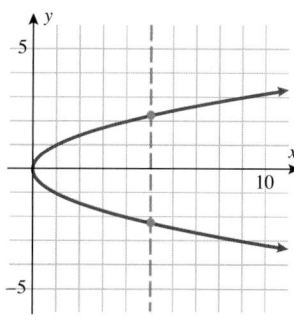   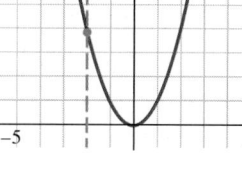   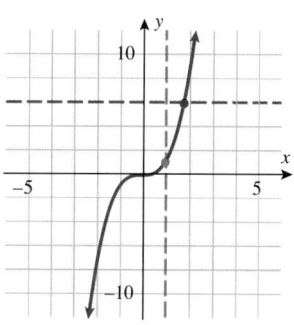

| Not a function | Function, but not one-to-one | One-to-one function |

(fails vertical line test)   (passes vertical line test, but fails horizontal line test)   (passes both vertical and horizontal line tests)

■ **YOUR TURN** Determine whether each of the functions is a one-to-one function.

  a. $f(x) = x + 2$ **b.** $f(x) = x^2 + 1$

■ **Answer:**
a. yes   **b.** no

Another way of writing the definition of a one-to-one function is

$$\text{If } f(x_1) = f(x_2), \text{ then } x_1 = x_2.$$

In the Your Turn following Example 2, we found (using the horizontal line test) that $f(x) = x + 2$ is a one-to-one function, but that $f(x) = x^2 + 1$ is not a one-to-one function. We can also use this alternative definition to determine algebraically whether a function is one-to-one.

WORDS	**MATH**
State the function.	$f(x) = x + 2$
Let there be two real numbers, x_1 and x_2, such that $f(x_1) = f(x_2)$.	$x_1 + 2 = x_2 + 2$
Subtract 2 from both sides of the equation.	$x_1 = x_2$

$f(x) = x + 2$ is a one-to-one function.

WORDS	**MATH**
State the function.	$f(x) = x^2 + 1$
Let there be two real numbers, x_1 and x_2, such that $f(x_1) = f(x_2)$.	$x_1^2 + 1 = x_2^2 + 1$
Subtract 1 from both sides of the equation.	$x_1^2 = x_2^2$
Solve for x_1.	$x_1 = \pm x_2$

$f(x) = x^2 + 2$ is *not* a one-to-one function.

 EXAMPLE 3 **Determining Algebraically Whether a Function Is One-to-One**

Determine algebraically whether the functions are one-to-one.

a. $f(x) = 5x^3 - 2$ **b.** $f(x) = |x + 1|$

Solution (a):

Find $f(x_1)$ and $f(x_2)$.	$f(x_1) = 5x_1^3 - 2$ and $f(x_2) = 5x_2^3 - 2$
Let $f(x_1) = f(x_2)$.	$5x_1^3 - 2 = 5x_2^3 - 2$
Add 2 to both sides of the equation.	$5x_1^3 = 5x_2^3$
Divide both sides of the equation by 5.	$x_1^3 = x_2^3$
Take the cube root of both sides of the equation.	$\left(x_1^3\right)^{1/3} = \left(x_2^3\right)^{1/3}$
Simplify.	$x_1 = x_2$

$f(x) = 5x^3 - 2$ is a one-to-one function.

Solution (b):

Find $f(x_1)$ and $f(x_2)$.	$f(x_1) = |x_1 + 1|$ and $f(x_2) = |x_2 + 1|$
Let $f(x_1) = f(x_2)$.	$|x_1 + 1| = |x_2 + 1|$
Solve the absolute value equation.	$(x_1 + 1) = (x_2 + 1)$ or $(x_1 + 1) = -(x_2 + 1)$
	$x_1 = x_2$ or $x_1 = -x_2 - 2$

$f(x) = |x + 1|$ is **not** a one-to-one function.

Inverse Functions

If a function is one-to-one, then the function maps each x to exactly one y, and no two x-values map to the same y-value. This implies that there is a one-to-one correspondence between the inputs (domain) and outputs (range) of a one-to-one function $f(x)$. In the special case of a one-to-one function, it would be possible to map from the output (range of f) back to the input (domain of f), and this mapping would also be a function. The function that maps the output back to the input of a function f is called the **inverse function** and is denoted $f^{-1}(x)$.

A one-to-one function f maps every x in the domain to a unique and distinct corresponding y in the range. Therefore, the inverse function f^{-1} maps every y back to a unique and distinct x.

The function notations $f(x) = y$ and $f^{-1}(y) = x$ indicate that if the point (x, y) satisfies the function, then the point (y, x) satisfies the inverse function.

For example, let the function $h(x) = \{(-1, 0), (1, 2), (3, 4)\}$.

$$h = \{(-1, 0), (1, 2), (3, 4)\}$$

Domain   Range

$$-1 \rightleftarrows 0$$

$$1 \rightleftarrows 2 \qquad h \text{ is a one-to-one function}$$

$$3 \rightleftarrows 4$$

Range   Domain

$$h^{-1} = \{(0, -1), (2, 1), (4, 3)\}$$

The inverse function undoes whatever the function does. For example, if $f(x) = 5x$, then the function f maps any value x in the domain to a value $5x$ in the range. If we want to map backwards or undo the $5x$, we develop a function called the inverse function that takes $5x$ as input and maps back to x as output. The inverse function is $f^{-1}(x) = \frac{1}{5}x$. Note that if we input $5x$ into the inverse function, the output is x: $f^{-1}(5x) = \frac{1}{5}(5x) = x$.

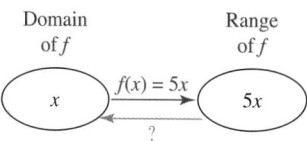

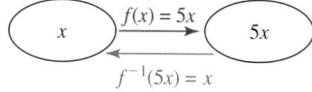

> **DEFINITION**   **Inverse Function**
>
> If f and g denote two one-to-one functions such that
>
> $$f(g(x)) = x \text{ for every } x \text{ in the domain of } g$$
> $$\text{and}$$
> $$g(f(x)) = x \text{ for every } x \text{ in the domain of } f,$$
>
> then g is the **inverse** of the function f. The function g is denoted by f^{-1} (read "f-inverse").

▼ **CAUTION**

$$f^{-1} \neq \frac{1}{f}$$

Note: f^{-1} is used to denote the inverse of f. The superscript -1 is not used as an exponent and, therefore, does not represent the reciprocal of f: $\dfrac{1}{f}$.

Two properties hold true relating one-to-one functions to their inverses: (1) The range of the function is the domain of the inverse, and the range of the inverse is the domain of the function, and (2) the composite function that results with a function and its inverse (and vice versa) is the identity function x.

> Domain of f = range of f^{-1} and range of f = domain of f^{-1}
>
> $$f^{-1}(f(x)) = x \quad \text{and} \quad f(f^{-1}(x)) = x$$

EXAMPLE 4 Verifying Inverse Functions

Verify that $f^{-1}(x) = \frac{1}{2}x - 2$ is the inverse of $f(x) = 2x + 4$.

Solution:

Show that $f^{-1}(f(x)) = x$ and $f(f^{-1}(x)) = x$.

Write f^{-1} using placeholder notation.

$$f^{-1}(\square) = \frac{1}{2}(\square) - 2$$

Substitute $f(x) = 2x + 4$ into f^{-1}.

$$f^{-1}(f(x)) = \frac{1}{2}(2x + 4) - 2$$

Simplify.

$$f^{-1}(f(x)) = x + 2 - 2 = x$$

$$f^{-1}(f(x)) = x$$

Write f using placeholder notation.

$$f(\square) = 2(\square) + 4$$

Substitute $f^{-1}(x) = \frac{1}{2}x - 2$ into f.

$$f(f^{-1}(x)) = 2\left(\frac{1}{2}x - 2\right) + 4$$

Simplify.

$$f(f^{-1}(x)) = x - 4 + 4 = x$$

$$f(f^{-1}(x)) = x$$

Note the relationship between the domain and range of f and f^{-1}.

	DOMAIN	**RANGE**
$f(x) = 2x + 4$	$(-\infty, \infty)$	$(-\infty, \infty)$
$f^{-1}(x) = \frac{1}{2}x - 2$	$(-\infty, \infty)$	$(-\infty, \infty)$

 ### EXAMPLE 5 Verifying Inverse Functions with Domain Restrictions

Verify that $f^{-1}(x) = x^2$, for $x \geq 0$, is the inverse of $f(x) = \sqrt{x}$.

Solution:

Show that $f^{-1}(f(x)) = x$ and $f(f^{-1}(x)) = x$.

Write f^{-1} using placeholder notation.

$$f^{-1}(\square) = (\square)^2$$

Substitute $f(x) = \sqrt{x}$ into f^{-1}.

$$f^{-1}(f(x)) = \left(\sqrt{x}\right)^2 = x$$

$$f^{-1}(f(x)) = x \text{ for } x \geq 0$$

Write f using placeholder notation.

$$f(\square) = \sqrt{(\square)}$$

Substitute $f^{-1}(x) = x^2$, $x \geq 0$ into f.

$$f(f^{-1}(x)) = \sqrt{x^2} = x, x \geq 0$$

$$f(f^{-1}(x)) = x \text{ for } x \geq 0$$

	DOMAIN	**RANGE**
$f(x) = \sqrt{x}$	$[0, \infty)$	$[0, \infty)$
$f^{-1}(x) = x^2, x \geq 0$	$[0, \infty)$	$[0, \infty)$

Graphical Interpretation of Inverse Functions

In Example 4, we showed that $f^{-1}(x) = \frac{1}{2}x - 2$ is the inverse of $f(x) = 2x + 4$. Let's now investigate the graphs that correspond to the function f and its inverse f^{-1}.

$f(x)$

x	y
-3	-2
-2	0
-1	2
0	4

$f^{-1}(x)$

x	y
-2	-3
0	-2
2	-1
4	0

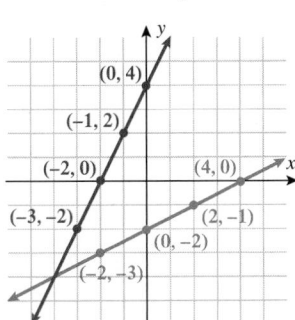

Note that the point $(-3, -2)$ lies on the function and the point $(-2, -3)$ lies on the inverse. In fact, every point (a, b) that lies on the function corresponds to a point (b, a) that lies on the inverse.

Draw the line $y = x$ on the graph. In general, the point (b, a) on the inverse $f^{-1}(x)$ is the reflection (about $y = x$) of the point (a, b) on the function $f(x)$.

In general, if the point (a, b) is on the graph of a function, then the point (b, a) is on the graph of its inverse.

EXAMPLE 6   Graphing the Inverse Function

Given the graph of the function $f(x)$, plot the graph of its inverse $f^{-1}(x)$.

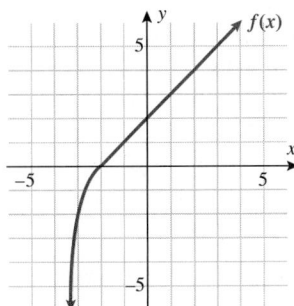

Solution:

Because the points $(-3, -2)$, $(-2, 0)$, $(0, 2)$, and $(2, 4)$ lie on the graph of f, then the points $(-2, -3)$, $(0, -2)$, $(2, 0)$, and $(4, 2)$ lie on the graph of f^{-1}.

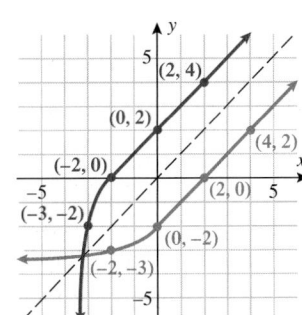

■ **Answer:**

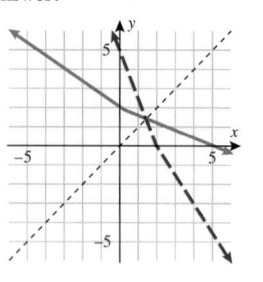

■ **YOUR TURN** Given the graph of a function f, plot the inverse function.

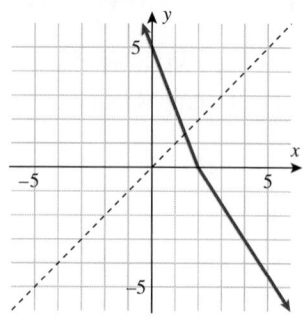

We have developed the definition of an inverse function, and properties of inverses. At this point, you should be able to determine whether two functions are inverses of one another. Let's turn our attention to another problem: How do you find the inverse of a function?

Finding the Inverse Function

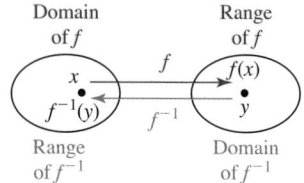

If the point (a, b) lies on the graph of a function, then the point (b, a) lies on the graph of the inverse function. The symmetry about the line $y = x$ tells us that the roles of x and y interchange. Therefore, if we start with every point (x, y) that lies on the graph of a function, then every point (y, x) lies on the graph of its inverse. Algebraically, this corresponds to interchanging x and y. Finding the inverse of a finite set of ordered pairs is easy: Simply interchange the x- and y-coordinates. Earlier, we found that if $h(x) = \{(-1, 0), (1, 2), (3, 4)\}$, then $h^{-1}(x) = \{(0, -1), (2, 1), (4, 3)\}$. But how do we find the inverse of a function defined by an equation?

Recall the mapping relationship if f is a one-to-one function. This relationship implies that $f(x) = y$ and $f^{-1}(y) = x$. Let's use these two identities to find the inverse. Now consider the function defined by $f(x) = 3x - 1$. To find f^{-1}, we let $f(x) = y$, which yields $y = 3x - 1$. Solve for the variable $x: x = \frac{1}{3}y + \frac{1}{3}$.

Recall that $f^{-1}(y) = x$, so we have found the inverse to be $f^{-1}(y) = \frac{1}{3}y + \frac{1}{3}$. It is customary to write the independent variable as x, so we write the inverse as $f^{-1}(x) = \frac{1}{3}x + \frac{1}{3}$. Now that we have found the inverse, let's confirm that the properties $f^{-1}(f(x)) = x$ and $f(f^{-1}(x)) = x$ hold.

$$f(f^{-1}(x)) = 3\left(\frac{1}{3}x + \frac{1}{3}\right) - 1 = x + 1 - 1 = x$$

$$f^{-1}(f(x)) = \frac{1}{3}(3x - 1) + \frac{1}{3} = x - \frac{1}{3} + \frac{1}{3} = x$$

FINDING THE INVERSE OF A FUNCTION

Let f be a one-to-one function, then the following procedure can be used to find the inverse function f^{-1} if the inverse exists.

STEP	PROCEDURE	EXAMPLE
1	Let $y = f(x)$.	$f(x) = -3x + 5$ $y = -3x + 5$
2	Solve the resulting equation for x in terms of y (if possible).	$3x = -y + 5$ $x = -\frac{1}{3}y + \frac{5}{3}$
3	Let $x = f^{-1}(y)$.	$f^{-1}(y) = -\frac{1}{3}y + \frac{5}{3}$
4	Let $y = x$ (interchange x and y).	$f^{-1}(x) = -\frac{1}{3}x + \frac{5}{3}$

The same result is found if we first interchange x and y and then solve for y in terms of x.

STEP	PROCEDURE	EXAMPLE
1	Let $y = f(x)$.	$f(x) = -3x + 5$ $y = -3x + 5$
2	Interchange x and y.	$x = -3y + 5$
3	Solve for y in terms of x.	$3y = -x + 5$ $y = -\frac{1}{3}x + \frac{5}{3}$
4	Let $y = f^{-1}(x)$	$f^{-1}(x) = -\frac{1}{3}x + \frac{5}{3}$

Note the following:

- Verify first that a function is one-to-one prior to finding an inverse (if it is not one-to-one, then the inverse does not exist).
- State the domain restrictions on the inverse function. The domain of f is the range of f^{-1} and vice versa.
- To verify that you have found the inverse, show that $f(f^{-1}(x)) = x$ for all x in the domain of f^{-1} and $f^{-1}(f(x)) = x$ for all x in the domain of f.

Technology Tip

Using a graphing utility, plot

$y_1 = f(x) = \sqrt{x + 2}$,

$y_2 = f^{-1}(x) = x^2 - 2$ for $x \geq 0$,

and $y_3 = x$.

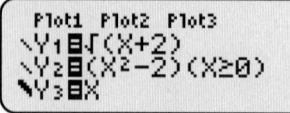

Note that the function $f(x)$ and its inverse $f^{-1}(x)$ are symmetric about the line $y = x$.

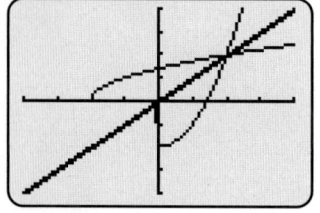

Study Tip

Had we ignored the domain and range in Example 7, we would have found the inverse function to be the square function $f(x) = x^2 - 2$, which is not a one-to-one function. It is only when we restrict the domain of the square function that we get a one-to-one function.

EXAMPLE 7   The Inverse of a Square Root Function

Find the inverse of the function $f(x) = \sqrt{x + 2}$ and state the domain and range of both f and f^{-1}.

Solution:

$f(x)$ is a one-to-one function because it passes the horizontal line test.

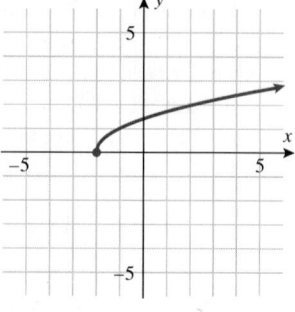

STEP 1   Let $y = f(x)$. $y = \sqrt{x + 2}$

STEP 2   Interchange x and y. $x = \sqrt{y + 2}$

STEP 3   Solve for y.

        Square both sides of the equation. $x^2 = y + 2$

        Subtract 2 from both sides. $x^2 - 2 = y$ or $y = x^2 - 2$

STEP 4   Let $y = f^{-1}(x)$. $f^{-1}(x) = x^2 - 2$

Note any domain restrictions. (State the domain and range of both f and f^{-1}.)

    f: Domain: $[-2, \infty)$ Range: $[0, \infty)$

    f^{-1}:   Domain: $[0, \infty)$ Range: $[-2, \infty)$

The inverse of $f(x) = \sqrt{x + 2}$ is $\boxed{f^{-1}(x) = x^2 - 2 \text{ for } x \geq 0}$.

Check.

$f^{-1}(f(x)) = x$ for all x in the domain of f.

$$f^{-1}(f(x)) = \left(\sqrt{x + 2}\right)^2 - 2$$
$$= x + 2 - 2 \text{ for } x \geq -2$$
$$= x$$

$f(f^{-1}(x)) = x$ for all x in the domain of f^{-1}.

$$f(f^{-1}(x)) = \sqrt{(x^2 - 2) + 2}$$
$$= \sqrt{x^2} \text{ for } x \geq 0$$
$$= x$$

Note that the function $f(x) = \sqrt{x + 2}$ and its inverse $f^{-1}(x) = x^2 - 2$ for $x \geq 0$ are symmetric about the line $y = x$.

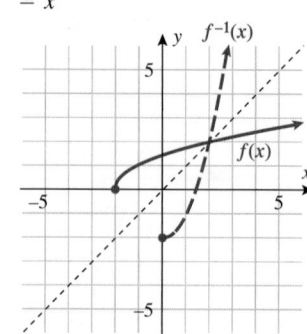

■ **Answers:**

a. $f^{-1}(x) = \dfrac{x + 3}{7}$, Domain:

    $(-\infty, \infty)$, Range: $(-\infty, \infty)$

b. $g^{-1}(x) = x^2 + 1$, Domain:

    $[0, \infty)$, Range: $[1, \infty)$

■ **YOUR TURN** Find the inverse of the given function and state the domain and range of the inverse function.

    a. $f(x) = 7x - 3$ **b.** $g(x) = \sqrt{x - 1}$

EXAMPLE 8   A Function That Does Not Have an Inverse Function

Find the inverse of the function $f(x) = |x|$ if it exists.

Solution:

The function $f(x) = |x|$ fails the horizontal
line test and therefore is not a one-to-one
function. Because f is not a one-to-one function,
its inverse function does not exist.

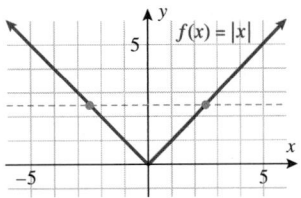

EXAMPLE 9   Finding the Inverse Function

The function $f(x) = \dfrac{2}{x + 3}$, $x \neq -3$, is a one-to-one function. Find its inverse.

Solution:

STEP 1 Let $y = f(x)$.	$y = \dfrac{2}{x + 3}$
STEP 2 Interchange x and y.	$x = \dfrac{2}{y + 3}$
STEP 3 Solve for y.	
Multiply the equation by $(y + 3)$.	$x(y + 3) = 2$
Eliminate the parentheses.	$xy + 3x = 2$
Subtract $3x$ from both sides.	$xy = -3x + 2$
Divide the equation by x.	$y = \dfrac{-3x + 2}{x} = -3 + \dfrac{2}{x}$
STEP 4 Let $y = f^{-1}(x)$.	$f^{-1}(x) = -3 + \dfrac{2}{x}$
Note any domain restrictions on $f^{-1}(x)$.	$x \neq 0$

The inverse of the function $f(x) = \dfrac{2}{x + 3}$, $x \neq -3$, is $\boxed{f^{-1}(x) = -3 + \dfrac{2}{x}, x \neq 0}$.

Check.

$$f^{-1}(f(x)) = -3 + \dfrac{2}{\left(\dfrac{2}{x + 3}\right)} = -3 + (x + 3) = x, x \neq -3$$

$$f(f^{-1}(x)) = \dfrac{2}{\left(-3 + \dfrac{2}{x}\right) + 3} = \dfrac{2}{\left(\dfrac{2}{x}\right)} = x, x \neq 0$$

■ YOUR TURN The function $f(x) = \dfrac{4}{x - 1}$, $x \neq 1$, is a one-to-one function.
Find its inverse.

Technology Tip

The graphs of $y_1 = f(x) = \dfrac{2}{x + 3}$,
$x \neq -3$, and $y_2 = f^{-1}(x) = -3 + \dfrac{2}{x}$,
$x \neq 0$, are shown.

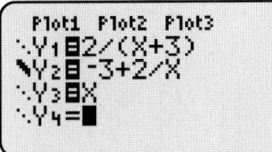

Note that the graphs of the function
$f(x)$ and its inverse $f^{-1}(x)$ are
symmetric about the line $y = x$.

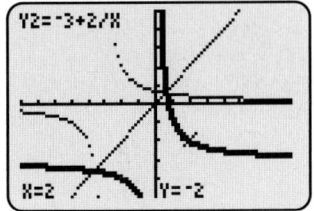

Study Tip

The range of the function is equal to
the domain of its inverse function.

■ Answer: $f^{-1}(x) = 1 + \dfrac{4}{x}, x \neq 0$

Note in Example 9 that the domain of f is $(-\infty, -3) \cup (-3, \infty)$ and the domain of f^{-1} is
$(-\infty, 0) \cup (0, \infty)$. Therefore, we know that the range of f is $(-\infty, 0) \cup (0, \infty)$, and the
range of f^{-1} is $(-\infty, -3) \cup (-3, \infty)$.

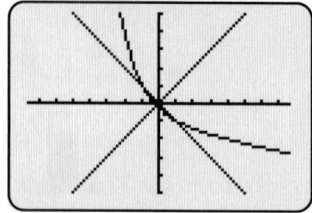

EXAMPLE 10 Finding the Inverse of a Piecewise-Defined Function

Determine whether the function $f(x) = \begin{cases} x^2 & x < 0 \\ -x & x \geq 0 \end{cases}$ is a one-to-one function. If it is a one-to-one function, find its inverse.

Solution:

The graph of the function f passes the horizontal line test and therefore f is a one-to-one function.

STEP 1 Let $y = f(x)$.

Let $y_1 = x_1^2$ for $x_1 < 0$ and $y_2 = -x_2$ for $x_2 \geq 0$ represent the two pieces of f. Note the domain and range for each piece.

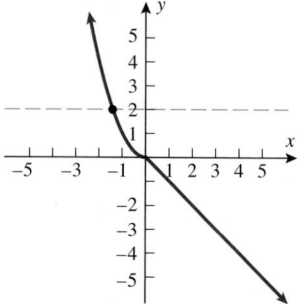

EQUATION	DOMAIN	RANGE
$y_1 = x_1^2$	$x_1 < 0$	$y_1 > 0$
$y_2 = -x_2$	$x_2 \geq 0$	$y_2 \leq 0$

STEP 2 Solve for x in terms of y.

Solve for x_1. $x_1 = \pm\sqrt{y_1}$

Select the negative root since $x_1 < 0$. $x_1 = -\sqrt{y_1}$

Solve for x_2. $x_2 = -y_2$

STEP 3 Let $x = f^{-1}(y)$.

$x_1 = f^{-1}(y_1) = -\sqrt{y_1}$	$x_1 < 0$	$y_1 > 0$
$x_2 = f^{-1}(y_2) = -y_2$	$x_2 \geq 0$	$y_2 \leq 0$

Express the two "pieces" in terms of a piecewise-defined function.

$$f^{-1}(y) = \begin{cases} -\sqrt{y} & y > 0 \\ -y & y \leq 0 \end{cases}$$

STEP 4 Let $y = x$ (interchange x and y).

$$f^{-1}(x) = \begin{cases} -\sqrt{x} & x > 0 \\ -x & x \leq 0 \end{cases}$$

SECTION
1.5 SUMMARY

One-to-One Functions

Each input in the domain corresponds to exactly one output in the range, and no two inputs map to the same output. There are three ways to test a function to determine whether it is a one-to-one function.

1. **Discrete points:** For the set of all points (a, b), verify that no y-values are repeated.
2. **Algebraic equations:** Let $f(x_1) = f(x_2)$; if it can be shown that $x_1 = x_2$, then the function is one-to-one.
3. **Graphs:** Use the horizontal line test; if any horizontal line intersects the graph of the function in more than one point, then the function is not one-to-one.

Properties of Inverse Functions

1. If f is a one-to-one function, then f^{-1} exists.
2. Domain and range
   - Domain of f = range of f^{-1}
   - Domain of f^{-1} = range of f
3. Composition of inverse functions
   - $f^{-1}(f(x)) = x$ for all x in the domain of f
   - $f(f^{-1}(x)) = x$ for all x in the domain of f^{-1}
4. The graphs of f and f^{-1} are symmetric with respect to the line $y = x$.

Procedure for Finding the Inverse of a Function

1. Let $y = f(x)$.
2. Interchange x and y.
3. Solve for y.
4. Let $y = f^{-1}(x)$.

SECTION
1.5 EXERCISES

■ SKILLS

In Exercises 1–10, determine whether the given relation is a function. If it is a function, determine whether it is a one-to-one function.

1. $\{(0, 0), (9, -3), (4, -2), (4, 2), (9, 3)\}$

2. $\{(0, 1), (1, 1), (2, 1), (3, 1)\}$

3. $\{(0, 1), (1, 0), (2, 1), (-2, 1), (5, 4), (-3, 4)\}$

4. $\{(0, 0), (-1, -1), (-2, -8), (1, 1), (2, 8)\}$

5.
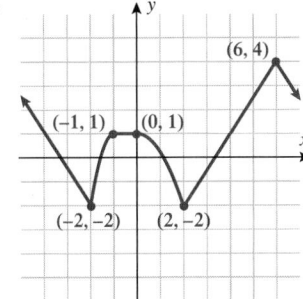

6.
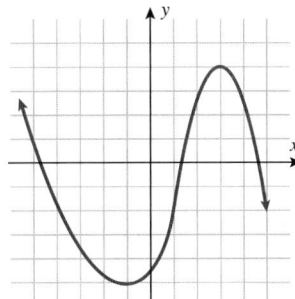

7.
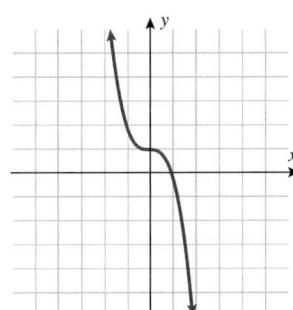

8.
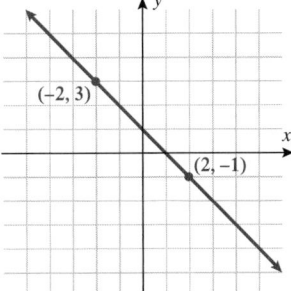

9.
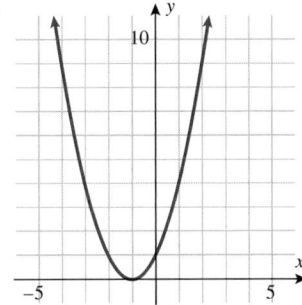

10.
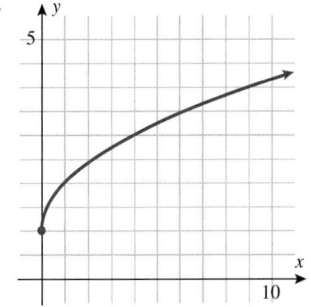

In Exercises 11–18, determine algebraically and graphically whether the function is one-to-one.

11. $f(x) = |x - 3|$

12. $f(x) = (x - 2)^2 + 1$

13. $f(x) = \dfrac{1}{x - 1}$

14. $f(x) = \sqrt[3]{x}$

15. $f(x) = x^2 - 4$

16. $f(x) = \sqrt{x + 1}$

17. $f(x) = x^3 - 1$

18. $f(x) = \dfrac{1}{x + 2}$

In Exercises 19–28, verify that the function $f^{-1}(x)$ is the inverse of $f(x)$ by showing that $f(f^{-1}(x)) = x$ and $f^{-1}(f(x)) = x$. Graph $f(x)$ and $f^{-1}(x)$ on the same axes to show the symmetry about the line $y = x$.

19. $f(x) = 2x + 1;\ f^{-1}(x) = \dfrac{x - 1}{2}$

20. $f(x) = \dfrac{x - 2}{3};\ f^{-1}(x) = 3x + 2$

21. $f(x) = \sqrt{x - 1}, x \ge 1;\ f^{-1}(x) = x^2 + 1, x \ge 0$

22. $f(x) = 2 - x^2, x \ge 0;\ f^{-1}(x) = \sqrt{2 - x}, x \le 2$

23. $f(x) = \dfrac{1}{x};\ f^{-1}(x) = \dfrac{1}{x}, x \ne 0$

24. $f(x) = (5 - x)^{1/3};\ f^{-1}(x) = 5 - x^3$

25. $f(x) = \dfrac{1}{2x + 6}, x \ne -3;\ f^{-1}(x) = \dfrac{1}{2x} - 3, x \ne 0$

26. $f(x) = \dfrac{3}{4 - x}, x \ne 4;\ f^{-1}(x) = 4 - \dfrac{3}{x}, x \ne 0$

27. $f(x) = \dfrac{x + 3}{x + 4}, x \ne -4;\ f^{-1}(x) = \dfrac{3 - 4x}{x - 1}, x \ne 1$

28. $f(x) = \dfrac{x - 5}{3 - x}, x \ne 3;\ f^{-1}(x) = \dfrac{3x + 5}{x + 1}, x \ne -1$

In Exercises 29–36, graph the inverse of the one-to-one function that is given.

29.

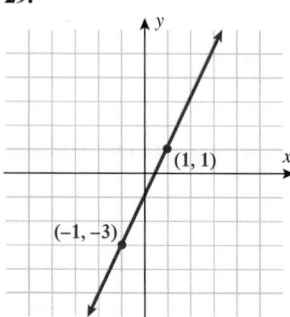

30.

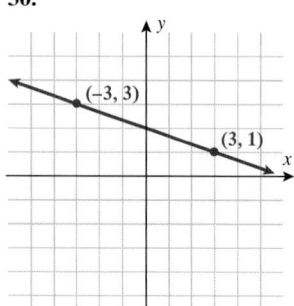

31.

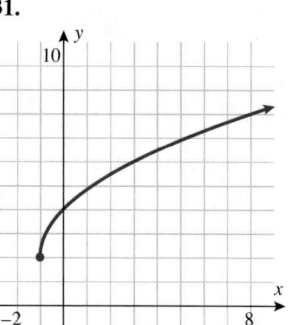

32.

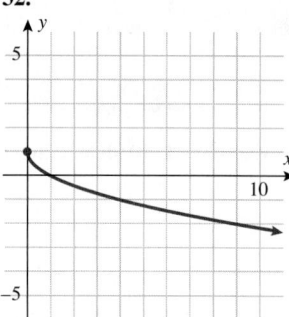

33.

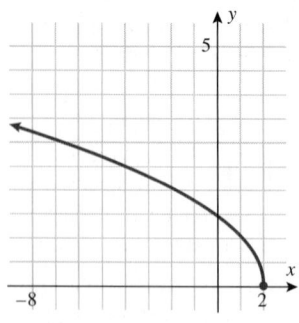

34.

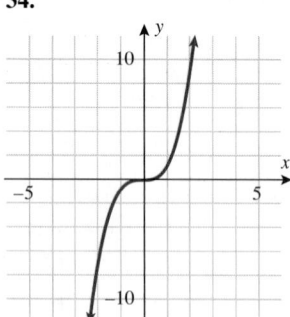

35.

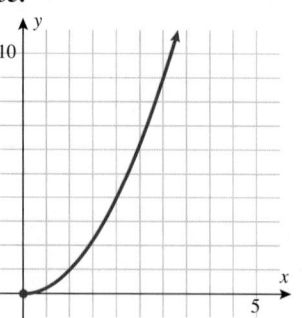

36.

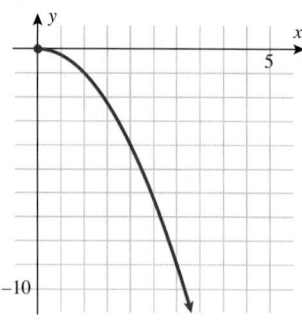

In Exercises 37–56, the function f is one-to-one. Find its inverse, and check your answer. State the domain and range of both f and f^{-1}.

37. $f(x) = -3x + 2$

38. $f(x) = 2x + 3$

39. $f(x) = x^3 + 1$

40. $f(x) = x^3 - 1$

41. $f(x) = \sqrt{x - 3}$

42. $f(x) = \sqrt{3 - x}$

43. $f(x) = x^2 - 1, x \geq 0$

44. $f(x) = 2x^2 + 1, x \geq 0$

45. $f(x) = (x + 2)^2 - 3, x \geq -2$

46. $f(x) = (x - 3)^2 - 2, x \geq 3$

47. $f(x) = \dfrac{2}{x}$

48. $f(x) = -\dfrac{3}{x}$

49. $f(x) = \dfrac{2}{3 - x}$

50. $f(x) = \dfrac{7}{x + 2}$

51. $f(x) = \dfrac{7x + 1}{5 - x}$

52. $f(x) = \dfrac{2x + 5}{7 + x}$

53. $f(x) = \dfrac{1}{\sqrt{x}}$

54. $f(x) = \dfrac{x}{\sqrt{x + 1}}$

55. $f(x) = \sqrt{\dfrac{x + 1}{x - 2}}$

56. $f(x) = \sqrt{x^2 - 1}, x \geq 1$

In Exercises 57–62, graph the piecewise-defined function to determine whether it is a one-to-one function. If it is a one-to-one function, find its inverse.

57. $G(x) = \begin{cases} 0 & x < 0 \\ \sqrt{x} & x \geq 0 \end{cases}$

58. $G(x) = \begin{cases} \dfrac{1}{x} & x < 0 \\ \sqrt{x} & x \geq 0 \end{cases}$

59. $f(x) = \begin{cases} \sqrt[3]{x} & x \leq -1 \\ x^2 + 2x & -1 < x \leq 1 \\ \sqrt{x} + 2 & x > 1 \end{cases}$

60. $f(x) = \begin{cases} -x & x < -2 \\ \sqrt{4 - x^2} & -2 \leq x \leq 0 \\ -\dfrac{1}{x} & x > 0 \end{cases}$

61. $f(x) = \begin{cases} x & x \leq -1 \\ x^3 & -1 < x < 1 \\ x & x \geq 1 \end{cases}$

62. $f(x) = \begin{cases} x + 3 & x \leq -2 \\ |x| & -2 < x < 2 \\ x^2 & x \geq 2 \end{cases}$

■ **APPLICATIONS**

63. Temperature. The equation used to convert from degrees Celsius to degrees Fahrenheit is $f(x) = \frac{9}{5}x + 32$. Determine the inverse function $f^{-1}(x)$. What does the inverse function represent?

64. Temperature. The equation used to convert from degrees Fahrenheit to degrees Celsius is $C(x) = \frac{5}{9}(x - 32)$. Determine the inverse function $C^{-1}(x)$. What does the inverse function represent?

65. Budget. The Richmond rowing club is planning to enter the Head of the Charles race in Boston and is trying to figure out how much money to raise. The entry fee is $250 per boat for the first 10 boats and $175 for each additional boat. Find the cost function $C(x)$ as a function of the number of boats x the club enters. Find the inverse function that will yield how many boats the club can enter as a function of how much money it will raise.

66. Long-Distance Calling Plans. A phone company charges $.39 per minute for the first 10 minutes of a long-distance phone call and $.12 per minute every minute after that. Find the cost function $C(x)$ as a function of length x of the phone call in minutes. Suppose you buy a "prepaid" phone card that is planned for a single call. Find the inverse function that determines how many minutes you can talk as a function of how much you prepaid.

67. Salary. A student works at Target making $7 per hour and the weekly number of hours worked per week x varies. If Target withholds 25% of his earnings for taxes and Social Security, write a function $E(x)$ that expresses the student's take-home pay each week. Find the inverse function $E^{-1}(x)$. What does the inverse function tell you?

68. Salary. A grocery store pays you $8 per hour for the first 40 hours per week and time and a half for overtime. Write a piecewise-defined function that represents your weekly earnings $E(x)$ as a function of the number of hours worked x. Find the inverse function $E^{-1}(x)$. What does the inverse function tell you?

69. ATM Charges. A bank charges $.60 each time that a client uses an ATM machine for the first 15 transactions of the month, and $.90 for every additional transaction. Find a function $M(x)$ as a function of the number of monthly transactions that describes the amount charged by the bank. Find the inverse function that will yield how many transactions per month a client can do as a function of the client's budget.

70. Truck Renting. For renting a truck, a company charges $19.95 per day plus $.80 per mile plus 10% in taxes. Find the cost of renting a truck for two days as a function of the miles driven x. Find the inverse function that determines how many miles you can drive as a function of the money you have available.

71. Depreciation. The value of a family car decreases $600 per year for the first 5 years; after that it depreciates $900 per year. Find the resale value function $V(x)$, where x is the number of years the family has owned the car whose original price was $20,000. Find the inverse function $V^{-1}(x)$. What does the inverse function tell you?

72. Production. The number of pounds of strawberries produced per square yard in a strawberry field depends on the number of plants per square yard. When $x < 25$ plants are planted per square yard, each square yard produces $\dfrac{50 - x}{5}$ pounds. Find a function P as a function of the number of plants x per square yard. Find the inverse function P^{-1}. What does the inverse function tell you?

■ CATCH THE MISTAKE

In Exercises 73–76, explain the mistake that is made.

73. Is $x = y^2$ a one-to-one function?

Solution:

Yes, this graph represents a one-to-one function because it passes the horizontal line test.

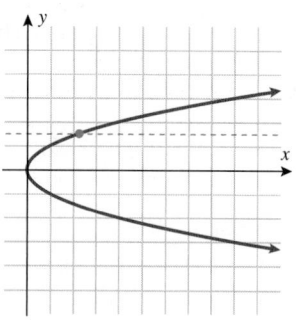

This is incorrect. What mistake was made?

74. A linear one-to-one function is graphed below. Draw its inverse.

Solution:

Note that the points $(3, 3)$ and $(0, -4)$ lie on the graph of the function.

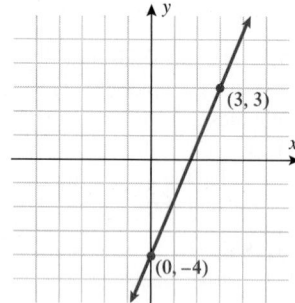

By symmetry, the points $(-3, -3)$ and $(0, 4)$ lie on the graph of the inverse.

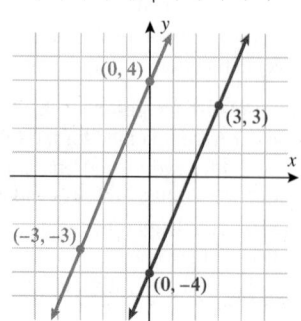

This is incorrect. What mistake was made?

75. Given the function $f(x) = x^2$, find the inverse function $f^{-1}(x)$.

Solution:

Step 1: Let $y = f(x)$.	$y = x^2$
Step 2: Solve for x.	$x = \sqrt{y}$
Step 3: Interchange x and y.	$y = \sqrt{x}$
Step 4: Let $y = f^{-1}(x)$.	$f^{-1}(x) = \sqrt{x}$

Check: $f(f^{-1}(x)) = (\sqrt{x})^2 = x$ and $f^{-1}(f(x)) = \sqrt{x^2} = x$.

The inverse of $f(x) = x^2$ is $f^{-1}(x) = \sqrt{x}$.

This is incorrect. What mistake was made?

76. Given the function $f(x) = \sqrt{x - 2}$, find the inverse function $f^{-1}(x)$, and state the domain restrictions on $f^{-1}(x)$.

Solution:

Step 1: Let $y = f(x)$.	$y = \sqrt{x - 2}$
Step 2: Interchange x and y.	$x = \sqrt{y - 2}$
Step 3: Solve for y.	$y = x^2 + 2$
Step 4: Let $f^{-1}(x) = y$.	$f^{-1}(x) = x^2 + 2$

Step 5: Domain restrictions $f(x) = \sqrt{x - 2}$ has the domain restriction that $x \geq 2$.

The inverse of $f(x) = \sqrt{x - 2}$ is $f^{-1}(x) = x^2 + 2$.

The domain of $f^{-1}(x)$ is $x \geq 2$.

This is incorrect. What mistake was made?

CONCEPTUAL

In Exercises 77–80, determine whether each statement is true or false.

77. Every even function is a one-to-one function.

78. Every odd function is a one-to-one function.

79. It is not possible that $f = f^{-1}$.

80. A function f has an inverse. If the function lies in quadrant II, then its inverse lies in quadrant IV.

81. If $(0, b)$ is the y-intercept of a one-to-one function f, what is the x-intercept of the inverse f^{-1}?

82. If $(a, 0)$ is the x-intercept of a one-to-one function f, what is the y-intercept of the inverse f^{-1}?

CHALLENGE

83. The unit circle is not a function. If we restrict ourselves to the semicircle that lies in quadrants I and II, the graph represents a function, but it is not a one-to-one function. If we further restrict ourselves to the quarter circle lying in quadrant I, the graph does represent a one-to-one function. Determine the equations of both the one-to-one function and its inverse. State the domain and range of both.

84. Find the inverse of $f(x) = \dfrac{c}{x}, c \neq 0$.

85. Under what conditions is the linear function $f(x) = mx + b$ a one-to-one function?

86. Assuming that the conditions found in Exercise 85 are met, determine the inverse of the linear function.

87. Determine the value of a that makes $f(x) = \dfrac{x - 2}{x^2 - a}$ a one-to-one function. Determine $f^{-1}(x)$ and its domain.

88. The point (a, b) lies on the graph of the one-to-one function $y = f(x)$. What other points are guaranteed to lie on the graph of $y = f^{-1}(x)$?

TECHNOLOGY

In Exercises 89–92, graph the following functions and determine whether they are one-to-one.

89. $f(x) = |4 - x^2|$

90. $f(x) = \dfrac{3}{x^3 + 2}$

91. $f(x) = x^{1/3} - x^5$

92. $f(x) = \dfrac{1}{x^{1/2}}$

In Exercises 93–96, graph the functions f and g and the line $y = x$ in the same screen. Do the two functions appear to be inverses of each other?

93. $f(x) = \sqrt{3x - 5}$; $g(x) = \dfrac{x^2}{3} + \dfrac{5}{3}$

94. $f(x) = \sqrt{4 - 3x}$; $g(x) = \dfrac{4}{3} - \dfrac{x^2}{3}, x \geq 0$

95. $f(x) = (x - 7)^{1/3} + 2$; $g(x) = x^3 - 6x^2 + 12x - 1$

96. $f(x) = \sqrt[3]{x + 3} - 2$; $g(x) = x^3 + 6x^2 + 12x + 6$

PREVIEW TO CALCULUS

For Exercises 97–100, refer to the following:

In calculus, the difference quotient $\dfrac{f(x + h) - f(x)}{h}$ of a function f is used to find the derivative f' of f, by allowing h to approach zero, $h \to 0$. The derivative of the inverse function $(f^{-1})'$ can be found using the formula

$$(f^{-1})'(x) = \dfrac{1}{f'(f^{-1}(x))}$$

provided that the denominator is not 0 and both f and f^{-1} are differentiable. For the following one-to-one function, find (a) f^{-1}, (b) f', (c) $(f^{-1})'$, and (d) verify the formula above. For (b) and (c), use the difference quotient.

97. $f(x) = 2x + 1$

98. $f(x) = x^2, x > 0$

99. $f(x) = \sqrt{x + 2}, x > -2$

100. $f(x) = \dfrac{1}{x + 1}, x > -1$

The U.S. National Oceanic and Atmospheric Association (NOAA) monitors temperature and carbon emissions at its observatory in Mauna Loa, Hawaii. NOAA's goal is to help foster an informed society that uses a comprehensive understanding of the role of the oceans, coasts, and atmosphere in the global ecosystem to make the best social and economic decisions. The data presented in this chapter are from the Mauna Loa Observatory, where historical atmospheric measurements have been recorded for the last 50 years. You will develop linear models based on this data to predict temperature and carbon emissions in the future.

The following table summarizes average yearly temperature in degrees Fahrenheit °F and carbon dioxide emissions in parts per million (ppm) for **Mauna Loa, Hawaii.**

Year	1960	1965	1970	1975	1980	1985	1990	1995	2000	2005
Temperature (°F)	44.45	43.29	43.61	43.35	46.66	45.71	45.53	47.53	45.86	46.23
CO_2 Emissions (ppm)	316.9	320.0	325.7	331.1	338.7	345.9	354.2	360.6	369.4	379.7

1. Plot the temperature data with time on the horizontal axis and temperature on the vertical axis. Let $t = 0$ correspond to 1960.

2. Find a *linear function* that models the temperature in Mauna Loa.
   a. Use data from 1965 and 1995.
   b. Use data from 1960 and 1990.
   c. Use linear regression and all data given.

3. Predict what the temperature will be in Mauna Loa in 2020.
   a. Apply the line found in Exercise 2(a).
   b. Apply the line found in Exercise 2(b).
   c. Apply the line found in Exercise 2(c).

4. Predict what the temperature will be in Mauna Loa in 2100.
   a. Apply the line found in Exercise 2(a).
   b. Apply the line found in Exercise 2(b).
   c. Apply the line found in Exercise 2(c).

5. Do you think your models support the claim of "global warming"? Explain.

6. Plot the carbon dioxide emissions data with time on the horizontal axis and carbon dioxide emissions on the vertical axis. Let $t = 0$ correspond to 1960.

7. Find a *linear function* that models the CO_2 emissions (ppm) in Mauna Loa.
   a. Use data from 1965 and 1995.
   b. Use data from 1960 and 1990.
   c. Use linear regression and all data given.

8. Predict the expected CO_2 emissions in Mauna Loa in 2020.
   a. Apply the line found in Exercise 7(a).
   b. Apply the line found in Exercise 7(b).
   c. Apply the line found in Exercise 7(c).

9. Predict the expected CO_2 emissions in Mauna Loa in 2100.
   a. Apply the line found in Exercise 7(a).
   b. Apply the line found in Exercise 7(b).
   c. Apply the line found in Exercise 7(c).

10. Do you think your models support the claim of the "greenhouse effect"? Explain.

SECTION	CONCEPT	PAGES	REVIEW EXERCISES	KEY IDEAS/POINT
1.1	**Functions**	106–117	1–28	
	Definition of a function	106–108	1–2	All functions are relations but not all relations are functions.
	Functions defined by equations	108–110	3–18	A vertical line can intersect a function in at most one point.
	Function notation	111–115	9–20	Placeholder notation Difference quotient: $\dfrac{f(x + h) - f(x)}{h}, h \neq 0$
	Domain of a function	115–117	21–26	Are there any restrictions on x?
1.2	**Graphs of functions**	123–135	29–48	
	Common functions	123–126		$f(x) = mx + b, f(x) = x, f(x) = x^2,$ $f(x) = x^3, f(x) = \sqrt{x}, f(x) = \sqrt[3]{x},$ $f(x) = \lvert x \rvert, f(x) = \dfrac{1}{x}$
	Even and odd functions	126–128	29–32	$f(-x) = f(x)$ Symmetry about y-axis $f(-x) = -f(x)$ Symmetry about origin
	Increasing and decreasing functions	128–129	33–36	■ Increasing: rises (left to right) ■ Decreasing: falls (left to right)
	Average rate of change	129–132	37–42	$\dfrac{f(x_2) - f(x_1)}{x_2 - x_1} \quad x_1 \neq x_2$
	Piecewise-defined functions	132–135	43–46	Points of discontinuity
1.3	**Graphing techniques: Transformations**	143–152	49–66	Shift the graph of $f(x)$.
	Horizontal and vertical shifts	143–147	49–66	$f(x + c)$ — c units to the left — where $c > 0$ $f(x - c)$ — c units to the right — where $c > 0$ $f(x) + c$ — c units upward — where $c > 0$ $f(x) - c$ — c units downward — where $c > 0$
	Reflection about the axes	147–149	49–66	$-f(x)$ — Reflection about the x-axis $f(-x)$ — Reflection about the y-axis
	Stretching and compressing	150–152	49–66	$cf(x)$ if $c > 1$ — stretch vertically $cf(x)$ if $0 < c < 1$ — compress vertically $f(cx)$ if $c > 1$ — compress horizontally $f(cx)$ if $0 < c < 1$ — stretch horizontally

SECTION	CONCEPT	PAGES	REVIEW EXERCISES	KEY IDEAS/POINT
1.4	**Combining functions**	157–163	67–90	
	Adding, subtracting, multiplying, and dividing functions	157–158	67–72	$(f + g)(x) = f(x) + g(x)$ $(f - g)(x) = f(x) - g(x)$ $(f \cdot g)(x) = f(x) \cdot g(x)$ Domain of the resulting function is the intersection of the individual domains. $\left(\dfrac{f}{g}\right)(x) = \dfrac{f(x)}{g(x)},\ g(x) \neq 0$ Domain of the quotient is the intersection of the domains of f and g, and any points when $g(x) = 0$ must be eliminated.
	Composition of functions	159–163	73–88	$(f \circ g)(x) = f(g(x))$ The domain of the composite function is a subset of the domain of $g(x)$. Values for x must be eliminated if their corresponding values $g(x)$ are not in the domain of f.
1.5	**One-to-one functions and inverse functions**	167–178	91–112	
	One-to-one functions	168–170	91–100	■ No two x-values map to the same y-value. If $f(x_1) = f(x_2)$, then $x_1 = x_2$. ■ A horizontal line may intersect a one-to-one function in at most one point.
	Inverse functions	171–172	101–112	■ Only one-to-one functions have inverses. ■ $f^{-1}(f(x)) = x$ for all x in the domain of f. ■ $f(f^{-1}(x)) = x$ for all x in the domain of f^{-1}. ■ Domain of f = range of f^{-1}. Range of f = domain of f^{-1}.
	Graphical interpretation of inverse functions	173–174	101–104	■ The graph of a function and its inverse are symmetric about the line $y = x$. ■ If the point (a, b) lies on the graph of a function, then the point (b, a) lies on the graph of its inverse.
	Finding the inverse function	174–178	105–112	1. Let $y = f(x)$. 2. Interchange x and y. 3. Solve for y. 4. Let $y = f^{-1}(x)$.

1.1 Functions

Determine whether each relation is a function. Assume that the coordinate pair (x, y) represents independent variable x and dependent variable y.

1. $\{(-2, 3), (1, -3), (0, 4), (2, 6)\}$

2. $\{(4, 7), (2, 6), (3, 8), (1, 7)\}$

3. $x^2 + y^2 = 36$

4. $x = 4$

5. $y = |x + 2|$

6. $y = \sqrt{x}$

7.

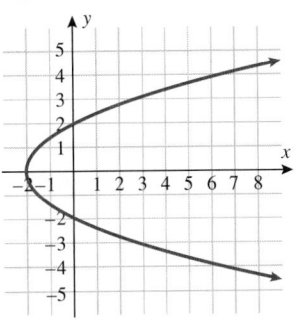

8.

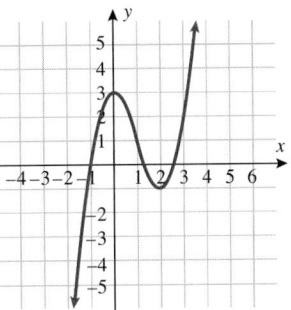

Use the graphs of the functions to find:

9.

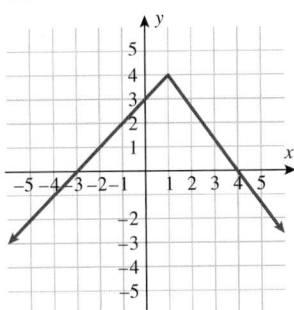

10.

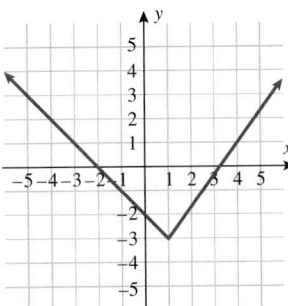

a. $f(-1)$ **b.** $f(1)$

c. x, where $f(x) = 0$

a. $f(-4)$ **b.** $f(0)$

c. x, where $f(x) = 0$

11.

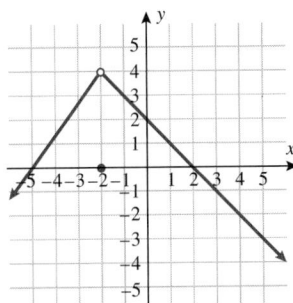

12.

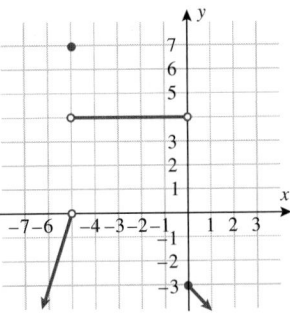

a. $f(-2)$ **b.** $f(4)$

c. x, where $f(x) = 0$

a. $f(-5)$ **b.** $f(0)$

c. x, where $f(x) = 0$

Evaluate the given quantities using the following three functions:

$$f(x) = 4x - 7 \qquad F(t) = t^2 + 4t - 3 \qquad g(x) = |x^2 + 2x + 4|$$

13. $f(3)$

14. $F(4)$

15. $f(-7) \cdot g(3)$

16. $\dfrac{F(0)}{g(0)}$

17. $\dfrac{f(2) - F(2)}{g(0)}$

18. $f(3 + h)$

19. $\dfrac{f(3 + h) - f(3)}{h}$

20. $\dfrac{F(t + h) - F(t)}{h}$

Find the domain of the given function. Express the domain in interval notation.

21. $f(x) = -3x - 4$

22. $g(x) = x^2 - 2x + 6$

23. $h(x) = \dfrac{1}{x + 4}$

24. $F(x) = \dfrac{7}{x^2 + 3}$

25. $G(x) = \sqrt{x - 4}$

26. $H(x) = \dfrac{1}{\sqrt{2x - 6}}$

Challenge

27. If $f(x) = \dfrac{D}{x^2 - 16}$, $f(4)$ and $f(-4)$ are undefined, and $f(5) = 2$, find D.

28. Construct a function that is undefined at $x = -3$ and $x = 2$ such that the point $(0, -4)$ lies on the graph of the function.

1.2 Graphs of Functions

Determine whether the function is even, odd, or neither.

29. $h(x) = x^3 - 7x$

30. $f(x) = x^4 + 3x^2$

31. $f(x) = \dfrac{1}{x^3} + 3x$

32. $f(x) = \dfrac{1}{x^2} + 3x^4 + |x|$

In Exercises 33–36, state the (a) domain, (b) range, and (c) x-interval(s), where the function is increasing, decreasing, or constant. Find the values of (d) $f(0)$, (e) $f(-3)$, and (f) $f(3)$.

33.

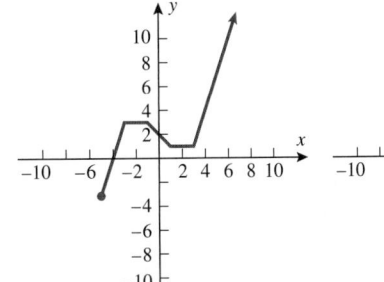

34.

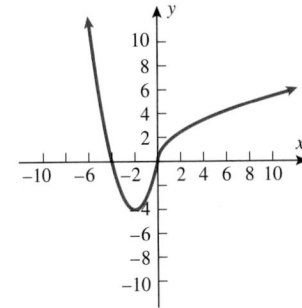

35.   **36.**

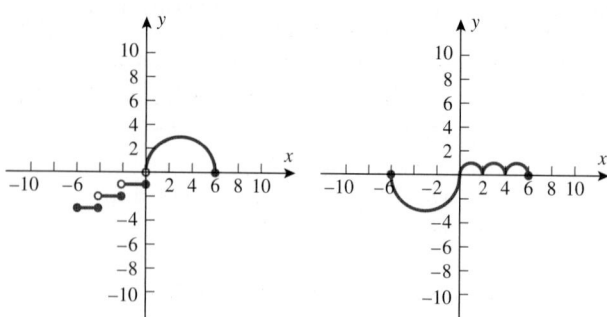

In Exercises 37–40, find the difference quotient $\dfrac{f(x+h)-f(x)}{h}$

for each function.

37. $f(x) = x^3 - 1$ **38.** $f(x) = \dfrac{x-1}{x+2}$

39. $f(x) = x + \dfrac{1}{x}$ **40.** $f(x) = \sqrt{\dfrac{x}{x+1}}$

41. Find the average rate of change of $f(x) = 4 - x^2$ from $x = 0$ to $x = 2$.

42. Find the average rate of change of $f(x) = |2x - 1|$ from $x = 1$ to $x = 5$.

Graph the piecewise-defined function. State the domain and range in interval notation.

43. $F(x) = \begin{cases} x^2 & x < 0 \\ 2 & x \geq 0 \end{cases}$

44. $f(x) = \begin{cases} -2x - 3 & x \leq 0 \\ 4 & 0 < x \leq 1 \\ x^2 + 4 & x > 1 \end{cases}$

45. $f(x) = \begin{cases} x^2 & x \leq 0 \\ -\sqrt{x} & 0 < x \leq 1 \\ |x + 2| & x > 1 \end{cases}$

46. $F(x) = \begin{cases} x^2 & x < 0 \\ x^3 & 0 < x < 1 \\ -|x| - 1 & x \geq 1 \end{cases}$

Applications

47. Housing Cost. In 2001 the market value of a house was $135,000; in 2006 the market price of the same house was $280,000. What is the average rate of the market price as a function of the time, where $t = 0$ corresponds to 2001.

48. Digital TV Conversion. A newspaper reported that by February 2009, only 38% of the urban population was ready for the conversion to digital TV. Ten weeks later, the newspaper reported that 64% of the population was prepared for the broadcasting change. Find the average rate of change of the population percent as a function of the time (in weeks).

1.3 Graphing Techniques: Transformations

Graph the following functions using graphing aids:

49. $y = -(x - 2)^2 + 4$ **50.** $y = |-x + 5| - 7$

51. $y = \sqrt[3]{x - 3} + 2$ **52.** $y = \dfrac{1}{x - 2} - 4$

53. $y = -\tfrac{1}{2}x^3$ **54.** $y = 2x^2 + 3$

Use the given graph to graph the following:

55.   **56.**

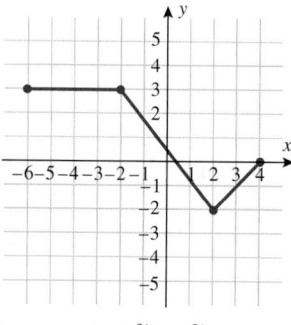 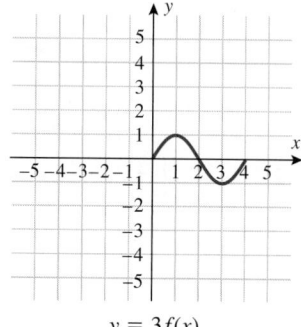

$y = f(x - 2)$ $y = 3f(x)$

57.   **58.**

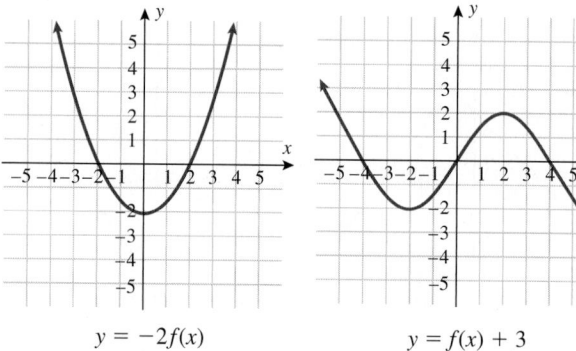

$y = -2f(x)$ $y = f(x) + 3$

Write the function whose graph is the graph of $y = \sqrt{x}$, but is transformed accordingly, and state the domain of the resulting function.

59. Shifted to the left 3 units

60. Shifted down 4 units

61. Shifted to the right 2 units and up 3 units

62. Reflected about the y-axis

63. Stretched by a factor of 5 and shifted vertically down 6 units

64. Compressed by a factor of 2 and shifted vertically up 3 units

Transform the function into the form $f(x) = c(x - h)^2 + k$ by completing the square and graph the resulting function using transformations.

65. $y = x^2 + 4x - 8$ **66.** $y = 2x^2 + 6x - 5$

1.4 Combining Functions

Given the functions g and h, find $g + h$, $g - h$, $g \cdot h$, and $\dfrac{g}{h}$, and state the domain.

67. $g(x) = -3x - 4$

$\quad h(x) = x - 3$

68. $g(x) = 2x + 3$

$\quad h(x) = x^2 + 6$

69. $g(x) = \dfrac{1}{x^2}$

$\quad h(x) = \sqrt{x}$

70. $g(x) = \dfrac{x + 3}{2x - 4}$

$\quad h(x) = \dfrac{3x - 1}{x - 2}$

71. $g(x) = \sqrt{x - 4}$

$\quad h(x) = \sqrt{2x + 1}$

72. $g(x) = x^2 - 4$

$\quad h(x) = x + 2$

For the given functions f and g, find the composite functions $f \circ g$ and $g \circ f$, and state the domains.

73. $f(x) = 3x - 4$

$\quad g(x) = 2x + 1$

74. $f(x) = x^3 + 2x - 1$

$\quad g(x) = x + 3$

75. $f(x) = \dfrac{2}{x + 3}$

$\quad g(x) = \dfrac{1}{4 - x}$

76. $f(x) = \sqrt{2x^2 - 5}$

$\quad g(x) = \sqrt{x + 6}$

77. $f(x) = \sqrt{x - 5}$

$\quad g(x) = x^2 - 4$

78. $f(x) = \dfrac{1}{\sqrt{x}}$

$\quad g(x) = \dfrac{1}{x^2 - 4}$

Evaluate $f(g(3))$ and $g(f(-1))$, if possible.

79. $f(x) = 4x^2 - 3x + 2$

$\quad g(x) = 6x - 3$

80. $f(x) = \sqrt{4 - x}$

$\quad g(x) = x^2 + 5$

81. $f(x) = \dfrac{x}{|2x - 3|}$

$\quad g(x) = |5x + 2|$

82. $f(x) = \dfrac{1}{x - 1}$

$\quad g(x) = x^2 - 1$

83. $f(x) = x^2 - x + 10$

$\quad g(x) = \sqrt[3]{x - 4}$

84. $f(x) = \dfrac{4}{x^2 - 2}$

$\quad g(x) = \dfrac{1}{x^2 - 9}$

Write the function as a composite $f(g(x))$ of two functions f and g.

85. $h(x) = 3(x - 2)^2 + 4(x - 2) + 7$

86. $h(x) = \dfrac{\sqrt[3]{x}}{1 - \sqrt[3]{x}}$

87. $h(x) = \dfrac{1}{\sqrt{x^2 + 7}}$

88. $h(x) = \sqrt{|3x + 4|}$

Applications

89. **Rain.** A rain drop hitting a lake makes a circular ripple. If the radius, in inches, grows as a function of time, in minutes, $r(t) = 25\sqrt{t + 2}$, find the area of the ripple as a function of time.

90. **Geometry.** Let the area of a rectangle be given by $42 = l \cdot w$, and let the perimeter be $36 = 2 \cdot l + 2 \cdot w$. Express the perimeter in terms of w.

1.5 One-to-One Functions and Inverse Functions

Determine whether the given function is a one-to-one function.

91. $\{(-2, 0), (4, 5), (3, 7)\}$

92. $\{(-8, -6), (-4, 2), (0, 3), (2, -8), (7, 4)\}$

93. $y = \sqrt{x}$ **94.** $y = x^2$ **95.** $f(x) = x^3$ **96.** $f(x) = \dfrac{1}{x^2}$

In Exercises 97–100, determine whether the function is one-to-one.

97.

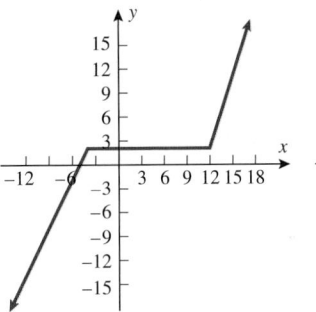

98.

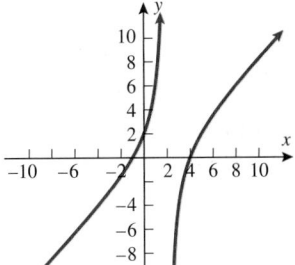

99.

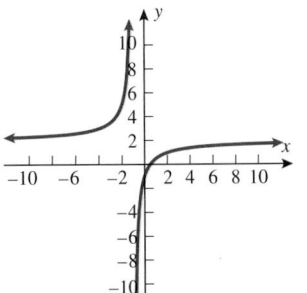

100.

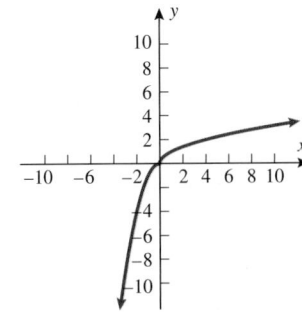

Verify that the function $f^{-1}(x)$ is the inverse of $f(x)$ by showing that $f(f^{-1}(x)) = x$. Graph $f(x)$ and $f^{-1}(x)$ on the same graph and show the symmetry about the line $y = x$.

101. $f(x) = 3x + 4$; $f^{-1}(x) = \dfrac{x - 4}{3}$

102. $f(x) = \dfrac{1}{4x - 7}$; $f^{-1}(x) = \dfrac{1 + 7x}{4x}$

103. $f(x) = \sqrt{x + 4}; f^{-1}(x) = x^2 - 4 \quad x \geq 0$

104. $f(x) = \dfrac{x + 2}{x - 7}; f^{-1}(x) = \dfrac{7x + 2}{x - 1}$

The function f is one-to-one. Find its inverse and check your answer. State the domain and range of both f and f^{-1}.

105. $f(x) = 2x + 1$

106. $f(x) = x^5 + 2$

107. $f(x) = \sqrt{x + 4}$

108. $f(x) = (x + 4)^2 + 3 \quad x \geq -4$

109. $f(x) = \dfrac{x + 6}{x + 3}$

110. $f(x) = 2\sqrt[3]{x - 5} - 8$

Applications

111. Salary. A pharmaceutical salesperson makes $22,000 base salary a year plus 8% of the total products sold. Write a function $S(x)$ that represents her yearly salary as a function of the total dollars x worth of products sold. Find $S^{-1}(x)$. What does this inverse function tell you?

112. Volume. Express the volume V of a rectangular box that has a square base of length s and is 3 feet high as a function of the square length. Find V^{-1}. If a certain volume is desired, what does the inverse tell you?

Technology Exercises

Section 1.1

113. Use a graphing utility to graph the function and find the domain. Express the domain in interval notation.

$$f(x) = \frac{1}{\sqrt{x^2 - 2x - 3}}$$

114. Use a graphing utility to graph the function and find the domain. Express the domain in interval notation.

$$f(x) = \frac{x^2 - 4x - 5}{x^2 - 9}$$

Section 1.2

115. Use a graphing utility to graph the function. State the (a) domain, (b) range, and (c) x intervals where the function is increasing, decreasing, and constant.

$$f(x) = \begin{cases} 1 - x & x < -1 \\ [[x]] & -1 \leq x < 2 \\ x + 1 & x > 2 \end{cases}$$

116. Use a graphing utility to graph the function. State the (a) domain, (b) range, and (c) x intervals where the function is increasing, decreasing, and constant.

$$f(x) = \begin{cases} |x^2 - 1| & -2 < x < 2 \\ \sqrt{x - 2} + 4 & x > 2 \end{cases}$$

Section 1.3

117. Use a graphing utility to graph $f(x) = x^2 - x - 6$ and $g(x) = x^2 - 5x$. Use transforms to describe the relationship between $f(x)$ and $g(x)$?

118. Use a graphing utility to graph $f(x) = 2x^2 - 3x - 5$ and $g(x) = -2x^2 - x + 6$. Use transforms to describe the relationship between $f(x)$ and $g(x)$?

Section 1.4

119. Using a graphing utility, plot $y_1 = \sqrt{2x + 3}$, $y_2 = \sqrt{4 - x}$, and $y_3 = \dfrac{y_1}{y_2}$. What is the domain of y_3?

120. Using a graphing utility, plot $y_1 = \sqrt{x^2 - 4}$, $y_2 = x^2 - 5$, and $y_3 = y_1^2 - 5$. If y_1 represents a function f and y_2 represents a function g, then y_3 represents the composite function $g \circ f$. The graph of y_3 is only defined for the domain of $g \circ f$. State the domain of $g \circ f$.

Section 1.5

121. Use a graphing utility to graph the function and determine whether it is one-to-one.

$$f(x) = \frac{6}{\sqrt[5]{x^3 - 1}}$$

122. Use a graphing utility to graph the functions f and g and the line $y = x$ in the same screen. Are the two functions inverses of each other?

$$f(x) = \sqrt[4]{x - 3} + 1, \, g(x) = x^4 - 4x^3 + 6x^2 - 4x + 3$$

Assuming that x represents the independent variable and y represents the dependent variable, classify the relationships as:

a. not a function
b. a function, but not one-to-one
c. a one-to-one function

1. $f(x) = |2x + 3|$ **2.** $x = y^2 + 2$ **3.** $y = \sqrt[3]{x + 1}$

Use $f(x) = \sqrt{x - 2}$ and $g(x) = x^2 + 11$, and determine the desired quantity or expression. In the case of an expression, state the domain.

4. $f(11) - 2g(-1)$

5. $\left(\dfrac{f}{g}\right)(x)$

6. $\left(\dfrac{g}{f}\right)(x)$

7. $g(f(x))$

8. $(f + g)(6)$

9. $f(g(\sqrt{7}))$

Determine whether the function is odd, even, or neither.

10. $f(x) = |x| - x^2$

11. $f(x) = 9x^3 + 5x - 3$

12. $f(x) = \dfrac{2}{x}$

Graph the functions. State the domain and range of each function.

13. $f(x) = -\sqrt{x - 3} + 2$

14. $f(x) = -2(x - 1)^2$

15. $f(x) = \begin{cases} -x & x < -1 \\ 1 & -1 < x < 2 \\ x^2 & x \geq 2 \end{cases}$

Use the graphs of the function to find:

16.

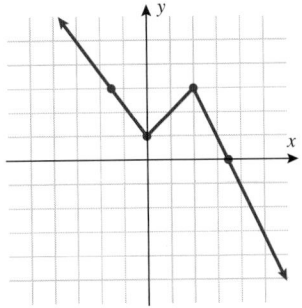

$y = f(x)$

a. $f(3)$ b. $f(0)$ c. $f(-4)$
d. x, where $f(x) = 3$ e. x, where $f(x) = 0$

17.

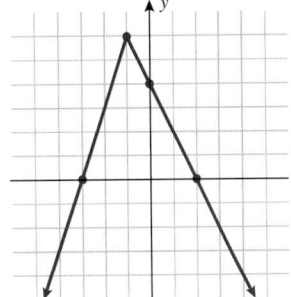

$y = g(x)$

a. $g(3)$ b. $g(0)$ c. $g(-4)$
d. x, where $g(x) = 0$

18.

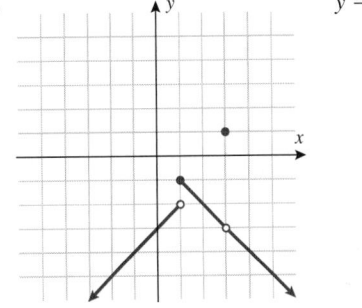

$y = f(x)$

a. $p(0)$ b. x, where $p(x) = 0$
c. $p(1)$ d. $p(3)$

Find $\dfrac{f(x + h) - f(x)}{h}$ for:

19. $f(x) = 3x^2 - 4x + 1$ **20.** $f(x) = x^3 - \dfrac{1}{\sqrt{x}}$

Find the average rate of change of the given functions.

21. $f(x) = 64 - 16x^2$ for $x = 0$ to $x = 2$

22. $f(x) = \sqrt{x - 1}$ for $x = 2$ to $x = 10$

In Exercises 23–26, given the function f, find the inverse if it exists. State the domain and range of both f and f^{-1}.

23. $f(x) = \sqrt{x - 5}$ **24.** $f(x) = x^2 + 5$

25. $f(x) = \dfrac{2x + 1}{5 - x}$ **26.** $f(x) = \begin{cases} -x & x \leq 0 \\ -x^2 & x > 0 \end{cases}$

27. What domain restriction can be made so that $f(x) = x^2$ has an inverse?

28. If the point $(-2, 5)$ lies on the graph of a function, what point lies on the graph of its inverse function?

29. Pressure. A mini-submarine descends at a rate of 5 feet per second. The pressure on the submarine structure is a linear function of the depth; when the submarine is on the surface, the pressure is 10 pounds per square inch, and when 100 ft underwater, the pressure is 28 pounds per square inch. Write a function that describes the pressure P as a function of the time t in seconds.

30. Geometry. Both the volume V and surface area S of a sphere are functions of the radius R. Write the volume as a function of the surface area.

31. Circles. If a quarter circle is drawn by tracing the unit circle in quadrant III, what does the inverse of that function look like? Where is it located?

32. Sprinkler. A sprinkler head malfunctions at midfield in an NFL football field. The puddle of water forms a circular pattern around the sprinkler head with a radius in yards that grows as a function of time, in hours: $r(t) = 10\sqrt{t}$. When will the puddle reach the sidelines? (A football field is 30 yards from sideline to sideline.)

33. Internet. The cost of airport Internet access is $15 for the first 30 minutes and $1 per minute for each minute after that. Write a function describing the cost of the service as a function of minutes used.

34. Temperature and CO_2 Emissions. The following table shows average yearly temperature in degrees Fahrenheit °F and carbon dioxide emissions in parts per million (ppm) for Mauna Loa, Hawaii. Scientists discovered that both temperature and CO_2 emissions are linear functions of the time. Write a function that describes the temperature T as a function of the CO_2 emissions x. Use this function to determine the temperature when the CO_2 reaches the level of 375 ppm.

Year	2000	2005
Temperature (°F)	45.86	46.23
CO_2 emissions (ppm)	369.4	379.7

2

Polynomial and Rational Functions

The gas mileage you achieve (in whatever vehicle you drive) is a function of speed, which can be modeled by a *polynomial function*. The number of turning points in the graph of a polynomial function is related to the degree of that polynomial.

We can approximate the above trends with a simple second-degree polynomial function, called a *quadratic function*, whose graph is a parabola. We see from the graph why hypermilers do not drive above posted speed limits (and often drive below them).

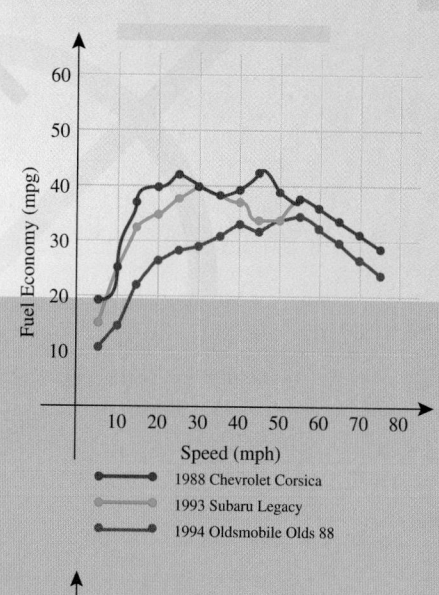

1988 Chevrolet Corsica
1993 Subaru Legacy
1994 Oldsmobile Olds 88

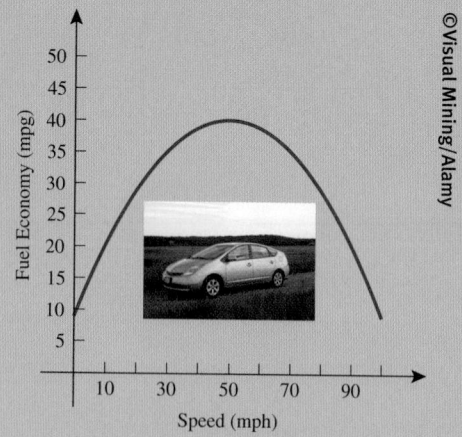

©Visual Mining/Alamy

IN THIS CHAPTER we will start by discussing quadratic functions (polynomial functions of degree 2), whose graphs are parabolas. We will find the vertex, which is the maximum or minimum point on the graph. Then we will expand our discussion to higher degree polynomial functions. We will discuss techniques to find zeros of polynomial functions and strategies for graphing polynomial functions. Lastly, we will discuss rational functions, which are ratios of polynomial functions.

POLYNOMIAL AND RATIONAL FUNCTIONS

2.1 Quadratic Functions	2.2 Polynomial Functions of Higher Degree	2.3 Dividing Polynomials	2.4 The Real Zeros of a Polynomial Function	2.5 Complex Zeros: The Fundamental Theorem of Algebra	2.6 Rational Functions
• Graphing Quadratic Functions in Standard Form • Graphing Quadratic Functions in General Form	• Graphs of Polynomial Functions • Real Zeros of a Polynomial Function • Graphing General Polynomial Functions	• Long Division of Polynomials • Synthetic Division of Polynomials	• The Remainder Theorem and the Factor Theorem • The Rational Zero Theorem and Descartes' Rule of Signs • Factoring Polynomials • The Intermediate Value Theorem • Graphing Polynomial Functions	• The Fundamental Theorem of Algebra • Complex Zeros • Factoring Polynomials	• Domain of Rational Functions • Vertical, Horizontal, and Slant Asymptotes • Graphing Rational Functions

CHAPTER OBJECTIVES

- Given a quadratic function in either standard or general form, find its vertex and sketch its graph (Section 2.1).
- Use multiplicity of zeros and end behavior as guides in sketching the graph of a polynomial function (Section 2.2).
- Divide polynomials using long division and understand when synthetic division can be used (Section 2.3).
- Find all real zeros of a polynomial function (x-intercepts) and use these as guides in sketching the graph of a polynomial function (Section 2.4).
- Factor a polynomial function completely over the set of complex numbers (Section 2.5).
- Use asymptotes and intercepts as guides in graphing rational functions (Section 2.6).

SKILLS OBJECTIVES

- Graph a quadratic function in standard form.
- Graph a quadratic function in general form.
- Find the equation of a parabola.
- Solve application problems that involve quadratic functions.

CONCEPTUAL OBJECTIVES

- Recognize characteristics of graphs of quadratic functions (parabolas):
  - whether the parabola opens up or down
  - whether the vertex is a maximum or minimum
  - the axis of symmetry

In Chapter 1, we studied functions in general. In this chapter, we will learn about a special group of functions called *polynomial functions*. Polynomial functions are simple functions; often, more complicated functions are approximated by polynomial functions. Polynomial functions model many real-world applications such as the stock market, football punts, business costs, revenues and profits, and the flight path of NASA's "vomit comet." Let's start by defining a polynomial function.

DEFINITION **Polynomial Function**

Let n be a nonnegative integer, and let $a_n, a_{n-1}, \ldots, a_2, a_1, a_0$ be real numbers with $a_n \neq 0$. The function

$$f(x) = a_n x^n + a_{n-1}x^{n-1} + \cdots + a_2 x^2 + a_1 x + a_0$$

is called a **polynomial function of x with degree n**. The coefficient a_n is called the **leading coefficient**, and a_0 is the constant.

Polynomials of particular degrees have special names. In Chapter 1, the library of functions included the constant function $f(x) = b$, which is a horizontal line; the linear function $f(x) = mx + b$, which is a line with slope m and y-intercept $(0, b)$; the square function $f(x) = x^2$; and the cube function $f(x) = x^3$. These are all special cases of a polynomial function.

In Section 1.3, we graphed functions using transformation techniques such as $F(x) = (x + 1)^2 - 2$, which can be graphed by starting with the square function $y = x^2$ and shifting 1 unit to the left and down 2 units. See the graph on the left.

Note that if we eliminate the parentheses in $F(x) = (x + 1)^2 - 2$ to get

$$\begin{aligned} F(x) &= x^2 + 2x + 1 - 2 \\ &= x^2 + 2x - 1 \end{aligned}$$

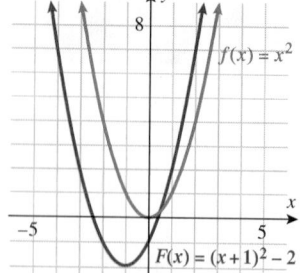

the result is a function defined by a second-degree polynomial (a polynomial with x^2 as the highest degree term), which is also called a *quadratic function*.

DEFINITION **Quadratic Function**

Let a, b, and c be real numbers with $a \neq 0$. The function

$$f(x) = ax^2 + bx + c$$

is called a **quadratic function**.

The graph of any quadratic function is a **parabola**. If the leading coefficient a is *positive*, then the parabola opens *upward*. If the leading coefficient a is *negative*, then the parabola opens *downward*. The **vertex** (or turning point) is the *minimum* point, or low point, on the graph if the parabola opens upward, whereas it is the *maximum* point, or high point, on the graph if the parabola opens downward. The vertical line that intersects the parabola at the vertex is called the **axis of symmetry**.

The axis of symmetry is the line $x = h$, and the vertex is located at the point (h, k), as shown in the following two figures:

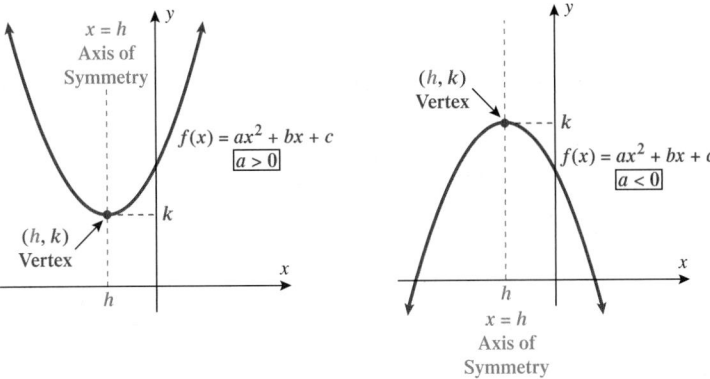

Graphing Quadratic Functions in Standard Form

In general, writing a quadratic function in the form

$$f(x) = a(x - h)^2 + k$$

allows the vertex (h, k) and the axis of symmetry $x = h$ to be determined by inspection. This form is a convenient way to express a quadratic function in order to quickly determine its corresponding graph. Hence, this form is called *standard form.*

QUADRATIC FUNCTION: STANDARD FORM

The quadratic function

$$f(x) = a(x - h)^2 + k$$

is in **standard form**. The graph of f is a parabola whose vertex is the point (h, k). The parabola is symmetric with respect to the line $x = h$. If $a > 0$, the parabola opens up. If $a < 0$, the parabola opens down.

Recall that graphing linear functions requires finding two points on the line, or a point and the slope of the line. However, for a quadratic function, simply knowing two points that lie on its graph is no longer sufficient. Below is a general step-by-step procedure for graphing quadratic functions given in standard form.

GRAPHING QUADRATIC FUNCTIONS

To graph $f(x) = a(x - h)^2 + k$

Step 1: Determine whether the parabola opens up or down.

$$a > 0 \quad \text{up}$$
$$a < 0 \quad \text{down}$$

Step 2: Determine the vertex (h, k).
Step 3: Find the y-intercept (by setting $x = 0$).
Step 4: Find any x-intercepts [by setting $f(x) = 0$ and solving for x].
Step 5: Plot the vertex and intercepts and connect them with a smooth curve.

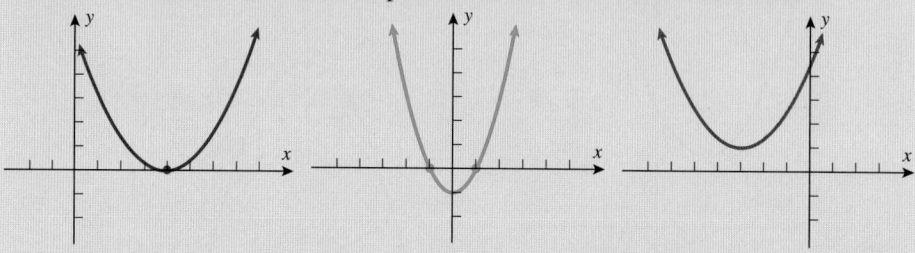

Note that Step 4 says to "find any x-intercepts." Parabolas opening up or down will always have a y-intercept. However, they can have **one**, **two**, or **no** x-intercepts. The figures above illustrate this for parabolas opening up, and the same can be said about parabolas opening down.

EXAMPLE 1   Graphing a Quadratic Function Given in Standard Form

Graph the quadratic function $f(x) = (x - 3)^2 - 1$.

Solution:

STEP 1 The parabola opens up. $a = 1$, so $a > 0$

STEP 2 Determine the vertex. $(h, k) = (3, -1)$

STEP 3 Find the y-intercept. $f(0) = (-3)^2 - 1 = 8$
                                        $(0, 8)$ corresponds to the y-intercept

STEP 4 Find any x-intercepts. $f(x) = (x - 3)^2 - 1 = 0$
                                        $(x - 3)^2 = 1$

　　　Use the square root method. $x - 3 = \pm 1$

　　　Solve. $x = 2 \quad \text{or} \quad x = 4$
                                        $(2, 0)$ and $(4, 0)$ correspond to the x-intercepts

STEP 5 Plot the vertex and intercepts
        $(3, -1)$, $(0, 8)$, $(2, 0)$, $(4, 0)$.

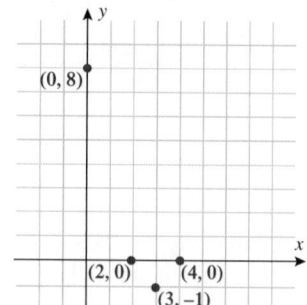

Connect the points with a smooth curve opening up.

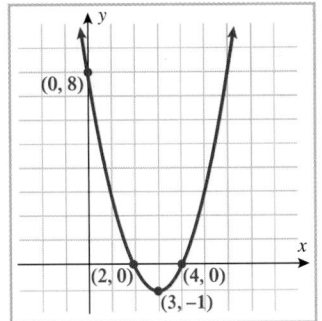

The graph in Example 1 could also have been found by shifting the square function to the right 3 units and down 1 unit.

■ **YOUR TURN** Graph the quadratic function $f(x) = (x - 1)^2 - 4$.

■ **Answer:**

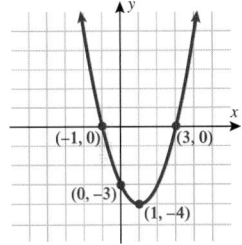

EXAMPLE 2 **Graphing a Quadratic Function Given in Standard Form with a Negative Leading Coefficient**

Graph the quadratic function $f(x) = -2(x - 1)^2 - 3$.

Solution:

STEP 1 The parabola opens down. $a = -2$, so $a < 0$

STEP 2 Determine the vertex. $(h, k) = (1, -3)$

STEP 3 Find the y-intercept. $f(0) = -2(-1)^2 - 3 = -2 - 3 = -5$
$(0, -5)$ corresponds to the y-intercept

STEP 4 Find any x-intercepts. $f(x) = -2(x - 1)^2 - 3 = 0$
$$-2(x - 1)^2 = 3$$
$$(x - 1)^2 = -\frac{3}{2}$$

A real quantity squared cannot be negative so there are no real solutions. There are no x-intercepts.

STEP 5 Plot the vertex $(1, -3)$ and y-intercept $(0, -5)$. Connect the points with a smooth curve.

Note that the axis of symmetry is $x = 1$. Because the point $(0, -5)$ lies on the parabola, then by symmetry with respect to $x = 1$, the point $(2, -5)$ also lies on the graph.

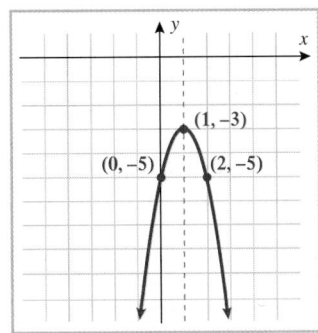

■ **Answer:**

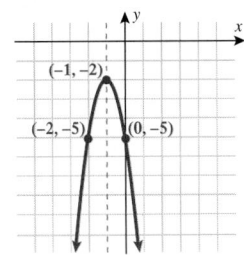

■ **YOUR TURN** Graph the quadratic function $f(x) = -3(x + 1)^2 - 2$.

When graphing quadratic functions (parabolas), have *at least 3 points* labeled on the graph.

- When there are *x*-intercepts (Example 1), label the vertex, *y*-intercept, and *x*-intercepts.

- When there are no *x*-intercepts (Example 2), label the vertex, *y*-intercept, and another point.

Graphing Quadratic Functions in General Form

A quadratic function is often written in one of two forms:

$$\text{Standard form: } f(x) = a(x - h)^2 + k$$
$$\text{General form: } f(x) = ax^2 + bx + c$$

When the quadratic function is expressed in standard form, the graph is easily obtained by identifying the vertex (h, k) and the intercepts and drawing a smooth curve that opens either up or down, depending on the sign of a.

Typically, quadratic functions are expressed in general form and a graph is the ultimate goal, so we must first express the quadratic function in standard form. One technique for transforming a quadratic function from general form to standard form was reviewed in Section 0.2 and is called *completing the square*.

EXAMPLE 3   Graphing a Quadratic Function Given in General Form

Write the quadratic function $f(x) = x^2 - 6x + 4$ in standard form and graph f.

Solution:

Express the quadratic function in standard form by completing the square.

Write the original function.	$f(x) = x^2 - 6x + 4$
Group the variable terms together.	$= (x^2 - 6x) + 4$

Complete the square.

Half of -6 is -3; -3 squared is 9.

Add and subtract 9 within the parentheses.	$= (x^2 - 6x + 9 - 9) + 4$
Write the -9 outside the parentheses.	$= (x^2 - 6x + 9) - 9 + 4$
Write the expression inside the parentheses as a perfect square and simplify.	$= (x - 3)^2 - 5$

Now that the quadratic function is written in standard form, $f(x) = (x - 3)^2 - 5$, we follow our step-by-step procedure for graphing a quadratic function in standard form.

STEP 1 The parabola opens up. $a = 1$, so $a > 0$

STEP 2 Determine the vertex. $(h, k) = (3, -5)$

Technology Tip

Use a graphing utility to graph the function $f(x) = x^2 - 6x + 4$ as y_1.

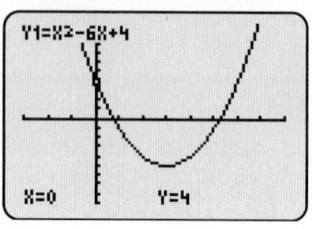

STEP 3 Find the *y*-intercept.

$$f(0) = (0)^2 - 6(0) + 4 = 4$$
$(0, 4)$ corresponds to the *y*-intercept

STEP 4 Find any *x*-intercepts.

$$f(x) = 0$$
$$f(x) = (x - 3)^2 - 5 = 0$$
$$(x - 3)^2 = 5$$
$$x - 3 = \pm\sqrt{5}$$
$$x = 3 \pm \sqrt{5}$$

$(3 + \sqrt{5}, 0)$ and $(3 - \sqrt{5}, 0)$ correspond to the *x*-intercepts.

STEP 5 Plot the vertex and intercepts
$(3, -5), (0, 4), (3 + \sqrt{5}, 0)$,
and $(3 - \sqrt{5}, 0)$.

Connect the points with a smooth
parabolic curve.

Note: $3 + \sqrt{5} \approx 5.24$ and
$3 - \sqrt{5} \approx 0.76$.

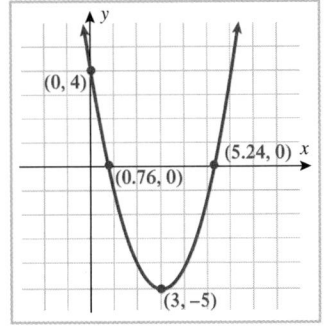

Study Tip

Although either form (standard or general) can be used to find the intercepts, it is often more convenient to use the general form when finding the *y*-intercept and the standard form when finding the *x*-intercept.

■ **Answer:**

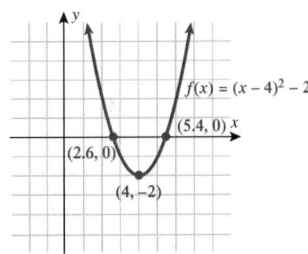

■ **YOUR TURN** Write the quadratic function $f(x) = x^2 - 8x + 14$ in standard form and graph f.

EXAMPLE 4 Graphing a Quadratic Function Given in General Form with a Negative Leading Coefficient

Graph the quadratic function $f(x) = -3x^2 + 6x + 2$.

Solution:

Express the function in standard form by completing the square.

Write the original function. $\qquad f(x) = -3x^2 + 6x + 2$

Group the variable terms together. $\qquad = (-3x^2 + 6x) + 2$

-3 is a common factor. $\qquad = -3(x^2 - 2x) + 2$

Add and subtract 1 inside the
parentheses to create a perfect square. $\qquad = -3(x^2 - 2x + 1 - 1) + 2$

Regroup the terms. $\qquad = -3(x^2 - 2x + 1) - 3(-1) + 2$

Write the expression inside the parentheses
as a perfect square and simplify. $\qquad = -3(x - 1)^2 + 5$

Now that the quadratic function is written in standard form, $f(x) = -3(x - 1)^2 + 5$, we follow our step-by-step procedure for graphing a quadratic function in standard form.

STEP 1 The parabola opens down. $\qquad a = -3$; therefore, $a < 0$

STEP 2 Determine the vertex. $\qquad (h, k) = (1, 5)$

STEP 3 Find the *y*-intercept using
the general form. $\qquad f(0) = -3(0)^2 + 6(0) + 2 = 2$
$(0, 2)$ corresponds to the *y*-intercept

Technology Tip

Use a graphing utility to graph the function $f(x) = -3x^2 + 6x + 2$ as y_1.

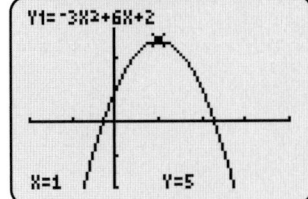

STEP 4 Find any x-intercepts using the standard form.

$$f(x) = -3(x - 1)^2 + 5 = 0$$
$$-3(x - 1)^2 = -5$$
$$(x - 1)^2 = \frac{5}{3}$$
$$x - 1 = \pm\sqrt{\frac{5}{3}}$$
$$x = 1 \pm \sqrt{\frac{5}{3}}$$
$$= 1 \pm \frac{\sqrt{15}}{3}$$

The x-intercepts are $\left(1 + \frac{\sqrt{15}}{3}, 0\right)$ and $\left(1 - \frac{\sqrt{15}}{3}, 0\right)$.

STEP 5 Plot the vertex and intercepts

$(1, 5), (0, 2), \left(1 + \frac{\sqrt{15}}{3}, 0\right)$, and

$\left(1 - \frac{\sqrt{15}}{3}, 0\right)$.

Connect the points with a smooth curve.

Note: $1 + \frac{\sqrt{15}}{3} \approx 2.3$ and

$1 - \frac{\sqrt{15}}{3} \approx -0.3.$

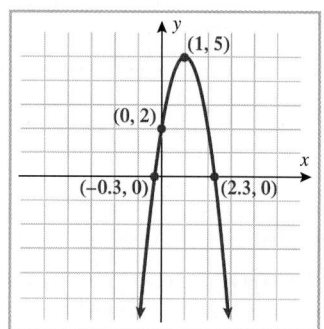

■ **Answer:**

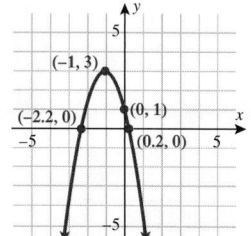

■ **YOUR TURN** Graph the quadratic function $f(x) = -2x^2 - 4x + 1$.

In Examples 3 and 4, the quadratic functions were given in general form and they were transformed into standard form by completing the square. It can be shown (by completing the square) that the vertex of a quadratic function in general form, $f(x) = ax^2 + bx + c$, is located at $x = -\frac{b}{2a}$.

VERTEX OF A PARABOLA

The graph of a quadratic function $f(x) = ax^2 + bx + c$ is a parabola with the **vertex** located at the point

$$\left(-\frac{b}{2a}, f\left(-\frac{b}{2a}\right)\right)$$

GRAPHING A QUADRATIC FUNCTION IN GENERAL FORM

Step 1: Find the vertex.
Step 2: Determine whether the parabola opens up or down.
  ■ If $a > 0$, the parabola opens up.
  ■ If $a < 0$, the parabola opens down.
Step 3: Find additional points near the vertex.
Step 4: Sketch the graph with a parabolic curve.

EXAMPLE 5 **Graphing a Quadratic Function Given in General Form**

Sketch the graph of $f(x) = -2x^2 + 4x + 5$.

Solution: Let $a = -2$, $b = 4$, and $c = 5$.

STEP 1 Find the vertex.

$$x = -\frac{b}{2a} = -\frac{4}{2(-2)} = 1$$

$$f(1) = -2(1)^2 + 4(1) + 5 = 7$$

Vertex: $(1, 7)$

STEP 2 The parabola opens down.

$a = -2$

STEP 3 Find additional points near the vertex.

x	-1	0	1	2	3
$f(x)$	$f(-1) = -1$	$f(0) = 5$	$f(1) = 7$	$f(2) = 5$	$f(3) = -1$

STEP 4 Label the vertex and additional points then sketch the graph.

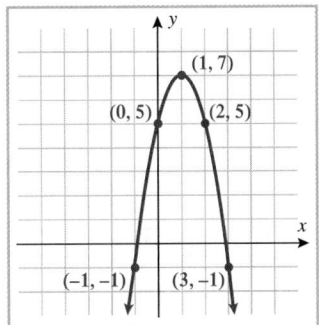

■ **YOUR TURN** Sketch the graph of $f(x) = 3x^2 - 6x + 4$.

■ **Answer:**

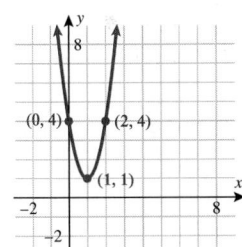

Finding the Equation of a Parabola

It is important to understand that the equation $y = x^2$ is equivalent to the quadratic function $f(x) = x^2$. Both have the same parabolic graph. Thus far, we have been given the function and then asked to find characteristics (vertex and intercepts) in order to graph. We now turn our attention to the problem of determining the function, given certain characteristics.

EXAMPLE 6 **Finding the Quadratic Function Given the Vertex and a Point That Lies on Its Graph**

Find the quadratic function whose graph has a vertex at $(3, 4)$ and which passes through the point $(2, 3)$. Express the quadratic function in both standard and general forms.

Solution:

Write the standard form of a quadratic function.

$$f(x) = a(x - h)^2 + k$$

Substitute the coordinates of the vertex $(h, k) = (3, 4)$.

$$f(x) = a(x - 3)^2 + 4$$

Use the point (2, 3) to find *a*.

The point (2, 3) implies $f(2) = 3$.

$$f(2) = a(2 - 3)^2 + 4 = 3$$

Solve for *a*.

$$a(2 - 3)^2 + 4 = 3$$
$$a(-1)^2 + 4 = 3$$
$$a + 4 = 3$$
$$a = -1$$

Write both forms of the quadratic function.

Standard form: $\boxed{f(x) = -(x - 3)^2 + 4}$ General form: $\boxed{f(x) = -x^2 + 6x - 5}$

■ **Answer:** Standard form:
$$f(x) = (x + 3)^2 - 5$$

■ **YOUR TURN** Find the standard form of the equation of a parabola whose graph has a vertex at $(-3, -5)$ and which passes through the point $(-2, -4)$.

As we have seen in Example 6, once the vertex is known, the leading coefficient *a* can be found from any point that lies on the parabola.

Application Problems That Involve Quadratic Functions

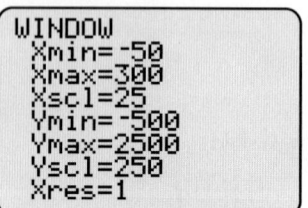

Technology Tip

Use a graphing utility to graph the cost function
$C(x) = 2000 - 15x + 0.05x^2$ as y_1.

```
WINDOW
 Xmin=-50
 Xmax=300
 Xscl=25
 Ymin=-500
 Ymax=2500
 Yscl=250
 Xres=1
```

```
Plot1 Plot2 Plot3
\Y1 ▄2000-15X+.05
X²
\Y2=■
```

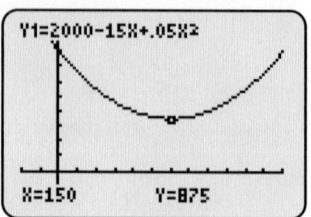

```
Y1=2000-15X+.05X²
```

X=150   Y=875

■ **Answer:** 250,000 bottles

Because the vertex of a parabola represents either the minimum or maximum value of the quadratic function, in application problems it often suffices simply to find the vertex.

EXAMPLE 7   Finding the Minimum Cost of Manufacturing a Motorcycle

A company that produces motorcycles has a per unit production cost of

$$C(x) = 2000 - 15x + 0.05x^2$$

where *C* is the cost in dollars to manufacture a motorcycle and *x* is the number of motorcycles produced. How many motorcycles should be produced in order to minimize the cost of each motorcycle? What is the corresponding minimum cost?

Solution:

The graph of the quadratic function is a parabola.

Rewrite the quadratic function in general form. $C(x) = 0.05x^2 - 15x + 2000$

The parabola opens up, because *a* is positive. $a = 0.05 > 0$

Because the parabola opens up, the vertex of the parabola is a *minimum*.

Find the *x*-coordinate of the vertex. $x = -\dfrac{b}{2a} = -\dfrac{(-15)}{2(0.05)} = 150$

The company keeps per unit cost to a minimum when 150 motorcycles are produced.

The minimum cost is $875 per motorcycle. $C(150) = 875$

■ **YOUR TURN** The revenue associated with selling vitamins is

$$R(x) = 500x - 0.001x^2$$

where *R* is the revenue in dollars and *x* is the number of bottles of vitamins sold. Determine how many bottles of vitamins should be sold to maximize the revenue.

EXAMPLE 8   Finding the Dimensions That Yield a Maximum Area

You have just bought a puppy and want to fence in an area in the backyard for her. You buy 100 linear feet of fence from Home Depot and have decided to make a rectangular fenced-in area using the back of your house as one side. Determine the dimensions of the rectangular pen that will maximize the area in which your puppy may roam. What is the maximum area of the rectangular pen?

Solution:

STEP 1 **Identify the question.**

Find the dimensions of the rectangular pen.

STEP 2 **Draw a picture.**

STEP 3 **Set up a function.**

If we let x represent the length of one side of the rectangle, then the opposite side is also of length x. Because there are 100 feet of fence, the remaining fence left for the side opposite the house is $100 - 2x$.

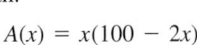

The area of a rectangle is equal to length times width:

$$A(x) = x(100 - 2x)$$

STEP 4 **Find the maximum value of the function.**

$$A(x) = x(100 - 2x) = -2x^2 + 100x$$

Find the maximum of the parabola that corresponds to the quadratic function for area $A(x) = -2x^2 + 100x$.

$a = -2$ and $b = 100$; therefore, the maximum occurs when

$$x = -\frac{b}{2a} = -\frac{100}{2(-2)} = 25$$

Replacing x with 25 in our original diagram:

The dimensions of the rectangle are

$\boxed{25 \text{ feet by } 50 \text{ feet}}$.

The maximum area $A(25) = 1250$ is

$\boxed{1250 \text{ square feet}}$.

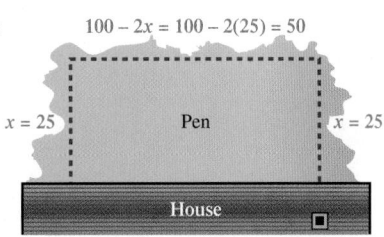

STEP 5 **Check the solution.**

Two sides are 25 feet and one side is 50 feet, and together they account for all 100 feet of fence.

■ **YOUR TURN** Suppose you have 200 linear feet of fence to enclose a rectangular garden. Determine the dimensions of the rectangle that will yield the greatest area.

Technology Tip

Use a graphing utility to graph the area function $A(x) = -2x^2 + 100x$.

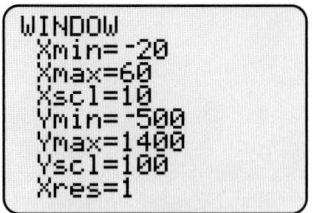

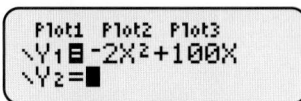

The maximum occurs when $x = 25$. The maximum area is $y = 1250$ square feet.

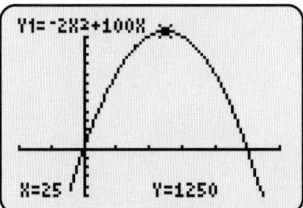

A table of values supports the solution.

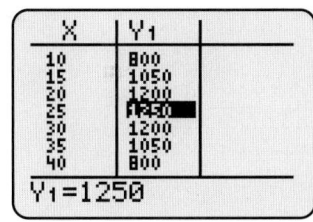

■ **Answer:** 50 feet by 50 feet

EXAMPLE 9   Path of a Punted Football

The path of a particular punt follows the quadratic function: $h(x) = -\frac{1}{8}(x - 5)^2 + 50$, where $h(x)$ is the height of the ball in yards and x corresponds to the horizontal distance in yards. Assume $x = 0$ corresponds to midfield (the 50 yard line). For example, $x = -20$ corresponds to the punter's own 30 yard line, whereas $x = 20$ corresponds to the other team's 30 yard line.

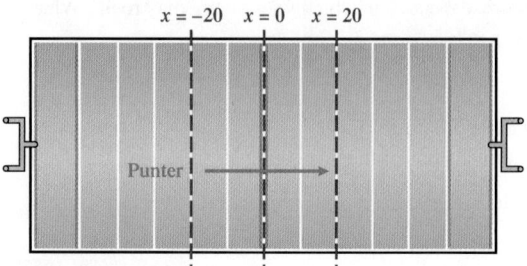

$x = -20$ $x = 0$ $x = 20$

Punter

a. Find the maximum height the ball achieves.
b. Find the horizontal distance the ball covers. Assume the height is zero when the ball is kicked and when the ball is caught.

Solution (a):

Identify the vertex since it is given in standard form. $\qquad\qquad (h, k) = (5, 50)$

The maximum height of the punt occurs at the other team's 45 yard line, and the height the ball achieves is 50 yards (150 feet) .

Solution (b):

The height when the ball is kicked
or caught is zero.

$$h(x) = -\frac{1}{8}(x - 5)^2 + 50 = 0$$

Solve for x.

$$\frac{1}{8}(x - 5)^2 = 50$$
$$(x - 5)^2 = 400$$
$$(x - 5) = \pm\sqrt{400}$$
$$x = 5 \pm 20$$
$$x = -15 \quad\text{and}\quad x = 25$$

The horizontal distance is the distance between these two points: $|25 - (-15)| =$ 40 yards .

SECTION 2.1   SUMMARY

All quadratic functions $f(x) = ax^2 + bx + c$ or $f(x) = a(x - h)^2 + k$ have graphs that are parabolas:

- If $a > 0$, the parabola opens up.
- If $a < 0$, the parabola opens down.
- The vertex is at the point

$$(h, k) = \left(-\frac{b}{2a}, f\left(-\frac{b}{2a}\right)\right) = \left(-\frac{b}{2a}, \frac{4ac - b^2}{4a}\right)$$

- When the quadratic function is given in general form, completing the square can be used to rewrite the function in standard form.
- At least three points are needed to graph a quadratic function:
  - vertex
  - y-intercept
  - x-intercept(s) or other point(s)

SECTION 2.1 EXERCISES

▪ SKILLS

In Exercises 1–4, match the quadratic function with its graph.

1. $f(x) = 3(x + 2)^2 - 5$ **2.** $f(x) = 2(x - 1)^2 + 3$ **3.** $f(x) = -\frac{1}{2}(x + 3)^2 + 2$ **4.** $f(x) = -\frac{1}{3}(x - 2)^2 + 3$

a. 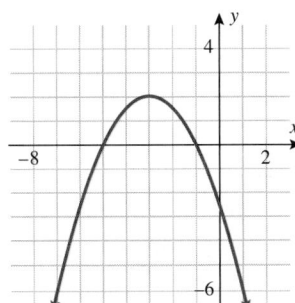 b. 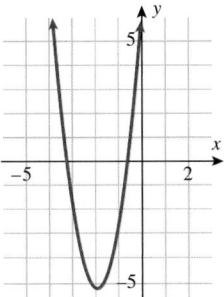 c. 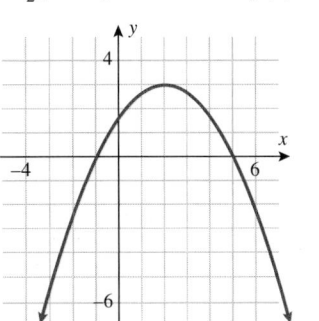 d.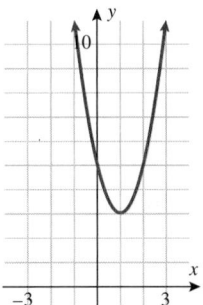

In Exercises 5–8, match the quadratic function with its graph.

5. $f(x) = 3x^2 + 5x - 2$ **6.** $f(x) = 3x^2 - x - 2$ **7.** $f(x) = -x^2 + 2x - 1$ **8.** $f(x) = -2x^2 - x + 3$

a. 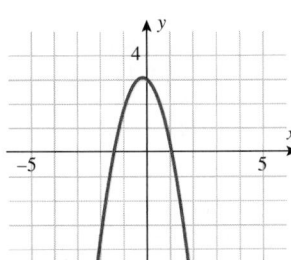 b. 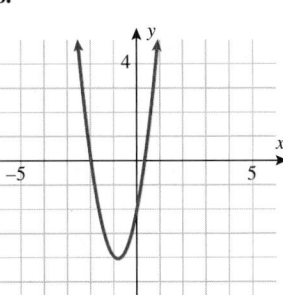 c. 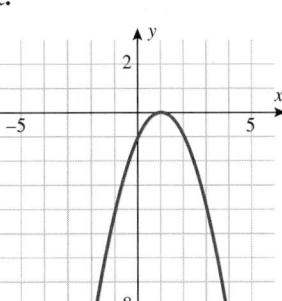 d.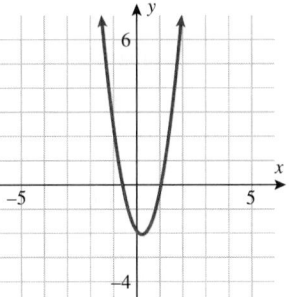

In Exercises 9–22, graph the quadratic function, which is given in standard form.

9. $f(x) = (x + 1)^2 - 2$ **10.** $f(x) = (x + 2)^2 - 1$ **11.** $f(x) = (x - 2)^2 - 3$

12. $f(x) = (x - 4)^2 + 2$ **13.** $f(x) = -(x - 3)^2 + 9$ **14.** $f(x) = -(x - 5)^2 - 4$

15. $f(x) = -(x + 1)^2 - 3$ **16.** $f(x) = -(x - 2)^2 + 6$ **17.** $f(x) = 2(x - 2)^2 + 2$

18. $f(x) = -3(x + 2)^2 - 15$ **19.** $f(x) = \left(x - \frac{1}{3}\right)^2 + \frac{1}{9}$ **20.** $f(x) = \left(x + \frac{1}{4}\right)^2 - \frac{1}{2}$

21. $f(x) = -0.5(x - 0.25)^2 + 0.75$ **22.** $f(x) = -0.2(x + 0.6)^2 + 0.8$

In Exercises 23–34, rewrite the quadratic function in standard form by completing the square.

23. $f(x) = x^2 + 6x - 3$ **24.** $f(x) = x^2 + 8x + 2$ **25.** $f(x) = -x^2 - 10x + 3$

26. $f(x) = -x^2 - 12x + 6$ **27.** $f(x) = 2x^2 + 8x - 2$ **28.** $f(x) = 3x^2 - 9x + 11$

29. $f(x) = -4x^2 + 16x - 7$ **30.** $f(x) = -5x^2 + 100x - 36$ **31.** $f(x) = x^2 + 10x$

32. $f(x) = -4x^2 + 12x - 2$ **33.** $f(x) = \frac{1}{2}x^2 - 4x + 3$ **34.** $f(x) = -\frac{1}{3}x^2 + 6x + 4$

In Exercises 35–44, graph the quadratic function.

35. $f(x) = x^2 + 6x - 7$ **36.** $f(x) = x^2 - 3x + 10$ **37.** $f(x) = -x^2 - 5x + 6$ **38.** $f(x) = -x^2 + 3x + 4$

39. $f(x) = 4x^2 - 5x + 10$ **40.** $f(x) = 3x^2 + 9x - 1$ **41.** $f(x) = -2x^2 - 12x - 16$ **42.** $f(x) = -3x^2 + 12x - 12$

43. $f(x) = \frac{1}{2}x^2 - \frac{1}{2}$ **44.** $f(x) = -\frac{1}{3}x^2 + \frac{4}{3}$

In Exercises 45–54, find the vertex of the parabola associated with each quadratic function.

45. $f(x) = 33x^2 - 2x + 15$ **46.** $f(x) = 17x^2 + 4x - 3$

47. $f(x) = \frac{1}{2}x^2 - 7x + 5$ **48.** $f(x) = -\frac{1}{3}x^2 + \frac{2}{5}x + 4$

49. $f(x) = -\frac{2}{5}x^2 + \frac{3}{7}x + 2$ **50.** $f(x) = -\frac{1}{7}x^2 - \frac{2}{3}x + \frac{1}{9}$

51. $f(x) = -0.002x^2 - 0.3x + 1.7$ **52.** $f(x) = 0.05x^2 + 2.5x - 1.5$

53. $f(x) = 0.06x^2 - 2.6x + 3.52$ **54.** $f(x) = -3.2x^2 + 0.8x - 0.14$

In Exercises 55–66, find the quadratic function that has the given vertex and goes through the given point.

55. vertex: $(-1, 4)$ point: $(0, 2)$ **56.** vertex: $(2, -3)$ point: $(0, 1)$ **57.** vertex: $(2, 5)$ point: $(3, 0)$

58. vertex: $(1, 3)$ point: $(-2, 0)$ **59.** vertex: $(-1, -3)$ point: $(-4, 2)$ **60.** vertex: $(0, -2)$ point: $(3, 10)$

61. vertex: $(-2, -4)$ point: $(-1, 6)$ **62.** vertex: $(5, 4)$ point: $(2, -5)$ **63.** vertex: $\left(\frac{1}{2}, -\frac{3}{4}\right)$ point: $\left(\frac{3}{4}, 0\right)$

64. vertex: $\left(-\frac{5}{6}, \frac{2}{3}\right)$ point: $(0, 0)$ **65.** vertex: $(2.5, -3.5)$ point: $(4.5, 1.5)$ **66.** vertex: $(1.8, 2.7)$ point: $(-2.2, -2.1)$

■ APPLICATIONS

Exercises 67 and 68 concern the path of a punted football. Refer to the diagram in Example 9.

67. Sports. The path of a particular punt follows the quadratic function

$$h(x) = -\frac{8}{125}(x + 5)^2 + 40$$

where $h(x)$ is the height of the ball in yards and x corresponds to the horizontal distance in yards. Assume $x = 0$ corresponds to midfield (the 50 yard line). For example, $x = -20$ corresponds to the punter's own 30 yard line, whereas $x = 20$ corresponds to the other team's 30 yard line.

a. Find the maximum height the ball achieves.

b. Find the horizontal distance the ball covers. Assume the height is zero when the ball is kicked and when the ball is caught.

68. Sports. The path of a particular punt follows the quadratic function

$$h(x) = -\frac{5}{40}(x - 30)^2 + 50$$

where $h(x)$ is the height of the ball in yards and x corresponds to the horizontal distance in yards. Assume $x = 0$ corresponds to midfield (the 50 yard line). For example, $x = -20$ corresponds to the punter's own 30 yard line, whereas $x = 20$ corresponds to the other team's 30 yard line.

a. Find the maximum height the ball achieves.

b. Find the horizontal distance the ball covers. Assume the height is zero when the ball is kicked and when the ball is caught.

69. Ranching. A rancher has 10,000 linear feet of fencing and wants to enclose a rectangular field and then divide it into two equal pastures with an internal fence parallel to one of the rectangular sides. What is the maximum area of each pasture? Round to the nearest square foot.

70. Ranching. A rancher has 30,000 linear feet of fencing and wants to enclose a rectangular field and then divide it into four equal pastures with three internal fences parallel to one of the rectangular sides. What is the maximum area of each pasture?

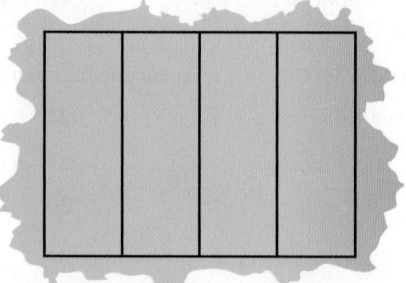

71. Gravity. A person standing near the edge of a cliff 100 feet above a lake throws a rock upward with an initial speed of 32 feet per second. The height of the rock above the lake at the bottom of the cliff is a function of time and is described by

$$h(t) = -16t^2 + 32t + 100$$

a. How many seconds will it take until the rock reaches its maximum height? What is that height?

b. At what time will the rock hit the water?

c. Over what time interval is the rock higher than the cliff?

100 feet

72. Gravity. A person holds a pistol straight upward and fires. The initial velocity of most bullets is around 1200 feet/second. The height of the bullet is a function of time and is described by

$$h(t) = -16t^2 + 1200t$$

How long, after the gun is fired, does the person have to get out of the way of the bullet falling from the sky?

73. Zero Gravity. As part of their training, astronauts ride the "vomit comet," NASA's reduced gravity KC 135A aircraft that performs parabolic flights to simulate weightlessness. The plane starts at an altitude of 20,000 feet and makes a steep climb at 52° with the horizon for 20–25 seconds and then dives at that same angle back down, repeatedly. The equation governing the altitude of the flight is

$$A(x) = -0.0003x^2 + 9.3x - 46,075$$

where $A(x)$ is altitude and x is horizontal distance in feet.

a. What is the maximum altitude the plane attains?

b. Over what horizontal distance is the entire maneuver performed? (Assume the starting and ending altitude is 20,000 feet.)

NASA's "Vomit Comet"

Courtesy NASA

74. Sports. A soccer ball is kicked from the ground at a 45° angle with an initial velocity of 40 feet per second. The height of the soccer ball above the ground is given by $H(x) = -0.0128x^2 + x$, where x is the horizontal distance the ball travels.

a. What is the maximum height the ball reaches?

b. What is the horizontal distance the ball travels?

75. Profit. A small company in Virginia Beach manufactures handcrafted surfboards. The profit of selling x boards is given by

$$P(x) = 20,000 + 80x - 0.4x^2$$

a. How many boards should be made to maximize the profit?

b. What is the maximum profit?

76. Environment: Fuel Economy. Gas mileage (miles per gallon, mpg) can be approximated by a quadratic function of speed. For a particular automobile, assume the vertex occurs when the speed is 50 mph (the mpg will be 30).

a. Write a quadratic function that models this relationship, assuming 70 mph corresponds to 25 mpg.

b. What gas mileage would you expect for this car driving 90 mph?

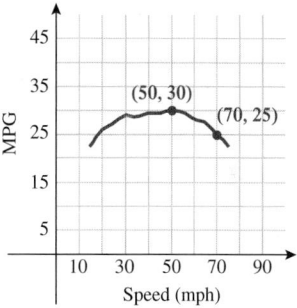

For Exercises 77 and 78, use the following information:

One function of particular interest in economics is the **profit function**. We denote this function by $P(x)$. It is defined to be the difference between revenue $R(x)$ and cost $C(x)$ so that

$$P(x) = R(x) - C(x)$$

The total revenue received from the sale of x goods at price p is given by

$$R(x) = px$$

The total cost function relates the cost of production to the level of output x. This includes both fixed costs C_f and variable costs C_v (costs per unit produced). The total cost in producing x goods is given by

$$C(x) = C_f + C_v x$$

Thus, the profit function is

$$P(x) = px - C_f - C_v x$$

Assume fixed costs are $1,000, variable costs per unit are $20, and the demand function is

$$p = 100 - x$$

77. Profit. How many units should the company produce to break even?

78. Profit. What is the maximum profit?

79. Cell Phones. The number of cell phones in the United States can be approximated by a quadratic function. In 1996 there were approximately 16 million cell phones, and in 2005 there were approximately 100 million. Let t be the number of years since 1996. The number of cell phones in 1996 is represented by $(0, 16)$, and the number in 2005 is $(9, 100)$. Let $(0, 16)$ be the vertex.

a. Find a quadratic function that represents the number of cell phones.

b. Based on this model, how many cell phones will be in use in 2010?

80. Underage Smoking. The number of underage cigarette smokers (ages 10–17) has declined in the United States. The peak percent was in 1998 at 49%. In 2006 this had dropped to 36%. Let t be time in years after 1998 ($t = 0$ corresponds to 1998).

a. Find a quadratic function that models the percent of underage smokers as a function of time. Let $(0, 49)$ be the vertex.

b. Now that you have the model, predict the percent of underage smokers in 2010.

81. Drug Concentration. The concentration of a drug in the bloodstream, measured in parts per million, can be modeled with a quadratic function. In 50 minutes the concentration is 93.75 parts per million. The maximum concentration of the drug in the bloodstream occurs in 225 minutes and is 400 parts per million.

a. Find a quadratic function that models the concentration of the drug as a function of time in minutes.

b. After the concentration peaks, eventually the drug will be eliminated from the body. How many minutes will it take until the concentration finally reaches 0?

82. Revenue. Jeff operates a mobile car washing business. When he charged $20 a car, he washed 70 cars a month. He raised the price to $25 a car and his business dropped to 50 cars a month.

a. Find a linear function that represents the demand equation (the price per car as a function of the number of cars washed).

b. Find the revenue function $R(x) = xp$.

c. How many cars should he wash to maximize the revenue?

d. What price should he charge to maximize revenue?

■ CATCH THE MISTAKE

In Exercises 83–86, explain the mistake that is made. There may be a single mistake or there may be more than one mistake.

83. Plot the quadratic function $f(x) = (x + 3)^2 - 1$.

Solution:

Step 1: The parabola opens up because $a = 1 > 0$.

Step 2: The vertex is $(3, -1)$.

Step 3: The y-intercept is $(0, 8)$.

Step 4: The x-intercepts are $(2, 0)$ and $(4, 0)$.

Step 5: Plot the vertex and intercepts, and connect the points with a smooth curve.

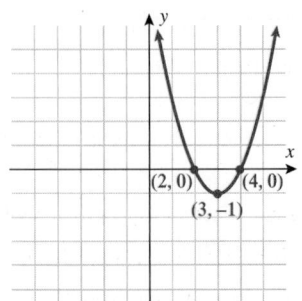

This is incorrect. What mistake(s) was made?

84. Determine the vertex of the quadratic function $f(x) = 2x^2 - 6x - 18$.

Solution:

Step 1: The vertex is given by $(h, k) = \left(-\dfrac{b}{2a}, f\left(-\dfrac{b}{2a}\right)\right)$.

In this case, $a = 2$ and $b = 6$.

Step 2: The x-coordinate of the vertex is

$$x = -\frac{6}{2(2)} = -\frac{6}{4} = -\frac{3}{2}$$

Step 3: The y-coordinate of the vertex is

$$f\left(-\frac{3}{2}\right) = -2\left(-\frac{3}{2}\right)^2 + 6\left(-\frac{3}{2}\right) - 18$$

$$= -2\left(\frac{9}{4}\right) - \frac{18}{2} - 18$$

$$= -\frac{9}{2} - 9 - 18$$

$$= -\frac{63}{2}$$

This is incorrect. What mistake(s) was made?

85. Rewrite the following quadratic function in standard form:

$$f(x) = -x^2 + 2x + 3$$

Solution:

Step 1: Group the variables together. $(-x^2 + 2x) + 3$

Step 2: Factor out a negative. $-(x^2 + 2x) + 3$

Step 3: Add and subtract 1 inside the parentheses. $-(x^2 + 2x + 1 - 1) + 3$

Step 4: Factor out the -1. $-(x^2 + 2x + 1) + 1 + 3$

Step 5: Simplify. $-(x + 1)^2 + 4$

This is incorrect. What mistake(s) was made?

86. Find the quadratic function whose vertex is $(2, -3)$ and whose graph passes through the point $(9, 0)$.

Solution:

Step 1: Write the quadratic function in standard form. $f(x) = a(x - h)^2 + k$

Step 2: Substitute $(h, k) = (2, -3)$. $f(x) = a(x - 2)^2 - 3$

Step 3: Substitute the point $(9, 0)$ and solve for a. $f(0) = a(0 - 2)^2 - 3 = 9$
$$4a - 3 = 9$$
$$4a = 12$$
$$a = 3$$

The quadratic function sought is $f(x) = 3(x - 2)^2 - 3$.

This is incorrect. What mistake(s) was made?

▪ CONCEPTUAL

In Exercises 87–90, determine whether each statement is true or false.

87. A quadratic function must have a y-intercept.

88. A quadratic function must have an x-intercept.

89. A quadratic function may have more than one y-intercept.

90. A quadratic function may have more than one x-intercept.

91. For the general quadratic equation, $f(x) = ax^2 + bx + c$, show that the vertex is $(h, k) = \left(-\dfrac{b}{2a}, f\left(-\dfrac{b}{2a}\right)\right)$.

92. Given the quadratic function $f(x) = a(x - h)^2 + k$, determine the x- and y-intercepts in terms of a, h, and k.

▪ CHALLENGE

93. A rancher has 1000 feet of fence to enclose a pasture.

   a. Determine the maximum area if a rectangular fence is used.

   b. Determine the maximum area if a circular fence is used.

94. A 600-room hotel in Orlando is filled to capacity every night when the rate is $90 per night. For every $5 increase in the rate, 10 fewer rooms are filled. How much should the hotel charge to produce the maximum income? What is its maximum income?

95. The speed of the river current is $\dfrac{1}{x + 4}$ mph. In quiet waters, the speed of a swimmer is $\dfrac{1}{x + 11}$ mph. When the swimmer swims down the river, her speed is $\frac{25}{144}$ mph. What is the value of x?

96. When a rectangle is reduced 25% (length and width each reduced by 25%), the new length equals the original width. Find the dimensions of the original rectangle given that the area of the reduced rectangle is 36 sq ft.

▪ TECHNOLOGY

97. On a graphing calculator, plot the quadratic function $f(x) = -0.002x^2 + 5.7x - 23$.

   a. Identify the vertex of this parabola.

   b. Identify the y-intercept.

   c. Identify the x-intercepts (if any).

   d. What is the axis of symmetry?

98. Determine the quadratic function whose vertex is $(-0.5, 1.7)$ and whose graph passes through the point $(0, 4)$.

   a. Write the quadratic function in general form.

   b. Plot this quadratic function with a graphing calculator.

   c. Zoom in on the vertex and y-intercept. Do they agree with the given values?

In Exercises 99 and 100, (a) use the calculator commands $\boxed{\text{STAT}}$ $\boxed{\text{QuadReg}}$ to model the data using a quadratic function; (b) write the quadratic function in standard form and identify the vertex; (c) plot this quadratic function with a graphing calculator and use the $\boxed{\text{TRACE}}$ key to highlight the given points. Do they agree with the given values?

99.

x	-2	2	5
y	-29.28	21.92	18.32

100.

x	-9	-2	4
y	-2.72	-16.18	6.62

▪ PREVIEW TO CALCULUS

Parabolas, ellipses, and hyperbolas form a family of curves called conic sections. These curves are studied later in Chapter 9 and in calculus. The general equation of each curve is given below:

Parabola: $(x - h)^2 = 4p(y - k)$

Ellipse: $\dfrac{(x - h)^2}{a^2} + \dfrac{(y - k)^2}{b^2} = 1$

Hyperbola: $\dfrac{(x - h)^2}{a^2} - \dfrac{(y - k)^2}{b^2} = 1$

In Exercises 101–104, write the general equation of each conic section and identify the curve.

101. $4x^2 + 9y^2 = 36y$

102. $x^2 + 16y = 4y^2 + 2x + 19$

103. $x^2 + 6x - 20y + 5 = 0$

104. $x^2 + 105 = 6x - 40y - 4y^2$

SECTION 2.2   POLYNOMIAL FUNCTIONS OF HIGHER DEGREE

SKILLS OBJECTIVES

- Graph polynomial functions using transformations.
- Identify real zeros of a polynomial function and their multiplicities.
- Determine the end behavior of a polynomial function.
- Sketch graphs of polynomial functions using intercepts and end behavior.

CONCEPTUAL OBJECTIVES

- Understand that real zeros of polynomial functions correspond to x-intercepts.
- Understand the intermediate value theorem and how it assists in graphing polynomial functions.
- Realize that end behavior is a result of the leading term dominating.

DEFINITION   Polynomial Function

Let n be a nonnegative integer and let $a_n, a_{n-1}, \ldots, a_2, a_1, a_0$ be real numbers with $a_n \neq 0$. The function

$$f(x) = a_n x^n + a_{n-1} x^{n-1} + \cdots + a_2 x^2 + a_1 x + a_0$$

is called a **polynomial function of x with degree n**. The coefficient a_n is called the leading coefficient.

EXAMPLE 1   Identifying Polynomials and Their Degree

For each of the functions given, determine whether the function is a polynomial function. If it is a polynomial function, then state the degree of the polynomial. If it is not a polynomial function, justify your answer.

a. $f(x) = 3 - 2x^5$ b. $F(x) = \sqrt{x} + 1$ c. $g(x) = 2$

d. $h(x) = 3x^2 - 2x + 5$ e. $H(x) = 4x^5(2x - 3)^2$ f. $G(x) = 2x^4 - 5x^3 - 4x^{-2}$

Solution:

a. $f(x)$ is a polynomial function of degree 5.

b. $F(x)$ is not a polynomial function. The variable x is raised to the power of $\frac{1}{2}$, which is not an integer.

c. $g(x)$ is a polynomial function of degree zero, also known as a constant function. Note that $g(x) = 2$ can also be written as $g(x) = 2x^0$ (assuming $x \neq 0$).

d. $h(x)$ is a polynomial function of degree 2. A polynomial function of degree 2 is called a quadratic function.

e. $H(x)$ is a polynomial function of degree 7. *Note:* $4x^5(4x^2 - 12x + 9) = 16x^7 - 48x^6 + 36x^5$.

f. $G(x)$ is not a polynomial function. $-4x^{-2}$ has an exponent that is negative.

■ **YOUR TURN** For each of the functions given, determine whether the function is a polynomial function. If it is a polynomial function, then state the degree of the polynomial. If it is not a polynomial function, justify your answer.

a. $f(x) = \dfrac{1}{x} + 2$ b. $g(x) = 3x^8(x - 2)^2(x + 1)^3$

■ **Answer:**
a. $f(x)$ is not a polynomial because x is raised to the power of -1, which is a negative integer.
b. $g(x)$ is a polynomial of degree 13.

Graphs of Polynomial Functions

Whenever we have discussed a particular polynomial function of degree 0, 1, or 2, we have graphed it too. These functions are summarized in the table below.

POLYNOMIAL	DEGREE	SPECIAL NAME	GRAPH
$f(x) = c$	0	Constant function	Horizontal line
$f(x) = mx + b$	1	Linear function	Line • Slope $= m$ • y-intercept: $(0, b)$
$f(x) = ax^2 + bx + c$	2	Quadratic function	Parabola • Opens up if $a > 0$. • Opens down if $a < 0$.

How do we graph polynomial functions that are of degree 3 or higher, and why do we care? Polynomial functions model real-world applications. One example is the percentage of fat in our bodies as we age. We can model the weight of a baby after it comes home from the hospital as a function of time. When a baby comes home from the hospital, it usually experiences weight loss. Then typically there is an increase in the percent of body fat when the baby is nursing. When infants start to walk, the increase in exercise is associated with a drop in the percentage of fat. Growth spurts in children are examples of the percent of body fat increasing and decreasing. Later in life, our metabolism slows down, and typically, the percent of body fat increases. We will model this with a polynomial function. Other examples are stock prices, the federal funds rate, and yo-yo dieting as functions of time.

Graphs of all polynomial functions are both *continuous* and *smooth*. A **continuous** graph is one you can draw completely without picking up your pencil (the graph has no jumps or holes). A **smooth** graph has no sharp corners. The following graphs illustrate what it means to be smooth (no sharp corners or cusps) and continuous (no holes or jumps).

The graph is *not continuous.*

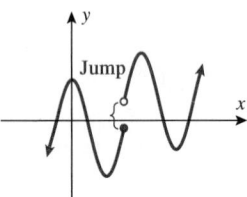

The graph is *not continuous.*

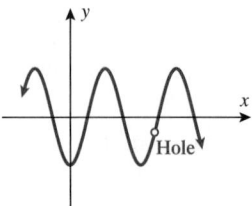

Study Tip

All polynomial functions have graphs that are both continuous and smooth.

The graph is *continuous* but *not smooth.*

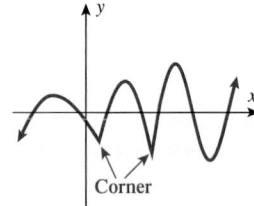

The graph is *continuous* and *smooth.*

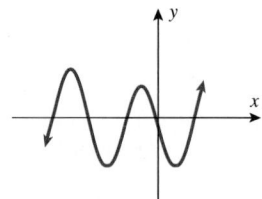

All polynomial functions have graphs that are both continuous and smooth. Recall from Chapter 1 that graphs of functions can be drawn by hand using graphing aids such as intercepts and symmetry. The graphs of polynomial functions can be graphed using these same aids. Let's start with the simplest types of polynomial functions, called **power functions**. Power functions are monomial functions (Appendix) of the form $f(x) = x^n$, where n is a positive integer.

DEFINITION   **Power Function**

Let n be a positive integer and the coefficient $a \neq 0$ be a real number. The function

$$f(x) = ax^n$$

is called a **power function of degree n.**

Power functions with *even* powers look similar to the square function.

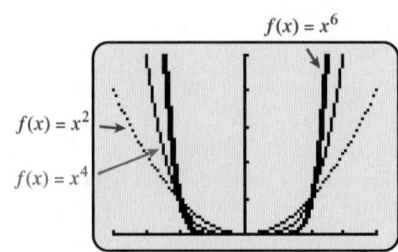

Power functions with *odd* powers (other than $n = 1$) look similar to the cube function.

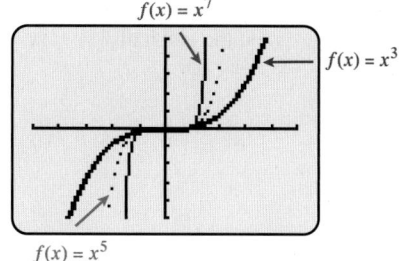

$$f(x) = x^7$$
$$f(x) = x^3$$
$$f(x) = x^5$$

All even power functions have similar characteristics to a quadratic function (parabola), and all odd ($n > 1$) power functions have similar characteristics to a cubic function. For example, all even functions are symmetric with respect to the *y*-axis, whereas all odd functions are symmetric with respect to the origin. This table summarizes their characteristics.

CHARACTERISTICS OF POWER FUNCTIONS: $f(x) = x^n$

	n EVEN	*n* ODD
Symmetry	*y*-axis	Origin
Domain	$(-\infty, \infty)$	$(-\infty, \infty)$
Range	$[0, \infty)$	$(-\infty, \infty)$
Some key points that lie on the graph	$(-1, 1)$, $(0, 0)$, and $(1, 1)$	$(-1, -1)$, $(0, 0)$, and $(1, 1)$
Increasing	$(0, \infty)$	$(-\infty, \infty)$
Decreasing	$(-\infty, 0)$	Nowhere

Graphing Polynomial Functions Using Transformations of Power Functions

We now have the tools to graph polynomial functions that are transformations of power functions. We will use the power functions combined with our graphing techniques such as horizontal and vertical shifting and reflection (Section 1.3).

EXAMPLE 2   Graphing Transformations of Power Functions

Graph the function $f(x) = (x - 1)^3$.

Solution:

STEP 1 Start with the graph of $y = x^3$.

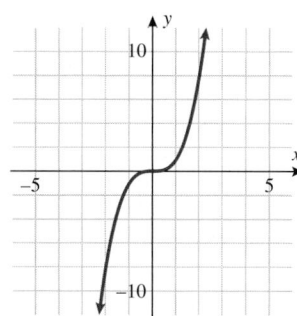

STEP 2 Shift $y = x^3$ to the right 1 unit to
yield the graph of $f(x) = (x - 1)^3$.

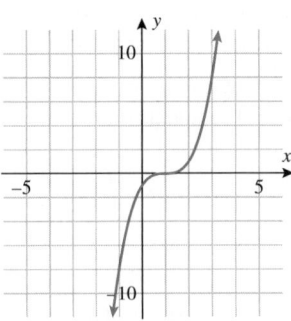

Answer: $f(x) = 1 - x^4$

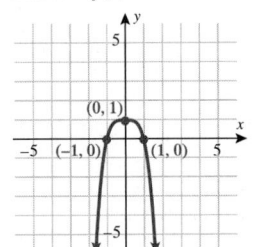

■ **YOUR TURN** Graph the function $f(x) = 1 - x^4$.

Real Zeros of a Polynomial Function

How do we graph general polynomial functions of degree greater than or equal to 3 if they cannot be written as transformations of power functions? We start by identifying the x-intercepts of the polynomial function. Recall that we determine the x-intercepts by setting the function equal to *zero* and solving for x. Therefore, an alternative name for an x-intercept of a function is a *zero* of the function. In our experience, to set a quadratic function equal to zero, the first step is to factor the quadratic expression into linear factors and then set each factor equal to zero. Therefore, there are four equivalent relationships that are summarized in the following box.

Study Tip

Real zeros correspond to x-intercepts.

REAL ZEROS OF POLYNOMIAL FUNCTIONS

If $f(x)$ is a polynomial function and a is a *real* number, then the following statements are equivalent.

- ■ $x = a$ is a **solution**, or **root**, of the equation $f(x) = 0$.
- ■ $(a, 0)$ is an **x-intercept** of the graph of $f(x)$.
- ■ $x = a$ is a **zero** of the function $f(x)$.
- ■ $(x - a)$ is a **factor** of $f(x)$.

Let's use a simple polynomial function to illustrate these four relationships. We'll focus on the quadratic function $f(x) = x^2 - 1$. The graph of this function is a parabola that opens up and has as its vertex the point $(0, -1)$.

SOLUTION	X-INTERCEPT		ZERO	FACTOR
$x = -1$ and $x = 1$ are solutions, or roots, of the equation $x^2 - 1 = 0$.	The x-intercepts correspond to the points $(-1, 0)$ and $(1, 0)$.	$f(x) = x^2 - 1$ $(-1, 0)$ $(1, 0)$ $(0, -1)$	$f(-1) = 0$ $f(1) = 0$	$f(x) = (x - 1)(x + 1)$

We have a good reason for wanting to know the x-intercepts, or zeros. When the value of a continuous function transitions from negative to positive and vice versa, it must pass through zero.

DEFINITION **Intermediate Value Theorem**

Let a and b be real numbers such that $a < b$ and let f be a polynomial function. If $f(a)$ and $f(b)$ have opposite signs, then there is at least one zero between a and b.

The **intermediate value theorem** will be used later in this chapter to assist us in finding the real zeros of a polynomial function. For now, it tells us that in order to change signs, the graph of a polynomial function must pass through the x-axis. In other words, once we know the zeros, then we know that between two consecutive zeros the graph of a polynomial function is either entirely above the x-axis or entirely below the x-axis. This enables us to break down the x-axis into intervals that we can test, which will assist us in graphing polynomial functions. Keep in mind, though, that the existence of a zero does not imply that the function will change signs—as you will see in the subsection on graphing general polynomial functions.

Technology Tip

Graph $y_1 = x^3 + x^2 - 2x$.

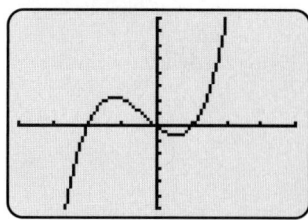

 EXAMPLE 3 **Identifying the Real Zeros of a Polynomial Function**

Find the zeros of the polynomial function $f(x) = x^3 + x^2 - 2x$.

Solution:

Set the function equal to zero.	$x^3 + x^2 - 2x = 0$
Factor out an x common to all three terms.	$x(x^2 + x - 2) = 0$
Factor the quadratic expression inside the parentheses.	$x(x + 2)(x - 1) = 0$
Apply the zero product property.	$x = 0$ or $(x + 2) = 0$ or $(x - 1) = 0$
Solve.	$x = -2, x = 0,$ and $x = 1$

The zeros are $\boxed{-2, 0, \text{ and } 1}$.

The zeros of the function -2, 0, and 1 correspond to the x-intercepts $(-2, 0)$, $(0, 0)$, and $(1, 0)$.

The table supports the real zeros shown by the graph.

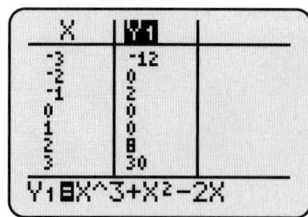

■ **YOUR TURN** Find the zeros of the polynomial function $f(x) = x^3 - 7x^2 + 12x$.

■ **Answer:** The zeros are 0, 3, and 4.

When factoring a quadratic equation, if the factor is raised to a power greater than 1, the corresponding root, or zero, is repeated. For example, the quadratic equation $x^2 - 2x + 1 = 0$ when factored is written as $(x - 1)^2 = 0$. The solution, or root, in this case is $x = 1$, and we say that it is a **repeated** root. Similarly, when determining zeros of higher order polynomial functions, if a factor is repeated, we say that the zero is a repeated, or **multiple**, zero of the function. The number of times that a zero repeats is called its *multiplicity*.

DEFINITION   **Multiplicity of a Zero**

If $(x - a)^n$ is a factor of a polynomial f, then a is called a **zero of multiplicity n** of f.

EXAMPLE 4   Finding the Multiplicities of Zeros of a Polynomial Function

Find the zeros, and state their multiplicities, of the polynomial function
$g(x) = (x - 1)^2 (x + \frac{3}{5})^7 (x + 5)$.

Solution:

> 1 is a zero of multiplicity 2.
>
> $-\frac{3}{5}$ is a zero of multiplicity 7.
>
> -5 is a zero of multiplicity 1.

Note: Adding the multiplicities yields the degree of the polynomial. The polynomial $g(x)$ is of degree 10, since $2 + 7 + 1 = 10$.

■ **Answer:**

   0 is a zero of multiplicity 2.
   2 is a zero of multiplicity 3.
   $-\frac{1}{2}$ is a zero of multiplicity 5.

■ **YOUR TURN**   For the polynomial $h(x)$, determine the zeros and state their multiplicities.

$$h(x) = x^2(x - 2)^3\left(x + \frac{1}{2}\right)^5$$

EXAMPLE 5   Finding a Polynomial from Its Zeros

Find a polynomial of degree 7 whose zeros are

$$-2 \text{ (multiplicity 2)} \qquad 0 \text{ (multiplicity 4)} \qquad 1 \text{ (multiplicity 1)}$$

Solution:

If $x = a$ is a zero, then $(x - a)$ is a factor.	$f(x) = (x + 2)^2(x - 0)^4(x - 1)^1$
Simplify.	$= x^4(x + 2)^2(x - 1)$
Square the binomial.	$= x^4(x^2 + 4x + 4)(x - 1)$
Multiply the two polynomials.	$= x^4(x^3 + 3x^2 - 4)$
Distribute x^4.	$= \boxed{x^7 + 3x^6 - 4x^4}$

Graphing General Polynomial Functions

Study Tip

It is not always possible to find x-intercepts. Sometimes there are no x-intercepts (for some even degree polynomial functions).

Let's develop a strategy for sketching an approximate graph of any polynomial function. First, we determine the x- and y-intercepts. Then we use the x-intercepts, or zeros, to divide the domain into intervals where the value of the polynomial is positive or negative so that we can find points in those intervals to assist in sketching a smooth and continuous graph. *Note:* It is not always possible to find x-intercepts. Some even degree polynomial functions have no x-intercepts on their graph.

EXAMPLE 6   Using a Strategy for Sketching the Graph of a Polynomial Function

Sketch the graph of $f(x) = (x + 2)(x - 1)^2$.

Solution:

STEP 1 Find the y-intercept.
(Let $x = 0$.)

$f(0) = (2)(-1)^2 = 2$
$(0, 2)$ is the y-intercept

STEP 2 Find any x-intercepts.
(Set $f(x) = 0$.)

$f(x) = (x + 2)(x - 1)^2 = 0$
$x = -2$ or $x = 1$
$(-2, 0)$ and $(1, 0)$ are the x-intercepts

STEP 3 Plot the intercepts.

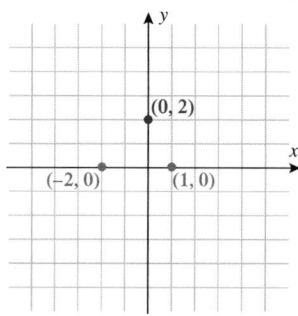

STEP 4 Divide the x-axis into intervals: $(-\infty, -2)$, $(-2, 1)$, and $(1, \infty)$

STEP 5 Select a number in each interval and test each interval. The function $f(x)$ either *crosses* the x-axis at an x-intercept or *touches* the x-axis at an x-intercept. Therefore, we need to check each of these intervals to determine whether the function is positive (above the x-axis) or negative (below the x-axis). We do so by selecting numbers in the intervals and determining the value of the function at the corresponding points.

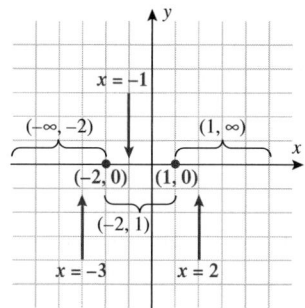

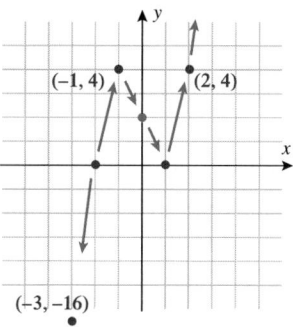

Interval	$(-\infty, -2)$	$(-2, 1)$	$(1, \infty)$
Number Selected in Interval	-3	-1	2
Value of Function	$f(-3) = -16$	$f(-1) = 4$	$f(2) = 4$
Point on Graph	$(-3, -16)$	$(-1, 4)$	$(2, 4)$
Interval Relation to x-Axis	Below x-axis	Above x-axis	Above x-axis

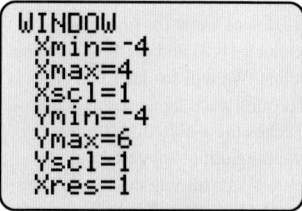
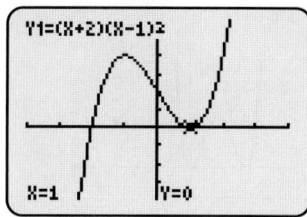
Study Tip

Although there may be up to n x-intercepts for the graph of a polynomial function of degree n, there will always be exactly one y-intercept.

From the table, we find three additional points on the graph: $(-3, -16)$, $(-1, 4)$, and $(2, 4)$. The point $(-2, 0)$ is an intercept where the function *crosses* the x-axis, because it is below the x-axis to the left of -2 and above the x-axis to the right of -2. The point $(1, 0)$ is an intercept where the function *touches* the x-axis, because it is above the x-axis on both sides of $x = 1$. Connecting these points with a smooth curve yields the graph.

STEP 6 Sketch a plot of the function.

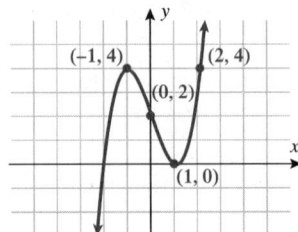

Study Tip

We do not know for sure that the points $(-1, 4)$ and $(1, 0)$ are turning points. We will see later that $(1, 0)$ is a turning point because the graph touches the x-axis at $(1, 0)$, but a graphing utility suggests that $(-1, 4)$ is a turning point, and later in calculus you will learn how to find relative maximum points and relative minimum points.

■ YOUR TURN Sketch the graph of $f(x) = x^2(x + 3)^2$.

■ Answer:

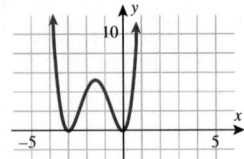

In Example 6, we found that the function crosses the x-axis at the point $(-2, 0)$. Note that -2 is a zero of multiplicity 1. We also found that the function touches the x-axis at the point $(1, 0)$. Note that 1 is a zero of multiplicity 2. In general, zeros with even multiplicity correspond to intercepts where the function touches the x-axis, and zeros with odd multiplicity correspond to intercepts where the function crosses the x-axis.

MULTIPLICITY OF A ZERO AND RELATION TO THE GRAPH OF A POLYNOMIAL FUNCTION

If a is a zero of $f(x)$, then

Study Tip

In general, zeros with *even* multiplicity correspond to intercepts where the function *touches* the x-axis and zeros with *odd* multiplicity correspond to intercepts where the function *crosses* the x-axis.

Multiplicity of a	$f(x)$ on Either Side of $x = a$	Graph of Function at the Intercept
Even	Does not change sign	Touches the x-axis (turns around) at point $(a, 0)$
Odd	Changes sign	Crosses the x-axis at point $(a, 0)$

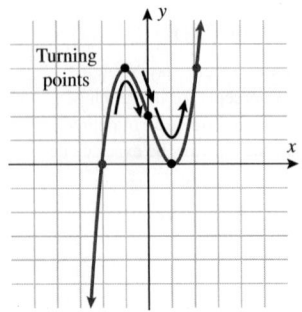

Also in Example 6, we know that somewhere in the interval $(-2, 1)$ the function must reach a relative or local maximum and then turn back toward the x-axis, because both points $(-2, 0)$ and $(1, 0)$ correspond to x-intercepts. When we sketch the graph, it "appears" that the point $(-1, 4)$ is a *turning point*. The point $(1, 0)$ also corresponds to a turning point. In general, if f is a polynomial of degree n, then the graph of f has at most $n - 1$ turning points.

Study Tip

If f is a polynomial of degree n, then the graph of f has at most $n - 1$ turning points.

The point $(-1, 4)$, which we call a turning point, is also a relative or local "high point" on the graph in the vicinity of the point $(-1, 4)$. Also note that the point $(1, 0)$, which we call a turning point, is a relative or local "low point" on the graph in the vicinity of the point $(1, 0)$. We call a "high point" on a graph a **local (relative) maximum** and a "low point" on a graph a **local (relative) minimum**. For quadratic functions we can find the maximum or minimum point by finding the vertex. However, for higher degree polynomial functions, we rely on graphing utilities to assist us in locating such points. Later in calculus, techniques will be developed for finding such points exactly. For now, we use the zoom and trace features to locate such points on a graph, and we can use the $\boxed{\text{table}}$ feature of a graphing utility to approximate relative minima or maxima.

Let us take the polynomial $f(x) = x^3 - 2x^2 - 5x + 6$. Using methods discussed thus far we can find that the x-intercepts of its graph are $(-2, 0)$, $(1, 0)$, and $(3, 0)$ and the y-intercept is the point $(0, 6)$. We can also find additional points that lie on the graph such as $(-1, 8)$ and $(2, -4)$. Plotting these points, we might "think" that the points $(-1, 8)$ and $(2, -4)$ might be turning points, but a graphing utility reveals an approximate relative maximum at the point $(-0.7863, 8.2088207)$ and an approximate relative minimum at the point $(2.1196331, -4.060673)$.

Intercepts and turning points assist us in sketching graphs of polynomial functions. Another piece of information that will assist us in graphing polynomial functions is knowledge of the *end behavior*. All polynomials eventually rise or fall without bound as x gets large in both the positive $(x \to \infty)$ and negative $(x \to -\infty)$ directions. The highest degree monomial within the polynomial dominates the *end behavior*. In other words, the highest power term is eventually going to overwhelm the other terms as x grows without bound.

Technology Tip

Use TI to graph the function $f(x) = x^3 - 2x^2 - 5x + 6$ as Y_1.

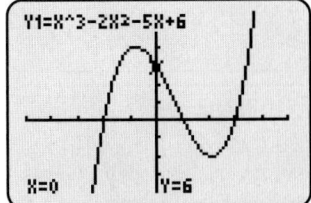

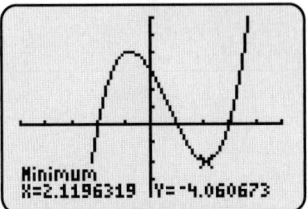

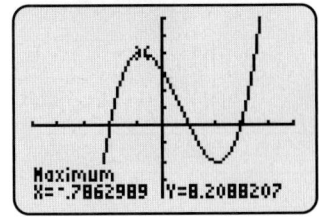

END BEHAVIOR

As x gets large in the positive $(x \to \infty)$ and negative $(x \to -\infty)$ directions, the graph of the polynomial

$$f(x) = a_n x^n + a_{n-1} x^{n-1} + \cdots + a_2 x^2 + a_1 x + a_0$$

has the same behavior as the power function

$$y = a_n x^n$$

Power functions behave much like a quadratic function (parabola) for even-degree polynomial functions and much like a cubic function for odd-degree polynomial functions. There are four possibilities because the leading coefficient can be positive or negative with either an odd or even power.

Let $y = a_n x^n$; then

n	Even	Even	Odd	Odd
a_n	Positive	Negative	Negative	Positive
$x \to -\infty$ (Left)	The graph of the function *rises*.	The graph of the function *falls*.	The graph of the function *rises*.	The graph of the function *falls*.
$x \to \infty$ (Right)	The graph of the function *rises*.	The graph of the function *falls*.	The graph of the function *falls*.	The graph of the function *rises*.
Graph	$a_n > 0$	$a_n < 0$	$a_n < 0$	$a_n > 0$

Technology Tip

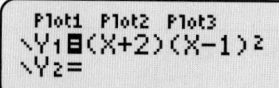

The graph of $f(x) = (x + 2)(x - 1)^2$ is shown.

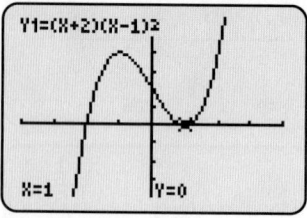

Note: The graph crosses the x-axis at the point $x = -2$ and touches the x-axis at the point $x = 1$. A table of values supports the graph.

EXAMPLE 7   Graphing a Polynomial Function

Sketch a graph of the polynomial function $f(x) = 2x^4 - 8x^2$.

Solution:

STEP 1 Determine the y-intercept: ($x = 0$).

$f(0) = 0$

The y-intercept corresponds to the point $(0, 0)$.

STEP 2 Find the zeros of the polynomial.

Factor out the common $2x^2$.

Factor the quadratic binomial.

Set $f(x) = 0$.

$$f(x) = 2x^4 - 8x^2$$
$$= 2x^2(x^2 - 4)$$
$$= 2x^2(x - 2)(x + 2)$$
$$= 2x^2(x - 2)(x + 2) = 0$$

0 is a zero of multiplicity 2. The graph will *touch* the x-axis.

2 is a zero of multiplicity 1. The graph will *cross* the x-axis.

-2 is a zero of multiplicity 1. The graph will *cross* the x-axis.

STEP 3 Determine the end behavior.

$f(x) = 2x^4 - 8x^2$ behaves like $y = 2x^4$.

$y = 2x^4$ is of even degree, and the leading coefficient is positive, so the graph rises without bound as x gets large in both the positive and negative directions.

STEP 4 Sketch the intercepts and end behavior.

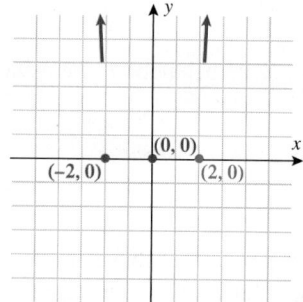

STEP 5 Find additional points.

x	-1	$-\frac{1}{2}$	$\frac{1}{2}$	1
$f(x)$	-6	$-\frac{15}{8}$	$-\frac{15}{8}$	-6

STEP 6 Sketch the graph.

■ estimate additional points

■ connect with a smooth curve

Note the symmetry about the y-axis. This function is an even function: $f(-x) = f(x)$.

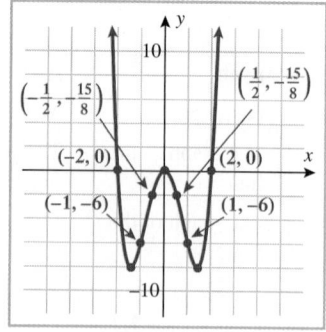

It is important to note that the absolute minimum occurs when $x = \pm\sqrt{2} \approx \pm1.14$, but at this time can only be illustrated using a graphing utility.

■ **Answer:**

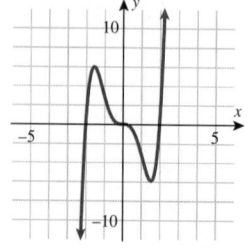

■ **YOUR TURN** Sketch a graph of the polynomial function $f(x) = x^5 - 4x^3$.

SECTION
2.2 SUMMARY

In general, polynomials can be graphed in one of two ways:

- Use graph-shifting techniques with power functions.
- General polynomial function.
  1. Identify intercepts.
  2. Determine each real zero and its multiplicity, and ascertain whether the graph crosses or touches the x-axis there.

3. x-intercepts (real zeros) divide the x-axis into intervals. Test points in the intervals to determine whether the graph is above or below the x-axis.
4. Determine the end behavior by investigating the end behavior of the highest degree monomial.
5. Sketch the graph with a smooth curve.

SECTION
2.2 EXERCISES

▪ SKILLS

In Exercises 1–10, determine which functions are polynomials, and for those that are, state their degree.

1. $g(x) = (x + 2)^3(x - \frac{3}{5})^2$ 2. $g(x) = \left(x - \frac{1}{4}\right)^4 \left(x + \sqrt{7}\right)^2$ 3. $g(x) = x^5(x + 2)(x - 6.4)$ 4. $g(x) = x^4(x - 1)^2(x + 2.5)^3$

5. $h(x) = \sqrt{x} + 1$ 6. $h(x) = (x - 1)^{1/2} + 5x$ 7. $F(x) = x^{1/3} + 7x^2 - 2$ 8. $F(x) = 3x^2 + 7x - \dfrac{2}{3x}$

9. $G(x) = \dfrac{x + 1}{x^2}$ 10. $H(x) = \dfrac{x^2 + 1}{2}$

In Exercises 11–18, match the polynomial function with its graph.

11. $f(x) = -3x + 1$ 12. $f(x) = -3x^2 - x$ 13. $f(x) = x^2 + x$ 14. $f(x) = -2x^3 + 4x^2 - 6x$

15. $f(x) = x^3 - x^2$ 16. $f(x) = 2x^4 - 18x^2$ 17. $f(x) = -x^4 + 5x^3$ 18. $f(x) = x^5 - 5x^3 + 4x$

a.

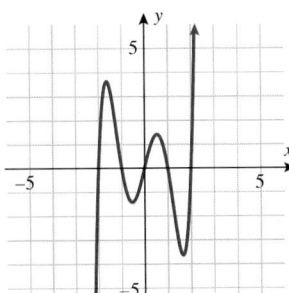

b.

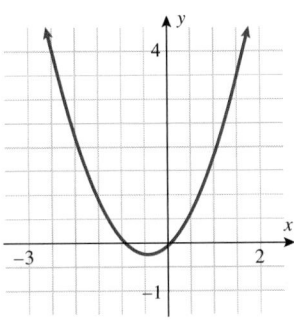

c.

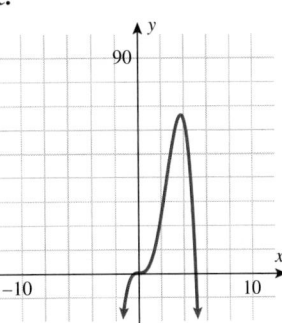

d.

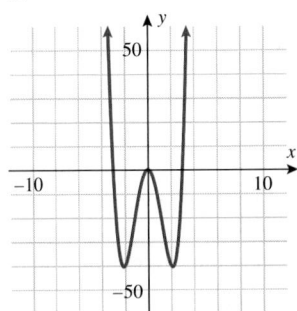

e.

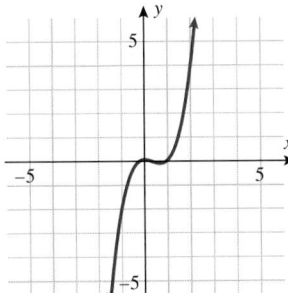

f.

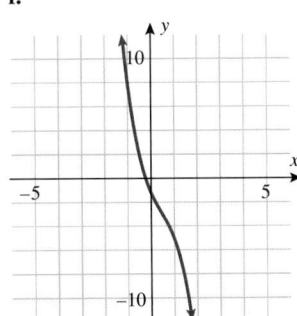

g.

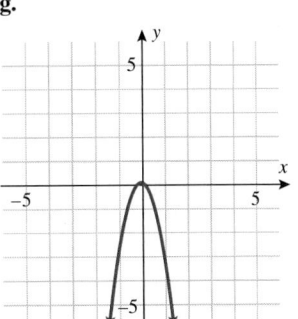

h.

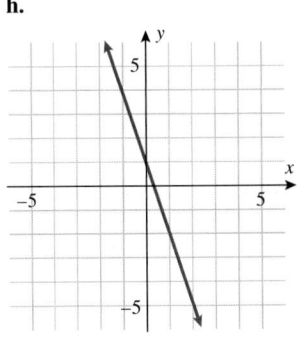

In Exercises 19–24, graph each function by transforming a power function $y = x^n$.

19. $f(x) = (x - 2)^4$

20. $f(x) = (x + 2)^5$

21. $f(x) = x^5 + 3$

22. $f(x) = -x^4 - 3$

23. $f(x) = 3 - (x + 1)^4$

24. $f(x) = (x - 3)^5 - 2$

In Exercises 25–36, find all the real zeros (and state their multiplicities) of each polynomial function.

25. $f(x) = 2(x - 3)(x + 4)^3$

26. $f(x) = -3(x + 2)^3(x - 1)^2$

27. $f(x) = 4x^2(x - 7)^2(x + 4)$

28. $f(x) = 5x^3(x + 1)^4(x - 6)$

29. $f(x) = 4x^2(x - 1)^2(x^2 + 4)$

30. $f(x) = 4x^2(x^2 - 1)(x^2 + 9)$

31. $f(x) = 8x^3 + 6x^2 - 27x$

32. $f(x) = 2x^4 + 5x^3 - 3x^2$

33. $f(x) = -2.7x^3 - 8.1x^2$

34. $f(x) = 1.2x^6 - 4.6x^4$

35. $f(x) = \frac{1}{3}x^6 + \frac{2}{5}x^4$

36. $f(x) = \frac{2}{7}x^5 - \frac{3}{4}x^4 + \frac{1}{2}x^3$

In Exercises 37–50, find a polynomial (there are many) of minimum degree that has the given zeros.

37. $-3, 0, 1, 2$

38. $-2, 0, 2$

39. $-5, -3, 0, 2, 6$

40. $0, 1, 3, 5, 10$

41. $-\frac{1}{2}, \frac{2}{3}, \frac{3}{4}$

42. $-\frac{3}{4}, -\frac{1}{3}, 0, \frac{1}{2}$

43. $1 - \sqrt{2}, 1 + \sqrt{2}$

44. $1 - \sqrt{3}, 1 + \sqrt{3}$

45. -2 (multiplicity 3), 0 (multiplicity 2)

46. -4 (multiplicity 2), 5 (multiplicity 3)

47. -3 (multiplicity 2), 7 (multiplicity 5)

48. 0 (multiplicity 1), 10 (multiplicity 3)

49. $-\sqrt{3}$ (multiplicity 2), -1 (multiplicity 1), 0 (multiplicity 2), $\sqrt{3}$ (multiplicity 2)

50. $-\sqrt{5}$ (multiplicity 2), 0 (multiplicity 1), 1 (multiplicity 2), $\sqrt{5}$ (multiplicity 2)

In Exercises 51–68, for each polynomial function given: (a) list each real zero and its multiplicity; (b) determine whether the graph touches or crosses at each x-intercept; (c) find the y-intercept and a few points on the graph; (d) determine the end behavior; and (e) sketch the graph.

51. $f(x) = (x - 2)^3$

52. $f(x) = -(x + 3)^3$

53. $f(x) = x^3 - 9x$

54. $f(x) = -x^3 + 4x^2$

55. $f(x) = -x^3 + x^2 + 2x$

56. $f(x) = x^3 - 6x^2 + 9x$

57. $f(x) = -x^4 - 3x^3$

58. $f(x) = x^5 - x^3$

59. $f(x) = 12x^6 - 36x^5 - 48x^4$

60. $f(x) = 7x^5 - 14x^4 - 21x^3$

61. $f(x) = 2x^5 - 6x^4 - 8x^3$

62. $f(x) = -5x^4 + 10x^3 - 5x^2$

63. $f(x) = x^3 - x^2 - 4x + 4$

64. $f(x) = x^3 - x^2 - x + 1$

65. $f(x) = -(x + 2)^2(x - 1)^2$

66. $f(x) = (x - 2)^3(x + 1)^3$

67. $f(x) = x^2(x - 2)^3(x + 3)^2$

68. $f(x) = -x^3(x - 4)^2(x + 2)^2$

In Exercises 69–72, for each graph given: (a) list each real zero and its smallest possible multiplicity; (b) determine whether the degree of the polynomial is even or odd; (c) determine whether the leading coefficient of the polynomial is positive or negative; (d) find the y-intercept; and (e) write an equation for the polynomial function (assume the least degree possible).

69.

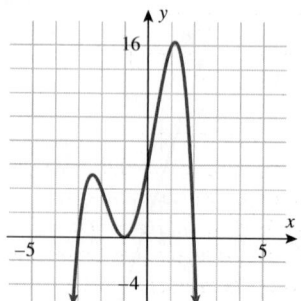

70.

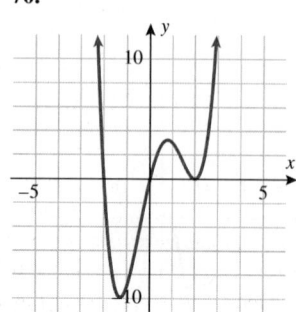

71.

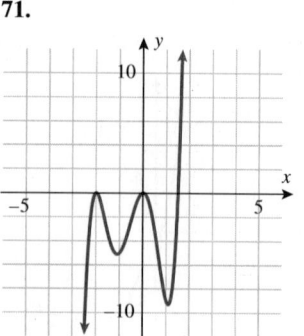

72.

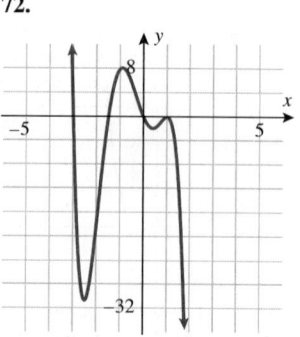

▪ APPLICATIONS

73. Weight. Jennifer has joined a gym to lose weight and feel better. She still likes to cheat a little and will enjoy the occasional bad meal with an ice cream dream dessert and then miss the gym for a couple of days. Given in the table is Jennifer's weight for a period of 8 months. Her weight can be modeled as a polynomial. Plot these data. How many turning points are there? Assuming these are the minimum number of turning points, what is the lowest degree polynomial that can represent Jennifer's weight?

MONTH	WEIGHT
1	169
2	158
3	150
4	161
5	154
6	159
7	148
8	153

74. Stock Value. A day trader checks the stock price of Coca-Cola during a 4-hour period (given below). The price of Coca-Cola stock during this 4-hour period can be modeled as a polynomial function. Plot these data. How many turning points are there? Assuming these are the minimum number of turning points, what is the lowest degree polynomial that can represent the Coca-Cola stock price?

PERIOD WATCHING STOCK MARKET	PRICE
1	$53.00
2	$56.00
3	$52.70
4	$51.50

75. Stock Value. The price of Tommy Hilfiger stock during a 4-hour period is given below. If a third-degree polynomial models this stock, do you expect the stock to go up or down in the fifth period?

PERIOD WATCHING STOCK MARKET	PRICE
1	$15.10
2	$14.76
3	$15.50
4	$14.85

76. Stock Value. The stock prices for Coca-Cola during a 4-hour period on another day yield the following results. If a third-degree polynomial models this stock, do you expect the stock to go up or down in the fifth period?

PERIOD WATCHING STOCK MARKET	PRICE
1	$52.80
2	$53.00
3	$56.00
4	$52.70

For Exercises 77 and 78, the following table graph illustrates the average federal funds rate in the month of January (2000 to 2008):

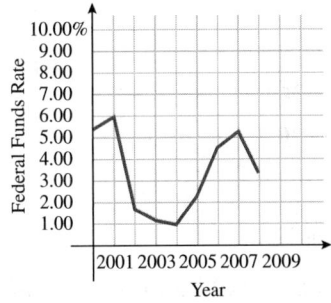

77. Finance. If a polynomial function is used to model the federal funds rate data shown in the graph, determine the degree of the lowest degree polynomial that can be used to model those data.

78. Finance. Should the leading coefficient in the polynomial found in Exercise 77 be positive or negative? Explain.

79. Air Travel. An airline has a daily flight Chicago–Miami. The number of passengers per flight is given in the table below. Which would be the minimum degree of a polynomial that models the number of passengers of the airline?

DAY	PASSENGERS
Monday	180
Tuesday	150
Wednesday	175
Thursday	160
Friday	100
Saturday	98
Sunday	120

80. Air Travel. The airline in Exercise 79 discovered that the information about the number of passengers corresponding to Monday and Sunday was mixed. On Sunday, they have 180 passengers, while on Monday, they have 120 passengers. Determine the degree of the lowest degree polynomial that can be used to model those data.

81. Temperature. The weather report indicates that the daily highest temperatures for next week can be described as a cubic polynomial function. The forecasting for Tuesday, Wednesday, and Thursday is 39°, 42°, and 35°, respectively. Thursday will be the coolest day of the week. What can you say about Monday's temperature T?

82. Sports. A basketball player scored more than 20 points on each of the past 9 games.

GAME	1	2	3	4	5	6	7	8	9
POINTS	25	27	30	28	25	24	26	27	25

a. If he scores 24 points in the next game, what is the degree of the polynomial function describing this data?

b. If he scores 26 points in the next game, what is the degree of the polynomial function describing this data?

▪ CATCH THE MISTAKE

In Exercises 83–86, explain the mistake that is made.

83. Find a fourth-degree polynomial function with zeros -2, $-1, 3, 4$.

Solution:
$$f(x) = (x - 2)(x - 1)(x + 3)(x + 4)$$

This is incorrect. What mistake was made?

84. Determine the end behavior of the polynomial function $f(x) = x(x - 2)^3$.

Solution:

This polynomial has similar end behavior to the graph of $y = x^3$.

End behavior falls to the left and rises to the right.

This is incorrect. What mistake was made?

85. Graph the polynomial function $f(x) = (x - 1)^2(x + 2)^3$.

Solution:

The zeros are -2 and 1, and therefore, the x-intercepts are $(-2, 0)$ and $(1, 0)$.

The y-intercept is $(0, 8)$.

Plotting these points and connecting with a smooth curve yield the graph on the right.

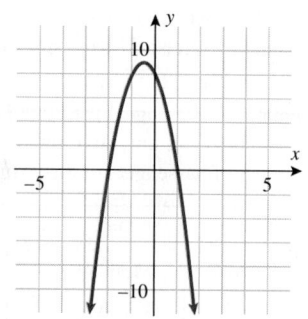

This graph is incorrect. What did we forget to do?

86. Graph the polynomial function $f(x) = (x + 1)^2(x - 1)^2$.

Solution:

The zeros are -1 and 1, so the x-intercepts are $(-1, 0)$ and $(1, 0)$.

The y-intercept is $(0, 1)$.

Plotting these points and connecting with a smooth curve yield the graph on the right.

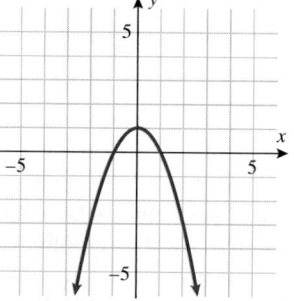

This graph is incorrect. What did we forget to do?

▪ CONCEPTUAL

In Exercises 87–90, determine whether each statement is true or false.

87. The graph of a polynomial function might not have any y-intercepts.

88. The graph of a polynomial function might not have any x-intercepts.

89. The domain of all polynomial functions is $(-\infty, \infty)$.

90. The range of all polynomial functions is $(-\infty, \infty)$.

91. What is the maximum number of zeros that a polynomial of degree n can have?

92. What is the maximum number of turning points a graph of an nth-degree polynomial can have?

■ CHALLENGE

93. Find a seventh-degree polynomial that has the following graph characteristics: The graph touches the x-axis at $x = -1$, and the graph crosses the x-axis at $x = 3$. Plot this polynomial function.

94. Find a fifth-degree polynomial that has the following graph characteristics: The graph touches the x-axis at $x = 0$ and crosses the x-axis at $x = 4$. Plot the polynomial function.

95. Determine the zeros of the polynomial
$f(x) = x^3 + (b - a)x^2 - abx$ for the positive real numbers a and b.

96. Graph the function $f(x) = x^2(x - a)^2(x - b)^2$ for the positive real numbers a, b, where $b > a$.

■ TECHNOLOGY

In Exercises 97 and 98, use a graphing calculator or computer to graph each polynomial. From that graph, estimate the x-intercepts (if any). Set the function equal to zero, and solve for the zeros of the polynomial. Compare the zeros with the x-intercepts.

97. $f(x) = x^4 + 2x^2 + 1$

98. $f(x) = 1.1x^3 - 2.4x^2 + 5.2x$

For each polynomial in Exercises 99 and 100, determine the power function that has similar end behavior. Plot this power function and the polynomial. Do they have similar end behavior?

99. $f(x) = -2x^5 - 5x^4 - 3x^3$

100. $f(x) = x^4 - 6x^2 + 9$

In Exercises 101 and 102, use a graphing calculator or a computer to graph each polynomial. From the graph, estimate the x-intercepts and state the zeros of the function and their multiplicities.

101. $f(x) = x^4 - 15.9x^3 + 1.31x^2 + 292.905x + 445.7025$

102. $f(x) = -x^5 + 2.2x^4 + 18.49x^3 - 29.878x^2 - 76.5x + 100.8$

In Exercises 103 and 104, use a graphing calculator or a computer to graph each polynomial. From the graph, estimate the coordinates of the relative maximum and minimum points. Round your answers to 2 decimal places.

103. $f(x) = 2x^4 + 5x^3 - 10x^2 - 15x + 8$

104. $f(x) = 2x^5 - 4x^4 - 12x^3 + 18x^2 + 16x - 7$

■ PREVIEW TO CALCULUS

In calculus we study the extreme values of functions; in order to find these values we need to solve different types of equations.

In Exercises 105–108, use the Intermediate Value Theorem to find all the zeros of the polynomial functions in the given interval. Round all your answers to three decimal places.

105. $x^3 + 3x - 5 = 0$, $[0, 2]$

106. $x^5 - x + 0.5 = 0$, $[0, 1]$

107. $x^4 - 3x^3 + 6x^2 - 7 = 0$, $[-2, 2]$

108. $x^3 + x^2 - 2x - 2 = 0$, $[1, 2]$

SECTION
2.3 DIVIDING POLYNOMIALS

SKILLS OBJECTIVES

- Divide polynomials with long division.
- Divide polynomials with synthetic division.
- Check the results now that you know the division algorithm.

CONCEPTUAL OBJECTIVES

- Extend long division of real numbers to polynomials.
- Understand *when* synthetic division can be used.

Long Division of Polynomials

Let's start with an example whose answer we already know. We know that a quadratic expression can be factored into the product of two linear factors: $x^2 + 4x - 5 = (x + 5)(x - 1)$. Therefore, if we divide both sides of the equation by $(x - 1)$, we get

$$\frac{x^2 + 4x - 5}{x - 1} = x + 5$$

We can state this by saying $x^2 + 4x - 5$ divided by $x - 1$ is equal to $x + 5$. Confirm this statement by long division:

$$x - 1 \overline{)x^2 + 4x - 5}$$

Note that although this is standard division notation, the **dividend**, $x^2 + 4x - 5$, and the **divisor**, $x - 1$, are both polynomials that consist of multiple terms. The *leading* terms of each algebraic expression will guide us.

WORDS	MATH
	$\boxed{x}$
Q: x times what quantity gives x^2?	$\boxed{x} - 1 \overline{)\boxed{x^2} + 4x - 5}$
A: x	
	x
	$x - 1 \overline{)x^2 + 4x - 5}$
Multiply $x(x - 1) = x^2 - x$.	$\underline{x^2 - x}$
	x
Subtract $(x^2 - x)$ from $x^2 + 4x - 5$.	$x - 1 \overline{)x^2 + 4x - 5}$
Note: $-(x^2 - x) = -x^2 + x$.	$\underline{-x^2 + x}$
Bring down the -5.	$5x - 5$
	$x + \boxed{5}$
Q: x times what quantity is $5x$?	$\boxed{x} - 1 \overline{)x^2 + 4x - 5}$
A: 5	$\underline{-x^2 + x}$
Multiply $5(x - 1) = 5x - 5$.	$\boxed{5x} - 5$
	$x + 5$
	$x - 1 \overline{)x^2 + 4x - 5}$
	$\underline{-x^2 + x}$
	$5x - 5$
Subtract $(5x - 5)$.	$\underline{-5x + 5}$
Note: $-(5x - 5) = -5x + 5$.	0

The **quotient** is $x + 5$, and as expected, the **remainder** is 0. By long division we have shown that

$$\boxed{\frac{x^2 + 4x - 5}{x - 1} = x + 5}$$

Check: Multiplying the equation by $x - 1$ yields $x^2 + 4x - 5 = (x + 5)(x - 1)$, which we knew to be true.

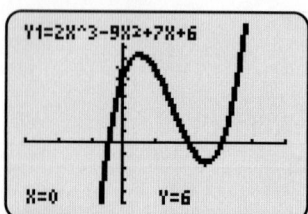
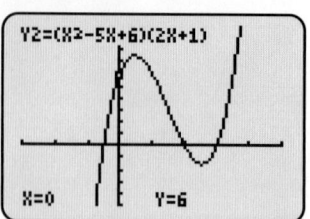

EXAMPLE 1 **Dividing Polynomials Using Long Division; Zero Remainder**

Divide $2x^3 - 9x^2 + 7x + 6$ by $2x + 1$.

Solution:

$$\begin{array}{r} x^2 - 5x + 6 \\ 2x+1\overline{)2x^3 - 9x^2 + 7x + 6} \end{array}$$

Multiply: $x^2(2x + 1)$.

$$-\left(2x^3 + x^2\right)$$

Subtract: Bring down the $7x$.

$$-10x^2 + 7x$$

Multiply: $-5x(2x + 1)$.

$$-\left(-10x^2 - 5x\right)$$

Subtract: Bring down the 6.

$$12x + 6$$

Multiply: $6(2x + 1)$.

$$-(12x + 6)$$

Subtract.

$$0$$

Quotient:

$$\boxed{x^2 - 5x + 6}$$

Check: $(2x + 1)\left(x^2 - 5x + 6\right) = 2x^3 - 9x^2 + 7x + 6$.

Note: The divisor cannot be equal to zero, $2x + 1 \neq 0$, so we say $x \neq -\frac{1}{2}$.

■ **YOUR TURN** Divide $4x^3 + 13x^2 - 2x - 15$ by $4x + 5$.

■ **Answer:** $x^2 + 2x - 3$, remainder 0.

Why are we interested in dividing polynomials? Because it helps us find zeros of polynomials. In Example 1, using long division, we found that

$$2x^3 - 9x^2 + 7x + 6 = (2x + 1)\left(x^2 - 5x + 6\right)$$

Factoring the quadratic expression enables us to write the cubic polynomial as a product of three linear factors:

$$2x^3 - 9x^2 + 7x + 6 = (2x + 1)\left(x^2 - 5x + 6\right) = (2x + 1)(x - 3)(x - 2)$$

Set the value of the polynomial equal to zero, $(2x + 1)(x - 3)(x - 2) = 0$, and solve for x. The zeros of the polynomial are $-\frac{1}{2}$, 2, and 3. In Example 1 and in the Your Turn, the remainder was 0. Sometimes there is a nonzero remainder (Example 2).

EXAMPLE 2 **Dividing Polynomials Using Long Division; Nonzero Remainder**

Divide $6x^2 - x - 2$ by $x + 1$.

Solution:

$$\begin{array}{r} 6x - 7 \\ x+1\overline{)6x^2 - x - 2} \end{array}$$

Multiply $6x(x + 1)$.

$$-\left(6x^2 + 6x\right)$$

Subtract and bring down -2.

$$-7x - 2$$

Multiply $-7(x + 1)$.

$$-(-7x - 7)$$

Subtract and identify the remainder.

$$+ 5$$

Dividend   Quotient   Remainder

$$\underset{\text{Divisor}}{\frac{6x^2 - x - 2}{x + 1}} = 6x - 7 + \underset{\text{Divisor}}{\frac{5}{x + 1}} \quad \boxed{x \neq -1}$$

Check: Multiply equation by $x + 1$. $6x^2 - x - 2 = (6x - 7)(x + 1) + 5$

$$= 6x^2 - x - 7 + 5$$

$$= 6x^2 - x - 2 \checkmark$$

■ **Answer:** $2x^2 + 3x - 1$ R: -4 or

$$2x^2 + 3x - 1 - \frac{4}{x - 1}$$

■ **YOUR TURN** Divide $2x^3 + x^2 - 4x - 3$ by $x - 1$.

In general, when a polynomial is divided by another polynomial, we express the result in the following form:

$$\frac{P(x)}{d(x)} = Q(x) + \frac{r(x)}{d(x)}$$

where $P(x)$ is the **dividend**, $d(x) \neq 0$ is the **divisor**, $Q(x)$ is the **quotient**, and $r(x)$ is the **remainder**. Multiplying this equation by the divisor $d(x)$ leads us to the division algorithm.

THE DIVISION ALGORITHM

If $P(x)$ and $d(x)$ are polynomials with $d(x) \neq 0$, and if the degree of $P(x)$ is greater than or equal to the degree of $d(x)$, then unique polynomials $Q(x)$ and $r(x)$ exist such that

$$P(x) = d(x) \cdot Q(x) + r(x)$$

If the remainder $r(x) = 0$, then we say that $d(x)$ divides $P(x)$ and that $d(x)$ and $Q(x)$ are factors of $P(x)$.

 EXAMPLE 3   Long Division of Polynomials with "Missing" Terms

Divide $3x^4 + 2x^3 + x^2 + 4$ by $x^2 + 1$.

Solution:

Insert $0x$ as a placeholder in both the divisor and the dividend.

Multiply $3x^2(x^2 + 0x + 1)$.

Subtract and bring down $0x$.

Multiply $2x(x^2 + 0x + 1)$.

Subtract and bring down 4.

Multiply $-2(x^2 - 2x + 1)$.

Subtract and get remainder $-2x + 6$.

$$\begin{array}{r} 3x^2 + 2x - 2 \\ x^2 + 0x + 1 \overline{)\ 3x^4 + 2x^3 + x^2 + 0x + 4} \\ \underline{-(3x^4 + 0x^3 + 3x^2)} \\ 2x^3 - 2x^2 + 0x \\ \underline{-(2x^3 + 0x^2 + 2x)} \\ -2x^2 - 2x + 4 \\ \underline{-(-2x^2 + 0x - 2)} \\ -2x + 6 \end{array}$$

$$\boxed{\frac{3x^4 + 2x^3 + x^2 + 4}{x^2 + 1} = 3x^2 + 2x - 2 + \frac{-2x + 6}{x^2 + 1}}$$

■ **Answer:**

$$2x^2 + 6 + \frac{11x^2 + 18x + 36}{x^3 - 3x - 4}$$

■ **YOUR TURN** Divide $2x^5 + 3x^2 + 12$ by $x^3 - 3x - 4$.

Synthetic Division of Polynomials

In the special case when the *divisor is a linear factor* of the form $x - a$ or $x + a$, there is another, more efficient way to divide polynomials. This method is called **synthetic division**. It is called synthetic because it is a contrived shorthand way of dividing a polynomial by a linear factor. A detailed step-by-step procedure is given below for synthetic division. Let's divide $x^4 - x^3 - 2x + 2$ by $x + 1$ using synthetic division.

Step 1: Write the division in synthetic form.
- List the coefficients of the dividend. **Remember to use 0 for a placeholder**.
- The divisor is $x + 1$, so $x = -1$ is used.

Study Tip

If $(x - a)$ is a divisor, then a is the number used in synthetic division.

Coefficients of Dividend

$$-1 \,\big|\; 1 \quad -1 \quad 0 \quad -2 \quad 2$$

Step 2: *Bring down* the first term (**1**) in the dividend.

$$-1 \,\big|\; 1 \quad -1 \quad 0 \quad -2 \quad 2$$

Bring down the 1

$$1$$

Step 3: *Multiply* the -1 by this leading coefficient (**1**), and place the product up and to the right in the second column.

$$-1 \,\big|\; 1 \quad -1 \quad 0 \quad -2 \quad 2$$
$$-1$$
$$1$$

Step 4: *Add* the values in the second column.

$$-1 \,\big|\; 1 \quad -1 \quad 0 \quad -2 \quad 2$$
$$-1 \ \text{ADD}$$
$$1 \quad -2$$

Step 5: Repeat Steps 3 and 4 until all columns are filled.

$$-1 \,\big|\; 1 \quad -1 \quad 0 \quad -2 \quad 2$$
$$-1 \quad 2\!\downarrow \ -2\!\downarrow \ 4\!\downarrow$$
$$1 \quad -2 \quad 2 \quad -4 \quad 6$$

Step 6: Identify the **quotient** by assigning powers of x in descending order, beginning with $x^{n-1} = x^{4-1} = x^3$. The last term is the **remainder**.

$$-1 \,\big|\; 1 \quad -1 \quad 0 \quad -2 \quad 2$$
$$-1 \quad 2 \quad -2 \quad 4$$
$$\underbrace{1 \quad -2 \quad 2 \quad -4}_{\text{Quotient Coefficients}} \quad \underset{\text{Remainder}}{6}$$
$$x^3 - 2x^2 + 2x - 4$$

Study Tip

Synthetic division can only be used when the divisor is of the form $x - a$. Realize that a may be negative, as in the divisor $x + 2$.

We know that the degree of the first term of the quotient is 3, because a fourth-degree polynomial was divided by a first-degree polynomial. Let's compare dividing $x^4 - x^3 - 2x + 2$ by $x + 1$ using both long division and synthetic division.

Long Division

$$
\begin{array}{r}
x^3 - 2x^2 + 2x - 4 \\
x + 1 \overline{)\, x^4 - x^3 + 0x^2 - 2x + 2} \\
\underline{x^4 + x^3 } \\
-2x^3 + 0x^2 \\
\underline{-(-2x^3 - 2x^2)} \\
2x^2 - 2x \\
\underline{-(2x^2 + 2x)} \\
-4x + 2 \\
\underline{-(-4x - 4)} \\
+ 6
\end{array}
$$

Synthetic Division

$$-1 \,\big|\; 1 \quad -1 \quad 0 \quad -2 \quad 2$$
$$-1 \quad 2 \quad -2 \quad 4$$
$$\underbrace{1 \quad -2 \quad 2 \quad -4}\quad 6$$
$$x^3 - 2x^2 + 2x - 4$$

Both long division and synthetic division yield the same answer.

$$\frac{x^4 - x^3 - 2x + 2}{x + 1} = x^3 - 2x^2 + 2x - 4 + \frac{6}{x + 1}$$

EXAMPLE 4   Synthetic Division

Use synthetic division to divide $3x^5 - 2x^3 + x^2 - 7$ by $x + 2$.

Solution:

STEP 1 Write the division in synthetic form.
- List the coefficients of the dividend. Remember to use 0 for a placeholder.
- The divisor of the original problem is $x + 2$. If we set $x + 2 = 0$, we find that $x = -2$, so -2 is the divisor for synthetic division.

$$-2 \begin{array}{|rrrrrr} 3 & 0 & -2 & 1 & 0 & -7 \end{array}$$

STEP 2 Perform the synthetic division steps.

$$-2 \begin{array}{|rrrrrr} 3 & 0 & -2 & 1 & 0 & -7 \\ & -6 & 12 & -20 & 38 & -76 \\ \hline 3 & -6 & 10 & -19 & 38 & -83 \end{array}$$

STEP 3 Identify the quotient and remainder.

$$-2 \begin{array}{|rrrrrr} 3 & 0 & -2 & 1 & 0 & -7 \\ & -6 & 12 & -20 & 38 & -76 \\ \hline 3 & -6 & 10 & -19 & 38 & \boxed{-83} \end{array}$$

$$\underbrace{3x^4 - 6x^3 + 10x^2 - 19x + 38}$$

$$\boxed{\dfrac{3x^5 - 2x^3 + x^2 - 7}{x + 2} = 3x^4 - 6x^3 + 10x^2 - 19x + 38 - \dfrac{83}{x + 2}}$$

■ **Answer:** $2x^2 + 2x + 1 + \dfrac{4}{x - 1}$

■ **YOUR TURN** Use synthetic division to divide $2x^3 - x + 3$ by $x - 1$.

Division of Polynomials
- Long division can always be used.
- Synthetic division is restricted to when the divisor is of the form $x - a$ or $x + a$.

Expressing Results
- $\dfrac{\text{Dividend}}{\text{Divisor}} = \text{quotient} + \dfrac{\text{remainder}}{\text{divisor}}$
- Dividend = (quotient)(divisor) + remainder

When Remainder Is Zero
- Dividend = (quotient)(divisor)
- Quotient and divisor are factors of the dividend.

▪ SKILLS

In Exercises 1–26, divide the polynomials using long division. Use exact values and express the answer in the form $Q(x) = \ ?, \ r(x) = \ ?$.

1. $\left(3x^2 - 9x - 5\right) \div (x - 2)$

2. $\left(x^2 + 4x - 3\right) \div (x - 1)$

3. $\left(3x^2 - 13x - 10\right) \div (x + 5)$

4. $\left(3x^2 - 13x - 10\right) \div (x + 5)$

5. $\left(x^2 - 4\right) \div (x + 4)$

6. $\left(x^2 - 9\right) \div (x - 2)$

7. $\left(9x^2 - 25\right) \div (3x - 5)$

8. $\left(5x^2 - 3\right) \div (x + 1)$

9. $\left(4x^2 - 9\right) \div (2x + 3)$

10. $\left(8x^3 + 27\right) \div (2x + 3)$

11. $\left(11x + 20x^2 + 12x^3 + 2\right) \div (3x + 2)$

12. $\left(12x^3 + 2 + 11x + 20x^2\right) \div (2x + 1)$

13. $\left(4x^3 - 2x + 7\right) \div (2x + 1)$

14. $\left(6x^4 - 2x^2 + 5\right) \div (-3x + 2)$

15. $\left(4x^3 - 12x^2 - x + 3\right) \div \left(x - \frac{1}{2}\right)$

16. $\left(12x^3 + 1 + 7x + 16x^2\right) \div \left(x + \frac{1}{3}\right)$

17. $\left(-2x^5 + 3x^4 - 2x^2\right) \div \left(x^3 - 3x^2 + 1\right)$

18. $\left(-9x^6 + 7x^4 - 2x^3 + 5\right) \div \left(3x^4 - 2x + 1\right)$

19. $\dfrac{x^4 - 1}{x^2 - 1}$

20. $\dfrac{x^4 - 9}{x^2 + 3}$

21. $\dfrac{40 - 22x + 7x^3 + 6x^4}{6x^2 + x - 2}$

22. $\dfrac{-13x^2 + 4x^4 + 9}{4x^2 - 9}$

23. $\dfrac{-3x^4 + 7x^3 - 2x + 1}{x - 0.6}$

24. $\dfrac{2x^5 - 4x^3 + 3x^2 + 5}{x - 0.9}$

25. $\left(x^4 + 0.8x^3 - 0.26x^2 - 0.168x + 0.0441\right) \div \left(x^2 + 1.4x + 0.49\right)$

26. $\left(x^5 + 2.8x^4 + 1.34x^3 - 0.688x^2 - 0.2919x + 0.0882\right) \div \left(x^2 - 0.6x + 0.09\right)$

In Exercises 27–46, divide the polynomial by the linear factor with synthetic division. Indicate the quotient $Q(x)$ and the remainder $r(x)$.

27. $\left(3x^2 + 7x + 2\right) \div (x + 2)$

28. $\left(2x^2 + 7x - 15\right) \div (x + 5)$

29. $\left(7x^2 - 3x + 5\right) \div (x + 1)$

30. $\left(4x^2 + x + 1\right) \div (x - 2)$

31. $\left(3x^2 + 4x - x^4 - 2x^3 - 4\right) \div (x + 2)$

32. $\left(3x^2 - 4 + x^3\right) \div (x - 1)$

33. $\left(x^4 + 1\right) \div (x + 1)$

34. $\left(x^4 + 9\right) \div (x + 3)$

35. $\left(x^4 - 16\right) \div (x + 2)$

36. $\left(x^4 - 81\right) \div (x - 3)$

37. $\left(2x^3 - 5x^2 - x + 1\right) \div \left(x + \frac{1}{2}\right)$

38. $\left(3x^3 - 8x^2 + 1\right) \div \left(x + \frac{1}{3}\right)$

39. $\left(2x^4 - 3x^3 + 7x^2 - 4\right) \div \left(x - \frac{2}{3}\right)$

40. $\left(3x^4 + x^3 + 2x - 3\right) \div \left(x - \frac{3}{4}\right)$

41. $\left(2x^4 + 9x^3 - 9x^2 - 81x - 81\right) \div (x + 1.5)$

42. $\left(5x^3 - x^2 + 6x + 8\right) \div (x + 0.8)$

43. $\dfrac{x^7 - 8x^4 + 3x^2 + 1}{x - 1}$

44. $\dfrac{x^6 + 4x^5 - 2x^3 + 7}{x + 1}$

45. $\left(x^6 - 49x^4 - 25x^2 + 1225\right) \div (x - \sqrt{5})$

46. $\left(x^6 - 4x^4 - 9x^2 + 36\right) \div (x - \sqrt{3})$

In Exercises 47–60, divide the polynomials by either long division or synthetic division.

47. $\left(6x^2 - 23x + 7\right) \div (3x - 1)$

48. $\left(6x^2 + x - 2\right) \div (2x - 1)$

49. $\left(x^3 - x^2 - 9x + 9\right) \div (x - 1)$

50. $\left(x^3 + 2x^2 - 6x - 12\right) \div (x + 2)$

51. $\left(x^3 + 6x^2 - 2x - 5\right) \div \left(x^2 - 1\right)$

52. $\left(3x^5 - x^3 + 2x^2 - 1\right) \div \left(x^3 + x^2 - x + 1\right)$

53. $\left(x^6 - 2x^5 + x^4 - 6x^3 + 7x^2 - 4x + 7\right) \div \left(x^2 + 1\right)$

54. $\left(x^6 - 1\right) \div \left(x^2 + x + 1\right)$

55. $\left(x^5 + 4x^3 + 2x^2 - 1\right) \div (x - 2)$

56. $\left(x^4 - x^2 + 3x - 10\right) \div (x + 5)$

57. $\left(x^4 - 25\right) \div \left(x^2 - 1\right)$

58. $\left(x^3 - 8\right) \div \left(x^2 - 2\right)$

59. $\left(x^7 - 1\right) \div (x - 1)$

60. $\left(x^6 - 27\right) \div (x - 3)$

▪ APPLICATIONS

61. Geometry. The area of a rectangle is $6x^4 + 4x^3 - x^2 - 2x - 1$ square feet. If the length of the rectangle is $2x^2 - 1$ feet, what is the width of the rectangle?

62. Geometry. If the rectangle in Exercise 61 is the base of a rectangular box with volume $18x^5 + 18x^4 + x^3 - 7x^2 - 5x - 1$ cubic feet, what is the height of the box?

63. Travel. If a car travels a distance of $x^3 + 60x^2 + x + 60$ miles at an average speed of $x + 60$ miles per hour, how long does the trip take?

64. Sports. If a quarterback throws a ball $-x^2 - 5x + 50$ yards in $5 - x$ seconds, how fast is the football traveling?

▪ CATCH THE MISTAKE

In Exercises 65–68, explain the mistake that is made.

65. Divide $x^3 - 4x^2 + x + 6$ by $x^2 + x + 1$.

Solution:

$$
\begin{array}{r}
x - 3 \\
x^2 + x + 1{\overline{\smash{\big)}\,x^3 - 4x^2 + x + 6}} \\
\underline{x^3 + x^2 + x} \\
-3x^2 + 2x + 6 \\
\underline{-3x^2 - 3x - 3} \\
-x + 3
\end{array}
$$

This is incorrect. What mistake was made?

66. Divide $x^4 - 3x^2 + 5x + 2$ by $x - 2$.

Solution:

$$
\begin{array}{r|rrrr}
-2 & 1 & -3 & 5 & 2 \\
 & & -2 & 10 & -30 \\
\hline
 & 1 & -5 & 15 & \boxed{-28}
\end{array}
$$
$$x^2 - 5x + 15$$

This is incorrect. What mistake was made?

67. Divide $x^3 + 4x - 12$ by $x - 3$.

Solution:

$$
\begin{array}{r|rrr}
3 & 1 & 4 & -12 \\
 & & 3 & 21 \\
\hline
 & 1 & 7 & \boxed{9}
\end{array}
$$
$$x + 7$$

This is incorrect. What mistake was made?

68. Divide $x^3 + 3x^2 - 2x + 1$ by $x^2 + 1$.

Solution:

$$
\begin{array}{r|rrrr}
-1 & 1 & 3 & -2 & 1 \\
 & & -1 & -2 & 4 \\
\hline
 & 1 & 2 & -4 & \boxed{5}
\end{array}
$$
$$x^2 - 2x - 4$$

This is incorrect. What mistake was made?

▪ CONCEPTUAL

In Exercises 69–74, determine whether each statement is true or false.

69. A fifth-degree polynomial divided by a third-degree polynomial will yield a quadratic quotient.

70. A third-degree polynomial divided by a linear polynomial will yield a linear quotient.

71. Synthetic division can be used whenever the degree of the dividend is exactly one more than the degree of the divisor.

72. When the remainder is zero, the divisor is a factor of the dividend.

73. When both the dividend and the divisor have the same degree, the quotient equals one.

74. Long division must be used whenever the degree of the divisor is greater than one.

▪ CHALLENGE

75. Is $x + b$ a factor of $x^3 + (2b - a)x^2 + (b^2 - 2ab)x - ab^2$?

76. Is $x + b$ a factor of $x^4 + (b^2 - a^2)x^2 - a^2b^2$?

77. Divide $x^{3n} + x^{2n} - x^n - 1$ by $x^n - 1$.

78. Divide $x^{3n} + 5x^{2n} + 8x^n + 4$ by $x^n + 1$.

79. Plot $\dfrac{2x^3 - x^2 + 10x - 5}{x^2 + 5}$. What type of function is it?

Perform this division using long division, and confirm that the graph corresponds to the quotient.

80. Plot $\dfrac{x^3 - 3x^2 + 4x - 12}{x - 3}$. What type of function is it?

Perform this division using synthetic division, and confirm that the graph corresponds to the quotient.

81. Plot $\dfrac{x^4 + 2x^3 - x - 2}{x + 2}$. What type of function is it?

Perform this division using synthetic division, and confirm that the graph corresponds to the quotient.

82. Plot $\dfrac{x^5 - 9x^4 + 18x^3 + 2x^2 - 5x - 3}{x^4 - 6x^3 + 2x + 1}$. What type of function is it? Perform this division using long division, and confirm that the graph corresponds to the quotient.

83. Plot $\dfrac{-6x^3 + 7x^2 + 14x - 15}{2x + 3}$. What type of function is it?

Perform this division using long division, and confirm that the graph corresponds to the quotient.

84. Plot $\dfrac{-3x^5 - 4x^4 + 29x^3 + 36x^2 - 18x}{3x^2 + 4x - 2}$. What type of function is it? Perform this division using long division, and confirm that the graph corresponds to the quotient.

▪ **PREVIEW TO CALCULUS**

For some of the operations in calculus it is convenient to write rational fractions $\dfrac{P(x)}{d(x)}$ in the form $Q(x) + \dfrac{r(x)}{d(x)}$, where $\dfrac{P(x)}{d(x)} = Q(x) + \dfrac{r(x)}{d(x)}$. In Exercises 85–88, write each rational function $\dfrac{P(x)}{d(x)}$ in the form $Q(x) + \dfrac{r(x)}{d(x)}$.

85. $\dfrac{2x^2 - x}{x + 2}$

86. $\dfrac{5x^3 + 2x^2 - 3x}{x - 3}$

87. $\dfrac{2x^4 + 3x^2 + 6}{x^2 + x + 1}$

88. $\dfrac{3x^5 - 2x^3 + x^2 + x - 6}{x^2 + x + 5}$

SECTION
2.4 THE REAL ZEROS OF A POLYNOMIAL FUNCTION

SKILLS OBJECTIVES

▪ Apply the remainder theorem to evaluate a polynomial function.
▪ Apply the factor theorem.
▪ Use the rational zero (root) theorem to list possible rational zeros.
▪ Apply Descartes' rule of signs to determine the possible combination of positive and negative real zeros.
▪ Utilize the upper and lower bound theorems to narrow the search for real zeros.
▪ Find the real zeros of a polynomial function.
▪ Factor a polynomial function.
▪ Employ the intermediate value theorem to approximate a real zero.

CONCEPTUAL OBJECTIVES

▪ Understand that a polynomial of degree n has at most n real zeros.
▪ Understand that a real zero can be either rational or irrational and that irrational zeros will not be listed as possible zeros through the rational zero test.
▪ Realize that rational zeros can be found exactly, whereas irrational zeros must be approximated.

The zeros of a polynomial function assist us in finding the x-intercepts of the graph of a polynomial function. How do we find the zeros of a polynomial function if we cannot factor them easily? For polynomial functions of degree 2, we have the quadratic formula, which allows us to find the two zeros. For polynomial functions whose degree is greater than 2, much more work is required.* In this section, we focus our attention on finding the *real* zeros of a polynomial function. Later, in Section 2.5, we expand our discussion to *complex* zeros of polynomial functions.

In this section, we start by listing possible rational zeros. As you will see, there are sometimes many possibilities. We can then narrow the search using Descartes' rule of signs, which tells us possible combinations of positive and negative real zeros. We can narrow the search even further with the upper and lower bound rules. Once we have tested possible values and determined a zero, we will employ synthetic division to divide the polynomial by the linear factor associated with the zero. We will continue the process until we have factored the polynomial function into a product of either linear factors or irreducible quadratic factors. Last, we will discuss how to find irrational real zeros using the intermediate value theorem.

The Remainder Theorem and the Factor Theorem

If we divide the polynomial function $f(x) = x^3 - 2x^2 + x - 3$ by $x - 2$ using synthetic division, we find the remainder is -1.

$$
\begin{array}{r|rrrr}
2 & 1 & -2 & 1 & -3 \\
  &   & 2 & 0 & 2 \\
\hline
  & 1 & 0 & 1 & -1
\end{array}
$$

Notice that if we evaluate the function at $x = 2$, the result is -1. $f(2) = -1$
This leads us to the *remainder theorem*.

REMAINDER THEOREM

If a polynomial $P(x)$ is divided by $x - a$, then the remainder is $r = P(a)$.

Technology Tip

A graphing utility can be used to evaluate $P(2)$. Enter $P(x) = 4x^5 - 3x^4 + 2x^3 - 7x^2 + 9x - 5$ as Y_1.

```
Plot1 Plot2 Plot3
\Y1■4X^5-3X^4+2X
^3-7X²+9X-5
\Y2=
```

To evaluate $P(2)$, press $\boxed{\text{VARS}}$

$\boxed{\text{Y-VARS}}$ $\boxed{\text{1:Function..}}$ $\boxed{\text{ENTER}}$

$\boxed{1:Y_1}$ $\boxed{\text{ENTER}}$ $\boxed{(\ 2\ )}$ $\boxed{\text{ENTER}}$

```
Y₁(2)
                    81
■
```

The remainder theorem tells you that polynomial division can be used to evaluate a polynomial function at a particular point.

EXAMPLE 1   Two Methods for Evaluating Polynomials

Let $P(x) = 4x^5 - 3x^4 + 2x^3 - 7x^2 + 9x - 5$ and evaluate $P(2)$ by

a. evaluating $P(2)$ directly.
b. the remainder theorem and synthetic division.

Solution:

a. $P(2) = 4(2)^5 - 3(2)^4 + 2(2)^3 - 7(2)^2 + 9(2) - 5$
$= 4(32) - 3(16) + 2(8) - 7(4) + 9(2) - 5$
$= 128 - 48 + 16 - 28 + 18 - 5$
$= \boxed{81}$

---

*There are complicated formulas for finding the zeros of polynomial functions of degree 3 and 4, but there are no such formulas for degree 5 and higher polynomials (according to the Abel–Ruffini theorem).

b.

$$
\begin{array}{r|rrrrrr}
2 & 4 & -3 & 2 & -7 & 9 & -5 \\
  &   & 8 & 10 & 24 & 34 & 86 \\
\hline
  & 4 & 5 & 12 & 17 & 43 & \boxed{81}
\end{array}
$$

■ **YOUR TURN** Let $P(x) = -x^3 + 2x^2 - 5x + 2$ and evaluate $P(-2)$ using the remainder theorem and synthetic division.

■ **Answer:** $P(-2) = 28$

Recall that when a polynomial is divided by $x - a$, if the remainder is zero, we say that $x - a$ is a factor of the polynomial. Through the remainder theorem, we now know that the remainder is related to evaluation of the polynomial at the point $x = a$. We are then led to the *factor theorem*.

FACTOR THEOREM

If $P(a) = 0$, then $x - a$ is a factor of $P(x)$. Conversely, if $x - a$ is a factor of $P(x)$, then $P(a) = 0$.

Technology Tip

EXAMPLE 2   Using the Factor Theorem to Factor a Polynomial

Determine whether $x + 2$ is a factor of $P(x) = x^3 - 2x^2 - 5x + 6$. If so, factor $P(x)$ completely.

Solution:

STEP 1 Divide $P(x) = x^3 - 2x^2 - 5x + 6$ by $x + 2$ using synthetic division.

$$
\begin{array}{r|rrrr}
-2 & 1 & -2 & -5 & 6 \\
   &   & -2 & 8 & -6 \\
\hline
   & 1 & -4 & 3 & \boxed{0} \\
\end{array}
$$
$$x^2 - 4x + 3$$

Since the remainder is zero, $P(-2) = 0$, $\boxed{x + 2 \text{ is a factor}}$ of $P(x) = x^3 - 2x^2 - 5x + 6$.

STEP 2 Write $P(x)$ as a product.

$$P(x) = (x + 2)(x^2 - 4x + 3)$$

STEP 3 Factor the quadratic polynomial.

$$\boxed{P(x) = (x + 2)(x - 3)(x - 1)}$$

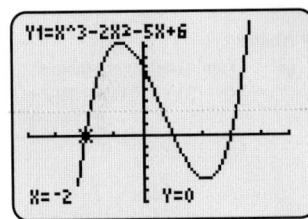

The three zeros of the function give the three factors $x + 2$, $x - 1$, and $x - 3$. A table of values supports the zeros of the graph.

■ **YOUR TURN** Determine whether $x - 1$ is a factor of $P(x) = x^3 - 4x^2 - 7x + 10$. If so, factor $P(x)$ completely.

■ **Answer:** $(x - 1)$ is a factor; $P(x) = (x - 5)(x - 1)(x + 2)$

Technology Tip

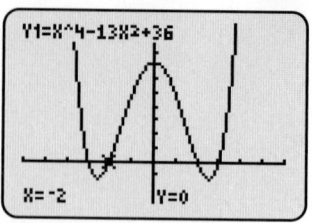

The zeros of the function at $x = -3$, $x = -2$, $x = 2$, and $x = 3$ are shown in the graph. A table of values supports the zeros of the graph.

X	Y1
-3	0
-2	0
-1	24
0	36
1	24
2	0
3	0

Y₁◼X^4-13X²+36

Answer:
$(x - 3)$ and $(x + 2)$ are factors; $P(x) = (x - 3)(x + 2)(x - 1)(x + 1)$

EXAMPLE 3   Using the Factor Theorem to Factor a Polynomial

Determine whether $x - 3$ and $x + 2$ are factors of $P(x) = x^4 - 13x^2 + 36$. If so, factor $P(x)$ completely.

Solution:

STEP 1   With synthetic division, divide $P(x) = x^4 - 13x^2 + 36$ by $x - 3$.

$$
\begin{array}{r|rrrrr}
3 & 1 & 0 & -13 & 0 & 36 \\
  &   & 3 & 9 & -12 & -36 \\
\hline
  & 1 & 3 & -4 & -12 & \boxed{0} \\
\end{array}
$$
$$x^3 + 3x^2 - 4x - 12$$

Because the remainder is 0, $\boxed{x - 3 \text{ is a factor}}$, and we can write the polynomial as

$$P(x) = (x - 3)(x^3 + 3x^2 - 4x - 12)$$

STEP 2   With synthetic division, divide the remaining cubic polynomial $(x^3 + 3x^2 - 4x - 12)$ by $x + 2$.

$$
\begin{array}{r|rrrr}
-2 & 1 & 3 & -4 & -12 \\
   &   & -2 & -2 & 12 \\
\hline
   & 1 & 1 & -6 & \boxed{0} \\
\end{array}
$$
$$x^2 + x - 6$$

Because the remainder is 0, $\boxed{x + 2 \text{ is a factor}}$, and we can now write the polynomial as

$$P(x) = (x - 3)(x + 2)(x^2 + x - 6)$$

STEP 3   Factor the quadratic polynomial: $x^2 + x - 6 = (x + 3)(x - 2)$.

STEP 4   Write $P(x)$ as a product of linear factors:

$$\boxed{P(x) = (x - 3)(x - 2)(x + 2)(x + 3)}$$

◼ YOUR TURN   Determine whether $x - 3$ and $x + 2$ are factors of $P(x) = x^4 - x^3 - 7x^2 + x + 6$. If so, factor $P(x)$ completely.

The Search for Real Zeros

In all of the examples thus far, the polynomial function and one or more real zeros (or linear factors) were given. Now, we will not be given any real zeros to start with. Instead, we will develop methods to search for them.

Each real zero corresponds to a linear factor and each linear factor is of degree 1. Therefore, the largest number of real zeros a polynomial function can have is equal to the degree of the polynomial.

THE NUMBER OF REAL ZEROS

A polynomial function cannot have more real zeros than its degree.

The following functions illustrate that a polynomial function of degree n can have at most n real zeros:

POLYNOMIAL FUNCTION	DEGREE	REAL ZEROS	COMMENTS
$f(x) = x^2 - 9$	2	$x = \pm 3$	**Two** real zeros
$f(x) = x^2 + 4$	2	None	**No** real zeros
$f(x) = x^3 - 1$	3	$x = 1$	**One** real zero
$f(x) = x^3 - x^2 - 6x$	3	$x = -2, 0, 3$	**Three** real zeros

Now that we know the *maximum* number of real zeros a polynomial function can have, let us discuss how to find these zeros.

The Rational Zero Theorem and Descartes' Rule of Signs

When the coefficients of a polynomial are integers, then the *rational zero theorem* (*rational root test*) gives us a list of possible rational zeros. We can then test these possible values to determine whether they really do correspond to actual zeros. *Descartes' rule of signs* tells us the possible combinations of *positive* real zeros and *negative* real zeros. Using Descartes' rule of signs will assist us in narrowing down the large list of possible zeros generated through the rational zero theorem to a (hopefully) shorter list of possible zeros. First, let's look at the rational zero theorem; then we'll turn to Descartes' rule of signs.

THE RATIONAL ZERO THEOREM (RATIONAL ROOT TEST)

If the polynomial function $P(x) = a_n x^n + a_{n-1} x^{n-1} + \cdots + a_2 x^2 + a_1 x + a_0$ has *integer* coefficients, then every rational zero of $P(x)$ has the form

$$\text{Rational zero} = \frac{\text{factors of } a_0}{\text{factors of } a_n} = \frac{\text{factors of constant term}}{\text{factors of leading coefficient}}$$

$$= \pm \frac{\text{positive factors of constant term}}{\text{positive factors of leading coefficient}}$$

To use this theorem, simply list all combinations of factors of both the constant term a_0 and the leading coefficient term a_n and take all appropriate combinations of ratios. This procedure is illustrated in Example 4. Notice that when the leading coefficient is 1, then the possible rational zeros will simply be the possible factors of the constant term.

EXAMPLE 4   Using the Rational Zero Theorem

Determine possible rational zeros for the polynomial $P(x) = x^4 - x^3 - 5x^2 - x - 6$ by the rational zero theorem. Test each one to find all rational zeros.

Solution:

STEP 1 List factors of the constant and leading coefficient terms.

$a_0 = -6 \qquad \pm 1, \pm 2, \pm 3, \pm 6$
$a_n = 1 \qquad \pm 1$

STEP 2 List possible rational zeros $\dfrac{a_0}{a_n}$.

$$\frac{\pm 1}{\pm 1}, \frac{\pm 2}{\pm 1}, \frac{\pm 3}{\pm 1}, \frac{\pm 6}{\pm 1} = \pm 1, \pm 2, \pm 3, \pm 6$$

There are three ways to test whether any of these are zeros: Substitute these values into the polynomial to see which ones yield zero, or use either polynomial division or synthetic division to divide the polynomial by these possible zeros, and look for a zero remainder.

STEP 3 Test possible zeros by looking for zero remainders.

1 is not a zero: $P(1) = (1)^4 - (1)^3 - 5(1)^2 - (1) - 6 = -12$

-1 is not a zero: $P(-1) = (-1)^4 - (-1)^3 - 5(-1)^2 - (-1) - 6 = -8$

We could continue testing with direct substitution, but let us now use synthetic division as an alternative.

Study Tip

The remainder can be found by evaluating the function or synthetic division. For simple values like $x = \pm 1$, it is easier to evaluate the polynomial function. For other values, it is often easier to use synthetic division.

2 is not a zero:

$$
\begin{array}{r|rrrrr}
2 & 1 & -1 & -5 & -1 & -6 \\
  &   & 2 & 2 & -6 & -14 \\
\hline
  & 1 & 1 & -3 & -7 & \boxed{-20}
\end{array}
$$

-2 is a zero:

$$
\begin{array}{r|rrrrr}
-2 & 1 & -1 & -5 & -1 & -6 \\
   &   & -2 & 6 & -2 & 6 \\
\hline
   & 1 & -3 & 1 & -3 & \boxed{0}
\end{array}
$$

Since -2 is a zero, then $x + 2$ is a factor of $P(x)$, and the remaining quotient is $x^3 - 3x^2 + x - 3$. Therefore, if there are any other real roots remaining, we can now use the simpler $x^3 - 3x^2 + x - 3$ for the dividend. Also note that the rational zero theorem can be applied to the new dividend and possibly shorten the list of possible rational zeros. In this case, the possible rational zeros of $F(x) = x^3 - 3x^2 + x - 3$ are ± 1 and ± 3.

3 is a zero:

$$
\begin{array}{r|rrrr}
3 & 1 & -3 & 1 & -3 \\
  &   & 3 & 0 & 3 \\
\hline
  & 1 & 0 & 1 & \boxed{0}
\end{array}
$$

We now know that $\boxed{-2}$ and $\boxed{3}$ are confirmed zeros. If we continue testing, we will find that the other possible zeros fail. This is a fourth-degree polynomial, and we have found two rational real zeros. We see in the graph on the right that these two real zeros correspond to the x-intercepts.

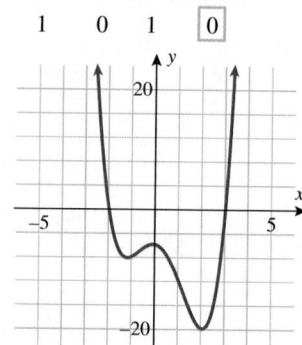

■ **Answer:** Possible rational zeros: ± 1 and ± 3. Rational real zeros: 1 and -3.

■ **YOUR TURN** List the possible rational zeros of the polynomial $P(x) = x^4 + 2x^3 - 2x^2 + 2x - 3$, and determine rational real zeros.

Notice in Example 4 that the polynomial function $P(x) = x^4 - x^3 - 5x^2 - x - 6$ had two rational real zeros, -2 and 3. This implies that $x + 2$ and $x - 3$ are factors of $P(x)$. Also note in the last step when we divided by the zero 3, the quotient was $x^2 + 1$. Therefore, we can write the polynomial in factored form as

$$
P(x) = \underbrace{(x + 2)}_{\substack{\text{linear} \\ \text{factor}}} \underbrace{(x - 3)}_{\substack{\text{linear} \\ \text{factor}}} \underbrace{(x^2 + 1)}_{\substack{\text{irreducible} \\ \text{quadratic} \\ \text{factor}}}
$$

Notice that the first two factors are of degree 1, so we call them **linear factors**. The third expression, $x^2 + 1$, is of degree 2 and cannot be factored in terms of real numbers.

We will discuss complex zeros in the next section. For now, we say that a quadratic expression, $ax^2 + bx + c$, is called **irreducible** if it cannot be factored over the real numbers.

EXAMPLE 5 Factoring a Polynomial Function

Write the following polynomial function as a product of linear and/or irreducible quadratic factors:

$$P(x) = x^4 - 4x^3 + 4x^2 - 36x - 45$$

Solution:

Use the rational zero theorem to list possible rational roots.

$$x = \pm 1, \pm 3, \pm 5, \pm 9, \pm 15, \pm 45$$

Test possible zeros by evaluating the function or by utilizing synthetic division.

$x = 1$ is not a zero. $\qquad P(1) = -80$

$x = -1$ is a zero. $\qquad P(-1) = 0$

Divide $P(x)$ by $x + 1$.

$$
\begin{array}{r|rrrrr}
-1 & 1 & -4 & 4 & -36 & -45 \\
   &   & -1 & 5 & -9 & 45 \\
\hline
   & 1 & -5 & 9 & -45 & \boxed{0}
\end{array}
$$

$x = 5$ is a zero.

$$
\begin{array}{r|rrrr}
5 & 1 & -5 & 9 & -45 \\
  &   & 5 & 0 & 45 \\
\hline
  & 1 & 0 & 9 & \boxed{0}
\end{array}
$$
$$\underbrace{}_{x^2 + 9}$$

The factor $x^2 + 9$ is irreducible.

Write the polynomial as a product of linear and/or irreducible quadratic factors.

$$\boxed{P(x) = (x - 5)(x + 1)(x^2 + 9)}$$

Notice that the graph of this polynomial function has x-intercepts at $x = -1$ and $x = 5$.

■ **YOUR TURN** Write the following polynomial function as a product of linear and/or irreducible quadratic factors:

$$P(x) = x^4 - 2x^3 - x^2 - 4x - 6$$

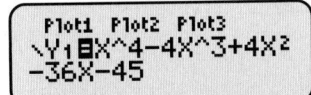

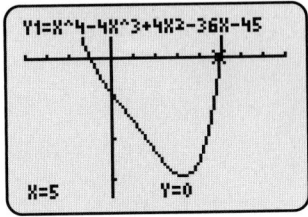

A table of values supports the real zeros of the graph.

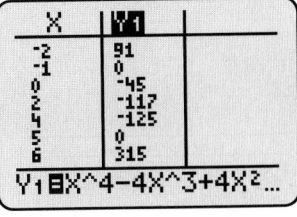

■ **Answer:**
$P(x) = (x + 1)(x - 3)(x^2 + 2)$

The rational zero theorem lists possible zeros. It would be helpful if we could narrow that list. Descartes' rule of signs determines the possible combinations of positive real zeros and negative real zeros through variations of sign. A *variation in sign* is a sign difference seen between consecutive coefficients.

$$P(x) = 2x^6 - 5x^5 - 3x^4 + 2x^3 - x^2 - x - 1$$

Sign Change − to +

Sign Change + to −

Sign Change + to −

This polynomial experiences three sign changes or variations in sign.

DESCARTES' RULE OF SIGNS

If the polynomial function $P(x) = a_n x^n + a_{n-1} x^{n-1} + \cdots + a_2 x^2 + a_1 x + a_0$ has real coefficients and $a_0 \neq 0$, then:

- The number of **positive** real zeros of the polynomial is either equal to the number of variations of sign of $P(x)$ or less than that number by an even integer.
- The number of **negative** real zeros of the polynomial is either equal to the number of variations of sign of $P(-x)$ or less than that number by an even integer.

Descartes' rule of signs narrows our search for real zeros, because we don't have to test all of the possible rational zeros. For example, if we know there is one positive real zero, then if we find a positive rational zero, we no longer need to continue to test possible positive zeros.

EXAMPLE 6 Using Descartes' Rule of Signs to Find Possible Combinations of Real Zeros

Determine the possible combinations of real zeros for

$$P(x) = x^4 - 2x^3 + x^2 + 2x - 2$$

Solution:

$P(x)$ has 3 variations in sign.

$$P(x) = x^4 - 2x^3 + x^2 + 2x - 2$$

Apply Descartes' rule of signs. $P(x)$ has *either* 3 or 1 **positive** real zero.

Find $P(-x)$.

$$P(-x) = (-x)^4 - 2(-x)^3 + (-x)^2 + 2(-x) - 2$$
$$= x^4 + 2x^3 + x^2 - 2x - 2$$

$P(-x)$ has 1 variation in sign. $P(-x) = x^4 + 2x^3 + x^2 - 2x - 2$

Apply Descartes' rule of signs. $\boxed{P(x) \text{ has } 1 \text{ \textbf{negative} real zero.}}$

Since $P(x) = x^4 - 2x^3 + x^2 + 2x - 2$ is a *fourth*-degree polynomial, there are at most 4 real zeros. One zero is a negative real number.

$\boxed{P(x) \text{ could have } 3 \text{ positive real zeros or } 1 \text{ positive real zero.}}$

Look at the graph in the Technology Tip to confirm 1 negative real zero and 1 positive real zero.

Technology Tip

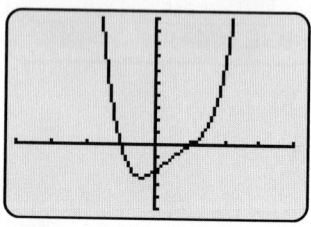

There are 1 negative real zero and 1 positive real zero.

The number of negative real zero is 1 and the number of positive real zero is 1. So, it has 2 complex conjugate zeros.

A table of values supports the zeros of the function.

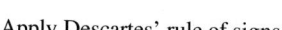

■ **Answer:** Positive real zeros: 1
Negative real zeros: 3 or 1

■ **YOUR TURN** Determine the possible combinations of zeros for

$$P(x) = x^4 + 2x^3 + x^2 + 8x - 12$$

Factoring Polynomials

Now let's draw on the tests discussed in this section thus far to help us in finding all real zeros of a polynomial function. Doing so will enable us to factor polynomials.

EXAMPLE 7   Factoring a Polynomial

Write the polynomial $P(x) = x^5 + 2x^4 - x - 2$ as a product of linear and/or irreducible quadratic factors.

Solution:

STEP 1 Determine variations in sign.

$P(x)$ has 1 sign change. $\qquad\qquad P(x) = x^5 + 2x^4 - x - 2$

$P(-x)$ has 2 sign changes. $\qquad\quad P(-x) = -x^5 + 2x^4 + x - 2$

STEP 2 Apply Descartes' rule of signs. $\quad$ Positive Real Zeros: $\qquad$ 1

$\qquad\qquad\qquad\qquad\qquad\qquad\qquad$ Negative Real Zeros: $\qquad$ 2 or 0

STEP 3 Use the rational zero theorem to
determine the possible rational zeros. $\quad \pm 1, \pm 2$

We know (Step 2) that there is one positive real zero, so test the possible
positive rational zeros first.

STEP 4 Test possible rational zeros.

$$
\begin{array}{r|rrrrrr}
1 & 1 & 2 & 0 & 0 & -1 & -2 \\
  &   & 1 & 3 & 3 & 3 & 2 \\
\hline
  & 1 & 3 & 3 & 3 & 2 & \boxed{0}
\end{array}
$$

1 is a zero:

Now that we have found *the* positive
zero, we can test the other two
possible negative zeros—because
either they both are zeros or neither
is a zero (or one is a double root).

$$
\begin{array}{r|rrrrr}
-1 & 1 & 3 & 3 & 3 & 2 \\
   &   & -1 & -2 & -1 & -2 \\
\hline
   & 1 & 2 & 1 & 2 & \boxed{0}
\end{array}
$$

-1 is a zero:

Let's now try the other possible
negative zero, -2.

$$
\begin{array}{r|rrrr}
-2 & 1 & 2 & 1 & 2 \\
   &   & -2 & 0 & -2 \\
\hline
   & 1 & 0 & 1 & \boxed{0}
\end{array}
$$

-2 is a zero: $\qquad\qquad\qquad\qquad\qquad\quad \underbrace{1 \quad 0 \quad 1}_{x^2+1}$

STEP 5 Three of the five have been found to be zeros: $-1, -2$, and 1.

STEP 6 Write the fifth-degree polynomial as a product of 3 linear factors and an irreducible
quadratic factor.

$$\boxed{P(x) = (x - 1)(x + 1)(x + 2)(x^2 + 1)}$$

■ YOUR TURN Write the polynomial $P(x) = x^5 - 2x^4 + x^3 - 2x^2 - 2x + 4$ as a
product of linear and/or irreducible quadratic factors.

■ **Answer:** $P(x) =$
$(x - 2)(x + 1)(x - 1)(x^2 + 2)$

The rational zero theorem gives us possible rational zeros of a polynomial, and
Descartes' rule of signs gives us possible combinations of positive and negative real zeros.
Additional aids that help eliminate possible zeros are the *upper* and *lower bound rules*.

Study Tip

If $f(x)$ has a common monomial factor, it should be canceled first before applying the bound rules.

These rules can give you an upper and lower bound on the real zeros of a polynomial function. If $f(x)$ has a common monomial factor, you should factor it out first, and then follow the upper and lower bound rules.

UPPER AND LOWER BOUND RULES

Let $f(x)$ be a polynomial with real coefficients and a positive leading coefficient. Suppose $f(x)$ is divided by $x - c$ using synthetic division.

1. If $c > 0$ and each number in the bottom row is either positive or zero, c is an **upper bound** for the real zeros of f.
2. If $c < 0$ and the numbers in the bottom row are alternately positive and negative (zero entries count as either positive or negative), c is a **lower bound** for the real zeros of f.

EXAMPLE 8   Using Upper and Lower Bounds to Eliminate Possible Zeros

Find the real zeros of $f(x) = 4x^3 - x^2 + 36x - 9$.

Solution:

STEP 1 The rational zero theorem gives possible rational zeros.

$$\frac{\text{Factors of } 9}{\text{Factors of } 4} = \frac{\pm 1, \pm 3, \pm 9}{\pm 1, \pm 2, \pm 4}$$

$$= \pm 1, \pm \frac{1}{2}, \pm \frac{1}{4}, \pm \frac{3}{4}, \pm \frac{3}{2}, \pm \frac{9}{4}, \pm 3, \pm \frac{9}{2}, \pm 9$$

STEP 2 Apply Descartes' rule of signs:

$f(x)$ has 3 sign variations.   3 or 1 positive real zeros

$f(-x)$ has no sign variations.   no negative real zeros

STEP 3 Try $x = 1$.

$$\begin{array}{r|rrrr} 1 & 4 & -1 & 36 & -9 \\ & & 4 & 3 & 39 \\ \hline & 4 & 3 & 39 & 30 \end{array}$$

$x = 1$ is not a zero, but because the last row contains all positive entries, $x = 1$ is an *upper* bound. Since we know there are no negative real zeros, we restrict our search to between 0 and 1.

Study Tip

In Example 8, Steps 3 and 4 long division can be used as well as evaluating the function at $x = 1$ and $x = \frac{1}{4}$ to determine if these are zeros.

STEP 4 Try $x = \frac{1}{4}$.

$$\begin{array}{r|rrrr} \frac{1}{4} & 4 & -1 & 36 & -9 \\ & & 1 & 0 & 9 \\ \hline & 4 & 0 & 36 & 0 \end{array}$$

$\frac{1}{4}$ is a zero and the quotient $4x^2 + 36$ has all positive coefficients; therefore, $\frac{1}{4}$ is an upper bound, so $\boxed{\frac{1}{4} \text{ is the only real zero}}$.

Note: If $f(x)$ has a common monomial factor, it should be factored out first before applying the bound rules.

The Intermediate Value Theorem

In our search for zeros, we sometimes encounter irrational zeros, as in, for example, the polynomial

$$f(x) = x^5 - x^4 - 1$$

Descartes' rule of signs tells us there is exactly one real positive zero. However, the rational zero test yields only $x = \pm 1$, neither of which is a zero. So if we know there is a real positive zero and we know it's not rational, it must be irrational. Notice that $f(1) = -1$ and $f(2) = 15$. Since polynomial functions are continuous and the function goes from negative to positive between $x = 1$ and $x = 2$, we expect a zero somewhere in that interval. Generating a graph with a graphing utility, we find that there is a zero around $x = 1.3$.

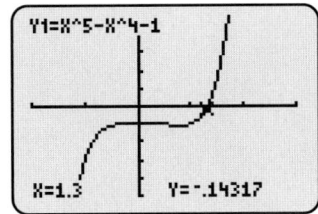

The *intermediate value theorem* is based on the fact that polynomial functions are continuous.

INTERMEDIATE VALUE THEOREM

Let a and b be real numbers such that $a < b$ and $f(x)$ is a polynomial function. If $f(a)$ and $f(b)$ have opposite signs, then there is at least one real zero between a and b.

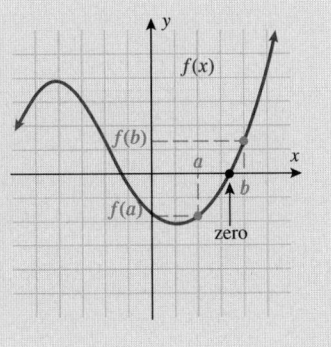

If the intermediate value theorem tells us that there is a real zero in the interval (a, b), how do we approximate that zero? The **bisection method*** is a root-finding algorithm that approximates the solution to the equation $f(x) = 0$. In the bisection method the interval is divided in half and then the subinterval that contains the zero is selected. This is repeated until the bisection method converges to an approximate root of f.

---

*In calculus you will learn Newton's method, which is a more efficient approximation technique for finding zeros.

Technology Tip

The graph of $f(x) = x^5 - x^4 - 1$ is shown. To find the zero of the function, press

$$\boxed{\text{2nd}}\ \boxed{\text{Calc}}\ \boxed{\blacktriangledown}\ \boxed{\text{2:zero}}\ \boxed{\text{ENTER}}$$

$$\boxed{1}\ \boxed{\text{ENTER}}\ \boxed{2}\ \boxed{\text{ENTER}}\ \boxed{\text{ENTER}}$$

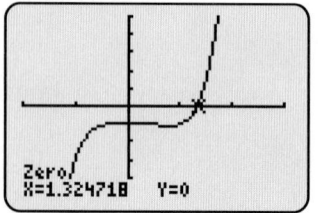

A table of values supports this zero of the function.

EXAMPLE 9   Approximating Real Zeros of a Polynomial Function

Approximate the real zero of $f(x) = x^5 - x^4 - 1$.

Note: Descartes' rule of signs tells us that there are no real negative zeros and there is exactly one real positive zero.

Solution:

Find two consecutive integer values for x that have corresponding function values opposite in sign.

x	1	2
$f(x)$	-1	15

Note that a graphing utility would have shown an x-intercept between $x = 1$ and $x = 2$.

Apply the bisection method, with $a = 1$ and $b = 2$.

$$c = \frac{a+b}{2} = \frac{1+2}{2} = \frac{3}{2}$$

Evaluate the function at $x = c$.

$$f(1.5) \approx 1.53$$

Compare the values of f at the endpoints and midpoint.

$$f(1) = -1, f(1.5) \approx 1.53, f(2) = 15$$

Select the subinterval corresponding to the *opposite* signs of f.

$$(1, 1.5)$$

Apply the bisection method again (repeat the algorithm).

$$\frac{1 + 1.5}{2} = 1.25$$

Evaluate the function at $x = 1.25$.

$$f(1.25) \approx -0.38965$$

Compare the values of f at the endpoints and midpoint.

$$f(1) = -1, f(1.25) \approx -0.38965, f(1.5) \approx 1.53$$

Select the subinterval corresponding to the *opposite* signs of f.

$$(1.25, 1.5)$$

Apply the bisection method again (repeat the algorithm).

$$\frac{1.25 + 1.5}{2} = 1.375$$

Evaluate the function at $x = 1.375$.

$$f(1.375) \approx 0.3404$$

Compare the values of f at the endpoints and midpoint.

$$f(1.25) \approx -0.38965, f(1.375) \approx 0.3404, f(1.5) \approx 1.53$$

Select the subinterval corresponding to the *opposite* signs of f.

$$(1.25, 1.375)$$

We can continue this procedure (*applying the bisection method*) to find that the zero is somewhere between $x = 1.32$ and $x = 1.33$, since $f(1.32) \approx -0.285$ and $f(1.33) \approx 0.0326$.

We find that, to three significant digits, $\boxed{1.32}$ is an approximation to the real zero.

Graphing Polynomial Functions

In Section 2.2, we graphed simple polynomial functions that were easily factored. Now that we have procedures for finding real zeros of polynomial functions (rational zero theorem, Descartes' rule of signs, and upper and lower bound rules for rational zeros, and the intermediate value theorem and the bisection method for irrational zeros), let us return to the topic of graphing polynomial functions. Since a real zero of a polynomial function corresponds to an x-intercept of its graph, we now have methods for finding (or estimating) any x-intercepts of the graph of any polynomial function.

EXAMPLE 10 **Graphing a Polynomial Function**

Graph the function $f(x) = 2x^4 - 2x^3 + 5x^2 + 17x - 22$.

Solution:

STEP 1 **Find the y-intercept.** $\qquad\qquad\qquad\qquad f(0) = -22$

STEP 2 **Find any x-intercepts (real zeros).**

Apply Descartes' rule of signs.

3 sign changes correspond to
3 or 1 positive real zeros. $\qquad\qquad f(x) = 2x^4 - 2x^3 + 5x^2 + 17x - 22$

1 sign change corresponds to
1 negative real zero. $\qquad\qquad f(-x) = 2x^4 + 2x^3 + 5x^2 - 17x - 22$

Apply the rational zero theorem.

Let $a_0 = -22$ and $a_n = 2$. $\qquad \dfrac{\text{Factors of } a_0}{\text{Factors of } a_n} = \pm\dfrac{1}{2}, \pm 1, \pm 2, \pm\dfrac{11}{2}, \pm 11, \pm 22$

Test the possible zeros.

$x = 1$ is a zero. $\qquad\qquad\qquad\qquad f(1) = 0$

There are no other rational zeros.

Apply the upper bound rule.

$$
\begin{array}{r|rrrr}
1 & 2 & -2 & 5 & 17 & -22 \\
  &   & 2 & 0 & 5 & 22 \\
\hline
  & 2 & 0 & 5 & 22 & \boxed{0}
\end{array}
$$

Since $x = 1$ is positive and all of the numbers in the bottom row are positive (or zero), $x = 1$ is an upper bound for the real zeros. We know there is exactly one negative real zero, but none of the possible zeros from the rational zero theorem is a zero. Therefore, the negative real zero is irrational.

Apply the intermediate value theorem and the bisection method.

f is positive at $x = -2$. $\qquad\qquad\qquad f(-2) = 12$

f is negative at $x = -1$. $\qquad\qquad\qquad f(-1) = -30$

Use the bisection method to find the negative
real zero between -2 and -1. $\qquad\qquad x \approx -1.85$

STEP 3 **Determine the end behavior.** $\qquad\qquad y = 2x^4$

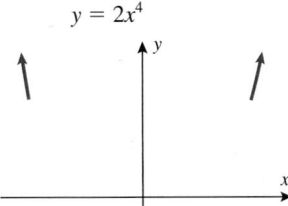

The graph of
$f(x) = 2x^4 - 2x^3 + 5x^2 + 17x - 22$
is shown.

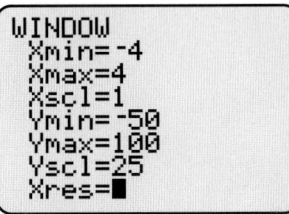

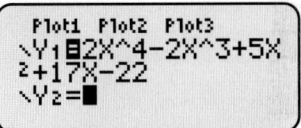

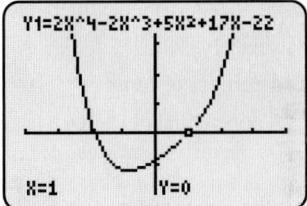

To find the zero of the function, press:

2nd | Calc | ▼ | 2:zero | ENTER

−3 | ENTER | −1 | ENTER | ENTER

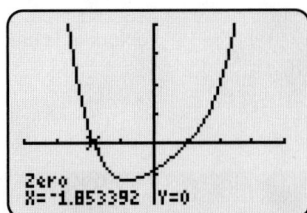

A table of values supports this zero of
the function and the graph.

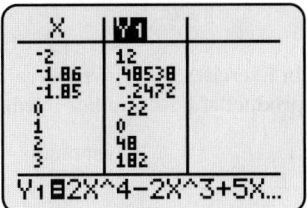

STEP 4 **Find additional points.**

x	-2	-1.85	-1	0	1	2
$f(x)$	12	0	-30	-22	0	48
Point	$(-2, 12)$	$(-1.85, 0)$	$(-1, -30)$	$(0, -22)$	$(1, 0)$	$(2, 48)$

STEP 5 **Sketch the graph.**

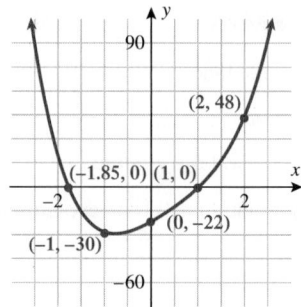

SECTION 2.4 SUMMARY

In this section, we discussed how to find the real zeros of a polynomial function. Once real zeros are known, it is possible to write the polynomial function as a product of linear and/or irreducible quadratic factors.

The Number of Zeros

- A polynomial of degree n has *at most n* real zeros.
- *Descartes' rule of signs* determines the possible combinations of positive and negative real zeros.
- *Upper and lower bounds* help narrow the search for zeros.

How to Find Zeros

- *Rational zero theorem:* List possible rational zeros:

$$\frac{\text{Factors of constant, } a_0}{\text{Factors of leading coefficient, } a_n}$$

- *Irrational zeros:* Approximate zeros by determining when the polynomial function changes sign (intermediate value theorem).

Procedure for Factoring a Polynomial Function

- List possible rational zeros (rational zero theorem).
- List possible combinations of positive and negative real zeros (Descartes' rule of signs).
- Test possible values until a zero is found.*
- Once a real zero is found, repeat testing on the quotient until linear and/or irreducible quadratic factors remain.
- If there is a real zero but all possible rational roots have failed, then approximate the zero using the *intermediate value theorem* and the *bisection method*.

*Depending on the form of the quotient, upper and lower bounds may eliminate possible zeros.

SECTION 2.4 EXERCISES

■ SKILLS

In Exercises 1–10, given a real zero of the polynomial, determine all other real zeros, and write the polynomial in terms of a product of linear and/or irreducible quadratic factors.

Polynomial	**Zero**	**Polynomial**	**Zero**
1. $P(x) = x^3 - 13x + 12$	1	**2.** $P(x) = x^3 + 3x^2 - 10x - 24$	3
3. $P(x) = 2x^3 + x^2 - 13x + 6$	$\frac{1}{2}$	**4.** $P(x) = 3x^3 - 14x^2 + 7x + 4$	$-\frac{1}{3}$

5. $P(x) = x^4 - 2x^3 - 11x^2 - 8x - 60$ $-3, 5$ **6.** $P(x) = x^4 - x^3 + 7x^2 - 9x - 18$ $-1, 2$

7. $P(x) = x^4 - 5x^2 + 10x - 6$ $1, -3$ **8.** $P(x) = x^4 - 4x^3 + x^2 + 6x - 40$ $4, -2$

9. $P(x) = x^4 + 6x^3 + 13x^2 + 12x + 4$ -2 (multiplicity 2)   **10.** $P(x) = x^4 + 4x^3 - 2x^2 - 12x + 9$ 1 (multiplicity 2)

In Exercises 11–18, use the rational zero theorem to list the *possible* rational zeros.

11. $P(x) = x^4 + 3x^2 - 8x + 4$ **12.** $P(x) = -x^4 + 2x^3 - 5x + 4$ **13.** $P(x) = x^5 - 14x^3 + x^2 - 15x + 12$

14. $P(x) = x^5 - x^3 - x^2 + 4x + 9$ **15.** $P(x) = 2x^6 - 7x^4 + x^3 - 2x + 8$ **16.** $P(x) = 3x^5 + 2x^4 - 5x^3 + x - 10$

17. $P(x) = 5x^5 + 3x^4 + x^3 - x - 20$ **18.** $P(x) = 4x^6 - 7x^4 + 4x^3 + x - 21$

In Exercises 19–22, list the possible rational zeros, and test to determine all rational zeros.

19. $P(x) = x^4 + 2x^3 - 9x^2 - 2x + 8$ **20.** $P(x) = x^4 + 2x^3 - 4x^2 - 2x + 3$

21. $P(x) = 2x^3 - 9x^2 + 10x - 3$ **22.** $P(x) = 3x^3 - 5x^2 - 26x - 8$

In Exercises 23–34, use Descartes' rule of signs to determine the possible number of positive real zeros and negative real zeros.

23. $P(x) = x^4 - 32$ **24.** $P(x) = x^4 + 32$ **25.** $P(x) = x^5 - 1$

26. $P(x) = x^5 + 1$ **27.** $P(x) = x^5 - 3x^3 - x + 2$ **28.** $P(x) = x^4 + 2x^2 - 9$

29. $P(x) = 9x^7 + 2x^5 - x^3 - x$ **30.** $P(x) = 16x^7 - 3x^4 + 2x - 1$ **31.** $P(x) = x^6 - 16x^4 + 2x^2 + 7$

32. $P(x) = -7x^6 - 5x^4 - x^2 + 2x + 1$ **33.** $P(x) = -3x^4 + 2x^3 - 4x^2 + x - 11$ **34.** $P(x) = 2x^4 - 3x^3 + 7x^2 + 3x + 2$

For each polynomial in Exercises 35–52: (a) use Descartes' rule of signs to determine the possible combinations of positive real zeros and negative real zeros; (b) use the rational zero test to determine possible rational zeros; (c) test for rational zeros; and (d) factor as a product of linear and/or irreducible quadratic factors.

35. $P(x) = x^3 + 6x^2 + 11x + 6$ **36.** $P(x) = x^3 - 6x^2 + 11x - 6$ **37.** $P(x) = x^3 - 7x^2 - x + 7$

38. $P(x) = x^3 - 5x^2 - 4x + 20$ **39.** $P(x) = x^4 + 6x^3 + 3x^2 - 10x$ **40.** $P(x) = x^4 - x^3 - 14x^2 + 24x$

41. $P(x) = x^4 - 7x^3 + 27x^2 - 47x + 26$ **42.** $P(x) = x^4 - 5x^3 + 5x^2 + 25x - 26$ **43.** $P(x) = 10x^3 - 7x^2 - 4x + 1$

44. $P(x) = 12x^3 - 13x^2 + 2x - 1$ **45.** $P(x) = 6x^3 + 17x^2 + x - 10$ **46.** $P(x) = 6x^3 + x^2 - 5x - 2$

47. $P(x) = x^4 - 2x^3 + 5x^2 - 8x + 4$ **48.** $P(x) = x^4 + 2x^3 + 10x^2 + 18x + 9$ **49.** $P(x) = x^6 + 12x^4 + 23x^2 - 36$

50. $P(x) = x^4 - x^2 - 16x^2 + 16$ **51.** $P(x) = 4x^4 - 20x^3 + 37x^2 - 24x + 5$ **52.** $P(x) = 4x^4 - 8x^3 + 7x^2 + 30x + 50$

In Exercises 53–56, use the information found in Exercises 37, 41, 45, and 51 to assist in sketching a graph of each polynomial function.

53. Exercise 37   **54.** Exercise 41   **55.** Exercise 45   **56.** Exercise 51

In Exercises 57–64, use the intermediate value theorem to approximate the real zero in the indicated interval. Approximate to two decimal places.

57. $f(x) = x^4 - 3x^3 + 4$ $[1, 2]$ **58.** $f(x) = x^5 - 3x^3 + 1$ $[0, 1]$

59. $f(x) = 7x^5 - 2x^2 + 5x - 1$ $[0, 1]$ **60.** $f(x) = -2x^3 + 3x^2 + 6x - 7$ $[-2, -1]$

61. $f(x) = x^3 - 2x^2 - 8x - 3$ $[-1, 0]$ **62.** $f(x) = x^4 + 4x^2 - 7x - 13$ $[-2, -1]$

63. $f(x) = x^5 + 2x^4 - 6x^3 - 25x^2 + 8x - 10$ $[2, 3]$ **64.** $f(x) = \frac{1}{2}x^6 + x^4 - 2x^2 - x + 1$ $[1, 2]$

▪ APPLICATIONS

65. Geometry. The distances (in inches) from one vertex of a rectangle to the other three vertices are x, $x + 2$, and $x + 4$. Find the dimensions of the rectangle.

66. Geometry. A box is constructed to contain a volume of 97.5 in^3. The length of the base is 3.5 in. larger than the width, and the height is 0.5 in. larger than the length. Find the dimensions of the box.

67. Agriculture. The weekly volume (in liters) of milk produced in a farm is given by $v(x) = x^3 + 21x^2 - 1480x$, where x is the number of cows. Find the number of cows that corresponds to a total production of 1500 liters of milk in a week.

68. Profit. A bakery uses the formula $f(x) = 2x^4 - 7x^3 + 3x^2 + 8x$ to determine the profit of selling x loaves of bread. How many loaves of bread must be sold to have a profit of $4? Assume $x \geq 1$.

▪ CATCH THE MISTAKE

In Exercises 69 and 70, explain the mistake that is made.

69. Use Descartes' rule of signs to determine the possible combinations of zeros of

$$P(x) = 2x^5 + 7x^4 + 9x^3 + 9x^2 + 7x + 2$$

Solution:

No sign changes, so no positive real zeros.

$$P(x) = 2x^5 + 7x^4 + 9x^3 + 9x^2 + 7x + 2$$

Five sign changes, so five negative real zeros.

$$P(-x) = -2x^5 + 7x^4 - 9x^3 + 9x^2 - 7x + 2$$

This is incorrect. What mistake was made?

70. Determine whether $x - 2$ is a factor of

$$P(x) = x^3 - 2x^2 - 5x + 6$$

Solution:

$$\begin{array}{r|rrrr} -2 & 1 & -2 & -5 & 6 \\ & & -2 & 8 & -6 \\ \hline & 1 & -4 & 3 & \boxed{0} \end{array}$$

Yes, $x - 2$ is a factor of $P(x)$.

This is incorrect. What mistake was made?

▪ CONCEPTUAL

In Exercises 71–74, determine whether each statement is true or false.

71. All real zeros of a polynomial correspond to x-intercepts.

72. A polynomial of degree n, $n > 0$, must have at least one zero.

73. A polynomial of degree n, $n > 0$, can be written as a product of n linear factors over real numbers.

74. The number of sign changes in a polynomial is equal to the number of positive real zeros of that polynomial.

75. A polynomial of degree n, $n > 0$, must have exactly n x-intercepts.

76. A polynomial with an odd number of zeros must have odd degree.

▪ CHALLENGE

77. Given that $x = a$ is a zero of $P(x) = x^3 - (a + b + c)x^2 + (ab + ac + bc)x - abc$, find the other two zeros, given that a, b, and c are real numbers and $a > b > c$.

78. Given that $x = a$ is a zero of $p(x) = x^3 + (-a + b - c)x^2 - (ab + bc - ac)x + abc$, find the other two real zeros, given that a, b, and c are real positive numbers.

79. Given that b is a zero of $P(x) = x^4 - (a + b)x^3 + (ab - c^2)x^2 + (a + b)c^2x - abc^2$, find the other three real zeros, given that a, b, and c are real positive numbers.

80. Given that a is a zero of $P(x) = x^4 + 2(b - a)x^3 + (a^2 - 4ab + b^2)x^2 + 2ab(a - b)x + a^2b^2$, find the other three real zeros, given that a and b are real positive numbers.

▪ TECHNOLOGY

In Exercises 81 and 82, determine all possible rational zeros of the polynomial. There are many possibilities. Instead of trying them all, use a graphing calculator or software to graph $P(x)$ to help find a zero to test.

81. $P(x) = x^3 - 2x^2 + 16x - 32$

82. $P(x) = x^3 - 3x^2 + 16x - 48$

In Exercises 83 and 84: (a) determine all possible rational zeros of the polynomial, using a graphing calculator or software to graph $P(x)$ to help find the zeros; and (b) factor as a product of linear and/or irreducible quadratic factors.

83. $P(x) = 12x^4 + 25x^3 + 56x^2 - 7x - 30$

84. $P(x) = -3x^3 - x^2 - 7x - 49$

▪ PREVIEW TO CALCULUS

In calculus we use the zeros of the derivative f' of a function f to determine whether the function f is increasing or decreasing around the zeros.

In Exercises 85–88, find the zeros of each polynomial function and determine the intervals over which $f(x) > 0$.

85. $f(x) = x^3 - 4x^2 - 7x + 10$

86. $f(x) = 6x^3 - 13x^2 - 11x + 8$

87. $f(x) = -2x^4 + 5x^3 + 7x^2 - 10x - 6$

88. $f(x) = -3x^4 + 14x^3 - 11x^2 + 14x - 8$

SECTION 2.5   COMPLEX ZEROS: THE FUNDAMENTAL THEOREM OF ALGEBRA

SKILLS OBJECTIVES

- Find the complex zeros of a polynomial function.
- Use the complex conjugate zeros theorem.
- Factor polynomial functions over the complex numbers.

CONCEPTUAL OBJECTIVES

- Extend the domain of polynomial functions to complex numbers.
- Understand how the fundamental theorem of algebra guarantees at least one zero.
- Understand why complex zeros occur in conjugate pairs.

In Section 2.4, we found the *real* zeros of a polynomial function. In this section, we find the *complex* zeros of a polynomial function. The domain of polynomial functions thus far has been the set of all real numbers. Now, we consider a more general case, where the domain of a polynomial function is the set of *complex numbers*. Note that the set of real numbers is a subset of the complex numbers. (Choose the imaginary part to be zero.)

It is important to note, however, that when we are discussing *graphs* of polynomial functions, we restrict the domain to the set of real numbers.

A *zero* of a polynomial $P(x)$ is the *solution* or *root* of the equation $P(x) = 0$. The *zeros of a polynomial can be complex numbers.* However, since the axes of the xy-plane represent real numbers, we interpret zeros as x-intercepts only when the zeros are real numbers.

We can illustrate the relationship between real and complex zeros of polynomial functions and their graphs with two similar examples. Let's take the two quadratic functions

Study Tip

The zeros of a polynomial can be complex numbers. Only when the zeros are real numbers do we interpret zeros as x-intercepts.

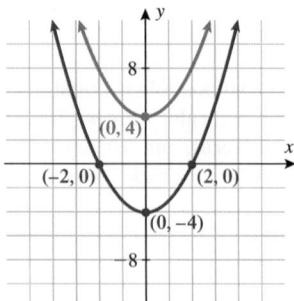

$f(x) = x^2 - 4$ and $g(x) = x^2 + 4$. The graphs of these two functions are parabolas that open upward with $f(x)$ shifted down 4 units and $g(x)$ shifted up 4 units as shown on the left. Setting each function equal to zero and solving for x, we find that the zeros for $f(x)$ are -2 and 2 and the zeros for $g(x)$ are $-2i$ and $2i$. Notice that the x-intercepts for $f(x)$ are $(-2, 0)$ and $(2, 0)$ and $g(x)$ has no x-intercepts.

The Fundamental Theorem of Algebra

In Section 2.4, we were able to write a polynomial function as a product of linear and/or irreducible quadratic factors. Now, we consider factors over complex numbers. Therefore, what were irreducible quadratic factors over real numbers will now be a product of two linear factors over the complex numbers.

What are the minimum and maximum number of zeros a polynomial can have? Every polynomial has *at least one zero* (provided the degree is greater than zero). The largest number of zeros a polynomial can have is equal to the degree of the polynomial.

THE FUNDAMENTAL THEOREM OF ALGEBRA

Every polynomial $P(x)$ of degree $n > 0$ has *at least one zero* in the complex number system.

The fundamental theorem of algebra and the factor theorem are used to prove the following *n* zeros theorem.

n ZEROS THEOREM

Every polynomial $P(x)$ of degree $n > 0$ can be expressed as the product of n linear factors in the complex number system. Hence, $P(x)$ has exactly n zeros, not necessarily distinct.

These two theorems are illustrated with five polynomials below:

a. The **first**-degree polynomial $f(x) = x + 3$ has exactly **one** zero: $x = -3$.

b. The **second**-degree polynomial $f(x) = x^2 + 10x + 25 = (x + 5)(x + 5)$ has exactly **two** zeros: $x = -5$ and $x = -5$. It is customary to write this as a single zero of multiplicity 2 or refer to it as a repeated root.

c. The **third**-degree polynomial $f(x) = x^3 + 16x = x(x^2 + 16) = x(x + 4i)(x - 4i)$ has exactly **three** zeros: $x = 0$, $x = -4i$, and $x = 4i$.

d. The **fourth**-degree polynomial $f(x) = x^4 - 1 = (x^2 - 1)(x^2 + 1)$
$= (x - 1)(x + 1)(x - i)(x + i)$ has exactly **four** zeros:
$x = 1, x = -1, x = i$, and $x = -i$.

e. The **fifth**-degree polynomial $f(x) = x^5 = x \cdot x \cdot x \cdot x \cdot x$ has exactly **five** zeros: $x = 0$, which has multiplicity 5.

The fundamental theorem of algebra and the *n* zeros theorem only tell you that the zeros *exist*—not how to find them. We must rely on techniques discussed in Section 2.4 and additional strategies discussed in this section to determine the zeros.

Complex Zeros

Often, at a grocery store or a drugstore, we see signs for special offers—"buy one, get one free." A similar phenomenon occurs for complex zeros of a polynomial function with real coefficients. If we restrict the coefficients of a polynomial to real numbers, complex zeros always come in conjugate pairs. In other words, if a zero of a polynomial function is a complex number, then another zero will always be its complex conjugate. Look at the third-degree polynomial in the above illustration, part (c), where two of the zeros were $-4i$ and $4i$, and in part (d), where two of the zeros were i and $-i$. In general, if we restrict the coefficients of a polynomial to real numbers, complex zeros always come in conjugate pairs.

COMPLEX CONJUGATE ZEROS THEOREM

If a polynomial $P(x)$ has real coefficients, and if $a + bi$ is a zero of $P(x)$, then its complex conjugate $a - bi$ is also a zero of $P(x)$.

We use the complex zeros theorem to assist us in factoring a higher degree polynomial.

EXAMPLE 1   Factoring a Polynomial with Complex Zeros

Factor the polynomial $P(x) = x^4 - x^3 - 5x^2 - x - 6$ given that i is a zero of $P(x)$.

Since $P(x)$ is a *fourth*-degree polynomial, we expect *four* zeros. The goal in this problem is to write $P(x)$ as a product of four linear factors: $P(x) = (x - a)(x - b)(x - c)(x - d)$, where a, b, c, and d are complex numbers and represent the zeros of the polynomial.

Solution:

Write known zeros and linear factors.

Since i is a zero, we know that $-i$ is a zero.
$$x = i \quad \text{and} \quad x = -i$$

We now know two linear factors of $P(x)$.
$$(x - i) \quad \text{and} \quad (x + i)$$

Write $P(x)$ as a product of four factors.
$$P(x) = (x - i)(x + i)(x - c)(x - d)$$

Multiply the two known factors.
$$(x + i)(x - i) = x^2 - i^2$$
$$= x^2 - (-1)$$
$$= x^2 + 1$$

Rewrite the polynomial.
$$P(x) = (x^2 + 1)(x - c)(x - d)$$

Divide both sides of the equation by $x^2 + 1$.
$$\frac{P(x)}{x^2 + 1} = (x - c)(x - d)$$

Divide $P(x)$ by $x^2 + 1$ using long division.

$$
\begin{array}{r}
x^2 - x - 6 \\
x^2 + 0x + 1 \overline{)x^4 - x^3 - 5x^2 - x - 6} \\
\underline{-(x^4 + 0x^3 + x^2)} \\
-x^3 - 6x^2 - x \\
\underline{-(-x^3 + 0x^2 - x)} \\
-6x^2 + 0x - 6 \\
\underline{-(-6x^2 + 0x - 6)} \\
0
\end{array}
$$

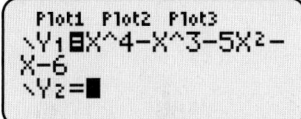
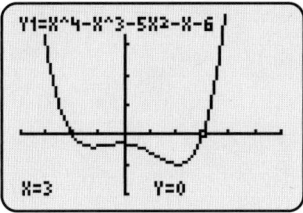

Since the remainder is 0, $x^2 - x - 6$ is a factor.

$$P(x) = \left(x^2 + 1\right)\left(x^2 - x - 6\right)$$

Factor the quotient $x^2 - x - 6$.

$$x^2 - x - 6 = (x - 3)(x + 2)$$

Write $P(x)$ as a product of four linear factors.

$$\boxed{P(x) = (x - i)(x + i)(x - 3)(x + 2)}$$

Check: $P(x)$ is a *fourth*-degree polynomial and we found *four* zeros, two of which are complex conjugates.

■ **Answer:** $P(x) =$
$(x - 2i)(x + 2i)(x - 1)(x - 2)$
Note: The zeros of $P(x)$ are 1, 2, 2i, and $-2i$.

■ **YOUR TURN** Factor the polynomial $P(x) = x^4 - 3x^3 + 6x^2 - 12x + 8$ given that $x - 2i$ is a factor.

EXAMPLE 2 **Factoring a Polynomial with Complex Zeros**

Factor the polynomial $P(x) = x^4 - 2x^3 + x^2 + 2x - 2$ given that $1 + i$ is a zero of $P(x)$.

Since $P(x)$ is a *fourth*-degree polynomial, we expect *four* zeros. The goal in this problem is to write $P(x)$ as a product of four linear factors: $P(x) = (x - a)(x - b)(x - c)(x - d)$, where a, b, c, and d are complex numbers and represent the zeros of the polynomial.

Solution:

STEP 1 Write known zeros and linear factors.

Since $1 + i$ is a zero, we know that $1 - i$ is a zero.

$$x = 1 + i \text{ and } x = 1 - i$$

We now know two linear factors of $P(x)$.

$$[x - (1 + i)] \text{ and } [x - (1 - i)]$$

STEP 2 Write $P(x)$ as a product of four factors.

$$P(x) = [x - (1 + i)][x - (1 - i)](x - c)(x - d)$$

STEP 3 Multiply the first two terms.

$$[x - (1 + i)][x - (1 - i)]$$

First regroup the expressions in each bracket.

$$[(x - 1) - i][(x - 1) + i]$$

Use the special product $(a - b)(a + b) = a^2 - b^2$, where a is $(x - 1)$ and b is i.

$$(x - 1)^2 - i^2$$
$$\left(x^2 - 2x + 1\right) - (-1)$$
$$x^2 - 2x + 2$$

STEP 4 Rewrite the polynomial.

$$P(x) = \left(x^2 - 2x + 2\right)(x - c)(x - d)$$

STEP 5 Divide both sides of the equation by $x^2 - 2x + 2$, and substitute in the original polynomial $P(x) = x^4 - 2x^3 + x^2 + 2x - 2$.

$$\frac{x^4 - 2x^3 + x^2 + 2x - 2}{x^2 - 2x + 2} = (x - c)(x - d)$$

STEP 6 Divide the left side of the equation using long division.

$$\frac{x^4 - 2x^3 + x^2 + 2x - 2}{x^2 - 2x + 2} = x^2 - 1$$

STEP 7 Factor $x^2 - 1$.

$$(x - 1)(x + 1)$$

STEP 8 Write $P(x)$ as a product of four linear factors.

$$\boxed{P(x) = [x - (1 + i)][x - (1 - i)][x - 1][x + 1]}$$

■ **Answer:** $P(x) = [x - (1 + 2i)] \cdot$
$[x - (1 - 2i)](x - 1)(x + 3)$
Note: The zeros of $P(x)$ are 1, -3, $1 + 2i$, and $1 - 2i$.

■ **YOUR TURN** Factor the polynomial $P(x) = x^4 - 2x^2 + 16x - 15$ given that $1 + 2i$ is a zero.

Because an n-degree polynomial function has exactly n zeros and since complex zeros always come in conjugate pairs, if the degree of the polynomial is **odd**, there is guaranteed to be **at least one zero that is a real number**. If the degree of the polynomial is even, there is no guarantee that a zero will be real—all the zeros could be complex.

EXAMPLE 3   Finding Possible Combinations of Real and Complex Zeros

List the possible combinations of real and complex zeros for the given polynomials.

a. $17x^5 + 2x^4 - 3x^3 + x^2 - 5$ **b.** $5x^4 + 2x^3 - x + 2$

Solution:

a.   Since this is a *fifth*-degree polynomial, there are *five* zeros. Because complex zeros come in conjugate pairs, the table describes the possible five zeros.

REAL ZEROS	COMPLEX ZEROS
1	4
3	2
5	0

Applying Descartes' rule of signs, we find that there are 3 or 1 positive real zeros and 2 or 0 negative real zeros.

POSITIVE REAL ZEROS	NEGATIVE REAL ZEROS	COMPLEX ZEROS
1	0	4
3	0	2
1	2	2
3	2	0

b.   Because this is a *fourth*-degree polynomial, there are *four* zeros. Since complex zeros come in conjugate pairs, the table describes the possible four zeros.

REAL ZEROS	COMPLEX ZEROS
0	4
2	2
4	0

Applying Descartes' rule of signs, we find that there are 2 or 0 positive real zeros and 2 or 0 negative real zeros.

POSITIVE REAL ZEROS	NEGATIVE REAL ZEROS	COMPLEX ZEROS
0	0	4
2	0	2
0	2	2
2	2	0

▪ **YOUR TURN** List the possible combinations of real and complex zeros for

$$P(x) = x^6 - 7x^5 + 8x^3 - 2x + 1$$

▪ **Answer:**

REAL ZEROS	COMPLEX ZEROS
0	6
2	4
4	2
6	0

Factoring Polynomials

Now let's draw on the tests discussed in this chapter to help us find all the zeros of a polynomial. Doing so will enable us to write polynomials as a product of linear factors. Before reading Example 4, reread Section 2.4, Example 7.

EXAMPLE 4 Factoring a Polynomial

Factor the polynomial $P(x) = x^5 + 2x^4 - x - 2$.

Solution:

STEP 1 Determine variations in sign.

$P(x)$ has 1 sign change. $\qquad\qquad$ $P(x) = x^5 + 2x^4 - x - 2$

$P(-x)$ has 2 sign changes. $\qquad\quad$ $P(-x) = -x^5 + 2x^4 + x - 2$

STEP 2 Apply Descartes' rule of signs and summarize the results in a table.

POSITIVE REAL ZEROS	NEGATIVE REAL ZEROS	COMPLEX ZEROS
1	2	2
1	0	4

STEP 3 Utilize the rational zero theorem to determine the possible rational zeros. $\qquad\qquad \pm 1, \pm 2$

STEP 4 Test possible rational zeros.

1 is a zero: $\qquad\qquad\qquad\qquad P(1) = (1)^5 + 2(1)^4 - (1) - 2 = 0$

Now that we have found *the* positive zero, we can test the two negative zeros.

-1 is a zero: $\qquad\qquad\qquad P(-1) = (-1)^5 + 2(-1)^4 - (-1) - 2 = 0$

-2 is a zero: $\qquad\qquad\qquad P(-2) = (-2)^5 + 2(-2)^4 - (-2) - 2$
$\qquad\qquad\qquad\qquad\qquad\qquad\quad = -32 + 32 + 2 - 2 = 0$

STEP 5 Three of the five zeros have been found: -1, -2, and 1.

STEP 6 Write the fifth-degree polynomial as a product of five linear factors.

$$P(x) = x^5 + 2x^4 - x - 2 = (x - 1)(x + 1)(x + 2)(x - c)(x - d)$$

We know that the remaining two zeros are complex according to Descartes' rule of signs and that complex roots come in conjugate pairs.

STEP 7 Multiply the three known linear factors.

$$(x - 1)(x + 1)(x + 2) = x^3 + 2x^2 - x - 2$$

STEP 8 Rewrite the polynomial.

$$P(x) = x^5 + 2x^4 - x - 2 = \left(x^3 + 2x^2 - x - 2\right)(x - c)(x - d)$$

STEP 9 Divide both sides of the equation by $x^3 + 2x^2 - x - 2$.

$$\frac{x^5 + 2x^4 - x - 2}{x^3 + 2x^2 - x - 2} = (x - c)(x - d)$$

STEP 10 Dividing the left side of the equation with long division yields

$$\frac{x^5 + 2x^4 - x - 2}{x^3 + 2x^2 - x - 2} = x^2 + 1 = (x - i)(x + i)$$

STEP 11 Write the remaining two zeros. $\qquad x = \pm i$

STEP 12 Write $P(x)$ as a product of linear factors.

$$\boxed{P(x) = (x - 1)(x + 1)(x + 2)(x - i)(x + i)}$$

Study Tip

From Step 2 we know there is one positive real zero, so test the positive possible rational zeros first in Step 4.

Study Tip

In Step 4 we could have used synthetic division to show that 1, -1, and -2 are all zeros of $P(x)$.

Technology Tip

The graph of $P(x) = x^5 + 2x^4 - x - 2$ is shown.

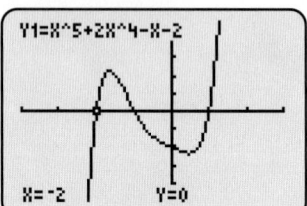

The real zeros of the function at $x = -2$, $x = -1$, and $x = 1$ give the factors of $x + 2$, $x + 1$, and $x - 1$. Use synthetic division to find the other factors.

A table of values supports the real zeros of the function and its factors.

X	Y1
-3	-80
-2	0
-1.5	2.0313
-1	0
-.5	-1.406
1	0
2	60

Y1◼X^5+2X^4-X-2

Study Tip

$P(x)$ is a *fifth*-degree polynomial, so we expect *five* zeros.

SECTION
2.5 SUMMARY

In this section, we discussed **complex zeros** of polynomial functions. A polynomial function $P(x)$ of degree n with real coefficients has the following properties:

- $P(x)$ has at least one zero (if $n > 0$) and no more than n zeros.
- If $a + bi$ is a zero, then $a - bi$ is also a zero.
- The polynomial can be written as a product of linear factors, not necessarily distinct.

SECTION
2.5 EXERCISES

▪ SKILLS

In Exercises 1–8, find all zeros (real and complex). Factor the polynomial as a product of linear factors.

1. $P(x) = x^2 + 4$ **2.** $P(x) = x^2 + 9$ **3.** $P(x) = x^2 - 2x + 2$ **4.** $P(x) = x^2 - 4x + 5$

5. $P(x) = x^4 - 16$ **6.** $P(x) = x^4 - 81$ **7.** $P(x) = x^4 - 25$ **8.** $P(x) = x^4 - 9$

In Exercises 9–16, a polynomial function is described. Find all remaining zeros.

9. Degree: 3   Zeros: $-1, i$ **10.** Degree: 3   Zeros: $1, -i$

11. Degree: 4   Zeros: $2i, 3 - i$ **12.** Degree: 4   Zeros: $3i, 2 + i$

13. Degree: 6   Zeros: 2 (multiplicity 2), $1 - 3i, 2 + 5i$ **14.** Degree: 6   Zeros: -2 (multiplicity 2), $1 - 5i, 2 + 3i$

15. Degree: 6   Zeros: $-i, 1 - i$ (multiplicity 2)   **16.** Degree: 6   Zeros: $2i, 1 + i$ (multiplicity 2)

In Exercises 17–28, given a zero of the polynomial, determine all other zeros (real and complex) and write the polynomial in terms of a product of linear factors.

	Polynomial	Zero		Polynomial	Zero
17.	$P(x) = x^4 - 2x^3 - 11x^2 - 8x - 60$	$-2i$	**18.**	$P(x) = x^4 - x^3 + 7x^2 - 9x - 18$	$3i$
19.	$P(x) = x^4 - 4x^3 + 4x^2 - 4x + 3$	i	**20.**	$P(x) = x^4 - x^3 + 2x^2 - 4x - 8$	$-2i$
21.	$P(x) = x^4 - 2x^3 + 10x^2 - 18x + 9$	$-3i$	**22.**	$P(x) = x^4 - 3x^3 + 21x^2 - 75x - 100$	$5i$
23.	$P(x) = x^4 - 9x^2 + 18x - 14$	$1 + i$	**24.**	$P(x) = x^4 - 4x^3 + x^2 + 6x - 40$	$1 - 2i$
25.	$P(x) = x^4 - 6x^3 + 6x^2 + 24x - 40$	$3 - i$	**26.**	$P(x) = x^4 - 4x^3 + 4x^2 + 4x - 5$	$2 + i$
27.	$P(x) = x^4 - 9x^3 + 29x^2 - 41x + 20$	$2 - i$	**28.**	$P(x) = x^4 - 7x^3 + 14x^2 + 2x - 20$	$3 + i$

In Exercises 29–52, factor each polynomial as a product of linear factors.

29. $P(x) = x^3 - x^2 + 9x - 9$ **30.** $P(x) = x^3 - 2x^2 + 4x - 8$ **31.** $P(x) = x^3 - 5x^2 + x - 5$

32. $P(x) = x^3 - 7x^2 + x - 7$ **33.** $P(x) = x^3 + x^2 + 4x + 4$ **34.** $P(x) = x^3 + x^2 - 2$

35. $P(x) = x^3 - x^2 - 18$ **36.** $P(x) = x^4 - 2x^3 - 2x^2 - 2x - 3$ **37.** $P(x) = x^4 - 2x^3 - 11x^2 - 8x - 60$

38. $P(x) = x^4 - x^3 + 7x^2 - 9x - 18$ **39.** $P(x) = x^4 - 4x^3 - x^2 - 16x - 20$ **40.** $P(x) = x^4 - 3x^3 + 11x^2 - 27x + 18$

41. $P(x) = x^4 - 7x^3 + 27x^2 - 47x + 26$ **42.** $P(x) = x^4 - 5x^3 + 5x^2 + 25x - 26$ **43.** $P(x) = -x^4 - 3x^3 + x^2 + 13x + 10$

44. $P(x) = -x^4 - x^3 + 12x^2 + 26x + 24$ **45.** $P(x) = x^4 - 2x^3 + 5x^2 - 8x + 4$ **46.** $P(x) = x^4 + 2x^3 + 10x^2 + 18x + 9$

47. $P(x) = x^6 + 12x^4 + 23x^2 - 36$ **48.** $P(x) = x^6 - 2x^5 + 9x^4 - 16x^3 + 24x^2 - 32x + 16$

49. $P(x) = 4x^4 - 20x^3 + 37x^2 - 24x + 5$ **50.** $P(x) = 4x^4 - 44x^3 + 145x^2 - 114x + 26$

51. $P(x) = 3x^5 - 2x^4 + 9x^3 - 6x^2 - 12x + 8$ **52.** $P(x) = 2x^5 - 5x^4 + 4x^3 - 26x^2 + 50x - 25$

▪ APPLICATIONS

In Exercises 53–56, assume the profit model is given by a polynomial function $P(x)$, where x is the number of units sold by the company per year.

53. Profit. If the profit function of a given company has all imaginary zeros and the leading coefficient is positive, would you invest in this company? Explain.

54. Profit. If the profit function of a given company has all imaginary zeros and the leading coefficient is negative, would you invest in this company? Explain.

55. Profit. If the profit function of a company is modeled by a third-degree polynomial with a negative leading coefficient and this polynomial has two complex conjugates as zeros and one positive real zero, would you invest in this company? Explain.

56. Profit. If the profit function of a company is modeled by a third-degree polynomial with a positive leading coefficient and this polynomial has two complex conjugates as zeros and one positive real zero, would you invest in this company? Explain.

▪ CATCH THE MISTAKE

In Exercises 57 and 58, explain the mistake that is made.

57. Given that 1 is a zero of $P(x) = x^3 - 2x^2 + 7x - 6$, find all other zeros.

Solution:

Step 1: $P(x)$ is a third-degree polynomial, so we expect three zeros.

Step 2: Because 1 is a zero, -1 is a zero, so two linear factors are $(x - 1)$ and $(x + 1)$.

Step 3: Write the polynomial as a product of three linear factors.

$$P(x) = (x - 1)(x + 1)(x - c)$$
$$P(x) = (x^2 - 1)(x - c)$$

Step 4: To find the remaining linear factor, we divide $P(x)$ by $x^2 - 1$.

$$\frac{x^3 - 2x^2 + 7x - 6}{x^2 - 1} = x - 2 + \frac{6x - 8}{x^2 - 1}$$

Which has a nonzero remainder? What went wrong?

58. Factor the polynomial $P(x) = 2x^3 + x^2 + 2x + 1$.

Solution:

Step 1: Since $P(x)$ is an odd-degree polynomial, we are guaranteed one real zero (since complex zeros come in conjugate pairs).

Step 2: Apply the rational zero test to develop a list of potential rational zeros.

Possible zeros: ± 1

Step 3: Test possible zeros.

1 is not a zero: $P(x) = 2(1)^3 + (1)^2 + 2(1) + 1$
$$= 6$$

-1 is not a zero: $P(x) = 2(-1)^3 + (-1)^2 + 2(-1) + 1$
$$= -2$$

Note: $-\frac{1}{2}$ is the real zero. Why did we not find it?

▪ CONCEPTUAL

In Exercises 59–62, determine whether each statement is true or false.

59. If $x = 1$ is a zero of a polynomial function, then $x = -1$ is also a zero of the polynomial function.

60. All zeros of a polynomial function correspond to x-intercepts.

61. A polynomial function of degree n, $n > 0$ must have at least one zero.

62. A polynomial function of degree n, $n > 0$ can be written as a product of n linear factors.

63. Is it possible for an odd-degree polynomial to have all imaginary complex zeros? Explain.

64. Is it possible for an even-degree polynomial to have all imaginary zeros? Explain.

■ CHALLENGE

In Exercises 65 and 66, assume a and b are nonzero real numbers.

65. Find a polynomial function that has degree 6, and for which bi is a zero of multiplicity 3.

66. Find a polynomial function that has degree 4, and for which $a + bi$ is a zero of multiplicity 2.

67. Find a polynomial function that has degree 6, and for which ai and bi are zeros, where ai has multiplicity 2. Assume $|a| \neq |b|$.

68. Assuming $|a| \neq |b|$, find a polynomial function of lowest degree for which ai and bi are zeros of equal multiplicity.

■ TECHNOLOGY

For Exercises 69 and 70, determine possible combinations of real and complex zeros. Plot $P(x)$ and identify any real zeros with a graphing calculator or software. Does this agree with your list?

69. $P(x) = x^4 + 13x^2 + 36$

70. $P(x) = x^6 + 2x^4 + 7x^2 - 130x - 288$

For Exercises 71 and 72, find all zeros (real and complex). Factor the polynomial as a product of linear factors.

71. $P(x) = -5x^5 + 3x^4 - 25x^3 + 15x^2 - 20x + 12$

72. $P(x) = x^5 + 2.1x^4 - 5x^3 - 5.592x^2 + 9.792x - 3.456$

■ PREVIEW TO CALCULUS

In Exercises 73–76, refer to the following:

In calculus we study the integration of rational functions by partial fractions.

a. Factor each polynomial into linear factors. Use complex numbers when necessary.

b. Factor each polynomial using only real numbers.

73. $f(x) = x^3 + x^2 + x + 1$

74. $f(x) = x^3 - 6x^2 + 21x - 26$

75. $f(x) = x^4 + 5x^2 + 4$

76. $f(x) = x^4 - 2x^3 - 7x^2 + 18x - 18$

SECTION
2.6 RATIONAL FUNCTIONS

SKILLS OBJECTIVES

- Find the domain of a rational function.
- Determine vertical, horizontal, and slant asymptotes of rational functions.
- Graph rational functions.

CONCEPTUAL OBJECTIVE

- Interpret the behavior of the graph of a rational function approaching an asymptote.

So far in this chapter, we have discussed polynomial functions. We now turn our attention to *rational functions*, which are *ratios* of polynomial functions. Ratios of integers are called *rational numbers*. Similarly, ratios of polynomial functions are called *rational functions*.

DEFINITION   **Rational Function**

A function $f(x)$ is a **rational function** if

$$f(x) = \frac{n(x)}{d(x)} \qquad d(x) \neq 0$$

where the numerator $n(x)$ and the denominator $d(x)$ are polynomial functions. The domain of $f(x)$ is the set of all real numbers x such that $d(x) \neq 0$.

Note: If $d(x)$ is a constant, then $f(x)$ is a polynomial function.

Domain of Rational Functions

The domain of any polynomial function is the set of all real numbers. When we divide two polynomial functions, the result is a *rational function*, and we must exclude any values of x that make the denominator equal to zero.

EXAMPLE 1   **Finding the Domain of a Rational Function**

Find the domain of the rational function $f(x) = \dfrac{x + 1}{x^2 - x - 6}$. Express the domain in interval notation.

Solution:

Set the denominator equal to zero. $x^2 - x - 6 = 0$

Factor. $(x + 2)(x - 3) = 0$

Solve for x. $x = -2 \quad \text{or} \quad x = 3$

Eliminate these values from the domain. $x \neq -2 \quad \text{or} \quad x \neq 3$

State the domain in interval notation. $\boxed{(-\infty, -2) \cup (-2, 3) \cup (3, \infty)}$

■ **Answer:** The domain is the set of all real numbers such that $x \neq -1$ or $x \neq 4$. Interval notation: $(-\infty, -1) \cup (-1, 4) \cup (4, \infty)$

■ **YOUR TURN** Find the domain of the rational function $f(x) = \dfrac{x - 2}{x^2 - 3x - 4}$. Express the domain in interval notation.

It is important to note that there are not always restrictions on the domain. For example, if the denominator is never equal to zero, the domain is the set of all real numbers.

EXAMPLE 2 When the Domain of a Rational Function Is the Set of All Real Numbers

Find the domain of the rational function $g(x) = \dfrac{3x}{x^2 + 9}$. Express the domain in interval notation.

Solution:

Set the denominator equal to zero.	$x^2 + 9 = 0$
Subtract 9 from both sides.	$x^2 = -9$
Solve for x.	$x = -3i$ or $x = 3i$
There are no *real* solutions; therefore, the domain has no restrictions.	$\mathbb{R}$, the set of all real numbers
State the domain in interval notation.	$\boxed{(-\infty, \infty)}$

■ **YOUR TURN** Find the domain of the rational function $g(x) = \dfrac{5x}{x^2 + 4}$. Express the domain in interval notation.

■ **Answer:** The domain is the set of all real numbers. Interval notation: $(-\infty, \infty)$

It is important to note that $f(x) = \dfrac{x^2 - 4}{x + 2}$, where $x \neq -2$, and $g(x) = x - 2$ are *not* the same function. Although $f(x)$ can be written in the factored form $f(x) = \dfrac{(x - 2)(x + 2)}{x + 2} = x - 2$, its domain is different. The domain of $g(x)$ is the set of all real numbers, whereas the domain of $f(x)$ is the set of all real numbers such that $x \neq -2$. If we were to plot $f(x)$ and $g(x)$, they would both look like the line $y = x - 2$. However, $f(x)$ would have a hole, or discontinuity, at the point $x = -2$.

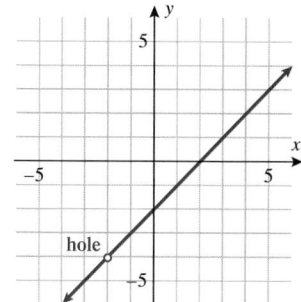

Vertical, Horizontal, and Slant Asymptotes

If a function is not defined at a point, then it is still useful to know how the function behaves near that point. Let's start with a simple rational function, the reciprocal function $f(x) = \dfrac{1}{x}$. This function is defined everywhere except at $x = 0$.

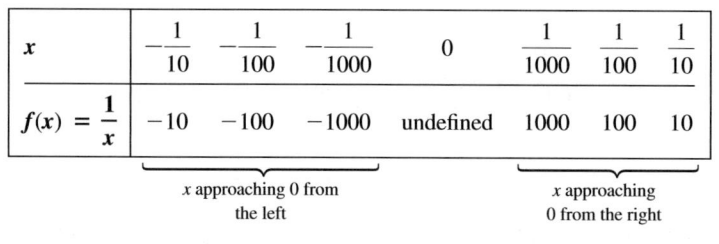

x	$-\dfrac{1}{10}$	$-\dfrac{1}{100}$	$-\dfrac{1}{1000}$	0	$\dfrac{1}{1000}$	$\dfrac{1}{100}$	$\dfrac{1}{10}$
$f(x) = \dfrac{1}{x}$	-10	-100	-1000	undefined	1000	100	10

x approaching 0 from the left
x approaching 0 from the right

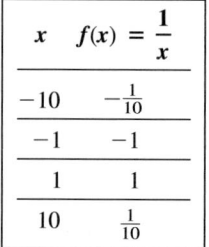

x	$f(x) = \dfrac{1}{x}$
-10	$-\dfrac{1}{10}$
-1	-1
1	1
10	$\dfrac{1}{10}$

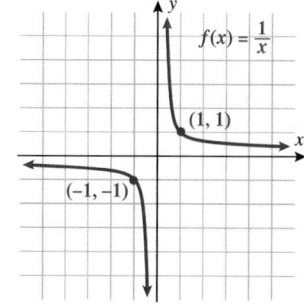

We cannot let $x = 0$, because that point is not in the domain of the function. We should, however, ask the question, "How does $f(x)$ behave as x *approaches* zero?" Let us take values that get closer and closer to $x = 0$, such as $\frac{1}{10}, \frac{1}{100}, \frac{1}{1000}, \dots$ (See the table above.) We use an *arrow* to represent the word *approach*, a *positive* superscript to represent from the *right*, and a *negative* superscript to represent from the *left*. A plot of this function can be generated using point-plotting techniques. The following are observations of the graph $f(x) = \dfrac{1}{x}$.

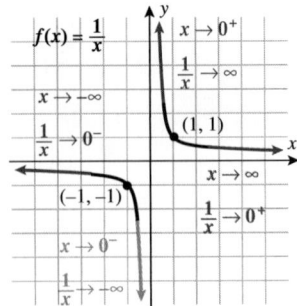

WORDS	MATH
As x approaches zero from the *right*, the function $f(x)$ increases without bound.	$x \to 0^+$ $\dfrac{1}{x} \to \infty$
As x approaches zero from the *left*, the function $f(x)$ decreases without bound.	$x \to 0^-$ $\dfrac{1}{x} \to -\infty$
As x approaches infinity (increases without bound), the function $f(x)$ approaches zero from *above*.	$x \to \infty$ $\dfrac{1}{x} \to 0^+$
As x approaches negative infinity (decreases without bound), the function $f(x)$ approaches zero from *below*.	$x \to -\infty$ $\dfrac{1}{x} \to 0^-$

The symbol ∞ does not represent an actual real number. This symbol represents growing without bound.

1. Notice that the function is not defined at $x = 0$. The y-axis, or the vertical line $x = 0$, represents the *vertical asymptote*.
2. Notice that the value of the function is never equal to zero. The x-axis is approached but not actually reached by the function. The x-axis, or $y = 0$, is a *horizontal asymptote*.

Asymptotes are lines that the graph of a function approaches. Suppose a football team's defense is its own 8 yard line and the team gets an "offsides" penalty that results in loss of "half the distance to the goal." Then the offense would get the ball on the 4 yard line. Suppose the defense gets another penalty on the next play that results in "half the distance to the goal." The offense would then get the ball on the 2 yard line. If the defense received 10 more penalties all resulting in "half the distance to the goal," would the referees *give* the offense a touchdown? No, because although the offense may appear to be snapping the ball from the goal line, technically it has not actually reached the goal line. Asymptotes utilize the same concept.

We will start with *vertical asymptotes*. Although the function $f(x) = \dfrac{1}{x}$ had one vertical asymptote, in general, rational functions can have *none*, *one*, or *several* vertical asymptotes. We will first formally define what a vertical asymptote is and then discuss how to find it.

DEFINITION   **Vertical Asymptotes**

The line $x = a$ is a **vertical asymptote** for the graph of a function if $f(x)$ either increases or decreases without bound as x approaches a from either the left or the right.

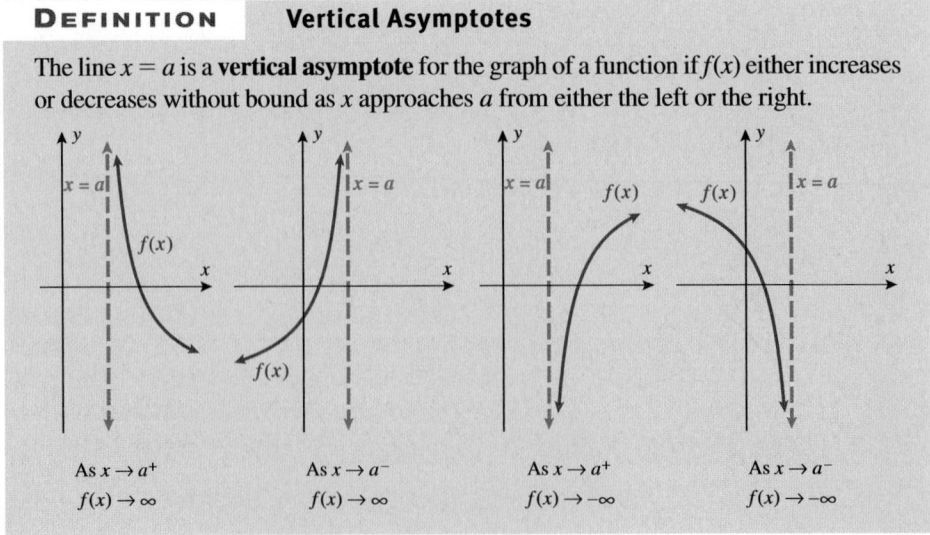

Vertical asymptotes assist us in graphing rational functions since they essentially "steer" the function in the vertical direction. How do we locate the vertical asymptotes of a rational function? Set the denominator equal to zero. If the numerator and denominator have no common factors, then any numbers that are excluded from the domain of a rational function locate vertical asymptotes.

A rational function $f(x) = \dfrac{n(x)}{d(x)}$ is said to be in **lowest terms** if the numerator $n(x)$ and denominator $d(x)$ have no common factors. Let $f(x) = \dfrac{n(x)}{d(x)}$ be a rational function in lowest terms; then any zeros of the numerator $n(x)$ correspond to x-intercepts of the graph of f, and any zeros of the denominator $d(x)$ correspond to vertical asymptotes of the graph of f. If a rational function does have a common factor (is not in lowest terms), then the common factor(s) should be canceled, resulting in an equivalent rational function $R(x)$ in lowest terms. If $(x - a)^p$ is a factor of the numerator and $(x - a)^q$ is a factor of the denominator, then there is a *hole* in the graph at $x = a$ provided $p \geq q$ and $x = a$ is a vertical asymptote if $p < q$.

LOCATING VERTICAL ASYMPTOTES

Let $f(x) = \dfrac{n(x)}{d(x)}$ be a rational function in lowest terms [that is, assume $n(x)$ and $d(x)$ are polynomials with no common factors]; then the graph of f has a vertical asymptote at any real zero of the denominator $d(x)$. That is, if $d(a) = 0$, then $x = a$ corresponds to a vertical asymptote on the graph of f.

Note: If f is a rational function that is not in lowest terms, then divide out the common factors, resulting in a rational function R that is in lowest terms. Any common factor $x - a$ of the function f corresponds to a hole in the graph of f at $x = a$ provided the multiplicity of a in the numerator is greater than or equal to the multiplicity of a in the denominator.

Study Tip

The vertical asymptotes of a rational function in *lowest terms* occur at x-values that make the denominator equal to zero.

EXAMPLE 3   Determining Vertical Asymptotes

Locate any vertical asymptotes of the rational function $f(x) = \dfrac{5x + 2}{6x^2 - x - 2}$.

Solution:

Factor the denominator.

$$f(x) = \frac{5x + 2}{(2x + 1)(3x - 2)}$$

The numerator and denominator have no common factors, which means that all zeros of the denominator correspond to vertical asymptotes.

Set the denominator equal to zero. $\qquad 2x + 1 = 0 \quad$ and $\quad 3x - 2 = 0$

Solve for x. $\qquad\qquad\qquad\qquad x = -\dfrac{1}{2} \quad$ and $\quad x = \dfrac{2}{3}$

The vertical asymptotes are $\boxed{x = -\tfrac{1}{2}}$ and $\boxed{x = \tfrac{2}{3}}$.

■ **YOUR TURN** Locate any vertical asymptotes of the following rational function:

$$f(x) = \frac{3x - 1}{2x^2 - x - 15}$$

■ **Answer:** $x = -\tfrac{5}{2}$ and $x = 3$

EXAMPLE 4 **Determining Vertical Asymptotes When the Rational Function Is Not in Lowest Terms**

Locate any vertical asymptotes of the rational function $f(x) = \dfrac{x+2}{x^3 - 3x^2 - 10x}$.

Solution:

Factor the denominator.

$$\begin{aligned} x^3 - 3x^2 - 10x &= x(x^2 - 3x - 10) \\ &= x(x-5)(x+2) \end{aligned}$$

Write the rational function in factored form.

$$f(x) = \frac{(x+2)}{x(x-5)(x+2)}$$

Cancel (divide out) the common factor $(x + 2)$.

$$R(x) = \frac{1}{x(x-5)} \quad x \neq -2$$

Find the values when the denominator of R is equal to zero.

$$x = 0 \quad \text{and} \quad x = 5$$

The vertical asymptotes are $\boxed{x = 0}$ and $\boxed{x = 5}$.

Note: $x = -2$ is not in the domain of $f(x)$, even though there is no vertical asymptote there. There is a "hole" in the graph at $x = -2$. Graphing calculators do not always show such "holes."

■ **Answer:** $x = 3$

■ **YOUR TURN** Locate any vertical asymptotes of the following rational function:

$$f(x) = \frac{x^2 - 4x}{x^2 - 7x + 12}$$

We now turn our attention to *horizontal asymptotes*. As we have seen, rational functions can have several vertical asymptotes. However, rational functions can have *at most* one horizontal asymptote. Horizontal asymptotes imply that a function approaches a constant value as x becomes large in the positive or negative direction. Another difference between vertical and horizontal asymptotes is that the graph of a function never touches a vertical asymptote but, as you will see in the next box, the graph of a function may cross a horizontal asymptote, just not at the "ends" $(x \rightarrow \pm\infty)$.

DEFINITION **Horizontal Asymptote**

The line $y = b$ is a **horizontal asymptote** of the graph of a function if $f(x)$ approaches b as x increases or decreases without bound. The following are three examples:

$$\text{As } x \rightarrow \infty, f(x) \rightarrow b$$

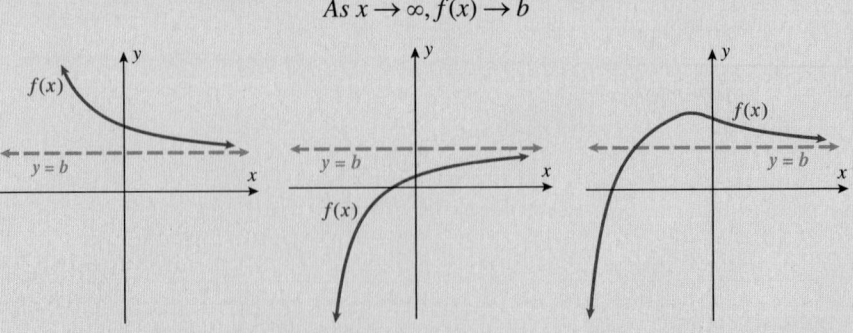

Note: A horizontal asymptote steers a function as x gets large. Therefore, when x is not large, the function may cross the asymptote.

How do we determine whether a horizontal asymptote exists? And, if it does, how do we locate it? We investigate the value of the rational function as $x \to \infty$ or as $x \to -\infty$. One of two things will happen: Either the rational function will increase or decrease without bound or the rational function will approach a constant value.

We say that a rational function is **proper** if the degree of the numerator is less than the degree of the denominator. Proper rational functions, like $f(x) = \dfrac{1}{x}$, approach zero as x gets large. Therefore, all proper rational functions have the specific horizontal asymptote, $y = 0$ (see Example 5a).

We say that a rational function is **improper** if the degree of the numerator is greater than or equal to the degree of the denominator. In this case, we can divide the numerator by the denominator and determine how the quotient behaves as x increases without bound.

- If the quotient is a constant (resulting when the degrees of the numerator and denominator are equal), then as $x \to \infty$ or as $x \to -\infty$, the rational function approaches the constant quotient (see Example 5b).
- If the quotient is a polynomial function of degree 1 or higher, then the quotient depends on x and does not approach a constant value as x increases (see Example 5c). In this case, we say that there is no horizontal asymptote.

We find horizontal asymptotes by comparing the degree of the numerator and the degree of the denominator. There are three cases to consider:

1. The degree of the numerator is less than the degree of the denominator.
2. The degree of the numerator is equal to the degree of the denominator.
3. The degree of the numerator is greater than the degree of the denominator.

LOCATING HORIZONTAL ASYMPTOTES

Let f be a rational function given by

$$f(x) = \frac{n(x)}{d(x)} = \frac{a_n x^n + a_{n-1} x^{n-1} + \cdots + a_1 x + a_0}{b_m x^m + b_{m-1} x^{m-1} + \cdots + b_1 x + b_0}$$

where $n(x)$ and $d(x)$ are polynomials.

1. When $n < m$, the x-axis ($y = 0$) is the horizontal asymptote.
2. When $n = m$, the line $y = \dfrac{a_n}{b_m}$ (ratio of leading coefficients) is the horizontal asymptote.
3. When $n > m$, there is no horizontal asymptote.

In other words:

1. When the degree of the numerator is less than the degree of the denominator, then $y = 0$ is the horizontal asymptote.
2. When the degree of the numerator is the same as the degree of the denominator, then the horizontal asymptote is the ratio of the leading coefficients.
3. If the degree of the numerator is greater than the degree of the denominator, then there is no horizontal asymptote.

Technology Tip

The following graphs correspond to the rational functions given in Example 5. The horizontal asymptotes are apparent, but are not drawn in the graph.

a. Graph $f(x) = \dfrac{8x + 3}{4x^2 + 1}$.

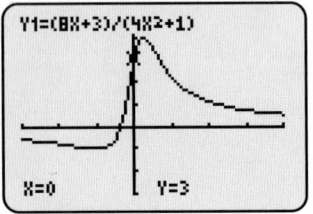

b. Graph $g(x) = \dfrac{8x^2 + 3}{4x^2 + 1}$.

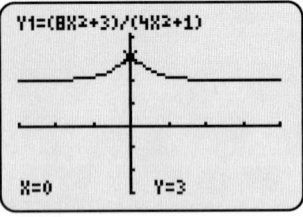

c. Graph $h(x) = \dfrac{8x^3 + 3}{4x^2 + 1}$.

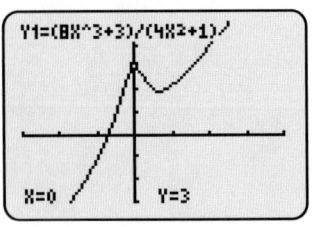

■ **Answer:** $y = -\frac{7}{4}$ is the horizontal asymptote.

Study Tip

There are three types of linear asymptotes: horizontal, vertical, and *slant*.

EXAMPLE 5   Finding Horizontal Asymptotes

Determine whether a horizontal asymptote exists for the graph of each of the given rational functions. If it does, locate the horizontal asymptote.

a. $f(x) = \dfrac{8x + 3}{4x^2 + 1}$ **b.** $g(x) = \dfrac{8x^2 + 3}{4x^2 + 1}$ **c.** $h(x) = \dfrac{8x^3 + 3}{4x^2 + 1}$

Solution (a):

The degree of the numerator $8x + 3$ is 1. $n = 1$

The degree of the denominator $4x^2 + 1$ is 2. $m = 2$

The degree of the numerator is less than the
degree of the denominator. $n < m$

The x-axis is the horizontal asymptote for the graph of $f(x)$. $y = 0$

The line $\boxed{y = 0}$ is the horizontal asymptote for the graph of $f(x)$.

Solution (b):

The degree of the numerator $8x^2 + 3$ is 2. $n = 2$

The degree of the denominator $4x^2 + 1$ is 2. $m = 2$

The degree of the numerator is equal to the
degree of the denominator. $n = m$

The ratio of the leading coefficients is the
horizontal asymptote for the graph of $g(x)$. $y = \dfrac{8}{4} = 2$

The line $\boxed{y = 2}$ is the horizontal asymptote for the graph of $g(x)$.

If we divide the numerator by the denominator,
the resulting quotient is the constant 2. $g(x) = \dfrac{8x^2 + 3}{4x^2 + 1} = 2 + \dfrac{1}{4x^2 + 1}$

Solution (c):

The degree of the numerator $8x^3 + 3$ is 3. $n = 3$

The degree of the denominator $4x^2 + 1$ is 2. $m = 2$

The degree of the numerator is greater than
the degree of the denominator. $n > m$

The graph of the rational function $h(x)$ has $\boxed{\text{no horizontal asymptote}}$.

If we divide the numerator by the denominator,
the resulting quotient is a linear function and
corresponds to the slant asymptote $y = 2x$. $h(x) = \dfrac{8x^3 + 3}{4x^2 + 1} = 2x + \dfrac{-2x + 3}{4x^2 + 1}$

■ **YOUR TURN** Find the horizontal asymptote (if one exists) for the graph of the rational function $f(x) = \dfrac{7x^3 + x - 2}{-4x^3 + 1}$.

There are three types of lines: horizontal (slope is zero), vertical (slope is undefined), and slant (nonzero slope). Similarly, there are three types of linear asymptotes: horizontal, vertical, and *slant*.

Recall that in dividing polynomials, the degree of the quotient is always the difference between the degree of the numerator and the degree of the denominator. For example, a cubic (third-degree) polynomial divided by a quadratic (second-degree) polynomial results in a linear (first-degree) polynomial. A fifth-degree polynomial divided by a fourth-degree polynomial results in a first-degree (linear) polynomial. When the degree of the numerator is exactly one more than the degree of the denominator, the quotient is linear and represents a *slant asymptote.*

SLANT ASYMPTOTES

Let f be a rational function given by $f(x) = \dfrac{n(x)}{d(x)}$, where $n(x)$ and $d(x)$ are polynomials and the degree of $n(x)$ is *one more than* the degree of $d(x)$. On dividing $n(x)$ by $d(x)$, the rational function can be expressed as

$$f(x) = mx + b + \frac{r(x)}{d(x)}$$

where the degree of the remainder $r(x)$ is less than the degree of $d(x)$ and the line $y = mx + b$ is a **slant asymptote** for the graph of f.

Note that as $x \to -\infty$ or $x \to \infty$, $f(x) \to mx + b$.

EXAMPLE 6   Finding Slant Asymptotes

Determine the slant asymptote of the rational function $f(x) = \dfrac{4x^3 + x^2 + 3}{x^2 - x + 1}$.

Solution:

Divide the numerator by the denominator with long division.

$$
\begin{array}{r}
4x + 5 \\
x^2 - x + 1 \overline{)\,4x^3 + x^2 + 0x + 3} \\
-(4x^3 - 4x^2 + 4x) \\
\hline
5x^2 - 4x + 3 \\
-(5x^2 - 5x + 5) \\
\hline
x - 2
\end{array}
$$

Note that as $x \to \pm\infty$, the rational expression approaches 0.

$$f(x) = 4x + 5 + \underbrace{\frac{x - 2}{x^2 - x + 1}}_{\substack{\to 0 \text{ as} \\ x \to \pm\infty}}$$

The quotient is the slant asymptote.

$$\boxed{y = 4x + 5}$$

■ **YOUR TURN** Find the slant asymptote of the rational function $f(x) = \dfrac{x^2 + 3x + 2}{x - 2}$.

■ **Answer:** $y = x + 5$

Graphing Rational Functions

We can now graph rational functions using asymptotes as graphing aids. The following box summarizes the six-step procedure for graphing rational functions.

Study Tip

Common factors need to be divided out first; then the remaining x-values corresponding to a denominator value of 0 are vertical asymptotes.

GRAPHING RATIONAL FUNCTIONS

Let f be a rational function given by $f(x) = \dfrac{n(x)}{d(x)}$.

Step 1: Find the domain of the rational function f.

Step 2: Find the **intercept(s)**.
- y-intercept: evaluate $f(0)$.
- x-intercept: solve the equation $n(x) = 0$ for x in the domain of f.

Step 3: Find any **holes**.
- Factor the numerator and denominator.
- Divide out common factors.
- A common factor $x - a$ corresponds to a hole in the graph of f at $x = a$ if the multiplicity of a in the numerator is greater than or equal to the multiplicity of a in the denominator.
- The result is an equivalent rational function $R(x) = \dfrac{p(x)}{q(x)}$ in lowest terms.

Step 4: Find any **asymptotes**.
- Vertical asymptotes: solve $q(x) = 0$.
- Compare the degree of the numerator and the degree of the denominator to determine whether either a horizontal or slant asymptote exists. If one exists, find it.

Step 5: Find **additional points** on the graph of f—particularly near asymptotes.

Step 6: Sketch the graph; draw the asymptotes, label the intercept(s) and additional points, and complete the graph with a smooth curve between and beyond the vertical asymptotes.

Study Tip

Any real number excluded from the domain of a rational function corresponds to either a vertical asymptote or a hole on its graph.

It is important to note that any real number eliminated from the domain of a rational function corresponds to either a vertical asymptote or a hole on its graph.

 EXAMPLE 7   Graphing a Rational Function

Graph the rational function $f(x) = \dfrac{x}{x^2 - 4}$.

Solution:

STEP 1 Find the **domain**.

Set the denominator equal to zero.	$x^2 - 4 = 0$
Solve for x.	$x = \pm 2$
State the domain.	$(-\infty, -2) \cup (-2, 2) \cup (2, \infty)$

STEP 2 Find the **intercepts**.

y-intercept: $\qquad f(0) = \dfrac{0}{-4} = 0 \qquad y = 0$

x-intercepts: $\qquad f(x) = \dfrac{x}{x^2 - 4} = 0 \qquad x = 0$

The only intercept is at the point $\boxed{(0, 0)}$.

STEP 3 Find any holes.

$$f(x) = \frac{x}{(x + 2)(x - 2)}$$

There are no common factors, so f is in lowest terms.
Since there are no common factors, there are no holes on the graph of f.

STEP 4 Find any **asymptotes**.

Vertical asymptotes:

$$d(x) = (x + 2)(x - 2) = 0$$

$$\boxed{x = -2} \quad \text{and} \quad \boxed{x = 2}$$

Horizontal asymptote:

$$\frac{\text{Degree of numerator}}{\text{Degree of denominator}} = \frac{1}{2}$$

Degree of numerator < Degree of denominator $\boxed{y = 0}$

STEP 5 Find **additional points** on the graph.

x	-3	-1	1	3
$f(x)$	$-\frac{3}{5}$	$\frac{1}{3}$	$-\frac{1}{3}$	$\frac{3}{5}$

STEP 6 **Sketch** the graph; label the intercepts, asymptotes, and additional points and complete with a smooth curve approaching the asymptotes.

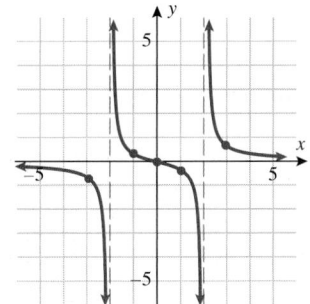

■ **Answer:**

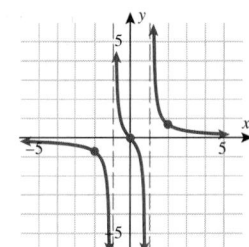

■ **YOUR TURN** Graph the rational function $f(x) = \dfrac{x}{x^2 - 1}$.

EXAMPLE 8   Graphing a Rational Function with No Horizontal or Slant Asymptotes

State the asymptotes (if there are any) and graph the rational function $f(x) = \dfrac{x^4 - x^3 - 6x^2}{x^2 - 1}$.

Solution:

STEP 1 Find the domain.

Set the denominator equal to zero. $x^2 - 1 = 0$

Solve for x. $x = \pm 1$

State the domain. $(-\infty, -1) \cup (-1, 1) \cup (1, \infty)$

STEP 2 Find the **intercepts**.

y-intercept: $f(0) = \dfrac{0}{-1} = 0$

x-intercepts: $n(x) = x^4 - x^3 - 6x^2 = 0$

Factor. $x^2(x - 3)(x + 2) = 0$

Solve. $x = 0, x = 3, \text{ and } x = -2$

The **intercepts** are the points $(0, 0)$, $(3, 0)$, and $(-2, 0)$.

Technology Tip

The behavior of each function as x approaches ∞ or $-\infty$ can be shown using tables of values.

Graph $f(x) = \dfrac{x^4 - x^3 - 6x^2}{x^2 - 1}$.

```
WINDOW
 Xmin=-4
 Xmax=5
 Xscl=1
 Ymin=-8
 Ymax=8
 Yscl=1
 Xres=1
```

```
Plot1 Plot2 Plot3
\Y1B(X^4-X^3-6X²
)/(X²-1)
\Y2=■
```

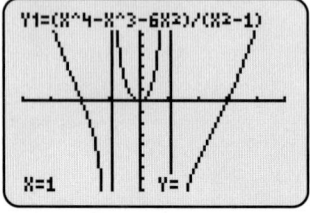

The graph of $f(x)$ shows that the vertical asymptotes are at $x = \pm1$ and there is no horizontal asymptote or slant asymptote.

STEP 3 Find any **holes**. $\qquad\qquad f(x) = \dfrac{x^2(x-3)(x+2)}{(x-1)(x+1)}$

There are no common factors, so f is in lowest terms.
Since there are no common factors, there are no holes on the graph of f.

STEP 4 Find the **asymptotes**.

Vertical asymptote:	$d(x) = x^2 - 1 = 0$
Factor.	$(x+1)(x-1) = 0$
Solve.	$x = -1$ and $x = 1$

No horizontal asymptote: degree of $n(x) >$ degree of $d(x)$ $\qquad$ [4 > 2]

No slant asymptote: degree of $n(x) -$ degree of $d(x) > 1$ $\qquad$ [4 − 2 = 2 > 1]

The **asymptotes** are $x = -1$ and $x = 1$.

STEP 5 Find **additional points** on the graph.

x	−3	−0.5	0.5	2	4
$f(x)$	6.75	1.75	2.08	−5.33	6.4

STEP 6 **Sketch** the graph; label the **intercepts** and **asymptotes**, and complete with a smooth curve between and beyond the vertical asymptote.

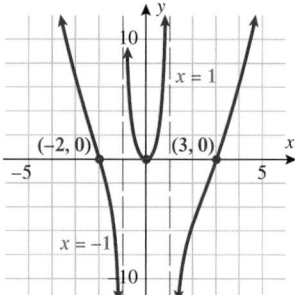

■ **Answer:** Vertical asymptote: $x = -2$. No horizontal or slant asymptotes.

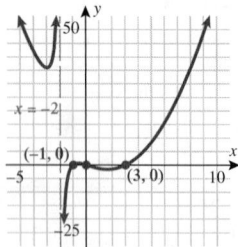

■ **YOUR TURN** State the asymptotes (if there are any) and graph the rational function:

$$f(x) = \frac{x^3 - 2x^2 - 3x}{x + 2}$$

EXAMPLE 9 **Graphing a Rational Function with a Horizontal Asymptote**

State the asymptotes (if there are any) and graph the rational function

$$f(x) = \frac{4x^3 + 10x^2 - 6x}{8 - x^3}$$

Solution:

STEP 1 Find the **domain**.

Set the denominator equal to zero.	$8 - x^3 = 0$
Solve for x.	$x = 2$
State the domain.	$(-\infty, 2) \cup (2, \infty)$

STEP 2 Find the **intercepts**.

y-intercept: $\quad\quad\quad\quad\quad\quad\quad\quad\quad$ $f(0) = \dfrac{0}{8} = 0$

x-intercepts: $\quad\quad\quad\quad\quad\quad\quad\quad$ $n(x) = 4x^3 + 10x^2 - 6x = 0$

Factor. $\quad\quad\quad\quad\quad\quad\quad\quad\quad\quad$ $2x(2x - 1)(x + 3) = 0$

Solve. $\quad\quad\quad\quad\quad\quad\quad\quad\quad\quad$ $x = 0,\ x = \dfrac{1}{2},\ \text{and } x = -3$

The **intercepts** are the points $(0, 0)$, $\left(\tfrac{1}{2}, 0\right)$, and $(-3, 0)$.

STEP 3 Find the **holes**. $\quad\quad\quad\quad$ $f(x) = \dfrac{2x(2x - 1)(x + 3)}{(2 - x)(x^2 + 2x + 4)}$

There are no common factors, so f is in lowest terms (no holes).

STEP 4 Find the **asymptotes**.

Vertical asymptote: $\quad\quad\quad\quad\quad$ $d(x) = 8 - x^3 = 0$

Solve. $\quad\quad\quad\quad\quad\quad\quad\quad\quad\quad$ $x = 2$

Horizontal asymptote: $\quad\quad\quad$ degree of $n(x)$ = degree of $d(x)$

Use leading coefficients. $\quad\quad\quad$ $y = \dfrac{4}{-1} = -4$

The **asymptotes** are $x = 2$ and $y = -4$.

STEP 5 Find **additional points** on the graph.

x	-4	-1	$\tfrac{1}{4}$	1	3
$f(x)$	-1	1.33	-0.10	1.14	-9.47

STEP 6 **Sketch** the graph; label the intercepts and asymptotes and complete with a smooth curve.

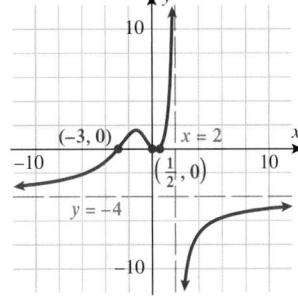

■ **YOUR TURN** Graph the rational function $f(x) = \dfrac{2x^2 - 7x + 6}{x^2 - 3x - 4}$. Give equations of the vertical and horizontal asymptotes and state the intercepts.

Technology Tip

The behavior of each function as x approaches ∞ or $-\infty$ can be shown using tables of values.

Graph $f(x) = \dfrac{4x^3 - 10x^2 - 6x}{8 - x^3}$.

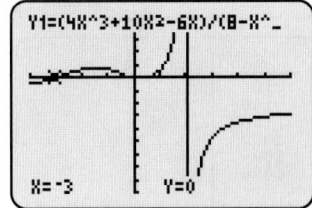

The graph of $f(x)$ shows that the vertical asymptote is at $x = 1$ and the horizontal asymptote is at $y = -4$.

■ **Answer:** Vertical asymptotes: $x = 4,\ x = -1$
Horizontal asymptote: $y = 2$
Intercepts: $\left(0, -\tfrac{3}{2}\right)$, $\left(\tfrac{3}{2}, 0\right)$, $(2, 0)$

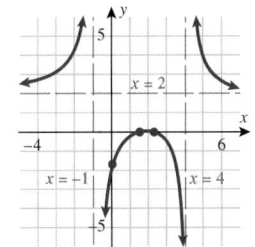

Technology Tip

The behavior of each function as x approaches ∞ or $-\infty$ can be shown using tables of values.

Graph $f(x) = \dfrac{x^2 - 3x - 4}{x + 2}$.

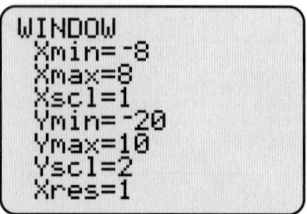

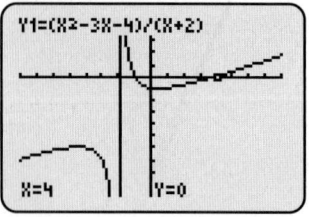

The graph of $f(x)$ shows that the vertical asymptote is at $x = -2$ and the slant asymptote is at $y = x - 5$.

Answer:
Vertical asymptote: $x = 3$
Slant asymptote: $y = x + 4$

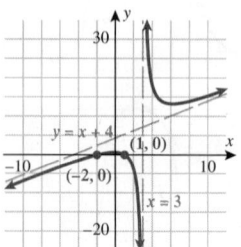

EXAMPLE 10   Graphing a Rational Function with a Slant Asymptote

Graph the rational function $f(x) = \dfrac{x^2 - 3x - 4}{x + 2}$.

Solution:

STEP 1 Find the **domain**.

Set the denominator equal to zero. $x + 2 = 0$

Solve for x. $x = -2$

State the domain. $(-\infty, -2) \cup (-2, \infty)$

STEP 2 Find the **intercepts**.

y-intercept: $f(0) = -\dfrac{4}{2} = -2$

x-intercepts: $n(x) = x^2 - 3x - 4 = 0$

Factor. $(x + 1)(x - 4) = 0$

Solve. $x = -1$ and $x = 4$

The **intercepts** are the points $(0, -2)$, $(-1, 0)$, and $(4, 0)$.

STEP 3 Find any **holes**. $f(x) = \dfrac{(x - 4)(x + 1)}{(x + 2)}$

There are no common factors, so f is in lowest terms.
Since there are no common factors, there are no holes on the graph of f.

STEP 4 Find the **asymptotes**.

Vertical asymptote: $d(x) = x + 2 = 0$

Solve. $x = -2$

Slant asymptote: degree of $n(x)$ − degree of $d(x) = 1$

Divide $n(x)$ by $d(x)$. $f(x) = \dfrac{x^2 - 3x - 4}{x + 2} = x - 5 + \dfrac{6}{x + 2}$

Write the equation of the asymptote. $y = x - 5$

The **asymptotes** are $x = -2$ and $y = x - 5$.

STEP 5 Find **additional points** on the graph.

x	-6	-5	-3	5	6
$f(x)$	-12.5	-12	-14	0.86	1.75

STEP 6 **Sketch** the graph; label the intercepts and asymptotes, and complete with a smooth curve between and beyond the vertical asymptote.

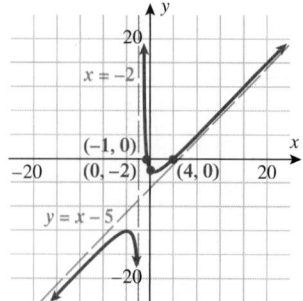

YOUR TURN For the function $f(x) = \dfrac{x^2 + x - 2}{x - 3}$, state the asymptotes (if any exist) and graph the function.

EXAMPLE 11   Graphing a Rational Function with a Hole in the Graph

Graph the rational function $f(x) = \dfrac{x^2 + x - 6}{x^2 - x - 2}$.

Solution:

STEP 1 Find the **domain**.

Set the denominator equal to zero. $x^2 - x - 2 = 0$

Solve for x. $(x - 2)(x + 1) = 0$

$x = -1$ or $x = 2$

State the domain. $(-\infty, -1) \cup (-1, 2) \cup (2, \infty)$

STEP 2 Find the **intercepts**.

y-intercept: $f(0) = \dfrac{-6}{-2} = 3 \qquad y = 3$

x-intercepts: $n(x) = x^2 + x - 6 = 0$

$(x + 3)(x - 2) = 0$

$x = -3$ or $x = 2$

The intercepts correspond to the points $\boxed{(0, 3)}$ and $\boxed{(-3, 0)}$. The point $(2, 0)$ appears to be an x-intercept; however, $x = 2$ is not in the domain of the function.

STEP 3 Find any **holes**.

$f(x) = \dfrac{(x - 2)(x + 3)}{(x - 2)(x + 1)}$

Since $x - 2$ is a common factor, there is a *hole* in the graph of f at $x = 2$.

Dividing out the common factor generates an equivalent rational function in lowest terms. $R(x) = \dfrac{(x + 3)}{(x + 1)}$

STEP 4 Find the **asymptotes**.

Vertical asymptotes: $x + 1 = 0$

$\boxed{x = -1}$

Horizontal asymptote: $\dfrac{\text{Degree of numerator}}{\text{Degree of denominator}} = \overset{f}{\dfrac{2}{2}} = \overset{R}{\dfrac{1}{1}}$

Since the degree of the numerator equals the degree of the denominator, use the leading coefficients. $\boxed{y = \dfrac{1}{1} = 1}$

STEP 5 Find **additional points** on the graph.

x	-4	-2	$-\tfrac{1}{2}$	1	3
$f(x)$ or $R(x)$	$\tfrac{1}{3}$	-1	5	2	$\tfrac{3}{2}$

STEP 6 **Sketch** the graph; label the intercepts, asymptotes, and additional points and complete with a smooth curve approaching asymptotes.

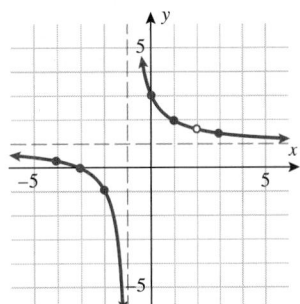

■ **YOUR TURN** Graph the rational function $f(x) = \dfrac{x^2 - x - 2}{x^2 + x - 6}$.

Technology Tip

The behavior of each function as x approaches ∞ or $-\infty$ can be shown using tables of values.

Graph $f(x) = \dfrac{x^2 + x - 6}{x^2 - x - 2}$.

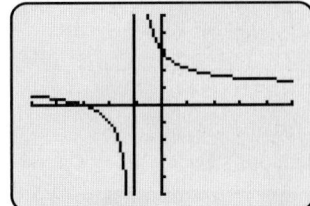

The graph of $f(x)$ shows that the vertical asymptote is at $x = -1$ and the horizontal asymptote is at $y = 1$.

Notice that the hole at $x = 2$ is not apparent in the graph. A table of values supports the graph.

■ **Answer:**

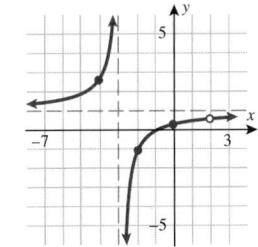

In this section, rational functions were discussed.

$$f(x) = \frac{n(x)}{d(x)}$$

- **Domain:** All real numbers except the x-values that make the denominator equal to zero, $d(x) = 0$.
- **Vertical Asymptotes:** Vertical lines, $x = a$, where $d(a) = 0$, after all common factors have been divided out. Vertical asymptotes steer the graph and are never touched.
- **Horizontal Asymptote:** Horizontal line, $y = b$, that steers the graph as $x \rightarrow \pm\infty$.
  1. If degree of the numerator $<$ degree of the denominator, then $y = 0$ is a horizontal asymptote.
  2. If degree of the numerator $=$ degree of the denominator, then $y = c$ is a horizontal asymptote, where c is the ratio of the leading coefficients of the numerator and denominator, respectively.
  3. If degree of the numerator $>$ degree of the denominator, then there is no horizontal asymptote.
- **Slant Asymptote:** Slant line, $y = mx + b$, that steers the graph as $x \rightarrow \pm\infty$.

1. If degree of the numerator $-$ degree of the denominator $= 1$, then there is a slant asymptote.
2. Divide the numerator by the denominator. The quotient corresponds to the equation of the line (slant asymptote).

Procedure for Graphing Rational Functions

1. Find the domain of the function.
2. Find the intercept(s).
   - y-intercept (does not exist if $x = 0$ is a vertical asymptote)
   - x-intercepts (if any)
3. Find any holes.
   - If $x - a$ is a common factor of the numerator and denominator, then $x = a$ corresponds to a hole in the graph of the rational function if the multiplicity of a in the numerator is greater than or equal to the multiplicity of a in the denominator. The result after the common factor is canceled is an equivalent rational function in lowest terms (no common factor).
4. Find any asymptotes.
   - Vertical asymptotes
   - Horizontal/slant asymptotes
5. Find additional points on the graph.
6. Sketch the graph: Draw the asymptotes and label the intercepts and points and connect with a smooth curve.

SKILLS

In Exercises 1–8, find the domain of each rational function.

1. $f(x) = \dfrac{x + 4}{x^2 + x - 12}$

2. $f(x) = \dfrac{x - 1}{x^2 + 2x - 3}$

3. $f(x) = \dfrac{x - 2}{x^2 - 4}$

4. $f(x) = \dfrac{x + 7}{2(x^2 - 49)}$

5. $f(x) = \dfrac{7x}{x^2 + 16}$

6. $f(x) = -\dfrac{2x}{x^2 + 9}$

7. $f(x) = -\dfrac{3(x^2 + x - 2)}{2(x^2 - x - 6)}$

8. $f(x) = \dfrac{5(x^2 - 2x - 3)}{(x^2 - x - 6)}$

In Exercises 9–16, find all vertical asymptotes and horizontal asymptotes (if there are any).

9. $f(x) = \dfrac{1}{x + 2}$

10. $f(x) = \dfrac{1}{5 - x}$

11. $f(x) = \dfrac{7x^3 + 1}{x + 5}$

12. $f(x) = \dfrac{2 - x^3}{2x - 7}$

13. $f(x) = \dfrac{6x^5 - 4x^2 + 5}{6x^2 + 5x - 4}$

14. $f(x) = \dfrac{6x^2 + 3x + 1}{3x^2 - 5x - 2}$

15. $f(x) = \dfrac{\frac{1}{3}x^2 + \frac{1}{3}x - \frac{1}{4}}{x^2 + \frac{1}{9}}$

16. $f(x) = \dfrac{\frac{1}{10}\left(x^2 - 2x + \frac{3}{10}\right)}{2x - 1}$

In Exercises 17–22, find the slant asymptote corresponding to the graph of each rational function.

17. $f(x) = \dfrac{x^2 + 10x + 25}{x + 4}$

18. $f(x) = \dfrac{x^2 + 9x + 20}{x - 3}$

19. $f(x) = \dfrac{2x^2 + 14x + 7}{x - 5}$

20. $f(x) = \dfrac{3x^3 + 4x^2 - 6x + 1}{x^2 - x - 30}$

21. $f(x) = \dfrac{8x^4 + 7x^3 + 2x - 5}{2x^3 - x^2 + 3x - 1}$

22. $f(x) = \dfrac{2x^6 + 1}{x^5 - 1}$

In Exercises 23–28, match the function to the graph.

23. $f(x) = \dfrac{3}{x - 4}$

24. $f(x) = \dfrac{3x}{x - 4}$

25. $f(x) = \dfrac{3x^2}{x^2 - 4}$

26. $f(x) = -\dfrac{3x^2}{x^2 + 4}$

27. $f(x) = \dfrac{3x^2}{4 - x^2}$

28. $f(x) = \dfrac{3x^2}{x + 4}$

a.

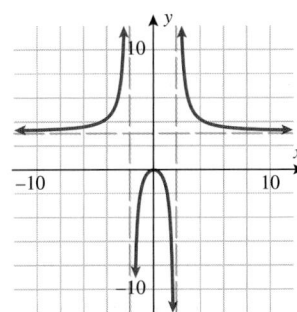

b.

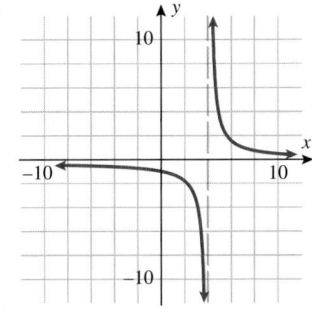

c.

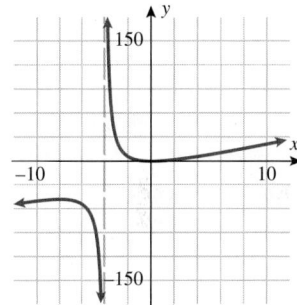

d.

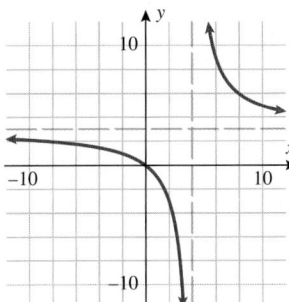

e.

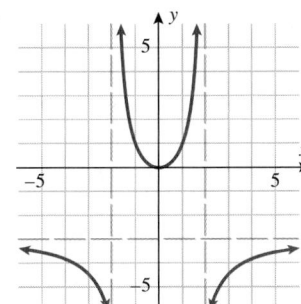

f.

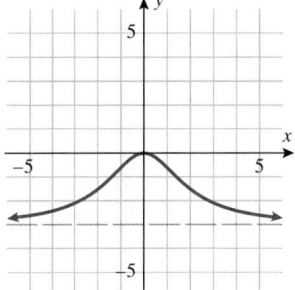

In Exercises 29–56, graph the rational functions. Locate any asymptotes on the graph.

29. $f(x) = \dfrac{2}{x + 1}$

30. $f(x) = \dfrac{4}{x - 2}$

31. $f(x) = \dfrac{2x}{x - 1}$

32. $f(x) = \dfrac{4x}{x + 2}$

33. $f(x) = \dfrac{x - 1}{x}$

34. $f(x) = \dfrac{2 + x}{x - 1}$

35. $f(x) = \dfrac{2(x^2 - 2x - 3)}{x^2 + 2x}$

36. $f(x) = \dfrac{3(x^2 - 1)}{x^2 - 3x}$

37. $f(x) = \dfrac{x^2}{x + 1}$

38. $f(x) = \dfrac{x^2 - 9}{x + 2}$

39. $f(x) = \dfrac{2x^3 - x^2 - x}{x^2 - 4}$

40. $f(x) = \dfrac{3x^3 + 5x^2 - 2x}{x^2 + 4}$

41. $f(x) = \dfrac{x^2 + 1}{x^2 - 1}$

42. $f(x) = \dfrac{1 - x^2}{x^2 + 1}$

43. $f(x) = \dfrac{7x^2}{(2x + 1)^2}$

44. $f(x) = \dfrac{12x^4}{(3x + 1)^4}$

45. $f(x) = \dfrac{1 - 9x^2}{(1 - 4x^2)^3}$

46. $f(x) = \dfrac{25x^2 - 1}{(16x^2 - 1)^2}$

47. $f(x) = 3x + \dfrac{4}{x}$

48. $f(x) = x - \dfrac{4}{x}$

49. $f(x) = \dfrac{(x - 1)^2}{(x^2 - 1)}$

50. $f(x) = \dfrac{(x + 1)^2}{(x^2 - 1)}$

51. $f(x) = \dfrac{(x - 1)(x^2 - 4)}{(x - 2)(x^2 + 1)}$

52. $f(x) = \dfrac{(x - 1)(x^2 - 9)}{(x - 3)(x^2 + 1)}$

53. $f(x) = \dfrac{3x(x - 1)}{x(x^2 - 4)}$

54. $f(x) = \dfrac{-2x(x - 3)}{x(x^2 + 1)}$

55. $f(x) = \dfrac{x^2(x + 5)}{2x(x^2 + 3)}$

56. $f(x) = \dfrac{4x(x - 1)(x + 2)}{x^2(x^2 - 4)}$

In Exercises 57–60, for each graph of the rational function given determine: (a) all intercepts, (b) all asymptotes, and (c) an equation of the rational function.

57.

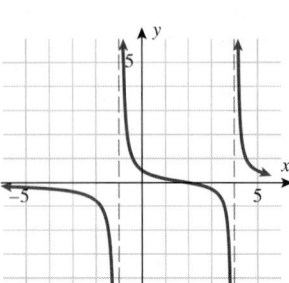

58.

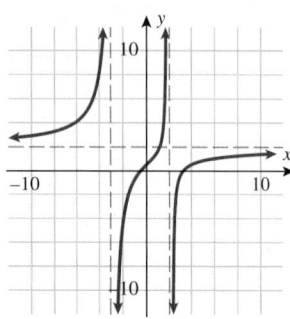

59.

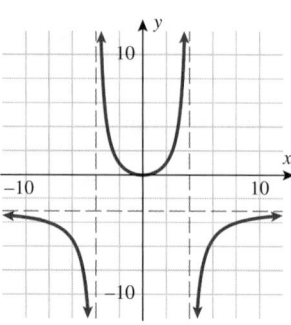

60.

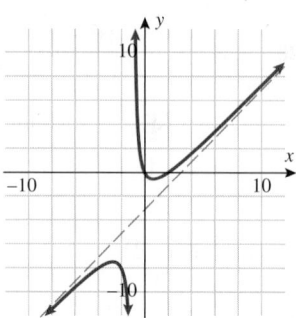

■ **APPLICATIONS**

61. Epidemiology. Suppose the number of individuals infected by a virus can be determined by the formula

$$n(t) = \frac{9500t - 2000}{4 + t}$$

where $t > 0$ is the time in months.

a. Find the number of infected people by the end of the fourth month.

b. After how many months are there 5500 infected people?

c. What happens with the number of infected people if the trend continues?

62. Investment. A financial institution offers to its investors a variable annual interest rate using the formula

$$r(x) = \frac{4x^2}{x^2 + 2x + 5}$$

where x is the amount invested in thousands of dollars.

a. What is the annual interest rate for an investment of $8,000?

b. What is the annual interest rate for an investment of $20,000?

c. What is the maximum annual interest rate offered by them?

63. Medicine. The concentration C of a particular drug in a person's bloodstream t minutes after injection is given by

$$C(t) = \frac{2t}{t^2 + 100}$$

a. What is the concentration in the bloodstream after 1 minute?

b. What is the concentration in the bloodstream after 1 hour?

c. What is the concentration in the bloodstream after 5 hours?

d. Find the horizontal asymptote of $C(t)$. What do you expect the concentration to be after several days?

64. Medicine. The concentration C of aspirin in the bloodstream t hours after consumption is given by $C(t) = \dfrac{t}{t^2 + 40}$.

a. What is the concentration in the bloodstream after $\frac{1}{2}$ hour?

b. What is the concentration in the bloodstream after 1 hour?

c. What is the concentration in the bloodstream after 4 hours?

d. Find the horizontal asymptote for $C(t)$. What do you expect the concentration to be after several days?

65. Typing. An administrative assistant is hired after graduating from high school and learns to type on the job. The number of words he can type per minute is given by

$$N(t) = \frac{130t + 260}{t + 5} \qquad t \geq 0$$

where t is the number of months he has been on the job.

a. How many words per minute can he type the day he starts?

b. How many words per minute can he type after 12 months?

c. How many words per minute can he type after 3 years?

d. How many words per minute would you expect him to type if he worked there until he retired?

66. Memorization. A professor teaching a large lecture course tries to learn students' names. The number of names she can remember $N(t)$ increases with each week in the semester t and is given by the rational function

$$N(t) = \frac{600t}{t + 20}$$

How many students' names does she know by the third week in the semester? How many students' names should she know by the end of the semester (16 weeks)? According to this function, what are the most names she can remember?

67. Food. The amount of food that cats typically eat increases as their weight increases. A rational function that describes this is $F(x) = \dfrac{10x^2}{x^2 + 4}$, where the amount of food $F(x)$ is given in ounces and the weight of the cat x is given in pounds. Calculate the horizontal asymptote. How many ounces of food will most adult cats eat?

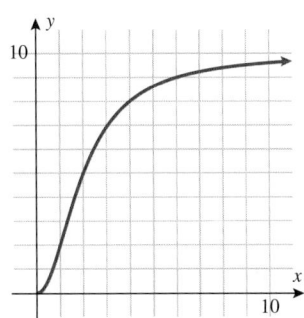

68. Memorization. The *Guinness Book of World Records, 2004* states that Dominic O'Brien (England) memorized on a single sighting a random sequence of 54 separate packs of cards all shuffled together (2808 cards in total) at Simpson's-In-The-Strand, London, England, on May 1, 2002. He memorized the cards in 11 hours 42 minutes, and then recited them in exact sequence in a time of 3 hours 30 minutes. With only a 0.5% margin of error allowed (no more than 14 errors), he broke the record with just 8 errors. If we let x represent the time (hours) it takes to memorize the cards and y represent the number of cards memorized, then a rational function that models this event is given by $y = \dfrac{2800x^2 + x}{x^2 + 2}$. According to this model, how many cards could be memorized in an hour? What is the greatest number of cards that can be memorized?

69. Gardening. A 500-square-foot rectangular garden will be enclosed with fencing. Write a rational function that describes how many linear feet of fence will be needed to enclose the garden as a function of the width of the garden w.

70. Geometry. A rectangular picture has an area of 414 square inches. A border (matting) is used when framing. If the top and bottom borders are each 4 inches and the side borders are 3.5 inches, write a function that represents the area $A(l)$ of the entire frame as a function of the length of the picture l.

■ CATCH THE MISTAKE

In Exercises 71–74, explain the mistake that is made.

71. Determine the vertical asymptotes of the function $f(x) = \dfrac{x - 1}{x^2 - 1}$.

Solution:

Set the denominator equal to zero. $x^2 - 1 = 0$

Solve for x. $x = \pm 1$

The vertical asymptotes are $x = -1$ and $x = 1$.

The following is a correct graph of the function:

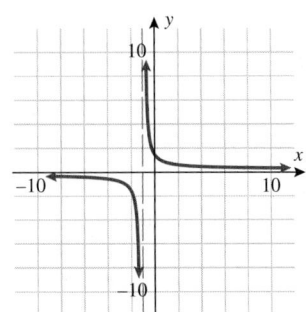

Note that only $x = -1$ is an asymptote. What went wrong?

72. Determine the vertical asymptotes of $f(x) = \dfrac{2x}{x^2 + 1}$.

Solution:

Set the denominator equal to zero. $x^2 + 1 = 0$

Solve for x. $x = \pm 1$

The vertical asymptotes are $x = -1$ and $x = 1$.

The following is a correct graph of the function:

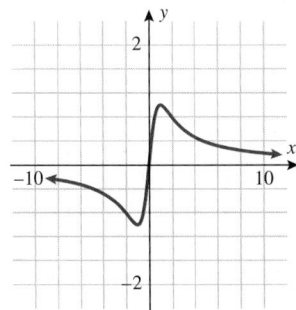

Note that there are no vertical asymptotes. What went wrong?

73. Determine whether a horizontal or a slant asymptote exists for the function $f(x) = \dfrac{9 - x^2}{x^2 - 1}$. If one does, find it.

Solution:

Step 1: The degree of the numerator equals the degree of the denominator, so there is a horizontal asymptote.

Step 2: The horizontal asymptote is the ratio of the lead coefficients: $y = \dfrac{9}{1} = 9$.

The horizontal asymptote is $y = 9$.

The following is a correct graph of the function.

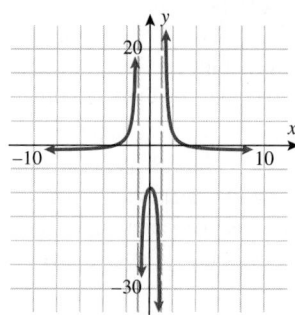

Note that there is no horizontal asymptote at $y = 9$. What went wrong?

74. Determine whether a horizontal or a slant asymptote exists for the function $f(x) = \dfrac{x^2 + 2x - 1}{3x^3 - 2x^2 - 1}$. If one does, find it.

Solution:

Step 1: The degree of the denominator is exactly one more than the degree of the numerator, so there is a slant asymptote.

Step 2: Divide.

$$
\begin{array}{r}
3x - 8 \\
x^2 + 2x - 1\overline{)3x^3 - 2x^2 + 0x - 1} \\
\underline{3x^3 + 6x^2 - 3x} \\
-8x^2 + 3x - 1 \\
\underline{8x^2 - 16x + 8} \\
19x - 9
\end{array}
$$

The slant asymptote is $y = 3x - 8$.

The following is the correct graph of the function.

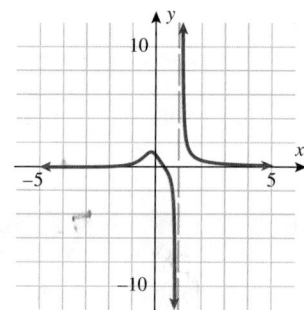

Note that $y = 3x - 8$ is not an asymptote. What went wrong?

CONCEPTUAL

For Exercises 75–78, determine whether each statement is true or false.

75. A rational function can have either a horizontal asymptote or an oblique asymptote, but not both.

76. A rational function can have at most one vertical asymptote.

77. A rational function can cross a vertical asymptote.

78. A rational function can cross a horizontal or an oblique asymptote.

79. Determine the asymptotes of the rational function
$$f(x) = \dfrac{(x - a)(x + b)}{(x - c)(x + d)}.$$

80. Determine the asymptotes of the rational function
$$f(x) = \dfrac{3x^2 + b^2}{x^2 + a^2}.$$

CHALLENGE

81. Write a rational function that has vertical asymptotes at $x = -3$ and $x = 1$ and a horizontal asymptote at $y = 4$.

82. Write a rational function that has no vertical asymptotes, approaches the x-axis as a horizontal asymptote, and has an x-intercept of $(3, 0)$.

83. Write a rational function that has no vertical asymptotes and oblique asymptote $y = x$, y-intercept $(0, 1)$, and x-intercept $(-1, 0)$. Round your answers to two decimal places.

84. Write a rational function that has vertical asymptotes at $x = -3$ and $x = 1$ and oblique asymptote $y = 3x$, y-intercept $(0, 2)$, and x-intercept $(2, 0)$. Round your answers to two decimal places.

▪ TECHNOLOGY

85. Determine the vertical asymptotes of $f(x) = \dfrac{x - 4}{x^2 - 2x - 8}$.

Graph this function utilizing a graphing utility. Does the graph confirm the asymptotes?

86. Determine the vertical asymptotes of $f(x) = \dfrac{2x + 1}{6x^2 + x - 1}$.

Graph this function utilizing a graphing utility. Does the graph confirm the asymptotes?

87. Find the asymptotes and intercepts of the rational function

$f(x) = \dfrac{1}{3x + 1} - \dfrac{2}{x}$. (*Note:* Combine the two expressions

into a single rational expression.) Graph this function utilizing a graphing utility. Does the graph confirm what you found?

88. Find the asymptotes and intercepts of the rational function

$f(x) = -\dfrac{1}{x^2 + 1} + \dfrac{1}{x}$. (*Note:* Combine the two expressions

into a single rational expression.) Graph this function utilizing a graphing utility. Does the graph confirm what you found?

For Exercises 89 and 90: **(a) Identify all asymptotes for each function. (b) Plot $f(x)$ and $g(x)$ in the same window. How does the end behavior of the function f differ from that of g? (c) Plot $g(x)$ and $h(x)$ in the same window. How does the end behavior of g differ from that of h? (d) Combine the two expressions into a single rational expression for the functions g and h. Does the strategy of finding horizontal and slant asymptotes agree with your findings in (b) and (c)?**

89. $f(x) = \dfrac{1}{x - 3}$, $g(x) = 2 + \dfrac{1}{x - 3}$, and

$h(x) = -3 + \dfrac{1}{x - 3}$

90. $f(x) = \dfrac{2x}{x^2 - 1}$, $g(x) = x + \dfrac{2x}{x^2 - 1}$, and

$h(x) = x - 3 + \dfrac{2x}{x^2 - 1}$

▪ PREVIEW TO CALCULUS

In calculus the integral of a rational function f on an interval $[a, b]$ might not exist if f has a vertical asymptote in $[a, b]$.

In Exercises 91–94, find the vertical asymptotes of each rational function.

91. $f(x) = \dfrac{x - 1}{x^3 - 2x^2 - 13x - 10}$ $[0, 3]$

92. $f(x) = \dfrac{x^2 + x + 2}{x^3 + 2x^2 - 25x - 50}$ $[-3, 2]$

93. $f(x) = \dfrac{5x + 2}{6x^2 - x - 2}$ $[-2, 0]$

94. $f(x) = \dfrac{6x - 2x^2}{x^3 + x}$ $[-1, 1]$

MODELING YOUR WORLD

The following table summarizes average yearly temperature in degrees Fahrenheit (°F) and carbon dioxide emissions in parts per million (ppm) for **Mauna Loa, Hawaii.**

YEAR	1960	1965	1970	1975	1980	1985	1990	1995	2000	2005
TEMPERATURE	44.45	43.29	43.61	43.35	46.66	45.71	45.53	47.53	45.86	46.23
CO$_2$ EMISSIONS (ppm)	316.9	320.0	325.7	331.1	338.7	345.9	354.2	360.6	369.4	379.7

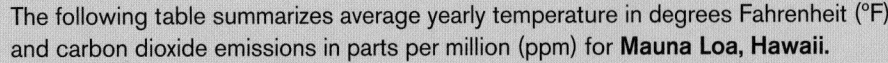

In the Modeling Your World in Chapter 1, the temperature and carbon emissions were modeled with *linear functions*. Now, let us model these same data using *polynomial functions*.

1. Plot the temperature data with time on the horizontal axis and temperature on the vertical axis. Let $t = 0$ correspond to 1960. Adjust the vertical range of the graph to (43, 48). How many turning points (local maxima and minima) do these data exhibit? What is the lowest degree polynomial function whose graph can pass through these data?

2. Find a *polynomial function* that models the temperature in Mauna Loa.

   a. Find a quadratic function: Let the data from 1995 correspond to the vertex of the graph and apply the 2005 data to determine the function.
   b. Find a quadratic function: Let the data from 2000 correspond to the vertex of the graph and apply the 2005 data to determine the function.
   c. Utilize regression and all data given to find a polynomial function whose degree is found in 1.

3. Predict what the temperature will be in Mauna Loa in 2020.

   a. Use the line found in Exercise 2(a).
   b. Use the line found in Exercise 2(b).
   c. Use the line found in Exercise 2(c).

4. Predict what the temperature will be in Mauna Loa in 2100.

   a. Use the line found in Exercise 2(a).
   b. Use the line found in Exercise 2(b).
   c. Use the line found in Exercise 2(c).

5. Do your models support the claim of "global warming"? Explain.

6. Plot the carbon dioxide emissions data with time on the horizontal axis and carbon dioxide emissions on the vertical axis. Let $t = 0$ correspond to 1960. Adjust the vertical range of the graph to (315, 380).

7. Find a *quadratic function* that models the CO_2 emissions (ppm) in Mauna Loa.

   a. Let the data from 1960 correspond to the vertex of the graph and apply the 2005 data to determine the function.
   b. Let the data from 1980 correspond to the vertex of the graph and apply the 2005 data to determine the function.
   c. Utilize regression and all data given.

8. Predict the expected CO_2 levels in Mauna Loa in 2020.

   a. Use the line found in Exercise 7(a).
   b. Use the line found in Exercise 7(b).
   c. Use the line found in Exercise 7(c).

9. Predict the expected CO_2 levels in Mauna Loa in 2100.

   a. Use the line found in Exercise 7(a).
   b. Use the line found in Exercise 7(b).
   c. Use the line found in Exercise 7(c).

10. Do your models support the claim of "global warming"? Explain. Do these models give similar predictions to the linear models found in Chapter 1?

11. Discuss differences in models and predictions found in parts (a), (b), and (c) and also discuss difference in linear and polynomial functions.

SECTION	CONCEPT	PAGES	REVIEW EXERCISES	KEY IDEAS/POINTS
2.1	**Quadratic functions**	196–206	1–28	
	Graphing quadratic functions in standard form	197–200	1–8	$f(x) = a(x - h)^2 + k$ ▪ Vertex: (h, k) ▪ Opens up: $a > 0$ ▪ Opens down: $a < 0$
	Graphing quadratic functions in general form	200–206	9–20	$f(x) = ax^2 + bx + c$, vertex is $(h, k) = \left(-\dfrac{b}{2a}, f\left(-\dfrac{b}{2a}\right)\right)$
2.2	**Polynomial functions of higher degree**	212–222	29–56	$P(x) = a_n x^n + a_{n-1} x^{n-1} + \cdots + a_2 x^2 + a_1 x + a_0$ is a polynomial of degree n.
	Graphs of polynomial functions	213–216	33–40	$y = x^n$ behave similar to ▪ $y = x^2$, when n is even. ▪ $y = x^3$, when n is odd.
	Real zeros of a polynomial function	216–218	41–50	$P(x) = (x - a)(x - b)^n = 0$ ▪ a is a zero of multiplicity 1. ▪ b is a zero of multiplicity n.
	Graphing general polynomial functions	218–222	51–54	Intercepts; zeros and multiplicities; end behavior
2.3	**Dividing polynomials**	227–232	57–70	Use zero placeholders for missing terms.
	Long division of polynomials	227–230	57–60	Can be used for all polynomial division.
	Synthetic division of polynomials	230–232	61–64	Can only be used when dividing by $(x \pm a)$.
2.4	**The real zeros of a polynomial function**	235–248	71–100	$P(x) = a_n x^n + a_{n-1} x^{n-1} + \cdots + a_2 x^2 + a_1 x + a_0$ If $P(c) = 0$, then c is a zero of $P(x)$.
	The remainder theorem and the factor theorem	236–239	71–74	If $P(x)$ is divided by $x - a$, then the remainder r is $r = P(a)$.
	The rational zero theorem and Descartes' rule of signs	239–242	79–94	Possible zeros $= \dfrac{\text{Factors of } a_0}{\text{Factors of } a_n}$ Number of positive or negative real zeros is related to the number of sign variations in $P(x)$ or $P(-x)$.
	Factoring polynomials	243–244	95–100	1. List possible rational zeros (rational zero theorem). 2. List possible combinations of positive and negative real zeros (Descartes' rule of signs). 3. Test possible values until a zero is found. 4. Once a real zero is found, use synthetic division. Then repeat testing on the quotient until linear and/or irreducible quadratic factors are reached. 5. If there is a real zero but all possible rational roots have failed, then approximate the real zero using the intermediate value theorem/bisection method.

SECTION	CONCEPT	PAGES	REVIEW EXERCISES	KEY IDEAS/POINTS
	The intermediate value theorem	245–246		Intermediate value theorem and the bisection method are used to approximate irrational zeros.
	Graphing polynomial functions	246–248	95–100	1. Find the intercepts. 2. Determine end behavior. 3. Find additional points. 4. Sketch a smooth curve.
2.5	**Complex zeros: The fundamental theorem of algebra**	251–256	101–116	$P(x) = a_n x^n + a_{n-1}x^{n-1} + \cdots + a_2 x^2 + a_1 x + a_0$ $P(x) = \underbrace{(x - c_1)(x - c_2) \cdots (x - c_n)}_{n \text{ factors}}$ where the c's represent complex (not necessarily distinct) zeros.
	The fundamental theorem of algebra	252	101–104	$P(x)$ of degree n has at least one zero and at most n zeros.
	Complex zeros	253–255	105–112	If $a + bi$ is a zero of $P(x)$, then $a - bi$ is also a zero.
	Factoring polynomials	255–256	109–116	The polynomial can be written as a product of linear factors, not necessarily distinct.
2.6	**Rational functions**	259–273	117–128	$f(x) = \dfrac{n(x)}{d(x)} \qquad d(x) \neq 0$
	Domain of rational functions	260–261	117–122	**Domain:** All real numbers except x-values that make the denominator equal to zero; that is, $d(x) = 0$. A rational function $f(x) = \dfrac{n(x)}{d(x)}$ is said to be in *lowest terms* if $n(x)$ and $d(x)$ have no common factors.
	Vertical, horizontal, and slant asymptotes	261–267	117–122	A rational function that has a common factor $x - a$ in both the numerator and denominator has a hole at $x = a$ in its graph if the multiplicity of a in the numerator is greater than or equal to the multiplicity of a in the denominator. *Vertical Asymptotes* A rational function in lowest terms has a vertical asymptote corresponding to any x-values that make the denominator equal to zero. *Horizontal Asymptote* ■ $y = 0$ if degree of $n(x) <$ degree of $d(x)$. ■ No horizontal asymptote if degree of $n(x) >$ degree of $d(x)$. ■ $y = \dfrac{\text{Leading coefficient of } n(x)}{\text{Leading coefficient of } d(x)}$ if degree of $n(x) =$ degree of $d(x)$.

Section	Concept	Pages	Review Exercises	Key Ideas/Points
				Slant Asymptote
				If degree of $n(x)$ − degree of $d(x) = 1$.
				Divide $n(x)$ by $d(x)$ and the quotient determines the slant asymptote; that is, $y = $ quotient.
	Graphing rational functions	267–273	123–128	1. Find the domain of the function.
				2. Find the intercept(s).
				3. Find any holes.
				4. Find any asymptote.
				5. Find additional points on the graph.
				6. *Sketch the graph:* Draw the asymptotes and label the intercepts and points and connect with a smooth curve.

2.1 Quadratic Functions

Match the quadratic function with its graph.

1. $f(x) = -2(x + 6)^2 + 3$ **2.** $f(x) = \frac{1}{4}(x - 4)^2 + 2$

3. $f(x) = x^2 + x - 6$ **4.** $f(x) = -3x^2 - 10x + 8$

a.

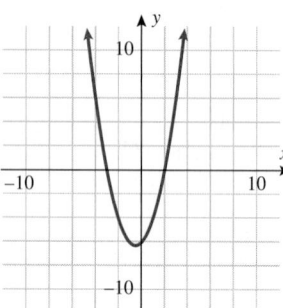

b.

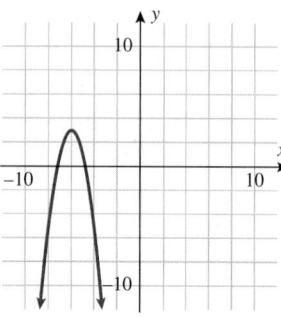

c.

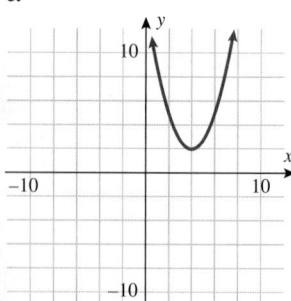

d.

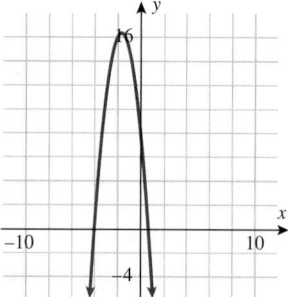

Graph the quadratic function given in standard form.

5. $f(x) = -(x - 7)^2 + 4$ **6.** $f(x) = (x + 3)^2 - 5$

7. $f(x) = -\frac{1}{2}(x - \frac{1}{3})^2 + \frac{2}{5}$ **8.** $f(x) = 0.6(x - 0.75)^2 + 0.5$

Rewrite the quadratic function in standard form by completing the square.

9. $f(x) = x^2 - 3x - 10$ **10.** $f(x) = x^2 - 2x - 24$

11. $f(x) = 4x^2 + 8x - 7$ **12.** $f(x) = -\frac{1}{4}x^2 + 2x - 4$

Graph the quadratic function given in general form.

13. $f(x) = x^2 - 3x + 5$

14. $f(x) = -x^2 + 4x + 2$

15. $f(x) = -4x^2 + 2x + 3$

16. $f(x) = -0.75x^2 + 2.5$

Find the vertex of the parabola associated with each quadratic function.

17. $f(x) = 13x^2 - 5x + 12$

18. $f(x) = \frac{2}{5}x^2 - 4x + 3$

19. $f(x) = -0.45x^2 - 0.12x + 3.6$

20. $f(x) = -\frac{3}{4}x^2 + \frac{2}{5}x + 4$

Find the quadratic function that has the given vertex and goes through the given point.

21. vertex: $(-2, 3)$ point: $(1, 4)$

22. vertex: $(4, 7)$ point: $(-3, 1)$

23. vertex: $(2.7, 3.4)$ point: $(3.2, 4.8)$

24. vertex: $\left(-\frac{5}{2}, \frac{7}{4}\right)$ point: $\left(\frac{1}{2}, \frac{3}{5}\right)$

Applications

25. Profit. The revenue and the cost of a local business are given below as functions of the number of units x in thousands produced and sold. Use the cost and the revenue to answer the questions that follow.

$$C(x) = \frac{1}{3}x + 2 \quad \text{and} \quad R(x) = -2x^2 + 12x - 12$$

  a. Determine the profit function.

  b. State the break-even points.

  c. Graph the profit function.

  d. What is the range of units to make and sell that will correspond to a profit?

26. Geometry. Given the length of a rectangle is $2x - 4$ and the width is $x + 7$, find the area of the rectangle. What dimensions correspond to the largest area?

27. Geometry. A triangle has a base of $x + 2$ units and a height of $4 - x$ units. Determine the area of the triangle. What dimensions correspond to the largest area?

28. Geometry. A person standing at a ridge in the Grand Canyon throws a penny upward and toward the pit of the canyon. The height of the penny is given by the function:

$$h(t) = -12t^2 + 80t$$

  a. What is the maximum height that the penny will reach?

  b. How many seconds will it take the penny to hit the ground below?

2.2 Polynomial Functions of Higher Degree

Determine which functions are polynomials, and for those, state their degree.

29. $f(x) = x^6 - 2x^5 + 3x^2 + 9x - 42$

30. $f(x) = (3x - 4)^3(x + 6)^2$

31. $f(x) = 3x^4 - x^3 + x^2 + \sqrt[4]{x} + 5$

32. $f(x) = 5x^3 - 2x^2 + \dfrac{4x}{7} - 3$

Match the polynomial function with its graph.

33. $f(x) = 2x - 5$

34. $f(x) = -3x^2 + x - 4$

35. $f(x) = x^4 - 2x^3 + x^2 - 6$

36. $f(x) = x^7 - x^5 + 3x^4 + 3x + 7$

a.

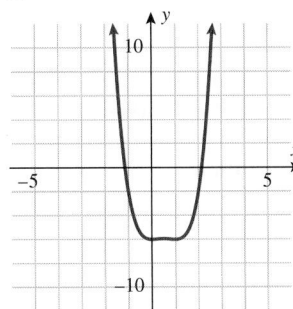

b.

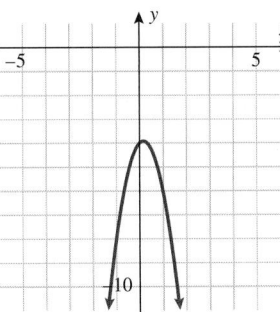

c.

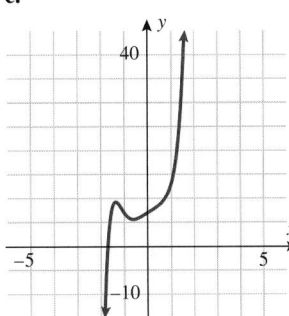

d.

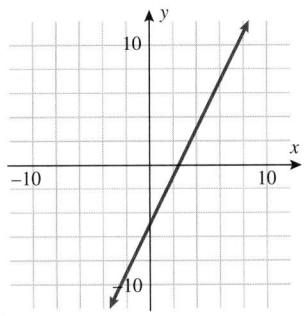

Graph each function by transforming a power function $y = x^n$.

37. $f(x) = -x^7$

38. $f(x) = (x - 3)^3$

39. $f(x) = x^4 - 2$

40. $f(x) = -6 - (x + 7)^5$

Find all the real zeros of each polynomial function, and state their multiplicities.

41. $f(x) = 3(x + 4)^2(x - 6)^5$

42. $f(x) = 7x(2x - 4)^3(x + 5)$

43. $f(x) = x^5 - 13x^3 + 36x$

44. $f(x) = 4.2x^4 - 2.6x^2$

Find a polynomial of minimum degree that has the given zeros.

45. $-3, 0, 4$

46. $2, 4, 6, -8$

47. $-\frac{2}{5}, \frac{3}{4}, 0$

48. $2 - \sqrt{5}, 2 + \sqrt{5}$

49. -2 (multiplicity of 2), 3 (multiplicity of 2)

50. 3 (multiplicity of 2), -1 (multiplicity of 2), 0 (multiplicity of 3)

For each polynomial function given: (a) list each real zero and its multiplicity; (b) determine whether the graph touches or crosses at each x-intercept; (c) find the y-intercept and a few points on the graph; (d) determine the end behavior; and (e) sketch the graph.

51. $f(x) = x^2 - 5x - 14$

52. $f(x) = -(x - 5)^5$

53. $f(x) = 6x^7 + 3x^5 - x^2 + x - 4$

54. $f(x) = -x^4(3x + 6)^3(x - 7)^3$

Applications

55. Salary. Tiffany has started tutoring students x hours per week. The tutoring job corresponds to the following additional income:

$$f(x) = (x - 1)(x - 3)(x - 7)$$

a. Graph the polynomial function.

b. Give any real zeros that occur.

c. How many hours of tutoring are financially beneficial to Tiffany?

56. Profit. The following function is the profit for Walt Disney World, where $P(x)$ represents profit in millions of dollars and x represents the month ($x = 1$ corresponds to January):

$$P(x) = 3(x - 2)^2(x - 5)^2(x - 10)^2 \qquad 1 \le x \le 12$$

Graph the polynomial. When are the peak seasons?

2.3 Dividing Polynomials

Divide the polynomials with long division. If you choose to use a calculator, do not round off. Keep the exact values instead. Express the answer in the form $Q(x) = ?, \quad r(x) = ?$.

57. $\left(x^2 + 2x - 6\right) \div (x - 2)$

58. $\left(2x^2 - 5x - 1\right) \div (2x - 3)$

59. $\left(4x^4 - 16x^3 + x - 9 + 12x^2\right) \div (2x - 4)$

60. $\left(6x^2 + 2x^3 - 4x^4 + 2 - x\right) \div \left(2x^2 + x - 4\right)$

Use synthetic division to divide the polynomial by the linear factor. Indicate the quotient $Q(x)$ and the remainder $r(x)$.

61. $\left(x^4 + 4x^3 + 5x^2 - 2x - 8\right) \div (x + 2)$

62. $\left(x^3 - 10x + 3\right) \div (2 + x)$

63. $\left(x^6 - 64\right) \div (x + 8)$

64. $\left(2x^5 + 4x^4 - 2x^3 + 7x + 5\right) \div \left(x - \frac{3}{4}\right)$

Divide the polynomials with either long division or synthetic division.

65. $\left(5x^3 + 8x^2 - 22x + 1\right) \div \left(5x^2 - 7x + 3\right)$

66. $\left(x^4 + 2x^3 - 5x^2 + 4x + 2\right) \div (x - 3)$

67. $\left(x^3 - 4x^2 + 2x - 8\right) \div (x + 1)$

68. $\left(x^3 - 5x^2 + 4x - 20\right) \div \left(x^2 + 4\right)$

Applications

69. Geometry. The area of a rectangle is given by the polynomial $6x^4 - 8x^3 - 10x^2 + 12x - 16$. If the width is $2x - 4$, what is the length of the rectangle?

70. Volume. A 10 inch by 15 inch rectangular piece of cardboard is used to make a box. Square pieces x inches on a side are cut out from the corners of the cardboard and then the sides are folded up. Find the volume of the box.

2.4 The Real Zeros of a Polynomial Function

Find the following values by applying synthetic division. Check by substituting the value into the function.

$$f(x) = 6x^5 + x^4 - 7x^2 + x - 1 \qquad g(x) = x^3 + 2x^2 - 3$$

71. $f(-2)$ **72.** $f(1)$ **73.** $g(1)$ **74.** $g(-1)$

Determine whether the number given is a zero of the polynomial.

75. $-3, P(x) = x^3 - 5x^2 + 4x + 2$

76. 2 and $-2, P(x) = x^4 - 16$

77. $1, P(x) = 2x^4 - 2x$

78. $4, P(x) = x^4 - 2x^3 - 8x$

Given a zero of the polynomial, determine all other real zeros, and write the polynomial in terms of a product of linear or irreducible factors.

Polynomial	Zero
79. $P(x) = x^4 - 6x^3 + 32x$	-2
80. $P(x) = x^3 - 7x^2 + 36$	3
81. $P(x) = x^5 - x^4 - 8x^3 + 12x^2$	0
82. $P(x) = x^4 - 32x^2 - 144$	6

Use Descartes' rule of signs to determine the possible number of positive real zeros and negative real zeros.

83. $P(x) = x^4 + 3x^3 - 16$

84. $P(x) = x^5 + 6x^3 - 4x - 2$

85. $P(x) = x^9 - 2x^7 + x^4 - 3x^3 + 2x - 1$

86. $P(x) = 2x^5 - 4x^3 + 2x^2 - 7$

Use the rational zero theorem to list the possible rational zeros.

87. $P(x) = x^3 - 2x^2 + 4x + 6$

88. $P(x) = x^5 - 4x^3 + 2x^2 - 4x - 8$

89. $P(x) = 2x^4 + 2x^3 - 36x^2 - 32x + 64$

90. $P(x) = -4x^5 - 5x^3 + 4x + 2$

List the possible rational zeros, and test to determine all rational zeros.

91. $P(x) = 2x^3 - 5x^2 + 1$

92. $P(x) = 12x^3 + 8x^2 - 13x + 3$

93. $P(x) = x^4 - 5x^3 + 20x - 16$

94. $P(x) = 24x^4 - 4x^3 - 10x^2 + 3x - 2$

For each polynomial: **(a)** use Descartes' rule of signs to determine the possible combinations of positive real zeros and negative real zeros; **(b)** use the rational zero test to determine possible rational zeros; **(c)** determine, if possible, the smallest value in the list of possible rational zeros for $P(x)$ that serves as a lower bound for all real zeros; **(d)** test for rational zeros; **(e)** factor as a product of linear and/or irreducible quadratic factors; and **(f)** graph the polynomial function.

95. $P(x) = x^3 + 3x - 5$

96. $P(x) = x^3 + 3x^2 - 6x - 8$

97. $P(x) = x^3 - 9x^2 + 20x - 12$

98. $P(x) = x^4 - x^3 - 7x^2 + x + 6$

99. $P(x) = x^4 - 5x^3 - 10x^2 + 20x + 24$

100. $P(x) = x^5 - 3x^3 - 6x^2 + 8x$

2.5 Complex Zeros: The Fundamental Theorem of Algebra

Find all zeros. Factor the polynomial as a product of linear factors.

101. $P(x) = x^2 + 25$ **102.** $P(x) = x^2 + 16$

103. $P(x) = x^2 - 2x + 5$ **104.** $P(x) = x^2 + 4x + 5$

A polynomial function is described. Find all remaining zeros.

105. Degree: 4   Zeros: $-2i, 3 + i$

106. Degree: 4   Zeros: $3i, 2 - i$

107. Degree: 6   Zeros: $i, 2 - i$ (multiplicity 2)

108. Degree: 6   Zeros: $2i, 1 - i$ (multiplicity 2)

Given a zero of the polynomial, determine all other zeros (real and complex) and write the polynomial in terms of a product of linear factors.

Polynomial	Zero
109. $P(x) = x^4 - 3x^3 - 3x^2 - 3x - 4$	i
110. $P(x) = x^4 - 4x^3 + x^2 + 16x - 20$	$2 - i$
111. $P(x) = x^4 - 2x^3 + 11x^2 - 18x + 18$	$-3i$
112. $P(x) = x^4 - 5x^2 + 10x - 6$	$1 + i$

Factor each polynomial as a product of linear factors.

113. $P(x) = x^4 - 81$

114. $P(x) = x^3 - 6x^2 + 12x$

115. $P(x) = x^3 - x^2 + 4x - 4$

116. $P(x) = x^4 - 5x^3 + 12x^2 - 2x - 20$

2.6 Rational Functions

Determine the vertical, horizontal, or slant asymptotes (if they exist) for the following rational functions.

117. $f(x) = \dfrac{7 - x}{x + 2}$

118. $f(x) = \dfrac{2 - x^2}{(x - 1)^3}$

119. $f(x) = \dfrac{4x^2}{x + 1}$

120. $f(x) = \dfrac{3x^2}{x^2 + 9}$

121. $f(x) = \dfrac{2x^2 - 3x + 1}{x^2 + 4}$

122. $f(x) = \dfrac{-2x^2 + 3x + 5}{x + 5}$

Graph the rational functions.

123. $f(x) = -\dfrac{2}{x - 3}$

124. $f(x) = \dfrac{5}{x + 1}$

125. $f(x) = \dfrac{x^2}{x^2 + 4}$

126. $f(x) = \dfrac{x^2 - 36}{x^2 + 25}$

127. $f(x) = \dfrac{x^2 - 49}{x + 7}$

128. $f(x) = \dfrac{2x^2 - 3x - 2}{2x^2 - 5x - 3}$

Technology

Section 2.1

129. On a graphing calculator, plot the quadratic function:

$$f(x) = 0.005x^2 - 4.8x - 59$$

    a. Identify the vertex of this parabola.

    b. Identify the y-intercept.

    c. Identify the x-intercepts (if any).

    d. What is the axis of symmetry?

130. Determine the quadratic function whose vertex is $(2.4, -3.1)$ and passes through the point $(0, 5.54)$.

    a. Write the quadratic function in general form.

    b. Plot this quadratic function with a graphing calculator.

    c. Zoom in on the vertex and y-intercept. Do they agree with the given values?

Section 2.2

Use a graphing calculator or a computer to graph each polynomial. From the graph, estimate the x-intercepts and state the zeros of the function and their multiplicities.

131. $f(x) = 5x^3 - 11x^2 - 10.4x + 5.6$

132. $f(x) = -x^3 - 0.9x^2 + 2.16x - 2.16$

Section 2.3

133. Plot $\dfrac{15x^3 - 47x^2 + 38x - 8}{3x^2 - 7x + 2}$. What type of function is it? Perform this division using long division, and confirm that the graph corresponds to the quotient.

134. Plot $\dfrac{-4x^3 + 14x^2 - x - 15}{x - 3}$. What type of function is it? Perform this division using synthetic division, and confirm that the graph corresponds to the quotient.

Section 2.4

(a) Determine all possible rational zeros of the polynomial. Use a graphing calculator or software to graph $P(x)$ to help find the zeros. (b) Factor as a product of linear and/or irreducible quadratic factors.

135. $P(x) = x^4 - 3x^3 - 12x^2 + 20x + 48$

136. $P(x) = -5x^5 - 18x^4 - 32x^3 - 24x^2 + x + 6$

Section 2.5

Find all zeros (real and complex). Factor the polynomial as a product of linear factors.

137. $P(x) = 2x^3 + x^2 - 2x - 91$

138. $P(x) = -2x^4 + 5x^3 + 37x^2 - 160x + 150$

Section 2.6

(a) Graph the function $f(x)$ utilizing a graphing utility to determine whether it is a one-to-one function. (b) If it is, find its inverse. (c) Graph both functions in the same viewing window.

139. $f(x) = \dfrac{2x - 3}{x + 1}$

140. $f(x) = \dfrac{4x + 7}{x - 2}$

1. Graph the quadratic function $f(x) = -(x - 4)^2 + 1$.

2. Write the quadratic function in standard form $f(x) = -x^2 + 4x - 1$.

3. Find the vertex of the parabola $f(x) = -\frac{1}{2}x^2 + 3x - 4$.

4. Find a quadratic function whose graph has a vertex at $(-3, -1)$ and whose graph passes through the point $(-4, 1)$.

5. Find a sixth-degree polynomial function with the given zeros:

   2 of multiplicity 3   1 of multiplicity 2   0 of multiplicity 1

6. For the polynomial function $f(x) = x^4 + 6x^3 - 7x$:

   a. List each real zero and its multiplicity.

   b. Determine whether the graph touches or crosses at each x-intercept.

   c. Find the y-intercept and a few points on the graph.

   d. Determine the end behavior.

   e. Sketch the graph.

7. Divide $-4x^4 + 2x^3 - 7x^2 + 5x - 2$ by $2x^2 - 3x + 1$.

8. Divide $17x^5 - 4x^3 + 2x - 10$ by $x + 2$.

9. Is $x - 3$ a factor of $x^4 + x^3 - 13x^2 - x + 12$?

10. Determine whether -1 is a zero of $P(x) = x^{21} - 2x^{18} + 5x^{12} + 7x^3 + 3x^2 + 2$.

11. Given that $x - 7$ is a factor of $P(x) = x^3 - 6x^2 - 9x + 14$, factor the polynomial in terms of linear factors.

12. Given that $3i$ is a zero of $P(x) = x^4 - 3x^3 + 19x^2 - 27x + 90$, find all other zeros.

13. Can a polynomial have zeros that are not x-intercepts? Explain.

14. Apply Descartes' rule of signs to determine the possible combinations of positive real zeros, negative real zeros, and complex zeros of $P(x) = 3x^5 + 2x^4 - 3x^3 + 2x^2 - x + 1$.

15. From the rational zero test, list all possible rational zeros of $P(x) = 3x^4 - 7x^2 + 3x + 12$.

In Exercises 16–18, determine all zeros of the polynomial function and graph.

16. $P(x) = -x^3 + 4x$

17. $P(x) = 2x^3 - 3x^2 + 8x - 12$

18. $P(x) = x^4 - 6x^3 + 10x^2 - 6x + 9$

19. **Sports.** A football player shows up in August at 300 pounds. After 2 weeks of practice in the hot sun, he is down to 285 pounds. Ten weeks into the season he is up to 315 pounds because of weight training. In the spring he does not work out, and he is back to 300 pounds by the next August. Plot these points on a graph. What degree polynomial could this be?

20. **Profit.** The profit of a company is governed by the polynomial $P(x) = x^3 - 13x^2 + 47x - 35$, where x is the number of units sold in thousands. How many units does the company have to sell to break even?

21. **Interest Rate.** The interest rate for a 30-year fixed mortgage fluctuates with the economy. In 1970 the mortgage interest rate was 8%, and in 1988 it peaked at 13%. In 2002 it dipped down to 4%, and in 2005 it was up to 6%. What is the lowest degree polynomial that can represent this function?

In Exercises 22–25, determine (if any) the:

   a. x- and y-intercepts

   b. vertical asymptotes

   c. horizontal asymptotes

   d. slant asymptotes

   e. graph

22. $f(x) = \dfrac{2x - 9}{x + 3}$

23. $g(x) = \dfrac{x}{x^2 - 4}$

24. $h(x) = \dfrac{3x^3 - 3}{x^2 - 4}$

25. $F(x) = \dfrac{x - 3}{x^2 - 2x - 8}$

26. **Food.** After a sugary snack, the glucose level of the average body almost doubles. The percentage increase in glucose level y can be approximated by the rational function $y = \dfrac{25x}{x^2 + 50}$, where x represents the number of minutes after eating the snack. Graph the function.

27. a. Use the calculator commands $\boxed{\text{STAT}}$ $\boxed{\text{QuadReg}}$ to model the data using a quadratic function.

   b. Write the quadratic function in standard form and identify the vertex.

   c. Find the x-intercepts.

   d. Plot this quadratic function with a graphing calculator. Do they agree with the given values?

x	-3	2.2	7.5
y	10.01	-9.75	25.76

28. Find the asymptotes and intercepts of the rational function $f(x) = \dfrac{x(2x - 3)}{x^2 - 3x} + 1$. (*Note:* Combine the two expressions into a single rational expression.) Graph this function utilizing a graphing utility. Does the graph confirm what you found?

1. If $f(x) = 4x - \dfrac{1}{\sqrt{x+2}}$, find $f(2), f(-1), f(1+h)$, and $f(-x)$.

2. If $f(x) = (x-1)^4 - \sqrt{2x+3}$, $f(1), f(3)$, and $f(x+h)$.

3. If $f(x) = \dfrac{3x-5}{2-x-x^2}$, find $f(-3), f(0), f(1)$, and $f(4)$.

4. If $f(x) = 4x^3 - 3x^2 + 5$, evaluate the difference quotient $\dfrac{f(x+h) - f(x)}{h}$.

5. If $f(x) = \sqrt{x} - \dfrac{1}{x^2}$, evaluate the difference quotient $\dfrac{f(x+h) - f(x)}{h}$.

6. If $f(x) = \begin{cases} 0 & x < 0 \\ 3x + x^2 & 0 \le x \le 4 \\ |2x - x^3| & x > 4 \end{cases}$
   find $f(-5), f(0), f(3), f(4)$, and $f(5)$.

In Exercises 7 and 8, (a) Graph the piecewise-defined functions. (b) State the domain and range in interval notation. (c) Determine the intervals where the function is increasing, decreasing, or constant.

7. $f(x) = \begin{cases} |6 - 2x| & x \le 8 \\ 10 & 8 < x < 10 \\ \dfrac{1}{x-10} & x > 10 \end{cases}$

8. $f(x) = \begin{cases} (x+5)^2 - 6 & x < -2 \\ \sqrt{x-1} + 3 & -2 \le x < 10 \\ 26 - 2x & 10 \le x \le 14 \end{cases}$

9. The position of a particle is described by the curve $y = \dfrac{2t}{t^2+3}$, where t is time (in seconds). What is the average rate of change of the position as a function of time from $t = 5$ to $t = 9$?

10. Express the domain of the function $f(x) = \sqrt{6x - 7}$ with interval notation.

11. Determine whether the function $g(x) = \sqrt{x+10}$ is even, odd, or neither.

12. For the function $y = -(x+1)^2 + 2$, identify all of the transformations of $y = x^2$.

13. Sketch the graph of $y = \sqrt{x-1} + 3$ and identify all transformations.

14. Find the composite function $f \circ g$ and state the domain for $f(x) = x^2 - 3$ and $g(x) = \sqrt{x+2}$.

15. Evaluate $g(f(-1))$ for $f(x) = 7 - 2x^2$ and $g(x) = 2x - 10$.

16. Find the inverse of the function $f(x) = (x-4)^2 + 2$, where $x \ge 4$.

17. Find a quadratic function whose graph has a vertex at $(-2, 3)$ and passes through the point $(-1, 4)$.

18. Find all of the real zeros and state the multiplicity of the function $f(x) = -3.7x^4 - 14.8x^3$.

19. Use long division to find the quotient $Q(x)$ and the remainder $r(x)$ of $(-20x^3 - 8x^2 + 7x - 5) \div (-5x + 3)$.

20. Use synthetic division to find the quotient $Q(x)$ and the remainder $r(x)$ of $(2x^3 + 3x^2 - 11x + 6) \div (x - 3)$.

21. List the possible rational zeros, and test to determine all rational zeros for $P(x) = 12x^3 + 29x^2 + 7x - 6$.

22. Given the real zero $x = 5$ of the polynomial $P(x) = 2x^3 - 3x^2 - 32x - 15$, determine all the other zeros and write the polynomial in terms of a product of linear factors.

23. Factor the polynomial $P(x) = x^3 - 5x^2 + 2x + 8$ completely.

24. Factor the polynomial $P(x) = x^5 + 7x^4 + 15x^3 + 5x^2 - 16x - 12$ completely.

25. Find all vertical and horizontal asymptotes of $f(x) = \dfrac{3x-5}{x^2-4}$.

26. Graph the function $f(x) = \dfrac{2x^3 - x^2 - x}{x^2 - 1}$.

27. Find the asymptotes and intercepts of the rational function $f(x) = \dfrac{5}{2x-3} - \dfrac{1}{x}$. (Note: Combine the two expressions into a single rational expression.) Graph this function utilizing a graphing utility. Does the graph confirm what you found?

28. Find the asymptotes and intercepts of the rational function $f(x) = \dfrac{6x}{3x+1} - \dfrac{6x}{4x-1}$. (Note: Combine the two expressions into a single rational expression.) Graph this function utilizing a graphing utility. Does the graph confirm what you found?

3

Exponential and Logarithmic Functions

Most populations initially grow *exponentially*, but then as resources become limited, the population reaches a *carrying capacity*. This exponential increase followed by a saturation at some carrying capacity is called *logistic growth*. Often when a particular species is placed on the endangered species list, it is protected from human predators and then its population size increases until naturally leveling off at some carrying capacity.

The U.S. Fish and Wildlife Service removed the gray wolf (*canis lupus*) from the Wisconsin list of endangered and threatened species in 2004, and placed it on the list of protected wild animals. About 537 to 564 wolves existed in Wisconsin in the late winter of 2008.

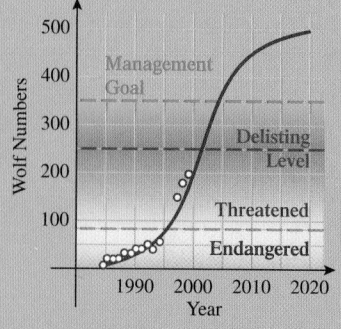

Wisconsin Wolf Population Growth if Carrying Capacity is 500 Wolves

IN THIS CHAPTER we will discuss exponential functions and their inverses, logarithmic functions. We will graph these functions and use their properties to solve exponential and logarithmic equations. We will then discuss particular exponential and logarithmic models that represent phenomena such as compound interest, world populations, conservation biology models, carbon dating, pH values in chemistry, and the bell curve that is fundamental in statistics for describing how quantities vary in the real world.

EXPONENTIAL AND LOGARITHMIC FUNCTIONS

3.1
Exponential Functions and Their Graphs

3.2
Logarithmic Functions and Their Graphs

3.3
Properties of Logarithms

3.4
Exponential and Logarithmic Equations

3.5
Exponential and Logarithmic Models

- Evaluating Exponential Functions
- Graphs of Exponential Functions
- The Natural Base *e*
- Applications of Exponential Functions

- Evaluating Logarithms
- Common and Natural Logarithms
- Graphs of Logarithmic Functions
- Applications of Logarithms

- Properties of Logarithms
- Change-of-Base Formula

- Solving Exponential Equations
- Solving Logarithmic Equations

- Exponential Growth Models
- Exponential Decay Models
- Gaussian (Normal) Distribution Models
- Logistic Growth Models
- Logarithmic Models

CHAPTER OBJECTIVES

- Evaluate exponential functions for particular values and understand the characteristics of the graph of an exponential function (Section 3.1).

- Evaluate logarithmic functions for particular values and understand the characteristics of the graph of a logarithmic function (Section 3.2).

- Understand that logarithmic functions are inverses of exponential functions and derive the properties of logarithms (Section 3.3).

- Solve exponential and logarithmic equations (Section 3.4).

- Use the exponential growth, exponential decay, logarithmic, logistic growth, and Gaussian distribution models to represent real-world phenomena (Section 3.5).

SKILLS OBJECTIVES

- Evaluate exponential functions.
- Graph exponential functions.
- Find the domain and range of exponential functions.
- Define the number e.
- Solve real-world problems using exponential functions.

CONCEPTUAL OBJECTIVES

- Understand the difference between algebraic and exponential functions.
- Understand that irrational exponents lead to approximations.

Evaluating Exponential Functions

Most of the functions (polynomial, rational, radical, etc.) we have studied thus far have been **algebraic functions**. Algebraic functions involve basic operations, powers, and roots. In this chapter, we discuss *exponential functions* and *logarithmic functions*. The following table illustrates the difference between algebraic functions and *exponential functions*:

FUNCTION	VARIABLE IS IN THE	CONSTANT IS IN THE	EXAMPLE	EXAMPLE
Algebraic	Base	Exponent	$f(x) = x^2$	$g(x) = x^{1/3}$
Exponential	Exponent	Base	$F(x) = 2^x$	$G(x) = \left(\dfrac{1}{3}\right)^x$

DEFINITION   **Exponential Function**

An **exponential function** with **base** b is denoted by

$$f(x) = b^x$$

where b and x are any real numbers such that $b > 0$ and $b \neq 1$.

Note:

- We eliminate $b = 1$ as a value for the base because it merely yields the constant function $f(x) = 1^x = 1$.

- We eliminate negative values for b because they would give non-real-number values such as $(-9)^{1/2} = \sqrt{-9} = 3i$.

- We eliminate $b = 0$ because 0^x corresponds to an undefined value when x is negative.

Sometimes the value of an exponential function for a specific argument can be found by inspection as an *exact* number.

x	-3	-1	0	1	3
$F(x) = 2^x$	$2^{-3} = \dfrac{1}{2^3} = \dfrac{1}{8}$	$2^{-1} = \dfrac{1}{2^1} = \dfrac{1}{2}$	$2^0 = 1$	$2^1 = 2$	$2^3 = 8$

If an exponential function cannot be evaluated exactly, then we find the decimal *approximation* using a calculator.

x	-2.7	$-\frac{4}{5}$	$\frac{5}{7}$	2.7
$F(x) = 2^x$	$2^{-2.7} \approx 0.154$	$2^{-4/5} \approx 0.574$	$2^{5/7} \approx 1.641$	$2^{2.7} \approx 6.498$

The domain of exponential functions, $f(x) = b^x$, is the set of all real numbers. All of the arguments discussed in the first two tables have been rational numbers. What happens if x is irrational? We can approximate the irrational number with a decimal approximation such as $b^\pi \approx b^{3.14}$ or $b^{\sqrt{2}} \approx b^{1.41}$.

Consider $7^{\sqrt{3}}$, and realize that the irrational number $\sqrt{3}$ is a decimal that never terminates or repeats: $\sqrt{3} \approx 1.7320508$. We can show in advanced mathematics that there is a number $7^{\sqrt{3}}$, and although we cannot write it exactly, we can approximate the number. In fact, the closer the exponent is to $\sqrt{3}$, the closer the approximation is to $7^{\sqrt{3}}$.

It is important to note that the properties of exponents (Appendix) hold when the exponent is any real number (rational or irrational).

$7^{1.7} \approx 27.3317$

$7^{1.73} \approx 28.9747$

$7^{1.732} \approx 29.0877$

$\cdots$

$7^{\sqrt{3}} \approx 29.0906$

EXAMPLE 1   Evaluating Exponential Functions

Let $f(x) = 3^x$, $g(x) = \left(\frac{1}{4}\right)^x$ and $h(x) = 10^{x-2}$. Find the following values:

a. $f(2)$ **b.** $f(\pi)$ **c.** $g\left(-\frac{3}{2}\right)$ **d.** $h(2.3)$ **e.** $f(0)$ **f.** $g(0)$

If an approximation is required, approximate to four decimal places.

Solution:

a. $f(2) = 3^2 = \boxed{9}$

b. $f(\pi) = 3^\pi \approx \boxed{31.5443}$ *

c. $g\left(-\frac{3}{2}\right) = \left(\frac{1}{4}\right)^{-3/2} = 4^{3/2} = \left(\sqrt{4}\right)^3 = 2^3 = \boxed{8}$

d. $h(2.3) = 10^{2.3-2} = 10^{0.3} \approx \boxed{1.9953}$

e. $f(0) = 3^0 = \boxed{1}$

f. $g(0) = \left(\frac{1}{4}\right)^0 = \boxed{1}$

Notice that parts (a) and (c) were evaluated exactly, whereas parts (b) and (d) required approximation using a calculator.

■ **YOUR TURN** Let $f(x) = 2^x$ and $g(x) = \left(\frac{1}{9}\right)^x$ and $h(x) = 5^{x-2}$. Find the following values:

   a. $f(4)$ **b.** $f(\pi)$ **c.** $g\left(-\frac{3}{2}\right)$ **d.** $h(2.9)$

   Evaluate exactly when possible, and round to four decimal places when a calculator is needed.

■ **Answer: a.** 16   **b.** 8.8250
            c. 27   **d.** 4.2567

---

*In part (b), the π button on the calculator is selected. If we instead approximate π by 3.14, we get a slightly different approximation for the function value:

$$f(\pi) = 3^\pi \approx 3^{3.14} \approx 31.4891$$

Graphs of Exponential Functions

Let's graph two exponential functions, $y = 2^x$ and $y = 2^{-x} = \left(\frac{1}{2}\right)^x$, by plotting points.

x	$y = 2^x$	(x, y)
-2	$2^{-2} = \dfrac{1}{2^2} = \dfrac{1}{4}$	$\left(-2, \dfrac{1}{4}\right)$
-1	$2^{-1} = \dfrac{1}{2^1} = \dfrac{1}{2}$	$\left(-1, \dfrac{1}{2}\right)$
0	$2^0 = 1$	$(0, 1)$
1	$2^1 = 2$	$(1, 2)$
2	$2^2 = 4$	$(2, 4)$
3	$2^3 = 8$	$(3, 8)$

x	$y = 2^{-x}$	(x, y)
-3	$2^{-(-3)} = 2^3 = 8$	$(-3, 8)$
-2	$2^{-(-2)} = 2^2 = 4$	$(-2, 4)$
-1	$2^{-(-1)} = 2^1 = 2$	$(-1, 2)$
0	$2^0 = 1$	$(0, 1)$
1	$2^{-1} = \dfrac{1}{2^1} = \dfrac{1}{2}$	$\left(1, \dfrac{1}{2}\right)$
2	$2^{-2} = \dfrac{1}{2^2} = \dfrac{1}{4}$	$\left(2, \dfrac{1}{4}\right)$

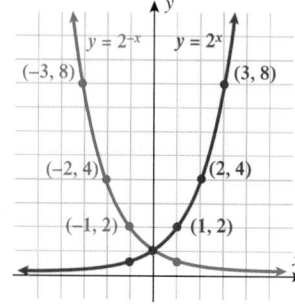

Notice that both graphs' y-intercept is $(0, 1)$ (as shown to the left) and neither graph has an x-intercept. The x-axis is a horizontal asymptote for both graphs. The following box summarizes general characteristics of the graphs of exponential functions.

CHARACTERISTICS OF GRAPHS OF EXPONENTIAL FUNCTIONS

$$f(x) = b^x, \qquad b > 0, \qquad b \neq 1$$

- Domain: $(-\infty, \infty)$
- Range: $(0, \infty)$
- x-intercepts: none
- y-intercept: $(0, 1)$
- Horizontal asymptote: x-axis
- The graph passes through $(1, b)$ and $\left(-1, \dfrac{1}{b}\right)$.
- As x increases, $f(x)$ increases if $b > 1$ and decreases if $0 < b < 1$.
- The function f is one-to-one.

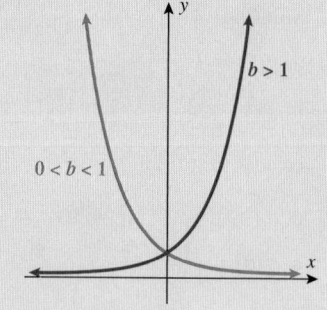

Since exponential functions, $f(x) = b^x$, all go through the point $(0, 1)$ and have the x-axis as a horizontal asymptote, we can find the graph by finding two additional points as outlined in the following procedure.

PROCEDURE FOR GRAPHING $f(x) = b^x$

Step 1: Label the point $(0, 1)$ corresponding to the y-intercept $f(0)$.

Step 2: Find and label two additional points corresponding to $f(-1)$ and $f(1)$.

Step 3: Connect the three points with a *smooth* curve with the x-axis as the horizontal asymptote.

EXAMPLE 2 **Graphing Exponential Functions for $b > 1$**

Graph the function $f(x) = 5^x$.

Solution:

STEP 1: Label the y-intercept $(0, 1)$.

$f(0) = 5^0 = 1$

STEP 2: Label the point $(1, 5)$.

$f(1) = 5^1 = 5$

Label the point $(-1, 0.2)$.

$f(-1) = 5^{-1} = \dfrac{1}{5} = 0.2$

STEP 3: Sketch a smooth curve through the three points with the x-axis as a horizontal asymptote.

Domain: $(-\infty, \infty)$

Range: $(0, \infty)$

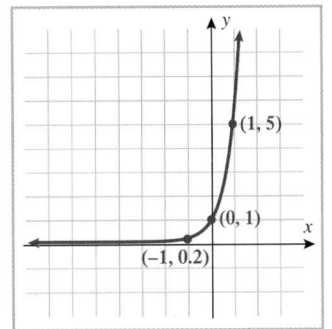

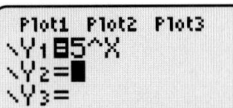
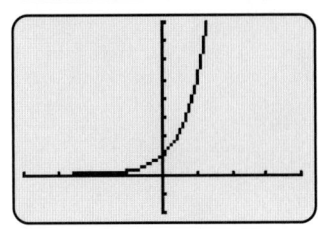
■ **YOUR TURN** Graph the function $f(x) = 5^{-x}$.

■ **Answer:**

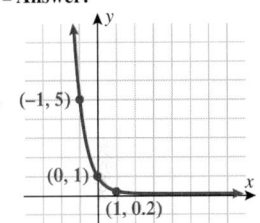

EXAMPLE 3 **Graphing Exponential Functions for $b < 1$**

Graph the function $f(x) = \left(\frac{2}{5}\right)^x$.

Solution:

STEP 1: Label the y-intercept $(0, 1)$.

$f(0) = \left(\dfrac{2}{5}\right)^0 = 1$

STEP 2: Label the point $(-1, 2.5)$.

$f(-1) = \left(\dfrac{2}{5}\right)^{-1} = \dfrac{5}{2} = 2.5$

Label the point $(1, 0.4)$.

$f(1) = \left(\dfrac{2}{5}\right)^1 = \dfrac{2}{5} = 0.4$

STEP 3: Sketch a smooth curve through the three points with the x-axis as a horizontal asymptote.

Domain: $(-\infty, \infty)$

Range: $(0, \infty)$

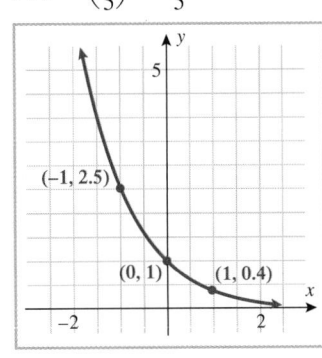

Exponential functions, like all functions, can be graphed by point-plotting. We can also use transformations (horizontal and vertical shifting and reflection; Section 1.3) to graph exponential functions.

EXAMPLE 4 **Graphing Exponential Functions Using a Horizontal or Vertical Shift**

a. Graph the function $F(x) = 2^{x-1}$. State the domain and range of F.
b. Graph the function $G(x) = 2^x + 1$. State the domain and range of G.

Solution (a):

Identify the base function.

$f(x) = 2^x$

Identify the base function y-intercept and horizontal asymptote.

$(0, 1)$ and $y = 0$

The graph of the function F is found by shifting the graph of the function f to the right 1 unit.

$F(x) = f(x - 1)$

Shift the y-intercept to the right 1 unit.

$(0, 1)$ shifts to $(1, 1)$

The horizontal asymptote is not altered by a horizontal shift.

$y = 0$

Find additional points on the graph.

$F(0) = 2^{0-1} = 2^{-1} = \dfrac{1}{2}$

$F(2) = 2^{2-1} = 2^1 = 2$

Sketch the graph of $F(x) = 2^{x-1}$ with a *smooth* curve.

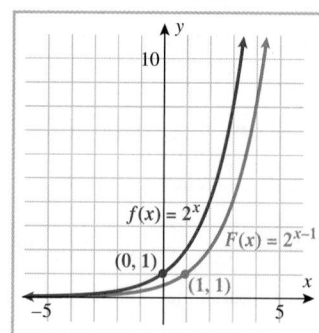

Domain: $(-\infty, \infty)$
Range: $(0, \infty)$

Solution (b):

Identify the base function.

$f(x) = 2^x$

Identify the base function y-intercept and horizontal asymptote.

$(0, 1)$ and $y = 0$

The graph of the function G is found by shifting the graph of the function f up 1 unit.

$G(x) = f(x) + 1$

Shift the y-intercept up 1 unit.

$(0, 1)$ shifts to $(0, 2)$

Shift the horizontal asymptote up 1 unit.

$y = 0$ shifts to $y = 1$

Find additional points on the graph.

$G(1) = 2^1 + 1 = 2 + 1 = 3$

$G(-1) = 2^{-1} + 1 = \dfrac{1}{2} + 1 = \dfrac{3}{2}$

Sketch the graph of $G(x) = 2^x + 1$ with a *smooth* curve.

Domain: $(-\infty, \infty)$
Range: $(1, \infty)$

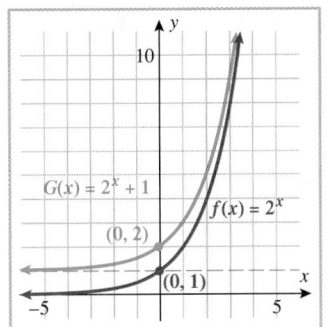

EXAMPLE 5   Graphing Exponential Functions Using Both Horizontal and Vertical Shifts

Graph the function $F(x) = 3^{x+1} - 2$. State the domain and range of F.

Solution:

Identify the base function.

$$f(x) = 3^x$$

Identify the base function y-intercept and horizontal asymptote.

$(0, 1)$ and $y = 0$

The graph of the function F is found by shifting the graph of the function f to the left 1 unit and down 2 units.

$$F(x) = f(x + 1) - 2$$

Shift the y-intercept to the left 1 unit and down 2 units.

$(0, 1)$ shifts to $(-1, -1)$

Shift the horizontal asymptote down 2 units.

$y = 0$ shifts to $y = -2$

Find additional points on the graph.

$$F(0) = 3^{0+1} - 2 = 3 - 2 = 1$$
$$F(1) = 3^{1+1} - 2 = 9 - 2 = 7$$

Sketch the graph of $F(x) = 3^{x+1} - 2$ with a *smooth* curve.

Domain: $(-\infty, \infty)$
Range: $(-2, \infty)$

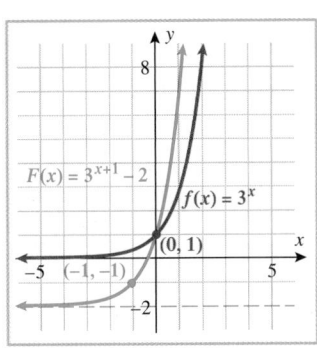

Technology Tip

The graph of $f(x) = 3^{x+1} - 2$ is shown.

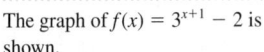

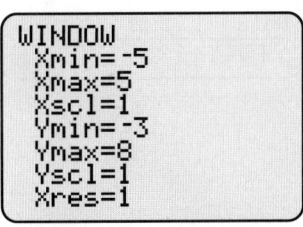

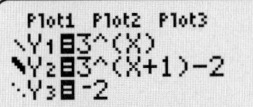

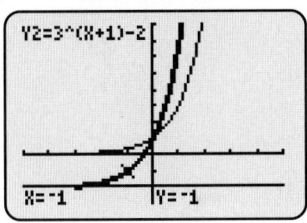

■ **Answer:** Domain: $(-\infty, \infty)$
                  Range: $(-1, \infty)$

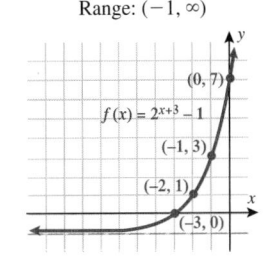

■ **YOUR TURN** Graph $f(x) = 2^{x+3} - 1$. State the domain and range of f.

The Natural Base e

Any positive real number can serve as the base for an exponential function. A particular irrational number, denoted by the letter e, appears as the base in many applications, as you will soon see when we discuss continuous compounded interest. Although you will see 2 and 10 as common bases, the base that appears most often is e, because e, as you will come to see in your further studies of mathematics, is the **natural base**. The exponential function with base e, $f(x) = e^x$, is called the **exponential function** or the **natural exponential function**. Mathematicians did not pull this irrational number out of a hat. The number e has many remarkable properties, but most simply, it comes from evaluating the expression $\left(1 + \dfrac{1}{m}\right)^m$ as m gets large (increases without bound).

m	$\left(1 + \dfrac{1}{m}\right)^m$
1	2
10	2.59374
100	2.70481
1000	2.71692
10,000	2.71815
100,000	2.71827
1,000,000	**2.71828**

$e \approx 2.71828$

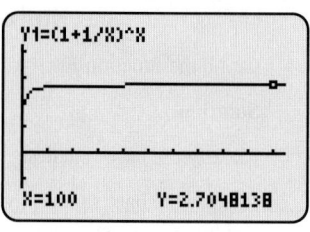

Calculators have an e^x button for approximating the natural exponential function.

EXAMPLE 6 Evaluating the Natural Exponential Function

Evaluate $f(x) = e^x$ for the given x-values. Round your answers to four decimal places.

a. $x = 1$ **b.** $x = -1$ **c.** $x = 1.2$ **d.** $x = -0.47$

Solution:

a. $f(1) = e^1 \approx 2.718281828 \approx \boxed{2.7183}$

b. $f(-1) = e^{-1} \approx 0.367879441 \approx \boxed{0.3679}$

c. $f(1.2) = e^{1.2} \approx 3.320116923 \approx \boxed{3.3201}$

d. $f(-0.47) = e^{-0.47} \approx 0.625002268 \approx \boxed{0.6250}$

Like all exponential functions of the form $f(x) = b^x$, $f(x) = e^x$ and $f(x) = e^{-x}$ have $(0, 1)$ as their y-intercept and the x-axis as a horizontal asymptote as shown in the figure on the right.

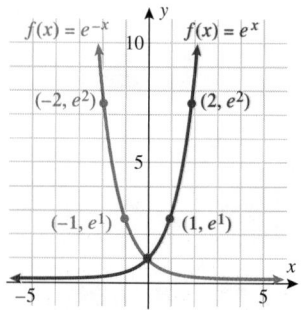

EXAMPLE 7 **Graphing Exponential Functions with Base *e***

Graph the function $f(x) = 3 + e^{2x}$.

Solution:

x	$f(x) = 3 + e^{2x}$	(x, y)
-2	3.02	$(-2, 3.02)$
-1	3.14	$(1, 3.14)$
0	4	$(0, 4)$
1	10.39	$(1, 10.39)$
2	57.60	$(2, 57.60)$

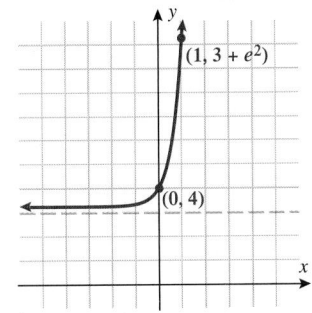

Note: The y-intercept is $(0, 4)$ and the line $y = 3$ is the horizontal asymptote.

■ **YOUR TURN** Graph the function $f(x) = e^{x+1} - 2$.

Technology Tip

The graph of $y_1 = 3 + e^{2x}$ is shown. To setup the split screen, press

MODE G-T ENTER GRAPH .

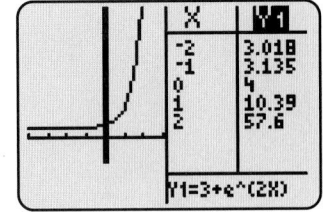

■ **Answer:**

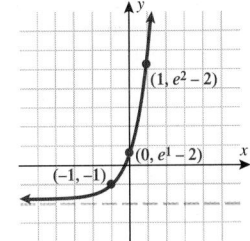

Applications of Exponential Functions

Exponential functions describe either *growth* or *decay*. Populations and investments are often modeled with exponential growth functions, while the declining value of a used car and the radioactive decay of isotopes are often modeled with exponential decay functions. In Section 3.5, various exponential models will be discussed. In this section, we discuss doubling time, half-life, and compound interest.

A successful investment program, growing at about 7.2% per year, will double in size every 10 years. Let's assume that you will retire at the age of 65. There is a saying: *It's not the first time your money doubles, it's the last time that makes such a difference.* As you may already know or as you will soon find, it is important to start investing early.

Suppose Maria invests $5,000 at age 25 and David invests $5,000 at age 35. Let's calculate how much will accrue from the initial $5,000 investment by the time they each retire, assuming their money doubles every 10 years and that they both retire at age 65.

AGE	MARIA	DAVID
25	$5,000	
35	$10,000	$5,000
45	$20,000	$10,000
55	$40,000	$20,000
65	**$80,000**	**$40,000**

They each made a one-time investment of $5,000. By investing 10 years sooner, Maria made twice what David made.

A measure of growth rate is the *doubling time*, the time it takes for something to double. Often doubling time is used to describe populations.

DOUBLING TIME GROWTH MODEL

The doubling time growth model is given by

$$P = P_0 2^{t/d}$$

where
    P = Population at time t
    P_0 = Population at time $t = 0$
    d = Doubling time

Note that when $t = d$, $P = 2P_0$ (population is equal to twice the original).

The units for P and P_0 are the same and can be any quantity (people, dollars, etc.). The units for t and d must be the same (years, weeks, days, hours, seconds, etc.).

In the investment scenario with Maria and David, $P_0 = \$5{,}000$ and $d = 10$ years, so the model used to predict how much money the original $5,000 investment yielded is $P = 5000(2)^{t/10}$. Maria retired 40 years after the original investment, $t = 40$, and David retired 30 years after the original investment, $t = 30$.

$$\text{Maria: } P = 5000(2)^{40/10} = 5000(2)^4 = 5000(16) = 80{,}000$$

$$\text{David: } P = 5000(2)^{30/10} = 5000(2)^3 = 5000(8) \;\; = 40{,}000$$

EXAMPLE 8   Doubling Time of Populations

In 2004 the population in Kazakhstan, a country in Asia, reached 15 million. It is estimated that the population doubles in 30 years. If the population continues to grow at the same rate, what will the population be in 2024? Round to the nearest million.

Solution:

Write the doubling model. $P = P_0 2^{t/d}$

Substitute $P_0 = 15$ million,
$d = 30$ years, and $t = 20$ years. $P = 15(2)^{20/30}$

Simplify. $P = 15(2)^{2/3} \approx 23.8110$

In 2024, there will be approximately $\boxed{24 \text{ million people}}$ in Kazakhstan.

■ **Answer:** 38 million

■ **YOUR TURN**   What will the approximate population in Kazakhstan be in 2044? Round to the nearest million.

We now turn our attention from exponential growth to exponential decay, or negative growth. Suppose you buy a brand-new car from a dealership for $24,000. The value of a car decreases over time according to an exponential decay function. The **half-life** of this particular car, or the time it takes for the car to depreciate 50%, is approximately 3 years. The exponential decay is described by

$$A = A_0 \left(\frac{1}{2}\right)^{t/h}$$

where A_0 is the amount the car is worth (in dollars) when new (that is, when $t = 0$), A is the amount the car is worth (in dollars) after t years, and h is the half-life in years. In our car scenario, $A_0 = 24{,}000$ and $h = 3$:

$$A = 24{,}000\left(\frac{1}{2}\right)^{t/3}$$

How much is the car worth after three years? Six years? Nine years? Twenty-four years?

$$t = 3: \qquad A = 24{,}000\left(\frac{1}{2}\right)^{3/3} = 24{,}000\left(\frac{1}{2}\right) = 12{,}000$$

$$t = 6: \qquad A = 24{,}000\left(\frac{1}{2}\right)^{6/3} = 24{,}000\left(\frac{1}{2}\right)^2 = 6000$$

$$t = 9: \qquad A = 24{,}000\left(\frac{1}{2}\right)^{9/3} = 24{,}000\left(\frac{1}{2}\right)^3 = 3000$$

$$t = 24: \qquad A = 24{,}000\left(\frac{1}{2}\right)^{24/3} = 24{,}000\left(\frac{1}{2}\right)^8 = 93.75 \approx 100$$

The car that was worth $24,000 new is worth $12,000 in 3 years, $6,000 in 6 years, $3,000 in 9 years, and about $100 in the junkyard in 24 years.

EXAMPLE 9   Radioactive Decay

The radioactive isotope of potassium ^{42}K, which is used in the diagnosis of brain tumors, has a half-life of 12.36 hours. If 500 milligrams of potassium 42 are taken, how many milligrams will remain after 24 hours? Round to the nearest milligram.

Solution:

Write the half-life formula.
$$A = A_0\left(\frac{1}{2}\right)^{t/h}$$

Substitute $A_0 = 500$ mg, $h = 12.36$ hours, $t = 24$ hours.
$$A = 500\left(\frac{1}{2}\right)^{24/12.36}$$

Simplify.
$$A \approx 500(0.2603) \approx 130.15$$

After 24 hours, there are approximately $\boxed{130 \text{ milligrams}}$ of potassium 42 left.

■ **YOUR TURN** How many milligrams of potassium 42 are expected to be left in the body after 1 week?

■ **Answer:** 0.04 milligram (less than 1 milligram)

In Section 0.1, *simple interest* was defined where the interest I is calculated based on the principal P, the annual interest rate r, and the time t in years, using the formula $I = Prt$.

If the interest earned in a period is then reinvested at the same rate, future interest is earned on both the principal and the reinvested interest during the next period. Interest paid on both the principal and interest is called *compound interest.*

COMPOUND INTEREST

If a **principal** P is invested at an annual **rate** r **compounded** n times a year, then the **amount** A in the account at the end of t years is given by

$$A = P\left(1 + \frac{r}{n}\right)^{nt}$$

The annual interest rate r is expressed as a decimal.

The following list shows the typical number of times interest is compounded:

Annually	$n = 1$	Monthly	$n = 12$
Semiannually	$n = 2$	Weekly	$n = 52$
Quarterly	$n = 4$	Daily	$n = 365$

EXAMPLE 10 **Compound Interest**

If $3,000 is deposited in an account paying 3% compounded quarterly, how much will you have in the account in 7 years?

Solution:

Write the compound interest formula.

$$A = P\left(1 + \frac{r}{n}\right)^{nt}$$

Substitute $P = 3000$, $r = 0.03$, $n = 4$, and $t = 7$.

$$A = 3000\left(1 + \frac{0.03}{4}\right)^{(4)(7)}$$

Simplify.

$$A = 3000(1.0075)^{28} \approx 3698.14$$

You will have $3,698.14 in the account.

■ **YOUR TURN** If $5,000 is deposited in an account paying 6% compounded annually, how much will you have in the account in 4 years?

Notice in the compound interest formula that as n increases the amount A also increases. In other words, the more times the interest is compounded per year, the more money you make. Ideally, your bank will compound your interest infinitely many times. This is called *compounding continuously*. We will now show the development of the continuous compounding formula, $A = Pe^{rt}$.

WORDS	MATH
Write the compound interest formula.	$A = P\left(1 + \dfrac{r}{n}\right)^{nt}$
Note that $\dfrac{r}{n} = \dfrac{1}{n/r}$ and $nt = \left(\dfrac{n}{r}\right)rt$.	$A = P\left(1 + \dfrac{1}{n/r}\right)^{(n/r)rt}$
Let $m = \dfrac{n}{r}$.	$A = P\left(1 + \dfrac{1}{m}\right)^{mrt}$
Use the exponential property: $x^{mrt} = (x^m)^{rt}$.	$A = P\left[\left(1 + \dfrac{1}{m}\right)^m\right]^{rt}$

Recall that as m increases, $\left(1 + \dfrac{1}{m}\right)^m$ approaches e. Therefore, as the number of times the interest is compounded approaches infinity, or as $n \to \infty$, the amount in an account $A = P\left(1 + \dfrac{r}{n}\right)^{nt}$ approaches $A = Pe^{rt}$.

CONTINUOUS COMPOUND INTEREST

If a **principal** P is invested at an annual **rate** r **compounded continuously**, then the **amount** A in the account at the end of t years is given by

$$A = Pe^{rt}$$

The annual interest rate r is expressed as a decimal.

It is important to note that for a given interest rate, the highest return you can earn is by compounding continuously.

EXAMPLE 11 Continuously Compounded Interest

If $3,000 is deposited in a savings account paying 3% a year compounded continuously, how much will you have in the account in 7 years?

Solution:

Write the continuous compound interest formula. $A = Pe^{rt}$

Substitute $P = 3000$, $r = 0.03$, and $t = 7$. $A = 3000e^{(0.03)(7)}$

Simplify. $A \approx 3701.034$

There will be $\boxed{\$3,701.03}$ in the account in 7 years.

Note: In Example 10, we worked this same problem compounding *quarterly,* and the result was $3,698.14.

If the number of times per year interest is compounded increases, then the total interest earned that year also increases.

▪ **YOUR TURN** If $5,000 is deposited in an account paying 6% compounded continuously, how much will be in the account in 4 years?

SECTION 3.1 SUMMARY

In this section, we discussed exponential functions (constant base, variable exponent).

General Exponential Functions: $f(x) = b^x$, $b \neq 1$, and $b > 0$

1. Evaluating exponential functions
   - Exact (by inspection): $f(x) = 2^x$ $f(3) = 2^3 = 8$.
   - Approximate (with the aid of a calculator): $f(x) = 2^x$ $f(\sqrt{3}) = 2^{\sqrt{3}} \approx 3.322$
2. Graphs of exponential functions
   - Domain: $(-\infty, \infty)$ and range: $(0, \infty)$.
   - The point $(0, 1)$ corresponds to the y-intercept.
   - The graph passes through the points $(1, b)$ and $\left(-1, \frac{1}{b}\right)$.
   - The x-axis is a horizontal asymptote.
   - The function f is one-to-one.

Procedure for Graphing: $f(x) = b^x$

Step 1: Label the point $(0, 1)$ corresponding to the y-intercept $f(0)$.
Step 2: Find and label two additional points corresponding to $f(-1)$ and $f(1)$.
Step 3: Connect the three points with a smooth curve with the x-axis as the horizontal asymptote.

The Natural Exponential Function: $f(x) = e^x$
   - The irrational number e is called the natural base.
   - $e = \left(1 + \frac{1}{m}\right)^m$ as $m \to \infty$
   - $e \approx 2.71828$

Applications of Exponential Functions (all variables expressed in consistent units)

1. Doubling time: $P = P_0 2^{t/d}$
   - d is doubling time.
   - P is population at time t.
   - P_0 is population at time $t = 0$.
2. Half-life: $A = A_0 \left(\frac{1}{2}\right)^{t/h}$
   - h is the half-life.
   - A is amount at time t.
   - A_0 is amount at time $t = 0$.
3. Compound interest (P = principal, A = amount after t years, r = interest rate)
   - Compounded n times a year: $A = P\left(1 + \frac{r}{n}\right)^{nt}$
   - Compounded continuously: $A = Pe^{rt}$

SECTION
3.1 EXERCISES

▪ SKILLS

In Exercises 1–6, evaluate *exactly* (without using a calculator). For rational exponents, consider converting to radical form first.

1. 5^{-2} **2.** 4^{-3} **3.** $8^{2/3}$ **4.** $27^{2/3}$ **5.** $\left(\frac{1}{9}\right)^{-3/2}$ **6.** $\left(\frac{1}{16}\right)^{-3/2}$

In Exercises 7–12, approximate with a calculator. Round your answer to four decimal places.

7. $5^{\sqrt{2}}$ **8.** $6^{\sqrt{3}}$ **9.** e^2 **10.** $e^{1/2}$ **11.** $e^{-\pi}$ **12.** $e^{-\sqrt{2}}$

In Exercises 13–20, for the functions $f(x) = 3^x$, $g(x) = \left(\frac{1}{16}\right)^x$, and $h(x) = 10^{x+1}$, find the function value at the indicated points.

13. $f(3)$ **14.** $h(1)$ **15.** $g(-1)$ **16.** $f(-2)$

17. $g\left(-\frac{1}{2}\right)$ **18.** $g\left(-\frac{3}{2}\right)$ **19.** $f(e)$ **20.** $g(\pi)$

In Exercises 21–26, match the graph with the function.

21. $y = 5^{x-1}$ **22.** $y = 5^{1-x}$ **23.** $y = -5^x$

24. $y = -5^{-x}$ **25.** $y = 1 - 5^{-x}$ **26.** $y = 5^x - 1$

a.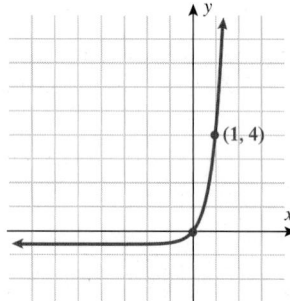

b.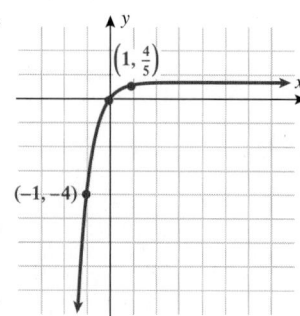

c.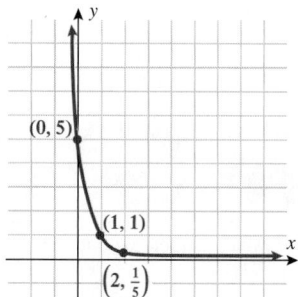

d.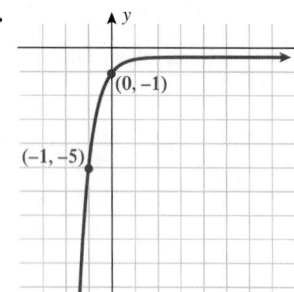

e.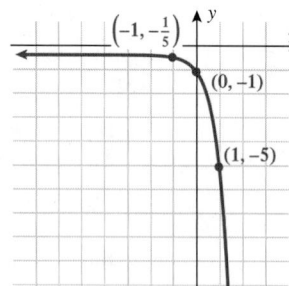

f.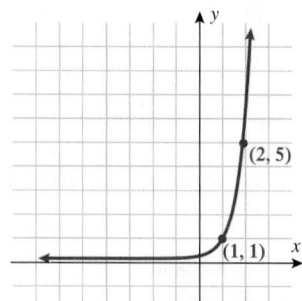

In Exercises 27–46, graph the exponential function using transformations. State the *y*-intercept, two additional points, the domain, the range, and the horizontal asymptote.

27. $f(x) = 6^x$ **28.** $f(x) = 7^x$ **29.** $f(x) = 10^{-x}$ **30.** $f(x) = 4^{-x}$ **31.** $f(x) = e^x$ **32.** $f(x) = -e^{-x}$

33. $f(x) = e^{-x}$ **34.** $f(x) = -e^x$ **35.** $f(x) = 2^x - 1$ **36.** $f(x) = 3^x - 1$ **37.** $f(x) = 2 - e^x$ **38.** $f(x) = 1 + e^{-x}$

39. $f(x) = 5 + 4^{-x}$ **40.** $f(x) = 5^x - 2$ **41.** $f(x) = e^{x+1} - 4$ **42.** $f(x) = e^{x-1} + 2$ **43.** $f(x) = 3e^{x/2}$ **44.** $f(x) = 2e^{-x}$

45. $f(x) = 1 + \left(\frac{1}{2}\right)^{x-2}$ **46.** $f(x) = 2 - \left(\frac{1}{3}\right)^{x+1}$

▪ APPLICATIONS

47. Population Doubling Time. In 2002 there were 7.1 million people living in London, England. If the population is expected to double by 2090, what is the expected population in London in 2050?

48. Population Doubling Time. In 2004 the population in Morganton, Georgia, was 43,000. The population in Morganton doubled by 2010. If the growth rate remains the same, what is the expected population in Morganton in 2020?

49. Investments. Suppose an investor buys land in a rural area for $1,500 an acre and sells some of it 5 years later at $3,000 an acre and the rest of it 10 years later at $6,000. Write a function that models the value of land in that area, assuming the growth rate stays the same. What would the expected cost per acre be 30 years after the initial investment of $1,500?

50. Salaries. Twin brothers, Collin and Cameron, get jobs immediately after graduating from college at the age of 22. Collin opts for the higher starting salary, $55,000, and stays with the same company until he retires at 65. His salary doubles every 15 years. Cameron opts for a lower starting salary, $35,000, but moves to a new job every 5 years; he doubles his salary every 10 years until he retires at 65. What is the annual salary of each brother upon retirement?

51. Radioactive Decay. A radioactive isotope of selenium, ^{75}Se, which is used in medical imaging of the pancreas, has a half-life of 119.77 days. If 200 milligrams are given to a patient, how many milligrams are left after 30 days?

52. Radioactive Decay. The radioactive isotope indium-111 (^{111}In), used as a diagnostic tool for locating tumors associated with prostate cancer, has a half-life of 2.807 days. If 300 milligrams are given to a patient, how many milligrams will be left after a week?

53. Radioactive Decay. A radioactive isotope of beryllium-11 decays to borom-11 with a half-life of 13.81 seconds. Beryllium is given to patients that suffer Chronic Beryllium Disease (CBD). If 800 mg are given to a CBD patient, how much beryllium is present after 2 minutes? Round your answer to the nearest milligram.

54. Radioactive Decay. If the CBD patient in Exercise 53 is given 1000 mg, how much beryllium is present after 1 minute? Round your answer to the nearest milligram.

55. Depreciation of Furniture. A couple buy a new bedroom set for $8,000 and 10 years later sell it for $4,000. If the depreciation continues at the same rate, how much would the bedroom set be worth in 4 more years?

56. Depreciation of a Computer. A student buys a new laptop for $1,500 when she arrives as a freshman. A year later, the computer is worth approximately $750. If the depreciation continues at the same rate, how much would she expect to sell her laptop for when she graduates 4 years after she bought it?

57. Compound Interest. If you put $3,200 in a savings account that earns 2.5% interest per year compounded quarterly, how much would you expect to have in that account in 3 years?

58. Compound Interest. If you put $10,000 in a savings account that earns 3.5% interest per year compounded annually, how much would you expect to have in that account in 5 years?

59. Compound Interest. How much money should you put in a savings account now that earns 5% a year compounded daily if you want to have $32,000 in 18 years?

60. Compound Interest. How much money should you put in a savings account now that earns 3.0% a year compounded weekly if you want to have $80,000 in 15 years?

61. Compound Interest. If you put $3,200 in a savings account that pays 2% a year compounded continuously, how much will you have in the account in 15 years?

62. Compound Interest. If you put $7,000 in a money market account that pays 4.3% a year compounded continuously, how much will you have in the account in 10 years?

63. Compound Interest. How much money should you deposit into a money market account that pays 5% a year compounded continuously to have $38,000 in the account in 20 years?

64. Compound Interest. How much money should you deposit into a certificate of deposit that pays 6% a year compounded continuously to have $80,000 in the account in 18 years?

▪ CATCH THE MISTAKE

In Exercises 65–68, explain the mistake that is made.

65. Evaluate the expression $4^{-1/2}$.

Solution: $4^{-1/2} = 4^2 = 16$

The correct value is $\frac{1}{2}$. What mistake was made?

66. Evaluate the function for the given x: $f(x) = 4^x$ for $x = \frac{3}{2}$.

Solution: $f\left(\frac{3}{2}\right) = 4^{3/2}$

$$= \frac{4^3}{4^2} = \frac{64}{16} = 4$$

The correct value is 8. What mistake was made?

67. If $2,000 is invested in a savings account that earns 2.5% interest compounding continuously, how much will be in the account in one year?

Solution:

Write the compound continuous interest
formula. $A = Pe^{rt}$

Substitute $P = 2000$, $r = 2.5$, and $t = 1$. $A = 2000e^{(2.5)(1)}$

Simplify. $A = 24{,}364.99$

This is incorrect. What mistake was made?

68. If $5,000 is invested in a savings account that earns 3% interest compounding continuously, how much will be in the account in 6 months?

Solution:

Write the compound continuous interest
formula. $A = Pe^{rt}$

Substitute $P = 5000$, $r = 0.03$, and $t = 6$. $A = 5000e^{(0.03)(6)}$

Simplify. $A = 5986.09$

This is incorrect. What mistake was made?

■ **CONCEPTUAL**

In Exercises 69–72, determine whether each statement is true or false.

69. The function $f(x) = -e^{-x}$ has the y-intercept $(0,1)$.

70. The function $f(x) = -e^{-x}$ has a horizontal asymptote along the x-axis.

71. The functions $y = 3^{-x}$ and $y = \left(\frac{1}{3}\right)^x$ have the same graphs.

72. $e = 2.718$.

73. Plot $f(x) = 3^x$ and its inverse on the same graph.

74. Plot $f(x) = e^x$ and its inverse on the same graph.

75. Graph $f(x) = e^{|x|}$.

76. Graph $f(x) = e^{-|x|}$.

■ **CHALLENGE**

77. Find the y-intercept and horizontal asymptote of
$f(x) = be^{-x+1} - a$.

78. Find the y-intercept and horizontal asymptote of
$f(x) = a + be^{x+1}$.

79. Graph $f(x) = b^{|x|}$, $b > 1$, and state the domain.

80. Graph the function $f(x) = \begin{cases} a^x & x < 0 \\ a^{-x} & x \geq 0 \end{cases}$ where $a > 1$.

81. Graph the function
$$f(x) = \begin{cases} -a^x & x < 0 \\ -a^{-x} & x \geq 0 \end{cases} \quad \text{where } 0 < a < 1.$$

82. Find the y-intercept and horizontal asymptote(s) of
$f(x) = 2^x + 3^x$.

■ **TECHNOLOGY**

83. Plot the function $y = \left(1 + \dfrac{1}{x}\right)^x$. What is the horizontal asymptote as x increases?

84. Plot the functions $y = 2^x$, $y = e^x$, and $y = 3^x$ in the same viewing screen. Explain why $y = e^x$ lies between the other two graphs.

85. Plot $y_1 = e^x$ and $y_2 = 1 + x + \dfrac{x^2}{2} + \dfrac{x^3}{6} + \dfrac{x^4}{24}$ in the same viewing screen. What do you notice?

86. Plot $y_1 = e^{-x}$ and $y_2 = 1 - x + \dfrac{x^2}{2} - \dfrac{x^3}{6} + \dfrac{x^4}{24}$ in the same viewing screen. What do you notice?

87. Plot the functions $f(x) = \left(1 + \dfrac{1}{x}\right)^x$, $g(x) = \left(1 + \dfrac{2}{x}\right)^x$, and $h(x) = \left(1 + \dfrac{2}{x}\right)^{2x}$ in the same viewing screen. Compare their horizontal asymptotes as x increases. What can you say about the function values of f, g, and h in terms of the powers of e as x increases?

88. Plot the functions $f(x) = \left(1 + \dfrac{1}{x}\right)^x$, $g(x) = \left(1 - \dfrac{1}{x}\right)^x$, and $h(x) = \left(1 - \dfrac{2}{x}\right)^x$ in the same viewing screen. Compare their horizontal asymptotes as x increases. What can you say about the function values of f, g, and h in terms of the powers of e as x increases?

▪ PREVIEW TO CALCULUS

In calculus the following two functions are studied:

$$\sinh x = \frac{e^x - e^{-x}}{2} \quad \text{and} \quad \cosh x = \frac{e^x + e^{-x}}{2}$$

89. Determine whether $f(x) = \sinh x$ is an even function or an odd function.

90. Determine whether $f(x) = \cosh x$ is an even function or an odd function.

91. Show that $\cosh^2 x - \sinh^2 x = 1$.

92. Show that $\cosh x + \sinh x = e^x$.

| SECTION 3.2 | LOGARITHMIC FUNCTIONS AND THEIR GRAPHS |

SKILLS OBJECTIVES

- Convert exponential expressions to logarithmic expressions.
- Convert logarithmic expressions to exponential expressions.
- Evaluate logarithmic expressions exactly by inspection.
- Approximate common and natural logarithms using a calculator.
- Graph logarithmic functions.
- Determine domain restrictions on logarithmic functions.

CONCEPTUAL OBJECTIVES

- Interpret logarithmic functions as inverses of exponential functions.
- Understand that logarithmic functions allow very large ranges of numbers in science and engineering applications to be represented on a smaller scale.

Evaluating Logarithms

In Section 3.1, we found that the graph of an exponential function, $f(x) = b^x$, passes through the point $(0, 1)$, with the x-axis as a horizontal asymptote. The graph passes both the vertical line test (for a function) and the horizontal line test (for a one-to-one function), and therefore an inverse exists. We will now apply the technique outlined in Section 1.5 to find the inverse of $f(x) = b^x$:

WORDS	MATH
Let $y = f(x)$.	$y = b^x$
Interchange x and y.	$x = b^y$
Solve for y.	$y = ?$

DEFINITION   **Logarithmic Function**

For $x > 0$, $b > 0$, and $b \neq 1$, the **logarithmic function with base b** is denoted $f(x) = \log_b x$, where

$$y = \log_b x \qquad \text{if and only if} \qquad x = b^y$$

We read $\log_b x$ as "log base b of x."

Study Tip

• $\log_b x = y$ is equivalent to $b^y = x$.
• The exponent y is called a logarithm (or "log" for short).

This definition says that $x = b^y$ (**exponential form**) and $y = \log_b x$ (**logarithmic form**) are equivalent. One way to remember this relationship is by adding arrows to the logarithmic form:

$$\log_b x = y \iff b^y = x$$

EXAMPLE 1   **Changing from Logarithmic Form to Exponential Form**

Express each equation in its equivalent exponential form.

a. $\log_2 8 = 3$ **b.** $\log_9 3 = \frac{1}{2}$ **c.** $\log_5 \left(\frac{1}{25}\right) = -2$

Solution:

a. $\log_2 8 = 3$ is equivalent to $\boxed{8 = 2^3}$

b. $\log_9 3 = \frac{1}{2}$ is equivalent to $\boxed{3 = 9^{1/2}} = \sqrt{9}$

c. $\log_5 \left(\frac{1}{25}\right) = -2$ is equivalent to $\boxed{\frac{1}{25} = 5^{-2}}$

■ **Answer: a.** $9 = 3^2$
       b. $4 = 16^{1/2}$
       c. $\frac{1}{8} = 2^{-3}$

■ **YOUR TURN**   Write each equation in its equivalent exponential form.

     a. $\log_3 9 = 2$ **b.** $\log_{16} 4 = \frac{1}{2}$ **c.** $\log_2 \left(\frac{1}{8}\right) = -3$

EXAMPLE 2   **Changing from Exponential Form to Logarithmic Form**

Write each equation in its equivalent logarithmic form.

a. $16 = 2^4$ **b.** $9 = \sqrt{81}$ **c.** $\frac{1}{9} = 3^{-2}$ **d.** $x^a = z$

Solution:

a. $16 = 2^4$ is equivalent to $\boxed{\log_2 16 = 4}$

b. $9 = \sqrt{81} = 81^{1/2}$ is equivalent to $\boxed{\log_{81} 9 = \frac{1}{2}}$

c. $\frac{1}{9} = 3^{-2}$ is equivalent to $\boxed{\log_3 \left(\frac{1}{9}\right) = -2}$

d. $x^a = z$ is equivalent to $\boxed{\log_x z = a \quad \text{for } x > 0}$

■ **Answer: a.** $\log_9 81 = 2$
       b. $\log_{144} 12 = \frac{1}{2}$
       c. $\log_7 \left(\frac{1}{49}\right) = -2$
       d. $\log_y w = b \quad \text{for } y > 0$

■ **YOUR TURN**   Write each equation in its equivalent logarithmic form.

     a. $81 = 9^2$ **b.** $12 = \sqrt{144}$ **c.** $\frac{1}{49} = 7^{-2}$ **d.** $y^b = w$

Some logarithms can be found exactly, while others must be approximated. Example 3 illustrates how to find the exact value of a logarithm. Example 4 illustrates approximating values of logarithms with a calculator.

EXAMPLE 3 **Finding the Exact Value of a Logarithm**

Find the exact value of

a. $\log_3 81$ **b.** $\log_{169} 13$ **c.** $\log_5 \left(\frac{1}{5}\right)$

Solution (a):

The logarithm has some value. Let's call it x.	$\log_3 81 = x$
Change from logarithmic to exponential form.	$3^x = 81$
3 raised to what power is 81?	$3^4 = 81$ $x = 4$
Change from exponential to logarithmic form.	$\boxed{\log_3 81 = 4}$

Solution (b):

The logarithm has some value. Let's call it x.	$\log_{169} 13 = x$
Change from logarithmic to exponential form.	$169^x = 13$
169 raised to what power is 13?	$169^{1/2} = \sqrt{169} = 13$ $x = \dfrac{1}{2}$
Change from exponential to logarithmic form.	$\boxed{\log_{169} 13 = \dfrac{1}{2}}$

Solution (c):

The logarithm has some value. Let's call it x.	$\log_5 \left(\dfrac{1}{5}\right) = x$
Change from logarithmic to exponential form.	$5^x = \dfrac{1}{5}$
5 raised to what power is $\frac{1}{5}$?	$5^{-1} = \dfrac{1}{5}$ $x = -1$
Change from exponential to logarithmic form.	$\boxed{\log_5 \left(\dfrac{1}{5}\right) = -1}$

■ **YOUR TURN** Evaluate the given logarithms exactly.

a. $\log_2 \frac{1}{2}$ **b.** $\log_{100} 10$ **c.** $\log_{10} 1000$

■ **Answer: a.** $\log_2 \frac{1}{2} = -1$
  b. $\log_{100} 10 = \frac{1}{2}$
  c. $\log_{10} 1000 = 3$

Common and Natural Logarithms

Two logarithmic bases that arise frequently are base 10 and base e. The logarithmic function of base 10 is called the **common logarithmic function**. Since it is common, $f(x) = \log_{10} x$ is often expressed as $f(x) = \log x$. Thus, if no explicit base is indicated, base 10 is implied. The logarithmic function of base e is called the **natural logarithmic function**. The natural logarithmic function $f(x) = \log_e x$ is often expressed as $f(x) = \ln x$. Both the LOG and LN buttons appear on scientific and graphing calculators. For the *logarithms* (not the functions), we say "the log" (for base 10) and "the natural log" (for base e).

Study Tip

• $\log_{10} x = \log x$. No explicit base implies base 10.
• $\log_e x = \ln x$

Earlier in this section, we evaluated logarithms exactly by converting to exponential form and identifying the exponent. For example, to evaluate $\log_{10} 100$, we ask the question, 10 raised to what power is 100? The answer is 2.

Calculators enable us to approximate logarithms. For example, evaluate $\log_{10} 233$. We are unable to evaluate this exactly by asking the question, 10 raised to what power is 233? Since $10^2 < 10^x < 10^3$, we know the answer x must lie between 2 and 3. Instead, we use a calculator to find an approximate value 2.367.

EXAMPLE 4 Using a Calculator to Evaluate Common and Natural Logarithms

Use a calculator to evaluate the common and natural logarithms. Round your answers to four decimal places.

a. $\log 415$ **b.** $\ln 415$ **c.** $\log 1$ **d.** $\ln 1$ **e.** $\log(-2)$ **f.** $\ln(-2)$

Solution:

a. $\log(415) \approx 2.618048097 \approx \boxed{2.6180}$ **b.** $\ln(415) \approx 6.02827852 \approx \boxed{6.0283}$

c. $\log(1) = \boxed{0}$ **d.** $\ln(1) = \boxed{0}$

e. $\log(-2)$ $\boxed{\text{undefined}}$ **f.** $\ln(-2)$ $\boxed{\text{undefined}}$

Study Tip

Logarithms can only be evaluated for positive arguments.

Parts (c) and (d) in Example 4 illustrate that all logarithmic functions pass through the point $(1, 0)$. Parts (e) and (f) in Example 4 illustrate that the domains of logarithmic functions are positive real numbers.

Graphs of Logarithmic Functions

The general logarithmic function $y = \log_b x$ is defined as the inverse of the exponential function $y = b^x$. Therefore, when these two functions are plotted on the same graph, they are symmetric about the line $y = x$. Notice the symmetry about the line $y = x$ when $y = b^x$ and $y = \log_b x$ are plotted on the same graph.

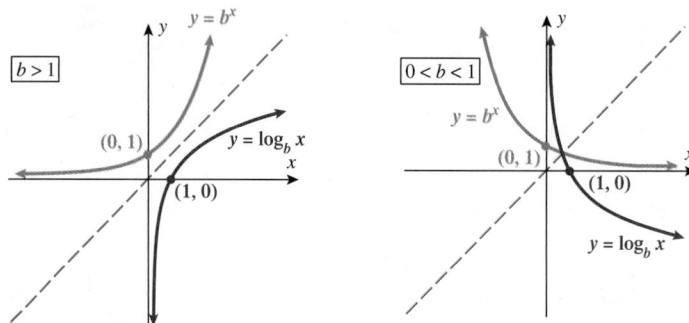

Comparison of Inverse Functions:
$f(x) = \log_b x$ and $f^{-1}(x) = b^x$

EXPONENTIAL FUNCTION	LOGARITHMIC FUNCTION
$y = b^x$	$y = \log_b x$
y-intercept $(0, 1)$	x-intercept $(1, 0)$
Domain $(-\infty, \infty)$	Domain $(0, \infty)$
Range $(0, \infty)$	Range $(-\infty, \infty)$
Horizontal asymptote: x-axis	Vertical asymptote: y-axis

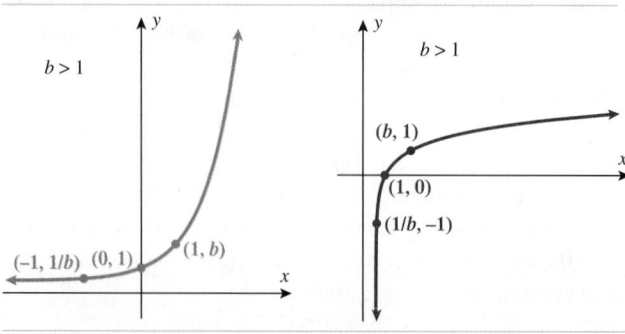

Additionally, the domain of one function is the range of the other, and vice versa. When dealing with logarithmic functions, special attention must be paid to the domain of the function. The domain of $y = \log_b x$ is $(0, \infty)$. In other words, you can only take the log of a positive real number, $x > 0$.

EXAMPLE 5 **Finding the Domain of a Shifted Logarithmic Function**

Find the domain of each of the given logarithmic functions.

a. $f(x) = \log_b(x - 4)$ **b.** $g(x) = \log_b(5 - 2x)$

Solution (a):

Set the argument greater than zero.	$x - 4 > 0$
Solve the inequality.	$x > 4$
Write the domain in interval notation.	$\boxed{(4, \infty)}$

Solution (b):

Set the argument greater than zero.	$5 - 2x > 0$
Solve the inequality.	$-2x > -5$
	$2x < 5$
	$x < \dfrac{5}{2}$
Write the domain in interval notation.	$\boxed{\left(-\infty, \dfrac{5}{2}\right)}$

■ **YOUR TURN** Find the domain of the given logarithmic functions.

a. $f(x) = \log_b(x + 2)$ **b.** $g(x) = \log_b(3 - 5x)$

■ **Answer: a.** $(-2, \infty)$
b. $\left(-\infty, \frac{3}{5}\right)$

It is important to note that when finding the domain of a logarithmic function, we set the argument strictly greater than zero and solve.

EXAMPLE 6 **Finding the Domain of a Logarithmic Function with a Complicated Argument**

Find the domain of each of the given logarithmic functions.

a. $\ln(x^2 - 9)$ **b.** $\log(|x + 1|)$

Solution (a):

Set the argument greater than zero.	$x^2 - 9 > 0$
Solve the inequality.	$\boxed{(-\infty, -3) \cup (3, \infty)}$

Solution (b):

Set the argument greater than zero.	$|x + 1| > 0$
Solve the inequality.	$x \neq -1$
Write the domain in interval notation.	$\boxed{(-\infty, -1) \cup (-1, \infty)}$

■ **YOUR TURN** Find the domain of each of the given logarithmic functions.

a. $\ln(x^2 - 4)$ **b.** $\log(|x - 3|)$

■ **Answer: a.** $(-\infty, -2) \cup (2, \infty)$
b. $(-\infty, 3) \cup (3, \infty)$

Recall from Section 1.3 that a technique for graphing general functions is transformations of known functions. For example, to graph $f(x) = (x - 3)^2 + 1$, we start with the known parabola $y = x^2$, whose vertex is at $(0, 0)$, and we shift that graph to the right 3 units and up 1 unit. We use the same techniques for graphing logarithmic functions. To graph $y = \log_b(x + 2) - 1$, we start with the graph of $y = \log_b(x)$ and shift the graph to the left 2 units and down 1 unit.

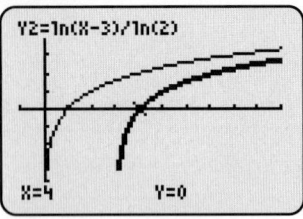

Technology Tip

a. To graph $\log_2(x - 3)$, use the change-base formula (see Section 3.3) to express it as $\dfrac{\ln(x - 3)}{\ln(2)}$. Graphs of $Y_1 = \log_2 x$ and $Y_2 = \log_2(x - 3)$ are shown.

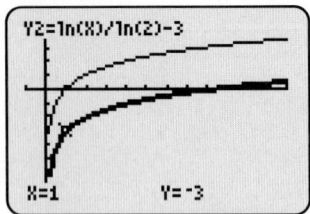

b. Graphs of $Y_1 = \log_2 x$ and $Y_2 = \log_2(x) - 3$ are shown.

EXAMPLE 7 Graphing Logarithmic Functions Using Horizontal and Vertical Shifts

Graph the functions, and state the domain and range of each.

a. $y = \log_2(x - 3)$ **b.** $\log_2(x) - 3$

Solution:

Identify the base function. $y = \log_2 x$

Label key features of $y = \log_2 x$.

x-intercept: $(1, 0)$

Vertical asymptote: $x = 0$

Additional points: $(2, 1)$, $(4, 2)$

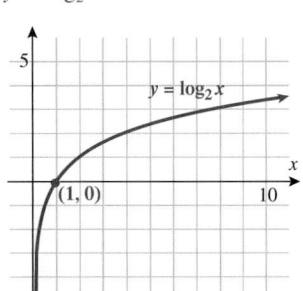

a. Shift base function to the *right* 3 units.

x-intercept: $(4, 0)$

Vertical asymptote: $x = 3$

Additional points: $(5, 1)$, $(7, 2)$

| Domain: $(3, \infty)$ | Range: $(-\infty, \infty)$ |

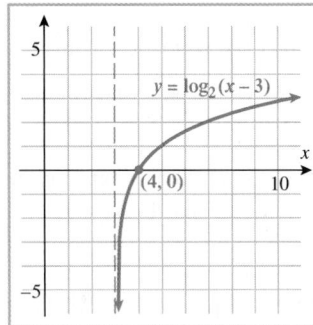

b. Shift base function *down* 3 units.

x-intercept: $(1, -3)$

Vertical asymptote: $x = 0$

Additional points: $(2, -2)$, $(4, -1)$

| Domain: $(0, \infty)$ | Range: $(-\infty, \infty)$ |

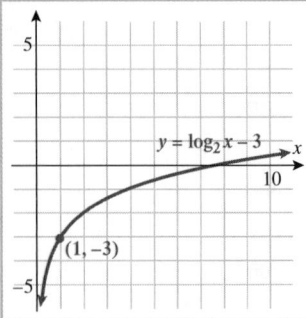

■ **Answer:**
a. Domain: $(0, \infty)$ Range: $(-\infty, \infty)$
b. Domain: $(-3, \infty)$ Range: $(-\infty, \infty)$
c. Domain: $(0, \infty)$ Range: $(-\infty, \infty)$

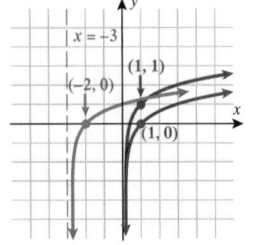

■ **YOUR TURN** Graph the functions and state the domain and range of each.

a. $y = \log_3 x$ **b.** $y = \log_3(x + 3)$ **c.** $\log_3(x) + 1$

All of the transformation techniques (shifting, reflection, and compression) discussed in Chapter 1 also apply to logarithmic functions. For example, the graphs of $-\log_2 x$ and $\log_2(-x)$ are found by reflecting the graph of $y = \log_2 x$ about the x-axis and y-axis, respectively.

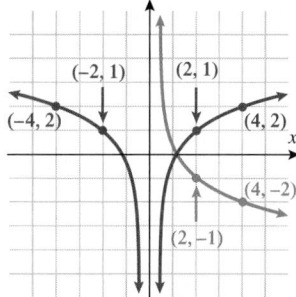

EXAMPLE 8   Graphing Logarithmic Functions Using Transformations

Graph the function $f(x) = -\log_2(x - 3)$ and state its domain and range.

Solution:

Graph $y = \log_2 x$.

x-intercept: $(1, 0)$

Vertical asymptote: $x = 0$

Additional points: $(2, 1)$, $(4, 2)$

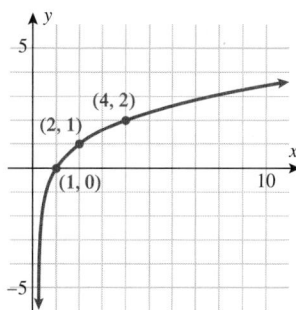

Graph $y = \log_2(x - 3)$ by shifting $y = \log_2 x$ to the *right* 3 units.

x-intercept: $(4, 0)$

Vertical asymptote: $x = 3$

Additional points: $(5, 1)$, $(7, 2)$

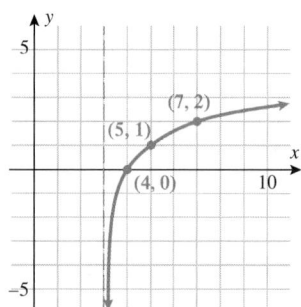

Graph $y = -\log_2(x - 3)$ by reflecting $y = \log_2(x - 3)$ about the x-axis.

x-intercept: $(4, 0)$

Vertical asymptote: $x = 3$

Additional points: $(5, -1)$, $(7, -2)$

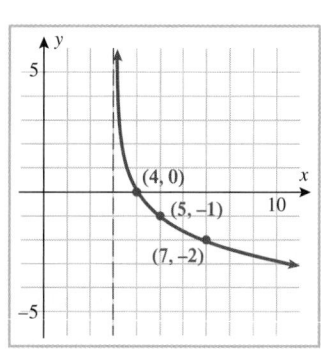

Domain: $(3, \infty)$ Range: $(-\infty, \infty)$

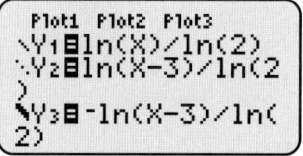
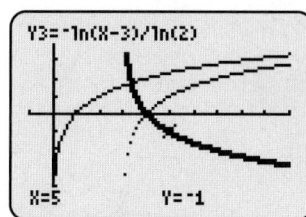

Applications of Logarithms

Logarithms are used to make a large range of numbers manageable. For example, to create a scale to measure a human's ability to hear, we must have a way to measure the sound intensity of an explosion, even though that intensity can be more than a trillion (10^{12}) times greater than that of a soft whisper. Decibels in engineering and physics, pH in chemistry, and the Richter scale for earthquakes are all applications of logarithmic functions.

The **decibel** is a logarithmic unit used to measure the magnitude of a physical quantity relative to a specified reference level. The *decibel* (dB) is employed in many engineering and science applications. The most common application is the intensity of sound.

DEFINITION   **Decibel (Sound)**

The **decibel** is defined as

$$D = 10 \log\left(\frac{I}{I_T}\right)$$

where D is the decibel level (dB), I is the intensity of the sound measured in watts per square meter, and I_T is the intensity threshold of the least audible sound a human can hear.

The human average threshold is $I_T = 1 \times 10^{-12}$ W/m².

Notice that when $I = I_T$, then $D = 10 \overset{0}{\log 1} = 0$ dB. People who work professionally with sound, such as acoustics engineers or medical hearing specialists, refer to this threshold level I_T as "0 dB." The following table illustrates typical sounds we hear and their corresponding decibel levels.

SOUND SOURCE	SOUND INTENSITY (W/m²)	DECIBELS (dB)
Threshold of hearing	1.0×10^{-12}	0
Vacuum cleaner	1.0×10^{-4}	80
iPod	1.0×10^{-2}	100
Jet engine	1.0×10^{3}	150

We add logarithms (decibels) as a substitute for multiplying numbers. For example, a whisper (approximately 0 dB) from someone standing next to a jet engine (150 dB) might go unheard because when these are added, we get approximately 150 dB (the jet engine).

 EXAMPLE 9   **Calculating Decibels of Sounds**

Suppose you have seats to a concert given by your favorite musical artist. Calculate the approximate decibel level associated with the typical sound intensity, given $I = 1 \times 10^{-2}$ W/m².

Solution:

Write the decibel-scale formula.

$$D = 10 \log\left(\frac{I}{I_T}\right)$$

Substitute $I = 1 \times 10^{-2}$ W/m² and $I_T = 1 \times 10^{-12}$ W/m².

$$D = 10 \log\left(\frac{1 \times 10^{-2}}{1 \times 10^{-12}}\right)$$

Simplify.

$$D = 10 \log\left(10^{10}\right)$$

Recall that the implied base for log is 10.

$$D = 10 \log_{10}\left(10^{10}\right)$$

Evaluate the right side. $\left[\log_{10}\left(10^{10}\right) = 10\right]$

$$D = 10 \cdot 10$$

$$D = 100$$

The typical sound level in the front row of a rock concert is $\boxed{100 \text{ dB}}$.

■ **YOUR TURN** Calculate the approximate decibels associated with a sound so loud it will cause instant perforation of the eardrums, $I = 1 \times 10^4 \text{ W/m}^2$.

■ **Answer:** 160 dB

The Richter scale (earthquakes) is another application of logarithms.

DEFINITION   **Richter Scale**

The magnitude M of an earthquake is measured using the **Richter scale**

$$M = \frac{2}{3} \log\left(\frac{E}{E_0}\right)$$

where

      M is the magnitude
      E is the seismic energy released by the earthquake (in joules)
      E_0 is the energy released by a reference earthquake $E_0 = 10^{4.4}$ joules

EXAMPLE 10   **Calculating the Magnitude of an Earthquake**

On October 17, 1989, just moments before game 3 of the World Series between the Oakland A's and the San Francisco Giants was about to start—with 60,000 fans in Candlestick Park—a devastating earthquake erupted. Parts of interstates and bridges collapsed, and President George H. W. Bush declared the area a disaster zone. The earthquake released approximately 1.12×10^{15} joules. Calculate the magnitude of the earthquake using the Richter scale.

Technology Tip

Enter the number 1.12×10^{15} using the scientific notation key $\boxed{\text{EXP}}$ or $\boxed{\text{EE}}$.

```
2/3log(1.12E15/1
0^4.4)
         7.099478682
```

Solution:

Write the Richter scale formula.

$$M = \frac{2}{3} \log\left(\frac{E}{E_0}\right)$$

Substitute $E = 1.12 \times 10^{15}$ and $E_0 = 10^{4.4}$.

$$M = \frac{2}{3} \log\left(\frac{1.12 \times 10^{15}}{10^{4.4}}\right)$$

Simplify.

$$M = \frac{2}{3} \log\left(1.12 \times 10^{10.6}\right)$$

Approximate the logarithm using a calculator.

$$M \approx \frac{2}{3}(10.65) \approx 7.1$$

The 1989 earthquake in California measured $\boxed{7.1}$ on the Richter scale.

■ **YOUR TURN** On May 3, 1996, Seattle experienced a moderate earthquake. The energy that the earthquake released was approximately 1.12×10^{12} joules. Calculate the magnitude of the 1996 Seattle earthquake using the Richter scale.

■ **Answer:** 5.1

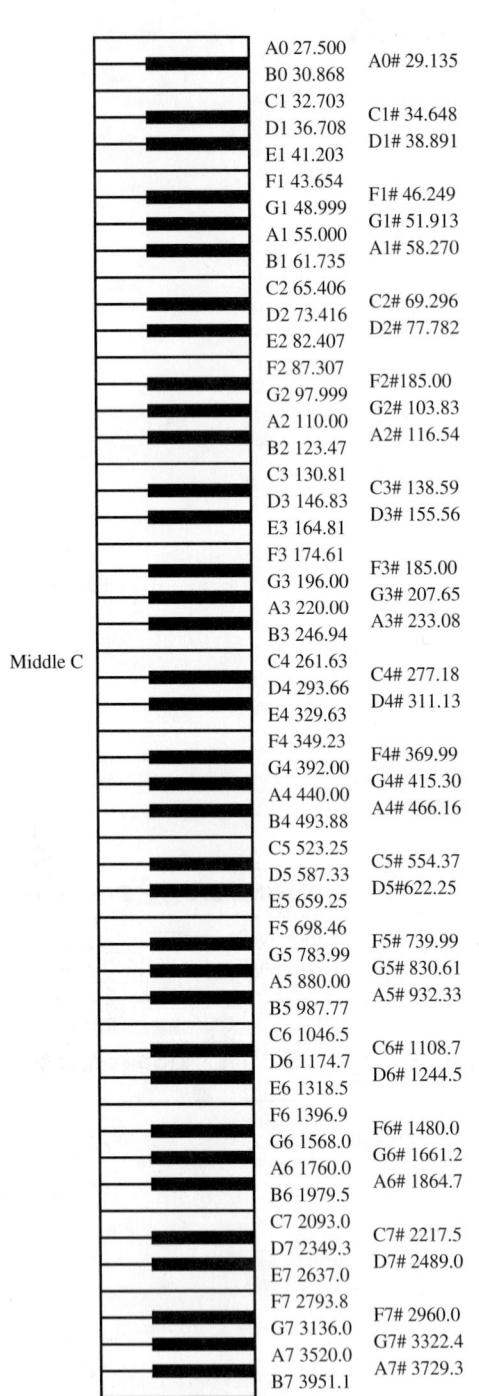

A0 27.500	
B0 30.868	A0# 29.135
C1 32.703	
D1 36.708	C1# 34.648
E1 41.203	D1# 38.891
F1 43.654	
G1 48.999	F1# 46.249
A1 55.000	G1# 51.913
B1 61.735	A1# 58.270

A **logarithmic scale** expresses the logarithm of a physical quantity instead of the quantity itself. In music, the pitch is the perceived fundamental frequency of sound. The note A above middle C on a piano has the pitch associated with a pure tone of 440 hertz (Hz). An octave is the interval between one musical pitch and another with either double or half its frequency. For example, if a note has a frequency of 440 Hz, then the note an octave above it has a frequency of 880 Hz, and the note an octave below it has a frequency of 220 Hz. Therefore, the ratio of two notes an octave apart is 2:1.

The following table lists the frequencies associated with A notes:

NOTE	A_1	A_2	A_3	A_4	A_5	A_6	A_7
Frequency (Hz)	55	110	220	440	880	1760	3520
Octave with respect to A_4	-3	-2	-1	0	$+1$	$+2$	$+3$

We can graph $\dfrac{\text{Frequency of note}}{440 \text{ Hz}}$ on the horizontal axis and the octave (with respect to A_4) on the vertical axis.

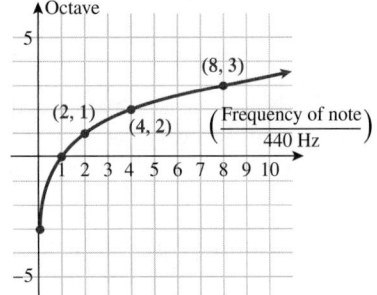

If we instead graph the logarithm of this quantity, $\log\left(\dfrac{\text{Frequency of note}}{440 \text{ Hz}}\right)$, we see that using a logarithmic scale expresses octaves linearly (up or down an octave). In other words, an "octave" is a purely logarithmic concept.

Semi-Log Plot

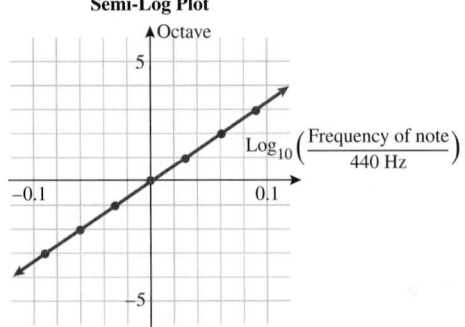

When a logarithmic scale is used, we typically classify a graph one of two ways:

- Log-log plot (both the horizontal and vertical axes use logarithmic scales)
- Semi-log plot (one of the axes uses a logarithmic scale)

The second graph with octaves on the vertical axis and the log of the ratio of frequencies on the horizontal axis is called a semi-log plot. An example of a log-log plot would be frequency versus wavelength in engineering. Frequency is inversely proportional to the wavelength. In a vacuum $f = \dfrac{c}{\lambda}$, where f is the frequency (in hertz), $c = 3.0 \times 10^8$ m/s^2 is the speed of light in a vacuum, and λ is the wavelength in meters.

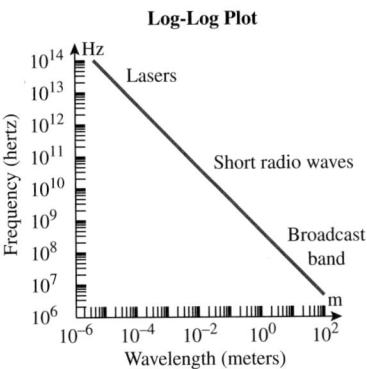

Log-Log Plot

The logarithmic scales allow us to represent a large range of numbers. In this graph, the x-axis ranges from microns, 10^{-6} meters, to hundreds of meters, and the y-axis ranges from megahertz (MHz), 10^6 hertz, to hundreds of terahertz (THz), 10^{12} hertz.

SECTION
3.2 SUMMARY

In this section, logarithmic functions were defined as inverses of exponential functions.

$$y = \log_b x \text{ is equivalent to } x = b^y$$

NAME	EXPLICIT BASE	IMPLICIT BASE
Common logarithm	$f(x) = \log_{10} x$	$f(x) = \log x$
Natural logarithm	$f(x) = \log_e x$	$f(x) = \ln x$

Evaluating Logarithms

- *Exact:* Convert to exponential form first, then evaluate.
- *Approximate:* Natural and common logarithms with calculators.

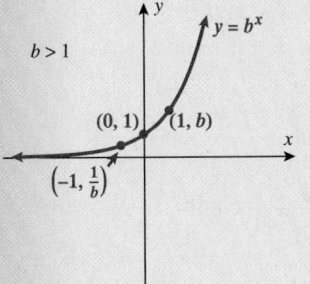

Graphs of Logarithmic Functions

EXPONENTIAL FUNCTION	LOGARITHMIC FUNCTION
$y = b^x$	$y = \log_b x$
y-intercept: $(0, 1)$	x-intercept: $(1, 0)$
Domain: $(-\infty, \infty)$	Domain: $(0, \infty)$
Range: $(0, \infty)$	Range: $(-\infty, \infty)$
Horizontal asymptote: x-axis	Vertical asymptote: y-axis

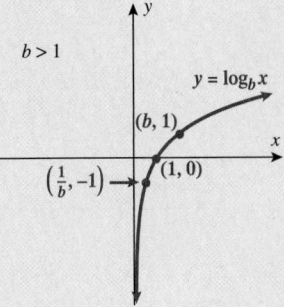

SKILLS

In Exercises 1–20, write each logarithmic equation in its equivalent exponential form.

1. $\log_{81} 3 = \frac{1}{4}$ **2.** $\log_{121} 11 = \frac{1}{2}$ **3.** $\log_2 \left(\frac{1}{32}\right) = -5$ **4.** $\log_3 \left(\frac{1}{81}\right) = -4$ **5.** $\log 0.01 = -2$

6. $\log 0.0001 = -4$ **7.** $\log 10,000 = 4$ **8.** $\log 1000 = 3$ **9.** $\log_{1/4} (64) = -3$ **10.** $\log_{1/6} (36) = -2$

11. $-1 = \ln \left(\frac{1}{e}\right)$ **12.** $1 = \ln e$ **13.** $\ln 1 = 0$ **14.** $\log 1 = 0$ **15.** $\ln 5 = x$

16. $\ln 4 = y$ **17.** $z = \log_x y$ **18.** $y = \log_x z$ **19.** $x = \log_y (x + y)$ **20.** $z = \ln x^y$

In Exercises 21–34, write each exponential equation in its equivalent logarithmic form.

21. $0.00001 = 10^{-5}$ **22.** $3^6 = 729$ **23.** $78,125 = 5^7$ **24.** $100,000 = 10^5$ **25.** $15 = \sqrt{225}$

26. $7 = \sqrt[3]{343}$ **27.** $\frac{8}{125} = \left(\frac{2}{5}\right)^3$ **28.** $\frac{8}{27} = \left(\frac{2}{3}\right)^3$ **29.** $3 = \left(\frac{1}{27}\right)^{-1/3}$ **30.** $4 = \left(\frac{1}{1024}\right)^{-1/5}$

31. $e^x = 6$ **32.** $e^{-x} = 4$ **33.** $x = y^z$ **34.** $z = y^x$

In Exercises 35–46, evaluate the logarithms exactly (if possible).

35. $\log_2 1$ **36.** $\log_5 1$ **37.** $\log_5 3125$ **38.** $\log_3 729$

39. $\log 10^7$ **40.** $\log 10^{-2}$ **41.** $\log_{1/4} 4096$ **42.** $\log_{1/7} 2401$

43. $\log 0$ **44.** $\ln 0$ **45.** $\log(-100)$ **46.** $\ln(-1)$

In Exercises 47–54, approximate (if possible) the common and natural logarithms using a calculator. Round to two decimal places.

47. $\log 29$ **48.** $\ln 29$ **49.** $\ln 380$ **50.** $\log 380$

51. $\log 0$ **52.** $\ln 0$ **53.** $\ln 0.0003$ **54.** $\log 0.0003$

In Exercises 55–66, state the domain of the logarithmic function in interval notation.

55. $f(x) = \log_2(x + 5)$ **56.** $f(x) = \log_2(4x - 1)$ **57.** $f(x) = \log_3(5 - 2x)$ **58.** $f(x) = \log_3(5 - x)$

59. $f(x) = \ln(7 - 2x)$ **60.** $f(x) = \ln(3 - x)$ **61.** $f(x) = \log|x|$ **62.** $f(x) = \log|x + 1|$

63. $f(x) = \log(x^2 + 1)$ **64.** $f(x) = \log(1 - x^2)$ **65.** $f(x) = \log(10 + 3x - x^2)$ **66.** $f(x) = \log_3(x^3 - 3x^2 + 3x - 1)$

In Exercises 67–72, match the graph with the function.

67. $y = \log_5 x$ **68.** $y = \log_5(-x)$ **69.** $y = -\log_5(-x)$

70. $y = \log_5(x + 3) - 1$ **71.** $y = \log_5(1 - x) - 2$ **72.** $y = -\log_5(3 - x) + 2$

a.
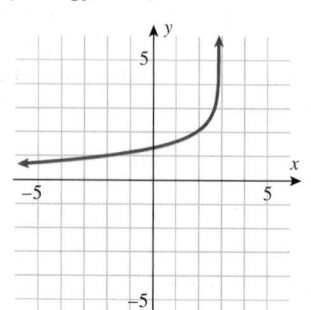

b.
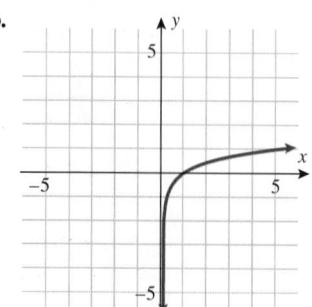

c.
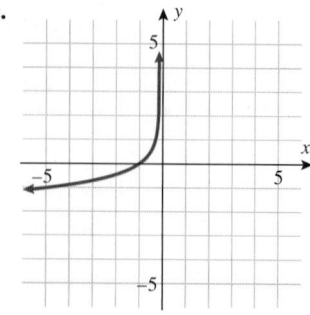

d. 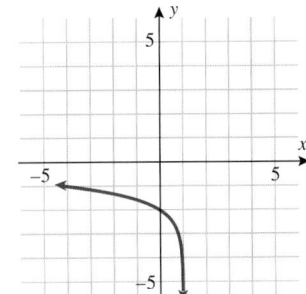   **e.** 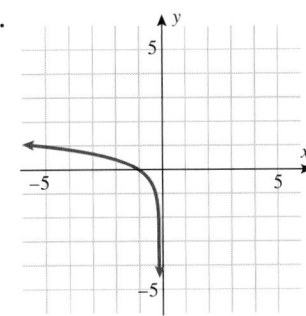   **f.**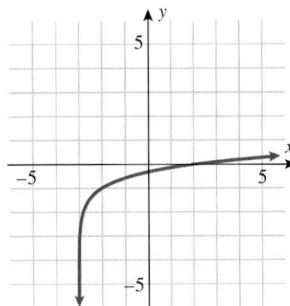

In Exercises 73–84, graph the logarithmic function using transformation techniques. State the domain and range of f.

73. $f(x) = \log(x - 1)$

74. $f(x) = \log(x + 2)$

75. $\ln(x + 2)$

76. $\ln(x - 1)$

77. $f(x) = \log_3(x + 2) - 1$

78. $f(x) = \log_3(x + 1) - 2$

79. $f(x) = -\log(x) + 1$

80. $f(x) = \log(-x) + 2$

81. $f(x) = \ln(x + 4)$

82. $f(x) = \ln(4 - x)$

83. $f(x) = \log(2x)$

84. $f(x) = 2\ln(-x)$

■ APPLICATIONS

85. Sound. Calculate the decibels associated with *normal conversation* if the intensity is $I = 1 \times 10^{-6}$ W/m².

86. Sound. Calculate the decibels associated with the *onset of pain* if the intensity is $I = 1 \times 10^{1}$ W/m².

87. Sound. Calculate the decibels associated with attending *a football game in a loud college stadium* if the intensity is $I = 1 \times 10^{-0.3}$ W/m².

88. Sound. Calculate the decibels associated with a *doorbell* if the intensity is $I = 1 \times 10^{-4.5}$ W/m².

89. Earthquakes. On Good Friday 1964, one of the most severe North American earthquakes ever recorded struck Alaska. The energy released measured 1.41×10^{17} joules. Calculate the magnitude of the 1964 Alaska earthquake using the Richter scale.

90. Earthquakes. On January 22, 2003, Colima, Mexico, experienced a major earthquake. The energy released measured 6.31×10^{15} joules. Calculate the magnitude of the 2003 Mexican earthquake using the Richter scale.

91. Earthquakes. On December 26, 2003, a major earthquake rocked southeastern Iran. In Bam, 30,000 people were killed, and 85% of buildings were damaged or destroyed. The energy released measured 2×10^{14} joules. Calculate the magnitude of the 2003 Iran earthquake with the Richter scale.

92. Earthquakes. On November 1, 1755, Lisbon was destroyed by an earthquake, which killed 90,000 people and destroyed 85% of the city. It was one of the most destructive earthquakes in history. The energy released measured 8×10^{17} joules. Calculate the magnitude of the 1755 Lisbon earthquake with the Richter scale.

For Exercises 93–98, refer to the following:

The pH of a solution is a measure of the molar concentration of hydrogen ions, H^+, in moles per liter, in the solution, which means that it is a measure of the acidity or basicity of the solution. The letters pH stand for "power of hydrogen," and the numerical value is defined as

$$\text{pH} = -\log_{10}\left[H^+\right]$$

Very acid corresponds to pH values near 1, neutral corresponds to a pH near 7 (pure water), and very basic corresponds to values near 14. In the next six exercises you will be asked to calculate the pH value of wine, Pepto-Bismol, normal rainwater, bleach, and fruit. List these six liquids and use your intuition to classify them as neutral, acidic, very acidic, basic, or very basic before you calculate their actual pH values.

93. Chemistry. If wine has an approximate hydrogen ion concentration of 5.01×10^{-4}, calculate its pH value.

94. Chemistry. Pepto-Bismol has a hydrogen ion concentration of about 5.01×10^{-11}. Calculate its pH value.

95. Chemistry. Normal rainwater is slightly acidic and has an approximate hydrogen ion concentration of $10^{-5.6}$. Calculate its pH value. Acid rain and tomato juice have similar approximate hydrogen ion concentrations of 10^{-4}. Calculate the pH value of acid rain and tomato juice.

96. Chemistry. Bleach has an approximate hydrogen ion concentration of 5.0×10^{-13}. Calculate its pH value.

97. Chemistry. An apple has an approximate hydrogen ion concentration of $10^{-3.6}$. Calculate its pH value.

98. Chemistry. An orange has an approximate hydrogen ion concentration of $10^{-4.2}$. Calculate its pH value.

99. Archaeology. Carbon dating is a method used to determine the age of a fossil or other organic remains. The age t in years is related to the mass C (in milligrams) of carbon 14 through a logarithmic equation:

$$t = -\frac{\ln\left(\dfrac{C}{500}\right)}{0.0001216}$$

How old is a fossil that contains 100 milligrams of carbon 14?

100. Archaeology. Repeat Exercise 99, only now the fossil contains 40 milligrams of carbon 14.

101. Broadcasting. Decibels are used to quantify losses associated with atmospheric interference in a communication system. The ratio of the power (watts) received to the power transmitted (watts) is often compared. Often, *watts* are transmitted, but losses due to the atmosphere typically correspond to *milliwatts* being received:

$$dB = 10 \log\left(\frac{\text{Power received}}{\text{Power transmitted}}\right)$$

If 1 W of power is transmitted and 3 mW is received, calculate the power loss in dB.

102. Broadcasting. Repeat Exercise 101, assuming 3 W of power is transmitted and 0.2 mW is received.

■ CATCH THE MISTAKE

In Exercises 103–106, explain the mistake that is made.

103. Evaluate the logarithm $\log_2 4$.

Solution:

Set the logarithm equal to x.	$\log_2 4 = x$
Write the logarithm in exponential form.	$x = 2^4$
Simplify.	$x = 16$
Answer:	$\log_2 4 = 16$

This is incorrect. The correct answer is $\log_2 4 = 2$. What went wrong?

104. Evaluate the logarithm $\log_{100} 10$.

Solution:

Set the logarithm equal to x.	$\log_{100} 10 = x$
Express the equation in exponential form.	$10^x = 100$
Solve for x.	$x = 2$
Answer:	$\log_{100} 10 = 2$

This is incorrect. The correct answer is $\log_{100} 10 = \frac{1}{2}$. What went wrong?

105. State the domain of the logarithmic function $f(x) = \log_2(x + 5)$ in interval notation.

Solution:

The domain of all logarithmic functions is $x > 0$.

Interval notation: $(0, \infty)$

This is incorrect. What went wrong?

106. State the domain of the logarithmic function $f(x) = \ln|x|$ in interval notation.

Solution:

Since the absolute value eliminates all negative numbers, the domain is the set of all real numbers.

Interval notation: $(-\infty, \infty)$

This is incorrect. What went wrong?

■ CONCEPTUAL

In Exercises 107–110, determine whether each statement is true or false.

107. The domain of the standard logarithmic function, $y = \ln x$, is the set of nonnegative real numbers.

108. The horizontal axis is the horizontal asymptote of the graph of $y = \ln x$.

109. The graphs of $y = \log x$ and $y = \ln x$ have the same x-intercept $(1, 0)$.

110. The graphs of $y = \log x$ and $y = \ln x$ have the same vertical asymptote, $x = 0$.

■ CHALLENGE

111. State the domain, range, and x-intercept of the function $f(x) = -\ln(x - a) + b$ for a and b real positive numbers.

112. State the domain, range, and x-intercept of the function $f(x) = \log(a - x) - b$ for a and b real positive numbers.

113. Graph the function $f(x) = \begin{cases} \ln(-x) & x < 0 \\ \ln(x) & x > 0 \end{cases}$.

114. Graph the function $f(x) = \begin{cases} -\ln(-x) & x < 0 \\ -\ln(x) & x > 0 \end{cases}$.

■ TECHNOLOGY

115. Apply a graphing utility to graph $y = e^x$ and $y = \ln x$ in the same viewing screen. What line are these two graphs symmetric about?

116. Apply a graphing utility to graph $y = 10^x$ and $y = \log x$ in the same viewing screen. What line are these two graphs symmetric about?

117. Apply a graphing utility to graph $y = \log x$ and $y = \ln x$ in the same viewing screen. What are the two common characteristics?

118. Using a graphing utility, graph $y = \ln|x|$. Is the function defined everywhere?

119. Apply a graphing utility to graph $f(x) = \ln(3x)$, $g(x) = \ln 3 + \ln x$, and $h(x) = (\ln 3)(\ln x)$ in the same viewing screen. Determine which two functions give the same graph, then state the domain of the function.

120. Apply a graphing utility to graph $f(x) = \ln(x^2 - 4)$, $g(x) = \ln(x + 2) + \ln(x - 2)$, and $h(x) = \ln(x + 2)\ln(x - 2)$ in the same viewing screen. Determine the domain where two functions give the same graph.

■ PREVIEW TO CALCULUS

In Exercises 121–124, refer to the following:

Recall that the derivative of f can be found by letting $h \to 0$ in the difference quotient $\dfrac{f(x + h) - f(x)}{h}$. In calculus we prove that $\dfrac{e^h - 1}{h} = 1$, when h approaches 0; that is, for really small values of h, $\dfrac{e^h - 1}{h}$ gets very close to 1.

121. Use this information to find the derivative of $f(x) = e^x$.

122. Use this information to find the derivative of $f(x) = e^{2x}$.
[*Hint:* $e^{2h} - 1 = (e^h - 1)(e^h + 1)$]

We also prove in calculus that the derivative of the inverse function f^{-1} is given by $(f^{-1})'(x) = \dfrac{1}{f'(f^{-1}(x))}$.

123. Given $f(x) = e^x$, find
    a. $f^{-1}(x)$ **b.** $(f^{-1})'(x)$

124. Given $f(x) = e^{2x}$, find
    a. $f^{-1}(x)$ **b.** $(f^{-1})'(x)$

SECTION
3.3 PROPERTIES OF LOGARITHMS

SKILLS OBJECTIVES

■ Write a single logarithm as a sum or difference of logarithms.
■ Write a logarithmic expression as a single logarithm.
■ Evaluate logarithms of a general base (other than base 10 or e).

CONCEPTUAL OBJECTIVES

■ Derive the seven basic logarithmic properties.
■ Derive the change-of-base formula.

Properties of Logarithms

Since exponential functions and logarithmic functions are inverses of one another, properties of exponents are related to properties of logarithms. We will start by reviewing properties of exponents (Appendix), and then proceed to properties of logarithms.

PROPERTIES OF EXPONENTS

Let a, b, m, and n be any real numbers and $m > 0$, $n > 0$, and $b \neq 0$; then the following are true:

1. $b^m \cdot b^n = b^{m+n}$

2. $b^{-m} = \dfrac{1}{b^m} = \left(\dfrac{1}{b}\right)^m$

3. $\dfrac{b^m}{b^n} = b^{m-n}$

4. $(b^m)^n = b^{mn}$

5. $(ab)^m = a^m \cdot b^m$

6. $b^0 = 1$

7. $b^1 = b$

From these properties of exponents, we can develop similar properties for logarithms. We list seven basic properties.

PROPERTIES OF LOGARITHMS

If b, M, and N are positive real numbers, where $b \neq 1$, and p and x are real numbers, then the following are true:

1. $\log_b 1 = 0$

2. $\log_b b = 1$

3. $\log_b b^x = x$

4. $b^{\log_b x} = x \qquad x > 0$

5. $\log_b MN = \log_b M + \log_b N$ *Product rule:* Log of a product is the sum of the logs.

6. $\log_b \left(\dfrac{M}{N}\right) = \log_b M - \log_b N$ *Quotient rule:* Log of a quotient is the difference of the logs.

7. $\log_b M^p = p \log_b M$ *Power rule:* Log of a number raised to an exponent is the exponent times the log of the number.

We will devote this section to proving and illustrating these seven properties.

The first two properties follow directly from the definition of a logarithmic function and properties of exponentials.

$$\text{Property (1):} \qquad \log_b 1 = 0 \text{ since } b^0 = 1$$

$$\text{Property (2):} \qquad \log_b b = 1 \text{ since } b^1 = b$$

The third and fourth properties follow from the fact that exponential functions and logarithmic functions are inverses of one another. Recall that inverse functions satisfy the relationship that $f^{-1}(f(x)) = x$ for all x in the domain of $f(x)$, and $f\left(f^{-1}(x)\right) = x$ for all x in the domain of f^{-1}. Let $f(x) = b^x$ and $f^{-1}(x) = \log_b x$.

Property (3):

Write the inverse identity. $f^{-1}(f(x)) = x$

Substitute $f^{-1}(x) = \log_b x$. $\log_b(f(x)) = x$
Substitute $f(x) = b^x$. $\log_b b^x = x$

Property (4):

Write the inverse identity. $f\left(f^{-1}(x)\right) = x$

Substitute $f(x) = b^x$. $b^{f^{-1}(x)} = x$
Substitute $f^{-1}(x) = \log_b x \qquad x > 0$. $b^{\log_b x} = x$

The first four properties are summarized below for common and natural logarithms:

COMMON AND NATURAL LOGARITHM PROPERTIES

Common Logarithm (base 10)	Natural Logarithm (base e)
1. $\log 1 = 0$	**1.** $\ln 1 = 0$
2. $\log 10 = 1$	**2.** $\ln e = 1$
3. $\log 10^x = x$	**3.** $\ln e^x = x$
4. $10^{\log x} = x \qquad x > 0$	**4.** $e^{\ln x} = x \qquad x > 0$

EXAMPLE 1   Using Logarithmic Properties

Use properties (1)–(4) to simplify the expressions.

a. $\log_{10} 10$ 　　 **b.** $\ln 1$ 　　 **c.** $10^{\log(x+8)}$

d. $e^{\ln(2x+5)}$ 　　 **e.** $\log 10^{x^2}$ 　　 **f.** $\ln e^{x+3}$

Solution:

a. Use property (2).　　　　　　 $\log_{10} 10 = \boxed{1}$

b. Use property (1).　　　　　　 $\ln 1 = \boxed{0}$

c. Use property (4).　　　　　 $10^{\log(x+8)} = \boxed{x + 8} \qquad x > -8$

d. Use property (4).　　　　　 $e^{\ln(2x+5)} = \boxed{2x + 5} \qquad x > -\frac{5}{2}$

e. Use property (3).　　　　　 $\log 10^{x^2} = \boxed{x^2}$

f. Use property (3).　　　　　 $\ln e^{x+3} = \boxed{x + 3}$

The fifth through seventh properties follow from the properties of exponents and the definition of logarithms. We will prove the product rule and leave the proofs of the quotient and power rules for the exercises.

$$\text{Property (5): } \quad \log_b MN = \log_b M + \log_b N$$

WORDS	MATH
Assume two logs that have the same base.	Let $u = \log_b M$ and $v = \log_b N$ $M > 0, N > 0$
Change to equivalent exponential forms.	$b^u = M$ and $b^v = N$
Write the log of a product.	$\log_b MN$
Substitute $M = b^u$ and $N = b^v$.	$= \log_b(b^u \, b^v)$
Use properties of exponents.	$= \log_b(b^{u+v})$
Apply property (3).	$= u + v$
Substitute $u = \log_b M, v = \log_b N$.	$= \log_b M + \log_b N$

$$\boxed{\log_b MN = \log_b M + \log_b N}$$

In other words, the log of a product is the sum of the logs. Let us illustrate this property with a simple example.

$$\underbrace{\log_2 8}_{3} + \underbrace{\log_2 4}_{2} = \underbrace{\log_2 32}_{5}$$

Notice that $\log_2 8 + \log_2 4 \neq \log_2 12$.

EXAMPLE 2   Writing a Logarithmic Expression as a Sum of Logarithms

Use the logarithmic properties to write the expression $\log_b\!\left(u^2\sqrt{v}\right)$ as a sum of simpler logarithms.

Solution:

Convert the radical to exponential form. $\log_b\!\left(u^2\sqrt{v}\right) = \log_b\!\left(u^2 v^{1/2}\right)$

Use the product property (5). $= \log_b u^2 + \log_b v^{1/2}$

Use the power property (7). $= \boxed{2\log_b u + \tfrac{1}{2}\log_b v}$

■ **Answer:**
$\log_b\!\left(x^4\sqrt[3]{y}\right) = 4\log_b x + \tfrac{1}{3}\log_b y$

■ **YOUR TURN** Use the logarithmic properties to write the expression $\log_b\!\left(x^4\sqrt[3]{y}\right)$ as a sum of simpler logarithms.

EXAMPLE 3   Writing a Sum of Logarithms as a Single Logarithmic Expression: The Right Way and the Wrong Way

Use properties of logarithms to write the expression $2\log_b 3 + 4\log_b u$ as a single logarithmic expression.

COMMON MISTAKE

A common mistake is to write the sum of the logs as a log of the sum.

$$\log_b M + \log_b N \neq \log_b(M + N)$$

▼ **CAUTION**

$\log_b M + \log_b N = \log_b(MN)$

$\log_b M + \log_b N \neq \log_b(M + N)$

★ **CORRECT**   ✖ **INCORRECT**

Use the power property (7).

$2\log_b 3 + 4\log_b u = \log_b 3^2 + \log_b u^4$

Simplify.

$\log_b 9 + \log_b u^4$ $\neq \log_b(9 + u^4)$ **ERROR**

Use the product property (5).

$\boxed{= \log_b(9u^4)}$

■ **Answer:** $\ln(x^2 y^3)$

■ **YOUR TURN** Express $2\ln x + 3\ln y$ as a single logarithm.

EXAMPLE 4   Writing a Logarithmic Expression as a Difference of Logarithms

Write the expression $\ln\!\left(\dfrac{x^3}{y^2}\right)$ as a difference of logarithms.

Solution:

Apply the quotient property (6). $\ln\!\left(\dfrac{x^3}{y^2}\right) = \ln(x^3) - \ln(y^2)$

Apply the power property (7). $= \boxed{3\ln x - 2\ln y}$

■ **Answer:** $4\log a - 5\log b$

■ **YOUR TURN** Write the expression $\log\!\left(\dfrac{a^4}{b^5}\right)$ as a difference of logarithms.

Another common mistake is misinterpreting the quotient rule.

EXAMPLE 5   Writing the Difference of Logarithms as a Logarithm of a Quotient

Write the expression $\frac{2}{3}\ln x - \frac{1}{2}\ln y$ as a logarithm of a quotient.

COMMON MISTAKE

$$\log_b M - \log_b N \neq \frac{\log_b M}{\log_b N}$$

⭐ **CORRECT**

Use the power property (7).

$$\frac{2}{3}\ln x - \frac{1}{2}\ln y = \ln x^{2/3} - \ln y^{1/2}$$

Use the quotient property (6).

$$\ln\left(\frac{x^{2/3}}{y^{1/2}}\right)$$

❌ **INCORRECT**

$$\frac{\ln x^{2/3}}{\ln y^{1/2}}\quad \textbf{ERROR}$$

■ **YOUR TURN** Write the expression $\frac{1}{2}\log a - 3\log b$ as a single logarithm.

■ **Answer:** $\log\left(\dfrac{a^{1/2}}{b^3}\right)$

EXAMPLE 6   Combining Logarithmic Expressions into a Single Logarithm

Write the expression $3\log_b x + \log_b(2x + 1) - 2\log_b 4$ as a single logarithm.

Solution:

Use the power property (7) on the first and third terms.

$$= \log_b x^3 + \log_b(2x + 1) - \log_b 4^2$$

Use the product property (5) on the first two terms.

$$= \log_b\left[x^3(2x + 1)\right] - \log_b 16$$

Use the quotient property (6).

$$= \log_b\left[\frac{x^3(2x + 1)}{16}\right]$$

■ **YOUR TURN** Write the expression $2\ln x - \ln(3y) + 3\ln z$ as a single logarithm.

■ **Answer:** $\ln\left(\dfrac{x^2 z^3}{3y}\right)$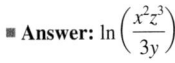

EXAMPLE 7   Expanding a Logarithmic Expression into a Sum or Difference of Logarithms

Write $\ln\left(\dfrac{x^2 - x - 6}{x^2 + 7x + 6}\right)$ as a sum or difference of logarithms.

Solution:

Factor the numerator and denominator.

$$= \ln\left[\frac{(x - 3)(x + 2)}{(x + 6)(x + 1)}\right]$$

Use the quotient property (6).

$$= \ln[(x - 3)(x + 2)] - \ln[(x + 6)(x + 1)]$$

Use the product property (5).

$$= \ln(x - 3) + \ln(x + 2) - [\ln(x + 6) + \ln(x + 1)]$$

Eliminate brackets.

$$= \ln(x - 3) + \ln(x + 2) - \ln(x + 6) - \ln(x + 1)$$

Change-of-Base Formula

Recall that in the last section, we were able to evaluate logarithms two ways: (1) exactly by writing the logarithm in exponential form and identifying the exponent and (2) using a calculator if the logarithms were base 10 or e. How do we evaluate a logarithm of general base if we cannot identify the exponent? We use the *change-of-base formula*.

EXAMPLE 8   Using Properties of Logarithms to Change the Base to Evaluate a General Logarithm

Evaluate $\log_3 8$. Round the answer to four decimal places.

Solution:

Let $y = \log_3 8$.	$y = \log_3 8$
Write the logarithm in exponential form.	$3^y = 8$
Take the log of both sides.	$\log 3^y = \log 8$
Use the power property (7).	$y \log 3 = \log 8$
Divide both sides by log 3.	$y = \dfrac{\log 8}{\log 3}$
Use a calculator to approximate.	$y \approx 1.892789261$
Substitute $y = \log_3 8$.	$\log_3 8 \approx \boxed{1.8928}$

Example 8 illustrated our ability to use properties of logarithms to change from base 3 to base 10, which our calculators can handle. This leads to the general change-of-base formula.

CHANGE-OF-BASE FORMULA

For any logarithmic bases a and b and any positive number M, the change-of-base formula says that

$$\log_b M = \frac{\log_a M}{\log_a b}$$

In the special case when a is either 10 or e, this relationship becomes

Common Logarithms		**Natural Logarithms**
$\log_b M = \dfrac{\log M}{\log b}$	or	$\log_b M = \dfrac{\ln M}{\ln b}$

It does not matter what base we select (10, e, or any other base); the ratio will be the same.

Proof of Change-of-Base Formula

WORDS	**MATH**
Let y be the logarithm we want to evaluate.	$y = \log_b M$
Write $y = \log_b M$ in exponential form.	$b^y = M$
Let a be any positive real number (where $a \neq 1$).	
Take the log of base a of both sides of the equation.	$\log_a b^y = \log_a M$
Use the power rule on the left side of the equation.	$y \log_a b = \log_a M$
Divide both sides of the equation by $\log_a b$.	$y = \dfrac{\log_a M}{\log_a b}$

EXAMPLE 9   Using the Change-of-Base Formula

Use the change-of-base formula to evaluate $\log_4 17$. Round to four decimal places.

Solution:

We will illustrate this two ways (choosing common and natural logarithms) using a scientific calculator.

Common Logarithms

Use the change-of-base formula with base 10. $\log_4 17 = \dfrac{\log 17}{\log 4}$

Approximate with a calculator. ≈ 2.043731421

$\approx \boxed{2.0437}$

Natural Logarithms

Use the change-of-base formula with base e. $\log_4 17 = \dfrac{\ln 17}{\ln 4}$

Approximate with a calculator. ≈ 2.043731421

$\approx \boxed{2.0437}$

■ **YOUR TURN** Use the change-of-base formula to approximate $\log_7 34$. Round to four decimal places.

■ **Answer:** $\log_7 34 \approx 1.8122$

SECTION
3.3 SUMMARY

Properties of Logarithms

If b, M, and N are positive real numbers, where $b \neq 1$, and p and x are real numbers, then the following are true:

■ Product Property: $\log_b MN = \log_b M + \log_b N$

■ Quotient Property: $\log_b\left(\dfrac{M}{N}\right) = \log_b M - \log_b N$

■ Power Property: $\log_b M^p = p\log_b M$

GENERAL LOGARITHM	COMMON LOGARITHM	NATURAL LOGARITHM
$\log_b 1 = 0$	$\log 1 = 0$	$\ln 1 = 0$
$\log_b b = 1$	$\log 10 = 1$	$\ln e = 1$
$\log_b b^x = x$	$\log 10^x = x$	$\ln e^x = x$
$b^{\log_b x} = x \quad x > 0$	$10^{\log x} = x \quad x > 0$	$e^{\ln x} = x \quad x > 0$
Change-of-base formula	$\log_b M = \dfrac{\log M}{\log b}$	$\log_b M = \dfrac{\ln M}{\ln b}$

▪ SKILLS

In Exercises 1–20, apply the properties of logarithms to simplify each expression. Do not use a calculator.

1. $\log_9 1$ **2.** $\log_{69} 1$ **3.** $\log_{1/2}\left(\frac{1}{2}\right)$ **4.** $\log_{3.3} 3.3$ **5.** $\log_{10} 10^8$

6. $\ln e^3$ **7.** $\log_{10} 0.001$ **8.** $\log_3 3^7$ **9.** $\log_2 \sqrt{8}$ **10.** $\log_5 \sqrt[3]{5}$

11. $8^{\log_8 5}$ **12.** $2^{\log_2 5}$ **13.** $e^{\ln(x+5)}$ **14.** $10^{\log(3x^2+2x+1)}$ **15.** $5^{3\log_5 2}$

16. $7^{2\log_7 5}$ **17.** $7^{-2\log_7 3}$ **18.** $e^{-2\ln 10}$ **19.** $7e^{-3\ln x}$ **20.** $-19e^{-2\ln x^2}$

In Exercises 21–36, write each expression as a sum or difference of logarithms.

$$\text{Example: } \log(m^2 n^5) = 2\log m + 5\log n$$

21. $\log_b(x^3 y^5)$ **22.** $\log_b(x^{-3} y^{-5})$ **23.** $\log_b(x^{1/2} y^{1/3})$ **24.** $\log_b(\sqrt{r}\,\sqrt[3]{t})$

25. $\log_b\left(\dfrac{r^{1/3}}{s^{1/2}}\right)$ **26.** $\log_b\left(\dfrac{r^4}{s^2}\right)$ **27.** $\log_b\left(\dfrac{x}{yz}\right)$ **28.** $\log_b\left(\dfrac{xy}{z}\right)$

29. $\log(x^2 \sqrt{x+5})$ **30.** $\log[(x-3)(x+2)]$ **31.** $\ln\left[\dfrac{x^3(x-2)^2}{\sqrt{x^2+5}}\right]$ **32.** $\ln\left[\dfrac{\sqrt{x+3}\,\sqrt[3]{x-4}}{(x+1)^4}\right]$

33. $\log\left(\dfrac{x^2-2x+1}{x^2-9}\right)$ **34.** $\log\left(\dfrac{x^2-x-2}{x^2+3x-4}\right)$ **35.** $\ln\sqrt{\dfrac{x^2+3x-10}{x^2-3x+2}}$ **36.** $\ln\left[\dfrac{\sqrt[3]{x-1}(3x-2)^4}{(x+1)\sqrt{x-1}}\right]^2$

In Exercises 37–48, write each expression as a single logarithm.

$$\text{Example: } 2\log m + 5\log n = \log(m^2 n^5)$$

37. $3\log_b x + 5\log_b y$ **38.** $2\log_b u + 3\log_b v$ **39.** $5\log_b u - 2\log_b v$ **40.** $3\log_b x - \log_b y$

41. $\frac{1}{2}\log_b x + \frac{2}{3}\log_b y$ **42.** $\frac{1}{2}\log_b x - \frac{2}{3}\log_b y$ **43.** $2\log u - 3\log v - 2\log z$ **44.** $3\log u - \log 2v - \log z$

45. $\ln(x+1) + \ln(x-1) - 2\ln(x^2+3)$ **46.** $\ln\sqrt{x-1} + \ln\sqrt{x+1} - 2\ln(x^2-1)$

47. $\frac{1}{2}\ln(x+3) - \frac{1}{3}\ln(x+2) - \ln(x)$ **48.** $\frac{1}{3}\ln(x^2+4) - \frac{1}{2}\ln(x^2-3) - \ln(x-1)$

In Exercises 49–58, evaluate the logarithms using the change-of-base formula. Round to four decimal places.

49. $\log_5 7$ **50.** $\log_4 19$ **51.** $\log_{1/2} 5$ **52.** $\log_5 \frac{1}{2}$ **53.** $\log_{2.7} 5.2$

54. $\log_{7.2} 2.5$ **55.** $\log_\pi 10$ **56.** $\log_\pi 2.7$ **57.** $\log_{\sqrt{3}} 8$ **58.** $\log_{\sqrt{2}} 9$

▪ APPLICATIONS

59. Sound. Sitting in the front row of a rock concert exposes us to a sound pressure (or sound level) of 1×10^{-1} W/m² (or 110 dB), and a normal conversation is typically around 1×10^{-6} W/m² (or 60 dB). How many decibels are you exposed to if a friend is talking in your ear at a rock concert? (*Note:* 160 dB causes perforation of the eardrums.) (*Hint:* Add the sound pressures and convert to dB.)

60. Sound. A whisper corresponds to 1×10^{-10} W/m² (or 20 dB) and a normal conversation is typically around 1×10^{-6} W/m² (or 60 dB). How many decibels are you exposed to if one friend is whispering in your ear, while the other one is talking at a normal level? (*Hint:* Add the sound pressures and convert to dB.)

For Exercises 61 and 62, refer to the following:

There are two types of waves associated with an earthquake: *compression* and *shear*. The compression, or longitudinal, waves displace material behind the earthquake's path. Longitudinal waves travel at great speeds and are often called "primary waves" or simply "P" waves. Shear, or transverse, waves displace material at right angles to the earthquake's path. Transverse waves do not travel as rapidly through the Earth's crust and mantle as do longitudinal waves, and they are called "secondary" or "S" waves.

61. Earthquakes. If a seismologist records the energy of P waves as 4.5×10^{12} joules and the energy of S waves as 7.8×10^8 joules, what is the total energy (sum the two energies)? What would the combined effect be on the Richter scale?

62. Earthquakes. Repeat Exercise 61, assuming the energy associated with the P waves is 5.2×10^{11} joules and the energy associated with the S waves is 4.1×10^9 joules.

63. Photography. In photographic quality assurance, logarithms are used to determine, for instance, the density. Density is the common logarithm of the opacity, which is the quotient of the amount of incident light and the amount of transmitted light. What is the density of a photographic material that only transmits 90% of the incident light?

64. pH Scale. The pH scale measures how acidic or basic a substance is. pH is defined as the negative logarithm of the hydrogen ion activity in an aqueous solution, a_H. Thus, if $a_H = 0.01$, then $pH = -\log 0.01 = 2$. Determine the pH of a liquid with $a_H = 0.00407$. Round your answer to the nearest hundredth.

65. pH Scale. How many times more acidic is a substance with $pH = 3.2$ than a substance with $pH = 4.4$? Round your answer to the nearest integer.

66. Information Theory. In information theory, logarithms in base 2 are often used. The capacity C of a noisy channel with bandwidth W and signal and noise powers S and N is $C = W \log_2\left(1 + \frac{S}{N}\right)$. The signal noise ratio R is given by $R = 10 \log\left(\frac{S}{N}\right)$. Assuming a channel with a bandwidth of 3 MHz and a signal noise ratio $R = 2$ dB, calculate the channel capacity.

■ **CATCH THE MISTAKE**

In Exercises 67–70, simplify if possible and explain the mistake that is made.

67. $3 \log 5 - \log 25$

Solution:

Apply the quotient property (6).	$\frac{3 \log 5}{\log 25}$
Write $25 = 5^2$.	$\frac{3 \log 5}{\log 5^2}$
Apply the power property (7).	$\frac{3 \log 5}{2 \log 5}$
Simplify.	$\frac{3}{2}$

This is incorrect. The correct answer is log 5. What mistake was made?

68. $\ln 3 + 2 \ln 4 - 3 \ln 2$

Solution:

Apply the power property (7).	$\ln 3 + \ln 4^2 - \ln 2^3$
Simplify.	$\ln 3 + \ln 16 - \ln 8$
Apply property (5).	$\ln(3 + 16 - 8)$
Simplify.	$\ln 11$

This is incorrect. The correct answer is ln 6. What mistake was made?

69. $\log_2 x + \log_3 y - \log_4 z$

Solution:

Apply the product property (5).	$\log_6 xy - \log_4 z$
Apply the quotient property (6).	$\log_{24} xyz$

This is incorrect. What mistake was made?

70. $2(\log 3 - \log 5)$

Solution:

Apply the quotient property (6).	$2\left(\log \frac{3}{5}\right)$
Apply the power property (7).	$\left(\log \frac{3}{5}\right)^2$
Apply a calculator to approximate.	≈ 0.0492

This is incorrect. What mistake was made?

▪ CONCEPTUAL

In Exercises 71–74, determine whether each statement is true or false.

71. $\log e = \dfrac{1}{\ln 10}$

72. $\ln e = \dfrac{1}{\log 10}$

73. $\ln(xy)^3 = (\ln x + \ln y)^3$

74. $\dfrac{\ln a}{\ln b} = \dfrac{\log a}{\log b}$

75. $\log 12x^3 = 36 \log x$

76. $e^{\ln x^2} = x^2$

▪ CHALLENGE

77. Prove the quotient rule: $\log_b\left(\dfrac{M}{N}\right) = \log_b M - \log_b N$.

  Hint: Let $u = \log_b M$ and $v = \log_b N$. Write both in exponential form and find the quotient $\log_b\left(\dfrac{M}{N}\right)$.

78. Prove the power rule: $\log_b M^p = p \log_b M$. *Hint:* Let $u = \log_b M$. Write this log in exponential form and find $\log_b M^p$.

79. Write in terms of simpler logarithmic forms.

$$\log_b\left(\sqrt{\dfrac{x^2}{y^3 z^{-5}}}\right)^6$$

80. Show that $\log_b\left(\dfrac{1}{x}\right) = -\log_b x$.

81. Show that $\log_b\left(\dfrac{a^2}{b^3}\right)^{-3} = 9 - \dfrac{6}{\log_a b}$.

82. Given that $\log_b 2 = 0.4307$ and $\log_b 3 = 0.6826$, find $\log_b \sqrt{48}$. Do not use a calculator.

▪ TECHNOLOGY

83. Use a graphing calculator to plot $y = \ln(2x)$ and $y = \ln 2 + \ln x$. Are they the same graph?

84. Use a graphing calculator to plot $y = \ln(2 + x)$ and $y = \ln 2 + \ln x$. Are they the same graph?

85. Use a graphing calculator to plot $y = \dfrac{\log x}{\log 2}$ and $y = \log x - \log 2$. Are they the same graph?

86. Use a graphing calculator to plot $y = \log\left(\dfrac{x}{2}\right)$ and $y = \log x - \log 2$. Are they the same graph?

87. Use a graphing calculator to plot $y = \ln\left(x^2\right)$ and $y = 2 \ln x$. Are they the same graph?

88. Use a graphing calculator to plot $y = (\ln x)^2$ and $y = 2 \ln x$. Are they the same graph?

89. Use a graphing calculator to plot $y = \ln x$ and $y = \dfrac{\log x}{\log e}$. Are they the same graph?

90. Use a graphing calculator to plot $y = \log x$ and $y = \dfrac{\ln x}{\ln 10}$. Are they the same graph?

▪ PREVIEW TO CALCULUS

In calculus we prove that the derivative of $f + g$ is $f' + g'$ and that the derivative of $f - g$ is $f' - g'$. It is also shown in calculus that if $f(x) = \ln x$ then $f'(x) = \dfrac{1}{x}$.

91. Use these properties to find the derivative of $f(x) = \ln x^2$.

92. Use these properties to find the derivative of $f(x) = \ln \dfrac{1}{x}$.

93. Find the derivative of $f(x) = \ln \dfrac{1}{x^2}$.

94. Find the derivative of $f(x) = \ln x^2 + \ln x^3$.

SECTION 3.4 EXPONENTIAL AND LOGARITHMIC EQUATIONS

SKILLS OBJECTIVES

- Solve exponential equations.
- Solve logarithmic equations.
- Solve application problems involving exponential and logarithmic equations.

CONCEPTUAL OBJECTIVE

- Understand how exponential and logarithmic equations are solved using properties of one-to-one functions and inverses.

To solve algebraic equations such as $x^2 - 9 = 0$, the goal is to solve for the variable, x, by finding the values of x that make the statement true. Exponential and logarithmic equations have the variable (x) buried within an exponent or a logarithm, but the goal is the same. Find the value(s) of x that makes the statement true.

$$\text{Exponential equation:} \quad e^{2x+1} = 5$$

$$\text{Logarithmic equation:} \quad \log(3x - 1) = 7$$

There are two methods for solving exponential and logarithmic equations that are based on the properties of one-to-one functions and inverses. To solve simple exponential and logarithmic equations, we will use one-to-one properties. To solve more complicated exponential and logarithmic equations, we will use properties of inverses. The following box summarizes the one-to-one and inverse properties that hold true when $b > 0$ and $b \neq 1$.

ONE-TO-ONE PROPERTIES

$b^x = b^y$	if and only if	$x = y$
$\log_b x = \log_b y$	if and only if	$x = y$

INVERSE PROPERTIES

$$b^{\log_b x} = x \qquad x > 0$$
$$\log_b b^x = x$$

We now outline a strategy for solving exponential and logarithmic equations using the one-to-one and inverse properties.

STRATEGIES FOR SOLVING EXPONENTIAL AND LOGARITHMIC EQUATIONS

Strategy for Solving Exponential Equations

TYPE OF EQUATION	STRATEGY	EXAMPLE
Simple	1. Rewrite both sides of the equation in terms of the same base.	$2^{x-3} = 32$ $2^{x-3} = 2^5$
	2. Use the one-to-one property to equate the exponents.	$x - 3 = 5$
	3. Solve for the variable.	$\boxed{x = 8}$
Complicated	1. Isolate the exponential expression.	$3e^{2x} - 2 = 7$ $3e^{2x} = 9$ $e^{2x} = 3$
	2. Take the same logarithm* of both sides.	$\ln e^{2x} = \ln 3$
	3. Simplify using the inverse properties.	$2x = \ln 3$
	4. Solve for the variable.	$\boxed{x = \frac{1}{2}\ln 3}$

Strategy for Solving Logarithmic Equations

TYPE OF EQUATION	STRATEGY	EXAMPLE
Simple	1. Combine logarithms on each side of the equation using properties.	$\log(x - 3) + \log x = \log 4$ $\log x(x - 3) = \log 4$
	2. Use the one-to-one property to equate the arguments.	$x(x - 3) = 4$
	3. Solve for the variable.	$x^2 - 3x - 4 = 0$ $(x - 4)(x + 1) = 0$ $x = -1, 4$
	4. Eliminate any extraneous solutions.	Eliminate $x = -1$ because $\log -1$ is undefined. $\boxed{x = 4}$
Complicated	1. Combine and isolate the logarithmic expressions.	$\log_5(x + 2) - \log_5 x = 2$ $\log_5\left(\dfrac{x + 2}{x}\right) = 2$
	2. Rewrite the equation in exponential form.	$\dfrac{x + 2}{x} = 5^2$
	3. Solve for the variable.	$x + 2 = 25x$ $24x = 2$ $\boxed{x = \frac{1}{12}}$
	4. Eliminate any extraneous solutions.	

*Take the logarithm with base that is equal to the base of the exponent and use the property $\log_b b^x = x$ or take the natural logarithm and use the property $\ln M^p = p \ln M$.

Solving Exponential Equations

EXAMPLE 1 **Solving a Simple Exponential Equation**

Solve the exponential equations using the one-to-one property.

a. $3^x = 81$ **b.** $5^{7-x} = 125$ **c.** $\left(\frac{1}{2}\right)^{4y} = 16$

Solution (a):

Substitute $81 = 3^4$. $3^x = 3^4$

Use the one-to-one property to identify x. $\boxed{x = 4}$

Solution (b):

Substitute $125 = 5^3$. $5^{7-x} = 5^3$

Use the one-to-one property. $7 - x = 3$

Solve for x. $\boxed{x = 4}$

Solution (c):

Substitute $\left(\frac{1}{2}\right)^{4y} = \left(\frac{1}{2^{4y}}\right) = 2^{-4y}$. $2^{-4y} = 16$

Substitute $16 = 2^4$. $2^{-4y} = 2^4$

Use the one-to-one property to identify y. $\boxed{y = -1}$

■ **Answer: a.** $x = 4$ **b.** $y = -3$

■ **YOUR TURN** Solve the following equations:

a. $2^{x-1} = 8$ **b.** $\left(\frac{1}{3}\right)^y = 27$

In Example 1, we were able to rewrite the equation in a form with the same bases so that we could use the one-to-one property. In Example 2, we will not be able to write both sides in a form with the same bases. Instead, we will use properties of inverses.

EXAMPLE 2 **Solving a More Complicated Exponential Equation with a Base Other Than 10 or e**

Solve the exponential equations and round the answers to four decimal places.

a. $5^{3x} = 16$ **b.** $4^{3x+2} = 71$

Solution (a):

Take the natural logarithm of both sides of the equation. $\ln 5^{3x} = \ln 16$

Use the power property on the left side of the equation. $3x \ln 5 = \ln 16$

Divide both sides of the equation by $3 \ln 5$. $x = \dfrac{\ln 16}{3 \ln 5}$

Use a calculator to approximate x to four decimal places. $\boxed{x \approx 0.5742}$

Technology Tip

```
((ln(71)/ln(4)-2
)/3
         .3582911866
```

Solution (b):

Rewrite in logarithmic form.

$$3x + 2 = \log_4 71$$

Subtract 2 from both sides.

$$3x = \log_4 71 - 2$$

Divide both sides by 3.

$$x = \frac{\log_4 71 - 2}{3}$$

Use the change-of-base formula, $\log_4 71 = \dfrac{\ln 71}{\ln 4}$.

$$x = \frac{\dfrac{\ln 71}{\ln 4} - 2}{3}$$

Use a calculator to approximate x to four decimal places.

$$x \approx \frac{3.0749 - 2}{3} \approx \boxed{0.3583}$$

We could have proceeded in an alternative way by taking either the natural log or the common log of both sides and using the power property (instead of using the change-of-base formula) to evaluate the logarithm with base 4.

Take the natural logarithm of both sides.

$$\ln\left(4^{3x+2}\right) = \ln 71$$

Use the power property (7).

$$(3x + 2)\ln 4 = \ln 71$$

Divide by $\ln 4$.

$$3x + 2 = \frac{\ln 71}{\ln 4}$$

Subtract 2 and divide by 3.

$$x = \frac{\dfrac{\ln 71}{\ln 4} - 2}{3}$$

Use a calculator to approximate x.

$$x \approx \frac{3.0749 - 2}{3} \approx \boxed{0.3583}$$

■ **Answer:** $y \approx \pm 1.4310$

■ **YOUR TURN** Solve the equation $5^{y^2} = 27$ and round the answer to four decimal places.

EXAMPLE 3 Solving a More Complicated Exponential Equation with Base 10 or e

Solve the exponential equation $4e^{x^2} = 64$ and round the answer to four decimal places.

Solution:

Divide both sides by 4.

$$e^{x^2} = 16$$

Take the natural logarithm (ln) of both sides.

$$\ln\left(e^{x^2}\right) = \ln 16$$

Simplify the left side with the property of inverses.

$$x^2 = \ln 16$$

Solve for x using the square-root method.

$$x = \pm\sqrt{\ln 16}$$

Use a calculator to approximate x to four decimal places.

$$\boxed{x \approx \pm 1.6651}$$

■ **Answer:** $x \approx 1.9225$

■ **YOUR TURN** Solve the equation $10^{2x-3} = 7$ and round the answer to four decimal places.

EXAMPLE 4   Solving an Exponential Equation Quadratic in Form

Solve the equation $e^{2x} - 4e^x + 3 = 0$ and round the answer to four decimal places.

Solution:

Let $u = e^x$. *(Note: $u^2 = e^x \cdot e^x = e^{2x}$.)*	$u^2 - 4u + 3 = 0$
Factor.	$(u - 3)(u - 1) = 0$
Solve for u.	$u = 3$ or $u = 1$
Substitute $u = e^x$.	$e^x = 3$ or $e^x = 1$
Take the natural logarithm (ln) of both sides.	$\ln(e^x) = \ln 3$ or $\ln(e^x) = \ln 1$
Simplify with the properties of logarithms.	$x = \ln 3$ or $x = \ln 1$
Approximate or evaluate exactly the right sides.	$\boxed{x \approx 1.0986}$ or $\boxed{x = 0}$

Technology Tip

The graph of the function $e^{2x} - 4e^x + 3$ is shown. The x-intercepts are $x \approx 1.10$ and $x = 0$.

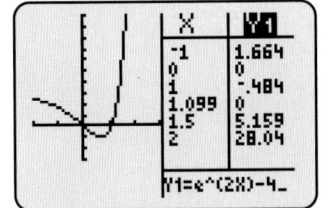

■ **Answer:** $x \approx 0.3010$

■ **YOUR TURN** Solve the equation $100^x - 10^x - 2 = 0$ and round the answer to four decimal places.

Solving Logarithmic Equations

We can solve simple logarithmic equations using the property of one-to-one functions. For more complicated logarithmic equations, we can employ properties of logarithms and properties of inverses. **Solutions must be checked to eliminate extraneous solutions**.

EXAMPLE 5   Solving a Simple Logarithmic Equation

Solve the equation $\log_4(2x - 3) = \log_4(x) + \log_4(x - 2)$.

Solution:

Apply the product property (5) on the right side.	$\log_4(2x - 3) = \log_4[x(x - 2)]$
Apply the property of one-to-one functions.	$2x - 3 = x(x - 2)$
Distribute and simplify.	$x^2 - 4x + 3 = 0$
Factor.	$(x - 3)(x - 1) = 0$
Solve for x.	$x = 3$ or $x = 1$

The possible solution $x = 1$ must be eliminated because it is not in the domain of two of the logarithmic functions.

$$x = 1: \log_4(\overbrace{-1}^{\text{undefined}}) \overset{?}{=} \log_4(1) + \log_4(\overbrace{-1}^{\text{undefined}})$$

$$\boxed{x = 3}$$

Study Tip

Solutions should be checked in the original equation to eliminate extraneous solutions.

■ **Answer:** $x = 2$

■ **YOUR TURN** Solve the equation $\ln(x + 8) = \ln(x) + \ln(x + 3)$.

EXAMPLE 6 **Solving a More Complicated Logarithmic Equation**

Solve the equation $\log_3(9x) - \log_3(x - 8) = 4$.

Solution:

Employ the quotient property (6) on the left side.

$$\log_3\left(\frac{9x}{x - 8}\right) = 4$$

Write in exponential form. $\log_b x = y \Rightarrow x = b^y$

$$\frac{9x}{x - 8} = 3^4$$

Simplify the right side.

$$\frac{9x}{x - 8} = 81$$

Multiply the equation by the LCD, $x - 8$.

$$9x = 81(x - 8)$$

Eliminate parentheses.

$$9x = 81x - 648$$

Solve for x.

$$-72x = -648$$

$$\boxed{x = 9}$$

Check: $\log_3[9 \cdot 9] - \log_3[9 - 8] = \log_3[81] - \log_3 1 = 4 - 0 = 4$

■ **Answer:** $x = 2$

■ **YOUR TURN** Solve the equation $\log_2(4x) - \log_2(2) = 2$.

EXAMPLE 7 **Solving a Logarithmic Equation with No Solution**

Solve the equation $\ln(3 - x^2) = 7$.

Solution:

Write in exponential form.

$$3 - x^2 = e^7$$

Simplify.

$$x^2 = 3 - e^7$$

$3 - e^7$ is negative.

$$x^2 = \text{negative real number}$$

There are no real numbers that when squared yield a negative real number. Therefore, there is $\boxed{\text{no real solution}}$.

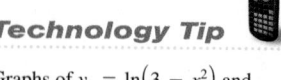
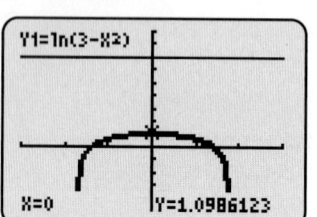
Applications

Archaeologists determine the age of a fossil by how much carbon 14 is present at the time of discovery. The number of grams of carbon 14 based on the radioactive decay of the isotope is given by

$$A = A_0 e^{-0.000124t}$$

where A is the number of grams of carbon 14 at the present time, A_0 is the number of grams of carbon 14 while alive, and t is the number of years since death. Using the inverse properties, we can isolate t.

WORDS	MATH
Divide by A_0.	$\dfrac{A}{A_0} = e^{-0.000124t}$
Take the natural logarithm of both sides.	$\ln\left(\dfrac{A}{A_0}\right) = \ln\left(e^{-0.000124t}\right)$
Simplify the right side utilizing properties of inverses.	$\ln\left(\dfrac{A}{A_0}\right) = -0.000124t$
Solve for t.	$t = -\dfrac{1}{0.000124}\ln\left(\dfrac{A}{A_0}\right)$

Let's assume that animals have approximately 1000 mg of carbon 14 in their bodies when they are alive. If a fossil has 200 mg of carbon 14, approximately how old is the fossil? Substituting $A = 200$ and $A_0 = 1000$ into our equation for t, we find

$$t = -\frac{1}{0.000124}\ln\left(\frac{1}{5}\right) \approx 12{,}979.338$$

The fossil is approximately 13,000 years old.

EXAMPLE 8 **Calculating How Many Years It Will Take for Money to Double**

You save $1,000 from a summer job and put it in a CD earning 5% compounding continuously. How many years will it take for your money to double? Round to the nearest year.

Solution:

Recall the compound continuous interest formula.	$A = Pe^{rt}$
Substitute $P = 1000$, $A = 2000$, and $r = 0.05$.	$2000 = 1000e^{0.05t}$
Divide by 1000.	$2 = e^{0.05t}$
Take the natural logarithm of both sides.	$\ln 2 = \ln\left(e^{0.05t}\right)$
Simplify with the property $\ln e^x = x$.	$\ln 2 = 0.05t$
Solve for t.	$t = \dfrac{\ln 2}{0.05} \approx 13.8629$

It will take almost $\boxed{14 \text{ years}}$ for your money to double.

■ YOUR TURN How long will it take $1,000 to triple (become $3,000) in a savings account earning 10% a year compounding continuously? Round your answer to the nearest year.

■ **Answer:** approximately 11 years

When an investment is compounded continuously, how long will it take for that investment to double?

WORDS	**MATH**
Write the interest formula for compounding continuously.	$A = Pe^{rt}$
Let $A = 2P$ (investment doubles).	$2P = Pe^{rt}$
Divide both sides of the equation by P.	$2 = e^{rt}$
Take the natural log of both sides of the equation.	$\ln 2 = \ln e^{rt}$
Simplify the right side by applying the property $\ln e^x = x$.	$\ln 2 = rt$
Divide both sides by r to get the exact value for t.	$t = \dfrac{\ln 2}{r}$
Now, approximate $\ln 2 \approx 0.7$.	$t \approx \dfrac{0.7}{r}$
Multiply the numerator and denominator by 100.	$t \approx \dfrac{70}{100r}$

This is the "rule of 70."

If we divide 70 by the interest rate (compounding continuously), we get the approximate time for an investment to double. In Example 8, the interest rate (compounding continuously) is 5%. Dividing 70 by 5 yields 14 years.

SECTION 3.4 SUMMARY

Strategy for Solving Exponential Equations

TYPE OF EQUATION	STRATEGY
Simple	1. Rewrite both sides of the equation in terms of the same base.
	2. Use the one-to-one property to equate the exponents.
	3. Solve for the variable.
Complicated	1. Isolate the exponential expression.
	2. Take the same logarithm* of both sides.
	3. Simplify using the inverse properties.
	4. Solve for the variable.

*Take the logarithm with the base that is equal to the base of the exponent and use the property $\log_b b^x = x$ or take the natural logarithm and use the property in $M^p = p \ln M$.

Strategy for Solving Logarithmic Equations

TYPE OF EQUATION	STRATEGY
Simple	1. Combine logarithms on each side of the equation using properties.
	2. Use the one-to-one property to equate the arguments.
	3. Solve for the variable.
	4. Eliminate any extraneous solutions.
Complicated	1. Combine and isolate the logarithmic expressions.
	2. Rewrite the equation in exponential form.
	3. Solve for the variable.
	4. Eliminate any extraneous solutions.

SECTION 3.4 EXERCISES

SKILLS

In Exercises 1–14, solve the exponential equations exactly for x.

1. $2^{x^2} = 16$

2. $169^x = 13$

3. $\left(\frac{2}{3}\right)^{x+1} = \frac{27}{8}$

4. $\left(\frac{3}{5}\right)^{x+1} = \frac{25}{9}$

5. $e^{2x+3} = 1$

6. $10^{x^2-1} = 1$

7. $7^{2x-5} = 7^{3x-4}$

8. $125^x = 5^{2x-3}$

9. $2^{x^2+12} = 2^{7x}$

10. $5^{x^2-3} = 5^{2x}$

11. $9^x = 3^{x^2-4x}$

12. $16^{x-1} = 2^{x^2}$

13. $e^{5x-1} = e^{x^2+3}$

14. $10^{x^2-8} = 100^x$

In Exercises 15–40, solve the exponential equations. Make sure to isolate the base to a power first. Round your answers to three decimal places.

15. $27 = 2^{3x-1}$

16. $15 = 7^{3-2x}$

17. $3e^x - 8 = 7$

18. $5e^x + 12 = 27$

19. $9 - 2e^{0.1x} = 1$

20. $21 - 4e^{0.1x} = 5$

21. $2(3^x) - 11 = 9$

22. $3(2^x) + 8 = 35$

23. $e^{3x+4} = 22$

24. $e^{x^2} = 73$

25. $3e^{2x} = 18$

26. $4(10^{3x}) = 20$

27. $4e^{2x+1} = 17$

28. $5(10^{x^2+2x+1}) = 13$

29. $3(4^{x^2-4}) = 16$

30. $7 \cdot \left(\frac{1}{4}\right)^{6-5x} = 3$

31. $e^{2x} + 7e^x - 3 = 0$

32. $e^{2x} - 4e^x - 5 = 0$

33. $(3^x - 3^{-x})^2 = 0$

34. $(3^x - 3^{-x})(3^x + 3^{-x}) = 0$

35. $\dfrac{2}{e^x - 5} = 1$

36. $\dfrac{17}{e^x + 4} = 2$

37. $\dfrac{20}{6 - e^{2x}} = 4$

38. $\dfrac{4}{3 - e^{3x}} = 8$

39. $\dfrac{4}{10^{2x} - 7} = 2$

40. $\dfrac{28}{10^x + 3} = 4$

In Exercises 41–56, solve the logarithmic equations exactly.

41. $\log_3(2x + 1) = 4$

42. $\log_2(3x - 1) = 3$

43. $\log_2(4x - 1) = -3$

44. $\log_4(5 - 2x) = -2$

45. $\ln x^2 - \ln 9 = 0$

46. $\log x^2 + \log x = 3$

47. $\log_5(x - 4) + \log_5 x = 1$

48. $\log_2(x - 1) + \log_2(x - 3) = 3$

49. $\log(x - 3) + \log(x + 2) = \log(4x)$

50. $\log_2(x + 1) + \log_2(4 - x) = \log_2(6x)$

51. $\log_4(4x) - \log_4\left(\dfrac{x}{4}\right) = 3$

52. $\log_3(7 - 2x) - \log_3(x + 2) = 2$

53. $\log(2x - 5) - \log(x - 3) = 1$

54. $\log_3(10 - x) - \log_3(x + 2) = 1$

55. $\log_4\left(x^2 + 5x + 4\right) - 2\log_4(x + 1) = 2$

56. $\log_2(x + 1) + \log_2(x + 5) - \log_2(2x + 5) = 2$

In Exercises 57–70, solve the logarithmic equations. Round your answers to three decimal places.

57. $\log(2x + 5) = 2$

58. $\ln(4x - 7) = 3$

59. $\ln\left(x^2 + 1\right) = 4$

60. $\log(x^2 + 4) = 2$

61. $\ln(2x + 3) = -2$

62. $\log(3x - 5) = -1$

63. $\log(2 - 3x) + \log(3 - 2x) = 1.5$

64. $\log_2(3 - x) + \log_2(1 - 2x) = 5$

65. $\ln(x) + \ln(x - 2) = 4$

66. $\ln(4x) + \ln(2 + x) = 2$

67. $\log_7(1 - x) - \log_7(x + 2) = \log_7 x$

68. $\log_5(x + 1) - \log_5(x - 1) = \log_5 x$

69. $\ln\sqrt{x + 4} - \ln\sqrt{x - 2} = \ln\sqrt{x + 1}$

70. $\log\left(\sqrt{1 - x}\right) - \log\left(\sqrt{x + 2}\right) = \log x$

■ APPLICATIONS

71. Money. If money is invested in a savings account earning 3.5% interest compounded yearly, how many years will pass until the money triples?

72. Money. If money is invested in a savings account earning 3.5% interest compounded monthly, how many years will pass until the money triples?

73. Money. If \$7,500 is invested in a savings account earning 5% interest compounded quarterly, how many years will pass until there is \$20,000?

74. Money. If \$9,000 is invested in a savings account earning 6% interest compounded continuously, how many years will pass until there is \$15,000?

75. Earthquakes. On September 25, 2003, an earthquake that measured 7.4 on the Richter scale shook Hokkaido, Japan. How much energy (joules) did the earthquake emit?

76. Earthquakes. Again, on that same day (September 25, 2003), a second earthquake that measured 8.3 on the Richter scale shook Hokkaido, Japan. How much energy (joules) did the earthquake emit?

77. Sound. Matt likes to drive around campus in his classic Mustang with the stereo blaring. If his boom stereo has a sound intensity of 120 dB, how many watts per square meter does the stereo emit?

78. Sound. The New York Philharmonic has a sound intensity of 100 dB. How many watts per square meter does the orchestra emit?

79. Anesthesia. When a person has a cavity filled, the dentist typically administers a local anesthetic. After leaving the dentist's office, one's mouth often remains numb for several more hours. If a shot of anesthesia is injected into the bloodstream at the time of the procedure ($t = 0$), and the amount of anesthesia still in the bloodstream t hours after the initial injection is given by $A = A_0 e^{-0.5t}$, in how many hours will only 10% of the original anesthetic still be in the bloodstream?

80. Investments. Money invested in an account that compounds interest continuously at a rate of 3% a year is modeled by $A = A_0 e^{0.03t}$, where A is the amount in the investment after t years and A_0 is the initial investment. How long will it take the initial investment to double?

81. Biology. The U.S. Fish and Wildlife Service is releasing a population of the endangered Mexican gray wolf in a protected area along the New Mexico and Arizona border. They estimate the population of the Mexican gray wolf to be approximated by

$$P(t) = \frac{200}{1 + 24e^{-0.2t}}$$

How many years will it take for the population to reach 100 wolves?

82. Introducing a New Car Model. If the number of new model Honda Accord hybrids purchased in North America is given by $N = \dfrac{100,000}{1 + 10e^{-2t}}$, where t is the number of weeks after Honda releases the new model, how many weeks will it take after the release until there are 50,000 Honda hybrids from that batch on the road?

83. Earthquakes. A P wave measures 6.2 on the Richter scale, and an S wave measures 3.3 on the Richter scale. What is their combined measure on the Richter scale?

84. Sound. You and a friend get front row seats to a rock concert. The music level is 100 dB, and your normal conversation is 60 dB. If your friend is telling you something during the concert, how many decibels are you subjecting yourself to?

■ CATCH THE MISTAKE

In Exercises 85–88, explain the mistake that is made.

85. Solve the equation: $4e^x = 9$.

Solution:

Take the natural log of both sides. $\quad \ln(4e^x) = \ln 9$

Apply the property of inverses. $\quad\quad 4x = \ln 9$

Solve for x. $\quad\quad x = \dfrac{\ln 9}{4} \approx 0.55$

This is incorrect. What mistake was made?

86. Solve the equation: $\log(x) + \log(3) = 1$.

Solution:

Apply the product property (5). $\quad \log(3x) = 1$

Exponentiate (base 10). $\quad 10^{\log(3x)} = 1$

Apply the properties of inverses. $\quad 3x = 1$

Solve for x. $\quad\quad x = \dfrac{1}{3}$

This is incorrect. What mistake was made?

87. Solve the equation: $\log(x) + \log(x + 3) = 1$ for x.

Solution:

Apply the product property (5). $\quad \log(x^2 + 3x) = 1$

Exponentiate both sides (base 10). $\quad 10^{\log(x^2+3x)} = 10^1$

Apply the property of inverses. $\quad x^2 + 3x = 10$

Factor. $\quad (x+5)(x-2) = 0$

Solve for x. $\quad x = -5$ and $x = 2$

This is incorrect. What mistake was made?

88. Solve the equation: $\log x + \log 2 = \log 5$.

Solution:

Combine the logarithms on the left. $\quad \log(x+2) = \log 5$

Apply the property of one-to-one functions. $\quad x + 2 = 5$

Solve for x. $\quad\quad x = 3$

This is incorrect. What mistake was made?

■ CONCEPTUAL

In Exercises 89–92, determine whether each statement is true or false.

89. The sum of logarithms with the same base is equal to the logarithm of the product.

90. A logarithm squared is equal to two times the logarithm.

91. $e^{\log x} = x$

92. $e^x = -2$ has no solution.

93. $\log_3(x^2 + x - 6) = 1$ has two solutions.

94. The division of two logarithms with the same base is equal to the logarithm of the subtraction.

■ CHALLENGE

95. Solve for x in terms of b:

$$\frac{1}{3}\log_b(x^3) + \frac{1}{2}\log_b(x^2 - 2x + 1) = 2$$

96. Solve exactly:

$$2\log_b(x) + 2\log_b(1-x) = 4$$

97. Solve $y = \dfrac{3000}{1 + 2e^{-0.2t}}$ for t in terms of y.

98. State the range of values of x that the following identity holds: $e^{\ln(x^2 - a)} = x^2 - a$.

99. A function called the hyperbolic cosine is defined as the average of exponential growth and exponential decay by $f(x) = \dfrac{e^x + e^{-x}}{2}$. If we restrict the domain of f to $[0, \infty)$, find its inverse.

100. A function called the hyperbolic sine is defined by $f(x) = \dfrac{e^x - e^{-x}}{2}$. Find its inverse.

■ TECHNOLOGY

101. Solve the equation $\ln 3x = \ln(x^2 + 1)$. Using a graphing calculator, plot the graphs $y = \ln(3x)$ and $y = \ln(x^2 + 1)$ in the same viewing rectangle. Zoom in on the point where the graphs intersect. Does this agree with your solution?

102. Solve the equation $10^{x^2} = 0.001^x$. Using a graphing calculator, plot the graphs $y = 10^{x^2}$ and $y = 0.001^x$ in the same viewing rectangle. Does this confirm your solution?

103. Use a graphing utility to help solve $3^x = 5x + 2$.

104. Use a graphing utility to help solve $\log x^2 = \ln(x - 3) + 2$.

105. Use a graphing utility to graph $y = \dfrac{e^x + e^{-x}}{2}$. State the domain. Determine whether there are any symmetry and asymptote.

106. Use a graphing utility to graph $y = \dfrac{e^x + e^{-x}}{e^x - e^{-x}}$. State the domain. Determine whether there are any symmetry and asymptote.

■ PREVIEW TO CALCULUS

107. The hyperbolic sine function is defined by $\sinh x = \dfrac{e^x - e^{-x}}{2}$. Find its inverse function $\sinh^{-1} x$.

108. The hyperbolic tangent is defined by $\tanh x = \dfrac{e^x - e^{-x}}{e^x + e^{-x}}$. Find its inverse function $\tanh^{-1} x$.

In Exercises 109–110, refer to the following:

In calculus, to find the derivative of a function of the form $y = k^x$, where k is a constant, we apply logarithmic differentiation. The first step in this process consists of writing $y = k^x$ in an equivalent form using the natural logarithm. Use the properties of this section to write an equivalent form of the following implicitly defined functions.

109. $y = 2^x$

110. $y = 4^x \cdot 3^{x+1}$

SKILLS OBJECTIVES

- Apply exponential growth and exponential decay models to biological, demographic, and economic phenomena.
- Represent distributions by means of a Gaussian model.
- Use logistic growth models to represent phenomena involving limits to growth.
- Solve problems such as species populations, credit card payoff, and wearoff of anesthesia through logarithmic models.

CONCEPTUAL OBJECTIVE

- Recognize exponential growth, exponential decay, Gaussian distributions, logistic growth, and logarithmic models.

The following table summarizes the five primary models that involve exponential and logarithmic functions:

NAME	MODEL	GRAPH	APPLICATIONS
Exponential growth	$f(t) = ce^{kt}$ $k > 0$	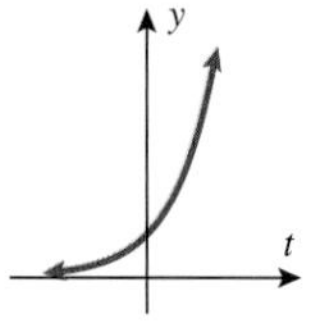	World populations, bacteria growth, appreciation, global spread of the HIV virus
Exponential decay	$f(t) = ce^{-kt}$ $k > 0$	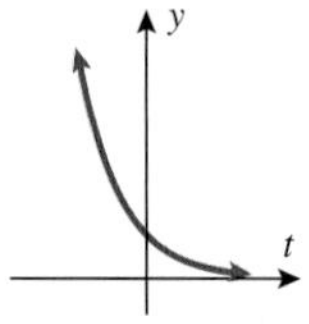	Radioactive decay, carbon dating, depreciation
Gaussian (normal) distribution	$f(x) = ce^{-(x-a)^2/k}$	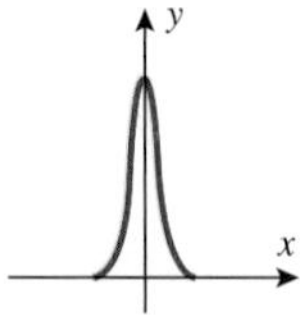	Bell curve (grade distribution), life expectancy, height/weight charts, intensity of a laser beam, IQ tests
Logistic growth	$f(t) = \dfrac{a}{1 + ce^{-kt}}$	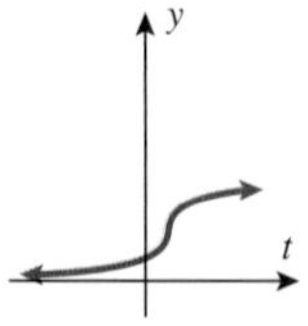	Conservation biology, learning curve, spread of virus on an island, carrying capacity
Logarithmic	$f(t) = a + c \log t$ $f(t) = a + c \ln t$	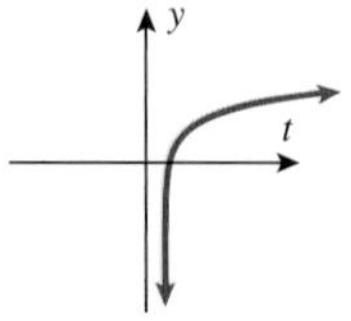	Population of species, anesthesia wearing off, time to pay off credit cards

Exponential Growth Models

Quite often one will hear that something "grows exponentially," meaning that it grows very fast and at increasing speed. In mathematics, the precise meaning of **exponential growth** is a *growth rate of a function that is proportional to its current size*. Let's assume you get a 5% raise every year in a government job. If your annual starting salary out of college is $40,000, then your first raise will be $2,000. Fifteen years later your annual salary will be approximately $83,000 and your next 5% raise will be around $4,150. The raise is always 5% of the current salary, so the larger the current salary, the larger the raise.

In Section 3.1, we saw that interest that is compounded continuously is modeled by $A = Pe^{rt}$. Here A stands for amount and P stands for principal. There are similar models for populations; these take the form $N(t) = N_0e^{rt}$, where N_0 represents the number of people at time $t = 0$, r is the annual growth rate, t is time in years, and N represents the number of people at time t. In general, any model of the form $f(x) = ce^{kx}$, $k > 0$, models exponential growth.

EXAMPLE 1   World Population Projections

The world population is the total number of humans on Earth at a given time. In 2000 the world population was 6.1 billion and in 2005 the world population was 6.5 billion. Find the annual growth rate and determine what year the population will reach 9 billion.

Solution:

Assume an exponential growth model. $N(t) = N_0e^{rt}$

Let $t = 0$ correspond to 2000. $N(0) = N_0 = 6.1$

In 2005, $t = 5$, the population was 6.5 billion. $6.5 = 6.1e^{5r}$

Solve for r. $$\frac{6.5}{6.1} = e^{5r}$$

$$\ln\left(\frac{6.5}{6.1}\right) = \ln\left(e^{5r}\right)$$

$$\ln\left(\frac{6.5}{6.1}\right) = 5r$$

$$r \approx 0.012702681$$

The annual growth rate is approximately $\boxed{1.3\%}$ per year.

Assuming the growth rate stays the
same, write a population model. $N(t) = 6.1e^{0.013}t$

Let $N(t) = 9$. $9 = 6.1e^{0.013t}$

Solve for t. $$e^{0.013t} = \frac{9}{6.1}$$

$$\ln(e^{0.013t}) = \ln\left(\frac{9}{6.1}\right)$$

$$0.013t = \ln\left(\frac{9}{6.1}\right)$$

$$t \approx 29.91813894$$

In $\boxed{2030}$ the world population will reach 9 billion if the same growth rate is maintained.

■ **Answer:** 1% per year; 2115

■ **YOUR TURN** The population of North America (United States and Canada) was 300 million in 1995, and in 2005 the North American population was 332 million. Find the annual growth rate (round to the nearest percent) and use that rounded growth rate to determine what year the population will reach 1 billion.

Exponential Decay Models

We mentioned radioactive decay briefly in Section 3.1. Radioactive decay is the process in which a radioactive isotope of an element (atoms) loses energy by emitting radiation in the form of particles. This results in loss of mass of the isotope, which we measure as a reduction in the rate of radioactive emission. This process is random, but given a large number of atoms, the decay rate is directly proportional to the mass of the radioactive substance. Since the mass is decreasing, we say this represents *exponential decay*, $m = m_0 e^{-rt}$, where m_0 represents the initial mass at time $t = 0$, r is the decay rate, t is time, and m represents the mass at time t. In general, any model of the form $f(x) = ce^{-kx}$, $k > 0$, models **exponential decay**.

Typically, the decay rate r is expressed in terms of the half-life h. Recall (Section 3.1) that half-life is the time it takes for a quantity to decrease by half.

WORDS	**MATH**
Write the radioactive decay model.	$m = m_0 e^{-rt}$
Divide both sides by m_0.	$\dfrac{m}{m_0} = e^{-rt}$
The remaining mass of the radioactive isotope is half of the initial mass when $t = h$.	$\dfrac{1}{2} = e^{-rh}$
Solve for r.	
Take the natural logarithm of both sides.	$\ln\left(\dfrac{1}{2}\right) = \ln\left(e^{-rh}\right)$
Simplify.	$\underbrace{\ln 1}_{0} - \ln 2 = -rh$
	$rh = \ln 2$
	$\boxed{r = \dfrac{\ln 2}{h}}$

Technology Tip

Graphs of $Y_1 = 500e^{-0.56x}$ with $x = t$ and $Y_2 = 5$ are shown.

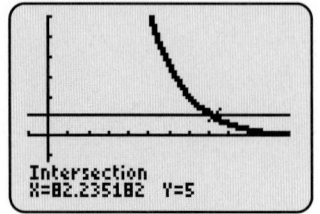

```
Plot1 Plot2 Plot3
\Y1■500e^(-.056X
)
\Y2■5
```

```
Intersection
X=82.235182   Y=5
```

EXAMPLE 2   Radioactive Decay

The radioactive isotope of potassium ^{42}K, which is vital in the diagnosis of brain tumors, has a half-life of 12.36 hours.

a. Determine the exponential decay model that represents the mass of ^{42}K.

b. If 500 milligrams of potassium-42 are taken, how many milligrams of this isotope will remain after 48 hours?

c. How long will it take for the original 500-milligram sample to decay to a mass of 5 milligrams?

Solution (a):

Write the relationship between rate of decay and half-life.	$r = \dfrac{\ln 2}{h}$
Let $h = 12.36$.	$r \approx 0.056$
Write the exponential decay model for the mass of ^{42}K.	$\boxed{m = m_0 e^{-0.056t}}$

Solution (b):

Let $m_0 = 500$ and $t = 48$.
$$m = 500e^{-(0.056)(48)} \approx 34.00841855$$

There are approximately 34 milligrams of ^{42}K still in the body after 48 hours.

Note: Had we used the full value of $r = 0.056079868$, the resulting mass would have been $m = 33.8782897$, which is approximately 34 mg.

Solution (c):

Write the exponential decay model for the mass of ^{42}K.
$$m = m_0 e^{-0.056t}$$

Let $m = 5$ and $m_0 = 500$.
$$5 = 500e^{-0.056t}$$

Solve for t.

Divide by 500.
$$e^{-0.056t} = \frac{5}{500} = \frac{1}{100}$$

Take the natural logarithm of both sides.
$$\ln\left(e^{-0.056t}\right) = \ln\left(\frac{1}{100}\right)$$

Simplify.
$$-0.056t = \ln\left(\frac{1}{100}\right)$$

Divide by -0.056 and approximate with a calculator.
$$t \approx 82.2352$$

It will take approximately 82 hours for the original 500-mg substance to decay to a mass of 5 mg.

■ **YOUR TURN** The radioactive element radon-222 has a half-life of 3.8 days.

    a. Determine the exponential decay model that represents the mass of radon-222.

    b. How much of a 64-gram sample of radon-222 will remain after 7 days? Round to the nearest gram.

    c. How long will it take for the original 64-gram sample to decay to a mass of 4 grams? Round to the nearest day.

■ **Answer: a.** $m = m_0 e^{-0.1824t}$
      b. 18 grams
      c. 15 days

Gaussian (Normal) Distribution Models

If your instructor plots the grades from the last test, typically you will see a **Gaussian (normal) distribution** of scores, otherwise known as the *bell-shaped curve*. Other examples of phenomena that tend to follow a Gaussian distribution are SAT scores, height distributions of adults, and standardized tests like IQ assessments.

The graph to the right represents a Gaussian distribution of IQ scores. The average score, which for IQ is 100, is always the maximum point on the curve. The typical probability distribution is

$$F(x) = \frac{1}{\sigma\sqrt{2\pi}} e^{-(x-\mu)^2/2\sigma^2}$$

where μ is the average or mean value and the variance is σ^2.

Any model of the form $f(x) = ce^{-(x-a)^2/k}$ is classified as a **Gaussian model**.

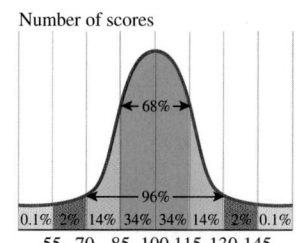

Number of scores

68%

96%

0.1% 2% 14% 34% 34% 14% 2% 0.1%

55 70 85 100 115 130 145
Intelligence quotient
(Score on Wechsler Adult Intelligence Scale)

EXAMPLE 3 **Weight Distributions**

Suppose each member of a Little League football team is weighed and the weight distribution follows the Gaussian model $f(x) = 10e^{-(x-100)^2/25}$.

a. Graph the weight distribution.
b. What is the average weight of a member of this team?
c. Approximately how many boys weigh 95 pounds?

Solution:

a.

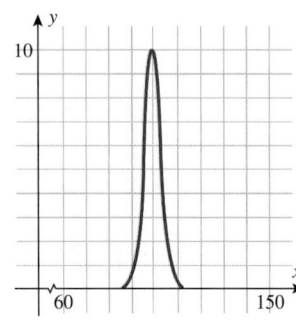

b. $\boxed{100 \text{ pounds}}$

c. $f(95) = 10e^{-(95-100)^2/25}$

$= 10e^{-25/25}$

$= 10e^{-1}$

≈ 3.6788

Approximately $\boxed{4 \text{ boys}}$ weigh 95 lb.

Logistic Growth Models

Technology Tip

Plot the graph of $Y_1 = \dfrac{500000}{1 + 5e^{-0.12x}}$ with $x = t$. Use the $\boxed{\text{TRACE}}$ key to find $f(20)$ and $f(30)$.

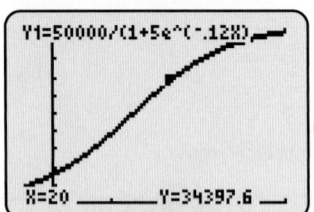

A table of values supports the solution at $t = 20$ and $t = 30$.

Earlier in this section, we discussed exponential growth models for populations that experience uninhibited growth. Now we will turn our attention to *logistic growth*, which models population growth when there are factors that impact the ability to grow, such as food and space. For example, if 10 rabbits are dropped off on an uninhabited island, they will reproduce and the population of rabbits on that island will experience rapid growth. The population will continue to increase rapidly until the rabbits start running out of space or food on the island. In other words, under favorable conditions the growth is not restricted, while under less favorable conditions the growth becomes restricted. This type of growth is represented by **logistic growth models**, $f(x) = \dfrac{a}{1 + ce^{-kx}}$. Ultimately, the population of rabbits reaches the island's *carrying capacity, a*.

 EXAMPLE 4 **Number of Students on a College Campus**

In 2008 the University of Central Florida was the sixth largest university in the country. The number of students can be modeled by the function

$f(t) = \dfrac{50,000}{1 + 5e^{-0.12t}}$, where t is time in years and $t = 0$ corresponds to 1970.

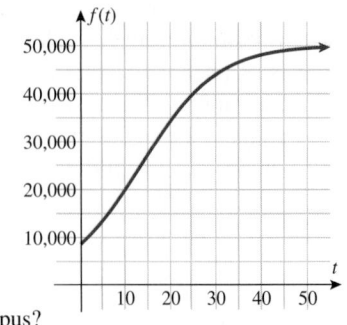

a. How many students attended UCF in 1990? Round to the nearest thousand.
b. How many students attended UCF in 2000?
c. What is the carrying capacity of the UCF main campus?

Round all answers to the nearest thousand.

Solution (a): Let $t = 20$. $\qquad f(20) = \dfrac{50{,}000}{1 + 5e^{-0.12(20)}} \approx \boxed{34{,}000}$

Solution (b): Let $t = 30$. $\qquad f(30) = \dfrac{50{,}000}{1 + 5e^{-0.12(30)}} \approx \boxed{44{,}000}$

Solution (c): As t increases, the UCF student population approaches $\boxed{50{,}000}$.

Logarithmic Models

Homeowners typically ask the question, "If I increase my payment, how long will it take to pay off my current mortgage?" In general, a loan over t years with an annual interest rate r with n periods per year corresponds to an interest rate per period of $i = \dfrac{r}{n}$. Typically, loans are paid in equal payments consisting of the principal P plus total interest divided by the total number of periods over the life of the loan nt. The periodic payment R is given by

$$R = P\,\dfrac{i}{1 - (1 + i)^{-nt}}$$

We can find the time (in years) it will take to pay off the loan as a function of periodic payment by solving for t.

WORDS	**MATH**
Multiply both sides by $1 - (1 + i)^{-nt}$.	$R\left[1 - (1 + i)^{-nt}\right] = Pi$
Eliminate the brackets.	$R - R(1 + i)^{-nt} = Pi$
Subtract R.	$-R(1 + i)^{-nt} = Pi - R$
Divide by $-R$.	$(1 + i)^{-nt} = 1 - \dfrac{Pi}{R}$
Take the natural log of both sides.	$\ln(1 + i)^{-nt} = \ln\left(1 - \dfrac{Pi}{R}\right)$
Use the power property for logarithms.	$-nt\ln(1 + i) = \ln\left(1 - \dfrac{Pi}{R}\right)$
Isolate t.	$t = -\dfrac{\ln\left(1 - \dfrac{Pi}{R}\right)}{n\ln(1 + i)}$
Let $i = \dfrac{r}{n}$.	$\boxed{t = -\dfrac{\ln\left(1 - \dfrac{Pr}{nR}\right)}{n\ln\left(1 + \dfrac{r}{n}\right)}}$

Technology Tip

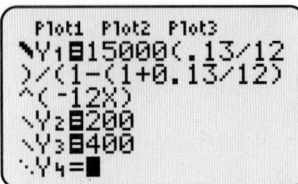

Use the keystrokes 2nd Calc 5:Intersect to find the points of intersection.

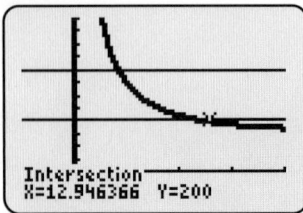

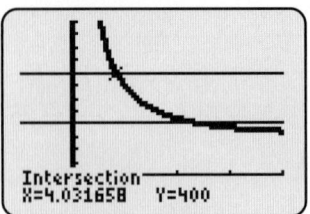

EXAMPLE 5   Paying Off Credit Cards

James owes $15,000 on his credit card. The annual interest rate is 13% compounded monthly.

a. Find the time it will take to pay off his credit card if he makes payments of $200 per month.

b. Find the time it will take to pay off his credit card if he makes payments of $400 per month.

Let $P = 15,000$, $r = 0.13$, and $n = 12$.

$$t = -\frac{\ln\left(1 - \dfrac{15,000(0.13)}{12R}\right)}{12\ln\left(1 + \dfrac{0.13}{12}\right)}$$

Solution (a): Let $R = 200$.

$$t = -\frac{\ln\left(1 - \dfrac{15,000(0.13)}{12(200)}\right)}{12\ln\left(1 + \dfrac{0.13}{12}\right)} \approx 13$$

$200 monthly payments will allow James to pay off his credit card in about ⟨ 13 years ⟩.

Solution (b): Let $R = 400$.

$$t = -\frac{\ln\left(1 - \dfrac{15,000(0.13)}{12(400)}\right)}{12\ln\left(1 + \dfrac{0.13}{12}\right)} \approx 4$$

$400 monthly payments will allow James to pay off the balance in approximately ⟨ 4 years ⟩. It is important to note that doubling the payment reduced the time to pay off the balance by less than a third.

SECTION

3.5 SUMMARY

In this section, we discussed five main types of models that involve exponential and logarithmic functions.

NAME	MODEL	APPLICATIONS
Exponential growth	$f(t) = ce^{kt}$, $k > 0$	Uninhibited growth (populations/inflation)
Exponential decay	$f(t) = ce^{-kt}$, $k > 0$	Carbon dating, depreciation
Gaussian (normal) distributions	$f(x) = ce^{-(x-a)^2/k}$	Bell curves (standardized tests, height/weight charts, distribution of power flux of laser beams)
Logistic growth	$f(t) = \dfrac{a}{1 + ce^{-kt}}$	Conservation biology (growth limited by factors like food and space), learning curve
Logarithmic	$f(t) = c\ln t$	Time to pay off credit cards, annuity planning

SECTION

3.5 EXERCISES

■ SKILLS

In Exercises 1–6, match the function with the graph (a to f) and the model name (i to v).

1. $f(t) = 5e^{2t}$

2. $N(t) = 28e^{-t/2}$

3. $T(x) = 4e^{-(x-80)^2/10}$

4. $P(t) = \dfrac{200}{1 + 5e^{-0.4t}}$

5. $D(x) = 4 + \log(x - 1)$

6. $h(t) = 2 + \ln(t + 3)$

Model Name

i. Logarithmic   **ii.** Logistic   **iii.** Gaussian   **iv.** Exponential growth   **v.** Exponential decay

Graphs

a.
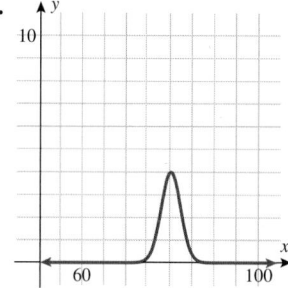

b.
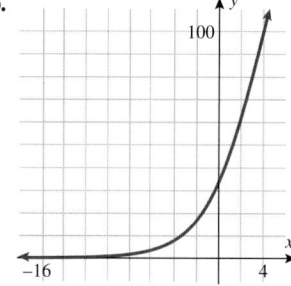

c.
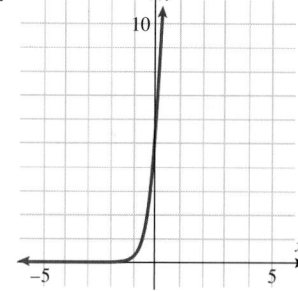

d.
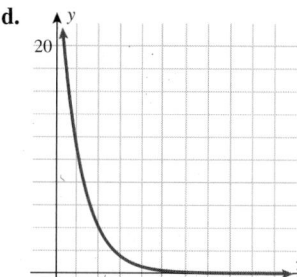

e.
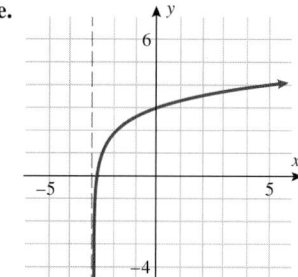

f.
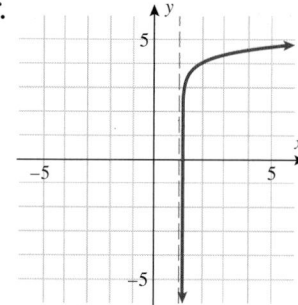

■ APPLICATIONS

7. Population Growth. The population of the Philippines in 2003 was 80 million. It increases 2.36% per year. What is the expected population of the Philippines in 2010? Apply the formula $N = N_0e^{rt}$, where N represents the number of people.

8. Population Growth. China's urban population is growing at 2.5% a year, compounding continuously. If there were 13.7 million people in Shanghai in 1996, approximately how many people will there be in 2016? Apply the formula $N = N_0e^{rt}$, where N represents the number of people.

9. Population Growth. Port St. Lucie, Florida, had the United States' fastest growth rate among cities with a population of 100,000 or more between 2003 and 2004. In 2003 the population was 103,800 and increasing at a rate of 12% per year. In what year should the population reach 200,000? (Let $t = 0$ correspond to 2003.) Apply the formula $N = N_0e^{rt}$, where N represents the number of people.

10. Population Growth. San Francisco's population has been declining since the "dot com" bubble burst. In 2002 the population was 776,000. If the population is declining at a rate of 1.5% per year, in what year will the population be 700,000? (Let $t = 0$ correspond to 2002.) Apply the formula $N = N_0e^{-rt}$, where N represents the number of people.

11. Cellular Phone Plans. The number of cell phones in China is exploding. In 2007 there were 487.4 million cell phone subscribers and the number is increasing at a rate of 16.5% per year. How many cell phone subscribers are expected in 2010? Use the formula $N = N_0 e^{rt}$, where N represents the number of cell phone subscribers. Let $t = 0$ correspond to 2007.

12. Bacteria Growth. A colony of bacteria is growing exponentially. Initially, 500 bacteria were in the colony. The growth rate is 20% per hour. (a) How many bacteria should be in the colony in 12 hours? (b) How many in 1 day? Use the formula $N = N_0 e^{rt}$, where N represents the number of bacteria.

13. Real Estate Appreciation. In 2004 the average house in Birmingham, AL cost $185,000, and real estate prices were increasing at an amazing rate of 30% per year. What was the expected cost of an average house in Birmingham in 2007? Use the formula $N = N_0 e^{rt}$, where N represents the average cost of a home. Round to the nearest thousand.

14. Real Estate Appreciation. The average cost of a single family home in Seattle, WA in 2004 was $230,000. In 2005 the average cost was $252,000. If this trend continued, what was the expected cost in 2007? Use the formula $N = N_0 e^{rt}$, where N represents the average cost of a home. Round to the nearest thousand.

15. Oceanography (Growth of Phytoplankton). Phytoplankton are microscopic plants that live in the ocean. Phytoplankton grow abundantly in oceans around the world and are the foundation of the marine food chain. One variety of phytoplankton growing in tropical waters is increasing at a rate of 20% per month. If it is estimated that there are 100 million in the water, how many will there be in 6 months? Utilize formula $N = N_0 e^{rt}$, where N represents the population of phytoplankton.

16. Oceanography (Growth of Phytoplankton). In Arctic waters there are an estimated 50,000,000 phytoplankton. The growth rate is 12% per month. How many phytoplankton will there be in 3 months? Utilize formula $N = N_0 e^{rt}$, where N represents the population of phytoplankton.

17. HIV/AIDS. In 2003 an estimated 1 million people had been infected with HIV in the United States. If the infection rate increases at an annual rate of 2.5% a year compounding continuously, how many Americans will be infected with the HIV virus by 2010?

18. HIV/AIDS. In 2003 there were an estimated 25 million people who have been infected with HIV in sub-Saharan Africa. If the infection rate increases at an annual rate of 9% a year compounding continuously, how many Africans will be infected with the HIV virus by 2010?

19. Anesthesia. When a person has a cavity filled, the dentist typically gives a local anesthetic. After leaving the dentist's office, one's mouth often is numb for several more hours. If 100 ml of anesthesia is injected into the local tissue at the time of the procedure ($t = 0$), and the amount of anesthesia still in the local tissue t hours after the initial injection is given by $A = 100e^{-0.5t}$, how much remains in the local tissue 4 hours later?

20. Anesthesia. When a person has a cavity filled, the dentist typically gives a local anesthetic. After leaving the dentist's office, one's mouth often is numb for several more hours. If 100 ml of anesthesia is injected into the local tissue at the time of the procedure ($t = 0$), and the amount of anesthesia still in the local tissue t hours after the initial injection is given by $A = 100e^{-0.5t}$, how much remains in the local tissue 12 hours later?

21. Radioactive Decay. Carbon-14 has a half-life of 5730 years. How long will it take 5 grams of carbon-14 to be reduced to 2 grams?

22. Radioactive Decay. Radium-226 has a half-life of 1600 years. How long will it take 5 grams of radium-226 to be reduced to 2 grams?

23. Radioactive Decay. The half-life of uranium-238 is 4.5 billion years. If 98% of uranium-238 remains in a fossil, how old is the fossil?

24. Decay Levels in the Body. A drug has a half-life of 12 hours. If the initial dosage is 5 milligrams, how many milligrams will be in the patient's body in 16 hours?

In Excercises 25–28, use the following formula for Newton's Law of Cooling:

If you take a hot dinner out of the oven and place it on the kitchen countertop, the dinner cools until it reaches the temperature of the kitchen. Likewise, a glass of ice set on a table in a room eventually melts into a glass of water at that room temperature. The rate at which the hot dinner cools or the ice in the glass melts at any given time is proportional to the difference between its temperature and the temperature of its surroundings (in this case, the room). This is called **Newton's law of cooling** (or warming) and is modeled by

$$T = T_S + (T_0 - T_S)e^{-kt}$$

where T is the temperature of an object at time t, T_s is the temperature of the surrounding medium, T_0 is the temperature of the object at time $t = 0$, t is the time, and k is a constant.

25. Newton's Law of Cooling. An apple pie is taken out of the oven with an internal temperature of 325 °F. It is placed on a rack in a room with a temperature of 72 °F. After 10 minutes, the temperature of the pie is 200 °F. What will the temperature of the pie be 30 minutes after coming out of the oven?

26. Newton's Law of Cooling. A cold drink is taken out of an ice chest with a temperature of 38 °F and placed on a picnic table with a surrounding temperature of 75 °F. After 5 minutes, the temperature of the drink is 45 °F. What will the temperature of the drink be 20 minutes after it is taken out of the chest?

27. Forensic Science (Time of Death). A body is discovered in a hotel room. At 7:00 A.M. a police detective found the body's temperature to be 85 °F. At 8:30 A.M. a medical examiner measures the body's temperature to be 82 °F. Assuming the room in which the body was found had a constant temperature of 74 °F, how long has the victim been dead? (Normal body temperature is 98.6 °F.)

28. Forensic Science (Time of Death). At 4 A.M. a body is found in a park. The police measure the body's temperature to be 90 °F. At 5 A.M. the medical examiner arrives and determines the temperature to be 86 °F. Assuming the temperature of the park was constant at 60 °F, how long has the victim been dead?

29. Depreciation of Automobile. A new Lexus IS250 has a book value of \$38,000, and after 1 year has a book value of \$32,000. What is the car's value in 4 years? Apply the formula $N = N_0 e^{-rt}$, where N represents the value of the car. Round to the nearest hundred.

30. Depreciation of Automobile. A new Hyundai Triburon has a book value of \$22,000, and after 2 years a book value of \$14,000. What is the car's value in 4 years? Apply the formula $N = N_0 e^{-rt}$, where N represents the value of the car. Round to the nearest hundred.

31. Automotive. A new model BMW convertible coupe is designed and produced in time to appear in North America in the fall. BMW Corporation has a limited number of new models available. The number of new model BMW convertible coupes purchased in North America is given by $N = \dfrac{100,000}{1 + 10e^{-2t}}$, where t is the number of weeks after the BMW is released.

a. How many new model BMW convertible coupes will have been purchased 2 weeks after the new model becomes available?

b. How many after 30 weeks?

c. What is the maximum number of new model BMW convertible coupes that will be sold in North America?

32. iPhone. The number of iPhones purchased is given by $N = \dfrac{2,000,000}{1 + 2e^{-4t}}$, where t is the time in weeks after they are made available for purchase.

a. How many iPhones are purchased within the first 2 weeks?

b. How many iPhones are purchased within the first month?

33. Spread of a Disease. The number of MRSA (methicillin-resistant *Staphylococcus aureus*) cases has been rising sharply in England and Wales since 1997. In 1997, 2422 cases were reported. The number of cases reported in 2003 was 7684. How many cases will be expected in 2010? (Let $t = 0$ correspond to 1997.) Use the formula $N = N_0 e^{-rt}$, where N represents the number of cases reported.

34. Spread of a Virus. Dengue fever, an illness carried by mosquitoes, is occurring in one of the worst outbreaks in decades across Latin America and the Caribbean. In 2004, 300,000 cases were reported, and 630,000 cases in 2007. How many cases might be expected in 2010? (Let $t = 0$ be 2004.) Use the formula $N = N_0 e^{-rt}$, where N represents the number of cases.

35. Carrying Capacity. The Virginia Department of Fish and Game stock a mountain lake with 500 trout. Officials believe the lake can support no more than 10,000 trout. The number of trout is given by $N = \dfrac{10,000}{1 + 19e^{-1.56t}}$, where t is time in years. How many years will it take for the trout population to reach 5000?

36. Carrying Capacity. The World Wildlife Fund has placed 1000 rare pygmy elephants in a conservation area in Borneo. They believe 1600 pygmy elephants can be supported in this environment. The number of elephants is given by $N = \dfrac{1600}{1 + 0.6e^{-0.14t}}$, where t is time in years. How many years will it take the herd to reach 1200 elephants?

37. Lasers. The intensity of a laser beam is given by the ratio of power to area. A particular laser beam has an intensity function given by $I = e^{-r^2}$ mW/cm^2, where r is the radius off the center axis given in centimeters. Where is the beam brightest (largest intensity)?

38. Lasers. The intensity of a laser beam is given by the ratio of power to area. A particular laser beam has an intensity function given by $I = e^{-r^2}$ mW/cm^2, where r is the radius off the center axis given in centimeters. What percentage of the on-axis intensity ($r = 0$) corresponds to $r = 2$ cm?

39. Grade Distribution. Suppose the first test in this class has a normal, or bell-shaped, grade distribution of test scores, with an average score of 75. An approximate function that models your class's grades on test 1 is $N(x) = 10e^{-(x-75)^2/25^2}$, where N represents the number of students who received the score x.

a. Graph this function.

b. What is the average grade?

c. Approximately how many students scored a 50?

d. Approximately how many students scored 100?

40. Grade Distribution. Suppose the final exam in this class has a normal, or bell-shaped, grade distribution of exam scores, with an average score of 80. An approximate function that models your class's grades on the exam is $N(x) = 10e^{-(x-80)^2/16^2}$, where N represents the number of students who received the score x.

a. Graph this function.

b. What is the average grade?

c. Approximately how many students scored a 60?

d. Approximately how many students scored 100?

41. Time to Pay Off Debt. Diana just graduated from medical school owing $80,000 in student loans. The annual interest rate is 9%.

a. Approximately how many years will it take to pay off her student loan if she makes a monthly payment of $750?

b. Approximately how many years will it take to pay off her loan if she makes a monthly payment of $1,000?

42. Time to Pay Off Debt. Victor owes $20,000 on his credit card. The annual interest rate is 17%.

a. Approximately how many years will it take him to pay off this credit card if he makes a monthly payment of $300?

b. Approximately how many years will it take him to pay off this credit card if he makes a monthly payment of $400?

■ CATCH THE MISTAKE

In Exercises 43 and 44, explain the mistake that is made.

43. The city of Orlando, Florida, has a population that is growing at 7% a year, compounding continuously. If there were 1.1 million people in greater Orlando in 2006, approximately how many people will there be in 2016? Apply the formula $N = N_0 e^{rt}$, where N represents the number of people.

Solution:

Use the population growth model. $\quad N = N_0 e^{rt}$

Let $N_0 = 1.1$, $r = 7$, and $t = 10$. $\quad N = 1.1e^{(7)(10)}$

Approximate with a calculator. $\quad 2.8 \times 10^{30}$

This is incorrect. What mistake was made?

44. The city of San Antonio, Texas, has a population that is growing at 5% a year, compounding continuously. If there were 1.3 million people in the greater San Antonio area in 2006, approximately how many people will there be in 2016? Apply the formula $N = N_0 e^{rt}$, where N represents the number of people.

Solution:

Use the population growth model. $\quad N = N_0 e^{rt}$

Let $N_0 = 1.3$, $r = 5$, and $t = 10$. $\quad N = 1.3e^{(5)(10)}$

Approximate with a calculator. $\quad 6.7 \times 10^{21}$

This is incorrect. What mistake was made?

■ CONCEPTUAL

In Exercises 45–48, determine whether each statement is true or false.

45. When a species gets placed on an endangered species list, the species begins to grow rapidly, and then reaches a carrying capacity. This can be modeled by logistic growth.

46. A professor has 400 students one semester. The number of names (of her students) she is able to memorize can be modeled by a logarithmic function.

47. The spread of lice at an elementary school can be modeled by exponential growth.

48. If you purchase a laptop computer this year ($t = 0$), then the value of the computer can be modeled with exponential decay.

■ CHALLENGE

In Exercises 49 and 50, refer to the logistic model $f(t) = \dfrac{a}{1 + ce^{-kt}}$**, where a is the carrying capacity.**

49. As c increases, does the model reach the carrying capacity in less time or more time?

50. As k increases, does the model reach the carrying capacity in less time or more time?

51. A culture of 100 bacteria grows at a rate of 20% every day. Two days later, 60 of the same type of bacteria are placed in a culture that allows a 30% daily growth rate. After how many days do both cultures have the same population?

52. Consider the quotient $Q = \dfrac{P_1 e^{r_1 t}}{P_2 e^{r_2 t}}$ of two models of exponential growth.

a. If $r_1 > r_2$, what can you say about Q?

b. If $r_1 < r_2$, what can you say about Q?

53. Consider the models of exponential decay $f(t) = (2 + c)e^{-k_1 t}$ and $g(t) = ce^{-k_2 t}$. Suppose that $f(1) = g(1)$, what is the relationship between k_1 and k_2?

54. Suppose that both logistic growth models $f(t) = \dfrac{a_1}{1 + c_1 e^{-k_1 t}}$ and $g(t) = \dfrac{a_2}{1 + c_2 e^{-k_2 t}}$ have horizontal asymptote $y = 100$. What can you say about the corresponding carrying capacities?

■ **TECHNOLOGY**

55. Wing Shan just graduated from dental school owing $80,000 in student loans. The annual interest is 6%. Her time t to pay off the loan is given by

$$t = -\frac{\ln\left[1 - \dfrac{80{,}000(0.06)}{nR}\right]}{n\ln\left(1 + \dfrac{0.06}{n}\right)}$$

where n is the number of payment periods per year and R is the periodic payment.

a. Use a graphing utility to graph

$$t_1 = -\frac{\ln\left[1 - \dfrac{80{,}000(0.06)}{12x}\right]}{12\ln\left(1 + \dfrac{0.06}{12}\right)} \text{ as } Y_1 \text{ and}$$

$$t_2 = -\frac{\ln\left[1 - \dfrac{80{,}000(0.06)}{26x}\right]}{26\ln\left(1 + \dfrac{0.06}{26}\right)} \text{ as } Y_2.$$

Explain the difference in the two graphs.

b. Use the ⬚TRACE⬚ key to estimate the number of years that it will take Wing Shan to pay off her student loan if she can afford a monthly payment of $800.

c. If she can make a biweekly payment of $400, estimate the number of years that it will take her to pay off the loan.

d. If she adds $200 more to her monthly or $100 more to her biweekly payment, estimate the number of years that it will take her to pay off the loan.

56. Amy has a credit card debt in the amount of $12,000. The annual interest is 18%. Her time t to pay off the loan is given by

$$t = -\frac{\ln\left[1 - \dfrac{12{,}000(0.18)}{nR}\right]}{n\ln\left(1 + \dfrac{0.18}{n}\right)}$$

where n is the number of payment periods per year and R is the periodic payment.

a. Use a graphing utility to graph

$$t_1 = -\frac{\ln\left[1 - \dfrac{12{,}000(0.18)}{12x}\right]}{12\ln\left(1 + \dfrac{0.18}{12}\right)} \text{ as } Y_1 \text{ and}$$

$$t_2 = -\frac{\ln\left[1 - \dfrac{12{,}000(0.18)}{26x}\right]}{26\ln\left(1 + \dfrac{0.18}{26}\right)} \text{ as } Y_2.$$

Explain the difference in the two graphs.

b. Use the ⬚TRACE⬚ key to estimate the number of years that it will take Amy to pay off her credit card if she can afford a monthly payment of $300.

c. If she can make a biweekly payment of $150, estimate the number of years that it will take her to pay off the credit card.

d. If Amy adds $100 more to her monthly or $50 more to her biweekly payment, estimate the number of years that it will take her to pay off the credit card.

■ **PREVIEW TO CALCULUS**

In Exercises 57–60, refer to the following:

In calculus, we find the derivative, $f'(x)$, of a function $f(x)$ by allowing h to approach 0 in the difference quotient $\dfrac{f(x + h) - f(x)}{h}$ of functions involving exponential functions.

57. Find the difference quotient of the exponential growth model $f(x) = Pe^{kx}$, where P and k are positive constants.

58. Find the difference quotient of the exponential decay model $f(x) = Pe^{-kx}$, where P and k are positive constants.

59. Use the fact that $\dfrac{e^h - 1}{h} = 1$ when h is close to zero to find the derivative of $f(x) = e^x + x$.

60. Find the difference quotient of $f(x) = \cosh x$ and use it to prove that $(\cosh x)' = \sinh x$.

The following table summarizes the average yearly temperature in degrees Fahrenheit (°F) and carbon dioxide emissions in parts per million (ppm) for **Mauna Loa, Hawaii.**

YEAR	1960	1965	1970	1975	1980	1985	1990	1995	2000	2005
TEMPERATURE	44.45	43.29	43.61	43.35	46.66	45.71	45.53	47.53	45.86	46.23
CO$_2$ EMISSIONS (PPM)	316.9	320.0	325.7	331.1	338.7	345.9	354.2	360.6	369.4	379.7

In the Modeling Your World in Chapters 1 and 2, the temperature and carbon emissions were modeled with *linear functions* and *polynomial functions*, respectively. Now, let us model these same data using *exponential* and *logarithmic functions*.

1. Plot the temperature data, with time on the horizontal axis and temperature on the vertical axis. Let $t = 1$ correspond to 1960.

2. Find a *logarithmic function* with base e, $f(t) = A \ln Bt$, that models the temperature in Mauna Loa.

   a. Apply data from 1965 and 2005.   **b.** Apply data from 2000 and 2005.
   c. Apply regression and all data given.

3. Predict what the temperature will be in Mauna Loa in 2020.

   a. Use the line found in Exercise 2(a).   **b.** Use the line found in Exercise 2(b).
   c. Use the line found in Exercise 2(c).

4. Predict what the temperature will be in Mauna Loa in 2100.

   a. Use the line found in Exercise 2(a).   **b.** Use the line found in Exercise 2(b).
   c. Use the line found in Exercise 2(c).

5. Do your models support the claim of "global warming"? Explain. Do these logarithmic models give similar predictions to the linear models found in Chapter 1 and the polynomial models found in Chapter 2?

6. Plot the carbon dioxide emissions data, with time on the horizontal axis and carbon dioxide emissions on the vertical axis. Let $t = 0$ correspond to 1960.

7. Find an *exponential function* with base e, $f(t) = Ae^{bt}$, that models the CO$_2$ emissions (ppm) in Mauna Loa.

   a. Apply data from 1960 and 2005.   **b.** Apply data from 1960 and 2000.
   c. Apply regression and all data given.

8. Predict the expected CO$_2$ levels in Mauna Loa in 2020.

   a. Use the line found in Exercise 7(a).   **b.** Use the line found in Exercise 7(b).
   c. Use the line found in Exercise 7(c).

9. Predict the expected CO$_2$ levels in Mauna Loa in 2100.

   a. Use the line found in Exercise 7(a).   **b.** Use the line found in Exercise 7(b).
   c. Use the line found in Exercise 7(c).

10. Do your models support the claim of "global warming"? Explain. Do these exponential models give similar predictions to the linear models found in Chapter 1 or the polynomial models found in Chapter 2?

11. Comparing the models developed in Chapters 1–3, do you believe that global temperatures are best modeled with a linear, polynomial, or logarithmic function?

12. Comparing the models developed in Chapters 1–3, do you believe that CO$_2$ emissions are best modeled by linear, polynomial, or exponential functions?

SECTION	CONCEPT	PAGES	REVIEW EXERCISES	KEY IDEAS/POINTS
3.1	**Exponential functions and their graphs**	292–303	1–28	
	Evaluating exponential functions	292–293	1–12	$f(x) = b^x$ $\quad b > 0, b \neq 1$
	Graphs of exponential functions	294–297	13–24	y-intercept $(0, 1)$ $\quad$ Horizontal asymptote: $y = 0$; the points $(1, b)$ and $(-1, 1/b)$
	The natural base e	298–299	5–8, 21–24	$f(x) = e^x$
	Applications of exponential functions	299–303	25–28	Doubling time: $P = P_0 2^{t/d}$ Compound interest: $A = P\left(1 + \dfrac{r}{n}\right)^{nt}$ Compounded continuously: $A = Pe^{rt}$
3.2	**Logarithmic functions and their graphs**	307–317	29–60	$y = \log_b x \quad x > 0$ $b > 0, b \neq 1$
	Evaluating logarithms	307–309	29–40	$y = \log_b x$ and $x = b^y$
	Common and natural logarithms	309–310	41–44	$y = \log x \quad$ Common (base 10) $y = \ln x \quad$ Natural (base e)
	Graphs of logarithmic functions	310–313	49–56	x-intercept $(1, 0)$ $\quad$ Vertical asymptote: $x = 0$; the points $(b, 1)$ and $(1/b, -1)$
	Applications of logarithms	314–317	57–60	Decibel scale: $D = 10 \log\left(\dfrac{I}{I_T}\right) \quad I_T = 1 \times 10^{-12}$ W/m^2 Richter scale: $M = \dfrac{2}{3} \log\left(\dfrac{E}{E_0}\right) \quad E_0 = 10^{4.4}$ joules
3.3	**Properties of logarithms**	321–327	61–74	
	Properties of logarithms	321–325	61–74	1. $\log_b 1 = 0$ 2. $\log_b b = 1$ 3. $\log_b b^x = x$ 4. $b^{\log_b x} = x \quad x > 0$ Product property: 5. $\log_b MN = \log_b M + \log_b N$ Quotient property: 6. $\log_b\left(\dfrac{M}{N}\right) = \log_b M - \log_b N$ Power property: 7. $\log_b M^p = p \log_b M$
	Change-of-base formula	326–327	71–74	$\log_b M = \dfrac{\log M}{\log b} \quad$ or $\quad \log_b M = \dfrac{\ln M}{\ln b}$

CHAPTER REVIEW

3.1 Exponential Functions and Their Graphs

Approximate each number using a calculator and round your answer to two decimal places.

1. $8^{4.7}$ 2. $\pi^{2/5}$ 3. $4 \cdot 5^{0.2}$ 4. $1.2^{1.2}$

Approximate each number using a calculator and round your answer to two decimal places.

5. $e^{3.2}$ 6. e^{π} 7. $e^{\sqrt{\pi}}$ 8. $e^{-2.5\sqrt{3}}$

Evaluate each exponential function for the given values.

9. $f(x) = 2^{4-x}$ $f(-2.2)$
10. $f(x) = -2^{x+4}$ $f(1.3)$
11. $f(x) = \left(\frac{2}{5}\right)^{1-6x}$ $f\left(\frac{1}{2}\right)$
12. $f(x) = \left(\frac{4}{7}\right)^{5x+1}$ $f\left(\frac{1}{5}\right)$

Match the graph with the function.

13. $y = 2^{x-2}$ 14. $y = -2^{2-x}$
15. $y = 2 + 3^{x+2}$ 16. $y = -2 - 3^{2-x}$

a.

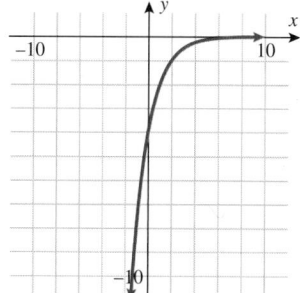

b.

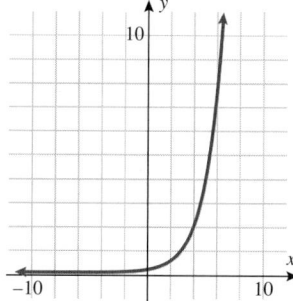

c.   d.

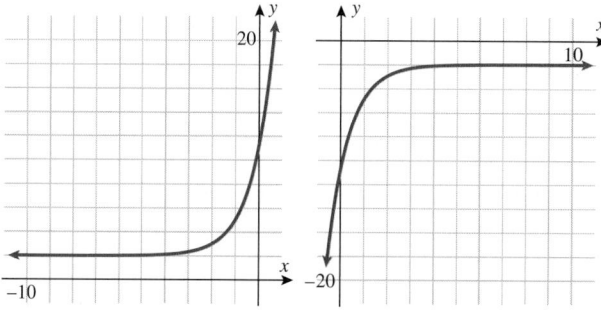

State the y-intercept and the horizontal asymptote, and graph the exponential function.

17. $y = -6^{-x}$ 18. $y = 4 - 3^x$

19. $y = 1 + 10^{-2x}$ 20. $y = 4^x - 4$

State the y-intercept and horizontal asymptote, and graph the exponential function.

21. $y = e^{-2x}$ 22. $y = e^{x-1}$

23. $y = 3.2e^{x/3}$ 24. $y = 2 - e^{1-x}$

Applications

25. **Compound Interest.** If $4,500 is deposited into an account paying 4.5% compounding semiannually, how much will you have in the account in 7 years?

26. **Compound Interest.** How much money should be put in a savings account now that earns 4.0% a year compounded quarterly if you want $25,000 in 8 years?

27. **Compound Interest.** If $13,450 is put in a money market account that pays 3.6% a year compounded continuously, how much will be in the account in 15 years?

28. **Compound Interest.** How much money should be invested today in a money market account that pays 2.5% a year compounded continuously if you desire $15,000 in 10 years?

3.2 Logarithmic Functions and Their Graphs

Write each logarithmic equation in its equivalent exponential form.

29. $\log_4 64 = 3$ 30. $\log_4 2 = \frac{1}{2}$

31. $\log\left(\frac{1}{100}\right) = -2$ 32. $\log_{16} 4 = \frac{1}{2}$

Write each exponential equation in its equivalent logarithmic form.

33. $6^3 = 216$ 34. $10^{-4} = 0.0001$

35. $\frac{4}{169} = \left(\frac{2}{13}\right)^2$ 36. $\sqrt[3]{512} = 8$

Evaluate the logarithms exactly.

37. $\log_7 1$ 38. $\log_4 256$

39. $\log_{1/6} 1296$ 40. $\log 10^{12}$

Approximate the common and natural logarithms utilizing a calculator. Round to two decimal places.

41. $\log 32$ 42. $\ln 32$

43. $\ln 0.125$ 44. $\log 0.125$

State the domain of the logarithmic function in interval notation.

45. $f(x) = \log_3(x + 2)$ 46. $f(x) = \log_2(2 - x)$

47. $f(x) = \log(x^2 + 3)$ 48. $f(x) = \log(3 - x^2)$

Match the graph with the function.

49. $y = \log_7 x$

50. $y = -\log_7(-x)$

51. $y = \log_7(x + 1) - 3$

52. $y = -\log_7(1 - x) + 3$

a.

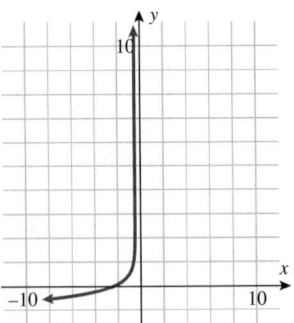

b.

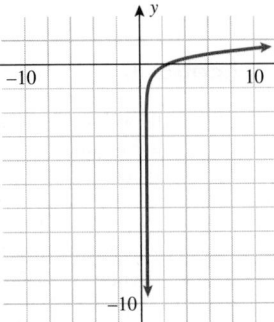

c.

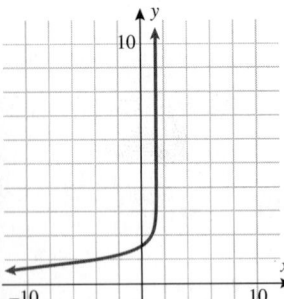

d.

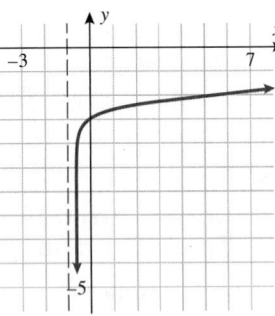

Graph the logarithmic function with transformation techniques.

53. $f(x) = \log_4(x - 4) + 2$

54. $f(x) = \log_4(x + 4) - 3$

55. $f(x) = -\log_4(x) - 6$

56. $f(x) = -2\log_4(-x) + 4$

Applications

57. Chemistry. Calculate the pH value of milk, assuming it has a concentration of hydrogen ions given by $H^+ = 3.16 \times 10^{-7}$.

58. Chemistry. Calculate the pH value of Coca-Cola, assuming it has a concentration of hydrogen ions given by $H^+ = 2.0 \times 10^{-3}$.

59. Sound. Calculate the decibels associated with a teacher speaking to a medium-sized class if the sound intensity is 1×10^{-7} W/m².

60. Sound. Calculate the decibels associated with an alarm clock if the sound intensity is 1×10^{-4} W/m².

3.3 Properties of Logarithms

Use the properties of logarithms to simplify each expression.

61. $\log_{2.5} 2.5$ **62.** $\log_2 \sqrt{16}$ **63.** $2.5^{\log_{2.5} 6}$ **64.** $e^{-3\ln 6}$

Write each expression as a sum or difference of logarithms.

65. $\log_c x^a y^b$

66. $\log_3 x^2 y^{-3}$

67. $\log_j\left(\dfrac{rs}{t^3}\right)$

68. $\log x^c \sqrt{x + 5}$

69. $\log\left[\dfrac{a^{1/2}}{b^{3/2} c^{2/5}}\right]$

70. $\log_7\left[\dfrac{c^3 d^{1/3}}{e^6}\right]^{1/3}$

Evaluate the logarithms using the change-of-base formula.

71. $\log_8 3$ **72.** $\log_5 \frac{1}{2}$ **73.** $\log_\pi 1.4$ **74.** $\log_{\sqrt{3}} 2.5$

3.4 Exponential and Logarithmic Equations

Solve the exponential equations exactly for x.

75. $4^x = \frac{1}{256}$ **76.** $3^{x^2} = 81$

77. $e^{3x-4} = 1$ **78.** $e^{\sqrt{x}} = e^{4.8}$

79. $\left(\frac{1}{3}\right)^{x+2} = 81$ **80.** $100^{x^2-3} = 10$

Solve the exponential equation. Round your answer to three decimal places.

81. $e^{2x+3} - 3 = 10$ **82.** $2^{2x-1} + 3 = 17$

83. $e^{2x} + 6e^x + 5 = 0$ **84.** $4e^{0.1x} = 64$

85. $(2^x - 2^{-x})(2^x + 2^{-x}) = 0$ **86.** $5(2^x) = 25$

Solve the logarithmic equations exactly.

87. $\log(3x) = 2$

88. $\log_3(x + 2) = 4$

89. $\log_4 x + \log_4 2x = 8$

90. $\log_6 x + \log_6(2x - 1) = \log_6 3$

Solve the logarithmic equations. Round your answers to three decimal places.

91. $\ln x^2 = 2.2$

92. $\ln(3x - 4) = 7$

93. $\log_3(2 - x) - \log_3(x + 3) = \log_3 x$

94. $4\log(x + 1) - 2\log(x + 1) = 1$

3.5 Exponential and Logarithmic Models

95. Compound Interest. If Tania needs $30,000 a year from now for a down payment on a new house, how much should she put in a 1-year CD earning 5% a year compounding continuously so that she will have exactly $30,000 a year from now?

96. Stock Prices. Jeremy is tracking the stock value of Best Buy (BBY on the NYSE). In 2003 he purchased 100 shares at $28 a share. The stock did not pay dividends because the company reinvested all earnings. In 2005 Jeremy cashed out and sold the stock for $4,000. What was the annual rate of return on BBY?

97. Compound Interest. Money is invested in a savings account earning 4.2% interest compounded quarterly. How many years will pass until the money doubles?

98. Compound Interest. If $9,000 is invested in an investment earning 8% interest compounded continuously, how many years will pass until there is $22,500?

99. Population. Nevada has the fastest-growing population according to the U.S. Census Bureau. In 2004 the population of Nevada was 2.62 million and increasing at an annual rate of 3.5%. What is the expected population in 2010? (Let $t = 0$ be 2004.) Apply the formula $N = N_0 e^{rt}$, where N is the population.

100. Population. The Hispanic population in the United States is the fastest growing of any ethnic group. In 1996 there were an estimated 28.3 million Hispanics in the United States, and in 2000 there were an estimated 32.5 million. What is the expected population of Hispanics in the United States in 2010? (Let $t = 0$ be 1996.) Apply the formula $N = N_0 e^{rt}$, where N is the population.

101. Bacteria Growth. Bacteria are growing exponentially. Initially, there were 1000 bacteria; after 3 hours there were 2500. How many bacteria should be expected in 6 hours? Apply the formula $N = N_0 e^{rt}$, where N is the number of bacteria.

102. Population. In 2003 the population of Phoenix, Arizona, was 1,388,215. In 2004 the population was 1,418,041. What is the expected population in 2010? (Let $t = 0$ be 2003.) Apply the formula $N = N_0 e^{rt}$, where N is the population.

103. Radioactive Decay. Strontium-90 has a half-life of 28 years. How long will it take for 20 grams of this to decay to 5 grams? Apply the formula $N = N_0 e^{-rt}$, where N is the number of grams.

104. Radioactive Decay. Plutonium-239 has a half-life of 25,000 years. How long will it take for 100 grams to decay to 20 grams? Apply the formula $N = N_0 e^{-rt}$, where N is the number of grams.

105. Wild Life Population. The *Boston Globe* reports that the fish population of the Essex River in Massachusetts is declining. In 2003 it was estimated there were 5600 fish in the river, and in 2004 there were only 2420 fish. How many fish should be expected in 2010 if this trend continues? Apply the formula $N = N_0 e^{-rt}$, where N is the number of fish.

106. Car Depreciation. A new Acura TSX costs $28,200. In 2 years the value will be $24,500. What is the expected value in 6 years? Apply the formula $N = N_0 e^{-rt}$, where N is the value of the car.

107. Carrying Capacity. The carrying capacity of a species of beach mice in St. Croix is given by $M = 1000\left(1 - e^{-0.035t}\right)$, where M is the number of mice and t is time in years ($t = 0$ corresponds to 1998). How many mice will there be in 2010?

108. Population. The city of Brandon, Florida, had 50,000 residents in 1970, and since the crosstown expressway was built, its population has increased 2.3% per year. If the growth continues at the same rate, how many residents will Brandon have in 2030?

Technology

Section 3.1

109. Use a graphing utility to graph the function
$f(x) = \left(1 + \dfrac{\sqrt{2}}{x}\right)^x$. Determine the horizontal asymptote as x increases.

110. Use a graphing utility to graph the functions $y = e^{-x+2}$ and $y = 3^x + 1$ in the same viewing screen. Estimate the coordinates of the point of intersection. Round your answers to three decimal places.

Section 3.2

111. Use a graphing utility to graph the functions $y = \log_{2.4}(3x - 1)$ and $y = \log_{0.8}(x - 1) + 3.5$ in the same viewing screen. Estimate the coordinates of the point of intersection. Round your answers to three decimal places.

112. Use a graphing utility to graph the functions $y = \log_{2.5}(x - 1) + 2$ and $y = 3.5^{x-2}$ in the same viewing screen. Estimate the coordinates of the point(s) of intersection. Round your answers to three decimal places.

Section 3.3

113. Use a graphing utility to graph $f(x) = \log_2\left(\dfrac{x^3}{x^2 - 1}\right)$ and $g(x) = 3 \log_2 x - \log_2(x + 1) - \log_2(x - 1)$ in the same viewing screen. Determine the domain where the two functions give the same graph.

114. Use a graphing utility to graph $f(x) = \ln\left(\dfrac{9 - x^2}{x^2 - 1}\right)$ and $g(x) = \ln(3 - x) + \ln(3 + x) - \ln(x + 1) - \ln(x - 1)$ in the same viewing screen. Determine the domain where the two functions give the same graph.

Section 3.4

115. Use a graphing utility to graph $y = \dfrac{e^x - e^{-x}}{e^x + e^{-x}}$. State the domain. Determine if there are any symmetry and asymptote.

116. Use a graphing utility to graph $y = \dfrac{1}{e^x - e^{-x}}$. State the domain. Determine if there are any symmetry and asymptote.

Section 3.5

117. A drug with initial dosage of 4 milligrams has a half-life of 18 hours. Let $(0, 4)$ and $(18, 2)$ be two points.

   a. Determine the equation of the dosage.

   b. Use $\boxed{\text{STAT}}$ $\boxed{\text{CALC}}$ $\boxed{\text{ExpReg}}$ to model the equation of the dosage.

   c. Are the equations in (a) and (b) the same?

118. In Exercise 105, let $t = 0$ be 2003 and $(0, 5600)$ and $(1, 2420)$ be the two points.

   a. Use $\boxed{\text{STAT}}$ $\boxed{\text{CALC}}$ $\boxed{\text{ExpReg}}$ to model the equation for the fish population.

   b. Using the equation found in (a), how many fish should be expected in 2010?

   c. Does the answer in (b) agree with the answer in Exercise 105?

CHAPTER 3 PRACTICE TEST

1. Simplify $\log 10^{x^3}$.

2. Use a calculator to evaluate $\log_5 326$ (round to two decimal places).

3. Find the exact value of $\log_{1/3} 81$.

4. Rewrite the expression $\ln\left[\dfrac{e^{5x}}{x(x^4 + 1)}\right]$ in a form with no logarithms of products, quotients, or powers.

In Exercises 5–20, solve for x, exactly if possible. If an approximation is required, round your answer to three decimal places.

5. $e^{x^2-1} = 42$

6. $e^{2x} - 5e^x + 6 = 0$

7. $27e^{0.2x+1} = 300$

8. $3^{2x-1} = 15$

9. $3\ln(x - 4) = 6$

10. $\log(6x + 5) - \log 3 = \log 2 - \log x$

11. $\ln(\ln x) = 1$

12. $\log_2(3x - 1) - \log_2(x - 1) = \log_2(x + 1)$

13. $\log_6 x + \log_6(x - 5) = 2$

14. $\ln(x + 2) - \ln(x - 3) = 2$

15. $\ln x + \ln(x + 3) = 1$

16. $\log_2\left(\dfrac{2x + 3}{x - 1}\right) = 3$

17. $\dfrac{12}{1 + 2e^x} = 6$

18. $\ln x + \ln(x - 3) = 2$

19. State the domain of the function $f(x) = \log\left(\dfrac{x}{x^2 - 1}\right)$.

20. State the range of x values for which the following is true: $10^{\log(4x-a)} = 4x - a$.

In Exercises 21–24, find all intercepts and asymptotes, and graph.

21. $f(x) = 3^{-x} + 1$

22. $f(x) = \left(\frac{1}{2}\right)^x - 3$

23. $f(x) = \ln(2x - 3) + 1$

24. $f(x) = \log(1 - x) + 2$

25. **Interest.** If $5,000 is invested at a rate of 6% a year, compounded quarterly, what is the amount in the account after 8 years?

26. **Interest.** If $10,000 is invested at a rate of 5%, compounded continuously, what is the amount in the account after 10 years?

27. **Sound.** A lawn mower's sound intensity is approximately 1×10^{-3} W/m². Assuming your threshold of hearing is 1×10^{-12} W/m², calculate the decibels associated with the lawn mower.

28. **Population.** The population in Seattle, Washington, has been increasing at a rate of 5% a year. If the population continues to grow at that rate, and in 2004 there are 800,000 residents, how many residents will there be in 2010? (*Hint:* $N = N_0e^{rt}$.)

29. **Earthquake.** An earthquake is considered moderate if it is between 5 and 6 on the Richter scale. What is the energy range in joules for a moderate earthquake?

30. **Radioactive Decay.** The mass $m(t)$ remaining after t hours from a 50-gram sample of a radioactive substance is given by the equation $m(t) = 50e^{-0.0578t}$. After how long will only 30 grams of the substance remain? Round your answer to the nearest hour.

31. **Bacteria Growth.** The number of bacteria in a culture is increasing exponentially. Initially, there were 200 in the culture. After 2 hours there are 500. How many should be expected in 8 hours? Round your answer to the nearest hundred.

32. **Carbon Decay.** Carbon-14 has a half-life of 5730 years. How long will it take for 100 grams to decay to 40 grams?

33. **Spread of a Virus.** The number of people infected by a virus is given by $N = \dfrac{2000}{1 + 3e^{-0.4t}}$, where t is time in days. In how many days will 1000 people be infected?

34. **Oil Consumtion.** The world consumption of oil was 76 million barrels per day in 2002. In 2004 the consumption was 83 million barrels per day. How many barrels are expected to be consumed in 2010?

35. Use a graphing utility to graph $y = \dfrac{e^x - e^{-x}}{2}$. State the domain. Determine if there are any symmetry and asymptote.

36. Use a graphing utility to help solve the equation $4^{3-x} = 2x - 1$. Round your answer to two decimal places.

1. Find the domain and range of the function $f(x) = \dfrac{3}{\sqrt{x^2 - 9}}$.

2. If $f(x) = 1 + 3x$ and $g(x) = x^2 - 1$, find

    a. $f + g$ **b.** $f - g$ **c.** $f \cdot g$ **d.** $\dfrac{f}{g}$

    and state the domain of each.

3. Write the function below as a composite of two functions f and g. (More than one answer is correct.)

$$f(g(x)) = \frac{1 - e^{2x}}{1 + e^{2x}}$$

4. Determine whether $f(x) = \sqrt[5]{x^3 + 1}$ is one-to-one. If f is one-to-one, find its inverse f^{-1}.

5. Find the quadratic function whose vertex is $(-2, 3)$ and goes through the point $(1, -1)$.

6. Write the polynomial $f(x) = 3x^3 + 6x^2 - 15x - 18$ as a product of linear factors.

7. Solve the equation $e^x + \sqrt{e^x} - 12 = 0$. Round your answer to three decimal places.

8. Using the function $f(x) = 4x - x^2$, evaluate the difference quotient $\dfrac{f(x + h) - f(x)}{h}$.

9. Given the piecewise-defined function

$$f(x) = \begin{cases} 5 & -2 < x \le 0 \\ 2 - \sqrt{x} & 0 < x < 4 \\ x - 3 & x \ge 4 \end{cases}$$

    find

    a. $f(4)$ **b.** $f(0)$ **c.** $f(1)$ **d.** $f(-4)$

    e. State the domain and range in interval notation.

    f. Determine the intervals where the function is increasing, decreasing, or constant.

10. Sketch the graph of the function $y = \sqrt{1 - x}$ and identify all transformations.

11. Determine whether the function $f(x) = \sqrt{x - 4}$ is one-to-one.

12. The volume of a cylinder with circular base is 400 in³. Its height is 10 in. Find its radius. Round your answer to three decimal places.

13. Find the vertex of the parabola associated with the quadratic function $f(x) = -4x^2 + 8x - 5$.

14. Find a polynomial of minimum degree (there are many) that has the zeros $x = -5$ (multiplicity 2) and $x = 9$ (multiplicity 4).

15. Use synthetic division to find the quotient $Q(x)$ and remainder $r(x)$ of $\left(3x^2 - 4x^3 - x^4 + 7x - 20\right) \div (x + 4)$.

16. Given the zero $x = 2 + i$ of the polynomial $P(x) = x^4 - 7x^3 + 13x^2 + x - 20$, determine all the other zeros and write the polynomial as the product of linear factors.

17. Find the vertical and slant asymptotes of $f(x) = \dfrac{x^2 + 7}{x - 3}$.

18. Graph the rational function $f(x) = \dfrac{3x}{x + 1}$. Give all asymptotes.

19. Graph the function $f(x) = 5x^2 (7 - x)^2(x + 3)$.

20. If \$5,400 is invested at 2.75% compounded monthly, how much is in the account after 4 years?

21. Give the exact value of $\log_3 243$.

22. Write the expression $\frac{1}{2} \ln(x + 5) - 2 \ln(x + 1) - \ln(3x)$ as a single logarithm.

23. Solve the logarithmic equation exactly: $10^{2 \log(4x+9)} = 121$.

24. Give an exact solution to the exponential equation $5^{x^2} = 625$.

25. If \$8,500 is invested at 4% compounded continuously, how many years will pass until there is \$12,000?

26. Use a graphing utility to help solve the equation $e^{3-2x} = 2^{x-1}$. Round your answer to two decimal places.

27. Strontium-90 with an initial amount of 6 grams has a half-life of 28 years.

    a. Use $\boxed{\text{STAT}}$ $\boxed{\text{CALC}}$ $\boxed{\text{ExpReg}}$ to model the equation of the amount remaining.

    b. How many grams will remain after 32 years? Round your answer to two decimal places.

4

Trigonometric Functions of Angles

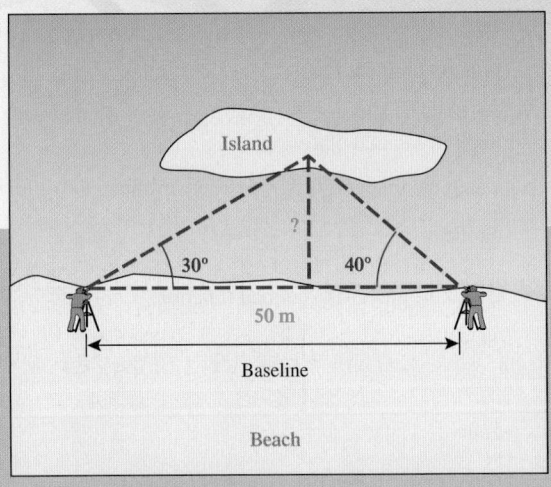

Surveyors use trigonometry to indirectly measure distances. Since angles are easier to measure than distances, surveyors set up a baseline between two stations and measure the distance between the two stations and the angles made by the baseline and some third station.

If there are two stations along the shoreline and the distance along the beach between the two stations is 50 meters and the angles between the baseline (beach) and the line of sight to the island are 30° and 40°, then the Law of Sines can be used to find the shortest distance from the beach to the island.*

*See Section 4.4, Exercises 47 and 48.

TRIGONOMETRIC FUNCTIONS OF ANGLES

4.1 Angle Measure	4.2 Right Triangle Trigonometry	4.3 Trigonometric Functions of Angles	4.4 The Law of Sines	4.5 The Law of Cosines
• Degrees and Radians • Coterminal Angles • Arc Length • Area of a Circular Sector • Linear and Angular Speeds	• Right Triangle Ratios • Evaluating Trigonometric Functions Exactly for Special Angle Measures • Solving Right Triangles	• Trigonometric Functions: The Cartesian Plane • Ranges of the Trigonometric Functions • Reference Angles and Reference Right Triangles • Evaluating Trigonometric Functions for Nonacute Angles	• Solving Oblique Triangles: Four Cases • The Law of Sines	• Solving Oblique Triangles Using the Law of Cosines • The Area of a Triangle

CHAPTER OBJECTIVES

- Understand angle measures in both degrees and radians, and convert between the two (Section 4.1).
- Find trigonometric function values for acute angles (Section 4.2).
- Find trigonometric function values for any (acute or nonacute) angle (Section 4.3).
- Use the Law of Sines to solve oblique triangles (Section 4.4).
- Use the Law of Cosines to solve oblique triangles (Section 4.5).

SKILLS OBJECTIVES

- Convert angle measure between degrees and radians.
- Find the complement or supplement of an angle.
- Identify coterminal angles.
- Calculate the length of an arc along a circle.
- Calculate the area of a circular sector.

CONCEPTUAL OBJECTIVES

- Understand that degrees and radians are both angle measures.
- Realize that radians are unitless (dimensionless).
- Understand the relationship between linear speed and angular speed.

An **angle** is formed when two **rays** share the same endpoint. The common endpoint is called the **vertex**.

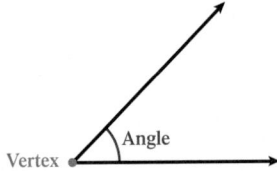

We say that an angle is formed when a ray is rotated around its endpoint. The ray in its original position is called the **initial ray** or the **initial side** of an angle. In the Cartesian plane, we assume the initial side of an angle is the positive *x*-axis. The ray after it is rotated is called the **terminal ray** or the **terminal side** of an angle. Rotation in a counterclockwise direction corresponds to a **positive angle**, whereas rotation in a clockwise direction corresponds to a **negative angle**.

Study Tip

Positive angle: counterclockwise
Negative angle: clockwise

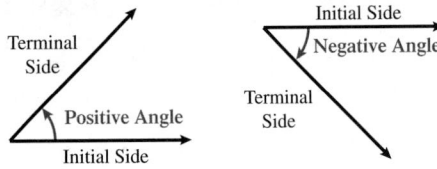

Degrees and Radians

Lengths, or distances, can be measured in different units: feet, miles, and meters are three common units. In order to compare angles of different sizes, we need a standard unit of measure. One way to measure the size of an angle is with **degree measure**.

DEFINITION   **Degree Measure of Angles**

An angle formed by one complete counterclockwise rotation has **measure 360 degrees**, denoted 360°.

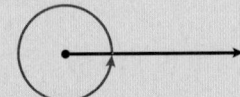

One complete counterclockwise revolution = 360°

WORDS   **MATH**

360° represents 1 complete counterclockwise rotation. $\dfrac{360°}{360°} = 1$

180° represents a $\frac{1}{2}$ counterclockwise rotation. $\dfrac{180°}{360°} = \dfrac{1}{2}$

90° represents a $\frac{1}{4}$ counterclockwise rotation. $\dfrac{90°}{360°} = \dfrac{1}{4}$

1° represents a $\frac{1}{360}$ counterclockwise rotation. $\dfrac{1°}{360°} = \dfrac{1}{360}$

The Greek letter θ (theta) is the most common name for an angle in mathematics. Other common names of angles are α (alpha), β (beta), and γ (gamma).

WORDS   **MATH**

An angle measuring exactly 90° is called a **right angle**.

A right angle is often represented by the adjacent sides of a rectangle, indicating that the two rays are *perpendicular*.

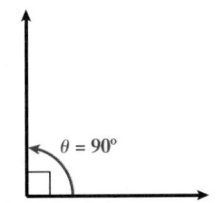

Right Angle: quarter rotation

An angle measuring exactly 180° is called a **straight angle**.

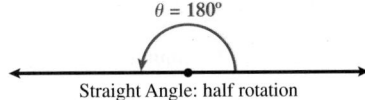

Straight Angle: half rotation

An angle measuring greater than 0°, but less than 90°, is called an **acute angle**.

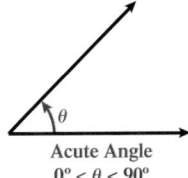

Acute Angle
$0° < \theta < 90°$

An angle measuring greater than 90°, but less than 180°, is called an **obtuse angle**.

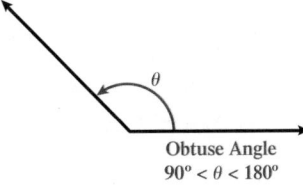

Obtuse Angle
$90° < \theta < 180°$

If the sum of the measures of two positive angles is 90°, the angles are called **complementary**. We say that α is the **complement** of β (and vice versa).

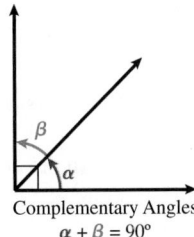

Complementary Angles
$\alpha + \beta = 90°$

If the sum of the measures of two positive angles is 180°, the angles are called **supplementary**. We say that α is the **supplement** of β (and vice versa).

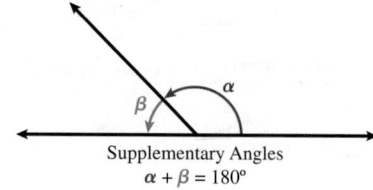

Supplementary Angles
$\alpha + \beta = 180°$

EXAMPLE 1 Finding Measures of Complementary and Supplementary Angles

Find the measure of each angle.

a. Find the complement of 50°.

b. Find the supplement of 110°.

c. Represent the complement of α in terms of α.

d. Find two supplementary angles such that the first angle is twice as large as the second angle.

Solution:

a. The sum of complementary angles is 90°. $\theta + 50° = 90°$

   Solve for θ. $\theta = \boxed{40°}$

b. The sum of supplementary angles is 180°. $\theta + 110° = 180°$

   Solve for θ. $\theta = \boxed{70°}$

c. Let β be the complement of α.

   The sum of complementary angles is 90°. $\alpha + \beta = 90°$

   Solve for β. $\beta = \boxed{90° - \alpha}$

d. The sum of supplementary angles is 180°. $\alpha + \beta = 180°$

   Let $\beta = 2\alpha$. $\alpha + 2\alpha = 180°$

   Solve for α. $3\alpha = 180°$

                                                        $\alpha = 60°$

   Substitute $\alpha = 60°$ into $\beta = 2\alpha$. $\beta = 120°$

   The angles have measures $\boxed{60°}$ and $\boxed{120°}$.

■ **Answer:** The angles have measures 45° and 135°.

■ **YOUR TURN** Find two supplementary angles such that the first angle is three times as large as the second angle.

It is important not to confuse an angle with its measure. In Example 1(d), angle α is a rotation and the measure of that rotation is 60°.

In geometry and most everyday applications, angles are measured in degrees. However, in calculus a more natural angle measure is *radian measure*. Using radian measure allows us to write trigonometric functions as functions of not only angles but also real numbers in general.

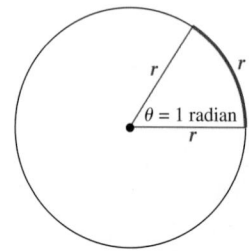

Now we think of the angle in the context of a circle. A **central angle** is an angle that has its vertex at the center of a circle. When the intercepted arc's length is equal to the radius, the measure of the central angle is 1 **radian**.

DEFINITION   Radian Measure

If a central angle θ in a circle with radius r intercepts
an arc on the circle of length s, then the measure of θ,
in **radians**, is given by

$$\theta(\text{in radians}) = \frac{s}{r}$$

Note: The formula is valid only if s (arc length) and r
(radius) are expressed in the same units.

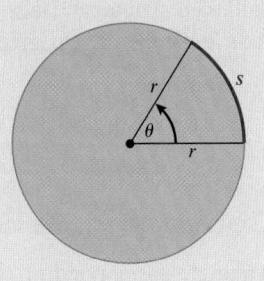

▼ **CAUTION**

To correctly calculate radians from

the formula $\theta = \frac{s}{r}$, the radius and

arc length must be expressed in the
same units.

Note that both s and r are measured in units of length. When both are given in the same units,
the units cancel, giving the number of radians as a *dimensionless* (unitless) real number. One
full rotation corresponds to an arc length equal to the circumference $2\pi r$ of the circle with
radius r. We see then that one full rotation is equal to 2π radians.

$$\theta_{\text{full rotation}} = \frac{2\pi r}{r} = 2\pi$$

EXAMPLE 2   Finding the Radian Measure of an Angle

What is the measure (in radians) of a central angle θ that intercepts an arc of length 6
centimeters on a circle with radius 2 meters?

COMMON MISTAKE

A common mistake is to forget to first put the radius and arc length in the same units.

⭐ **CORRECT**

Write the formula relating radian
measure to arc length and radius.

$$\theta(\text{in radians}) = \frac{s}{r}$$

Substitute $s = 6$ cm and $r = 2$ m
into the radian expression.

$$\theta = \frac{6\,\text{cm}}{2\,\text{m}}$$

Convert the radius (2) meters to
centimeters: 2 m = 200 cm.

$$\theta = \frac{6\,\text{cm}}{200\,\text{cm}}$$

The units, cm, cancel and the
result is a unitless real number.

$$\theta = 0.03\,\text{rad}$$

❌ **INCORRECT**

Substitute $s = 6$ cm and $r = 2$ m into
the radian expression.

$$\theta = \frac{6\,\text{cm}}{2\,\text{m}}$$

Simplify. $\theta = 3\,\text{rad}$

ERROR

▼ **CAUTION**

Units for arc length and radius must

be the same to use $\theta = \frac{s}{r}$.

■ **Answer:** 0.3 rad

■ **YOUR TURN**   What is the measure (in radians) of a central angle θ that intercepts
an arc of length 12 mm on a circle with radius 4 cm?

In the above example, the units, cm, canceled, therefore correctly giving *radians* as a unitless real number. Because radians are unitless, the word radians (or rad) is often omitted. If an angle measure is given simply as a real number, then radians are implied.

WORDS	MATH
The measure of θ is 4 degrees.	$\theta = 4°$
The measure of θ is 4 radians.	$\theta = 4$

Converting Between Degrees and Radians

An angle corresponding to one full rotation is said to have measure 360° or 2π radians. Therefore, $180° = \pi$ rad.

- To convert degrees to radians, multiply the degree measure by $\dfrac{\pi}{180°}$.

- To convert radians to degrees, multiply the radian measure by $\dfrac{180°}{\pi}$.

EXAMPLE 3   Converting Between Degrees and Radians

Convert:

a. 45° to radians   **b.** 472° to radians   **c.** $\dfrac{2\pi}{3}$ to degrees

Solution (a):

Multiply 45° by $\dfrac{\pi}{180°}$.

$$(45°)\left(\dfrac{\pi}{180°}\right) = \dfrac{45°\pi}{180°}$$

Simplify.

$$= \dfrac{\pi}{4} \text{ radians}$$

Note: $\dfrac{\pi}{4}$ is the exact value. A calculator can be used to approximate this expression.

Scientific and graphing calculators have a π button (on most scientific calculators, it requires using a shift or second command). The decimal approximation rounded to three decimal places is 0.785.

Exact value: $\boxed{\dfrac{\pi}{4}}$

Approximate value: $\boxed{0.785}$

Solution (b):

Multiply 472° by $\dfrac{\pi}{180°}$.

$$472°\left(\dfrac{\pi}{180°}\right)$$

Simplify (factor out the common 4).

$$= \boxed{\dfrac{118}{45}\pi}$$

Approximate with a calculator.

$$\boxed{\approx 8.238}$$

Solution (c):

Multiply $\dfrac{2\pi}{3}$ by $\dfrac{180°}{\pi}$.

$$\dfrac{2\pi}{3} \cdot \dfrac{180°}{\pi}$$

Simplify.

$$= \boxed{120°}$$

■ **Answer:**

a. $\dfrac{\pi}{3}$ or approximately 1.047

b. $\dfrac{23}{9}\pi$ or approximately 8.029

c. 270°

■ **YOUR TURN** Convert:

a. 60° to radians   **b.** 460° to radians   **c.** $\dfrac{3\pi}{2}$ to degrees

Coterminal Angles

Angles in Standard Position

If the *initial side* of an angle is aligned along the *positive x-axis* and the *vertex* of the angle is positioned at the *origin*, then the angle is said to be in *standard position*.

DEFINITION	**Standard Position**

An angle is said to be in **standard position** if its initial side is along the positive *x*-axis and its vertex is at the origin.

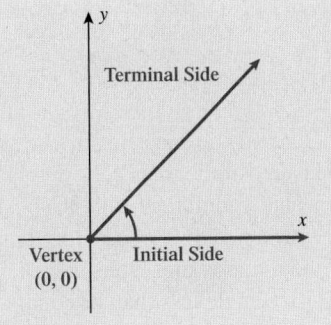

We say that an angle lies in the quadrant in which its terminal side lies. Angles in standard position with terminal sides along the *x*-axis or *y*-axis ($90°, 180°, 270°, 360°$, etc.) are called **quadrantal angles**.

$$90° \text{ or } \frac{\pi}{2}$$

$90° < \theta < 180°$	$0° < \theta < 90°$
QII	QI
$\frac{\pi}{2} < \theta < \pi$	$0 < \theta < \frac{\pi}{2}$

$180° \text{ or } \pi$

$0° \text{ or } 0;$
$360° \text{ or } 2\pi$

$180° < \theta < 270°$	$270° < \theta < 360°$
QIII	QIV
$\pi < \theta < \frac{3\pi}{2}$	$\frac{3\pi}{2} < \theta < 2\pi$

$$270° \text{ or } \frac{3\pi}{2}$$

Coterminal Angles

DEFINITION	**Coterminal Angles**

Two angles in standard position with the same terminal side are called **coterminal angles**.

For example, $-40°$ and $320°$ are measures of coterminal angles; their terminal rays are identical even though they are formed by rotations in opposite directions. The angles $60°$ and $420°$ are also coterminal; angles larger than $360°$ or less than $-360°$ are generated by continuing the rotation beyond one full circle. Thus, all coterminal angles have the same initial side (positive *x*-axis) and the same terminal side, just different amounts and/or direction of rotation.

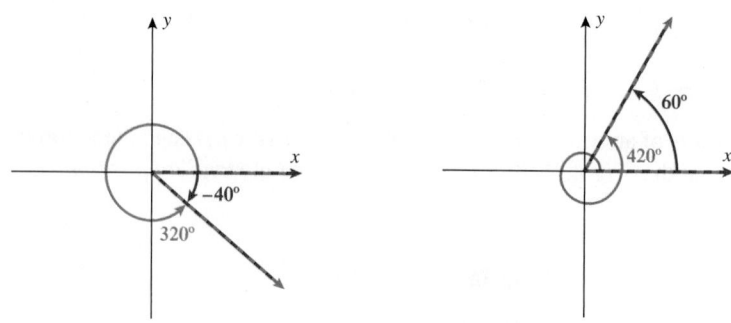

To find the measure of the smallest nonnegative coterminal angle of a given angle measured in degrees follow this procedure. If the given angle is positive, subtract 360° repeatedly until the result is a positive angle less than or equal to 360°. If the given angle is negative, add 360° repeatedly until the result is a positive angle less than or equal to 360°. If the angle is measured in radians, subtract or add equivalently 2π until your result is a positive angle less than or equal to 2π.

EXAMPLE 4   Finding Measures of Coterminal Angles

Determine the angle with the smallest possible positive measure that is coterminal with each of the following angles:

a. $830°$ **b.** $-520°$ **c.** $\dfrac{11\pi}{3}$

Solution (a):

Since $830°$ is positive, subtract $360°$.	$830° - 360° = 470°$
Subtract $360°$ again.	$470° - 360° = \boxed{110°}$

Solution (b):

Since $-520°$ is negative, add $360°$.	$-520° + 360° = -160°$
Add $360°$ again.	$-160° + 360° = \boxed{200°}$

Solution (c):

Since $\dfrac{11\pi}{3}$ is positive, subtract 2π.	$\dfrac{11\pi}{3} - 2\pi = \boxed{\dfrac{5\pi}{3}}$

■ **Answer: a.** $180°$
   b. $290°$
   c. $\dfrac{11\pi}{6}$

■ **YOUR TURN** Determine the angle with the smallest possible positive measure that is coterminal with each of the following angles:

a. $900°$ **b.** $-430°$ **c.** $-\dfrac{13\pi}{6}$

Applications of Radian Measure

We now look at applications of radian measure that involve calculating *arc lengths*, *areas of circular sectors*, and *angular and linear speeds*. All of these applications are related to the definition of radian measure.

Arc Length

Study Tip

To use the relationship

$$s = r\theta$$

the angle θ must be in radians.

DEFINITION   **Arc Length**

If a central angle θ in a circle with radius r intercepts an arc on the circle of length s, then the **arc length** s is given by

$$s = r\theta \qquad\qquad \theta \text{ is given in radians}$$

EXAMPLE 5   **Finding Arc Length When the Angle Has Degree Measure**

The International Space Station (ISS) is in an approximately circular orbit 400 km above the surface of the Earth. If the ground station tracks the space station when it is within a 45° central angle of this circular orbit above the tracking antenna, how many kilometers does the ISS cover while it is being tracked by the ground station? Assume that the radius of the Earth is 6400 kilometers. Round to the nearest kilometer.

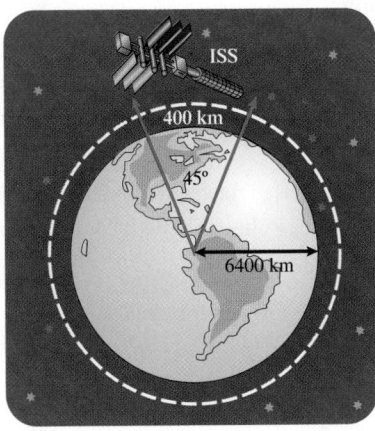

ISS
400 km
45°
6400 km

Technology Tip

Use the TI to evaluate the expression for s.

$$s = (6800 \text{ km})(45°)\left(\frac{\pi}{180°}\right)$$

Press 2nd ∧ for π. Type
6800 × 45 × 2nd ∧ ÷
180 ENTER .

```
6800*45*π/180
        5340.707511
```

Solution:

Write the formula for arc length when the angle has degree measure.	$s = r\theta_d\left(\dfrac{\pi}{180°}\right)$
Substitute $r = 6400 + 400 = 6800$ km and $\theta_d = 45°$.	$s = (6800 \text{ km})(45°)\left(\dfrac{\pi}{180°}\right)$
Evaluate with a calculator.	$s \approx 5340.708$ km
Round to the nearest kilometer.	$s \approx \boxed{5341 \text{ km}}$

The ISS travels approximately 5341 km during the ground station tracking.

Study Tip

When the angle is given in degrees, the arc length formula becomes

$$s = r \cdot \theta_d\left(\frac{\pi}{180°}\right)$$

■ **YOUR TURN** If the ground station in Example 5 could track the ISS within a 60° central angle, how far would the ISS travel during the tracking?

■ **Answer:** 7121 km

Area of a Circular Sector

DEFINITION   **Area of a Circular Sector**

The **area of a sector** of a circle with radius r and central angle θ is given by

$$A = \frac{1}{2}r^2\theta \qquad\qquad \theta \text{ is given in radians}$$

Study Tip

To use the relationship

$$A = \frac{1}{2}r^2\theta$$

the angle θ must be in radians.

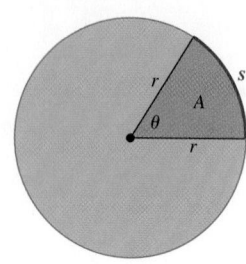

WORDS	MATH
Write the ratio of the area of the sector to the area of the entire circle.	$\dfrac{A}{\pi r^2}$
Write the ratio of the central angle θ to the measure of one full rotation.	$\dfrac{\theta}{2\pi}$
The ratios must be equal (proportionality of sector to circle).	$\dfrac{A}{\pi r^2} = \dfrac{\theta}{2\pi}$
Multiply both sides of the equation by πr^2.	$\pi r^2 \cdot \dfrac{A}{\pi r^2} = \dfrac{\theta}{2\pi} \cdot \pi r^2$
Simplify.	$A = \dfrac{1}{2}r^2\theta$

EXAMPLE 6 **Finding the Area of a Sector When the Angle Has Degree Measure**

Sprinkler heads come in all different sizes depending on the angle of rotation desired. If a sprinkler head rotates 90° and has enough pressure to keep a constant 25-foot spray, what is the area of the sector of the lawn that gets watered? Round to the nearest square foot.

Solution:

Write the formula for circular sector area in degrees.	$A = \dfrac{1}{2}r^2\theta_d\left(\dfrac{\pi}{180°}\right)$
Substitute $r = 25$ ft and $\theta_d = 90°$ into the area equation.	$A = \dfrac{1}{2}(25 \text{ ft})^2(90°)\left(\dfrac{\pi}{180°}\right)$
Simplify.	$A = \left(\dfrac{625\pi}{4}\right) \text{ ft}^2 \approx 490.87 \text{ ft}^2$
Round to the nearest square foot.	$A \approx \boxed{491 \text{ ft}^2}$

■ **Answer:** 1414 ft²

■ **YOUR TURN** If a sprinkler head rotates 180° and has enough pressure to keep a constant 30-foot spray, what is the area of the sector of the lawn it can water? Round to the nearest square foot.

Linear and Angular Speeds

Recall the relationship between distance, rate (assumed to be constant), and time: $d = rt$. Rate is speed, and in words this formula can be rewritten as

$$\text{speed} = \frac{\text{distance}}{\text{time}}$$

It is important to note that we assume speed is constant. If we think of a car driving around a circular track, the distance it travels is the arc length s; and if we let v represent speed and t represent time, we have the formula for speed along a circular path (*linear speed*):

$$v = \frac{s}{t}$$

DEFINITION   Linear Speed

If a point P moves along the circumference of a circle at a constant speed, then the **linear speed** v is given by

$$v = \frac{s}{t}$$

where s is the arc length and t is the time.

EXAMPLE 7   Linear Speed

A car travels at a constant speed around a circular track with circumference equal to 2 miles. If the car records a time of 15 minutes for 9 laps, what is the linear speed of the car in miles per hour?

Solution:

Calculate the distance traveled around the circular track.

$$s = (9 \, \text{laps})\left(\frac{2 \, \text{miles}}{\text{lap}}\right) = 18 \, \text{miles}$$

Substitute $t = 15$ minutes and $s = 18$ miles into $v = \frac{s}{t}$.

$$v = \frac{18 \, \text{miles}}{15 \, \text{minutes}}$$

Convert the linear speed from miles per minute to miles per hour.

$$v = \left(\frac{18 \, \text{miles}}{15 \, \text{minutes}}\right)\left(\frac{60 \, \text{minutes}}{1 \, \text{hour}}\right)$$

Simplify.

$$v = \boxed{72 \, \text{miles per hour}}$$

■ **Answer:** 105 mph

■ **YOUR TURN** A car travels at a constant speed around a circular track with circumference equal to 3 miles. If the car records a time of 12 minutes for 7 laps, what is the linear speed of the car in miles per hour?

To calculate linear speed, we find how fast a position along the circumference of a circle is changing. To calculate *angular speed*, we find how fast the central angle is changing.

DEFINITION   Angular Speed

If a point P moves along the circumference of a circle at a constant speed, then the central angle θ that is formed with the terminal side passing through point P also changes over some time t at a constant speed. The **angular speed** ω (omega) is given by

$$\omega = \frac{\theta}{t} \qquad \text{where } \theta \text{ is given in radians}$$

EXAMPLE 8   Angular Speed

A lighthouse in the middle of a channel rotates its light in a circular motion with constant speed. If the beacon of light completes 1 rotation every 10 seconds, what is the angular speed of the beacon in radians per minute?

Solution:

Calculate the angle measure in radians associated with 1 rotation.

$$\theta = 2\pi$$

Substitute $\theta = 2\pi$ and $t = 10$ seconds into $\omega = \dfrac{\theta}{t}$.

$$\omega = \frac{2\pi \,(\text{radians})}{10\ \text{seconds}}$$

Convert the angular speed from radians per second to radians per minute.

$$\omega = \frac{2\pi \,(\text{radians})}{10\ \text{seconds}} \cdot \frac{60\ \text{seconds}}{1\ \text{minute}}$$

Simplify.

$$\omega = \boxed{12\pi\ \text{rad/min}}$$

■ **Answer:** $\omega = 3\pi\,\text{rad/min}$

■ **YOUR TURN** If the lighthouse in Example 8 is adjusted so that the beacon rotates 1 time every 40 seconds, what is the angular speed of the beacon in radians per minute?

Relationship Between Linear and Angular Speeds

Angular speed and *linear speed* are related through the *radius*.

WORDS	MATH
Write the definition of radian measure.	$\theta = \dfrac{s}{r}$
Write the definition of arc length (θ in radians).	$s = r\theta$
Divide both sides by t.	$\dfrac{s}{t} = \dfrac{r\theta}{t}$
Rewrite the right side of the equation.	$\dfrac{s}{t} = r\dfrac{\theta}{t}$
Recall the definitions of **linear** and **angular** speeds.	$v = \dfrac{s}{t}$ and $\omega = \dfrac{\theta}{t}$
Substitute $v = \dfrac{s}{t}$ and $\omega = \dfrac{\theta}{t}$ into $\dfrac{s}{t} = r\dfrac{\theta}{t}$.	$v = r\omega$

Study Tip

This relationship between linear and angular speed assumes the angle is given in radians.

RELATING LINEAR AND ANGULAR SPEEDS

If a point P moves at a constant speed along the circumference of a circle with radius r, then the **linear speed** v and the **angular speed** ω are related by

$$v = r\omega \qquad \text{or} \qquad \omega = \frac{v}{r}$$

Note: This relationship is true only when θ is given in radians.

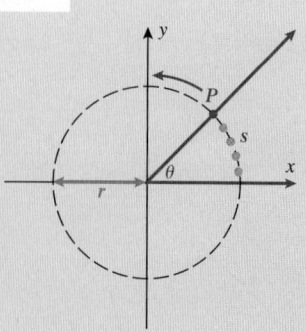

Notice that tires of two different radii with the same angular speed have different linear speeds. The larger tire has the faster linear speed.

 EXAMPLE 9 **Relating Linear and Angular Speeds**

A Ford F-150 truck comes standard with tires that have a diameter of 25.7 inches (17″ rims). If the owner decides to upgrade to tires with a diameter of 28.2 inches (19″ rims) without having the onboard computer updated, how fast will the truck *actually* be traveling when the speedometer reads 75 mph?

14.1 in.
12.85 in.

Solution:

The computer onboard the F-150 "thinks" the tires are 25.7 inches in diameter and knows the angular speed. Use the programmed tire diameter and speedometer reading to calculate the angular speed. Then use that angular speed and the upgraded tire diameter to get the actual speed (linear speed).

Write the formula for the angular speed.
$$\omega = \frac{v}{r}$$

Substitute $v = 75\,\dfrac{\text{miles}}{\text{hour}}$ and
$r = \dfrac{25.7}{2} = 12.85$ inches into the formula.
$$\omega = \frac{75\,\dfrac{\text{miles}}{\text{hour}}}{12.85\text{ inches}}$$

1 mile $= 5280$ feet $= 63,360$ inches.
$$\omega = \frac{75(63,360)\,\dfrac{\text{inches}}{\text{hour}}}{12.85\text{ inches}}$$

Simplify.
$$\omega \approx 369,805\,\frac{\text{radians}}{\text{hour}}$$

Write the linear speed formula.
$$v = r\omega$$

Substitute $r = \dfrac{28.2}{2} = 14.1$ inches
and $\omega \approx 369,805\,\dfrac{\text{radians}}{\text{hour}}$.
$$v = (14.1\text{ inches})\left(369,805\,\frac{\text{radians}}{\text{hour}}\right)$$

Simplify.
$$v \approx 5,214,251\,\frac{\text{inches}}{\text{hour}}$$

1 mile $= 5280$ feet $= 63,360$ inches.
$$v \approx \frac{5,214,251\,\dfrac{\text{inches}}{\text{hour}}}{63,360\,\dfrac{\text{inches}}{\text{mile}}}$$

$$v \approx \boxed{82.296\,\frac{\text{miles}}{\text{hour}}}$$

Study Tip

We could have solved Example 9 the following way:
$$\frac{75\text{ mph}}{25.7} = \frac{x}{28.2}$$
$$x = \left(\frac{28.2}{25.7}\right)(75)$$
$$\approx 82.296\text{ mph}$$

Although the speedometer indicates a speed of 75 mph, the actual speed is approximately $\boxed{82\text{ mph}}$.

■ Answer: approximately 62 mph

■ YOUR TURN Suppose the owner of the Ford F-150 truck in Example 9 decides to downsize the tires from their original 25.7-inch diameter to a 24.4-inch diameter. If the speedometer indicates a speed of 65 mph, what is the actual speed of the truck?

SECTION 4.1 SUMMARY

Angle measures can be converted between degrees and radians in the following way:

- To convert degrees to radians, multiply the degree measure by $\frac{\pi}{180°}$.
- To convert radians to degrees, multiply the radian measure by $\frac{180°}{\pi}$.

(Remember that $\pi = 180°$.)

Coterminal angles in standard position have terminal sides that coincide.

The length of a circular arc is given by $s = r\theta$, where θ is the central angle given in radians and r is the radius of the circle.

The area of a circular sector is given by $A = \frac{1}{2}r^2\theta$, where θ is the central angle given in radians and r is the radius of the circle.

Linear speed, $v = \frac{s}{t}$, and angular speed, $\omega = \frac{\theta}{t}$, are related through the radius: $v = r\omega$.

SECTION 4.1 EXERCISES

SKILLS

In Exercises 1–6, find (a) the complement and (b) the supplement of the given angles.

1. 18°　　**2.** 39°　　**3.** 42°　　**4.** 57°　　**5.** 89°　　**6.** 75°

In Exercises 7–12, find the measure (in radians) of a central angle θ that intercepts an arc of length s on a circle with radius r.

7. $r = 22$ in., $s = 4$ in.　　**8.** $r = 6$ in., $s = 1$ in.　　**9.** $r = 100$ cm, $s = 20$ mm

10. $r = 1$ m, $s = 2$ cm　　**11.** $r = \frac{1}{4}$ in., $s = \frac{1}{32}$ in.　　**12.** $r = \frac{3}{4}$ cm, $s = \frac{3}{14}$ cm

In Exercises 13–28, convert from degrees to radians. Leave the answers in terms of π.

13. 30°　**14.** 60°　**15.** 45°　**16.** 90°　**17.** 315°　**18.** 270°

19. 75°　**20.** 100°　**21.** 170°　**22.** 340°　**23.** 780°　**24.** 540°

25. −210°　**26.** −320°　**27.** −3600°　**28.** 1800°

In Exercises 29–42, convert from radians to degrees.

29. $\frac{\pi}{6}$　**30.** $\frac{\pi}{4}$　**31.** $\frac{3\pi}{4}$　**32.** $\frac{7\pi}{6}$　**33.** $\frac{3\pi}{8}$　**34.** $\frac{11\pi}{9}$　**35.** $\frac{5\pi}{12}$

36. $\frac{7\pi}{3}$　**37.** 9π　**38.** -6π　**39.** $\frac{19\pi}{20}$　**40.** $\frac{13\pi}{36}$　**41.** $-\frac{7\pi}{15}$　**42.** $-\frac{8\pi}{9}$

In Exercises 43–50, convert from radians to degrees. Round your answers to the nearest hundredth of a degree.

43. 4　　**44.** 3　　**45.** 0.85　　**46.** 3.27

47. −2.7989　　**48.** −5.9841　　**49.** $2\sqrt{3}$　　**50.** $5\sqrt{7}$

In Exercises 51–56, convert from degrees to radians. Round your answers to three significant digits.

51. 47°　**52.** 65°　**53.** 112°　**54.** 172°　**55.** 56.5°　**56.** 298.7°

In Exercises 57–68, state in which quadrant or on which axis each angle with the given measure in standard position would lie.

57. 145°　**58.** 175°　**59.** 270°　**60.** 180°　**61.** −540°　**62.** −450°

63. $\frac{2\pi}{5}$　**64.** $\frac{4\pi}{7}$　**65.** $\frac{13\pi}{4}$　**66.** $\frac{18\pi}{11}$　**67.** 2.5　**68.** 11.4

In Exercises 69–80, determine the angle of the smallest possible positive measure that is coterminal with each of the angles whose measure is given. Use degree or radian measures accordingly.

69. $412°$ **70.** $379°$ **71.** $-92°$ **72.** $-187°$ **73.** $-390°$ **74.** $945°$

75. $\dfrac{29\pi}{3}$ **76.** $\dfrac{47\pi}{7}$ **77.** $-\dfrac{313\pi}{9}$ **78.** $-\dfrac{217\pi}{4}$ **79.** -30 **80.** 42

In Exercises 81–88, find the exact length of the arc made by the indicated central angle and radius of each circle.

81. $\theta = \dfrac{\pi}{12}, r = 8$ ft   **82.** $\theta = \dfrac{\pi}{8}, r = 6$ yd   **83.** $\theta = \frac{1}{2}, r = 5$ in.   **84.** $\theta = \frac{3}{4}, r = 20$ m

85. $\theta = 22°, r = 18\ \mu$m   **86.** $\theta = 14°, r = 15\ \mu$m   **87.** $\theta = 8°, r = 1500$ km   **88.** $\theta = 3°, r = 1800$ km

In Exercises 89–94, find the area of the circular sector given the indicated radius and central angle. Round your answers to three significant digits.

89. $\theta = \dfrac{3\pi}{8}, r = 2.2$ km   **90.** $\theta = \dfrac{5\pi}{6}, r = 13$ mi   **91.** $\theta = 56°, r = 4.2$ cm

92. $\theta = 27°, r = 2.5$ mm   **93.** $\theta = 1.2°, r = 1.5$ ft   **94.** $\theta = 14°, r = 3.0$ ft

In Exercises 95–98, find the linear speed of a point that moves with constant speed in a circular motion if the point travels along the circle of arc length s in time t.

95. $s = 2$ m, $t = 5$ sec   **96.** $s = 12$ ft, $t = 3$ min   **97.** $s = 68,000$ km, $t = 250$ hr   **98.** $s = 7524$ mi, $t = 12$ days

In Exercises 99–102, find the distance traveled (arc length) of a point that moves with constant speed v along a circle in time t.

99. $v = 2.8$ m/sec, $t = 3.5$ sec   **100.** $v = 6.2$ km/hr, $t = 4.5$ hr

101. $v = 4.5$ mi/hr, $t = 20$ min   **102.** $v = 5.6$ ft/sec, $t = 2$ min

In Exercises 103–106, find the angular speed (radians/second) associated with rotating a central angle θ in time t.

103. $\theta = 25\pi, t = 10$ sec   **104.** $\theta = \dfrac{3\pi}{4}, t = \dfrac{1}{6}$ sec   **105.** $\theta = 200°, t = 5$ sec   **106.** $\theta = 60°, t = 0.2$ sec

In Exercises 107–110, find the linear speed of a point traveling at a constant speed along the circumference of a circle with radius r and angular speed ω.

107. $\omega = \dfrac{2\pi\ \text{rad}}{3\ \text{sec}}, r = 9$ in.   **108.** $\omega = \dfrac{3\pi\ \text{rad}}{4\ \text{sec}}, r = 8$ cm   **109.** $\omega = \dfrac{\pi\ \text{rad}}{20\ \text{sec}}, r = 5$ mm   **110.** $\omega = \dfrac{5\pi\ \text{rad}}{16\ \text{sec}}, r = 24$ ft

In Exercises 111–114, find the distance a point travels along a circle over a time t, given the angular speed ω and radius r of the circle. Round your answers to three significant digits.

111. $r = 5$ cm, $\omega = \dfrac{\pi\ \text{rad}}{6\ \text{sec}}, t = 10$ sec   **112.** $r = 2$ mm, $\omega = 6\pi\dfrac{\text{rad}}{\text{sec}}, t = 11$ sec

113. $r = 5.2$ in., $\omega = \dfrac{\pi\ \text{rad}}{15\ \text{sec}}, t = 10$ min   **114.** $r = 3.2$ ft, $\omega = \dfrac{\pi\ \text{rad}}{4\ \text{sec}}, t = 3$ min

▪ APPLICATIONS

For Exercises 115 and 116, refer to the following:

A common school locker combination lock is shown. The lock has a dial with 40 calibration marks numbered 0 to 39. A combination consists of three of these numbers (e.g., 5-35-20). To open the lock, the following steps are taken:

iStockphoto

- Turn the dial clockwise two full turns.
- Continue turning clockwise until the first number of the combination.
- Turn the dial counterclockwise one full turn.
- Continue turning counterclockwise until the 2nd number is reached.
- Turn the dial clockwise again until the 3rd number is reached.
- Pull the shank and the lock will open.

115. Combination Lock. Given that the initial position of the dial is at zero (shown in the illustration), how many degrees is the dial rotated in total (sum of clockwise and counterclockwise rotations) in opening the lock if the combination is 35-5-20?

116. Combination Lock. Given that the initial position of the dial is at zero (shown in the illustration), how many degrees is the dial rotated in total (sum of clockwise and counterclockwise rotations) in opening the lock if the combination is 20-15-5?

117. Tires. A car owner decides to upgrade from tires with a diameter of 24.3 inches to tires with a diameter of 26.1 inches. If she doesn't update the onboard computer, how fast will she actually be traveling when the speedometer reads 65 mph? Round to the nearest mph.

118. Tires. A car owner decides to upgrade from tires with a diameter of 24.8 inches to tires with a diameter of 27.0 inches. If she doesn't update the onboard computer, how fast will she actually be traveling when the speedometer reads 70 mph? Round to the nearest mph.

For Exercises 119 and 120, refer to the following:

NASA explores artificial gravity as a way to counter the physiologic effects of extended weightlessness for future space exploration. NASA's centrifuge has a 58-foot-diameter arm.

Courtesy NASA

119. NASA. If two humans are on opposite (red and blue) ends of the centrifuge and their linear speed is 200 miles per hour, how fast is the arm rotating? Express the answer in radians per second to two significant digits.

120. NASA. If two humans are on opposite (red and blue) ends of the centrifuge and they rotate one full rotation every second, what is their linear speed in feet per second?

▪ CATCH THE MISTAKE

In Exercises 121 and 122, explain the mistake that is made.

121. If the radius of a set of tires on a car is 15 inches and the tires rotate 180° per second, how fast is the car traveling (linear speed) in miles per hour?

Solution:

Write the formula for linear speed.

$$v = r\omega$$

Let $r = 15$ inches and $\omega = 180°$ per second

$$v = (15 \text{ in.})\left(\frac{180°}{\text{sec}}\right)$$

Simplify.

$$v = 2700 \frac{\text{in.}}{\text{sec}}$$

Let 1 mile = 5280 feet
= 63,360 inches and
1 hour = 3600 seconds.

$$v = \left(\frac{2700 \cdot 3600}{63,360}\right) \text{mph}$$

Simplify.

$$v \approx 153.4 \text{ mph}$$

This is incorrect. The correct answer is approximately 2.7 mph. What mistake was made?

122. If a bicycle has tires with radius 10 inches and the tires rotate 90° per $\frac{1}{2}$ second, how fast is the bicycle traveling (linear speed) in miles per hour?

Solution:

Write the formula for linear speed. $v = r\omega$

Let $r = 10$ inches and $\omega = 180°$ per second. $v = (10 \text{ in.})\left(\dfrac{180°}{\sec}\right)$

Simplify. $v = \dfrac{1800 \text{ in.}}{\sec}$

Let 1 mile = 5280 feet = 63,360 inches and 1 hour = 3600 seconds. $v = \left(\dfrac{1800 \cdot 3600}{63{,}360}\right)\text{mph}$

Simplify. $v \approx 102.3$ mph

This is incorrect. The correct answer is approximately 1.8 mph. What mistake was made?

■ CONCEPTUAL

In Exercises 123–126, determine whether each statement is true or false.

123. If the radius of a circle doubles, then the arc length (associated with a fixed central angle) doubles.

124. If the radius of a circle doubles, then the area of the sector (associated with a fixed central angle) doubles.

125. If the angular speed doubles, then the number of revolutions doubles.

126. If the central angle of a sector doubles, then the area corresponding to the sector is double the area of the original sector.

■ CHALLENGE

127. What is the measure (in degrees) of the smaller angle the hour and minute hands make when the time is 12:20?

128. What is the measure (in degrees) of the smaller angle the hour and minute hands make when the time is 9:10?

129. Find the area of the shaded region below:

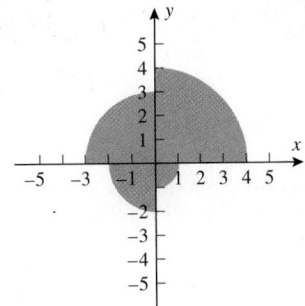

130. Find the perimeter of the shaded region in Exercise 129.

▪ TECHNOLOGY

In Exercises 131 and 132, find the measure (in degrees, minutes, and nearest seconds) of a central angle θ that intercepts an arc on a circle with indicated radius r and arc length s. With the TI calculator commands ANGLE and DMS, change to degrees, minutes, and seconds.

131. $r = 78.6$ cm, $s = 94.4$ cm

132. $r = 14.2$ inches, $s = 23.8$ inches

▪ PREVIEW TO CALCULUS

In calculus we work with real numbers; thus, the measure of an angle must be in radians.

133. What is the measure (in radians) of a central angle θ that intercepts an arc of length 2π cm on a circle of radius 10 cm?

134. Determine the angle of the smallest possible positive measure (in radians) that is coterminal with the angle 750°.

135. The area of a sector of a circle with radius 3 in. and central angle θ is $\dfrac{3\pi}{2}$ in.². What is the radian measure of θ?

136. An object is rotating at 600° per second, find the central angle θ, in radians, when $t = 3$ sec.

SECTION
4.2 RIGHT TRIANGLE TRIGONOMETRY

SKILLS OBJECTIVES

- Learn the trigonometric functions as ratios of sides of a right triangle.
- Evaluate trigonometric functions exactly for special angles.
- Evaluate trigonometric functions using a calculator.

CONCEPTUAL OBJECTIVES

- Understand that right triangle ratios are based on the properties of similar triangles.
- Understand the difference between evaluating trigonometric functions exactly and using a calculator.

The word **trigonometry** stems from the Greek words *trigonon*, which means triangle, and *metrein*, which means to measure. Trigonometry began as a branch of geometry and was utilized extensively by early Greek mathematicians to determine unknown distances. The major *trigonometric functions*, including *sine*, *cosine*, and *tangent*, were first defined as ratios of sides in a right triangle. This is the way we will define them in this section. Since the two angles, besides the right angle, in a right triangle have to be acute, a second kind of definition was needed to extend the domain of trigonometric functions to nonacute angles in the Cartesian plane (Section 4.3). Starting in the eighteenth century, broader definitions of the trigonometric functions came into use, under which the functions are associated with points on the unit circle (Section 5.1).

Right Triangle Ratios

Similar Triangles

The word *similar* in mathematics means identical in shape, although not necessarily the same size. It is important to note that two triangles can have the exact same shape (same angles) but have different sizes.

DEFINITION **Similar Triangles**

Similar triangles are triangles with equal corresponding angle measures (equal angles).

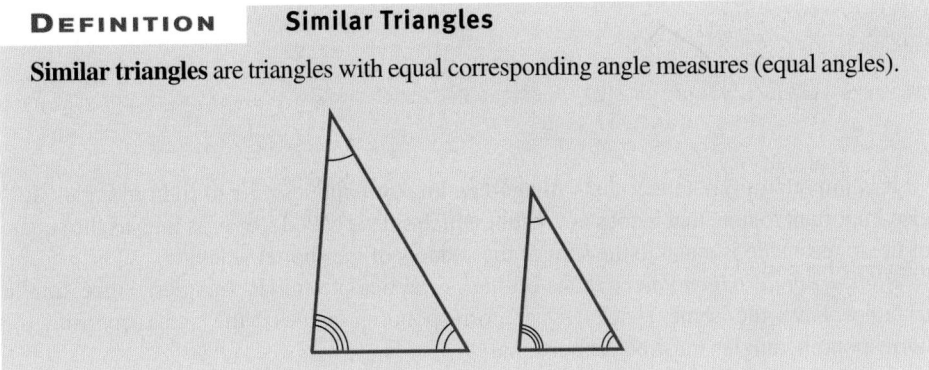

Right Triangles

A **right triangle** is a triangle in which one of the angles is a right angle 90°. Since one angle is 90°, the other two angles must be complementary (sum to 90°), so that the sum of all three angles is 180°. The longest side of a right triangle, called the **hypotenuse**, is opposite the right angle. The other two sides are called the **legs** of the right triangle.

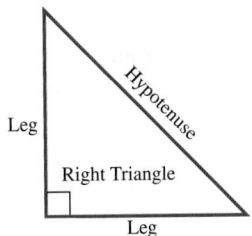

The *Pythagorean theorem* relates the sides of a right triangle. It says that the sum of the squares of the lengths of the two legs is equal to the square of the length of the hypotenuse. It is important to note that length (a synonym of distance) is always positive.

PYTHAGOREAN THEOREM

In any right triangle, the square of the length of the longest side (hypotenuse) is equal to the sum of the squares of the lengths of the other two sides (legs).

$$a^2 + b^2 = c^2$$

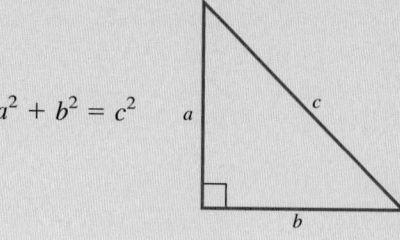

Study Tip

The Pythagorean theorem applies only to *right* triangles.

It is important to note that the Pythagorean theorem applies *only* to right triangles. It is also important to note that it does not matter which side is called a or b, as long as the square of the longest side is equal to the sum of the squares of the shorter sides.

Right-triangle trigonometry relies on the properties of similar triangles. Since similar triangles have the same shape (equal corresponding angles), the sides opposite the corresponding angles must be proportional.

Right Triangle Ratios

The concept of similar triangles, one of the basic insights in trigonometry, allows us to determine the length of a side of one triangle if we know the length of certain sides of a similar triangle. Consider the following similar triangles:

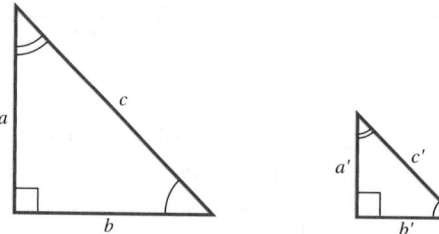

In similar triangles, the sides opposite corresponding angles must be proportional so the following ratios hold true:

$$\frac{a}{a'} = \frac{b}{b'} = \frac{c}{c'}$$

Separate the common ratios into three equations:

$$\frac{a}{a'} = \frac{b}{b'} \qquad \frac{b}{b'} = \frac{c}{c'} \qquad \frac{a}{a'} = \frac{c}{c'}$$

WORDS	MATH
Start with the first ratio.	$\dfrac{a}{a'} = \dfrac{b}{b'}$
Cross multiply.	$ab' = a'b$
Divide both sides by bb'.	$\dfrac{ab'}{bb'} = \dfrac{a'b}{bb'}$
Simplify.	$\dfrac{a}{b} = \dfrac{a'}{b'}$

Similarly, it can be shown that $\dfrac{b}{c} = \dfrac{b'}{c'}$ and $\dfrac{a}{c} = \dfrac{a'}{c'}$.

Notice that even though the sizes of the triangles are different, since the corresponding angles are equal, the ratio of the two legs of the large triangle is equal to the corresponding ratio of the legs of the small triangle, or $\dfrac{a}{b} = \dfrac{a'}{b'}$. Similarly, the ratios of a leg and the hypotenuse of the large triangle and the corresponding leg and hypotenuse of the small triangle are also equal; that is, $\dfrac{b}{c} = \dfrac{b'}{c'}$ and $\dfrac{a}{c} = \dfrac{a'}{c'}$.

For any right triangle, there are six possible ratios of sides that can be calculated for each acute angle θ:

$$\dfrac{b}{c} \quad \dfrac{a}{c} \quad \dfrac{b}{a}$$

$$\dfrac{c}{b} \quad \dfrac{c}{a} \quad \dfrac{a}{b}$$

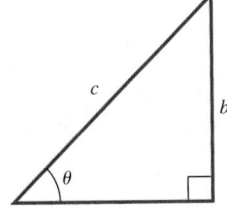

These ratios are referred to as **trigonometric ratios** or **trigonometric functions**, since they depend on the angle θ, and each is given a name:

FUNCTION NAME	ABBREVIATION
Sine	sin
Cosine	cos
Tangent	tan
Cosecant	csc
Secant	sec
Cotangent	cot

WORDS	MATH
The sine of θ	$\sin\theta$
The cosine of θ	$\cos\theta$
The tangent of θ	$\tan\theta$
The cosecant of θ	$\csc\theta$
The secant of θ	$\sec\theta$
The cotangent of θ	$\cot\theta$

Sine, cosine, tangent, cotangent, secant, and cosecant are names given to specific ratios of lengths of sides of right triangles.

DEFINITION   **Trigonometric Functions**

Let θ be an acute angle in a right triangle, then

$$\sin\theta = \frac{b}{c} \qquad \cos\theta = \frac{a}{c} \qquad \tan\theta = \frac{b}{a}$$

$$\csc\theta = \frac{c}{b} \qquad \sec\theta = \frac{c}{a} \qquad \cot\theta = \frac{a}{b}$$

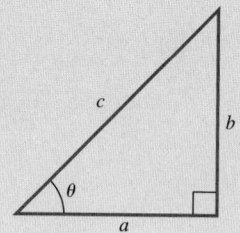

The following terminology will be used throughout this text (refer to the right triangle above):

- The **hypotenuse** is always opposite the right angle.
- One leg (b) is **opposite** the angle θ.
- One leg (a) is **adjacent** to the angle θ.

Also notice that since $\sin\theta = \dfrac{b}{c}$ and $\cos\theta = \dfrac{a}{c}$, then $\tan\theta = \dfrac{\sin\theta}{\cos\theta} = \dfrac{\frac{b}{c}}{\frac{a}{c}} = \dfrac{b}{a}$.

Using this terminology, we arrive at an alternative definition that is easier to remember.

DEFINITION   **Trigonometric Functions (Alternate Form)**

For an acute angle θ in a right triangle:

$$\sin\theta = \frac{\text{opposite}}{\text{hypotenuse}} \qquad \cos\theta = \frac{\text{adjacent}}{\text{hypotenuse}} \qquad \tan\theta = \frac{\text{opposite}}{\text{adjacent}}$$

and their reciprocals:

$$\csc\theta = \frac{1}{\sin\theta} = \frac{\text{hypotenuse}}{\text{opposite}}$$

$$\sec\theta = \frac{1}{\cos\theta} = \frac{\text{hypotenuse}}{\text{adjacent}}$$

$$\cot\theta = \frac{1}{\tan\theta} = \frac{\text{adjacent}}{\text{opposite}}$$

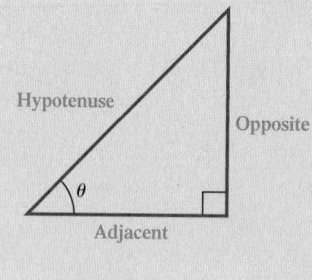

Reciprocal Identities

The three main trigonometric functions should be learned in terms of the following ratios:

$$\sin\theta = \frac{\text{opposite}}{\text{hypotenuse}} \qquad \cos\theta = \frac{\text{adjacent}}{\text{hypotenuse}} \qquad \tan\theta = \frac{\text{opposite}}{\text{adjacent}}$$

The remaining three trigonometric functions can be derived from $\sin\theta$, $\cos\theta$, and $\tan\theta$ using the *reciprocal identities*. Recall that the **reciprocal** of x is $\dfrac{1}{x}$ for $x \neq 0$.

RECIPROCAL IDENTITIES

$$\csc\theta = \frac{1}{\sin\theta} \qquad \sec\theta = \frac{1}{\cos\theta} \qquad \cot\theta = \frac{1}{\tan\theta}$$

Evaluating Trigonometric Functions Exactly for Special Angle Measures

There are three special acute angles that are very important in trigonometry: $30°, 45°,$ and $60°$. We can combine the relationships governing their side lengths

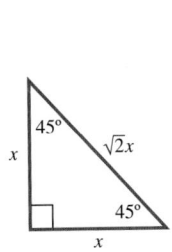 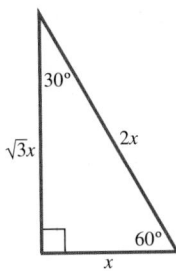

with the trigonometric ratios developed in this section to evaluate the trigonometric functions for the special angle measures of $30°, 45°,$ and $60°$.

Consider a $45°$-$45°$-$90°$ triangle.

WORDS	MATH
A $45°$-$45°$-$90°$ triangle is an isosceles (two legs are equal) right triangle.	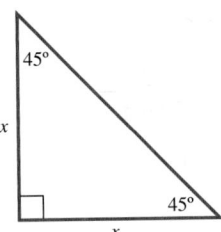

Use the Pythagorean theorem.	$x^2 + x^2 = \text{hypotenuse}^2$
Simplify the left side of the equation.	$2x^2 = \text{hypotenuse}^2$
Solve for the hypotenuse.	$\text{hypotenuse} = \pm\sqrt{2x^2} = \pm\sqrt{2}\,|x|$
x and the hypotenuse are lengths and must be positive.	$\text{hypotenuse} = \sqrt{2}\,x$

The hypotenuse of a $45°$-$45°$-$90°$ triangle is $\sqrt{2}$ times the length of either leg.

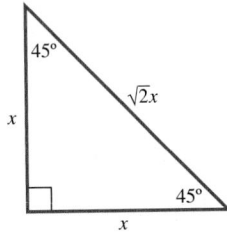

Let us now determine the relationship of the sides of a 30°-60°-90° triangle. We start with an equilateral triangle (equal sides and equal angles of measure 60°).

WORDS	**MATH**
Draw an equilateral triangle with sides $2x$.	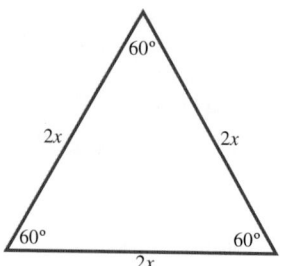

Draw a line segment from one vertex that is perpendicular to the opposite side; this line segment represents the height of the triangle, h, and bisects the base. There are now two identical 30°-60°-90° triangles.

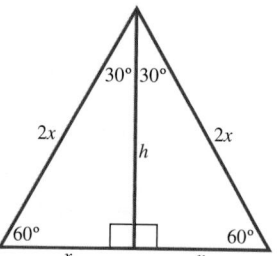

Notice that in each triangle the hypotenuse is twice the shortest leg, which is opposite the 30° angle.

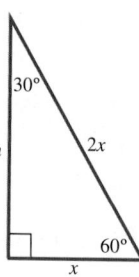

To find the length h, use the Pythagorean theorem.

$$h^2 + x^2 = (2x)^2$$
$$h^2 + x^2 = 4x^2$$

Solve for h.

$$h^2 = 3x^2$$
$$h = \pm\sqrt{3x^2} = \pm\sqrt{3}|x|$$

h and x are lengths and must be positive.

$$h = \sqrt{3}x$$

The hypotenuse of a 30°-60°-90° is twice the length of the leg opposite the 30° angle, the shortest leg.

The leg opposite the 60° angle is $\sqrt{3}$ times the length of the leg opposite the 30° angle, the shortest leg.

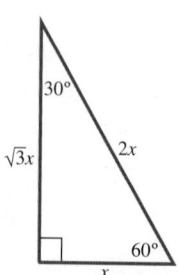

EXAMPLE 1 **Evaluating the Trigonometric Functions Exactly for 30°**

Evaluate the six trigonometric functions for an angle that measures 30°.

Solution:

Label the sides (opposite, adjacent, and hypotenuse) of the 30°-60°-90° triangle with respect to the 30° angle.

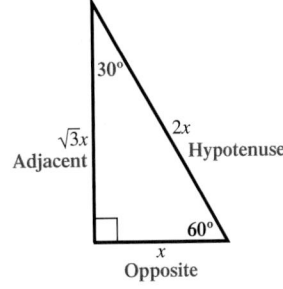

Use the right triangle ratio definitions of sine, cosine, and tangent.

$$\sin 30° = \frac{\text{opposite}}{\text{hypotenuse}} = \frac{x}{2x} = \frac{1}{2}$$

$$\cos 30° = \frac{\text{adjacent}}{\text{hypotenuse}} = \frac{\sqrt{3}x}{2x} = \frac{\sqrt{3}}{2}$$

$$\tan 30° = \frac{\text{opposite}}{\text{adjacent}} = \frac{x}{\sqrt{3}x} = \frac{1}{\sqrt{3}} = \frac{1}{\sqrt{3}} \cdot \frac{\sqrt{3}}{\sqrt{3}} = \frac{\sqrt{3}}{3}$$

Use the reciprocal identities to obtain cosecant, secant, and cotangent.

$$\csc 30° = \frac{1}{\sin 30°} = \frac{1}{\frac{1}{2}} = 2$$

$$\sec 30° = \frac{1}{\cos 30°} = \frac{1}{\frac{\sqrt{3}}{2}} = \frac{2}{\sqrt{3}} = \frac{2}{\sqrt{3}} \cdot \frac{\sqrt{3}}{\sqrt{3}} = \frac{2\sqrt{3}}{3}$$

$$\cot 30° = \frac{1}{\tan 30°} = \frac{1}{\frac{\sqrt{3}}{3}} = \frac{3}{\sqrt{3}} = \frac{3}{\sqrt{3}} \cdot \frac{\sqrt{3}}{\sqrt{3}} = \sqrt{3}$$

The six trigonometric functions evaluated for an angle measuring 30° are

$$\boxed{\sin 30° = \frac{1}{2}} \qquad \boxed{\cos 30° = \frac{\sqrt{3}}{2}} \qquad \boxed{\tan 30° = \frac{\sqrt{3}}{3}}$$

$$\boxed{\csc 30° = 2} \qquad \boxed{\sec 30° = \frac{2\sqrt{3}}{3}} \qquad \boxed{\cot 30° = \sqrt{3}}$$

■ **YOUR TURN** Evaluate the six trigonometric functions for an angle that measures 60°.

In comparing our answers in Example 1 and Your Turn, we see that the following cofunction relationships are true. We call these *cofunction* relationships.

$$\sin 30° = \cos 60° \qquad \sec 30° = \csc 60° \qquad \tan 30° = \cot 60°$$

$$\sin 60° = \cos 30° \qquad \sec 60° = \csc 30° \qquad \tan 60° = \cot 30°$$

Notice that 30° and 60° are complementary angles.

Technology Tip

Set calculator in degree mode. Use directions from Technology Tip on the next page, but with 30° instead of 45°.

```
sin(30)
            .5
cos(30)
       .8660254038
√(3)/2
       .8660254038
```

```
tan(30)
       .5773502692
√(3)/3
       .5773502692
■
```

To find the values of csc 30°, sec 30°, and cot 30°, find the reciprocal of three main trigonometric functions.

```
sin(30)⁻¹
            2
cos(30)⁻¹
       1.154700538
2*√(3)/3
       1.154700538
■
```

```
tan(30)⁻¹
       1.732050808
√(3)
       1.732050808
■
```

■ **Answer:**

$$\sin 60° = \frac{\sqrt{3}}{2} \qquad \cos 60° = \frac{1}{2}$$

$$\tan 60° = \sqrt{3} \qquad \csc 60° = \frac{2\sqrt{3}}{3}$$

$$\sec 60° = 2 \qquad \cot 60° = \frac{\sqrt{3}}{3}$$

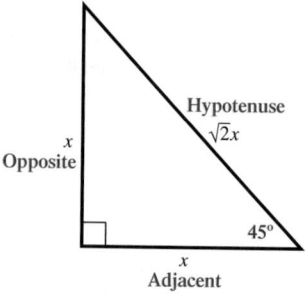
EXAMPLE 2   Evaluating the Trigonometric Functions Exactly for 45°

Evaluate the six trigonometric functions for an angle that measures 45°.

Solution:

Label the sides of the 45°-45°-90° triangle as opposite, adjacent, or hypotenuse with respect to one of the 45° angles.

Use the right triangle ratio definitions of sine, cosine, and tangent.

$$\sin 45° = \frac{\text{opposite}}{\text{hypotenuse}} = \frac{x}{\sqrt{2}x} = \frac{1}{\sqrt{2}} = \frac{1}{\sqrt{2}} \cdot \frac{\sqrt{2}}{\sqrt{2}} = \frac{\sqrt{2}}{2}$$

$$\cos 45° = \frac{\text{adjacent}}{\text{hypotenuse}} = \frac{x}{\sqrt{2}x} = \frac{1}{\sqrt{2}} = \frac{1}{\sqrt{2}} \cdot \frac{\sqrt{2}}{\sqrt{2}} = \frac{\sqrt{2}}{2}$$

$$\tan 45° = \frac{\text{opposite}}{\text{adjacent}} = \frac{x}{x} = 1$$

Use the reciprocal identities to obtain cosecant, secant, and cotangent.

$$\csc 45° = \frac{1}{\sin 45°} = \frac{1}{\frac{\sqrt{2}}{2}} = \frac{2}{\sqrt{2}} = \frac{2}{\sqrt{2}} \cdot \frac{\sqrt{2}}{\sqrt{2}} = \sqrt{2}$$

$$\sec 45° = \frac{1}{\cos 45°} = \frac{1}{\frac{\sqrt{2}}{2}} = \frac{2}{\sqrt{2}} = \frac{2}{\sqrt{2}} \cdot \frac{\sqrt{2}}{\sqrt{2}} = \sqrt{2}$$

$$\cot 45° = \frac{1}{\tan 45°} = \frac{1}{1} = 1$$

The six trigonometric functions evaluated for an angle measuring 45° are

$\sin 45° = \dfrac{\sqrt{2}}{2}$	$\cos 45° = \dfrac{\sqrt{2}}{2}$	$\tan 45° = 1$
$\csc 45° = \sqrt{2}$	$\sec 45° = \sqrt{2}$	$\cot 45° = 1$

We see that the following *cofunction* relationships are true

$$\sin 45° = \cos 45° \qquad \sec 45° = \csc 45° \qquad \tan 45° = \cot 45°$$

since 45° and 45° are complementary angles.

The trigonometric function values for the three special angle measures, 30°, 45°, and 60°, are summarized in the following table.

Trigonometric Function Values for Special Angles

Degrees	Radians	$\sin\theta$	$\cos\theta$	$\tan\theta$	$\cot\theta$	$\sec\theta$	$\csc\theta$
30°	$\dfrac{\pi}{6}$	$\dfrac{1}{2}$	$\dfrac{\sqrt{3}}{2}$	$\dfrac{\sqrt{3}}{3}$	$\sqrt{3}$	$\dfrac{2\sqrt{3}}{3}$	2
45°	$\dfrac{\pi}{4}$	$\dfrac{\sqrt{2}}{2}$	$\dfrac{\sqrt{2}}{2}$	1	1	$\sqrt{2}$	$\sqrt{2}$
60°	$\dfrac{\pi}{3}$	$\dfrac{\sqrt{3}}{2}$	$\dfrac{1}{2}$	$\sqrt{3}$	$\dfrac{\sqrt{3}}{3}$	2	$\dfrac{2\sqrt{3}}{3}$

It is important to **learn** the special values in **red** for sine and cosine. All other values in the table can be found through reciprocals or quotients of these two functions. Notice that tangent is the ratio of sine to cosine.

$$\sin\theta = \frac{\text{opposite}}{\text{hypotenuse}} \qquad \cos\theta = \frac{\text{adjacent}}{\text{hypotenuse}} \qquad \tan\theta = \frac{\sin\theta}{\cos\theta} = \frac{\frac{\text{opposite}}{\text{hypotenuse}}}{\frac{\text{adjacent}}{\text{hypotenuse}}} = \frac{\text{opposite}}{\text{adjacent}}$$

Using Calculators to Evaluate Trigonometric Functions

We now turn our attention to using calculators to evaluate trigonometric functions, which often results in an approximation. Scientific and graphing calculators have buttons for sine (sin), cosine (cos), and tangent (tan) functions.

EXAMPLE 3 **Evaluating Trigonometric Functions with a Calculator**

Use a calculator to find the values of

a. $\sin 75°$ **b.** $\tan 67°$ **c.** $\sec 52°$ **d.** $\cos\left(\dfrac{\pi}{6}\right)$ **e.** $\tan\left(\dfrac{\pi}{8}\right)$

Round your answers to four decimal places.

Solution:

a. 0.965925826 ≈ 0.9659

b. 2.355852366 ≈ 2.3559

c. $\cos 52° \approx 0.615661475$ $1/x\ (\text{or } x^{-1})$ 1.624269245 ≈ 1.6243

d. 0.866025403 ≈ 0.8660

e. 0.414213562 ≈ 0.4142

Note: We know $\cos\left(\dfrac{\pi}{6}\right) = \dfrac{\sqrt{3}}{2} \approx 0.8660$.

■ **YOUR TURN** Use a calculator to find the values of

    a. $\cos 22°$ **b.** $\tan 81°$ **c.** $\csc 37°$ **d.** $\sin\left(\dfrac{\pi}{4}\right)$ **e.** $\cot\left(\dfrac{\pi}{12}\right)$

Round your answers to four decimal places.

Study Tip

In calculating secant, cosecant, and cotangent function values with a calculator, it is important not to round the number until after using the reciprocal function key $1/x$.

■ **Answer: a.** 0.9272   **b.** 6.3138   **c.** 1.6616   **d.** 0.7071   **e.** 3.7321

When calculating secant, cosecant, and cotangent function values with a calculator, it is important not to round the number until after using the reciprocal function key $1/x$ or x^{-1} in order to be as accurate as possible.

Solving Right Triangles

A triangle has three angles and three sides. To *solve a triangle* means to find the length of all three sides and the measures of all three angles.

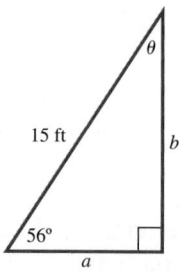

Technology Tip

```
15cos(56)
       8.387893552
```

```
15sin(56)
      12.43556359
```

EXAMPLE 4 Solving a Right Triangle Given an Angle and a Side

Solve the right triangle—find a, b, and θ.

Solution:

STEP 1 **Solve for θ.**

The two acute angles in a right triangle are complementary.

$$\theta + 56° = 90°$$

Solve for θ.

$$\boxed{\theta = 34°}$$

STEP 2 **Solve for a.**

Cosine of an angle is equal to the adjacent side over the hypotenuse.

$$\cos 56° = \frac{a}{15}$$

Solve for a.

$$a = 15 \cos 56°$$

Evaluate the right side of the expression using a calculator.

$$a \approx 8.38789$$

Round a to two significant digits.

$$\boxed{a \approx 8.4 \text{ ft}}$$

STEP 3 **Solve for b.**

Notice that there are two ways to solve for b: trigonometric functions or the Pythagorean theorem. Although it is tempting to use the Pythagorean theorem, it is better to use the given information with trigonometric functions than to use a value that has already been rounded, which could make results less accurate.

Sine of an angle is equal to the opposite side over the hypotenuse.

$$\sin 56° = \frac{b}{15}$$

Solve for b.

$$b = 15 \sin 56°$$

Evaluate the right side of the expression using a calculator.

$$b \approx 12.43556$$

Round b to two significant digits.

$$\boxed{b \approx 12 \text{ ft}}$$

STEP 4 **Check.**

Check the trigonometric values of the specific angles by calculating the trigonometric ratios.

$$\sin 34° \overset{?}{=} \frac{8.4}{15} \qquad \cos 34° \overset{?}{=} \frac{12}{15} \qquad \tan 34° \overset{?}{=} \frac{8.4}{12}$$

$$0.5592 \approx 0.56 \qquad 0.8290 \approx 0.80 \qquad 0.6745 \approx 0.70$$

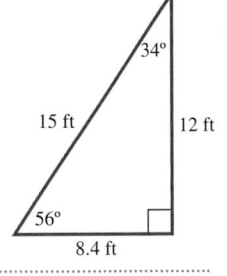

■ **YOUR TURN** Solve the right triangle— find a, b, and θ.

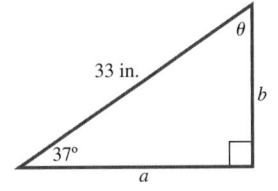

■ **Answer:** $\theta = 53°, a \approx 26$ in., $b \approx 20$ in.

Sometimes in solving a right triangle, the side lengths are given and we need to find the angles. To do this, we can work backwards from the table of known values. For example, if $\cos\theta = \frac{1}{2}$, what is θ? We see from the table of exact values that $\theta = 60°$ or $\frac{\pi}{3}$. To find an angle that is not listed in the table, we use the inverse cosine function, $\cos^{-1}$, key on a calculator. In a later section, we will discuss inverse trigonometric functions, but for now we will simply use the inverse trigonometric function keys on a calculator to determine the unknown angle.

EXAMPLE 5 **Solving a Right Triangle Given Two Sides**

Solve the right triangle—find a, α, and β.

Solution:

STEP 1 **Solve for α.**

Cosine of an angle is equal to the adjacent side over hypotenuse. $\cos\alpha = \dfrac{19.67 \text{ cm}}{37.21 \text{ cm}}$

Evaluate the right side using a calculator. $\cos\alpha \approx 0.528621338$

Write the angle α in terms of the inverse cosine function. $\alpha \approx \cos^{-1} 0.528621338$

Use a calculator to evaluate the inverse cosine function. $\alpha \approx 58.08764854°$

Round α to the nearest hundredth of a degree. $\boxed{\alpha \approx 58.09°}$

STEP 2 **Solve for β.**

The two acute angles in a right triangle are complementary. $\alpha + \beta = 90°$

Substitute $\alpha \approx 58.09°$. $58.09 + \beta \approx 90°$

Solve for β. $\boxed{\beta \approx 31.91°}$

The answer is already rounded to the nearest hundredth of a degree.

Technology Tip

To find $\cos^{-1}\left(\dfrac{19.67}{37.21}\right)$, first press

19.67 ÷ 37.21 ENTER.

Next, press 2nd COS for $\cos^{-1}$ and 2nd (−) for the answer, ANS

COS⁻¹ ANS) ENTER.

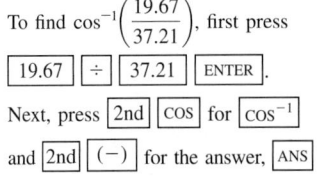

Or enter the entire expression as $\cos^{-1}\left(\dfrac{19.67}{37.21}\right)$. Press 2nd COS

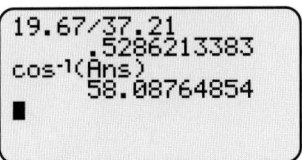

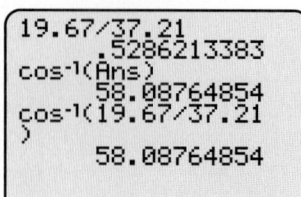

Using a scientific calculator, press

$\boxed{\cos^{-1}}$ $\boxed{(}$ $\boxed{19.67}$ $\boxed{\div}$ $\boxed{37.21}$

$\boxed{)}$ $\boxed{\text{ENTER}}$.

■ **Answer:** $a \approx 16.0\,\text{mi}$, $\alpha \approx 43.0°$, $\beta \approx 47.0°$

Study Tip

To find a length in a right triangle, use the sine, cosine, or tangent function. To find an angle measure in a right triangle, given the proper ratio of side lengths, use the inverse sine, inverse cosine, or inverse tangent function.

Technology Tip

To calculate $400\tan 0.01°$, press

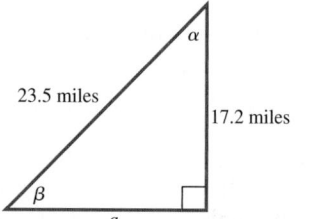

STEP 3 Solve for a.

Use the Pythagorean theorem. $a^2 + b^2 = c^2$

Substitute the given values for b and c. $a^2 + 19.67^2 = 37.21^2$

Solve for a. $a \approx 31.5859969$

Round a to four significant digits. $\boxed{a \approx 31.59\,\text{cm}}$

STEP 4 Check.

Check the trigonometric values of the specific angles by calculating the trigonometric ratio.

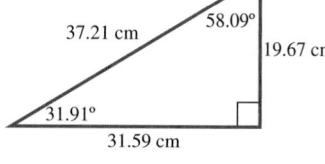

$$\sin 31.91° \overset{?}{=} \frac{19.67}{37.21} \qquad \sin 58.09° \overset{?}{=} \frac{31.59}{37.21}$$

$$0.5286 \approx 0.5286 \qquad\qquad 0.8489 \approx 0.8490$$

■ **YOUR TURN** Solve the right triangle— find a, α, and β.

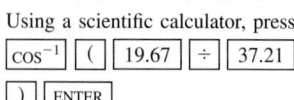

Applications

Suppose NASA wants to talk with the International Space Station (ISS), which is traveling at a speed of 17,700 mph (7900 meters per second), 400 km (250 miles) above the surface of Earth. If the antennas at the ground station in Houston have a pointing error of even 1/100 of a degree, that is, $0.01°$, the ground station will miss the chance to talk with the astronauts.

EXAMPLE 6   Pointing Error

Assume that the ISS (which is 108 meters long and 73 meters wide) is in a 400-km low earth orbit. If the communications antennas have a $0.01°$ pointing error, how many meters off will the communications link be?

Solution:

Draw a right triangle that depicts this scenario.

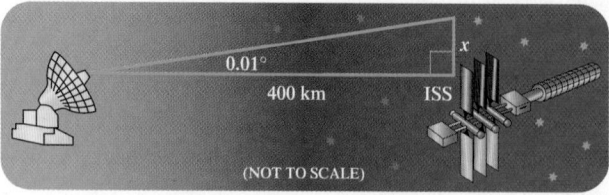

(NOT TO SCALE)

Identify the tangent ratio. $\tan 0.01° = \dfrac{x}{400}$

Solve for x. $x = (400\,\text{km}) \tan 0.01°$

Evaluate the expression on the right. $x \approx 0.06981317\,\text{km}$

400 km is accurate to three significant digits, so we express the answer to three significant digits.

The pointing error causes the signal to be off by $\boxed{69.8\,\text{meters}}$. Since the ISS is only 108 meters long, it is possible that the signal will be missed by the astronaut crew.

In navigation, the word **bearing** means the direction in which a vessel is pointed. **Heading** is the direction in which the vessel is actually traveling. Heading and bearing are only synonyms when there is no wind. Direction is often given as a bearing, which is the measure of an acute angle with respect to the north–south vertical line. "The plane has a bearing of N 20° E" means that the plane is pointed 20° to the east of due north.

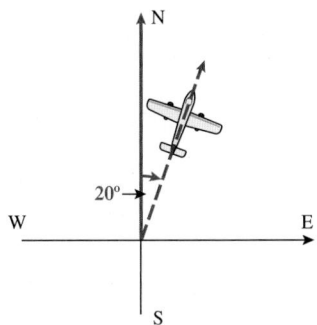

 EXAMPLE 7 **Bearing (Navigation)**

A jet takes off bearing N 28° E and flies 5 miles, and then makes a left (90°) turn and flies 12 miles further. If the control tower operator wants to locate the plane, what bearing should she use?

Solution:

Draw a picture that represents this scenario.

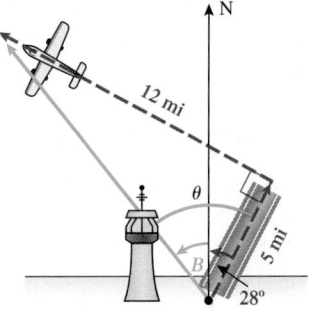

Identify the tangent ratio.

$$\tan\theta = \frac{12}{5}$$

Use the inverse tangent function to solve for θ.

$$\theta = \tan^{-1}\left(\frac{12}{5}\right) \approx 67.4°$$

Subtract 28° from θ to find the bearing.

$$B \approx 67.4° - 28° \approx 39.4°$$

Round to the nearest degree.

$$\boxed{B \approx \text{N } 39° \text{ W}}$$

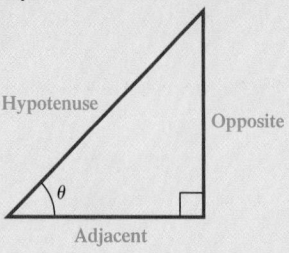

SECTION 4.2 SUMMARY

The trigonometric functions defined in terms of ratios of side lengths of right triangles are given by

$$\sin \theta = \frac{\text{opposite}}{\text{hypotenuse}}$$

$$\cos \theta = \frac{\text{adjacent}}{\text{hypotenuse}}$$

$$\tan \theta = \frac{\text{opposite}}{\text{adjacent}}$$

And the remaining three trigonometric functions can be found using the reciprocal identities:

$$\csc \theta = \frac{1}{\sin \theta} \quad \sin \theta \neq 0$$

$$\sec \theta = \frac{1}{\cos \theta} \quad \cos \theta \neq 0$$

$$\cot \theta = \frac{1}{\tan \theta} \quad \tan \theta \neq 0$$

The following table lists the values of the sine and cosine functions for special acute angles:

θ (DEGREES)	θ (RADIANS)	SIN θ	COS θ
30°	$\frac{\pi}{6}$	$\frac{1}{2}$	$\frac{\sqrt{3}}{2}$
45°	$\frac{\pi}{4}$	$\frac{\sqrt{2}}{2}$	$\frac{\sqrt{2}}{2}$
60°	$\frac{\pi}{3}$	$\frac{\sqrt{3}}{2}$	$\frac{1}{2}$

SECTION 4.2 EXERCISES

■ SKILLS

In Exercises 1–6, refer to the triangle in the drawing to find the indicated trigonometric function values. Rationalize any denominators containing radicals that you encounter in the answers.

1. $\cos \theta$
2. $\sin \theta$
3. $\sec \theta$
4. $\csc \theta$
5. $\tan \theta$
6. $\cot \theta$

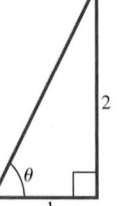

For Exercises 7–12, refer to the triangle in the drawing to find the indicated trigonometric function values. Rationalize any denominators containing radicals that you encounter in the answers.

7. $\sin \theta$
8. $\cos \theta$
9. $\sec \theta$
10. $\csc \theta$
11. $\cot \theta$
12. $\tan \theta$

In Exercises 13–18, match the trigonometric function values.

a. $\dfrac{1}{2}$ b. $\dfrac{\sqrt{3}}{2}$ c. $\dfrac{\sqrt{2}}{2}$

13. $\sin 30°$ **14.** $\sin 60°$ **15.** $\cos\left(\dfrac{\pi}{6}\right)$ **16.** $\cos\left(\dfrac{\pi}{3}\right)$ **17.** $\sin 45°$ **18.** $\cos\left(\dfrac{\pi}{4}\right)$

In Exercises 19–21, use the results in Exercises 13–18 and the trigonometric quotient identity $\tan\theta = \dfrac{\sin\theta}{\cos\theta}$ to calculate the following values:

19. $\tan 30°$ **20.** $\tan\left(\dfrac{\pi}{4}\right)$ **21.** $\tan 60°$

In Exercises 22–30, use the results in Exercises 13–21 and the reciprocal identities $\csc\theta = \dfrac{1}{\sin\theta}$, $\sec\theta = \dfrac{1}{\cos\theta}$, and $\cot\theta = \dfrac{1}{\tan\theta}$ to calculate the following values:

22. $\csc 30°$ **23.** $\sec 30°$ **24.** $\cot\left(\dfrac{\pi}{6}\right)$ **25.** $\csc\left(\dfrac{\pi}{3}\right)$ **26.** $\sec 60°$ **27.** $\cot 60°$

28. $\csc 45°$ **29.** $\sec\left(\dfrac{\pi}{4}\right)$ **30.** $\cot\left(\dfrac{\pi}{4}\right)$

In Exercises 31–46, use a calculator to evaluate the trigonometric functions for the indicated angle values. Round your answers to four decimal places.

31. $\sin 37°$ **32.** $\sin 17.8°$ **33.** $\cos 82°$ **34.** $\cos 21.9°$ **35.** $\sin\left(\dfrac{\pi}{12}\right)$ **36.** $\sin\left(\dfrac{5\pi}{9}\right)$

37. $\cos\left(\dfrac{6\pi}{5}\right)$ **38.** $\cos\left(\dfrac{13\pi}{7}\right)$ **39.** $\tan 54°$ **40.** $\tan 43.2°$ **41.** $\tan\left(\dfrac{\pi}{8}\right)$ **42.** $\cot\left(\dfrac{3\pi}{5}\right)$

43. $\csc\left(\dfrac{10\pi}{19}\right)$ **44.** $\sec\left(\dfrac{4\pi}{9}\right)$ **45.** $\cot 55°$ **46.** $\cot 29°$

In Exercises 47–54, refer to the right triangle diagram and the given information to find the indicated measure. Write your answers for angle measures in decimal degrees.

47. $\alpha = 55°$, $c = 22$ ft; find a.

48. $\alpha = 55°$, $c = 22$ ft; find b.

49. $\alpha = 20.5°$, $b = 14.7$ mi; find a.

50. $\beta = 69.3°$, $a = 0.752$ mi; find b.

51. $\beta = 25°$, $a = 11$ km; find c.

52. $\beta = 75°$, $b = 26$ km; find c.

53. $b = 2.3$ m, $c = 4.9$ m; find α.

54. $b = 7.8$ m, $c = 13$ m; find β.

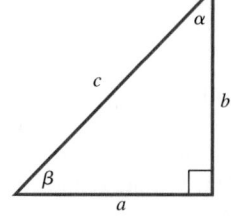

In Exercises 55–66, refer to the right triangle diagram and the given information to solve the right triangle. Write your answers for angle measures in decimal degrees.

55. $\alpha = 32°$ and $c = 12$ ft

56. $\alpha = 65°$ and $c = 37$ ft

57. $\beta = 72°$ and $c = 9.7$ mm

58. $\beta = 45°$ and $c = 7.8$ mm

59. $\alpha = 54.2°$ and $a = 111$ mi

60. $\beta = 47.2°$ and $a = 9.75$ mi

61. $a = 42.5$ ft and $b = 28.7$ ft

62. $a = 19.8$ ft and $c = 48.7$ ft

63. $a = 35,236$ km and $c = 42,766$ km

64. $b = 0.1245$ mm and $c = 0.8763$ mm

65. $\beta = 25.4°$ and $b = 11.6$ in.

66. $\beta = 39.21°$ and $b = 6.3$ m

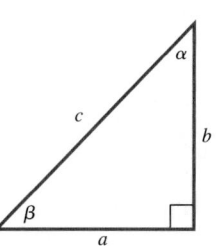

■ APPLICATIONS

Exercises 67 and 68 illustrate a mid-air refueling scenario that military aircraft often enact. Assume the elevation angle that the hose makes with the plane being fueled is $\theta = 36°$.

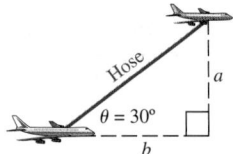

67. Mid-Air Refueling. If the hose is 150 feet long, what should be the altitude difference a between the two planes? Round to the nearest foot.

68. Mid-Air Refueling. If the smallest acceptable altitude difference a between the two planes is 100 feet, how long should the hose be? Round to the nearest foot.

Exercises 69–72 are based on the idea of a glide slope (the angle the flight path makes with the ground).

Precision Approach Path Indicator (PAPI) lights are used as a visual approach slope aid for pilots landing aircraft. A typical glide path for commercial jet airliners is 3°. The space shuttle has an outer glide approach of 18°–20°. PAPI lights are typically configured as a row of four lights. All four lights are on, but in different combinations of red or white. If all four lights are white, then the angle of descent is too high; if all four lights are red, then the angle of descent is too low; and if there are two white and two red, then the approach is perfect.

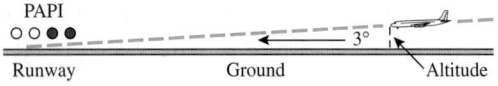

69. Glide Path of a Commercial Jet Airliner. If a commercial jetliner is 5000 feet (about 1 mile) ground distance from the runway, what should the altitude of the plane be to achieve two red and two white PAPI lights? (Assume this corresponds to a 3° glide path.)

70. Glide Path of a Commercial Jet Airliner. If a commercial jetliner is at an altitude of 450 feet when it is 5200 feet from the runway (approximately 1 mile ground distance), what is the glide slope angle? Will the pilot see white lights, red lights, or both?

71. Glide Path of the Space Shuttle *Orbiter*. If the pilot of the space shuttle *Orbiter* is at an altitude of 3000 feet when she is 15,500 ft (approximately 3 miles) from the shuttle landing facility (ground distance), what is her glide slope angle (round to the nearest degree)? Is she too high or too low?

72. Glide Path of the Space Shuttle *Orbiter*. If the same pilot in Exercise 71 raises the nose of the gliding shuttle so that she drops only 500 feet by the time she is 7800 feet from the shuttle landing strip (ground distance), what is her glide angle then (round to the nearest degree)? Is she within the specs to land the shuttle?

In Exercises 73 and 74, refer to the illustration below, which shows a search and rescue helicopter with a 30° field of view with a searchlight.

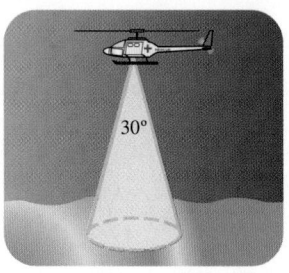

73. Search and Rescue. If the search and rescue helicopter is flying at an altitude of 150 feet above sea level, what is the diameter of the circle illuminated on the surface of the water?

74. Search and Rescue. If the search and rescue helicopter is flying at an altitude of 500 feet above sea level, what is the diameter of the circle illuminated on the surface of the water?

For Exercises 75–78, refer to the following:

Geostationary orbits are useful because they cause a satellite to appear stationary with respect to a fixed point on the rotating Earth. As a result, an antenna (dish TV) can point in a fixed direction and maintain a link with the satellite. The satellite orbits in the direction of Earth's rotation at an altitude of approximately 35,000 km.

75. Dish TV. If your dish TV antenna has a pointing error of 0.000278°, how long would the satellite have to be in order to maintain a link? Round your answer to the nearest meter.

76. Dish TV. If your dish TV antenna has a pointing error of 0.000139°, how long would the satellite have to be in order to maintain a link? Round your answer to the nearest meter.

77. Dish TV. If the satellite in a geostationary orbit (at 35,000 km) was only 10 meters long, about how accurate would the pointing of the dish have to be? Give the answer in degrees to two significant digits.

78. Dish TV. If the satellite in a geostationary orbit (at 35,000 km) was only 30 meters long, about how accurate would the pointing of the dish have to be? Give the answer in degrees to two significant digits.

79. Angle of Inclination (Skiing). The angle of inclination of a mountain with triple black diamond ski trails is 65°. If a skier at the top of the mountain is at an elevation of 4000 feet, how long is the ski run from the top to the base of the mountain? Round to the nearest foot.

80. Bearing (Navigation). If a plane takes off bearing N 33° W and flies 6 miles and then makes a right (90°) turn and flies 10 miles further, what bearing will the traffic controller use to locate the plane?

For Exercises 81 and 82, refer to the following:

The structure of molecules is critical to the study of materials science and organic chemistry, and has countless applications to a variety of interesting phenomena. Trigonometry plays a critical role in determining the bonding angles of molecules. For instance, the structure of the $(FeCl_4Br_2)^{-3}$ ion (dibromatetrachlorideferrate III) is shown in the figure below.

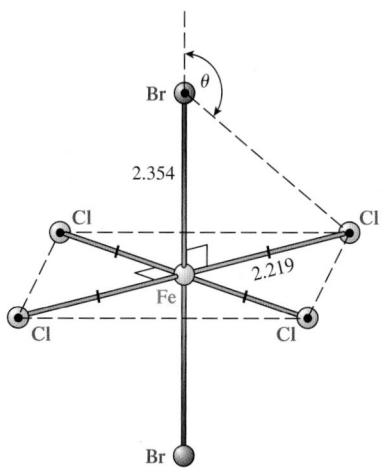

81. Chemistry. Determine the angle θ [i.e., the angle between the axis containing the apical bromide atom (Br) and the segment connecting Br to Cl].

82. Chemistry. Now, suppose one of the chlorides (Cl) is removed. The resulting structure is triagonal in nature, resulting in the following structure. Does the angle θ change? If so, what is its new value?

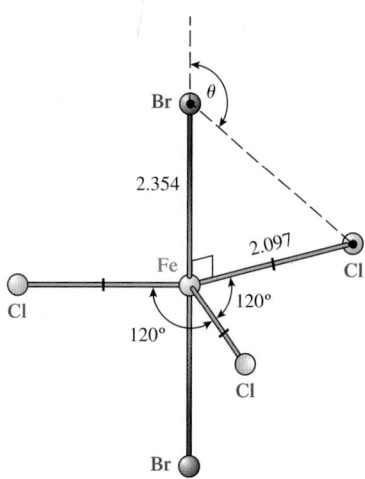

83. Construction. Two neighborhood kids are planning to build a treehouse in Tree I, and connect it with a zipline to Tree II that is 40 yards away. The base of the treehouse will be 20 feet above the ground, and a platform will be nailed into Tree II, 3 feet above the ground. The plan is to connect the base of the treehouse on Tree I to an anchor 2 feet above the platform on Tree II.

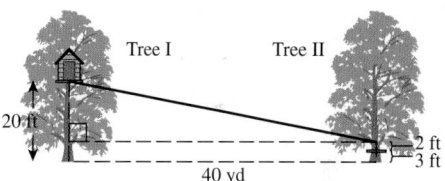

How much zipline (in feet) will they need? Round your answer to the nearest foot.

84. Construction. In Exercise 83, what is the angle of depression β that the zipline makes with Tree I? Express your answer in two significant digits.

85. Construction. A pool that measures 5 feet above the ground is to be placed between the trees directly in the path of the zipline in Exercise 83. Assuming that a rider of the zipline dangles at most 3 feet below the wire anywhere in route, what is the closest the edge of the pool can be placed to tree 2 so that the pool will not impede a rider's trip down the zipline?

86. Construction. If the treehouse is to be built so that its base is now 22 feet above the base of tree 1, where should the anchor on tree 2 (to which the zipline is connected) be placed in order to ensure the same angle of depression found in Exercise 84?

▪ CATCH THE MISTAKE

For Exercises 87–90, explain the mistake that is made.

For the triangle in the drawing, calculate the indicated trigonometric function values.

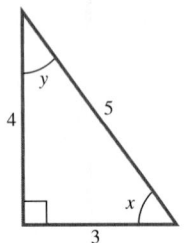

87. Calculate $\sin y$.

Solution:

| Formulate sine in terms of trigonometric ratios. | $\sin y = \dfrac{\text{opposite}}{\text{hypotenuse}}$ |

| The opposite side is 4, and the hypotenuse is 5. | $\sin y = \dfrac{4}{5}$ |

This is incorrect. What mistake was made?

88. Calculate $\tan x$.

Solution:

| Formulate tangent in terms of trigonometric ratios. | $\tan x = \dfrac{\text{adjacent}}{\text{opposite}}$ |

| The adjacent side is 3, and the opposite side is 4. | $\tan x = \dfrac{3}{4}$ |

This is incorrect. What mistake was made?

89. Calculate $\sec x$.

Solution:

| Formulate sine in terms of trigonometric ratios. | $\sin x = \dfrac{\text{opposite}}{\text{hypotenuse}}$ |

| The opposite side is 4, and the hypotenuse is 5. | $\sin x = \dfrac{4}{5}$ |

| Write secant as the reciprocal of sine. | $\sec x = \dfrac{1}{\sin x}$ |

| Simplify. | $\sec x = \dfrac{1}{\frac{4}{5}} = \dfrac{5}{4}$ |

This is incorrect. What mistake was made?

90. Calculate $\csc y$.

Solution:

| Formulate cosine in terms of trigonometric ratios. | $\cos y = \dfrac{\text{adjacent}}{\text{hypotenuse}}$ |

| The adjacent side is 4, and the hypotenuse is 5. | $\cos y = \dfrac{4}{5}$ |

| Write cosecant as the reciprocal of cosine. | $\csc y = \dfrac{1}{\cos y}$ |

| Simplify. | $\csc y = \dfrac{1}{\frac{4}{5}} = \dfrac{5}{4}$ |

This is incorrect. What mistake was made?

▪ CONCEPTUAL

In Exercises 91–94, determine whether each statement is true or false.

91. If you are given the measures of two sides of a right triangle, you can solve the right triangle.

92. If you are given the measures of one side and one acute angle of a right triangle, you can solve the right triangle.

93. If you are given the two acute angles of a right triangle, you can solve the right triangle.

94. If you are given the hypotenuse of a right triangle and the angle opposite the hypotenuse, you can solve the right triangle.

In Exercises 95–98, use trignometric ratios and the assumption that a is much larger than b.

Thus far, in this text we have discussed trigonometric values only for acute angles, or for $0° < \theta < 90°$. How do we determine these values when θ is approximately $0°$ or $90°$? We will formally consider these cases in the next section, but for now, draw and label a right triangle that has one angle very close to $0°$, so that the opposite side is very small compared to the adjacent side. Then the hypotenuse and the adjacent side will be very close to the same length.

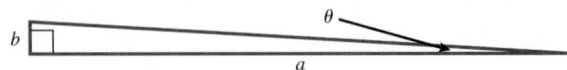

95. Approximate $\sin 0°$ without using a calculator.

96. Approximate $\cos 0°$ without using a calculator.

97. Approximate $\cos 90°$ without using a calculator.

98. Approximate $\sin 90°$ without using a calculator.

■ CHALLENGE

For Exercises 99 and 100, consider the following diagram:

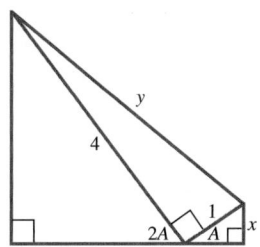

99. Determine x.

100. Determine y.

■ TECHNOLOGY

101. Calculate $\sec 70°$ in the following two ways:

  a. Find $\cos 70°$ to three decimal places and then divide 1 by that number. Write that number to five decimal places.

  b. With a calculator in degree mode, enter 70, cos, 1/x, and round the result to five decimal places.

102. Calculate $\csc 40°$ in the following two ways:

  a. Find $\sin 40°$ to three decimal places and then divide 1 by that number. Write this last result to five decimal places.

  b. With a calculator in degree mode, enter 40, sin, 1/x, and round the result to five decimal places.

103. Calculate $\cot 54.9°$ in the following two ways:

  a. Find $\tan 54.9°$ to three decimal places and then divide 1 by that number. Write that number to five decimal places.

  b. With a calculator in degree mode, enter 54.9, tan, 1/x, and round the result to five decimal places.

104. Calculate $\sec 18.6°$ in the following two ways:

  a. Find $\cos 18.6°$ to three decimal places and then divide 1 by that number. Write that number to five decimal places.

  b. With a calculator in degree mode, enter 18.6, cos, 1/x, and round the result to five decimal places.

■ PREVIEW TO CALCULUS

In calculus, the value of $F(b) - F(a)$ of a function $F(x)$ at $x = a$ and $x = b$ plays an important role in the calculation of definite integrals. In Exercises 105–108, find the exact value of $F(b) - F(a)$.

105. $F(x) = \sec x, a = \dfrac{\pi}{6}, b = \dfrac{\pi}{3}$

106. $F(x) = \sin^3 x, a = 0, b = \dfrac{\pi}{4}$

107. $F(x) = \tan x + 2\cos x, a = 0, b = \dfrac{\pi}{3}$

108. $F(x) = \dfrac{\cot x - 4\sin x}{\cos x}, a = \dfrac{\pi}{4}, b = \dfrac{\pi}{3}$

SKILLS OBJECTIVES

- Calculate trigonometric function values for nonacute angles.
- Determine the reference angle of a nonacute angle.
- Calculate the trigonometric function values for quadrantal angles.

CONCEPTUAL OBJECTIVES

- Understand that right triangle ratio definitions of trigonometric functions for acute angles are consistent with definitions of trigonometric functions for all angles in the Cartesian plane.
- Understand why some trigonometric functions are undefined for quadrantal angles.
- Understand that the ranges of the sine and cosine functions are bounded, whereas the ranges for the other four trigonometric functions are unbounded.

In Section 4.2, we defined trigonometric functions as ratios of side lengths of right triangles. This definition holds only for acute $(0° < \theta < 90°)$ angles, since the two angles in a right triangle other than the right angle must be acute. We now define trigonometric functions as ratios of x- and y-coordinates and distances in the Cartesian plane, which for acute angles is consistent with right triangle trigonometry. However, this second approach also enables us to formulate trigonometric functions for quadrantal angles (whose terminal side lies along an axis) and nonacute angles.

Trigonometric Functions: The Cartesian Plane

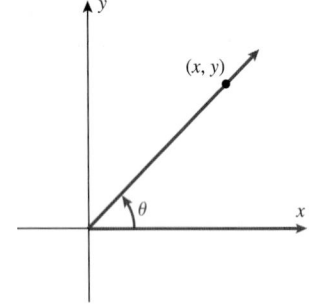

To define the trigonometric functions in the Cartesian plane, let us start with an acute angle θ in standard position. Choose any point (x, y) on the terminal side of the angle as long as it is not the vertex (the origin).

A right triangle can be drawn so that the right angle is made when a perpendicular segment connects the point (x, y) to the x-axis. Notice that the side opposite θ has length y and the other leg of the right triangle has length x.

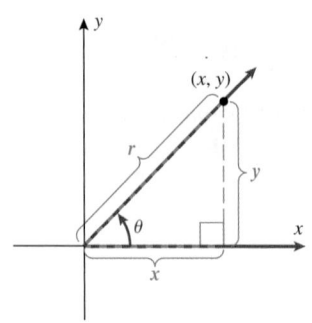

The distance r from the origin $(0, 0)$ to the point (x, y) can be found using the distance formula.

$$r = \sqrt{(x - 0)^2 + (y - 0)^2}$$
$$r = \sqrt{x^2 + y^2}$$

Since r is a distance and x and y are not both zero, r is always positive.

$$r > 0$$

Using our first definition of trigonometric functions in terms of right triangle ratios (Section 4.2), we know that $\sin \theta = \dfrac{\text{opposite}}{\text{hypotenuse}}$. From this picture, we see that sine can also be defined by the relation $\sin \theta = \dfrac{y}{r}$. Similar reasoning holds for all six trigonometric functions and leads us to the second definition of the trigonometric functions, in terms of ratios of coordinates of a point and distances in the Cartesian plane.

DEFINITION   **Trigonometric Functions**

Let (x, y) be a point, other than the origin, on the terminal side of an angle θ in standard position. Let r be the distance from the point (x, y) to the origin. Then the six trigonometric functions are defined as

$$\sin \theta = \frac{y}{r} \qquad\qquad \cos \theta = \frac{x}{r} \qquad\qquad \tan \theta = \frac{y}{x} \quad (x \neq 0)$$

$$\csc \theta = \frac{r}{y} \ \ (y \neq 0) \qquad \sec \theta = \frac{r}{x} \ \ (x \neq 0) \qquad \cot \theta = \frac{x}{y} \ \ (y \neq 0)$$

where $r = \sqrt{x^2 + y^2}$, or $x^2 + y^2 = r^2$. The distance r is positive: $r > 0$.

EXAMPLE 1   **Calculating Trigonometric Function Values for Acute Angles**

The terminal side of an angle θ in standard position passes through the point $(2, 5)$. Calculate the values of the six trigonometric functions for angle θ.

Solution:

STEP 1 Draw the angle and label the point **(2, 5)**.

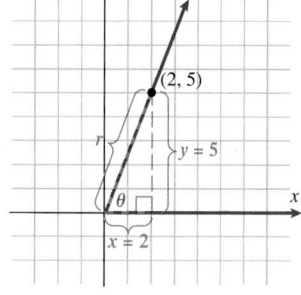

STEP 2 Calculate the distance *r*.

$$r = \sqrt{2^2 + 5^2} = \sqrt{29}$$

STEP 3 Formulate the trigonometric functions in terms of *x, y,* and *r*.

Let $x = 2, y = 5, r = \sqrt{29}$.

$$\sin \theta = \frac{y}{r} = \frac{5}{\sqrt{29}} \qquad\qquad \cos \theta = \frac{x}{r} = \frac{2}{\sqrt{29}} \qquad\qquad \tan \theta = \frac{y}{x} = \frac{5}{2}$$

$$\csc \theta = \frac{r}{y} = \frac{\sqrt{29}}{5} \qquad\qquad \sec \theta = \frac{r}{x} = \frac{\sqrt{29}}{2} \qquad\qquad \cot \theta = \frac{x}{y} = \frac{2}{5}$$

STEP 4 **Rationalize any denominators containing a radical.**

$$\sin\theta = \frac{5}{\sqrt{29}}\cdot\frac{\sqrt{29}}{\sqrt{29}} = \frac{5\sqrt{29}}{29} \qquad \cos\theta = \frac{2}{\sqrt{29}}\cdot\frac{\sqrt{29}}{\sqrt{29}} = \frac{2\sqrt{29}}{29}$$

STEP 5 **Write the values of the six trigonometric functions for θ.**

$$\sin\theta = \frac{5\sqrt{29}}{29} \qquad \cos\theta = \frac{2\sqrt{29}}{29} \qquad \tan\theta = \frac{5}{2}$$

$$\csc\theta = \frac{\sqrt{29}}{5} \qquad \sec\theta = \frac{\sqrt{29}}{2} \qquad \cot\theta = \frac{2}{5}$$

Note: In Example 1, we could have used the values of the sine, cosine, and tangent functions along with the reciprocal identities to calculate the cosecant, secant, and cotangent function values.

■ Study Tip

There is no need to memorize definitions for secant, cosecant, and cotangent functions, since their values can be derived from the reciprocals of the sine, cosine, and tangent function values.

■ Answer:

$$\sin\theta = \frac{7\sqrt{58}}{58} \qquad \cos\theta = \frac{3\sqrt{58}}{58}$$

$$\tan\theta = \frac{7}{3} \qquad \csc\theta = \frac{\sqrt{58}}{7}$$

$$\sec\theta = \frac{\sqrt{58}}{3} \qquad \cot\theta = \frac{3}{7}$$

■ YOUR TURN The terminal side of an angle θ in standard position passes through the point $(3, 7)$. Calculate the values of the six trigonometric functions for angle θ.

We can now find values for nonacute angles (angles with measure greater than or equal to 90°) as well as negative angles.

EXAMPLE 2 **Calculating Trigonometric Function Values for Nonacute Angles**

The terminal side of an angle θ in standard position passes through the point $(-4, -7)$. Calculate the values of the six trigonometric functions for angle θ.

Solution:

STEP 1 **Draw the angle and label the point $(-4, -7)$.**

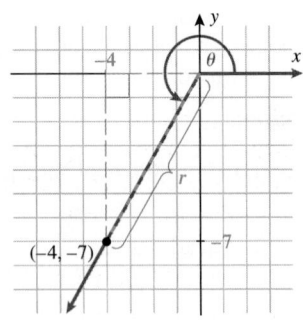

STEP 2 **Calculate the distance r.** $\qquad r = \sqrt{(-4)^2 + (-7)^2} = \sqrt{65}$

STEP 3 **Formulate the trigonometric functions in terms of x, y, and r.**

Let $x = -4, y = -7$, and $r = \sqrt{65}$.

$$\sin\theta = \frac{y}{r} = \frac{-7}{\sqrt{65}} \qquad \cos\theta = \frac{x}{r} = \frac{-4}{\sqrt{65}} \qquad \tan\theta = \frac{y}{x} = \frac{-7}{-4} = \frac{7}{4}$$

$$\csc\theta = \frac{r}{y} = \frac{\sqrt{65}}{-7} \qquad \sec\theta = \frac{r}{x} = \frac{\sqrt{65}}{-4} \qquad \cot\theta = \frac{x}{y} = \frac{-4}{-7} = \frac{4}{7}$$

STEP 4 **Rationalize the radical denominators in the sine and cosine functions.**

$$\sin\theta = \frac{y}{r} = \frac{-7}{\sqrt{65}} \cdot \frac{\sqrt{65}}{\sqrt{65}} = -\frac{7\sqrt{65}}{65}$$

$$\cos\theta = \frac{x}{r} = \frac{-4}{\sqrt{65}} \cdot \frac{\sqrt{65}}{\sqrt{65}} = -\frac{4\sqrt{65}}{65}$$

STEP 5 **Write the values of the six trigonometric functions for θ.**

$$\sin\theta = -\frac{7\sqrt{65}}{65}$$

$$\cos\theta = -\frac{4\sqrt{65}}{65}$$

$$\tan\theta = \frac{7}{4}$$

$$\csc\theta = -\frac{\sqrt{65}}{7}$$

$$\sec\theta = -\frac{\sqrt{65}}{4}$$

$$\cot\theta = \frac{4}{7}$$

■ **YOUR TURN** The terminal side of an angle θ in standard position passes through the point $(-3, -5)$. Calculate the values of the six trigonometric functions for angle θ.

■ **Answer:**

$$\sin\theta = -\frac{5\sqrt{34}}{34} \qquad \cos\theta = -\frac{3\sqrt{34}}{34}$$

$$\tan\theta = \frac{5}{3} \qquad \csc\theta = -\frac{\sqrt{34}}{5}$$

$$\sec\theta = -\frac{\sqrt{34}}{3} \qquad \cot\theta = \frac{3}{5}$$

Algebraic Signs of Trigonometric Functions

We have defined trigonometric functions as ratios of x, y, and r. Since r is the distance from the origin to the point (x, y) and distance is never negative, r is always taken as the positive solution to $r^2 = x^2 + y^2$, so $r = \sqrt{x^2 + y^2}$.

The x-coordinate is positive in quadrants **I** and **IV** and negative in quadrants **II** and **III**. The y-coordinate is positive in quadrants **I** and **II** and negative in quadrants **III** and **IV**. Recall the definition of the six trigonometric functions in the Cartesian plane:

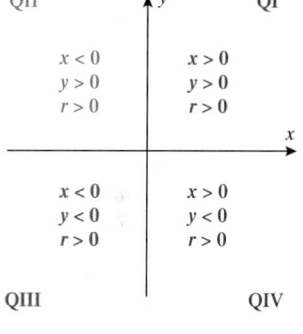

$$\sin\theta = \frac{y}{r} \qquad \cos\theta = \frac{x}{r} \qquad \tan\theta = \frac{y}{x} \ (x \neq 0)$$

$$\csc\theta = \frac{r}{y} \ (y \neq 0) \qquad \sec\theta = \frac{r}{x} \ (x \neq 0) \qquad \cot\theta = \frac{x}{y} \ (y \neq 0)$$

Therefore, the algebraic sign, $+$ or $-$, of each trigonometric function will depend on which quadrant contains the terminal side of angle θ. Let us look at the three main trigonometric functions: sine, cosine, and tangent. In quadrant I, all three functions are positive since x, y, and r are all positive. However, in quadrant II, only sine is positive since y and r are both positive. In quadrant III, only tangent is positive, and in quadrant IV, only cosine is positive. The expression "**A**ll **S**tudents **T**ake **C**alculus" helps us remember which of the three main trigonometric functions are positive in each quadrant.

Study Tip

All Students Take Calculus is an expression that helps us remember which of the three (sine, cosine, tangent) functions are positive in quadrants I, II, III, and IV.

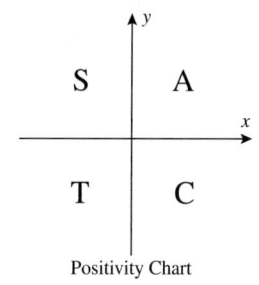

Positivity Chart

PHRASE	QUADRANT	POSITIVE TRIGONOMETRIC FUNCTION
All	I	All three: sine, cosine, and tangent
Students	II	Sine
Take	III	Tangent
Calculus	IV	Cosine

The following table indicates the algebraic sign of all six trigonometric functions according to the quadrant in which the terminal side of an angle θ lies. Notice that the reciprocal functions have the same sign.

TERMINAL SIDE OF θ IN QUADRANT	$\sin\theta$	$\cos\theta$	$\tan\theta$	$\cot\theta$	$\sec\theta$	$\csc\theta$
I	+	+	+	+	+	+
II	+	−	−	−	−	+
III	−	−	+	+	−	−
IV	−	+	−	−	+	−

Technology Tip

To draw the terminal side of the angle θ in quadrant III through the point $(-3, -4)$, press GRAPH
2nd: DRAW ▼ 2:Line(
ENTER .

Now use arrows to move the cursor to the origin and press ENTER .

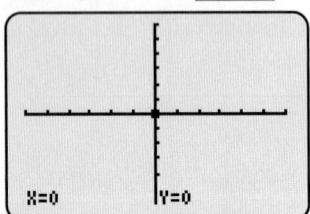

X=0   Y=0

Use ◄ to move the cursor to the left to about $x = -3$ and press
ENTER . Use ▼ to move the cursor down to about $y = -4$ and press
ENTER .

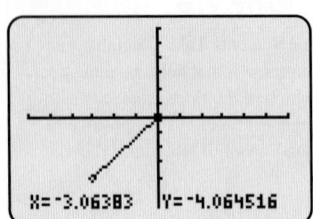

X=-3.06383   Y=-4.064516

■ **Answer:** $\cos\theta = -\dfrac{\sqrt{7}}{4}$

EXAMPLE 3   Evaluating a Trigonometric Function When One Trigonometric Function Value and the Quadrant of the Terminal Side Is Known

If $\cos\theta = -\frac{3}{5}$ and the terminal side of angle θ lies in quadrant III, find $\sin\theta$.

Solution:

STEP 1 Draw some angle θ in QIII.

STEP 2 Identify known quantities from the information given.

Recall that $\cos\theta = \dfrac{x}{r}$ and $r > 0$.

$\cos\theta = -\dfrac{3}{5} = \dfrac{-3}{5} = \dfrac{x}{r}$

Identify x and r. $x = -3$ and $r = 5$

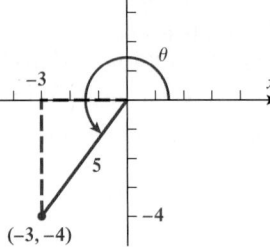

STEP 3 Since x and r are known, find y.

Substitute $x = -3$ and $r = 5$ into $x^2 + y^2 = r^2$. $(-3)^2 + y^2 = 5^2$

Solve for y. $9 + y^2 = 25$

$$y^2 = 16$$

$$y = \pm 4$$

STEP 4 Select the sign of y based on quadrant information.

Since the terminal side of angle θ lies in quadrant III, $y < 0$. $y = -4$

STEP 5 Find $\sin\theta$.

$$\sin\theta = \dfrac{y}{r} = \dfrac{-4}{5}$$

$$\boxed{\sin\theta = -\dfrac{4}{5}}$$

■ **YOUR TURN** If $\sin\theta = -\frac{3}{4}$ and the terminal side of angle θ lies in quadrant III, find $\cos\theta$.

We can also make a table showing the values of the trigonometric functions when the terminal side of angle θ lies along each axis (i.e., when θ is any of the quadrantal angles).

When the terminal side lies along the x-axis, then $y = 0$. When $y = 0$, notice that $r = \sqrt{x^2 + y^2} = \sqrt{x^2} = |x|$. When the terminal side lies along the positive x-axis, $x > 0$; and when the terminal side lies along the negative x-axis, $x < 0$. Therefore, when the terminal side of the angle lies on the positive x-axis, then $y = 0$, $x > 0$, and $r = x$; and when the terminal side lies along the negative x-axis, then $y = 0$, $x < 0$, and $r = |x|$. A similar argument can be made when the terminal side lies along the y-axis, which results in $r = |y|$.

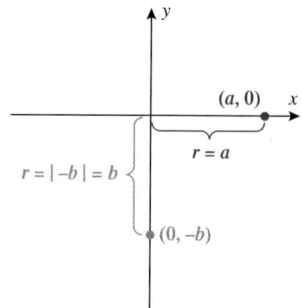

TERMINAL SIDE OF θ LIES ALONG THE	$\sin \theta$	$\cos \theta$	$\tan \theta$	$\cot \theta$	$\sec \theta$	$\csc \theta$
Positive x-axis (e.g., 0° or 360° or 0 or 2π)	0	1	0	undefined	1	undefined
Positive y-axis $\left(\text{e.g., }90° \text{ or } \dfrac{\pi}{2}\right)$	1	0	undefined	0	undefined	1
Negative x-axis (e.g., 180° or π)	0	-1	0	undefined	-1	undefined
Negative y-axis $\left(\text{e.g., }270° \text{ or } \dfrac{3\pi}{2}\right)$	-1	0	undefined	0	undefined	-1

EXAMPLE 4 Working with Values of the Trigonometric Functions for Quadrantal Angles

Evaluate each of the following expressions, if possible:

a. $\cos 540° + \sin 270°$ **b.** $\cot\left(\dfrac{\pi}{2}\right) + \tan\left(-\dfrac{\pi}{2}\right)$

Solution (a):

The terminal side of an angle with measure 540° lies along the negative x-axis.

$$540° - 360° = 180°$$

Evaluate cosine of an angle whose terminal side lies along the negative x-axis.

$$\cos 540° = -1$$

Evaluate sine of an angle whose terminal side lies along the negative y-axis.

$$\sin 270° = -1$$

Sum the sine and cosine values.

$$\cos 540° + \sin 270° = -1 + (-1)$$

$$\cos 540° + \sin 270° = \boxed{-2}$$

Check: Evaluate this expression with a calculator.

Solution (b):

Evaluate cotangent of an angle whose terminal side lies along the positive y-axis.

$$\cot\left(\dfrac{\pi}{2}\right) = 0$$

The terminal side of an angle with measure $-\dfrac{\pi}{2}$ lies along the negative y-axis.

$$\tan\left(-\dfrac{\pi}{2}\right) = \tan\left(\dfrac{3\pi}{2}\right)$$

The tangent function is undefined for an angle whose terminal side lies along the negative y-axis.

$\tan\left(-\dfrac{\pi}{2}\right)$ is $\boxed{\text{undefined}}$

Even though $\cot\left(\dfrac{\pi}{2}\right)$ is defined, since $\tan\left(-\dfrac{\pi}{2}\right)$ is undefined, the sum of the two expressions is also undefined.

■ **Answer: a.** 0   **b.** undefined

■ **YOUR TURN** Evaluate each of the following expressions, if possible:

     a. $\csc\left(\dfrac{\pi}{2}\right) + \sec\pi$ **b.** $\csc(-630°) + \sec(-630°)$

Ranges of the Trigonometric Functions

Thus far, we have discussed what the algebraic sign of a trigonometric function value for an angle in a particular quadrant, but we haven't discussed how to find actual values of the trigonometric functions for nonacute angles. We will need to define *reference angles* and reference right triangles. However, before we proceed, let's get a feel for the ranges (set of values of the functions) we will expect.

Let us start with an angle θ in quadrant I and the sine function defined as the ratio $\sin\theta = \dfrac{y}{r}$.

$$\sin\theta = \frac{y}{r}$$

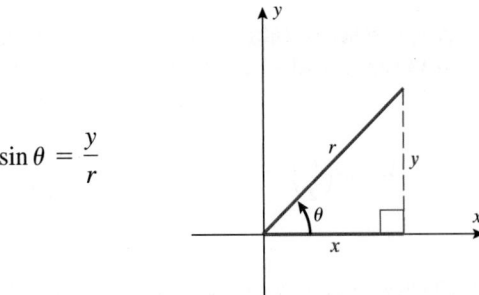

If we keep the value of r constant, then as the measure of θ increases toward 90° or $\dfrac{\pi}{2}$, y increases. Notice that the value of y approaches the value of r until they are equal when $\theta = 90°\left(\text{or } \dfrac{\pi}{2}\right)$, and y can never be larger than r.

$$y \le r$$

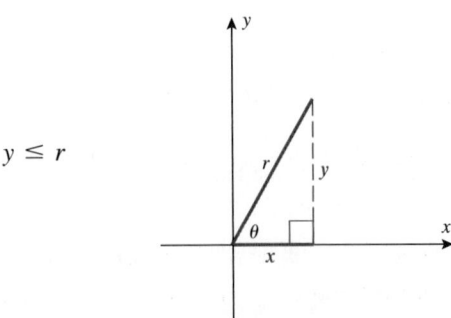

A similar analysis can be conducted in quadrant IV as θ approaches $-90°$ from $0°$ (note that y is negative in quadrant IV). A result that is valid in all four quadrants is $|y| \leq r$.

WORDS	MATH
Write the absolute value inequality as a double inequality.	$-r \leq y \leq r$
Divide both sides by r.	$-1 \leq \dfrac{y}{r} \leq 1$
Let $\sin \theta = \dfrac{y}{r}$.	$-1 \leq \sin \theta \leq 1$

Similarly, by allowing θ to approach $0°$ and $180°$, we can show that $|x| \leq r$, which leads to the range of the cosine function: $-1 \leq \cos\theta \leq 1$. Sine and cosine values range between -1 and 1 and since secant and cosecant are reciprocals of the cosine and sine functions, respectively, their ranges are stated as

$$\sec\theta \leq -1 \text{ or } \sec\theta \geq 1 \qquad \csc\theta \leq -1 \text{ or } \csc\theta \geq 1$$

Since $\tan\theta = \dfrac{y}{x}$ and $\cot\theta = \dfrac{x}{y}$ and since $x < y, x = y,$ and $x > y$ are all possible, the values of the tangent and cotangent functions can be any real numbers (positive, negative, or zero). The following box summarizes the ranges of the trigonometric functions.

RANGES OF THE TRIGONOMETRIC FUNCTIONS

For any angle θ for which the trigonometric functions are defined, the six trigonometric functions have the following ranges:

- $-1 \leq \sin\theta \leq 1$
- $-1 \leq \cos\theta \leq 1$
- $\tan\theta$ and $\cot\theta$ can equal any real number.
- $\sec\theta \leq -1$ or $\sec\theta \geq 1$
- $\csc\theta \leq -1$ or $\csc\theta \geq 1$

EXAMPLE 5   **Determining Whether a Value Is Within the Range of a Trigonometric Function**

Determine whether each statement is possible or not.

a. $\cos\theta = 1.001$

b. $\cot\theta = 0$

c. $\sec\theta = \dfrac{\sqrt{3}}{2}$

Solution (a): Not possible, because $1.001 > 1$.

Solution (b): Possible, because $\cot 90° = 0$.

Solution (c): Not possible, because $\dfrac{\sqrt{3}}{2} \approx 0.866 < 1$.

▪ **YOUR TURN** Determine whether each statement is possible or not.

    a. $\sin\theta = -1.1$ **b.** $\tan\theta = 2$ **c.** $\csc\theta = \sqrt{3}$

▪ **Answer: a.** not possible
    b. possible
    c. possible

Reference Angles and Reference Right Triangles

Now that we know the trigonometric function ranges and their algebraic signs in each of the four quadrants, we can evaluate the trigonometric functions of nonacute angles. Before we do that, however, we first must discuss *reference angles* and *reference right triangles*.

Every nonquadrantal angle in standard position has a corresponding *reference angle* and *reference right triangle*. We have already calculated the trigonometric function values for quadrantal angles.

DEFINITION   **Reference Angle**

For angle θ, $0° < \theta < 360°$ or $0 < \theta < 2\pi$, in standard position whose terminal side lies in one of the four quadrants, there exists a **reference angle** α, which is the acute angle formed by the terminal side of angle θ and the *x*-axis.

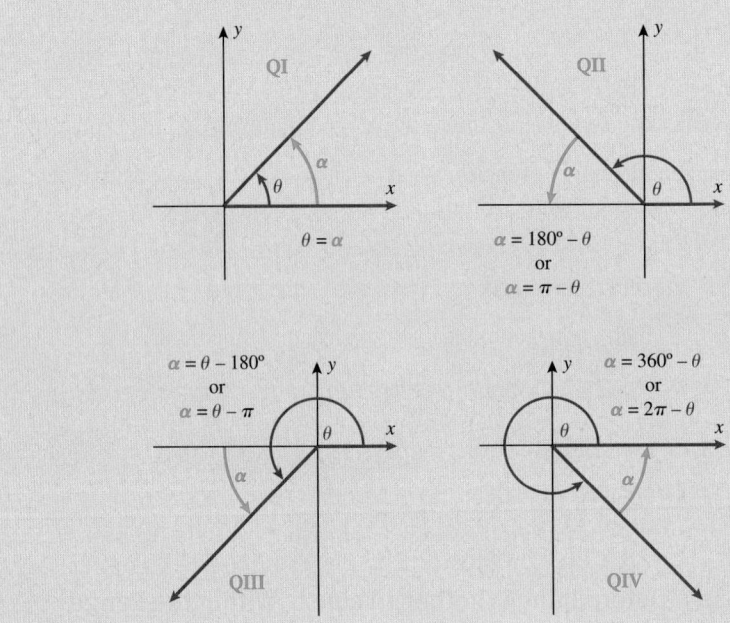

The reference angle is the positive, acute angle that the terminal side makes with the *x*-axis.

EXAMPLE 6   Finding Reference Angles

Find the reference angle for each angle given.

a. $210°$ **b.** $\dfrac{3\pi}{4}$ **c.** $422°$

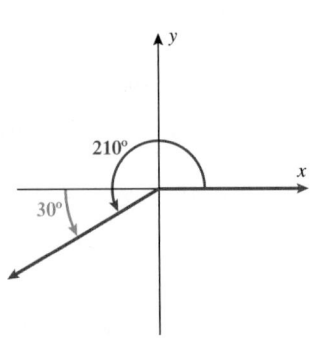

Solution (a):

The terminal side of angle θ lies in quadrant III.

The reference angle is formed by the terminal side and the negative *x*-axis.

$210° - 180° = \boxed{30°}$

Solution (b):

The terminal side of angle θ lies in quadrant II.

The reference angle is formed by the terminal side and the negative x-axis.

$$\pi - \frac{3\pi}{4} = \boxed{\frac{\pi}{4}}$$

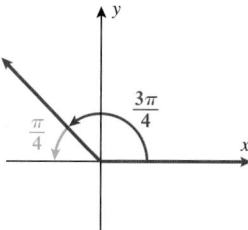

Solution (c):

The terminal side of angle θ lies in quadrant I.

The reference angle is formed by the terminal side and the positive x-axis.

$$422° - 360° = \boxed{62°}$$

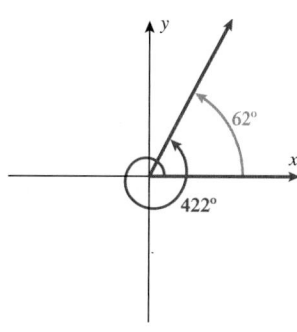

■ **YOUR TURN** Find the reference angle for each angle given.

    a. $160°$ **b.** $\dfrac{7\pi}{4}$ **c.** $600°$

■ **Answer:**

  a. $20°$ **b.** $\dfrac{\pi}{4}$ **c.** $60°$

DEFINITION   **Reference Right Triangle**

To form a **reference right triangle** for angle θ, where $0° < \theta < 360°$ or $0 < \theta < 2\pi$, drop a perpendicular line from the terminal side of the angle to the x-axis. The right triangle now has reference angle α as one of its acute angles.

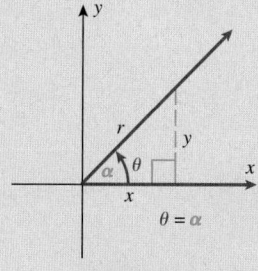

$\theta = \alpha$

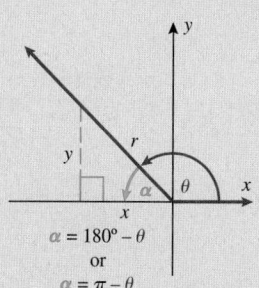

$\alpha = 180° - \theta$
or
$\alpha = \pi - \theta$

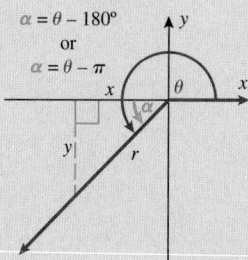

$\alpha = \theta - 180°$
or
$\alpha = \theta - \pi$

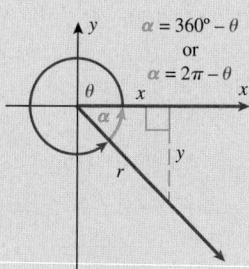

$\alpha = 360° - \theta$
or
$\alpha = 2\pi - \theta$

In Section 4.2, we first defined the trigonometric functions of an acute angle as ratios of lengths of sides of a right triangle. For example, $\sin\theta = \dfrac{\text{opposite}}{\text{hypotenuse}}$. The lengths of the sides of triangles are always positive.

In this section, we defined the sine function of any angle as $\sin\theta = \dfrac{y}{r}$. Notice in the above box that for a nonacute angle θ, $\sin\theta = \dfrac{y}{r}$; and for the acute reference angle α, $\sin\alpha = \dfrac{|y|}{r}$. The only difference between these two expressions is the algebraic sign, since r is always positive and y is positive or negative depending on the quadrant.

Therefore, to calculate the trigonometric function values for a nonacute angle, simply find the trigonometric values for the reference angle and determine the correct algebraic sign according to the quadrant in which the terminal side lies.

Evaluating Trigonometric Functions for Nonacute Angles

Let's look at a specific example before we generalize a procedure for evaluating trigonometric function values for nonacute angles.

Suppose we have the angles in standard position with measure $60°, 120°, 240°$, and $300°$ or $\dfrac{\pi}{3}, \dfrac{2\pi}{3}, \dfrac{4\pi}{3}$, or $\dfrac{5\pi}{3}$, respectively. Notice that the reference angle for all these angles is $60°$ or $\dfrac{\pi}{3}$.

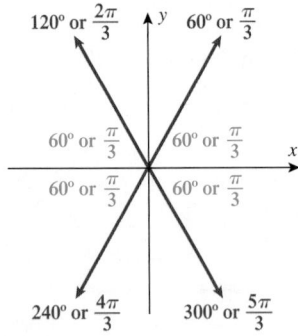

If we draw reference triangles and let the shortest leg have length 1, we find that the other leg has length $\sqrt{3}$ and the hypotenuse has length 2. (Recall the relationships for side lengths of a $30°$–$60°$–$90°$ triangle.)

Notice that the legs of the triangles have lengths (always positive) 1 and $\sqrt{3}$; however, the coordinates are $(\pm1, \pm\sqrt{3})$. Therefore, when we calculate the trigonometric functions for any of the angles, $60°\left(\dfrac{\pi}{3}\right)$, $120°\left(\dfrac{2\pi}{3}\right)$, $240°\left(\dfrac{4\pi}{3}\right)$, and $300°\left(\dfrac{5\pi}{3}\right)$, we can simply calculate the trigonometric functions for the reference angle, $60°\left(\dfrac{\pi}{3}\right)$, and determine the algebraic sign $(+$ or $-)$ for the particular trigonometric function and quadrant.

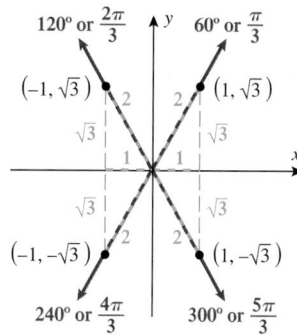

Study Tip

The value of a trigonometric function of an angle is the same as the trigonometric value of its reference angle, except there may be an algebraic sign (+ or −) difference between the two values.

To find the value of $\cos 120°$, we first recognize that the terminal side of an angle with 120° measure lies in quadrant II. We also know that cosine is negative in quadrant II. We then calculate the cosine of the reference angle, $60°$.

$$\cos 60° = \frac{\text{adjacent}}{\text{hypotenuse}} = \frac{1}{2}$$

Since we know $\cos 120°$ is negative because it lies in quadrant II, we know that

$$\cos 120° = -\frac{1}{2}$$

Similarly, we know that $\cos 240° = -\frac{1}{2}$ and $\cos 300° = \frac{1}{2}$.

For any angle whose terminal side lies along one of the axes, we consult the table in this section for the values of the trigonometric functions for quadrantal angles. If the terminal side lies in one of the four quadrants, then the angle is said to be nonquadrantal and the following procedure can be used.

PROCEDURE FOR EVALUATING FUNCTION VALUES FOR ANY NONQUADRANTAL ANGLE θ

Step 1: ■ If $0° < \theta < 360°$ or $0 < \theta < 2\pi$, proceed to Step 2.
 ■ If $\theta < 0°$, add 360° as many times as needed to get a coterminal angle with measure between 0° and 360°. Similarly, if $\theta < 0$, add 2π as many times as needed to get a coterminal angle with measure between 0 and 2π.
 ■ If $\theta > 360°$, subtract 360° as many times as needed to get a coterminal angle with measure between 0° and 360°. Similarly, if $\theta > 2\pi$, subtract 2π as many times as needed to get a coterminal angle with measure between 0 and 2π.

Step 2: Find the quadrant in which the terminal side of the angle in Step 1 lies.

Step 3: Find the reference angle α of the angle found in Step 1.

Step 4: Find the trigonometric function values for the reference angle α.

Step 5: Determine the correct algebraic signs (+ or −) for the trigonometric function values based on the quadrant identified in Step 2.

Step 6: Combine the trigonometric values found in Step 4 with the algebraic signs in Step 5 to give the trigonometric function values of θ.

We follow the above procedure for all angles except when we get to Step 4. In Step 4, we evaluate exactly if possible the special angles $\left(30°, \ 45°, \ 60° \text{ or } \dfrac{\pi}{6}, \dfrac{\pi}{4}, \dfrac{\pi}{3}\right)$; otherwise, we use a calculator to approximate.

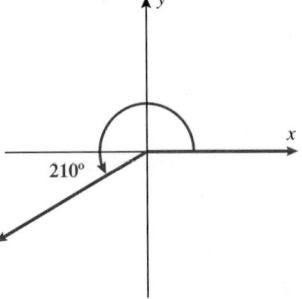

EXAMPLE 7 Evaluating the Cosine Function of a Special Angle Exactly

Find the exact value of $\cos 210°$.

Solution:

The terminal side of $\theta = 210°$ lies in quadrant III.

Find the reference angle for $\theta = 210°$. $210° - 180° = 30°$

Find the value of the cosine of the reference angle. $\cos 30° = \dfrac{\sqrt{3}}{2}$

Determine the algebraic sign for the cosine in quadrant III.   Negative $(-)$

Combine the algebraic sign of the cosine in quadrant III with the value of the cosine of the reference angle.

$$\cos 210° = \boxed{-\dfrac{\sqrt{3}}{2}}$$

 Answer: $-\dfrac{1}{2}$

■ **YOUR TURN** Find the exact value of $\sin 330°$.

EXAMPLE 8 Evaluating the Cosecant Function of a Special Angle Exactly

Find the exact value of $\csc\left(-\dfrac{7\pi}{6}\right)$.

Solution:

Add 2π to get a coterminal angle between 0 and 2π.

$$-\frac{7\pi}{6} + 2\pi = \frac{5\pi}{6}$$

The terminal side of the angle lies in quadrant II.

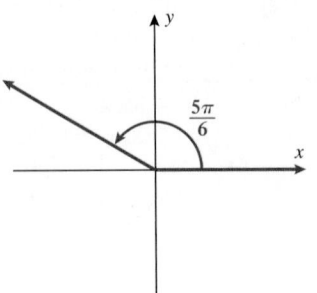

Find the reference angle for the angle with measure $\dfrac{5\pi}{6}$.

$$\pi - \dfrac{5\pi}{6} = \dfrac{\pi}{6}$$

Find the value of the cosecant of the reference angle.

$$\csc\left(\dfrac{\pi}{6}\right) = \dfrac{1}{\sin\left(\dfrac{\pi}{6}\right)} = \dfrac{1}{\dfrac{1}{2}} = 2$$

Determine the algebraic sign for the cosecant in quadrant II.

Positive $(+)$

Combine the algebraic sign of the cosecant in quadrant II with the value of the cosecant of the reference angle.

$$\csc\left(-\dfrac{7\pi}{6}\right) = \boxed{2}$$

▪ **YOUR TURN** Find the exact value of $\sec\left(-\dfrac{11\pi}{6}\right)$.

▪ **Answer:** $\dfrac{2\sqrt{3}}{3}$

EXAMPLE 9   Finding Exact Angle Measures Given Trigonometric Function Values

Find all values of θ where $0° \leq \theta \leq 360°$, when $\sin\theta = -\dfrac{\sqrt{3}}{2}$.

Solution:

Determine in which quadrants sine is negative.

QIII and QIV

Since the absolute value of $\sin\theta$ is $\dfrac{\sqrt{3}}{2}$, the reference angle has measure $60°$.

$$\sin 60° = \dfrac{\sqrt{3}}{2}$$

Determine the angles between $180°$ and $360°$ in QIII and QIV with reference angle $60°$.

Quadrant III: $180° + 60° = 240°$

Quadrant IV: $360° - 60° = 300°$

The two angles are $\boxed{240°}$ and $\boxed{300°}$.

▪ **YOUR TURN** Find all values of θ where $0° \leq \theta \leq 360°$, when $\cos\theta = -\dfrac{\sqrt{3}}{2}$.

▪ **Answer:** $150°$ and $210°$

EXAMPLE 10 Finding Approximate Angle Measures Given Trigonometric Function Values

Find the measure of an angle θ (rounded to the nearest degree) if $\sin \theta = -0.6293$ and the terminal side of θ (in standard position) lies in quadrant III, where $0° \le \theta \le 360°$.

Solution:

The sine of the reference angle is 0.6293.	$\sin \alpha = 0.6293$
Find the reference angle.	$\alpha = \sin^{-1}(0.6293) \approx 38.998°$
Round the reference angle to the nearest degree.	$\alpha \approx 39°$
Find θ, which lies in quadrant III.	$\theta \approx 180° + 39° \approx 219°$
	$\boxed{\theta \approx 219°}$
Check with a calculator.	$\sin 219° \approx -0.6293$

■ **Answer:** 122°

■ **YOUR TURN** Find the measure of θ, the smallest positive angle (rounded to the nearest degree), if $\cos \theta = -0.5299$ and the terminal side of θ (in standard position) lies in quadrant II.

SECTION 4.3 SUMMARY

The trigonometric functions are defined in the Cartesian plane for any angle as follows:

Let (x, y) be a point, other than the origin, on the terminal side of an angle θ in standard position. Let r be the distance from the point (x, y) to the origin. Then the sine, cosine, and tangent functions are defined as

$$\sin \theta = \frac{y}{r} \qquad \cos \theta = \frac{x}{r} \qquad \tan \theta = \frac{y}{x} \quad (x \ne 0)$$

The range of the sine and cosine functions is $[-1, 1]$, whereas the range of the secant and cosecant functions is $(-\infty, -1] \cup [1, \infty)$.

Reference angles and reference right triangles can be used to evaluate trigonometric functions for nonacute angles.

SECTION 4.3 EXERCISES

SKILLS

In Exercises 1–14, the terminal side of an angle θ in standard position passes through the indicated point. Calculate the values of the six trigonometric functions for angle θ.

1. $(3, 6)$
2. $(8, 4)$
3. $\left(\frac{1}{2}, \frac{2}{5}\right)$
4. $\left(\frac{4}{7}, \frac{2}{3}\right)$
5. $(-2, 4)$

6. $(-1, 3)$
7. $(-4, -7)$
8. $(-9, -5)$
9. $\left(-\sqrt{2}, \sqrt{3}\right)$
10. $\left(-\sqrt{3}, \sqrt{2}\right)$

11. $\left(-\sqrt{5}, -\sqrt{3}\right)$
12. $\left(-\sqrt{6}, -\sqrt{5}\right)$
13. $\left(-\frac{10}{3}, -\frac{4}{3}\right)$
14. $\left(-\frac{2}{9}, -\frac{1}{3}\right)$

In Exercises 15–24, indicate the quadrant in which the terminal side of θ must lie in order for the information to be true.

15. $\cos \theta$ is positive and $\sin \theta$ is negative.
16. $\cos \theta$ is negative and $\sin \theta$ is positive.

17. $\tan \theta$ is negative and $\sin \theta$ is positive.
18. $\tan \theta$ is positive and $\cos \theta$ is negative.

19. $\sec\theta$ and $\csc\theta$ are both positive.

20. $\sec\theta$ and $\csc\theta$ are both negative.

21. $\cot\theta$ and $\cos\theta$ are both positive.

22. $\cot\theta$ and $\sin\theta$ are both negative.

23. $\tan\theta$ is positive and $\sec\theta$ is negative.

24. $\cot\theta$ is negative and $\csc\theta$ is positive.

In Exercises 25–36, find the indicated trigonometric function values.

25. If $\cos\theta = -\dfrac{3}{5}$, and the terminal side of θ lies in quadrant III, find $\sin\theta$.

26. If $\tan\theta = -\dfrac{5}{12}$, and the terminal side of θ lies in quadrant II, find $\cos\theta$.

27. If $\sin\theta = \dfrac{60}{61}$, and the terminal side of θ lies in quadrant II, find $\tan\theta$.

28. If $\cos\theta = \dfrac{40}{41}$, and the terminal side of θ lies in quadrant IV, find $\tan\theta$.

29. If $\tan\theta = \dfrac{84}{13}$, and the terminal side of θ lies in quadrant III, find $\sin\theta$.

30. If $\sin\theta = -\dfrac{7}{25}$, and the terminal side of θ lies in quadrant IV, find $\cos\theta$.

31. If $\sec\theta = -2$, and the terminal side of θ lies in quadrant III, find $\tan\theta$.

32. If $\cot\theta = 1$, and the terminal side of θ lies in quadrant I, find $\sin\theta$.

33. If $\csc\theta = \dfrac{2}{\sqrt{3}}$ and the terminal side of θ lies in quadrant II, find $\cot\theta$.

34. If $\sec\theta = -\dfrac{13}{5}$ and the terminal side of θ lies in quadrant II, find $\csc\theta$.

35. If $\cot\theta = -\sqrt{3}$ and the terminal side of θ lies in quadrant IV, find $\sec\theta$.

36. If $\cot\theta = -\dfrac{13}{84}$ and the terminal side of θ lies in quadrant II, find $\csc\theta$.

In Exercises 37–46, evaluate each expression, if possible.

37. $\cos(-270°) + \sin 450°$

38. $\sin(-270°) + \cos 450°$

39. $\sin 630° + \tan(-540°)$

40. $\cos(-720°) + \tan 720°$

41. $\cos(3\pi) - \sec(-3\pi)$

42. $\sin\left(-\dfrac{5\pi}{2}\right) + \csc\left(\dfrac{3\pi}{2}\right)$

43. $\csc\left(-\dfrac{7\pi}{2}\right) - \cot\left(\dfrac{7\pi}{2}\right)$

44. $\sec(-3\pi) + \tan(3\pi)$

45. $\tan 720° + \sec 720°$

46. $\cot 450° - \cos(-450°)$

In Exercises 47–56, determine whether each statement is possible or not.

47. $\sin\theta = -0.999$

48. $\cos\theta = 1.0001$

49. $\cos\theta = \dfrac{2\sqrt{6}}{3}$

50. $\sin\theta = \dfrac{\sqrt{2}}{10}$

51. $\tan\theta = 4\sqrt{5}$

52. $\cot\theta = -\dfrac{\sqrt{6}}{7}$

53. $\sec\theta = -\dfrac{4}{\sqrt{7}}$

54. $\csc\theta = \dfrac{\pi}{2}$

55. $\cot\theta = 500$

56. $\sec\theta = 0.9996$

In Exercises 57–68, evaluate the following expressions *exactly*:

57. $\cos 240°$

58. $\cos 120°$

59. $\sin\left(\dfrac{5\pi}{3}\right)$

60. $\sin\left(\dfrac{7\pi}{4}\right)$

61. $\tan 210°$

62. $\sec 135°$

63. $\tan(-315°)$

64. $\sec(-330°)$

65. $\csc\left(\dfrac{11\pi}{6}\right)$

66. $\csc\left(-\dfrac{4\pi}{3}\right)$

67. $\cot(-315°)$

68. $\cot 150°$

In Exercises 69–76, find all possible values of θ, where $0° \le \theta \le 360°$.

69. $\cos\theta = \dfrac{\sqrt{3}}{2}$ **70.** $\sin\theta = \dfrac{\sqrt{3}}{2}$ **71.** $\sin\theta = -\dfrac{1}{2}$ **72.** $\cos\theta = -\dfrac{1}{2}$

73. $\cos\theta = 0$ **74.** $\sin\theta = 0$ **75.** $\sin\theta = -1$ **76.** $\cos\theta = -1$

In Exercises 77–90, find the smallest positive measure of θ (rounded to the nearest degree) if the indicated information is true.

77. $\sin\theta = 0.9397$ and the terminal side of θ lies in quadrant II.

78. $\cos\theta = 0.7071$ and the terminal side of θ lies in quadrant IV.

79. $\cos\theta = -0.7986$ and the terminal side of θ lies in quadrant II.

80. $\sin\theta = -0.1746$ and the terminal side of θ lies in quadrant III.

81. $\tan\theta = -0.7813$ and the terminal side of θ lies in quadrant IV.

82. $\cos\theta = -0.3420$ and the terminal side of θ lies in quadrant III.

83. $\tan\theta = -0.8391$ and the terminal side of θ lies in quadrant II.

84. $\tan\theta = 11.4301$ and the terminal side of θ lies in quadrant III.

85. $\sin\theta = -0.3420$ and the terminal side of θ lies in quadrant IV.

86. $\sin\theta = -0.4226$ and the terminal side of θ lies in quadrant III.

87. $\sec\theta = 1.0001$ and the terminal side of θ lies in quadrant I.

88. $\sec\theta = -3.1421$ and the terminal side of θ lies in quadrant II.

89. $\csc\theta = -2.3604$ and the terminal side of θ lies in quadrant IV.

90. $\csc\theta = -1.0001$ and the terminal side of θ lies in quadrant III.

■ APPLICATIONS

In Exercises 91–94, refer to the following:

When light passes from one substance to another, such as from air to water, its path bends. This is called refraction and is what is seen in eyeglass lenses, camera lenses, and gems. The rule governing the change in the path is called *Snell's law*, named after a Dutch astronomer: $n_1 \sin\theta_1 = n_2 \sin\theta_2$, where n_1 and n_2 are the indices of refraction of the different substances and θ_1 and θ_2 are the respective angles that light makes with a line perpendicular to the surface at the boundary between substances. The figure shows the path of light rays going from air to water. Assume that the index of refraction in air is 1.

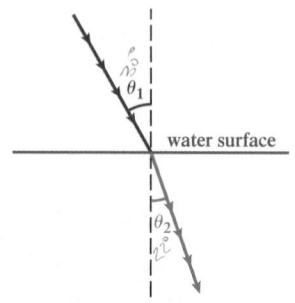

91. If light rays hit the water's surface at an angle of $30°$ from the perpendicular and are refracted to an angle of $22°$ from the perpendicular, then what is the refraction index for water? Round the answer to two significant digits.

92. If light rays hit a glass surface at an angle of $30°$ from the perpendicular and are refracted to an angle of $18°$ from the perpendicular, then what is the refraction index for that glass? Round the answer to two significant digits.

93. If the refraction index for a diamond is 2.4, then to what angle is light refracted if it enters the diamond at an angle of $30°$? Round the answer to two significant digits.

94. If the refraction index for a rhinestone is 1.9, then to what angle is light refracted if it enters the rhinestone at an angle of $30°$? Round the answer to two significant digits.

CATCH THE MISTAKE

In Exercises 95 and 96, explain the mistake that is made.

95. Evaluate the expression $\sec 120°$ exactly.

Solution:

$120°$ lies in quadrant II. The reference angle is $30°$.

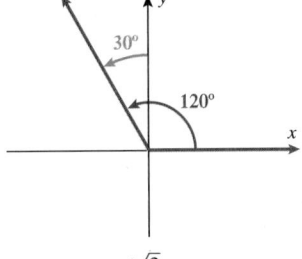

Find the cosine of the reference angle.

$$\cos 30° = \frac{\sqrt{3}}{2}$$

Cosine is negative in quadrant II.

$$\cos 120° = -\frac{\sqrt{3}}{2}$$

Secant is the reciprocal of cosine.

$$\sec 120° = -\frac{2}{\sqrt{3}} = -\frac{2\sqrt{3}}{3}$$

This is incorrect. What mistake was made?

96. Find the measure of the smallest positive angle θ (rounded to the nearest degree) if $\cos\theta = -0.2388$ and the terminal side of θ (in standard position) lies in quadrant III.

Solution:

Evaluate with a calculator.

$$\theta = \cos^{-1}(-0.2388) = 103.8157°$$

Approximate to the nearest degree.

$$\theta \approx 104°$$

This is incorrect. What mistake was made?

CONCEPTUAL

In Exercises 97–104, determine whether each statement is true or false.

97. It is possible for all six trigonometric functions of the same angle to have positive values.

98. It is possible for all six trigonometric functions of the same angle to have negative values.

99. The trigonometric function value for any angle with negative measure must be negative.

100. The trigonometric function value for any angle with positive measure must be positive.

101. $\sec^2\theta - 1$ can be negative for some value of θ.

102. $(\sec\theta)(\csc\theta)$ is negative only when the terminal side of θ lies in quadrant II or IV.

103. $\cos\theta = \cos(\theta + 360°n)$, where n is an integer.

104. $\sin\theta = \sin(\theta + 2\pi n)$, where n is an integer.

CHALLENGE

105. If the terminal side of angle θ passes through the point $(-3a, 4a)$, find $\cos\theta$. Assume $a > 0$.

106. If the terminal side of angle θ passes through the point $(-3a, 4a)$, find $\sin\theta$. Assume $a > 0$.

107. Find the equation of the line with negative slope that passes through the point $(a, 0)$ and makes an acute angle θ with the x-axis. The equation of the line will be in terms of x, a, and a trigonometric function of θ. Assume $a > 0$.

108. Find the equation of the line with positive slope that passes through the point $(a, 0)$ and makes an acute angle θ with the x-axis. The equation of the line will be in terms of x, a, and a trigonometric function of θ. Assume $a > 0$.

109. If $\tan\theta = \frac{a}{b}$, where a and b are positive, and if θ lies in quadrant III, find $\sin\theta$.

110. If $\tan\theta = -\frac{a}{b}$, where a and b are positive, and if θ lies in quadrant II, find $\cos\theta$.

111. If $\csc\theta = -\frac{a}{b}$, where a and b are po~~~~ in quadrant IV, find $\cot\theta$

112. If $\sec\theta = -\frac{a}{~}$ in quad~

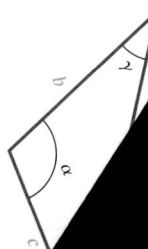

TECHNOLOGY

In Exercises 113–120, use a calculator to evaluate the following expressions. If you get an error, explain why.

113. $\cos 270°$

114. $\tan 270°$

115. $\cot 270°$

117. $\cos(-270°)$

119. $\sec(-270°)$

In calculus, the value $F(b) - F(a)$ of a function $F(x)$ at $x = a$ and $x = b$ plays an important role in the calculation of definite integrals.

In Exercises 121–124, find the exact value of $F(b) - F(a)$.

121. $F(x) = 2\tan x + \cos x, a = -\dfrac{\pi}{6}, b = \dfrac{\pi}{4}$

122. $F(x) = \sin^2 x + \cos^2 x, a = \dfrac{3\pi}{4}, b = \dfrac{7\pi}{6}$

123. $F(x) = \sec^2 x + 1, a = \dfrac{5\pi}{6}, b = \dfrac{4\pi}{3}$

124. $F(x) = \cot x - \csc^2 x, a = \dfrac{7\pi}{6}, b = \dfrac{7\pi}{4}$

SECTION
4.4 THE LAW OF SINES

SKILLS OBJECTIVES

■ Classify an oblique triangle as one of four cases.
■ Solve AAS or ASA triangle cases.
■ Solve ambiguous SSA triangle cases.

CONCEPTUAL OBJECTIVES

■ Understand the derivation of the Law of Sines.
■ Understand that the ambiguous case can yield no triangle, one triangle, or two triangles.
■ Understand why an AAA case cannot be solved.

Solving Oblique Triangles: Four Cases

Thus far we have discussed only *right* triangles. There are, however, two types of triangles, right and *oblique*. An **oblique triangle** is any triangle that does not have a right angle. An oblique triangle is either an **acute triangle**, having three acute angles, or an **obtuse triangle**, having one obtuse (between 90° and 180°) angle.

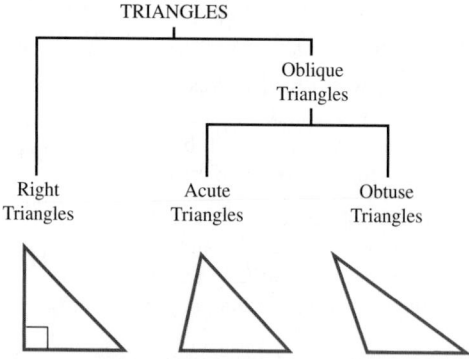

It is customary to label oblique triangles in the following way:

■ angle α (alpha) opposite side a.
■ angle β (beta) opposite side b.
■ angle γ (gamma) opposite side c.

$\alpha + \beta + \gamma = 180°$

Remember that the sum of the three angles of any triangle must equal 180°. In Section 4.3, we solved right triangles. In this section, we solve oblique triangles, which means we find the lengths of all three sides and the measures of all three angles. In order to solve an oblique triangle, *we need to know the length of one side* and one of the following three:

- two angles
- one angle and another side
- the other two sides

This requirement leads to the following four possible cases to consider:

REQUIRED INFORMATION TO SOLVE OBLIQUE TRIANGLES

CASE	WHAT'S GIVEN	EXAMPLES/NAMES
Case 1	Measures of one side and two angles	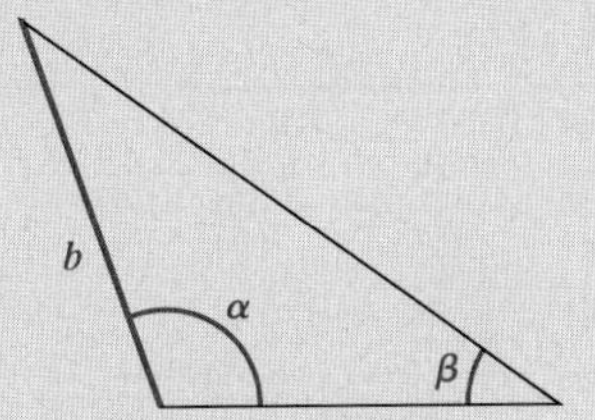 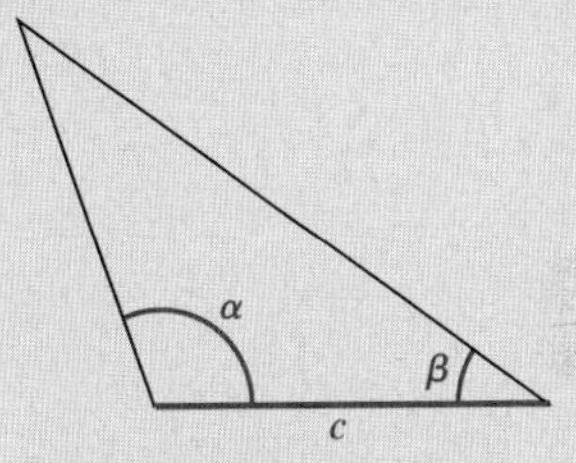 AAS: Angle-Angle-Side ASA: Angle-Side-Angle
Case 2	Measures of two sides and the angle opposite one of them	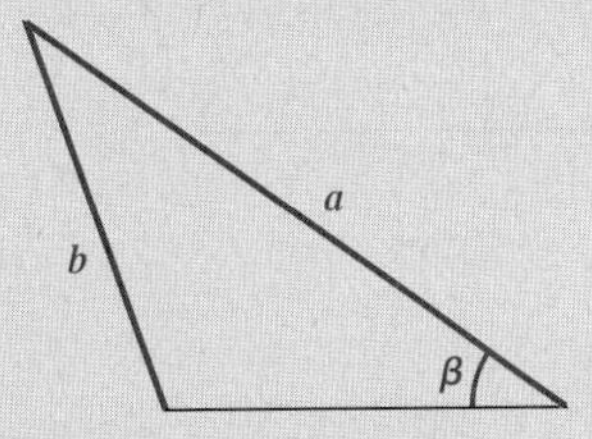 SSA: Side-Side-Angle
Case 3	Measures of two sides and the angle between them	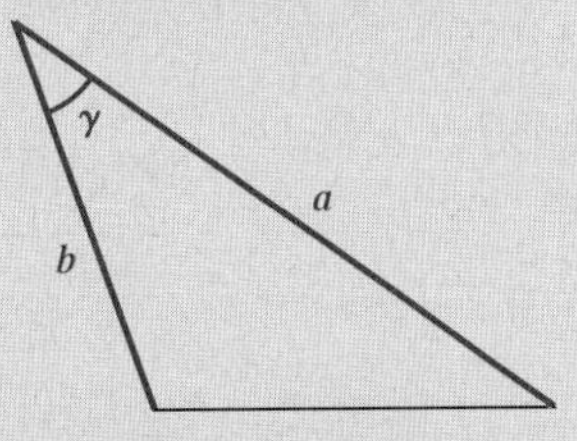 SAS: Side-Angle-Side
Case 4	Measures of three sides	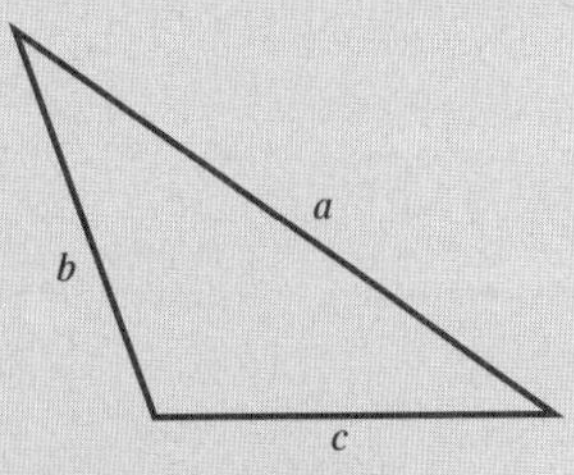 SSS: Side-Side-Side

Notice that there is no AAA case, because two similar triangles can have the same angle measures but different side lengths. That is why at least the length of one side must be known.

In this section, we will derive the Law of Sines, which will enable us to solve Case 1 and Case 2 problems. In the next section, we will derive the Law of Cosines, which will enable us to solve Case 3 and Case 4 problems.

The Law of Sines

Let us start with two oblique triangles, an acute triangle and an obtuse triangle.

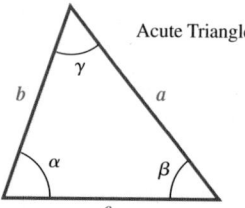

Acute Triangle

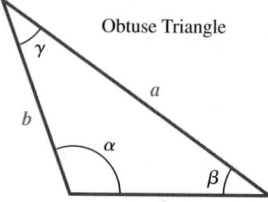

Obtuse Triangle

The following discussion applies to both triangles. First, construct an altitude (perpendicular) h from the vertex at angle γ to the side (or its extension) opposite γ.

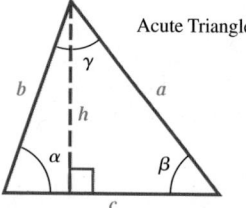

Acute Triangle

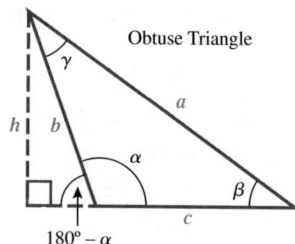

Obtuse Triangle

WORDS	MATH
Formulate sine ratios for the acute triangle.	$\sin\alpha = \dfrac{h}{b}$ and $\sin\beta = \dfrac{h}{a}$
Formulate sine ratios for the obtuse triangle.	$\sin(180° - \alpha) = \dfrac{h}{b}$ and $\sin\beta = \dfrac{h}{a}$
For the obtuse triangle, apply the sine difference identity.*	$\sin(180° - \alpha) = \sin 180° \cos\alpha - \cos 180° \sin\alpha$ $= 0 \cdot \cos\alpha - (-1)\sin\alpha$ $= \sin\alpha$
Therefore, in either triangle we find the same equation.	$\sin\alpha = \dfrac{h}{b}$ and $\sin\beta = \dfrac{h}{a}$
Solve for h in both equations.	$h = b\sin\alpha$ and $h = a\sin\beta$
Since h is equal to itself, equate the expressions for h.	$b\sin\alpha = a\sin\beta$
Divide both sides by ab.	$\dfrac{b\sin\alpha}{ab} = \dfrac{a\sin\beta}{ab}$
Divide out common factors.	$\boxed{\dfrac{\sin\alpha}{a} = \dfrac{\sin\beta}{b}}$

---

* The sine difference identify, $\sin(x - y) = \sin x \cdot \cos y - \cos x \cdot \sin y$ is derived in Section 6.2.

In a similar manner, we can extend an altitude (perpendicular) from angle α, and we

will find that $\boxed{\dfrac{\sin \gamma}{c} = \dfrac{\sin \beta}{b}}$. Equating these two expressions leads us to the third ratio

of the *Law of Sines*: $\boxed{\dfrac{\sin \alpha}{a} = \dfrac{\sin \gamma}{c}}$.

THE LAW OF SINES

For a triangle with sides of lengths a, b, and c, and opposite angles of measures α, β, and γ, the following is true:

$$\frac{\sin \alpha}{a} = \frac{\sin \beta}{b} = \frac{\sin \gamma}{c}$$

In other words, the ratio of the sine of an angle in a triangle to its opposite side is equal to the ratios of the sines of the other two angles to their opposite sides.

Study Tip

Remember that the longest side is opposite the largest angle; the shortest side is opposite the smallest angle.

Some things to note before we begin solving oblique triangles are:

- The angles and sides share the same progression of magnitude:
  - The longest side of a triangle is opposite the largest angle.
  - The shortest side of a triangle is opposite the smallest angle.
- Draw the triangle and label the angles and sides.
- If two angle measures are known, start by determining the third angle.
- Whenever possible, in successive steps always return to given values rather than refer to calculated (approximate) values.

Study Tip

Always use given values rather than calculated (approximated) values for better accuracy.

Keeping these pointers in mind will help you determine whether your answers are reasonable.

Case 1: Two Angles and One Side (AAS or ASA)

EXAMPLE 1   Using the Law of Sines to Solve a Triangle (AAS)

Solve the triangle.

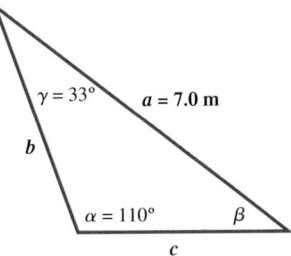

$\gamma = 33°$ $a = 7.0$ m

b

$\alpha = 110°$ β

c

Solution:

This is an AAS (angle-angle-side) case because two angles and a side are given and the side is opposite one of the angles.

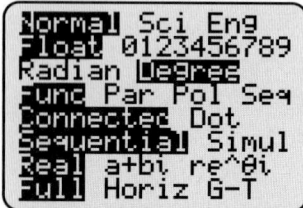
Study Tip

Notice in Step 3 that we used a that is given, as opposed to b that has been calculated (approximated).

STEP 1 Find β.

The sum of the measures of the angles in a triangle is 180°. $\alpha + \beta + \gamma = 180°$

Let $\alpha = 110°$ and $\gamma = 33°$. $110° + \beta + 33° = 180°$

Solve for β. $\boxed{\beta = 37°}$

STEP 2 Find b.

Use the Law of Sines with the known side a. $\dfrac{\sin \alpha}{a} = \dfrac{\sin \beta}{b}$

Isolate b. $b = \dfrac{a\sin \beta}{\sin \alpha}$

Let $\alpha = 110°$, $\beta = 37°$, and $a = 7$ m. $b = \dfrac{7\sin 37°}{\sin 110°}$

Use a calculator to approximate b. $b \approx 4.483067$ m

Round b to two significant digits. $\boxed{b \approx 4.5 \text{ m}}$

STEP 3 Find c.

Use the Law of Sines with the known side a. $\dfrac{\sin \alpha}{a} = \dfrac{\sin \gamma}{c}$

Isolate c. $c = \dfrac{a\sin \gamma}{\sin \alpha}$

Let $\alpha = 110°$, $\gamma = 33°$, and $a = 7$ m. $c = \dfrac{(7 \text{ m})\sin 33°}{\sin 110°}$

Use a calculator to approximate c. $c \approx 4.057149$ m

Round c to two significant digits. $\boxed{c \approx 4.1 \text{ m}}$

STEP 4 Draw and label the triangle.

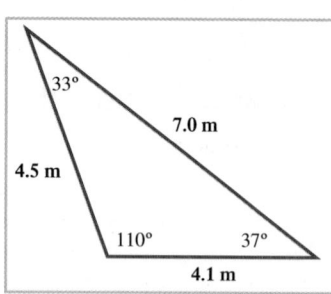

■ **Answer:** $\gamma = 32°$, $a \approx 42$ ft, $c \approx 23$ ft

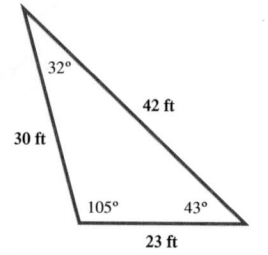

■ **YOUR TURN** Solve the triangle.

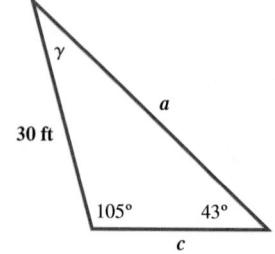

EXAMPLE 2 **Using the Law of Sines to Solve a Triangle (ASA)**

Solve the triangle.

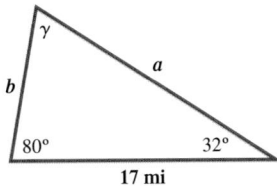

Solution:

This is an ASA (angle-side-angle) case because two angles and a side are given and the side is not opposite one of the angles.

STEP 1 Find γ.

The sum of the measures of the angles in a triangle is $180°$.	$\alpha + \beta + \gamma = 180°$
Let $\alpha = 80°$ and $\beta = 32°$.	$80° + 32° + \gamma = 180°$
Solve for γ.	$\boxed{\gamma = 68°}$

STEP 2 Find b.

Write the Law of Sines to include the known side c.	$\dfrac{\sin\beta}{b} = \dfrac{\sin\gamma}{c}$
Isolate b.	$b = \dfrac{c\sin\beta}{\sin\gamma}$
Let $\beta = 32°$, $\gamma = 68°$, and $c = 17$ miles.	$b = \dfrac{(17\text{ mi})\sin 32°}{\sin 68°}$
Use a calculator to approximate b.	$b \approx 9.7161177$ mi
Round b to two significant digits.	$\boxed{b \approx 9.7 \text{ miles}}$

STEP 3 Find a.

Write the Law of Sines again incorporating the known side c.	$\dfrac{\sin\alpha}{a} = \dfrac{\sin\gamma}{c}$
Isolate a.	$a = \dfrac{c\sin\alpha}{\sin\gamma}$
Let $\alpha = 80°$, $\gamma = 68°$, and $c = 17$ miles.	$a = \dfrac{(17\text{ mi})\sin 80°}{\sin 68°}$
Use a calculator to approximate a.	$a \approx 18.056539$ mi
Round a to two significant digits.	$\boxed{a \approx 18 \text{ mi}}$

STEP 4 Draw and label the triangle.

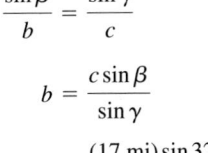

Technology Tip

Step 2: Use the calculator to find $b = \dfrac{17\sin 32°}{\sin 68°}$.

Step 3: Use the calculator to find $a = \dfrac{17\sin 80°}{\sin 68°}$.

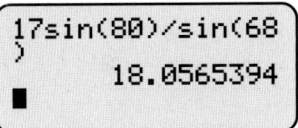

■ **Answer:** $\alpha = 35°$, $b \approx 21$ in., $c \approx 18$ in.

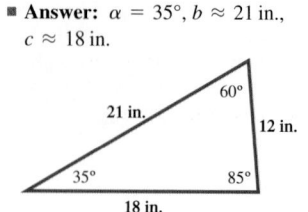

■ **YOUR TURN** Solve the triangle.

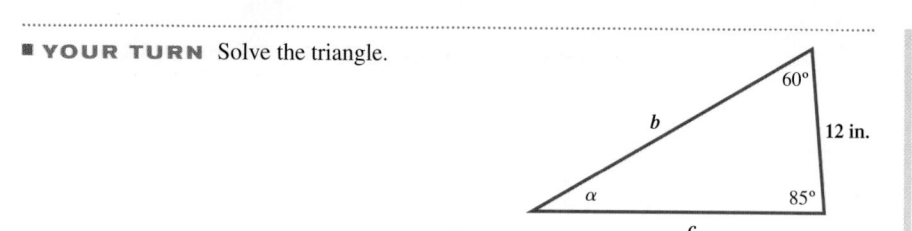

Case 2 (Ambiguous Case): Two Sides and One Angle (SSA)

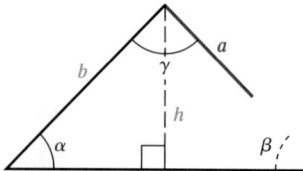

If we are given the measures of two sides and an angle opposite one of the sides, then we call that Case 2, SSA (side-side-angle). This case is called the ambiguous case, because the given information by itself can represent one triangle, two triangles, or no triangle at all. If the angle given is acute, then the possibilities are zero, one, or two triangles. If the angle given is obtuse, then the possibilities are zero or one triangle. The possibilities come from the fact that $\sin\alpha = k$, where $0 < k < 1$, has two solutions for α: one in quadrant I (acute angle) and one in quadrant II (obtuse angle). In the figure on the left, note that $h = b\sin\alpha$ by the definition of the sine ratio, and a given value of a may turn out to be smaller than, equal to, or larger than h. In addition, since $0 < \sin\alpha < 1$, then $h < b$.

Given Angle (α) Is Acute

CONDITION	PICTURE	NUMBER OF TRIANGLES
$0 < a < h$, in this case, $\sin\beta > 1$ (impossible)	No Triangle	0
$a = h$, in this case, $\sin\beta = 1$	Right Triangle	1
$h < a < b$, in this case, $0 < \sin\beta < 1$	Acute Triangle / Obtuse Triangle	2
$a \geq b$, in this case, $0 < \sin\beta < 1$	Acute Triangle	1

Given Angle (α) Is Obtuse

CONDITION	PICTURE	NUMBER OF TRIANGLES
	No Triangle	
$a \leq b$, in this case, $\sin \beta \geq 1$ (impossible)		0
	One Triangle	
$a > b$, in this case, $0 < \sin \beta < 1$		1

Study Tip

If an angle given is obtuse, then the side opposite that angle must be longer than the other sides (longest side opposite largest angle).

EXAMPLE 3   Solving the Ambiguous Case (SSA)—One Triangle

Solve the triangle $a = 23$ ft, $b = 11$ ft, and $\alpha = 122°$.

Solution:

This is the ambiguous case because the measures of two sides and an angle opposite one of those sides are given. Since the given angle α is obtuse and $a > b$, we expect one triangle.

STEP 1 Find β.

Use the Law of Sines.

$$\frac{\sin \alpha}{a} = \frac{\sin \beta}{b}$$

Isolate $\sin \beta$.

$$\sin \beta = \frac{b \sin \alpha}{a}$$

Let $a = 23$ ft, $b = 11$ ft, and $\alpha = 122°$.

$$\sin \beta = \frac{(11 \text{ ft}) \sin 122°}{23 \text{ ft}}$$

Use a calculator to evaluate the right side.

$$\sin \beta \approx 0.40558822$$

Use a calculator to approximate β.

$$\beta \approx \sin^{-1}(0.40558822) \approx \boxed{24°}$$

STEP 2 Find γ.

The measures of angles in a triangle sum to 180°.

$$\alpha + \beta + \gamma = 180°$$

Substitute $\alpha = 122°$ and $\beta \approx 24°$.

$$122° + 24° + \gamma \approx 180°$$

Solve for γ.

$$\boxed{\gamma \approx 34°}$$

STEP 3 Find c.

Use the Law of Sines.

$$\frac{\sin \alpha}{a} = \frac{\sin \gamma}{c}$$

Isolate c.

$$c = \frac{a \sin \gamma}{\sin \alpha}$$

Substitute $a = 23$ ft, $\alpha = 122°$, and $\gamma \approx 34°$.

$$c \approx \frac{(23 \text{ ft}) \sin 34°}{\sin 122°}$$

Use a calculator to evaluate c.

$$\boxed{c \approx 15 \text{ ft}}$$

STEP 4 Draw and label the triangle.

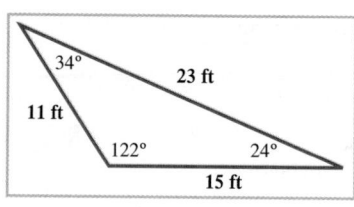

■ **Answer:** $\beta \approx 32°$, $\gamma \approx 15°$, $b \approx 35$ mm

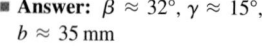

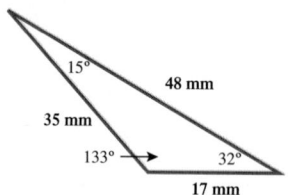

■ **YOUR TURN** Solve the triangle $\alpha = 133°$, $a = 48$ mm, and $c = 17$ mm.

EXAMPLE 4 Solving the Ambiguous Case (SSA)— Two Triangles

Solve the triangle $a = 8.1$ m, $b = 8.3$ m, and $\alpha = 72°$.

Solution:

This is the ambiguous case because the measures of two sides and an angle opposite one of those sides are given. Since the given angle α is acute and $a < b$, we expect two triangles.

STEP 1 Find β.

Write the Law of Sines for the given information.	$\dfrac{\sin \alpha}{a} = \dfrac{\sin \beta}{b}$
Isolate $\sin \beta$.	$\sin \beta = \dfrac{b \sin \alpha}{a}$
Let $a = 8.1$ m, $b = 8.3$ m, and $\alpha = 72°$.	$\sin \beta = \dfrac{(8.3 \text{ m}) \sin 72°}{8.1 \text{ m}}$
Use a calculator to evaluate the right side.	$\sin \beta \approx 0.974539393$
Use a calculator to approximate β. Note that β can be acute or obtuse.	$\beta \approx \sin^{-1}(0.974539393) \approx 77°$
This is the quadrant I solution (β is acute).	$\boxed{\beta_1 \approx 77°}$
The quadrant II solution is $\beta_2 = 180 - \beta_1$.	$\boxed{\beta_2 \approx 103°}$

STEP 2 Find γ.

The measures of the angles in a triangle sum to 180°.	$\alpha + \beta + \gamma = 180°$
Substitute $\alpha = 72°$ and $\beta_1 \approx 77°$.	$72° + 77° + \gamma_1 \approx 180°$
Solve for γ_1.	$\boxed{\gamma_1 \approx 31°}$
Substitute $\alpha = 72°$ and $\beta_2 \approx 103°$.	$72° + 103° + \gamma_2 \approx 180°$
Solve for γ_2.	$\boxed{\gamma_2 \approx 5°}$

STEP 3 Find c.

Use the Law of Sines.

$$\frac{\sin \alpha}{a} = \frac{\sin \gamma}{c}$$

Isolate c.

$$c = \frac{a \sin \gamma}{\sin \alpha}$$

Substitute $a = 8.1$ m, $\alpha = 72°$, and $\gamma_1 \approx 31°$.

$$c_1 \approx \frac{(8.1 \text{ m}) \sin 31°}{\sin 72°}$$

Use a calculator to evaluate c_1.

$$\boxed{c_1 \approx 4.4 \text{ m}}$$

Substitute $a = 8.1$ m, $\alpha = 72°$, and $\gamma_2 \approx 5°$.

$$c_2 \approx \frac{(8.1 \text{ m}) \sin 5°}{\sin 72°}$$

Use a calculator to evaluate c_2.

$$\boxed{c_2 \approx 0.74 \text{ m}}$$

STEP 4 Draw and label the two triangles.

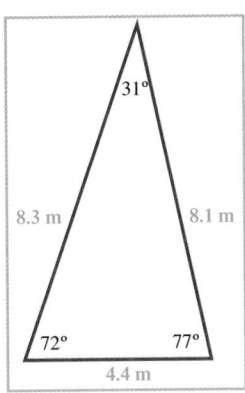

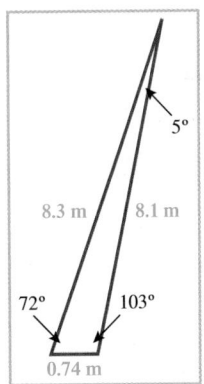

Study Tip

Notice that when there are two solutions in the SSA case, one triangle will be obtuse.

EXAMPLE 5   Solving the Ambiguous Case (SSA)—No Triangle

Solve the triangle $\alpha = 107°, a = 6$, and $b = 8$.

Solution:

This is the ambiguous case because the measures of two sides and an angle opposite one of those sides are given. Since the given angle α is obtuse and $a < b$, we expect no triangle since the longer side is not opposite the largest angle.

Write the Law of Sines.

$$\frac{\sin \alpha}{a} = \frac{\sin \beta}{b}$$

Isolate $\sin \beta$.

$$\sin \beta = \frac{b \sin \alpha}{a}$$

Let $\alpha = 107°, a = 6$, and $b = 8$.

$$\sin \beta = \frac{8 \sin 107°}{6}$$

Use a calculator to evaluate the right side.

$$\sin \beta \approx 1.28$$

Since the range of the sine function is $[-1, 1]$, there is no angle β such that $\sin \beta \approx 1.28$. Therefore, there is $\boxed{\text{no triangle}}$ with the given measurements.

Note: Had the geometric contradiction not been noticed, your work analytically will show a contradiction of $\sin \beta > 1$.

SECTION 4.4 SUMMARY

In this section, we solved oblique triangles. When given the measures of three parts of a triangle, we classify the triangle according to the given data (sides and angles). Four cases arise:

- one side and two angles (AAS or ASA)
- two sides and the angle opposite one of the sides (SSA)
- two sides and the angle between sides (SAS)
- three sides (SSS)

The Law of Sines

$$\frac{\sin \alpha}{a} = \frac{\sin \beta}{b} = \frac{\sin \gamma}{c}$$

can be used to solve the first two cases (AAS or ASA, and SSA). It is important to note that the SSA case is called the ambiguous case because any one of three results is possible: no triangle, one triangle, or two triangles.

SECTION 4.4 EXERCISES

SKILLS

In Exercises 1–6, classify each triangle problem as cases AAS, ASA, SAS, SSA, or SSS on the basis of the given information.

1. $c, a,$ and α SSA
2. $c, a,$ and γ SSA
3. $a, b,$ and c SSS
4. $a, b,$ and γ ASA
5. $\alpha, \beta,$ and c ASA
6. $\beta, \gamma,$ and a SAS

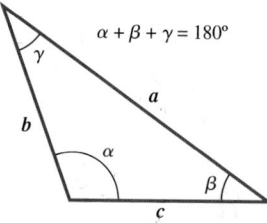

$\alpha + \beta + \gamma = 180°$

In Exercises 7–16, solve each of the following triangles with the given measures.

7. $\alpha = 45°, \beta = 60°, a = 10$ m
8. $\beta = 75°, \gamma = 60°, b = 25$ in.
9. $\alpha = 46°, \gamma = 72°, b = 200$ cm
10. $\gamma = 100°, \beta = 40°, a = 16$ ft
11. $\alpha = 16.3°, \gamma = 47.6°, c = 211$ yd
12. $\beta = 104.2°, \gamma = 33.6°, a = 26$ in.
13. $\alpha = 30°, \beta = 30°, c = 12$ m
14. $\alpha = 45°, \gamma = 75°, c = 9$ in.
15. $\beta = 26°, \gamma = 57°, c = 100$ yd
16. $\alpha = 80°, \gamma = 30°, b = 3$ ft

In Exercises 17–34, the measures of two sides and an angle are given. Determine whether a triangle (or two) exist, and if so, solve the triangle(s).

17. $a = 4, b = 5, \alpha = 16°$
18. $b = 30, c = 20, \beta = 70°$
19. $a = 12, c = 12, \gamma = 40°$
20. $b = 111, a = 80, \alpha = 25°$
21. $a = 21, b = 14, \beta = 100°$
22. $a = 13, b = 26, \alpha = 120°$
23. $\alpha = 30°, b = 18, a = 9$
24. $\alpha = 45°, b = \sqrt{2}, a = 1$
25. $\alpha = 34°, b = 7, a = 10$
26. $\alpha = 71°, b = 5.2, a = 5.2$
27. $\alpha = 21.3°, b = 6.18, a = 6.03$
28. $\alpha = 47.3°, b = 7.3, a = 5.32$
29. $\alpha = 116°, b = 4\sqrt{3}, a = 5\sqrt{2}$
30. $\alpha = 51°, b = 4\sqrt{3}, a = 4\sqrt{5}$
31. $b = 500, c = 330, \gamma = 40°$
32. $b = 16, a = 9, \beta = 137°$
33. $a = \sqrt{2}, b = \sqrt{7}, \beta = 106°$
34. $b = 15.3, c = 27.2, \gamma = 11.6°$

■ **APPLICATIONS**

For Exercises 35 and 36, refer to the following:

NASA Kennedy Space Center

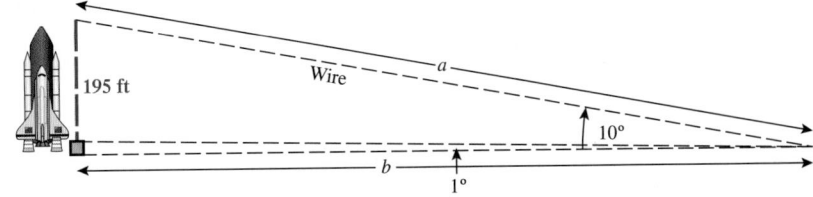

On the launch pad at Kennedy Space Center, there is an escape basket that can hold four astronauts. The basket slides down a wire that is attached 195 feet high, above the base of the launch pad. The angle of inclination measured from where the basket would touch the ground to the base of the launch pad is 1°, and the angle of inclination from that same point to where the wire is attached is 10°.

35. NASA. How long is the wire a?

36. NASA. How far from the launch pad does the basket touch the ground? That is, find b.

37. Hot-Air Balloon. A hot-air balloon is sighted at the same time by two friends who are 1.0 mile apart on the same side of the balloon. The angles of elevation from the two friends are 20.5° and 25.5°. How high is the balloon?

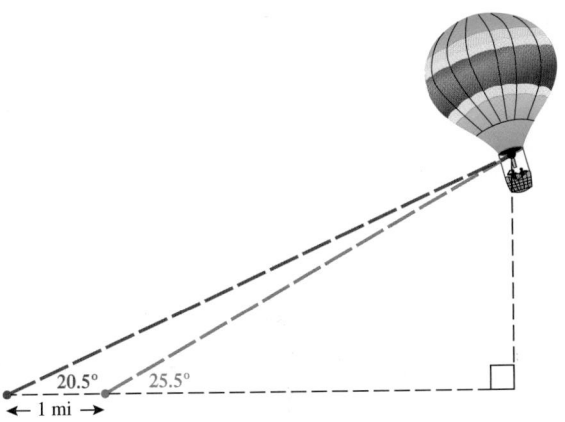

38. Hot-Air Balloon. A hot-air balloon is sighted at the same time by two friends who are 2 miles apart on the same side of the balloon. The angles of elevation from the two friends are 10° and 15°. How high is the balloon?

39. Rocket Tracking. A tracking station has two telescopes that are 1.0 mile apart. The telescopes can lock onto a rocket after it is launched and record the angles of elevation to the

rocket. If the angles of elevation from telescopes A and B are 30° and 80°, respectively, then how far is the rocket from telescope A?

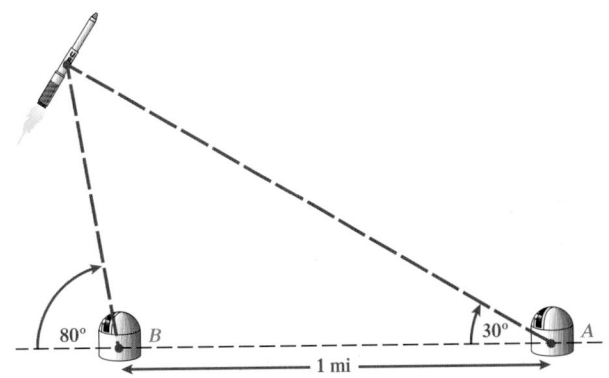

40. Rocket Tracking. Given the data in Exercise 39, how far is the rocket from telescope B?

41. Distance Across River. An engineer wants to construct a bridge across a fast-moving river. Using a straight-line segment between two points that are 100 feet apart along his side of the river, he measures the angles formed when sighting the point on the other side where he wants to have the bridge end. If the angles formed at points A and B are 65° and 15°, respectively, how far is it from point A to the point on the other side of the river? Round to the nearest foot.

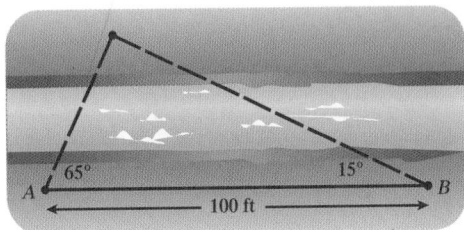

42. Distance Across River. Given the data in Exercise 41, how far is it from point B to the point on the other side of the river? Round to the nearest foot.

43. Lifeguard Posts. Two lifeguard chairs, labeled P and Q, are located 400 feet apart. A troubled swimmer is spotted by both lifeguards. If the lifeguard at P reports the swimmer at angle 35° (with respect to the line segment connecting P and Q) and the lifeguard at Q reports the swimmer at angle 41°, how far is the swimmer from P?

44. Rock Climbing. A rock climbing enthusiast is creating a climbing route rated as 5.8 level (i.e., medium difficulty) on the wall at the local rock gym. Given the difficulty of the route, he wants to avoid placing any two holds on the same vertical or horizontal line on the wall. If he places holds at P, Q, and R such that $\angle QPR = 40°$, $QR = 6$ feet, and $QP = 4.5$ feet, how far is the hold at P from the hold at R?

45. Tennis. After a long rally between two friends playing tennis, Player 2 lobs the ball into Player 1's court, enabling him to hit an overhead smash such that the angle between the racquet head (at the point of contact with the ball) and his body is 56°. The ball travels 20.3 feet to the other side of the court where Player 2 volleys the ball off the ground at an angle α such that the ball travels directly toward Player 1. The ball travels 19.4 feet during this return. Find angle α.

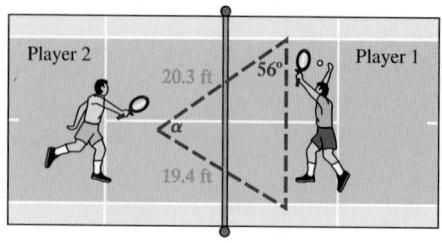

46. Tennis. Shocked by the move Player 1 made in Exercise 45, Player 2 is forced to quickly deflect the ball straight back to Player 1. Player 1 reaches behind himself and is able to contact the ball at the same height above the ground with which Player 1 initially hit it. If the angle between Player 1's racquet position at the end of the previous shot and its current position at the point of contact of this shot is 130°, and the angle with which it contacts the ball is 25°, how far has the ball traveled horizontally as a result of Player 2's hit?

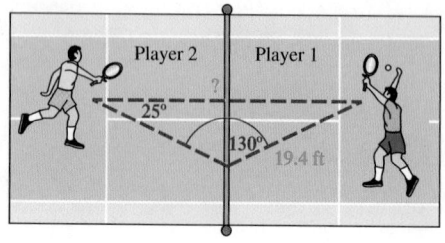

47. Surveying. There are two stations along the shoreline and the distance along the beach between the two stations is 50 meters. The angles between the baseline (beach) and the line of sight to the island are 30° and 40°. Find the shortest distance from the beach to the island. Round to the nearest meter.

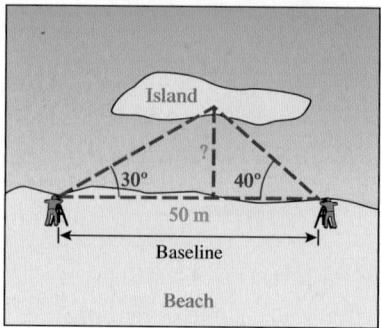

48. Surveying. There are two stations along a shoreline and the distance along the beach between the two stations is 200 feet. The angles between the baseline (beach) and the line of sight to the island are 30° and 50°. Find the shortest distance from the beach to the island. Round to the nearest foot.

49. Bowling. The 6-8 split is common in bowling. To make this split, a bowler stands dead center and throws the ball hard and straight directly toward the right of the 6 pin. The distance from the ball at the point of release to the 8 pin is 63.2 feet. See the diagram. How far does the ball travel from the bowler to the 6 pin?

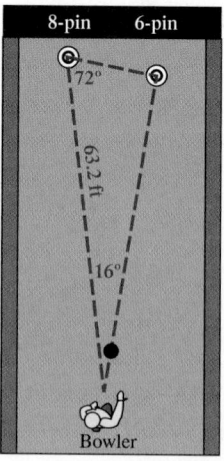

50. Bowling. A bowler is said to get a strike on the "Brooklyn side" of the head pin if he hits the head pin on the side opposite the pocket. (For a right-handed bowler, the pocket is to the right of the head pin.) There is a small range for the angle at which the ball must contact the head pin in order to convert all of the pins. If the measurements are as shown, how far does the ball travel (assuming it is thrown straight with no hook) before it contacts the head pin?

■ CATCH THE MISTAKE

In Exercises 51 and 52, explain the mistake that is made.

51. Solve the triangle $\alpha = 120°$, $a = 7$, and $b = 9$.

Solution:

Use the Law of Sines to find β.

$$\frac{\sin\alpha}{a} = \frac{\sin\beta}{b}$$

Let $\alpha = 120°$, $a = 7$, and $b = 9$.

$$\frac{\sin 120°}{7} = \frac{\sin\beta}{9}$$

Solve for $\sin\beta$.

$$\sin\beta = 1.113$$

Solve for β.

$$\beta = 42°$$

Sum the angle measures to 180°.

$$120° + 42° + \gamma = 180°$$

Solve for γ.

$$\gamma = 18°$$

Use the Law of Sines to find c.

$$\frac{\sin\alpha}{a} = \frac{\sin\gamma}{c}$$

Let $\alpha = 120°$, $a = 7$, and $\gamma = 18°$.

$$\frac{\sin 120°}{7} = \frac{\sin 18°}{c}$$

Solve for c.

$$c = 2.5$$

$\alpha = 120°$, $\beta = 42°$, $\gamma = 18°$, $a = 7$, $b = 9$, and $c = 2.5$.

This is incorrect. The longest side is not opposite the longest angle. There is no triangle that makes the original measurements work. What mistake was made?

52. Solve the triangle $\alpha = 40°$, $a = 7$, and $b = 9$.

Solution:

Use the Law of Sines to find β.

$$\frac{\sin\alpha}{a} = \frac{\sin\beta}{b}$$

Let $\alpha = 40°$, $a = 7$, and $b = 9$.

$$\frac{\sin 40°}{7} = \frac{\sin\beta}{9}$$

Solve for $\sin\beta$.

$$\sin\beta = 0.826441212$$

Solve for β.

$$\beta = 56°$$

Find γ.

$$40° + 56° + \gamma = 180°$$
$$\gamma = 84°$$

Use the Law of Sines to find c.

$$\frac{\sin\alpha}{a} = \frac{\sin\gamma}{c}$$

Let $\alpha = 40°$, $a = 7$, and $\gamma = 84°$.

$$\frac{\sin 40°}{7} = \frac{\sin 84°}{c}$$

Solve for c.

$$c = 11$$

$\alpha = 40°$, $\beta = 56°$, $\gamma = 84°$, $a = 7$, $b = 9$ and $c = 11$.

This is incorrect. What mistake was made?

■ CONCEPTUAL

In Exercises 53–58, determine whether each statment is true or false.

53. The Law of Sines applies only to right triangles.

54. If you are given the measures of two sides and any angle, there is a unique solution for the triangle.

55. An acute triangle is an oblique triangle.

56. An obtuse triangle is an oblique triangle.

57. If you are given two sides that have the same length in a triangle, then there can be at most one triangle.

58. If α is obtuse and $\beta = \dfrac{\alpha}{2}$, then the situation is unambiguous.

■ CHALLENGE

The following identities are useful in Exercises 59 and 60, and will be derived in Chapter 6.

$$\sin x + \sin y = 2\sin\left(\frac{x+y}{2}\right)\cos\left(\frac{x-y}{2}\right)$$

$$\sin(2x) = 2\sin x\cos x$$

$$\sin(x \pm y) = \sin x\cos y \pm \cos x\sin y$$

$$\cos(x \pm y) = \cos x\cos y \pm \sin x\sin y$$

59. Mollweide's Identity. For any triangle, the following identity is true. It is often used to check the solution of a triangle since all six pieces of information (three sides and three angles) are involved. Derive the identity using the Law of Sines.

$$(a + b)\sin(\tfrac{1}{2}\gamma) = c\cos\left[\tfrac{1}{2}(\alpha - \beta)\right]$$

60. The Law of Tangents. Use the Law of Sines and trigonometric identities to show that for any triangle, the following is true:

$$\frac{a - b}{a + b} = \frac{\tan\left[\tfrac{1}{2}(\alpha - \beta)\right]}{\tan\left[\tfrac{1}{2}(\alpha + \beta)\right]}$$

61. Use the Law of Sines to prove that all angles in an equilateral triangle must have the same measure.

62. Suppose that you have a triangle with side lengths a, b, and c, and angles α, β, and γ, respectively, directly across from them. If it is known that $a = \frac{1}{\sqrt{2}} b$, $c = 2$, α is an acute angle, and $\beta = 2\alpha$, solve the triangle.

▪ TECHNOLOGY

For Exercises 63–68, let A, B, and C be the lengths of the three sides with X, Y, and Z as the opposite corresponding angles. Write a program to solve the given triangle with a calculator.

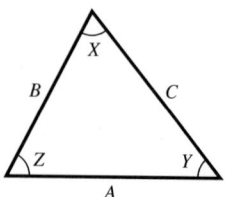

63. $A = 10$, $Y = 40°$, and $Z = 72°$

64. $B = 42.8$, $X = 31.6°$, and $Y = 82.2°$

65. $A = 22$, $B = 17$, and $X = 105°$

66. $B = 16.5$, $C = 9.8$, and $Z = 79.2°$

67. $A = 25.7$, $C = 12.2$, and $X = 65°$

68. $A = 54.6$, $B = 12.9$, and $Y = 23°$

▪ PREVIEW TO CALCULUS

In calculus, some applications of the derivative require the solution of triangles. In Exercises 69–72, solve each triangle using the Law of Sines.

69. In an oblique triangle ABC, $\beta = 45°$, $\gamma = 60°$, and $b = 20$ in. Find the length of a. Round your answer to the nearest unit.

70. In an oblique triangle ABC, $\beta = \frac{2\pi}{9}$, $\gamma = \frac{5\pi}{9}$, and $a = 200$ ft. Find the length of c. Round your answer to the nearest unit.

71. In an oblique triangle ABC, $b = 14$ m, $c = 14$ m, and $\alpha = \frac{4\pi}{7}$. Find the length of a. Round your answer to the nearest unit.

72. In an oblique triangle ABC, $b = 30$ cm, $c = 45$ cm, and $\gamma = 35°$. Find the length of a. Round your answer to the nearest unit.

SECTION
4.5 THE LAW OF COSINES

SKILLS OBJECTIVES

- Solve SAS triangles.
- Solve SSS triangles.
- Find the area of triangles in the SAS case.
- Find the area of triangles in the SSS case.

CONCEPTUAL OBJECTIVES

- Understand the derivation of the Law of Cosines.
- Develop a strategy for which angles (larger or smaller) and which method (the Law of Sines or the Law of Cosines) to select to solve oblique triangles.

Solving Oblique Triangles Using the Law of Cosines

In Section 4.4, we learned that to solve oblique triangles means to find all three side lengths and angle measures, and that at least one side length must be known. We need two additional pieces of information to solve an oblique triangle (combinations of side lengths and/or angles). We found that there are four cases:

- Case 1: AAS or ASA (measures of two angles and a side are given)
- Case 2: SSA (measures of two sides and an angle opposite one of the sides are given)
- Case 3: SAS (measures of two sides and the angle between them are given)
- Case 4: SSS (measures of three sides are given)

We used the Law of Sines to solve Case 1 and Case 2 triangles. Now, we need the *Law of Cosines* to solve Case 3 and Case 4 triangles.

WORDS	**MATH**
Start with an oblique (acute) triangle.	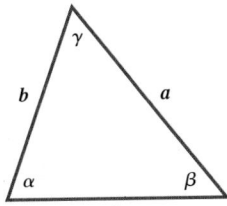

Drop a perpendicular line segment from γ to side c with height h.

The result is two right triangles within the larger triangle.

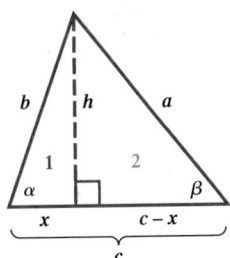

Use the Pythagorean theorem to write the relationship between the side lengths in both right triangles.

Triangle 1: $x^2 + h^2 = b^2$

Triangle 2: $(c - x)^2 + h^2 = a^2$

Solve for h^2 in both equations.

Triangle 1: $h^2 = b^2 - x^2$

Triangle 2: $h^2 = a^2 - (c - x)^2$

Since the segment of length h is shared, set $h^2 = h^2$, for the two triangles. $b^2 - x^2 = a^2 - (c - x)^2$

Multiply out the squared binomial on the right. $b^2 - x^2 = a^2 - (c^2 - 2cx + x^2)$

Eliminate the parentheses. $b^2 - x^2 = a^2 - c^2 + 2cx - x^2$

Add x^2 to both sides. $b^2 = a^2 - c^2 + 2cx$

Isolate a^2. $a^2 = b^2 + c^2 - 2cx$

Notice that $\cos\alpha = \dfrac{x}{b}$. Let $x = b\cos\alpha$. $\boxed{a^2 = b^2 + c^2 - 2bc\cos\alpha}$

Note: If we instead drop the perpendicular line segment with length h from the angle α or the angle β, we can derive the other two parts of the Law of Cosines:

$$\boxed{b^2 = a^2 + c^2 - 2ac\cos\beta} \quad \text{and} \quad \boxed{c^2 = a^2 + b^2 - 2ab\cos\gamma}$$

THE LAW OF COSINES

For a triangle with sides of length a, b, and c, and opposite angle measures α, β, and γ, the following equations are true:

$$a^2 = b^2 + c^2 - 2bc \cos \alpha$$

$$b^2 = a^2 + c^2 - 2ac \cos \beta$$

$$c^2 = a^2 + b^2 - 2ab \cos \gamma$$

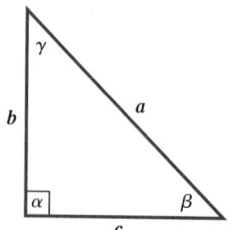

It is important to note that the Law of Cosines can be used to find side lengths or angles in any triangle in cases SAS or SSS, as long as three of the four variables in any of the equations are known, the fourth can be calculated.

Notice that in the special case of a right triangle (say, $\alpha = 90°$),

$$a^2 = b^2 + c^2 - 2bc \underbrace{\cos 90°}_{0}$$

one of the equations of the Law of Cosines reduces to the Pythagorean theorem:

$$\underbrace{a^2}_{\text{hyp}} = \underbrace{b^2}_{\text{leg}} + \underbrace{c^2}_{\text{leg}}$$

The Pythagorean theorem can thus be regarded as a special case of the Law of Cosines.

Case 3: Solving Oblique Triangles (SAS)

Study Tip

The Pythagorean theorem is a special case of the Law of Cosines.

We now solve SAS triangle problems where the measures of two sides and the angle between them are given. We start by using the Law of Cosines to solve for the length of the side opposite the given angle. We then can apply either the Law of Sines or the Law of Cosines to find the second angle measure.

Technology Tip

Step 1: Use the calculator to find the value of b.

Step 2: Use the calculator to find γ.

EXAMPLE 1 Using the Law of Cosines to Solve a Triangle (SAS)

Solve the triangle $a = 13$, $c = 6$, and $\beta = 20°$.

Solution:

The measures of two sides and the angle between them are given (SAS).

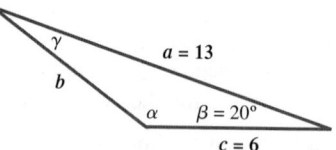

Notice that the Law of Sines can't be used, because it requires the measures of at least one angle and the side opposite that angle.

STEP 1 Find b.

Apply the Law of Cosines that involves β.	$b^2 = a^2 + c^2 - 2ac \cos \beta$
Let $a = 13$, $c = 6$, and $\beta = 20°$.	$b^2 = 13^2 + 6^2 - 2(13)(6) \cos 20°$
Evaluate the right side with a calculator.	$b^2 \approx 58.40795$
Solve for b.	$b \approx \pm 7.6425$
Round to two significant digits; b can only be positive.	$\boxed{b \approx 7.6}$

STEP 2 Find γ.

Use the Law of Sines to find the smaller angle measure, γ.

$$\frac{\sin \gamma}{c} = \frac{\sin \beta}{b}$$

Isolate $\sin \gamma$.

$$\sin \gamma = \frac{c \sin \beta}{b}$$

Let $b \approx 7.6$, $c = 6$, and $\beta = 20°$.

$$\sin \gamma \approx \frac{6 \sin 20°}{7.6}$$

Apply the inverse sine function.

$$\gamma \approx \sin^{-1}\left(\frac{6 \sin 20°}{7.6}\right)$$

Evaluate the right side with a calculator.

$$\gamma \approx 15.66521°$$

Round to the nearest degree.

$$\boxed{\gamma \approx 16°}$$

STEP 3 Find α.

The angle measures must sum to 180°.

$$\alpha + 20° + 16° \approx 180°$$

Solve for α.

$$\boxed{\alpha \approx 144°}$$

■ **YOUR TURN** Solve the triangle $b = 4.2$, $c = 1.8$, and $\alpha = 35°$.

■ **Answer:** $a \approx 2.9$, $\gamma \approx 21°$, $\beta \approx 124°$

Notice the steps we took in solving an SAS triangle:

1. Find the length of the side opposite the given angle using the Law of Cosines.

2. Solve for the smaller angle using the Law of Sines.

3. Solve for the larger angle using the fact that angles of a triangle sum to 180°.

You may be thinking, "Would it matter if we had solved for α before solving for γ?" Yes, it does matter—in this problem you cannot solve for α by the Law of Sines before finding γ. The Law of Sines can be used only on the smaller angle (opposite the shortest side). If we had tried to use the Law of Sines with the obtuse angle α, the inverse sine would have resulted in $\alpha = 36°$. Since the sine function is positive in QI and QII, we would not know whether that angle was $\alpha = 36°$ or its supplementary angle $\alpha = 144°$. Notice that $c < a$; therefore, the angles opposite those sides must have the same relationship, $\gamma < \alpha$. We choose the smaller angle first. Alternatively, if we want to solve for the obtuse angle first, we can use the Law of Cosines to solve for α. If you use the Law of Cosines to find the second angle, you can choose either angle. The Law of Cosines can be used to find the measure of either acute or obtuse angles.

Study Tip

Although the Law of Sines is sometimes ambiguous, the Law of Cosines is never ambiguous.

Case 4: Solving Oblique Triangles (SSS)

We now solve oblique triangles when all three side lengths are given (the SSS case). In this case, start by finding the largest angle (opposite the largest side) using the Law of Cosines. Then apply the Law of Sines to find either of the remaining two angles. Lastly, find the third angle with the triangle angle sum identity.

EXAMPLE 2   Using the Law of Cosines to Solve a Triangle (SSS)

Solve the triangle $a = 8, b = 6,$ and $c = 7$.

Solution:

STEP 1 Identify the largest angle, which is α.

Write the equation of the Law of Cosines that involves α.

$$a^2 = b^2 + c^2 - 2bc\cos\alpha$$

Let $a = 8, b = 6,$ and $c = 7$.

$$8^2 = 6^2 + 7^2 - 2(6)(7)\cos\alpha$$

Simplify and isolate $\cos\alpha$.

$$\cos\alpha = \frac{6^2 + 7^2 - 8^2}{2(6)(7)} = 0.25$$

Approximate with a calculator.

$$\alpha = \cos^{-1}(0.25) \approx \boxed{75.5°}$$

STEP 2 Find either of the remaining angles. We will solve for β.

Write the Law of Sines.

$$\frac{\sin\alpha}{a} = \frac{\sin\beta}{b}$$

Isolate $\sin\beta$.

$$\sin\beta = \frac{b\sin\alpha}{a}$$

Let $a = 8, b = 6,$ and $\alpha = 75.5°$.

$$\sin\beta \approx \frac{6\sin 75.5°}{8}$$

Approximate with a calculator.

$$\beta \approx \sin^{-1}\left(\frac{6\sin 75.5°}{8}\right) \approx \boxed{46.6°}$$

STEP 3 Find the third angle, γ.

The sum of the angle measures is 180°.

$$75.5° + 46.6° + \gamma \approx 180°$$

Solve for γ.

$$\boxed{\gamma \approx 57.9°}$$

■ **Answer:** $\alpha \approx 38.2°, \beta \approx 60.0°,$ $\gamma \approx 81.8°$

■ **YOUR TURN** Solve the triangle $a = 5, b = 7,$ and $c = 8$.

The Area of a Triangle

The general formula for the area of a triangle and the sine function together can be used to develop a formula for the area of a triangle when the measures of two sides and the angle between them are given.

WORDS

Start with an acute triangle, given b, c, and α.

MATH

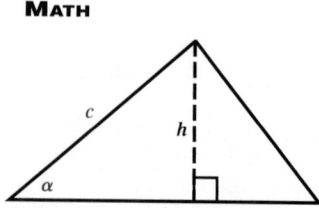

Write the sine ratio in the right triangle for the acute angle α.

$$\sin\alpha = \frac{h}{c}$$

Solve for h.

$$h = c\sin\alpha$$

Write the formula for area of a triangle.

$$A_{\text{triangle}} = \frac{1}{2}bh$$

Substitute $h = c\sin\alpha$.

$$\boxed{A_{\text{SAS}} = \frac{1}{2}bc\sin\alpha}$$

Now we can calculate the area of this triangle with the given information (the measures of two sides and the angle between them: b, c, and α). Similarly, it can be shown that the other formulas for SAS triangles are

$$\boxed{A_{\text{SAS}} = \frac{1}{2}ab\sin\gamma} \quad \text{and} \quad \boxed{A_{\text{SAS}} = \frac{1}{2}ac\sin\beta}$$

AREA OF A TRIANGLE (SAS)

For any triangle where the measures of two sides and the angle between them are known, the area for that triangle is given by one of the following formulas (depending on which angle and side measures are given):

$$A_{\text{SAS}} = \frac{1}{2}bc\sin\alpha \qquad \text{when } b, c, \text{ and } \alpha \text{ are known}$$

$$A_{\text{SAS}} = \frac{1}{2}ab\sin\gamma \qquad \text{when } a, b, \text{ and } \gamma \text{ are known}$$

$$A_{\text{SAS}} = \frac{1}{2}ac\sin\beta \qquad \text{when } a, c, \text{ and } \beta \text{ are known}$$

In other words, the area of a triangle equals one-half the product of two of its sides and the sine of the angle between them.

EXAMPLE 3   Finding the Area of a Triangle (SAS Case)

Find the area of the triangle $a = 7.0$ ft, $b = 9.3$ ft, and $\gamma = 86°$.

Solution:

Apply the area formula where a, b, and γ are given.

$$A = \frac{1}{2}ab\sin\gamma$$

Substitute $a = 7.0$ ft, $b = 9.3$ ft, and $\gamma = 86°$.

$$A = \frac{1}{2}(7.0\,\text{ft})(9.3\,\text{ft})\sin 86°$$

Approximate with a calculator.

$$A \approx 32.47071\,\text{ft}^2$$

Round to two significant digits.

$$\boxed{A \approx 32\,\text{ft}^2}$$

■ **YOUR TURN** Find the area of the triangle $a = 3.2$ m, $c = 5.1$ m, and $\beta = 49°$.

Technology Tip

Use the calculator to find A.

```
1/2*7*9.3*sin(86
)
          32.47070984
```

■ **Answer:** 6.2 m^2

Study Tip

The Pythagorean Identity
$$\sin^2\theta + \cos^2\theta = 1$$
can be proven using any of the three trigonometric definitions:
$$\left(\frac{y}{r}\right)^2 + \left(\frac{x}{r}\right)^2 = \frac{y^2 + x^2}{r^2} = \frac{r^2}{r^2} = 1$$

The Law of Cosines can be used to develop a formula for the area of an SSS triangle, called **Heron's formula**.

WORDS	MATH
Start with any of the formulas for SAS triangles.	$A = \frac{1}{2}ab\sin\gamma$
Square both sides.	$A^2 = \frac{1}{4}a^2b^2\sin^2\gamma$
Isolate $\sin^2\gamma$.	$\frac{4A^2}{a^2b^2} = \sin^2\gamma$
Apply the Pythagorean identity.	$\frac{4A^2}{a^2b^2} = 1 - \cos^2\gamma$
Factor the difference of the two squares on the right.	$\frac{4A^2}{a^2b^2} = (1 - \cos\gamma)(1 + \cos\gamma)$
Solve the Law of Cosines, $c^2 = a^2 + b^2 - 2ab\cos\gamma$, for $\cos\gamma$.	$\cos\gamma = \frac{a^2 + b^2 - c^2}{2ab}$
Substitute $\cos\gamma = \frac{a^2 + b^2 - c^2}{2ab}$ into $\frac{4A^2}{a^2b^2} = (1 - \cos\gamma)(1 + \cos\gamma)$.	$\frac{4A^2}{a^2b^2} = \left[1 - \frac{a^2 + b^2 - c^2}{2ab}\right]\left[1 + \frac{a^2 + b^2 - c^2}{2ab}\right]$
Combine the expressions in brackets.	$\frac{4A^2}{a^2b^2} = \left[\frac{2ab - a^2 - b^2 + c^2}{2ab}\right]\left[\frac{2ab + a^2 + b^2 - c^2}{2ab}\right]$
Group the terms in the numerators on the right.	$\frac{4A^2}{a^2b^2} = \left[\frac{-(a^2 - 2ab + b^2) + c^2}{2ab}\right]\left[\frac{(a^2 + 2ab + b^2) - c^2}{2ab}\right]$
Write the numerators on the right as the difference of two squares.	$\frac{4A^2}{a^2b^2} = \left[\frac{c^2 - (a - b)^2}{2ab}\right]\left[\frac{(a + b)^2 - c^2}{2ab}\right]$
Factor the numerators on the right. Recall: $x^2 - y^2 = (x - y)(x + y)$.	$\frac{4A^2}{a^2b^2} = \left[\frac{(c - [a - b])(c + [a - b])}{2ab}\right]\left[\frac{([a + b] - c)([a + b] + c)}{2ab}\right]$
Simplify.	$\frac{4A^2}{a^2b^2} = \left[\frac{(c - a + b)(c + a - b)}{2ab}\right]\left[\frac{(a + b - c)(a + b + c)}{2ab}\right]$
	$\frac{4A^2}{a^2b^2} = \frac{(c - a + b)(c + a - b)(a + b - c)(a + b + c)}{4a^2b^2}$
Solve for A^2 by multiplying both sides by $\frac{a^2b^2}{4}$.	$A^2 = \frac{1}{16}(c - a + b)(c + a - b)(a + b - c)(a + b + c)$

The semiperimeter s is half the perimeter of the triangle.	$$s = \frac{a + b + c}{2}$$
Manipulate each of the four factors:	$c - a + b = a + b + c - 2a = 2s - 2a = 2(s - a)$ $c + a - b = a + b + c - 2b = 2s - 2b = 2(s - b)$ $a + b - c = a + b + c - 2c = 2s - 2c = 2(s - c)$ $a + b + c = 2s$
Substitute in these values for the four factors.	$$A^2 = \frac{1}{16} \cdot 2(s - a) \cdot 2(s - b) \cdot 2(s - c) \cdot 2s$$
Simplify.	$$A^2 = s(s - a)(s - b)(s - c)$$
Solve for A (area is always positive).	$$\boxed{A = \sqrt{s(s - a)(s - b)(s - c)}}$$

AREA OF A TRIANGLE (SSS CASE—HERON'S FORMULA)

For any triangle where the lengths of the three sides are known, the area for that triangle is given by the following formula:

$$A_{SSS} = \sqrt{s(s - a)(s - b)(s - c)}$$

where a, b, and c are the lengths of the sides of the triangle and s is half the perimeter of the triangle, called the semiperimeter.

$$s = \frac{a + b + c}{2}$$

EXAMPLE 4   Finding the Area of a Triangle (SSS Case)

Find the area of the triangle $a = 5$, $b = 6$, and $c = 9$.

Solution:

Find the semiperimeter s.	$$s = \frac{a + b + c}{2}$$
Substitute $a = 5, b = 6$, and $c = 9$.	$$s = \frac{5 + 6 + 9}{2}$$
Simplify.	$s = 10$
Write the formula for the area of a triangle in the SSS case (Heron's formula).	$$A = \sqrt{s(s - a)(s - b)(s - c)}$$
Substitute $a = 5, b = 6, c = 9$, and $s = 10$.	$$A = \sqrt{10(10 - 5)(10 - 6)(10 - 9)}$$
Simplify the radicand.	$$A = \sqrt{10 \cdot 5 \cdot 4 \cdot 1}$$
Evaluate the radical.	$$\boxed{A = 10\sqrt{2} \approx 14 \text{ sq units}}$$

■ **YOUR TURN** Find the area of the triangle $a = 3$, $b = 5$, and $c = 6$.

Technology Tip

Use the calculator to find A.

```
√(10(10-5)(10-6)
(10-9)
              14.14213562
```

■ **Answer:** $2\sqrt{14} \approx 7.5$ sq units

We can solve any triangle given three measures, as long as one of the measures is a side length. Depending on the information given, we either apply the **Law of Sines**

$$\frac{\sin \alpha}{a} = \frac{\sin \beta}{b} = \frac{\sin \gamma}{c}$$

and the angle sum identity, or we apply a combination of the **Law of Cosines**,

$$a^2 = b^2 + c^2 - 2bc \cos \alpha \qquad b^2 = a^2 + c^2 - 2ac \cos \beta \qquad c^2 = a^2 + b^2 - 2ab \cos \gamma$$

the Law of Sines, and the angle sum identity. The table below summarizes the strategies for solving oblique triangles covered in Sections 4.4 and 4.5.

OBLIQUE TRIANGLE	WHAT'S KNOWN	PROCEDURE FOR SOLVING
AAS or ASA	Two angles and a side	Step 1: Find the remaining angle with $\alpha + \beta + \gamma = 180°$. Step 2: Find the remaining sides with the Law of Sines.
SSA	Two sides and an angle opposite one of the sides	This is the ambiguous case, so there is either no triangle, one triangle, or two triangles. If the given angle is obtuse, then there is either one or no triangle. If the given angle is acute, then there is no triangle, one triangle, or two triangles. Step 1: Apply the Law of Sines to find one of the angles. Step 2: Find the remaining angle with $\alpha + \beta + \gamma = 180°$. Step 3: Find the remaining side with the Law of Sines. If two triangles exist, then the angle found in Step 1 can be either acute or obtuse, and Step 2 and Step 3 must be performed for each triangle.
SAS	Two sides and an angle between the sides	Step 1: Find the third side with the Law of Cosines. Step 2: Find the smaller angle with the Law of Sines. Step 3: Find the remaining angle with $\alpha + \beta + \gamma = 180°$.
SSS	Three sides	Step 1: Find the largest angle with the Law of Cosines. Step 2: Find either remaining angle with the Law of Sines. Step 3: Find the last remaining angle with $\alpha + \beta + \gamma = 180°$.

Formulas for calculating the areas of triangles (SAS and SSS cases) were derived. The three area formulas for the SAS case depend on which angles and sides are given.

$$A_{\text{SAS}} = \frac{1}{2} bc \sin \alpha \qquad A_{\text{SAS}} = \frac{1}{2} ab \sin \gamma \qquad A_{\text{SAS}} = \frac{1}{2} ac \sin \beta$$

The Law of Cosines was instrumental in developing a formula for the area of a triangle (SSS case) when all three sides are given.

$$(Heron's\ formula) \quad A_{\text{SSS}} = \sqrt{s(s-a)(s-b)(s-c)} \quad \text{where} \quad s = \frac{a+b+c}{2}$$

▪ SKILLS

In Exercises 1–28, solve each triangle.

1. $a = 4$, $c = 3$, $\beta = 100°$

2. $a = 6$, $b = 10$, $\gamma = 80°$

3. $b = 7$, $c = 2$, $\alpha = 16°$

4. $b = 5$, $a = 6$, $\gamma = 170°$

5. $b = 5$, $c = 5$, $\alpha = 20°$

6. $a = 4.2$, $b = 7.3$, $\gamma = 25°$

7. $a = 9, c = 12, \beta = 23°$

8. $b = 6, c = 13, \alpha = 16°$

9. $a = 4, c = 8, \beta = 60°$

10. $b = 3, c = \sqrt{18}, \alpha = 45°$

11. $a = 8, b = 5, c = 6$

12. $a = 6, b = 9, c = 12$

13. $a = 4, b = 4, c = 5$

14. $a = 17, b = 20, c = 33$

15. $a = 8.2, b = 7.1, c = 6.3$

16. $a = 1492, b = 2001, c = 1776$

17. $a = 4, b = 5, c = 10$

18. $a = 1.3, b = 2.7, c = 4.2$

19. $a = 12, b = 5, c = 13$

20. $a = 4, b = 5, c = \sqrt{41}$

21. $\alpha = 40°, \beta = 35°, a = 6$

22. $b = 11.2, a = 19.0, \gamma = 13.3°$

23. $\alpha = 31°, b = 5, a = 12$

24. $a = 11, c = 12, \gamma = 60°$

25. $a = \sqrt{7}, b = \sqrt{8}, c = \sqrt{3}$

26. $\beta = 106°, \gamma = 43°, a = 1$

27. $b = 11, c = 2, \beta = 10°$

28. $\alpha = 25°, a = 6, c = 9$

In Exercises 29–50, find the area of each triangle with measures given.

29. $a = 8, c = 16, \beta = 60°$

30. $b = 6, c = 4\sqrt{3}, \alpha = 30°$

31. $a = 1, b = \sqrt{2}, \alpha = 45°$

32. $b = 2\sqrt{2}, c = 4, \beta = 45°$

33. $a = 6, b = 8, \gamma = 80°$

34. $b = 9, c = 10, \alpha = 100°$

35. $a = 4, c = 7, \beta = 27°$

36. $a = 6.3, b = 4.8, \gamma = 17°$

37. $b = 100, c = 150, \alpha = 36°$

38. $c = 0.3, a = 0.7, \beta = 145°$

39. $a = 15, b = 15, c = 15$

40. $a = 1, b = 1, c = 1$

41. $a = 7, b = \sqrt{51}, c = 10$

42. $a = 9, b = 40, c = 41$

43. $a = 6, b = 10, c = 9$

44. $a = 40, b = 50, c = 60$

45. $a = 14.3, b = 15.7, c = 20.1$

46. $a = 146.5, b = 146.5, c = 100$

47. $a = 14,000, b = 16,500, c = 18,700$

48. $a = \sqrt{2}, b = \sqrt{3}, c = \sqrt{5}$

49. $a = 80, b = 75, c = 160$

50. $a = 19, b = 23, c = 3$

■ **APPLICATIONS**

51. Aviation. A plane flew due north at 500 mph for 3 hours. A second plane, starting at the same point and at the same time, flew southeast at an angle 150° clockwise from due north at 435 mph for 3 hours. At the end of the 3 hours, how far apart were the two planes? Round to the nearest mile.

52. Aviation. A plane flew due north at 400 mph for 4 hours. A second plane, starting at the same point and at the same time, flew southeast at an angle 120° clockwise from due north at 300 mph for 4 hours. At the end of the 4 hours, how far apart were the two planes? Round to the nearest mile.

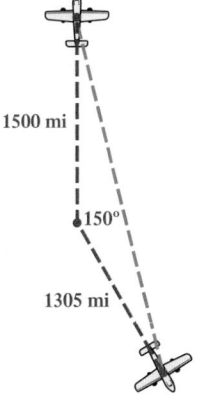

1500 mi

150°

1305 mi

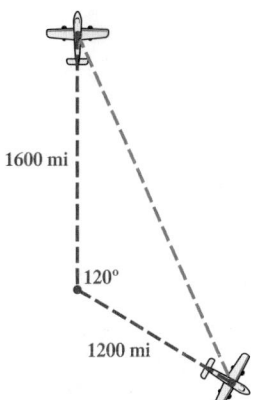

1600 mi

120°

1200 mi

53. Aviation. A plane flew N 30° W at 350 mph for 2.5 hours. A second plane, starting at the same point and at the same time, flew 35° at an angle clockwise from due north at 550 mph for 2.5 hours. At the end of 2.5 hours, how far apart were the two planes? Round to the nearest mile.

54. Aviation. A plane flew N 30° W at 350 mph for 3 hours. A second plane starts at the same point and takes off at the same time. It is known that after 3 hours, the two planes are 2100 miles apart. Find the original bearing of the second plane, to the nearest hundredth of a degree.

55. Sliding Board. A 40-foot slide leaning against the bottom of a building's window makes a 55° angle with the building. The angle formed with the building by the line of sight from the top of the window to the point on the ground where the slide ends is 40°. How tall is the window?

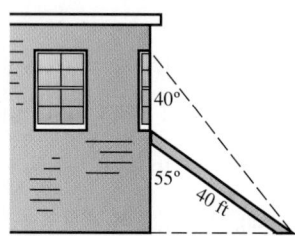

56. Airplane Slide. An airplane door is 6 feet high. If a slide attached to the bottom of the open door is at an angle of 40° with the ground, and the angle formed by the line of sight from where the slide touches the ground to the top of the door is 45°, how long is the slide?

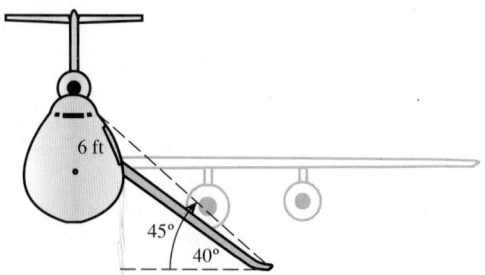

57. Law Enforcement. Two members of a SWAT team and the thief they are to apprehend are positioned as follows:

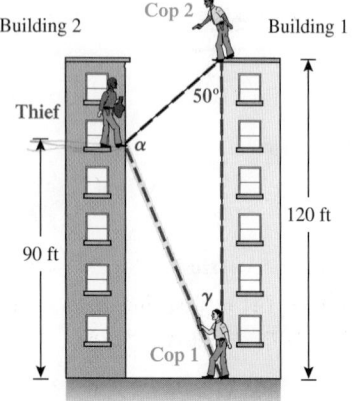

When the signal is given, Cop 2 shoots a zipline across to the window where the thief is spotted (at a 50° angle to Building 1) and Cop 1 shines a very bright light directly at the thief. Find the angle γ at which Cop 1 holds the light to shine it directly at the thief. Round to the nearest hundredth degree.

58. Law Enforcement. In reference to Exercise 57, what angle does the zipline make with respect to Building 2?

59. Surveying. A glaciologist needs to determine the length across a certain crevice on Mendenhall glacier in order to circumvent it with his team. He has the following measurements:

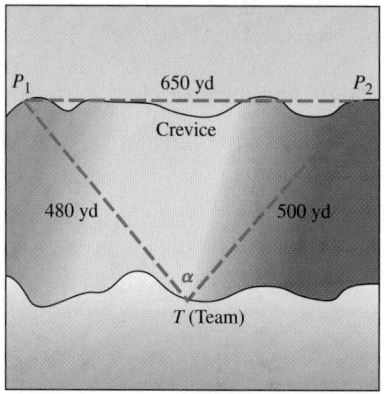

Find α.

60. Surveying. A glaciologist needs to determine the length across a certain crevice on Mendenhall glacier in order to circumvent it with her team. She has the following measurements:

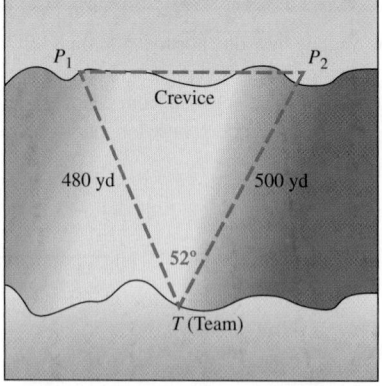

Find the approximate length across the crevice.

61. Parking Lot. A parking lot is to have the shape of a parallelogram that has adjacent sides measuring 200 feet and 260 feet. The acute angle between two adjacent sides is 65°. What is the area of the parking lot?

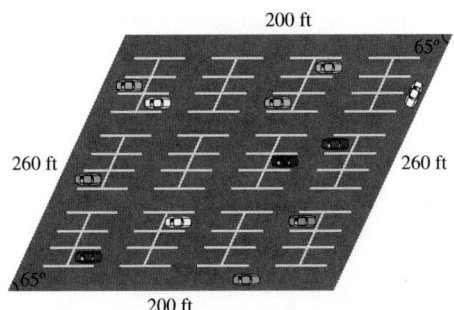

200 ft

65°

260 ft   260 ft

65°

200 ft

62. Parking Lot. A parking lot is to have the shape of a parallelogram that has adjacent sides measuring 250 feet and 300 feet. The acute angle between two adjacent sides is 55°. What is the area of the parking lot?

63. Regular Hexagon. A regular hexagon has sides measuring 3 feet. What is its area? Recall that the measure of an angle of a regular n-gon is given by the formula angle $= \dfrac{180°(n-2)}{n}$.

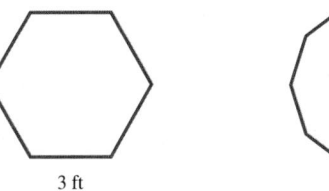

3 ft   5 in.

64. Regular Decagon. A regular decagon has sides measuring 5 inches. What is its area?

65. Geometry. A quadrilateral $ABCD$ has sides of lengths $AB = 2$, $BC = 3$, $CD = 4$, and $DA = 5$. The angle between AB and BC is 135°. Find the area of $ABCD$.

66. Geometry. A quadrilateral $ABCD$ has sides of lengths $AB = 5$, $BC = 6$, $CD = 7$, and $DA = 8$. The angle between AB and BC is 135°. Find the area of $ABCD$.

■ CATCH THE MISTAKE

In Exercises 67 and 68, explain the mistake that is made.

67. Solve the triangle $b = 3$, $c = 4$, and $\alpha = 30°$.

Solution:

Step 1: Find a.

Apply the Law of Cosines. $\qquad a^2 = b^2 + c^2 - 2bc\cos\alpha$

Let $b = 3$, $c = 4$, and $\alpha = 30°$. $\qquad a^2 = 3^2 + 4^2 - 2(3)(4)\cos 30°$

Solve for a. $\qquad\qquad a \approx 2.1$

Step 2: Find γ.

Apply the Law of Sines. $\qquad \dfrac{\sin\alpha}{a} = \dfrac{\sin\gamma}{c}$

Solve for $\sin\gamma$. $\qquad \sin\gamma = \dfrac{c\sin\alpha}{a}$

Solve for γ. $\qquad \gamma = \sin^{-1}\left(\dfrac{c\sin\alpha}{a}\right)$

Let $a = 2.1$, $c = 4$, and $\alpha = 30°$. $\qquad \gamma \approx 72°$

Step 3: Find β.

$\alpha + \beta + \gamma = 180° \qquad 30° + \beta + 72° = 180°$

Solve for β. $\qquad\qquad \beta \approx 78°$

$a \approx 2.1$, $b = 3$, $c = 4$, $\alpha = 30°$, $\beta \approx 78°$, and $\gamma \approx 72°$

This is incorrect. The longest side is not opposite the largest angle. What mistake was made?

68. Solve the triangle $a = 6$, $b = 2$, and $c = 5$.

Solution:

Step 1: Find β.

Apply the Law of Cosines. $\qquad b^2 = a^2 + c^2 - 2ac\cos\beta$

Solve for β. $\qquad \beta = \cos^{-1}\left(\dfrac{a^2 + c^2 - b^2}{2ac}\right)$

Let $a = 6$, $b = 2$, $c = 5$. $\qquad \beta \approx 18°$

Step 2: Find α.

Apply the Law of Sines. $\qquad \dfrac{\sin\alpha}{a} = \dfrac{\sin\beta}{b}$

Solve for α. $\qquad \alpha = \sin^{-1}\left(\dfrac{a\sin\beta}{b}\right)$

Let $a = 6$, $b = 2$, and $\beta = 18°$. $\qquad \alpha \approx 68°$

Step 3: Find γ.

$\alpha + \beta + \gamma = 180°$

$68° + 18° + \gamma = 180°$

$\gamma \approx 94°$

$a = 6$, $b = 2$, $c = 5$, $\alpha \approx 68°$, $\beta \approx 18°$, and $\gamma \approx 94°$

This is incorrect. The longest side is not opposite the largest angle. What mistake was made?

CONCEPTUAL

In Exercises 69–74, determine whether each statement is true or false.

69. Given the lengths of all three sides of a triangle, there is insufficient information to solve the triangle.

70. Given three angles of a triangle, there is insufficient information to solve the triangle.

71. The Pythagorean theorem is a special case of the Law of Cosines.

72. The Law of Cosines is a special case of the Pythagorean theorem.

73. If an obtuse triangle is isosceles, then knowing the measure of the obtuse angle and a side adjacent to it is sufficient to solve the triangle.

74. All acute triangles can be solved using the Law of Cosines.

CHALLENGE

75. Show that $\dfrac{\cos\alpha}{a} + \dfrac{\cos\beta}{b} + \dfrac{\cos\gamma}{c} = \dfrac{a^2 + b^2 + c^2}{2abc}$.
(*Hint:* Use the Law of Cosines.)

76. Show that $a = c\cos\beta + b\cos\gamma$. (*Hint:* Use the Law of Cosines.)

The following half-angle identities are useful in Exercises 77 and 78, and will be derived in Chapter 6.

$$\cos\left(\frac{x}{2}\right) = \sqrt{\frac{1 + \cos x}{2}} \qquad \tan\left(\frac{x}{2}\right) = \sqrt{\frac{1 - \cos x}{1 + \cos x}}$$

77. Consider the following diagram and express $\cos\left(\dfrac{X}{2}\right)$ in terms of a.

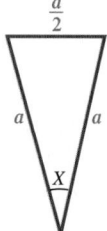

78. Using the diagram in Exercise 77, express $\tan\left(\dfrac{X}{2}\right)$ in terms of a.

79. Show that the area for an SAA triangle is given by

$$A = \frac{a^2 \sin\beta \sin\gamma}{2\sin\alpha}$$

Assume that $\alpha, \beta,$ and a are given.

TECHNOLOGY

For Exercises 83–88, let A, B, and C be the lengths of the three sides with X, Y, and Z as the corresponding angle measures in a triangle. Write a program using a TI calculator to solve each triangle with the given measures.

80. Show that the area of an isosceles triangle with equal sides of length s is given by

$$A_{\text{isosceles}} = \frac{1}{2} s^2 \sin\theta$$

where θ is the angle between the two equal sides.

81. Find the area of the shaded region.

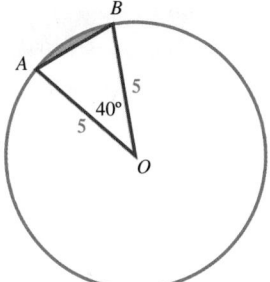

82. Find the area of the shaded region.

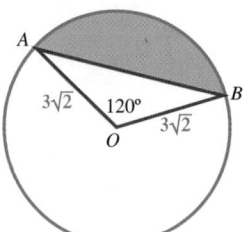

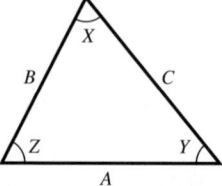

83. $B = 45$, $C = 57$, and $X = 43°$

84. $B = 24.5$, $C = 31.6$, and $X = 81.5°$

85. $A = 29.8$, $B = 37.6$, and $C = 53.2$

86. $A = 100$, $B = 170$, and $C = 250$

87. $A = \sqrt{12}$, $B = \sqrt{21}$, and $Z = 62.8°$

88. $A = 1235$, $B = 987$, and $C = 1456$

▪ PREVIEW TO CALCULUS

In calculus, some applications of the derivative require the solution of triangles. In Exercises 89–92, solve each triangle using the Law of Cosines.

89. Two ships start moving from the same port at the same time. One moves north at 40 mph, while the other moves southeast at 50 mph. Find the distance between the ships 4 hours later. Round your answer to the nearest mile.

90. An airport radar detects two planes approaching. The distance between the planes is 80 miles; the closest plane is 60 miles from the airport and the other plane is 70 miles from the airport. What is the angle (in degrees) formed by the planes and the airport?

91. An athlete runs along a circular track, of radius 100 m, runs from A to B and then decides to take a shortcut to go to C. If the measure of angle BAC is $\dfrac{2\pi}{9}$, find the distance

covered by the athlete if the distance from A to B is 153 m. Round your answer to the nearest integer.

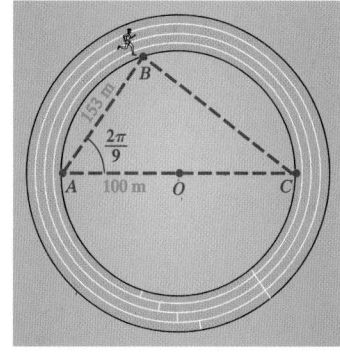

92. A regular pentagon is inscribed in a circle of radius 10 ft. Find its perimeter. Round your answer to the nearest tenth.

MODELING YOUR WORLD

The Intergovernmental Panel on Climate Change (IPCC) claims that carbon dioxide (CO_2) production from increased industrial activity (such as fossil fuel burning and other human activities) has increased the CO_2 concentrations in the atmosphere. Because it is a greenhouse gas, elevated CO_2 levels will increase global mean (average) temperature. In this section, we will examine the increasing rate of carbon emissions on Earth.

In 1955 there were (globally) 2 billion tons of carbon emitted per year. In 2005 the carbon emission had more than tripled, reaching approximately 7 billion tons of carbon emitted per year. Currently, we are on the path to doubling our current carbon emissions in the next 50 years.

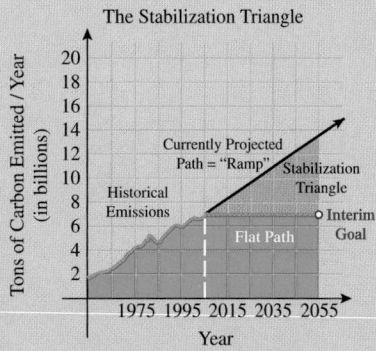

Two Princeton professors* (Stephen Pacala and Rob Socolow) introduced the Climate Carbon Wedge concept. A "wedge" is a strategy to reduce carbon emissions that grow in a 50-year time period from 0 to 1.0 GtC/yr (gigatons of carbon per year).

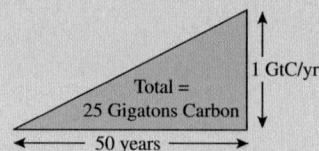

1. Consider eight scenarios (staying on path of one of the seven wedges). A check is done at the 10-year mark. What total GtC per year would we have to measure to correspond to the following projected paths?

   a. Flat path (no increase) over 50 years (2005 to 2055)
   b. Increase of 1 GtC over 50 years (2005 to 2055)
   c. Increase of 2 GtC over 50 years (2005 to 2055)
   d. Increase of 3 GtC over 50 years (2005 to 2055)
   e. Increase of 4 GtC over 50 years (2005 to 2055)
   f. Increase of 5 GtC over 50 years (2005 to 2055)
   g. Increase of 6 GtC over 50 years (2005 to 2055)
   h. Increase of 7 GtC over 50 years (2005 to 2055) (projected path)

2. Consider the angle θ in each wedge. For each of the seven wedges (and the flat path), find the GtC/yr rate in terms of $\tan\theta$.

   a. Flat path
   b. Increase of 1 GtC/50 years
   c. Increase of 2 GtC/50 years
   d. Increase of 3 GtC/50 years
   e. Increase of 4 GtC/50 years
   f. Increase of 5 GtC/50 years
   g. Increase of 6 GtC/50 years
   h. Increase of 7 GtC/50 years (projected path)

3. Research the "climate carbon wedge" concept and discuss the types of changes (transportation efficiency, transportation conservation, building efficiency, efficiency in electricity production, alternate energies, etc.) the world would have to make that would correspond to each of the seven wedges.

   a. Flat path
   b. Wedge 1
   c. Wedge 2
   d. Wedge 3
   e. Wedge 4
   f. Wedge 5
   g. Wedge 6
   h. Wedge 7

---

*S. Pacala and R. Socolow, "Stabilization Wedges: Solving the Climate Problem for the Next 50 Years with Current Technologies," *Science*, Vol. 305 (2004).

SECTION	CONCEPT	PAGES	REVIEW EXERCISES	KEY IDEAS/POINTS
4.1	**Angle measure**	364–375	1–26	
	Degrees and radians	364–368	1–24	Converting between degrees and radians (Remember that $\pi = 180°$.) ■ Degrees to radians: Multiply by $\frac{\pi}{180°}$ ■ Radians to degrees: Multiply by $\frac{180°}{\pi}$
	Coterminal angles	369–370		Two angles in standard position with the same terminal side
	Arc length	371		$s = r\theta$ θ is in radians.
	Area of a circular sector	371–372		$A = \frac{1}{2}r^2\theta$ θ is in radians.
	Linear and angular speeds	372–375	25–26	**Linear speed** v is given by $$v = \frac{s}{t}$$ where s is the arc length (or distance along the arc) and t is time. **Angular speed** ω is given by $$\omega = \frac{\theta}{t}$$ where θ is given in radians. Linear and angular speeds are related through the radius of the circle: $$v = r\omega \quad \text{or} \quad \omega = \frac{v}{r}$$
4.2	**Right triangle trigonometry**	380–393	27–48	
	Right triangle ratios	381–385	27–32	$\sin\theta = \frac{\text{opposite}}{\text{hypotenuse}}$ (SOH) $\cos\theta = \frac{\text{adjacent}}{\text{hypotenuse}}$ (CAH) $\tan\theta = \frac{\text{opposite}}{\text{adjacent}}$ (TOA)

SECTION	CONCEPT	PAGES	REVIEW EXERCISES	KEY IDEAS/POINTS

Reciprocal identities

$$\cot\theta = \frac{1}{\tan\theta} \qquad \csc\theta = \frac{1}{\sin\theta} \qquad \sec\theta = \frac{1}{\cos\theta}$$

Evaluating trigonometric functions exactly for special angle measures	385–390	33–38	

θ	$\sin\theta$	$\cos\theta$
30°	$\dfrac{1}{2}$	$\dfrac{\sqrt{3}}{2}$
45°	$\dfrac{\sqrt{2}}{2}$	$\dfrac{\sqrt{2}}{2}$
60°	$\dfrac{\sqrt{3}}{2}$	$\dfrac{1}{2}$

The other trigonometric functions can be found for these values using $\tan\theta = \dfrac{\sin\theta}{\cos\theta}$ and the reciprocal identities.

Solving right triangles	390–393		

4.3	**Trigonometric functions of angles**	400–414	49–72	

Trigonometric functions: The Cartesian plane	400–406	49–72	

$$\sin\theta = \frac{y}{r} \qquad \cos\theta = \frac{x}{r} \qquad \tan\theta = \frac{y}{x}$$

$$\csc\theta = \frac{r}{y} \qquad \sec\theta = \frac{r}{x} \qquad \cot\theta = \frac{x}{y}$$

where $x^2 + y^2 = r^2 \Rightarrow r = \sqrt{x^2 + y^2}$

The distance r is positive: $r > 0$.

Algebraic signs of trigonometric functions

θ	QI	QII	QIII	QIV
$\sin\theta$	+	+	−	−
$\cos\theta$	+	−	−	+
$\tan\theta$	+	−	+	−

Trigonometric function values for quadrantal angles.

θ	0°	90°	180°	270°
$\sin\theta$	0	1	0	−1
$\cos\theta$	1	0	−1	0
$\tan\theta$	0	undefined	0	undefined
$\cot\theta$	undefined	0	undefined	0
$\sec\theta$	1	undefined	−1	undefined
$\csc\theta$	undefined	1	undefined	−1

SECTION	CONCEPT	PAGES	REVIEW EXERCISES	KEY IDEAS/POINTS
	Ranges of the trigonometric functions	406–407		$\sin\theta$ and $\cos\theta$: $[-1, 1]$ $\tan\theta$ and $\cot\theta$: $(-\infty, \infty)$ $\sec\theta$ and $\csc\theta$: $(-\infty, -1] \cup [1, \infty)$
	Reference angles and reference right triangles	408–410	59–72	The reference angle α for angle θ (between $0°$ and $360°$) is given by ■ QI: $\alpha = \theta$ ■ QII: $\alpha = 180° - \theta$ or $\pi - \theta$ ■ QIII: $\alpha = \theta - 180°$ or $\theta - \pi$ ■ QIV: $\alpha = 360° - \theta$ or $2\pi - \theta$
	Evaluating trigonometric functions for nonacute angles	410–414	59–72	
4.4	**The Law of Sines**	418–427	73–90	
	Solving oblique triangles: Four cases	418–420	73–90	Oblique (Nonright) Triangles
	The Law of Sines	420–427	73–90	$\dfrac{\sin\alpha}{a} = \dfrac{\sin\beta}{b} = \dfrac{\sin\gamma}{c}$ Use for: ■ AAS (or ASA) triangles ■ SSA triangles (ambiguous case)
4.5	**The Law of Cosines**	432–439	91–120	
	Solving oblique triangles using the Law of Cosines	432–436	91–110	$a^2 = b^2 + c^2 - 2bc\cos\alpha$ $b^2 = a^2 + c^2 - 2ac\cos\beta$ $c^2 = a^2 + b^2 - 2ab\cos\gamma$ Use for: ■ SAS triangles ■ SSS triangles
	The area of a triangle	436–439	111–120	**The area of a triangle (SAS case)** $A_{SAS} = \frac{1}{2}bc\sin\alpha$ when b, c, and α are known. $A_{SAS} = \frac{1}{2}ab\sin\gamma$ when a, b, and γ are known. $A_{SAS} = \frac{1}{2}ac\sin\beta$ when a, c, and β are known. **The area of a triangle (SSS case)** Use Heron's formula for the SSS case: $$A_{SSS} = \sqrt{s(s-a)(s-b)(s-c)}$$ where a, b, and c are the lengths of the sides of the triangle and s is half the perimeter of the triangle, called the semiperimeter. $$s = \frac{a+b+c}{2}$$

4.1 Angle Measure

Find (a) the complement and (b) the supplement of the given angles.

1. $28°$ 2. $17°$ 3. $35°$

4. $78°$ 5. $89.01°$ 6. $0.013°$

Convert from degrees to radians. Leave your answers in terms of π.

7. $135°$ 8. $240°$ 9. $330°$ 10. $180°$

11. $216°$ 12. $108°$ 13. $1620°$ 14. $900°$

Convert from radians to degrees.

15. $\dfrac{\pi}{3}$ 16. $\dfrac{11\pi}{6}$ 17. $\dfrac{5\pi}{4}$ 18. $\dfrac{2\pi}{3}$

19. $\dfrac{5\pi}{9}$ 20. $\dfrac{17\pi}{10}$ 21. 10π 22. $\dfrac{31\pi}{2}$

Applications

23. **Clock.** What is the measure (in degrees) of the angle that the minute hand sweeps in exactly 25 minutes?

24. **Clock.** What is the measure (in degrees) of the angle that the second hand sweeps in exactly 15 seconds?

25. A ladybug is clinging to the outer edge of a child's spinning disk. The disk is 4 inches in diameter and is spinning at 60 revolutions per minute. How fast is the ladybug traveling in inches/minute?

26. How fast is a motorcyclist traveling in miles per hour if his tires are 30 inches in diameter and the angular speed of the tire is 10π radians per second?

4.2 Right Triangle Trigonometry

Use the following triangle to find the indicated trigonometric functions. Rationalize any denominators that you encounter in the answers.

27. $\cos\theta$

28. $\sin\theta$

29. $\sec\theta$

30. $\csc\theta$

31. $\tan\theta$

32. $\cot\theta$

Label each trigonometric function value with the corresponding value (a–c).

a. $\dfrac{\sqrt{3}}{2}$ b. $\dfrac{1}{2}$ c. $\dfrac{\sqrt{2}}{2}$

33. $\sin 30°$ 34. $\cos 30°$ 35. $\cos 60°$

36. $\sin 60°$ 37. $\sin 45°$ 38. $\cos 45°$

Use a calculator to approximate the following trigonometric function values. Round the answers to four decimal places.

39. $\sin 42°$ 40. $\cos 57°$ 41. $\cos 17.3°$ 42. $\tan 25.2°$

43. $\cot 33°$ 44. $\sec 16.8°$ 45. $\csc 40.25°$ 46. $\cot 19.76°$

The following exercises illustrate a mid-air refueling scenario that U.S. military aircraft often use. Assume the elevation angle that the hose makes with the plane being fueled is $\theta = 30°$.

47. **Mid-Air Refueling.** If the hose is 150 feet long, what should the altitude difference a be between the two planes?

48. **Mid-Air Refueling.** If the smallest acceptable altitude difference, a, between the two planes is 100 feet, how long should the hose be?

4.3 Trigonometric Functions of Angles

In the following exercises, the terminal side of an angle θ in standard position passes through the indicated point. Calculate the values of the six trigonometric functions for angle θ.

49. $(6, -8)$ 50. $(-24, -7)$ 51. $(-6, 2)$ 52. $(-40, 9)$

53. $(\sqrt{3}, 1)$ 54. $(-9, -9)$ 55. $\left(\frac{1}{2}, -\frac{1}{4}\right)$ 56. $\left(-\frac{3}{4}, \frac{5}{6}\right)$

57. $(-1.2, -2.4)$ 58. $(0.8, -2.4)$

Evaluate the following expressions exactly:

59. $\sin 330°$ 60. $\cos(-300°)$ 61. $\tan 150°$

62. $\cot 315°$ 63. $\sec(-150°)$ 64. $\csc 210°$

65. $\sin\left(\dfrac{7\pi}{4}\right)$ 66. $\cos\left(\dfrac{7\pi}{6}\right)$ 67. $\tan\left(-\dfrac{2\pi}{3}\right)$

68. $\cot\left(\dfrac{4\pi}{3}\right)$ 69. $\sec\left(\dfrac{5\pi}{4}\right)$ 70. $\csc\left(-\dfrac{8\pi}{3}\right)$

71. $\sec\left(\dfrac{5\pi}{6}\right)$ 72. $\cos\left(-\dfrac{11\pi}{6}\right)$

4.4 The Law of Sines

Solve the given triangles.

73. $\alpha = 10°, \beta = 20°, a = 4$

74. $\beta = 40°, \gamma = 60°, b = 10$

75. $\alpha = 5°, \beta = 45°, c = 10$

76. $\beta = 60°, \gamma = 70°, a = 20$

77. $\gamma = 11°, \alpha = 11°, c = 11$

78. $\beta = 20°, \gamma = 50°, b = 8$

79. $\alpha = 45°, \gamma = 45°, b = 2$

80. $\alpha = 60°, \beta = 20°, c = 17$

81. $\alpha = 12°, \gamma = 22°, a = 99$

82. $\beta = 102°, \gamma = 27°, a = 24$

Two sides and an angle are given. Determine whether a triangle (or two) exist and, if so, solve the triangle.

83. $a = 7, b = 9, \alpha = 20°$

84. $b = 24, c = 30, \beta = 16°$

85. $a = 10, c = 12, \alpha = 24°$

86. $b = 100, c = 116, \beta = 12°$

87. $a = 40, b = 30, \beta = 150°$

88. $b = 2, c = 3, \gamma = 165°$

89. $a = 4, b = 6, \alpha = 10°$

90. $c = 25, a = 37, \gamma = 4°$

4.5 The Law of Cosines

Solve each triangle.

91. $a = 40, b = 60, \gamma = 50°$

92. $b = 15, c = 12, \alpha = 140°$

93. $a = 24, b = 25, c = 30$

94. $a = 6, b = 6, c = 8$

95. $a = \sqrt{11}, b = \sqrt{14}, c = 5$

96. $a = 22, b = 120, c = 122$

97. $b = 7, c = 10, \alpha = 14°$

98. $a = 6, b = 12, \gamma = 80°$

99. $b = 10, c = 4, \alpha = 90°$

100. $a = 4, b = 5, \gamma = 75°$

101. $a = 10, b = 11, c = 12$

102. $a = 22, b = 24, c = 25$

103. $b = 16, c = 18, \alpha = 100°$

104. $a = 25, c = 25, \beta = 9°$

105. $b = 12, c = 40, \alpha = 10°$

106. $a = 26, b = 20, c = 10$

107. $a = 26, b = 40, c = 13$

108. $a = 1, b = 2, c = 3$

109. $a = 6.3, b = 4.2, \alpha = 15°$

110. $b = 5, c = 6, \beta = 35°$

Find the area of each triangle described.

111. $b = 16, c = 18, \alpha = 100°$

112. $a = 25, c = 25, \beta = 9°$

113. $a = 10, b = 11, c = 12$

114. $a = 22, b = 24, c = 25$

115. $a = 26, b = 20, c = 10$

116. $a = 24, b = 32, c = 40$

117. $b = 12, c = 40, \alpha = 10°$

118. $a = 21, c = 75, \beta = 60°$

Applications

119. Area of Inscribed Triangle. The area of a triangle inscribed in a circle can be found if you know the lengths of the sides of the triangle and the radius of the circle: $A = \dfrac{abc}{4r}$. Find the radius of the circle that circumscribes the triangle if all the sides of the triangle measure 9.0 inches and the area of the triangle is 35 square inches.

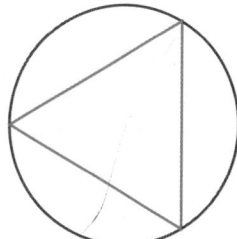

120. Area of Inscribed Triangle. The area of a triangle inscribed in a circle can be found if you know the lengths of the sides of the triangle and the radius of the circle: $A = \dfrac{abc}{4r}$. Find the radius of the circle that circumscribes the triangle if the sides of the triangle measure 9, 12, and 15 inches and the area of the triangle is 54 square inches.

1. A 5-foot girl is standing *in* the Grand Canyon, and she wants to estimate the depth of the canyon. The sun casts her shadow 6 inches along the ground. To measure the shadow cast by the top of the canyon, she walks the length of the shadow. She takes 200 steps and estimates that each step is roughly 3 feet. Approximately how tall is the Grand Canyon?

2. Fill in the values in the table.

θ	$\sin \theta$	$\cos \theta$	$\tan \theta$	$\cot \theta$	$\sec \theta$	$\csc \theta$
30°						
45°						
60°						

3. What is the difference between $\cos \theta = \frac{2}{3}$ and $\cos \theta \approx 0.6\overline{6}$?

4. Fill in the table with exact values for the quadrantal angles and the algebraic signs for the quadrants.

	0°	QI	90°	QII	180°	QIII	270°	QIV	360°
$\sin \theta$									
$\cos \theta$									

5. If $\cot \theta < 0$ and $\sec \theta > 0$, in which quadrant does the terminal side of θ lie?

6. Evaluate $\sin 210°$ exactly.

7. Convert $\dfrac{13\pi}{4}$ to degree measure.

8. Convert 260° to radian measure. Leave the answer in terms of π.

9. What is the area of the sector swept by the second hand of a clock in 25 seconds? Assume the radius of the sector is 3 inches.

10. What is the measure in radians of the smaller angle between the hour and minute hands at 10:10?

Solve the triangles if possible.

11. $\alpha = 30°, \beta = 40°, b = 10$

12. $\alpha = 47°, \beta = 98°, \gamma = 35°$

13. $a = 7, b = 9, c = 12$

14. $\alpha = 45°, a = 8, b = 10$

15. $a = 1, b = 1, c = 2$

16. $a = \dfrac{23}{7}, c = \dfrac{5}{7}, \beta = 61.2°$

17. $\alpha = 110°, \beta = 20°, a = 5$

18. $b = \dfrac{\sqrt{5}}{2}, c = 3\sqrt{5}, \alpha = 45°$

In Exercises 19 and 20, find the areas of the given triangles.

19. $\gamma = 72°, a = 10, b = 12$

20. $a = 7, b = 10, c = 13$

1. Find the average rate of change for $f(x) = \dfrac{5}{x}$ from $x = 2$ to $x = 4$.

2. Use interval notation to express the domain of the function $f(x) = \sqrt{x^2 - 25}$.

3. Using the function $f(x) = 5 - x^2$, evaluate the difference quotient $\dfrac{f(x + h) - f(x)}{h}$.

4. Given the piecewise-defined function
$$f(x) = \begin{cases} x^2 & x < 0 \\ 2x - 1 & 0 \le x < 5 \\ 5 - x & x \ge 5 \end{cases}$$
find:
   a. $f(0)$ b. $f(4)$ c. $f(5)$ d. $f(-4)$
   e. State the domain and range in interval notation.
   f. Determine the intervals where the function is increasing, decreasing, or constant.

5. Evaluate $g(f(-1))$ for $f(x) = \sqrt[3]{x - 7}$ and $g(x) = \dfrac{5}{3 - x}$.

6. Find the inverse of the function $f(x) = \dfrac{5x + 2}{x - 3}$.

7. Find the quadratic function that has the vertex $(0, 7)$ and goes through the point $(2, -1)$.

8. Find all of the real zeros and state the multiplicity of each for the function $f(x) = \frac{1}{7}x^5 + \frac{2}{9}x^3$.

9. Graph the rational function $f(x) = \dfrac{x^2 + 3}{x - 2}$. Give all asymptotes.

10. Factor the polynomial $P(x) = 4x^4 - 4x^3 + 13x^2 + 18x + 5$ as a product of linear factors.

11. How much money should be put in a savings account now that earns 5.5% a year compounded continuously, if you want to have $85,000 in 15 years?

12. Evaluate $\log_{4.7} 8.9$ using the change-of-base formula. Round the answer to three decimal places.

13. Solve the equation $5\left(10^{2x}\right) = 37$ for x. Round the answer to three decimal places.

14. Solve for x: $\ln\sqrt{6 - 3x} - \frac{1}{2}\ln(x + 2) = \ln(x)$.

15. In a 45°-45°-90° triangle, if the two legs have a length of 15 feet, how long is the hypotenuse?

16. **Height of a tree.** The shadow of a tree measures $15\frac{1}{3}$ feet. At the same time of day the shadow of a 6-foot pole measures 2.3 feet. How tall is the tree?

17. Convert 432° to radians.

18. Convert $\dfrac{5\pi}{9}$ to degrees.

19. Find the exact value of $\tan\left(\dfrac{4\pi}{3}\right)$.

20. Find the exact value of $\sec\left(-\dfrac{7\pi}{6}\right)$.

21. Use a calculator to find the value of $\csc 37°$. Round your answer to four decimal places.

22. In the right triangle below, find a, b, and θ. Round each to the nearest tenth.

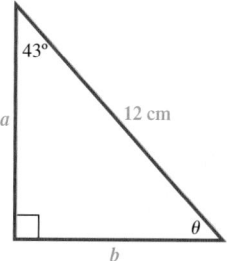

23. Solve the triangle below. Round the side lengths to the nearest centimeter.

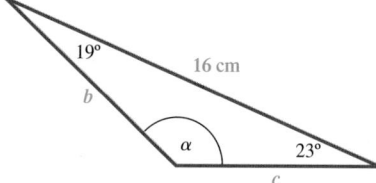

24. Solve the triangle $a = 2$, $b = 4$, and $c = 5$. Round your answer to the nearest degree.

5

Trigonometric Functions of Real Numbers

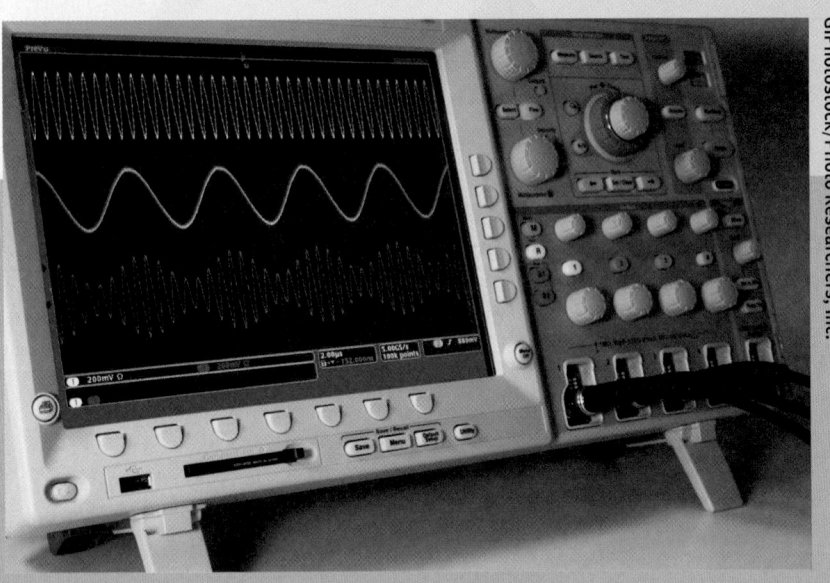

An oscilloscope displays voltage (vertical axis) as a function of time (horizontal axis) of an electronic signal. The electronic signal is an electric representation of some periodic process that occurs in the real world, such as a human pulse or a sound wave. Oscilloscopes are used in medicine, the sciences, and engineering, and allow the shape of a signal to be displayed, which allows the amplitude and frequency of the repetitive signal to then be determined. The oscilloscope above displays a *sine* wave.

IN THIS CHAPTER we will use the unit circle approach to define trigonometric functions. We will graph the sine and cosine functions and find periods, amplitudes, and phase shifts. Applications such as harmonic motion will be discussed. Combinations of sinusoidal functions will be discussed through a technique called the addition of ordinates. Lastly, we will discuss the graphs of the other trigonometric functions (tangent, cotangent, secant, and cosecant).

TRIGONOMETRIC FUNCTIONS OF REAL NUMBERS

5.1
Trigonometric Functions:
The Unit Circle Approach

- Trigonometric Functions and the Unit Circle
- Circular Functions
- Properties of Circular Functions

5.2
Graphs of Sine and Cosine Functions

- The Graph of $f(x) = \sin x$
- The Graph of $f(x) = \cos x$
- The Amplitude and Period of Sinusoidal Graphs
- Graphing a Shifted Sinusoidal Function: $y = A \sin(Bx + C) + D$ and $y = A \cos(Bx + C) + D$
- Harmonic Motion
- Graphing Sums of Functions: Addition of Ordinates

5.3
Graphs of Other Trigonometric Functions

- The Tangent Function
- The Cotangent Function
- The Secant Function
- The Cosecant Function
- Graphing Tangent, Cotangent, Secant, and Cosecant Functions
- Translations of Trigonometric Functions

CHAPTER OBJECTIVES

- Define trigonometric functions using the unit circle approach (Section 5.1).
- Graph a sinusoidal function and determine its amplitude, period, and phase shift (Section 5.2).
- Graph tangent, cotangent, secant, and cosecant functions (Section 5.3).

SKILLS OBJECTIVES

- Draw the unit circle showing the special angles, and label cosine and sine values.
- Determine the domain and range of circular functions.
- Classify trigonometric functions as even or odd.

CONCEPTUAL OBJECTIVES

- Understand that the definition of trigonometric functions using the unit circle approach is consistent with both of the previous definitions (right triangle trigonometry and trigonometric functions of nonacute angles in the Cartesian plane).
- Relate x-coordinates and y-coordinates of points on the unit circle to the values of cosine and sine functions.
- Visualize the periodic properties of circular functions.

Trigonometric Functions and the Unit Circle

Recall that the first definition of trigonometric functions we developed was in terms of ratios of sides of right triangles (Section 4.2). Then in Section 4.3 we superimposed right triangles on the Cartesian plane, which led to a second definition of trigonometric functions (for any angle) in terms of ratios of x- and y-coordinates of a point and the distance from the origin to that point. In this section, we inscribe the right triangles into the unit circle in the Cartesian plane, which will yield a third definition of trigonometric functions. It is important to note that all three definitions are consistent with one another.

Recall that the equation for the **unit circle** centered at the origin is given by $x^2 + y^2 = 1$. We will use the term *circular function* later in this section, but it is important to note that a circle is not a function (it does not pass the vertical line test).

If we form a central angle θ in the unit circle such that the terminal side lies in quadrant I, we can use the previous two definitions of sine and cosine when $r = 1$ (i.e., in the unit circle).

Note: In radians, $\theta = \dfrac{s}{r}$, and since $r = 1$, we know that $\theta = s$.

TRIGONOMETRIC FUNCTION	RIGHT TRIANGLE TRIGONOMETRY	CARTESIAN PLANE
$\sin\theta$	$\dfrac{\text{opposite}}{\text{hypotenuse}} = \dfrac{y}{1} = y$	$\dfrac{y}{r} = \dfrac{y}{1} = y$
$\cos\theta$	$\dfrac{\text{adjacent}}{\text{hypotenuse}} = \dfrac{x}{1} = x$	$\dfrac{x}{r} = \dfrac{x}{1} = x$

Notice that the point (x, y) on the unit circle can be written as $(\cos\theta, \sin\theta)$. We can now summarize the exact values for **sine** and **cosine** in the illustration on the following page.

The following observations are consistent with properties of trigonometric functions we've studied already:

- $\sin\theta > 0$ in QI and QII.
- $\cos\theta > 0$ in QI and QIV.
- The unit circle equation $x^2 + y^2 = 1$ leads to the Pythagorean identity $\cos^2\theta + \sin^2\theta = 1$.

Study Tip

$(\cos\theta, \sin\theta)$ represents a point (x, y) on the unit circle.

$(x, y) = (\cos \theta, \sin \theta)$, where θ is the central angle whose terminal side intersects the unit circle at (x, y).

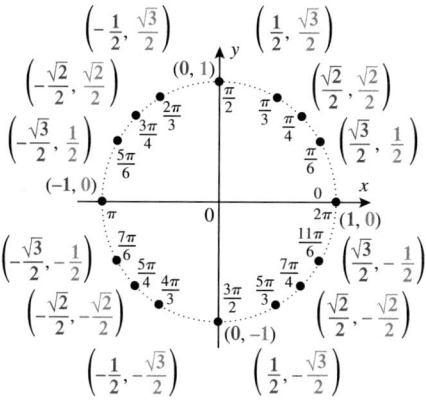

Circular Functions

Using the unit circle relationship, $(x, y) = (\cos \theta, \sin \theta)$, where θ is the central angle whose terminal side intersects the unit circle at the point (x, y), we can now define the remaining trigonometric functions using this unit circle approach and the quotient and reciprocal identities. Because the trigonometric functions are defined in terms of the unit *circle*, the trigonometric functions are often called **circular functions**.

DEFINITION   **Trigonometric Functions**

Unit Circle Approach

Let (x, y) be any point on the unit circle. If θ is a real number that represents the distance from the point $(1, 0)$ along the circumference to the point (x, y), then

$$\sin \theta = y \qquad\qquad \cos \theta = x \qquad\qquad \tan \theta = \frac{y}{x} \quad x \neq 0$$

$$\csc \theta = \frac{1}{y} \quad y \neq 0 \qquad \sec \theta = \frac{1}{x} \quad x \neq 0 \qquad \cot \theta = \frac{x}{y} \quad y \neq 0$$

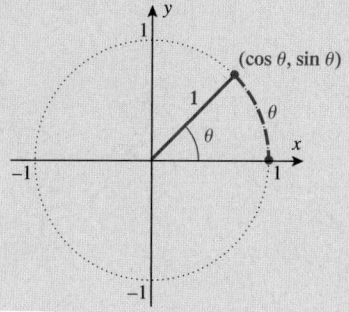

Technology Tip

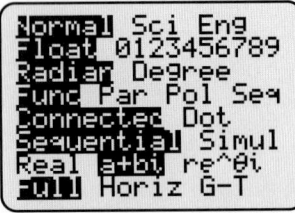

Use a TI/scientific calculator to check the values for $\sin\left(\dfrac{7\pi}{4}\right)$, $\cos\left(\dfrac{5\pi}{6}\right)$,

and $\tan\left(\dfrac{3\pi}{2}\right)$. Be sure to set the TI in radian mode.

```
Normal Sci Eng
Float 0123456789
Radian Degree
Func Par Pol Seq
Connected Dot
Sequential Simul
Real a+bi re^θi
Full Horiz G-T
```

```
sin(7π/4)
         -.7071067812
-√(2)/2
         -.7071067812
```

```
cos(5π/6)
         -.8660254038
-√(3)/2
         -.8660254038
■
```

Since $\tan\left(\dfrac{3\pi}{2}\right)$ is undefined, TI will display an error message.

```
tan(3π/2)
■
```

```
ERR:DOMAIN
1⬛Quit
2:Goto
```

EXAMPLE 1   Finding Exact Circular Function Values

Find the exact values for

a. $\sin\left(\dfrac{7\pi}{4}\right)$ **b.** $\cos\left(\dfrac{5\pi}{6}\right)$ **c.** $\tan\left(\dfrac{3\pi}{2}\right)$

Solution (a):

The angle $\dfrac{7\pi}{4}$ corresponds to the coordinates $\left(\dfrac{\sqrt{2}}{2}, -\dfrac{\sqrt{2}}{2}\right)$ on the unit circle.

The value of the sine function is the y-coordinate. $\boxed{\sin\left(\dfrac{7\pi}{4}\right) = -\dfrac{\sqrt{2}}{2}}$

Solution (b):

The angle $\dfrac{5\pi}{6}$ corresponds to the coordinate $\left(-\dfrac{\sqrt{3}}{2}, \dfrac{1}{2}\right)$ on the unit circle.

The value of the cosine function is the x-coordinate. $\boxed{\cos\left(\dfrac{5\pi}{6}\right) = -\dfrac{\sqrt{3}}{2}}$

Solution (c):

The angle $\dfrac{3\pi}{2}$ corresponds to the coordinate $(0, -1)$ on the unit circle.

The value of the cosine function is the x-coordinate. $\cos\left(\dfrac{3\pi}{2}\right) = 0$

The value of the sine function is the y-coordinate. $\sin\left(\dfrac{3\pi}{2}\right) = -1$

Tangent is the ratio of sine to cosine. $\tan\left(\dfrac{3\pi}{2}\right) = \dfrac{\sin\left(\dfrac{3\pi}{2}\right)}{\cos\left(\dfrac{3\pi}{2}\right)}$

Let $\cos\left(\dfrac{3\pi}{2}\right) = 0$ and $\sin\left(\dfrac{3\pi}{2}\right) = -1$. $\tan\left(\dfrac{3\pi}{2}\right) = \dfrac{-1}{0}$

$\boxed{\tan\left(\dfrac{3\pi}{2}\right) \text{ is undefined.}}$

■ **Answer: a.** $\dfrac{1}{2}$ **b.** $\dfrac{\sqrt{2}}{2}$ **c.** $-\sqrt{3}$

■ **YOUR TURN** Find the exact values for

a. $\sin\left(\dfrac{5\pi}{6}\right)$ **b.** $\cos\left(\dfrac{7\pi}{4}\right)$ **c.** $\tan\left(\dfrac{2\pi}{3}\right)$

EXAMPLE 2 **Solving Equations Involving Circular Functions**

Use the unit circle to find all values of θ, $0 \le \theta \le 2\pi$, for which $\sin \theta = -\frac{1}{2}$.

Solution:

The value of sine is the y-coordinate.

Since the value of sine is negative, θ must lie in QIII or QIV.

There are two values for θ that are greater than or equal to zero and less than or equal to 2π that correspond to $\sin\theta = -\frac{1}{2}$.

$$\boxed{\theta = \frac{7\pi}{6}, \frac{11\pi}{6}}$$

■ **YOUR TURN** Find all values of θ, $0 \le \theta \le 2\pi$, for which $\cos \theta = -\frac{1}{2}$.

■ **Answer:** $\theta = \dfrac{2\pi}{3}, \dfrac{4\pi}{3}$

Properties of Circular Functions

WORDS	MATH
For a point (x, y) that lies on the unit circle, $x^2 + y^2 = 1$.	$-1 \le x \le 1$ and $-1 \le y \le 1$
Since $(x, y) = (\cos \theta, \sin \theta)$, the following holds.	$-1 \le \cos \theta \le 1$ and $-1 \le \sin \theta \le 1$
State the **domain and range of the cosine and sine functions**.	Domain: $(-\infty, \infty)$ Range: $[-1, 1]$
Since $\cot \theta = \dfrac{\cos \theta}{\sin \theta}$ and $\csc \theta = \dfrac{1}{\sin \theta}$, the values for θ that make $\sin \theta = 0$ must be eliminated from the **domain of the cotangent and cosecant functions**.	Domain: $\theta \ne n\pi$, n an integer
Since $\tan \theta = \dfrac{\sin \theta}{\cos \theta}$ and $\sec \theta = \dfrac{1}{\cos \theta}$, the values for θ that make $\cos \theta = 0$ must be eliminated from the **domain of the tangent and secant functions**.	Domain: $\theta \ne \dfrac{(2n + 1)\pi}{2}$, n an integer

The following box summarizes the domains and ranges of the circular functions.

DOMAINS AND RANGES OF THE CIRCULAR FUNCTIONS

For any real number θ and integer n:

FUNCTION	DOMAIN	RANGE
$\sin \theta$	$(-\infty, \infty)$	$[-1,1]$
$\cos \theta$	$(-\infty, \infty)$	$[-1,1]$
$\tan \theta$	all real numbers such that $\theta \neq \dfrac{(2n + 1)\pi}{2}$	$(-\infty, \infty)$
$\cot \theta$	all real numbers such that $\theta \neq n\pi$	$(-\infty, \infty)$
$\sec \theta$	all real numbers such that $\theta \neq \dfrac{(2n + 1)\pi}{2}$	$(-\infty, -1] \cup [1, \infty)$
$\csc \theta$	all real numbers such that $\theta \neq n\pi$	$(-\infty, -1] \cup [1, \infty)$

Recall from algebra that **even functions** are functions for which $f(-x) = f(x)$ and **odd functions** are functions for which $f(-x) = -f(x)$.

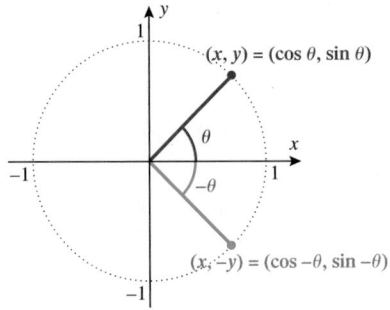

The cosine function is an even function. $\boxed{\cos \theta = \cos(-\theta)}$

The sine function is an odd function. $\boxed{\sin(-\theta) = -\sin \theta}$

Technology Tip

Use a TI/scientific calculator to check the value of $\cos\left(-\dfrac{5\pi}{6}\right)$.

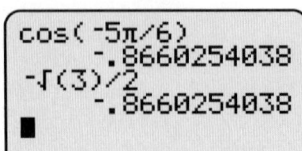

■ **Answer:** $-\dfrac{1}{2}$

EXAMPLE 3 Using Properties of Circular Functions

Evaluate $\cos\left(-\dfrac{5\pi}{6}\right)$.

Solution:

The cosine function is an even function. $\qquad \cos\left(-\dfrac{5\pi}{6}\right) = \cos\left(\dfrac{5\pi}{6}\right)$

Use the unit circle to evaluate cosine. $\qquad \cos\left(\dfrac{5\pi}{6}\right) = -\dfrac{\sqrt{3}}{2}$

$$\cos\left(-\dfrac{5\pi}{6}\right) = \boxed{-\dfrac{\sqrt{3}}{2}}$$

■ **YOUR TURN** Evaluate $\sin\left(-\dfrac{5\pi}{6}\right)$.

It is important to note that although circular functions can be evaluated exactly for some special angles, a calculator can be used to approximate circular functions for any angle. It is important to set the calculator to radian mode first, since θ is a real number.

 EXAMPLE 4 **Evaluating Circular Functions with a Calculator**

Use a calculator to evaluate $\sin\left(\dfrac{7\pi}{12}\right)$. Round the answer to four decimal places.

⭐ **CORRECT**

Evaluate with a calculator.

$$0.965925826$$

Round to four decimal places.

$$\sin\left(\dfrac{7\pi}{12}\right) \approx \boxed{0.9659}$$

❌ **INCORRECT**

Evaluate with a calculator.

$$0.031979376 \quad \textbf{ERROR}$$

(Calculator in degree mode)

Many calculators automatically reset to degree mode after every calculation, so make sure to always check what mode the calculator indicates.

■ **YOUR TURN** Use a calculator to evaluate $\tan\left(\dfrac{9\pi}{5}\right)$. Round the answer to four decimal places.

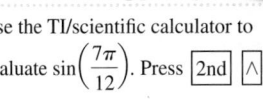

■ **Answer:** -0.7265

EXAMPLE 5 **Even and Odd Circular Functions**

Show that the secant function is an even function.

Solution:

Show that $\sec(-\theta) = \sec\theta$.

Secant is the reciprocal of cosine.

$$\sec(-\theta) = \dfrac{1}{\cos(-\theta)}$$

Cosine is an even function, $\cos(-\theta) = \cos\theta$.

$$\sec(-\theta) = \dfrac{1}{\cos\theta}$$

Secant is the reciprocal of cosine, $\sec\theta = \dfrac{1}{\cos\theta}$.

$$\sec(-\theta) = \dfrac{1}{\cos\theta} = \sec\theta$$

Since $\sec(-\theta) = \sec\theta$, the secant function is an even function .

SECTION

5.1 SUMMARY

In this section, we have defined trigonometric functions as circular functions. Any point (x, y) that lies on the unit circle satisfies the equation $x^2 + y^2 = 1$. The Pythagorean identity $\cos^2\theta + \sin^2\theta = 1$ can also be represented on the unit circle where $(x, y) = (\cos\theta, \sin\theta)$, and where θ is the central angle whose terminal side intersects the unit circle at the point (x, y). The cosine function is an even function, $\cos(-\theta) = \cos\theta$; the sine function is an odd function, $\sin(-\theta) = -\sin\theta$.

SECTION
5.1 EXERCISES

▪ SKILLS

In Exercises 1–14, find the exact values of the indicated trigonometric functions using the unit circle.

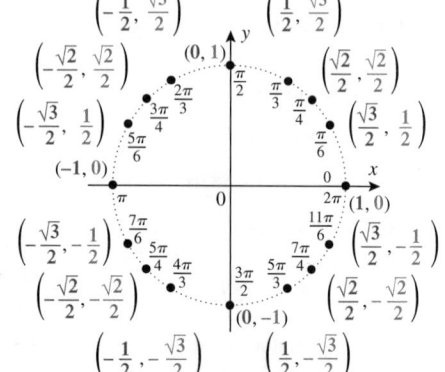

1. $\sin\left(\dfrac{5\pi}{3}\right)$

2. $\cos\left(\dfrac{5\pi}{3}\right)$

3. $\cos\left(\dfrac{7\pi}{6}\right)$

4. $\sin\left(\dfrac{7\pi}{6}\right)$

5. $\sin\left(\dfrac{3\pi}{4}\right)$

6. $\cos\left(\dfrac{3\pi}{4}\right)$

7. $\tan\left(\dfrac{7\pi}{4}\right)$

8. $\cot\left(\dfrac{7\pi}{4}\right)$

9. $\sec\left(\dfrac{5\pi}{4}\right)$

10. $\csc\left(\dfrac{5\pi}{3}\right)$

11. $\tan\left(\dfrac{4\pi}{3}\right)$

12. $\cot\left(\dfrac{11\pi}{6}\right)$

13. $\csc\left(\dfrac{5\pi}{6}\right)$

14. $\cot\left(\dfrac{2\pi}{3}\right)$

In Exercises 15–34, use the unit circle and the fact that sine is an odd function and cosine is an even function to find the exact values of the indicated functions.

15. $\sin\left(-\dfrac{2\pi}{3}\right)$

16. $\sin\left(-\dfrac{5\pi}{4}\right)$

17. $\sin\left(-\dfrac{\pi}{3}\right)$

18. $\sin\left(-\dfrac{7\pi}{6}\right)$

19. $\cos\left(-\dfrac{3\pi}{4}\right)$

20. $\cos\left(-\dfrac{5\pi}{3}\right)$

21. $\cos\left(-\dfrac{5\pi}{6}\right)$

22. $\cos\left(-\dfrac{7\pi}{4}\right)$

23. $\sin\left(-\dfrac{5\pi}{4}\right)$

24. $\sin(-\pi)$

25. $\sin\left(-\dfrac{3\pi}{2}\right)$

26. $\sin\left(-\dfrac{\pi}{3}\right)$

27. $\cos\left(-\dfrac{\pi}{4}\right)$

28. $\cos\left(-\dfrac{3\pi}{4}\right)$

29. $\cos\left(-\dfrac{\pi}{2}\right)$

30. $\cos\left(-\dfrac{7\pi}{6}\right)$

31. $\csc\left(-\dfrac{5\pi}{6}\right)$

32. $\sec\left(-\dfrac{7\pi}{4}\right)$

33. $\tan\left(-\dfrac{11\pi}{6}\right)$

34. $\cot\left(-\dfrac{11\pi}{6}\right)$

In Exercises 35–54, use the unit circle to find all of the exact values of θ that make the equation true in the indicated interval.

35. $\cos\theta = \dfrac{\sqrt{3}}{2}, 0 \le \theta \le 2\pi$

36. $\cos\theta = -\dfrac{\sqrt{3}}{2}, 0 \le \theta \le 2\pi$

37. $\sin\theta = -\dfrac{\sqrt{3}}{2}, 0 \le \theta \le 2\pi$

38. $\sin\theta = \dfrac{\sqrt{3}}{2}, 0 \le \theta \le 2\pi$

39. $\sin\theta = 0, 0 \le \theta \le 4\pi$

40. $\sin\theta = -1, 0 \le \theta \le 4\pi$

41. $\cos\theta = -1, 0 \le \theta \le 4\pi$

42. $\cos\theta = 0, 0 \le \theta \le 4\pi$

43. $\tan\theta = -1, 0 \le \theta \le 2\pi$

44. $\cot\theta = 1, 0 \le \theta \le 2\pi$

45. $\sec\theta = -\sqrt{2}, 0 \le \theta \le 2\pi$

46. $\csc\theta = \sqrt{2}, 0 \le \theta \le 2\pi$

47. $\csc\theta$ is undefined, $0 \le \theta \le 2\pi$

48. $\sec\theta$ is undefined, $0 \le \theta \le 2\pi$

49. $\tan\theta$ is undefined, $0 \le \theta \le 2\pi$

50. $\cot\theta$ is undefined, $0 \le \theta \le 2\pi$

51. $\csc\theta = -2, 0 \le \theta \le 2\pi$

52. $\cot\theta = -\sqrt{3}, 0 \le \theta \le 2\pi$

53. $\sec\theta = \dfrac{2\sqrt{3}}{3}, 0 \le \theta \le 2\pi$

54. $\tan\theta = \dfrac{\sqrt{3}}{3}, 0 \le \theta \le 2\pi$

▪ APPLICATIONS

For Exercises 55 and 56, refer to the following:

The average daily temperature in Peoria, Illinois, can be predicted by the formula $T = 50 - 28\cos\left[\dfrac{2\pi(x - 31)}{365}\right]$, where x is the number of the day in the year (January 1 = 1, February 1 = 32, etc.) and T is in degrees Fahrenheit.

55. **Atmospheric Temperature.** What is the expected temperature on February 15?

56. **Atmospheric Temperature.** What is the expected temperature on August 15? (Assume it is not a leap year.)

For Exercises 57 and 58, refer to the following:

The human body temperature normally fluctuates during the day. A person's body temperature can be predicted by the formula $T = 99.1 - 0.5 \sin\left(x + \dfrac{\pi}{12}\right)$, where x is the number of hours since midnight and T is in degrees Fahrenheit.

57. Body Temperature. What is the person's temperature at 6:00 A.M.?

58. Body Temperature. What is the person's temperature at 9:00 P.M.?

For Exercises 59 and 60, refer to the following:

The height of the water in a harbor changes with the tides. The height of the water at a particular hour during the day can be determined by the formula $h(x) = 5 + 4.8 \sin\left[\dfrac{\pi}{6}(x + 4)\right]$, where x is the number of hours since midnight and h is the height of the tide in feet.

59. Tides. What is the height of the tide at 3:00 P.M.?

Bill Brooks/Alamy

Bill Brooks/Alamy

60. Tides. What is the height of the tide at 5:00 A.M.?

61. Yo-Yo Dieting. A woman has been yo-yo dieting for years. Her weight changes throughout the year as she gains and loses weight. Her weight in a particular month can be determined by the formula $w(x) = 145 + 10 \cos\left(\dfrac{\pi}{6}x\right)$, where x is the month and w is in pounds. If $x = 1$ corresponds to January, how much does she weigh in June?

62. Yo-Yo Dieting. How much does the woman in Exercise 61 weigh in December?

63. Seasonal Sales. The average number of guests visiting the Magic Kingdom at Walt Disney World per day is given by $n(x) = 30,000 + 20,000 \sin\left[\dfrac{\pi}{2}(x + 1)\right]$, where n is the number of guests and x is the month. If January corresponds to $x = 1$, how many people on average are visiting the Magic Kingdom per day in February?

64. Seasonal Sales. How many guests are visiting the Magic Kingdom in Exercise 63 in December?

■ CATCH THE MISTAKE

In Exercises 65 and 66, explain the mistake that is made.

65. Use the unit circle to evaluate $\tan\left(\dfrac{5\pi}{6}\right)$ exactly.

Solution:

Tangent is the ratio of sine to cosine.
$$\tan\left(\frac{5\pi}{6}\right) = \frac{\sin\left(\dfrac{5\pi}{6}\right)}{\cos\left(\dfrac{5\pi}{6}\right)}$$

Use the unit circle to identify sine and cosine.
$$\sin\left(\frac{5\pi}{6}\right) = -\frac{\sqrt{3}}{2} \quad \text{and} \quad \cos\left(\frac{5\pi}{6}\right) = \frac{1}{2}$$

Substitute values for sine and cosine.
$$\tan\left(\frac{5\pi}{6}\right) = \frac{-(\sqrt{3}/2)}{1/2}$$

Simplify.
$$\tan\left(\frac{5\pi}{6}\right) = -\sqrt{3}$$

This is incorrect. What mistake was made?

66. Use the unit circle to evaluate $\sec\left(\dfrac{11\pi}{6}\right)$ exactly.

Solution:

Secant is the reciprocal of cosine.
$$\sec\left(\frac{11\pi}{6}\right) = \frac{1}{\cos\left(\dfrac{11\pi}{6}\right)}$$

Use the unit circle to evaluate cosine.
$$\cos\left(\frac{11\pi}{6}\right) = -\frac{1}{2}$$

Substitute the value for cosine.
$$\sec\left(\frac{11\pi}{6}\right) = \frac{1}{-\dfrac{1}{2}}$$

Simplify.
$$\sec\left(\frac{11\pi}{6}\right) = -2$$

This is incorrect. What mistake was made?

■ **CONCEPTUAL**

In Exercises 67–72, determine whether each statement is true or false.

67. $\sin(2n\pi + \theta) = \sin\theta$, n an integer.

68. $\cos(2n\pi + \theta) = \cos\theta$, n an integer.

69. $\sin\theta = 1$ when $\theta = \dfrac{(2n+1)\pi}{2}$, n an integer.

70. $\cos\theta = 1$ when $\theta = n\pi$, n an integer.

71. $\tan(\theta + 2n\pi) = \tan\theta$, n an integer.

72. $\tan\theta = 0$ if and only if $\theta = \dfrac{(2n+1)\pi}{2}$, n an integer.

73. Is cosecant an even or an odd function? Justify your answer.

74. Is tangent an even or an odd function? Justify your answer.

■ **CHALLENGE**

75. Find all the values of θ, $0 \le \theta \le 2\pi$, for which the equation $\sin\theta = \cos\theta$ is true.

76. Find all the values of θ (θ is any real number) for which the equation $\sin\theta = \cos\theta$ is true.

77. Find all the values of θ, $0 \le \theta \le 2\pi$, for which the equation $2\sin\theta = \csc\theta$ is true.

78. Find all the values of θ, $0 \le \theta \le 2\pi$, for which the equation $\cos\theta = \frac{1}{4}\sec\theta$ is true.

79. Find all the values of θ (θ is any real number) for which the equation $3\csc\theta = 4\sin\theta$ is true.

80. Find all the values of θ (θ is any real number) for which the equation $4\cos\theta = 3\sec\theta$ is true.

81. Does there exist an angle $0 \le \theta < 2\pi$ such that $\tan\theta = \cot\theta$?

82. Does there exist an angle $0 \le \theta < 2\pi$ such that $\sec\theta = \csc(-\theta)$?

■ **TECHNOLOGY**

83. Use a calculator to approximate $\sin 423°$. What do you expect $\sin(-423°)$ to be? Verify your answer with a calculator.

84. Use a calculator to approximate $\cos 227°$. What do you expect $\cos(-227°)$ to be? Verify your answer with a calculator.

85. Use a calculator to approximate $\tan 81°$. What do you expect $\tan(-81°)$ to be? Verify your answer with a calculator.

86. Use a calculator to approximate $\csc 211°$. What do you expect $\csc(-211°)$ to be? Verify your answer with a calculator.

For Exercises 87–90, refer to the following:

Set the calculator in parametric and radian modes and let
$$X_1 = \cos T$$
$$Y_1 = \sin T$$

Set the window so that $0 \le T \le 2\pi$, step $= \dfrac{\pi}{15}$, $-2 \le X \le 2$, and $-2 \le Y \le 2$. To approximate the sine or cosine of a T value, use the $\boxed{TRACE}$ key, type in the T value, and read the corresponding coordinates from the screen.

87. Approximate $\cos\left(\dfrac{\pi}{3}\right)$, take 5 steps of $\dfrac{\pi}{15}$ each, and read the x-coordinate.

88. Approximate $\sin\left(\dfrac{\pi}{3}\right)$, take 5 steps of $\dfrac{\pi}{15}$ each, and read the y-coordinate.

89. Approximate $\sin\left(\dfrac{2\pi}{3}\right)$ to four decimal places.

90. Approximate $\cos\left(\dfrac{5\pi}{4}\right)$ to four decimal places.

■ **PREVIEW TO CALCULUS**

The Fundamental Theorem of Calculus establishes that the definite integral $\int_a^b f(x)dx$ equals $F(b) - F(a)$, where F is any antiderivative of a continuous function f.

In Exercises 91–94, use the information below to find the exact value of each definite integral.

91. $\int_0^{\pi} \sin x\,dx$

92. $\int_{\pi/4}^{5\pi/6} \cos x\,dx$

93. $\int_{7\pi/6}^{5\pi/4} \sec^2 x\,dx$

94. $\int_{5\pi/3}^{11\pi/6} \csc x\cot x\,dx$

FUNCTION	$\sin x$	$\cos x$	$\sec^2 x$	$\csc x\cot x$
ANTIDERIVATIVE	$-\cos x$	$\sin x$	$\tan x$	$-\csc x$

SECTION 5.2 GRAPHS OF SINE AND COSINE FUNCTIONS

SKILLS OBJECTIVES

- Graph the sine and cosine functions.
- Determine the domain and range of the sine and cosine functions.
- Determine the amplitude and period of sinusoidal functions.
- Determine the phase shift of a sinusoidal function.
- Solve harmonic motion problems.
- Graph sums of functions.

CONCEPTUAL OBJECTIVES

- Understand why the graphs of the sine and cosine functions are called sinusoidal graphs.
- Understand the cyclic nature of periodic functions.
- Visualize harmonic motion as a sinusoidal function.

The following are examples of things that repeat in a predictable way (are roughly periodic):

- heartbeat
- tide levels
- time of sunrise
- average outdoor temperature for the time of year

The trigonometric functions are *strictly* periodic. In the unit circle, the value of any of the trigonometric functions is the same for any coterminal angle (same initial and terminal sides no matter how many full rotations the angle makes). For example, if we add (or subtract) multiples of 2π to (from) the angle θ, the values for sine and cosine are unchanged.

$$\sin(\theta + 2n\pi) = \sin\theta \qquad \text{or} \qquad \cos(\theta + 2n\pi) = \cos\theta \quad (n \text{ is any integer})$$

DEFINITION **Periodic Function**

A function f is called a **periodic function** if there is a positive number p such that

$$f(x + p) = f(x) \qquad \text{for all } x \text{ in the domain of } f$$

If p is the smallest such number for which this equation holds, then p is called the **fundamental period**.

You will see in this chapter that sine, cosine, secant, and cosecant have fundamental period 2π, but that tangent and cotangent have fundamental period π.

The Graph of $f(x) = \sin x$

Let us start by point-plotting the sine function. We select special values for the sine function that we already know.

x	$f(x) = \sin x$	(x, y)
0	$\sin 0 = 0$	$(0, 0)$
$\dfrac{\pi}{4}$	$\sin\left(\dfrac{\pi}{4}\right) = \dfrac{\sqrt{2}}{2}$	$\left(\dfrac{\pi}{4}, \dfrac{\sqrt{2}}{2}\right)$
$\dfrac{\pi}{2}$	$\sin\left(\dfrac{\pi}{2}\right) = 1$	$\left(\dfrac{\pi}{2}, 1\right)$
$\dfrac{3\pi}{4}$	$\sin\left(\dfrac{3\pi}{4}\right) = \dfrac{\sqrt{2}}{2}$	$\left(\dfrac{3\pi}{4}, \dfrac{\sqrt{2}}{2}\right)$
π	$\sin \pi = 0$	$(\pi, 0)$
$\dfrac{5\pi}{4}$	$\sin\left(\dfrac{5\pi}{4}\right) = -\dfrac{\sqrt{2}}{2}$	$\left(\dfrac{5\pi}{4}, -\dfrac{\sqrt{2}}{2}\right)$
$\dfrac{3\pi}{2}$	$\sin\left(\dfrac{3\pi}{2}\right) = -1$	$\left(\dfrac{3\pi}{2}, -1\right)$
$\dfrac{7\pi}{4}$	$\sin\left(\dfrac{7\pi}{4}\right) = -\dfrac{\sqrt{2}}{2}$	$\left(\dfrac{7\pi}{4}, -\dfrac{\sqrt{2}}{2}\right)$
2π	$\sin(2\pi) = 0$	$(2\pi, 0)$

By plotting the above coordinates (x, y), we can obtain the graph of one **period**, or **cycle**, of the graph of $y = \sin x$. Note that $\dfrac{\sqrt{2}}{2} \approx 0.7$.

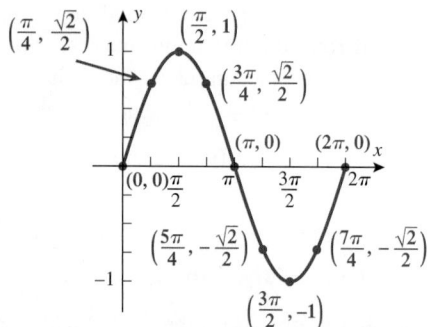

We can extend the graph horizontally in both directions (left and right) since the domain of the sine function is the set of all real numbers.

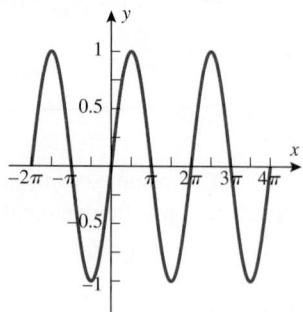

From here on, we are no longer showing angles on the unit circle but are now showing angles as *real numbers* in radians on the *x*-axis of the *Cartesian* graph. Therefore, we no longer illustrate a "terminal side" to an angle—the physical arcs and angles no longer exist; only their measures exist, as values of the *x*-coordinate.

If we graph the function $f(x) = \sin x$, the x-intercepts correspond to values of x at which the sine function is equal to zero.

x	$f(x) = \sin x$	(x, y)	
0	$\sin 0 = 0$	$(0, 0)$	
π	$\sin \pi = 0$	$(\pi, 0)$	
2π	$\sin(2\pi) = 0$	$(2\pi, 0)$	
3π	$\sin(3\pi) = 0$	$(3\pi, 0)$	
4π	$\sin(4\pi) = 0$	$(4\pi, 0)$	
...			
$n\pi$	$\sin(n\pi) = 0$	$(n\pi, 0)$	where n is an integer.

Notice that the point $(0, 0)$ is both a y-intercept and an x-intercept but all x-intercepts have the form $(n\pi, 0)$. The maximum value of the sine function is 1, and the minimum value of the sine function is -1, which occurs at odd multiples of $\dfrac{\pi}{2}$.

x	$f(x) = \sin x$	(x, y)
$\dfrac{\pi}{2}$	$\sin\left(\dfrac{\pi}{2}\right) = 1$	$\left(\dfrac{\pi}{2}, 1\right)$
$\dfrac{3\pi}{2}$	$\sin\left(\dfrac{3\pi}{2}\right) = -1$	$\left(\dfrac{3\pi}{2}, -1\right)$
$\dfrac{5\pi}{2}$	$\sin\left(\dfrac{5\pi}{2}\right) = 1$	$\left(\dfrac{5\pi}{2}, 1\right)$
$\dfrac{7\pi}{2}$	$\sin\left(\dfrac{7\pi}{2}\right) = -1$	$\left(\dfrac{7\pi}{2}, -1\right)$
...	...	...
$\dfrac{(2n+1)\pi}{2}$	$\sin\left(\dfrac{(2n+1)\pi}{2}\right) = \pm 1$	$\left(\dfrac{(2n+1)\pi}{2}, \pm 1\right)$

The following box summarizes the sine function:

SINE FUNCTION $f(x) = \sin x$

- Domain: $(-\infty, \infty)$ or $-\infty < x < \infty$
- Range: $[-1, 1]$ or $-1 \le y \le 1$
- The sine function is an odd function:
  - symmetric about the origin
  - $f(-x) = -f(x)$
- The sine function is a periodic function with fundamental period 2π.
- The x-intercepts, $0, \pm\pi, \pm 2\pi, \ldots$, are of the form $n\pi$, where n is an integer.
- The maximum (1) and minimum (-1) values of the sine function correspond to x-values of the form $\dfrac{(2n+1)\pi}{2}$, such as $\pm\dfrac{\pi}{2}, \pm\dfrac{3\pi}{2}, \pm\dfrac{5\pi}{2}, \ldots$.

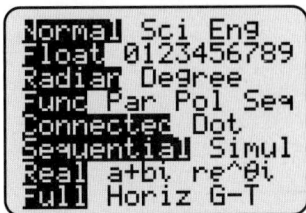
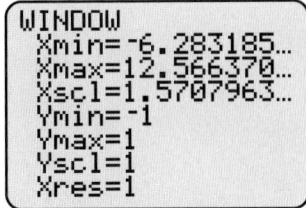
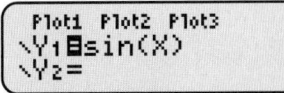
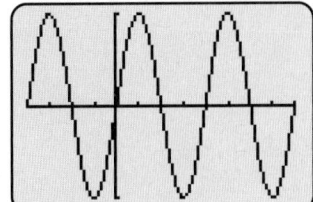

The Graph of $f(x) = \cos x$

Let us start by point-plotting the cosine function.

x	$f(x) = \cos x$	(x, y)
0	$\cos 0 = 1$	$(0, 1)$
$\dfrac{\pi}{4}$	$\cos\left(\dfrac{\pi}{4}\right) = \dfrac{\sqrt{2}}{2}$	$\left(\dfrac{\pi}{4}, \dfrac{\sqrt{2}}{2}\right)$
$\dfrac{\pi}{2}$	$\cos\left(\dfrac{\pi}{2}\right) = 0$	$\left(\dfrac{\pi}{2}, 0\right)$
$\dfrac{3\pi}{4}$	$\cos\left(\dfrac{3\pi}{4}\right) = -\dfrac{\sqrt{2}}{2}$	$\left(\dfrac{3\pi}{4}, -\dfrac{\sqrt{2}}{2}\right)$
π	$\cos \pi = -1$	$(\pi, -1)$
$\dfrac{5\pi}{4}$	$\cos\left(\dfrac{5\pi}{4}\right) = -\dfrac{\sqrt{2}}{2}$	$\left(\dfrac{5\pi}{4}, -\dfrac{\sqrt{2}}{2}\right)$
$\dfrac{3\pi}{2}$	$\cos\left(\dfrac{3\pi}{2}\right) = 0$	$\left(\dfrac{3\pi}{2}, 0\right)$
$\dfrac{7\pi}{4}$	$\cos\left(\dfrac{7\pi}{4}\right) = \dfrac{\sqrt{2}}{2}$	$\left(\dfrac{7\pi}{4}, \dfrac{\sqrt{2}}{2}\right)$
2π	$\cos(2\pi) = 1$	$(2\pi, 1)$

By plotting the above coordinates (x, y), we can obtain the graph of one period, or cycle, of the graph of $y = \cos x$. Note that $\dfrac{\sqrt{2}}{2} \approx 0.7$.

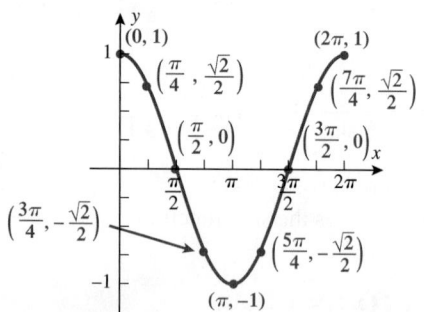

We can extend the graph horizontally in both directions (left and right) since the domain of the cosine function is all real numbers.

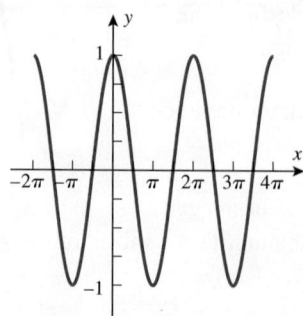

If we graph the function $f(x) = \cos x$, the x-intercepts correspond to values of x at which the cosine function is equal to zero.

x	$f(x) = \cos x$	(x, y)
$\dfrac{\pi}{2}$	$\cos\left(\dfrac{\pi}{2}\right) = 0$	$\left(\dfrac{\pi}{2}, 0\right)$
$\dfrac{3\pi}{2}$	$\cos\left(\dfrac{3\pi}{2}\right) = 0$	$\left(\dfrac{3\pi}{2}, 0\right)$
$\dfrac{5\pi}{2}$	$\cos\left(\dfrac{5\pi}{2}\right) = 0$	$\left(\dfrac{5\pi}{2}, 0\right)$
$\dfrac{7\pi}{2}$	$\cos\left(\dfrac{7\pi}{2}\right) = 0$	$\left(\dfrac{7\pi}{2}, 0\right)$
$\cdots$	$\cdots$	$\cdots$
$\dfrac{(2n+1)\pi}{2}$	$\cos\left(\dfrac{(2n+1)\pi}{2}\right) = 0$	$\left(\dfrac{(2n+1)\pi}{2}, 0\right)$ where n is an integer.

The point $(0, 1)$ is the y-intercept, and there are several x-intercepts of the form $\left(\dfrac{2n+1}{\pi}, 0\right)$. The maximum value of the cosine function is 1 and the minimum value of the cosine function is -1; these values occur at alternating integer multiples of π.

x	$f(x) = \cos x$	(x, y)
0	$\cos 0 = 1$	$(0, 1)$
π	$\cos \pi = -1$	$(\pi, -1)$
2π	$\cos(2\pi) = 1$	$(2\pi, 1)$
3π	$\cos(3\pi) = -1$	$(3\pi, -1)$
4π	$\cos(4\pi) = 1$	$(4\pi, 1)$
$\cdots$		
$n\pi$	$\cos(n\pi) = \pm 1$	$(n\pi, \pm 1)$

The following box summarizes the cosine function:

Technology Tip

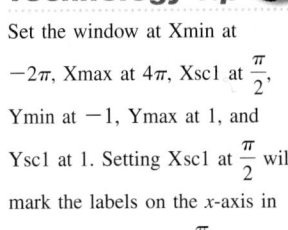

Set the window at Xmin at -2π, Xmax at 4π, Xscl at $\dfrac{\pi}{2}$, Ymin at -1, Ymax at 1, and Yscl at 1. Setting Xscl at $\dfrac{\pi}{2}$ will mark the labels on the x-axis in terms of multiples of $\dfrac{\pi}{2}$.

```
WINDOW
Xmin=-6.283185...
Xmax=12.566370...
Xscl=1.5707963...
Ymin=-1
Ymax=1
Yscl=1
Xres=1
```

Use Y= to enter the function $\cos(X)$.

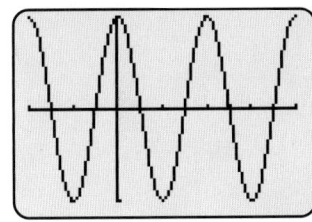

COSINE FUNCTION $f(x) = \cos x$

- Domain: $(-\infty, \infty)$ or $-\infty < x < \infty$
- Range: $[-1, 1]$ or $-1 \le y \le 1$
- The cosine function is an even function:
  - symmetric about the y-axis
  - $f(-x) = f(x)$
- The cosine function is a periodic function with fundamental period 2π.
- The x-intercepts, $\pm\dfrac{\pi}{2}, \pm\dfrac{3\pi}{2}, \pm\dfrac{5\pi}{2}, \ldots$, are

  odd integer multiples of $\dfrac{\pi}{2}$ that have the form $\dfrac{(2n+1)\pi}{2}$, where n is an integer.
- The maximum (1) and minimum (-1) values of the cosine function correspond to x-values of the form $n\pi$, such as $0, \pm\pi, \pm2\pi, \ldots$.

The Amplitude and Period of Sinusoidal Graphs

In mathematics, the word **sinusoidal** means "resembling the sine function." Let us start by graphing $f(x) = \sin x$ and $f(x) = \cos x$ on the same graph. Notice that they have similar characteristics (domain, range, period, and shape).

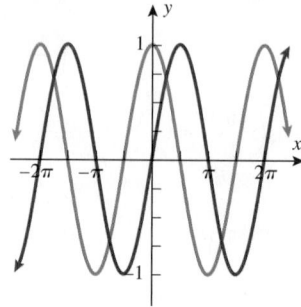

In fact, if we were to shift the cosine graph to the right $\dfrac{\pi}{2}$ units, the two graphs would be identical. For that reason, we refer to any graphs of the form $y = \cos x$ or $y = \sin x$ as **sinusoidal functions**.

We now turn our attention to graphs of the form $y = A \sin Bx$ and $y = A \cos Bx$, which are graphs like $y = \sin x$ and $y = \cos x$ that have been stretched or compressed vertically and horizontally.

EXAMPLE 1   Vertical Stretching and Compressing

Plot the functions $y = 2\sin x$ and $y = \frac{1}{2}\sin x$ on the same graph with $y = \sin x$ on the interval $-4\pi \leq x \leq 4\pi$.

Solution:

STEP 1   Make a table with the coordinate values of the graphs.

x	0	$\dfrac{\pi}{2}$	π	$\dfrac{3\pi}{2}$	2π
$\sin x$	**0**	**1**	**0**	**−1**	**0**
$2\sin x$	0	2	0	−2	0
$\frac{1}{2}\sin x$	0	$\frac{1}{2}$	0	$-\frac{1}{2}$	0

STEP 2   Label the points on the graph and connect with a smooth curve over one period, $0 \leq x \leq 2\pi$.

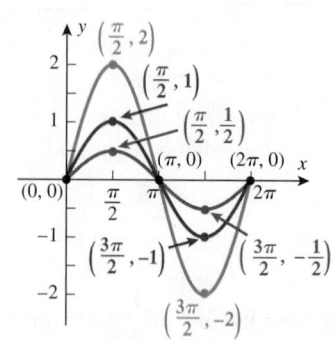

STEP 3   Extend the graph in both directions (repeat every 2π).

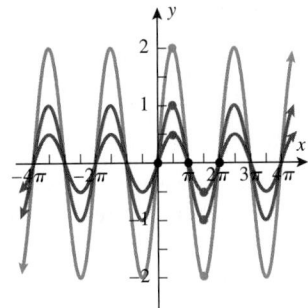

 ■ YOUR TURN   Plot the functions $y = 3\cos x$ and $y = \frac{1}{3}\cos x$ on the same graph with $y = \cos x$ on the interval $-2\pi \leq x \leq 2\pi$.

■ Answer:

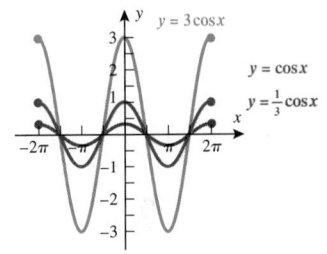

Notice in Example 1 and the corresponding Your Turn that:

■ $y = 2\sin x$ has the shape and period of $y = \sin x$ but is stretched vertically.
■ $y = \frac{1}{2}\sin x$ has the shape and period of $y = \sin x$ but is compressed vertically.
■ $y = 3\cos x$ has the shape and period of $y = \cos x$ but is stretched vertically.
■ $y = \frac{1}{3}\cos x$ has the shape and period of $y = \cos x$ but is compressed vertically.

In general, functions of the form $y = A\sin x$ and $y = A\cos x$ are stretched in the vertical direction when $|A| > 1$ and compressed in the vertical direction when $|A| < 1$.

The **amplitude** of a periodic function is half the difference between the maximum value of the function and the minimum value of the function. For the functions $y = \sin x$ and $y = \cos x$, the maximum value is 1 and the minimum value is -1. Therefore, the amplitude of each of these two functions is $|A| = \frac{1}{2}|1 - (-1)| = 1$.

AMPLITUDE OF SINUSOIDAL FUNCTIONS

For sinusoidal functions of the form $y = A\sin Bx$ and $y = A\cos Bx$, the **amplitude** is $|A|$. When $|A| < 1$, the graph is compressed in the vertical direction, and when $|A| > 1$, the graph is stretched in the vertical direction.

EXAMPLE 2   Finding the Amplitude of Sinusoidal Functions

State the amplitude of

a. $f(x) = -4\cos x$

b. $g(x) = \frac{1}{5}\sin x$

Solution (a): The amplitude is the magnitude of -4. $A = |-4| = \boxed{4}$

Solution (b): The amplitude is the magnitude of $\frac{1}{5}$. $A = \left|\frac{1}{5}\right| = \boxed{\dfrac{1}{5}}$

Technology Tip

Set the window at Xmin at

-2π, Xmax at 2π, Xsc1 at $\frac{\pi}{2}$,

Ymin at -1, Ymax at 1, and Ysc1

at 1. Setting Xsc1 at $\frac{\pi}{2}$ will mark

the labels on the x-axis in terms of

multiples of $\frac{\pi}{2}$.

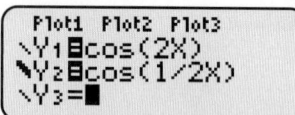

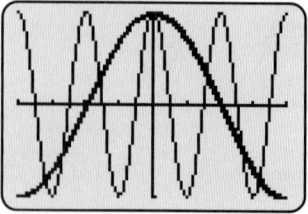

EXAMPLE 3   Horizontal Stretching and Compressing

Plot the functions $y = \cos(2x)$ and $y = \cos\left(\frac{1}{2}x\right)$ on the same graph with $y = \cos x$ on the interval $-2\pi \le x \le 2\pi$.

Solution:

STEP 1   Make a table with the coordinate values of the graphs. It is necessary only to select the points that correspond to x-intercepts, $(y = 0)$, and maximum and minimum points, $(y = \pm 1)$. Usually, the period is divided into four subintervals (which you will see in Examples 5–7).

x	0	$\frac{\pi}{4}$	$\frac{\pi}{2}$	$\frac{3\pi}{4}$	π	$\frac{5\pi}{4}$	$\frac{3\pi}{2}$	$\frac{7\pi}{4}$	2π
$\cos x$	1		0		-1		0		1
$\cos(2x)$	1	0	-1	0	1	0	-1	0	1
$\cos\left(\frac{1}{2}x\right)$	1				0				-1

STEP 2   Label the points on the graph and connect with a smooth curve.

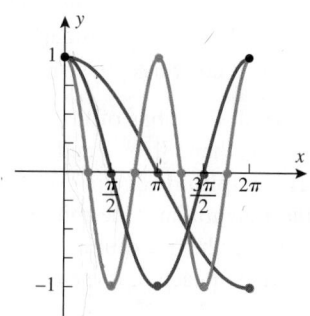

STEP 3   Extend the graph to cover the entire interval: $-2\pi \le x \le 2\pi$.

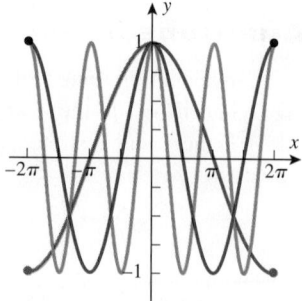

■ **Answer:**

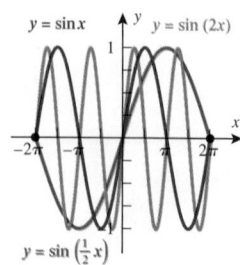

■ **YOUR TURN**   Plot the functions $y = \sin(2x)$ and $y = \sin\left(\frac{1}{2}x\right)$ on the same graph with $y = \sin x$ on the interval $-2\pi \le x \le 2\pi$.

Notice in Example 3 and the corresponding Your Turn that:

- $y = \cos(2x)$ has the shape and amplitude of $y = \cos x$ but is compressed horizontally.
- $y = \cos\left(\frac{1}{2}x\right)$ has the shape and amplitude of $y = \cos x$ but is stretched horizontally.
- $y = \sin(2x)$ has the shape and amplitude of $y = \sin x$ but is compressed horizontally.
- $y = \sin\left(\frac{1}{2}x\right)$ has the shape and amplitude of $y = \sin x$ but is stretched horizontally.

In general, functions of the form $y = \sin Bx$ and $y = \cos Bx$, with $B > 0$, are compressed in the horizontal direction when $B > 1$ and stretched in the horizontal direction when $0 < B < 1$. Negative arguments $(B < 0)$ are included in the context of *reflections*.

The period of the functions $y = \sin x$ and $y = \cos x$ is 2π. To find the period of a function of the form $y = A \sin Bx$ or $y = A \cos Bx$, set Bx equal to 2π and solve for x.

$$Bx = 2\pi$$

$$x = \frac{2\pi}{B}$$

PERIOD OF SINUSOIDAL FUNCTIONS

For sinusoidal functions of the form $y = A \sin Bx$ and $y = A \cos Bx$, the **period** is $\dfrac{2\pi}{B}$.

When $0 < B < 1$, the graph is stretched in the horizontal direction, and when $B > 1$, the graph is compressed in the horizontal direction.

Study Tip

When B is negative, the period is $\dfrac{2\pi}{|B|}$.

EXAMPLE 4   **Finding the Period of a Sinusoidal Function**

State the period of

a. $y = \cos(4x)$

b. $y = \sin\left(\frac{1}{3}x\right)$

Solution (a):

Compare $\cos(4x)$ with $\cos Bx$ to identify B. $B = 4$

Calculate the period of $\cos(4x)$, using $p = \dfrac{2\pi}{B}$. $p = \dfrac{2\pi}{4} = \dfrac{\pi}{2}$

The period of $\cos(4x)$ is $\boxed{p = \dfrac{\pi}{2}}$.

Solution (b):

Compare $\sin\left(\frac{1}{3}x\right)$ with $\sin Bx$ to identify B. $B = \dfrac{1}{3}$

Calculate the period of $\sin\left(\frac{1}{3}x\right)$, using $p = \dfrac{2\pi}{B}$. $p = \dfrac{2\pi}{\dfrac{1}{3}} = 6\pi$

The period of $\sin\left(\frac{1}{3}x\right)$ is $\boxed{p = 6\pi}$.

■ YOUR TURN   State the period of

a. $y = \sin(3x)$ **b.** $y = \cos\left(\frac{1}{2}x\right)$

■ **Answer:**

a. $p = \dfrac{2\pi}{3}$ **b.** $p = 4\pi$

Study Tip

Divide the period by 4 to get the key values along the *x*-axis for graphing.

Now that you know the basic graphs of $y = \sin x$ and $y = \cos x$, you can sketch one cycle (period) of these graphs with the following *x*-values: $0, \frac{\pi}{2}, \pi, \frac{3\pi}{2}, 2\pi$. For a period of 2π, we used steps of $\frac{\pi}{2}$. Therefore, for functions of the form $y = A \sin Bx$ or $y = A \cos Bx$, when we start at the origin and as long as we include these four basic values during one period, we are able to sketch the graphs.

Technology Tip

Use a TI calculator to check the graph of $y = 3 \sin(2x)$.

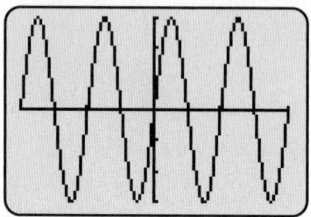

STRATEGY FOR SKETCHING GRAPHS OF SINUSOIDAL FUNCTIONS

To graph $y = A \sin Bx$ or $y = A \cos Bx$ with $B > 0$:

Step 1: Find the amplitude $|A|$ and period $\frac{2\pi}{B}$.

Step 2: Divide the period into four subintervals of equal lengths.

Step 3: Make a table and evaluate the function for *x*-values from Step 2 starting at $x = 0$.

Step 4: Draw the *xy*-plane (label the *y*-axis from $-|A|$ to $|A|$) and plot the points found in Step 3.

Step 5: Connect the points with a sinusoidal curve (with amplitude $|A|$).

Step 6: Extend the graph over one or two additional periods in both directions (left and right).

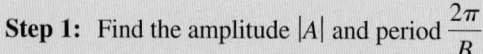

EXAMPLE 5 **Graphing Sinusoidal Functions of the Form $y = A \sin Bx$**

Use the strategy for graphing a sinusoidal function to graph $y = 3 \sin(2x)$.

Solution:

STEP 1 Find the amplitude and period for $A = 3$ and $B = 2$. $\quad |A| = |3| = 3 \quad$ and $\quad p = \dfrac{2\pi}{B} = \dfrac{2\pi}{2} = \pi$

STEP 2 Divide the period π into four equal steps. $\quad \dfrac{\pi}{4}$

STEP 3 Make a table starting at $x = 0$ to the period $x = \pi$ in steps of $\dfrac{\pi}{4}$.

x	$y = 3\sin(2x)$	(x, y)
0	$3[\sin 0] = 3[0] = 0$	$(0, 0)$
$\dfrac{\pi}{4}$	$3\left[\sin\left(\dfrac{\pi}{2}\right)\right] = 3[1] = 3$	$\left(\dfrac{\pi}{4}, 3\right)$
$\dfrac{\pi}{2}$	$3[\sin \pi] = 3[0] = 0$	$\left(\dfrac{\pi}{2}, 0\right)$
$\dfrac{3\pi}{4}$	$3\left[\sin\left(\dfrac{3\pi}{2}\right)\right] = 3[-1] = -3$	$\left(\dfrac{3\pi}{4}, -3\right)$
π	$3[\sin(2\pi)] = 3[0] = 0$	$(\pi, 0)$

STEP 4 Draw the xy-plane and label the points in the table.

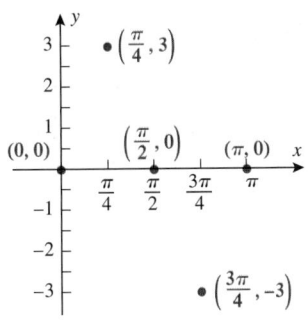

STEP 5 Connect the points with a sinusoidal curve.

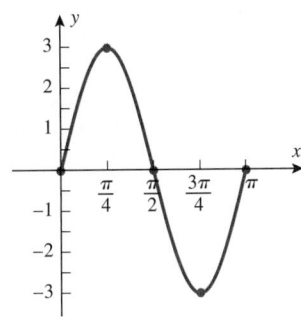

STEP 6 Repeat over several periods (to the left and right).

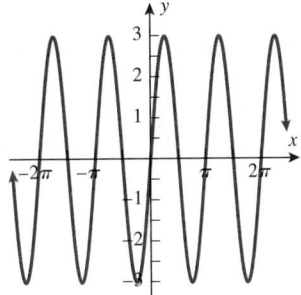

■ **Answer:**

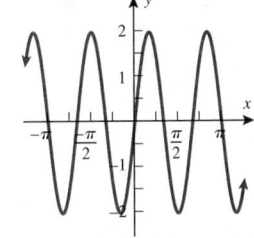

■ **YOUR TURN** Use the strategy for graphing sinusoidal functions to graph $y = 2\sin(3x)$.

EXAMPLE 6 **Graphing Sinusoidal Functions of the Form $y = A\cos Bx$**

Use the strategy for graphing a sinusoidal function to graph $y = -2\cos(\frac{1}{3}x)$.

Solution:

STEP 1 Find the amplitude and period for $A = -2$ and $B = \frac{1}{3}$.

$|A| = |-2| = 2$ and $p = \dfrac{2\pi}{B} = \dfrac{2\pi}{\dfrac{1}{3}} = 6\pi$

STEP 2 Divide the period 6π into four equal steps.

$\dfrac{6\pi}{4} = \dfrac{3\pi}{2}$

Technology Tip

Use a TI calculator to check the graph of $y = -2\cos(\frac{1}{3}x)$. Set the window at Xmin at -12π, Xmax at 12π, Xscl at $\dfrac{3\pi}{2}$, Ymin at -2, Ymax at 2, and Yscl at 1. Setting Xscl at $\dfrac{3\pi}{2}$ will mark the labels on the x-axis in terms of multiples of $\dfrac{3\pi}{2}$.

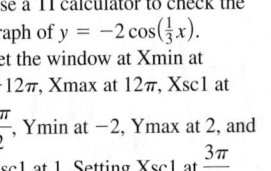

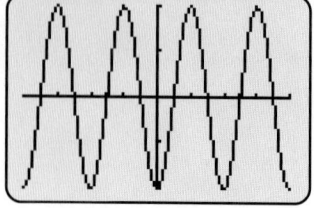

STEP 3 Make a table starting at $x = 0$ and completing one period of 6π in steps of $\dfrac{3\pi}{2}$.

x	$y = -2\cos(\frac{1}{3}x)$	(x, y)
0	$-2[\cos 0] = -2[1] = -2$	$(0, -2)$
$\dfrac{3\pi}{2}$	$-2\left[\cos\left(\dfrac{\pi}{2}\right)\right] = -2[0] = 0$	$\left(\dfrac{3\pi}{2}, 0\right)$
3π	$-2[\cos \pi] = -2[-1] = 2$	$(3\pi, 2)$
$\dfrac{9\pi}{2}$	$-2\left[\cos\left(\dfrac{3\pi}{2}\right)\right] = -2[0] = 0$	$\left(\dfrac{9\pi}{2}, 0\right)$
6π	$-2[\cos(2\pi)] = -2[1] = -2$	$(6\pi, -2)$

STEP 4 Draw the xy-plane and label the points in the table.

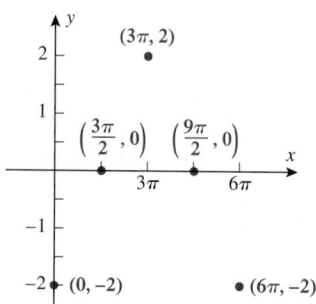

STEP 5 Connect the points with a sinusoidal curve.

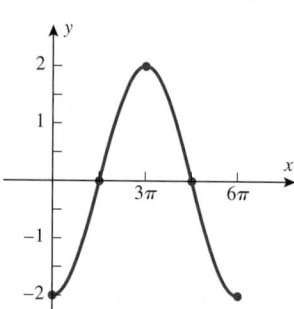

STEP 6 Repeat over several periods (to the left and right).

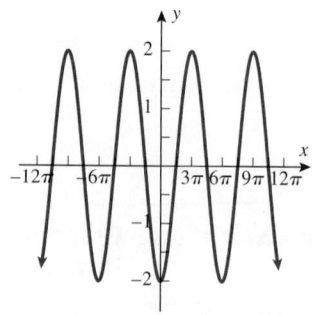

■ **Answer:**

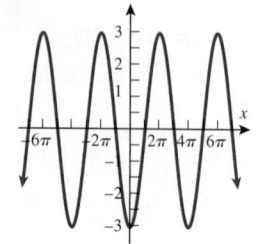

■ **YOUR TURN** Use the strategy for graphing a sinusoidal function to graph $y = -3\cos(\frac{1}{2}x)$.

Notice in Example 6 and the corresponding Your Turn that when A is negative, the result is a reflection of the original function (sine or cosine) about the x-axis.

EXAMPLE 7   **Finding an Equation for a Sinusoidal Graph**

Find an equation for the graph.

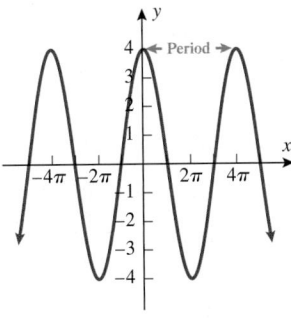

Solution:

This graph represents a cosine function.	$y = A \cos Bx$
The amplitude is 4 (half the maximum spread).	$|A| = 4$
The period $\dfrac{2\pi}{B}$ is equal to 4π.	$\dfrac{2\pi}{B} = 4\pi$
Solve for B.	$B = \dfrac{1}{2}$
Substitute $A = 4$ and $B = \frac{1}{2}$ into $y = A \cos Bx$.	$\boxed{y = 4 \cos \left(\dfrac{1}{2}x\right)}$

■ YOUR TURN   Find an equation for the graph.

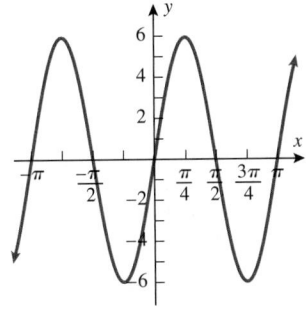

■ **Answer:** $y = 6 \sin(2x)$

Graphing a Shifted Sinusoidal Function: $y = A\sin(Bx + C) + D$ and $y = A\cos(Bx + C) + D$

Recall from Section 1.3 that we graph functions using horizontal and vertical translations (shifts) in the following way ($c > 0$):

- To graph $f(x + c)$, shift $f(x)$ to the **left** c units.
- To graph $f(x - c)$, shift $f(x)$ to the **right** c units.
- To graph $f(x) + c$, shift $f(x)$ **up** c units.
- To graph $f(x) - c$, shift $f(x)$ **down** c units.

To graph functions of the form $y = A\sin(Bx + C) + D$ and $y = A\cos(Bx + C) + D$, utilize the strategy below.

STRATEGY FOR GRAPHING $y = A\sin(Bx + C) + D$ AND $y = A\cos(Bx + C) + D$

A strategy for graphing $y = A\sin(Bx + C) + D$ is outlined below. The same strategy can be used to graph $y = A\cos(Bx + C) + D$.

Step 1: Find the amplitude $|A|$.

Step 2: Find the period $\dfrac{2\pi}{B}$ and **phase shift** $-\dfrac{C}{B}$.

Step 3: Graph $y = A\sin(Bx + C)$ over one period $\left(\text{from } -\dfrac{C}{B} \text{ to } -\dfrac{C}{B} + \dfrac{2\pi}{B}\right)$.

Step 4: Extend the graph over several periods.

Step 5: Shift the graph of $y = A\sin(Bx + C)$ vertically D units.

Note: If we rewrite the function in **standard form**, we get

$$y = A\sin\left[B\left(x + \dfrac{C}{B}\right)\right] + D$$

which makes it easier to identify the **phase shift**.

If $B < 0$, we can use properties of even and odd functions:

$$\sin(-x) = -\sin x \qquad \cos(-x) = \cos x$$

Study Tip

Rewriting in standard form

$$y = A\sin\left[B\left(x + \dfrac{C}{B}\right)\right]$$

makes identifying the phase shift easier.

Study Tip

An alternative method for finding the period and phase shift is to first write the function in standard form.

$$y = 5\cos\left[4\left(x + \dfrac{\pi}{4}\right)\right]$$

$$B = 4$$

$$\text{Period} = \dfrac{2\pi}{B} = \dfrac{2\pi}{4} = \dfrac{\pi}{2}.$$

Phase shift $= \dfrac{\pi}{4}$ units to the left.

EXAMPLE 8   **Graphing Functions of the Form $y = A\cos(Bx \pm C)$**

Graph $y = 5\cos(4x + \pi)$ over one period.

Solution:

STEP 1   State the amplitude. $|A| = |5| = 5$

STEP 2   Calculate the period and phase shift.

The interval for one period is from 0 to 2π. $4x + \pi = 0$ to $4x + \pi = 2\pi$

Solve for x. $x = -\dfrac{\pi}{4}$ to $x = -\dfrac{\pi}{4} + \dfrac{\pi}{2}$

Identify the phase shift. $-\dfrac{C}{B} = -\dfrac{\pi}{4}$

Identify the period $\dfrac{2\pi}{B}$. $\dfrac{2\pi}{B} = \dfrac{2\pi}{4} = \dfrac{\pi}{2}$

STEP 3   Graph.

Draw a cosine function starting at $x = -\dfrac{\pi}{4}$ with period $\dfrac{\pi}{2}$ and amplitude 5.

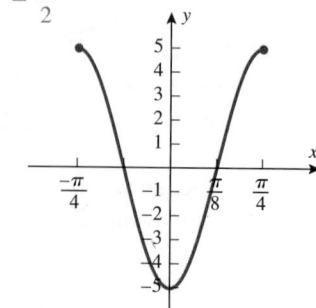

■ **Answer:**

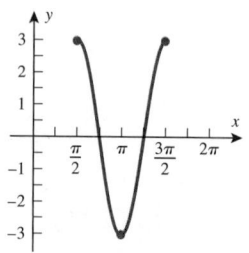

■ **YOUR TURN**   Graph $y = 3\cos(2x - \pi)$ over one period.

EXAMPLE 9   **Graphing Sinusoidal Functions**

Graph $y = -3 + 2\cos(2x - \pi)$.

Solution:

STEP 1   Find the amplitude. $|A| = |2| = 2$

STEP 2   Find the phase shift and period.

Set $2x - \pi$ equal to 0 and 2π. $2x - \pi = 0$ to $2x - \pi = 2\pi$

Solve for x. $x = \dfrac{\pi}{2}$ to $x = \dfrac{\pi}{2} + \pi$

Identify phase shift and period. $-\dfrac{C}{B} = \dfrac{\pi}{2}$.

$$\dfrac{2\pi}{B} = \pi$$

STEP 3   Graph $y = 2\cos(2x - \pi)$ starting
at $x = \dfrac{\pi}{2}$ over one period, π.

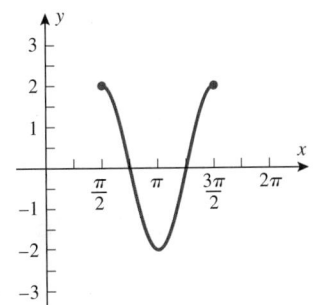

STEP 4   Extend the graph of $y = 2\cos(2x - \pi)$
over several periods.

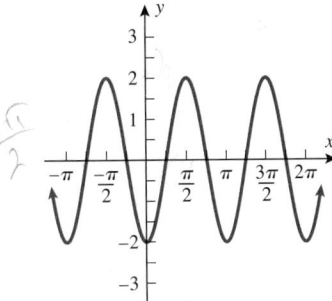

STEP 5   Shift the graph of $y = 2\cos(2x - \pi)$
down 3 units to arrive at the graph of
$y = -3 + 2\cos(2x - \pi)$.

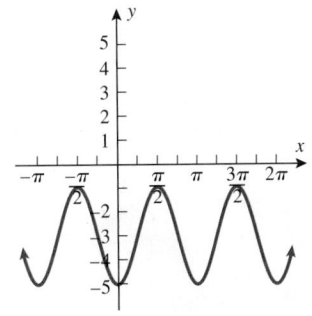

■ YOUR TURN   Graph $y = -2 + 3\sin(2x + \pi)$.

Technology Tip

Use a TI calculator to check the
graph of $y = -3 + 2\cos(2x - \pi)$.

Set the window at Xmin at $-\dfrac{\pi}{2}$,

Xmax at $\dfrac{3\pi}{2}$, Xscl at $\dfrac{\pi}{4}$, Ymin at

-5, Ymax at 1, and Yscl at 1.

Setting Xscl at $\dfrac{\pi}{4}$ will mark the

labels on the x-axis in terms of

multiples of $\dfrac{\pi}{4}$.

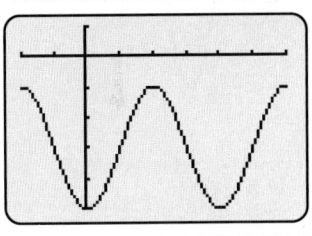

■ Answer:

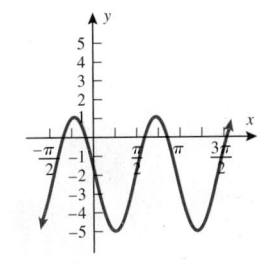

Harmonic Motion

One of the most important applications of sinusoidal functions is in describing *harmonic motion*, which we define as the symmetric periodic movement of an object or quantity about a center (equilibrium) position or value. The oscillation of a pendulum is a form of harmonic motion. Other examples are the recoil of a spring balance scale when a weight is placed on the tray and the variation of current or voltage within an AC circuit.

There are three types of harmonic motion: **simple harmonic motion**, **damped harmonic motion**, and **resonance**.

Simple Harmonic Motion

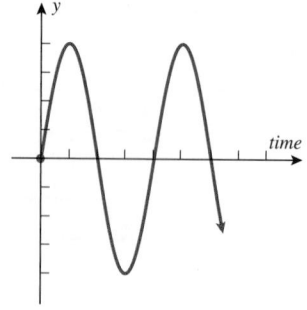

Simple harmonic motion is the kind of *unvarying* periodic motion that would occur in an ideal situation in which no resistive forces, such as friction, cause the amplitude of oscillation to decrease over time: the amplitude stays in exactly the same range in each period as time—the variable on the horizontal axis—increases. It will also occur if energy is being supplied at the correct rate to overcome resistive forces. Simple harmonic motion occurs, for example, in an AC electric circuit when a power source is consistently supplying energy. When you are swinging on a swing and "pumping" energy into the swing to keep it in motion at a constant period and amplitude, you are sustaining simple harmonic motion.

Damped Harmonic Motion

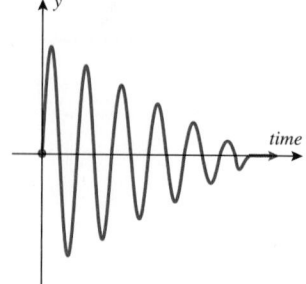

In damped harmonic motion, the amplitude of the periodic motion decreases as time increases. If you are on a moving swing and stop "pumping" new energy into the swing, the swing will continue moving with a constant period, but the amplitude—the height to which the swing will rise—will diminish with each cycle as the swing is slowed down by friction with the air or between its own moving parts.

Resonance

Resonance occurs when the amplitude of periodic motion increases as time increases. It is caused when the energy applied to an oscillating object or system is more than what is needed to oppose friction or other forces and sustain simple harmonic motion. Instead, the applied energy *increases* the amplitude of harmonic motion with each cycle. With resonance, eventually, the amplitude becomes unbounded and the result is disastrous. Bridges have collapsed because of resonance. On the previous page are pictures of the Tacoma Narrows Bridge (near Seattle, Washington) that collapsed due to high winds resulting in resonance. Soldiers know that when they march across a bridge, they must break cadence to prevent resonance.

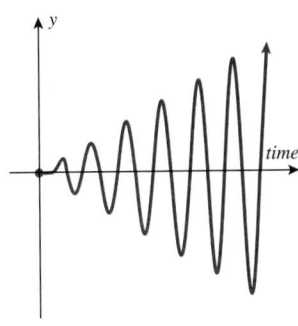

Examples of Harmonic Motion

If we hang a weight from a spring, then while the resulting "system" is at rest, we say it is in the equilibrium position.

If we then pull down on the weight and release it, the elasticity in the spring pulls the weight up and causes it to start oscillating up and down.

If we neglect friction and air resistance, we can imagine that the combination of the weight and the spring will oscillate indefinitely; the height of the weight with respect to the equilibrium position can be modeled by a simple sinusoidal function. This is an example of **simple harmonic motion**.

SIMPLE HARMONIC MOTION

The position of a point oscillating around an equilibrium position at time t is modeled by the sinusoidal function

$$y = A \sin \omega t \qquad \text{or} \qquad y = A \cos \omega t$$

Here $|A|$ is the amplitude and the period is $\dfrac{2\pi}{\omega}$, where $\omega > 0$.

EXAMPLE 10   Simple Harmonic Motion

Let the height of the seat of a swing be equal to zero when the swing is at rest. Assume that a child starts swinging until she reaches the highest she can swing and keeps her effort constant. Suppose the height $h(t)$ of the seat can be given by

$$h(t) = 8\sin\left(\frac{\pi}{2}t\right)$$

where t is time in seconds and h is the height in feet. Note that positive h indicates height reached swinging forward and negative h indicates height reached swinging backward. Assume that $t = 0$ is when the child passes through the equilibrium position swinging forward.

a. Graph the height function $h(t)$ for $0 \le t \le 4$.

b. What is the maximum height above ground reached by the seat of the swing?

c. What is the period of the swinging child?

Solution (a):

Make a table with integer values of t. $0 \le t \le 4$

t (seconds)	$y = h(t) = 8\sin\left(\frac{\pi}{2}t\right)$ (feet)	(t, y)
0	$8\sin 0 = 0$	$(0, 0)$
1	$8\sin\left(\dfrac{\pi}{2}\right) = 8$	$(1, 8)$
2	$8\sin \pi = 0$	$(2, 0)$
3	$8\sin\left(\dfrac{3\pi}{2}\right) = -8$	$(3, -8)$
4	$8\sin(2\pi) = 0$	$(4, 0)$

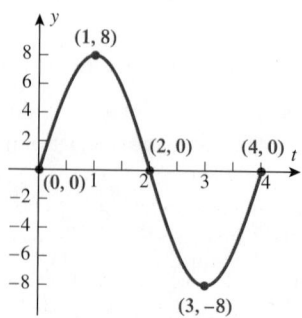

Labeling the time and height on the original diagram, we see that the maximum height is 8 feet and the period is 4 seconds.

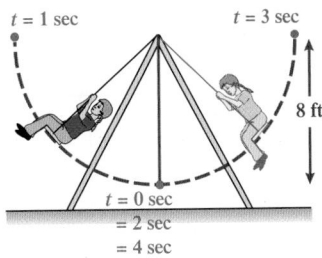

$t = 1$ sec $t = 3$ sec

8 ft

$t = 0$ sec
$= 2$ sec
$= 4$ sec

Solutions (b) and (c):

$$h(t) = 8 \sin\left(\frac{\pi}{2} t\right)$$

The maximum height above the equilibrium is the amplitude.

$$A = 8 \text{ ft}$$

The period is $\dfrac{2\pi}{\omega}$.

$$p = \frac{2\pi}{\dfrac{\pi}{2}} = 4 \text{ sec}$$

Damped harmonic motion can be modeled by a sinusoidal function whose amplitude decreases as time increases. If we again hang a weight from a spring so that it is suspended at rest and then pull down on the weight and release, the weight will oscillate about the equilibrium point. This time we will not neglect friction and air resistance: The weight will oscillate closer and closer to the equilibrium point over time until the weight eventually comes to rest at the equilibrium point. This is an example of damped harmonic motion.

The product of any decreasing function and the original periodic function will describe damped oscillatory motion. Here are two examples of functions that describe damped harmonic motion:

$$y = \frac{1}{t} \sin \omega t \qquad y = e^{-t} \cos \omega t$$

where e^{-t} is a decreasing exponential function (exponential decay).

EXAMPLE 11   Damped Harmonic Motion

Assume that the child in Example 10 decides to stop pumping and allows the swing to continue until she eventually comes to rest. Assume that

$$h(t) = \frac{8}{t} \cos\left(\frac{\pi}{2} t\right)$$

where t is time in seconds and h is the height in feet above the resting position. Note that positive h indicates height reached swinging forward and negative h indicates height reached swinging backward, assuming that $t = 1$ is when the child passes through the equilibrium position swinging backward and stops "pumping."

a. Graph the height function $h(t)$ for $1 \le t \le 8$.
b. What is the height above ground at 4 seconds? At 8 seconds? After 1 minute?

Technology Tip

To set up a table for $y = \dfrac{8}{t}\cos\left(\dfrac{\pi}{2}t\right)$,

enter $Y_1 = \dfrac{8}{x}\cos\left(\dfrac{\pi}{2}x\right)$ and select

TBLSET .

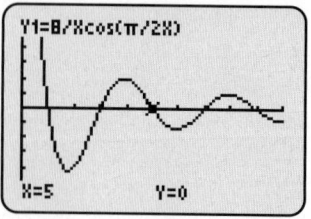

```
Plot1 Plot2 Plot3
\Y₁∎8/Xcos(π/2X)
■
```

```
TABLE SETUP
 TblStart=1
 ∆Tbl=1
Indpnt: Auto Ask
Depend: Auto Ask
```

Press 2nd TABLE .

```
  X  | Y1
  1  | 0
  2  | -4
  3  | 0
  4  | 2
  5  | 0
  6  | -1.333
  7  | -3E-13
Y₁∎8/Xcos(π/2X)
```

Now graph the function in the [0, 10] by [−6, 6] viewing rectangle.

```
Y1=8/Xcos(π/2X)
X=5          Y=0
```

Solution (a):

Make a table with integer values of t. $\qquad 1 \le t \le 8$

t (seconds)	$y = h(t) = \dfrac{8}{t}\cos\left(\dfrac{\pi}{2}t\right)$ (feet)	(t, y)
1	$\dfrac{8}{1}\cos\left(\dfrac{\pi}{2}\right) = 0$	$(1, 0)$
2	$\dfrac{8}{2}\cos\pi = -4$	$(2, -4)$
3	$\dfrac{8}{3}\cos\left(\dfrac{3\pi}{2}\right) = 0$	$(3, 0)$
4	$\dfrac{8}{4}\cos(2\pi) = 2$	$(4, 2)$
5	$\dfrac{8}{5}\cos\left(\dfrac{5\pi}{2}\right) = 0$	$(5, 0)$
6	$\dfrac{8}{6}\cos(3\pi) = -\dfrac{4}{3}$	$\left(6, -\dfrac{4}{3}\right)$
7	$\dfrac{8}{7}\cos\left(\dfrac{7\pi}{2}\right) = 0$	$(7, 0)$
8	$\dfrac{8}{8}\cos(4\pi) = 1$	$(8, 1)$

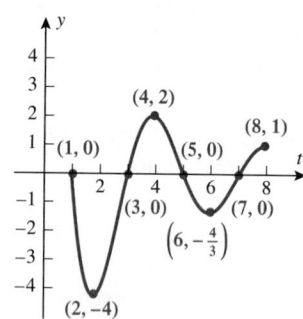

Solution (b):

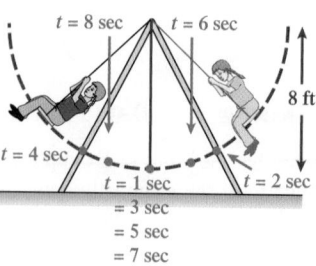

The height is 2 feet when t is 4 seconds. $\qquad \dfrac{8}{4}\cos(2\pi) = 2$

The height is 1 foot when t is 8 seconds. $\qquad \dfrac{8}{8}\cos(4\pi) = 1$

The height is 0.13 feet when t is 1 minute (60 seconds). $\qquad \dfrac{8}{60}\cos(30\pi) = 0.1333$

Resonance can be represented by the product of any increasing function and the original sinusoidal function. Here are two examples of functions that result in resonance as time increases:

$$y = t \cos \omega t \qquad y = e^t \sin \omega t$$

Graphing Sums of Functions: Addition of Ordinates

Since you have the ability to graph sinusoidal functions, let us now consider graphing sums of functions such as

$$y = x - \sin\left(\frac{\pi x}{2}\right) \qquad y = \sin x + \cos x \qquad y = 3 \sin x + \cos(2x)$$

The method for graphing these sums is called the **addition of ordinates**, because we add the corresponding y-values (ordinates). The following table illustrates the ordinates (y-values) of the two sinusoidal functions $\sin x$ and $\cos x$; adding the corresponding ordinates leads to the y-values of $y = \sin x + \cos x$.

x	$\sin x$	$\cos x$	$y = \sin x + \cos x$
0	0	1	1
$\dfrac{\pi}{4}$	$\dfrac{\sqrt{2}}{2}$	$\dfrac{\sqrt{2}}{2}$	$\sqrt{2}$
$\dfrac{\pi}{2}$	1	0	1
$\dfrac{3\pi}{4}$	$\dfrac{\sqrt{2}}{2}$	$-\dfrac{\sqrt{2}}{2}$	0
π	0	-1	-1
$\dfrac{5\pi}{4}$	$-\dfrac{\sqrt{2}}{2}$	$-\dfrac{\sqrt{2}}{2}$	$-\sqrt{2}$
$\dfrac{3\pi}{2}$	-1	0	-1
$\dfrac{7\pi}{4}$	$-\dfrac{\sqrt{2}}{2}$	$\dfrac{\sqrt{2}}{2}$	0
2π	0	1	1

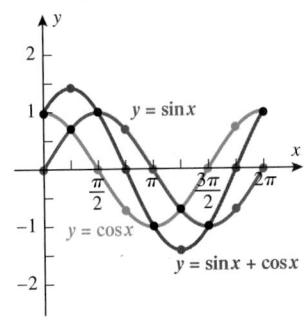

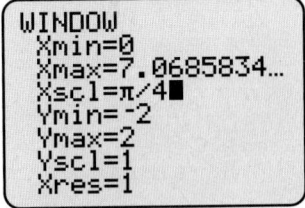

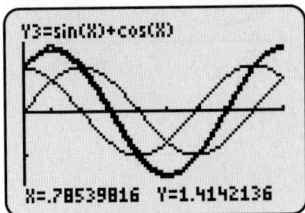
Using a graphing utility, we can graph $Y_1 = \sin X$, $Y_2 = \cos X$, and $Y_3 = Y_1 + Y_2$.

To display the graphs of x,

$-\sin\left(\dfrac{\pi x}{2}\right)$, and $x - \sin\left(\dfrac{\pi x}{2}\right)$ in the

same $[0, 4]$ by $[-1, 5]$ viewing

window, enter $Y_1 = x$,

$Y_2 = -\sin\left(\dfrac{\pi x}{2}\right)$, and

$Y_3 = x - \sin\left(\dfrac{\pi x}{2}\right)$.

To graph Y_3 using a thicker line, use
the ◄ key to go to the left of Y_3,
press ENTER , and select the
thicker line.

```
WINDOW
 Xmin=0
 Xmax=4
 Xscl=1
 Ymin=-1
 Ymax=5
 Yscl=1
 Xres=1
```

```
Plot1 Plot2 Plot3
\Y1■X
\Y2■-sin(πX/2)
\Y3■X-sin(πX/2)
```

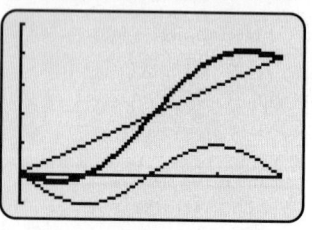

EXAMPLE 12   Graphing Sums of Functions

Graph $y = x - \sin\left(\dfrac{\pi x}{2}\right)$ on the interval $0 \le x \le 4$.

Solution:

Let $y_1 = x$ and $y_2 = -\sin\left(\dfrac{\pi x}{2}\right)$.

State the amplitude and period
of the graph of y_2. $|A| = |-1| = 1, p = 4$

Make a table of x- and
y-values of y_1, y_2, and
$y = y_1 + y_2$.

x	$y_1 = x$	$y_2 = -\sin\left(\dfrac{\pi x}{2}\right)$	$y = x + \left[-\sin\left(\dfrac{\pi x}{2}\right)\right]$
0	0	0	0
1	1	-1	0
2	2	0	2
3	3	1	4
4	4	0	4

Graph $y_1 = x$, $y_2 = -\sin\left(\dfrac{\pi x}{2}\right)$,

and $y = x - \sin\left(\dfrac{\pi x}{2}\right)$.

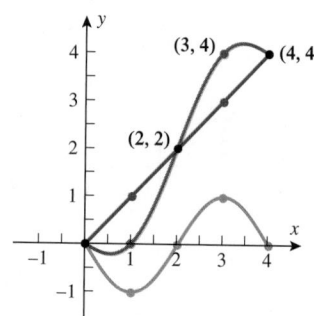

EXAMPLE 13 **Graphing Sums of Sine and Cosine Functions**

Graph $y = 3\sin x + \cos(2x)$ on the interval $0 \le x \le 2\pi$.

Solution:

Let $y_1 = 3\sin x$, and state the amplitude and period of its graph.

$$|A| = 3, p = 2\pi$$

Let $y_2 = \cos(2x)$, and state the amplitude and period of its graph.

$$|A| = 1, p = \pi$$

Make a table of x- and y-values of y_1, y_2, and $y = y_1 + y_2$.

x	$y_1 = 3\sin x$	$y_2 = \cos(2x)$	$y = 3\sin x + \cos(2x)$
0	0	1	1
$\dfrac{\pi}{4}$	$\dfrac{3\sqrt{2}}{2}$	0	$\dfrac{3\sqrt{2}}{2}$
$\dfrac{\pi}{2}$	3	-1	2
$\dfrac{3\pi}{4}$	$\dfrac{3\sqrt{2}}{2}$	0	$\dfrac{3\sqrt{2}}{2}$
π	0	1	1
$\dfrac{5\pi}{4}$	$-\dfrac{3\sqrt{2}}{2}$	0	$-\dfrac{3\sqrt{2}}{2}$
$\dfrac{3\pi}{2}$	-3	-1	-4
$\dfrac{7\pi}{4}$	$-\dfrac{3\sqrt{2}}{2}$	0	$-\dfrac{3\sqrt{2}}{2}$
2π	0	1	1

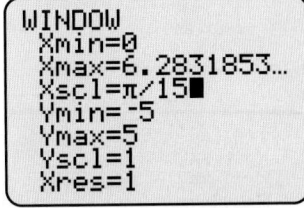

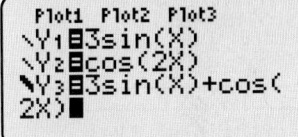

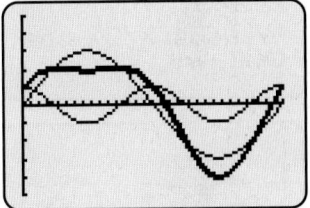

Graph $y_1 = 3\sin x$, $y_2 = \cos(2x)$, and $y = 3\sin x + \cos(2x)$.

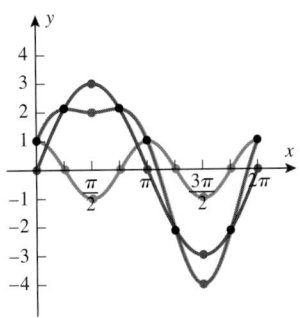

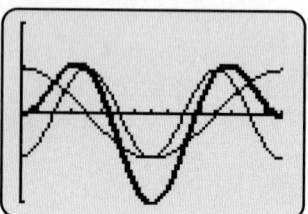
EXAMPLE 14 Graphing Sums of Cosine Functions

Graph $y = \cos\left(\dfrac{x}{2}\right) - \cos x$ on the interval $0 \leq x \leq 4\pi$.

Solution:

Let $y_1 = \cos\left(\dfrac{x}{2}\right)$ and
state the amplitude
and period of its graph. $|A| = 1, p = 4\pi$

Let $y_2 = -\cos x$ and
state the amplitude
and period of its graph. $|A| = |-1| = 1, p = 2\pi$

Make a table of x- and
y-values of y_1, y_2, and
$y = y_1 + y_2$.

x	$y_1 = \cos\left(\dfrac{x}{2}\right)$	$y_2 = -\cos x$	$y = \cos\left(\dfrac{x}{2}\right) + (-\cos x)$
0	1	-1	0
$\dfrac{\pi}{2}$	$\dfrac{\sqrt{2}}{2}$	0	$\dfrac{\sqrt{2}}{2}$
π	0	1	1
$\dfrac{3\pi}{2}$	$-\dfrac{\sqrt{2}}{2}$	0	$-\dfrac{\sqrt{2}}{2}$
2π	-1	-1	-2
$\dfrac{5\pi}{2}$	$-\dfrac{\sqrt{2}}{2}$	0	$-\dfrac{\sqrt{2}}{2}$
3π	0	1	1
$\dfrac{7\pi}{2}$	$\dfrac{\sqrt{2}}{2}$	0	$\dfrac{\sqrt{2}}{2}$
4π	1	-1	0

Graph $y_1 = \cos\left(\dfrac{x}{2}\right)$, $y_2 = -\cos x$,

and $y = \cos\left(\dfrac{x}{2}\right) - \cos x$.

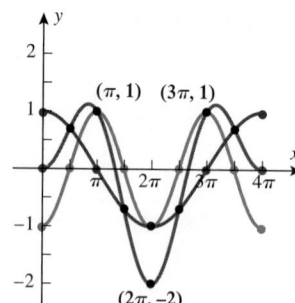

SECTION 5.2 SUMMARY

The sine function is an odd function, and its graph is symmetric about the origin. The cosine function is an even function, and its graph is symmetric about the y-axis. Graphs of the form $y = A \sin Bx$ and $y = A \cos Bx$ have amplitude $|A|$ and period $\dfrac{2\pi}{B}$.

In order to graph sinusoidal functions, you can start by point-plotting. A more efficient way is to first determine the amplitude and period. Divide the period into four equal parts and choose the values of the division points starting at 0 for x. Make a table of those four points and graph them (this is the graph of one period) by labeling the four coordinates and drawing a smooth sinusoidal curve. Extend the graph to the left and right.

To find the equation of a sinusoidal function given its graph, start by first finding the amplitude (half the distance between the maximum and minimum values) so you can find A. Then determine the period so you can find B. Graphs of the form $y = A \sin(Bx + C) + D$ and $y = A \cos(Bx + C) + D$ can be graphed using graph-shifting techniques.

Harmonic motion is one of the major applications of sinusoidal functions. To graph combinations of trigonometric functions, add the corresponding y-values of the individual functions.

SECTION 5.2 EXERCISES

▪ SKILLS

In Exercises 1–10, match the function with its graph (a–j).

1. $y = -\sin x$

2. $y = \sin x$

3. $y = \cos x$

4. $y = -\cos x$

5. $y = 2\sin x$

6. $y = 2\cos x$

7. $y = \sin\left(\tfrac{1}{2}x\right)$

8. $y = \cos\left(\tfrac{1}{2}x\right)$

9. $y = -2\cos\left(\tfrac{1}{2}x\right)$

10. $y = -2\sin\left(\tfrac{1}{2}x\right)$

a.

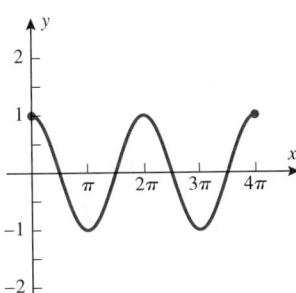

b.

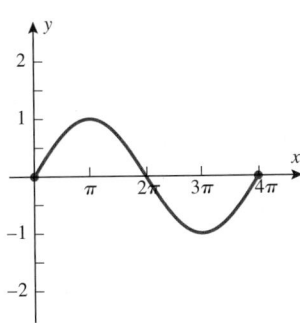

c.

d.

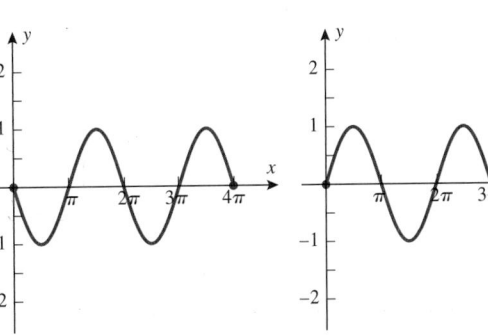

e.

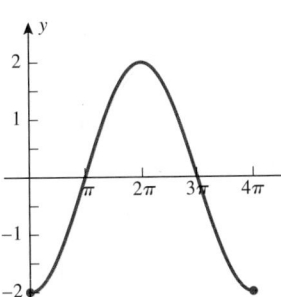

f.

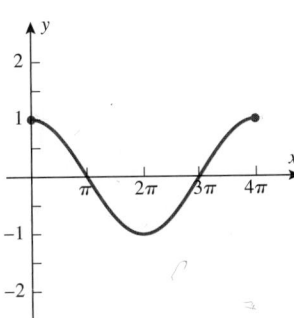

g.

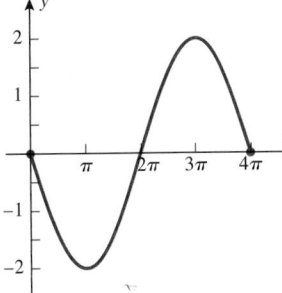

h.

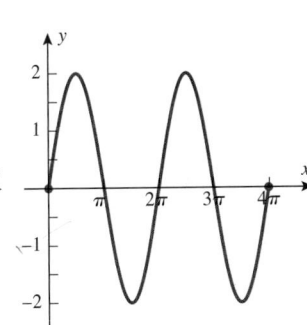

i.

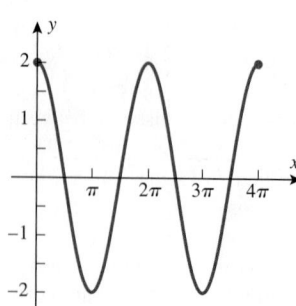

j.

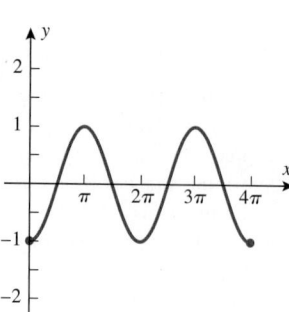

In Exercises 11–20, state the amplitude and period of each function.

11. $y = \dfrac{3}{2}\cos(3x)$ **12.** $y = \dfrac{2}{3}\sin(4x)$ **13.** $y = -\sin(5x)$ **14.** $y = -\cos(7x)$ **15.** $y = \dfrac{2}{3}\cos\left(\dfrac{3}{2}x\right)$

16. $y = \dfrac{3}{2}\sin\left(\dfrac{2}{3}x\right)$ **17.** $y = -3\cos(\pi x)$ **18.** $y = -2\sin(\pi x)$ **19.** $y = 5\sin\left(\dfrac{\pi}{3}x\right)$ **20.** $y = 4\cos\left(\dfrac{\pi}{4}x\right)$

In Exercises 21–32, graph the given function over one period.

21. $y = 8\cos x$ **22.** $y = 7\sin x$ **23.** $y = \sin(4x)$ **24.** $y = \cos(3x)$

25. $y = -3\cos\left(\dfrac{1}{2}x\right)$ **26.** $y = -2\sin\left(\dfrac{1}{4}x\right)$ **27.** $y = -3\sin(\pi x)$ **28.** $y = -2\cos(\pi x)$

29. $y = 5\cos(2\pi x)$ **30.** $y = 4\sin(2\pi x)$ **31.** $y = -3\sin\left(\dfrac{\pi}{4}x\right)$ **32.** $y = -4\sin\left(\dfrac{\pi}{2}x\right)$

In Exercises 33–40, graph the given function over the interval $[-2p, 2p]$, where p is the period of the function.

33. $y = -4\cos\left(\dfrac{1}{2}x\right)$ **34.** $y = -5\sin\left(\dfrac{1}{2}x\right)$ **35.** $y = -\sin(6x)$ **36.** $y = -\cos(4x)$

37. $y = 3\cos\left(\dfrac{\pi}{4}x\right)$ **38.** $y = 4\sin\left(\dfrac{\pi}{4}x\right)$ **39.** $y = \sin(4\pi x)$ **40.** $y = \cos(6\pi x)$

In Exercises 41–48, find the equation for each graph.

41.

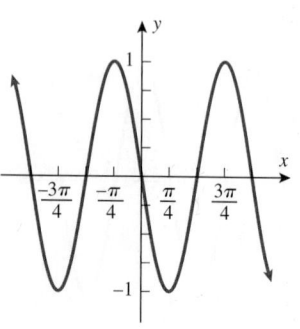

42.

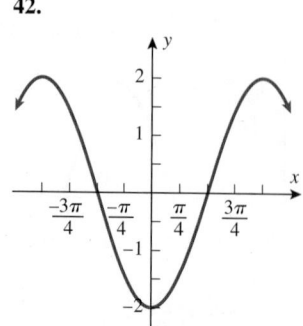

43.

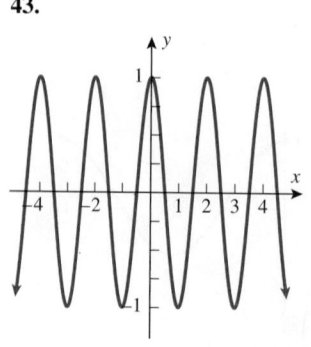

44.

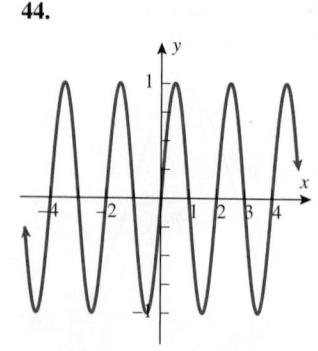

45.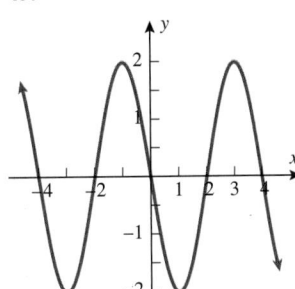

46.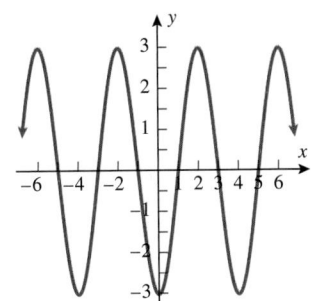

47.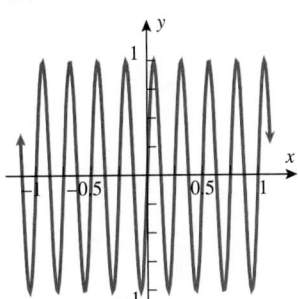

48.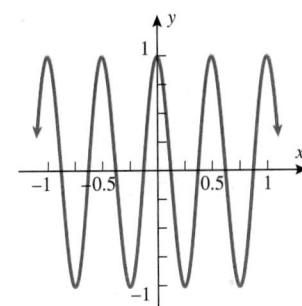

In Exercises 49–60, state the amplitude, period, and phase shift (including direction) of the given function.

49. $y = 2\sin(\pi x - 1)$

50. $y = 4\cos(x + \pi)$

51. $y = -5\cos(3x + 2)$

52. $y = -7\sin(4x - 3)$

53. $y = 6\sin[-\pi(x + 2)]$

54. $y = 3\sin\left[-\dfrac{\pi}{2}(x - 1)\right]$

55. $y = 3\sin(2x + \pi)$

56. $y = -4\cos(2x - \pi)$

57. $y = -\dfrac{1}{4}\cos\left(\dfrac{1}{4}x - \dfrac{\pi}{2}\right)$

58. $y = \dfrac{1}{2}\sin\left(\dfrac{1}{3}x + \pi\right)$

59. $y = 2\cos\left[\dfrac{\pi}{2}(x - 4)\right]$

60. $y = -5\sin[-\pi(x + 1)]$

In Exercises 61–66, sketch the graph of the function over the indicated interval.

61. $y = \dfrac{1}{2} + \dfrac{3}{2}\cos(2x + \pi), \left[-\dfrac{3\pi}{2}, \dfrac{3\pi}{2}\right]$

62. $y = \dfrac{1}{3} + \dfrac{2}{3}\sin(2x - \pi), \left[-\dfrac{3\pi}{2}, \dfrac{3\pi}{2}\right]$

63. $y = \dfrac{1}{2} - \dfrac{1}{2}\sin\left(\dfrac{1}{2}x - \dfrac{\pi}{4}\right), \left[-\dfrac{7\pi}{2}, \dfrac{9\pi}{2}\right]$

64. $y = -\dfrac{1}{2} + \dfrac{1}{2}\cos\left(\dfrac{1}{2}x + \dfrac{\pi}{4}\right), \left[-\dfrac{9\pi}{2}, \dfrac{7\pi}{2}\right]$

65. $y = -3 + 4\sin[\pi(x - 2)], [0, 4]$

66. $y = 4 - 3\cos[\pi(x + 1)], [-1, 3]$

In Exercises 67–94, add the ordinates of the individual functions to graph each summed function on the indicated interval.

67. $y = 2x - \cos(\pi x), 0 \le x \le 4$

68. $y = 3x - 2\cos(\pi x), 0 \le x \le 4$

69. $y = \dfrac{1}{3}x + 2\cos(2x), 0 \le x \le 2\pi$

70. $y = \dfrac{1}{4}x + 3\cos\left(\dfrac{x}{2}\right), 0 \le x \le 4\pi$

71. $y = x - \cos\left(\dfrac{3\pi}{2}x\right), 0 \le x \le 6$

72. $y = -2x + 2\sin\left(\dfrac{\pi}{2}x\right), -2 \le x \le 2$

73. $y = \dfrac{1}{4}x - \dfrac{1}{2}\cos[\pi(x - 1)], 2 \le x \le 6$

74. $y = -\dfrac{1}{3}x + \dfrac{1}{3}\sin\left[\dfrac{\pi}{6}(x + 2)\right], -2 \le x \le 10$

75. $y = \sin x - \cos x, 0 \le x \le 2\pi$

76. $y = \cos x - \sin x, 0 \le x \le 2\pi$

77. $y = 3\cos x + \sin x, 0 \le x \le 2\pi$

78. $y = 3\sin x - \cos x, 0 \le x \le 2\pi$

79. $y = 4\cos x - \sin(2x), 0 \le x \le 2\pi$

80. $y = \dfrac{1}{2}\sin x + 2\cos(4x), -\pi \le x \le \pi$

81. $y = 2\sin[\pi(x - 1)] - 2\cos[\pi(x + 1)], -1 \le x \le 2$

82. $y = \sin\left[\dfrac{\pi}{4}(x + 2)\right] + 3\cos\left[\dfrac{3\pi}{3}(x - 1)\right], 1 \le x \le 5$

83. $y = \cos\left(\dfrac{x}{2}\right) + \cos(2x), 0 \le x \le 4\pi$

84. $y = \sin(2x) + \sin(3x), -\pi \le x \le \pi$

85. $y = \sin\left(\dfrac{x}{2}\right) + \sin(2x), 0 \le x \le 4\pi$

86. $y = -\sin\left(\dfrac{\pi}{4}x\right) - 3\sin\left(\dfrac{5\pi}{4}x\right), 0 \le x \le 4$

87. $y = -\dfrac{1}{3}\sin\left(\dfrac{\pi}{6}x\right) + \dfrac{2}{3}\sin\left(\dfrac{5\pi}{6}x\right), 0 \le x \le 3$

88. $y = 8\cos x - 6\cos\left(\dfrac{1}{2}x\right), -2\pi \le x \le 2\pi$

89. $y = -\dfrac{1}{4}\cos\left(\dfrac{\pi}{6}x\right) - \dfrac{1}{2}\cos\left(\dfrac{\pi}{3}x\right), 0 \le x \le 12$

90. $y = 2\cos\left(\dfrac{3}{2}x\right) - \cos\left(\dfrac{1}{2}x\right), -2\pi \le x \le 2\pi$

91. $y = 2\sin\left(\dfrac{x}{2}\right) - \cos(2x), 0 \le x \le 4\pi$

92. $y = 2\cos\left(\dfrac{x}{2}\right) + \sin(2x), 0 \le x \le 4\pi$

93. $y = 2\sin[\pi(x-1)] + 3\sin\left[2\pi\left(x+\dfrac{1}{2}\right)\right], -2 \le x \le 2$

94. $y = -\dfrac{1}{2}\cos\left(x + \dfrac{\pi}{3}\right) - 2\cos\left(x - \dfrac{\pi}{6}\right), -\pi \le x \le \pi$

■ APPLICATIONS

For Exercises 95–98, refer to the following:

A weight hanging on a spring will oscillate up and down about its equilibrium position after it is pulled down and released.

This is an example of simple harmonic motion. This motion would continue forever if there were not any friction or air resistance. Simple harmonic motion can be described with the function $y = A\cos\left(t\sqrt{\dfrac{k}{m}}\right)$, where $|A|$ is the amplitude, t is the time in seconds, m is the mass of the weight, and k is a constant particular to the spring.

95. Simple Harmonic Motion. If the height of the spring is measured in centimeters and the mass in grams, then what are the amplitude and mass if $y = 4\cos\left(\dfrac{t\sqrt{k}}{2}\right)$?

96. Simple Harmonic Motion. If a spring is measured in centimeters and the mass in grams, then what are the amplitude and mass if $y = 3\cos\left(3t\sqrt{k}\right)$?

97. Frequency of Oscillations. The frequency of the oscillations in cycles per second is determined by $f = \dfrac{1}{p}$, where p is the period. What is the frequency for the oscillation modeled by $y = 3\cos\left(\dfrac{t}{2}\right)$?

98. Frequency of Oscillations. The frequency of the oscillations f is given by $f = \dfrac{1}{p}$, where p is the period. What is the frequency of oscillation modeled by $y = 3.5\cos(3t)$?

99. Sound Waves. A pure tone created by a vibrating tuning fork shows up as a sine wave on an oscilloscope's screen. A tuning fork vibrating at 256 hertz (Hz) gives the tone middle C and can have the equation $y = 0.005\sin[(2\pi)(256t)]$, where the amplitude is in centimeters (cm) and the time t in seconds. What are the amplitude and frequency of the wave where the frequency is $\dfrac{1}{p}$ in cycles per second? (*Note:* 1 hertz = 1 cycle per second.)

100. Sound Waves. A pure tone created by a vibrating tuning fork shows up as a sine wave on an oscilloscope's screen. A tuning fork vibrating at 288 Hz gives the tone D and can have the equation $y = 0.005\sin[(2\pi)(288t)]$, where the amplitude is in centimeters (cm) and the time t in seconds. What are the amplitude and frequency of the wave where the frequency is $\dfrac{1}{p}$ in cycles per second?

101. Sound Waves. If a sound wave is represented by $y = 0.008\sin(750\pi t)$ cm, what are its amplitude and frequency? See Exercise 99.

102. Sound Waves. If a sound wave is represented by $y = 0.006\cos(1000\pi t)$ cm, what are its amplitude and frequency? See Exercise 99.

For Exercises 103–106, refer to the following:

When an airplane flies faster than the speed of sound, the sound waves that are formed take on a cone shape, and where the cone hits the ground, a sonic boom is heard. If θ is the angle of the vertex of the cone, then $\sin\left(\dfrac{\theta}{2}\right) = \dfrac{330\,\text{m/sec}}{V} = \dfrac{1}{M}$, where V is the speed of the plane and M is the Mach number.

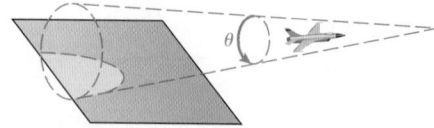

103. Sonic Booms. What is the speed of the plane if the plane is flying at Mach 2?

104. Sonic Booms. What is the Mach number if the plane is flying at 990 m/sec?

105. Sonic Booms. What is the speed of the plane if the cone angle is 60°?

106. Sonic Booms. What is the speed of the plane if the cone angle is 30°?

For Exercises 107 and 108, refer to the following:

With the advent of summer come fireflies. They are intriguing because they emit a flashing luminescence that beckons their mate to them. It is known that the speed and intensity of the flashing are related to the temperature—the higher the temperature, the quicker and more intense the flashing becomes. If you ever watch a single firefly, you will see that the intensity of the flashing is periodic with time. The intensity of light emitted is measured in *candelas per square meter* (of firefly). To give an idea of this unit of measure, the intensity of a picture on a typical TV screen is about 450 candelas per square meter.

The measurement for the intensity of the light emitted by a typical firefly at its brightest moment is about 50 candelas per square meter. Assume that a typical cycle of this flashing is 4 seconds and that the intensity is essentially zero candelas at the beginning and ending of a cycle.

107. Bioluminescence in Fireflies. Find an equation that describes this flashing. What is the intensity of the flashing at 4 minutes?

108. Bioluminescence in Fireflies. Graph the equation from Exercise 107 for a period of 30 seconds.

■ CATCH THE MISTAKE

In Exercises 109 and 110, explain the mistake that is made.

109. Graph the function $y = -2\cos x$.

Solution:

Find the amplitude. $|A| = |-2| = 2$

The graph of $y = -2\cos x$ is similar to the graph of $y = \cos x$ with amplitude 2.

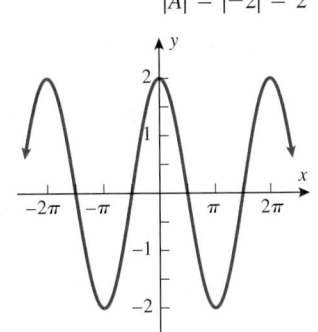

This is incorrect. What mistake was made?

110. Graph the function $y = -\sin(2x)$.

Solution:

Make a table with values.

x	$y = -\sin(2x)$	(x, y)
0	$y = -\sin 0 = 0$	$(0, 0)$
$\dfrac{\pi}{2}$	$y = -\sin \pi = 0$	$(0, 0)$
π	$y = -\sin(2\pi) = 0$	$(0, 0)$
$\dfrac{3\pi}{2}$	$y = -\sin(3\pi) = 0$	$(0, 0)$
2π	$y = -\sin(4\pi) = 0$	$(0, 0)$

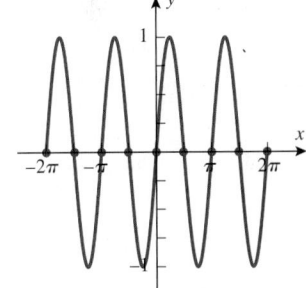

Graph the function by plotting these points and connecting them with a sinusoidal curve.

This is incorrect. What mistake was made?

■ CONCEPTUAL

In Exercises 111–114, determine whether each statement is true or false. (A and B are positive real numbers.)

111. The graph of $y = -A\cos Bx$ is the graph of $y = A\cos Bx$ reflected about the x-axis.

112. The graph of $y = A\sin(-Bx)$ is the graph of $y = A\sin Bx$ reflected about the x-axis.

113. The graph of $y = -A\cos(-Bx)$ is the graph of $y = A\cos Bx$.

114. The graph of $y = -A\sin(-Bx)$ is the graph of $y = A\sin Bx$.

In Exercises 115–118, A and B are positive real numbers.

115. Find the y-intercept of the function $y = A \cos Bx$.

116. Find the y-intercept of the function $y = A \sin Bx$.

117. Find the x-intercepts of the function $y = A \sin Bx$.

118. Find the x-intercepts of the function $y = A \cos Bx$.

■ CHALLENGE

119. Find the y-intercept of $y = -A \sin\left(Bx + \dfrac{\pi}{6}\right)$.

120. Find the y-intercept of $y = A \cos(Bx - \pi) + C$.

121. Find the x-intercept(s) of $y = A \sin Bx + A$.

122. Find an expression involving C and A that describes the values of C for which the graph of $y = A \cos Bx + C$ does not cross the x-axis. (Assume that $A > 0$.)

123. What is the range of $y = 2A \sin(Bx + C) - \dfrac{A}{2}$?

124. Can the y-coordinate of a point on the graph of
$$y = A \sin Bx + 3A \cos\left(\dfrac{B}{2}x\right)$$
exceed $4A$? Explain. (Assume that $A > 0$.)

■ TECHNOLOGY

125. Use a graphing calculator to graph $Y_1 = 5 \sin x$ and $Y_2 = \sin(5x)$. Is the following statement true based on what you see? $y = \sin cx$ has the same graph as $y = c \sin x$.

126. Use a graphing calculator to graph $Y_1 = 3 \cos x$ and $Y_2 = \cos(3x)$. Is the following statement true based on what you see? $y = \cos cx$ has the same graph as $y = c \cos x$.

127. Use a graphing calculator to graph $Y_1 = \sin x$ and $Y_2 = \cos\left(x - \dfrac{\pi}{2}\right)$. What do you notice?

128. Use a graphing calculator to graph $Y_1 = \cos x$ and $Y_2 = \sin\left(x + \dfrac{\pi}{2}\right)$. What do you notice?

129. Use a graphing calculator to graph $Y_1 = \cos x$ and $Y_2 = \cos(x + c)$, where

  a. $c = \dfrac{\pi}{3}$, and explain the relationship between Y_2 and Y_1.

  b. $c = -\dfrac{\pi}{3}$, and explain the relationship between Y_2 and Y_1.

130. Use a graphing calculator to graph $Y_1 = \sin x$ and $Y_2 = \sin(x + c)$, where

  a. $c = \dfrac{\pi}{3}$, and explain the relationship between Y_2 and Y_1.

  b. $c = -\dfrac{\pi}{3}$, and explain the relationship between Y_2 and Y_1.

For Exercises 131 and 132, refer to the following:

Damped oscillatory motion, or *damped oscillation*, occurs when things in oscillatory motion experience friction or resistance.

The friction causes the amplitude to decrease as a function of time. Mathematically, we can use a negative exponential function to damp the oscillations in the form of

$$f(t) = e^{-t} \sin t$$

131. Damped Oscillation. Graph the functions $Y_1 = e^{-t}$, $Y_2 = \sin t$, and $Y_3 = e^{-t} \sin t$ in the same viewing window (let t range from 0 to 2π). What happens as t increases?

132. Damped Oscillation. Graph $Y_1 = e^{-t} \sin t$, $Y_2 = e^{-2t} \sin t$, and $Y_3 = e^{-4t} \sin t$ in the same viewing window. What happens to $Y = e^{-kt} \sin t$ as k increases?

133. Use a graphing calculator to graph $Y_1 = \sin x$ and $Y_2 = \sin x + c$, where

  a. $c = 1$, and explain the relationship between Y_2 and Y_1.

  b. $c = -1$, and explain the relationship between Y_2 and Y_1.

134. Use a graphing calculator to graph $Y_1 = \cos x$ and $Y_2 = \cos x + c$, where

  a. $c = \dfrac{1}{2}$, and explain the relationship between Y_2 and Y_1.

  b. $c = -\dfrac{1}{2}$, and explain the relationship between Y_2 and Y_1.

135. What is the amplitude of the function $y = 3 \cos x + 4 \sin x$? Use a graphing calculator to graph $Y_1 = 3 \cos x$, $Y_2 = 4 \sin x$, and $Y_3 = 3 \cos x + 4 \sin x$ in the same viewing window.

136. What is the amplitude of the function $y = \sqrt{3} \cos x - \sin x$? Use a graphing calculator to graph $Y_1 = \sqrt{3} \cos x$, $Y_2 = \sin x$, and $Y_3 = \sqrt{3} \cos x - \sin x$ in the same viewing window.

■ PREVIEW TO CALCULUS

In calculus, the definite integral $\int_a^b f(x)\,dx$ is used to find the area below the graph of f, above the x-axis, between $x = a$ and $x = b$. For example, $\int_0^2 x\,dx = 2$, as you can see in the following figure:

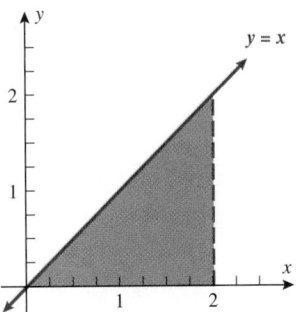

The Fundamental Theorem of Calculus establishes that the definite integral $\int_a^b f(x)\,dx$ equals $F(b) - F(a)$, where F is any antiderivative of a continuous function f.

In Exercises 137–140, first shade the area corresponding to the definite integral and then use the information below to find the exact value of the area.

FUNCTION	$\sin x$	$\cos x$
ANTIDERIVATIVE	$-\cos x$	$\sin x$

137. $\displaystyle\int_0^{\pi} \sin x\,dx$

138. $\displaystyle\int_{-\pi/2}^{\pi/2} \cos x\,dx$

139. $\displaystyle\int_0^{\pi/2} \cos x\,dx$

140. $\displaystyle\int_0^{\pi/2} \sin x\,dx$

SECTION 5.3 GRAPHS OF OTHER TRIGONOMETRIC FUNCTIONS

SKILLS OBJECTIVES

- Determine the domain and range of the tangent, cotangent, secant, and cosecant functions.
- Graph basic tangent, cotangent, secant, and cosecant functions.
- Determine the period of tangent, cotangent, secant, and cosecant functions.
- Graph translated tangent, cotangent, secant, and cosecant functions.

CONCEPTUAL OBJECTIVES

- Relate domain restrictions to vertical asymptotes.
- Understand the pattern that vertical asymptotes follow.
- Understand the relationships between the graphs of the cosine and secant functions and the sine and cosecant functions.

Section 5.2 focused on graphing sinusoidal functions (sine and cosine). We now turn our attention to graphing the other trigonometric functions: tangent, cotangent, secant, and cosecant. We know the graphs of the sine and cosine functions, and we can get the graphs of the other trigonometric functions from the sinusoidal functions. Recall the reciprocal and quotient identities:

$$\tan x = \frac{\sin x}{\cos x} \qquad \cot x = \frac{\cos x}{\sin x} \qquad \sec x = \frac{1}{\cos x} \qquad \csc x = \frac{1}{\sin x}$$

Recall that in graphing rational functions, a *vertical asymptote* corresponds to a denominator equal to zero (as long as the numerator and denominator have no common factors). A **vertical**

asymptote is a vertical (dashed) line that represents a value of x for which the function is not defined. As you will see in this section, tangent and secant functions have graphs with vertical asymptotes at the x-values where cosine is equal to zero, and cotangent and cosecant functions have graphs with vertical asymptotes at the x-values where sine is equal to zero.

One important difference between the sinusoidal functions, $y = \sin x$ and $y = \cos x$, and the other four trigonometric functions ($y = \tan x$, $y = \sec x$, $y = \csc x$, and $y = \cot x$) is that the sinusoidal functions have defined amplitudes, whereas the other four trigonometric functions do not (since they are unbounded vertically).

The Tangent Function

Since the tangent function is a quotient that relies on the sine and cosine functions, let us start with a table of values for the quadrantal angles.

x	$\sin x$	$\cos x$	$\tan x = \dfrac{\sin x}{\cos x}$	(x, y) OR ASYMPTOTE
0	0	1	0	$(0, 0)$
$\dfrac{\pi}{2}$	1	0	undefined	vertical asymptote: $x = \dfrac{\pi}{2}$
π	0	-1	0	$(\pi, 0)$
$\dfrac{3\pi}{2}$	-1	0	undefined	vertical asymptote: $x = \dfrac{3\pi}{2}$
2π	0	1	0	$(2\pi, 0)$

Notice that the x-intercepts correspond to integer multiples of π and vertical asymptotes correspond to odd integer multiples of $\dfrac{\pi}{2}$.

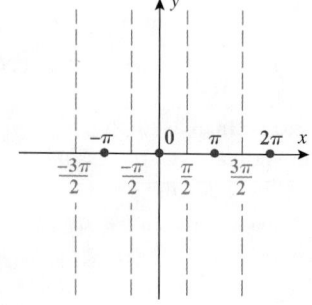

We know that the graph of the tangent function is undefined at the odd integer multiples of $\dfrac{\pi}{2}$, so its graph cannot cross the vertical asymptotes. The question is, what happens between the asymptotes? We know the x-intercepts, so let us now make a table for special values of x.

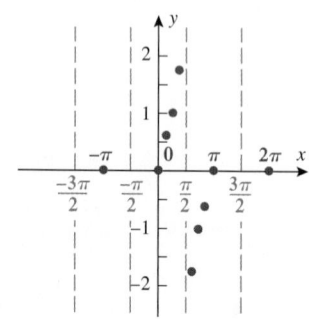

x	$\sin x$	$\cos x$	$\tan x = \dfrac{\sin x}{\cos x}$	(x, y)
$\dfrac{\pi}{6}$	$\dfrac{1}{2}$	$\dfrac{\sqrt{3}}{2}$	$\dfrac{1}{\sqrt{3}} = \dfrac{\sqrt{3}}{3} \approx 0.577$	$\left(\dfrac{\pi}{6}, 0.577\right)$
$\dfrac{\pi}{4}$	$\dfrac{\sqrt{2}}{2}$	$\dfrac{\sqrt{2}}{2}$	1	$\left(\dfrac{\pi}{4}, 1\right)$
$\dfrac{\pi}{3}$	$\dfrac{\sqrt{3}}{2}$	$\dfrac{1}{2}$	$\sqrt{3} \approx 1.732$	$\left(\dfrac{\pi}{3}, 1.732\right)$
$\dfrac{2\pi}{3}$	$\dfrac{\sqrt{3}}{2}$	$-\dfrac{1}{2}$	$-\sqrt{3} \approx -1.732$	$\left(\dfrac{2\pi}{3}, -1.732\right)$
$\dfrac{3\pi}{4}$	$\dfrac{\sqrt{2}}{2}$	$-\dfrac{\sqrt{2}}{2}$	-1	$\left(\dfrac{3\pi}{4}, -1\right)$
$\dfrac{5\pi}{6}$	$\dfrac{1}{2}$	$-\dfrac{\sqrt{3}}{2}$	$\dfrac{1}{-\sqrt{3}} = -\dfrac{\sqrt{3}}{3} \approx -0.577$	$\left(\dfrac{5\pi}{6}, -0.577\right)$

WORDS	MATH
What happens as x approaches $\frac{\pi}{2}$?	$\lim\limits_{x \to \frac{\pi}{2}} \tan x$

We must consider which way x approaches $\frac{\pi}{2}$:

x approaches $\frac{\pi}{2}$ from the left.	$\lim\limits_{x \to \left(\frac{\pi}{2}\right)^-} \tan x$
x approaches $\frac{\pi}{2}$ from the right.	$\lim\limits_{x \to \left(\frac{\pi}{2}\right)^+} \tan x$
The numerical approximation to $\frac{\pi}{2}$ is 1.571.	$\frac{\pi}{2} \approx 1.571$

As x approaches $\frac{\pi}{2}$ from the left, the value of the tangent function increases toward a large positive number.

$\lim\limits_{x \to \left(\frac{\pi}{2}\right)^-}$	$(\tan x)$
1.5	14.1
1.55	48.1
1.57	1255.8

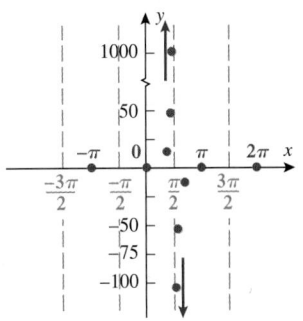

Thus, as x approaches $\frac{\pi}{2}$ from the left, $\tan x$ approaches infinity.

$$\lim\limits_{x \to \left(\frac{\pi}{2}\right)^-} \tan x = \infty$$

As x approaches $\frac{\pi}{2}$ from the right, the value of the tangent function decreases toward a large negative number.

$\lim\limits_{x \to \left(\frac{\pi}{2}\right)^+}$	$(\tan x)$
1.65	-12.6
1.59	-52.1
1.58	-108.6

Thus, as x approaches $\frac{\pi}{2}$ from the right, $\tan x$ is negative and the absolute value of the tangent function gets larger and larger, so we say, $\tan x$ approaches negative infinity.

$$\lim\limits_{x \to \left(\frac{\pi}{2}\right)^+} \tan x = -\infty$$

Technology Tip

```
Plot1 Plot2 Plot3
\Y1 ■ tan(X)
\Y2=■
```

```
WINDOW
 Xmin=-6.283185...
 Xmax=6.2831853...
 Xscl=.78539816...
 Ymin=-5
 Ymax=5
 Yscl=1
 Xres=1
```

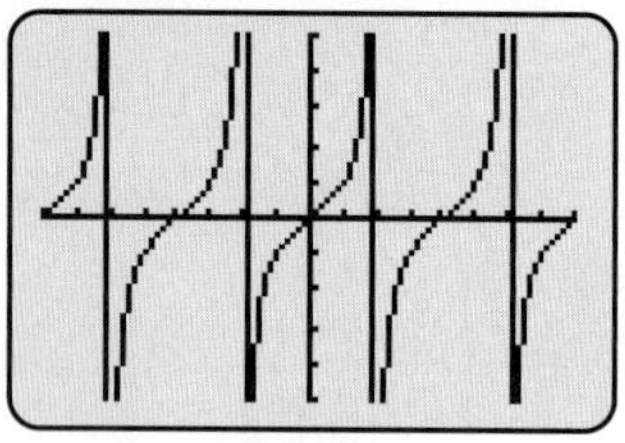

GRAPH OF $y = \tan x$

1. The x-intercepts occur at multiples of π. $\quad x = n\pi$
2. Vertical asymptotes occur at odd integer multiples of $\frac{\pi}{2}$. $\quad x = \frac{(2n + 1)\pi}{2}$
3. The domain is the set of all real numbers except odd integer multiples of $\frac{\pi}{2}$. $\quad x \neq \frac{(2n + 1)\pi}{2}$
4. The range is the set of all real numbers. $\quad (-\infty, \infty)$
5. $y = \tan x$ has period π.
6. $y = \tan x$ is an odd function (symmetric about the origin). $\quad \tan(-x) = -\tan x$
7. The graph has no defined amplitude, since the funtion is unbounded.

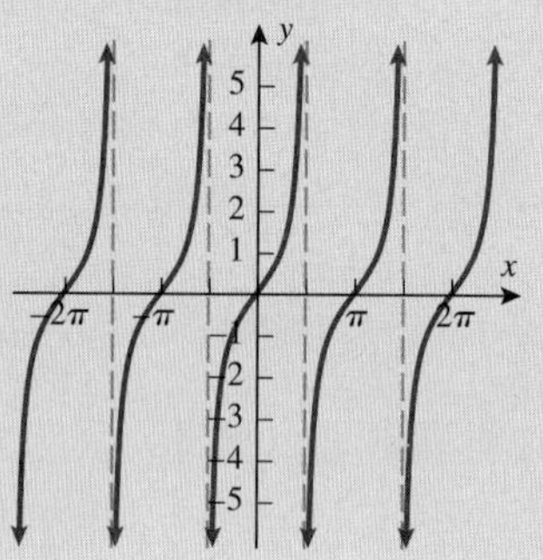

Note: n is an integer.

The Cotangent Function

The cotangent function is similar to the tangent function in that it is a quotient involving the sine and cosine functions. The difference is that cotangent has cosine in the numerator and sine in the denominator: $\cot x = \frac{\cos x}{\sin x}$. The graph of $y = \tan x$ has x-intercepts corresponding to integer multiples of π and vertical asymptotes corresponding to odd integer multiples of $\frac{\pi}{2}$. The graph of the cotangent function is the reverse in that it has x-intercepts corresponding to odd integer multiples of $\frac{\pi}{2}$ and vertical asymptotes corresponding to integer multiples of π. This is because the x-intercepts occur when the numerator, $\cos x$, is equal to 0 and the vertical asymptotes occur when the denominator, $\sin x$, is equal to 0.

Study Tip

The graphs of $y = \tan x$ and $y = \cot x$ both have period π, and neither of them has defined amplitude.

GRAPH OF $y = \cot x$

1. The x-intercepts occur at odd integer multiples of $\dfrac{\pi}{2}$.

$$x = \dfrac{(2n + 1)\pi}{2}$$

2. Vertical asymptotes occur at integer multiples of π.

$$x = n\pi$$

3. The domain is the set of all real numbers except integer multiples of π.

$$x \ne n\pi$$

4. The range is the set of all real numbers.

$$(-\infty, \infty)$$

5. $y = \cot x$ has period π.

6. $y = \cot x$ is an odd function (symmetric about the origin).

$$\cot(-x) = -\cot x$$

7. The graph has no defined amplitude, since the funtion is unbounded.

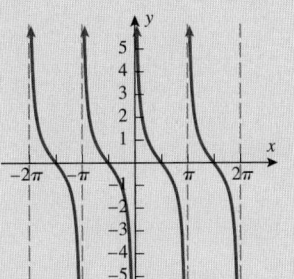

Note: n is an integer.

Technology Tip

To graph $\cot x$, use the reciprocal property to enter $(\tan x)^{-1}$.

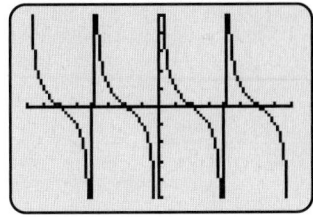

The Secant Function

Since $y = \cos x$ has period 2π, the secant function, which is the reciprocal of the cosine function, $\sec x = \dfrac{1}{\cos x}$, also has period 2π. We now illustrate values of the secant function with a table.

x	$\cos x$	$\sec x = \dfrac{1}{\cos x}$	(x, y) OR ASYMPTOTE
0	1	1	$(0, 1)$
$\dfrac{\pi}{2}$	0	undefined	vertical asymptote: $x = \dfrac{\pi}{2}$
π	-1	-1	$(\pi, -1)$
$\dfrac{3\pi}{2}$	0	undefined	vertical asymptote: $x = \dfrac{3\pi}{2}$
2π	1	1	$(2\pi, 1)$

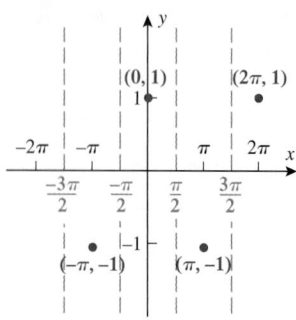

Again, we ask the same question: What happens as x approaches the vertical asymptotes? The same asymptotic behavior occurs that we found with the tangent function.

If we graph $y = \cos x$ (the "guide" function) and $y = \sec x$ on the same graph, we notice the following:

- The x-intercepts of $y = \cos x$ correspond to the vertical asymptotes of $y = \sec x$.
- The range of cosine is $[-1, 1]$ and the range of secant is $(-\infty, -1] \cup [1, \infty)$.
- When cosine is positive, secant is positive, and when one is negative, the other is negative.

The cosine function is used as the guide function to graph the secant function.

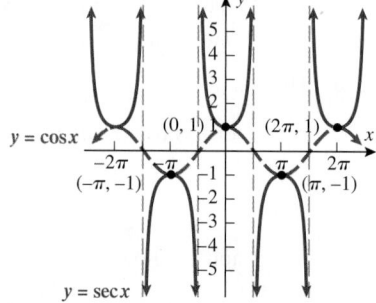

GRAPH OF $y = \sec x$

Technology Tip

To graph $\sec x$, enter as $(\cos x)^{-1}$.

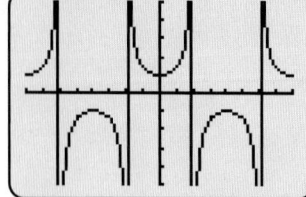

1. There are no x-intercepts.
2. Vertical asymptotes occur at odd integer multiples of $\dfrac{\pi}{2}$.
   $$x = \frac{(2n + 1)\pi}{2}$$
3. The domain is the set of all real numbers except odd integer multiples of $\dfrac{\pi}{2}$.
   $$x \neq \frac{(2n + 1)\pi}{2}$$
4. The range is $(-\infty, -1] \cup [1, \infty)$.
5. $y = \sec x$ has period 2π.
6. $y = \sec x$ is an even function (symmetric about the y-axis).
   $$\sec(-x) = \sec x$$
7. The graph has no defined amplitude, since the function is unbounded.

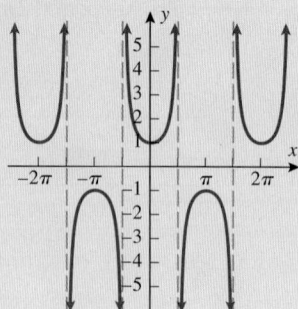

Note: n is an integer.

The Cosecant Function

Since $y = \sin x$ has period 2π, the cosecant function, which is the reciprocal of the sine function, $\csc x = \dfrac{1}{\sin x}$, also has period 2π. We now illustrate values of cosecant with a table.

x	$\sin x$	$\csc x = \dfrac{1}{\sin x}$	(x, y) OR ASYMPTOTE
0	0	undefined	vertical asymptote: $x = 0$
$\dfrac{\pi}{2}$	1	1	$\left(\dfrac{\pi}{2}, 1\right)$
π	0	undefined	vertical asymptote: $x = \pi$
$\dfrac{3\pi}{2}$	-1	-1	$\left(\dfrac{3\pi}{2}, -1\right)$
2π	0	undefined	vertical asymptote: $x = 2\pi$

Again, we ask the same question: What happens as x approaches the vertical asymptotes? The same asymptotic behavior occurs that we found with the cotangent function.

If we graph $y = \sin x$ (the "guide" function) and $y = \csc x$ on the same graph, we notice the following:

- The x-intercepts of $y = \sin x$ correspond to the vertical asymptotes of $y = \csc x$.
- The range of sine is $[-1, 1]$ and the range of cosecant is $(-\infty, -1] \cup [1, \infty)$.
- When sine is positive, cosecant is positive, and when one is negative, the other is negative.

The sine function is used as the guide function to graph the cosecant function.

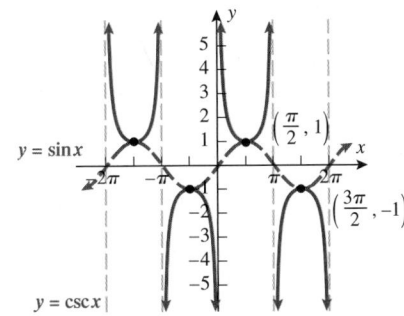

GRAPH OF $y = \csc x$

1. There are no x-intercepts.
2. Vertical asymptotes occur at integer multiples of π. $x = n\pi$
3. The domain is the set of all real numbers
   except integer multiples of π. $x \neq n\pi$
4. The range is $(-\infty, -1] \cup [1, \infty)$.
5. $y = \csc x$ has period 2π.
6. $y = \csc x$ is an odd function (symmetric
   about the origin). $\csc(-x) = -\csc x$
7. The graph has no defined amplitude, since
   the function is unbounded.

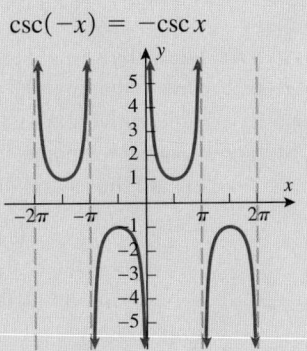

Note: n is an integer.

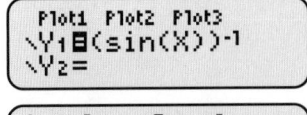
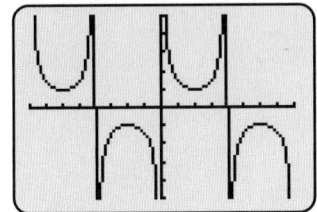

Graphing Tangent, Cotangent, Secant, and Cosecant Functions

FUNCTION	$y = \sin x$	$y = \cos x$	$y = \tan x$	$y = \cot x$	$y = \sec x$	$y = \csc x$
Graph of One Period						
Domain	$\mathbb{R}$	$\mathbb{R}$	$x \neq \dfrac{(2n+1)\pi}{2}$	$x \neq n\pi$	$x \neq \dfrac{(2n+1)\pi}{2}$	$x \neq n\pi$
Range	$[-1, 1]$	$[-1, 1]$	$\mathbb{R}$	$\mathbb{R}$	$(-\infty, -1] \cup [1, \infty)$	$(-\infty, -1] \cup [1, \infty)$
Amplitude	1	1	none	none	none	none
Period	2π	2π	π	π	2π	2π
x-intercepts	$x = n\pi$	$x = \dfrac{(2n+1)\pi}{2}$	$x = n\pi$	$x = \dfrac{(2n+1)\pi}{2}$	none	none
Vertical Asymptotes	none	none	$x = \dfrac{(2n+1)\pi}{2}$	$x = n\pi$	$x = \dfrac{(2n+1)\pi}{2}$	$x = n\pi$

Note: n is an integer.

We use these basic functions as the starting point for graphing general tangent, cotangent, secant, and cosecant functions.

GRAPHING TANGENT AND COTANGENT FUNCTIONS

Graphs of $y = A \tan Bx$ and $y = A \cot Bx$ can be obtained using the following steps (assume $B > 0$):

Step 1: Calculate the period $\dfrac{\pi}{B}$.

Step 2: Find two neighboring vertical asymptotes.

For $y = A \tan Bx$: $\quad Bx = -\dfrac{\pi}{2}$ and $Bx = \dfrac{\pi}{2}$

For $y = A \cot Bx$: $\quad Bx = 0$ and $Bx = \pi$

Step 3: Find the x-intercept between the two asymptotes.

For $y = A \tan Bx$: $\quad Bx = 0$

For $y = A \cot Bx$: $\quad Bx = \dfrac{\pi}{2}$

Step 4: Draw the vertical asymptotes and label the x-intercept.

Step 5: Divide the interval between the asymptotes into four equal parts. Set up a table with coordinates corresponding to the points in the interval.

Step 6: Connect the points with a smooth curve. Use arrows to indicate the behavior toward the asymptotes.

- If $A > 0$
  - $y = A \tan Bx$ increases from left to right.
  - $y = A \cot Bx$ decreases from left to right.
- If $A < 0$
  - $y = A \tan Bx$ decreases from left to right.
  - $y = A \cot Bx$ increases from left to right.

EXAMPLE 1 **Graphing $y = A \tan Bx$**

Graph $y = -3 \tan(2x)$ on the interval $-\dfrac{\pi}{2} \leq x \leq \dfrac{\pi}{2}$.

Solution: $A = -3$, $B = 2$

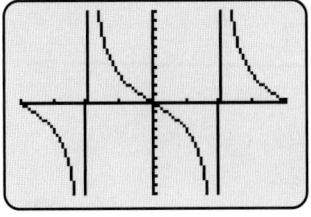
STEP 1 Calculate the period.

$$\dfrac{\pi}{B} = \dfrac{\pi}{2}$$

STEP 2 Find two vertical asymptotes.

$$Bx = -\dfrac{\pi}{2} \quad \text{and} \quad Bx = \dfrac{\pi}{2}$$

Substitute $B = 2$ and solve for x.

$$x = -\dfrac{\pi}{4} \quad \text{and} \quad x = \dfrac{\pi}{4}$$

STEP 3 Find the x-intercept between the asymptotes.

$$Bx = 0$$
$$x = 0$$

STEP 4 Draw the vertical asymptotes $x = -\dfrac{\pi}{4}$ and $x = \dfrac{\pi}{4}$ and label the x-intercept $(0, 0)$.

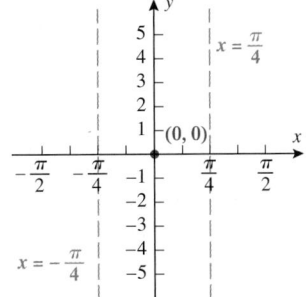

STEP 5 Divide the period $\dfrac{\pi}{2}$ into four equal parts, in steps of $\dfrac{\pi}{8}$. Set up a table with coordinates corresponding to values of $y = -3 \tan(2x)$.

x	$y = -3 \tan(2x)$	(x, y)
$-\dfrac{\pi}{4}$	undefined	vertical asymptote, $x = -\dfrac{\pi}{4}$
$-\dfrac{\pi}{8}$	3	$\left(-\dfrac{\pi}{8}, 3\right)$
0	0	$(0, 0)$
$\dfrac{\pi}{8}$	-3	$\left(\dfrac{\pi}{8}, -3\right)$
$\dfrac{\pi}{4}$	undefined	vertical asymptote, $x = \dfrac{\pi}{4}$

STEP 6 Graph the points from the table and connect with a smooth curve. Repeat to the right and left until you reach the interval endpoints.

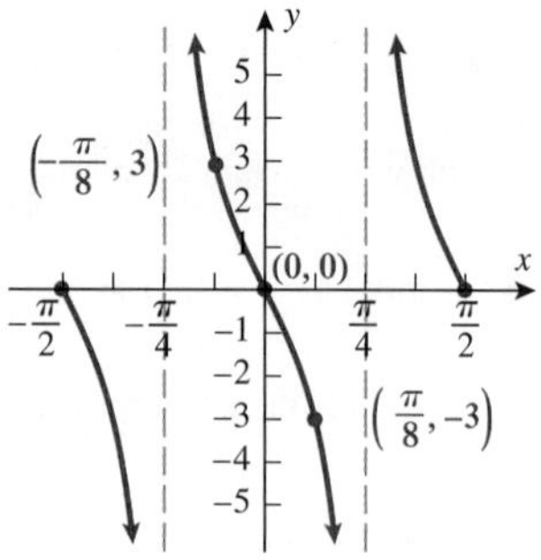

■ **YOUR TURN** Graph $y = \frac{1}{3}\tan\left(\frac{1}{2}x\right)$ on the interval $-\pi \leq x \leq \pi$.

■ **Answer:**

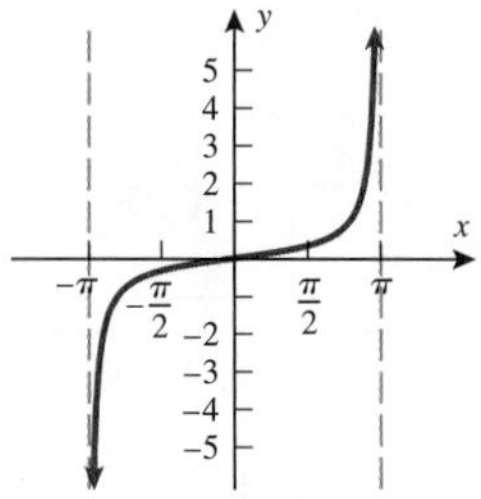

Technology Tip

To graph $y = 4\cot\left(\frac{1}{2}x\right)$ on the interval $-2\pi \leq x \leq 2\pi$, enter $y = 4\left[\tan\left(\frac{1}{2}x\right)\right]^{-1}$.

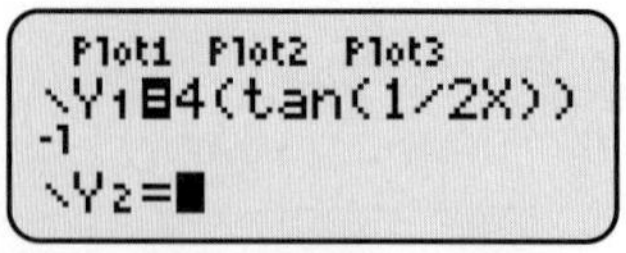

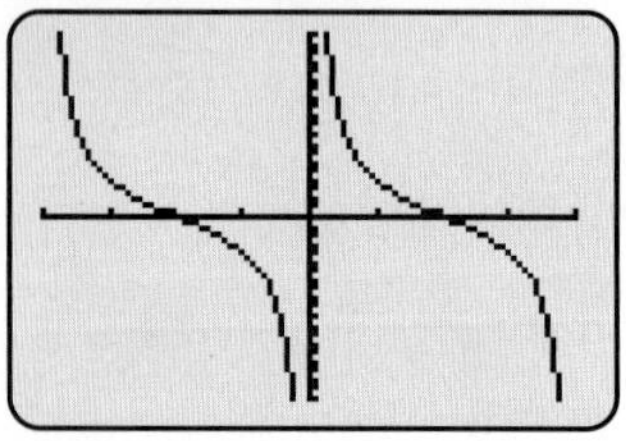

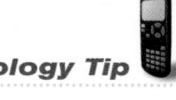

EXAMPLE 2 Graphing $y = A\cot Bx$

Graph $y = 4\cot\left(\frac{1}{2}x\right)$ on the interval $-2\pi \leq x \leq 2\pi$.

Solution: $A = 4,\ B = \frac{1}{2}$

STEP 1 Calculate the period. $\qquad \frac{\pi}{B} = 2\pi$

STEP 2 Find two vertical asymptotes. $\qquad Bx = 0$ and $Bx = \pi$

Substitute $B = \frac{1}{2}$ and solve for x. $\qquad x = 0$ and $x = 2\pi$

STEP 3 Find the x-intercept between the asymptotes. $\qquad Bx = \frac{\pi}{2}$

$x = \pi$

STEP 4 Draw the vertical asymptotes $x = 0$ and $x = 2\pi$ and label the x-intercept $(\pi, 0)$.

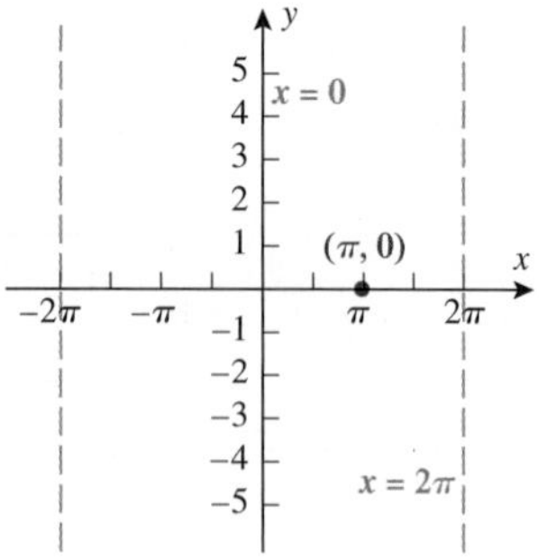

STEP 5 Divide the period 2π into four equal parts, in steps of $\dfrac{\pi}{2}$. Set up a table with coordinates corresponding to values of $y = 4\cot\left(\tfrac{1}{2}x\right)$.

x	$y = 4\cot\left(\tfrac{1}{2}x\right)$	(x, y)
0	undefined	vertical asymptote, $x = 0$
$\dfrac{\pi}{2}$	4	$\left(\dfrac{\pi}{2}, 4\right)$
π	0	$(\pi, 0)$
$\dfrac{3\pi}{2}$	-4	$\left(\dfrac{3\pi}{2}, -4\right)$
2π	undefined	vertical asymptote, $x = 2\pi$

STEP 6 Graph the points from the table and connect with a smooth curve. Repeat to the right and left until you reach the interval endpoints.

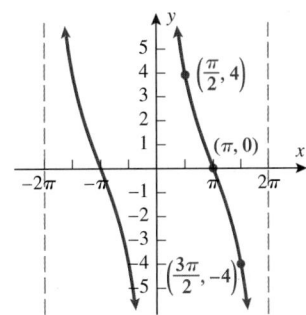

■ **YOUR TURN** Graph $y = 2\cot(2x)$ on the interval $-\pi \leq x \leq \pi$.

■ **Answer:**

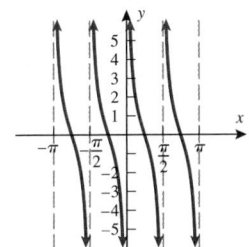

GRAPHING SECANT AND COSECANT FUNCTIONS

Graphs of $y = A\sec Bx$ and $y = A\csc Bx$ can be obtained using the following steps:

Step 1: Graph the corresponding guide function with a dashed curve.

For $y = A\sec Bx$, use $y = A\cos Bx$ as a guide.
For $y = A\csc Bx$, use $y = A\sin Bx$ as a guide.

Step 2: Draw the asymptotes, which correspond to the x-intercepts of the guide function.

Step 3: Draw the U shape between the asymptotes. If the guide function has a positive value between the asymptotes, the U opens upward; and if the guide function has a negative value, the U opens downward.

Technology Tip

To graph $y = 2\sec(\pi x)$ on the interval $-2 \le x \le 2$, enter $y = [2\cos(\pi x)]^{-1}$.

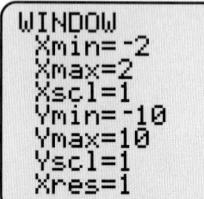

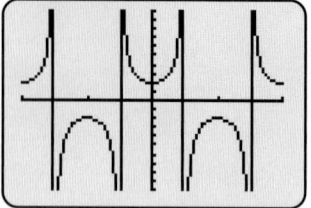

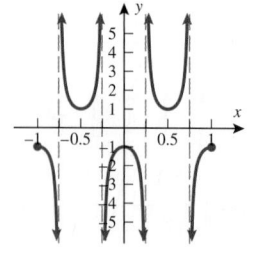

■ Answer:

EXAMPLE 3 Graphing $y = A\sec Bx$

Graph $y = 2\sec(\pi x)$ on the interval $-2 \le x \le 2$.

Solution:

STEP 1 Graph the corresponding guide function with a dashed curve.

For $y = 2\sec(\pi x)$, use $y = 2\cos(\pi x)$ as a guide.

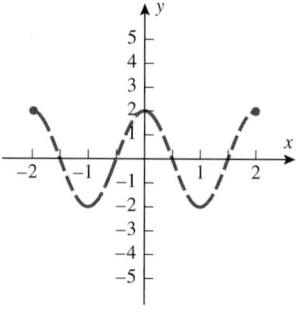

STEP 2 Draw the asymptotes, which correspond to the x-intercepts of the guide function.

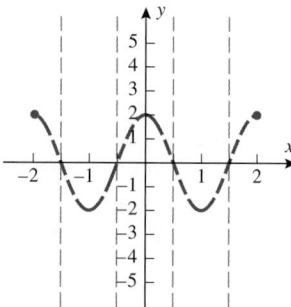

STEP 3 Use the U shape between the asymptotes. If the guide function is positive, the U opens upward, and if the guide function is negative, the U opens downward.

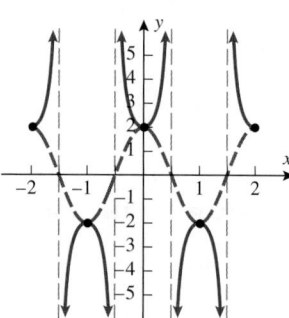

■ YOUR TURN Graph $y = -\sec(2\pi x)$ on the interval $-1 \le x \le 1$.

EXAMPLE 4 **Graphing $y = A \csc Bx$**

Graph $y = -3\csc(2\pi x)$ on the interval $-1 \le x \le 1$.

Solution:

STEP 1 Graph the corresponding guide function
with a dashed curve.

For $y = -3\csc(2\pi x)$, use
$y = -3\sin(2\pi x)$ as a guide.

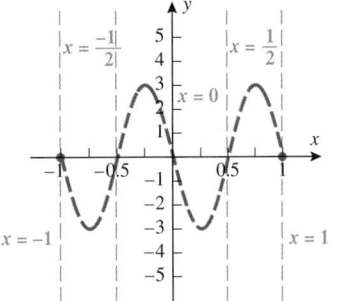

Technology Tip

To graph $y = -3\csc(2\pi x)$ on the
interval $-1 \le x \le 1$, enter
$y = -3[\sin(2\pi x)]^{-1}$.

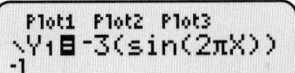

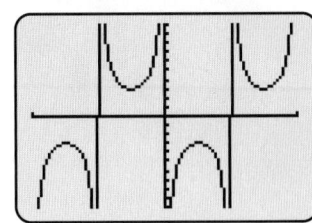

STEP 2 Draw the asymptotes, which
correspond to the x-intercepts
of the guide function.

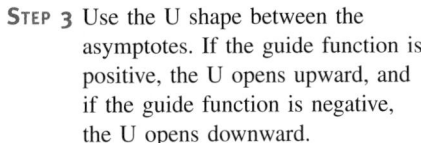

STEP 3 Use the U shape between the
asymptotes. If the guide function is
positive, the U opens upward, and
if the guide function is negative,
the U opens downward.

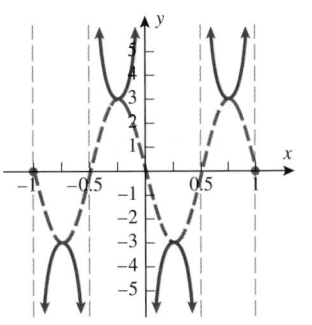

■ **YOUR TURN** Graph $y = \frac{1}{2}\csc(\pi x)$ on the interval $-1 \le x \le 1$.

■ **Answer:**

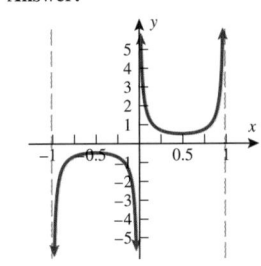

Translations of Trigonometric Functions

Vertical translations and horizontal translations (phase shifts) of the tangent, cotangent,
secant, and cosecant functions are graphed the same way as vertical and horizontal
translations of sinusoidal graphs are drawn. For tangent and cotangent functions, we
follow the same procedure as we did with sinusoidal functions. For secant and cotangent
functions, we graph the guide function first and then translate up or down depending on
the sign of the vertical shift.

EXAMPLE 5   Graphing $y = A \tan(Bx + C) + D$

Graph $y = 1 - \tan\left(x - \dfrac{\pi}{2}\right)$ on $-\pi \le x \le \pi$. State the domain and range on the interval.

There are two ways to approach graphing this function. Both will be illustrated.

Solution (1):

Plot $y = \tan x$, and then do the following:

- Shift the curve to the right $\dfrac{\pi}{2}$ units. $\qquad\qquad y = \tan\left(x - \dfrac{\pi}{2}\right)$

- Reflect the curve about the x-axis (because of the negative sign). $\qquad y = -\tan\left(x - \dfrac{\pi}{2}\right)$

- Shift the entire graph up 1 unit. $\qquad\qquad y = 1 - \tan\left(x - \dfrac{\pi}{2}\right)$

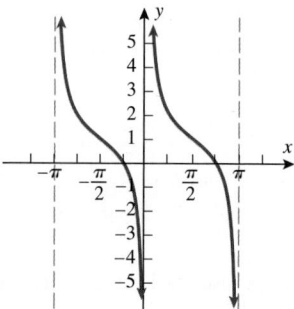

Solution (2): Graph $y = -\tan\left(x - \dfrac{\pi}{2}\right)$, and then shift the entire graph up 1 unit, because $D = 1$.

STEP 1 Calculate the period. $\qquad\qquad \dfrac{\pi}{B} = \pi$

STEP 2 Find two vertical asymptotes. $\qquad x - \dfrac{\pi}{2} = -\dfrac{\pi}{2}$ and $x - \dfrac{\pi}{2} = \dfrac{\pi}{2}$

Solve for x. $\qquad\qquad\qquad\qquad x = 0$ and $x = \pi$

STEP 3 Find the x-intercept between the asymptotes. $\qquad x - \dfrac{\pi}{2} = 0$

$\qquad\qquad\qquad\qquad\qquad\qquad x = \dfrac{\pi}{2}$

STEP 4 Draw the vertical asymptotes $x = 0$ and $x = \pi$ and label the x-intercept $\left(\dfrac{\pi}{2}, 0\right)$.

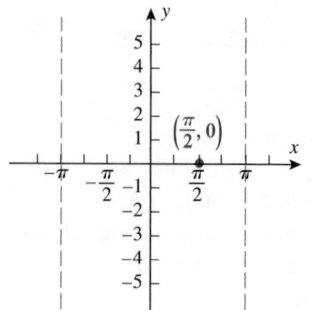

STEP 5 Divide the period π into four equal parts, in steps of $\dfrac{\pi}{4}$. Set up a table with coordinates corresponding to values of $y = -\tan\left(x - \dfrac{\pi}{2}\right)$ between the two asymptotes.

x	$y = -\tan\left(x - \dfrac{\pi}{2}\right)$	(x, y)
$x = 0$	undefined	vertical asymptote, $x = 0$
$\dfrac{\pi}{4}$	1	$\left(\dfrac{\pi}{4}, 1\right)$
$\dfrac{\pi}{2}$	0	$\left(\dfrac{\pi}{2}, 0\right)$
$\dfrac{3\pi}{4}$	-1	$\left(\dfrac{3\pi}{4}, -1\right)$
$x = \pi$	undefined	vertical asymptote, $x = \pi$

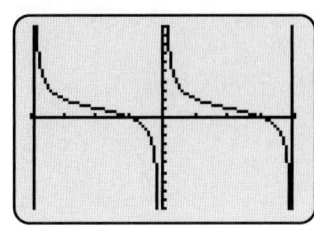
STEP 6 Graph the points from the table and connect with a smooth curve. Repeat to the right and left until reaching the interval endpoints.

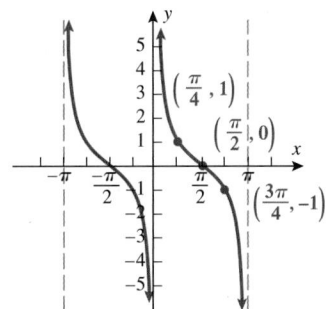

STEP 7 Shift the entire graph up 1 unit to arrive at the graph of

$$y = 1 - \tan\left(x - \dfrac{\pi}{2}\right).$$

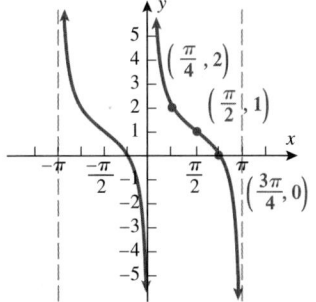

■ **Answer:**
Domain:

$$\left[-\pi, -\dfrac{\pi}{2}\right) \cup \left(-\dfrac{\pi}{2}, \dfrac{\pi}{2}\right) \cup \left(\dfrac{\pi}{2}, \pi\right]$$

Range: $(-\infty, \infty)$

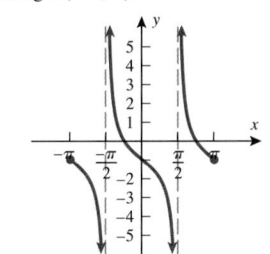

STEP 8 State the domain and range on the interval.

Domain: $(-\pi, 0) \cup (0, \pi)$
Range: $(-\infty, \infty)$

■ **YOUR TURN** Graph $y = -1 + \cot\left(x + \dfrac{\pi}{2}\right)$ on $-\pi \le x \le \pi$. State the domain and range on the interval.

EXAMPLE 6   Graphing $y = A \csc(Bx + C) + D$

Graph $y = 1 - \csc(2x - \pi)$ on $-\pi \le x \le \pi$. State the domain and range on the interval.

Solution:

Graph $y = -\csc(2x - \pi)$, and shift the entire graph up 1 unit to arrive at the graph of $y = 1 - \csc(2x - \pi)$.

STEP 1 Draw the guide function,
$$y = -\sin(2x - \pi).$$

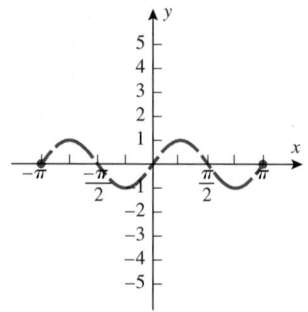

STEP 2 Draw the vertical asymptotes of
$y = -\csc(2x - \pi)$ that correspond
to the x-intercepts of
$y = -\sin(2x - \pi)$.

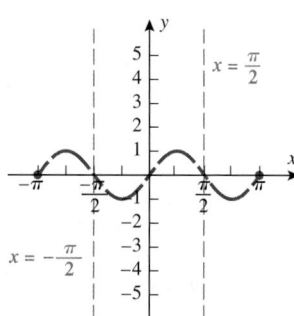

STEP 3 Use the U shape between the
asymptotes. If the guide function is
positive, the U opens upward, and if
the guide function is negative, the U
opens downward.

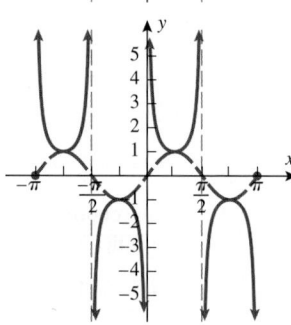

STEP 4 Shift the entire graph up 1 unit
to arrive at the graph of
$y = 1 - \csc(2x - \pi)$.

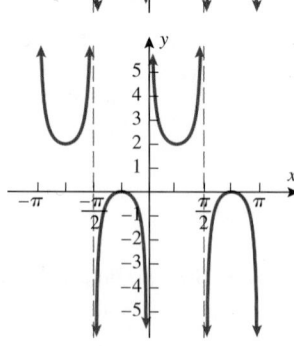

■ **Answer:**
Domain:

$$\left[-1, -\tfrac{1}{2}\right) \cup \left(-\tfrac{1}{2}, \tfrac{1}{2}\right) \cup \left(\tfrac{1}{2}, 1\right]$$

Range: $(-\infty, -3] \cup [-1, \infty)$

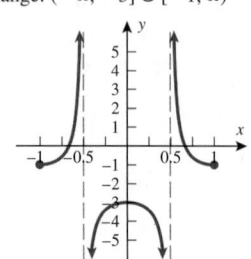

STEP 5 State the domain and
range on the interval.

Domain: $\left(-\pi, -\dfrac{\pi}{2}\right) \cup \left(-\dfrac{\pi}{2}, 0\right) \cup \left(0, \dfrac{\pi}{2}\right) \cup \left(\dfrac{\pi}{2}, \pi\right)$

Range: $(-\infty, 0] \cup [2, \infty)$

■ **YOUR TURN** Graph $y = -2 + \sec(\pi x - \pi)$ on $-1 \le x \le 1$. State the domain
and range on the interval.

SECTION

5.3 SUMMARY

The tangent and cotangent functions have period π, whereas the secant and cosecant functions have period 2π. To graph the tangent and cotangent functions, first identify the vertical asymptotes and x-intercepts, and then find values of the function within a period (i.e., between the asymptotes). To find graphs of secant and cosecant functions, first graph their guide functions (cosine and sine, respectively), and then label vertical asymptotes that correspond to x-intercepts of the guide function. The graphs of the secant and cosecant functions resemble the letter U opening up or down. The secant and cosecant functions are positive when their guide function is positive and negative when their guide function is negative.

SECTION

5.3 EXERCISES

■ SKILLS

In Exercises 1–8, match the graphs to the functions (a–h).

1. $y = -\tan x$

2. $y = -\csc x$

3. $y = \sec(2x)$

4. $y = \csc(2x)$

5. $y = \cot(\pi x)$

6. $y = -\cot(\pi x)$

7. $y = 3 \sec x$

8. $y = 3 \csc x$

a.

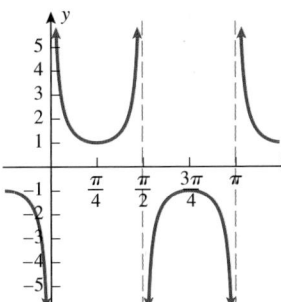

b.

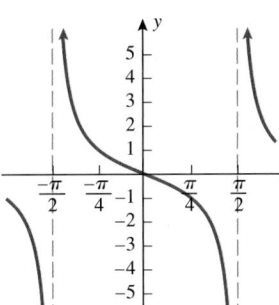

c.

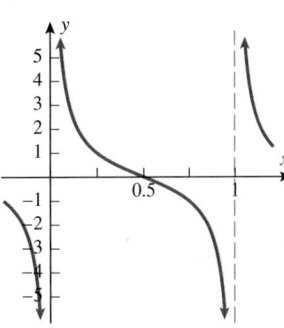

d.

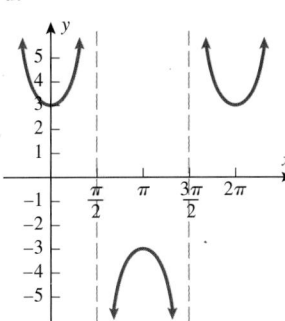

e.

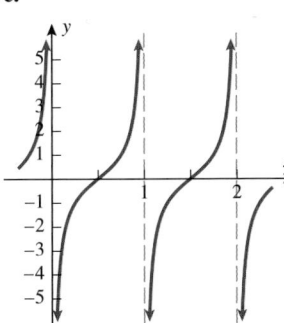

f.

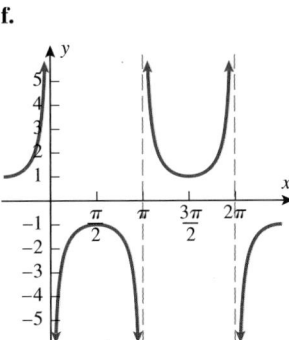

g.

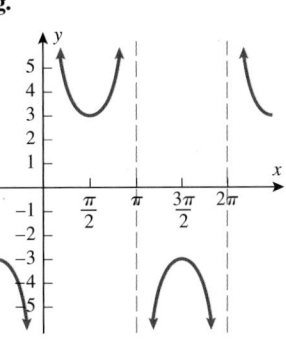

h.

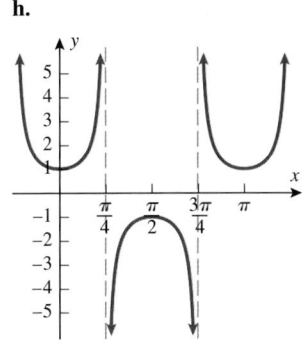

In Exercises 9–28, graph the functions over the indicated intervals.

9. $y = \tan(\tfrac{1}{2}x)$, $-2\pi \le x \le 2\pi$

10. $y = \cot(\tfrac{1}{2}x)$, $-2\pi \le x \le 2\pi$

11. $y = -\cot(2\pi x)$, $-1 \le x \le 1$

12. $y = -\tan(2\pi x)$, $-1 \le x \le 1$

13. $y = 2\tan(3x)$, $-\pi \le x \le \pi$

14. $y = 2\tan(\tfrac{1}{3}x)$, $-3\pi \le x \le 3\pi$

15. $y = -\dfrac{1}{4}\cot\left(\dfrac{x}{2}\right)$, $-2\pi \le x \le 2\pi$

16. $y = -\dfrac{1}{2}\tan\left(\dfrac{x}{4}\right)$, $-4\pi \le x \le 4\pi$

17. $y = -\tan\left(x - \dfrac{\pi}{2}\right)$, $-\pi \le x \le \pi$

18. $y = \tan\left(x + \dfrac{\pi}{4}\right)$, $-\pi \le x \le \pi$

19. $y = 2\tan\left(x + \dfrac{\pi}{6}\right)$, $-\pi \le x \le \pi$

20. $y = -\tfrac{1}{2}\tan(x + \pi)$, $-\pi \le x \le \pi$

21. $y = \cot\left(x - \dfrac{\pi}{4}\right)$, $-\pi \le x \le \pi$

22. $y = -\cot\left(x + \dfrac{\pi}{2}\right)$, $-\pi \le x \le \pi$

23. $y = -\dfrac{1}{2}\cot\left(x + \dfrac{\pi}{3}\right)$, $-\pi \le x \le \pi$

24. $y = 3\cot\left(x - \dfrac{\pi}{6}\right)$, $-\pi \le x \le \pi$

25. $y = \tan(2x - \pi)$, $-2\pi \le x \le 2\pi$

26. $y = \cot(2x - \pi)$, $-2\pi \le x \le 2\pi$

27. $y = \cot\left(\dfrac{x}{2} + \dfrac{\pi}{4}\right)$, $-\pi \le x \le \pi$

28. $y = \tan\left(\dfrac{x}{3} - \dfrac{\pi}{3}\right)$, $-\pi \le x \le \pi$

In Exercises 29–46, graph the functions over the indicated intervals.

29. $y = \sec(\tfrac{1}{2}x)$, $-2\pi \le x \le 2\pi$

30. $y = \csc(\tfrac{1}{2}x)$, $-2\pi \le x \le 2\pi$

31. $y = -\csc(2\pi x)$, $-1 \le x \le 1$

32. $y = -\sec(2\pi x)$, $-1 \le x \le 1$

33. $y = \dfrac{1}{3}\sec\left(\dfrac{\pi}{2}x\right)$, $-4 \le x \le 4$

34. $y = \dfrac{1}{2}\csc\left(\dfrac{\pi}{3}x\right)$, $-6 \le x \le 6$

35. $y = -3\csc\left(\dfrac{x}{3}\right)$, $-6\pi \le x \le 0$

36. $y = -4\sec\left(\dfrac{x}{2}\right)$, $-4\pi \le x \le 4\pi$

37. $y = 2\sec(3x)$, $0 \le x \le 2\pi$

38. $y = 2\csc(\tfrac{1}{3}x)$, $-3\pi \le x \le 3\pi$

39. $y = -3\csc\left(x - \dfrac{\pi}{2}\right)$, over at least one period

40. $y = 5\sec\left(x + \dfrac{\pi}{4}\right)$, over at least one period

41. $y = \tfrac{1}{2}\sec(x - \pi)$, over at least one period

42. $y = -4\csc(x + \pi)$, over at least one period

43. $y = 2\sec(2x - \pi)$, $-2\pi \le x \le 2\pi$

44. $y = 2\csc(2x + \pi)$, $-2\pi \le x \le 2\pi$

45. $y = -\tfrac{1}{4}\sec(3x + \pi)$

46. $y = -\dfrac{2}{3}\csc\left(4x - \dfrac{\pi}{2}\right)$, $-\pi \le x \le \pi$

In Exercises 47–56, graph the functions over at least one period.

47. $y = 3 - 2\sec\left(x - \dfrac{\pi}{2}\right)$

48. $y = -3 + 2\csc\left(x + \dfrac{\pi}{2}\right)$

49. $y = \dfrac{1}{2} + \dfrac{1}{2}\tan\left(x - \dfrac{\pi}{2}\right)$

50. $y = \dfrac{3}{4} - \dfrac{1}{4}\cot\left(x + \dfrac{\pi}{2}\right)$

51. $y = -2 + 3\csc(2x - \pi)$

52. $y = -1 + 4\sec(2x + \pi)$

53. $y = -1 - \sec\left(\dfrac{1}{2}x - \dfrac{\pi}{4}\right)$

54. $y = -2 + \csc\left(\dfrac{1}{2}x + \dfrac{\pi}{4}\right)$

55. $y = -2 - 3\cot\left(2x - \dfrac{\pi}{4}\right)$, $-\pi \le x \le \pi$

56. $y = -\dfrac{1}{4} + \dfrac{1}{2}\sec\left(\pi x + \dfrac{\pi}{4}\right)$, $-2 \le x \le 2$

In Exercises 57–66, state the domain and range of the functions.

57. $y = \tan\left(\pi x - \dfrac{\pi}{2}\right)$

58. $y = \cot\left(x - \dfrac{\pi}{2}\right)$

59. $y = 2\sec(5x)$

60. $y = -4\sec(3x)$

61. $y = 2 - \csc\left(\tfrac{1}{2}x - \pi\right)$

62. $y = 1 - 2\sec\left(\tfrac{1}{2}x + \pi\right)$

63. $y = -3\tan\left(\dfrac{\pi}{4}x - \pi\right) + 1$

64. $y = \dfrac{1}{4}\cot\left(2\pi x + \dfrac{\pi}{3}\right) - 3$

65. $y = -2 + \dfrac{1}{2}\sec\left(\pi x + \dfrac{\pi}{2}\right)$

66. $y = \dfrac{1}{2} - \dfrac{1}{3}\csc\left(3x - \dfrac{\pi}{2}\right)$

■ APPLICATIONS

67. Tower of Pisa. The angle between the ground and the Tower of Pisa is about 85°. Its inclination measured at the base is 4.2 m. What is the vertical distance from the top of the tower to the ground?

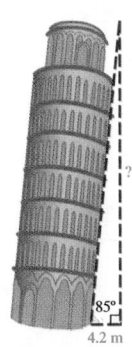

68. Architecture. The angle of elevation from the top of a building 40 ft tall to the top of another building 75 ft tall is $\dfrac{\pi}{6}$. What is the distance between the buildings?

69. Lighthouse. A lighthouse is located on a small island 3 miles offshore. The distance x is given by $x = 3\tan(\pi t)$,

where t is the time measured in seconds. Suppose that at midnight the light beam forms a straight angle with the shoreline. Find x at

a. $t = \tfrac{2}{3}$ s   **b.** $t = \tfrac{3}{4}$ s   **c.** 1 s

d. $t = \tfrac{5}{4}$ s   **e.** $t = \tfrac{4}{3}$ s

Round to the nearest length.

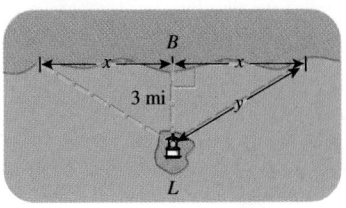

70. Lighthouse. If the length of the light beam is determined by $y = 3|\sec(\pi t)|$, find y at

a. $t = \tfrac{2}{3}$ s   **b.** $t = \tfrac{3}{4}$ s   **c.** 1 s

d. $t = \tfrac{5}{4}$ s   **e.** $t = \tfrac{4}{3}$ s

Round to the nearest length.

■ CATCH THE MISTAKE

In Exercises 71 and 72, explain the mistake that is made.

71. Graph $y = 3\csc(2x)$.

Solution:

Graph the guide function, $y = \sin(2x)$.

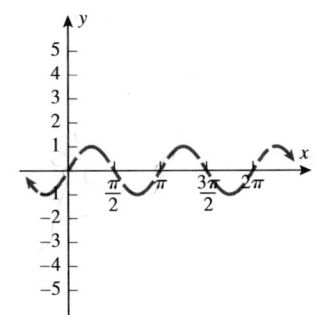

Draw vertical asymptotes at x-values that correspond to x-intercepts of the guide function.

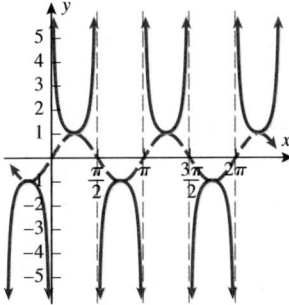

Draw the cosecant function.

This is incorrect. What mistake was made?

72. Graph $y = \tan(4x)$.

Solution:

Step 1: Calculate the period. $\dfrac{\pi}{B} = \dfrac{\pi}{4}$

Step 2: Find two vertical asymptotes. $4x = 0$ and $4x = \pi$

Solve for x. $x = 0$ and $x = \dfrac{\pi}{4}$

Step 3: Find the x-intercept between the asymptotes. $4x = \dfrac{\pi}{2}$

$x = \dfrac{\pi}{8}$

Step 4: Draw the vertical asymptotes $x = 0$ and $x = \dfrac{\pi}{4}$

and label the x-intercept $\left(\dfrac{\pi}{8}, 0\right)$.

Step 5: Graph.

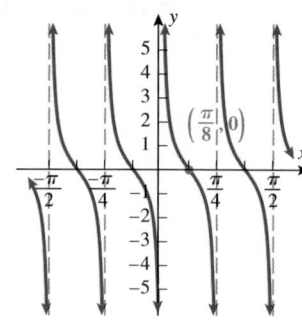

This is incorrect. What mistake was made?

CONCEPTUAL

In Exercises 73 and 74, determine whether each statement is true or false.

73. $\sec\left(x - \dfrac{\pi}{2}\right) = \csc x$

74. $\csc\left(x - \dfrac{\pi}{2}\right) = \sec x$

75. For what values of n do $y = \tan x$ and $y = \tan(x - n\pi)$ have the same graph?

76. For what values of n do $y = \csc x$ and $y = \csc(x - n\pi)$ have the same graph?

77. Solve the equation $\tan(2x - \pi) = 0$ for x in the interval $[-\pi, \pi]$ by graphing.

78. Solve the equation $\csc(2x + \pi) = 0$ for x in the interval $[-\pi, \pi]$ by graphing.

79. Find the x-intercepts of $y = A\tan(Bx + C)$.

80. For what x-values does the graph of $y = -A\sec\left(\dfrac{\pi}{2}x\right)$ lie above the x-axis? (Assume $A > 0$.)

81. How many solutions are there to the equation $\tan x = x$? Explain.

82. For what values of A do the graphs of $y = A\sin(Bx + C)$ and $y = -2\csc\left(\dfrac{\pi}{6}x - \pi\right)$ never intersect?

TECHNOLOGY

83. What is the amplitude of the function $y = \cos x + \sin x$? Use a graphing calculator to graph $Y_1 = \cos x$, $Y_2 = \sin x$, and $Y_3 = \cos x + \sin x$ in the same viewing window.

84. Graph $Y_1 = \cos x + \sin x$ and $Y_2 = \sec x + \csc x$ in the same viewing window. Based on what you see, is $Y_1 = \cos x + \sin x$ the guide function for $Y_2 = \sec x + \csc x$?

85. What is the period of the function $y = \tan x + \cot x$? Use a graphing calculator to graph $Y_1 = \tan x + \cot x$ and $Y_3 = 2\csc(2x)$ in the same viewing window.

86. What is the period of the function $y = \tan\left(2x + \dfrac{\pi}{2}\right)$? Use a graphing calculator to graph $Y_1 = \tan\left(2x + \dfrac{\pi}{2}\right)$, $Y_2 = \tan\left(2x - \dfrac{\pi}{2}\right)$, and $Y_3 = \tan\left(-2x + \dfrac{\pi}{2}\right)$ in the same viewing window. Describe the relationships of Y_1 and Y_2 and Y_2 and Y_3.

▪ PREVIEW TO CALCULUS

In calculus, the definite integral $\int_a^b f(x)\,dx$ is used to find the area below the graph of a continuous function f, above the x-axis, and between $x = a$ and $x = b$. The Fundamental Theorem of Calculus establishes that the definite integral $\int_a^b f(x)\,dx$ equals $F(b) - F(a)$, where F is any antiderivative of a continuous function f.

In Exercises 87–90, first shade the area corresponding to the definite integral and then use the information below to find the exact value of the area.

FUNCTION	$\tan x$	$\cot x$	$\sec x$	$\csc x$
ANTIDERIVATIVE	$-\ln|\cos x|$	$\ln|\sin x|$	$\ln|\sec x + \tan x|$	$-\ln|\csc x + \cot x|$

87. $\displaystyle\int_0^{\pi/4} \tan x\,dx$ **88.** $\displaystyle\int_{\pi/4}^{\pi/2} \cot x\,dx$ **89.** $\displaystyle\int_0^{\pi/4} \sec x\,dx$ **90.** $\displaystyle\int_{\pi/4}^{\pi/2} \csc x\,dx$

MODELING YOUR WORLD

Some would argue that temperatures are oscillatory in nature: that we are not experiencing global warming at all, but instead a natural cycle in Earth's temperature. In the Modeling Your World features in Chapters 1–3, you modeled mean temperatures with linear, polynomial, and logarithmic models—all of which modeled increasing temperatures. Looking at the data another way, we can demonstrate—over a short period of time—how it might appear to be oscillatory, which we can model with a sinusoidal curve. Be warned, however, that even the stock market, which increases gradually over time, looks sinusoidal over a short period.

The following table summarizes average yearly temperature in degree Fahrenheit (°F) and carbon dioxide emissions in parts per million (ppm) for the Mauna Loa Observatory in Hawaii.

YEAR	1960	1965	1970	1975	1980	1985	1990	1995	2000	2005
Temperature	44.45	43.29	43.61	43.35	46.66	45.71	45.53	47.53	45.86	46.23
CO_2 Emissions (ppm)	316.9	320.0	325.7	331.1	338.7	345.9	354.2	360.6	369.4	379.7

1. Find a *sinusoidal function* of the form $f(t) = k + A\sin Bt$ that models the temperature in Mauna Loa. Assume that the peak amplitude occurs in 1980 and again in 2010 (at 46.66°). Let $t = 0$ correspond to 1960.

2. Do your models support the claim of global warming? Explain.

3. You have modeled these same data with different types of models—oscillatory and nonoscillatory. Over the short 30-year period from 1980 to 2010, you have a sinusoidal model that can fit the data, but does it prove anything in the long term?

4. Would a resonance-type sinusoidal function of the form $f(t) = k + At\sin Bt$ perhaps be a better fit for 1960 to 2005? Develop a function of this form that models the data.

CHAPTER 5 REVIEW

SECTION	CONCEPT	PAGES	REVIEW EXERCISES	KEY IDEAS/POINTS
5.1	**Trigonometric functions: The unit circle approach**	456–461	1–20	
	Trigonometric functions and the unit circle	456–457	1–20	
	Circular functions	457–459	1–20	
	Properties of circular functions	459–461	13–20	Cosine is an even function. $\cos(-\theta) = \cos\theta$ Sine is an odd function. $\sin(-\theta) = -\sin\theta$

SECTION	CONCEPT	PAGES	REVIEW EXERCISES	KEY IDEAS/POINTS
5.2	**Graphs of sine and cosine functions**	465–488	21–44	
	The graph of $f(x) = \sin x$	465–467	21–26	Odd function: $f(-x) = -f(x)$
	The graph of $f(x) = \cos x$	468–469	21–26	Even function: $f(-x) = f(x)$
	The amplitude and period of sinusoidal graphs	470–477	27–34	$y = A\sin Bx$ or $y = A\cos Bx$, $B > 0$ Amplitude $= \|A\|$ ■ $\|A\| > 1$ stretch vertically. ■ $\|A\| < 1$ compress vertically. Period $= \dfrac{2\pi}{B}$ ■ $B > 1$ compress horizontally. ■ $B < 1$ stretch horizontally.
	Graphing a shifted sinusoidal function: $y = A\sin(Bx + C) + D$ and $y = A\cos(Bx + C) + D$	477–479	35–40	■ $y = A\sin(Bx \pm C) = A\sin\left[B\left(x \pm \dfrac{C}{B}\right)\right]$ has period $\dfrac{2\pi}{B}$ and a phase shift of $\dfrac{C}{B}$ units to the left $(+)$ or the right $(-)$. ■ $y = A\cos(Bx \pm C) = A\cos\left[B\left(x \pm \dfrac{C}{B}\right)\right]$ has period $\dfrac{2\pi}{B}$ and a phase shift of $\dfrac{C}{B}$ units to the left $(+)$ or the right $(-)$. ■ To graph $y = A\sin(Bx + C) + D$ or $y = A\cos(Bx + C) + D$, start with the graph of $y = A\sin(Bx + C)$ or $y = A\cos(Bx + C)$ and shift up or down D units.

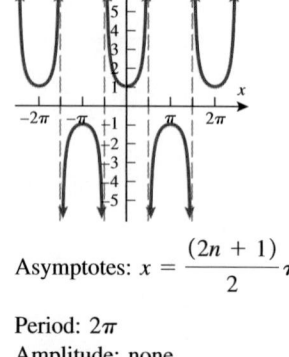

SECTION	CONCEPT	PAGES	REVIEW EXERCISES	KEY IDEAS/POINTS
	The cosecant function	500–501	45–56	Asymptotes: $x = n\pi$ Period: 2π Amplitude: none x-intercepts: none
	Graphing tangent, cotangent, secant, and cosecant functions	502–507	45–56	
	Translations of trigonometric functions	507–510	45–56	$y = A\tan(Bx + C)$ or $y = A\cot(Bx + C)$ To find asymptotes, set $Bx + C$ equal to ■ $-\dfrac{\pi}{2}$ and $\dfrac{\pi}{2}$ for tangent. ■ 0 and π for cotangent. To find x-intercepts, set $Bx + C$ equal to ■ 0 for tangent. ■ $\dfrac{\pi}{2}$ for cotangent. $y = A\sec(Bx + C)$ or $y = A\csc(Bx + C)$ To graph $y = A\sec(Bx + C)$, use $y = A\cos(Bx + C)$ as the guide. To graph $y = A\csc(Bx + C)$, use $y = A\sin(Bx + C)$ as the guide. Intercepts on the guide function correspond to vertical asymptotes of secant or cosecant functions.

5.1 Trigonometric Functions: The Unit Circle Approach

Find each trigonometric function value in *exact* form.

1. $\tan\left(\dfrac{5\pi}{6}\right)$
2. $\cos\left(\dfrac{5\pi}{6}\right)$
3. $\sin\left(\dfrac{11\pi}{6}\right)$

4. $\sec\left(\dfrac{11\pi}{6}\right)$
5. $\cot\left(\dfrac{5\pi}{4}\right)$
6. $\csc\left(\dfrac{5\pi}{4}\right)$

7. $\sin\left(\dfrac{3\pi}{2}\right)$
8. $\cos\left(\dfrac{3\pi}{2}\right)$
9. $\cos\pi$

10. $\tan\left(\dfrac{7\pi}{4}\right)$
11. $\cos\left(\dfrac{\pi}{3}\right)$
12. $\sin\left(\dfrac{11\pi}{6}\right)$

13. $\sin\left(-\dfrac{7\pi}{4}\right)$
14. $\tan\left(-\dfrac{2\pi}{3}\right)$
15. $\csc\left(-\dfrac{3\pi}{2}\right)$

16. $\cot\left(-\dfrac{5\pi}{6}\right)$
17. $\cos\left(-\dfrac{7\pi}{6}\right)$
18. $\sec\left(-\dfrac{3\pi}{4}\right)$

19. $\tan\left(-\dfrac{13\pi}{6}\right)$
20. $\cos\left(-\dfrac{14\pi}{3}\right)$

5.2 Graphs of Sine and Cosine Functions

Refer to the graph of the sinusoidal function to answer the questions.

21. Determine the period of the function.

22. Determine the amplitude of the function.

23. Write an equation for the sinusoidal function.

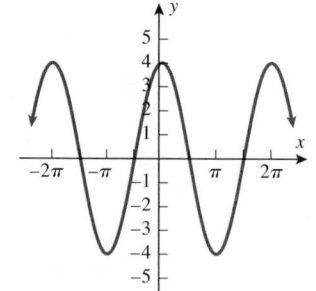

Refer to the graph of the sinusoidal function to answer the questions.

24. Determine the period of the function.

25. Determine the amplitude of the function.

26. Write an equation for the sinusoidal function.

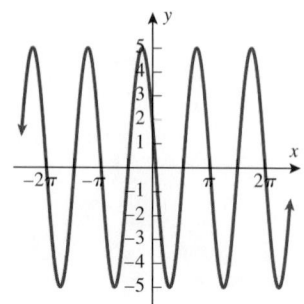

Determine the amplitude and period of each function.

27. $y = -2\cos(2\pi x)$
28. $y = \dfrac{1}{3}\sin\left(\dfrac{\pi}{2}x\right)$

29. $y = \dfrac{1}{5}\sin(3x)$
30. $y = -\dfrac{7}{6}\cos(6x)$

Graph each function from -2π to 2π.

31. $y = -2\sin\left(\dfrac{x}{2}\right)$
32. $y = 3\sin(3x)$

33. $y = \dfrac{1}{2}\cos(2x)$
34. $y = -\dfrac{1}{4}\cos\left(\dfrac{x}{2}\right)$

State the amplitude, period, phase shift, and vertical shift of each function.

35. $y = 2 + 3\sin\left(x - \dfrac{\pi}{2}\right)$

36. $y = 3 - \dfrac{1}{2}\sin\left(x + \dfrac{\pi}{4}\right)$

37. $y = -2 - 4\cos\left[3\left(x + \dfrac{\pi}{4}\right)\right]$

38. $y = -1 + 2\cos\left[2\left(x - \dfrac{\pi}{3}\right)\right]$

39. $y = -\dfrac{1}{2} + \dfrac{1}{3}\cos\left(\pi x - \dfrac{1}{2}\right)$

40. $y = \dfrac{3}{4} - \dfrac{1}{6}\sin\left(\dfrac{\pi}{6}x + \dfrac{\pi}{3}\right)$

Graph each function from $-\pi$ to π.

41. $y = 3x - \cos(2x)$

42. $y = -\dfrac{1}{2}\cos(4x) + \dfrac{1}{2}\cos(2x)$

43. $y = 2\sin\left(\dfrac{1}{3}x\right) - 3\sin(3x)$

44. $y = 5\cos x + 3\sin\left(\dfrac{x}{2}\right)$

5.3 Graphs of Other Trigonometric Functions

State the domain and range of each function.

45. $y = 4\tan\left(x + \dfrac{\pi}{2}\right)$

46. $y = \cot 2\left(x - \dfrac{\pi}{2}\right)$

47. $y = 3\sec(2x)$

48. $y = 1 + 2\csc x$

49. $y = -\dfrac{1}{2} + \dfrac{1}{4}\sec\left(\pi x - \dfrac{2\pi}{3}\right)$

50. $y = 3 - \dfrac{1}{2}\csc(2x - \pi)$

Graph each function on the interval $[-2\pi, 2\pi]$.

51. $y = -\tan\left(x - \dfrac{\pi}{4}\right)$

52. $y = 1 + \cot(2x)$

53. $y = 2 + \sec(x - \pi)$

54. $y = -\csc\left(x + \dfrac{\pi}{4}\right)$

55. $y = \dfrac{1}{2} + 2\csc\left(2x - \dfrac{\pi}{2}\right)$

56. $y = -1 - \dfrac{1}{2}\sec\left(\pi x - \dfrac{3\pi}{4}\right)$

Technology Exercises

Section 5.1

In Exercises 57 and 58, refer to the following:

A graphing calculator can be used to graph the unit circle with parametric equations (these will be covered in more detail in Section 8.7). For now, set the calculator in parametric and radian modes and let

$$X_1 = \cos T$$
$$Y_1 = \sin T$$

Set the window so that $0 \le T \le 2\pi$, step $= \dfrac{\pi}{15}$, $-2 \le X \le 2$, and $-2 \le Y \le 2$. To approximate the sine or cosines of a T value, use the $\boxed{\text{TRACE}}$ key, enter the T value, and read the corresponding coordinates from the screen.

57. Use the above steps to approximate $\cos\left(\dfrac{13\pi}{12}\right)$ to four decimal places.

58. Use the above steps to approximate $\sin\left(\dfrac{5\pi}{6}\right)$ to four decimal places.

Section 5.2

59. Use a graphing calculator to graph $Y_1 = \cos x$ and $Y_2 = \cos(x + c)$, where

  a. $c = \dfrac{\pi}{6}$, and explain the relationship between Y_2 and Y_1.

  b. $c = -\dfrac{\pi}{6}$, and explain the relationship between Y_2 and Y_1.

60. Use a graphing calculator to graph $Y_1 = \sin x$ and $Y_2 = \sin x + c$, where

  a. $c = \dfrac{1}{2}$, and explain the relationship between Y_2 and Y_1.

  b. $c = -\dfrac{1}{2}$, and explain the relationship between Y_2 and Y_1.

Section 5.3

61. What is the amplitude of the function $y = 4\cos x - 3\sin x$? Use a graphing calculator to graph $Y_1 = 4\cos x$, $Y_2 = 3\sin x$, and $Y_3 = 4\cos x - 3\sin x$ in the same viewing window.

62. What is the amplitude of the function $y = \sqrt{3}\sin x + \cos x$? Use a graphing calculator to graph $Y_1 = \sqrt{3}\sin x$, $Y_2 = \cos x$, and $Y_3 = \sqrt{3}\sin x + \cos x$ in the same viewing window.

REVIEW EXERCISES

1. State the amplitude and period of $y = -5\sin(3x)$.

2. Graph $y = -2\cos(\frac{1}{2}x)$ over $-4\pi \le x \le 4\pi$.

3. Graph $y = 1 + 3\sin(x + \pi)$ over $-3\pi \le x \le 3\pi$.

4. Graph $y = 4 - \sin\left(x - \frac{\pi}{2}\right)$ over $-6\pi \le x \le 6\pi$.

5. Graph $y = -2 - \cos\left(x + \frac{\pi}{2}\right)$ over $-4\pi \le x \le 4\pi$.

6. Graph $y = 3 + 2\cos\left(x + \frac{3\pi}{2}\right)$ over $-5\pi \le x \le 5\pi$.

7. Graph $y = \tan\left(\pi x - \frac{\pi}{2}\right)$ over two periods.

8. The vertical asymptotes of $y = 2\csc(3x - \pi)$ correspond to the _____ of $y = 2\sin(3x - \pi)$.

9. State the x-intercepts of $y = \tan(2x)$ for all x.

10. State the phase shift and vertical shift for
$y = -\cot\left(\frac{\pi}{3}x - \pi\right)$.

11. State the range of $y = -3\sec\left(2x + \frac{\pi}{3}\right) - 1$.

12. State the domain of $y = \tan\left(2x - \frac{\pi}{6}\right) + 3$.

13. Graph $y = -2\csc\left(x + \frac{\pi}{2}\right)$ over two periods.

14. Find the x-intercept(s) of $y = \frac{6}{\sqrt{3}} - 3\sec\left(6x - \frac{5\pi}{6}\right)$.

15. True or false: The equation $2\sin\theta = 2.0001$ has no solution.

16. On what x-intervals does the graph of $y = \cos(2x)$ lie below the x-axis?

17. Write the equation of a sine function that has amplitude 4, vertical shift $\frac{1}{2}$ down, phase shift $\frac{3}{2}$ to the left, and period π.

18. Write the equation of a cotangent function that has period π, vertical shift 0.01 up, and no phase shift.

19. Graph $y = \cos(3x) - \frac{1}{2}\sin(3x)$ for $0 \le x \le \pi$.

20. $y = -\frac{1}{5}\cos\left(\frac{x}{3}\right)$
    a. Graph the function over one period.
    b. Determine the amplitude, period, and phase shift.

21. $y = 4\sin(2\pi x)$
    a. Graph the function over one period.
    b. Determine the amplitude, period, and phase shift.

22. $y = -2\sin(3x + 4\pi) + 1$
    a. Write the sinusoidal function in standard form.
    b. Determine its amplitude, period, and phase shift.
    c. Graph the function over one period.

23. $y = 6 + 5\cos(2x - \pi)$
    a. Write the sinusoidal function in standard form.
    b. Determine its amplitude, period, and phase shift.
    c. Graph the function over one period.

24. Graph the function $y = 2\cos x - \sin x$ on the interval $[-\pi, \pi]$ by adding the ordinates of each individual function.

1. Find the domain of $f(x) = \dfrac{4}{\sqrt{15 + 3x}}$. Express the domain in interval notation.

2. On February 24, the gasoline price (per gallon) was $1.94; on May 24, its price per gallon was $2.39. Find the average rate of change per month in the gasoline price from February 24 to May 24.

3. Write the function whose graph is the graph of $y = |x|$, but stretched by a factor of 2, shifted up 4 units, and shifted to the left 6 units. Graph the function on the interval $[-10, 10]$.

4. Given $f(x) = 2x - 5$ and $g(x) = x^2 + 7$, find

   a. $f + g$

   b. $f - g$

   c. $f \cdot g$

   d. f/g

   e. $f \circ g$

5. Find the inverse function of the one-to-one function $f(x) = \dfrac{x - 2}{3x + 5}$. Estimate the domain and range of both f and f^{-1}.

6. For $f(x) = -x^2(2x - 6)^3(x + 5)^4$

   a. list each zero and its multiplicity.

   b. sketch the graph.

7. Divide $(6x^4 - 5x^3 + 6x^2 + 7x - 4) \div (2x^2 - 1)$ using long division. Express the answer in the form $Q(x) = ?$ and $r(x) = ?$.

8. For the polynomial function $f(x) = x^4 + x^3 - 7x^2 - x + 6$

   a. factor as a product of linear and/or irreducible quadratic factors.

   b. graph the function.

9. Factor the polynomial $P(x) = x^4 - 4x^2 - 5$ as a product of linear factors.

10. Given $f(x) = \dfrac{x^2 - 5x - 14}{2x^2 + 14x + 20}$,

   a. determine the vertical, horizontal, or slant asymptotes (if they exist).

   b. graph the function.

11. If $3,000 is deposited into an account paying 1.2% compounding quarterly, how much will you have in the account in 10 years?

12. Approximate $\log_2 19$ utilizing a calculator. Round to two decimal places.

13. Write $\ln\left(\dfrac{a^3}{b^2 c^5}\right)$ as a sum or difference of logarithms.

14. Solve the exponential equation $4^{3x-2} + 5 = 23$. Round your answer to three decimal places.

15. Solve the exponential equation $\log(x + 2) + \log(x + 3) = \log(2x + 10)$. Round your answer to three decimal places.

16. Find the area of a circular sector with radius $r = 6.5$ cm and central angle $\theta = \dfrac{4\pi}{5}$. Round your answer to the nearest integer.

17. Solve the right triangle $\beta = 27°$, $c = 14$ in. Round your answer to the nearest hundredth.

18. The terminal side of an angle θ in standard position passes through the point $(3, -2)$. Calculate the exact value of the six trigonometric functions for angle θ.

19. Solve the triangle $\alpha = 68°$, $a = 24$ m, and $b = 24.5$ m.

20. Solve the triangle $a = 5$, $b = 6$, and $c = 7$.

21. Given that $\sin \theta = \dfrac{1}{2}$ and $\dfrac{\pi}{2} < \theta < \pi$, find the exact value of all the other trigonometric functions.

22. Graph the function $y = 4\cos(2x + \pi)$ over one period.

23. The frequency of the oscillations f is given by $f = \dfrac{1}{p}$, where p is the period. What is the frequency of the oscillations modeled by $y = 1.14 \sin(4t)$?

24. Graph the function $y = -2 + 5\csc\left(4x - \dfrac{\pi}{2}\right)$ over two periods. State the range of the function.

6

Analytic Trigonometry

When you press a touch-tone button to dial a phone number, how does the phone system know which key you have pressed? Dual Tone Multi-Frequency (DTMF), also known as Touch-Tone dialing, was developed by Bell Labs in the 1960s. The Touch-Tone system also introduced a standardized *keypad* layout.

The keypad is laid out in a 4×3 matrix, with each row representing a low frequency and each column representing a high frequency.

FREQUENCY	1209 Hz	1336 Hz	1477 Hz
697 Hz	1	2	3
770 Hz	4	5	6
852 Hz	7	8	9
941 Hz	*	0	#

When you press the number 8, the phone sends a sinusoidal tone that combines a low-frequency tone of 852 Hz and a high-frequency tone of 1336 Hz. The result can be found using sum-to-product *trigonometric identities*.

IN THIS CHAPTER we will verify trigonometric identities. Specific identities that we will discuss are sum and difference, double-angle and half-angle, and product-to-sum and sum-to-product. Inverse trigonometric functions will be defined. Trigonometric identities and inverse trigonometric functions will be used to solve trigonometric equations.

ANALYTIC TRIGONOMETRY

6.1 Verifying Trigonometric Identities	6.2 Sum and Difference Identities	6.3 Double-Angle and Half-Angle Identities	6.4 Product-to-Sum and Sum-to-Product Identities	6.5 Inverse Trigonometric Functions	6.6 Trigonometric Equations
• Fundamental Identities • Simplifying Trigonometric Expressions Using Identities • Verifying Identities	• Sum and Difference Identities for the Cosine Function • Sum and Difference Identities for the Sine Function • Sum and Difference Identities for the Tangent Function	• Double-Angle Identities • Half-Angle Identities	• Product-to-Sum Identities • Sum-to-Product Identities	• Inverse Sine Function • Inverse Cosine Function • Inverse Tangent Function • Remaining Inverse Trigonometric Functions • Finding Exact Values for Expressions Involving Inverse Trigonometric Functions	• Solving Trigonometric Equations by Inspection • Solving Trigonometric Equations Using Algebraic Techniques • Solving Trigonometric Equations That Require the Use of Inverse Functions • Using Trigonometric Identities to Solve Trigonometric Equations

CHAPTER OBJECTIVES

- Verify trigonometric identities (Section 6.1).
- Use the sum and difference identities to simplify trigonometric expressions (Section 6.2).
- Use the double-angle and half-angle identities to simplify trigonometric expressions (Section 6.3).
- Use the product-to-sum and sum-to-product identities to simplify trigonometric expressions (Section 6.4).
- Evaluate the inverse trigonometric functions for specific values (Section 6.5).
- Solve trigonometric equations (Section 6.6).

SKILLS OBJECTIVES

- Apply fundamental identities.
- Simplify trigonometric expressions using identities.
- Verify trigonometric identities.

CONCEPTUAL OBJECTIVES

- Understand that there is more than one way to verify an identity.
- Understand that identities must hold for all values in the domain of the functions that are related by the identities.

Fundamental Identities

Study Tip

Just because an equation is true for *some* values of *x* does not mean it is an identity.

In mathematics, an **identity** is an equation that is true for *all* values of the variable for which the expressions in the equation are defined. If an equation is true for only *some* values of the variable, it is a **conditional equation**.

The following boxes summarize trigonometric identities that have been discussed in Chapters 4 and 5.

RECIPROCAL IDENTITIES

RECIPROCAL IDENTITIES	EQUIVALENT FORMS	DOMAIN RESTRICTIONS
$\csc x = \dfrac{1}{\sin x}$	$\sin x = \dfrac{1}{\csc x}$	$x \neq n\pi \quad n = $ integer
$\sec x = \dfrac{1}{\cos x}$	$\cos x = \dfrac{1}{\sec x}$	$x \neq \dfrac{n\pi}{2} \quad n = $ odd integer
$\cot x = \dfrac{1}{\tan x}$	$\tan x = \dfrac{1}{\cot x}$	$x \neq \dfrac{n\pi}{2} \quad n = $ integer

QUOTIENT IDENTITIES

QUOTIENT IDENTITIES	DOMAIN RESTRICTIONS
$\tan x = \dfrac{\sin x}{\cos x}$	$\cos x \neq 0 \quad x \neq \dfrac{n\pi}{2} \quad n = $ odd integer
$\cot x = \dfrac{\cos x}{\sin x}$	$\sin x \neq 0 \quad x \neq n\pi \quad n = $ integer

PYTHAGOREAN IDENTITIES

$$\sin^2 x + \cos^2 x = 1 \qquad \tan^2 x + 1 = \sec^2 x \qquad 1 + \cot^2 x = \csc^2 x$$

In Chapter 5, we discussed even and odd trigonometric functions, which like even and odd functions in general have these respective properties:

TYPE OF FUNCTION	ALGEBRAIC IDENTITY	GRAPH
Even	$f(-x) = f(x)$	Symmetry about the y-axis
Odd	$f(-x) = -f(x)$	Symmetry about the origin

We already learned in Chapter 5 that the sine function is an odd function and the cosine function is an even function. Combining this knowledge with the reciprocal and quotient identities, we arrive at the *even-odd identities*, which we can add to our list of basic identities.

EVEN-ODD IDENTITIES

Odd $\begin{cases} \sin(-x) = -\sin x \\ \csc(-x) = -\csc x \\ \tan(-x) = -\tan x \\ \cot(-x) = -\cot x \end{cases}$ Even $\begin{cases} \cos(-x) = \cos x \\ \sec(-x) = \sec x \end{cases}$

Technology Tip

Graphs of $y_1 = \sin(-x)$ and $y_2 = -\sin x$.

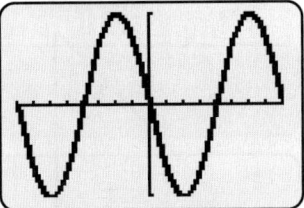

Cofunctions

Recall complementary angles (Section 4.1). Notice the *co* in *co*sine, *co*secant, and *co*tangent functions. These *cofunctions* are based on the relationship of *co*mplementary angles. Let us look at a right triangle with labled sides and angles.

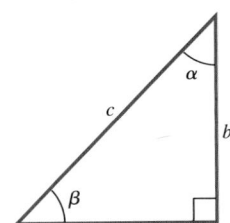

$$\sin\beta = \frac{\text{opposite of }\beta}{\text{hypotenuse}} = \frac{b}{c}$$
$$\cos\alpha = \frac{\text{adjacent to }\alpha}{\text{hypotenuse}} = \frac{b}{c}$$
$$\left.\right\} \sin\beta = \cos\alpha$$

Recall that the sum of the measures of the three angles in a triangle is 180°. In a right triangle, one angle is 90°; therefore, the two acute angles are complementary angles (the measures sum to 90°). You can see in the triangle above that β and α are *co*mplementary angles. In other words, the sine of an angle is the same as the *co*sine of the *co*mplement of that angle. This is true for all *trigonometric confunction* pairs.

COFUNCTION THEOREM

A trigonometric function of an angle is always equal to the cofunction of the complement of the angle. If $\alpha + \beta = 90°$ $\left(\text{or } \alpha + \beta = \dfrac{\pi}{2} \right)$, then

$$\sin \beta = \cos \alpha$$
$$\sec \beta = \csc \alpha$$
$$\tan \beta = \cot \alpha$$

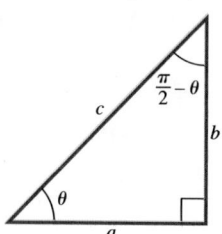

COFUNCTION IDENTITIES

$$\sin\theta = \cos(90° - \theta) \qquad \cos\theta = \sin(90° - \theta)$$
$$\tan\theta = \cot(90° - \theta) \qquad \cot\theta = \tan(90° - \theta)$$
$$\sec\theta = \csc(90° - \theta) \qquad \csc\theta = \sec(90° - \theta)$$

EXAMPLE 1 **Writing Trigonometric Function Values in Terms of Their Cofunctions**

Write each function or function value in terms of its cofunction.

a. $\sin 30°$ **b.** $\tan x$ **c.** $\csc 40°$

Solution (a):

Cosine is the cofunction of sine. $\sin\theta = \cos(90° - \theta)$

Substitute $\theta = 30°$. $\sin 30° = \cos(90° - 30°)$

Simplify. $\boxed{\sin 30° = \cos 60°}$

Solution (b):

Cotangent is the cofunction of tangent. $\tan\theta = \cot(90° - \theta)$

Substitute $\theta = x$. $\boxed{\tan x = \cot(90° - x)}$

Solution (c):

Cosecant is the cofunction of secant. $\csc\theta = \sec(90° - \theta)$

Substitute $\theta = 40°$. $\csc 40° = \sec(90° - 40°)$

Simplify. $\boxed{\csc 40° = \sec 50°}$

■ **Answer: a.** $\sin 45°$ **b.** $\sec(90° - y)$

■ **YOUR TURN** Write each function or function value in terms of its cofunction.

     a. $\cos 45°$ **b.** $\csc y$

Simplifying Trigonometric Expressions Using Identities

We can use the fundamental identities and algebraic manipulation to simplify more complicated trigonometric expressions. In simplifying trigonometric expressions, one approach is to first convert all expressions into sines and cosines and then simplify.

EXAMPLE 2 **Simplifying Trigonometric Expressions**

Simplify $\tan x \sin x + \cos x$.

Solution:

Write the tangent function in terms of the sine and cosine functions: $\tan x = \dfrac{\sin x}{\cos x}$.

$\tan x \cdot \sin x + \cos x$

$= \dfrac{\sin x}{\cos x}\sin x + \cos x$

Simplify. $= \dfrac{\sin^2 x}{\cos x} + \cos x$

Write as a fraction with a single quotient by finding a common denominator, $\cos x$.	$= \dfrac{\sin^2 x + \cos^2 x}{\cos x}$
Use the Pythagorean identity: $\sin^2 x + \cos^2 x = 1$.	$= \dfrac{1}{\cos x}$
Use the reciprocal identity $\sec x = \dfrac{1}{\cos x}$.	$= \boxed{\sec x}$

■ **Answer:** $\csc x$

■ **YOUR TURN** Simplify $\cot x \cos x + \sin x$.

In Example 2, $\tan x$ and $\sec x$ are not defined for odd integer multiples of $\dfrac{\pi}{2}$. In the Your Turn, $\cot x$ and $\csc x$ are not defined for integer multiples of π. Both the original expression and the simplified form are governed by the same restrictions. There are times when the original expression is subject to more domain restrictions than the simplified form and thus special attention must be given to domain restrictions.

For example, the algebraic expression $\dfrac{x^2 - 1}{x + 1}$ is under the domain restriction $x \neq -1$ because that value for x makes the value of the denominator equal to zero. If we forget to state the domain restrictions, we might simplify the algebraic expression as $\dfrac{x^2 - 1}{x + 1} = \dfrac{(x - 1)(x + 1)}{(x + 1)} = x - 1$ and assume this is true for all values of x. The correct simplification is $\dfrac{x^2 - 1}{x + 1} = x - 1$ for $x \neq -1$. In fact, if we were to graph both the original expression $y = \dfrac{x^2 - 1}{x + 1}$ and the line $y = x - 1$, they would coincide, except that the graph of the original expression would have a "hole" or discontinuity at $x = -1$. In this chapter, it is assumed that the domain of the simplified expression is the same as the domain of the original expression.

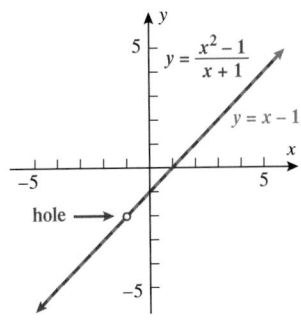

EXAMPLE 3 Simplifying Trigonometric Expressions

Simplify $\dfrac{1}{\csc^2 x} + \dfrac{1}{\sec^2 x}$.

Solution:

Rewrite the expression in terms of quotients squared.	$\dfrac{1}{\csc^2 x} + \dfrac{1}{\sec^2 x} = \left(\dfrac{1}{\csc x}\right)^2 + \left(\dfrac{1}{\sec x}\right)^2$
Use the reciprocal identities to write the cosecant and secant functions in terms of sines and cosines: $\sin x = \dfrac{1}{\csc x}$ and $\cos x = \dfrac{1}{\sec x}$.	$= \sin^2 x + \cos^2 x$
Use the Pythagorean identity: $\sin^2 x + \cos^2 x = 1$.	$= \boxed{1}$

■ **Answer:** $\tan^2 x$

■ **YOUR TURN** Simplify $\dfrac{1}{\cos^2 x} - 1$.

Verifying Identities

We will now use the trigonometric identities to verify, or establish, other trigonometric identities. For example, verify that

$$(\sin x - \cos x)^2 - 1 = -2\sin x \cos x$$

The good news is that we will know we are done when we get there, since we know the desired identity. But how do we get there? How do we verify that the identity is true? Remember that it must be true for *all* x, not just some x. Therefore, it is not enough to simply select values for x and show it is true for those specific values.

WORDS	MATH
Start with one side of the equation (the more complicated side).	$(\sin x - \cos x)^2 - 1$
Remember that $(a - b)^2 = a^2 - 2ab + b^2$ and expand $(\sin x - \cos x)^2$.	$= \sin^2 x - 2\sin x \cos x + \cos^2 x - 1$
Group the $\sin^2 x$ and $\cos^2 x$ terms and use the Pythagorean identity.	$= -2\sin x \cos x + \underbrace{(\sin^2 x + \cos^2 x)}_{1} - 1$
Simplify.	$= -2\sin x \cos x$

When we arrive at the right side of the equation, then we have succeeded in verifying the identity. In verifying trigonometric identities, there is no one procedure that works for all identities. You must manipulate one side of the equation until it looks like the other side. Here are two suggestions that are generally helpful:

1. Convert all trigonometric expressions to sines and cosines.
2. Write all sums or differences of fractions (quotients) as a single fraction (quotient).

The following suggestions help guide the way in verifying trigonometric identities.

GUIDELINES FOR VERIFYING TRIGONOMETRIC IDENTITIES

- Start with the more complicated side of the equation.
- Combine all sums and differences of fractions (quotients) into a single fraction (quotient).
- Use fundamental trigonometric identities.
- Use algebraic techniques to manipulate one side of the equation until the other side of the equation is achieved.
- Sometimes it is helpful to convert all trigonometric functions into sines and cosines.

It is important to note that trigonometric identities must be valid for all values of the independent variable (usually, x or θ) for which the expressions in the equation are defined (domain of the equation).

EXAMPLE 4   Verifying Trigonometric Identities

Verify the identity $\dfrac{\tan x - \cot x}{\tan x + \cot x} = \sin^2 x - \cos^2 x$.

Solution:

Start with the more complicated side of the equation.

$$\frac{\tan x - \cot x}{\tan x + \cot x}$$

Use the quotient identity to write the tangent and cotangent functions in terms of the sine and cosine functions.

$$= \frac{\dfrac{\sin x}{\cos x} - \dfrac{\cos x}{\sin x}}{\dfrac{\sin x}{\cos x} + \dfrac{\cos x}{\sin x}}$$

Multiply by $\dfrac{\sin x \cos x}{\sin x \cos x}$.

$$= \frac{\left(\dfrac{\sin x}{\cos x} - \dfrac{\cos x}{\sin x}\right)}{\left(\dfrac{\sin x}{\cos x} + \dfrac{\cos x}{\sin x}\right)}\left(\dfrac{\sin x \cos x}{\sin x \cos x}\right)$$

Simplify.

$$= \frac{\sin^2 x - \cos^2 x}{\sin^2 x + \cos^2 x}$$

Use the Pythagorean identity: $\sin^2 x + \cos^2 x = 1$.

$$= \boxed{\sin^2 x - \cos^2 x}$$

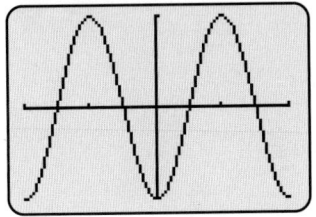
EXAMPLE 5   Determining Whether a Trigonometric Equation Is an Identity

Determine whether $\left(1 - \cos^2 x\right)\left(1 + \cot^2 x\right) = 0$.

Solution:

Use the quotient identity to write the cotangent function in terms of the sine and cosine functions.

$$\left(1 - \cos^2 x\right)\left(1 + \cot^2 x\right) = \left(1 - \cos^2 x\right)\left(1 + \frac{\cos^2 x}{\sin^2 x}\right)$$

Combine the expression in the second parentheses so that it is a single quotient.

$$= \left(1 - \cos^2 x\right)\left(\frac{\sin^2 x + \cos^2 x}{\sin^2 x}\right)$$

Use the Pythagorean identity.

$$= \underbrace{\left(1 - \cos^2 x\right)}_{\sin^2 x}\left(\frac{\overbrace{\sin^2 x + \cos^2 x}^{1}}{\sin^2 x}\right)$$

Eliminate the parentheses.

$$= \frac{\sin^2 x}{\sin^2 x}$$

Simplify.

$$= 1$$

Since $1 \neq 0$, this is not an identity.

EXAMPLE 6 **Verifying Trigonometric Identities**

Verify that $\dfrac{\sin(-x)}{\cos(-x)\tan(-x)} = 1$.

Solution:

Start with the left side of the equation.

$$\dfrac{\sin(-x)}{\cos(-x)\tan(-x)}$$

Use the even-odd identities.

$$= \dfrac{-\sin x}{-\cos x \tan x}$$

Simplify.

$$= \dfrac{\sin x}{\cos x \tan x}$$

Use the quotient identity to write the tangent function in terms of the sine and cosine functions.

$$= \dfrac{\sin x}{\cos x \left(\dfrac{\sin x}{\cos x}\right)}$$

Divide out the cosine term in the denominator.

$$= \dfrac{\sin x}{\sin x}$$

Simplify.

$$= 1$$

> **Study Tip**
>
> Start with the more complicated expression (side) and manipulate until reaching the simpler expression (on the other side).

We have verified that $\dfrac{\sin(-x)}{\cos(-x)\tan(-x)} = 1$.

So far we have discussed working with only one side of the identity until arriving at the other side. Another method for verifying identities is to work with (simplify) each side separately and use identities and algebraic techniques to arrive at the same result on both sides.

EXAMPLE 7 **Verifying an Identity by Simplifying Both Sides Separately**

Verify that $\dfrac{\sin x + 1}{\sin x} = -\dfrac{\cot^2 x}{1 - \csc x}$.

Solution:

Left-hand side:

$$\dfrac{\sin x + 1}{\sin x} = \dfrac{\sin x}{\sin x} + \dfrac{1}{\sin x} = 1 + \csc x$$

Right-hand side:

$$\dfrac{-\cot^2 x}{1 - \csc x} = \dfrac{1 - \csc^2 x}{1 - \csc x} = \dfrac{(1 - \csc x)(1 + \csc x)}{(1 - \csc x)} = 1 + \csc x$$

Since the left-hand side equals the right-hand side, the equation is an identity.

SECTION

6.1 **SUMMARY**

We combined the fundamental trigonometric identities—reciprocal, quotient, Pythagorean, even-odd, and cofunction—with algebraic techniques to simplify trigonometric expressions and verify more complex trigonometric identities. Two steps that we often use in both simplifying trigonometric expressions and verifying trigonometric identities are: (1) writing all trigonometric functions in terms of the sine and cosine functions, and (2) combining sums or differences of quotients into a single quotient.

When verifying trigonometric identities, we typically work with the more complicated side (keeping the other side in mind as our goal). Another approach to verifying trigonometric identities is to work on each side separately and arrive at the same result.

■ SKILLS

In Exercises 1–6, use the cofunction identities to fill in the blanks.

1. $\sin 60° = \cos\underline{\hspace{1cm}}$

2. $\sin 45° = \cos\underline{\hspace{1cm}}$

3. $\cos x = \sin\underline{\hspace{1cm}}$

4. $\cot A = \tan\underline{\hspace{1cm}}$

5. $\csc 30° = \sec\underline{\hspace{1cm}}$

6. $\sec B = \csc\underline{\hspace{1cm}}$

In Exercises 7–14, write the trigonometric function values in terms of its cofunction.

7. $\sin(x + y)$

8. $\sin(60° - x)$

9. $\cos(20° + A)$

10. $\cos(A + B)$

11. $\cot(45° - x)$

12. $\sec(30° - \theta)$

13. $\csc(60° - \theta)$

14. $\tan(40° + \theta)$

In Exercises 15–38, simplify each of the trigonometric expressions.

15. $\sin x \csc x$

16. $\tan x \cot x$

17. $\sec(-x)\cot x$

18. $\tan(-x)\cos(-x)$

19. $\csc(-x)\sin x$

20. $\cot(-x)\tan x$

21. $\sec x \cos(-x) + \tan^2 x$

22. $\sec(-x)\tan(-x)\cos(-x)$

23. $(\sin^2 x)(\cot^2 x + 1)$

24. $(\cos^2 x)(\tan^2 x + 1)$

25. $(\sin x - \cos x)(\sin x + \cos x)$

26. $(\sin x + \cos x)^2$

27. $\dfrac{\csc x}{\cot x}$

28. $\dfrac{\sec x}{\tan x}$

29. $\dfrac{1 - \cot(-x)}{1 + \cot x}$

30. $\sec^2 x - \tan^2(-x)$

31. $\dfrac{1 - \cos^4 x}{1 + \cos^2 x}$

32. $\dfrac{1 - \sin^4 x}{1 + \sin^2 x}$

33. $\dfrac{1 - \cot^4 x}{1 - \cot^2 x}$

34. $\dfrac{1 - \tan^4(-x)}{1 - \tan^2 x}$

35. $1 - \dfrac{\sin^2 x}{1 - \cos x}$

36. $1 - \dfrac{\cos^2 x}{1 + \sin x}$

37. $\dfrac{\tan x - \cot x}{\tan x + \cot x} + 2\cos^2 x$

38. $\dfrac{\tan x - \cot x}{\tan x + \cot x} + \cos^2 x$

In Exercises 39–64, verify each of the trigonometric identities.

39. $(\sin x + \cos x)^2 + (\sin x - \cos x)^2 = 2$

40. $(1 - \sin x)(1 + \sin x) = \cos^2 x$

41. $(\csc x + 1)(\csc x - 1) = \cot^2 x$

42. $(\sec x + 1)(\sec x - 1) = \tan^2 x$

43. $\tan x + \cot x = \csc x \sec x$

44. $\csc x - \sin x = \cot x \cos x$

45. $\dfrac{2 - \sin^2 x}{\cos x} = \sec x + \cos x$

46. $\dfrac{2 - \cos^2 x}{\sin x} = \csc x + \sin x$

47. $[\cos(-x) - 1][1 + \cos x] = -\sin^2 x$

48. $\tan(-x)\cot x = -1$

49. $\dfrac{\sec(-x)\cot x}{\csc(-x)} = -1$

50. $\csc(-x) - 1 = \dfrac{\cot^2 x}{\csc(-x) + 1}$

51. $\dfrac{1}{\csc^2 x} + \dfrac{1}{\sec^2 x} = 1$

52. $\dfrac{1}{\cot^2 x} - \dfrac{1}{\tan^2 x} = \sec^2 x - \csc^2 x$

53. $\dfrac{1}{1 - \sin x} + \dfrac{1}{1 + \sin x} = 2\sec^2 x$

54. $\dfrac{1}{1 - \cos x} + \dfrac{1}{1 + \cos x} = 2\csc^2 x$

55. $\dfrac{\sin^2 x}{1 - \cos x} = 1 + \cos x$

56. $\dfrac{\cos^2 x}{1 - \sin x} = 1 + \sin x$

57. $\sec x + \tan x = \dfrac{1}{\sec x - \tan x}$

58. $\csc x + \cot x = \dfrac{1}{\csc x - \cot x}$

59. $\dfrac{\csc x - \tan x}{\sec x + \cot x} = \dfrac{\cos x - \sin^2 x}{\sin x + \cos^2 x}$

60. $\dfrac{\sec x + \tan x}{\csc x + 1} = \tan x$

61. $\dfrac{\cos^2 x + 1 + \sin x}{\cos^2 x + 3} = \dfrac{1 + \sin x}{2 + \sin x}$

62. $\dfrac{\sin x + 1 - \cos^2 x}{\cos^2 x} = \dfrac{\sin x}{1 - \sin x}$

63. $\sec x(\tan x + \cot x) = \dfrac{\csc x}{\cos^2 x}$

64. $\tan x(\csc x - \sin x) = \cos x$

In Exercises 65–78, determine whether each equation is a conditional equation or an identity.

65. $\cos^2 x(\tan x - \sec x)(\tan x + \sec x) = 1$

66. $\cos^2 x(\tan x - \sec x)(\tan x + \sec x) = \sin^2 x - 1$

67. $\dfrac{\csc x \cot x}{\sec x \tan x} = \cot^3 x$

68. $\sin x \cos x = 0$

69. $\sin x + \cos x = \sqrt{2}$

70. $\sin^2 x + \cos^2 x = 1$

71. $\tan^2 x - \sec^2 x = 1$

72. $\sec^2 x - \tan^2 x = 1$

73. $\sin x = \sqrt{1 - \cos^2 x}$

74. $\csc x = \sqrt{1 + \cot^2 x}$

75. $\sqrt{\sin^2 x + \cos^2 x} = 1$

76. $\sqrt{\sin^2 x + \cos^2 x} = \sin x + \cos x$

77. $(\sin x - \cos x)^2 = \sin^2 x - \cos^2 x$

78. $[\sin(-x) - 1][\sin(-x) + 1] = \cos^2 x$

■ APPLICATIONS

79. Area of a Circle. Show that the area of a circle with radius $r = \sec x$ is equal to $\pi + \pi(\tan x)^2$.

80. Area of a Triangle. Show that the area of a triangle with base $b = \cos x$ and height $h = \sec x$ is equal to $\frac{1}{2}$.

81. Pythagorean Theorem. Find the length of the hypotenuse of a right triangle whose legs have lengths 1 and $\tan\theta$.

82. Pythagorean Theorem. Find the length of the hypotenuse of a right triangle whose legs have lengths 1 and $\cot\theta$.

■ CATCH THE MISTAKE

In Exercises 83–86, explain the mistake that is made.

83. Verify the identity $\dfrac{\cos x}{1 - \tan x} + \dfrac{\sin x}{1 - \cot x} = \sin x + \cos x$.

Solution:

Start with the left side of the equation.
$$\frac{\cos x}{1 - \tan x} + \frac{\sin x}{1 - \cot x}$$

Write the tangent and cotangent functions in terms of sines and cosines.
$$= \frac{\cos x}{1 - \dfrac{\sin x}{\cos x}} + \frac{\sin x}{1 - \dfrac{\cos x}{\sin x}}$$

Cancel the common cosine in the first term and sine in the second term.
$$= \frac{1}{1 - \sin x} + \frac{1}{1 - \cos x}$$

This is incorrect. What mistake was made?

84. Verify the identity $\dfrac{\cos^3 x \sec x}{1 - \sin x} = 1 + \sin x$.

Solution:

Start with the equation on the left.
$$\frac{\cos^3 x \sec x}{1 - \sin x}$$

Rewrite secant in terms of sine.
$$= \frac{\cos^3 x \dfrac{1}{\sin x}}{1 - \sin x}$$

Simplify.
$$= \frac{\cos^3 x}{1 - \sin^2 x}$$

Use the Pythagorean identity.
$$= \frac{\cos^3 x}{\cos^2 x}$$

Simplify.
$$= \cos x$$

This is incorrect. What mistakes were made?

85. Determine whether the equation is a conditional equation or an identity: $\dfrac{\tan x}{\cot x} = 1$.

Solution:

Start with the left side.
$$\frac{\tan x}{\cot x}$$

Rewrite the tangent and cotangent functions in terms of sines and cosines.
$$= \frac{\dfrac{\sin x}{\cos x}}{\dfrac{\cos x}{\sin x}}$$

Simplify.
$$= \frac{\sin^2 x}{\cos^2 x} = \tan^2 x$$

Let $x = \dfrac{\pi}{4}$. *Note:* $\tan\left(\dfrac{\pi}{4}\right) = 1$. $= 1$

Since $\dfrac{\tan x}{\cot x} = 1$, this equation is an identity.

This is incorrect. What mistake was made?

86. Determine whether the equation is a conditional equation or an identity: $|\sin x| - \cos x = 1$.

Solution:

Start with the left side of the equation.
$$|\sin x| - \cos x$$

Let $x = \dfrac{n\pi}{2}$, where n is an odd integer.
$$\left|\sin\left(\frac{n\pi}{2}\right)\right| - \cos\left(\frac{n\pi}{2}\right)$$

Simplify.
$$|\pm 1| - 0 = 1$$

Since $|\sin x| - \cos x = 1$, this is an identity.

This is incorrect. What mistake was made?

■ CONCEPTUAL

In Exercises 87 and 88, determine whether each statement is true or false.

87. If an equation is true for some values (but not all values), then it is still an identity.

88. If an equation has an infinite number of solutions, then it is an identity.

89. In which quadrants is the equation $\cos\theta = \sqrt{1 - \sin^2\theta}$ true?

90. In which quadrants is the equation $-\cos\theta = \sqrt{1 - \sin^2\theta}$ true?

91. In which quadrants is the equation $\csc\theta = -\sqrt{1 + \cot^2\theta}$ true?

92. In which quadrants is the equation $\sec\theta = \sqrt{1 + \tan^2\theta}$ true?

93. Do you think that $\sin(A + B) = \sin A + \sin B$? Why?

94. Do you think that $\cos(\frac{1}{2}A) = \frac{1}{2}\cos A$? Why?

95. Do you think $\tan(2A) = 2\tan A$? Why?

96. Do you think $\cot(A^2) = (\cot A)^2$? Why?

■ CHALLENGE

97. Simplify $(a\sin x + b\cos x)^2 + (b\sin x - a\cos x)^2$.

98. Simplify $\dfrac{1 + \cot^3 x}{1 + \cot x} + \cot x$.

99. Show that $\csc\left(\dfrac{\pi}{2} + \theta + 2n\pi\right) = \sec\theta$, n an integer.

100. Show that $\sec\left(\dfrac{\pi}{2} - \theta - 2n\pi\right) = \csc\theta$, n an integer.

101. Simplify $\csc\left(2\pi - \dfrac{\pi}{2} - \theta\right)\cdot\sec\left(\theta - \dfrac{\pi}{2}\right)\cdot\sin(-\theta)$.

102. Simplify $\tan\theta\cdot\cot(2\pi - \theta)$.

■ TECHNOLOGY

In the next section, you will learn the sum and difference identities. In Exercises 103–106, we illustrate these identities with graphing calculators.

103. Determine the correct sign (+ or −) for $\cos(A + B) = \cos A\cos B \underset{?}{\pm} \sin A\sin B$ by graphing $Y_1 = \cos(A + B)$, $Y_2 = \cos A\cos B + \sin A\sin B$, and $Y_2 = \cos A\cos B - \sin A\sin B$ in the same viewing rectangle for several values of A and B.

104. Determine the correct sign (+ or −) for $\cos(A - B) = \cos A\cos B \underset{?}{\pm} \sin A\sin B$ by graphing $Y_1 = \cos(A - B)$, $Y_2 = \cos A\cos B + \sin A\sin B$, and $Y_2 = \cos A\cos B - \sin A\sin B$ in the same viewing rectangle for several values of A and B.

105. Determine the correct sign (+ or −) for $\sin(A + B) = \sin A\cos B \underset{?}{\pm} \cos A\sin B$ by graphing $Y_1 = \sin(A + B)$, $Y_2 = \sin A\cos B + \cos A\sin B$, and $Y_2 = \sin A\cos B - \cos A\sin B$ in the same viewing rectangle for several values of A and B.

106. Determine the correct sign (+ or −) for $\sin(A - B) = \sin A\cos B \underset{?}{\pm} \cos A\sin B$ by graphing $Y_1 = \sin(A - B)$, $Y_2 = \sin A\cos B + \cos A\sin B$, and $Y_2 = \sin A\cos B - \cos A\sin B$ in the same viewing rectangle for several values of A and B.

■ PREVIEW TO CALCULUS

For Exercises 107–110, refer to the following:

In calculus, when integrating expressions such as $\sqrt{a^2 - x^2}$, $\sqrt{a^2 + x^2}$, and $\sqrt{x^2 - a^2}$, trigonometric functions are used as "dummy" functions to eliminate the radical. Once the integration is performed, the trigonometric function is "unsubstituted." These trigonometric substitutions (and corresponding trigonometric identities) are used to simplify these types of expressions.

When simplifying, it is important to remember that

$$|x| = \begin{cases} x & \text{if } x \geq 0 \\ -x & \text{if } x < 0 \end{cases}$$

536 CHAPTER 6 Analytic Trigonometry

EXPRESSIONS	SUBSTITUTION		TRIGONOMETRIC IDENTITY
$\sqrt{a^2 - x^2}$	$x = a\sin\theta$	$-\dfrac{\pi}{2} \le \theta \le \dfrac{\pi}{2}$	$1 - \sin^2\theta = \cos^2\theta$
$\sqrt{a^2 + x^2}$	$x = a\tan\theta$	$-\dfrac{\pi}{2} \le \theta \le \dfrac{\pi}{2}$	$1 + \tan^2\theta = \sec^2\theta$
$\sqrt{x^2 - a^2}$	$x = a\sec\theta$	$0 \le \theta < \dfrac{\pi}{2}$ or $\pi \le \theta < \dfrac{3\pi}{2}$	$\sec^2\theta - 1 = \tan^2\theta$

107. Start with the expression $\sqrt{a^2 - x^2}$ and let $x = a\sin\theta$, assuming $-\dfrac{\pi}{2} \le \theta \le \dfrac{\pi}{2}$. Simplify the original expression so that it contains no radicals.

108. Start with the expression $\sqrt{a^2 + x^2}$ and let $x = a\tan\theta$, assuming $-\dfrac{\pi}{2} < \theta < \dfrac{\pi}{2}$. Simplify the original expression so that it contains no radicals.

109. Start with the expression $\sqrt{x^2 - a^2}$ and let $x = a\sec\theta$, assuming $0 \le \theta < \dfrac{\pi}{2}$. Simplify the original expression so that it contains no radicals.

110. Use a trigonometric substitution to simplify the expression $\sqrt{9 - x^2}$ so that it contains no radicals.

SECTION 6.2 SUM AND DIFFERENCE IDENTITIES

SKILLS OBJECTIVES

- Find exact values of trigonometric functions of certain rational multiples of π by using the sum and difference identities.
- Develop new identities from the sum and difference identities.

CONCEPTUAL OBJECTIVE

- Understand that a trigonometric function of a sum is not the sum of the trigonometric functions.

In this section, we will consider trigonometric functions with arguments that are sums and differences. In general, $f(A + B) \ne f(A) + f(B)$. First, it is important to note that function notation is not distributive:

$$\cos(A + B) \ne \cos A + \cos B$$

This principle is easy to prove. Let $A = \pi$ and $B = 0$; then

$$\cos(A + B) = \cos(\pi + 0) = \cos(\pi) = -1$$
$$\cos A + \cos B = \cos\pi + \cos 0 = -1 + 1 = 0$$

In this section, we will derive some new and important identities.

- Sum and difference identities for the cosine, sine, and tangent functions
- Cofunction identities

We begin with the familiar distance formula, from which we can derive the sum and difference identities for the cosine function. From there we can derive the sum and difference formulas for the sine and tangent functions.

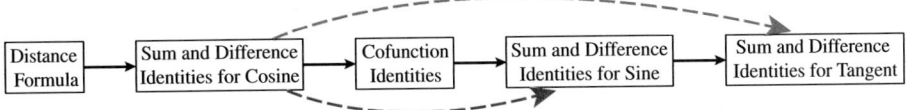

Before we start deriving and working with trigonometric sum and difference identities, let us first discuss why these are important. Sum and difference (and later product-to-sum and sum-to-product) identities are important because they allow calculation in functional (analytic) form and often lead to evaluating expressions *exactly* (as opposed to approximating them with calculators). The identities developed in this chapter are useful in such applications as musical sound, where they allow the determination of the "beat" frequency. In calculus, these identities will simplify the integration and differentiation processes.

Sum and Difference Identities for the Cosine Function

Recall from Section 5.1 that the unit circle approach gave the relationship between the coordinates along the unit circle and the sine and cosine functions. Specifically, the x-coordinate corresponded to the value of the cosine function and the y-coordinate corresponded to the value of the sine function.

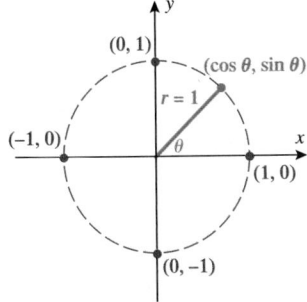

Let us now draw the unit circle with two angles α and β, realizing that the two terminal sides of these angles form a third angle, $\alpha - \beta$.

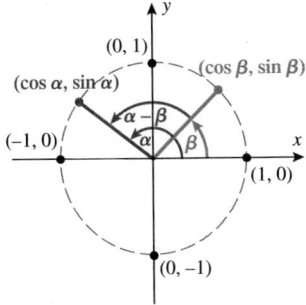

If we label the points $P_1 = (\cos\alpha, \sin\alpha)$ and $P_2 = (\cos\beta, \sin\beta)$, we can then draw a segment connecting points P_1 and P_2.

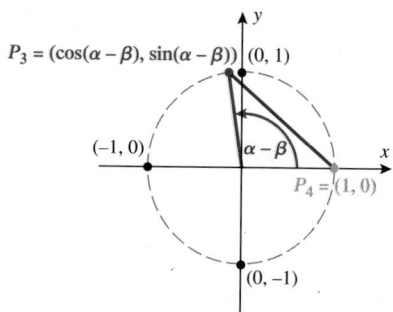

If we rotate the angle clockwise so the central angle $\alpha - \beta$ is in standard position, then the two points where the initial and terminal sides intersect the unit circle are $P_4 = (1, 0)$ and $P_3 = (\cos(\alpha - \beta), \sin(\alpha - \beta))$, respectively.

Study Tip

The distance from point $P_1 = (x_1, y_1)$ to $P_2 = (x_2, y_2)$ is given by the distance formula

$$d(P_1, P_2) = \sqrt{(x_2 - x_1)^2 + (y_2 - y_1)^2}.$$

The distance from P_1 to P_2 is equal to the length of the segment joining the points. Similarly, the distance from P_3 to P_4 is equal to the length of the segment joining the points. Since the lengths of the segments are equal, we say that the distances are equal: $d(P_1, P_2) = d(P_3, P_4)$.

WORDS	MATH
Set the distances (segment lengths) equal.	$d(P_1, P_2) = d(P_3, P_4)$
Apply the distance formula	$\sqrt{(x_2 - x_1)^2 + (y_2 - y_1)^2} = \sqrt{(x_4 - x_3)^2 + (y_4 - y_3)^2}$

Substitute $P_1 = (x_1, y_1) = (\cos\alpha, \sin\alpha)$ and $P_2 = (x_2, y_2) = (\cos\beta, \sin\beta)$ into the left side of the equation and $P_3 = (x_3, y_3) = (\cos(\alpha - \beta), \sin(\alpha - \beta))$ and $P_4 = (x_4, y_4) = (1, 0)$ into the right side of the equation.

$$\sqrt{[\cos\beta - \cos\alpha]^2 + [\sin\beta - \sin\alpha]^2} = \sqrt{[1 - \cos(\alpha - \beta)]^2 + [0 - \sin(\alpha - \beta)]^2}$$

Square both sides of the equation.	$[\cos\beta - \cos\alpha]^2 + [\sin\beta - \sin\alpha]^2 = [1 - \cos(\alpha - \beta)]^2 + [0 - \sin(\alpha - \beta)]^2$
Eliminate the brackets.	$\cos^2\beta - 2\cos\beta\,\cos\alpha + \cos^2\alpha + \sin^2\beta - 2\sin\beta\,\sin\alpha + \sin^2\alpha$ $= 1 - 2\cos(\alpha - \beta) + \cos^2(\alpha - \beta) + \sin^2(\alpha - \beta)$
Regroup terms on each side and use the Pythagorean identity.	$\underbrace{\cos^2\alpha + \sin^2\alpha}_{1} - 2\cos\alpha\,\cos\beta - 2\sin\alpha\,\sin\beta + \underbrace{\cos^2\beta + \sin^2\beta}$ $= 1 - 2\cos(\alpha - \beta) + \underbrace{\cos^2(\alpha - \beta) + \sin^2(\alpha - \beta)}_{1}$
Simplify.	$2 - 2\cos\alpha\,\cos\beta - 2\sin\alpha\,\sin\beta = 2 - 2\cos(\alpha - \beta)$
Subtract 2 from both sides.	$-2\cos\alpha\,\cos\beta - 2\sin\alpha\,\sin\beta = -2\cos(\alpha - \beta)$
Divide by -2.	$\cos\alpha\,\cos\beta + \sin\alpha\,\sin\beta = \cos(\alpha - \beta)$
Write the **difference identity for the cosine function.**	$\boxed{\cos(\alpha - \beta) = \cos\alpha\cos\beta + \sin\alpha\sin\beta}$

We can now derive the sum identity for the cosine function from the difference identity for the cosine function and the properties of even and odd functions.

WORDS	MATH
Apply the difference identity.	$\cos(\alpha + \beta) = \cos[\alpha - (-\beta)]$ $\cos(\alpha + \beta) = \cos\alpha\cos(-\beta) + \sin\alpha\sin(-\beta)$
Simplify the left side and use properties of even and odd functions on the right side.	$\cos(\alpha + \beta) = \cos\alpha(\cos\beta) + \sin\alpha(-\sin\beta)$
Write the **sum identity for the cosine function**.	$\boxed{\cos(\alpha + \beta) = \cos\alpha\cos\beta - \sin\alpha\sin\beta}$

SUM AND DIFFERENCE IDENTITIES FOR THE COSINE FUNCTION

Sum $\qquad \cos(A + B) = \cos A\cos B - \sin A\sin B$

Difference $\quad \cos(A - B) = \cos A\cos B + \sin A\sin B$

EXAMPLE 1 Finding Exact Values for the Cosine Function

Evaluate each of the following cosine expressions exactly:

a. $\cos\left(\dfrac{7\pi}{12}\right)$

b. $\cos 15°$

Solution (a):

Write $\dfrac{7\pi}{12}$ as a sum of known "special" angles.	$\cos\left(\dfrac{7\pi}{12}\right) = \cos\left(\dfrac{4\pi}{12} + \dfrac{3\pi}{12}\right)$
Simplify.	$\cos\left(\dfrac{7\pi}{12}\right) = \cos\left(\dfrac{\pi}{3} + \dfrac{\pi}{4}\right)$
Write the sum identity for the cosine function.	$\cos(A + B) = \cos A\cos B - \sin A\sin B$
Substitute $A = \dfrac{\pi}{3}$ and $B = \dfrac{\pi}{4}$.	$\cos\left(\dfrac{7\pi}{12}\right) = \cos\left(\dfrac{\pi}{3}\right)\cos\left(\dfrac{\pi}{4}\right) - \sin\left(\dfrac{\pi}{3}\right)\sin\left(\dfrac{\pi}{4}\right)$
Evaluate the expressions on the right exactly.	$\cos\left(\dfrac{7\pi}{12}\right) = \dfrac{1}{2}\dfrac{\sqrt{2}}{2} - \dfrac{\sqrt{3}}{2}\dfrac{\sqrt{2}}{2}$
Simplify.	$\boxed{\cos\left(\dfrac{7\pi}{12}\right) = \dfrac{\sqrt{2} - \sqrt{6}}{4}}$

Technology Tip

a. Use a calculator to check the values for $\cos\left(\dfrac{7\pi}{12}\right)$ and $\dfrac{\sqrt{2} - \sqrt{6}}{4}$.

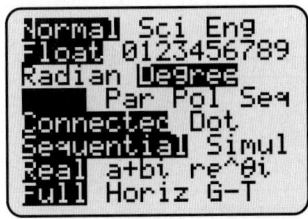

b. Use a calculator to check the values of $\cos 15°$ and $\dfrac{\sqrt{2} + \sqrt{6}}{4}$. Be sure the calculator is set in degree mode.

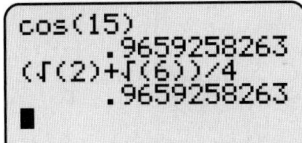

Solution (b):

Write 15° as a difference of known "special" angles.

$$\cos 15° = \cos(45° - 30°)$$

Write the difference identity for the cosine function.

$$\cos(A - B) = \cos A \cos B + \sin A \sin B$$

Substitute $A = 45°$ and $B = 30°$.

$$\cos 15° = \cos 45° \cos 30° + \sin 45° \sin 30°$$

Evaluate the expressions on the right exactly.

$$\cos 15° = \frac{\sqrt{2}}{2} \frac{\sqrt{3}}{2} + \frac{\sqrt{2}}{2} \frac{1}{2}$$

Simplify.

$$\boxed{\cos 15° = \frac{\sqrt{6} + \sqrt{2}}{4}}$$

■ **Answer: a.** $\dfrac{\sqrt{6} - \sqrt{2}}{4}$

    b. $\dfrac{\sqrt{6} - \sqrt{2}}{4}$

■ **YOUR TURN** Use the sum or difference identities for the cosine function to evaluate each cosine expression exactly.

    a. $\cos\left(\dfrac{5\pi}{12}\right)$ **b.** $\cos 75°$

Example 1 illustrates an important characteristic of the sum and difference identities: that we can now find the exact trigonometric function value of angles that are multiples of $15°$ (or, equivalently, $\dfrac{\pi}{12}$), since each of these can be written as a sum or difference of angles for which we know the trigonometric function values exactly.

Technology Tip

a.

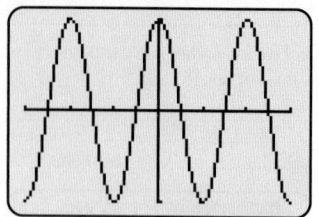

b.

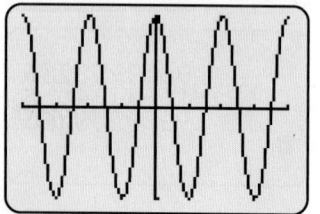

EXAMPLE 2 **Writing a Sum or Difference as a Single Cosine Expression**

Use the sum or the difference identity for the cosine function to write each of the following expressions as a single cosine expression:

a. $\sin(5x)\sin(2x) + \cos(5x)\cos(2x)$

b. $\cos x \cos(3x) - \sin x \sin(3x)$

Solution (a):

Because of the positive sign, this will be a cosine of a difference.

Reverse the expression and write the formula.

$$\cos A \cos B + \sin A \sin B = \cos(A - B)$$

Identify A and B.

$$A = 5x \quad \text{and} \quad B = 2x$$

Substitute $A = 5x$ and $B = 2x$ into the difference identity.

$$\cos(5x)\cos(2x) + \sin(5x)\sin(2x) = \cos(5x - 2x)$$

Simplify.

$$\cos(5x)\cos(2x) + \sin(5x)\sin(2x) = \boxed{\cos(3x)}$$

Notice that if we had selected $A = 2x$ and $B = 5x$ instead, the result would have been $\cos(-3x)$, but since the cosine function is an even function, this would have simplified to $\cos(3x)$.

Solution (b):

Because of the negative sign, this will be a cosine of a sum.

Reverse the expression and write the formula.	$\cos A \cos B - \sin A \sin B = \cos(A + B)$
Identify A and B.	$A = x$ and $B = 3x$
Substitute $A = x$ and $B = 3x$ into the sum identity.	$\cos x \cos(3x) - \sin x \sin(3x) = \cos(x + 3x)$
Simplify.	$\cos x \cos(3x) - \sin x \sin(3x) = \boxed{\cos(4x)}$

■ **Answer:** $\cos(3x)$

■ **YOUR TURN** Write as a single cosine expression.

$$\cos(4x)\cos(7x) + \sin(4x)\sin(7x)$$

Sum and Difference Identities for the Sine Function

We can now use the cofunction identities (Section 6.1) together with the sum and difference identities for the cosine function to develop the sum and difference identities for the sine function.

WORDS	**MATH**
Start with the cofunction identity.	$\sin\theta = \cos\left(\dfrac{\pi}{2} - \theta\right)$
Let $\theta = A + B$.	$\sin(A + B) = \cos\left[\dfrac{\pi}{2} - (A + B)\right]$
Regroup the terms in the cosine expression.	$\sin(A + B) = \cos\left[\left(\dfrac{\pi}{2} - A\right) - B\right]$
Use the difference identity for the cosine function.	$\sin(A + B) = \cos\left(\dfrac{\pi}{2} - A\right)\cos B + \sin\left(\dfrac{\pi}{2} - A\right)\sin B$
Use the cofunction identities.	$\sin(A + B) = \underbrace{\cos\left(\dfrac{\pi}{2} - A\right)}_{\sin A}\cos B + \underbrace{\sin\left(\dfrac{\pi}{2} - A\right)}_{\cos A}\sin B$
Simplify.	$\boxed{\sin(A + B) = \sin A \cos B + \cos A \sin B}$

Now we can derive the difference identity for the sine function using the sum identity for the sine function and the properties of even and odd functions.

WORDS	**MATH**
Replace B with $-B$ in the sum identity.	$\sin(A + (-B)) = \sin A \cos(-B) + \cos A \sin(-B)$
Simplify using even and odd identities.	$\boxed{\sin(A - B) = \sin A \cos B - \cos A \sin B}$

SUM AND DIFFERENCE IDENTITIES FOR THE SINE FUNCTION

Sum	$\sin(A + B) = \sin A \cos B + \cos A \sin B$
Difference	$\sin(A - B) = \sin A \cos B - \cos A \sin B$

EXAMPLE 3 **Finding Exact Values for the Sine Function**

Use the sum or the difference identity for the sine function to evaluate each sine expression exactly.

a. $\sin\left(\dfrac{5\pi}{12}\right)$ **b.** $\sin 75°$

Solution (a):

Write $\dfrac{5\pi}{12}$ as a sum of known "special" angles.

$$\sin\left(\frac{5\pi}{12}\right) = \sin\left(\frac{2\pi}{12} + \frac{3\pi}{12}\right)$$

Simplify.

$$\sin\left(\frac{5\pi}{12}\right) = \sin\left(\frac{\pi}{6} + \frac{\pi}{4}\right)$$

Write the sum identity for the sine function.

$$\sin(A + B) = \sin A \cos B + \cos A \sin B$$

Substitute $A = \dfrac{\pi}{6}$ and $B = \dfrac{\pi}{4}$.

$$\sin\left(\frac{5\pi}{12}\right) = \sin\left(\frac{\pi}{6}\right)\cos\left(\frac{\pi}{4}\right) + \cos\left(\frac{\pi}{6}\right)\sin\left(\frac{\pi}{4}\right)$$

Evaluate the expressions on the right exactly.

$$\sin\left(\frac{5\pi}{12}\right) = \left(\frac{1}{2}\right)\left(\frac{\sqrt{2}}{2}\right) + \left(\frac{\sqrt{3}}{2}\right)\left(\frac{\sqrt{2}}{2}\right)$$

Simplify.

$$\boxed{\sin\left(\frac{5\pi}{12}\right) = \frac{\sqrt{2} + \sqrt{6}}{4}}$$

Solution (b):

Write 75° as a sum of known "special" angles.

$$\sin 75° = \sin(45° + 30°)$$

Write the sum identity for the sine function.

$$\sin(A + B) = \sin A \cos B + \cos A \sin B$$

Substitute $A = 45°$ and $B = 30°$.

$$\sin 75° = \sin 45° \cos 30° + \cos 45° \sin 30°$$

Evaluate the expressions on the right exactly.

$$\sin 75° = \left(\frac{\sqrt{2}}{2}\right)\left(\frac{\sqrt{3}}{2}\right) + \left(\frac{\sqrt{2}}{2}\right)\left(\frac{1}{2}\right)$$

Simplify.

$$\boxed{\sin 75° = \frac{\sqrt{6} + \sqrt{2}}{4}}$$

Technology Tip

Use a calculator to check the values of $\sin\left(\dfrac{5\pi}{12}\right)$ and $\dfrac{\sqrt{2} + \sqrt{6}}{4}$. Be sure the calculator is in radian mode.

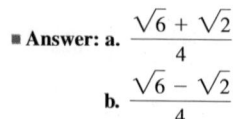

```
sin(5π/12)
        .9659258263
(√(2)+√(6))/4
        .9659258263
```

■ **Answer: a.** $\dfrac{\sqrt{6} + \sqrt{2}}{4}$

b. $\dfrac{\sqrt{6} - \sqrt{2}}{4}$

■ **YOUR TURN** Use the sum or the difference identity for the sine function to evaluate each sine expression exactly.

a. $\sin\left(\dfrac{7\pi}{12}\right)$ **b.** $\sin 15°$

We see in Example 3 that the sum and difference identities allow us to calculate exact values for trigonometric functions of angles that are multiples of 15° $\left(\text{or, equivalently, }\dfrac{\pi}{12}\right)$, as we saw with the cosine function.

EXAMPLE 4 **Writing a Sum or Difference as a Single Sine Expression**

Graph $y = 3\sin x \cos(3x) + 3\cos x \sin(3x)$.

Solution:

Use the sum identity for the sine function to write the expression as a single sine expression.

Factor out the common 3.
$$y = 3[\sin x \cos(3x) + \cos x \sin(3x)]$$

Write the sum identity for the sine function.
$$\sin A \cos B + \cos A \sin B = \sin(A + B)$$

Identify A and B.
$$A = x \text{ and } B = 3x$$

Substitute $A = x$ and $B = 3x$ into the sum identity.
$$\sin x \cos(3x) + \cos x \sin(3x) = \sin(x + 3x) = \sin(4x)$$

Simplify.
$$y = 3[\underbrace{\sin x \cos(3x) + \cos x \sin(3x)}_{\sin(4x)}]$$

Graph $y = 3\sin(4x)$.

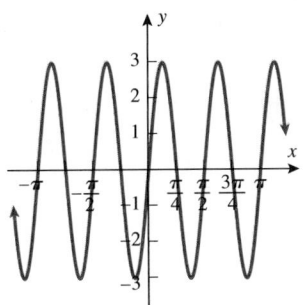

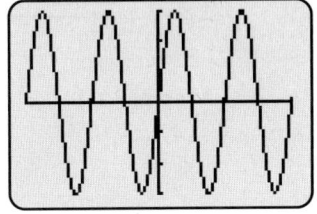
Sum and Difference Identities for the Tangent Function

We now develop the sum and difference identities for the tangent function.

WORDS	MATH
Start with the quotient identity.	$\tan x = \dfrac{\sin x}{\cos x}$
Let $x = A + B$.	$\tan(A + B) = \dfrac{\sin(A + B)}{\cos(A + B)}$
Use the sum identities for the sine and cosine functions.	$\tan(A + B) = \dfrac{\sin A \cos B + \cos A \sin B}{\cos A \cos B - \sin A \sin B}$
Multiply the numerator and denominator by $\dfrac{1}{\cos A \cos B}$.	$\tan(A + B) = \dfrac{\dfrac{\sin A \cos B + \cos A \sin B}{\cos A \cos B}}{\dfrac{\cos A \cos B - \sin A \sin B}{\cos A \cos B}} = \dfrac{\dfrac{\sin A \cos B}{\cos A \cos B} + \dfrac{\cos A \sin B}{\cos A \cos B}}{\dfrac{\cos A \cos B}{\cos A \cos B} - \dfrac{\sin A \sin B}{\cos A \cos B}}$
Simplify.	$\tan(A + B) = \dfrac{\left(\dfrac{\sin A}{\cos A}\right) + \left(\dfrac{\sin B}{\cos B}\right)}{1 - \left(\dfrac{\sin A}{\cos A}\right)\left(\dfrac{\sin B}{\cos B}\right)}$

Write the expressions inside the parentheses in terms of the tangent function.

$$\tan(A + B) = \frac{\tan A + \tan B}{1 - \tan A \tan B}$$

Replace B with $-B$.

$$\tan(A - B) = \frac{\tan A + \tan(-B)}{1 - \tan A \tan(-B)}$$

Since the tangent function is an odd function, $\tan(-B) = -\tan B$.

$$\tan(A - B) = \frac{\tan A - \tan B}{1 + \tan A \tan B}$$

SUM AND DIFFERENCE IDENTITIES FOR THE TANGENT FUNCTION

Sum $\qquad \tan(A + B) = \dfrac{\tan A + \tan B}{1 - \tan A \tan B}$

Difference $\qquad \tan(A - B) = \dfrac{\tan A - \tan B}{1 + \tan A \tan B}$

Technology Tip

If $\sin \alpha = -\frac{1}{3}$ and α is in QIII, then $\alpha = \pi + \sin^{-1}\left(\frac{1}{3}\right)$. If $\cos \beta = -\frac{1}{4}$ and β is in QII, then $\beta = \pi - \cos^{-1}\left(\frac{1}{4}\right)$. Now use the graphing calculator to find $\tan(\alpha + \beta)$ by entering

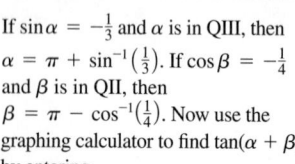

In Step 4, use the graphing calculator to evaluate both expressions

$$\frac{\dfrac{\sqrt{2}}{4} - \sqrt{15}}{1 - \left(\dfrac{\sqrt{2}}{4}\right)\left(-\sqrt{15}\right)} \text{ and}$$

$$\frac{\sqrt{2} - 4\sqrt{15}}{4 + \sqrt{30}} \text{ by entering}$$

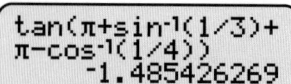

EXAMPLE 5   Finding Exact Values for the Tangent Function

Find the exact value of $\tan(\alpha + \beta)$ if $\sin \alpha = -\frac{1}{3}$ and $\cos \beta = -\frac{1}{4}$ and the terminal side of α lies in QIII and the terminal side of β lies in QII.

Solution:

STEP 1   Write the sum identity for the tangent function.

$$\tan(\alpha + \beta) = \frac{\tan \alpha + \tan \beta}{1 - \tan \alpha \tan \beta}$$

STEP 2   Find $\tan \alpha$.

The terminal side of α lies in QIII.

$$\sin \alpha = \frac{y}{r} = -\frac{1}{3}.$$

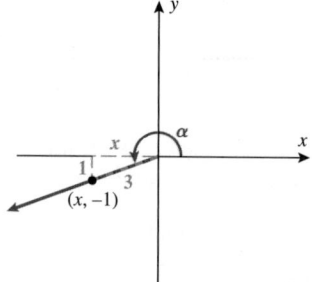

Solve for x. $\left(\text{Recall } x^2 + y^2 = r^2.\right)$

$$x^2 + (-1)^2 = 3^2$$
$$x = \pm\sqrt{8}$$

Take the negative sign since x is negative in QIII.

$$x = -2\sqrt{2}$$

Find $\tan \alpha$.

$$\tan \alpha = \frac{y}{x} = \frac{-1}{-2\sqrt{2}} = \frac{1}{2\sqrt{2}} \cdot \frac{\sqrt{2}}{\sqrt{2}} = \frac{\sqrt{2}}{4}$$

STEP 3 Find $\tan\beta$.

The terminal side of β lies in QII.

$$\cos\beta = -\frac{1}{4} = \frac{x}{r}.$$

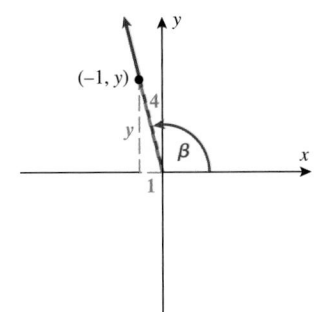

Solve for y. $\left(\text{Recall } x^2 + y^2 = r^2.\right)$

$$(-1)^2 + y^2 = 4^2$$
$$y = \pm\sqrt{15}$$

Take the positive sign since y is positive in QII.

$$y = \sqrt{15}$$

Find $\tan\beta$.

$$\tan\beta = \frac{y}{x} = \frac{\sqrt{15}}{-1} = -\sqrt{15}$$

STEP 4 Substitute $\tan\alpha = \dfrac{\sqrt{2}}{4}$ and $\tan\beta = -\sqrt{15}$ into the sum identity for the tangent function.

$$\tan(\alpha + \beta) = \frac{\dfrac{\sqrt{2}}{4} - \sqrt{15}}{1 - \left(\dfrac{\sqrt{2}}{4}\right)\left(-\sqrt{15}\right)}$$

Multiply the numerator and the denominator by 4.

$$\tan(\alpha + \beta) = \frac{4\left(\dfrac{\sqrt{2}}{4} - \sqrt{15}\right)}{4\left(1 + \dfrac{\sqrt{30}}{4}\right)}$$

$$= \frac{\sqrt{2} - 4\sqrt{15}}{4 + \sqrt{30}}$$

The expression $\boxed{\tan(\alpha + \beta) = \dfrac{\sqrt{2} - 4\sqrt{15}}{4 + \sqrt{30}}}$ can be simplified further if we rationalize the denominator.

It is important to note in Example 5 that right triangles have been superimposed in the Cartesian plane. The coordinate pair (x, y) can have positive or negative values, but the radius r is always positive. When right triangles are superimposed, with one vertex at the point (x, y) and another vertex at the origin, it is important to understand that triangles have positive side lengths.

SECTION 6.2 SUMMARY

In this section, we derived the sum and difference identities for the cosine function using the distance formula. The cofunction identities and sum and difference identities for the cosine function were used to derive the sum and difference identities for the sine function. We combined the sine and cosine sum and difference identities to determine the tangent sum and difference identities. The sum and difference identities enabled us to evaluate a trigonometric expression exactly for any multiple of $15°\left(\text{i.e., } \dfrac{\pi}{12}\right).$

$$\cos(A + B) = \cos A \cos B - \sin A \sin B$$
$$\cos(A - B) = \cos A \cos B + \sin A \sin B$$
$$\sin(A + B) = \sin A \cos B + \cos A \sin B$$
$$\sin(A - B) = \sin A \cos B - \cos A \sin B$$
$$\tan(A + B) = \frac{\tan A + \tan B}{1 - \tan A \tan B}$$
$$\tan(A - B) = \frac{\tan A - \tan B}{1 + \tan A \tan B}$$

SECTION
6.2 EXERCISES

SKILLS

In Exercises 1–16, find the exact value for each trigonometric expression.

1. $\sin\left(\dfrac{\pi}{12}\right)$
2. $\cos\left(\dfrac{\pi}{12}\right)$
3. $\cos\left(-\dfrac{5\pi}{12}\right)$
4. $\sin\left(-\dfrac{5\pi}{12}\right)$
5. $\tan\left(-\dfrac{\pi}{12}\right)$
6. $\tan\left(\dfrac{13\pi}{12}\right)$

7. $\sin 105°$
8. $\cos 195°$
9. $\tan(-105°)$
10. $\tan 165°$
11. $\cot\left(\dfrac{\pi}{12}\right)$
12. $\cot\left(-\dfrac{5\pi}{12}\right)$

13. $\sec\left(-\dfrac{11\pi}{12}\right)$
14. $\sec\left(-\dfrac{13\pi}{12}\right)$
15. $\csc(-255°)$
16. $\csc(-15°)$

In Exercises 17–28, write each expression as a single trigonometric function.

17. $\sin(2x)\sin(3x) + \cos(2x)\cos(3x)$
18. $\sin x \sin(2x) - \cos x \cos(2x)$

19. $\sin x \cos(2x) - \cos x \sin(2x)$
20. $\sin(2x)\cos(3x) + \cos(2x)\sin(3x)$

21. $\cos(\pi - x)\sin x + \sin(\pi - x)\cos x$
22. $\sin\left(\dfrac{\pi}{3}x\right)\cos\left(-\dfrac{\pi}{2}x\right) - \cos\left(\dfrac{\pi}{3}x\right)\sin\left(-\dfrac{\pi}{2}x\right)$

23. $(\sin A - \sin B)^2 + (\cos A - \cos B)^2 - 2$
24. $(\sin A + \sin B)^2 + (\cos A + \cos B)^2 - 2$

25. $2 - (\sin A + \cos B)^2 - (\cos A + \sin B)^2$
26. $2 - (\sin A - \cos B)^2 - (\cos A + \sin B)^2$

27. $\dfrac{\tan 49° - \tan 23°}{1 + \tan 49° \tan 23°}$
28. $\dfrac{\tan 49° + \tan 23°}{1 - \tan 49° \tan 23°}$

In Exercises 29–34, find the exact value of the indicated expression using the given information and identities.

29. Find the exact value of $\cos(\alpha + \beta)$ if $\cos\alpha = -\frac{1}{3}$ and $\cos\beta = -\frac{1}{4}$ and the terminal side of α lies in QIII and the terminal side of β lies in QII.

30. Find the exact value of $\cos(\alpha - \beta)$ if $\cos\alpha = \frac{1}{3}$ and $\cos\beta = -\frac{1}{4}$ and the terminal side of α lies in QIV and the terminal side of β lies in QII.

31. Find the exact value of $\sin(\alpha - \beta)$ if $\sin\alpha = -\frac{3}{5}$ and $\sin\beta = \frac{1}{5}$ and the terminal side of α lies in QIII and the terminal side of β lies in QI.

32. Find the exact value of $\sin(\alpha + \beta)$ if $\sin\alpha = -\frac{3}{5}$ and $\sin\beta = \frac{1}{5}$ and the terminal side of α lies in QIII and the terminal side of β lies in QII.

33. Find the exact value of $\tan(\alpha + \beta)$ if $\sin\alpha = -\frac{3}{5}$ and $\cos\beta = -\frac{1}{4}$ and the terminal side of α lies in QIII and the terminal side of β lies in QII.

34. Find the exact value of $\tan(\alpha - \beta)$ if $\sin\alpha = -\frac{3}{5}$ and $\cos\beta = -\frac{1}{4}$ and the terminal side of α lies in QIII and the terminal side of β lies in QII.

In Exercises 35–52, determine whether each equation is a conditional equation or an identity.

35. $\sin(A + B) + \sin(A - B) = 2\sin A \cos B$
36. $\cos(A + B) + \cos(A - B) = 2\cos A \cos B$

37. $\sin\left(x - \dfrac{\pi}{2}\right) = \cos\left(x + \dfrac{\pi}{2}\right)$
38. $\sin\left(x + \dfrac{\pi}{2}\right) = \cos\left(x + \dfrac{\pi}{2}\right)$

39. $\dfrac{\sqrt{2}}{2}(\sin x + \cos x) = \sin\left(x + \dfrac{\pi}{4}\right)$
40. $\sqrt{3}\cos x + \sin x = 2\cos\left(x + \dfrac{\pi}{3}\right)$

41. $\sin^2 x = \dfrac{1 - \cos(2x)}{2}$

42. $\cos^2 x = \dfrac{1 + \cos(2x)}{2}$

43. $\sin(2x) = 2 \sin x \cos x$

44. $\cos(2x) = \cos^2 x - \sin^2 x$

45. $\sin(A + B) = \sin A + \sin B$

46. $\cos(A + B) = \cos A + \cos B$

47. $\tan(\pi + B) = \tan B$

48. $\tan(A - \pi) = \tan A$

49. $\cot(3\pi + x) = \dfrac{1}{\tan x}$

50. $\csc(2x) = 2 \sec x \csc x$

51. $\dfrac{1 + \tan x}{1 - \tan x} = \tan\left(x - \dfrac{\pi}{4}\right)$

52. $\cot\left(x + \dfrac{\pi}{4}\right) = \dfrac{1 - \tan x}{1 + \tan x}$

In Exercises 53–62, graph each of the functions by first rewriting it as a sine, cosine, or tangent of a difference or sum.

53. $y = \cos\left(\dfrac{\pi}{3}\right) \sin x + \cos x \sin\left(\dfrac{\pi}{3}\right)$

54. $y = \cos\left(\dfrac{\pi}{3}\right) \sin x - \cos x \sin\left(\dfrac{\pi}{3}\right)$

55. $y = \sin x \sin\left(\dfrac{\pi}{4}\right) + \cos x \cos\left(\dfrac{\pi}{4}\right)$

56. $y = \sin x \sin\left(\dfrac{\pi}{4}\right) - \cos x \cos\left(\dfrac{\pi}{4}\right)$

57. $y = -\sin x \cos(3x) - \cos x \sin(3x)$

58. $y = \sin x \sin(3x) + \cos x \cos(3x)$

59. $y = \dfrac{1 + \tan x}{1 - \tan x}$

60. $y = \dfrac{\sqrt{3} - \tan x}{1 + \sqrt{3}\tan x}$

61. $y = \dfrac{1 + \sqrt{3}\tan x}{\sqrt{3} - \tan x}$

62. $y = \dfrac{1 - \tan x}{1 + \tan x}$

■ APPLICATIONS

In Exercises 63 and 64, refer to the following:

Sum and difference identities can be used to simplify more complicated expressions. For instance, the sine and cosine function can be represented by infinite polynomials called power series.

$$\cos x = 1 - \frac{x^2}{2!} + \frac{x^4}{4!} - \frac{x^6}{6!} + \frac{x^8}{8!} - \cdots$$

$$\sin x = x - \frac{x^3}{3!} + \frac{x^5}{5!} - \frac{x^7}{7!} + \frac{x^9}{9!} - \cdots$$

63. Power Series. Find the power series that represents $\cos\left(x - \dfrac{\pi}{4}\right)$.

64. Power Series. Find the power series that represents $\sin\left(x + \dfrac{3\pi}{2}\right)$.

For Exercises 65 and 66, use the following:

A nonvertical line makes an angle with the x-axis. In the figure, we see that the line L_1 makes an acute angle θ_1 with the x-axis. Similarly, the line L_2 makes an acute angle θ_2 with the x-axis.

$\tan \theta_1 = $ slope of $L_1 = m_1$

$\tan \theta_2 = $ slope of $L_2 = m_2$

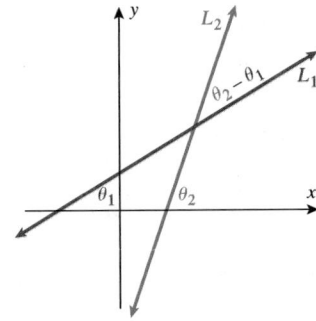

65. Angle Between Two Lines. Show that

$$\tan(\theta_2 - \theta_1) = \frac{m_2 - m_1}{1 + m_1 m_2}$$

66. Relating Tangent and Slope. Show that

$$\tan(\theta_1 - \theta_2) = \frac{m_1 - m_2}{1 + m_1 m_2}$$

For Exercises 67 and 68, refer to the following:

An electric field E of a wave with constant amplitude A, propagating a distance z, is given by

$$E = A\cos(kz - ct)$$

where k is the propagation wave number, which is related to the wavelength λ by $k = \dfrac{2\pi}{\lambda}$, and where $c = 3.0 \times 10^8$ m/s is the speed of light in a vacuum, and t is time in seconds.

67. Electromagnetic Wave Propagation. Use the cosine difference identity to express the electric field in terms of both sine and cosine functions. When the quotient of the propagation distance z and the wavelength λ are equal to an integer, what do you notice?

68. Electromagnetic Wave Propagation. Use the cosine difference identity to express the electric field in terms of both sine and cosine functions. When $t = 0$, what do you notice?

■ **CATCH THE MISTAKE**

In Exercises 69 and 70, explain the mistake that is made.

69. Find the exact value of $\tan\left(\dfrac{5\pi}{12}\right)$.

Solution:

Write $\dfrac{5\pi}{12}$ as a sum. $\qquad \tan\left(\dfrac{\pi}{4}+\dfrac{\pi}{6}\right)$

Distribute. $\qquad\qquad\quad \tan\left(\dfrac{\pi}{4}\right)+\tan\left(\dfrac{\pi}{6}\right)$

Evaluate the tangent
function for $\dfrac{\pi}{4}$ and $\dfrac{\pi}{6}$. $\qquad 1+\dfrac{\sqrt{3}}{3}$

This is incorrect. What mistake was made?

70. Find the exact value of $\tan\left(-\dfrac{7\pi}{6}\right)$.

Solution:

The tangent function is
an even function. $\qquad\qquad\qquad \tan\left(\dfrac{7\pi}{6}\right)$

Write $\dfrac{7\pi}{6}$ as a sum. $\qquad\qquad \tan\left(\pi+\dfrac{\pi}{6}\right)$

Use the tangent sum identity, $\qquad \dfrac{\tan\pi+\tan\left(\dfrac{\pi}{6}\right)}{1-\tan\pi\tan\left(\dfrac{\pi}{6}\right)}$
$\tan(A+B)=\dfrac{\tan A+\tan B}{1-\tan A\tan B}$.

Evaluate the tangent functions
on the right. $\qquad\qquad\qquad \dfrac{0+\dfrac{1}{\sqrt{3}}}{1-0}$

Simplify. $\qquad\qquad\qquad\qquad \dfrac{\sqrt{3}}{3}$

This is incorrect. What mistake was made?

■ **CONCEPTUAL**

**In Exercises 71–74, determine whether each statement is
true or false.**

71. $\cos 15° = \cos 45° - \cos 30°$

72. $\sin\left(\dfrac{\pi}{2}\right) = \sin\left(\dfrac{\pi}{3}\right) + \sin\left(\dfrac{\pi}{6}\right)$

73. $\tan\left(x+\dfrac{\pi}{4}\right) = 1 + \tan x$

74. $\cot\left(\dfrac{\pi}{4}-x\right) = \dfrac{1+\tan x}{1-\tan x}$

■ **CHALLENGE**

75. Verify that $\sin(A + B + C) = \sin A\cos B\cos C +$
$\cos A\sin B\cos C + \cos A\cos B\sin C - \sin A\sin B\sin C$.

76. Verify that $\cos(A + B + C) = \cos A\cos B\cos C$
$- \sin A\sin B\cos C - \sin A\cos B\sin C - \cos A\sin B\sin C$.

77. Although in general the statement $\sin(A - B) = \sin A - \sin B$
is not true, it is true for some values. Determine some
values of A and B that make this statement true.

78. Although in general the statement $\sin(A + B) = \sin A + \sin B$
is not true, it is true for some values. Determine some values
of A and B that make this statement true.

■ **TECHNOLOGY**

79. In Exercise 63, you showed that the difference quotient for
$f(x) = \sin x$ is $\cos x\left(\dfrac{\sin h}{h}\right) - \sin x\left(\dfrac{1-\cos h}{h}\right)$.

Plot $Y_1 = \cos x\left(\dfrac{\sin h}{h}\right) - \sin x\left(\dfrac{1-\cos h}{h}\right)$ for

a. $h = 1$ **b.** $h = 0.1$ **c.** $h = 0.01$

What function does the difference quotient for $f(x) = \sin x$
resemble when h approaches zero?

80. Show that the difference quotient for $f(x) = \cos x$ is
$-\sin x\left(\dfrac{\sin h}{h}\right) - \cos x\left(\dfrac{1-\cos h}{h}\right)$.

Plot $Y_1 = -\sin x\left(\dfrac{\sin h}{h}\right) - \cos x\left(\dfrac{1-\cos h}{h}\right)$ for

a. $h = 1$ **b.** $h = 0.1$ **c.** $h = 0.01$

What function does the difference quotient for $f(x) = \cos x$
resemble when h approaches zero?

81. Show that the difference quotient for $f(x) = \sin(2x)$ is

$$\cos(2x)\left[\frac{\sin(2h)}{h}\right] - \sin(2x)\left[\frac{1 - \cos(2h)}{h}\right].$$

Plot $Y_1 = \cos(2x)\left[\dfrac{\sin(2h)}{h}\right] - \sin(2x)\left[\dfrac{1 - \cos(2h)}{h}\right]$ for

a. $h = 1$ **b.** $h = 0.1$ **c.** $h = 0.01$

What function does the difference quotient for $f(x) = \sin(2x)$ resemble when h approaches zero?

82. Show that the difference quotient for $f(x) = \cos(2x)$ is

$$-\sin(2x)\left[\frac{\sin(2h)}{h}\right] - \cos(2x)\left[\frac{1 - \cos(2h)}{h}\right].$$

Plot $Y_1 = -\sin(2x)\left[\dfrac{\sin(2h)}{h}\right] - \cos(2x)\left[\dfrac{1 - \cos(2h)}{h}\right]$ for

a. $h = 1$ **b.** $h = 0.1$ **c.** $h = 0.01$

What function does the difference quotient for $f(x) = \cos(2x)$ resemble when h approaches zero?

▪ PREVIEW TO CALCULUS

In calculus, one technique used to solve differential equations consists of the separation of variables. For example, consider the equation $x^2 + 3y\dfrac{f(y)}{g(x)} = 0$, which is equivalent to $3yf(y) = -x^2 g(x)$. Here each side of the equation contains only one type of variable, either x or y.

In Exercises 83–86, use the sum and difference identities to separate the variables in each equation.

83. $\sin(x + y) = 0$

84. $\cos(x - y) = 0$

85. $\tan(x + y) = 2$

86. $\cos(x + y) = \sin y$

SECTION 6.3 DOUBLE-ANGLE AND HALF-ANGLE IDENTITIES

SKILLS OBJECTIVES

▪ Use the double-angle identities to find exact values of certain trigonometric functions.
▪ Use the double-angle identities to help in verifying identities.
▪ Use the half-angle identities to find exact values of certain trigonometric functions.
▪ Use half-angle identities to help in verifying identities.

CONCEPTUAL OBJECTIVE

▪ Understand that the double-angle identities are derived from the sum identities.
▪ Understand that the half-angle identities are derived from the double-angle identities.

Double-Angle Identities

In previous chapters, we could only evaluate trigonometric functions exactly for reference angles of $30°$, $45°$, and $60°$ or $\dfrac{\pi}{6}, \dfrac{\pi}{4}$, and $\dfrac{\pi}{3}$; note that as of the previous section, we now can include multiples of $\dfrac{\pi}{12}$ among these "special" angles. Now, we can use *double-angle identities* to also evaluate the trigonometric function values for other angles that are even integer multiples of the special angles or to verify other trigonometric identities. One important distinction now is that we will be able to find exact values of many functions using the double-angle identities without needing to know the value of the angle.

Derivation of Double-Angle Identities

To derive the double-angle identities, we let $A = B$ in the sum identities.

WORDS	MATH
Write the identity for the sine of a sum.	$\sin(A + B) = \sin A \cos B + \cos A \sin B$
Let $B = A$.	$\sin(A + A) = \sin A \cos A + \cos A \sin A$
Simplify.	$\boxed{\sin(2A) = 2 \sin A \cos A}$
Write the identity for the cosine of a sum.	$\cos(A + B) = \cos A \cos B - \sin A \sin B$
Let $B = A$.	$\cos(A + A) = \cos A \cos A - \sin A \sin A$
Simplify.	$\boxed{\cos(2A) = \cos^2 A - \sin^2 A}$

We can write the double-angle identity for the cosine function two other ways if we use the Pythagorean identity:

1. Write the identity for the cosine function of a double angle.

$$\cos(2A) = \cos^2 A - \sin^2 A$$

Use the Pythagorean identity for the cosine function.

$$\cos(2A) = \underbrace{\cos^2 A}_{1-\sin^2 A} - \sin^2 A$$

Simplify.

$$\boxed{\cos(2A) = 1 - 2\sin^2 A}$$

2. Write the identity for the cosine function of a double angle.

$$\cos(2A) = \cos^2 A - \sin^2 A$$

Use the Pythagorean identity for the sine function.

$$\cos(2A) = \cos^2 A - \underbrace{\sin^2 A}_{1-\cos^2 A}$$

Simplify.

$$\boxed{\cos(2A) = 2\cos^2 A - 1}$$

The tangent function can always be written as a quotient, $\tan(2A) = \dfrac{\sin(2A)}{\cos(2A)}$, if $\sin(2A)$ and $\cos(2A)$ are known. Here we write the double-angle identity for the tangent function in terms of only the tangent function.

Write the tangent of a sum identity.	$\tan(A + B) = \dfrac{\tan A + \tan B}{1 - \tan A \tan B}$
Let $B = A$.	$\tan(A + A) = \dfrac{\tan A + \tan A}{1 - \tan A \tan A}$
Simplify.	$\boxed{\tan(2A) = \dfrac{2\tan A}{1 - \tan^2 A}}$

DOUBLE-ANGLE IDENTITIES

SINE	COSINE	TANGENT
$\sin(2A) = 2\sin A \cos A$	$\cos(2A) = \cos^2 A - \sin^2 A$	$\tan(2A) = \dfrac{2\tan A}{1 - \tan^2 A}$
	$\cos(2A) = 1 - 2\sin^2 A$	
	$\cos(2A) = 2\cos^2 A - 1$	

Applying Double-Angle Identities

EXAMPLE 1   Finding Exact Values Using Double-Angle Identities

If $\cos x = \frac{2}{3}$, find $\sin(2x)$ given $\sin x < 0$.

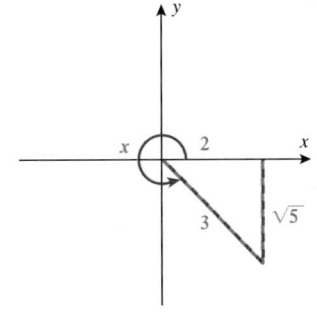

Technology Tip

If $\cos x = \frac{2}{3}$ and $\sin x < 0$, then x is in QIV and one value for x is $x = 2\pi - \cos^{-1}\left(\frac{2}{3}\right)$. Now use the graphing calculator to find $\sin(2x)$, entering $\sin\left\{2\left[2\pi - \cos^{-1}\left(\frac{2}{3}\right)\right]\right\}$, and compare that value to $-\dfrac{4\sqrt{5}}{9}$.

```
sin(2(2π-cos-1(2/
3)))
           -.99380799
-4√(5)/9
           -.99380799
```

Solution:

Find $\sin x$.

Use the Pythagorean identity. $\sin^2 x + \cos^2 x = 1$

Substitute $\cos x = \frac{2}{3}$. $\sin^2 x + \left(\dfrac{2}{3}\right)^2 = 1$

Solve for $\sin x$, which is negative. $\sin x = -\sqrt{1 - \dfrac{4}{9}}$

Simplify. $\sin x = -\dfrac{\sqrt{5}}{3}$

Find $\sin(2x)$.

Use the double-angle formula for the sine function. $\sin(2x) = 2\sin x \cos x$

Substitute $\sin x = -\dfrac{\sqrt{5}}{3}$ and $\cos x = \dfrac{2}{3}$. $\sin(2x) = 2\left(-\dfrac{\sqrt{5}}{3}\right)\left(\dfrac{2}{3}\right)$

Simplify. $\boxed{\sin(2x) = -\dfrac{4\sqrt{5}}{9}}$

▪ **YOUR TURN** If $\cos x = -\frac{1}{3}$, find $\sin(2x)$ given $\sin x < 0$.

▪ **Answer:** $\sin(2x) = \dfrac{4\sqrt{2}}{9}$

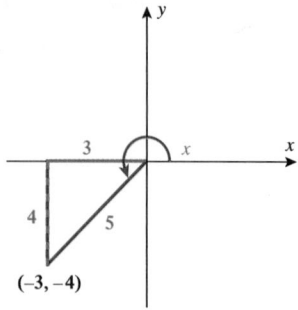

(-3, -4)

If $\sin x = -\frac{4}{5}$ and $\cos x < 0$, then x is in QIII and a value for x is $x = \pi + \sin^{-1}(\frac{4}{5})$. Now use the graphing calculator to find $\sin(2x), \cos(2x)$, and $\tan(2x)$.

```
sin(2(π+sin⁻¹(4/5
)))
                .96
Ans▸Frac
              24/25
```

```
cos(2(π+sin⁻¹(4/5
)))
               -.28
Ans▸Frac
              -7/25
```

```
tan(2(π+sin⁻¹(4/5
)))
        -3.428571429
Ans▸Frac
               -24/7
```

■ **Answer:** $\sin(2x) = -\frac{24}{25}$,
$\cos(2x) = -\frac{7}{25}, \tan(2x) = \frac{24}{7}$

EXAMPLE 2 Finding Exact Values Using Double-Angle Identities

If $\sin x = -\frac{4}{5}$ and $\cos x < 0$, find $\sin(2x), \cos(2x)$, and $\tan(2x)$.

Solution:

Solve for $\cos x$.

Use the Pythagorean identity. $\qquad\qquad \sin^2 x + \cos^2 x = 1$

Substitute $\sin x = -\frac{4}{5}$. $\qquad\qquad \left(-\frac{4}{5}\right)^2 + \cos^2 x = 1$

Simplify. $\qquad\qquad \cos^2 x = \frac{9}{25}$

Solve for $\cos x$, which is negative. $\qquad \cos x = -\sqrt{\frac{9}{25}} = -\frac{3}{5}$

Find $\sin(2x)$.

Use the double-angle identity for the sine function. $\qquad \sin(2x) = 2\sin x \cos x$

Substitute $\sin x = -\frac{4}{5}$ and $\cos x = -\frac{3}{5}$. $\qquad \sin(2x) = 2\left(-\frac{4}{5}\right)\left(-\frac{3}{5}\right)$

Simplify. $\qquad\qquad \boxed{\sin(2x) = \frac{24}{25}}$

Find $\cos(2x)$.

Use the double-angle identity for the cosine function. $\qquad \cos(2x) = \cos^2 x - \sin^2 x$

Substitute $\sin x = -\frac{4}{5}$ and $\cos x = -\frac{3}{5}$. $\qquad \cos(2x) = \left(-\frac{3}{5}\right)^2 - \left(-\frac{4}{5}\right)^2$

Simplify. $\qquad\qquad \boxed{\cos(2x) = -\frac{7}{25}}$

Find $\tan(2x)$.

Use the quotient identity. $\qquad\qquad \tan\theta = \frac{\sin\theta}{\cos\theta}$

Let $\theta = 2x$. $\qquad\qquad \tan(2x) = \frac{\sin(2x)}{\cos(2x)}$

Substitute $\sin(2x) = \frac{24}{25}$ and $\cos(2x) = -\frac{7}{25}$. $\qquad \tan(2x) = \frac{\frac{24}{25}}{-\frac{7}{25}}$

Simplify. $\qquad\qquad \boxed{\tan(2x) = -\frac{24}{7}}$

Note: We could also have found $\tan(2x)$ first by finding $\tan x = \frac{\sin x}{\cos x}$ and then using the value for $\tan x$ in the double-angle identity, $\tan(2A) = \frac{2\tan A}{1 - \tan^2 A}$.

■ **YOUR TURN** If $\cos x = \frac{3}{5}$ and $\sin x < 0$, find $\sin(2x), \cos(2x)$, and $\tan(2x)$.

EXAMPLE 3   Verifying Trigonometric Identities Using Double-Angle Identities

Verify the identity $(\sin x - \cos x)^2 = 1 - \sin(2x)$.

Solution:

Start with the left side of the equation. $\boxed{(\sin x - \cos x)^2}$

Expand by squaring. $= \sin^2 x - 2\sin x \cos x + \cos^2 x$

Group the $\sin^2 x$ and $\cos^2 x$ terms. $= \sin^2 x + \cos^2 x - 2\sin x \cos x$

Apply the Pythagorean identity. $= \underbrace{\sin^2 x + \cos^2 x}_{1} - 2\sin x \cos x$

Apply the sine double-angle identity. $= 1 - \underbrace{2\sin x \cos x}_{\sin(2x)}$

Simplify. $= \boxed{1 - \sin(2x)}$

EXAMPLE 4   Verifying Multiple-Angle Identities

Verify the identity $\cos(3x) = \left(1 - 4\sin^2 x\right)\cos x$.

Solution:

Write the cosine of a sum identity. $\cos(A + B) = \cos A \cos B - \sin A \sin B$

Let $A = 2x$ and $B = x$. $\cos(2x + x) = \cos(2x)\cos x - \sin(2x)\sin x$

Apply the double-angle identities. $\cos(3x) = \underbrace{\cos(2x)}_{1 - 2\sin^2 x} \cos x - \underbrace{\sin(2x)}_{2\sin x \cos x} \sin x$

Simplify. $\cos(3x) = \cos x - 2\sin^2 x \cos x - 2\sin^2 x \cos x$

$\cos(3x) = \cos x - 4\sin^2 x \cos x$

Factor out the common cosine term. $\boxed{\cos(3x) = \left(1 - 4\sin^2 x\right)\cos x}$

EXAMPLE 5   Simplifying Trigonometric Expressions Using Double-Angle Identities

Graph $y = \dfrac{\cot x - \tan x}{\cot x + \tan x}$.

Solution:

Simplify $y = \dfrac{\cot x - \tan x}{\cot x + \tan x}$ first.

Write the cotangent and tangent functions in terms of the sine and cosine functions.

$y = \dfrac{\dfrac{\cos x}{\sin x} - \dfrac{\sin x}{\cos x}}{\dfrac{\cos x}{\sin x} + \dfrac{\sin x}{\cos x}}$

Multiply the numerator and the denominator by $\sin x \cos x$.

$y = \dfrac{\left(\dfrac{\cos x}{\sin x} - \dfrac{\sin x}{\cos x}\right)}{\left(\dfrac{\cos x}{\sin x} + \dfrac{\sin x}{\cos x}\right)}\left(\dfrac{\sin x \cos x}{\sin x \cos x}\right)$

Simplify.

$y = \dfrac{\cos^2 x - \sin^2 x}{\cos^2 x + \sin^2 x}$

Technology Tip

Graphs of $y = \dfrac{\cot x - \tan x}{\cot x + \tan x}$ and $y = \cos(2x)$.

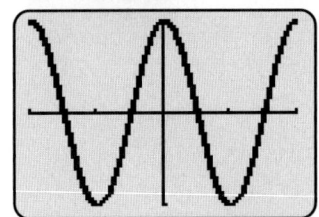

Use the double-angle and Pythagorean identities.

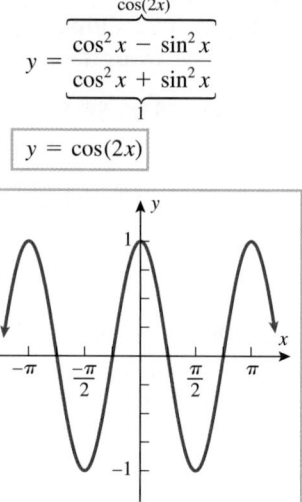

$$y = \dfrac{\overbrace{\cos^2 x - \sin^2 x}^{\cos(2x)}}{\underbrace{\cos^2 x + \sin^2 x}_{1}}$$

$$\boxed{y = \cos(2x)}$$

Graph $y = \cos(2x)$.

Half-Angle Identities

We now use the *double-angle identities* to develop the *half-angle identities*. Like the double-angle identities, the half-angle identities will allow us to find certain exact values of trigonometric functions and to verify other trigonometric identities. The *half-angle identities* come directly from the double-angle identities. We start by rewriting the second and third forms of the cosine double-angle identity to obtain identities for the square of the sine and cosine functions, $\sin^2$ and $\cos^2$.

WORDS	**MATH**
Write the second form of the cosine double-angle identity.	$\cos(2A) = 1 - 2\sin^2 A$
Find $\sin^2 A$.	
Isolate the $2\sin^2 A$ term on one side of the equation.	$2\sin^2 A = 1 - \cos(2A)$
Divide both sides by 2.	$\boxed{\sin^2 A = \dfrac{1 - \cos(2A)}{2}}$
Find $\cos^2 A$.	
Write the third form of the cosine double-angle identity.	$\cos(2A) = 2\cos^2 A - 1$
Isolate the $2\cos^2 A$ term on one side of the equation.	$2\cos^2 A = 1 + \cos^2 A$
Divide both sides by 2.	$\boxed{\cos^2 A = \dfrac{1 + \cos(2A)}{2}}$
Find $\tan^2 A$.	
Taking the quotient of these leads us to another identity.	$\tan^2 A = \dfrac{\sin^2 A}{\cos^2 A} = \dfrac{\dfrac{1 - \cos(2A)}{2}}{\dfrac{1 + \cos(2A)}{2}}$
Simplify.	$\boxed{\tan^2 A = \dfrac{1 - \cos(2A)}{1 + \cos(2A)}}$

These three identities for the squared functions—really, alternative forms of the double-angle identities—are used in calculus as power reduction formulas (identities that allow us to reduce the power of the trigonometric function from 2 to 1):

$$\sin^2 A = \frac{1 - \cos(2A)}{2}$$

$$\cos^2 A = \frac{1 + \cos(2A)}{2}$$

$$\tan^2 A = \frac{1 - \cos(2A)}{1 + \cos(2A)}$$

We can now use these forms of the double-angle identities to derive the *half-angle identities*.

WORDS	MATH
For the *sine half-angle identity*, start with the double-angle formula involving both the sine and cosine functions, $\cos(2x) = 1 - 2\sin^2 x$, and solve for $\sin^2 x$.	$\sin^2 x = \dfrac{1 - \cos(2x)}{2}$
Solve for $\sin x$.	$\sin x = \pm\sqrt{\dfrac{1 - \cos(2x)}{2}}$
Let $x = \dfrac{A}{2}$.	$\sin\left(\dfrac{A}{2}\right) = \pm\sqrt{\dfrac{1 - \cos 2\left(\dfrac{A}{2}\right)}{2}}$
Simplify.	$\sin\left(\dfrac{A}{2}\right) = \pm\sqrt{\dfrac{1 - \cos A}{2}}$
For the *cosine half-angle identity*, start with the double-angle formula involving only the cosine function, $\cos(2x) = 2\cos^2 x - 1$, and solve for $\cos^2 x$.	$\cos^2 x = \dfrac{1 + \cos(2x)}{2}$
Solve for $\cos x$.	$\cos x = \pm\sqrt{\dfrac{1 + \cos(2x)}{2}}$
Let $x = \dfrac{A}{2}$.	$\cos\left(\dfrac{A}{2}\right) = \pm\sqrt{\dfrac{1 + \cos 2\left(\dfrac{A}{2}\right)}{2}}$
Simplify.	$\cos\left(\dfrac{A}{2}\right) = \pm\sqrt{\dfrac{1 + \cos A}{2}}$
For the *tangent half-angle identity*, start with the quotient identity.	$\tan\left(\dfrac{A}{2}\right) = \dfrac{\sin\left(\dfrac{A}{2}\right)}{\cos\left(\dfrac{A}{2}\right)}$
Substitute half-angle identities for the sine and cosine functions.	$\tan\left(\dfrac{A}{2}\right) = \dfrac{\pm\sqrt{\dfrac{1 - \cos A}{2}}}{\pm\sqrt{\dfrac{1 + \cos A}{2}}}$
Simplify.	$\tan\left(\dfrac{A}{2}\right) = \pm\sqrt{\dfrac{1 - \cos A}{1 + \cos A}}$

Note: We can also find $\tan\left(\dfrac{A}{2}\right)$ by starting with the identity $\tan^2 x = \dfrac{1 - \cos(2x)}{1 + \cos(2x)}$, solving for $\tan x$, and letting $x = \dfrac{A}{2}$. The tangent function also has two other similar forms for $\tan\left(\dfrac{A}{2}\right)$ (see Exercises 127 and 128).

HALF-ANGLE IDENTITIES

Study Tip

The sign $+$ or $-$ is determined by what quadrant contains $\dfrac{A}{2}$ and what the sign of the particular trigonometric function is in that quadrant.

SINE	COSINE	TANGENT
$\sin\left(\dfrac{A}{2}\right) = \pm\sqrt{\dfrac{1 - \cos A}{2}}$	$\cos\left(\dfrac{A}{2}\right) = \pm\sqrt{\dfrac{1 + \cos A}{2}}$	$\tan\left(\dfrac{A}{2}\right) = \pm\sqrt{\dfrac{1 - \cos A}{1 + \cos A}}$
		$\tan\left(\dfrac{A}{2}\right) = \dfrac{\sin A}{1 + \cos A}$
		$\tan\left(\dfrac{A}{2}\right) = \dfrac{1 - \cos A}{\sin A}$

It is important to note that these identities hold for any real number A or any angle with either degree measure or radian measure A as long as both sides of the equation are defined. The sign ($+$ or $-$) is determined by the sign of the trigonometric function in the quadrant that contains $\dfrac{A}{2}$.

EXAMPLE 6   Finding Exact Values Using Half-Angle Identities

Use a half-angle identity to find $\cos 15°$.

Technology Tip

Use a TI calculator to check the values of $\cos 15°$ and $\sqrt{\dfrac{2 + \sqrt{3}}{4}}$. Be sure the calculator is in degree mode.

```
cos(15)
        .9659258263
√((2+√(3))/4)
        .9659258263
```

Solution:

Write $\cos 15°$ in terms of a half angle.
$$\cos 15° = \cos\left(\dfrac{30°}{2}\right)$$

Write the half-angle identity for the cosine function.
$$\cos\left(\dfrac{A}{2}\right) = \pm\sqrt{\dfrac{1 + \cos A}{2}}$$

Substitute $A = 30°$.
$$\cos\left(\dfrac{30°}{2}\right) = \pm\sqrt{\dfrac{1 + \cos 30°}{2}}$$

Simplify.
$$\cos 15° = \pm\sqrt{\dfrac{1 + \dfrac{\sqrt{3}}{2}}{2}}$$

$15°$ is in QI, where the cosine function is positive.
$$\cos 15° = \sqrt{\dfrac{2 + \sqrt{3}}{4}} = \boxed{\dfrac{\sqrt{2 + \sqrt{3}}}{2}}$$

■ **Answer:** $\sin 22.5° = \dfrac{\sqrt{2 - \sqrt{2}}}{2}$

■ **YOUR TURN** Use a half-angle identity to find $\sin 22.5°$.

EXAMPLE 7   Finding Exact Values Using Half-Angle Identities

Use a half-angle identity to find $\tan\left(\dfrac{11\pi}{12}\right)$.

Solution:

Write $\tan\left(\dfrac{11\pi}{12}\right)$ in terms of a half angle.

$$\tan\left(\frac{11\pi}{12}\right) = \tan\left(\frac{\frac{11\pi}{6}}{2}\right)$$

Write the half-angle identity for the tangent function.*

$$\tan\left(\frac{A}{2}\right) = \frac{1 - \cos A}{\sin A}$$

Substitute $A = \dfrac{11\pi}{6}$.

$$\tan\left(\frac{\frac{11\pi}{6}}{2}\right) = \frac{1 - \cos\left(\frac{11\pi}{6}\right)}{\sin\left(\frac{11\pi}{6}\right)}$$

Simplify.

$$\tan\left(\frac{11\pi}{12}\right) = \frac{1 - \frac{\sqrt{3}}{2}}{-\frac{1}{2}}$$

$$\tan\left(\frac{11\pi}{12}\right) = \boxed{\sqrt{3} - 2}$$

$\dfrac{11\pi}{12}$ is in QII, where tangent is negative. Notice that if we approximate $\tan\left(\dfrac{11\pi}{12}\right)$ with a calculator, we find that $\tan\left(\dfrac{11\pi}{12}\right) \approx -0.2679$ and $\sqrt{3} - 2 \approx -0.2679$.

*This form of the tangent half-angle identity was selected because of mathematical simplicity. If we had selected either of the other forms, we would have obtained an expression that had a square root within a square root or a radical in the denominator (requiring rationalization).

■ **YOUR TURN** Use a half-angle identity to find $\tan\left(\dfrac{\pi}{8}\right)$.

Technology Tip

Use a TI calculator to check the values of $\tan\left(\dfrac{11\pi}{12}\right)$ and $\sqrt{3} - 2$. Be sure the calculator is in radian mode.

```
tan(11π/12)
           -.2679491924
√(3)-2
           -.2679491924
■
```

■ **Answer:** $\dfrac{\sqrt{2}}{2 + \sqrt{2}}$ or $\sqrt{2} - 1$

EXAMPLE 8   Finding Exact Values Using Half-Angle Identities

If $\cos x = \dfrac{3}{5}$ and $\dfrac{3\pi}{2} < x < 2\pi$, find $\sin\left(\dfrac{x}{2}\right)$, $\cos\left(\dfrac{x}{2}\right)$, and $\tan\left(\dfrac{x}{2}\right)$.

Solution:

Determine in which quadrant $\dfrac{x}{2}$ lies.

Since $\dfrac{3\pi}{2} < x < 2\pi$, we divide by 2.

$$\frac{3\pi}{4} < \frac{x}{2} < \pi$$

$\dfrac{x}{2}$ lies in QII; therefore, the sine function is positive and both the cosine and tangent functions are negative.

Technology Tip

If $\cos x = \dfrac{3}{5}$ and $\dfrac{3\pi}{2} < x < 2\pi$, then x is in QIV, $\dfrac{x}{2}$ is in QII, and $x = 2\pi - \cos^{-1}\left(\frac{3}{5}\right)$. Now use the graphing calculator to find $\sin\left(\dfrac{x}{2}\right)$, $\cos\left(\dfrac{x}{2}\right)$, and $\tan\left(\dfrac{x}{2}\right)$.

```
sin((2π-cos⁻¹(3/5
))/2)
         .4472135955
√(5)/5
         .4472135955
```

```
cos((2π-cos⁻¹(3/5
))/2)
        -.894427191
-2√(5)/5
        -.894427191
■
```

```
tan((2π-cos⁻¹(3/5
))/2)
              -.5
Ans►Frac
              -1/2
```

Write the half-angle identity for the sine function.

$$\sin\left(\frac{x}{2}\right) = \pm\sqrt{\frac{1 - \cos x}{2}}$$

Substitute $\cos x = \frac{3}{5}$.

$$\sin\left(\frac{x}{2}\right) = \pm\sqrt{\frac{1 - \frac{3}{5}}{2}}$$

Simplify.

$$\sin\left(\frac{x}{2}\right) = \pm\sqrt{\frac{1}{5}} = \pm\frac{\sqrt{5}}{5}$$

Since $\frac{x}{2}$ lies in QII, choose the positive value for the sine function.

$$\boxed{\sin\left(\frac{x}{2}\right) = \frac{\sqrt{5}}{5}}$$

Write the half-angle identity for the cosine function.

$$\cos\left(\frac{x}{2}\right) = \pm\sqrt{\frac{1 + \cos x}{2}}$$

Substitute $\cos x = \frac{3}{5}$.

$$\cos\left(\frac{x}{2}\right) = \pm\sqrt{\frac{1 + \frac{3}{5}}{2}}$$

Simplify.

$$\cos\left(\frac{x}{2}\right) = \pm\sqrt{\frac{4}{5}} = \pm\frac{2\sqrt{5}}{5}$$

Since $\frac{x}{2}$ lies in QII, choose the negative value for the cosine function.

$$\boxed{\cos\left(\frac{x}{2}\right) = -\frac{2\sqrt{5}}{5}}$$

Use the quotient identity for tangent.

$$\tan\left(\frac{x}{2}\right) = \frac{\sin\left(\frac{x}{2}\right)}{\cos\left(\frac{x}{2}\right)}$$

Substitute $\sin\left(\frac{x}{2}\right) = \frac{\sqrt{5}}{5}$ and $\cos\left(\frac{x}{2}\right) = -\frac{2\sqrt{5}}{5}$.

$$\tan\left(\frac{x}{2}\right) = \frac{\frac{\sqrt{5}}{5}}{-\frac{2\sqrt{5}}{5}}$$

Simplify.

$$\boxed{\tan\left(\frac{x}{2}\right) = -\frac{1}{2}}$$

■ **Answer:** $\sin\left(\frac{x}{2}\right) = \frac{2\sqrt{5}}{5}$,

$\cos\left(\frac{x}{2}\right) = -\frac{\sqrt{5}}{5}$, $\tan\left(\frac{x}{2}\right) = -2$

■ **YOUR TURN** If $\cos x = -\frac{3}{5}$ and $\pi < x < \frac{3\pi}{2}$, find $\sin\left(\frac{x}{2}\right)$, $\cos\left(\frac{x}{2}\right)$, and $\tan\left(\frac{x}{2}\right)$.

EXAMPLE 9 **Using Half-Angle Identities to Verify Other Identities**

Verify the identity $\cos^2\left(\frac{x}{2}\right) = \frac{\tan x + \sin x}{2\tan x}$.

Solution:

Write the cosine half-angle identity.

$$\cos\left(\frac{x}{2}\right) = \pm\sqrt{\frac{1 + \cos x}{2}}$$

Square both sides of the equation.

$$\cos^2\left(\frac{x}{2}\right) = \frac{1 + \cos x}{2}$$

Multiply the numerator and denominator on the right side by $\tan x$.

$$\cos^2\left(\frac{x}{2}\right) = \left(\frac{1 + \cos x}{2}\right)\left(\frac{\tan x}{\tan x}\right)$$

Simplify.

$$\cos^2\left(\frac{x}{2}\right) = \frac{\tan x + \cos x \tan x}{2\tan x}$$

Note that $\cos x \tan x = \sin x$.

$$\boxed{\cos^2\left(\frac{x}{2}\right) = \frac{\tan x + \sin x}{2\tan x}}$$

An alternative solution is to start with the right-hand side.

Solution (alternative):

Start with the right-hand side.

$$\frac{\tan x + \sin x}{2\tan x}$$

Write this expression as the sum of two expressions.

$$= \frac{\tan x}{2\tan x} + \frac{\sin x}{2\tan x}$$

Simplify.

$$= \frac{1}{2} + \frac{1}{2}\frac{\sin x}{\tan x}$$

Write $\tan x = \dfrac{\sin x}{\cos x}$.

$$= \frac{1}{2} + \frac{1}{2}\frac{\sin x}{\dfrac{\sin x}{\cos x}}$$

$$= \frac{1}{2}(1 + \cos x)$$

$$= \cos^2\left(\frac{x}{2}\right)$$

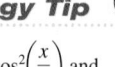

Technology Tip

Graphs of $y_1 = \cos^2\left(\dfrac{x}{2}\right)$ and

$y_2 = \dfrac{\tan x + \sin x}{2\tan x}$.

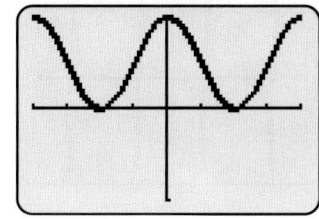

EXAMPLE 10 **Using Half-Angle Identities to Verify Other Trigonometric Identities**

Verify the identity $\tan x = \csc(2x) - \cot(2x)$.

Solution:

Write the third half-angle formula for the tangent function.

$$\tan\left(\frac{A}{2}\right) = \frac{1 - \cos A}{\sin A}$$

Write the right side as a difference of two expressions having the same denominator.

$$\tan\left(\frac{A}{2}\right) = \frac{1}{\sin A} - \frac{\cos A}{\sin A}$$

Substitute the reciprocal and quotient identities, respectively, on the right.

$$\tan\left(\frac{A}{2}\right) = \csc A - \cot A$$

Let $A = 2x$.

$$\boxed{\tan x = \csc(2x) - \cot(2x)}$$

Notice in Example 10 that we started with the third half-angle identity for the tangent function. In Example 11 we will start with the second half-angle identity for the tangent function. In general, you select the form that appears to lead to the desired expression.

Technology Tip

Graphs of $y_1 = \dfrac{\sin(2\pi x)}{1 + \cos(2\pi x)}$ and $y_2 = \tan(\pi x)$.

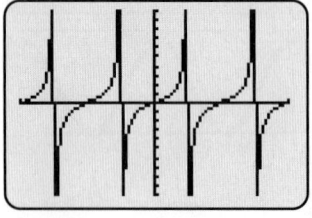

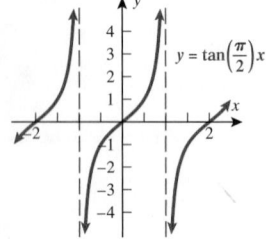

■ **Answer:**

EXAMPLE 11 Using Half-Angle Identities to Simplify Trigonometric Expressions

Graph $y = \dfrac{\sin(2\pi x)}{1 + \cos(2\pi x)}$.

Solution:

Simplify the trigonometric expression using a half-angle identity for the tangent function.

Write the second half-angle identity for the tangent function.

$$\tan\left(\frac{A}{2}\right) = \frac{\sin A}{1 + \cos A}$$

Let $A = 2\pi x$.

$$\tan(\pi x) = \frac{\sin(2\pi x)}{1 + \cos(2\pi x)}$$

Graph $y = \tan(\pi x)$.

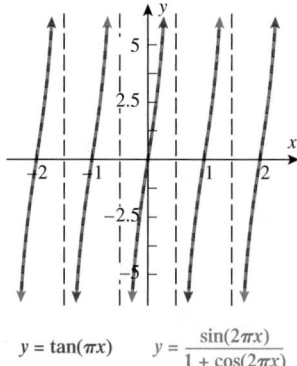

$$y = \tan(\pi x) \qquad y = \frac{\sin(2\pi x)}{1 + \cos(2\pi x)}$$

■ **YOUR TURN** Graph $y = \dfrac{1 - \cos(\pi x)}{\sin(\pi x)}$.

In this section, we derived the double-angle identities from the sum identities. We then used the double-angle identities to find exact values of trigonometric functions, to verify other trigonometric identities, and to simplify trigonometric expressions.

$$\sin(2A) = 2\sin A \cos A$$

$$\cos(2A) = \cos^2 A - \sin^2 A$$

$$= 1 - 2\sin^2 A$$

$$= 2\cos^2 A - 1$$

$$\tan(2A) = \frac{2\tan A}{1 - \tan^2 A}$$

The double-angle identities were used to derive the half-angle identities. We then used the half-angle identities to find certain exact values of trigonometric functions, verify other trigonometric identities, and simplify trigonometric expressions.

$$\sin\left(\frac{A}{2}\right) = \pm\sqrt{\frac{1 - \cos A}{2}} \qquad \cos\left(\frac{A}{2}\right) = \pm\sqrt{\frac{1 + \cos A}{2}}$$

$$\tan\left(\frac{A}{2}\right) = \pm\sqrt{\frac{1 - \cos A}{1 + \cos A}}$$

We determine the sign, $+$ or $-$, by first deciding which quadrant contains $\dfrac{A}{2}$ and then finding the sign of the indicated trigonometric function in that quadrant.

Recall that there are three forms of the tangent half-angle identity. There is no need to memorize the other forms of the tangent half-angle identity, since they can be derived by first using the Pythagorean identity and algebraic manipulation.

SECTION
6.3 EXERCISES

■ SKILLS

In Exercises 1–12, use the double-angle identities to answer the following questions:

1. If $\sin x = \dfrac{1}{\sqrt{5}}$ and $\cos x < 0$, find $\sin(2x)$.

2. If $\sin x = \dfrac{1}{\sqrt{5}}$ and $\cos x < 0$, find $\cos(2x)$.

3. If $\cos x = \dfrac{5}{13}$ and $\sin x < 0$, find $\tan(2x)$.

4. If $\cos x = -\dfrac{5}{13}$ and $\sin x < 0$, find $\tan(2x)$.

5. If $\tan x = \dfrac{12}{5}$ and $\pi < x < \dfrac{3\pi}{2}$, find $\sin(2x)$.

6. If $\tan x = \dfrac{12}{5}$ and $\pi < x < \dfrac{3\pi}{2}$, find $\cos(2x)$.

7. If $\sec x = \sqrt{5}$ and $\sin x > 0$, find $\tan(2x)$.

8. If $\sec x = \sqrt{3}$ and $\sin x < 0$, find $\tan(2x)$.

9. If $\csc x = -2\sqrt{5}$ and $\cos x < 0$, find $\sin(2x)$.

10. If $\csc x = -\sqrt{13}$ and $\cos x > 0$, find $\sin(2x)$.

11. If $\cos x = -\dfrac{12}{13}$ and $\csc x < 0$, find $\cot(2x)$.

12. If $\sin x = \dfrac{12}{13}$ and $\cot x < 0$, find $\csc(2x)$.

In Exercises 13–24, simplify each expression. Evaluate the resulting expression exactly, if possible.

13. $\dfrac{2\tan 15°}{1 - \tan^2 15°}$

14. $\dfrac{2\tan\left(\dfrac{\pi}{8}\right)}{1 - \tan^2\left(\dfrac{\pi}{8}\right)}$

15. $\sin\left(\dfrac{\pi}{8}\right)\cos\left(\dfrac{\pi}{8}\right)$

16. $\sin 15° \cos 15°$

17. $\cos^2(2x) - \sin^2(2x)$

18. $\cos^2(x + 2) - \sin^2(x + 2)$

19. $\dfrac{2\tan\left(\dfrac{5\pi}{12}\right)}{1 - \tan^2\left(\dfrac{5\pi}{12}\right)}$

20. $\dfrac{2\tan\left(\dfrac{x}{2}\right)}{1 - \tan^2\left(\dfrac{x}{2}\right)}$

21. $1 - 2\sin^2\left(\dfrac{7\pi}{12}\right)$

22. $2\sin^2\left(-\dfrac{5\pi}{8}\right) - 1$

23. $2\cos^2\left(-\dfrac{7\pi}{12}\right) - 1$

24. $1 - 2\cos^2\left(-\dfrac{\pi}{8}\right)$

In Exercises 25–40, use the double-angle identities to verify each identity.

25. $\csc(2A) = \tfrac{1}{2}\csc A \sec A$

26. $\cot(2A) = \tfrac{1}{2}(\cot A - \tan A)$

27. $(\sin x - \cos x)(\cos x + \sin x) = -\cos(2x)$

28. $(\sin x + \cos x)^2 = 1 + \sin(2x)$

29. $\cos^2 x = \dfrac{1 + \cos(2x)}{2}$

30. $\sin^2 x = \dfrac{1 - \cos(2x)}{2}$

31. $\cos^4 x - \sin^4 x = \cos(2x)$

32. $\cos^4 x + \sin^4 x = 1 - \tfrac{1}{2}\sin^2(2x)$

33. $8\sin^2 x\cos^2 x = 1 - \cos(4x)$

34. $[\cos(2x) - \sin(2x)][\sin(2x) + \cos(2x)] = \cos(4x)$

35. $-\tfrac{1}{2}\sec^2 x = -2\sin^2 x\csc^2(2x)$

36. $4\csc(4x) = \dfrac{\sec x \csc x}{\cos(2x)}$

37. $\sin(3x) = \sin x(4\cos^2 x - 1)$

38. $\tan(3x) = \dfrac{\tan x(3 - \tan^2 x)}{(1 - 3\tan^2 x)}$

39. $\tfrac{1}{2}\sin(4x) = 2\sin x\cos x - 4\sin^3 x\cos x$

40. $\cos(4x) = [\cos(2x) - \sin(2x)][(\cos(2x) + \sin(2x)]$

In Exercises 41–50, graph the functions.

41. $y = \dfrac{\sin(2x)}{1 - \cos(2x)}$

42. $y = \dfrac{2\tan x}{2 - \sec^2 x}$

43. $y = \dfrac{\cot x + \tan x}{\cot x - \tan x}$

44. $y = \frac{1}{2}\tan x \cot x \sec x \csc x$

45. $y = \sin(2x)\cos(2x)$

46. $y = 3\sin(3x)\cos(-3x)$

47. $y = 1 - \dfrac{\tan x \cot x}{\sec x \csc x}$

48. $y = 3 - 2\dfrac{\sec(2x)}{\csc(2x)}$

49. $y = \dfrac{\sin(2x)}{\cos x} - 3\cos(2x)$

50. $y = 2 + \dfrac{\sin(2x)}{\cos x} - 3\cos(2x)$

In Exercises 51–66, use the half-angle identities to find the exact values of the trigonometric expressions.

51. $\sin 15°$

52. $\cos 22.5°$

53. $\cos\left(\dfrac{11\pi}{12}\right)$

54. $\sin\left(\dfrac{\pi}{8}\right)$

55. $\cos 75°$

56. $\sin 75°$

57. $\tan 67.5°$

58. $\tan 202.5°$

59. $\sec\left(-\dfrac{9\pi}{8}\right)$

60. $\csc\left(\dfrac{9\pi}{8}\right)$

61. $\cot\left(\dfrac{13\pi}{8}\right)$

62. $\cot\left(\dfrac{7\pi}{8}\right)$

63. $\sec\left(\dfrac{5\pi}{8}\right)$

64. $\csc\left(-\dfrac{5\pi}{8}\right)$

65. $\cot(-135°)$

66. $\cot 105°$

In Exercises 67–82, use the half-angle identities to find the desired function values.

67. If $\cos x = \dfrac{5}{13}$ and $\sin x < 0$, find $\sin\left(\dfrac{x}{2}\right)$.

68. If $\cos x = -\dfrac{5}{13}$ and $\sin x < 0$, find $\cos\left(\dfrac{x}{2}\right)$.

69. If $\tan x = \dfrac{12}{5}$ and $\pi < x < \dfrac{3\pi}{2}$, find $\sin\left(\dfrac{x}{2}\right)$.

70. If $\tan x = \dfrac{12}{5}$ and $\pi < x < \dfrac{3\pi}{2}$, find $\cos\left(\dfrac{x}{2}\right)$.

71. If $\sec x = \sqrt{5}$ and $\sin x > 0$, find $\tan\left(\dfrac{x}{2}\right)$.

72. If $\sec x = \sqrt{3}$ and $\sin x < 0$, find $\tan\left(\dfrac{x}{2}\right)$.

73. If $\csc x = 3$ and $\cos x < 0$, find $\sin\left(\dfrac{x}{2}\right)$.

74. If $\csc x = -3$ and $\cos x > 0$, find $\cos\left(\dfrac{x}{2}\right)$.

75. If $\cos x = -\dfrac{1}{4}$ and $\csc x < 0$, find $\cot\left(\dfrac{x}{2}\right)$.

76. If $\cos x = \dfrac{1}{4}$ and $\cot x < 0$, find $\csc\left(\dfrac{x}{2}\right)$.

77. If $\cot x = -\dfrac{24}{5}$ and $\dfrac{\pi}{2} < x < \pi$, find $\cos\left(\dfrac{x}{2}\right)$.

78. If $\cot x = -\dfrac{24}{5}$ and $\dfrac{\pi}{2} < x < \pi$, find $\sin\left(\dfrac{x}{2}\right)$.

79. If $\sin x = -0.3$ and $\sec x > 0$, find $\tan\left(\dfrac{x}{2}\right)$.

80. If $\sin x = -0.3$ and $\sec x < 0$, find $\cot\left(\dfrac{x}{2}\right)$.

81. If $\sec x = 2.5$ and $\tan x > 0$, find $\cot\left(\dfrac{x}{2}\right)$.

82. If $\sec x = -3$ and $\cot x < 0$, find $\tan\left(\dfrac{x}{2}\right)$.

In Exercises 83–88, simplify each expression using half-angle identities. Do not evaluate.

83. $\sqrt{\dfrac{1 + \cos\left(\dfrac{5\pi}{6}\right)}{2}}$

84. $\sqrt{\dfrac{1 - \cos\left(\dfrac{\pi}{4}\right)}{2}}$

85. $\dfrac{\sin 150°}{1 + \cos 150°}$

86. $\dfrac{1 - \cos 150°}{\sin 150°}$

87. $\sqrt{\dfrac{1 - \cos\left(\dfrac{5\pi}{4}\right)}{1 + \cos\left(\dfrac{5\pi}{4}\right)}}$

88. $\sqrt{\dfrac{1 - \cos 15°}{1 + \cos 15°}}$

In Exercises 89–100, use the half-angle identities to verify the identities.

89. $\sin^2\left(\dfrac{x}{2}\right) + \cos^2\left(\dfrac{x}{2}\right) = 1$

90. $\cos^2\left(\dfrac{x}{2}\right) - \sin^2\left(\dfrac{x}{2}\right) = \cos x$

91. $\sin(-x) = -2\sin\left(\dfrac{x}{2}\right)\cos\left(\dfrac{x}{2}\right)$

92. $2\cos^2\left(\dfrac{x}{4}\right) = 1 + \cos\left(\dfrac{x}{2}\right)$

93. $\tan^2\left(\dfrac{x}{2}\right) = \dfrac{1 - \cos x}{1 + \cos x}$

94. $\tan^2\left(\dfrac{x}{2}\right) = (\csc x - \cot x)^2$

95. $\tan\left(\dfrac{A}{2}\right) + \cot\left(\dfrac{A}{2}\right) = 2\csc A$

96. $\cot\left(\dfrac{A}{2}\right) - \tan\left(\dfrac{A}{2}\right) = 2\cot A$

97. $\csc^2\left(\dfrac{A}{2}\right) = \dfrac{2(1 + \cos A)}{\sin^2 A}$

98. $\sec^2\left(\dfrac{A}{2}\right) = \dfrac{2(1 - \cos A)}{\sin^2 A}$

99. $\csc\left(\dfrac{A}{2}\right) = \pm|\csc A|\sqrt{2 + 2\cos A}$

100. $\sec\left(\dfrac{A}{2}\right) = \pm|\csc A|\sqrt{2 - 2\cos A}$

In Exercises 101–108, graph the functions.

101. $y = 4\cos^2\left(\dfrac{x}{2}\right)$

102. $y = -6\sin^2\left(\dfrac{x}{2}\right)$

103. $y = \dfrac{1 - \tan^2\left(\dfrac{x}{2}\right)}{1 + \tan^2\left(\dfrac{x}{2}\right)}$

104. $y = 1 - \left[\sin\left(\dfrac{x}{2}\right) + \cos\left(\dfrac{x}{2}\right)\right]^2$

105. $y = 4\sin^2\left(\dfrac{x}{2}\right) - 1$

106. $y = -\dfrac{1}{6}\cos^2\left(\dfrac{x}{2}\right) + 2$

107. $y = \sqrt{\dfrac{1 - \cos(2x)}{1 + \cos(2x)}} \quad 0 \le x < \pi$

108. $y = \sqrt{\dfrac{1 + \cos(3x)}{2}} + 3 \quad 0 \le x \le \dfrac{\pi}{3}$

■ APPLICATIONS

For Exercises 109 and 110, refer to the following:

An ore-crusher wheel consists of a heavy disk spinning on its axle. The normal (crushing) force F, in pounds, between the wheel and the inclined track is determined by

$$F = W\sin\theta + \dfrac{1}{2}\psi^2\left[\dfrac{C}{R}(1 - \cos 2\theta) + \dfrac{A}{l}\sin 2\theta\right]$$

where W is the weight of the wheel in pounds, θ is the angle of the axis, C and A are moments of inertia, R is the radius of the wheel, l is the distance from the wheel to the pin where the axle is attached, and ψ is the speed in rpm that the wheel is spinning. The optimum crushing force occurs when the angle θ is between 45° and 90°.

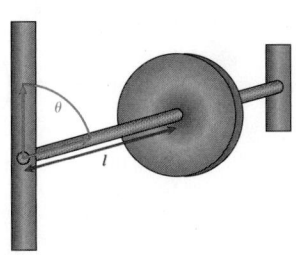

109. Ore-Crusher Wheel. Find F if the angle is 60°, W is 500 lb, ψ is 200 rpm, $\dfrac{C}{R} = 750$, and $\dfrac{A}{l} = 3.75$.

110. Ore-Crusher Wheel. Find F if the angle is 75°, W is 500 lb, ψ is 200 rpm, $\dfrac{C}{R} = 750$, and $\dfrac{A}{l} = 3.75$.

111. Area of an Isosceles Triangle. Consider the triangle below, where the vertex angle measures θ, the equal sides measure a, the height is h, and half the base is b. (In an isosceles triangle, the perpendicular dropped from the vertex angle divides the triangle into two congruent triangles.) The two triangles formed are right triangles.

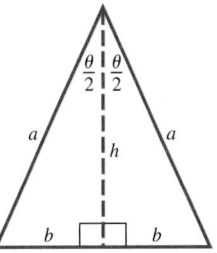

In the right triangles, $\sin\left(\dfrac{\theta}{2}\right) = \dfrac{b}{a}$ and $\cos\left(\dfrac{\theta}{2}\right) = \dfrac{h}{a}$. Multiply each side of each equation by a to get $b = a\sin\left(\dfrac{\theta}{2}\right), h = a\cos\left(\dfrac{\theta}{2}\right)$.

The area of the entire isosceles triangle is $A = \frac{1}{2}(2b)h = bh$.

Substitute the values for b and h into the area formula. Show that the area is equivalent to $\left(\dfrac{a^2}{2}\right)\sin\theta$.

112. Area of an Isosceles Triangle. Use the results from Exercise 111 to find the area of an isosceles triangle whose equal sides measure 7 inches and whose base angles each measure 75°.

113. With the information given in the diagram below, compute y.

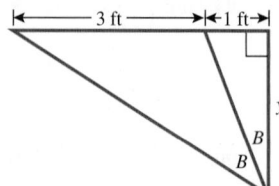

114. With the information given in the diagram below, compute x.

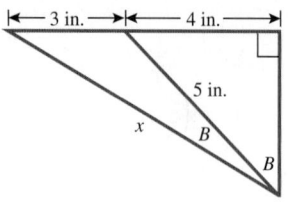

■ CATCH THE MISTAKE

In Exercises 115–118, explain the mistake that is made.

115. If $\cos x = \frac{1}{3}$, find $\sin(2x)$ given $\sin x < 0$.

Solution:

Write the double-angle identity for the sine function.

$$\sin(2x) = 2\sin x \cos x$$

Solve for $\sin x$ using the Pythagorean identity.

$$\sin^2 x + \left(\frac{1}{3}\right)^2 = 1$$

$$\sin x = \frac{2\sqrt{2}}{3}$$

Substitute $\cos x = \frac{1}{3}$ and $\sin x = \frac{2\sqrt{2}}{3}$.

$$\sin(2x) = 2\left(\frac{2\sqrt{2}}{3}\right)\left(\frac{1}{3}\right)$$

Simplify.

$$\sin(2x) = \frac{4\sqrt{2}}{9}$$

This is incorrect. What mistake was made?

116. If $\sin x = \frac{1}{3}$, find $\tan(2x)$ given $\cos x < 0$.

Solution:

Use the quotient identity.

$$\tan(2x) = \frac{\sin(2x)}{\cos x}$$

Use the double-angle formula for the sine function.

$$\tan(2x) = \frac{2\sin x \cos x}{\cos x}$$

Cancel the common cosine factors.

$$\tan(2x) = 2\sin x$$

Substitute $\sin x = \frac{1}{3}$.

$$\tan(2x) = \frac{2}{3}$$

This is incorrect. What mistake was made?

117. If $\cos x = -\frac{1}{3}$, find $\sin\left(\frac{x}{2}\right)$ given $\pi < x < \frac{3\pi}{2}$.

Solution:

Write the half-angle identity for the sine function.

$$\sin\left(\frac{x}{2}\right) = \pm\sqrt{\frac{1 - \cos x}{2}}$$

Substitute $\cos x = -\frac{1}{3}$.

$$\sin\left(\frac{x}{2}\right) = \pm\sqrt{\frac{1 + \frac{1}{3}}{2}}$$

Simplify.

$$\sin\left(\frac{x}{2}\right) = \pm\sqrt{\frac{2}{3}} = \pm\frac{\sqrt{2}}{\sqrt{3}}$$

The sine function is negative.

$$\sin\left(\frac{x}{2}\right) = -\frac{\sqrt{2}}{\sqrt{3}}$$

This is incorrect. What mistake was made?

118. If $\cos x = \frac{1}{3}$, find $\tan^2\left(\frac{x}{2}\right)$.

Solution:

Use the quotient identity.

$$\tan^2\left(\frac{x}{2}\right) = \frac{\sin^2\left(\frac{x}{2}\right)}{\cos^2 x}$$

Use the half-angle identity for the sine function.

$$\tan^2\left(\frac{x}{2}\right) = \frac{\frac{1 - \cos x}{2}}{\cos^2 x}$$

Simplify.

$$\tan^2\left(\frac{x}{2}\right) = \frac{1}{2}\left(\frac{1}{\cos^2 x} - \frac{\cos x}{\cos^2 x}\right)$$

$$\tan^2\left(\frac{x}{2}\right) = \frac{1}{2}\left(\frac{1}{\cos^2 x} - \frac{1}{\cos x}\right)$$

Substitute $\cos x = \frac{1}{3}$.

$$\tan^2\left(\frac{x}{2}\right) = \frac{1}{2}\left(\frac{1}{\frac{1}{9}} - \frac{1}{\frac{1}{3}}\right)$$

$$\tan^2\left(\frac{x}{2}\right) = 3$$

This is incorrect. What mistake was made?

■ CONCEPTUAL

For Exercises 119–126, determine whether each statement is true or false.

119. $\sin(2A) + \sin(2A) = \sin(4A)$

120. $\cos(4A) - \cos(2A) = \cos(2A)$

121. If $\tan x > 0$, then $\tan(2x) > 0$.

122. If $\sin x > 0$, then $\sin(2x) > 0$.

123. $\sin\left(\dfrac{A}{2}\right) + \sin\left(\dfrac{A}{2}\right) = \sin A$

124. $\cos\left(\dfrac{A}{2}\right) + \cos\left(\dfrac{A}{2}\right) = \cos A$

125. If $\tan x > 0$, then $\tan\left(\dfrac{x}{2}\right) > 0$.

126. If $\sin x > 0$, then $\sin\left(\dfrac{x}{2}\right) > 0$.

127. Given $\tan\left(\dfrac{A}{2}\right) = \pm\sqrt{\dfrac{1 - \cos A}{1 + \cos A}}$, verify $\tan\left(\dfrac{A}{2}\right) = \dfrac{\sin A}{1 + \cos A}$. Substitute $A = \pi$ into the identity and explain your results.

128. Given $\tan\left(\dfrac{A}{2}\right) = \pm\sqrt{\dfrac{1 - \cos A}{1 + \cos A}}$, verify $\tan\left(\dfrac{A}{2}\right) = \dfrac{1 - \cos A}{\sin A}$. Substitute $A = \pi$ into the identity and explain your results.

■ CHALLENGE

129. Is the identity $\tan(2x) = \dfrac{2\tan x}{1 - \tan^2 x}$ true for $x = \dfrac{\pi}{4}$? Explain.

130. Is the identity $2\csc(2x) = \dfrac{1 + \tan^2 x}{\tan x}$ true for $x = \dfrac{\pi}{2}$? Explain.

131. Prove that $\cot\left(\dfrac{A}{4}\right) = \pm\sqrt{\dfrac{1 + \cos\left(\dfrac{A}{2}\right)}{1 - \cos\left(\dfrac{A}{2}\right)}}$.

132. Prove that $\cot\left(-\dfrac{A}{2}\right)\sec\left(\dfrac{A}{2}\right)\csc\left(-\dfrac{A}{2}\right)\tan\left(\dfrac{A}{2}\right) = 2\csc A$.

133. Find the values of x in the interval $[0, 2\pi]$ for which $\tan\left(\dfrac{x}{2}\right) > 0$.

134. Find the values of x in the interval $[0, 2\pi]$ for which $\cot\left(\dfrac{x}{2}\right) \leq 0$.

■ TECHNOLOGY

One cannot *prove* that an equation is an identity using technology, but rather one uses it as a first step to see whether or not the equation *seems* to be an identity.

135. With a graphing calculator, plot $Y_1 = 1 - \dfrac{(2x)^2}{2!} + \dfrac{(2x)^4}{4!}$ and $Y_2 = \cos(2x)$ for x range $[-1, 1]$. Is Y_1 a good approximation to Y_2?

136. With a graphing calculator, plot $Y_1 = (2x) - \dfrac{(2x)^3}{3!} + \dfrac{(2x)^5}{5!}$ and $Y_2 = \sin(2x)$ for x range $[-1, 1]$. Is Y_1 a good approximation to Y_2?

137. With a graphing calculator, plot $Y_1 = \left(\dfrac{x}{2}\right) - \dfrac{\left(\dfrac{x}{2}\right)^3}{3!} + \dfrac{\left(\dfrac{x}{2}\right)^5}{5!}$ and $Y_2 = \sin\left(\dfrac{x}{2}\right)$ for x range $[-1, 1]$. Is Y_1 a good approximation to Y_2?

138. Using a graphing calculator, plot $Y_1 = 1 - \dfrac{\left(\dfrac{x}{2}\right)^2}{2!} + \dfrac{\left(\dfrac{x}{2}\right)^4}{4!}$ and $Y_2 = \cos\left(\dfrac{x}{2}\right)$ for x range $[-1, 1]$. Is Y_1 a good approximation to Y_2?

▪ PREVIEW TO CALCULUS

In calculus, we work with the derivative of expressions containing trigonometric functions. Usually, it is better to work with a simplified version of these expressions.

In Exercises 139–142, simplify each expression using the double-angle and half-angle identities.

139. $\dfrac{\dfrac{2\sin x}{\cos x}}{\dfrac{\cos^2 x - \sin^2 x}{\cos^2 x}}$

140. $\cos^4 x - 6\sin^2 x \cos^2 x + \sin^4 x$

141. $3\sin x \cos^2 x - \sin^3 x$

142. $\sqrt{\dfrac{1 - \sqrt{\dfrac{1 + \cos x}{2}}}{1 + \sqrt{\dfrac{1 + \cos x}{2}}}}$

SECTION 6.4 PRODUCT-TO-SUM AND SUM-TO-PRODUCT IDENTITIES

SKILLS OBJECTIVES

▪ Express products of trigonometric functions as sums of trigonometric functions.
▪ Express sums of trigonometric functions as products of trigonometric functions.

CONCEPTUAL OBJECTIVES

▪ Understand that the sum and difference identities are used to derive product-to-sum identities.
▪ Understand that the product-to-sum identities are used to derive the sum-to-product identities.

In calculus, often it is helpful to write products of trigonometric functions as sums of other trigonometric functions, and vice versa. In this section, we discuss the *product-to-sum identities*, which convert products to sums, and *sum-to-product identities*, which convert sums to products.

Product-to-Sum Identities

The *product-to-sum identities* are derived from the sum and difference identities.

WORDS	MATH
Write the identity for the cosine of a sum.	$\cos A \cos B - \sin A \sin B = \cos(A + B)$
Write the identity for the cosine of a difference.	$\cos A \cos B + \sin A \sin B = \cos(A - B)$
Add the two identities.	$2\cos A \cos B = \cos(A + B) + \cos(A - B)$
Divide both sides by 2.	$\cos A \cos B = \dfrac{1}{2}\left[\cos(A + B) + \cos(A - B)\right]$
	$\cos A \cos B + \sin A \sin B = \cos(A - B)$
Subtract the sum identity from the difference identity.	$-\cos A \cos B + \sin A \sin B = -\cos(A + B)$
	$2\sin A \sin B = \cos(A - B) - \cos(A + B)$
Divide both sides by 2.	$\sin A \sin B = \dfrac{1}{2}\left[\cos(A - B) - \cos(A + B)\right]$

Write the identity for the sine of a sum.

Write the identity for the sine of a difference.

Add the two identities.

$$\sin A \cos B + \cos A \sin B = \sin(A + B)$$
$$\underline{\sin A \cos B - \cos A \sin B = \sin(A - B)}$$
$$2 \sin A \cos B = \sin(A + B) + \sin(A - B)$$

Divide both sides by 2.

$$\sin A \cos B = \frac{1}{2}\left[\sin(A + B) + \sin(A - B)\right]$$

PRODUCT-TO-SUM IDENTITIES

1. $\cos A \cos B = \frac{1}{2}\left[\cos(A + B) + \cos(A - B)\right]$

2. $\sin A \sin B = \frac{1}{2}\left[\cos(A - B) - \cos(A + B)\right]$

3. $\sin A \cos B = \frac{1}{2}\left[\sin(A + B) + \sin(A - B)\right]$

EXAMPLE 1   Illustrating a Product-to-Sum Identity for Specific Values

Show that product-to-sum identity (3) is true when $A = 30°$ and $B = 90°$.

Solution:

Write product-to-sum identity (3).

$$\sin A \cos B = \frac{1}{2}\left[\sin(A + B) + \sin(A - B)\right]$$

Let $A = 30°$ and $B = 90°$.

$$\sin 30° \cos 90° = \frac{1}{2}\left[\sin(30° + 90°) + \sin(30° - 90°)\right]$$

Simplify.

$$\sin 30° \cos 90° = \frac{1}{2}\left[\sin 120° + \sin(-60°)\right]$$

Evaluate the trigonometric functions.

$$\frac{1}{2} \cdot 0 = \frac{1}{2}\left[\frac{\sqrt{3}}{2} - \frac{\sqrt{3}}{2}\right]$$

Simplify.

$$0 = 0$$

EXAMPLE 2   Convert a Product to a Sum

Convert the product $\cos(4x)\cos(3x)$ to a sum.

Solution:

Write product-to-sum identity (1).

$$\cos A \cos B = \frac{1}{2}[\cos(A + B) + \cos(A - B)]$$

Let $A = 4x$ and $B = 3x$.

$$\cos(4x)\cos(3x) = \frac{1}{2}[\cos(4x + 3x) + \cos(4x - 3x)]$$

Simplify.

$$\cos(4x)\cos(3x) = \frac{1}{2}[\cos(7x) + \cos x]$$

■ **YOUR TURN**   Convert the product $\cos(2x)\cos(5x)$ to a sum.

Technology Tip

Graphs of $y_1 = \cos(4x)\cos(3x)$ and $y_2 = \frac{1}{2}[\cos(7x) + \cos x]$.

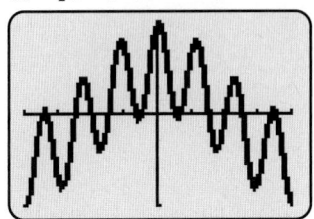

■ **Answer:** $\frac{1}{2}[\cos(7x) + \cos(3x)]$

EXAMPLE 3 **Converting Products to Sums**

Express $\sin(2x)\sin(3x)$ in terms of cosines.

COMMON MISTAKE

A common mistake that is often made is calling the product of two sines the square of a sine.

★ **CORRECT**

Write product-to-sum identity (2).

$$\sin A \sin B$$
$$= \frac{1}{2}[\cos(A - B) - \cos(A + B)]$$

Let $A = 2x$ and $B = 3x$.

$$\sin(2x)\sin(3x)$$
$$= \frac{1}{2}[\cos(2x - 3x) - \cos(2x + 3x)]$$

Simplify.

$$\sin(2x)\sin(3x)$$
$$= \frac{1}{2}[\cos(-x) - \cos(5x)]$$

The cosine function is an even function; thus,
$$\sin(2x)\sin(3x) = \frac{1}{2}[\cos x - \cos(5x)].$$

✖ **INCORRECT**

Multiply the two sine functions.

$$\sin(2x)\sin(3x) = \sin^2(6x^2) \quad \textbf{ERROR}$$

▼ **CAUTION**

1. $\sin A \sin B \neq \sin^2(AB)$.

2. The argument must be the same in order to use the identity:
$\sin A \sin A = (\sin A)^2 = \sin^2 A$.

■ **Answer:** $\frac{1}{2}[\cos x - \cos(3x)]$

■ **YOUR TURN** Express $\sin x \sin(2x)$ in terms of cosines.

Sum-to-Product Identities

The *sum-to-product identities* can be obtained from the product-to-sum identities.

WORDS	MATH
Write the identity for the product of the sine and cosine functions.	$\frac{1}{2}[\sin(x + y) + \sin(x - y)] = \sin x \cos y$
Let $x + y = A$ and $x - y = B$, then $x = \dfrac{A + B}{2}$ and $y = \dfrac{A - B}{2}$.	
Substitute these values into the identity.	$\frac{1}{2}[\sin A + \sin B] = \sin\left(\dfrac{A + B}{2}\right)\cos\left(\dfrac{A - B}{2}\right)$
Multiply by 2.	$\sin A + \sin B = 2\sin\left(\dfrac{A + B}{2}\right)\cos\left(\dfrac{A - B}{2}\right)$

The other three *sum-to-product* identities can be found similarly. All are summarized in the box below.

SUM-TO-PRODUCT IDENTITIES

4. $\sin A + \sin B = 2\sin\left(\dfrac{A+B}{2}\right)\cos\left(\dfrac{A-B}{2}\right)$

5. $\sin A - \sin B = 2\sin\left(\dfrac{A-B}{2}\right)\cos\left(\dfrac{A+B}{2}\right)$

6. $\cos A + \cos B = 2\cos\left(\dfrac{A+B}{2}\right)\cos\left(\dfrac{A-B}{2}\right)$

7. $\cos A - \cos B = -2\sin\left(\dfrac{A+B}{2}\right)\sin\left(\dfrac{A-B}{2}\right)$

EXAMPLE 4 Illustrating a Sum-to-Product Identity for Specific Values

Show that sum-to-product identity (7) is true when $A = 30°$ and $B = 90°$.

Solution:

Write the sum-to-product identity (7).

$$\cos A - \cos B = -2\sin\left(\frac{A+B}{2}\right)\sin\left(\frac{A-B}{2}\right)$$

Let $A = 30°$ and $B = 90°$.

$$\cos 30° - \cos 90° = -2\sin\left(\frac{30° + 90°}{2}\right)\sin\left(\frac{30° - 90°}{2}\right)$$

Simplify.

$$\cos 30° - \cos 90° = -2\sin 60° \sin(-30°)$$

The sine function is an odd function.

$$\cos 30° - \cos 90° = 2\sin 60° \sin 30°$$

Evaluate the trigonometric functions.

$$\frac{\sqrt{3}}{2} - 0 = 2\left(\frac{\sqrt{3}}{2}\right)\left(\frac{1}{2}\right)$$

Simplify.

$$\frac{\sqrt{3}}{2} = \frac{\sqrt{3}}{2}$$

EXAMPLE 5 Convert a Sum to a Product

Convert $-9[\sin(2x) - \sin(10x)]$, a trigonometric expression containing a sum, to a product.

Solution:

The expression inside the brackets is in the form of identity (5).

$$\sin A - \sin B = 2\sin\left(\frac{A-B}{2}\right)\cos\left(\frac{A+B}{2}\right)$$

Let $A = 2x$ and $B = 10x$.

$$\sin(2x) - \sin(10x) = 2\sin\left(\frac{2x - 10x}{2}\right)\cos\left(\frac{2x + 10x}{2}\right)$$

Simplify.

$$\sin(2x) - \sin(10x) = 2\sin(-4x)\cos(6x)$$

The sine function is an odd function.

$$\sin(2x) - \sin(10x) = -2\sin(4x)\cos(6x)$$

Multiply both sides by -9.

$$\boxed{-9[\sin(2x) - \sin(10x)] = 18\sin(4x)\cos(6x)}$$

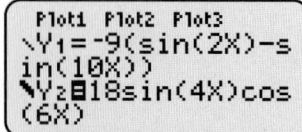

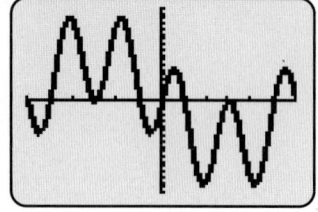

EXAMPLE 6   Simplifying a Trigonometric Expression

Simplify the expression $\sin\left(\dfrac{x+y}{2}\right)\cos\left(\dfrac{x-y}{2}\right) + \sin\left(\dfrac{x-y}{2}\right)\cos\left(\dfrac{x+y}{2}\right)$.

Solution:

Use identities (4) and (5).

$$\underbrace{\sin\left(\frac{x+y}{2}\right)\cos\left(\frac{x-y}{2}\right)}_{\frac{1}{2}[\sin x + \sin y]} + \underbrace{\sin\left(\frac{x-y}{2}\right)\cos\left(\frac{x+y}{2}\right)}_{\frac{1}{2}[\sin x - \sin y]}$$

$$= \frac{1}{2}\sin x + \frac{1}{2}\sin y + \frac{1}{2}\sin x - \frac{1}{2}\sin y$$

Simplify. $= \boxed{\sin x}$

Applications

In music, a tone is a fixed pitch (frequency) that is given a name. If two notes are sounded simultaneously, then they interfere and produce another tone, often called a "beat." The beat frequency is the difference of the two frequencies. The more rapid the beat, the further apart the two frequencies of the tones are. When musicians tune their instruments, they use a tuning fork to sound a tone and then tune the instrument until the beat is eliminated; then the tuning fork and instrument are in tune with each other. Mathematically, a tone is represented as $A\cos(2\pi ft)$, where A is the amplitude (loudness), f is the frequency in Hz, and t is time in seconds. The following table summarizes common tones and frequencies.

C	D	E	F	G	A	B
262 Hz	294 Hz	330 Hz	349 Hz	392 Hz	440 Hz	494 Hz

EXAMPLE 7   Music

Express the musical tone that is heard when a C and G are simultaneously struck (assume they have the same loudness).

Find the beat frequency, $f_2 - f_1$. Assume uniform amplitude (loudness), $A = 1$.

Solution:

Write the mathematical description of the C tone. $\cos(2\pi f_1 t)$, $f_1 = 262$ Hz

Write the mathematical description of the G tone. $\cos(2\pi f_2 t)$, $f_2 = 392$ Hz

Add the two notes. $\cos(524\pi t) + \cos(784\pi t)$

Use sum-to-product identities. $\cos(524\pi t) + \cos(784\pi t)$

$$= 2\cos\left(\frac{524\pi t + 784\pi t}{2}\right)\cos\left(\frac{524\pi t - 784\pi t}{2}\right)$$

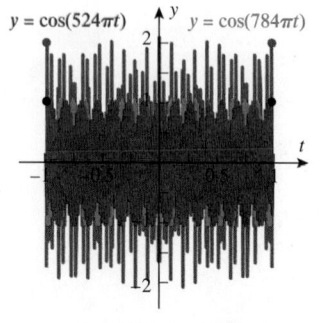

$y = \cos(524\pi t)$ $y = \cos(784\pi t)$

$y = 2\cos(654\pi t)\cos(130\pi t)$

Simplify.

$$= 2\cos(654\pi t)\cos(-130\pi t)$$

$$= 2\cos(2 \cdot 327 \cdot \pi \cdot t)\cos(130\,\pi t)$$

Identify the beat frequency. $f_2 - f_1 = 392 - 262 = 130\,\text{Hz}$

Therefore, the tone of average frequency, 327 Hz, has a beat of 130 Hz (beats/sec).

Notice that an average frequency results, $\cos(2 \cdot \underbrace{327}_{\substack{\text{average}\\\text{frequency}}} \cdot \pi \cdot t)$, and that frequency is modulated by a beat frequency, 130 Hz.

Note: In actual musical performance, the original two tones, C (262 Hz) and G (392 Hz) do not completely fuse with each other. You will hear the average frequency (327 Hz) and the beat frequency (130 Hz) *together with* the original C and G tones.

SECTION 6.4 SUMMARY

In this section, we used the sum and difference identities to derive the product-to-sum identities. The product-to-sum identities allowed us to express products as sums.

$$\cos A \cos B = \frac{1}{2}[\cos(A + B) + \cos(A - B)]$$

$$\sin A \sin B = \frac{1}{2}[\cos(A - B) - \cos(A + B)]$$

$$\sin A \cos B = \frac{1}{2}[\sin(A + B) + \sin(A - B)]$$

We then used the product-to-sum identities to derive the sum-to-product identities. The sum-to-product identities allow us to express sums as products.

$$\sin A + \sin B = 2\sin\left(\frac{A + B}{2}\right)\cos\left(\frac{A - B}{2}\right)$$

$$\sin A - \sin B = 2\sin\left(\frac{A - B}{2}\right)\cos\left(\frac{A + B}{2}\right)$$

$$\cos A + \cos B = 2\cos\left(\frac{A + B}{2}\right)\cos\left(\frac{A - B}{2}\right)$$

$$\cos A - \cos B = -2\sin\left(\frac{A + B}{2}\right)\sin\left(\frac{A - B}{2}\right)$$

SECTION 6.4 EXERCISES

SKILLS

In Exercises 1–14, write each product as a sum or difference of sines and/or cosines.

1. $\sin(2x)\cos x$

2. $\cos(10x)\sin(5x)$

3. $5\sin(4x)\sin(6x)$

4. $-3\sin(2x)\sin(4x)$

5. $4\cos(-x)\cos(2x)$

6. $-8\cos(3x)\cos(5x)$

7. $\sin\left(\frac{3x}{2}\right)\sin\left(\frac{5x}{2}\right)$

8. $\sin\left(\frac{\pi x}{2}\right)\sin\left(\frac{5\pi x}{2}\right)$

9. $\cos\left(\frac{2x}{3}\right)\cos\left(\frac{4x}{3}\right)$

10. $\sin\left(-\frac{\pi}{4}x\right)\cos\left(-\frac{\pi}{2}x\right)$

11. $-3\cos(0.4x)\cos(1.5x)$

12. $2\sin(2.1x)\sin(3.4x)$

13. $4\sin\left(-\sqrt{3}x\right)\cos\left(3\sqrt{3}x\right)$

14. $-5\cos\left(-\frac{\sqrt{2}}{3}x\right)\sin\left(\frac{5\sqrt{2}}{3}x\right)$

In Exercises 15–28, write each expression as a product of sines and/or cosines.

15. $\cos(5x) + \cos(3x)$

16. $\cos(2x) - \cos(4x)$

17. $\sin(3x) - \sin x$

18. $\sin(10x) + \sin(5x)$

19. $\sin\left(\dfrac{x}{2}\right) - \sin\left(\dfrac{5x}{2}\right)$

20. $\cos\left(\dfrac{x}{2}\right) - \cos\left(\dfrac{5x}{2}\right)$

21. $\cos\left(\dfrac{2}{3}x\right) + \cos\left(\dfrac{7}{3}x\right)$

22. $\sin\left(\dfrac{2}{3}x\right) + \sin\left(\dfrac{7}{3}x\right)$

23. $\sin(0.4x) + \sin(0.6x)$

24. $\cos(0.3x) - \cos(0.5x)$

25. $\sin(\sqrt{5}x) - \sin(3\sqrt{5}x)$

26. $\cos(-3\sqrt{7}x) - \cos(2\sqrt{7}x)$

27. $\cos\left(-\dfrac{\pi}{4}x\right) + \cos\left(\dfrac{\pi}{6}x\right)$

28. $\sin\left(\dfrac{3\pi}{4}x\right) + \sin\left(\dfrac{5\pi}{4}x\right)$

In Exercises 29–34, simplify the trigonometric expressions.

29. $\dfrac{\cos(3x) - \cos x}{\sin(3x) + \sin x}$

30. $\dfrac{\sin(4x) + \sin(2x)}{\cos(4x) - \cos(2x)}$

31. $\dfrac{\cos x - \cos(3x)}{\sin(3x) - \sin x}$

32. $\dfrac{\sin(4x) + \sin(2x)}{\cos(4x) + \cos(2x)}$

33. $\dfrac{\cos(5x) + \cos(2x)}{\sin(5x) - \sin(2x)}$

34. $\dfrac{\sin(7x) - \sin(2x)}{\cos(7x) - \cos(2x)}$

In Exercises 35–42, verify the identities.

35. $\dfrac{\sin A + \sin B}{\cos A + \cos B} = \tan\left(\dfrac{A + B}{2}\right)$

36. $\dfrac{\sin A - \sin B}{\cos A + \cos B} = \tan\left(\dfrac{A - B}{2}\right)$

37. $\dfrac{\cos A - \cos B}{\sin A + \sin B} = -\tan\left(\dfrac{A - B}{2}\right)$

38. $\dfrac{\cos A - \cos B}{\sin A - \sin B} = -\tan\left(\dfrac{A + B}{2}\right)$

39. $\dfrac{\sin A + \sin B}{\sin A - \sin B} = \tan\left(\dfrac{A + B}{2}\right)\cot\left(\dfrac{A - B}{2}\right)$

40. $\dfrac{\cos A - \cos B}{\cos A + \cos B} = -\tan\left(\dfrac{A + B}{2}\right)\tan\left(\dfrac{A - B}{2}\right)$

41. $\dfrac{\cos(A + B) + \cos(A - B)}{\sin(A + B) + \sin(A - B)} = \cot A$

42. $\dfrac{\cos(A - B) - \cos(A + B)}{\sin(A + B) + \sin(A - B)} = \tan B$

■ **APPLICATIONS**

43. Music. Write a mathematical description of a tone that results from simultaneously playing a G and a B. What is the beat frequency? What is the average frequency?

44. Music. Write a mathematical description of a tone that results from simultaneously playing an F and an A. What is the beat frequency? What is the average frequency?

45. Optics. Two optical signals with uniform $(A = 1)$ intensities and wavelengths of 1.55 μm and 0.63 μm are "beat" together. What is the resulting sum if their individual signals are given by $\sin\left(\dfrac{2\pi t c}{1.55\,\mu\text{m}}\right)$ and $\sin\left(\dfrac{2\pi t c}{0.63\,\mu\text{m}}\right)$, where $c = 3.0 \times 10^8$ m/s? (*Note:* 1 μm = 10^{-6} m.)

46. Optics. The two optical signals in Exercise 45 are beat together. What are the average frequency and the beat frequency?

For Exercises 47 and 48, refer to the following:

Touch-tone keypads have the following simultaneous low and high frequencies.

FREQUENCY	1209 Hz	1336 Hz	1477 Hz
697 Hz	1	2	3
770 Hz	4	5	6
852 Hz	7	8	9
941 Hz	*	0	#

The signal given when a key is pressed is $\sin(2\pi f_1 t) + \sin(2\pi f_2 t)$, where f_1 is the low frequency and f_2 is the high frequency.

47. Touch-Tone Dialing. What is the mathematical function that models the sound of dialing 4?

48. Touch-Tone Dialing. What is the mathematical function that models the sound of dialing 3?

49. Area of a Triangle. A formula for finding the area of a triangle when given the measures of the angles and one side is Area $= \dfrac{a^2 \sin B \sin C}{2 \sin A}$, where a is the side opposite angle A. If the measures of angles B and C are 52.5° and 7.5°, respectively, and if $a = 10$ feet, use the appropriate product-to-sum identity to change the formula so that you can solve for the area of the triangle exactly.

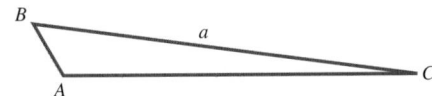

50. Area of a Triangle. If the measures of angles B and C in Exercise 49 are 75° and 45°, respectively, and if $a = 12$ inches, use the appropriate product-to-sum identity to change the formula so that you can solve for the area of the triangle exactly.

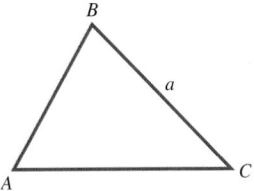

■ CATCH THE MISTAKE

In Exercises 51 and 52, explain the mistake that is made.

51. Simplify the expression $(\cos A - \cos B)^2 + (\sin A - \sin B)^2$.

Solution:

Expand by squaring.

$\cos^2 A - 2\cos A \cos B + \cos^2 B + \sin^2 A - 2\sin A \sin B + \sin^2 B$

Group terms.

$\cos^2 A + \sin^2 A - 2\cos A \cos B - 2\sin A \sin B + \cos^2 B + \sin^2 B$

Simplify using the Pythagorean identity.

$\underbrace{\cos^2 A + \sin^2 A}_{1} - 2\cos A \cos B - 2\sin A \sin B + \underbrace{\cos^2 B + \sin^2 B}_{1}$

Factor the common 2. $2(1 - \cos A \cos B - \sin A \sin B)$

Simplify. $2(1 - \cos AB - \sin AB)$

This is incorrect. What mistakes were made?

52. Simplify the expression $(\sin A - \sin B)(\cos A + \cos B)$.

Solution:

Multiply the expressions using the distributive property.

$\sin A \cos A + \sin A \cos B - \sin B \cos A - \sin B \cos B$

Cancel the second and third terms.

$\sin A \cos A - \sin B \cos B$

Use the product-to-sum identity.

$\underset{\frac{1}{2}[\sin(A + A) + \sin(A - A)]}{\underline{\sin A \cos A}} \qquad - \qquad \underset{\frac{1}{2}[\sin(B + B) + \sin(B - B)]}{\underline{\sin B \cos B}}$

Simplify. $= \dfrac{1}{2}\sin(2A) - \dfrac{1}{2}\sin(2B)$

This is incorrect. What mistake was made?

■ CONCEPTUAL

In Exercises 53–56, determine whether each statement is true or false.

53. $\cos A \cos B = \cos AB$

54. $\sin A \sin B = \sin AB$

55. The product of two cosine functions is a sum of two other cosine functions.

56. The product of two sine functions is a difference of two cosine functions.

57. Write $\sin A \sin B \sin C$ as a sum or difference of sines and cosines.

58. Write $\cos A \cos B \cos C$ as a sum or difference of sines and cosines.

■ CHALLENGE

59. Prove the addition formula
$\cos(A + B) = \cos A \cos B - \sin A \sin B$
using the identities of this section.

60. Prove the difference formula
$\sin(A - B) = \sin A \cos B - \sin B \cos A$
using the identities of this section.

61. Graph $y = 1 - 3\sin(\pi x)\sin\left(-\dfrac{\pi}{6}x\right)$.

62. Graph $y = 4\sin(2x - 1)\cos(2 - x)$.

63. Graph $y = -\cos\left(\dfrac{2\pi}{3}x\right)\cos\left(\dfrac{5\pi}{6}x\right)$.

64. Graph $y = x - \cos(2x)\sin(3x)$.

▪ TECHNOLOGY

65. Suggest an identity $4\sin x\cos x\cos(2x) = $ _____ by graphing $Y_1 = 4\sin x\cos x\cos(2x)$ and determining the function based on the graph.

66. Suggest an identity $1 + \tan x\tan(2x) = $ _____ by graphing $Y_1 = 1 + \tan x\tan(2x)$ and determining the function based on the graph.

67. With a graphing calculator, plot $Y_1 = \sin(4x)\sin(2x)$, $Y_2 = \sin(6x)$, and $Y_3 = \frac{1}{2}[\cos(2x) - \cos(6x)]$ in the same viewing rectangle $[0, 2\pi]$ by $[-1, 1]$. Which graphs are the same?

68. With a graphing calculator, plot $Y_1 = \cos(4x)\cos(2x)$, $Y_2 = \cos(6x)$, and $Y_3 = \frac{1}{2}[\cos(6x) + \cos(2x)]$ in the same viewing rectangle $[0, 2\pi]$ by $[-1, 1]$. Which graphs are the same?

▪ PREVIEW TO CALCULUS

In calculus, the method of separation of variables is used to solve certain differential equations. Given an equation with two variables, the method consists of writing the equation in such a way that each side of the equation contains only one type of variable.

In Exercises 69–72, use the product-to-sum and sum-to-product identities to separate the variables x and y in each equation.

69. $\sin\left(\dfrac{x + y}{2}\right)\sin\left(\dfrac{x - y}{2}\right) = \dfrac{1}{5}$

70. $\dfrac{1}{2} = \sin\left(\dfrac{x + y}{2}\right)\cos\left(\dfrac{x - y}{2}\right)$

71. $\sin(x + y) = 1 + \sin(x - y)$

72. $2 + \cos(x + y) = \cos(x - y)$

SECTION
6.5 INVERSE TRIGONOMETRIC FUNCTIONS

SKILLS OBJECTIVES

- Develop inverse trigonometric functions.
- Find values of inverse trigonometric functions.
- Graph inverse trigonometric functions.

CONCEPTUAL OBJECTIVES

- Understand the different notations for inverse trigonometric functions.
- Understand why domain restrictions on trigonometric functions are needed for inverse trigonometric functions to exist.
- Extend properties of inverse functions to develop inverse trigonometric identities.

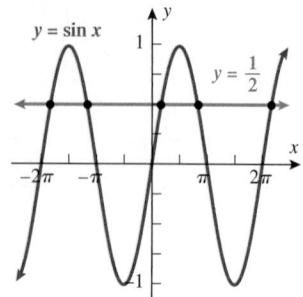

In Section 1.5, we discussed one-to-one functions and inverse functions. Here we present a summary of that section. A function is one-to-one if it passes the horizontal line test: No two x-values map to the same y-value.

Notice that the sine function does not pass the horizontal line test. However, if we restrict the domain to $-\dfrac{\pi}{2} \le x \le \dfrac{\pi}{2}$, then the restricted function is one-to-one.

Recall that if $y = f(x)$, then $x = f^{-1}(y)$.

The following are the properties of inverse functions:

1. If f is a one-to-one function, then the inverse function f^{-1} exists.
2. The domain of $f^{-1} = $ the range of f.
   The range of $f^{-1} = $ the domain of f.
3. $f^{-1}(f(x)) = x$ for all x in the domain of f.
   $f(f^{-1}(x)) = x$ for all x in the domain of f^{-1}.
4. The graph of f^{-1} is the reflection of the graph of f about the line $y = x$. If the point (a, b) lies on the graph of a function, then the point (b, a) lies on the graph of its inverse.

Inverse Sine Function

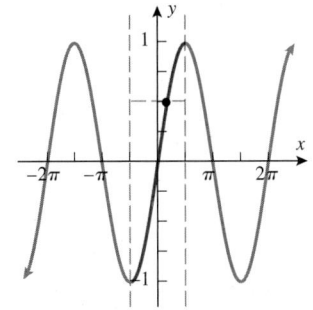

Let us start with the sine function with the restricted domain $\left[-\dfrac{\pi}{2}, \dfrac{\pi}{2}\right]$.

$$y = \sin x \qquad \text{Domain: } \left[-\frac{\pi}{2}, \frac{\pi}{2}\right] \qquad \text{Range: } [-1, 1]$$

x	y
$-\dfrac{\pi}{2}$	-1
$-\dfrac{\pi}{4}$	$-\dfrac{\sqrt{2}}{2}$
0	0
$\dfrac{\pi}{4}$	$\dfrac{\sqrt{2}}{2}$
$\dfrac{\pi}{2}$	1

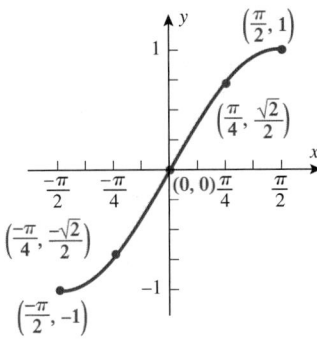

Study Tip

The inverse sine function gives an angle on the right half of the unit circle (QI and QIV).

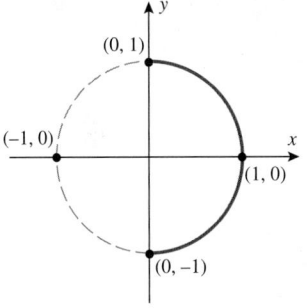

By the properties of inverse functions, the inverse sine function will have a domain of $[-1, 1]$ and a range of $\left[-\dfrac{\pi}{2}, \dfrac{\pi}{2}\right]$. To find the inverse sine function, we interchange the x- and y-values of $y = \sin x$.

$$y = \sin^{-1} x \qquad \text{Domain: } [-1, 1] \qquad \text{Range: } \left[-\frac{\pi}{2}, \frac{\pi}{2}\right]$$

x	y
-1	$-\dfrac{\pi}{2}$
$-\dfrac{\sqrt{2}}{2}$	$-\dfrac{\pi}{4}$
0	0
$\dfrac{\sqrt{2}}{2}$	$\dfrac{\pi}{4}$
1	$\dfrac{\pi}{2}$

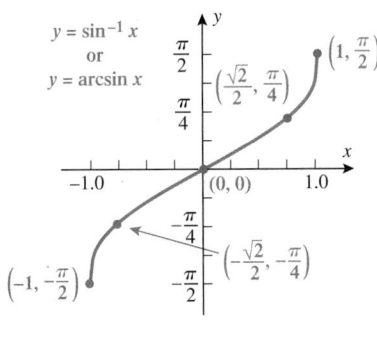

Technology Tip

To graph $y = \sin^{-1} x$, use $[-1, 1]$ as the domain and $\left[-\dfrac{\pi}{2}, \dfrac{\pi}{2}\right]$ as the range.

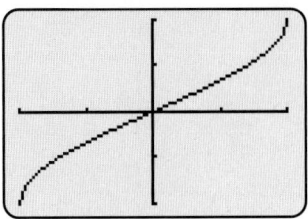

Notice that the inverse sine function, like the sine function, is an odd function (symmetric about the origin).

If the sine of an angle is known, what is the measure of that angle? The inverse sine function determines that angle measure. Another notation for the inverse sine function is $\arcsin x$.

INVERSE SINE FUNCTION

$$\underbrace{y = \sin^{-1} x \text{ or } y = \arcsin x}_{\text{"}y\text{ is the inverse sine of }x\text{"}} \qquad \text{means} \qquad \underbrace{x = \sin y}_{\substack{\text{"}y\text{ is the angle measure} \\ \text{whose sine equals }x\text{"}}}$$

$$\text{where } -1 \le x \le 1 \text{ and } -\frac{\pi}{2} \le y \le \frac{\pi}{2}$$

Study Tip

Trigonometric functions take angle measures and return real numbers. Inverse trigonometric functions take real numbers and return angle measures.

Study Tip

$$\sin^{-1}x \neq \frac{1}{\sin x}$$

It is important to note that the -1 as the superscript indicates an inverse function. Therefore, the inverse sine function should not be interpreted as a reciprocal:

$$\sin^{-1}x \neq \frac{1}{\sin x}$$

EXAMPLE 1 Finding Exact Values of an Inverse Sine Function

Find the exact value of each of the following expressions:

a. $\sin^{-1}\left(\dfrac{\sqrt{3}}{2}\right)$ **b.** $\arcsin\left(-\dfrac{1}{2}\right)$

Technology Tip

Set the TI/scientific calculator to degree mode by typing MODE .

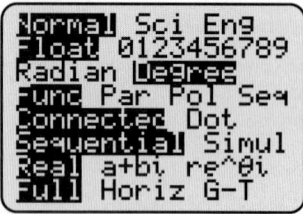

a. Using the TI calculator to find $\sin^{-1}\left(\dfrac{\sqrt{3}}{2}\right)$, type 2nd SIN for SIN^{-1} and 2nd x^2 for $\sqrt{}$.

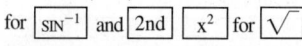
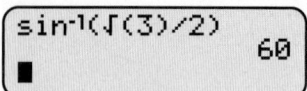

b. $\arcsin\left(-\dfrac{1}{2}\right)$

Solution (a):

Let $\theta = \sin^{-1}\left(\dfrac{\sqrt{3}}{2}\right)$. $\sin\theta = \dfrac{\sqrt{3}}{2}$ when $-\dfrac{\pi}{2} \le \theta \le \dfrac{\pi}{2}$

Which value of θ, in the range $-\dfrac{\pi}{2} \le \theta \le \dfrac{\pi}{2}$, corresponds to a sine value of $\dfrac{\sqrt{3}}{2}$?

- The range $-\dfrac{\pi}{2} \le \theta \le \dfrac{\pi}{2}$ corresponds to quadrants I and IV.
- The sine function is positive in quadrant I.
- We look for a value of θ in quadrant I that has a sine value of $\dfrac{\sqrt{3}}{2}$. $\theta = \dfrac{\pi}{3}$

$\sin\dfrac{\pi}{3} = \dfrac{\sqrt{3}}{2}$ and $\dfrac{\pi}{3}$ is in the interval $\left[-\dfrac{\pi}{2}, \dfrac{\pi}{2}\right]$. $\boxed{\sin^{-1}\left(\dfrac{\sqrt{3}}{2}\right) = \dfrac{\pi}{3}}$

Calculator Confirmation: Since $\dfrac{\pi}{3} = 60°$, if our calculator is set in degree mode, we should find that $\sin^{-1}\left(\dfrac{\sqrt{3}}{2}\right)$ is equal to 60°.

Solution (b):

Let $\theta = \arcsin\left(-\dfrac{1}{2}\right)$. $\sin\theta = -\dfrac{1}{2}$ when $-\dfrac{\pi}{2} \le \theta \le \dfrac{\pi}{2}$

Which value of θ, in the range $-\dfrac{\pi}{2} \le \theta \le \dfrac{\pi}{2}$, corresponds to a sine value of $-\dfrac{1}{2}$?

- The range $-\dfrac{\pi}{2} \le \theta \le \dfrac{\pi}{2}$ corresponds to quadrants I and IV.
- The sine function is negative in quadrant IV.
- We look for a value of θ in quadrant IV that has a sine value of $-\dfrac{1}{2}$. $\theta = -\dfrac{\pi}{6}$

$\sin\left(-\dfrac{\pi}{6}\right) = -\dfrac{1}{2}$ and $-\dfrac{\pi}{6}$ is in the interval $\left[-\dfrac{\pi}{2}, \dfrac{\pi}{2}\right]$. $\boxed{\arcsin\left(-\dfrac{1}{2}\right) = -\dfrac{\pi}{6}}$

Calculator Confirmation: Since $-\dfrac{\pi}{6} = -30°$, if our calculator is set in degree mode, we should find that $\sin^{-1}\left(-\dfrac{1}{2}\right)$ is equal to $-30°$.

■ **Answer: a.** $-\dfrac{\pi}{3}$ **b.** $\dfrac{\pi}{6}$

■ **YOUR TURN** Find the exact value of each of the following expressions:

a. $\sin^{-1}\left(-\dfrac{\sqrt{3}}{2}\right)$ **b.** $\arcsin\left(\dfrac{1}{2}\right)$

In Example 1, it is important to note that in part (a), both $60°$ and $120°$ correspond to the sine function equal to $\dfrac{\sqrt{3}}{2}$ and only one of them is valid, which is why the domain restrictions are necessary for inverse functions except for quadrantal angles. There are always two angles (values) from 0 to $360°$ or 0 to 2π (except that only $90°$, or $\dfrac{\pi}{2}$, and $270°$, or $\dfrac{3\pi}{2}$, correspond to 1 and -1, respectively) that correspond to the sine function equal to a particular value.

It is important to note that the inverse sine function has a domain $[-1, 1]$. For example, $\sin^{-1}3$ does not exist because 3 is not in the domain. Notice that calculator evaluation of $\sin^{-1}3$ says *error*. Calculators can be used to evaluate inverse sine functions when an exact evaluation is not feasible, just as they are for the basic trigonometric functions. For example, $\sin^{-1}0.3 \approx 17.46°$, or 0.305 radians.

We now state the properties relating the sine function and the inverse sine function that follow directly from properties of inverse functions.

SINE-INVERSE SINE IDENTITIES

$$\sin^{-1}(\sin x) = x \qquad \text{for} \qquad -\frac{\pi}{2} \le x \le \frac{\pi}{2}$$

$$\sin(\sin^{-1}x) = x \qquad \text{for} \qquad -1 \le x \le 1$$

For example, $\sin^{-1}\left[\sin\left(\dfrac{\pi}{12}\right)\right] = \dfrac{\pi}{12}$, since $\dfrac{\pi}{12}$ is in the interval $\left[-\dfrac{\pi}{2}, \dfrac{\pi}{2}\right]$. However, you must be careful not to overlook the domain restriction for which these identities hold, as illustrated in the next example.

EXAMPLE 2   Using Inverse Identities to Evaluate Expressions Involving Inverse Sine Functions

Find the exact value of each of the following trigonometric expressions:

a. $\sin\left[\sin^{-1}\left(\dfrac{\sqrt{2}}{2}\right)\right]$ **b.** $\sin^{-1}\left[\sin\left(\dfrac{3\pi}{4}\right)\right]$

Solution (a):

Write the appropriate identity. $\sin(\sin^{-1}x) = x$ for $-1 \le x \le 1$

Let $x = \dfrac{\sqrt{2}}{2}$, which is in the interval $[-1, 1]$.

Since the domain restriction is met, the identity can be used.

$$\sin\left[\sin^{-1}\left(\frac{\sqrt{2}}{2}\right)\right] = \frac{\sqrt{2}}{2}$$

Technology Tip

Use a TI calculator to find $\sin^{-1}3$ and $\sin^{-1}0.3$. Be sure to set the calculator in radian mode.

```
sin⁻¹(3)
■
```

```
ERR:DOMAIN
1∎Quit
2:Goto
```

```
sin⁻¹(3)
sin⁻¹(0.3)
          .304692654
■
```

Technology Tip

a. Check the answer of

$\sin\left[\sin^{-1}\left(\dfrac{\sqrt{2}}{2}\right)\right]$ with a

calculator.

```
sin(sin-1(√(2)/2)
)
         .7071067812
√(2)/2
         .7071067812
■
```

b. Check the answer of

$\sin^{-1}\left[\sin\left(\dfrac{3\pi}{4}\right)\right]$ with a calculator in

radian mode.

```
sin-1(sin(3π/4))
         .7853981634
π/4
         .7853981634
■
```

Answer: a. $-\dfrac{1}{2}$ **b.** $\dfrac{\pi}{6}$

Solution (b):

COMMON MISTAKE

Ignoring the domain restrictions on inverse identities.

⭐ CORRECT	❌ INCORRECT

CORRECT

Write the appropriate identity.

$\sin^{-1}(\sin x) = x$ for $-\dfrac{\pi}{2} \le x \le \dfrac{\pi}{2}$

Let $x = \dfrac{3\pi}{4}$, which is *not* in the

interval $\left[-\dfrac{\pi}{2}, \dfrac{\pi}{2}\right]$.

Since the domain restriction is not met, the identity cannot be used. Instead, we look for a value in the domain that corresponds to the same value of sine.

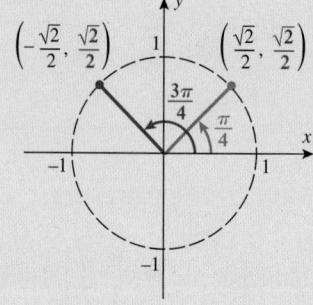

Substitute $\sin\dfrac{3\pi}{4} = \sin\dfrac{\pi}{4}$ into the expression.

$\sin^{-1}\left[\sin\left(\dfrac{3\pi}{4}\right)\right] = \sin^{-1}\left[\sin\left(\dfrac{\pi}{4}\right)\right]$

Since $\dfrac{\pi}{4}$ is in the interval $\left[-\dfrac{\pi}{2}, \dfrac{\pi}{2}\right]$,

we can use the identity.

$\sin^{-1}\left[\sin\left(\dfrac{3\pi}{4}\right)\right] = \sin^{-1}\left[\sin\left(\dfrac{\pi}{4}\right)\right] = \boxed{\dfrac{\pi}{4}}$

INCORRECT

$\sin^{-1}(\sin x) = x$ **ERROR**

Let $x = \dfrac{3\pi}{4}$.

(Forgot the domain restriction.)

$\sin^{-1}\left[\sin\left(\dfrac{3\pi}{4}\right)\right] = \dfrac{3\pi}{4}$

INCORRECT

YOUR TURN Find the exact value of each of the following trigonometric expressions:

a. $\sin\left[\sin^{-1}\left(-\dfrac{1}{2}\right)\right]$ **b.** $\sin^{-1}\left[\sin\left(\dfrac{5\pi}{6}\right)\right]$

Inverse Cosine Function

The cosine function is also not a one-to-one function, so we must restrict the domain in order to develop the inverse cosine function.

$$y = \cos x \qquad \text{Domain: } [0, \pi] \qquad \text{Range: } [-1, 1]$$

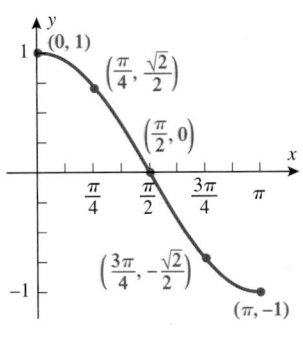

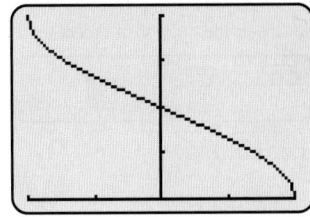
x	y
0	1
$\dfrac{\pi}{4}$	$\dfrac{\sqrt{2}}{2}$
$\dfrac{\pi}{2}$	0
$\dfrac{3\pi}{4}$	$-\dfrac{\sqrt{2}}{2}$
π	-1

By the properties of inverses, the inverse cosine function will have a domain of $[-1, 1]$ and a range of $[0, \pi]$. To find the inverse cosine function, we interchange the x- and y-values of $y = \cos x$.

$$y = \cos^{-1} x \qquad \text{Domain: } [-1, 1] \qquad \text{Range: } [0, \pi]$$

x	y
-1	π
$-\dfrac{\sqrt{2}}{2}$	$\dfrac{3\pi}{4}$
0	$\dfrac{\pi}{2}$
$\dfrac{\sqrt{2}}{2}$	$\dfrac{\pi}{4}$
1	0

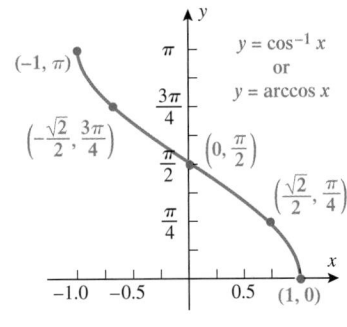

Study Tip

The inverse cosine function gives an angle on the top half of the unit circle (QI and QII).

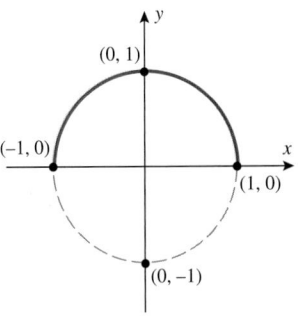

Notice that the inverse cosine function, unlike the cosine function, is not symmetric about the y-axis or the origin. Although the inverse sine and inverse cosine functions have the same domain, they behave differently. The inverse sine function increases on its domain (from left to right), whereas the inverse cosine function decreases on its domain (from left to right).

If the cosine of an angle is known, what is that measure of that angle? The inverse cosine function determines that angle measure. Another notation for the inverse cosine function is $\arccos x$.

INVERSE COSINE FUNCTION

$$\underbrace{y = \cos^{-1} x \text{ or } y = \arccos x}_{\text{"}y\text{ is the inverse cosine of }x\text{"}} \qquad \text{means} \qquad \underbrace{x = \cos y}_{\substack{\text{"}y\text{ is the angle measure} \\ \text{whose cosine equals }x\text{"}}}$$

$$\text{where } -1 \le x \le 1 \text{ and } 0 \le y \le \pi$$

Study Tip

$$\cos^{-1} x \ne \dfrac{1}{\cos x}$$

EXAMPLE 3   Finding Exact Values of an Inverse Cosine Function

Find the exact value of each of the following expressions:

a. $\cos^{-1}\left(-\dfrac{\sqrt{2}}{2}\right)$ **b.** $\arccos 0$

Solution (a):

Let $\theta = \cos^{-1}\left(-\dfrac{\sqrt{2}}{2}\right)$. $\qquad\qquad\qquad$ $\cos\theta = -\dfrac{\sqrt{2}}{2}$ when $0 \le \theta \le \pi$

Which value of θ, in the range $0 \le \theta \le \pi$, corresponds to a cosine value of $-\dfrac{\sqrt{2}}{2}$?

- The range $0 \le \theta \le \pi$ corresponds to quadrants I and II.
- The cosine function is negative in quadrant II.
- We look for a value of θ in quadrant II that has a cosine value of $-\dfrac{\sqrt{2}}{2}$. $\theta = \dfrac{3\pi}{4}$

$\cos\left(\dfrac{3\pi}{4}\right) = -\dfrac{\sqrt{2}}{2}$ and $\dfrac{3\pi}{4}$ is in the interval $[0, \pi]$. $\boxed{\cos^{-1}\left(-\dfrac{\sqrt{2}}{2}\right) = \dfrac{3\pi}{4}}$

Calculator Confirmation: Since $\dfrac{3\pi}{4} = 135°$, if our calculator is set in degree mode, we should find that $\cos^{-1}\left(-\dfrac{\sqrt{2}}{2}\right)$ is equal to $135°$.

Solution (b):

Let $\theta = \arccos 0$. $\qquad\qquad\qquad$ $\cos\theta = 0$ when $0 \le \theta \le \pi$

Which value of θ, in the range $0 \le \theta \le \pi$, corresponds to a cosine value of 0? $\qquad$ $\theta = \dfrac{\pi}{2}$

$\cos\left(\dfrac{\pi}{2}\right) = 0$ and $\dfrac{\pi}{2}$ is in the interval $[0, \pi]$. $\boxed{\arccos 0 = \dfrac{\pi}{2}}$

Calculator Confirmation: Since $\dfrac{\pi}{2} = 90°$, if our calculator is set in degree mode, we should find that $\cos^{-1} 0$ is equal to $90°$.

■ **Answer: a.** $\dfrac{\pi}{4}$ **b.** 0

■ **YOUR TURN** Find the exact value of each of the following expressions:

$$\textbf{a. } \cos^{-1}\left(\dfrac{\sqrt{2}}{2}\right) \qquad \textbf{b. } \arccos 1$$

We now state the properties relating the cosine function and the inverse cosine function that follow directly from the properties of inverses.

COSINE-INVERSE COSINE IDENTITIES

$$\cos^{-1}(\cos x) = x \qquad \text{for} \qquad 0 \le x \le \pi$$

$$\cos(\cos^{-1} x) = x \qquad \text{for} \qquad -1 \le x \le 1$$

As was the case with inverse identities for the sine function, you must be careful not to overlook the domain restrictions governing when each of these identities hold.

EXAMPLE 4 **Using Inverse Identities to Evaluate Expressions Involving Inverse Cosine Functions**

Find the exact value of each of the following trigonometric expressions:

a. $\cos\left[\cos^{-1}\left(-\dfrac{1}{2}\right)\right]$ **b.** $\cos^{-1}\left[\cos\left(\dfrac{7\pi}{4}\right)\right]$

Solution (a):

Write the appropriate identity. $\cos(\cos^{-1}x) = x$ for $-1 \le x \le 1$

Let $x = -\dfrac{1}{2}$, which is in the interval $[-1, 1]$.

Since the domain restriction is met, the identity can be used. $\cos\left[\cos^{-1}\left(-\dfrac{1}{2}\right)\right] = \boxed{-\dfrac{1}{2}}$

Solution (b):

Write the appropriate identity. $\cos^{-1}(\cos x) = x$ for $0 \le x \le \pi$

Let $x = \dfrac{7\pi}{4}$, which is *not* in the interval $[0, \pi]$.

Since the domain restriction is not met, the identity cannot be used.

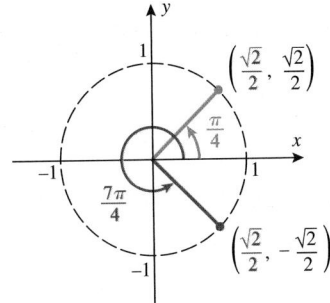

Instead, we find another angle in the interval that has the same cosine value. $\cos\left(\dfrac{7\pi}{4}\right) = \cos\left(\dfrac{\pi}{4}\right)$

Substitute $\cos\left(\dfrac{7\pi}{4}\right) = \cos\left(\dfrac{\pi}{4}\right)$ into the expression. $\cos^{-1}\left[\cos\left(\dfrac{7\pi}{4}\right)\right] = \cos^{-1}\left[\cos\left(\dfrac{\pi}{4}\right)\right]$

Since $\dfrac{\pi}{4}$ is in the interval $[0, \pi]$, we can use the identity. $= \boxed{\dfrac{\pi}{4}}$

▪ **YOUR TURN** Find the exact value of each of the following trigonometric expressions:

a. $\cos\left[\cos^{-1}\left(\dfrac{1}{2}\right)\right]$ **b.** $\cos^{-1}\left[\cos\left(-\dfrac{\pi}{6}\right)\right]$

■ **Answer: a.** $\dfrac{1}{2}$ **b.** $\dfrac{\pi}{6}$

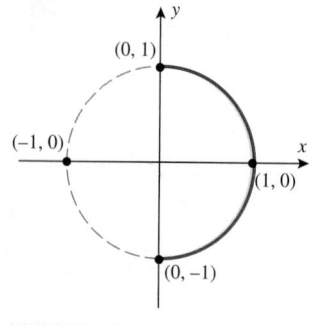

Inverse Tangent Function

The tangent function, too, is not a one-to-one function (it fails the horizontal line test). Let us start with the tangent function with a restricted domain:

$$y = \tan x \quad \text{Domain: } \left(-\frac{\pi}{2}, \frac{\pi}{2}\right) \quad \text{Range: } (-\infty, \infty)$$

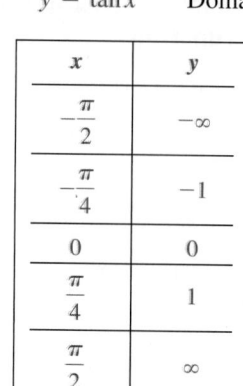

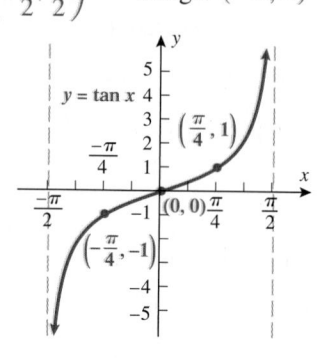

x	y
$-\dfrac{\pi}{2}$	$-\infty$
$-\dfrac{\pi}{4}$	-1
0	0
$\dfrac{\pi}{4}$	1
$\dfrac{\pi}{2}$	∞

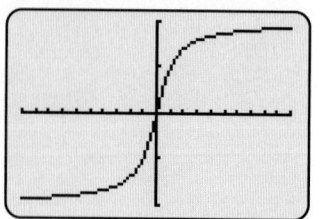

By the properties of inverse functions, the inverse tangent function will have a domain of $(-\infty, \infty)$ and a range of $\left(-\dfrac{\pi}{2}, \dfrac{\pi}{2}\right)$. To find the inverse tangent function, interchange x and y values.

$$y = \tan^{-1}x \quad \text{Domain: } (-\infty, \infty) \quad \text{Range: } \left(-\frac{\pi}{2}, \frac{\pi}{2}\right)$$

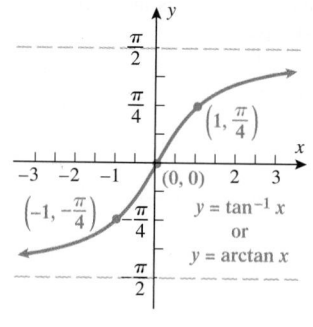

x	y
$-\infty$	$-\dfrac{\pi}{2}$
-1	$-\dfrac{\pi}{4}$
0	0
1	$\dfrac{\pi}{4}$
∞	$\dfrac{\pi}{2}$

Notice that the inverse tangent function, like the tangent function, is an odd function (it is symmetric about the origin).

The inverse tangent function allows us to answer the question: If the tangent of an angle is known, what is the measure of that angle? Another notation for the inverse tangent function is arctan x.

INVERSE TANGENT FUNCTION

$$\underbrace{y = \tan^{-1}x \text{ or } y = \arctan x}_{\text{"}y\text{ is the inverse tangent of }x\text{"}} \quad \text{means} \quad \underbrace{x = \tan y}_{\substack{\text{"}y\text{ is the angle measure whose} \\ \text{tangent equals }x\text{"}}}$$

$$\text{where } -\frac{\pi}{2} < y < \frac{\pi}{2}$$

EXAMPLE 5   Finding Exact Values of an Inverse Tangent Function

Find the exact value of each of the following expressions:

a. $\tan^{-1}(\sqrt{3})$ **b.** $\arctan 0$

Solution (a):

Let $\theta = \tan^{-1}(\sqrt{3})$. $\tan \theta = \sqrt{3}$ when $-\dfrac{\pi}{2} < \theta < \dfrac{\pi}{2}$

Which value of θ, in the range $-\dfrac{\pi}{2} < \theta < \dfrac{\pi}{2}$, $\theta = \dfrac{\pi}{3}$

corresponds to a tangent value of $\sqrt{3}$?

$\tan\left(\dfrac{\pi}{3}\right) = \sqrt{3}$ and $\dfrac{\pi}{3}$ is in the interval $\left(-\dfrac{\pi}{2}, \dfrac{\pi}{2}\right)$. $\boxed{\tan^{-1}(\sqrt{3}) = \dfrac{\pi}{3}}$

Calculator Confirmation: Since $\dfrac{\pi}{3} = 60°$, if our calculator is set in degree mode, we should find that $\tan^{-1}(\sqrt{3})$ is equal to $60°$.

Solution (b):

Let $\theta = \arctan 0$. $\tan \theta = 0$ when $-\dfrac{\pi}{2} < \theta < \dfrac{\pi}{2}$

Which value of θ, in the range $-\dfrac{\pi}{2} < \theta < \dfrac{\pi}{2}$, $\theta = 0$

corresponds to a tangent value of 0?

$\tan 0 = 0$, and 0 is in the interval $\left(-\dfrac{\pi}{2}, \dfrac{\pi}{2}\right)$. $\boxed{\arctan 0 = 0}$

Calculator Confirmation: $\tan^{-1} 0$ is equal to 0.

We now state the properties relating the tangent function and the inverse tangent function that follow directly from the properties of inverses.

TANGENT-INVERSE TANGENT IDENTITIES

$$\tan^{-1}(\tan x) = x \qquad \text{for} \qquad -\dfrac{\pi}{2} < x < \dfrac{\pi}{2}$$

$$\tan(\tan^{-1} x) = x \qquad \text{for} \qquad -\infty < x < \infty$$

EXAMPLE 6   Using Inverse Identities to Evaluate Expressions Involving Inverse Tangent Functions

Find the exact value of each of the following trigonometric expressions:

a. $\tan(\tan^{-1} 17)$ **b.** $\tan^{-1}\left[\tan\left(\dfrac{2\pi}{3}\right)\right]$

Solution (a):

Write the appropriate identity. $\tan(\tan^{-1} x) = x$ for $-\infty < x < \infty$

Let $x = 17$, which is in the interval $(-\infty, \infty)$.

Since the domain restriction is met, the identity can be used. $\boxed{\tan(\tan^{-1} 17) = 17}$

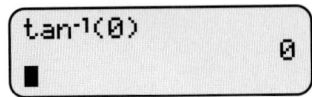
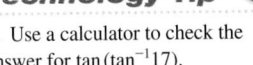
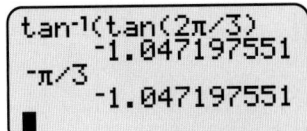

Solution (b):

Write the appropriate identity.

$$\tan^{-1}(\tan x) = x \text{ for } -\frac{\pi}{2} < x < \frac{\pi}{2}$$

Let $x = \frac{2\pi}{3}$, which is *not* in the interval $\left(-\frac{\pi}{2}, \frac{\pi}{2}\right)$.

Since the domain restriction is not met, the identity cannot be used.

Instead, we find another angle in the interval that has the same tangent value.

$$\tan\left(\frac{2\pi}{3}\right) = \tan\left(-\frac{\pi}{3}\right)$$

Substitute $\tan\left(\frac{2\pi}{3}\right) = \tan\left(-\frac{\pi}{3}\right)$ into the expression.

$$\tan^{-1}\left[\tan\left(\frac{2\pi}{3}\right)\right] = \tan^{-1}\left[\tan\left(-\frac{\pi}{3}\right)\right]$$

Since $-\frac{\pi}{3}$ is in the interval $\left(-\frac{\pi}{2}, \frac{\pi}{2}\right)$, we can use the identity.

$$\tan^{-1}\left[\tan\left(\frac{2\pi}{3}\right)\right] = \boxed{-\frac{\pi}{3}}$$

■ **Answer:** $\frac{\pi}{6}$

■ **YOUR TURN** Find the exact value of $\tan^{-1}\left[\tan\left(\frac{7\pi}{6}\right)\right]$.

Remaining Inverse Trigonometric Functions

The remaining three inverse trigonometric functions are defined similarly to the previous ones.

- ■ Inverse cotangent function: $\cot^{-1}x$ or $\operatorname{arccot} x$
- ■ Inverse secant function: $\sec^{-1}x$ or $\operatorname{arcsec} x$
- ■ Inverse cosecant function: $\csc^{-1}x$ or $\operatorname{arccsc} x$

A table summarizing all six of the inverse trigonometric functions is given below:

INVERSE FUNCTION	$y = \sin^{-1}x$	$y = \cos^{-1}x$	$y = \tan^{-1}x$	$y = \cot^{-1}x$	$y = \sec^{-1}x$	$y = \csc^{-1}x$
DOMAIN	$[-1, 1]$	$[-1, 1]$	$(-\infty, \infty)$	$(-\infty, \infty)$	$(-\infty, -1] \cup [1, \infty)$	$(-\infty, -1] \cup [1, \infty)$
RANGE	$\left[-\frac{\pi}{2}, \frac{\pi}{2}\right]$	$[0, \pi]$	$\left(-\frac{\pi}{2}, \frac{\pi}{2}\right)$	$(0, \pi)$	$\left[0, \frac{\pi}{2}\right) \cup \left(\frac{\pi}{2}, \pi\right]$	$\left[-\frac{\pi}{2}, 0\right) \cup \left(0, \frac{\pi}{2}\right]$
GRAPH						

EXAMPLE 7 Finding the Exact Value of Inverse Trigonometric Functions

Find the exact value of the following expressions:

a. $\cot^{-1}(\sqrt{3})$ **b.** $\csc^{-1}(\sqrt{2})$ **c.** $\sec^{-1}(-\sqrt{2})$

Solution (a):

Let $\theta = \cot^{-1}(\sqrt{3})$. $\cot\theta = \sqrt{3}$ when $0 < \theta < \pi$

Which value of θ, in the range $0 < \theta < \pi$, corresponds to a cotangent value of $\sqrt{3}$? $\theta = \dfrac{\pi}{6}$

$\cot\left(\dfrac{\pi}{6}\right) = \sqrt{3}$ and $\dfrac{\pi}{6}$ is in the interval $(0, \pi)$. $\boxed{\cot^{-1}(\sqrt{3}) = \dfrac{\pi}{6}}$

Solution (b):

Let $\theta = \csc^{-1}(\sqrt{2})$. $\csc\theta = \sqrt{2}$

Which value of θ, in the range $\left[-\dfrac{\pi}{2}, 0\right) \cup \left(0, \dfrac{\pi}{2}\right]$, corresponds to a cosecant value of $\sqrt{2}$? $\theta = \dfrac{\pi}{4}$

$\csc\left(\dfrac{\pi}{4}\right) = \sqrt{2}$ and $\dfrac{\pi}{4}$ is in the interval $\left[-\dfrac{\pi}{2}, 0\right) \cup \left(0, \dfrac{\pi}{2}\right]$. $\boxed{\csc^{-1}(\sqrt{2}) = \dfrac{\pi}{4}}$

Solution (c):

Let $\theta = \sec^{-1}(-\sqrt{2})$. $\sec\theta = -\sqrt{2}$

Which value of θ, in the range $\left[0, \dfrac{\pi}{2}\right) \cup \left(\dfrac{\pi}{2}, \pi\right]$, corresponds to a secant value of $-\sqrt{2}$? $\theta = \dfrac{3\pi}{4}$

$\sec\left(\dfrac{3\pi}{4}\right) = -\sqrt{2}$ and $\dfrac{3\pi}{4}$ is in the interval $\left[0, \dfrac{\pi}{2}\right) \cup \left(\dfrac{\pi}{2}, \pi\right]$. $\boxed{\sec^{-1}(-\sqrt{2}) = \dfrac{3\pi}{4}}$

Technology Tip

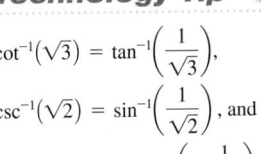

$\cot^{-1}(\sqrt{3}) = \tan^{-1}\left(\dfrac{1}{\sqrt{3}}\right)$,

$\csc^{-1}(\sqrt{2}) = \sin^{-1}\left(\dfrac{1}{\sqrt{2}}\right)$, and

$\sec^{-1}(-\sqrt{2}) = \cos^{-1}\left(-\dfrac{1}{\sqrt{2}}\right)$.

```
tan⁻¹(1/√(3))
          .5235987756
π/6
          .5235987756
```

```
sin⁻¹(1/√(2))
          .7853981634
π/4
          .7853981634
```

```
cos⁻¹(-1/√(2))
          2.35619449
3π/4
          2.35619449
■
```

How do we approximate the inverse secant, inverse cosecant, and inverse cotangent functions with a calculator? Scientific calculators have keys ($\sin^{-1}$, $\cos^{-1}$, and $\tan^{-1}$) for three of the inverse trigonometric functions but not for the other three. Recall that we find the cosecant, secant, and cotangent function values by taking sine, cosine, or tangent, and finding the reciprocal.

$$\csc x = \dfrac{1}{\sin x} \qquad \sec x = \dfrac{1}{\cos x} \qquad \cot x = \dfrac{1}{\tan x}$$

However, *the reciprocal approach cannot be used for inverse functions.* The three inverse trigonometric functions $\csc^{-1}x$, $\sec^{-1}x$, and $\cot^{-1}x$ cannot be found by finding the reciprocal of $\sin^{-1}x$, $\cos^{-1}x$, or $\tan^{-1}x$.

$$\csc^{-1}x \neq \dfrac{1}{\sin^{-1}x} \qquad \sec^{-1}x \neq \dfrac{1}{\cos^{-1}x} \qquad \cot^{-1}x \neq \dfrac{1}{\tan^{-1}x}$$

Study Tip

$\sec^{-1}x \neq \dfrac{1}{\cos^{-1}x}$

$\csc^{-1}x \neq \dfrac{1}{\sin^{-1}x}$

$\cot^{-1}x \neq \dfrac{1}{\tan^{-1}x}$

Instead, we seek the equivalent $\sin^{-1}x$, $\cos^{-1}x$, or $\tan^{-1}x$ values by algebraic means, always remembering to look within the correct domain and range.

Start with the inverse secant function.

$$y = \sec^{-1}x \qquad \text{for} \qquad x \le -1 \qquad \text{or} \qquad x \ge 1$$

Write the equivalent secant expression.

$$\sec y = x \qquad \text{for} \qquad 0 \le y < \frac{\pi}{2} \qquad \text{or} \qquad \frac{\pi}{2} < y \le \pi$$

Apply the reciprocal identity.

$$\frac{1}{\cos y} = x$$

Simplify using algebraic techniques.

$$\cos y = \frac{1}{x}$$

Write the result in terms of the inverse cosine function.

$$y = \cos^{-1}\left(\frac{1}{x}\right)$$

Therefore, we have the relationship:

$$\sec^{-1}x = \cos^{-1}\left(\frac{1}{x}\right) \qquad \text{for} \qquad x \le -1 \qquad \text{or} \qquad x \ge 1$$

The other relationships will be found in the exercises and are summarized below:

INVERSE SECANT, INVERSE COSECANT, AND INVERSE COTANGENT IDENTITIES

$$\sec^{-1}x = \cos^{-1}\left(\frac{1}{x}\right) \qquad \text{for} \qquad x \le -1 \text{ or } x \ge 1$$

$$\csc^{-1}x = \sin^{-1}\left(\frac{1}{x}\right) \qquad \text{for} \qquad x \le -1 \text{ or } x \ge 1$$

$$\cot^{-1}x = \begin{cases} \tan^{-1}\left(\dfrac{1}{x}\right) & \text{for} \quad x > 0 \\ \pi + \tan^{-1}\left(\dfrac{1}{x}\right) & \text{for} \quad x < 0 \end{cases}$$

Technology Tip

Use the inverse trigonometry function identities to find

a. $\sec^{-1}2 = \cos^{-1}\left(\frac{1}{2}\right)$

b. $\cot^{-1}7 = \tan^{-1}\left(\frac{1}{7}\right)$

```
cos-1(1/2)
         1.047197551
π/3
         1.047197551
```

```
tan-1(1/7)
         .1418970546
Ans*180/π
         8.130102354
```

EXAMPLE 8 Using Inverse Identities

a. Find the exact value of $\sec^{-1}2$.
b. Use a calculator to find the value of $\cot^{-1}7$.

Solution (a):

Let $\theta = \sec^{-1}2$. $\qquad\qquad \sec\theta = 2$ on $\left[0, \dfrac{\pi}{2}\right) \cup \left(\dfrac{\pi}{2}, \pi\right]$

Substitute the reciprocal identity. $\qquad \dfrac{1}{\cos\theta} = 2$

Solve for $\cos\theta$. $\qquad\qquad \cos\theta = \dfrac{1}{2}$

The restricted interval $\left[0, \dfrac{\pi}{2}\right) \cup \left(\dfrac{\pi}{2}, \pi\right]$ corresponds to quadrants I and II.

The cosine function is positive in quadrant I. $\qquad \theta = \dfrac{\pi}{3}$

$$\boxed{\sec^{-1}2 = \cos^{-1}\left(\frac{1}{2}\right) = \frac{\pi}{3}}$$

Solution (b):

Since we do not know an exact value that would correspond to the cotangent function equal to 7, we proceed using identities and a calculator.

Select the correct identity, given that $x = 7 > 0$. $\qquad \cot^{-1}x = \tan^{-1}\left(\dfrac{1}{x}\right)$

Let $x = 7$. $\qquad\qquad\qquad\qquad\qquad\qquad \cot^{-1}7 = \tan^{-1}\left(\dfrac{1}{7}\right)$

Evaluate the right side with a calculator. $\qquad \boxed{\cot^{-1}7 \approx 8.13°}$

Finding Exact Values for Expressions Involving Inverse Trigonometric Functions

We will now find exact values of trigonometric expressions that involve inverse trigonometric functions.

EXAMPLE 9   Finding Exact Values of Trigonometric Expressions Involving Inverse Trigonometric Functions

Find the exact value of $\cos\left[\sin^{-1}\left(\frac{2}{3}\right)\right]$.

Solution:

STEP 1 Let $\theta = \sin^{-1}\left(\frac{2}{3}\right)$. $\qquad\qquad \sin\theta = \dfrac{2}{3}$ when $-\dfrac{\pi}{2} \le \theta \le \dfrac{\pi}{2}$

The range $-\dfrac{\pi}{2} \le \theta \le \dfrac{\pi}{2}$ corresponds to quadrants I and IV.

The sine function is positive in quadrant I.

STEP 2 Draw angle θ in quadrant I.

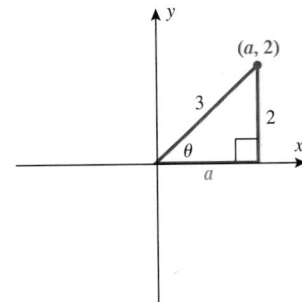

Label the sides known from the sine value. $\qquad \sin\theta = \dfrac{2}{3} = \dfrac{\text{opposite}}{\text{hypotenuse}}$

STEP 3 Find the unknown side length a. $\qquad a^2 + 2^2 = 3^2$

Solve for a. $\qquad\qquad\qquad\qquad a = \pm\sqrt{5}$

Since θ is in quadrant I, a is positive. $\qquad a = \sqrt{5}$

STEP 4 Find $\cos\left[\sin^{-1}\left(\frac{2}{3}\right)\right]$.

Substitute $\theta = \sin^{-1}\left(\frac{2}{3}\right)$.

$$\cos\left[\sin^{-1}\left(\frac{2}{3}\right)\right] = \cos\theta$$

Find $\cos\theta$.

$$\cos\theta = \frac{\text{adjacent}}{\text{hypotenuse}} = \frac{\sqrt{5}}{3}$$

$$\boxed{\cos\left[\sin^{-1}\left(\frac{2}{3}\right)\right] = \frac{\sqrt{5}}{3}}$$

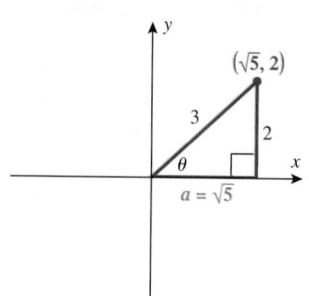

■ **Answer:** $\dfrac{2\sqrt{2}}{3}$

■ **YOUR TURN** Find the exact value of $\sin\left[\cos^{-1}\left(\frac{1}{3}\right)\right]$.

EXAMPLE 10 **Finding Exact Values of Trigonometric Expressions Involving Inverse Trigonometric Functions**

Find the exact value of $\tan\left[\cos^{-1}\left(-\frac{7}{12}\right)\right]$.

Solution:

STEP 1 Let $\theta = \cos^{-1}\left(-\frac{7}{12}\right)$.

$$\cos\theta = -\frac{7}{12} \text{ when } 0 \leq \theta \leq \pi$$

The range $0 \leq \theta \leq \pi$ corresponds to quadrants I and II.

The cosine function is negative in quadrant II.

STEP 2 Draw angle θ in quadrant II.

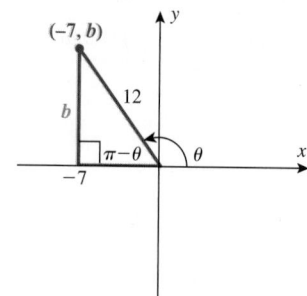

Label the sides known from the cosine value.

$$\cos\theta = -\frac{7}{12} = \frac{\text{adjacent}}{\text{hypotenuse}}$$

STEP 3 Find the length of the unknown side b.

$$b^2 + (-7)^2 = 12^2$$

Solve for b.

$$b = \pm\sqrt{95}$$

Since θ is in quadrant II, b is positive.

$$b = \sqrt{95}$$

STEP 4 Find $\tan\left[\cos^{-1}\left(-\frac{7}{12}\right)\right]$.

Substitute $\theta = \cos^{-1}\left(-\frac{7}{12}\right)$.

$$\tan\left[\cos^{-1}\left(-\frac{7}{12}\right)\right] = \tan\theta$$

Find $\tan\theta$.

$$\tan\theta = \frac{\text{opposite}}{\text{adjacent}} = \frac{\sqrt{95}}{-7}$$

$$\boxed{\tan\left[\cos^{-1}\left(-\frac{7}{12}\right)\right] = -\frac{\sqrt{95}}{7}}$$

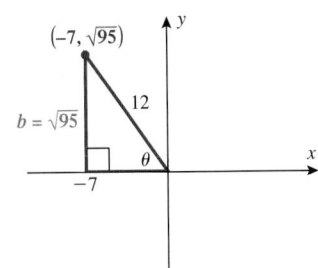

■ **YOUR TURN** Find the exact value of $\tan\left[\sin^{-1}\left(-\frac{3}{7}\right)\right]$.

■ **Answer:** $-\dfrac{3\sqrt{10}}{20}$

SECTION 6.5 SUMMARY

If a trigonometric function value of an angle or of a real number is known, what is that number, as defined by the domain restriction? Inverse trigonometric functions determine the angle measure (or the value of the argument). To define the inverse trigonometric relations as functions, we first restrict the trigonometric functions to domains in which they are one-to-one functions. Exact values for inverse trigonometric functions can be found when the function values are those of the special angles. Inverse trigonometric functions also provide a means for evaluating one trigonometric function when we are given the value of another. It is important to note that the -1 as a superscript indicates an inverse function, not a reciprocal.

INVERSE FUNCTION	$y = \sin^{-1}x$	$y = \cos^{-1}x$	$y = \tan^{-1}x$	$y = \cot^{-1}x$	$y = \sec^{-1}x$	$y = \csc^{-1}x$
DOMAIN	$[-1, 1]$	$[-1, 1]$	$(-\infty, \infty)$	$(-\infty, \infty)$	$(-\infty, -1] \cup [1, \infty)$	$(-\infty, -1] \cup [1, \infty)$
RANGE	$\left[-\frac{\pi}{2}, \frac{\pi}{2}\right]$	$[0, \pi]$	$\left(-\frac{\pi}{2}, \frac{\pi}{2}\right)$	$(0, \pi)$	$\left[0, \frac{\pi}{2}\right) \cup \left(\frac{\pi}{2}, \pi\right]$	$\left[-\frac{\pi}{2}, 0\right) \cup \left(0, \frac{\pi}{2}\right]$
GRAPH						

SECTION 6.5 EXERCISES

■ SKILLS

In Exercises 1–16, find the exact value of each expression. Give the answer in radians.

1. $\arccos\left(\dfrac{\sqrt{2}}{2}\right)$

2. $\arccos\left(-\dfrac{\sqrt{2}}{2}\right)$

3. $\arcsin\left(-\dfrac{\sqrt{3}}{2}\right)$

4. $\arcsin\left(\dfrac{1}{2}\right)$

5. $\cot^{-1}(-1)$

6. $\tan^{-1}\left(\dfrac{\sqrt{3}}{3}\right)$

7. $\text{arcsec}\left(\dfrac{2\sqrt{3}}{3}\right)$

8. $\text{arccsc}(-1)$

9. $\csc^{-1} 2$ **10.** $\sec^{-1}(-2)$ **11.** $\arctan\left(-\sqrt{3}\right)$ **12.** $\operatorname{arccot}\left(\sqrt{3}\right)$

13. $\sin^{-1} 0$ **14.** $\tan^{-1} 1$ **15.** $\sec^{-1}(-1)$ **16.** $\cot^{-1} 0$

In Exercises 17–32, find the exact value of each expression. Give the answer in degrees.

17. $\cos^{-1}\left(\dfrac{1}{2}\right)$ **18.** $\cos^{-1}\left(-\dfrac{\sqrt{3}}{2}\right)$ **19.** $\sin^{-1}\left(\dfrac{\sqrt{2}}{2}\right)$ **20.** $\sin^{-1} 0$

21. $\cot^{-1}\left(-\dfrac{\sqrt{3}}{3}\right)$ **22.** $\tan^{-1}\left(-\sqrt{3}\right)$ **23.** $\arctan\left(\dfrac{\sqrt{3}}{3}\right)$ **24.** $\operatorname{arccot} 1$

25. $\operatorname{arccsc}(-2)$ **26.** $\csc^{-1}\left(-\dfrac{2\sqrt{3}}{3}\right)$ **27.** $\operatorname{arcsec}\left(-\sqrt{2}\right)$ **28.** $\operatorname{arccsc}\left(-\sqrt{2}\right)$

29. $\sin^{-1}(-1)$ **30.** $\arctan(-1)$ **31.** $\operatorname{arccot} 0$ **32.** $\operatorname{arcsec}(-1)$

In Exercises 33–42, use a calculator to evaluate each expression. Give the answer in degrees and round it to two decimal places.

33. $\cos^{-1}(0.5432)$ **34.** $\sin^{-1}(0.7821)$ **35.** $\tan^{-1}(1.895)$ **36.** $\tan^{-1}(3.2678)$

37. $\sec^{-1}(1.4973)$ **38.** $\sec^{-1}(2.7864)$ **39.** $\csc^{-1}(-3.7893)$ **40.** $\csc^{-1}(-6.1324)$

41. $\cot^{-1}(-4.2319)$ **42.** $\cot^{-1}(-0.8977)$

In Exercises 43–52, use a calculator to evaluate each expression. Give the answer in radians and round it to two decimal places.

43. $\sin^{-1}(-0.5878)$ **44.** $\sin^{-1}(0.8660)$ **45.** $\cos^{-1}(0.1423)$ **46.** $\tan^{-1}(-0.9279)$

47. $\tan^{-1}(1.3242)$ **48.** $\cot^{-1}(2.4142)$ **49.** $\cot^{-1}(-0.5774)$ **50.** $\sec^{-1}(-1.0422)$

51. $\csc^{-1}(3.2361)$ **52.** $\csc^{-1}(-2.9238)$

In Exercises 53–76, evaluate each expression exactly, if possible. If not possible, state why.

53. $\sin^{-1}\left[\sin\left(\dfrac{5\pi}{12}\right)\right]$ **54.** $\sin^{-1}\left[\sin\left(-\dfrac{5\pi}{12}\right)\right]$ **55.** $\sin[\sin^{-1}(1.03)]$ **56.** $\sin[\sin^{-1}(1.1)]$

57. $\sin^{-1}\left[\sin\left(-\dfrac{7\pi}{6}\right)\right]$ **58.** $\sin^{-1}\left[\sin\left(\dfrac{7\pi}{6}\right)\right]$ **59.** $\cos^{-1}\left[\cos\left(\dfrac{4\pi}{3}\right)\right]$ **60.** $\cos^{-1}\left[\cos\left(-\dfrac{5\pi}{3}\right)\right]$

61. $\cot\left[\cot^{-1}\left(\sqrt{3}\right)\right]$ **62.** $\cot^{-1}\left[\cot\left(\dfrac{5\pi}{4}\right)\right]$ **63.** $\sec^{-1}\left[\sec\left(-\dfrac{\pi}{3}\right)\right]$ **64.** $\sec\left[\sec^{-1}\left(\dfrac{1}{2}\right)\right]$

65. $\csc\left[\csc^{-1}\left(\dfrac{1}{2}\right)\right]$ **66.** $\csc^{-1}\left[\csc\left(\dfrac{7\pi}{6}\right)\right]$ **67.** $\cot(\cot^{-1} 0)$ **68.** $\cot^{-1}\left[\cot\left(-\dfrac{\pi}{4}\right)\right]$

69. $\tan^{-1}\left[\tan\left(-\dfrac{\pi}{4}\right)\right]$ **70.** $\tan^{-1}\left[\tan\left(\dfrac{\pi}{4}\right)\right]$ **71.** $\sec(\sec^{-1} 0)$ **72.** $\csc^{-1}(\csc \pi)$

73. $\cot^{-1}\left[\cot\left(\dfrac{8\pi}{3}\right)\right]$ **74.** $\tan^{-1}[\tan(8\pi)]$ **75.** $\csc^{-1}\left[\csc\left(\dfrac{15\pi}{4}\right)\right]$ **76.** $\sec^{-1}\left[\sec\left(\dfrac{17\pi}{2}\right)\right]$

In Exercises 77–88, evaluate each expression exactly.

77. $\cos\left[\sin^{-1}\left(\dfrac{3}{4}\right)\right]$ **78.** $\sin\left[\cos^{-1}\left(\dfrac{2}{3}\right)\right]$ **79.** $\sin\left[\tan^{-1}\left(\dfrac{12}{5}\right)\right]$ **80.** $\cos\left[\tan^{-1}\left(\dfrac{7}{24}\right)\right]$

81. $\tan\left[\sin^{-1}\left(\dfrac{3}{5}\right)\right]$ **82.** $\tan\left[\cos^{-1}\left(\dfrac{2}{5}\right)\right]$ **83.** $\sec\left[\sin^{-1}\left(\dfrac{\sqrt{2}}{5}\right)\right]$ **84.** $\sec\left[\cos^{-1}\left(\dfrac{\sqrt{7}}{4}\right)\right]$

85. $\csc\left[\cos^{-1}\left(\dfrac{1}{4}\right)\right]$ **86.** $\csc\left[\sin^{-1}\left(\dfrac{1}{4}\right)\right]$ **87.** $\cot\left[\sin^{-1}\left(\dfrac{60}{61}\right)\right]$ **88.** $\cot\left[\sec^{-1}\left(\dfrac{41}{9}\right)\right]$

■ **APPLICATIONS**

89. Alternating Current. Alternating electrical current in amperes (A) is modeled by the equation $i = I \sin(2\pi ft)$, where i is the current, I is the maximum current, t is time in seconds, and f is the frequency in hertz (Hz is the number of cycles per second). If the frequency is 5 Hz and maximum current is 115 A, what time t corresponds to a current of 85 A? Find the smallest positive value of t.

90. Alternating Current. If the frequency is 100 Hz and maximum current is 240 A, what time t corresponds to a current of 100 A? Find the smallest positive value of t.

91. Hours of Daylight. The number of hours of daylight in San Diego, California, can be modeled with $H(t) = 12 + 2.4 \sin(0.017t - 1.377)$, where t is the day of the year (January 1, $t = 1$, etc.). For what value of t is the number of hours of daylight equal to 14.4? If May 31 is the 151st day of the year, what month and day correspond to that value of t?

92. Hours of Daylight. Repeat Exercise 91. For what value of t is the number of hours of daylight equal to 9.6? What month and day correspond to the value of t? (You may have to count backwards.)

93. Money. A young couple get married and immediately start saving money. They renovate a house and are left with less and less saved money. They have children after 10 years and are in debt until their children are in college. They then save until retirement. A formula that represents the percentage of their annual income that they either save (positive) or are in debt (negative) is given by $P(t) = 12.5 \cos(0.157t) + 2.5$, where $t = 0$ corresponds to the year they were married. How many years into their marriage do they first accrue debt?

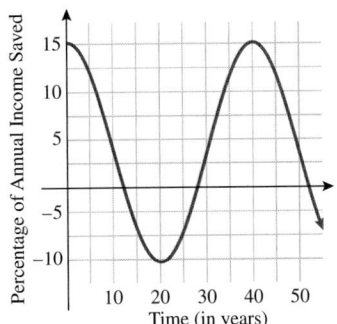

94. Money. For the couple in Exercise 93, how many years into their marriage are they back to saving 15% of their annual income?

95. Viewing Angle of Painting. A museum patron whose eye level is 5 feet above the floor is studying a painting that is 8 feet in height and mounted on the wall 4 feet above the floor. If the patron is x feet from the wall, use $\tan(\alpha + \beta)$ to express $\tan\theta$, where θ is the angle that the patron's eye sweeps from the top to the bottom of the painting.

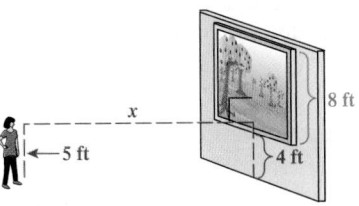

96. Viewing Angle of Painting. Using the equation for $\tan\theta$ in Exercise 95, solve for θ using the inverse tangent. Then find the measure of the angles θ for $x = 10$ and $x = 20$ (to the nearest degree).

97. Earthquake Movement. The horizontal movement of a point that is k kilometers away from an earthquake's fault line can be estimated with

$$M = \frac{f}{2}\left[1 - \frac{2\tan^{-1}\left(\frac{k}{d}\right)}{\pi}\right]$$

where M is the movement of the point in meters, f is the total horizontal displacement occurring along the fault line, k is the distance of the point from the fault line, and d is the depth in kilometers of the focal point of the earthquake. If an earthquake produces a displacement f of 2 meters and the depth of the focal point is 4 kilometers, then what is the movement M of a point that is 2 kilometers from the fault line? of a point 10 kilometers from the fault line?

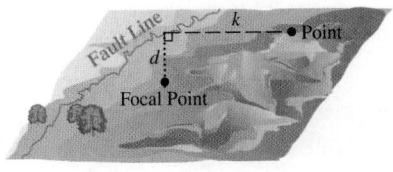

98. Earthquake Movement. Repeat Exercise 97. If an earthquake produces a displacement f of 3 meters and the depth of the focal point is 2.5 kilometers, then what is the movement M of a point that is 5 kilometers from the fault line? of a point 10 kilometers from the fault line?

99. Laser Communication. A laser communication system depends on a narrow beam, and a direct line of sight is necessary for communication links. If a transmitter/receiver for a laser system is placed between two buildings (see the figure) and the other end of the system is located on a low-earth-orbit satellite, then the link is operational only when the satellite and the ground system have a line of sight (when the buildings are not in the way). Find the angle θ that corresponds to the system being operational (i.e., find the maximum value of θ that permits the system to be operational). Express θ in terms of inverse tangent functions and the distance from the shorter building.

100. Laser Communication. Repeat Exercise 99, assuming that the ground system is on top of a 20-foot tower.

■ CATCH THE MISTAKE

In Exercises 101–104, explain the mistake that is made.

101. Evaluate the expression exactly: $\sin^{-1}\left[\sin\left(\dfrac{3\pi}{5}\right)\right]$.

Solution:

Use the identity $\sin^{-1}(\sin x) = x$ on $0 \le x \le \pi$.

Since $\dfrac{3\pi}{5}$ is in the interval

$[0, \pi]$, the identity

can be used. $\sin^{-1}\left[\sin\left(\dfrac{3\pi}{5}\right)\right] = \dfrac{3\pi}{5}$

This is incorrect. What mistake was made?

102. Evaluate the expression exactly: $\cos^{-1}\left[\cos\left(-\dfrac{\pi}{5}\right)\right]$.

Solution:

Use the identity $\cos^{-1}(\cos x) = x$ on $-\dfrac{\pi}{2} \le x \le \dfrac{\pi}{2}$.

Since $-\dfrac{\pi}{5}$ is in the interval

$\left[-\dfrac{\pi}{2}, \dfrac{\pi}{2}\right]$, the identity

can be used. $\cos^{-1}\left[\cos\left(-\dfrac{\pi}{5}\right)\right] = -\dfrac{\pi}{5}$

This is incorrect. What mistake was made?

103. Evaluate the expression exactly: $\cot^{-1}(2.5)$.

Solution:

Use the reciprocal identity. $\cot^{-1}(2.5) = \dfrac{1}{\tan^{-1}(2.5)}$

Evaluate $\tan^{-1}(2.5) = 1.19$. $\cot^{-1}(2.5) = \dfrac{1}{1.19}$

Simplify. $\cot^{-1}(2.5) = 0.8403$

This is incorrect. What mistake was made?

104. Evaluate the expression exactly: $\csc^{-1}\left(\dfrac{1}{4}\right)$.

Solution:

Use the reciprocal identity. $\csc^{-1}\left(\dfrac{1}{4}\right) = \dfrac{1}{\sin^{-1}\left(\dfrac{1}{4}\right)}$

Evaluate $\sin^{-1}\left(\dfrac{1}{4}\right) = 14.478$. $\csc^{-1}\left(\dfrac{1}{4}\right) = \dfrac{1}{14.478}$

Simplify. $\csc^{-1}\left(\dfrac{1}{4}\right) = 0.0691$

This is incorrect. What mistake was made?

■ CONCEPTUAL

In Exercises 105–108, determine whether each statement is true or false.

105. The inverse secant function is an even function.

106. The inverse cosecant function is an odd function.

107. $\csc^{-1}(\csc \theta) = \theta$, for all θ in the domain of cosecant.

108. $\sin^{-1}(2x) \cdot \csc^{-1}(2x) = 1$, for all x for which both functions are defined.

109. Explain why $\sec^{-1}\left(\dfrac{1}{2}\right)$ does not exist.

110. Explain why $\csc^{-1}\left(\dfrac{1}{2}\right)$ does not exist.

■ CHALLENGE

111. Evaluate exactly: $\sin\left[\cos^{-1}\left(\dfrac{\sqrt{2}}{2}\right) + \sin^{-1}\left(-\dfrac{1}{2}\right)\right]$.

112. Determine the x-values for which

$$\sin^{-1}\left[2\sin\left(\dfrac{3x}{2}\right)\cos\left(\dfrac{3x}{2}\right)\right] = 3x$$

113. Evaluate exactly: $\sin(2\sin^{-1}1)$.

114. Let $f(x) = 2 - 4\sin\left(x - \dfrac{\pi}{2}\right)$.

   a. State an accepted domain of $f(x)$ so that $f(x)$ is a one-to-one function.

   b. Find $f^{-1}(x)$ and state its domain.

115. Let $f(x) = 3 + \cos\left(x - \dfrac{\pi}{4}\right)$.

   a. State an accepted domain of $f(x)$ so that $f(x)$ is a one-to-one function.

   b. Find $f^{-1}(x)$ and state its dómain.

116. Let $f(x) = 1 - \tan\left(x + \dfrac{\pi}{3}\right)$.

   a. State an accepted domain of $f(x)$ so that $f(x)$ is a one-to-one function.

   b. Find $f^{-1}(x)$ and state its domain.

117. Let $f(x) = 2 + \dfrac{1}{4}\cot\left(2x - \dfrac{\pi}{6}\right)$.

   a. State an accepted domain of $f(x)$ so that $f(x)$ is a one-to-one function.

   b. Find $f^{-1}(x)$ and state its domain.

118. Let $f(x) = -\csc\left(\dfrac{\pi}{4}x - 1\right)$.

   a. State an accepted domain of $f(x)$ so that $f(x)$ is a one-to-one function.

   b. Find $f^{-1}(x)$ and state its domain.

■ TECHNOLOGY

119. Use a graphing calculator to plot $Y_1 = \sin(\sin^{-1}x)$ and $Y_2 = x$ for the domain $-1 \le x \le 1$. If you then increase the domain to $-3 \le x \le 3$, you get a different result. Explain the result.

120. Use a graphing calculator to plot $Y_1 = \cos(\cos^{-1}x)$ and $Y_2 = x$ for the domain $-1 \le x \le 1$. If you then increase the domain to $-3 \le x \le 3$, you get a different result. Explain the result.

121. Use a graphing calculator to plot $Y_1 = \csc^{-1}(\csc x)$ and $Y_2 = x$. Determine the domain for which the following statement is true: $\csc^{-1}(\csc x) = x$. Give the domain in terms of π.

122. Use a graphing calculator to plot $Y_1 = \sec^{-1}(\sec x)$ and $Y_2 = x$. Determine the domain for which the following statement is true: $\sec^{-1}(\sec x) = x$. Give the domain in terms of π.

123. Given $\tan x = \dfrac{40}{9}$ and $\pi < x < \dfrac{3\pi}{2}$:

   a. Find $\sin(2x)$ using the double-angle identity.

   b. Use the inverse of tangent to find x in QIII and use a calculator to find $\sin(2x)$. Round to five decimal places.

   c. Are the results in (a) and (b) the same?

124. Given $\sin x = -\dfrac{1}{\sqrt{10}}$ and $\dfrac{3\pi}{2} < x < 2\pi$:

   a. Find $\tan(2x)$ using the double-angle identity.

   b. Use the inverse of sine to find x in QIV and find $\tan(2x)$.

   c. Are the results in (a) and (b) the same?

■ PREVIEW TO CALCULUS

In calculus, we study the derivatives of inverse trigonometric functions. In order to obtain these formulas, we use the definitions of the functions and a right triangle. Thus, if $y = \sin^{-1}x$, then $\sin y = x$; the right triangle associated with this equation is given below, where we can see that $\cos y = \sqrt{1 - x^2}$.

In Exercises 125–128, use this idea to find the indicated expression.

125. If $y = \tan^{-1}x$, find $\sec^2 y$.

126. If $y = \cos^{-1}x$, find $\sin y$.

127. If $y = \sec^{-1}x$, find $\sec y \tan y$.

128. If $y = \cot^{-1}x$, find $\csc^2 y$.

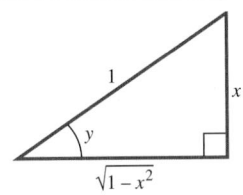

Skills Objectives

- Solve trigonometric equations by inspection.
- Solve trigonometric equations using algebraic techniques.
- Solve trigonometric equations using inverse functions.
- Solve trigonometric equations (involving more than one trigonometric function) using trigonometric identities.

Conceptual Objectives

- Understand that solving trigonometric equations is similar to solving algebraic equations.
- Realize that the goal in solving trigonometric equations is to find the value(s) for the independent variable that make(s) the equation true.

Solving Trigonometric Equations by Inspection

The goal in solving equations in one variable is to find the values for that variable which make the equation true. For example, $9x = 72$ can be solved by inspection by asking the question, "9 times what is 72?" The answer is $x = 8$. We approach simple trigonometric equations the same way we approach algebraic equations: We inspect the equation and determine the solution.

EXAMPLE 1 Solving a Trigonometric Equation by Inspection

Solve each of the following equations over $[0, 2\pi]$:

a. $\sin x = \frac{1}{2}$ **b.** $\cos(2x) = \frac{1}{2}$

Solution (a):

Ask the question, "sine of what angles is $\frac{1}{2}$?"

$$x = \frac{\pi}{6} \quad \text{or} \quad x = \frac{5\pi}{6}$$

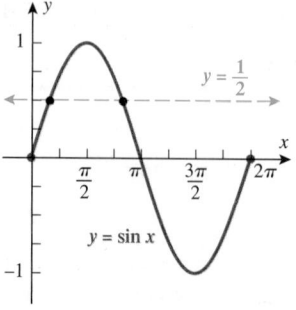

Solution (b):

Ask the question, "cosine of what angles is $\frac{1}{2}$?"

In this case, the angle is equal to $2x$.

$$2x = \frac{\pi}{3} \quad \text{or} \quad 2x = \frac{5\pi}{3}$$

Solve for x: $\boxed{x = \dfrac{\pi}{6} \quad \text{or} \quad x = \dfrac{5\pi}{6}}$.

Quadrant III and IV solutions

correspond to $\boxed{\dfrac{7\pi}{6} \quad \text{or} \quad \dfrac{11\pi}{6}}$.

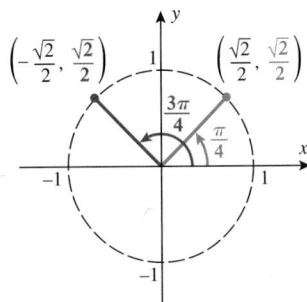

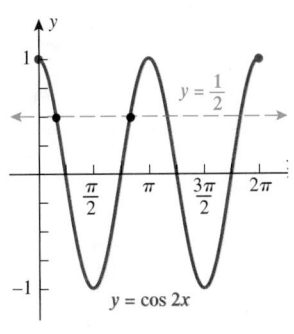

■ **Answer: a.** $x = \dfrac{\pi}{3}, \dfrac{5\pi}{3}$

b. $x = \dfrac{\pi}{12}, \dfrac{5\pi}{12}, \dfrac{13\pi}{12}, \dfrac{17\pi}{12}$

■ **YOUR TURN** Solve each of the following equations over $[0, 2\pi]$:

     a. $\cos x = \frac{1}{2}$ **b.** $\sin(2x) = \frac{1}{2}$

EXAMPLE 2 Solving a Trigonometric Equation by Inspection

Solve the equation $\sin x = \dfrac{\sqrt{2}}{2}$.

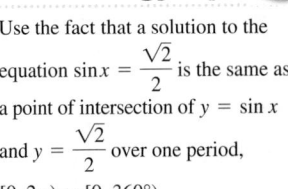

Technology Tip

Use the fact that a solution to the equation $\sin x = \dfrac{\sqrt{2}}{2}$ is the same as a point of intersection of $y = \sin x$ and $y = \dfrac{\sqrt{2}}{2}$ over one period, $[0, 2\pi)$ or $[0, 360°)$.

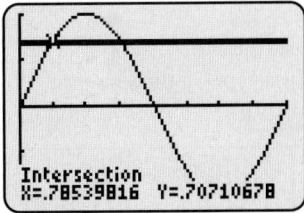

Solution:

STEP 1 Solve over one period, $[0, 2\pi)$.

    Ask the question, "sine of what angles is $\dfrac{\sqrt{2}}{2}$?"

DEGREES	$x = 45°$ or $x = 135°$
RADIANS	$x = \dfrac{\pi}{4}$ or $x = \dfrac{3\pi}{4}$

    The sine function is positive in quadrants I and II.

STEP 2 Solve over all real numbers.

    Since the sine function has a period of $360°$ or 2π, adding integer multiples of $360°$ or 2π will give the other solutions.

DEGREES	$x = 45° + 360°n$ or $x = 135° + 360°n$
RADIANS	$x = \dfrac{\pi}{4} + 2n\pi$ or $x = \dfrac{3\pi}{4} + 2n\pi$, where n is any integer

Study Tip

Find *all* solutions unless the domain is restricted.

■ **Answer:**

DEGREES	$x = 60° + 360°n$ or $x = 300° + 360°n$
RADIANS	$x = \dfrac{\pi}{3} + 2n\pi$ or $x = \dfrac{5\pi}{3} + 2n\pi$, where n is any integer

■ **YOUR TURN** Solve the equation $\cos x = \frac{1}{2}.$ $= 2x^\circ = \frac{\pi}{6}$

Notice that the equations in Example 2 and the Your Turn have an infinite number of solutions. Unless the domain is restricted, you must find *all* solutions.

$= \frac{\pi}{8} + 2n\pi \cdot \text{ or } x = \frac{5\pi}{8} + 2n\pi$

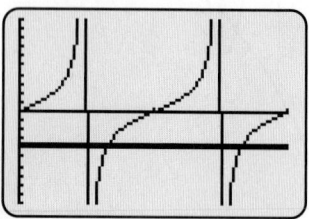
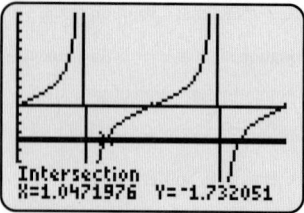
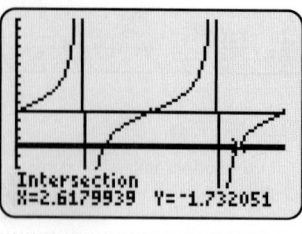
EXAMPLE 3 Solving a Trigonometric Equation by Inspection

Solve the equation $\tan(2x) = -\sqrt{3}$.

Solution:

STEP 1 Solve over one period, $[0, \pi)$.

Ask the question, "tangent of what angles is $-\sqrt{3}$?" Note that the angle in this case is $2x$.

DEGREES	$2x = 120°$
RADIANS	$2x = \dfrac{2\pi}{3}$

The tangent function is negative in quadrants II and IV. Since $[0, \pi)$ includes quadrants I and II, we find only the angle in quadrant II. (The solution corresponding to quadrant IV will be found when we extend the solution over all real numbers.)

STEP 2 Solve over all x.

Since the tangent function has a period of 180°, or π, adding integer multiples of 180° or π will give all of the other solutions.

DEGREES	$2x = 120° + 180°n$
RADIANS	$2x = \dfrac{2\pi}{3} + n\pi,$ where n is any integer

Solve for x by dividing by 2.

DEGREES	$x = 60° + 90°n$
RADIANS	$x = \dfrac{\pi}{3} + \dfrac{n}{2}\pi,$ where n is any integer

Note:

- There are infinitely many solutions. If we graph $y = \tan(2x)$ and $y = -\sqrt{3}$, we see that there are infinitely many points of intersection.

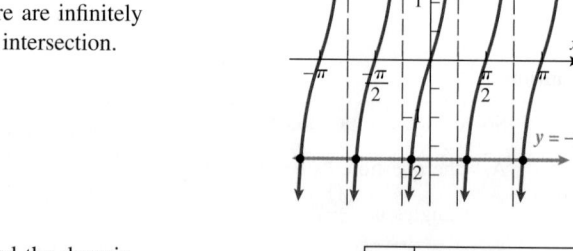

- Had we restricted the domain to $0 \leq x < 2\pi$, the solutions (in radians) would be the values given to the right in the table.

n	$x = \dfrac{\pi}{3} + \dfrac{n}{2}\pi$
0	$x = \dfrac{\pi}{3}$
1	$x = \dfrac{5\pi}{6}$
2	$x = \dfrac{4\pi}{3}$
3	$x = \dfrac{11\pi}{6}$

Notice that only $n = 0, 1, 2, 3$ yield x-values in the domain $0 \leq x < 2\pi$.

Notice that in Step 2 of Example 2, $2n\pi$ was added to get all of the solutions, whereas in Step 2 of Example 3, we added $n\pi$ to the argument of the tangent function. The reason that we added $2n\pi$ in Example 2 and $n\pi$ in Example 3 is because the sine function has period 2π, whereas the tangent function has period π.

Solving Trigonometric Equations Using Algebraic Techniques

We now will use algebraic techniques to solve trigonometric equations. Let us first start with linear and quadratic equations. For linear equations, we solve for the variable by isolating it. For quadratic equations, we often employ factoring or the quadratic formula. If we can let x represent the trigonometric function and the resulting equation is either linear or quadratic, then we use techniques learned in solving algebraic equations.

TYPE	EQUATION	SUBSTITUTION	ALGEBRAIC EQUATION
Linear trigonometric equation	$4\sin\theta - 2 = -4$	$x = \sin\theta$	$4x - 2 = -4$
Quadratic trigonometric equation	$2\cos^2\theta + \cos\theta - 1 = 0$	$x = \cos\theta$	$2x^2 + x - 1 = 0$

It is not necessary to make the substitution, though it is often convenient. Frequently, one can see how to factor a quadratic trigonometric equation without first converting it to an algebraic equation. In Example 4 we will not use the substitution. However, in Example 5 we will illustrate the use of a substitution.

EXAMPLE 4   Solving a Linear Trigonometric Equation

Solve $4\sin\theta - 2 = -4$ on $0 \le \theta < 2\pi$.

Solution:

STEP 1 Solve for $\sin\theta$. $\qquad\qquad\qquad\qquad 4\sin\theta - 2 = -4$

Add 2. $\qquad\qquad\qquad\qquad\qquad\qquad\quad 4\sin\theta = -2$

Divide by 4. $\qquad\qquad\qquad\qquad\qquad\quad \sin\theta = -\dfrac{1}{2}$

STEP 2 Find the values of θ on $0 \le \theta < 2\pi$ that satisfy the equation $\sin\theta = -\frac{1}{2}$.

The sine function is negative in quadrants III and IV.

$\sin\left(\dfrac{7\pi}{6}\right) = -\dfrac{1}{2}$ and $\sin\left(\dfrac{11\pi}{6}\right) = -\dfrac{1}{2}.$ $\qquad$ $\boxed{\theta = \dfrac{7\pi}{6}}$ or $\boxed{\theta = \dfrac{11\pi}{6}}$

■ **YOUR TURN** Solve $2\cos\theta + 1 = 2$ on $0 \le \theta < 2\pi$.

■ **Answer:** $\theta = \dfrac{\pi}{3}$ or $\dfrac{5\pi}{3}$

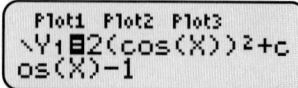

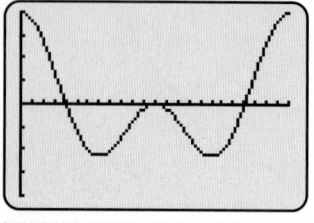

EXAMPLE 5   Solving a Quadratic Trigonometric Equation

Solve $2\cos^2\theta + \cos\theta - 1 = 0$ on $0 \le \theta < 2\pi$.

Solution:

STEP 1 Solve for $\cos\theta$. $2\cos^2\theta + \cos\theta - 1 = 0$

Let $x = \cos\theta$. $2x^2 + x - 1 = 0$

Factor the quadratic equation. $(2x - 1)(x + 1) = 0$

Set each factor equal to 0. $2x - 1 = 0$ or $x + 1 = 0$

Solve each for x. $x = \dfrac{1}{2}$ or $x = -1$

Substitute $x = \cos\theta$. $\cos\theta = \dfrac{1}{2}$ or $\cos\theta = -1$

STEP 2 Find the values of θ on $0 \le \theta < 2\pi$ which satisfy the equation $\cos\theta = \frac{1}{2}$.

The cosine function is positive in quadrants I and IV.

$$\cos\left(\frac{\pi}{3}\right) = \frac{1}{2} \text{ and } \cos\left(\frac{5\pi}{3}\right) = \frac{1}{2}.$$
$\boxed{\theta = \dfrac{\pi}{3}}$ or $\boxed{\theta = \dfrac{5\pi}{3}}$

STEP 3 Find the values of θ on $0 \le \theta < 2\pi$ that satisfy the equation $\cos\theta = -1$.

$\cos\pi = -1$. $\boxed{\theta = \pi}$

The solutions to $2\cos^2\theta + \cos\theta - 1 = 0$ on $0 \le \theta < 2\pi$ are $\theta = \dfrac{\pi}{3}$, $\theta = \dfrac{5\pi}{3}$, and $\theta = \pi$.

▪ **Answer:** $\theta = \dfrac{\pi}{2}, \dfrac{7\pi}{6}$, or $\dfrac{11\pi}{6}$

▪ **YOUR TURN** Solve $2\sin^2\theta - \sin\theta - 1 = 0$ on $0 \le \theta < 2\pi$.

If Example 5 asked for the solution to the trigonometric equation over all real numbers, then the solutions would be $\theta = \dfrac{\pi}{3} \pm 2n\pi, \dfrac{5\pi}{3} \pm 2n\pi$, and $\pi \pm 2n\pi$.

Solving Trigonometric Equations That Require the Use of Inverse Functions

Thus far, we have been able to solve the trigonometric equations exactly. Now we turn our attention to situations that require using a calculator and inverse functions to approximate a solution to a trigonometric equation.

EXAMPLE 6 Solving a Trigonometric Equation That Requires the Use of Inverse Functions

Solve $\tan^2\theta - \tan\theta = 6$ on $0° \le \theta < 180°$.

Solution:

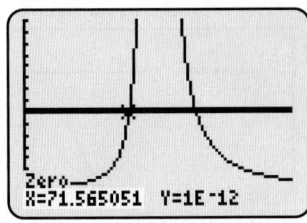

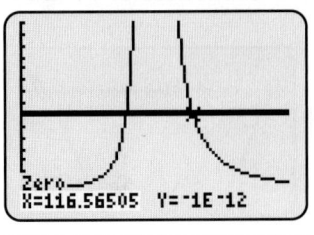

STEP 1 Solve for $\tan\theta$.

Subtract 6. $\tan^2\theta - \tan\theta - 6 = 0$

Factor the quadratic trigonometric
expression on the left. $(\tan\theta - 3)(\tan\theta + 2) = 0$

Set the factors equal to 0. $\tan\theta - 3 = 0$ or $\tan\theta + 2 = 0$

Solve for $\tan\theta$. $\tan\theta = 3$ or $\tan\theta = -2$

STEP 2 Solve $\tan\theta = 3$ on $0° \le \theta < 180°$.

The tangent function is positive on $0° \le \theta < 180°$ only in quadrant I.

Write the equivalent inverse notation to $\tan\theta = 3$. $\theta = \tan^{-1}3$

Use a calculator to evaluate (approximate) θ. $\boxed{\theta \approx 71.6°}$

STEP 3 Solve $\tan\theta = -2$ on $0° \le \theta < 180°$.

The tangent function is negative on $0° \le \theta < 180°$ only in quadrant II.

A calculator gives values of the inverse tangent in quadrants I and IV.

We will call the reference angle in quadrant IV "α."

Write the equivalent inverse notation
to $\tan\alpha = -2$. $\alpha = \tan^{-1}(-2)$

Use a calculator to evaluate (approximate) α. $\alpha \approx -63.4°$

To find the value of θ in quadrant II, add $180°$. $\theta = \alpha + 180°$

$\boxed{\theta \approx 116.6°}$

The solutions to $\tan^2\theta - \tan\theta = 6$ on $0° \le \theta < 180°$ are $\theta = 71.6°$ and $\theta = 116.6°$.

■ YOUR TURN Solve $\tan^2\theta + \tan\theta = 6$ on $0° \le \theta < 180°$.

■ Answer: $\theta \approx 63.4°$ or $108.4°$

Recall that in solving algebraic quadratic equations, one method (when factoring is not obvious or possible) is to use the Quadratic Formula.

$$ax^2 + bx + c = 0 \text{ has solutions } x = \frac{-b \pm \sqrt{b^2 - 4ac}}{2a}$$

Technology Tip

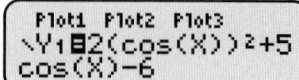

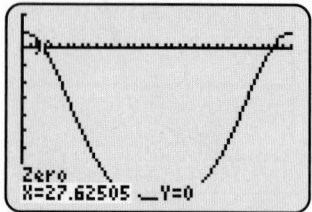

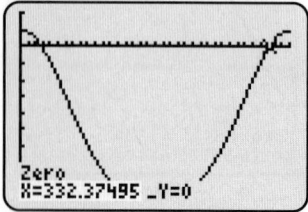

EXAMPLE 7 **Solving a Quadratic Trigonometric Equation That Requires the Use of the Quadratic Formula and Inverse Functions**

Solve $2\cos^2\theta + 5\cos\theta - 6 = 0$ on $0° \le \theta < 360°$.

Solution:

STEP 1 Solve for $\cos\theta$. $\qquad\qquad\qquad\qquad 2\cos^2\theta + 5\cos\theta - 6 = 0$

Let $x = \cos\theta$. $\qquad\qquad\qquad\qquad\quad 2x^2 + 5x - 6 = 0$

Use the Quadratic Formula, $\qquad x = \dfrac{-5 \pm \sqrt{5^2 - 4(2)(-6)}}{2(2)}$
$a = 2, b = 5, c = -6$.

Simplify. $\qquad\qquad\qquad\qquad\qquad x = \dfrac{-5 \pm \sqrt{73}}{4}$

Use a calculator to approximate
the solution. $\qquad\qquad\qquad\qquad x \approx -3.3860 \quad$ or $\quad x \approx 0.8860$

Let $x = \cos\theta$. $\qquad\qquad\qquad \cos\theta \approx -3.3860 \quad$ or $\quad \cos\theta \approx 0.8860$

STEP 2 Solve $\cos\theta = -3.3860$ on $0° \le \theta < 360°$.

Recall that the range of the cosine function is $[-1,1]$; therefore, the cosine function can never equal a number outside that range $(-3.3860 < -1)$.

The equation $\cos\theta = -3.3860$ has *no solution*.

STEP 3 Solve $\cos\theta = 0.8860$ on $0° \le \theta < 360°$.

The cosine function is positive in quadrants I and IV. Since a calculator gives inverse cosine values only in quadrants I and II, we will have to use a reference angle to get the quadrant IV solution.

Write the equivalent inverse
notation for $\cos\theta = 0.8860$. $\qquad\qquad \theta = \cos^{-1} 0.8860$

Use a calculator to evaluate
(approximate) the solution. $\qquad\qquad \boxed{\theta \approx 27.6°}$

To find the second solution
(in quadrant IV), subtract the $\qquad\qquad \theta = 360° - 27.6°$
reference angle from 360°. $\qquad\qquad \boxed{\theta \approx 332.4°}$

The solutions to $2\cos^2\theta + 5\cos\theta - 6 = 0$ on $0° \le \theta < 360°$ are $\theta \approx 27.6°$ and $\theta \approx 332.4°$.

■ **Answer:** $\theta \approx 242.4°$ or $297.6°$

■ **YOUR TURN** Solve $2\sin^2\theta - 5\sin\theta - 6 = 0$ on $0° \le \theta < 360°$.

Using Trigonometric Identities to Solve Trigonometric Equations

We now consider trigonometric equations that involve more than one trigonometric function. Trigonometric identities are an important part of solving these types of equations.

EXAMPLE 8 **Using Trigonometric Identities to Solve Trigonometric Equations**

Solve $\sin x + \cos x = 1$ on $0 \le x < 2\pi$.

Solution:

Square both sides.

$$\sin^2 x + 2\sin x \cos x + \cos^2 x = 1$$

Label the Pythagorean identity.

$$\underbrace{\sin^2 x + \cos^2 x}_{1} + 2\sin x \cos x = 1$$

Subtract 1 from both sides.

$$2\sin x \cos x = 0$$

Use the Zero Product Property.

$$\sin x = 0 \quad \text{or} \quad \cos x = 0$$

Solve for x on $0 \le x < 2\pi$.

$$x = 0 \quad \text{or} \quad x = \pi \quad \text{or} \quad x = \frac{\pi}{2} \quad \text{or} \quad x = \frac{3\pi}{2}$$

Because we squared the equation, we have to check for extraneous solutions.

Check $x = 0$.

$$\sin 0 + \cos 0 = 0 + 1 = 1 \quad \checkmark$$

Check $x = \pi$.

$$\sin \pi + \cos \pi = 0 - 1 = -1 \quad \text{X}$$

Check $x = \dfrac{\pi}{2}$.

$$\sin\left(\frac{\pi}{2}\right) + \cos\left(\frac{\pi}{2}\right) = 1 + 0 = 1 \quad \checkmark$$

Check $x = \dfrac{3\pi}{2}$.

$$\sin\left(\frac{3\pi}{2}\right) + \cos\left(\frac{3\pi}{2}\right) = -1 + 0 = -1 \quad \text{X}$$

The solutions to $\sin x + \cos x = 1$ on $0 \le x < 2\pi$ are $\boxed{x = 0}$ and $\boxed{x = \dfrac{\pi}{2}}$.

■ **YOUR TURN** Solve $\sin x - \cos x = 1$ on $0 \le x < 2\pi$.

Technology Tip

Find the points of intersection of $y_1 = \sin x + \cos x$ and $y_2 = 1$.

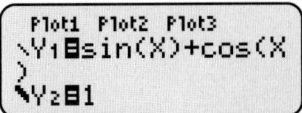

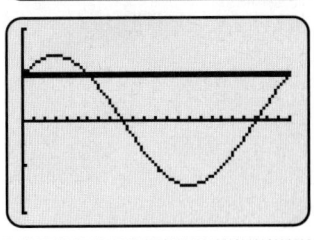

■ **Answer:** $x = \dfrac{\pi}{2}$ or $x = \pi$

EXAMPLE 9 **Using Trigonometric Identities to Solve Trigonometric Equations**

Solve $\sin(2x) = \sin x$ on $0 \le x < 2\pi$.

COMMON MISTAKE

Dividing by a trigonometric function (which could be equal to zero).

⭐ **CORRECT**

Use the double-angle formula for sine.

$$\underbrace{\sin(2x)}_{2\sin x \cos x} = \sin x$$

Subtract $\sin x$.

$$2\sin x \cos x - \sin x = 0$$

Factor the common $\sin x$.

$$(\sin x)(2\cos x - 1) = 0$$

Set each factor equal to 0.

$$\sin x = 0 \quad \text{or} \quad 2\cos x - 1 = 0$$

$$\sin x = 0 \quad \text{or} \quad \cos x = \frac{1}{2}$$

Solve $\sin x = 0$ for x on $0 \le x < 2\pi$.

$$\boxed{x = 0} \quad \text{or} \quad \boxed{x = \pi}$$

Solve $\cos x = \frac{1}{2}$ for x on $0 \le x < 2\pi$.

$$\boxed{x = \frac{\pi}{3}} \quad \text{or} \quad \boxed{x = \frac{5\pi}{3}}$$

❌ **INCORRECT**

$$2\sin x \cos x = \sin x$$

Divide by $\sin x$. **ERROR**

$$2\cos x = 1$$

Solve for $\cos x$.

$$\cos x = \frac{1}{2}$$

Missing solutions from $\sin x = 0$.

The solutions to $\sin(2x) = \sin x$ are $x = 0, \dfrac{\pi}{3}, \pi,$ and $\dfrac{5\pi}{3}$.

▼ **CAUTION**

Do not divide equations by trigonometric functions, as they can sometimes equal zero.

■ **Answer:** $x = \dfrac{\pi}{2}, \dfrac{3\pi}{2}, \dfrac{\pi}{6},$ or $\dfrac{5\pi}{6}$

■ **YOUR TURN** Solve $\sin(2x) = \cos x$ on $0 \le x < 2\pi$.

EXAMPLE 10   **Using Trigonometric Identities to Solve Trigonometric Equations**

Solve $\sin x + \csc x = -2$.

Solution:

Use the reciprocal identity.

$$\sin x + \underbrace{\csc x}_{\frac{1}{\sin x}} = -2$$

Add 2.

$$\sin x + 2 + \frac{1}{\sin x} = 0$$

Multiply by $\sin x$. (Note $\sin x \neq 0$.)

$$\sin^2 x + 2\sin x + 1 = 0$$

Factor as a perfect square.

$$(\sin x + 1)^2 = 0$$

Solve for $\sin x$.

$$\sin x = -1$$

Solve for x on one period of the sine function, $[0, 2\pi)$.

$$x = \frac{3\pi}{2}$$

Add integer multiples of 2π to obtain all solutions.

$$\boxed{x = \frac{3\pi}{2} + 2n\pi}$$

EXAMPLE 11   **Using Trigonometric Identities and Inverse Functions to Solve Trigonometric Equations**

Solve $3\cos^2\theta + \sin\theta = 3$ on $0° \leq \theta < 360°$.

Solution:

Use the Pythagorean identity.

$$3\underbrace{\cos^2\theta}_{1-\sin^2\theta} + \sin\theta = 3$$

Subtract 3.

$$3(1 - \sin^2\theta) + \sin\theta - 3 = 0$$

Eliminate the parentheses.

$$3 - 3\sin^2\theta + \sin\theta - 3 = 0$$

Simplify.

$$-3\sin^2\theta + \sin\theta = 0$$

Factor the common $\sin\theta$.

$$\sin\theta(1 - 3\sin\theta) = 0$$

Set each factor equal to 0.

$$\sin\theta = 0 \quad \text{or} \quad 1 - 3\sin\theta = 0$$

Solve for $\sin\theta$.

$$\sin\theta = 0 \quad \text{or} \quad \sin\theta = \frac{1}{3}$$

Solve $\sin\theta = 0$ for x on $0° \leq \theta < 360°$.

$$\boxed{\theta = 0°} \quad \text{or} \quad \boxed{\theta = 180°}$$

Solve $\sin\theta = \frac{1}{3}$ for x on $0° \leq \theta < 360°$.

The sine function is positive in quadrants I and II.

A calculator gives inverse values only in quadrant I.

Write the equivalent inverse notation for $\sin\theta = \frac{1}{3}$.

$$\theta = \sin^{-1}\left(\frac{1}{3}\right)$$

Use a calculator to approximate the quadrant I solution.

$$\boxed{\theta \approx 19.5°}$$

To find the quadrant II solution, subtract the reference angle from 180°.

$$\theta \approx 180° - 19.5°$$
$$\boxed{\theta \approx 160.5°}$$

Technology Tip

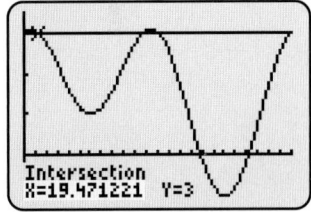

Intersection
X=19.471221 Y=3

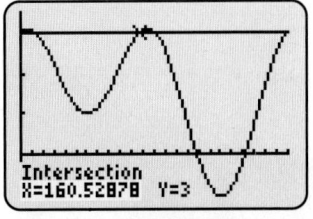

Intersection
X=160.52878 Y=3

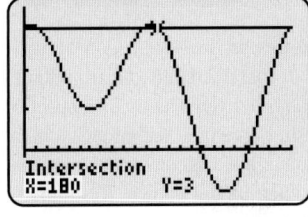

Intersection
X=180   Y=3

Applications

 EXAMPLE 12 **Applications Involving Trigonometric Equations**

Light bends (refracts) according to Snell's law, which states

$$n_i \sin(\theta_i) = n_r \sin(\theta_r)$$

where

- n_i is the refractive index of the medium the light is leaving.
- θ_i is the incident angle between the light ray and the normal (perpendicular) to the interface between mediums.
- n_r is the refractive index of the medium the light is entering.
- θ_r is the refractive angle between the light ray and the normal (perpendicular) to the interface between mediums.

Janis Christie/Getty Images, Inc.

Assume that light is going from air into a diamond. Calculate the refractive angle θ_r if the incidence angle is $\theta_i = 32°$ and the index of refraction values for air and diamond are $n_i = 1.00$ and $n_r = 2.417$, respectively.

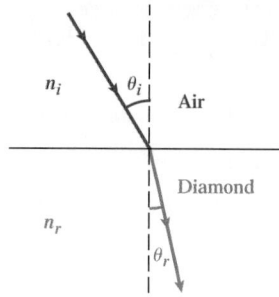

Solution:

Write Snell's law.	$n_i \sin(\theta_i) = n_r \sin(\theta_r)$
Substitute $\theta_i = 32°$, $n_i = 1.00$, and $n_r = 2.417$.	$\sin 32° = 2.417 \sin \theta_r$
Isolate $\sin \theta_r$ and simplify.	$\sin \theta_r = \dfrac{\sin 32°}{2.417} \approx 0.21925$
Solve for θ_r using the inverse sine function.	$\theta_r \approx \sin^{-1}(0.21925) \approx 12.665°$
Round to the nearest degree.	$\boxed{\theta_r \approx 13°}$

SECTION
6.6 SUMMARY

In this section, we began by solving basic trigonometric equations that contained only one trigonometric function. Some such equations can be solved exactly by inspection, and others can be solved exactly using algebraic techniques similar to those of linear and quadratic equations. Calculators and inverse functions are needed when exact values are not known. It is important to note that calculators give the inverse function in only one of the two relevant quadrants. The other quadrant solutions must be found using reference angles. Trigonometric identities are useful for solving equations that involve more than one trigonometric function. With trigonometric identities we can transform such equations into equations involving only one trigonometric function, and then we can apply algebraic techniques.

SECTION
6.6 EXERCISES

■ SKILLS

In Exercises 1–20, solve the given trigonometric equation exactly over the indicated interval.

1. $\cos\theta = -\dfrac{\sqrt{2}}{2}, 0 \le \theta < 2\pi$

2. $\sin\theta = -\dfrac{\sqrt{2}}{2}, 0 \le \theta < 2\pi$

3. $\csc\theta = -2, 0 \le \theta < 4\pi$

4. $\sec\theta = -2, 0 \le \theta < 4\pi$

5. $\tan\theta = 0$, all real numbers

6. $\cot\theta = 0$, all real numbers

7. $\sin(2\theta) = -\dfrac{1}{2}, 0 \le \theta < 2\pi$

8. $\cos(2\theta) = \dfrac{\sqrt{3}}{2}, 0 \le \theta < 2\pi$

9. $\sin\left(\dfrac{\theta}{2}\right) = -\dfrac{1}{2}$, all real numbers

10. $\cos\left(\dfrac{\theta}{2}\right) = -1$, all real numbers

11. $\tan(2\theta) = \sqrt{3}, -2\pi \le \theta < 2\pi$

12. $\tan(2\theta) = -\sqrt{3}$, all real numbers

13. $\sec\theta = -2, -2\pi \le \theta < 0$

14. $\csc\theta = \dfrac{2\sqrt{3}}{3}, -\pi \le \theta < \pi$

15. $\cot(4\theta) = -\dfrac{\sqrt{3}}{3}$, all real numbers

16. $\tan(5\theta) = 1$, all real numbers

17. $\sec(3\theta) = -1, -2\pi \le \theta \le 0$

18. $\sec(4\theta) = \sqrt{2}, 0 \le \theta \le \pi$

19. $\csc(3\theta) = 1, -2\pi \le \theta \le 0$

20. $\csc(6\theta) = -\dfrac{2\sqrt{3}}{3}, 0 \le \theta \le \pi$

In Exercises 21–40, solve the given trigonometric equation exactly on $0 \le \theta < 2\pi$.

21. $2\sin(2\theta) = \sqrt{3}$

22. $2\cos\left(\dfrac{\theta}{2}\right) = -\sqrt{2}$

23. $3\tan(2\theta) - \sqrt{3} = 0$

24. $4\tan\left(\dfrac{\theta}{2}\right) - 4 = 0$

25. $2\cos(2\theta) + 1 = 0$

26. $4\csc(2\theta) + 8 = 0$

27. $\sqrt{3}\cot\left(\dfrac{\theta}{2}\right) - 3 = 0$

28. $\sqrt{3}\sec(2\theta) + 2 = 0$

29. $\tan^2\theta - 1 = 0$

30. $\sin^2\theta + 2\sin\theta + 1 = 0$

31. $2\cos^2\theta - \cos\theta = 0$

32. $\tan^2\theta - \sqrt{3}\tan\theta = 0$

33. $\csc^2\theta + 3\csc\theta + 2 = 0$

34. $\cot^2\theta = 1$

35. $\sin^2\theta + 2\sin\theta - 3 = 0$

36. $2\sec^2\theta + \sec\theta - 1 = 0$

37. $\sec^2\theta - 1 = 0$

38. $\csc^2\theta - 1 = 0$

39. $\sec^2(2\theta) - \dfrac{4}{3} = 0$

40. $\csc^2(2\theta) - 4 = 0$

In Exercises 41–60, solve the given trigonometric equation on $0° \le \theta < 360°$ and express the answer in degrees to two decimal places.

41. $\sin(2\theta) = -0.7843$

42. $\cos(2\theta) = 0.5136$

43. $\tan\left(\dfrac{\theta}{2}\right) = -0.2343$

44. $\sec\left(\dfrac{\theta}{2}\right) = 1.4275$

45. $5\cot\theta - 9 = 0$

46. $5\sec\theta + 6 = 0$

47. $4\sin\theta + \sqrt{2} = 0$

48. $3\cos\theta - \sqrt{5} = 0$

49. $4\cos^2\theta + 5\cos\theta - 6 = 0$

50. $6\sin^2\theta - 13\sin\theta - 5 = 0$

51. $6\tan^2\theta - \tan\theta - 12 = 0$

52. $6\sec^2\theta - 7\sec\theta - 20 = 0$

53. $15\sin^2(2\theta) + \sin(2\theta) - 2 = 0$

54. $12\cos^2\left(\dfrac{\theta}{2}\right) - 13\cos\left(\dfrac{\theta}{2}\right) + 3 = 0$

55. $\cos^2\theta - 6\cos\theta + 1 = 0$

56. $\sin^2\theta + 3\sin\theta - 3 = 0$

57. $2\tan^2\theta - \tan\theta - 7 = 0$

58. $3\cot^2\theta + 2\cot\theta - 4 = 0$

59. $\csc^2(3\theta) - 2 = 0$

60. $\sec^2\left(\dfrac{\theta}{2}\right) - 2 = 0$

In Exercises 61–88, solve the trigonometric equations exactly on the indicated interval, $0 \le x < 2\pi$.

61. $\sin x = \cos x$

62. $\sin x = -\cos x$

63. $\sec x + \cos x = -2$

64. $\sin x + \csc x = 2$

65. $\sec x - \tan x = \dfrac{\sqrt{3}}{3}$

66. $\sec x + \tan x = 1$

67. $\csc x + \cot x = \sqrt{3}$

68. $\csc x - \cot x = \dfrac{\sqrt{3}}{3}$

69. $2 \sin x - \csc x = 0$

70. $2 \sin x + \csc x = 3$

71. $\sin(2x) = 4 \cos x$

72. $\sin(2x) = \sqrt{3} \sin x$

73. $\sqrt{2} \sin x = \tan x$

74. $\cos(2x) = \sin x$

75. $\tan(2x) = \cot x$

76. $3 \cot(2x) = \cot x$

77. $\sqrt{3} \sec x = 4 \sin x$

78. $\sqrt{3} \tan x = 2 \sin x$

79. $\sin^2 x - \cos(2x) = -\dfrac{1}{4}$

80. $\sin^2 x - 2 \sin x = 0$

81. $\cos^2 x + 2 \sin x + 2 = 0$

82. $2 \cos^2 x = \sin x + 1$

83. $2 \sin^2 x + 3 \cos x = 0$

84. $4 \cos^2 x - 4 \sin x = 5$

85. $\cos(2x) + \cos x = 0$

86. $2 \cot x = \csc x$

87. $\dfrac{1}{4} \sec(2x) = \sin(2x)$

88. $-\dfrac{1}{4} \csc\left(\dfrac{1}{2}x\right) = \cos\left(\dfrac{1}{2}x\right)$

In Exercises 89–98, solve each trigonometric equation on $0° \le \theta < 360°$. Express solutions in degrees and round to two decimal places.

89. $\cos(2x) + \dfrac{1}{2} \sin x = 0$

90. $\sec^2 x = \tan x + 1$

91. $6 \cos^2 x + \sin x = 5$

92. $\sec^2 x = 2 \tan x + 4$

93. $\cot^2 x - 3 \csc x - 3 = 0$

94. $\csc^2 x + \cot x = 7$

95. $2 \sin^2 x + 2 \cos x - 1 = 0$ **96.** $\sec^2 x + \tan x - 2 = 0$

97. $\dfrac{1}{16} \csc^2\left(\dfrac{x}{4}\right) - \cos^2\left(\dfrac{x}{4}\right) = 0$ **98.** $-\dfrac{1}{4} \sec^2\left(\dfrac{x}{8}\right) + \sin^2\left(\dfrac{x}{8}\right) = 0$

▪ APPLICATIONS

99. Sales. Monthly sales of soccer balls are approximated by $S = 400 \sin\left(\dfrac{\pi}{6}x\right) + 2000$, where x is the number of the month (January is $x = 1$, etc.). During which month do sales reach 2400?

100. Sales. Monthly sales of soccer balls are approximated by $S = 400 \sin\left(\dfrac{\pi}{6}x\right) + 2000$, where x is the number of the month (January is $x = 1$, etc.). During which two months do sales reach 1800?

101. Home Improvement. A rain gutter is constructed from a single strip of sheet metal by bending as shown below and on the right, so that the base and sides are the same length. Express the area of the cross section of the rain gutter as a function of the angle θ (note that the expression will also involve x).

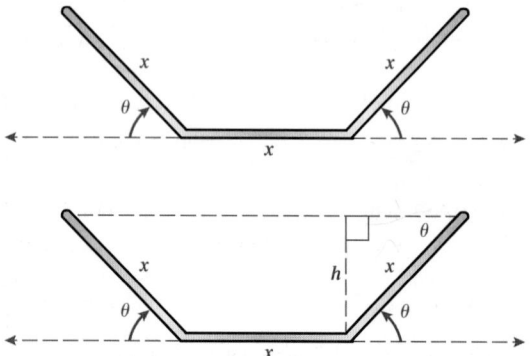

102. Home Improvement. A rain gutter is constructed from a single strip of sheet metal by bending as shown above, so that the base and sides are the same length. When the area of the cross section of the rain gutter is expressed as a function of the angle θ, you can then determine the value of θ that produces the cross section with the greatest possible area. The angle is found by solving the equation $\cos^2\theta - \sin^2\theta + \cos\theta = 0$. Which angle gives the maximum area?

103. Deer Population. The number of deer on an island is given by $D = 200 + 100 \sin\left(\dfrac{\pi}{2}x\right)$, where x is the number of years since 2000. Which is the first year after 2000 that the number of deer reaches 300?

104. Deer Population. The number of deer on an island is given by $D = 200 + 100 \sin\left(\dfrac{\pi}{6}x\right)$, where x is the number of years since 2000. Which is the first year after 2000 that the number of deer reaches 150?

105. Optics. Assume that light is going from air into a diamond. Calculate the refractive angle θ_r if the incidence angle is $\theta_i = 75°$ and the index of refraction values for air and diamond are $n_i = 1.00$ and $n_r = 2.417$, respectively. Round to the nearest degree. (See Example 12 for Snell's law.)

106. Optics. Assume that light is going from a diamond into air. Calculate the refractive angle θ_r if the incidence angle is $\theta_i = 15°$ and the index of refraction values for diamond and air are $n_i = 2.417$ and $n_r = 1.00$, respectively. Round to the nearest degree. (See Example 12 for Snell's law.)

107. Air in Lungs. If a person breathes in and out every 3 seconds, the volume of air in the lungs can be modeled by $A = 2\sin\left(\dfrac{\pi}{3}x\right)\cos\left(\dfrac{\pi}{3}x\right) + 3$, where A is in liters of air and x is in seconds. How many seconds into the cycle is the volume of air equal to 4 liters?

108. Air in Lungs. For the function given in Exercise 107, how many seconds into the cycle is the volume of air equal to 2 liters?

For Exercises 109 and 110, refer to the following:

The figure below shows the graph of $y = 2\cos x - \cos(2x)$ between -2π and 2π. The maximum and minimum values of the curve occur at the *turning points* and are found in the solutions of the equation $-2\sin x + 2\sin(2x) = 0$.

109. Finding Turning Points. Solve for the coordinates of the turning points of the curve between 0 and 2π.

110. Finding Turning Points. Solve for the coordinates of the turning points of the curve between -2π and 0.

■ CATCH THE MISTAKE

In Exercises 111–114, explain the mistake that is made.

111. Solve $\sqrt{2 + \sin\theta} = \sin\theta$ on $0 \le \theta \le 2\pi$.

Solution:

Square both sides.	$2 + \sin\theta = \sin^2\theta$
Gather all terms to one side.	$\sin^2\theta - \sin\theta - 2 = 0$
Factor.	$(\sin\theta - 2)(\sin\theta + 1) = 0$
Set each factor equal to zero.	$\sin\theta - 2 = 0$ or $\sin\theta + 1 = 0$
Solve for $\sin\theta$.	$\sin\theta = 2$ or $\sin\theta = -1$
Solve $\sin\theta = 2$ for θ.	no solution
Solve $\sin\theta = -1$ for θ.	$\theta = \dfrac{3\pi}{2}$

This is incorrect. What mistake was made?

112. Solve $\sqrt{3\sin\theta - 2} = -\sin\theta$ on $0 \le \theta \le 2\pi$.

Solution:

Square both sides.	$3\sin\theta - 2 = \sin^2\theta$
Gather all terms to one side.	$\sin^2\theta - 3\sin\theta + 2 = 0$
Factor.	$(\sin\theta - 2)(\sin\theta - 1) = 0$
Set each factor equal to zero.	$\sin\theta - 2 = 0$ or $\sin\theta - 1 = 0$
Solve for $\sin\theta$.	$\sin\theta = 2$ or $\sin\theta = 1$
Solve $\sin\theta = 2$ for θ.	no solution
Solve $\sin\theta = 1$ for θ.	$\theta = \dfrac{\pi}{2}$

This is incorrect. What mistake was made?

113. Solve $3\sin(2x) = 2\cos x$ on $0° \le \theta \le 180°$.

Solution:

Use the double-angle identity for the sine function.	$3\underbrace{\sin(2x)}_{2\sin x\cos x} = 2\cos x$
Simplify.	$6\sin x\cos x = 2\cos x$
Divide by $2\cos x$.	$3\sin x = 1$
Divide by 3.	$\sin x = \dfrac{1}{3}$
Write the equivalent inverse notation.	$x = \sin^{-1}\left(\dfrac{1}{3}\right)$
Use a calculator to approximate the solution.	$x \approx 19.47°$, QI solution
The QII solution is:	$x \approx 180° - 19.47° \approx 160.53°$

This is incorrect. What mistake was made?

114. Solve $\sqrt{1 + \sin x} = \cos x$ on $0 \le x \le 2\pi$.

Solution:

Square both sides.	$1 + \sin x = \cos^2 x$
Use the Pythagorean identity.	$1 + \sin x = \underbrace{\cos^2 x}_{1 - \sin^2 x}$
Simplify.	$\sin^2 x + \sin x = 0$
Factor.	$\sin x(\sin x + 1) = 0$
Set each factor equal to zero.	$\sin x = 0$ or $\sin x + 1 = 0$
Solve for $\sin x$.	$\sin x = 0$ or $\sin x = -1$
Solve for x.	$x = 0, \pi, \dfrac{3\pi}{2}, 2\pi$

This is incorrect. What mistake was made?

▪ CONCEPTUAL

In Exercises 115–118, determine whether each statement is true or false.

115. Linear trigonometric equations always have one solution on $[0, 2\pi]$.

116. Quadratic trigonometric equations always have two solutions on $[0, 2\pi]$.

117. If a trigonometric equation has all real numbers as its solution, then it is an identity.

118. If a trigonometric equation has an infinite number of solutions, then it is an identity.

▪ CHALLENGE

119. Solve $16\sin^4\theta - 8\sin^2\theta = -1$ over $0 \le \theta \le 2\pi$.

120. Solve $\left|\cos\left(\theta + \dfrac{\pi}{4}\right)\right| = \dfrac{\sqrt{3}}{2}$ over all real numbers.

121. Solve for the smallest positive x that makes this statement true:
$$\sin\left(x + \frac{\pi}{4}\right) + \sin\left(x - \frac{\pi}{4}\right) = \frac{\sqrt{2}}{2}$$

122. Solve for the smallest positive x that makes this statement true:
$$\cos x \cos 15° + \sin x \sin 15° = 0.7$$

123. Find all real numbers x such that $\dfrac{1 - \cos\left(\dfrac{x}{3}\right)}{1 + \cos\left(\dfrac{x}{3}\right)} + 1 = 0$.

124. Find all real numbers θ such that $\sec^4\left(\dfrac{1}{3}\theta\right) - 1 = 0$.

125. Find all real numbers θ such that $\csc^4\left(\dfrac{\pi}{4}\theta - \pi\right) - 4 = 0$.

126. Find all real numbers x such that
$$2\tan(3x) = \sqrt{3} - \sqrt{3}\tan^2(3x).$$

▪ TECHNOLOGY

Graphing calculators can be used to find approximate solutions to trigonometric equations. For the equation $f(x) = g(x)$, let $Y_1 = f(x)$ and $Y_2 = g(x)$. The x-values that correspond to points of intersections represent solutions.

127. With a graphing utility, solve the equation $\sin\theta = \cos(2\theta)$ on $0 \le \theta \le \pi$.

128. With a graphing utility, solve the equation $\csc\theta = \sec\theta$ on $0 \le \theta \le \dfrac{\pi}{2}$.

129. With a graphing utility, solve the equation $\sin\theta = \sec\theta$ on $0 \le \theta \le \pi$.

130. With a graphing utility, solve the equation $\cos\theta = \csc\theta$ on $0 \le \theta \le \pi$.

131. With a graphing utility, find all of the solutions to the equation $\sin\theta = e^\theta$ for $\theta \ge 0$.

132. With a graphing utility, find all of the solutions to the equation $\cos\theta = e^\theta$ for $\theta \ge 0$.

Find the smallest positive values of x that make the statement true. Give the answer in degrees and round to two decimal places.

133. $\sec(3x) + \csc(2x) = 5$

134. $\cot(5x) + \tan(2x) = -3$

135. $e^x - \tan x = 0$ **136.** $e^x + 2\sin x = 1$

137. $\ln x - \sin x = 0$ **138.** $\ln x - \cos x = 0$

▪ PREVIEW TO CALCULUS

In calculus, the definite integral is used to find the area between two intersecting curves (functions). When the curves correspond to trigonometric functions, we need to solve trigonometric equations.

In Exercises 139–142, solve each trigonometric equation within the indicated interval.

139. $\cos x = 2 - \cos x$, $0 \le x \le 2\pi$

140. $\cos(2x) = \sin x$, $-\dfrac{\pi}{2} \le x \le \dfrac{\pi}{6}$

141. $2\sin x = \tan x$, $-\dfrac{\pi}{3} \le x \le \dfrac{\pi}{3}$

142. $\sin(2x) - \cos(2x) = 0$, $0 < x \le \pi$

MODELING YOUR WORLD

In the Modeling Your World feature in Chapter 5, you modeled mean *temperatures* with sinusoidal models. Now we consider *carbon emissions*, which are greenhouse gases that have been shown to negatively affect the ozone layer. Over the last 50 years, we have increased our global carbon emissions at an alarming rate. Recall that the graph of the inverse tangent function increases rapidly and then levels off at the horizontal

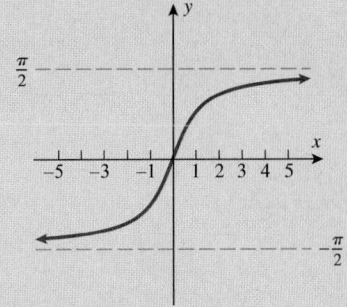

asymptote, $y = \dfrac{\pi}{2} \approx 1.57$. To achieve a similar plateau

with carbon emissions, drastic environmental regulations will need to be enacted.

The carbon emissions data over the last 50 years suggest an almost linear climb (similar to the inverse tangent graph from $x = 0$ to $x = 1$). If the world started reducing carbon emissions, they might possibly reach a plateau level.

The following table summarizes average yearly temperature in degrees Fahrenheit (°F) and carbon dioxide emissions in parts per million (ppm) for Mauna Loa, Hawaii.

Year	1960	1965	1970	1975	1980	1985	1990	1995	2000	2005
Temperature	44.45	43.29	43.61	43.35	46.66	45.71	45.53	47.53	45.86	46.23
CO_2 emissions (ppm)	316.9	320.0	325.7	331.1	338.7	345.9	354.2	360.6	369.4	379.7

1. Plot the carbon emissions data with time on the horizontal axis and CO_2 emissions (in ppm) on the vertical axis. Let $t = 0$ correspond to 1960.

2. Find an *inverse tangent function* of the form $f(x) = A\tan^{-1}(Bx) + k$ that models the carbon emissions in Mauna Loa.

   a. Use the data from 1960, 1985, and 2005.
   b. Use the data from 1960, 1965, and 1995.

3. According to your model, what are the expected carbon emissions in 2050?

4. Describe the ways by which the world might be able to reach a plateau level of carbon emissions instead of the predicted increased rates.

SECTION	CONCEPT	PAGES	REVIEW EXERCISES	KEY IDEAS/POINTS
6.1	**Verifying trigonometric identities**	526–532	1–26	Identities must hold for *all* values of *x* (not just some values of *x*) for which both sides of the equation are defined.
	Fundamental identities	526–528	1–26	*Reciprocal identities* $\csc\theta = \dfrac{1}{\sin\theta} \quad \sec\theta = \dfrac{1}{\cos\theta} \quad \cot\theta = \dfrac{1}{\tan\theta}$ *Quotient identities* $\tan\theta = \dfrac{\sin\theta}{\cos\theta} \quad \cot\theta = \dfrac{\cos\theta}{\sin\theta}$ *Pythagorean identities* $\sin^2\theta + \cos^2\theta = 1$ $\tan^2\theta + 1 = \sec^2\theta$ $1 + \cot^2\theta = \csc^2\theta$ *Cofunction identities* $\sin\theta = \cos\left(\dfrac{\pi}{2} - \theta\right) \quad \csc\theta = \sec\left(\dfrac{\pi}{2} - \theta\right)$ $\cos\theta = \sin\left(\dfrac{\pi}{2} - \theta\right) \quad \sec\theta = \csc\left(\dfrac{\pi}{2} - \theta\right)$ $\tan\theta = \cot\left(\dfrac{\pi}{2} - \theta\right) \quad \cot\theta = \tan\left(\dfrac{\pi}{2} - \theta\right)$
	Simplifying trigonometric expressions using identities	528–529	7–14	Use the reciprocal, quotient, or Pythagorean identities to simplify trigonometric expressions.
	Verifying identities	530–532	15–20	■ Convert all trigonometric expressions to sines and cosines. ■ Write all sums or differences of fractions as a single fraction.
6.2	**Sum and difference identities**	536–545	27–44	$f(A \pm B) \neq f(A) \pm f(B)$ For trigonometric functions, we have the sum and difference identities.
	Sum and difference identities for the cosine function	537–541	27–44	$\cos(A + B) = \cos A \cos B - \sin A \sin B$ $\cos(A - B) = \cos A \cos B + \sin A \sin B$
	Sum and difference identities for the sine function	541–543	27–44	$\sin(A + B) = \sin A \cos B + \cos A \sin B$ $\sin(A - B) = \sin A \cos B - \cos A \sin B$
	Sum and difference identities for the tangent function	543–545	27–44	$\tan(A + B) = \dfrac{\tan A + \tan B}{1 - \tan A \tan B}$ $\tan(A - B) = \dfrac{\tan A - \tan B}{1 + \tan A \tan B}$

Section	Concept	Pages	Review Exercises	Key Ideas/Points
6.3	**Double-angle and half-angle identities**	549–560	45–80	
	Double-angle identities	549–554	45–60	$\sin(2A) = 2\sin A\cos A$ $\cos(2A) = \cos^2 A - \sin^2 A$ $\quad = 1 - 2\sin^2 A = 2\cos^2 A - 1$ $\tan(2A) = \dfrac{2\tan A}{1 - \tan^2 A}$
	Half-angle identities	554–560	61–80	$\sin\left(\dfrac{A}{2}\right) = \pm\sqrt{\dfrac{1 - \cos A}{2}}$ $\cos\left(\dfrac{A}{2}\right) = \pm\sqrt{\dfrac{1 + \cos A}{2}}$ $\tan\left(\dfrac{A}{2}\right) = \pm\sqrt{\dfrac{1 - \cos A}{1 + \cos A}} = \dfrac{\sin A}{1 + \cos A} = \dfrac{1 - \cos A}{\sin A}$
6.4	**Product-to-sum and sum-to-product identities**	566–571	81–92	
	Product-to-sum identities	566–568	81–92	$\cos A\cos B = \tfrac{1}{2}[\cos(A + B) + \cos(A - B)]$ $\sin A\sin B = \tfrac{1}{2}[\cos(A - B) - \cos(A + B)]$ $\sin A\cos B = \tfrac{1}{2}[\sin(A + B) + \sin(A - B)]$
	Sum-to-product identities	568–571	81–92	$\sin A + \sin B = 2\sin\left(\dfrac{A + B}{2}\right)\cos\left(\dfrac{A - B}{2}\right)$ $\sin A - \sin B = 2\sin\left(\dfrac{A - B}{2}\right)\cos\left(\dfrac{A + B}{2}\right)$ $\cos A + \cos B = 2\cos\left(\dfrac{A + B}{2}\right)\cos\left(\dfrac{A - B}{2}\right)$ $\cos A - \cos B = -2\sin\left(\dfrac{A + B}{2}\right)\sin\left(\dfrac{A - B}{2}\right)$
6.5	**Inverse trigonometric functions**	574–589	93–124	$\sin^{-1}x$ or arcsin x $\qquad\cos^{-1}x$ or arccos x $\tan^{-1}x$ or arctan x $\qquad\cot^{-1}x$ or arccot x $\sec^{-1}x$ or arcsec x $\qquad\csc^{-1}x$ or arccsc x
	Inverse sine function	575–578	93–124	**Definition** $y = \sin^{-1}x$ means $x = \sin y$ $-1 \le x \le 1$ and $-\dfrac{\pi}{2} \le y \le \dfrac{\pi}{2}$ **Identities** $\sin^{-1}(\sin x) = x$ for $-\dfrac{\pi}{2} \le x \le \dfrac{\pi}{2}$ $\sin(\sin^{-1}x) = x$ for $-1 \le x \le 1$

Section	Concept	Pages	Review Exercises	Key Ideas/Points
	Inverse cosine function	579–581	93–124	**Definition** $y = \cos^{-1}x$ means $x = \cos y$ $-1 \leq x \leq 1$ and $0 \leq y \leq \pi$ **Identities** $\cos^{-1}(\cos x) = x$ for $0 \leq x \leq \pi$ $\cos(\cos^{-1}x) = x$ for $-1 \leq x \leq 1$

Section	Concept	Pages	Review Exercises	Key Ideas/Points
	Inverse tangent function	582–584	93–124	**Definition** $y = \tan^{-1}x$ means $x = \tan y$ $-\infty < x < \infty$ and $-\dfrac{\pi}{2} < y < \dfrac{\pi}{2}$ **Identities** $\tan^{-1}(\tan x) = x$ for $-\dfrac{\pi}{2} < x < \dfrac{\pi}{2}$ $\tan(\tan^{-1}x) = x$ for $-\infty < x < \infty$

Inverse cotangent function

Definition

$y = \cot^{-1}x$ means $x = \cot y$

$-\infty < x < \infty$ and $0 < y < \pi$

Identity

$$\cot^{-1}x = \begin{cases} \tan^{-1}\left(\dfrac{1}{x}\right), x > 0 \\ \pi + \tan^{-1}\left(\dfrac{1}{x}\right), x < 0 \end{cases}$$

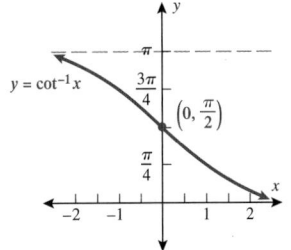

Inverse secant function

Definition

$y = \sec^{-1}x$ means $x = \sec y$

$x \leq -1$ or $x \geq 1$ and

$0 \leq y < \dfrac{\pi}{2}$ or $\dfrac{\pi}{2} < y \leq \pi$

Identity

$\sec^{-1}x = \cos^{-1}\left(\dfrac{1}{x}\right)$ for $x \leq -1$ or $x \geq 1$

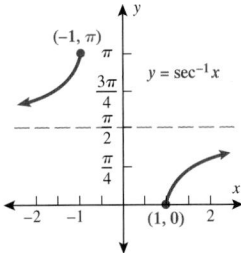

SECTION	CONCEPT	PAGES	REVIEW EXERCISES	KEY IDEAS/POINTS
				Inverse cosecant function **Definition** $y = \csc^{-1}x$ means $x = \csc y$ $x \leq -1$ or $x \geq 1$ and $-\dfrac{\pi}{2} \leq y < 0$ or $0 < y \leq \dfrac{\pi}{2}$ **Identity** $\csc^{-1}x = \sin^{-1}\left(\dfrac{1}{x}\right)$ for $x \leq -1$ or $x \geq 1$
	Finding exact values for expressions involving inverse trigonometric functions	587–589	113–124	
6.6	**Trigonometric equations**	594–604	125–160	Goal: Find the values of the variable that make the equation true.
	Solving trigonometric equations by inspection	594–597	125–130	Solve: $\sin\theta = \dfrac{\sqrt{2}}{2}$ on $0 \leq \theta \leq 2\pi$. Answer: $\theta = \dfrac{\pi}{4}$ or $\theta = \dfrac{3\pi}{4}$. Solve: $\sin\theta = \dfrac{\sqrt{2}}{2}$ on all real numbers. Answer: $\theta = \begin{cases} \dfrac{\pi}{4} + 2n\pi \\ \dfrac{3\pi}{4} + 2n\pi \end{cases}$ where n is an integer.
	Solving trigonometric equations using algebraic techniques	597–598	131–136	Transform trigonometric equations into linear or quadratic algebraic equations by making a substitution such as $x = \sin\theta$. Then use algebraic methods for solving linear and quadratic equations. If an expression is squared, always check for extraneous solutions.
	Solving trigonometric equations that require the use of inverse functions	598–600	137–160	Follow the same procedures outlined by inspection or algebraic methods. Finding the solution requires the use of inverse functions and a calculator. Be careful: Calculators only give one solution (the one in the range of the inverse function).
	Using trigonometric identities to solve trigonometric equations	601–604	143–160	Use trigonometric identities to transform an equation with multiple trigonometric functions into an equation with only one trigonometric function. Then use the methods outlined above.

6.1 Verifying Trigonometric Identities

Use the cofunction identities to fill in the blanks.

1. $\sin 30° = \cos$ ___60°___

2. $\cos A = \sin$ ___B___

3. $\tan 45° = \cot$ ___45°___

4. $\csc 60° = \sec$ ___30°___

5. $\sec 30° = \csc$ ___60°___

6. $\cot 60° = \tan$ ___30°___

Simplify the following trigonometric expressions:

7. $\tan x(\cot x + \tan x)$

8. $(\sec x + 1)(\sec x - 1)$

9. $\dfrac{\tan^4 x - 1}{\tan^2 x - 1}$

10. $\sec^2 x(\cot^2 x - \cos^2 x)$

11. $\cos x[\cos(-x) - \tan(-x)] - \sin x$

12. $\dfrac{\tan^2 x + 1}{2\sec^2 x}$

13. $\dfrac{\csc^3(-x) + 8}{\csc x - 2}$

14. $\dfrac{\csc^2 x - 1}{\cot x}$

Verify the trigonometric identities.

15. $(\tan x + \cot x)^2 - 2 = \tan^2 x + \cot^2 x$

16. $\csc^2 x - \cot^2 x = 1$

17. $\dfrac{1}{\sin^2 x} - \dfrac{1}{\tan^2 x} = 1$

18. $\dfrac{1}{\csc x + 1} + \dfrac{1}{\csc x - 1} = \dfrac{2\tan x}{\cos x}$

19. $\dfrac{\tan^2 x - 1}{\sec^2 x + 3\tan x + 1} = \dfrac{\tan x - 1}{\tan x + 2}$

20. $\cot x(\sec x - \cos x) = \sin x$

Determine whether each of the following equations is a conditional equation or an identity:

21. $2\tan^2 x + 1 = \dfrac{1 + \sin^2 x}{\cos^2 x}$

22. $\sin x - \cos x = 0$

23. $\cot^2 x - 1 = \tan^2 x$

24. $\cos^2 x(1 + \cot^2 x) = \cot^2 x$

25. $\left(\cot x - \dfrac{1}{\tan x}\right)^2 = 0$

26. $\csc x + \sec x = \dfrac{1}{\sin x + \cos x}$

6.2 Sum and Difference Identities

Find the exact value for each trigonometric expression.

27. $\cos\left(\dfrac{7\pi}{12}\right)$ 28. $\sin\left(\dfrac{\pi}{12}\right)$

29. $\tan(-15°)$ 30. $\cot 105°$

Write each expression as a single trigonometric function.

31. $\sin(4x)\cos(3x) - \cos(4x)\sin(3x)$

32. $\sin(-x)\sin(-2x) + \cos(-x)\cos(-2x)$

33. $\dfrac{\tan(5x) - \tan(4x)}{1 + \tan(5x)\tan(4x)}$

34. $\dfrac{\tan\left(\dfrac{\pi}{4}\right) + \tan\left(\dfrac{\pi}{3}\right)}{1 - \tan\left(\dfrac{\pi}{4}\right)\tan\left(\dfrac{\pi}{3}\right)}$

Find the exact value of the indicated expression using the given information and identities.

35. Find the exact value of $\tan(\alpha - \beta)$ if $\sin\alpha = -\dfrac{3}{5}$, $\sin\beta = -\dfrac{24}{25}$, the terminal side of α lies in QIV, and the terminal side of β lies in QIII.

36. Find the exact value of $\cos(\alpha + \beta)$ if $\cos\alpha = -\dfrac{5}{13}$, $\sin\beta = \dfrac{7}{25}$, the terminal side of α lies in QII, and the terminal side of β also lies in QII.

37. Find the exact value of $\cos(\alpha - \beta)$ if $\cos\alpha = \dfrac{9}{41}$, $\cos\beta = \dfrac{7}{25}$, the terminal side of α lies in QIV, and the terminal side of β lies in QI.

38. Find the exact value of $\sin(\alpha - \beta)$ if $\sin\alpha = -\dfrac{5}{13}$, $\cos\beta = -\dfrac{4}{5}$, the terminal side of α lies in QIII, and the terminal side of β lies in QII.

Determine whether each of the following equations is a conditional equation or an identity:

39. $2\cos A\cos B = \cos(A + B) + \cos(A - B)$

40. $2\sin A\sin B = \cos(A - B) - \cos(A + B)$

Graph the following functions:

41. $y = \cos\left(\dfrac{\pi}{2}\right)\cos x - \sin\left(\dfrac{\pi}{2}\right)\sin x$

42. $y = \sin\left(\dfrac{2\pi}{3}\right)\cos x + \cos\left(\dfrac{2\pi}{3}\right)\sin x$

43. $y = \dfrac{2\tan\left(\dfrac{x}{3}\right)}{1 - \tan^2\left(\dfrac{x}{3}\right)}$

44. $y = \dfrac{\tan(\pi x) - \tan x}{1 + \tan(\pi x)\tan x}$

6.3 Double-Angle and Half-Angle Identities

Use double-angle identities to answer the following questions:

45. If $\sin x = \dfrac{3}{5}$ and $\dfrac{\pi}{2} < x < \pi$, find $\cos(2x)$.

46. If $\cos x = \dfrac{7}{25}$ and $\dfrac{3\pi}{2} < x < 2\pi$, find $\sin(2x)$.

47. If $\cot x = -\dfrac{11}{61}$ and $\dfrac{3\pi}{2} < x < 2\pi$, find $\tan(2x)$.

48. If $\tan x = -\dfrac{12}{5}$ and $\dfrac{\pi}{2} < x < \pi$, find $\cos(2x)$.

49. If $\sec x = \dfrac{25}{24}$ and $0 < x < \dfrac{\pi}{2}$, find $\sin(2x)$.

50. If $\csc x = \dfrac{5}{4}$ and $\dfrac{\pi}{2} < x < \pi$, find $\tan(2x)$.

Simplify each of the following expressions. Evaluate exactly, if possible.

51. $\cos^2 15° - \sin^2 15°$

52. $\dfrac{2\tan\left(-\dfrac{\pi}{12}\right)}{1 - \tan^2\left(-\dfrac{\pi}{12}\right)}$

53. $6\sin\left(\dfrac{\pi}{12}\right)\cos\left(\dfrac{\pi}{12}\right)$

54. $1 - 2\sin^2\left(\dfrac{\pi}{8}\right)$

Verify the following identities:

55. $\sin^3 A - \cos^3 A = (\sin A - \cos A)\left[1 + \tfrac{1}{2}\sin(2A)\right]$

56. $2\sin A\cos^3 A - 2\sin^3 A\cos A = \cos(2A)\sin(2A)$

57. $\tan A = \dfrac{\sin(2A)}{1 + \cos(2A)}$

58. $\tan A = \dfrac{1 - \cos(2A)}{\sin(2A)}$

59. Launching a Missile. When launching a missile for a given range, the minimum velocity needed is related to the angle θ of the launch, and the velocity is determined by $V = \dfrac{2\cos(2\theta)}{1 + \cos(2\theta)}$. Show that V is equivalent to $1 - \tan^2\theta$.

60. Launching a Missile. When launching a missile for a given range, the minimum velocity needed is related to the angle θ of the launch, and the velocity is determined by $V = \dfrac{2\cos(2\theta)}{1 + \cos(2\theta)}$. Find the value of V when $\theta = \dfrac{\pi}{6}$.

Use half-angle identities to find the exact value of each of the following trigonometric expressions:

61. $\sin(-22.5°)$

62. $\cos 67.5°$

63. $\cot\left(\dfrac{3\pi}{8}\right)$

64. $\csc\left(-\dfrac{7\pi}{8}\right)$

65. $\sec(-165°)$

66. $\tan(-75°)$

Use half-angle identities to find each of the following values:

67. If $\sin x = -\dfrac{7}{25}$ and $\pi < x < \dfrac{3\pi}{2}$, find $\sin\left(\dfrac{x}{2}\right)$.

68. If $\cos x = -\dfrac{4}{5}$ and $\dfrac{\pi}{2} < x < \pi$, find $\cos\left(\dfrac{x}{2}\right)$.

69. If $\tan x = \dfrac{40}{9}$ and $\pi < x < \dfrac{3\pi}{2}$, find $\tan\left(\dfrac{x}{2}\right)$.

70. If $\sec x = \dfrac{17}{15}$ and $\dfrac{3\pi}{2} < x < 2\pi$, find $\sin\left(\dfrac{x}{2}\right)$.

Simplify each expression using half-angle identities. Do not evaluate.

71. $\sqrt{\dfrac{1 - \cos\left(\dfrac{\pi}{6}\right)}{2}}$

72. $\sqrt{\dfrac{1 - \cos\left(\dfrac{11\pi}{6}\right)}{1 + \cos\left(\dfrac{11\pi}{6}\right)}}$

Verify each of the following identities:

73. $\left[\sin\left(\dfrac{A}{2}\right) + \cos\left(\dfrac{A}{2}\right)\right]^2 = 1 + \sin A$

74. $\sec^2\left(\dfrac{A}{2}\right) + \tan^2\left(\dfrac{A}{2}\right) = \dfrac{3 - \cos A}{1 + \cos A}$

75. $\csc^2\left(\dfrac{A}{2}\right) + \cot^2\left(\dfrac{A}{2}\right) = \dfrac{3 + \cos A}{1 - \cos A}$

76. $\tan^2\left(\dfrac{A}{2}\right) + 1 = \sec^2\left(\dfrac{A}{2}\right)$

Graph each of the following functions:

77. $y = \sqrt{\dfrac{1 - \cos\left(\dfrac{\pi}{12}x\right)}{2}}$

78. $y = \cos^2\left(\dfrac{x}{2}\right) - \sin^2\left(\dfrac{x}{2}\right)$

79. $y = -\sqrt{\dfrac{1 - \cos x}{1 + \cos x}}$

80. $y = \sqrt{\dfrac{1 + \cos(3x - 1)}{2}}$

6.4 Product-to-Sum and Sum-to-Product Identities

Write each product as a sum or difference of sines and/or cosines.

81. $6\sin(5x)\cos(2x)$

82. $3\sin(4x)\sin(2x)$

Write each expression as a product of sines and/or cosines.

83. $\cos(5x) - \cos(3x)$

84. $\sin\left(\dfrac{5x}{2}\right) + \sin\left(\dfrac{3x}{2}\right)$

85. $\sin\left(\dfrac{4x}{3}\right) - \sin\left(\dfrac{2x}{3}\right)$

86. $\cos(7x) + \cos x$

Simplify each trigonometric expression.

87. $\dfrac{\cos(8x) + \cos(2x)}{\sin(8x) - \sin(2x)}$

88. $\dfrac{\sin(5x) + \sin(3x)}{\cos(5x) + \cos(3x)}$

Verify the identities.

89. $\dfrac{\sin A + \sin B}{\cos A - \cos B} = -\cot\left(\dfrac{A - B}{2}\right)$

90. $\dfrac{\sin A - \sin B}{\cos A - \cos B} = -\cot\left(\dfrac{A + B}{2}\right)$

91. $\csc\left(\dfrac{A - B}{2}\right) = \dfrac{2\sin\left(\dfrac{A + B}{2}\right)}{\cos B - \cos A}$

92. $\sec\left(\dfrac{A + B}{2}\right) = \dfrac{2\sin\left(\dfrac{A - B}{2}\right)}{\sin A - \sin B}$

6.5 Inverse Trigonometric Functions

Find the exact value of each expression. Give the answer in radians.

93. $\arctan 1$

94. $\operatorname{arccsc}(-2)$

95. $\cos^{-1} 0$

96. $\sin^{-1}(\ 1)$

97. $\sec^{-1}\left(\dfrac{2}{\sqrt{3}}\right)$

98. $\cot^{-1}(-\sqrt{3})$

Find the exact value of each expression. Give the answer in degrees.

99. $\csc^{-1}(-1)$

100. $\arctan(-1)$

101. $\operatorname{arccot}\left(\dfrac{\sqrt{3}}{3}\right)$

102. $\cos^{-1}\left(\dfrac{\sqrt{2}}{2}\right)$

103. $\sin^{-1}\left(-\dfrac{\sqrt{3}}{2}\right)$

104. $\sec^{-1} 1$

Use a calculator to evaluate each expression. Give the answer in degrees and round to two decimal places.

105. $\sin^{-1}(-0.6088)$

106. $\tan^{-1}(1.1918)$

107. $\sec^{-1}(1.0824)$

108. $\cot^{-1}(-3.7321)$

Use a calculator to evaluate each expression. Give the answer in radians and round to two decimal places.

109. $\cos^{-1}(-0.1736)$

110. $\tan^{-1}(0.1584)$

111. $\csc^{-1}(-10.0167)$

112. $\sec^{-1}(-1.1223)$

Evaluate each expression exactly, if possible. If not possible, state why.

113. $\sin^{-1}\left[\sin\left(-\dfrac{\pi}{4}\right)\right]$

114. $\cos\left[\cos^{-1}\left(-\dfrac{\sqrt{2}}{2}\right)\right]$

115. $\tan\left[\tan^{-1}(-\sqrt{3})\right]$

116. $\cot^{-1}\left[\cot\left(\dfrac{11\pi}{6}\right)\right]$

117. $\csc^{-1}\left[\csc\left(\dfrac{2\pi}{3}\right)\right]$

118. $\sec\left[\sec^{-1}\left(-\dfrac{2\sqrt{3}}{3}\right)\right]$

Evaluate each expression exactly.

119. $\sin\left[\cos^{-1}\left(\dfrac{11}{61}\right)\right]$

120. $\cos\left[\tan^{-1}\left(\dfrac{40}{9}\right)\right]$

121. $\tan\left[\cot^{-1}\left(\dfrac{6}{7}\right)\right]$

122. $\cot\left[\sec^{-1}\left(\dfrac{25}{7}\right)\right]$

123. $\sec\left[\sin^{-1}\left(\dfrac{1}{6}\right)\right]$

124. $\csc\left[\cot^{-1}\left(\dfrac{5}{12}\right)\right]$

6.6 Trigonometric Equations

Solve the given trigonometric equation over the indicated interval.

125. $\sin(2\theta) = -\dfrac{\sqrt{3}}{2}, 0 \le \theta \le 2\pi$

126. $\sec\left(\dfrac{\theta}{2}\right) = 2, -2\pi \le \theta \le 2\pi$

127. $\sin\left(\dfrac{\theta}{2}\right) = -\dfrac{\sqrt{2}}{2}, -2\pi \le \theta \le 2\pi$

128. $\csc(2\theta) = 2, 0 \le \theta \le 2\pi$

129. $\tan\left(\dfrac{1}{3}\theta\right) = -1, 0 \le \theta \le 6\pi$

130. $\cot(4\theta) = -\sqrt{3}, -\pi \le \theta \le \pi$

Solve each trigonometric equation exactly on $0 \le \theta \le 2\pi$.

131. $4\cos(2\theta) + 2 = 0$

132. $\sqrt{3}\tan\left(\dfrac{\theta}{2}\right) - 1 = 0$

133. $2\tan(2\theta) + 2 = 0$

134. $2\sin^2\theta + \sin\theta - 1 = 0$

135. $\tan^2\theta + \tan\theta = 0$

136. $\sec^2\theta - 3\sec\theta + 2 = 0$

Solve the given trigonometric equations on $0° \le \theta \le 360°$ and express the answer in degrees to two decimal places.

137. $\tan(2\theta) = -0.3459$ **138.** $6\sin\theta - 5 = 0$

139. $4\cos^2\theta + 3\cos\theta = 0$ **140.** $12\cos^2\theta - 7\cos\theta + 1 = 0$

141. $\csc^2\theta - 3\csc\theta - 1 = 0$ **142.** $2\cot^2\theta + 5\cot\theta - 4 = 0$

Solve each trigonometric equation exactly on the interval $0 \le \theta \le 2\pi$.

143. $\sec x = 2\sin x$ **144.** $3\tan x + \cot x = 2\sqrt{3}$

145. $\sqrt{3}\tan x - \sec x = 1$ **146.** $2\sin(2x) = \cot x$

147. $\sqrt{3}\tan x = 2\sin x$ **148.** $2\sin x = 3\cot x$

149. $\cos^2 x + \sin x + 1 = 0$ **150.** $2\cos^2 x - \sqrt{3}\cos x = 0$

151. $\cos(2x) + 4\cos x + 3 = 0$ **152.** $\sin(2x) + \sin x = 0$

153. $\tan^2\left(\frac{1}{2}x\right) - 1 = 0$ **154.** $\cot^2\left(\frac{1}{3}x\right) - 1 = 0$

Solve each trigonometric equation on $0° \le \theta \le 360°$. Give the answers in degrees and round to two decimal places.

155. $\csc^2 x + \cot x = 1$ **156.** $8\cos^2 x + 6\sin x = 9$

157. $\sin^2 x + 2 = 2\cos x$ **158.** $\cos(2x) = 3\sin x - 1$

159. $\cos x - 1 = \cos(2x)$ **160.** $12\cos^2 x + 4\sin x = 11$

Technology Exercises

Section 6.1

161. Is $\cos 73° = \sqrt{1 - \sin^2 73°}$? Use a calculator to find each of the following:

   a. $\cos 73°$

   b. $1 - \sin 73°$

   c. $\sqrt{1 - \sin^2 73°}$

   Which results are the same?

162. Is $\csc 28° = \sqrt{1 + \cot^2 28°}$? Use a calculator to find each of the following:

   a. $\cot 28°$

   b. $1 + \cot 28°$

   c. $\sqrt{1 + \cot^2 28°}$

   Which results are the same?

Section 6.2

Recall that the difference quotient for a function f is given by
$$\frac{f(x + h) + f(x)}{h}.$$

163. Show that the difference quotient for $f(x) = \sin(3x)$ is
$$[\cos(3x)]\left[\frac{\sin(3h)}{h}\right] - [\sin(3x)]\left[\frac{1 - \cos(3h)}{h}\right].$$

Plot $Y_1 = [\cos(3x)]\left[\dfrac{\sin(3h)}{h}\right] - [\sin(3x)]\left[\dfrac{1 - \cos(3h)}{h}\right]$

for

   a. $h = 1$

   b. $h = 0.1$

   c. $h = 0.01$

What function does the difference quotient for $f(x) = \sin(3x)$ resemble when h approaches zero?

164. Show that the difference quotient for $f(x) = \cos(3x)$ is
$$[-\sin(3x)]\left[\frac{\sin(3h)}{h}\right] - [\cos(3x)]\left[\frac{1 - \cos(3h)}{h}\right].$$

Plot $Y_1 = [-\sin(3x)]\left[\dfrac{\sin(3h)}{h}\right] - [\cos(3x)]\left[\dfrac{1 - \cos(3h)}{h}\right]$

for

   a. $h = 1$

   b. $h = 0.1$

   c. $h = 0.01$

What function does the difference quotient for $f(x) = \cos(3x)$ resemble when h approaches zero?

Section 6.3

165. With a graphing calculator, plot $Y_1 = \tan(2x)$, $Y_2 = 2\tan x$, and $Y_3 = \dfrac{2\tan x}{1 - \tan^2 x}$ in the same viewing rectangle $[-2\pi, 2\pi]$ by $[-10, 10]$. Which graphs are the same?

166. With a graphing calculator, plot $Y_1 = \cos(2x)$, $Y_2 = 2\cos x$, and $Y_3 = 1 - 2\sin^2 x$ in the same viewing rectangle $[-2\pi, 2\pi]$ by $[-2, 2]$. Which graphs are the same?

167. With a graphing calculator, plot $Y_1 = \cos\left(\dfrac{x}{2}\right)$, $Y_2 = \dfrac{1}{2}\cos x$, and $Y_3 = -\sqrt{\dfrac{1 + \cos x}{2}}$ in the same viewing rectangle $[\pi, 2\pi]$ by $[-1, 1]$. Which graphs are the same?

168. With a graphing calculator, plot $Y_1 = \sin\left(\dfrac{x}{2}\right)$, $Y_2 = \dfrac{1}{2}\sin x$, and $Y_3 = -\sqrt{\dfrac{1 - \cos x}{2}}$ in the same viewing rectangle $[2\pi, 4\pi]$ by $[-1, 1]$. Which graphs are the same?

Section 6.4

169. With a graphing calculator, plot $Y_1 = \sin(5x)\cos(3x)$, $Y_2 = \sin(4x)$, and $Y_3 = \dfrac{1}{2}[\sin(8x) + \sin(2x)]$ in the same viewing rectangle $[0, 2\pi]$ by $[-1, 1]$. Which graphs are the same?

170. With a graphing calculator, plot $Y_1 = \sin(3x)\cos(5x)$, $Y_2 = \cos(4x)$, and $Y_3 = \dfrac{1}{2}[\sin(8x) - \sin(2x)]$ in the same viewing rectangle $[0, 2\pi]$ by $[-1, 1]$. Which graphs are the same?

Section 6.5

171. Given $\cos x = -\dfrac{1}{\sqrt{5}}$ and $\dfrac{\pi}{2} < x < \pi$:

    a. Find $\cos(2x)$ using the double-angle identity.
    b. Use the inverse of cosine to find x in QII and to find $\cos(2x)$.
    c. Are the results in (a) and (b) the same?

172. Given $\cos x = \dfrac{5}{12}$ and $\dfrac{3\pi}{2} < x < 2\pi$:

    a. Find $\cos\left(\dfrac{1}{2}x\right)$ using the half-angle identity.
    b. Use the inverse of cosine to find x in QIV and to find $\cos\left(\dfrac{1}{2}x\right)$. Round to five decimal places.
    c. Are the results in (a) and (b) the same?

Section 6.6

Find the smallest positive value of x that makes each statement true. Give the answer in radians and round to four decimal places.

173. $\ln x + \sin x = 0$

174. $\ln x + \cos x = 0$

1. For what values of x does the quotient identity $\tan x = \dfrac{\sin x}{\cos x}$ not hold?

2. Is the equation $\sqrt{\sin^2 x + \cos^2 x} = \sin x + \cos x$ a conditional equation or an identity?

3. Evaluate $\sin\left(-\dfrac{\pi}{8}\right)$ exactly.

4. Evaluate $\tan\left(\dfrac{7\pi}{12}\right)$ exactly.

5. If $\cos x = \dfrac{2}{5}$ and $\dfrac{3\pi}{2} < x < 2\pi$, find $\sin\left(\dfrac{x}{2}\right)$.

6. If $\sin x = -\dfrac{1}{5}$ and $\pi < x < \dfrac{3\pi}{2}$, find $\cos(2x)$.

7. Write $\cos(7x)\cos(3x) - \sin(3x)\sin(7x)$ as a cosine or sine of a sum or difference.

8. Write $-\dfrac{2\tan x}{1 - \tan^2 x}$ as a single tangent function.

9. Write $\sqrt{\dfrac{1 + \cos(a + b)}{2}}$ as a single cosine function if $a + b$ is an angle in QII. Assume $\dfrac{\pi}{2} < a + b < \pi$.

10. Write $2\sin\left(\dfrac{x + 3}{2}\right)\cos\left(\dfrac{x - 3}{2}\right)$ as a sum of two sine functions.

11. Write $10\cos(3 - x) + 10\cos(x + 3)$ as a product of two cosine functions.

12. In the expression $\sqrt{9 - u^2}$, let $u = 3\sin x$. What is the resulting expression?

Solve the trigonometric equations exactly, if possible. Otherwise, use a calculator to approximate solution(s).

13. $2\sin\theta = -\sqrt{3}$ on all real numbers.

14. $2\cos^2\theta + \cos\theta - 1 = 0$ on $0 \le \theta \le 2\pi$

15. $\sin 2\theta = \dfrac{1}{2}\cos\theta$ over $0 \le \theta \le 360°$

16. $\sqrt{\sin x + \cos x} = -1$ over $0 \le \theta \le 2\pi$

17. Determine whether $(1 + \cot x)^2 = \csc^2 x$ is a conditional or an identity.

18. Evaluate $\csc\left(-\dfrac{\pi}{12}\right)$ exactly.

19. If $\sin x = -\dfrac{5}{13}$ and $\pi < x < \dfrac{3\pi}{2}$, find $\cos\left(\dfrac{x}{2}\right)$.

20. If $\cos x = -0.26$ and $\dfrac{\pi}{2} < x < \pi$, find $\sin(2x)$.

21. Express $y = \sqrt{\dfrac{1 + \dfrac{\sqrt{2}}{2}\left[\cos\left(\dfrac{\pi}{3}x\right) + \sin\left(\dfrac{\pi}{3}x\right)\right]}{1 - \dfrac{\sqrt{2}}{2}\left[\cos\left(\dfrac{\pi}{3}x\right) + \sin\left(\dfrac{\pi}{3}x\right)\right]}}$ as a cotangent function.

22. Calculate $\csc\left(\csc^{-1}\sqrt{2}\right)$.

23. Determine an interval on which $f(x) = a + b\csc(\pi x + c)$ is one-to-one, and determine the inverse of $f(x)$ on this interval. Assume that a, b, and c are all positive.

24. Find the range of $y = -\dfrac{\pi}{4} + \arctan(2x - 3)$.

25. Solve $\cos\left(\dfrac{\pi}{4}\theta\right) = -\dfrac{1}{2}$, for all real numbers.

26. Solve $\sqrt{\dfrac{1 - \cos(2\pi x)}{1 + \cos(2\pi x)}} = -\dfrac{1}{\sqrt{3}}$, for all real numbers.

27. Solve $\dfrac{\sqrt{3}}{\csc\left(\dfrac{x}{3}\right)} = \cos\left(\dfrac{x}{3}\right)$, for all real numbers.

28. Show that the difference quotient for $f(x) = \cos\left(\tfrac{1}{2}x\right)$ is
$$-\sin\left(\dfrac{1}{2}x\right)\left[\dfrac{\sin\left(\tfrac{1}{2}h\right)}{h}\right] - \cos\left(\dfrac{1}{2}x\right)\left[\dfrac{1 - \cos\left(\tfrac{1}{2}h\right)}{h}\right].$$

Plot $Y_1 = -\sin\left(\dfrac{1}{2}x\right)\left[\dfrac{\sin\left(\tfrac{1}{2}h\right)}{h}\right] - \cos(2x)\left[\dfrac{1 - \cos\left(\tfrac{1}{2}h\right)}{h}\right]$

for
a. $h = 1$
b. $h = 0.1$
c. $h = 0.01$

What function does the difference quotient for $f(x) = \cos\left(\tfrac{1}{2}x\right)$ resemble when h approaches zero?

29. Given $\tan x = \dfrac{3}{4}$ and $\pi < x < \dfrac{3\pi}{2}$:
a. Find $\sin\left(\tfrac{1}{2}x\right)$ using the half-angle identity.
b. Use the inverse of tangent to find x in QIII and to find $\sin\left(\tfrac{1}{2}x\right)$. Round to five decimal places.
c. Are the results in (a) and (b) the same?

1. Find the exact value of the following trigonometric functions:

   a. $\sin\left(\dfrac{7\pi}{3}\right)$

   b. $\tan\left(-\dfrac{5\pi}{3}\right)$

   c. $\csc\left(\dfrac{11\pi}{6}\right)$

2. Find the exact value of the following trigonometric functions:

   a. $\sec\left(\dfrac{5\pi}{6}\right)$

   b. $\cos\left(-\dfrac{3\pi}{4}\right)$

   c. $\cot\left(\dfrac{7\pi}{6}\right)$

3. Find the exact value of the following inverse trigonometric functions:

   a. $\cos^{-1}\left(-\dfrac{1}{2}\right)$

   b. $\csc^{-1}(-2)$

   c. $\cot(-\sqrt{3})$

4. Determine whether the relation $x^2 - y^2 = 25$ is a function.

5. Determine whether the function $g(x) = \sqrt{2 - x^2}$ is odd or even.

6. For the function $y = 5(x - 4)^2$, identify all of the transformations of $y = x^2$.

7. Find the composite function, $f \circ g$, and state the domain for $f(x) = x^3 - 1$ and $g(x) = \dfrac{1}{x}$.

8. Find the inverse of the function $f(x) = \sqrt[3]{x} - 1$.

9. Find the vertex of the parabola associated with the quadratic function $f(x) = \frac{1}{4}x^2 + \frac{3}{5}x - \frac{6}{25}$.

10. Find a polynomial of minimum degree (there are many) that has the zeros $x = -\sqrt{7}$ (multiplicity 2), $x = 0$ (multiplicity 3), $x = \sqrt{7}$ (multiplicity 2).

11. Use long division to find the quotient $Q(x)$ and the remainder $r(x)$ of $(5x^3 - 4x^2 + 3) \div (x^2 + 1)$.

12. Given the zero $x = 4i$ of the polynomial $P(x) = x^4 + 2x^3 + x^2 + 32x - 240$, determine all the other zeros and write the polynomial in terms of a product of linear factors.

13. Find the vertical and horizontal asymptotes of the function $f(x) = \dfrac{0.7x^2 - 5x + 11}{x^2 - x - 6}$.

14. If \$5,400 is invested at 2.25% compounded continuously, how much is in the account after 4 years?

15. Use interval notation to express the domain of the function $f(x) = \log_4 (x + 3)$.

16. Use properties of logarithms to simplify the expression $\log_\pi 1$.

17. Give an exact solution to the logarithmic equation $\log_5(x + 2) + \log_5(6 - x) = \log_5(3x)$.

18. If money is invested in a savings account earning 4% compounded continuously, how many years will it take for the money to triple?

19. Use a calculator to evaluate $\cos 62°$. Round the answer to four decimal places.

20. **Angle of Inclination (Skiing).** The angle of inclination of a mountain with triple black diamond ski trails is 63°. If a skier at the top of the mountain is at an elevation of 4200 feet, how long is the ski run from the top to the base of the mountain?

21. Convert $-105°$ to radians. Leave the answer in terms of π.

22. Find all of the exact values of θ, when $\tan\theta = 1$ and $0 \le \theta \le 2\pi$.

23. Determine whether the equation $\cos^2 x - \sin^2 x = 1$ is a conditional equation or an identity.

24. Simplify $\dfrac{2\tan\left(-\dfrac{\pi}{8}\right)}{1 - \tan^2\left(-\dfrac{\pi}{8}\right)}$ and evaluate exactly.

25. Evaluate exactly the expression $\tan\left[\sin^{-1}\left(\dfrac{5}{13}\right)\right]$.

26. With a graphing calculator, plot $Y_1 = \sin x \cos(3x)$, $Y_2 = \cos(4x)$, and $Y_3 = \frac{1}{2}[\sin(4x) - \sin(2x)]$ in the same viewing rectangle $[0, 2\pi]$ by $[-1, 1]$. Which graphs are the same?

27. Find the smallest positive value of x that makes the statement true. Give the answer in radians and round to four decimal places.

$$\ln x - \sin(2x) = 0$$

7

Vectors, the Complex Plane, and Polar Coordinates

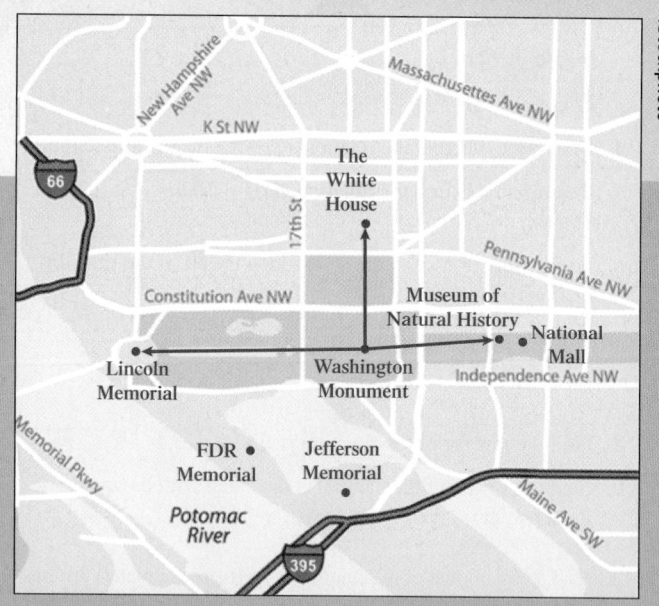

iStockphoto

A coordinate system is used to locate a point in a plane. In the Cartesian plane, rectangular coordinates (x, y) are used to describe the location of a point. For example, we can say that the Museum of Natural History in Washington D.C. is at the corner of Constitution Avenue and 12th Street. But we can also describe the location of the Museum of Natural History as being $\frac{1}{4}$ mile east-northeast of the Washington Monument. Instead of using a grid of streets running east-west and north-south, it is sometimes more convenient to give a location with respect to a distance and direction from a fixed point. In the *polar coordinate system*, the location of a point is given in *polar coordinates* as (r, θ), where r is the distance and θ is the direction angle of the point from a fixed reference point (origin).

IN THIS CHAPTER vectors will be defined and combined with the Law of Sines and the Law of Cosines to find resulting velocity and force vectors. The dot product (product of two vectors) is defined and used in physical problems like calculating work. Trigonometric functions are then used to define polar coordinates. Lastly, we define polar coordinates and examine polar equations and their corresponding graphs.

VECTORS, THE COMPLEX PLANE, AND POLAR COORDINATES

7.1 Vectors	7.2 The Dot Product	7.3 Polar (Trigonometric) Form of Complex Numbers	7.4 Products, Quotients, Powers, and Roots of Complex Numbers	7.5 Polar Coordinates and Graphs of Polar Equations
• Magnitude and Direction of Vectors • Vector Operations • Horizontal and Vertical Components of a Vector • Unit Vectors • Resultant Vectors	• Multiplying Two Vectors • Angle Between Two Vectors • Work	• Complex Numbers in Rectangular Form • Complex Numbers in Polar Form • Converting Complex Numbers Between Rectangular and Polar Forms	• Products of Complex Numbers • Quotients of Complex Numbers • Powers of Complex Numbers • Roots of Complex Numbers	• Polar Coordinates • Converting Between Polar and Rectangular Coordinates • Graphs of Polar Equations

CHAPTER OBJECTIVES

- Find the direction and magnitude of a vector (Section 7.1).
- Find the dot product of two vectors (Section 7.2).
- Express complex numbers in polar form (Section 7.3).
- Use De Moivre's theorem to find a complex number raised to a power (Section 7.4).
- Convert between rectangular and polar coordinates (Section 7.5).

SKILLS OBJECTIVES

- Represent vectors geometrically and algebraically.
- Find the magnitude and direction of a vector.
- Add and subtract vectors.
- Perform scalar multiplication of a vector.
- Find unit vectors.
- Express a vector in terms of its horizontal and vertical components.

CONCEPTUAL OBJECTIVES

- Understand the difference between scalars and vectors.
- Relate the geometric and algebraic representations of vectors.

Magnitude and Direction of Vectors

What is the difference between velocity and speed? Speed has only *magnitude*, whereas velocity has both *magnitude* and *direction*. We use **scalars**, which are real numbers, to denote magnitudes such as speed and mass. We use **vectors**, which have magnitude *and* direction, to denote quantities such as velocity (speed in a certain direction) and force (weight in a certain direction).

A vector quantity is geometrically denoted by a **directed line segment**, which is a line segment with an arrow representing direction. There are many ways to denote a vector. For example, the vector shown in the margin can be denoted as $\mathbf{u}$, $\vec{u}$, or $\overrightarrow{AB}$, where A is the **initial point** and B is the **terminal point**.

It is customary in books to use the bold letter to represent a vector and when handwritten (as in your class notes and homework) to use the arrow on top to denote a vector.

In this book, we will limit our discussion to vectors in a plane (two-dimensional). It is important to note that geometric representation can be extended to three dimensions and algebraic representation can be extended to any higher dimension, as you will see in the exercises.

Study Tip

The magnitude of a vector is the distance between the initial and terminal points of the vector.

Geometric Interpretation of Vectors

The *magnitude* of a vector can be denoted one of two ways: $|\mathbf{u}|$ or $\|\mathbf{u}\|$. We will use the former notation.

MAGNITUDE: $|\mathbf{u}|$

The **magnitude** of a vector $\mathbf{u}$, denoted $|\mathbf{u}|$, is the length of the directed line segment, that is the distance between the initial and terminal points of the vector.

Two vectors have the **same direction** if they are parallel and point in the same direction. Two vectors have **opposite direction** if they are parallel and point in opposite directions.

EQUAL VECTORS: $\mathbf{u} = \mathbf{v}$

Two vectors $\mathbf{u}$ and $\mathbf{v}$ are **equal** ($\mathbf{u} = \mathbf{v}$) if they have the same magnitude ($|\mathbf{u}| = |\mathbf{v}|$) and the same direction.

Equal Vectors $\mathbf{u} = \mathbf{v}$	Same Magnitude but Opposite Direction $\mathbf{u} = -\mathbf{v}$	Same Magnitude $|\mathbf{u}| = |\mathbf{v}|$	Different Magnitude	Same Direction Different Magnitude

It is important to note that vectors do not have to coincide to be equal.

Study Tip

Equal vectors can be translated (shifted) so that they coincide.

VECTOR ADDITION: U + V

Two vectors, **u** and **v**, can be added together using the **tail-to-tip** (or head-to-tail) rule.

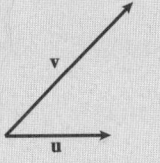

Translate **v** so that its tail end (the initial point) is located at the tip (head) end (the terminal point) of **u**.

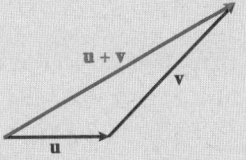

The **sum**, **u** + **v**, is the **resultant** vector from the tail end of **u** to the tip end of **v**.

The difference, **u** − **v** or **u** + (−**v**), is the resultant vector from the tip of **v** to the tip of **u** when the vectors are placed tail to tail.

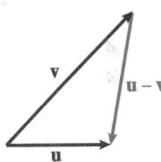

Algebraic Interpretation of Vectors

Since vectors that have the same direction and magnitude are equal, any vector can be translated to an equal vector with its initial point located at the origin in the Cartesian plane. Therefore, we now consider vectors in a rectangular coordinate system.

A vector with its initial point at the origin is called a **position vector**, or a vector in standard position. A position vector **u** with its terminal point at (a, b) is denoted

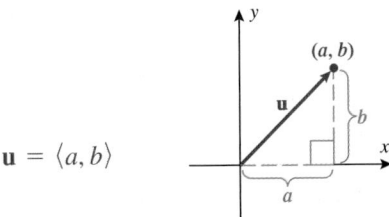

$$\mathbf{u} = \langle a, b \rangle$$

where the real numbers a and b are called the **components** of vector **u**.

Notice the subtle difference between coordinate notation and vector notation. The point is denoted with parentheses, (a, b), whereas the vector is denoted with angled brackets, $\langle a, b \rangle$. The notation $\langle a, b \rangle$ denotes a vector whose initial point is $(0, 0)$ and terminal point is (a, b).

The vector with initial point $(3, 4)$ and terminal point $(8, 10)$ is equal to the vector $\langle 5, 6 \rangle$, which has initial point $(0, 0)$ and terminal point $(5, 6)$.

Recall that the geometric definition of the *magnitude* of a vector is the *length* of the vector.

MAGNITUDE: |u|

The **magnitude** (or norm) of a vector, $\mathbf{u} = \langle a, b \rangle$, is

$$|\mathbf{u}| = \sqrt{a^2 + b^2}$$

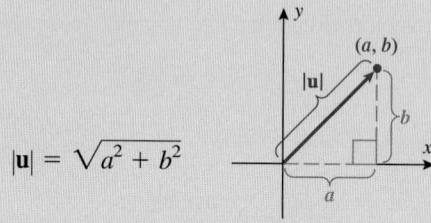

EXAMPLE 1 Finding the Magnitude of a Vector

Find the magnitude of the vector $\mathbf{u} = \langle 3, -4 \rangle$.

Solution:

Write the formula for magnitude of a vector. $|\mathbf{u}| = \sqrt{a^2 + b^2}$

Let $a = 3$ and $b = -4$. $|\mathbf{u}| = \sqrt{3^2 + (-4)^2}$

Simplify. $|\mathbf{u}| = \boxed{\sqrt{25} = 5}$

Note: If we graph the vector $\mathbf{u} = \langle 3, -4 \rangle$, we see that the distance from the origin to the point $(3, -4)$ is 5 units.

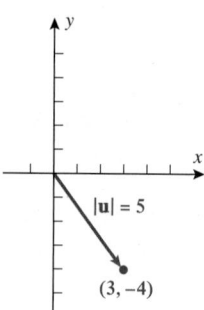

■ **Answer:** $\sqrt{26}$

■ **YOUR TURN** Find the magnitude of the vector $\mathbf{v} = \langle -1, 5 \rangle$.

DIRECTION ANGLE OF A VECTOR

The positive angle between the x-axis and a position vector is called the **direction angle**, denoted θ.

$$\tan \theta = \frac{b}{a}, \text{ where } a \neq 0$$

EXAMPLE 2 Finding the Direction Angle of a Vector

Find the direction angle of the vector $\mathbf{v} = \langle -1, 5 \rangle$.

Solution:

Start with $\tan\theta = \dfrac{b}{a}$ and let $a = -1$ and $b = 5$.
$\qquad \tan\theta = \dfrac{5}{-1}$

With a calculator, find $\tan^{-1}(-5)$.
$\qquad \tan^{-1}(-5) = -78.7°$

The calculator gives a QIV angle.

The point $(-1, 5)$ lies in QII.

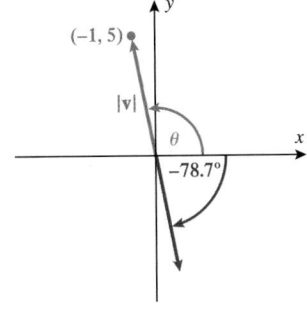

Add 180°.
$\qquad \theta = -78.7° + 180° = 101.3°$

$$\boxed{\theta = 101.3°}$$

Technology Tip

Use the calculator to find θ.

```
tan⁻¹(-5)
        -78.69006753
```

■ **YOUR TURN** Find the direction angle of the vector $\mathbf{u} = \langle 3, -4 \rangle$.

■ **Answer:** 306.9°

Recall that two vectors are equal if they have the same magnitude and direction. Algebraically, this corresponds to their corresponding vector components (a and b) being equal.

EQUAL VECTORS: U = V

The vectors $\mathbf{u} = \langle a, b \rangle$ and $\mathbf{v} = \langle c, d \rangle$ are **equal** (that is, $\mathbf{u} = \mathbf{v}$) if $a = c$ and $b = d$.

Vector addition is done geometrically with the tail-to-tip rule. Algebraically, vector addition is performed component by component.

VECTOR ADDITION: U + V

If $\mathbf{u} = \langle a, b \rangle$ and $\mathbf{v} = \langle c, d \rangle$, then $\mathbf{u} + \mathbf{v} = \langle a + c, b + d \rangle$.

EXAMPLE 3 Adding Vectors

Let $\mathbf{u} = \langle 2, -7 \rangle$ and $\mathbf{v} = \langle -3, 4 \rangle$. Find $\mathbf{u} + \mathbf{v}$.

Solution:

Let $\mathbf{u} = \langle 2, -7 \rangle$ and $\mathbf{v} = \langle -3, 4 \rangle$ in the addition formula.
$\qquad \mathbf{u} + \mathbf{v} = \langle 2 + (-3), -7 + 4 \rangle$

Simplify.
$\qquad \mathbf{u} + \mathbf{v} = \boxed{\langle -1, -3 \rangle}$

■ **YOUR TURN** Let $\mathbf{u} = \langle 1, 2 \rangle$ and $\mathbf{v} = \langle -5, -4 \rangle$. Find $\mathbf{u} + \mathbf{v}$.

■ **Answer:** $\mathbf{u} + \mathbf{v} = \langle -4, -2 \rangle$

Vector Operations

We now summarize vector operations. As we have seen, addition and subtraction are performed component by component. Multiplication, however, is not as straightforward. To perform **scalar multiplication** of a vector (to multiply a vector by a real number), we multiply each component by the scalar. In Section 7.2, we will study a form of multiplication for two vectors that is defined as long as the vectors have the same number of components; it gives a result known as the *dot product*, and is useful in solving common problems in physics.

SCALAR MULTIPLICATION: *k*u

If k is a scalar (real number) and $\mathbf{u} = \langle a, b \rangle$, then

$$k\mathbf{u} = k\langle a, b \rangle = \langle ka, kb \rangle$$

Scalar multiplication corresponds to

- Increasing the length of the vector: $|k| > 1$
- Decreasing the length of the vector: $|k| < 1$
- Changing the direction of the vector: $k < 0$

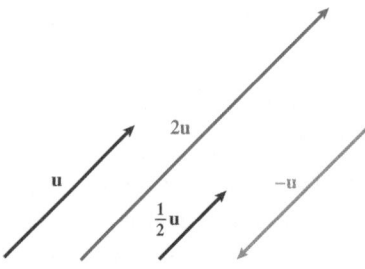

The following box is a summary of vector operations:

VECTOR OPERATIONS

If $\mathbf{u} = \langle a, b \rangle$, $\mathbf{v} = \langle c, d \rangle$, and k is a scalar, then

$$\mathbf{u} + \mathbf{v} = \langle a + c, b + d \rangle$$

$$\mathbf{u} - \mathbf{v} = \langle a - c, b - d \rangle$$

$$k\mathbf{u} = k\langle a, b \rangle = \langle ka, kb \rangle$$

The zero vector, $\mathbf{0} = \langle 0, 0 \rangle$, is a vector in any direction with a magnitude equal to zero. We now can state the algebraic properties (associative, commutative, and distributive) of vectors.

DEFINITION   **Algebraic Properties of Vectors**

$$\mathbf{u} + \mathbf{v} = \mathbf{v} + \mathbf{u}$$

$$(\mathbf{u} + \mathbf{v}) + \mathbf{w} = \mathbf{u} + (\mathbf{v} + \mathbf{w})$$

$$(k_1 k_2)\mathbf{u} = k_1(k_2\mathbf{u})$$

$$k(\mathbf{u} + \mathbf{v}) = k\mathbf{u} + k\mathbf{v}$$

$$(k_1 + k_2)\mathbf{u} = k_1\mathbf{u} + k_2\mathbf{u}$$

$$0\mathbf{u} = \mathbf{0} \qquad 1\mathbf{u} = \mathbf{u} \qquad -1\mathbf{u} = -\mathbf{u}$$

$$\mathbf{u} + (-\mathbf{u}) = \mathbf{0}$$

Horizontal and Vertical Components of a Vector

The **horizontal component** a and **vertical component** b of a vector $\mathbf{u}$ are related to the magnitude of the vector, $|\mathbf{u}|$, through the sine and cosine functions.

$$\cos\theta = \frac{a}{|\mathbf{u}|}$$

$$\sin\theta = \frac{b}{|\mathbf{u}|}$$

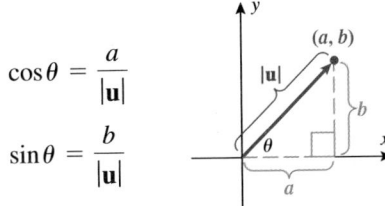

HORIZONTAL AND VERTICAL COMPONENTS OF A VECTOR

The horizontal and vertical components of vector $\mathbf{u}$, with magnitude $|\mathbf{u}|$ and direction angle θ, are given by

horizontal component: $a = |\mathbf{u}|\cos\theta$

vertical component: $b = |\mathbf{u}|\sin\theta$

The vector $\mathbf{u}$ can then be written as $\mathbf{u} = \langle a, b \rangle = \langle |\mathbf{u}|\cos\theta, |\mathbf{u}|\sin\theta \rangle$.

EXAMPLE 4   Finding the Horizontal and Vertical Components of a Vector

Find the vector that has a magnitude of 6 and a direction angle of 15°.

Solution:

Write the horizontal and vertical components of a vector $\mathbf{u}$.

$a = |\mathbf{u}|\cos\theta$ and $b = |\mathbf{u}|\sin\theta$

Let $|\mathbf{u}| = 6$ and $\theta = 15°$.

$a = 6\cos 15°$ and $b = 6\sin 15°$

Evaluate the sine and cosine functions of 15°.

$a = 5.8$ and $b = 1.6$

Let $\mathbf{u} = \langle a, b \rangle$.

$\mathbf{u} = \boxed{\langle 5.8, 1.6 \rangle}$

■ **Answer:** $\mathbf{u} = \langle 0.78, 2.9 \rangle$

■ **YOUR TURN** Find the vector that has a magnitude of 3 and direction angle of 75°.

Unit Vectors

A **unit vector** is any vector with magnitude equal to 1 or $|\mathbf{u}| = 1$. It is often useful to be able to find a unit vector in the same direction of some vector $\mathbf{v}$. A unit vector can be formed from any nonzero vector as follows:

FINDING A UNIT VECTOR

If $\mathbf{v}$ is a nonzero vector, then

$$\mathbf{u} = \frac{\mathbf{v}}{|\mathbf{v}|} = \frac{1}{|\mathbf{v}|} \cdot \mathbf{v}$$

is a **unit vector** in the same direction as $\mathbf{v}$. In other words, multiplying any nonzero vector by the reciprocal of its magnitude results in a unit vector.

Study Tip

Multiplying a nonzero vector by the reciprocal of its magnitude results in a unit vector.

It is important to notice that since the magnitude is always a scalar, then the reciprocal of the magnitude is always a scalar. A scalar times a vector is a vector.

 EXAMPLE 5   Finding a Unit Vector

Find a unit vector in the same direction as $\mathbf{v} = \langle -3, -4 \rangle$.

Solution:

Find the magnitude of the vector
$\mathbf{v} = \langle -3, -4 \rangle$.

$$|\mathbf{v}| = \sqrt{(-3)^2 + (-4)^2}$$

Simplify.

$$|\mathbf{v}| = 5$$

Multiply $\mathbf{v}$ by the reciprocal of its magnitude.

$$\frac{1}{|\mathbf{v}|} \cdot \mathbf{v}$$

Let $|\mathbf{v}| = 5$ and $\mathbf{v} = \langle -3, -4 \rangle$.

$$\frac{1}{5} \langle -3, -4 \rangle$$

Simplify.

$$\boxed{\left\langle -\frac{3}{5}, -\frac{4}{5} \right\rangle}$$

Check: The unit vector, $\left\langle -\dfrac{3}{5}, -\dfrac{4}{5} \right\rangle$, should have a magnitude of 1.

$$\sqrt{\left(-\frac{3}{5}\right)^2 + \left(-\frac{4}{5}\right)^2} = \sqrt{\frac{25}{25}} = 1$$

■ **Answer:** $\left\langle \frac{5}{13}, -\frac{12}{13} \right\rangle$

■ **YOUR TURN** Find a unit vector in the same direction as $\mathbf{v} = \langle 5, -12 \rangle$.

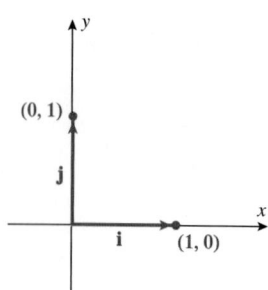

Two important unit vectors are the horizontal and vertical unit vectors **i** and **j**. The unit vector **i** has an initial point at the origin and terminal point at $(1, 0)$. The unit vector **j** has an initial point at the origin and terminal point at $(0, 1)$. We can use these unit vectors to represent vectors algebraically. For example, the vector $\langle 3, -4 \rangle = 3\mathbf{i} - 4\mathbf{j}$.

Resultant Vectors

Vectors arise in many applications. **Velocity vectors** and **force vectors** are two that we will discuss. For example, suppose that you are at the beach and you think you are swimming straight out at a certain speed (magnitude and direction). This is your **apparent velocity** with respect to the water. After a few minutes you turn around to look at the shore, and you are farther out than you thought and appear to have drifted down the beach. This is because of the **current** of the water. When the **current velocity** and the apparent velocity are added together, the result is the **actual** or **resultant velocity**.

EXAMPLE 6   Resultant Velocities

A boat's speedometer reads 25 mph (which is relative to the water) and sets a course due east (90° from due north). If the river is moving 10 mph due north, what is the resultant (actual) velocity of the boat?

Solution:

Draw a picture.

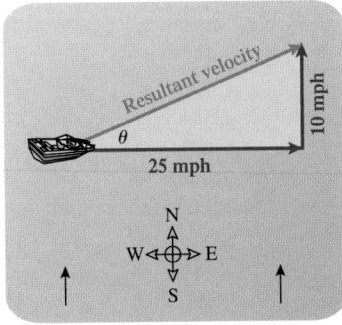

Label the horizontal and vertical components of the resultant vector.

$\langle 25, 10 \rangle$

Determine the magnitude of the resultant vector.

$$\sqrt{25^2 + 10^2} = 5\sqrt{29} \approx 27 \text{ mph}$$

Determine the direction angle.

$$\tan \theta = \frac{10}{25}$$

Solve for θ.

$$\theta = 21.8°$$

The actual velocity of the boat has magnitude ⎿27 mph⏋ and the boat is headed

⎿21.8° north of east or 68.2° east of north⏋.

In Example 6, the three vectors formed a right triangle. In Example 7, the three vectors form an oblique triangle.

EXAMPLE 7   Resultant Velocities

A speedboat traveling 30 mph has a compass heading of 100° east of north. The current velocity has a magnitude of 15 mph and its heading is 22° east of north. Find the resultant (actual) velocity of the boat.

Solution:

Draw a picture.

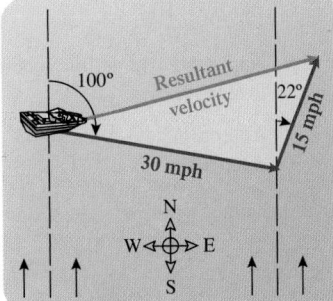

Label the supplementary angles to 100°.

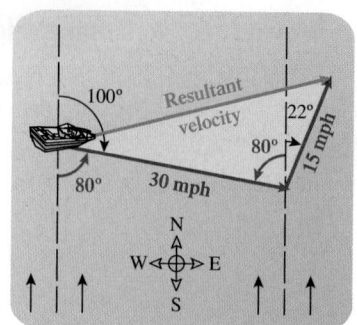

Draw and label the oblique triangle.

The magnitude of the actual (resultant) velocity is b.

The heading of the actual (resultant) velocity is $100° - \alpha$.

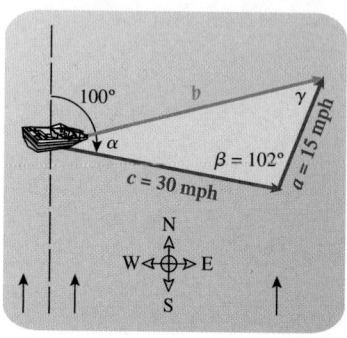

Use the Law of Sines and the Law of Cosines to solve for α and b.

Find b: Apply the Law of Cosines. $b^2 = a^2 + c^2 - 2ac \cos \beta$

    Let $a = 15$, $c = 30$, and
    $\beta = 102°$. $b^2 = 15^2 + 30^2 - 2(15)(30) \cos 102°$

    Solve for b. $\boxed{b \approx 36 \text{ mph}}$

Find α: Apply the Law of Sines. $\dfrac{\sin \alpha}{a} = \dfrac{\sin \beta}{b}$

    Isolate $\sin \alpha$. $\sin \alpha = \dfrac{a}{b} \sin \beta$

    Let $a = 15$, $b = 36$,
    and $\beta = 102°$. $\sin \alpha = \dfrac{15}{36} \sin 102°$

    Apply the inverse sine function
    to solve for α. $\alpha = \sin^{-1}\left(\dfrac{15}{36} \sin 102°\right)$

    Approximate α with a
    calculator. $\alpha \approx 24°$

Actual heading: $100° - \alpha = 100° - 24° = \boxed{76°}$

The actual velocity vector of the boat has magnitude $\boxed{36 \text{ mph}}$ and the boat is headed $\boxed{76° \text{ east of north}}$.

Two vectors combine to yield a resultant vector. The opposite vector to the resultant vector is called the **equilibrant**.

EXAMPLE 8   Finding an Equilibrant

A skier is being pulled up a slope by a handle lift. Let F_1 represent the vertical force due to gravity and F_2 represent the force of the skier pushing against the side of the mountain, at an angle of 35° to the horizontal. If the weight of the skier is 145 pounds, that is, $|F_1| = 145$, find the magnitude of the equilibrant force F_3 required to hold the skier in place (i.e., to keep the skier from sliding down the mountain). Assume that the side of the mountain is a frictionless surface.

Technology Tip

Use a TI calculator to find

$|F_3| = 145 \sin 35°$.

```
145sin(35)
        83.16858327
```

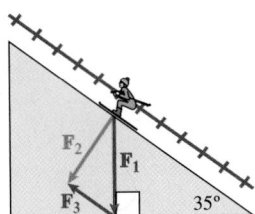

Solution:

The angle between vectors F_1 and F_2 is 35°.

The magnitude of vector F_3 is the force required to hold the skier in place.

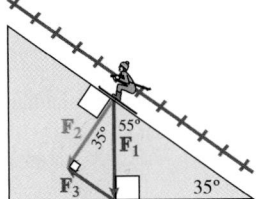

Relate the magnitudes (side lengths) to the given angle using the sine ratio.

$$\sin 35° = \frac{|F_3|}{|F_1|}$$

Solve for $|F_3|$.

$$|F_3| = |F_1| \sin 35°$$

Let $|F_1| = 145$.

$$|F_3| = 145 \sin 35°$$

$$|F_3| = 83.16858$$

A force of approximately $\boxed{83 \text{ pounds}}$ is required to keep the skier from sliding down the hill.

EXAMPLE 9   Resultant Forces

A barge runs aground outside the channel. A single tugboat cannot generate enough force to pull the barge off the sandbar. A second tugboat comes to assist. The following diagram illustrates the force vectors, F_1 and F_2, from the tugboats. What is the resultant force vector of the two tugboats?

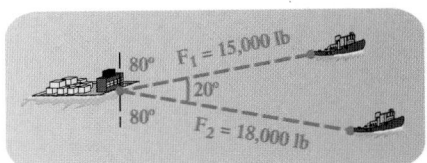

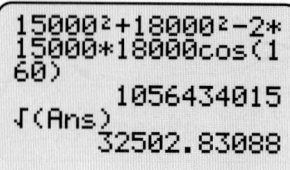
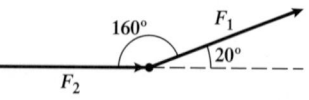
Solution:

Using the tail-to-tip rule, we can add these two vectors and form a triangle below:

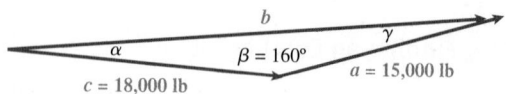

Find *b*: Apply the Law of Cosines.

$$b^2 = a^2 + c^2 - 2ac\cos\beta$$

Let $a = 15{,}000$, $c = 18{,}000$, and $\beta = 160°$.

$$b^2 = 15{,}000^2 + 18{,}000^2 - 2(15{,}000)(18{,}000)\cos 160°$$

Solve for *b*.

$$\boxed{b = 32{,}503 \text{ lb}}$$

Find *α*: Apply the Law of Sines.

$$\frac{\sin\alpha}{a} = \frac{\sin\beta}{b}$$

Isolate $\sin\alpha$.

$$\sin\alpha = \frac{a}{b}\sin\beta$$

Let $a = 15{,}000$, $b = 32{,}503$, and $\beta = 160°$.

$$\sin\alpha = \frac{15{,}000}{32{,}503}\sin 160°$$

Apply the inverse sine function to solve for *α*.

$$\alpha = \sin^{-1}\left(\frac{15{,}000}{32{,}503}\sin 160°\right)$$

Approximate *α* with a calculator.

$$\boxed{\alpha \approx 9.08°}$$

The resulting force is $\boxed{32{,}503 \text{ lb}}$ at an angle of $\boxed{9° \text{ from the tug pulling with a force of } 18{,}000 \text{ lb}}$.

SECTION 7.1 SUMMARY

In this section, we discussed scalars (real numbers) and vectors. Scalars have only magnitude, whereas vectors have both magnitude and direction.

$$\text{Vector:} \quad \mathbf{u} = \langle a, b\rangle$$

$$\text{Magnitude:} \quad |\mathbf{u}| = \sqrt{a^2 + b^2}$$

$$\text{Direction } (\theta): \quad \tan\theta = \frac{b}{a}$$

We defined vectors both algebraically and geometrically and gave interpretations of magnitude and vector addition in both methods.

Vector addition is performed algebraically component by component.

$$\langle a, b\rangle + \langle c, d\rangle = \langle a + c, b + d\rangle.$$

The trigonometric functions are used to express the horizontal and vertical components of a vector.

$$\text{Horizontal component:} \quad a = |\mathbf{u}|\cos\theta$$

$$\text{Vertical component:} \quad b = |\mathbf{u}|\sin\theta$$

Velocity and force vectors illustrate applications of the Law of Sines and the Law of Cosines.

SECTION 7.1 EXERCISES

■ SKILLS

In Exercises 1–6, find the magnitude of the vector AB.

1. $A = (2, 7)$ and $B = (5, 9)$

2. $A = (-2, 3)$ and $B = (3, -4)$

3. $A = (4, 1)$ and $B = (-3, 0)$

4. $A = (-1, -1)$ and $B = (2, -5)$

5. $A = (0, 7)$ and $B = (-24, 0)$

6. $A = (-2, 1)$ and $B = (4, 9)$

In Exercises 7–16, find the magnitude and direction angle of the given vector.

7. $u = \langle 3, 8 \rangle$ **8.** $u = \langle 4, 7 \rangle$ **9.** $u = \langle 5, -1 \rangle$ **10.** $u = \langle -6, -2 \rangle$

11. $u = \langle -4, 1 \rangle$ **12.** $u = \langle -6, 3 \rangle$ **13.** $u = \langle -8, 0 \rangle$ **14.** $u = \langle 0, 7 \rangle$

15. $u = \langle \sqrt{3}, 3 \rangle$ **16.** $u = \langle -5, -5 \rangle$

In Exercises 17–24, perform the indicated vector operation, given $u = \langle -4, 3 \rangle$ and $v = \langle 2, -5 \rangle$.

17. $u + v$ **18.** $u - v$ **19.** $3u$ **20.** $-2u$

21. $2u + 4v$ **22.** $5(u + v)$ **23.** $6(u - v)$ **24.** $2u - 3v + 4u$

In Exercises 25–34, find the vector, given its magnitude and direction angle.

25. $|u| = 7, \theta = 25°$ **26.** $|u| = 5, \theta = 75°$ **27.** $|u| = 16, \theta = 100°$ **28.** $|u| = 8, \theta = 200°$

29. $|u| = 4, \theta = 310°$ **30.** $|u| = 8, \theta = 225°$ **31.** $|u| = 9, \theta = 335°$ **32.** $|u| = 3, \theta = 315°$

33. $|u| = 2, \theta = 120°$ **34.** $|u| = 6, \theta = 330°$

In Exercises 35–44, find a unit vector in the direction of the given vector.

35. $v = \langle -5, -12 \rangle$ **36.** $v = \langle 3, 4 \rangle$ **37.** $v = \langle 60, 11 \rangle$ **38.** $v = \langle -7, 24 \rangle$

39. $v = \langle 24, -7 \rangle$ **40.** $v = \langle -10, 24 \rangle$ **41.** $v = \langle -9, -12 \rangle$ **42.** $v = \langle 40, -9 \rangle$

43. $v = \langle \sqrt{2}, 3\sqrt{2} \rangle$ **44.** $v = \langle -4\sqrt{3}, -2\sqrt{3} \rangle$

In Exercises 45–50, express the vector in terms of unit vectors i and j.

45. $\langle 7, 3 \rangle$ **46.** $\langle -2, 4 \rangle$ **47.** $\langle 5, -3 \rangle$ **48.** $\langle -6, -2 \rangle$ **49.** $\langle -1, 0 \rangle$ **50.** $\langle 0, 2 \rangle$

In Exercises 51–56, perform the indicated vector operation.

51. $(5i - 2j) + (-3i + 2j)$ **52.** $(4i - 2j) + (3i - 5j)$ **53.** $(-3i + 3j) - (2i - 2j)$

54. $(i - 3j) - (-2i + j)$ **55.** $(5i + 3j) + (2i - 3j)$ **56.** $(-2i + j) + (2i - 4j)$

APPLICATIONS

57. Bullet Speed. A bullet is fired from ground level at a speed of 2200 feet per second at an angle of 30° from the horizontal. Find the magnitude of the horizontal and vertical components of the velocity vector.

58. Weightlifting. A 50-pound weight lies on an inclined bench that makes an angle of 40° with the horizontal. Find the component of the weight directed perpendicular to the bench and also the component of the weight parallel to the inclined bench.

59. Weight of a Boat. A force of 630 pounds is needed to pull a speedboat and its trailer up a ramp that has an incline of 13°. What is the combined weight of the boat and its trailer?

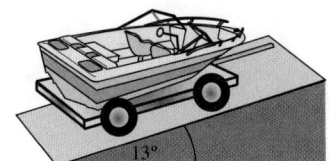

60. Weight of a Boat. A force of 500 pounds is needed to pull a speedboat and its trailer up a ramp that has an incline of 16°. What is the weight of the boat and its trailer?

61. Speed and Direction of a Ship. A ship's captain sets a course due north at 10 mph. The water is moving at 6 mph due west. What is the actual velocity of the ship and in what direction is it traveling?

62. Speed and Direction of a Ship. A ship's captain sets a course due west at 12 mph. The water is moving at 3 mph due north. What is the actual velocity of the ship and in what direction is it traveling?

63. Heading and Airspeed. A plane has a compass heading of 60° east of due north and an airspeed of 300 mph. The wind is blowing at 40 mph with a heading of 30° west of due north. What are the plane's actual heading and airspeed?

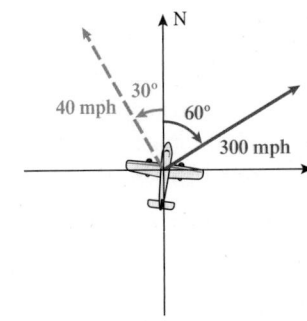

64. Heading and Airspeed. A plane has a compass heading of 30° east of due north and an airspeed of 400 mph. The wind is blowing at 30 mph with a heading of 60° west of due north. What are the plane's actual heading and airspeed?

65. Sliding Box. A box weighing 500 pounds is held in place on an inclined plane that has an angle of 30°. What force is required to hold it in place?

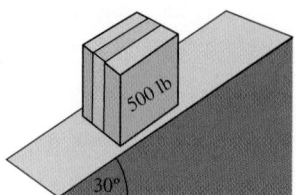

66. Sliding Box. A box weighing 500 pounds is held in place on an inclined plane that has an angle of 10°. What force is required to hold it in place?

67. Baseball. A baseball player throws a ball with an initial velocity of 80 feet per second at an angle of 40° with the horizontal. What are the vertical and horizontal components of the velocity?

68. Baseball. A baseball pitcher throws a ball with an initial velocity of 100 feet per second at an angle of 5° with the horizontal. What are the vertical and horizontal components of the velocity?

For Exercises 69 and 70, refer to the following:

In a post pattern in football, the receiver in motion runs past the quarterback parallel to the line of scrimmage (A), runs perpendicular to the line of scrimmage (B), and then cuts toward the goal post (C).

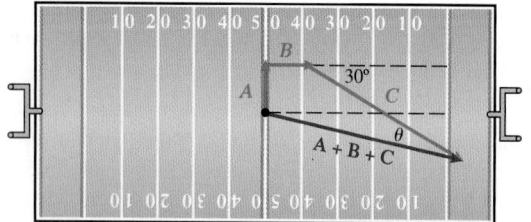

69. Football. A receiver runs the post pattern. If the magnitudes of the vectors are $|A| = 4$ yards, $|B| = 12$ yards, and $|C| = 20$ yards, find the magnitude of the resultant vector $A + B + C$.

70. Football. A receiver runs the post pattern. If the magnitudes of the vectors are $|A| = 4$ yards, $|B| = 12$ yards, and $|C| = 20$ yards, find the direction angle θ.

71. Resultant Force. A force with a magnitude of 100 pounds and another with a magnitude of 400 pounds are acting on an object. The two forces have an angle of 60° between them. What is the direction of the resultant force with respect to the force of 400 pounds?

72. Resultant Force. A force with a magnitude of 100 pounds and another with a magnitude of 400 pounds are acting on an object. The two forces have an angle of 60° between them. What is the magnitude of the resultant force?

73. Resultant Force. A force of 1000 pounds is acting on an object at an angle of 45° from the horizontal. Another force of 500 pounds is acting at an angle of −40° from the horizontal. What is the magnitude of the resultant force?

74. Resultant Force. A force of 1000 pounds is acting on an object at an angle of 45° from the horizontal. Another force of 500 pounds is acting at an angle of −40° from the horizontal. What is the angle of the resultant force?

75. Resultant Force. Forces with magnitudes of 200 N and 180 N act on a hook. The angle between these two forces is 45°. Find the direction and magnitude of the resultant of these forces.

76. Resultant Force. Forces with magnitudes of 100 N and 50 N act on a hook. The angle between these two forces is 30°. Find the direction and magnitude of the resultant of these forces.

77. Exercise Equipment. A tether ball weighing 5 pounds is pulled outward from a pole by a horizontal force u until the rope makes a 45° angle with the pole. Determine the resulting tension (in pounds) on the rope and magnitude of u.

78. Exercise Equipment. A tether ball weighing 8 pounds is pulled outward from a pole by a horizontal force u until the rope makes a 60° angle with the pole. Determine the resulting tension (in pounds) on the rope and magnitude of u.

79. Recreation. A freshman wishes to sign up for four different clubs during orientation. Each club is positioned at a different table in the gym and the clubs of interest to him are positioned at A, B, C, and D, as pictured below. He starts at the entrance way O and walks directly toward A, followed by B and C, then to D, and then back to O.

a. Find the resultant vector of all his movement.

b. How far did he walk during this sign-up adventure?

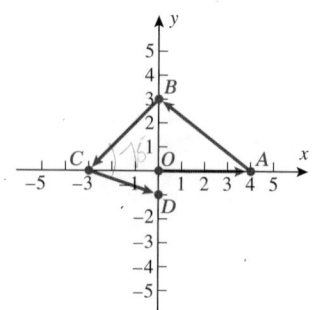

80. Recreation. A freshman wishes to sign up for three different clubs during orientation. Each club is positioned at a different table in the gym and the clubs of interest to him are positioned at A, B, and C, as pictured below. He starts at the entrance way O at the far end of the gym, walks directly toward A, followed by B and C, and then exits the gym through the exit P at the opposite end.

a. Find the resultant vector of all his movement.

b. How far did he walk during this sign-up adventure?

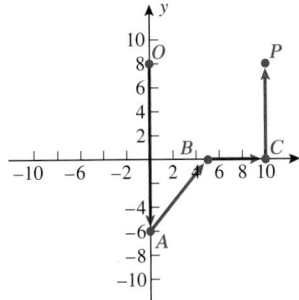

81. Torque. *Torque* is the tendency for an arm to rotate about a pivot point. If a force $\mathbf{F}$ is applied at an angle θ to turn an arm of length L, as pictured below, then the magnitude of the torque $= L|\mathbf{F}|\sin\theta$.

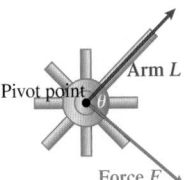

Assume that a force of 45 N is applied to a bar 0.2 meters wide on a sewer shut-off valve at an angle 85°. What is the magnitude of the torque in N-m?

82. Torque. You walk through a swinging mall door to enter a department store. You exert a force of 40 N applied perpendicular to the door. The door is 0.85 meters wide. Assuming that you pushed the door at its edge and the hinge is the pivot point, find the magnitude of the torque.

83. Torque. You walk through a swinging mall door to enter a department store. You exert a force of 40 N applied at an angle 110° to the door. The door is 0.85 meters wide. Assuming that you pushed the door at its edge and the hinge is the pivot point, find the magnitude of the torque.

84. Torque. Suppose that within the context of Exercises 82 and 83, the magnitude of the torque turned out to be 0 N-m. When can this occur?

85. Resultant Force. A person is walking two dogs fastened to separate leashes that meet in a connective hub, leading to a single leash that she is holding. Dog 1 applies a force N 60° W with a magnitude of 8, and Dog 2 applies a force of N 45° E with a magnitude of 6. Find the magnitude and direction of the force $\mathbf{w}$ that the walker applies to the leash in order to counterbalance the total force exerted by the dogs.

86. Resultant Force. A person is walking three dogs fastened to separate leashes that meet in a connective hub, leading to a single leash that she is holding. Dog 1 applies a force N 60° W with a magnitude of 8, Dog 2 applies a force of N 45° E with a magnitude of 6, and Dog 3 moves directly N with a magnitude of 12. Find the magnitude and direction of the force $\mathbf{w}$ that the walker applies to the leash in order to counterbalance the total force exerted by the dogs.

■ CATCH THE MISTAKE

In Exercises 87 and 88, explain the mistake that is made.

87. Find the magnitude of the vector $\langle -2, -8 \rangle$.

Solution:

Factor the -1.

$$-\langle 2, 8 \rangle$$

Find the magnitude of $\langle 2, 8 \rangle$.

$$|\langle 2, 8 \rangle| = \sqrt{2^2 + 8^2} = \sqrt{68} = 2\sqrt{17}$$

Write the magnitude of $\langle -2, -8 \rangle$.

$$|\langle -2, -8 \rangle| = -2\sqrt{17}$$

This is incorrect. What mistake was made?

88. Find the direction angle of the vector $\langle -2, -8 \rangle$.

Solution:

Write the formula for the direction angle of $\langle a, b \rangle$.

$$\tan\theta = \frac{b}{a}$$

Let $a = -2$ and $b = -8$.

$$\tan\theta = \frac{-8}{-2}$$

Apply the inverse tangent function.

$$\theta = \tan^{-1}4$$

Evaluate with a calculator.

$$\theta = 76°$$

This is incorrect. What mistake was made?

■ CONCEPTUAL

In Exercises 89–92, determine whether each statment is true or false.

89. The magnitude of the vector **i** is the imaginary number i.

90. The arrow components of equal vectors must coincide.

91. The magnitude of a vector is always greater than or equal to the magnitude of its horizontal component.

92. The magnitude of a vector is always greater than or equal to the magnitude of its vertical component.

93. Would a scalar or a vector represent the following? *The car is driving 72 mph due east (90° with respect to north).*

94. Would a scalar or vector represent the following? *The granite has a mass of 131 kg.*

95. Find the magnitude of the vector $\langle -a, b \rangle$ if $a > 0$ and $b > 0$.

96. Find the direction angle of the vector $\langle -a, b \rangle$ if $a > 0$ and $b > 0$.

■ CHALLENGE

97. Show that if **u** is a unit vector in the direction of **v**, then $\mathbf{v} = |\mathbf{u}|\,\mathbf{u}$.

98. Show that if $\mathbf{u} = a\mathbf{i} + b\mathbf{j}$ is a unit vector, then (a, b) lies on the unit circle.

99. A vector **u** is a *linear combination* of **p** and **q** if there exist constants c_1 and c_2 such that $\mathbf{u} = c_1\mathbf{p} + c_2\mathbf{q}$. Show that $\langle -6, 4 \rangle$ is a linear combination of $\langle -8, 4 \rangle$ and $\langle 1, -1 \rangle$.

100. Show that $\left\langle -\frac{2}{9}a, \frac{8}{9}b \right\rangle$ is a linear combination of $\langle a, 3b \rangle$ and $\langle -a, -b \rangle$, for any real constants a and b.

101. Prove that $\mathbf{u} + 3(2\mathbf{v} - \mathbf{u}) = 6\mathbf{v} - 2\mathbf{u}$, showing carefully how all relevant properties and definitions enter the proof.

102. Let $\mathbf{u} = \langle 2a, a \rangle$, $\mathbf{v} = \langle -a, -2a \rangle$. Compute $\left| \dfrac{2\mathbf{u}}{|\mathbf{v}|} - \dfrac{3\mathbf{v}}{|\mathbf{u}|} \right|$.

■ TECHNOLOGY

For Exercises 103–108, refer to the following:

Vectors can be represented as column matrices. For example, the vector $\mathbf{u} = \langle 3, -4 \rangle$ can be represented as a 2×1 column matrix $\begin{bmatrix} 3 \\ -4 \end{bmatrix}$. With a TI-83 calculator, vectors can be entered as matrices in two ways, directly or via MATRIX .

Directly:

```
[[3][-4]]
         [[3 ]
          [-4]]
■
```

Matrix:

```
[A]
         [[3 ]
          [-4]]
```

Use a calculator to perform the vector operation given $\mathbf{u} = \langle 8, -5 \rangle$ and $\mathbf{v} = \langle -7, 11 \rangle$.

103. $\mathbf{u} + 3\mathbf{v}$

104. $-9(\mathbf{u} - 2\mathbf{v})$

Use a calculator to find a unit vector in the direction of the given vector.

105. $\mathbf{u} = \langle 10, -24 \rangle$

106. $\mathbf{u} = \langle -9, -40 \rangle$

Use the graphing calculator SUM command to find the magnitude of the given vector. Also, find the direction angle to the nearest degree.

107. $\langle -33, 180 \rangle$

108. $\langle -20, -30\sqrt{5} \rangle$

■ PREVIEW TO CALCULUS

There is a branch of calculus devoted to the study of vector-valued functions; these are functions that map real numbers onto vectors. For example, $\mathbf{v}(t) = \langle t, 2t \rangle$.

109. Find the magnitude of the vector-valued function $\mathbf{v}(t) = \langle \cos t, \sin t \rangle$.

110. Find the direction of the vector-valued function $\mathbf{v}(t) = \langle -3t, -4t \rangle$.

The difference quotient for the vector-valued function $\mathbf{v}(t)$ is defined as $\dfrac{\mathbf{v}(t + h) - \mathbf{v}(t)}{h}$. In Exercises 111 and 112, find the difference quotient of the vector-valued function.

111. $\mathbf{v}(t) = \langle t, t^2 \rangle$

112. $\mathbf{v}(t) = \langle t^2 + 1, t^3 \rangle$

SKILLS OBJECTIVES

- Find the dot product of two vectors.
- Use the dot product to find the angle between two vectors.
- Determine whether two vectors are parallel or perpendicular.
- Use the dot product to calculate the amount of work associated with a physical problem.

CONCEPTUAL OBJECTIVES

- Understand that the dot products of parallel or perpendicular vectors are respective limiting cases of the dot product of two vectors.
- Project a vector onto another vector.

Multiplying Two Vectors

In this course, we apply two types of multiplication defined for vectors: scalar multiplication and the dot product. Scalar multiplication (which we already demonstrated in Section 7.1) is multiplication of a scalar by a vector; the result is a vector. Now we discuss the *dot product* of two vectors. In this case, there are two important things to note: (1) The dot product of two vectors is defined only if the vectors have the same number of components and (2) if the dot product does exist, then the result is a scalar.

DOT PRODUCT

The **dot product** of two vectors $\mathbf{u} = \langle a, b \rangle$ and $\mathbf{v} = \langle c, d \rangle$ is given by

$$\mathbf{u} \cdot \mathbf{v} = ac + bd$$

$\mathbf{u} \cdot \mathbf{v}$ is pronounced "u dot v."

EXAMPLE 1 **Finding the Dot Product of Two Vectors**

Find the dot product $\langle -7, 3 \rangle \cdot \langle 2, 5 \rangle$.

Solution:

Sum the products of the first components and the products of the second components.

$$\langle -7, 3 \rangle \cdot \langle 2, 5 \rangle = (-7)(2) + (3)(5)$$

Simplify.

$$\boxed{\langle -7, 3 \rangle \cdot \langle 2, 5 \rangle = 1}$$

Study Tip

The dot product of two vectors is a scalar.

- **Answer:** -9

- **YOUR TURN** Find the dot product $\langle 6, 1 \rangle \cdot \langle -2, 3 \rangle$.

The following box summarizes the properties of the dot product:

PROPERTIES OF THE DOT PRODUCT

1. $\mathbf{u} \cdot \mathbf{v} = \mathbf{v} \cdot \mathbf{u}$
2. $\mathbf{u} \cdot \mathbf{u} = |\mathbf{u}|^2$
3. $\mathbf{0} \cdot \mathbf{u} = 0$
4. $k(\mathbf{u} \cdot \mathbf{v}) = (k\mathbf{u}) \cdot \mathbf{v} = \mathbf{u} \cdot (k\mathbf{v})$
5. $(\mathbf{u} + \mathbf{v}) \cdot \mathbf{w} = \mathbf{u} \cdot \mathbf{w} + \mathbf{v} \cdot \mathbf{w}$
6. $\mathbf{u} \cdot (\mathbf{v} + \mathbf{w}) = \mathbf{u} \cdot \mathbf{v} + \mathbf{u} \cdot \mathbf{w}$

These properties are verified in the exercises.

Angle Between Two Vectors

We can use these properties to develop an equation that relates the angle between two vectors and the dot product of the vectors.

WORDS	MATH
Let $\mathbf{u}$ and $\mathbf{v}$ be two vectors with the same initial point, and let θ be the angle between them.	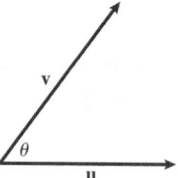
The vector $\mathbf{u} - \mathbf{v}$ is opposite angle θ.	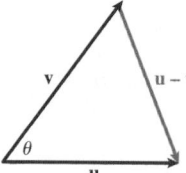
A triangle is formed with side lengths equal to the magnitudes of the three vectors.	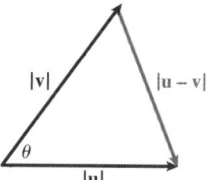

Apply the Law of Cosines.

$$|\mathbf{u} - \mathbf{v}|^2 = |\mathbf{u}|^2 + |\mathbf{v}|^2 - 2|\mathbf{u}||\mathbf{v}|\cos\theta$$

Use properties of the dot product to rewrite the left side of equation.

Property (2): $\quad |\mathbf{u} - \mathbf{v}|^2 = (\mathbf{u} - \mathbf{v}) \cdot (\mathbf{u} - \mathbf{v})$

Property (6): $\quad\quad\quad\quad\quad = \mathbf{u} \cdot (\mathbf{u} - \mathbf{v}) - \mathbf{v} \cdot (\mathbf{u} - \mathbf{v})$

Property (6): $\quad\quad\quad\quad\quad = \mathbf{u} \cdot \mathbf{u} - \mathbf{u} \cdot \mathbf{v} - \mathbf{v} \cdot \mathbf{u} + \mathbf{v} \cdot \mathbf{v}$

Property (2): $\quad\quad\quad\quad\quad = |\mathbf{u}|^2 - \mathbf{u} \cdot \mathbf{v} - \mathbf{v} \cdot \mathbf{u} + |\mathbf{v}|^2$

Property (1): $\quad\quad\quad\quad\quad = |\mathbf{u}|^2 - 2(\mathbf{u} \cdot \mathbf{v}) + |\mathbf{v}|^2$

Substitute this last expression for the left side of the original Law of Cosines equation.

$$|\mathbf{u}|^2 - 2(\mathbf{u} \cdot \mathbf{v}) + |\mathbf{v}|^2 = |\mathbf{u}|^2 + |\mathbf{v}|^2 - 2|\mathbf{u}||\mathbf{v}|\cos\theta$$

Simplify.

$$-2(\mathbf{u} \cdot \mathbf{v}) = -2|\mathbf{u}||\mathbf{v}|\cos\theta$$

Isolate $\cos\theta$.

$$\cos\theta = \frac{\mathbf{u} \cdot \mathbf{v}}{|\mathbf{u}||\mathbf{v}|}$$

Notice that **u** and **v** have to be nonzero vectors, since we divided by them in the last step.

ANGLE BETWEEN TWO VECTORS

If θ is the angle between two nonzero vectors **u** and **v**, where $0° \leq \theta \leq 180°$, then

$$\cos\theta = \frac{\mathbf{u} \cdot \mathbf{v}}{|\mathbf{u}||\mathbf{v}|}$$

sidebar
Study Tip

The angle between two vectors

$\theta = \cos^{-1}\left(\dfrac{\mathbf{u} \cdot \mathbf{v}}{|\mathbf{u}||\mathbf{v}|}\right)$ is an angle

between 0° and 180° (the range of the inverse cosine function).

In the Cartesian plane, there are two angles between two vectors, θ and $360° - \theta$. **We assume that θ is the "smaller" angle.**

EXAMPLE 2   Finding the Angle Between Two Vectors

Find the angle between $\langle 2, -3 \rangle$ and $\langle -4, 3 \rangle$.

Solution:

Let $\mathbf{u} = \langle 2, -3 \rangle$ and $\mathbf{v} = \langle -4, 3 \rangle$.

STEP 1 Find $\mathbf{u} \cdot \mathbf{v}$.

$$\begin{aligned}\mathbf{u} \cdot \mathbf{v} &= \langle 2, -3 \rangle \cdot \langle -4, 3 \rangle \\ &= (2)(-4) + (-3)(3) = -17\end{aligned}$$

STEP 2 Find $|\mathbf{u}|$.

$$|\mathbf{u}| = \sqrt{\mathbf{u} \cdot \mathbf{u}} = \sqrt{2^2 + (-3)^2} = \sqrt{13}$$

STEP 3 Find $|\mathbf{v}|$.

$$|\mathbf{v}| = \sqrt{\mathbf{v} \cdot \mathbf{v}} = \sqrt{(-4)^2 + 3^2} = \sqrt{25} = 5$$

STEP 4 Find θ.

$$\cos\theta = \frac{\mathbf{u} \cdot \mathbf{v}}{|\mathbf{u}||\mathbf{v}|} = \frac{-17}{5\sqrt{13}}$$

Approximate θ with a calculator.

$$\theta = \cos^{-1}\left(-\frac{17}{5\sqrt{13}}\right) \approx 160.559965°$$

$$\boxed{\theta \approx 161°}$$

STEP 5 Draw a picture to confirm the answer.

Draw the vectors $\langle 2, -3 \rangle$ and $\langle -4, 3 \rangle$.

161° appears to be correct.

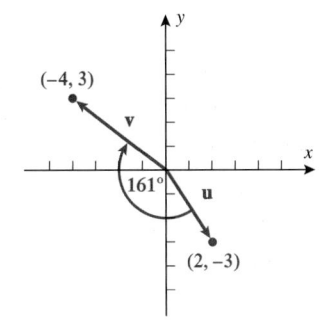

sidebar
Technology Tip

Use a TI calculator to find θ,

$$\cos\theta = \frac{-17}{5\sqrt{13}}.$$

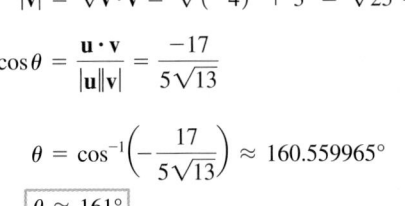

■ **Answer:** 38°

■ **YOUR TURN** Find the angle between $\langle 1, 5 \rangle$ and $\langle -2, 4 \rangle$.

When two vectors are **parallel**, the angle between them is 0° or 180°.

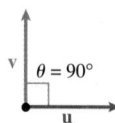

$\theta = 0°$ **u** $\theta = 180°$

     v   **v**   **u**

When two vectors are **perpendicular (orthogonal)**, the angle between them is 90°.

V $\theta = 90°$

   u

Note: We did not include 270° because the angle $0° \le \theta \le 180°$ between two vectors is taken to be the smaller angle.

WORDS	MATH
When two vectors **u** and **v** are perpendicular, $\theta = 90°$.	$\cos 90° = \dfrac{\mathbf{u} \cdot \mathbf{v}}{|\mathbf{u}||\mathbf{v}|}$
Substitute $\cos 90° = 0$.	$0 = \dfrac{\mathbf{u} \cdot \mathbf{v}}{|\mathbf{u}||\mathbf{v}|}$
Therefore, the dot product of **u** and **v** must be zero.	$\mathbf{u} \cdot \mathbf{v} = \mathbf{0}$

ORTHOGONAL VECTORS

Two vectors **u** and **v** are **orthogonal** (perpendicular) if and only if their dot product is zero.

$$\mathbf{u} \cdot \mathbf{v} = \mathbf{0}$$

EXAMPLE 3   **Determining Whether Vectors Are Orthogonal**

Determine whether each pair of vectors is orthogonal.

a. $\mathbf{u} = \langle 2, -3 \rangle$ and $\mathbf{v} = \langle 3, 2 \rangle$ **b.** $\mathbf{u} = \langle -7, -3 \rangle$ and $\mathbf{v} = \langle 7, 3 \rangle$

Solution (a):

Find the dot product $\mathbf{u} \cdot \mathbf{v}$. $\mathbf{u} \cdot \mathbf{v} = (2)(3) + (-3)(2)$

Simplify. $\mathbf{u} \cdot \mathbf{v} = 0$

> Vectors **u** and **v** are orthogonal, since $\mathbf{u} \cdot \mathbf{v} = 0$.

Solution (b):

Find the dot product $\mathbf{u} \cdot \mathbf{v}$. $\mathbf{u} \cdot \mathbf{v} = (-7)(7) + (-3)(3)$

Simplify. $\mathbf{u} \cdot \mathbf{v} = -58$

> Vectors **u** and **v** are not orthogonal, since $\mathbf{u} \cdot \mathbf{v} \neq 0$.

Work

If you had to carry either of the following items—barbells or pillows—for 1 mile, which would you choose?

You would probably pick the pillows over the barbell and weights, because the pillows are lighter. It requires less work to carry the pillows than it does to carry the weights. If asked to carry either of them 1 mile or 10 miles, you would probably pick 1 mile, because it's a shorter distance and requires less work. **Work** is done when a *force causes an object to move a certain distance.*

The simplest case is when the force is in the same direction as the displacement—for example, a stagecoach (the horses pull with a force in the same direction). In this case the work is defined as the magnitude of the force times the magnitude of the displacement, distance d.

$$W = |\mathbf{F}|\,d$$

Notice that the magnitude of the force is a scalar, the distance d is a scalar, and hence the product is a scalar.

If the horses pull with a force of 1000 pounds and they move the stagecoach 100 feet, the work done by the force is

$$W = (1000\,\text{lb})(100\,\text{ft}) = 100{,}000\,\text{ft-lb}$$

In many physical applications, however, the force is not in the same direction as the displacement, and hence vectors (not just their magnitudes) are required.

We often want to know how much of a vector is applied in a certain direction. For example, when your car runs out of gasoline and you try to push it, some of the force vector $\mathbf{F}_1$ you generate from pushing translates into the horizontal direction $\mathbf{F}_2$ (hence the car moves). If we let θ be the angle between the vectors $\mathbf{F}_1$ and $\mathbf{F}_2$, then the horizontal component of $\mathbf{F}_1$ is $\mathbf{F}_2$, where $|F_2| = |F_1|\cos\theta$.

If the couple in the picture push at an angle of $25°$ with a force of 150 pounds, then the horizontal component of the force vector $\mathbf{F}_1$ is

$$(150\,\text{lb})(\cos 25°) \approx 136\,\text{lb}$$

To develop a generalized formula when the force exerted and the displacement are not in the same direction, we start with the formula, $\cos\theta = \dfrac{\mathbf{u}\cdot\mathbf{v}}{|\mathbf{u}||\mathbf{v}|}$.

Isolate the dot product $\mathbf{u}\cdot\mathbf{v}$. $\mathbf{u}\cdot\mathbf{v} = |\mathbf{u}||\mathbf{v}|\cos\theta$

Let $\mathbf{u} = \mathbf{F}$ and $\mathbf{v} = \mathbf{d}$. $W = \mathbf{F}\cdot\mathbf{d} = |\mathbf{F}||\mathbf{d}|\cos\theta = \underbrace{|\mathbf{F}|\cos\theta}_{\substack{\text{magnitude of force} \\ \text{in direction of displacement}}} \cdot \underbrace{|\mathbf{d}|}_{\text{distance}}$

WORK

If an object is moved from point A to point B by a constant force, then the work associated with this displacement is

$$W = \mathbf{F}\cdot\mathbf{d}$$

where $\mathbf{d}$ is the displacement vector and $\mathbf{F}$ is the force vector.

Work is typically expressed in one of two units:

SYSTEM	FORCE	DISTANCE	WORK
U.S. customary	pound	foot	ft-lb
SI	newton	meter	N-m

EXAMPLE 4   Calculating Work

How much work is done when a force (in pounds) $\mathbf{F} = \langle 2, 4 \rangle$ moves an object from $(0, 0)$ to $(5, 9)$ (the distance is in feet)?

Solution:

Find the displacement vector $\mathbf{d}$. $\mathbf{d} = \langle 5, 9 \rangle$

Apply the work formula, $W = \mathbf{F}\cdot\mathbf{d}$. $W = \langle 2, 4 \rangle \cdot \langle 5, 9 \rangle$

Calculate the dot product. $W = (2)(5) + (4)(9)$

Simplify. $\boxed{W = 46 \text{ ft-lb}}$

■ **Answer:** 25 N-m

■ **YOUR TURN** How much work is done when a force (in newtons) $\mathbf{F} = \langle 1, 3 \rangle$ moves an object from $(0, 0)$ to $(4, 7)$ (the distance is in meters)?

SECTION 7.2 SUMMARY

In this section, we defined a dot product as a form of multiplication of two vectors. A scalar times a vector results in a vector, whereas the dot product of two vectors is a scalar.

$$\langle a, b \rangle \cdot \langle c, d \rangle = ac + bc$$

We developed a formula that determines the angle θ between two vectors.

$$\cos\theta = \frac{\mathbf{u}\cdot\mathbf{v}}{|\mathbf{u}||\mathbf{v}|}$$

Orthogonal (perpendicular) vectors have an angle of 90° between them, and consequently, the dot product of two orthogonal vectors is equal to zero. Work is the result of a force displacing an object. When the force and displacement are in the same direction, the work is equal to the product of the magnitude of the force and the distance (magnitude of the displacement). When the force and displacement are not in the same direction, work is the dot product of the force vector and displacement vector, $\mathbf{W} = \mathbf{F}\cdot\mathbf{d}$.

SECTION
7.2 EXERCISES

■ SKILLS

In Exercises 1–12, find the indicated dot product.

1. $\langle 4, -2 \rangle \cdot \langle 3, 5 \rangle$

2. $\langle 7, 8 \rangle \cdot \langle 2, -1 \rangle$

3. $\langle -5, 6 \rangle \cdot \langle 3, 2 \rangle$

4. $\langle 6, -3 \rangle \cdot \langle 2, 1 \rangle$

5. $\langle -7, -4 \rangle \cdot \langle -2, -7 \rangle$

6. $\langle 5, -2 \rangle \cdot \langle -1, -1 \rangle$

7. $\langle \sqrt{3}, -2 \rangle \cdot \langle 3\sqrt{3}, -1 \rangle$

8. $\langle 4\sqrt{2}, \sqrt{7} \rangle \cdot \langle -\sqrt{2}, -\sqrt{7} \rangle$

9. $\langle 5, a \rangle \cdot \langle -3a, 2 \rangle$

10. $\langle 4x, 3y \rangle \cdot \langle 2y, -5x \rangle$

11. $\langle 0.8, -0.5 \rangle \cdot \langle 2, 6 \rangle$

12. $\langle -18, 3 \rangle \cdot \langle 10, -300 \rangle$

In Exercises 13–24, find the angle (round to the nearest degree) between each pair of vectors.

13. $\langle -4, 3 \rangle$ and $\langle -5, -9 \rangle$

14. $\langle 2, -4 \rangle$ and $\langle 4, -1 \rangle$

15. $\langle -2, -3 \rangle$ and $\langle -3, 4 \rangle$

16. $\langle 6, 5 \rangle$ and $\langle 3, -2 \rangle$

17. $\langle -4, 6 \rangle$ and $\langle -6, 8 \rangle$

18. $\langle 1, 5 \rangle$ and $\langle -3, -2 \rangle$

19. $\langle -2, 2\sqrt{3} \rangle$ and $\langle -\sqrt{3}, 1 \rangle$

20. $\langle -3\sqrt{3}, -3 \rangle$ and $\langle -2\sqrt{3}, 2 \rangle$

21. $\langle -5\sqrt{3}, -5 \rangle$ and $\langle \sqrt{2}, -\sqrt{2} \rangle$

22. $\langle -5, -5\sqrt{3} \rangle$ and $\langle 2, -\sqrt{2} \rangle$

23. $\langle 4, 6 \rangle$ and $\langle -6, -9 \rangle$

24. $\langle 2, 8 \rangle$ and $\langle -12, 3 \rangle$

In Exercises 25–36, determine whether each pair of vectors is orthogonal.

25. $\langle -6, 8 \rangle$ and $\langle -8, 6 \rangle$

26. $\langle 5, -2 \rangle$ and $\langle -5, 2 \rangle$

27. $\langle 6, -4 \rangle$ and $\langle -6, -9 \rangle$

28. $\langle 8, 3 \rangle$ and $\langle -6, 16 \rangle$

29. $\langle 0.8, 4 \rangle$ and $\langle 3, -6 \rangle$

30. $\langle -7, 3 \rangle$ and $\langle \frac{1}{7}, -\frac{1}{3} \rangle$

31. $\langle 5, -0.4 \rangle$ and $\langle 1.6, 20 \rangle$

32. $\langle 12, 9 \rangle$ and $\langle 3, -4 \rangle$

33. $\langle \sqrt{3}, \sqrt{6} \rangle$ and $\langle -\sqrt{2}, 1 \rangle$

34. $\langle \sqrt{7}, -\sqrt{3} \rangle$ and $\langle 3, 7 \rangle$

35. $\langle \frac{4}{3}, \frac{8}{15} \rangle$ and $\langle -\frac{1}{12}, \frac{5}{24} \rangle$

36. $\langle \frac{5}{6}, \frac{6}{7} \rangle$ and $\langle \frac{36}{25}, -\frac{49}{36} \rangle$

■ APPLICATIONS

37. Lifting Weights. How much work does it take to lift 100 pounds vertically 4 feet?

38. Lifting Weights. How much work does it take to lift 150 pounds vertically 3.5 feet?

39. Raising Wrecks. How much work is done by a crane to lift a 2-ton car to a level of 20 feet?

40. Raising Wrecks. How much work is done by a crane to lift a 2.5-ton car to a level of 25 feet?

41. Work. To slide a crate across the floor, a force of 50 pounds at a 30° angle is needed. How much work is done if the crate is dragged 30 feet?

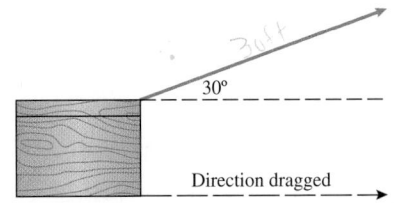

42. Work. To slide a crate across the floor, a force of 800 pounds at a 20° angle is needed. How much work is done if the crate is dragged 50 feet?

43. Close a Door. A sliding door is closed by pulling a cord with a constant force of 35 pounds at a constant angle of 45°. The door is moved 6 feet to close it. How much work is done?

44. Close a Door. A sliding door is closed by pulling a cord with a constant force of 45 pounds at a constant angle of 55°. The door is moved 6 feet to close it. How much work is done?

45. Braking Power. A car that weighs 2500 pounds is parked on a hill in San Francisco with a slant of 40° from the horizontal. How much force will keep it from rolling down the hill?

46. Towing Power. A car that weighs 2500 pounds is parked on a hill in San Francisco with a slant of 40° from the horizontal. A tow truck has to remove the car from its parking spot and move it 120 feet up the hill. How much work is required?

47. Towing Power. A semi trailer truck that weighs 40,000 pounds is parked on a hill in San Francisco with a slant of 10° from the horizontal. A tow truck has to remove the truck from its parking spot and move it 100 feet up the hill. How much work is required?

48. Braking Power. A truck that weighs 40,000 pounds is parked on a hill in San Francisco with a slant of 10° from the horizontal. How much force will keep it from rolling down the hill?

49. Business. Suppose that $\mathbf{u} = \langle 2000, 5000 \rangle$ represents the number of units of battery A and B, respectively, produced by a company and $\mathbf{v} = \langle 8.40, 6.50 \rangle$ represents the price (in dollars) of a 10-pack of battery A and B, respectively. Compute and interpret $\mathbf{u} \cdot \mathbf{v}$.

50. Demographics. Suppose that $\mathbf{u} = \langle 120, 80 \rangle$ represents the number of males and females in a high school class, and $\mathbf{v} = \langle 7.2, 5.3 \rangle$ represents the average number of minutes it takes a male and female, respectively, to register. Compute and interpret $\mathbf{u} \cdot \mathbf{v}$.

51. Geometry. Use vector methods to show that the diagonals of a rhombus are perpendicular to each other.

52. Geometry. Let $\mathbf{u}$ be a unit vector, and consider the following diagram:

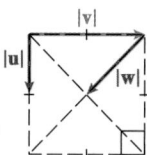

Compute $\mathbf{u} \cdot \mathbf{v}$ and $\mathbf{u} \cdot \mathbf{w}$.

53. Geometry. Consider the following diagram:

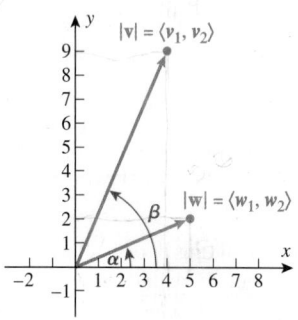

a. Compute $\cos\beta$, $\sin\beta$, $\cos\alpha$, and $\sin\alpha$.

b. Use (a) to show that $\cos(\alpha - \beta) = \dfrac{\mathbf{v} \cdot \mathbf{w}}{\sqrt{\mathbf{v} \cdot \mathbf{v}} \sqrt{\mathbf{w} \cdot \mathbf{w}}}$.

■ CATCH THE MISTAKE

In Exercises 61 and 62, explain the mistake that is made.

61. Find the dot product $\langle -3, 2 \rangle \cdot \langle 2, 5 \rangle$.

Solution:

Multiply component by component. $\quad \langle -3, 2 \rangle \cdot \langle 2, 5 \rangle = \langle (-3)(2), (2)(5) \rangle$

Simplify. $\quad \langle -3, 2 \rangle \cdot \langle 2, 5 \rangle = \langle -6, 10 \rangle$

This is incorrect. What mistake was made?

---

54. Geometry. Consider the diagram in Exercise 53.

a. Compute $\cos\beta$, $\sin\beta$, $\cos\alpha$, and $\sin\alpha$.

b. Use (a) to show that $\cos(\alpha + \beta) = \dfrac{\mathbf{v} \cdot \langle w_1, -w_2 \rangle}{\sqrt{\mathbf{v} \cdot \mathbf{v}} \sqrt{\mathbf{w} \cdot \mathbf{w}}}$.

55. Tennis. A player hits an overhead smash at full arm extension at the top of his racquet, which is 7 feet from the ground. The ball travels 16.3 feet (ignore the effects of gravity). Consult the following diagram:

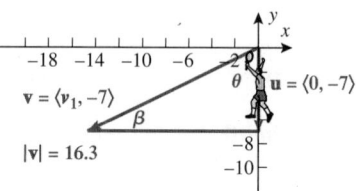

a. Determine v_1.

b. Find the angle θ with which the player hits this smash.

56. Tennis. In Exercise 55, use the dot product to determine the angle β with which the ball hits the ground.

57. Optimization. Let $\mathbf{u} = \langle a, b \rangle$ be a given vector and suppose that the head of $\mathbf{n} = \langle n_1, n_2 \rangle$ lies on the circle $x^2 + y^2 = r^2$. Find the vector $\mathbf{n}$ such that $\mathbf{u} \cdot \mathbf{n}$ is as big as possible. Find the actual value of $\mathbf{u} \cdot \mathbf{n}$ in this case.

58. Optimization. Let $\mathbf{u} = \langle a, b \rangle$ be a given vector and suppose that the head of $\mathbf{n} = \langle n_1, n_2 \rangle$ lies on the circle $x^2 + y^2 = r^2$. Find the vector $\mathbf{n}$ such that $\mathbf{u} \cdot \mathbf{n}$ is as small as possible. Find the actual value of $\mathbf{u} \cdot \mathbf{n}$ in this case.

59. Pursuit Theory. Assume that the head of $\mathbf{u}$ is restricted so that its tail is at the origin and its head is on the unit circle in quadrant II or quadrant III. A vector $\mathbf{v}$ has its tail at the origin and its head must lie on the line $y = 2 - x$ in quadrant I. Find the least value of $\mathbf{u} \cdot \mathbf{v}$.

60. Pursuit Theory. Assume that the head of $\mathbf{u}$ is restricted so that its tail is at the origin and its head is on the unit circle in quadrant I or quadrant IV. A vector $\mathbf{v}$ has its tail at the origin and its head must lie on the line $y = 2 - x$ in quadrant I. Find the largest value of $\mathbf{u} \cdot \mathbf{v}$.

62. Find the dot product $\langle 11, 12 \rangle \cdot \langle -2, 3 \rangle$.

Solution:

Multiply the outer and inner components.

$$\langle 11, 12 \rangle \cdot \langle -2, 3 \rangle = (11)(3) + (12)(-2)$$

Simplify. $\quad \langle 11, 12 \rangle \cdot \langle -2, 3 \rangle = 9$

This is incorrect. What mistake was made?

■ CONCEPTUAL

In Exercises 63–66, determine whether each statement is true or false.

63. A dot product of two vectors is a vector.

64. A dot product of two vectors is a scalar.

65. Orthogonal vectors have a dot product equal to zero.

66. If the dot product of two nonzero vectors is equal to zero, then the vectors must be perpendicular.

For Exercises 67 and 68, refer to the following to find the dot product:

The dot product of vectors with n components is

$$\langle a_1, a_2, \ldots, a_n \rangle \cdot \langle b_1, b_2, \ldots, b_n \rangle = a_1 b_1 + a_2 b_2 + \cdots + a_n b_n.$$

67. $\langle 3, 7, -5 \rangle \cdot \langle -2, 4, 1 \rangle$

68. $\langle 1, 0, -2, 3 \rangle \cdot \langle 5, 2, 3, 1 \rangle$

In Exercises 69–72, given $u = \langle a, b \rangle$ and $v = \langle c, d \rangle$, show that the following properties are true:

69. $\mathbf{u} \cdot \mathbf{v} = \mathbf{v} \cdot \mathbf{u}$

70. $\mathbf{u} \cdot \mathbf{u} = |\mathbf{u}|^2$

71. $\mathbf{0} \cdot \mathbf{u} = 0$

72. $k(\mathbf{u} \cdot \mathbf{v}) = (k\mathbf{u}) \cdot \mathbf{v} = \mathbf{u} \cdot (k\mathbf{v})$, k is a scalar

■ CHALLENGE

73. Show that $\mathbf{u} \cdot (\mathbf{v} + \mathbf{w}) = \mathbf{u} \cdot \mathbf{v} + \mathbf{u} \cdot \mathbf{w}$.

74. Show that $|\mathbf{u} - \mathbf{v}|^2 = |\mathbf{u}|^2 + |\mathbf{v}|^2 - 2(\mathbf{u} \cdot \mathbf{v})$.

75. The *projection of* **v** *onto* **u** is defined by $\text{proj}_\mathbf{u} \, \mathbf{v} = \left(\dfrac{\mathbf{u} \cdot \mathbf{v}}{|\mathbf{u}|^2} \right) \mathbf{u}$.

This vector is depicted below. Heuristically, this is the "shadow" of **v** on **u**.

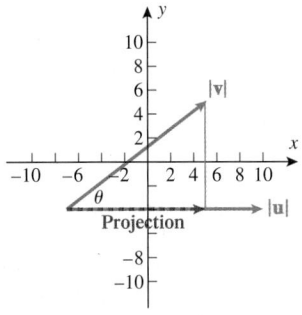

a. Compute $\text{proj}_\mathbf{u} \, 2\mathbf{u}$.

b. What is $\text{proj}_\mathbf{u} \, c\mathbf{u}$ for any $c > 0$?

76. a. Compute $\text{proj}_\mathbf{u} \, 2\mathbf{u}$.

b. What is $\text{proj}_\mathbf{u} \, c\mathbf{u}$ for any $c > 0$?

77. Suppose that you are given a vector **u**. For what vectors **v** does $\text{proj}_\mathbf{u} \, \mathbf{v} = \mathbf{0}$?

78. True or false: $\text{proj}_\mathbf{u} (\mathbf{v} + \mathbf{w}) = \text{proj}_\mathbf{u} \, \mathbf{v} + \text{proj}_\mathbf{u} \, \mathbf{w}$.

79. If **u** and **v** are unit vectors, determine the maximum and minimum value of $(-2\mathbf{u}) \cdot (3\mathbf{v})$.

80. Assume that the angle between **u** and **v** is $\theta = \dfrac{\pi}{3}$. Show that

$$\frac{(\mathbf{u} \cdot \mathbf{v})\mathbf{u}}{|\mathbf{v}|} - \frac{(\mathbf{v} \cdot \mathbf{u})\mathbf{v}}{|\mathbf{u}|} = \frac{|\mathbf{u}|\mathbf{u} - |\mathbf{v}|\mathbf{v}}{2}.$$

■ TECHNOLOGY

For Exercises 81 and 82, find the indicated dot product with a calculator.

81. $\langle -11, 34 \rangle \cdot \langle 15, -27 \rangle$

82. $\langle 23, -350 \rangle \cdot \langle 45, 202 \rangle$

83. A rectangle has sides with lengths 18 units and 11 units. Find the angle to one decimal place between the diagonal and the side with length of 18 units. (*Hint:* Set up a rectangular coordinate system, and use vectors $\langle 18, 0 \rangle$ to represent the side of length 18 units and $\langle 18, 11 \rangle$ to represent the diagonal.)

84. The definition of a dot product and the formula to find the angle between two vectors can be extended and applied to vectors with more than two components. A rectangular box has sides with lengths 12 feet, 7 feet, and 9 feet. Find the angle, to the nearest degree, between the diagonal and the side with length 7 feet.

Use the graphing calculator $\boxed{\text{SUM}}$ command to find the angle (round to the nearest degree) between each pair of vectors.

85. $\langle -25, 42 \rangle, \langle 10, 35 \rangle$

86. $\langle -12, 9 \rangle, \langle -21, -13 \rangle$

There is a branch of calculus devoted to the study of vector-valued functions; these are functions that map real numbers onto vectors. For example, $\mathbf{v}(t) = \langle t, 2t \rangle$.

87. Calculate the dot product of the vector-valued functions $\mathbf{u}(t) = \langle 2t, t^2 \rangle$ and $\mathbf{v}(t) = \langle t, -3t \rangle$.

88. Calculate the dot product of the vector-valued functions $\mathbf{u}(t) = \langle \cos t, \sin t \rangle$ and $\mathbf{v}(t) = \langle \cos t, -\sin t \rangle$.

89. Find the angle between the vector-valued functions
$$\mathbf{u}(t) = \langle \sin t, \cos t \rangle \text{ and } \mathbf{v}(t) = \langle \csc t, -\cos t \rangle \text{ when } t = \frac{\pi}{6}.$$

90. Find the values of t that make the vector-valued functions $\mathbf{u}(t) = \langle \sin t, \sin t \rangle$ and $\mathbf{v}(t) = \langle \cos t, -\sin t \rangle$ orthogonal.

SECTION
7.3
POLAR (TRIGONOMETRIC) FORM OF COMPLEX NUMBERS

SKILLS OBJECTIVES

- Graph a point in the complex plane.
- Convert complex numbers from rectangular form to polar form.
- Convert complex numbers from polar form to rectangular form.

CONCEPTUAL OBJECTIVES

- Understand that a complex number can be represented either in rectangular or polar form.
- Relate the horizontal axis in the complex plane to the real component of a complex number.
- Relate the vertical axis in the complex plane to the imaginary component of a complex number.

Complex Numbers in Rectangular Form

We are already familiar with the **rectangular coordinate system**, where the horizontal axis is called the x-axis and the vertical axis is called the y-axis. In our study of complex numbers, we refer to the **standard (rectangular) form** as $a + bi$, where a represents the real part and b represents the imaginary part. If we let the horizontal axis be the **real axis** and the vertical axis be the **imaginary axis**, the result is the **complex plane**. The point $a + bi$ is located in the complex plane by finding the coordinates (a, b).

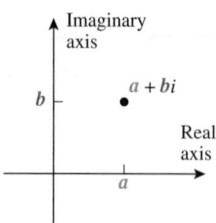

When $b = 0$, the result is a real number, and therefore any numbers along the horizontal axis are real numbers. When $a = 0$, the result is an imaginary number, so any numbers along the vertical axis are imaginary numbers.

The variable z is often used to represent a complex number: $z = x + iy$. Complex numbers are analogous to vectors. Suppose we define a vector $\mathbf{z} = \langle x, y \rangle$, whose initial point is the origin and whose terminal point is (x, y); then the magnitude of that vector would be $|\mathbf{z}| = \sqrt{x^2 + y^2}$. Similarly, the magnitude, or *modulus*, of a complex number is defined like the magnitude of a position vector in the xy-plane, as the distance from the origin $(0, 0)$ to the point (x, y) in the complex plane.

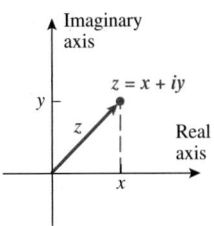

DEFINITION   Modulus of a Complex Number

The **modulus**, or magnitude, of a complex number $z = x + iy$ is the distance from the origin to the point (x, y) in the complex plane given by

$$|z| = \sqrt{x^2 + y^2}$$

Recall that a complex number $z = x + iy$ has a complex conjugate $\bar{z} = x - iy$. The bar above a complex number denotes its conjugate. Notice that

$$z\bar{z} = (x + iy)(x - iy) = x^2 - i^2 y^2 = x^2 + y^2$$

and therefore the modulus can also be written as

$$|z| = \sqrt{z\bar{z}}$$

EXAMPLE 1   Finding the Modulus of a Complex Number

Find the modulus of $z = -3 + 2i$.

COMMON MISTAKE

Including the i in the imaginary part.

★ **CORRECT**

Let $x = -3$ and $y = 2$ in
$|z| = \sqrt{x^2 + y^2}$.

$$|-3 + 2i| = \sqrt{(-3)^2 + 2^2}$$

Eliminate the parentheses.

$$|-3 + 2i| = \sqrt{9 + 4}$$

Simplify.

$$\boxed{|z| = |-3 + 2i| = \sqrt{13}}$$

✖ **INCORRECT**

Let $x = -3$ and $y = 2i$ **ERROR**

$$|-3 + 2i| = \sqrt{(-3)^2 + (2i)^2}$$

The i is not included in the formula. Only the imaginary part (the coefficient of i) is used.

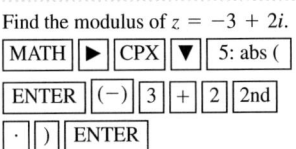
■ **YOUR TURN** Find the modulus of $z = 2 - 5i$.

■ Answer: $|z| = |2 - 5i| = \sqrt{29}$

Complex Numbers in Polar Form

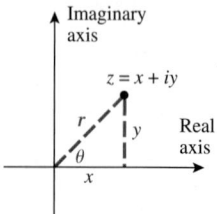

Imaginary
axis

$z = x + iy$

Real
axis

We say that a complex number $z = x + iy$ is in *rectangular* form because it is located at the point (x, y), which is expressed in rectangular coordinates, in the complex plane. Another convenient way of expressing complex numbers is in *polar* form. Recall from our study of vectors (Section 7.1) that vectors have both magnitude and a direction angle. The same is true of numbers in the complex plane. Let r represent the magnitude, or distance from the origin to the point (x, y), and θ represent the direction angle; then we have the following relationships:

$$r = \sqrt{x^2 + y^2}$$

$$\sin\theta = \frac{y}{r} \qquad \cos\theta = \frac{x}{r} \qquad \text{and} \qquad \tan\theta = \frac{y}{x} \quad (x \neq 0)$$

Isolating x and y in the sinusoidal functions, we find

$$x = r\cos\theta \qquad y = r\sin\theta$$

When we use these expressions for x and y, a complex number can be written in *polar* form.

$$z = x + yi = (r\cos\theta) + (r\sin\theta)i = r(\cos\theta + i\sin\theta)$$

POLAR (TRIGONOMETRIC) FORM OF COMPLEX NUMBERS

The following expression is the **polar form** of a complex number:

$$z = r(\cos\theta + i\sin\theta)$$

where r represents the **modulus** (magnitude) of the complex number and θ represents the **argument** of z.

The following is standard notation for modulus and argument:

$$r = \text{mod}\,z = |z|$$
$$\theta = \text{Arg}\,z, \qquad 0 \leq \theta < 2\pi$$

Converting Complex Numbers Between Rectangular and Polar Forms

We can convert back and forth between rectangular and polar (trigonometric) forms of complex numbers using the modulus and trigonometric ratios.

$$r = \sqrt{x^2 + y^2} \qquad \sin\theta = \frac{y}{r} \qquad \cos\theta = \frac{x}{r} \qquad \text{and} \qquad \tan\theta = \frac{y}{x} \quad (x \neq 0)$$

CONVERTING COMPLEX NUMBERS FROM RECTANGULAR FORM TO POLAR FORM

Step 1: Plot the point $z = x + iy$ in the complex plane (note the quadrant).

Step 2: Find r. Use $r = \sqrt{x^2 + y^2}$.

Step 3: Find θ. Apply $\tan\theta = \frac{y}{x}$, $x \neq 0$, where θ is in the quadrant found in Step 1.

Step 4: Write the complex number in polar form: $z = r(\cos\theta + i\sin\theta)$.

Realize that imaginary numbers, $x = 0$ and $z = bi$, lie on the imaginary axis. Therefore, $\theta = 90°$ if $b > 0$ and $\theta = 270°$ if $b < 0$.

EXAMPLE 2 Converting from Rectangular to Polar Form

Express the complex number $z = \sqrt{3} - i$ in polar form.

Solution:

STEP 1 Plot the point.

The point lies in QIV.

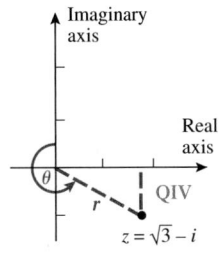

STEP 2 Find r.

Let $x = \sqrt{3}$ and $y = -1$
in $r = \sqrt{x^2 + y^2}$. $\qquad r = \sqrt{(\sqrt{3})^2 + (-1)^2}$

Eliminate the parentheses. $\qquad r = \sqrt{3 + 1}$

Simplify. $\qquad \boxed{r = 2}$

STEP 3 Find θ.

Let $x = \sqrt{3}$ and $y = -1$
in $\tan\theta = \dfrac{y}{x}$. $\qquad \tan\theta = -\dfrac{1}{\sqrt{3}}$

Solve for θ $\qquad \theta = \tan^{-1}\left(-\dfrac{1}{\sqrt{3}}\right) = -\dfrac{\pi}{6}$

Find the reference angle. $\qquad$ reference angle $= \dfrac{\pi}{6}$

The complex number lies in QIV. $\qquad \boxed{\theta = \dfrac{11\pi}{6}}$

STEP 4 Write the complex number in polar form. $\qquad \boxed{z = 2\left[\cos\left(\dfrac{11\pi}{6}\right) + i\sin\left(\dfrac{11\pi}{6}\right)\right]}$
$z = r(\cos\theta + i\sin\theta)$

Note: An alternative form is in degrees: $z = 2(\cos 330° + i\sin 330°)$.

....

■ **YOUR TURN** Express the complex number $z = 1 - i\sqrt{3}$ in polar form.

Technology Tip

Express the complex number $z = \sqrt{3} - i$ in polar form.

Method I: Use $\tan^{-1}\left(\dfrac{1}{\sqrt{3}}\right)$ to find the reference angle for θ, which is in QIV.

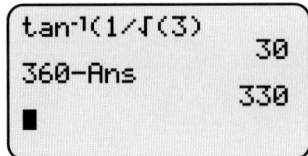

Method II: Use the $\boxed{\text{angle (}}$ feature on the calculator to find θ. You still have to find the actual angle in QIV. Press

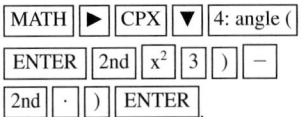

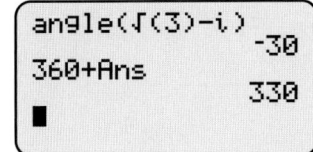

■ **Answer:**

$z = 2\left[\cos\left(\dfrac{5\pi}{3}\right) + i\sin\left(\dfrac{5\pi}{3}\right)\right]$ or

$2(\cos 300° + i\sin 300°)$

You must be very careful in converting from rectangular to polar form. Remember that the inverse tangent function is a one-to-one function and will yield values in quadrants I and IV. If the point lies in quadrant II or III, add 180° to the angle found through the inverse tangent function.

EXAMPLE 3   Converting from Rectangular to Polar Form

Technology Tip

Express the complex number $z = -2 + i$ in polar form.

```
abs(-2+i)
         2.236067977
√(5)
         2.236067977
angle(-2+i)
         153.4349488
```

COMMON MISTAKE

Forgetting to confirm the quadrant, which results in using the reference angle instead of the actual angle.

Express the complex number $z = -2 + i$ in polar form.

⭐ **CORRECT**   ❌ **INCORRECT**

Step 1: Plot the point.

The point lies in QII.

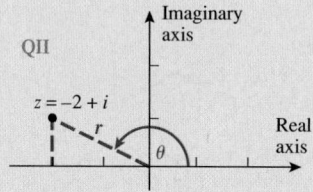

Step 2: Find r.

Let $x = -2$ and $y = 1$ in
$r = \sqrt{x^2 + y^2}$.

$$r = \sqrt{(-2)^2 + 1^2}$$

Simplify.

$$\boxed{r = \sqrt{5}}$$

Step 3: Find θ.

Let $x = -2$ and $y = 1$ in
$\tan \theta = \dfrac{y}{x}$.

$$\tan \theta = -\frac{1}{2}$$

$$\theta = \tan^{-1}\left(-\frac{1}{2}\right)$$

$$= -26.565°$$

The complex number lies in
QII.

$$\boxed{\theta = -26.6° + 180° = 153.4°}$$

Evaluate the inverse function with a calculator.

$$\theta = \tan^{-1}\left(-\frac{1}{2}\right) = -26.565°$$

Write the complex number in polar form.

$z = r(\cos \theta + i \sin \theta)$
$z = \sqrt{5}[\cos(-26.6°) + i \sin(-26.6°)]$

$\theta = -26.565°$ lies in QIV, whereas the original point lies in QII. Therefore, we should have added 180° to θ in order to arrive at a point in QII.

Step 4: Write the complex number in
polar form $z = r(\cos \theta + i \sin \theta)$.

$$\boxed{z = \sqrt{5}(\cos 153.4° + i \sin 153.4°)}$$

■ **Answer:**
$z = \sqrt{5}(\cos 116.6° + i \sin 116.6°)$

■ **YOUR TURN** Express the complex number $z = -1 + 2i$ in polar form.

To convert from polar to rectangular form, simply evaluate the trigonometric functions.

EXAMPLE 4 Converting from Polar to Rectangular Form

Express $z = 4(\cos 120° + i\sin 120°)$ in rectangular form.

Solution:

Evaluate the trigonometric functions exactly.

$$z = 4\left(\underbrace{\cos 120°}_{-\frac{1}{2}} + i\underbrace{\sin 120°}_{\frac{\sqrt{3}}{2}}\right)$$

Distribute the 4.

$$z = 4\left(-\frac{1}{2}\right) + 4\left(\frac{\sqrt{3}}{2}\right)i$$

Simplify.

$$\boxed{z = -2 + 2\sqrt{3}i}$$

■ **YOUR TURN** Express $z = 2(\cos 210° + i\sin 210°)$ in rectangular form.

Technology Tip

Express $z = 4(\cos 120° + i\sin 120°)$ in rectangular form.

```
4cos(120)
                    -2
4sin(120)
          3.464101615
2√(3)
          3.464101615
```

■ **Answer:** $z = -\sqrt{3} - i$

EXAMPLE 5 Using a Calculator to Convert from Polar to Rectangular Form

Express $z = 3(\cos 109° + i\sin 109°)$ in rectangular form. Round to four decimal places.

Solution:

Use a calculator to evaluate the trigonometric functions.

$$z = 3\left(\underbrace{\cos 109°}_{-0.325568} + i\underbrace{\sin 109°}_{0.945519}\right)$$

Simplify.

$$\boxed{z = -0.9767 + 2.8366i}$$

■ **YOUR TURN** Express $z = 7(\cos 217° + i\sin 217°)$ in rectangular form. Round to four decimal places.

Technology Tip

Express $z = 4(\cos 109° + i\sin 109°)$ in rectangular form.

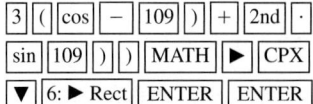

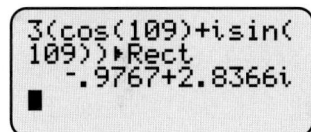

```
3(cos(109)+isin(
109))▶Rect
     -.9767+2.8366i
■
```

■ **Answer:** $z = -5.5904 - 4.2127i$

SECTION 7.3 SUMMARY

In the complex plane, the horizontal axis is the real axis and the vertical axis is the imaginary axis. We can express complex numbers in either rectangular or polar form:

rectangular form: $z = x + iy$

or

polar form: $z = r(\cos\theta + i\sin\theta)$

The modulus of a complex number, $z = x + iy$, is given by

$$|z| = \sqrt{x^2 + y^2}$$

To convert from rectangular to polar form, we use the relationships

$$r = \sqrt{x^2 + y^2} \quad \text{and} \quad \tan\theta = \frac{y}{x}, x \neq 0 \text{ and } 0 \leq \theta < 2\pi$$

It is important to note in which quadrant the point lies. To convert from polar to rectangular form, simply evaluate the trigonometric functions.

$$x = r\cos\theta \quad \text{and} \quad y = r\sin\theta$$

SECTION 7.3 EXERCISES

■ SKILLS

In Exercises 1–8, graph each complex number in the complex plane.

1. $7 + 8i$

2. $3 + 5i$

3. $-2 - 4i$

4. $-3 - 2i$

5. 2

6. 7

7. $-3i$

8. $-5i$

In Exercises 9–24, express each complex number in polar form.

9. $1 - i$

10. $2 + 2i$

11. $1 + \sqrt{3}i$

12. $-3 - \sqrt{3}i$

13. $-4 + 4i$

14. $\sqrt{5} - \sqrt{5}i$

15. $\sqrt{3} - 3i$

16. $-\sqrt{3} + i$

17. $3 + 0i$

18. $-2 + 0i$

19. $-\frac{1}{2} - \frac{1}{2}i$

20. $\frac{1}{6} - \frac{1}{6}i$

21. $-\sqrt{6} - \sqrt{6}i$

22. $\frac{1}{3} - \frac{1}{3}i$

23. $-5 + 5i$

24. $3 + 3i$

In Exercises 25–40, use a calculator to express each complex number in polar form.

25. $3 - 7i$

26. $2 + 3i$

27. $-6 + 5i$

28. $-4 - 3i$

29. $-5 + 12i$

30. $24 + 7i$

31. $8 - 6i$

32. $-3 + 4i$

33. $-\frac{1}{2} + \frac{3}{4}i$

34. $-\frac{5}{8} - \frac{11}{4}i$

35. $5.1 + 2.3i$

36. $1.8 - 0.9i$

37. $-2\sqrt{3} - \sqrt{5}i$

38. $-\frac{4\sqrt{5}}{3} + \frac{\sqrt{5}}{2}i$

39. $4.02 - 2.11i$

40. $1.78 - 0.12i$

In Exercises 41–52, express each complex number in rectangular form.

41. $5(\cos 180° + i\sin 180°)$

42. $2(\cos 135° + i\sin 135°)$

43. $2(\cos 315° + i\sin 315°)$

44. $3(\cos 270° + i\sin 270°)$

45. $-4(\cos 60° + i\sin 60°)$

46. $-4(\cos 210° + i\sin 210°)$

47. $\sqrt{3}(\cos 150° + i\sin 150°)$

48. $\sqrt{3}(\cos 330° + i\sin 330°)$

49. $\sqrt{2}\left[\cos\left(\frac{\pi}{4}\right) + i\sin\left(\frac{\pi}{4}\right)\right]$

50. $2\left[\cos\left(\frac{5\pi}{6}\right) + i\sin\left(\frac{5\pi}{6}\right)\right]$

51. $6\left[\cos\left(\frac{3\pi}{4}\right) + i\sin\left(\frac{3\pi}{4}\right)\right]$

52. $4\left[\cos\left(\frac{11\pi}{6}\right) + i\sin\left(\frac{11\pi}{6}\right)\right]$

In Exercises 53–64, use a calculator to express each complex number in rectangular form.

53. $5(\cos 295° + i\sin 295°)$

54. $4(\cos 35° + i\sin 35°)$

55. $3(\cos 100° + i\sin 100°)$

56. $6(\cos 250° + i\sin 250°)$

57. $-7(\cos 140° + i\sin 140°)$

58. $-5(\cos 320° + i\sin 320°)$

59. $3\left[\cos\left(\frac{11\pi}{12}\right) + i\sin\left(\frac{11\pi}{12}\right)\right]$

60. $2\left[\cos\left(\frac{4\pi}{7}\right) + i\sin\left(\frac{4\pi}{7}\right)\right]$

61. $-2\left[\cos\left(\frac{3\pi}{5}\right) + i\sin\left(\frac{3\pi}{5}\right)\right]$

62. $-4\left[\cos\left(\frac{15\pi}{11}\right) + i\sin\left(\frac{15\pi}{11}\right)\right]$

63. $-5\left[\cos\left(\frac{4\pi}{9}\right) + i\sin\left(\frac{4\pi}{9}\right)\right]$

64. $6\left[\cos\left(\frac{13\pi}{8}\right) + i\sin\left(\frac{13\pi}{8}\right)\right]$

■ **APPLICATIONS**

65. Road Construction. Engineers are planning the construction of a bypass in a north-south highway to connect cities B and C.

a. What is the distance from A to C?

b. Write the vector AC as a complex number in polar form. (Use degrees for the angle.)

c. What is the angle BAC?

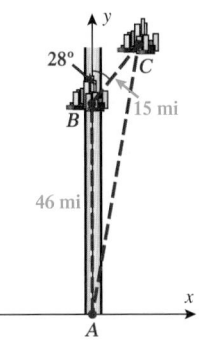

66. Road Construction. Engineers are planning the construction of a bypass in a north-south highway to connect cities B and C.

a. What is the distance from A to C?

b. Write the vector AC as a complex number in polar form. (Use degrees for the angle.)

c. What is the angle BAC?

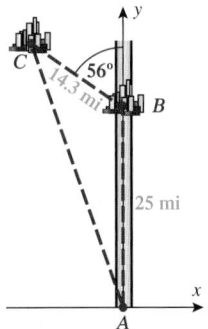

67. City Map Barcelona. The city of Barcelona is crossed by the Diagonal Avenue as shown in the map.

a. Which complex numbers, in rectangular form, represent the street segment A-B, B-C, and C-D?

b. Which complex number, in polar form, represents A-D?

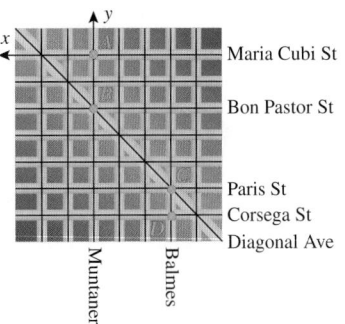

68. City Map Washington, D.C. A simplified map of Washington, D.C., is shown below.

a. Which complex numbers, in rectangular form, represent the street segment A-B, B-C, and C-D?

b. Which complex number, in polar form, represents A-D?

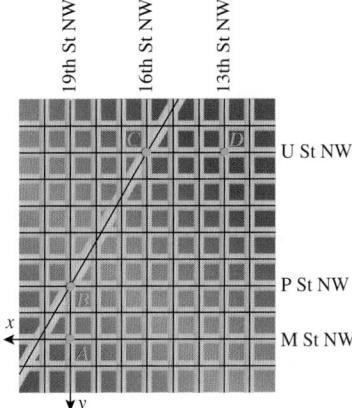

For Exercises 69 and 70, refer to the following:

In the design of AC circuits, the voltage across a resistance is regarded as a real number. When the voltage goes across an inductor or a capacitor, it is considered an imaginary number: positive in the inductor case and negative in the capacitor case. The impedance results from the combination of the voltages in the circuit and is given by the formula

$$z = |z|(\cos\theta + i\sin\theta), \text{ where } |z| = \sqrt{R^2 + I^2}$$

$$\text{and } \theta = \tan^{-1}\left(\frac{I}{R}\right)$$

where z is impedance, R is resistance, and I is inductance.

69. AC Circuits. Find the impedance of a circuit with resistance 4 ohms and inductor of 6 ohms. Write your answer in polar form.

70. AC Circuits. Find the impedance of a circuit with resistance 7 ohms and capacitor of 5 ohms. Write your answer in polar form.

■ CATCH THE MISTAKE

In Exercises 71 and 72, explain the mistake that is made.

71. Express $z = -3 - 8i$ in polar form.

Solution:

Find r. $r = \sqrt{x^2 + y^2} = \sqrt{9 + 64} = \sqrt{73}$

Find θ. $\tan\theta = \dfrac{8}{3}$

$$\theta = \tan^{-1}\left(\dfrac{8}{3}\right) = 69.44°$$

Write the complex number in polar form.

$$z = \sqrt{73}(\cos 69.44° + i\sin 69.44°)$$

This is incorrect. What mistake was made?

72. Express $z = -3 + 8i$ in polar form.

Solution:

Find r. $r = \sqrt{x^2 + y^2} = \sqrt{9 + 64} = \sqrt{73}$

Find θ. $\tan\theta = -\dfrac{8}{3}$

$$\theta = \tan^{-1}\left(-\dfrac{8}{3}\right) = -69.44°$$

Write the complex number in polar form.

$$z = \sqrt{73}[\cos(-69.44°) + i\sin(-69.44°)]$$

This is incorrect. What mistake was made?

■ CONCEPTUAL

In Exercises 73–76, determine whether each statement is true or false.

73. In the complex plane, any point that lies along the horizontal axis is a real number.

74. In the complex plane, any point that lies along the vertical axis is an imaginary number.

75. The modulus of z and the modulus of $\bar{z}$ are equal.

76. The argument of z and the argument of $\bar{z}$ are equal.

77. Find the argument of $z = a$, where a is a positive real number.

78. Find the argument of $z = bi$, where b is a positive real number.

79. Find the modulus of $z = bi$, where b is a negative real number.

80. Find the modulus of $z = a$, where a is a negative real number.

In Exercises 81 and 82, use a calculator to express the complex number in polar form.

81. $a - 2ai$, where $a > 0$

82. $-3a - 4ai$, where $a > 0$

■ CHALLENGE

83. Suppose that a complex number z lies on the circle $x^2 + y^2 = \pi^2$. If $\cos\left(\dfrac{\theta}{2}\right) = \dfrac{1}{2}$ and $\sin\theta < 0$, find the rectangular form of z.

84. Suppose that a complex number z lies on the circle $x^2 + y^2 = 8$. If $\sin\left(\dfrac{\theta}{2}\right) = -\dfrac{\sqrt{3}}{2}$ and $\cos\theta < 0$, find the rectangular form of z.

85. Consider the following diagram:

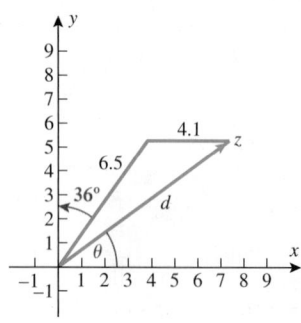

Find z in trigonometric form.
(*Hint:* Use the Law of Cosines.)

86. Consider the following diagram:

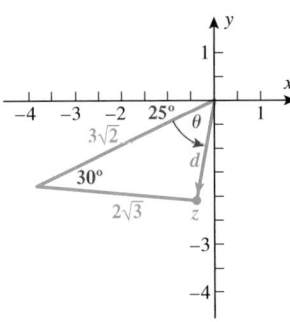

Find z in trigonometric form.
(*Hint:* Use the Law of Cosines.)

87. Consider the complex number in polar form
$z = r\cos\theta + i\sin\theta$. What is the polar form of $-z$?

88. Consider the complex number in polar form
$z = r\cos\theta + i\sin\theta$. What is the polar form of $\bar{z}$?

▪ **TECHNOLOGY**

Graphing calculators are able to convert complex numbers from rectangular to polar form using the $\boxed{\text{abs}}$ command to find the modulus and the angle command to find the angle.

89. Find abs $(1 + i)$. Find angle $(1 + i)$. Write $1 + i$ in polar form.

90. Find abs $(1 - i)$. Find angle $(1 - i)$. Write $1 - i$ in polar form.

A second way of using a graphing calculator to convert between rectangular and polar coordinates is with the $\boxed{\text{Pol}}$ and $\boxed{\text{Rec}}$ commands.

91. Find Pol(2, 1). Write $2 + i$ in polar form.

92. Find Rec(345°). Write $3(\cos 45° + i\sin 45°)$ in rectangular form.

Another way of using a graphing calculator to represent complex numbers in rectangular form is to enter the real and imaginary parts as a list of two numbers and use the $\boxed{\text{SUM}}$ command to find the modulus.

93. Write $28 - 21i$ in polar form using the $\boxed{\text{SUM}}$ command to find its modulus, and round the angle to the nearest degree.

94. Write $-\sqrt{21} + 10i$ in polar form using the $\boxed{\text{SUM}}$ command to find its modulus, and round the angle to the nearest degree.

▪ **PREVIEW TO CALCULUS**

In Exercises 95–98, refer to the following:

The use of a different system of coordinates simplifies many mathematical expressions and some calculations are performed in an easier way. The rectangular coordinates (x, y) are transformed into polar coordinates by the equations

$$x = r\cos\theta \quad \text{and} \quad y = r\sin\theta$$

where $r = \sqrt{x^2 + y^2}\,(r \neq 0)$ and $\tan\theta = \dfrac{y}{x}\,(x \neq 0)$. In polar coordinates, the equation of the unit circle $x^2 + y^2 = 1$ is just $r = 1$.

In calculus, we use polar coordinates extensively. Transform the rectangular equation to polar form.

95. $x^2 + y^2 = 25$

96. $x^2 + y^2 = 4x$

97. $y^2 - 2y = -x^2$

98. $(x^2 + y^2)^2 - 16(x^2 - y^2) = 0$

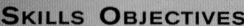

SKILLS OBJECTIVES

- Find the product of two complex numbers.
- Find the quotient of two complex numbers.
- Raise a complex number to an integer power.
- Find the nth root of a complex number.
- Solve an equation by finding complex roots.

CONCEPTUAL OBJECTIVES

- Derive the identities for products and quotients of complex numbers.
- Relate De Moivre's theorem (the power rule) for complex numbers to the product rule for complex numbers.

In this section, we will multiply complex numbers, divide complex numbers, raise complex numbers to powers, and find roots of complex numbers.

Products of Complex Numbers

First, we will derive a formula for the product of two complex numbers.

WORDS	MATH
Start with two complex numbers z_1 and z_2.	$z_1 = r_1(\cos\theta_1 + i\sin\theta_1)$ $z_2 = r_2(\cos\theta_2 + i\sin\theta_2)$
Multiply z_1 and z_2.	$z_1 z_2 = r_1 r_2(\cos\theta_1 + i\sin\theta_1)(\cos\theta_2 + i\sin\theta_2)$
Use the FOIL method to multiply the expressions in parentheses.	$z_1 z_2 = r_1 r_2(\cos\theta_1\cos\theta_2 + i\cos\theta_1\sin\theta_2 + i\sin\theta_1\cos\theta_2 + \underset{-1}{i^2}\sin\theta_1\sin\theta_2)$
Group the real parts and the imaginary parts.	$z_1 z_2 = r_1 r_2[(\cos\theta_1\cos\theta_2 - \sin\theta_1\sin\theta_2) + i(\cos\theta_1\sin\theta_2 + \sin\theta_1\cos\theta_2)]$
Apply the cosine and sine sum identities.	$z_1 z_2 = r_1 r_2\left[\underset{\cos(\theta_1+\theta_2)}{(\cos\theta_1\cos\theta_2 - \sin\theta_1\sin\theta_2)} + i\underset{\sin(\theta_1+\theta_2)}{(\cos\theta_1\sin\theta_2 + \sin\theta_1\cos\theta_2)}\right]$
Simplify.	$z_1 z_2 = r_1 r_2[\cos(\theta_1 + \theta_2) + i\sin(\theta_1 + \theta_2)]$

Study Tip

When two complex numbers are multiplied, the magnitudes are multiplied and the arguments are added.

PRODUCT OF TWO COMPLEX NUMBERS

Let $z_1 = r_1(\cos\theta_1 + i\sin\theta_1)$ and $z_2 = r_2(\cos\theta_2 + i\sin\theta_2)$ be two complex numbers. The complex product $z_1 z_2$ is given by

$$z_1 z_2 = r_1 r_2[\cos(\theta_1 + \theta_2) + i\sin(\theta_1 + \theta_2)]$$

In other words, *when multiplying two complex numbers, multiply the magnitudes and add the arguments.*

EXAMPLE 1   Multiplying Complex Numbers

Find the product of $z_1 = 3(\cos 35° + i\sin 35°)$ and $z_2 = 2(\cos 10° + i\sin 10°)$.

Solution:

Set up the product.

$$z_1 z_2 = 3(\cos 35° + i\sin 35°) \cdot 2(\cos 10° + i\sin 10°)$$

Multiply the magnitudes and add the arguments.

$$z_1 z_2 = 3 \cdot 2[\cos(35° + 10°) + i\sin(35° + 10°)]$$

Simplify.

$$z_1z_2 = 6(\cos 45° + i\sin 45°)$$

The product is in polar form. To express the product in rectangular form, evaluate the trigonometric functions.

$$z_1z_2 = 6\left(\frac{\sqrt{2}}{2} + i\frac{\sqrt{2}}{2}\right) = 3\sqrt{2} + 3i\sqrt{2}$$

Product in polar form:

$$\boxed{z_1z_2 = 6(\cos 45° + i\sin 45°)}$$

Product in rectangular form:

$$\boxed{z_1z_2 = 3\sqrt{2} + 3i\sqrt{2}}$$

■ **YOUR TURN** Find the product of $z_1 = 2(\cos 55° + i\sin 55°)$ and $z_2 = 5(\cos 65° + i\sin 65°)$. Express the answer in both polar and rectangular form.

Quotients of Complex Numbers

We now derive a formula for the quotient of two complex numbers.

WORDS	MATH
Start with two complex numbers z_1 and z_2.	$z_1 = r_1(\cos\theta_1 + i\sin\theta_1)$ $z_2 = r_2(\cos\theta_2 + i\sin\theta_2)$
Divide z_1 by z_2.	$\dfrac{z_1}{z_2} = \dfrac{r_1(\cos\theta_1 + i\sin\theta_1)}{r_2(\cos\theta_2 + i\sin\theta_2)} = \left(\dfrac{r_1}{r_2}\right)\left(\dfrac{\cos\theta_1 + i\sin\theta_1}{\cos\theta_2 + i\sin\theta_2}\right)$
Multiply the numerator and the denominator of the second expression in parentheses by the conjugate of the denominator, $\cos\theta_2 - i\sin\theta_2$.	$\dfrac{z_1}{z_2} = \left(\dfrac{r_1}{r_2}\right)\left(\dfrac{\cos\theta_1 + i\sin\theta_1}{\cos\theta_2 + i\sin\theta_2}\right)\left(\dfrac{\cos\theta_2 - i\sin\theta_2}{\cos\theta_2 - i\sin\theta_2}\right)$
Use the FOIL method to multiply the expressions in parentheses in the last two expressions.	$\dfrac{z_1}{z_2} = \left(\dfrac{r_1}{r_2}\right)\left(\dfrac{\cos\theta_1\cos\theta_2 - i^2\sin\theta_1\sin\theta_2 + i\sin\theta_1\cos\theta_2 - i\sin\theta_2\cos\theta_1}{\cos^2\theta_2 - i^2\sin^2\theta_2}\right)$
Substitute $i^2 = -1$ and group the real parts and the imaginary parts.	$\dfrac{z_1}{z_2} = \left(\dfrac{r_1}{r_2}\right)\left[\dfrac{(\cos\theta_1\cos\theta_2 + \sin\theta_1\sin\theta_2) + i(\sin\theta_1\cos\theta_2 - \sin\theta_2\cos\theta_1)}{\underbrace{\cos^2\theta_2 + \sin^2\theta_2}_{1}}\right]$
Simplify.	$\dfrac{z_1}{z_2} = \left(\dfrac{r_1}{r_2}\right)[(\cos\theta_1\cos\theta_2 + \sin\theta_1\sin\theta_2) + i(\sin\theta_1\cos\theta_2 - \sin\theta_2\cos\theta_1)]$
Use the cosine and sine difference identities.	$\dfrac{z_1}{z_2} = \left(\dfrac{r_1}{r_2}\right)\underbrace{(\cos\theta_1\cos\theta_2 + \sin\theta_1\sin\theta_2)}_{\cos(\theta_1-\theta_2)} + i\underbrace{(\sin\theta_1\cos\theta_2 - \sin\theta_2\cos\theta_1)}_{\sin(\theta_1-\theta_2)}$
Simplify.	$z_1z_2 = \dfrac{r_1}{r_2}[\cos(\theta_1 - \theta_2) + i\sin(\theta_1 - \theta_2)]$

It is important to notice that the argument difference is the argument of the numerator minus the argument of the denominator.

QUOTIENT OF TWO COMPLEX NUMBERS

Let $z_1 = r_1(\cos\theta_1 + i\sin\theta_1)$ and $z_2 = r_2(\cos\theta_2 + i\sin\theta_2)$ be two complex numbers. The complex quotient $\dfrac{z_1}{z_2}$ is given by

$$\frac{z_1}{z_2} = \frac{r_1}{r_2}[\cos(\theta_1 - \theta_2) + i\sin(\theta_1 - \theta_2)]$$

In other words, *when dividing two complex numbers, divide the magnitudes and subtract the arguments. It is important to note that the argument difference is the argument of the complex number in the numerator minus the argument of the complex number in the denominator.*

Technology Tip

Let $z_1 = 6(\cos 125° + i\sin 125°)$
and $z_2 = 3(\cos 65° + i\sin 65°)$.

Find $\dfrac{z_1}{z_2}$. Be sure to include
parentheses for z_1 and z_2.

```
(6(cos(125)+isin
(125)))/(3(cos(6
5)+isin(65)))
    1.0000+1.7321i
√(3)
           1.7321
■
```

EXAMPLE 2   Dividing Complex Numbers

Let $z_1 = 6(\cos 125° + i\sin 125°)$ and $z_2 = 3(\cos 65° + i\sin 65°)$. Find $\dfrac{z_1}{z_2}$.

Solution:

Set up the quotient.

$$\frac{z_1}{z_2} = \frac{6(\cos 125° + i\sin 125°)}{3(\cos 65° + i\sin 65°)}$$

Divide the magnitudes and subtract the arguments.

$$\frac{z_1}{z_2} = \frac{6}{3}[\cos(125° - 65°) + i\sin(125° - 65°)]$$

Simplify.

$$\frac{z_1}{z_2} = 2(\cos 60° + i\sin 60°)$$

The quotient is in polar form. To express the quotient in rectangular form, evaluate the trigonometric functions.

$$\frac{z_1}{z_2} = 2\left(\frac{1}{2} + i\frac{\sqrt{3}}{2}\right) = 1 + i\sqrt{3}$$

The quotient in polar form:

$$\boxed{\dfrac{z_1}{z_2} = 2(\cos 60° + i\sin 60°)}$$

The quotient in rectangular form:

$$\boxed{\dfrac{z_1}{z_2} = 1 + i\sqrt{3}}$$

■ **Answer:**

$\dfrac{z_1}{z_2} = 2(\cos 210° + i\sin 210°)$ or

$\dfrac{z_1}{z_2} = -\sqrt{3} - i$

■ **YOUR TURN** Let $z_1 = 10(\cos 275° + i\sin 275°)$ and $z_2 = 5(\cos 65° + i\sin 65°)$.

Find $\dfrac{z_1}{z_2}$. Express the answers in both polar and rectangular form.

When multiplying or dividing complex numbers, we have considered only those values of θ such that $0° \leq \theta \leq 360°$. When the value of θ is negative or greater than $360°$, find the coterminal angle in the interval $[0°, 360°]$.

Powers of Complex Numbers

Raising a number to a positive integer power is the same as multiplying that number by itself repeated times.

$$x^3 = x \cdot x \cdot x \qquad (a + b)^2 = (a + b)(a + b)$$

Therefore, raising a complex number to a power that is a positive integer is the same as multiplying the complex number by itself multiple times. Let us illustrate this with the complex number $z = r(\cos\theta + i\sin\theta)$, which we will raise to positive integer powers (n).

WORDS	**MATH**
Take the case $n = 2$.	$z^2 = [r(\cos\theta + i\sin\theta)][r(\cos\theta + i\sin\theta)]$
Apply the product rule (multiply the magnitudes and add the arguments).	$z^2 = r^2(\cos 2\theta + i\sin 2\theta)$
Take the case $n = 3$.	$z^3 = z^2 z = [r^2(\cos 2\theta + i\sin 2\theta)][r(\cos\theta + i\sin\theta)]$
Apply the product rule (multiply the magnitudes and add the arguments).	$z^3 = r^3(\cos 3\theta + i\sin 3\theta)$
Take the case $n = 4$.	$z^4 = z^3 z = [r^3(\cos 3\theta + i\sin 3\theta)][r(\cos\theta + i\sin\theta)]$
Apply the product rule (multiply the magnitudes and add the arguments).	$z^4 = r^4(\cos 4\theta + i\sin 4\theta)$
The pattern observed for any n is	$z^n = r^n(\cos n\theta + i\sin n\theta)$

Although we will not prove this generalized representation of a complex number raised to a power, it was proved by Abraham De Moivre, and hence its name.

DE MOIVRE'S THEOREM

If $z = r(\cos\theta + i\sin\theta)$ is a complex number, then

$$z^n = r^n(\cos n\theta + i\sin n\theta)$$

when n is a positive integer ($n \geq 1$).

In other words, when raising a complex number to a power n, raise the magnitude to the same power n and multiply the argument by n.

Although De Moivre's theorem was proved for all real numbers n, we will only use it for positive integer values of n and their reciprocals (nth roots). This is a very powerful theorem. For example, if asked to find $(\sqrt{3} - i)^{10}$, you have two choices: (1) Multiply out the expression algebraically, which we will call the long way or (2) convert to polar coordinates and use De Moivre's theorem, which we will call the short way. We will use De Moivre's theorem.

Technology Tip

Find $(\sqrt{3} + i)^{10}$ and express the answer in rectangular form.

```
(√(3)+i)^10
512-886.8100135i
(2(cos(30)+isin(
30)))^10
512-886.8100135i
```

EXAMPLE 3 **Finding a Power of a Complex Number**

Find $(\sqrt{3} + i)^{10}$ and express the answer in rectangular form.

Solution:

STEP 1 Convert to polar form. $(\sqrt{3} + i)^{10} = [2(\cos 30° + i\sin 30°)]^{10}$

STEP 2 Apply De Moivre's theorem $(\sqrt{3} + i)^{10} = [2(\cos 30° + i\sin 30°)]^{10}$
with $n = 10$.
$$= 2^{10}[\cos(10 \cdot 30°) + i\sin(10 \cdot 30°)]$$

STEP 3 Simplify. $(\sqrt{3} + i)^{10} = 2^{10}(\cos 300° + i\sin 300°)$

Evaluate 2^{10} and the sine
and cosine functions.
$$= 1024\left(\frac{1}{2} - i\frac{\sqrt{3}}{2}\right)$$

Simplify. $\boxed{= 512 - 512i\sqrt{3}}$

■ **Answer:** $-512 - 512i\sqrt{3}$

■ **YOUR TURN** Find $(1 + i\sqrt{3})^{10}$ and express the answer in rectangular form.

Roots of Complex Numbers

De Moivre's theorem is the basis for the *nth root theorem*. Before we proceed, let us motivate it with a problem: Solve $x^3 - 1 = 0$.

WORDS	MATH
Add 1 to both sides of the equation.	$x^3 = 1$
Raise both sides to the $\frac{1}{3}$ power.	$(x^3)^{1/3} = 1^{1/3}$
Simplify.	$x = 1^{1/3}$

You may think the answer is $x = 1$. There are, however, two complex solutions in addition to this real solution. Recall that a polynomial of degree n has n solutions (roots) in the complex number system. So the polynomial $P(x) = x^3 - 1$, which is degree three, has three solutions (roots). How do we find the other two solutions? We use the *nth root theorem* to find the additional cube roots of 1.

WORDS	MATH
Let z be a complex number.	$z = r(\cos\theta + i\sin\theta)$
For k = integer, θ and $\theta + 2k\pi$ are coterminal.	$z = r[\cos(\theta + 2k\pi) + i\sin(\theta + 2k\pi)]$
Apply De Moivre's theorem for $1/n$.	$z^{1/n} = r^{1/n}\left[\cos\left(\dfrac{\theta + 2k\pi}{n}\right) + i\sin\left(\dfrac{\theta + 2k\pi}{n}\right)\right]$
Simplify.	$z^{1/n} = r^{1/n}\left[\cos\left(\dfrac{\theta}{n} + \dfrac{2k\pi}{n}\right) + i\sin\left(\dfrac{\theta}{n} + \dfrac{2k\pi}{n}\right)\right]$

Notice that when $k = n$, the arguments $\dfrac{\theta}{n} + 2\pi$ and $\dfrac{\theta}{n}$ are coterminal. Therefore, to get distinct roots, $k = 0, 1, \ldots, n - 1$.

If we let z be a given complex number and w be any complex number that satisfies the relationship $z^{1/n} = w$ or $z = w^n$ where $n \geq 2$, we say that w is a **complex *n*th root** of z.

nTH ROOT THEOREM

The **nth roots** of the complex number $z = r(\cos\theta + i\sin\theta)$ are given by

$$w_k = r^{1/n}\left[\cos\left(\frac{\theta}{n} + \frac{2k\pi}{n}\right) + i\sin\left(\frac{\theta}{n} + \frac{2k\pi}{n}\right)\right] \qquad \theta \text{ in radians}$$

or

$$w_k = r^{1/n}\left[\cos\left(\frac{\theta}{n} + \frac{k \cdot 360°}{n}\right) + i\sin\left(\frac{\theta}{n} + \frac{k \cdot 360°}{n}\right)\right] \qquad \theta \text{ in degrees}$$

where $k = 0, 1, 2, \ldots, n - 1$.

EXAMPLE 4   Finding Roots of Complex Numbers

Find the three distinct cube roots of $-4 - 4i\sqrt{3}$, and plot the roots in the complex plane.

Solution:

STEP 1 Write $-4 - 4i\sqrt{3}$ in polar form. $\qquad\qquad\qquad 8(\cos 240° + i\sin 240°)$

STEP 2 Find the three cube roots.

$$w_k = r^{1/n}\left[\cos\left(\frac{\theta}{n} + \frac{k \cdot 360°}{n}\right) + i\sin\left(\frac{\theta}{n} + \frac{k \cdot 360°}{n}\right)\right]$$

$$\theta = 240°, \ r = 8, \ n = 3, \ k = 0, 1, 2$$

$k = 0$: $\qquad w_0 = 8^{1/3}\left[\cos\left(\frac{240°}{3} + \frac{0 \cdot 360°}{3}\right) + i\sin\left(\frac{240°}{3} + \frac{0 \cdot 360°}{3}\right)\right]$

Simplify. $\qquad \boxed{w_0 = 2(\cos 80° + i\sin 80°)}$

$k = 1$: $\qquad w_1 = 8^{1/3}\left[\cos\left(\frac{240°}{3} + \frac{1 \cdot 360°}{3}\right) + i\sin\left(\frac{240°}{3} + \frac{1 \cdot 360°}{3}\right)\right]$

Simplify. $\qquad \boxed{w_1 = 2(\cos 200° + i\sin 200°)}$

$k = 2$: $\qquad w_2 = 8^{1/3}\left[\cos\left(\frac{240°}{3} + \frac{2 \cdot 360°}{3}\right) + i\sin\left(\frac{240°}{3} + \frac{2 \cdot 360°}{3}\right)\right]$

Simplify. $\qquad \boxed{w_2 = 2(\cos 320° + i\sin 320°)}$

Technology Tip

Find the three distinct roots of $-4 - 4i\sqrt{3}$.

Caution: If you use a TI calculator to find $(-4 - 4i\sqrt{3})^{1/3}$, the calculator will return only one root.

```
(-4-4i√(3))^(1/3
)
         1.53-1.29i
2(cos(320)+isin(
320))
         1.53-1.29i
```

To find all three distinct roots, you need to change to polar form and apply the nth root theorem.

STEP 3 Plot the three cube roots in the complex plane.

Notice the following:

- The roots all have a magnitude of 2.

- The roots lie on a circle of radius 2.

- The roots are equally spaced around the circle ($120°$ apart).

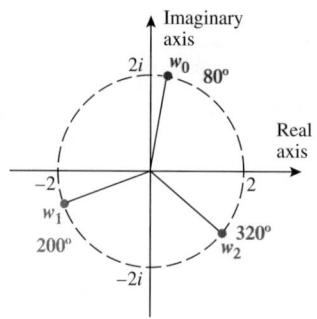

■ **Answer:**

$w_0 = 2(\cos 100° + i \sin 100°)$
$w_1 = 2(\cos 220° + i \sin 220°)$
$w_2 = 2(\cos 340° + i \sin 340°)$

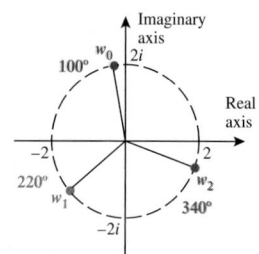

■ **YOUR TURN** Find the three distinct cube roots of $4 - 4i\sqrt{3}$, and plot the roots in the complex plane.

Solving Equations Using Roots of Complex Numbers

Let us return to solving the equation $x^3 - 1 = 0$. As stated, $x = 1$ is the real solution to this cubic equation. However, there are two additional (complex) solutions. Since we are finding the zeros of a third-degree polynomial, we expect three solutions. Furthermore, when complex solutions arise in finding the roots of polynomials, they come in conjugate pairs.

EXAMPLE 5 Solving Equations Using Complex Roots

Find all complex solutions to $x^3 - 1 = 0$.

Solution: $x^3 = 1$

STEP 1 Write 1 in polar form. $\qquad\qquad 1 = 1 + 0i = \cos 0° + i \sin 0°$

STEP 2 Find the three cube roots of 1.

$$w_k = r^{1/n}\left[\cos\left(\frac{\theta}{n} + \frac{k \cdot 360°}{n}\right) + i \sin\left(\frac{\theta}{n} + \frac{k \cdot 360°}{n}\right)\right]$$

$$r = 1, \theta = 0°, n = 3, k = 0, 1, 2$$

$k = 0:$ $\qquad w_0 = 1^{1/3}\left[\cos\left(\frac{0°}{3} + \frac{0 \cdot 360°}{3}\right) + i \sin\left(\frac{0°}{3} + \frac{0 \cdot 360°}{3}\right)\right]$

Simplify. $\qquad w_0 = \cos 0° + i \sin 0°$

$k = 1:$ $\qquad w_1 = 1^{1/3}\left[\cos\left(\frac{0°}{3} + \frac{1 \cdot 360°}{3}\right) + i \sin\left(\frac{0°}{3} + \frac{1 \cdot 360°}{3}\right)\right]$

Simplify. $\qquad w_1 = \cos 120° + i \sin 120°$

$k = 2:$ $\qquad w_2 = 1^{1/3}\left[\cos\left(\frac{0°}{3} + \frac{2 \cdot 360°}{3}\right) + i \sin\left(\frac{0°}{3} + \frac{2 \cdot 360°}{3}\right)\right]$

Simplify. $\qquad w_2 = \cos 240° + i \sin 240°$

STEP 3 Write the roots in rectangular form.

w_0:

$$w_0 = \underbrace{\cos 0°}_{1} + i \underbrace{\sin 0°}_{0} = 1$$

w_1:

$$w_1 = \underbrace{\cos 120°}_{-\frac{1}{2}} + i \underbrace{\sin 120°}_{\frac{\sqrt{3}}{2}} = -\frac{1}{2} + i\frac{\sqrt{3}}{2}$$

w_2:

$$w_2 = \underbrace{\cos 240°}_{-\frac{1}{2}} + i \underbrace{\sin 240°}_{-\frac{\sqrt{3}}{2}} = -\frac{1}{2} - i\frac{\sqrt{3}}{2}$$

STEP 4 Write the solutions to the equation $x^3 - 1 = 0$.

$$\boxed{x = 1} \qquad \boxed{x = -\frac{1}{2} + i\frac{\sqrt{3}}{2}} \qquad \boxed{x = -\frac{1}{2} - i\frac{\sqrt{3}}{2}}$$

Notice that there is one real solution and there are two (nonreal) complex solutions and that the two complex solutions are complex conjugates.

It is always a good idea to check that the solutions indeed satisfy the equation. The equation $x^3 - 1 = 0$ can also be written as $x^3 = 1$, so the check in this case is to cube the three solutions and confirm that the result is 1.

$x = 1$:

$$1^3 = 1$$

$x = -\dfrac{1}{2} + \dfrac{i\sqrt{3}}{2}$:

$$\left(-\frac{1}{2} + i\frac{\sqrt{3}}{2}\right)^3 = \left(-\frac{1}{2} + i\frac{\sqrt{3}}{2}\right)^2 \left(-\frac{1}{2} + i\frac{\sqrt{3}}{2}\right)$$

$$= \left(-\frac{1}{2} - i\frac{\sqrt{3}}{2}\right)\left(-\frac{1}{2} + i\frac{\sqrt{3}}{2}\right)$$

$$= \frac{1}{4} + \frac{3}{4}$$

$$= 1$$

$x = -\dfrac{1}{2} - \dfrac{i\sqrt{3}}{2}$:

$$\left(-\frac{1}{2} - i\frac{\sqrt{3}}{2}\right)^3 = \left(-\frac{1}{2} - i\frac{\sqrt{3}}{2}\right)^2 \left(-\frac{1}{2} - i\frac{\sqrt{3}}{2}\right)$$

$$= \left(-\frac{1}{2} + i\frac{\sqrt{3}}{2}\right)\left(-\frac{1}{2} - i\frac{\sqrt{3}}{2}\right)$$

$$= \frac{1}{4} + \frac{3}{4}$$

$$= 1$$

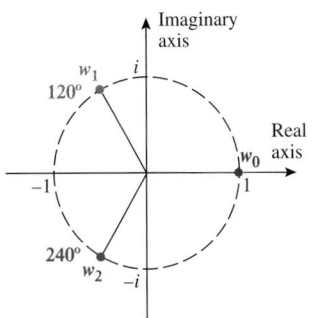

Technology Tip

The solution to the equation $x^3 - 1 = 0$ is $x = (1 + 0i)^{1/3}$.

```
abs(1)
              1.00
angle(1)
              0.00
```

```
(cos(0)+isin(0))
              1.00
(cos(0+360/3)+is
in(0+360/3))
         -.50+.87i
■
```

```
cos(0)+isin(0)
              1.00
cos(120)+isin(12
0)
         -.50+.87i
```

```
(cos(0+2*360/3)+
isin(0+2*360/3))
         -.50-.87i
■
```

```
cos(240)+isin(24
0)
         -.50-.87i
```

SECTION
7.4　SUMMARY

In this section, we multiplied and divided complex numbers and, using De Moivre's theorem, raised complex numbers to integer powers and found the nth roots of complex numbers, as follows:

Let $z_1 = r_1(\cos\theta_1 + i\sin\theta_1)$ and

$z_2 = r_2(\cos\theta_2 + i\sin\theta_2)$ be two complex numbers.

The **product** $z_1 z_2$ is given by

$$z_1 z_2 = r_1 r_2[\cos(\theta_1 + \theta_2) + i\sin(\theta_1 + \theta_2)]$$

The **quotient** $\dfrac{z_1}{z_2}$ is given by

$$\frac{z_1}{z_2} = \frac{r_1}{r_2}[\cos(\theta_1 - \theta_2) + i\sin(\theta_1 - \theta_2)]$$

Let $z = r(\cos\theta + i\sin\theta)$ be a complex number. Then for a positive integer n,

z raised to a **power** n is given by

$$z^n = r^n(\cos n\theta + i\sin n\theta)$$

The **nth roots** of z are given by

$$w_k = r^{1/n}\left[\cos\left(\frac{\theta}{n} + \frac{k \cdot 360°}{n}\right) + i\sin\left(\frac{\theta}{n} + \frac{k \cdot 360°}{n}\right)\right]$$

where $k = 0, 1, 2, \ldots, n - 1$.

SECTION
7.4　EXERCISES

▪ SKILLS

In Exercises 1–10, find the product $z_1 z_2$ and express it in rectangular form.

1. $z_1 = 4(\cos 40° + i\sin 40°)$ and $z_2 = 3(\cos 80° + i\sin 80°)$

2. $z_1 = 2(\cos 100° + i\sin 100°)$ and $z_2 = 5(\cos 50° + i\sin 50°)$

3. $z_1 = 4(\cos 80° + i\sin 80°)$ and $z_2 = 2(\cos 145° + i\sin 145°)$

4. $z_1 = 3(\cos 130° + i\sin 130°)$ and $z_2 = 4(\cos 170° + i\sin 170°)$

5. $z_1 = 2(\cos 10° + i\sin 10°)$ and $z_2 = 4(\cos 80° + i\sin 80°)$

6. $z_1 = 3(\cos 190° + i\sin 190°)$ and $z_2 = 5(\cos 80° + i\sin 80°)$

7. $z_1 = \sqrt{3}\left[\cos\left(\dfrac{\pi}{12}\right) + i\sin\left(\dfrac{\pi}{12}\right)\right]$ and $z_2 = \sqrt{27}\left[\cos\left(\dfrac{\pi}{6}\right) + i\sin\left(\dfrac{\pi}{6}\right)\right]$

8. $z_1 = \sqrt{5}\left[\cos\left(\dfrac{\pi}{15}\right) + i\sin\left(\dfrac{\pi}{15}\right)\right]$ and $z_2 = \sqrt{5}\left[\cos\left(\dfrac{4\pi}{15}\right) + i\sin\left(\dfrac{4\pi}{15}\right)\right]$

9. $z_1 = 4\left[\cos\left(\dfrac{3\pi}{8}\right) + i\sin\left(\dfrac{3\pi}{8}\right)\right]$ and $z_2 = 3\left[\cos\left(\dfrac{\pi}{8}\right) + i\sin\left(\dfrac{\pi}{8}\right)\right]$

10. $z_1 = 6\left[\cos\left(\dfrac{2\pi}{9}\right) + i\sin\left(\dfrac{2\pi}{9}\right)\right]$ and $z_2 = 5\left[\cos\left(\dfrac{\pi}{9}\right) + i\sin\left(\dfrac{\pi}{9}\right)\right]$

In Exercises 11–20, find the quotient $\dfrac{z_1}{z_2}$ and express it in rectangular form.

11. $z_1 = 6(\cos 100° + i \sin 100°)$ and $z_2 = 2(\cos 40° + i \sin 40°)$

12. $z_1 = 8(\cos 80° + i \sin 80°)$ and $z_2 = 2(\cos 35° + i \sin 35°)$

13. $z_1 = 10(\cos 200° + i \sin 200°)$ and $z_2 = 5(\cos 65° + i \sin 65°)$

14. $z_1 = 4(\cos 280° + i \sin 280°)$ and $z_2 = 4(\cos 55° + i \sin 55°)$

15. $z_1 = \sqrt{12}(\cos 350° + i \sin 350°)$ and $z_2 = \sqrt{3}(\cos 80° + i \sin 80°)$

16. $z_1 = \sqrt{40}(\cos 110° + i \sin 110°)$ and $z_2 = \sqrt{10}(\cos 20° + i \sin 20°)$

17. $z_1 = 9\left[\cos\left(\dfrac{5\pi}{12}\right) + i \sin\left(\dfrac{5\pi}{12}\right)\right]$ and $z_2 = 3\left[\cos\left(\dfrac{\pi}{12}\right) + i \sin\left(\dfrac{\pi}{12}\right)\right]$

18. $z_1 = 8\left[\cos\left(\dfrac{5\pi}{8}\right) + i \sin\left(\dfrac{5\pi}{8}\right)\right]$ and $z_2 = 4\left[\cos\left(\dfrac{3\pi}{8}\right) + i \sin\left(\dfrac{3\pi}{8}\right)\right]$

19. $z_1 = 45\left[\cos\left(\dfrac{22\pi}{15}\right) + i \sin\left(\dfrac{22\pi}{15}\right)\right]$ and $z_2 = 9\left[\cos\left(\dfrac{2\pi}{15}\right) + i \sin\left(\dfrac{2\pi}{15}\right)\right]$

20. $z_1 = 22\left[\cos\left(\dfrac{11\pi}{18}\right) + i \sin\left(\dfrac{11\pi}{18}\right)\right]$ and $z_2 = 11\left[\cos\left(\dfrac{5\pi}{18}\right) + i \sin\left(\dfrac{5\pi}{18}\right)\right]$

In Exercises 21–30, find the result of each expression using De Moivre's theorem. Write the answer in rectangular form.

21. $(-1 + i)^5$ **22.** $(1 - i)^4$ **23.** $(-\sqrt{3} + i)^6$ **24.** $(\sqrt{3} - i)^8$ **25.** $(1 - \sqrt{3}i)^4$

26. $(-1 + \sqrt{3}i)^5$ **27.** $(4 - 4i)^8$ **28.** $(-3 + 3i)^{10}$ **29.** $(4\sqrt{3} + 4i)^7$ **30.** $(-5 + 5\sqrt{3}i)^7$

In Exercises 31–40, find all *n*th roots of *z*. Write the answers in polar form, and plot the roots in the complex plane.

31. $2 - 2i\sqrt{3}$, $n = 2$ **32.** $2 + 2\sqrt{3}i$, $n = 2$ **33.** $\sqrt{18} - \sqrt{18}i$, $n = 2$ **34.** $-\sqrt{2} + \sqrt{2}i$, $n = 2$

35. $4 + 4\sqrt{3}i$, $n = 3$ **36.** $-\dfrac{27}{2} + \dfrac{27\sqrt{3}}{2}i$, $n = 3$ **37.** $\sqrt{3} - i$, $n = 3$ **38.** $4\sqrt{2} + 4\sqrt{2}i$, $n = 3$

39. $8\sqrt{2} - 8\sqrt{2}i$, $n = 4$ **40.** $-\sqrt{128} + \sqrt{128}i$, $n = 4$

In Exercises 41–56, find all complex solutions to the given equations.

41. $x^4 - 16 = 0$ **42.** $x^3 - 8 = 0$ **43.** $x^3 + 8 = 0$ **44.** $x^3 + 1 = 0$

45. $x^4 + 16 = 0$ **46.** $x^6 + 1 = 0$ **47.** $x^6 - 1 = 0$ **48.** $4x^2 + 1 = 0$

49. $x^2 + i = 0$ **50.** $x^2 - i = 0$ **51.** $x^4 - 2i = 0$ **52.** $x^4 + 2i = 0$

53. $x^5 + 32 = 0$ **54.** $x^5 - 32 = 0$ **55.** $x^7 - \pi^{14}i = 0$ **56.** $x^7 + \pi^{14} = 0$

■ APPLICATIONS

57. Complex Pentagon. When you graph the five fifth roots of $-\dfrac{\sqrt{2}}{2} - \dfrac{\sqrt{2}}{2}i$ and connect the points, you form a pentagon. Find the roots and draw the pentagon.

58. Complex Square. When you graph the four fourth roots of $16i$ and connect the points, you form a square. Find the roots and draw the square.

59. Hexagon. Compute the six sixth roots of $\dfrac{1}{2} - \dfrac{\sqrt{3}}{2}i$, and form a hexagon by connecting successive roots.

60. Octagon. Compute the eight eighth roots of $2i$, and form an octagon by connecting successive roots.

▪ CATCH THE MISTAKE

In Exercises 61–64, explain the mistake that is made.

61. Let $z_1 = 6(\cos 65° + i \sin 65°)$ and

$z_2 = 3(\cos 125° + i \sin 125°)$. Find $\dfrac{z_1}{z_2}$.

Solution:

Use the quotient formula.

$$\frac{z_1}{z_2} = \frac{r_1}{r_2}[\cos(\theta_1 - \theta_2) + i \sin(\theta_1 - \theta_2)]$$

Substitute values.

$$\frac{z_1}{z_2} = \frac{6}{3}[\cos(125° - 65°) + i \sin(125° - 65°)]$$

Simplify. $\dfrac{z_1}{z_2} = 2(\cos 60° + i \sin 60°)$

Evaluate the trigonometric functions.

$$\frac{z_1}{z_2} = 2\left(\frac{1}{2} + i\frac{\sqrt{3}}{2}\right) = 1 + i\sqrt{3}$$

This is incorrect. What mistake was made?

62. Let $z_1 = 6(\cos 65° + i \sin 65°)$ and
$z_2 = 3(\cos 125° + i \sin 125°)$. Find $z_1 z_2$.

Solution:

Write the product.

$$z_1 z_2 = 6(\cos 65° + i \sin 65°) \cdot 3(\cos 125° + i \sin 125°)$$

Multiply the magnitudes.

$$z_1 z_2 = 18(\cos 65° + i \sin 65°)(\cos 125° + i \sin 125°)$$

Multiply the cosine terms and sine terms (add the arguments).

$$z_1 z_2 = 18[\cos(65° + 125°) + i^2 \sin(65° + 125°)]$$

Simplify ($i^2 = -1$).

$$z_1 z_2 = 18(\cos 190° - \sin 190°)$$

This is incorrect. What mistake was made?

63. Find $(\sqrt{2} + i\sqrt{2})^6$.

Solution:

Raise each term to the sixth power. $\qquad (\sqrt{2})^6 + i^6(\sqrt{2})^6$

Simplify. $\qquad\qquad\qquad\qquad\qquad 8 + 8i^6$

Let $i^6 = i^4 \cdot i^2 = -1$. $\qquad\qquad 8 - 8 = 0$

This is incorrect. What mistake was made?

64. Find all complex solutions to $x^5 - 1 = 0$.

Solution:

Add 1 to both sides. $\qquad\qquad\qquad\qquad x^5 = 1$

Raise both sides to the fifth power. $\qquad x = 1^{1/5}$

Simplify. $\qquad\qquad\qquad\qquad\qquad\qquad x = 1$

This is incorrect. What mistake was made?

▪ CONCEPTUAL

In Exercises 65–70, determine whether the statement is true or false.

65. The product of two complex numbers is a complex number.

66. The quotient of two complex numbers is a complex number.

67. There are always n distinct real solutions of the equation $x^n - a = 0$, where a is not zero.

68. There are always n distinct complex solutions of the equation $x^n - a = 0$, where a is not zero.

69. There are n distinct complex zeros of $\dfrac{1}{a + bi}$, where a and b are positive real numbers.

70. There exists a complex number for which there is no complex square root.

71. The distance between any consecutive pair of the n complex roots of a number is a constant.

72. If $2\left[\cos\left(\dfrac{\pi}{2}\right) + i \sin\left(\dfrac{\pi}{2}\right)\right]$ is one of the n complex roots of a number, then n is even.

▪ CHALLENGE

In Exercises 73–76, use the following identity:

In calculus you will see an identity called Euler's formula or identity, $e^{i\theta} = \cos\theta + i\sin\theta$. Notice that when $\theta = \pi$, the identity reduces to $e^{i\pi} + 1 = 0$, which is a beautiful identity in that it relates the five fundamental numbers (e, π, 1, i, and 0) and the fundamental operations (multiplication, addition, exponents, and equality) in mathematics.

73. Let $z_1 = r_1(\cos\theta_1 + i\sin\theta_1) = r_1 e^{i\theta_1}$ and
$z_2 = r_2(\cos\theta_2 + i\sin\theta_2) = r_2 e^{i\theta_2}$ be two complex numbers. Use the properties of exponentials to show that
$z_1 z_2 = r_1 r_2[\cos(\theta_1 + \theta_2) + i\sin(\theta_1 + \theta_2)]$.

74. Let $z_1 = r_1(\cos\theta_1 + i\sin\theta_1) = r_1 e^{i\theta_1}$ and
$z_2 = r_2(\cos\theta_2 + i\sin\theta_2) = r_2 e^{i\theta_2}$ be two complex numbers.
Use the properties of exponentials to show that
$$\frac{z_1}{z_2} = \frac{r_1}{r_2}[\cos(\theta_1 - \theta_2) + i\sin(\theta_1 - \theta_2)].$$

75. Let $z = r(\cos\theta + i\sin\theta) = re^{i\theta}$. Use the properties of exponents to show that $z^n = r^n(\cos n\theta + i\sin n\theta)$.

76. Let $z = r(\cos\theta + i\sin\theta) = re^{i\theta}$. Use the properties of exponents to show that
$$w_k = r^{1/n}\left[\cos\left(\frac{\theta}{n} + \frac{2k\pi}{n}\right) + i\sin\left(\frac{\theta}{n} + \frac{2k\pi}{n}\right)\right].$$

77. Use De Moivre's theorem to prove the identity
$\cos 2\theta = \cos^2\theta - \sin^2\theta$.

78. Use De Moivre's theorem to derive an expression for $\sin 3\theta$.

79. Use De Moivre's theorem to derive an expression for $\cos 3\theta$.

80. Calculate $\dfrac{\left(\dfrac{1}{2} + \dfrac{\sqrt{3}}{2}i\right)^{14}}{\left(\dfrac{1}{2} - \dfrac{\sqrt{3}}{2}i\right)^{20}}$.

81. Calculate $(1 - i)^n \cdot (1 + i)^m$, where n and m are positive integers.

82. Calculate $\dfrac{(1 + i)^n}{(1 - i)^m}$, where n and m are positive integers.

■ **TECHNOLOGY**

For Exercises 83–88, refer to the following:

According to the nth root theorem, the first of the nth roots of the complex number $z = r(\cos\theta + i\sin\theta)$ is given by

$$w_1 = r^{1/n}\left[\cos\left(\frac{\theta}{n} + \frac{2\pi}{n}\right) + i\sin\left(\frac{\theta}{n} + \frac{2\pi}{n}\right)\right], \text{ with } \theta \text{ in radians}$$

or $w_1 = r^{1/n}\left[\cos\left(\dfrac{\theta}{n} + \dfrac{360°}{n}\right) + i\sin\left(\dfrac{\theta}{n} + \dfrac{360°}{n}\right)\right]$,

with θ in degrees.

Using the graphing calculator to plot the n roots of a complex number z, enter $r_1 = r$, $\theta \min = \dfrac{\theta}{n}$, $\theta \max = 2\pi + \dfrac{\theta}{n}$ or

$360° + \dfrac{\theta}{n}$, $\theta \text{ step} = \dfrac{2\pi}{n}$ or $\dfrac{360°}{n}$, $x\min = -r$, $x\max = r$,

$y\min = -r$, $y\max = r$, and $\boxed{\text{MODE}}$ in radians or degrees.

83. Find the fifth roots of $\dfrac{\sqrt{3}}{2} - \dfrac{1}{2}i$, and plot the roots with a calculator.

84. Find the fourth roots of $-\dfrac{\sqrt{2}}{2} + \dfrac{\sqrt{2}}{2}i$, and plot the roots with a calculator.

85. Find the sixth roots of $-\dfrac{1}{2} - \dfrac{\sqrt{3}}{2}i$, and draw the complex hexagon with a calculator.

86. Find the fifth roots of $-4 + 4i$, and draw the complex pentagon with a calculator.

87. Find the cube roots of $\dfrac{27\sqrt{2}}{2} + \dfrac{27\sqrt{2}}{2}i$, and draw the complex triangle with a calculator.

88. Find the fifth roots of $8\sqrt{2}\left(\sqrt{3} - 1\right) + 8\sqrt{2}\left(\sqrt{3} - 1\right)i$, and draw the complex triangle with a calculator.

■ **PREVIEW TO CALCULUS**

In advanced calculus, complex numbers in polar form are used extensively. Use De Moivre's formula to show that

89. $\cos(2\theta) = \cos^2\theta - \sin^2\theta$

90. $\sin(2\theta) = 2\sin\theta\cos\theta$

91. $\cos(3\theta) = 4\cos^3\theta - 3\cos\theta$

92. $\sin(3\theta) = 3\sin\theta - 4\sin^3\theta$

SKILLS OBJECTIVES

- Plot points in the polar coordinate system.
- Convert between rectangular and polar coordinates.
- Convert equations between polar form and rectangular form.
- Graph polar equations.

CONCEPTUAL OBJECTIVES

- Relate the rectangular coordinate system to the polar coordinate system.
- Classify common shapes that arise from plotting certain types of polar equations.

We have discussed the rectangular and the trigonometric (polar) form of complex numbers in the complex plane. We now turn our attention back to the familiar Cartesian plane, where the horizontal axis represents the x-variable, the vertical axis represents the y-variable, and points in this plane represent pairs of real numbers. It is often convenient to instead represent real-number plots in the *polar coordinate system*.

Polar Coordinates

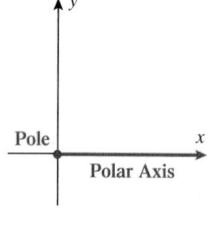

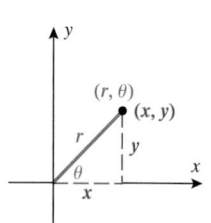

The **polar coordinate system** is anchored by a point, called the **pole** (taken to be the **origin**), and a ray with a vertex at the pole, called the **polar axis**. The polar axis is normally shown where we expect to find the positive x-axis in Cartesian coordinates.

If you align the pole with the origin on the rectangular graph and the polar axis with the positive x-axis, you can label a point either with rectangular coordinates (x, y) or with an ordered pair (r, θ) in **polar coordinates**.

Typically, polar graph paper is used that gives the angles and radii. The graph below gives the angles in radians (the angle also can be given in degrees) and shows the radii from 0 through 5.

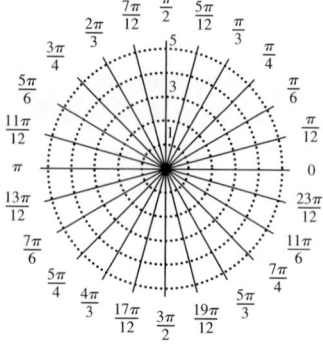

When plotting points in the polar coordinate system, $|r|$ represents the distance from the origin to the point. The following procedure guides us in plotting points in the polar coordinate system.

POINT-PLOTTING POLAR COORDINATES

To plot a point (r, θ):

1. Start on the polar axis and rotate the terminal side of an angle to the value θ.

2. If $r > 0$, the point is r units from the origin in the *same direction* of the terminal side of θ.

3. If $r < 0$, the point is $|r|$ units from the origin in the *opposite direction* of the terminal side of θ.

EXAMPLE 1   Plotting Points in the Polar Coordinate System

Plot the points in a polar coordinate system.

a. $\left(3, \dfrac{3\pi}{4}\right)$ b. $(-2, 60°)$

Solution (a):

Start by placing a pencil along the polar axis.

Rotate the pencil to the angle $\dfrac{3\pi}{4}$.

Go out (in the direction of the pencil) 3 units.

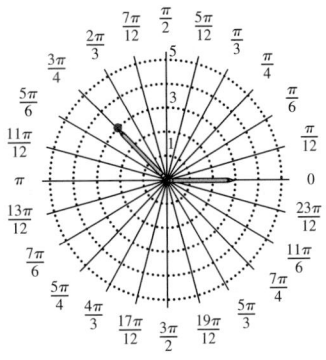

Solution (b):

Start by placing a pencil along the polar axis.

Rotate the pencil to the angle 60°.

Go out (opposite the direction of the pencil) 2 units.

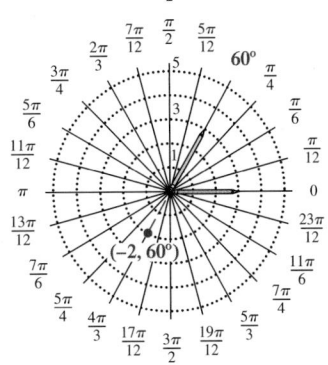

■ **Answer:**

■ **YOUR TURN**   Plot the points in the polar coordinate system.

a. $\left(-4, \dfrac{3\pi}{2}\right)$ b. $(3, 330°)$

In polar form it is important to note that (r, θ), the name of the point, is not unique, whereas in rectangular form (x, y) it is unique. For example, $(2, 30°) = (-2, 210°)$.

Converting Between Polar and Rectangular Coordinates

The relationships between polar and rectangular coordinates are the familiar ones:

$$\sin\theta = \frac{y}{r}$$

$$\cos\theta = \frac{x}{r}$$

$$\tan\theta = \frac{y}{x} \quad (x \neq 0)$$

$$r^2 = x^2 + y^2$$

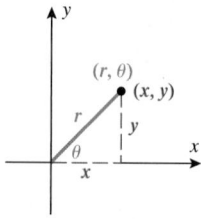

CONVERTING BETWEEN POLAR AND RECTANGULAR COORDINATES

FROM	TO	IDENTITIES	
Polar (r, θ)	Rectangular (x, y)	$x = r\cos\theta$	$y = r\sin\theta$
Rectangular (x, y)	Polar (r, θ)	$r = \sqrt{x^2 + y^2}$	$\tan\theta = \dfrac{y}{x} \quad (x \neq 0)$

EXAMPLE 2 Converting Between Polar and Rectangular Coordinates

a. Convert $\left(-1, \sqrt{3}\right)$ to polar coordinates.

b. Convert $\left(6\sqrt{2}, 135°\right)$ to rectangular coordinates.

Solution (a): $\left(-1, \sqrt{3}\right)$ lies in quadrant II.

Identify x and y. $\qquad\qquad x = -1, \qquad y = \sqrt{3}$

Find r. $\qquad\qquad r = \sqrt{x^2 + y^2} = \sqrt{(-1)^2 + \left(\sqrt{3}\right)^2} = \sqrt{4} = 2$

Find θ. $\qquad\qquad \tan\theta = \dfrac{\sqrt{3}}{-1} \quad (\theta \text{ lies in quadrant II})$

Identify θ from the unit circle. $\qquad\qquad \theta = \dfrac{2\pi}{3}$

Write the point in polar coordinates. $\qquad \boxed{\left(2, \dfrac{2\pi}{3}\right)}$

Solution (b): $\left(6\sqrt{2}, 135°\right)$ lies in quadrant II.

Identify r and θ. $\qquad\qquad r = 6\sqrt{2} \qquad \theta = 135°$

Find x. $\qquad x = r\cos\theta = 6\sqrt{2}\cos 135° = 6\sqrt{2}\left(-\dfrac{\sqrt{2}}{2}\right) = -6$

Find y. $\qquad y = r\sin\theta = 6\sqrt{2}\sin 135° = 6\sqrt{2}\left(\dfrac{\sqrt{2}}{2}\right) = 6$

Write the point in rectangular coordinates. $\qquad \boxed{(-6, 6)}$

Graphs of Polar Equations

We are familiar with equations in rectangular form such as

$$y = 3x + 5 \qquad y = x^2 + 2 \qquad x^2 + y^2 = 9$$
$$\text{(line)} \qquad\quad \text{(parabola)} \qquad\quad \text{(circle)}$$

We now discuss equations in polar form (known as **polar equations**) such as

$$r = 5\theta \qquad r = 2\cos\theta \qquad r = \sin(5\theta)$$

which you will learn to recognize in this section as typical equations whose plots are some general shapes.

Our first example deals with two of the simplest forms of polar equations: when r or θ is constant. The results are a circle centered at the origin and a line that passes through the origin, respectively.

EXAMPLE 3 **Graphing a Polar Equation of the Form $r =$ Constant or $\theta =$ Constant**

Graph the polar equations.

a. $r = 3$ **b.** $\theta = \dfrac{\pi}{4}$

Solution (a): Constant value of r

Approach 1 (polar coordinates): $\qquad\qquad r = 3$ (θ can take on any value)

> Plot points for arbitrary θ and $r = 3$.
>
> Connect the points; a circle with radius 3.

Approach 2 (rectangular coordinates): $\qquad\qquad r = 3$

> Square both sides. $\qquad\qquad r^2 = 9$
>
> Remember that in rectangular coordinates $r^2 = x^2 + y^2$. $\qquad\qquad x^2 + y^2 = 3^2$
>
> This is a circle, centered at the origin, with radius 3.

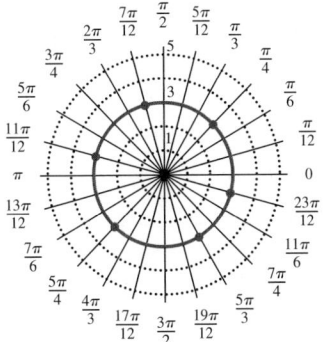

Solution (b): Constant value of θ

Approach 1: $\theta = \dfrac{\pi}{4}$ (r can take on any value, positive or negative.)

> Plot points for $\theta = \dfrac{\pi}{4}$ at several arbitrary values of r.
>
> Connect the points. The result is a line passing through the origin with slope $= 1$.

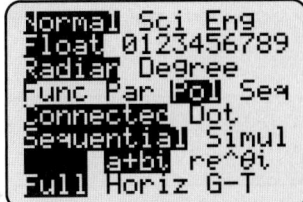
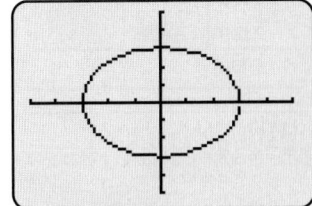
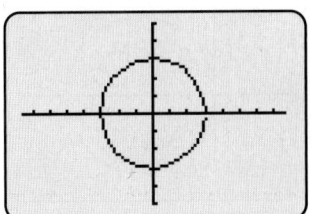

Approach 2: $\theta = \dfrac{\pi}{4}$

Take the tangent of both sides. $\tan\theta = \underbrace{\tan\left(\dfrac{\pi}{4}\right)}_{1}$

Use the identity $\tan\theta = \dfrac{y}{x}$. $\dfrac{y}{x} = 1$

Multiply by x. $y = x$

The result is a line passing through the origin with slope $= 1$.

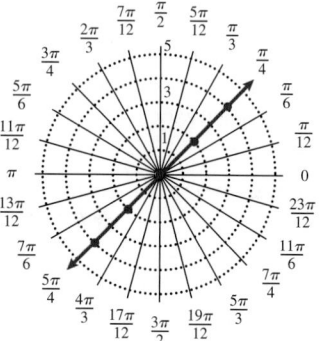

Technology Tip

Graph $r = 4\cos\theta$.

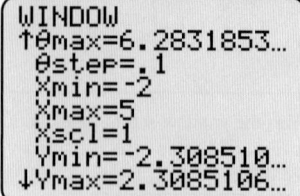

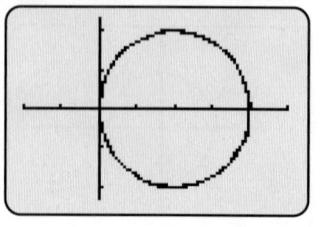

Rectangular equations that depend on varying (not constant) values of x or y can be graphed by point-plotting (making a table and plotting the points). We will use this same procedure for graphing polar equations that depend on varying (not constant) values of r or θ.

EXAMPLE 4   Graphing a Polar Equation of the Form $r = c \cdot \cos\theta$ or $r = c \cdot \sin\theta$

Graph $r = 4\cos\theta$.

Solution:

STEP 1 Make a table and find several key values.

θ	$r = 4\cos\theta$	(r, θ)
0	$4(1) = 4$	$(4, 0)$
$\dfrac{\pi}{4}$	$4\left(\dfrac{\sqrt{2}}{2}\right) \approx 2.8$	$\left(2.8, \dfrac{\pi}{4}\right)$
$\dfrac{\pi}{2}$	$4(0) = 0$	$\left(0, \dfrac{\pi}{2}\right)$
$\dfrac{3\pi}{4}$	$4\left(-\dfrac{\sqrt{2}}{2}\right) \approx -2.8$	$\left(-2.8, \dfrac{3\pi}{4}\right)$
π	$4(-1) = -4$	$(-4, \pi)$

STEP 2 Plot the points in polar coordinates.

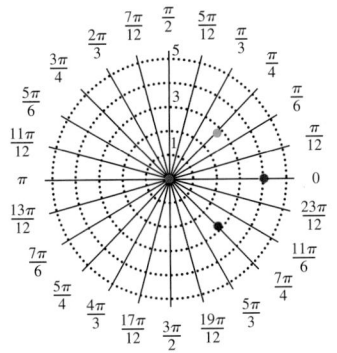

STEP 3 Connect the points with a smooth curve.

Notice that $(4, 0)$ and $(-4, \pi)$ correspond to the same point. There is no need to continue with angles beyond π, because the result would be to go around the same circle again.

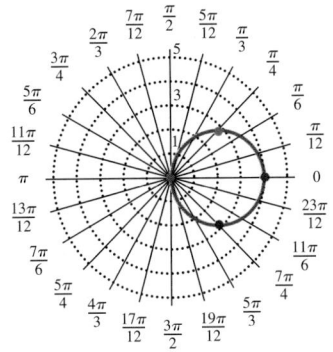

■ **Answer:**

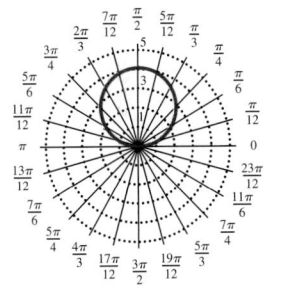

■ **YOUR TURN** Graph $r = 4 \sin \theta$.

Compare the result of Example 4, the graph of $r = 4 \cos \theta$, with the result of the Your Turn, the graph of $r = 4 \sin \theta$. Notice that they are 90° out of phase (we simply rotate one graph 90° about the pole to get the other graph).

In general, graphs of polar equations of the form $r = a \sin \theta$ and $r = a \cos \theta$ are circles.

Study Tip

Graphs of $r = a \sin \theta$ and $r = a \cos \theta$ are circles.

WORDS	**MATH**	
Start with the polar form.	$r = a \sin \theta$	$r = a \cos \theta$
Apply trigonometric ratios: $\sin \theta = \dfrac{y}{r}$ and $\cos \theta = \dfrac{x}{r}$.	$r = a \dfrac{y}{r}$	$r = a \dfrac{x}{r}$
Multiply the equations by r.	$r^2 = ay$	$r^2 = ax$
Let $r^2 = x^2 + y^2$.	$x^2 + y^2 = ay$	$x^2 + y^2 = ax$
Group x terms together and y terms together.	$x^2 + (y^2 - ay) = 0$	$(x^2 - ax) + y^2 = 0$
Complete the square on the expressions in parentheses.	$x^2 + \left[y^2 - ay + \left(\dfrac{a}{2} \right)^2 \right] = \left(\dfrac{a}{2} \right)^2$ $x^2 + \left(y - \dfrac{a}{2} \right)^2 = \left(\dfrac{a}{2} \right)^2$	$\left[x^2 - ax + \left(\dfrac{a}{2} \right)^2 \right] + y^2 = \left(\dfrac{a}{2} \right)^2$ $\left(x - \dfrac{a}{2} \right)^2 + y^2 = \left(\dfrac{a}{2} \right)^2$
Identify the center and radius.	Center: $\left(0, \dfrac{a}{2} \right)$ Radius: $\dfrac{a}{2}$	Center: $\left(\dfrac{a}{2}, 0 \right)$ Radius: $\dfrac{a}{2}$

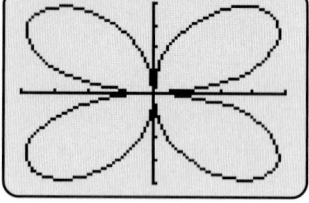
EXAMPLE 5　Graphing a Polar Equation of the Form $r = c \cdot \sin(2\theta)$ or $r = c \cdot \cos(2\theta)$

Graph $r = 5\sin(2\theta)$.

Solution:

STEP 1　Make a table and find key values. Since the argument of the sine function is doubled, the period is halved. Therefore, instead of steps of $\dfrac{\pi}{4}$, take steps of $\dfrac{\pi}{8}$.

θ	$r = 5\sin(2\theta)$	(r, θ)
0	$5(0) = 0$	$(0, 0)$
$\dfrac{\pi}{8}$	$5\left(\dfrac{\sqrt{2}}{2}\right) = 3.5$	$\left(3.5, \dfrac{\pi}{8}\right)$
$\dfrac{\pi}{4}$	$5(1) = 5$	$\left(5, \dfrac{\pi}{4}\right)$
$\dfrac{3\pi}{8}$	$5\left(\dfrac{\sqrt{2}}{2}\right) = 3.5$	$\left(3.5, \dfrac{3\pi}{8}\right)$
$\dfrac{\pi}{2}$	$5(0) = 0$	$(0, \pi)$

STEP 2　Label the polar coordinates.

The values in the table represent what happens in quadrant I. The same pattern repeats in the other three quadrants. The result is a **four-leaved rose**.

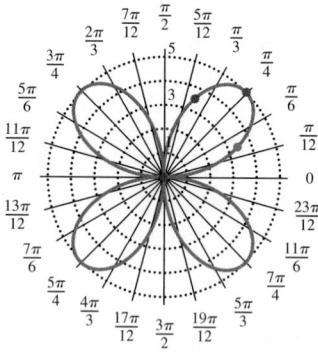

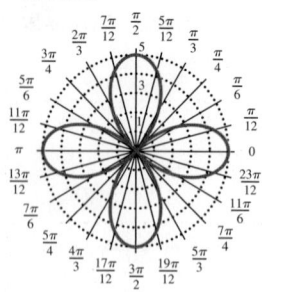
STEP 3　Connect the points with smooth curves.

■ **YOUR TURN**　Graph $r = 5\cos(2\theta)$.

Compare the result of Example 5, the graph of $r = 5\sin(2\theta)$, with the result of the Your Turn, the graph of $r = 5\cos(2\theta)$. Notice that they are 45° out of phase (we rotate one graph 45° about the pole to get the other graph).

In general, for $r = a\sin(n\theta)$ or $r = a\cos(n\theta)$, the graph has n leaves if n is odd and $2n$ leaves if n is even.

The next class of graphs are called **limaçons**, which have equations of the form $r = a \pm b\cos\theta$ or $r = a \pm b\sin\theta$. When $a = b$, the result is a **cardioid** (heart shape).

EXAMPLE 6   The Cardioid as a Polar Equation

Graph $r = 2 + 2\cos\theta$.

Solution:

STEP 1 Make a table and find key values.

This behavior repeats in QIII and QIV, because the cosine function has corresponding values in QI and QIV and in QII and QIII.

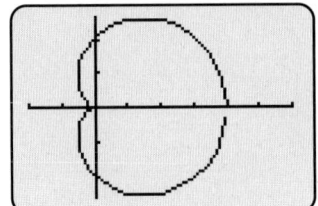
θ	$r = 2 + 2\cos\theta$	(r, θ)
0	$2 + 2(1) = 4$	$(4, 0)$
$\dfrac{\pi}{4}$	$2 + 2\left(\dfrac{\sqrt{2}}{2}\right) = 3.4$	$\left(3.4, \dfrac{\pi}{4}\right)$
$\dfrac{\pi}{2}$	$2 + 2(0) = 2$	$\left(2, \dfrac{\pi}{2}\right)$
$\dfrac{3\pi}{4}$	$2 + 2\left(-\dfrac{\sqrt{2}}{2}\right) \approx 0.6$	$\left(0.6, \dfrac{3\pi}{4}\right)$
π	$2 + 2(-1) = 0$	$(0, \pi)$

STEP 2 Plot the points in polar coordinates.

STEP 3 Connect the points with a smooth curve. The curve is a *cardioid*, a term formed from Greek roots meaning "heart-shaped."

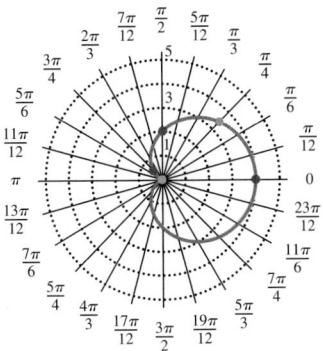

EXAMPLE 7   Graphing a Polar Equation of the Form $r = c \cdot \theta$

Graph $r = 0.5\theta$.

Solution:

STEP 1 Make a table and find key values.

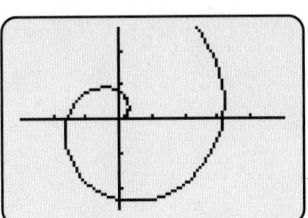
θ	$r = 0.5\theta$	(r, θ)
0	$0.5(0) = 0$	$(0, 0)$
$\dfrac{\pi}{2}$	$0.5\left(\dfrac{\pi}{2}\right) = 0.8$	$\left(0.8, \dfrac{\pi}{2}\right)$
π	$0.5(\pi) = 1.6$	$(1.6, \pi)$
$\dfrac{3\pi}{2}$	$0.5\left(\dfrac{3\pi}{2}\right) = 2.4$	$\left(2.4, \dfrac{3\pi}{2}\right)$
2π	$0.5(2\pi) = 3.1$	$(3.1, 2\pi)$

STEP 2 Plot the points in polar coordinates.

STEP 3 Connect the points with a smooth curve.
The curve is a *spiral*.

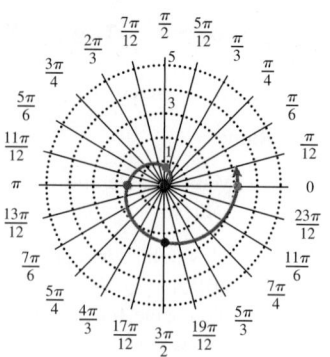

EXAMPLE 8 Graphing a Polar Equation of the Form $r^2 = c \cdot \sin(2\theta)$ or $r^2 = c \cdot \cos(2\theta)$

Graph $r^2 = 4\cos(2\theta)$.

Solution:

STEP 1 Make a table and find key values.

Solving for r yields $r = \pm 2\sqrt{\cos(2\theta)}$. All coordinates $(-r, \theta)$ can be expressed as $(r, \theta + \pi)$. The following table does not have values for $\dfrac{\pi}{4} < \theta < \dfrac{3\pi}{4}$, because the corresponding values of $\cos(2\theta)$ are negative and hence r is an imaginary number. The table also does not have values for $\theta > \pi$, because $2\theta > 2\pi$ and the corresponding points are repeated.

θ	$\cos(2\theta)$	$r = \pm 2\sqrt{\cos(2\theta)}$	(r, θ)	
0	1	$r = \pm 2$	$(2, 0)$ and	$(-2, 0) = (2, \pi)$
$\dfrac{\pi}{6}$	0.5	$r = \pm 1.4$	$\left(1.4, \dfrac{\pi}{6}\right)$ and $\left(-1.4, \dfrac{\pi}{6}\right) =$	$\left(1.4, \dfrac{7\pi}{6}\right)$
$\dfrac{\pi}{4}$	0	$r = 0$	$\left(0, \dfrac{\pi}{4}\right)$	
$\dfrac{3\pi}{4}$	0	$r = 0$	$\left(0, \dfrac{3\pi}{4}\right)$	
$\dfrac{5\pi}{6}$	0.5	$r = \pm 1.4$	$\left(1.4, \dfrac{5\pi}{6}\right)$ and $\left(-1.4, \dfrac{5\pi}{6}\right) =$	$\left(1.4, \dfrac{11\pi}{6}\right)$
π	1	$r = \pm 2$	$(2, \pi)$ and	$(-2, \pi) = (2, 2\pi)$

STEP 2 Plot the points in polar coordinates.

STEP 3 Connect the points with a smooth curve.
The resulting curve is known as a
lemniscate.

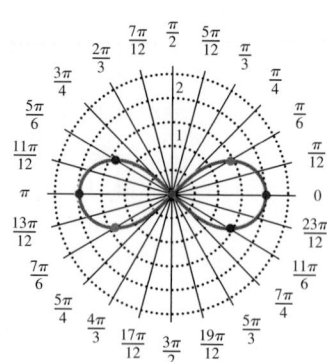

Converting Equations Between Polar and Rectangular Form

It is not always advantageous to plot an equation in the form in which it is given. It is sometimes easier to first convert to rectangular form and then plot. For example, to plot $r = \dfrac{2}{\cos\theta + \sin\theta}$, we could make a table of values. However, as you will see in Example 9, it is much easier to convert this equation to rectangular coordinates.

EXAMPLE 9   Converting an Equation from Polar Form to Rectangular Form

Graph $r = \dfrac{2}{\cos\theta + \sin\theta}$.

Solution:

Multiply the equation by $\cos\theta + \sin\theta$.
$$r(\cos\theta + \sin\theta) = 2$$

Eliminate parentheses.
$$r\cos\theta + r\sin\theta = 2$$

Convert the result to rectangular form.
$$\underset{x}{\underline{r\cos\theta}} + \underset{y}{\underline{r\sin\theta}} = 2$$

Simplify. The result is a straight line.
$$\boxed{y = -x + 2}$$

Graph the line.

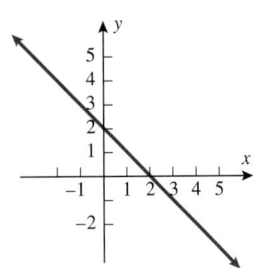

■ **YOUR TURN** Graph $r = \dfrac{2}{\cos\theta - \sin\theta}$.

Technology Tip

Graph $r = \dfrac{2}{\cos\theta + \sin\theta}$.

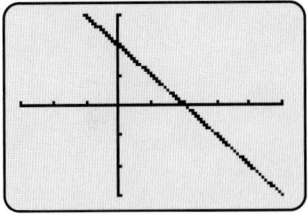

■ **Answer:** $y = x - 2$

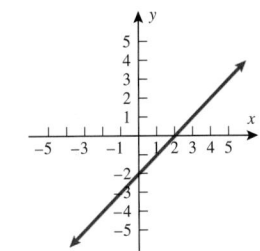

SECTION 7.5 SUMMARY

Graph polar coordinates (r, θ) in the polar coordinate system first by rotating a ray to get the terminal side of the angle. Then if r is positive, go out r units from the origin in the direction of the terminal side. If r is negative, go out $|r|$ units in the opposite direction of the terminal side. Conversions between polar and rectangular forms are given by

FROM	TO	IDENTITIES	
Polar (r, θ)	Rectangular (x, y)	$x = r\cos\theta$	$y = r\sin\theta$
Rectangular (x, y)	Polar (r, θ)	$r = \sqrt{x^2 + y^2}$	$\tan\theta = \dfrac{y}{x}, x \neq 0$

We can graph polar equations by point-plotting. Common shapes that arise are given in the following table. Sine and cosine curves have the same shapes (just rotated). If more than one equation is given, then the top equation corresponds to the actual graph. In this table, a and b are assumed to be positive.

CLASSIFICATION	DESCRIPTION	POLAR EQUATIONS	GRAPH
Line	Radial line	$\theta = a$	
Circle	Centered at origin	$r = a$	
Circle	Touches pole/center on polar axis	$r = a\cos\theta$	
Circle	Touches pole/center on line $\theta = \dfrac{\pi}{2}$	$r = a\sin\theta$	
Limaçon	Cardioid **a = b**	$r = a + b\cos\theta$ $r = a + b\sin\theta$	
Limaçon	Without inner loop $a > b$	$r = -a - b\cos\theta$ $r = a + b\sin\theta$	
Limaçon	With inner loop $a < b$	$r = a + b\sin\theta$ $r = a + b\cos\theta$	
Lemniscate		$r^2 = a^2\cos(2\theta)$ $r^2 = a^2\sin(2\theta)$	
Rose	Three* petals	$r = a\sin(3\theta)$ $r = a\cos(3\theta)$	

CLASSIFICATION	DESCRIPTION	POLAR EQUATIONS	GRAPH
Rose	Four* petals	$r = a\sin(2\theta)$ $r = a\cos(2\theta)$	
Spiral		$r = a\theta$	

*In the argument $n\theta$, if n is odd, there are n petals (leaves), and if n is even, there are $2n$ petals (leaves).

SECTION
7.5 EXERCISES

■ SKILLS

In Exercises 1–10, plot each indicated point in a polar coordinate system.

1. $\left(3, \dfrac{5\pi}{6}\right)$ **2.** $\left(2, \dfrac{5\pi}{4}\right)$ **3.** $\left(4, \dfrac{11\pi}{6}\right)$ **4.** $\left(1, \dfrac{2\pi}{3}\right)$ **5.** $\left(-2, \dfrac{\pi}{6}\right)$

6. $\left(-4, \dfrac{7\pi}{4}\right)$ **7.** $(-4, 270°)$ **8.** $(3, 135°)$ **9.** $(4, 225°)$ **10.** $(-2, 60°)$

In Exercises 11–20, convert each point to exact polar coordinates. Assume that $0 \le \theta < 2\pi$.

11. $(2, 2\sqrt{3})$ **12.** $(3, -3)$ **13.** $(-1, -\sqrt{3})$ **14.** $(6, 6\sqrt{3})$ **15.** $(-4, 4)$

16. $(0, \sqrt{2})$ **17.** $(3, 0)$ **18.** $(-7, -7)$ **19.** $(-\sqrt{3}, -1)$ **20.** $(2\sqrt{3}, -2)$

In Exercises 21–30, convert each point to exact rectangular coordinates.

21. $\left(4, \dfrac{5\pi}{3}\right)$ **22.** $\left(2, \dfrac{3\pi}{4}\right)$ **23.** $\left(-1, \dfrac{5\pi}{6}\right)$ **24.** $\left(-2, \dfrac{7\pi}{4}\right)$ **25.** $\left(0, \dfrac{11\pi}{6}\right)$

26. $(6, 0)$ **27.** $(2, 240°)$ **28.** $(-3, 150°)$ **29.** $(-1, 135°)$ **30.** $(5, 315°)$

In Exercises 31–34, match the polar graphs with their corresponding equations.

31. $r = 4\cos\theta$ **32.** $r = 2\theta$ **33.** $r = 3 + 3\sin\theta$ **34.** $r = 3\sin(2\theta)$

a.

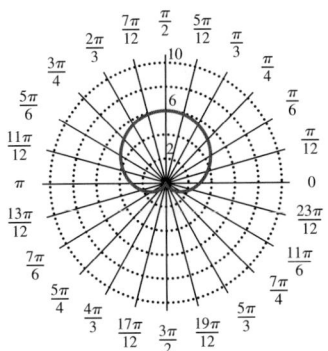

b.

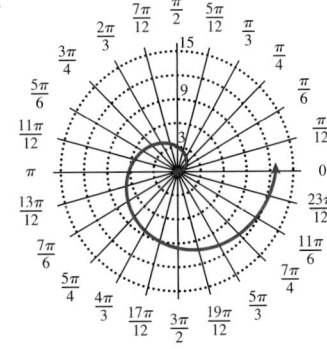

c.

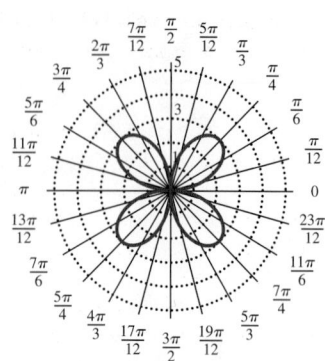

d.

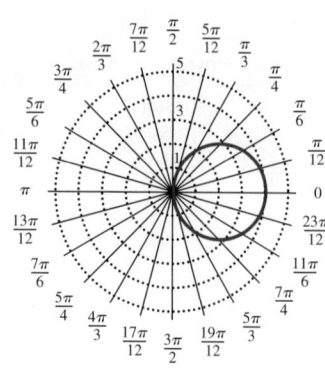

In Exercises 35–50, graph each equation.

35. $r = 5$

36. $\theta = -\dfrac{\pi}{3}$

37. $r = 2\cos\theta$

38. $r = 3\sin\theta$

39. $r = 4\sin(2\theta)$

40. $r = 5\cos(2\theta)$

41. $r = 3\sin(3\theta)$

42. $r = 4\cos(3\theta)$

43. $r^2 = 9\cos(2\theta)$

44. $r^2 = 16\sin(2\theta)$

45. $r = -2\cos\theta$

46. $r = -3\sin(3\theta)$

47. $r = 4\theta$

48. $r = -2\theta$

49. $r = -3 + 2\cos\theta$

50. $r = 2 + 3\sin\theta$

In Exercises 51–54, convert the equation from polar to rectangular form. Identify the resulting equation as a line, parabola, or circle.

51. $r(\sin\theta + 2\cos\theta) = 1$

52. $r(\sin\theta - 3\cos\theta) = 2$

53. $r^2\cos^2\theta - 2r\cos\theta + r^2\sin^2\theta = 8$

54. $r^2\cos^2\theta - r\sin\theta = -2$

In Exercises 55–60, graph the polar equation.

55. $r = -\frac{1}{3}\theta$

56. $r = \frac{1}{4}\theta$

57. $r = 4\sin(5\theta)$

58. $r = -3\cos(4\theta)$

59. $r = -2 - 3\cos\theta$

60. $r = 4 - 3\sin\theta$

■ APPLICATIONS

61. Halley's Comet. Halley's comet travels an elliptical path that can be modeled with the polar equation

$$r = \frac{0.587(1 + 0.967)}{1 - 0.967\cos\theta}.$$ Sketch the graph of the path of Halley's comet.

62. Dwarf Planet Pluto. The planet Pluto travels in an elliptical orbit that can be modeled with the polar equation

$$r = \frac{29.62(1 + 0.249)}{1 - 0.249\cos\theta}.$$ Sketch the graph of Pluto's orbit.

For Exercises 63 and 64, refer to the following:

Spirals are seen in nature, as in the swirl of a pine cone; they are also used in machinery to convert motions. An Archimedes spiral has the general equation $r = a\theta$. A more general form for the equation of a spiral is $r = a\theta^{1/n}$, where n is a constant that determines how tightly the spiral is wrapped.

63. Archimedes Spiral. Compare the Archimedes spiral $r = \theta$ with the spiral $r = \theta^{1/2}$ by graphing both on the same polar graph.

64. Archimedes Spiral. Compare the Archimedes spiral $r = \theta$ with the spiral $r = \theta^{4/3}$ by graphing both on the same polar graph.

For Exercises 65 and 66, refer to the following:

The *lemniscate motion* occurs naturally in the flapping of birds' wings. The bird's vertical lift and wing sweep create the distinctive figure-eight pattern. The patterns vary with the different wing profiles.

65. Flapping Wings of Birds. Compare the following two possible lemniscate patterns by graphing them on the same polar graph: $r^2 = 4\cos(2\theta)$ and $r^2 = \frac{1}{4}\cos(2\theta)$.

66. Flapping Wings of Birds. Compare the following two possible lemniscate patterns by graphing them on the same polar graph: $r^2 = 4\cos(2\theta)$ and $r^2 = 4\cos(2\theta + 2)$.

For Exercises 67 and 68, refer to the following:

Many microphone manufacturers advertise that their microphones' exceptional pickup capabilities isolate the sound source and minimize background noise. These microphones are described as cardioid microphones because of the pattern formed by the range of the pickup.

67. Cardioid Pickup Pattern. Graph the cardioid curve $r = 2 + 2\sin\theta$ to see what the range looks like.

68. Cardioid Pickup Pattern. Graph the cardioid curve $r = -4 - 4\sin\theta$ to see what the range looks like.

For Exercises 69 and 70, refer to the following:

The sword artistry of the Samurai is legendary in Japanese folklore and myth. The elegance with which a samurai could wield a sword rivals the grace exhibited by modern figure skaters. In more modern times, such legends have been rendered digitally in many different video games (e.g., *Onimusha*). In order to make the characters realistically move across the screen, and in particular, wield various sword motions true to the legends, trigonometric functions are extensively used in constructing the underlying graphics module. One famous movement is a figure eight, swept out with two hands on the sword. The actual path of the tip of the blade as the movement progresses in this figure-eight motion depends essentially on the length L of the sword and the speed with which it is swept out. Such a path is modeled using a polar equation of the form

$$r^2\theta = L\cos(A\theta) \quad \text{or} \quad r^2\theta = L\sin(A\theta), \quad \theta_1 \le \theta \le \theta_2$$

whose graphs are called *lemniscates*.

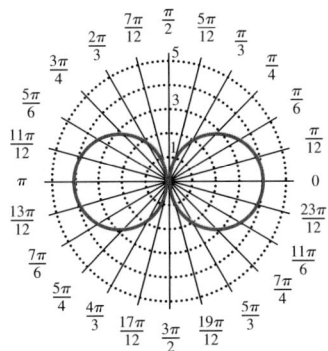

69. Video Games. Graph the following equations:

a. $r^2\theta = 5\cos\theta, 0 \le \theta \le 2\pi$

b. $r^2\theta = 5\cos(2\theta), 0 \le \theta \le \pi$

c. $r^2\theta = 5\cos(4\theta), 0 \le \theta \le \dfrac{\pi}{2}$

What do you notice about all of these graphs? Suppose that the movement of the tip of the sword in a game is governed by these graphs. Describe what happens if you change the domain in (b) and (c) to $0 \le \theta \le 2\pi$.

70. Video Games. Write a polar equation that would describe the motion of a sword 12 units long that makes 8 complete motions in $[0, 2\pi]$.

71. Home Improvement. The owner of a garden maze has decided to replace the square central plot of the maze, which is 60 feet × 60 feet, with a section comprised of a spiral to make the participants literally feel as though they are going in circles as they reach the tall slide in the center of the maze (that leads to the exit). Assume that the center of the maze is at the origin. If the walkway must be 3 feet wide to accommodate a participant, and the bushes on either side of the walkway are 1.5 feet thick, how many times can the spiral wrap around before the center is reached? Graph the resulting spiral.

72. Home Improvement. Consider the garden maze in Exercise 71, but now suppose that the owner wants the spiral to wrap around exactly 3 times. How wide can the walkway then be throughout the spiral? Graph it.

73. Magnetic Pendulum. A magnetic bob is affixed to an arm of length L, which is fastened to a pivot point. Three magnets of equal strength are positioned on a plane 8 inches from the center; one is placed on the x-axis, one at 120° with respect to the positive x-axis, and the other at 240° with respect to the positive x-axis. The path swept out is a three-petal rose, as shown below:

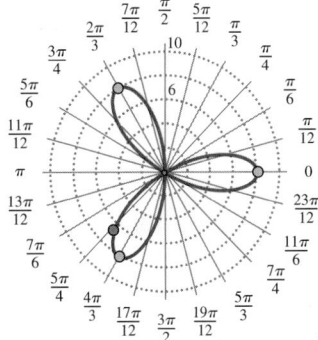

a. Find the equation of this path.

b. How many times does the path retrace itself on the interval $[0, 100\pi]$?

74. Magnetic Pendulum. In reference to the context of Exercise 73, now position 8 magnets, each 8 units from the origin and at the vertices of a regular octagon, one being on the x-axis. Assume that the path of the pendulum is an eight-petal rose.

a. Find the equation of this path.

b. Graph this equation.

▪ CATCH THE MISTAKE

In Exercises 75 and 76, explain the mistake that is made.

75. Convert $(-2, -2)$ to polar coordinates.

Solution:

Label x and y. $x = -2, y = -2$

Find r. $r = \sqrt{x^2 + y^2} = \sqrt{4 + 4} = \sqrt{8} = 2\sqrt{2}$

Find θ. $\tan\theta = \dfrac{-2}{-2} = 1$

$$\theta = \tan^{-1}(1) = \frac{\pi}{4}$$

Write the point in polar coordinates. $\left(2\sqrt{2}, \dfrac{\pi}{4}\right)$

This is incorrect. What mistake was made?

76. Convert $\left(-\sqrt{3}, 1\right)$ to polar coordinates.

Solution:

Label x and y. $x = -\sqrt{3}, y = 1$

Find r. $r = \sqrt{x^2 + y^2} = \sqrt{3 + 1} = \sqrt{4} = 2$

Find θ. $\tan\theta = \dfrac{1}{-\sqrt{3}} = -\dfrac{1}{\sqrt{3}}$

$$\theta = \tan^{-1}\left(-\frac{1}{\sqrt{3}}\right) = -\frac{\pi}{4}$$

Write the point in polar coordinates. $\left(2, -\dfrac{\pi}{4}\right)$

This is incorrect. What mistake was made?

▪ CONCEPTUAL

In Exercises 77 and 78, determine whether each statement is true or false.

77. All cardioids are limaçons, but not all limaçons are cardioids.

78. All limaçons are cardioids, but not all cardioids are limaçons.

79. Find the polar equation that is equivalent to a vertical line, $x = a$.

80. Find the polar equation that is equivalent to a horizontal line, $y = b$.

81. Give another pair of polar coordinates for the point (a, θ).

82. Convert $(-a, b)$ to polar coordinates. Assume that $a > 0, b > 0$.

▪ CHALLENGE

83. Determine the values of θ at which $r = 4\cos\theta$ and $r\cos\theta = 1$ intersect. Graph both equations.

84. Find the Cartesian equation for $r = a\sin\theta + b\cos\theta$, where a and b are positive. Identify the type of graph.

85. Find the Cartesian equation for $r = \dfrac{a\sin(2\theta)}{\cos^3\theta - \sin^3\theta}$.

86. Identify an equation for the following graph:

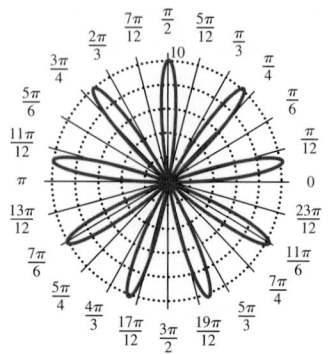

87. Consider the equation $r = 2a\cos(\theta - b)$. Sketch the graph for various values of a and b, and then give a general description of the graph.

88. Consider the equation $r = a\sin(b\theta)$, where $a, b > 0$. Determine the smallest number M for which the graph starts to repeat.

■ TECHNOLOGY

89. Given $r = \cos\left(\dfrac{\theta}{2}\right)$, find the θ-intervals for the inner loop above the x-axis.

90. Given $r = 2\cos\left(\dfrac{3\theta}{2}\right)$, find the θ-intervals for the petal in the first quadrant.

91. Given $r = 1 + 3\cos\theta$, find the θ-intervals for the inner loop.

92. Given $r = 1 + \sin(2\theta)$ and $r = 1 - \cos(2\theta)$, find all points of intersection.

93. Given $r = 2 + \sin(4\theta)$ and $r = 1$, find the angles of all points of intersection.

94. Given $r = 2 - \cos(3\theta)$ and $r = 1.5$, find the angles of all points of intersection.

■ PREVIEW TO CALCULUS

In calculus, when we need to find the area enclosed by two polar curves, the first step consists of finding the points where the curves coincide.

In Exercises 95–98, find the points of intersection of the given curves.

95. $r = 4\sin\theta$ and $r = 4\cos\theta$

96. $r = \cos\theta$ and $r = 2 + 3\cos\theta$

97. $r = 1 - \sin\theta$ and $r = 1 + \cos\theta$

98. $r = 1 + \sin\theta$ and $r = 1 + \cos\theta$

MODELING YOUR WORLD

In the summer of 2008, the average price of unleaded gasoline in the United States topped $4 per gallon. This meant that considering hybrid alternative automobiles was no longer just a "green" pursuit for the environmentally conscious, but an attractive financial option. When hybrids were first introduced, gasoline was roughly $3 per gallon and hybrid owners had to drive three to five years (depending on how many miles per year they drove) before their savings at the pump equaled the initial additional cost of the purchase. With gasoline surpassing the $4 per gallon price (with a projected $5 per gallon by 2010), the rising price of gasoline allows a consumer to make up the initial additional cost of a hybrid automobile in as little as one to three years, depending on driving habits.

The following table illustrates the approximate gross vehicle weight of both a large SUV and a small hybrid and the approximate fuel economy rates in miles per gallon:

AUTOMOBILE	WEIGHT	MPG
Ford Expedition	7100 lb	18
Toyota Prius	2800 lb	45

Recall that the amount of work to push an object that weighs F pounds a distance of d feet along a horizontal is $W = \mathbf{F \cdot d}$.

1. Calculate how much work it would take to move a Ford Expedition 100 feet.

2. Calculate how much work it would take to move a Toyota Prius 100 feet.

3. Compare the values you calculated in Questions 1 and 2. What is the ratio of work to move the Expedition to work required to move the Prius?

4. Compare the result in Question 3 with the ratio of fuel economy (mpg) for these two vehicles. What can you conclude about the relationship between weight of an automobile and fuel economy?

5. Calculate the work required to move both the Ford Expedition and Toyota Prius 100 feet along an incline that makes a 45° angle with the ground (horizontal).

6. Based on your results in Question 5, do you expect the fuel economy ratios to be the same in this inclined scenario compared with the horizontal? In other words, should consumers in Florida (flat) be guided by the same "numbers" as consumers in the Appalachian Mountains (North Carolina)?

SECTION	CONCEPT	PAGES	REVIEW EXERCISES	KEY IDEAS/POINTS
7.1	**Vectors**	624–634	1–20	Vector **u** or $\overrightarrow{AB}$
	Magnitude and direction of vectors	624–627	1–8 13–16	$\|\mathbf{u}\| = \sqrt{a^2 + b^2}$ $\tan\theta = \dfrac{b}{a}$ Magnitude (length of a vector): $\mathbf{u} = \langle a, b\rangle$ Geometric: tail-to-tip $\mathbf{u} = \langle a, b\rangle$ and $\mathbf{v} = \langle c, d\rangle$ $\mathbf{u} + \mathbf{v} = \langle a + c, b + d\rangle$
	Vector operations	628	9–12 19–20	Scalar multiplication: $k\langle a, b\rangle = \langle ka, kb\rangle$
	Horizontal and vertical components of a vector	629		Horizontal component: $a = \|\mathbf{u}\|\cos\theta$ Vertical component: $b = \|\mathbf{u}\|\sin\theta$
	Unit vectors	629–630	17–20	$\mathbf{u} = \dfrac{\mathbf{v}}{\|\mathbf{v}\|}$
	Resultant vectors	630–634		■ Resultant velocities ■ Resultant forces
7.2	**The dot product**	639–644	21–40	■ The product of a scalar and a vector is a vector. ■ The dot product of two vectors is a scalar.
	Multiplying two vectors	639–640	21–26	$\mathbf{u} = \langle a, b\rangle$ and $\mathbf{v} = \langle c, d\rangle$ $\mathbf{u}\cdot\mathbf{v} = ac + bd$
	Angle between two vectors	640–642	27–40	If θ is the angle between two nonzero vectors **u** and **v**, where $0° \leq \theta \leq 180°$, then $$\cos\theta = \dfrac{\mathbf{u}\cdot\mathbf{v}}{\|\mathbf{u}\|\|\mathbf{v}\|}$$ Orthogonal (perpendicular) vectors: $\mathbf{u}\cdot\mathbf{v} = 0$
	Work	643–644		When force and displacement are in the same direction: $W = \|\mathbf{F}\|\|\mathbf{d}\|$. When force and displacement are not in the same direction: $W = \mathbf{F}\cdot\mathbf{d}$.

CHAPTER REVIEW

Section	Concept	Pages	Review Exercises	Key Ideas/Points
7.5	**Polar coordinates and graphs of polar equations**	670–679	77–92	
	Polar coordinates	670–671	77–88	To plot a point (r, θ): ■ Start on the polar axis and rotate a ray to form the terminal side of an angle θ. ■ If $r > 0$, the point is r units from the origin in the *same direction* as the terminal side of θ. ■ If $r < 0$, the point is $\lvert r \rvert$ units from the origin in the *opposite direction* of the terminal side of θ.
	Converting between polar and rectangular coordinates	672	77–88	From polar (r, θ) to rectangular (x, y): $$x = r\cos\theta \qquad y = r\sin\theta$$ From rectangular (x, y) to polar (r, θ): $$r = \sqrt{x^2 + y^2} \qquad \tan\theta = \frac{y}{x} \quad x \neq 0$$
	Graphs of polar equations	673–679	89–92	Radial line, circle, spiral, rose petals, lemniscate, and limaçon

7.1 Vectors

Find the magnitude of vector AB.

1. $A = (4, -3)$ and $B = (-8, 2)$

2. $A = (-2, 11)$ and $B = (2, 8)$

3. $A = (0, -3)$ and $B = (5, 9)$

4. $A = (3, -11)$ and $B = (9, -3)$

Find the magnitude and direction angle of the given vector.

5. $\mathbf{u} = \langle -10, 24 \rangle$
6. $\mathbf{u} = \langle -5, -12 \rangle$

7. $\mathbf{u} = \langle 16, -12 \rangle$
8. $\mathbf{u} = \langle 0, 3 \rangle$

Perform the vector operation, given that $\mathbf{u} = \langle 7, -2 \rangle$ and $\mathbf{v} = \langle -4, 5 \rangle$.

9. $2\mathbf{u} + 3\mathbf{v}$
10. $\mathbf{u} - \mathbf{v}$

11. $6\mathbf{u} + \mathbf{v}$
12. $-3(\mathbf{u} + 2\mathbf{v})$

Find the vector, given its magnitude and direction angle.

13. $|\mathbf{u}| = 10, \theta = 75°$
14. $|\mathbf{u}| = 8, \theta = 225°$

15. $|\mathbf{u}| = 12, \theta = 105°$
16. $|\mathbf{u}| = 20, \theta = 15°$

Find a unit vector in the direction of the given vector.

17. $\mathbf{v} = \langle \sqrt{6}, -\sqrt{6} \rangle$
18. $\mathbf{v} = \langle -11, 60 \rangle$

Perform the indicated vector operation.

19. $(3\mathbf{i} - 4\mathbf{j}) + (2\mathbf{i} + 5\mathbf{j})$
20. $(-6\mathbf{i} + \mathbf{j}) - (9\mathbf{i} - \mathbf{j})$

7.2 The Dot Product

Find the indicated dot product.

21. $\langle 6, -3 \rangle \cdot \langle 1, 4 \rangle$
22. $\langle -6, 5 \rangle \cdot \langle -4, 2 \rangle$

23. $\langle 3, 3 \rangle \cdot \langle 3, -6 \rangle$
24. $\langle -2, -8 \rangle \cdot \langle -1, 1 \rangle$

25. $\langle 0, 8 \rangle \cdot \langle 1, 2 \rangle$
26. $\langle 4, -3 \rangle \cdot \langle -1, 0 \rangle$

Find the angle (round to the nearest degree) between each pair of vectors.

27. $\langle 3, 4 \rangle$ and $\langle -5, 12 \rangle$
28. $\langle -4, 5 \rangle$ and $\langle 5, -4 \rangle$

29. $\langle 1, \sqrt{2} \rangle$ and $\langle -1, 3\sqrt{2} \rangle$
30. $\langle 7, -24 \rangle$ and $\langle -6, 8 \rangle$

31. $\langle 3, 5 \rangle$ and $\langle -4, -4 \rangle$
32. $\langle -1, 6 \rangle$ and $\langle 2, -2 \rangle$

Determine whether each pair of vectors is orthogonal.

33. $\langle 8, 3 \rangle$ and $\langle -3, 12 \rangle$
34. $\langle -6, 2 \rangle$ and $\langle 4, 12 \rangle$

35. $\langle 5, -6 \rangle$ and $\langle -12, -10 \rangle$
36. $\langle 1, 1 \rangle$ and $\langle -4, 4 \rangle$

37. $\langle 0, 4 \rangle$ and $\langle 0, -4 \rangle$
38. $\langle -7, 2 \rangle$ and $\langle \frac{1}{7}, -\frac{1}{2} \rangle$

39. $\langle 6z, a - b \rangle$ and $\langle a + b, -6z \rangle$

40. $\langle a - b, -1 \rangle$ and $\langle a + b, a^2 - b^2 \rangle$

7.3 Polar (Trigonometric) Form of Complex Numbers

Graph each complex number in the complex plane.

41. $-6 + 2i$
42. $5i$

Express each complex number in polar form.

43. $\sqrt{2} - \sqrt{2}i$
44. $\sqrt{3} + i$

45. $-8i$
46. $-8 - 8i$

With a calculator, express each complex number in polar form.

47. $-60 + 11i$
48. $9 - 40i$

49. $15 + 8i$
50. $-10 - 24i$

Express each complex number in rectangular form.

51. $6(\cos 300° + i \sin 300°)$

52. $4(\cos 210° + i \sin 210°)$

53. $\sqrt{2}(\cos 135° + i \sin 135°)$

54. $4(\cos 150° + i \sin 150°)$

With a calculator, express each complex number in rectangular form.

55. $4(\cos 200° + i \sin 200°)$

56. $3(\cos 350° + i \sin 350°)$

7.4 Products, Quotients, Powers, and Roots of Complex Numbers

Find the product $z_1 z_2$.

57. $3(\cos 200° + i \sin 200°)$ and $4(\cos 70° + i \sin 70°)$

58. $3(\cos 20° + i \sin 20°)$ and $4(\cos 220° + i \sin 220°)$

59. $7(\cos 100° + i \sin 100°)$ and $3(\cos 140° + i \sin 140°)$

60. $(\cos 290° + i \sin 290°)$ and $4(\cos 40° + i \sin 40°)$

Find the quotient $\frac{z_1}{z_2}$.

61. $\sqrt{6}(\cos 200° + i \sin 200°)$ and $\sqrt{6}(\cos 50° + i \sin 50°)$

62. $18(\cos 190° + i \sin 190°)$ and $2(\cos 100° + i \sin 100°)$

63. $24(\cos 290° + i \sin 290°)$ and $4(\cos 110° + i \sin 110°)$

64. $\sqrt{200}(\cos 93° + i \sin 93°)$ and $\sqrt{2}(\cos 48° + i \sin 48°)$

Find the result of each expression using De Moivre's theorem. Write the answer in rectangular form.

65. $(3 + 3i)^4$
66. $(3 + \sqrt{3}i)^4$

67. $(1 + \sqrt{3}i)^5$
68. $(-2 - 2i)^7$

Find all *n*th roots of *z*. Write the answers in polar form, and plot the roots in the complex plane.

69. $2 + 2\sqrt{3}i$, $n = 2$ **70.** $-8 + 8\sqrt{3}i$, $n = 4$

71. -256, $n = 4$ **72.** $-18i$, $n = 2$

Find all complex solutions to the given equations.

73. $x^3 + 216 = 0$ **74.** $x^4 - 1 = 0$

75. $x^4 + 1 = 0$ **76.** $x^3 - 125 = 0$

7.5 Polar Coordinates and Graphs of Polar Equations

Convert each point to exact polar coordinates (assuming that $0 \le \theta < 2\pi$), and then graph the point in the polar coordinate system.

77. $(-2, 2)$ **78.** $(4, -4\sqrt{3})$

79. $(-5\sqrt{3}, -5)$ **80.** $(\sqrt{3}, \sqrt{3})$

81. $(0, -2)$ **82.** $(11, 0)$

Convert each point to exact rectangular coordinates.

83. $\left(-3, \dfrac{5\pi}{3}\right)$ **84.** $\left(4, \dfrac{5\pi}{4}\right)$

85. $\left(2, \dfrac{\pi}{3}\right)$ **86.** $\left(6, \dfrac{7\pi}{6}\right)$

87. $\left(1, \dfrac{4\pi}{3}\right)$ **88.** $\left(-3, \dfrac{7\pi}{4}\right)$

Graph each equation.

89. $r = 4\cos(2\theta)$ **90.** $r = \sin(3\theta)$

91. $r = -\theta$ **92.** $r = 4 - 3\sin\theta$

Technology Exercises

Section 7.1

With the graphing calculator [SUM] command, find the magnitude of the given vector. Also, find the direction angle to the nearest degree.

93. $\langle 25, -60 \rangle$ **94.** $\langle -70, 10\sqrt{15} \rangle$

Section 7.2

With the graphing calculator [SUM] command, find the angle (round to the nearest degree) between each pair of vectors.

95. $\langle 14, 37 \rangle$, $\langle 9, -26 \rangle$

96. $\langle -23, -8 \rangle$, $\langle 18, -32 \rangle$

Section 7.3

Another way of using a graphing calculator to represent complex numbers in rectangular form is to enter the real and imaginary parts as a list of two numbers and use the [SUM] command to find the modulus.

97. Write $-\sqrt{23} - 11i$ in polar form using the [SUM] command to find its modulus, and round the angle to the nearest degree.

98. Write $11 + \sqrt{23}i$ in polar form using the [SUM] command to find its modulus, and round the angle to the nearest degree.

Section 7.4

99. Find the fourth roots of $-8 + 8\sqrt{3}i$, and draw the complex rectangle with the calculator.

100. Find the fourth roots of $8\sqrt{3} + 8i$, and draw the complex rectangle with the calculator.

Section 7.5

101. Given $r = 1 - 2\sin 3\theta$, find the angles of all points of intersection (where $r = 0$).

102. Given $r = 1 + 2\cos 3\theta$, find the angles of all points of intersection (where $r = 0$).

1. Find the magnitude and direction angle of the vector $\mathbf{u} = \langle -5, 12 \rangle$.

2. Find a unit vector pointing in the same direction as $\mathbf{v} = \langle -3, -4 \rangle$.

3. Perform the indicated operations:
   a. $2\langle -1, 4 \rangle - 3\langle 4, 1 \rangle$
   b. $\langle -7, -1 \rangle \cdot \langle 2, 2 \rangle$

4. In a post pattern in football, the receiver in motion runs past the quarterback parallel to the line of scrimmage (A), runs 12 yards perpendicular to the line of scrimmage (B), and then cuts toward the goal post (C).

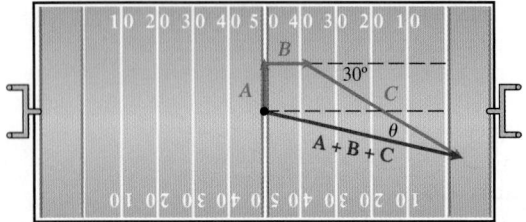

A receiver runs the post pattern. If the magnitudes of the vectors are $|A| = 3$ yards, $|B| = 12$ yards, and $|C| = 18$ yards, find the magnitude of the resultant vector $A + B + C$ and the direction angle θ.

5. Find the dot product $\langle 4, -51 \rangle \cdot \langle -2, -\frac{1}{3} \rangle$.

6. If the dot product $\langle a, -2a \rangle \cdot \langle 4, 5 \rangle = 18$, find the value of a.

For Exercises 7 and 8, use the complex number
$z = 16(\cos 120° + i \sin 120°)$.

7. Find z^4.

8. Find the four distinct fourth roots of z.

9. Convert the point $(3, 210°)$ to rectangular coordinates.

10. Convert the point $\left(4, \frac{5\pi}{4} \right)$ to rectangular coordinates.

11. Convert the point $(30, -15)$ to polar coordinates.

12. Graph $r = 6\sin(2\theta)$.

13. Graph $r^2 = 9\cos(2\theta)$.

14. Find x such that $\langle x, 1 \rangle$ is perpendicular to $3\mathbf{i} - 4\mathbf{j}$.

15. Prove that $\mathbf{u} \cdot (\mathbf{v} - \mathbf{w}) = \mathbf{u} \cdot \mathbf{v} - \mathbf{u} \cdot \mathbf{w}$.

16. Construct a unit vector in the opposite direction of $\langle 3, 5 \rangle$.

17. Compute $\mathbf{u} \cdot \mathbf{v}$ if $|\mathbf{u}| = 4$, $|\mathbf{v}| = 10$, and $\theta = \frac{2\pi}{3}$.

18. Determine whether $\mathbf{u}$ and $\mathbf{v}$ are parallel, perpendicular, or neither: $\mathbf{u} = \langle \sin\theta, \cos\theta \rangle$ $\mathbf{v} = \langle -\cos\theta, \sin\theta \rangle$.

19. Find the magnitude of $-\mathbf{i} - \mathbf{j}$.

20. Determine θ, when a streetlight is formed as follows:

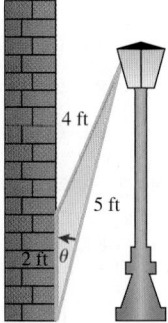

21. Two tugboats pull a cruiser off the port of Miami. The first one pulls with a force of 25,000 lb and the second one pulls with a force of 27,000 lb. If the angle between the lines connecting the cruiser with the tugboats is 25°, what is the resultant force vector of the two tugboats?

22. True or false: If $\mathbf{u} + \mathbf{v}$ is perpendicular to $\mathbf{u} - \mathbf{v}$, then $|\mathbf{u}| = |\mathbf{v}|$.

23. Solve $z^4 + 256i = 0$.

24. Convert to a Cartesian equation: $r^2 = \tan\theta$.

25. With the graphing calculator $\boxed{\text{SUM}}$ command, find the angle (round to the nearest degree) between each pair of vectors: $\langle -8, -11 \rangle$ and $\langle -16, 26 \rangle$.

26. Find the fourth roots of $-8\sqrt{3} - 8i$, and draw the complex rectangle with the calculator.

1. Given $f(x) = x^2 - 4$ and $g(x) = \dfrac{1}{\sqrt{3x + 5}}$, find $(f \circ g)(x)$ and the domain of f, g, and $f \circ g$.

2. Determine whether the function $f(x) = |x^3|$ is even, odd, or neither.

3. Find the quadratic function whose graph has a vertex at $(-1, 2)$ and that passes through the point $(2, -1)$. Express the quadratic function in both standard and general forms.

4. For the polynomial function
   $$f(x) = x^5 - 4x^4 + x^3 + 10x^2 - 4x - 8$$
   a. List each real zero and its multiplicity.
   b. Determine whether the graph touches or crosses at each x-intercept.

5. Find all vertical and horizontal or slant asymptotes (if any) in the following:
   $$f(x) = \frac{x^3 - 3x^2 + 2x - 1}{x^2 - 2x + 1}$$

6. How much money should be invested today in a money market account that pays 1.4% a year compounded continuously if you desire $5,000 in 8 years?

7. Write the exponential equation $\sqrt[4]{625} = 5$ in its equivalent logarithmic form.

8. What is the radian measure of an angle of $305°$? Express your answer in terms of π.

9. Find all trigonometric functions of the angle θ. Rationalize any denominators containing radicals that you encounter in your answers.

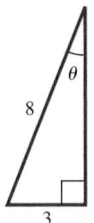

10. Solve the triangle $\alpha = 30°$, $\beta = 30°$, and $c = 4$ in.

11. Solve the triangle $a = 14.2$ m, $b = 16.5$ m, and $\gamma = 50°$.

12. Find the exact value of each trigonometric function:
   a. $\sin\left(\dfrac{3\pi}{2}\right)$
   b. $\cos 0$
   c. $\tan\left(\dfrac{5\pi}{4}\right)$
   d. $\cot\left(\dfrac{11\pi}{6}\right)$
   e. $\sec\left(\dfrac{2\pi}{3}\right)$
   f. $\csc\left(\dfrac{5\pi}{6}\right)$

13. State the amplitude, period, phase shift, and vertical shift of the function $y = 4 - \frac{1}{3}\sin(4x - \pi)$.

14. If $\sin x = \dfrac{1}{\sqrt{3}}$ and $\cos x < 0$, find $\cos(2x)$.

15. Find the exact value of $\tan\left[\cos^{-1}\left(-\frac{3}{5}\right) + \sin^{-1}\left(\frac{1}{2}\right)\right]$.

16. Find $\left(1 + \sqrt{3}i\right)^8$. Express the answer in rectangular form.

8

Systems of Linear Equations and Inequalities

Cryptography is the practice and study of encryption and decryption—encoding data so that it can be decoded only by specific individuals. In other words, it turns a message into gibberish so that only the person who has the deciphering tools can turn that gibberish back into the original message. ATM cards, online shopping sites, and secure military communications all depend on coding and decoding of information. Matrices are used extensively in cryptography. A *matrix* is used as the "key" to encode the data, and then its *inverse matrix* is used as the key to decode the data.*

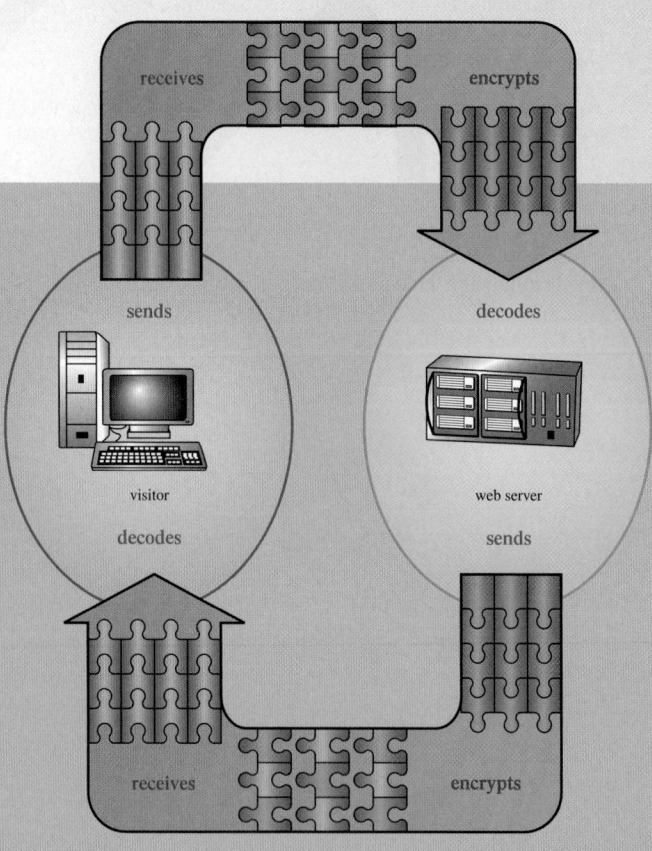

---

*Section 8.4, Exercises 83–88.

IN THIS CHAPTER we will solve systems of linear equations using the elimination and substitution methods. We will then solve systems of linear equations using matrices three different ways: using augmented matrices (Gauss–Jordan elimination), matrix algebra (inverse matrices), and determinants (Cramer's rule). We will then discuss an application of systems of linear equations that is useful in calculus called partial–fraction decomposition. Finally, we will solve systems of linear inequalities.

SYSTEMS OF LINEAR EQUATIONS AND INEQUALITIES

8.1 Systems of Linear Equations in Two Variables	8.2 Systems of Linear Equations in Three Variables	8.3 Systems of Linear Equations and Matrices	8.4 Matrix Algebra	8.5 The Determinant of a Square Matrix and Cramer's Rule	8.6 Partial Fractions	8.7 Systems of Linear Inequalities in Two Variables
• Substitution Method • Elimination Method • Graphing Method • Three Methods and Three Types of Solutions	• Solving Systems of Linear Equations in Three Variables • Types of Solutions	• Augmented Matrices • Row Operations on a Matrix • Row–Echelon Form of a Matrix • Gaussian Elimination with Back-Substitution • Gauss–Jordan Elimination • Inconsistent and Dependent Systems	• Equality of Matrices • Matrix Addition and Subtraction • Scalar and Matrix Multiplication • Matrix Equations • Finding the Inverse of a Square Matrix • Solving Systems of Linear Equations Using Matrix Algebra and Inverses of Square Matrices	• Determinant of a 2×2 Matrix • Determinant of an $n \times n$ Matrix • Cramer's Rule: Systems of Linear Equations in Two Variables • Cramer's Rule: Systems of Linear Equations in Three Variables	• Distinct Linear Factors • Repeated Linear Factors • Distinct Irreducible Quadratic Factors • Repeated Irreducible Quadratic Factors	• Linear Inequalities in Two Variables • Systems of Linear Inequalities in Two Variables • The Linear Programming Model

CHAPTER OBJECTIVES

- Solve systems of linear equations in two variables using elimination and substitution methods (Section 8.1).
- Solve systems of linear equations in three variables using elimination and substitution methods (Section 8.2).
- Use Gauss–Jordan elimination (augmented matrices) to solve systems of linear equations in more than two variables (Section 8.3).
- Use matrix algebra and inverse matrices to solve systems of linear equations (Section 8.4).
- Use Cramer's rule to solve systems of linear equations (Section 8.5).
- Perform partial–fraction decomposition on rational expressions (Section 8.6).
- Solve systems of linear inequalities in two variables (Section 8.7).

SKILLS OBJECTIVES

- Solve systems of linear equations in two variables using the substitution method.
- Solve systems of linear equations in two variables using the elimination method.
- Solve systems of linear equations in two variables by graphing.
- Solve applications involving systems of linear equations.

CONCEPTUAL OBJECTIVES

- Understand that a system of linear equations has either one solution, no solution, or infinitely many solutions.
- Visualize two lines that intersect at one point, no points (parallel lines), or infinitely many points (same line).

A linear equation in two variables is given in standard form by

$$Ax + By = C$$

and the graph of this linear equation is a line, provided that A and B are not both equal to zero. In this section, we discuss **systems of linear equations**, which can be thought of as simultaneous equations. To **solve** a system of linear equations in two variables means to find the solution that satisfies *both* equations. Suppose we are given the following system of equations:

$$x + 2y = 6$$
$$3x - y = 11$$

We can interpret the solution to this system of equations both algebraically and graphically.

	ALGEBRAIC	GRAPHICAL
Solution	$x = 4$ and $y = 1$	$(4, 1)$
Check	**Equation 1** **Equation 2** $x + 2y = 6$ $3x - y = 11$ $(4) + 2(1) = 6$ ✓ $3(4) - 1 = 11$ ✓	
Interpretation	$x = 4$ and $y = 1$ satisfy both equations.	The point $(4, 1)$ lies on both lines.

This particular example had *one solution*. There are systems of equations that have *no solution* or *infinitely many solutions*. We give these systems special names: **independent**, **inconsistent**, and **dependent**, respectively.

INDEPENDENT SYSTEM	INCONSISTENT SYSTEM	DEPENDENT SYSTEM
One solution	No solution	Infinitely many solutions
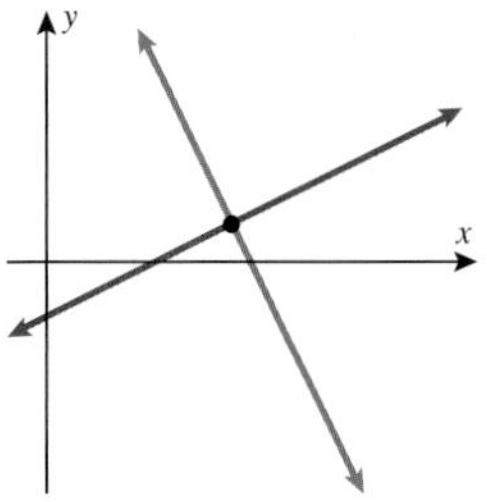	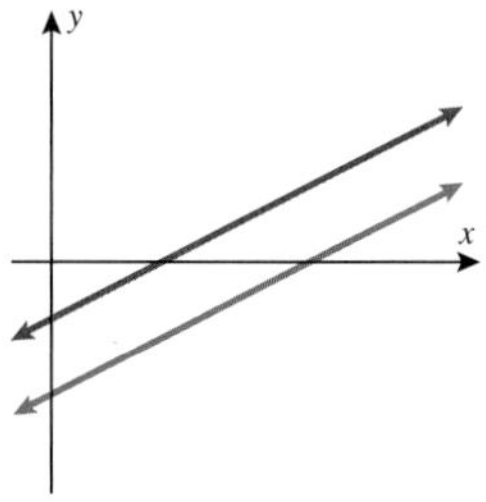	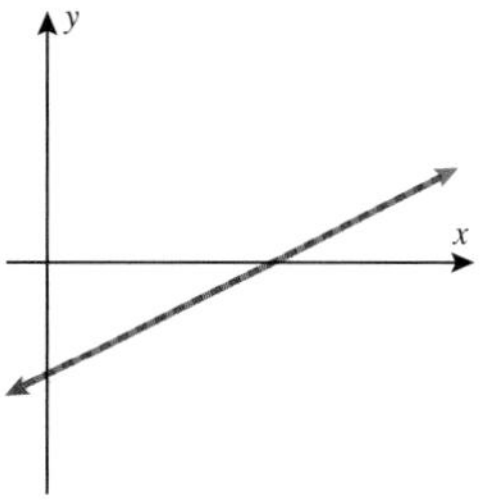
Lines have different slopes.	Lines are parallel (same slope and different y-intercepts).	Lines coincide (same slope and same y-intercept).

In this section, we discuss three methods for solving systems of two linear equations in two variables: *substitution*, *elimination*, and *graphing*. We use the algebraic methods—substitution and elimination—to find solutions exactly; we then look at a graphical interpretation of the solution (two lines that intersect at one point, parallel lines, or coinciding lines).

We will illustrate each method with the same example given earlier:

$$x + 2y = 6 \qquad \text{Equation (1)}$$

$$3x - y = 11 \qquad \text{Equation (2)}$$

Substitution Method

The following box summarizes the substitution method for solving systems of two linear equations in two variables:

SUBSTITUTION METHOD

Step 1: **Solve** one of the equations for one variable in terms of the other variable.

Equation (2): $y = 3x - 11$

Step 2: **Substitute** the expression found in Step 1 into the *other* equation. The result is an equation in one variable.

Equation (1): $x + 2(3x - 11) = 6$

Step 3: **Solve** the equation obtained in Step 2.

$x + 6x - 22 = 6$
$7x = 28$
$\boxed{x = 4}$

Step 4: **Back-substitute** the value found in Step 3 into the expression found in Step 1.

$y = 3(4) - 11$
$\boxed{y = 1}$

Step 5: **Check** that the solution satisfies *both* equations. Substitute (4, 1) into both equations.

Equation (1): $x + 2y = 6$
$(4) + 2(1) = 6 \checkmark$

Equation (2): $3x - y = 11$
$3(4) - 1 = 11 \checkmark$

EXAMPLE 1   Determining by Substitution That a System Has One Solution

Use the substitution method to solve the following system of linear equations:

$$x + y = 8 \qquad \text{Equation (1)}$$
$$3x - y = 4 \qquad \text{Equation (2)}$$

Solution:

STEP 1 Solve Equation (2) for y in terms of x. $\qquad\qquad y = 3x - 4$

STEP 2 Substitute $y = 3x - 4$ into Equation (1). $\qquad x + (3x - 4) = 8$

STEP 3 Solve for x. $\qquad\qquad\qquad\qquad\qquad\qquad\qquad x + 3x - 4 = 8$
$$4x = 12$$
$$\boxed{x = 3}$$

STEP 4 Back-substitute $x = 3$ into Equation (1). $\qquad\qquad 3 + y = 8$
$$\boxed{y = 5}$$

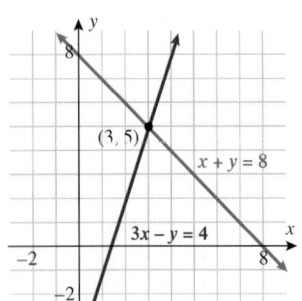

STEP 5 Check that (3, 5) satisfies *both* equations.

$$\text{Equation (1):} \qquad x + y = 8$$
$$3 + 5 = 8$$

$$\text{Equation (2):} \qquad 3x - y = 4$$
$$3(3) - 5 = 4$$

Note: The graphs of the two equations are two lines that intersect at the point (3, 5).

EXAMPLE 2   Determining by Substitution That a System Has No Solution

Use the substitution method to solve the following system of linear equations:

$$x - y = 2 \qquad \text{Equation (1)}$$
$$2x - 2y = 10 \qquad \text{Equation (2)}$$

Solution:

STEP 1 Solve Equation (1) for y in terms of x. $\qquad\qquad y = x - 2$

STEP 2 Substitute $y = x - 2$ into Equation (2). $\qquad 2x - 2(x - 2) = 10$

STEP 3 Solve for x. $\qquad\qquad\qquad\qquad\qquad\qquad\qquad 2x - 2x + 4 = 10$
$$4 = 10$$

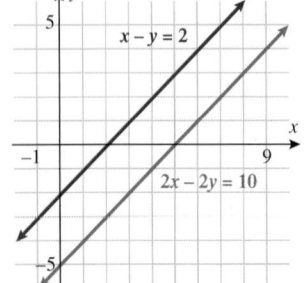

$4 = 10$ is never true, so this is called an inconsistent system. There is $\boxed{\text{no solution}}$ to this system of linear equations.

Note: The graphs of the two equations are parallel lines.

EXAMPLE 3   **Determining by Substitution That a System Has Infinitely Many Solutions**

Use the substitution method to solve the following system of linear equations:

$$x - y = 2 \qquad \text{Equation (1)}$$
$$-x + y = -2 \qquad \text{Equation (2)}$$

Solution:

STEP 1 Solve Equation (1) for y in terms of x. $\qquad\qquad y = x - 2$

STEP 2 Substitute $y = x - 2$ into Equation (2). $\qquad -x + (x - 2) = -2$

STEP 3 Solve for x. $\qquad\qquad\qquad\qquad\qquad\qquad -x + x - 2 = -2$
$$-2 = -2$$

$-2 = -2$ is always true, so this is called a dependent system. Notice, for instance, that the points $(2, 0)$, $(4, 2)$, and $(7, 5)$ all satisfy both equations. In fact, there are $\boxed{\text{infinitely many solutions}}$ to this system of linear equations. All solutions are in the form (x, y), where $\boxed{y = x - 2}$.

Note: The graphs of these two equations are the same line.

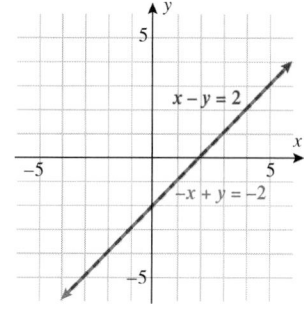

■ **YOUR TURN**   Use the substitution method to solve each system of linear equations.

a. $2x + y = 3$
$4x + 2y = 4$

b. $x - y = 2$
$4x - 3y = 10$

c. $x + 2y = 1$
$2x + 4y = 2$

Answer: a. no solution
b. $(4, 2)$
c. infinitely many solutions where $y = -\frac{1}{2}x + \frac{1}{2}$

Elimination Method

We now turn our attention to another method, *elimination*, which is often preferred over substitution and will later be used in higher order systems. In a system of two linear equations in two variables, the equations can be combined, resulting in a third equation in one variable, thus *eliminating* one of the variables. The following is an example of when elimination would be preferred because the y terms sum to zero when the two equations are added together:

$$\begin{array}{rcr} 2x - y = & 5 \\ -x + y = & -2 \\ \hline x = & 3 \end{array}$$

When you cannot eliminate a variable simply by *adding* the two equations, multiply one equation by a constant that will cause the coefficients of some variable in the two equations to match and be opposite in sign.

The following box summarizes the *elimination method*, also called the *addition method*, for solving systems of two linear equations in two variables using the same example given earlier:

$$x + 2y = 6 \qquad \text{Equation (1)}$$
$$3x - y = 11 \qquad \text{Equation (2)}$$

ELIMINATION METHOD

Step 1*: **Multiply** the coefficients of one (or both) of the equations so that one of the variables will be eliminated when the two equations are added.

Multiply Equation (2) by 2:
$$6x - 2y = 22$$

Step 2: **Eliminate** one of the variables by adding the equation found in Step 1 to the *other* original equation. The result is an equation in one variable.

$$
\begin{array}{r}
x + 2y = 6 \\
\underline{6x - 2y = 22} \\
7x \quad\;\; = 28
\end{array}
$$

Step 3: **Solve** the equation obtained in Step 2.

$$7x = 28$$
$$\boxed{x = 4}$$

Step 4: **Back-substitute** the value found in Step 3 into either of the two original equations.

$$(4) + 2y = 6$$
$$2y = 2$$
$$\boxed{y = 1}$$

Step 5: **Check** that the solution satisfies *both* equations. Substitute $(4, 1)$ into both equations.

Equation (1):
$$x + \quad 2y = 6$$
$$(4) + 2(1) = 6 \checkmark$$

Equation (2):
$$3x - y = 11$$
$$3(4) - 1 = 11 \checkmark$$

*Step 1 is not necessary in cases where a pair of corresponding terms already sum to zero.

EXAMPLE 4 **Applying the Elimination Method When One Variable Is Eliminated by Adding the Two Original Equations**

Use the elimination method to solve the following system of linear equations:

$$2x - y = -5 \qquad \text{Equation (1)}$$
$$4x + y = \;\;11 \qquad \text{Equation (2)}$$

Solution:

STEP 1 Not necessary.

STEP 2 Eliminate y by adding Equation (1) to Equation (2).

$$
\begin{array}{r}
2x - y = -5 \\
\underline{4x + y = \;\;11} \\
6x \quad\;\; = \;\;6
\end{array}
$$

STEP 3 Solve for x.

$$\boxed{x = 1}$$

STEP 4 Back-substitute $x = 1$ into Equation (2). Solve for y.

$$4(1) + y = 11$$
$$\boxed{y = 7}$$

STEP 5 Check that $(1, 7)$ satisfies both equations.

Equation (1): $\quad 2x - \quad y = -5$
$$2(1) - (7) = -5 \checkmark$$

Equation (2): $\quad 4x + \quad y = 11$
$$4(1) + (7) = 11 \checkmark$$

Note: The graphs of the two given equations correspond to two lines that intersect at the point $(1, 7)$.

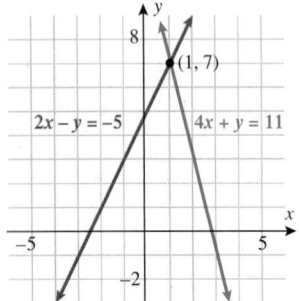

In Example 4, we eliminated the variable y simply by adding the two equations. Sometimes it is necessary to multiply one (Example 5) or both (Example 6) equations by constants prior to adding.

 EXAMPLE 5   **Applying the Elimination Method When Multiplying One Equation by a Constant Is Necessary**

Use the elimination method to solve the following system of linear equations:

$$-4x + 3y = 23 \qquad \text{Equation (1)}$$
$$12x + 5y = 1 \qquad \text{Equation (2)}$$

Solution:

STEP 1 Multiply Equation (1) by 3.

$$-12x + 9y = 69$$

STEP 2 Eliminate x by adding the modified Equation (1) to Equation (2).

$$\begin{array}{r} -12x + 9y = 69 \\ 12x + 5y = 1 \\ \hline 14y = 70 \end{array}$$

STEP 3 Solve for y.

$$\boxed{y = 5}$$

STEP 4 Back-substitute $y = 5$ into Equation (2).

Solve for x.

$$12x + 5(5) = 1$$
$$12x + 25 = 1$$
$$12x = -24$$
$$\boxed{x = -2}$$

> **Study Tip**
>
> Be sure to multiply the **entire** equation by the constant.

STEP 5 Check that $(-2, 5)$ satisfies both equations.

Equation (1):
$$-4(-2) + 3(5) = 23$$
$$8 + 15 = 23 \checkmark$$

Equation (2):
$$12(-2) + 5(5) = 1$$
$$-24 + 25 = 1 \checkmark$$

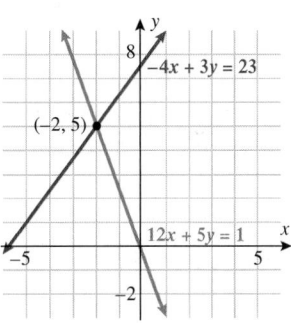

Note: The graphs of the two given equations correspond to two lines that intersect at the point $(-2, 5)$.

In Example 5, we eliminated x simply by multiplying the first equation by a constant and adding the result to the second equation. In order to eliminate either of the variables in Example 6, we will have to multiply *both* equations by constants prior to adding.

EXAMPLE 6   Applying the Elimination Method When Multiplying Both Equations by Constants Is Necessary

Use the elimination method to solve the following system of linear equations:

$$3x + 2y = 1 \qquad \text{Equation (1)}$$
$$5x + 7y = 9 \qquad \text{Equation (2)}$$

Solution:

STEP 1 Multiply Equation (1) by 5 and Equation (2) by -3.

$$15x + 10y = 5$$
$$-15x - 21y = -27$$

STEP 2 Eliminate x by adding the modified Equation (1) to the modified Equation (2).

$$\begin{array}{r} 15x + 10y = 5 \\ -15x - 21y = -27 \\ \hline -11y = -22 \end{array}$$

STEP 3 Solve for y.

$$\boxed{y = 2}$$

STEP 4 Back-substitute $y = 2$ into Equation (1). Solve for x.

$$3x + 2(2) = 1$$
$$3x = -3$$
$$\boxed{x = -1}$$

STEP 5 Check that $(-1, 2)$ satisfies both equations.

Equation (1): $\qquad\qquad 3x + 2y = 1$
$$3(-1) + 2(2) = 1 \checkmark$$

Equation (2): $\qquad\qquad 5x + 7y = 9$
$$5(-1) + 7(2) = 9 \checkmark$$

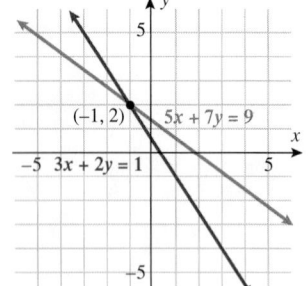

Note: The graphs of the two given equations correspond to two lines that intersect at the point $(-1, 2)$.

Notice in Example 6 that we could have also eliminated y by multiplying the first equation by 7 and the second equation by -2. Typically, the choice is dictated by which approach will keep the coefficients as simple as possible. In the event that the original coefficients contain fractions or decimals, first rewrite the equations in standard form with integer coefficients and then make the decision.

EXAMPLE 7   Determining by the Elimination Method That a System Has No Solution

Use the elimination method to solve the following system of linear equations:

$$-x + y = 7 \qquad \text{Equation (1)}$$
$$2x - 2y = 4 \qquad \text{Equation (2)}$$

Solution:

STEP 1 Multiply Equation (1) by 2.

$$-2x + 2y = 14$$

STEP 2 Eliminate y by adding the modified Equation (1) found in Step 1 to Equation (2).

$$\begin{array}{r} -2x + 2y = 14 \\ 2x - 2y = 4 \\ \hline 0 = 18 \end{array}$$

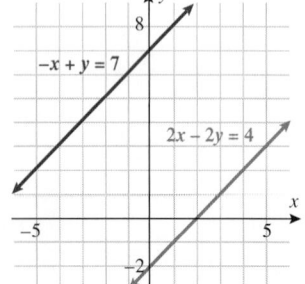

This system is inconsistent since $0 = 18$ is never true. Therefore, there are no values of x and y that satisfy both equations. We say that there is $\boxed{\text{no solution}}$ to this system of linear equations.

Note: The graphs of the two equations are two parallel lines.

Study Tip

Systems of linear equations in two variables have either one solution, no solution, or infinitely many solutions.

EXAMPLE 8 **Determining by the Elimination Method That a System Has Infinitely Many Solutions**

Use the elimination method to solve the following system of linear equations:

$$7x + y = 2 \qquad \text{Equation (1)}$$
$$-14x - 2y = -4 \qquad \text{Equation (2)}$$

Solution:

STEP 1 Multiply Equation (1) by 2. $\qquad\qquad\qquad\qquad 14x + 2y = 4$

STEP 2 Add the modified Equation (1) found in Step 1 to Equation (2).

$$\begin{array}{r} 14x + 2y = 4 \\ -14x - 2y = -4 \\ \hline 0 = 0 \end{array}$$

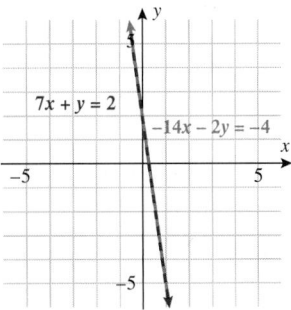

This system is dependent since $0 = 0$ is always true. We say that there are

infinitely many solutions to this system of linear equations of the form $y = -7x + 2$.

Note: The graphs of the two equations are the same line.

YOUR TURN Apply the elimination method to solve each system of linear equations.

a. $\begin{aligned} 2x + 3y &= 1 \\ 4x - 3y &= -7 \end{aligned}$ **b.** $\begin{aligned} x - 5y &= 2 \\ -10x + 50y &= -20 \end{aligned}$ **c.** $\begin{aligned} x - y &= 14 \\ -x + y &= 9 \end{aligned}$

■ **Answer:**
a. $(-1, 1)$
b. infinitely many solutions of the form $y = \frac{1}{5}x - \frac{2}{5}$
c. no solution

Graphing Method

A third way to solve a system of linear equations in two variables is to graph the two lines. If the two lines intersect, then the point of intersection is the solution. Graphing is the most labor-intensive method for solving systems of linear equations in two variables. The graphing method is typically not used to solve systems of linear equations when an exact solution is desired. Instead, it is used to interpret or confirm the solution(s) found by the other two methods (substitution and elimination). If you are using a graphing calculator, however, you will get as accurate an answer using the graphing method as you will when applying the other methods.

The following box summarizes the graphing method for solving systems of linear equations in two variables using the same example given earlier:

$$x + 2y = 6 \qquad \text{Equation (1)}$$
$$3x - y = 11 \qquad \text{Equation (2)}$$

GRAPHING METHOD

Step 1*: **Write** the equations in slope–intercept form.

Equation (1):   Equation (2):

$$y = -\frac{1}{2}x + 3 \qquad\qquad y = 3x - 11$$

Step 2: **Graph** the two lines.

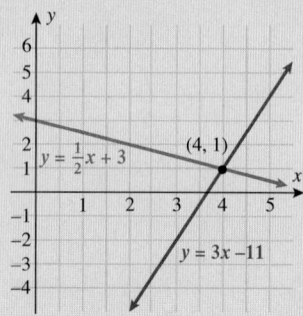

Step 3: **Identify** the point of intersection.   (4, 1)

Step 4: **Check** that the solution satisfies *both* equations.

Equation (1):   Equation (2):

$$x + 2y = 6 \qquad\qquad 3x - y = 11$$
$$(4) + 2(1) = 6 \checkmark \qquad 3(4) - 1 = 11 \checkmark$$

*Step 1 is not necessary when the lines are already in slope–intercept form.

EXAMPLE 9   **Determining by Graphing That a System Has One Solution**

Use graphing to solve the following system of linear equations:

$$x + y = 2 \qquad \text{Equation (1)}$$
$$3x - y = 2 \qquad \text{Equation (2)}$$

Solution:

STEP 1 Write each equation in slope–intercept form.

$$y = -x + 2 \qquad \text{Equation (1)}$$
$$y = 3x - 2 \qquad \text{Equation (2)}$$

STEP 2 Plot both lines on the same graph.

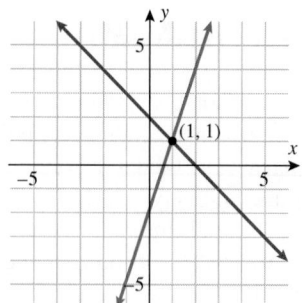

STEP 3 Identify the point of intersection.   (1, 1)

STEP 4 Check that the point $(1, 1)$ satisfies both equations.

$$x + y = 2$$
$$1 + 1 = 2 \checkmark \qquad \text{Equation (1)}$$

$$3x - y = 2$$
$$3(1) - (1) = 2 \checkmark \qquad \text{Equation (2)}$$

Note: There is one solution, because the two lines intersect at one point.

EXAMPLE 10 **Determining by Graphing That a System Has No Solution**

Technology Tip

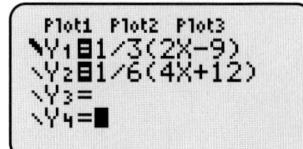

Use graphing to solve the following system of linear equations:

$$2x - 3y = 9 \qquad \text{Equation (1)}$$
$$-4x + 6y = 12 \qquad \text{Equation (2)}$$

First solve each equation for y.
The graphs of $Y_1 = \frac{1}{3}(2x - 9)$ and $Y_2 = \frac{1}{6}(4x + 12)$ are shown.

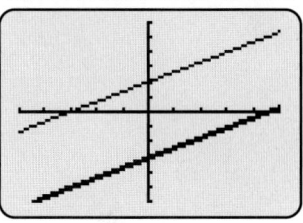

Solution:

STEP 1 Write each equation in slope–intercept form.

$$y = \frac{2}{3}x - 3 \qquad \text{Equation (1)}$$
$$y = \frac{2}{3}x + 2 \qquad \text{Equation (2)}$$

STEP 2 Plot both lines on the same graph.

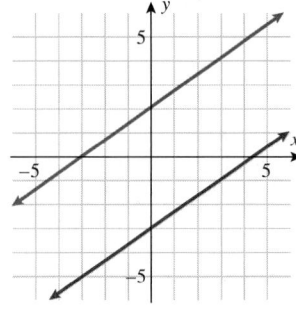

STEP 3 Identify the point of intersection. None

There is no solution, because two parallel lines never intersect.

Note: Both lines have the same slope, but different y-intercepts.

The graphs and table show that there is no solution to the system.

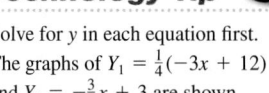

EXAMPLE 11 **Determining by Graphing That a System Has Infinitely Many Solutions**

Use graphing to solve the following system of linear equations:

$$3x + 4y = 12 \qquad \text{Equation (1)}$$
$$\frac{3}{4}x + y = 3 \qquad \text{Equation (2)}$$

Technology Tip

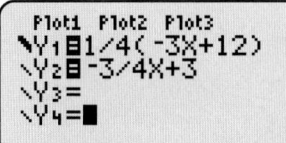

Solve for y in each equation first.
The graphs of $Y_1 = \frac{1}{4}(-3x + 12)$ and $Y_2 = -\frac{3}{4}x + 3$ are shown.

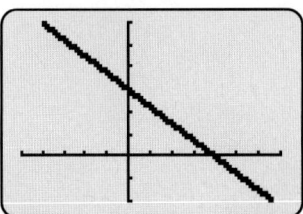

Solution:

STEP 1 Write each equation in slope–intercept form.

$$y = -\frac{3}{4}x + 3 \qquad y = -\frac{3}{4}x + 3$$

STEP 2 Plot both lines on the same graph.

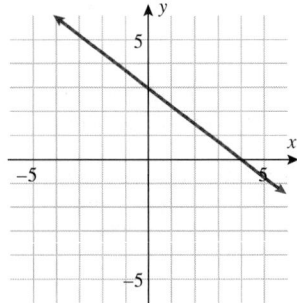

Both lines are on the same graph.

STEP 3 Identify the point of intersection.　　　　　　Infinitely many points

There are $\boxed{\text{infinitely many solutions, } y = -\frac{3}{4}x + 3}$, since the two lines are identical and coincide.

■ **Answer:**

a. infinitely many solutions of the form $y = \frac{1}{2}x - \frac{1}{2}$

b. $(3, 1)$

c. no solution

■ **YOUR TURN** Utilize graphing to solve each system of linear equations.

a. $\begin{aligned} x - 2y &= 1 \\ 2x - 4y &= 2 \end{aligned}$　　b. $\begin{aligned} x - 2y &= 1 \\ 2x + y &= 7 \end{aligned}$　　c. $\begin{aligned} 2x + y &= 3 \\ 2x + y &= 7 \end{aligned}$

Three Methods and Three Types of Solutions

Given any system of two linear equations in two variables, any of the three methods (substitution, elimination, or graphing) can be utilized. If you find that it is easy to eliminate a variable by adding multiples of the two equations, then elimination is the preferred choice. If you do not see an obvious elimination, then solve the system by substitution. For exact solutions, choose one of these two algebraic methods. You should typically use graphing to confirm the solution(s) you have found by applying the other two methods or when you are using a graphing utility.

EXAMPLE 12　Identifying Which Method to Use

State which of the two algebraic methods (elimination or substitution) would be the preferred method to solve each system of linear equations.

a. $\begin{aligned} x - 2y &= 1 \\ -x + y &= 2 \end{aligned}$　　b. $\begin{aligned} x &= 2y - 1 \\ 2x - y &= 4 \end{aligned}$　　c. $\begin{aligned} 7x - 20y &= 1 \\ 5x + 3y &= 18 \end{aligned}$

Solution:

a. **Elimination:** Because the x variable is eliminated when the two equations are added.

b. **Substitution:** Because the first equation is easily substituted into the second equation (for x).

c. **Either:** There is no preferred method, as both elimination and substitution require substantial work.

Regardless of which method is used to solve systems of two linear equations in two variables, in general, we can summarize the three types of solutions both algebraically and graphically.

THREE TYPES OF SOLUTIONS TO SYSTEMS OF LINEAR EQUATIONS

NUMBER OF SOLUTIONS	GRAPHICAL INTERPRETATION
One solution	The two lines intersect at one point.
No solution	The two lines are parallel. (*Same* slope/*different* y-intercepts.)
Infinitely many solutions	The two lines coincide. (*Same* slope/*same* y-intercept.)

Applications

Suppose you have two job offers that require sales. One pays a higher base, while the other pays a higher commission. Which job do you take?

EXAMPLE 13   Deciding Which Job to Take

Suppose that upon graduation you are offered a job selling biomolecular devices to laboratories studying DNA. The Beckman-Coulter Company offers you a job selling its DNA sequencer with an annual base salary of $20,000 plus 5% commission on total sales. The MJ Research Corporation offers you a job selling its PCR Machine that makes copies of DNA with an annual base salary of $30,000 plus 3% commission on sales. Determine what the total sales would have to be to make the Beckman-Coulter job the better offer.

Solution:

STEP 1 **Identify the question.**

When would these two jobs have equal compensations?

STEP 2 **Make notes.**

Beckman-Coulter salary	20,000 + 5% of sales
MJ Research salary	30,000 + 3% of sales

STEP 3 **Set up the equations.**

Let x = total sales and y = compensation.

Equation (1)   Beckman-Coulter: $y = 20{,}000 + 0.05x$

Equation (2)   MJ Research: $y = 30{,}000 + 0.03x$

STEP 4 **Solve the system of equations.**
Substitution method*

Substitute Equation (1)
into Equation (2). $20{,}000 + 0.05x = 30{,}000 + 0.03x$

Solve for x. $0.02x = 10{,}000$
                                                    $x = 500{,}000$

If you make $500,000 worth of sales per year, the jobs will yield equal compensations. If you sell less than $500,000, the MJ Research job is the better offer, and more than $500{,}000$, the Beckman-Coulter job is the better offer.

*The elimination method could also have been used.

STEP 5 **Check the solution.**

Equation (1)   Beckman-Coulter: $y = 20{,}000 + 0.05(500{,}000) = \$45{,}000$

Equation (2)   MJ Research: $y = 30{,}000 + 0.03(500{,}000) = \$45{,}000$

Technology Tip

The graphs of $Y_1 = 20{,}000 + 0.05x$ and $Y_2 = 30{,}000 + 0.03x$ are shown.

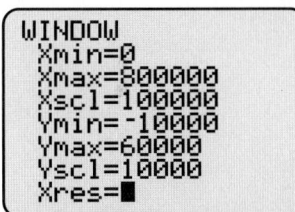

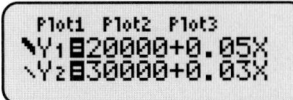

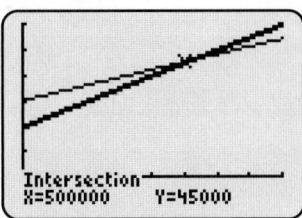

The graphs and table support the solution to the system.

In this section, we discussed two algebraic techniques for solving systems of two linear equations in two variables:

- Substitution method
- Elimination method

The algebraic methods are preferred for exact solutions, and the graphing method is typically used to give a visual interpretation and confirmation of the solution. There are three types of solutions to systems of two linear equations in two variables: one solution, no solution, or infinitely many solutions.

INDEPENDENT SYSTEM	INCONSISTENT SYSTEM	DEPENDENT SYSTEM
One solution	No solution	Infinitely many solutions
Lines have different slopes.	Lines are parallel (same slope and different y-intercepts).	Lines coincide (same slope and same y-intercept).

SKILLS

In Exercises 1–20, solve each system of linear equations by substitution.

1. $x + y = 7$
$x - y = 9$

2. $x - y = -10$
$x + y = 4$

3. $2x - y = 3$
$x - 3y = 4$

4. $4x + 3y = 3$
$2x + y = 1$

5. $3x + y = 5$
$2x - 5y = -8$

6. $6x - y = -15$
$2x - 4y = -16$

7. $2u + 5v = 7$
$3u - v = 5$

8. $m - 2n = 4$
$3m + 2n = 1$

9. $2x + y = 7$
$-2x - y = 5$

10. $3x - y = 2$
$3x - y = 4$

11. $4r - s = 1$
$8r - 2s = 2$

12. $-3p + q = -4$
$6p - 2q = 8$

13. $5r - 3s = 15$
$-10r + 6s = -30$

14. $-5p - 3q = -1$
$10p + 6q = 2$

15. $2x - 3y = -7$
$3x + 7y = 24$

16. $4x - 5y = -7$
$3x + 8y = 30$

17. $\frac{1}{3}x - \frac{1}{4}y = 0$
$-\frac{2}{3}x + \frac{3}{4}y = 2$

18. $\frac{1}{5}x + \frac{2}{3}y = 10$
$-\frac{1}{2}x - \frac{1}{6}y = -7$

19. $-3.9x + 4.2y = 15.3$
$-5.4x + 7.9y = 16.7$

20. $6.3x - 7.4y = 18.6$
$2.4x + 3.5y = 10.2$

In Exercises 21–40, solve each system of linear equations by elimination.

21. $x - y = -3$
$x + y = 7$

22. $x - y = -10$
$x + y = 8$

23. $5x + 3y = -3$
$3x - 3y = -21$

24. $-2x + 3y = 1$
$2x - y = 7$

25. $2x - 7y = 4$
$5x + 7y = 3$

26. $3x + 2y = 6$
$-3x + 6y = 18$

27. $2x + 5y = 7$
$3x - 10y = 5$

28. $6x - 2y = 3$
$-3x + 2y = -2$

29. $2x + 5y = 5$
$-4x - 10y = -10$

30. $11x + 3y = 3$
$22x + 6y = 6$

31. $3x - 2y = 12$
$4x + 3y = 16$

32. $5x - 2y = 7$
$3x + 5y = 29$

33. $6x - 3y = -15$
$7x + 2y = -12$

34. $7x - 4y = -1$
$3x - 5y = 16$

35. $4x - 5y = 22$
$3x + 4y = 1$

36. $6x - 5y = 32$
$2x - 6y = 2$

37. $\frac{1}{3}x + \frac{1}{2}y = 1$
$\frac{1}{5}x + \frac{7}{2}y = 2$

38. $\frac{1}{2}x - \frac{1}{3}y = 0$
$\frac{3}{2}x + \frac{1}{2}y = \frac{3}{4}$

39. $3.4x + 1.7y = 8.33$
$-2.7x - 7.8y = 15.96$

40. $-0.04x + 1.12y = 9.815$
$2.79x + 1.19y = -0.165$

In Exercises 41–44, match the systems of equations with the graphs.

41. $3x - y = 1$
$3x + y = 5$

42. $-x + 2y = -1$
$2x + y = 7$

43. $2x + y = 3$
$2x + y = 7$

44. $x - 2y = 1$
$2x - 4y = 2$

a.

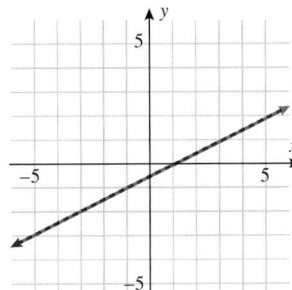

b.

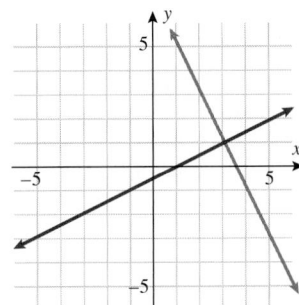

c.

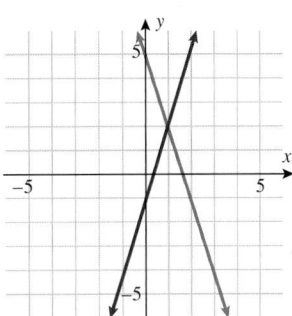

d.

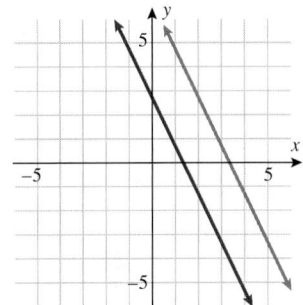

In Exercises 45–52, solve each system of linear equations by graphing.

45. $y = -x$
$y = x$

46. $x - 3y = 0$
$x + 3y = 0$

47. $2x + y = -3$
$x + y = -2$

48. $x - 2y = -1$
$-x - y = -5$

49. $\frac{1}{2}x - \frac{2}{3}y = 4$
$\frac{1}{4}x - y = 6$

50. $\frac{1}{5}x - \frac{5}{2}y = 10$
$\frac{1}{15}x - \frac{5}{6}y = \frac{10}{3}$

51. $1.6x - y = 4.8$
$-0.8x + 0.5y = 1.5$

52. $1.1x - 2.2y = 3.3$
$-3.3x + 6.6y = -6.6$

■ APPLICATIONS

53. Mixture. In chemistry lab, Stephanie has to make a 37 milliliter solution that is 12% HCl. All that is in the lab is 8% and 15% HCl. How many milliliters of each solution should she use to obtain the desired mix?

54. Mixture. A mechanic has 340 gallons of gasoline and 10 gallons of oil to make gas/oil mixtures. He wants one mixture to be 4% oil and the other mixture to be 2.5% oil. If he wants to use all of the gas and oil, how many gallons of gas and oil are in each of the resulting mixtures?

55. Salary Comparison. Upon graduation with a degree in management of information systems (MIS), you decide to work for a company that buys data from states' departments of motor vehicles and sells to banks and car dealerships customized reports detailing how many cars at each dealership are financed through particular banks. Autocount Corporation offers you a $15,000 base salary and 10% commission on your total annual sales. Polk Corporation offers you a base salary of $30,000 plus a 5% commission on your total annual sales. How many total sales would you have to make per year to earn more money at Autocount?

56. Salary Comparison. Two types of residential real estate agents are those who sell existing houses (resale) and those who sell new homes for developers. Resale of existing homes typically earns 6% commission on every sale, and representing developers in selling new homes typically earns a base salary of $15,000 per year plus an additional 1.5% commission, because agents are required to work 5 days a week on site in a new development. Find the total value (dollars) an agent would have to sell per year to make more money in resale than in new homes?

57. Gas Mileage. A Honda Accord gets approximately 26 mpg on the highway and 19 mpg in the city. You drove 349.5 miles on a full tank (16 gallons) of gasoline. Approximately how many miles did you drive in the city and how many on the highway?

58. Wireless Plans. AT&T is offering a 600-minute peak plan with free mobile-to-mobile and weekend minutes at $59 per month plus $0.13 per minute for every minute over 600. The next plan up is the 800-minute plan that costs $79 per month. You think you may go over 600 minutes, but are not sure you need 800 minutes. How many minutes would you have to talk for the 800-minute plan to be the better deal?

59. Distance/Rate/Time. A direct flight on Delta Air Lines from Atlanta to Paris is 4000 miles and takes approximately 8 hours going east (Atlanta to Paris) and 10 hours going west (Paris to Atlanta). Although the plane averages the same airspeed, there is a headwind while traveling west and a tailwind while traveling east, resulting in different air speeds. What is the average air speed of the plane, and what is the average wind speed?

60. Distance/Rate/Time. A private pilot flies a Cessna 172 on a trip that is 500 miles each way. It takes her approximately 3 hours to get there and 4 hours to return. What is the approximate average air speed of the Cessna, and what is the approximate wind speed?

61. Investment Portfolio. Leticia has been tracking two volatile stocks. Stock A over the last year has increased 10%, and stock B has increased 14% (using a simple interest model). She has $10,000 to invest and would like to split it between these two stocks. If the stocks continue to perform at the same rate, how much should she invest in each for one year to result in a balance of $11,260?

62. Investment Portfolio. Toby split his savings into two different investments, one earning 5% and the other earning 7%. He put twice as much in the investment earning the higher rate. In one year, he earned $665 in interest. How much money did he invest in each account?

63. Break-Even Analysis. A company produces CD players for a unit cost of $15.00 per CD player. The company has fixed costs of $120.00. If each CD player can be sold for $30.00, how many CD players must be sold to break even? Determine the cost equation first. Next, determine the revenue equation. Use the two equations you have found to determine the break-even point.

64. Managing a Lemonade Stand. An elementary-school-age child wants to have a lemonade stand. She would sell each glass of lemonade for $0.25. She has determined that each glass of lemonade costs about $0.10 to make (for lemons and sugar). It costs her $15.00 for materials to make the lemonade stand. How many glasses of lemonade must she sell to break even?

65. Meal Cost. An airline is deciding which meals to buy from its provider. If the airline orders the same number of meals of types I and II totalling 150 meals, the cost is $1275; if they order 60% of type I and 40% of type II, the cost is $1260. What is the cost of each type of meal?

66. Meal Cost. In a school district, the board of education has decided on two menus to serve in the school cafeterias. The annual budget for the meal plan is $1.2 million and one of the menus is 5% more expensive than the other. What is the annual cost of each menu? Round your answer to the nearest integer.

67. Population. The U.S. Census Bureau reports that Florida's population in the year 2008 was 18,328,340 habitants. The number of females exceeded the number of males by 329,910. What is the number of habitants, by gender, in Florida in 2008?

68. Population. According to the U.S. Census Bureau, in 2000, the U.S. population was 281,420,906 habitants. Some projections indicate that by 2020 there will be 341,250,007 habitants. The number of senior citizens will increase 30%, while the number of citizens under the age of 65 will increase 20%. Find the number of senior citizens and nonsenior citizens in the year 2000. Round your answer to the nearest integer.

■ CATCH THE MISTAKE

In Exercises 69–72, explain the mistake that is made.

69. Solve the system of equations by elimination.

$$2x + y = -3$$
$$3x + y = 8$$

Solution:

Multiply Equation (1) by -1.	$2x - y = -3$
Add the result to Equation (2).	$3x + y = 8$
	$\overline{5x = 5}$
Solve for x.	$x = 1$
Substitute $x = 1$ into Equation (2).	$3(1) + y = 8$
	$y = 5$

The answer $(1, 5)$ is incorrect. What mistake was made?

70. Solve the system of equations by elimination.

$$4x - y = 12$$
$$4x - y = 24$$

Solution:

Multiply Equation (1) by -1.	$-4x + y = -12$
Add the result to Equation (2).	$-4x + y = -12$
	$4x - y = 24$
	$\overline{0 = 12}$

Answer: Infinitely many solutions.

This is incorrect. What mistake was made?

71. Solve the system of equations by substitution.

$$x + 3y = -4$$
$$-x + 2y = -6$$

Solution:

Solve Equation (1) for x.	$x = -3y - 4$
Substitute $x = -3y - 4$ into Equation (2).	$-(-3y - 4) + 2y = -6$
Solve for y.	$3y - 4 + 2y = -6$
	$5y = -2$
	$y = -\dfrac{2}{5}$

Substitute $y = -\frac{2}{5}$ into Equation (1). $x + 3\left(-\dfrac{2}{5}\right) = -4$

Solve for x. $x = -\dfrac{14}{5}$

The answer $\left(-\frac{2}{5}, -\frac{14}{5}\right)$ is incorrect. What mistake was made?

72. Solve the system of equations by graphing.

$$2x + 3y = 5$$
$$4x + 6y = 10$$

Solution:

Write both equations in slope–intercept form.

$$y = -\dfrac{2}{3}x + \dfrac{5}{3}$$
$$y = -\dfrac{2}{3}x + \dfrac{5}{3}$$

Since these lines have the same slope, they are parallel lines. Parallel lines do not intersect, so there is no solution.

This is incorrect. What mistake was made?

■ CONCEPTUAL

In Exercises 73–76, determine whether each statement is true or false on the xy-plane.

73. A system of equations represented by a graph of two lines with the same slope always has no solution.

74. A system of equations represented by a graph of two lines with slopes that are negative reciprocals always has one solution.

75. If two lines do not have exactly one point of intersection, then they must be parallel.

76. The system of equations, $Ax - By = 1$ and $-Ax + By = -1$, has no solution.

77. The point $(2, -3)$ is a solution to the system of equations

$$Ax + By = -29$$
$$Ax - By = 13$$

Find A and B.

78. If you graph the lines

$$x - 50y = 100$$
$$x - 48y = -98$$

they appear to be parallel lines. However, there is a unique solution. Explain how this might be possible.

■ CHALLENGE

79. Energy Drinks. A nutritionist wishes to market a new vitamin-enriched fruit drink and is preparing two versions of it to distribute at a local health club. She has 100 cups of pineapple juice and 4 cups of super vitamin-enriched pomegranate concentrate. One version of the drink is to contain 2% pomegranate and the other version 4% pomegranate. How much of each drink can she create if drinks are 1 cup and she uses all of the ingredients?

80. Easter Eggs. A family is coloring Easter eggs and wants to make 2 shades of purple, "light purple" and "deep purple." They have 30 tablespoons of deep red solution and 2 tablespoons of blue solution. If "light purple" consists of 2% blue solution and "deep purple" consists of 10% blue solution, how much of each version of purple solution can be created?

81. The line $y = mx + b$ connects the points $(-2, 4)$ and $(4, -2)$. Find the values of m and b.

82. Find b and c such that the parabola $y = x^2 + bx + c$ goes through the points $(2, 7)$ and $(-6, 7)$.

83. Find b and c such that the parabola $y = bx^2 + bx + c$ goes through the points $(4, 46)$ and $(-2, 10)$.

84. The system of equations

$$x^2 + y^2 = 4$$
$$x^2 - y^2 = 2$$

can be solved by a change of variables. Taking $u = x^2$ and $v = y^2$, we can transform the system into

$$u + v = 4$$
$$u - v = 2$$

Find the solutions of the original system.

85. The system of equations

$$x^2 + 2y^2 = 11$$
$$4x^2 + y^2 = 16$$

can be solved by a change of variables. Taking $u = x^2$ and $v = y^2$, we can transform the system into

$$u + 2v = 11$$
$$4u + v = 16$$

Find the solutions of the original system.

86. The parabola $y = bx^2 - 2x - a$ goes through the points $(-2, a)$ and $(-1, b - 2)$. Find a and b.

■ TECHNOLOGY

87. Apply a graphing utility to graph the two equations $y = -1.25x + 17.5$ and $y = 2.3x - 14.1$. Approximate the solution to this system of linear equations.

88. Apply a graphing utility to graph the two equations $y = 14.76x + 19.43$ and $y = 2.76x + 5.22$. Approximate the solution to this system of linear equations.

89. Apply a graphing utility to graph the two equations $23x + 15y = 7$ and $46x + 30y = 14$. Approximate the solution to this system of linear equations.

90. Apply a graphing utility to graph the two equations $-3x + 7y = 2$ and $6x - 14y = 3$. Approximate the solution to this system of linear equations.

91. Apply a graphing utility to graph the two equations $\frac{1}{3}x - \frac{5}{12}y = \frac{5}{6}$ and $\frac{3}{7}x + \frac{1}{14}y = \frac{29}{28}$. Approximate the solution to this system of linear equations.

92. Apply a graphing utility to graph the two equations $\frac{5}{9}x + \frac{11}{13}y = 2$ and $\frac{3}{4}x + \frac{5}{7}y = \frac{13}{14}$. Approximate the solution to this system of linear equations.

■ PREVIEW TO CALCULUS

For Exercises 93–96, refer to the following:

In calculus, when integrating rational functions, we decompose the function into partial fractions. This technique involves the solution of systems of equations. For example, suppose

$$\frac{1}{x^2 + x - 2} = \frac{1}{(x - 1)(x + 2)}$$

$$= \frac{A}{x - 1} + \frac{B}{x + 2}$$

$$= \frac{A(x + 2) + B(x - 1)}{(x - 1)(x + 2)}$$

and we want to find A and B such that $1 = A(x + 2) + B(x - 1)$, which is equivalent to $1 = (A + B)x + (2A - B)$. From this equation, we obtain the system of equations

$$A + B = 0$$
$$2A - B = 1$$

which solution is $\left(\frac{1}{3}, -\frac{1}{3}\right)$.

Find the values of A and B that make each equation true.

93. $x + 5 = A(x + 2) + B(x - 4)$

94. $6x = A(x + 1) + B(x - 2)$

95. $x + 1 = A(x + 2) + B(x - 3)$

96. $5 = A(x - 2) + B(2x + 1)$

SYSTEMS OF LINEAR EQUATIONS IN THREE VARIABLES

SKILLS OBJECTIVES

- Solve systems of linear equations in three variables using a combination of both the elimination method and the substitution method.
- Solve application problems using systems of linear equations in three variables.

CONCEPTUAL OBJECTIVES

- Understand that a graph of a linear equation in three variables corresponds to a plane.
- Identify three types of solutions: one solution (point), no solution, or infinitely many solutions (a single line in three-dimensional space or a plane).

In Section 8.1, we solved systems of two linear equations in two variables. Graphs of linear equations in two variables correspond to lines. Now we turn our attention to linear equations in *three* variables. A **linear equation in three variables**, x, y, and z, is given by

$$Ax + By + Cz = D$$

where A, B, C, and D are real numbers that are not all equal to zero. All three variables have degree equal to one, which is why this is called a linear equation in three variables. The graph of any equation in three variables requires a three-dimensional coordinate system.

The x-axis, y-axis, and z-axis are each perpendicular to the other two. For the three-dimensional coordinate system on the right, a point $(x, y, z) = (2, 3, 1)$ is found by starting at the origin, moving 2 units to the right, 3 units up, and 1 unit out toward you.

In two variables, the graph of a linear equation is a line. In three variables, however, the graph of a linear equation is a **plane**. A plane can be thought of as an infinite sheet of paper. When solving systems of linear equations in three variables, we find one of three possibilities: one solution, no solution, or infinitely many solutions.

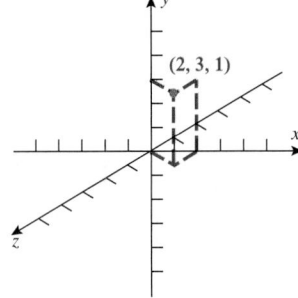
$(2, 3, 1)$

Study Tip

If all three planes are coplaner (the same plane) there are infinitely many solutions.

One Solution	No Solution	Infinitely Many Solutions
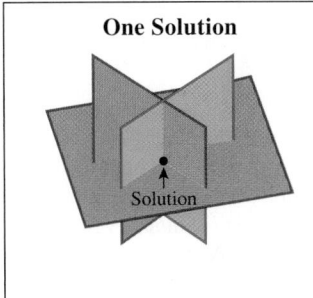 Solution	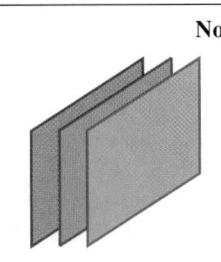 or 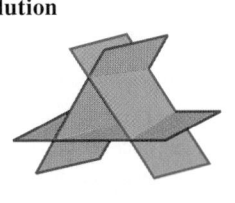	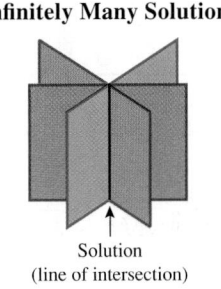 Solution (line of intersection)

Solving Systems of Linear Equations in Three Variables

There are many ways to solve systems of linear equations in more than two variables. One method is to combine the elimination and substitution methods, which will be discussed in this section. Other methods involve matrices, which will be discussed in Sections 8.3–8.5. We now outline a procedure for solving systems of linear equations in three variables, which can be extended to solve systems of more than three variables. Solutions are sometimes given as ordered triples of the form (x, y, z).

SOLVING SYSTEMS OF LINEAR EQUATIONS IN THREE VARIABLES USING ELIMINATION AND SUBSTITUTION

Step 1: Reduce the system of three equations in three variables to two equations in two (of the same) variables by applying elimination.

Step 2: Solve the resulting system of two linear equations in two variables by applying elimination or substitution.

Step 3: Substitute the solutions in Step 2 into *any* one of the original equations and solve for the third variable.

Step 4: Check that the solution satisfies *all* three original equations.

Technology Tip

The ⟦ref⟧ under the ⟦MATRIX⟧ menu will be used to solve the system of equations by entering the coefficients of x, y, z, and the constant.

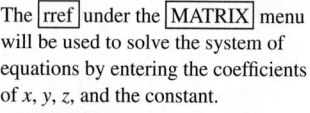

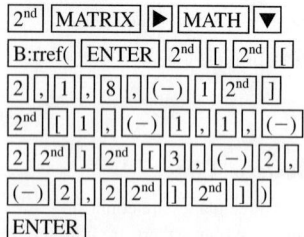

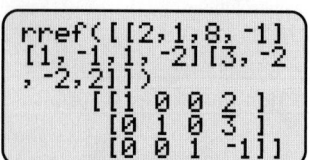

First row gives $x = 2$, second row gives $y = 3$, and third row gives $z = -1$.

Note: The TI function ⟦rref⟧ stands for reduced row echelon form.

EXAMPLE 1 Solving a System of Linear Equations in Three Variables

Solve the system:

$$2x + y + 8z = -1 \qquad \text{Equation (1)}$$
$$x - y + z = -2 \qquad \text{Equation (2)}$$
$$3x - 2y - 2z = 2 \qquad \text{Equation (3)}$$

Solution:

Inspecting the three equations, we see that y is easily eliminated when Equations (1) and (2) are added, because the coefficients of y, $+1$ and -1, are equal in magnitude and opposite in sign. We can also eliminate y from Equation (3) by adding Equation (3) to *either* 2 times Equation (1) *or* -2 times Equation (2). Therefore, our plan of attack is to eliminate y from the system of equations, so the result will be two equations in two variables x and z.

Study Tip

First eliminate the *same* variable
from two different pairs of equations.

STEP 1 Eliminate y in Equation (1) and Equation (2).

Equation (1):
Equation (2):
Add.

$$2x + y + 8z = -1$$
$$\underline{x - y + z = -2}$$
$$3x \quad + 9z = -3$$

Eliminate y in Equation (2) and Equation (3).

Multiply Equation (2) by -2.
Equation (3):
Add.

$$-2x + 2y - 2z = 4$$
$$\underline{3x - 2y - 2z = 2}$$
$$x \quad - 4z = 6$$

STEP 2 Solve the system of two linear equations
in two variables.

$$3x + 9z = -3$$
$$x - 4z = 6$$

Substitution* method: $x = 4z + 6$

$$3(4z + 6) + 9z = -3$$

Distribute.

$$12z + 18 + 9z = -3$$

Combine like terms.

$$21z = -21$$

Solve for z.

$$\boxed{z = -1}$$

Substitute $z = -1$ into $x = 4z + 6$.

$$x = 4(-1) + 6 = 2$$

$\boxed{x = 2}$ and $\boxed{z = -1}$ are the solutions to the system of two equations.

STEP 3 Substitute $x = 2$ and $z = -1$ into any one of the three original equations and
solve for y.

Substitute $x = 2$ and $z = -1$ into Equation (2).

$$2 - y - 1 = -2$$

Solve for y.

$$\boxed{y = 3}$$

STEP 4 Check that $x = 2$, $y = 3$, and $z = -1$ satisfy all three equations.

Equation (1): $2(2) + 3 + 8(-1) = 4 + 3 - 8 = -1$

Equation (2): $2 - 3 - 1 = -2$

Equation (3): $3(2) - 2(3) - 2(-1) = 6 - 6 + 2 = 2$

The solution is $\boxed{x = 2, y = 3, z = -1, \text{ or } (2, 3, -1)}$.

*Elimination method could also be used.

■ **YOUR TURN** Solve the system:
$$2x - y + 3z = -1$$
$$x + y - z = 0$$
$$3x + 3y - 2z = 1$$

In Example 1 and the Your Turn, the variable y was eliminated by adding the first and
second equations. In practice, any of the three variables can be eliminated, but typically we
select the most convenient variable to eliminate. If a variable is missing from one of the
equations (has a coefficient of 0), then we eliminate that variable from the other two equations.

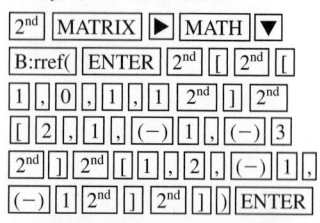
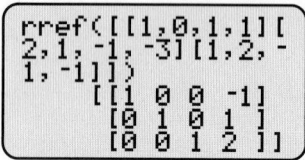
EXAMPLE 2   **Solving a System of Linear Equations in Three Variables When One Variable Is Missing**

Solve the system:

$$
\begin{aligned}
x \quad\quad + z &= 1 && \text{Equation (1)} \\
2x + y - z &= -3 && \text{Equation (2)} \\
x + 2y - z &= -1 && \text{Equation (3)}
\end{aligned}
$$

Solution:

Since y is missing from Equation (1), y is the variable to be eliminated in Equation (2) and Equation (3).

STEP 1   Eliminate y.

Multiply Equation (2) by -2.
Equation (3):

$$
\begin{aligned}
-4x - 2y + 2z &= 6 \\
x + 2y - z &= -1
\end{aligned}
$$

Add.

$$-3x \quad\quad + z = 5$$

STEP 2   Solve the system of two equations.

$$x + z = 1$$

Equation (1) and the resulting equation in Step 1.

$$-3x + z = 5$$

Multiply the second equation by (-1) and add it to the first equation.

$$
\begin{aligned}
x + z &= 1 \\
3x - z &= -5 \\
\hline
4x \quad\quad &= -4
\end{aligned}
$$

Solve for x.

$$x = -1$$

Substitute $x = -1$ into Equation (1).

$$-1 + z = 1$$

Solve for z.

$$z = 2$$

STEP 3   Substitute $x = -1$ and $z = 2$ into one of the original equations [Equation (2) or Equation (3)] and solve for y.

Substitute $x = -1$ and $z = 2$ into $x + 2y - z = -1$.

$$(-1) + 2y - 2 = -1$$

Gather like terms.

$$2y = 2$$

Solve for y.

$$y = 1$$

STEP 4   Check that $x = -1$, $y = 1$, and $z = 2$ satisfy all three equations.

Equation (1): $(-1) + 2 = 1$

Equation (2): $2(-1) + (1) - (2) = -3$

Equation (3): $(-1) + 2(1) - (2) = -1$

The solution is $\boxed{x = -1, y = 1, z = 2}$.

■ **Answer:** $x = 1, y = 2, z = -3$

■ **YOUR TURN**   Solve the system:

$$
\begin{aligned}
x + y + z &= 0 \\
2x \quad\quad + z &= -1 \\
x - y - z &= 2
\end{aligned}
$$

Types of Solutions

Systems of linear equations in three variables have three possible solutions: one solution, infinitely many solutions, or no solution. Examples 1 and 2 each had one solution. Examples 3 and 4 illustrate systems with infinitely many solutions and no solution, respectively.

EXAMPLE 3 **A Dependent System of Linear Equations in Three Variables (Infinitely Many Solutions)**

Solve the system:

$2x + y - z = 4$	Equation (1)
$x \quad + y = 2$	Equation (2)
$3x + 2y - z = 6$	Equation (3)

Solution:

Since z is missing from Equation (2), z is the variable to be eliminated from Equation (1) and Equation (3).

STEP 1 Eliminate z.

Multiply Equation (1) by (-1).
Equation (3):
Add.

$$-2x - y + z = -4$$
$$3x + 2y - z = 6$$
$$\overline{x + y = 2}$$

STEP 2 Solve the system of two equations:
Equation (2) and the resulting
equation in Step 1.

$$x + y = 2$$
$$x + y = 2$$

Multiply the first equation by (-1)
and add it to the second equation.

$$-x - y = -2$$
$$\overline{x + y = 2}$$
$$0 = 0$$

This statement is always true; therefore, there are infinitely many solutions. The original system has been reduced to a system of two identical linear equations. Therefore, the equations are dependent (share infinitely many solutions). Typically, to define those infinitely many solutions, we let $z = a$, where a stands for any real number, and then find x and y in terms of a. The resulting ordered triple showing the three variables in terms of a is called a **parametric representation** of a line in three dimensions.

STEP 3 Let $z = a$ and find x and y in terms of a.

Solve Equation (2) for y.

$$y = 2 - x$$

Let $y = 2 - x$ and $z = a$ in Equation (1).

$$2x + (2 - x) - a = 4$$

Solve for x.

$$2x + 2 - x - a = 4$$
$$x - a = 2$$
$$\boxed{x = a + 2}$$

Let $x = a + 2$ in Equation (2).

$$(a + 2) + y = 2$$

Solve for y.

$$\boxed{y = -a}$$

The infinitely many solutions are written as $(a + 2, -a, a)$.

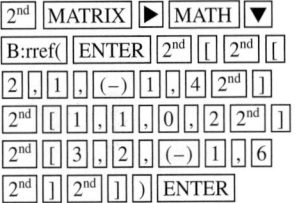
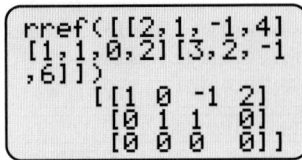

Step 4 Check that $x = a + 2$, $y = -a$, and $z = a$ satisfy all three equations.

Equation (1): $2(a + 2) + (-a) - a = 2a + 4 - a - a = 4$ ✓

Equation (2): $(a + 2) + (-a) = a + 2 - a = 2$ ✓

Equation (3): $3(a + 2) + 2(-a) - a = 3a + 6 - 2a - a = 6$ ✓

▪ **Answer:** $(a - 1, a + 1, a)$

▪ **YOUR TURN** Solve the system:
$$\begin{aligned} x + y - 2z &= 0 \\ x \quad\;\;\; - z &= -1 \\ x - 2y + z &= -3 \end{aligned}$$

Technology Tip

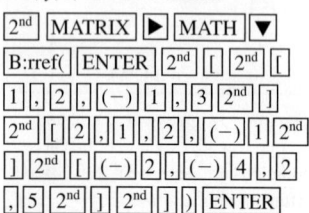

The $\boxed{\text{rref}}$ under the $\boxed{\text{MATRIX}}$ menu will be used to solve the system of equations by entering the coefficients of x, y, z, and the constant.

$\boxed{2^{\text{nd}}}$ $\boxed{\text{MATRIX}}$ $\boxed{\blacktriangleright}$ $\boxed{\text{MATH}}$ $\boxed{\blacktriangledown}$
$\boxed{\text{B:rref(}}$ $\boxed{\text{ENTER}}$ $\boxed{2^{\text{nd}}}$ $\boxed{[}$ $\boxed{2^{\text{nd}}}$ $\boxed{[}$
$\boxed{1}$ $\boxed{,}$ $\boxed{2}$ $\boxed{,}$ $\boxed{(-)}$ $\boxed{1}$ $\boxed{,}$ $\boxed{3}$ $\boxed{2^{\text{nd}}}$ $\boxed{]}$
$\boxed{2^{\text{nd}}}$ $\boxed{[}$ $\boxed{2}$ $\boxed{,}$ $\boxed{1}$ $\boxed{,}$ $\boxed{2}$ $\boxed{,}$ $\boxed{(-)}$ $\boxed{1}$ $\boxed{2^{\text{nd}}}$
$\boxed{]}$ $\boxed{2^{\text{nd}}}$ $\boxed{[}$ $\boxed{(-)}$ $\boxed{2}$ $\boxed{,}$ $\boxed{(-)}$ $\boxed{4}$ $\boxed{,}$ $\boxed{2}$
$\boxed{,}$ $\boxed{5}$ $\boxed{2^{\text{nd}}}$ $\boxed{]}$ $\boxed{2^{\text{nd}}}$ $\boxed{]}$ $\boxed{)}$ $\boxed{\text{ENTER}}$

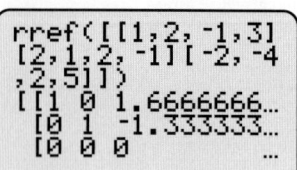

```
rref([[1,2,-1,3]
[2,1,2,-1][-2,-4
,2,5]])
[[1 0 1.6666666…
[0 1 -1.333333…
[0 0 0
                …
```

To change to fractions, press:

$\boxed{\text{MATH}}$ $\boxed{\text{1:Frac}}$ $\boxed{\text{ENTER}}$ $\boxed{\text{ENTER}}$

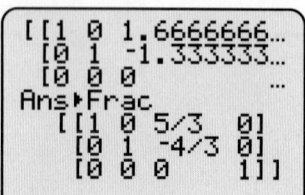

```
[[1 0 1.6666666…
[0 1 -1.333333…
[0 0 0
                …
Ans▶Frac
[[1 0 5/3  0]
[0 1 -4/3 0]
[0 0 0    1]]
```

The third row gives $0 = 1$, which is a contradiction. Therefore, there is no solution.

EXAMPLE 4 An Inconsistent System of Linear Equations in Three Variables (No Solution)

Solve the system:
$$\begin{aligned} x + 2y - z &= 3 && \text{Equation (1)} \\ 2x + y + 2z &= -1 && \text{Equation (2)} \\ -2x - 4y + 2z &= 5 && \text{Equation (3)} \end{aligned}$$

Solution:

Step 1 Eliminate x.

Multiply Equation (1) by -2.
Equation (2):
Add.

$$\begin{aligned} -2x - 4y + 2z &= -6 \\ 2x + y + 2z &= -1 \\ \hline -3y + 4z &= -7 \end{aligned}$$

Equation (2):
Equation (3):
Add.

$$\begin{aligned} 2x + y + 2z &= -1 \\ -2x - 4y + 2z &= 5 \\ \hline -3y + 4z &= 4 \end{aligned}$$

Step 2 Solve the system of two equations:

$$\begin{aligned} -3y + 4z &= -7 \\ -3y + 4z &= 4 \end{aligned}$$

Multiply the top equation by (-1) and add it to the second equation.

$$\begin{aligned} 3y - 4z &= 7 \\ -3y + 4z &= 4 \\ \hline 0 &= 11 \end{aligned}$$

This is a contradiction, or inconsistent statement, and therefore, there is $\boxed{\text{no solution}}$.

So far in this section, we have discussed only systems of *three* linear equations in *three* variables. What happens if we have a system of *two* linear equations in *three* variables? The two linear equations in three variables will always correspond to two planes in three dimensions. The possibilities are no solution (the two planes are parallel) or infinitely many solutions (the two planes intersect in a line or two planes are coplanar).

NO SOLUTION	INFINITELY MANY SOLUTIONS (LINE)
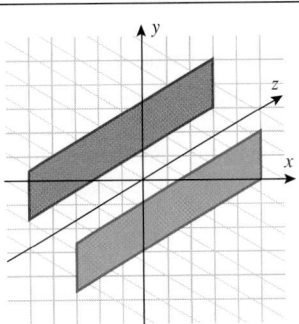	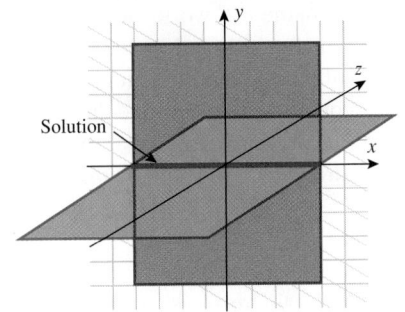

EXAMPLE 5   Solving a System of Two Linear Equations in Three Variables

Solve the system of linear equations: $x - y + z = 7$ Equation (1)

$\qquad\qquad\qquad\qquad\qquad\qquad\qquad x + y + 2z = 2$ Equation (2)

Solution:

Eliminate y by adding the two equations.

$$\begin{array}{r} x - y + z = 7 \\ x + y + 2z = 2 \\ \hline 2x + 3z = 9 \end{array}$$

Therefore, Equation (1) and Equation (2) are both true if $2x + 3z = 9$. Since we know there is a solution, it must be a line. To define the line of intersection, we again turn to parametric representation.

Let $\boxed{z = a}$, where a is any real number.

$$2x + 3a = 9$$

Solve for x.

$$\boxed{x = \frac{9}{2} - \frac{3}{2}a}$$

Substitute $z = a$ and $x = \frac{9}{2} - \frac{3}{2}a$ into Equation (1).

$$\left(\frac{9}{2} - \frac{3}{2}a\right) - y + a = 7$$

Solve for y.

$$\boxed{y = -\frac{1}{2}a - \frac{5}{2}}$$

The solution is the line in three dimensions given by $\boxed{\left(\frac{9}{2} - \frac{3}{2}a, -\frac{1}{2}a - \frac{5}{2}, a\right)}$, where a is any real number.

Note: Every real number a corresponds to a point on the line of intersection.

a	$\left(\frac{9}{2} - \frac{3}{2}a, -\frac{1}{2}a - \frac{5}{2}, a\right)$
-1	$(6, -2, -1)$
0	$\left(\frac{9}{2}, -\frac{5}{2}, 0\right)$
1	$(3, -3, 1)$

Modeling with a System of Three Linear Equations

Many times in the real world we see a relationship that looks like a particular function such as a quadratic function and we know particular data points, but we do not know the function. We start with the general function, fit the curve to particular data points, and solve a system of linear equations to determine the specific function parameters.

Suppose you want to model a stock price as a function of time and based on the data you feel a quadratic model would be the best fit. Therefore, the model is given by

$$P(t) = at^2 + bt + c$$

where $P(t)$ is the price of the stock at time t. If we have data corresponding to three distinct points $[t, P(t)]$, the result is a system of three linear equations in three variables a, b, and c. We can solve the resulting system of linear equations, which determines the coefficients a, b, and c of the quadratic model for stock price.

EXAMPLE 6 Stock Value

The Oracle Corporation's stock (ORCL) over 3 days (Wednesday, October 13, to Friday, October 15, 2004) can be approximately modeled by a quadratic function: $f(t) = at^2 + bt + c$. If Wednesday corresponds to $t - 1$, where t is in days, then the following data points approximately correspond to the stock value:

t	$f(t)$	DAYS
1	$12.20	Wednesday
2	$12.00	Thursday
3	$12.20	Friday

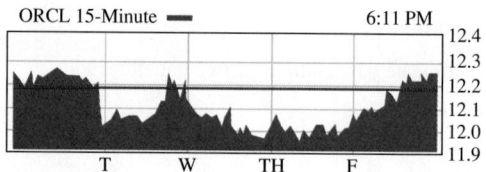

Determine the function that models this behavior.

Solution:

Substitute the points (1, 12.20), (2, 12.00), and (3, 12.20) into $f(t) = at^2 + bt + c$.

$$a(1)^2 + b(1) + c = 12.20$$

$$a(2)^2 + b(2) + c = 12.00$$

$$a(3)^2 + b(3) + c = 12.20$$

Simplify to a system of three equations in three variables (a, b, and c).

$$a + b + c = 12.20 \quad \text{Equation (1)}$$
$$4a + 2b + c = 12.00 \quad \text{Equation (2)}$$
$$9a + 3b + c = 12.20 \quad \text{Equation (3)}$$

Solve for a, b, and c by applying the technique of this section.

STEP 1 Eliminate c.

Multiply Equation (1) by (-1).	$-a - b - c = -12.20$
Equation (2):	$4a + 2b + c = 12.20$
Add.	$3a + b = -0.20$

Multiply Equation (1) by -1.	$-a - b - c = -12.20$
Equation (3):	$9a + 3b + c = 12.20$
Add.	$8a + 2b = 0$

Step 2 Solve the system of two equations.

$$3a + b = -0.20$$
$$8a + 2b = 0$$

Multiply the first equation by -2 and add to the second equation.

$$-6a - 2b = 0.40$$
$$8a + 2b = 0$$

Add.

$$2a = 0.4$$

Solve for a.

$$a = 0.2$$

Substitute $a = 0.2$ into $8a + 2b = 0$.

$$8(0.2) + 2b = 0$$

Simplify.

$$2b = -1.6$$

Solve for b.

$$b = -0.8$$

Step 3 Substitute $a = 0.2$ and $b = -0.8$ into one of the original three equations.

Substitute $a = 0.2$ and $b = -0.8$ into $a + b + c = 12.20$.

$$0.2 - 0.8 + c = 12.20$$

Gather like terms.

$$-0.6 + c = 12.20$$

Solve for c.

$$c = 12.80$$

Step 4 Check that $a = 0.2$, $b = -0.8$, and $c = 12.80$ satisfy all three equations.

Equation (1): $\quad a + b + c = 0.2 - 0.8 + 12.8 = 12.20$

Equation (2): $\quad 4a + 2b + c = 4(0.2) + 2(-0.8) + 12.80$
$$= 0.8 - 1.6 + 12.8 = 12.00$$

Equation (3): $\quad 9a + 3b + c = 9(0.2) + 3(-0.8) + 12.80$
$$= 1.8 - 2.4 + 12.8 = 12.20$$

The model is given by $\boxed{f(t) = 0.2t^2 - 0.8t + 12.80}$.

SECTION 8.2 SUMMARY

Graphs of linear equations in *two* variables are *lines*, whereas graphs of linear equations in *three* variables are *planes*. Systems of linear equations in three variables have one of three solutions:

- One solution (point)
- No solution (no intersection of all three planes)
- Infinitely many solutions (planes intersect along a line)

When the solution to a system of three linear equations is a line in three dimensions, we use parametric representation to express the solution.

SECTION 8.2 EXERCISES

SKILLS

In Exercises 1–32, solve each system of linear equations.

1. $\quad x - y + z = 6$
$\quad -x + y + z = 3$
$\quad -x - y - z = 0$

2. $\quad -x - y + z = -1$
$\quad -x + y - z = 3$
$\quad x - y - z = 5$

3. $\quad x + y - z = 2$
$\quad -x - y - z = -3$
$\quad -x + y - z = 6$

4. $\quad x + y + z = -1$
$\quad -x + y - z = 3$
$\quad -x - y + z = 8$

5. $-x + y - z = -1$
$x - y - z = 3$
$x + y - z = 9$

6. $x - y - z = 2$
$-x - y + z = 4$
$-x + y - z = 6$

7. $2x - 3y + 4z = -3$
$-x + y + 2z = 1$
$5x - 2y - 3z = 7$

8. $x - 2y + z = 0$
$-2x + y - z = -5$
$13x + 7y + 5z = 6$

9. $3y - 4x + 5z = 2$
$2x - 3y - 2z = -3$
$3z + 4y - 2x = 1$

10. $2y + z - x = 5$
$2x + 3z - 2y = 0$
$-2z + y - 4x = 3$

11. $x - y + z = -1$
$y - z = -1$
$-x + y + z = 1$

12. $-y + z = 1$
$x - y + z = -1$
$x - y - z = -1$

13. $3x - 2y - 3z = -1$
$x - y + z = -4$
$2x + 3y + 5z = 14$

14. $3x - y + z = 2$
$x - 2y + 3z = 1$
$2x + y - 3z = -1$

15. $-3x - y - z = 2$
$x + 2y - 3z = 4$
$2x - y + 4z = 6$

16. $2x - 3y + z = 1$
$x + 4y - 2z = 2$
$3x - y + 4z = -3$

17. $3x + 2y + z = 4$
$-4x - 3y - z = -15$
$x - 2y + 3z = 12$

18. $3x - y + 4z = 13$
$-4x - 3y - z = -15$
$x - 2y + 3z = 12$

19. $-x + 2y + z = -2$
$3x - 2y + z = 4$
$2x - 4y - 2z = 4$

20. $2x - y = 1$
$-x + z = -2$
$-2x + y = -1$

21. $x - z - y = 10$
$2x - 3y + z = -11$
$y - x + z = -10$

22. $2x + z + y = -3$
$2y - z + x = 0$
$x + y + 2z = 5$

23. $3x_1 + x_2 - x_3 = 1$
$x_1 - x_2 + x_3 = -3$
$2x_1 + x_2 + x_3 = 0$

24. $2x_1 + x_2 + x_3 = -1$
$x_1 + x_2 - x_3 = 5$
$3x_1 - x_2 - x_3 = 1$

25. $2x + 5y = 9$
$x + 2y - z = 3$
$-3x - 4y + 7z = 1$

26. $x - 2y + 3z = 1$
$-2x + 7y - 9z = 4$
$x + z = 9$

27. $2x_1 - x_2 + x_3 = 3$
$x_1 - x_2 + x_3 = 2$
$-2x_1 + 2x_2 - 2x_3 = -4$

28. $x_1 - x_2 - 2x_3 = 0$
$-2x_1 + 5x_2 + 10x_3 = -3$
$3x_1 + x_2 = 0$

29. $2x + y - z = 2$
$x - y - z = 6$

30. $3x + y - z = 0$
$x + y + 7z = 4$

31. $4x + 3y - 3z = 5$
$6x + 2z = 10$

32. $x + 2y + 4z = 12$
$-3x - 4y + 7z = 21$

▪ APPLICATIONS

Exercises 33 and 34 rely on a selection of Subway sandwiches whose nutrition information is given in the table.

Suppose you are going to eat only Subway sandwiches for a week (seven days) for lunch and dinner (total of fourteen meals).

SANDWICH	CALORIES	FAT (GRAMS)
Mediterranean chicken	350	18
Six-inch tuna	430	19
Six-inch roast beef	290	5

www.subway.com

33. Diet. Your goal is a total of 4840 calories and 190 grams of fat. How many of each sandwich would you eat that week to obtain this goal?

34. Diet. Your goal is a total of 4380 calories and 123 grams of fat. How many of each sandwich would you eat that week to obtain this goal?

Exercises 35 and 36 involve vertical motion and the effect of gravity on an object.

Because of gravity, an object that is projected upward will eventually reach a maximum height and then fall to the ground. The equation that determines the height h of a projectile t seconds after it is shot upward is given by

$$h = \frac{1}{2}at^2 + v_0t + h_0$$

where a is the acceleration due to gravity, h_0 is the initial height of the object at time $t = 0$, and v_0 is the initial velocity of the object at time $t = 0$. Note that a projectile follows the path of a parabola opening down, so $a < 0$.

35. Vertical Motion. An object is thrown upward and the following table depicts the height of the ball t seconds after the projectile is released. Find the initial height, initial velocity, and acceleration due to gravity.

t SECONDS	HEIGHT (FEET)
1	36
2	40
3	12

36. Vertical Motion. An object is thrown upward and the following table depicts the height of the ball t seconds after the projectile is released. Find the initial height, initial velocity, and acceleration due to gravity.

t SECONDS	HEIGHT (FEET)
1	84
2	136
3	156

37. Data Curve-Fitting. The number of minutes that an average person of age x spends driving a car can be modeled by a quadratic function $y = ax^2 + bx + c$, where $a < 0$ and $18 \leq x \leq 65$. The following table gives the average number of minutes per day that a person spends driving a car. Determine the quadratic function that models this quantity.

AGE	AVERAGE DAILY MINUTES DRIVING
20	30
40	60
60	40

38. Data Curve-Fitting. The average age when a woman gets married began increasing during the last century. In 1930 the average age was 18.6, in 1950 the average age was 20.2, and in 2002 the average age was 25.3. Find a quadratic function $y = ax^2 + bx + c$, where $a > 0$ and $18 < y < 35$, that models the average age y when a woman gets married as a function of the year x ($x = 0$ corresponds to 1930). What will the average age be in 2010?

39. Money. Tara and Lamar decide to place $20,000 of their savings into investments. They put some in a money market account earning 3% interest, some in a mutual fund that has been averaging 7% a year, and some in a stock that rose 10% last year. If they put $6,000 more in the money market than in the mutual fund and the mutual fund and stocks experience the same growth the next year as they did the previous year, they will earn $1,180 in a year. How much money did Tara and Lamar put in each of the three investments?

40. Money. Tara talks Lamar into putting less money in the money market and more money in the stock (see Exercise 39). They place $20,000 of their savings into investments. They put some in a money market account earning 3% interest, some in a mutual fund that has been averaging 7% a year, and some in a stock that rose 10% last year. If they put $6,000 more in the stock than in the mutual fund and the mutual fund and stock experience the same growth the next year as they did the previous year, they will earn $1,680 in a year. How much money did Tara and Lamar put in each of the three investments?

41. Ski Production. A company produces three types of skis: regular model, trick ski, and slalom ski. They need to fill a customer order of 110 pairs of skis. There are two major production divisions within the company: labor and finishing. Each regular model of skis requires 2 hours of labor and 1 hour of finishing. Each trick ski model requires 3 hours of labor and 2 hours of finishing. Finally, each slalom ski model requires 3 hours of labor and 5 hours of finishing. Suppose the company has only 297 labor hours and 202 finishing hours. How many of each type ski can be made under these restrictions?

42. Automobile Production. An automobile manufacturing company produces three types of automobiles: compact car, intermediate, and luxury model. The company has the capability of producing 500 automobiles. Suppose that each compact-model car requires 200 units of steel and 30 units of rubber, each intermediate model requires 300 units of steel and 20 units of rubber, and each luxury model requires 250 units of steel and 45 units of rubber. The number of units of steel available is 128,750, and the number of units of rubber available is 15,625. How many of each type of automobile can be produced with these restraints?

43. Computer versus Man. *The Seattle Times* reported a story on November 18, 2006, about a game of Scrabble played between a human and a computer. The best Scrabble player in the United States was pitted against a computer program designed to play the game. Remarkably, the human beat the computer in the best of two out of three games competition. The total points scored by both computer and the man for all three games was 2591. The difference between the first game's total and second game's total was 62 points. The difference between the first game's total and the third game's total was only 2 points. Determine the total number of points scored by both computer and the man for each of the three contests.

44. Brain versus Computer. Can the human brain perform more calculations per second than a supercomputer? The calculating speed of the three top supercomputers, IBM's Blue Gene/L, IBM's BGW, and IBM's ASC Purple, has been determined. The speed of IBM's Blue Gene/L is 245 teraflops more than that of IBM's BGW. The computing speed of IBM's BGW is 22 teraflops more than that of IBM's ASC Purple. The combined speed of all three top supercomputers is 568 teraflops. Determine the computing speed (in teraflops) of each supercomputer. A **teraflop** is a measure of a computer's speed and can be expressed as 1 trillion floating-point operations per second. By comparison, it is estimated that the human brain can perform 10 quadrillion calculations per second.

45. Production. A factory manufactures three types of golf balls: Eagle, Birdie, and Bogey. The daily production is 10,000 balls. The number of Eagle and Birdie balls combined equals the number of Bogey balls produced. If the factory makes three times more Birdie than Eagle balls, find the daily production of each type of ball.

46. Pizza. Three-cheese pizzas are made with a mixture of three types of cheese. The cost of a pizza containing 2 parts of each cheese is $2.40. A pizza made with 2 parts of cheese A, 1 part of cheese B, and 2 parts of cheese C costs $2.20, while a pizza made with 2 parts of cheese A, 2 parts of cheese B, and 3 parts of cheese C costs $2.70. Determine the cost, per part, of each cheese.

47. TV Commercials. A TV station sells intervals of time for commercials of 10 seconds for $100, 20 seconds for $180, and 40 seconds for $320. It has 2 minutes for publicity during a game with a total revenue of $1,060 for six commercials shown. Find the number of commercials of each length sold by the TV station if there are twice as many 10 second commercials as 40 second commercials.

48. Airline. A commercial plane has 270 seats divided into three classes: first class, business, and coach. The first-class seats are a third of the business-class seats. There are 250 more coach seats than first-class seats. Find the number of seats of each class in the airplane.

▪ CATCH THE MISTAKE

In Exercises 49 and 50, explain the mistake that is made.

49. Solve the system of equations.

Equation (1):	$2x - y + z = 2$
Equation (2):	$x - y\quad = 1$
Equation (3):	$x\quad + z = 1$

Solution:

Equation (2):	$x - y\quad = 1$
Equation (3):	$x\quad + z = 1$

Add Equation (2) and Equation (3). $\quad -y + z = 2$

Multiply Equation (1) by (-1). $\quad -2x + y - z = -2$

Add. $\quad -2x \quad = 0$

Solve for x. $\qquad x = 0$

Substitute $x = 0$ into Equation (2). $\quad 0 - y = 1$

Solve for y. $\qquad y = -1$

Substitute $x = 0$ into Equation (3). $\quad 0 + z = 1$

Solve for z. $\qquad z = 1$

The answer is $x = 0$, $y = -1$, and $z = 1$.

This is incorrect. Although $x = 0$, $y = -1$, and $z = 1$ does satisfy the three original equations, it is only one of infinitely many solutions. What mistake was made?

50. Solve the system of equations.

Equation (1):	$x + 3y + 2z = 4$
Equation (2):	$3x + 10y + 9z = 17$
Equation (3):	$2x + 7y + 7z = 17$

Solution:

Multiply Equation (1) by -3. $\quad -3x - 9y - 6z = -12$

Equation (2): $\quad \underline{3x + 10y + 9z = \quad 17}$

Add. $\qquad y + 3z = 5$

Multiply Equation (1) by -2. $\quad -2x - 6y - 4z = -8$

Equation (3): $\quad \underline{2x + 7y + 7z = 17}$

Add. $\qquad y + 3z = 9$

Solve the system of two equations. $\quad y + 3z = 5$

$\qquad y + 3z = 9$

Infinitely many solutions.

Let $z = a$, then $y = 5 - 3a$.

Substitute $z = a$ and $y = 5 - 3a$ into Equation (1).

$x + 3y + 2z = 4$

$x + 3(5 - 3a) + 2a = 4$

Eliminate parentheses. $\quad x + 15 - 9a + 2a = 4$

Solve for x. $\qquad x = 7a - 11$

The answer is $x = 7a - 11$, $y = 5 - 3a$, and $z = a$.

This is incorrect. There is no solution. What mistake was made?

▪ CONCEPTUAL

In Exercises 51–54, determine whether each statement is true or false.

51. A system of linear equations that has more variables than equations cannot have a unique solution.

52. A system of linear equations that has the same number of equations as variables always has a unique solution.

53. The linear equation $Ax + By = C$ always represents a straight line.

54. If the system of linear equations

$$x + 2y + 3z = a$$
$$2x + 3y + z = b$$
$$3x + y + 2z = c$$

has a unique solution $\left(\frac{1}{6}, \frac{1}{6}, \frac{1}{6}\right)$, then the system of equations

$$x + 2y + 3z = 2a$$
$$2x + 3y + z = 2b$$
$$3x + y + 2z = 2c$$

has a unique solution $\left(\frac{1}{3}, \frac{1}{3}, \frac{1}{3}\right)$.

55. The circle given by the equation $x^2 + y^2 + ax + by + c = 0$ passes through the points $(-2, 4)$, $(1, 1)$, and $(-2, -2)$. Find a, b, and c.

56. The circle given by the equation $x^2 + y^2 + ax + by + c = 0$ passes through the points $(0, 7)$, $(6, 1)$, and $(5, 4)$. Find a, b, and c.

■ CHALLENGE

57. A fourth-degree polynomial,

$$f(x) = ax^4 + bx^3 + cx^2 + dx + e, \text{ with } a < 0,$$

can be used to represent the following data on the number of deaths per year due to lightning strikes. Assume 1999 corresponds to $x = -2$ and 2003 corresponds to $x = 2$. Use the data to determine a, b, c, d, and e.

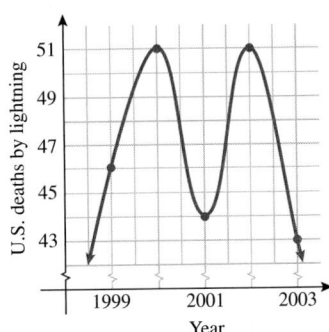

58. A copy machine accepts nickels, dimes, and quarters. After 1 hour, there are 30 coins total and their value is $4.60. If there are four more quarters than nickels, how many nickels, quarters, and dimes are in the machine?

In Exercises 59–62, solve the system of linear equations.

59.
$$\begin{aligned} 2y + z &= 3 \\ 4x \quad - z &= -3 \\ 7x - 3y - 3z &= 2 \\ x - y - z &= -2 \end{aligned}$$

60.
$$\begin{aligned} -2x - y + 2z &= 3 \\ 3x \quad - 4z &= 2 \\ 2x + y \quad &= -1 \\ -x + y - z &= -8 \end{aligned}$$

61.
$$\begin{aligned} 3x_1 - 2x_2 + x_3 + 2x_4 &= -2 \\ -x_1 + 3x_2 + 4x_3 + 3x_4 &= 4 \\ x_1 + x_2 + x_3 + x_4 &= 0 \\ 5x_1 + 3x_2 + x_3 + 2x_4 &= -1 \end{aligned}$$

62.
$$\begin{aligned} 5x_1 + 3x_2 + 8x_3 + x_4 &= 1 \\ x_1 + 2x_2 + 5x_3 + 2x_4 &= 3 \\ 4x_1 \quad + x_3 - 2x_4 &= -3 \\ x_2 + x_3 + x_4 &= 0 \end{aligned}$$

63. Find the values of A, B, C, and D such that the following equation is true:

$$x^3 + x^2 + 2x + 3 = (Ax + B)(x^2 + 3) + (Cx + D)(x^2 + 2)$$

64. Find the values of A, B, C, D, and E such that the following equation is true:

$$\begin{aligned} Ax^3(x + 1) &+ Bx^2(x + 1) + Cx(x + 1) + D(x + 1) + Ex^4 \\ &= 4x^4 + x + 1 \end{aligned}$$

■ TECHNOLOGY

In Exercises 65 and 66, employ a graphing calculator to solve the system of linear equations (most graphing calculators have the capability of solving linear systems with the user entering the coefficients).

65.
$$\begin{aligned} x - z - y &= 10 \\ 2x - 3y + z &= -11 \\ y - x + z &= -10 \end{aligned}$$

66.
$$\begin{aligned} 2x + z + y &= -3 \\ 2y - z + x &= 0 \\ x + y + 2z &= 5 \end{aligned}$$

67. Graphing calculators and graphing utilities have the ability to graph in three dimensions (3D) as opposed to the traditional two dimensions (2D). The line must be given in the form $z = ax + by + c$. Rewrite the system of equations in Exercise 65 in this form and graph the three lines in 3D. What is the point of intersection? Compare that with your answer in Exercise 65.

68. Graphing calculators and graphing utilities have the ability to graph in three dimensions (3D) as opposed to the traditional two dimensions (2D). The line must be given in the form $z = ax + by + c$. Rewrite the system of equations in Exercise 64 in this form and graph the three lines in 3D. What is the point of intersection? Compare that with your answer in Exercise 66.

In Exercises 69 and 70, employ a graphing calculator to solve the system of equations.

69.
$$\begin{aligned} 0.2x - 0.7y + 0.8z &= 11.2 \\ -1.2x + 0.3y - 1.5z &= 0 \\ 0.8x - 0.1y + 2.1z &= 6.4 \end{aligned}$$

70.
$$\begin{aligned} 1.8x - 0.5y + 2.4z &= 1.6 \\ 0.3x \quad - 0.6z &= 0.2 \end{aligned}$$

■ PREVIEW TO CALCULUS

In calculus, when integrating rational functions, we decompose the function into partial fractions. This technique involves the solution of systems of equations.

In Exercises 71–74, find the values of A, B, and C that make each equation true.

71. $5x^2 + 6x + 2 = A(x^2 + 2x + 5) + (Bx + C)(x + 2)$

72. $2x^2 - 3x + 2 = A(x^2 + 1) + (Bx + C)x$

73. $3x + 8 = A(x^2 + 5x + 6) + B(x^2 + 3x) + C(x^2 + 2x)$

74. $x^2 + x + 1 = A(x^2 + 5x + 6) + B(x^2 + 4x + 3)$
$\qquad\qquad + C(x^2 + 3x + 2)$

SKILLS OBJECTIVES

- Write a system of linear equations as an augmented matrix.
- Perform row operations on an augmented matrix.
- Write a matrix in row–echelon form.
- Solve systems of linear equations using Gaussian elimination with back-substitution.
- Write a matrix in reduced row–echelon form.
- Solve systems of linear equations using Gauss–Jordan elimination.

CONCEPTUAL OBJECTIVES

- Visualize an augmented matrix as a system of linear equations.
- Understand that solving systems with augmented matrices is equivalent to solving by the method of elimination.
- Recognize matrices that correspond to inconsistent and dependent systems.

Some information is best displayed in a table. For example, the number of calories burned per half hour of exercise depends on the person's weight, as illustrated in the following table. Note that the rows correspond to activities and the columns correspond to weight.

ACTIVITY	127–137 LB	160–170 LB	180–200 LB
Walking/4 mph	156	183	204
Volleyball	267	315	348
Jogging/5 mph	276	345	381

Another example is the driving distance in miles from cities in Arizona (columns) to cities outside the state (rows).

CITY	FLAGSTAFF	PHOENIX	TUCSON	YUMA
Albuquerque, NM	325	465	440	650
Las Vegas, NV	250	300	415	295
Los Angeles, CA	470	375	490	285

If we selected only the numbers in each of the preceding tables and placed brackets around them, the result would be a *matrix*.

$$\text{Calories:} \begin{bmatrix} 156 & 183 & 204 \\ 267 & 315 & 348 \\ 276 & 345 & 381 \end{bmatrix} \qquad \text{Miles:} \begin{bmatrix} 325 & 465 & 440 & 650 \\ 250 & 300 & 415 & 295 \\ 470 & 375 & 490 & 285 \end{bmatrix}$$

A *matrix* is a rectangular array of numbers written within brackets.

$$\begin{bmatrix} a_{11} & a_{12} & \cdots & a_{1j} & \cdots & a_{1n} \\ a_{21} & a_{22} & \cdots & a_{2j} & \cdots & a_{2n} \\ \vdots & \vdots & \cdots & \vdots & \cdots & \vdots \\ a_{i1} & a_{i2} & \cdots & a_{ij} & \cdots & a_{in} \\ \vdots & \vdots & \cdots & \vdots & \cdots & \vdots \\ a_{m1} & a_{m2} & \cdots & a_{mj} & \cdots & a_{mn} \end{bmatrix}$$

Each number a_{ij} in the matrix is called an **entry** (or **element**) of the matrix. The first subscript i is the **row index**, and the second subscript j is the **column index**. This matrix contains m rows and n columns, and is said to be of **order** $m \times n$.

When the number of rows equals the number of columns (i.e., when $m = n$), the matrix is a **square matrix** of order n. In a square matrix, the entries $a_{11}, a_{22}, a_{33}, \ldots, a_{nn}$ are the **main diagonal** entries.

The matrix

$$A_{4\times3} = \begin{bmatrix} * & * & * \\ * & * & * \\ * & a_{32} & * \\ * & * & * \end{bmatrix}$$

has order (dimensions) 4×3, since there are four rows and three columns. The entry a_{32} is in the third row and second column.

EXAMPLE 1 Finding the Order of a Matrix

Determine the order of each matrix given.

a. $\begin{bmatrix} 2 & 1 \\ 3 & 0 \end{bmatrix}$

b. $\begin{bmatrix} 1 & -2 & 5 \\ -1 & 3 & 4 \end{bmatrix}$

c. $\begin{bmatrix} -2 & 5 & 4 \\ 1 & -\frac{1}{3} & 0 \\ 3 & 8 & 1 \end{bmatrix}$

d. $\begin{bmatrix} 4 & 9 & -\frac{1}{2} & 3 \end{bmatrix}$

e. $\begin{bmatrix} 3 & -2 \\ 5 & 1 \\ 0 & -\frac{2}{3} \\ 7 & 6 \end{bmatrix}$

Solution:

a. This matrix has **2** rows and **2** columns, so the order of the matrix is $\boxed{2 \times 2}$.

b. This matrix has **2** rows and **3** columns, so the order of the matrix is $\boxed{2 \times 3}$.

c. This matrix has **3** rows and **3** columns, so the order of the matrix is $\boxed{3 \times 3}$ or 3 since it is a square matrix.

d. This matrix has **1** row and **4** columns, so the order of the matrix is $\boxed{1 \times 4}$.

e. This matrix has **4** rows and **2** columns, so the order of the matrix is $\boxed{4 \times 2}$.

A matrix with only one column is called a **column matrix**, and a matrix that has only one row is called a **row matrix**. Notice that in Example 1 the matrices given in parts (a) and (c) are square matrices and the matrix given in part (d) is a row matrix.

You can use matrices as a shorthand way of writing systems of linear equations. There are two ways we can represent systems of linear equations with matrices: as *augmented matrices* or with *matrix equations*. In this section, we will discuss *augmented matrices* and solve systems of linear equations using two methods: *Gaussian elimination with back-substitution* and *Gauss–Jordan elimination*.

Augmented Matrices

A particular type of matrix that is used in representing a system of linear equations is an **augmented matrix**. It resembles a matrix with an additional vertical line and column of numbers, hence the name *augmented*. The following table illustrates examples of augmented matrices that represent systems of linear equations:

SYSTEM OF LINEAR EQUATIONS	AUGMENTED MATRIX
$3x + 4y = 1$ $x - 2y = 7$	$\begin{bmatrix} 3 & 4 & 1 \\ 1 & -2 & 7 \end{bmatrix}$
$x - y + z = 2$ $2x + 2y - 3z = -3$ $x + y + z = 6$	$\begin{bmatrix} 1 & -1 & 1 & 2 \\ 2 & 2 & -3 & -3 \\ 1 & 1 & 1 & 6 \end{bmatrix}$
$x + y + z = 0$ $3x + 2y - z = 2$	$\begin{bmatrix} 1 & 1 & 1 & 0 \\ 3 & 2 & -1 & 2 \end{bmatrix}$

Note the following:

- Each row represents an equation.
- The vertical line represents the equal sign.
- The first column represents the coefficients of the variable x.
- The second column represents the coefficients of the variable y.
- The third column (in the second and third systems) represents the coefficients of the variable z.
- The coefficients of the variables are on the left of the equal sign (vertical line) and the constants are on the right.
- Any variable that does not appear in an equation has an implied coefficient of 0.

EXAMPLE 2   Writing a System of Linear Equations as an Augmented Matrix

Write each system of linear equations as an augmented matrix.

a. $2x - y = 5$
$-x + 2y = 3$

b. $3x - 2y + 4z = 5$
$y - 3z = -2$
$7x \quad\quad - z = 1$

c. $x_1 - x_2 + 2x_3 - 3 = 0$
$x_1 + x_2 - 3x_3 + 5 = 0$
$x_1 - x_2 + x_3 - 2 = 0$

Solution:

a.

$$\left[\begin{array}{rr|r} 2 & -1 & 5 \\ -1 & 2 & 3 \end{array}\right]$$

b. Note that all missing terms have a 0 coefficient.

$3x - 2y + 4z = 5$
$0x + y - 3z = -2$
$7x + 0y - z = 1$

$$\left[\begin{array}{rrr|r} 3 & -2 & 4 & 5 \\ 0 & 1 & -3 & -2 \\ 7 & 0 & -1 & 1 \end{array}\right]$$

c. Write the constants on the right side of the vertical line in the matrix.

$x_1 - x_2 + 2x_3 = 3$
$x_1 + x_2 - 3x_3 = -5$
$x_1 - x_2 + x_3 = 2$

$$\left[\begin{array}{rrr|r} 1 & -1 & 2 & 3 \\ 1 & 1 & -3 & -5 \\ 1 & -1 & 1 & 2 \end{array}\right]$$

■ **YOUR TURN** Write each system of linear equations as an augmented matrix.

a. $2x + y - 3 = 0$
$x - y = 5$

b. $y - x + z = 7$
$x - y - z = 2$
$z - y = -1$

■ **Answer:**

a. $\left[\begin{array}{rr|r} 2 & 1 & 3 \\ 1 & -1 & 5 \end{array}\right]$

b. $\left[\begin{array}{rrr|r} -1 & 1 & 1 & 7 \\ 1 & -1 & -1 & 2 \\ 0 & -1 & 1 & -1 \end{array}\right]$

Row Operations on a Matrix

Row operations on a matrix are used to solve a system of linear equations when the system is written as an augmented matrix. Recall from the elimination method in Sections 8.1 and 8.2 that we could interchange equations, multiply an entire equation by a nonzero constant, and add a multiple of one equation to another equation to produce equivalent systems. Because each row in a matrix represents an equation, the operations that produced equivalent systems of equations that were used in the elimination method will also produce equivalent augmented matrices.

Study Tip

Each missing term in an equation of the system of linear equations is represented with a zero in the augmented matrix.

ROW OPERATIONS

The following operations on an augmented matrix will yield an equivalent matrix:

1. Interchange any two rows.
2. Multiply a row by a nonzero constant.
3. Add a multiple of one row to another row.

The following symbols describe these row operations:

1. $R_i \leftrightarrow R_j$ Interchange row i with row j.
2. $cR_i \rightarrow R_i$ Multiply row i by the constant c.
3. $cR_i + R_j \rightarrow R_j$ Multiply row i by the constant c and add to row j, writing the results in row j.

EXAMPLE 3 Applying a Row Operation to an Augmented Matrix

For each matrix, perform the given operation.

a. $\begin{bmatrix} 2 & -1 & | & 3 \\ 0 & 2 & | & 1 \end{bmatrix}$ $R_1 \leftrightarrow R_2$

b. $\begin{bmatrix} -1 & 0 & 1 & | & -2 \\ 3 & -1 & 2 & | & 3 \\ 0 & 1 & 3 & | & 1 \end{bmatrix}$ $2R_3 \to R_3$

c. $\begin{bmatrix} 1 & 2 & 0 & 2 & | & 2 \\ 0 & 1 & 2 & 3 & | & 5 \end{bmatrix}$ $R_1 - 2R_2 \to R_1$

Solution:

a. Interchange the first row with the second row.

$\begin{bmatrix} 2 & -1 & | & 3 \\ 0 & 2 & | & 1 \end{bmatrix}$ $R_1 \leftrightarrow R_2$ $\begin{bmatrix} 0 & 2 & | & 1 \\ 2 & -1 & | & 3 \end{bmatrix}$

b. Multiply the third row by 2.

$\begin{bmatrix} -1 & 0 & 1 & | & -2 \\ 3 & -1 & 2 & | & 3 \\ 0 & 1 & 3 & | & 1 \end{bmatrix}$ $2R_3 \to R_3$ $\begin{bmatrix} -1 & 0 & 1 & | & -2 \\ 3 & -1 & 2 & | & 3 \\ 0 & 2 & 6 & | & 2 \end{bmatrix}$

c. From row 1 subtract 2 times row 2, and write the answer in row 1. Note that finding row 1 minus 2 times row 2 is the same as adding row 1 to the product of -2 with row 2.

$R_1 - 2R_2 \to R_1$ $\begin{bmatrix} 1-2(0) & 2-2(1) & 0-2(2) & 2-2(3) & | & 2-2(5) \\ 0 & 1 & 2 & 3 & | & 5 \end{bmatrix}$

$\begin{bmatrix} 1 & 0 & -4 & -4 & | & -8 \\ 0 & 1 & 2 & 3 & | & 5 \end{bmatrix}$

■ **Answer:**

$\begin{bmatrix} 1 & 0 & 0 & | & 1 \\ 0 & 1 & 2 & | & 3 \\ 0 & 0 & 1 & | & 2 \end{bmatrix}$

■ **YOUR TURN** Perform the operation $R_1 + 2R_3 \to R_1$ on the matrix.

$\begin{bmatrix} 1 & 0 & -2 & | & -3 \\ 0 & 1 & 2 & | & 3 \\ 0 & 0 & 1 & | & 2 \end{bmatrix}$

Row–Echelon Form of a Matrix

We can solve systems of linear equations using augmented matrices with two procedures: *Gaussian elimination with back-substitution*, which uses row operations to transform a matrix into *row–echelon form*, and *Gauss–Jordan elimination*, which uses row operations to transform a matrix into *reduced row–echelon form*.

Row–Echelon Form

A matrix is in **row–echelon** form if it has all three of the following properties:

1. Any rows consisting entirely of 0s are at the bottom of the matrix.
2. For each row that does not consist entirely of 0s, the first (leftmost) nonzero entry is 1 (called the leading 1).
3. For two successive nonzero rows, the leading 1 in the higher row is farther to the left than the leading 1 in the lower row.

Reduced Row–Echelon Form

If a matrix in row–echelon form has the following additional property, then the matrix is in **reduced row–echelon form**:

4. Every column containing a leading 1 has zeros in every position above and below the leading 1.

EXAMPLE 4   **Determining Whether a Matrix Is in Row–Echelon Form**

Determine whether each matrix is in row–echelon form. If it is in row–echelon form, determine whether it is in reduced row–echelon form.

a. $\begin{bmatrix} 1 & 3 & 2 & | & 3 \\ 0 & 1 & 4 & | & 2 \\ 0 & 0 & 1 & | & -1 \end{bmatrix}$ **b.** $\begin{bmatrix} 1 & 3 & 2 & | & 3 \\ 0 & 1 & 1 & | & 3 \\ 0 & 0 & 0 & | & 0 \end{bmatrix}$ **c.** $\begin{bmatrix} 1 & 0 & 3 & | & 2 \\ 0 & 1 & -1 & | & 5 \end{bmatrix}$

d. $\begin{bmatrix} 1 & 0 & | & 1 \\ 0 & 3 & | & 1 \end{bmatrix}$ **e.** $\begin{bmatrix} 1 & 0 & 0 & | & 3 \\ 0 & 1 & 0 & | & 5 \\ 0 & 0 & 1 & | & 7 \end{bmatrix}$ **f.** $\begin{bmatrix} 1 & 3 & 2 & | & 3 \\ 0 & 0 & 1 & | & 2 \\ 0 & 1 & 0 & | & -3 \end{bmatrix}$

Solution:

The matrices in (a), (b), (c), and (e) are in row–echelon form. The matrix in (d) is not in row–echelon form, by condition 2; the leading nonzero entry is not a 1 in each row. If the "3" were a "1," the matrix would be in reduced row–echelon form. The matrix in (f) is not in row–echelon form, because of condition 3; the leading 1 in row 2 is not to the left of the leading 1 in row 3. The matrices in (c) and (e) are in reduced row–echelon form, because in the columns containing the leading 1s there are zeros in every position above and below the leading 1.

Gaussian Elimination with Back-Substitution

Gaussian elimination with back-substitution is a method that uses row operations to transform an augmented matrix into row–echelon form and then uses back-substitution to find the solution to the system of linear equations.

GAUSSIAN ELIMINATION WITH BACK-SUBSTITUTION

Step 1: Write the system of linear equations as an augmented matrix.

Step 2: Use row operations to rewrite the augmented matrix in row–echelon form.

Step 3: Write the system of linear equations that corresponds to the matrix in row–echelon form found in Step 2.

Step 4: Use the system of linear equations found in Step 3 together with back-substitution to find the solution of the system.

Study Tip

For row–echelon form, get 1s along the main diagonal and 0s below these 1s.

The order in which we perform row operations is important. You should move from left to right. Here is an example of Step 2 in the procedure:

$$\begin{bmatrix} 1 & * & * & | & * \\ * & * & * & | & * \\ * & * & * & | & * \end{bmatrix} \rightarrow \begin{bmatrix} 1 & * & * & | & * \\ 0 & * & * & | & * \\ 0 & * & * & | & * \end{bmatrix} \rightarrow \begin{bmatrix} 1 & * & * & | & * \\ 0 & 1 & * & | & * \\ 0 & * & * & | & * \end{bmatrix} \rightarrow \begin{bmatrix} 1 & * & * & | & * \\ 0 & 1 & * & | & * \\ 0 & 0 & * & | & * \end{bmatrix} \rightarrow \begin{bmatrix} 1 & * & * & | & * \\ 0 & 1 & * & | & * \\ 0 & 0 & 1 & | & * \end{bmatrix}$$

Matrices are not typically used for systems of linear equations in two variables because the methods from Section 8.1 (substitution and elimination) are more efficient. Example 5 illustrates this procedure with a simple system of linear equations in two variables.

EXAMPLE 5   Using Gaussian Elimination with Back-Substitution to Solve a System of Two Linear Equations in Two Variables

Apply Gaussian elimination with back-substitution to solve the system of linear equations.

$$2x + \ y = -8$$
$$x + 3y = \ \ \ 6$$

Solution:

STEP 1 Write the system of linear equations as an augmented matrix.
$$\begin{bmatrix} 2 & 1 & -8 \\ 1 & 3 & 6 \end{bmatrix}$$

STEP 2 Use row operations to rewrite the matrix in row–echelon form.

Get a 1 in the top left. Interchange rows 1 and 2.
$$\begin{bmatrix} 2 & 1 & -8 \\ 1 & 3 & 6 \end{bmatrix} \ R_1 \leftrightarrow R_2 \ \begin{bmatrix} 1 & 3 & 6 \\ 2 & 1 & -8 \end{bmatrix}$$

Get a 0 below the leading 1 in row 1.
$$\begin{bmatrix} 1 & 3 & 6 \\ 2 & 1 & -8 \end{bmatrix} \ R_2 - 2R_1 \rightarrow R_2 \ \begin{bmatrix} 1 & 3 & 6 \\ 0 & -5 & -20 \end{bmatrix}$$

Get a leading 1 in row 2. Make the "−5" a "1" by dividing by −5. Dividing by −5 is the same as multiplying by its reciprocal $-\frac{1}{5}$.
$$\begin{bmatrix} 1 & 3 & 6 \\ 0 & -5 & -20 \end{bmatrix} \ -\tfrac{1}{5}R_2 \rightarrow R_2 \ \begin{bmatrix} 1 & 3 & 6 \\ 0 & 1 & 4 \end{bmatrix}$$

The resulting matrix is in row–echelon form.

STEP 3 Write the system of linear equations corresponding to the row–echelon form of the matrix resulting in Step 2.
$$\begin{bmatrix} 1 & 3 & 6 \\ 0 & 1 & 4 \end{bmatrix} \rightarrow \begin{array}{r} x + 3y = 6 \\ y = 4 \end{array}$$

STEP 4 Use back-substitution to find the solution to the system.

Let $y = 4$ in the first equation $x + 3y = 6$. $x + 3(4) = 6$

Solve for x. $x = -6$

The solution to the system of linear equations is $\boxed{x = -6, y = 4}$.

EXAMPLE 6   Using Gaussian Elimination with Back-Substitution to Solve a System of Three Linear Equations in Three Variables

Use Gaussian elimination with back-substitution to solve the system of linear equations.

$$2x + \ y + 8z = -1$$
$$x - \ y + \ z = -2$$
$$3x - 2y - 2z = \ \ \ 2$$

Solution:

STEP 1 Write the system of linear equations as an augmented matrix.
$$\begin{bmatrix} 2 & 1 & 8 & -1 \\ 1 & -1 & 1 & -2 \\ 3 & -2 & -2 & 2 \end{bmatrix}$$

STEP 2 Use row operations to rewrite the matrix in row–echelon form.

Get a 1 in the top left.
Interchange rows 1 and 2.

$$R_1 \leftrightarrow R_2 \quad \begin{bmatrix} 1 & -1 & 1 & | & -2 \\ 2 & 1 & 8 & | & -1 \\ 3 & -2 & -2 & | & 2 \end{bmatrix}$$

Get 0s below the leading 1
in row 1.

$$R_2 - 2R_1 \rightarrow R_2 \quad \begin{bmatrix} 1 & -1 & 1 & | & -2 \\ 0 & 3 & 6 & | & 3 \\ 3 & -2 & -2 & | & 2 \end{bmatrix}$$

$$R_3 - 3R_1 \rightarrow R_3 \quad \begin{bmatrix} 1 & -1 & 1 & | & -2 \\ 0 & 3 & 6 & | & 3 \\ 0 & 1 & -5 & | & 8 \end{bmatrix}$$

Get a leading 1 in row 2. Make the
"3" a "1" by dividing by 3.

$$\tfrac{1}{3}R_2 \rightarrow R_2 \quad \begin{bmatrix} 1 & -1 & 1 & | & -2 \\ 0 & 1 & 2 & | & 1 \\ 0 & 1 & -5 & | & 8 \end{bmatrix}$$

Get a zero below the leading
1 in row 2.

$$R_3 - R_2 \rightarrow R_3 \quad \begin{bmatrix} 1 & -1 & 1 & | & -2 \\ 0 & 1 & 2 & | & 1 \\ 0 & 0 & -7 & | & 7 \end{bmatrix}$$

Get a leading 1 in row 3. Make the
"−7" a "1" by dividing by −7.

$$-\tfrac{1}{7}R_3 \rightarrow R_3 \quad \begin{bmatrix} 1 & -1 & 1 & | & -2 \\ 0 & 1 & 2 & | & 1 \\ 0 & 0 & 1 & | & -1 \end{bmatrix}$$

STEP 3 Write the system of linear equations corresponding
to the row–echelon form of the matrix resulting in Step 2.

$$\begin{aligned} x - y + z &= -2 \\ y + 2z &= 1 \\ z &= -1 \end{aligned}$$

STEP 4 Use back-substitution to find the solution to the system.

Let $\boxed{z = -1}$ in the second equation $y + 2z = 1$. $\qquad y + 2(-1) = 1$

Solve for y. $\boxed{y = 3}$

Let $y = 3$ and $z = -1$ in the first equation
$x - y + z = -2$. $\qquad x - (3) + (-1) = -2$

Solve for x. $\boxed{x = 2}$

The solution to the system of linear equations is $\boxed{x = 2, y = 3, \text{ and } z = -1}$.

■ Answer: $x = -1, y = 2, z = 1$

■ YOUR TURN Use Gaussian elimination with back-substitution to solve the system
of linear equations.

$$\begin{aligned} x + y - z &= 0 \\ 2x + y + z &= 1 \\ 2x - y + 3z &= -1 \end{aligned}$$

Gauss–Jordan Elimination

In Gaussian elimination with back-substitution, we used row operations to rewrite the matrix in an equivalent row–echelon form. If we continue using row operations until the matrix is in *reduced* row–echelon form, this eliminates the need for back-substitution, and we call this process *Gauss–Jordan elimination*.

GAUSS–JORDAN ELIMINATION

Step 1: Write the system of linear equations as an augmented matrix.

Step 2: Use row operations to rewrite the augmented matrix in *reduced* row–echelon form.

Step 3: Write the system of linear equations that corresponds to the matrix in reduced row–echelon form found in Step 2. The result is the solution to the system.

Study Tip

For reduced row–echelon form, get 1s along the main diagonal and 0s above and below these 1s.

The order in which we perform row operations is important. You should move from left to right. Think of this process as climbing *down* a set of stairs first and then back up the stairs second. On the way *down* the stairs always use operations with rows *above* where you currently are, and on the way back *up* the stairs always use rows *below* where you currently are.

Down the stairs:

$$
\begin{bmatrix} 1 & * & * & | & * \\ * & * & * & | & * \\ * & * & * & | & * \end{bmatrix}
\rightarrow
\begin{bmatrix} 1 & * & * & | & * \\ 0 & * & * & | & * \\ 0 & * & * & | & * \end{bmatrix}
\rightarrow
\begin{bmatrix} 1 & * & * & | & * \\ 0 & 1 & * & | & * \\ 0 & * & * & | & * \end{bmatrix}
\rightarrow
\begin{bmatrix} 1 & * & * & | & * \\ 0 & 1 & * & | & * \\ 0 & 0 & * & | & * \end{bmatrix}
\rightarrow
\begin{bmatrix} 1 & * & * & | & * \\ 0 & 1 & * & | & * \\ 0 & 0 & 1 & | & * \end{bmatrix}
$$

Up the stairs:

$$
\begin{bmatrix} 1 & * & * & | & * \\ 0 & 1 & * & | & * \\ 0 & 0 & 1 & | & * \end{bmatrix}
\rightarrow
\begin{bmatrix} 1 & * & * & | & * \\ 0 & 1 & 0 & | & * \\ 0 & 0 & 1 & | & * \end{bmatrix}
\rightarrow
\begin{bmatrix} 1 & * & 0 & | & * \\ 0 & 1 & 0 & | & * \\ 0 & 0 & 1 & | & * \end{bmatrix}
\rightarrow
\begin{bmatrix} 1 & 0 & 0 & | & * \\ 0 & 1 & 0 & | & * \\ 0 & 0 & 1 & | & * \end{bmatrix}
$$

EXAMPLE 7   **Using Gauss–Jordan Elimination to Solve a System of Linear Equations in Three Variables**

Apply Gauss–Jordan elimination to solve the system of linear equations.

$$x - y + 2z = -1$$
$$3x + 2y - 6z = 1$$
$$2x + 3y + 4z = 8$$

Solution:

STEP 1 Write the system as an augmented matrix.

$$\begin{bmatrix} 1 & -1 & 2 & | & -1 \\ 3 & 2 & -6 & | & 1 \\ 2 & 3 & 4 & | & 8 \end{bmatrix}$$

STEP 2 Utilize row operations to rewrite the matrix in reduced row–echelon form.

There is already a 1 in the first row/first column.

Get 0s below the leading 1 in row 1.

$R_2 - 3R_1 \rightarrow R_2$
$R_3 - 2R_1 \rightarrow R_3$
$$\begin{bmatrix} 1 & -1 & 2 & | & -1 \\ 0 & 5 & -12 & | & 4 \\ 0 & 5 & 0 & | & 10 \end{bmatrix}$$

Get a 1 in row 2/column 2.

$R_2 \leftrightarrow R_3$
$$\begin{bmatrix} 1 & -1 & 2 & | & -1 \\ 0 & 5 & 0 & | & 10 \\ 0 & 5 & -12 & | & 4 \end{bmatrix}$$

$\frac{1}{5}R_2 \rightarrow R_2$
$$\begin{bmatrix} 1 & -1 & 2 & | & -1 \\ 0 & 1 & 0 & | & 2 \\ 0 & 5 & -12 & | & 4 \end{bmatrix}$$

Get a 0 in row 3/column 2.

$R_3 - 5R_2 \rightarrow R_3$
$$\begin{bmatrix} 1 & -1 & 2 & | & -1 \\ 0 & 1 & 0 & | & 2 \\ 0 & 0 & -12 & | & -6 \end{bmatrix}$$

Get a 1 in row 3/column 3.

$-\frac{1}{12}R_3 \rightarrow R_3$
$$\begin{bmatrix} 1 & -1 & 2 & | & -1 \\ 0 & 1 & 0 & | & 2 \\ 0 & 0 & 1 & | & \frac{1}{2} \end{bmatrix}$$

Now, go back up the stairs.

Get 0s above the 1 in row 3/column 3.

$R_1 - 2R_3 \rightarrow R_1$
$$\begin{bmatrix} 1 & -1 & 0 & | & -2 \\ 0 & 1 & 0 & | & 2 \\ 0 & 0 & 1 & | & \frac{1}{2} \end{bmatrix}$$

Get a 0 in row 1/column 2.

$R_1 + R_2 \rightarrow R_1$
$$\begin{bmatrix} 1 & 0 & 0 & | & 0 \\ 0 & 1 & 0 & | & 2 \\ 0 & 0 & 1 & | & \frac{1}{2} \end{bmatrix}$$

STEP 3 Identify the solution.

$$\boxed{x = 0, \, y = 2, \, z = \tfrac{1}{2}}$$

■ Answer: $x = -1, y = 2, z = 3$

■ YOUR TURN Use an augmented matrix and Gauss–Jordan elimination to solve the system of equations.

$$x + y - z = -2$$
$$3x + y - z = -4$$
$$2x - 2y + 3z = 3$$

EXAMPLE 8 **Solving a System of Four Linear Equations in Four Variables**

Solve the system of equations with Gauss–Jordan elimination.

$$x_1 + x_2 - x_3 + 3x_4 = 3$$
$$3x_2 - 2x_4 = 4$$
$$2x_1 - 3x_3 = -1$$
$$4x_4 + 2x_1 = -6$$

Solution:

STEP 1 Write the system as an augmented matrix.

$$\begin{bmatrix} 1 & 1 & -1 & 3 & 3 \\ 0 & 3 & 0 & -2 & 4 \\ 2 & 0 & -3 & 0 & -1 \\ 2 & 0 & 0 & 4 & -6 \end{bmatrix}$$

STEP 2 Use row operations to rewrite the matrix in reduced row–echelon form.

There is already a 1 in the first row/first column.

Get 0s below the 1 in row 1/column 1.

$$\begin{matrix} \\ R_3 - 2R_1 \to R_3 \\ R_4 - 2R_1 \to R_4 \end{matrix} \quad \begin{bmatrix} 1 & 1 & -1 & 3 & 3 \\ 0 & 3 & 0 & -2 & 4 \\ 0 & -2 & -1 & -6 & -7 \\ 0 & -2 & 2 & -2 & -12 \end{bmatrix}$$

Get a 1 in row 2/column 2.

$$R_2 \leftrightarrow R_4 \quad \begin{bmatrix} 1 & 1 & -1 & 3 & 3 \\ 0 & -2 & 2 & -2 & -12 \\ 0 & -2 & -1 & -6 & -7 \\ 0 & 3 & 0 & -2 & 4 \end{bmatrix}$$

$$-\tfrac{1}{2}R_2 \leftrightarrow R_2 \quad \begin{bmatrix} 1 & 1 & -1 & 3 & 3 \\ 0 & 1 & -1 & 1 & 6 \\ 0 & -2 & -1 & -6 & -7 \\ 0 & 3 & 0 & -2 & 4 \end{bmatrix}$$

Get 0s below the 1 in row 2/column 2.

$$\begin{matrix} R_3 + 2R_2 \to R_3 \\ R_4 - 3R_2 \to R_4 \end{matrix} \quad \begin{bmatrix} 1 & 1 & -1 & 3 & 3 \\ 0 & 1 & -1 & 1 & 6 \\ 0 & 0 & -3 & -4 & 5 \\ 0 & 0 & 3 & -5 & -14 \end{bmatrix}$$

Get a 1 in row 3/column 3.

$$-\tfrac{1}{3}R_3 \to R_3 \quad \begin{bmatrix} 1 & 1 & -1 & 3 & 3 \\ 0 & 1 & -1 & 1 & 6 \\ 0 & 0 & 1 & \tfrac{4}{3} & -\tfrac{5}{3} \\ 0 & 0 & 3 & -5 & -14 \end{bmatrix}$$

Get a 0 in row 4/column 3.

$$R_4 - 3R_3 \to R_4 \quad \begin{bmatrix} 1 & 1 & -1 & 3 & 3 \\ 0 & 1 & -1 & 1 & 6 \\ 0 & 0 & 1 & \tfrac{4}{3} & -\tfrac{5}{3} \\ 0 & 0 & 0 & -9 & -9 \end{bmatrix}$$

Get a 1 in row 4/column 4.

$$-\tfrac{1}{9}R_4 \to R_4 \qquad \begin{bmatrix} 1 & 1 & -1 & 3 & | & 3 \\ 0 & 1 & -1 & 1 & | & 6 \\ 0 & 0 & 1 & \tfrac{4}{3} & | & -\tfrac{5}{3} \\ 0 & 0 & 0 & 1 & | & 1 \end{bmatrix}$$

Now go back up the stairs.

Get 0s above the 1 in row 4/column 4.

$$\begin{matrix} R_3 - \tfrac{4}{3}R_4 \to R_3 \\ R_2 - R_4 \to R_2 \\ R_1 - 3R_4 \to R_1 \end{matrix} \qquad \begin{bmatrix} 1 & 1 & -1 & 0 & | & 0 \\ 0 & 1 & -1 & 0 & | & 5 \\ 0 & 0 & 1 & 0 & | & -3 \\ 0 & 0 & 0 & 1 & | & 1 \end{bmatrix}$$

Get 0s above the 1 in row 3/column 3.

$$\begin{matrix} R_2 + R_3 \to R_2 \\ R_1 + R_3 \to R_1 \end{matrix} \qquad \begin{bmatrix} 1 & 1 & 0 & 0 & | & -3 \\ 0 & 1 & 0 & 0 & | & 2 \\ 0 & 0 & 1 & 0 & | & -3 \\ 0 & 0 & 0 & 1 & | & 1 \end{bmatrix}$$

Get a 0 in row 1/column 2.

$$R_1 - R_2 \to R_1 \qquad \begin{bmatrix} 1 & 0 & 0 & 0 & | & -5 \\ 0 & 1 & 0 & 0 & | & 2 \\ 0 & 0 & 1 & 0 & | & -3 \\ 0 & 0 & 0 & 1 & | & 1 \end{bmatrix}$$

STEP 3 Identify the solution.
$$\boxed{x_1 = -5, x_2 = 2, x_3 = -3, x_4 = 1}$$

Inconsistent and Dependent Systems

Recall from Section 8.1 that systems of linear equations can be independent, inconsistent, or dependent systems and therefore have *one solution, no solution,* or *infinitely many solutions.* All of the systems we have solved so far in this section have been independent systems (unique solution). When solving a system of linear equations using Gaussian elimination or Gauss–Jordan elimination, the following will indicate the three possible types of solutions.

SYSTEM	TYPE OF SOLUTION	MATRIX DURING GAUSS–JORDAN ELIMINATION	EXAMPLE	
Independent	One (unique) solution	Diagonal entries are all 1s, and the 0s occupy all other coefficient positions.	$\begin{bmatrix} 1 & 0 & 0 & \vert & 1 \\ 0 & 1 & 0 & \vert & -3 \\ 0 & 0 & 1 & \vert & 2 \end{bmatrix}$	or $\begin{matrix} x = 1 \\ y = -3 \\ z = 2 \end{matrix}$
Inconsistent	No solution	One row will have only zero entries for coefficients and a nonzero entry for the constant.	$\begin{bmatrix} 1 & 0 & 0 & \vert & 1 \\ 0 & 1 & 0 & \vert & -3 \\ 0 & 0 & 0 & \vert & 2 \end{bmatrix}$	or $\begin{matrix} x = 1 \\ y = -3 \\ 0 = 2 \end{matrix}$
Dependent	Infinitely many solutions	One row will be entirely 0s when the number of equations equals the number of variables.	$\begin{bmatrix} 1 & 0 & -2 & \vert & 1 \\ 0 & 1 & 1 & \vert & -3 \\ 0 & 0 & 0 & \vert & 0 \end{bmatrix}$	or $\begin{matrix} x - 2z = 1 \\ y + z = -3 \\ 0 = 0 \end{matrix}$

EXAMPLE 9   **Determining That a System Is Inconsistent: No Solution**

Solve the system of equations.

$$\begin{aligned}
x + 2y - z &= 3 \\
2x + y + 2z &= -1 \\
-2x - 4y + 2z &= 5
\end{aligned}$$

Solution:

STEP 1 Write the system of equations as an augmented matrix.

$$\begin{bmatrix} 1 & 2 & -1 & | & 3 \\ 2 & 1 & 2 & | & -1 \\ -2 & -4 & 2 & | & 5 \end{bmatrix}$$

STEP 2 Apply row operations to rewrite the matrix in row–echelon form.

Get 0s below the 1 in column 1.

$$\begin{matrix} R_2 - 2R_1 \rightarrow R_2 \\ R_3 + 2R_1 \rightarrow R_3 \end{matrix} \quad \begin{bmatrix} 1 & 2 & -1 & | & 3 \\ 0 & -3 & 4 & | & -7 \\ 0 & 0 & 0 & | & 11 \end{bmatrix}$$

There is no need to continue because row 3 is a contradiction. $0x + 0y + 0z = 11$ or $0 = 11$

Since this is inconsistent, there is *no solution* to this system of equations.

EXAMPLE 10   **Determining That a System Is Dependent: Infinitely Many Solutions**

Solve the system of equations.

$$\begin{aligned}
x + z &= 3 \\
2x + y + 4z &= 8 \\
3x + y + 5z &= 11
\end{aligned}$$

Solution:

STEP 1 Write the system of equations as an augmented matrix.

$$\begin{bmatrix} 1 & 0 & 1 & | & 3 \\ 2 & 1 & 4 & | & 8 \\ 3 & 1 & 5 & | & 11 \end{bmatrix}$$

STEP 2 Use row operations to rewrite the matrix in reduced row–echelon form.

Get the 0s below the 1 in column 1.

$$\begin{matrix} R_2 - 2R_1 \rightarrow R_2 \\ R_3 - 3R_1 \rightarrow R_3 \end{matrix} \quad \begin{bmatrix} 1 & 0 & 1 & | & 3 \\ 0 & 1 & 2 & | & 2 \\ 0 & 1 & 2 & | & 2 \end{bmatrix}$$

Get a 0 in row 3/column 2.

$$R_3 - R_2 \rightarrow R_3 \quad \begin{bmatrix} 1 & 0 & 1 & | & 3 \\ 0 & 1 & 2 & | & 2 \\ 0 & 0 & 0 & | & 0 \end{bmatrix}$$

Study Tip

In a system with three variables, say, x, y, and z, we typically let $z = a$ (where a is called a parameter) and then solve for x and y in terms of a.

This matrix is in reduced row–echelon form. This matrix corresponds to a dependent system of linear equations and has infinitely many solutions.

STEP 3 Write the augmented matrix as a system of linear equations.

$$\begin{aligned} x + z &= 3 \\ y + 2z &= 2 \end{aligned}$$

Let $z = a$, where a is any real number, and substitute this into the two equations.

We find that $x = 3 - a$ and $y = 2 - 2a$. The answer is written as

$$\boxed{x = 3 - a, \, y = 2 - 2a, \, z = a}$$ for a any real number.

A common mistake that is made is to identify a unique solution as no solution when one of the variables is equal to zero. For example, what is the difference between the following two matrices?

$$\begin{bmatrix} 1 & 0 & 2 & | & 1 \\ 0 & 1 & 3 & | & 2 \\ 0 & 0 & 3 & | & 0 \end{bmatrix} \quad \text{and} \quad \begin{bmatrix} 1 & 0 & 2 & | & 1 \\ 0 & 1 & 3 & | & 2 \\ 0 & 0 & 0 & | & 3 \end{bmatrix}$$

The first matrix has a *unique solution*, whereas the second matrix has *no solution*. The third row of the first matrix corresponds to the equation $3z = 0$, which implies that $z = 0$. The third row of the second matrix corresponds to the equation $0x + 0y + 0z = 3$ or $0 = 3$, which is inconsistent, and therefore the system has no solution.

EXAMPLE 11 **Determining That a System Is Dependent: Infinitely Many Solutions**

Solve the system of linear equations.

$$2x + y + z = 8$$
$$x + y - z = -3$$

Solution:

STEP 1 Write the system of equations as an augmented matrix. $\quad \begin{bmatrix} 2 & 1 & 1 & | & 8 \\ 1 & 1 & -1 & | & -3 \end{bmatrix}$

STEP 2 Use row operations to rewrite the matrix in reduced row–echelon form.

Get a 1 in row 1/column 1. $\qquad R_1 \leftrightarrow R_2 \quad \begin{bmatrix} 1 & 1 & -1 & | & -3 \\ 2 & 1 & 1 & | & 8 \end{bmatrix}$

Get a 0 in row 2/column 1. $\qquad R_2 - 2R_1 \rightarrow R_2 \quad \begin{bmatrix} 1 & 1 & -1 & | & -3 \\ 0 & -1 & 3 & | & 14 \end{bmatrix}$

Get a 1 in row 2/column 2. $\qquad -R_2 \rightarrow R_2 \quad \begin{bmatrix} 1 & 1 & -1 & | & -3 \\ 0 & 1 & -3 & | & -14 \end{bmatrix}$

Get a 0 in row 1/column 2. $\qquad R_1 - R_2 \rightarrow R_1 \quad \begin{bmatrix} 1 & 0 & 2 & | & 11 \\ 0 & 1 & -3 & | & -14 \end{bmatrix}$

This matrix is in reduced row–echelon form.

STEP 3 Identify the solution.

$$x + 2z = 11$$
$$y - 3z = -14$$

Let $z = a$, where a is any real number. Substituting $z = a$ into these two equations gives the infinitely many solutions $\boxed{x = 11 - 2a,\ y = 3a - 14,\ z = a}$.

■ **YOUR TURN** Solve the system of equations using an augmented matrix.

$$x + y + z = 0$$
$$3x + 2y - z = 2$$

■ **Answer:** $x = 3a + 2$, $y = -4a - 2$, $z = a$, where a is any real number.

Applications

Remember Jared who lost all that weight eating at Subway and is still keeping it off 10 years later? He ate Subway sandwiches for lunch and dinner for one year and lost 235 pounds! The following table gives nutritional information for Subway's 6-inch sandwiches advertised with 6 grams of fat or less.

SANDWICH	CALORIES	FAT (g)	CARBOHYDRATES (g)	PROTEIN (g)
Veggie Delight	350	18	17	36
Oven-roasted chicken breast	430	19	46	20
Ham (Black Forest without cheese)	290	5	45	19

EXAMPLE 12 Subway Diet

Suppose you are going to eat only Subway 6-inch sandwiches for a week (seven days) for both lunch and dinner (total of 14 meals). If your goal is to eat 388 grams of protein and 4900 calories in those 14 sandwiches, how many of each sandwich should you eat that week?

Solution:

STEP 1 Determine the system of linear equations.

Let three variables represent the number of each type of sandwich you eat in a week.

x = number of Veggie Delight sandwiches
y = number of chicken breast sandwiches
z = number of ham sandwiches

The total number of sandwiches eaten is 14. $\qquad x + y + z = 14$

The total number of calories consumed is 4900. $\qquad 350x + 430y + 290z = 4900$

The total number of grams of protein
consumed is 388. $\qquad 36x + 20y + 19z = 388$

Write an augmented matrix representing this system of linear equations.

$$\begin{bmatrix} 1 & 1 & 1 & | & 14 \\ 350 & 430 & 290 & | & 4900 \\ 36 & 20 & 19 & | & 388 \end{bmatrix}$$

STEP 2 Utilize row operations to rewrite the matrix in reduced row–echelon form.

$$\begin{matrix} R_2 - 350R_1 \rightarrow R_2 \\ R_3 - 36R_1 \rightarrow R_3 \end{matrix} \quad \begin{bmatrix} 1 & 1 & 1 & | & 14 \\ 0 & 80 & -60 & | & 0 \\ 0 & -16 & -17 & | & -116 \end{bmatrix}$$

$$\tfrac{1}{80}R_2 \rightarrow R_2 \quad \begin{bmatrix} 1 & 1 & 1 & | & 14 \\ 0 & 1 & -\tfrac{3}{4} & | & 0 \\ 0 & -16 & -17 & | & -116 \end{bmatrix}$$

$$R_3 + 16R_2 \rightarrow R_3 \quad \begin{bmatrix} 1 & 1 & 1 & | & 14 \\ 0 & 1 & -\tfrac{3}{4} & | & 0 \\ 0 & 0 & -29 & | & -116 \end{bmatrix}$$

$$-\frac{1}{29}R_3 \rightarrow R_3 \quad \begin{bmatrix} 1 & 1 & 1 & | & 14 \\ 0 & 1 & -\frac{3}{4} & | & 0 \\ 0 & 0 & 1 & | & 4 \end{bmatrix}$$

$$\begin{array}{c} R_2 + \frac{3}{4}R_3 \rightarrow R_2 \\ R_1 - R_3 \rightarrow R_1 \end{array} \quad \begin{bmatrix} 1 & 1 & 0 & | & 10 \\ 0 & 1 & 0 & | & 3 \\ 0 & 0 & 1 & | & 4 \end{bmatrix}$$

$$R_1 - R_2 \rightarrow R_1 \quad \begin{bmatrix} 1 & 0 & 0 & | & 7 \\ 0 & 1 & 0 & | & 3 \\ 0 & 0 & 1 & | & 4 \end{bmatrix}$$

STEP 3 Identify the solution. $\boxed{x = 7, \ y = 3, \ z = 4}$

You should eat $\boxed{\text{7 Veggie Delights, 3 oven-roasted chicken breast, and 4 ham sandwiches}}$.

EXAMPLE 13 Fitting a Curve to Data

The amount of money awarded in medical malpractice suits is rising. This can be modeled with a quadratic function $y = at^2 + bt + c$, where $t > 0$ and $a > 0$. Determine a quadratic function that passes through the three points shown on the graph. Based on this trend, how much money will be spent on malpractice in 2011?

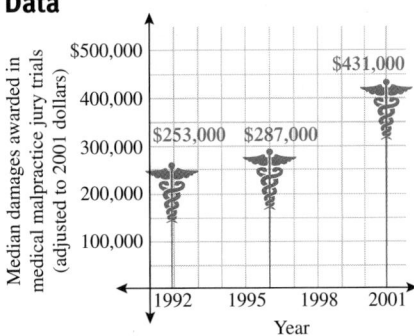

Solution:

Let 1991 correspond to $t = 0$ and y represent the number of dollars awarded for malpractice suits. The following data are reflected in the illustration above:

YEAR	t	y (THOUSANDS OF DOLLARS)	(t, y)
1992	1	253	(1, 253)
1996	5	287	(5, 287)
2001	10	431	(10, 431)

Substitute the three points (1, 253), (5, 287), and (10, 431) into the general quadratic equation: $y = at^2 + bt + c$.

POINT	$y = at^2 + bt + c$	SYSTEM OF EQUATIONS
(1, 253)	$253 = a(1)^2 + b(1) + c$	$a + b + c = 253$
(5, 287)	$287 = a(5)^2 + b(5) + c$	$25a + 5b + c = 287$
(10, 431)	$431 = a(10)^2 + b(10) + c$	$100a + 10b + c = 431$

Technology Tip

Enter the matrix into the TI. Press $\boxed{2^{nd}}$ $\boxed{\text{MATRIX}}$. Use $\boxed{\blacktriangleright}$ $\boxed{\text{EDIT}}$ $\boxed{\text{ENTER}}$ $\boxed{\text{ENTER}}$.

Now enter the size by typing the number of rows first and the number of columns second. Enter the elements of the matrix one row at a time by pressing the number and $\boxed{\text{ENTER}}$ each time.

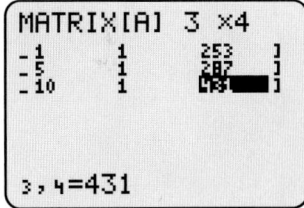

When done, press $\boxed{2^{nd}}$ $\boxed{\text{QUIT}}$. To show the matrix A, press $\boxed{2^{nd}}$ $\boxed{\text{MATRIX}}$ $\boxed{\text{ENTER}}$ $\boxed{\text{ENTER}}$.

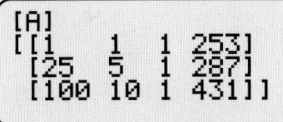

To find the reduced row–echelon form of the original matrix directly, use $\boxed{\text{rref}}$ (matrix) command. Press $\boxed{2^{nd}}$ $\boxed{\text{MATRIX}}$ $\boxed{\blacktriangleright}$ $\boxed{\text{MATH}}$ $\boxed{\blacktriangledown}$ $\boxed{\text{B:rref(}}$ $\boxed{\text{ENTER}}$ $\boxed{2^{nd}}$ $\boxed{\text{MATRIX}}$ $\boxed{\text{ENTER}}$ $\boxed{)}$ $\boxed{\blacktriangleright}$ $\boxed{\text{MATH}}$ $\boxed{\text{1:Frac}}$ $\boxed{\text{ENTER}}$.

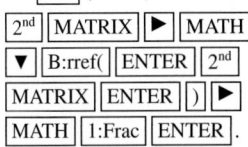

The solution to the system is $a = \frac{203}{90}, b = -\frac{151}{30}, c = \frac{2302}{9}$.

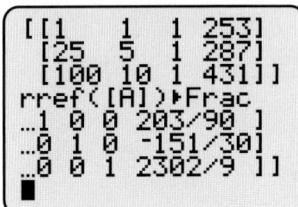

STEP 1 Write this system of linear equations as an augmented matrix.

$$\begin{bmatrix} 1 & 1 & 1 & | & 253 \\ 25 & 5 & 1 & | & 287 \\ 100 & 10 & 1 & | & 431 \end{bmatrix}$$

STEP 2 Apply row operations to rewrite the matrix in reduced row–echelon form.

$$R_2 - 25R_1 \rightarrow R_2 \quad \begin{bmatrix} 1 & 1 & 1 & | & 253 \\ 0 & -20 & -24 & | & -6038 \\ 100 & 10 & 1 & | & 431 \end{bmatrix}$$

$$R_3 - 100R_1 \rightarrow R_3 \quad \begin{bmatrix} 1 & 1 & 1 & | & 253 \\ 0 & -20 & -24 & | & -6038 \\ 0 & -90 & -99 & | & -24,869 \end{bmatrix}$$

$$-\tfrac{1}{20}R_2 \rightarrow R_2 \quad \begin{bmatrix} 1 & 1 & 1 & | & 253 \\ 0 & 1 & \frac{6}{5} & | & \frac{3019}{10} \\ 0 & -90 & -99 & | & -24,869 \end{bmatrix}$$

$$R_3 + 90R_2 \rightarrow R_3 \quad \begin{bmatrix} 1 & 1 & 1 & | & 253 \\ 0 & 1 & \frac{6}{5} & | & \frac{3019}{10} \\ 0 & 0 & 9 & | & 2302 \end{bmatrix}$$

$$\tfrac{1}{9}R_3 \rightarrow R_3 \quad \begin{bmatrix} 1 & 1 & 1 & | & 253 \\ 0 & 1 & \frac{6}{5} & | & \frac{3019}{10} \\ 0 & 0 & 1 & | & \frac{2302}{9} \end{bmatrix}$$

$$R_2 - \tfrac{6}{5}R_3 \rightarrow R_2 \quad \begin{bmatrix} 1 & 1 & 1 & | & 253 \\ 0 & 1 & 0 & | & -\frac{151}{30} \\ 0 & 0 & 1 & | & \frac{2302}{9} \end{bmatrix}$$

$$R_1 - R_3 \rightarrow R_1 \quad \begin{bmatrix} 1 & 1 & 0 & | & -\frac{25}{9} \\ 0 & 1 & 0 & | & -\frac{151}{30} \\ 0 & 0 & 1 & | & \frac{2302}{9} \end{bmatrix}$$

$$R_1 - R_2 \rightarrow R_1 \quad \begin{bmatrix} 1 & 0 & 0 & | & \frac{203}{90} \\ 0 & 1 & 0 & | & -\frac{151}{30} \\ 0 & 0 & 1 & | & \frac{2302}{9} \end{bmatrix}$$

STEP 3 Identify the solution. $\quad a = \dfrac{203}{90}, \quad b = -\dfrac{151}{30}, \quad c = \dfrac{2302}{9}$

Substituting $a = \frac{203}{90}, b = -\frac{151}{30}, c = \frac{2302}{9}$ into $y = at^2 + bt + c$, we find that the thousands of dollars spent on malpractice suits as a function of year is given by

$$\boxed{y = \dfrac{203}{90}t^2 - \dfrac{151}{30}t + \dfrac{2302}{9}} \qquad 1991 \text{ is } t = 0$$

Notice that all three points lie on this curve.

For 2011, we let $t = 20$, which results in approximately $\boxed{\$1.06\text{ M}}$ in malpractice.

SECTION
8.3 SUMMARY

In this section, we used augmented matrices to represent a system of linear equations.

$$\begin{aligned} a_1x + b_1y + c_1z &= d_1 \\ a_2x + b_2y + c_2z &= d_2 \\ a_3x + b_3y + c_3z &= d_3 \end{aligned} \iff \begin{bmatrix} a_1 & b_1 & c_1 & d_1 \\ a_2 & b_2 & c_2 & d_2 \\ a_3 & b_3 & c_3 & d_3 \end{bmatrix}$$

Any missing terms correspond to a 0 in the matrix. A matrix is in **row–echelon** form if it has all three of the following properties:

1. Any rows consisting entirely of 0s are at the bottom of the matrix.
2. For each row that does not consist entirely of 0s, the first (leftmost) nonzero entry is 1 (called the leading 1).

3. For two successive nonzero rows, the leading 1 in the higher row is farther to the left than the leading 1 in the lower row.

If a matrix in row–echelon form has the following additional property, then the matrix is in **reduced row–echelon form**:

4. Every column containing a leading 1 has zeros in every position above and below the leading 1.

The two methods used for solving systems of linear equations represented as augmented matrices are Gaussian elimination with back-substitution and Gauss–Jordan elimination. In both cases, we represent the system of linear equations as an augmented matrix and then use row operations to rewrite in row–echelon form. With Gaussian elimination we then stop and perform back-substitution to solve the system, and with Gauss–Jordan elimination we continue with row operations until the matrix is in reduced row–echelon form and then identify the solution to the system.

SECTION
8.3 EXERCISES

■ SKILLS

In Exercises 1–6, determine the order of each matrix.

1. $\begin{bmatrix} -1 & 3 & 4 \\ 2 & 7 & 9 \end{bmatrix}$
2. $\begin{bmatrix} 0 & 1 \\ 3 & 9 \\ 7 & 8 \end{bmatrix}$
3. $\begin{bmatrix} 1 & 2 & 3 & 4 \end{bmatrix}$
4. $\begin{bmatrix} 3 \\ 7 \\ -1 \\ 10 \end{bmatrix}$
5. $[0]$
6. $\begin{bmatrix} -1 & 3 & 6 & 8 \\ 2 & 9 & 7 & 3 \\ 5 & 4 & -2 & -10 \\ 6 & 3 & 1 & 5 \end{bmatrix}$

In Exercises 7–14, write the augmented matrix for each system of linear equations.

7. $\begin{aligned} 3x - 2y &= 7 \\ -4x + 6y &= -3 \end{aligned}$

8. $\begin{aligned} -x + y &= 2 \\ x - y &= -4 \end{aligned}$

9. $\begin{aligned} 2x - 3y + 4z &= -3 \\ -x + y + 2z &= 1 \\ 5x - 2y - 3z &= 7 \end{aligned}$

10. $\begin{aligned} x - 2y + z &= 0 \\ -2x + y - z &= -5 \\ 13x + 7y + 5z &= 6 \end{aligned}$

11. $\begin{aligned} x + y &= 3 \\ x - z &= 2 \\ y + z &= 5 \end{aligned}$

12. $\begin{aligned} x - y &= -4 \\ y + z &= 3 \end{aligned}$

13. $\begin{aligned} 3y - 4x + 5z - 2 &= 0 \\ 2x - 3y - 2z &= -3 \\ 3z + 4y - 2x - 1 &= 0 \end{aligned}$

14. $\begin{aligned} 2y + z - x - 3 &= 2 \\ 2x + 3z - 2y &= 0 \\ -2z + y - 4x - 3 &= 0 \end{aligned}$

In Exercises 15–20, write the system of linear equations represented by the augmented matrix. Utilize the variables x, y, and z.

15. $\begin{bmatrix} -3 & 7 & 2 \\ 1 & 5 & 8 \end{bmatrix}$

16. $\begin{bmatrix} -1 & 2 & 4 & 4 \\ 7 & 9 & 3 & -3 \\ 4 & 6 & -5 & 8 \end{bmatrix}$

17. $\begin{bmatrix} -1 & 0 & 0 & 4 \\ 7 & 9 & 3 & -3 \\ 4 & 6 & -5 & 8 \end{bmatrix}$

18. $\begin{bmatrix} 2 & 3 & -4 & 6 \\ 7 & -1 & 5 & 9 \end{bmatrix}$

19. $\begin{bmatrix} 1 & 0 & a \\ 0 & 1 & b \end{bmatrix}$

20. $\begin{bmatrix} 3 & 0 & 5 & 1 \\ 0 & -4 & 7 & -3 \\ 2 & -1 & 0 & 8 \end{bmatrix}$

In Exercises 21–30, indicate whether each matrix is in row–echelon form. If it is, determine whether it is in reduced row–echelon form.

21. $\begin{bmatrix} 1 & 0 & 3 \\ 1 & 1 & 2 \end{bmatrix}$

22. $\begin{bmatrix} 0 & 1 & 3 \\ 1 & 0 & 2 \end{bmatrix}$

23. $\begin{bmatrix} 1 & 0 & -1 & -3 \\ 0 & 1 & 3 & 14 \end{bmatrix}$

24. $\begin{bmatrix} 1 & 0 & 0 & -3 \\ 0 & 1 & 3 & 14 \end{bmatrix}$

25. $\begin{bmatrix} 1 & 0 & 1 & 3 \\ 0 & 0 & 0 & 0 \\ 0 & 1 & 2 & 2 \end{bmatrix}$

26. $\begin{bmatrix} 1 & 0 & 1 & 3 \\ 0 & 1 & 2 & 2 \\ 0 & 0 & 0 & 0 \end{bmatrix}$

27. $\begin{bmatrix} 1 & 0 & 0 & 3 \\ 0 & 1 & 0 & 2 \\ 0 & 0 & 1 & 5 \end{bmatrix}$

28. $\begin{bmatrix} -1 & 0 & 0 & 3 \\ 0 & -1 & 0 & 2 \\ 0 & 0 & -1 & 5 \end{bmatrix}$

29. $\begin{bmatrix} 1 & 0 & 0 & 1 & 3 \\ 0 & 1 & 0 & 3 & 2 \\ 0 & 0 & 1 & 0 & 5 \\ 0 & 0 & 0 & 1 & 0 \end{bmatrix}$

30. $\begin{bmatrix} 1 & 0 & 0 & 1 & 3 \\ 0 & 1 & 0 & 3 & 2 \\ 0 & 0 & 1 & 0 & 5 \\ 0 & 0 & 0 & 0 & 0 \end{bmatrix}$

In Exercises 31–40, perform the indicated row operations on each augmented matrix.

31. $\begin{bmatrix} 1 & -2 & -3 \\ 2 & 3 & -1 \end{bmatrix}$ $R_2 - 2R_1 \rightarrow R_2$

32. $\begin{bmatrix} 2 & -3 & -4 \\ 1 & 2 & 5 \end{bmatrix}$ $R_1 \leftrightarrow R_2$

33. $\begin{bmatrix} 1 & -2 & -1 & 3 \\ 2 & 1 & -3 & 6 \\ 3 & -2 & 5 & -8 \end{bmatrix}$ $R_2 - 2R_1 \rightarrow R_2$

34. $\begin{bmatrix} 1 & -2 & 1 & 3 \\ 0 & 1 & -2 & 6 \\ -3 & 0 & -1 & -5 \end{bmatrix}$ $R_3 + 3R_1 \rightarrow R_3$

35. $\begin{bmatrix} 1 & -2 & 5 & -1 & 2 \\ 0 & 3 & 0 & -1 & -2 \\ 0 & -2 & 1 & -2 & 5 \\ 0 & 0 & 1 & -1 & -6 \end{bmatrix}$ $R_3 + R_2 \rightarrow R_2$

36. $\begin{bmatrix} 1 & 0 & 5 & -10 & 15 \\ 0 & 1 & 2 & -3 & 4 \\ 0 & 2 & -3 & 0 & -1 \\ 0 & 0 & 1 & -1 & -3 \end{bmatrix}$ $R_2 - \frac{1}{2}R_3 \rightarrow R_3$

37. $\begin{bmatrix} 1 & 0 & 5 & -10 & -5 \\ 0 & 1 & 2 & -3 & -2 \\ 0 & 2 & -3 & 0 & -1 \\ 0 & -3 & 2 & -1 & -3 \end{bmatrix}$ $\begin{aligned} R_3 - 2R_2 &\rightarrow R_3 \\ R_4 + 3R_2 &\rightarrow R_4 \end{aligned}$

38. $\begin{bmatrix} 1 & 0 & 4 & 0 & 1 \\ 0 & 1 & 2 & 0 & -2 \\ 0 & 0 & 1 & 0 & 0 \\ 0 & 0 & 0 & 1 & -3 \end{bmatrix}$ $\begin{aligned} R_2 - 2R_3 &\rightarrow R_2 \\ R_1 - 4R_3 &\rightarrow R_1 \end{aligned}$

39. $\begin{bmatrix} 1 & 0 & 4 & 8 & 3 \\ 0 & 1 & 2 & -3 & -2 \\ 0 & 0 & 1 & 6 & 3 \\ 0 & 0 & 0 & 1 & -3 \end{bmatrix}$ $\begin{aligned} R_3 - 6R_4 &\rightarrow R_3 \\ R_2 + 3R_4 &\rightarrow R_2 \\ R_1 - 8R_4 &\rightarrow R_1 \end{aligned}$

40. $\begin{bmatrix} 1 & 0 & -1 & 5 & 2 \\ 0 & 1 & 2 & 3 & -5 \\ 0 & 0 & 1 & -2 & 2 \\ 0 & 0 & 0 & 1 & 1 \end{bmatrix}$ $\begin{aligned} R_3 + 2R_4 &\rightarrow R_3 \\ R_2 - 3R_4 &\rightarrow R_2 \\ R_1 - 5R_4 &\rightarrow R_1 \end{aligned}$

In Exercises 41–50, use row operations to transform each matrix to reduced row–echelon form.

41. $\begin{bmatrix} 1 & 2 & 4 \\ 2 & 3 & 2 \end{bmatrix}$

42. $\begin{bmatrix} 1 & -1 & 3 \\ -3 & 2 & 2 \end{bmatrix}$

43. $\begin{bmatrix} 1 & -1 & 1 & -1 \\ 0 & 1 & -1 & -1 \\ -1 & 1 & 1 & 1 \end{bmatrix}$

44. $\begin{bmatrix} 0 & -1 & 1 & 1 \\ 1 & -1 & 1 & -1 \\ 1 & -1 & -1 & -1 \end{bmatrix}$

45. $\begin{bmatrix} 3 & -2 & -3 & -1 \\ 1 & -1 & 1 & -4 \\ 2 & 3 & 5 & 14 \end{bmatrix}$

46. $\begin{bmatrix} 3 & -1 & 1 & 2 \\ 1 & -2 & 3 & 1 \\ 2 & 1 & -3 & -1 \end{bmatrix}$

47. $\begin{bmatrix} 2 & 1 & -6 & 4 \\ 1 & -2 & 2 & -3 \end{bmatrix}$

48. $\begin{bmatrix} -3 & -1 & 2 & -1 \\ -1 & -2 & 1 & -3 \end{bmatrix}$

49. $\begin{bmatrix} -1 & 2 & 1 & -2 \\ 3 & -2 & 1 & 4 \\ 2 & -4 & -2 & 4 \end{bmatrix}$

50. $\begin{bmatrix} 2 & -1 & 0 & 1 \\ -1 & 0 & 1 & -2 \\ -2 & 1 & 0 & -1 \end{bmatrix}$

In Exercises 51–70, solve the system of linear equations using Gaussian elimination with back-substitution.

51. $\begin{aligned} 2x + 3y &= 1 \\ x + y &= -2 \end{aligned}$

52. $\begin{aligned} 3x + 2y &= 11 \\ x - y &= 12 \end{aligned}$

53. $\begin{aligned} -x + 2y &= 3 \\ 2x - 4y &= -6 \end{aligned}$

54. $\begin{aligned} 3x - y &= -1 \\ 2y + 6x &= 2 \end{aligned}$

55. $\begin{aligned} \tfrac{2}{3}x + \tfrac{1}{3}y &= \tfrac{8}{9} \\ \tfrac{1}{2}x + \tfrac{1}{4}y &= \tfrac{3}{4} \end{aligned}$

56. $\begin{aligned} 0.4x - 0.5y &= 2.08 \\ -0.3x + 0.7y &= 1.88 \end{aligned}$

57.
$$x - z - y = 10$$
$$2x - 3y + z = -11$$
$$y - x + z = -10$$

58.
$$2x + z + y = -3$$
$$2y - z + x = 0$$
$$x + y + 2z = 5$$

59.
$$3x_1 + x_2 - x_3 = 1$$
$$x_1 - x_2 + x_3 = -3$$
$$2x_1 + x_2 + x_3 = 0$$

60.
$$2x_1 + x_2 + x_3 = -1$$
$$x_1 + x_2 - x_3 = 5$$
$$3x_1 - x_2 - x_3 = 1$$

61.
$$2x + 5y = 9$$
$$x + 2y - z = 3$$
$$-3x - 4y + 7z = 1$$

62.
$$x - 2y + 3z = 1$$
$$-2x + 7y - 9z = 4$$
$$x + z = 9$$

63.
$$2x_1 - x_2 + x_3 = 3$$
$$x_1 - x_2 + x_3 = 2$$
$$-2x_1 + 2x_2 - 2x_3 = -4$$

64.
$$x_1 - x_2 - 2x_3 = 0$$
$$-2x_1 + 5x_2 + 10x_3 = -3$$
$$3x_1 + x_2 = 0$$

65.
$$2x + y - z = 2$$
$$x - y - z = 6$$

66.
$$3x + y - z = 0$$
$$x + y + 7z = 4$$

67.
$$2y + z = 3$$
$$4x - z = -3$$
$$7x - 3y - 3z = 2$$
$$x - y - z = -2$$

68.
$$-2x - y + 2z = 3$$
$$3x - 4z = 2$$
$$2x + y = -1$$
$$-x + y - z = -8$$

69.
$$3x_1 - 2x_2 + x_3 + 2x_4 = -2$$
$$-x_1 + 3x_2 + 4x_3 + 3x_4 = 4$$
$$x_1 + x_2 + x_3 + x_4 = 0$$
$$5x_1 + 3x_2 + x_3 + 2x_4 = -1$$

70.
$$5x_1 + 3x_2 + 8x_3 + x_4 = 1$$
$$x_1 + 2x_2 + 5x_3 + 2x_4 = 3$$
$$4x_1 + x_3 - 2x_4 = -3$$
$$x_2 + x_3 + x_4 = 0$$

In Exercises 71–86, solve the system of linear equations using Gauss–Jordan elimination.

71.
$$x + 3y = -5$$
$$-2x - y = 0$$

72.
$$5x - 4y = 31$$
$$3x + 7y = -19$$

73.
$$x + y = 4$$
$$-3x - 3y = 10$$

74.
$$3x - 4y = 12$$
$$-6x + 8y = -24$$

75.
$$x - 2y + 3z = 5$$
$$3x + 6y - 4z = -12$$
$$-x - 4y + 6z = 16$$

76.
$$x + 2y - z = 6$$
$$2x - y + 3z = -13$$
$$3x - 2y + 3z = -16$$

77.
$$x + y + z = 3$$
$$x - z = 1$$
$$y - z = -4$$

78.
$$x - 2y + 4z = 2$$
$$2x - 3y - 2z = -3$$
$$\frac{1}{2}x + \frac{1}{4}y + z = -2$$

79.
$$x + 2y + z = 3$$
$$2x - y + 3z = 7$$
$$3x + y + 4z = 5$$

80.
$$x + 2y + z = 3$$
$$2x - y + 3z = 7$$
$$3x + y + 4z = 10$$

81.
$$3x - y + z = 8$$
$$x + y - 2z = 4$$

82.
$$x - 2y + 3z = 10$$
$$-3x + z = 9$$

83.
$$4x - 2y + 5z = 20$$
$$x + 3y - 2z = 6$$

84.
$$y + z = 4$$
$$x + y = 8$$

85.
$$x - y - z - w = 1$$
$$2x + y + z + 2w = 3$$
$$x - 2y - 2z - 3w = 0$$
$$3x - 4y + z + 5w = -3$$

86.
$$x - 3y + 3z - 2w = 4$$
$$x + 2y - z = -3$$
$$x + 3z + 2w = 3$$
$$y + z + 5w = 6$$

■ APPLICATIONS

87. Astronomy. Astronomers have determined the number of stars in a small region of the universe to be 2,880,968 classified as red dwarfs, yellow, and blue stars. For every blue star there are 120 red dwarfs; for every red dwarf there are 3000 yellow stars. Determine the number of stars by type in that region of the universe.

88. Orange Juice. Orange juice producers use three varieties of oranges: Hamlin, Valencia, and navel. They want to make a juice mixture to sell at $3.00 per gallon. The price per gallon of each variety of juice is $2.50, $3.40, and $2.80, respectively. To maintain their quality standards, they use the same amount of Valencia and navel oranges. Determine the quantity of each juice used to produce 1 gallon of mixture.

Exercises 89 and 90 rely on a selection of Subway sandwiches whose nutrition information is given in the table below. Suppose you are going to eat only Subway sandwiches for a week (seven days) for lunch and dinner (a total of 14 meals).

SANDWICH	CALORIES	FAT (g)	CARBOHYDRATES (g)	PROTEIN (g)
Mediterranean chicken	350	18	17	36
6-inch tuna	430	19	46	20
6-inch roast beef	290	5	45	19
Turkey–bacon wrap	430	27	20	34

www.subway.com

89. Diet. Your goal is a low-fat diet consisting of 526 grams of carbohydrates, 168 grams of fat, and 332 grams of protein. How many of each sandwich would you eat that week to obtain this goal?

90. Diet. Your goal is a low-carb diet consisting of 5180 calories, 335 grams of carbohydrates, and 263 grams of fat. How many of each sandwich would you eat that week to obtain this goal?

Exercises 91 and 92 involve vertical motion and the effect of gravity on an object.

Because of gravity, an object that is projected upward will eventually reach a maximum height and then fall to the ground. The equation that relates the height h of a projectile t seconds after it is projected upward is given by

$$h = \frac{1}{2}at^2 + v_0 t + h_0$$

where a is the acceleration due to gravity, h_0 is the initial height of the object at time $t = 0$, and v_0 is the initial velocity of the object at time $t = 0$. Note that a projectile follows the path of a parabola opening down, so $a < 0$.

91. Vertical Motion. An object is thrown upward, and the table below depicts the height of the ball t seconds after the projectile is released. Find the initial height, initial velocity, and acceleration due to gravity.

t (SECONDS)	HEIGHT (FEET)
1	34
2	36
3	6

92. Vertical Motion. An object is thrown upward, and the table below depicts the height of the ball t seconds after the projectile is released. Find the initial height, initial velocity, and acceleration due to gravity.

t (SECONDS)	HEIGHT (FEET)
1	54
2	66
3	46

93. Data Curve-Fitting. The average number of minutes that a person spends driving a car can be modeled by a quadratic function $y = ax^2 + bx + c$, where $a < 0$ and $15 < x < 65$. The table below gives the average number of minutes a day that a person spends driving a car. Determine a quadratic function that models this quantity.

AGE	AVERAGE DAILY MINUTES DRIVING
16	25
40	64
65	40

94. Data Curve-Fitting. The average age when a woman gets married has been increasing during the last century. In 1920 the average age was 18.4, in 1960 the average age was 20.3, and in 2002 the average age was 25.30. Find a quadratic function $y = ax^2 + bx + c$, where $a > 0$ and $18 < x < 35$, that models the average age y when a woman gets married as a function of the year x ($x = 0$ corresponds to 1920). What will the average age be in 2010?

95. Money. Gary and Ginger decide to place $10,000 of their savings into investments. They put some in a money market account earning 3% interest, some in a mutual fund that has been averaging 7% a year, and some in a stock that rose 10% last year. If they put $3,000 more in the money market than in the mutual fund and the mutual fund and stocks have the same growth in the next year as they did in the previous year, they will earn $540 in a year. How much money did they put in each of the three investments?

96. Money. Ginger talks Gary into putting less money in the money market and more money in the stock (see Exercise 95). They place $10,000 of their savings into investments. They put some in a money market account earning 3% interest, some in a mutual fund that has been averaging 7% a year, and some in a stock that rose 10% last year. If they put $3,000 more in the stock than in the mutual fund and the mutual fund and stock have the same growth in the next year as they did in the previous year, they will earn $840 in a year. How much money did they put in each of the three investments?

97. Manufacturing. A company produces three products x, y, and z. Each item of product x requires 20 units of steel, 2 units of plastic, and 1 unit of glass. Each item of product y requires 25 units of steel, 5 units of plastic, and no units of glass. Each item of product z requires 150 units of steel, 10 units of plastic, and 0.5 units of glass. The available amounts of steel, plastic, and glass are 2400, 310, and 28, respectively. How many items of each type can the company produce and utilize all the available raw materials?

98. Geometry. Find the values of a, b, and c such that the graph of the quadratic function $y = ax^2 + bx + c$ passes through the points $(1, 5)$, $(-2, -10)$, and $(0, 4)$.

99. Ticket Sales. One hundred students decide to buy tickets to a football game. There are three types of tickets: general admission, reserved, and end zone. Each general admission ticket costs $20.00, each reserved ticket costs $40.00, and each end zone ticket costs $15.00. The students spend a total of $2,375.00 for all the tickets. There are five more reserved tickets than general admission tickets, and 20 more end zone tickets than general admission tickets. How many of each type of ticket were purchased by the students?

100. Exercise and Nutrition. Ann would like to exercise one hour per day to burn calories and lose weight. She would like to engage in three activities: walking, step-up exercise, and weight training. She knows she can burn 85 calories walking at a certain pace in 15 minutes, 45 calories doing the step-up exercise in 10 minutes, and 137 calories by weight training for 20 minutes.

a. Determine the number of calories per minute she can burn doing each activity.

b. Suppose she has time to exercise for only one hour (60 minutes). She sets a goal of burning 358 calories in one hour and would like to weight train twice as long as walking. How many minutes must she engage in each exercise to burn the required number of calories in one hour?

101. Geometry. The circle given by the equation $x^2 + y^2 + ax + by + c = 0$ passes through the points $(4, 4)$, $(-3, -1)$, and $(1, -3)$. Find a, b, and c.

102. Geometry. The circle given by the equation $x^2 + y^2 + ax + by + c = 0$ passes through the points $(0, 7)$, $(6, 1)$, and $(5, 4)$. Find a, b, and c.

▪ CATCH THE MISTAKE

In Exercises 103–106, explain the mistake that is made.

103. Solve the system of equations using an augmented matrix.

$$\begin{aligned} y - x + z &= 2 \\ x - 2z + y &= -3 \\ x + y + z &= 6 \end{aligned}$$

Solution:

Step 1: Write as an augmented matrix.

$$\begin{bmatrix} 1 & -1 & 1 & | & 2 \\ 1 & -2 & 1 & | & -3 \\ 1 & 1 & 1 & | & 6 \end{bmatrix}$$

Step 2: Reduce the matrix using Gaussian elimination.

$$\begin{bmatrix} 1 & -1 & 1 & | & 2 \\ 0 & 1 & 0 & | & 5 \\ 0 & 0 & 0 & | & -6 \end{bmatrix}$$

Step 3: Identify the solution. Row 3 is inconsistent, so there is no solution.

This is incorrect. The correct answer is $x = 1$, $y = 2$, $z = 3$. What mistake was made?

104. Perform the indicated row operations on the matrix.

$$\begin{bmatrix} 1 & -1 & 1 & | & 2 \\ 2 & -3 & 1 & | & 4 \\ 3 & 1 & 2 & | & -6 \end{bmatrix}$$

a. $R_2 - 2R_1 \rightarrow R_2$
b. $R_3 - 3R_1 \rightarrow R_3$

Solution:

a. $\begin{bmatrix} 1 & -1 & 1 & | & 2 \\ 0 & -3 & 1 & | & 4 \\ 3 & 1 & 2 & | & -6 \end{bmatrix}$

b. $\begin{bmatrix} 1 & -1 & 1 & | & 2 \\ 2 & -3 & 1 & | & 4 \\ 0 & 1 & 2 & | & -6 \end{bmatrix}$

This is incorrect. What mistake was made?

105. Solve the system of equations using an augmented matrix.

$$\begin{aligned} 3x - 2y + z &= -1 \\ x + y - z &= 3 \\ 2x - y + 3z &= 0 \end{aligned}$$

Solution:

Step 1: Write the system as an augmented matrix.

$$\begin{bmatrix} 3 & -2 & 1 & | & -1 \\ 1 & 1 & -1 & | & 3 \\ 2 & -1 & 3 & | & 0 \end{bmatrix}$$

Step 2: Reduce the matrix using Gaussian elimination.

$$\begin{bmatrix} 1 & 0 & 0 & | & 1 \\ 0 & 1 & 0 & | & 2 \\ 0 & 0 & 1 & | & 0 \end{bmatrix}$$

Step 3: Identify the answer: Row 3 is inconsistent $1 = 0$, therefore there is no solution.

This is incorrect. What mistake was made?

106. Solve the system of equations using an augmented matrix.

$$\begin{aligned} x + 3y + 2z &= 4 \\ 3x + 10y + 9z &= 17 \\ 2x + 7y + 7z &= 17 \end{aligned}$$

Solution:

Step 1: Write the system as an augmented matrix.

$$\begin{bmatrix} 1 & 3 & 2 & | & 4 \\ 3 & 10 & 9 & | & 17 \\ 2 & 7 & 7 & | & 17 \end{bmatrix}$$

Step 2: Reduce the matrix using Gaussian elimination.

$$\begin{bmatrix} 1 & 0 & -7 & | & -11 \\ 0 & 1 & 3 & | & 5 \\ 0 & 0 & 0 & | & 4 \end{bmatrix}$$

Step 3: Identify the answer:
Infinitely many solutions. $\begin{aligned} x &= 7t - 11 \\ y &= -3t + 5 \\ z &= t \end{aligned}$

This is incorrect. What mistake was made?

In Exercises 107–114, determine whether each of the following statements is true or false:

107. A system of equations represented by a nonsquare coefficient matrix cannot have a unique solution.

108. The procedure for Gaussian elimination can be used only for a system of linear equations represented by a square matrix.

109. A system of linear equations represented by a square coefficient matrix that has a unique solution has a reduced matrix with 1s along the main diagonal and 0s above and below the 1s.

110. A system of linear equations represented by a square coefficient matrix with an all-zero row has infinitely many solutions.

111. When a system of linear equations is represented by a square augmented matrix, the system of equations always has a unique solution.

112. Gauss–Jordan elimination produces a matrix in reduced row–echelon form.

113. An inconsistent system of linear equations has infinitely many solutions.

114. Every system of linear equations with a unique solution is represented by an augmented matrix of order $n \times (n + 1)$. (Assume no two rows are identical.)

115. A fourth-degree polynomial $f(x) = ax^4 + bx^3 + cx^2 + dx + k$, with $a < 0$, can be used to represent the data on the number of deaths per year due to lightning strikes (assume 1999 corresponds to $x = 0$).

Use the data below to determine a, b, c, d, and k.

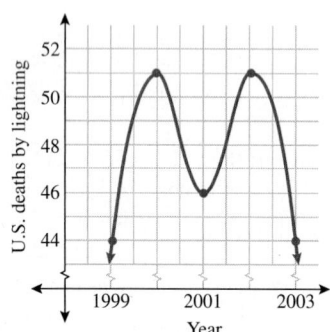

116. A copy machine accepts nickels, dimes, and quarters. After one hour, it holds 30 coins total, and their value is $4.60. How many nickels, quarters, and dimes are in the machine?

117. A ferry goes down a river from city A to city B in 5 hours. The return trip takes 7 hours. How long will a canoe take to make the trip from A to B if it moves at the river speed?

118. Solve the system of equations.

$$\frac{3}{x} - \frac{4}{y} + \frac{6}{z} = 1$$
$$\frac{9}{x} + \frac{8}{y} - \frac{12}{z} = 3$$
$$\frac{9}{x} - \frac{4}{y} + \frac{12}{z} = 4$$

119. The sides of a triangle are formed by the lines $x - y = -3$, $3x + 4y = 5$, and $6x + y = 17$. Find the vertices of the triangle.

120. A winery has three barrels, A, B, and C, containing mixtures of three different wines, w_1, w_2, and w_3. In barrel A, the wines are in the ratio 1:2:3. In barrel B, the wines are in the ratio 3:5:7. In barrel C, the wines are in the ratio 3:7:9. How much wine must be taken from each barrel to get a mixture containing 17 liters of w_1, 35 liters of w_2, and 47 liters of w_3?

121. In Exercise 57, you were asked to solve this system of equations using an augmented matrix.

$$x - z - y = 10$$
$$2x - 3y + z = -11$$
$$y - x + z = -10$$

A graphing calculator or graphing utility can be used to solve systems of linear equations by entering the coefficients of the matrix. Solve this system and confirm your answer with the calculator's answer.

122. In Exercise 58, you were asked to solve this system of equations using an augmented matrix.

$$2x + z + y = -3$$
$$2y - z + x = 0$$
$$x + y + 2z = 5$$

A graphing calculator or graphing utility can be use to solve systems of linear equations by entering the coefficients of the matrix. Solve this system and confirm your answer with the calculator's answer.

In Exercises 123 and 124, you are asked to model a set of three points with a quadratic function $y = ax^2 + bx + c$ and determine the quadratic function.

a. Set up a system of equations, use a graphing utility or graphing calculator to solve the system by entering the coefficients of the augmented matrix.

b. Use the graphing calculator commands $\boxed{\text{STAT}}$ $\boxed{\text{QuadReg}}$ to model the data using a quadratic function. Round your answers to two decimal places.

123. $(-6, -8), (2, 7), (7, 1)$

124. $(-9, 20), (2, -18), (11, 16)$

■ PREVIEW TO CALCULUS

In calculus, when solving systems of linear differential equations with initial conditions, the solution of a system of linear equations is required.

In Exercises 125–128, solve each system of equations.

125.
$$\begin{aligned} c_1 + c_2 &= 0 \\ c_1 + 5c_2 &= -3 \end{aligned}$$

126.
$$\begin{aligned} 3c_1 + 3c_2 &= 0 \\ 2c_1 + 3c_2 &= 0 \end{aligned}$$

127.
$$\begin{aligned} 2c_1 + 2c_2 + 2c_3 &= 0 \\ 2c_1 \quad\quad - 2c_3 &= 2 \\ c_1 - c_2 + c_3 &= 6 \end{aligned}$$

128.
$$\begin{aligned} c_1 + c_4 &= 1 \\ c_3 &= 1 \\ c_2 + 3c_4 &= 1 \\ c_1 - 2c_3 &= 1 \end{aligned}$$

SECTION
8.4 MATRIX ALGEBRA

SKILLS OBJECTIVES

- Add and subtract matrices.
- Perform scalar multiplication.
- Multiply two matrices.
- Write a system of linear equations as a matrix equation.
- Find the inverse of a square matrix.
- Solve systems of linear equations using inverse matrices.

CONCEPTUAL OBJECTIVES

- Understand what is meant by equal matrices.
- Understand why multiplication of some matrices is undefined.
- Realize that matrix multiplication is *not* commutative.
- Visualize a system of linear equations as a matrix equation.
- Understand that only a square matrix can have an inverse.
- Realize that not every square matrix has an inverse.

In Section 8.3, we defined a matrix with m rows and n columns to have order $m \times n$.

$$A = \begin{bmatrix} a_{11} & a_{12} & \cdots & a_{1n} \\ a_{21} & a_{22} & \cdots & a_{2n} \\ \vdots & \vdots & \cdots & \vdots \\ a_{m1} & a_{m2} & \cdots & a_{mn} \end{bmatrix}$$

Capital letters are used to represent (or name) a matrix, and lowercase letters are used to represent the entries (elements) of the matrix. The subscripts are used to denote the location (row/column) of each entry. The order of a matrix is often written as a subscript of the matrix name: $A_{m \times n}$. Other words like "size" and "dimension" are used as synonyms of "order." Matrices are a convenient way to represent data.

There is an entire field of study called **matrix algebra** that treats matrices similarly to functions and variables in traditional algebra. This section serves as an introduction to matrix algebra. It is important to pay special attention to the *order* of a matrix, because it determines whether certain operations are defined.

Equality of Matrices

Two matrices are equal if and only if they have the same order, $m \times n$, and all of their corresponding entries are equal.

DEFINITION **Equality of Matrices**

Two matrices, A and B, are **equal**, written as $A = B$, if and only if *both* of the following are true:

- A and B have the same order $m \times n$.
- Every pair of corresponding entries is equal: $a_{ij} = b_{ij}$ for all $i = 1, 2, \ldots, m$ and $j = 1, 2, \ldots, n$.

EXAMPLE 1 **Equality of Matrices**

Referring to the definition of equality of matrices, find the indicated entries.

$$\begin{bmatrix} a_{11} & a_{12} & a_{13} \\ a_{21} & a_{22} & a_{23} \\ a_{31} & a_{32} & a_{33} \end{bmatrix} = \begin{bmatrix} 2 & -7 & 1 \\ 0 & 5 & -3 \\ -1 & 8 & 9 \end{bmatrix}$$

Find the main diagonal entries: a_{11}, a_{22}, and a_{33}.

Solution:

Since the matrices are equal, their corresponding entries are equal.

$\boxed{a_{11} = 2}$ $\boxed{a_{22} = 5}$ $\boxed{a_{33} = 9}$

Matrix Addition and Subtraction

Two matrices, A and B, can be added or subtracted only if they have the *same order*. Suppose A and B are both of order $m \times n$; then the *sum* $A + B$ is found by adding corresponding entries, or taking $a_{ij} + b_{ij}$. The *difference* $A - B$ is found by subtracting the entries in B from the corresponding entries in A, or finding $a_{ij} - b_{ij}$.

DEFINITION **Matrix Addition and Matrix Subtraction**

If A is an $m \times n$ matrix and B is an $m \times n$ matrix, then their **sum** $A + B$ is an $m \times n$ matrix whose entries are given by

$$a_{ij} + b_{ij}$$

and their **difference** $A - B$ is an $m \times n$ matrix whose entries are given by

$$a_{ij} - b_{ij}$$

EXAMPLE 2   Adding and Subtracting Matrices

Given that $A = \begin{bmatrix} -1 & 3 & 4 \\ -5 & 2 & 0 \end{bmatrix}$ and $B = \begin{bmatrix} 2 & 1 & -3 \\ 0 & -5 & 4 \end{bmatrix}$, find:

a. $A + B$ **b.** $A - B$

Solution:

Since $A_{2\times3}$ and $B_{2\times3}$ have the same order, they can be added or subtracted.

a. Write the sum.

$$A + B = \begin{bmatrix} -1 & 3 & 4 \\ -5 & 2 & 0 \end{bmatrix} + \begin{bmatrix} 2 & 1 & -3 \\ 0 & -5 & 4 \end{bmatrix}$$

Add the corresponding entries.

$$= \begin{bmatrix} -1+2 & 3+1 & 4+(-3) \\ -5+0 & 2+(-5) & 0+4 \end{bmatrix}$$

Simplify.

$$= \begin{bmatrix} 1 & 4 & 1 \\ -5 & -3 & 4 \end{bmatrix}$$

b. Write the difference.

$$A - B = \begin{bmatrix} -1 & 3 & 4 \\ -5 & 2 & 0 \end{bmatrix} - \begin{bmatrix} 2 & 1 & -3 \\ 0 & -5 & 4 \end{bmatrix}$$

Subtract the corresponding entries.

$$= \begin{bmatrix} -1-2 & 3-1 & 4-(-3) \\ -5-0 & 2-(-5) & 0-4 \end{bmatrix}$$

Simplify.

$$= \begin{bmatrix} -3 & 2 & 7 \\ -5 & 7 & -4 \end{bmatrix}$$

■ **YOUR TURN** Perform the indicated matrix operations, if possible.

$$A = \begin{bmatrix} -4 & 0 \\ 1 & 2 \end{bmatrix} \quad B = \begin{bmatrix} 2 & 3 \\ -4 & 0 \end{bmatrix} \quad C = \begin{bmatrix} 2 & 9 & 5 & -1 \end{bmatrix} \quad D = \begin{bmatrix} 0 \\ -3 \\ 4 \\ 2 \end{bmatrix}$$

a. $B - A$ **b.** $C + D$ **c.** $A + B$ **d.** $A + D$

It is important to note that only matrices of the same order can be added or subtracted. For example, if $A = \begin{bmatrix} -1 & 3 & 4 \\ -5 & 2 & 0 \end{bmatrix}$ and $B = \begin{bmatrix} 5 & -3 \\ 12 & 1 \end{bmatrix}$, the sum and difference of these matrices are undefined because $A_{2\times3}$ and $B_{2\times2}$ do not have the same order.

A matrix whose entries are all equal to 0 is called a **zero matrix**, denoted **0**. The following are examples of zero matrices:

2×2 square zero matrix $\begin{bmatrix} 0 & 0 \\ 0 & 0 \end{bmatrix}$

3×2 zero matrix $\begin{bmatrix} 0 & 0 \\ 0 & 0 \\ 0 & 0 \end{bmatrix}$

1×4 zero matrix $\begin{bmatrix} 0 & 0 & 0 & 0 \end{bmatrix}$

If A, an $m \times n$ matrix, is added to the $m \times n$ zero matrix, the result is A.

$$A + \mathbf{0} = A$$

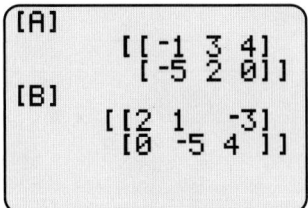

For example,

$$\begin{bmatrix} 1 & -3 \\ 2 & 5 \end{bmatrix} + \begin{bmatrix} 0 & 0 \\ 0 & 0 \end{bmatrix} = \begin{bmatrix} 1 & -3 \\ 2 & 5 \end{bmatrix}$$

Because of this result, an $m \times n$ zero matrix is called the **additive identity** for $m \times n$ matrices. Similarly, for any matrix A, there exists an **additive inverse**, $-A$, such that each entry of $-A$ is the negative of the corresponding entry of A.

For example, $A = \begin{bmatrix} 1 & -3 \\ 2 & 5 \end{bmatrix}$ and $-A = \begin{bmatrix} -1 & 3 \\ -2 & -5 \end{bmatrix}$, and adding these two matrices results in a zero matrix: $A + (-A) = \mathbf{0}$.

The same properties that hold for adding real numbers also hold for adding matrices, provided that addition of matrices is defined.

PROPERTIES OF MATRIX ADDITION

If A, B, and C are all $m \times n$ matrices and $\mathbf{0}$ is the $m \times n$ zero matrix, then the following are true:

Commutative property: $A + B = B + A$

Associative property: $(A + B) + C = A + (B + C)$

Additive identity property: $A + \mathbf{0} = A$

Additive inverse property: $A + (-A) = \mathbf{0}$

Scalar and Matrix Multiplication

There are two types of multiplication involving matrices: *scalar multiplication* and *matrix multiplication*. A **scalar** is any real number. *Scalar multiplication* is the multiplication of a matrix by a scalar, or real number, and is defined for all matrices. *Matrix multiplication* is the multiplication of two matrices and is defined only for certain pairs of matrices, depending on the order of each matrix.

Scalar Multiplication

To multiply a matrix A by a scalar k, multiply every entry in A by k.

$$3\begin{bmatrix} -1 & 0 & 4 \\ 7 & 5 & -2 \end{bmatrix} = \begin{bmatrix} 3(-1) & 3(0) & 3(4) \\ 3(7) & 3(5) & 3(-2) \end{bmatrix} = \begin{bmatrix} -3 & 0 & 12 \\ 21 & 15 & -6 \end{bmatrix}$$

Here, the scalar is $k = 3$.

DEFINITION   **Scalar Multiplication**

If A is an $m \times n$ matrix and k is any real number, then their product kA is an $m \times n$ matrix whose entries are given by

$$ka_{ij}$$

In other words, every entry a_{ij} of A is multiplied by k.

In general, uppercase letters are used to denote a matrix and lowercase letters are used to denote scalars. Notice that the elements of each matrix are also represented with lowercase letters, since they are real numbers.

EXAMPLE 3 Multiplying a Matrix by a Scalar

Given that $A = \begin{bmatrix} -1 & 2 \\ -3 & 4 \end{bmatrix}$ and $B = \begin{bmatrix} 0 & 1 \\ -2 & 3 \end{bmatrix}$, perform:

a. $2A$ **b.** $-3B$ **c.** $2A - 3B$

Solution (a):

Write the scalar multiplication.

$$2A = 2\begin{bmatrix} -1 & 2 \\ -3 & 4 \end{bmatrix}$$

Multiply all entries of A by 2.

$$2A = \begin{bmatrix} 2(-1) & 2(2) \\ 2(-3) & 2(4) \end{bmatrix}$$

Simplify.

$$2A = \begin{bmatrix} -2 & 4 \\ -6 & 8 \end{bmatrix}$$

Solution (b):

Write the scalar multiplication.

$$-3B = -3\begin{bmatrix} 0 & 1 \\ -2 & 3 \end{bmatrix}$$

Multiply all entries of B by -3.

$$-3B = \begin{bmatrix} -3(0) & -3(1) \\ -3(-2) & -3(3) \end{bmatrix}$$

Simplify.

$$-3B = \begin{bmatrix} 0 & -3 \\ 6 & -9 \end{bmatrix}$$

Solution (c):

Add the results of parts (a) and (b).

$$2A - 3B = 2A + (-3B)$$

$$2A - 3B = \begin{bmatrix} -2 & 4 \\ -6 & 8 \end{bmatrix} + \begin{bmatrix} 0 & -3 \\ 6 & -9 \end{bmatrix}$$

Add the corresponding entries.

$$2A - 3B = \begin{bmatrix} -2 + 0 & 4 + (-3) \\ -6 + 6 & 8 + (-9) \end{bmatrix}$$

Simplify.

$$2A - 3B = \begin{bmatrix} -2 & 1 \\ 0 & -1 \end{bmatrix}$$

■ **YOUR TURN** For the matrices A and B given in Example 3, find $-5A + 2B$.

Technology Tip

Enter matrices as A and B.

```
[A]
          [[-1 2]
           [-3 4]]
[B]
          [[0  1]
           [-2 3]]
```

Now enter $2A$, $-3B$, $2A - 3B$.

```
2[A]
          [[-2 4]
           [-6 8]]
-3[B]
          [[0 -3]
           [6 -9]]
■
```

```
2[A]-3[B]
          [[-2 1 ]
           [0  -1]]
■
```

■ **Answer:**

$$-5A + 2B = \begin{bmatrix} 5 & -8 \\ 11 & -14 \end{bmatrix}$$

Matrix Multiplication

Scalar multiplication is straightforward in that it is defined for all matrices and is performed by multiplying every entry in the matrix by the scalar. Addition of matrices is also an entry-by-entry operation. *Matrix multiplication*, on the other hand, is not as straightforward in that we *do not multiply the corresponding entries* and it is not defined for all matrices. Matrices are multiplied using a row-by-column method.

Study Tip

When we multiply matrices, we *do not* multiply corresponding entries.

Study Tip

For the product *AB* of two matrices *A* and *B* to be defined, the number of columns in the first matrix must equal the number of rows in the second matrix.

Before we even try to find the product *AB* of two matrices *A* and *B*, we first have to determine whether the product is defined. For the product *AB* to exist, **the number of columns in the first matrix *A* must equal the number of rows in the second matrix *B*.** In other words, if the matrix $A_{m \times n}$ has *m* rows and *n* columns and the matrix $B_{n \times p}$ has *n* rows and *p* columns, then the product $(AB)_{m \times p}$ is defined and has *m* rows and *p* columns.

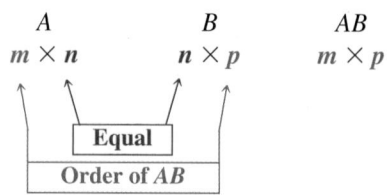

Matrix:	*A*	*B*	*AB*
Order:	$m \times n$	$n \times p$	$m \times p$

Equal

Order of *AB*

EXAMPLE 4 Determining Whether the Product of Two Matrices Is Defined

Given the matrices

$$A = \begin{bmatrix} 1 & -2 & 0 \\ 5 & -1 & 3 \end{bmatrix} \quad B = \begin{bmatrix} 2 & 3 \\ 0 & 7 \\ 4 & 9 \end{bmatrix} \quad C = \begin{bmatrix} 6 & -1 \\ 5 & 2 \end{bmatrix} \quad D = \begin{bmatrix} -3 & -2 \end{bmatrix}$$

state whether each of the following products exists. If the product exists, state the order of the product matrix.

a. *AB* **b.** *AC* **c.** *BC* **d.** *CD* **e.** *DC*

Solution:

Label the order of each matrix: $A_{2 \times 3}$, $B_{3 \times 2}$, $C_{2 \times 2}$, and $D_{1 \times 2}$.

a. *AB* is defined, because *A* has 3 columns and *B* has 3 rows.
$A_{2 \times 3} B_{3 \times 2}$
AB is order $\boxed{2 \times 2}$.
$(AB)_{2 \times 2}$

b. *AC* is $\boxed{\text{not defined}}$, because *A* has 3 columns and *C* has 2 rows.

c. *BC* is defined, because *B* has 2 columns and *C* has 2 rows.
$B_{3 \times 2} C_{2 \times 2}$
BC is order $\boxed{3 \times 2}$.
$(BC)_{3 \times 2}$

d. *CD* is $\boxed{\text{not defined}}$, because *C* has 2 columns and *D* has 1 row.

e. *DC* is defined, because *D* has 2 columns and *C* has 2 rows.
$D_{1 \times 2} C_{2 \times 2}$
DC is order $\boxed{1 \times 2}$.
$(DC)_{1 \times 2}$

Notice that in part (d) we found that *CD* is not defined, but in part (e) we found that *DC* is defined. **Matrix multiplication is not commutative.** Therefore, the order in which matrices are multiplied is important in determining whether the product is defined or undefined. For the product of two matrices to exist, the number of *columns* in the *first* matrix *A* must equal the number of *rows* in the *second* matrix *B*.

■ **Answer:**

a. *DA* exists and is order 1×3.
b. *CB* does not exist.
c. *BA* exists and is order 3×3

■ **YOUR TURN** For the matrices given in Example 4, state whether the following products exist. If the product exists, state the order of the product matrix.

a. *DA* **b.** *CB* **c.** *BA*

Now that we can determine whether a product of two matrices is defined and, if so, what the order of the resulting product is, let us turn our attention to how to multiply two matrices.

DEFINITION   Matrix Multiplication

If A is an $m \times n$ matrix and B is an $n \times p$ matrix, then their product AB is an $m \times p$ matrix whose entries are given by

$$(ab)_{ij} = a_{i1}b_{1j} + a_{i2}b_{2j} + \ldots + a_{in}b_{nj}$$

In other words, the entry $(ab)_{ij}$, which is in the ith row and jth column of AB, is the sum of the products of the corresponding entries in the ith row of A and the jth column of B. Multiply *across* the row and *down* the column.

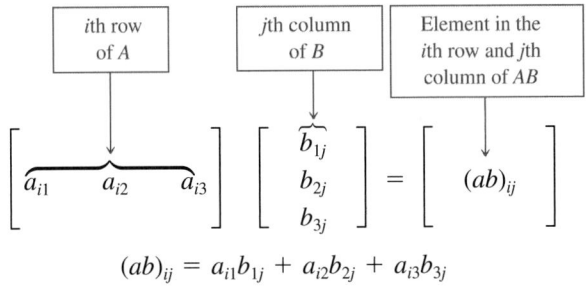

$$(ab)_{ij} = a_{i1}b_{1j} + a_{i2}b_{2j} + a_{i3}b_{3j}$$

EXAMPLE 5   Multiplication of Two 2 × 2 Matrices

Given $A = \begin{bmatrix} 1 & 2 \\ 3 & 4 \end{bmatrix}$ and $B = \begin{bmatrix} 5 & 6 \\ 7 & 8 \end{bmatrix}$, find AB.

COMMON MISTAKE

Do not multiply entry by entry.

⭐ CORRECT

Write the product of the two matrices A and B.

$$AB = \begin{bmatrix} 1 & 2 \\ 3 & 4 \end{bmatrix}\begin{bmatrix} 5 & 6 \\ 7 & 8 \end{bmatrix}$$

Perform the row-by-column multiplication.

$$AB = \begin{bmatrix} (1)(5) + (2)(7) & (1)(6) + (2)(8) \\ (3)(5) + (4)(7) & (3)(6) + (4)(8) \end{bmatrix}$$

Simplify.

$$AB = \begin{bmatrix} 19 & 22 \\ 43 & 50 \end{bmatrix}$$

❌ INCORRECT

Multiply the corresponding entries.

ERROR

$$AB \neq \begin{bmatrix} (1)(5) & (2)(6) \\ (3)(7) & (4)(8) \end{bmatrix}$$

Technology Tip

Enter the matrices as A and B and calculate AB.

```
[A]
                [[1 2]
                 [3 4]]
[B]
                [[5 6]
■                [7 8]]
```

```
[A] [B]
                [[19 22]
                 [43 50]]
```

■ **Answer:**

$$BA = \begin{bmatrix} 5 & 6 \\ 7 & 8 \end{bmatrix}\begin{bmatrix} 1 & 2 \\ 3 & 4 \end{bmatrix}$$

$$= \begin{bmatrix} 23 & 34 \\ 31 & 46 \end{bmatrix}$$

■ **YOUR TURN** For matrices A and B given in Example 5, find BA.

Compare the products obtained in Example 5 and the preceding Your Turn. Note that $AB \neq BA$. Therefore, there is **no commutative property for matrix multiplication**.

■ **Answer:** $AB = \begin{bmatrix} 0 & -5 \\ -1 & -7 \end{bmatrix}$

EXAMPLE 6   Multiplying Matrices

For $A = \begin{bmatrix} -1 & 2 & -3 \\ -2 & 0 & 4 \end{bmatrix}$ and $B = \begin{bmatrix} 2 & 0 \\ 1 & 3 \\ -1 & -2 \end{bmatrix}$, find AB.

Solution:

Since A is order 2×3 and B is order 3×2, the product AB is defined and has order 2×2.
$$A_{2\times3}B_{3\times2} = (AB)_{2\times2}$$

Write the product of the two matrices.
$$AB = \begin{bmatrix} -1 & 2 & -3 \\ -2 & 0 & 4 \end{bmatrix}\begin{bmatrix} 2 & 0 \\ 1 & 3 \\ -1 & -2 \end{bmatrix}$$

Perform the row-by-column multiplication.

$$AB = \begin{bmatrix} (-1)(2) + (2)(1) + (-3)(-1) & (-1)(0) + (2)(3) + (-3)(-2) \\ (-2)(2) + (0)(1) + (4)(-1) & (-2)(0) + (0)(3) + (4)(-2) \end{bmatrix}$$

Simplify.
$$AB = \begin{bmatrix} 3 & 12 \\ -8 & -8 \end{bmatrix}$$

■ **YOUR TURN** For $A = \begin{bmatrix} 1 & 0 & 2 \\ -3 & -1 & 4 \end{bmatrix}$ and $B = \begin{bmatrix} 0 & -1 \\ 1 & 2 \\ 0 & -2 \end{bmatrix}$, find AB.

EXAMPLE 7   Multiplying Matrices

For $A = \begin{bmatrix} 1 & 0 & 3 \\ -2 & 5 & -1 \end{bmatrix}$ and $B = \begin{bmatrix} -2 & 0 & 1 \\ -3 & -1 & 4 \\ 0 & 2 & 5 \end{bmatrix}$, find AB.

Solution:

Since A is order 2×3 and B is order 3×3, the product AB is defined and has order 2×3.
$$A_{2\times3}B_{3\times3} = (AB)_{2\times3}$$

Write the product of the two matrices.
$$AB = \begin{bmatrix} 1 & 0 & 3 \\ -2 & 5 & -1 \end{bmatrix}\begin{bmatrix} -2 & 0 & 1 \\ -3 & -1 & 4 \\ 0 & 2 & 5 \end{bmatrix}$$

Perform the row-by-column multiplication.

$$AB = \begin{bmatrix} (1)(-2) + (0)(-3) + (3)(0) & (1)(0) + (0)(-1) + (3)(2) & (1)(1) + (0)(4) + (3)(5) \\ (-2)(-2) + (5)(-3) + (-1)(0) & (-2)(0) + (5)(-1) + (-1)(2) & (-2)(1) + (5)(4) + (-1)(5) \end{bmatrix}$$

Simplify.
$$AB = \begin{bmatrix} -2 & 6 & 16 \\ -11 & -7 & 13 \end{bmatrix}$$

■ **Answer: a.** $AB = \begin{bmatrix} 4 & 5 \\ 8 & 10 \\ 12 & 15 \end{bmatrix}$
b. does not exist

■ **YOUR TURN** Given $A = \begin{bmatrix} 1 \\ 2 \\ 3 \end{bmatrix}$ and $B = [4 \quad 5]$, find:

a. AB, if it exists   **b.** BA, if it exists

Although we have shown repeatedly that there is no commutative property of multiplication for matrices, matrices do have an associative property of multiplication, as well as a distributive property of multiplication similar to real numbers.

PROPERTIES OF MATRIX MULTIPLICATION

If A, B, and C are all matrices for which AB, AC, BC, $A + B$, and $B + C$ are all defined, then the following properties are true:

Associative property: $A(BC) = (AB)C$

Distributive property: $A(B + C) = AB + AC$ or $(A + B)C = AC + BC$

EXAMPLE 8   Application of Matrix Multiplication

The following table gives fuel and electric requirements per mile associated with gasoline and electric automobiles:

	NUMBER OF GALLONS/MILE	NUMBER OF kW-hr/MILE
Gas car	0.05	0
Hybrid car	0.02	0.1
Electric car	0	0.25

The following table gives an average cost for gasoline and electricity:

Cost per gallon of gasoline	$3.00
Cost per kW-hr of electricity	$0.05

a. Let matrix A represent the gasoline and electricity consumption and matrix B represent the costs of gasoline and electricity.

b. Find AB and describe what the entries of the product matrix represent.

c. Assume you drive 12,000 miles per year. What are the yearly costs associated with driving the three types of cars?

Solution (a):

A has order 3×2.

$$A = \begin{bmatrix} 0.05 & 0 \\ 0.02 & 0.1 \\ 0 & 0.25 \end{bmatrix}$$

B has order 2×1.

$$B = \begin{bmatrix} \$3.00 \\ \$0.05 \end{bmatrix}$$

Solution (b):

Find the order of the product matrix AB.

$$A_{3\times2}B_{2\times1} = (AB)_{3\times1}$$

$$AB = \begin{bmatrix} 0.05 & 0 \\ 0.02 & 0.1 \\ 0 & 0.25 \end{bmatrix} \begin{bmatrix} \$3.00 \\ \$0.05 \end{bmatrix}$$

Calculate AB.

$$= \begin{bmatrix} (0.05)(\$3.00) + (0)(\$0.05) \\ (0.02)(\$3.00) + (0.1)(\$0.05) \\ (0)(\$3.00) + (0.25)(\$0.05) \end{bmatrix}$$

$$AB = \begin{bmatrix} \$0.15 \\ \$0.065 \\ \$0.0125 \end{bmatrix}$$

Technology Tip

Enter the matrices as A and B and calculate AB.

```
[A]
         [[.05 0   ]
          [.02 .1  ]
          [0   .25]]
[B]
         [[3   ]
          [.05]]
```

```
[A] [B]
         [[.15   ]
          [.065  ]
          [.0125]]
```

Interpret the product matrix.

$$AB = \begin{bmatrix} \text{Cost per mile to drive the gas car} \\ \text{Cost per mile to drive the hybrid car} \\ \text{Cost per mile to drive the electric car} \end{bmatrix}$$

Solution (c):

Find $12,000AB$.

$$12,000 \begin{bmatrix} \$0.15 \\ \$0.065 \\ \$0.0125 \end{bmatrix} = \begin{bmatrix} \$1,800 \\ \$780 \\ \$150 \end{bmatrix}$$

GAS/ELECTRIC COSTS PER YEAR ($)	
Gas car	1,800
Hybrid car	780
Electric car	150

Matrix Equations

Matrix equations are another way of writing systems of linear equations.

WORDS	MATH
Start with a matrix equation.	$\begin{bmatrix} 2 & -3 \\ 1 & 5 \end{bmatrix}\begin{bmatrix} x \\ y \end{bmatrix} = \begin{bmatrix} -7 \\ 9 \end{bmatrix}$
Multiply the two matrices on the left.	$\begin{bmatrix} 2x - 3y \\ x + 5y \end{bmatrix} = \begin{bmatrix} -7 \\ 9 \end{bmatrix}$
Apply equality of two matrices.	$\begin{aligned} 2x - 3y &= -7 \\ x + 5y &= 9 \end{aligned}$

Let A be a matrix with m rows and n columns, which represents the coefficients in the system. Also, let X be a column matrix of order $n \times 1$ that represents the variables in the system and let B be a column matrix of order $m \times 1$ that represents the constants in the system. Then, a system of linear equations can be written as $AX = B$.

System of Linear Equations	A	X	B	Matrix Equation: AX = B
$\begin{aligned} 3x + 4y &= 1 \\ x - 2y &= 7 \end{aligned}$	$\begin{bmatrix} 3 & 4 \\ 1 & -2 \end{bmatrix}$	$\begin{bmatrix} x \\ y \end{bmatrix}$	$\begin{bmatrix} 1 \\ 7 \end{bmatrix}$	$\begin{bmatrix} 3 & 4 \\ 1 & -2 \end{bmatrix}\begin{bmatrix} x \\ y \end{bmatrix} = \begin{bmatrix} 1 \\ 7 \end{bmatrix}$
$\begin{aligned} x - y + z &= 2 \\ 2x + 2y - 3z &= -3 \\ x + y + z &= 6 \end{aligned}$	$\begin{bmatrix} 1 & -1 & 1 \\ 2 & 2 & -3 \\ 1 & 1 & 1 \end{bmatrix}$	$\begin{bmatrix} x \\ y \\ z \end{bmatrix}$	$\begin{bmatrix} 2 \\ -3 \\ 6 \end{bmatrix}$	$\begin{bmatrix} 1 & -1 & 1 \\ 2 & 2 & -3 \\ 1 & 1 & 1 \end{bmatrix}\begin{bmatrix} x \\ y \\ z \end{bmatrix} = \begin{bmatrix} 2 \\ -3 \\ 6 \end{bmatrix}$
$\begin{aligned} x + y + z &= 0 \\ 3x + 2y - z &= 2 \end{aligned}$	$\begin{bmatrix} 1 & 1 & 1 \\ 3 & 2 & -1 \end{bmatrix}$	$\begin{bmatrix} x \\ y \\ z \end{bmatrix}$	$\begin{bmatrix} 0 \\ 2 \end{bmatrix}$	$\begin{bmatrix} 1 & 1 & 1 \\ 3 & 2 & -1 \end{bmatrix}\begin{bmatrix} x \\ y \\ z \end{bmatrix} = \begin{bmatrix} 0 \\ 2 \end{bmatrix}$

EXAMPLE 9   Writing a System of Linear Equations as a Matrix Equation

Write each system of linear equations as a matrix equation.

a. $2x - y = 5$
$-x + 2y = 3$

b. $3x - 2y + 4z = 5$
$y - 3z = -2$
$7x \quad - z = 1$

c. $x_1 - x_2 + 2x_3 - 3 = 0$
$x_1 + x_2 - 3x_3 + 5 = 0$
$x_1 - x_2 + x_3 - 2 = 0$

Solution:

a.
$$\begin{bmatrix} 2 & -1 \\ -1 & 2 \end{bmatrix}\begin{bmatrix} x \\ y \end{bmatrix} = \begin{bmatrix} 5 \\ 3 \end{bmatrix}$$

b. Note that all missing terms have 0 coefficients.

$3x - 2y + 4z = 5$
$0x + y - 3z = -2$
$7x + 0y - z = 1$

$$\begin{bmatrix} 3 & -2 & 4 \\ 0 & 1 & -3 \\ 7 & 0 & -1 \end{bmatrix}\begin{bmatrix} x \\ y \\ z \end{bmatrix} = \begin{bmatrix} 5 \\ -2 \\ 1 \end{bmatrix}$$

c. Write the constants on the right side of the equal sign.

$x_1 - x_2 + 2x_3 = 3$
$x_1 + x_2 - 3x_3 = -5$
$x_1 - x_2 + x_3 = 2$

$$\begin{bmatrix} 1 & -1 & 2 \\ 1 & 1 & -3 \\ 1 & -1 & 1 \end{bmatrix}\begin{bmatrix} x_1 \\ x_2 \\ x_3 \end{bmatrix} = \begin{bmatrix} 3 \\ -5 \\ 2 \end{bmatrix}$$

■ **YOUR TURN** Write each system of linear equations as a matrix equation.

a. $2x + y - 3 = 0$
$x - y = 5$

b. $y - x + z = 7$
$x - y - z = 2$
$z - y = -1$

■ **Answer:**

a. $\begin{bmatrix} 2 & 1 \\ 1 & -1 \end{bmatrix}\begin{bmatrix} x \\ y \end{bmatrix} = \begin{bmatrix} 3 \\ 5 \end{bmatrix}$

b. $\begin{bmatrix} -1 & 1 & 1 \\ 1 & -1 & -1 \\ 0 & -1 & 1 \end{bmatrix}\begin{bmatrix} x \\ y \\ z \end{bmatrix} = \begin{bmatrix} 7 \\ 2 \\ -1 \end{bmatrix}$

Finding the Inverse of a Square Matrix

Before we discuss solving systems of linear equations in the form $AX = B$, let us first recall how we solve $ax = b$, where a and b are real numbers (not matrices).

WORDS	MATH
Write the linear equation in one variable.	$ax = b$
Multiply both sides by a^{-1} (same as dividing by a), provided $a \neq 0$.	$a^{-1}ax = a^{-1}b$
Simplify.	$\underset{1}{\underline{a^{-1}a}}x = a^{-1}b$
	$x = a^{-1}b$

Recall that a^{-1}, or $\frac{1}{a}$, is the *multiplicative inverse* of a because $a^{-1}a = 1$. And we call 1 the *multiplicative identity*, because any number multiplied by 1 is itself. Before we solve matrix equations, we need to define the *multiplicative identity matrix* and the *multiplicative inverse matrix*.

A square matrix of order $n \times n$ with 1s along the **main diagonal** (a_{ii}) and 0s for all other elements is called the **multiplicative identity matrix I_n**.

$$I_2 = \begin{bmatrix} 1 & 0 \\ 0 & 1 \end{bmatrix} \qquad I_3 = \begin{bmatrix} 1 & 0 & 0 \\ 0 & 1 & 0 \\ 0 & 0 & 1 \end{bmatrix} \qquad I_4 = \begin{bmatrix} 1 & 0 & 0 & 0 \\ 0 & 1 & 0 & 0 \\ 0 & 0 & 1 & 0 \\ 0 & 0 & 0 & 1 \end{bmatrix}$$

Since a real number multiplied by 1 is itself $(a \cdot 1 = a)$, we expect that a matrix multiplied by the appropriate identity matrix should result in itself. Remember, the order in which matrices are multiplied makes a difference. Notice the appropriate identity matrix may differ, depending on the order of multiplication, but the identity matrix will always be square.

$$A_{m \times n} I_n = A_{m \times n} \qquad \text{and} \qquad I_m A_{m \times n} = A_{m \times n}$$

EXAMPLE 10   Multiplying a Matrix by the Multiplicative Identity Matrix I_n

For $A = \begin{bmatrix} -2 & 4 & 1 \\ 3 & 7 & -1 \end{bmatrix}$, find $I_2 A$.

Solution:

Write the two matrices. $A = \begin{bmatrix} -2 & 4 & 1 \\ 3 & 7 & -1 \end{bmatrix} \qquad I_2 = \begin{bmatrix} 1 & 0 \\ 0 & 1 \end{bmatrix}$

Find the product $I_2 A$. $I_2 A = \begin{bmatrix} 1 & 0 \\ 0 & 1 \end{bmatrix} \begin{bmatrix} -2 & 4 & 1 \\ 3 & 7 & -1 \end{bmatrix}$

$I_2 A = \begin{bmatrix} (1)(-2)+(0)(3) & (1)(4)+(0)(7) & (1)(1)+(0)(-1) \\ (0)(-2)+(1)(3) & (0)(4)+(1)(7) & (0)(1)+(1)(-1) \end{bmatrix}$

$I_2 A = \boxed{\begin{bmatrix} -2 & 4 & 1 \\ 3 & 7 & -1 \end{bmatrix}} = A$

■ **Answer:**

$AI_3 = \begin{bmatrix} -2 & 4 & 1 \\ 3 & 7 & 1 \end{bmatrix} = A$

■ **YOUR TURN** For A in Example 10, find AI_3.

The identity matrix I_n will assist us in developing the concept of an *inverse of a square matrix*.

Study Tip

• Only a *square* matrix can have an inverse.
• Not all square matrices have inverses.

DEFINITION   Inverse of a Square Matrix

Let A be a square $n \times n$ matrix. If there exists a square $n \times n$ matrix A^{-1} such that

$$AA^{-1} = I_n \qquad \text{and} \qquad A^{-1}A = I_n$$

then A^{-1}, stated as "A inverse," is the **inverse** of A.

It is important to note that only a square matrix can have an inverse. Even then, not all square matrices have inverses.

EXAMPLE 11   Multiplying a Matrix by Its Inverse

Verify that the inverse of $A = \begin{bmatrix} 1 & 3 \\ 2 & 5 \end{bmatrix}$ is $A^{-1} = \begin{bmatrix} -5 & 3 \\ 2 & -1 \end{bmatrix}$.

Solution:

Show that $AA^{-1} = I_2$ and $A^{-1}A = I_2$.

Find the product AA^{-1}.

$$AA^{-1} = \begin{bmatrix} 1 & 3 \\ 2 & 5 \end{bmatrix}\begin{bmatrix} -5 & 3 \\ 2 & -1 \end{bmatrix}$$

$$= \begin{bmatrix} (1)(-5) + (3)(2) & (1)(3) + (3)(-1) \\ (2)(-5) + (5)(2) & (2)(3) + (5)(-1) \end{bmatrix}$$

$$= \begin{bmatrix} 1 & 0 \\ 0 & 1 \end{bmatrix} = I_2$$

Find the product $A^{-1}A$.

$$A^{-1}A = \begin{bmatrix} -5 & 3 \\ 2 & -1 \end{bmatrix}\begin{bmatrix} 1 & 3 \\ 2 & 5 \end{bmatrix}$$

$$= \begin{bmatrix} (-5)(1) + (3)(2) & (-5)(3) + (3)(5) \\ (2)(1) + (-1)(2) & (2)(3) + (-1)(5) \end{bmatrix}$$

$$= \begin{bmatrix} 1 & 0 \\ 0 & 1 \end{bmatrix} = I_2$$

■ **YOUR TURN** Verify that the inverse of $A = \begin{bmatrix} 1 & 4 \\ 2 & 9 \end{bmatrix}$ is $A^{-1} = \begin{bmatrix} 9 & -4 \\ -2 & 1 \end{bmatrix}$.

Technology Tip

Enter the matrices as A, and A^{-1} as B.

```
[A]
        [[1 3]
         [2 5]]
[B]
        [[-5 3 ]
         [2  -1]]
■
```

Enter AA^{-1} as AB, and A^{-1} as BA.

```
[A] [B]
        [[1 0]
         [0 1]]
[B] [A]
        [[1 0]
         [0 1]]
```

■ **Answer:** $AA^{-1} = A^{-1}A = I_2$

Now that we can show that two matrices are inverses of one another, let us describe the process for finding an inverse, if it exists. If an inverse A^{-1} exists, then the matrix A is said to be **nonsingular**. If the inverse does not exist, then the matrix A is said to be **singular**.

Let $A = \begin{bmatrix} 1 & -1 \\ 2 & -3 \end{bmatrix}$ and the inverse be $A^{-1} = \begin{bmatrix} w & x \\ y & z \end{bmatrix}$, where w, x, y, and z are variables to be determined. A matrix and its inverse must satisfy the identity $AA^{-1} = I_2$.

WORDS	MATH
The product of a matrix and its inverse is the identity matrix.	$\begin{bmatrix} 1 & -1 \\ 2 & -3 \end{bmatrix}\begin{bmatrix} w & x \\ y & z \end{bmatrix} = \begin{bmatrix} 1 & 0 \\ 0 & 1 \end{bmatrix}$
Multiply the two matrices on the left.	$\begin{bmatrix} w - y & x - z \\ 2w - 3y & 2x - 3z \end{bmatrix} = \begin{bmatrix} 1 & 0 \\ 0 & 1 \end{bmatrix}$
Equate corresponding matrix elements.	$\begin{aligned} w - y &= 1 \\ 2w - 3y &= 0 \end{aligned} \quad \text{and} \quad \begin{aligned} x - z &= 0 \\ 2x - 3z &= 1 \end{aligned}$

Notice that there are two systems of equations, both of which can be solved by several methods (elimination, substitution, or augmented matrices). We will find that $w = 3$, $x = -1$, $y = 2$, and $z = -1$. Therefore, we know the inverse is $A^{-1} = \begin{bmatrix} 3 & -1 \\ 2 & -1 \end{bmatrix}$. But, instead, let us use augmented matrices in order to develop the general procedure.

Write the two systems of equations as two augmented matrices:

$$\begin{matrix} w & y \\ \begin{bmatrix} 1 & -1 & | & 1 \\ 2 & -3 & | & 0 \end{bmatrix} \end{matrix} \qquad \begin{matrix} x & z \\ \begin{bmatrix} 1 & -1 & | & 0 \\ 2 & -3 & | & 1 \end{bmatrix} \end{matrix}$$

Since the left side is the same for each augmented matrix, we can combine these two matrices into one matrix, thereby simultaneously solving both systems of equations.

$$\begin{bmatrix} 1 & -1 & | & 1 & 0 \\ 2 & -3 & | & 0 & 1 \end{bmatrix}$$

Notice that the right side of the vertical line is the identity matrix I_2.

Using Gauss–Jordan elimination, transform the matrix on the left to the identity matrix.

$$\begin{bmatrix} 1 & -1 & | & 1 & 0 \\ 2 & -3 & | & 0 & 1 \end{bmatrix}$$

$$R_2 - 2R_1 \rightarrow R_2 \quad \begin{bmatrix} 1 & -1 & | & 1 & 0 \\ 0 & -1 & | & -2 & 1 \end{bmatrix}$$

$$-R_2 \rightarrow R_2 \quad \begin{bmatrix} 1 & -1 & | & 1 & 0 \\ 0 & 1 & | & 2 & -1 \end{bmatrix}$$

$$R_1 + R_2 \rightarrow R_1 \quad \begin{bmatrix} 1 & 0 & | & 3 & -1 \\ 0 & 1 & | & 2 & -1 \end{bmatrix}$$

The matrix on the right of the vertical line is the inverse $A^{-1} = \begin{bmatrix} 3 & -1 \\ 2 & -1 \end{bmatrix}$.

FINDING THE INVERSE OF A SQUARE MATRIX

To find the inverse of an $n \times n$ matrix A:

Step 1: Form the matrix $[A \mid I_n]$.

Step 2: Use row operations to transform this entire augmented matrix to $[I_n \mid A^{-1}]$. This is done by applying Gauss–Jordan elimination to reduce A to the identity matrix I_n. If this is not possible, then A is a singular matrix and no inverse exists.

Step 3: Verify the result by showing that $AA^{-1} = I_n$ and $A^{-1}A = I_n$.

Technology Tip

A graphing calculator can be used to find the inverse of A. Enter the matrix A.

```
[A]
        [[1 2]
         [3 5]]
■
```

To find A^{-1}, press [2nd] [MATRIX]
[1:[A]] [ENTER] [x^{-1}] [ENTER].

```
[A]
        [[1 2]
         [3 5]]
[A]-1
        [[-5 2 ]
         [3  -1]]
```

EXAMPLE 12 Finding the Inverse of a 2 × 2 Matrix

Find the inverse of $A = \begin{bmatrix} 1 & 2 \\ 3 & 5 \end{bmatrix}$.

Solution:

STEP 1 Form the matrix $[A \mid I_2]$.

$$\begin{bmatrix} 1 & 2 & | & 1 & 0 \\ 3 & 5 & | & 0 & 1 \end{bmatrix}$$

STEP 2 Use row operations to transform A into I_2.

$$R_2 - 3R_1 \rightarrow R_2 \quad \begin{bmatrix} 1 & 2 & | & 1 & 0 \\ 0 & -1 & | & -3 & 1 \end{bmatrix}$$

$$-R_2 \rightarrow R_2 \quad \begin{bmatrix} 1 & 2 & | & 1 & 0 \\ 0 & 1 & | & 3 & -1 \end{bmatrix}$$

$$R_1 - 2R_2 \rightarrow R_1 \quad \begin{bmatrix} 1 & 0 & | & -5 & 2 \\ 0 & 1 & | & 3 & -1 \end{bmatrix}$$

Identify the inverse.

$$A^{-1} = \begin{bmatrix} -5 & 2 \\ 3 & -1 \end{bmatrix}$$

STEP 3 Check.

$$AA^{-1} = \begin{bmatrix} 1 & 2 \\ 3 & 5 \end{bmatrix}\begin{bmatrix} -5 & 2 \\ 3 & -1 \end{bmatrix} = \begin{bmatrix} 1 & 0 \\ 0 & 1 \end{bmatrix} = I_2$$

$$A^{-1}A = \begin{bmatrix} -5 & 2 \\ 3 & -1 \end{bmatrix}\begin{bmatrix} 1 & 2 \\ 3 & 5 \end{bmatrix} = \begin{bmatrix} 1 & 0 \\ 0 & 1 \end{bmatrix} = I_2$$

■ **YOUR TURN** Find the inverse of $A = \begin{bmatrix} 2 & 3 \\ 5 & 8 \end{bmatrix}$.

■ **Answer:** $A^{-1} = \begin{bmatrix} 8 & -3 \\ -5 & 2 \end{bmatrix}$

This procedure for finding an inverse of a square matrix is used for all square matrices of order $n \times n$. For the special case of a 2×2 matrix, there is a formula (that will be derived in Exercises 107 and 108) for finding the inverse.

Let $A = \begin{bmatrix} a & b \\ c & d \end{bmatrix}$ represent any 2×2 matrix; then the inverse matrix is given by

$$A^{-1} = \frac{1}{ad - bc}\begin{bmatrix} d & -b \\ -c & a \end{bmatrix} \qquad ad - bc \neq 0$$

The denominator $ad - bc$ is called the *determinant* of the matrix A and will be discussed in Section 8.5.

We found the inverse of $A = \begin{bmatrix} 1 & 2 \\ 3 & 5 \end{bmatrix}$ in Example 12. Let us now find the inverse using this formula.

WORDS

Write the formula for A^{-1}.

Substitute $a = 1, b = 2, c = 3$, and $d = 5$ into the formula.

Simplify.

MATH

$$A^{-1} = \frac{1}{ad - bc}\begin{bmatrix} d & -b \\ -c & a \end{bmatrix}$$

$$A^{-1} = \frac{1}{(1)(5) - (2)(3)}\begin{bmatrix} 5 & -2 \\ -3 & 1 \end{bmatrix}$$

$$A^{-1} = (-1)\begin{bmatrix} 5 & -2 \\ -3 & 1 \end{bmatrix}$$

$$A^{-1} = \begin{bmatrix} -5 & 2 \\ 3 & -1 \end{bmatrix}$$

The result is the same as that we found in Example 12.

EXAMPLE 13   Finding That No Inverse Exists: Singular Matrix

Find the inverse of $A = \begin{bmatrix} 1 & -5 \\ -1 & 5 \end{bmatrix}$.

Study Tip

If the determinant of a 2×2 matrix is equal to 0, then its inverse does not exist.

Solution:

STEP 1 Form the matrix $[A \mid I_2]$.

$$\begin{bmatrix} 1 & -5 & | & 1 & 0 \\ -1 & 5 & | & 0 & 1 \end{bmatrix}$$

STEP 2 Apply row operations to transform A into I_2.

$$R_2 + R_1 \rightarrow R_2 \qquad \begin{bmatrix} 1 & -5 & | & 1 & 0 \\ 0 & 0 & | & 1 & 1 \end{bmatrix}$$

We cannot convert the left-hand side of the augmented matrix to I_2 because of the all-zero row on the left-hand side. Therefore, $\boxed{A \text{ is not invertible}}$; that is, A has no inverse, or A^{-1} does not exist. We say that A is **singular**.

Technology Tip

A graphing calculator can be used to find the inverse of A. Enter the matrix A.

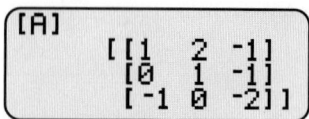

[A]
[[1   2   -1]
 [0   1   -1]
 [-1   0   -2]]

Press $\boxed{2^{\text{nd}}}$ $\boxed{\text{MATRIX}}$ $\boxed{1:[A]}$
$\boxed{\text{ENTER}}$ $\boxed{X^{-1}}$ $\boxed{\text{ENTER}}$.

[A]⁻¹
[[2   -4   1]
 [-1   3 -1]
 [-1   2 -1]]

■ **Answer:** $A^{-1} = \begin{bmatrix} 0 & 1 & 1 \\ 1 & -1 & -1 \\ 0 & 2 & 1 \end{bmatrix}$

EXAMPLE 14 Finding the Inverse of a 3 × 3 Matrix

Find the inverse of $A = \begin{bmatrix} 1 & 2 & -1 \\ 0 & 1 & -1 \\ -1 & 0 & -2 \end{bmatrix}$.

Solution:

STEP 1 Form the matrix $[A \mid I_3]$.

$$\left[\begin{array}{ccc|ccc} 1 & 2 & -1 & 1 & 0 & 0 \\ 0 & 1 & -1 & 0 & 1 & 0 \\ -1 & 0 & -2 & 0 & 0 & 1 \end{array}\right]$$

STEP 2 Apply row operations to transform A into I_3.

$$R_3 + R_1 \to R_3 \quad \left[\begin{array}{ccc|ccc} 1 & 2 & -1 & 1 & 0 & 0 \\ 0 & 1 & -1 & 0 & 1 & 0 \\ 0 & 2 & -3 & 1 & 0 & 1 \end{array}\right]$$

$$R_3 - 2R_2 \to R_3 \quad \left[\begin{array}{ccc|ccc} 1 & 2 & -1 & 1 & 0 & 0 \\ 0 & 1 & -1 & 0 & 1 & 0 \\ 0 & 0 & -1 & 1 & -2 & 1 \end{array}\right]$$

$$-R_3 \to R_3 \quad \left[\begin{array}{ccc|ccc} 1 & 2 & -1 & 1 & 0 & 0 \\ 0 & 1 & -1 & 0 & 1 & 0 \\ 0 & 0 & 1 & -1 & 2 & -1 \end{array}\right]$$

$$\begin{array}{c} R_2 + R_3 \to R_2 \\ R_1 + R_3 \to R_1 \end{array} \quad \left[\begin{array}{ccc|ccc} 1 & 2 & 0 & 0 & 2 & -1 \\ 0 & 1 & 0 & -1 & 3 & -1 \\ 0 & 0 & 1 & -1 & 2 & -1 \end{array}\right]$$

$$R_1 - 2R_2 \to R_1 \quad \left[\begin{array}{ccc|ccc} 1 & 0 & 0 & 2 & -4 & 1 \\ 0 & 1 & 0 & -1 & 3 & -1 \\ 0 & 0 & 1 & -1 & 2 & -1 \end{array}\right]$$

Identify the inverse.

$$A^{-1} = \begin{bmatrix} 2 & -4 & 1 \\ -1 & 3 & -1 \\ -1 & 2 & -1 \end{bmatrix}$$

STEP 3 Check. $AA^{-1} = \begin{bmatrix} 1 & 2 & -1 \\ 0 & 1 & -1 \\ -1 & 0 & -2 \end{bmatrix}\begin{bmatrix} 2 & -4 & 1 \\ -1 & 3 & -1 \\ -1 & 2 & -1 \end{bmatrix} = \begin{bmatrix} 1 & 0 & 0 \\ 0 & 1 & 0 \\ 0 & 0 & 1 \end{bmatrix} = I_3$

$$A^{-1}A = \begin{bmatrix} 2 & -4 & 1 \\ -1 & 3 & -1 \\ -1 & 2 & -1 \end{bmatrix}\begin{bmatrix} 1 & 2 & -1 \\ 0 & 1 & -1 \\ -1 & 0 & -2 \end{bmatrix} = \begin{bmatrix} 1 & 0 & 0 \\ 0 & 1 & 0 \\ 0 & 0 & 1 \end{bmatrix} = I_3$$

■ **YOUR TURN** Find the inverse of $A = \begin{bmatrix} 1 & 1 & 0 \\ -1 & 0 & 1 \\ 2 & 0 & -1 \end{bmatrix}$.

Solving Systems of Linear Equations Using Matrix Algebra and Inverses of Square Matrices

We can solve systems of linear equations using matrix algebra. We will use a system of three equations and three variables to demonstrate the procedure. However, it can be extended to any square system.

Linear System of Equations	**Matrix Form of the System**

$$a_1x + b_1y + c_1z = d_1$$
$$a_2x + b_2y + c_2z = d_2$$
$$a_3x + b_3y + c_3z = d_3$$

$$\underbrace{\begin{bmatrix} a_1 & b_1 & c_1 \\ a_2 & b_2 & c_2 \\ a_3 & b_3 & c_3 \end{bmatrix}}_{A} \underbrace{\begin{bmatrix} x \\ y \\ z \end{bmatrix}}_{X} = \underbrace{\begin{bmatrix} d_1 \\ d_2 \\ d_3 \end{bmatrix}}_{B}$$

Recall that a system of linear equations has a unique solution, no solution, or infinitely many solutions. If a system of n equations in n variables has a unique solution, it can be found using the following procedure:

WORDS	**MATH**
Write the system of linear equations as a matrix equation.	$A_{n\times n}X_{n\times 1} = B_{n\times 1}$
Multiply both sides of the equation by A^{-1}.	$A^{-1}AX = A^{-1}B$
A matrix times its inverse is the identity matrix.	$I_nX = A^{-1}B$
A matrix times the identity matrix is equal to itself.	$X = A^{-1}B$

Notice the order in which the right side is multiplied, $X_{n\times 1} = A^{-1}_{n\times n}B_{n\times 1}$, and remember that matrix multiplication is not commutative. Therefore, you multiply both sides of the matrix equation in the same order.

SOLVING A SYSTEM OF LINEAR EQUATIONS USING MATRIX ALGEBRA: UNIQUE SOLUTION

If a system of linear equations is represented by the matrix equation $AX = B$, where A is a nonsingular square matrix, then the system has a unique solution given by

$$X = A^{-1}B$$

Technology Tip

Use a T1 to find the inverse of A and X. Enter the matrices A and B.

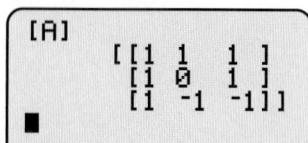

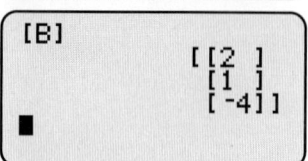

Now use the graphing calculator to find the inverse of A, A^{-1}. Press

2nd | MATRIX | 1: [A] | ENTER

X^{-1} | ENTER .

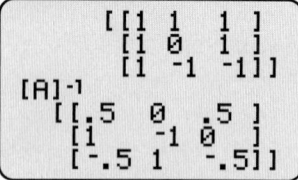

To show elements using fractions, press 2nd | MATRIX | 1:[A]

ENTER | X^{-1} | MATH | 1: ► Frac

ENTER | ENTER .

To enter $A^{-1}B$, press

2nd | ANS | 2nd | MATRIX | 2:[B]

ENTER | ENTER .

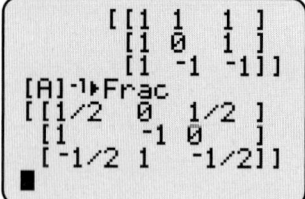

The solution to the system is $x = -1$, $y = 1$, and $z = 2$.

Solve the system of equations using matrix algebra.

$$\begin{aligned} x + y + z &= 2 \\ x \quad\;\; + z &= 1 \\ x - y - z &= -4 \end{aligned}$$

Solution:

Write the system in matrix form.

$$AX = B$$

$$A = \begin{bmatrix} 1 & 1 & 1 \\ 1 & 0 & 1 \\ 1 & -1 & -1 \end{bmatrix} \quad X = \begin{bmatrix} x \\ y \\ z \end{bmatrix} \quad B = \begin{bmatrix} 2 \\ 1 \\ -4 \end{bmatrix}$$

Find the inverse of A.

Form the matrix $[A \mid I_3]$.

$$\left[\begin{array}{ccc|ccc} 1 & 1 & 1 & 1 & 0 & 0 \\ 1 & 0 & 1 & 0 & 1 & 0 \\ 1 & -1 & -1 & 0 & 0 & 1 \end{array}\right]$$

$$\begin{array}{c} R_2 - R_1 \rightarrow R_2 \\ R_3 - R_1 \rightarrow R_3 \end{array} \left[\begin{array}{ccc|ccc} 1 & 1 & 1 & 1 & 0 & 0 \\ 0 & -1 & 0 & -1 & 1 & 0 \\ 0 & -2 & -2 & -1 & 0 & 1 \end{array}\right]$$

$$-R_2 \rightarrow R_2 \left[\begin{array}{ccc|ccc} 1 & 1 & 1 & 1 & 0 & 0 \\ 0 & 1 & 0 & 1 & -1 & 0 \\ 0 & -2 & -2 & -1 & 0 & 1 \end{array}\right]$$

$$R_3 + 2R_2 \rightarrow R_3 \left[\begin{array}{ccc|ccc} 1 & 1 & 1 & 1 & 0 & 0 \\ 0 & 1 & 0 & 1 & -1 & 0 \\ 0 & 0 & -2 & 1 & -2 & 1 \end{array}\right]$$

$$-\tfrac{1}{2}R_3 \rightarrow R_3 \left[\begin{array}{ccc|ccc} 1 & 1 & 1 & 1 & 0 & 0 \\ 0 & 1 & 0 & 1 & -1 & 0 \\ 0 & 0 & 1 & -\tfrac{1}{2} & 1 & -\tfrac{1}{2} \end{array}\right]$$

$$R_1 - R_3 \rightarrow R_1 \left[\begin{array}{ccc|ccc} 1 & 1 & 0 & \tfrac{3}{2} & -1 & \tfrac{1}{2} \\ 0 & 1 & 0 & 1 & -1 & 0 \\ 0 & 0 & 1 & -\tfrac{1}{2} & 1 & -\tfrac{1}{2} \end{array}\right]$$

$$R_1 - R_2 \rightarrow R_1 \left[\begin{array}{ccc|ccc} 1 & 0 & 0 & \tfrac{1}{2} & 0 & \tfrac{1}{2} \\ 0 & 1 & 0 & 1 & -1 & 0 \\ 0 & 0 & 0 & -\tfrac{1}{2} & 1 & -\tfrac{1}{2} \end{array}\right]$$

Identify the inverse.

$$A^{-1} = \begin{bmatrix} \tfrac{1}{2} & 0 & \tfrac{1}{2} \\ 1 & -1 & 0 \\ -\tfrac{1}{2} & 1 & -\tfrac{1}{2} \end{bmatrix}$$

The solution to the system is $X = A^{-1}B$.

$$X = A^{-1}B = \begin{bmatrix} \tfrac{1}{2} & 0 & \tfrac{1}{2} \\ 1 & -1 & 0 \\ -\tfrac{1}{2} & 1 & -\tfrac{1}{2} \end{bmatrix} \begin{bmatrix} 2 \\ 1 \\ -4 \end{bmatrix}$$

Simplify.

$$X = \begin{bmatrix} x \\ y \\ z \end{bmatrix} = \begin{bmatrix} -1 \\ 1 \\ 2 \end{bmatrix}$$

$$\boxed{x = -1, y = 1, z = 2}$$

■ **YOUR TURN** Solve the system of equations using matrix algebra.

$$\begin{aligned} x + y - z &= 3 \\ y + z &= 1 \\ 2x + 3y + z &= 5 \end{aligned}$$

■ **Answer:** $x = 0, y = 2, z = -1$

Cryptography Applications

Cryptography is the practice of hiding information, or secret communication. Let's assume you want to send your ATM PIN code over the Internet, but you don't want hackers to be able to retrieve it. You can represent the PIN code in a matrix and then multiply that PIN matrix by a "key" matrix so that it is encrypted. If the person you send it to has the "inverse key" matrix, he can multiply the encrypted matrix he receives by the inverse key matrix and the result will be the original PIN matrix. Although PIN numbers are typically four digits, we will assume two digits to illustrate the process.

WORDS	**MATH**
Suppose the two-digit ATM PIN is 13.	$P = \begin{bmatrix} 1 & 3 \end{bmatrix}$
Apply any 2×2 nonsingular matrix as the "key" (encryption) matrix.	$K = \begin{bmatrix} 2 & 3 \\ 5 & 8 \end{bmatrix}$
Multiply the PIN and encryption matrices.	$PK = \begin{bmatrix} 1 & 3 \end{bmatrix}\begin{bmatrix} 2 & 3 \\ 5 & 8 \end{bmatrix}$ $= \begin{bmatrix} 1(2) + 3(5) & 1(3) + 3(8) \end{bmatrix}$ $= \begin{bmatrix} 17 & 27 \end{bmatrix}$
The receiver of the encrypted matrix sees only $\begin{bmatrix} 17 & 27 \end{bmatrix}$.	
The decoding "key" is the inverse matrix K^{-1}.	$K^{-1} = \begin{bmatrix} 8 & -3 \\ -5 & 2 \end{bmatrix}$

Study Tip

$$K = \begin{bmatrix} 2 & 3 \\ 5 & 8 \end{bmatrix}$$

$$K^{-1} = \frac{1}{(2)(8) - (3)(5)}\begin{bmatrix} 8 & -3 \\ -5 & 2 \end{bmatrix}$$

$$= \begin{bmatrix} 8 & -3 \\ -5 & 2 \end{bmatrix}$$

Any receiver who has the decoding key can multiply the received encrypted matrix by the decoding "key" matrix. The result is the original transmitted PIN number.

$$\begin{bmatrix} 17 & 27 \end{bmatrix}\begin{bmatrix} 8 & -3 \\ -5 & 2 \end{bmatrix} = \begin{bmatrix} 17(8) + 27(-5) & 17(-3) + 27(2) \end{bmatrix} = \begin{bmatrix} 1 & 3 \end{bmatrix}$$

8.4 SUMMARY

Matrices can be used to represent data. Operations such as equality, addition, subtraction, and scalar multiplication are performed entry by entry. Two matrices can be added or subtracted only if they have the same order. Matrix multiplication, however, requires that the number of columns in the first matrix is equal to the number of rows in the second matrix and is performed using a row-by-column procedure.

Matrix Multiplication Is Not Commutative: $AB \neq BA$

OPERATION	ORDER REQUIREMENT
Equality	Same: $A_{m \times n} = B_{m \times n}$
Addition	Same: $A_{m \times n} + B_{m \times n}$
Subtraction	Same: $A_{m \times n} - B_{m \times n}$
Scalar multiplication	None: $kA_{m \times n}$
Matrix multiplication	$A_{m \times n} B_{n \times p} = (AB)_{m \times p}$

Systems of linear equations can be solved using matrix equations.

SYSTEM OF LINEAR EQUATIONS	A	X	B	MATRIX EQUATION: $AX = B$
$\begin{aligned} x - y + z &= 2 \\ 2x + 2y - 3z &= -3 \\ x + y + z &= 6 \end{aligned}$	$\begin{bmatrix} 1 & -1 & 1 \\ 2 & 2 & -3 \\ 1 & 1 & 1 \end{bmatrix}$	$\begin{bmatrix} x \\ y \\ z \end{bmatrix}$	$\begin{bmatrix} 2 \\ -3 \\ 6 \end{bmatrix}$	$\begin{bmatrix} 1 & -1 & 1 \\ 2 & 2 & -3 \\ 1 & 1 & 1 \end{bmatrix} \begin{bmatrix} x \\ y \\ z \end{bmatrix} = \begin{bmatrix} 2 \\ -3 \\ 6 \end{bmatrix}$

If this system of linear equations has a unique solution, then the solution is represented by

$$X = A^{-1}B$$

A^{-1} is the inverse of A, that is, $AA^{-1} = A^{-1}A = I$, and is found by

$$[A_{n \times n} | I_n] \rightarrow [I_n | A_{n \times n}^{-1}]$$

SECTION
8.4 **EXERCISES**

▪ SKILLS

In Exercises 1–8, state the order of each matrix.

1. $\begin{bmatrix} -1 & 2 & 4 \\ 7 & -3 & 9 \end{bmatrix}$

2. $\begin{bmatrix} 3 & 5 \\ 2 & 6 \\ -1 & -4 \end{bmatrix}$

3. $\begin{bmatrix} -4 & 5 \\ 0 & 1 \end{bmatrix}$

4. $\begin{bmatrix} -4 & 5 & 3 & 7 \end{bmatrix}$

5. $\begin{bmatrix} -3 & 4 & 1 \\ 10 & 8 & 0 \\ -2 & 5 & 7 \end{bmatrix}$

6. $\begin{bmatrix} 1 \\ 2 \\ 3 \\ 4 \end{bmatrix}$

7. $\begin{bmatrix} -3 & 6 & 0 & 5 \\ 4 & -9 & 2 & 7 \\ 1 & 8 & 3 & 6 \\ 5 & 0 & -4 & 11 \end{bmatrix}$

8. $\begin{bmatrix} -1 & 3 & 6 & 9 \\ 2 & 5 & -7 & 8 \end{bmatrix}$

In Exercises 9–14, solve for the indicated variables.

9. $\begin{bmatrix} 2 & x \\ y & 3 \end{bmatrix} = \begin{bmatrix} 2 & -5 \\ 1 & 3 \end{bmatrix}$

10. $\begin{bmatrix} -3 & 17 \\ x & y \end{bmatrix} = \begin{bmatrix} -3 & 17 \\ 10 & 12 \end{bmatrix}$

11. $\begin{bmatrix} x+y & 3 \\ x-y & 9 \end{bmatrix} = \begin{bmatrix} -5 & z \\ -1 & 9 \end{bmatrix}$

12. $\begin{bmatrix} x & -4 \\ y & 7 \end{bmatrix} = \begin{bmatrix} 2+y & -4 \\ 5 & 7 \end{bmatrix}$

13. $\begin{bmatrix} 3 & 4 \\ 0 & 12 \end{bmatrix} = \begin{bmatrix} x-y & 4 \\ 0 & 2y+x \end{bmatrix}$

14. $\begin{bmatrix} 9 & 2b+1 \\ -5 & 16 \end{bmatrix} = \begin{bmatrix} a^2 & 9 \\ 2a+1 & b^2 \end{bmatrix}$

In Exercises 15–24, perform the indicated operations for each expression, if possible.

$$A = \begin{bmatrix} -1 & 3 & 0 \\ 2 & 4 & 1 \end{bmatrix} \quad B = \begin{bmatrix} 0 & 2 & 1 \\ 3 & -2 & 4 \end{bmatrix} \quad C = \begin{bmatrix} 0 & 1 \\ 2 & -1 \\ 3 & 1 \end{bmatrix} \quad D = \begin{bmatrix} 2 & -3 \\ 0 & 1 \\ 4 & -2 \end{bmatrix}$$

15. $A + B$

16. $C + D$

17. $C - D$

18. $A - B$

19. $B + C$

20. $A + D$

21. $D - B$

22. $C - A$

23. $2A + 3B$

24. $2B - 3A$

In Exercises 25–44, perform the indicated operations for each expression, if possible.

$$A = \begin{bmatrix} 1 & 2 & -1 \\ 0 & 3 & 1 \\ 5 & 0 & -2 \end{bmatrix} \quad B = \begin{bmatrix} 2 & 0 & -3 \end{bmatrix} \quad C = \begin{bmatrix} -1 & 7 & 2 \\ 3 & 0 & 1 \end{bmatrix} \quad D = \begin{bmatrix} 3 & 0 \\ 1 & -1 \\ 2 & 5 \end{bmatrix}$$

$$E = \begin{bmatrix} -1 & 0 & 1 \\ 2 & 1 & 4 \\ -3 & 1 & 5 \end{bmatrix} \quad F = \begin{bmatrix} 1 \\ 0 \\ -1 \end{bmatrix} \quad G = \begin{bmatrix} 1 & 2 \\ 3 & 4 \end{bmatrix}$$

25. CD

26. BF

27. DC

28. $(A + E)D$

29. DG

30. $2A + 3E$

31. GD

32. $ED + C$

33. $-4BD$

34. $-3ED$

35. $B(A + E)$

36. $GC + 5C$

37. $FB + 5A$

38. A^2

39. $G^2 + 5G$

40. $C \cdot (2E)$

41. $(2E) \cdot F$

42. $CA + 5C$

43. DF

44. AE

In Exercises 45–50, determine whether B is the multiplicative inverse of A using $AA^{-1} = I$.

45. $A = \begin{bmatrix} 8 & -11 \\ -5 & 7 \end{bmatrix}$ $B = \begin{bmatrix} 7 & 11 \\ 5 & 8 \end{bmatrix}$

46. $A = \begin{bmatrix} 7 & -9 \\ -3 & 4 \end{bmatrix}$ $B = \begin{bmatrix} 4 & 9 \\ 3 & 7 \end{bmatrix}$

47. $A = \begin{bmatrix} 3 & 1 \\ 1 & -2 \end{bmatrix}$ $B = \begin{bmatrix} \frac{2}{7} & \frac{1}{7} \\ \frac{1}{7} & -\frac{3}{7} \end{bmatrix}$

48. $A = \begin{bmatrix} 2 & 3 \\ 1 & -1 \end{bmatrix}$ $B = \begin{bmatrix} \frac{1}{5} & \frac{3}{5} \\ \frac{1}{5} & -\frac{2}{5} \end{bmatrix}$

49. $A = \begin{bmatrix} 1 & -1 & 1 \\ 1 & 0 & -1 \\ 0 & 1 & -1 \end{bmatrix}$ $B = \begin{bmatrix} 1 & 0 & 1 \\ 1 & -1 & 2 \\ 1 & -1 & 1 \end{bmatrix}$

50. $A = \begin{bmatrix} -1 & 0 & -1 \\ -1 & 1 & -2 \\ -1 & 1 & -1 \end{bmatrix}$ $B = \begin{bmatrix} -1 & 1 & -1 \\ -1 & 0 & 1 \\ 0 & -1 & 1 \end{bmatrix}$

In Exercises 51–62, find A^{-1}, if possible.

51. $A = \begin{bmatrix} 2 & 1 \\ -1 & 0 \end{bmatrix}$

52. $A = \begin{bmatrix} 3 & 1 \\ 2 & 1 \end{bmatrix}$

53. $A = \begin{bmatrix} \frac{1}{3} & 2 \\ 5 & \frac{3}{4} \end{bmatrix}$

54. $A = \begin{bmatrix} \frac{1}{4} & 2 \\ \frac{1}{3} & \frac{2}{3} \end{bmatrix}$

55. $A = \begin{bmatrix} 1 & 1 & 1 \\ 1 & -1 & -1 \\ -1 & 1 & -1 \end{bmatrix}$

56. $A = \begin{bmatrix} 1 & -1 & 1 \\ 1 & 1 & 1 \\ -1 & 2 & -3 \end{bmatrix}$

57. $A = \begin{bmatrix} 1 & 0 & 1 \\ 0 & 1 & 1 \\ 1 & -1 & 0 \end{bmatrix}$

58. $A = \begin{bmatrix} 1 & 2 & -3 \\ 1 & -1 & -1 \\ 1 & 0 & -4 \end{bmatrix}$

59. $A = \begin{bmatrix} 2 & 4 & 1 \\ 1 & 1 & -1 \\ 1 & 1 & 0 \end{bmatrix}$

60. $A = \begin{bmatrix} 1 & 0 & 1 \\ 1 & 1 & -1 \\ 2 & 1 & -1 \end{bmatrix}$

61. $A = \begin{bmatrix} 1 & 1 & -1 \\ 1 & -1 & 1 \\ 2 & -1 & -1 \end{bmatrix}$

62. $A = \begin{bmatrix} 1 & -1 & -1 \\ 1 & 1 & -3 \\ 3 & -5 & 1 \end{bmatrix}$

In Exercises 63–74, apply matrix algebra to solve the system of linear equations.

63. $2x - y = 5$
  $x + y = 1$

64. $2x - 3y = 12$
  $x + y = 1$

65. $4x - 9y = -1$
  $7x - 3y = \frac{5}{2}$

66. $7x - 3y = 1$
  $4x - 5y = -\frac{7}{5}$

67. $x + y + z = 1$
  $x - y - z = -1$
  $-x + y - z = -1$

68. $x - y + z = 0$
  $x + y + z = 2$
  $-x + 2y - 3z = 1$

69. $x \quad\ + z = 3$
  $y + z = 1$
  $x - y \quad = 2$

70. $x + 2y - 3z = 1$
  $x - y - z = 3$
  $x \quad\ - 4z = 0$

71. $2x + 4y + z = -5$
  $x + y - z = 7$
  $x + y \quad = 0$

72. $x \quad\ + z = 3$
  $x + y - z = -3$
  $2x + y - z = -5$

73. $x + y - z = 4$
  $x - y + z = 2$
  $2x - y - z = -3$

74. $x - y - z = 0$
  $x + y - 3z = 2$
  $3x - 5y + z = 4$

▪ APPLICATIONS

75. Smoking. On January 6 and 10, 2000, the Harris Poll conducted a survey of adult smokers in the United States. When asked, "Have you ever tried to quit smoking?", 70% said yes and 30% said no. Write a 2×1 matrix—call it A—that represents those smokers. When asked what consequences smoking would have on their lives, 89% believed it would increase their chance of getting lung cancer and 84% believed smoking would shorten their lives. Write a 2×1 matrix—call it B—that represents those smokers. If there are 46 million adult smokers in the United States

a. What does $46A$ tell us?
b. What does $46B$ tell us?

76. Women in Science. According to the study of science and engineering indicators by the National Science Foundation (www.nsf.gov), the number of female graduate students in science and engineering disciplines has increased over the last 30 years. In 1981, 24% of mathematics graduate students were female and 23% of graduate students in computer science were female. In 1991, 32% of mathematics graduate students and 21% of computer science graduate students were female. In 2001, 38% of mathematics graduate students and 30% of computer science graduate students were female. Write three 2×1 matrices representing the percentage of female graduate students.

$$A = \begin{bmatrix} \% \text{ female} - \text{math} - 1981 \\ \% \text{ female} - \text{C.S.} - 1981 \end{bmatrix}$$

$$B = \begin{bmatrix} \% \text{ female} - \text{math} - 1991 \\ \% \text{ female} - \text{C.S.} - 1991 \end{bmatrix}$$

$$C = \begin{bmatrix} \% \text{ female} - \text{math} - 2001 \\ \% \text{ female} - \text{C.S.} - 2001 \end{bmatrix}$$

What does $C - B$ tell us? What does $B - A$ tell us? What can you conclude about the number of women pursuing mathematics and computer science graduate degrees?

Note: C.S. = computer science.

77. Registered Voters. According to the U.S. Census Bureau (www.census.gov), in the 2000 national election, 58.9% of men over the age of 18 were registered voters, but only 41.4% voted; and 62.8% of women over 18 were registered voters, but only 43% actually voted. Write a 2×2 matrix with the following data:

$$A = \begin{bmatrix} \text{Percentage of registered} & \text{Percentage of registered} \\ \text{male voters} & \text{female voters} \\ \text{Percent of males} & \text{Percent of females} \\ \text{who voted} & \text{who voted} \end{bmatrix}$$

If we let B be a 2×1 matrix representing the total population of males and females over the age of 18 in the United States, or

$$B = \begin{bmatrix} 100 \text{ M} \\ 110 \text{ M} \end{bmatrix}, \text{ what does } AB \text{ tell us?}$$

78. Job Application. A company has two rubrics for scoring job applicants based on weighting education, experience, and the interview differently.

Matrix A	Rubric 1	Rubric 2
Education	0.5	0.6
Experience	0.3	0.1
Interview	0.2	0.3

Applicants receive a score from 1 to 10 in each category (education, experience, and interview). Two applicants are shown in matrix B.

Matrix B	Education	Experience	Interview
Applicant 1	8	7	5
Applicant 2	6	8	8

What is the order of BA? What does each entry in BA tell us?

79. Taxes. The IRS allows an individual to deduct business expenses in the following way: $0.45 per mile driven, 50% of entertainment costs, and 100% of actual expenses. Represent these deductions in the given order as a row matrix A. In 2006, Jamie had the following business expenses: $2,700 in entertainment, $15,200 actual expenses, and he drove 7523 miles. Represent Jamie's expenses in the given order as a column matrix B. Multiply these two matrices to find the total amount of business expenses Jamie can claim on his 2006 federal tax form: AB.

80. Tips on Service. Marilyn decides to go to the Safety Harbor Spa for a day of pampering. She is treated to a hot stone massage ($85), a manicure and pedicure ($75), and a haircut ($100). Represent the costs of the individual services as a row matrix A (in the given order). She decides to tip her masseur 25%, her nail tech 20%, and her hair stylist 15%. Represent the tipping percentages as a column matrix B (in the given order). Multiply these matrices to find the total amount in tips AB she needs to add to her final bill.

Use the following tables for exercises 81 and 82:

The following table gives fuel and electric requirements per mile associated with gasoline and electric automobiles:

	NUMBER OF GALLONS/MILE	NUMBER OF KW-HR/MILE
SUV full size	0.06	0
Hybrid car	0.02	0.1
Electric car	0	0.3

The following table gives an average cost for gasoline and electricity:

Cost per gallon of gasoline	$3.80
Cost per kW-hr of electricity	$0.05

81. Environment. Let matrix A represent the gasoline and electricity consumption and matrix B represent the costs of gasoline and electricity. Find AB and describe what the elements of the product matrix represent. (*Hint:* A has order 3×2 and B has order 2×1.)

82. Environment. Assume you drive 12,000 miles per year. What are the yearly costs associated with driving the three types of cars in Exercise 81?

For Exercises 83–88, apply the following decoding scheme:

1	A	10	J	19	S
2	B	11	K	20	T
3	C	12	L	21	U
4	D	13	M	22	V
5	E	14	N	23	W
6	F	15	O	24	X
7	G	16	P	25	Y
8	H	17	Q	26	Z
9	I	18	R		

The encoding matrix is $\begin{bmatrix} 1 & 1 & 0 \\ -1 & 0 & 1 \\ 2 & 0 & -1 \end{bmatrix}$. The encrypted matrices are given below. For each of the following, determine the 3-letter word that is originally transmitted. (*Hint:* All six words are parts of the body.)

83. Cryptography. $[55 \quad 10 \quad -22]$

84. Cryptography. $[31 \quad 8 \quad -7]$

85. Cryptography. $[21 \quad 12 \quad -2]$

86. Cryptography. $[9 \quad 1 \quad 5]$

87. Cryptography. $[-10 \quad 5 \quad 20]$

88. Cryptography. $[40 \quad 5 \quad -17]$

■ CATCH THE MISTAKE

In Exercises 89–92, explain the mistake that is made.

89. Multiply $\begin{bmatrix} 3 & 2 \\ 1 & 4 \end{bmatrix}\begin{bmatrix} -1 & 3 \\ -2 & 5 \end{bmatrix}$.

Solution:

Multiply corresponding elements.

$$\begin{bmatrix} 3 & 2 \\ 1 & 4 \end{bmatrix}\begin{bmatrix} -1 & 3 \\ -2 & 5 \end{bmatrix} = \begin{bmatrix} (3)(-1) & (2)(3) \\ (1)(-2) & (4)(5) \end{bmatrix}$$

Simplify.

$$\begin{bmatrix} 3 & 2 \\ 1 & 4 \end{bmatrix}\begin{bmatrix} -1 & 3 \\ -2 & 5 \end{bmatrix} = \begin{bmatrix} -3 & 6 \\ -2 & 20 \end{bmatrix}$$

This is incorrect. What mistake was made?

90. Multiply $\begin{bmatrix} 3 & 2 \\ 1 & 4 \end{bmatrix}\begin{bmatrix} -1 & 3 \\ -2 & 5 \end{bmatrix}$.

Solution:

Multiply using column-by-row method.

$$\begin{bmatrix} 3 & 2 \\ 1 & 4 \end{bmatrix}\begin{bmatrix} -1 & 3 \\ -2 & 5 \end{bmatrix} = \begin{bmatrix} (3)(-1) + (1)(3) & (2)(-1) + (4)(3) \\ (3)(-2) + (1)(5) & (2)(-2) + (4)(5) \end{bmatrix}$$

Simplify.

$$\begin{bmatrix} 3 & 2 \\ 1 & 4 \end{bmatrix}\begin{bmatrix} -1 & 3 \\ -2 & 5 \end{bmatrix} = \begin{bmatrix} 0 & 10 \\ -1 & 16 \end{bmatrix}$$

This is incorrect. What mistake was made?

91. Find the inverse of $A = \begin{bmatrix} 1 & 0 & 1 \\ -1 & 0 & -1 \\ 1 & 2 & 0 \end{bmatrix}$.

Solution:

Write the matrix $[A \,|\, I_3]$. $\begin{bmatrix} 1 & 0 & 1 & | & 1 & 0 & 0 \\ -1 & 0 & -1 & | & 0 & 1 & 0 \\ 1 & 2 & 0 & | & 0 & 0 & 1 \end{bmatrix}$

Use Gaussian elimination to reduce A.

$$\begin{matrix} R_2 + R_1 \rightarrow R_2 \\ R_3 - R_1 \rightarrow R_3 \end{matrix} \quad \begin{bmatrix} 1 & 0 & 1 & | & 1 & 0 & 0 \\ 0 & 0 & 0 & | & 1 & 1 & 0 \\ 0 & 2 & -1 & | & -1 & 0 & 1 \end{bmatrix}$$

$$R_2 \leftrightarrow R_3 \quad \begin{bmatrix} 1 & 0 & 1 & | & 1 & 0 & 0 \\ 0 & 2 & -1 & | & -1 & 0 & 1 \\ 0 & 0 & 0 & | & 1 & 1 & 0 \end{bmatrix}$$

$$\tfrac{1}{2}R_2 \rightarrow R_2 \quad \begin{bmatrix} 1 & 0 & 1 & | & 1 & 0 & 0 \\ 0 & 1 & -\frac{1}{2} & | & -\frac{1}{2} & 0 & \frac{1}{2} \\ 0 & 0 & 0 & | & 1 & 1 & 0 \end{bmatrix}$$

$A^{-1} = \begin{bmatrix} 1 & 0 & 0 \\ -\frac{1}{2} & 0 & \frac{1}{2} \\ 1 & 1 & 0 \end{bmatrix}$ is incorrect because $AA^{-1} \neq I_3$.

What mistake was made?

92. Find the inverse of A given that $A = \begin{bmatrix} 2 & 5 \\ 3 & 10 \end{bmatrix}$.

Solution:

$$A^{-1} = \frac{1}{A} \qquad A^{-1} = \frac{1}{\begin{bmatrix} 2 & 5 \\ 3 & 10 \end{bmatrix}}$$

Simplify.

$$A^{-1} = \begin{bmatrix} \frac{1}{2} & \frac{1}{5} \\ \frac{1}{3} & \frac{1}{10} \end{bmatrix}$$

This is incorrect. What mistake was made?

■ CONCEPTUAL

In Exercises 93–98, determine whether the statements are true or false.

93. If $A = \begin{bmatrix} a_{11} & a_{12} \\ a_{21} & a_{22} \end{bmatrix}$ and $B = \begin{bmatrix} b_{11} & b_{12} \\ b_{21} & b_{22} \end{bmatrix}$, then

$$AB = \begin{bmatrix} a_{11}b_{11} & a_{12}b_{12} \\ a_{21}b_{21} & a_{22}b_{22} \end{bmatrix}.$$

94. If AB is defined, then $AB = BA$.

95. AB is defined only if the number of columns in A equals the number of rows in B.

96. $A + B$ is defined only if A and B have the same order.

97. If $A = \begin{bmatrix} a_{11} & a_{12} \\ a_{21} & a_{22} \end{bmatrix}$, then $A^{-1} = \begin{bmatrix} \dfrac{1}{a_{11}} & \dfrac{1}{a_{12}} \\ \dfrac{1}{a_{21}} & \dfrac{1}{a_{22}} \end{bmatrix}$.

98. All square matrices have inverses.

99. For $A = \begin{bmatrix} a_{11} & a_{12} \\ a_{21} & a_{22} \end{bmatrix}$, find A^2.

100. In order for $A^2_{m \times n}$ to be defined, what condition (with respect to m and n) must be met?

101. For what values of x does the inverse of A not exist, given
$A = \begin{bmatrix} x & 6 \\ 3 & 2 \end{bmatrix}$?

102. Let $A = \begin{bmatrix} a & 0 & 0 \\ 0 & b & 0 \\ 0 & 0 & c \end{bmatrix}$. Find A^{-1}. Assume $abc \neq 0$.

■ CHALLENGE

103. For $A = \begin{bmatrix} 1 & 1 \\ 1 & 1 \end{bmatrix}$ find $A, A^2, A^3, \ldots$. What is A^n?

104. For $A = \begin{bmatrix} 1 & 0 \\ 0 & 1 \end{bmatrix}$ find $A, A^2, A^3, \ldots$. What is A^n?

105. If $A_{m \times n}B_{n \times p}$ is defined, explain why $(A_{m \times n}B_{n \times p})^2$ is not defined for $m \neq p$.

106. Given $C_{n \times m}$ and $A_{m \times n} = B_{m \times n}$, explain why $AC \neq CB$, if $m \neq n$.

107. Verify that $A^{-1} = \dfrac{1}{ad - bc}\begin{bmatrix} d & -b \\ -c & a \end{bmatrix}$ is the inverse of

$A = \begin{bmatrix} a & b \\ c & d \end{bmatrix}$, provided $ad - bc \neq 0$.

108. Let $A = \begin{bmatrix} a & b \\ c & d \end{bmatrix}$ and form the matrix $[A \mid I_2]$. Apply row operations to transform into $[I_2 \mid A^{-1}]$. Show

$$A^{-1} = \dfrac{1}{ad - bc}\begin{bmatrix} d & -b \\ -c & a \end{bmatrix} \text{ such that } ad - bc \neq 0.$$

109. Why does the square matrix $A = \begin{bmatrix} 2 & 3 \\ 4 & 6 \end{bmatrix}$ not have an inverse?

110. Why does the square matrix $A = \begin{bmatrix} 1 & 2 & -1 \\ 2 & 4 & -2 \\ 0 & 1 & 3 \end{bmatrix}$ not have an inverse?

■ TECHNOLOGY

In Exercises 111–116, apply a graphing utility to perform the indicated matrix operations, if possible.

$$A = \begin{bmatrix} 1 & 7 & 9 & 2 \\ -3 & -6 & 15 & 11 \\ 0 & 3 & 2 & 5 \\ 9 & 8 & -4 & 1 \end{bmatrix} \qquad B = \begin{bmatrix} 7 & 9 \\ 8 & 6 \\ -4 & -2 \\ 3 & 1 \end{bmatrix}$$

111. AB **112.** BA **113.** BB **114.** AA

$$A = \begin{bmatrix} 2 & 1 & 1 \\ -3 & 0 & 2 \\ 4 & -6 & 0 \end{bmatrix}$$

115. A^2 **116.** A^5

In Exercises 117 and 118, apply a graphing utility to perform the indicated matrix operations.

$$A = \begin{bmatrix} 1 & 7 & 9 & 2 \\ -3 & -6 & 15 & 11 \\ 0 & 3 & 2 & 5 \\ 9 & 8 & -4 & 1 \end{bmatrix}$$

117. Find A^{-1}.   **118.** Find AA^{-1}.

■ **PREVIEW TO CALCULUS**

In calculus, when finding the inverse of a vector function, it is fundamental that the matrix of partial derivatives is not singular. In Exercises 119–122, find the inverse of each matrix.

119. $\begin{bmatrix} 2x & 2y \\ 2x & -2y \end{bmatrix}$ **120.** $\begin{bmatrix} 1 & 1 \\ uy & ux \end{bmatrix}$

121. $\begin{bmatrix} \cos\theta & \sin\theta \\ -\sin\theta & \cos\theta \end{bmatrix}$

122. $\begin{bmatrix} \cos\theta & -r\sin\theta & 0 \\ \sin\theta & r\cos\theta & 0 \\ 0 & 0 & 1 \end{bmatrix}$

SECTION
8.5
THE DETERMINANT OF A SQUARE MATRIX AND CRAMER'S RULE

SKILLS OBJECTIVES

- Find the determinant of a 2 × 2 matrix.
- Find the determinant of an $n \times n$ matrix.
- Use Cramer's rule to solve a square system of linear equations.

CONCEPTUAL OBJECTIVES

- Derive Cramer's rule.
- Understand that if a determinant of a matrix is equal to zero, then that matrix does not have an inverse.
- Understand that Cramer's rule can be used to find only a unique solution.

In Section 8.3, we discussed Gauss–Jordan elimination as a way to solve systems of linear equations using augmented matrices. Then in Section 8.4, we employed matrix algebra and inverses to solve systems of linear equations that are square (same number of equations as variables). In this section, we will describe another method, called Cramer's rule, for solving systems of linear equations. Cramer's rule is applicable only to square systems. *Determinants* of square matrices play a vital role in Cramer's rule and indicate whether a matrix has an inverse.

Determinant of a 2 × 2 Matrix

Every square matrix A has a number associated with it called its *determinant*, denoted $\det(A)$ or $|A|$.

DEFINITION   **Determinant of a 2 × 2 Matrix**

The **determinant** of the 2 × 2 matrix $A = \begin{bmatrix} a & b \\ c & d \end{bmatrix}$ is given by

$$\det(A) = |A| = \begin{vmatrix} a & b \\ c & d \end{vmatrix} = ad - bc$$

Although the symbol for determinant, | |, looks like absolute value bars, the determinant can be any real number (positive, negative, or zero). The determinant of a 2 × 2 matrix is found by finding the product of the main diagonal entries (top left to bottom right) and subtracting the product of the entries along the other diagonal (bottom left to top right).

$$\begin{vmatrix} a & b \\ c & d \end{vmatrix} = ad - bc$$

Study Tip

The determinant of a 2 × 2 matrix is found by finding the product of the main diagonal entries and subtracting the product of the other diagonal entries.

EXAMPLE 1   Finding the Determinant of a 2 × 2 Matrix

Find the determinant of each matrix.

a. $\begin{bmatrix} 2 & -5 \\ -1 & 3 \end{bmatrix}$ **b.** $\begin{bmatrix} 0.5 & 0.2 \\ -3.0 & -4.2 \end{bmatrix}$ **c.** $\begin{bmatrix} \frac{2}{3} & 1 \\ 2 & 3 \end{bmatrix}$

Solution:

a. $\begin{vmatrix} 2 & -5 \\ -1 & 3 \end{vmatrix} = (2)(3) - (-1)(-5) = 6 - 5 = \boxed{1}$

b. $\begin{vmatrix} 0.5 & 0.2 \\ -3 & -4.2 \end{vmatrix} = (0.5)(-4.2) - (-3)(0.2) = -2.1 + 0.6 = \boxed{-1.5}$

c. $\begin{vmatrix} \frac{2}{3} & 1 \\ 2 & 3 \end{vmatrix} = \left(\frac{2}{3}\right)(3) - (2)(1) = 2 - 2 = \boxed{0}$

In Example 1, we see that determinants are real numbers that can be positive, negative, or zero. Although evaluating determinants of 2 × 2 matrices is a simple process, one **common mistake** is reversing the difference: $\begin{vmatrix} a & b \\ c & d \end{vmatrix} \neq bc - ad.$

■ **YOUR TURN** Evaluate the determinant $\begin{vmatrix} -2 & 1 \\ -3 & 2 \end{vmatrix}.$

Technology Tip

A graphing calculator can be used to find the determinant of each matrix.

Enter the matrix $\begin{bmatrix} 2 & -5 \\ -1 & 3 \end{bmatrix}$ as A.

Press $\boxed{2^{nd}}$ $\boxed{\text{MATRIX}}$. Use $\boxed{\blacktriangleright}$ to highlight $\boxed{\text{MATH}}$ $\boxed{1:\det(}$ $\boxed{\text{ENTER}}$ $\boxed{2^{nd}}$ $\boxed{\text{MATH}}$ $\boxed{\text{ENTER}}$ $\boxed{)}$ $\boxed{\text{ENTER}}$.

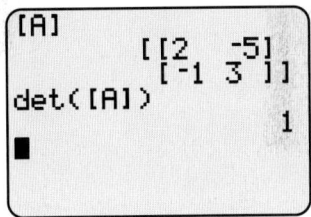

■ **Answer:** −1

Determinant of an *n* × *n* Matrix

In order to define the *determinant* of a 3 × 3 or a general *n* × *n* (where *n* ≥ 3) matrix, we first define *minors* and *cofactors* of a square matrix.

DEFINITION   **Minor and Cofactor**

Let *A* be a square matrix of order *n* × *n*. Then:

■ The **minor** M_{ij} of the entry a_{ij} is the determinant of the $(n - 1) \times (n - 1)$ matrix obtained when the *i*th row and *j*th column of *A* are deleted.
■ The **cofactor** C_{ij} of the entry a_{ij} is given by $C_{ij} = (-1)^{i+j}M_{ij}.$

The following table illustrates entries, minors, and cofactors of the matrix:

$$A = \begin{bmatrix} 1 & -3 & 2 \\ 4 & -1 & 0 \\ 5 & -2 & 3 \end{bmatrix}$$

Entry a_{ij}	Minor M_{ij}	Cofactor C_{ij}
$a_{11} = 1$	For M_{11}, delete the first row and first column:	$C_{11} = (-1)^{1+1}M_{11}$
	$$\begin{bmatrix} 1 & 3 & 2 \\ 4 & -1 & 0 \\ 5 & -2 & 3 \end{bmatrix}$$	$= (1)(-3)$
		$= -3$
	$M_{11} = \begin{vmatrix} -1 & 0 \\ -2 & 3 \end{vmatrix} = -3 - 0 = -3$	
$a_{32} = -2$	For M_{32}, delete the third row and second column:	$C_{32} = (-1)^{3+2}M_{32}$
	$$\begin{bmatrix} 1 & 3 & 2 \\ 4 & -1 & 0 \\ 5 & -2 & 3 \end{bmatrix}$$	$= (-1)(-8)$
		$= 8$
	$M_{32} = \begin{vmatrix} 1 & 2 \\ 4 & 0 \end{vmatrix} = 0 - 8 = -8$	

Notice that the cofactor is simply the minor multiplied by either 1 or -1, depending on whether $i + j$ is even or odd. Therefore, we can make the following sign pattern for 3×3 and 4×4 matrices and obtain the cofactor by multiplying the minor with the appropriate sign ($+1$ or -1):

$$\begin{bmatrix} + & - & + \\ - & + & - \\ + & - & + \end{bmatrix} \qquad \begin{bmatrix} + & - & + & - \\ - & + & - & + \\ + & - & + & - \\ - & + & - & + \end{bmatrix}$$

DEFINITION   **Determinant of an $n \times n$ Matrix**

Let A be an $n \times n$ matrix. Then the **determinant** of A is found by summing the entries in *any* row of A (or column of A) multiplied by each entries' respective cofactor.

If A is a 3×3 matrix, the determinant can be given by

$$\det(A) = a_{11}C_{11} + a_{12}C_{12} + a_{13}C_{13}$$

this is called **expanding the determinant by the first row**. It is important to note that any row or column can be used. Typically, the row or column with the most zeros is selected because it makes the arithmetic simpler.

Combining the definitions of minors, cofactors, and determinants, we now give a general definition for the determinant of a 3×3 matrix.

Row 1 expansion: $\begin{vmatrix} a_1 & b_1 & c_1 \\ a_2 & b_2 & c_2 \\ a_3 & b_3 & c_3 \end{vmatrix} = a_1 \begin{vmatrix} b_2 & c_2 \\ b_3 & c_3 \end{vmatrix} - b_1 \begin{vmatrix} a_2 & c_2 \\ a_3 & c_3 \end{vmatrix} + c_1 \begin{vmatrix} a_2 & b_2 \\ a_3 & b_3 \end{vmatrix}$

Column 1 expansion: $\begin{vmatrix} a_1 & b_1 & c_1 \\ a_2 & b_2 & c_2 \\ a_3 & b_3 & c_3 \end{vmatrix} = a_1 \begin{vmatrix} b_2 & c_2 \\ b_3 & c_3 \end{vmatrix} - a_2 \begin{vmatrix} b_1 & c_1 \\ b_3 & c_3 \end{vmatrix} + a_3 \begin{vmatrix} b_1 & c_1 \\ b_2 & c_2 \end{vmatrix}$

Whichever row or column is expanded, an alternating sign scheme is used (see sign arrays above). Notice that in either of the expansions above, each 2×2 determinant obtained is found by crossing out the row and column containing the entry that is multiplying the determinant.

EXAMPLE 2   Finding the Determinant of a 3 × 3 Matrix

For the given matrix, expand the determinant by the *first row*.

$$\begin{bmatrix} 2 & 1 & 3 \\ -1 & 5 & -2 \\ -3 & 7 & 4 \end{bmatrix}$$

Solution:

Expand the determinant by the first row. Remember the alternating sign.

$$\begin{vmatrix} 2 & 1 & 3 \\ -1 & 5 & -2 \\ -3 & 7 & 4 \end{vmatrix} = +2 \begin{vmatrix} 5 & -2 \\ 7 & 4 \end{vmatrix} - 1 \begin{vmatrix} -1 & -2 \\ -3 & 4 \end{vmatrix} + 3 \begin{vmatrix} -1 & 5 \\ -3 & 7 \end{vmatrix}$$

Evaluate the resulting 2 × 2 determinants.

$$= 2[(5)(4) - (7)(-2)] - 1[(-1)(4) - (-3)(-2)] + 3[(-1)(7) - (-3)(5)]$$

$$= 2[20 + 14] - [-4 - 6] + 3[-7 + 15]$$

Simplify. $$= 2(34) - (-10) + 3(8)$$

$$= 68 + 10 + 24$$

$$= \boxed{102}$$

Study Tip

The determinant by the third column is also 102. It does not matter on which row or column the expansion occurs.

■ **Answer:** 156

■ **YOUR TURN** For the given matrix, expand the determinant by the first row.

$$\begin{bmatrix} 1 & 3 & -2 \\ 2 & 5 & 4 \\ 7 & -1 & 6 \end{bmatrix}$$

Determinants can be expanded by any row *or* column. Typically, the row or column with the most zeros is selected to simplify the arithmetic.

EXAMPLE 3   Finding the Determinant of a 3 × 3 Matrix

Find the determinant of the matrix $\begin{vmatrix} -1 & 2 & 0 \\ 4 & 7 & 1 \\ 5 & 3 & 0 \end{vmatrix}$.

Solution:

Since there are two 0s in the third column, expand the determinant by the third column. Recall the sign array.

$$\begin{bmatrix} + & - & + \\ - & + & - \\ + & - & + \end{bmatrix}$$

$$\begin{vmatrix} -1 & 2 & 0 \\ 4 & 7 & 1 \\ 5 & 3 & 0 \end{vmatrix} = +0 \begin{vmatrix} 4 & 7 \\ 5 & 3 \end{vmatrix} - 1 \begin{vmatrix} -1 & 2 \\ 5 & 3 \end{vmatrix} + 0 \begin{vmatrix} -1 & 2 \\ 4 & 7 \end{vmatrix}$$

There is no need to calculate the two determinants that are multiplied by 0s, since 0 times any real number is zero.

$$\begin{vmatrix} -1 & 2 & 0 \\ 4 & 7 & 1 \\ 5 & 3 & 0 \end{vmatrix} = 0 - 1 \underset{-3-10}{\begin{vmatrix} -1 & 2 \\ 5 & 3 \end{vmatrix}} + 0$$

Simplify. $$= -1(-13) = \boxed{13}$$

■ **YOUR TURN** Evaluate the determinant $\begin{vmatrix} 1 & -2 & 1 \\ -1 & 0 & 3 \\ -4 & 0 & 2 \end{vmatrix}$.

Technology Tip

Enter the matrix as *A* and find the determinant.

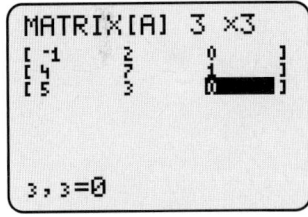

Press [2nd] [MATRIX]. Use [▶] to highlight [MATH] [1:det(] [ENTER] [2nd] [MATH] [ENTER] [)] [ENTER].

■ **Answer:** 20

EXAMPLE 4   Finding the Determinant of a 4 × 4 Matrix

Find the determinant of the matrix $\begin{vmatrix} 1 & -2 & 3 & 4 \\ -4 & 0 & -1 & 0 \\ -3 & 9 & 6 & 5 \\ -5 & 7 & 2 & 1 \end{vmatrix}$.

Solution:

Since there are two 0s in the second row, expand the determinant by the second row. Recall the sign array for a 4 × 4 matrix.

$$\begin{bmatrix} + & - & + & - \\ - & + & - & + \\ + & - & + & - \\ - & + & - & + \end{bmatrix}$$

$$\begin{bmatrix} 1 & -2 & 3 & 4 \\ -4 & 0 & -1 & 0 \\ -3 & 9 & 6 & 5 \\ -5 & 7 & 2 & 1 \end{bmatrix} = -\,-(-4)\begin{vmatrix} -2 & 3 & 4 \\ 9 & 6 & 5 \\ 7 & 2 & 1 \end{vmatrix} + 0 - (-1)\begin{vmatrix} 1 & -2 & 4 \\ -3 & 9 & 5 \\ -5 & 7 & 1 \end{vmatrix} + 0$$

Evaluate the two 3 × 3 determinants.

$$\begin{vmatrix} -2 & 3 & 4 \\ 9 & 6 & 5 \\ 7 & 2 & 1 \end{vmatrix} = -2\begin{vmatrix} 6 & 5 \\ 2 & 1 \end{vmatrix} - 3\begin{vmatrix} 9 & 5 \\ 7 & 1 \end{vmatrix} + 4\begin{vmatrix} 9 & 6 \\ 7 & 2 \end{vmatrix}$$

$$= -2(6 - 10) - 3(9 - 35) + 4(18 - 42)$$
$$= -2(-4) - 3(-26) + 4(-24)$$
$$= 8 + 78 - 96$$
$$= -10$$

$$\begin{vmatrix} 1 & -2 & 4 \\ -3 & 9 & 5 \\ -5 & 7 & 1 \end{vmatrix} = 1\begin{vmatrix} 9 & 5 \\ 7 & 1 \end{vmatrix} - (-2)\begin{vmatrix} -3 & 5 \\ -5 & 1 \end{vmatrix} + 4\begin{vmatrix} -3 & 9 \\ -5 & 7 \end{vmatrix}$$

$$= 1(9 - 35) + 2(-3 + 25) + 4(-21 + 45)$$
$$= -26 + 2(22) + 4(24)$$
$$= -26 + 44 + 96$$
$$= 114$$

$$\begin{bmatrix} 1 & -2 & 3 & 4 \\ -4 & 0 & -1 & 0 \\ -3 & 9 & 6 & 5 \\ -5 & 7 & 2 & 1 \end{bmatrix} = 4\underbrace{\begin{vmatrix} -2 & 3 & 4 \\ 9 & 6 & 5 \\ 7 & 2 & 1 \end{vmatrix}}_{-10} + \underbrace{\begin{vmatrix} 1 & -2 & 4 \\ -3 & 9 & 5 \\ -5 & 7 & 1 \end{vmatrix}}_{114} = 4(-10) + 114 = \boxed{74}$$

Cramer's Rule: Systems of Linear Equations in Two Variables

Let's now apply determinants of 2×2 matrices to solve systems of linear equations in two variables. We begin by solving the general system of two linear equations in two variables:

$$(1) \quad a_1x + b_1y = c_1$$

$$(2) \quad a_2x + b_2y = c_2$$

Solve for x using elimination (eliminate y).

Multiply (1) by b_2. $\qquad\qquad\qquad\qquad\qquad b_2a_1x + b_2b_1y = b_2c_1$

Multiply (2) by $-b_1$. $\qquad\qquad\qquad\qquad \underline{-b_1a_2x - b_1b_2y = -b_1c_2}$

Add the two new equations to eliminate y. $\qquad (a_1b_2 - a_2b_1)x = (b_2c_1 - b_1c_2)$

Divide both sides by $(a_1b_2 - a_2b_1)$. $\qquad\qquad\qquad x = \dfrac{(b_2c_1 - b_1c_2)}{(a_1b_2 - a_2b_1)}$

Write both the numerator and the denominator as determinants. $\qquad\qquad x = \dfrac{\begin{vmatrix} c_1 & b_1 \\ c_2 & b_2 \end{vmatrix}}{\begin{vmatrix} a_1 & b_1 \\ a_2 & b_2 \end{vmatrix}}$

Solve for y using elimination (eliminate x).

Multiply (1) by $-a_2$. $\qquad\qquad\qquad\qquad\qquad -a_2a_1x - a_2b_1y = -a_2c_1$

Multiply (2) by a_1. $\qquad\qquad\qquad\qquad\quad \underline{a_1a_2x + a_1b_2y = a_1c_2}$

Add the two new equations to eliminate x. $\qquad (a_1b_2 - a_2b_1)y = (a_1c_2 - a_2c_1)$

Divide both sides by $(a_1b_2 - a_2b_1)$. $\qquad\qquad\qquad y = \dfrac{(a_1c_2 - a_2c_1)}{(a_1b_2 - a_2b_1)}$

Write both the numerator and the denominator as determinants. $\qquad\qquad y = \dfrac{\begin{vmatrix} a_1 & c_1 \\ a_2 & c_2 \end{vmatrix}}{\begin{vmatrix} a_1 & b_1 \\ a_2 & b_2 \end{vmatrix}}$

Notice that the solutions for x and y involve three determinants. If we let

$$D = \begin{vmatrix} a_1 & b_1 \\ a_2 & b_2 \end{vmatrix} \qquad D_x = \begin{vmatrix} c_1 & b_1 \\ c_2 & b_2 \end{vmatrix} \qquad D_y = \begin{vmatrix} a_1 & c_1 \\ a_2 & c_2 \end{vmatrix},$$

$$\text{then} \quad x = \frac{D_x}{D} \quad \text{and} \quad y = \frac{D_y}{D}.$$

Notice that the matrix D is the determinant of the coefficient matrix of the system and cannot equal zero ($D \neq 0$) or there will be no unique solution. These formulas for solving a system of two linear equations in two variables are known as *Cramer's rule*.

CRAMER'S RULE FOR SOLVING SYSTEMS OF TWO LINEAR EQUATIONS IN TWO VARIABLES

For the system of linear equations

$$a_1x + b_1y = c_1$$
$$a_2x + b_2y = c_2$$

let

$$D = \begin{vmatrix} a_1 & b_1 \\ a_2 & b_2 \end{vmatrix} \qquad D_x = \begin{vmatrix} c_1 & b_1 \\ c_2 & b_2 \end{vmatrix} \qquad D_y = \begin{vmatrix} a_1 & c_1 \\ a_2 & c_2 \end{vmatrix}$$

If $D \neq 0$, then the solution to the system of linear equations is

$$x = \frac{D_x}{D} \qquad y = \frac{D_y}{D}$$

If $D = 0$, then the system of linear equations has either no solution or infinitely many solutions.

Notice that the determinants D_x and D_y are similar to the determinant D. A three-step procedure is outlined for setting up the three determinants for a system of two linear equations in two variables:

$$a_1x + b_1y = c_1$$
$$a_2x + b_2y = c_2$$

Step 1: Set up D.

Apply the coefficients of x and y. $\qquad D = \begin{vmatrix} a_1 & b_1 \\ a_2 & b_2 \end{vmatrix}$

Step 2: Set up D_x.

Start with D and replace the coefficients of x (column 1) with the constants on the right side of the equal sign. $\qquad D_x = \begin{vmatrix} c_1 & b_1 \\ c_2 & b_2 \end{vmatrix}$

Step 3: Set up D_y.

Start with D and replace the coefficients of y (column 2) with the constants on the right side of the equal sign. $\qquad D_y = \begin{vmatrix} a_1 & c_1 \\ a_2 & c_2 \end{vmatrix}$

EXAMPLE 5   Using Cramer's Rule to Solve a System of Two Linear Equations

Apply Cramer's rule to solve the system.

$$x + 3y = 1$$
$$2x + y = -3$$

Solution:

Set up the three determinants.

$$D = \begin{vmatrix} 1 & 3 \\ 2 & 1 \end{vmatrix}$$

$$D_x = \begin{vmatrix} 1 & 3 \\ -3 & 1 \end{vmatrix}$$

$$D_y = \begin{vmatrix} 1 & 1 \\ 2 & -3 \end{vmatrix}$$

Evaluate the determinants.

$$D = 1 - 6 = -5$$
$$D_x = 1 - (-9) = 10$$
$$D_y = -3 - 2 = -5$$

Solve for x and y.

$$x = \frac{D_x}{D} = \frac{10}{-5} = -2$$

$$y = \frac{D_y}{D} = \frac{-5}{-5} = 1$$

$$\boxed{x = -2, y = 1}$$

```
[A]
        [[1   3]
         [-3  1]]
[B]
        [[1  1]
         [2  -3]]
```
```
[C]
        [[1  3]
         [2  1]]
```

To solve for x and y, enter D_x/C as A/C for x and D_y/C as B/C for y.

```
det([A])/det([C]
)
               -2
det([B])/det([C]
)
                1
```

■ **Answer:** $x = 5, y = -6$

■ **YOUR TURN** Apply Cramer's rule to solve the system.

$$5x + 4y = 1$$
$$-3x - 2y = -3$$

Recall from Section 8.1 that systems of two linear equations in two variables led to one of three possible outcomes: a unique solution, no solution, and infinitely many solutions. When $D = 0$, Cramer's rule does not apply and the system is either inconsistent (has no solution) or contains dependent equations (has infinitely many solutions).

Cramer's Rule: Systems of Linear Equations in Three Variables

Cramer's rule can also be used to solve higher order systems of linear equations. The following box summarizes Cramer's rule for solving a system of three equations in three variables:

CRAMER'S RULE: SOLUTION FOR SYSTEMS OF THREE EQUATIONS IN THREE VARIABLES

The system of linear equations

$$a_1x + b_1y + c_1z = d_1$$
$$a_2x + b_2y + c_2z = d_2$$
$$a_3x + b_3y + c_3z = d_3$$

has the solution

$$x = \frac{D_x}{D} \qquad y = \frac{D_y}{D} \qquad z = \frac{D_z}{D} \qquad D \neq 0$$

where the determinants are given as follows:

Display the coefficients of x, y, and z.

$$D = \begin{vmatrix} a_1 & b_1 & c_1 \\ a_2 & b_2 & c_2 \\ a_3 & b_3 & c_3 \end{vmatrix}$$

Replace the coefficients of x (column 1) in D with the constants on the right side of the equal sign.

$$D_x = \begin{vmatrix} d_1 & b_1 & c_1 \\ d_2 & b_2 & c_2 \\ d_3 & b_3 & c_3 \end{vmatrix}$$

Replace the coefficients of y (column 2) in D with the constants on the right side of the equal sign.

$$D_y = \begin{vmatrix} a_1 & d_1 & c_1 \\ a_2 & d_2 & c_2 \\ a_3 & d_3 & c_3 \end{vmatrix}$$

Replace the coefficients of z (column 3) in D with the constants on the right side of the equal sign.

$$D_z = \begin{vmatrix} a_1 & b_1 & d_1 \\ a_2 & b_2 & d_2 \\ a_3 & b_3 & d_3 \end{vmatrix}$$

EXAMPLE 6   Using Cramer's Rule to Solve a System of Three Linear Equations

Use Cramer's rule to solve the system.

$$\begin{aligned} 3x - 2y + 3z &= -3 \\ 5x + 3y + 8z &= -2 \\ x + y + 3z &= 1 \end{aligned}$$

Solution:

Set up the four determinants.

D contains the coefficients of x, y, and z.

$$D = \begin{vmatrix} 3 & -2 & 3 \\ 5 & 3 & 8 \\ 1 & 1 & 3 \end{vmatrix}$$

Replace a column with constants on the right side of the equation.

$$D_x = \begin{vmatrix} -3 & -2 & 3 \\ -2 & 3 & 8 \\ 1 & 1 & 3 \end{vmatrix} \quad D_y = \begin{vmatrix} 3 & -3 & 3 \\ 5 & -2 & 8 \\ 1 & 1 & 3 \end{vmatrix} \quad D_z = \begin{vmatrix} 3 & -2 & -3 \\ 5 & 3 & -2 \\ 1 & 1 & 1 \end{vmatrix}$$

Evaluate the determinants.

$$\begin{aligned} D &= 3(9-8) - (-2)(15-8) + 3(5-3) = 23 \\ D_x &= -3(9-8) - (-2)(-6-8) + 3(-2-3) = -46 \\ D_y &= 3(-6-8) - (-3)(15-8) + 3(5+2) = 0 \\ D_z &= 3(3+2) - (-2)(5+2) - 3(5-3) = 23 \end{aligned}$$

Solve for x, y, and z.

$$x = \frac{D_x}{D} = \frac{-46}{23} = -2 \qquad y = \frac{D_y}{D} = \frac{0}{23} = 0 \qquad z = \frac{D_z}{D} = \frac{23}{23} = 1$$

$$\boxed{x = -2, y = 0, z = 1}$$

■ **Answer:** $x = 1, y = -2, z = 3$

■ **YOUR TURN** Use Cramer's rule to solve the system.

$$\begin{aligned} 2x + 3y + z &= -1 \\ x - y - z &= 0 \\ -3x - 2y + 3z &= 10 \end{aligned}$$

As was the case in two equations, when $D = 0$, Cramer's rule does not apply and the system of three equations is either inconsistent (no solution) or contains dependent equations (infinitely many solutions).

SECTION
8.5 SUMMARY

In this section, **determinants** were discussed for square matrices.

ORDER	DETERMINANT	ARRAY
2×2	$\det(A) = \|A\| = \begin{vmatrix} a & b \\ c & d \end{vmatrix} = ad - bc$	
3×3	$\begin{vmatrix} a_1 & b_1 & c_1 \\ a_2 & b_2 & c_2 \\ a_3 & b_3 & c_3 \end{vmatrix} = a_1 \begin{vmatrix} b_2 & c_2 \\ b_3 & c_3 \end{vmatrix} - b_1 \begin{vmatrix} a_2 & c_2 \\ a_3 & c_3 \end{vmatrix} + c_1 \begin{vmatrix} a_2 & b_2 \\ a_3 & b_3 \end{vmatrix}$ Expansion by first row (any row or column can be used)	$\begin{bmatrix} + & - & + \\ - & + & - \\ + & - & + \end{bmatrix}$

Cramer's rule was developed for 2×2 and 3×3 matrices, but it can be extended to general $n \times n$ matrices. When the coefficient determinant is equal to zero ($D = 0$), then the system is either inconsistent (and has no solution) or represents dependent equations (and has infinitely many solutions), and Cramer's rule does not apply.

SYSTEM	ORDER	SOLUTION	DETERMINANTS
$\begin{aligned} a_1 x + b_1 y &= c_1 \\ a_2 x + b_2 y &= c_2 \end{aligned}$	2×2	$x = \dfrac{D_x}{D} \qquad y = \dfrac{D_y}{D}$	$D = \begin{vmatrix} a_1 & b_1 \\ a_2 & b_2 \end{vmatrix} \neq 0$ $D_x = \begin{vmatrix} c_1 & b_1 \\ c_2 & b_2 \end{vmatrix}$ $D_y = \begin{vmatrix} a_1 & c_1 \\ a_2 & c_2 \end{vmatrix}$
$\begin{aligned} a_1 x + b_1 y + c_1 z &= d_1 \\ a_2 x + b_2 y + c_2 z &= d_2 \\ a_3 x + b_3 y + c_3 z &= d_3 \end{aligned}$	3×3	$x = \dfrac{D_x}{D} \quad y = \dfrac{D_y}{D} \quad z = \dfrac{D_z}{D}$	$D = \begin{vmatrix} a_1 & b_1 & c_1 \\ a_2 & b_2 & c_2 \\ a_3 & b_3 & c_3 \end{vmatrix} \neq 0$ $D_x = \begin{vmatrix} d_1 & b_1 & c_1 \\ d_2 & b_2 & c_2 \\ d_3 & b_3 & c_3 \end{vmatrix}$ $D_y = \begin{vmatrix} a_1 & d_1 & c_1 \\ a_2 & d_2 & c_2 \\ a_3 & d_3 & c_3 \end{vmatrix}$ $D_z = \begin{vmatrix} a_1 & b_1 & d_1 \\ a_2 & b_2 & d_2 \\ a_3 & b_3 & d_3 \end{vmatrix}$

SECTION
8.5 EXERCISES

■ SKILLS

In Exercises 1–10, evaluate each 2 × 2 determinant.

1. $\begin{vmatrix} 1 & 2 \\ 3 & 4 \end{vmatrix}$ **2.** $\begin{vmatrix} 1 & -2 \\ -3 & -4 \end{vmatrix}$ **3.** $\begin{vmatrix} 7 & 9 \\ -5 & -2 \end{vmatrix}$ **4.** $\begin{vmatrix} -3 & -11 \\ 7 & 15 \end{vmatrix}$ **5.** $\begin{vmatrix} 0 & 7 \\ 4 & -1 \end{vmatrix}$

6. $\begin{vmatrix} 0 & 0 \\ 1 & 0 \end{vmatrix}$ **7.** $\begin{vmatrix} -1.2 & 2.4 \\ -0.5 & 1.5 \end{vmatrix}$ **8.** $\begin{vmatrix} -1.0 & 1.4 \\ 1.5 & -2.8 \end{vmatrix}$ **9.** $\begin{vmatrix} \frac{3}{4} & \frac{1}{3} \\ 2 & \frac{8}{9} \end{vmatrix}$ **10.** $\begin{vmatrix} -\frac{1}{2} & \frac{1}{4} \\ \frac{2}{3} & -\frac{8}{9} \end{vmatrix}$

In Exercises 11–30, use Cramer's rule to solve each system of equations, if possible.

11. $x + y = -1$
$x - y = 11$

12. $x + y = -1$
$x - y = -9$

13. $3x + 2y = -4$
$-2x + y = 5$

14. $5x + 3y = 1$
$4x - 7y = -18$

15. $3x - 2y = -1$
$5x + 4y = -31$

16. $x - 4y = -7$
$3x + 8y = 19$

17. $7x - 3y = -29$
$5x + 2y = 0$

18. $6x - 2y = 24$
$4x + 7y = 41$

19. $3x + 5y = 16$
$y - x = 0$

20. $-2x - 3y = 15$
$7y + 4x = -33$

21. $3x - 5y = 7$
$-6x + 10y = -21$

22. $3x - 5y = 7$
$6x - 10y = 14$

23. $2x - 3y = 4$
$-10x + 15y = -20$

24. $2x - 3y = 2$
$10x - 15y = 20$

25. $3x + \frac{1}{2}y = 1$
$4x + \frac{1}{3}y = \frac{5}{3}$

26. $\frac{3}{2}x + \frac{9}{4}y = \frac{9}{8}$
$\frac{1}{3}x + \frac{1}{4}y = \frac{1}{12}$

27. $0.3x - 0.5y = -0.6$
$0.2x + y = 2.4$

28. $0.5x - 0.4y = -3.6$
$10x + 3.6y = -14$

29. $y = 17x + 7$
$y = -15x + 7$

30. $9x = -45 - 2y$
$4x = -3y - 20$

In Exercises 31–42, evaluate each 3 × 3 determinant.

31. $\begin{vmatrix} 3 & 1 & 0 \\ 2 & 0 & -1 \\ -4 & 1 & 0 \end{vmatrix}$ **32.** $\begin{vmatrix} 1 & 1 & 0 \\ 0 & 2 & -1 \\ 0 & -3 & 5 \end{vmatrix}$ **33.** $\begin{vmatrix} 2 & 1 & -5 \\ 3 & 0 & -1 \\ 4 & 0 & 7 \end{vmatrix}$ **34.** $\begin{vmatrix} 2 & 1 & -5 \\ 3 & -7 & 0 \\ 4 & -6 & 0 \end{vmatrix}$

35. $\begin{vmatrix} 1 & 1 & -5 \\ 3 & -7 & -4 \\ 4 & -6 & 9 \end{vmatrix}$ **36.** $\begin{vmatrix} -3 & 2 & -5 \\ 1 & 8 & 2 \\ 4 & -6 & 9 \end{vmatrix}$ **37.** $\begin{vmatrix} 1 & 3 & 4 \\ 2 & -1 & 1 \\ 3 & -2 & 1 \end{vmatrix}$ **38.** $\begin{vmatrix} -7 & 2 & 5 \\ \frac{7}{8} & 3 & 4 \\ -1 & 4 & 6 \end{vmatrix}$

39. $\begin{vmatrix} -3 & 1 & 5 \\ 2 & 0 & 6 \\ 4 & 7 & -9 \end{vmatrix}$ **40.** $\begin{vmatrix} 1 & -1 & 5 \\ 3 & -3 & 6 \\ 4 & 9 & 0 \end{vmatrix}$ **41.** $\begin{vmatrix} -2 & 1 & -7 \\ 4 & -2 & 14 \\ 0 & 1 & 8 \end{vmatrix}$ **42.** $\begin{vmatrix} 5 & -2 & -1 \\ 4 & -9 & -3 \\ 2 & 8 & -6 \end{vmatrix}$

In Exercises 43–58, apply Cramer's rule to solve each system of equations, if possible.

43. $x + y - z = 0$
$x - y + z = 4$
$x + y + z = 10$

44. $-x + y + z = -4$
$x + y - z = 0$
$x + y + z = 2$

45. $3x + 8y + 2z = 28$
$-2x + 5y + 3z = 34$
$4x + 9y + 2z = 29$

46. $7x + 2y - z = -1$
$6x + 5y + z = 16$
$-5x - 4y + 3z = -5$

47. $3x + 5z = 11$
$4y + 3z = -9$
$2x - y = 7$

48. $3x - 2z = 7$
$4x + z = 24$
$6x - 2y = 10$

49. $x + y - z = 5$
$x - y + z = -1$
$-2x - 2y + 2z = -10$

50. $x + y - z = 3$
$x - y + z = -2$
$-2x - 2y + 2z = -6$

51.
$$x + y + z = 9$$
$$x - y + z = 3$$
$$-x + y - z = 5$$

52.
$$x + y + z = 6$$
$$x - y - z = 0$$
$$-x + y + z = 7$$

53.
$$x + 2y + 3z = 11$$
$$-2x + 3y + 5z = 29$$
$$4x - y + 8z = 19$$

54.
$$8x - 2y + 5z = 36$$
$$3x + y - z = 17$$
$$2x - 6y + 4z = -2$$

55.
$$x - 4y + 7z = 49$$
$$-3x + 2y - z = -17$$
$$5x + 8y - 2z = -24$$

56.
$$\tfrac{1}{2}x - 2y + 7z = 25$$
$$x + \tfrac{1}{4}y - 4z = -2$$
$$-4x + 5y = -56$$

57.
$$2x + 7y - 4z = -5.5$$
$$-x - 4y - 5z = -19$$
$$4x - 2y - 9z = -38$$

58.
$$4x - 2y + z = -15$$
$$3x + y - 2z = -20$$
$$-6x + y + 5z = 51$$

▪ APPLICATIONS

In Exercises 59 and 60, three points, (x_1, y_1), (x_2, y_2), and (x_3, y_3), are collinear if and only if

$$\begin{vmatrix} x_1 & y_1 & 1 \\ x_2 & y_2 & 1 \\ x_3 & y_3 & 1 \end{vmatrix} = 0$$

59. Geometry. Apply determinants to determine whether the points, $(-2, -1)$, $(1, 5)$, and $(3, 9)$, are collinear.

60. Geometry. Apply determinants to determine whether the points, $(2, -6)$, $(-7, 30)$, and $(5, -18)$, are collinear.

For Exercises 61–64, the area of a triangle with vertices, (x_1, y_1), (x_2, y_2), and (x_3, y_3), is given by

$$\text{Area} = \pm\frac{1}{2}\begin{vmatrix} x_1 & y_1 & 1 \\ x_2 & y_2 & 1 \\ x_3 & y_3 & 1 \end{vmatrix}$$

where the sign is chosen so that the area is positive.

61. Geometry. Apply determinants to find the area of a triangle with vertices, $(3, 2)$, $(5, 2)$, and $(3, -4)$. Check your answer by plotting these vertices in a Cartesian plane and using the formula for area of a right triangle.

62. Geometry. Apply determinants to find the area of a triangle with vertices, $(2, 3)$, $(7, 3)$, and $(7, 7)$. Check your answer by plotting these vertices in a Cartesian plane and using the formula for area of a right triangle.

63. Geometry. Apply determinants to find the area of a triangle with vertices, $(1, 2)$, $(3, 4)$, and $(-2, 5)$.

64. Geometry. Apply determinants to find the area of a triangle with vertices, $(-1, -2)$, $(3, 4)$, and $(2, 1)$.

65. Geometry. An equation of a line that passes through two points (x_1, y_1) and (x_2, y_2) can be expressed as a determinant equation as follows:

$$\begin{vmatrix} x & y & 1 \\ x_1 & y_1 & 1 \\ x_2 & y_2 & 1 \end{vmatrix} = 0$$

Apply the determinant to write an equation of the line passing through the points $(1, 2)$ and $(2, 4)$. Expand the determinant and express the equation of the line in slope–intercept form.

66. Geometry. If three points (x_1, y_1), (x_2, y_2), and (x_3, y_3) are collinear (lie on the same line), then the following determinant equation must be satisfied:

$$\begin{vmatrix} x_1 & y_1 & 1 \\ x_2 & y_2 & 1 \\ x_3 & y_3 & 1 \end{vmatrix} = 0$$

Determine whether $(0, 5)$, $(2, 0)$, and $(1, 2)$ are collinear.

67. Electricity: Circuit Theory. The following equations come from circuit theory. Find the currents I_1, I_2, and I_3.

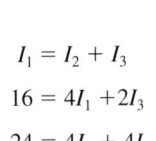

$$I_1 = I_2 + I_3$$
$$16 = 4I_1 + 2I_3$$
$$24 = 4I_1 + 4I_2$$

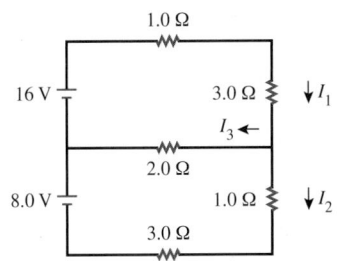

68. Electricity: Circuit Theory. The following equations come from circuit theory. Find the currents I_1, I_2, and I_3.

$$I_1 = I_2 + I_3$$
$$24 = 6I_1 + 3I_3$$
$$36 = 6I_1 + 6I_2$$

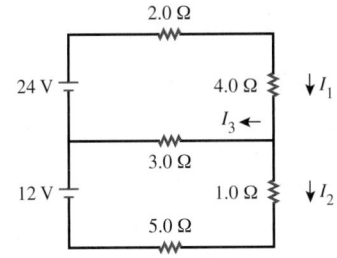

▪ CATCH THE MISTAKE

In Exercises 69–72, explain the mistake that is made.

69. Evaluate the determinant $\begin{vmatrix} 2 & 1 & 3 \\ -3 & 0 & 2 \\ 1 & 4 & -1 \end{vmatrix}$.

Solution:

Expand the 3×3 determinant in terms of the 2×2 determinants.

$$\begin{vmatrix} 2 & 1 & 3 \\ -3 & 0 & 2 \\ 1 & 4 & -1 \end{vmatrix} = 2\begin{vmatrix} 0 & 2 \\ 4 & -1 \end{vmatrix} + 1\begin{vmatrix} -3 & 2 \\ 1 & -1 \end{vmatrix} + 3\begin{vmatrix} -3 & 0 \\ 1 & 4 \end{vmatrix}$$

Expand the 2×2 determinants. $= 2(0 - 8) + 1(3 - 2) + 3(-12 - 0)$

Simplify. $= -16 + 1 - 36 = -51$

This is incorrect. What mistake was made?

70. Evaluate the determinant $\begin{vmatrix} 2 & 1 & 3 \\ -3 & 0 & 2 \\ 1 & 4 & -1 \end{vmatrix}$.

Solution:

Expand the 3×3 determinant in terms of the 2×2 determinants.

$$\begin{vmatrix} 2 & 1 & 3 \\ -3 & 0 & 2 \\ 1 & 4 & -1 \end{vmatrix} = 2\begin{vmatrix} 0 & 2 \\ 4 & -1 \end{vmatrix} - 1\begin{vmatrix} -3 & 2 \\ 1 & -1 \end{vmatrix} + 3\begin{vmatrix} -3 & 2 \\ 1 & -1 \end{vmatrix}$$

Expand the 2×2 determinants. $= 2(0 - 8) - 1(3 - 2) + 3(3 - 2)$

Simplify. $= -16 - 1 + 3 = -14$

This is incorrect. What mistake was made?

71. Solve the system of linear equations.

$$\begin{aligned} 2x + 3y &= 6 \\ -x - y &= -3 \end{aligned}$$

Solution:

Set up the determinants.

$$D = \begin{vmatrix} 2 & 3 \\ -1 & -1 \end{vmatrix}, D_x = \begin{vmatrix} 2 & 6 \\ -1 & -3 \end{vmatrix}, \text{ and } D_y = \begin{vmatrix} 6 & 3 \\ -3 & -1 \end{vmatrix}$$

Evaluate the determinants. $D = 1, D_x = 0, \text{ and } D_y = 3$

Solve for x and y. $x = \dfrac{D_x}{D} = \dfrac{0}{1} = 0 \text{ and } y = \dfrac{D_y}{D} = \dfrac{3}{1} = 3$

$x = 0, y = 3$ is incorrect. What mistake was made?

72. Solve the system of linear equations.

$$\begin{aligned} 4x - 6y &= 0 \\ 4x + 6y &= 4 \end{aligned}$$

Solution:

Set up the determinants.

$$D = \begin{vmatrix} 4 & -6 \\ 4 & 6 \end{vmatrix}, D_x = \begin{vmatrix} 0 & -6 \\ 4 & 6 \end{vmatrix}, \text{ and } D_y = \begin{vmatrix} 4 & 0 \\ 4 & 4 \end{vmatrix}$$

Evaluate the determinants. $D = 48, D_x = 24, \text{ and } D_y = 16$

Solve for x and y. $x = \dfrac{D}{D_x} = \dfrac{48}{24} = 2 \text{ and } y = \dfrac{D_y}{D} = \dfrac{48}{16} = 3$

$x = 2, y = 3$ is incorrect. What mistake was made?

▪ CONCEPTUAL

In Exercises 73–76, determine whether each statement is true or false.

73. The value of a determinant changes sign if any two rows are interchanged.

74. If all the entries in any column are equal to zero, the value of the determinant is 0.

75. $\begin{vmatrix} 2 & 6 & 4 \\ 0 & 2 & 8 \\ 4 & 0 & 10 \end{vmatrix} = 2\begin{vmatrix} 1 & 3 & 2 \\ 0 & 1 & 4 \\ 2 & 0 & 5 \end{vmatrix}$

76. $\begin{vmatrix} 3 & 1 & 2 \\ 0 & 2 & 8 \\ 3 & 1 & 2 \end{vmatrix} = 0$

77. Calculate the determinant $\begin{vmatrix} a & 0 & 0 \\ 0 & b & 0 \\ 0 & 0 & c \end{vmatrix}$.

78. Calculate the determinant $\begin{vmatrix} a_1 & b_1 & c_1 \\ 0 & b_2 & c_2 \\ 0 & 0 & c_3 \end{vmatrix}$.

■ **CHALLENGE**

79. Evaluate the determinant:

$$\begin{vmatrix} 1 & -2 & -1 & 3 \\ 4 & 0 & 1 & 2 \\ 0 & 3 & 2 & 4 \\ 1 & -3 & 5 & -4 \end{vmatrix}$$

80. For the system of equations

$$3x + 2y = 5$$
$$ax - 4y = 1$$

find a that guarantees no unique solution.

81. Show that

$$\begin{vmatrix} a_1 & b_1 & c_1 \\ a_2 & b_2 & c_2 \\ a_3 & b_3 & c_3 \end{vmatrix} = \begin{matrix} a_1b_2c_3 + b_1c_2a_3 + c_1a_2b_3 \\ - a_3b_2c_1 - b_3c_2a_1 - b_1a_2c_3 \end{matrix}$$

by expanding down the second column.

82. Show that

$$\begin{vmatrix} a_1 & b_1 & c_1 \\ a_2 & b_2 & c_2 \\ a_3 & b_3 & c_3 \end{vmatrix} = \begin{matrix} a_1b_2c_3 + b_1c_2a_3 + c_1a_2b_3 \\ - a_3b_2c_1 - b_3c_2a_1 - c_3b_1a_2 \end{matrix}$$

by expanding across the third row.

83. Show that

$$\begin{vmatrix} a^2 & a & 1 \\ b^2 & b & 1 \\ c^2 & c & 1 \end{vmatrix} = (a - b)(a - c)(b - c)$$

84. For the system of equations

$$x + 3y + 2z = 0$$
$$x + ay + 4z = 0$$
$$2y + az = 0$$

find the value(s) of a that guarantees no unique solution.

■ **TECHNOLOGY**

In Exercises 85–88, apply a graphing utility to evaluate the determinants.

85. $\begin{vmatrix} 1 & 1 & -5 \\ 3 & -7 & -4 \\ 4 & -6 & 9 \end{vmatrix}$ Compare with your answer to Exercise 35.

86. $\begin{vmatrix} -3 & 2 & -5 \\ 1 & 8 & 2 \\ 4 & -6 & 9 \end{vmatrix}$ Compare with your answer to Exercise 36.

87. $\begin{vmatrix} -3 & 2 & -1 & 3 \\ 4 & 1 & 5 & 2 \\ 17 & 2 & 2 & 8 \\ 13 & -4 & 10 & -11 \end{vmatrix}$

88. $\begin{vmatrix} -3 & 21 & 19 & 3 \\ 4 & 1 & 16 & 2 \\ 17 & 31 & 2 & 5 \\ 13 & -4 & 10 & 2 \end{vmatrix}$

In Exercises 89 and 90, apply Cramer's rule to solve each system of equations and a graphing utility to evaluate the determinants.

89. $3.1x + 1.6y - 4.8z = -33.76$
$5.2x - 3.4y + 0.5z = -36.68$
$0.5x - 6.4y + 11.4z = 25.96$

90. $-9.2x + 2.7y + 5.1z = -89.20$
$4.3x - 6.9y - 7.6z = 38.89$
$2.8x - 3.9y - 3.5z = 34.08$

■ **PREVIEW TO CALCULUS**

In calculus, determinants are used when evaluating double and triple integrals through a change of variables. In these cases, the elements of the determinant are functions.

In Exercises 91–94, find each determinant.

91. $\begin{vmatrix} \cos\theta & -r\sin\theta \\ \sin\theta & r\cos\theta \end{vmatrix}$

92. $\begin{vmatrix} 2x & 2y \\ 2x & 2y - 2 \end{vmatrix}$

93. $\begin{vmatrix} \sin\phi\cos\theta & -\rho\sin\phi\sin\theta & \rho\cos\phi\cos\theta \\ \sin\phi\sin\theta & \rho\sin\phi\cos\theta & \rho\cos\phi\sin\theta \\ \cos\phi & 0 & -\rho\sin\phi \end{vmatrix}$

94. $\begin{vmatrix} \cos\theta & -r\sin\theta & 0 \\ \sin\theta & r\cos\theta & 0 \\ 0 & 0 & 1 \end{vmatrix}$

SKILLS OBJECTIVES

- Decompose rational expressions into sums of partial fractions when the denominators contain:
  - Distinct linear factors
  - Repeated linear factors
  - Distinct irreducible quadratic factors
  - Repeated irreducible quadratic factors

CONCEPTUAL OBJECTIVE

- Understand the connection between partial-fraction decomposition and systems of linear equations.

In Chapter 2, we studied polynomial functions, and in Section 2.6, we discussed ratios of polynomial functions, called rational functions. Rational expressions are of the form

$$\frac{n(x)}{d(x)} \qquad d(x) \neq 0$$

where the numerator $n(x)$ and the denominator $d(x)$ are polynomials. Examples of rational expressions are

$$\frac{4x - 1}{2x + 3} \qquad \frac{2x + 5}{x^2 - 1} \qquad \frac{3x^4 - 2x + 5}{x^2 + 2x + 4}$$

Suppose we are asked to add two rational expressions: $\dfrac{2}{x + 1} + \dfrac{5}{x - 3}$.

We already possess the skills to accomplish this. We first identify the least common denominator $(x + 1)(x - 3)$ and combine the fractions into a single expression.

$$\frac{2}{x + 1} + \frac{5}{x - 3} = \frac{2(x - 3) + 5(x + 1)}{(x + 1)(x - 3)} = \frac{2x - 6 + 5x + 5}{(x + 1)(x - 3)} = \frac{7x - 1}{x^2 - 2x - 3}$$

How do we do this in reverse? For example, how do we start with $\dfrac{7x - 1}{x^2 - 2x - 3}$ and write this expression as a sum of two simpler expressions?

$$\underbrace{\frac{7x - 1}{x^2 - 2x - 3}}_{} = \overbrace{\underbrace{\frac{2}{x + 1}}_{\text{Partial Fraction}} + \underbrace{\frac{5}{x - 3}}_{\text{Partial Fraction}}}^{\text{Partial-Fraction Decomposition}}$$

Each of the two expressions on the right is called a **partial fraction**. The sum of these fractions is called the **partial-fraction decomposition** of $\dfrac{7x - 1}{x^2 - 2x - 3}$.

Partial-fraction decomposition is an important tool in calculus. Calculus operations such as differentiation and integration are often made simpler if you apply partial fractions. The reason partial fractions were not discussed until now is because partial-fraction decomposition *requires the ability to solve systems of linear equations*. Since partial-fraction decomposition is made possible by the techniques of solving systems of linear equations, we consider partial fractions an important application of systems of linear equations.

As mentioned earlier, a rational expression is the ratio of two polynomial expressions $n(x)/d(x)$ and we assume that $n(x)$ and $d(x)$ are polynomials with no common factors other than 1. If the degree of $n(x)$ is less than the degree of $d(x)$, then the rational expression

$n(x)/d(x)$ is said to be **proper**. If the degree of $n(x)$ is greater than or equal to the degree of $d(x)$, the rational expression is said to be **improper**. If the rational expression is improper, it should first be divided using long division.

$$\frac{n(x)}{d(x)} = Q(x) + \frac{r(x)}{d(x)}$$

The result is the sum of a quotient $Q(x)$ and a rational expression, which is the ratio of the remainder $r(x)$ and the divisor $d(x)$. The rational expression $r(x)/d(x)$ is proper, and the techniques outlined in this section can be applied to its partial-fraction decomposition.

Partial-fraction decomposition of proper rational expressions always begins with factoring the denominator $d(x)$. The goal is to write $d(x)$ as a product of distinct linear factors, but that may not always be possible. Sometimes $d(x)$ can be factored into a product of linear factors, where one or more are repeated. And, sometimes the factored form of $d(x)$ contains irreducible quadratic factors, such as $x^2 + 1$. There are times when the irreducible quadratic factors are repeated, such as $(x^2 + 1)^2$. A procedure is now outlined for partial-fraction decomposition.

PARTIAL-FRACTION DECOMPOSITION

To write a rational expression $\dfrac{n(x)}{d(x)}$ as a sum of partial fractions:

Step 1: Determine whether the rational expression is proper or improper.

- Proper: degree of $n(x) <$ degree of $d(x)$
- Improper: degree of $n(x) \geq$ degree of $d(x)$

Step 2: If proper, proceed to Step 3.

If improper, divide $\dfrac{n(x)}{d(x)}$ using polynomial (long) division and write

the result as $\dfrac{n(x)}{d(x)} = Q(x) + \dfrac{r(x)}{d(x)}$ and proceed to Step 3 with $\dfrac{r(x)}{d(x)}$.

Step 3: Factor $d(x)$. One of four possible cases will arise:

Case 1 Distinct (nonrepeated) *linear* factors: $(ax + b)$

Example: $d(x) = (3x - 1)(x + 2)$

Case 2 One or more repeated linear factors: $(ax + b)^m$ $m \geq 2$

Example: $d(x) = (x + 5)^2(x - 3)$

Case 3 One or more distinct irreducible $\left(ax^2 + bx + c = 0 \text{ has no real roots}\right)$ quadratic factors: $\left(ax^2 + bx + c\right)$

Example: $d(x) = \left(x^2 + 4\right)(x + 1)(x - 2)$

Case 4 One or more repeated irreducible quadratic factors:
$\left(ax^2 + bx + c\right)^m$

Example: $d(x) = \left(x^2 + x + 1\right)^2(x + 1)(x - 2)$

Step 4: Decompose the rational expression into a sum of partial fractions according to the procedure outlined in each case in this section.

Step 4 depends on which cases, or types of factors, arise. It is important to note that these four cases are not exclusive and combinations of different types of factors will appear.

Distinct Linear Factors

> ### CASE 1: $d(x)$ HAS ONLY DISTINCT (NONREPEATED) LINEAR FACTORS
>
> If $d(x)$ is a polynomial of degree p, and it can be factored into p linear factors
>
> $$d(x) = \underbrace{(ax + b)(cx + d)\ldots}_{p \text{ linear factors}}$$
>
> where no two factors are the same, then the partial-fraction decomposition of $\dfrac{n(x)}{d(x)}$ can be written as
>
> $$\frac{n(x)}{d(x)} = \frac{A}{(ax + b)} + \frac{B}{(cx + d)} + \cdots$$
>
> where the numerators, A, B, and so on are constants to be determined.

The goal is to write a proper rational expression as the sum of proper rational expressions. Therefore, if the denominator is a linear factor (degree 1), then the numerator is a constant (degree 0).

Technology Tip

Use a TI to check the graph of $Y_1 = \dfrac{5x + 13}{x^2 + 4x - 5}$ and its partial-fraction decomposition $Y_2 = \dfrac{3}{x - 1} + \dfrac{2}{x + 5}$. The graphs and tables of values are shown.

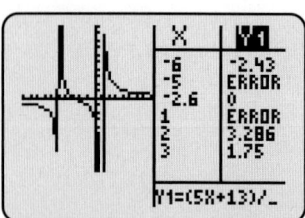

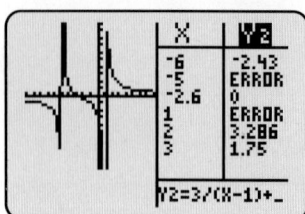

■ Answer:
$$\frac{4x - 13}{x^2 - 3x - 10} = \frac{3}{x + 2} + \frac{1}{x - 5}$$

EXAMPLE 1 Partial-Fraction Decomposition with Distinct Linear Factors

Find the partial-fraction decomposition of $\dfrac{5x + 13}{x^2 + 4x - 5}$.

Solution:

Factor the denominator.
$$\frac{5x + 13}{(x - 1)(x + 5)}$$

Express as a sum of two partial fractions.
$$\frac{5x + 13}{(x - 1)(x + 5)} = \frac{A}{(x - 1)} + \frac{B}{(x + 5)}$$

Multiply the two sides of the equation by the LCD $(x - 1)(x + 5)$.
$$5x + 13 = A(x + 5) + B(x - 1)$$

Eliminate the parentheses.
$$5x + 13 = Ax + 5A + Bx - B$$

Group the x's and constants on the right.
$$5x + 13 = (A + B)x + (5A - B)$$

Identify like terms.
$$5x + 13 = (A + B)x + (5A - B)$$

Equate the coefficients of x.
$$5 = A + B$$

Equate the constant terms.
$$13 = 5A - B$$

Solve the system of two linear equations using any method to solve for A and B.
$$A = 3, B = 2$$

Substitute $A = 3$, $B = 2$ into the partial-fraction decomposition.
$$\boxed{\frac{5x + 13}{(x - 1)(x + 5)} = \frac{3}{(x - 1)} + \frac{2}{(x + 5)}}$$

Check by adding the partial fractions.
$$\frac{3}{(x - 1)} + \frac{2}{(x + 5)} = \frac{3(x + 5) + 2(x - 1)}{(x - 1)(x + 5)} = \frac{5x + 13}{x^2 + 4x - 5}$$

■ YOUR TURN Find the partial-fraction decomposition of $\dfrac{4x - 13}{x^2 - 3x - 10}$.

In Example 1, we started with a rational expression that had a numerator of degree 1 and a denominator of degree 2. Partial-fraction decomposition enabled us to write that rational expression as a sum of two rational expressions with degree 0 numerators and degree 1 denominators.

Repeated Linear Factors

CASE 2: $d(x)$ HAS AT LEAST ONE REPEATED LINEAR FACTOR

If $d(x)$ can be factored into a product of linear factors, then the partial-fraction decomposition will proceed as in Case 1, with the exception of a repeated factor $(ax + b)^m$, $m \geq 2$. Any linear factor repeated m times will result in the sum of m partial fractions

$$\frac{A}{(ax + b)} + \frac{B}{(ax + b)^2} + \frac{C}{(ax + b)^3} + \cdots + \frac{M}{(ax + b)^m}$$

where the numerators, $A, B, C, \ldots, M$ are constants to be determined.

Note that if $d(x)$ is of degree p, the general form of the decomposition will have p partial fractions. If some numerator constants turn out to be zero, then the final decomposition may have fewer than p partial fractions.

EXAMPLE 2 **Partial-Fraction Decomposition with a Repeated Linear Factor**

Find the partial-fraction decomposition of $\dfrac{-3x^2 + 13x - 12}{x^3 - 4x^2 + 4x}$.

Solution:

Factor the denominator.

$$\frac{-3x^2 + 13x - 12}{x(x - 2)^2}$$

Express as a sum of three partial fractions.

$$\frac{-3x^2 + 13x - 12}{x(x - 2)^2} = \frac{A}{x} + \frac{B}{(x - 2)} + \frac{C}{(x - 2)^2}$$

Multiply both sides by the LCD $x(x - 2)^2$.

$$-3x^2 + 13x - 12 = A(x - 2)^2 + Bx(x - 2) + Cx$$

Eliminate the parentheses.

$$-3x^2 + 13x - 12 = Ax^2 - 4Ax + 4A + Bx^2 - 2Bx + Cx$$

Group like terms on the right.

$$-3x^2 + 13x - 12 = (A + B)x^2 + (-4A - 2B + C)x + 4A$$

Identify like terms on both sides.

$$-3x^2 + 13x - 12 = (A + B)x^2 + (-4A - 2B + C)x + 4A$$

Equate the coefficients of x^2. $-3 = A + B$ (1)

Equate the coefficients of x. $13 = -4A - 2B + C$ (2)

Equate the constant terms. $-12 = 4A$ (3)

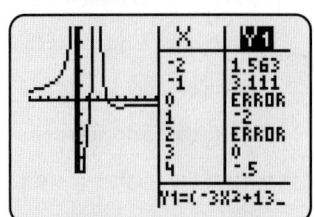
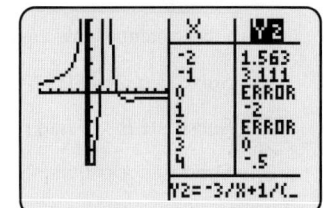

Solve the system of three equations for A, B, and C.

Solve (3) for A. $\hspace{3cm} A = -3$

Substitute $A = -3$ into (1). $\hspace{2cm} B = 0$

Substitute $A = -3$ and $B = 0$ into (2). $\hspace{0.5cm} C = 1$

Substitute $A = -3$, $B = 0$, $C = 1$ into the partial-fraction decomposition.

$$\frac{-3x^2 + 13x - 12}{x(x - 2)^2} = \frac{-3}{x} + \frac{0}{(x - 2)} + \frac{1}{(x - 2)^2}$$

$$\boxed{\frac{-3x^2 + 13x - 12}{x^3 - 4x^2 + 4x} = \frac{-3}{x} + \frac{1}{(x - 2)^2}}$$

Check by adding the partial fractions.

$$\frac{-3}{x} + \frac{1}{(x - 2)^2} = \frac{-3(x - 2)^2 + 1(x)}{x(x - 2)^2} = \frac{-3x^2 + 13x - 12}{x^3 - 4x^2 + 4x}$$

■ **Answer:**
$$\frac{x^2 + 1}{x^3 + 2x^2 + x} = \frac{1}{x} - \frac{2}{(x + 1)^2}$$

■ **YOUR TURN** Find the partial-fraction decomposition of $\dfrac{x^2 + 1}{x^3 + 2x^2 + x}$.

EXAMPLE 3 Partial-Fraction Decomposition with Multiple Repeated Linear Factors

Find the partial-fraction decomposition of $\dfrac{2x^3 + 6x^2 + 6x + 9}{x^4 + 6x^3 + 9x^2}$.

Solution:

Factor the denominator.
$$\frac{2x^3 + 6x^2 + 6x + 9}{x^2(x + 3)^2}$$

Express as a sum of four partial fractions.
$$\frac{2x^3 + 6x^2 + 6x + 9}{x^2(x + 3)^2} = \frac{A}{x} + \frac{B}{x^2} + \frac{C}{(x + 3)} + \frac{D}{(x + 3)^2}$$

Multiply both sides by the LCD $x^2(x + 3)^2$.
$$2x^3 + 6x^2 + 6x + 9 = Ax(x + 3)^2 + B(x + 3)^2 + Cx^2(x + 3) + Dx^2$$

Eliminate the parentheses.
$$2x^3 + 6x^2 + 6x + 9 = Ax^3 + 6Ax^2 + 9Ax + Bx^2 + 6Bx + 9B + Cx^3 + 3Cx^2 + Dx^2$$

Group like terms on the right.
$$2x^3 + 6x^2 + 6x + 9 = (A + C)x^3 + (6A + B + 3C + D)x^2 + (9A + 6B)x + 9B$$

Identify like terms on both sides.
$$2x^3 + 6x^2 + 6x + 9 = (A + C)x^3 + (6A + B + 3C + D)x^2 + (9A + 6B)x + 9B$$

Equate the coefficients of x^3. $\hspace{2cm} 2 = A + C \hspace{3cm}$ (1)

Equate the coefficients of x^2. $\hspace{2cm} 6 = 6A + B + 3C + D \hspace{1.5cm}$ (2)

Equate the coefficients of x. $\hspace{2.2cm} 6 = 9A + 6B \hspace{3cm}$ (3)

Equate the constant terms. $\hspace{2.5cm} 9 = 9B \hspace{3.5cm}$ (4)

Solve the system of four equations for A, B, C, and D.

Solve Equation (4) for B. $\hspace{4cm} B = 1$

Substitute $B = 1$ into Equation (3) and solve for A. $\hspace{1cm} A = 0$

Substitute $A = 0$ into Equation (1) and solve for C. $\hspace{1cm} C = 2$

Substitute $A = 0$, $B = 1$, and $C = 2$ into Equation (2) and solve for D. $\hspace{0.5cm} D = -1$

Substitute $A = 0$, $B = 1$, $C = 2$, $D = -1$ into the partial-fraction decomposition.

$$\frac{2x^3 + 6x^2 + 6x + 9}{x^2(x + 3)^2} = \frac{0}{x} + \frac{1}{x^2} + \frac{2}{(x + 3)} + \frac{-1}{(x + 3)^2}$$

$$\boxed{\frac{2x^3 + 6x^2 + 6x + 9}{x^2(x + 3)^2} = \frac{1}{x^2} + \frac{2}{(x + 3)} - \frac{1}{(x + 3)^2}}$$

Check by adding the partial fractions.

$$\frac{1}{x^2} + \frac{2}{(x + 3)} - \frac{1}{(x + 3)^2} = \frac{(x + 3)^2 + 2x^2(x + 3) - 1(x^2)}{x^2(x + 3)^2}$$

$$= \frac{2x^3 + 6x^2 + 6x + 9}{x^4 + 6x^3 + 9x^2}$$

■ **YOUR TURN** Find the partial-fraction decomposition of $\dfrac{2x^3 + 2x + 1}{x^4 + 2x^3 + x^2}$.

■ **Answer:** $\dfrac{2x^3 + 2x + 1}{x^4 + 2x^3 + x^2}$

$$= \frac{1}{x^2} + \frac{2}{(x + 1)} - \frac{3}{(x + 1)^2}$$

Distinct Irreducible Quadratic Factors

There will be times when a polynomial cannot be factored into a product of linear factors with real coefficients. For example, $x^2 + 4$, $x^2 + x + 1$, and $9x^2 + 3x + 2$ are all examples of *irreducible quadratic* expressions. The general form of an **irreducible quadratic factor** is given by

$$ax^2 + bx + c \qquad \text{where } ax^2 + bx + c = 0 \text{ has no real roots}$$

CASE 3: $d(x)$ HAS A DISTINCT IRREDUCIBLE QUADRATIC FACTOR

If the factored form of $d(x)$ contains an irreducible quadratic factor $ax^2 + bx + c$, then the partial-fraction decomposition will contain a term of the form

$$\frac{Ax + B}{ax^2 + bx + c}$$

where A and B are constants to be determined.

Recall that for a proper rational expression, the degree of the numerator is less than the degree of the denominator. For irreducible quadratic (degree 2) denominators we assume a linear (degree 1) numerator. For example,

$$\frac{7x^2 + 2}{(2x + 1)(x^2 + 1)} = \underbrace{\frac{A}{(2x + 1)}}_{\substack{\text{Constant numerator} \\ \text{Linear factor}}} + \underbrace{\frac{Bx + C}{(x^2 + 1)}}_{\substack{\text{Linear numerator} \\ \text{Quadratic factor}}}$$

Study Tip

In a partial-fraction decomposition, the degree of the numerator is always 1 less than the degree of the denominator.

A constant is used in the numerator when the denominator consists of a linear expression and a linear expression is used in the numerator when the denominator consists of a quadratic expression.

Use a TI to check the graph of

$$Y_1 = \frac{7x^2 + 2}{(2x + 1)(x^2 + 1)}$$ and its

partial-fraction decomposition

$$Y_2 = \frac{3}{2x + 1} + \frac{2x - 1}{x^2 + 1}.$$ The graphs

and tables of values are shown.

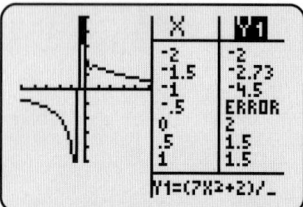

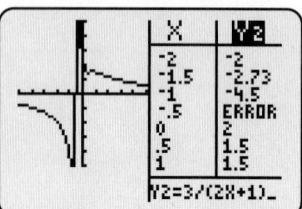

EXAMPLE 4 **Partial-Fraction Decomposition with an Irreducible Quadratic Factor**

Find the partial-fraction decomposition of $\dfrac{7x^2 + 2}{(2x + 1)(x^2 + 1)}$.

Solution:

The denominator is already in factored form. $\dfrac{7x^2 + 2}{(2x + 1)(x^2 + 1)}$

Express as a sum of two partial fractions. $\dfrac{7x^2 + 2}{(2x + 1)(x^2 + 1)} = \dfrac{A}{(2x + 1)} + \dfrac{Bx + C}{(x^2 + 1)}$

Multiply both sides by the LCD $(2x + 1)(x^2 + 1)$. $7x^2 + 2 = A(x^2 + 1) + (Bx + C)(2x + 1)$

Eliminate the parentheses. $7x^2 + 2 = Ax^2 + A + 2Bx^2 + Bx + 2Cx + C$

Group like terms on the right. $7x^2 + 2 = (A + 2B)x^2 + (B + 2C)x + (A + C)$

Identify like terms on both sides. $7x^2 + 0x + 2 = (A + 2B)x^2 + (B + 2C)x + (A + C)$

Equate the coefficients of x^2. $7 = A + 2B$

Equate the coefficients of x. $0 = B + 2C$

Equate the constant terms. $2 = A + C$

Solve the system of three equations for A, B, and C. $A = 3, B = 2, C = -1$

Substitute $A = 3$, $B = 2$, $C = -1$ into the partial-fraction decomposition.

$$\boxed{\dfrac{7x^2 + 2}{(2x + 1)(x^2 + 1)} = \dfrac{3}{(2x + 1)} + \dfrac{2x - 1}{(x^2 + 1)}}$$

Check by adding the partial fractions.

$$\dfrac{3}{(2x + 1)} + \dfrac{2x - 1}{(x^2 + 1)} = \dfrac{3(x^2 + 1) + (2x - 1)(2x + 1)}{(2x + 1)(x^2 + 1)} = \dfrac{7x^2 + 2}{(2x + 1)(x^2 + 1)}$$

■ **Answer:** $\dfrac{-2x^2 + x + 6}{(x - 1)(x^2 + 4)}$

$= \dfrac{1}{x - 1} - \dfrac{3x + 2}{x^2 + 4}$

■ **YOUR TURN** Find the partial-fraction decomposition of $\dfrac{-2x^2 + x + 6}{(x - 1)(x^2 + 4)}$.

Repeated Irreducible Quadratic Factors

CASE 4: *d*(*x*) HAS A REPEATED IRREDUCIBLE QUADRATIC FACTOR

If the factored form of $d(x)$ contains an irreducible quadratic factor $\left(ax^2 + bx + c\right)^m$, where $b^2 - 4ac < 0$, then the partial-fraction decomposition will contain a series of terms of the form

$$\dfrac{A_1x + B_1}{ax^2 + bx + c} + \dfrac{A_2x + B_2}{\left(ax^2 + bx + c\right)^2} + \dfrac{A_3x + B_3}{\left(ax^2 + bx + c\right)^3} + \cdots + \dfrac{A_mx + B_m}{\left(ax^2 + bx + c\right)^m}$$

where A_i and B_i with $i = 1, 2, \ldots, m$, are constants to be determined.

EXAMPLE 5 **Partial-Fraction Decomposition with a Repeated Irreducible Quadratic Factor**

Find the partial-fraction decomposition of $\dfrac{x^3 - x^2 + 3x + 2}{\left(x^2 + 1\right)^2}$.

Solution:

The denominator is already in factored form.

$$\dfrac{x^3 - x^2 + 3x + 2}{\left(x^2 + 1\right)^2}$$

Express as a sum of two partial fractions.

$$\dfrac{x^3 - x^2 + 3x + 2}{\left(x^2 + 1\right)^2} = \dfrac{Ax + B}{x^2 + 1} + \dfrac{Cx + D}{\left(x^2 + 1\right)^2}$$

Multiply both sides by the LCD $\left(x^2 + 1\right)^2$.

$$x^3 - x^2 + 3x + 2 = (Ax + B)\left(x^2 + 1\right) + Cx + D$$

Eliminate the parentheses.

$$x^3 - x^2 + 3x + 2 = Ax^3 + Bx^2 + Ax + B + Cx + D$$

Group like terms on the right.

$$x^3 - x^2 + 3x + 2 = Ax^3 + Bx^2 + (A + C)x + (B + D)$$

Identify like terms on both sides.

$$x^3 - x^2 + 3x + 2 = Ax^3 + Bx^2 + (A + C)x + (B + D)$$

Equate the coefficients of x^3.

$$1 = A \tag{1}$$

Equate the coefficients of x^2.

$$-1 = B \tag{2}$$

Equate the coefficients of x.

$$3 = A + C \tag{3}$$

Equate the **constant** terms.

$$2 = B + D \tag{4}$$

Substitute $A = 1$ into Equation (3) and solve for C.

$$C = 2$$

Substitute $B = -1$ into Equation (4) and solve for D.

$$D = 3$$

Substitute $A = 1$, $B = -1$, $C = 2$, $D = 3$ into the partial-fraction decomposition.

$$\boxed{\dfrac{x^3 - x^2 + 3x + 2}{\left(x^2 + 1\right)^2} = \dfrac{x - 1}{x^2 + 1} + \dfrac{2x + 3}{\left(x^2 + 1\right)^2}}$$

Check by adding the partial fractions.

$$\dfrac{x - 1}{x^2 + 1} + \dfrac{2x + 3}{\left(x^2 + 1\right)^2} = \dfrac{(x - 1)\left(x^2 + 1\right) + (2x + 3)}{\left(x^2 + 1\right)^2} = \dfrac{x^3 - x^2 + 3x + 2}{\left(x^2 + 1\right)^2}$$

■ **YOUR TURN** Find the partial-fraction decomposition of $\dfrac{3x^3 + x^2 + 4x - 1}{\left(x^2 + 4\right)^2}$.

■ **Answer:** $\dfrac{3x^3 + x^2 + 4x - 1}{\left(x^2 + 4\right)^2}$

$$= \dfrac{3x + 1}{x^2 + 4} - \dfrac{8x + 5}{\left(x^2 + 4\right)^2}$$

Combinations of All Four Cases

As you probably can imagine, there are rational expressions that have combinations of all four cases, which can lead to a system of several equations when solving for the unknown constants in the numerators of the partial fractions.

EXAMPLE 6   Partial-Fraction Decomposition

Find the partial-fraction decomposition of $\dfrac{x^5 + x^4 + 4x^3 - 3x^2 + 4x - 8}{x^2(x^2 + 2)^2}$.

Solution:

The denominator is already in factored form.

$$\frac{x^5 + x^4 + 4x^3 - 3x^2 + 4x - 8}{x^2(x^2 + 2)^2}$$

Express as a sum of partial fractions.

There are repeated linear and irreducible quadratic factors.

$$\frac{x^5 + x^4 + 4x^3 - 3x^2 + 4x - 8}{x^2(x^2 + 2)^2} = \frac{A}{x} + \frac{B}{x^2} + \frac{Cx + D}{(x^2 + 2)} + \frac{Ex + F}{(x^2 + 2)^2}$$

Multiply both sides by the LCD $x^2(x^2 + 2)^2$.

$$\begin{aligned} x^5 + x^4 &+ 4x^3 - 3x^2 + 4x - 8 \\ &= Ax(x^2 + 2)^2 + B(x^2 + 2)^2 + (Cx + D)x^2(x^2 + 2) + (Ex + F)x^2 \end{aligned}$$

Eliminate the parentheses.

$$\begin{aligned} x^5 + x^4 &+ 4x^3 - 3x^2 + 4x - 8 \\ &= Ax^5 + 4Ax^3 + 4Ax + Bx^4 + 4Bx^2 + 4B + Cx^5 + 2Cx^3 + Dx^4 + 2Dx^2 + Ex^3 + Fx^2 \end{aligned}$$

Group like terms on the right.

$$\begin{aligned} x^5 + x^4 &+ 4x^3 - 3x^2 + 4x - 8 \\ &= (A + C)x^5 + (B + D)x^4 + (4A + 2C + E)x^3 + (4B + 2D + F)x^2 + 4Ax + 4B \end{aligned}$$

Equating the coefficients of like terms leads to six equations.

$$A + C = 1$$
$$B + D = 1$$
$$4A + 2C + E = 4$$
$$4B + 2D + F = -3$$
$$4A = 4$$
$$4B = -8$$

Solve this system of equations.

$$A = 1, \quad B = -2, \quad C = 0, \quad D = 3, \quad E = 0, \quad F = -1$$

Substitute $A = 1, B = -2, C = 0, D = 3, E = 0, F = -1$ into the partial-fraction decomposition.

$$\frac{x^5 + x^4 + 4x^3 - 3x^2 + 4x - 8}{x^2(x^2 + 2)^2} = \frac{1}{x} + \frac{-2}{x^2} + \frac{0x + 3}{(x^2 + 2)} + \frac{0x + -1}{(x^2 + 2)^2}$$

$$\boxed{\frac{x^5 + x^4 + 4x^3 - 3x^2 + 4x - 8}{x^2(x^2 + 2)^2} = \frac{1}{x} - \frac{2}{x^2} + \frac{3}{(x^2 + 2)} - \frac{1}{(x^2 + 2)^2}}$$

Check by adding the partial fractions.

SECTION 8.6 SUMMARY

A rational expression $\dfrac{n(x)}{d(x)}$ is

- **Proper:** If the degree of the numerator is less than the degree of the denominator.

- **Improper:** If the degree of the numerator is equal to or greater than the degree of the denominator.

Partial-Fraction Decomposition of Proper Rational Expressions

1. Distinct (nonrepeated) linear factors

Example: $\dfrac{3x - 10}{(x - 5)(x + 4)} = \dfrac{A}{x - 5} + \dfrac{B}{x + 4}$

2. Repeated linear factors

Example: $\dfrac{2x + 5}{(x - 3)^2(x + 1)} = \dfrac{A}{x - 3} + \dfrac{B}{(x - 3)^2} + \dfrac{C}{x + 1}$

3. Distinct irreducible quadratic factors

Example: $\dfrac{1 - x}{(x^2 + 1)(x^2 + 8)} = \dfrac{Ax + B}{x^2 + 1} + \dfrac{Cx + D}{x^2 + 8}$

4. Repeated irreducible quadratic factors

Example: $\dfrac{4x^2 - 3x + 2}{(x^2 + 1)^2} = \dfrac{Ax + B}{x^2 + 1} + \dfrac{Cx + D}{(x^2 + 1)^2}$

SECTION 8.6 EXERCISES

SKILLS

In Exercises 1–6, match the rational expression (1–6) with the form of the partial-fraction decomposition (a–f).

1. $\dfrac{3x + 2}{x(x^2 - 25)}$ **2.** $\dfrac{3x + 2}{x(x^2 + 25)}$ **3.** $\dfrac{3x + 2}{x^2(x^2 + 25)}$ **4.** $\dfrac{3x + 2}{x^2(x^2 - 25)}$ **5.** $\dfrac{3x + 2}{x(x^2 + 25)^2}$ **6.** $\dfrac{3x + 2}{x^2(x^2 + 25)^2}$

a. $\dfrac{A}{x} + \dfrac{B}{x^2} + \dfrac{Cx + D}{x^2 + 25}$

b. $\dfrac{A}{x} + \dfrac{Bx + C}{x^2 + 25} + \dfrac{Dx + E}{(x^2 + 25)^2}$

c. $\dfrac{A}{x} + \dfrac{Bx + C}{x^2 + 25}$

d. $\dfrac{A}{x} + \dfrac{B}{x + 5} + \dfrac{C}{x - 5}$

e. $\dfrac{A}{x} + \dfrac{B}{x^2} + \dfrac{Cx + D}{x^2 + 25} + \dfrac{Ex + F}{(x^2 + 25)^2}$

f. $\dfrac{A}{x} + \dfrac{B}{x^2} + \dfrac{C}{x + 5} + \dfrac{D}{x - 5}$

In Exercises 7–14, write the form of the partial-fraction decomposition. Do not solve for the constants.

7. $\dfrac{9}{x^2 - x - 20}$

8. $\dfrac{8}{x^2 - 3x - 10}$

9. $\dfrac{2x + 5}{x^3 - 4x^2}$

10. $\dfrac{x^2 + 2x - 1}{x^4 - 9x^2}$

11. $\dfrac{2x^3 - 4x^2 + 7x + 3}{(x^2 + x + 5)}$

12. $\dfrac{2x^3 + 5x^2 + 6}{(x^2 - 3x + 7)}$

13. $\dfrac{3x^3 - x + 9}{(x^2 + 10)^2}$

14. $\dfrac{5x^3 + 2x^2 + 4}{(x^2 + 13)^2}$

In Exercises 15–40, find the partial-fraction decomposition for each rational function.

15. $\dfrac{1}{x(x + 1)}$

16. $\dfrac{1}{x(x - 1)}$

17. $\dfrac{x}{x(x - 1)}$

18. $\dfrac{x}{x(x + 1)}$

19. $\dfrac{9x - 11}{(x - 3)(x + 5)}$

20. $\dfrac{8x - 13}{(x - 2)(x + 1)}$

21. $\dfrac{3x + 1}{(x - 1)^2}$

22. $\dfrac{9y - 2}{(y - 1)^2}$

23. $\dfrac{4x - 3}{x^2 + 6x + 9}$

24. $\dfrac{3x + 1}{x^2 + 4x + 4}$

25. $\dfrac{4x^2 - 32x + 72}{(x + 1)(x - 5)^2}$

26. $\dfrac{4x^2 - 7x - 3}{(x + 2)(x - 1)^2}$

27. $\dfrac{5x^2 + 28x - 6}{(x + 4)(x^2 + 3)}$

28. $\dfrac{x^2 + 5x + 4}{(x - 2)(x^2 + 2)}$

29. $\dfrac{-2x^2 - 17x + 11}{(x - 7)(3x^2 - 7x + 5)}$

30. $\dfrac{14x^2 + 8x + 40}{(x + 5)(2x^2 - 3x + 5)}$

31. $\dfrac{x^3}{(x^2 + 9)^2}$

32. $\dfrac{x^2}{(x^2 + 9)^2}$

33. $\dfrac{2x^3 - 3x^2 + 7x - 2}{(x^2 + 1)^2}$

34. $\dfrac{-x^3 + 2x^2 - 3x + 15}{(x^2 + 8)^2}$

35. $\dfrac{3x + 1}{x^4 - 1}$

36. $\dfrac{2 - x}{x^4 - 81}$

37. $\dfrac{5x^2 + 9x - 8}{(x - 1)(x^2 + 2x - 1)}$

38. $\dfrac{10x^2 - 5x + 29}{(x - 3)(x^2 + 4x + 5)}$

39. $\dfrac{3x}{x^3 - 1}$

40. $\dfrac{5x + 2}{x^3 - 8}$

■ APPLICATIONS

41. Optics. The relationship between the distance of an object to a lens d_o, the distance to the image d_i, and the focal length f of the lens is given by

$$\frac{f(d_i + d_o)}{d_i d_o} - 1$$

Use partial-fraction decomposition to write the lens law in terms of sums of fractions. What does each term represent?

42. Sums. Find the partial-fraction decomposition of $\dfrac{1}{n(n + 1)}$, and apply it to find the sum of

$$\frac{1}{1 \cdot 2} + \frac{1}{2 \cdot 3} + \frac{1}{3 \cdot 4} + \cdots + \frac{1}{999 \cdot 1000}$$

In Exercises 43 and 44, refer to the following:

Laplace transforms are used to solve differential equations. The Laplace transform of $f(t)$ is denoted by $L\{f(t)\}$; thus, $L\{e^{3t}\}$ is the Laplace transform of $f(t) = e^{3t}$. It is known that

$$L\{e^{kt}\} = \frac{1}{s - k} \text{ and } L\{e^{-kt}\} = \frac{1}{s + k}. \text{ Then the inverse Laplace}$$

transform of $g(s) = \dfrac{1}{s - k}$ is $L^{-1}\left\{\dfrac{1}{s - k}\right\} = e^{kt}$. Inverse Laplace transforms are linear:

$$L^{-1}\{f(t) + g(t)\} = L^{-1}\{f(t)\} + L^{-1}\{g(t)\}$$

43. Laplace Transform. Use partial fractions to find the inverse Laplace transform of $\dfrac{9 + s}{4 - s^2}$.

44. Laplace Transform. Use partial fractions to find the inverse Laplace transform of $\dfrac{2s^2 + 3s - 2}{s(s + 1)(s - 2)}$.

■ CATCH THE MISTAKE

In Exercises 45 and 46, explain the mistake that is made.

45. Find the partial-fraction decomposition of $\dfrac{3x^2 + 3x + 1}{x(x^2 + 1)}$.

Solution:

Write the partial-fraction decomposition form.
$$\frac{3x^2 + 3x + 1}{x(x^2 + 1)} = \frac{A}{x} + \frac{B}{x^2 + 1}$$

Multiply both sides by the LCD $x(x^2 + 1)$.
$$3x^2 + 3x + 1 = A(x^2 + 1) + Bx$$

Eliminate the parentheses.
$$3x^2 + 3x + 1 = Ax^2 + Bx + A$$

Matching like terms leads to three equations.
$$A = 3, B = 3, \text{ and } A = 1$$

This is incorrect. What mistake was made?

46. Find the partial-fraction decomposition of $\dfrac{3x^4 - x - 1}{x(x - 1)}$.

Solution:

Write the partial-fraction decomposition form.
$$\frac{3x^4 - x - 1}{x(x - 1)} = \frac{A}{x} + \frac{B}{x - 1}$$

Multiply both sides by the LCD $x(x - 1)$.
$$3x^4 - x - 1 = A(x - 1) + Bx$$

Eliminate the parentheses and group like terms.
$$3x^4 - x - 1 = (A + B)x - A$$

Compare like coefficients.
$$A = 1, B = -2$$

This is incorrect. What mistake was made?

▪ CONCEPTUAL

In Exercises 47–52, determine whether each statement is true or false.

47. Partial-fraction decomposition can be employed only when the degree of the numerator is greater than the degree of the denominator.

48. The degree of the denominator of a proper rational expression is equal to the number of partial fractions in its decomposition.

49. Partial-fraction decomposition depends on the factors of the denominator.

50. A rational function can always be decomposed into partial fractions with linear or irreducible quadratic factors in each denominator.

51. The partial-fraction decomposition of a rational function $\dfrac{f(x)}{(x-a)^n}$ has the form $\dfrac{A_1}{(x-a)} + \dfrac{A_2}{(x-a)^2} + \cdots + \dfrac{A_n}{(x-a)^n}$, where all the numbers A_i are nonzero.

52. The rational function $\dfrac{1}{x^3 + 1}$ cannot be decomposed into partial fractions.

▪ CHALLENGE

For Exercises 53–58, find the partial-fraction decomposition.

53. $\dfrac{x^2 + 4x - 8}{x^3 - x^2 - 4x + 4}$

54. $\dfrac{ax + b}{x^2 - c^2}$ a, b, c are real numbers.

55. $\dfrac{2x^3 + x^2 - x - 1}{x^4 + x^3}$

56. $\dfrac{-x^3 + 2x - 2}{x^5 - x^4}$

57. $\dfrac{x^5 + 2}{\left(x^2 + 1\right)^3}$

58. $\dfrac{x^2 - 4}{\left(x^2 + 1\right)^3}$

▪ TECHNOLOGY

59. Apply a graphing utility to graph $y_1 = \dfrac{5x + 4}{x^2 + x - 2}$ and $y_2 = \dfrac{3}{x - 1} + \dfrac{2}{x + 2}$ in the same viewing rectangle. Is y_2 the partial-fraction decomposition of y_1?

60. Apply a graphing utility to graph $y_1 = \dfrac{2x^2 + 2x - 5}{x^3 + 5x}$ and $y_2 = \dfrac{3x + 2}{x^2 + 5} - \dfrac{1}{x}$ in the same viewing rectangle. Is y_2 the partial-fraction decomposition of y_1?

61. Apply a graphing utility to graph $y_1 = \dfrac{x^9 + 8x - 1}{x^5\left(x^2 + 1\right)^3}$ and $y_2 = \dfrac{4}{x} - \dfrac{1}{x^5} + \dfrac{2}{x^2 + 1} - \dfrac{3x + 2}{\left(x^2 + 1\right)^2}$ in the same viewing rectangle. Is y_2 the partial-fraction decomposition of y_1?

62. Apply a graphing utility to graph $y_1 = \dfrac{x^3 + 2x + 6}{(x + 3)\left(x^2 - 4\right)^3}$ and $y_2 = \dfrac{2}{x + 3} + \dfrac{x + 3}{\left(x^2 - 4\right)^3}$ in the same viewing rectangle. Is y_2 the partial-fraction decomposition of y_1?

63. Apply a graphing utility to graph $y_1 = \dfrac{2x^3 - 8x + 16}{(x - 2)^2\left(x^2 + 4\right)}$ and $y_2 = \dfrac{1}{x - 2} + \dfrac{2}{(x - 2)^2} + \dfrac{x + 4}{x^2 + 4}$ in the same viewing rectangle. Is y_2 the partial-fraction decomposition of y_1?

64. Apply a graphing utility to graph $y_1 = \dfrac{3x^3 + 14x^2 + 6x + 51}{\left(x^2 + 3x - 4\right)\left(x^2 + 2x + 5\right)}$ and $y_2 = \dfrac{2}{x - 1} - \dfrac{1}{x + 4} + \dfrac{2x - 3}{x^2 + 2x + 5}$ in the same viewing rectangle. Is y_2 the partial-fraction decomposition of y_1?

▪ PREVIEW TO CALCULUS

In calculus, partial fractions are used to calculate the sums of infinite series. In Exercises 65–68, find the partial-fraction decomposition of the summand.

65. $\displaystyle\sum_{k=1}^{\infty} \dfrac{9}{k(k + 3)}$

66. $\displaystyle\sum_{k=1}^{\infty} \dfrac{1}{k(k + 1)}$

67. $\displaystyle\sum_{k=1}^{\infty} \dfrac{2k + 1}{k^2(k + 1)^2}$

68. $\displaystyle\sum_{k=1}^{\infty} \dfrac{4}{k(k + 1)(k + 2)}$

SKILLS OBJECTIVES

- Graph a linear inequality in two variables.
- Graph a system of linear inequalities in two variables.
- Solve an optimization problem using linear programming.

CONCEPTUAL OBJECTIVES

- Interpret the difference between solid and dashed lines.
- Interpret an overlapped shaded region as a solution.

Linear Inequalities in Two Variables

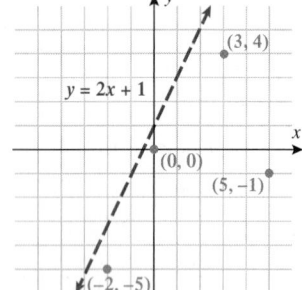

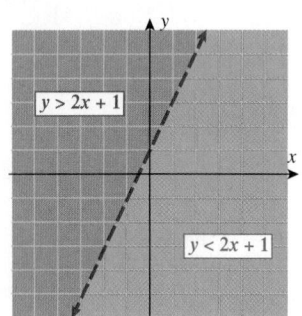

Recall in Section 0.6 that $y = 2x + 1$ is an *equation in two variables* whose graph is a line in the xy-plane. We now turn our attention to **linear inequalities in two variables**. For example, if we change the $=$ in $y = 2x + 1$ to $<$, we get $y < 2x + 1$. The solution to this inequality in two variables is the set of all points (x, y) that make this inequality true. Some solutions to this inequality are $(-2, -5)$, $(0, 0)$, $(3, 4)$, $(5, -1)$, ...

In fact, the entire region *below* the line $y = 2x + 1$ satisfies the inequality $y < 2x + 1$. If we reverse the sign of the inequality to get $y > 2x + 1$, then the entire region *above* the line $y = 2x + 1$ represents the solution to the inequality.

Any line divides the xy-plane into two **half-planes**. For example, the line $y = 2x + 1$ divides the xy-plane into two half-planes represented as $y > 2x + 1$ and $y < 2x + 1$. Recall that with inequalities in one variable we used the notation of parentheses and brackets to denote the type of inequality (strict or nonstrict). We use a similar notation with linear inequalities in two variables. If the inequality is a strict inequality, $<$ or $>$, then the line is *dashed*, and, if the inequality includes the equal sign, $\le$ or $\ge$, then a *solid* line is used. The following box summarizes the procedure for graphing a linear inequality in two variables.

Study Tip

A dashed line means that the points that lie on the line are not included in the solution of the linear inequality.

GRAPHING A LINEAR INEQUALITY IN TWO VARIABLES

Step 1: Change the inequality sign, $<$, $\le$, $\ge$, or $>$, to an equal sign, $=$.

Step 2: Draw the line that corresponds to the resulting equation in Step 1.
- If the inequality is strict, $<$ or $>$, use a **dashed** line.
- If the inequality is not strict, $\le$ or $\ge$, use a **solid** line.

Step 3: Test a point.
- Select a point in one half-plane and test to see whether it satisfies the inequality. If it does, then so do all the points in that region (half-plane). If not, then none of the points in that half-plane satisfy the inequality.
- Repeat this step for the other half-plane.

Step 4: Shade the half-plane that satisfies the inequality.

EXAMPLE 1   Graphing a Strict Linear Inequality in Two Variables

Graph the inequality $3x + y < 2$.

Solution:

STEP 1   Change the inequality sign to an
equal sign.

$$3x + y = 2$$

STEP 2   Draw the line.

Convert from standard form to
slope–intercept form.

$$y = -3x + 2$$

Since the inequality $<$ is a strict
inequality, use a **dashed** line.

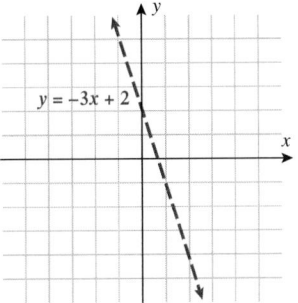

STEP 3   Test points in each half-plane.

Substitute $(3, 0)$ into $3x + y < 2$. $3(3) + 0 < 2$

The point $(3, 0)$ does not satisfy the inequality. $9 < 2$

Substitute $(-2, 0)$ into $3x + y < 2$. $3(-2) + 0 < 2$

The point $(-2, 0)$ does satisfy the inequality. $-6 < 2$

STEP 4   Shade the region containing
the point $(-2, 0)$.

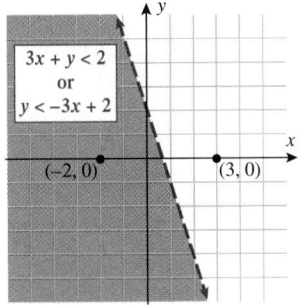

▪ **YOUR TURN**   Graph the inequality $-x + y > -1$.

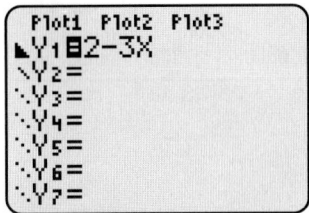
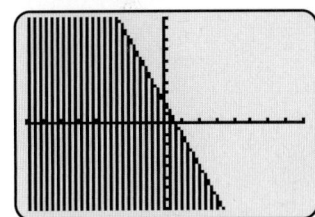
▪ **Answer:**

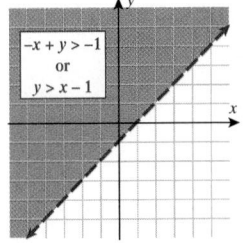

Technology Tip

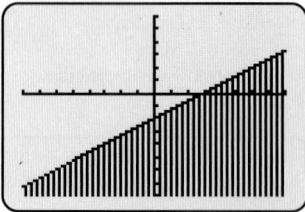

EXAMPLE 2 Graphing a Nonstrict Linear Inequality in Two Variables

Graph the inequality $2x - 3y \geq 6$.

Solution:

STEP 1 Change the inequality sign to an equal sign. $\qquad 2x - 3y = 6$

STEP 2 Draw the line.

Convert from standard form to slope–intercept form. $\qquad y = \dfrac{2}{3}x - 2$

Since the inequality $\geq$ is not a strict inequality, use a **solid** line.

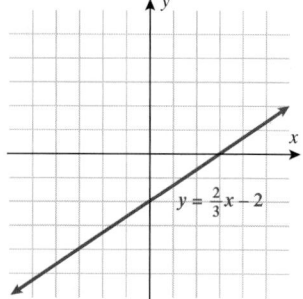

STEP 3 Test points in each half-plane.

Substitute $(5, 0)$ into $2x - 3y \geq 6$. $\qquad 2(5) - 3(0) \geq 6$

The point $(5, 0)$ satisfies the inequality. $\qquad 10 \geq 6$

Substitute $(0, 0)$ into $2x - 3y \geq 6$. $\qquad 2(0) - 3(0) \geq 6$

$\qquad\qquad\qquad\qquad\qquad\qquad\qquad\qquad\qquad 0 \geq 6$

The point $(0, 0)$ does not satisfy the inequality.

STEP 4 Shade the region containing the point $(5, 0)$.

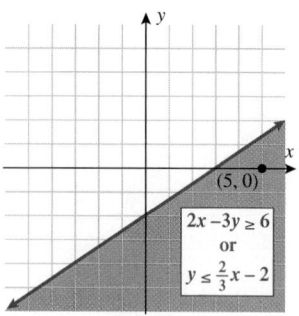

■ **Answer:**

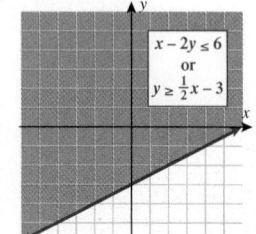

■ **YOUR TURN** Graph the inequality $x - 2y \leq 6$.

Systems of Linear Inequalities in Two Variables

Systems of linear inequalities are similar to *systems of linear equations.* In systems of linear equations we sought the points that satisfied *all* of the equations. The **solution set of a system of inequalities** contains the points that satisfy *all* of the inequalities. The graph of a system of inequalities can be obtained by simultaneously graphing each individual inequality and finding where the shaded regions intersect (or overlap), if at all.

EXAMPLE 3   Solving a System of Two Linear Inequalities

Graph the system of inequalities: $x + y \geq -2$
$x + y \leq 2$

Solution:

STEP 1 Change the inequality signs to equal signs. $x + y = -2$
$x + y = 2$

STEP 2 Draw the two lines.

Because the inequality signs are
not strict, use solid lines.

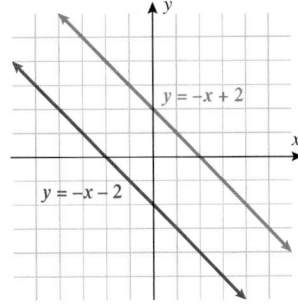

STEP 3 Test points for each inequality.

$x + y \geq -2$

Substitute $(-4, 0)$ into $x + y \geq -2$. $-4 \geq -2$
The point $(-4, 0)$ does not satisfy
the inequality.
Substitute $(0, 0)$ into $x + y \geq -2$. $0 \geq -2$
The point $(0, 0)$ does satisfy the inequality.

$x + y \leq 2$

Substitute $(0, 0)$ into $x + y \leq 2$. $0 \leq 2$
The point $(0, 0)$ does satisfy the inequality.
Substitute $(4, 0)$ into $x + y \leq 2$. $4 \leq 2$
The point $(4, 0)$ does not satisfy the inequality.

STEP 4 For $x + y \geq -2$, shade the region   For $x + y \leq 2$, shade the region
above that includes $(0, 0)$.   *below* that includes $(0, 0)$.

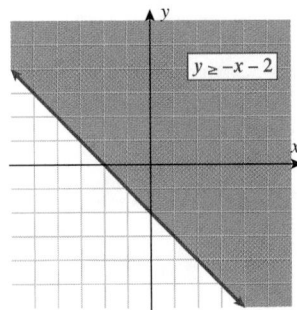

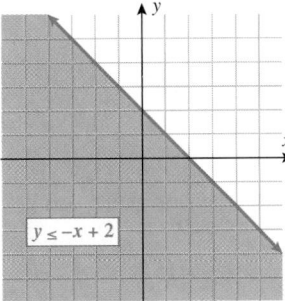

STEP 5 All of the points in the overlapping
region and on the lines constitute
the solution.

Notice that three sample points $(0, 0)$,
$(-1, 1)$, and $(1, -1)$ all lie in the
shaded region and all three satisfy
both inequalities.

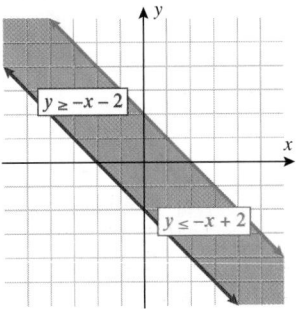

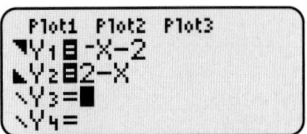
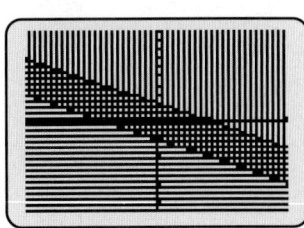

Technology Tip

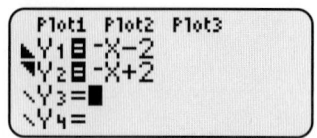

Solve for y in each inequality first. Enter $y_1 \leq -x - 2$ and $y_2 \geq -x + 2$.

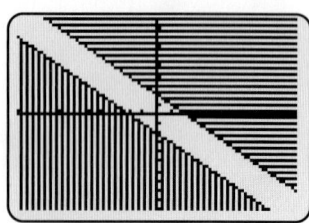

There is no overlapping region. Therefore, there is no solution to the system of inequalities.

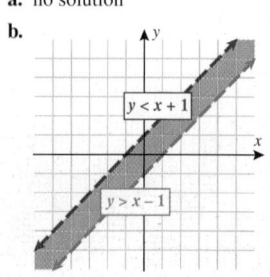

EXAMPLE 4   Solving a System of Two Linear Inequalities with No Solution

Graph the system of inequalities: $x + y \leq -2$
                                       $x + y \geq 2$

Solution:

STEP 1   Change the inequality signs to equal signs. $x + y = -2$
                                                             $x + y = 2$

STEP 2   Draw the two lines.

Because the inequality signs are not strict, use solid lines.

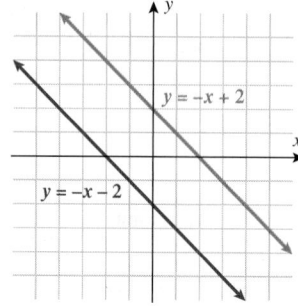

STEP 3   Test points for each inequality.

$x + y \leq -2$

Substitute $(-4, 0)$ into $x + y \leq -2$. $-4 \leq -2$

The point $(-4, 0)$ does satisfy the inequality.

Substitute $(0, 0)$ into $x + y \leq -2$. $0 \leq -2$

The point $(0, 0)$ does not satisfy the inequality.

$x + y \geq 2$

Substitute $(0, 0)$ into $x + y \geq 2$. $0 \geq 2$

The point $(0, 0)$ does not satisfy the inequality.

Substitute $(4, 0)$ into $x + y \geq 2$. $4 \geq 2$

The point $(4, 0)$ does satisfy the inequality.

STEP 4   For $x + y \leq -2$, shade the region *below* that includes $(-4, 0)$.

For $x + y \leq 2$, shade the region *above* that includes $(4, 0)$.

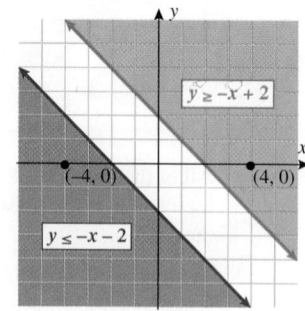

STEP 5   There is no overlapping region. Therefore, no points satisfy both inequalities. We say there is $\boxed{\text{no solution}}$.

■ **Answer:**

a. no solution

b.

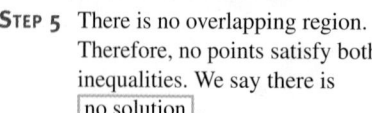

▪ **YOUR TURN**   Graph the solution to the system of inequalities.

a. $y > x + 1$ **b.** $y < x + 1$
 $y < x - 1$ $y > x - 1$

Thus far we have addressed only systems of two linear inequalities. Systems with more than two inequalities are treated in a similar manner. The solution is the set of all points that satisfy *all* of the inequalities. When there are more than two linear inequalities, the solution may be a **bounded** region. We can algebraically determine where the lines intersect by setting the *y*-values equal to each other.

EXAMPLE 5   Solving a System of Multiple Linear Inequalities

Solve the system of inequalities:
$$y \leq x$$
$$y \geq -x$$
$$y < 3$$

Solution:

STEP 1 Change the inequalities to equal signs.
$$y = x$$
$$y = -x$$
$$y = 3$$

STEP 2 Draw the three lines.

To determine the points of intersection, set the *y*-values equal.

Point where $y = x$ and $y = -x$ intersect:
$$x = -x$$
$$x = 0$$
$$(0, 0)$$

Substitute $x = 0$ into $y = x$.

Point where $y = -x$ and $y = 3$ intersect:
$$-x = 3$$
$$x = -3$$
$$(-3, 3)$$

Point where $y = 3$ and $y = x$ intersect:
$$x = 3$$
$$(3, 3)$$

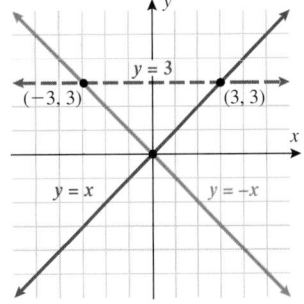

STEP 3 Test points to determine the shaded half-planes corresponding to $y \leq x$, $y \geq -x$, and $y < 3$.

STEP 4 All of the points in the overlapping region (orange) and along the boundaries of the region corresponding to the lines $y = -x$ and $y = x$ constitute the solution.

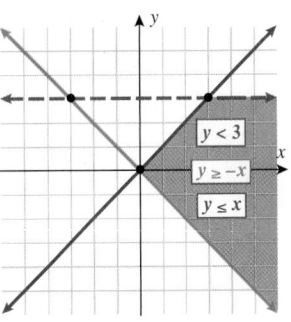

Technology Tip

Solve for *y* in each inequality first. Enter $y_1 \leq x$, $y_2 \geq -x$, and $y_3 < 3$.

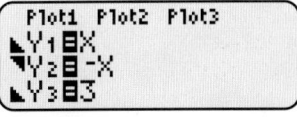

The overlapping region is the solution to the system of inequalities.

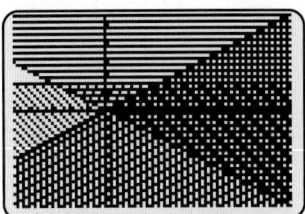

Applications

In economics, the point where the supply and demand curves intersect is called the **equilibrium point. Consumer surplus** is a measure of the amount that consumers benefit by being able to purchase a product for a price less than the maximum they would be willing to pay. **Producer surplus** is a measure of the amount that producers benefit by selling at a market price that is higher than the least they would be willing to sell for.

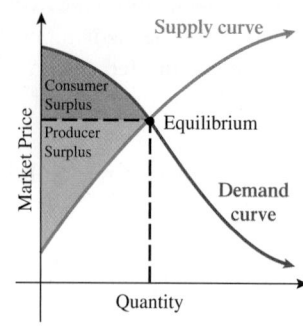

EXAMPLE 6   Consumer Surplus and Producer Surplus

The Tesla Motors Roadster is the first electric car that will be able to travel 220 miles on a single charge. The price of a 2010 model is approximately $90,000.

Courtesy Tesla Motors

Suppose the supply and demand equations for this electric car are given by

$$P = 90{,}000 - 0.1x \quad \text{(Demand)}$$
$$P = 10{,}000 + 0.3x \quad \text{(Supply)}$$

where P is the price in dollars and x is the number of cars produced. Calculate the consumer surplus and the producer surplus for these two equations.

Solution:

Find the equilibrium point.

$$90{,}000 - 0.1x = 10{,}000 + 0.3x$$
$$0.4x = 80{,}000$$
$$x = 200{,}000$$

Let $x = 200{,}000$ in either the supply or demand equation.

$$P = 90{,}000 - 0.1(200{,}000) = 70{,}000$$
$$P = 10{,}000 + 0.3(200{,}000) = 70{,}000$$

According to these models, if the price of a Tesla Motors Roadster is $70,000, then 200,000 cars will be sold and there will be no surplus.

Write the systems of linear inequalities that correspond to consumer surplus and producer surplus.

CONSUMER SURPLUS	PRODUCER SURPLUS
$P \leq 90{,}000 - 0.1x$	$P \geq 10{,}000 + 0.3x$
$P \geq 70{,}000$	$P \leq 70{,}000$
$x \geq 0$	$x \geq 0$

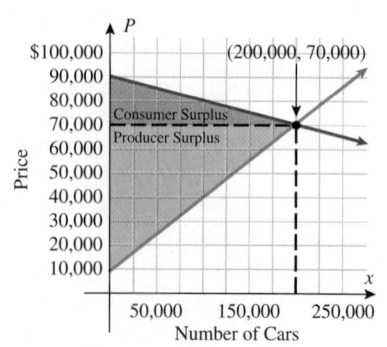

The consumer surplus is the area of the red triangle.

$$A = \frac{1}{2}bh$$
$$= \frac{1}{2}(200,000)(20,000)$$

The consumer surplus is \$2B.

$$= 2,000,000,000$$

The producer surplus is the area of the blue triangle.

$$A = \frac{1}{2}bh$$
$$= \frac{1}{2}(200,000)(60,000)$$

The producer surplus is \$6B.

$$= 6,000,000,000$$

The graph of the systems of linear inequalities in Example 6 are said to be **bounded**, whereas the graphs of the systems of linear inequalities in Examples 3–5 are said to be **unbounded**. Any points that correspond to boundary lines intersecting are called **corner points** or **vertices**. In Example 6, the vertices corresponding to the consumer surplus are the points (0, 90,000), (0, 70,000), and (200,000, 70,000), and the vertices corresponding to the producer surplus are the points (0, 70,000), (0, 10,000), and (200,000, 70,000).

The Linear Programming Model

Often we seek to maximize or minimize a function subject to constraints. This process is called **optimization**. When the function we seek to minimize or maximize is linear and the constraints are given in terms of linear inequalities, a graphing approach to such problems is called **linear programming**. In linear programming, we start with a linear equation, called the **objective function**, that represents the quantity that is to be maximized or minimized.

The goal is to minimize or maximize the objective function $z = Ax + By$ subject to *constraints*. In other words, find the points (x, y) that make the value of z the largest (or smallest). The **constraints** are a system of linear inequalities, and the common shaded region represents the **feasible (possible) solutions**.

If the constraints form a bounded region, the maximum or minimum value of the objective function will occur using the coordinates of one of the vertices. If the region is not bounded, then if an optimal solution exists, it will occur at a vertex. A procedure for solving linear programming problems is outlined below:

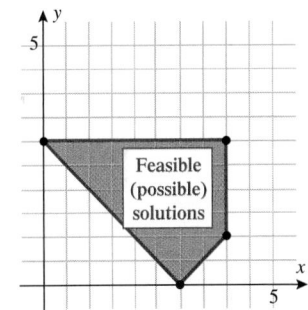

Feasible (possible) solutions

SOLVING AN OPTIMIZATION PROBLEM USING LINEAR PROGRAMMING

Step 1: Write the objective function. This expression represents the quantity that is to be minimized or maximized.

Step 2: Write the constraints. This is a system of linear inequalities.

Step 3: Graph the constraints. Graph the system of linear inequalities and shade the common region, which contains the feasible solutions.

Step 4: Identify the vertices. The corner points (vertices) of the shaded region represent possible solutions for maximizing or minimizing the objective function.

Step 5: Evaluate the objective function for each vertex. For each corner point of the shaded region, substitute the coordinates into the objective function and list the value of the objective function.

Step 6: Identify the optimal solution. The largest (maximum) or smallest (minimum) value of the objective function in Step 5 is the optimal solution.

EXAMPLE 7   Maximizing an Objective Function

Find the maximum value of $z = 2x + y$ subject to the constraints:

$$x \geq 1 \qquad x \leq 4 \qquad x + y \leq 5 \qquad y \geq 0$$

Solution:

STEP 1 Write the objective function. $z = 2x + y$

STEP 2 Write the constraints.

$$x \geq 1$$
$$x \leq 4$$
$$y \leq -x + 5$$
$$y \geq 0$$

STEP 3 Graph the constraints.

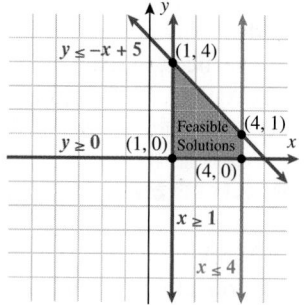

Study Tip

The bounded region is the region that satisfies *all* of the constraints. Only vertices of the bounded region correspond to possible solutions. Even though $y = -x + 5$ and $y = 0$ intersect at $x = 5$, that point of intersection is outside the shaded region and therefore is *not* one of the vertices.

STEP 4 Identify the vertices. $(1, 4), (4, 1), (1, 0), (4, 0)$

STEP 5 Evaluate the objective function for each vertex.

VERTEX	x	y	OBJECTIVE FUNCTION: $z = 2x + y$
$(1, 4)$	1	4	$2(1) + 4 = 6$
$(4, 1)$	4	1	$2(4) + 1 = 9$
$(1, 0)$	1	0	$2(1) + 0 = 2$
$(4, 0)$	4	0	$2(4) + 0 = 8$

STEP 6 The maximum value of z is **9**, subject to the given constraints when $x = 4$ and $y = 1$.

■ **Answer:** The maximum value of z is **7**, which occurs when $x = 1$ and $y = 2$.

■ **YOUR TURN** Find the maximum value of $z = x + 3y$ subject to the constraints:

$$x \geq 1 \qquad x \leq 3 \qquad y \leq -x + 3 \qquad y \geq 0$$

EXAMPLE 8   Minimizing an Objective Function

Find the minimum value of $z = 4x + 5y$ subject to the constraints:

$$x \geq 0 \qquad 2x + y \leq 6 \qquad x + y \leq 5 \qquad y \geq 0$$

Solution:

STEP 1 Write the objective function.

$z = 4x + 5y$

STEP 2 Write the constraints.

$x \geq 0$
$y \leq -2x + 6$
$y \leq -x + 5$
$y \geq 0$

STEP 3 Graph the constraints.

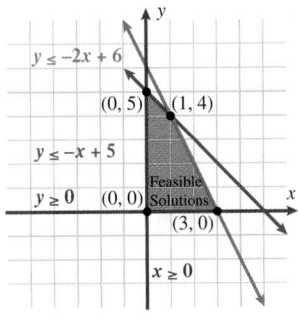

STEP 4 Identify the vertices.

$(0, 0), (0, 5), (1, 4), (3, 0)$

STEP 5 Evaluate the objective function for each vertex.

VERTEX	x	y	OBJECTIVE FUNCTION: $z = 4x + 5y$
$(0, 0)$	0	0	$4(0) + 5(0) = \mathbf{0}$
$(0, 5)$	0	5	$4(0) + 5(5) = \mathbf{25}$
$(1, 4)$	1	4	$4(1) + 5(4) = \mathbf{24}$
$(3, 0)$	3	0	$4(3) + 5(0) = \mathbf{12}$

STEP 6   The minimum value of z is **0**, which occurs when $x = 0$ and $y = 0$.

■ **YOUR TURN** Find the minimum value of $z = 2x + 3y$ subject to the constraints:

$$x \geq 1 \qquad 2x + y \leq 8 \qquad x + y \geq 4$$

Study Tip

Maxima or minima of objective functions only occur at the vertices of the shaded region corresponding to the constraints.

■ **Answer:** The minimum value of z is **8**, which occurs when $x = 4$ and $y = 0$.

EXAMPLE 9   Solving an Optimization Problem Using Linear Programming: Unbounded Region

Find the maximum value and minimum value of $z = 7x + 3y$ subject to the constraints:

$$y \geq 0 \qquad -2x + y \leq 0 \qquad -x + y \geq -4$$

Solution:

STEP 1 Write the objective function. $z = 7x + 3y$

STEP 2 Write the constraints.

$$y \geq 0$$
$$y \leq 2x$$
$$y \geq x - 4$$

STEP 3 Graph the constraints.

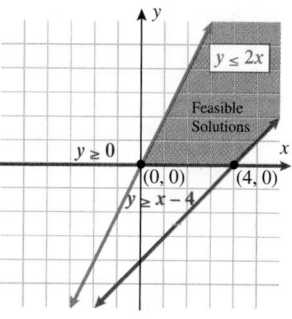

STEP 4 Identify the vertices. $(0, 0), (4, 0)$

STEP 5 Evaluate the objective function for each vertex.

VERTEX	x	y	OBJECTIVE FUNCTION: z = 7x + 3y
$(0, 0)$	0	0	$7(0) + 3(0) = \mathbf{0}$
$(4, 0)$	4	0	$7(4) + 3(0) = \mathbf{28}$

STEP 6   The minimum value of z is **0**, which occurs when $x = 0$ and $y = 0$.

 There is no maximum value , because if we select a point in the shaded region, say, $(3, 3)$, the objective function at $(3, 3)$ is equal to 30, which is greater than 28.

When the feasible solutions are contained in a bounded region, then a maximum and a minimum exist and are each located at one of the vertices. If the feasible solutions are contained in an unbounded region, then if a maximum or minimum exists, it is located at one of the vertices.

SECTION
8.7 SUMMARY

Graphing a Linear Inequality

1. Change the inequality sign to an equal sign.
2. Draw the line $y = mx + b$. (Dashed for strict inequalities and solid for nonstrict inequalities.)
3. Test a point. (Select a point in one-half plane and test the inequality. Repeat this step for the other half-plane.)
4. Shade the half-plane of the overlapping region that satisfies the linear inequality.

Graphing a System of Linear Inequalities

- Draw the individual linear inequalities.
- The overlapped shaded region, if it exists, is the solution.

Linear Programming Model

1. Write the objective function.
2. Write the constraints.
3. Graph the constraints.
4. Identify the vertices.
5. Evaluate the objective function for each vertex.
6. Identify the optimal solution.

SECTION
8.7 EXERCISES

■ SKILLS

In Exercises 1–4, match the linear inequality with the correct graph.

1. $y > x$ **2.** $y \geq x$ **3.** $y < x$ **4.** $y \leq x$

a.
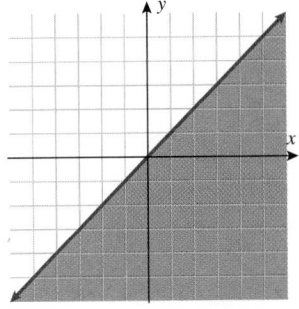

b.
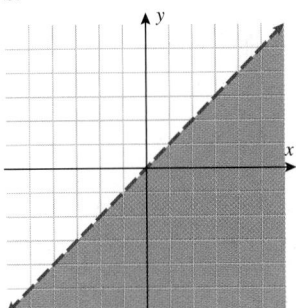

c.
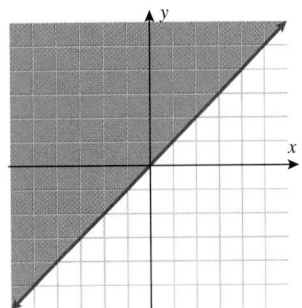

d.
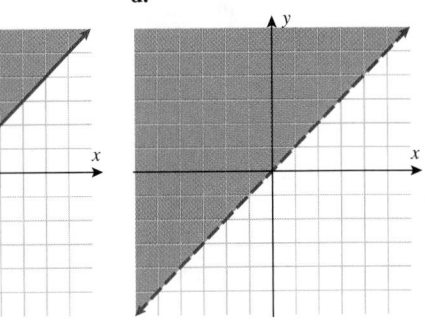

In Exercises 5–20, graph each linear inequality.

5. $y > x - 1$ **6.** $y \geq -x + 1$ **7.** $y \leq -x$ **8.** $y > -x$

9. $y \leq -3x + 2$ **10.** $y < 2x + 3$ **11.** $y \leq -2x + 1$ **12.** $y > 3x - 2$

13. $3x + 4y < 2$ **14.** $2x + 3y > -6$ **15.** $5x + 3y < 15$ **16.** $4x - 5y \leq 20$

17. $4x - 2y \geq 6$ **18.** $6x - 3y \geq 9$ **19.** $6x + 4y \leq 12$ **20.** $5x - 2y \geq 10$

In Exercises 21–50, graph each system of inequalities or indicate that the system has no solution.

21. $y \geq x - 1$
$y \leq x + 1$

22. $y > x + 1$
$y < x - 1$

23. $y > 2x + 1$
$y < 2x - 1$

24. $y \leq 2x - 1$
$y \geq 2x + 1$

25. $y \geq 2x$
$y \leq 2x$

26. $y > 2x$
$y < 2x$

27. $x > -2$
$x < 4$

28. $y < 3$
$y > 0$

29. $x \geq 2$
$y \leq x$

30. $y \leq 3$
$y \geq x$

31. $y > x$
$x < 0$
$y < 4$

32. $y \le x$
$x \ge 0$
$y \le 1$

33. $x + y > 2$
$y < 1$
$x > 0$

34. $x + y < 4$
$x > 0$
$y \ge 1$

35. $-x + y > 1$
$y < 3$
$x > 0$

36. $x - y > 2$
$y < 4$
$x \ge 0$

37. $x + 3y > 6$
$y < 1$
$x \ge 1$

38. $x + 2y > 4$
$y < 1$
$x \ge 0$

39. $y \ge \ \ x - 1$
$y \le -x + 3$
$y < \ \ x + 2$

40. $y < \ \ 4 - x$
$y > \ \ x - 4$
$y > -x - 4$

41. $\ \ x + y > -4$
$-x + y < \ \ 2$
$y \ge -1$
$y \le \ \ 1$

42. $y < \ \ x + 2$
$y > \ \ x - 2$
$y < -x + 2$
$y > -x - 2$

43. $\ \ \ \ y < x + 3$
$x + y \ge 1$
$y \ge 1$
$y \le 3$

44. $\ \ \ \ y \le -x + 2$
$y - x \ge -3$
$y \ge -2$
$y \le \ \ 1$

45. $y + x < \ \ 2$
$y + x \ge \ \ 4$
$y \ge -2$
$y \le \ \ 1$

46. $y - x < \ \ 3$
$y + x > \ \ 3$
$y \le -2$
$y \ge -4$

47. $2x - y < 2$
$2x + y > 2$
$y < 2$

48. $3x - y > \ \ 3$
$3x + y < \ \ 3$
$y < -2$

49. $x + 4y > 5$
$x - 4y < 5$
$x > 6$

50. $2x - 3y < 6$
$2x + 3y > 6$
$x < 4$

In Exercises 51–54, find the value of the objective function at each of the vertices. What is the maximum value of the objective function? What is the minimum value of the objective function?

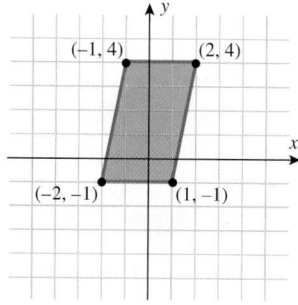

51. Objective function: $z = 2x + 3y$

52. Objective function: $z = 3x + 2y$

53. Objective function: $z = 1.5x + 4.5y$

54. Objective function: $z = \frac{2}{3}x + \frac{3}{5}y$

In Exercises 55–62, minimize or maximize each objective function subject to the constraints.

55. Minimize $z = 7x + 4y$ subject to

$x \ge 0 \qquad y \ge 0 \qquad -x + y \le 4$

56. Maximize $z = 3x + 5y$ subject to

$x \ge 0 \qquad y \ge 0 \qquad -x + y \ge 4$

57. Maximize $z = 4x + 3y$ subject to

$x \ge 0 \qquad y \le -x + 4 \qquad y \ge -x$

58. Minimize $z = 4x + 3y$ subject to

$x \ge \ \ 0 \qquad \quad y \ge 0$
$x + y \le 10 \qquad x + y \ge 0$

59. Minimize $z = 2.5x + 3.1y$ subject to

$x \ge 0 \qquad y \ge 0 \qquad x \le 4$
$-x + y \le 2 \qquad x + y \le 6$

60. Maximize $z = 2.5x - 3.1y$ subject to

$x \ge 1 \qquad y \le 7 \qquad x \le 3$
$-x + y \ge 2 \qquad x + y \ge 6$

61. Maximize $z = \frac{1}{4}x + \frac{2}{5}y$ subject to

$x + y \ge 5 \qquad x + y \le 7$
$-x + y \le 5 \qquad -x + y \ge 3$

62. Minimize $z = \frac{1}{3}x - \frac{2}{5}y$ subject to

$x + y \ge 6 \qquad x + y \le 8$
$-x + y \le 6 \qquad -x + y \ge 4$

▪ APPLICATIONS

For Exercises 63–66, employ the following supply and demand equations:

Demand: $\quad P = 80 - 0.01x$
Supply: $\quad P = 20 + 0.02x$

where P is the price in dollars when x units are produced.

63. Consumer Surplus. Write a system of linear inequalities corresponding to the consumer surplus.

64. Producer Surplus. Write a system of linear inequalities corresponding to the producer surplus.

65. Consumer Surplus. Calculate the consumer surplus given the supply and demand equations.

66. Producer Surplus. Calculate the producer surplus given the supply and demand equations.

67. Hurricanes. After back-to-back-to-back-to-back hurricanes (Charley, Frances, Ivan, and Jeanne) in Florida in the summer of 2004, FEMA sent disaster relief trucks to Florida. Floridians mainly needed drinking water and generators. Each truck could carry no more than 6000 pounds of cargo or 2400 cubic feet of cargo. Each case of bottled water takes up 1 cubic foot of space and weighs 25 pounds. Each generator takes up 20 cubic feet and weighs 150 pounds. Let x represent the number of cases of water and y represent the number of generators, and write a system of linear inequalities that describes the number of generators and cases of water each truck can haul to Florida.

68. Hurricanes. Repeat Exercise 67 with a smaller truck and different supplies. Suppose the smaller trucks that can haul 2000 pounds and 1500 cubic feet of cargo are used to haul plywood and tarps. A case of plywood is 60 cubic feet and weighs 500 pounds. A case of tarps is 10 cubic feet and weighs 50 pounds. Letting x represent the number of cases of plywood and y represent the number of cases of tarps, write a system of linear inequalities that describes the number of cases of tarps and plywood each truck can haul to Florida. Graph the system of linear inequalities.

69. Hurricanes. After the 2004 hurricanes in Florida, a student at Valencia Community College decided to create two T-shirts to sell. One T-shirt said, "I survived Charley on Friday the Thirteenth," and the second said, "I survived Charley, Frances, Ivan, and Jeanne." The Charley T-shirt costs him $7 to make and he sold it for $13. The other T-shirt cost him $5 to make and he sold it for $10. He did not want to invest more than $1,000. He estimated that the total demand would not exceed 180 T-shirts. Find the number of each type of T-shirt he needed to make to yield maximum profit.

70. Hurricanes. After Hurricane Charley devastated central Florida unexpectedly, Orlando residents prepared for Hurricane Frances by boarding up windows and filling up their cars with gas. It took 5 hours of standing in line to get plywood, and lines for gas were just as time-consuming. A student at Seminole Community College decided to do a spoof of the "Got Milk" ads and created two T-shirts: "Got Plywood" showing a line of people in a home improvement store, and "Got Gas" showing a street lined with cars waiting to pump gasoline. The "Got Plywood" shirts cost $8 to make, and she sold them for $13. The "Got Gas" shirts cost $6 to make, and she sold them for $10. She decided to limit her costs to $1,400. She estimated that demand for these T-shirts would not exceed 200 T-shirts. Find the number of each type of T-shirt she should have made to yield maximum profit.

71. Computer Business. A computer science major and a business major decide to start a small business that builds and sells desktop computers and laptop computers. They buy the parts, assemble them, load the operating system, and sell the computers to other students. The costs for parts, time to assemble each computer, and profit are summarized in the following table:

	DESKTOP	LAPTOP
Cost of parts	$700	$400
Time to assemble (hours)	5	3
Profit	$500	$300

They were able to get a small business loan in the amount of $10,000 to cover costs. They plan on making these computers over the summer and selling them the first day of class. They can dedicate at most only 90 hours to assembling these computers. They estimate that the demand for laptops will be at least three times as great as the demand for desktops. How many of each type of computer should they make to maximize profit?

72. Computer Business. Repeat Exercise 71 if the two students are able to get a loan for $30,000 to cover costs and they can dedicate at most 120 hours to assembling the computers.

73. Passenger Ratio. The Eurostar is a high-speed train that travels between London, Brussels, and Paris. There are 30 cars on each departure. Each train car is designated first-class or second-class. Based on demand for each type of fare, there should always be at least two but no more than four first-class train cars. The management wants to claim that the ratio of first-class to second-class cars never exceeds 1:8. If the profit on each first-class train car is twice as much as the profit on each second-class train car, find the number of each class of train car that will generate a maximum profit.

74. Passenger Ratio. Repeat Exercise 73. This time, assume that there has to be at least one first-class train car and that the profit from each first-class train car is 1.2 times as much as the profit from each second-class train car. The ratio of first-class to second-class cannot exceed 1:10.

75. Production. A manufacturer of skis produces two models: a regular ski and a slalom ski. A set of regular skis produces a $25.00 profit and a set of slalom skis produces a profit of $50.00. The manufacturer expects a customer demand of at least 200 pairs of regular skis and at least 80 pair of slalom skis. The maximum number of pairs of skis that can be produced by this company is 400. How many of each model of skis should be produced to maximize profits?

76. Donut Inventory. A well-known donut store makes two popular types of donuts: crème-filled and jelly-filled. The manager knows from past statistics that the number of dozens of donuts sold is at least 10, but no more than 30. To prepare the donuts for frying, the baker needs (on the average) 3 minutes for a dozen crème-filled and 2 minutes for jelly-filled. The baker has at most two hours available per day to prepare the donuts. How many dozens of each type should be prepared to maximize the daily profit if there is a $1.20 profit for each dozen crème-filled and $1.80 profit for each dozen jelly-filled donuts?

▪ CATCH THE MISTAKE

In Exercises 77 and 78, explain the mistake that is made.

77. Graph the inequality
$y \geq 2x + 1$.

Solution:

Graph the line
$y = 2x + 1$ with
a solid line.

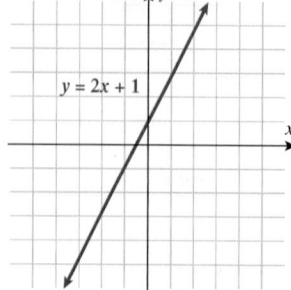

Since the inequality
is $\geq$, shade to the *right*.

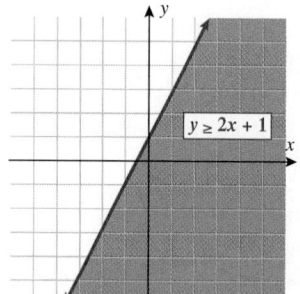

This is incorrect. What
mistake was made?

78. Graph the inequality $y < 2x + 1$.

Solution:

Graph the line
$y = 2x + 1$ with
a solid line.

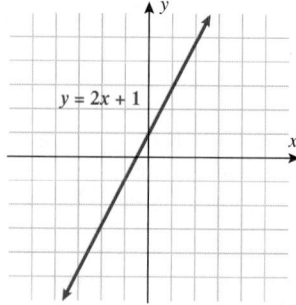

Since the inequality
is $<$, shade *below*.

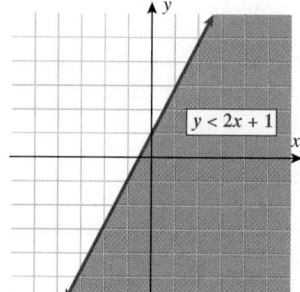

This is incorrect. What
mistake was made?

▪ CONCEPTUAL

In Exercises 79–84, determine whether each statement is true or false.

79. A linear inequality always has a solution that is a half-plane.

80. A dashed curve is used for strict inequalities.

81. A solid curve is used for strict inequalities.

82. A system of linear inequalities always has a solution.

83. An objective function always has a maximum or minimum.

84. An objective function subject to constraints that correspond to a bounded region always has a maximum and a minimum.

▪ CHALLENGE

In Exercises 85 and 86, for the system of linear inequalities, assume $a, b, c,$ and d are real numbers.

$$x \geq a$$
$$x < b$$
$$y > c$$
$$y \leq d$$

85. Describe the solution when $a < b$ and $c < d$.

86. What will the solution be if $a > b$ and $c > d$?

For Exercises 87 and 88, use the following system of linear inequalities:

$$y \leq \ ax + b$$
$$y \geq -ax + b$$

87. If a and b are positive real numbers, graph the solution.

88. If a and b are negative real numbers, graph the solution.

89. Maximize the objective function $z = 2x + y$ subject to the conditions, where $a > 2$.

$$ax + y \geq -a$$
$$-ax + y \leq \ a$$
$$ax + y \leq \ a$$
$$-ax + y \geq -a$$

90. Maximize the objective function $z = x + 2y$ subject to the conditions, where $a > b > 0$.

$$x + y \geq a$$
$$-x + y \leq a$$
$$x + y \leq a + b$$
$$-x + y \geq a - b$$

▪ TECHNOLOGY

In Exercises 91 and 92, apply a graphing utility to graph the following inequalities.

91. $4x - 2y \geq 6$ (Check with your answer to Exercise 17.)

92. $6x - 3y \geq 9$ (Check with your answer to Exercise 18.)

In Exercises 93 and 94, use a graphing utility to graph each system of inequalities or indicate that the system has no solution.

93. $-0.05x + 0.02y \geq 0.12$
$0.01x + 0.08y \leq 0.08$

94. $y \leq 2x + 3$
$y > -0.5x + 5$

▪ PREVIEW TO CALCULUS

In calculus, the first steps when solving the problem of finding the area enclosed by a set of curves are similar to those for finding the feasible region in a linear programming problem.

In Exercises 95–98, graph the system of inequalities and identify the vertices, that is, the points of intersection of the given curves.

95. $y \leq x + 2$
$y \geq x^2$

96. $x \leq 25$
$x \geq y^2$

97. $y \leq x^2$
$y \geq x^3$
$x \geq 0$

98. $y \geq x^3$
$y \leq -x$
$y \geq x + 6$

MODELING YOUR WORLD

In 2005 hybrid vehicles were introduced in the U.S. market. The demand for hybrids, which are typically powered by a combination of gasoline and electric batteries, was based on popular recognition of petroleum as an increasingly scarce nonrenewable resource, as well as consumers' need to combat rising prices at the gas pumps. In addition to achieving greater fuel economy than conventional internal combustion engine vehicles (ICEVs), their use also results in reduced emissions.

An online "Gas Mileage Impact Calculator," created by the American Council for an Energy-Efficient Economy (www.aceee.org), was used to generate the following tables comparing a conventional sedan (four-door) and an SUV versus their respective hybrid counterparts.

Gas Mileage Impact Calculator

	TOYOTA CAMRY 2.4L 4, AUTO $3.75/GALLON 15,000 MI/YEAR	TOYOTA CAMRY HYBRID 2.4L 4, AUTO $3.75/GALLON 15,000 MI/YEAR
Gas consumption	611 gallons	449 gallons
Gas cost	$2,289.75	$1,681.99
Fuel economy	25 mpg	33 mpg
EMISSIONS		
Carbon dioxide (greenhouse gas)	11,601 pounds	8,522 pounds
Carbon monoxide (poisonous gas)	235 pounds	169 pounds
Nitrogen oxides (lung irritant and smog)	10 pounds	7 pounds
Particulate matter (soot)	255 grams	255 grams
Hydrocarbons (smog)	6 pounds	8 pounds

	TOYOTA HIGHLANDER 3.5L 6, AUTO STK $3.75/GALLON 15,000 MI/YEAR	TOYOTA HIGHLANDER HYBRID 3.3L 6, AUTO AWD $3.75/GALLON 15,000 MI/YEAR
Gas consumption	740 gallons	576 gallons
Gas cost	$2,773.44	$2,158.33
Fuel economy	20 mpg	26 mpg
EMISSIONS		
Carbon dioxide (greenhouse gas)	14,052 pounds	10,936 pounds
Carbon monoxide (poisonous gas)	229 pounds	187 pounds
Nitrogen oxides (lung irritant and smog)	11 pounds	8 pounds
Particulate matter (soot)	320 grams	399 grams
Hydrocarbons (smog)	7 pounds	16 pounds

The MSRP and mileage comparisons for the 2008 models are given below:

	CAMRY	CAMRY HYBRID	HIGHLANDER	HIGHLANDER HYBRID
MSRP	$19,435	$26,065	$28,035	$34,435
Miles per gallon in city	21	33	18	27
Miles per gallon on highway	31	34	24	25

For the following questions, assume that you drive 15,000 miles per year (all in the city) and the price of gasoline is $3.75 per gallon.

1. Write a linear equation that models the total cost of owning and operating each vehicle y as a function of the number of years of ownership x.

   a. Camry
   c. Highlander
   b. Camry Hybrid
   d. Highlander Hybrid

2. Write a linear equation that models the total number of pounds of carbon dioxide each vehicle emits y as a function of the number of years of ownership x.

   a. Camry
   c. Highlander
   b. Camry Hybrid
   d. Highlander Hybrid

3. How many years would you have to own and drive the vehicle for the hybrid to be the better deal?

   a. Camry Hybrid versus Camry
   b. Highlander Hybrid versus Highlander

4. How many years would you have to own and drive the vehicle for the hybrid to emit 50% less carbon dioxide than its conventional counterpart?

   a. Camry Hybrid versus Camry
   b. Highlander Hybrid versus Highlander

SECTION	CONCEPT	PAGES	REVIEW EXERCISES	KEY IDEAS/POINTS
8.1	**Systems of linear equations in two variables**	696–707	1–22	$A_1x + B_1y = C_1$ $A_2x + B_2y = C_2$
	Substitution method	697–699	1–16	Solve for one variable in terms of the other and substitute that expression into the other equation.
	Elimination method	699–703	1–16	Eliminate a variable by adding multiples of the equations.
	Graphing method	703–706	17–20	Graph the two lines. The solution is the point of intersection. Parallel lines have no solution and identical lines have infinitely many solutions.
	Three methods and three types of solutions	706–707	1–22	One solution, no solution, infinitely many solutions
8.2	**Systems of linear equations in three variables**	713–721	23–28	Planes in a three-dimensional coordinate system
	Solving systems of linear equations in three variables	714–716	23–28	Step 1: Reduce the system to two equations and two unknowns. Step 2: Solve the resulting system from Step 1. Step 3: Substitute solutions found in Step 2 into any of the equations to find the third variable. Step 4: Check.
	Types of solutions	717–721	23–28	One solution (point), no solution, or infinitely many solutions (line or the same plane)
8.3	**Systems of linear equations and matrices**	726–742	29–56	
	Augmented matrices	728–729	29–32	$a_1x + b_1y + c_1z = d_1$ $a_2x + b_2y + c_2z = d_2 \Rightarrow$ $a_3x + b_3y + c_3z = d_3$ $\begin{bmatrix} a_1 & b_1 & c_1 & \vert & d_1 \\ a_2 & b_2 & c_2 & \vert & d_2 \\ a_3 & b_3 & c_3 & \vert & d_3 \end{bmatrix}$
	Row operations on a matrix	729–730	37–44	1. $R_i \leftrightarrow R_j$ — Interchange row i with row j. 2. $cR_i \to R_i$ — Multiply row i by the constant c. 3. $cR_i + R_j \to R_j$ — Multiply row i by the constant c and add to row j, writing the results in row j.
	Row–echelon form of a matrix	730–731	33–36	A matrix is in **row–echelon form** if it has all three of the following properties: 1. Any rows consisting entirely of 0s are at the bottom of the matrix. 2. For each row that does not consist entirely of 0s, the first (leftmost) nonzero entry is 1 (called the leading 1). 3. For two successive nonzero rows, the leading 1 in the higher row is farther to the left than the leading 1 in the lower row. If a matrix in row–echelon form has the following additional property, then the matrix is in **reduced row–echelon form**: 4. Every column containing a leading 1 has zeros in every position above and below the leading 1.

$$\begin{array}{c}
\qquad\quad \text{Column 1} \quad \text{Column 2} \quad \cdots \quad \text{Column } j \quad \cdots \quad \text{Column } n \\
\begin{array}{c}
\text{Row 1} \\ \text{Row 2} \\ \vdots \\ \text{Row } i \\ \vdots \\ \text{Row } m
\end{array}
\begin{bmatrix}
a_{11} & a_{12} & \cdots & a_{1j} & \cdots & a_{1n} \\
a_{21} & a_{22} & \cdots & a_{2j} & \cdots & a_{2n} \\
\vdots & \vdots & \cdots & \vdots & \cdots & \vdots \\
a_{i1} & a_{i2} & \cdots & a_{ij} & \cdots & a_{in} \\
\vdots & \vdots & \cdots & \vdots & \cdots & \vdots \\
a_{m1} & a_{m2} & \cdots & a_{mj} & \cdots & a_{mn}
\end{bmatrix}
\end{array}$$

CHAPTER REVIEW

Determinant of a 2 × 2 matrix — 774–775 — 93–96

$$\begin{vmatrix} a & b \\ c & d \end{vmatrix} = ad - bc$$

Determinant of an n × n matrix — 775–778 — 103–106

Let A be a square matrix of order $n \times n$. Then:

■ The **minor** M_{ij} of the element a_{ij} is the determinant of the $(n-1) \times (n-1)$ matrix obtained when the ith row and jth column of A are deleted.

■ The **cofactor** C_{ij} of the element a_{ij} is given by $C_{ij} = (-1)^{i+j}M_{ij}$.

$$\begin{bmatrix} 1 & -3 & 2 \\ 4 & -1 & 0 \\ 5 & -2 & 3 \end{bmatrix}$$

$$M_{11} = \begin{vmatrix} -1 & 0 \\ -2 & 3 \end{vmatrix} = -3 - 0 = -3$$

$$C_{11} = (-1)^{1+1}M_{11} = (1)(-3) = -3$$

Sign pattern of cofactors for the determinant of a 3 × 3 matrix:

$$\begin{bmatrix} + & - & + \\ - & + & - \\ + & - & + \end{bmatrix}$$

If A is a 3 × 3 matrix, the determinant can be given by $\det(A) = a_{11}C_{11} + a_{12}C_{12} + a_{13}C_{13}$. This is called **expanding the determinant by the first row**. (Note that any row or column can be used.)

$$\begin{vmatrix} a_1 & b_1 & c_1 \\ a_2 & b_2 & c_2 \\ a_3 & b_3 & c_3 \end{vmatrix} = a_1 \begin{vmatrix} b_2 & c_2 \\ b_3 & c_3 \end{vmatrix} - b_1 \begin{vmatrix} a_2 & c_2 \\ a_3 & c_3 \end{vmatrix} + c_1 \begin{vmatrix} a_2 & b_2 \\ a_3 & b_3 \end{vmatrix}$$

Cramer's rule: Systems of linear equations in two variables — 779–781 — 97–102

The system

$$a_1 x + b_1 y = c_1$$
$$a_2 x + b_2 y = c_2$$

has the solution

$$x = \frac{D_x}{D} \qquad y = \frac{D_y}{D} \quad \text{if } D \neq 0$$

where

$$D = \begin{vmatrix} a_1 & b_1 \\ a_2 & b_2 \end{vmatrix} \qquad D_x = \begin{vmatrix} c_1 & b_1 \\ c_2 & b_2 \end{vmatrix} \qquad D_y = \begin{vmatrix} a_1 & c_1 \\ a_2 & c_2 \end{vmatrix}$$

SECTION	CONCEPT	PAGES	REVIEW EXERCISES	KEY IDEAS/POINTS
	Cramer's rule: Systems of linear equations in three variables	781−783	107–110	The system $$a_1x + b_1y + c_1z = d_1$$ $$a_2x + b_2y + c_2z = d_2$$ $$a_3x + b_3y + c_3z = d_3$$ has the solution $$x = \frac{D_x}{D} \qquad y = \frac{D_y}{D} \qquad z = \frac{D_z}{D} \quad \text{if } D \neq 0$$ where $$D = \begin{vmatrix} a_1 & b_1 & c_1 \\ a_2 & b_2 & c_2 \\ a_3 & b_3 & c_3 \end{vmatrix} \qquad D_x = \begin{vmatrix} d_1 & b_1 & c_1 \\ d_2 & b_2 & c_2 \\ d_3 & b_3 & c_3 \end{vmatrix}$$ $$D_y = \begin{vmatrix} a_1 & d_1 & c_1 \\ a_2 & d_2 & c_2 \\ a_3 & d_3 & c_3 \end{vmatrix} \qquad D_z = \begin{vmatrix} a_1 & b_1 & d_1 \\ a_2 & b_2 & d_2 \\ a_3 & b_3 & d_3 \end{vmatrix}$$
8.6	**Partial fractions**	788−796	111–126	$\dfrac{n(x)}{d(x)} \qquad$ Factor $d(x)$
	Distinct linear factors	790−791		$$\frac{n(x)}{d(x)} = \frac{A}{(ax + b)} + \frac{B}{(cx + d)} + \cdots$$
	Repeated linear factors	791−793		$$\frac{n(x)}{d(x)} = \frac{A}{(ax + b)} + \frac{B}{(ax + b)^2} + \cdots + \frac{M}{(ax + b)^m}$$
	Distinct irreducible quadratic factors	793−794		$$\frac{n(x)}{d(x)} = \frac{Ax + B}{ax^2 + bx + c}$$
	Repeated irreducible quadratic factors	794−796		$$\frac{n(x)}{d(x)} = \frac{A_1x + B_1}{ax^2 + bx + c} + \frac{A_2x + B_2}{\left(ax^2 + bx + c\right)^2} +$$ $$\frac{A_3x + B_3}{\left(ax^2 + bx + c\right)^3} + \cdots + \frac{A_mx + B_m}{\left(ax^2 + bx + c\right)^m}$$
8.7	**Systems of linear inequalities in two variables**	800−810	127–146	
	Linear inequalities in two variables	800−802		■ $\leq$ or $\geq$ use solid lines. ■ $<$ or $>$ use dashed lines.
	Systems of linear inequalities in two variables	802−807		Solutions are determined graphically by finding the common shaded regions.
	The linear programming model	807−810		Finding optimal solutions Minimizing or maximizing a function subject to constraints (linear inequalities)

8.1 Systems of Linear Equations in Two Variables

Solve each system of linear equations.

1. $r - s = 3$
$r + s = 3$

2. $3x + 4y = 2$
$x - y = 6$

3. $-4x + 2y = 3$
$4x - y = 5$

4. $0.25x - 0.5y = 0.6$
$0.5x + 0.25y = 0.8$

5. $x + y = 3$
$x - y = 1$

6. $3x + y = 4$
$2x + y = 1$

7. $4c - 4d = 3$
$c + d = 4$

8. $5r + 2s = 1$
$r - s = -3$

9. $y = -\frac{1}{2}x$
$y = \frac{1}{2}x + 2$

10. $2x + 4y = -2$
$4x - 2y = 3$

11. $1.3x - 2.4y = 1.6$
$0.7x - 1.2y = 1.4$

12. $\frac{1}{4}x - \frac{3}{4}y = 12$
$\frac{1}{2}y + \frac{1}{4}x = \frac{1}{2}$

13. $5x - 3y = 21$
$-2x + 7y = -20$

14. $6x - 2y = -2$
$4x + 3y = 16$

15. $10x - 7y = -24$
$7x + 4y = 1$

16. $\frac{1}{3}x - \frac{2}{9}y = \frac{2}{9}$
$\frac{4}{5}x + \frac{3}{4}y = -\frac{3}{4}$

Match each system of equations with its graph.

17. $2x - 3y = 4$
$x + 4y = 3$

18. $5x - y = 2$
$5x - y = -2$

19. $x + 2y = -6$
$2x + 4y = -12$

20. $5x + 2y = 3$
$4x - 2y = 6$

a.

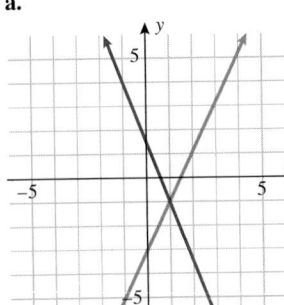

b.

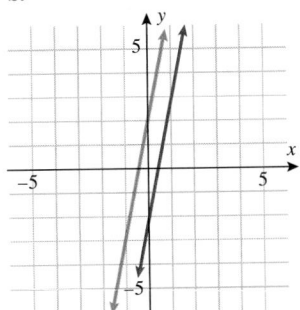

c.

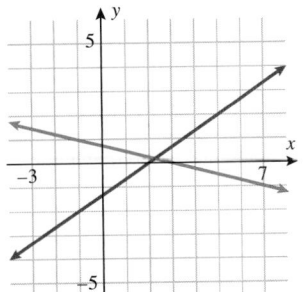

d.

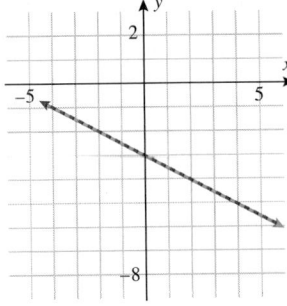

Applications

21. Chemistry. In chemistry lab, Alexandra needs to make a 42-milliliter solution that is 15% NaCl. All that is in the lab is 6% and 18% NaCl. How many milliliters of each solution should she use to obtain the desired mix?

22. Gas Mileage. A Nissan Sentra gets approximately 32 mpg on the highway and 18 mpg in the city. Suppose 265 miles were driven on a full tank (12 gallons) of gasoline. Approximately how many miles were driven in the city and how many on the highway?

8.2 Systems of Linear Equations in Three Variables

Solve each system of linear equations.

23. $x + y + z = 1$
$x - y - z = -3$
$-x + y + z = 3$

24. $x - 2y + z = 3$
$2x - y + z = -4$
$3x - 3y - 5z = 2$

25. $x + y + z = 7$
$x - y - z = 17$
$y + z = 5$

26. $x + z = 3$
$-x + y - z = -1$
$x + y + z = 5$

Applications

27. Fitting a Curve to Data. The average number of flights on a commercial plane that a person takes per year can be modeled by a quadratic function $y = ax^2 + bx + c$, where $a < 0$, and x represents age: $16 \le x \le 65$. The following table gives the average number of flights per year that a person takes on a commercial airline. Determine a quadratic function that models this quantity. (*Note:* Coefficients will be approximate.)

AGE	NUMBER OF FLIGHTS PER YEAR
16	2
40	6
65	4

28. Investment Portfolio. Danny and Paula decide to invest $20,000 of their savings. They put some in an IRA account earning 4.5% interest, some in a mutual fund that has been averaging 8% a year, and some in a stock that earned 12% last year. If they put $4,000 more in the IRA than in the mutual fund, and the mutual fund and stock have the same growth in the next year as they did in the previous year, they will earn $1,525 in a year. How much money did they put in each of the three investments?

8.3 Systems of Linear Equations and Matrices

Write the augmented matrix for each system of linear equations.

29. $5x + 7y = 2$
$3x - 4y = -2$

30. $2.3x - 4.5y = 6.8$
$-0.4x + 2.1y = -9.1$

31. $2x - z = 3$
$y - 3z = -2$
$x + 4z = -3$

32. $2y - x + 3z = 1$
$4z - 2y + 3x = -2$
$x - y - 4z = 0$

Indicate whether each matrix is in row–echelon form. If it is, state whether it is in *reduced* row–echelon form.

33. $\begin{bmatrix} 1 & 1 & | & 0 \\ 0 & 1 & | & 2 \end{bmatrix}$

34. $\begin{bmatrix} 1 & 2 & | & 0 \\ 0 & 0 & | & 1 \end{bmatrix}$

35. $\begin{bmatrix} 2 & 0 & 1 & | & 1 \\ 0 & -2 & 0 & | & 2 \\ 0 & 0 & 2 & | & 3 \end{bmatrix}$

36. $\begin{bmatrix} 1 & 0 & 1 & 0 & | & 2 \\ 0 & 0 & 1 & 1 & | & -3 \\ 0 & 1 & 0 & 0 & | & 2 \\ 0 & 0 & 0 & 1 & | & 1 \end{bmatrix}$

Perform the indicated row operations on each matrix.

37. $\begin{bmatrix} 1 & -2 & | & 1 \\ 0 & -2 & | & 2 \end{bmatrix}$ $\frac{-1}{2}R_2 \to R_2$

38. $\begin{bmatrix} 1 & 4 & | & 1 \\ 2 & -2 & | & 3 \end{bmatrix}$ $R_2 - 2R_1 \to R_2$

39. $\begin{bmatrix} 1 & -2 & 0 & | & 1 \\ 0 & -2 & 3 & | & -2 \\ 0 & 1 & -4 & | & 8 \end{bmatrix}$ $R_2 + R_1 \to R_1$

40. $\begin{bmatrix} 1 & 1 & 1 & 6 & | & 0 \\ 0 & 2 & -2 & 3 & | & -2 \\ 0 & 0 & 1 & -2 & | & 4 \\ 0 & -1 & 3 & -3 & | & 3 \end{bmatrix}$ $\begin{matrix} -2R_1 + R_2 \to R_1 \\ R_4 + R_3 \to R_4 \end{matrix}$

Apply row operations to transform each matrix to reduced row–echelon form.

41. $\begin{bmatrix} 1 & 3 & | & 0 \\ 3 & 4 & | & 1 \end{bmatrix}$

42. $\begin{bmatrix} 1 & 2 & -1 & | & 0 \\ 0 & 1 & -1 & | & -1 \\ -2 & 0 & 1 & | & -2 \end{bmatrix}$

43. $\begin{bmatrix} 4 & 1 & -2 & | & 0 \\ 1 & 0 & -1 & | & 0 \\ -2 & 1 & 1 & | & 12 \end{bmatrix}$

44. $\begin{bmatrix} 2 & 3 & 2 & | & 1 \\ 0 & -1 & 1 & | & -2 \\ 1 & 1 & -1 & | & 6 \end{bmatrix}$

Solve the system of linear equations using augmented matrices.

45. $3x - 2y = 2$
$-2x + 4y = 1$

46. $2x - 7y = 22$
$x + 5y = -23$

47. $5x - y = 9$
$x + 4y = 6$

48. $8x + 7y = 10$
$-3x + 5y = 42$

49. $x - 2y + z = 3$
$2x - y + z = -4$
$3x - 3y - 5z = 2$

50. $3x - y + 4z = 18$
$5x + 2y - z = -20$
$x + 7y - 6z = -38$

51. $x - 4y + 10z = -61$
$3x - 5y + 8z = -52$
$-5x + y - 2z = 8$

52. $4x - 2y + 5z = 17$
$x + 6y - 3z = -\frac{17}{2}$
$-2x + 5y + z = 2$

53. $3x + y + z = -4$
$x - 2y + z = -6$

54. $2x - y + 3z = 6$
$3x + 2y - z = 12$

Applications

55. Fitting a Curve to Data. The average number of flights on a commercial plane that a person takes a year can be modeled by a quadratic function $y = ax^2 + bx + c$, where $a < 0$ and x represents age: $16 < x < 65$. The table below gives the average number of flights per year that a person takes on a commercial airline. Determine a quadratic function that models this quantity by solving for a, b, and c using matrices and compare with Exercise 27. (*Note:* Coefficients will be approximate.)

AGE	NUMBER OF FLIGHTS PER YEAR
16	2
40	6
65	4

56. Investment Portfolio. Danny and Paula decide to invest $20,000 of their savings in investments. They put some in an IRA account earning 4.5% interest, some in a mutual fund that has been averaging 8% a year, and some in a stock that earned 12% last year. If they put $3,000 more in the mutual fund than in the IRA, and the mutual fund and stock have the same growth in the next year as they did in the previous year, they will earn $1,877.50 in a year. How much money did they put in each of the three investments?

8.4 Matrix Algebra

Calculate the given expression, if possible.

$A = \begin{bmatrix} 2 & -3 \\ 0 & 1 \end{bmatrix}$ $B = \begin{bmatrix} 1 & 5 & -1 \\ 3 & 7 & 2 \end{bmatrix}$ $C = \begin{bmatrix} 5 & 0 & 1 \\ 2 & -1 & 4 \\ 0 & 3 & 6 \end{bmatrix}$

$D = \begin{bmatrix} 5 & 2 \\ 9 & 7 \end{bmatrix}$ $E = \begin{bmatrix} 2 & 0 & 3 \\ 4 & 1 & -1 \end{bmatrix}$

57. $A + C$ **58.** $B + A$ **59.** $B + E$

60. $A + D$ **61.** $2A + D$ **62.** $3E + B$

63. $2D - 3A$ **64.** $3B - 4E$ **65.** $5A - 2D$

66. $5B - 4E$ **67.** AB **68.** BC

69. DA **70.** AD **71.** $BC + E$

72. DB **73.** EC **74.** CE

Determine whether B is the multiplicative inverse of A using $AA^{-1} = I$.

75. $A = \begin{bmatrix} 6 & 4 \\ 4 & 2 \end{bmatrix}$ $B = \begin{bmatrix} -0.5 & 1 \\ 1 & -1.5 \end{bmatrix}$

76. $A = \begin{bmatrix} 1 & -2 \\ 2 & -4 \end{bmatrix}$ $B = \begin{bmatrix} 1 & 2 \\ 2 & -2 \end{bmatrix}$

77. $A = \begin{bmatrix} 1 & -2 & 6 \\ 2 & 3 & -2 \\ 0 & -1 & 1 \end{bmatrix}$ $B = \begin{bmatrix} -\frac{1}{7} & \frac{4}{7} & 2 \\ \frac{2}{7} & -\frac{1}{7} & -2 \\ \frac{2}{7} & -\frac{1}{7} & -1 \end{bmatrix}$

78. $A = \begin{bmatrix} 0 & 7 & 6 \\ 1 & 0 & -4 \\ -2 & 1 & 0 \end{bmatrix}$ $B = \begin{bmatrix} 1 & 1 & 1 \\ -2 & -2 & -2 \\ 2 & 0 & 6 \end{bmatrix}$

Find A^{-1}, if it exists.

79. $A = \begin{bmatrix} 1 & 2 \\ -3 & 4 \end{bmatrix}$ **80.** $A = \begin{bmatrix} -2 & 7 \\ -4 & 6 \end{bmatrix}$

81. $A = \begin{bmatrix} 0 & 1 \\ -2 & 0 \end{bmatrix}$ **82.** $A = \begin{bmatrix} 3 & -1 \\ -2 & 2 \end{bmatrix}$

83. $A = \begin{bmatrix} 1 & 3 & -2 \\ 2 & 1 & -1 \\ 0 & 1 & -3 \end{bmatrix}$ **84.** $A = \begin{bmatrix} 0 & 1 & 0 \\ 4 & 1 & 2 \\ -3 & -2 & 1 \end{bmatrix}$

85. $A = \begin{bmatrix} -1 & 1 & 0 \\ -2 & 1 & 2 \\ 1 & 2 & 4 \end{bmatrix}$ **86.** $A = \begin{bmatrix} -4 & 4 & 3 \\ 1 & 2 & 2 \\ 3 & -1 & 6 \end{bmatrix}$

Solve the system of linear equations using matrix algebra.

87. $\begin{aligned} 3x - y &= 11 \\ 5x + 2y &= 33 \end{aligned}$ **88.** $\begin{aligned} 6x + 4y &= 15 \\ -3x - 2y &= -1 \end{aligned}$

89. $\begin{aligned} \frac{5}{8}x - \frac{2}{3}y &= -3 \\ \frac{3}{4}x + \frac{5}{6}y &= 16 \end{aligned}$ **90.** $\begin{aligned} x + y - z &= 0 \\ 2x - y + 3z &= 18 \\ 3x - 2y + z &= 17 \end{aligned}$

91. $\begin{aligned} 3x - 2y + 4z &= 11 \\ 6x + 3y - 2z &= 6 \\ x - y + 7z &= 20 \end{aligned}$ **92.** $\begin{aligned} 2x + 6y - 4z &= 11 \\ -x - 3y + 2z &= -\frac{11}{2} \\ 4x + 5y + 6z &= 20 \end{aligned}$

8.5 The Determinant of a Square Matrix and Cramer's Rule

Evaluate each 2×2 determinant.

93. $\begin{vmatrix} 2 & 4 \\ 3 & 2 \end{vmatrix}$ **94.** $\begin{vmatrix} -2 & -4 \\ -3 & 2 \end{vmatrix}$

95. $\begin{vmatrix} 2.4 & -2.3 \\ 3.6 & -1.2 \end{vmatrix}$ **96.** $\begin{vmatrix} -\frac{1}{4} & 4 \\ \frac{3}{4} & -4 \end{vmatrix}$

Employ Cramer's rule to solve each system of equations, if possible.

97. $\begin{aligned} x - y &= 2 \\ x + y &= 4 \end{aligned}$ **98.** $\begin{aligned} 3x - y &= -17 \\ -x + 5y &= 43 \end{aligned}$

99. $\begin{aligned} 2x + 4y &= 12 \\ x - 2y &= 6 \end{aligned}$ **100.** $\begin{aligned} -x + y &= 4 \\ 2x - 6y &= -5 \end{aligned}$

101. $\begin{aligned} -3x &= 40 - 2y \\ 2x &= 25 + y \end{aligned}$ **102.** $\begin{aligned} 3x &= 20 + 4y \\ y - x &= -6 \end{aligned}$

Evaluate each 3×3 determinant.

103. $\begin{vmatrix} 1 & 2 & 2 \\ 0 & 1 & 3 \\ 2 & -1 & 0 \end{vmatrix}$ **104.** $\begin{vmatrix} 0 & -2 & 1 \\ 0 & -3 & 7 \\ 1 & -10 & -3 \end{vmatrix}$

105. $\begin{vmatrix} a & 0 & -b \\ -a & b & c \\ 0 & 0 & -d \end{vmatrix}$ **106.** $\begin{vmatrix} -2 & -4 & 6 \\ 2 & 0 & 3 \\ -1 & 2 & \frac{3}{4} \end{vmatrix}$

Employ Cramer's rule to solve each system of equations, if possible.

107. $\begin{aligned} x + y - 2z &= -2 \\ 2x - y + z &= 3 \\ x + y + z &= 4 \end{aligned}$ **108.** $\begin{aligned} -x - y + z &= 3 \\ x + 2y - 2z &= 8 \\ 2x + y + 4z &= -4 \end{aligned}$

109. $\begin{aligned} 3x \quad + 4z &= -1 \\ x + y + 2z &= -3 \\ y - 4z &= -9 \end{aligned}$ **110.** $\begin{aligned} x + y + z &= 0 \\ -x - 3y + 5z &= -2 \\ 2x + y - 3z &= -4 \end{aligned}$

8.6 Partial Fractions

Write the form of each partial-fraction decomposition. Do not solve for the constants.

111. $\dfrac{4}{(x - 1)^2(x + 3)(x - 5)}$ **112.** $\dfrac{7}{(x - 9)(3x + 5)^2(x + 4)}$

113. $\dfrac{12}{x(4x + 5)(2x + 1)^2}$ **114.** $\dfrac{2}{(x + 1)(x - 5)(x - 9)^2}$

115. $\dfrac{3}{x^2 + x - 12}$ **116.** $\dfrac{x^2 + 3x - 2}{x^3 + 6x^2}$

117. $\dfrac{3x^3 + 4x^2 + 56x + 62}{(x^2 + 17)^2}$ **118.** $\dfrac{x^3 + 7x^2 + 10}{(x^2 + 13)^2}$

Find the partial-fraction decomposition for each rational function.

119. $\dfrac{9x + 23}{(x - 1)(x + 7)}$

120. $\dfrac{12x + 1}{(3x + 2)(2x - 1)}$

121. $\dfrac{13x^2 + 90x - 25}{2x^3 - 50x}$

122. $\dfrac{5x^2 + x + 24}{x^3 + 8x}$

123. $\dfrac{2}{x^2 + x}$

124. $\dfrac{x}{x(x + 3)}$

125. $\dfrac{5x - 17}{x^2 + 4x + 4}$

126. $\dfrac{x^3}{(x^2 + 64)^2}$

8.7 Systems of Linear Inequalities in Two Variables

Graph each linear inequality.

127. $y \geq -2x + 3$

128. $y < x - 4$

129. $2x + 4y > 5$

130. $5x + 2y \leq 4$

131. $y \geq -3x + 2$

132. $y < x - 2$

133. $3x + 8y \leq 16$

134. $4x - 9y \leq 18$

Graph each system of inequalities or indicate that the system has no solution.

135. $y \geq x + 2$
$y \leq x - 2$

136. $y \geq 3x$
$y \leq 3x$

137. $x \leq -2$
$y > x$

138. $x + 3y \geq 6$
$2x - y \leq 8$

139. $3x - 4y \leq 16$
$5x + 3y > 9$

140. $x + y > -4$
$x - y < 3$
$y \geq -2$
$x \leq 8$

Minimize or maximize the objective function subject to the constraints.

141. Minimize $z = 2x + y$ subject to

$x \geq 0 \quad y \geq 0 \quad x + y \leq 3$

142. Maximize $z = 2x + 3y$ subject to

$x \geq 0 \quad y \geq 0$
$-x + y \leq 0 \quad x \leq 3$

143. Minimize $z = 3x - 5y$ subject to

$2x + y > 6 \quad 2x - y < 6 \quad x > 0$

144. Maximize $z = -2x + 7y$ subject to

$3x + y < 7 \quad x - 2y > 1 \quad x \geq 0$

Applications

For Exercises 145 and 146, refer to the following:

An art student decides to hand-paint coasters and sell sets at a flea market. She decides to make two types of coaster sets: an ocean watercolor and black-and-white geometric shapes. The cost, profit, and time it takes her to paint each set are summarized in the table below.

	OCEAN WATERCOLOR	GEOMETRIC SHAPES
Cost	$4	$2
Profit	$15	$8
Hours	3	2

145. Profit. If the student's costs cannot exceed $100 and she can spend only 90 hours total painting the coasters, determine the number of each type she should make to maximize her profit.

146. Profit. If the student's costs cannot exceed $300 and she can spend only 90 hours painting, determine the number of each type she should make to maximize her profit.

Technology Exercises

Section 8.1

147. Apply a graphing utility to graph the two equations $0.4x + 0.3y = -0.1$ and $0.5x - 0.2y = 1.6$. Find the solution to this system of linear equations.

148. Apply a graphing utility to graph the two equations $\frac{1}{2}x + \frac{3}{10}y = \frac{1}{5}$ and $-\frac{5}{3}x + \frac{1}{2}y = \frac{4}{3}$. Find the solution to this system of linear equations.

Section 8.2

Employ a graphing calculator to solve the system of equations.

149. $5x - 3y + 15z = 21$
$-2x + 0.8y - 4z = -8$
$2.5x - y + 7.5z = 12$

150. $2x - 1.5y + 3z = 9.5$
$0.5x - 0.375y + 0.75z = 1.5$

Section 8.3

In Exercises 151 and 152, refer to the following:

You are asked to model a set of three points with a quadratic function $y = ax^2 + bx + c$ and determine the quadratic function.

a. Set up a system of equations; use a graphing utility or graphing calculator to solve the system by entering the coefficients of the augmented matrix.

b. Use the graphing calculator commands STAT QuadReg to model the data using a quadratic function. Round your answers to two decimal places.

151. $(-10, 12.5), (3, -2.8), (9, 8.5)$

152. $(-4, 10), (2.5, -9.5), (13.5, 12.6)$

Section 8.4

Apply a graphing utility to perform the indicated matrix operations, if possible.

$$A = \begin{bmatrix} -6 & 0 & 4 \\ 1 & 3 & 5 \\ 2 & -1 & 0 \end{bmatrix} \quad B = \begin{bmatrix} 5 & 1 \\ 0 & 2 \\ -8 & 4 \end{bmatrix} \quad C = \begin{bmatrix} 4 & -3 & 0 \\ 1 & 2 & 5 \end{bmatrix}$$

153. ABC

154. CAB

Apply a graphing utility and matrix algebra to solve the system of linear equations.

155. $\begin{aligned} 6.1x - 14.2y &= 75.495 \\ -2.3x + 7.2y &= -36.495 \end{aligned}$

156. $\begin{aligned} 7.2x + 3.2y - 1.7z &= 5.53 \\ -1.3x + 4.1y + 2.8z &= -23.949 \end{aligned}$

Section 8.5

Apply Cramer's rule to solve each system of equations and a graphing utility to evaluate the determinants.

157. $\begin{aligned} 4.5x - 8.7y &= -72.33 \\ -1.4x + 5.3y &= 31.32 \end{aligned}$

158. $\begin{aligned} 1.4x + 3.6y + 7.5z &= 42.08 \\ 2.1x - 5.7y - 4.2z &= 5.37 \\ 1.8x - 2.8y - 6.2z &= -9.86 \end{aligned}$

Section 8.6

159. Apply a graphing utility to graph $y_1 = \dfrac{x^2 + 4}{x^4 - x^2}$ and

$y_2 = -\dfrac{4}{x^2} + \dfrac{5/2}{x - 1} - \dfrac{5/2}{x + 1}$ in the same viewing rectangle. Is y_2 the partial-fraction decomposition of y_1?

160. Apply a graphing utility to graph

$y_1 = \dfrac{x^3 + 6x^2 + 27x + 38}{(x^2 + 8x + 17)(x^2 + 6x + 13)}$ and

$y_2 = \dfrac{2x + 1}{x^2 + 8x + 17} - \dfrac{x - 3}{x^2 + 6x + 13}$ in the same viewing rectangle. Is y_2 the partial-fraction decomposition of y_1?

Section 8.7

In Exercises 161 and 162, use a graphing utility to graph each system of inequalities or indicate that the system has no solution.

161. $\begin{aligned} 2x + 5y &\geq -15 \\ y &\leq -\tfrac{2}{3}x - 1 \end{aligned}$

162. $\begin{aligned} y &\leq 0.5x \\ y &> -1.5x + 6 \end{aligned}$

163. Maximize $z = 6.2x + 1.5y$ subject to

$$\begin{aligned} 4x - 3y &\leq 5.4 \\ 2x + 4.5y &\leq 6.3 \\ 3x - y &\geq -10.7 \end{aligned}$$

164. Minimize $z = 1.6x - 2.8y$ subject to

$$\begin{aligned} y &\geq 3.2x - 4.8 & x &\geq -2 \\ y &\leq 3.2x + 4.8 & x &\leq 4 \end{aligned}$$

CHAPTER 8 PRACTICE TEST

Solve each system of linear equations using elimination and/or substitution methods.

1. $\begin{aligned} x - 2y &= 1 \\ -x + 3y &= 2 \end{aligned}$

2. $\begin{aligned} 3x + 5y &= -2 \\ 7x + 11y &= -6 \end{aligned}$

3. $\begin{aligned} x - y &= 2 \\ -2x + 2y &= -4 \end{aligned}$

4. $\begin{aligned} 3x - 2y &= 5 \\ 6x - 4y &= 0 \end{aligned}$

5. $\begin{aligned} x + y + z &= -1 \\ 2x + y + z &= 0 \\ -x + y + 2z &= 0 \end{aligned}$

6. $\begin{aligned} 6x + 9y + z &= 5 \\ 2x - 3y + z &= 3 \\ 10x + 12y + 2z &= 9 \end{aligned}$

In Exercises 7 and 8, write the system of linear equations as an augmented matrix.

7. $\begin{aligned} 6x + 9y + z &= 5 \\ 2x - 3y + z &= 3 \\ 10x + 12y + 2z &= 9 \end{aligned}$

8. $\begin{aligned} 3x + 2y - 10z &= 2 \\ x + y - z &= 5 \end{aligned}$

9. Perform the following row operations.

$$\begin{bmatrix} 1 & 3 & 5 \\ 2 & 7 & -1 \\ -3 & -2 & 0 \end{bmatrix} \begin{aligned} R_2 - 2R_1 &\rightarrow R_2 \\ R_3 + 3R_1 &\rightarrow R_3 \end{aligned}$$

10. Rewrite the following matrix in reduced row–echelon form.

$$\left[\begin{array}{ccc|c} 2 & -1 & 1 & 3 \\ 1 & 1 & -1 & 0 \\ 3 & 2 & -2 & 1 \end{array}\right]$$

In Exercises 11 and 12, solve the systems of linear equations using augmented matrices.

11. $6x + 9y + z = 5$
$2x - 3y + z = 3$
$10x + 12y + 2z = 9$

12. $3x + 2y - 10z = 2$
$x + y - z = 5$

13. Multiply the matrices, if possible.

$$\begin{bmatrix} 1 & -2 & 5 \\ 0 & -1 & 3 \end{bmatrix} \begin{bmatrix} 0 & 4 \\ 3 & -5 \\ -1 & 1 \end{bmatrix}$$

14. Add the matrices, if possible.

$$\begin{bmatrix} 1 & -2 & 5 \\ 0 & -1 & 3 \end{bmatrix} + \begin{bmatrix} 0 & 4 \\ 3 & -5 \\ -1 & 1 \end{bmatrix}$$

15. Find the inverse of $\begin{bmatrix} 4 & 3 \\ 5 & -1 \end{bmatrix}$, if it exists.

16. Find the inverse of $\begin{bmatrix} 1 & -3 & 2 \\ 4 & 2 & 0 \\ -1 & 2 & 5 \end{bmatrix}$, if it exists.

17. Solve the system of linear equations with matrix algebra (inverses).

$$3x - y + 4z = 18$$
$$x + 2y + 3z = 20$$
$$-4x + 6y - z = 11$$

Calculate the determinant.

18. $\begin{vmatrix} 7 & -5 \\ 2 & -1 \end{vmatrix}$

19. $\begin{vmatrix} 1 & -2 & -1 \\ 3 & -5 & 2 \\ 4 & -1 & 0 \end{vmatrix}$

In Exercises 20 and 21, solve the system of linear equations using Cramer's rule.

20. $x - 2y = 1$
$-x + 3y = 2$

21. $3x + 5y - 2z = -6$
$7x + 11y + 3z = 2$
$x - y + z = 4$

22. A company has two rubrics for scoring job applicants based on weighting education, experience, and the interview differently.

Matrix A:

	Rubric 1	Rubric 2
Education	0.4	0.6
Experience	0.5	0.1
Interview	0.1	0.3

Applicants receive a score from 1 to 10 in each category (education, experience, and interview). Two applicants are shown in the matrix B.

Matrix B:

	Education	Experience	Interview
Applicant 1	4	7	3
Applicant 2	6	5	4

What is the order of BA? What does each entry in BA tell us?

Write each rational expression as a sum of partial fractions.

23. $\dfrac{2x + 5}{x^2 + x}$

24. $\dfrac{3x - 13}{(x - 5)^2}$

25. $\dfrac{5x - 3}{x(x^2 - 9)}$

26. $\dfrac{1}{2x^2 + 5x - 3}$

Graph the inequalities.

27. $-2x + y < 6$

28. $4x - y \geq 8$

In Exercises 29 and 30, graph the system of inequalities.

29. $x + y \leq 4$
$-x + y \geq -2$

30. $x + 3y \leq 6$
$2x - y \leq 4$

31. Minimize the function $z = 5x + 7y$ subject to the constraints
$x \geq 0 \quad y \geq 0 \quad x + y \leq 3 \quad -x + y \geq 1$

32. Find the maximum value of the objective function $z = 3x + 6y$ given the constraints

$$x \geq 0 \qquad y \geq 0$$
$$x + y \leq 6 \qquad -x + 2y \leq 4$$

33. Apply a graphing utility and matrix algebra to solve the system of linear equations.

$$5.6x - 2.7y = 87.28$$
$$-4.2x + 8.4y = -106.26$$

34. You are asked to model a set of three points with a quadratic function $y = ax^2 + bx + c$.

  a. Set up a system of equations; use a graphing utility or graphing calculator to solve the system by entering the coefficients of the augmented matrix.

  b. Use the graphing calculator commands $\boxed{\text{STAT}}$ $\boxed{\text{QuadReg}}$ to model the data using a quadratic function.

$$(-3, 6), (1, 12), (5, 7)$$

1. Evaluate $g[f(-1)]$, with $f(x) = \sqrt{2x + 11}$ and $g(x) = x^3$.

2. Use interval notation to express the domain of the function
$$G(x) = \frac{9}{\sqrt{1 - 5x}}.$$

3. Using the function $f(x) = x^2 - 3x + 2$, evaluate the difference quotient $\dfrac{f(x + h) - f(x)}{h}$.

4. Find all the real zeros (and state the multiplicity) of
$f(x) = -4x(x - 7)^2(x + 13)^3$.

5. Find the vertex of the parabola $f(x) = -0.04x^2 + 1.2x - 3$.

6. Factor the polynomial $P(x) = x^4 + 8x^2 - 9$ as a product of linear factors.

7. Find the vertical and horizontal asymptotes of the function
$$f(x) = \frac{5x - 7}{3 - x}.$$

8. Approximate e^π using a calculator. Round your answer to two decimal places.

9. Evaluate $\log_5 0.2$ exactly.

10. Solve $5^{2x-1} = 11$ for x. Round the answer to three decimal places.

11. Evaluate $\log_2 6$ using the change-of-base formula. Round your answer to three decimal places.

12. Solve $\ln(5x - 6) = 2$. Round your answer to three decimal places.

13. Give the exact value of $\cos 30°$.

14. How much money should be put in a savings account now that earns 4.7% a year compounded weekly, if you want to have $65,000 in 17 years?

15. The terminal side of angle θ in standard position passes through the point $(-5, 2)$. Calculate the exact values of the six trigonometric function for angle θ.

16. Find all values of θ, where $0° \leq \theta \leq 360°$, when
$$\cos\theta = -\frac{\sqrt{3}}{2}.$$

17. Graph the function $y = \tan\left(\frac{1}{4}x\right)$ over the interval $-2\pi \leq x \leq 2\pi$.

18. Verify the identity $\cos(3x) = \cos x(1 - 4\sin^2 x)$.

19. State the domain and range of the function $y = 5\tan\left(x - \dfrac{\pi}{2}\right)$.

20. Simplify the trigonometric expression $\dfrac{\sec^4 x - 1}{\sec^2 x + 1}$.

21. Use the half-angle identities to find the exact value of
$$\tan\left(-\frac{3\pi}{8}\right).$$

22. Write the product $7\sin(-2x)\sin(5x)$ as a sum or difference of sines and/or cosines.

23. Solve the triangle $\beta = 106.3°$, $\gamma = 37.4°$, $a = 76.1$ m.

24. Find the angle (rounded to the nearest degree) between the vectors $\langle 2, 3 \rangle$ and $\langle -4, -5 \rangle$.

25. Find all complex solutions to $x^3 - 27 = 0$.

26. Graph $\theta = -\dfrac{\pi}{4}$.

27. Given
$$A = \begin{bmatrix} 3 & 4 & -7 \\ 0 & 1 & 5 \end{bmatrix} \quad B = \begin{bmatrix} 8 & -2 & 6 \\ 9 & 0 & -1 \end{bmatrix} \quad C = \begin{bmatrix} 9 & 0 \\ 1 & 2 \end{bmatrix}$$
find CB.

28. Solve the system using Gauss–Jordan elimination.
$$\begin{aligned} x - 2y + 3z &= 11 \\ 4x + 5y - z &= -8 \\ 3x + y - 2z &= 1 \end{aligned}$$

29. Use Cramer's rule to solve the system of equations.
$$\begin{aligned} 7x + 5y &= 1 \\ -x + 4y &= -1 \end{aligned}$$

30. Write the matrix equation, find the inverse of the coefficient matrix, and solve the system using matrix algebra.
$$\begin{aligned} 2x + 5y &= -1 \\ -x + 4y &= 7 \end{aligned}$$

31. Graph the system of linear inequalities.
$$\begin{aligned} y &> -x \\ y &\geq -3 \\ x &\leq 3 \end{aligned}$$

9

Conics, Systems of Nonlinear Equations and Inequalities, and Parametric Equations

We will now study three types of conic sections or conics: the parabola, the ellipse, and the hyperbola. The trajectory of a basketball is a *parabola*, the Earth's orbit around the Sun is an *ellipse*, and the shape of a cooling tower is a *hyperbola*.

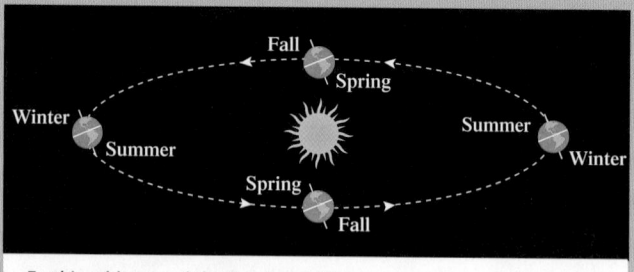

Earth's orbit around the Sun is an *ellipse*.

IN THIS CHAPTER we define the three conic sections: the parabola, the ellipse, and the hyperbola. Algebraic equations and the graphs of these conics are discussed. We solve systems of nonlinear equations and inequalities involving parabolas, ellipses, and hyperbolas. We then will determine how rotating the axes changes the equation of a conic, and with our results we will be able to identify the graph of a general second-degree equation as one of the three conics. We will discuss the equations of the conics first in rectangular coordinates and then in polar coordinates. Finally, we will look at parametric equations, which give orientation along a plane curve.

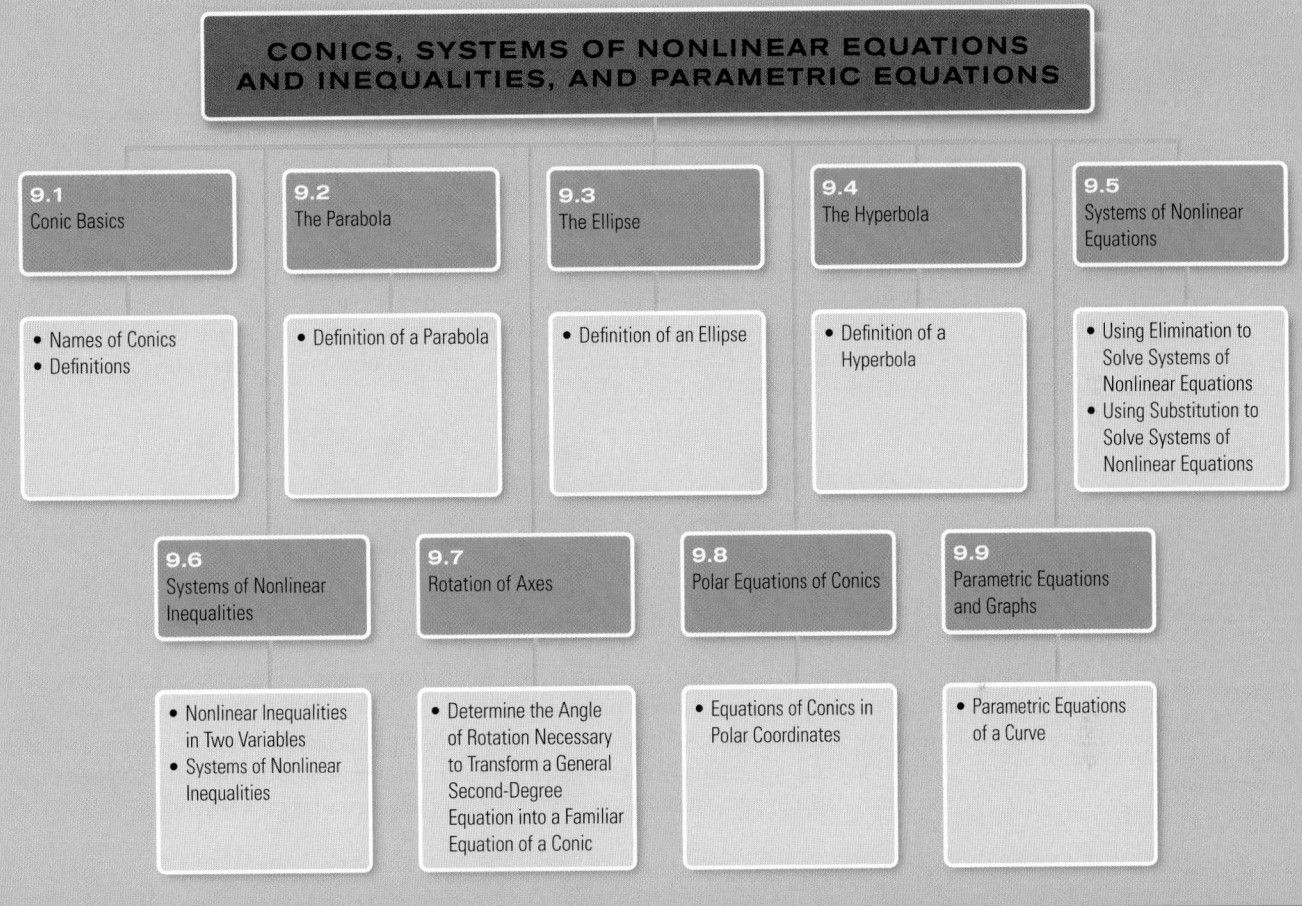

CONICS, SYSTEMS OF NONLINEAR EQUATIONS AND INEQUALITIES, AND PARAMETRIC EQUATIONS

9.1 Conic Basics
- Names of Conics
- Definitions

9.2 The Parabola
- Definition of a Parabola

9.3 The Ellipse
- Definition of an Ellipse

9.4 The Hyperbola
- Definition of a Hyperbola

9.5 Systems of Nonlinear Equations
- Using Elimination to Solve Systems of Nonlinear Equations
- Using Substitution to Solve Systems of Nonlinear Equations

9.6 Systems of Nonlinear Inequalities
- Nonlinear Inequalities in Two Variables
- Systems of Nonlinear Inequalities

9.7 Rotation of Axes
- Determine the Angle of Rotation Necessary to Transform a General Second-Degree Equation into a Familiar Equation of a Conic

9.8 Polar Equations of Conics
- Equations of Conics in Polar Coordinates

9.9 Parametric Equations and Graphs
- Parametric Equations of a Curve

CHAPTER OBJECTIVES

- Identify if a second-degree equation in two variables corresponds to a parabola, an ellipse, or a hyperbola (Section 9.1).
- Graph parabolas whose vertex is at the point (h, k) (Section 9.2).
- Graph an ellipse whose center is at the point (h, k) (Section 9.3).
- Graph a hyperbola whose center is at the point (h, k) (Section 9.4).
- Solve systems of nonlinear equations (Section 9.5).
- Graph systems of nonlinear inequalities (Section 9.6).
- Transform general second-degree equations into recognizable equations of conics by analyzing the rotation of axes (Section 9.7).
- Express equations of conics in polar coordinates (Section 9.8).
- Express projectile motion using parametric equations (Section 9.9).

SKILLS OBJECTIVES

- Learn the name of each conic section.
- Define conics.
- Recognize the algebraic equation associated with each conic.

CONCEPTUAL OBJECTIVES

- Understand each conic as an intersection of a plane and a cone.
- Understand how the three equations of the conic sections are related to the general form of a second-degree equation in two variables.

Names of Conics

The word *conic* is derived from the word *cone*. Let's start with a (right circular) **double cone** (see the figure on the left).

Conic sections are curves that result from the intersection of a plane and a double cone. The four conic sections are a **circle**, an **ellipse**, a **parabola**, and a **hyperbola**. **Conics** is an abbreviation for conic sections.

Circle

Ellipse

Parabola

Hyperbola

Study Tip

A circle is a special type of ellipse. All circles are ellipses, but not all ellipses are circles.

In Section 0.5, circles were discussed, and we will show that a circle is a particular type of an ellipse. Now we will discuss parabolas, ellipses, and hyperbolas. There are two ways in which we usually describe conics: graphically and algebraically. An entire section will be devoted to each of the three conics, but here we will summarize the definitions of a parabola, an ellipse, and a hyperbola and show how to identify the equations of these conics.

Definitions

You already know that a circle consists of all points equidistant (at a distance equal to the radius) from a point (the center). Ellipses, parabolas, and hyperbolas have similar definitions in that they all have a constant distance (or a sum or difference of distances) to some reference point(s).

A **parabola** is the set of all points that are **equidistant from both a line and a point**. An **ellipse** is the set of all points, the **sum of whose distances to two fixed points is constant**. A **hyperbola** is the set of all points, the **difference of whose distances to two fixed points is a constant**.

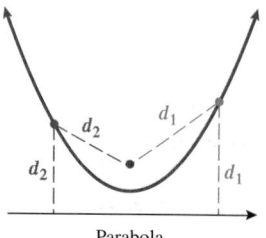

Parabola

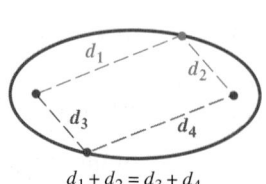

$d_1 + d_2 = d_3 + d_4$
Ellipse

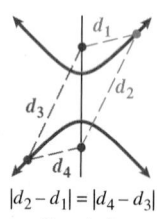

$|d_2 - d_1| = |d_4 - d_3|$
Hyperbola

The **general form of a second-degree equation in two variables**, x and y, is given by

$$Ax^2 + Bxy + Cy^2 + Dx + Ey + F = 0$$

If we let $A = 1, B = 0, C = 1, D = 0, E = 0$, and $F = -r^2$, this general equation reduces to the equation of a circle centered at the origin: $x^2 + y^2 = r^2$. In fact, all three conics (parabolas, ellipses, and hyperbolas) are special cases of the general second-degree equation.

Recall from Section 0.2 (Quadratic Equations) that the discriminant, $b^2 - 4ac$, determines what types of solutions result from solving a second-degree equation in one variable. If the discriminant is positive, the solutions are two distinct real roots. If the discriminant is zero, the solution is a real repeated root. If the discriminant is negative, the solutions are two complex conjugate roots.

The concept of discriminant is also applicable to second-degree equations in two variables. The discriminant $B^2 - 4AC$ determines the *shape* of the conic section.

CONIC	DISCRIMINANT
Ellipse	$B^2 - 4AC < 0$
Parabola	$B^2 - 4AC = 0$
Hyperbola	$B^2 - 4AC > 0$

Study Tip

All circles are ellipses since $B^2 - 4AC < 0$.

Using the discriminant to identify the shape of the conic will not work for degenerate cases (when the polynomial factors). For example,

$$2x^2 - xy - y^2 = 0$$

At first glance, one may think this is a hyperbola because $B^2 - 4AC > 0$, but this is a degenerate case.

$$(2x + y)(x - y) = 0$$

$$2x + y = 0 \qquad \text{or} \qquad x - y = 0$$
$$y = -2x \qquad \text{or} \qquad y = x$$

The graph is two intersecting lines.

We now identify conics from the general form of a second-degree equation in two variables.

EXAMPLE 1   Determining the Type of Conic

Determine what type of conic corresponds to each of the following equations:

a. $\dfrac{x^2}{a^2} + \dfrac{y^2}{b^2} = 1$ **b.** $y = x^2$ **c.** $\dfrac{x^2}{a^2} - \dfrac{y^2}{b^2} = 1$

Solution:

Write the general form of the second-degree equation:

$$Ax^2 + Bxy + Cy^2 + Dx + Ey + F = 0$$

a. Identify $A, B, C, D, E,$ and F. $A = \dfrac{1}{a^2}, B = 0, C = \dfrac{1}{b^2}, D = 0, E = 0, F = -1$

Calculate the discriminant. $B^2 - 4AC = -\dfrac{4}{a^2 b^2} < 0$

Since the discriminant is negative, the equation $\dfrac{x^2}{a^2} + \dfrac{y^2}{b^2} = 1$ is that of an **ellipse**.

Notice that if $a = b = r$, then this equation of an ellipse reduces to the general equation of a circle, $x^2 + y^2 = r^2$, centered at the origin, with radius r.

b. Identify A, B, C, D, E, and F. $A = 1, B = 0, C = 0, \; D = 0, E = -1, F = 0$

Calculate the discriminant. $B^2 - 4AC = 0$

Since the discriminant is zero, the equation $y = x^2$ is a **parabola**.

c. Identify A, B, C, D, E, and F. $A = \dfrac{1}{a^2}, B = 0, C = -\dfrac{1}{b^2}, D = 0, E = 0, F = -1$

Calculate the discriminant. $B^2 - 4AC = \dfrac{4}{a^2 b^2} > 0$

Since the discriminant is positive, the equation $\dfrac{x^2}{a^2} - \dfrac{y^2}{b^2} = 1$ is a **hyperbola**.

■ **Answer: a.** ellipse
       b. hyperbola
       c. parabola

■ **YOUR TURN**   Determine what type of conic corresponds to each of the following equations:

    a. $2x^2 + y^2 = 4$ **b.** $2x^2 = y^2 + 4$ **c.** $2y^2 = x$

In the next three sections, we will discuss the standard forms of equations and the graphs of parabolas, ellipses, and hyperbolas.

SECTION 9.1 SUMMARY

In this section, we defined the three conic sections and determined their general equations with respect to the general form of a second-degree equation in two variables:

$$Ax^2 + Bxy + Cy^2 + Dx + Ey + F = 0$$

The following table summarizes the three conics: ellipse, parabola, and hyperbola.

CONIC	GEOMETRIC DEFINITION: THE SET OF ALL POINTS	DISCRIMINANT
Ellipse	the sum of whose distances to two fixed points is constant	Negative: $B^2 - 4AC < 0$
Parabola	equidistant to both a line and a point	Zero: $B^2 - 4AC = 0$
Hyperbola	the difference of whose distances to two fixed points is a constant	Positive: $B^2 - 4AC > 0$

It is important to note that a circle is a special type of ellipse.

SECTION 9.1 EXERCISES

■ SKILLS

In Exercises 1–12, identify the conic section as a parabola, ellipse, circle, or hyperbola.

1. $x^2 + xy - y^2 + 2x = -3$ **2.** $x^2 + xy + y^2 + 2x = -3$ **3.** $2x^2 + 2y^2 = 10$ **4.** $x^2 - 4x + y^2 + 2y = 4$

5. $2x^2 - y^2 = 4$ **6.** $2y^2 - x^2 = 16$ **7.** $5x^2 + 20y^2 = 25$ **8.** $4x^2 + 8y^2 = 30$

9. $x^2 - y = 1$ **10.** $y^2 - x = 2$ **11.** $x^2 + y^2 = 10$ **12.** $x^2 + y^2 = 100$

SKILLS OBJECTIVES

- Graph a parabola given the focus, directrix, and vertex.
- Find the equation of a parabola whose vertex is at the origin.
- Find the equation of a parabola whose vertex is at the point (h, k).
- Solve applied problems that involve parabolas.

CONCEPTUAL OBJECTIVES

- Derive the general equation of a parabola.
- Identify, draw, and use the focus, directrix, and axis of symmetry.

Definition of a Parabola

Recall from Section 2.1 that the graphs of quadratic functions such as

$$f(x) = a(x - h)^2 + k \quad \text{or} \quad y = ax^2 + bx + c$$

were *parabolas* that opened either upward or downward. We now expand our discussion to *parabolas* that open to the right or left. We did not discuss these types of parabolas before because they are not functions (they fail the vertical line test).

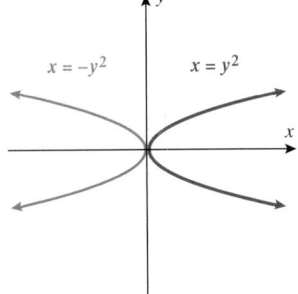

DEFINITION   Parabola

A **parabola** is the set of all points in a plane that are equidistant from a fixed line, the **directrix**, and a fixed point not on the line, the **focus**. The line through the focus and perpendicular to the directrix is the **axis of symmetry**. The **vertex** of the parabola is located at the midpoint between the directrix and the focus along the axis of symmetry.

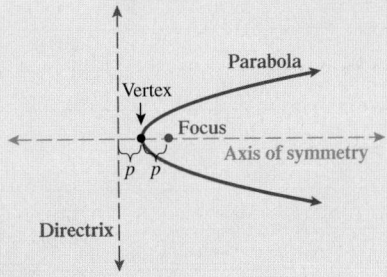

Here p is the distance along the axis of symmetry from the directrix to the vertex and from the vertex to the focus.

Let's consider a parabola with the vertex at the origin and the focus on the positive x-axis. Let the distance from the vertex to the focus be p. Therefore, the focus is located at the point $(p, 0)$. Since the distance from the vertex to the focus is p, the distance from the vertex to the directrix must also be p. Since the axis of symmetry is the x-axis, the directrix must be perpendicular to the x-axis. Therefore, the directrix is given by $x = -p$. Any point, (x, y), must have the same distance to the focus, $(p, 0)$, as does the directrix, $(-p, y)$.

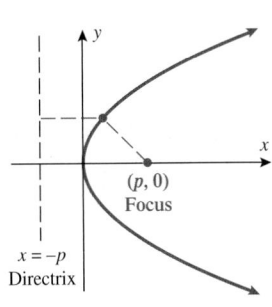

Derivation of the Equation of a Parabola

WORDS	MATH
Calculate the distance from (x, y) to $(p, 0)$ with the distance formula.	$\sqrt{(x - p)^2 + y^2}$
Calculate the distance from (x, y) to $(-p, y)$ with the distance formula.	$\sqrt{(x - (-p))^2 + 0^2}$
Set the two distances equal to one another.	$\sqrt{(x - p)^2 + y^2} = \sqrt{(x + p)^2}$
Recall that $\sqrt{x^2} = |x|$.	$\sqrt{(x - p)^2 + y^2} = |x + p|$
Square both sides of the equation.	$(x - p)^2 + y^2 = (x + p)^2$
Square the binomials inside the parentheses.	$x^2 - 2px + p^2 + y^2 = x^2 + 2px + p^2$
Simplify.	$\boxed{y^2 = 4px}$

The equation $y^2 = 4px$ represents a parabola opening right ($p > 0$) with the vertex at the origin. The following box summarizes parabolas that have a vertex at the origin and a focus along either the x-axis or the y-axis:

EQUATION OF A PARABOLA WITH VERTEX AT THE ORIGIN

The standard (conic) form of the equation of a **parabola** with vertex at the origin is given by

EQUATION	$y^2 = 4px$	$x^2 = 4py$
VERTEX	$(0, 0)$	$(0, 0)$
FOCUS	$(p, 0)$	$(0, p)$
DIRECTRIX	$x = -p$	$y = -p$
AXIS OF SYMMETRY	x-axis	y-axis
$p > 0$	opens to the right	opens upward
$p < 0$	opens to the left	opens downward
GRAPH ($p > 0$)	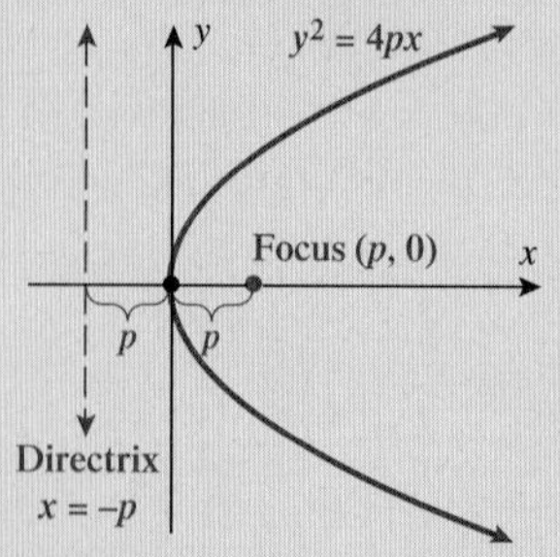	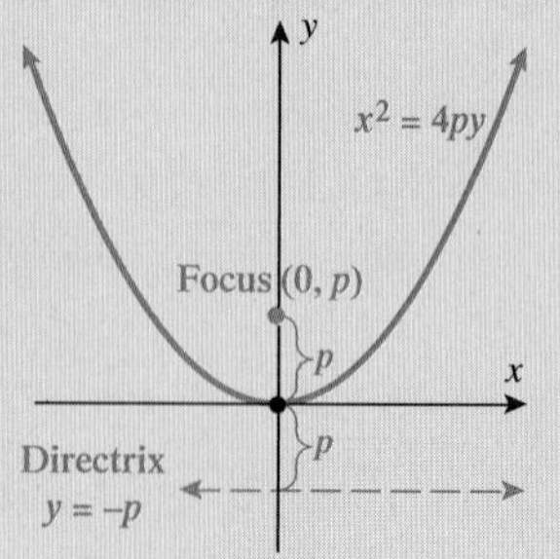

EXAMPLE 1   Finding the Focus and Directrix of a Parabola Whose Vertex Is Located at the Origin

Find the focus and directrix of a parabola whose equation is $y^2 = 8x$.

Solution:

Compare this parabola with the general equation of a parabola.

$$y^2 = 4px$$
$$y^2 = 8x$$

Let $y^2 = 8x$.

$$4px = 8x$$

Solve for p (assume $x \neq 0$).

$$4p = 8$$
$$p = 2$$

The focus of a parabola of the form $y^2 = 4px$ is $(p, 0)$.

The directrix of a parabola of the form $y^2 = 4px$ is $x = -p$.

The focus is $\boxed{(2, 0)}$ and the directrix is $\boxed{x = -2}$.

■ **YOUR TURN** Find the focus and directrix of a parabola whose equation is $y^2 = 16x$.

■ **Answer:** The focus is $(4, 0)$ and the directrix is $x = -4$.

Study Tip

The focus and directrix define a parabola, but do not appear on its graph.

Graphing a Parabola with a Vertex at the Origin

When a seamstress starts with a pattern for a custom-made suit, the pattern is used as a guide. The pattern is not sewn into the suit, but rather removed once it is used to determine the exact shape and size of the fabric to be sewn together. The focus and directrix of a parabola are similar to the pattern used by a seamstress. Although the focus and directrix define a parabola, they do not appear on the graph of a parabola.

We can draw an approximate sketch of a parabola whose vertex is at the origin with three pieces of information. We know that the vertex is located at $(0, 0)$. Additional information that we seek is the direction in which the parabola opens and approximately how wide or narrow to draw the parabolic curve. The direction toward which the parabola opens is found from the equation. An equation of the form $y^2 = 4px$ opens either left or right. It opens right if $p > 0$ and opens left if $p < 0$. An equation of the form $x^2 = 4py$ opens either up or down. It opens up if $p > 0$ and opens down if $p < 0$. How narrow or wide should we draw the parabolic curve? If we select a few points that satisfy the equation, we can use those as graphing aids.

In Example 1, we found that the focus of that parabola is located at $(2, 0)$. If we select the x-coordinate of the focus $x = 2$, and substitute that value into the equation of the parabola $y^2 = 8x$, we find the corresponding y values to be $y = -4$ and $y = 4$. If we plot the three points $(0, 0)$, $(2, -4)$, and $(2, 4)$ and then connect the points with a parabolic curve, we get the graph on the right.

The line segment that passes through the focus $(2, 0)$ is parallel to the directrix $x = -2$, and whose endpoints are on the parabola is called the **latus rectum**. The latus rectum in this case has length 8. The latus rectum is a graphing aid that assists us in determining how wide or how narrow to draw the parabola.

In general, the points on a parabola of the form $y^2 = 4px$ that lie above and below the focus $(p, 0)$ satisfy the equation $y^2 = 4p^2$ and are located at $(p, -2p)$ and $(p, 2p)$. The latus rectum will have length $4|p|$. Similarly, a parabola of the form $x^2 = 4py$ will have a horizontal latus rectum of length $4|p|$.

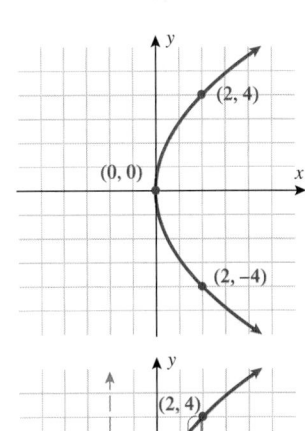

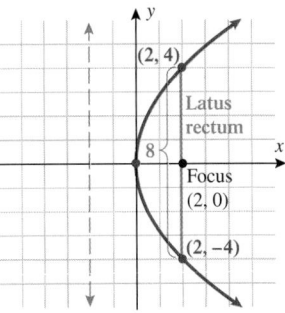

Technology Tip

To graph $x^2 = -12y$ with a graphing calculator, solve for y first. That is, $y = -\frac{1}{12}x^2$.

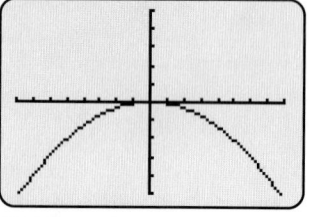

EXAMPLE 2 Graphing a Parabola Whose Vertex Is at the Origin, Using the Focus, Directrix, and Latus Rectum as Graphing Aids

Determine the focus, directrix, and length of the latus rectum of the parabola $x^2 = -12y$. Employ these to assist in graphing the parabola.

Solution:

Compare this parabola with the general equation of a parabola.

$$x^2 = 4py \qquad x^2 = -12y$$

Solve for p.

$$4p = -12$$
$$p = -3$$

A parabola of the form $x^2 = 4py$ has focus $(0, p)$, directrix $y = -p$, and a latus rectum of length $4|p|$. For this parabola, $p = -3$; therefore, the focus is $\boxed{(0, -3)}$, the directrix is $\boxed{y = 3}$, and the length of the latus rectum is $\boxed{12}$.

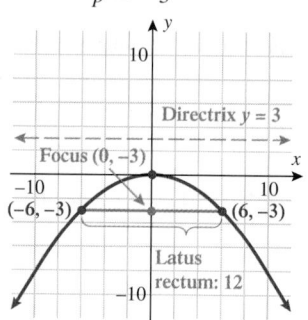

■ **Answer:**
The focus is $(-2, 0)$.
The directrix is $x = 2$.
The length of the latus rectum is 8.

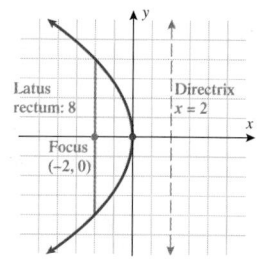

■ **YOUR TURN** Find the focus, directrix, and length of the latus rectum of the parabola $y^2 = -8x$, and use these to graph the parabola.

Finding the Equation of a Parabola with a Vertex at the Origin

Thus far we have started with the equation of a parabola and then determined its focus and directrix. Let's now reverse the process. For example, if we know the focus and directrix of a parabola, how do we find the equation of the parabola? If we are given the focus and directrix, then we can find the vertex, which is the midpoint between the focus and the directrix. If the vertex is at the origin, then we know the general equation of the parabola that corresponds to the focus.

EXAMPLE 3 Finding the Equation of a Parabola Given the Focus and Directrix When the Vertex Is at the Origin

Find the standard form of the equation of a parabola whose focus is at the point $\left(0, \frac{1}{2}\right)$ and whose directrix is $y = -\frac{1}{2}$. Graph the equation.

Solution:

The midpoint of the segment joining the focus and the directrix along the axis of symmetry is the vertex.

Calculate the midpoint between $\left(0, \frac{1}{2}\right)$ and $\left(0, -\frac{1}{2}\right)$.

$$\text{Vertex} = \left(\frac{0 + 0}{2}, \frac{\frac{1}{2} - \frac{1}{2}}{0}\right) = (0, 0).$$

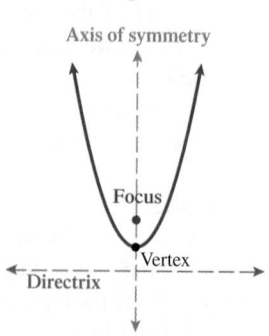

A parabola with vertex at $(0, 0)$, focus at $(0, p)$, and directrix $y = -p$ corresponds to the equation $x^2 = 4py$.

Identify p given that the focus is $(0, p) = \left(0, \frac{1}{2}\right)$.

$$p = \frac{1}{2}$$

Substitute $p = \frac{1}{2}$ into the standard equation of a parabola with vertex at the origin $x^2 = 4py$.

$$x^2 = 2y$$

Now that the equation is known, a few points can be selected, and the parabola can be point-plotted. Alternatively, the length of the latus rectum can be calculated to sketch the approximate width of the parabola.

To graph $x^2 = 2y$, first calculate the latus rectum.

$$4|p| = 4\left(\frac{1}{2}\right) = 2$$

Label the focus, directrix, and latus rectum, and draw a parabolic curve whose vertex is at the origin that intersects with the latus rectum's endpoints.

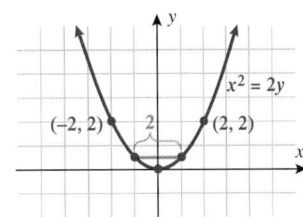

■ **YOUR TURN**   Find the equation of a parabola whose focus is at the point $(-5, 0)$ and whose directrix is $x = 5$.

■ **Answer:** $y^2 = -20x$

Before we proceed to parabolas with general vertices, let's first make a few observations: The larger the latus rectum, the more rapidly the parabola widens. An alternative approach for graphing the parabola is to plot a few points that satisfy the equation of the parabola, which is the approach in most textbooks.

Parabola with a Vertex at (h, k)

Recall (Section 0.5) that the graph of $x^2 + y^2 = r^2$ is a circle with radius r centered at the origin, whereas the graph of $(x - h)^2 + (y - k)^2 = r^2$ is a circle with radius r centered at the point (h, k). In other words, the center is shifted from the origin to the point (h, k). This same translation (shift) can be used to describe parabolas whose vertex is at the point (h, k).

Study Tip

When $(h, k) = (0, 0)$, the vertex of the parabola is located at the origin.

EQUATION OF A PARABOLA WITH VERTEX AT THE POINT (h, k)

The standard (conic) form of the equation of a parabola with vertex at the point (h, k) is given by

EQUATION	$(y - k)^2 = 4p(x - h)$	$(x - h)^2 = 4p(y - k)$
VERTEX	(h, k)	(h, k)
FOCUS	$(p + h, k)$	$(h, p + k)$
DIRECTRIX	$x = -p + h$	$y = -p + k$
AXIS OF SYMMETRY	$y = k$	$x = h$
$p > 0$	opens to the right	opens upward
$p < 0$	opens to the left	opens downward

In order to find the vertex of a parabola given a general second-degree equation, first complete the square (Section 0.2) in order to identify (h, k). Then determine whether the parabola opens up, down, left, or right. Identify points that lie on the graph of the parabola. It is important to note that intercepts are often the easiest points to find, since they are the points where one of the variables is set equal to zero.

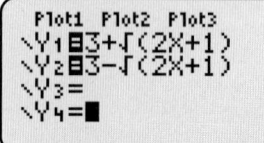
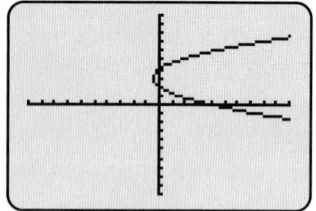
■ **Answer:**
Vertex: $\left(-\frac{9}{5}, -2\right)$
x-intercept: $x = -1$
y-intercepts: $y = -5$ and $y = 1$

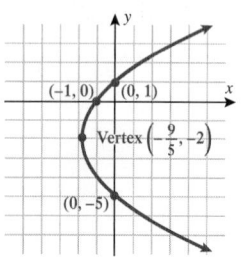

EXAMPLE 4 Graphing a Parabola with Vertex (h, k)

Graph the parabola given by the equation $y^2 - 6y - 2x + 8 = 0$.

Solution:

Transform this equation into the form $(y - k)^2 = 4p(x - h)$, since this equation is of degree 2 in y and degree 1 in x. We know this parabola opens either to the left or right.

Complete the square on y: $\qquad\qquad y^2 - 6y - 2x + 8 = 0$

    Isolate the y terms. $\qquad\qquad\qquad\qquad y^2 - 6y = 2x - 8$

    Add 9 to both sides to complete the square. $\quad y^2 - 6y + 9 = 2x - 8 + 9$

    Write the left side as a perfect square. $\qquad (y - 3)^2 = 2x + 1$

    Factor out a 2 on the right side. $\qquad\qquad (y - 3)^2 = 2\left(x + \frac{1}{2}\right)$

Compare with $(y - k)^2 = 4p(x - h)$ and identify (h, k) and p.

$$(h, k) = \left(-\frac{1}{2}, 3\right)$$

$$4p = 2 \Rightarrow p = \frac{1}{2}$$

The vertex is at the point $\left(-\frac{1}{2}, 3\right)$, and since $p = \frac{1}{2}$ is positive, the parabola opens to the right. Since the parabola's vertex lies in quadrant II and it opens to the right, we know there are two y-intercepts and one x-intercept. Apply the general equation $y^2 - 6y - 2x + 8 = 0$ to find the intercepts.

Find the y-intercepts (set $x = 0$). $\qquad\qquad y^2 - 6y + 8 = 0$

    Factor. $\qquad\qquad\qquad\qquad\qquad (y - 2)(y - 4) = 0$

    Solve for y. $\qquad\qquad\qquad\qquad y = 2 \quad \text{or} \quad y = 4$

Find the x-intercept (set $y = 0$). $\qquad\qquad -2x + 8 = 0$

    Solve for x. $\qquad\qquad\qquad\qquad\qquad x = 4$

Label the following points and connect them with a smooth curve:

    Vertex: $\qquad\left(-\frac{1}{2}, 3\right)$

    y-intercepts: $(0, 2)$ and $(0, 4)$

    x-intercept: $(4, 0)$

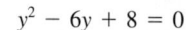

■ **YOUR TURN** For the equation $y^2 + 4y - 5x - 5 = 0$, identify the vertex and the intercepts, and graph.

EXAMPLE 5 Graphing a Parabola with Vertex (*h*, *k*)

Graph the parabola given by the equation $x^2 - 2x - 8y - 7 = 0$.

Solution:

Transform this equation into the form $(x - h)^2 = 4p(y - k)$, since this equation is degree 2 in *x* and degree 1 in *y*. We know this parabola opens either upward or downward.

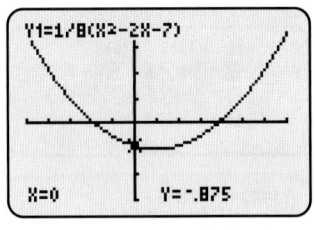

Complete the square on *x*: $\qquad\qquad\qquad\qquad x^2 - 2x - 8y - 7 = 0$

    Isolate the *x* terms. $\qquad\qquad\qquad\qquad\quad x^2 - 2x = 8y + 7$

    Add 1 to both sides to complete the square. $\quad x^2 - 2x + 1 = 8y + 7 + 1$

    Write the left side as a perfect square. $\qquad (x - 1)^2 = 8y + 8$

    Factor out the 8 on the right side. $\qquad\quad (x - 1)^2 = 8(y + 1)$

Compare with $(x - h)^2 = 4p(y - k)$ and identify (h, k) and *p*.

$$(h, k) = (1, -1)$$

$$4p = 8 \Rightarrow p = 2$$

The vertex is at the point $(1, -1)$, and since $p = 2$ is positive, the parabola opens upward. Since the parabola's vertex lies in quadrant IV and it opens upward, we know there are two *x*-intercepts and one *y*-intercept. Use the general equation $x^2 - 2x - 8y - 7 = 0$ to find the intercepts.

Find the *y*-intercept (set $x = 0$). $\qquad\qquad\qquad\qquad -8y - 7 = 0$

    Solve for *y*. $\qquad\qquad\qquad\qquad\qquad\qquad\qquad y = -\dfrac{7}{8}$

Find the *x*-intercepts (set $y = 0$). $\qquad\qquad\qquad x^2 - 2x - 7 = 0$

    Solve for *x*. $\qquad x = \dfrac{2 \pm \sqrt{4 + 28}}{2} = \dfrac{2 \pm \sqrt{32}}{2} = \dfrac{2 \pm 4\sqrt{2}}{2} = 1 \pm 2\sqrt{2}$

Label the following points and connect with a smooth curve:

    Vertex: $\qquad\quad (1, -1)$

    y-intercept: $\quad \left(0, -\dfrac{7}{8}\right)$

    x-intercepts: $\quad \left(1 - 2\sqrt{2}, 0\right)$ and $\left(1 + 2\sqrt{2}, 0\right)$

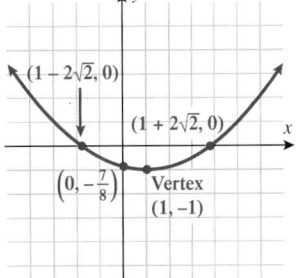

■ Answer:

Vertex: $(-1, 1)$

x-intercepts: $x = -1 \pm 2\sqrt{2}$

y-intercept: $y = \dfrac{7}{8}$

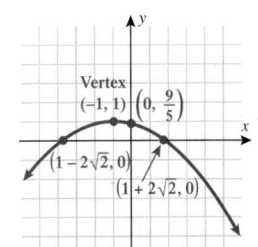

■ YOUR TURN For the equation $x^2 + 2x + 8y - 7 = 0$, identify the vertex and the intercepts, and graph.

Technology Tip

Use a TI to check the graph of $y^2 + 6y - 12x + 33 = 0$. Use $y^2 + 6y - (12x - 33) = 0$ to solve for y first. That is, $y_1 = -3 + 2\sqrt{3x - 6}$ or $y_2 = -3 - 2\sqrt{3x - 6}$.

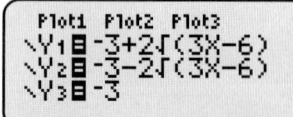

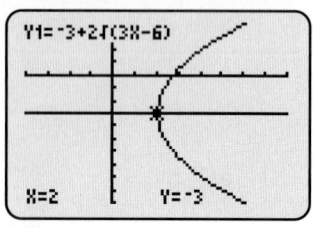

■ **Answer:** $y^2 + 6y + 8x - 7 = 0$

EXAMPLE 6 Finding the Equation of a Parabola with Vertex (h, k)

Find the general form of the equation of a parabola whose vertex is located at the point $(2, -3)$ and whose focus is located at the point $(5, -3)$.

Solution:

Draw a Cartesian plane and label the vertex and focus. The vertex and focus share the same axis of symmetry $y = -3$, and indicate a parabola opening to the right.

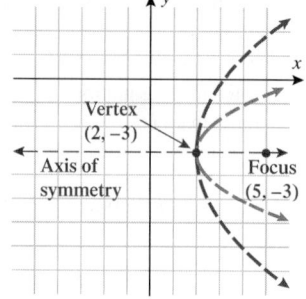

Write the standard (conic) equation of a parabola opening to the right.

$$(y - k)^2 = 4p(x - h) \qquad p > 0$$

Substitute the vertex $(h, k) = (2, -3)$, into the standard equation.

$$[y - (-3)]^2 = 4p(x - 2)$$

Find p.

The general form of the vertex is (h, k) and the focus is $(h + p, k)$.

For this parabola, the vertex is $(2, -3)$ and the focus is $(5, -3)$.

Find p by taking the difference of the x-coordinates. $\qquad p = 3$

Substitute $p = 3$ into $[y - (-3)]^2 = 4p(x - 2)$. $\qquad (y + 3)^2 = 4(3)(x - 2)$

Eliminate the parentheses. $\qquad y^2 + 6y + 9 = 12x - 24$

Simplify. $\qquad \boxed{y^2 + 6y - 12x + 33 = 0}$

■ **YOUR TURN** Find the equation of the parabola whose vertex is located at $(2, -3)$ and whose focus is located at $(0, -3)$.

Applications

If we start with a parabola in the xy-plane and rotate it around its axis of symmetry, the result will be a three-dimensional paraboloid. Solar cookers illustrate the physical property that the rays of light coming into a parabola should be reflected to the focus. A flashlight reverses this process in that its light source at the focus illuminates a parabolic reflector to direct the beam outward.

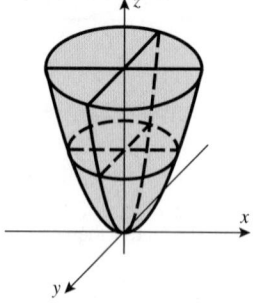

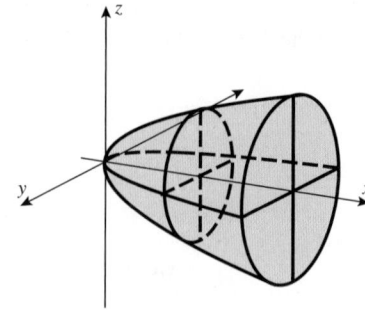

Satellite dish

Digital Vision

A satellite dish is in the shape of a paraboloid. Functioning as an antenna, the parabolic dish collects all of the incoming signals and reflects them to a single point, the focal point, which is where the receiver is located. In Examples 7 and 8, and in the Applications Exercises, the intention is not to find the three-dimensional equation of the paraboloid, but rather the equation of the plane parabola that's rotated to generate the paraboloid.

EXAMPLE 7   Finding the Location of the Receiver in a Satellite Dish

A satellite dish is 24 feet in diameter at its opening and 4 feet deep in its center. Where should the receiver be placed?

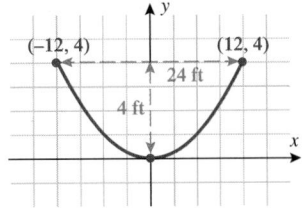

Solution:

Draw a parabola with a vertex at the origin representing the center cross section of the satellite dish.

Write the standard equation of a parabola opening upward with vertex at (0, 0).

$$x^2 = 4py$$

The point (12, 4) lies on the parabola, so substitute (12, 4) into $x^2 = 4py$.

$$(12)^2 = 4p(4)$$

Simplify.

$$144 = 16p$$

Solve for p.

$$p = 9$$

Substitute $p = 9$ into the focus $(0, p)$.

focus: (0, 9)

The receiver should be placed 9 feet from the vertex of the dish.

Parabolic antennas work for sound in addition to light. Have you ever wondered how the sound of the quarterback calling audible plays is heard by the sideline crew? The crew holds a parabolic system with a microphone at the focus. All of the sound in the direction of the parabolic system is reflected toward the focus, where the microphone amplifies and records the sound.

Francis Specker/Icon SMI/NewsCom

EXAMPLE 8   Finding the Equation of a Parabolic Sound Dish

If the parabolic sound dish the sideline crew is holding has a 2-foot diameter at the opening and the microphone is located 6 inches from the vertex, find the equation that governs the center cross section of the parabolic sound dish.

Solution:

Write the standard equation of a parabola opening to the right with the vertex at the origin (0, 0).

$$x = 4py^2$$

The focus is located 6 inches $\left(\frac{1}{2} \text{ foot}\right)$ from the vertex.

$$(p, 0) = \left(\frac{1}{2}, 0\right)$$

Solve for p.

$$p = \frac{1}{2}$$

Let $p = \frac{1}{2}$ in $x = 4py^2$.

$$x = 4\left(\frac{1}{2}\right)y^2$$

Simplify.

$$x = 2y^2$$

SECTION
9.2 **SUMMARY**

In this section, we discussed parabolas whose vertex is at the origin.

EQUATION	$y^2 = 4px$	$x^2 = 4py$
VERTEX	$(0, 0)$	$(0, 0)$
FOCUS	$(p, 0)$	$(0, p)$
DIRECTRIX	$x = -p$	$y = -p$
AXIS OF SYMMETRY	x-axis	y-axis
$p > 0$	opens to the right	opens upward
$p < 0$	opens to the left	opens downward
GRAPH		

For parabolas whose vertex is at the point, (h, k):

EQUATION	$(y - k)^2 = 4p(x - h)$	$(x - h)^2 = 4p(y - k)$
VERTEX	(h, k)	(h, k)
FOCUS	$(p + h, k)$	$(h, p + k)$
DIRECTRIX	$x = -p + h$	$y = -p + k$
AXIS OF SYMMETRY	$y = k$	$x = h$
$p > 0$	opens to the right	opens upward
$p < 0$	opens to the left	opens downward

SECTION
9.2 EXERCISES

▪ SKILLS

In Exercises 1–4, match the parabola to the equation.

1. $y^2 = 4x$

2. $y^2 = -4x$

3. $x^2 = -4y$

4. $x^2 = 4y$

a.

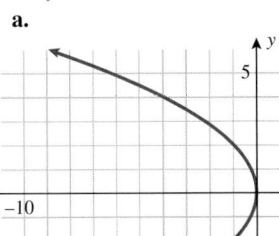

b.

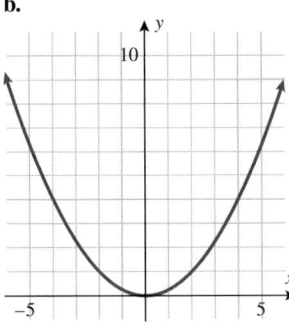

c.

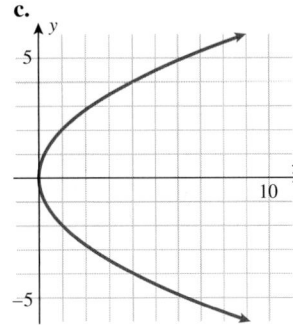

d.

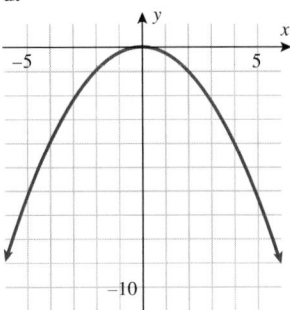

In Exercises 5–8, match the parabola to the equation.

5. $(y - 1)^2 = 4(x - 1)$

6. $(y + 1)^2 = -4(x - 1)$

7. $(x + 1)^2 = -4(y + 1)$

8. $(x - 1)^2 = 4(y - 1)$

a.

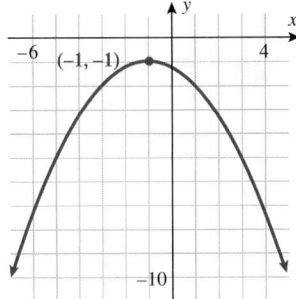

b.

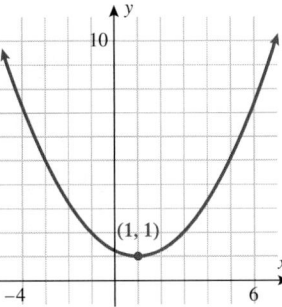

c.

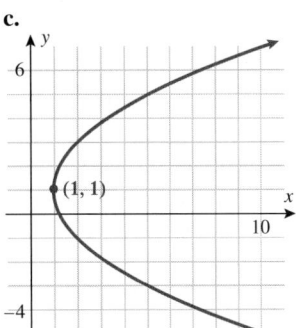

d.

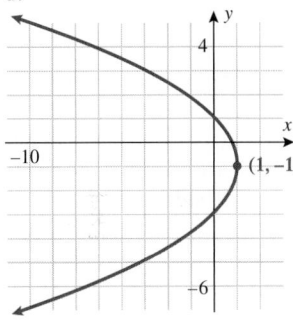

In Exercises 9–20, find an equation for the parabola described.

9. Vertex at $(0, 0)$; focus at $(0, 3)$

10. Vertex at $(0, 0)$; focus at $(2, 0)$

11. Vertex at $(0, 0)$; focus at $(-5, 0)$

12. Vertex at $(0, 0)$; focus at $(0, -4)$

13. Vertex at $(3, 5)$; focus at $(3, 7)$

14. Vertex at $(3, 5)$; focus at $(7, 5)$

15. Vertex at $(2, 4)$; focus at $(0, 4)$

16. Vertex at $(2, 4)$; focus at $(2, -1)$

17. Focus at $(2, 4)$; directrix at $y = -2$

18. Focus at $(2, -2)$; directrix at $y = 4$

19. Focus at $(3, -1)$; directrix at $x = 1$

20. Focus at $(-1, 5)$; directrix at $x = 5$

In Exercises 21–24, write an equation for each parabola.

21.

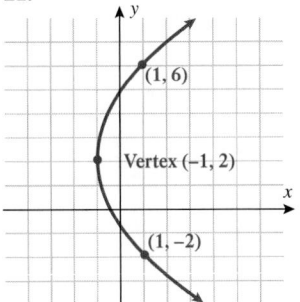

22.

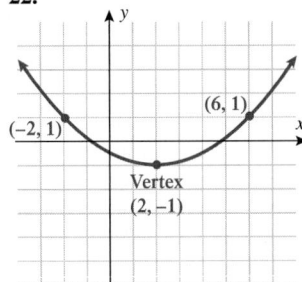

23.

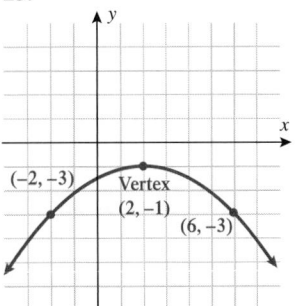

24.

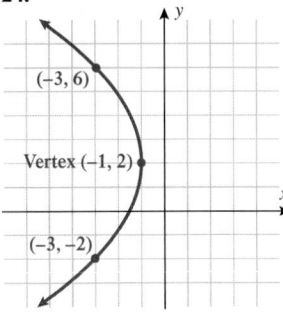

In Exercises 25–32, find the focus, vertex, directrix, and length of latus rectum and graph the parabola.

25. $x^2 = 8y$ **26.** $x^2 = -12y$ **27.** $y^2 = -2x$ **28.** $y^2 = 6x$

29. $x^2 = 16y$ **30.** $x^2 = -8y$ **31.** $y^2 = 4x$ **32.** $y^2 = -16x$

In Exercises 33–44, find the vertex and graph the parabola.

33. $(y - 2)^2 = 4(x + 3)$ **34.** $(y + 2)^2 = -4(x - 1)$ **35.** $(x - 3)^2 = -8(y + 1)$ **36.** $(x + 3)^2 = -8(y - 2)$

37. $(x + 5)^2 = -2y$ **38.** $y^2 = -16(x + 1)$ **39.** $y^2 - 4y - 2x + 4 = 0$ **40.** $x^2 - 6x + 2y + 9 = 0$

41. $y^2 + 2y - 8x - 23 = 0$ **42.** $x^2 - 6x - 4y + 10 = 0$ **43.** $x^2 - x + y - 1 = 0$ **44.** $y^2 + y - x + 1 = 0$

▪ APPLICATIONS

45. Satellite Dish. A satellite dish measures 8 feet across its opening and 2 feet deep at its center. The receiver should be placed at the focus of the parabolic dish. Where is the focus?

46. Satellite Dish. A satellite dish measures 30 feet across its opening and 5 feet deep at its center. The receiver should be placed at the focus of the parabolic dish. Where is the focus?

47. Eyeglass Lens. Eyeglass lenses can be thought of as very wide parabolic curves. If the focus occurs 2 centimeters from the center of the lens and the lens at its opening is 5 centimeters, find an equation that governs the shape of the center cross section of the lens.

48. Optical Lens. A parabolic lens focuses light onto a focal point 3 centimeters from the vertex of the lens. How wide is the lens 0.5 centimeter from the vertex?

Exercises 49 and 50 are examples of solar cookers. Parabolic shapes are often used to generate intense heat by collecting sun rays and focusing all of them at a focal point.

49. Solar Cooker. The parabolic cooker MS-ST10 is delivered as a kit, handily packed in a single carton, with complete assembly instructions and even the necessary tools.

Solar cooker, Ubuntu Village,
Johannesburg, South Africa

Thanks to the reflector diameter of 1 meter, it develops an immense power: 1 liter of water boils in significantly less than half an hour. If the rays are focused 40 centimeters from the vertex, find the equation for the parabolic cooker.

50. Le Four Solaire at Font-Romeur "Mirrors of the Solar Furnace." There is a reflector in the Pyrenees Mountains that is eight stories high. It cost $2 million and took 10 years to build. Made of 9000 mirrors arranged in a parabolic formation, it can reach 6000 °F just from the Sun hitting it!

Solar furnace, Odellio, France

If the diameter of the parabolic mirror is 100 meters and the sunlight is focused 25 meters from the vertex, find the equation for the parabolic dish.

51. Sailing Under a Bridge. A bridge with a parabolic shape has an opening 80 feet wide at the base (where the bridge meets the water), and the height in the center of the bridge is 20 feet. A sailboat whose mast reaches 17 feet above the water is traveling under the bridge 10 feet from the center of the bridge. Will it clear the bridge without scraping its mast? Justify your answer.

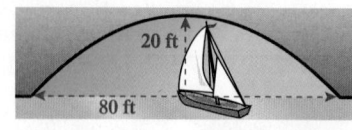

52. Driving Under a Bridge. A bridge with a parabolic shape reaches a height of 25 feet in the center of the road, and the width of the bridge opening at ground level is 20 feet combined (both lanes). If an RV is 10 feet tall and 8 feet wide, it won't make it under the bridge if it hugs the center line. Will it clear the bridge if it straddles the center line? Justify your answer.

53. Parabolic Telescope. The Arecibo radio telescope in Puerto Rico has an enormous reflecting surface, or radio mirror. The huge "dish" is 1000 feet in diameter and 167 feet deep and covers an area of about 20 acres. Using these dimensions, determine the focal length of the telescope. Find the equation for the dish portion of the telescope.

54. Suspension Bridge. If one parabolic segment of a suspension bridge is 300 feet and if the cables at the vertex are suspended 10 feet above the bridge, whereas the height of the cables 150 feet from the vertex reaches 60 feet, find the equation of the parabolic path of the suspension cables.

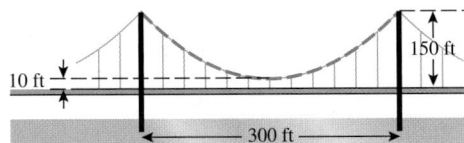

55. Health. In a meditation state, the pulse rate (pulses per minute) can be modeled by $p(t) = 0.18t^2 - 5.4t + 95.5$, where t is in minutes. What is the minimum pulse rate according to this model?

56. Health. In a distress situation, the pulse rate (pulses per minutes) can be modeled by $p(t) = -1.1t^2 + 22t + 80$, where t is the time in seconds. What is the maximum pulse rate according to this model?

■ CATCH THE MISTAKE

In Exercises 57 and 58, explain the mistake that is made.

57. Find an equation for a parabola whose vertex is at the origin and whose focus is at the point $(3, 0)$.

Solution:

Write the general equation for a parabola whose vertex is at the origin. $\qquad x^2 = 4py$

The focus of this parabola is $(p, 0) = (3, 0)$. $\quad p = 3$

Substitute $p = 3$ into $x^2 = 4py$. $\qquad x^2 = 12y$

This is incorrect. What mistake was made?

58. Find an equation for a parabola whose vertex is at the point $(3, 2)$ and whose focus is located at $(5, 2)$.

Solution:

Write the equation associated with a parabola whose vertex is $(3, 2)$. $\quad (x - h)^2 = 4p(y - k)$

Substitute $(3, 2)$ into $(x - h)^2 = 4p(y - k)$. $\qquad (x - 3)^2 = 4p(y - 2)$

The focus is located at $(5, 2)$; therefore, $p = 5$.

Substitute $p = 5$ into $(x - 3)^2 = 4p(y - 2)$. $\qquad (x - 3)^2 = 20(y - 2)$

This is incorrect. What mistake(s) was made?

■ CONCEPTUAL

In Exercises 59–62, determine whether each statement is true or false.

59. The vertex lies on the graph of a parabola.

60. The focus lies on the graph of a parabola.

61. The directrix lies on the graph of a parabola.

62. The endpoints of the latus rectum lie on the graph of a parabola.

In Exercises 63 and 64, use the following equation:

$$\frac{(y - k)^2}{(x - h)} = 4$$

63. Find the directrix of the parabola.

64. Determine whether the parabola opens to the right or to the left.

In Exercises 65 and 66, use the following information about the graph of the parabola:

> Axis of symmetry: $x = 6$
> Directrix: $y = 4$
> Focus: $(6, 9)$

65. Find the vertex of the parabola.

66. Find the equation of the parabola.

▪ CHALLENGE

67. Derive the standard equation of a parabola with its vertex at the origin, opening upward $x^2 = 4py$. [Calculate the distance d_1 from any point on the parabola (x, y) to the focus $(0, p)$. Calculate the distance d_2 from any point on the parabola (x, y) to the directrix $(-p, y)$. Set $d_1 = d_2$.]

68. Derive the standard equation of a parabola opening right, $y^2 = 4px$. [Calculate the distance d_1 from any point on the parabola (x, y) to the focus $(p, 0)$. Calculate the distance d_2 from any point on the parabola (x, y) to the directrix $(x, -p)$. Set $d_1 = d_2$.]

69. Two parabolas with the same axis of symmetry, $y = 6$, intersect at the point (4, 2). If the directrix of one of these parabolas is the y-axis and the directrix of the other parabola is $x = 8$, find the equations of the parabolas.

70. Two parabolas with the same axis of symmetry, $x = 9$, intersect at the point $(6, -5)$. If the directrix of one of these parabolas is $y = -11$ and the directrix of the other parabola is $y = 1$, find the equations of the parabolas.

71. Find the points of intersection of the parabolas with foci $\left(0, \frac{3}{2}\right)$ and $\left(0, -\frac{3}{4}\right)$, and directresses $y = \frac{1}{2}$ and $y = -\frac{5}{4}$, respectively.

72. Find two parabolas with focus $(1, 2p)$ and vertices $(1, p)$ and $(1, -p)$ that intersect each other.

▪ TECHNOLOGY

73. With a graphing utility, plot the parabola $x^2 - x + y - 1 = 0$. Compare with the sketch you drew for Exercise 43.

74. With a graphing utility, plot the parabola $y^2 + y - x + 1 = 0$. Compare with the sketch you drew for Exercise 44.

75. In your mind, picture the parabola given by $(y + 3.5)^2 = 10(x - 2.5)$. Where is the vertex? Which way does this parabola open? Now plot the parabola with a graphing utility.

76. In your mind, picture the parabola given by $(x + 1.4)^2 = -5(y + 1.7)$. Where is the vertex? Which way does this parabola open? Now plot the parabola with a graphing utility.

77. In your mind, picture the parabola given by $(y - 1.5)^2 = -8(x - 1.8)$. Where is the vertex? Which way does this parabola open? Now plot the parabola with a graphing utility.

78. In your mind, picture the parabola given by $(x + 2.4)^2 = 6(y - 3.2)$. Where is the vertex? Which way does this parabola open? Now plot the parabola with a graphing utility.

▪ PREVIEW TO CALCULUS

In calculus, to find the area between two curves, first we need to find the point of intersection of the two curves. In Exercises 79–82, find the points of intersection of the two parabolas.

79. Parabola I: vertex: $(0, -1)$; directrix: $y = -\frac{5}{4}$

Parabola II: vertex: $(0, 7)$; directrix: $y = \frac{29}{4}$

80. Parabola I: vertex: $(0, 0)$; focus: $(0, 1)$

Parabola II: vertex: $(1, 0)$; focus: $(1, 1)$

81. Parabola I: vertex: $\left(5, \frac{5}{3}\right)$; focus: $\left(5, \frac{29}{12}\right)$

Parabola II: vertex: $\left(\frac{13}{2}, \frac{289}{24}\right)$; focus: $\left(\frac{13}{2}, \frac{253}{24}\right)$

82. Parabola I: focus: $\left(-2, -\frac{35}{4}\right)$; directrix: $y = -\frac{37}{4}$

Parabola II: focus: $\left(2, -\frac{101}{4}\right)$; directrix: $y = -\frac{99}{4}$

SKILLS OBJECTIVES

- Graph an ellipse given the center, major axis, and minor axis.
- Find the equation of an ellipse centered at the origin.
- Find the equation of an ellipse centered at the point (h, k).
- Solve applied problems that involve ellipses.

CONCEPTUAL OBJECTIVES

- Derive the general equation of an ellipse.
- Understand the meaning of major and minor axes and foci.
- Understand the properties of an ellipse that result in a circle.
- Interpret eccentricity in terms of the shape of the ellipse.

Definition of an Ellipse

If we were to take a piece of string, tie loops at both ends, and tack the ends down so that the string had lots of slack, we would have the picture on the right. If we then took a pencil and pulled the string taut and traced our way around for one full rotation, the result would be an ellipse. See the second figure on the right.

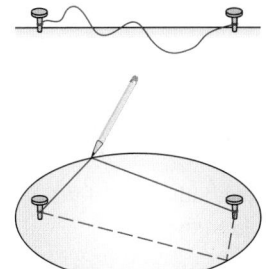

DEFINITION · Ellipse

An **ellipse** is the set of all points in a plane the sum of whose distances from two fixed points is constant. These two fixed points are called **foci** (plural of focus). A line segment through the foci called the **major axis** intersects the ellipse at the **vertices**. The midpoint of the line segment joining the vertices is called the **center**. The line segment that intersects the center and joins two points on the ellipse and is perpendicular to the major axis is called the **minor axis**.

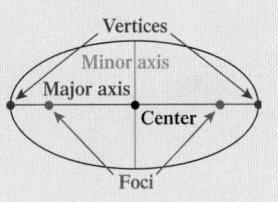

Let's start with an ellipse whose center is located at the origin. Using graph-shifting techniques, we can later extend the characteristics of an ellipse centered at a point other than the origin. Ellipses can vary from the shape of circles to something quite elongated, either horizontally or vertically, that resembles the shape of a racetrack. We say that the ellipse has either greater (elongated) or lesser (circular) *eccentricity*; as we will see, there is a simple mathematical definition of *eccentricity*. It can be shown that the standard equation of an ellipse with its center at the origin is given by one of two forms, depending on whether the orientation of the major axis of the ellipse is horizontal or vertical. For $a > b > 0$, if the major axis is horizontal, then the equation is given by $\dfrac{x^2}{a^2} + \dfrac{y^2}{b^2} = 1$, and if the major axis is vertical, then the equation is given by $\dfrac{x^2}{b^2} + \dfrac{y^2}{a^2} = 1$.

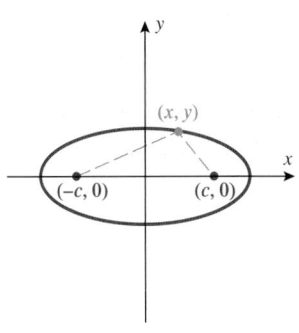

Let's consider an ellipse with its center at the origin and the foci on the x-axis. Let the distance from the center to the focus be c. Therefore, the foci are located at the points $(-c, 0)$ and $(c, 0)$. The line segment containing the foci is called the major axis, and it lies along the x-axis. The sum of the two distances from the foci to any point (x, y) must be constant.

Derivation of the Equation of an Ellipse

WORDS	MATH
Calculate the distance from (x, y) to $(-c, 0)$ by applying the distance formula.	$\sqrt{[x - (-c)]^2 + y^2}$
Calculate the distance from (x, y) to $(c, 0)$ by applying the distance formula.	$\sqrt{(x - c)^2 + y^2}$
The sum of these two distances is equal to a constant ($2a$ for convenience).	$\sqrt{[x - (-c)]^2 + y^2} + \sqrt{(x - c)^2 + y^2} = 2a$
Isolate one radical.	$\sqrt{[x - (-c)]^2 + y^2} = 2a - \sqrt{(x - c)^2 + y^2}$
Square both sides of the equation.	$(x + c)^2 + y^2 = 4a^2 - 4a\sqrt{(x - c)^2 + y^2} + (x - c)^2 + y^2$
Square the binomials inside the parentheses.	$x^2 + 2cx + c^2 + y^2 = 4a^2 - 4a\sqrt{(x - c)^2 + y^2}$ $+ x^2 - 2cx + c^2 + y^2$
Simplify.	$4cx - 4a^2 = -4a\sqrt{(x - c)^2 + y^2}$
Divide both sides of the equation by -4.	$a^2 - cx = a\sqrt{(x - c)^2 + y^2}$
Square both sides of the equation.	$\left(a^2 - cx\right)^2 = a^2\left[(x - c)^2 + y^2\right]$
Square the binomials inside the parentheses.	$a^4 - 2a^2cx + c^2x^2 = a^2(x^2 - 2cx + c^2 + y^2)$
Distribute the a^2 term.	$a^4 - 2a^2cx + c^2x^2 = a^2x^2 - 2a^2cx + a^2c^2 + a^2y^2$
Group the x and y terms together, respectively, on one side and constants on the other side.	$c^2x^2 - a^2x^2 - a^2y^2 = a^2c^2 - a^4$
Factor out the common factors.	$\left(c^2 - a^2\right)x^2 - a^2y^2 = a^2\left(c^2 - a^2\right)$
Multiply both sides of the equation by -1.	$\left(a^2 - c^2\right)x^2 + a^2y^2 = a^2\left(a^2 - c^2\right)$
We can make the argument that $a > c$ in order for a point to be on the ellipse (and not on the x-axis). Thus, since a and c represent distances and therefore are positive, we know that $a^2 > c^2$, or $a^2 - c^2 > 0$. Hence, we can divide both sides of the equation by $a^2 - c^2$, since $a^2 - c^2 \neq 0$.	$x^2 + \dfrac{a^2y^2}{\left(a^2 - c^2\right)} = a^2$
Let $b^2 = a^2 - c^2$.	$x^2 + \dfrac{a^2y^2}{b^2} = a^2$
Divide both sides of the equation by a^2.	$\boxed{\dfrac{x^2}{a^2} + \dfrac{y^2}{b^2} = 1}$

The equation $\dfrac{x^2}{a^2} + \dfrac{y^2}{b^2} = 1$ represents an ellipse with its center at the origin with the foci along the x-axis, since $a > b$. The following box summarizes ellipses that have their center at the origin and foci along either the x-axis or y-axis:

EQUATION OF AN ELLIPSE WITH CENTER AT THE ORIGIN

The **standard form of the equation of an ellipse** with its center at the origin is given by

ORIENTATION OF MAJOR AXIS	Horizontal (along the x-axis)	Vertical (along the y-axis)
EQUATION	$\dfrac{x^2}{a^2} + \dfrac{y^2}{b^2} = 1 \quad a > b > 0$	$\dfrac{x^2}{b^2} + \dfrac{y^2}{a^2} = 1 \quad a > b > 0$
FOCI	$(-c, 0)$ and $(c, 0)$ where $c^2 = a^2 - b^2$	$(0, -c)$ and $(0, c)$ where $c^2 = a^2 - b^2$
VERTICES	$(-a, 0)$ and $(a, 0)$	$(0, -a)$ and $(0, a)$
OTHER INTERCEPTS	$(0, b)$ and $(0, -b)$	$(b, 0)$ and $(-b, 0)$
GRAPH		

In both cases, the value of c, the distance along the major axis from the center to the focus, is given by $c^2 = a^2 - b^2$. The length of the major axis is $2a$ and the length of the minor axis is $2b$.

It is important to note that the vertices correspond to intercepts when an ellipse is centered at the origin. One of the first things we notice about an ellipse is its *eccentricity*. The **eccentricity**, denoted e, is given by $e = \dfrac{c}{a}$, where $0 < e < 1$. The circle is a limiting form of an ellipse, $c = 0$. In other words, if the eccentricity is close to 0, then the ellipse resembles a circle, whereas if the eccentricity is close to 1, then the ellipse is quite elongated, or eccentric.

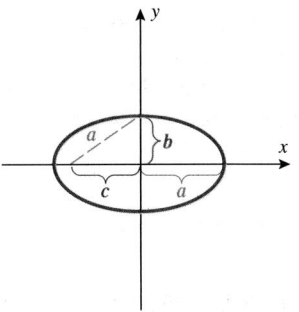

Graphing an Ellipse with Center at the Origin

The equation of an ellipse in standard form can be used to graph an ellipse. Although an ellipse is defined in terms of the foci, the foci are not part of the graph. It is important to note that if the divisor of the term with x^2 is larger than the divisor of the term with y^2, then the ellipse is elongated horizontally.

Technology Tip

Use a graphing calculator to check the graph of $\dfrac{x^2}{25} + \dfrac{y^2}{9} = 1$.

Solve for y first. That is,

$$y_1 = 3\sqrt{1 - \dfrac{x^2}{25}} \text{ and}$$

$$y_2 = -3\sqrt{1 - \dfrac{x^2}{25}}.$$

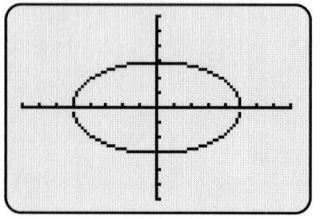

EXAMPLE 1 Graphing an Ellipse with a Horizontal Major Axis

Graph the ellipse given by $\dfrac{x^2}{25} + \dfrac{y^2}{9} = 1$.

Solution:

Since $25 > 9$, the major axis is horizontal. $\qquad a^2 = 25 \quad$ and $\quad b^2 = 9$

Solve for a and b. $\qquad\qquad\qquad\qquad\qquad a = 5 \quad$ and $\quad b = 3$

Identify the vertices: $(-a, 0)$ and $(a, 0)$. $\qquad (-5, 0) \quad$ and $\quad (5, 0)$

Identify the endpoints (y-intercepts) on
the minor axis: $(0, -b)$ and $(0, b)$. $\qquad (0, -3) \quad$ and $\quad (0, 3)$

Graph by labeling the points $(-5, 0)$,
$(5, 0)$, $(0, -3)$, and $(0, 3)$ and connecting
them with a smooth curve.

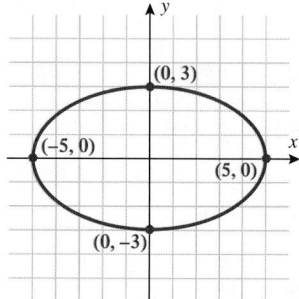

If the divisor of x^2 is larger than the divisor of y^2, then the major axis is horizontal along the x-axis, as in Example 1. If the divisor of y^2 is larger than the divisor of x^2, then the major axis is vertical along the y-axis, as you will see in Example 2.

Study Tip

If the divisor of x^2 is larger than the divisor of y^2, then the major axis is horizontal along the x-axis, as in Example 1. If the divisor of y^2 is larger than the divisor of x^2, then the major axis is vertical along the y-axis, as you will see in Example 2.

EXAMPLE 2 Graphing an Ellipse with a Vertical Major Axis

Graph the ellipse given by $16x^2 + y^2 = 16$.

Solution:

Write the equation in standard form by dividing by 16. $\qquad \dfrac{x^2}{1} + \dfrac{y^2}{16} = 1$

Since $16 > 1$, this ellipse is elongated vertically. $\qquad a^2 = 16 \quad$ and $\quad b^2 = 1$

Solve for a and b. $\qquad\qquad\qquad\qquad\qquad a = 4 \quad$ and $\quad b = 1$

Identify the vertices: $(0, -a)$ and $(0, a)$. $\qquad (0, -4) \quad$ and $\quad (0, 4)$

Identify the x-intercepts on the minor
axis: $(-b, 0)$ and $(b, 0)$. $\qquad (-1, 0) \quad$ and $\quad (1, 0)$

Graph by labeling the points $(0, -4)$,
$(0, 4)$, $(-1, 0)$, and $(1, 0)$ and connecting
them with a smooth curve.

■ **Answer:**

a.

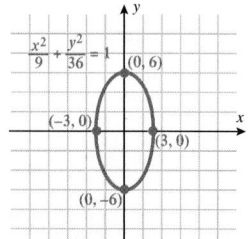

b.

■ **YOUR TURN** Graph the ellipses:

a. $\dfrac{x^2}{9} + \dfrac{y^2}{4} = 1$ $\qquad$ **b.** $\dfrac{x^2}{9} + \dfrac{y^2}{36} = 1$

Finding the Equation of an Ellipse with Center at the Origin

What if we know the vertices and the foci of an ellipse and want to find the equation to which it corresponds? The axis on which the foci and vertices are located is the major axis. Therefore, we will have the standard equation of an ellipse, and a will be known (from the vertices). Since c is known from the foci, we can use the relation $c^2 = a^2 - b^2$ to determine the unknown b.

EXAMPLE 3 Finding the Equation of an Ellipse Centered at the Origin

Find the standard form of the equation of an ellipse with foci at $(-3, 0)$ and $(3, 0)$ and vertices $(-4, 0)$ and $(4, 0)$.

Solution:

The major axis lies along the x-axis, since it contains the foci and vertices.

Write the corresponding general equation of an ellipse. $\qquad \dfrac{x^2}{a^2} + \dfrac{y^2}{b^2} = 1$

Identify a from the vertices:

Match vertices $(-4, 0) = (-a, 0)$ and $(4, 0) = (a, 0)$. $\qquad a = 4$

Identify c from the foci:

Match foci $(-3, 0) = (-c, 0)$ and $(3, 0) = (c, 0)$. $\qquad c = 3$

Substitute $a = 4$ and $c = 3$ into $b^2 = a^2 - c^2$. $\qquad b^2 = 4^2 - 3^2$

Simplify. $\qquad b^2 = 7$

Substitute $a^2 = 16$ and $b^2 = 7$ into $\dfrac{x^2}{a^2} + \dfrac{y^2}{b^2} = 1$. $\qquad \dfrac{x^2}{16} + \dfrac{y^2}{7} = 1$

The equation of the ellipse is $\boxed{\dfrac{x^2}{16} + \dfrac{y^2}{7} = 1}$.

■ **YOUR TURN** Find the standard form of the equation of an ellipse with vertices at $(0, -6)$ and $(0, 6)$ and foci $(0, -5)$ and $(0, 5)$.

■ **Answer:** $\dfrac{x^2}{11} + \dfrac{y^2}{36} = 1$

An Ellipse with Center at the Point (h, k)

We can use graph-shifting techniques to graph ellipses that are centered at a point other than the origin. For example, to graph $\dfrac{(x - h)^2}{a^2} + \dfrac{(y - k)^2}{b^2} = 1$ (assuming h and k are positive constants), start with the graph of $\dfrac{x^2}{a^2} + \dfrac{y^2}{b^2} = 1$ and shift to the right h units and up k units. The center, the vertices, the foci, and the major and minor axes all shift. In other words, the two ellipses are identical in shape and size, except that the ellipse $\dfrac{(x - h)^2}{a^2} + \dfrac{(y - k)^2}{b^2} = 1$ is centered at the point (h, k).

The following table summarizes the characteristics of ellipses centered at a point other than the origin:

EQUATION OF AN ELLIPSE WITH CENTER AT THE POINT (h, k)

The **standard form of the equation of an ellipse** with its center at the point (h, k) is given by

ORIENTATION OF MAJOR AXIS	Horizontal (parallel to the x-axis)	Vertical (parallel to the y-axis)
EQUATION	$\dfrac{(x-h)^2}{a^2} + \dfrac{(y-k)^2}{b^2} = 1$	$\dfrac{(x-h)^2}{b^2} + \dfrac{(y-k)^2}{a^2} = 1$
FOCI	$(h-c, k)$ and $(h+c, k)$	$(h, k-c)$ and $(h, k+c)$
GRAPH		

In both cases, $a > b > 0$, $c^2 = a^2 - b^2$, the length of the major axis is $2a$, and the length of the minor axis is $2b$.

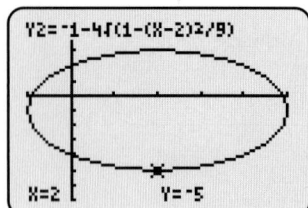

Technology Tip

Use a TI to check the graph of $\dfrac{(x-2)^2}{9} + \dfrac{(y+1)^2}{16} = 1$.

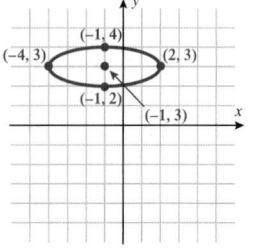

■ **Answer:**

EXAMPLE 4 Graphing an Ellipse with Center (h, k) Given the Equation in Standard Form

Graph the ellipse given by $\dfrac{(x-2)^2}{9} + \dfrac{(y+1)^2}{16} = 1$.

Solution:

Write the equation in the form

$\dfrac{(x-h)^2}{b^2} + \dfrac{(y-k)^2}{a^2} = 1$. $\qquad$ $\dfrac{(x-2)^2}{3^2} + \dfrac{[y-(-1)]^2}{4^2} = 1$

Identify a, b, and the center (h, k). $\qquad$ $a = 4$, $b = 3$, and $(h, k) = (2, -1)$

Draw a graph and label the center: $(2, -1)$.

Since $a = 4$, the vertices are up 4 and down 4 units from the center: $(2, -5)$ and $(2, 3)$.

Since $b = 3$, the endpoints of the minor axis are to the left and right 3 units: $(-1, -1)$ and $(5, -1)$.

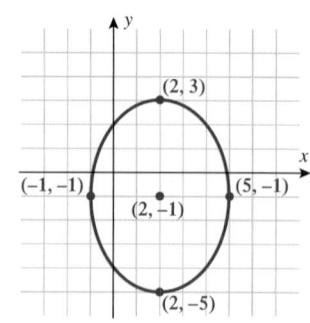

■ **YOUR TURN** Graph the ellipse given by $\dfrac{(x+1)^2}{9} + \dfrac{(y-3)^2}{1} = 1$.

All active members of the Lambda Chi fraternity are college students, but not all college students are members of the Lambda Chi fraternity. Similarly, all circles are ellipses, but not all ellipses are circles. When $a = b$, the standard equation of an ellipse simplifies to a standard equation of a circle. Recall that when we are given the equation of a circle in general form, we first complete the square in order to express the equation in standard form, which allows the center and radius to be identified. We use that same approach when the equation of an ellipse is given in a general form.

EXAMPLE 5 Graphing an Ellipse with Center (h, k) Given an Equation in General Form

Graph the ellipse given by $4x^2 + 24x + 25y^2 - 50y - 39 = 0$.

Solution:

Transform the general equation into standard form.

Group x terms together and y terms together and add 39 to both sides.
$$\left(4x^2 + 24x\right) + \left(25y^2 - 50y\right) = 39$$

Factor out the 4 common to the x terms and the 25 common to the y terms.
$$4\left(x^2 + 6x\right) + 25\left(y^2 - 2y\right) = 39$$

Complete the square on x and y.
$$4\left(x^2 + 6x + 9\right) + 25\left(y^2 - 2y + 1\right) = 39 + 4(9) + 25(1)$$

Simplify.
$$4(x + 3)^2 + 25(y - 1)^2 = 100$$

Divide by 100.
$$\frac{(x + 3)^2}{25} + \frac{(y - 1)^2}{4} = 1$$

Since $25 > 4$, this is an ellipse with a horizontal major axis.

Now that the equation of the ellipse is in standard form, compare to $\frac{(x - h)^2}{a^2} + \frac{(y - k)^2}{b^2} = 1$ and identify a, b, h, k.
$$a = 5, b = 2, \text{ and } (h, k) = (-3, 1)$$

Since $a = 5$, the vertices are 5 units to left and right of the center.
$$(-8, 1) \text{ and } (2, 1)$$

Since $b = 2$, the endpoints of the minor axis are up and down 2 units from the center.
$$(-3, -1) \text{ and } (-3, 3)$$

Graph.

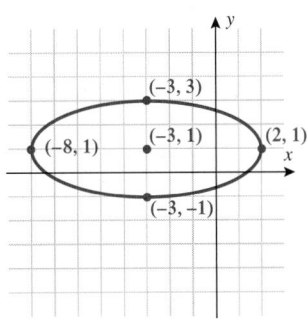

■ YOUR TURN Write the equation $4x^2 + 32x + y^2 - 2y + 61 = 0$ in standard form. Identify the center, vertices, and endpoints of the minor axis, and graph.

Technology Tip

Use a TI to check the graph of $4x^2 + 24x + 25y^2 - 50y - 39 = 0$.

Use $\frac{(x + 3)^2}{25} + \frac{(y - 1)^2}{4} = 1$ to solve for y first. That is,

$$y_1 = 1 + 2\sqrt{1 - \frac{(x + 3)^2}{25}} \text{ or}$$

$$y_2 = 1 - 2\sqrt{1 - \frac{(x + 3)^2}{25}}.$$

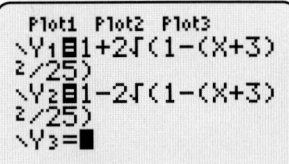

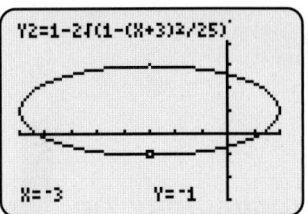

■ **Answer:**
$$\frac{(x + 4)^2}{1} + \frac{(y - 1)^2}{4} = 1$$

Center: $(-4, 1)$
Vertices: $(-4, -1)$ and $(-4, 3)$
Endpoints of minor axis: $(-5, 1)$ and $(-3, 1)$

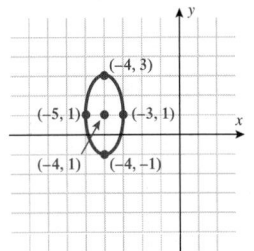

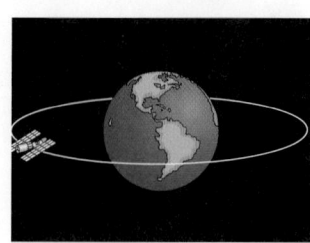

Applications

There are many examples of ellipses all around us. On Earth we have racetracks, and in our solar system, the planets travel in elliptical orbits with the Sun as a focus. Satellites are in elliptical orbits around Earth. Most communications satellites are in a *geosynchronous* (GEO) orbit—they orbit Earth once each day. In order to stay over the same spot on Earth, a *geostationary* satellite has to be directly above the equator; it circles Earth in exactly the time it takes Earth to turn once on its axis, and its orbit has to follow the path of the equator as Earth rotates. Otherwise, from Earth the satellite would appear to move in a north–south line every day.

If we start with an ellipse in the *xy*-plane and rotate it around its major axis, the result is a three-dimensional ellipsoid.

 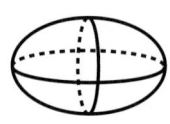

A football and a blimp are two examples of ellipsoids. The ellipsoidal shape allows for a more aerodynamic path.

PhotoDisc, Inc.

Peter Phipp/Age Fotostock America, Inc.

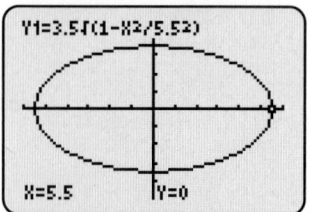
EXAMPLE 6 An Official NFL Football

A longitudinal section (that includes the two vertices and the center) of an official Wilson NFL football is an ellipse. The longitudinal section is approximately 11 inches long and 7 inches wide. Write an equation governing the elliptical longitudinal section.

Solution:

Locate the center of the ellipse at the origin and orient the football horizontally.

Write the general equation of an ellipse centered at the origin. $\dfrac{x^2}{a^2} + \dfrac{y^2}{b^2} = 1$

The length of the major axis is 11 inches. $2a = 11$

Solve for *a*. $a = 5.5$

The length of the minor axis is 7 inches. $2b = 7$

Solve for *b*. $b = 3.5$

Substitute $a = 5.5$ and $b = 3.5$ into $\dfrac{x^2}{a^2} + \dfrac{y^2}{b^2} = 1$. $\boxed{\dfrac{x^2}{5.5^2} + \dfrac{y^2}{3.5^2} = 1}$

SECTION
9.3 SUMMARY

In this section, we first analyzed ellipses that are centered at the origin.

ORIENTATION OF MAJOR AXIS	Horizontal along the x-axis	Vertical along the y-axis
EQUATION	$\dfrac{x^2}{a^2} + \dfrac{y^2}{b^2} = 1 \qquad a > b > 0$	$\dfrac{x^2}{b^2} + \dfrac{y^2}{a^2} = 1 \qquad a > b > 0$
FOCI*	$(-c, 0)$ and $(c, 0)$	$(0, -c)$ and $(0, c)$
VERTICES	$(-a, 0)$ and $(a, 0)$	$(0, -a)$ and $(0, a)$
OTHER INTERCEPTS	$(0, -b)$ and $(0, b)$	$(-b, 0)$ and $(b, 0)$
GRAPH		

$*c^2 = a^2 - b^2$

For ellipses centered at the origin, we can graph an ellipse by finding all four intercepts.
For ellipses centered at the point (h, k), the major and minor axes and endpoints of the ellipse all shift accordingly.
It is important to note that when $a = b$, the ellipse is a circle.

SECTION
9.3 EXERCISES

■ SKILLS

In Exercises 1–4, match the equation to the ellipse.

1. $\dfrac{x^2}{36} + \dfrac{y^2}{16} = 1$

2. $\dfrac{x^2}{16} + \dfrac{y^2}{36} = 1$

3. $\dfrac{x^2}{8} + \dfrac{y^2}{72} = 1$

4. $4x^2 + y^2 = 1$

a.

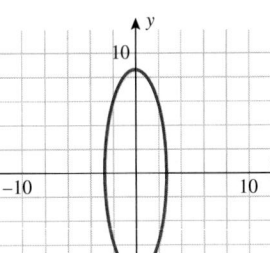

b.

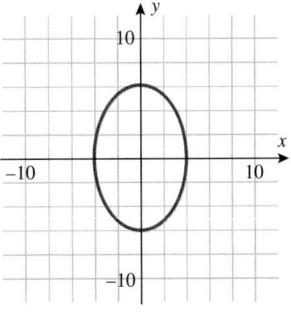

c.

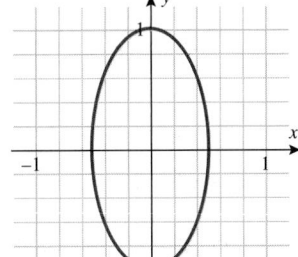

d.

In Exercises 5–16, graph each ellipse. Label the center and vertices.

5. $\dfrac{x^2}{25} + \dfrac{y^2}{16} = 1$

6. $\dfrac{x^2}{49} + \dfrac{y^2}{9} = 1$

7. $\dfrac{x^2}{16} + \dfrac{y^2}{64} = 1$

8. $\dfrac{x^2}{25} + \dfrac{y^2}{144} = 1$

9. $\dfrac{x^2}{100} + y^2 = 1$

10. $9x^2 + 4y^2 = 36$

11. $\dfrac{4}{9}x^2 + 81y^2 = 1$

12. $\dfrac{4}{25}x^2 + \dfrac{100}{9}y^2 = 1$

13. $4x^2 + y^2 = 16$

14. $x^2 + y^2 = 81$

15. $8x^2 + 16y^2 = 32$

16. $10x^2 + 25y^2 = 50$

In Exercises 17–24, find the standard form of the equation of an ellipse with the given characteristics.

17. Foci: $(-4, 0)$ and $(4, 0)$ Vertices: $(-6, 0)$ and $(6, 0)$

18. Foci: $(-1, 0)$ and $(1, 0)$ Vertices: $(-3, 0)$ and $(3, 0)$

19. Foci: $(0, -3)$ and $(0, 3)$ Vertices: $(0, -4)$ and $(0, 4)$

20. Foci: $(0, -1)$ and $(0, 1)$ Vertices: $(0, -2)$ and $(0, 2)$

21. Major axis vertical with length of 8, minor axis length of 4, and centered at $(0, 0)$

22. Major axis horizontal with length of 10, minor axis length of 2, and centered at $(0, 0)$

23. Vertices $(0, -7)$ and $(0, 7)$ and endpoints of minor axis $(-3, 0)$ and $(3, 0)$

24. Vertices $(-9, 0)$ and $(9, 0)$ and endpoints of minor axis $(0, -4)$ and $(0, 4)$

In Exercises 25–28, match each equation with the ellipse.

25. $\dfrac{(x-3)^2}{4} + \dfrac{(y+2)^2}{25} = 1$

26. $\dfrac{(x+3)^2}{4} + \dfrac{(y-2)^2}{25} = 1$

27. $\dfrac{(x-3)^2}{25} + \dfrac{(y+2)^2}{4} = 1$

28. $\dfrac{(x+3)^2}{25} + \dfrac{(y-2)^2}{4} = 1$

a.

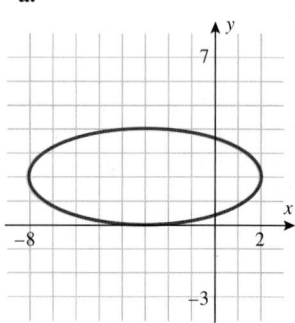

b.

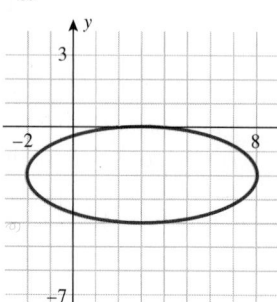

c.

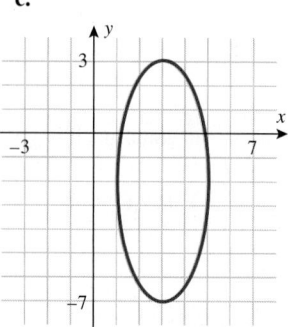

d.

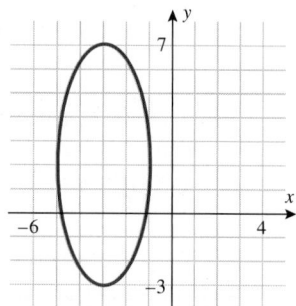

In Exercises 29–38, graph each ellipse. Label the center and vertices.

29. $\dfrac{(x-1)^2}{16} + \dfrac{(y-2)^2}{4} = 1$

30. $\dfrac{(x+1)^2}{36} + \dfrac{(y+2)^2}{9} = 1$

31. $10(x+3)^2 + (y-4)^2 = 80$

32. $3(x+3)^2 + 12(y-4)^2 = 36$

33. $x^2 + 4y^2 - 24y + 32 = 0$

34. $25x^2 + 2y^2 - 4y - 48 = 0$

35. $x^2 - 2x + 2y^2 - 4y - 5 = 0$

36. $9x^2 - 18x + 4y^2 - 27 = 0$

37. $5x^2 + 20x + y^2 + 6y - 21 = 0$

38. $9x^2 + 36x + y^2 + 2y + 36 = 0$

In Exercises 39–46, find the standard form of the equation of an ellipse with the given characteristics.

39. Foci: $(-2, 5)$ and $(6, 5)$ Vertices: $(-3, 5)$ and $(7, 5)$

40. Foci: $(2, -2)$ and $(4, -2)$ Vertices: $(0, -2)$ and $(6, -2)$

41. Foci: $(4, -7)$ and $(4, -1)$ Vertices: $(4, -8)$ and $(4, 0)$

42. Foci: $(2, -6)$ and $(2, -4)$ Vertices:$(2, -7)$ and $(2, -3)$

43. Major axis vertical with length of 8, minor axis length of 4, and centered at $(3, 2)$

44. Major axis horizontal with length of 10, minor axis length of 2, and centered at $(-4, 3)$

45. Vertices $(-1, -9)$ and $(-1, 1)$ and endpoints of minor axis $(-4, -4)$ and $(2, -4)$

46. Vertices $(-2, 3)$ and $(6, 3)$ and endpoints of minor axis $(2, 1)$ and $(2, 5)$

■ **APPLICATIONS**

47. Carnival Ride. The Zipper, a favorite carnival ride, maintains an elliptical shape with a major axis of 150 feet and a minor axis of 30 feet. Assuming it is centered at the origin, find an equation for the ellipse.

Zipper

48. Carnival Ride. A Ferris wheel traces an elliptical path with both a major and minor axis of 180 feet. Assuming it is centered at the origin, find an equation for the ellipse (circle).

Ferris wheel, Barcelona, Spain

For Exercises 49 and 50, refer to the following information:

A high school wants to build a football field surrounded by an elliptical track. A regulation football field must be 120 yards long and 30 yards wide.

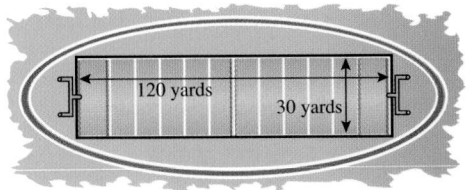

49. Sports Field. Suppose the elliptical track is centered at the origin and has a horizontal major axis of length 150 yards and a minor axis length of 40 yards.

a. Write an equation for the ellipse.
b. Find the width of the track at the end of the field. Will the track completely enclose the football field?

50. Sports Field. Suppose the elliptical track is centered at the origin and has a horizontal major axis of length 150 yards. How long should the minor axis be in order to enclose the field?

For Exercises 51 and 52, refer to orbits in our solar system:

The planets have elliptical orbits with the Sun as one of the foci. Pluto (orange), the planet furthest from the Sun, has a very elongated, or flattened, elliptical orbit, whereas Earth (royal blue) has an almost circular orbit. Because of Pluto's flattened path, it is not always the planet furthest from the Sun.

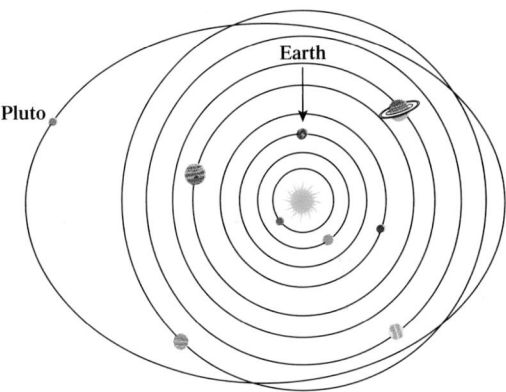

51. Planetary Orbits. The orbit of the dwarf planet Pluto has approximately the following characteristics (assume the Sun is the focus):

■ The length of the semimajor axis $2a$ is approximately 11,827,000,000 km.
■ The perihelion distance from the dwarf planet to the Sun is 4,447,000,000 km.

Determine the equation for Pluto's elliptical orbit around the Sun.

52. Planetary Orbits. Earth's orbit has approximately the following characteristics (assume the Sun is the focus):

■ The length of the semimajor axis $2a$ is approximately 299,700,000 km.
■ The perihelion distance from Earth to the Sun is 147,100,000 km.

Determine the equation for Earth's elliptical orbit around the Sun.

For Exercises 53 and 54, refer to the following information:

Asteroids orbit the Sun in elliptical patterns and often cross paths with Earth's orbit, making life a little tense now and again. A few asteroids have orbits that cross Earth's orbit—called "Apollo asteroids" or "Earth-crossing asteroids." In recent years, asteroids have passed within 100,000 kilometers of Earth!

53. Asteroids. Asteroid 433, or Eros, is the second largest near-Earth asteroid. The semimajor axis is 150 million kilometers and the eccentricity is 0.223, where eccentricity is defined as $e = \sqrt{1 - \dfrac{b^2}{a^2}}$, where a is the semimajor axis or $2a$ is the major axis, and b is the semiminor axis or $2b$ is the minor axis. Find the equation of Eros's orbit. Round a and b to the nearest million kilometers.

54. Asteroids. The asteroid Toutatis is the largest near-Earth asteroid. The semimajor axis is 350 million kilometers and the eccentricity is 0.634, where eccentricity is defined as $e = \sqrt{1 - \dfrac{b^2}{a^2}}$, where a is the semimajor axis or $2a$ is the major axis, and b is the semimajor axis or $2b$ is the minor axis. On September 29, 2004, it missed Earth by 961,000 miles. Find the equation of Toutatis's orbit.

55. Halley's Comet. The eccentricity of Halley's Comet is approximately 0.967. If a comet had e almost equal to 1, what would its orbit appear to be from Earth?

56. Halley's Comet. The length of the semimajor axis is 17.8 AU (astronomical units) and the eccentricity is approximately 0.967. Find the equation of Halley's Comet. (Assume 1 AU = 150 million km.)

57. Medicine. The second time that a drug is given to a patient, the relationship between the drug concentration c (in milligrams per cm^3) and the time t (in hours) is given by

$$t^2 + 9c^2 - 6t - 18c + 9 = 0$$

Determine the highest concentration of drug in the patient's bloodstream.

58. Fuel Transportation. Tanks built to transport fuel and other hazardous materials have an elliptical cross section, which makes them more stable for the transportation of these materials. If the lengths of the major and minor axes of the elliptical cross section are 8 ft and 6 ft, respectively, and the tank is 30 ft long, find the volume of the tank. Round your answer to the nearest integer. (*Hint:* The area of an ellipse is $\pi \cdot a \cdot b$.)

■ CATCH THE MISTAKE

In Exercises 59 and 60, explain the mistake that is made.

59. Graph the ellipse given by $\dfrac{x^2}{6} + \dfrac{y^2}{4} = 1$.

Solution:

Write the standard form of the equation of an ellipse.
$$\frac{x^2}{a^2} + \frac{y^2}{b^2} = 1$$

Identify a and b.
$$a = 6, b = 4$$

Label the vertices and the endpoints of the minor axis, $(-6, 0)$, $(6, 0)$, $(0, -4)$, $(0, 4)$, and connect with an elliptical curve.

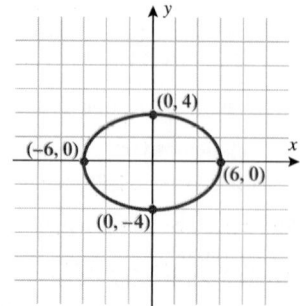

This is incorrect. What mistake was made?

60. Determine the foci of the ellipse $\dfrac{x^2}{16} + \dfrac{y^2}{9} = 1$.

Solution:

Write the general equation of a horizontal ellipse.
$$\frac{x^2}{a^2} + \frac{y^2}{b^2} = 1$$

Identify a and b.
$$a = 4, b = 3$$

Substitute $a = 4$, $b = 3$ into $c^2 = a^2 + b^2$.
$$c^2 = 4^2 + 3^2$$

Solve for c.
$$c = 5$$

Foci are located at $(-5, 0)$ and $(5, 0)$.

The points $(-5, 0)$ and $(5, 0)$ are located outside of the ellipse.

This is incorrect. What mistake was made?

■ CONCEPTUAL

In Exercises 61–64, determine whether each statement is true or false.

61. If you know the vertices of an ellipse, you can determine the equation for the ellipse.

62. If you know the foci and the endpoints of the minor axis, you can determine the equation for the ellipse.

63. Ellipses centered at the origin have symmetry with respect to the x-axis, y-axis, and the origin.

64. All ellipses are circles, but not all circles are ellipses.

65. How many ellipses, with major and minor axes parallel to the coordinate axes, have focus $(-2, 0)$ and pass through the point $(-2, 2)$?

66. How many ellipses have vertices $(-3, 0)$ and $(3, 0)$?

67. If two ellipses intersect each other, what is the minimum number of intersection points?

68. If two ellipses intersect each other, what is the maximum number of intersection points?

■ CHALLENGE

69. The eccentricity of an ellipse is defined as $e = \dfrac{c}{a}$. Compare the eccentricity of the orbit of Pluto to that of Earth (refer to Exercises 51 and 52).

70. The eccentricity of an ellipse is defined as $e = \dfrac{c}{a}$. Since $a > c > 0$, then $0 < e < 1$. Describe the shape of an ellipse when

  a. e is close to zero
  b. e is close to one
  c. $e = 0.5$

71. Find the equation of an ellipse centered at the origin containing the points $(1, 3)$ and $(4, 2)$.

72. Find the equation of an ellipse centered at the origin containing the points $\left(1, \dfrac{6\sqrt{5}}{5}\right)$ and $\left(-\dfrac{5}{3}, 2\right)$.

73. Find the equation of an ellipse centered at $(2, -3)$ that passes through the points $\left(1, -\dfrac{1}{3}\right)$ and $(5, -3)$.

74. Find the equation of an ellipse centered at $(1, -2)$ that passes through the points $(1, -4)$ and $(2, -2)$.

■ TECHNOLOGY

75. Graph the following three ellipses: $x^2 + y^2 = 1$, $x^2 + 5y^2 = 1$, and $x^2 + 10y^2 = 1$. What can be said to happen to the ellipse $x^2 + cy^2 = 1$ as c increases?

76. Graph the following three ellipses: $x^2 + y^2 = 1$, $5x^2 + y^2 = 1$, and $10x^2 + y^2 = 1$. What can be said to happen to the ellipse $cx^2 + y^2 = 1$ as c increases?

77. Graph the following three ellipses: $x^2 + y^2 = 1$, $5x^2 + 5y^2 = 1$, and $10x^2 + 10y^2 = 1$. What can be said to happen to the ellipse $cx^2 + cy^2 = 1$ as c increases?

78. Graph the equation $\dfrac{x^2}{9} - \dfrac{y^2}{16} = 1$. Notice what a difference the sign makes. Is this an ellipse?

79. Graph the following three ellipses: $x^2 + y^2 = 1$, $0.5x^2 + y^2 = 1$, and $0.05x^2 + y^2 = 1$. What can be said to happen to ellipse $cx^2 + y^2 = 1$ as c decreases?

80. Graph the following three ellipses: $x^2 + y^2 = 1$, $x^2 + 0.5y^2 = 1$, and $x^2 + 0.05y^2 = 1$. What can be said to happen to ellipse $x^2 + cy^2 = 1$ as c decreases?

■ PREVIEW TO CALCULUS

In calculus, the derivative of a function is used to find its maximum and minimum values. In the case of an ellipse, with major and minor axes parallel to the coordinate axes, the maximum and minimum values correspond to the y-coordinate of the vertices that lie on its vertical axis of symmetry.

  In Exercises 81–84, find the maximum and minimum values of each ellipse.

81. $4x^2 + y^2 - 24x + 10y + 57 = 0$

82. $9x^2 + 4y^2 + 72x + 16y + 124 = 0$

83. $81x^2 + 100y^2 - 972x + 1600y + 1216 = 0$

84. $25x^2 + 16y^2 + 200x + 256y - 176 = 0$

SKILLS OBJECTIVES

- Find a hyperbola's foci and vertices.
- Find the equation of a hyperbola centered at the origin.
- Graph a hyperbola using asymptotes as graphing aids.
- Find the equation of a hyperbola centered at the point (h, k).
- Solve applied problems that involve hyperbolas.

CONCEPTUAL OBJECTIVES

- Derive the general equation of a hyperbola.
- Identify, apply, and graph the transverse axis, vertices, and foci.
- Use asymptotes to determine the shape of a hyperbola.

The definition of a hyperbola is similar to the definition of an ellipse. An ellipse is the set of all points, the *sum* of whose distances from two points (the foci) is constant. A *hyperbola* is the set of all points, the *difference* of whose distances from two points (the foci) is constant. What distinguishes their equations is a minus sign.

Ellipse centered at the origin: $\quad \dfrac{x^2}{a^2} + \dfrac{y^2}{b^2} = 1$

Hyperbola centered at the origin: $\quad \dfrac{x^2}{a^2} - \dfrac{y^2}{b^2} = 1$

DEFINITION   Hyperbola

A **hyperbola** is the set of all points in a plane the difference of whose distances from two fixed points is a positive constant. These two fixed points are called **foci**. The hyperbola has two separate curves called **branches**. The two points where the hyperbola intersects the line joining the foci are called **vertices**. The line segment joining the vertices is called the **transverse axis of the hyperbola**. The midpoint of the transverse axis is called the **center**.

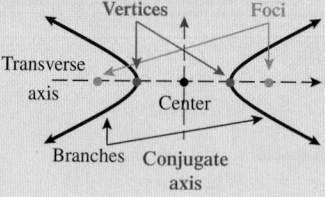

Definition of a Hyperbola

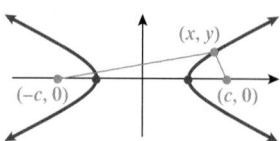

Let's consider a hyperbola with the center at the origin and the foci on the x-axis. Let the distance from the center to the focus be c. Therefore, the foci are located at the points $(-c, 0)$ and $(c, 0)$. The difference of the two distances from the foci to any point (x, y) must be constant. We then can follow a similar analysis as done with an ellipse.

Derivation of the Equation of a Hyperbola

WORDS	MATH
The difference of these two distances is equal to a constant ($2a$ for convenience).	$$\sqrt{[x - (-c)]^2 + y^2} - \sqrt{(x - c)^2 + y^2} = \pm 2a$$
Following the same procedure that we did with an ellipse leads to:	$$(c^2 - a^2)x^2 - a^2y^2 = a^2(c^2 - a^2)$$
We can make the argument that $c > a$ in order for a point to be on the hyperbola (and not on the x-axis). Therefore, since a and c represent distances and therefore are positive, we know that $c^2 > a^2$, or $c^2 - a^2 > 0$. Hence, we can divide both sides of the equation by $c^2 - a^2$, since $c^2 - a^2 \neq 0$.	$$x^2 - \frac{a^2y^2}{(c^2 - a^2)} = a^2$$
Let $b^2 = c^2 - a^2$.	$$x^2 - \frac{a^2y^2}{b^2} = a^2$$
Divide both sides of the equation by a^2.	$$\frac{x^2}{a^2} - \frac{y^2}{b^2} = 1$$

The equation $\dfrac{x^2}{a^2} - \dfrac{y^2}{b^2} = 1$ represents a hyperbola with its center at the origin, with the foci along the x-axis. The following box summarizes hyperbolas that have their center at the origin and foci along either the x-axis or y-axis:

EQUATION OF A HYPERBOLA WITH CENTER AT THE ORIGIN

The **standard form of the equation of a hyperbola** with its center at the origin is given by

ORIENTATION OF TRANSVERSE AXIS	Horizontal (along the x-axis)	Vertical (along the y-axis)
EQUATION	$\dfrac{x^2}{a^2} - \dfrac{y^2}{b^2} = 1$	$\dfrac{y^2}{a^2} - \dfrac{x^2}{b^2} = 1$
FOCI	$(-c, 0)$ and $(c, 0)$ where $c^2 = a^2 + b^2$	$(0, -c)$ and $(0, c)$ where $c^2 = a^2 + b^2$
ASYMPTOTES	$y = \dfrac{b}{a}x$ and $y = -\dfrac{b}{a}x$	$y = \dfrac{a}{b}x$ and $y = -\dfrac{a}{b}x$
VERTICES	$(-a, 0)$ and $(a, 0)$	$(0, -a)$ and $(0, a)$
TRANSVERSE AXIS	Horizontal length $2a$	Vertical length $2a$
GRAPH	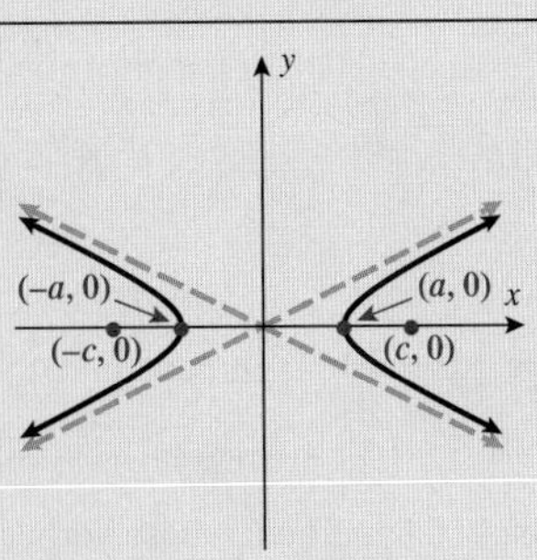	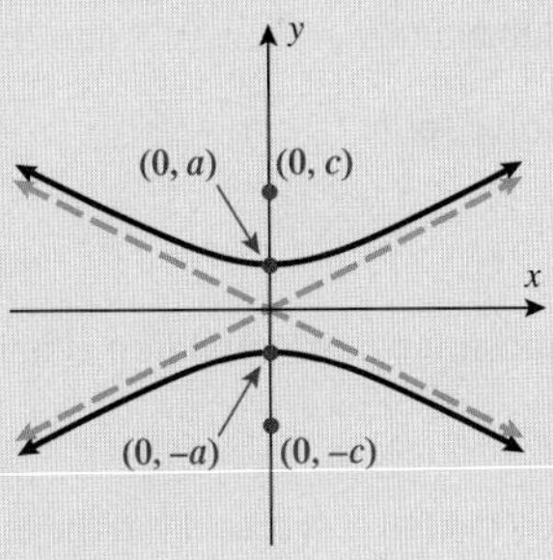

Note that for $\dfrac{x^2}{a^2} - \dfrac{y^2}{b^2} = 1$, if $x = 0$, then $-\dfrac{y^2}{b^2} = 1$, which yields an imaginary number for y. However, when $y = 0$, $\dfrac{x^2}{a^2} = 1$, and therefore $x = \pm a$. The vertices for this hyperbola are $(-a, 0)$ and $(a, 0)$.

Technology Tip

Use a TI to check the graph of $\dfrac{x^2}{9} - \dfrac{y^2}{4} = 1$ Solve for y first. That is,

$$y_1 = 2\sqrt{\dfrac{x^2}{9} - 1} \text{ or } y_2 = -2\sqrt{\dfrac{x^2}{9} - 1}.$$

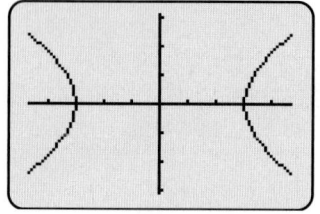

■ **Answer:** Vertices: $(0, -4)$ and $(0, 4)$
Foci: $(0, -6)$ and $(0, 6)$

EXAMPLE 1 Finding the Foci and Vertices of a Hyperbola Given the Equation

Find the foci and vertices of the hyperbola given by $\dfrac{x^2}{9} - \dfrac{y^2}{4} = 1$.

Solution:

Compare to the standard equation of a hyperbola, $\dfrac{x^2}{a^2} - \dfrac{y^2}{b^2} = 1$. $\qquad a^2 = 9, b^2 = 4$

Solve for a and b. $\qquad\qquad\qquad\qquad\qquad\qquad\qquad a = 3, b = 2$

Substitute $a = 3$ into the vertices, $(-a, 0)$ and $(a, 0)$. $\qquad (-3, 0)$ and $(3, 0)$

Substitute $a = 3, b = 2$ into $c^2 = a^2 + b^2$. $\qquad\qquad\qquad c^2 = 3^2 + 2^2$

Solve for c. $\qquad\qquad\qquad\qquad\qquad\qquad\qquad\qquad c^2 = 13$

$$c = \sqrt{13}$$

Substitute $c = \sqrt{13}$ into the foci, $(-c, 0)$ and $(c, 0)$. $\qquad \left(-\sqrt{13}, 0\right)$ and $\left(\sqrt{13}, 0\right)$

The vertices are $\boxed{(-3, 0)}$ and $\boxed{(3, 0)}$, and the foci are $\boxed{\left(-\sqrt{13}, 0\right)}$ and $\boxed{\left(\sqrt{13}, 0\right)}$.

■ **YOUR TURN** Find the vertices and foci of the hyperbola $\dfrac{y^2}{16} - \dfrac{x^2}{20} = 1$.

Technology Tip

Use a TI to check the graph of $\dfrac{y^2}{16} - \dfrac{x^2}{9} = 1$. Solve for y first. That is, $y_1 = 4\sqrt{1 + \dfrac{x^2}{9}}$ or

$$y_2 = -4\sqrt{1 - \dfrac{x^2}{9}}.$$

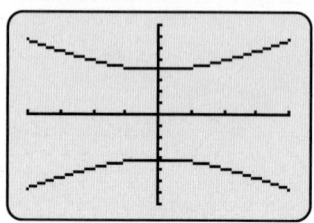

■ **Answer:** $\dfrac{x^2}{4} - \dfrac{y^2}{12} = 1$

EXAMPLE 2 Finding the Equation of a Hyperbola Given Foci and Vertices

Find the standard form of the equation of a hyperbola whose vertices are located at $(0, -4)$ and $(0, 4)$ and whose foci are located at $(0, -5)$ and $(0, 5)$.

Solution:

The center is located at the midpoint of the segment joining the vertices.

$$\left(\dfrac{0 + 0}{2}, \dfrac{-4 + 4}{2}\right) = (0, 0)$$

Since the foci and vertices are located on the y-axis, the standard equation is given by:

$$\dfrac{y^2}{a^2} - \dfrac{x^2}{b^2} = 1$$

The vertices $(0, \pm a)$ and the foci $(0, \pm c)$ can be used to identify a and c. $\qquad a = 4, c = 5$

Substitute $a = 4, c = 5$ into $b^2 = c^2 - a^2$. $\qquad b^2 = 5^2 - 4^2$

Solve for b. $\qquad\qquad\qquad\qquad\qquad\qquad b^2 = 25 - 16 = 9$

$$b = 3$$

Substitute $a = 4$ and $b = 3$ into $\dfrac{y^2}{a^2} - \dfrac{x^2}{b^2} = 1$. $\qquad \boxed{\dfrac{y^2}{16} - \dfrac{x^2}{9} = 1}$

■ **YOUR TURN** Find the equation of a hyperbola whose vertices are located at $(-2, 0)$ and $(2, 0)$ and whose foci are located at $(-4, 0)$ and $(4, 0)$.

Graphing a Hyperbola

To graph a hyperbola, we use the vertices and asymptotes. The asymptotes are found by the equations $y = \pm\dfrac{b}{a}x$ or $y = \pm\dfrac{a}{b}x$, depending on whether the transverse axis is horizontal or vertical. An easy way to draw these graphing aids is to first draw the rectangular box that passes through the vertices and the points $(0, \pm b)$ or $(\pm b, 0)$. The **conjugate axis** is perpendicular to the transverse axis and has length $2b$. The asymptotes pass through the center of the hyperbola and the corners of the rectangular box.

$$\frac{x^2}{a^2} - \frac{y^2}{b^2} = 1 \qquad\qquad \frac{y^2}{a^2} - \frac{x^2}{b^2} = 1$$

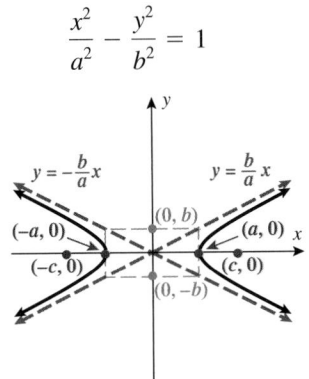
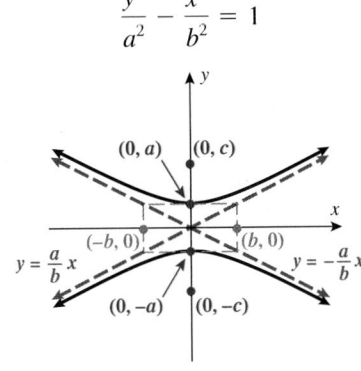

 EXAMPLE 3 **Graphing a Hyperbola Centered at the Origin with a Horizontal Transverse Axis**

Graph the hyperbola given by $\dfrac{x^2}{4} - \dfrac{y^2}{9} = 1$.

Solution:

Compare $\dfrac{x^2}{2^2} - \dfrac{y^2}{3^2} = 1$ to the general equation $\dfrac{x^2}{a^2} - \dfrac{y^2}{b^2} = 1$.

Identify a and b. $a = 2 \quad$ and $\quad b = 3$

The transverse axis of this hyperbola lies on the x-axis.

Label the vertices $(-a, 0) = (-2, 0)$
and $(a, 0) = (2, 0)$ and the points
$(0, -b) = (0, -3)$ and $(0, b) = (0, 3)$.
Draw the rectangular box that passes
through those points. Draw the asymptotes
that pass through the center and the corners
of the rectangle.

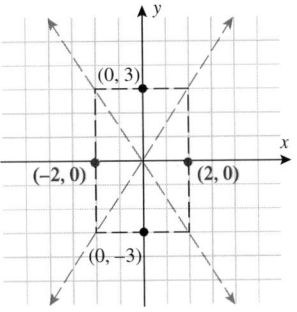

Draw the two branches of the hyperbola,
each passing through a vertex and guided
by the asymptotes.

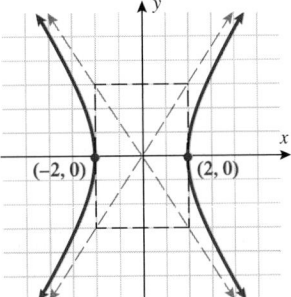

Technology Tip

Use a TI to check the graph of
$\dfrac{x^2}{4} - \dfrac{y^2}{9} = 1$. Solve for y first.

That is, $y_1 = 3\sqrt{\dfrac{x^2}{4} - 1}$ or
$y_2 = -3\sqrt{\dfrac{x^2}{4} - 1}$.

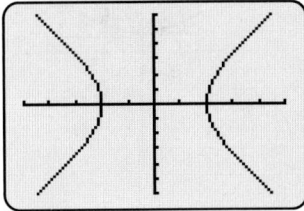

To add the two asymptotes, enter
$y_3 = \frac{3}{2}x$ or $y_4 = -\frac{3}{2}x$.

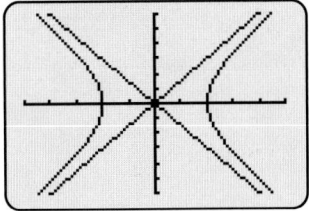

In Example 3, if we let $y = 0$, then $\dfrac{x^2}{4} = 1$ or $x = \pm 2$. Thus the vertices are $(-2, 0)$ and $(2, 0)$, and the transverse axis lies along the x-axis. Note that if $x = 0$, $y = \pm 3i$.

Technology Tip

Use a TI to check the graph of $\dfrac{y^2}{16} - \dfrac{x^2}{4} = 1$. Solve for y first.

That is, $y_1 = 4\sqrt{1 + \dfrac{x^2}{4}}$ or $y_2 = -4\sqrt{1 - \dfrac{x^2}{4}}$.

To add the two asymptotes, enter $y_3 = 2x$ or $y_4 = -2x$.

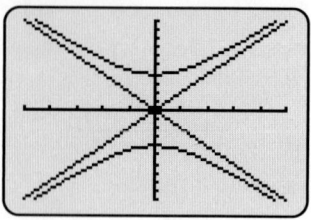

■ Answer:

a.

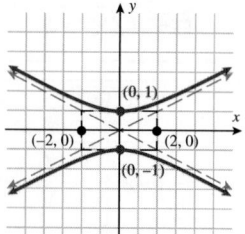

b.

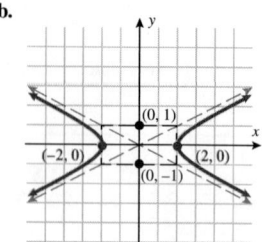

EXAMPLE 4 **Graphing a Hyperbola Centered at the Origin with a Vertical Transverse Axis**

Graph the hyperbola given by $\dfrac{y^2}{16} - \dfrac{x^2}{4} = 1$.

Solution:

Compare $\dfrac{y^2}{4^2} - \dfrac{x^2}{2^2} = 1$ to the general equation $\dfrac{y^2}{a^2} - \dfrac{x^2}{b^2} = 1$.

Identify a and b. $a = 4$ and $b = 2$

The transverse axis of this hyperbola lies along the y-axis.

Label the vertices $(0, -a) = (0, -4)$ and $(0, a) = (0, 4)$, and the points $(-b, 0) = (-2, 0)$ and $(b, 0) = (2, 0)$. Draw the rectangular box that passes through those points. Draw the asymptotes that pass through the center and the corners of the rectangle.

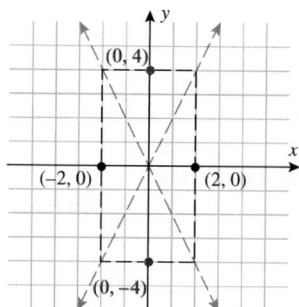

Draw the two branches of the hyperbola, each passing through a vertex and guided by the asymptotes.

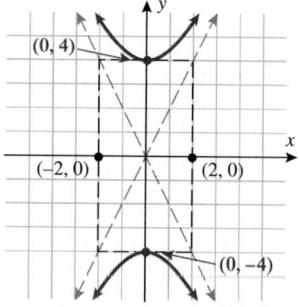

■ YOUR TURN Graph the hyperbolas:

a. $\dfrac{y^2}{1} - \dfrac{x^2}{4} = 1$ **b.** $\dfrac{x^2}{4} - \dfrac{y^2}{1} = 1$

Hyperbolas Centered at the Point (*h*, *k*)

We can use graph-shifting techniques to graph hyperbolas that are centered at a point other than the origin—say, (h, k). For example, to graph $\dfrac{(x - h)^2}{a^2} - \dfrac{(y - k)^2}{b^2} = 1$, start with the graph of $\dfrac{x^2}{a^2} - \dfrac{y^2}{b^2} = 1$ and shift to the right h units and up k units. The center, the vertices, the foci, the transverse and conjugate axes, and the asymptotes all shift. The following table summarizes the characteristics of hyperbolas centered at a point other than the origin:

EQUATION OF A HYPERBOLA WITH CENTER AT THE POINT (*h*, *k*)

The **standard form of the equation of a hyperbola** with its center at the point (h, k) is given by

ORIENTATION OF TRANSVERSE AXIS	Horizontal (parallel to the *x*-axis)	Vertical (parallel to the *y*-axis)
EQUATION	$\dfrac{(x - h)^2}{a^2} - \dfrac{(y - k)^2}{b^2} = 1$	$\dfrac{(y - k)^2}{a^2} - \dfrac{(x - h)^2}{b^2} = 1$
VERTICES	$(h - a, k)$ and $(h + a, k)$	$(h, k - a)$ and $(h, k + a)$
FOCI	$(h - c, k)$ and $(h + c, k)$ where $c^2 = a^2 + b^2$	$(h, k - c)$ and $(h, k + c)$ where $c^2 = a^2 + b^2$
GRAPH	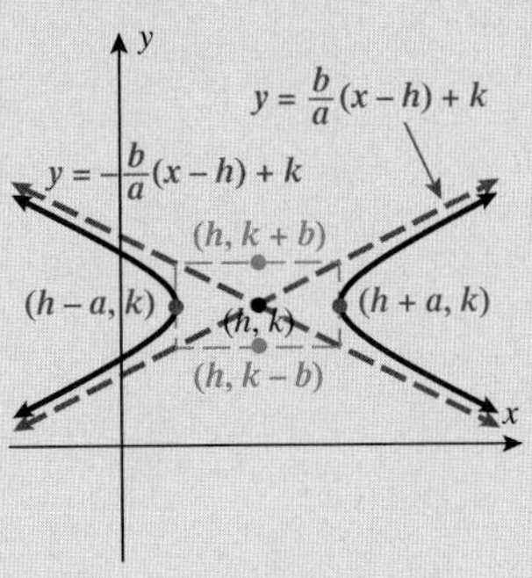	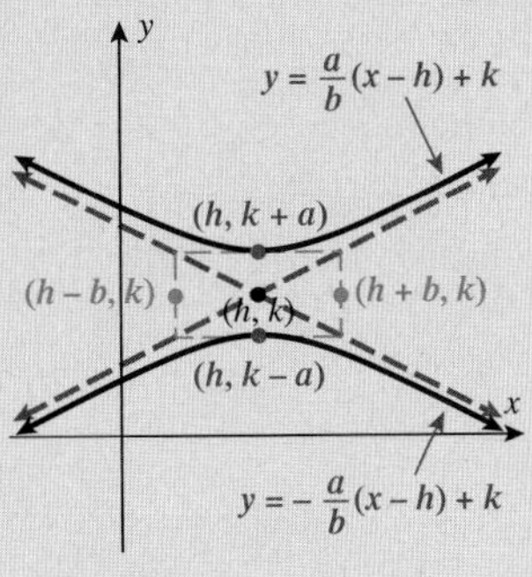

Technology Tip

Use a TI to check the graph of
$$\frac{(y-2)^2}{16} - \frac{(x-1)^2}{9} = 1.$$
Solve for y first. That is,

$$y_1 = 2 + 4\sqrt{1 + \frac{(x-1)^2}{9}} \text{ or}$$

$$y_2 = 2 - 4\sqrt{1 + \frac{(x-1)^2}{4}}.$$

To add the two asymptotes, enter

$$y_3 = \frac{4}{3}(x-1) + 2 \text{ or}$$

$$y_4 = -\frac{4}{3}(x-1) + 2.$$

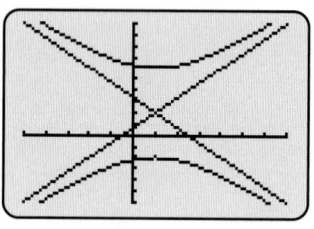

EXAMPLE 5   Graphing a Hyperbola with Center Not at the Origin

Graph the hyperbola $\dfrac{(y-2)^2}{16} - \dfrac{(x-1)^2}{9} = 1$.

Solution:

Compare $\dfrac{(y-2)^2}{4^2} - \dfrac{(x-1)^2}{3^2} = 1$ to the general equation $\dfrac{(y-k)^2}{a^2} - \dfrac{(x-h)^2}{b^2} = 1$.

Identify a, b, and (h, k). $a = 4$, $b = 3$, and $(h, k) = (1, 2)$

The transverse axis of this hyperbola lies along $x = 2$, which is parallel to the y-axis.

Label the vertices $(h, k - a) = (1, -2)$ and
$(h, k + a) = (1, 6)$ and the points $(h - b, k) = (-2, 2)$
and $(h + b, k) = (4, 2)$. Draw the rectangular box that
passes through those points. Draw the asymptotes that
pass through the center $(h, k) = (1, 2)$ and the corners
of the rectangle. Draw the two branches of the hyperbola,
each passing through a vertex and guided by the
asymptotes.

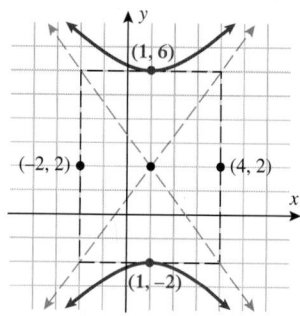

EXAMPLE 6   Transforming an Equation of a Hyperbola to Standard Form

Graph the hyperbola $9x^2 - 16y^2 - 18x + 32y - 151 = 0$.

Solution:

Complete the square on the x
terms and y terms, respectively.

$$9(x^2 - 2x) - 16(y^2 - 2y) = 151$$
$$9(x^2 - 2x + 1) - 16(y^2 - 2y + 1) = 151 + 9 - 16$$
$$9(x-1)^2 - 16(y-1)^2 = 144$$
$$\frac{(x-1)^2}{16} - \frac{(y-1)^2}{9} = 1$$

Compare $\dfrac{(x-1)^2}{16} - \dfrac{(y-1)^2}{9} = 1$ to the general form $\dfrac{(x-h)^2}{a^2} - \dfrac{(y-k)^2}{b^2} = 1$.

Identify a, b, and (h, k). $a = 4$, $b = 3$, and $(h, k) = (1, 1)$

The transverse axis of this hyperbola lies along $y = 1$.

Label the vertices $(h - a, k) = (-3, 1)$ and
$(h + a, k) = (5, 1)$ and the points
$(h, k - b) = (1, -2)$ and $(h, k + b) = (1, 4)$.
Draw the rectangular box that passes through these
points. Draw the asymptotes that pass through the
center $(1, 1)$ and the corners of the box. Draw the
two branches of the hyperbola, each passing through
a vertex and guided by the asymptotes.

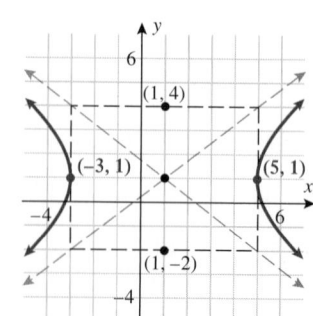

Applications

Nautical navigation is assisted by hyperbolas. For example, suppose that two radio stations on a coast are emitting simultaneous signals. If a boat is at sea, it will be slightly closer to one station than the other station, which results in a small time difference between the received signals from the two stations. If the boat follows the path associated with a constant time difference, that path will be hyperbolic.

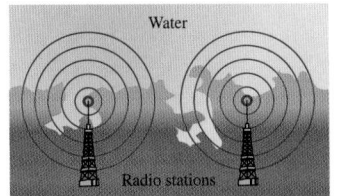

The synchronized signals would intersect one another in associated hyperbolas. Each time difference corresponds to a different path. The radio stations are the foci of the hyperbolas. This principle forms the basis of a hyperbolic radio navigation system known as *loran* (**LO**ng-**RA**nge **N**avigation).

There are navigational charts that correspond to different time differences. A ship selects the hyperbolic path that will take it to the desired port, and the loran chart lists the corresponding time difference.

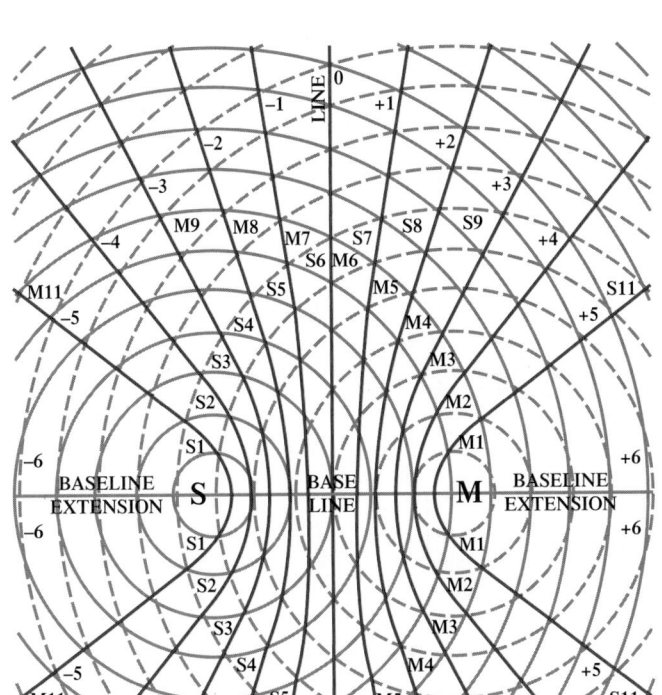

EXAMPLE 7 Nautical Navigation Using Loran

Two loran stations are located 200 miles apart along a coast. If a ship records a time difference of 0.00043 second and continues on the hyperbolic path corresponding to that difference, where does it reach shore?

Solution:

Draw the xy-plane and the two stations corresponding to the foci at $(-100, 0)$ and $(100, 0)$. Draw the ship somewhere in quadrant I.

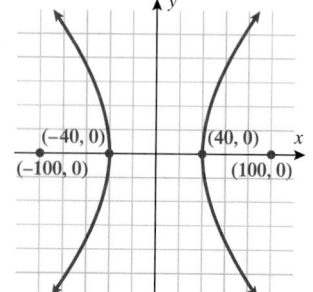

The hyperbola corresponds to a path where the difference of the distances between the ship and the respective stations remains constant. The constant is $2a$, where $(a, 0)$ is a vertex. Find that difference by using $d = rt$. Assume that the speed of the radio signal is 186,000 miles per second.

Substitute $r = 186,000$ miles/second and $t = 0.00043$ second into $d = rt$.

$$d = (186{,}000 \text{ miles/second})(0.00043 \text{ second}) \approx 80 \text{ miles}$$

Set the constant equal to $2a$. $\qquad\qquad 2a = 80$

Find a vertex $(a, 0)$. $\qquad\qquad\qquad (40, 0)$

The ship reaches shore between the two stations, 60 miles from station B and 140 miles from station A.

SECTION 9.4 SUMMARY

In this section, we discussed hyperbolas centered at the origin.

EQUATION	$\dfrac{x^2}{a^2} - \dfrac{y^2}{b^2} = 1$	$\dfrac{y^2}{a^2} - \dfrac{x^2}{b^2} = 1$
TRANSVERSE AXIS	Horizontal (x-axis), length $2a$	Vertical (y-axis), length $2a$
CONJUGATE AXIS	Vertical (y-axis), length $2b$	Horizontal (x-axis), length $2b$
VERTICES	$(-a, 0)$ and $(a, 0)$	$(0, -a)$ and $(0, a)$
FOCI	$(-c, 0)$ and $(c, 0)$ where $c^2 = a^2 + b^2$	$(0, -c)$ and $(0, c)$ where $c^2 = a^2 + b^2$
ASYMPTOTE	$y = \dfrac{b}{a}x$ and $y = -\dfrac{b}{a}x$	$y = \dfrac{a}{b}x$ and $y = -\dfrac{a}{b}x$
GRAPH	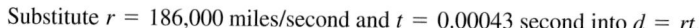	

For a hyperbola centered at (h, k), the vertices, foci, and asymptotes all shift accordingly.

SECTION
9.4 EXERCISES

■ SKILLS

In Exercises 1–4, match each equation with the corresponding hyperbola.

1. $\dfrac{x^2}{36} - \dfrac{y^2}{16} = 1$ **2.** $\dfrac{y^2}{36} - \dfrac{x^2}{16} = 1$ **3.** $\dfrac{x^2}{8} - \dfrac{y^2}{72} = 1$ **4.** $4y^2 - x^2 = 1$

a.　　　　　　　　　**b.**　　　　　　　　　**c.**　　　　　　　　　**d.**

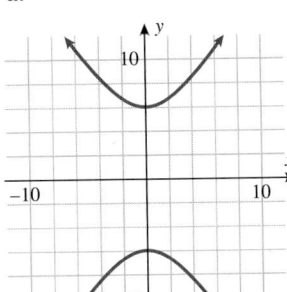　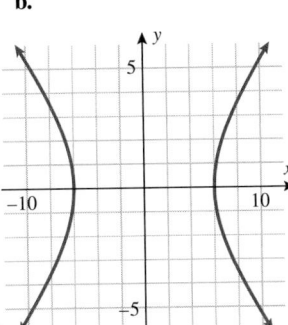　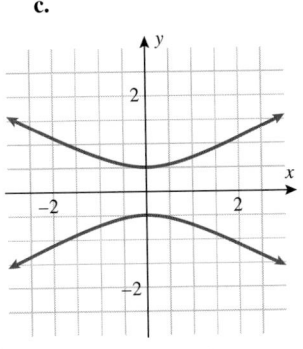　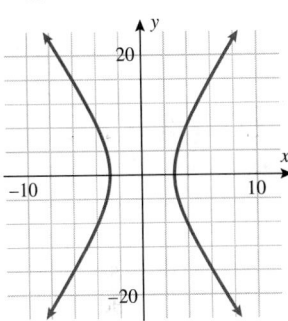

In Exercises 5–16, graph each hyperbola.

5. $\dfrac{x^2}{25} - \dfrac{y^2}{16} = 1$ **6.** $\dfrac{x^2}{49} - \dfrac{y^2}{9} = 1$ **7.** $\dfrac{y^2}{16} - \dfrac{x^2}{64} = 1$ **8.** $\dfrac{y^2}{144} - \dfrac{x^2}{25} = 1$

9. $\dfrac{x^2}{100} - y^2 = 1$ **10.** $9y^2 - 4x^2 = 36$ **11.** $\dfrac{4y^2}{9} - 81x^2 = 1$ **12.** $\dfrac{4}{25}x^2 - \dfrac{100}{9}y^2 = 1$

13. $4x^2 - y^2 = 16$ **14.** $y^2 - x^2 = 81$ **15.** $8y^2 - 16x^2 = 32$ **16.** $10x^2 - 25y^2 = 50$

In Exercises 17–24, find the standard form of an equation of the hyperbola with the given characteristics.

17. Vertices: $(-4, 0)$ and $(4, 0)$ Foci: $(-6, 0)$ and $(6, 0)$　　**18.** Vertices: $(-1, 0)$ and $(1, 0)$ Foci: $(-3, 0)$ and $(3, 0)$

19. Vertices: $(0, -3)$ and $(0, 3)$ Foci: $(0, -4)$ and $(0, 4)$　　**20.** Vertices: $(0, -1)$ and $(0, 1)$ Foci: $(0, -2)$ and $(0, 2)$

21. Center: $(0, 0)$; transverse: x-axis; asymptotes: $y = x$ and $y = -x$　　**22.** Center: $(0, 0)$; transverse: y-axis; asymptotes: $y = x$ and $y = -x$

23. Center: $(0, 0)$; transverse axis: y-axis; asymptotes: $y = 2x$ and $y = -2x$　　**24.** Center: $(0, 0)$; transverse axis: x-axis; asymptotes: $y = 2x$ and $y = -2x$

In Exercises 25–28, match each equation with the hyperbola.

25. $\dfrac{(x-3)^2}{4} - \dfrac{(y+2)^2}{25} = 1$ **26.** $\dfrac{(x+3)^2}{4} - \dfrac{(y-2)^2}{25} = 1$ **27.** $\dfrac{(y-3)^2}{25} - \dfrac{(x+2)^2}{4} = 1$ **28.** $\dfrac{(y+3)^2}{25} - \dfrac{(x-2)^2}{4} = 1$

a.　　　　　　　　　**b.**　　　　　　　　　**c.**　　　　　　　　　**d.**

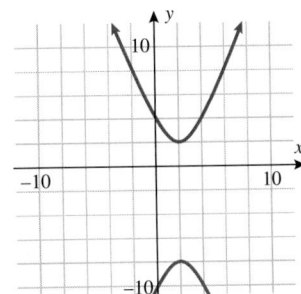　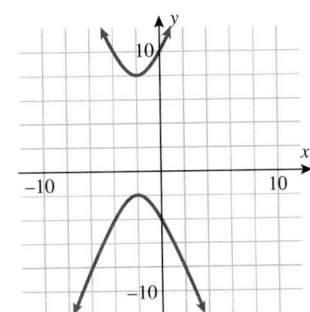　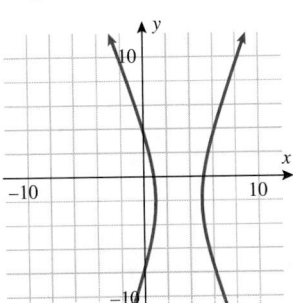　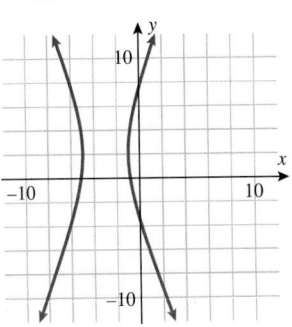

In Exercises 29–38, graph each hyperbola.

29. $\dfrac{(x-1)^2}{16} - \dfrac{(y-2)^2}{4} = 1$

30. $\dfrac{(y+1)^2}{36} - \dfrac{(x+2)^2}{9} = 1$

31. $10(y+3)^2 - (x-4)^2 = 80$

32. $3(x+3)^2 - 12(y-4)^2 = 36$

33. $x^2 - 4x - 4y^2 = 0$

34. $-9x^2 + y^2 + 2y - 8 = 0$

35. $-9x^2 - 18x + 4y^2 - 8y - 41 = 0$

36. $25x^2 - 50x - 4y^2 - 8y - 79 = 0$

37. $x^2 - 6x - 4y^2 - 16y - 8 = 0$

38. $-4x^2 - 16x + y^2 - 2y - 19 = 0$

In Exercises 39–42, find the standard form of the equation of a hyperbola with the given characteristics.

39. Vertices: $(-2, 5)$ and $(6, 5)$ Foci: $(-3, 5)$ and $(7, 5)$

40. Vertices: $(1, -2)$ and $(3, -2)$ Foci: $(0, -2)$ and $(4, -2)$

41. Vertices: $(4, -7)$ and $(4, -1)$ Foci: $(4, -8)$ and $(4, 0)$

42. Vertices: $(2, -6)$ and $(2, -4)$ Foci: $(2, -7)$ and $(2, -3)$

■ APPLICATIONS

43. Ship Navigation. Two loran stations are located 150 miles apart along a coast. If a ship records a time difference of 0.0005 second and continues on the hyperbolic path corresponding to that difference, where will it reach shore?

44. Ship Navigation. Two loran stations are located 300 miles apart along a coast. If a ship records a time difference of 0.0007 second and continues on the hyperbolic path corresponding to that difference, where will it reach shore? Round to the nearest mile.

45. Ship Navigation. If the captain of the ship in Exercise 43 wants to reach shore between the stations and 30 miles from one of them, what time difference should he look for?

46. Ship Navigation. If the captain of the ship in Exercise 44 wants to reach shore between the stations and 50 miles from one of them, what time difference should he look for?

47. Light. If the light from a lamp casts a hyperbolic pattern on the wall due to its lampshade, calculate the equation of the hyperbola if the distance between the vertices is 2 feet and the foci are half a foot from the vertices.

48. Special Ops. A military special ops team is calibrating its recording devices used for passive ascertaining of enemy location. They place two recording stations, alpha and bravo, 3000 feet apart (alpha is due east of bravo). The team detonates small explosives 300 feet west of alpha and records the time it takes each station to register an explosion. The team also sets up a second set of explosives directly north of the alpha station. How many feet north of alpha should the team set off the explosives if it wants to record the same times as on the first explosion?

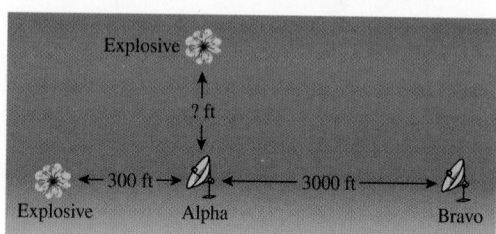

In Exercises 49–52, refer to the following:

The navigation system loran (long-range navigation) uses the reflection properties of a hyperbola. Two synchronized radio signals are transmitted at a constant speed by two distant radio stations (foci of the hyperbola). Based on the order of arrival and the interval between the signals, the location of the craft along a branch of a hyperbola can be determined. The distance between the radio stations and the craft remains constant. With the help of a third station, the location of the craft can be determined exactly as the intersection of the branches of two hyperbolas.

49. Loran Navigation System. Two radio stations, located at the same latitude, are separated by 200 km. A vessel navigates following a trajectory parallel to the line connecting A and B, 50 km north of this line. The radio signal transmitted travels at 320 m/μs. The vessel receives the signal from B, 400 μs after receiving the signal from A. Find the location of the vessel.

50. Loran Navigation System. Two radio stations, located at the same latitude, are separated by 300 km. A vessel navigates following a trajectory parallel to the line connecting A and B, 80 km north of this line. The radio signals transmitted travel at 350 m/μs. The vessel receives the signal from B, 380 μs after receiving the signal from A. Find the location of the vessel.

51. Loran Navigation System. Two radio stations, located at the same latitude, are separated by 460 km. A vessel navigates following a trajectory parallel to the line connecting A and B, 60 km north of this line. The radio signals transmitted travel at 420 m/μs. The vessel receives the signal from B, 500 μs after receiving the signal from A. Find the location of the vessel.

52. Loran Navigation System. Two radio stations, located at the same latitude, are separated by 520 km. A vessel navigates following a trajectory parallel to the line connecting A and B, 40 km north of this line. The radio signals transmitted travel at 500 m/μs. The vessel receives the signal from B, 450 μs after receiving the signal from A. Find the location of the vessel.

■ CATCH THE MISTAKE

In Exercises 53 and 54, explain the mistake that is made.

53. Graph the hyperbola $\dfrac{y^2}{4} - \dfrac{x^2}{9} = 1$.

Solution:

Compare the equation to the standard form and solve for a and b. $a = 2, b = 3$

Label the vertices $(-a, 0)$ and $(a, 0)$. $(-2, 0)$ and $(2, 0)$

Label the points $(0, -b)$ and $(0, b)$. $(0, -3)$ and $(0, 3)$

Draw the rectangle connecting these four points, and align the asymptotes so that they pass through the center and the corner of the boxes. Then draw the hyperbola using the vertices and asymptotes.

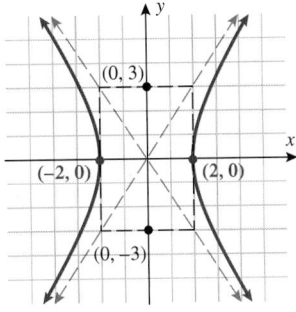

This is incorrect. What mistake was made?

54. Graph the hyperbola $\dfrac{x^2}{1} - \dfrac{y^2}{4} = 1$.

Solution:

Compare the equation to the general form and solve for a and b. $a = 2, b = 1$

Label the vertices $(-a, 0)$ and $(a, 0)$. $(-2, 0)$ and $(2, 0)$

Label the points $(0, -b)$ and $(0, b)$. $(0, -1)$ and $(0, 1)$

Draw the rectangle connecting these four points, and align the asymptotes so that they pass through the center and the corner of the boxes. Then draw the hyperbola using the vertices and asymptotes.

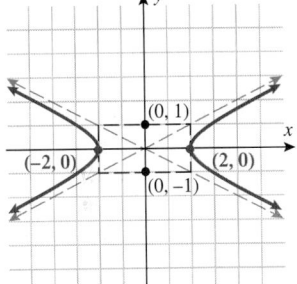

This is incorrect. What mistake was made?

■ CONCEPTUAL

In Exercises 55–58, determine whether each statement is true or false.

55. If you know the vertices of a hyperbola, you can determine the equation for the hyperbola.

56. If you know the foci and vertices, you can determine the equation for the hyperbola.

57. Hyperbolas centered at the origin have symmetry with respect to the x-axis, y-axis, and the origin.

58. The center and foci are part of the graph of a hyperbola.

59. If the point (p, q) lies on the hyperbola $\dfrac{x^2}{a^2} - \dfrac{y^2}{b^2} = 1$, find three other points that lie on the hyperbola.

60. Given the hyperbola $\dfrac{x^2}{4} - \dfrac{y^2}{b^2} = 1$, find b such that the asymptotes are perpendicular to each other.

61. A vertical line intersects the hyperbola $\dfrac{x^2}{9} - \dfrac{y^2}{4} = 1$, at the point (p, q), and intersects the hyperbola $\dfrac{x^2}{9} - \dfrac{y^2}{16} = 1$ at the point (p, r). Determine the relationship between q and r. Assume p, r, and q are positive real numbers.

62. Does the line $y = \dfrac{2b}{a}x$ intersect the hyperbola $\dfrac{x^2}{a^2} - \dfrac{y^2}{b^2} = 1$?

■ **CHALLENGE**

63. Find the general equation of a hyperbola whose asymptotes are perpendicular.

64. Find the general equation of a hyperbola whose vertices are $(3, -2)$ and $(-1, -2)$ and whose asymptotes are the lines $y = 2x - 4$ and $y = -2x$.

65. Find the asymptotes of the graph of the hyperbola given by $9y^2 - 16x^2 - 36y - 32x - 124 = 0$.

66. Find the asymptotes of the graph of the hyperbola given by $5x^2 - 4y^2 + 20x + 8y - 4 = 0$.

67. If the line $3x + 5y - 7 = 0$ is perpendicular to one of the asymptotes of the graph of the hyperbola given by $\dfrac{x^2}{a^2} - \dfrac{y^2}{b^2} = 1$ with vertices at $(\pm 3, 0)$, find the foci.

68. If the line $2x - y + 9 = 0$ is perpendicular to one of the asymptotes of the graph of the hyperbola given by $\dfrac{y^2}{a^2} - \dfrac{x^2}{b^2} = 1$ with vertices at $(0, \pm 1)$, find the foci.

■ **TECHNOLOGY**

69. Graph the following three hyperbolas: $x^2 - y^2 = 1$, $x^2 - 5y^2 = 1$, and $x^2 - 10y^2 = 1$. What can be said to happen to the hyperbola $x^2 - cy^2 = 1$ as c increases?

70. Graph the following three hyperbolas: $x^2 - y^2 = 1$, $5x^2 - y^2 = 1$, and $10x^2 - y^2 = 1$. What can be said to happen to the hyperbola $cx^2 - y^2 = 1$ as c increases?

71. Graph the following three hyperbolas: $x^2 - y^2 = 1$, $0.5x^2 - y^2 = 1$, and $0.05x^2 - y^2 = 1$. What can be said to happen to the hyperbola $cx^2 - y^2 = 1$ as c decreases?

72. Graph the following three hyperbolas: $x^2 - y^2 = 1$, $x^2 - 0.5y^2 = 1$, and $x^2 - 0.05y^2 = 1$. What can be said to happen to the hyperbola $x^2 - cy^2 = 1$ as c decreases?

■ **PREVIEW TO CALCULUS**

In Exercises 73 and 74, refer to the following:

In calculus, we study hyperbolic functions. The hyperbolic sine is defined by $\sinh u = \dfrac{e^u - e^{-u}}{2}$; the hyperbolic cosine is defined by $\cosh u = \dfrac{e^u + e^{-u}}{2}$.

73. If $x = \cosh u$ and $y = \sinh u$, show that $x^2 - y^2 = 1$.

74. If $x = \dfrac{e^u + e^{-u}}{e^u - e^{-u}}$ and $y = \dfrac{2}{e^u - e^{-u}}$, show that $x^2 - y^2 = 1$.

In Exercises 75 and 76, refer to the following:

In calculus, we use the difference quotient $\dfrac{f(x + h) - f(x)}{h}$ to find the derivative of the function f.

75. Find the derivative of $y = f(x)$, where $y^2 - x^2 = 1$ and $y < 0$.

76. Find the difference quotient of $y = f(x)$, where $4x^2 + y^2 = 1$ and $y > 0$.

SKILLS OBJECTIVES

- Solve a system of nonlinear equations with elimination.
- Solve a system of nonlinear equations with substitution.
- Eliminate extraneous solutions.

CONCEPTUAL OBJECTIVES

- Interpret the algebraic solution graphically.
- Understand the types of solutions: distinct number of solutions, no solution, and infinitely many solutions.
- Understand that equations of conic sections are nonlinear equations.

In Chapter 8, we discussed solving systems of *linear* equations. We applied elimination and substitution to solve systems of linear equations in two variables, and we employed matrices to solve systems of linear equations in three or more variables. Recall that a system of linear equations in two variables has one of three types of solutions:

One solution	Two lines that intersect at one point	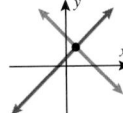
No solution	Two parallel lines (never intersect)	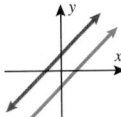
Infinitely many solutions	Two lines that coincide (same line)	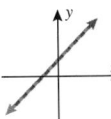

Notice that systems of *linear* equations in two variables always corresponded to *lines*. Now we turn our attention to systems of *nonlinear* equations in two variables. If any of the equations in a system of equations is nonlinear, then the system is a nonlinear system. The following are systems of nonlinear equations:

$$\begin{cases} y = x^2 + 1 \text{ (Parabola)} \\ y = 2x + 2 \text{ (Line)} \end{cases} \quad \begin{cases} x^2 + y^2 = 25 \text{ (Circle)} \\ y = x \quad \text{(Line)} \end{cases} \quad \begin{cases} \dfrac{x^2}{9} + \dfrac{y^2}{4} = 1 \text{ (Ellipse)} \\ \dfrac{y^2}{16} - \dfrac{x^2}{25} = 1 \text{ (Hyperbola)} \end{cases}$$

To find the solution to these systems, we ask the question, "At what point(s)—if any—do the graphs of these equations intersect?" Since some nonlinear equations represent conics, this is a convenient time to discuss systems of nonlinear equations.

How many points of intersection do a line and a parabola have? The answer depends on which line and which parabola. As we see in the following graphs, the answer can be one, two, or none.

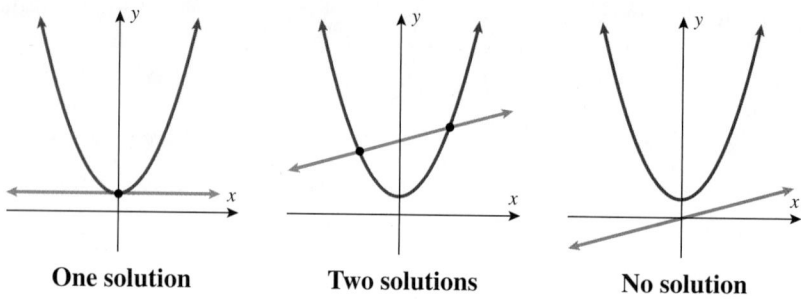

One solution	**Two solutions**	**No solution**

How many points of intersection do a parabola and an ellipse have? One, two, three, four, or no points of intersection correspond to one solution, two solutions, three solutions, four solutions, or no solution, respectively.

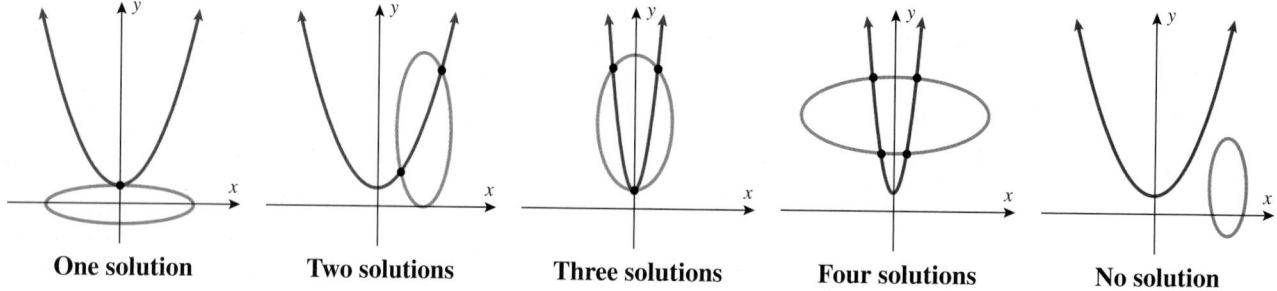

One solution	**Two solutions**	**Three solutions**	**Four solutions**	**No solution**

How many points of intersection do a parabola and a hyperbola have? The answer depends on which parabola and which hyperbola. As we see in the following graphs, the answer can be one, two, three, four, or none.

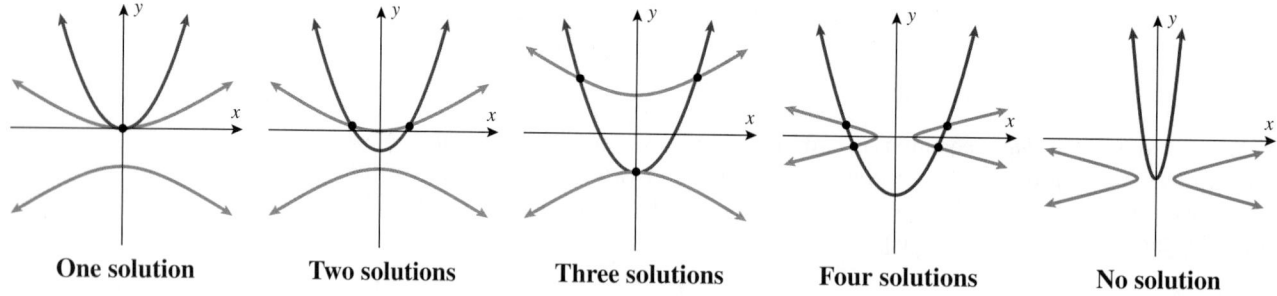

One solution	**Two solutions**	**Three solutions**	**Four solutions**	**No solution**

Using Elimination to Solve Systems of Nonlinear Equations

The first three examples in this section use elimination to solve systems of two nonlinear equations. In linear systems, we can eliminate either variable. In nonlinear systems, the variable to eliminate is the one that is raised to the same power in both equations.

EXAMPLE 1   Solving a System of Two Nonlinear Equations by Elimination: One Solution

Solve the system of equations, and graph the corresponding line and parabola to verify the answer.

$$\text{Equation (1):} \quad 2x - y = 3$$

$$\text{Equation (2):} \quad x^2 - y = 2$$

Solution:

Equation (1): $\qquad\qquad\qquad\qquad\qquad\qquad\qquad 2x - y = 3$

Multiply both sides of Equation (2) by -1. $\qquad\qquad -x^2 + y = -2$

Add. $\qquad\qquad\qquad\qquad\qquad\qquad\qquad\qquad\quad 2x - x^2 = 1$

Gather all terms to one side. $\qquad\qquad\qquad\quad x^2 - 2x + 1 = 0$

Factor. $\qquad\qquad\qquad\qquad\qquad\qquad\qquad\quad (x - 1)^2 = 0$

Solve for x. $\qquad\qquad\qquad\qquad\qquad\qquad\qquad\qquad x = 1$

Substitute $x = 1$ into original Equation (1). $\qquad 2(1) - y = 3$

Solve for y. $\qquad\qquad\qquad\qquad\qquad\qquad\qquad\qquad y = -1$

The solution is $x = 1$, $y = -1$, or $(1, -1)$.

Graph the line $y = 2x - 3$ and the parabola $y = x^2 - 2$ and confirm that the point of intersection is $(1, -1)$.

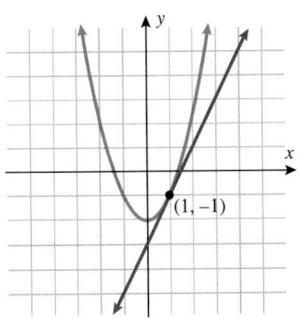

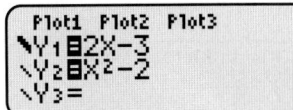
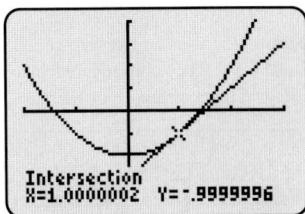

Technology Tip

Use a TI to solve the system of equations. Solve for y in each equation first; that is, $y_1 = x^2 - 7$, $y_2 = \sqrt{9 - x^2}$, and $y_3 = -\sqrt{9 - x^2}$.

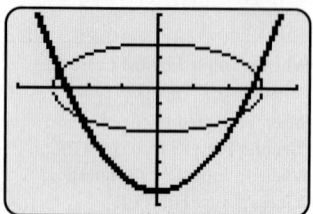

Note: The circle appears elliptical because the x- and y-axes are not of equal scale. The zoom square feature can give the appearance of a circle.

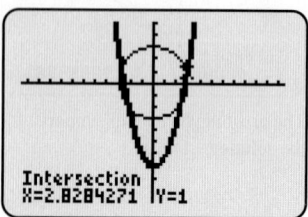

EXAMPLE 2   Solving a System of Two Nonlinear Equations with Elimination: More Than One Solution

Solve the system of equations, and graph the corresponding parabola and circle to verify the answer.

$$\text{Equation (1):}\qquad -x^2 + y = -7$$
$$\text{Equation (2):}\qquad x^2 + y^2 = 9$$

Solution:

Equation (1):	$-x^2 + y = -7$
Equation (2):	$x^2 + y^2 = 9$
Add.	$y^2 + y = 2$
Gather all terms to one side.	$y^2 + y - 2 = 0$
Factor.	$(y + 2)(y - 1) = 0$
Solve for y.	$y = -2 \quad \text{or} \quad y = 1$
Substitute $y = -2$ into Equation (2).	$x^2 + (-2)^2 = 9$
Solve for x.	$x = \pm\sqrt{5}$
Substitute $y = 1$ into Equation (2).	$x^2 + (1)^2 = 9$
Solve for x.	$x = \pm\sqrt{8} = \pm2\sqrt{2}$

There are four solutions: $\boxed{\left(-\sqrt{5}, -2\right), \left(\sqrt{5}, -2\right), \left(-2\sqrt{2}, 1\right), \text{and } \left(2\sqrt{2}, 1\right)}$

Graph the parabola $y = x^2 - 7$ and the circle $x^2 + y^2 = 9$ and confirm the four points of intersection.

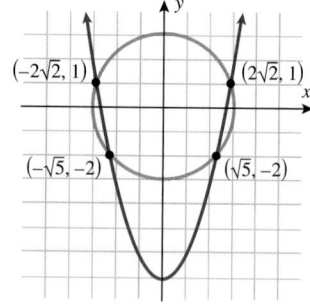

EXAMPLE 3 **Solving a System of Two Nonlinear Equations with Elimination: No Solution**

Solve the system of equations, and graph the corresponding parabolas to verify the answer.

$$\text{Equation (1):} \quad x^2 + y = 3$$
$$\text{Equation (2):} \quad -x^2 + y = 5$$

Solution:

Equation (1):	$x^2 + y = 3$
Equation (2):	$-x^2 + y = 5$
Add.	$2y = 8$
Solve for y.	$y = 4$
Substitute $y = 4$ into Equation (1).	$x^2 + 4 = 3$
Simplify.	$x^2 = -1$

$x^2 = -1$ has no real solution.

There is $\boxed{\text{no solution}}$ to this system of nonlinear equations.

Graph the parabola $x^2 + y = 3$ and the parabola $y = x^2 + 5$ and confirm there are no points of intersection.

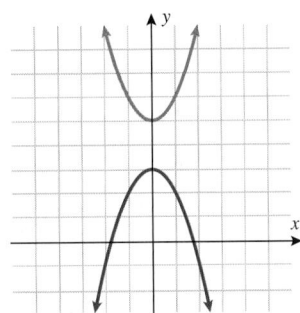

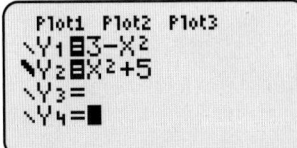
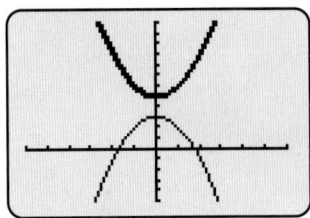

EXAMPLE 4   Solving a System of Nonlinear Equations with Elimination

Solve the system of nonlinear equations with elimination.

$$\text{Equation (1):} \quad \frac{x^2}{4} + y^2 = 1$$

$$\text{Equation (2):} \quad x^2 - y^2 = 1$$

Solution:

Add Equations (1) and (2) to eliminate y^2.

$$\frac{x^2}{4} + y^2 = 1$$
$$\underline{x^2 - y^2 = 1}$$
$$\frac{5}{4}x^2 = 2$$

Solve for x.

$$x^2 = \frac{8}{5}$$

$$x = \pm\sqrt{\frac{8}{5}}$$

Let $x = \pm\sqrt{\frac{8}{5}}$ in Equation (2).

$$\left(\pm\sqrt{\frac{8}{5}}\right)^2 - y^2 = 1$$

Solve for y.

$$y^2 = \frac{8}{5} - 1 = \frac{3}{5}$$

$$y = \pm\sqrt{\frac{3}{5}}$$

There are four solutions:

$$\left(-\sqrt{\frac{8}{5}}, -\sqrt{\frac{3}{5}}\right), \left(-\sqrt{\frac{8}{5}}, \sqrt{\frac{3}{5}}\right), \left(\sqrt{\frac{8}{5}}, -\sqrt{\frac{3}{5}}\right), \text{ and } \left(\sqrt{\frac{8}{5}}, \sqrt{\frac{3}{5}}\right)$$

A calculator can be used to approximate these solutions:

$$\sqrt{\frac{8}{5}} \approx 1.26$$

$$\sqrt{\frac{3}{5}} \approx 0.77$$

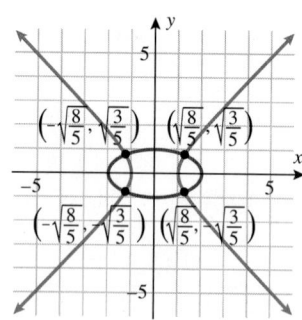

■ **Answer:**

a. $(-1, 2)$ and $(2, 5)$

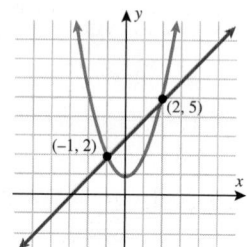

b. no solution

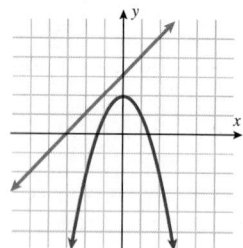

c. no solution

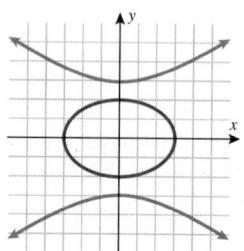

···

■ **YOUR TURN**   Solve the following systems of nonlinear equations:

a. $-x + y = 3$
    $x^2 - y = -1$

b. $x^2 + y = 2$
    $-x + y = 5$

c. $\dfrac{x^2}{9} + \dfrac{y^2}{4} = 1$
    $\dfrac{y^2}{9} - \dfrac{x^2}{16} = 1$

Using Substitution to Solve Systems of Nonlinear Equations

Elimination is based on the idea of eliminating one of the variables and solving the remaining equation in one variable. This is not always possible with nonlinear systems. For example, a system consisting of a circle and a line

$$x^2 + y^2 = 5$$
$$-x + y = 1$$

cannot be solved with elimination, because both variables are raised to different powers in each equation. We now turn to the substitution method. It is important to always check solutions, because extraneous solutions are possible.

Study Tip

Extraneous solutions are possible when you have one equation with a linear (or odd) power and one equation with a second-degree (or even) power.

EXAMPLE 5 **Solving a System of Nonlinear Equations with Substitution**

Solve the system of equations, and graph the corresponding circle and line to verify the answer.

Equation (1): $x^2 + y^2 = 5$

Equation (2): $-x + y = 1$

Solution:

Rewrite Equation (2) with y isolated.

Equation (1):	$x^2 + y^2 = 5$
Equation (2):	$y = x + 1$
Substitute Equation (2), $y = x + 1$, into Equation (1).	$x^2 + (x + 1)^2 = 5$
Eliminate the parentheses.	$x^2 + x^2 + 2x + 1 = 5$
Gather like terms.	$2x^2 + 2x - 4 = 0$
Divide by 2.	$x^2 + x - 2 = 0$
Factor.	$(x + 2)(x - 1) = 0$
Solve for x.	$x = -2$ or $x = 1$
Substitute $x = -2$ into Equation (1).	$(-2)^2 + y^2 = 5$
Solve for y.	$y = -1$ or $y = 1$
Substitute $x = 1$ into Equation (1).	$(1)^2 + y^2 = 5$
Solve for y.	$y = -2$ or $y = 2$

There appear to be four solutions: $(-2, -1)$, $(-2, 1)$, $(1, -2)$, and $(1, 2)$, but a line can intersect a circle in no more than two points. Therefore, at least two solutions are *extraneous*. All four points satisfy Equation (1), but only $(-2, -1)$ and $(1, 2)$ also satisfy Equation (2).

The answer is $(-2, -1)$ and $(1, 2)$.

Graph the circle $x^2 + y^2 = 5$ and the line $y = x + 1$ and confirm the two points of intersection.

Note: After solving for x, had we substituted back into the linear Equation (2) instead of Equation (1), extraneous solutions would not have appeared. In general, **substitute back into the lowest-degree equation, and always check solutions**.

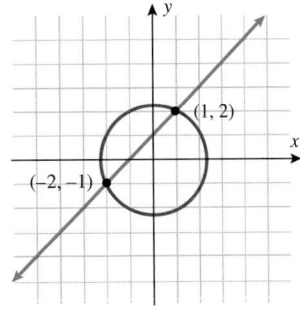

Technology Tip

Use a TI to solve the system of equations. Solve for y in each equation first; that is, $y_1 = \sqrt{5 - x^2}$, $y_2 = -\sqrt{5 - x^2}$, and $y_3 = x + 1$.

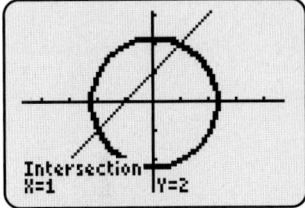

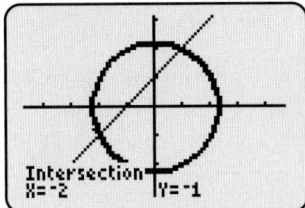

Note: The graphs support the solutions to the system.

Study Tip

Substitute back into the lowest-degree equation, and always check solutions.

■ YOUR TURN Solve the system of equations $x^2 + y^2 = 13$ and $x + y = 5$.

■ Answer: (2, 3) and (3, 2)

In Example 6, the equation $xy = 2$ can also be shown to be a rotated hyperbola (Section 9.7). For now, we can express this equation in terms of a reciprocal function $y = \dfrac{2}{x}$, a topic we discussed in Section 1.2.

EXAMPLE 6 Solving a System of Nonlinear Equations with Substitution

Solve the system of equations.

$$\text{Equation (1):} \qquad x^2 + y^2 = 5$$
$$\text{Equation (2):} \qquad xy = 2$$

Solution:

Since Equation (2) tells us that $xy = 2$, we know that neither x nor y can be zero.

Solve Equation (2) for y. $\qquad\qquad y = \dfrac{2}{x}$

Substitute $y = \dfrac{2}{x}$ into Equation (1). $\qquad x^2 + \left(\dfrac{2}{x}\right)^2 = 5$

Eliminate the parentheses. $\qquad x^2 + \dfrac{4}{x^2} = 5$

Multiply by x^2. $\qquad x^4 + 4 = 5x^2$

Collect the terms to one side. $\qquad x^4 - 5x^2 + 4 = 0$

Factor. $\qquad (x^2 - 4)(x^2 - 1) = 0$

Solve for x. $\qquad x = \pm 2 \quad \text{or} \quad x = \pm 1$

Substitute $x = -2$ into Equation (2), $xy = 2$, and solve for y. $\qquad y = -1$

Substitute $x = 2$ into Equation (2), $xy = 2$, and solve for y. $\qquad y = 1$

Substitute $x = -1$ into Equation (2), $xy = 2$, and solve for y. $\qquad y = -2$

Substitute $x = 1$ into Equation (2), $xy = 2$, and solve for y. $\qquad y = 2$

Check to see that there are four solutions: $(-2, -1), (-1, -2), (2, 1),$ and $(1, 2)$.

Note: It is important to check the solutions either algebraically or graphically (see the graph on the left).

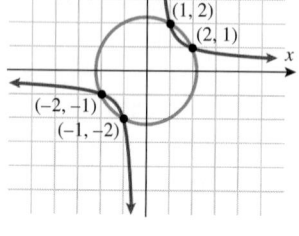

■ **Answer:** $(-1, -1)$ and $(1, 1)$

■ **YOUR TURN** Solve the system of equations $x^2 + y^2 = 2$ and $xy = 1$.

Applications

EXAMPLE 7 Calculating How Much Fence to Buy

A couple buy a rectangular piece of property advertised as 10 acres (approximately 400,000 square feet). They want two fences to divide the land into an internal grazing area and a surrounding riding path. If they want the riding path to be 20 feet wide, one fence will enclose the property and one internal fence will sit 20 feet inside the outer fence. If the internal grazing field is 237,600 square feet, how many linear feet of fencing should they buy?

Solution:

Use the five-step procedure for solving word problems from Section 0.1, and use two variables.

STEP 1 Identify the question.

How many linear feet of fence should they buy? Or, what is the sum of the perimeters of the two fences?

STEP 2 Make notes or draw a sketch.

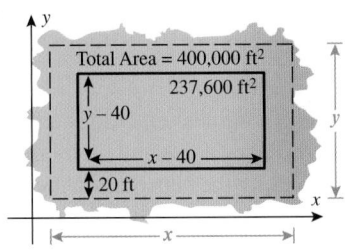

STEP 3 Set up the equations.

$x =$ length of property $\qquad\qquad x - 40 =$ length of internal field

$y =$ width of property $\qquad\qquad y - 40 =$ width of internal field

$$\text{Equation (1):} \qquad xy = 400{,}000$$

$$\text{Equation (2):} \qquad (x - 40)(y - 40) = 237{,}600$$

STEP 4 Solve the system of equations.

Substitution Method

Since Equation (1) tells us that $xy = 400{,}000$, we know that neither x nor y can be zero.

Solve Equation (1) for y. $\qquad\qquad\qquad\qquad\qquad y = \dfrac{400{,}000}{x}$

Substitute $y = \dfrac{400{,}000}{x}$ into Equation (2). $\qquad (x - 40)\left(\dfrac{400{,}000}{x} - 40\right) = 237{,}600$

Eliminate the parentheses. $\qquad 400{,}000 - 40x - \dfrac{16{,}000{,}000}{x} + 1600 = 237{,}600$

Multiply by the LCD, x. $\qquad 400{,}000x - 40x^2 - 16{,}000{,}000 + 1600x = 237{,}600x$

Collect like terms on one side. $\qquad 40x^2 - 164{,}000x + 16{,}000{,}000 = 0$

Divide by 40. $\qquad\qquad x^2 - 4100x + 400{,}000 = 0$

Factor. $\qquad\qquad (x - 4000)(x - 100) = 0$

Solve for x. $\qquad\qquad x = 4000 \quad \text{or} \quad x = 100$

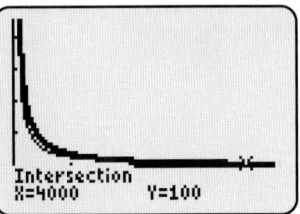
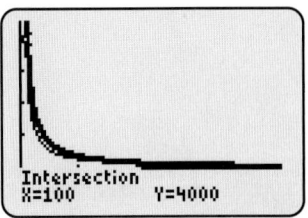

Substitute $x = 4000$ into the original Equation (1).	$4000y = 400{,}000$
Solve for y.	$y = 100$
Substitute $x = 100$ into the original Equation (1).	$100y = 400{,}000$
Solve for y.	$y = 4000$

The two solutions yield the same dimensions: 4000×100. The inner field has the dimensions 3960×60. Therefore, the sum of the perimeters of the two fences is

$$2(4000) + 2(100) + 2(3960) + 2(60) = 8000 + 200 + 7920 + 120 = 16{,}240$$

> The couple should buy 16,240 linear feet of fencing.

STEP 5 **Check the solution.**

The point $(4000, 100)$ satisfies both Equation (1) and Equation (2).

It is important to note that some nonlinear equations are not conic sections (they could be exponential, logarithmic, or higher-degree polynomial equations). These systems of linear equations are typically solved by the substitution method (see the exercises).

SECTION 9.5 SUMMARY

In this section, systems of two equations were discussed when at least one of the equations is nonlinear (e.g., conics). The substitution method and elimination method can *sometimes* be applied to nonlinear systems. When graphing the two equations, the points of intersection are the solutions of the system. Systems of nonlinear equations can have more than one solution. Also, extraneous solutions can appear, so it is important to always check solutions.

SECTION 9.5 EXERCISES

SKILLS

In Exercises 1–12, solve the system of equations by applying the elimination method.

1. $x^2 - y = -2$
$-x + y = 4$

2. $x^2 + y = 2$
$2x + y = -1$

3. $x^2 + y = 1$
$2x + y = 2$

4. $x^2 - y = 2$
$-2x + y = -3$

5. $x^2 + y = -5$
$-x + y = 3$

6. $x^2 - y = -7$
$x + y = -2$

7. $x^2 + y^2 = 1$
$x^2 - y = -1$

8. $x^2 + y^2 = 1$
$x^2 + y = -1$

9. $x^2 + y^2 = 3$
$4x^2 + y = 0$

10. $x^2 + y^2 = 6$
$-7x^2 + y = 0$

11. $x^2 + y^2 = -6$
$-2x^2 + y = 7$

12. $x^2 + y^2 = 5$
$3x^2 + y = 9$

In Exercises 13–24, solve the system of equations by applying the substitution method.

13. $x + y = 2$
$x^2 + y^2 = 2$

14. $x - y = -2$
$x^2 + y^2 = 2$

15. $xy = 4$
$x^2 + y^2 = 10$

16. $xy = -3$
$x^2 + y^2 = 12$

17. $y = x^2 - 3$
$y = -4x + 9$

18. $y = -x^2 + 5$
$y = 3x - 4$

19. $x^2 + xy - y^2 = 5$
$x - y = -1$

20. $x^2 + xy + y^2 = 13$
$x + y = -1$

21. $2x - y = 3$
$x^2 + y^2 - 2x + 6y = -9$

22. $x^2 + y^2 - 2x - 4y = 0$
$-2x + y = -3$

23. $4x^2 + 12xy + 9y^2 = 25$
$-2x + y = 1$

24. $-4xy + 4y^2 = 8$
$3x + y = 2$

In Exercises 25–40, solve the system of equations by applying any method.

25. $x^3 - y^3 = 63$
$x - y = 3$

26. $x^3 + y^3 = -26$
$x + y = -2$

27. $4x^2 - 3xy = -5$
$-x^2 + 3xy = 8$

28. $2x^2 + 5xy = 2$
$x^2 - xy = 1$

29. $2x^2 - xy = 28$
$4x^2 - 9xy = 28$

30. $-7xy + 2y^2 = -3$
$-3xy + y^2 = 0$

31. $4x^2 + 10y^2 = 26$
$-2x^2 + 2y^2 = -6$

32. $x^3 + y^3 = 19$
$x^3 - y^3 = -35$

33. $\log_x(2y) = 3$
$\log_x(y) = 2$

34. $\log_x(y) = 1$
$\log_x(2y) = \frac{1}{2}$

35. $\dfrac{1}{x^3} + \dfrac{1}{y^2} = 17$
$\dfrac{1}{x^3} - \dfrac{1}{y^2} = -1$

36. $\dfrac{2}{x^2} + \dfrac{3}{y^2} = \dfrac{5}{6}$
$\dfrac{4}{x^2} - \dfrac{9}{y^2} = 0$

37. $2x^2 + 4y^4 = -2$
$6x^2 + 3y^4 = -1$

38. $x^2 + y^2 = -2$
$x^2 + y^2 = -1$

39. $2x^2 - 5y^2 + 8 = 0$
$x^2 - 7y^2 + 4 = 0$

40. $x^2 + y^2 = 4x + 6y - 12$
$9x^2 + 4y^2 = 36x + 24y - 36$

In Exercises 41 and 42, graph each equation and find the point(s) of intersection.

41. The parabola $y = x^2 - 6x + 11$ and the line $y = -x + 7$

42. The circle $x^2 + y^2 - 4x - 2y + 5 = 0$ and the line $-x + 3y = 6$

■ APPLICATIONS

43. Numbers. The sum of two numbers is 10, and the difference of their squares is 40. Find the numbers.

44. Numbers. The difference of two numbers is 3, and the difference of their squares is 51. Find the numbers.

45. Numbers. The product of two numbers is equal to the reciprocal of the difference of their reciprocals. The product of the two numbers is 72. Find the numbers.

46. Numbers. The ratio of the sum of two numbers to the difference of the two numbers is 9. The product of the two numbers is 80. Find the numbers.

47. Geometry. A rectangle has a perimeter of 36 centimeters and an area of 80 square centimeters. Find the dimensions of the rectangle.

48. Geometry. Two concentric circles have perimeters that add up to 16π and areas that add up to 34π. Find the radii of the two circles.

49. Horse Paddock. An equestrian buys a 5-acre rectangular parcel (approximately 200,000 square feet) and is going to fence in the entire property and then divide the parcel into two halves with a fence. If 2200 linear feet of fencing is required, what are the dimensions of the parcel?

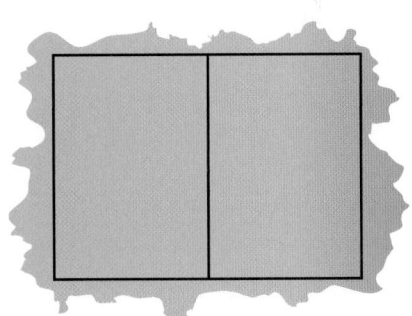

50. Dog Run. A family moves into a new home and decides to fence in the yard to give its dog room to roam. If the area that will be fenced in is rectangular and has an area of 11,250 square feet, and the length is twice as much as the width, how many linear feet of fence should the family buy?

51. Footrace. Your college algebra professor and Jeremy Wariner (2004 Olympic Gold Medalist in the men's 400 meter) decided to race. The race was 400 meters and Jeremy gave your professor a 1-minute head start, and still crossed the finish line 1 minute 40 seconds before your professor. If Jeremy ran five times faster than your professor, what was each person's average speed?

52. Footrace. You decided to race Jeremy Wariner for 800 meters. At that distance, Jeremy runs approximately twice as fast as you. He gave you a 1-minute head start and crossed the finish line 20 seconds before you. What were each of your average speeds?

53. Velocity. Two cars start moving simultaneously in the same direction. The first car moves at 50 mph; the speed of the second car is 40 mph. A half-hour later, another car starts moving in the same direction. The third car reaches the first one 1.5 hours after it reached the second car. Find the speed of the third car.

54. Design. Two boxes are constructed to contain the same volume. In the first box, the width is 16 cm larger than the depth and the length is five times the depth. In the second box, both the length and width are 4 cm shorter and the depth is 25% larger than in the first box. Find the dimensions of the second box.

55. Numbers. Find a number consisting of four digits such that

- the sum of the squares of the thousands and the units is 13.
- the sum of the squares of the hundreds and tens is 85.
- the hundreds is one more than the tens.
- the thousands is one more than the units.
- when 1089 is subtracted from the number, the result has the same digits but in inverse order.

56. Numbers. Find a number consisting of three digits such that

- the sum of the cubes of the hundreds and units is 9.
- the tens is one more than twice the hundreds.
- the hundreds is one more than the units.

■ CATCH THE MISTAKE

In Exercises 57 and 58, explain the mistake that is made.

57. Solve the system of equations: $x^2 + y^2 = 4$
$x + y = 2$

Solution:

Multiply the second equation by (-1) and add to the first equation. $x^2 - x = 2$

Subtract 2. $x^2 - x - 2 = 0$

Factor. $(x + 1)(x - 2) = 0$

Solve for x. $x = -1$ and $x = 2$

Substitute $x = -1$ and $x = 2$ into $x + y = 2$. $-1 + y = 2$ and $2 + y = 2$

Solve for y. $y = 3$ and $y = 0$

The answer is $(-1, 3)$ and $(2, 0)$.

This is incorrect. What mistake was made?

58. Solve the system of equations: $x^2 + y^2 = 5$
$2x - y = 0$

Solution:

Solve the second equation for y. $y = 2x$

Substitute $y = 2x$ into the first equation. $x^2 + (2x)^2 = 5$

Eliminate the parentheses. $x^2 + 4x^2 = 5$

Gather like terms. $5x^2 = 5$

Solve for x. $x = -1$ and $x = 1$

Substitute $x = -1$ into the first equation. $(-1)^2 + y^2 = 5$

Solve for y. $y = -2$ and $y = 2$

Substitute $x = 1$ into the first equation. $(1)^2 + y^2 = 5$

Solve for y. $y = -2$ and $y = 2$

The answers are $(-1, -2)$, $(-1, 2)$, $(1, -2)$, and $(1, 2)$.

This is incorrect. What mistake was made?

CONCEPTUAL

In Exercises 59–62, determine whether each statement is true or false.

59. A system of equations representing a line and a parabola can intersect in at most three points.

60. A system of equations representing a line and a cubic function can intersect in at most three places.

61. The elimination method can always be used to solve systems of two nonlinear equations.

62. The substitution method always works for solving systems of nonlinear equations.

63. A circle and a line have at most two points of intersection. A circle and a parabola have at most four points of intersection. What is the greatest number of points of intersection that a circle and an nth-degree polynomial can have?

64. A line and a parabola have at most two points of intersection. A line and a cubic function have at most three points of intersection. What is the greatest number of points of intersection that a line and an nth-degree polynomial can have?

CHALLENGE

65. Find a system of equations representing a line and a parabola that has only one real solution.

66. Find a system of equations representing a circle and a parabola that has only one real solution.

In Exercises 67–70, solve each system of equations.

67. $x^4 + 2x^2y^2 + y^4 = 25$
$x^4 - 2x^2y^2 + y^4 = 9$

68. $x^4 + 2x^2y^2 + y^4 = 169$
$x^4 - 2x^2y^2 + y^4 = 25$

69. $x^4 + 2x^2y^2 + y^4 = -25$
$x^4 - 2x^2y^2 + y^4 = -9$

70. $x^4 + 2x^2y^2 + y^4 = -169$
$x^4 - 2x^2y^2 + y^4 = -25$

TECHNOLOGY

In Exercises 71–76, use a graphing utility to solve the systems of equations.

71. $y = e^x$
$y = \ln x$

72. $y = 10^x$
$y = \log x$

73. $2x^3 + 4y^2 = 3$
$xy^3 = 7$

74. $3x^4 - 2xy + 5y^2 = 19$
$x^4y = 5$

75. $5x^3 + 2y^2 = 40$
$x^3y = 5$

76. $4x^4 + 2xy + 3y^2 = 60$
$x^4y = 8 - 3x^4$

PREVIEW TO CALCULUS

In calculus, when finding the derivative of equations in two variables, we typically use implicit differentiation. A more direct approach is used when an equation can be solved for one variable in terms of the other variable.

In Exercises 77–80, solve each equation for y in terms of x.

77. $x^2 + 4y^2 = 8, y < 0$

78. $y^2 + 2xy + 4 = 0, y > 0$

79. $x^3y^3 = 9y, y > 0$

80. $3xy = -x^3y^2, y < 0$

SKILLS OBJECTIVES

- Graph a nonlinear inequality in two variables.
- Graph a system of nonlinear inequalities in two variables.

CONCEPTUAL OBJECTIVES

- Understand that a nonlinear inequality in two variables may be represented by either a bounded or an unbounded region.
- Interpret an overlapping shaded region as a solution.

Nonlinear Inequalities in Two Variables

Linear inequalities are expressed in the form $Ax + By \leq C$. Specific expressions can involve either of the strict or either of the nonstrict inequalities. Examples of **nonlinear inequalities in two variables** are

$$9x^2 + 16y^2 \geq 1 \qquad x^2 + y^2 > 1 \qquad y \leq -x^2 + 3 \qquad \text{and} \qquad \frac{x^2}{20} - \frac{y^2}{81} < 1$$

We follow the same procedure as we did with linear inequalities. We change the inequality to an equal sign, graph the resulting nonlinear equation, test points from the two regions, and shade the region that makes the inequality true. For strict inequalities, $<$ or $>$, we use dashed curves, and for nonstrict inequalities, $\leq$ or $\geq$, we use solid curves.

EXAMPLE 1 **Graphing a Strict Nonlinear Inequality in Two Variables**

Graph the inequality $x^2 + y^2 > 1$.

Solution:

STEP 1 Change the inequality sign to an equal sign. $\qquad\qquad x^2 + y^2 = 1$

The equation is the equation of a circle.

STEP 2 Draw the graph of the circle.

The center is $(0, 0)$ and the radius is 1.

Since the inequality $>$ is a strict inequality, draw the circle as a **dashed** curve.

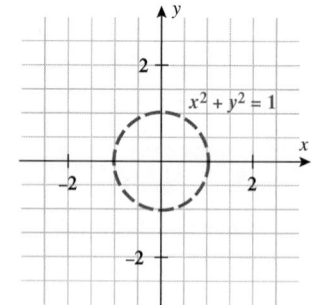

STEP 3 Test points in each region (outside the circle and inside the circle).

Substitute $(2, 0)$ into $x^2 + y^2 > 1$. $\qquad\qquad 4 \geq 1$

The point $(2, 0)$ satisfies the inequality.

Substitute $(0, 0)$ into $x^2 + y^2 > 1$. $\qquad\qquad 0 \geq 1$

The point $(0, 0)$ does not satisfy the inequality.

STEP 4 Shade the region containing the point $(2, 0)$.

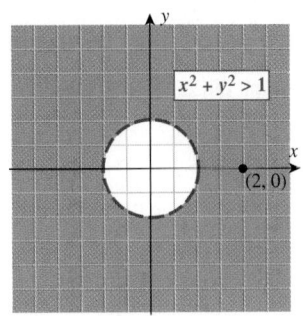

$x^2 + y^2 > 1$

$(2, 0)$

Technology Tip

Use a TI to graph the inequality $y \le -x^2 + 3$. Enter $y_1 = -x^2 + 3$. For $\le$, use the arrow key to move the cursor to the left of Y_1 and type $\boxed{\text{ENTER}}$ until you see ◣.

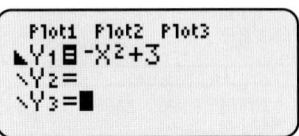

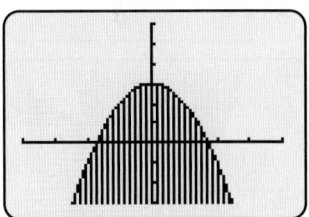

Note: The parabola should be drawn solid.

EXAMPLE 2 **Graphing a Nonstrict Nonlinear Inequality in Two Variables**

Graph the inequality $y \le -x^2 + 3$.

Solution:

STEP 1 Change the inequality sign to an equal sign. $\qquad\qquad y = -x^2 + 3$

The equation is that of a parabola.

STEP 2 Graph the parabola.

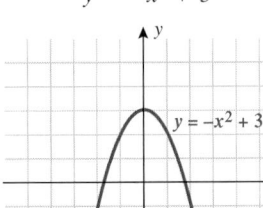

$y = -x^2 + 3$

Reflect the base function, $f(x) = x^2$, about the x-axis and shift up 3 units. Since the inequality $\le$ is a nonstrict inequality, draw the parabola as a **solid** curve.

STEP 3 Test points in each region (inside the parabola and outside the parabola).

Substitute $(3, 0)$ into $y \le -x^2 + 3$. $\qquad\qquad\qquad 0 \le -6$

The point $(3, 0)$ does not satisfy the inequality.

Substitute $(0, 0)$ into $y \le -x^2 + 3$. $\qquad\qquad\qquad 0 \le 3$

The point $(0, 0)$ does satisfy the inequality.

STEP 4 Shade the region containing the point $(0, 0)$.

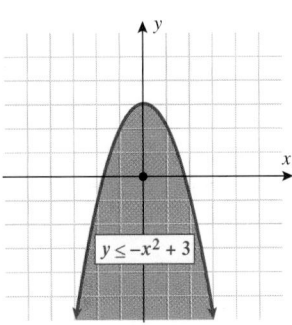

$y \le -x^2 + 3$

■ **Answer:**

a.

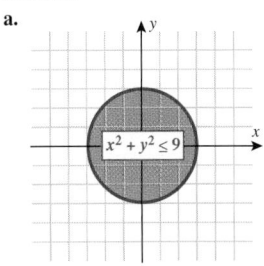

$x^2 + y^2 \le 9$

b.

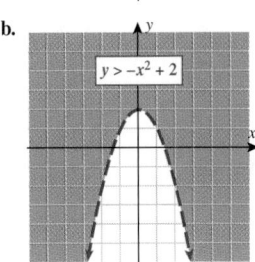

$y > -x^2 + 2$

■ **YOUR TURN** Graph the following inequalities:

$\qquad$ **a.** $x^2 + y^2 \le 9$ $\qquad$ **b.** $y > -x^2 + 2$

Systems of Nonlinear Inequalities

To solve a system of inequalities, first graph the inequalities and shade the region containing the points that satisfy each inequality. The overlap of all the shaded regions is the solution.

Technology Tip

Use a TI to graph the solution to the system of inequalities $y \geq x^2 - 1$ and $y < x + 1$. First enter $y_1 = x^2 - 1$. For $\geq$, use the arrow key to move the cursor to the left of Y_1 and type ENTER until you see ◥. Next enter $y_2 = x + 1$. For $<$, use the arrow key to move the cursor to the left of Y_2 and type ENTER until you see ◣.

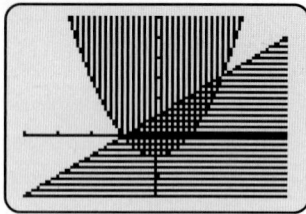

Note: The parabola should be drawn solid and the line should be drawn dashed.

■ **Answer:**

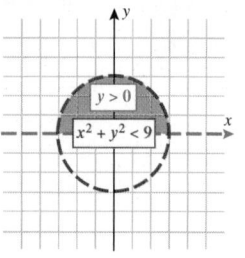

EXAMPLE 3 Graphing a System of Inequalities

Graph the solution to the system of inequalities: $\quad y \geq x^2 - 1$
$$y < x + 1$$

Solution:

STEP 1 Change the inequality signs to equal signs.
$$y = x^2 - 1$$
$$y = x + 1$$

STEP 2 The resulting equations represent a parabola (to be drawn solid) and a line (to be drawn dashed). Graph the two equations.

To determine the points of intersection, set the y-values equal to each other. $\qquad x^2 - 1 = x + 1$

Write the quadratic equation in standard form. $\qquad x^2 - x - 2 = 0$

Factor. $\qquad (x - 2)(x + 1) = 0$

Solve for x. $\qquad x = 2 \quad \text{or} \quad x = -1$

Substitute $x = 2$ into $y = x + 1$. $\qquad (2, 3)$

Substitute $x = -1$ into $y = x + 1$. $\qquad (-1, 0)$

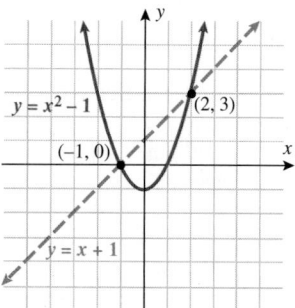

STEP 3 Test points and shade the regions.

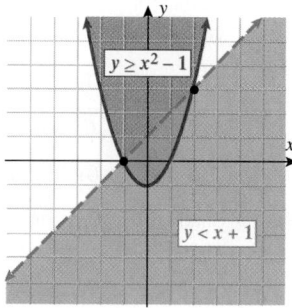

STEP 4 Shade the common region.

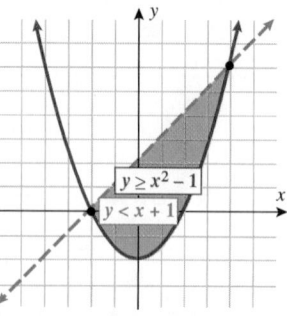

■ **YOUR TURN** Graph the solution to the system of inequalities: $\quad x^2 + y^2 < 9$
$$y > 0$$

EXAMPLE 4   Solving a System of Nonlinear Inequalities

Solve the system of inequalities: $x^2 + y^2 < 2$
$\qquad\qquad\qquad\qquad\qquad\qquad\quad y \geq x^2$

Solution:

STEP 1 Change the inequality signs to equal signs.
$\qquad\qquad\qquad\qquad\qquad\qquad\qquad\qquad\quad x^2 + y^2 = 2$
$\qquad\qquad\qquad\qquad\qquad\qquad\qquad\qquad\qquad\quad y = x^2$

STEP 2 The resulting equations correspond to a circle (to be drawn dashed) and a parabola (to be drawn solid). Graph the two inequalities.

To determine the points of intersection, solve the system of equations by substitution.
$$x^2 + \underbrace{(x^2)^2}_{y} = 2$$
$$x^4 + x^2 - 2 = 0$$

Factor.
$$(x^2 + 2)(x^2 - 1) = 0$$

Solve for x.
$$\underbrace{x^2 = -2}_{\text{no solution}} \quad \text{or} \quad \underbrace{x^2 = 1}_{x = \pm 1}$$

The points of intersection are $(-1, 1)$ and $(1, 1)$.

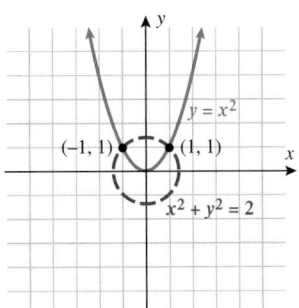

STEP 3 Test points and shade the region.

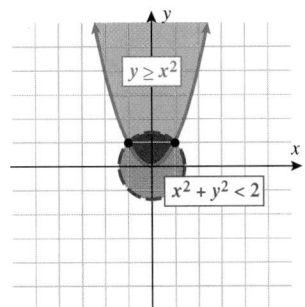

STEP 4 Identify the common region as the solution.

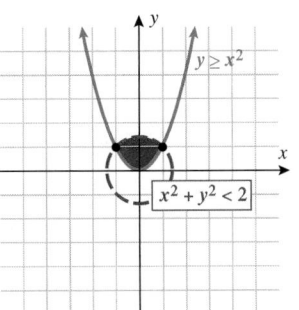

■ **YOUR TURN** Solve the system of inequalities: $x^2 + y^2 < 2$
$\qquad\qquad\qquad\qquad\qquad\qquad\qquad\qquad\qquad\quad y < x^2$

Technology Tip

Use a TI to graph the solution to the system of inequalities $x^2 + y^2 < 2$ and $y \geq x^2$. First enter
$$y_1 < \sqrt{2 - x^2} \text{ and}$$
$$y_2 > -\sqrt{2 - x^2}.$$

Use the arrow key to move the cursor to the left of Y_1 and Y_2 and type
ENTER until you see ▲ for $<$
and ▼ for $>$. Next enter $y_3 = x^2$.

For $\geq$, use the arrow key to move the cursor to the left of Y_3 and type
ENTER until you see ▼.

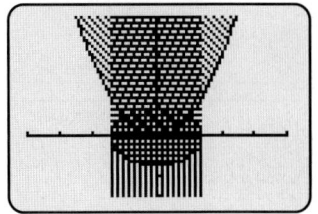

■ **Answer:**

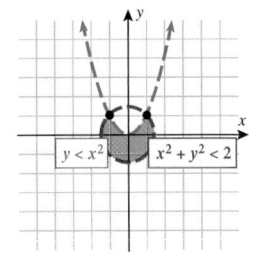

It is important to note that any inequality based on an equation whose graph is not a line is considered a nonlinear inequality.

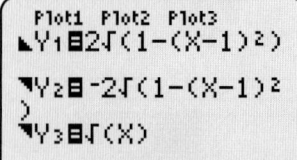
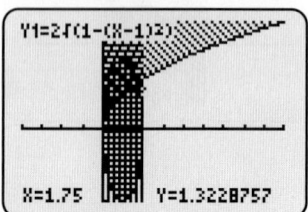
EXAMPLE 5 Solving a System of Nonlinear Inequalities

Solve the system of inequalities:

$$(x - 1)^2 + \frac{y^2}{4} < 1$$

$$y \geq \sqrt{x}$$

Solution:

STEP 1 Change the inequality signs to equal signs.

$$(x - 1)^2 + \frac{y^2}{4} = 1$$
$$y = \sqrt{x}$$

STEP 2 The resulting equations correspond to an ellipse (to be drawn dashed) and the square-root function (to be drawn solid). Graph the two inequalities.

To determine the points of intersection, solve the system of equations by substitution.

$$(x - 1)^2 + \frac{(\sqrt{x})^2}{4} = 1$$

Multiply by 4.

$$4(x - 1)^2 + x = 4$$

Expand the binomial squared.

$$4(x^2 - 2x + 1) + x = 4$$

Distribute.

$$4x^2 - 8x + 4 + x = 4$$

Combine like terms and gather terms to one side.

$$4x^2 - 7x = 0$$

Factor.

$$x(4x - 7) = 0$$

Solve for x.

$$x = 0 \quad \text{and} \quad x = \frac{7}{4}$$

The points of intersection are $(0, 0)$ and $\left(\dfrac{7}{4}, \sqrt{\dfrac{7}{4}}\right)$.

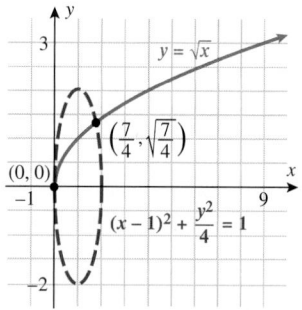

STEP 3 Shade the solution.

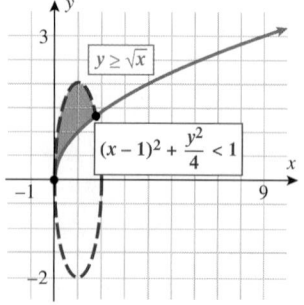

SECTION
9.6 SUMMARY

SECTION 9.6 SUMMARY

In this section, we discussed nonlinear inequalities in two variables. Sometimes these result in bounded regions (e.g., $x^2 + y^2 \leq 1$), and sometimes these result in unbounded regions (e.g., $x^2 + y^2 > 1$).

When solving systems of inequalities, we first graph each of the inequalities separately and then look for the intersection (overlap) of all shaded regions.

SECTION 9.6 EXERCISES

SKILLS

In Exercises 1–12, match the nonlinear inequality with the correct graph.

1. $x^2 + y^2 < 25$

2. $x^2 + y^2 \leq 9$

3. $\dfrac{x^2}{9} + \dfrac{y^2}{16} \geq 1$

4. $\dfrac{x^2}{4} + \dfrac{y^2}{9} > 1$

5. $y \geq x^2 - 3$

6. $x^2 \geq 16y$

7. $x \geq y^2 - 4$

8. $\dfrac{x^2}{9} + \dfrac{y^2}{25} \geq 1$

9. $9x^2 + 9y^2 < 36$

10. $(x - 2)^2 + (y + 3)^2 \leq 9$

11. $\dfrac{x^2}{4} - \dfrac{y^2}{9} \geq 1$

12. $\dfrac{y^2}{16} - \dfrac{x^2}{9} < 1$

a.

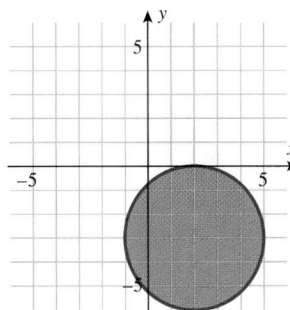

b.

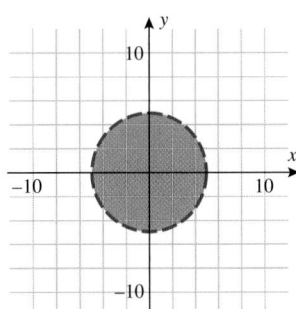

c.

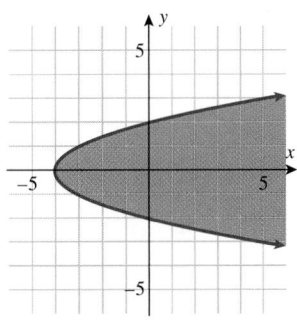

d.

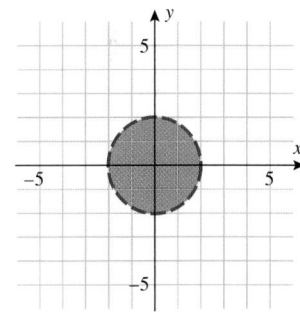

e.

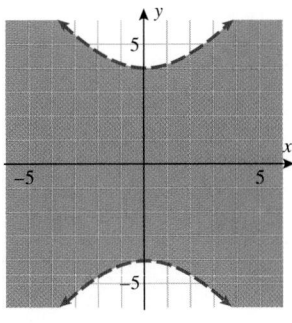

f.

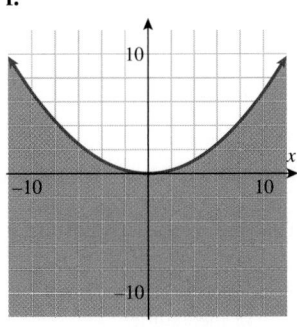

g.

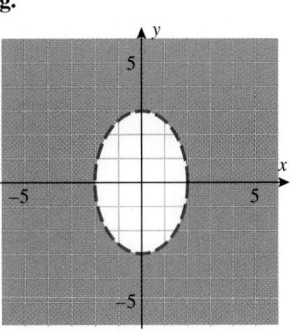

h.

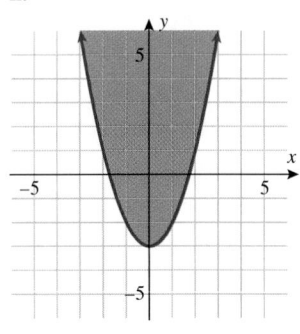

i. 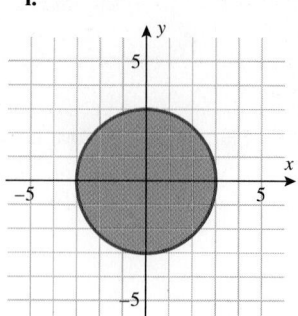 **j.** 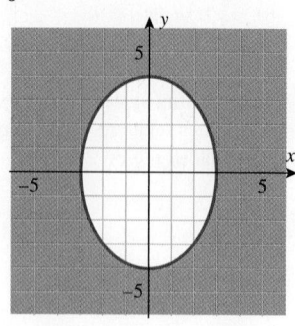 **k.** 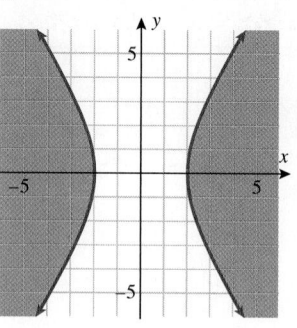 **l.**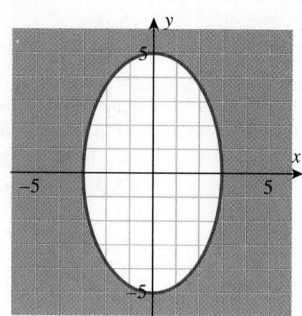

In Exercises 13–30, graph the nonlinear inequality.

13. $y \leq x^2 - 2$

14. $y \geq -x^2 + 3$

15. $x^2 + y^2 > 4$

16. $x^2 + y^2 < 16$

17. $x^2 + y^2 - 2x + 4y + 4 \geq 0$

18. $x^2 + y^2 + 2x - 2y - 2 \leq 0$

19. $3x^2 + 4y^2 \leq 12$

20. $\dfrac{(x-2)^2}{9} + \dfrac{(y+1)^2}{25} > 1$

21. $9x^2 + 16y^2 - 18x + 96y + 9 > 0$

22. $\dfrac{(x-2)^2}{4} - \dfrac{(y+3)^2}{1} \geq 1$

23. $9x^2 - 4y^2 \geq 26$

24. $\dfrac{(y+1)^2}{9} - \dfrac{(x+2)^2}{16} < 1$

25. $36x^2 - 9y^2 \geq 324$

26. $25x^2 - 36y^2 + 200x + 144y - 644 \geq 0$

27. $y \geq e^x$

28. $y \leq \ln x$

29. $y < -x^3$

30. $y > -x^4$

In Exercises 31–50, graph each system of inequalities or indicate that the system has no solution.

31. $y < x + 1$
$y \leq x^2$

32. $y < x^2 + 4x$
$y \leq 3 - x$

33. $y \geq 2 + x$
$y \leq 4 - x^2$

34. $y \geq (x - 2)^2$
$y \leq 4 - x$

35. $y \leq -(x + 2)^2$
$y > -5 + x$

36. $y \geq (x - 1)^2 + 2$
$y \leq 10 - x$

37. $-x^2 + y > -1$
$x^2 + y < 1$

38. $x < -y^2 + 1$
$x > y^2 - 1$

39. $y \geq x^2$
$x \geq y^2$

40. $y < x^2$
$x > y^2$

41. $x^2 + y^2 < 36$
$2x + y > 3$

42. $x^2 + y^2 < 36$
$y > 6$

43. $x^2 + y^2 < 25$
$y \geq 6 + x$

44. $(x - 1)^2 + (y + 2)^2 \leq 36$
$y \geq x - 3$

45. $x^2 + y^2 \leq 9$
$y \geq 1 + x^2$

46. $x^2 + y^2 \geq 16$
$x^2 + (y - 3)^2 \leq 9$

47. $x^2 - y^2 < 4$
$y > 1 - x^2$

48. $\dfrac{x^2}{4} - \dfrac{y^2}{9} \leq 1$
$y \geq x - 5$

49. $y < e^x$
$y > \ln x \quad x > 0$

50. $y < 10^x$
$y > \log x \quad x > 0$

■ **APPLICATIONS**

51. Find the area enclosed by the system of inequalities.
$x^2 + y^2 < 9$
$x > 0$

52. Find the area enclosed by the system of inequalities.
$x^2 + y^2 \leq 5$
$x \leq 0$
$y \geq 0$

In Exercises 53 and 54, refer to the following:

The area enclosed by the ellipse $\dfrac{x^2}{a^2} + \dfrac{y^2}{b^2} = 1$ is given by $ab\pi$.

53. Find the area enclosed by the system of inequalities.
$4x^2 + y^2 \leq 16$
$x \leq 0$
$y \geq 0$

54. Find the area enclosed by the system of inequalities.
$$9x^2 + 4y^2 \geq 36$$
$$x^2 + y^2 \leq 9$$

55. Find the area enclosed by the system of inequalities.
$$y \leq x^2$$
$$x \geq 0$$
$$x \leq 6$$
$$y \geq x - 6$$

In Exercises 55 and 56, refer to the following:

The area below $y = x^2$, above $y = 0$, and between $x = 0$ and $x = a$ is $\dfrac{a^3}{3}$.

56. Find the area enclosed by the system of inequalities.
$$y \leq x^2 + 4$$
$$y \geq x$$
$$x \geq -3$$
$$x \leq 3$$

■ **CATCH THE MISTAKE**

In Exercises 57 and 58, explain the mistake that is made.

57. Graph the system of inequalities:
$$x^2 + y^2 < 1$$
$$x^2 + y^2 > 4$$

Solution:

Draw the circles
$x^2 + y^2 = 1$ and
$x^2 + y^2 = 4$.

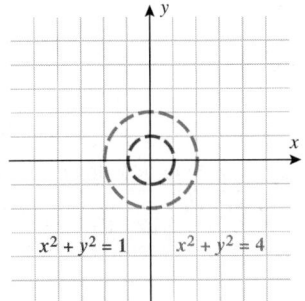

Shade outside $x^2 + y^2 = 1$
and inside $x^2 + y^2 = 4$.

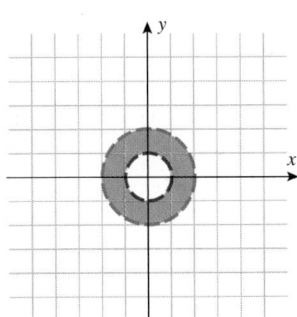

This is incorrect. What mistake was made?

58. Graph the system of inequalities:
$$x > -y^2 + 1$$
$$x < y^2 - 1$$

Solution:

Draw the parabolas
$x = -y^2 + 1$ and
$x = y^2 - 1$.

Shade the region
between the curves.

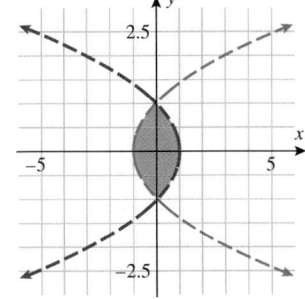

This is incorrect. What mistake was made?

■ **CONCEPTUAL**

In Exercises 59–66, determine whether each statement is true or false.

59. A nonlinear inequality always represents a bounded region.

60. A system of inequalities always has a solution.

61. The solution to the following system of equations is symmetric with respect to the y-axis:
$$\frac{x^2}{a^2} - \frac{y^2}{b^2} \leq 1$$
$$y \geq x^2 - 2a$$

62. The solution to the following system of equations is symmetric with respect to the origin:
$$\frac{x^2}{a^2} - \frac{y^2}{b^2} \leq 1$$
$$y \geq \sqrt[3]{x}$$

63. The solution to the following system of equations is symmetric with respect to the origin:
$$\frac{x^2}{a^2} - \frac{y^2}{b^2} \geq 1$$
$$\frac{x^2}{4a^2} + \frac{y^2}{a^2} \leq 1$$

64. The solution to the following system of equations is bounded:

$$\frac{x^2}{a^2} - \frac{y^2}{b^2} \leq 1$$
$$x \geq -2a$$
$$x \leq 2a$$

65. The following systems of inequalities have the same solution:

$$16x^2 - 25y^2 \geq 400 \qquad \frac{y^2}{25} - \frac{x^2}{16} \geq 1$$
$$x \geq -6 \qquad\qquad y \geq -6$$
$$x \leq 6 \qquad\qquad y \leq 6$$

66. The solution to the following system of inequalities is unbounded:

$$\frac{x^2}{a^2} - \frac{y^2}{b^2} \leq 1$$
$$\frac{y^2}{a^2} - \frac{x^2}{b^2} \leq 1$$

■ CHALLENGE

67. For the system of nonlinear inequalities $\begin{array}{l} x^2 + y^2 \geq a^2 \\ x^2 + y^2 \leq b^2 \end{array}$, what restriction must be placed on the values of a and b for this system to have a solution? Assume that a and b are real numbers.

68. Can $x^2 + y^2 < -1$ ever have a real solution? What types of numbers would x and/or y have to be to satisfy this inequality?

69. Find a positive real number a such that the area enclosed by the curves is the same.

$$x^2 + y^2 = 144 \quad \text{and} \quad \frac{x^2}{a^2} + \frac{y^2}{4^2} = 1$$

70. If the area of the regions enclosed by $x^2 + y^2 = 1$ and $\frac{x^2}{a^2} + \frac{y^2}{b^2} = 1$ are equal, what can you say about a and b?

71. If the solution to

$$4x^2 + 9y^2 \leq 36$$
$$(x - h)^2 \leq 4y$$

is symmetric with respect to the y-axis, what can you say about h?

72. The solution to

$$\frac{x^2}{a^2} + \frac{y^2}{b^2} \leq 1$$
$$y \geq x$$

is located in quadrants I, II, and III. If the sections in quadrants I and III have the same area, what can you say about a and b?

■ TECHNOLOGY

In Exercises 73–80, use a graphing utility to graph the inequalities.

73. $x^2 + y^2 - 2x + 4y + 4 \geq 0$

74. $x^2 + y^2 + 2x - 2y - 2 \leq 0$

75. $y \geq e^x$

76. $y \leq \ln x$

77. $y < e^x$
$y > \ln x \qquad x > 0$

78. $y < 10^x$
$y > \log x \qquad x > 0$

79. $x^2 - 4y^2 + 5x - 6y + 18 \geq 0$

80. $x^2 - 2xy + 4y^2 + 10x - 25 \leq 0$

■ PREVIEW TO CALCULUS

In calculus, the problem of finding the area enclosed by a set of curves can be seen as the problem of finding the area enclosed by a system of inequalities.
In Exercises 81–84, graph the system of inequalities.

81. $y \leq x^3 - x$
$y \geq x^2 - 1$

82. $y \leq x^3$
$y \geq 2x - x^2$

83. $y \geq x^3 - 2x^2$
$y \leq x^2$
$y \leq 5$

84. $y \leq \sqrt{1 - x^2}$
$y \geq x^2$
$y \leq 2x$

SKILLS OBJECTIVES

- Transform general second-degree equations into recognizable equations of conics by analyzing rotation of axes.
- Determine the angle of rotation that will transform a general second-degree equation into a familiar equation of a conic section.
- Graph a rotated conic.

CONCEPTUAL OBJECTIVE

- Understand how the equation of a conic section is altered by rotation of axes.

In Sections 9.1 through 9.4, we learned to recognize equations of parabolas, ellipses, and hyperbolas that were centered at any point in the Cartesian plane and whose vertices and foci were aligned either along or parallel to either the x-axis or the y-axis. We learned, for example, that the equation of an ellipse centered at the origin takes the form

$$\frac{x^2}{a^2} + \frac{y^2}{b^2} = 1$$

where the major and minor axes are, respectively, either the x- or the y-axis depending on whether a is greater than or less than b. Now let us look at an equation of a conic section whose graph is *not* aligned with the x- or y-axis: the equation $5x^2 - 8xy + 5y^2 - 9 = 0$.

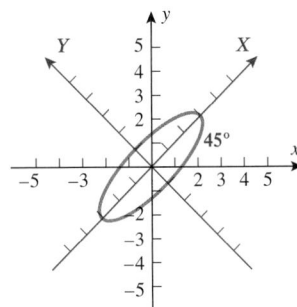

This graph can be thought of as an ellipse that started with the major axis along the x-axis and the minor axis along the y-axis and then was rotated counterclockwise 45°. A new XY-coordinate system can be introduced that has the same origin but is rotated by a certain amount from the standard xy-coordinate system. In this example, the major axis of the ellipse lies along the new X-axis and the minor axis lies along the new Y-axis. We will see that we can write the equation of this ellipse as

$$\frac{X^2}{9} + \frac{Y^2}{1} = 1$$

We will now develop the *rotation of axes formulas*, which allow us to transform the generalized second-degree equation in xy, that is, $Ax^2 + Bxy + Cy^2 + Dx + Ey + F = 0$, into an equation in XY of a conic that is familiar to us.

WORDS	**MATH**
Let the new XY-coordinate system be displaced from the xy-coordinate system by rotation through an angle θ. Let P represent some point a distance r from the origin.	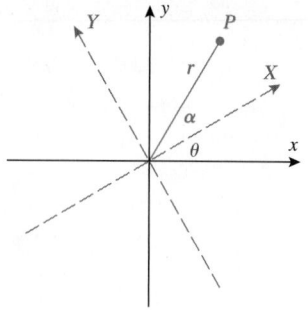
We can represent the point P as either the point (x, y) or the point (X, Y). We define the angle α as the angle r makes with the X-axis and $\alpha + \theta$ as the angle r makes with the x-axis.	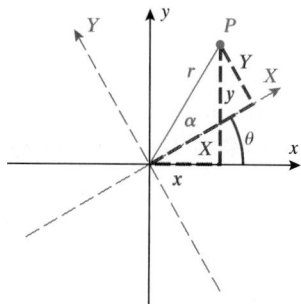
We can represent the point P in polar coordinates using the following relationships:	$x = r\cos(\alpha + \theta)$ $y = r\sin(\alpha + \theta)$ $X = r\cos\alpha$ $Y = r\sin\alpha$

Let us now derive the relationships between the two coordinate systems.

WORDS	**MATH**
Start with the x-term and write the cosine identity for a sum.	$x = r\cos(\alpha + \theta)$ $\quad = r(\cos\alpha\,\cos\theta - \sin\alpha\,\sin\theta)$
Eliminate the parentheses and group r with the α-terms.	$x = (r\cos\alpha)\cos\theta - (r\sin\alpha)\sin\theta$
Substitute according to the relationships $X = r\cos\alpha$ and $Y = r\sin\alpha$.	$\boxed{x = X\cos\theta - Y\sin\theta}$
Start with the y-term and write the sine identity for a sum.	$y = r\sin(\alpha + \theta)$ $\quad = r(\sin\alpha\,\cos\theta + \cos\alpha\,\sin\theta)$
Eliminate the parentheses and group r with the α-terms.	$y = (r\sin\alpha)\cos\theta + (r\cos\alpha)\sin\theta$
Substitute according to the relationships $X = r\cos\alpha$ and $Y = r\sin\alpha$.	$\boxed{y = Y\cos\theta + X\sin\theta}$

By treating the highlighted equations for x and y as a system of linear equations in X and Y, we can then solve for X and Y in terms of x and y. The results are summarized in the following box:

ROTATION OF AXES FORMULAS

Suppose that the x- and y-axes in the rectangular coordinate plane are rotated through an acute angle θ to produce the X- and Y-axes. Then, the coordinates (x, y) and (X, Y) are related according to the following equations:

$$x = X\cos\theta - Y\sin\theta$$
$$y = X\sin\theta + Y\cos\theta$$
or
$$X = x\cos\theta + y\sin\theta$$
$$Y = -x\sin\theta + y\cos\theta$$

EXAMPLE 1 Rotating the Axes

If the xy-coordinate axes are rotated $60°$, find the XY-coordinates of the point $(x, y) = (-3, 4)$.

Solution:

Start with the rotation formulas.

$$X = x\cos\theta + y\sin\theta$$
$$Y = -x\sin\theta + y\cos\theta$$

Let $x = -3$, $y = 4$, and $\theta = 60°$.

$$X = -3\cos 60° + 4\sin 60°$$
$$Y = -(-3)\sin 60° + 4\cos 60°$$

Simplify.

$$X = -3\underbrace{\cos 60°}_{\frac{1}{2}} + 4\underbrace{\sin 60°}_{\frac{\sqrt{3}}{2}}$$

$$Y = 3\underbrace{\sin 60°}_{\frac{\sqrt{3}}{2}} + 4\underbrace{\cos 60°}_{\frac{1}{2}}$$

$$X = -\frac{3}{2} + 2\sqrt{3}$$

$$Y = \frac{3\sqrt{3}}{2} + 2$$

The XY-coordinates are $\boxed{\left(-\frac{3}{2} + 2\sqrt{3}, \frac{3\sqrt{3}}{2} + 2\right)}$.

■ YOUR TURN: If the xy-coordinate axes are rotated $30°$, find the XY-coordinates of the point $(x, y) = (3, -4)$.

■ **Answer:**
$\left(\frac{3\sqrt{3}}{2} - 2, -\frac{3}{2} - 2\sqrt{3}\right)$

EXAMPLE 2   Rotating an Ellipse

Show that the graph of the equation $5x^2 - 8xy + 5y^2 - 9 = 0$ is an ellipse aligning with coordinate axes that are rotated by 45°.

Solution:

Start with the rotation formulas.

$$x = X\cos\theta - Y\sin\theta$$
$$y = X\sin\theta + Y\cos\theta$$

Let $\theta = 45°$.

$$x = \underset{\frac{\sqrt{2}}{2}}{X\cos 45°} - \underset{\frac{\sqrt{2}}{2}}{Y\sin 45°}$$
$$y = \underset{\frac{\sqrt{2}}{2}}{X\sin 45°} + \underset{\frac{\sqrt{2}}{2}}{Y\cos 45°}$$

Simplify.

$$x = \frac{\sqrt{2}}{2}(X - Y)$$
$$y = \frac{\sqrt{2}}{2}(X + Y)$$

Substitute $x = \dfrac{\sqrt{2}}{2}(X - Y)$ and $y = \dfrac{\sqrt{2}}{2}(X + Y)$ into $5x^2 - 8xy + 5y^2 - 9 = 0$.

$$5\left[\frac{\sqrt{2}}{2}(X - Y)\right]^2 - 8\left[\frac{\sqrt{2}}{2}(X - Y)\right]\left[\frac{\sqrt{2}}{2}(X + Y)\right] + 5\left[\frac{\sqrt{2}}{2}(X + Y)\right]^2 - 9 = 0$$

Simplify.
$$\frac{5}{2}(X^2 - 2XY + Y^2) - 4(X^2 - Y^2) + \frac{5}{2}(X^2 + 2XY + Y^2) - 9 = 0$$

$$\frac{5}{2}X^2 - 5XY + \frac{5}{2}Y^2 - 4X^2 + 4Y^2 + \frac{5}{2}X^2 + 5XY + \frac{5}{2}Y^2 = 9$$

Combine like terms.

$$X^2 + 9Y^2 = 9$$

Divide by 9.

$$\boxed{\dfrac{X^2}{9} + \dfrac{Y^2}{1} = 1}$$

This (as discussed earlier) is an ellipse whose major axis is along the X-axis.

The vertices are at the points $(X, Y) = (\pm 3, 0)$.

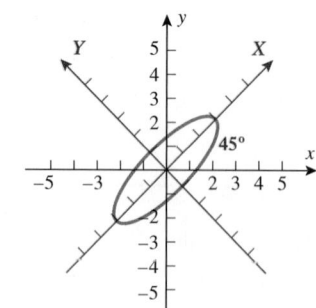

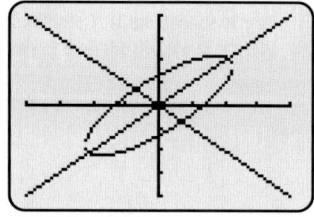

Determine the Angle of Rotation Necessary to Transform a General Second-Degree Equation into a Familiar Equation of a Conic

In Section 9.1, we stated that the general second-degree equation

$$Ax^2 + Bxy + Cy^2 + Dx + Ey + F = 0$$

corresponds to a graph of a conic. Which type of conic it is depends on the value of the discriminant, $B^2 - 4AC$. In Sections 9.2–9.4, we discussed graphs of parabolas, ellipses, and hyperbolas with vertices along either the axes or lines parallel (or perpendicular) to the axes. In all cases the value of B was taken to be zero. When the value of B is nonzero, the result is a conic with vertices along the new XY-axes (or, respectively, parallel and perpendicular to them), which are the original xy-axes rotated through an angle θ. If given θ, we can determine the rotation equations as illustrated in Example 2, but how do we find the angle θ that represents the *angle of rotation*?

To find the angle of rotation, let us start with a general second-degree polynomial equation:

$$Ax^2 + Bxy + Cy^2 + Dx + Ey + F = 0$$

We want to transform this equation into an equation in X and Y that does not contain an XY-term. Suppose we rotate our coordinates by an angle θ and use the rotation equations

$$x = X\cos\theta - Y\sin\theta \qquad y = X\sin\theta + Y\cos\theta$$

in the general second-degree polynomial equation; then the result is

$$A(X\cos\theta - Y\sin\theta)^2 + B(X\cos\theta - Y\sin\theta)(X\sin\theta + Y\cos\theta)$$
$$+ C(X\cos\theta + Y\sin\theta)^2 + D(X\cos\theta - Y\sin\theta) + E(X\sin\theta + Y\cos\theta) + F = 0$$

If we expand these expressions and collect like terms, the result is an equation of the form

$$aX^2 + bXY + cY^2 + dX + eY + f = 0$$

where

$$a = A\cos^2\theta + B\sin\theta\cos\theta + C\sin^2\theta$$
$$b = B(\cos^2\theta - \sin^2\theta) + 2(C - A)\sin\theta\cos\theta$$
$$c = A\sin^2\theta - B\sin\theta\cos\theta + C\cos^2\theta$$
$$d = D\cos\theta + E\sin\theta$$
$$e = -D\sin\theta + E\cos\theta$$
$$f = F$$

WORDS	MATH
We do not want this new equation to have an XY-term, so we set $b = 0$.	$B(\cos^2\theta - \sin^2\theta) + 2(C - A)\sin\theta\cos\theta = 0$
We can use the double-angle formulas to simplify.	$\underbrace{B(\cos^2\theta - \sin^2\theta)}_{\cos(2\theta)} + \underbrace{(C - A)2\sin\theta\cos\theta}_{\sin(2\theta)} = 0$
Subtract the $\sin(2\theta)$ term from both sides of the equation.	$B\cos(2\theta) = (A - C)\sin(2\theta)$
Divide by $B\sin(2\theta)$.	$\dfrac{B\cos(2\theta)}{B\sin(2\theta)} = \dfrac{(A - C)\sin(2\theta)}{B\sin(2\theta)}$
Simplify.	$\boxed{\cot(2\theta) = \dfrac{A - C}{B}}$

ANGLE OF ROTATION FORMULA

To transform the equation of a conic

$$Ax^2 + Bxy + Cy^2 + Dx + Ey + F = 0$$

into an equation in X and Y without an XY-term, rotate the xy-axes by an acute angle θ that satisfies the equation

$$\cot(2\theta) = \frac{A - C}{B} \quad \text{or} \quad \tan(2\theta) = \frac{B}{A - C}$$

Notice that the trigonometric equation $\cot(2\theta) = \dfrac{A - C}{B}$ or $\tan(2\theta) = \dfrac{B}{A - C}$ can be solved exactly for some values of θ (Example 3) and will have to be approximated with a calculator for other values of θ (Example 4).

EXAMPLE 3   Determining the Angle of Rotation I: The Value of the Cotangent Function Is That of a Known (Special) Angle

Determine the angle of rotation necessary to transform the following equation into an equation in X and Y with no XY-term:

$$3x^2 + 2\sqrt{3}xy + y^2 + 2x - 2\sqrt{3}y = 0$$

Solution:

Identify the A, B, and C parameters in the equation.

$$\underset{A}{3x^2} + \underset{B}{2\sqrt{3}\,xy} + \underset{C}{1y^2} + 2x - 2\sqrt{3}y = 0$$

Write the rotation formula.

$$\cot(2\theta) = \frac{A - C}{B}$$

Let $A = 3, B = 2\sqrt{3}$, and $C = 1$.

$$\cot(2\theta) = \frac{3 - 1}{2\sqrt{3}}$$

Simplify.

$$\cot(2\theta) = \frac{1}{\sqrt{3}}$$

Apply the reciprocal identity.

$$\tan(2\theta) = \sqrt{3}$$

From our knowledge of trigonometric exact values, we know that $2\theta = 60°$ or $\boxed{\theta = 30°}$.

Technology Tip

To find the angle of rotation, use

$$\theta = \frac{1}{2}\tan^{-1}\left(\frac{B}{A - C}\right)$$ and set the

TI calculator to degree mode. Substitute $A = 3, B = 2\sqrt{3}$, $C = 1$.

```
1/2tan-1(2√(3)/(3
-1))
              30
```

EXAMPLE 4   **Determining the Angle of Rotation II: The Argument of the Cotangent Function Needs to Be Approximated with a Calculator**

Determine the angle of rotation necessary to transform the following equation into an equation in X and Y with no XY-term. Round to the nearest tenth of a degree.

$$4x^2 + 2xy - 6y^2 - 5x + y - 2 = 0$$

Solution:

Identify the A, B, and C parameters in the equation.

$$\underset{A}{4x^2} + \underset{B}{2xy} - \underset{C}{6y^2} - 5x + y - 2 = 0$$

Write the rotation formula.

$$\cot(2\theta) = \frac{A - C}{B}$$

Let $A = 4$, $B = 2$, and $C = -6$.

$$\cot(2\theta) = \frac{4 - (-6)}{2}$$

Simplify.

$$\cot(2\theta) = 5$$

Apply the reciprocal identity.

$$\tan(2\theta) = \frac{1}{5} = 0.2$$

Write the result as an inverse tangent function.

$$2\theta = \tan^{-1}(0.2)$$

With a calculator evaluate the right side of the equation.

$$2\theta \approx 11.31°$$

Solve for θ and round to the nearest tenth of a degree.

$$\boxed{\theta = 5.7°}$$

Technology Tip

To find the angle of rotation, use

$$\theta = \frac{1}{2}\tan^{-1}\left(\frac{B}{A - C}\right)$$ and set

the TI calculator to degree mode. Substitute $A = 4$, $B = 2$, and $C = -6$.

```
1/2tan-1(2/(4-(-6
)))
        5.654966237
■
```

Special attention must be given when evaluating the inverse tangent function on a calculator, as the result is always in quadrant I or IV. If 2θ turns out to be negative, then $180°$ must be added so that 2θ is in quadrant II (as opposed to quadrant IV). Then θ will be an acute angle lying in quadrant I.

Recall that we stated (without proof) in Section 9.1 that we can identify a general equation

$$Ax^2 + Bxy + Cy^2 + Dx + Ey + F = 0$$

as that of a particular conic depending on the discriminant.

Parabola	$B^2 - 4AC = 0$
Ellipse	$B^2 - 4AC < 0$
Hyperbola	$B^2 - 4AC > 0$

EXAMPLE 5   Graphing a Rotated Conic

For the equation $x^2 + 2xy + y^2 - \sqrt{2}x - 3\sqrt{2}y + 6 = 0$:

a. Determine which conic the equation represents.

b. Find the rotation angle required to eliminate the XY-term in the new coordinate system.

c. Transform the equation in x and y into an equation in X and Y.

d. Graph the resulting conic.

Technology Tip

To find the angle of rotation, use

$$\theta = \frac{1}{2}\tan^{-1}\left(\frac{B}{A - C}\right)$$ and set the

calculator to degree mode. Substitute $A = 1$, $B = 2$, and $C = 1$.

```
1/2tan⁻¹(2/(1-1))
■
```

```
ERR:DIVIDE BY 0
1⯑Quit
2:Goto
```

The calculator displays an error message, which means that $\cos(2\theta) = 0$. Therefore, $2\theta = 90°$ or $\theta = 45°$.

To graph the equation $x^2 + 2xy + y^2 - \sqrt{2}x - 3\sqrt{2}y + 6 = 0$ with a TI-83 or TI-83 Plus calculator, you need to solve for y using the quadratic formula.

$y^2 + y(2x - 3\sqrt{2}) +$
$(x^2 - \sqrt{2}x + 6) = 0$

$a = 1$, $b = 2x - 3\sqrt{2}$,

$c = x^2 - \sqrt{2}x + 6$

$$y = \frac{-b \pm \sqrt{b^2 - 4ac}}{2a}$$

$$y = \frac{-2x + 3\sqrt{2} \pm \sqrt{-8\sqrt{2}x - 6}}{2}$$

Solution (a):

Identify A, B, and C.

$$\underset{A}{1x^2} + \underset{B}{2xy} + \underset{C}{1y^2} - \sqrt{2}x - 3\sqrt{2}y + 6 = 0$$

$$A = 1, B = 2, C = 1$$

Compute the discriminant. $B^2 - 4AC = 2^2 - 4(1)(1) = 0$

Since the discriminant equals zero, the equation represents a **parabola**.

Solution (b):

Write the rotation formula. $\cot(2\theta) = \dfrac{A - C}{B}$

Let $A = 1, B = 2,$ and $C = 1$. $\cot(2\theta) = \dfrac{1 - 1}{2}$

Simplify. $\cot(2\theta) = 0$

Write the cotangent function in terms $\dfrac{\cos(2\theta)}{\sin(2\theta)} = 0$
of the sine and cosine functions.

The numerator must equal zero. $\cos(2\theta) = 0$

From our knowledge of trigonometric exact values, we know that $2\theta = 90°$ or $\boxed{\theta = 45°}$.

Solution (c):

Start with the equation
$x^2 + 2xy + y^2 - \sqrt{2}x - 3\sqrt{2}y + 6 = 0$,
and use the rotation formulas
with $\theta = 45°$.

$$x = X\cos 45° - Y\sin 45° = \frac{\sqrt{2}}{2}(X - Y)$$

$$y = X\sin 45° + Y\cos 45° = \frac{\sqrt{2}}{2}(X + Y)$$

Find x^2, xy, and y^2.

$$x^2 = \left[\frac{\sqrt{2}}{2}(X - Y)\right]^2 = \frac{1}{2}(X^2 - 2XY + Y^2)$$

$$xy = \left[\frac{\sqrt{2}}{2}(X - Y)\right]\left[\frac{\sqrt{2}}{2}(X + Y)\right] = \frac{1}{2}(X^2 - Y^2)$$

$$y^2 = \left[\frac{\sqrt{2}}{2}(X + Y)\right]^2 = \frac{1}{2}(X^2 + 2XY + Y^2)$$

Substitute the values
for x, y, x^2, xy,
and y^2 into the
original equation.

$$x^2 + 2xy + y^2 - \sqrt{2}x - 3\sqrt{2}y + 6 = 0$$

$$\frac{1}{2}(X^2 - 2XY + Y^2) + 2\frac{1}{2}(X^2 - Y^2)$$

$$+ \frac{1}{2}(X^2 + 2XY + Y^2) - \sqrt{2}\left[\frac{\sqrt{2}}{2}(X - Y)\right]$$

$$- 3\sqrt{2}\left[\frac{\sqrt{2}}{2}(X + Y)\right] + 6 = 0$$

Eliminate the parentheses
and combine like terms.

$$2X^2 - 4X - 2Y + 6 = 0$$

Divide by 2.

$$X^2 - 2X - Y + 3 = 0$$

Add Y.

$$Y = (X^2 - 2X) + 3$$

Complete the square on X.

$$Y = (X - 1)^2 + 2$$

Solution (d):

This is a parabola opening
upward in the XY-coordinate
system shifted to the right
1 unit and up 2 units.

Now enter

$$y1 = \frac{-2x + 3\sqrt{2} + \sqrt{-8\sqrt{2}x - 6}}{2}$$

and

$$y2 = \frac{-2x + 3\sqrt{2} - \sqrt{-8\sqrt{2}x - 6}}{2}.$$

To graph the X- and Y-axes, enter
$y3 = \tan(45)x$ and

$$y4 = -\frac{1}{\tan(45)}x.$$

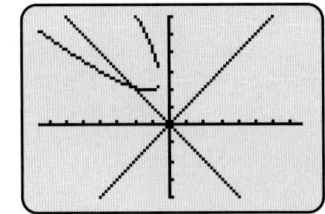

SECTION
9.7 SUMMARY

In this section, we found that the graph of the general second-degree equation

$$Ax^2 + Bxy + Cy^2 + Dx + Ey + F = 0$$

can represent conics in a system of rotated axes.

The following are the rotation formulas relating the xy-coordinate system to a rotated coordinate system with axes X and Y

$$x = X\cos\theta - Y\sin\theta$$

$$y = X\sin\theta + Y\cos\theta$$

where the rotation angle θ is found from the equation

$$\cot(2\theta) = \frac{A - C}{B} \quad \text{or} \quad \tan(2\theta) = \frac{B}{A - C}$$

SECTION
9.7 EXERCISES

▪ SKILLS

In Exercises 1–8, the coordinates of a point in the xy-coordinate system are given. Assuming that the XY-axes are found by rotating the xy-axes by the given angle θ, find the corresponding coordinates for the point in the XY-system.

1. $(2, 4)$, $\theta = 45°$ **2.** $(5, 1)$, $\theta = 60°$ **3.** $(-3, 2)$, $\theta = 30°$ **4.** $(-4, 6)$, $\theta = 45°$

5. $(-1, -3)$, $\theta = 60°$ **6.** $(4, -4)$, $\theta = 45°$ **7.** $(0, 3)$, $\theta = 60°$ **8.** $(-2, 0)$, $\theta = 30°$

In Exercises 9–24, (a) identify the type of conic from the discriminant, (b) transform the equation in x and y into an equation in X and Y (without an XY-term) by rotating the x- and y-axes by the indicated angle θ to arrive at the new X- and Y-axes, and (c) graph the resulting equation (showing both sets of axes).

9. $xy - 1 = 0$, $\theta = 45°$ **10.** $xy - 4 = 0$, $\theta = 45°$

11. $x^2 + 2xy + y^2 + \sqrt{2}x - \sqrt{2}y - 1 = 0$, $\theta = 45°$ **12.** $2x^2 - 4xy + 2y^2 - \sqrt{2}x + 1 = 0$, $\theta = 45°$

13. $y^2 - \sqrt{3}xy + 3 = 0$, $\theta = 30°$ **14.** $x^2 - \sqrt{3}xy - 3 = 0$, $\theta = 60°$

15. $7x^2 - 2\sqrt{3}xy + 5y^2 - 8 = 0$, $\theta = 60°$ **16.** $4x^2 + \sqrt{3}xy + 3y^2 - 45 = 0$, $\theta = 30°$

17. $3x^2 + 2\sqrt{3}xy + y^2 + 2x - 2\sqrt{3}y - 2 = 0$, $\theta = 30°$ **18.** $x^2 + 2\sqrt{3}xy + 3y^2 - 2\sqrt{3}x + 2y - 4 = 0$, $\theta = 60°$

19. $7x^2 + 4\sqrt{3}xy + 3y^2 - 9 = 0$, $\theta = \dfrac{\pi}{6}$ **20.** $37x^2 + 42\sqrt{3}xy + 79y^2 - 400 = 0$, $\theta = \dfrac{\pi}{3}$

21. $7x^2 - 10\sqrt{3}xy - 3y^2 + 24 = 0$, $\theta = \dfrac{\pi}{3}$ **22.** $9x^2 + 14\sqrt{3}xy - 5y^2 + 48 = 0$, $\theta = \dfrac{\pi}{6}$

23. $x^2 - 2xy + y^2 - \sqrt{2}x - \sqrt{2}y - 8 = 0$, $\theta = \dfrac{\pi}{4}$ **24.** $x^2 + 2xy + y^2 + 3\sqrt{2}x + \sqrt{2}y = 0$, $\theta = \dfrac{\pi}{4}$

In Exercises 25–38, determine the angle of rotation necessary to transform the equation in x and y into an equation in X and Y with no XY-term.

25. $x^2 + 4xy + y^2 - 4 = 0$ **26.** $3x^2 + 5xy + 3y^2 - 2 = 0$

27. $2x^2 + \sqrt{3}xy + 3y^2 - 1 = 0$ **28.** $4x^2 + \sqrt{3}xy + 3y^2 - 1 = 0$

29. $2x^2 + \sqrt{3}xy + y^2 - 5 = 0$ **30.** $2\sqrt{3}x^2 + xy + 3\sqrt{3}y^2 + 1 = 0$

31. $\sqrt{2}x^2 + xy + \sqrt{2}y^2 - 1 = 0$ **32.** $x^2 + 10xy + y^2 + 2 = 0$

33. $12\sqrt{3}x^2 + 4xy + 8\sqrt{3}y^2 - 1 = 0$ **34.** $4x^2 + 2xy + 2y^2 - 7 = 0$

35. $5x^2 + 6xy + 4y^2 - 1 = 0$ **36.** $x^2 + 2xy + 12y^2 + 3 = 0$

37. $3x^2 + 10xy + 5y^2 - 1 = 0$ **38.** $10x^2 + 3xy + 2y^2 + 3 = 0$

In Exercises 39–48, graph the second-degree equation. (*Hint:* Transform the equation into an equation that contains no xy-term.)

39. $21x^2 + 10\sqrt{3}xy + 31y^2 - 144 = 0$ **40.** $5x^2 + 6xy + 5y^2 - 8 = 0$

41. $8x^2 - 20xy + 8y^2 + 18 = 0$ **42.** $3y^2 - 26\sqrt{3}xy - 23x^2 - 144 = 0$

43. $3x^2 + 2\sqrt{3}xy + y^2 + 2x - 2\sqrt{3}y - 12 = 0$ **44.** $3x^2 - 2\sqrt{3}xy + y^2 - 2x - 2\sqrt{3}y - 4 = 0$

45. $37x^2 - 42\sqrt{3}xy + 79y^2 - 400 = 0$ **46.** $71x^2 - 58\sqrt{3}xy + 13y^2 + 400 = 0$

47. $x^2 + 2xy + y^2 + 5\sqrt{2}x + 3\sqrt{2}y = 0$ **48.** $7x^2 - 4\sqrt{3}xy + 3y^2 - 9 = 0$

▪ CONCEPTUAL

In Exercises 49–52, determine whether each statement is true or false.

49. The graph of the equation $x^2 + kxy + 9y^2 = 5$, where k is any positive constant less than 6, is an ellipse.

50. The graph of the equation $x^2 + kxy + 9y^2 = 5$, where k is any constant greater than 6, is a parabola.

51. The reciprocal function is a rotated hyperbola.

52. The equation $\sqrt{x} + \sqrt{y} = 3$ can be transformed into the equation $X^2 + Y^2 = 9$.

▪ **CHALLENGE**

53. Determine the equation in X and Y that corresponds to $\dfrac{x^2}{a^2} + \dfrac{y^2}{b^2} = 1$ when the axes are rotated through

a. $90°$ **b.** $180°$

54. Determine the equation in X and Y that corresponds to $\dfrac{x^2}{a^2} - \dfrac{y^2}{b^2} = 1$ when the axes are rotated through

a. $90°$ **b.** $180°$

55. Identify the conic section with equation $y^2 + ax^2 = x$ for $a < 0, a > 0, a = 0$, and $a = 1$.

56. Identify the conic section with equation $x^2 - ay^2 = y$ for $a < 0, a > 0, a = 0$, and $a = 1$.

▪ **TECHNOLOGY**

For Exercises 57–62, refer to the following:

To use a TI-83 or TI-83 Plus (function-driven software or graphing utility) to graph a general second-degree equation, you need to solve for y. Let us consider a general second-degree equation $Ax^2 + Bxy + Cy^2 + Dx + Ey + F = 0$.

Group y^2 terms together, y terms together, and the remaining terms together.

$$Ax^2 + \underline{Bxy} + Cy^2 + Dx + \underset{=}{Ey} + F = 0$$

$$Cy^2 + (Bxy + Ey) + (Ax^2 + Dx + F) = 0$$

Factor out the common y in the first set of parentheses.

$$Cy^2 + y(Bx + E) + (Ax^2 + Dx + F) = 0$$

Now this is a quadratic equation in y: $ay^2 + by + c = 0$.

Use the quadratic formula to solve for y.

$$Cy^2 + y(Bx + E) + (Ax^2 + Dx + F) = 0$$

$$a = C, b = Bx + E, c = Ax^2 + Dx + F$$

$$y = \frac{-b \pm \sqrt{b^2 - 4ac}}{2a} \qquad y = \frac{-(Bx + E) \pm \sqrt{(Bx + E)^2 - 4(C)(Ax^2 + Dx + F)}}{2(C)}$$

$$y = \frac{-(Bx + E) \pm \sqrt{B^2x^2 + 2BEx + E^2 - 4ACx^2 - 4CDx - 4CF}}{2C}$$

$$y = \frac{-(Bx + E) \pm \sqrt{(B^2 - 4AC)x^2 + (2BE - 4CD)x + (E^2 - 4CF)}}{2C}$$

Case I: $B^2 - 4AC = 0 \to$ The second-degree equation $Ax^2 + Bxy + Cy^2 + Dx + Ey + F = 0$ is a parabola.

$$y = \frac{-(Bx + E) \pm \sqrt{(2BE - 4CD)x + (E^2 - 4CF)}}{2C}$$

Case II: $B^2 - 4AC < 0 \to$ The second-degree equation $Ax^2 + Bxy + Cy^2 + Dx + Ey + F = 0$ is an ellipse.

$$y = \frac{-(Bx + E) \pm \sqrt{(B^2 - 4AC)x^2 + (2BE - 4CD)x + (E^2 - 4CF)}}{2C}$$

Case III: $B^2 - 4AC > 0 \to$ The second-degree equation $Ax^2 + Bxy + Cy^2 + Dx + Ey + F = 0$ is a hyperbola.

$$y = \frac{-(Bx + E) \pm \sqrt{(B^2 - 4AC)x^2 + (2BE - 4CD)x + (E^2 - 4CF)}}{2C}$$

57. Use a graphing utility to explore the second-degree equation $3x^2 + 2\sqrt{3}xy + y^2 + Dx + Ey + F = 0$ for the following values of D, E, and F:

a. $D = 1, E = 3, F = 2$

b. $D = -1, E = -3, F = 2$

Show the angle of rotation to the nearest degree. Explain the differences.

58. Use a graphing utility to explore the second-degree equation $x^2 + 3xy + 3y^2 + Dx + Ey + F = 0$ for the following values of D, E, and F:

a. $D = 2$, $E = 6$, $F = -1$

b. $D = 6$, $E = 2$, $F = -1$

Show the angle of rotation to the nearest degree. Explain the differences.

59. Use a graphing utility to explore the second-degree equation $2x^2 + 3xy + y^2 + Dx + Ey + F = 0$ for the following values of D, E, and F:

a. $D = 2$, $E = 1$, $F = -2$

b. $D = 2$, $E = 1$, $F = 2$

Show the angle of rotation to the nearest degree. Explain the differences.

60. Use a graphing utility to explore the second-degree equation $2\sqrt{3}x^2 + xy + \sqrt{3}y^2 + Dx + Ey + F = 0$ for the following values of D, E, and F:

a. $D = 2$, $E = 1$, $F = -1$

b. $D = 2$, $E = 6$, $F = -1$

Show the angle of rotation to the nearest degree. Explain the differences.

61. Use a graphing utility to explore the second-degree equation $Ax^2 + Bxy + Cy^2 + 2x + y - 1 = 0$ for the following values of A, B, and C:

a. $A = 4$, $B = -4$, $C = 1$

b. $A = 4$, $B = 4$, $C = -1$

c. $A = 1$, $B = -4$, $C = 4$

Show the angle of rotation to the nearest degree. Explain the differences.

62. Use a graphing utility to explore the second-degree equation $Ax^2 + Bxy + Cy^2 + 3x + 5y - 2 = 0$ for the following values of A, B, and C:

a. $A = 1$, $B = -4$, $C = 4$

b. $A = 1$, $B = 4$, $D = -4$

Show the angle of rotation to the nearest degree. Explain the differences.

■ PREVIEW TO CALCULUS

In calculus, when finding the area between two curves, we need to find the points of intersection of the curves.

In Exercises 63–66, find the points of intersection of the rotated conic sections.

63. $\quad x^2 + 2xy = 10$
$\quad\;\; 3x^2 - xy = 2$

64. $\quad x^2 - 3xy + 2y^2 = 0$
$\quad\;\; x^2 + xy \quad\quad = 6$

65. $\quad 2x^2 - 7xy + 2y^2 = -1$
$\quad\;\;\; x^2 - 3xy + y^2 = 1$

66. $\quad\;\; 4x^2 + xy + 4y^2 = 22$
$\quad -3x^2 + 2xy - 3y^2 = -11$

SECTION
9.8 POLAR EQUATIONS OF CONICS

SKILLS OBJECTIVES

- Define conics in terms of eccentricity.
- Express equations of conics in polar form.
- Graph the polar equations of conics.

CONCEPTUAL OBJECTIVE

- Define all conics in terms of a focus and a directrix.

In Section, 9.1, we discussed parabolas, ellipses, and hyperbolas in terms of geometric definitions. Then in Sections 9.2–9.4, we examined the rectangular equations of these conics. The equations for ellipses and hyperbolas when their centers are at the origin were simpler than when they were not (when the conics were shifted). In Section 7.5, we discussed polar coordinates and graphing of polar equations. In this section, we develop a more unified definition of the three conics in terms of a single focus and a directrix. You will see in this section that if the *focus* is located at the origin, then equations of conics are simpler when written in polar coordinates.

Alternative Definition of Conics

Recall that when we work with rectangular coordinates, we define a parabola (Sections 9.1 and 9.2) in terms of a fixed point (focus) and a line (directrix), whereas we define an ellipse and hyperbola (Sections 9.1, 9.3, and 9.4) in terms of two fixed points (the foci). However, it is possible to define all three conics in terms of a single focus and a directrix.

The following alternative representation of conics depends on a parameter called *eccentricity*.

ALTERNATIVE DESCRIPTION OF CONICS

Let D be a fixed line (the **directrix**), F be a fixed point (a **focus**) not on D, and e be a fixed positive number (**eccentricity**). The set of all points P such that the ratio of the distance from P to F to the distance from P to D equals the constant e defines a conic section.

$$\frac{d(P, F)}{d(P, D)} = e$$

- If $e = 1$, the conic is a **parabola**.
- If $e < 1$, the conic is an **ellipse**.
- If $e > 1$, the conic is a **hyperbola**.

When $e = 1$, the result is a parabola, described by the same definition we used previously in Section 9.1. When $e \neq 1$, the result is either an ellipse or a hyperbola. The major axis of an ellipse passes through the focus and is perpendicular to the directrix. The transverse axis of a hyperbola also passes through the focus and is perpendicular to the directrix. If we let c represent the distance from the focus to the center and a represent the distance from the vertex to the center, then eccentricity is given by

$$e = \frac{c}{a}$$

Equations of Conics in Polar Coordinates

In polar coordinates, if we locate the focus of a conic at the pole and the directrix is either perpendicular or parallel to the polar axis, then we have four possible scenarios:

- The directrix is *perpendicular* to the polar axis and p units to the *right* of the pole.
- The directrix is *perpendicular* to the polar axis and p units to the *left* of the pole.
- The directrix is *parallel* to the polar axis and p units *above* the pole.
- The directrix is *parallel* to the polar axis and p units *below* the pole.

Let us take the case in which the directrix is perpendicular to the polar axis and p units to the right of the pole.

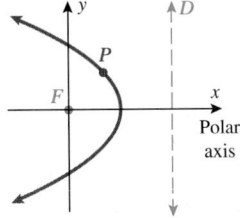

In polar coordinates (r, θ), we see that the distance from the focus to a point P is equal to r, that is, $d(P, F) = r$, and the distance from P to the closest point on the directrix is $d(P, D) = p - r\cos\theta$.

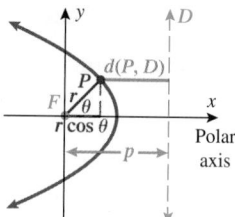

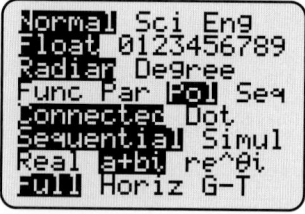
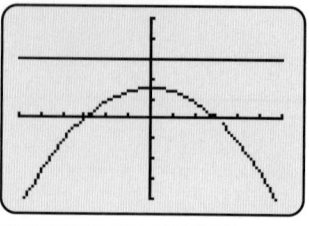
WORDS	MATH
Substitute $d(P, F) = r$ and $d(P, D) = p - r\cos\theta$ into the formula for eccentricity, $\dfrac{d(P, F)}{d(P, D)} = e$.	$\dfrac{r}{p - r\cos\theta} = e$
Multiply the result by $p - r\cos\theta$.	$r = e(p - r\cos\theta)$
Eliminate the parentheses.	$r = ep - er\cos\theta$
Add $er\cos\theta$ to both sides of the equation.	$r + er\cos\theta = ep$
Factor out the common r.	$r + (1 + e\cos\theta) = ep$
Divide both sides by $1 + e\cos\theta$.	$\boxed{r = \dfrac{ep}{1 + e\cos\theta}}$

We need not derive the other three cases here, but note that if the directrix is perpendicular to the polar axis and p units to the *left* of the pole, the resulting polar equation is

$$r = \frac{ep}{1 - e\cos\theta}$$

If the directrix is parallel to the polar axis, the directrix is either above ($y = p$) or below ($y = -p$) the polar axis and we get the sine function instead of the cosine function, as summarized in the following box:

POLAR EQUATIONS OF CONICS

The following polar equations represent conics with one focus at the origin and with eccentricity e. It is assumed that the positive x-axis represents the polar axis.

EQUATION	DESCRIPTION
$r = \dfrac{ep}{1 + e\cos\theta}$	The directrix is *vertical* and p units to the *right* of the pole.
$r = \dfrac{ep}{1 - e\cos\theta}$	The directrix is *vertical* and p units to the *left* of the pole.
$r = \dfrac{ep}{1 + e\sin\theta}$	The directrix is *horizontal* and p units *above* the pole.
$r = \dfrac{ep}{1 - e\sin\theta}$	The directrix is *horizontal* and p units *below* the pole.

ECCENTRICITY	THE CONIC IS A	THE ____ IS PERPENDICULAR TO THE DIRECTRIX
$e = 1$	Parabola	Axis of symmetry
$e < 1$	Ellipse	Major axis
$e > 1$	Hyperbola	Transverse axis

EXAMPLE 1 Finding the Polar Equation of a Conic

Find a polar equation for a parabola that has its focus at the origin and whose directrix is the line $y = 3$.

Solution:

The directrix is horizontal and above the pole.

$$r = \frac{ep}{1 + e\sin\theta}$$

A parabola has eccentricity $e = 1$, and we know that $p = 3$.

$$r = \frac{3}{1 + \sin\theta}$$

■ **YOUR TURN** Find a polar equation for a parabola that has its focus at the origin and whose directrix is the line $x = -3$.

■ **Answer:** $r = \dfrac{3}{1 - 6\cos\theta}$

EXAMPLE 2 Identifying a Conic from Its Equation

Identify the type of conic represented by the equation $r = \dfrac{10}{3 + 2\cos\theta}$.

Solution:

To identify the type of conic, we need to rewrite the equation in the form:

$$r = \frac{ep}{1 \pm e\cos\theta}$$

Divide the numerator and denominator by 3.

$$r = \frac{\dfrac{10}{3}}{\left(1 + \dfrac{2}{3}\cos\theta\right)}$$

Identify e in the denominator.

$$= \frac{\dfrac{10}{3}}{\left(1 + \underset{e}{\dfrac{2}{3}}\cos\theta\right)}$$

The numerator is equal to ep.

$$= \frac{\overset{p}{\dfrac{10}{3}} \cdot \overset{e}{\dfrac{2}{3}}}{\left(1 + \underset{e}{\dfrac{2}{3}}\cos\theta\right)}$$

Since $e = \frac{2}{3} < 1$, the conic is an $\boxed{\text{ellipse}}$. The directrix is $x = 5$, so the major axis is along the x-axis (perpendicular to the directrix).

■ **YOUR TURN** Identify the type of conic represented by the equation

$$r = \frac{10}{2 - 10\sin\theta}$$

Use $\boxed{Y=}$ to enter the polar equation $r = \dfrac{10}{3 + 2\cos\theta}$.

$\boxed{r1 =}\ \boxed{10}\ \boxed{\div}\ \boxed{(}\ \boxed{3}\ \boxed{+}\ \boxed{2}$
$\boxed{\cos}\ \boxed{X, T, \theta, n}\ \boxed{)}\ \boxed{)}$

To enter the equation of the directrix $x = 5$, use its polar form.

$$x = 5 \quad r\cos\theta = 5 \quad r = \frac{5}{\cos\theta}$$

$\boxed{r2 =}\ \boxed{5}\ \boxed{\div}\ \boxed{\cos}\ \boxed{X, T, \theta, n}\ \boxed{)}$

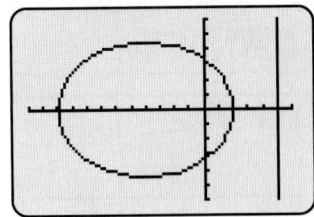

■ **Answer:** hyperbola, $e = 5$, with transverse axis along the y-axis

In Example 2 we found that the polar equation $r = \dfrac{10}{3 + 2\cos\theta}$ is an ellipse with its major axis along the x-axis. We will graph this ellipse in Example 3.

Technology Tip

Use $\boxed{Y =}$ to enter the polar

equation $r = \dfrac{10}{3 + 2\cos\theta}$.

$\boxed{r1} = \boxed{10} \boxed{\div} \boxed{(} \boxed{(} \boxed{3} \boxed{+} \boxed{2}$

$\boxed{\cos} \boxed{X, T, \theta, n} \boxed{)} \boxed{)}$

Use the $\boxed{\text{TRACE}}$ key to trace the
vertices of the ellipse.

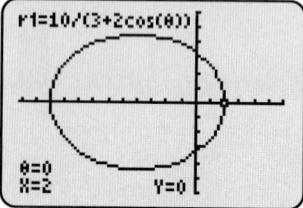

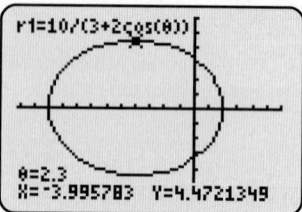

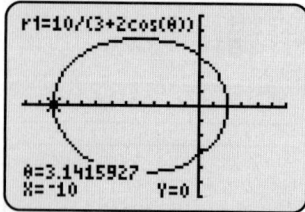

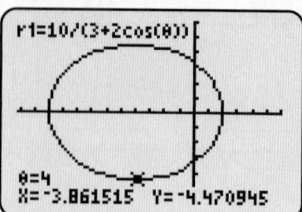

EXAMPLE 3   Graphing a Conic from Its Equation

The graph of the polar equation $r = \dfrac{10}{3 + 2\cos\theta}$ is an ellipse.

a. Find the vertices.

b. Find the center of the ellipse.

c. Find the lengths of the major and minor axes.

d. Graph the ellipse.

Solution (a):

From Example 2 we see that $e = \frac{2}{3}$, which corresponds to an ellipse, and $x = 5$ is the directrix.
     The major axis is perpendicular to the directrix. Therefore, the major axis lies along
the polar axis. To find the vertices (which lie along the major axis), let $\theta = 0$ and $\theta = \pi$.

$\theta = 0$:
$$r = \frac{10}{3 + 2\cos\theta} = \frac{10}{5} = 2$$

$\theta = \pi$:
$$r = \frac{10}{3 + 2\cos\pi} = \frac{10}{1} = 10$$

The vertices are the points $\boxed{V_1 = (2, 0)}$ and $\boxed{V_2 = (10, \pi)}$.

Solution (b):

The vertices in rectangular coordinates are $V_1 = (2, 0)$ and $V_2 = (-10, 0)$.

The midpoint (in rectangular coordinates) between the two vertices is the point $(-4, 0)$,
which corresponds to the point $\boxed{(4, \pi)}$ in polar coordinates.

Solution (c):

The length of the major axis, $2a$, is
the distance between the vertices. $\boxed{2a = 12}$

The length $a = 6$ corresponds to the distance from the center to a vertex.

Apply the formula $e = \dfrac{c}{a}$ with $a = 6$
and $e = \frac{2}{3}$ to find c. $c = ae = 6\left(\dfrac{2}{3}\right) = 4$

Let $a = 6$ and $c = 4$ in $b^2 = a^2 - c^2$. $b^2 = 6^2 - 4^2 = 20$

Solve for b. $b = 2\sqrt{5}$

The length of the minor axis is $\boxed{2b = 4\sqrt{5}}$.

Solution (d):

Graph the ellipse.

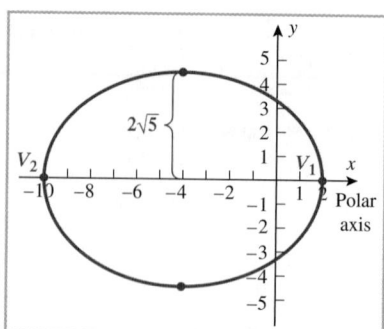

EXAMPLE 4   Identifying and Graphing a Conic from Its Equation

Identify and graph the conic defined by the equation $r = \dfrac{2}{2 + 3\sin\theta}$.

Solution:

Rewrite the equation in
the form $r = \dfrac{ep}{1 + e\sin\theta}$.

$$r = \frac{2}{2 + 3\sin\theta} = \frac{\overset{p}{\left(\frac{2}{3}\right)}\overset{e}{\left(\frac{3}{2}\right)}}{1 + \underset{e}{\left(\frac{3}{2}\right)}\sin\theta}$$

The conic is a *hyperbola* since $e = \frac{3}{2} > 1$.

The directrix is horizontal and $\frac{2}{3}$ unit above the pole (origin).

To find the vertices, let $\theta = \dfrac{\pi}{2}$ and $\theta = \dfrac{3\pi}{2}$.

$\theta = \dfrac{\pi}{2}$:

$$r = \frac{2}{2 + 3\sin\left(\dfrac{\pi}{2}\right)} = \frac{2}{5}$$

$\theta = \dfrac{3\pi}{2}$:

$$r = \frac{2}{2 + 3\sin\left(\dfrac{3\pi}{2}\right)} = \frac{2}{-1} = -2$$

The vertices in polar coordinates are $\left(\dfrac{2}{5}, \dfrac{\pi}{2}\right)$ and $\left(-2, \dfrac{3\pi}{2}\right)$.

The vertices in rectangular coordinates are $V_1 = \left(0, \frac{2}{5}\right)$ and $V_2 = (0, 2)$.

The center is the midpoint between the vertices: $\left(0, \frac{6}{5}\right)$.

The distance from the center to a focus is $c = \frac{6}{5}$.

Apply the formula $e = \dfrac{c}{a}$ with $c = \dfrac{6}{5}$ and
$e = \dfrac{3}{2}$ to find a.

$$a = \frac{c}{e} = \frac{\dfrac{6}{5}}{\dfrac{3}{2}} = \frac{4}{5}$$

Let $a = \frac{4}{5}$ and $c = \frac{6}{5}$ in $b^2 = c^2 - a^2$.

$$b^2 = \left(\frac{6}{5}\right)^2 - \left(\frac{4}{5}\right)^2 = \frac{20}{25}$$

Solve for b.

$$b = \frac{2\sqrt{5}}{5}$$

The asymptotes are given by
$y = \pm\dfrac{a}{b}(x - h) + k$, where $a = \dfrac{4}{5}$,
$b = \dfrac{2\sqrt{5}}{5}$, and $(h, k) = \left(0, \dfrac{6}{5}\right)$.

$$y = \pm\frac{2}{\sqrt{5}}x + \frac{6}{5}$$

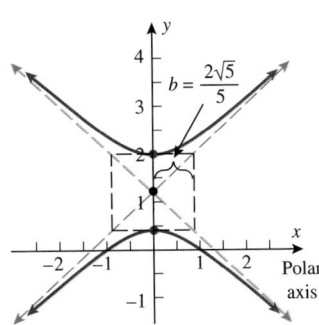

Technology Tip

Use $\boxed{Y =}$ to enter the polar

equation $r = \dfrac{2}{2 + 3\sin\theta}$.

$\boxed{\text{r1}} = \boxed{2}\boxed{\div}\boxed{(}\boxed{2}\boxed{+}\boxed{3}$
$\boxed{\sin}\boxed{\text{X, T, }\theta\text{, n}}\boxed{)}\boxed{)}$

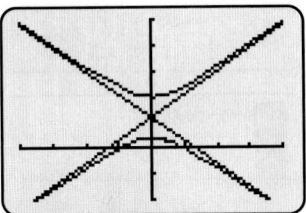

Use the $\boxed{\text{TRACE}}$ key to trace the
vertices of the hyperbola.

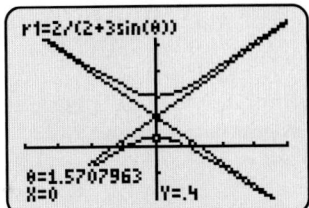

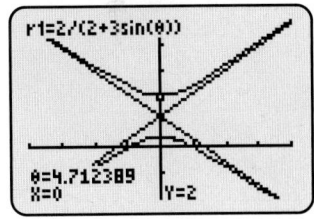

It is important to note that although we relate specific points (vertices, foci, etc.) to rectangular coordinates, another approach to finding a rough sketch is to simply point-plot the equation in polar coordinates.

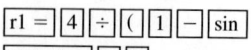

Technology Tip

Use $\boxed{Y =}$ to enter the polar equation $r = \dfrac{4}{1 - \sin\theta}$.

$\boxed{r1} = \boxed{4}\,\boxed{\div}\,\boxed{(}\,\boxed{(}\,\boxed{1}\,\boxed{-}\,\boxed{\sin}$
$\boxed{X, T, \theta, n}\,\boxed{)}\,\boxed{)}$

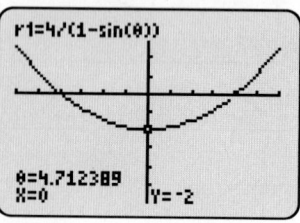

```
Plot1  Plot2  Plot3
\r1■4/(1-sin(θ))
```

```
r1=4/(1-sin(θ))
```

```
θ=4.712389
X=0            Y=-2
```

EXAMPLE 5   Graphing a Conic by Point-Plotting in Polar Coordinates

Sketch a graph of the conic $r = \dfrac{4}{1 - \sin\theta}$.

Solution:

STEP 1 The conic is a parabola because the equation is in the form

$$r = \frac{(4)(1)}{1 - (1)\sin\theta}$$

Make a table with key values for θ and r.

θ	$r = \dfrac{4}{1 - \sin\theta}$	(r, θ)
0	$r = \dfrac{4}{1 - \sin 0} = \dfrac{4}{1} = 4$	$(4, 0)$
$\dfrac{\pi}{2}$	$r = \dfrac{4}{1 - \sin\dfrac{\pi}{2}} = \dfrac{4}{1 - 1} = \dfrac{4}{0}$	undefined
π	$r = \dfrac{4}{1 - \sin\pi} = \dfrac{4}{1} = 4$	$(4, \pi)$
$\dfrac{3\pi}{2}$	$r = \dfrac{4}{1 - \sin\dfrac{3\pi}{2}} = \dfrac{4}{1 - (-1)} = \dfrac{4}{2} = 2$	$\left(2, \dfrac{3\pi}{2}\right)$
2π	$r = \dfrac{4}{1 - \sin 2\pi} = \dfrac{4}{1} = 4$	$(4, 2\pi)$

STEP 2 Plot the points on a polar graph and connect them with a smooth parabolic curve.

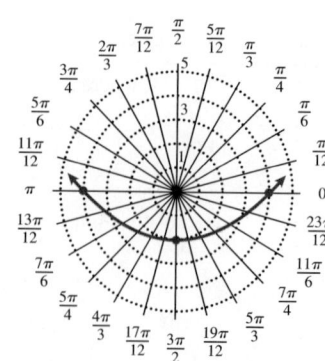

SECTION
9.8 SUMMARY

In this section, we found that we could graph polar equations of conics by identifying a single focus and the directrix. There are four possible equations in terms of eccentricity e:

EQUATION	DESCRIPTION
$r = \dfrac{ep}{1 + e\cos\theta}$	The directrix is *vertical* and p units to the *right* of the pole.
$r = \dfrac{ep}{1 - e\cos\theta}$	The directrix is *vertical* and p units to the *left* of the pole.
$r = \dfrac{ep}{1 + e\sin\theta}$	The directrix is *horizontal* and p units *above* the pole.
$r = \dfrac{ep}{1 - e\sin\theta}$	The directrix is *horizontal* and p units *below* the pole.

SECTION
9.8 EXERCISES

▪ **SKILLS**

In Exercises 1–14, find the polar equation that represents the conic described (assume that a focus is at the origin).

Conic	Eccentricity	Directrix		Conic	Eccentricity	Directrix
1. Ellipse	$e = \frac{1}{2}$	$y = -5$		**2.** Ellipse	$e = \frac{1}{3}$	$y = 3$
3. Hyperbola	$e = 2$	$y = 4$		**4.** Hyperbola	$e = 3$	$y = -2$
5. Parabola	$e = 1$	$x = 1$		**6.** Parabola	$e = 1$	$x = -1$
7. Ellipse	$e = \frac{3}{4}$	$x = 2$		**8.** Ellipse	$e = \frac{2}{3}$	$x = -4$
9. Hyperbola	$e = \frac{4}{3}$	$x = -3$		**10.** Hyperbola	$e = \frac{3}{2}$	$x = 5$
11. Parabola	$e = 1$	$y = -3$		**12.** Parabola	$e = 1$	$y = 4$
13. Ellipse	$e = \frac{3}{5}$	$y = 6$		**14.** Hyperbola	$e = \frac{8}{5}$	$y = 5$

In Exercises 15–26, identify the conic (parabola, ellipse, or hyperbola) that each polar equation represents.

15. $r = \dfrac{4}{1 + \cos\theta}$

16. $r = \dfrac{3}{2 - 3\sin\theta}$

17. $r = \dfrac{2}{3 + 2\sin\theta}$

18. $r = \dfrac{3}{2 - 2\cos\theta}$

19. $r = \dfrac{2}{4 + 8\cos\theta}$

20. $r = \dfrac{1}{4 - \cos\theta}$

21. $r = \dfrac{7}{3 + \cos\theta}$

22. $r = \dfrac{4}{5 + 6\sin\theta}$

23. $r = \dfrac{40}{5 + 5\sin\theta}$

24. $r = \dfrac{5}{5 - 4\sin\theta}$

25. $r = \dfrac{1}{1 - 6\cos\theta}$

26. $r = \dfrac{5}{3 - 3\sin\theta}$

In Exercises 27–42, for the given polar equations: (a) identify the conic as either a parabola, an ellipse, or a hyperbola; (b) find the eccentricity and vertex (or vertices); and (c) graph.

27. $r = \dfrac{2}{1 + \sin\theta}$

28. $r = \dfrac{4}{1 - \cos\theta}$

29. $r = \dfrac{4}{1 - 2\sin\theta}$

30. $r = \dfrac{3}{3 + 8\cos\theta}$

31. $r = \dfrac{2}{2 + \sin\theta}$

32. $r = \dfrac{1}{3 - \sin\theta}$

33. $r = \dfrac{1}{2 - 2\sin\theta}$

34. $r = \dfrac{1}{1 - 2\sin\theta}$

35. $r = \dfrac{4}{3 + \cos\theta}$ **36.** $r = \dfrac{2}{5 + 4\sin\theta}$ **37.** $r = \dfrac{6}{2 + 3\sin\theta}$ **38.** $r = \dfrac{6}{1 + \cos\theta}$

39. $r = \dfrac{2}{5 + 5\cos\theta}$ **40.** $r = \dfrac{10}{6 - 3\cos\theta}$ **41.** $r = \dfrac{6}{3\cos\theta + 1}$ **42.** $r = \dfrac{15}{3\sin\theta + 5}$

■ APPLICATIONS

For Exercises 43 and 44, refer to the following:

Planets travel in elliptical orbits around a single focus, the Sun. Pluto (orange), the dwarf planet furthest from the Sun, has a pronounced elliptical orbit, whereas Earth (royal blue) has an almost circular orbit. The polar equation of a planet's orbit can be expressed as

$$r = \frac{a(1 - e^2)}{(1 - e\cos\theta)}$$

where e is the eccentricity and $2a$ is the length of the major axis. It can also be shown that the perihelion distance (minimum distance from the Sun to a planet) and the aphelion distance (maximum distance from the Sun to the planet) can be represented by $r = a(1 - e)$ and $r = a(1 + e)$, respectively.

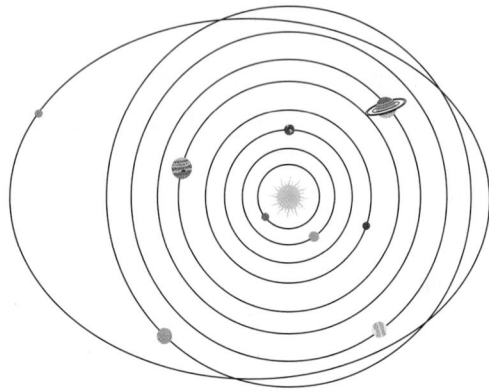

43. Planetary Orbits. Pluto's orbit is summarized in the picture below. Find the eccentricity of Pluto's orbit. Find the polar equation that governs Pluto's orbit.

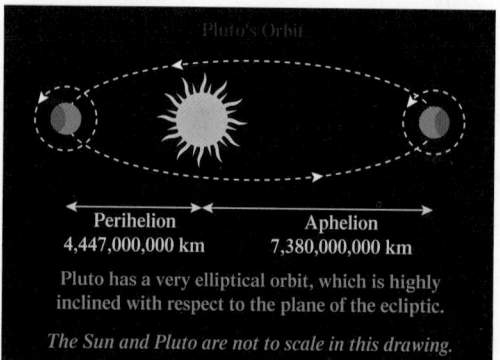

Perihelion
4,447,000,000 km

Aphelion
7,380,000,000 km

Pluto has a very elliptical orbit, which is highly inclined with respect to the plane of the ecliptic.

The Sun and Pluto are not to scale in this drawing.

44. Planetary Orbits. Earth's orbit is summarized in the picture below. Find the eccentricity of Earth's orbit. Find the polar equation that governs Earth's orbit.

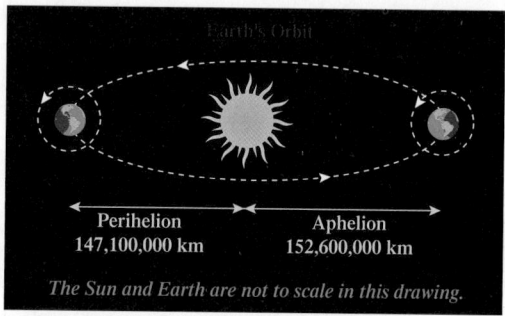

Perihelion
147,100,000 km

Aphelion
152,600,000 km

The Sun and Earth are not to scale in this drawing.

For Exercises 45 and 46, refer to the following:

Asteroids, meteors, and comets all orbit the Sun in elliptical patterns and often cross paths with Earth's orbit, making life a little tense now and again. Asteroids are large rocks (bodies under 1000 km across), meteors range from sand particles to rocks, and comets are masses of debris. A few asteroids have orbits that cross Earth's orbits—called Apollos or Earth-crossing asteroids. In recent years, asteroids have passed within 100,000 km of Earth!

45. Asteroids. The asteroid 433 or Eros is the second largest near-Earth asteroid. The semimajor axis of its orbit is 150 million km and the eccentricity is 0.223. Find the polar equation of Eros's orbit.

46. Asteroids. The asteroid Toutatis is the largest near-Earth asteroid. The semimajor axis of its orbit is 350 million km and the eccentricity is 0.634. On September 29, 2004, it missed Earth by 961,000 miles. Find the polar equation of Toutatis's orbit.

47. Earth's Orbit. A simplified model of Earth's orbit around the Sun is given by $r = \dfrac{1}{1 + 0.0167\cos\theta}$. Find the center of the orbit in

a. rectangular coordinates
b. polar coordinates

48. Uranus's Orbit. A simplified model of Uranus's orbit around the Sun is given by $r = \dfrac{1}{1 + 0.0461\cos\theta}$. Find the center of the orbit in

a. rectangular coordinates
b. polar coordinates

49. Orbit of Halley's Comet. A simplified model of the orbit of Halley's Comet around the Sun is given by $r = \dfrac{1}{1 + 0.967\sin\theta}$. Find the center of the orbit in rectangular coordinates.

50. Orbit of the Hale–Bopp Comet. A simplified model of the orbit of the Hale–Bopp Comet around the Sun is given by $r = \dfrac{1}{1 + 0.995\sin\theta}$. Find the center of the orbit in rectangular coordinates.

■ CONCEPTUAL

51. When $0 < e < 1$, the conic is an ellipse. Does the conic become more elongated or elliptical as e approaches 1 or as e approaches 0?

52. Show that $r = \dfrac{ep}{1 - e\sin\theta}$ is the polar equation of a conic with a horizontal directrix that is p units *below* the pole.

53. Convert from rectangular to polar coordinates to show that the equation of a hyperbola, $\dfrac{x^2}{a^2} - \dfrac{y^2}{b^2} = 1$, in polar form is $r^2 = -\dfrac{b^2}{1 - e^2\cos^2\theta}$.

54. Convert from rectangular to polar coordinates to show that the equation of an ellipse, $\dfrac{x^2}{a^2} + \dfrac{y^2}{b^2} = 1$, in polar form is $r^2 = \dfrac{b^2}{1 - e^2\cos^2\theta}$.

■ CHALLENGE

55. Find the major diameter of the ellipse with polar equation $r = \dfrac{ep}{1 + e\cos\theta}$ in terms of e and p.

56. Find the minor diameter of the ellipse with polar equation $r = \dfrac{ep}{1 + e\cos\theta}$ in terms of e and p.

57. Find the center of the ellipse with polar equation $r = \dfrac{ep}{1 + e\cos\theta}$ in terms of e and p.

58. Find the length of the latus rectum of the parabola with polar equation $r = \dfrac{p}{1 + \cos\theta}$. Assume that the focus is at the origin.

■ TECHNOLOGY

59. Let us consider the polar equations $r = \dfrac{ep}{1 + e\cos\theta}$ and $r = \dfrac{ep}{1 - e\cos\theta}$ with eccentricity $e = 1$. With a graphing utility, explore the equations with $p = 1, 2,$ and 6. Describe the behavior of the graphs as $p \to \infty$ and also the difference between the two equations.

60. Let us consider the polar equations $r = \dfrac{ep}{1 + e\sin\theta}$ and $r = \dfrac{ep}{1 - e\sin\theta}$ with eccentricity $e = 1$. With a graphing utility, explore the equations with $p = 1, 2,$ and 6. Describe the behavior of the graphs as $p \to \infty$ and also the difference between the two equations.

61. Let us consider the polar equations $r = \dfrac{ep}{1 + e\cos\theta}$ and $r = \dfrac{ep}{1 - e\cos\theta}$ with $p = 1$. With a graphing utility, explore the equations with $e = 1.5, 3,$ and 6. Describe the behavior of the graphs as $e \to \infty$ and also the difference between the two equations.

62. Let us consider the polar equations $r = \dfrac{ep}{1 + e\sin\theta}$ and $r = \dfrac{ep}{1 - e\sin\theta}$ with $p = 1$. With a graphing utility, explore the equations with $e = 1.5, 3,$ and 6. Describe the behavior of the graphs as $e \to \infty$ and also the difference between the two equations.

63. Let us consider the polar equations $r = \dfrac{ep}{1 + e\cos\theta}$ and $r = \dfrac{ep}{1 - e\cos\theta}$ with $p = 1$. With a graphing utility, explore the equations with $e = 0.001, 0.5, 0.9,$ and 0.99. Describe the behavior of the graphs as $e \to 1$ and also the difference between the two equations. Be sure to set the window parameters properly.

64. Let us consider the polar equations $r = \dfrac{ep}{1 + e\sin\theta}$ and $r = \dfrac{ep}{1 - e\sin\theta}$ with $p = 1$. With a graphing utility, explore the equations with $e = 0.001, 0.5, 0.9,$ and 0.99. Describe the behavior of the graphs as $e \to 1$ and also the difference between the two equations. Be sure to set the window parameters properly.

65. Let us consider the polar equation $r = \dfrac{5}{5 + 2\sin\theta}$.

Explain why the graphing utility gives the following graphs with the specified window parameters:

a. $[-2, 2]$ by $[-2, 2]$ with θ step $= \dfrac{\pi}{2}$

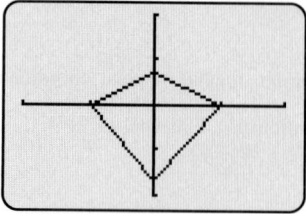

b. $[-2, 2]$ by $[-2, 2]$ with θ step $= \dfrac{\pi}{3}$

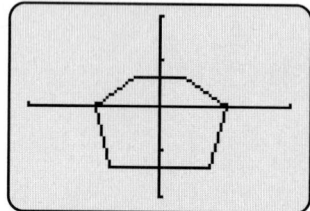

66. Let us consider the polar equation $r = \dfrac{2}{1 + \cos\theta}$. Explain why a graphing utility gives the following graphs with the specified window parameters:

a. $[-2, 2]$ by $[-4, 4]$ with θ step $= \dfrac{\pi}{2}$

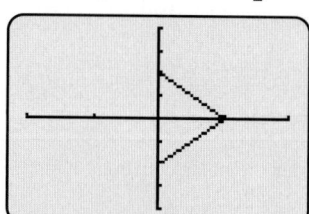

b. $[-2, 2]$ by $[-4, 4]$ with θ step $= \dfrac{\pi}{3}$

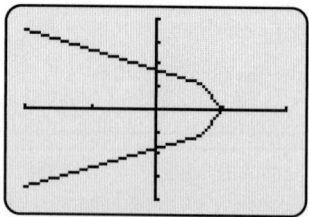

67. Let us consider the polar equation $r = \dfrac{6}{1 + 3\sin\theta}$.

Explain why a graphing utility gives the following graphs with the specified window parameters:

a. $[-8, 8]$ by $[-2, 4]$ with θ step $= \dfrac{\pi}{2}$

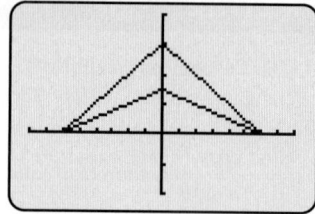

b. $[-4, 8]$ by $[-2, 6]$ with θ step $= 0.4\pi$

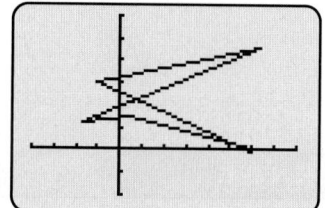

68. Let us consider the polar equation $r = \dfrac{2}{1 - \sin\theta}$. Explain why a graphing utility gives the following graphs with the specified window parameters:

a. $[-4, 4]$ by $[-2, 4]$ with θ step $= \dfrac{\pi}{3}$

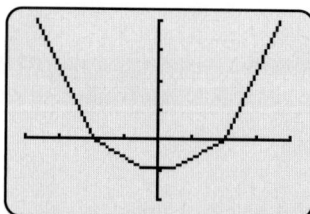

b. $[-4, 4]$ by $[-2, 6]$ with θ step $= 0.8\pi$

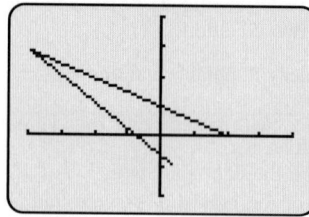

■ **PREVIEW TO CALCULUS**

In calculus, when finding the area between two polar curves, we need to find the points of intersection of the two curves.

In Exercises 69–72, find the values of θ where the two conic sections intersect on $[0, 2\pi]$.

69. $r = \dfrac{2}{2 + \sin\theta}, r = \dfrac{2}{2 + \cos\theta}$

70. $r = \dfrac{1}{3 + 2\sin\theta}, r = \dfrac{1}{3 - 2\sin\theta}$

71. $r = \dfrac{1}{4 - 3\sin\theta}, r = \dfrac{1}{-1 + 7\sin\theta}$

72. $r = \dfrac{1}{5 + 2\cos\theta}, r = \dfrac{1}{10 - 8\cos\theta}$

SKILLS OBJECTIVES

- Graph parametric equations.
- Find an equation (in rectangular form) that corresponds to a graph defined parametrically.
- Find parametric equations for a graph that is defined by an equation in rectangular form.

CONCEPTUAL OBJECTIVES

- Understand that the results of increasing the value of the parameter reveal the orientation of a curve or the direction of motion along it.
- Use time as a parameter in parametric equations.

Parametric Equations of a Curve

Thus far we have talked about graphs in planes. For example, the equation $x^2 + y^2 = 1$ when graphed in a plane is the unit circle. Similarly, the function $f(x) = \sin x$ when graphed in a plane is a sinusoidal curve. Now, we consider the **path along a curve**. For example, if a car is being driven on a circular racetrack, we want to see the movement along the circle. We can determine where (position) along the circle the car is at some time t using *parametric equations*. Before we define *parametric equations* in general, let us start with a simple example.

Let $x = \sin t$ and $y = \cos t$ and $t \geq 0$. We then can make a table of some corresponding values.

t SECONDS	$x = \cos t$	$y = \sin t$	(x, y)
0	$x = \cos 0 = 1$	$y = \sin 0 = 0$	$(1, 0)$
$\dfrac{\pi}{2}$	$x = \cos\left(\dfrac{\pi}{2}\right) = 0$	$y = \sin\left(\dfrac{\pi}{2}\right) = 1$	$(0, 1)$
π	$x = \cos \pi = -1$	$y = \sin \pi = 0$	$(-1, 0)$
$\dfrac{3\pi}{2}$	$x = \cos\left(\dfrac{3\pi}{2}\right) = 0$	$y = \sin\left(\dfrac{3\pi}{2}\right) = -1$	$(0, -1)$
2π	$x = \cos(2\pi) = 1$	$y = \sin(2\pi) = 0$	$(1, 0)$

If we plot these points and note the correspondence to time (by converting all numbers to decimals), we will be tracing a *path* counterclockwise along the unit circle.

TIME (SECONDS)	$t = 0$	$t = 1.57$	$t = 3.14$	$t = 4.71$
POSITION	$(1, 0)$	$(0, 1)$	$(-1, 0)$	$(0, -1)$

Notice that at time $t = 6.28$ seconds we are back to the point $(1, 0)$.

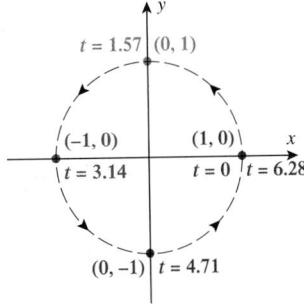

We can see that the path represents the unit circle, since $x^2 + y^2 = \cos^2 t + \sin^2 t = 1$.

DEFINITION   **Parametric Equations**

Let $x = f(t)$ and $y = g(t)$ be functions defined for t on some interval. The set of points $(x, y) = [f(t), g(t)]$ represents a **plane curve**. The equations

$$x = f(t) \qquad \text{and} \qquad y = g(t)$$

are called **parametric equations** of the curve. The variable t is called the **parameter**.

Parametric equations are useful for showing movement along a curve. We insert arrows in the graph to show **direction**, or **orientation**, along the curve as t increases.

EXAMPLE 1   **Graphing a Curve Defined by Parametric Equations**

Graph the curve defined by the parametric equations

$$x = t^2 \qquad y = (t - 1) \qquad t \text{ in } [-2, 2]$$

Indicate the orientation with arrows.

Solution:

STEP 1 Make a table and find values for t, x, and y.

t	$x = t^2$	$y = (t - 1)$	(x, y)
$t = -2$	$x = (-2)^2 = 4$	$y = (-2 - 1) = -3$	$(4, -3)$
$t = -1$	$x = (-1)^2 = 1$	$y = (-1 - 1) = -2$	$(1, -2)$
$t = 0$	$x = 0^2 = 0$	$y = (0 - 1) = -1$	$(0, -1)$
$t = 1$	$x = 1^2 = 1$	$y = (1 - 1) = 0$	$(1, 0)$
$t = 2$	$x = 2^2 = 4$	$y = (2 - 1) = 1$	$(4, 1)$

STEP 2 Plot the points in the xy-plane.

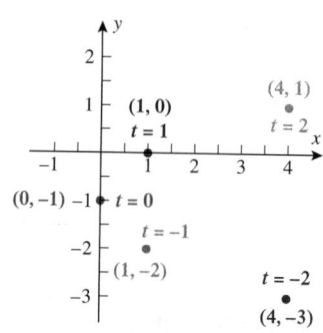

STEP 3 Connect the points with a
smooth curve and use arrows
to indicate direction.

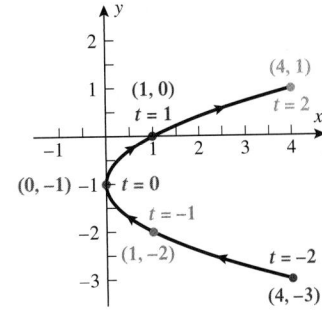

The shape of the graph appears to be a parabola. The parametric equations are $x = t^2$ and $y = (t - 1)$. If we solve the second equation for t, getting $t = y + 1$, and substitute this expression into $x = t^2$, the result is $x = (y + 1)^2$. The graph of $x = (y + 1)^2$ is a parabola with vertex at the point $(0, -1)$ and opening to the right.

■ **YOUR TURN** Graph the curve defined by the parametric equations

$$x = t + 1 \qquad y = t^2 \qquad t \text{ in } [-2, 2]$$

Indicate the orientation with arrows.

■ **Answer:**

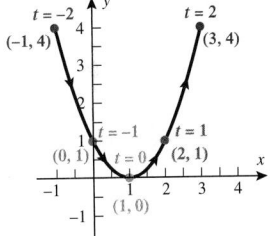

Sometimes it is easier to show the rectangular equivalent of the curve and eliminate the parameter.

EXAMPLE 2 **Graphing a Curve Defined by Parametric Equations by First Finding an Equivalent Rectangular Equation**

Graph the curve defined by the parametric equations

$$x = 4\cos t \qquad y = 3\sin t \qquad t \text{ is any real number}$$

Indicate the orientation with arrows.

Solution:

One approach is to point-plot as in Example 1. A second approach is to find the equivalent rectangular equation that represents the curve.

We apply the Pythagorean identity. $\qquad\qquad \sin^2 t + \cos^2 t = 1$

Find $\sin^2 t$ from the parametric equation for y. $\qquad y = 3\sin t$

  Square both sides. $\qquad\qquad\qquad\qquad\qquad y^2 = 9\sin^2 t$

  Divide by 9. $\qquad\qquad\qquad\qquad\qquad\qquad \sin^2 t = \dfrac{y^2}{9}$

Similarly, find $\cos^2 t$. $\qquad\qquad\qquad\qquad x = 4\cos t$

  Square both sides. $\qquad\qquad\qquad\qquad\qquad x^2 = 16\cos^2 t$

  Divide by 16. $\qquad\qquad\qquad\qquad\qquad\qquad \cos^2 t = \dfrac{x^2}{16}$

Substitute $\sin^2 t = \dfrac{y^2}{9}$ and $\cos^2 t = \dfrac{x^2}{16}$ into $\sin^2 t + \cos^2 t = 1$. $\qquad \dfrac{y^2}{9} + \dfrac{x^2}{16} = 1$

The curve is an ellipse.

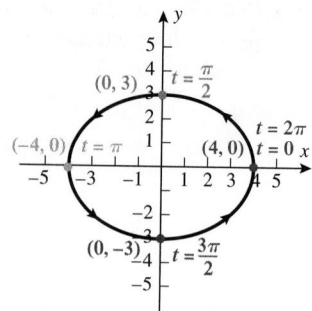

The orientation is counterclockwise. For example, when $t = 0$, the position is $(4, 0)$; when $t = \dfrac{\pi}{2}$, the position is $(0, 3)$; and when $t = \pi$, the position is $(-4, 0)$.

Applications of Parametric Equations

Parametric equations can be used to describe motion in many applications. Two that we will discuss are the *cycloid* and a *projectile*. Suppose that you paint a red X on a bicycle tire. As the bicycle moves in a straight line, if you watch the motion of the red X, you will see that it follows the path of a **cycloid**.

The parametric equations that define a cycloid are

$$x = a(t - \sin t) \qquad \text{and} \qquad y = a(1 - \cos t)$$

where t is any real number.

EXAMPLE 3   Graphing a Cycloid

Graph the cycloid given by $x = 2(t - \sin t)$ and $y = 2(1 - \cos t)$ for t in $[0, 4\pi]$.

Solution:

STEP 1 Make a table and find key values for t, x, and y.

t	$x = 2(t - \sin t)$	$y = 2(1 - \cos t)$	(x, y)
$t = 0$	$x = 2(0 - 0) = 0$	$y = 2(1 - 1) = 0$	$(0, 0)$
$t = \pi$	$x = 2(\pi - 0) = 2\pi$	$y = 2[1 - (-1)] = 4$	$(2\pi, 4)$
$t = 2\pi$	$x = 2(2\pi - 0) = 4\pi$	$y = 2(1 - 1) = 0$	$(4\pi, 0)$
$t = 3\pi$	$x = 2(3\pi - 0) = 6\pi$	$y = 2[1 - (-1)] = 4$	$(6\pi, 4)$
$t = 4\pi$	$x = 2(4\pi - 0) = 8\pi$	$y = 2(1 - 1) = 0$	$(8\pi, 0)$

STEP 2 Plot points in a plane and connect them with a smooth curve.

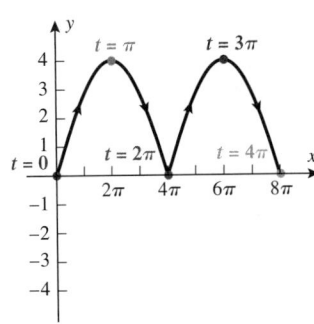

Another example of parametric equations describing real-world phenomena is projectile motion. The accompanying photo of Michelle Wie hitting a golf ball illustrates an example of a projectile.

Jeff Topping/Corbis Images; (inset photo) Getty Images, Inc.

Let v_0 be the initial velocity of an object, θ be the angle of inclination with the horizontal, and h be the initial height above the ground. Then the parametric equations describing the **projectile motion** (which will be developed in calculus) are

$$x = (v_0 \cos \theta) t \qquad \text{and} \qquad y = -\frac{1}{2} g t^2 + (v_0 \sin \theta) t + h$$

where t is time and g is the constant acceleration due to gravity $\left(9.8 \text{ m/sec}^2 \text{ or } 32 \text{ ft/sec}^2\right)$.

EXAMPLE 4   Graphing Projectile Motion

Suppose that Michelle Wie hits her golf ball with an initial velocity of 160 feet per second at an angle of 30° with the ground. How far is her drive, assuming that the length of the drive is from the tee to the point where the ball first hits the ground? Graph the curve representing the path of the golf ball. Assume that she hits the ball straight off the tee and down the fairway.

Solution:

STEP 1 Find the parametric equations that describe the golf ball that Michelle Wie drove.

Write the parametric equations for projectile motion.

$$x = (v_0 \cos \theta) t \quad \text{and} \quad y = -\frac{1}{2} g t^2 + (v_0 \sin \theta) t + h$$

Let $g = 32 \text{ ft/sec}^2$, $v_0 = 160 \text{ ft/sec}$, $h = 0$, and $\theta = 30°$.

$$x = (160 \cdot \cos 30°) t \quad \text{and} \quad y = -16 t^2 + (160 \cdot \sin 30°) t$$

Evaluate the sine and cosine functions and simplify.

$$x = 80\sqrt{3}\, t \quad \text{and} \quad y = -16 t^2 + 80 t$$

STEP 2 Graph the projectile motion.

t	$x = 80\sqrt{3}\,t$	$y = -16t^2 + 80t$	(x, y)
$t = 0$	$x = 80\sqrt{3}(0) = 0$	$y = -16(0)^2 + 80(0) = 0$	$(0, 0)$
$t = 1$	$x = 80\sqrt{3}(1) \approx 139$	$y = -16(1)^2 + 80(1) = 64$	$(139, 64)$
$t = 2$	$x = 80\sqrt{3}(2) = 277$	$y = -16(2)^2 + 80(2) = 96$	$(277, 96)$
$t = 3$	$x = 80\sqrt{3}(3) \approx 416$	$y = -16(3)^2 + 80(3) = 96$	$(416, 96)$
$t = 4$	$x = 80\sqrt{3}(4) \approx 554$	$y = -16(4)^2 + 80(4) = 64$	$(554, 64)$
$t = 5$	$x = 80\sqrt{3}(5) \approx 693$	$y = -16(5)^2 + 80(5) = 0$	$(693, 0)$

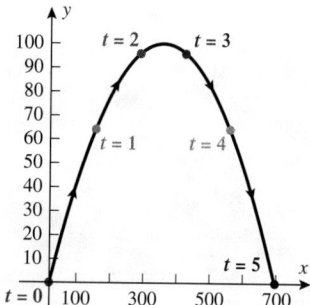

We can see that we selected our time increments well [the last point, $(693, 0)$, corresponds to the ball hitting the ground 693 feet from the tee].

STEP 3 Identify the horizontal distance from the tee to the point where the ball first hits the ground. Algebraically, we can determine the distance of the tee shot by setting the height y equal to zero.

Factor (divide) out $-16t$.

$$y = -16t^2 + 80t = 0$$
$$-16t(t - 5) = 0$$

Solve for t.

$$t = 0 \text{ or } t = 5$$

The ball hits the ground after 5 seconds.

Let $t = 5$ in the horizontal distance, $x = 80\sqrt{3}\,t$.

$$x = 80\sqrt{3}(5) \approx 693$$

> The ball hits the ground 693 feet from the tee.

SECTION 9.9 SUMMARY

Parametric equations are a way of describing as a function of t, the parameter, the path an object takes along a curve in the xy-plane. Parametric equations have equivalent rectangular equations. Typically, the method of graphing a set of parametric equations is to eliminate t and graph the corresponding rectangular equation. Once the curve is found, orientation along the curve can be determined by finding points corresponding to different t-values. Two important applications are cycloids and projectiles, whose paths we can trace using parametric equations.

SECTION
9.9 EXERCISES

▪ SKILLS

In Exercises 1–30, graph the curve defined by the parametric equations.

1. $x = t + 1,\ y = \sqrt{t},\ t \geq 0$

2. $x = 3t,\ y = t^2 - 1,\ t \text{ in } [0, 4]$

3. $x = -3t,\ y = t^2 + 1,\ t \text{ in } [0, 4]$

4. $x = t^2 - 1,\ y = t^2 + 1,\ t \text{ in } [-3, 3]$

5. $x = t^2,\ y = t^3,\ t \text{ in } [-2, 2]$

6. $x = t^3 + 1,\ y = t^3 - 1,\ t \text{ in } [-2, 2]$

7. $x = \sqrt{t},\ y = t,\ t \text{ in } [0, 10]$

8. $x = t,\ y = \sqrt{t^2 + 1},\ t \text{ in } [0, 10]$

9. $x = (t + 1)^2,\ y = (t + 2)^3,\ t \text{ in } [0, 1]$

10. $x = (t - 1)^3,\ y = (t - 2)^2,\ t \text{ in } [0, 4]$

11. $x = e^t,\ y = e^{-t},\ -\ln 3 \leq t \leq \ln 3$

12. $x = e^{-2t},\ y = e^{2t} + 4,\ -\ln 2 \leq t \leq \ln 3$

13. $x = 2t^4 - 1,\ y = t^8 + 1,\ 0 \leq t \leq 4$

14. $x = 3t^6 - 1,\ y = 2t^3,\ -1 \leq t \leq 1$

15. $x = t(t - 2)^3,\ y = t(t - 2)^3,\ 0 \leq t \leq 4$

16. $x = -t\sqrt[3]{t},\ y = -5t^8 - 2,\ -3 \leq t \leq 3$

17. $x = 3\sin t,\ y = 2\cos t,\ t \text{ in } [0, 2\pi]$

18. $x = \cos(2t),\ y = \sin t,\ t \text{ in } [0, 2\pi]$

19. $x = \sin t + 1,\ y = \cos t - 2,\ t \text{ in } [0, 2\pi]$

20. $x = \tan t,\ y = 1,\ t \text{ in } \left[-\dfrac{\pi}{4}, \dfrac{\pi}{4}\right]$

21. $x = 1,\ y = \sin t,\ t \text{ in } [-2\pi, 2\pi]$

22. $x = \sin t,\ y = 2,\ t \text{ in } [0, 2\pi]$

23. $x = \sin^2 t,\ y = \cos^2 t,\ t \text{ in } [0, 2\pi]$

24. $x = 2\sin^2 t,\ y = 2\cos^2 t,\ t \text{ in } [0, 2\pi]$

25. $x = 2\sin(3t),\ y = 3\cos(2t),\ t \text{ in } [0, 2\pi]$

26. $x = 4\cos(2t),\ y = t,\ t \text{ in } [0, 2\pi]$

27. $x = \cos\left(\dfrac{t}{2}\right) - 1,\ y = \sin\left(\dfrac{t}{2}\right) + 1,\ -2\pi \leq t \leq 2\pi$

28. $x = \sin\left(\dfrac{t}{3}\right) + 3,\ y = \cos\left(\dfrac{t}{3}\right) - 1,\ 0 \leq t \leq 6\pi$

29. $x = 2\sin\left(t + \dfrac{\pi}{4}\right),\ y = -2\cos\left(t + \dfrac{\pi}{4}\right),\ -\dfrac{\pi}{4} \leq t \leq \dfrac{7\pi}{4}$

30. $x = -3\cos^2(3t),\ y = 2\cos(3t),\ -\dfrac{\pi}{3} \leq t \leq \dfrac{\pi}{3}$

In Exercises 31–40, the given parametric equations define a plane curve. Find an equation in rectangular form that also corresponds to the plane curve.

31. $x = \dfrac{1}{t},\ y = t^2$

32. $x = t^2 - 1,\ y = t^2 + 1$

33. $x = t^3 + 1,\ y = t^3 - 1$

34. $x = 3t,\ y = t^2 - 1$

35. $x = t,\ y = \sqrt{t^2 + 1}$

36. $x = \sin^2 t,\ y = \cos^2 t$

37. $x = 2\sin^2 t,\ y = 2\cos^2 t$

38. $x = \sec^2 t,\ y = \tan^2 t$

39. $x = 4(t^2 + 1),\ y = 1 - t^2$

40. $x = \sqrt{t - 1},\ y = \sqrt{t}$

▪ APPLICATIONS

For Exercises 41–50, recall that the flight of a projectile can be modeled with the parametric equations

$$x = (v_0 \cos\theta)t \qquad y = -16t^2 + (v_0 \sin\theta)t + h$$

where t is in seconds, v_0 is the initial velocity, θ is the angle with the horizontal, and x and y are in feet.

41. Flight of a Projectile. A projectile is launched from the ground at a speed of 400 ft/sec at an angle of 45° with the horizontal. After how many seconds does the projectile hit the ground?

42. Flight of a Projectile. A projectile is launched from the ground at a speed of 400 ft/sec at an angle of 45° with the horizontal. How far does the projectile travel (what is the horizontal distance), and what is its maximum altitude?

43. Flight of a Baseball. A baseball is hit at an initial speed of 105 mph and an angle of 20° at a height of 3 feet above the ground. If home plate is 420 feet from the back fence, which is 15 feet tall, will the baseball clear the back fence for a home run?

44. Flight of a Baseball. A baseball is hit at an initial speed of 105 mph and an angle of 20° at a height of 3 feet above the ground. If there is no back fence or other obstruction, how far does the baseball travel (horizontal distance), and what is its maximum height?

45. Bullet Fired. A gun is fired from the ground at an angle of 60°, and the bullet has an initial speed of 700 ft/sec. How high does the bullet go? What is the horizontal (ground) distance between the point where the gun is fired and the point where the bullet hits the ground?

46. Bullet Fired. A gun is fired from the ground at an angle of 60°, and the bullet has an initial speed of 2000 ft/sec. How high does the bullet go? What is the horizontal (ground) distance between the point where the gun is fired and the point where the bullet hits the ground?

47. Missile Fired. A missile is fired from a ship at an angle of 30°, an initial height of 20 feet above the water's surface, and a speed of 4000 ft/sec. How long will it be before the missile hits the water?

48. Missile Fired. A missile is fired from a ship at an angle of 40°, an initial height of 20 feet above the water's surface, and a speed of 5000 ft/sec. Will the missile be able to hit a target that is 2 miles away?

49. Path of a Projectile. A projectile is launched at a speed of 100 ft/sec at an angle of 35° with the horizontal. Plot the path of the projectile on a graph. Assume that $h = 0$.

50. Path of a Projectile. A projectile is launched at a speed of 150 ft/sec at an angle of 55° with the horizontal. Plot the path of the projectile on a graph. Assume that $h = 0$.

For Exercises 51 and 52, refer to the following:

Modern amusement park rides are often designed to push the envelope in terms of speed, angle, and ultimately g's, and usually take the form of gargantuan roller coasters or skyscraping towers.

However, even just a couple of decades ago, such creations were depicted only in fantasy-type drawings, with their creators never truly believing their construction would become a reality. Nevertheless, thrill rides still capable of nauseating any would-be rider were still able to be constructed; one example is the *Calypso*. This ride is a not-too-distant cousin of the more well-known *Scrambler*. It consists of four rotating arms (instead of three like the Scrambler), and on each of these arms, four cars (equally spaced around the circumference of a circular frame) are attached. Once in motion, the main piston to which the four arms are connected rotates clockwise, while each of the four arms themselves rotates counterclockwise. The combined motion appears as a blur to any onlooker from the crowd, but the motion of a single rider is much less chaotic. In fact, a single rider's path can be modeled by the following graph:

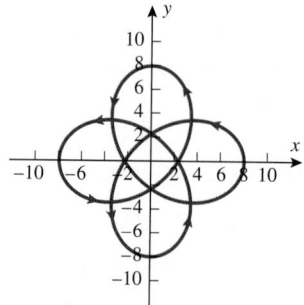

The equation of this graph is defined parametrically by

$$x(t) = A\cos t + B\cos(-3t)$$
$$y(t) = A\sin t + B\sin(-3t) \quad 0 \le t \le 2\pi$$

51. Amusement Rides. What is the location of the rider at $t = 0$, $t = \dfrac{\pi}{2}$, $t = \pi$, $t = \dfrac{3\pi}{2}$, and $t = 2\pi$?

52. Amusement Rides. Suppose that the ride conductor was rather sinister and speeded up the ride to twice the speed. How would you modify the parametric equations to model such a change? Now vary the values of A and B. What do you think these parameters are modeling in this problem?

■ CATCH THE MISTAKE

In Exercises 53 and 54, explain the mistake that is made.

53. Find the rectangular equation that corresponds to the plane curve defined by the parametric equations $x = t + 1$ and $y = \sqrt{t}$. Describe the plane curve.

Solution:

Square $y = \sqrt{t}$. $y^2 = t$

Substitute $t = y^2$ into $x = t + 1$. $x = y^2 + 1$

The graph of $x = y^2 + 1$ is a parabola opening to the right with its vertex at $(1, 0)$.

This is incorrect. What mistake was made?

54. Find the rectangular equation that corresponds to the plane curve defined by the parametric equations $x = \sqrt{t}$ and $y = t - 1$. Describe the plane curve.

Solution:

Square $x = \sqrt{t}$. $x^2 = t$

Substitute $t = x^2$ into $y = t - 1$. $y = x^2 - 1$

The graph of $y = x^2 - 1$ is a parabola opening up with its vertex at $(0, -1)$.

This is incorrect. What mistake was made?

■ CONCEPTUAL

In Exercises 55 and 56, determine whether each statement is true or false.

55. Curves given by equations in rectangular form have orientation.

56. Curves given by parametric equations have orientation.

57. Determine what type of curve the parametric equations $x = \sqrt{t}$ and $y = \sqrt{1 - t}$ define.

58. Determine what type of curve the parametric equations $x = \ln t$ and $y = t$ define.

■ CHALLENGE

59. Prove that $x = a \tan t$, $y = b \sec t$, $0 \le t \le 2\pi$, $t \ne \dfrac{\pi}{2}, \dfrac{3\pi}{2}$ are parametric equations for a hyperbola. Assume that a and b are nonzero constants.

60. Prove that $x = a \csc\left(\dfrac{t}{2}\right)$, $y = b \cot\left(\dfrac{t}{2}\right)$, $0 \le t \le 4\pi$, $t \ne \pi, 3\pi$ are parametric equations for a hyperbola. Assume that a and b are nonzero constants.

61. Consider the parametric curve $x = a \sin^2 t - b \cos^2 t$, $y = b \cos^2 t + a \sin^2 t$, $0 \le t \le \dfrac{\pi}{2}$. Assume that a and b are nonzero constants. Find the Cartesian equation for this curve.

62. Consider the parametric curve $x = a \sin t + a \cos t$, $y = a \cos t - a \sin t$, $0 \le t \le 2\pi$. Assume that a is not zero. Find the Cartesian equation for this curve.

63. Consider the parametric curve $x = e^{at}$, $y = be^t$, $t > 0$. Assume that a is a positive integer and b is a positive real number. Determine the Cartesian equation.

64. Consider the parametric curve $x = a \ln t$, $y = \ln(bt)$, $t > 0$. Assume that b is a positive integer and a is a positive real number. Determine the Cartesian equation.

■ TECHNOLOGY

65. Consider the parametric equations: $x = a \sin t - \sin(at)$ and $y = a \cos t + \cos(at)$. With a graphing utility, explore the graphs for $a = 2, 3$, and 4.

66. Consider the parametric equations: $x = a \cos t - b \cos(at)$ and $y = a \sin t + \sin(at)$. With a graphing utility, explore the graphs for $a = 3$ and $b = 1$, $a = 4$ and $b = 2$, and $a = 6$ and $b = 2$. Find the t-interval that gives one cycle of the curve.

67. Consider the parametric equations: $x = \cos(at)$ and $y = \sin(bt)$. With a graphing utility, explore the graphs for $a = 2$ and $b = 4$, $a = 4$ and $b = 2$, $a = 1$ and $b = 3$, and $a = 3$ and $b = 1$. Find the t-interval that gives one cycle of the curve.

68. Consider the parametric equations: $x = a \sin(at) - \sin t$ and $y = a \cos(at) - \cos t$. With a graphing utility, explore the graphs for $a = 2$ and 3. Describe the t-interval for each case.

69. Consider the parametric equations $x = a \cos(at) - \sin t$ and $y = a \sin(at) - \cos t$. With a graphing utility, explore the graphs for $a = 2$ and 3. Describe the t-interval for each case.

70. Consider the parametric equations $x = a \sin(at) - \cos t$ and $y = a \cos(at) - \sin t$. With a graphing utility, explore the graphs for $a = 2$ and 3. Describe the t-interval for each case.

■ PREVIEW TO CALCULUS

In calculus, some operations can be simplified by using parametric equations. Finding the points of intersection (if they exist) of two curves given by parametric equations is a standard procedure.

In Exercises 71–74, find the points of intersection of the given curves given s and t are any real numbers.

71. Curve I: $x = t$, $y = t^2 - 1$
Curve II: $x = s + 1$, $y = 4 - s$

72. Curve I: $x = t^2 + 3$, $y = t$
Curve II: $x = s + 2$, $y = 1 - s$

73. Curve I: $x = 100t$, $y = 80t - 16t^2$
Curve II: $x = 100 - 200t$, $y = -16t^2 + 144t - 224$

74. Curve I: $x = t^2$, $y = t + 1$
Curve II: $x = 2 + s$, $y = 1 - s$

In the Modeling Your World features in Chapters 1–3, you used the average yearly temperature in degrees Fahrenheit (°F) and carbon dioxide emissions in parts per million (ppm) collected by NOAA in Mauna Loa, Hawaii, to develop linear (Chapter 1) and nonlinear (Chapters 2 and 3) models. In the following exercises, you will determine when these different models actually predict the same temperatures and carbon emissions. It is important to realize that not only can different models be used to predict trends, but also the choice of data those models are fitted to also affects the models and hence the predicted values.

YEAR	1960	1965	1970	1975	1980	1985	1990	1995	2000	2005
TEMPERATURE	44.45	43.29	43.61	43.35	46.66	45.71	45.53	47.53	45.86	46.23
CO₂ EMISSIONS (PPM)	316.9	320.0	325.7	331.1	338.7	345.9	354.2	360.6	369.4	379.7

1. Solve the system of nonlinear equations governing mean temperature that was found by using two data points:

   Equation (1): Use the linear model developed in Modeling Your World, Chapter 1, Exercise 2(a).

   Equation (2): Use the quadratic model found in Modeling Your World, Chapter 2, Exercise 2(a).

2. For what year do the models used in Exercise 1 agree? Compare the value given by the models that year to the actual data for the year.

3. Solve the system of nonlinear equations governing mean temperature that was found by applying regression (all data points):

   Equation (1): Use the linear model developed in Modeling Your World, Chapter 1, Exercise 2(c).

   Equation (2): Use the quadratic model found in Modeling Your World, Chapter 2, Exercise 2(c).

4. For what year do the models used in Exercise 3 agree? Compare the value given by the models that year to the actual data for the year.

5. Solve the system of nonlinear equations governing carbon dioxide emissions that was found by using two data points:

   Equation (1): Use the linear model developed in Modeling Your World, Chapter 1, Exercise 7(a).

   Equation (2): Use the quadratic model found in Modeling Your World, Chapter 2, Exercise 7(a).

6. For what year do the models used in Exercise 5 agree? Compare the value given by the models that year to the actual data for the year.

7. Solve the system of nonlinear equations governing carbon emissions that was found by applying regression (all data points):

   Equation (1): Use the linear model developed in Modeling Your World, Chapter 1, Exercise 7(c).

   Equation (2): Use the quadratic model found in Modeling Your World, Chapter 2, Exercise 7(c).

8. For what year do the models used in Exercise 7 agree? Compare the value given by the models that year to the actual data for the year.

SECTION	CONCEPT	PAGES	REVIEW EXERCISES	KEY IDEAS/POINTS
9.1	**Conic basics**	830–832	1–4	
	Names of conics	830	1–4	Parabola, ellipse, and hyperbola
	Definitions	830–832	1–4	**Parabola:** Distances from a point to a reference point (focus) and a reference line (directrix) are equal. **Ellipse:** Sum of the distances between the point and two reference points (foci) is constant. **Hyperbola:** Difference of the distances between the point and two reference points (foci) is constant.
9.2	**The parabola**	833–841	5–24	
	Definition of a parabola	833–841	5–24	

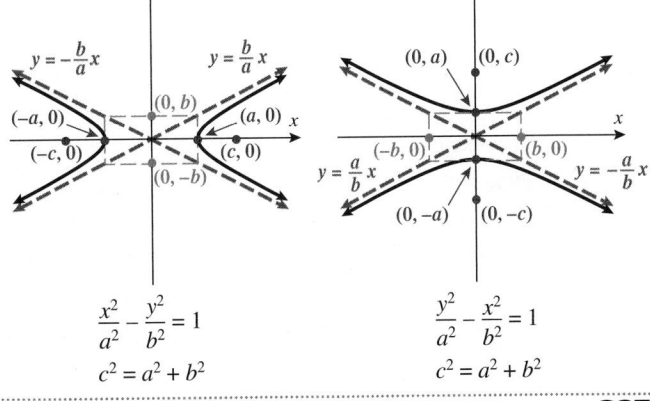

Up: $p > 0$ Down: $p < 0$ Right: $p > 0$ Left: $p < 0$

9.3	**The ellipse**	847–854	25–40	
	Definition of an ellipse	847–854	25–40	

$$\frac{x^2}{a^2} + \frac{y^2}{b^2} = 1$$
$$c^2 = a^2 - b^2$$

$$\frac{x^2}{b^2} + \frac{y^2}{a^2} = 1$$
$$c^2 = a^2 - b^2$$

9.4	**The hyperbola**	860–868	41–56	
	Definition of a hyperbola	860–868	41–56	

$$\frac{x^2}{a^2} - \frac{y^2}{b^2} = 1$$
$$c^2 = a^2 + b^2$$

$$\frac{y^2}{a^2} - \frac{x^2}{b^2} = 1$$
$$c^2 = a^2 + b^2$$

SECTION	CONCEPT	PAGES	REVIEW EXERCISES	KEY IDEAS/POINTS
9.5	**Systems of nonlinear equations**	873–882	57–68	There is no procedure guaranteed to solve nonlinear equations.
	Using elimination to solve systems of nonlinear equations	874–878	57–60	Eliminate a variable by either adding one equation to or subtracting one equation from the other.
	Using substitution to solve systems of nonlinear equations	879–882	61–64	Solve for one variable in terms of the other and substitute into the second equation.
9.6	**Systems of nonlinear inequalities**	886–890	69–80	Solutions are determined graphically by finding the common shaded regions. ■ $\leq$ or $\geq$ use solid curves. ■ $<$ or $>$ use dashed curves.
	Nonlinear inequalities in two variables	886–887	69–74	Step 1: Rewrite the inequality as an equation. Step 2: Graph the equation. Step 3: Test points. Step 4: Shade.
	Systems of nonlinear inequalities	887–890	75–80	Graph the individual inequalities and the solution in the common (overlapping) shaded region.
9.7	**Rotation of axes**	895–903	81–88	$x = X\cos\theta - Y\sin\theta$ $y = X\sin\theta + Y\cos\theta$
	Determine the angle of rotation necessary to transform a general second-degree equation into a familiar equation of a conic	899–903	81–88	$\cot(2\theta) = \dfrac{A - C}{B}$ or $\tan(2\theta) = \dfrac{B}{A - C}$
9.8	**Polar equations of conics**	906–912	89–94	All three conics (parabolas, ellipses, and hyperbolas) are defined in terms of a single focus and a directrix.
	Equations of conics in polar coordinates	907–912	89–94	The directrix is *vertical* and p units to the *right* of the pole. $$r = \frac{ep}{1 + e\cos\theta}$$ The directrix is *vertical* and p units to the *left* of the pole. $$r = \frac{ep}{1 - e\cos\theta}$$ The directrix is *horizontal* and p units *above* the pole. $$r = \frac{ep}{1 + e\sin\theta}$$ The directrix is *horizontal* and p units *below* the pole. $$r = \frac{ep}{1 - e\sin\theta}$$
9.9	**Parametric equations and graphs**	917–922	95–102	
	Parametric equations of a curve	917–922	95–102	Parametric equations: $x = f(t)$ and $y = g(t)$ Plane curve: $(x, y) = (f(t), g(t))$

9.1 Conic Basics

Determine whether each statement is true or false.

1. The focus is a point on the graph of the parabola.

2. The graph of $y^2 = 8x$ is a parabola that opens upward.

3. $\dfrac{x^2}{9} - \dfrac{y^2}{1} = 1$ is the graph of a hyperbola that has a horizontal transverse axis.

4. $\dfrac{(x + 1)^2}{9} + \dfrac{(y - 3)^2}{16} = 1$ is a graph of an ellipse whose center is $(1, 3)$.

9.2 The Parabola

Find an equation for the parabola described.

5. Vertex at $(0, 0)$; Focus at $(3, 0)$

6. Vertex at $(0, 0)$; Focus at $(0, 2)$

7. Vertex at $(0, 0)$; Directrix at $x = 5$

8. Vertex at $(0, 0)$; Directrix at $y = 4$

9. Vertex at $(2, 3)$; Focus at $(2, 5)$

10. Vertex at $(-1, -2)$; Focus at $(1, -2)$

11. Focus at $(1, 5)$; Directrix at $y = 7$

12. Focus at $(2, 2)$; Directrix at $x = 0$

Find the focus, vertex, directrix, and length of the latus rectum, and graph the parabola.

13. $x^2 = -12y$

14. $x^2 = 8y$

15. $y^2 = x$

16. $y^2 = -6x$

17. $(y + 2)^2 = 4(x - 2)$

18. $(y - 2)^2 = -4(x + 1)$

19. $(x + 3)^2 = -8(y - 1)$

20. $(x - 3)^2 = -8(y + 2)$

21. $x^2 + 5x + 2y + 25 = 0$

22. $y^2 + 2y - 16x + 1 = 0$

Applications

23. **Satellite Dish.** A satellite dish measures 10 feet across its opening and 2 feet deep at its center. The receiver should be placed at the focus of the parabolic dish. Where should the receiver be placed?

24. **Clearance Under a Bridge.** A bridge with a parabolic shape reaches a height of 40 feet in the center of the road, and the width of the bridge opening at ground level is 30 feet combined (both lanes). If an RV is 14 feet tall and 8 feet wide, will it make it through the tunnel?

9.3 The Ellipse

Graph each ellipse.

25. $\dfrac{x^2}{9} + \dfrac{y^2}{64} = 1$

26. $\dfrac{x^2}{81} + \dfrac{y^2}{49} = 1$

27. $25x^2 + y^2 = 25$

28. $4x^2 + 8y^2 = 64$

Find the standard form of an equation of the ellipse with the given characteristics.

29. Foci: $(-3, 0)$ and $(3, 0)$ Vertices: $(-5, 0)$ and $(5, 0)$

30. Foci: $(0, -2)$ and $(0, 2)$ Vertices: $(0, -3)$ and $(0, 3)$

31. Major axis vertical with length of 16, minor axis length of 6, and centered at $(0, 0)$

32. Major axis horizontal with length of 30, minor axis length of 20, and centered at $(0, 0)$

Graph each ellipse.

33. $\dfrac{(x - 7)^2}{100} + \dfrac{(y + 5)^2}{36} = 1$

34. $20(x + 3)^2 + (y - 4)^2 = 120$

35. $4x^2 - 16x + 12y^2 + 72y + 123 = 0$

36. $4x^2 - 8x + 9y^2 - 72y + 147 = 0$

Find the standard form of an equation of the ellipse with the given characteristics.

37. Foci: $(-1, 3)$ and $(7, 3)$ Vertices: $(-2, 3)$ and $(8, 3)$

38. Foci: $(1, -3)$ and $(1, -1)$ Vertices: $(1, -4)$ and $(1, 0)$

Applications

39. **Planetary Orbits.** Jupiter's orbit is summarized in the picture. Utilize the fact that the Sun is a focus to determine an equation for Jupiter's elliptical orbit around the Sun. Round to the nearest hundred thousand kilometers.

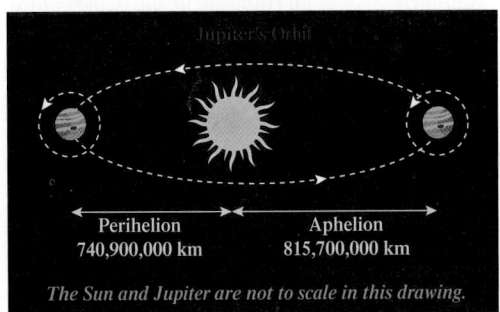

Perihelion 740,900,000 km   Aphelion 815,700,000 km

The Sun and Jupiter are not to scale in this drawing.

40. Planetary Orbits. Mars's orbit is summarized in the picture that follows. Utilize the fact that the Sun is a focus to determine an equation for Mars's elliptical orbit around the Sun. Round to the nearest million kilometers.

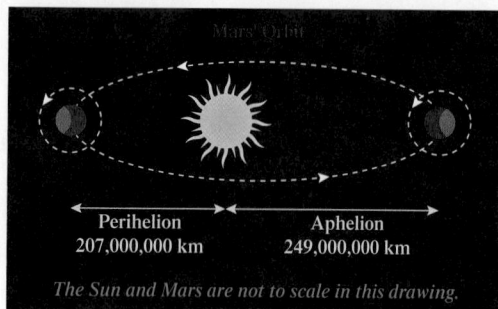

Perihelion
207,000,000 km

Aphelion
249,000,000 km

The Sun and Mars are not to scale in this drawing.

9.4 The Hyperbola

Graph each hyperbola.

41. $\dfrac{x^2}{9} - \dfrac{y^2}{64} = 1$ **42.** $\dfrac{x^2}{81} - \dfrac{y^2}{49} = 1$

43. $x^2 - 25y^2 = 25$ **44.** $8y^2 - 4x^2 = 64$

Find the standard form of an equation of the hyperbola with the given characteristics.

45. Vertices: $(-3, 0)$ and $(3, 0)$ Foci: $(-5, 0)$ and $(5, 0)$

46. Vertices: $(0, -1)$ and $(0, 1)$ Foci: $(0, -3)$ and $(0, 3)$

47. Center: $(0, 0)$; Transverse: y-axis; Asymptotes: $y = 3x$ and $y = -3x$

48. Center: $(0, 0)$; Transverse axis: y-axis; Asymptotes: $y = \frac{1}{2}x$ and $y = -\frac{1}{2}x$

Graph each hyperbola.

49. $\dfrac{(y - 1)^2}{36} - \dfrac{(x - 2)^2}{9} = 1$

50. $3(x + 3)^2 - 12(y - 4)^2 = 72$

51. $8x^2 - 32x - 10y^2 - 60y - 138 = 0$

52. $2x^2 + 12x - 8y^2 + 16y + 6 = 0$

Find the standard form of an equation of the hyperbola with the given characteristics.

53. Vertices: $(0, 3)$ and $(8, 3)$ Foci: $(-1, 3)$ and $(9, 3)$

54. Vertices: $(4, -2)$ and $(4, 0)$ Foci: $(4, -3)$ and $(4, 1)$

Applications

55. Ship Navigation. Two loran stations are located 220 miles apart along a coast. If a ship records a time difference of 0.00048 second and continues on the hyperbolic path corresponding to that difference, where would it reach shore? Assume that the speed of radio signals is 186,000 miles per second.

56. Ship Navigation. Two loran stations are located 400 miles apart along a coast. If a ship records a time difference of 0.0008 second and continues on the hyperbolic path corresponding to that difference, where would it reach shore?

9.5 Systems of Nonlinear Equations

Solve the system of equations with the elimination method.

57. $\begin{aligned} x^2 + y &= -3 \\ x - y &= 5 \end{aligned}$ **58.** $\begin{aligned} x^2 + y^2 &= 4 \\ x^2 + y &= 2 \end{aligned}$

59. $\begin{aligned} x^2 + y^2 &= 5 \\ 2x^2 - y &= 0 \end{aligned}$ **60.** $\begin{aligned} x^2 + y^2 &= 16 \\ 6x^2 + y^2 &= 16 \end{aligned}$

Solve the system of equations with the substitution method.

61. $\begin{aligned} x + y &= 3 \\ x^2 + y^2 &= 4 \end{aligned}$ **62.** $\begin{aligned} xy &= 4 \\ x^2 + y^2 &= 16 \end{aligned}$

63. $\begin{aligned} x^2 + xy + y^2 &= -12 \\ x - y &= 2 \end{aligned}$ **64.** $\begin{aligned} 3x + y &= 3 \\ x - y^2 &= -9 \end{aligned}$

Solve the system of equations by applying any method.

65. $\begin{aligned} x^3 - y^3 &= -19 \\ x - y &= -1 \end{aligned}$ **66.** $\begin{aligned} 2x^2 + 4xy &= 9 \\ x^2 - 2xy &= 0 \end{aligned}$

67. $\begin{aligned} \dfrac{2}{x^2} + \dfrac{1}{y^2} &= 15 \\ \dfrac{1}{x^2} - \dfrac{1}{y^2} &= -3 \end{aligned}$ **68.** $\begin{aligned} x^2 + y^2 &= 2 \\ x^2 + y^2 &= 4 \end{aligned}$

9.6 Systems of Nonlinear Inequalities

Graph the nonlinear inequality.

69. $y \geq x^2 + 3$ **70.** $x^2 + y^2 > 16$

71. $y \leq e^x$ **72.** $y < -x^3 + 2$

73. $y \geq \ln(x - 1)$ **74.** $9x^2 + 4y^2 \leq 36$

Solve each system of inequalities and shade the region on a graph, or indicate that the system has no solution.

75. $\begin{aligned} y &\geq x^2 - 2 \\ y &\leq -x^2 + 2 \end{aligned}$ **76.** $\begin{aligned} x^2 + y^2 &\leq 4 \\ y &\leq x \end{aligned}$

77. $\begin{aligned} y &\geq (x + 1)^2 - 2 \\ y &\leq 10 - x \end{aligned}$ **78.** $\begin{aligned} 3x^2 + 3y^2 &\leq 27 \\ y &\geq x - 1 \end{aligned}$

79. $\begin{aligned} 4y^2 - 9x^2 &\leq 36 \\ y &\geq x + 1 \end{aligned}$ **80.** $\begin{aligned} 9x^2 + 16y^2 &\leq 144 \\ y &\geq 1 - x^2 \end{aligned}$

9.7 Rotation of Axes

The coordinates of a point in the xy-coordinate system are given. Assuming the X- and Y-axes are found by rotating the x- and y-axes by the indicated angle θ, find the corresponding coordinates for the point in the XY-system.

81. $(-3, 2)$, $\theta = 60°$ **82.** $(4, -3)$, $\theta = 45°$

Transform the equation of the conic into an equation in X and Y (without an XY-term) by rotating the x- and y-axes through the indicated angle θ. Then graph the resulting equation.

83. $2x^2 + 4\sqrt{3}xy - 2y^2 - 16 = 0$, $\theta = 30°$

84. $25x^2 + 14xy + 25y^2 - 288 = 0$, $\theta = \dfrac{\pi}{4}$

Determine the angle of rotation necessary to transform the equation in x and y into an equation in X and Y with no XY-term.

85. $4x^2 + 2\sqrt{3}xy + 6y^2 - 9 = 0$

86. $4x^2 + 5xy + 4y^2 - 11 = 0$

Graph the second-degree equation.

87. $x^2 + 2xy + y^2 + \sqrt{2}x - \sqrt{2}y + 8 = 0$

88. $76x^2 + 48\sqrt{3}xy + 28y^2 - 100 = 0$

9.8 Polar Equations of Conics

Find the polar equation that represents the conic described.

89. An ellipse with eccentricity $e = \frac{3}{7}$ and directrix $y = -7$

90. A parabola with directrix $x = 2$

Identify the conic (parabola, ellipse, or hyperbola) that each polar equation represents.

91. $r = \dfrac{6}{4 - 5\cos\theta}$

92. $r = \dfrac{2}{5 + 3\sin\theta}$

For the given polar equations, find the eccentricity and vertex (or vertices), and graph the curve.

93. $r = \dfrac{4}{2 + \cos\theta}$

94. $r = \dfrac{6}{1 - \sin\theta}$

9.9 Parametric Equations and Graphs

Graph the curve defined by the parametric equations.

95. $x = \sin t, y = 4\cos t$ for t in $[-\pi, \pi]$

96. $x = 5\sin^2 t, y = 2\cos^2 t$ for t in $[-\pi, \pi]$

97. $x = 4 - t^2, y = t^2$ for t in $[-3, 3]$

98. $x = t + 3, y = 4$ for t in $[-4, 4]$

The given parametric equations define a plane curve. Find an equation in rectangular form that also corresponds to the plane curve.

99. $x = 4 - t^2, y = t$

100. $x = 5\sin^2 t, y = 2\cos^2 t$

101. $x = 2\tan^2 t, y = 4\sec^2 t$

102. $x = 3t^2 + 4, y = 3t^2 - 5$

Technology Exercises

Section 9.2

103. In your mind, picture the parabola given by $(x - 0.6)^2 = -4(y + 1.2)$. Where is the vertex? Which way does this parabola open? Now plot the parabola with a graphing utility.

104. In your mind, picture the parabola given by $(y - 0.2)^2 = 3(x - 2.8)$. Where is the vertex? Which way does this parabola open? Now plot the parabola with a graphing utility.

105. Given is the parabola $y^2 + 2.8y + 3x - 6.85 = 0$.
  a. Solve the equation for y, and use a graphing utility to plot the parabola.
  b. Transform the equation into the form $(y - k)^2 = 4p(x - h)$. Find the vertex. Which way does the parabola open?
  c. Do (a) and (b) agree with each other?

106. Given is the parabola $x^2 - 10.2x - y + 24.8 = 0$.
  a. Solve the equation for y, and use a graphing utility to plot the parabola.
  b. Transform the equation into the form $(x - h)^2 = 4p(y - k)$. Find the vertex. Which way does the parabola open?
  c. Do (a) and (b) agree with each other?

Section 9.3

107. Graph the following three ellipses: $4x^2 + y^2 = 1$, $4(2x)^2 + y^2 = 1$, and $4(3x)^2 + y^2 = 1$. What can be said to happen to ellipse $4(cx)^2 + y^2 = 1$ as c increases?

108. Graph the following three ellipses: $x^2 + 4y^2 = 1$, $x^2 + 4(2y)^2 = 1$, and $x^2 + 4(3y)^2 = 1$. What can be said to happen to ellipse $x^2 + 4(cy)^2 = 1$ as c increases?

Section 9.4

109. Graph the following three hyperbolas: $4x^2 - y^2 = 1$, $4(2x)^2 - y^2 = 1$, and $4(3x)^2 - y^2 = 1$. What can be said to happen to hyperbola $4(cx)^2 - y^2 = 1$ as c increases?

110. Graph the following three hyperbolas: $x^2 - 4y^2 = 1$, $x^2 - 4(2y)^2 = 1$, and $x^2 - 4(3y)^2 = 1$. What can be said to happen to hyperbola $x^2 - 4(cy)^2 = 1$ as c increases?

Section 9.5

With a graphing utility, solve the following systems of equations:

111. $7.5x^2 + 1.5y^2 = 12.25$
$x^2y = 1$

112. $4x^2 + 2xy + 3y^2 = 12$
$x^3y = 3 - 3x^3$

Section 9.6

With a graphing utility, graph the following systems of nonlinear inequalities:

113. $y \geq 10^x - 1$
$y \leq 1 - x^2$

114. $x^2 + 4y^2 \leq 36$
$y \geq e^x$

REVIEW EXERCISES

Section 9.7

115. With a graphing utility, explore the second-degree equation $Ax^2 + Bxy + Cy^2 + 10x - 8y - 5 = 0$ for the following values of A, B, and C:

a. $A = 2, B = -3, C = 5$

b. $A = 2, B = 3, C = -5$

Show the angle of rotation to one decimal place. Explain the differences.

116. With a graphing utility, explore the second-degree equation $Ax^2 + Bxy + Cy^2 + 2x - y = 0$ for the following values of A, B, and C:

a. $A = 1, B = -2, C = -1$

b. $A = 1, B = 2, D = 1$

Show the angle of rotation to the nearest degree. Explain the differences.

Section 9.8

117. Let us consider the polar equation $r = \dfrac{8}{4 + 5\sin\theta}$. Explain why a graphing utility gives the following graph with the specified window parameters:

$$[-6, 6] \text{ by } [-3, 9] \text{ with } \theta \text{ step } = \frac{\pi}{4}$$

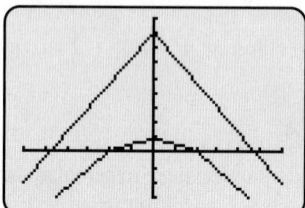

118. Let us consider the polar equation $r = \dfrac{9}{3 - 2\sin\theta}$. Explain why a graphing utility gives the following graph with the specified window parameters:

$$[-6, 6] \text{ by } [-3, 9] \text{ with } \theta \text{ step } = \frac{\pi}{2}$$

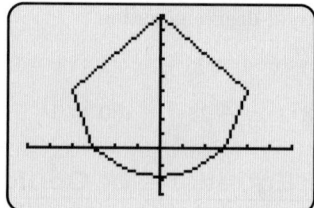

Section 9.9

119. Consider the parametric equations $x = a\cos at + b\sin bt$ and $y = a\sin at + b\cos bt$. Use a graphing utility to explore the graphs for $(a, b) = (2, 3)$ and $(a, b) = (3, 2)$. Describe the t-interval for each case.

120. Consider the parametric equations $x = a\sin at - b\cos bt$ and $y = a\cos at - b\sin bt$. Use a graphing utility to explore the graphs for $(a, b) = (1, 2)$ and $(a, b) = (2, 1)$. Describe the t-interval for each case.

Match the equation to the graph.

1. $x = 16y^2$

2. $y = 16x^2$

3. $x^2 + 16y^2 = 1$

4. $x^2 - 16y^2 = 1$

5. $16x^2 + y^2 = 1$

6. $16y^2 - x^2 = 1$

a.

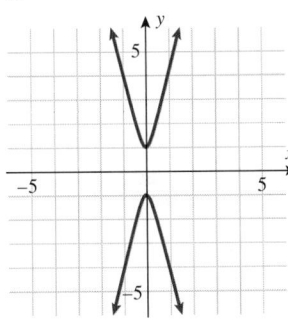

b.

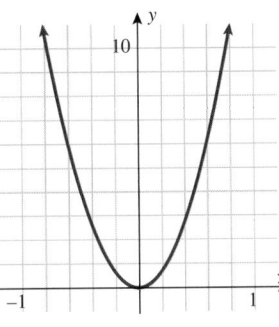

c.

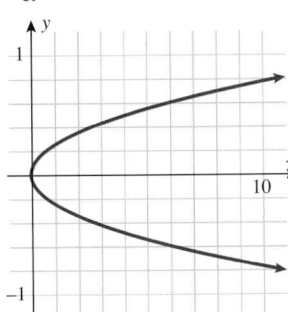

d.

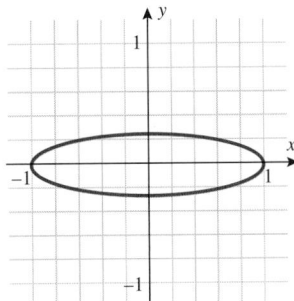

e.

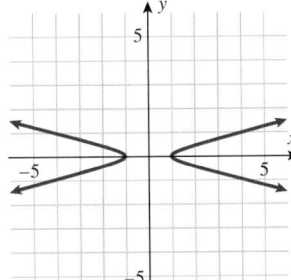

f.

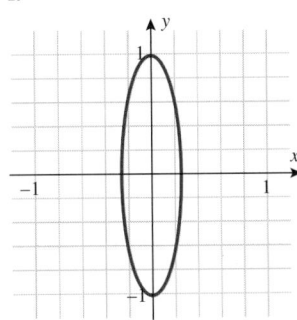

Find the equation of the conic with the given characteristics.

7. Parabola   vertex: $(0, 0)$ focus: $(-4, 0)$

8. Parabola   vertex: $(0, 0)$ directrix: $y = 2$

9. Parabola   vertex: $(-1, 5)$ focus: $(-1, 2)$

10. Parabola   vertex: $(2, -3)$ directrix: $x = 0$

11. Ellipse   center: $(0, 0)$
vertices: $(0, -4), (0, 4)$
foci: $(0, -3), (0, 3)$

12. Ellipse   center: $(0, 0)$
vertices: $(-3, 0), (3, 0)$
foci: $(-1, 0), (1, 0)$

13. Ellipse   vertices: $(2, -6), (2, 6)$
foci: $(2, -4), (2, 4)$

14. Ellipse   vertices: $(-7, -3), (-4, -3)$
foci: $(-6, -3), (-5, -3)$

15. Hyperbola   vertices: $(-1, 0)$ and $(1, 0)$
asymptotes: $y = -2x$ and $y = 2x$

16. Hyperbola   vertices: $(0, -1)$ and $(0, 1)$
asymptotes: $y = -\frac{1}{3}x$ and $y = \frac{1}{3}x$

17. Hyperbola   foci: $(2, -6), (2, 6)$
vertices: $(2, -4), (2, 4)$

18. Hyperbola   foci: $(-7, -3), (-4, -3)$
vertices: $(-6, -3), (-5, -3)$

Graph the following equations:

19. $9x^2 + 18x - 4y^2 + 16y - 43 = 0$

20. $4x^2 - 8x + y^2 + 10y + 28 = 0$

21. $y^2 + 4y - 16x + 20 = 0$

22. $x^2 - 4x + y + 1 = 0$

23. Eyeglass Lens. Eyeglass lenses can be thought of as very wide parabolic curves. If the focus occurs 1.5 centimeters from the center of the lens, and the lens at its opening is 4 centimeters across, find an equation that governs the shape of the lens.

24. Planetary Orbits. The planet Uranus's orbit is described in the following picture with the Sun as a focus of the elliptical orbit. Write an equation for the orbit.

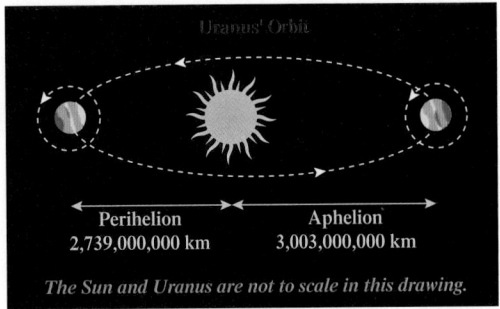

Uranus' Orbit

Perihelion
2,739,000,000 km

Aphelion
3,003,000,000 km

The Sun and Uranus are not to scale in this drawing.

Graph the following nonlinear inequalities:

25. $y < x^3 + 1$

26. $y^2 \geq 16x$

Graph the following systems of nonlinear inequalities:

27.
$$y \le 4 - x^2$$
$$16x^2 + 25y^2 \le 400$$

28.
$$y \le e^{-x}$$
$$y \ge x^2 - 4$$

29. Identify the conic represented by the equation

$$r = \frac{12}{3 + 2\sin\theta}.$$ State the eccentricity.

30. Use rotation of axes to transform the equation in x and y into an equation in X and Y that has no XY-term:
$6\sqrt{3}x^2 + 6xy + 4\sqrt{3}y^2 = 21\sqrt{3}$. State the rotation angle.

31. A golf ball is hit with an initial speed of 120 feet per second at an angle of $45°$ with the ground. How long will the ball stay in the air? How far will the ball travel (horizontal distance) before it hits the ground?

32. Describe (classify) the plane curve defined by the parametric equations $x = \sqrt{1 - t}$ and $y = \sqrt{t}$ for t in $[0, 1]$.

33. Use a graphing utility to graph the following nonlinear inequality:

$$x^2 + 4xy - 9y^2 - 6x + 8y + 28 \le 0$$

34. Use a graphing utility to solve the following systems of equations:

$$0.1225x^2 + 0.0289y^2 = 1$$
$$y^3 = 11x$$

Round your answers to three decimal places.

35. Given is the parabola $x^2 + 4.2x - y + 5.61 = 0$.
   a. Solve the equation for y and use a graphing utility to plot the parabola.
   b. Transform the equation into the form $(x - h)^2 = 4p(y - k)$. Find the vertex. Which way does the parabola open?
   c. Do (a) and (b) agree with each other?

36. With a graphing utility, explore the second-degree equation $Ax^2 + Bxy + Cy^2 + 10x - 8y - 5 = 0$ for the following values of A, B, and C:
   a. $A = 2, B = -\sqrt{3}, C = 1$
   b. $A = 2, B = \sqrt{3}, C = -1$

Show the angle of rotation to one decimal place. Explain the differences.

1. Solve for x: $(x + 2)^2 - (x + 2) - 20 = 0$.

2. Find an equation of a circle centered at $(5, 1)$ and passing through the point $(6, -2)$.

3. Evaluate the difference quotient $\dfrac{f(x + h) - f(x)}{h}$ for the function $f(x) = 8 - 7x$.

4. Write an equation that describes the following variation: I is directly proportional to both P and t, and $I = 90$ when $P = 1500$ and $t = 2$.

5. Find the quadratic function that has vertex $(7, 7)$ and goes through the point $(10, 10)$.

6. **Compound Interest.** How much money should you put in a savings account now that earns 4.7% interest a year compounded weekly if you want to have \$65,000 in 17 years?

7. Solve the logarithmic equation exactly: $\log x^2 - \log 16 = 0$.

8. In a 30°-60°-90° triangle, if the shortest leg has length 8 inches, what are the lengths of the other leg and the hypotenuse?

9. Use a calculator to evaluate $\cot(-27°)$. Round your answer to four decimal places.

10. **Sound Waves.** If a sound wave is represented by $y = 0.007 \sin 850\pi t$ cm, what are its amplitude and frequency?

11. For the trigonometric expression $\tan\theta(\csc\theta + \cos\theta)$, perform the operations and simplify. Write the answer in terms of $\sin\theta$ and $\cos\theta$.

12. Find the exact value of $\cos\left(-\dfrac{11\pi}{12}\right)$.

13. Solve the trigonometric equation $4\cos^2 x + 4\cos 2x + 1 = 0$ exactly over the interval $0 \le \theta \le 2\pi$.

14. **Airplane Speed.** A plane flew due north at 450 mph for 2 hours. A second plane, starting at the same point and at the same time, flew southeast at an angle of 135° clockwise from due north at 375 mph for 2 hours. At the end of 2 hours, how far apart were the two planes? Round to the nearest mile.

15. Find the vector with magnitude $|\mathbf{u}| = 15$ and direction angle $\theta = 110°$.

16. Given $z_1 = 5(\cos 15° + i\sin 15°)$ and $z_2 = 2(\cos 75° + i\sin 75°)$, find the product $z_1 z_2$ and express it in rectangular form.

17. At a food court, 3 medium sodas and 2 soft pretzels cost \$6.77. A second order of 5 medium sodas and 4 soft pretzels costs \$12.25. Find the cost of a soda and the cost of a soft pretzel.

18. Find the partial-fraction decomposition for the rational expression $\dfrac{3x + 5}{(x - 3)(x^2 + 5)}$.

19. Graph the system of inequalities or indicate that the system has no solution.
$$y \ge 3x - 2$$
$$y \le 3x + 2$$

20. Solve the system using Gauss–Jordan elimination.
$$x - 2y + z = 7$$
$$-3x + y + 2z = -11$$

21. Given $A = \begin{bmatrix} 3 & 4 & -7 \\ 0 & 1 & 5 \end{bmatrix}$, $B = \begin{bmatrix} 8 & -2 & 6 \\ 9 & 0 & -1 \end{bmatrix}$, and $C = \begin{bmatrix} 9 & 0 \\ 1 & 2 \end{bmatrix}$, find $2B - 3A$.

22. Use Cramer's rule to solve the system of equations.
$$25x + 40y = -12$$
$$75x - 105y = 69$$

23. Find the standard form of the equation of an ellipse with foci $(6, 2)$ and $(6, -6)$ and vertices $(6, 3)$ and $(6, -7)$.

24. Find the standard form of the equation of a hyperbola with vertices $(5, -2)$ and $(5, 0)$ and foci $(5, -3)$ and $(5, 1)$.

25. Solve the system of equations.
$$x + y = 6$$
$$x^2 + y^2 = 20$$

26. Use a graphing utility to graph the following equation:
$$x^2 - 3xy + 10y^2 - 1 = 0$$

27. Use a graphing utility to graph the following system of nonlinear inequalities:
$$y \ge e^{-0.3x} - 3.5$$
$$y \le 4 - x^2$$

10
Sequences and Series

Gaillardia Flower (55), Kenneth M. Highfil/Photo Researchers, Inc.; Michaelmas Daisy (89), Maxine Adcock/Photo Researchers, Inc.; Yellow Iris (3), Edward Kinsman/Photo Researchers, Inc.; Blue Columbine (5), Jeffrey Lepore/Photo Researchers, Inc.; Gerbera Daisy (34), Bonnie Sue Rauch/Photo Researchers, Inc.; Erect Dayflower (2), Michael Lustbader/Photo Researchers, Inc.; Calla Lilies (1), Adam Jones/Photo Researchers, Inc.; Cosmos Flower (8), Maria Mosolova/Photo Researchers, Inc.; Lemon Symphony (21), Bonnie Sue Rauch/Photo Researchers, Inc.; Black-Eyed Susan (13), Rod Planck/Photo Researchers, Inc.

A famous sequence that appears throughout nature is the *Fibonacci sequence*, where each term in the sequence is the sum of the previous two numbers:

$$1, 1, 2, 3, 5, 8, 13, 21, 34, 55, 89, \ldots$$

The number of petals in certain flowers are numbers in the Fibonacci sequence.

IN THIS CHAPTER we will first define a sequence and a series and then discuss two particular kinds of sequences and series called arithmetic and geometric. We will then discuss mathematical proof by induction. Lastly, we will discuss the Binomial theorem, which allows us an efficient way to perform binomial expansions.

SEQUENCES AND SERIES

10.1 Sequences and Series	10.2 Arithmetic Sequences and Series	10.3 Geometric Sequences and Series	10.4 Mathematical Induction	10.5 The Binomial Theorem
• Sequences • Factorial Notation • Recursion Formulas • Sums and Series	• Arithmetic Sequences • The General (*n*th) Term of an Arithmetic Sequence • The Sum of an Arithmetic Sequence	• Geometric Sequences • The General (*n*th) Term of a Geometric Sequence • Geometric Series	• Proof by Mathematical Induction	• Binomial Coefficients • Binomial Expansion • Pascal's Triangle • Finding a Particular Term of a Binomial Expansion

CHAPTER OBJECTIVES

- Use sigma notation to represent a series (Section 10.1).
- Find the sum of an arithmetic sequence (Section 10.2).
- Determine whether an infinite geometric series converges (Section 10.3).
- Prove mathematical statements using mathematical induction (Section 10.4).
- Perform binomial expansions (Section 10.5).

SKILLS OBJECTIVES

- Find terms of a sequence given the general term.
- Look for a pattern in a sequence and find the general term.
- Apply factorial notation.
- Apply recursion formulas.
- Use summation (sigma) notation to represent a series.
- Evaluate a series.

CONCEPTUAL OBJECTIVES

- Understand the difference between a sequence and a series.
- Understand the difference between a finite series and an infinite series.

Sequences

The word *sequence* means an order in which one thing follows another in succession. A sequence is an ordered list. For example, if we write $x, 2x^2, 3x^3, 4x^4, 5x^5, ?$, what would the next term in the *sequence* be, the one where the question mark now stands? The answer is $6x^6$.

Study Tip

A sequence is a set of terms written in a specific order

$$a_1, a_2, a_3, \ldots a_n, \ldots$$

where a_1 is called the first term, a_2 is called the second term, and a_n is called the nth term.

DEFINITION   **A Sequence**

A **sequence** is a function whose domain is a set of positive integers. The function values, or **terms**, of the sequence are written as

$$a_1, a_2, a_3, \ldots, a_n, \ldots$$

Rather than using function notation, sequences are usually written with subscript (or index) notation, $a_{subscript}$.

A **finite sequence** has the domain $\{1, 2, 3, \ldots, n\}$ for some positive integer n. An **infinite sequence** has the domain of all positive integers $\{1, 2, 3, \ldots\}$. There are times when it is convenient to start the indexing at 0 instead of 1:

$$a_0, a_1, a_2, a_3, \ldots, a_n, \ldots$$

Sometimes a pattern in the sequence can be obtained and the sequence can be written using a *general term*. In the previous example, $x, 2x^2, 3x^3, 4x^4, 5x^5, 6x^6, \ldots$, each term has the same exponent and coefficient. We can write this sequence as $a_n = nx^n$, $n = 1, 2, 3, 4, 5, 6, \ldots$, where a_n is called the **general** or **nth term**.

EXAMPLE 1   **Finding Several Terms of a Sequence, Given the General Term**

Find the first four ($n = 1, 2, 3, 4$) terms of each of the following sequences, given the general term.

a. $a_n = 2n - 1$

b. $b_n = \dfrac{(-1)^n}{n + 1}$

Solution (a):

$$a_n = 2n - 1$$

Find the first term, $n = 1$.

$$a_1 = 2(1) - 1 = 1$$

Find the second term, $n = 2$.

$$a_2 = 2(2) - 1 = 3$$

Find the third term, $n = 3$.

$$a_3 = 2(3) - 1 = 5$$

Find the fourth term, $n = 4$.

$$a_4 = 2(4) - 1 = 7$$

The first four terms of the sequence are $\boxed{1, 3, 5, 7}$.

Solution (b):

$$b_n = \frac{(-1)^n}{n + 1}$$

Find the first term, $n = 1$.

$$b_1 = \frac{(-1)^1}{1 + 1} = -\frac{1}{2}$$

Find the second term, $n = 2$.

$$b_2 = \frac{(-1)^2}{2 + 1} = \frac{1}{3}$$

Find the third term, $n = 3$.

$$b_3 = \frac{(-1)^3}{3 + 1} = -\frac{1}{4}$$

Find the fourth term, $n = 4$.

$$b_4 = \frac{(-1)^4}{4 + 1} = \frac{1}{5}$$

The first four terms of the sequence are $\boxed{-\frac{1}{2}, \frac{1}{3}, -\frac{1}{4}, \frac{1}{5}}$.

■ **YOUR TURN** Find the first four terms of the sequence $a_n = \dfrac{(-1)^n}{n^2}$.

■ **Answer:** $-1, \frac{1}{4}, -\frac{1}{9}, \frac{1}{16}$

EXAMPLE 2 **Finding the General Term, Given Several Terms of the Sequence**

Find the general term of the sequence, given the first five terms.

a. $1, \frac{1}{4}, \frac{1}{9}, \frac{1}{16}, \frac{1}{25}, \ldots$ **b.** $-1, 4, -9, 16, -25, \ldots$

Solution (a):

Write 1 as $\frac{1}{1}$.

$$\frac{1}{1}, \frac{1}{4}, \frac{1}{9}, \frac{1}{16}, \frac{1}{25}, \ldots$$

Notice that each denominator is an integer squared.

$$\frac{1}{1^2}, \frac{1}{2^2}, \frac{1}{3^2}, \frac{1}{4^2}, \frac{1}{5^2}, \ldots$$

Identify the general term.

$$\boxed{a_n = \frac{1}{n^2}} \qquad n = 1, 2, 3, 4, 5, \ldots$$

Solution (b):

Notice that each term includes an integer squared.

$$-1^2, 2^2, -3^2, 4^2, -5^2, \ldots$$

Identify the general term.

$$\boxed{b_n = (-1)^n n^2} \qquad n = 1, 2, 3, 4, 5, \ldots$$

Study Tip

$(-1)^n$ or $(-1)^{n+1}$ is a way to represent an alternating sequence.

■ **YOUR TURN** Find the general term of the sequence, given the first five terms.

a. $-\frac{1}{2}, \frac{1}{4}, -\frac{1}{6}, \frac{1}{8}, -\frac{1}{10}, \ldots$ **b.** $\frac{1}{2}, \frac{1}{4}, \frac{1}{8}, \frac{1}{16}, \frac{1}{32}, \ldots$

■ **Answer:**

a. $a_n = \dfrac{(-1)^n}{2n}$ **b.** $a_n = \dfrac{1}{2^n}$

Parts (b) in both Example 1 and Example 2 are called **alternating** sequences, because the terms alternate signs (positive and negative). If the odd-indexed terms, $a_1, a_3, a_5, \ldots$, are negative and the even-indexed terms, $a_2, a_4, a_6, \ldots$, are positive, we include $(-1)^n$ in the general term. If the opposite is true, and the odd-indexed terms are positive and the even-indexed terms are negative, we include $(-1)^{n+1}$ in the general term.

Factorial Notation

Find 0!, 1!, 2!, 3!, 4!, and 5!.

Scientific calculators:

Press	Display
0 ! =	1
1 ! =	1
2 ! =	2
3 ! =	6
4 ! =	24
5 ! =	120

Graphing calculators:

Press	Display
0 MATH PRB 4:! ENTER	1
1 MATH PRB 4:! ENTER	1
2 MATH PRB 4:! ENTER	2
3 MATH PRB 4:! ENTER	6
4 MATH PRB 4:! ENTER	24
5 MATH PRB 4:! ENTER	120

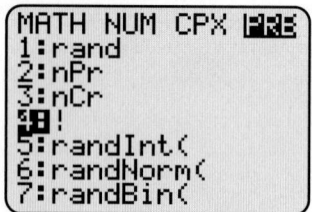

```
MATH NUM CPX PRB
1:rand
2:nPr
3:nCr
4:!
5:randInt(
6:randNorm(
7:randBin(
```

```
0!
              1
1!
              1
2!
              2
```

```
3!
              6
4!
             24
5!
            120
```

Many important sequences that arise in mathematics involve terms that are defined with products of consecutive positive integers. The products are expressed in *factorial notation*.

DEFINITION — Factorial

If n is a positive integer, then $n!$ (stated as "n factorial") is the product of all positive integers from n down to 1.

$$n! = n(n-1)(n-2)\cdots 3\cdot 2\cdot 1 \qquad n \geq 2$$

and $0! = 1$ and $1! = 1$.

The values of $n!$ for the first six nonnegative integers are

$$0! = 1$$
$$1! = 1$$
$$2! = 2\cdot 1 = 2$$
$$3! = 3\cdot 2\cdot 1 = 6$$
$$4! = 4\cdot 3\cdot 2\cdot 1 = 24$$
$$5! = 5\cdot 4\cdot 3\cdot 2\cdot 1 = 120$$

Notice that $4! = 4\cdot 3\cdot 2\cdot 1 = 4\cdot 3!$. In general, we can apply the formula $n! = n[(n-1)!]$. Often the brackets are not used, and the notation $n! = n(n-1)!$ implies calculating the factorial $(n-1)!$ and then multiplying that quantity by n. For example, to find $6!$, we employ the relationship $n! = n(n-1)!$ and set $n = 6$:

$$6! = 6\cdot 5! = 6\cdot 120 = 720$$

EXAMPLE 3 Finding the Terms of a Sequence Involving Factorials

Find the first four terms of the sequence, given the general term $a_n = \dfrac{x^n}{n!}$.

Solution:

Find the first term, $n = 1$.

$$a_1 = \frac{x^1}{1!} = x$$

Find the second term, $n = 2$.

$$a_2 = \frac{x^2}{2!} = \frac{x^2}{2\cdot 1} = \frac{x^2}{2}$$

Find the third term, $n = 3$.

$$a_3 = \frac{x^3}{3!} = \frac{x^3}{3\cdot 2\cdot 1} = \frac{x^3}{6}$$

Find the fourth term, $n = 4$.

$$a_4 = \frac{x^4}{4!} = \frac{x^4}{4\cdot 3\cdot 2\cdot 1} = \frac{x^4}{24}$$

The first four terms of the sequence are $\boxed{x, \dfrac{x^2}{2}, \dfrac{x^3}{6}, \dfrac{x^4}{24}}$.

EXAMPLE 4 Evaluating Expressions with Factorials

Evaluate each factorial expression.

a. $\dfrac{6!}{2! \cdot 3!}$ b. $\dfrac{(n + 1)!}{(n - 1)!}$

Solution (a):

Expand each factorial in the numerator and denominator.

$$\frac{6!}{2! \cdot 3!} = \frac{6 \cdot 5 \cdot 4 \cdot \cancel{3} \cdot \cancel{2} \cdot \cancel{1}}{2 \cdot 1 \cdot \cancel{3} \cdot \cancel{2} \cdot \cancel{1}}$$

Cancel the $3 \cdot 2 \cdot 1$ in both the numerator and denominator.

$$= \frac{6 \cdot 5 \cdot 4}{2 \cdot 1}$$

Simplify.

$$= \frac{6 \cdot 5 \cdot 2}{1} = 60$$

$$\boxed{\frac{6!}{2! \cdot 3!} = 60}$$

Solution (b):

Expand each factorial in the numerator and denominator.

$$\frac{(n + 1)!}{(n - 1)!} = \frac{(n + 1)(n)(n - 1)(n - 2) \cdots 3 \cdot 2 \cdot 1}{(n - 1)(n - 2) \cdots 3 \cdot 2 \cdot 1}$$

Cancel the $(n - 1)(n - 2) \cdots 3 \cdot 2 \cdot 1$ in both the numerator and denominator.

$$\frac{(n + 1)!}{(n - 1)!} = (n + 1)(n)$$

Alternatively,

$$\frac{(n + 1)!}{(n - 1)!} = \frac{(n + 1)(n)\cancel{(n - 1)!}}{\cancel{(n - 1)!}}$$

$$\boxed{\frac{(n + 1)!}{(n - 1)!} = n^2 + n}$$

COMMON MISTAKE

In Example 4 we found $\dfrac{6!}{2! \cdot 3!} = 60$. It is important to note that $2! \cdot 3! \neq 6!$

■ **YOUR TURN** Evaluate each factorial expression.

a. $\dfrac{3! \cdot 4!}{2! \cdot 6!}$ b. $\dfrac{(n + 2)!}{n!}$

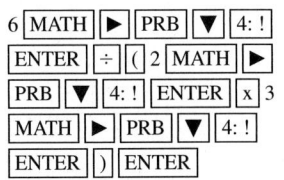
■ **Answer:**

a. $\frac{1}{10}$ b. $(n + 2)(n + 1)$

Recursion Formulas

Another way to define a sequence is **recursively**, or using a **recursion formula**. The first few terms are listed, and the recursion formula determines the remaining terms based on previous terms. For example, the famous Fibonacci sequence is 1, 1, 2, 3, 5, 8, 13, 21, 34, 55, 89, ... Each term in the Fibonacci sequence is found by adding the previous two terms.

$$1 + 1 = 2 \qquad 1 + 2 = 3 \qquad 2 + 3 = 5$$
$$3 + 5 = 8 \qquad 5 + 8 = 13 \qquad 8 + 13 = 21$$
$$13 + 21 = 34 \qquad 21 + 34 = 55 \qquad 34 + 55 = 89$$

We can define the Fibonacci sequence using a general term:

$$a_1 = 1, \quad a_2 = 1, \quad \text{and} \quad a_n = a_{n-2} + a_{n-1} \qquad n \geq 3$$

The Fibonacci sequence is found in places we least expect it (e.g., pineapples, broccoli, and flowers). The number of petals in certain flowers is a Fibonacci number. For example, a wild rose has 5 petals, lilies and irises have 3 petals, and daisies have 34, 55, or even 89 petals. The number of spirals in an Italian broccoli is a Fibonacci number (13).

Study Tip

If $a_n = a_{n-1} + a_{n-3}$, then $a_{100} = a_{99} + a_{98}$.

EXAMPLE 5 **Using a Recursion Formula to Find a Sequence**

Find the first four terms of the sequence a_n if $a_1 = 2$ and $a_n = 2a_{n-1} - 1$, $n \geq 2$.

Solution:

Write the first term, $n = 1$.	$a_1 = 2$
Find the second term, $n = 2$.	$a_2 = 2a_1 - 1 = 2(2) - 1 = 3$
Find the third term, $n = 3$.	$a_3 = 2a_2 - 1 = 2(3) - 1 = 5$
Find the fourth term, $n = 4$.	$a_4 = 2a_3 - 1 = 2(5) - 1 = 9$

The first four terms of the sequence are $\boxed{2, 3, 5, 9}$.

■ **Answer:** $1, \frac{1}{2}, \frac{1}{12}, \frac{1}{288}$

■ **YOUR TURN** Find the first four terms of the sequence

$$a_1 = 1 \quad \text{and} \quad a_n = \frac{a_{n-1}}{n!} \quad n \geq 2$$

Sums and Series

When we add the terms in a sequence, the result is a *series*.

DEFINITION **Series**

Given the infinite sequence $a_1, a_2, a_3, \ldots, a_n, \ldots$, the sum of all of the terms in the infinite sequence is called an **infinite series** and is denoted by

$$a_1 + a_2 + a_3 + \cdots + a_n + \cdots$$

and the sum of only the first n terms is called a **finite series**, or **nth partial sum**, and is denoted by

$$S_n = a_1 + a_2 + a_3 + \cdots + a_n$$

Study Tip

Σ is a regular Greek letter, but when used to represent the mathematical sum operation, we oversize it.

The capital Greek letter Σ (sigma) corresponds to the capital S in our alphabet. Therefore, we use Σ as a shorthand way to represent a sum (series). For example, the sum of the first five terms of the sequence $1, 4, 9, 16, 25, \ldots, n^2, \ldots$ can be represented using **sigma (or summation) notation**:

Study Tip

We often use the following notation in running text:

$$\sum_{n=1}^{5} = \sum_{n=1}^{5}$$

$$\sum_{n=1}^{5} n^2 = (1)^2 + (2)^2 + (3)^2 + (4)^2 + (5)^2$$

$$= 1 + 4 + 9 + 16 + 25$$

This is read "the sum as n goes from 1 to 5 of n^2." The letter n is called the **index of summation**, and often other letters are used instead of n. It is important to note that the sum can start at other integers besides 1.

Study Tip

The index of a summation (series) can start at any integer (not just 1).

If we wanted the sum of all of the terms in the sequence, we would represent that infinite series using summation notation as

$$\sum_{n=1}^{\infty} n^2 = 1 + 4 + 9 + 16 + 25 + \cdots$$

EXAMPLE 6   **Writing a Series Using Sigma Notation**

Represent each of the following series using sigma notation:

a. $1 + 1 + \frac{1}{2} + \frac{1}{6} + \frac{1}{24} + \frac{1}{120}$ **b.** $8 + 27 + 64 + 125 + \cdots$

Solution (a):

Write 1 as $\frac{1}{1}$.

$$\frac{1}{1} + \frac{1}{1} + \frac{1}{2} + \frac{1}{6} + \frac{1}{24} + \frac{1}{120}$$

Notice that we can write the denominators using factorials.

$$= \frac{1}{1} + \frac{1}{1} + \frac{1}{2!} + \frac{1}{3!} + \frac{1}{4!} + \frac{1}{5!}$$

Recall that $0! = 1$ and $1! = 1$.

$$= \frac{1}{0!} + \frac{1}{1!} + \frac{1}{2!} + \frac{1}{3!} + \frac{1}{4!} + \frac{1}{5!}$$

Identify the general term.

$$a_n = \frac{1}{n!} \qquad n = 0, 1, 2, 3, 4, 5$$

Write the finite series using sigma notation.

$$\boxed{\sum_{n=0}^{5} \frac{1}{n!}}$$

Solution (b):

Write the infinite series as a sum of terms cubed.

$$8 + 27 + 64 + 125 + \cdots$$
$$= 2^3 + 3^3 + 4^3 + 5^3 + \cdots$$

Identify the general term of the series.

$$a_n = n^3 \qquad n \geq 2$$

Write the infinite series using sigma notation.

$$\boxed{\sum_{n=2}^{\infty} n^3}$$

▪ **YOUR TURN**   Represent each of the following series using sigma notation:

a. $1 - \frac{1}{2} + \frac{1}{6} - \frac{1}{24} + \frac{1}{120} - \cdots$

b. $4 + 8 + 16 + 32 + 64 + \cdots$

▪ **Answer:**

a. $\displaystyle\sum_{n=1}^{\infty} \frac{(-1)^{n+1}}{n!}$ **b.** $\displaystyle\sum_{n=2}^{\infty} 2^n$

Now that we are comfortable with sigma (summation) notation, let's turn our attention to evaluating a series (calculating the sum). You can always evaluate a finite series. However, you cannot always evaluate an infinite series.

EXAMPLE 7   **Evaluating a Finite Series**

Evaluate the series $\sum_{i=0}^{4} (2i + 1)$.

Solution:

Write out the partial sum.

$$\sum_{i=0}^{4} (2i + 1) = \underset{\underset{(i=0)}{\uparrow}}{1} + \underset{\underset{(i=1)}{\downarrow}}{3} + \underset{\underset{(i=2)}{\uparrow}}{5} + \underset{\underset{(i=3)}{\downarrow}}{7} + \underset{\underset{(i=4)}{\uparrow}}{9}$$

Simplify.

$$= 25$$

$$\boxed{\sum_{i=0}^{4} (2i + 1) = 25}$$

▪ **YOUR TURN**   Evaluate the series $\sum_{n=1}^{5} (-1)^n n$.

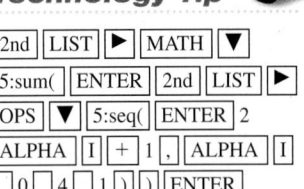

2nd LIST ▶ MATH ▼
5:sum(ENTER 2nd LIST ▶
OPS ▼ 5:seq(ENTER 2
ALPHA I + 1 , ALPHA I
, 0 , 4 , 1)) ENTER

```
sum(seq(2I+1,I,0
,4,1))
              25
```

▪ **Answer:** -3

Infinite series may or may not have a finite sum. For example, if we keep adding $1 + 1 + 1 + 1 + \cdots$, then there is no single real number that the series sums to because the sum continues to grow without bound. However, if we add $0.9 + 0.09 + 0.009 + 0.0009 + \cdots$, this sum is $0.9999\ldots = 0.\overline{9}$, which is a rational number, and it can be proven that $0.\overline{9} = 1$.

Technology Tip

a.

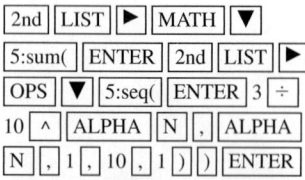

b.

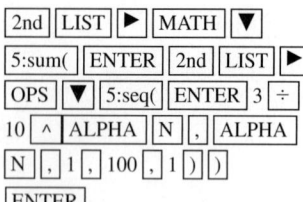

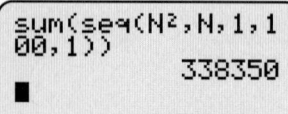

■ **Answer: a.** Series diverges.
  b. Series converges to $\frac{2}{3}$.

EXAMPLE 8   Evaluating an Infinite Series, If Possible

Evaluate the following infinite series, if possible.

a. $\displaystyle\sum_{n=1}^{\infty} \frac{3}{10^n}$ **b.** $\displaystyle\sum_{n=1}^{\infty} n^2$

Solution (a):

Expand the series.

$$\sum_{n=1}^{\infty} \frac{3}{10^n} = \frac{3}{10} + \frac{3}{100} + \frac{3}{1000} + \frac{3}{10,000} + \cdots$$

Write in decimal form.

$$\sum_{n=1}^{\infty} \frac{3}{10^n} = 0.3 + 0.03 + 0.003 + 0.0003 + \cdots$$

Calculate the sum.

$$\sum_{n=1}^{\infty} \frac{3}{10^n} = 0.333333\overline{3} = \frac{1}{3}$$

$$\boxed{\sum_{n=1}^{\infty} \frac{3}{10^n} = \frac{1}{3}}$$

Solution (b):

Expand the series.

$$\sum_{n=1}^{\infty} n^2 = 1 + 4 + 9 + 16 + 25 + 36 + \cdots$$

This sum is infinite since it continues to grow without any bound.

In part (a) we say that the series **converges** to $\frac{1}{3}$ and in part (b) we say that the series **diverges**.

■ **YOUR TURN** Evaluate the following infinite series, if possible.

a. $\displaystyle\sum_{n=1}^{\infty} 2n$ **b.** $\displaystyle\sum_{n=1}^{\infty} 6\left(\frac{1}{10}\right)^n$

Applications

The average stock price for Home Depot Inc (HD) was $37 in 2007, $28 in 2008, and $23 in 2009. If a_n is the yearly average stock price, where $n = 0$ corresponds to 2007, then $\frac{1}{3}\sum_{n=0}^{2} a_n$ tells us the average yearly stock price over the three year period 2007 to 2009.

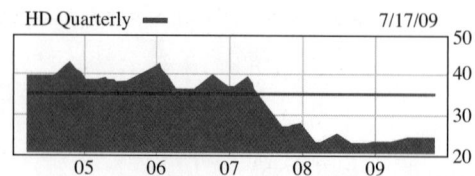

SECTION
10.1 SUMMARY

In this section, we discussed finite and infinite sequences and series. When the terms of a sequence are added together, the result is a series.

Finite sequence: $a_1, a_2, a_3, \ldots, a_n$
Infinite sequence: $a_1, a_2, a_3, \ldots, a_n, \ldots$

Finite series: $a_1 + a_2 + a_3 + \cdots + a_n$
Infinite series: $a_1 + a_2 + a_3 + \cdots + a_n + \cdots$

Factorial notation was also introduced:

$n! = n(n-1)\cdots 3 \cdot 2 \cdot 1 \quad n \geq 2$

and $0! = 1$ and $1! = 1$.

The sum of a finite series is always finite.

The sum of an infinite series is either convergent or divergent.

Sigma notation is used to express a series.

■ **Finite series:** $\displaystyle\sum_{i=1}^{n} a_i = a_1 + a_2 + \cdots + a_n$

■ **Infinite series:** $\displaystyle\sum_{n=1}^{\infty} a_n = a_1 + a_2 + a_3 + \cdots$

SECTION
10.1 EXERCISES

■ **SKILLS**

In Exercises 1–12, write the first four terms of each sequence. Assume n starts at 1.

1. $a_n = n$

2. $a_n = n^2$

3. $a_n = 2n - 1$

4. $a_n = x^n$

5. $a_n = \dfrac{n}{(n+1)}$

6. $a_n = \dfrac{(n+1)}{n}$

7. $a_n = \dfrac{2^n}{n!}$

8. $a_n = \dfrac{n!}{(n+1)!}$

9. $a_n = (-1)^n x^{n+1}$

10. $a_n = (-1)^{n+1} n^2$

11. $a_n = \dfrac{(-1)^n}{(n+1)(n+2)}$

12. $a_n = \dfrac{(n-1)^2}{(n+1)^2}$

In Exercises 13–20, find the indicated term of each sequence given.

13. $a_n = \left(\dfrac{1}{2}\right)^n \quad a_9 = ?$

14. $a_n = \dfrac{n}{(n+1)^2} \quad a_{15} = ?$

15. $a_n = \dfrac{(-1)^n n!}{(n+2)!} \quad a_{19} = ?$

16. $a_n = \dfrac{(-1)^{n+1}(n-1)(n+2)}{n} \quad a_{13} = ?$

17. $a_n = \left(1 + \dfrac{1}{n}\right)^2 \quad a_{100} = ?$

18. $a_n = 1 - \dfrac{1}{n^2} \quad a_{10} = ?$

19. $a_n = \log 10^n \quad a_{23} = ?$

20. $a_n = e^{\ln n} \quad a_{49} = ?$

In Exercises 21–28, write an expression for the nth term of the given sequence. Assume n starts at 1.

21. $2, 4, 6, 8, 10, \ldots$

22. $3, 6, 9, 12, 15, \ldots$

23. $\dfrac{1}{2 \cdot 1}, \dfrac{1}{3 \cdot 2}, \dfrac{1}{4 \cdot 3}, \dfrac{1}{5 \cdot 4}, \dfrac{1}{6 \cdot 5}, \ldots$

24. $\dfrac{1}{2}, \dfrac{1}{4}, \dfrac{1}{8}, \dfrac{1}{16}, \dfrac{1}{32}, \ldots$

25. $-\dfrac{2}{3}, \dfrac{4}{9}, -\dfrac{8}{27}, \dfrac{16}{81}, \ldots$

26. $\dfrac{1}{2}, \dfrac{3}{4}, \dfrac{9}{8}, \dfrac{27}{16}, \dfrac{81}{32}, \ldots$

27. $1, -1, 1, -1, 1, \ldots$

28. $\dfrac{1}{3}, -\dfrac{2}{4}, \dfrac{3}{5}, -\dfrac{4}{6}, \dfrac{5}{7}, \ldots$

In Exercises 29–40, simplify each ratio of factorials.

29. $\dfrac{9!}{7!}$ **30.** $\dfrac{4!}{6!}$ **31.** $\dfrac{29!}{27!}$ **32.** $\dfrac{32!}{30!}$ **33.** $\dfrac{75!}{77!}$ **34.** $\dfrac{100!}{103!}$

35. $\dfrac{97!}{93!}$ **36.** $\dfrac{101!}{98!}$ **37.** $\dfrac{(n-1)!}{(n+1)!}$ **38.** $\dfrac{(n+2)!}{n!}$ **39.** $\dfrac{(2n+3)!}{(2n+1)!}$ **40.** $\dfrac{(2n+2)!}{(2n-1)!}$

In Exercises 41–50, write the first four terms of the sequence defined by each recursion formula. Assume the sequence begins at $n = 1$.

41. $a_1 = 7 \qquad a_n = a_{n-1} + 3$ **42.** $a_1 = 2 \qquad a_n = a_{n-1} + 1$ **43.** $a_1 = 1 \qquad a_n = n \cdot a_{n-1}$

44. $a_1 = 2 \qquad a_n = (n+1) \cdot a_{n-1}$ **45.** $a_1 = 100 \qquad a_n = \dfrac{a_{n-1}}{n!}$ **46.** $a_1 = 20 \qquad a_n = \dfrac{a_{n-1}}{n^2}$

47. $a_1 = 1, a_2 = 2 \qquad a_n = a_{n-1} \cdot a_{n-2}$ **48.** $a_1 = 1, a_2 = 2 \qquad a_n = \dfrac{a_{n-2}}{a_{n-1}}$

49. $a_1 = 1, a_2 = -1 \qquad a_n = (-1)^n \left[a_{n-1}^2 + a_{n-2}^2 \right]$ **50.** $a_1 = 1, a_2 = -1 \qquad a_n = (n-1)\,a_{n-1} + (n-2)a_{n-2}$

In Exercises 51–64, evaluate each finite series.

51. $\displaystyle\sum_{n=1}^{5} 2$ **52.** $\displaystyle\sum_{n=1}^{5} 7$ **53.** $\displaystyle\sum_{n=0}^{4} n^2$ **54.** $\displaystyle\sum_{n=1}^{4} \dfrac{1}{n}$ **55.** $\displaystyle\sum_{n=1}^{6} (2n-1)$

56. $\displaystyle\sum_{n=1}^{6} (n+1)$ **57.** $\displaystyle\sum_{n=0}^{4} 1^n$ **58.** $\displaystyle\sum_{n=0}^{4} 2^n$ **59.** $\displaystyle\sum_{n=0}^{3} (-x)^n$ **60.** $\displaystyle\sum_{n=0}^{3} (-x)^{n+1}$

61. $\displaystyle\sum_{k=0}^{5} \dfrac{2^k}{k!}$ **62.** $\displaystyle\sum_{k=0}^{5} \dfrac{(-1)^k}{k!}$ **63.** $\displaystyle\sum_{k=0}^{4} \dfrac{x^k}{k!}$ **64.** $\displaystyle\sum_{k=0}^{4} \dfrac{(-1)^k x^k}{k!}$

In Exercises 65–68, evaluate each infinite series, if possible.

65. $\displaystyle\sum_{j=0}^{\infty} 2 \cdot (0.1)^j$ **66.** $\displaystyle\sum_{j=0}^{\infty} 5 \cdot \left(\tfrac{1}{10}\right)^j$ **67.** $\displaystyle\sum_{j=0}^{\infty} n^j \qquad n \ge 1$ **68.** $\displaystyle\sum_{j=0}^{\infty} 1^j$

In Exercises 69–76, use sigma notation to represent each sum.

69. $1 - \dfrac{1}{2} + \dfrac{1}{4} - \dfrac{1}{8} + \cdots + \dfrac{1}{64}$ **70.** $1 + \dfrac{1}{2} + \dfrac{1}{4} + \dfrac{1}{8} + \cdots + \dfrac{1}{64} + \cdots$

71. $1 - 2 + 3 - 4 + 5 - 6 + \cdots$ **72.** $1 + 2 + 3 + 4 + 5 + \cdots + 21 + 22 + 23$

73. $\dfrac{2 \cdot 1}{1} + \dfrac{3 \cdot 2 \cdot 1}{1} + \dfrac{4 \cdot 3 \cdot 2 \cdot 1}{2 \cdot 1} + \dfrac{5 \cdot 4 \cdot 3 \cdot 2 \cdot 1}{3 \cdot 2 \cdot 1} + \dfrac{6 \cdot 5 \cdot 4 \cdot 3 \cdot 2 \cdot 1}{4 \cdot 3 \cdot 2 \cdot 1}$

74. $1 + \dfrac{2}{1} + \dfrac{2^2}{2 \cdot 1} + \dfrac{2^3}{3 \cdot 2 \cdot 1} + \dfrac{2^4}{4 \cdot 3 \cdot 2 \cdot 1} + \cdots$

75. $1 - x + \dfrac{x^2}{2} - \dfrac{x^3}{6} + \dfrac{x^4}{24} - \dfrac{x^5}{120} + \cdots$

76. $x + x^2 + \dfrac{x^3}{2} + \dfrac{x^4}{6} + \dfrac{x^5}{24} + \dfrac{x^6}{120}$

▪ APPLICATIONS

77. Money. Upon graduation Jessica receives a commission from the U.S. Navy to become an officer and a $20,000 signing bonus for selecting aviation. She puts the entire bonus in an account that earns 6% interest compounded monthly. The balance in the account after n months is

$$A_n = 20,000\left(1 + \frac{0.06}{12}\right)^n \qquad n = 1, 2, 3, \ldots$$

Her commitment to the Navy is 6 years. Calculate A_{72}. What does A_{72} represent?

78. Money. Dylan sells his car during his freshman year and puts $7,000 in an account that earns 5% interest compounded quarterly. The balance in the account after n quarters is

$$A_n = 7000\left(1 + \frac{0.05}{4}\right)^n \qquad n = 1, 2, 3, \ldots$$

Calculate A_{12}. What does A_{12} represent?

79. Salary. An attorney is trying to calculate the costs associated with going into private practice. If she hires a paralegal to assist her, she will have to pay the paralegal $20.00 per hour. To be competitive with most firms, she will have to give her paralegal a $2 per hour raise each year. Find a general term of a sequence a_n that would represent the hourly salary of a paralegal with n years of experience. What will be the paralegal's salary with 20 years of experience?

80. NFL Salaries. A player in the NFL typically has a career that lasts 3 years. The practice squad makes the league minimum of $275,000 (2004) in the first year, with a $75,000 raise per year. Write the general term of a sequence a_n that represents the salary of an NFL player making the league minimum during his entire career. Assuming $n = 1$ corresponds to the first year, what does $\sum_{n=1}^{3} a_n$ represent?

81. Salary. Upon graduation Sheldon decides to go to work for a local police department. His starting salary is $30,000 per year, and he expects to get a 3% raise per year. Write the recursion formula for a sequence that represents his annual salary after n years on the job. Assume $n = 0$ represents his first year making $30,000.

82. *Escherichia coli.* A single cell of bacteria reproduces through a process called binary fission. *Escherichia coli* cells divide into two every 20 minutes. Suppose the same rate of division is maintained for 12 hours after the original cell enters the body. How many *E. coli* bacteria cells would be in the body 12 hours later? Suppose there is an infinite nutrient source so that the *E. coli* bacteria cells maintain the same rate of division for 48 hours after the original cell enters the body. How many *E. coli* bacteria cells would be in the body 48 hours later?

83. AIDS/HIV. A typical person has 500 to 1500 T cells per drop of blood in the body. HIV destroys the T cell count at a rate of 50–100 cells per drop of blood per year, depending on how aggressive it is in the body. Generally, the onset of AIDS occurs once the body's T cell count drops below 200. Write a sequence that represents the total number of T cells in a person infected with HIV. Assume that before infection the person has a 1000 T

cell count ($a_0 = 1000$) and the rate at which the infection spreads corresponds to a loss of 75 T cells per drop of blood per year. How much time will elapse until this person has full-blown AIDS?

84. Company Sales. Lowe's reported total sales from 2003 through 2004 in the billions. The sequence $a_n = 3.8 + 1.6n$ represents the total sales in billions of dollars. Assuming $n = 3$ corresponds to 2003, what were the reported sales in 2003 and 2004? What does $\frac{1}{2} \cdot \sum_{n=3}^{4} a_n$ represent?

85. Cost of Eating Out. A college student tries to save money by bringing a bag lunch instead of eating out. He will be able to save $100 per month. He puts the money into his savings account, which draws 1.2% interest and is compounded monthly. The balance in his account after n months of bagging his lunch is

$$A_n = 100,000[(1.001)^n - 1] \qquad n = 1, 2, \ldots$$

Calculate the first four terms of this sequence. Calculate the amount after 3 years (36 months).

86. Cost of Acrylic Nails. A college student tries to save money by growing her own nails out and not spending $50 per month on acrylic fills. She will be able to save $50 per month. She puts the money into her savings account, which draws 1.2% interest and is compounded monthly. The balance in her account after n months of natural nails is

$$A_n = 50,000[(1.001)^n - 1] \qquad n = 1, 2, \ldots$$

Calculate the first four terms of this sequence. Calculate the amount after 4 years (48 months).

87. Math and Engineering. The formula $\sum_{n=0}^{\infty} \frac{x^n}{n!}$ can be used to approximate the function $y = e^x$. Estimate e^2 by using the sum of the first five terms and compare this result with the calculator value of e^2.

88. Home Prices. If the inflation rate is 3.5% per year and the average price of a home is $195,000, the average price of a home after n years is given by $A_n = 195,000(1.035)^n$. Find the average price of the home after 6 years.

89. Approximating Functions. Polynomials can be used to approximate transcendental functions such as $\ln(x)$ and e^x, which are found in advanced mathematics and engineering. For example,

$$\sum_{n=0}^{\infty} (-1)^n \frac{(x - 1)^{n+1}}{n + 1}$$ can be used to approximate $\ln(x)$, when

x is close to 1. Use the first five terms of the series to approximate $\ln(1.1)$ and compare with the value indicated by your calculator for $\ln(1.1)$.

90. Future Value of an Annuity. The future value of an ordinary annuity is given by the formula $FV = PMT[((1 + i)^n - 1)/i]$, where PMT = amount paid into the account at the end of each period, i = interest rate per period, and n = number of compounding periods. If you invest $5,000 at the end of each year for 5 years, you will have an accumulated value of FV as given in the above formula at the end of the nth year. Determine how much is in the account at the end of each year for the next 5 years if $i = 0.06$.

■ CATCH THE MISTAKE

In Exercises 91–94, explain the mistake that is made.

91. Simplify the ratio of factorials $\dfrac{(3!)(5!)}{6!}$.

Solution:

Express 6! in factored form.

$$\frac{(3!)(5!)}{(3!)(2!)}$$

Cancel the 3! in the numerator and denominator.

$$\frac{(5!)}{(2!)}$$

Write out the factorials.

$$\frac{5\cdot4\cdot3\cdot2\cdot1}{2\cdot1}$$

Simplify.

$$5\cdot4\cdot3 = 60$$

$\dfrac{(3!)(5!)}{(3!)(2!)} \neq 60$. What mistake was made?

92. Simplify the factorial expression $\dfrac{2n(2n-2)!}{(2n+2)!}$.

Solution:

Express factorials in factored form.

$$\frac{2n(2n-2)(2n-4)(2n-6)\cdots}{(2n+2)(2n)(2n-2)(2n-4)(2n-6)\cdots}$$

Cancel common terms.

$$\frac{1}{2n+2}$$

This is incorrect. What mistake was made?

93. Find the first four terms of the sequence defined by $a_n = (-1)^{n+1}n^2$.

Solution:

Find the $n = 1$ term. $a_1 = -1$

Find the $n = 2$ term. $a_2 = 4$

Find the $n = 3$ term. $a_3 = -9$

Find the $n = 4$ term. $a_4 = 16$

The sequence $-1, 4, -9, 16, \ldots$ is incorrect. What mistake was made?

94. Evaluate the series $\sum_{k=0}^{3}(-1)^{k+1}k^2$.

Solution:

Write out the sum.

$$\sum_{k=0}^{3}(-1)^{k+1}k^2 = -1 + 4 - 9$$

Simplify the sum.

$$\sum_{k=0}^{3}(-1)^{k+1}k^2 = -6$$

This is incorrect. What mistake was made?

■ CONCEPTUAL

In Exercises 95–100, determine whether each statement is true or false.

95. $\sum_{k=0}^{n}cx^k = c\sum_{k=0}^{n}x^k$

96. $\sum_{i=1}^{n}(a_i + b_i) = \sum_{i=1}^{n}a_i + \sum_{i=1}^{n}b_i$

97. $\sum_{k=1}^{n}a_kb_k = \sum_{k=1}^{n}a_k \cdot \sum_{k=1}^{n}b_k$

98. $(a!)(b!) = (ab)!$

99. $\sum_{k=1}^{\infty}a_k = \infty$

100. If $m! < n!$, then $m < n$.

■ CHALLENGE

101. Write the first four terms of the sequence defined by the recursion formula

$$a_1 = C \qquad a_n = a_{n-1} + D \qquad D \neq 0$$

102. Write the first four terms of the sequence defined by the recursion formula

$$a_1 = C \qquad a_n = Da_{n-1} \qquad D \neq 0$$

103. Fibonacci Sequence. An explicit formula for the nth term of the Fibonacci sequence is

$$F_n = \frac{(1 + \sqrt{5})^n - (1 - \sqrt{5})^n}{2^n\sqrt{5}}$$

Apply algebra (not your calculator) to find the first two terms of this sequence and verify that these are indeed the first two terms of the Fibonacci sequence.

104. Let $a_n = \sqrt{a_{n-1}}$ for $n \geq 2$ and $a_1 = 7$. Find the first five terms of this sequence and make a generalization for the nth term.

105. The sequence defined by $a_n = \left(1 + \dfrac{1}{n}\right)^n$ approaches the number e as n gets large. Use a graphing calculator to find a_{100}, a_{1000}, $a_{10,000}$, and keep increasing n until the terms in the sequence approach 2.7183.

106. The Fibonacci sequence is defined by $a_1 = 1$, $a_2 = 1$, and $a_n = a_{n-2} + a_{n-1}$ for $n \geq 3$. The ratio $\dfrac{a_{n+1}}{a_n}$ is an approximation of the golden ratio. The ratio approaches a constant ϕ (phi) as n gets large. Find the golden ratio using a graphing utility.

107. Use a graphing calculator "SUM" to sum $\sum_{k=0}^{5} \dfrac{2^k}{k!}$. Compare it with your answer to Exercise 61.

108. Use a graphing calculator "SUM" to sum $\sum_{k=0}^{5} \dfrac{(-1)^k}{k!}$. Compare it with your answer to Exercise 62.

■ **PREVIEW TO CALCULUS**

In calculus, we study the convergence of sequences. A sequence is *convergent* when its terms approach a limiting value. For example, $a_n = \dfrac{1}{n}$ is convergent because its terms approach zero. If the terms of a sequence satisfy $a_1 \leq a_2 \leq a_3 \leq \ldots \leq a_n \leq \ldots$, the sequence is *monotonic nondecreasing*. If $a_1 \geq a_2 \geq a_3 \geq \ldots \geq a_n \geq \ldots$, the sequence is *monotonic nonincreasing*.

In Exercises 109–112, classify each sequence as monotonic or not monotonic. If the sequence is monotonic, determine whether it is nondecreasing or nonincreasing.

109. $a_n = \dfrac{4n}{n+5}$

110. $a_n = \sin\left(\dfrac{n\pi}{4}\right)$

111. $a_n = \dfrac{2 + (-1)^n}{n+4}$

112. $a_n = \dfrac{3n^2}{5n^2 + 1}$

SECTION 10.2 ARITHMETIC SEQUENCES AND SERIES

SKILLS OBJECTIVES

- Recognize an arithmetic sequence.
- Find the general, or *n*th, term of an arithmetic sequence.
- Evaluate a finite arithmetic series.
- Use arithmetic sequences and series to model real-world problems.

CONCEPTUAL OBJECTIVE

- Understand the difference between an arithmetic sequence and an arithmetic series.

Arithmetic Sequences

The word *arithmetic* (with emphasis on the third syllable) often implies adding or subtracting of numbers. *Arithmetic sequences* are sequences whose terms are found by adding a constant to each previous term. The sequence 1, 3, 5, 7, 9, . . . is arithmetic because each successive term is found by adding 2 to the previous term.

DEFINITION **Arithmetic Sequences**

A sequence is **arithmetic** if the difference of any two consecutive terms is constant: $a_{n+1} - a_n = d$, where the number d is called the **common difference**. Each term in the sequence is found by adding the same real number d to the previous term, so that $a_{n+1} = a_n + d$.

EXAMPLE 1 **Identifying the Common Difference in Arithmetic Sequences**

Determine whether each sequence is arithmetic. If so, find the common difference for each of the arithmetic sequences.

a. $5, 9, 13, 17, \ldots$ **b.** $18, 9, 0, -9, \ldots$ **c.** $\frac{1}{2}, \frac{5}{4}, 2, \frac{11}{4}, \ldots$

Solution (a):

Label the terms.

$$a_1 = 5, a_2 = 9, a_3 = 13, a_4 = 17, \ldots$$

Find the difference $d = a_{n+1} - a_n$.

$$d = a_2 - a_1 = 9 - 5 = \boxed{4}$$

Check that the difference of the next two successive pairs of terms is also 4.

$$d = a_3 - a_2 = 13 - 9 = 4$$
$$d = a_4 - a_3 = 17 - 13 = 4$$

There is a common difference of $\boxed{4}$. Therefore, this sequence is arithmetic and each successive term is found by adding 4 to the previous term.

Solution (b):

Label the terms.

$$a_1 = 18, a_2 = 9, a_3 = 0, a_4 = -9, \ldots$$

Find the difference $d = a_{n+1} - a_n$.

$$d = a_2 - a_1 = 9 - 18 = \boxed{-9}$$

Check that the difference of the next two successive pairs of terms is also -9.

$$d = a_3 - a_2 = 0 - 9 = -9$$
$$d = a_4 - a_3 = -9 - 0 = -9$$

There is a common difference of $\boxed{-9}$. Therefore, this sequence is arithmetic and each successive term is found by subtracting 9 from (i.e., adding -9 to) the previous term.

Solution (c):

Label the terms.

$$a_1 = \frac{1}{2}, a_2 = \frac{5}{4}, a_3 = 2, a_4 = \frac{11}{4}, \ldots$$

Find the difference $d = a_{n+1} - a_n$.

$$d = a_2 - a_1 = \frac{5}{4} - \frac{1}{2} = \boxed{\frac{3}{4}}$$

Check that the difference of the next two successive pairs of terms is also $\frac{3}{4}$.

$$d = a_3 - a_2 = 2 - \frac{5}{4} = \frac{3}{4}$$

$$d = a_4 - a_3 = \frac{11}{4} - 2 = \frac{3}{4}$$

There is a common difference of $\boxed{\frac{3}{4}}$. Therefore, this sequence is arithmetic and each successive term is found by adding $\frac{3}{4}$ to the previous term.

■ **Answer: a.** -5 **b.** $\frac{2}{3}$

■ **YOUR TURN** Find the common difference for each of the arithmetic sequences.

a. $7, 2, -3, -8, \ldots$ **b.** $1, \frac{5}{3}, \frac{7}{3}, 3, \ldots$

The General (nth) Term of an Arithmetic Sequence

To find a formula for the general, or nth, term of an arithmetic sequence, write out the first several terms and look for a pattern.

First term, $n = 1$. a_1

Second term, $n = 2$. $a_2 = a_1 + d$

Third term, $n = 3$. $a_3 = a_2 + d = (a_1 + d) + d = a_1 + 2d$

Fourth term, $n = 4$. $a_4 = a_3 + d = (a_1 + 2d) + d = a_1 + 3d$

In general, the nth term is given by $a_n = a_1 + (n - 1)d$.

THE nTH TERM OF AN ARITHMETIC SEQUENCE

The **nth term** of an arithmetic sequence with common difference d is given by

$$a_n = a_1 + (n - 1)d \qquad \text{for } n \geq 1$$

Technology Tip

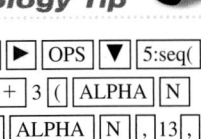

2nd | LIST | ► | OPS | ▼ | 5:seq(
ENTER | 2 | + | 3 | ((| ALPHA | N
− | 1 |)) | , | ALPHA | N | , | 13 | ,
13 | , | 1 |) | ENTER

```
seq(2+3(N−1),N,1
3,13,1)
            {38}
```

EXAMPLE 2   Finding the nth Term of an Arithmetic Sequence

Find the 13th term of the arithmetic sequence 2, 5, 8, 11, ...

Solution:

Identify the common difference. $d = 5 - 2 = 3$

Identify the first ($n = 1$) term. $a_1 = 2$

Substitute $a_1 = 2$ and $d = 3$ into $a_n = a_1 + (n - 1)d$. $a_n = 2 + 3(n - 1)$

Substitute $n = 13$ into $a_n = 2 + 3(n - 1)$. $a_{13} = 2 + 3(13 - 1) = \boxed{38}$

■ **Answer:** 66

■ **YOUR TURN**   Find the 10th term of the arithmetic sequence 3, 10, 17, 24, ...

EXAMPLE 3   Finding the Arithmetic Sequence

The 4th term of an arithmetic sequence is 16, and the 21st term is 67. Find a_1 and d and construct the sequence.

Solution:

Write the 4th and 21st terms. $a_4 = 16$ and $a_{21} = 67$

Adding d 17 times to a_4 results in a_{21}. $a_{21} = a_4 + 17d$

Substitute $a_4 = 16$ and $a_{21} = 67$. $67 = 16 + 17d$

Solve for d. $\boxed{d = 3}$

Substitute $d = 3$ into $a_n = a_1 + (n - 1)d$. $a_n = a_1 + 3(n - 1)$

Let $a_4 = 16$. $16 = a_1 + 3(4 - 1)$

Solve for a_1. $\boxed{a_1 = 7}$

The arithmetic sequence that starts at 7 and has a common difference of 3 is $\boxed{7, 10, 13, 16, \ldots}$.

■ **Answer:** 2, 6, 10, 14, . . .

■ **YOUR TURN**   Construct the arithmetic sequence whose 7th term is 26 and whose 13th term is 50.

The Sum of an Arithmetic Sequence

What is the sum of the first 100 counting numbers

$$1 + 2 + 3 + 4 + \cdots + 99 + 100 = \, ?$$

If we write this sum twice (one in ascending order and one in descending order) and add, we get 100 pairs of 101.

$$\begin{array}{rrrrrr}
1 + & 2 + & 3 + & 4 + \cdots + & 99 + & 100 \\
100 + & 99 + & 98 + & 97 + \cdots + & 2 + & 1 \\
\hline
101 + & 101 + & 101 + & 101 + \cdots + & 101 + & 101 = 100\,(101)
\end{array}$$

Since we added twice the sum, we divide by 2.

$$1 + 2 + 3 + 4 + \cdots + 99 + 100 = \frac{(101)(100)}{2} = 5050$$

Now, let us develop the sum of a general arithmetic series.

The sum of the first n terms of an arithmetic sequence is called the **nth partial sum**, or **finite arithmetic series**, and is denoted by S_n. An arithmetic sequence can be found by starting at the first term and adding the common difference to each successive term, and so the nth partial sum, or finite series, can be found the same way, but terminating the sum at the nth term:

$$S_n = a_1 + a_2 + a_3 + a_4 + \cdots$$
$$S_n = a_1 + (a_1 + d) + (a_1 + 2d) + (a_1 + 3d) + \cdots + (a_n)$$

Similarly, we can start with the nth term and find terms going backward by subtracting the common difference until we arrive at the first term:

$$S_n = a_n + a_{n-1} + a_{n-2} + a_{n-3} + \cdots$$
$$S_n = a_n + (a_n - d) + (a_n - 2d) + (a_n - 3d) + \cdots + (a_1)$$

Add these two representations of the nth partial sum. Notice that the d terms are eliminated:

$$\begin{array}{l}
S_n = a_1 + (a_1 + d) + (a_1 + 2d) + (a_1 + 3d) + \cdots + (a_n) \\
S_n = a_n + (a_n - d) + (a_n - 2d) + (a_n - 3d) + \cdots + (a_1) \\
\hline
2S_n = \underbrace{(a_1 + a_n) + (a_1 + a_n) + (a_1 + a_n) + \cdots + (a_1 + a_n)}_{n(a_1 + a_n)}
\end{array}$$

$$2S_n = n(a_1 + a_n) \qquad \text{or} \qquad S_n = \frac{n}{2}(a_1 + a_n)$$

> **DEFINITION**   **Evaluating a Finite Arithmetic Series**
>
> The sum of the first n terms of an arithmetic sequence (nth partial sum), called a **finite arithmetic series**, is given by the formula
>
> $$S_n = \frac{n}{2}(a_1 + a_n) \quad n \geq 2$$

Study Tip

S_n can also be written as

$$S_n = \frac{n}{2}[2a_1 + (n - 1)d]$$

For an arithmetic sequence, let $a_n = a_1 + (n - 1)d$.

$$S_n = \frac{n}{2}[a_1 + a_1 + (n - 1)d]$$

$$= \frac{n}{2}[2a_1 + (n - 1)d] = na_1 + \frac{n(n - 1)d}{2}$$

EXAMPLE 4   Evaluating a Finite Arithmetic Series

Evaluate the finite arithmetic series $\sum_{k=1}^{100} k$.

Solution:

Expand the arithmetic series.

$$\sum_{k=1}^{100} k = 1 + 2 + 3 + \cdots + 99 + 100$$

This is the sum of an arithmetic sequence of numbers with a common difference of 1.

Identify the parameters of the arithmetic sequence.

$$a_1 = 1, \, a_n = 100, \text{ and } n = 100$$

Substitute these values into

$S_n = \dfrac{n}{2}(a_1 + a_n)$.

$$S_{100} = \dfrac{100}{2}(1 + 100)$$

Simplify.

$$\boxed{S_{100} = 5050}$$

The sum of the first 100 natural numbers is 5050.

■ **YOUR TURN** Evaluate the following finite arithmetic series:

     a. $\displaystyle\sum_{k=1}^{30} k$ **b.** $\displaystyle\sum_{k=1}^{20} (2k + 1)$

EXAMPLE 5   Finding the *n*th Partial Sum of an Arithmetic Sequence

Find the sum of the first 20 terms of the arithmetic sequence 3, 8, 13, 18, 23, ...

Solution:

Recall the partial sum formula.

$$S_n = \dfrac{n}{2}(a_1 + a_n)$$

Find the 20th partial sum of this arithmetic sequence.

$$S_{20} = \dfrac{20}{2}(a_1 + a_{20})$$

Recall that the general *n*th term of an arithmetic sequence is given by

$$a_n = a_1 + (n - 1)d$$

Note that the first term of the arithmetic sequence is 3.

$$a_1 = 3$$

This is an arithmetic sequence with a common difference of 5.

$$d = 5$$

Substitute $a_1 = 3$ and $d = 5$ into $a_n = a_1 + (n - 1)d$.

$$a_n = 3 + (n - 1)5$$

Substitute $n = 20$ to find the 20th term.

$$a_{20} = 3 + (20 - 1)5 = 98$$

Substitute $a_1 = 3$ and $a_{20} = 98$ into the partial sum.

$$\boxed{S_{20} = 10(3 + 98) = 1010}$$

The sum of the first 20 terms of this arithmetic sequence is 1010.

■ **YOUR TURN** Find the sum of the first 25 terms of the arithmetic sequence 2, 6, 10, 14, 18, ...

Applications

EXAMPLE 6 Marching Band Formation

Suppose a band has 18 members in the first row, 22 members in the second row, and 26 members in the third row and continues with that pattern for a total of nine rows. How many marchers are there all together?

UC Berkeley marching band

David Young-Wolff/PhotoEdit

Solution:

The number of members in each row forms an arithmetic sequence with a common difference of 4, and the first row has 18 members.

$a_1 = 18 \quad d = 4$

Calculate the nth term of the sequence $a_n = a_1 + (n - 1)d$.

$a_n = 18 + (n - 1)4$

Find the 9th term, $n = 9$.

$a_9 = 18 + (9 - 1)4 = 50$

Calculate the sum $S_n = \dfrac{n}{2}(a_1 + a_n)$ of the nine rows.

$S_9 = \dfrac{9}{2}(a_1 + a_9)$

$= \dfrac{9}{2}(18 + 50)$

$= \dfrac{9}{2}(68)$

$= \boxed{306}$

There are 306 members in the marching band.

■ **Answer:** 328

■ **YOUR TURN** Suppose a bed of tulips is arranged in a garden so that there are 20 tulips in the first row, 26 tulips in the second row, and 32 tulips in the third row and the rows continue with that pattern for a total of 8 rows. How many tulips are there all together?

SECTION
10.2 SUMMARY

In this section, arithmetic sequences were defined as sequences of which each successive term is found by adding the same constant d to the previous term. Formulas were developed for the general, or nth, term of an arithmetic sequence, and for the nth partial sum of an arithmetic sequence, also called a finite arithmetic series.

$$a_n = a_1 + (n - 1)d \qquad n \geq 1$$

$$S_n = \frac{n}{2}(a_1 + a_n) = na_1 + \frac{n(n - 1)}{2}d$$

SECTION
10.2 EXERCISES

▪ SKILLS

In Exercises 1–10, determine whether each sequence is arithmetic. If it is, find the common difference.

1. $2, 5, 8, 11, 14, \ldots$

2. $9, 6, 3, 0, -3, -6, \ldots$

3. $1^2 + 2^2 + 3^2 + \cdots$

4. $1! + 2! + 3! + \cdots$

5. $3.33, 3.30, 3.27, 3.24, \ldots$

6. $0.7, 1.2, 1.7, 2.2, \ldots$

7. $4, \frac{14}{3}, \frac{16}{3}, 6, \ldots$

8. $2, \frac{7}{3}, \frac{8}{3}, 3, \ldots$

9. $10^1, 10^2, 10^3, 10^4, \ldots$

10. $120, 60, 30, 15, \ldots$

In Exercises 11–20, find the first four terms of each sequence described. Determine whether the sequence is arithmetic, and if so, find the common difference.

11. $a_n = -2n + 5$

12. $a_n = 3n - 10$

13. $a_n = n^2$

14. $a_n = \frac{n^2}{n!}$

15. $a_n = 5n - 3$

16. $a_n = -4n + 5$

17. $a_n = 10(n - 1)$

18. $a_n = 8n - 4$

19. $a_n = (-1)^n n$

20. $a_n = (-1)^{n+1} 2n$

In Exercises 21–28, find the general, or nth, term of each arithmetic sequence given the first term and the common difference.

21. $a_1 = 11 \qquad d = 5$

22. $a_1 = 5 \qquad d = 11$

23. $a_1 = -4 \qquad d = 2$

24. $a_1 = 2 \qquad d = -4$

25. $a_1 = 0 \qquad d = \frac{2}{3}$

26. $a_1 = -1 \qquad d = -\frac{3}{4}$

27. $a_1 = 0 \qquad d = e$

28. $a_1 = 1.1 \qquad d = -0.3$

In Exercises 29–32, find the specified term for each arithmetic sequence given.

29. The 10th term of the sequence $7, 20, 33, 46, \ldots$

30. The 19th term of the sequence $7, 1, -5, -11, \ldots$

31. The 100th term of the sequence $9, 2, -5, -12, \ldots$

32. The 90th term of the sequence $13, 19, 25, 31, \ldots$

In Exercises 33–38, for each arithmetic sequence described, find a_1 and d and construct the sequence by stating the general, or nth, term.

33. The 5th term is 44 and the 17th term is 152.

34. The 9th term is -19 and the 21st term is -55.

35. The 7th term is -1 and the 17th term is -41.

36. The 8th term is 47 and the 21st term is 112.

37. The 4th term is 3 and the 22nd term is 15.

38. The 11th term is -3 and the 31st term is -13.

In Exercises 39–50, find each sum given.

39. $\displaystyle\sum_{k=1}^{23} 2k$ **40.** $\displaystyle\sum_{k=0}^{20} 5k$ **41.** $\displaystyle\sum_{n=1}^{30} (-2n+5)$ **42.** $\displaystyle\sum_{n=0}^{17} (3n-10)$ **43.** $\displaystyle\sum_{j=3}^{14} 0.5j$ **44.** $\displaystyle\sum_{j=1}^{33} \frac{j}{4}$

45. $2+7+12+17+\cdots+62$ **46.** $1-3-7-\cdots-75$ **47.** $4+7+10+\cdots+151$

48. $2+0-2-\cdots-56$ **49.** $\frac{1}{6}-\frac{1}{6}-\frac{1}{2}-\cdots-\frac{13}{2}$ **50.** $\frac{11}{12}+\frac{7}{6}+\frac{17}{12}+\cdots+\frac{14}{3}$

In Exercises 51–56, find the indicated partial sum of each arithmetic series.

51. The first 18 terms of $1+5+9+13+\cdots$

52. The first 21 terms of $2+5+8+11+\cdots$

53. The first 43 terms of $1+\frac{1}{2}+0-\frac{1}{2}-\cdots$

54. The first 37 terms of $3+\frac{3}{2}+0-\frac{3}{2}-\cdots$

55. The first 18 terms of $-9+1+11+21+31+\cdots$

56. The first 21 terms of $-2+8+18+28+\cdots$

■ **APPLICATIONS**

57. Comparing Salaries. Colin and Camden are twin brothers graduating with B.S. degrees in biology. Colin takes a job at the San Diego Zoo making $28,000 for his first year with a $1,500 raise per year every year after that. Camden accepts a job at Florida Fish and Wildlife making $25,000 with a guaranteed $2,000 raise per year. How much will each of the brothers have made in a total of 10 years?

58. Comparing Salaries. On graduating with a Ph.D. in optical sciences, Jasmine and Megan choose different career paths. Jasmine accepts a faculty position at the University of Arizona making $80,000 with a guaranteed $2,000 raise every year. Megan takes a job with the Boeing Corporation making $90,000 with a guaranteed $5,000 raise each year. Calculate how much each woman will have made after 15 years.

59. Theater Seating. You walk into the premiere of Brad Pitt's new movie, and the theater is packed, with almost every seat filled. You want to estimate the number of people in the theater. You quickly count to find that there are 22 seats in the front row, and there are 25 rows in the theater. Each row appears to have one more seat than the row in front of it. How many seats are in that theater?

60. Field of Tulips. Every spring the Skagit County Tulip Festival plants more than 100,000 bulbs. In honor of the Tri-Delta sorority that has sent 120 sisters from the University of Washington to volunteer for the festival, Skagit County has planted tulips in the shape of $\Delta\Delta\Delta$. In each of the triangles there are 20 rows of tulips, each row having one less than the row before. How many tulips are planted in each delta if there is one tulip in the first row?

61. World's Largest Champagne Fountain. From December 28 to 30, 1999, Luuk Broos, director of Maison Luuk-Chalet Fontain, constructed a 56-story champagne fountain at the Steigenberger Kurhaus Hotel, Scheveningen, Netherlands. The fountain consisted of 30,856 champagne glasses. Assuming there was one glass at the top and the number of glasses in each row forms an arithmetic sequence, how many were on the bottom row (story)? How many glasses less did each successive row (story) have? Assume each story is one row.

62. Stacking of Logs. If 25 logs are laid side by side on the ground, and 24 logs are placed on top of those, and 23 logs are placed on the 3rd row, and the pattern continues until there is a single log on the 25th row, how many logs are in the stack?

63. Falling Object. When a skydiver jumps out of an airplane, she falls approximately 16 feet in the 1st second, 48 feet during the 2nd second, 80 feet during the 3rd second, 112 feet during the 4th second, and 144 feet during the 5th second, and this pattern continues. If she deploys her parachute after 10 seconds have elapsed, how far will she have fallen during those 10 seconds?

64. Falling Object. If a penny is dropped out of a plane, it falls approximately 4.9 meters during the 1st second, 14.7 meters during the 2nd second, 24.5 meters during the 3rd second, and 34.3 meters during the 4th second. Assuming this pattern continues, how many meters will the penny have fallen after 10 seconds?

65. Grocery Store. A grocer has a triangular display of oranges in a window. There are 20 oranges in the bottom row and the number of oranges decreases by one in each row above this row. How many oranges are in the display?

66. Salary. Suppose your salary is $45,000 and you receive a $1,500 raise for each year you work for 35 years.
a. How much will you earn during the 35th year?
b. What is the total amount you earned over your 35-year career?

67. Theater Seating. At a theater, seats are arranged in a triangular pattern of rows with each succeeding row having one more seat than the previous row. You count the number of seats in the fourth row and determine that there are 26 seats.
a. How many seats are in the first row?
b. Now, suppose there are 30 rows of seats. How many total seats are there in the theater?

68. Mathematics. Find the exact sum of
$$\frac{1}{e}+\frac{3}{e}+\frac{5}{e}+\cdots+\frac{23}{e}$$

CATCH THE MISTAKE

In Exercises 69–72, explain the mistake that is made.

69. Find the general, or nth, term of the arithmetic sequence $3, 4, 5, 6, 7, \ldots$.

Solution:

The common difference of this sequence is 1.	$d = 1$
The first term is 3.	$a_1 = 3$
The general term is $a_n = a_1 + nd$.	$a_n = 3 + n$

This is incorrect. What mistake was made?

70. Find the general, or nth, term of the arithmetic sequence $10, 8, 6, \ldots$.

Solution:

The common difference of this sequence is 2.	$d = 2$
The first term is 10.	$a_1 = 10$
The general term is $a_n = a_1 + (n-1)d$.	$a_n = 10 + 2(n-1)$

This is incorrect. What mistake was made?

71. Find the sum $\sum_{k=0}^{10}(2n + 1)$.

Solution:

The sum is given by $S_n = \dfrac{n}{2}(a_1 + a_n)$, where $n = 10$.

Identify the 1st and 10th terms. $a_1 = 1, a_{10} = 21$

Substitute $a_1 = 1$, $a_{10} = 21$, and $n = 10$ into $S_n = \dfrac{n}{2}(a_1 + a_n)$. $S_{10} = \dfrac{10}{2}(1 + 21) = 110$

This is incorrect. What mistake was made?

72. Find the sum $3 + 9 + 15 + 21 + 27 + 33 + \cdots + 87$.

Solution:

This is an arithmetic sequence with common difference of 6.	$d = 6$
The general term is given by $a_n = a_1 + (n-1)d$.	$a_n = 3 + (n-1)6$
87 is the 15th term of the series.	$a_{15} = 3 + (15-1)6 = 87$
The sum of the series is $S_n = \dfrac{n}{2}(a_n - a_1)$.	$S_{15} = \dfrac{15}{2}(87 - 3) = 630$

This is incorrect. What mistake was made?

CONCEPTUAL

In Exercises 73–76, determine whether each statement is true or false.

73. An arithmetic sequence and a finite arithmetic series are the same.

74. The sum of all infinite and finite arithmetic series can always be found.

75. An alternating sequence cannot be an arithmetic sequence.

76. The common difference of an arithmetic sequence is always positive.

CHALLENGE

77. Find the sum $a + (a + b) + (a + 2b) + \cdots + (a + nb)$.

78. Find the sum $\sum_{k=-29}^{30} \ln e^k$.

79. The wave number λ (reciprocal of wave length) of certain light waves in the spectrum of light emitted by hydrogen is given by $\lambda = R\left(\dfrac{1}{k^2} - \dfrac{1}{n^2}\right)$, $n > k$, where $R = 109{,}678$. A series of lines is given by holding k constant and varying the value of n. Suppose $k = 2$ and $n = 3, 4, 5, \ldots$. Find what value the wave number of the series approaches as n increases.

80. In a certain arithmetic sequence $a_1 = -4$ and $d = 6$. If $S_n = 570$, find the value of n.

TECHNOLOGY

81. Use a graphing calculator "SUM" to sum the natural numbers from 1 to 100.

82. Use a graphing calculator to sum the even natural numbers from 1 to 100.

83. Use a graphing calculator to sum the odd natural numbers from 1 to 100.

84. Use a graphing calculator to find $\sum_{n=1}^{30}(-2n + 5)$. Compare it with your answer to Exercise 41.

85. Use a graphing calculator to find $\sum_{n=1}^{100}[-59 + 5(n-1)]$.

86. Use a graphing calculator to find $\sum_{n=1}^{200}\left[-18 + \frac{4}{5}(n-1)\right]$.

■ **PREVIEW TO CALCULUS**

In calculus, when estimating certain integrals, we use sums of the form $\sum_{i=1}^{n} f(x_i)\Delta x$, where f is a function and Δx is a constant.

In Exercises 87–90, find the indicated sum.

87. $\sum_{i=1}^{100} f(x_i)\Delta x$, where $f(x_i) = 2i$ and $\Delta x = 0.1$

88. $\sum_{i=1}^{50} f(x_i)\Delta x$, where $f(x_i) = 4i - 2$ and $\Delta x = 0.01$

89. $\sum_{i=1}^{43} f(x_i)\Delta x$, where $f(x_i) = 6 + i$ and $\Delta x = 0.001$

90. $\sum_{i=1}^{85} f(x_i)\Delta x$, where $f(x_i) = 6 - 7i$ and $\Delta x = 0.2$

SECTION 10.3 GEOMETRIC SEQUENCES AND SERIES

SKILLS OBJECTIVES

■ Recognize a geometric sequence.
■ Find the general, or nth, term of a geometric sequence.
■ Evaluate a finite geometric series.
■ Evaluate an infinite geometric series, if it exists.
■ Use geometric sequences and series to model real-world problems.

CONCEPTUAL OBJECTIVES

■ Understand the difference between a geometric sequence and a geometric series.
■ Distinguish between an arithmetic sequence and a geometric sequence.
■ Understand why it is not possible to evaluate all infinite geometric series.

Geometric Sequences

In Section 10.2, we discussed *arithmetic* sequences, where successive terms had a *common difference*. In other words, each term was found by adding the same constant to the previous term. In this section, we discuss *geometric* sequences, where successive terms have a *common ratio*. In other words, each term is found by multiplying the previous term by the same constant. The sequence 4, 12, 36, 108, ... is geometric because each successive term is found by multiplying the previous term by 3.

DEFINITION **Geometric Sequences**

A sequence is **geometric** if each term in the sequence is found by multiplying the previous term by a number r, so that $a_{n+1} = r \cdot a_n$. Because $\dfrac{a_{n+1}}{a_n} = r$, the number r is called the **common ratio**.

EXAMPLE 1   **Identifying the Common Ratio in Geometric Sequences**

Find the common ratio for each of the geometric sequences.

a. $5, 20, 80, 320, \ldots$ **b.** $1, -\frac{1}{2}, \frac{1}{4}, -\frac{1}{8}, \ldots$ **c.** $\$5{,}000, \$5{,}500, \$6{,}050, \$6{,}655, \ldots$

Solution (a):

Label the terms.

$$a_1 = 5, a_2 = 20, a_3 = 80, a_4 = 320, \ldots$$

Find the ratio $r = \dfrac{a_{n+1}}{a_n}$.

$$r = \frac{a_2}{a_1} = \frac{20}{5} = 4$$

$$r = \frac{a_3}{a_2} = \frac{80}{20} = 4$$

$$r = \frac{a_4}{a_3} = \frac{320}{80} = 4$$

$$\boxed{\text{The common ratio is 4.}}$$

Solution (b):

Label the terms.

$$a_1 = 1, a_2 = -\frac{1}{2}, a_3 = \frac{1}{4}, a_4 = -\frac{1}{8}, \ldots$$

Find the ratio $r = \dfrac{a_{n+1}}{a_n}$.

$$r = \frac{a_2}{a_1} = \frac{-1/2}{1} = -\frac{1}{2}$$

$$r = \frac{a_3}{a_2} = \frac{1/4}{-1/2} = -\frac{1}{2}$$

$$r = \frac{a_4}{a_3} = \frac{-1/8}{1/4} = -\frac{1}{2}$$

$$\boxed{\text{The common ratio is } -\tfrac{1}{2}.}$$

Solution (c):

Label the terms.

$$a_1 = \$5{,}000, a_2 = \$5{,}500, a_3 = \$6{,}050, a_4 = \$6{,}655, \ldots$$

Find the ratio $r = \dfrac{a_{n+1}}{a_n}$.

$$r = \frac{a_2}{a_1} = \frac{\$5{,}500}{\$5{,}000} = 1.1$$

$$r = \frac{a_3}{a_2} = \frac{\$6{,}050}{\$5{,}500} = 1.1$$

$$r = \frac{a_4}{a_3} = \frac{\$6{,}655}{\$6{,}050} = 1.1$$

$$\boxed{\text{The common ratio is 1.1.}}$$

■ **YOUR TURN** Find the common ratio of each geometric series.

a. $1, -3, 9, -27, \ldots$ **b.** $320, 80, 20, 5, \ldots$

■ **Answer: a.** -3 **b.** $\frac{1}{4}$ or 0.25

The General (*n*th) Term of a Geometric Sequence

To find a formula for the general, or *n*th, term of a geometric sequence, write out the first several terms and look for a pattern.

WORDS	MATH
First term, $n = 1$.	a_1
Second term, $n = 2$.	$a_2 = a_1 \cdot r$
Third term, $n = 3$.	$a_3 = a_2 \cdot r = (a_1 \cdot r) \cdot r = a_1 \cdot r^2$
Fourth term, $n = 4$.	$a_4 = a_3 \cdot r = (a_1 \cdot r^2) \cdot r = a_1 \cdot r^3$

In general, the nth term is given by $a_n = a_1 \cdot r^{n-1}$.

THE nTH TERM OF A GEOMETRIC SEQUENCE

The **nth term** of a geometric sequence with common ratio r is given by

$$a_n = a_1 \cdot r^{n-1} \qquad \text{for } n \geq 1$$

Technology Tip

Use $\boxed{\text{seq}}$ to find the nth term of the sequence. To find the 7th term of the geometric sequence $a_n = 2 \cdot 5^{n-1}$, press

$\boxed{\text{2}^{\text{nd}}}\ \boxed{\text{LIST}}\ \boxed{\blacktriangleright}\ \boxed{\text{OPS}}\ \boxed{\blacktriangledown}\ \boxed{\text{5:seq(}}$

$\boxed{\text{ENTER}}\ \boxed{2}\ \boxed{\text{x}}\ \boxed{5}\ \boxed{\wedge}\ \boxed{(}\ \boxed{\text{ALPHA}}$

$\boxed{\text{N}}\ \boxed{-}\ \boxed{1}\ \boxed{)}\ \boxed{,}\ \boxed{\text{ALPHA}}\ \boxed{\text{N}}\ \boxed{,}$

$\boxed{7}\ \boxed{,}\ \boxed{7}\ \boxed{,}\ \boxed{1}\ \boxed{)}\ \boxed{\text{ENTER}}$.

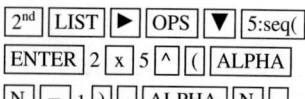

```
seq(2*5^(N-1),N,
7,7,1)
            {31250}
```

■ **Answer:** 49,152

EXAMPLE 2   Finding the nth Term of a Geometric Sequence

Find the 7th term of the sequence 2, 10, 50, 250, ...

Solution:

Identify the common ratio.

$$r = \frac{10}{2} = \frac{50}{10} = \frac{250}{50} = 5$$

Identify the first ($n = 1$) term.

$$a_1 = 2$$

Substitute $a_1 = 2$ and $r = 5$ into $a_n = a_1 \cdot r^{n-1}$.

$$a_n = 2 \cdot 5^{n-1}$$

Substitute $n = 7$ into $a_n = 2 \cdot 5^{n-1}$.

$$a_7 = 2 \cdot 5^{7-1} = 2 \cdot 5^6 = 31{,}250$$

> The 7th term of the geometric sequence is 31,250.

■ **YOUR TURN** Find the 8th term of the sequence 3, 12, 48, 192, ...

EXAMPLE 3   Finding the Geometric Sequence

Find the geometric sequence whose 5th term is 0.01 and whose common ratio is 0.1.

Solution:

Label the common ratio and 5th term.

$$a_5 = 0.01 \text{ and } r = 0.1$$

Substitute $a_5 = 0.01$, $n = 5$, and $r = 0.1$ into $a_n = a_1 \cdot r^{n-1}$.

$$0.01 = a_1 \cdot (0.1)^{5-1}$$

Solve for a_1.

$$a_1 = \frac{0.01}{(0.1)^4} = \frac{0.01}{0.0001} = 100$$

> The geometric sequence that starts at 100 and has a common ratio of 0.1 is 100, 10, 1, 0.1, 0.01, ...

■ **Answer:** 81, 27, 9, 3, 1, . . .

■ **YOUR TURN** Find the geometric sequence whose 4th term is 3 and whose common ratio is $\frac{1}{3}$.

Geometric Series

The sum of the terms of a geometric sequence is called a **geometric series**.

$$a_1 + a_1 \cdot r + a_1 \cdot r^2 + a_1 \cdot r^3 + \cdots$$

If we only sum the first n terms of a geometric sequence, the result is a **finite geometric series** given by

$$S_n = a_1 + a_1 \cdot r + a_1 \cdot r^2 + a_1 \cdot r^3 + \cdots + a_1 \cdot r^{n-1}$$

To develop a formula for this nth partial sum, we multiply the above equation by r:

$$r \cdot S_n = a_1 \cdot r + a_1 \cdot r^2 + a_1 \cdot r^3 + \cdots + a_1 \cdot r^{n-1} + a_1 \cdot r^n$$

Subtracting the second equation from the first equation, we find that all of the terms on the right side drop out except the *first* term in the first equation and the *last* term in the second equation:

$$
\begin{aligned}
S_n &= a_1 + a_1 \cdot r + a_1 \cdot r^2 + \cdots + a_1 r^{n-1} \\
-rS_n &= \qquad - a_1 \cdot r - a_1 \cdot r^2 - \cdots - a_1 r^{n-1} - a_1 r^n \\
\hline
S_n - rS_n &= a_1 \qquad\qquad\qquad\qquad\qquad\qquad\quad -a_1 r^n
\end{aligned}
$$

Factor the S_n out of the left side and the a_1 out of the right side:

$$S_n(1 - r) = a_1(1 - r^n)$$

Divide both sides by $(1 - r)$, assuming $r \neq 1$. The result is a general formula for the sum of a finite geometric series:

$$S_n = a_1 \frac{(1 - r^n)}{(1 - r)} \qquad r \neq 1$$

EVALUATING A FINITE GEOMETRIC SERIES

The sum of the first n terms of a geometric sequence, called a **finite geometric series**, is given by the formula

$$S_n = a_1 \frac{(1 - r^n)}{(1 - r)} \qquad r \neq 1$$

It is important to note that a finite geometric series can also be written in sigma (summation) notation:

$$S_n = \sum_{k=1}^{n} a_1 \cdot r^{k-1} = a_1 + a_1 \cdot r + a_1 \cdot r^2 + a_1 \cdot r^3 + \cdots + a_1 \cdot r^{n-1}$$

Study Tip

The underscript $k = 1$ applies only when the summation starts at the a_1 term. It is important to note which term is the starting term.

EXAMPLE 4   Evaluating a Finite Geometric Series

Evaluate each finite geometric series.

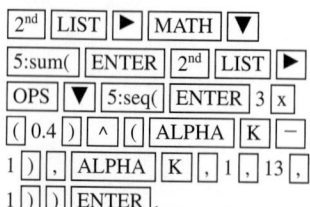

a. $\displaystyle\sum_{k=1}^{13} 3 \cdot (0.4)^{k-1}$

b. The first nine terms of the series $1 + 2 + 4 + 8 + 16 + 32 + 64 + \cdots$

Solution (a):

Identify a_1, n, and r. $\qquad a_1 = 3, n = 13,$ and $r = 0.4$

Substitute $a_1 = 3$, $n = 13$, and $r = 0.4$

into $S_n = a_1 \dfrac{(1 - r^n)}{(1 - r)}$. $\qquad S_{13} = 3\dfrac{\left(1 - 0.4^{13}\right)}{(1 - 0.4)}$

Simplify. $\qquad \boxed{S_{13} \approx 4.99997}$

Solution (b):

Identify the first term and common ratio. $\qquad a_1 = 1$ and $r = 2$

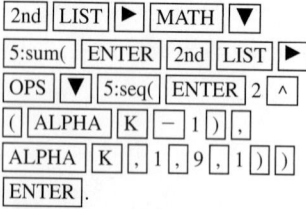
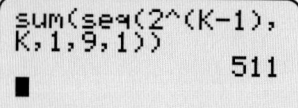
Substitute $a_1 = 1$ and $r = 2$ into $S_n = a_1\dfrac{(1 - r^n)}{(1 - r)}$. $\qquad S_n = \dfrac{(1 - 2^n)}{(1 - 2)}$

To sum the first nine terms, let $n = 9$. $\qquad S_9 = \dfrac{\left(1 - 2^9\right)}{(1 - 2)}$

Simplify. $\qquad \boxed{S_9 = 511}$

The sum of an infinite geometric sequence is called an **infinite geometric series**. Some infinite geometric series *converge* (yield a finite sum) and some *diverge* (do not have a finite sum). For example,

$$\frac{1}{2} + \frac{1}{4} + \frac{1}{8} + \frac{1}{16} + \frac{1}{32} + \cdots + \frac{1}{2^n} + \cdots = 1 \; \text{(converges)}$$

$$2 + 4 + 8 + 16 + 32 + \cdots + 2^n + \cdots \; \text{(diverges)}$$

For infinite geometric series that converge, the partial sum S_n approaches a single number as n gets large. The formula used to evaluate a finite geometric series

$$S_n = a_1 \frac{(1 - r^n)}{(1 - r)}$$

can be extended to an infinite geometric series for certain values of r. If $|r| < 1$, then when r is raised to a power, it continues to get smaller, approaching 0. For those values of r, the infinite geometric series converges to a finite sum.

$$\text{Let } n \to \infty; \text{ then } a_1\frac{(1 - r^n)}{(1 - r)} \to a_1\frac{(1 - 0)}{(1 - r)} = \frac{a_1}{1 - r}, \text{ if } |r| < 1.$$

EVALUATING AN INFINITE GEOMETRIC SERIES

The **sum of an infinite geometric series** is given by the formula

$$\sum_{n=1}^{\infty} a_1 r^{n-1} = \sum_{n=0}^{\infty} a_1 \cdot r^n = a_1 \frac{1}{(1 - r)} \qquad |r| < 1$$

EXAMPLE 5   Determining Whether the Sum of an Infinite Series Exists

Determine whether the sum exists for each of the geometric series.

a. $3 + 15 + 75 + 375 + \cdots$ **b.** $8 + 4 + 2 + 1 + \frac{1}{2} + \frac{1}{4} + \frac{1}{8} + \cdots$

Solution (a):

Identify the common ratio. $r = 5$

Since 5 is greater than 1, the $\boxed{\text{sum does not exist}}$. $|r| = 5 > 1$

Solution (b):

Identify the common ratio. $r = \dfrac{1}{2}$

Since $\frac{1}{2}$ is less than 1, the $\boxed{\text{sum exists}}$. $|r| = \dfrac{1}{2} < 1$

■ **Answer: a.** yes   **b.** no

■ **YOUR TURN** Determine whether the sum exists for each of the geometric series.

a. $81 + 9 + 1 + \frac{1}{9} + \cdots$ **b.** $1 + 5 + 25 + 125 + \cdots$

Do you expect $\frac{1}{4} + \frac{1}{12} + \frac{1}{36} + \frac{1}{64} + \cdots$ and $\frac{1}{4} - \frac{1}{12} + \frac{1}{36} - \frac{1}{64} + \cdots$ to sum to the same number? The answer is no, because the second series is an alternating series and terms are both added and subtracted. Hence, we would expect the second series to sum to a smaller number than the first series sums to.

EXAMPLE 6   Evaluating an Infinite Geometric Series

Evaluate each infinite geometric series.

a. $1 + \frac{1}{3} + \frac{1}{9} + \frac{1}{27} + \cdots$ **b.** $1 - \frac{1}{3} + \frac{1}{9} - \frac{1}{27} + \cdots$

Solution (a):

Identify the first term and the common ratio. $a_1 = 1 \qquad r = \dfrac{1}{3}$

Since $|r| = \frac{1}{3} < 1$, the sum of the series exists.

Substitute $a_1 = 1$ and $r = \frac{1}{3}$ into

$$\sum_{n=0}^{\infty} a_1 \cdot r^n = \frac{a_1}{(1 - r)}.$$

$$\frac{1}{1 - \dfrac{1}{3}}$$

Simplify.

$$= \frac{1}{\dfrac{2}{3}} = \frac{3}{2}$$

$$\boxed{1 + \frac{1}{3} + \frac{1}{9} + \frac{1}{27} + \cdots = \frac{3}{2}}$$

Solution (b):

Identify the first term and the common ratio.

$$a_1 = 1, r = -\frac{1}{3}$$

Since $|r| = \left|-\frac{1}{3}\right| = \frac{1}{3} < 1$, the sum of the series exists.

Substitute $a_1 = 1$ and $r = -\frac{1}{3}$ into $\sum_{n=0}^{\infty} a_1 \cdot r^n = \frac{a_1}{(1-r)}$.

Simplify.

$$= \frac{1}{1-(-1/3)} = \frac{1}{1+(1/3)} = \frac{1}{4/3} = \frac{3}{4}$$

$$\boxed{1 - \frac{1}{3} + \frac{1}{9} - \frac{1}{27} + \cdots = \frac{3}{4}}$$

Notice that the alternating series summed to $\frac{3}{4}$, whereas the positive series summed to $\frac{3}{2}$.

■ **Answer: a.** $\frac{3}{8}$ **b.** $\frac{3}{16}$

■ **YOUR TURN** Find the sum of each infinite geometric series.

a. $\frac{1}{4} + \frac{1}{12} + \frac{1}{36} + \frac{1}{64} + \cdots$ **b.** $\frac{1}{4} - \frac{1}{12} + \frac{1}{36} - \frac{1}{64} + \cdots$

It is important to note the restriction on the common ratio r. The absolute value of the common ratio has to be strictly less than 1 for an infinite geometric series to converge. Otherwise, the infinite geometric series diverges.

EXAMPLE 7 Evaluating an Infinite Geometric Series

Evaluate the infinite geometric series, if possible.

a. $\sum_{n=0}^{\infty} 2\left(-\frac{1}{4}\right)^n$ **b.** $\sum_{n=1}^{\infty} 3 \cdot 2^{n-1}$

Solution (a):

Identify a_1 and r.

$$\sum_{n=0}^{\infty} 2\left(-\frac{1}{4}\right)^n = \underset{a_1}{2} - \frac{1}{2} + \frac{1}{8} - \frac{1}{32} + \frac{1}{128} - \cdots$$

with $r = -\frac{1}{4}$.

Since $|r| = \left|-\frac{1}{4}\right| = \frac{1}{4} < 1$, the infinite geometric series converges.

$$\sum_{n=0}^{\infty} a_1 \cdot r^n = \frac{a_1}{(1-r)}$$

Let $a_1 = 2$ and $r = -\frac{1}{4}$.

$$= \frac{2}{\left[1-(-1/4)\right]}$$

Simplify.

$$= \frac{2}{1+(1/4)} = \frac{2}{5/4} = \frac{8}{5}$$

This infinite geometric series converges.

$$\boxed{\sum_{n=0}^{\infty} 2\left(-\frac{1}{4}\right)^n = \frac{8}{5}}$$

Solution (b):

Identify a_1 and r.

$$\sum_{n=1}^{\infty} 3 \cdot (2)^{n-1} = \underset{a_1}{3} + 6 + 12 + 24 + 48 + \cdots$$

with $r = 2$.

$$\boxed{\text{Since } r = 2 > 1, \text{ this infinite geometric series diverges.}}$$

Applications

Suppose you are given a job offer with a guaranteed percentage raise per year. What will your annual salary be 10 years from now? That answer can be obtained using a geometric sequence. Suppose you want to make voluntary contributions to a retirement account directly debited from your paycheck every month. Suppose the account earns a fixed percentage rate: How much will you have in 30 years if you deposit $50 a month? What is the difference in the total you will have in 30 years if you deposit $100 a month instead? These important questions about your personal finances can be answered using geometric sequences and series.

EXAMPLE 8 Future Salary: Geometric Sequence

Suppose you are offered a job as an event planner for the PGA Tour. The starting salary is $45,000, and employees are given a 5% raise per year. What will your annual salary be during the 10th year with the PGA Tour?

Solution:

Every year the salary is 5% more than the previous year.

Label the year 1 salary.	$a_1 = 45,000$
Calculate the year 2 salary.	$a_2 = 1.05 \cdot a_1$
Calculate the year 3 salary.	$a_3 = 1.05 \cdot a_2$
	$= 1.05(1.05 \cdot a_1) = (1.05)^2 a_1$
Calculate the year 4 salary.	$a_4 = 1.05 \cdot a_3$
	$= 1.05(1.05)^2 a_1 = (1.05)^3 a_1$
Identify the year n salary.	$a_n = 1.05^{n-1} a_1$
Substitute $n = 10$ and $a_1 = 45,000$.	$a_{10} = (1.05)^9 \cdot 45,000$
Simplify.	$a_{10} \approx 69,809.77$

During your 10th year with the company, your salary will be $69,809.77.

■ **YOUR TURN** Suppose you are offered a job with AT&T at $37,000 per year with a guaranteed raise of 4% every year. What will your annual salary be after 15 years with the company?

■ **Answer:** $64,072.03

EXAMPLE 9 Savings Growth: Geometric Series

Karen has maintained acrylic nails by paying for them with money earned from a part-time job. After hearing a lecture from her economics professor on the importance of investing early in life, she decides to remove the acrylic nails, which cost $50 per month, and do her own manicures. She has that $50 automatically debited from her checking account on the first of every month and put into a money market account that earns 3% interest compounded monthly. What will the balance be in the money market account exactly 2 years from the day of her initial $50 deposit?

Technology Tip

Use a calculator to find
$$S_{24} = 50(1.0025)\frac{(1 - 1.0025^{24})}{(1 - 1.0025)}.$$

Scientific calculators:

Press	Display
50 × 1.0025 ×	1238.23
(1 − 1.0025	
x^y 24) ÷ (	
1 − 1.0025) =	

Graphing calculators:

50 × 1.0025 × ((1 − 1.0025
^ 24) ÷ ((1 − 1.0025)

ENTER

```
50*1.0025*(1-1.0
025^24)/(1-1.002
5)
        1238.228737
```

■ **Answer:** $5,105.85

Solution:

Recall the compound interest formula.

$$A = P\left(1 + \frac{r}{n}\right)^{nt}$$

Substitute $r = 0.03$ and $n = 12$ into the compound interest formula.

$$A = P\left(1 + \frac{0.03}{12}\right)^{12t}$$

$$= P(1.0025)^{12t}$$

Let $t = \frac{n}{12}$, where n is the number of months of the investment.

$$A_n = P(1.0025)^n$$

The first deposit of $50 will gain interest for 24 months. $A_{24} = 50(1.0025)^{24}$

The second deposit of $50 will gain interest for 23 months. $A_{23} = 50(1.0025)^{23}$

The third deposit of $50 will gain interest for 22 months. $A_{22} = 50(1.0025)^{22}$

The last deposit of $50 will gain interest for 1 month. $A_1 = 50(1.0025)^1$

Sum the amounts accrued from the 24 deposits.

$$A_1 + A_2 + \cdots + A_{24} = 50(1.0025) + 50(1.0025)^2 + 50(1.0025)^3 + \cdots + 50(1.0025)^{24}$$

Identify the first term and common ratio. $a_1 = 50(1.0025)$ and $r = 1.0025$

Sum the first n terms of a geometric series. $S_n = a_1\frac{(1 - r^n)}{(1 - r)}$

Substitute $n = 24$, $a_1 = 50(1.0025)$, and $r = 1.0025$.

$$S_{24} = 50(1.0025)\frac{(1 - 1.0025^{24})}{(1 - 1.0025)}$$

Simplify. $S_{24} \approx 1238.23$

Karen will have $1,238.23 saved in her money market account in 2 years.

■ **YOUR TURN** Repeat Example 9 with Karen putting $100 (instead of $50) in the same money market account. Assume she does this for 4 years (instead of 2 years).

SECTION 10.3 SUMMARY

In this section, we discussed geometric sequences, in which each successive term is found by multiplying the previous term by a constant, so that $a_{n+1} = r \cdot a_n$. That constant, r, is called the common ratio. The nth term of a geometric sequence is given by $a_n = a_1 r^{n-1}, n \geq 1$ or $a_{n+1} = a_1 r^n, n \geq 0$. The sum of the terms of a geometric sequence is called a geometric series. Finite geometric series converge to a number. Infinite geometric series converge to a number if the absolute value of the common ratio is less than 1. If the absolute value of the common ratio is greater than or equal to 1, the infinite geometric series diverges and the sum does not exist. Many real-world applications involve geometric sequences and series, such as growth of salaries and annuities through percentage increases.

Finite Geometric Series: $\sum_{k=1}^{n} a_1 r^{k-1} = a_1\frac{(1 - r^n)}{(1 - r)}$ $r \neq 1$

Infinite Geometric Series: $\sum_{k=1}^{\infty} a_1 r^{k-1} = a_1\frac{1}{(1 - r)}$ $|r| < 1$

SECTION
10.3 EXERCISES

■ SKILLS

In Exercises 1–8, determine whether each sequence is geometric. If it is, find the common ratio.

1. $1, 3, 9, 27, \ldots$

2. $2, 4, 8, 16, \ldots$

3. $1, 4, 9, 16, 25, \ldots$

4. $1, \frac{1}{4}, \frac{1}{9}, \frac{1}{16}, \ldots$

5. $8, 4, 2, 1, \ldots$

6. $8, -4, 2, -1, \ldots$

7. $800, 1360, 2312, 3930.4, \ldots$

8. $7, 15.4, 33.88, 74.536, \ldots$

In Exercises 9–16, write the first five terms of each geometric series.

9. $a_1 = 6 \qquad r = 3$

10. $a_1 = 17 \qquad r = 2$

11. $a_1 = 1 \qquad r = -4$

12. $a_1 = -3 \qquad r = -2$

13. $a_1 = 10,000 \qquad r = 1.06$

14. $a_1 = 10,000 \qquad r = 0.8$

15. $a_1 = \frac{2}{3} \qquad r = \frac{1}{2}$

16. $a_1 = \frac{1}{10} \qquad r = -\frac{1}{5}$

In Exercises 17–24, write the formula for the nth term of each geometric series.

17. $a_1 = 5 \qquad r = 2$

18. $a_1 = 12 \qquad r = 3$

19. $a_1 = 1 \qquad r = -3$

20. $a_1 = -4 \qquad r = -2$

21. $a_1 = 1000 \qquad r = 1.07$

22. $a_1 = 1000 \qquad r = 0.5$

23. $a_1 = \frac{16}{3} \qquad r = -\frac{1}{4}$

24. $a_1 = \frac{1}{200} \qquad r = 5$

In Exercises 25–30, find the indicated term of each geometric sequence.

25. 7th term of the sequence $-2, 4, -8, 16, \ldots$

26. 10th term of the sequence $1, -5, 25, -225, \ldots$

27. 13th term of the sequence $\frac{1}{3}, \frac{2}{3}, \frac{4}{3}, \frac{8}{3}, \ldots$

28. 9th term of the sequence $100, 20, 4, 0.8, \ldots$

29. 15th term of the sequence $1000, 50, 2.5, 0.125, \ldots$

30. 8th term of the sequence $1000, -800, 640, -512, \ldots$

In Exercises 31–40, find the sum of each finite geometric series.

31. $\dfrac{1}{3} + \dfrac{2}{3} + \dfrac{2^2}{3} + \cdots + \dfrac{2^{12}}{3}$

32. $1 + \dfrac{1}{3} + \dfrac{1}{3^2} + \dfrac{1}{3^3} + \cdots + \dfrac{1}{3^{10}}$

33. $2 + 6 + 18 + 54 + \cdots + 2(3^9)$

34. $1 + 4 + 16 + 64 + \cdots + 4^9$

35. $\displaystyle\sum_{n=0}^{10} 2(0.1)^n$

36. $\displaystyle\sum_{n=0}^{11} 3(0.2)^n$

37. $\displaystyle\sum_{n=1}^{8} 2(3)^{n-1}$

38. $\displaystyle\sum_{n=13}^{9} \frac{2}{3}(5)^{n-1}$

39. $\displaystyle\sum_{k=0}^{13} 2^k$

40. $\displaystyle\sum_{k=0}^{13} \left(\frac{1}{2}\right)^k$

In Exercises 41–54, find the sum of each infinite geometric series, if possible.

41. $\displaystyle\sum_{n=0}^{\infty} \left(\frac{1}{2}\right)^n$

42. $\displaystyle\sum_{n=1}^{\infty} \left(\frac{1}{3}\right)^n$

43. $\displaystyle\sum_{n=1}^{\infty} \left(-\frac{1}{3}\right)^n$

44. $\displaystyle\sum_{n=0}^{\infty} \left(-\frac{1}{2}\right)^n$

45. $\displaystyle\sum_{n=0}^{\infty} 1^n$

46. $\displaystyle\sum_{n=0}^{\infty} 1.01^n$

47. $\displaystyle\sum_{n=0}^{\infty} -9\left(\frac{1}{3}\right)^n$

48. $\displaystyle\sum_{n=0}^{\infty} -8\left(-\frac{1}{2}\right)^n$

49. $\displaystyle\sum_{n=0}^{\infty} 10,000(0.05)^n$

50. $\displaystyle\sum_{n=0}^{\infty} 200(0.04)^n$

51. $\displaystyle\sum_{n=1}^{\infty} 0.4^n$

52. $0.3 + 0.03 + 0.003 + 0.0003 + \cdots$

53. $\displaystyle\sum_{n=0}^{\infty} 0.99^n$

54. $\displaystyle\sum_{n=0}^{\infty} \left(\frac{5}{4}\right)^n$

■ APPLICATIONS

55. Salary. Jeremy is offered a government job with the Department of Commerce. He is hired on the "GS" scale at a base rate of $34,000 with a 2.5% increase in his salary per year. Calculate what his salary will be after he has been with the Department of Commerce for 12 years.

56. Salary. Alison is offered a job with a small start-up company that wants to promote loyalty to the company with incentives for employees to stay with the company. The company offers her a starting salary of $22,000 with a guaranteed 15% raise per year. What will her salary be after she has been with the company for 10 years?

57. Depreciation. Brittany, a graduating senior in high school, receives a laptop computer as a graduation gift from her Aunt Jeanine so that she can use it when she gets to the University of Alabama. If the laptop costs $2,000 new and depreciates 50% per year, write a formula for the value of the laptop n years after it was purchased. How much will the laptop be worth when Brittany graduates from college (assuming she will graduate in 4 years)? How much will it be worth when she finishes graduate school? Assume graduate school is another 3 years.

58. Depreciation. Derek is deciding between a new Honda Accord and the BMW 325 series. The BMW costs $35,000 and the Honda costs $25,000. If the BMW depreciates at 20% per year and the Honda depreciates at 10% per year, find formulas for the value of each car n years after it is purchased. Which car is worth more in 10 years?

59. Bungee Jumping. A bungee jumper rebounds 70% of the height jumped. Assuming the bungee jump is made with a cord that stretches to 100 feet, how far will the bungee jumper travel upward on the fifth rebound?

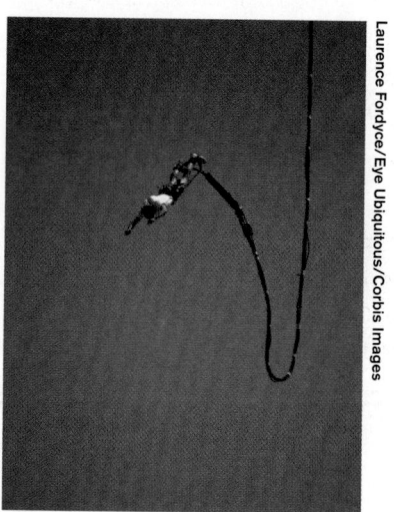

Laurence Fordyce/Eye Ubiquitous/Corbis Images

60. Bungee Jumping. A bungee jumper rebounds 65% of the height jumped. Assuming the bungee cord stretches 200 feet, how far will the bungee jumper travel upward on the eighth rebound?

61. Population Growth. One of the fastest-growing universities in the country is the University of Central Florida. The student populations each year starting in 2000 were 36,000, 37,800, 39,690, 41,675, ... Assuming this rate has continued, how many students are enrolled at UCF in 2010?

62. Web Site Hits. The Web site for Matchbox 20 (www.matchboxtwenty.com) has noticed that every week the number of hits to its Web site increases 5%. If there were 20,000 hits this week, how many will there be exactly 52 weeks from now if this rate continues?

63. Rich Man's Promise. A rich man promises that he will give you $1,000 on January 1, and every day after that, he will pay you 90% of what he paid you the day before. How many days will it take before you are making less than $1? How much will the rich man pay out for the entire month of January? Round to the nearest dollar.

64. Poor Man's Clever Deal. A poor man promises to work for you for $.01 the first day, $.02 on the second day, $.04 on the third day; his salary will continue to double each day. If he started on January 1 how much would he be paid to work on January 31? How much in total would he make during the month? Round to the nearest dollar.

65. Investing Lunch. A newlywed couple decides to stop going out to lunch every day and instead bring their lunch to work. They estimate it will save them $100 per month. They invest that $100 on the first of every month in an account that is compounded monthly and pays 5% interest. How much will be in the account at the end of 3 years?

66. Pizza as an Investment. A college freshman decides to stop ordering late-night pizzas (for both health and cost reasons). He realizes that he has been spending $50 a week on pizzas. Instead, he deposits $50 into an account that compounds weekly and pays 4% interest. (Assume 52 weeks annually.) How much money will be in the account after 52 weeks?

67. Tax-Deferred Annuity. Dr. Schober contributes $500 from her paycheck (weekly) to a tax-deferred investment account. Assuming the investment earns 6% and is compounded weekly, how much will be in the account after 26 weeks? 52 weeks?

68. Saving for a House. If a new graduate decides she wants to save for a house and she is able to put $300 every month into an account that earns 5% compounded monthly, how much will she have in the account after 5 years?

69. House Values. In 2008 you buy a house for $195,000. The value of the house appreciates 6.5% per year, on the average. How much is the house worth after 15 years?

70. The Bouncing Ball Problem. A ball is dropped from a height of 9 feet. Assume that on each bounce, the ball rebounds to one-third of its previous height. Find the total distance that the ball travels.

71. Probability. A fair coin is tossed repeatedly. The probability that the first head occurs on the nth toss is given by the function $p(n) = \left(\frac{1}{2}\right)^n$, where $n \geq 1$. Show that

$$\sum_{n=1}^{\infty} \left(\frac{1}{2}\right)^n = 1.0$$

72. Salary. Suppose you work for a supervisor who gives you two different options to choose from for your monthly pay. Option 1: The company pays you 1 cent for the first day of work, 2 cents the second day, 4 cents for the third day, 8 cents for the fourth day, and so on for 30 days. Option 2: You can receive a check right now for $10,000,000. Which pay option is better? How much better is it?

▪ CATCH THE MISTAKE

In Exercises 73–76, explain the mistake that is made.

73. Find the nth term of the geometric sequence

$$-1, \tfrac{1}{3}, -\tfrac{1}{9}, \tfrac{1}{27}, \ldots.$$

Solution:

Identify the first term and common ratio. $a_1 = -1$ and $r = \dfrac{1}{3}$

Substitute $a_1 = -1$ and $r = \tfrac{1}{3}$ into $a_n = a_1 \cdot r^{n-1}$. $a_n = (-1) \cdot \left(\dfrac{1}{3}\right)^{n-1}$

Simplify. $a_n = \dfrac{-1}{3^{n-1}}$

This is incorrect. What mistake was made?

74. Find the sum of the first n terms of the finite geometric series

$$2, 4, 8, 16, \ldots.$$

Solution:

Write the sum in sigma notation. $\displaystyle\sum_{k=1}^{n} (2)^k$

Identify the first term and common ratio. $a_1 = 1$ and $r = 2$

Substitute $a_1 = 1$ and $r = 2$ into $S_n = a_1 \dfrac{(1 - r^n)}{(1 - r)}$. $S_n = 1\dfrac{(1 - 2^n)}{(1 - 2)}$

Simplify. $S_n = 2^n - 1$

This is incorrect. What mistake was made?

75. Find the sum of the finite geometric series $\displaystyle\sum_{n=1}^{8} 4(-3)^n$.

Solution:

Identify the first term and common ratio. $a_1 = 4$ and $r = -3$

Substitute $a_1 = 4$ and $r = -3$ into $S_n = a_1 \dfrac{(1 - r^n)}{(1 - r)}$. $S_n = 4\dfrac{\left[1 - (-3)^n\right]}{[1 - (-3)]}$

$= 4\dfrac{\left[1 - (-3)^n\right]}{4}$

Simplify. $S_n = \left[1 - (-3)^n\right]$

Substitute $n = 8$. $S_8 = \left[1 - (-3)^8\right] = -6560$

This is incorrect. What mistake was made?

76. Find the sum of the infinite geometric series $\displaystyle\sum_{n=1}^{\infty} 2 \cdot 3^{n-1}$.

Solution:

Identify the first term and common ratio. $a_1 = 2$ and $r = 3$

Substitute $a_1 = 2$ and $r = 3$ into $S_\infty = a_1 \dfrac{1}{(1 - r)}$. $S_\infty = 2\dfrac{1}{(1 - 3)}$

Simplify. $S_\infty = -1$

This is incorrect. The series does not sum to -1. What mistake was made?

▪ CONCEPTUAL

In Exercises 77–80, determine whether each statement is true or false.

77. An alternating sequence cannot be a geometric sequence.

78. All finite and infinite geometric series can always be evaluated.

79. The common ratio of a geometric sequence can be positive or negative.

80. An infinite geometric series can be evaluated if the common ratio is less than or equal to 1.

▪ CHALLENGE

81. State the conditions for the sum

$$a + a \cdot b + a \cdot b^2 + \cdots + a \cdot b^n + \cdots$$

to exist. Assuming those conditions are met, find the sum.

82. Find the sum of $\displaystyle\sum_{k=0}^{20} \log 10^{2^k}$.

83. Represent the repeating decimal $0.474747\ldots$ as a fraction (ratio of two integers).

84. Suppose the sum of an infinite geometric series is

$$S = \dfrac{2}{1 - x}, \text{ where } x \text{ is a variable.}$$

a. Write out the first five terms of the series.

b. For what values of x will the series converge?

▪ TECHNOLOGY

85. Sum the series $\sum_{k=1}^{50} (-2)^{k-1}$. Apply a graphing utility to confirm your answer.

86. Does the sum of the infinite series $\sum_{n=0}^{\infty} \left(\frac{1}{3}\right)^n$ exist? Use a graphing calculator to find it.

87. Apply a graphing utility to plot $y_1 = 1 + x + x^2 + x^3 + x^4$ and $y_2 = \dfrac{1}{1 - x}$. Based on what you see, what do you expect the geometric series $\sum_{n=0}^{\infty} x^n$ to sum to?

88. Apply a graphing utility to plot $y_1 = 1 - x + x^2 - x^3 + x^4$ and $y_2 = \dfrac{1}{1 + x}$. Based on what you see, what do you expect the geometric series $\sum_{n=0}^{\infty} (-1)^n x^n$ to sum to?

▪ PREVIEW TO CALCULUS

In calculus, we study the convergence of geometric series. A geometric series with ratio r diverges if $|r| \geq 1$. If $|r| < 1$, then the geometric series converges to the sum

$$\sum_{n=0}^{\infty} ar^n = \frac{a}{1 - r}$$

In Exercises 89–92, determine the convergence or divergence of the series. If the series is convergent, find its sum.

89. $1 + \dfrac{2}{3} + \dfrac{4}{9} + \dfrac{8}{27} + \cdots$

90. $1 + \dfrac{5}{4} + \dfrac{25}{16} + \dfrac{125}{64} + \cdots$

91. $\dfrac{3}{8} + \dfrac{3}{32} + \dfrac{3}{128} + \dfrac{3}{512} + \cdots$

92. $\sum_{n=0}^{\infty} \dfrac{\pi}{3} \left(-\dfrac{8}{9}\right)^n$

SECTION
10.4 MATHEMATICAL INDUCTION

SKILLS OBJECTIVES

- Know the steps required to prove a statement by mathematical induction.
- Prove mathematical statements using mathematical induction.

CONCEPTUAL OBJECTIVES

- Understand that just because there appears to be a pattern, the pattern is not necessarily true for all values.
- Understand that when mathematical ideas are accepted, it is because they can be proved.

n	$n^2 - n + 41$	PRIME?
1	41	Yes
2	43	Yes
3	47	Yes
4	53	Yes
5	61	Yes

Is the expression $n^2 - n + 41$ *always* a prime number if n is a natural number? Your instinct may lead you to try a few values for n.

It appears that the statement might be true for all natural numbers. However, what about when $n = 41$?

$$n^2 - n + 41 = (41)^2 - 41 + 41 = 41^2$$

We find that when $n = 41$, $n^2 - n + 41$ is not prime. The moral of the story is that just because a pattern seems to exist for *some* values, the pattern is not necessarily true for *all* values. We must look for a way to show whether a statement is true for all values. In this section, we talk about *mathematical induction*, which is a way to show a statement is true for all values.

Mathematics is based on logic and proof (not assumptions or belief). One of the most famous mathematical statements was Fermat's Last Theorem. Pierre de Fermat (1601–1665) conjectured that there are no positive integer values for x, y, and z such that $x^n + y^n = z^n$, if $n \geq 3$. Although mathematicians *believed* that this theorem was true, no one was able to *prove* it until 350 years after the assumption was made. Professor Andrew Wiles at Princeton University received a $50,000 prize for successfully proving Fermat's Last Theorem in 1994.

Proof by Mathematical Induction

Mathematical induction is a technique used in precalculus and even in very advanced mathematics to prove many kinds of mathematical statements. In this section, you will use it to prove statements like "If $x > 1$, then $x^n > 1$ for all natural numbers n."

The principle of mathematical induction can be illustrated by a row of standing dominos, as in the image here. We make two assumptions:

1. The first domino is knocked down.
2. If a domino is knocked down, then the domino immediately following it will also be knocked down.

If both of these assumptions are true, then it is also true that all of the dominos will fall.

PRINCIPLE OF MATHEMATICAL INDUCTION

Let S_n be a statement involving the positive integer n. To prove that S_n is true for all positive integers, the following steps are required:

Step 1: Show that S_1 is true.
Step 2: Assume S_k is true and show that S_{k+1} is true ($k =$ positive integer).

Combining Steps 1 and 2 proves the statement is true for all positive integers (natural numbers).

EXAMPLE 1   Using Mathematical Induction

Apply the principle of mathematical induction to prove this statement:

$$\text{If } x > 1, \text{ then } x^n > 1 \text{ for all natural numbers } n.$$

Solution:

STEP 1   Show the statement is true for $n = 1$. $x^1 > 1$ because $x > 1$

STEP 2   Assume the statement is true for $n = k$. $x^k > 1$

   Show the statement is true for $k + 1$.

      Multiply both sides by x. $x^k \cdot x > 1 \cdot x$

      (Since $x > 1$, this step does not reverse the inequality sign.)

      Simplify. $x^{k+1} > x$

      Recall that $x > 1$. $x^{k+1} > x > 1$

Therefore, we have shown that $x^{k+1} > 1$.

This completes the induction proof. Thus, the following statement is true:

$$\text{"If } x > 1, \text{ then } x^n > 1 \text{ for } \textbf{all} \text{ natural numbers } n.\text{"}$$

EXAMPLE 2   Using Mathematical Induction

Use mathematical induction to prove that $n^2 + n$ is divisible by 2 for all natural numbers (positive integers) n.

Solution:

STEP 1   Show the statement we are
testing is true for $n = 1$. $1^2 + 1 = 2$

   2 is divisible by 2. $\dfrac{2}{2} = 1$

STEP 2 Assume the statement is true for $n = k$. $\dfrac{k^2 + k}{2} =$ an integer

Show it is true for $k + 1$ where $k \geq 1$. $\dfrac{(k + 1)^2 + (k + 1)}{2} \overset{?}{=}$ an integer

$$\dfrac{k^2 + 2k + 1 + k + 1}{2} \overset{?}{=} \text{an integer}$$

Regroup terms. $\dfrac{\left(k^2 + k\right) + 2(k + 1)}{2} \overset{?}{=}$ an integer

$$\dfrac{\left(k^2 + k\right)}{2} + \dfrac{2(k + 1)}{2} \overset{?}{=} \text{an integer}$$

We assumed $\dfrac{k^2 + k}{2} =$ an integer.   an integer $+ (k + 1) \overset{?}{=}$ an integer

Since k is a natural number (integer).   an integer $+$ an integer $=$ an integer

This completes the induction proof. The following statement is true:

"$n^2 + n$ is divisible by 2 for all natural numbers n."

Mathematical induction is often used to prove formulas for partial sums.

EXAMPLE 3   Proving a Partial–Sum Formula with Mathematical Induction

Apply mathematical induction to prove the following partial–sum formula:

$$1 + 2 + 3 + \cdots + n = \dfrac{n(n + 1)}{2} \text{ for all positive integers } n$$

Solution:

STEP 1 Show the formula is true for $n = 1$. $1 = \dfrac{1(1 + 1)}{2} = \dfrac{2}{2} = 1$

STEP 2 Assume the formula is true for $n = k$. $1 + 2 + 3 + \cdots + k = \dfrac{k(k + 1)}{2}$

Show it is true for $n = k + 1$.

$$1 + 2 + 3 + \cdots + k + (k + 1) \overset{?}{=} \dfrac{(k + 1)(k + 2)}{2}$$

$$\underset{\displaystyle \frac{k(k + 1)}{2}}{\underline{1 + 2 + 3 + \cdots + k}} + (k + 1) \overset{?}{=} \dfrac{(k + 1)(k + 2)}{2}$$

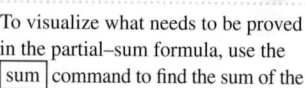

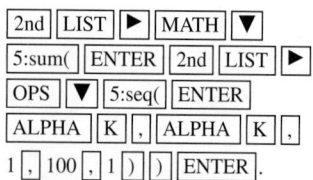

$$\dfrac{k(k + 1)}{2} + (k + 1) \overset{?}{=} \dfrac{(k + 1)(k + 2)}{2}$$

$$\dfrac{k(k + 1) + 2(k + 1)}{2} \overset{?}{=} \dfrac{(k + 1)(k + 2)}{2}$$

$$\dfrac{k^2 + 3k + 2}{2} \overset{?}{=} \dfrac{(k + 1)(k + 2)}{2}$$

$$\dfrac{(k + 1)(k + 2)}{2} = \dfrac{(k + 1)(k + 2)}{2}$$

This completes the induction proof. The following statement is true:

"$1 + 2 + 3 + \cdots + n = \dfrac{n(n + 1)}{2}$ for all positive integers n."

SECTION

10.4 SUMMARY

Just because we believe something is true does not mean that it is. In mathematics we rely on proof. In this section, we discussed *mathematical induction*, a process of proving some kinds of mathematical statements. The two-step procedure for mathematical induction is to (1) show the statement is true for $n = 1$, then (2) assume the statement is true for $n = k$ (any positive integer) and show the statement must be true for $n = k + 1$. The combination of Steps 1 and 2 proves the statement.

SECTION

10.4 EXERCISES

■ SKILLS

In Exercises 1–24, prove each statement using mathematical induction for all positive integers n.

1. $n^2 \le n^3$

2. If $0 < x < 1$, then $0 < x^n < 1$.

3. $2n \le 2^n$

4. $5^n < 5^{n+1}$

5. $n! > 2^n$ $n \ge 4$ (Show it is true for $n = 4$, instead of $n = 1$.)

6. $(1 + c)^n \ge nc$ $c > 1$

7. $n(n + 1)(n - 1)$ is divisible by 3.

8. $n^3 - n$ is divisible by 3.

9. $n^2 + 3n$ is divisible by 2.

10. $n(n + 1)(n + 2)$ is divisible by 6.

11. $2 + 4 + 6 + 8 + \cdots + 2n = n(n + 1)$

12. $1 + 3 + 5 + 7 + \cdots + (2n - 1) = n^2$

13. $1 + 3 + 3^2 + 3^3 + \cdots + 3^n = \dfrac{3^{n+1} - 1}{2}$

14. $2 + 4 + 8 + \cdots + 2^n = 2^{n+1} - 2$

15. $1^2 + 2^2 + 3^2 + \cdots + n^2 = \dfrac{n(n + 1)(2n + 1)}{6}$

16. $1^3 + 2^3 + 3^3 + \cdots + n^3 = \dfrac{n^2(n + 1)^2}{4}$

17. $\dfrac{1}{1 \cdot 2} + \dfrac{1}{2 \cdot 3} + \dfrac{1}{3 \cdot 4} + \cdots + \dfrac{1}{n(n + 1)} = \dfrac{n}{n + 1}$

18. $\dfrac{1}{2 \cdot 3} + \dfrac{1}{3 \cdot 4} + \cdots + \dfrac{1}{(n + 1)(n + 2)} = \dfrac{n}{2(n + 2)}$

19. $(1 \cdot 2) + (2 \cdot 3) + (3 \cdot 4) + \cdots + n(n + 1) = \dfrac{n(n + 1)(n + 2)}{3}$

20. $(1 \cdot 3) + (2 \cdot 4) + (3 \cdot 5) + \cdots + n(n + 2) = \dfrac{n(n + 1)(2n + 7)}{6}$

21. $1 + x + x^2 + x^3 + \cdots + x^{n-1} = \dfrac{1 - x^n}{1 - x}$ $x \ne 1$

22. $\dfrac{1}{2} + \dfrac{1}{4} + \dfrac{1}{8} + \cdots + \dfrac{1}{2^n} = 1 - \dfrac{1}{2^n}$

23. The sum of an arithmetic sequence: $a_1 + (a_1 + d) + (a_1 + 2d) + \cdots + [a_1 + (n - 1)d] = \dfrac{n}{2}[2a_1 + (n - 1)d]$.

24. The sum of a geometric sequence: $a_1 + a_1r + a_1r^2 + \cdots + a_1r^{n-1} = a_1\left(\dfrac{1 - r^n}{1 - r}\right), r \ne 1$.

■ **APPLICATIONS**

The Tower of Hanoi. This is a game with three pegs and n disks (largest on the bottom and smallest on the top). The goal is to move this entire tower of disks to another peg (in the same order). The challenge is that you may move only one disk at a time, and at no time can a larger disk be resting on a smaller disk. You may want to first go online to www.mazeworks.com/hanoi/index/htm and play the game.

Tower of Hanoi

■ **CONCEPTUAL**

In Exercises 31 and 32, determine whether each statement is true or false.

31. Assume S_k is true. If it can be shown that S_{k+1} is true, then S_n is true for all n, where n is any positive integer.

32. Assume S_1 is true. If it can be shown that S_2 and S_3 are true, then S_n is true for all n, where n is any positive integer.

■ **CHALLENGE**

33. Apply mathematical induction to prove

$$\sum_{k=1}^{n} k^4 = \frac{n(n+1)(2n+1)\left(3n^2 + 3n - 1\right)}{30}$$

34. Apply mathematical induction to prove

$$\sum_{k=1}^{n} k^5 = \frac{n^2(n+1)^2\left(2n^2 + 2n - 1\right)}{12}$$

35. Apply mathematical induction to prove

$$\left(1 + \frac{1}{1}\right)\left(1 + \frac{1}{2}\right)\left(1 + \frac{1}{3}\right)\cdots\left(1 + \frac{1}{n}\right) = n + 1$$

36. Apply mathematical induction to prove that $x + y$ is a factor of $x^{2n} - y^{2n}$.

25. What is the smallest number of moves needed if there are three disks?

26. What is the smallest number of moves needed if there are four disks?

27. What is the smallest number of moves needed if there are five disks?

28. What is the smallest number of moves needed if there are n disks? Prove it by mathematical induction.

29. **Telephone Infrastructure.** Suppose there are n cities that are to be connected with telephone wires. Apply mathematical induction to prove that the number of telephone wires required to connect the n cities is given by $\dfrac{n(n-1)}{2}$. Assume each city has to connect directly with any other city.

30. **Geometry.** Prove, with mathematical induction, that the sum of the measures of the interior angles in degrees of a regular polygon of n sides is given by the formula $(n-2)(180°)$ for $n \geq 3$. (*Hint:* Divide a polygon into triangles. For example, a four-sided polygon can be divided into two triangles. A five-sided polygon can be divided into three triangles. A six-sided polygon can be divided into four triangles, and so on.)

37. Apply mathematical induction to prove that $x - y$ is a factor of $x^{2n} - y^{2n}$.

38. Apply mathematical induction to prove

$$\ln(c_1 \cdot c_2 \cdot c_3 \cdots c_n) = \ln c_1 + \ln c_2 + \cdots + \ln c_n$$

39. Use a graphing calculator to sum the series

$$\frac{1}{2} + \frac{1}{4} + \frac{1}{8} + \cdots + \frac{1}{2^n} \text{ and evaluate the expression } 1 - \frac{1}{2^n}$$

for $n = 8$. Do they agree with each other? Do your answers confirm the proof for Exercise 22?

Andy Washnik

40. Use a graphing calculator to sum the series
$(1 \cdot 2) + (2 \cdot 3) + (3 \cdot 4) + \cdots + n(n + 1)$ and evaluate
the expression $\dfrac{n(n + 1)(n + 2)}{3}$ for $n = 200$. Do they
agree with each other? Do your answers confirm the proof
for Exercise 19?

■ PREVIEW TO CALCULUS

Several of the results studied in calculus must be proved by mathematical induction. In Exercises 41–44, apply mathematical induction to prove each formula.

41. $(\pi + 1) + (\pi + 2) + (\pi + 3) + \cdots + (\pi + n)$
$= \dfrac{n(2\pi + n + 1)}{2}$

42. $(1 + 1) + (2 + 4) + (3 + 9) + \cdots + (n + n^2)$
$= \dfrac{n(n + 1)(n + 2)}{3}$

43. $(1 + 1) + (2 + 8) + (3 + 27) + \cdots + (n + n^3)$
$= \dfrac{n(n + 1)(n^2 + n + 2)}{4}$

44. $(1 + 1) + (4 + 8) + (9 + 27) + \cdots + (n^2 + n^3)$
$= \dfrac{n(n + 1)(n + 2)(3n + 1)}{12}$

SECTION
10.5 THE BINOMIAL THEOREM

SKILLS OBJECTIVES

- Evaluate a binomial coefficient with the Binomial theorem.
- Evaluate a binomial coefficient with Pascal's triangle.
- Expand a binomial raised to a positive integer power.
- Find a particular term of a binomial expansion.

CONCEPTUAL OBJECTIVE

- Recognize patterns in binomial expansions.

A **binomial** is a polynomial that has two terms. The following are all examples of binomials:
$$x^2 + 2y \qquad a + 3b \qquad 4x^2 + 9$$

In this section, we will develop a formula for the expression for raising a binomial to a power n, where n is a positive integer.
$$(x^2 + 2y)^6 \qquad (a + 3b)^4 \qquad (4x^2 + 9)^5$$

To begin, let's start by writing out the expansions of $(a + b)^n$ for several values of n.

$$(a + b)^1 = a + b$$
$$(a + b)^2 = a^2 + 2ab + b^2$$
$$(a + b)^3 = a^3 + 3a^2b + 3ab^2 + b^3$$
$$(a + b)^4 = a^4 + 4a^3b + 6a^2b^2 + 4ab^3 + b^4$$
$$(a + b)^5 = a^5 + 5a^4b + 10a^3b^2 + 10a^2b^3 + 5ab^4 + b^5$$

There are several *patterns* that all of the **binomial expansions** have:

1. The number of terms in each resulting polynomial is always *one more* than the power of the binomial n. Thus, there are $n + 1$ terms in each expansion.

$$n = 3: \quad (a + b)^3 = \underbrace{a^3 + 3a^2b + 3ab^2 + b^3}_{\text{four terms}}$$

2. Each expansion has symmetry. For example, a and b can be interchanged and you will arrive at the same expansion. Furthermore, the powers of a decrease by 1 in each successive term, and the powers of b increase by 1 in each successive term.

$$(a + b)^3 = a^3b^0 + 3a^2b^1 + 3a^1b^2 + a^0b^3$$

3. The sum of the powers of each term in the expansion is n.

$$n = 3: \quad (a + b)^3 = a^3b^0 \overset{3+0=3}{} + 3a^2b^1 \overset{2+1=3}{} + 3a^1b^2 \overset{1+2=3}{} + a^0b^3 \overset{0+3=3}{}$$

4. The coefficients increase and decrease in a symmetric manner.

$$(a + b)^5 = 1a^5 + 5a^4b + 10a^3b^2 + 10a^2b^3 + 5ab^4 + 1b^5$$

Using these patterns, we can develop a generalized formula for $(a + b)^n$.

$$(a + b)^n = \square a^n + \square a^{n-1}b + \square a^{n-2}b^2 + \cdots + \square a^2b^{n-2} + \square ab^{n-1} + \square b^n$$

We know that there are $n + 1$ terms in the expansion. We also know that the sum of the powers of each term must equal n. The powers increase and decrease by 1 in each successive term, and if we interchanged a and b, the result would be the same expansion. The question that remains is, what coefficients go in the blanks?

Binomial Coefficients

We know that the coefficients must increase and then decrease in a symmetric order (similar to walking up and then down a hill). It turns out that the *binomial coefficients* are represented by a symbol that we will now define.

> **DEFINITION** **Binomial Coefficients**
>
> For nonnegative integers n and k, where $n \geq k$, the symbol $\binom{n}{k}$ is called the **binomial coefficient** and is defined by
>
> $$\binom{n}{k} = \frac{n!}{(n - k)!k!} \qquad \binom{n}{k} \text{ is read "}n \text{ choose } k\text{."}$$

You will see in the following sections that "n choose k" comes from counting combinations of n things taken k at a time.

EXAMPLE 1 **Evaluating a Binomial Coefficient**

Evaluate the following binomial coefficients:

a. $\binom{6}{4}$ **b.** $\binom{5}{5}$ **c.** $\binom{4}{0}$ **d.** $\binom{10}{9}$

Solution:

Select the top number as n and the bottom number as k and substitute into the binomial coefficient formula $\binom{n}{k} = \frac{n!}{(n - k)!k!}$.

a. $\dbinom{6}{4} = \dfrac{6!}{(6-4)!4!} = \dfrac{6!}{2!4!} = \dfrac{6 \cdot 5 \cdot 4 \cdot 3 \cdot 2 \cdot 1}{(2 \cdot 1)(4 \cdot 3 \cdot 2 \cdot 1)} = \dfrac{6 \cdot 5}{2} = \boxed{15}$

b. $\dbinom{5}{5} = \dfrac{5!}{(5-5)!5!} = \dfrac{5!}{0!5!} = \dfrac{1}{0!} = \dfrac{1}{1} = \boxed{1}$

c. $\dbinom{4}{0} = \dfrac{4!}{(4-0)!0!} = \dfrac{4!}{4!0!} = \dfrac{1}{0!} = \boxed{1}$

d. $\dbinom{10}{9} = \dfrac{10!}{(10-9)!9!} = \dfrac{10!}{1!9!} = \dfrac{10 \cdot 9!}{9!} = \boxed{10}$

■ **YOUR TURN** Evaluate the following binomial coefficients:

a. $\dbinom{9}{6}$ **b.** $\dbinom{8}{6}$

■ **Answer: a.** 84   **b.** 28

Parts (b) and (c) of Example 1 lead to the general formulas

$$\dbinom{n}{n} = 1 \qquad \text{and} \qquad \dbinom{n}{0} = 1$$

Binomial Expansion

Let's return to the question of the binomial expansion and how to determine the coefficients:

$$(a+b)^n = \square a^n + \square a^{n-1}b + \square a^{n-2}b^2 + \cdots + \square a^2 b^{n-2} + \square a b^{n-1} + \square b^n$$

The symbol $\dbinom{n}{k}$ is called a binomial coefficient because the coefficients in the blanks in the binomial expansion are equivalent to this symbol.

THE BINOMIAL THEOREM

Let a and b be real numbers; then for any positive integer n,

$$(a+b)^n = \dbinom{n}{0}a^n + \dbinom{n}{1}a^{n-1}b + \dbinom{n}{2}a^{n-2}b^2 + \cdots + \dbinom{n}{n-2}a^2 b^{n-2} + \dbinom{n}{n-1}a b^{n-1} + \dbinom{n}{n}b^n$$

or in sigma (summation) notation as

$$(a+b)^n = \sum_{k=0}^{n} \dbinom{n}{k} a^{n-k} b^k$$

EXAMPLE 2   Applying the Binomial Theorem

Expand $(x+2)^3$ with the Binomial theorem.

Solution:

Substitute $a = x$, $b = 2$, and $n = 3$ into the equation of the Binomial theorem.

$$(x+2)^3 = \sum_{k=0}^{3} \dbinom{3}{k} x^{3-k} 2^k$$

Expand the summation.

$$= \dbinom{3}{0}x^3 + \dbinom{3}{1}x^2 \cdot 2 + \dbinom{3}{2}x \cdot 2^2 + \dbinom{3}{3}2^3$$

Find the binomial coefficients.

$$= x^3 + 3x^2 \cdot 2 + 3x \cdot 2^2 + 2^3$$

Simplify.

$$= \boxed{x^3 + 6x^2 + 12x + 8}$$

■ **YOUR TURN** Expand $(x+5)^4$ with the Binomial theorem.

■ **Answer:**
$x^4 + 20x^3 + 150x^2 + 500x + 625$

EXAMPLE 3   Applying the Binomial Theorem

Expand $(2x - 3)^4$ with the Binomial theorem.

Solution:

Substitute $a = 2x$, $b = -3$, and $n = 4$ into the equation of the Binomial theorem.

$$(2x - 3)^4 = \sum_{k=0}^{4} \binom{4}{k}(2x)^{4-k}(-3)^k$$

Expand the summation.

$$= \binom{4}{0}(2x)^4 + \binom{4}{1}(2x)^3(-3) + \binom{4}{2}(2x)^2(-3)^2 + \binom{4}{3}(2x)(-3)^3 + \binom{4}{4}(-3)^4$$

Find the binomial coefficients.

$$= (2x)^4 + 4(2x)^3(-3) + 6(2x)^2(-3)^2 + 4(2x)(-3)^3 + (-3)^4$$

Simplify.

$$\boxed{= 16x^4 - 96x^3 + 216x^2 - 216x + 81}$$

■ **Answer:**
$81x^4 - 216x^3 + 216x^2 - 96x + 16$

■ **YOUR TURN** Expand $(3x - 2)^4$ with the Binomial theorem.

Pascal's Triangle

Instead of writing out the Binomial theorem and calculating the binomial coefficients using factorials every time you want to do a binomial expansion, we now present an alternative, more convenient way of remembering the binomial coefficients, called **Pascal's triangle**.

Notice that the first and last number in every row is 1. Each of the other numbers is found by adding the two numbers directly above it. For example,

$$3 = 2 + 1 \qquad 4 = 1 + 3 \qquad 10 = 6 + 4$$

Pascal's triangle

```
            1
          1   1
        1   2   1
      1   3   3   1
    1   4   6   4   1
  1   5   10   10   5   1
```

Let's arrange values of $\binom{n}{k}$ in a triangular pattern. Notice that the *value* of the binomial coefficients below are given in the margin.

$$\binom{0}{0}$$

$$\binom{1}{0} \quad \binom{1}{1}$$

$$\binom{2}{0} \quad \binom{2}{1} \quad \binom{2}{2}$$

$$\binom{3}{0} \quad \binom{3}{1} \quad \binom{3}{2} \quad \binom{3}{3}$$

$$\binom{4}{0} \quad \binom{4}{1} \quad \binom{4}{2} \quad \binom{4}{3} \quad \binom{4}{4}$$

$$\binom{5}{0} \quad \binom{5}{1} \quad \binom{5}{2} \quad \binom{5}{3} \quad \binom{5}{4} \quad \binom{5}{5}$$

It turns out that the numbers in Pascal's triangle are exactly the coefficients in a binomial expansion.

$$1$$
$$1a + 1b$$
$$1a^2 + 2ab + 1b^2$$
$$1a^3 + 3a^2b + 3ab^2 + 1b^3$$
$$1a^4 + 4a^3b + 6a^2b^2 + 4ab^3 + 1b^4$$
$$1a^5 + 5a^4b + 10a^3b^2 + 10a^2b^3 + 5ab^4 + 1b^5$$

The top row is called the *zero row* because it corresponds to the binomial raised to the zero power, $n = 0$. Since each row in Pascal's triangle starts and ends with a 1 and all other values are found by adding the two numbers directly above it, we can now easily calculate the sixth row.

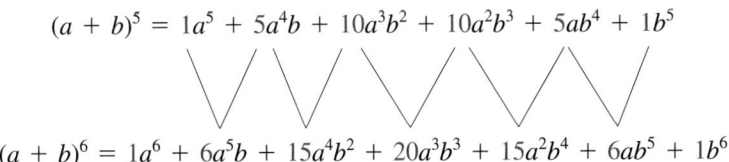

$$(a + b)^5 = 1a^5 + 5a^4b + 10a^3b^2 + 10a^2b^3 + 5ab^4 + 1b^5$$

$$(a + b)^6 = 1a^6 + 6a^5b + 15a^4b^2 + 20a^3b^3 + 15a^2b^4 + 6ab^5 + 1b^6$$

EXAMPLE 4   Applying Pascal's Triangle in a Binomial Expansion

Use Pascal's triangle to determine the binomial expansion of $(x + 2)^5$.

Solution:

Write the binomial expansion with blanks for coefficients.

$$(x + 2)^5 = \Box x^5 + \Box x^4 \cdot 2 + \Box x^3 \cdot 2^2 + \Box x^2 \cdot 2^3 + \Box x \cdot 2^4 + \Box 2^5$$

Write the binomial coefficients in the *fifth* row of Pascal's triangle.

$$1, 5, 10, 10, 5, 1$$

Substitute these coefficients into the blanks of the binomial expansion.

$$(x + 2)^5 = 1x^5 + 5x^4 \cdot 2 + 10x^3 \cdot 2^2 + 10x^2 \cdot 2^3 + 5x \cdot 2^4 + 1 \cdot 2^5$$

Simplify. $$\boxed{(x + 2)^5 = x^5 + 10x^4 + 40x^3 + 80x^2 + 80x + 32}$$

■ **YOUR TURN** Apply Pascal's triangle to determine the binomial expansion of $(x + 3)^4$.

■ **Answer:**
$x^4 + 12x^3 + 54x^2 + 108x + 81$

EXAMPLE 5   Applying Pascal's Triangle in a Binomial Expansion

Use Pascal's triangle to determine the binomial expansion of $(2x + 5)^4$.

Solution:

Write the binomial expansion with blanks for coefficients.

$$(2x + 5)^4 = \Box (2x)^4 + \Box (2x)^3 \cdot 5 + \Box (2x)^2 \cdot 5^2 + \Box (2x) \cdot 5^3 + \Box 5^4$$

Write the binomial coefficients in the *fourth* row of Pascal's triangle.

$$1, 4, 6, 4, 1$$

Substitute these coefficients into the blanks of the binomial expansion.

$$(2x + 5)^4 = 1(2x)^4 + 4(2x)^3 \cdot 5 + 6(2x)^2 \cdot 5^2 + 4(2x) \cdot 5^3 + 1 \cdot 5^4$$

Simplify. $\boxed{(2x + 5)^4 = 16x^4 + 160x^3 + 600x^2 + 1000x + 625}$

■ **Answer:**
a. $27x^3 + 54x^2 + 36x + 8$
b. $243x^5 - 810x^4 + 1080x^3 - 720x^2 + 240x - 32$

■ **YOUR TURN** Use Pascal's triangle to determine the binomial expansion of
a. $(3x + 2)^3$ **b.** $(3x - 2)^5$

Finding a Particular Term of a Binomial Expansion

What if we don't want to find the entire expansion, but instead want just a single term? For example, what is the fourth term of $(a + b)^5$?

WORDS	**MATH**
Recall the sigma notation.	$(a + b)^n = \sum_{k=0}^{n} \binom{n}{k} a^{n-k} b^k$
Let $n = 5$.	$(a + b)^5 = \sum_{k=0}^{5} \binom{5}{k} a^{5-k} b^k$
Expand.	$(a + b)^5 = \binom{5}{0}a^5 + \binom{5}{1}a^4b + \binom{5}{2}a^3b^2 + \underbrace{\binom{5}{3}a^2b^3}_{\text{fourth term}} + \binom{5}{4}ab^4 + \binom{5}{5}b^5$
Simplify the fourth term.	$10a^2b^3$

> **FINDING A PARTICULAR TERM OF A BINOMIAL EXPANSION**
>
> The $(r + 1)$ term of the expansion $(a + b)^n$ is $\binom{n}{r}a^{n-r}b^r$.

Technology Tip

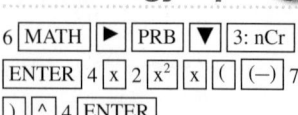

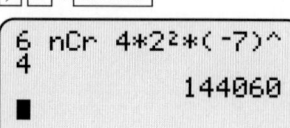

■ **Answer:** $1080x^3$

EXAMPLE 6   Finding a Particular Term of a Binomial Expansion

Find the fifth term of the binomial expansion of $(2x - 7)^6$.

Solution:

Recall that the $r + 1$ term of $(a + b)^n$ is $\binom{n}{r}a^{n-r}b^r$.

For the fifth term, let $r = 4$. $\qquad\qquad \binom{n}{4}a^{n-4}b^4$

For this expansion, let $a = 2x$, $b = -7$, and $n = 6$. $\qquad \binom{6}{4}(2x)^{6-4}(-7)^4$

Note that $\binom{6}{4} = 15$. $\qquad\qquad\qquad 15(2x)^2(-7)^4$

Simplify. $\qquad\qquad\qquad\qquad \boxed{144{,}060x^2}$

■ **YOUR TURN** What is the third term of the binomial expansion of $(3x - 2)^5$?

SECTION 10.5 SUMMARY

In this section, we developed a formula for expanding a binomial raised to a non-negative integer power, n. The patterns that surfaced were

- that the expansion displays symmetry between the two terms
- every expansion has $n + 1$ terms
- the powers sum to n
- the coefficients, called binomial coefficients, are ratios of factorials

$$(a + b)^n = \sum_{k=0}^{n} \binom{n}{k} a^{n-k} b^k$$

$$\binom{n}{k} = \frac{n!}{(n-k)!k!}$$

Also, Pascal's triangle, a shortcut method for evaluating the binomial coefficients, was discussed. The patterns in the triangle are that every row begins and ends with 1 and all other numbers are found by adding the two numbers in the row above the entry.

$$
\begin{array}{ccccccccccc}
 & & & & & 1 & & & & & \\
 & & & & 1 & & 1 & & & & \\
 & & & 1 & & 2 & & 1 & & & \\
 & & 1 & & 3 & & 3 & & 1 & & \\
 & 1 & & 4 & & 6 & & 4 & & 1 & \\
1 & & 5 & & 10 & & 10 & & 5 & & 1
\end{array}
$$

Lastly, a formula was given for finding a particular term of a binomial expansion; the $(r + 1)$ term of $(a + b)^n$ is $\binom{n}{r} a^{n-r} b^r$.

SECTION 10.5 EXERCISES

▪ SKILLS

In Exercises 1–10, evaluate each binomial coefficient.

1. $\binom{7}{3}$ **2.** $\binom{8}{2}$ **3.** $\binom{10}{8}$ **4.** $\binom{23}{21}$ **5.** $\binom{17}{0}$

6. $\binom{100}{0}$ **7.** $\binom{99}{99}$ **8.** $\binom{52}{52}$ **9.** $\binom{48}{45}$ **10.** $\binom{29}{26}$

In Exercises 11–32, expand each expression using the Binomial theorem.

11. $(x + 2)^4$ **12.** $(x + 3)^5$ **13.** $(y - 3)^5$ **14.** $(y - 4)^4$

15. $(x + y)^5$ **16.** $(x - y)^6$ **17.** $(x + 3y)^3$ **18.** $(2x - y)^3$

19. $(5x - 2)^3$ **20.** $(a - 7b)^3$ **21.** $\left(\dfrac{1}{x} + 5y\right)^4$ **22.** $\left(2x + \dfrac{3}{y}\right)^4$

23. $\left(x^2 + y^2\right)^4$ **24.** $\left(r^3 - s^3\right)^3$ **25.** $(ax + by)^5$ **26.** $(ax - by)^5$

27. $\left(\sqrt{x} + 2\right)^6$ **28.** $\left(3 + \sqrt{y}\right)^4$ **29.** $\left(a^{3/4} + b^{1/4}\right)^4$ **30.** $\left(x^{2/3} + y^{1/3}\right)^3$

31. $\left(x^{1/4} + 2\sqrt{y}\right)^4$ **32.** $\left(\sqrt{x} - 3y^{1/4}\right)^8$

In Exercises 33–36, expand each expression using Pascal's triangle.

33. $(r - s)^4$ **34.** $\left(x^2 + y^2\right)^7$ **35.** $(ax + by)^6$ **36.** $(x + 3y)^4$

In Exercises 37–44, find the coefficient C of the given term in each binomial expansion.

Binomial	Term	Binomial	Term	Binomial	Term
37. $(x + 2)^{10}$	Cx^6	**38.** $(3 + y)^9$	Cy^5	**39.** $(y - 3)^8$	Cy^4
40. $(x - 1)^{12}$	Cx^5	**41.** $(2x + 3y)^7$	Cx^3y^4	**42.** $(3x - 5y)^9$	Cx^2y^7
43. $\left(x^2 + y\right)^8$	Cx^8y^4	**44.** $\left(r - s^2\right)^{10}$	Cr^6s^8		

▪ APPLICATIONS

45. Lottery. In a state lottery in which 6 numbers are drawn from a possible 40 numbers, the number of possible 6-number combinations is equal to $\binom{40}{6}$. How many possible combinations are there?

46. Lottery. In a state lottery in which 6 numbers are drawn from a possible 60 numbers, the number of possible 6-number combinations is equal to $\binom{60}{6}$. How many possible combinations are there?

47. Poker. With a deck of 52 cards, 5 cards are dealt in a game of poker. There are a total of $\binom{52}{5}$ different 5-card poker hands that can be dealt. How many possible hands are there?

48. Canasta. In the card game Canasta, two decks of cards including the jokers are used and 11 cards are dealt to each person. There are a total of $\binom{108}{11}$ different 11-card Canasta hands that can be dealt. How many possible hands are there?

▪ CATCH THE MISTAKE

In Exercises 49 and 50, explain the mistake that is made.

49. Evaluate the expression $\binom{7}{5}$.

Solution:

Write out the binomial coefficient in terms of factorials.
$$\binom{7}{5} = \frac{7!}{5!}$$

Write out the factorials.
$$\binom{7}{5} = \frac{7!}{5!} = \frac{7 \cdot 6 \cdot 5 \cdot 4 \cdot 3 \cdot 2 \cdot 1}{5 \cdot 4 \cdot 3 \cdot 2 \cdot 1}$$

Simplify.
$$\binom{7}{5} = \frac{7!}{5!} = \frac{7 \cdot 6}{1} = 42$$

This is incorrect. What mistake was made?

50. Expand $(x + 2y)^4$.

Solution:

Write out the expansion with blanks.
$$(x + 2y)^4 - \square x^4 + \square x^3 y + \square x^2 y^2 + \square xy^3 + \square y^4$$

Write out the terms from the fifth row of Pascal's triangle.
$$1, 4, 6, 4, 1$$

Substitute these coefficients into the expansion.
$$(x + 2y)^4 = x^4 + 4x^3 y + 6x^2 y^2 + 4xy^3 + y^4$$

This is incorrect. What mistake was made?

▪ CONCEPTUAL

In Exercises 51–56, determine whether each statement is true or false.

51. The binomial expansion of $(x + y)^{10}$ has 10 terms.

52. The binomial expansion of $(x^2 + y^2)^{15}$ has 16 terms.

53. $\binom{n}{n} = 1$

54. $\binom{n}{-n} = -1$

55. The coefficient of x^8 in the expansion of $(2x - 1)^{12}$ is 126,720.

56. The sixth term of the binomial expansion of $(x^2 + y)^{10}$ is $252x^5 y^5$.

▪ CHALLENGE

57. Show that $\binom{n}{k} = \binom{n}{n - k}$, if $0 \le k \le n$.

58. Show that if n is a positive integer, then
$$\binom{n}{0} + \binom{n}{1} + \binom{n}{2} + \cdots + \binom{n}{n} = 2^n$$

Hint: Let $2^n = (1 + 1)^n$ and use the Binomial theorem to expand.

59. Show that $\binom{n}{k} = \binom{n - 1}{k - 1} + \binom{n - 1}{k}$ for $k < n$ and $n \ge 2$.

60. Show that $(n - 2k)\binom{n}{k} = n\left[\binom{n - 1}{k} - \binom{n - 1}{k - 1}\right]$ for $n > k$ and $n \ge 2$.

■ TECHNOLOGY

61. With a graphing utility, plot $y_1 = 1 - 3x + 3x^2 - x^3$, $y_2 = -1 + 3x - 3x^2 + x^3$, and $y_3 = (1 - x)^3$ in the same viewing screen. Which is the binomial expansion of $(1 - x)^3$, y_1 or y_2?

62. With a graphing utility, plot $y_1 = (x + 3)^4$, $y_2 = x^4 + 4x^3 + 6x^2 + 4x + 1$, and $y_3 = x^4 + 12x^3 + 54x^2 + 108x + 81$. Which is the binomial expansion of $(x + 3)^4$, y_2 or y_3?

63. With a graphing utility, plot $y_1 = 1 - 3x$, $y_2 = 1 - 3x + 3x^2$, $y_3 = 1 - 3x + 3x^2 - x^3$, and $y_4 = (1 - x)^3$ for $-1 < x < 1$. What do you notice happening each time an additional term is added to the series? Now, let $1 < x < 2$. Does the same thing happen?

64. With a graphing utility, plot $y_1 = 1 - \dfrac{3}{x}$, $y_2 = 1 - \dfrac{3}{x} + \dfrac{3}{x^2}$, $y_3 = 1 - \dfrac{3}{x} + \dfrac{3}{x^2} - \dfrac{1}{x^3}$, and $y_4 = \left(1 - \dfrac{1}{x}\right)^3$ for $1 < x < 2$.

What do you notice happening each time an additional term is added to the series? Now, let $0 < x < 1$. Does the same thing happen?

65. With a graphing utility, plot $y_1 = 1 + \dfrac{3}{x}$, $y_2 = 1 + \dfrac{3}{x} + \dfrac{3}{x^2}$, $y_3 = 1 + \dfrac{3}{x} + \dfrac{3}{x^2} - \dfrac{1}{x^3}$, and $y_4 = \left(1 + \dfrac{1}{x}\right)^3$ for $1 < x < 2$. What do you notice happening each time an additional term is added to the series? Now, let $0 < x < 1$. Does the same thing happen?

66. With a graphing utility, plot $y_1 = 1 + \dfrac{x}{1!}$, $y_2 = 1 + \dfrac{x}{1!} + \dfrac{x^2}{2!}$, $y_3 = 1 + \dfrac{x}{1!} + \dfrac{x^2}{2!} - \dfrac{x^3}{3!}$, and $y_4 = e^x$ for $-1 < x < 1$. What do you notice happening each time an additional term is added to the series? Now, let $1 < x < 2$. Does the same thing happen?

■ PREVIEW TO CALCULUS

In calculus, the difference quotient $\dfrac{f(x + h) - f(x)}{h}$ of a function f is used to find the derivative of the function f.

In Exercises 67 and 68, use the Binomial theorem to find the difference quotient of each function.

67. $f(x) = x^n$

68. $f(x) = (2x)^n$

69. In calculus, we learn that the derivative of $f(x) = (x + 1)^n$ is $f'(x) = n(x + 1)^{n-1}$. Using the Binomial theorem, find an expression for f' using sigma notation.

70. In calculus, we learn that the antiderivative of $f(x) = (x + 1)^n$ is $F(x) = \dfrac{(x + 1)^{n+1}}{n + 1}$ for $n \neq -1$. Using the Binomial theorem, find an expression for $F(x)$ using sigma notation.

In 2005 the world was producing 7 billion tons of carbon emissions per year. In 2055 this number is projected to double with the worldwide production of carbon emissions equaling 14 billion tons per year.

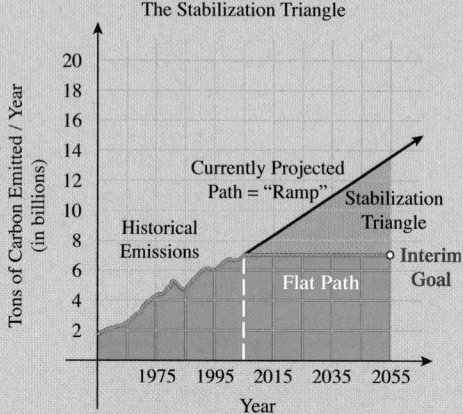

1. Determine the equation of the line for the projected path: an increase of 7 gigatons of carbon (GtC) over 50 years (2005–2055). Calculate the slope of the line.

2. What is the increase (per year) in the rate of carbon emissions per year based on the projected path model?

3. Develop a model in terms of a finite series that yields the total additional billions of tons of carbon emitted over the 50-year period (2005–2055) for the projected path over the flat path.

4. Calculate the total additional billions of tons of carbon of the projected path over the flat path [i.e., sum the series in (3)].

5. Discuss possible ways to provide the reduction between the projected path and the flat path based on the proposals given by Pacala and Socolow (professors at Princeton).*

*S. Pacala and R. Socolow, "Stabilization Wedges: Solving the Climate Problem for the Next 50 Years with Current Technologies," *Science*, Vol. 305 (2004).

SECTION	CONCEPT	PAGES	REVIEW EXERCISES	KEY IDEAS/POINTS
10.1	**Sequences and series**	938–944	1–30	
	Sequences	938–939	1–12	$a_1, a_2, a_3, \ldots, a_n, \ldots$ a_n is the general term.
	Factorial notation	940–941	13–16	$n! = n(n-1)(n-2) \cdot \cdots 3 \cdot 2 \cdot 1$ $n \geq 2$ $0! = 1$ and $1! = 1$
	Recursion formulas	941–942	17–20	When a_n is defined by previous terms a_{n-i} $(i = 1, 2, \ldots)$.
	Sums and series	942–944	21–28	Infinite series: $$\sum_{n=1}^{\infty} a_n = a_1 + a_2 + a_3 + \cdots + a_n + \cdots$$ Finite series, or nth partial sum, S_n: $$S_n = a_1 + a_2 + a_3 + \cdots + a_n$$
10.2	**Arithmetic sequences and series**	949–954	31–50	
	Arithmetic sequences	949–950	31–44	$a_{n+1} = a_n + d$ or $a_{n+1} - a_n = d$ d is called the **common difference**.
	The general (nth) term of an arithmetic sequence	951	37–44	$a_n = a_1 + (n-1)d$ for $n \geq 1$
	The sum of an arithmetic sequence	952–954	45–48	$S_n = \dfrac{n}{2}(a_1 + a_n)$
10.3	**Geometric sequences and series**	958–966	51–74	
	Geometric sequences	958–959	51–58	$a_{n+1} = r \cdot a_n$ or $\dfrac{a_{n+1}}{a_n} = r$ r is called the **common ratio**.
	The general (nth) term of a geometric sequence	959–960	59–66	$a_n = a_1 \cdot r^{n-1}$ for $n \geq 1$
	Geometric series	961–966	67–72	Finite series: $S_n = a_1 \dfrac{(1-r^n)}{(1-r)}$ $r \neq 1$ Infinite series: $\sum_{n=0}^{\infty} a_1 r^n = a_1 \dfrac{1}{(1-r)}$ $\lvert r \rvert < 1$
10.4	**Mathematical induction**	970–972	75–78	
	Proof by mathematical induction	971–972		Prove that S_n is true for all positive integers: Step 1: Show that S_1 is true. Step 2: Assume S_n is true for S_k and show it is true for S_{k+1} (k = positive integer).

10.1 Sequences and Series

Write the first four terms of each sequence. Assume n starts at 1.

1. $a_n = n^3$

2. $a_n = \dfrac{n!}{n}$

3. $a_n = 3n + 2$

4. $a_n = (-1)^n x^{n+2}$

Find the indicated term of each sequence.

5. $a_n = \left(\dfrac{2}{3}\right)^n \qquad a_5 = ?$

6. $a_n = \dfrac{n^2}{3^n} \qquad a_8 = ?$

7. $a_n = \dfrac{(-1)^n(n-1)!}{n(n+1)!} \qquad a_{15} = ?$

8. $a_n = 1 + \dfrac{1}{n} \qquad a_{10} = ?$

Write an expression for the nth term of each given sequence.

9. $3, -6, 9, -12, \ldots$

10. $1, \frac{1}{2}, 3, \frac{1}{4}, 5, \frac{1}{6}, 7, \frac{1}{8}, \ldots$

11. $-1, 1, -1, 1, \ldots$

12. $1, 10, 10^2, 10^3, \ldots$

Simplify each ratio of factorials.

13. $\dfrac{8!}{6!}$

14. $\dfrac{20!}{23!}$

15. $\dfrac{n(n-1)!}{(n+1)!}$

16. $\dfrac{(n-2)!}{n!}$

Write the first four terms of each sequence defined by the recursion formula.

17. $a_1 = 5 \qquad a_n = a_{n-1} - 2$

18. $a_1 = 1 \qquad a_n = n^2 \cdot a_{n-1}$

19. $a_1 = 1, a_2 = 2 \qquad a_n = (a_{n-1})^2 \cdot (a_{n-2})$

20. $a_1 = 1, a_2 = 2 \qquad a_n = \dfrac{a_{n-2}}{(a_{n-1})^2}$

Evaluate each finite series.

21. $\displaystyle\sum_{n=1}^{5} 3$

22. $\displaystyle\sum_{n=1}^{4} \dfrac{1}{n^2}$

23. $\displaystyle\sum_{n=1}^{6} (3n + 1)$

24. $\displaystyle\sum_{k=0}^{5} \dfrac{2^{k+1}}{k!}$

Use sigma (summation) notation to represent each sum.

25. $-1 + \frac{1}{2} - \frac{1}{4} + \frac{1}{8} + \cdots - \frac{1}{64}$

26. $2 + 4 + 6 + 8 + 10 + \cdots + 20$

27. $1 + x + \dfrac{x^2}{2} + \dfrac{x^3}{6} + \dfrac{x^4}{24} + \cdots$

28. $x - x^2 + \dfrac{x^3}{2} - \dfrac{x^4}{6} + \dfrac{x^5}{24} - \dfrac{x^6}{120} + \cdots$

Applications

29. A Marine's Investment. With the prospect of continued fighting in Iraq, in December 2004, the Marine Corps offered bonuses of as much as $30,000—in some cases, tax-free—to persuade enlisted personnel with combat experience and training to reenlist. Suppose a Marine put his entire $30,000 reenlistment bonus in an account that earned 4% interest compounded monthly. The balance in the account after n months would be

$$A_n = 30,000\left(1 + \dfrac{0.04}{12}\right)^n \qquad n = 1, 2, 3, \ldots$$

His commitment to the Marines is 5 years. Calculate A_{60}. What does A_{60} represent?

30. Sports. The NFL minimum salary for a rookie is $180,000. Suppose a rookie enters the league making the minimum and gets a $30,000 raise each year. Write the general term a_n of a sequence that represents the salary of an NFL player making the league minimum during his entire career. Assuming $n = 1$ corresponds to the first year, what does $\displaystyle\sum_{n=1}^{4} a_n$ represent?

10.2 Arithmetic Sequences and Series

Determine whether each sequence is arithmetic. If it is, find the common difference.

31. $7, 5, 3, 1, -1, \ldots$

32. $1^3 + 2^3 + 3^3 + \cdots$

33. $1, \frac{3}{2}, 2, \frac{5}{2}, \ldots$

34. $a_n = -n + 3$

35. $a_n = \dfrac{(n+1)!}{n!}$

36. $a_n = 5(n-1)$

Find the general, or nth, term of each arithmetic sequence given the first term and the common difference.

37. $a_1 = -4 \qquad d = 5$

38. $a_1 = 5 \qquad d = 6$

39. $a_1 = 1 \qquad d = -\frac{2}{3}$

40. $a_1 = 0.001 \qquad d = 0.01$

For each arithmetic sequence described below, find a_1 and d and construct the sequence by stating the general, or nth, term.

41. The 5th term is 13 and the 17th term is 37.

42. The 7th term is -14 and the 10th term is -23.

43. The 8th term is 52 and the 21st term is 130.

44. The 11th term is -30 and the 21st term is -80.

Find each sum.

45. $\displaystyle\sum_{k=1}^{20} 3k$

46. $\displaystyle\sum_{n=1}^{15} (n + 5)$

47. $2 + 8 + 14 + 20 + \cdots + 68$

48. $\frac{1}{4} - \frac{1}{4} - \frac{3}{4} - \cdots - \frac{31}{4}$

Applications

49. Salary. Upon graduating with MBAs, Bob and Tania opt for different career paths. Bob accepts a job with the U.S. Department of Transportation making $45,000 with a guaranteed $2,000 raise every year. Tania takes a job with Templeton Corporation making $38,000 with a guaranteed $4,000 raise every year. Calculate how many total dollars both Bob and Tania will have each made after 15 years.

50. Gravity. When a skydiver jumps out of an airplane, she falls approximately 16 feet in the 1st second, 48 feet during the 2nd second, 80 feet during the 3rd second, 112 feet during the 4th second, and 144 feet during the 5th second, and this pattern continues. If she deploys her parachute after 5 seconds have elapsed, how far will she have fallen during those 5 seconds?

10.3 Geometric Sequences and Series

Determine whether each sequence is geometric. If it is, find the common ratio.

51. $2, -4, 8, -16, \ldots$

52. $1, \dfrac{1}{2^2}, \dfrac{1}{3^2}, \dfrac{1}{4^2}, \ldots$

53. $20, 10, 5, \dfrac{5}{2}, \ldots$

54. $\dfrac{1}{100}, \dfrac{1}{10}, 1, 10, \ldots$

Write the first five terms of each geometric series.

55. $a_1 = 3 \qquad r = 2$

56. $a_1 = 10 \qquad r = \dfrac{1}{4}$

57. $a_1 = 100 \qquad r = -4$

58. $a_1 = -60 \qquad r = -\dfrac{1}{2}$

Write the formula for the nth term of each geometric series.

59. $a_1 = 7 \qquad r = 2$

60. $a_1 = 12 \qquad r = \dfrac{1}{3}$

61. $a_1 = 1 \qquad r = -2$

62. $a_1 = \dfrac{32}{5} \qquad r = -\dfrac{1}{4}$

Find the indicated term of each geometric sequence.

63. 25th term of the sequence $2, 4, 8, 16, \ldots$

64. 10th term of the sequence $\dfrac{1}{2}, 1, 2, 4, \ldots$

65. 12th term of the sequence $100, -20, 4, -0.8, \ldots$

66. 11th term of the sequence $1000, -500, 250, -125, \ldots$

Evaluate each geometric series, if possible.

67. $\dfrac{1}{2} + \dfrac{3}{2} + \dfrac{3^2}{2} + \cdots + \dfrac{3^8}{2}$

68. $1 + \dfrac{1}{2} + \dfrac{1}{2^2} + \dfrac{1}{2^3} + \cdots + \dfrac{1}{2^{10}}$

69. $\displaystyle\sum_{n=1}^{8} 5(3)^{n-1}$

70. $\displaystyle\sum_{n=1}^{7} \dfrac{2}{3}(5)^n$

71. $\displaystyle\sum_{n=0}^{\infty} \left(\dfrac{2}{3}\right)^n$

72. $\displaystyle\sum_{n=1}^{\infty} \left(-\dfrac{1}{5}\right)^{n+1}$

Applications

73. Salary. Murad is fluent in four languages and is offered a job with the U.S. government as a translator. He is hired on the "GS" scale at a base rate of $48,000 with a 2% increase in his salary per year. Calculate what his salary will be *after* he has been with the U.S. government for 12 years.

74. Boat Depreciation. Upon graduating from Auburn University, Philip and Steve get jobs at Disney Ride and Show Engineering and decide to buy a ski boat together. If the boat costs $15,000 new, and depreciates 20% per year, write a formula for the value of the boat n years after it was purchased. How much will the boat be worth when Philip and Steve have been working at Disney for 3 years?

10.4 Mathematical Induction

Prove each statement using mathematical induction for all positive integers n.

75. $3n \le 3^n$

76. $4^n < 4^{n+1}$

77. $2 + 7 + 12 + 17 + \cdots + (5n - 3) = \dfrac{n}{2}(5n - 1)$

78. $2n^2 > (n + 1)^2 \qquad n \ge 3$

10.5 The Binomial Theorem

Evaluate each binomial coefficient.

79. $\dbinom{11}{8}$

80. $\dbinom{10}{0}$

81. $\dbinom{22}{22}$

82. $\dbinom{47}{45}$

Expand each expression using the Binomial theorem.

83. $(x - 5)^4$

84. $(x + y)^5$

85. $(2x - 5)^3$

86. $\left(x^2 + y^3\right)^4$

87. $\left(\sqrt{x} + 1\right)^5$

88. $\left(x^{2/3} + y^{1/3}\right)^6$

Expand each expression using Pascal's triangle.

89. $(r - s)^5$

90. $(ax + by)^4$

Find the coefficient C of the term in each binomial expansion.

	Binomial	Term
91.	$(x - 2)^8$	Cx^6
92.	$(3 + y)^7$	Cy^4
93.	$(2x + 5y)^6$	Cx^2y^4
94.	$\left(r^2 - s\right)^8$	Cr^8s^4

Applications

95. Lottery. In a state lottery in which 6 numbers are drawn from a possible 53 numbers, the number of possible 6-number combinations is equal to $\dbinom{53}{6}$. How many possible combinations are there?

96. Canasta. In the card game Canasta, two decks of cards including the jokers are used, and 13 cards are dealt to each person. A total of $\binom{108}{13}$ different 13-card Canasta hands can be dealt. How many possible hands are there?

Technology

Section 10.1

97. Use a graphing calculator "SUM" to find the sum of the series $\sum_{n=1}^{6} \frac{1}{n^2}$.

98. Use a graphing calculator "SUM" to find the sum of the infinite series $\sum_{n=1}^{\infty} \frac{1}{n}$, if possible.

Section 10.2

99. Use a graphing calculator to sum $\sum_{n=1}^{75} \left[\frac{3}{2} + \frac{6}{7}(n-1) \right]$.

100. Use a graphing calculator to sum $\sum_{n=1}^{264} \left[-19 + \frac{1}{3}(n-1) \right]$.

Section 10.3

101. Apply a graphing utility to plot
$y_1 = 1 - 2x + 4x^2 - 8x^3 + 16x^4$ and $y_2 = \frac{1}{1+2x}$, and let x range from $[-0.3, 0.3]$. Based on what you see, what do you expect the geometric series $\sum_{n=0}^{\infty} (-1)^n(2x)^n$ to sum to in this range of x-values?

102. Does the sum of the infinite series $\sum_{n=0}^{\infty} \left(\frac{e}{\pi} \right)^n$ exist? Use a graphing calculator to find it and round to four decimal places.

Section 10.4

103. Use a graphing calculator to sum the series $2 + 7 + 12 + 17 + \cdots + (5n - 3)$ and evaluate the expression $\frac{n}{2}(5n - 1)$ for $n = 200$. Do they agree with each other? Do your answers confirm the proof for Exercise 77?

104. Use a graphing calculator to plot the graphs of $y_1 = 2x^2$ and $y_2 = (x + 1)^2$ in the $[100, 1000]$ by $[10{,}000, 2{,}500{,}000]$ viewing rectangle. Do your results confirm the proof for Exercise 78?

Section 10.5

105. With a graphing utility, plot $y_1 = 1 + 8x$, $y_2 = 1 + 8x + 24x^2$, $y_3 = 1 + 8x + 24x^2 + 32x^3$, $y_4 = 1 + 8x + 24x^2 + 32x^3 + 16x^4$, and $y_5 = (1 + 2x)^4$ for $-0.1 < x < 0.1$. What do you notice happening each time an additional term is added to the series? Now, let $0.1 < x < 1$. Does the same thing happen?

106. With a graphing utility, plot $y_1 = 1 - 8x$, $y_2 = 1 - 8x + 24x^2$, $y_3 = 1 - 8x + 24x^2 - 32x^3$, $y_4 = 1 - 8x + 24x^2 - 32x^3 + 16x^4$, and $y_5 = (1 - 2x)^4$ for $-0.1 < x < 0.1$. What do you notice happening each time an additional term is added to the series? Now, let $0.1 < x < 1$. Does the same thing happen?

For Exercises 1–5, use the sequence $1, x, x^2, x^3, \ldots$

1. Write the nth term of the sequence.

2. Classify this sequence as arithmetic, geometric, or neither.

3. Find the nth partial sum of the series S_n.

4. Assuming this sequence is infinite, write its series using sigma notation.

5. Assuming this sequence is infinite, what condition would have to be satisfied in order for the sum to exist?

6. Find the following sum: $\frac{1}{3} + \frac{1}{9} + \frac{1}{27} + \frac{1}{81} + \cdots$.

7. Find the following sum: $\sum_{n=1}^{10} 3 \cdot \left(\frac{1}{4}\right)^n$.

8. Find the following sum: $\sum_{k=1}^{50} (2k + 1)$.

9. Write the following series using sigma notation, then find its sum: $2 + 7 + 12 + 17 + \cdots + 497$.

10. Use mathematical induction to prove that
$2 + 4 + 6 + \cdots + 2n = n^2 + n$.

11. Evaluate $\dfrac{7!}{2!}$.

12. Find the third term of $(2x + y)^5$.

In Exercises 13 and 14, evaluate each expression.

13. $\binom{15}{12}$ 14. $\binom{k}{k}$

15. Simplify the expression $\dfrac{42!}{7! \cdot 37!}$.

16. Simplify the expression $\dfrac{(n + 2)!}{(n - 2)!}$ for $n \geq 2$.

17. Expand the expression $\left(x^2 + \dfrac{1}{x}\right)^5$.

18. Use the Binomial theorem to expand the binomial $(3x - 2)^4$.

Prove each statement using mathematical induction for all positive integers n.

19. $3 + 7 + 11 + \cdots + 4n - 1 = 2n^2 + n$

20. $2^1 + 1^3 + 2^2 + 2^3 + 2^3 + 3^3 + 2^4 + 4^3 + \cdots + 2^n + n^3$
$= 2^{n+1} - 2 + \dfrac{n^2(n + 1)^2}{4}$

Find the sum of each infinite geometric series, if possible.

21. $\sum_{n=0}^{\infty} \left(\dfrac{2}{5}\right)^n$

22. $1 + \dfrac{3}{4} + \dfrac{9}{16} + \dfrac{27}{64} + \cdots$

Apply sigma notation to represent each series.

23. $\dfrac{5}{8} + \dfrac{5}{10} + \dfrac{5}{12} + \dfrac{5}{14} + \cdots$

24. $\dfrac{1}{4} + \dfrac{5}{8} + \dfrac{7}{16} + \dfrac{17}{32} + \dfrac{31}{64} + \dfrac{65}{128} + \cdots$

In Exercises 25 and 26, expand the expression using the Binomial theorem.

25. $(2 - 3x)^7$ 26. $(x^2 + y^3)^6$

27. Find the constant term in the expression $\left(x^3 + \dfrac{1}{x^3}\right)^{20}$.

28. Use a graphing calculator to sum $\sum_{n=1}^{125} \left[-\dfrac{11}{4} + \dfrac{5}{6}(n - 1) \right]$.

1. Find the difference quotient $\dfrac{f(x+h)-f(x)}{h}$ of the function $f(x) = \dfrac{x^2}{x+1}$.

2. Given $f(x) = x^2 + x$ and $g(x) = \dfrac{1}{x+3}$, find $(f \circ g)(x)$.

3. Write the polynomial function $f(x) = x^4 - 4x^3 - 4x^2 - 4x - 5$ as a product of linear factors.

4. Find the inverse of the function $f(x) = 5x - 4$.

5. Use long division to divide the polynomials: $(-6x^5 + 3x^3 + 2x^2 - 7) \div (x^2 + 3)$.

6. Write the logarithmic equation $\log 0.001 = -3$ in its equivalent exponential form.

7. Solve for x: $\ln(5x - 6) = 2$. Round to three decimal places.

8. **Sprinkler Coverage.** A sprinkler has a 21-foot spray and it rotates through an angle of 50°. What is the area that the sprinkler covers?

9. Find the exact value of $\sin\left(-\dfrac{7\pi}{4}\right)$.

10. If $\tan x = \dfrac{7}{24}$ and $\pi < x < \dfrac{3\pi}{2}$, find $\cos(2x)$.

11. Solve the trigonometric equation exactly $2\cos^2\theta - \cos\theta - 1 = 0$ over $0 \le \theta \le 2\pi$.

12. Given $\gamma = 53°$, $a = 18$, and $c = 17$, determine if a triangle (or two) exist and if so solve the triangles.

13. Express the complex number $\sqrt{2} - \sqrt{2}i$ in polar form.

14. Solve the system of linear equations.
$$8x - 5y = 15$$
$$y = \dfrac{8}{5}x + 10$$

15. Solve the system of linear equations.
$$2x - y + z = 1$$
$$x - y + 4z = 3$$

16. Maximize the objective function $z = 4x + 5y$, subject to the constraints $x + y \le 5$, $x \ge 1$, $y \ge 2$.

17. Solve the system using Gauss–Jordan elimination.
$$x + 5y - 2z = 3$$
$$3x + y + 2z = -3$$
$$2x - 4y + 4z = 10$$

18. Given
$$A = \begin{bmatrix} 3 & 4 & -7 \\ 0 & 1 & 5 \end{bmatrix} \quad B = \begin{bmatrix} 8 & -2 & 6 \\ 9 & 0 & -1 \end{bmatrix} \quad C = \begin{bmatrix} 9 & 0 \\ 1 & 2 \end{bmatrix}$$
find $C(A + B)$.

19. Calculate the determinant.
$$\begin{vmatrix} 2 & 5 & -1 \\ 1 & 4 & 0 \\ -2 & 1 & 3 \end{vmatrix}$$

20. Find the equation of a parabola with vertex $(3, 5)$ and directrix $x = 7$.

21. Graph $x^2 + y^2 < 4$.

22. The parametric equations $x = 2\sin t$, $y = 3\cos t$ define a plane curve. Find an equation in rectangular form that also corresponds to the plane curve.

23. Find the sum of the finite series $\sum_{n=1}^{4} \dfrac{2^{n-1}}{n!}$.

24. Classify the sequence as arithmetic, geometric, or neither.
$$5, 15, 45, 135, \ldots$$

25. The number of subsets with k elements of a set with n elements is given by $\dfrac{n!}{k!(n-k)!}$. Find the number of subsets with 3 elements of a set with 7 elements.

26. Find the binomial expansion of $(x - x^2)^5$.

Appendix

Prerequisites: Fundamentals of Algebra

IN THIS APPENDIX you will first review real numbers. Integer exponents and scientific notation will be discussed, followed by rational exponents and radicals. Simplification of radicals and rationalization of denominators will be reviewed. Basic operations such as addition, subtraction, and multiplication of polynomials will be discussed followed by a review of how to factor polynomials. Rational expressions will be discussed and a brief overview of solving simple algebraic equations will be given. After reviewing all these aspects of real numbers, we will conclude with a review of complex numbers.

SKILLS OBJECTIVES

■ Classify real numbers as rational or irrational.
■ Round or truncate real numbers.
■ Evaluate algebraic expressions.
■ Apply properties of real numbers.

CONCEPTUAL OBJECTIVES

■ Understand that rational and irrational numbers are mutually exclusive and complementary subsets of real numbers.
■ Learn the order of operations for real numbers.

The Set of Real Numbers

A **set** is a group or collection of objects that are called **members** or **elements** of the set. If *every* member of set B is also a member of set A, then we say B is a **subset** of A and denote it as $B \subset A$.

For example, the starting lineup on a baseball team is a subset of the entire team. The set of **natural numbers**, $\{1, 2, 3, 4, \ldots\}$, is a subset of the set of **whole numbers**, $\{0, 1, 2, 3, 4, \ldots\}$, which is a subset of the set of **integers**, $\{\ldots, -4, -3, -2, -1, 0, 1, 2, 3, \ldots\}$, which is a subset of the set of *rational numbers*, which is a subset of the set of *real numbers*. The three dots, called an **ellipsis**, indicate that the pattern continues indefinitely.

If a set has no elements, it is called the **empty set**, or **null set**, and is denoted by the symbol $\varnothing$. The **set of real numbers** consists of two main subsets: *rational* and *irrational* numbers.

DEFINITION   **Rational Number**

A **rational number** is a number that can be expressed as a quotient (ratio) of two integers, $\dfrac{a}{b}$, where the integer a is called the **numerator** and the integer b is called the **denominator** and where $b \neq 0$.

Rational numbers include all integers or all fractions that are ratios of integers. Note that any integer can be written as a ratio whose denominator is equal to 1. In decimal form, the rational numbers are those that terminate or are nonterminating with a repeated decimal pattern, which we represent with an overbar. Those decimals that do not repeat and do not terminate are **irrational numbers**. The numbers

$$5, \quad -17, \quad \frac{1}{3}, \quad \sqrt{2}, \quad \pi, \quad 1.37, \quad 0, \quad -\frac{19}{17}, \quad 3.66\overline{6}, \quad 3.2179\ldots$$

are examples of **real** numbers, where 5, -17, $\frac{1}{3}$, 1.37, 0, $-\frac{19}{17}$, and $3.66\overline{6}$ are rational numbers, and $\sqrt{2}$, π, and $3.2179\ldots$ are irrational numbers. It is important to note that the ellipsis following the last decimal digit denotes continuing in an irregular fashion, whereas the absence of such dots to the right of the last decimal digit implies the decimal expansion terminates.

RATIONAL NUMBER (FRACTION)	CALCULATOR DISPLAY	DECIMAL REPRESENTATION	DESCRIPTION
$\frac{7}{2}$	3.5	3.5	*Terminates*
$\frac{15}{12}$	1.25	1.25	*Terminates*
$\frac{2}{3}$	0.666666666	$0.\overline{6}$	*Repeats*
$\frac{1}{11}$	0.09090909	$0.\overline{09}$	*Repeats*

Notice that the overbar covers the entire repeating pattern. The following figure and table illustrate the subset relationship and examples of different types of real numbers.

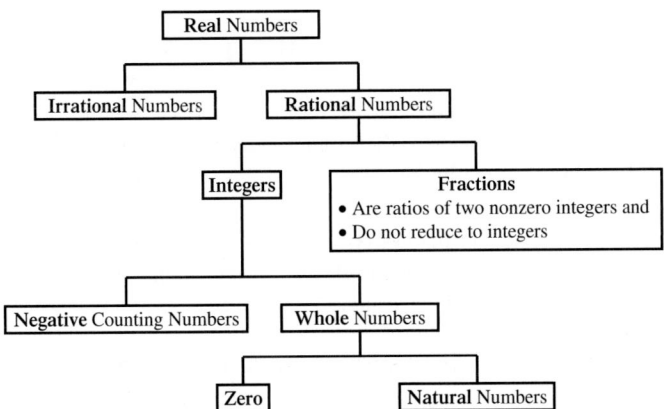

Study Tip

Every real number is either a rational number *or* an irrational number.

SYMBOL	NAME	DESCRIPTION	EXAMPLES
$\mathbb{N}$	Natural numbers	Counting numbers	$1, 2, 3, 4, 5, \ldots$
$\mathbb{W}$	Whole numbers	Natural numbers and zero	$0, 1, 2, 3, 4, 5, \ldots$
$\mathbb{Z}$	Integers	Whole numbers and negative natural numbers	$\ldots, -5, -4, -3, -2, -1, 0, 1, 2, 3, 4, 5, \ldots$
$\mathbb{Q}$	Rational numbers	Ratios of integers: $\dfrac{a}{b}$ $(b \neq 0)$ • Decimal representation terminates, or • Decimal representation repeats	$-17, -\frac{19}{7}, 0, \frac{1}{3}, 1.37, 3.66\overline{6}, 5$
$\mathbb{I}$	Irrational numbers	Numbers whose decimal representation does *not* terminate or repeat	$\sqrt{2}, 1.2179\ldots, \pi$
$\mathbb{R}$	Real numbers	Rational and irrational numbers	$\pi, 5, -\frac{2}{3}, 17.25, \sqrt{7}$

Since the set of real numbers can be formed by combining the set of rational numbers and the set of irrational numbers, then every real number is either rational or irrational. We say that the set of rational numbers and the set of irrational numbers are mutually exclusive (no shared elements) and complementary sets. The **real number line** is a graph used to represent the set of all real numbers.

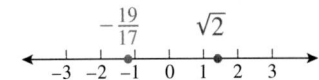

EXAMPLE 1   Classifying Real Numbers

Classify the following real numbers as rational or irrational:

$$-3, \quad 0, \quad \tfrac{1}{4}, \quad \sqrt{3}, \quad \pi, \quad 7.51, \quad \tfrac{1}{3}, \quad -\tfrac{8}{5}, \quad 6.66666$$

Solution:

Rational: $-3, \quad 0, \quad \tfrac{1}{4}, \quad 7.51, \quad \tfrac{1}{3}, \quad -\tfrac{8}{5}, \quad -6.66666$ Irrational: $\sqrt{3}, \pi$

■ **YOUR TURN** Classify the following real numbers as rational or irrational:

$$-\tfrac{7}{3}, \quad 5.999\overline{9}, \quad 12, \quad 0, \quad -5.27, \quad \sqrt{5}, \quad 2.010010001\ldots$$

Approximations: Rounding and Truncation

Every real number can be represented by a decimal. When a real number is in decimal form, we often approximate the number by either *rounding off* or *truncating* to a given decimal place. **Truncation** is "cutting off" or eliminating everything to the right of a certain decimal place. **Rounding** means looking to the right of the specified decimal place and making a judgment. If the digit to the right is greater than or equal to 5, then the specified digit is rounded up, or increased by 1 unit. If the digit to the right is less than 5, then the specified digit stays the same. In both of these cases, all decimal places to the right of the specified place are removed.

EXAMPLE 2   Approximating Decimals to Two Places

Approximate 17.368204 to two decimal places by

a. truncation   **b.** rounding

Solution:

a. To truncate, eliminate all digits to the right of the 6. $\boxed{17.36}$

b. To round, look to the right of the 6.
Because "8" is greater than 5, we round up (add 1 to the 6). $\boxed{17.37}$

■ **YOUR TURN** Approximate 23.02492 to two decimal places by

a. truncation   **b.** rounding

 ### EXAMPLE 3   Approximating Decimals to Four Places

Approximate 7.293516 to four decimal places by

a. truncation   **b.** rounding

Solution:

The "5" is in the fourth decimal place.

a. To truncate, eliminate all digits to the right of 5. $\boxed{7.2935}$

b. To round, look to the right of the 5.
Because "1" is less than 5, the 5 remains the same. $\boxed{7.2935}$

■ **YOUR TURN** Approximate −2.381865 to four decimal places by

a. truncation   **b.** rounding

In Examples 2 and 3, we see that *rounding and truncation sometimes yield the same approximation, but not always.*

Order of Operations

Addition, subtraction, multiplication, and division are called arithmetic operations. The results of these operations are called the sum, difference, product, and quotient, respectively. These four operations are summarized in the following table:

OPERATION	NOTATION	RESULT
Addition	$a + b$	Sum
Subtraction	$a - b$	Difference
Multiplication	$a \cdot b$ or ab or $(a)(b)$	Product
Division	$\dfrac{a}{b}$ or a/b $(b \neq 0)$	Quotient (Ratio)

Since algebra involves *variables* such as x, in order to avoid confusion, we avoid the traditional multiplication sign $\times$ when working with symbols; three alternatives are shown in the preceding table. Similarly, the arithmetic sign for division, $\div$, is often represented by vertical or slanted fractions.

The symbol $=$ is called the **equal sign**, and is read as "equals" or "is." It implies that the expression on one side of the equal sign is equivalent to (has the same value as) the expression on the other side of the equal sign.

WORDS	MATH
The sum of seven and eleven equals eighteen.	$7 + 11 = 18$
Three times five is fifteen.	$3 \cdot 5 = 15$
Four times six equals twenty-four.	$4(6) = 24$
Eight divided by two is four.	$\dfrac{8}{2} = 4$
Three subtracted from five is two.	$5 - 3 = 2$

When evaluating expressions involving real numbers, it is important to remember the correct *order of operations*. For example, how do we simplify the expression $3 + 2 \cdot 5$? Do we multiply first and then add, or add first and then multiply? In mathematics, conventional order implies multiplication first, and then addition: $3 + 2 \cdot 5 = 3 + 10 = 13$. Parentheses imply grouping of terms, and the necessary operations should be performed inside them first. If there are nested parentheses, start with the innermost parentheses and work your way out. Within parentheses follow the conventional order of operations. Exponents are an important part of order of operations and will be discussed in Section A.2.

ORDER OF OPERATIONS

1. Start with the innermost parentheses (grouping symbols) and work outward.
2. Perform all indicated multiplications and divisions, working from left to right.
3. Perform all additions and subtractions, working from left to right.

EXAMPLE 4   Simplifying Expressions Using the Correct Order of Operations

Simplify the expressions.

a. $4 + 3 \cdot 2 - 7 \cdot 5 + 6$ **b.** $\dfrac{7 - 6}{2 \cdot 3 + 8}$

Solution (a):

Perform multiplication first.

$$4 + \underbrace{3 \cdot 2}_{6} - \underbrace{7 \cdot 5}_{35} + 6$$

Then perform the indicated additions and subtractions.

$$= 4 + 6 - 35 + 6 = \boxed{-19}$$

Solution (b):

The numerator and the denominator are similar to expressions in parentheses. Simplify these separately first, following the correct order of operations.

Perform multiplication in the denominator first.

$$\dfrac{7 - 6}{\underbrace{2 \cdot 3}_{6} + 8}$$

Then perform subtraction in the numerator and addition in the denominator.

$$= \dfrac{7 - 6}{6 + 8} = \boxed{\dfrac{1}{14}}$$

■ **Answer: a.** 10   **b.** $\frac{3}{16}$

■ **YOUR TURN** Simplify the expressions.

a. $-7 + 4 \cdot 5 - 2 \cdot 6 + 9$ **b.** $\dfrac{9 - 6}{2 \cdot 5 + 6}$

Parentheses ( ), brackets [ ], and braces { } are the typical notations for grouping and are often used interchangeably. When nested (groups within groups), we generally use parentheses on the innermost, then brackets or braces on the outermost.

EXAMPLE 5   Simplifying Expressions That Involve Grouping Signs Using the Correct Order of Operations

Simplify the expression $3[5 \cdot (4 - 2) - 2 \cdot 7]$.

Solution:

Simplify the inner parentheses: $(4 - 2) = 2$.

$$= 3[5 \cdot 2 - 2 \cdot 7]$$

Inside the brackets, perform the multiplication $5 \cdot 2 = 10$ and $2 \cdot 7 = 14$.

$$= 3[10 - 14]$$

Inside the brackets, perform the subtraction.

$$= 3[-4]$$

Multiply.

$$= \boxed{-12}$$

■ **Answer:** -24

■ **YOUR TURN** Simplify the expression $2[-3 \cdot (13 - 5) + 4 \cdot 3]$.

Algebraic Expressions

Everything we have discussed until now has involved real numbers (explicitly). In algebra, however, numbers are often represented by letters (such as x and y), which we call **variables**. A **constant** is a fixed (known) number such as 5. A **coefficient** is the constant that is multiplied by a variable. Quantities within the *algebraic expression* that are separated by addition or subtraction are referred to as **terms**.

DEFINITION   **Algebraic Expression**

An **algebraic expression** is the combination of variables and constants using basic operations such as addition, subtraction, multiplication, and division. Each term is separated by addition or subtraction.

Algebraic Expression	Variable Term	Constant Term	Coefficient
$5x + 3$	$5x$	3	5

When we know the value of the variables, we can **evaluate an algebraic expression** using the **substitution principle**:

Algebraic expression: $5x + 3$
Value of the variable: $x = 2$
Substitute $x = 2$: $5(2) + 3 = 10 + 3 = 13$

EXAMPLE 6   **Evaluating Algebraic Expressions**

Evaluate the algebraic expression $7x + 2$ for $x = 3$.

Solution:

Start with the algebraic expression.	$7x + 2$
Substitute $x = 3$.	$7(3) + 2$
Perform multiplication.	$= 21 + 2$
Perform addition.	$= \boxed{23}$

YOUR TURN Evaluate the algebraic expression $6y + 4$ for $y = 2$.

■ **Answer:** 16

In Example 6, the value for the variable was specified in order for us to evaluate the algebraic expression. What if the value of the variable is not specified; can we simplify an expression like $3(2x - 5y)$? In this case, we cannot subtract $5y$ from $2x$. Instead, we rely on the *basic properties of real numbers*, or the *basic rules of algebra*.

Properties of Real Numbers

You probably already know many properties of real numbers. For example, if you add up four numbers, it does not matter in which order you add them. If you multiply five numbers, it does not matter in what order you multiply them. If you add 0 to a real number or multiply a real number by 1, the result yields the original real number. **Basic properties of real numbers** are summarized in the following table. Because these properties are true for variables and algebraic expressions, these properties are often called the **basic rules of algebra**.

PROPERTIES OF REAL NUMBERS (BASIC RULES OF ALGEBRA)

NAME	DESCRIPTION	MATH (LET a, b, AND c EACH BE ANY REAL NUMBER)	EXAMPLE
Commutative property of addition	Two real numbers can be added in any order.	$a + b = b + a$	$3x + 5 = 5 + 3x$
Commutative property of multiplication	Two real numbers can be multiplied in any order.	$ab = ba$	$y \cdot 3 = 3y$
Associative property of addition	When three real numbers are added, it does not matter which two numbers are added first.	$(a + b) + c = a + (b + c)$	$(x + 5) + 7 = x + (5 + 7)$
Associative property of multiplication	When three real numbers are multiplied, it does not matter which two numbers are multiplied first.	$(ab)c = a(bc)$	$(-3x)y = -3(xy)$
Distributive property	Multiplication is distributed over *all* the terms of the sums or differences within the parentheses.	$a(b + c) = ab + ac$	$5(x + 2) = 5x + 5 \cdot 2$
Additive identity property	Adding zero to any real number yields the same real number.	$a + 0 = a$ $0 + a = a$	$7y + 0 = 7y$
Multiplicative identity property	Multiplying any real number by 1 yields the same real number.	$a \cdot 1 = a$ $1 \cdot a = a$	$(8x)(1) = 8x$
Additive inverse property	The sum of a real number and its additive inverse (opposite) is zero.	$a + (-a) = 0$	$4x + (-4x) = 0$
Multiplicative inverse property	The product of a nonzero real number and its multiplicative inverse (reciprocal) is 1.	$a \cdot \dfrac{1}{a} = 1 \qquad a \neq 0$	$(x + 2) \cdot \left(\dfrac{1}{x + 2}\right) = 1$ $x \neq -2$

The properties in the previous table govern addition and multiplication. Subtraction can be defined in terms of addition of the *additive inverse*, and division can be defined in terms of multiplication by *the multiplicative inverse (reciprocal)*.

SUBTRACTION AND DIVISION

Let a and b be real numbers.

	MATH	TYPE OF INVERSE	WORDS
Subtraction	$a - b = a + (-b)$	$-b$ is the **additive inverse** or **opposite** of b.	Subtracting a real number is equal to adding its opposite.
Division	$a \div b = a \cdot \dfrac{1}{b}$, $b \neq 0$	$\dfrac{1}{b}$ is the **multiplicative inverse** or **reciprocal** of b.	Dividing by a real number is equal to multiplying by its reciprocal.

In the fractional form $\dfrac{a}{b}$, a is called the **numerator** and b is called the **denominator**.

It is important to note that the subtraction form of the distributive property is

$$a(b - c) = ab - ac$$

EXAMPLE 7 **Using the Distributive Property**

Use the distributive property to eliminate the parentheses.

a. $3(x + 5)$ **b.** $2(y - 6)$

Solution (a):

Use the distributive property.	$3(x + 5) = 3(x) + 3(5)$
Perform the multiplication.	$= \boxed{3x + 15}$

Solution (b):

Use the distributive property.	$2(y - 6) = 2(y) - 2(6)$
Perform the multiplication.	$= \boxed{2y - 12}$

■ **Answer: a.** $2x + 6$ **b.** $5y - 15$

■ **YOUR TURN** Use the distributive property to eliminate the parentheses.

a. $2(x + 3)$ **b.** $5(y - 3)$

You also probably know the rules that apply when multiplying a negative real number. For example, "a negative times a negative is a positive."

PROPERTIES OF NEGATIVES

DESCRIPTION	MATH (LET a AND b BE POSITIVE REAL NUMBERS)	EXAMPLE
A *negative* quantity *times* a *positive* quantity is a *negative* quantity.	$(-a)(b) = -ab$	$(-8)(3) = -24$
A *negative* quantity *divided* by a *positive* quantity is a *negative* quantity. or A *positive* quantity *divided* by a *negative* quantity is a *negative* quantity.	$\dfrac{-a}{b} = -\dfrac{a}{b}$ or $\dfrac{a}{-b} = -\dfrac{a}{b}$	$\dfrac{-16}{4} = -4$ or $\dfrac{15}{-3} = -5$
A *negative* quantity *times* a *negative* quantity is a *positive* quantity.	$(-a)(-b) = ab$	$(-2x)(-5) = 10x$
A *negative* quantity *divided* by a *negative* quantity is a *positive* quantity.	$\dfrac{-a}{-b} = \dfrac{a}{b}$	$\dfrac{-12}{-3} = 4$
The opposite of a negative quantity is a positive quantity (subtracting a negative quantity is equivalent to adding a positive quantity).	$-(-a) = a$	$-(-9) = 9$
A negative sign preceding an expression is distributed throughout the expression.	$-(a + b) = -a - b$ $-(a - b) = -a + b$	$-3(x + 5) = -3x - 15$ $-3(x - 5) = -3x + 15$

EXAMPLE 8 **Using Properties of Negatives**

Eliminate the parentheses and perform the operations.

a. $-5 + 7 - (-2)$ **b.** $-(-3)(-4)(-6)$

Solution:

a. Distribute the negative.

$$-5 + 7 \underbrace{- (-2)}_{2}$$

$$= -5 + 7 + 2$$

Combine the three quantities. $= \boxed{4}$

b. Multiply left to right. $[-(-3)][(-4)(-6)]$

Perform the multiplication inside the [ ]. $= [3][24]$

Multiply. $= \boxed{72}$

Technology Tip

a.

```
-5+7--2
                    4
```

b. Here are the calculator keystrokes for $-(-3)(-4)(-6)$.

```
--3*-4*-6
                   72
■
```

We use properties of negatives to define the *absolute value* of any real number. The **absolute value** of a real number a, denoted $|a|$, is its magnitude. On a number line this is the distance from the origin 0 to the point. For example, algebraically, the absolute value of 5 is 5, that is, $|5| = 5$; and the absolute value of -5 is 5, or $|-5| = 5$. Graphically, the distance on the real number line from 0 to either -5 or 5 is 5.

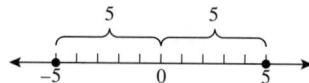

Notice that the absolute value does not change a positive real number, but changes a negative real number to a positive number. A negative number becomes a positive number if it is multiplied by -1.

IF a IS A ...	$|a|$	EXAMPLE
Positive real number	$|a| = a$	$|5| = 5$
Negative real number	$|a| = -a$	$|-5| = -(-5) = 5$
Zero	$|a| = a$	$|0| = 0$

EXAMPLE 9   Finding the Absolute Value of a Real Number

Evaluate the expressions.

a. $|-3 + 7|$ **b.** $|2 - 8|$

Solution:

a. $|-3 + 7| = |4| = \boxed{4}$

b. $|2 - 8| = |-6| = -(-6) = \boxed{6}$

Properties of the absolute value are discussed in Section 0.3.

 EXAMPLE 10   Using Properties of Negatives and the Distributive Property

Eliminate the parentheses $-(2x - 3y)$.

COMMON MISTAKE

A common mistake is applying a negative only to the first term.

⭐ CORRECT	❌ INCORRECT
$-(2x - 3y)$	**Error:** $-2x - 3y$
$= -(2x) - (-3y)$	
$= \boxed{-2x + 3y}$	The negative $(-)$ was not distributed through the second term.

■ **Answer: a.** $-2x - 10y$
     b. $-3 + 2b$

■ **YOUR TURN** Eliminate the parentheses.

a. $-2(x + 5y)$ **b.** $-(3 - 2b)$

What is the product of any real number and zero? The answer is zero. This property also leads to the zero product property, which is used together with factoring to solve quadratic equations. This is discussed in Section 0.3.

PROPERTIES OF ZERO

DESCRIPTION	MATH (LET *a* BE A REAL NUMBER)	EXAMPLE
A real number multiplied by zero is zero.	$a \cdot 0 = 0$	$0 \cdot x = 0$
Zero divided by a nonzero real number is zero.	$\dfrac{0}{a} = 0 \qquad a \neq 0$	$\dfrac{0}{3-x} = 0 \qquad x \neq 3$
A real number divided by zero is undefined.	$\dfrac{a}{0}$ is undefined $\qquad a \neq 0$	$\dfrac{x+2}{0}$ is undefined.

ZERO PRODUCT PROPERTY

DESCRIPTION	MATH	EXAMPLE
If the product of two real numbers is zero, then at least one of those numbers has to be zero.	If $ab = 0$, then $a = 0$ or $b = 0$ (or both).	If $x(x + 2) = 0$, then $x = 0$ or $x + 2 = 0$.

Note: If *a* and *b* are *both* equal to zero, then the product is still zero.

Fractions always seem to intimidate students. In fact, many instructors teach students to eliminate fractions in algebraic equations.

It is important to realize that you can never divide by zero. Therefore, in the following table of fractional properties it is assumed that no denominators are zero.

FRACTIONAL PROPERTIES

DESCRIPTION	MATH	ZERO CONDITION	EXAMPLE
Equivalent fractions	$\dfrac{a}{b} = \dfrac{c}{d}$ if and only if $ad = bc$	$b \neq 0$ and $d \neq 0$	$\dfrac{y}{2} = \dfrac{6y}{12}$ since $12y = 12y$
Multiplying two fractions	$\dfrac{a}{b} \cdot \dfrac{c}{d} = \dfrac{ac}{bd}$	$b \neq 0$ and $d \neq 0$	$\dfrac{3}{5} \cdot \dfrac{x}{7} = \dfrac{3x}{35}$
Adding fractions that have the same denominator	$\dfrac{a}{b} + \dfrac{c}{b} = \dfrac{a+c}{b}$	$b \neq 0$	$\dfrac{x}{3} + \dfrac{2}{3} = \dfrac{x+2}{3}$
Subtracting fractions that have the same denominator	$\dfrac{a}{b} - \dfrac{c}{b} = \dfrac{a-c}{b}$	$b \neq 0$	$\dfrac{7}{3} - \dfrac{5}{3} = \dfrac{2}{3}$
Adding fractions with different denominators using a least common denominator	$\dfrac{a}{b} + \dfrac{c}{d} = \dfrac{ad}{bd} + \dfrac{cb}{bd} = \dfrac{ad+bc}{bd}$	$b \neq 0$ and $d \neq 0$	$\dfrac{1}{2} + \dfrac{5}{3} = \dfrac{(1)(3)+(5)(2)}{6} = \dfrac{13}{6}$
Subtracting fractions with different denominators using a least common denominator	$\dfrac{a}{b} - \dfrac{c}{d} = \dfrac{ad}{bd} - \dfrac{cb}{bd} = \dfrac{ad-bc}{bd}$	$b \neq 0$ and $d \neq 0$	$\dfrac{1}{3} - \dfrac{1}{4} = \dfrac{(1)(4)-(1)(3)}{12} = \dfrac{1}{12}$
Dividing by a fraction is equivalent to multiplying by its reciprocal.	$\dfrac{a}{b} \div \dfrac{c}{d} = \dfrac{a}{b} \cdot \dfrac{d}{c}$	$b \neq 0, c \neq 0,$ and $d \neq 0$	$\dfrac{x}{3} \div \dfrac{2}{7} = \dfrac{x}{3} \cdot \dfrac{7}{2} = \dfrac{7x}{6}$

The **least common multiple** of two or more integers is the smallest integer that is evenly divisible by each of the integers. For example, the least common multiple (LCM) of 3 and 4 is 12. The LCM of 8 and 6 is 24. The reason the LCM of 8 and 6 is not 48 is that 8 and 6 have a common factor of 2. When adding and subtracting fractions, a common denominator can be found by multiplying the denominators. When there are common factors in the denominators, the *least* **common denominator** (LCD) is the LCM of the original denominators.

EXAMPLE 11   Performing Operations with Fractions

Perform the indicated operations involving fractions and simplify.

a. $\dfrac{2}{3} - \dfrac{1}{4}$ **b.** $\dfrac{2}{3} \div 4$ **c.** $\dfrac{x}{2} + \dfrac{3}{5}$

Solution (a):

Determine the least common denominator.	$3 \cdot 4 = 12$
Rewrite fractions applying the least common denominator.	$\dfrac{2}{3} - \dfrac{1}{4} = \dfrac{2 \cdot 4}{3 \cdot 4} - \dfrac{1 \cdot 3}{4 \cdot 3}$
	$= \dfrac{2(4) - 1(3)}{3(4)}$
Eliminate the parentheses.	$= \dfrac{8 - 3}{12}$
Combine terms in the numerator.	$= \boxed{\dfrac{5}{12}}$

Solution (b):

Rewrite 4 with an understood 1 in the denominator.	$\dfrac{2}{3} \div 4 = \dfrac{2}{3} \div \dfrac{4}{1}$
Dividing by a fraction is equivalent to multiplying by its reciprocal.	$= \dfrac{2}{3} \cdot \dfrac{1}{4}$
Multiply numerators and denominators, respectively.	$= \dfrac{2}{12}$
Reduce the fraction to simplest form.	$= \boxed{\dfrac{1}{6}}$

Solution (c):

Determine the least common denominator.	$2 \cdot 5 = 10$
Rewrite fractions in terms of the least common denominator.	$\dfrac{x}{2} + \dfrac{3}{5} = \dfrac{5x + 3(2)}{10}$
Simplify the numerator.	$= \boxed{\dfrac{5x + 6}{10}}$

Technology Tip

To change the decimal number to a fraction, press MATH 1: ▶ Frac ENTER ENTER.

a.

```
2/3-1/4
          .4166666667
Ans▶Frac
              5/12
■
```

b.

```
2/3/4
          .1666666667
Ans▶Frac
               1/6
■
```

■ **Answer:**

a. $\dfrac{11}{10}$ **b.** $\dfrac{2}{3}$ **c.** $\dfrac{10 - 3x}{15}$

■ **YOUR TURN** Perform the indicated operations involving fractions.

a. $\dfrac{3}{5} + \dfrac{1}{2}$ **b.** $\dfrac{1}{5} \div \dfrac{3}{10}$ **c.** $\dfrac{2}{3} - \dfrac{x}{5}$

SECTION A.1 SUMMARY

In this section, real numbers were defined as the set of all rational and irrational numbers. Decimals are approximated by either truncating or rounding.

- *Truncating*: Eliminate all values after a particular digit.
- *Rounding*: Look to the right of a particular digit. If the number is 5 or greater, increase the digit by 1; otherwise, leave it as is and eliminate all digits to the right.

The *order* in which we perform operations is

1. parentheses (grouping); work from inside outward.
2. multiplication/division; work from left to right.
3. addition/subtraction; work from left to right.

The *properties of real numbers* are employed as the basic rules of algebra when dealing with algebraic expressions.

- Commutative property of addition: $a + b = b + a$
- Commutative property of multiplication: $ab = ba$
- Associative property of addition: $(a + b) + c = a + (b + c)$
- Associative property of multiplication: $(ab)c = a(bc)$
- Distributive property: $a(b + c) = ab + ac$ or $a(b - c) = ab - ac$
- Additive identity: $a + 0 = a$
- Multiplicative identity: $a \cdot 1 = a$
- Additive inverse (opposite): $a + (-a) = 0$
- Multiplicative inverse (reciprocal): $a \cdot \dfrac{1}{a} = 1$ $a \neq 0$

Subtraction and division can be defined in terms of addition and multiplication.

- *Subtraction*: $a - b = a + (-b)$ (add the opposite)
- *Division*: $a \div b = a \cdot \dfrac{1}{b}$, where $b \neq 0$

(multiply by the reciprocal)

Properties of negatives were reviewed. If a and b are positive real numbers, then

- $(-a)(b) = -ab$
- $(-a)(-b) = ab$
- $-(-a) = a$
- $-(a + b) = -a - b$ and $-(a - b) = -a + b$
- $\dfrac{-a}{b} = -\dfrac{a}{b}$
- $\dfrac{-a}{-b} = \dfrac{a}{b}$

Absolute value of real numbers: $|a| = a$ if a is nonnegative, and $|a| = -a$ if a is negative. Properties of zero were reviewed.

- $a \cdot 0 = 0$ and $\dfrac{0}{a} = 0$ $a \neq 0$
- $\dfrac{a}{0}$ is undefined.
- Zero product property: If $ab = 0$, then $a = 0$ or $b = 0$ (or both).

Properties of fractions were also reviewed.

- $\dfrac{a}{b} \pm \dfrac{c}{d} = \dfrac{ad \pm bc}{bd}$ $b \neq 0$ and $d \neq 0$
- $\dfrac{a}{b} \div \dfrac{c}{d} = \dfrac{a}{b} \cdot \dfrac{d}{c}$ $b \neq 0$, $c \neq 0$, and $d \neq 0$

SECTION A.1 EXERCISES

SKILLS

In Exercises 1–8, classify the following real numbers as rational or irrational:

1. $\frac{11}{3}$

2. $\frac{22}{3}$

3. $2.07172737\ldots$

4. π

5. $2.7766\overline{776677}$

6. $5.22222\overline{2}$

7. $\sqrt{5}$

8. $\sqrt{17}$

In Exercises 9–16, approximate the real number to three decimal places by (a) rounding and (b) truncation.

9. 7.3471

10. 9.2549

11. 2.9949

12. 6.9951

13. 0.234492

14. 1.327491

15. 5.238473

16. 2.118465

In Exercises 17–40, perform the indicated operations in the correct order.

17. $5 + 2 \cdot 3 - 7$

18. $2 + 5 \cdot 4 + 3 \cdot 6$

19. $2 \cdot (5 + 7 \cdot 4 - 20)$

20. $-3 \cdot (2 + 7) + 8 \cdot (7 - 2 \cdot 1)$

21. $2 - 3[4(2 \cdot 3 + 5)]$

22. $4 \cdot 6(5 - 9)$

23. $8 - (-2) + 7$

24. $-10 - (-9)$

25. $-3 - (-6)$

26. $-5 + 2 - (-3)$

27. $x - (-y) - z$

28. $-a + b - (-c)$

29. $-(3x + y)$

30. $-(4a - 2b)$

31. $\dfrac{-3}{(5)(-1)}$

32. $-\dfrac{12}{(-3)(-4)}$

33. $-4 - 6[(5 - 8)(4)]$

34. $\dfrac{-14}{5 - (-2)}$

35. $-(6x - 4y) - (3x + 5y)$

36. $\dfrac{-4x}{6 - (-2)}$

37. $-(3 - 4x) - (4x + 7)$

38. $2 - 3[(4x - 5) - 3x - 7]$

39. $\dfrac{-4(5) - 5}{-5}$

40. $-6(2x + 3y) - [3x - (2 - 5y)]$

In Exercises 41–56, write as a single fraction and simplify.

41. $\dfrac{1}{3} + \dfrac{5}{4}$

42. $\dfrac{1}{2} - \dfrac{1}{5}$

43. $\dfrac{5}{6} - \dfrac{1}{3}$

44. $\dfrac{7}{3} - \dfrac{1}{6}$

45. $\dfrac{3}{2} + \dfrac{5}{12}$

46. $\dfrac{1}{3} + \dfrac{5}{9}$

47. $\dfrac{1}{9} - \dfrac{2}{27}$

48. $\dfrac{3}{7} - \dfrac{(-4)}{3} - \dfrac{5}{6}$

49. $\dfrac{x}{5} + \dfrac{2x}{15}$

50. $\dfrac{y}{3} - \dfrac{y}{6}$

51. $\dfrac{x}{3} - \dfrac{2x}{7}$

52. $\dfrac{y}{10} - \dfrac{y}{15}$

53. $\dfrac{4y}{15} - \dfrac{(-3y)}{4}$

54. $\dfrac{6x}{12} - \dfrac{7x}{20}$

55. $\dfrac{3}{40} + \dfrac{7}{24}$

56. $\dfrac{-3}{10} - \left(\dfrac{-7}{12}\right)$

In Exercises 57–68, perform the indicated operation and simplify, if possible.

57. $\dfrac{2}{7} \cdot \dfrac{14}{3}$

58. $\dfrac{2}{3} \cdot \dfrac{9}{10}$

59. $\dfrac{2}{7} \div \dfrac{10}{3}$

60. $\dfrac{4}{5} \div \dfrac{7}{10}$

61. $\dfrac{4b}{9} \div \dfrac{a}{27} \qquad a \neq 0$

62. $\dfrac{3a}{7} \div \dfrac{b}{21} \qquad b \neq 0$

63. $\dfrac{3x}{10} \div \dfrac{6x}{15} \qquad x \neq 0$

64. $4\dfrac{1}{5} \div 7\dfrac{1}{20}$

65. $\dfrac{3x}{4} \div \dfrac{9}{16y} \qquad y \neq 0$

66. $\dfrac{14m}{2} \cdot \dfrac{4}{7}$

67. $\dfrac{6x}{7} \div \dfrac{3y}{28} \qquad y \neq 0$

68. $2\dfrac{1}{3} \cdot 7\dfrac{5}{6}$

In Exercises 69–72, evaluate the algebraic expression for the specified values.

69. $\dfrac{-c}{2d}$ for $c = -4, d = 3$

70. $2l + 2w$ for $l = 5, w = 10$

71. $\dfrac{m_1 \cdot m_2}{r^2}$ for $m_1 = 3, m_2 = 4, r = 10$

72. $\dfrac{x - \mu}{\sigma}$ for $x = 100, \mu = 70, \sigma = 15$

▪ APPLICATIONS

On December 16, 2007, U.S. debt was estimated at $9,176,366,494,947, and at that time the estimated population was 303,818,361 citizens.

73. U.S. National Debt. Round the debt to the nearest million.

74. U.S. Population. Round the number of citizens to the nearest thousand.

75. U.S. Debt. If the debt is distributed evenly to all citizens, what is the national debt per citizen? Round your answer to the nearest dollar.

76. U.S. Debt. If the debt is distributed evenly to all citizens, what is the national debt per citizen? Round your answer to the nearest cent.

■ CATCH THE MISTAKE

In Exercises 77–80, explain the mistake that is made.

77. Round 13.2749 to two decimal places.

Solution:

The 9, to the right of the 4, causes the 4 to round to 5.	13.275
The 5, to the right of the 7, causes the 7 to be rounded to 8.	13.28

This is incorrect. What mistake was made?

78. Simplify the expression $\frac{2}{3} + \frac{1}{9}$.

Solution:

Add the numerators and denominators.
$$\frac{2+1}{3+9} = \frac{3}{12}$$

Reduce.
$$= \frac{1}{4}$$

This is incorrect. What mistake was made?

79. Simplify the expression $3(x + 5) - 2(4 + y)$.

Solution:

Eliminate parentheses.	$3x + 15 - 8 + y$
Simplify.	$3x + 7 + y$

This is incorrect. What mistake was made?

80. Simplify the expression $-3(x + 2) - (1 - y)$.

Solution:

Eliminate parentheses.	$-3x - 6 - 1 - y$
Simplify.	$-3x - 7 - y$

This is incorrect. What mistake was made?

■ CONCEPTUAL

In Exercises 81–84, determine whether each statement is true or false.

81. The entire student population is a subset of the student athletes.

82. The students who are members of fraternities or sororities are a subset of the entire student population.

83. Every integer is a rational number.

84. A real number can be both rational and irrational.

85. What restrictions are there on x for the following to be true?
$$\frac{3}{x} \div \frac{5}{x} = \frac{3}{5}$$

86. What restrictions are there on x for the following to be true?
$$\frac{x}{2} \div \frac{x}{6} = 3$$

■ CHALLENGE

In Exercises 87 and 88, simplify the expressions.

87. $-2[3(x - 2y) + 7] + [3(2 - 5x) + 10] - 7[-2(x - 3) + 5]$

88. $-2\{-5(y - x) - 2[3(2x - 5) + 7(2) - 4] + 3\} + 7$

■ TECHNOLOGY

89. Use your calculator to evaluate $\sqrt{1260}$. Does the answer appear to be a rational or an irrational number? Why?

90. Use your calculator to evaluate $\sqrt{\dfrac{144}{25}}$. Does the answer appear to be a rational or an irrational number? Why?

91. Use your calculator to evaluate $\sqrt{4489}$. Does the answer appear to be a rational or an irrational number? Why?

92. Use your calculator to evaluate $\sqrt{\dfrac{882}{49}}$. Does the answer appear to be a rational or an irrational number? Why?

SKILLS OBJECTIVES

- Evaluate expressions involving integer exponents.
- Apply properties of exponents.
- Use scientific notation.

CONCEPTUAL OBJECTIVES

- Visualize negative exponents as reciprocals.
- Understand that scientific notation is an effective way to represent very large or very small real numbers.

Integer Exponents

Exponents represent repeated multiplication. For example, $2 \cdot 2 \cdot 2 \cdot 2 \cdot 2 = 2^5$. The 2 that is repeatedly multiplied is called the *base*, and the small number 5 above and to the right of the 2 is called the *exponent*.

Study Tip

a^n: "a raised to the nth power"
a^2: "a squared"
a^3: "a cubed"

DEFINITION   **Natural-Number Exponent**

Let a be a real number and n be a natural number (positive integer); then a^n is defined as

$$a^n = \underbrace{a \cdot a \cdot a \cdots a}_{n \text{ factors}} \qquad (a \text{ appears as a factor } n \text{ times})$$

where n is the **exponent**, or **power**, and a is the **base**.

Technology Tip

```
4^3
              64
8^1
               8
```

```
5^4
             625
(1/2)^5
          .03125
Ans►Frac
            1/32
```

■ Answer: a. 216   b. $\frac{1}{81}$

EXAMPLE 1   **Evaluating Expressions Involving Natural-Number Exponents**

Evaluate the expressions.

a. 4^3 b. 8^1 c. 5^4 d. $\left(\frac{1}{2}\right)^5$

Solution:

a. $4^3 = 4 \cdot 4 \cdot 4 = \boxed{64}$

b. $8^1 = \boxed{8}$

c. $5^4 = 5 \cdot 5 \cdot 5 \cdot 5 = \boxed{625}$

d. $\left(\frac{1}{2}\right)^5 = \frac{1}{2} \cdot \frac{1}{2} \cdot \frac{1}{2} \cdot \frac{1}{2} \cdot \frac{1}{2} = \boxed{\frac{1}{32}}$

■ **YOUR TURN** Evaluate the expressions.

a. 6^3 b. $\left(\frac{1}{3}\right)^4$

We now include exponents in our order of operations:

1. Parentheses
2. Exponents
3. Multiplication/division
4. Addition/subtraction

EXAMPLE 2   **Evaluating Expressions Involving Natural-Number Exponents**

Evaluate the expressions.

a. $(-3)^4$ **b.** -3^4 **c.** $(-2)^3 \cdot 5^2$

Solution:

a. $(-3)^4 = (-3)(-3)(-3)(-3) = \boxed{81}$

b. $-3^4 = \underbrace{-3 \cdot 3 \cdot 3 \cdot 3}_{81} = \boxed{-81}$

c. $(-2)^3 \cdot 5^2 = \underbrace{(-2)(-2)(-2)}_{-8} \cdot \underbrace{5 \cdot 5}_{25} = \boxed{-200}$

■ **YOUR TURN**   Evaluate the expression $-4^3 \cdot 2^3$.

So far, we have discussed only exponents that are natural numbers (positive integers). When the exponent is a negative integer, we use the following property:

NEGATIVE-INTEGER EXPONENT PROPERTY

Let a be any nonzero real number and n be a natural number (positive integer); then

$$a^{-n} = \frac{1}{a^n} \qquad a \neq 0$$

In other words, a base raised to a negative-integer exponent is equivalent to the reciprocal of the base raised to the opposite- (positive-) integer exponent.

EXAMPLE 3   **Evaluating Expressions Involving Negative-Integer Exponents**

Evaluate the expressions.

a. 2^{-4} **b.** $\dfrac{1}{3^{-3}}$ **c.** $4^{-3} \cdot \dfrac{1}{2^{-4}}$ **d.** $-2^3 \cdot \dfrac{1}{(-6)^{-2}}$

Solution:

a. $2^{-4} = \dfrac{1}{2^4} = \boxed{\dfrac{1}{16}}$

b. $\dfrac{1}{3^{-3}} = \dfrac{1}{\left(\frac{1}{3^3}\right)} = 1 \div \left(\dfrac{1}{3^3}\right) = 1 \cdot \dfrac{3^3}{1} = 3^3 = \boxed{27}$

c. $4^{-3} \cdot \dfrac{1}{2^{-4}} = \dfrac{1}{4^3} \cdot 2^4 = \dfrac{16}{64} = \boxed{\dfrac{1}{4}}$

d. $-2^3 \cdot \dfrac{1}{(-6)^{-2}} = \underbrace{-2^3}_{-8} \cdot \underbrace{(-6)^2}_{36} = (-8)(36) = \boxed{-288}$

■ **YOUR TURN**   Evaluate the expressions.

a. $-\dfrac{1}{5^{-2}}$ **b.** $\dfrac{1}{3^{-2}} \cdot 6^{-2}$

Technology Tip

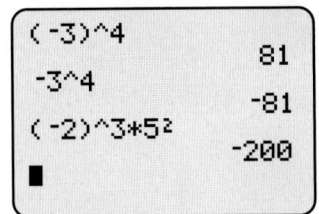

```
(-3)^4
                  81
-3^4
                 -81
(-2)^3*5²
                -200
■
```

■ **Answer:** -512

Study Tip

A negative exponent indicates a reciprocal.

Technology Tip

```
2^(-4)
               .0625
Ans▶Frac
                1/16
1/3^(-3)
                  27
■
```

```
4^(-3)*1/2^(-4)
                 .25
Ans▶Frac
                 1/4
-2^3*1/(-6)^(-2)
                -288
■
```

■ **Answer: a.** -25 **b.** $\frac{1}{4}$

Now we can evaluate expressions involving positive and negative exponents. How do we evaluate an expression with a zero exponent? We define any nonzero real number raised to the zero power as 1.

ZERO-EXPONENT PROPERTY

Let a be any nonzero real number; then

$$a^0 = 1 \qquad a \neq 0$$

EXAMPLE 4   Evaluating Expressions Involving Zero Exponents

Evaluate the expressions.

a. 5^0 **b.** $\dfrac{1}{2^0}$ **c.** $(-3)^0$ **d.** -4^0

Solution:

a. $5^0 = \boxed{1}$ **b.** $\dfrac{1}{2^0} = \dfrac{1}{1} = \boxed{1}$ **c.** $(-3)^0 = \boxed{1}$ **d.** $\dfrac{-4^0}{1} = \boxed{-1}$

We now can evaluate expressions involving integer (positive, negative, or zero) exponents. What about when expressions involving integer exponents are multiplied, divided, or raised to a power?

WORDS

When expressions with the *same base* are *multiplied*, the exponents are *added*.

MATH

$$2^3 \cdot 2^4 = \underbrace{2 \cdot 2 \cdot 2}_{2 \cdot 2 \cdot 2} \cdot \overbrace{2 \cdot 2 \cdot 2 \cdot 2}^{2 \cdot 2 \cdot 2 \cdot 2} = 2^{3+4} = 2^7$$

When expressions with the *same base* are *divided*, the exponents are *subtracted*.

$$\frac{2^5}{2^3} = \frac{2 \cdot 2 \cdot 2 \cdot 2 \cdot 2}{2 \cdot 2 \cdot 2} = \frac{2 \cdot 2}{1} = 2^2$$
or $2^{5-3} = 2^2$

When an expression involving an exponent is *raised to a power*, the exponents are *multiplied*.

$$(2^3)^2 = (8)^2 = 64$$
or $(2^3)^2 = 2^{3 \cdot 2} = 2^6 = 64$

The following table summarizes the *properties of integer exponents*:

PROPERTIES OF INTEGER EXPONENTS

NAME	DESCRIPTION	MATH (LET a AND b BE NONZERO REAL NUMBERS AND m AND n BE INTEGERS)	EXAMPLE
Product property	When multiplying exponentials with the same base, **add** exponents.	$a^m \cdot a^n = a^{m+n}$	$x^2 \cdot x^5 = x^{2+5} = x^7$
Quotient property	When dividing exponentials with the same base, **subtract** the exponents (numerator − denominator).	$\dfrac{a^m}{a^n} = a^{m-n}$	$\dfrac{x^5}{x^3} = x^{5-3} = x^2 \qquad x \neq 0$
Power property	When raising an exponential to a power, **multiply** exponents.	$(a^m)^n = a^{mn}$	$(x^2)^4 = x^{2 \cdot 4} = x^8$
Product-to-a-power property	A product raised to a power is equal to the product of each factor raised to the power.	$(ab)^n = a^n b^n$	$(2x)^3 = 2^3 \cdot x^3 = 8x^3$
Quotient-to-a-power property	A quotient raised to a power is equal to the quotient of the factors raised to the power.	$\left(\dfrac{a}{b}\right)^n = \dfrac{a^n}{b^n}$	$\left(\dfrac{x}{y}\right)^4 = \dfrac{x^4}{y^4} \qquad y \neq 0$

Common Errors Made Using Properties of Exponents

INCORRECT	CORRECT	ERROR
$x^4 \cdot x^3 = x^{12}$	$x^4 \cdot x^3 = x^7$	Exponents should be added (not multiplied).
$\dfrac{x^{18}}{x^6} = x^3$	$\dfrac{x^{18}}{x^6} = x^{12}; x \neq 0$	Exponents should be subtracted (not divided).
$(x^2)^3 = x^8$	$(x^2)^3 = x^6$	Exponents should be multiplied (not raised to power).
$(2x)^3 = 2x^3$	$(2x)^3 = 8x^3$	Both factors (the 2 and the x) should be cubed.
$2^3 \cdot 2^4 = 4^7$	$2^3 \cdot 2^4 = 2^7$	The original common base should be retained.
$2^3 \cdot 3^5 = 6^8$	$2^3 \cdot 3^5$	The properties of integer exponents require the *same* base.

We will now use properties of integer exponents to *simplify exponential expressions*. An exponential expression is **simplified** when:

- All parentheses (groupings) have been eliminated.
- A base appears only once.
- No powers are raised to other powers.
- All exponents are positive.

EXAMPLE 5 Simplifying Exponential Expressions

Simplify the expressions (assume all variables are nonzero).

a. $\left(-2x^2y^3\right)\left(5x^3y\right)$ **b.** $\left(2x^2yz^3\right)^3$ **c.** $\dfrac{25x^3y^6}{-5x^5y^4}$

Solution (a):

Parentheses imply multiplication.

Group the same bases together. $\left(-2x^2y^3\right)\left(5x^3y\right) = (-2)(5)x^2x^3y^3y$

Apply the product property. $= (-2)(5)\underbrace{x^2x^3}_{x^{2+3}}\underbrace{y^3y}_{y^{3+1}}$

Multiply the constants. $= \boxed{-10x^5y^4}$

Solution (b):

Apply the product to a power property. $\left(2x^2yz^3\right)^3 = (2)^3(x^2)^3(y)^3(z^3)^3$

Apply the power property. $= 8x^{2\cdot3}y^{1\cdot3}z^{3\cdot3}$

Simplify. $= \boxed{8x^6y^3z^9}$

Solution (c):

Group the same bases together. $\dfrac{25x^3y^6}{-5x^5y^4} = \left(\dfrac{25}{-5}\right)\left(\dfrac{x^3}{x^5}\right)\left(\dfrac{y^6}{y^4}\right)$

Apply the quotient property. $= (-5)x^{3-5}y^{6-4}$

$= -5x^{-2}y^2$

Apply the negative exponent property. $= \boxed{\dfrac{-5y^2}{x^2}}$

Study Tip

It is customary not to leave negative exponents. Instead, we use the negative exponent property to write exponential expressions with only positive exponents.

■ **YOUR TURN** Simplify the expressions (assume all variables are nonzero).

a. $\left(-3x^3y^2\right)\left(4xy^3\right)$ **b.** $\left(-3xy^3z^2\right)^3$ **c.** $\dfrac{-16x^4y^3}{4xy^7}$

■ **Answer:**

a. $-12x^4y^5$ **b.** $-27x^3y^9z^6$ **c.** $\dfrac{-4x^3}{y^4}$

EXAMPLE 6 **Simplifying Exponential Expressions**

Write each expression so that all exponents are positive (assume all variables are nonzero).

a. $\left(3x^2z^{-4}\right)^{-3}$ **b.** $\dfrac{\left(x^2y^{-3}\right)^2}{\left(x^{-1}y^4\right)^{-3}}$ **c.** $\dfrac{\left(-2xy^2\right)^3}{-\left(6xz^3\right)^2}$

Solution (a):

Apply the product-to-a-power property.

$\left(3x^2z^{-4}\right)^{-3} = (3)^{-3}\left(x^2\right)^{-3}\left(z^{-4}\right)^{-3}$

Apply the power property.

$= 3^{-3}x^{-6}z^{12}$

Apply the negative-integer exponent property.

$= \dfrac{z^{12}}{3^3x^6}$

Evaluate 3^3.

$= \boxed{\dfrac{z^{12}}{27x^6}}$

Solution (b):

Apply the product-to-a-power property.

$\dfrac{\left(x^2y^{-3}\right)^2}{\left(x^{-1}y^4\right)^{-3}} = \dfrac{x^4y^{-6}}{x^3y^{-12}}$

Apply the quotient property.

$= x^{4-3}y^{-6-(-12)}$

Simplify.

$= \boxed{xy^6}$

Solution (c):

Apply the product-to-a-power property on both the numerator and denominator.

$\dfrac{\left(-2xy^2\right)^3}{-\left(6xz^3\right)^2} = \dfrac{(-2)^3(x)^3(y^2)^3}{-(6)^2(x)^2(z^3)^2}$

Apply the power property.

$= \dfrac{-8x^3y^6}{-36x^2z^6}$

Group constant terms and x terms.

$= \left(\dfrac{-8}{-36}\right)\left(\dfrac{x^3}{x^2}\right)\left(\dfrac{y^6}{z^6}\right)$

Apply the quotient property.

$= \left(\dfrac{8}{36}\right)\left(x^{3-2}\right)\left(\dfrac{y^6}{z^6}\right)$

Simplify.

$= \boxed{\dfrac{2xy^6}{9z^6}}$

■ **Answer:** $\dfrac{2t}{s^3}$

■ **YOUR TURN** Simplify the exponential expression and express it in terms of positive exponents $\dfrac{\left(ts^2\right)^{-3}}{\left(2t^4s^3\right)^{-1}}$.

Scientific Notation

You are already familiar with base 10 raised to positive-integer powers. However, it can be inconvenient to write all the zeros out, so we give certain powers of 10 particular names: thousand, million, billion, trillion, and so on. For example, we say there are 300 million U.S. citizens as opposed to writing out 300,000,000 citizens. Or, we say that the national debt is $9 trillion as opposed to writing out $9,000,000,000,000. The following table contains scientific notation for positive exponents and examples of some common prefixes and abbreviations. One of the fundamental applications of scientific notation is *measurement*.

Exponential Form	Decimal Form	Number of Zeros Following the 1	Prefix	Abbreviation	Example
10^1	10	1			
10^2	100	2			
10^3	1,000 (one thousand)	3	kilo-	k	The relay-for-life team ran a total of 80 km (kilometers).
10^4	10,000	4			
10^5	100,000	5			
10^6	1,000,000 (one million)	6	mega-	M	Modern high-powered diesel–electric railroad locomotives typically have a peak power output of 3 to 5 MW (megawatts).
10^7	10,000,000	7			
10^8	100,000,000	8			
10^9	1,000,000,000 (one billion)	9	giga-	G	A flash drive typically has 1 to 4 GB (gigabytes) of storage.
10^{10}	10,000,000,000	10			
10^{11}	100,000,000,000	11			
10^{12}	1,000,000,000,000 (one trillion)	12	tera-	T	Laser systems offer higher frequencies on the order of THz (terahertz).

Notice that 10^8 is a 1 followed by 8 zeros; alternatively, you can start with 1.0 and move the decimal point 8 places to the right (insert zeros). The same type of table can be made for negative-integer powers with base 10. To find the real number associated with exponential form, start with 1.0 and move the decimal a certain number of places to the left (fill in missing decimal places with zeros).

Exponential Form	Decimal Form	Number of Places Decimal (1.0) Moves to the Left	Prefix	Abbreviation	Example
10^{-1}	0.1	1			
10^{-2}	0.01	2			
10^{-3}	0.001 (one thousandth)	3	milli-	m	Excedrin Extra Strength tablets each have 250 mg (milligrams) of acetaminophen.
10^{-4}	0.0001	4			
10^{-5}	0.00001	5			
10^{-6}	0.000001 (one millionth)	6	micro-	μ	A typical laser has a wavelength of 1.55 μm (micrometers[*]).
10^{-7}	0.0000001	7			
10^{-8}	0.00000001	8			
10^{-9}	0.000000001 (one billionth)	9	nano-	n	PSA levels less than 4 ng/ml (nanogram per milliliter of blood) represent low risk for prostate cancer.
10^{-10}	0.0000000001	10			
10^{-11}	0.00000000001	11			
10^{-12}	0.000000000001 (one trillionth)	12	pico-	p	A single yeast cell weighs 44 pg (picograms).

[*]In optics a micrometer is called a micron.

SCIENTIFIC NOTATION

A positive real number can be written in **scientific notation** with the form $c \times 10^n$, where $1 \leq c < 10$ and n is an integer.

Note that c is a real number between 1 and 10. Therefore, 22.5×10^3 is not in scientific notation, but we can convert it to scientific notation: 2.25×10^4.

For example, there are approximately 50 trillion cells in the human body. We write 50 trillion as 50 followed by 12 zeros: 50,000,000,000,000. An efficient way of writing such a large number is using **scientific notation**. Notice that 50,000,000,000,000 is **5** followed by **13** zeros, or in scientific notation, $\mathbf{5 \times 10^{13}}$. Very small numbers can also be written using scientific notation. For example, in laser communications a pulse width is 2 femtoseconds, or 0.000000000000002 second. Notice that if we start with **2.0** and move the decimal point **15** places to the left (adding zeros in between), the result is 0.000000000000002, or in scientific notation, $\mathbf{2 \times 10^{-15}}$.

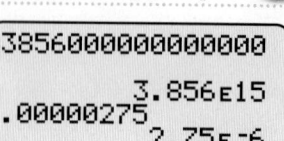

Technology Tip

```
3856000000000000
         3.856E15
.00000275
          2.75E-6
■
```

■ **Answer: a.** 4.52×10^9
       b. 4.3×10^{-7}

EXAMPLE 7 Expressing a Positive Real Number in Scientific Notation

Express the numbers in scientific notation.

a. 3,856,000,000,000,000   **b.** 0.00000275

Solution:

a. Rewrite the number with the implied decimal point.   3,856,000,000,000,000.

   Move the decimal point to the left 15 places. $= \boxed{3.856 \times 10^{15}}$

b. Move the decimal point to the right 6 places. $0.00000275 = \boxed{2.75 \times 10^{-6}}$

■ **YOUR TURN** Express the numbers in scientific notation.

   a. 4,520,000,000   **b.** 0.00000043

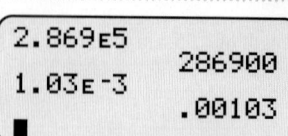

Technology Tip

```
2.869E5
          286900
1.03E-3
          .00103
■
```

■ **Answer: a.** 81,000
       b. 0.000000037

EXAMPLE 8 Converting from Scientific Notation to Decimals

Write each number as a decimal.

a. 2.869×10^5 **b.** 1.03×10^{-3}

Solution:

a. Move the decimal point 5 places to the right (add zeros in between).   286,900 or $\boxed{286,900}$

b. Move the decimal point 3 places to the left (add zeros in between). $\boxed{0.00103}$

■ **YOUR TURN** Write each number as a decimal.

   a. 8.1×10^4 **b.** 3.7×10^{-8}

In this section, we discussed properties of exponents.

Integer Exponents

The following table summarizes integer exponents. Let a be any real number and n be a natural number.

NAME	DESCRIPTION	MATH
Natural-number exponent	Multiply n factors of a.	$a^n = \underbrace{a \cdot a \cdot a \cdots a}_{n \text{ factors}}$
Negative-integer exponent property	A negative exponent implies a reciprocal.	$a^{-n} = \dfrac{1}{a^n} \quad a \neq 0$
Zero-exponent property	Any nonzero real number raised to the zero power is equal to one.	$a^0 = 1 \quad a \neq 0$

Properties of Integer Exponents

The following table summarizes properties of integer exponents. Let a and b be nonzero real numbers and m and n be integers.

NAME	DESCRIPTION	MATH
Product property	When multiplying exponentials with the same base, **add** exponents.	$a^m \cdot a^n = a^{m+n}$
Quotient property	When dividing exponentials with the same base, **subtract** the exponents (numerator − denominator).	$\dfrac{a^m}{a^n} = a^{m-n}$
Power property	When raising an exponential to a power, **multiply** exponents.	$(a^m)^n = a^{mn}$
Product to a power property	A product raised to a power is equal to the product of each factor raised to the power.	$(ab)^n = a^n b^n$
Quotient to a power property	A quotient raised to a power is equal to the quotient of the factors raised to the power.	$\left(\dfrac{a}{b}\right)^n = \dfrac{a^n}{b^n}$

Scientific Notation

Scientific notation is a convenient way of using exponents to represent either very small or very large numbers. Real numbers greater than 1 correspond to positive exponents in scientific notation, whereas real numbers greater than 0 but less than 1 correspond to negative exponents in scientific notation. Scientific notation offers the convenience of multiplying and dividing real numbers by applying properties of exponents.

REAL NUMBER (DECIMAL FORM)	PROCESS	SCIENTIFIC NOTATION
2,357,000,000	Move the implied decimal point to the *left* 9 places.	2.357×10^9
0.00000465	Move the decimal point to the *right* 6 places.	4.65×10^{-6}

▪ SKILLS

In Exercises 1–20, evaluate each expression.

1. 4^4

2. 5^3

3. $(-3)^5$

4. $(-4)^2$

5. -5^2

6. -7^2

7. $-2^2 \cdot 4$

8. $-3^2 \cdot 5$

9. 9^0

10. $-8x^0$

11. 10^{-1}

12. a^{-1}

13. 8^{-2}

14. 3^{-4}

15. $-6 \cdot 5^2$

16. $-2 \cdot 4^2$

17. $8 \cdot 2^{-3} \cdot 5$

18. $5 \cdot 2^{-4} \cdot 32$

19. $-6 \cdot 3^{-2} \cdot 81$

20. $6 \cdot 4^2 \cdot 4^{-4}$

In Exercises 21–50, simplify and write the resulting expression with only positive exponents.

21. $x^2 \cdot x^3$

22. $y^3 \cdot y^5$

23. $x^2 x^{-3}$

24. $y^3 \cdot y^{-7}$

25. $(x^2)^3$

26. $(y^3)^2$

27. $(4a)^3$

28. $(4x^2)^3$

29. $(-2t)^3$

30. $(-3b)^4$

31. $(5xy^2)^2(3x^3y)$

32. $(4x^2y)(2xy^3)^2$

33. $\dfrac{x^5 y^3}{x^7 y}$

34. $\dfrac{y^5 x^2}{y^{-2} x^{-5}}$

35. $\dfrac{(2xy)^2}{(-2xy)^3}$

36. $\dfrac{(-3x^3y)}{-4(x^2y^3)^3}$

37. $\left(\dfrac{b}{2}\right)^{-4}$

38. $\left(\dfrac{c}{3}\right)^{-2}$

39. $(9a^{-2}b^3)^{-2}$

40. $(-9x^{-3}y^2)^{-4}$

41. $\dfrac{a^{-2}b^3}{a^4 b^5}$

42. $\dfrac{x^{-3}y^2}{y^{-4}x^5}$

43. $\dfrac{(x^3 y^{-1})^2}{(xy^2)^{-2}}$

44. $\dfrac{(x^3 y^{-2})^2}{(x^4 y^3)^{-3}}$

45. $\dfrac{3(x^2y)^3}{12(x^{-2}y)^4}$

46. $\dfrac{(-4x^{-2})^2 y^3 z}{(2x^3)^{-2}(y^{-1}z)^4}$

47. $\dfrac{(x^{-4}y^5)^{-2}}{[-2(x^3)^2 y^{-4}]^5}$

48. $-2x^2(-2x^3)^5$

49. $\left[\dfrac{a^2(-xy^4)^3}{x^4(-a^3y^2)^2}\right]^3$

50. $\left[\dfrac{b^{-3}(-x^3y^2)^4}{y^2(-b^2x^5)^3}\right]^5$

51. Write $2^8 \cdot 16^3 \cdot (64)$ as a power of 2.

52. Write $3^9 \cdot 81^5 \cdot (9)$ as a power of 3.

In Exercises 53–60, express the given number in scientific notation.

53. 27,600,000

54. 144,000,000,000

55. 93,000,000

56. 1,234,500,000

57. 0.0000000567

58. 0.00000828

59. 0.000000123

60. 0.000000005

In Exercises 61–66, write the number as a decimal.

61. 4.7×10^7

62. 3.9×10^5

63. 2.3×10^4

64. 7.8×10^{-3}

65. 4.1×10^{-5}

66. 9.2×10^{-8}

▪ APPLICATIONS

67. Astronomy. The distance from Earth to Mars on a particular day can be 200 million miles. Express this distance in scientific notation.

68. Astronomy. The distance from Mars to the Sun on a particular day can be 142 million miles. Express this distance in scientific notation.

69. Lasers. The wavelength of a typical laser used for communications systems is 1.55 microns (or 1.55×10^{-6} meters). Express the wavelength in decimal representation in terms of meters.

70. Lasers. A ruby-red laser has a wavelength of 694 nanometers (or 6.93×10^{-7} meters). Express the wavelength in decimal representation in terms of meters.

▪ CATCH THE MISTAKE

In Exercises 71–74, explain the mistake that is made.

71. Simplify $(-2y^3)(3x^2y^2)$.

Group like factors together.	$(-2)(3)x^2 y^3 y^2$
Use the product property.	$-6x^2 y^6$

This is incorrect. What mistake was made?

72. Simplify $(2xy^2)^3$.

Eliminate the parentheses.	$(2xy^2)^3 = 2x^3 y^6$

This is incorrect. What mistake was made?

73. Simplify $(-2xy^3)^2(5x^2y)^2$.

Apply the product to a power property.	$= (-2)^2 x^2 (y^3)^2 (5)^2 (x^2)^2 y^2$
Apply the power rule.	$= 4x^2 y^9 \cdot 25 x^4 y^2$
Group like factors.	$= (4)(25) x^2 x^4 y^9 y^2$
Apply the product property.	$= 100 x^6 y^{11}$

This is incorrect. What mistake was made?

74. Simplify $\dfrac{-4x^{16}y^9}{8x^2 y^3}$.

Group like factors.	$= \left(\dfrac{-4}{8}\right)\left(\dfrac{x^{16}}{x^2}\right)\left(\dfrac{y^9}{y^3}\right)$
Use the quotient property.	$= -\dfrac{1}{2}x^8 y^3$

This is incorrect. What mistake was made?

▪ CONCEPTUAL

In Exercises 75–78, determine whether each statement is true or false

75. $-2^n = (-2)^n$, if n is an integer.

76. Any nonzero real number raised to the zero power is one.

77. $\dfrac{x^{n+1}}{x^n} = x$ for $x =$ any nonzero real number and n is an integer.

78. $x^{-1} + x^{-2} = x^{-3}$

▪ CHALLENGE

85. Earth's population is approximately 6.6×10^9 people, and there are approximately 1.5×10^8 square kilometers of land on the surface of Earth. If 1 square kilometer is approximately 247 acres, how many acres per person are there on Earth? Round to the nearest tenth of an acre.

86. The population of the United States is approximately 3.0×10^8 people, and there are approximately 3.79×10^6 square miles of land in the United States. If 1 square mile is approximately 640 acres, how many acres per person are there in the United States? Round to the nearest tenth of an acre.

▪ TECHNOLOGY

Scientific calculators have an EXP button that is used for scientific notation. For example, 2.5×10^3 can be input into the calculator by pressing 2.5 EXP 3.

89. Repeat Exercise 85 and confirm your answer with a calculator.

90. Repeat Exercise 86 and confirm your answer with a calculator.

79. Simplify $\left[(a^m)^n\right]^k$.

80. Simplify $\left[(a^{-m})^{-n}\right]^{-k}$.

In Exercises 81–84, evaluate the expression for the given value.

81. $-a^2 + 2ab$ for $a = -2, b = 3$

82. $2a^3 - 7a^2$ for $a = 4$

83. $-16t^2 + 100t$ for $t = 3$

84. $\dfrac{a^3 - 27}{a - 4}$ for $a = -2$

87. Evaluate $\dfrac{(4 \times 10^{-23})(3 \times 10^{12})}{(6 \times 10^{-10})}$. Express your answer in both scientific and decimal notation.

88. Evaluate $\dfrac{(2 \times 10^{-17})(5 \times 10^{13})}{(1 \times 10^{-6})}$. Express your answer in both scientific and decimal notation.

In Exercises 91 and 92, use a graphing utility or scientific calculator to evaluate the expression. Express your answer in scientific notation.

91. $\dfrac{(7.35 \times 10^{-26})(2.19 \times 10^{19})}{(3.15 \times 10^{-21})}$

92. $\dfrac{(1.6849 \times 10^{32})}{(8.12 \times 10^{16})(3.32 \times 10^{-9})}$

SECTION
A.3

POLYNOMIALS: BASIC OPERATIONS

SKILLS OBJECTIVES

▪ Add and subtract polynomials.
▪ Multiply polynomials.

CONCEPTUAL OBJECTIVES

▪ Recognize like terms.
▪ Learn formulas for special products.

The expressions

$$3x^2 - 7x - 1 \qquad 4y^3 - y \qquad 5z$$

are all examples of *polynomials* in one variable. A monomial in one variable, ax^k, is the product of a constant and a variable raised to a nonnegative-integer power. The constant a

is called the **coefficient** of the monomial, and k is called the **degree** of the monomial. A **polynomial** is a sum of monomials. The monomials that are part of a polynomial are called **terms**.

DEFINITION | **Polynomial**

A **polynomial in x** is an algebraic expression of the form

$$a_n x^n + a_{n-1} x^{n-1} + a_{n-2} x^{n-2} + \cdots + a_2 x^2 + a_1 x + a_0$$

where $a_0, a_1, a_2, \ldots, a_n$ are real numbers, with $a_n \neq 0$, and n is a nonnegative integer. The polynomial is of **degree** n, a_n is the **leading coefficient**, and a_0 is the **constant term**.

Polynomials with one, two, and three terms are called **monomials**, **binomials**, and **trinomials**, respectively. Polynomials in one variable are typically written in **standard form** in order of decreasing degrees, and the **degree** of the polynomial is determined by the highest degree (exponent) of any single term.

POLYNOMIAL	STANDARD FORM	SPECIAL NAME	DEGREE	DESCRIPTION
$4x^3 - 5x^7 + 2x - 6$	$-5x^7 + 4x^3 + 2x - 6$	Polynomial	7	A *seventh*-degree polynomial in x
$5 + 2y^3 - 4y$	$2y^3 - 4y + 5$	Trinomial	3	A *third*-degree trinomial in y
$7z^2 + 2$	$7z^2 + 2$	Binomial	2	A *second*-degree binomial in z
$-17x^5$	$-17x^5$	Monomial	5	A *fifth*-degree monomial in x

EXAMPLE 1 **Writing Polynomials in Standard Form**

Write the polynomials in standard form and state their degree, leading coefficient, and constant term.

a. $4x - 9x^5 + 2$
b. $3 - x^2$
c. $3x^2 - 8 + 14x^3 - 20x^8 + x$
d. $-7x^3 + 25x$

Solution:

	Standard Form	Degree	Leading Coefficient	Constant Term
a.	$-9x^5 + 4x + 2$	5	-9	2
b.	$-x^2 + 3$	2	-1	3
c.	$-20x^8 + 14x^3 + 3x^2 + x - 8$	8	-20	-8
d.	$-7x^3 + 25x$	3	-7	0

■ **Answer:** $-4x^3 + 17x^2 - x + 5$
Degree: 3
Leading coefficient: -4
Constant term: 5

■ **YOUR TURN** Write the polynomial in standard form and state its degree, leading coefficient, and constant term.

$$17x^2 - 4x^3 + 5 - x$$

Adding and Subtracting Polynomials

Polynomials are added and subtracted by combining *like terms*. **Like terms** are terms having the same variables and exponents. Like terms can be combined by adding their coefficients.

WORDS	MATH
Identify like terms.	$\underline{3x^2} + 2x + \underline{4x^2} + 5$
Add coefficients of like terms.	$7x^2 + 2x + 5$

Note: The $2x$ and 5 could not be combined because they are *not* like terms.

EXAMPLE 2 **Adding Polynomials**

Find the sum and simplify $\left(5x^2 - 2x + 3\right) + \left(3x^3 - 4x^2 + 7\right)$.

Solution:

Eliminate parentheses.	$5x^2 - 2x + 3 + 3x^3 - 4x^2 + 7$
Identify like terms.	$\underline{5x^2} - 2x + \underline{\underline{3}} + 3x^3 - \underline{4x^2} + \underline{\underline{7}}$
Combine like terms.	$x^2 - 2x + 10 + 3x^3$
Write in standard form.	$\boxed{3x^3 + x^2 - 2x + 10}$

- - - - -

■ **YOUR TURN** Find the sum and simplify.

$$\left(3x^2 + 5x - 2x^5\right) + \left(6x^3 - x^2 + 11\right)$$

■ **Answer:**
$-2x^5 + 6x^3 + 2x^2 + 5x + 11$

EXAMPLE 3 **Subtracting Polynomials**

Find the difference and simplify $\left(3x^3 - 2x + 1\right) - \left(x^2 + 5x - 9\right)$.

COMMON MISTAKE

Distributing the negative to only the first term in the second polynomial.

✪ CORRECT	✖ INCORRECT
Eliminate the parentheses.	**ERROR:**
$3x^3 - 2x + 1 - x^2 - 5x + 9$	$3x^3 - 2x + 1 - x^2 + 5x - 9$
Identify like terms.	Don't forget to distribute the negative through the entire second polynomial.
$3x^3 - \underline{2x} + \underline{\underline{1}} - x^2 - \underline{5x} + \underline{\underline{9}}$	
Combine like terms.	
$\boxed{3x^3 - x^2 - 7x + 10}$	

Study Tip

When subtracting polynomials, it is important to distribute the negative through *all* of the terms in the *second* polynomial.

- - - - -

■ **YOUR TURN** Find the difference and simplify.

$$\left(-7x^2 - x + 5\right) - \left(2 - x^3 + 3x\right)$$

■ **Answer:** $x^3 - 7x^2 - 4x + 3$

Multiplying Polynomials

The product of two monomials is found by using the properties of exponents (Section A.2). For example,

$$\left(-5x^3\right)\left(9x^2\right) = (-5)(9)x^{3+2} = -45x^5$$

To multiply a monomial and a polynomial, we use the distributive property (Section A.1).

EXAMPLE 4   Multiplying a Monomial and a Polynomial

Find the product and simplify $5x^2\left(3x^5 - x^3 + 7x - 4\right)$.

Solution:

Use the distributive property.

$$5x^2\left(3x^5 - x^3 + 7x - 4\right)$$

$$= 5x^2\left(3x^5\right) - 5x^2\left(x^3\right) + 5x^2(7x) - 5x^2(4)$$

Multiply each set of monomials.

$$= \boxed{15x^7 - 5x^5 + 35x^3 - 20x^2}$$

■ **Answer:** $12x^4 - 6x^3 + 3x^2$

■ **YOUR TURN** Find the product and simplify $3x^2\left(4x^2 - 2x + 1\right)$.

How do we multiply two polynomials if neither one is a monomial? For example, how do we find the product of a binomial and a trinomial such as $(2x - 5)\left(x^2 - 2x + 3\right)$? Notice that the binomial is a combination of two monomials. Therefore, we treat each monomial, $2x$ and -5, separately and then combine our results. In other words, use the distributive property repeatedly.

WORDS	**MATH**
Apply the distributive property.	$(2x - 5)\left(x^2 - 2x + 3\right) = 2x\left(x^2 - 2x + 3\right) - 5\left(x^2 - 2x + 3\right)$
Apply the distributive property.	$= (2x)\left(x^2\right) + (2x)(-2x) + (2x)(3) - 5\left(x^2\right) - 5(-2x) - 5(3)$
Multiply the monomials.	$= 2x^3 - 4x^2 + 6x - 5x^2 + 10x - 15$
Combine like terms.	$= 2x^3 - 9x^2 + 16x - 15$

EXAMPLE 5   Multiplying Two Polynomials

Multiply and simplify $\left(2x^2 - 3x + 1\right)\left(x^2 - 5x + 7\right)$.

Solution:

Multiply each term of the first trinomial by the entire second trinomial.

$$= 2x^2\left(x^2 - 5x + 7\right) - 3x\left(x^2 - 5x + 7\right) + 1\left(x^2 - 5x + 7\right)$$

Identify like terms.

$$= 2x^4 - 10x^3 + 14x^2 - 3x^3 + 15x^2 - 21x + x^2 - 5x + 7$$

Combine like terms.

$$= \boxed{2x^4 - 13x^3 + 30x^2 - 26x + 7}$$

■ **Answer:**
$-3x^5 + x^4 + x^3 - 14x^2 + 14x - 20$

■ **YOUR TURN** Multiply and simplify $\left(-x^3 + 2x - 4\right)\left(3x^2 - x + 5\right)$.

Special Products

The method outlined for multiplying polynomials works for *all* products of polynomials. For the special case when both polynomials are binomials, the **FOIL method** can also be used.

Study Tip

When the binomials are of the form $(ax + b)(cx + d)$, the outer and inner terms will be like terms and can be combined.

WORDS	MATH
Apply the distributive property.	$(5x - 1)(2x + 3) = 5x(2x + 3) - 1(2x + 3)$
Apply the distributive property.	$= 5x(2x) + 5x(3) - 1(2x) - 1(3)$
Multiply each set of monomials.	$= 10x^2 + 15x - 2x - 3$
Combine like terms.	$= 10x^2 + 13x - 3$

The FOIL method finds the products of the **F**irst terms, **O**uter terms, **I**nner terms, and **L**ast terms.

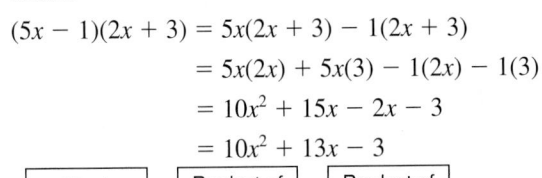

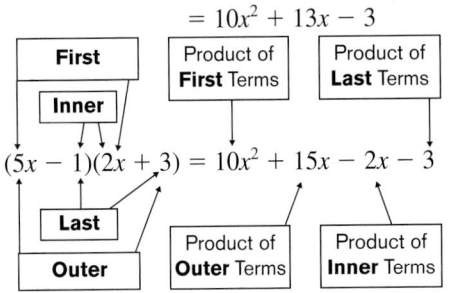

$(5x - 1)(2x + 3) = 10x^2 + 15x - 2x - 3$

EXAMPLE 6 Multiplying Binomials Using the FOIL Method

Multiply $(3x + 1)(2x - 5)$ using the FOIL method.

Solution:

Multiply the **first** terms.	$(3x)(2x) = 6x^2$
Multiply the **outer** terms.	$(3x)(-5) = -15x$
Multiply the **inner** terms.	$(1)(2x) = 2x$
Multiply the **last** terms.	$(1)(-5) = -5$
Add the first, outer, inner, and last terms, and identify the like terms.	$(3x + 1)(2x - 5) = 6x^2 - 15x + 2x - 5$
Combine like terms.	$= \boxed{6x^2 - 13x - 5}$

■ **YOUR TURN** Multiply $(2x - 3)(5x - 2)$.

■ **Answer:** $10x^2 - 19x + 6$

There are some products of binomials that occur frequently in algebra and are given special names. Example 7 illustrates the *difference of two squares* and *perfect squares*.

EXAMPLE 7 Multiplying Binomials Resulting in Special Products

Find: **a.** $(x - 5)(x + 5)$ **b.** $(x + 5)^2$ **c.** $(x - 5)^2$

Solution:

a. $(x - 5)(x + 5) = x^2 + 5x - 5x - 5^2 = x^2 - 5^2 = \boxed{x^2 - 25}$

b. $(x + 5)^2 = (x + 5)(x + 5) = x^2 + 5x + 5x + 5^2 = x^2 + 2(5x) + 5^2 = \boxed{x^2 + 10x + 25}$

c. $(x - 5)^2 = (x - 5)(x - 5) = x^2 - 5x - 5x + 5^2 = x^2 - 2(5x) + 5^2 = \boxed{x^2 - 10x + 25}$

Let a and b be any real number, variable, or algebraic expression in the following special products:

DIFFERENCE OF TWO SQUARES

$$(a + b)(a - b) = a^2 - b^2$$

PERFECT SQUARES

Square of a binomial sum: $(a + b)^2 = (a + b)(a + b) = a^2 + 2ab + b^2$

Square of a binomial difference: $(a - b)^2 = (a - b)(a - b) = a^2 - 2ab + b^2$

EXAMPLE 8   Finding the Square of a Binomial Sum

Find $(x + 3)^2$.

COMMON MISTAKE

Forgetting the middle term, which is *twice* the product of the two terms in the binomial.

CORRECT

$(x + 3)^2 = (x + 3)(x + 3)$

$\qquad = x^2 + 3x + 3x + 9$

$\qquad = x^2 + 6x + 9$

INCORRECT

ERROR:

$(x + 3)^2 \neq x^2 + 9$

Don't forget the middle term, which is twice the product of the two terms in the binomial.

EXAMPLE 9   Using Special Product Formulas

Find:

a. $(2x - 1)^2$ **b.** $(3 + 2y)^2$ **c.** $(4x + 3)(4x - 3)$

Solution (a):

Write the square of a binomial difference formula.

$$(a - b)^2 = a^2 - 2ab + b^2$$

Let $a = 2x$ and $b = 1$.

$$(2x - 1)^2 = (2x)^2 - 2(2x)(1) + 1^2$$

Simplify.

$$= \boxed{4x^2 - 4x + 1}$$

Solution (b):

Write the square of a binomial sum formula.

$$(a + b)^2 = a^2 + 2ab + b^2$$

Let $a = 3$ and $b = 2y$.

$$(3 + 2y)^2 = (3)^2 + 2(3)(2y) + (2y)^2$$

Simplify.

$$= 9 + 12y + 4y^2$$

Write in standard form.

$$= \boxed{4y^2 + 12y + 9}$$

Solution (c):

Write the difference of two squares formula.

$$(a + b)(a - b) = a^2 - b^2$$

Let $a = 4x$ and $b = 3$.

$$(4x + 3)(4x - 3) = (4x)^2 - 3^2$$

Simplify.

$$= \boxed{16x^2 - 9}$$

■ **YOUR TURN** Find:

a. $(3x + 1)^2$ b. $(1 - 3y)^2$ c. $(3x + 2)(3x - 2)$

■ **Answer: a.** $9x^2 + 6x + 1$
b. $9y^2 - 6y + 1$
c. $9x^2 - 4$

EXAMPLE 10   Cubing a Binomial

Find:

a. $(x + 2)^3$ b. $(x - 2)^3$

Study Tip

$$(x + 2)^3 \neq x^3 + 8$$
$$(x - 2)^3 \neq x^3 - 8$$

Solution (a):

Write the cube as a product of three binomials.

$$(x + 2)^3 = (x + 2)\underbrace{(x + 2)(x + 2)}_{(x + 2)^2}$$

Apply the perfect square formula.

$$= (x + 2)(x^2 + 4x + 4)$$

Apply the distributive property.

$$= x(x^2 + 4x + 4) + 2(x^2 + 4x + 4)$$

Apply the distributive property.

$$= x^3 + 4x^2 + 4x + 2x^2 + 8x + 8$$

Combine like terms.

$$= \boxed{x^3 + 6x^2 + 12x + 8}$$

Solution (b):

Write the cube as a product of three binomials.

$$(x - 2)^3 = (x - 2)\underbrace{(x - 2)(x - 2)}_{(x - 2)^2}$$

Apply the perfect square formula.

$$= (x - 2)(x^2 - 4x + 4)$$

Apply the distributive property.

$$= x(x^2 - 4x + 4) - 2(x^2 - 4x + 4)$$

Apply the distributive property.

$$= x^3 - 4x^2 + 4x - 2x^2 + 8x - 8$$

Combine like terms.

$$= \boxed{x^3 - 6x^2 + 12x - 8}$$

The formulas are for *cubes of binomials* (*perfect cubes*), as well as a *sum* or *difference* of *cubes*.

PERFECT CUBES

Cube of a binomial sum: $(a + b)^3 = a^3 + 3a^2b + 3ab^2 + b^3$
Cube of a binomial difference: $(a - b)^3 = a^3 - 3a^2b + 3ab^2 - b^3$

EXAMPLE 11   **Applying the Special Product Formulas**

Find:

a. $(2x + 1)^3$ **b.** $(2x - 5)^3$

Solution (a):

Write the cube of a binomial sum formula.

$$(a + b)^3 = a^3 + 3a^2b + 3ab^2 + b^3$$

Let $a = 2x$ and $b = 1$.

$$(2x + 1)^3 = (2x)^3 + 3(2x)^2(1) + 3(2x)(1)^2 + 1^3$$

Simplify.

$$= \boxed{8x^3 + 12x^2 + 6x + 1}$$

Solution (b):

Write the cube of a binomial difference formula.

$$(a - b)^3 = a^3 - 3a^2b + 3ab^2 - b^3$$

Let $a = 2x$ and $b = 5$.

$$(2x - 5)^3 = (2x)^3 - 3(2x)^2(5) + 3(2x)(5)^2 - 5^3$$

Simplify.

$$= \boxed{8x^3 - 60x^2 + 150x - 125}$$

■ **Answer:**
$27x^3 - 108x^2 + 144x + 64$

■ **YOUR TURN** Find $(3x - 4)^3$.

EXAMPLE 12   **Applying the Special Product Formulas for Binomials in Two Variables**

Find $(2x - 3y)^2$.

Solution:

Write the square of a binomial difference formula.

$$(a - b)^2 = (a - b)(a - b) = a^2 - 2ab + b^2$$

Let $a = 2x$ and $b = 3y$.

$$(2x - 3y)^2 = (2x)^2 - 2(2x)(3y) + (3y)^2$$

Simplify.

$$= \boxed{4x^2 - 12xy + 9y^2}$$

■ **Answer:** $9x^2 - 12xy + 4y^2$

■ **YOUR TURN** Find $(3x - 2y)^2$.

In this section, polynomials were defined. Polynomials with one, two, and three terms are called monomials, binomials, and trinomials, respectively. Polynomials are added and subtracted by combining like terms. Polynomials are multiplied by distributing the monomials in the first polynomial throughout the second polynomial. In the special case of the product of two binomials, the FOIL method can also be used. The following are special products of binomials.

Difference of Two Squares

$$(a + b)(a - b) = a^2 - b^2$$

Perfect Squares

Square of a binomial sum.

$$(a + b)^2 = (a + b)(a + b) = a^2 + 2ab + b^2$$

Square of a binomial difference.

$$(a - b)^2 = (a - b)(a - b) = a^2 - 2ab + b^2$$

Perfect Cubes

Cube of a binomial sum.

$$(a + b)^3 = a^3 + 3a^2b + 3ab^2 + b^3$$

Cube of a binomial difference.

$$(a - b)^3 = a^3 - 3a^2b + 3ab^2 - b^3$$

SECTION
A.3 EXERCISES

■ SKILLS

In Exercises 1–8, write the polynomial in standard form and state the degree of the polynomial.

1. $5x^2 - 2x^3 + 16 - 7x^4$

2. $7x^3 - 9x^2 + 5x - 4$

3. $4x + 3 - 6x^3$

4. $5x^5 - 7x^3 + 8x^4 - x^2 + 10$

5. 15

6. -14

7. $y - 2$

8. $x - 5$

In Exercises 9–24, add or subtract the polynomials, gather like terms, and write the simplified expression in standard form.

9. $\left(2x^2 - x + 7\right) + \left(-3x^2 + 6x - 2\right)$

10. $\left(3x^2 + 5x + 2\right) + \left(2x^2 - 4x - 9\right)$

11. $\left(-7x^2 - 5x - 8\right) - \left(-4x - 9x^2 + 10\right)$

12. $\left(8x^3 - 7x^2 - 10\right) - \left(7x^3 + 8x^2 - 9x\right)$

13. $\left(2x^4 - 7x^2 + 8\right) - \left(3x^2 - 2x^4 + 9\right)$

14. $\left(4x^2 - 9x - 2\right) - \left(5 - 3x - 5x^2\right)$

15. $\left(7z^2 - 2\right) - \left(5z^2 - 2z + 1\right)$

16. $\left(25y^3 - 7y^2 + 9y\right) - \left(14y^2 - 7y + 2\right)$

17. $\left(3y^3 - 7y^2 + 8y - 4\right) - \left(14y^3 - 8y + 9y^2\right)$

18. $\left(2x^2 + 3xy\right) - \left(x^2 + 8xy - 7y^2\right)$

19. $(6x - 2y) - 2(5x - 7y)$

20. $3a - \left[2a^2 - \left(5a - 4a^2 + 3\right)\right]$

21. $\left(2x^2 - 2\right) - (x + 1) - \left(x^2 - 5\right)$

22. $\left(3x^3 + 1\right) - \left(3x^2 - 1\right) - (5x - 3)$

23. $\left(4t - t^2 - t^3\right) - \left(3t^2 - 2t + 2t^3\right) + \left(3t^3 - 1\right)$

24. $\left(-z^3 - 2z^2\right) + \left(z^2 - 7z + 1\right) - \left(4z^3 + 3z^2 - 3z + 2\right)$

In Exercises 25–64, multiply the polynomials and write the expressions in standard form.

25. $5xy^2(7xy)$

26. $6z\left(4z^3\right)$

27. $2x^3\left(1 - x + x^2\right)$

28. $-4z^2\left(2 + z - z^2\right)$

29. $-2x^2\left(5 + x - 5x^2\right)$

30. $-\frac{1}{2}z\left(2z + 4z^2 - 10\right)$

31. $\left(x^2 + x - 2\right)2x^3$

32. $\left(x^2 - x + 2\right)3x^3$

33. $2ab^2\left(a^2 + 2ab - 3b^2\right)$

34. $bc^3d^2\left(b^2c + cd^3 - b^2d^4\right)$

35. $(2x + 1)(3x - 4)$

36. $(3z - 1)(4z + 7)$

37. $(x + 2)(x - 2)$

38. $(y - 5)(y + 5)$

39. $(2x + 3)(2x - 3)$

40. $(5y + 1)(5y - 1)$

41. $(2x - 1)(1 - 2x)$

42. $(4b - 5y)(4b + 5y)$

43. $\left(2x^2 - 3\right)\left(2x^2 + 3\right)$

44. $(4xy - 9)(4xy + 9)$

45. $\left(7y - 2y^2\right)\left(y - y^2 + 1\right)$

46. $\left(4 - t^2\right)\left(6t + 1 - t^2\right)$

47. $(x + 1)\left(x^2 - 2x + 3\right)$

48. $(x + 3)\left(x^2 - 3x + 9\right)$

49. $(t - 2)^2$

50. $(t - 3)^2$

51. $(z + 2)^2$

52. $(z + 3)^2$

53. $[(x + y) - 3]^2$

54. $\left(2x^2 + 3y\right)^2$

55. $(5x - 2)^2$

56. $(x + 1)\left(x^2 + x + 1\right)$

57. $y(3y + 4)(2y - 1)$

58. $p^2(p + 1)(p - 2)$

59. $\left(x^2 + 1\right)\left(x^2 - 1\right)$

60. $(t - 5)^2(t + 5)^2$

61. $(b - 3a)(a + 2b)(b + 3a)$

62. $(x - 2y)\left(x^2 + 2xy + 4y^2\right)$

63. $(x + y - z)(2x - 3y + 5z)$

64. $\left(5b^2 - 2b + 1\right)\left(3b - b^2 + 2\right)$

■ APPLICATIONS

In Exercises 65–68, profit is equal to revenue minus cost:
$P = R - C.$

65. Profit. Donna decides to sell fabric cord covers on eBay for $20 a piece. The material for each cord cover costs $9 and it costs her $100 a month to advertise on eBay. Let x be the number of cord covers sold. Write a polynomial representing her monthly profit.

66. Profit. Calculators are sold for $25.00 each. Advertising costs are $75.00 per month. Let x be the number of calculators sold. Write a polynomial representing the monthly profit earned by selling x calculators.

67. Profit. If the revenue associated with selling x units of a product is $R = -x^2 + 100x$, and the cost associated with producing x units of the product is $C = -100x + 7500$, find the polynomial that represents the profit of making and selling x units.

68. Profit. A business sells a certain quantity x of items. The revenue generated by selling x items is given by the equation $R = -\frac{1}{2}x^2 + 50x$. The costs are given by $C = 8000 - 150x$. Find a polynomial representing the net profit of this business when x items are sold.

69. Volume of a Box. A rectangular sheet of cardboard is to be used in the construction of a box by cutting out squares of side length x from each corner and turning up the sides. Suppose the dimensions of the original rectangle are 15 inches by 8 inches. Determine a polynomial in x that would give the volume of the box.

70. Volume of a Box. Suppose a box is to be constructed from a square piece of material of side length x by cutting out a 2-inch square from each corner and turning up the sides. Express the volume of the box as a polynomial in the variable x.

71. Geometry. Suppose a running track is constructed of a rectangular portion that measures $2x$ feet wide by $2x + 5$ feet long. Each end of the rectangular portion consists of a semicircle whose diameter is $2x$. Write a polynomial that determines the

a. perimeter of the track in terms of the variable x.
b. area of the track in terms of x.

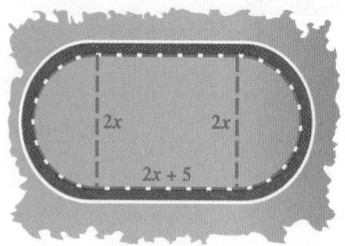

72. Geometry. A right circular cylinder whose radius is r and whose height is $2r$ is surmounted by a hemisphere of radius r.

a. Find a polynomial in the variable r that represents the volume of the "silo" shape.
b. Find a polynomial in r that represents the total surface area of the silo.

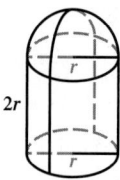

73. Engineering. The force of an electrical field is given by the equation $F = k\dfrac{q_1 q_2}{r^2}$. Suppose $q_1 = x$, $q_2 = 3x$, and $r = 10x$. Find a polynomial representing the force of the electrical field in terms of the variable x.

74. Engineering. If a football (or other projectile) is thrown upward, its height above the ground is given by the equation $s = 16t^2 + v_0 t + s_0$, where v_0 and s_0 are the initial velocity and initial height of the football, respectively, and t is the time in seconds. Suppose the football is thrown from the top of a building that is 192 feet tall, with an initial speed of 96 feet per second.

a. Write the polynomial that gives the height of the football in terms of the variable t (time).
b. What is the height of the football after 2 seconds have elapsed? Will the football hit the ground after 2 seconds?

■ CATCH THE MISTAKE

In Exercises 75 and 76, explain the mistake that is made.

75. Subtract and simplify $(2x^2 - 5) - (3x - x^2 + 1)$.

Solution:

Eliminate the parentheses. $2x^2 - 5 - 3x - x^2 + 1$

Collect like terms. $x^2 - 3x - 4$

This is incorrect. What mistake was made?

76. Simplify $(2 + x)^2$.

Solution:

Write the square of the binomial as the sum of the squares. $(2 + x)^2 = 2^2 + x^2$

Simplify. $= x^2 + 4$

This is incorrect. What mistake was made?

▪ CONCEPTUAL

In Exercises 77–80, determine whether each statement is true or false.

77. All binomials are polynomials.

78. The product of two monomials is a binomial.

79. $(x + y)^3 = x^3 - y^3$

80. $(x - y)^2 = x^2 + y^2$

In Exercises 81 and 82, let m and n be real numbers and $m > n$.

81. What degree is the *product* of a polynomial of degree n and a polynomial of degree m?

82. What degree is the *sum* of a polynomial of degree n and a polynomial of degree m?

▪ CHALLENGE

In Exercises 83–86, perform the indicated operations and simplify.

83. $\left(7x - 4y^2\right)^2 \left(7x + 4y^2\right)^2$

84. $\left(3x - 5y^2\right)^2 \left(3x + 5y^2\right)^2$

85. $(x - a)\left(x^2 + ax + a^2\right)$

86. $(x + a)\left(x^2 - ax + a^2\right)$

▪ TECHNOLOGY

87. Use a graphing utility to plot the graphs of the three expression $(2x + 3)(x - 4)$, $2x^2 + 5x - 12$, and $2x^2 - 5x - 12$. Which two graphs agree with each other?

88. Use a graphing utility to plot the graphs of the three expressions $(x + 5)^2$, $x^2 + 25$, and $x^2 + 10x + 25$. Which two graphs agree with each other?

SECTION
A.4 FACTORING POLYNOMIALS

SKILLS OBJECTIVES

- Factor out the greatest common factor.
- Factor the difference of two squares.
- Factor perfect squares.
- Factor the sum or difference of two cubes.
- Factor a trinomial as a product of binomials.
- Factor by grouping.

CONCEPTUAL OBJECTIVES

- Identify common factors.
- Understand that factoring has its basis in the distributive property.
- Identify prime (irreducible) polynomials.
- Develop a general strategy for factoring polynomials.

In Section A.3, we discussed multiplying polynomials. In this section, we examine the reverse of that process, which is called *factoring*. Consider the following product:

$$(x + 3)(x + 1) = x^2 + 4x + 3$$

To *factor* the resulting polynomial, you reverse the process to undo the multiplication:

$$x^2 + 4x + 3 = (x + 3)(x + 1)$$

The polynomials $(x + 3)$ and $(x + 1)$ are called **factors** of the polynomial $x^2 + 4x + 3$. The process of writing a polynomial as a product is called **factoring**. In Chapter 0, we solved linear and quadratic equations by factoring.

In this section, we will restrict our discussion to factoring polynomials with integer coefficients, which is called **factoring over the integers**. If a polynomial cannot be factored using integer coefficients, then it is **prime** or irreducible over the integers. When a polynomial is written as a product of prime polynomials, then the polynomial is said to be **factored completely**.

Greatest Common Factor

The simplest type of factoring of polynomials occurs when there is a factor common to every term of the polynomial. This **common factor** is a monomial that can be "factored out" by applying the distributive property in reverse:

$$ab + ac = a(b + c)$$

For example, $4x^2 - 6x$ can be written as $2x(x) - 2x(3)$. Notice that $2x$ is a common factor to both terms, so the distributive property tells us we can factor this polynomial to yield $2x(x - 3)$. Although 2 is a common factor and x is a common factor, the monomial $2x$ is called the *greatest common factor*.

GREATEST COMMON FACTOR

The monomial ax^k is called the **greatest common factor (GCF)** of a polynomial in x with integer coefficients if *both* of the following are true:

- a is the *greatest* integer factor common to all of the polynomial coefficients.
- k is the *smallest* exponent on x found among all of the terms of the polynomial.

POLYNOMIAL	GCF	WRITE EACH TERM AS A PRODUCT OF GCF AND REMAINING FACTOR	FACTORED FORM
$7x + 21$	7	$7(x) + 7(3)$	$7(x + 3)$
$3x^2 + 12x$	$3x$	$3x(x) + 3x(4)$	$3x(x + 4)$
$4x^3 + 2x + 6$	2	$2(2x^3) + 2x + 2(3)$	$2(2x^3 + x + 3)$
$6x^4 - 9x^3 + 12x^2$	$3x^2$	$3x^2(2x^2) - 3x^2(3x) + 3x^2(4)$	$3x^2(2x^2 - 3x + 4)$
$-5x^4 + 25x^3 - 20x^2$	$-5x^2$	$-5x^2x^2 - 5x^2(-5x) - 5x^2(4)$	$-5x^2(x^2 - 5x + 4)$

EXAMPLE 1   Factoring Polynomials by Extracting the Greatest Common Factor

Factor:

a. $6x^5 - 18x^4$ **b.** $6x^5 - 10x^4 - 8x^3 + 12x^2$

Solution (a):

Identify the greatest common factor. $6x^4$

Write each term as a product with the GCF as a factor. $6x^5 - 18x^4 = 6x^4(x) - 6x^4(3)$

Factor out the GCF. $= \boxed{6x^4(x - 3)}$

Solution (b):

Identify the greatest common factor. $\qquad$ $2x^2$

Write each term as a product with the GCF as a factor.

$$6x^5 - 10x^4 - 8x^3 + 12x^2 = 2x^2(3x^3) - 2x^2(5x^2) - 2x^2(4x) + 2x^2(6)$$

Factor out the GCF. $\qquad = \boxed{2x^2(3x^3 - 5x^2 - 4x + 6)}$

■ **YOUR TURN** Factor:

$\qquad$ **a.** $12x^3 - 4x$ $\qquad$ **b.** $3x^5 - 9x^4 + 12x^3 - 6x^2$

■ **Answer: a.** $4x(3x^2 - 1)$
$\qquad$ **b.** $3x^2(x^3 - 3x^2 + 4x - 2)$

Factoring Formulas: Special Polynomial Forms

The first step in factoring polynomials is to look for a common factor. If there is no common factor, then we look for special polynomial forms.

Difference of two squares:	$a^2 - b^2 = (a + b)(a - b)$
Perfect squares:	$a^2 + 2ab + b^2 = (a + b)^2$
	$a^2 - 2ab + b^2 = (a - b)^2$
Sum of two cubes:	$a^3 + b^3 = (a + b)(a^2 - ab + b^2)$
Difference of two cubes:	$a^3 - b^3 = (a - b)(a^2 + ab + b^2)$

EXAMPLE 2 **Factoring the Difference of Two Squares**

Factor:

a. $x^2 - 9$ $\qquad$ **b.** $4x^2 - 25$ $\qquad$ **c.** $x^4 - 16$

Solution (a):

Rewrite as the difference of two squares. $\qquad x^2 - 9 = x^2 - 3^2$

Let $a = x$ and $b = 3$ in $a^2 - b^2 = (a + b)(a - b)$. $\qquad = \boxed{(x + 3)(x - 3)}$

Solution (b):

Rewrite as the difference of two squares. $\qquad 4x^2 - 25 = (2x)^2 - 5^2$

Let $a = 2x$ and $b = 5$ in $a^2 - b^2 = (a + b)(a - b)$. $\qquad = \boxed{(2x + 5)(2x - 5)}$

Solution (c):

Rewrite as the difference of two squares. $\qquad x^4 - 16 = (x^2)^2 - 4^2$

Let $a = x^2$ and $b = 4$ in $a^2 - b^2 = (a + b)(a - b)$. $\qquad = (x^2 + 4)(x^2 - 4)$

Note that $x^2 - 4$ is also a difference of two squares. $\qquad = \boxed{(x + 2)(x - 2)(x^2 + 4)}$

■ **YOUR TURN** Factor:

$\qquad$ **a.** $x^2 - 4$ $\qquad$ **b.** $9x^2 - 16$ $\qquad$ **c.** $x^4 - 81$

■ **Answer: a.** $(x + 2)(x - 2)$
$\qquad$ **b.** $(3x + 4)(3x - 4)$
$\qquad$ **c.** $(x^2 + 9)(x - 3)(x + 3)$

A trinomial is a perfect square if it has the form $a^2 \pm 2ab + b^2$. Notice that:

- The first term and third term are perfect squares.
- The middle term is twice the product of the bases of these two perfect squares.
- The sign of the middle term determines the sign of the factored form:

$$a^2 \pm 2ab + b^2 = (a \pm b)^2$$

EXAMPLE 3 Factoring Trinomials That Are Perfect Squares

Factor:

a. $x^2 + 6x + 9$ **b.** $x^2 - 10x + 25$ **c.** $9x^2 - 12x + 4$

Solution (a):

Rewrite the trinomial so that the first and third terms are perfect squares.

$x^2 + 6x + 9 = x^2 + 6x + 3^2$

Notice that if we let $a = x$ and $b = 3$ in $a^2 + 2ab + b^2 = (a + b)^2$, then the middle term $6x$ is $2ab$.

$x^2 + 6x + 9 = x^2 + 2(3x) + 3^2$
$= \boxed{(x + 3)^2}$

Solution (b):

Rewrite the trinomial so that the first and third terms are perfect squares.

$x^2 - 10x + 25 = x^2 - 10x + 5^2$

Notice that if we let $a = x$ and $b = 5$ in $a^2 - 2ab + b^2 = (a - b)^2$, then the middle term $-10x$ is $-2ab$.

$x^2 - 10x + 25 = x^2 - 2(5x) + 5^2$
$= \boxed{(x - 5)^2}$

Solution (c):

Rewrite the trinomial so that the first and third terms are perfect squares.

$9x^2 - 12x + 4 = (3x)^2 - 2(3x)(2) + 2^2$

Notice that if we let $a = 3x$ and $b = 2$ in $a^2 - 2ab + b^2 = (a - b)^2$, then the middle term $-12x$ is $-2ab$.

$9x^2 - 12x + 4 = (3x)^2 - 2(3x)(2) + 2^2$
$= \boxed{(3x - 2)^2}$

- **Answer: a.** $(x + 4)^2$
  b. $(x - 2)^2$
  c. $(5x - 2)^2$

■ YOUR TURN Factor:

a. $x^2 + 8x + 16$ **b.** $x^2 - 4x + 4$ **c.** $25x^2 - 20x + 4$

EXAMPLE 4 Factoring the Sum of Two Cubes

Factor $x^3 + 27$.

Solution:

Rewrite as the sum of two cubes. $x^3 + 27 = x^3 + 3^3$

Write the sum of two cubes formula. $a^3 + b^3 = (a + b)(a^2 - ab + b^2)$

Let $a = x$ and $b = 3$. $x^3 + 27 = x^3 + 3^3 = \boxed{(x + 3)(x^2 - 3x + 9)}$

EXAMPLE 5 **Factoring the Difference of Two Cubes**

Factor $x^3 - 125$.

Solution:

Rewrite as the difference
of two cubes.

$$x^3 - 125 = x^3 - 5^3$$

Write the difference of two
cubes formula.

$$a^3 - b^3 = (a - b)(a^2 + ab + b^2)$$

Let $a = x$ and $b = 5$.

$$x^3 - 125 = x^3 - 5^3 = \boxed{(x - 5)(x^2 + 5x + 25)}$$

■ **YOUR TURN** Factor:

   a. $x^3 + 8$ **b.** $x^3 - 64$

■ **Answer: a.** $(x + 2)(x^2 - 2x + 4)$
      b. $(x - 4)(x^2 + 4x + 16)$

Factoring a Trinomial as a Product of Two Binomials

The first step in factoring is to look for a common factor. If there is no common factor, look to see whether the polynomial is a special form for which we know the factoring formula. If it is not of such a special form, then if it is a trinomial, we proceed with a general factoring strategy.

We know that $(x + 3)(x + 2) = x^2 + 5x + 6$, so we say the **factors** of $x^2 + 5x + 6$ are $(x + 3)$ and $(x + 2)$. In factored form we have $x^2 + 5x + 6 = (x + 3)(x + 2)$. Recall the FOIL method from Section A.3. The product of the last terms (3 and 2) is 6 and the sum of the products of the inner terms ($3x$) and the outer terms ($2x$) is $5x$. Let's pretend for a minute that we didn't know this factored form but had to work with the general form:

$$x^2 + 5x + 6 = (x + a)(x + b)$$

The goal is to find a and b. We start by multiplying the two binomials on the right.

$$x^2 + 5x + 6 = (x + a)(x + b) = x^2 + ax + bx + ab = x^2 + (a + b)x + ab$$

Compare the expression we started with on the left with the expression on the far right $x^2 + 5x + 6 = x^2 + (a + b)x + ab$. We see that $ab = 6$ and $(a + b) = 5$. *Start* with the possible combinations of a and b whose product is 6, and *then* look among those for the combination whose sum is 5.

$ab = 6$	a, b:	1, 6	$-1, -6$	2, 3	$-2, -3$
$a + b$		7	-7	**5**	-5

All of the possible a, b combinations in the first row have a product equal to 6, but only one of those has a sum equal to 5. Therefore, the factored form is

$$x^2 + 5x + 6 = (x + a)(x + b) = (x + 2)(x + 3)$$

EXAMPLE 6   **Factoring a Trinomial Whose Leading Coefficient Is 1**

Factor $x^2 + 10x + 9$.

Solution:

Write the trinomial as a product of two binomials in general form.

$$x^2 + 10x + 9 = (x + \square)(x + \square)$$

Write all of the integers whose product is 9.

Integers whose product is 9	1, 9	−1, −9	3, 3	−3, −3

Determine the sum of the integers.

Integers whose product is 9	**1, 9**	−1, −9	3, 3	−3, −3
Sum	**10**	−10	9	−9

Select 1, 9 because the product is 9 (last term of the trinomial) and the sum is 10 (middle term coefficient of the trinomial).

$$x^2 + 10x + 9 = (x + 9)(x + 1)$$

Check: $(x + 9)(x + 1) = x^2 + 9x + 1x + 9 = x^2 + 10x + 9$ ✓

■ **Answer:** $(x + 4)(x + 5)$

■ **YOUR TURN**   Factor $x^2 + 9x + 20$.

In Example 6, all terms in the trinomial are positive. When the constant term is negative, then (regardless of whether the middle term is positive or negative) the factors will be opposite in sign, as illustrated in Example 7.

EXAMPLE 7   **Factoring a Trinomial Whose Leading Coefficient Is 1**

Factor $x^2 - 3x - 28$.

Solution:

Write the trinomial as a product of two binomials in general form.

$$x^2 - 3x + -28 = (x + \square)(x - \square)$$

Write all of the integers whose product is −28.

Integers whose product is −28	1, −28	−1, 28	2, −14	−2, 14	4, −7	−4, 7

Determine the sum of the integers.

Integers whose product is −28	1, −28	−1, 28	2, −14	−2, 14	**4, −7**	−4, 7
Sum	−27	27	−12	12	**−3**	3

Select 4, −7 because the product is −28 (last term of the trinomial) and the sum is −3 (middle-term coefficient of the trinomial).

$$x^2 - 3x - 28 = (x + 4)(x - 7)$$

Check: $(x + 4)(x - 7) = x^2 - 7x + 4x + -28 = x^2 - 3x + -28$ ✓

■ **Answer:** $(x + 6)(x - 3)$

■ **YOUR TURN**   Factor $x^2 + 3x - 18$.

When the leading coefficient of the trinomial is not equal to 1, then we consider all possible factors using the following procedure, which is based on the FOIL method in reverse.

FACTORING A TRINOMIAL WHOSE LEADING COEFFICIENT IS NOT 1

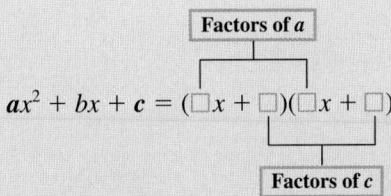

$$ax^2 + bx + c = (\square x + \square)(\square x + \square)$$

Step 1: Find two **F**irst terms whose product is the first term of the trinomial.

Step 2: Find two **L**ast terms whose product is the last term of the trinomial.

Step 3: Consider all possible combinations found in Steps 1 and 2 until the sum of the **O**uter and **I**nner products is equal to the middle term of the trinomial.

EXAMPLE 8   Factoring a Trinomial Whose Leading Coefficient Is Not 1

Factor $5x^2 + 9x - 2$.

Solution:

STEP 1 Start with the *first* term. Note that $5x \cdot x = 5x^2$. $(5x \pm \square)(x \pm \square)$

STEP 2 The product of the *last* terms should yield -2. $-1, 2$ or $1, -2$

STEP 3 Consider all possible factors based on Steps 1 and 2.

$$(5x - 1)(x + 2)$$
$$(5x + 1)(x - 2)$$
$$(5x + 2)(x - 1)$$
$$(5x - 2)(x + 1)$$

Since the *outer* ans *inner* products must sum to $9x$, the factored form must be $\boxed{5x^2 + 9x - 2 = (5x - 1)(x + 2)}$

Check: $(5x - 1)(x + 2) = 5x^2 + 10x - 1x - 2 = 5x^2 + 9x - 2$ ✓

■ YOUR TURN Factor $2t^2 + t - 3$.

■ Answer: $(2t + 3)(t - 1)$

EXAMPLE 9   Factoring a Trinomial Whose Leading Coefficient Is Not 1

Factor $15x^2 - x - 6$.

Solution:

STEP 1 Start with the *first* term. $(5x \pm \square)(3x \pm \square)$ or $(15x \pm \square)(x \pm \square)$

STEP 2 The product of the *last* terms should yield -6. $-1, 6$ or $1, -6$ or $2, -3$ or $-2, 3$

Study Tip

In Example 9, Step 3, we can eliminate any factors that have a common factor since there is no common factor to the terms in the trinomial.

STEP 3 Consider all possible factors based on Steps 1 and 2.

$(5x - 1)(3x + 6)$	$(15x - 1)(x + 6)$
$(5x + 6)(3x - 1)$	$(15x + 6)(x - 1)$
$(5x + 1)(3x - 6)$	$(15x + 1)(x - 6)$
$(5x - 6)(3x + 1)$	$(15x - 6)(x + 1)$
$(5x + 2)(3x - 3)$	$(15x + 2)(x - 3)$
$(5x - 3)(3x + 2)$	$(15x - 3)(x + 2)$
$(5x - 2)(3x + 3)$	$(15x - 2)(x + 3)$
$(5x + 3)(3x - 2)$	$(15x + 3)(x - 2)$

Since the *outer* and *inner* products must sum to $-x$, the factored form must be

$$\boxed{15x^2 - x - 6 = (5x + 3)(3x - 2)}$$

Check: $(5x + 3)(3x - 2) = 15x^2 - 10x + 9x - 6 = 15x^2 - x - 6$ ✓

■ **Answer:** $(3x - 4)(2x + 3)$

■ **YOUR TURN** Factor $6x^2 + x - 12$.

EXAMPLE 10 **Identifying Prime (Irreducible) Polynomials**

Factor $x^2 + x - 8$.

Solution:

Write the trinomial as a product of two binomials in general form.

$$x^2 + x - 8 = (x + \square)(x - \square)$$

Write all of the integers whose product is -8.

Integers whose product is -8	1, -8	$-1, 8$	4, -2	$-4, 2$

Determine the sum of the integers.

Integers whose product is -8	1, -8	$-1, 8$	4, -2	$-4, 2$
Sum	-7	7	2	-2

The middle term of the trinomial is $-x$, so we look for the sum of the integers that equals -1. Since no sum exists for the given combinations, we say that this polynomial is *prime* (irreducible) over the integers.

Factoring by Grouping

Much of our attention in this section has been on factoring trinomials. For polynomials with more than three terms, we first look for a common factor to all terms. If there is no common factor to all terms of the polynomial, we look for a group of terms that have a common factor. This strategy is called **factoring by grouping**.

EXAMPLE 11 **Factoring a Polynomial by Grouping**

Factor $x^3 - x^2 + 2x - 2$.

Solution:

Group the terms that have a common factor.

$$= (x^3 - x^2) + (2x - 2)$$

Factor out the common factor in each pair of parentheses.

$$= x^2(x - 1) + 2(x - 1)$$

Use the distributive property.

$$= \boxed{(x^2 + 2)(x - 1)}$$

EXAMPLE 12 Factoring a Polynomial by Grouping

Factor $2x^2 + 2x - x - 1$.

Solution:

Group the terms that have a common factor.	$= \left(2x^2 + 2x\right) + (-x - 1)$
Factor out the common factor in each pair of parentheses.	$= 2x(x + 1) - 1(x + 1)$
Use the distributive property.	$= \boxed{(2x - 1)(x + 1)}$

■ **YOUR TURN** Factor $x^3 + x^2 - 3x - 3$.

■ **Answer:** $(x + 1)\left(x^2 - 3\right)$

A Strategy for Factoring Polynomials

The first step in factoring a polynomial is to look for the greatest common factor. When specifically factoring trinomials look for special known forms: a perfect square or a difference of two squares. A general approach to factoring a trinomial uses the FOIL method in reverse. Finally, we look for factoring by grouping. The following strategy for factoring polynomials is based on the techniques discussed in this section.

STRATEGY FOR FACTORING POLYNOMIALS

1. Factor out the greatest common factor (monomial).
2. Identify any special polynomial forms and apply factoring formulas.
3. Factor a trinomial into a product of two binomials: $(ax + b)(cx + d)$.
4. Factor by grouping.

Study Tip

When factoring, always start by factoring out the GCF.

EXAMPLE 13 Factoring Polynomials

Factor:

a. $3x^2 - 6x + 3$ **b.** $-4x^3 + 2x^2 + 6x$
c. $15x^2 + 7x - 2$ **d.** $x^3 - x + 2x^2 - 2$

Solution (a):

Factor out the greatest common factor.	$3x^2 - 6x + 3 = 3\left(x^2 - 2x + 1\right)$
The trinomial is a perfect square.	$= \boxed{3(x - 1)^2}$

Solution (b):

Factor out the greatest common factor.	$-4x^3 + 2x^2 + 6x = -2x\left(2x^2 - x - 3\right)$
Use the FOIL method in reverse to factor the trinomial.	$= \boxed{-2x(2x - 3)(x + 1)}$

Solution (c):

There is no common factor. Use the FOIL method in reverse to factor the trinomial.	$15x^2 + 7x - 2 = \boxed{(3x + 2)(5x - 1)}$

Solution (d):

Factor by grouping.	$= \left(x^3 - x\right) + \left(2x^2 - 2\right)$
	$= x\left(x^2 - 1\right) + 2\left(x^2 - 1\right)$
	$= (x + 2)\left(x^2 - 1\right)$
	$= \boxed{(x + 2)(x - 1)(x + 1)}$

SECTION A.4 SUMMARY

In this section, we discussed factoring polynomials, which is the reverse process of multiplying polynomials. Four main techniques were discussed.

Greatest Common Factor: ax^k

- a is the greatest common factor for all coefficients of the polynomial.
- k is the smallest exponent on x found among all of the terms in the polynomial.

Factoring Formulas: Special Polynomial Forms

Difference of two squares: $a^2 - b^2 = (a + b)(a - b)$
Perfect squares: $a^2 + 2ab + b^2 = (a + b)^2$
$a^2 - 2ab + b^2 = (a - b)^2$
Sum of two cubes: $a^3 + b^3 = (a + b)(a^2 - ab + b^2)$
Difference of two cubes: $a^3 - b^3 = (a - b)(a^2 + ab + b^2)$

Factoring a Trinomial as a Product of Two Binomials

- $x^2 + bx + c = (x + ?)(x + ?)$

1. Find all possible combinations of factors whose product is c.
2. Of the combinations in Step 1, look for the sum of factors that equals b.

- $ax^2 + bx + c = (?x + ?)(?x + ?)$

1. Find all possible combinations of the first terms whose product is ax^2.
2. Find all possible combinations of the last terms whose product is c.
3. Consider all possible factors based on Steps 1 and 2.

Factoring by Grouping

- Group terms that have a common factor.
- Use the distributive property.

SECTION A.4 EXERCISES

SKILLS

In Exercises 1–12, factor each expression. Start by finding the greatest common factor (GCF).

1. $5x + 25$
2. $x^2 + 2x$
3. $4t^2 - 2$
4. $16z^2 - 20z$
5. $2x^3 - 50x$
6. $4x^2y - 8xy^2 + 16x^2y^2$
7. $3x^3 - 9x^2 + 12x$
8. $14x^4 - 7x^2 + 21x$
9. $x^3 - 3x^2 - 40x$
10. $-9y^2 + 45y$
11. $4x^2y^3 + 6xy$
12. $3z^3 - 6z^2 + 18z$

In Exercises 13–20, factor the difference of two squares.

13. $x^2 - 9$
14. $x^2 - 25$
15. $4x^2 - 9$
16. $1 - x^4$
17. $2x^2 - 98$
18. $144 - 81y^2$
19. $225x^2 - 169y^2$
20. $121y^2 - 49x^2$

In Exercises 21–32, factor the perfect squares.

21. $x^2 + 8x + 16$
22. $y^2 - 10y + 25$
23. $x^4 - 4x^2 + 4$
24. $1 - 6y + 9y^2$
25. $4x^2 + 12xy + 9y^2$
26. $x^2 - 6xy + 9y^2$
27. $9 - 6x + x^2$
28. $25x^2 - 20xy + 4y^2$
29. $x^4 + 2x^2 + 1$
30. $x^6 - 6x^3 + 9$
31. $p^2 + 2pq + q^2$
32. $p^2 - 2pq + q^2$

In Exercises 33–42, factor the sum or difference of two cubes.

33. $t^3 + 27$
34. $z^3 + 64$
35. $y^3 - 64$
36. $x^3 - 1$
37. $8 - x^3$
38. $27 - y^3$
39. $y^3 + 125$
40. $64x - x^4$
41. $27 + x^3$
42. $216x^3 - y^3$

In Exercises 43–52, factor each trinomial into a product of two binomials.

43. $x^2 - 6x + 5$ **44.** $t^2 - 5t - 6$ **45.** $y^2 - 2y - 3$ **46.** $y^2 - 3y - 10$

47. $2y^2 - 5y - 3$ **48.** $2z^2 - 4z - 6$ **49.** $3t^2 + 7t + 2$ **50.** $4x^2 - 2x - 12$

51. $-6t^2 + t + 2$ **52.** $-6x^2 - 17x + 10$

In Exercises 53–60, factor by grouping.

53. $x^3 - 3x^2 + 2x - 6$ **54.** $x^5 + 5x^3 - 3x^2 - 15$ **55.** $a^4 + 2a^3 - 8a - 16$ **56.** $x^4 - 3x^3 - x + 3$

57. $3xy - 5rx - 10rs + 6sy$ **58.** $6x^2 - 10x + 3x - 5$ **59.** $20x^2 + 8xy - 5xy - 2y^2$ **60.** $9x^5 - a^2x^3 - 9x^2 + a^2$

In Exercises 61–92, factor each of the polynomials completely, if possible. If the polynomial cannot be factored, state that it is prime.

61. $x^2 - 4y^2$ **62.** $a^2 + 5a + 6$ **63.** $3a^2 + a - 14$ **64.** $ax + b + bx + a$

65. $x^2 + 16$ **66.** $x^2 + 49$ **67.** $4z^2 + 25$ **68.** $\frac{1}{16} - b^4$

69. $6x^2 + 10x + 4$ **70.** $x^2 + 7x + 5$ **71.** $6x^2 + 13xy - 5y^2$ **72.** $15x + 15xy$

73. $36s^2 - 9t^2$ **74.** $3x^3 - 108x$ **75.** $a^2b^2 - 25c^2$ **76.** $2x^3 + 54$

77. $4x^2 - 3x - 10$ **78.** $10x - 25 - x^2$ **79.** $3x^3 - 5x^2 - 2x$ **80.** $2y^3 + 3y^2 - 2y$

81. $x^3 - 9x$ **82.** $w^3 - 25w$ **83.** $xy - x - y + 1$ **84.** $a + b + ab + b^2$

85. $x^4 + 5x^2 + 6$ **86.** $x^6 - 7x^3 - 8$ **87.** $x^2 - 2x - 24$ **88.** $25x^2 + 30x + 9$

89. $x^4 + 125x$ **90.** $x^4 - 1$ **91.** $x^4 - 81$ **92.** $10x^2 - 31x + 15$

■ APPLICATIONS

93. Geometry. A rectangle has a length of $2x + 4$ and a width of x. Express the perimeter of the rectangle as a factored polynomial in x.

94. Geometry. The volume of a box is given by the expression $x^3 + 7x^2 + 12x$. Express the volume as a factored polynomial in the variable x.

95. Business. The profit of a business is given by the expression $P = 2x^2 - 15x + 4x - 30$. Express the profit as a factored polynomial in the variable x.

96. Business. The break-even point for a company is given by solving the equation $3x^2 + 9x - 4x - 12 = 0$. Factor the polynomial on the left side of the equation.

97. Engineering. The height of a projectile is given by the equation $s = -16t^2 - 78t + 10$. Factor the expression on the right side of the equal sign.

98. Engineering. The electrical field at a point P between two charges is given by $k = \dfrac{10x - x^2}{100}$. Factor the numerator of this expression.

■ CATCH THE MISTAKE

In Exercises 99 and 100, explain the mistake that is made.

99. Factor $x^3 - x^2 - 9x + 9$.

   Solution:

   Group terms with common factors. $(x^3 - x^2) + (-9x + 9)$

   Factor out common factors. $x^2(x - 1) - 9(x - 1)$

   Distributive property. $(x - 1)(x^2 - 9)$

   Factor $x^2 - 9$. $(x - 1)(x - 3)^2$

   This is incorrect. What mistake was made?

   $(x+3)(x-3)$

100. Factor $4x^2 + 12x - 40$.

   Solution: $4(x^2 + 3x - 10)$

   Factor the trinomial into a product of binomials. $(2x - 4)(2x + 10)$

   Factor out a 2. $= 2(x - 2)(x + 5)$

   This is incorrect. What mistake was made?

■ CONCEPTUAL

In Exercises 101–104, determine whether each statement is true or false.

101. All trinomials can be factored into a product of two binomials.

102. All polynomials can be factored into prime factors with respect to the integers.

103. $x^2 - y^2 = (x - y)(x + y)$

104. $x^2 + y^2 = (x + y)^2$

■ CHALLENGE

105. Factor $a^{2n} - b^{2n}$ completely, assuming a, b, and n are positive integers.

106. Find all the values of c such that the trinomial $x^2 + cx - 14$ can be factored.

■ TECHNOLOGY

107. Use a graphing utility to plot the graphs of the three expressions $8x^3 + 1$, $(2x + 1)(4x^2 - 2x + 1)$, and $(2x - 1)(4x^2 + 2x + 1)$. Which two graphs agree with each other?

108. Use a graphing utility to plot the graphs of the three expressions $27x^3 - 1$, $(3x - 1)^3$, and $(3x - 1)(9x^2 + 3x + 1)$. Which two graphs agree with each other?

SECTION
A.5 RATIONAL EXPRESSIONS

SKILLS OBJECTIVES

- Find the domain of an algebraic expression.
- Reduce a rational expression to lowest terms.
- Multiply and divide rational expressions.
- Add and subtract rational expressions.
- Simplify complex rational expressions.

CONCEPTUAL OBJECTIVES

- Understand why rational expressions have domain restrictions.
- Understand the least common denominator method for rational expressions.

Rational Expressions and Domain Restrictions

Recall that a rational number is the ratio of two integers with the denominator not equal to zero. Similarly, the ratio, or quotient, of two polynomials is a **rational expression**.

Rational Numbers: $\qquad \dfrac{3}{7} \qquad \dfrac{5}{9} \qquad \dfrac{9}{11}$

Rational Expressions: $\qquad \dfrac{3x + 2}{x - 5} \qquad \dfrac{5x^2}{x^2 + 1} \qquad \dfrac{9}{3x - 2}$

As with rational numbers, the denominators of rational expressions are never equal to zero. In the above first and third rational expressions, there are values of the variable that would correspond to a denominator equal to zero; these values are not permitted:

$$\frac{3x + 2}{x - 5} \qquad x \neq 5 \qquad\qquad \frac{9}{3x - 2} \qquad x \neq \frac{2}{3}$$

In the second rational expression, $\dfrac{5x^2}{x^2 + 1}$, there are no real numbers that will correspond to a zero denominator.

The set of real numbers for which an algebraic expression is *defined* is called the **domain**. Since a rational expression is not defined if its denominator is zero, we must eliminate from the domain those values of the variable that would result in a zero denominator.

To find the domain of an algebraic expression, we ask the question, "What can x (the variable) be?" For rational expressions the answer in general is "any values except those that make the denominator equal to zero."

EXAMPLE 1   Finding the Domain of an Algebraic Expression

Find the domain of the expressions.

a. $2x^2 - 5x + 3$ **b.** $\dfrac{2x + 1}{x - 4}$ **c.** $\dfrac{x}{x^2 + 1}$ **d.** $\dfrac{3x + 1}{x}$

Solution:

ALGEBRAIC EXPRESSION	TYPE	DOMAIN	NOTE
a. $2x^2 - 5x + 3$	Polynomial	All real numbers	The domain of all polynomials is the set of all real numbers.
b. $\dfrac{2x + 1}{x - 4}$	Rational expression	All real numbers except $x = 4$	When $x = 4$, the rational expression is undefined.
c. $\dfrac{x}{x^2 + 1}$	Rational expression	All real numbers	There are no real numbers that will result in the denominator being equal to zero.
d. $\dfrac{3x + 1}{x}$	Rational expression	All real numbers except $x = 0$	When $x = 0$, the rational expression is undefined.

■ **YOUR TURN** Find the domain of the expressions.

a. $\dfrac{3x - 1}{x + 1}$ **b.** $\dfrac{5x - 1}{x}$ **c.** $3x^2 + 2x - 7$ **d.** $\dfrac{2x + 5}{x^2 + 4}$

■ **Answer: a.** $x \neq -1$
   b. $x \neq 0$
   c. all real numbers
   d. all real numbers

In this text, it will be assumed that the domain is the set of all real numbers except the real numbers shown to be excluded.

EXAMPLE 2   Excluding Values from the Domain of Rational Expressions

Determine what real numbers must be excluded from the domain of the following rational expressions:

a. $\dfrac{7x + 5}{x^2 - 4}$ **b.** $\dfrac{3x + 2}{x^2 - 5x}$

Solution (a):

Factor the denominator.

$$\dfrac{7x + 5}{x^2 - 4} = \dfrac{7x + 5}{(x + 2)(x - 2)}$$

Determine the values of x that will make the denominator equal to zero.

$\boxed{x = -2}$ and $\boxed{x = 2}$ must be excluded from the domain.

Solution (b):

Factor the denominator.

$$\frac{3x + 2}{x^2 - 5x} = \frac{3x + 2}{x(x - 5)}$$

Determine the values of x that will make the denominator equal to zero.

$\boxed{x = 0}$ and $\boxed{x = 5}$ must be excluded from the domain.

In this section, we will simplify rational expressions and perform operations on rational expressions such as multiplication, division, addition, and subtraction. The resulting expressions may not have *explicit* domain restrictions, but it is important to note that there are *implicit* domain restrictions, because the domain restrictions on the original rational expression still apply.

Simplifying Rational Expressions

Recall that a fraction is *reduced* when it is written with no common factors.

$$\frac{16}{12} = \frac{4 \cdot 4}{4 \cdot 3} = \left(\frac{4}{4}\right) \cdot \left(\frac{4}{3}\right) = (1) \cdot \left(\frac{4}{3}\right) = \frac{4}{3} \quad \text{or} \quad \frac{16}{12} = \frac{\cancel{4} \cdot 4}{\cancel{4} \cdot 3} = \frac{4}{3}$$

Similarly, rational expressions are **reduced to lowest terms**, or **simplified**, if the numerator and denominator have no common factors other than ± 1. As with real numbers, the ability to write fractions in reduced form is dependent on your ability to factor.

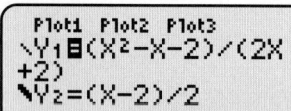
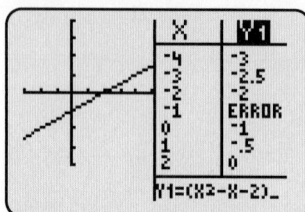
REDUCING A RATIONAL EXPRESSION TO LOWEST TERMS (SIMPLIFYING)

1. Factor the numerator and denominator completely.
2. State any domain restrictions.
3. Cancel (divide out) the common factors in the numerator and denominator.

EXAMPLE 3   Reducing a Rational Expression to Lowest Terms

Simplify $\dfrac{x^2 - x - 2}{2x + 2}$ and state any domain restrictions.

Solution:

Factor the numerator and denominator.

$$\frac{x^2 - x - 2}{2x + 2} = \frac{(x - 2)(x + 1)}{2(x + 1)}$$

State any domain restrictions.

$$x \neq -1$$

Cancel (divide out) the common factor, $x + 1$.

$$= \frac{(x - 2)\cancel{(x + 1)}}{2\cancel{(x + 1)}}$$

The rational expression is now in lowest terms (simplified).

$$= \boxed{\frac{x - 2}{2} \qquad x \neq -1}$$

■ **YOUR TURN** Simplify $\dfrac{x^2 + x - 2}{2x - 2}$ and state any domain restrictions.

The following table summarizes two **common mistakes** made with rational expressions:

CORRECT	INCORRECT	COMMENT
$\dfrac{x+5}{y+5}$ is already simplified.	**Error:** $\dfrac{x+\cancel{5}}{y+\cancel{5}} = \dfrac{x}{y}$	*Factors* can be divided out (canceled). Terms or parts of terms cannot be divided out. Remember to factor the numerator and denominator first, and then divide out common factors.
$\dfrac{x}{x^2+x}$ $= \dfrac{x}{x(x+1)} \qquad x \neq 0, x \neq -1$ $= \boxed{\dfrac{1}{x+1}} \qquad x \neq 0, x \neq -1$	**Error:** $\dfrac{x}{x^2+x} = \dfrac{1}{x+1} \qquad x \neq -1$ *Note:* Missing $x \neq 0$.	Determine the domain restrictions *before* dividing out common factors.

EXAMPLE 4 Simplifying Rational Expressions

Reduce $\dfrac{x^2-x-6}{x^2+x-2}$ to lowest terms and state any domain restrictions.

Solution:

Factor the numerator and denominator.	$\dfrac{x^2-x-6}{x^2+x-2} = \dfrac{(x-3)(x+2)}{(x-1)(x+2)}$
State domain restrictions.	$x \neq -2, x \neq 1$
Divide out the common factor, $x+2$.	$= \dfrac{(x-3)\cancel{(x+2)}}{(x-1)\cancel{(x+2)}}$
Simplify.	$= \boxed{\dfrac{x-3}{x-1}} \qquad x \neq -2, x \neq 1$

■ **YOUR TURN** Reduce $\dfrac{x^2+x-6}{x^2+2x-3}$ to lowest terms and state any domain restrictions.

■ **Answer:** $\dfrac{x-2}{x-1} \qquad x \neq -3, x \neq 1$

EXAMPLE 5   Simplifying Rational Expressions

Reduce $\dfrac{x^2 - 4}{2 - x}$ to lowest terms and state any domain restrictions.

Solution:

Factor the numerator and denominator.	$\dfrac{x^2 - 4}{2 - x} = \dfrac{(x - 2)(x + 2)}{(2 - x)}$
State domain restrictions.	$x \neq 2$
Factor out -1 in the denominator.	$= \dfrac{(x - 2)(x + 2)}{-(x - 2)}$
Cancel (divide out) the common factor, $x - 2$.	$= \dfrac{\cancel{(x - 2)}(x + 2)}{-\cancel{(x - 2)}}$
Simplify.	$= \boxed{-(x + 2) \qquad x \neq 2}$

■ **Answer:** $-(x + 5) \qquad x \neq 5$

■ **YOUR TURN** Reduce $\dfrac{x^2 - 25}{5 - x}$ to lowest terms and state any domain restrictions.

Multiplying and Dividing Rational Expressions

The same rules that apply to multiplying and dividing rational numbers also apply to rational expressions.

PROPERTY	RESTRICTION	DESCRIPTION
$\dfrac{a}{b} \cdot \dfrac{c}{d} = \dfrac{ac}{bd}$	$b \neq 0, d \neq 0$	Multiply numerators and denominators, respectively.
$\dfrac{a}{b} \div \dfrac{c}{d} = \dfrac{a}{b} \cdot \dfrac{d}{c}$	$b \neq 0, d \neq 0, c \neq 0$	Dividing is equivalent to multiplying by a reciprocal.

Multiplying Rational Expressions

1. Factor all numerators and denominators completely.
2. State any domain restrictions.
3. Divide the numerators and denominators by any common factors.
4. Multiply the remaining numerators and denominators, respectively.

Dividing Rational Expressions

1. Factor all numerators and denominators completely.
2. State any domain restrictions.
3. Rewrite division as multiplication by a reciprocal.
4. State any additional domain restrictions.
5. Divide the numerators and denominators by any common factors.
6. Multiply the remaining numerators and denominators, respectively.

EXAMPLE 6 Multiplying Rational Expressions

Multiply and simplify $\dfrac{3x+1}{4x^2+4x}\cdot\dfrac{x^3+3x^2+2x}{9x+3}$.

Solution:

Factor the numerators and denominators.

$$=\dfrac{(3x+1)}{4x(x+1)}\cdot\dfrac{x(x+1)(x+2)}{3(3x+1)}$$

State any domain restrictions.

$$x\neq 0, x\neq -1, x\neq -\dfrac{1}{3}$$

Divide the numerators and denominators by common factors.

$$=\dfrac{\cancel{(3x+1)}}{4\cancel{x}\cancel{(x+1)}}\cdot\dfrac{\cancel{x}\cancel{(x+1)}(x+2)}{3\cancel{(3x+1)}}$$

Simplify.

$$=\boxed{\dfrac{x+2}{12}\qquad x\neq 0, x\neq -1, x\neq -\dfrac{1}{3}}$$

....

■ **YOUR TURN** Multiply and simplify $\dfrac{2x+1}{3x^2-3x}\cdot\dfrac{x^3+2x^2-3x}{8x+4}$.

■ **Answer:**

$\dfrac{x+3}{12}\qquad x\neq 0, x\neq 1, x\neq -\dfrac{1}{2}$

EXAMPLE 7 Dividing Rational Expressions

Divide and simplify $\dfrac{x^2-4}{x}\div\dfrac{3x^3-12x}{5x^3}$.

Solution:

Factor numerators and denominators.

$$\dfrac{(x-2)(x+2)}{x}\div\dfrac{3x(x-2)(x+2)}{5x^3}$$

State any domain restrictions.

$$x\neq 0$$

Write the quotient as a product.

$$=\dfrac{(x-2)(x+2)}{x}\cdot\dfrac{5x^3}{3x(x-2)(x+2)}$$

State any additional domain restrictions.

$$x\neq -2, x\neq 2$$

Divide out the common factors.

$$=\dfrac{\cancel{(x-2)}\cancel{(x+2)}}{\cancel{x}}\cdot\dfrac{5x^3}{3\cancel{x}\cancel{(x-2)}\cancel{(x+2)}}$$

Simplify.

$$=\boxed{\dfrac{5x}{3}\qquad x\neq -2, x\neq 0, x\neq 2}$$

....

■ **YOUR TURN** Divide and simplify $\dfrac{x^2-9}{x}\div\dfrac{2x^3-18x}{7x^4}$.

■ **Answer:**

$\dfrac{7x^2}{2}\qquad x\neq -3, x\neq 0, x\neq 3$

Adding and Subtracting Rational Expressions

The same rules that apply to adding and subtracting rational numbers also apply to rational expressions.

PROPERTY	RESTRICTION	DESCRIPTION
$\dfrac{a}{b} \pm \dfrac{c}{b} = \dfrac{a \pm c}{b}$	$b \neq 0$	Adding or subtracting rational expressions when the denominators are the same
$\dfrac{a}{b} \pm \dfrac{c}{d} = \dfrac{ad \pm bc}{bd}$	$b \neq 0$ and $d \neq 0$	Adding or subtracting rational expressions when the denominators are different

EXAMPLE 8 **Adding and Subtracting Rational Expressions: Equal Denominators**

Perform the indicated operation and simplify.

a. $\dfrac{x + 7}{(x + 2)^2} + \dfrac{3x + 1}{(x + 2)^2}$ **b.** $\dfrac{6x + 7}{2x - 1} - \dfrac{2x + 9}{2x - 1}$

Solution (a):

Write as a single expression. $\qquad = \dfrac{x + 7 + 3x + 1}{(x + 2)^2}$

State any domain restrictions. $\qquad x \neq -2$

Combine like terms in the numerator. $\qquad = \dfrac{4x + 8}{(x + 2)^2}$

Factor out the common factor in the numerator. $\qquad = \dfrac{4(x + 2)}{(x + 2)^2}$

Cancel (divide out) the common factor, $x + 2$. $\qquad = \boxed{\dfrac{4}{x + 2}} \quad x \neq -2$

Solution (b):

Write as a single expression. Use parentheses around the second numerator to ensure that the negative will be distributed throughout all terms. $\qquad = \dfrac{6x + 7 - (2x + 9)}{2x - 1}$

State any domain restrictions. $\qquad x \neq \dfrac{1}{2}$

Eliminate parentheses. Distribute the negative. $\qquad = \dfrac{6x + 7 - 2x - 9}{2x - 1}$

Combine like terms in the numerator. $\qquad = \dfrac{4x - 2}{2x - 1}$

Factor out the common factor in the numerator. $\qquad = \dfrac{2(2x - 1)}{2x - 1}$

Divide out (cancel) the common factor, $2x - 1$. $\qquad = \boxed{2 \quad x \neq \dfrac{1}{2}}$

Study Tip

When subtracting a rational expression, distribute the negative of the quantity to be subtracted over all terms in the numerator.

EXAMPLE 9 **Adding and Subtracting Rational Expressions: No Common Factors in Denominators**

Perform the indicated operation and simplify.

a. $\dfrac{3-x}{2x+1} + \dfrac{x}{x-1}$ **b.** $\dfrac{1}{x^2} - \dfrac{2}{x+1}$

Solution (a):

The common denominator is the product of the denominators.

$$\frac{3-x}{2x+1} + \frac{x}{x-1} = \frac{(3-x)(x-1)}{(2x+1)(x-1)} + \frac{x(2x+1)}{(x-1)(2x+1)}$$

$$= \frac{(3-x)(x-1) + x(2x+1)}{(2x+1)(x-1)}$$

Eliminate parentheses in the numerator.

$$= \frac{3x-3-x^2+x+2x^2+x}{(2x+1)(x-1)}$$

Combine like terms in the numerator.

$$= \boxed{\frac{x^2+5x-3}{(2x+1)(x-1)} \qquad x \neq -\frac{1}{2}, x \neq 1}$$

Solution (b):

The common denominator is the product of the denominators.

$$\frac{1}{x^2} - \frac{2}{x+1} = \frac{(1)(x+1) - 2\left(x^2\right)}{x^2(x+1)}$$

Eliminate parentheses in the numerator.

$$= \frac{x+1-2x^2}{x^2(x+1)} = \frac{-2x^2+x+1}{x^2(x+1)} = \frac{-\left(2x^2-x-1\right)}{x^2(x+1)}$$

Write the numerator in factored form to ensure no further simplification is possible.

$$= \boxed{\frac{-(2x+1)(x-1)}{x^2(x+1)} \qquad x \neq -1, x \neq 0}$$

Study Tip

When adding or subtracting rational expressions whose denominators have no common factors, the least common denominator is the product of the two denominators.

■ **YOUR TURN** Perform the indicated operation and simplify.

a. $\dfrac{2x-1}{x+3} + \dfrac{x}{2x+1}$ **b.** $\dfrac{2}{x^2+1} - \dfrac{1}{x}$

■ **Answer:**

a. $\dfrac{5x^2+3x-1}{(x+3)(2x+1)}$ $x \neq -3, x \neq -\frac{1}{2}$

b. $\dfrac{-(x-1)^2}{x\left(x^2+1\right)}$ $x \neq 0$

When combining two or more fractions through addition or subtraction, recall that the **least common multiple (LCM)**, or **least common denominator (LCD)**, is the smallest real number that all of the denominators divide into evenly (i.e., the smallest of which all are factors). For example,

$$\frac{2}{3} + \frac{1}{6} - \frac{4}{9}$$

To find the LCD of these three fractions, factor the denominators into prime factors:

$$
\begin{aligned}
3 &= \mathbf{3} \\
6 &= \mathbf{3 \cdot 2} \\
\underline{9 = \mathbf{3} \quad \cdot \mathbf{3}} \\
\mathbf{3 \cdot 2 \cdot 3 = 18}
\end{aligned}
\qquad
\frac{2}{3} + \frac{1}{6} - \frac{4}{9} = \frac{12 + 3 - 8}{18} = \frac{7}{18}
$$

Rational expressions follow this same procedure, only now variables are also considered:

$$
\frac{1}{2x} + \frac{1}{x^3} = \frac{1}{2x} + \frac{1}{x \cdot x \cdot x}
\qquad\qquad
\text{LCD} = 2x^3
$$

$$
\frac{2}{x+1} - \frac{x}{2x+1} + \frac{3-x}{(x+1)^2} = \frac{2}{(x+1)} - \frac{x}{(2x+1)} + \frac{3-x}{(x+1)(x+1)}
\qquad
\text{LCD} = (x+1)^2(2x+1)
$$

The following box summarizes the LCD method for adding and subtracting rational expressions whose denominators have common factors.

THE LCD METHOD FOR ADDING AND SUBTRACTING RATIONAL EXPRESSIONS

1. Factor each of the denominators completely.
2. The LCD is the product of each of these distinct factors raised to the highest power to which that factor appears in any of the denominators.
3. Write each rational expression using the LCD for each denominator.
4. Add or subtract the resulting numerators.
5. Factor the resulting numerator to check for common factors.

 EXAMPLE 10 **Subtracting Rational Expressions: Common Factors in Denominators (LCD)**

Perform the indicated operation and write in simplified form.

$$
\frac{5x}{2x-6} - \frac{7x-2}{x^2-x-6}
$$

Solution:

Factor the denominators.
$$
= \frac{5x}{2(x-3)} - \frac{7x-2}{(x-3)(x+2)}
$$

Identify the least common denominator.
$$
\text{LCD} = 2(x-3)(x+2)
$$

Write each expression using the LCD as the denominator.
$$
= \frac{5x(x+2)}{2(x-3)(x+2)} - \frac{2(7x-2)}{2(x-3)(x+2)}
$$

Combine into one expression. Distribute the negative through the entire second numerator.
$$
= \frac{5x^2 + 10x - 14x + 4}{2(x-3)(x+2)}
$$

Simplify.
$$
\boxed{= \frac{5x^2 - 4x + 4}{2(x-3)(x+2)} \qquad x \neq -2,\ x \neq 3}
$$

■ Answer:
$$
\frac{2x^2 - 7x - 3}{3(x-2)(x+4)} \qquad x \neq -4,\ x \neq 2
$$

■ YOUR TURN Perform the indicated operation and write it in simplified form.

$$
\frac{2x}{3x-6} - \frac{5x+1}{x^2+2x-8}
$$

Complex Rational Expressions

A rational expression that contains another rational expression in either its numerator or denominator is called a **complex rational expression**. The following are examples of complex rational expressions:

$$\frac{\frac{1}{x} - 5}{2 + x} \qquad \frac{2 - x}{4 + \frac{3}{x - 1}} \qquad \frac{\frac{3}{x} - 7}{\frac{6}{2x - 5} - 1}$$

TWO METHODS FOR SIMPLIFYING COMPLEX RATIONAL EXPRESSIONS

Write a sum or difference of rational expressions that appear in either the numerator or denominator as a single rational expression. Once the complex rational expression contains a single rational expression in the numerator and one in the denominator, then rewrite the division as multiplication by the reciprocal.

OR

Find the LCD of all rational expressions contained in both the numerator and denominator. Multiply the numerator and denominator by this LCD and simplify.

EXAMPLE 11 Simplifying a Complex Rational Expression

Write the rational expression in simplified form.

$$\frac{\frac{2}{x} + 1}{1 + \frac{1}{x + 1}}$$

Solution:

State the domain restrictions.

$$\frac{\boxed{\dfrac{2}{x}} + 1}{1 + \boxed{\dfrac{1}{x + 1}}} \qquad \boxed{x \neq 0}, \boxed{x \neq -1},$$
$$\text{and } \boxed{x \neq -2}$$

Procedure 1:

Add the expressions in both the numerator and denominator.

$$= \frac{\dfrac{2}{x} + \dfrac{x}{x}}{\dfrac{x + 1}{x + 1} + \dfrac{1}{x + 1}} = \frac{\dfrac{2 + x}{x}}{\dfrac{(x + 1) + 1}{x + 1}}$$

Simplify.

$$= \frac{\dfrac{2 + x}{x}}{\dfrac{x + 2}{x + 1}}$$

Express the quotient as a product.

$$= \frac{2 + x}{x} \cdot \frac{x + 1}{x + 2}$$

Divide out the common factors.

$$= \frac{\cancel{2 + x}}{x} \cdot \frac{x + 1}{\cancel{x + 2}}$$

Write in simplified form.

$$= \boxed{\frac{x + 1}{x} \qquad x \neq -2, x \neq -1, x \neq 0}$$

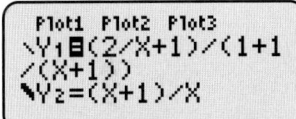
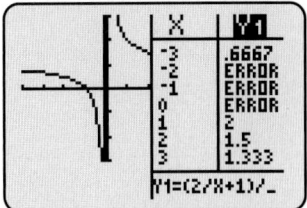
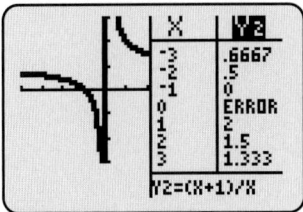

Procedure 2:

Find the LCD of the numerator and denominator.

$$\frac{\dfrac{2}{x} + 1}{1 + \dfrac{1}{x + 1}}$$

Identify the LCDs.

Numerator LCD: x
Denominator LCD: $x + 1$
Combined LCD: $x(x + 1)$

Multiply both numerator and denominator by their combined LCD.

$$= \frac{\dfrac{2}{x} + 1}{1 + \dfrac{1}{x + 1}} \cdot \frac{x(x + 1)}{x(x + 1)}$$

Multiply the numerators and denominators, respectively, applying the distributive property.

$$= \frac{\dfrac{2}{x} \cdot x(x + 1) + 1x(x + 1)}{1 \cdot x(x + 1) + \dfrac{1}{x + 1} \cdot x(x + 1)}$$

Divide out common factors.

$$= \frac{\dfrac{2}{\cancel{x}} \cdot \cancel{x}(x + 1) + 1x(x + 1)}{x(x + 1) + \dfrac{1}{\cancel{x + 1}} \cdot x(\cancel{x + 1})}$$

Simplify.

$$= \frac{2(x + 1) + x(x + 1)}{x(x + 1) + x}$$

Apply the distributive property.

$$= \frac{2x + 2 + x^2 + x}{x^2 + x + x}$$

Combine like terms.

$$= \frac{x^2 + 3x + 2}{x^2 + 2x}$$

Factor the numerator and denominator.

$$= \frac{(x + 2)(x + 1)}{x(x + 2)}$$

Divide out the common factor.

$$= \frac{\cancel{(x + 2)}(x + 1)}{x\cancel{(x + 2)}}$$

Write in simplified form.

$$= \boxed{\frac{x + 1}{x} \qquad x \neq -2, x \neq -1, x \neq 0}$$

EXAMPLE 12   Simplifying a Complex Rational Expression

Write the rational expression in simplified form.

$$\frac{\dfrac{1}{x^2 - 9} + 3}{1 - \dfrac{x}{2x + 6}} \qquad x \neq -6, -3, 3$$

Solution:

Factor the respective denominators.

$$\frac{\dfrac{1}{(x - 3)(x + 3)} + 3}{1 - \dfrac{x}{2(x + 3)}}$$

Identify the LCDs.

Numerator LCD: $(x - 3)(x + 3)$
Denominator LCD: $2(x + 3)$
Combined LCD: $2(x - 3)(x + 3)$

Multiply both the numerator and the denominator by the combined LCD.

$$= \frac{\dfrac{1}{(x - 3)(x + 3)} + 3}{1 - \dfrac{x}{2(x + 3)}} \cdot \frac{2(x + 3)(x - 3)}{2(x + 3)(x - 3)}$$

Multiply the numerators and denominators, respectively, applying the distributive property.

$$= \frac{\dfrac{2(x + 3)(x - 3)}{(x - 3)(x + 3)} + 3 \cdot 2(x + 3)(x - 3)}{1 \cdot 2(x + 3)(x - 3) - \dfrac{x \cdot 2(x + 3)(x - 3)}{2(x + 3)}}$$

Simplify.

$$= \frac{2 + 6(x + 3)(x - 3)}{1 \cdot 2(x + 3)(x - 3) - x(x - 3)}$$

Eliminate the parentheses.

$$= \frac{2 + 6x^2 - 54}{2x^2 - 18 - x^2 + 3x}$$

Combine like terms.

$$= \frac{6x^2 - 52}{x^2 + 3x - 18}$$

Factor the numerator and denominator to make sure there are no common factors.

$$= \frac{2(3x^2 - 26)}{(x + 6)(x - 3)} \qquad x \neq -6, x \neq -3, x \neq 3$$

In this section, rational expressions were defined as quotients of polynomials. The domain of any polynomial is the set of all real numbers. Since rational expressions are ratios of polynomials, the domain of rational expressions is the set of all real numbers *except those values that make the denominator equal to zero*. In this section, rational expressions were simplified (written with no common factors), multiplied, divided, added, and subtracted.

OPERATION	EXAMPLE	NOTE
Multiplying rational expressions	$\dfrac{2}{x+1} \cdot \dfrac{3x}{x-1} = \dfrac{6x}{x^2-1}$ $x \neq \pm 1$	State domain restrictions.
Dividing rational expressions	$\dfrac{2}{x+1} \div \dfrac{3x}{x-1}$ $x \neq \pm 1$ $= \dfrac{2}{x+1} \cdot \dfrac{x-1}{3x}$ $x \neq 0$ $= \dfrac{2(x-1)}{3x(x+1)}$ $x \neq 0, x \neq \pm 1$	When dividing rational expressions, remember to check for additional domain restrictions once the division is rewritten as multiplication by a reciprocal.
Adding/subtracting rational expressions with no common factors	$\dfrac{2}{x+1} + \dfrac{3x}{x-1}$ $x \neq \pm 1$ $\text{LCD} = (x+1)(x-1)$ $= \dfrac{2(x-1) + 3x(x+1)}{(x+1)(x-1)}$ $= \dfrac{2x-2+3x^2+3x}{(x+1)(x-1)}$ $= \dfrac{3x^2+5x-2}{(x+1)(x-1)}$ $= \dfrac{(3x-1)(x+2)}{(x+1)(x-1)}$ $x \neq \pm 1$	The least common denominator (LCD) is the product of the two denominators.
Adding/subtracting rational expressions with common factors	$\dfrac{3}{x(x+1)} - \dfrac{2}{x(x+2)}$ $x \neq -2, -1, 0$ $\text{LCD} = x(x+1)(x+2)$ $= \dfrac{3(x+2) - 2(x+1)}{x(x+1)(x+2)} = \dfrac{3x+6-2x-2}{x(x+1)(x+2)}$ $= \dfrac{x+4}{x(x+1)(x+2)}$ $x \neq -2, -1, 0$	The LCD is the product of each of the distinct factors raised to the highest power that appears in any of the denominators.

Complex rational expressions are simplified in one of two ways:

1. Combine the sum or difference of rational expressions in a numerator or denominator into a single rational expression. The result is a rational expression in the numerator and a rational expression in the denominator. Then write the division as multiplication by the reciprocal.

2. Multiply the numerator and denominator by the overall LCD (LCD for all rational expressions that appear). The result is a single rational expression. Then simplify, if possible.

SECTION
A.5 EXERCISES

SKILLS

In Exercises 1–10, state any real numbers that must be excluded from the domain of each rational expression.

1. $\dfrac{3}{x}$ **2.** $\dfrac{5}{x}$ **3.** $\dfrac{3}{x-1}$ **4.** $\dfrac{6}{y-1}$ **5.** $\dfrac{5x-1}{x+1}$

6. $\dfrac{2x}{3-x}$ **7.** $\dfrac{2p^2}{p^2-1}$ **8.** $\dfrac{3t}{t^2-9}$ **9.** $\dfrac{3p-1}{p^2+1}$ **10.** $\dfrac{2t-2}{t^2+4}$

In Exercises 11–30, reduce the rational expression to lowest terms and state any real numbers that must be excluded from the domain.

11. $\dfrac{(x+3)(x-9)}{2(x+3)(x+9)}$ **12.** $\dfrac{4y(y-8)(y+7)}{8y(y+7)(y+8)}$ **13.** $\dfrac{(x-3)(x+1)}{2(x+1)}$ **14.** $\dfrac{(2x+1)(x-3)}{3(x-3)}$

15. $\dfrac{2(3y+1)(2y-1)}{3(2y-1)(3y)}$ **16.** $\dfrac{7(2y+1)(3y-1)}{5(3y-1)(2y)}$ **17.** $\dfrac{(5y-1)(y+1)}{25y-5}$ **18.** $\dfrac{(2t-1)(t+2)}{4t+8}$

19. $\dfrac{(3x+7)(x-4)}{4x-16}$ **20.** $\dfrac{(t-7)(2t+5)}{3t-21}$ **21.** $\dfrac{x^2-4}{x-2}$ **22.** $\dfrac{t^3-t}{t-1}$

23. $\dfrac{x+7}{x+7}$ **24.** $\dfrac{2y+9}{2y+9}$ **25.** $\dfrac{x^2+9}{2x+9}$ **26.** $\dfrac{x^2+4}{2x+4}$

27. $\dfrac{x^2+5x+6}{x^2-3x-10}$ **28.** $\dfrac{x^2+19x+60}{x^2+8x+16}$ **29.** $\dfrac{6x^2-x-1}{2x^2+9x-5}$ **30.** $\dfrac{15x^2-x-2}{5x^2+13x-6}$

In Exercises 31–48, multiply the rational expressions and simplify. State any real numbers that must be excluded from the domain.

31. $\dfrac{x-2}{x+1}\cdot\dfrac{3x+5}{x-2}$ **32.** $\dfrac{4x+5}{x-2}\cdot\dfrac{3x+4}{4x+5}$ **33.** $\dfrac{5x+6}{x}\cdot\dfrac{2x}{5x-6}$

34. $\dfrac{4(x-2)(x+5)}{8x}\cdot\dfrac{16x}{(x-5)(x+5)}$ **35.** $\dfrac{2x-2}{3x}\cdot\dfrac{x^2+x}{x^2-1}$ **36.** $\dfrac{5x-5}{10x}\cdot\dfrac{x^2+x}{x^2-1}$

37. $\dfrac{3x^2-12}{x}\cdot\dfrac{x^2+5x}{x^2+3x-10}$ **38.** $\dfrac{4x^2-32x}{x}\cdot\dfrac{x^2+3x}{x^2-5x-24}$ **39.** $\dfrac{t+2}{3t-9}\cdot\dfrac{t^2-6t+9}{t^2+4t+4}$

40. $\dfrac{y+3}{3y+9}\cdot\dfrac{y^2-10y+25}{y^2+3y-40}$ **41.** $\dfrac{t^2+4}{t-3}\cdot\dfrac{3t}{t+2}$ **42.** $\dfrac{7a^2+21a}{14(a^2-9)}\cdot\dfrac{a+3}{7}$

43. $\dfrac{y^2-4}{y-3}\cdot\dfrac{3y}{y+2}$ **44.** $\dfrac{t^2+t-6}{t^2-4}\cdot\dfrac{8t}{2t^2}$ **45.** $\dfrac{3x^2-15x}{2x^3-50x}\cdot\dfrac{2x^2-7x-15}{3x^2+15x}$

46. $\dfrac{5t-1}{4t}\cdot\dfrac{4t^2+3t}{16t^2-9}$ **47.** $\dfrac{6x^2-11x-35}{8x^2-22x-21}\cdot\dfrac{4x^2-49}{9x^2-25}$ **48.** $\dfrac{3x^2-2x}{12x^3-8x^2}\cdot\dfrac{x^2-7x-18}{2x^2-162}$

In Exercises 49–66, divide the rational expressions and simplify. State any real numbers that must be excluded from the domain.

49. $\dfrac{3}{x}\div\dfrac{12}{x^2}$ **50.** $\dfrac{5}{x^2}\div\dfrac{10}{x^3}$ **51.** $\dfrac{6}{x-2}\div\dfrac{12}{(x-2)(x+2)}$

52. $\dfrac{5(x+6)}{10(x-6)}\div\dfrac{20(x+6)}{8}$ **53.** $\dfrac{1}{x-1}\div\dfrac{5}{x^2-1}$ **54.** $\dfrac{5}{3x-4}\div\dfrac{10}{9x^2-16}$

55. $\dfrac{2-p}{p^2-1} \div \dfrac{2p-4}{p+1}$

56. $\dfrac{4-x}{x^2-16} \div \dfrac{12-3x}{x-4}$

57. $\dfrac{36-n^2}{n^2-9} \div \dfrac{n+6}{n+3}$

58. $\dfrac{49-y^2}{y^2-25} \div \dfrac{7+y}{2y+10}$

59. $\dfrac{3t^3-6t^2-9t}{5t-10} \div \dfrac{6+6t}{4t-8}$

60. $\dfrac{x^3+8x^2+12x}{5x^2-10x} \div \dfrac{4x+8}{x^2-4}$

61. $\dfrac{w^2-w}{w} \div \dfrac{w^3-w}{5w^3}$

62. $\dfrac{y^2-3y}{2y} \div \dfrac{y^3-3y^2}{8y}$

63. $\dfrac{x^2+4x-21}{x^2+3x-10} \div \dfrac{x^2-2x-63}{x^2+x-20}$

64. $\dfrac{2y^2-5y-3}{2y^2-9y-5} \div \dfrac{3y-9}{y^2-5y}$

65. $\dfrac{20x^2-3x-2}{25x^2-4} \div \dfrac{12x^2+23x+5}{3x^2+5x}$

66. $\dfrac{x^2-6x-27}{2x^2+13x-7} \div \dfrac{2x^2-15x-27}{2x^2+9x-5}$

In Exercises 67–82, add or subtract the rational expression and simplify. State any real numbers that must be excluded from the domain.

67. $\dfrac{3}{x} - \dfrac{2}{5x}$

68. $\dfrac{5}{7x} - \dfrac{3}{x}$

69. $\dfrac{3}{p-2} + \dfrac{5p}{p+1}$

70. $\dfrac{4}{9+x} - \dfrac{5x}{x-2}$

71. $\dfrac{2x+1}{5x-1} - \dfrac{3-2x}{1-5x}$

72. $\dfrac{7}{2x-1} - \dfrac{5}{1-2x}$

73. $\dfrac{3y^2}{y+1} + \dfrac{1-2y}{y-1}$

74. $\dfrac{3}{1-x} + \dfrac{4}{x-1}$

75. $\dfrac{3x}{x^2-4} + \dfrac{3+x}{x+2}$

76. $\dfrac{x-1}{4-x^2} - \dfrac{x+1}{2+x}$

77. $\dfrac{x-1}{x-2} + \dfrac{x-6}{x^2-4}$

78. $\dfrac{2}{y-3} + \dfrac{7}{y+2}$

79. $\dfrac{5a}{a^2-b^2} - \dfrac{7}{b-a}$

80. $\dfrac{1}{y} + \dfrac{4}{y^2-4} - \dfrac{2}{y^2-2y}$

81. $7 + \dfrac{1}{x-3}$

82. $\dfrac{3}{5y+6} - \dfrac{4}{y-2} + \dfrac{y^2-y}{5y^2-4y-12}$

In Exercises 83–90, simplify the complex rational expressions. State any real numbers that must be excluded from the domain.

83. $\dfrac{\dfrac{1}{x}-1}{1-\dfrac{2}{x}}$

84. $\dfrac{\dfrac{3}{y}-5}{4-\dfrac{2}{y}}$

85. $\dfrac{3+\dfrac{1}{x}}{9-\dfrac{1}{x^2}}$

86. $\dfrac{\dfrac{1}{x}+\dfrac{2}{x^2}}{\dfrac{9}{x}-\dfrac{5}{x^2}}$

87. $\dfrac{\dfrac{1}{x-1}+1}{1-\dfrac{1}{x+1}}$

88. $\dfrac{\dfrac{7}{y+7}}{\dfrac{1}{y+7}-\dfrac{1}{y}}$

89. $\dfrac{\dfrac{1}{x-1}+1}{\dfrac{1}{x+1}+1}$

90. $\dfrac{\dfrac{3}{x+1}-\dfrac{3}{x-1}}{\dfrac{5}{x^2-1}}$

■ APPLICATIONS

91. Finance. The amount of payment made on a loan is given by the formula $A = \dfrac{pi}{1-1/(1+i)^n}$, where p is the principal (amount borrowed), and $i = \dfrac{r}{n}$, where r is the interest rate expressed as a decimal and n is the number of payments per year. Suppose $n = 5$. Simplify the formula as much as possible.

92. Finance. Use the formula $A = \dfrac{pi}{1-\dfrac{1}{(1+i)^{nt}}}$ to calculate the amount your monthly payment will be on a loan of $150,000 at an interest rate of 6.5% for 30 years ($nt = 360$).

93. Circuits. If two resistances are connected in parallel, the combined resistance is given by the formula $= \dfrac{1}{1/R_1+1/R_2}$, where R_1 and R_2 are the individual resistances. Simplify the formula.

94. Optics. The focal length of a lens can be calculated by applying the formula $f = \dfrac{1}{1/p+1/q}$, where p is the distance that the object is from the lens and q is the distance that the image is from the lens. Simplify the formula.

▪ CATCH THE MISTAKE

In Exercises 95 and 96, explain the mistake that is made.

95. Simplify $\dfrac{x^2 + 2x + 1}{x + 1}$.

Solution:

Factor the numerator. $\qquad \dfrac{x^2 + 2x + 1}{x + 1} = \dfrac{(x + 1)(x + 1)}{(x + 1)}$

Cancel the common factor, $x + 1$. $\qquad = \dfrac{(x + 1)\cancel{(x + 1)}}{\cancel{(x + 1)}}$

Write in simplified form. $\qquad\qquad = x + 1$

This is incorrect. What mistake was made?

96. Simplify $\dfrac{x + 1}{x^2 + 2x + 1}$.

Solution:

Cancel the common 1s. $\qquad \dfrac{x \cancel{+ 1}}{x^2 + 2x \cancel{+ 1}}$

Factor the denominator. $\qquad = \dfrac{x}{x(x + 2)}$

Cancel the common x. $\qquad = \dfrac{\cancel{x}}{\cancel{x}(x + 2)}$

Write in simplified form. $\qquad = \dfrac{1}{x + 2} \qquad x \neq -2$

This is incorrect. What mistake was made?

▪ CONCEPTUAL

In Exercises 97–100, determine whether each statement is true or false.

97. $\dfrac{x^2 - 81}{x - 9} = x + 9$

98. $\dfrac{x - 9}{x^2 - 81} = \dfrac{1}{x + 9} \qquad x \neq -9, 9$

99. When adding or subtracting rational expressions, the LCD is always the product of all the denominators.

100. $\dfrac{x - c}{c - x} = -1$ for all values of x.

▪ CHALLENGE

101. Perform the operation and simplify (remember to state domain restrictions).

$$\dfrac{x + a}{x + b} \div \dfrac{x + c}{x + d}$$

102. Write the numerator as the product of two binomials. Divide out any common factors of the numerator and denominator.

$$\dfrac{a^{2n} - b^{2n}}{a^n - b^n}$$

▪ TECHNOLOGY

103. Utilizing a graphing technology, plot the expression $y = \dfrac{x + 7}{x + 7}$. Zoom in near $x = -7$. Does this agree with what you found in Exercise 23?

104. Utilizing a graphing technology, plot the expression $y = \dfrac{x^2 - 4}{x - 2}$. Zoom in near $x = 2$. Does this agree with what you found in Exercise 21?

In Exercises 105 and 106, for each given expression: (a) simplify the expression, (b) use a graphing utility to plot the expression and the answer in (a) in the same viewing window, and (c) determine the domain restriction(s) where the graphs will agree with each other.

105. $\dfrac{1 + \dfrac{1}{x - 2}}{1 - \dfrac{1}{x + 2}}$

106. $\dfrac{1 - \dfrac{2}{x + 3}}{1 + \dfrac{1}{x + 4}}$

SKILLS OBJECTIVES

- Simplify square roots.
- Simplify radicals.
- Add and subtract radicals.
- Rationalize denominators containing radicals.
- Apply properties of rational exponents.

CONCEPTUAL OBJECTIVES

- Understand that radicals are equivalent to rational exponents.
- Understand that a radical implies one number (the principal root), not two ($\pm$ the principal root).

In Section A.2, we discussed integer exponents and their properties. For example, $4^2 = 16$ and $x^2 \cdot x^3 = x^5$. In this section, we expand our discussion of exponents to include any rational numbers. For example, $16^{1/2} = ?$ and $\left(x^{1/2}\right)^{3/4} = ?$. We will first start with a more familiar notation (*roots*) and discuss operations on *radicals*, and then *rational exponents* will be discussed.

Square Roots

> **DEFINITION** **Principal Square Root**
>
> Let a be any nonnegative real number; then the *nonnegative* real number b is called the **principal square root of** a, denoted $b = \sqrt{a}$, if $b^2 = a$. The symbol $\sqrt{}$ is called a **radical sign**, and a is called the **radicand**.

It is important to note that the principal square root b is nonnegative. "The principal square root of 16 is 4" implies $4^2 = 16$. Although it is also true that $(-4)^2 = 16$, the principal square root is defined to be nonnegative.

It is also important to note that negative real numbers do not have real square roots. For example, $\sqrt{-9}$ is not a real number because there are no real numbers that when squared yield -9. Since principal square roots are defined to be nonnegative, this means they must be zero or positive. The square root of zero is equal to zero: $\sqrt{0} = 0$. All other nonnegative principal square roots are positive.

EXAMPLE 1 **Evaluating Square Roots**

Evaluate the square roots, if possible.

a. $\sqrt{169}$ b. $\sqrt{\dfrac{4}{9}}$ c. $\sqrt{-36}$

Solution:

a. What positive real number squared results in 169? $\sqrt{169} = \boxed{13}$

   Check: $13^2 = 169$

b. What positive real number squared results in $\frac{4}{9}$? $\sqrt{\dfrac{4}{9}} = \boxed{\dfrac{2}{3}}$

   Check: $\left(\frac{2}{3}\right)^2 = \frac{4}{9}$

c. What positive real number squared results in -36? $\boxed{\text{No real number}}$

SQUARE ROOTS OF PERFECT SQUARES

Let a be any real number. Then

$$\sqrt{a^2} = |a|$$

EXAMPLE 2 Finding Square Roots of Perfect Squares

Evaluate:

a. $\sqrt{6^2}$ **b.** $\sqrt{(-7)^2}$ **c.** $\sqrt{x^2}$

Solution:

a. $\sqrt{6^2} = \sqrt{36} = \boxed{6}$ **b.** $\sqrt{(-7)^2} = \sqrt{49} = \boxed{7}$ **c.** $\sqrt{x^2} = \boxed{|x|}$

Simplifying Square Roots

So far only square roots of perfect squares have been discussed. Now we consider how to simplify square roots such as $\sqrt{12}$. We rely on the following properties:

PROPERTIES OF SQUARE ROOTS

Let a and b be nonnegative real numbers. Then

Property	Description	Example
$\sqrt{a \cdot b} = \sqrt{a} \cdot \sqrt{b}$	The square root of a product is the product of the square roots.	$\sqrt{20} = \sqrt{4} \cdot \sqrt{5} = 2\sqrt{5}$
$\sqrt{\dfrac{a}{b}} = \dfrac{\sqrt{a}}{\sqrt{b}} \quad b \neq 0$	The square root of a quotient is the quotient of the square roots.	$\sqrt{\dfrac{40}{49}} = \dfrac{\sqrt{40}}{\sqrt{49}} = \dfrac{\sqrt{4} \cdot \sqrt{10}}{7} = \dfrac{2\sqrt{10}}{7}$

EXAMPLE 3 Simplifying Square Roots

Simplify:

a. $\sqrt{48x^2}$ **b.** $\sqrt{28x^3}$ **c.** $\sqrt{12x} \cdot \sqrt{6x}$ **d.** $\dfrac{\sqrt{45x^3}}{\sqrt{5x}}$

Solution:

a. $\sqrt{48x^2} = \sqrt{48} \cdot \sqrt{x^2} = \sqrt{16 \cdot 3} \cdot \sqrt{x^2} = \underset{4}{\underline{\sqrt{16}}} \cdot \sqrt{3} \cdot \underset{|x|}{\underline{\sqrt{x^2}}} = \boxed{4\,|x|\,\sqrt{3}}$

b. $\sqrt{28x^3} = \sqrt{28} \cdot \sqrt{x^3} = \sqrt{4 \cdot 7} \cdot \sqrt{x^2 \cdot x} = \underset{2}{\underline{\sqrt{4}}} \cdot \sqrt{7} \cdot \underset{|x|}{\underline{\sqrt{x^2}}} \cdot \sqrt{x}$

$= 2|x|\sqrt{7}\,\sqrt{x} = 2|x|\sqrt{7x} = \boxed{2x\sqrt{7x}}$ since $x \geq 0$

c. $\sqrt{12x} \cdot \sqrt{6x} = \sqrt{72x^2} = \sqrt{36 \cdot 2 \cdot x^2} = \underset{6}{\underline{\sqrt{36}}} \cdot \sqrt{2} \cdot \underset{|x|}{\underline{\sqrt{x^2}}} = 6|x|\sqrt{2}$

$= \boxed{6x\sqrt{2}}$ since $x \geq 0$

d. $\dfrac{\sqrt{45x^3}}{\sqrt{5x}} = \sqrt{\dfrac{45x^3}{5x}} = \sqrt{9x^2} = \underset{3}{\underline{\sqrt{9}}} \cdot \underset{|x|}{\underline{\sqrt{x^2}}} = 3|x| = \boxed{3x}$ since $x > 0$

Note: $x \neq 0, \quad x > 0$

> **Study Tip**
>
> If you have a single odd power under the square root, like $\sqrt{x^3}$, the variable is forced to be nonnegative so the absolute value is not necessary.

■ **YOUR TURN** Simplify: **a.** $\sqrt{60x^3}$ **b.** $\dfrac{\sqrt{125x^5}}{\sqrt{25x^3}}$ given that $x > 0$

■ **Answer:**
a. $2x\sqrt{15x}$ **b.** $x\sqrt{5}$

Other (*n*th) Roots

We now expand our discussion from square roots to other *n*th roots.

DEFINITION	Principal *n*th Root

Let a be a real number and n be a positive integer. Then the real number b is called the **principal *n*th root of *a***, denoted $b = \sqrt[n]{a}$, if $b^n = a$. If n is even, then a and b are nonnegative real numbers. The positive integer n is called the **index**. The square root (where no index is shown) corresponds to $n = 2$, and the **cube root** corresponds to $n = 3$.

n	*a*	*b*	EXAMPLE
Even	Positive	Positive	$\sqrt[4]{16} = 2$ because $2^4 = 16$
Even	Negative	Not a real number	$\sqrt[4]{-16}$ is not a real number
Odd	Positive	Positive	$\sqrt[3]{27} = 3$ because $3^3 = 27$
Odd	Negative	Negative	$\sqrt[3]{-125} = -5$ because $(-5)^3 = -125$

A radical sign $\sqrt[n]{}$ combined with a radicand is called a **radical**.

PROPERTIES OF RADICALS

Let a and b be real numbers. Then

PROPERTY	DESCRIPTION	EXAMPLE
$\sqrt[n]{ab} = \sqrt[n]{a} \cdot \sqrt[n]{b}$ if $\sqrt[n]{a}$ and $\sqrt[n]{b}$ both exist.	The *n*th root of a product is the product of the *n*th roots.	$\sqrt[3]{16} = \sqrt[3]{8} \cdot \sqrt[3]{2} = 2\sqrt[3]{2}$
$\sqrt[n]{\dfrac{a}{b}} = \dfrac{\sqrt[n]{a}}{\sqrt[n]{b}} \quad b \neq 0$ if $\sqrt[n]{a}$ and $\sqrt[n]{b}$ both exist.	The *n*th root of a quotient is the quotient of the *n*th roots.	$\sqrt[4]{\dfrac{81}{16}} = \dfrac{\sqrt[4]{81}}{\sqrt[4]{16}} = \dfrac{3}{2}$
$\sqrt[n]{a^m} = \left(\sqrt[n]{a}\right)^m$	The *n*th root of a power is the power of the *n*th root.	$\sqrt[3]{8^2} = \left(\sqrt[3]{8}\right)^2 = (2)^2 = 4$
$\sqrt[n]{a^n} = a \quad n$ is odd.	When *n* is odd, the *n*th root of *a* raised to the *n*th power is *a*.	$\sqrt[3]{x^5} = \underset{x}{\sqrt[3]{x^3}} \cdot \sqrt[3]{x^2} = x\sqrt[3]{x^2}$
$\sqrt[n]{a^n} = |a| \quad n$ is even.	When *n* is even, the *n*th root of *a* raised to the *n*th power is the absolute value of *a*.	$\sqrt[4]{x^5} = \underset{|x|}{\sqrt[4]{x^4}} \cdot \sqrt[4]{x} = |x|\sqrt[4]{x}$

EXAMPLE 4 Simplifying Radicals

Simplify:

a. $\sqrt[3]{-24x^5}$ b. $\sqrt[4]{32x^5}$

Solution:

a. $\sqrt[3]{-24x^5} = \sqrt[3]{(-8)(3)x^3x^2} = \underbrace{\sqrt[3]{-8}}_{-2} \cdot \sqrt[3]{3} \cdot \underbrace{\sqrt[3]{x^3}}_{x} \cdot \sqrt[3]{x^2} = \boxed{-2x\sqrt[3]{3x^2}}$

b. $\sqrt[4]{32x^5} = \sqrt[4]{16 \cdot 2 \cdot x^4 \cdot x} = \underbrace{\sqrt[4]{16}}_{2} \cdot \sqrt[4]{2} \cdot \underbrace{\sqrt[4]{x^4}}_{|x|} \cdot \sqrt[4]{x} = 2|x|\sqrt[4]{2x} = \boxed{2x\sqrt[4]{2x}}$ since $x \geq 0$

Combining Like Radicals

We have already discussed properties for multiplying and dividing radicals. Now we focus on combining (adding or subtracting) radicals. Radicals with the same index and radicand are called **like radicals**. Only like radicals can be added or subtracted.

EXAMPLE 5 Combining Like Radicals

Combine the radicals, if possible.

a. $4\sqrt{3} - 6\sqrt{3} + 7\sqrt{3}$ b. $2\sqrt{5} - 3\sqrt{7} + 6\sqrt{3}$

c. $3\sqrt{5} + \sqrt{20} - 2\sqrt{45}$ d. $\sqrt[4]{10} - 2\sqrt[3]{10} + 3\sqrt{10}$

Solution (a):

Use the distributive property. $4\sqrt{3} - 6\sqrt{3} + 7\sqrt{3} = (4 - 6 + 7)\sqrt{3}$

Eliminate the parentheses. $= \boxed{5\sqrt{3}}$

Solution (b):

None of these radicals are alike.
The expression is in simplified form. $\boxed{2\sqrt{5} - 3\sqrt{7} + 6\sqrt{3}}$

Solution (c):

Write the radicands as
products with a factor of 5. $3\sqrt{5} + \sqrt{20} - 2\sqrt{45} = 3\sqrt{5} + \sqrt{4 \cdot 5} - 2\sqrt{9 \cdot 5}$

The square root of a product is
the product of square roots. $= 3\sqrt{5} + \sqrt{4} \cdot \sqrt{5} - 2\sqrt{9} \cdot \sqrt{5}$

Simplify the square roots of perfect squares. $= 3\sqrt{5} + 2\sqrt{5} - \underbrace{2(3)}_{6}\sqrt{5}$

All three radicals are now like radicals. $= 3\sqrt{5} + 2\sqrt{5} - 6\sqrt{5}$

Use the distributive property. $= (3 + 2 - 6)\sqrt{5}$

Simplify. $= \boxed{-\sqrt{5}}$

Solution (d):

None of these radicals are alike because
they have different indices. The expression
is in simplified form. $\boxed{\sqrt[4]{10} - 2\sqrt[3]{10} + 3\sqrt{10}}$

■ **YOUR TURN** Combine the radicals.

a. $4\sqrt[3]{7} - 6\sqrt[3]{7} + 9\sqrt[3]{7}$ b. $5\sqrt{24} - 2\sqrt{54}$

■ **Answer:**

a. $7\sqrt[3]{7}$ b. $4\sqrt{6}$

Rationalizing Denominators

When radicals appear in a quotient, it is customary to write the quotient with no radicals in the denominator whenever possible. This process is called **rationalizing the denominator** and involves multiplying by an expression that will eliminate the radical in the denominator.

For example, the expression $\dfrac{1}{\sqrt{3}}$ contains a single radical in the denominator. In a case like this, multiply the numerator and denominator by an appropriate radical expression, so that the resulting denominator will be radical-free:

$$\frac{1}{\sqrt{3}} \cdot \underbrace{\frac{\sqrt{3}}{\sqrt{3}}}_{1} = \frac{\sqrt{3}}{\sqrt{3}\cdot\sqrt{3}} = \frac{\sqrt{3}}{3}$$

If the denominator contains a sum of the form $a + \sqrt{b}$, multiply both the numerator and the denominator by the **conjugate** of the denominator, $a - \sqrt{b}$, which uses the difference of two squares to eliminate the radical term. Similarly, if the denominator contains a difference of the form $a - \sqrt{b}$, multiply both the numerator and the denominator by the conjugate of the denominator, $a + \sqrt{b}$. For example, to rationalize $\dfrac{1}{3 - \sqrt{5}}$, take the conjugate of the denominator, which is $3 + \sqrt{5}$:

$$\frac{1}{(3 - \sqrt{5})} \cdot \frac{(3 + \sqrt{5})}{(3 + \sqrt{5})} = \frac{3 + \sqrt{5}}{3^2 \underbrace{+\ 3\sqrt{5} - 3\sqrt{5}}_{\text{like terms}} - (\sqrt{5})^2} = \frac{3 + \sqrt{5}}{9 - 5} = \frac{3 + \sqrt{5}}{4}$$

In general, we apply the difference of two squares:

$$(\sqrt{a} + \sqrt{b})(\sqrt{a} - \sqrt{b}) = (\sqrt{a})^2 - (\sqrt{b})^2 = a - b$$

Notice that the product does not contain a radical. Therefore, to simplify the expression

$$\frac{1}{(\sqrt{a} + \sqrt{b})}$$

multiply the numerator and denominator by $(\sqrt{a} - \sqrt{b})$:

$$\frac{1}{(\sqrt{a} + \sqrt{b})} \cdot \frac{(\sqrt{a} - \sqrt{b})}{(\sqrt{a} - \sqrt{b})}$$

The denominator now contains no radicals:

$$\frac{(\sqrt{a} - \sqrt{b})}{(a - b)}$$

EXAMPLE 6 **Rationalizing Denominators**

Rationalize the denominators and simplify.

a. $\dfrac{2}{3\sqrt{10}}$ **b.** $\dfrac{5}{3-\sqrt{2}}$ **c.** $\dfrac{\sqrt{5}}{\sqrt{2}-\sqrt{7}}$

Solution (a):

Multiply the numerator and
denominator by $\sqrt{10}$.

$= \dfrac{2}{3\sqrt{10}}\cdot\dfrac{\sqrt{10}}{\sqrt{10}}$

Simplify.

$= \dfrac{2\sqrt{10}}{3\left(\sqrt{10}\right)^2} = \dfrac{2\sqrt{10}}{3(10)} = \dfrac{2\sqrt{10}}{30}$

Divide out the common 2 in the
numerator and denominator.

$= \boxed{\dfrac{\sqrt{10}}{15}}$

Solution (b):

Multiply the numerator and denominator
by the conjugate, $3+\sqrt{2}$.

$= \dfrac{5}{\left(3-\sqrt{2}\right)}\cdot\dfrac{\left(3+\sqrt{2}\right)}{\left(3+\sqrt{2}\right)}$

$= \dfrac{5\left(3+\sqrt{2}\right)}{\left(3-\sqrt{2}\right)\left(3+\sqrt{2}\right)}$

The denominator now contains no radicals.

$= \dfrac{15+5\sqrt{2}}{9-2}$

Simplify.

$= \boxed{\dfrac{15+5\sqrt{2}}{7}}$

Solution (c):

Multiply the numerator and denominator
by the conjugate, $\sqrt{2}+\sqrt{7}$.

$= \dfrac{\sqrt{5}}{\left(\sqrt{2}-\sqrt{7}\right)}\cdot\dfrac{\left(\sqrt{2}+\sqrt{7}\right)}{\left(\sqrt{2}+\sqrt{7}\right)}$

Multiply the numerators and
denominators, respectively.

$= \dfrac{\sqrt{5}\left(\sqrt{2}+\sqrt{7}\right)}{\left(\sqrt{2}-\sqrt{7}\right)\left(\sqrt{2}+\sqrt{7}\right)}$

The denominator now contains no radicals.

$= \dfrac{\sqrt{10}+\sqrt{35}}{2-7}$

Simplify.

$= \boxed{-\dfrac{\sqrt{10}+\sqrt{35}}{5}}$

■ **YOUR TURN** Write the expression $\dfrac{7}{1-\sqrt{3}}$ in simplified form.

■ **Answer:** $-\dfrac{7\left(1+\sqrt{3}\right)}{2}$

SIMPLIFIED FORM OF A RADICAL EXPRESSION

A radical expression is in **simplified form** if
■ No factor in the radicand is raised to a power greater than or equal to the
  index.
■ The power of the radicand does not share a common factor with the index.
■ The denominator does not contain a radical.
■ The radical does not contain a fraction.

EXAMPLE 7   Expressing a Radical Expression in Simplified Form

Express the radical expression in simplified form: $\sqrt[3]{\dfrac{16x^5}{81y^7}}$ $x \geq 0, y > 0$.

Solution:

Rewrite the expression so that the radical does not contain a fraction.

$$\sqrt[3]{\dfrac{16x^5}{81y^7}} = \dfrac{\sqrt[3]{16x^5}}{\sqrt[3]{81y^7}}$$

Substitute $16 = 2^4$ and $81 = 3^4$.

$$= \dfrac{\sqrt[3]{2^4 \cdot x^5}}{\sqrt[3]{3^4 \cdot y^7}}$$

Factors in both radicands are raised to powers greater than the index (3). Rewrite the expression so that each power remaining in the radicand is less than the index.

$$= \dfrac{2x\sqrt[3]{2x^2}}{3y^2\sqrt[3]{3y}}$$

The denominator contains a radical. In order to eliminate the radical in the denominator, we multiply the numerator and denominator by $\sqrt[3]{9y^2}$.

$$= \dfrac{2x\sqrt[3]{2x^2}}{3y^2\sqrt[3]{3y}} \cdot \dfrac{\sqrt[3]{9y^2}}{\sqrt[3]{9y^2}}$$

$$= \dfrac{2x\sqrt[3]{18x^2y^2}}{3y^2\sqrt[3]{27y^3}}$$

$$= \dfrac{2x\sqrt[3]{18x^2y^2}}{9y^3}$$

The radical expression now satisfies the conditions for simplified form.

$$\boxed{= \dfrac{2x\sqrt[3]{18x^2y^2}}{9y^3}}$$

Rational Exponents

We now use radicals to define rational exponents.

RATIONAL EXPONENTS: $\dfrac{1}{n}$

Let a be any real number and n be a positive integer. Then

$$a^{1/n} = \sqrt[n]{a}$$

where $\dfrac{1}{n}$ is the **rational exponent** of a.

- When n is even and a is negative, then $a^{1/n}$ and $\sqrt[n]{a}$ are not real numbers.
- Furthermore, if m is a positive integer with m and n having no common factors, then

$$a^{m/n} = \left(a^{1/n}\right)^m = \left(a^m\right)^{1/n} = \sqrt[n]{a^m}$$

Note: Any of the four notations can be used.

EXAMPLE 8 Simplifying Expressions with Rational Exponents

Simplify:

a. $16^{3/2}$ **b.** $(-8)^{2/3}$

Solution:

a. $16^{3/2} = \left(16^{1/2}\right)^3 = \left(\sqrt{16}\right)^3 = 4^3 = 64$

b. $(-8)^{2/3} = \left[(-8)^{1/3}\right]^2 = (-2)^2 = 4$

■ **YOUR TURN** Simplify $27^{2/3}$.

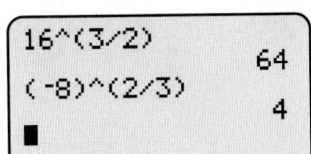

```
16^(3/2)
                    64
(-8)^(2/3)
                     4
■
```

■ **Answer:** 9

The properties of exponents that hold for integers also hold for rational numbers:

$$a^{-1/n} = \frac{1}{a^{1/n}} \quad \text{and} \quad a^{-m/n} = \frac{1}{a^{m/n}} \qquad a \neq 0$$

EXAMPLE 9 Simplifying Expressions with Negative Rational Exponents

Simplify $\dfrac{(9x)^{-1/2}}{4x^{-3/2}}$ $x > 0$.

Solution:

Negative exponents correspond to positive exponents in the reciprocal.

$$\frac{(9x)^{-1/2}}{4x^{-3/2}} = \frac{x^{3/2}}{4 \cdot (9x)^{1/2}}$$

Eliminate the parentheses.

$$= \frac{x^{3/2}}{4 \cdot 9^{1/2} x^{1/2}}$$

Apply the quotient property on x.

$$= \frac{x^{3/2 - 1/2}}{\underset{3}{4 \cdot 9^{1/2}}}$$

Simplify.

$$= \boxed{\frac{x}{12}}$$

■ **Answer:** $18x^2$

■ **YOUR TURN** Simplify $\dfrac{9x^{3/2}}{(4x)^{-1/2}}$ $x > 0$.

EXAMPLE 10 Simplifying Algebraic Expressions with Rational Exponents

Simplify $\dfrac{\left(-8x^2y\right)^{1/3}}{\left(9xy^4\right)^{1/2}}$ $x > 0, y > 0$.

Solution:

$$\frac{\left(-8x^2y\right)^{1/3}}{\left(9xy^4\right)^{1/2}} = \frac{(-8)^{1/3}\left(x^2\right)^{1/3}y^{1/3}}{9^{1/2}x^{1/2}\left(y^4\right)^{1/2}} = \frac{(-8)^{1/3}x^{2/3}y^{1/3}}{9^{1/2}x^{1/2}y^2} = \left(\frac{-2}{3}\right)x^{2/3-1/2}y^{1/3-2} = -\frac{2}{3}x^{1/6}y^{-5/3}$$

Write in terms of positive exponents.

$$= \boxed{-\frac{2x^{1/6}}{3y^{5/3}}}$$

■ **YOUR TURN** Simplify $\dfrac{\left(16x^3y\right)^{1/2}}{\left(27x^2y^3\right)^{1/3}}$ and write your answer with only positive exponents.

■ **Answer:** $\dfrac{4x^{5/6}}{3y^{1/2}}$

EXAMPLE 11   **Factoring Expressions with Rational Exponents**

Factor completely $x^{8/3} - 5x^{5/3} - 6x^{2/3}$.

Solution:

Factor out the greatest common factor $x^{2/3}$.

$$\underbrace{x^{8/3}}_{x^{2/3} \cdot x^{6/3}} - \underbrace{5x^{5/3}}_{x^{2/3} \cdot x^{3/3}} - 6x^{2/3} = x^{2/3}\left(x^2 - 5x - 6\right)$$

Factor the trinomial.

$$= \boxed{x^{2/3}(x - 6)(x + 1)}$$

■ **Answer:** $x^{1/3}(x - 2)(x + 1)$

■ **YOUR TURN**   Factor completely $x^{7/3} - x^{4/3} - 2x^{1/3}$.

SECTION A.6 SUMMARY

In this section, we defined radicals as "$b = \sqrt[n]{a}$ means $a = b^n$" for a and b positive real numbers when n is a positive even integer, and a and b any real numbers when n is a positive odd integer.

Properties of Radicals

PROPERTY		EXAMPLE
$\sqrt[n]{ab} = \sqrt[n]{a} \cdot \sqrt[n]{b}$		$\sqrt[3]{16} = \sqrt[3]{8} \cdot \sqrt[3]{2} = 2\sqrt[3]{2}$
$\sqrt[n]{\dfrac{a}{b}} = \dfrac{\sqrt[n]{a}}{\sqrt[n]{b}}$	$b \neq 0$	$\sqrt[4]{\dfrac{81}{16}} = \dfrac{\sqrt[4]{81}}{\sqrt[4]{16}} = \dfrac{3}{2}$
$\sqrt[n]{a^m} = \left(\sqrt[n]{a}\right)^m$		$\sqrt[3]{8^2} = \left(\sqrt[3]{8}\right)^2 = (2)^2 = 4$
$\sqrt[n]{a^n} = a$	n is odd	$\sqrt[3]{x^5} = \underbrace{\sqrt[3]{x^3}}_{x} \cdot \sqrt[3]{x^2} = x\sqrt[3]{x^2}$
$\sqrt[n]{a^n} = \lvert a \rvert$	n is even	$\sqrt[4]{x^6} = \sqrt[4]{x^4} \cdot \sqrt[4]{x^2} = \lvert x \rvert \sqrt[4]{x^2}$

Radicals can be combined only if they are like radicals (same radicand and index). Quotients with radicals in the denominator are usually rewritten with no radicals in the denominator.

Rational exponents were defined in terms of radicals: $a^{1/n} = \sqrt[n]{a}$. The properties for integer exponents we learned in Section A.2 also hold true for rational exponents:

$$a^{m/n} = \left(a^{1/n}\right)^m = \left(\sqrt[n]{a}\right)^m \quad \text{and} \quad a^{m/n} = (a^m)^{1/n} = \sqrt[n]{a^m}$$

Negative rational exponents: $a^{-m/n} = \dfrac{1}{a^{m/n}}$, for m and n positive integers with no common factors.

SECTION A.6 EXERCISES

■ SKILLS

In Exercises 1–24, evaluate each expression or state that it is not a real number.

1. $\sqrt{100}$

2. $\sqrt{121}$

3. $-\sqrt{144}$

4. $\sqrt{-169}$

5. $\sqrt[3]{-216}$

6. $\sqrt[3]{-125}$

7. $\sqrt[3]{343}$

8. $-\sqrt[3]{-27}$

9. $\sqrt[8]{1}$

10. $\sqrt[7]{-1}$

11. $\sqrt[3]{0}$

12. $\sqrt[5]{0}$

13. $\sqrt{-16}$

14. $\sqrt[5]{-1}$

15. $(-27)^{1/3}$

16. $(-64)^{1/3}$

17. $8^{2/3}$

18. $(-64)^{2/3}$

19. $(-32)^{1/5}$

20. $(-243)^{1/3}$

21. $(-1)^{1/3}$

22. $1^{5/2}$

23. $9^{3/2}$

24. $(27)^{2/3}$

In Exercises 25–40, simplify, if possible, the radical expressions.

25. $\sqrt{2} - 5\sqrt{2}$

26. $3\sqrt{5} - 7\sqrt{5}$

27. $3\sqrt{5} - 2\sqrt{5} + 7\sqrt{5}$

28. $6\sqrt{7} + 7\sqrt{7} - 10\sqrt{7}$

29. $\sqrt{12} \cdot \sqrt{2}$

30. $2\sqrt{5} \cdot 3\sqrt{40}$

31. $\sqrt[3]{12} \cdot \sqrt[3]{4}$

32. $\sqrt[4]{8} \cdot \sqrt[4]{4}$

33. $\sqrt{3}\sqrt{7}$

34. $\sqrt{5}\sqrt{2}$

35. $8\sqrt{25x^2}$

36. $16\sqrt{36y^4}$

37. $\sqrt{4x^2y}$

38. $\sqrt{16x^3y}$

39. $\sqrt[3]{-81x^6y^8}$

40. $\sqrt[5]{-32x^{10}y^8}$

In Exercises 41–56, rationalize the denominators.

41. $\sqrt{\dfrac{1}{3}}$

42. $\sqrt{\dfrac{2}{5}}$

43. $\dfrac{2}{3\sqrt{11}}$

44. $\dfrac{5}{3\sqrt{2}}$

45. $\dfrac{3}{1 - \sqrt{5}}$

46. $\dfrac{2}{1 + \sqrt{3}}$

47. $\dfrac{1 + \sqrt{2}}{1 - \sqrt{2}}$

48. $\dfrac{3 - \sqrt{5}}{3 + \sqrt{5}}$

49. $\dfrac{3}{\sqrt{2} - \sqrt{3}}$

50. $\dfrac{5}{\sqrt{2} + \sqrt{5}}$

51. $\dfrac{4}{3\sqrt{2} + 2\sqrt{3}}$

52. $\dfrac{7}{2\sqrt{3} + 3\sqrt{2}}$

53. $\dfrac{4 + \sqrt{5}}{3 + 2\sqrt{5}}$

54. $\dfrac{6}{3\sqrt{2} + 4}$

55. $\dfrac{\sqrt{7} + 3}{\sqrt{2} - \sqrt{5}}$

56. $\dfrac{\sqrt{y}}{\sqrt{x} - \sqrt{y}}$

In Exercises 57–64, simplify by applying the properties of rational exponents. Express your answers in terms of positive exponents.

57. $\left(x^{1/2} y^{2/3}\right)^6$

58. $\left(y^{2/3} \cdot y^{1/4}\right)^{12}$

59. $\dfrac{\left(x^{1/3} y^{1/2}\right)^{-3}}{\left(x^{-1/2} y^{1/4}\right)^2}$

60. $\dfrac{\left(x^{-2/3} \cdot y^{-3/4}\right)^{-2}}{\left(x^{1/3} \cdot y^{1/4}\right)^4}$

61. $\dfrac{x^{1/2} y^{1/5}}{x^{-2/3} y^{-9/5}}$

62. $\dfrac{\left(y^{-3/4} \cdot x^{-2/3}\right)^{12}}{\left(y^{1/4} \cdot x^{7/3}\right)^{24}}$

63. $\dfrac{\left(2x^{2/3}\right)^3}{\left(4x^{-1/3}\right)^2}$

64. $\dfrac{\left(2x^{-2/3}\right)^3}{\left(4x^{-4/3}\right)^2}$

In Exercises 65–68, factor each expression completely.

65. $x^{7/3} - x^{4/3} - 2x^{1/3}$

66. $8x^{1/4} + 4x^{5/4}$

67. $7x^{3/7} - 14x^{6/7} + 21x^{10/7}$

68. $7x^{-1/3} + 70x$

■ **APPLICATIONS**

69. **Gravity.** If a penny is dropped off a building, the time it takes (seconds) to fall d feet is given by $\sqrt{\dfrac{d}{16}}$. If a penny is dropped off a 1280-foot-tall building, how long will it take until it hits the ground? Round to the nearest second.

70. **Gravity.** If a ball is dropped off a building, the time it takes (seconds) to fall d meters is approximately given by $\sqrt{\dfrac{d}{5}}$. If a ball is dropped off a 600-meter-tall building, how long will it take until it hits the ground? Round to the nearest second.

71. **Kepler's Law.** The square of the period p (in years) of a planet's orbit around the Sun is equal to the cube of the planet's maximum distance from the Sun, d (in astronomical units or AU). This relationship can be expressed mathematically as $p^2 = d^3$. If this formula is solved for d, the resulting equation is $d = p^{2/3}$. If Saturn has an orbital period of 29.46 Earth years, calculate Saturn's maximum distance from the Sun to the nearest hundredth of an AU.

72. **Period of a Pendulum.** The period (in seconds) of a pendulum of length L (in meters) is given by $P = 2 \cdot \pi \cdot \left(\dfrac{L}{9.8}\right)^{1/2}$. If a certain pendulum has a length of 19.6 meters, determine the period P of this pendulum to the nearest tenth of a second.

CATCH THE MISTAKE

In Exercises 73 and 74, explain the mistake that is made.

73. Simplify $\left(4x^{1/2}\,y^{1/4}\right)^2$.

Solution:

Use the properties of exponents. $\qquad 4\left(x^{1/2}\right)^2\left(y^{1/4}\right)^2$

Simplify. $\qquad\qquad\qquad\qquad 4xy^{1/2}$

This is incorrect. What mistake was made?

74. Simplify $\dfrac{2}{5 - \sqrt{11}}$.

Solution:

Multiply the numerator and denominator by $5 - \sqrt{11}$.

$$\frac{2}{5 - \sqrt{11}} \cdot \frac{\left(5 - \sqrt{11}\right)}{\left(5 - \sqrt{11}\right)}$$

Multiply numerators and denominators.

$$\frac{2\left(5 - \sqrt{11}\right)}{25 - 11}$$

Simplify.

$$\frac{2\left(5 - \sqrt{11}\right)}{14} = \frac{5 - \sqrt{11}}{7}$$

This is incorrect. What mistake was made?

CONCEPTUAL

In Exercises 75–78, determine whether each statement is true or false.

75. $\sqrt{121} = \pm 11$

76. $\sqrt{x^2} = x$, where x is any real number.

77. $\sqrt{a^2 + b^2} = \sqrt{a} + \sqrt{b}$

78. $\sqrt{-4} = -2$

In Exercises 79 and 80, a, m, n, and k are any positive real numbers.

79. Simplify $\left(\left(a^m\right)^n\right)^k$.

80. Simplify $\left(a^{-k}\right)^{-1/k}$.

In Exercises 81 and 82, evaluate each algebraic expression for the specified values.

81. $\dfrac{\sqrt{b^2 - 4ac}}{2a}$ for $a = 1$, $b = 7$, $c = 12$

82. $\sqrt{b^2 - 4ac}$ for $a = 1$, $b = 7$, $c = 12$

CHALLENGE

83. Rationalize the denominator and simplify: $\dfrac{1}{\left(\sqrt{a} + \sqrt{b}\right)^2}$.

84. Rationalize the denominator and simplify: $\dfrac{\sqrt{a + b} - \sqrt{a}}{\sqrt{a + b} + \sqrt{a}}$.

TECHNOLOGY

85. Use a calculator to approximate $\sqrt{11}$ to three decimal places.

86. Use a calculator to approximate $\sqrt[3]{7}$ to three decimal places.

87. Given $\dfrac{4}{5\sqrt{2} + 4\sqrt{3}}$,

  a. Rationalize the denominator.

  b. Use a graphing utility to evaluate the expression and the answer.

  c. Do the expression and answer agree?

88. Given $\dfrac{2}{4\sqrt{5} - 3\sqrt{6}}$,

  a. Rationalize the denominator.

  b. Use a graphing utility to evaluate the expression and the answer.

  c. Do the expression and answer agree?

SKILLS OBJECTIVES

- Write radicals with negative radicands as imaginary numbers.
- Add and subtract complex numbers.
- Multiply complex numbers.
- Divide complex numbers.
- Raise complex numbers to powers.

CONCEPTUAL OBJECTIVES

- Understand that real numbers and imaginary numbers are subsets of complex numbers.
- Understand how to eliminate imaginary numbers in denominators.

In Section 0.2, we will be reviewing equations whose solutions sometimes involve the square roots of negative numbers. In Section A.6, when asked to evaluate the square root of a negative number, like $\sqrt{-16}$, we said, "It is not a real number," because there is no real number such that $x^2 = -16$. To include such roots in the number system, mathematicians created a new expanded set of numbers, called the *complex numbers*. The foundation of this new set of numbers is the imaginary unit i.

> **DEFINITION** **The Imaginary Unit i**
>
> The **imaginary unit** is denoted by the letter i and is defined as
>
> $$i = \sqrt{-1}$$
>
> where $i^2 = -1$.

Recall that for positive real numbers a and b, we defined the principal square root as

$$b = \sqrt{a} \quad \text{that means} \quad b^2 = a$$

Similarly, we define the *principal square root of a negative number* as $\sqrt{-a} = i\sqrt{a}$, since $\left(i\sqrt{a}\right)^2 = i^2 a = -a$.

> If $-a$ is a negative real number, then the **principal square root** of $-a$ is
>
> $$\sqrt{-a} = i\sqrt{a}$$
>
> where i is the imaginary unit and $i^2 = -1$.

We write $i\sqrt{a}$ instead of $\sqrt{a}\,i$ to avoid any confusion.

EXAMPLE 1 Using Imaginary Numbers to Simplify Radicals

Simplify using imaginary numbers.

a. $\sqrt{-9}$ **b.** $\sqrt{-8}$

Solution:

a. $\sqrt{-9} = i\sqrt{9} = \boxed{3i}$ **b.** $\sqrt{-8} = i\sqrt{8} = i \cdot 2\sqrt{2} = \boxed{2i\sqrt{2}}$

■ **YOUR TURN** Simplify $\sqrt{-144}$.

Technology Tip

Be sure to put the graphing calculator in $\boxed{a + bi}$ mode.

a. $\sqrt{-9}$ **b.** $\sqrt{-8}$

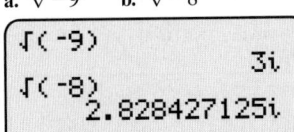

```
√(-9)
                    3i
√(-8)
         2.828427125i
```

■ **Answer:** $12i$

DEFINITION **Complex Number**

A **complex number** in standard form is defined as

$$a + bi$$

where a and b are real numbers and i is the imaginary unit. We call a the **real part** of the complex number and b the **imaginary part** of the complex number.

A complex number written as $a + bi$ is said to be in **standard form**. If $a = 0$ and $b \neq 0$, then the resulting complex number bi is called a **pure imaginary number**. If $b = 0$, then $a + bi$ is a real number. The set of all real numbers and the set of all imaginary numbers are both subsets of the set of complex numbers.

```
                 Complex Numbers
                      a + bi

     Real Numbers                  Imaginary Numbers
         a                               bi
      (b = 0)                          (a = 0)
```

The following are all examples of complex numbers:

$$17 \qquad 2 - 3i \qquad -5 + i \qquad 3 - i\sqrt{11} \qquad -9i$$

DEFINITION **Equality of Complex Numbers**

The complex numbers $a + bi$ and $c + di$ are **equal** if and only if $a = c$ and $b = d$. In other words, two complex numbers are equal if and only if both real parts are equal *and* both imaginary parts are equal.

Adding and Subtracting Complex Numbers

Complex numbers in the standard form $a + bi$ are treated in much the same way as binomials of the form $a + bx$. We can add, subtract, and multiply complex numbers the same way we performed these operations on binomials. When adding or subtracting complex numbers, combine real parts with real parts and combine imaginary parts with imaginary parts.

Technology Tip

Be sure to put the graphing calculator in $\boxed{a + bi}$ mode.

a. $(3 - 2i) + (-1 + i)$
b. $(2 - i) - (3 - 4i)$

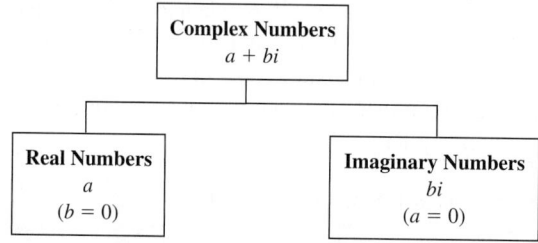

EXAMPLE 2 **Adding and Subtracting Complex Numbers**

Perform the indicated operation and simplify.

a. $(3 - 2i) + (-1 + i)$ **b.** $(2 - i) - (3 - 4i)$

Solution (a):

Eliminate the parentheses. $= 3 - 2i - 1 + i$

Group real and imaginary numbers, respectively. $= (3 - 1) + (-2i + i)$

Simplify. $= \boxed{2 - i}$

Solution (b):

Eliminate the parentheses (distribute the negative).	$= 2 - i - 3 + 4i$
Group real and imaginary numbers, respectively.	$= (2 - 3) + (-i + 4i)$
Simplify.	$= \boxed{-1 + 3i}$

■ **YOUR TURN** Perform the indicated operation and simplify: $(4 + i) - (3 - 5i)$.

Multiplying Complex Numbers

When multiplying complex numbers, you apply all the same methods as you did when multiplying binomials. It is important to remember that $i^2 = -1$.

WORDS	**MATH**
Multiply the complex numbers.	$(5 - i)(3 - 4i)$
Multiply using the distributive property.	$= 5(3) + 5(-4i) - i(3) - (i)(-4i)$
Eliminate the parentheses.	$= 15 - 20i - 3i + 4i^2$
Substitute $i^2 = -1$.	$= 15 - 20i - 3i + 4(-1)$
Simplify	$= 15 - 20i - 3i - 4$
Combine real parts and imaginary parts, respectively.	$= 11 - 23i$

EXAMPLE 3 Multiplying Complex Numbers

Multiply the complex numbers and express the result in standard form, $a + bi$.

a. $(3 - i)(2 + i)$ **b.** $i(-3 + i)$

Solution (a):

Use the distributive property.	$(3 - i)(2 + i) = 3(2) + 3(i) - i(2) - i(i)$
Eliminate the parentheses.	$= 6 + 3i - 2i - i^2$
Substitute $i^2 = -1$.	$= 6 + 3i - 2i - (-1)$
Group like terms.	$= (6 + 1) + (3i - 2i)$
Simplify.	$= \boxed{7 + i}$

Solution (b):

Use the distributive property.	$i(-3 + i) = -3i + i^2$
Substitute $i^2 = -1$.	$= -3i - 1$
Write in standard form.	$= \boxed{-1 - 3i}$

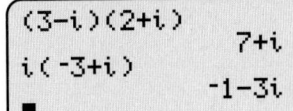

■ **YOUR TURN** Multiply the complex numbers and express the result in standard form, $a + bi$: $(4 - 3i)(-1 + 2i)$.

Dividing Complex Numbers

Recall the special product that produces a difference of two squares, $(a + b)(a - b) = a^2 - b^2$. This special product has only first and last terms because the outer and inner terms sum to zero. Similarly, if we multiply complex numbers in the same manner, the result is a real number because the imaginary terms sum to zero.

COMPLEX CONJUGATE

The product of a complex number, $z = a + bi$, and its **complex conjugate**, $\bar{z} = a - bi$, is a real number.

$$z\bar{z} = (a + bi)(a - bi) = a^2 - b^2i^2 = a^2 - b^2(-1) = a^2 + b^2$$

In order to write a quotient of complex numbers in standard form, $a + bi$, multiply the numerator and the denominator by the complex conjugate of the denominator. It is important to note that if i is present in the denominator, then the complex number is *not* in standard form.

Technology Tip

Be sure to put the graphing calculator in $\boxed{a + bi}$ mode.

$$\frac{2 - i}{1 + 3i}$$

To change the answer to the fraction form, press $\boxed{\text{MATH}}$, highlight $\boxed{1: \blacktriangleright \text{Frac}}$, press $\boxed{\text{ENTER}}$ and $\boxed{\text{ENTER}}$.

```
(2-i)/(1+3i)
        -.1-.7i
Ans▶Frac
     -1/10-7/10i
■
```

Answer: $\frac{10}{17} + \frac{11}{17}i$

EXAMPLE 4 **Dividing Complex Numbers**

Write the quotient in standard form: $\dfrac{2 - i}{1 + 3i}$.

Solution:

Multiply the numerator and denominator by the complex conjugate of the denominator, $1 - 3i$.

$$\frac{2 - i}{1 + 3i} = \left(\frac{2 - i}{1 + 3i}\right)\left(\frac{1 - 3i}{1 - 3i}\right)$$

Multiply the numerators and denominators, respectively.

$$= \frac{(2 - i)(1 - 3i)}{(1 + 3i)(1 - 3i)}$$

Use the FOIL method (or distributive property).

$$= \frac{2 - 6i - i + 3i^2}{1 - 3i + 3i - 9i^2}$$

Combine imaginary parts.

$$= \frac{2 - 7i + 3i^2}{1 - 9i^2}$$

Substitute $i^2 = -1$.

$$= \frac{2 - 7i - 3}{1 - 9(-1)}$$

Simplify the numerator and denominator.

$$= \frac{-1 - 7i}{10}$$

Write in standard form.

Recall that $\dfrac{a + b}{c} = \dfrac{a}{c} + \dfrac{b}{c}$.

$$= \boxed{-\frac{1}{10} - \frac{7}{10}i}$$

■**YOUR TURN** Write the quotient in standard form: $\dfrac{3 + 2i}{4 - i}$.

Raising Complex Numbers to Integer Powers

Note that i raised to the fourth power is 1. In simplifying imaginary numbers, we factor out i raised to the largest multiple of 4.

$$i = \sqrt{-1}$$
$$i^2 = -1$$
$$i^3 = i^2 \cdot i = (-1)i = -i$$
$$\underline{i^4 = i^2 \cdot i^2 = (-1)(-1) = 1}$$
$$i^5 = i^4 \cdot i = (1)(i) = i$$
$$i^6 = i^4 \cdot i^2 = (1)(-1) = -1$$
$$i^7 = i^4 \cdot i^3 = (1)(-i) = -i$$
$$i^8 = \left(i^4\right)^2 = 1$$

EXAMPLE 5 Raising the Imaginary Unit to Integer Powers

Simplify:

a. i^7 **b.** i^{13} **c.** i^{100}

Solution:

a. $i^7 = i^4 \cdot i^3 = (1)(-i) = \boxed{-i}$

b. $i^{13} = i^{12} \cdot i = \left(i^4\right)^3 \cdot i = 1^3 \cdot i = \boxed{i}$

c. $i^{100} = \left(i^4\right)^{25} = 1^{25} = \boxed{1}$

■ **YOUR TURN** Simplify i^{27}.

Technology Tip

```
i^7
            -3ᴇ-13-i
```

This is due to rounding off error in the programming. Since -3×10^{-13} can be approximated as 0, $i^7 = -i$.

```
i^13
            -3ᴇ-13+i
i^100
                  1
```

■ **Answer:** $-i$

EXAMPLE 6 Raising a Complex Number to an Integer Power

Write $(2 - i)^3$ in standard form.

Solution:

Recall the formula for cubing a binomial.

$$(a - b)^3 = a^3 - 3a^2b + 3ab^2 - b^3$$

Let $a = 2$ and $b = i$.

$$(2 - i)^3 = 2^3 - 3(2)^2(i) + 3(2)(i)^2 - i^3$$

Let $i^2 = -1$ and $i^3 = -i$.

$$= 2^3 - 3(2)^2(i) + (3)(2)(-1) - (-i)$$

Eliminate parentheses.

$$= 8 - 6 - 12i + i$$

Combine the real and imaginary parts, respectively.

$$= \boxed{2 - 11i}$$

■ **YOUR TURN** Write $(2 + i)^3$ in standard form.

Technology Tip

Be sure to put the graphing calculator in $\boxed{a + bi}$ mode.

$$(2 - i)^3$$

```
(2-i)^3
              2-11i
```

■ **Answer:** $2 + 11i$

The Imaginary Unit i

- $i = \sqrt{-1}$
- $i^2 = -1$

Complex Numbers

- *Standard Form*: $a + bi$, where a is the real part and b is the imaginary part.
- The set of real numbers and the set of pure imaginary numbers are subsets of the set of complex numbers.

Adding and Subtracting Complex Numbers

- $(a + bi) + (c + di) = (a + c) + (b + d)i$
- $(a + bi) - (c + di) = (a - c) + (b - d)i$
- To add or subtract complex numbers, add or subtract the real parts and imaginary parts, respectively.

Multiplying Complex Numbers

- $(a + bi)(c + di) = (ac - bd) + (ad + bc)i$
- Apply the same methods for multiplying binomials. It is important to remember that $i^2 = -1$.

Dividing Complex Numbers

- Complex conjugate of $a + bi$ is $a - bi$.
- In order to write a quotient of complex numbers in standard form, multiply the numerator and the denominator by the complex conjugate of the denominator:

$$\frac{a + bi}{c + di} \cdot \frac{(c - di)}{(c - di)}$$

SKILLS

In Exercises 1–12, write each expression as a complex number in standard form. Some expressions simplify to either a real number or a pure imaginary number.

1. $\sqrt{-16}$ **2.** $\sqrt{-100}$ **3.** $\sqrt{-20}$ **4.** $\sqrt{-24}$

5. $\sqrt[3]{-64}$ **6.** $\sqrt[3]{-27}$ **7.** $\sqrt{-64}$ **8.** $\sqrt{-27}$

9. $3 - \sqrt{-100}$ **10.** $4 - \sqrt{-121}$ **11.** $-10 - \sqrt{-144}$ **12.** $7 - \sqrt[3]{-125}$

In Exercises 13–40, perform the indicated operation, simplify, and express in standard form.

13. $(3 - 7i) + (-1 - 2i)$ **14.** $(1 + i) + (9 - 3i)$ **15.** $(3 - 4i) + (7 - 10i)$ **16.** $(5 + 7i) + (-10 - 2i)$

17. $(4 - 5i) - (2 - 3i)$ **18.** $(-2 + i) - (1 - i)$ **19.** $(-3 + i) - (-2 - i)$ **20.** $(4 + 7i) - (5 + 3i)$

21. $3(4 - 2i)$ **22.** $4(7 - 6i)$ **23.** $12(8 - 5i)$ **24.** $-3(16 + 4i)$

25. $-3(16 - 9i)$ **26.** $5(-6i + 3)$ **27.** $-6(17 - 5i)$ **28.** $-12(8 + 3i)$

29. $(1 - i)(3 + 2i)$ **30.** $(-3 + 2i)(1 - 3i)$ **31.** $(5 - 7i)(-3 + 4i)$ **32.** $(16 - 5i)(-2 - i)$

33. $(7 - 5i)(6 + 9i)$ **34.** $(-3 - 2i)(7 - 4i)$ **35.** $(12 - 18i)(-2 + i)$ **36.** $(-4 + 3i)(-4 - 3i)$

37. $\left(\frac{1}{2} + 2i\right)\left(\frac{4}{9} - 3i\right)$ **38.** $\left(-\frac{3}{4} + \frac{9}{16}i\right)\left(\frac{2}{3} + \frac{4}{9}i\right)$ **39.** $(-i + 17)(2 + 3i)$ **40.** $(-3i - 2)(-2 - 3i)$

For Exercises 41–48, for each complex number z, write the complex conjugate $\bar{z}$ and find $z\bar{z}$.

41. $z = 4 + 7i$ **42.** $z = 2 + 5i$ **43.** $z = 2 - 3i$ **44.** $z = 5 - 3i$

45. $z = 6 + 4i$ **46.** $z = -2 + 7i$ **47.** $z = -2 - 6i$ **48.** $z = -3 - 9i$

For Exercises 49–64, write each quotient in standard form.

49. $\dfrac{2}{i}$ **50.** $\dfrac{3}{i}$ **51.** $\dfrac{1}{3 - i}$ **52.** $\dfrac{2}{7 - i}$ **53.** $\dfrac{1}{3 + 2i}$ **54.** $\dfrac{1}{4 - 3i}$ **55.** $\dfrac{2}{7 + 2i}$ **56.** $\dfrac{8}{1 + 6i}$

57. $\dfrac{1 - i}{1 + i}$ **58.** $\dfrac{3 - i}{3 + i}$ **59.** $\dfrac{2 + 3i}{3 - 5i}$ **60.** $\dfrac{2 + i}{3 - i}$ **61.** $\dfrac{4 - 5i}{7 + 2i}$ **62.** $\dfrac{7 + 4i}{9 - 3i}$ **63.** $\dfrac{8 + 3i}{9 - 2i}$ **64.** $\dfrac{10 - i}{12 + 5i}$

For Exercises 65–76, simplify.

65. i^{15} **66.** i^{99} **67.** i^{40} **68.** i^{18} **69.** $(5 - 2i)^2$ **70.** $(3 - 5i)^2$

71. $(2 + 3i)^2$ **72.** $(4 - 9i)^2$ **73.** $(3 + i)^3$ **74.** $(2 + i)^3$ **75.** $(1 - i)^3$ **76.** $(4 - 3i)^3$

■ APPLICATIONS

Electrical impedance is the ratio of voltage to current in *AC* circuits. Let Z represent the total impedance of an electrical circuit. If there are two resistors in a circuit, let $Z_1 = 3 - 6i$ ohms and $Z_2 = 5 + 4i$ ohms.

77. Electrical Circuits in Series. When the resistors in the circuit are placed in series, the total impedance is the sum of the two impedances $Z = Z_1 + Z_2$. Find the total impedance of the electrical circuit in series.

78. Electrical Circuits in Parallel. When the resistors in the circuit are placed in parallel, the total impedance is given by $\dfrac{1}{Z} = \dfrac{1}{Z_1} + \dfrac{1}{Z_2}$. Find the total impedance of the electrical circuit in parallel.

■ CATCH THE MISTAKE

In Exercises 79 and 80, explain the mistake that is made.

79. Write the quotient in standard form: $\dfrac{2}{4 - i}$.

Solution:

Multiply the numerator and the denominator by $4 - i$. $\dfrac{2}{4 - i} = \dfrac{2}{4 - i} \cdot \dfrac{4 - i}{4 - i}$

Multiply the numerator using the distributive property and the denominator using the FOIL method. $= \dfrac{8 - 2i}{16 - 1}$

Simplify. $= \dfrac{8 - 2i}{15}$

Write in standard form. $= \dfrac{8}{15} - \dfrac{2}{15}i$

This is incorrect. What mistake was made?

80. Write the product in standard form: $(2 - 3i)(5 + 4i)$.

Solution:

Use the FOIL method to multiply the complex numbers.

$$(2 - 3i)(5 + 4i) = 10 - 7i - 12i^2$$

Simplify.

$$= -2 - 7i$$

This is incorrect. What mistake was made?

■ CONCEPTUAL

In Exercises 81–84, determine whether each statement is true or false.

81. The product is a real number: $(a + bi)(a - bi)$.

82. Imaginary numbers are a subset of the complex numbers.

83. Real numbers are a subset of the complex numbers.

84. There is no complex number that equals its conjugate.

■ CHALLENGE

85. Factor completely over the complex numbers: $x^4 + 2x^2 + 1$.

86. Factor completely over the complex numbers: $x^4 + 18x^2 + 81$.

■ TECHNOLOGY

In Exercises 87–90, apply a graphing utility to simplify the expression. Write your answer in standard form.

87. $(1 + 2i)^5$ 88. $(3 - i)^6$

89. $\dfrac{1}{(2 - i)^3}$ 90. $\dfrac{1}{(4 + 3i)^2}$

SECTION	CONCEPT	PAGES	REVIEW EXERCISES	KEY IDEAS/POINTS
A.1	**Real numbers**	994–1004	1–10	
	The set of real numbers	994–996	1–10	*Rational:* $\dfrac{a}{b}$, where a and b are integers or a decimal that terminates or repeats *Irrational:* nonrepeating/nonterminating decimal
	Approximations: Rounding and truncation	996–997	1–2	*Rounding:* Examine the digit to the right of the last desired digit. Digit $<$ 5: Keep last desired digit as is. Digit $\geq$ 5: Round the last desired digit up 1. *Truncating:* Eliminate all digits to the right of the desired digit.
	Order of operations	997–999	3–6	1. Parentheses 2. Multiplication and division 3. Addition and subtraction
	Properties of real numbers	999–1004	3–10	▪ $a(b + c) = ab + ac$ ▪ If $xy = 0$, then $x = 0$ or $y = 0$. ▪ $\dfrac{a}{b} \pm \dfrac{c}{d} = \dfrac{ad \pm bc}{bd}$ $b \neq 0$ and $d \neq 0$ ▪ $\dfrac{a}{b} \div \dfrac{c}{d} = \dfrac{a}{b} \cdot \dfrac{d}{c}$ $b \neq 0, c \neq 0$, and $d \neq 0$
A.2	**Integer exponents and scientific notation**	1008–1014	11–16	$a^n = \underbrace{a \cdot a \cdot a \cdot \cdots \cdot a}_{n \text{ factors}}$
	Integer exponents	1008–1012	11–14	$a^m \cdot a^n = a^{m+n}$ $\dfrac{a^m}{a^n} = a^{m-n}$ $a^0 = 1$ $(a^m)^n = a^{mn}$ $a^{-n} = \dfrac{1}{a^n} = \left(\dfrac{1}{a}\right)^n$ $a \neq 0$
	Scientific notation	1012–1014	15–16	$c \times 10^n$, where c is a positive real number $1 \leq c < 10$ and n is an integer.
A.3	**Polynomials: Basic operations**	1017–1024	17–28	
	Adding and subtracting polynomials	1019	17–20	Combine like terms.
	Multiplying polynomials	1020	21–28	Distributive property
	Special products	1021–1024	25–28	$(x + a)(x + b) = x^2 + (ax + bx) + ab$ Perfect Squares $(a + b)^2 = (a + b)(a + b) = a^2 + 2ab + b^2$ $(a - b)^2 = (a - b)(a - b) = a^2 - 2ab + b^2$ Difference of Two Squares $(a + b)(a - b) = a^2 - b^2$ Perfect Cubes $(a + b)^3 = a^3 + 3a^2b + 3ab^2 + b^3$ $(a - b)^3 = a^3 - 3a^2b + 3ab^2 - b^3$

SECTION	CONCEPT	PAGES	REVIEW EXERCISES	KEY IDEAS/POINTS
A.6	**Rational exponents and radicals**	1054–1062	53–56	
	Square roots	1054–1055	53–60	$\sqrt{25} = 5$
	Other (nth) roots	1056–1060	63–66	$b = \sqrt[n]{a}$ means $a = b^n$ for a and b positive real numbers and n a positive even integer, or for a and b any real numbers and n a positive odd integer. $$\sqrt[n]{ab} = \sqrt[n]{a} \cdot \sqrt[n]{b} \qquad \sqrt[n]{\frac{a}{b}} = \frac{\sqrt[n]{a}}{\sqrt[n]{b}} \qquad b \neq 0$$ $$\sqrt[n]{a^m} = \left(\sqrt[n]{a}\right)^m$$ $$\sqrt[n]{a^n} = a \qquad n \text{ is odd}$$ $$\sqrt[n]{a^n} = |a| \qquad n \text{ is even}$$
	Rational exponents	1060–1062	63–66	$a^{1/n} = \sqrt[n]{a}$ $a^{m/n} = \left(a^{1/n}\right)^m = \left(\sqrt[n]{a}\right)^m$ $a^{-m/n} = \dfrac{1}{a^{m/n}}$, for m and n positive integers with no common factors $a \neq 0$.
A.7	**Complex numbers**	1065–1069	67–80	$a + bi$, where a and b are real numbers.
	Adding and subtracting complex numbers	1066–1067	71–74	Combine real parts with real parts and imaginary parts with imaginary parts.
	Multiplying complex numbers	1067	71–74	Use the FOIL method and $i^2 = -1$ to simplify.
	Dividing complex numbers	1068	75–80	If $a + bi$ is in the denominator, then multiply the numerator and the denominator by $a - bi$. The result is a real number in the denominator.
	Raising complex numbers to integer powers	1069	69–70	$i = \sqrt{-1} \qquad i^2 = -1 \qquad i^3 = -i \qquad i^4 = 1$

APPENDIX REVIEW

APPENDIX REVIEW EXERCISES

A.1 Real Numbers

Approximate to two decimal places by (a) rounding and (b) truncating.

1. 5.21597

2. 7.3623

Simplify.

3. $7 - 2\cdot5 + 4\cdot3 - 5$

4. $-2(5+3) + 7(3 - 2\cdot5)$

5. $-\dfrac{16}{(-2)(-4)}$

6. $-3(x-y) + 4(3x - 2y)$

Perform the indicated operation and simplify.

7. $\dfrac{x}{4} - \dfrac{x}{3}$

8. $\dfrac{y}{3} + \dfrac{y}{5} - \dfrac{y}{6}$

9. $\dfrac{12}{7}\cdot\dfrac{21}{4}$

10. $\dfrac{a^2}{b^3} \div \dfrac{2a}{b^2}$

A.2 Integer Exponents and Scientific Notation

Simplify using properties of exponents.

11. $(-2z)^3$

12. $(-4z^2)^3$

13. $\dfrac{(3x^3y^2)^2}{2(x^2y)^4}$

14. $\dfrac{(2x^2y^3)^2}{(4xy)^3}$

15. Express 0.00000215 in scientific notation.

16. Express 7.2×10^9 as a real number.

A.3 Polynomials: Basic Operations

Perform the indicated operation and write the results in standard form.

17. $(14z^2 + 2) + (3z - 4)$

18. $(27y^2 - 6y + 2) - (y^2 + 3y - 7)$

19. $(36x^2 - 4x - 5) - (6x - 9x^2 + 10)$

20. $[2x - (4x^2 - 7x)] - [3x - (2x^2 + 5x - 4)]$

21. $5xy^2(3x - 4y)$

22. $-2st^2(-t + s - 2st)$

23. $(x-7)(x+9)$

24. $(2x+1)(3x-2)$

25. $(2x-3)^2$

26. $(5x-7)(5x+7)$

27. $(x^2+1)^2$

28. $(1-x^2)^2$

A.4 Factoring Polynomials

Factor out the common factor.

29. $14x^2y^2 - 10xy^3$

30. $30x^4 - 20x^3 + 10x^2$

Factor the trinomial into a product of two binomials.

31. $2x^2 + 9x - 5$

32. $6x^2 - 19x - 7$

33. $16x^2 - 25$

34. $9x^2 - 30x + 25$

Factor the sum or difference of two cubes.

35. $x^3 + 125$

36. $1 - 8x^3$

Factor into a product of three polynomials.

37. $2x^3 + 4x^2 - 30x$

38. $6x^3 - 5x^2 + x$

Factor into a product of two binomials by grouping.

39. $x^3 + x^2 - 2x - 2$

40. $2x^3 - x^2 + 6x - 3$

A.5 Rational Expressions

State the domain restrictions on each of the rational expressions.

41. $\dfrac{4x^2 - 3}{x^2 - 9}$

42. $\dfrac{1}{x^2 + 1}$

Simplify.

43. $\dfrac{x^2 - 4}{x - 2}$

44. $\dfrac{x - 5}{x - 5}$

45. $\dfrac{t^2 + t - 6}{t^2 - t - 2}$

46. $\dfrac{z^3 - z}{z^2 + z}$

Perform the indicated operation and simplify.

47. $\dfrac{x^2 + 3x - 10}{x^2 + 2x - 3}\cdot\dfrac{x^2 + x - 2}{x^2 + x - 6}$

48. $\dfrac{x^2 - x - 2}{x^3 + 3x^2} \div \dfrac{x + 1}{x^2 + 2x}$

49. $\dfrac{1}{x + 1} - \dfrac{1}{x + 3}$

50. $\dfrac{1}{x} - \dfrac{1}{x + 1} + \dfrac{1}{x + 2}$

Simplify.

51. $\dfrac{2 + \dfrac{1}{x - 3}}{\dfrac{1}{5x - 15} + 4}$

52. $\dfrac{\dfrac{1}{x} + \dfrac{2}{x^2}}{3 - \dfrac{1}{x^2}}$

A.6 Rational Exponents and Radicals

Simplify.

53. $\sqrt{20}$

54. $\sqrt{80}$

55. $\sqrt[3]{-125x^5y^4}$

56. $\sqrt[5]{32x^4y^5}$

57. $3\sqrt{20} + 5\sqrt{80}$

58. $4\sqrt{27x} - 8\sqrt{12x}$

59. $(2 + \sqrt{5})(1 - \sqrt{5})$

60. $(3 + \sqrt{x})(4 - \sqrt{x})$

61. $\dfrac{1}{2 - \sqrt{3}}$

62. $\dfrac{1}{3 - \sqrt{x}}$

63. $\dfrac{\left(3x^{2/3}\right)^2}{\left(4x^{1/3}\right)^2}$

64. $\dfrac{\left(4x^{3/4}\right)^2}{\left(2x^{-1/3}\right)^2}$

65. $\dfrac{5^{1/2}}{5^{1/3}}$

66. $\left(x^{-2/3}y^{1/4}\right)^{12}$

A.7 Complex Numbers

Simplify.

67. $\sqrt{-169}$

68. $\sqrt{-32}$

69. i^{19}

70. i^9

Express the product in standard form.

71. $(2 + 2i)(3 - 3i)$

72. $(1 + 6i)(1 + 5i)$

73. $(4 + 7i)^2$

74. $(7 - i)^2$

Express the quotient in standard form.

75. $\dfrac{1}{2 - i}$

76. $\dfrac{1}{3 + i}$

77. $\dfrac{7 + 2i}{4 + 5i}$

78. $\dfrac{6 - 5i}{3 - 2i}$

79. $\dfrac{10}{3i}$

80. $\dfrac{7}{2i}$

Technology

Section A.1

81. Use your calculator to evaluate $\sqrt{272.25}$. Does the answer appear to be a rational or an irrational number? Why?

82. Use your calculator to evaluate $\sqrt{\dfrac{1053}{81}}$. Does the answer appear to be a rational or an irrational number? Why?

Section A.2

Use a graphing utility to evaluate the expression. Express your answer in scientific notation.

83. $\dfrac{\left(8.2 \times 10^{11}\right)\left(1.167 \times 10^{-35}\right)}{\left(4.92 \times 10^{-18}\right)}$

84. $\dfrac{\left(1.4805 \times 10^{21}\right)}{\left(5.64 \times 10^{26}\right)\left(1.68 \times 10^{-9}\right)}$

Section A.3

85. Use a graphing utility to plot the graphs of the three expressions $(2x + 3)^3$, $8x^3 + 27$, and $8x^3 + 36x^2 + 54x + 27$. Which two graphs agree with each other?

86. Use a graphing utility to plot the graphs of the three expressions $(x - 3)^2$, $8x^2 + 9$, and $x^2 - 6x + 9$. Which two graphs agree with each other?

Section A.4

87. Use a graphing utility to plot the graphs of the three expressions $x^2 - 3x + 18$, $(x + 6)(x - 3)$, and $(x - 6)(x + 3)$. Which two graphs agree with each other?

88. Use a graphing utility to plot the graphs of the three expressions $x^2 - 8x + 16$, $(x + 4)^2$, and $(x - 4)^2$. Which two graphs agree with each other?

Section A.5

For each given expression: (a) simplify the expression, (b) use a graphing utility to plot the expression and the answer in (a) in the same viewing window, and (c) determine the domain restriction(s) where the graphs will agree with each other.

89. $\dfrac{1 - 4/x}{1 - 4/x^2}$

90. $\dfrac{1 - 3/x}{1 + 9/x^2}$

Section A.6

91. Given $\dfrac{6}{\sqrt{5} - \sqrt{2}}$,

    a. Rationalize the denominator.

    b. Use a graphing utility to evaluate the expression and the answer.

    c. Do the expression and answer agree?

92. Given $\dfrac{11}{2\sqrt{6} + \sqrt{13}}$,

    a. Rationalize the denominator.

    b. Use a graphing utility to evaluate the expression and the answer.

    c. Do the expression and answer agree?

Section A.7

In Exercises 93 and 94, apply a graphing utility to simplify the expression. Write your answer in standard form.

93. $(3 + 5i)^5$

94. $\dfrac{1}{(1 + 3i)^4}$

95. Use a graphing utility to evaluate the expression. Express your answer in scientific notation.

$$\dfrac{\left(5.6 \times 10^{-12}\right)\left(1.2 \times 10^{21}\right)}{\left(4.2 \times 10^{-5}\right)}$$

96. Apply a graphing utility to simplify the expression and write your answer in standard form.

$$\dfrac{1}{\left(6 + 2i\right)^4}$$

APPENDIX PRACTICE TEST

Simplify.

1. $\sqrt{16}$

2. $\sqrt[3]{54x^6}$

3. $-3(2 + 5^2) + 2(3 - 7) - (3^2 - 1)$

4. $\sqrt[5]{-32}$

5. $\sqrt{-12x^2}$

6. i^{17}

7. $\dfrac{(x^2y^{-3}z^{-1})^{-2}}{(x^{-1}y^2z^3)^{1/2}}$

8. $3\sqrt{x} - 4\sqrt{x} + 5\sqrt{x}$

9. $3\sqrt{18} - 4\sqrt{32}$

10. $(5\sqrt{6} - 2\sqrt{2})(\sqrt{6} + 3\sqrt{2})$

Perform the indicated operation and simplify.

11. $(3y^2 - 5y + 7) - (y^2 + 7y - 13)$

12. $(2x - 3)(5x + 7)$

Factor.

13. $x^2 - 16$

14. $3x^2 + 15x + 18$

15. $4x^2 + 12xy + 9y^2$

16. $x^4 - 2x^2 + 1$

17. $2x^2 - x - 1$

18. $6y^2 - y - 1$

19. $2t^3 - t^2 - 3t$

20. $2x^3 - 5x^2 - 3x$

21. $x^2 - 3yx + 4yx - 12y^2$

22. $x^4 + 5x^2 - 3x^2 - 15$

23. $81 + 3x^3$

24. $27x - x^4$

Perform the indicated operations and simplify.

25. $\dfrac{2}{x} + \dfrac{3}{x - 1}$

26. $\dfrac{5x}{x^2 - 7x + 10} - \dfrac{4}{x^2 - 25}$

27. $\dfrac{x - 1}{x^2 - 1} \cdot \dfrac{x^2 + x + 1}{x^3 - 1}$

28. $\dfrac{4x^2 - 9}{x^2 - 11x - 60} \cdot \dfrac{x^2 - 16}{2x + 3}$

29. $\dfrac{x - 3}{2x - 5} \div \dfrac{x^2 - 9}{5 - 2x}$

30. $\dfrac{1 - t}{3t + 1} \div \dfrac{t^2 - 2t + 1}{7t + 21t^2}$

In Exercises 31 and 32, write the resulting expression in standard form.

31. $(1 - 3i)(7 - 5i)$

32. $\dfrac{2 - 11i}{4 + i}$

33. Rationalize the denominator: $\dfrac{7 - 2\sqrt{3}}{4 - 5\sqrt{3}}$.

34. Represent 0.0000155 in scientific notation.

35. Simplify $\dfrac{\dfrac{1}{x} - \dfrac{2}{x + 1}}{x - 1}$ and state any domain restrictions.

36. For the given expression:

$$\dfrac{1 + \dfrac{5}{x}}{1 - \dfrac{25}{x^2}}$$

   a. Simplify the expression.
   b. Use a graphing utility to plot the expression and the answer in (a) in the same viewing window.
   c. Determine the domain restriction(s) where the graphs will agree with each other.

37. Apply a graphing utility to evaluate the expression. Round your answer to three decimal places.

$$\dfrac{\sqrt{5}}{\sqrt{13} - \sqrt{7}}$$

ANSWERS TO ODD NUMBERED EXERCISES*

CHAPTER 0

Section 0.1

1. $m = 2$ 　　**3.** $t = \frac{7}{5}$ 　　**5.** $x = -10$

7. $n = 2$ 　　**9.** $x = 12$ 　　**11.** $t = -\frac{15}{2}$

13. $x = -1$ 　　**15.** $p = -\frac{9}{2}$ 　　**17.** $x = \frac{1}{4}$

19. $x = -\frac{3}{2}$ 　　**21.** $a = -8$ 　　**23.** $x = -15$

25. $c = -\frac{35}{13}$ 　　**27.** $m = \frac{60}{11}$ 　　**29.** $x = 36$

31. $p = 8$ 　　**33.** $y = -2$ 　　**35.** $p = 2$

37. no solution 　　**39.** 12 miles 　　**41.** 270 units

43. $r_1 = 3$ feet, $r_2 = 6$ feet

45. The body is 63 inches or 5.25 feet.

47. $20,000 at 4% and $100,000 at 7%

49. $3,000 at 10%, $5,500 at 2%, and $5,500 at 40%

51. 70 ml of 5%, 30 ml of 15%

53. $\approx$ 9 minutes 　　**55.** $\approx$ 2.3 mph

57. Jogger: 6 mph 　　**59.** Bicyclist: 6 minutes
Walker: 4 mph 　　　　　Walker: 18 minutes

61. 22.5 hours 　　**63.** 2.4 hours

65. 2 field goals and 6 touchdowns

67. 3.5 feet from center

69. Fulcrum is 0.4 feet from Maria and 0.6 feet from Max.

71. The error is forgetting to do to one side of the equation what is done to the other side. The correct answer is $x = 5$.

73. $x = \dfrac{c - b}{a}$ 　　**75.** $w = \dfrac{P - 2l}{2}$

77. $h = \dfrac{2A}{b}$ 　　**79.** $w = \dfrac{A}{l}$ 　　**81.** $h = \dfrac{V}{lw}$

83. Janine: 58 mph; Tricia: 70 mph

85. $x = 2$

87. all real numbers

89. $191,983.35

91. Plan B is better for 5 or fewer plays/month. Plan A is better for 6 or more plays/month.

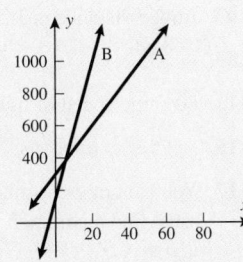

Section 0.2

1. $x = 2, 3$ 　　**3.** $p = 3, 5$ 　　**5.** $x = -4, 3$

7. $x = -\frac{1}{4}$ 　　**9.** $y = \frac{1}{3}$ 　　**11.** $y = 0, 2$

13. $p = \frac{2}{3}$ 　　**15.** $x = -3, 3$ 　　**17.** $x = -6, 2$

19. $p = -5, 5$ 　　**21.** $x = -2, 2$ 　　**23.** $p = -2\sqrt{2}, 2\sqrt{2}$

25. $x = -3i, 3i$ 　　　　　　　**27.** $x = -3, 9$

29. $x = \dfrac{-3 \pm 2i}{2}$ 　　**31.** $x = \dfrac{2 \pm 3\sqrt{3}}{5}$

33. $x = -2, 4$ 　　**35.** $x = -3, 1$

37. $t = 1, 5$ 　　**39.** $y = 1, 3$

41. $p = \dfrac{-4 \pm \sqrt{10}}{2}$ 　　**43.** $x = \frac{1}{2}, 3$

45. $x = \dfrac{4 \pm 3\sqrt{2}}{2}$ 　　**47.** $t = \dfrac{-3 \pm \sqrt{13}}{2}$

49. $s = \dfrac{-1 \pm \sqrt{3}i}{2}$ 　　**51.** $x = \dfrac{3 \pm \sqrt{57}}{6}$

53. $x = 1 \pm 4i$ 　　**55.** $x = \dfrac{-7 \pm \sqrt{109}}{10}$

57. $x = \dfrac{-4 \pm \sqrt{34}}{3}$ 　　**59.** $v = -2, 10$

61. $t = -6, 1$ 　　**63.** $x = -7, 1$

65. $p = 4 \pm 2\sqrt{3}$ 　　**67.** $w = \dfrac{-1 \pm \sqrt{167}i}{8}$

69. $p = \dfrac{9 \pm \sqrt{69}}{6}$ 　　**71.** $t = \dfrac{10 \pm \sqrt{130}}{10}$

73. $x = 0.4$ and $x = -0.3$

75. $t = 8$ (August) and $t = 12$ (December) 2003

77. $\approx$ 20 inches 　　**79.** 17 and 18 　　**81.** 9 ft $\times$ 15 ft

83. base is 6 units and height is 20 units

85. 2.5 seconds after it is dropped

87. $\approx$ 21.2 feet 　　**89.** 5 ft $\times$ 5 ft

91. border is 2.3 feet wide 　　**93.** 10 days

95. The problem is factored incorrectly. The correction would be $t = 6$ and $t = -1$.

97. When taking the square root of both sides, the i is missing from the right side. The correction would be $a = \pm\frac{3}{4}i$.

99. false 　　**101.** true

103. $x^2 - 2ax + a^2 = 0$ 　　**105.** $x^2 - 7x + 10 = 0$

107. $t = \pm\sqrt{\dfrac{2s}{g}}$ 　　**109.** $c = \pm\sqrt{a^2 + b^2}$

---

*Answers that require a proof, graph, or otherwise lengthy solution are not included.

111. $x = 0, \pm 2$

113. $x = \pm 2, -1$

115. $\dfrac{-b + \sqrt{b^2 - 4ac}}{2a} + \dfrac{-b - \sqrt{b^2 - 4ac}}{2a} = \dfrac{-2b}{2a} = \dfrac{-b}{a}$

117. $x^2 - 6x + 4 = 0$

119. 250 mph

121. $ax^2 - bx + c = 0$

123. Small jet: 300 mph; 757: 400 mph

125. $x = -1, 2$

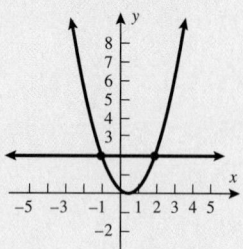

127. a. $x = -2, 4$

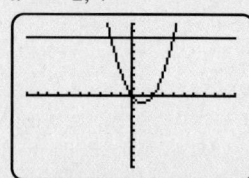

b. $b = -3, x = 1 \pm \sqrt{2}i$

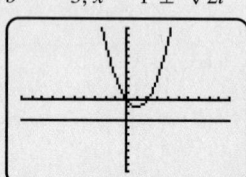

$b = -1, x = 1$

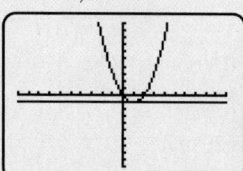

$b = 0, x = 0, 2$

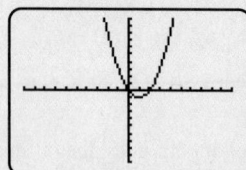

$b = 5, x = 1 \pm \sqrt{6}$

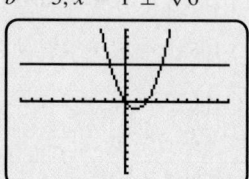

Section 0.3

1. no solution; $x \neq 2$

3. no solution; $p \neq 1$

5. $x = -10, x \neq -2$

7. no solution; $n \neq -1, 0$

9. no solution; $a \neq -3, 0$

11. $n = \frac{53}{11}; n \neq 1$

13. $x = -3; x \neq \frac{1}{2}, -\frac{1}{5}$

15. no solution; $t \neq 1$

17. $x = 3, 4$

19. $x = -\frac{3}{4}, 2$

21. no solution

23. $x = 5$

25. $y = -\frac{1}{2}$

27. $x = 4$

29. $y = 0, y = 25$

31. $s = 3, s = 6$

33. $x = -3, x = -1$

35. $x = 0$

37. $x = \frac{5}{2}$

39. no solution

41. $x = 1, x = 5$

43. $x = 7$

45. $x = 4$

47. $x = -8, x = 0$

49. $x = \pm\sqrt{2}, x = \pm 1$

51. $x = \pm\dfrac{\sqrt{6}}{2}i, x = \pm i\sqrt{2}$

53. $t = \frac{5}{4}, t = 3$

55. $x = \pm 1, \pm i, \pm\frac{1}{2}, \pm\frac{1}{2}i$

57. $y = -\frac{3}{4}, y = 1$

59. $z = 1$

61. $t = -27, t = 8$

63. $x = -\frac{4}{3}, x = 0$

65. $u = \pm 1, u = \pm 8$

67. $x = -3, 0, 4$

69. $p = 0, \pm\frac{3}{2}$

71. $u = 0, \pm 2, \pm 2i$

73. $x = \pm 3, 5$

75. $y = -2, 5, 7$

77. $x = 0, 3$

79. $t = \pm 5$

81. $y = 2, 3$

83. $p = 10$ or $p = 4$

85. $y = 3$ or $y = 5$

87. $t = 4$ or $t = 2$

89. $x = -1$ or $x = 8$

91. $y = 0$ or $y = \frac{2}{3}$

93. $x = \frac{47}{14}$ or $x = -\frac{23}{14}$

95. $x = -3$ or $x = 13$

97. $p = -13$ or $p = 7$

99. $y = -5$ or $y = 9$

101. $x = \pm\sqrt{3}$ and $x = \pm\sqrt{5}$

103. $x = \pm 2, \pm i\sqrt{6}$

105. $d_0 = \frac{15}{2}$ or 7.5 cm (in front of the lens)

107. Image distance is 3 cm, object distance is 6 cm.

109. $\approx$ 132 ft

111. 25 cm

113. 80% the speed of light

115. $t = 5$ is extraneous. No solution.

117. You cannot cross multiply with the -3 in the problem. You must find a common denominator first. The correct solution is $p = \frac{6}{5}$.

119. false

121. $x = \dfrac{a-b}{c} \quad x \neq 0$ **123.** $x = -2$

125. $x = \dfrac{by}{a-y-cy}, x \neq 0, -\dfrac{b}{c-1}$

127. $x = \frac{313}{64} \approx 4.891$ **129.** $x = 81$

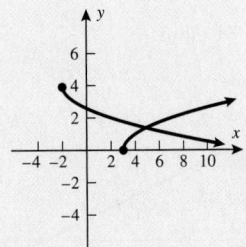

Section 0.4

1. [−2, 3)

3. (−3, 5]

5. [4, 6]

7. [−8, −6]

9. empty set

11. [1, 4)

13. [−1, 2)

15. $(-\infty, 4) \cup (4, \infty)$

17. $(-\infty, -3] \cup [3, \infty)$

19. (−3, 2]

21. $t > -3, (-3, \infty)$ **23.** $x < 6, (-\infty, 6)$

25. $y < 1, (-\infty, 1)$ **27.** $-8 \leq x < 4, [-8, 4)$

29. $-6 < y < 6, (-6, 6)$ **31.** $\frac{1}{2} \leq y \leq \frac{5}{4}, \left[\frac{1}{2}, \frac{5}{4}\right]$

33. $\left[-1, \frac{3}{2}\right]$ **35.** $\left(\frac{1}{3}, \frac{1}{2}\right)$

37. $\left(-\infty, -\frac{1}{2}\right] \cup [3, \infty)$

39. $\left(-\infty, -1 - \sqrt{5}\right] \cup \left[-1 + \sqrt{5}, \infty\right)$

41. $(2 - \sqrt{10}, 2 + \sqrt{10})$ **43.** $(-\infty, 0] \cup [3, \infty)$

45. $(-\infty, -3) \cup (3, \infty)$ **47.** $(-\infty, -2] \cup [0, 1]$

49. $(0, 1) \cup (1, \infty)$ **51.** $(-\infty, -2) \cup [-1, 2)$

53. $(-\infty, -5] \cup (-2, 0]$ **55.** $(-2, 2)$

57. $\mathbb{R}$ **59.** $[-3, 3) \cup (3, \infty)$

61. $(-3, -1] \cup (3, \infty)$ **63.** $(-\infty, -4) \cup (2, 5]$

65. $(-\infty, -2) \cup (2, \infty)$ **67.** $(-\infty, 2) \cup (6, \infty)$

69. [3, 5]   **71.** $\mathbb{R}$

73. $(-\infty, 2] \cup [5, \infty)$ **75.** $\mathbb{R}$

77. $\left(-\infty, -\frac{3}{2}\right] \cup \left[\frac{3}{2}, \infty\right)$ **79.** $(-\infty, -3) \cup (3, \infty)$

81. [−3, 3]   **83.** $0.9r_T \leq r_R \leq 1.1r_T$

85. $\$4,386.25 \leq T \leq \$15,698.75$

87. $30 < x < 100$ (30–100 orders will yield a profit)

89. For years 3–5, the car is worth more than you owe. In the first 3 years, you owe more than the car is worth.

91. 75 seconds   **93.** $d = 4$ (tie), $d < 4$ (win)

95. You must reverse the inequality.

97. You cannot divide by x.

99. true   **101.** false

103. $\mathbb{R}$ **105.** no solution

107. a. $-2 < x < 5$ **b.**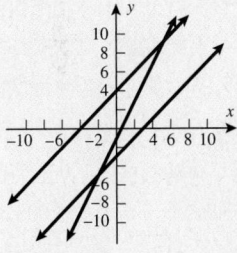
     c. agree

109. $\left(-\frac{1}{2}, \infty\right)$

Section 0.5

1. $d = 4, (3, 3)$ **3.** $d = 4\sqrt{2}, (1, 2)$

5. $d = 3\sqrt{10}, \left(-\frac{17}{2}, \frac{7}{2}\right)$ **7.** $d = 5, \left(-5, \frac{1}{2}\right)$

9. $d = 4\sqrt{2}, (-4, -6)$ **11.** $d = 5, \left(\frac{3}{2}, \frac{11}{6}\right)$

13. 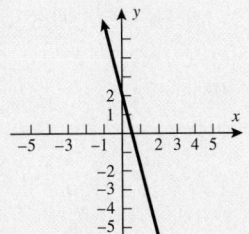   **15.**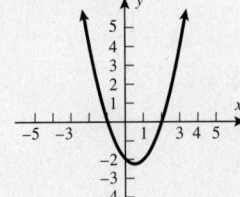

17.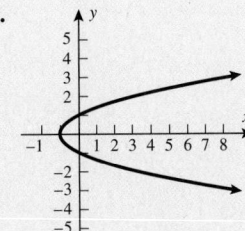

19. $(3, 0)$ and $(0, -6)$

21. $(4, 0)$ (*Note:* No y-intercept.)

23. $(\pm 2, 0)$ and $(0, \pm 4)$

25. x-axis

27. x-axis

29. y-axis

31.

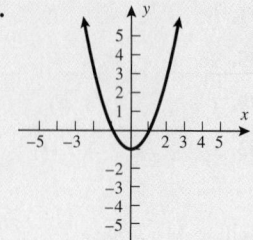

33.

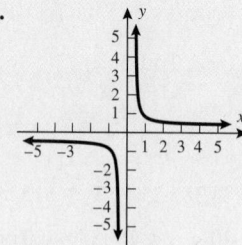

35.

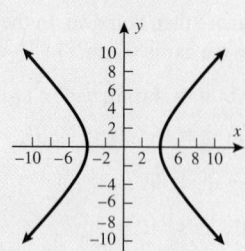

37. $(x - 5)^2 + (y - 7)^2 = 81$

39. $(x + 11)^2 + (y - 12)^2 = 169$

41. $(x - 5)^2 + (y + 3)^2 = 12$

43. $\left(x - \frac{2}{3}\right)^2 + \left(y + \frac{3}{5}\right)^2 = \frac{1}{16}$

45. $C = (2, -5)$ $r = 7$

47. $C = (4, 9)$ $r = 2\sqrt{5}$ **49.** $C = \left(\frac{2}{5}, \frac{1}{7}\right)$ $r = \frac{2}{3}$

51. $C = (5, 7)$ $r = 9$ **53.** $C = (1, 3)$ $r = 3$

55. $C = (5, -3)$ $r = 2\sqrt{3}$ **57.** $C = (3, 2)$ $r = 2\sqrt{3}$

59. $C = \left(\frac{1}{2}, -\frac{1}{2}\right)$ $r = \frac{1}{2}$ **61.** distance ≈ 268 miles

63. midpoint = (2003, 330) $330m in revenue

65. $x^2 + y^2 = 2{,}250{,}000$ **67.** $x^2 + y^2 = 40{,}000$

69.

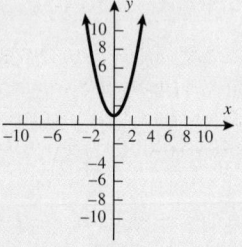

In general, more than 2 points are needed.

71. The standard form of an equation of a circle is $(x - h)^2 + (y - k)^2 = 25$, which would lead to a center of $(4, -3)$, not $(4, 3)$.

73. false   **75.** true

77. the point $(-5, 3)$ **79.** origin

81. $(x - 3)^2 + (y + 2)^2 = 20$

83. $4c = a^2 + b^2$ **85.** y-axis

87. a. $(x - 5.5)^2 + (y + 1.5)^2 = 39.69$, $(5.5, -1.5)$, $r = 6.3$

b. $y = -1.5 \pm \sqrt{39.69 - (x - 5.5)^2}$

c.

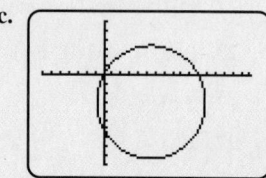

d. The graphs are the same.

Section 0.6

1. slope $= m = 3$ **3.** slope $= m = -2$

5. slope $= m = -\frac{19}{10}$ **7.** slope $= m \approx 2.379$

9. $m = -3$

11. $(0.5, 0)$ and $(0, -1)$, slope $= m = 2$ rising

13. $(1, 0)$ and $(0, 1)$, slope $= m = -1$ falling

15. $(0, 1)$, slope $= m = 0$ horizontal

17.

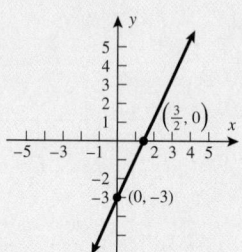

19.

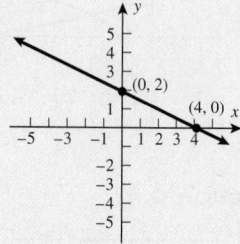

21.

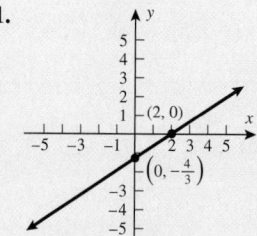

23.

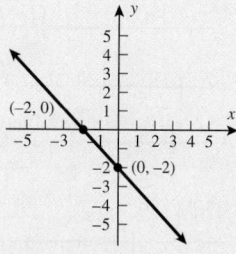

25.

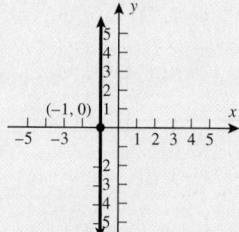

27.

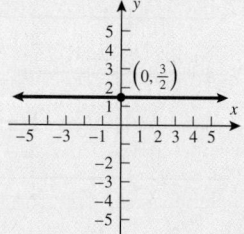

29.

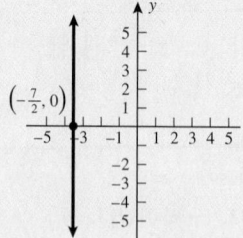

31. $y = \frac{2}{5}x - 2$, $m = \frac{2}{5}$, y-intercept: $(0, -2)$

33. $y = -\frac{1}{3}x + 2$, $m = -\frac{1}{3}$, y-intercept: $(0, 2)$

35. $y = 4x - 3$, $m = 4$, y-intercept: $(0, -3)$

37. $y = -2x + 4$, $m = -2$, y-intercept: $(0, 4)$

39. $y = \frac{2}{3}x - 2$, $m = \frac{2}{3}$, y-intercept: $(0, -2)$

41. $y = -\frac{3}{4}x + 6$, $m = -\frac{3}{4}$, y-intercept: $(0, 6)$

43. $y = 2x + 3$ **45.** $y = -\frac{1}{3}x$ **47.** $y = 2$

49. $x = \frac{3}{2}$ **51.** $y = 5x + 2$ **53.** $y = -3x - 4$

55. $y = \frac{3}{4}x - \frac{7}{4}$ **57.** $y = 4$ **59.** $x = -1$

61. $y = \frac{3}{5}x + \frac{1}{5}$ **63.** $y = -5x - 16$ **65.** $y = \frac{1}{6}x - \frac{121}{3}$

67. $y = -3x + 1$ **69.** $y = \frac{3}{2}x$ **71.** $x = 3$

73. $y = 7$ **75.** $y = \frac{6}{5}x + 6$ **77.** $x = -6$

79. $x = \frac{2}{5}$ **81.** $y = x - 1$ **83.** $y = -2x + 3$

85. $y = -\frac{1}{2}x + 1$ **87.** $y = 2x + 7$ **89.** $y = \frac{3}{2}x$

91. $y = 5$ **93.** $y = 2$ **95.** $y = \frac{3}{2}x - 4$

97. $C(h) = 1200 + 25h$; A 32-hour job will cost $2,000.

99. $375   **101.** $F = \frac{9}{5}C + 32$, $-40\,°C = -40\,°F$

103. The rate of change in inches per year is $\frac{1}{50}$.

105. 0.06 ounces per year. In 2040 we expect a baby will weigh 6 lb 12.4 oz.

107. y-intercept $(0, 35)$ is the flat monthly charge of $35.

109. -0.35 inches/year; 2.75 inches

111. 2.4 billion plastic bags/year; 404 billion

113. The correction that needs to be made is that for the x-intercept, $y = 0$, and for the y-intercept, $x = 0$.

115. Values for the numerator and denominator reversed.

117. true   **119.** false

121. The line perpendicular is vertical and has undefined slope.

123. $y = -\dfrac{A}{B}x + 1$ **125.** $y = \dfrac{B}{A}x + 2B - 1$

129. perpendicular   **131.** perpendicular

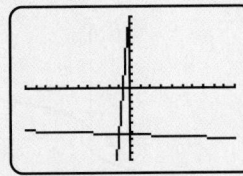 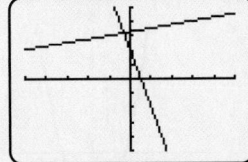

133. neither

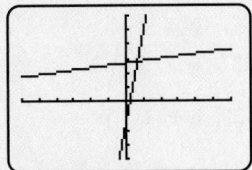

Section 0.7

1. $y = kx$ **3.** $V = kx^3$ **5.** $z = km$

7. $f = \dfrac{k}{\lambda}$ **9.** $F = k\dfrac{w}{L}$ **11.** $v = kgt$

13. $R = \dfrac{k}{PT}$ **15.** $y = k\sqrt{x}$ **17.** $d = rt$

19. $V = lwh$ **21.** $A = \pi r^2$ **23.** $V = \dfrac{\pi}{16}r^2 h$

25. $V = \dfrac{400{,}000}{P}$ **27.** $F = \dfrac{2\pi}{\lambda L}$ **29.** $t = \dfrac{19.2}{s}$

31. $R = \dfrac{4.9}{I^2}$ **33.** $R = \dfrac{0.01L}{A}$ **35.** $F = \dfrac{0.025 m_1 m_2}{d^2}$

37. $W = 7.5H$ **39.** 1292 mph   **41.** $F = 1.618H$

43. 24 cm   **45.** 20,000   **47.** 600 w/m^2

49. Bank of America 1.5%; Navy Federal Credit Union: 3%

51. $\frac{11}{12}$ or 0.92 atm   **53.** y varies *inversely* with x

55. true   **57.** b   **59.** $\sigma_{pl}^2 = 1.23 C_n^2 k^{7/6} L^{11/6}$

61. a. The least squares regression line is $y = 2.93x + 201.72$.

     b. The variation constant is 120.07, and the equation of the direct variation is $y = 120.074x^{0.26}$.

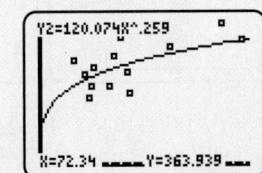

    c. The oil price per barrel was $72.70 in September 2006, so the predicted stock index from the least squares regression line is 415 and from the equation of direct variation is 364. The least squares regression line gives a closer approximation to the actual value, 417.

63. a. The least squares regression line is $y = -141.73x + 2419.35$.

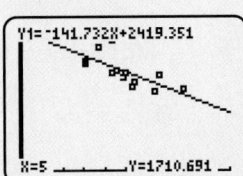

b. The variation constant is 3217.69 and the equation of the inverse variation is $y = \dfrac{3217.69}{x^{0.41}}$.

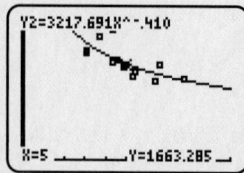

c. The 5-year maturity rate was 5.02% in September 2006, so the predicted number of housing units from the least squares regression line is 1708, and from the equation of inverse variation is 1661. The equation of the least squares regression line gives a closer approximation to the actual value, 1861.

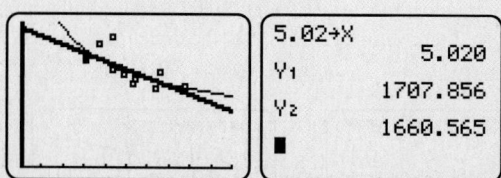

65. a. $y = 0.218x + 0.898$ **b.** $2.425 per gallon. Yes, it is very close to the actual price at $2.425 per gallon. **c.** $3.083

Review Exercises

1. $x = \dfrac{16}{7}$ **3.** $p = -\dfrac{8}{25}$ **5.** $x = 27$

7. $y = -\dfrac{17}{5}$ **9.** $b = \dfrac{6}{7}$ **11.** $x = -\dfrac{6}{17}$

13. $20,000 at 8% and $5,000 at 20%

15. 60 ml of 5% and 90 ml of 10%

17. $b = 7$ and $b = -3$ **19.** $x = 0$ and $x = 8$

21. $q = \pm 13$ **23.** $x = 2 \pm 4i$

25. $x = 6$ and $x = -2$ **27.** $x = \dfrac{1 \pm \sqrt{33}}{2}$

29. $t = \dfrac{7}{3}$ and $t = -1$ **31.** $f = \dfrac{1 \pm \sqrt{337}}{48}$

33. $q = \dfrac{3 \pm \sqrt{69}}{10}$ **35.** $x = \dfrac{5}{2}$ and $x = -1$

37. $x = \dfrac{2}{7}$ and $x = -3$ **39.** base = 4 ft, height = 1 ft

41. $x = \dfrac{6 \pm \sqrt{39}}{3}, x \neq 0$ **43.** $t = -\dfrac{34}{5}, t \neq 0, t \neq -4$

45. $x = -\dfrac{1}{2}, x \neq 0$ **47.** $x = 6$

49. $x = 125$ **51.** no solution

53. $x \approx -0.6$ **55.** $y = \frac{1}{4}, 1$

57. $x = -\dfrac{125}{8}, 1$ **59.** $x = -1, -\dfrac{1}{8}$

61. $x = \pm 2$ and $x = \pm 3i$ **63.** $x = -8, 0, 4$

65. $p = -2, 3, 2$ **67.** $p = -\dfrac{1}{2}, \dfrac{5}{2}, 3$

69. $y = \pm 9$ **71.** no solution

73. $x = 1.7$ and $x \approx 0.9667$

75. $(4, \infty)$

77. $[8, 12]$

79. $\left(-\infty, \dfrac{5}{3}\right)$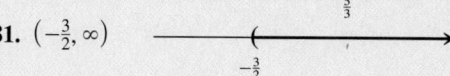

81. $\left(-\dfrac{3}{2}, \infty\right)$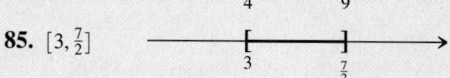

83. $(4, 9]$

85. $\left[3, \dfrac{7}{2}\right]$

87. $[-6, 6]$ **89.** $(-\infty, 0] \cup [4, \infty)$

91. $\left(-\infty, -\dfrac{3}{4}\right) \cup (4, \infty)$ **93.** $(0, 3)$

95. $(-\infty, -6] \cup [9, \infty)$ **97.** $(-\infty, 2) \cup (4, 5]$

99. $(-\infty, -11) \cup (3, \infty)$ **101.** $(-\infty, -3) \cup (3, \infty)$

103. $\mathbb{R}$ **105.** $3\sqrt{5}$

107. $\sqrt{205}$ **109.** $\left(\dfrac{5}{2}, 6\right)$

111. $(3.85, 5.3)$

113. $(\pm 2, 0)$ and $(0, \pm 1)$ **115.** $(\pm 3, 0)$

117. y-axis   **119.** origin

121. 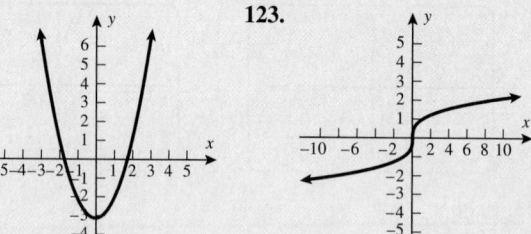   **123.**

125. 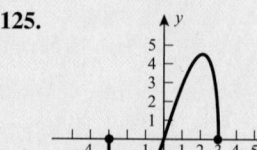   **127.** $C: (-2, -3)$ $r = 9$

129. not a circle

131. $y = -2x - 2$

133. $y = 6$

135. $y = \dfrac{5}{6}x + \dfrac{4}{3}$

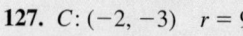

137. $y = -2x - 1$ **139.** $y = \frac{2}{3}x + \frac{1}{3}$

141. $C = 2\pi r$ **143.** $A = \pi r^2$

Practice Test

1. $p = -3$ **3.** $t = -4, t = 7$

5. $x = \frac{8}{3}$ and $x = -\frac{1}{2}$ **7.** $y = -8$

9. $x = 4$ **11.** $y = 1$

13. $x = 0, x = 2, x = 6$ **15.** $(-\infty, 17]$

17. $\left(-\frac{32}{5}, -6\right]$ **19.** $(-\infty, -1] \cup \left[\frac{4}{3}, \infty\right)$

21. $\left(-\frac{1}{2}, 3\right]$ **23.** $\sqrt{82}$

25. $2x^2 + y^2 = 8$ **27.** $(6, 0)$ and $(0, -2)$

29. $y = \frac{8}{3}x - 8$

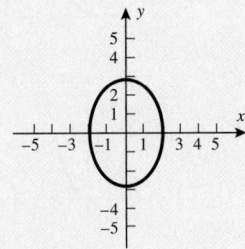

31. $y = x + 5$ **33.** $y = -2x + 3$

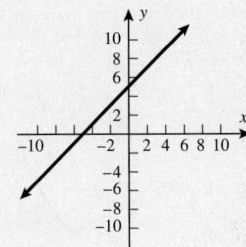   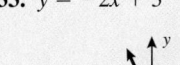

35. $F = \dfrac{30m}{p}$

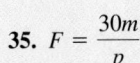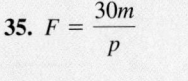

CHAPTER 1

Section 1.1

1. not a function   **3.** not a function

5. function   **7.** not a function

9. not a function   **11.** function

13. not a function   **15.** function

17. not a function

19. a. 5 **b.** 1 **c.** -3 **21. a.** 3 **b.** 2 **c.** 5

23. a. -5 **b.** -5 **c.** -5 **25. a.** 2 **b.** -8 **c.** -5

27. $x = 1$ **29.** $x = -3, 1$ **31.** $-4 \le x \le 4$

33. $x = 6$ **35.** -7 **37.** 6

39. -1 **41.** -33 **43.** $-\frac{7}{6}$

45. $\frac{2}{3}$ **47.** 4   **49.** $8 - x - a$

51. 2   **53.** 1   **55.** 2

57. 1   **59.** $(-\infty, \infty)$ **61.** $(-\infty, \infty)$

63. $(-\infty, 5) \cup (5, \infty)$ **65.** $(-\infty, -2) \cup (-2, 2) \cup (2, \infty)$

67. $(-\infty, \infty)$ **69.** $(-\infty, 7]$

71. $\left[-\frac{5}{2}, \infty\right)$ **73.** $(-\infty, -2] \cup [2, \infty)$

75. $(3, \infty)$ **77.** $(-\infty, \infty)$

79. $(-\infty, -4) \cup (-4, \infty)$ **81.** $\left(-\infty, \frac{3}{2}\right)$

83. $(-\infty, -2) \cup (3, \infty)$ **85.** $(-\infty, -4] \cup [4, \infty)$

87. $\left(-\infty, \frac{3}{2}\right)$ **89.** $(-\infty, \infty)$

91. $x = -2, x = 4$ **93.** $x = -1, x = 5, x = 6$

95. $T(6) = 64.8\,°\text{F}$ $T(12) = 90\,°\text{F}$

97. $h(2) = 27$ ft   domain: $[0, 2.8]$

99. $V(x) = x(10 - 2x)^2$ domain: $(0, 5)$

101. Yes, because every input (year) corresponds to exactly one output (federal funds rate).

103.

YEAR	TOTAL HEALTH-CARE COST FOR FAMILY PLANS	ORDERED PAIRS
1989	$4,000	(1989, 4000)
1993	$6,000	(1993, 6000)
1997	$6,000	(1997, 6000)
2001	$8,000	(2001, 8000)
2005	$11,000	(2005, 11,000)

105. a. 0   **b.** 1000   **c.** 2000

107. You must apply the vertical line test instead of the horizontal line test. Applying the vertical line test would show that the graph given is actually a function.

109. $f(x + 1) \ne f(x) + f(1)$

111. $G(-1 + h) \ne G(-1) + G(h)$

113. false   **115.** true

117. $A = 2$ **119.** $C = -5, D = -2$

121. $(-\infty, -a) \cup (-a, a) \cup (a, \infty)$

123. Warmest at noon: 90 °F. Outside the interval $[6, 18]$, the temperatures are too low.

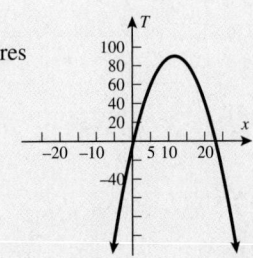

125. Graph of y_2 can be obtained by shifting the graph of y_1 two units to the right.

127. $f'(x) = 3x^2 + 1$ **129.** $f'(x) = \dfrac{8}{(x+3)^2}$

Section 1.2

1. neither   **3.** odd   **5.** even

7. even   **9.** neither   **11.** neither

13. neither   **15.** neither

17. a. $(-\infty, \infty)$ **b.** $[-1, \infty)$

    c. increasing: $(-1, \infty)$, decreasing: $(-3, -2)$,
      constant: $(-\infty, -3) \cup (-2, -1)$

    d. 0   **e.** -1 **f.** 2

19. a. $[-7, 2]$ **b.** $[-5, 4]$

    c. increasing: $(-4, 0)$, decreasing: $(-7, -4) \cup (0, 2)$,
      constant: none

    d. 4   **e.** 1   **f.** -5

21. a. $(-\infty, \infty)$ **b.** $(-\infty, \infty)$

    c. increasing: $(-\infty, -3) \cup (4, \infty)$, decreasing: none,
      constant: $(-3, 4)$

    d. 2   **e.** 2   **f.** 2

23. a. $(-\infty, \infty)$ **b.** $[-4, \infty)$

    c. increasing: $(0, \infty)$, decreasing: $(-\infty, 0)$,
      constant: none

    d. -4 **e.** 0   **f.** 0

25. a. $(-\infty, 0) \cup (0, \infty)$ **b.** $(-\infty, 0) \cup (0, \infty)$

    c. increasing: $(-\infty, 0) \cup (0, \infty)$, decreasing: none,
      constant: none

    d. undefined   **e.** 3   **f.** -3

27. a. $(-\infty, 0) \cup (0, \infty)$

    b. $(-\infty, 5) \cup [7, 7]$ or $(-\infty, 5) \cup \{7\}$

    c. increasing: $(-\infty, 0)$, decreasing: $(5, \infty)$, constant $(0, 5)$

    d. undefined   **e.** 3   **f.** 7

29. $2x + h - 1$ **31.** $2x + h + 3$

33. $2x + h - 3$ **35.** $-6x - 3h + 5$

37. $3x^2 + 3xh + h^2 + 2x + h$

39. $-\dfrac{2}{(x+h-2)(x-2)}$

41. $\dfrac{\sqrt{1-2x-2h} - \sqrt{1-2x}}{h} = -\dfrac{2}{\sqrt{1-2(x+h)} + \sqrt{1-2x}}$

43. $\dfrac{4}{h\sqrt{x+h}} - \dfrac{4}{h\sqrt{x}} = -\dfrac{4}{\sqrt{x}(x+h)(\sqrt{x} + \sqrt{x+h})}$

45. 13   **47.** 1   **49.** -2 **51.** -1

53. domain: $(-\infty, \infty)$
range: $(-\infty, 2]$
increasing: $(-\infty, 2)$
constant: $(2, \infty)$

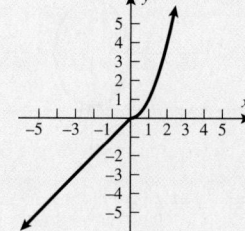

55. domain: $(-\infty, \infty)$
range: $[0, \infty)$
increasing: $(0, \infty)$
decreasing: $(-1, 0)$
constant: $(-\infty, -1)$

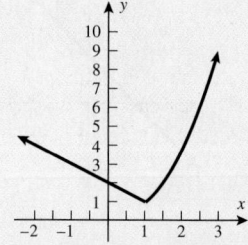

57. domain: $(-\infty, \infty)$
range: $(-\infty, \infty)$
increasing: $(-\infty, \infty)$

59. domain: $(-\infty, \infty)$
range: $[1, \infty)$
increasing: $(1, \infty)$
decreasing: $(-\infty, 1)$

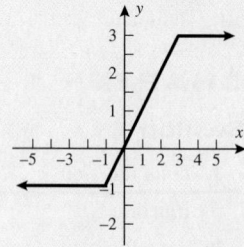

61. domain: $(-\infty, \infty)$
range: $[-1, 3]$
increasing: $(-1, 3)$
constant: $(-\infty, -1) \cup (3, \infty)$

63. domain: $(-\infty, \infty)$
range: $[1, 4]$
increasing: $(1, 2)$
constant: $(-\infty, 1) \cup (2, \infty)$

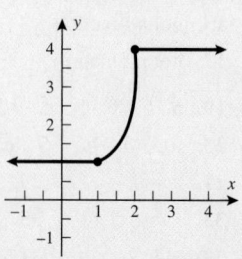

65. domain: $(-\infty, -2) \cup (-2, \infty)$
range: $(-\infty, \infty)$
increasing: $(-2, 1)$
decreasing: $(-\infty, -2) \cup (1, \infty)$

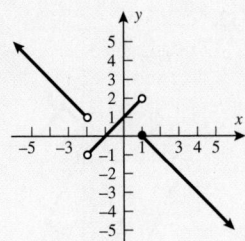

67. domain: $(-\infty, \infty)$
range: $[0, \infty)$
increasing: $(0, \infty)$
constant: $(-\infty, 0)$

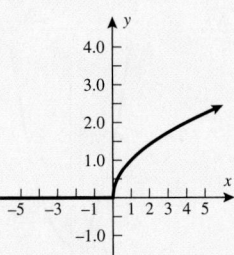

69. domain: $(-\infty, \infty)$
range: $(-\infty, \infty)$
decreasing: $(-\infty, 0) \cup (0, \infty)$

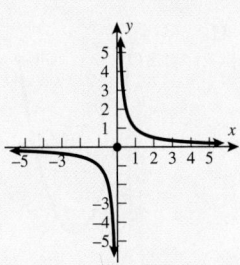

71. domain: $(-\infty, 1) \cup (1, \infty)$
range: $(-\infty, -1) \cup (-1, \infty)$
increasing: $(-1, 1)$
decreasing: $(-\infty, -1) \cup (1, \infty)$

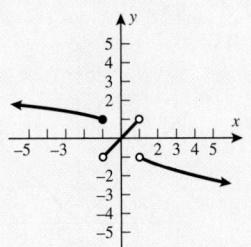

73. domain: $(-\infty, \infty)$
range: $(-\infty, 2) \cup [4, \infty)$
increasing:
$(-\infty, -2) \cup (0, 2) \cup (2, \infty)$
decreasing: $(-2, 0)$

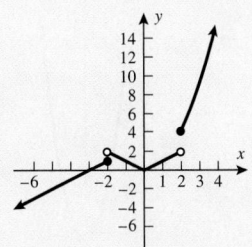

75. domain: $(-\infty, 1) \cup (1, \infty)$
range: $(-\infty, 1) \cup (1, \infty)$
increasing: $(-\infty, 1) \cup (1, \infty)$

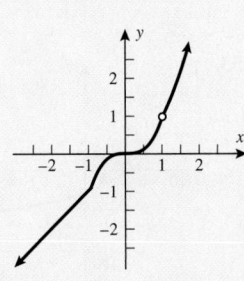

77. $C(x) = \begin{cases} 10x & 0 \le x \le 50 \\ 9x & 50 < x < 100 \\ 8x & x \ge 100 \end{cases}$

79. $C(x) = \begin{cases} 250x & 0 \le x \le 10 \\ 175x + 750 & x > 10 \end{cases}$

81. $C(x) = \begin{cases} 1000 + 35x & 0 \le x \le 100 \\ 2000 + 25x & x > 100 \end{cases}$

83. $R(x) = \begin{cases} 50{,}000 + 3x & 0 \le x \le 100{,}000 \\ 4x - 50{,}000 & x > 100{,}000 \end{cases}$

85. $P(x) = 65x - 800$

87. $C(x) = 0.88 + 0.17[[x]]$, where x represents the weight of the letter in ounces.

89. $f(t) = 3(-1)^{[[t]]}$

91. a. 20 million tons per year

   b. 110 million tons per year

93. 0 feet per second

95. The domain is incorrect. It should be $(-\infty, 0) \cup (0, \infty)$. The range is also incorrect and should be $(0, \infty)$.

97. $C(x) = \begin{cases} 15 & x \le 30 \\ 15 + 1(x - 30) & x > 30 \end{cases} = \begin{cases} 15 & x \le 30 \\ -15 + x & x > 30 \end{cases}$

99. false   **101.** yes, if $a = 2b$

103. yes, if $a = -4, b = -5$ **105.** odd

107. odd   **109.** domain: all real numbers
range: set of integers

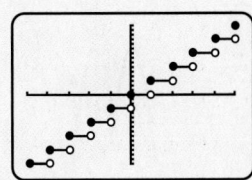

111. $f'(x) = 0$ **113.** $f'(x) = 2ax + b$

Section 1.3

1. $y = |x| + 3$ **3.** $y = |-x| = |x|$ **5.** $y = 3|x|$

7. $y = x^3 - 4$ **9.** $y = (x + 1)^3 + 3$ **11.** $y = (-x)^3 = -x^3$

13.

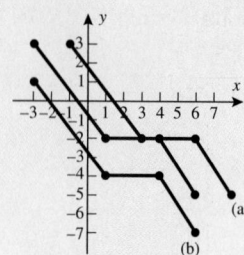

a. $y = f(x - 2)$
b. $y = f(x) - 2$

15.

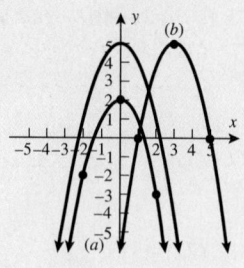

a. $y = f(x) - 3$
b. $y = f(x - 3)$

33.

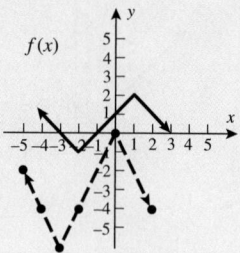

35.

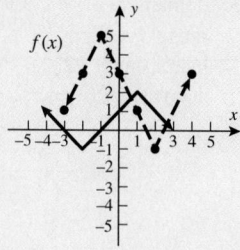

17.

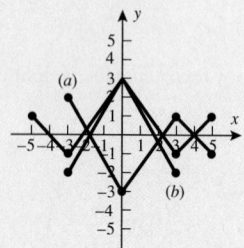

a. $y = -f(x)$
b. $y = f(-x)$

19.

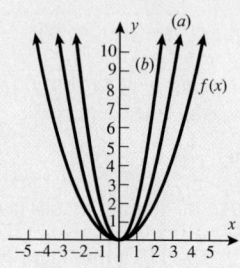

a. $y = 2f(x)$
b. $y = f(2x)$

37.

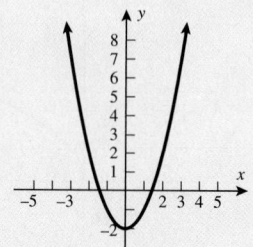

39.

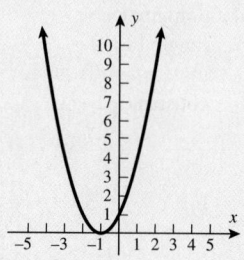

41.

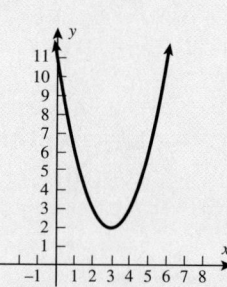

43.

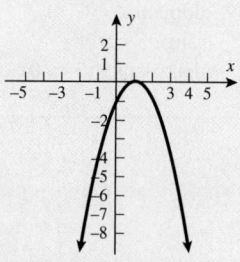

21.

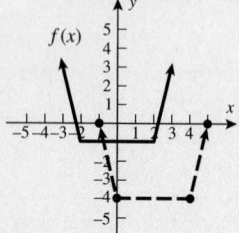

23.

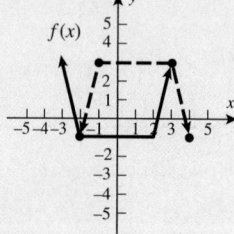

45.

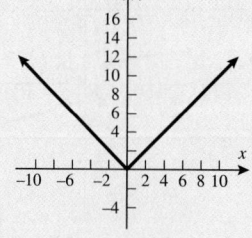

47.

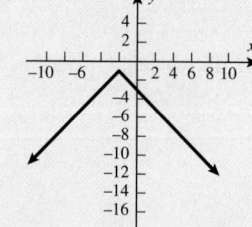

25.

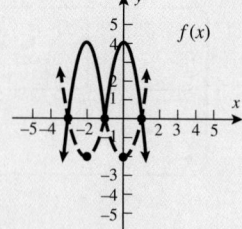

27.

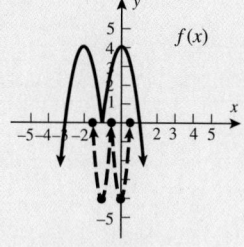

49.

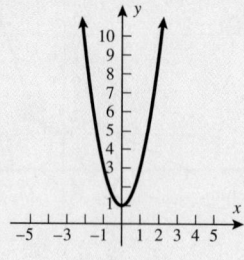

51.

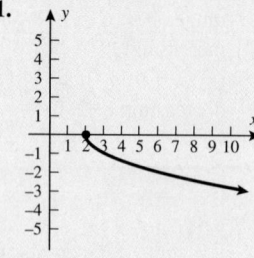

29.

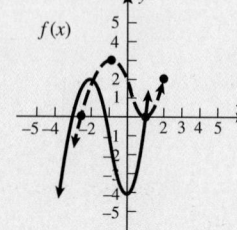

31.

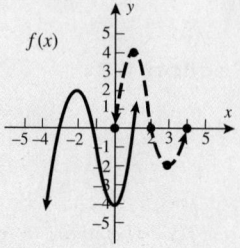

53.

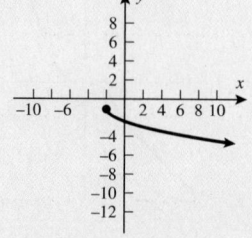

55.

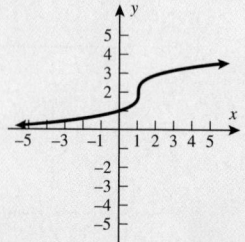

57.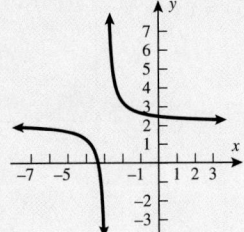

59.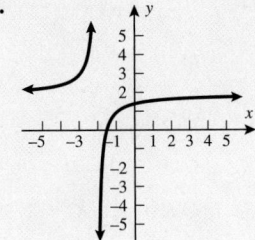

61.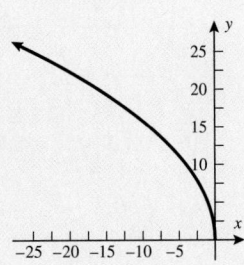

63.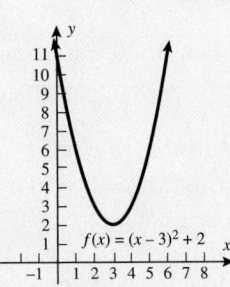

$f(x) = (x - 3)^2 + 2$

65.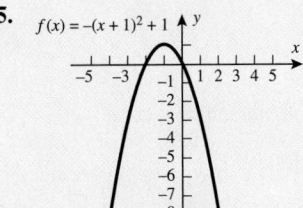

$f(x) = -(x + 1)^2 + 1$

67.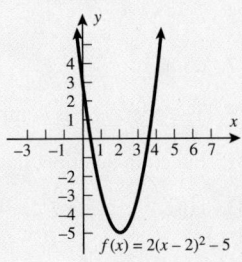

$f(x) = 2(x - 2)^2 - 5$

69. $S(x) = 10x$ and $S(x) = 10x + 50$

71. $T(x) = 0.33(x - 6500)$

73. $C(x) = 7 + 0.3(x - 1)$
$C(x + 1) = 7 + 0.3x$

75. $Q(t) = P(t + 50) = (t + 50)^3 - (t + 50)^2 + (t + 50) - 1$

77. **b.** shifted to the *right* 3 units

79. $|3 - x| = |x - 3|$; therefore, shift to the *right* 3 units.

81. true   **83.** true   **85.** true

87. $(a + 3, b + 2)$ **89.** $(a - 1, 2b - 1)$

91. Any part of the graph of $f(x)$ that is below the *x*-axis is reflected above it for $|f(x)|$.

a. 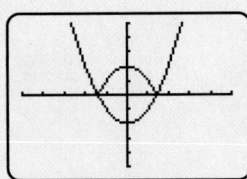   **b.**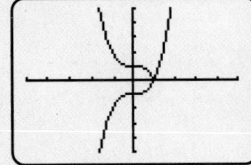

93. If $0 < a < 1$, you have a horizontal stretch. If $a > 1$, the graph is a horizontal compression.

a. 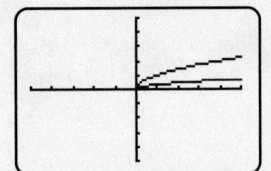   **b.**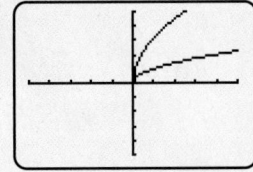

95. Each horizontal line in the graph of $y = [[x]]$ is stretched twice in length. There is a vertical shift of 1 unit up.

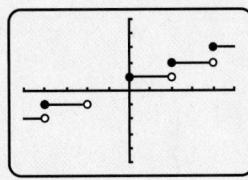

97. $f'(x) = 2x$, $g'(x) = 2(x - 1)$. The graph of g' is a horizontal shift of the graph of f'.

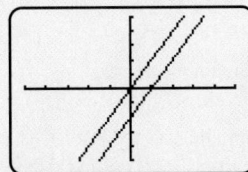

99. $f'(x) = 2$, $g'(x) = 2$. The graph of g' is the same as the graph of f'.

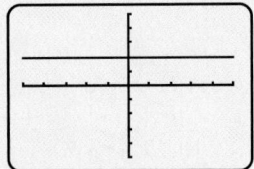

Section 1.4

1. $f(x) + g(x) = x + 2$
$f(x) - g(x) = 3x$ domain: $(-\infty, \infty)$
$f(x) \cdot g(x) = -2x^2 + x + 1$

$\dfrac{f(x)}{g(x)} = \dfrac{2x + 1}{1 - x}$ domain: $(-\infty, 1) \cup (1, \infty)$

3. $f(x) + g(x) = 3x^2 - x - 4$
$f(x) - g(x) = x^2 - x + 4$ domain: $(-\infty, \infty)$
$f(x) \cdot g(x) = 2x^4 - x^3 - 8x^2 + 4x$

$\dfrac{f(x)}{g(x)} = \dfrac{2x^2 - x}{x^2 - 4}$ domain: $(-\infty, -2) \cup (-2, 2) \cup (2, \infty)$

5. $f(x) + g(x) = \dfrac{1 + x^2}{x}$

$f(x) - g(x) = \dfrac{1 - x^2}{x}$ $\Bigg\}$ domain: $(-\infty, 0) \cup (0, \infty)$

$f(x) \cdot g(x) = 1$

$\dfrac{f(x)}{g(x)} = \dfrac{1/x}{x} = \dfrac{1}{x^2}$

7. $f(x) + g(x) = 3\sqrt{x}$ $\Big\}$ domain: $[0, \infty)$

$f(x) - g(x) = -\sqrt{x}$

$f(x) \cdot g(x) = 2x$ domain: $[0, \infty)$

$\dfrac{f(x)}{g(x)} = \dfrac{1}{2}$ domain: $(0, \infty)$

9. $f(x) + g(x) = \sqrt{4 - x} + \sqrt{x + 3}$

$f(x) - g(x) = \sqrt{4 - x} - \sqrt{x + 3}$ $\Big\}$ domain: $[-3, 4]$

$f(x) \cdot g(x) = \sqrt{4 - x}\sqrt{x + 3}$

$\dfrac{f(x)}{g(x)} = \dfrac{\sqrt{4 - x}}{\sqrt{x + 3}} = \dfrac{\sqrt{4 - x}\sqrt{x + 3}}{x + 3}$ domain: $(-3, 4]$

11. $(f \circ g)(x) = 2x^2 - 5$ domain: $(-\infty, \infty)$

$(g \circ f)(x) = 4x^2 + 4x - 2$ domain: $(-\infty, \infty)$

13. $(f \circ g)(x) = \dfrac{1}{x + 1}$ domain: $(-\infty, -1) \cup (-1, \infty)$

$(g \circ f)(x) = \dfrac{1}{x - 1} + 2$ domain: $(-\infty, 1) \cup (1, \infty)$

15. $(f \circ g)(x) = \dfrac{1}{|x - 1|}$ domain: $(-\infty, 1) \cup (1, \infty)$

$(g \circ f)(x) = \dfrac{1}{|x| - 1}$ domain: $(-\infty, -1) \cup (-1, 1) \cup (1, \infty)$

17. $(f \circ g)(x) = \sqrt{x + 4}$ domain: $[-4, \infty)$

$(g \circ f)(x) = \sqrt{x - 1} + 5$ domain: $[1, \infty)$

19. $(f \circ g)(x) = x$ domain: $(-\infty, \infty)$

$(g \circ f)(x) = x$ domain: $(-\infty, \infty)$

21. 15

23. 13

25. $26\sqrt{3}$

27. $\frac{110}{3}$

29. 11

31. $3\sqrt{2}$

33. undefined

35. undefined

37. 13

39. $f(g(1)) = \frac{1}{3}$ $g(f(2)) = 2$

41. undefined

43. $f(g(1)) = \frac{1}{3}$ $g(f(2)) = 4$

45. $f(g(1)) = \sqrt{5}$

$g(f(2)) = 6$

49. $f(g(1)) = \sqrt[3]{3}$

$g(f(2)) = 4$

61. $f(x) = 2x^2 + 5x$

$g(x) = 3x - 1$

63. $f(x) = \dfrac{2}{|x|}$

$g(x) = x - 3$

65. $f(x) = \dfrac{3}{\sqrt{x} - 2}$

$g(x) = x + 1$

67. $F(C(K)) = \frac{9}{5}(K - 273.15) + 32$

69. a. $A(x) = \left(\dfrac{x}{4}\right)^2$; x is the number of linear feet of fence.

b. $A(100) = 625$ square feet

c. $A(200) = 2500$ square feet

71. a. $C(p) = 62{,}000 - 20p$

b. $R(p) = 600{,}000 - 200p$

c. $P(p) = R(p) - C(p) = 538{,}000 - 180p$

73. area $= 150^2 \pi t = 22{,}500\pi t$

75. $d(h) = \sqrt{h^2 + 4}$

77. domain: $x \neq -2$

79. The operation is composition, *not* multiplication.

81. Function notation, not multiplication.

83. false   **85.** true

87. $(g \circ f)(x) = \dfrac{1}{x}$ $x \neq 0$

89. $(g \circ f)(x) = x$ $x \geq -a$

91.

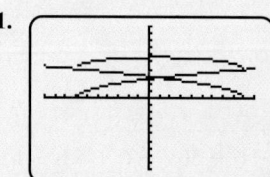

93.

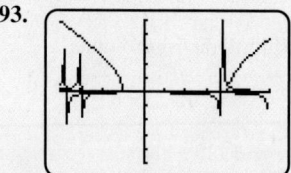

$(-\infty, -3) \cup (-3, -1] \cup [4, 6) \cup (6, \infty)$

95. $F'(x) = 1, G'(x) = 2x, H'(x) = 1 + 2x.\ H' = F' + G'$

97. $F'(x) = 0, G'(x) = \dfrac{1}{2\sqrt{x - 1}}, H'(x) = \dfrac{5}{2\sqrt{x - 1}}.$

$H' \neq F' \cdot G'$

Section 1.5

Exercise #	Function	One-to-One
1.	no	
3.	yes	no
5.	yes	no
7.	yes	yes
9.	yes	no

11. no **13.** yes **15.** no **17.** yes

19.
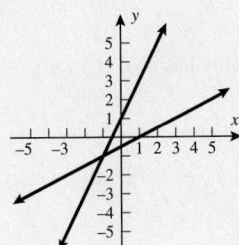

21.
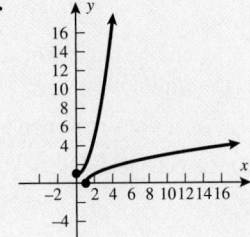

23.
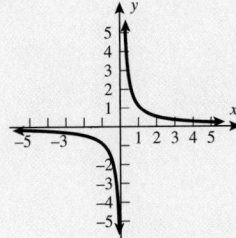

25.
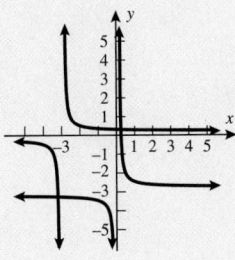

27.
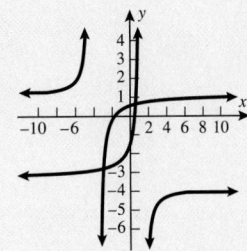

29.
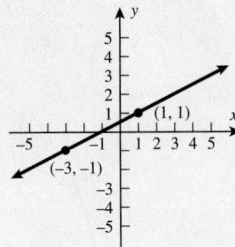

31.
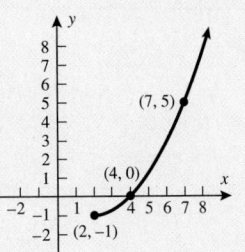

33.
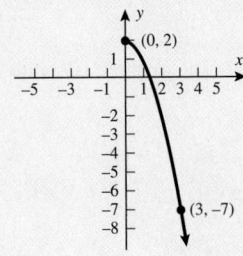

35.
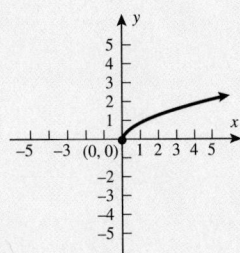

Exercise #	f^{-1}	Domain of f Range of f^{-1}	Range of f Domain of f^{-1}
37.	$f^{-1}(x) = -\frac{1}{3}x + \frac{2}{3}$	$(-\infty, \infty)$	$(-\infty, \infty)$
39.	$f^{-1}(x) = \sqrt[3]{x-1}$	$(-\infty, \infty)$	$(-\infty, \infty)$
41.	$f^{-1}(x) = x^2 + 3$	$[3, \infty)$	$[0, \infty)$
43.	$f^{-1}(x) = \sqrt{x+1}$	$[0, \infty)$	$[-1, \infty)$
45.	$f^{-1}(x) = -2 + \sqrt{x+3}$	$[-2, \infty)$	$[-3, \infty)$
47.	$f^{-1}(x) = \dfrac{2}{x}$	$(-\infty, 0) \cup (0, \infty)$	$(-\infty, 0) \cup (0, \infty)$
49.	$f^{-1}(x) = 3 - \dfrac{2}{x}$	$(-\infty, 3) \cup (3, \infty)$	$(-\infty, 0) \cup (0, \infty)$
51.	$f^{-1}(x) = \dfrac{5x - 1}{x + 7}$	$(-\infty, 5) \cup (5, \infty)$	$(-\infty, -7) \cup (-7, \infty)$
53.	$f^{-1}(x) = \dfrac{1}{x^2}$	$(0, \infty)$	$(0, \infty)$
55.	$f^{-1}(x) = \dfrac{2x^2 + 1}{x^2 - 1}$	$(-\infty, -1] \cup (2, \infty)$	$(-\infty, -1) \cup (-1, 1) \cup (1, \infty)$

57. The function is NOT one-to-one.

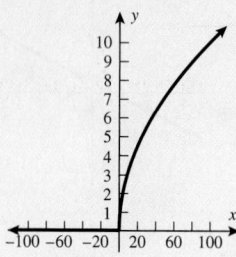

59. The function is one-to-one.

$$f^{-1}(x) = \begin{cases} x^3 & x \le -1 \\ -1 + \sqrt{x+1} & -1 < x \le 3 \\ (x-2)^2 & x > 3 \end{cases}$$

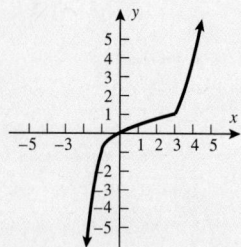

61. The function is one-to-one.

$$f^{-1}(x) = \begin{cases} x & x \le -1 \\ \sqrt[3]{x} & -1 < x < 1 \\ x & x \ge 1 \end{cases}$$

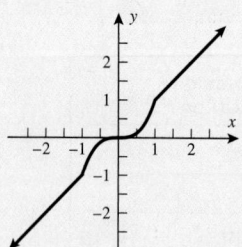

63. $f^{-1}(x) = \frac{5}{9}(x - 32)$; this now represents degrees Fahrenheit being turned into degrees Celsius.

65. $C(x) = \begin{cases} 250x & x \le 10 \\ 175x + 750 & x > 10 \end{cases}$

$$C^{-1}(x) = \begin{cases} \dfrac{x}{250} & x \le 2500 \\ \dfrac{x - 750}{175} & x > 2500 \end{cases}$$

67. $E(x) = 5.25x$; $E^{-1}(x) = \dfrac{x}{5.25}$ tells you how many hours the student will have to work to bring home x dollars.

69. $M(x) = \begin{cases} 0.60x & 0 \le x \le 15 \\ 0.9x - 4.5 & x > 15 \end{cases}$

$$M^{-1}(x) = \begin{cases} \dfrac{5x}{3} & 0 \le x \le 9 \\ \dfrac{10x}{9} + 5 & x > 9 \end{cases}$$

71. $V(x) = \begin{cases} 20{,}000 - 600x & 0 \le x \le 5 \\ 21{,}500 - 900x & x > 5 \end{cases}$

$$V^{-1}(x) = \begin{cases} \dfrac{21{,}500 - x}{900} & x < 17{,}000 \\ \dfrac{20{,}000 - x}{600} & 17{,}000 \le x \le 20{,}000 \end{cases}$$

It tells the number of years a family has owned a car that is worth x dollars.

73. No, it's not a function because it fails the vertical line test.

75. f is not one-to-one, so it does not have an inverse function.

77. false   **79.** false   **81.** $(b, 0)$

83. $f(x) = \sqrt{1 - x^2}$ $0 \le x \le 1$
$f^{-1}(x) = \sqrt{1 - x^2}$ $0 \le x \le 1$

85. $m \ne 0$

87. $a = 4$

$$f^{-1}(x) = \frac{1 - 2x}{x}$$

domain of f^{-1}: $(-\infty, 0) \cup (0, \infty)$

89. no   **91.** no

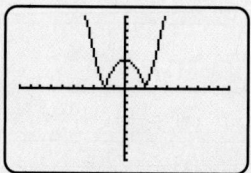 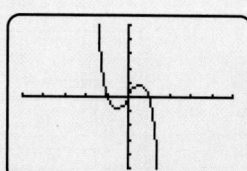

93. 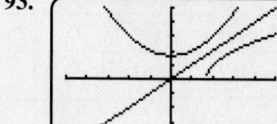   **95.**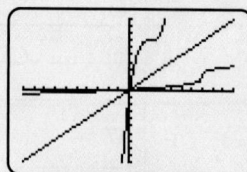

No, the functions are not inverses of each other. Had we restricted the domain of g to $x \ge 0$, then they would be inverses.

Yes, the functions are inverses of each other.

97. **a.** $f^{-1}(x) = \dfrac{x - 1}{2}$ **b.** $f'(x) = 2$

c. $(f^{-1})' = \dfrac{1}{2}$ **d.** $\dfrac{1}{2} = \dfrac{1}{2}$

99. a. $f^{-1}(x) = x^2 - 2 \quad x \geq 0$ **b.** $f'(x) = \dfrac{1}{2\sqrt{x+2}}$

   c. $(f^{-1})'(x) = 2x$ **d.** $2x = \dfrac{1}{\dfrac{1}{2\sqrt{(x^2-2)+2}}}$

Review Exercises

1. yes   **3.** no

5. yes   **7.** no

9. a. 2   **b.** 4   **c.** $x = -3, x = 4$

11. a. 0   **b.** -2 **c.** $x = -5, x = 2$

13. 5   **15.** -665 **17.** -2

19. 4   **21.** domain: $(-\infty, \infty)$ **23.** $(-\infty, -4) \cup (-4, \infty)$

25. $[4, \infty)$ **27.** $D = 18$

29. odd   **31.** odd

33. a. $[-5, \infty)$ **b.** $[-3, \infty)$

   c. increasing: $(-5, -3) \cup (3, \infty)$, decreasing: $(-1, 1)$, constant: $(-3, -1) \cup (1, 3)$

   d. 2   **e.** 3   **f.** 1

35. a. $[-6, 6]$ **b.** $[0, 3] \cup \{-3, -2, -1\}$

   c. increasing: $(0, 3)$, decreasing: $(3, 6)$, constant: $(-6, -4)$, $(-4, -2)$, $(-2, 0)$

   d. -1 **e.** -2 **f.** 3

37. $3x^2 + 3xh + h^2$ **39.** $1 - \dfrac{1}{x(x+h)}$

41. -2

43. domain: $(-\infty, \infty)$ **45.** domain: $(-\infty, \infty)$
range: $(0, \infty)$ range: $[-1, \infty)$

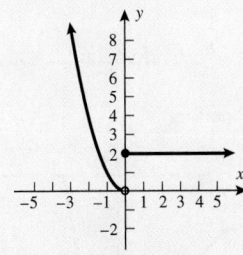 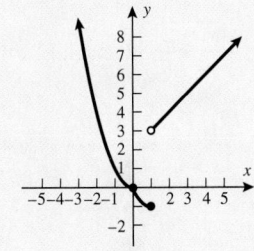

47. \$29,000 per year

49. 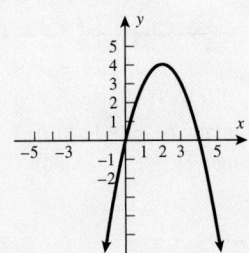   **51.**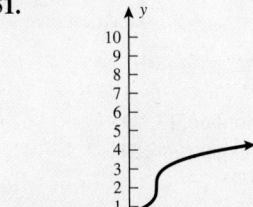

53.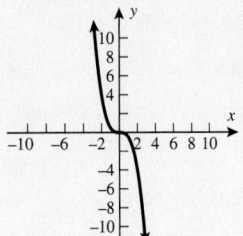

55.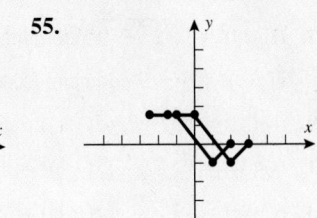

57.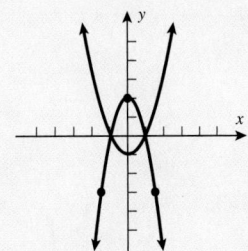

59. $y = \sqrt{x+3}$
domain: $[-3, \infty)$

61. $y = \sqrt{x-2} + 3$
domain: $[2, \infty)$

63. $y = 5\sqrt{x} - 6$
domain: $[0, \infty)$

65. $y = (x+2)^2 - 12$

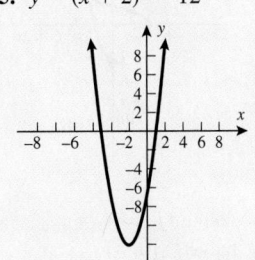

67. $g(x) + h(x) = -2x - 7$
domain: $(-\infty, \infty)$

$g(x) - h(x) = -4x - 1$
domain: $(-\infty, \infty)$

$g(x) \cdot h(x) = -3x^2 + 5x + 12$
domain: $(-\infty, \infty)$

$\dfrac{g(x)}{h(x)} = \dfrac{-3x - 4}{x - 3}$
domain: $(-\infty, 3) \cup (3, \infty)$

69. $g(x) + h(x) = \dfrac{1}{x^2} + \sqrt{x}$, domain: $(0, \infty)$

$g(x) - h(x) = \dfrac{1}{x^2} - \sqrt{x}$, domain: $(0, \infty)$

$g(x) \cdot h(x) = \dfrac{1}{x^{3/2}}$, domain: $(0, \infty)$

$\dfrac{g(x)}{h(x)} = \dfrac{1}{x^{5/2}}$, domain: $(0, \infty)$

71. $\left. \begin{array}{l} g(x) + h(x) = \sqrt{x-4} + \sqrt{2x+1} \\ g(x) - h(x) = \sqrt{x-4} - \sqrt{2x+1} \\ g(x) \cdot h(x) = \sqrt{x-4}\sqrt{2x+1} \\ \dfrac{g(x)}{h(x)} = \dfrac{\sqrt{x-4}}{\sqrt{2x+1}} \end{array} \right\}$ domain: $[4, \infty)$

73. $(f \circ g)(x) = 6x - 1$ domain: $(-\infty, \infty)$

$(g \circ f)(x) = 6x - 7$ domain: $(-\infty, \infty)$

75. $(f \circ g)(x) = \dfrac{8 - 2x}{13 - 3x}$
domain: $(-\infty, 4) \cup \left(4, \frac{13}{3}\right) \cup \left(\frac{13}{3}, \infty\right)$

$(g \circ f)(x) = \dfrac{x + 3}{4x + 10}$
domain: $(-\infty, -3) \cup \left(-3, -\frac{5}{2}\right) \cup \left(-\frac{5}{2}, \infty\right)$

77. $f(g(x)) = \sqrt{x^2 - 9}$ domain: $(-\infty, -3] \cup [3, \infty)$
$g(f(x)) = x - 9$ domain: $[5, \infty)$

79. $f(g(3)) = 857$ $g(f(-1)) = 51$

81. $f(g(3)) = \frac{17}{31}$ $g(f(-1)) = 1$

83. $f(g(3)) = 12$ $g(f(-1)) = 2$

85. $f(x) = 3x^2 + 4x + 7$ and $g(x) = x - 2$

87. $f(x) = \dfrac{1}{\sqrt{x}}$ and $g(x) = x^2 + 7$

89. $A(t) = 625\pi(t + 2)$ sq. in.

91. yes   **93.** yes   **95.** yes

97. no   **99.** yes

101. 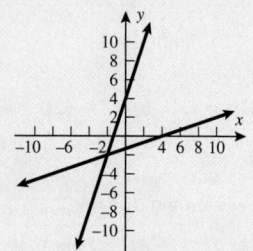   **103.**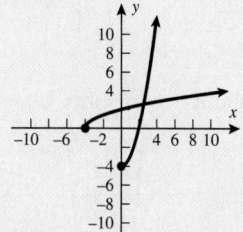

105. $f^{-1}(x) = \dfrac{x - 1}{2}$ **107.** $f^{-1}(x) = x^2 - 4$
domain f: $(-\infty, \infty)$
range f: $(-\infty, \infty)$
domain f^{-1}: $(-\infty, \infty)$
range f^{-1}: $(-\infty, \infty)$

domain f: $[-4, \infty)$
range f: $[0, \infty)$
domain f^{-1}: $[0, \infty)$
range f^{-1}: $[-4, \infty)$

109. $f^{-1}(x) = \dfrac{6 - 3x}{x - 1}$
domain f: $(-\infty, -3) \cup (-3, \infty)$
range f: $(-\infty, 1) \cup (1, \infty)$
domain f^{-1}: $(-\infty, 1) \cup (1, \infty)$
range f^{-1}: $(-\infty, -3) \cup (-3, \infty)$

111. $S(x) = 22,000 + 0.08x$ $S^{-1}(x) = \dfrac{(x - 22,000)}{0.08}$
Sales required to earn desired income.

113. $(-\infty, -1) \cup (3, \infty)$

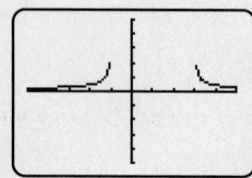

115.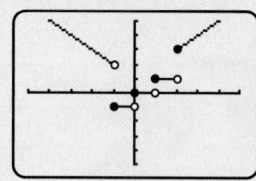

a. $(-\infty, 2) \cup (2, \infty)$ **b.** $\{-1, 0, 1\} \cup (2, \infty)$
c. increasing: $(2, \infty)$, decreasing: $(-\infty, -1)$,
    constant: $(-1, 0) \cup (0, 1) \cup (1, 2)$

117. The graph of f can be obtained by shifting the graph of g
two units to the left; that is, $f(x) = g(x + 2)$.

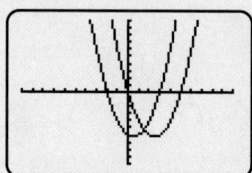

119. $[-1.5, 4)$ **121.** yes

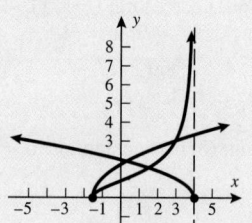   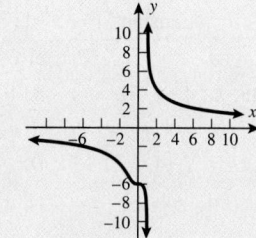

Practice Test

1. b   **3.** c

5. $\dfrac{\sqrt{x - 2}}{x^2 + 11}$ domain: $[2, \infty)$ **7.** $x + 9$ domain: $[2, \infty)$

9. 4   **11.** neither

13. 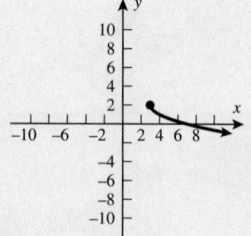   **15.**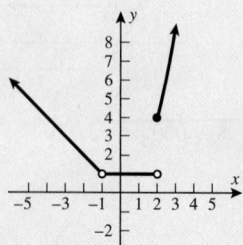

domain: $[3, \infty)$ domain: $(-\infty, -1) \cup (-1, \infty)$
range: $(-\infty, 2]$ range: $[1, \infty)$

17. a. -2 **b.** 4   **c.** -3 **d.** $x = 2, x = -3$

19. $6x + 3h - 4$ **21.** -32

23. $f^{-1}(x) = x^2 + 5, x \geq 0$
    f: domain: $[5, \infty)$, range: $[0, \infty)$
    f^{-1}: domain: $[0, \infty)$, range: $[5, \infty)$

25. $f^{-1}(x) = \dfrac{5x - 1}{x + 2}$

    f: domain: $(-\infty, 5) \cup (5, \infty)$, range: $(-\infty, -2) \cup (-2, \infty)$

    f^{-1}: domain: $(-\infty, -2) \cup (-2, \infty)$, range: $(-\infty, 5) \cup (5, \infty)$

27. $x \geq 0$ (*Note:* There is more than one answer.)

29. $P(t) = \frac{9}{10}t + 10$

31. quadrant III, "quarter of unit circle"

33. $f(x) = \begin{cases} 15 & 0 \leq x \leq 30 \\ -15 + x & x > 30 \end{cases}$

CHAPTER 2

Section 2.1

1. b   **3.** a   **5.** b   **7.** c

9.
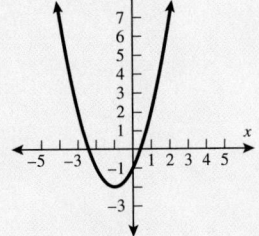

11.
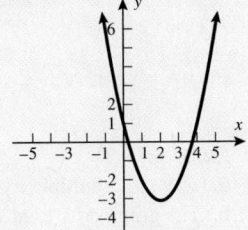

13.
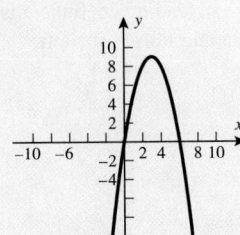

15.
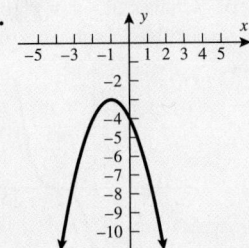

17.
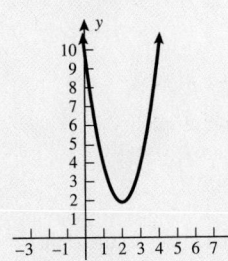

19.
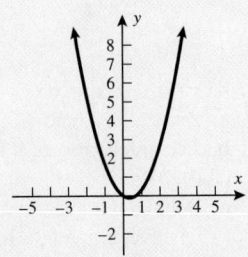

21.
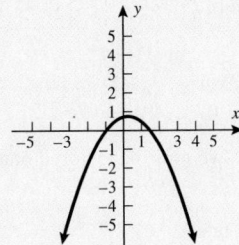

23. $f(x) = (x + 3)^2 - 12$

25. $f(x) = -(x + 5)^2 + 28$

27. $f(x) = 2(x + 2)^2 - 10$

29. $f(x) = -4(x - 2)^2 + 9$

31. $f(x) = (x + 5)^2 - 25$

33. $f(x) = \frac{1}{2}(x - 4)^2 - 5$

35.
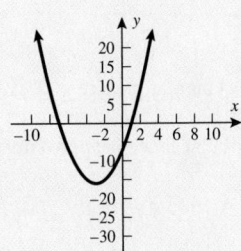

37.
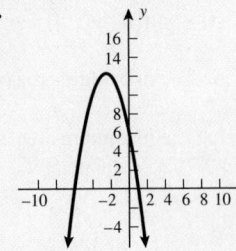

39.
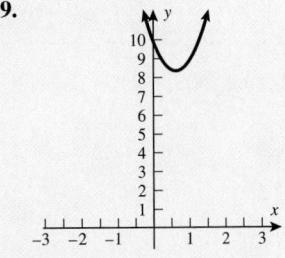

41.
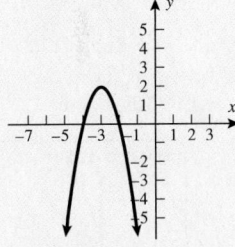

43.
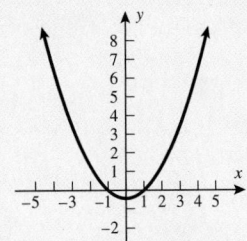

45. vertex $\left(\frac{1}{33}, \frac{494}{33}\right)$

47. vertex $\left(7, -\frac{39}{2}\right)$

49. vertex $\left(\frac{15}{28}, \frac{829}{392}\right)$

51. vertex $(-75, 12.95)$

53. $(21.67, -24.65)$

55. $y = -2(x + 1)^2 + 4$

57. $y = -5(x - 2)^2 + 5$ **59.** $y = \frac{5}{9}(x + 1)^2 - 3$

61. $f(x) = 10(x + 2)^2 - 4$ **63.** $y = 12\left(x - \frac{1}{2}\right)^2 - \frac{3}{4}$

65. $y = \frac{5}{4}(x - 2.5)^2 - 3.5$

67. a. 40 yards (120 feet)   **b.** 50 yards

69. 2,083,333 square feet

71. a. $t = 1$ second is when the rock reaches the maximum
     height of 116 feet.
    b. The rock hits the ground in approximately 3.69 seconds.
    c. Between 0 and 2 seconds: $0 < t < 2$.

73. Altitude is 26,000 feet; horizontal distance is 8944 feet.

75. a. 100 boards   **b.** $24,000

77. Between 15 and 16 units or 64 and 65 units to break even.

79. a. $f(t) = \frac{28}{27}t^2 + 16$ **b.** 219 million cell phones

81. a. $C(t) = -0.01(t - 225)^2 + 400$
  b. 425 minutes

83. The graph is incorrect. The vertex is $(-3, -1)$ and the x-intercepts are $(-2, 0)$ and $(-4, 0)$.

85. $f(x) = -(x - 1)^2 + 4$. The negative must be factored out of the x^2 and x terms.

87. true   **89.** false

91. $f(x) = a\left(x + \dfrac{b}{2a}\right)^2 + \dfrac{4ac - b^2}{4a}$

93. a. The maximum area of the rectangular pasture is 62,500 square feet.
  b. The maximum area of the circular fence is approximately 79,577 square feet.

95. 5

97. a. $(1425, 4038.25)$ **b.** $(0, -23)$
  c. $(4.04, 0)$ $(2845.96, 0)$ **d.** $x = 1425$

99. a. $y = -2x^2 + 12.8x + 4.32$
  b. $y = -2(x - 3.2)^2 + 24.8$, $(3.2, 24.8)$
  c. Yes, they agree.

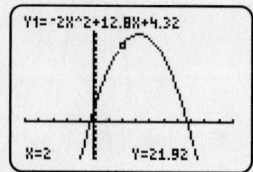

101. $\dfrac{x^2}{9} + \dfrac{(y - 2)^2}{4} = 1$, ellipse

103. $(x + 3)^2 = 20\left(y + \frac{1}{5}\right)$, parabola

Section 2.2

1. polynomial; degree 5   **3.** polynomial, degree 7

5. not a polynomial   **7.** not a polynomial

9. not a polynomial   **11.** h

13. b   **15.** e   **17.** c

19.   **21.**

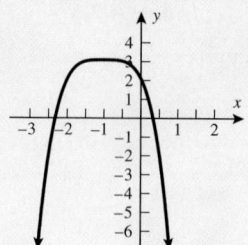

23.

25. zero at 3, multiplicity of 1;
  zero at -4, multiplicity of 3

27. zero at 0, multiplicity of 2;
  zero at 7, multiplicity of 2;
  zero at -4, multiplicity of 1

29. zero at 0, multiplicity of 2;
  zero at 1, multiplicity of 2

31. $f(x) = x(4x + 9)(2x - 3)$

  zero at 0, multiplicity of 1

  zero at $-\frac{9}{4}$, multiplicity of 1

  zero at $\frac{3}{2}$, multiplicity of 1

33. zero at 0, multiplicity 2; zero at -3, multiplicity 1

35. zero at 0, multiplicity 4

37. $P(x) = x(x + 3)(x - 1)(x - 2)$

39. $P(x) = x(x + 5)(x + 3)(x - 2)(x - 6)$

41. $P(x) = (2x + 1)(3x - 2)(4x - 3)$

43. $P(x) = x^2 - 2x - 1$

45. $P(x) = x^2(x + 2)^3$

47. $P(x) = (x + 3)^2(x - 7)^5$

49. $P(x) = x^2\left(x + \sqrt{3}\right)^2(x + 1)\left(x - \sqrt{3}\right)^2$
  $= x^2(x + 1)(x^2 - 3)^2$

51. $f(x) = (x - 2)^3$
  a. zero at 2; multiplicity of 3
  b. The graph crosses at $x = 2$.
  c. $(0, -8)$
  d. Behaves like $y = x^3$ (odd degree); leading coefficient is positive, graph rises to the right and falls to the left.
  e.

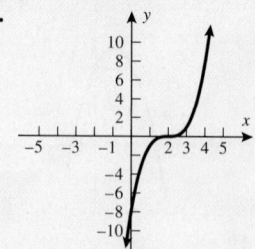

53. $f(x) = x^3 - 9x = x(x - 3)(x + 3)$
  a. zeros at 0, 3, and -3; each has multiplicity 1
  b. The graph crosses at all three zeros.
  c. $(0, 0)$
  d. Behaves like $y = x^3$ (odd degree); leading coefficient is positive, graph falls to the left and rises to the right.

e.

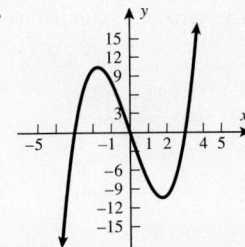

55. $f(x) = -x^3 + x^2 + 2x = -x(x - 2)(x + 1)$
   a. zeros at 0, 2, −1; each has multiplicity 1
   b. The graph crosses at all three zeros.
   c. (0, 0)
   d. Behaves like $y = -x^3$ (odd degree); leading coefficient is negative, graph rises to the left and falls to the right.
   e.

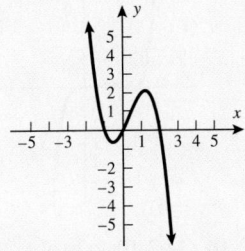

57. $f(x) = -x^4 - 3x^3 = -x^3(x + 3)$
   a. zero at 0, multiplicity 3; zero at −3, multiplicity 1
   b. The graph crosses at $x = 0$ and $x = -3$.
   c. (0, 0)
   d. Behaves like $y = -x^4$ (even degree); leading coefficient is negative.
   e.

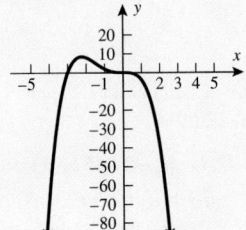

59. $f(x) = 12x^4(x - 4)(x + 1)$
   a. zero at 0, multiplicity 4; zero at 4, multiplicity 1; zero at −1, multiplicity 1
   b. The graph crosses at $x = -1$ and $x = 4$ while the graph touches at 0.
   c. (0, 0)
   d. Behaves like $y = x^6$ (even degree); leading coefficient is positive.
   e.

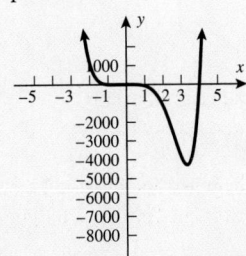

61. $f(x) = 2x^5 - 6x^4 - 8x^3 = 2x^3(x - 4)(x + 1)$
   a. zeros at −1, 0, and 4; 0 has multiplicity 3; −1 and 4 each have multiplicity 1
   b. The graph crosses at $x = 0$, $x = -1$, and $x = 4$.
   c. (0, 0)
   d. Behaves like $y = x^5$ (odd degree); leading coefficient is positive, graph falls to the left and rises to the right.
   e.

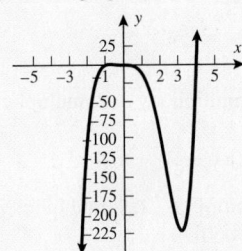

63. $f(x) = x^3 - x^2 - 4x + 4 = (x - 2)(x + 2)(x - 1)$
   a. zeros at −2, 1, and 2; each has multiplicity 1
   b. The graph crosses at $x = -2$, $x = 1$, and $x = 2$.
   c. (0, 4)
   d. Behaves like $y = x^3$ (odd degree); leading coefficient is positive, graph falls to the left and rises to the right.
   e.

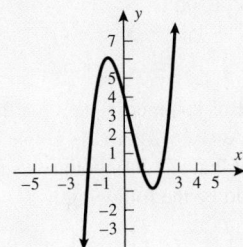

65. $f(x) = -(x + 2)^2(x - 1)^2$
   a. zeros at −2 and 1; with both having multiplicity of 2
   b. The graph touches at each one of the zeros.
   c. (0, −4)
   d. Behaves like $y = x^4$ (even degree); leading coefficient is negative.
   e.

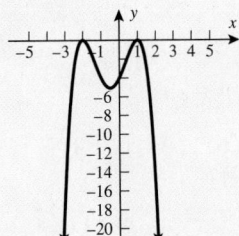

67. $f(x) = x^2(x - 2)^3(x + 3)^2$
   a. zeros at 0, 2, and −3; having multiplicities of 2, 3, and 2
   b. The graph touches at −3 and 0 but crosses at 2.
   c. (0, 0)
   d. Behaves like $y = x^7$; leading coefficient is positive.

e.

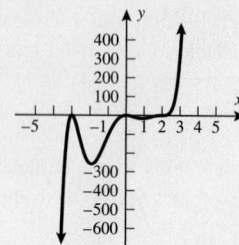

69. a. -3, multiplicity 1; -1, multiplicity 2; 2, multiplicity 1
b. even **c.** negative **d.** $(0, 6)$
e. $f(x) = -(x + 1)^2(x - 2)(x + 3)$

71. a. -2, multiplicity 2; 0, multiplicity 2; $\frac{3}{2}$, multiplicity 1
b. odd **c.** positive **d.** $(0, 0)$
e. $f(x) = x^2(2x - 3)(x + 2)^2$

73. Sixth-degree polynomial, because there are five known turning points.

75. down **77.** fourth-degree

79. 4 **81.** $35° < T < 39°$

83. If h is a zero then $(x - h)$ is a factor.
In this case: $P(x) = (x + 2)(x + 1)(x - 3)(x - 4)$.

85. $f(x) = (x - 1)^2(x + 2)^3$
Yes, the zeros are -2 and 1. But you must remember that this is a fifth-degree polynomial. At the -2 zero the graph crosses, but it should be noted that at 1 it only touches at this value. The correct graph would be the following:

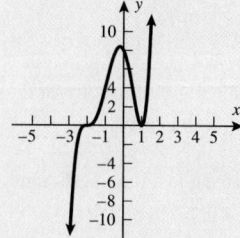

87. false **89.** true **91.** n

93. There are three possibilities.

a. $f(x) = (x + 1)^2(x - 3)^5$ **b.** $g(x) = (x + 1)^4(x - 3)^3$

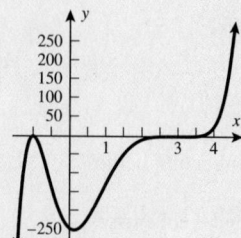 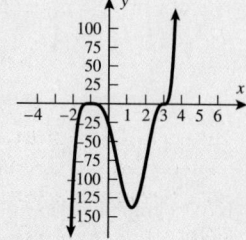

c. $h(x) = (x + 1)^6(x - 3)$

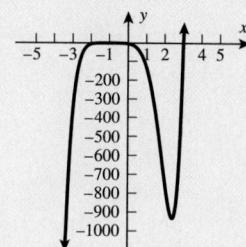

95. The zeros of the polynomial are 0, a, and $-b$.

97. no x-intercepts **99.** $y = -2x^5$
yes

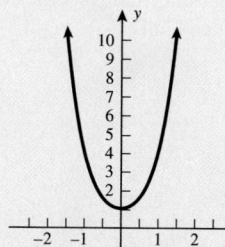 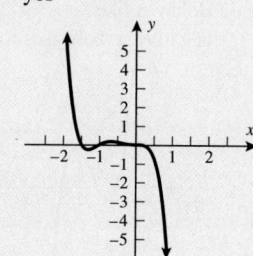

101. x-intercepts: $(-2.25, 0)$, $(6.2, 0)$, $(14.2, 0)$; zero at -2.25, multiplicity 2; zero at 6.2, multiplicity 1; zero at 14.2, multiplicity 1

103.

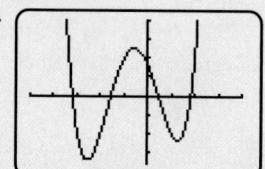

relative minimum at $(-2.56, -17.12)$ and $(1.27, -11.73)$; relative maximum at $(-0.58, 12.59)$

105. $x \approx 1.154$ **107.** $x \approx -0.865$ and $x \approx 1.363$

Section 2.3

1. $Q(x) = 3x - 3$ $r(x) = -11$

3. $Q(x) = 3x - 28$ $r(x) = 130$

5. $Q(x) = x - 4$ $r(x) = 12$

7. $Q(x) = 3x + 5$ $r(x) = 0$

9. $Q(x) = 2x - 3$ $r(x) = 0$

11. $Q(x) = 4x^2 + 4x + 1$ $r(x) = 0$

13. $Q(x) = 2x^2 - x - \frac{1}{2}$ $r(x) = \frac{15}{2}$

15. $Q(x) = 4x^2 - 10x - 6$ $r(x) = 0$

17. $Q(x) = -2x^2 - 3x - 9$ $r(x) = -27x^2 + 3x + 9$

19. $Q(x) = x^2 + 1$ $r(x) = 0$

21. $Q(x) = x^2 + x + \frac{1}{6}$ $r(x) = -\frac{121}{6}x + \frac{121}{3}$

23. $Q(x) = -3x^3 + 5.2x^2 + 3.12x - 0.128$ $r(x) = 0.9232$

25. $Q(x) = x^2 - 0.6x + 0.09$ $r(x) = 0$

27. $Q(x) = 3x + 1$ $r(x) = 0$

29. $Q(x) = 7x - 10$ $r(x) = 15$

31. $Q(x) = -x^3 + 3x - 2$ $r(x) = 0$

33. $Q(x) = x^3 - x^2 + x - 1$ $r(x) = 2$

35. $Q(x) = x^3 - 2x^2 + 4x - 8$ $r(x) = 0$

37. $Q(x) = 2x^2 - 6x + 2$ $r(x) = 0$

39. $Q(x) = 2x^3 - \frac{5}{3}x^2 + \frac{53}{9}x + \frac{106}{27}$ $r(x) = -\frac{112}{81}$

41. $Q(x) = 2x^3 + 6x^2 - 18x - 54$ $r(x) = 0$

43. $Q(x) = x^6 + x^5 + x^4 - 7x^3 - 7x^2 - 4x - 4$ $r(x) = -3$

45. $Q(x) = x^5 + \sqrt{5}x^4 - 44x^3 - 44\sqrt{5}x^2 - 245x - 245\sqrt{5}$
$r(x) = 0$

47. $Q(x) = 2x - 7$ $r(x) = 0$

49. $Q(x) = x^2 - 9$ $r(x) = 0$

51. $Q(x) = x + 6$ $r(x) = -x + 1$

53. $Q(x) = x^4 - 2x^3 - 4x + 7$ $r(x) = 0$

55. $Q(x) = x^4 + 2x^3 + 8x^2 + 18x + 36$ $r(x) = 71$

57. $Q(x) = x^2 + 1$ $r(x) = -24$

59. $Q(x) = x^6 + x^5 + x^4 + x^3 + x^2 + x + 1$ $r(x) = 0$

61. The width is $3x^2 + 2x + 1$ feet.

63. The trip takes $x^2 + 1$ hours.

65. In long division, you must subtract (not add) each term.

67. Forgot "0" placeholder.

69. true   **71.** false

73. false   **75.** yes

77. $x^{2n} + 2x^n + 1$

79. $2x - 1$ **81.** $x^3 - 1$

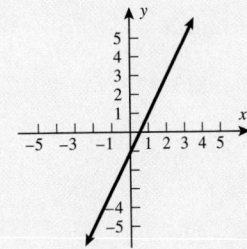   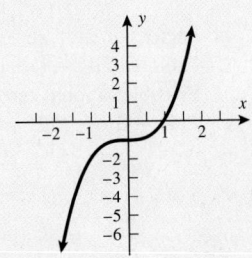

83. A quadratic function.

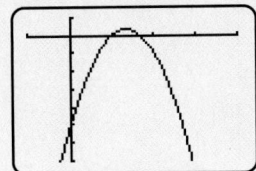

85. $2x - 5 + \dfrac{10}{x + 2}$

87. $2x^2 - 2x + 3 - \dfrac{x - 3}{x^2 + x + 1}$

Section 2.4

1. $1, 3, -4; P(x) = (x - 1)(x - 3)(x + 4)$

3. $\frac{1}{2}, -3, 2; P(x) = (2x - 1)(x + 3)(x - 2)$

5. $5, -3; P(x) = (x - 5)(x + 3)(x^2 + 4)$

7. $-3, 1; P(x) = (x + 3)(x - 1)(x^2 - 2x + 2)$

9. $-1, -2; P(x) = (x + 1)^2(x + 2)^2$

11. $P(x) = x^4 + 3x^2 - 8x + 4$
Factors of $a_0 = 4$: $\pm 1, \pm 2, \pm 4$
Factors of $a_n = 1$: ± 1
Possible rational zeros: $\pm 1, \pm 2, \pm 4$

13. $P(x) = x^5 - 14x^3 + x^2 - 15x + 12$
Factors of $a_0 = 12$: $\pm 1, \pm 2, \pm 3, \pm 4, \pm 6, \pm 12$
Factors of $a_n = 1$: ± 1
Possible rational zeros: $\pm 1, \pm 2, \pm 3, \pm 4, \pm 6, \pm 12$

15. $P(x) = 2x^6 - 7x^4 + x^3 - 2x + 8$
Factors of $a_0 = 8$: $\pm 1, \pm 2, \pm 4, \pm 8$
Factors of $a_n = 2$: $\pm 1, \pm 2$
Possible rational zeros: $\pm 1, \pm 2, \pm 4, \pm 8, \pm \frac{1}{2}$

17. $P(x) = 5x^5 + 3x^4 + x^3 - x - 20$
Factors of $a_0 = -20$: $\pm 1, \pm 2, \pm 4, \pm 5, \pm 10, \pm 20$
Factors of $a_n = 5$: $\pm 1, \pm 5$
Possible rational zeros:
$\pm 1, \pm 2, \pm 4, \pm 5, \pm 10, \pm 20, \pm \frac{1}{5}, \pm \frac{2}{5}, \pm \frac{4}{5}$

19. $P(x) = x^4 + 2x^3 - 9x^2 - 2x + 8$
Factors of $a_0 = 8$: $\pm 1, \pm 2, \pm 4, \pm 8$
Factors of $a_n = 1$: ± 1
Possible rational zeros: $\pm 1, \pm 2, \pm 4, \pm 8$
Testing the zeros:
$P(1) = 0$ $P(-1) = 0$ $P(2) = 0$ $P(-4) = 0$

21. $P(x) = 2x^3 - 9x^2 + 10x - 3$
Factors of $a_0 = -3$: $\pm 1, \pm 3$
Factors of $a_n = 2$: $\pm 1, \pm 2$
Possible rational zeros: $\pm 1, \pm 3, \pm \frac{1}{2}, \pm \frac{3}{2}$
$P(1) = 0$ $P(3) = 0$ $P\left(\frac{1}{2}\right) = 0$

23.

POSITIVE REAL ZEROS	NEGATIVE REAL ZEROS
1	1

25.

POSITIVE REAL ZEROS	NEGATIVE REAL ZEROS
1	0

27.

POSITIVE REAL ZEROS	NEGATIVE REAL ZEROS
2	1
0	1

29.

POSITIVE REAL ZEROS	NEGATIVE REAL ZEROS
1	1

Note that $x = 0$ is also a zero.

31.

POSITIVE REAL ZEROS	NEGATIVE REAL ZEROS
2	2
0	2
2	0
0	0

33.

POSITIVE REAL ZEROS	NEGATIVE REAL ZEROS
4	0
2	0
0	0

35. $P(x) = x^3 + 6x^2 + 11x + 6$

a. No sign changes: 0 positive real zeros
$P(-x) = (-x)^3 + 6(-x)^2 + 11(-x) + 6$
$\qquad = -x^3 + 6x^2 - 11x + 6$
3 or 1 negative real zeros

POSITIVE	NEGATIVE
0	3
0	1

b. Factors of $a_0 = 6$: $\pm1, \pm2, \pm3, \pm6$
Factors of $a_n = 1$: ±1
Possible rational zeros: $\pm1, \pm2, \pm3, \pm6$
c. $P(-1) = 0 \qquad P(-2) = 0 \qquad P(-3) = 0$
d. $x = -1 \qquad x = -2 \qquad x = -3$
$P(x) = (x + 1)(x + 2)(x + 3)$

37. $P(x) = x^3 - 7x^2 - x + 7$

a. 2 or 0 positive real zeros
$P(-x) = (-x)^3 - 7(-x)^2 - (-x) + 7$
$\qquad = -x^3 - 7x^2 + x + 7$
1 negative real zero

POSITIVE	NEGATIVE
2	1
0	1

b. Factors of $a_0 = 7$: $\pm1, \pm7$
Factors of $a_n = 1$: ±1
Possible rational zeros: $\pm1, \pm7$
c. $P(1) = 0 \qquad P(-1) = 0 \qquad P(7) = 0$
d. $P(x) = (x - 1)(x + 1)(x - 7)$

39. $P(x) = x^4 + 6x^3 + 3x^2 - 10x$

a. 1 positive real zero
$P(-x) = (-x)^4 + 6(-x)^3 + 3(-x)^2 - 10(-x)$
$\qquad = x^4 - 6x^3 + 3x^2 + 10x$
$\qquad = x(x^3 - 6x^2 + 3x + 10)$
2 or 0 negative real zeros

POSITIVE	NEGATIVE
1	2
1	0

b. Factors of $a_0 = -10$: $\pm1, \pm2, \pm5, \pm10$
Factors of $a_n = 1$: ±1
Possible rational zeros: $\pm1, \pm2, \pm5, \pm10$
c. $P(0) = 0 \qquad P(1) = 0 \qquad P(-2) = 0 \qquad P(-5) = 0$
d. $P(x) = x(x - 1)(x + 2)(x + 5)$

41. $P(x) = x^4 - 7x^3 + 27x^2 - 47x + 26$

a. 4, 2, or 0 positive real zeros
$P(-x) = (-x)^4 - 7(-x)^3 + 27(-x)^2 - 47(-x) + 26$
$\qquad = x^4 + 7x^3 + 27x^2 + 47x + 26$
no negative real zeros

POSITIVE	NEGATIVE
4	0
2	0
0	0

b. Factors of $a_0 = 26$: $\pm1, \pm2, \pm13, \pm26$
Factors of $a_n = 1$: ±1
Possible rational zeros: $\pm1, \pm2, \pm13, \pm26$
c. $P(1) = 0 \qquad P(2) = 0 \qquad (x - 1)(x - 2) = x^2 - 3x + 2$

$$
\begin{array}{r}
x^2 - 4x + 13 \\
\end{array}
$$

d. $x^2 - 3x + 2\overline{\smash{\big)}\ x^4 - 7x^3 + 27x^2 - 47x + 26}$
$$
\begin{array}{r}
\underline{-(x^4 - 3x^3 + 2x^2)} \\
-4x^3 + 25x^2 - 47x \\
\underline{-(-4x^3 + 12x^2 - 8x)} \\
13x^2 - 39x + 26 \\
\underline{-(13x^2 - 39x + 26)} \\
0
\end{array}
$$

$P(x) = (x - 1)(x - 2)(x^2 - 4x + 13)$

43. $P(x) = 10x^3 - 7x^2 - 4x + 1$
  a. 2 or 0 positive real zeros
$$P(-x) = 10(-x)^3 - 7(-x)^2 - 4(-x) + 1$$
$$= -10x^3 - 7x^2 + 4x + 1$$
  1 negative real zero

POSITIVE	NEGATIVE
2	1
0	1

  b. Factors of $a_0 = 1$: ± 1
    Factors of $a_n = 10$: $\pm 1, \pm 2, \pm 5, \pm 10$
    Possible rational zeros: $\pm 1, \pm\frac{1}{2}, \pm\frac{1}{5}, \pm\frac{1}{10}$
  c. $P(1) = 0 \qquad P\left(-\frac{1}{2}\right) = 0 \qquad P\left(\frac{1}{5}\right) = 0$
  d. $(x - 1)\left(x + \frac{1}{2}\right)\left(x - \frac{1}{5}\right) = 0$
$$P(x) = (x - 1)(2x + 1)(5x + 1)$$

45. $P(x) = 6x^3 + 17x^2 + x - 10$
  a. 1 positive real zero
$$P(-x) = 6(-x)^3 + 17(-x)^2 + (-x) - 10$$
$$= -6x^3 + 17x^2 - x - 10$$
  2 or 0 negative real zeros

POSITIVE	NEGATIVE
1	2
1	0

  b. Factors of $a_0 = -10$: $\pm 1, \pm 2, \pm 5, \pm 10$
    Factors of $a_n = 6$: $\pm 1, \pm 2, \pm 3, \pm 6$
    Possible rational zeros: $\pm 1, \pm\frac{1}{2}, \pm\frac{1}{3}, \pm\frac{1}{6}, \pm 2, \pm\frac{2}{3},$
    $\pm 5, \pm\frac{5}{2}, \pm\frac{5}{3}, \pm\frac{5}{6}, \pm 10, \pm\frac{10}{3}$
  c. $P(-1) = 0 \qquad P\left(-\frac{5}{2}\right) = 0 \qquad P\left(\frac{2}{3}\right) = 0$
  d. $(x + 1)\left(x + \frac{5}{2}\right)\left(x - \frac{2}{3}\right) = 0$
$$P(x) = (x + 1)(2x + 5)(3x - 2)$$

47. $P(x) = x^4 - 2x^3 + 5x^2 - 8x + 4$
  a. 4, 2, or 0 positive real zeros
$$P(-x) = x^4 + 2x^3 + 5x^2 + 8x + 4$$
  0 negative real zero

POSITIVE	NEGATIVE
4	0
2	0
0	0

  b. Factors of $a_0 = 4$: $\pm 1, \pm 2, \pm 4$
    Factors of $a_n = 1$: ± 1
    Possible rational zeros: $\pm 1, \pm 2, \pm 4$
  c. $P(1) = 0 \qquad x = 1$ (multiplicity 2)
  d. $P(x) = (x - 1)^2(x^2 + 4)$

49. $P(x) = x^6 + 12x^4 + 23x^2 - 36$
  a. 1 positive real zero
$$P(-x) = x^6 + 12x^4 + 23x^2 - 36$$
  1 negative real zero

POSITIVE	NEGATIVE
1	1

  b. Factors of $a_0 = -36$: $\pm 1, \pm 2, \pm 3, \pm 4, \pm 6, \pm 9, \pm 12,$
                   $\pm 18, \pm 36$
    Factors of $a_n = 1$: ± 1
    Possible rational zeros: $\pm 1, \pm 2, \pm 3, \pm 4, \pm 6, \pm 9, \pm 12,$
                       $\pm 18, \pm 36$
  c. $P(1) = 0 \qquad P(-1) = 0$
  d. $P(x) = (x + 1)(x - 1)(x^2 + 4)(x^2 + 9)$

51. $P(x) = 4x^4 - 20x^3 + 37x^2 - 24x + 5$
  a. 4, 2, or 0 positive real zeros
$$P(-x) = 4x^4 + 20x^3 + 37x^2 + 24x + 5$$
  0 negative real zero

POSITIVE	NEGATIVE
4	0
2	0
0	0

  b. Factors of $a_0 = 5$: $\pm 1, \pm 5$
    Factors of $a_n = 4$: $\pm 1, \pm 2, \pm 4$
    Possible rational zeros: $\pm 1, \pm\frac{1}{2}, \pm\frac{1}{4}, \pm 5, \pm\frac{5}{2}, \pm\frac{5}{4}$
  c. $P\left(\frac{1}{2}\right) = 0 \qquad \frac{1}{2}$ has multiplicity 2
  d. $P(x) = (2x - 1)^2(x^2 - 4x + 5)$

53.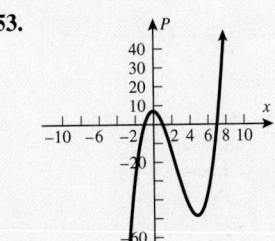

55.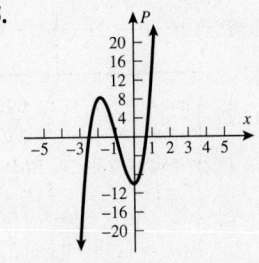

57. $x \approx 1.34$ **59.** $x \approx 0.22$ **61.** $x \approx -0.43$

63. $x \approx 2.88$ **65.** 6×8 in.   **67.** 30 cows

69. Yes, you can arrive at five negative zeros, but there may be 3 or 1.

POSITIVE	NEGATIVE
0	5
0	3
0	1

71. true   **73.** false   **75.** false

77. $x = b$ and $x = c$ **79.** $x = a, x = c$, and $x = -c$

81. $x = 2$ **83. a.** $-\frac{3}{4}, \frac{2}{3}$

b. $(3x - 2)(4x + 3)(x^2 + 2x + 5)$

85. $-2, 1, 5$ **87.** $\pm\sqrt{2}, -\frac{1}{2}, 3$

$(-2, 1) \cup (5, \infty)$ $\left(-\sqrt{2}, -\frac{1}{2}\right) \cup \left(\sqrt{2}, 3\right)$

Section 2.5

1. $x = \pm 2i; P(x) = (x - 2i)(x + 2i)$

3. $x = 1 \pm i; P(x) = [x - (1 + i)][x - (1 - i)]$

5. $x = \pm 2, \pm 2i; P(x) = (x - 2)(x + 2)(x - 2i)(x + 2i)$

7. $x = \pm\sqrt{5}, \pm i\sqrt{5};$
$P(x) = (x - \sqrt{5})(x + \sqrt{5})(x - i\sqrt{5})(x + i\sqrt{5})$

9. $-i$ **11.** $-2i, 3 + i$

13. $1 + 3i, 2 - 5i$ **15.** $i, 1 + i$ (multiplicity 2)

17. $2i, -3, 5; P(x) = (x - 5)(x + 3)(x - 2i)(x + 2i)$

19. $-i, 1, 3; P(x) = (x - 3)(x - 1)(x - i)(x + i)$

21. $3i, 1$ (multiplicity 2); $P(x) = (x - 1)^2(x - 3i)(x + 3i)$

23. $1 - i, -1 \pm 2\sqrt{2};$
$P(x) = [x - (1 - i)][x - (1 + i)]\left[x - (-1 - 2\sqrt{2})\right]\left[x - (-1 + 2\sqrt{2})\right]$

25. $3 + i, \pm 2; P(x) = (x - 2)(x + 2)[x - (3 + i)][x - (3 - i)]$

27. $1, 4, 2 + i; P(x) = (x - 4)(x - 1)[x - (2 - i)][x - (2 + i)]$

29. $P(x) = (x - 1)(x - 3i)(x + 3i)$

31. $P(x) = (x - 5)(x - i)(x + i)$

33. $P(x) = (x + 1)(x - 2i)(x + 2i)$

35. $P(x) = (x - 3)\left[x - (-1 - i\sqrt{5})\right]\left[x - (-1 + i\sqrt{5})\right]$

37. $P(x) = (x - 5)(x + 3)(x - 2i)(x + 2i)$

39. $P(x) = (x - 5)(x + 1)(x - 2i)(x + 2i)$

41. $P(x) = (x - 1)(x - 2)[x - (2 - 3i)][x - (2 + 3i)]$

43. $P(x) = -(x - 2)(x + 1)[x - (-2 - i)][x - (-2 + i)]$

45. $P(x) = (x - 1)^2(x - 2i)(x + 2i)$

47. $P(x) = (x - 1)(x + 1)(x - 2i)(x + 2i)(x - 3i)(x + 3i)$

49. $P(x) = (2x - 1)^2[x - (2 - i)][x - (2 + i)]$

51. $P(x) = (3x - 2)(x - 1)(x + 1)(x - 2i)(x + 2i)$

53. Yes—since there are no real zeros, this function is either always positive or always negative; the end behavior is similar to an even power function with a positive leading coefficient (rising in both ends). We expect profit to eventually rise (without bound!).

55. No—since there is only one real zero (and it corresponds to producing a positive number of units) then that is the "break-even point;" the end behavior is similar to an odd power function with a negative leading coefficient (falling to the right). We expect profit to fall (without bound!) after the break-even point.

57. Just because 1 is a zero does not mean -1 is a zero (holds only for complex conjugates). Should have used synthetic division with 1.

59. false   **61.** true

63. No—if there is an imaginary zero, then its conjugate is also a zero. Therefore, all imaginary zeros can only correspond to an even-degree polynomial.

65. $P(x) = x^6 + 3b^2x^4 + 3b^4x^2 + b^6$

67. $P(x) = x^6 + (2a^2 + b^2)x^4 + a^2(a^2 + 2b^2)x^2 + a^4b^2$

69.

REAL ZEROS	COMPLEX ZEROS
4	0
2	2
0	4

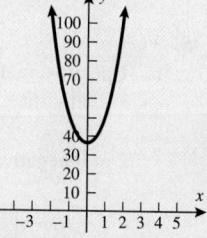

There are no real zeros; therefore, there are 4 complex zeros.

71. $\frac{3}{5}, \pm i, \pm 2i; (3 - 5x)(x + i)(x - i)(x + 2i)(x - 2i)$

73. a. $f(x) = (x + 1)(x - i)(x + i)$

b. $f(x) = (x + 1)(x^2 + 1)$

75. a. $f(x) = (x - i)(x + i)(x - 2i)(x + 2i)$

b. $f(x) = (x^2 + 1)(x^2 + 4)$

Section 2.6

1. $(-\infty, -4) \cup (-4, 3) \cup (3, \infty)$

3. $(-\infty, -2) \cup (-2, 2) \cup (2, \infty)$

5. $(-\infty, \infty)$

7. $(-\infty, -2) \cup (-2, 3) \cup (3, \infty)$

9. HA: $y = 0$ VA: $x = -2$

11. HA: none   VA: $x = -5$

13. HA: none   VA: $x = \frac{1}{2}$ and $x = -\frac{4}{3}$

15. HA: $y = \frac{1}{3}$ VA: none

17. $y = x + 6$ **19.** $y = 2x + 24$ **21.** $y = 4x + \frac{11}{2}$

23. b   **25.** a   **27.** e

29. 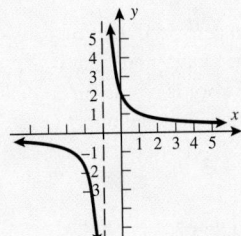   **31.**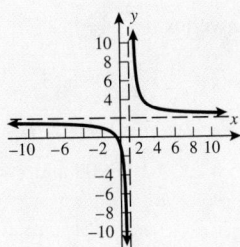

33. 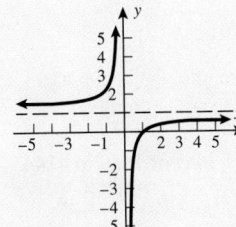   **35.**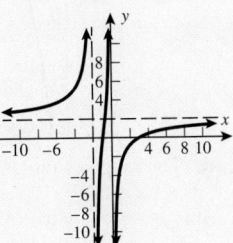

37. 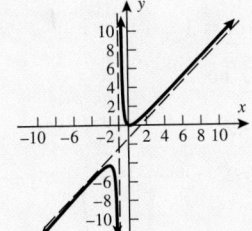   **39.**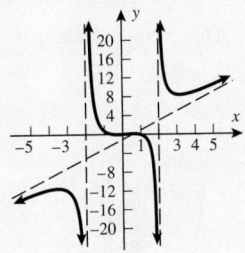

41. 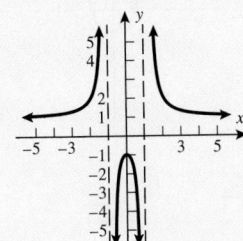   **43.**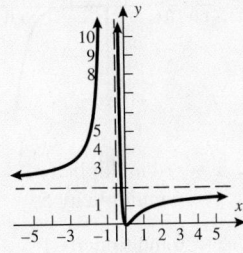

45. 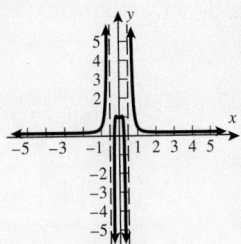   **47.**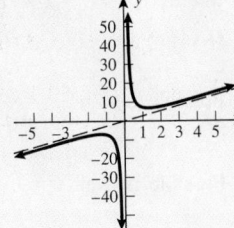

49. 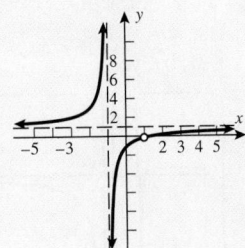   **51.**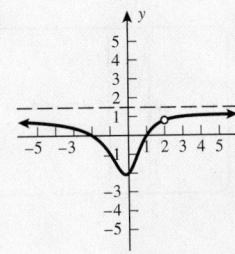

53. 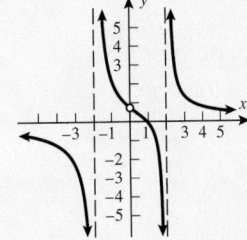   **55.**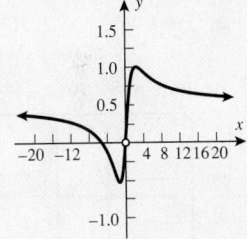

57. **a.** x-intercept: $(2, 0)$; y-intercept: $(0, 0.5)$

   b. HA: $y = 0$ VA: $x = -1, x = 4$

   c. $f(x) = \dfrac{x - 2}{(x + 1)(x - 4)}$

59. **a.** x-intercept: $(0, 0)$; y-intercept: $(0, 0)$

   b. HA: $y = -3$ VA: $x = -4, x = 4$

   c. $f(x) = \dfrac{-3x^2}{(x + 4)(x - 4)}$

61. **a.** 4500

   b. 6 months

   c. The number of infected people stabilizes around 9500.

63. **a.** $C(1) \approx 0.0198$

   b. $C(60) \approx 0.0324$

   c. $C(300) \approx 0.007 \approx 0$

   d. $y = 0$, after several days $C(t) \approx 0$.

65. **a.** $N(0) = 52$ wpm   **b.** $N(12) \approx 107$ wpm

   c. $N(36) \approx 120$ wpm   **d.** 130 wpm

67. 10 ounces   **69.** $\dfrac{2w^2 + 1000}{w}$

71. There is a common factor in the numerator and denominator. There is a hole at $x = 1$.

73. The horizontal asymptote was found incorrectly. The correct horizontal asymptote is $y = -1$.

75. true **77.** false

79. HA: $y = 1$ VA: $x = c$ and $x = -d$

81. Possible answer: $y = \dfrac{4x^2}{(x+3)(x-1)}$

83. Possible answer: $y = \dfrac{x^3 + 1}{x^2 + 1}$

85. $x = -2$; yes

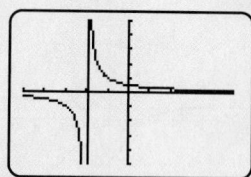

87. VA: $x = 0$, $x = -\frac{1}{3}$
HA: $y = 0$
$\left(-\frac{2}{5}, 0\right)$; yes

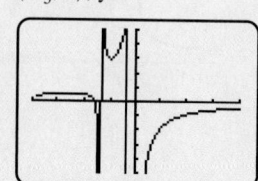

89. a. For f, VA: $x = 3$, HA: $y = 0$; for g, VA: $x = 3$, HA: $y = 2$; for h, VA: $x = 3$, HA: $y = -3$

b. As $x \to -\infty$ or $x \to \infty$, $f(x) \to 0$ and $g(x) \to 2$

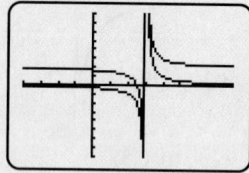

c. As $x \to -\infty$ or $x \to \infty$, $g(x) \to 2$ and $h(x) \to -3$

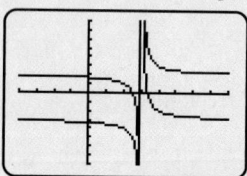

d. $g(x) = \dfrac{2x-5}{x-3}$, $h(x) = \dfrac{-3x+10}{x-3}$. Yes, if the degree of the numerator is the same as the degree of the denominator, then the horizontal asymptote is the ratio of the leading coefficients for both g and h.

91. VA: $x = -2$, $x = -1$, $x = 5$

93. VA: $x = -\frac{1}{2}$, $x = \frac{2}{3}$

Review Exercises

1. b   **3.** a

5.

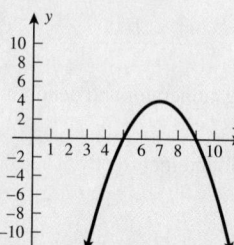

7.

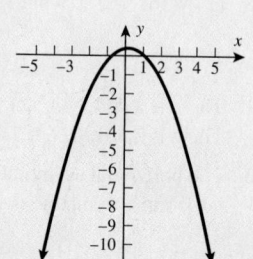

9. $f(x) = \left(x - \frac{3}{2}\right)^2 - \frac{49}{4}$

11. $f(x) = 4(x+1)^2 - 11$

13.

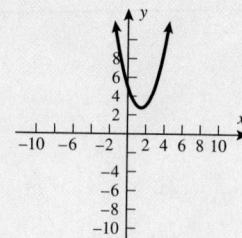

15.

17. vertex: $\left(\frac{5}{26}, \frac{599}{52}\right)$

19. vertex: $\left(-\frac{2}{15}, \frac{451}{125}\right)$

21. $y = \frac{1}{9}(x+2)^2 + 3$

23. $y = 5.6(x - 2.7)^2 + 3.4$

25. a. $P(x) = -2x^2 + \frac{35}{3}x - 14$

b. $x \approx 1.68909$ and $x \approx 4.1442433$

c.

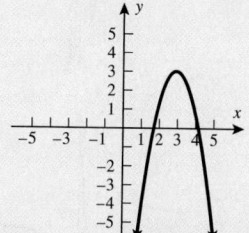

d. The range of units is $(1.689, 4.144)$ or 1689 to 4144.

27. $A(x) = -\frac{1}{2}x^2 + x + 4$
$= -\frac{1}{2}(x-1)^2 + \frac{9}{2}$

$x = 1$ is a maximum (4.5 square units).

Dimensions: base $= 3$ height $= 3$

29. yes, 6   **31.** no

33. d   **35.** a

37.

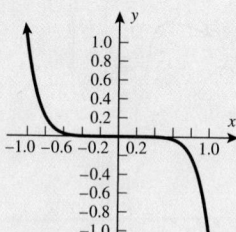

39.

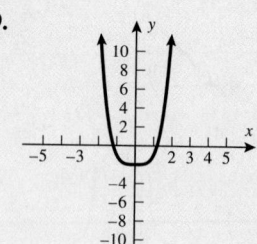

41. $x = -4$ multiplicity 2
$x = 6$ multiplicity 5

43. $x = 0$ multiplicity 1
$x = 3$ multiplicity 1
$x = -3$ multiplicity 1
$x = 2$ multiplicity 1
$x = -2$ multiplicity 1

45. $f(x) = x(x+3)(x-4)$

47. $f(x) = x(5x + 2)(4x - 3)$

49. $f(x) = (x + 2)^2(x - 3)^2$
$\quad = x^4 - 2x^3 - 11x^2 + 12x + 36$

51. $f(x) = (x - 7)(x + 2)$

a. 7 multiplicity 1
   -2 multiplicity 1
b. The graph crosses at
   $(-2, 0)$ and $(7, 0)$.
c. $(0, -14)$
d. rises right and left
e.

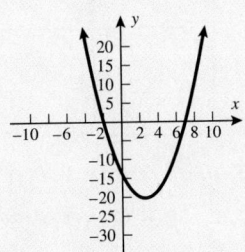

53. $f(x) = 6x^7 + 3x^5 - x^2 + x - 4$

a. 0.8748 multiplicity 1
b. The graph crosses at
   $(0.8748, 0)$.
c. $(0, -4)$
d. falls left; rises right
e.

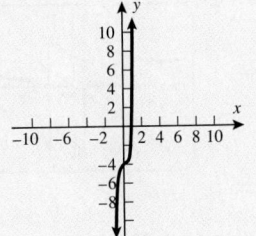

55. a.

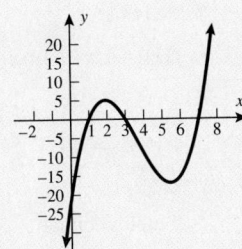

b. The real zeros occur at
   $x = 1$, $x = 3$, and $x = 7$.
c. Between 1 and 3 hours
   or more than 7 hours is
   financially beneficial.

57. $Q(x) = x + 4 \qquad r(x) = 2$

59. $Q(x) = 2x^3 - 4x^2 - 2x - \frac{7}{2} \qquad r(x) = -23$

61. $Q(x) = x^3 + 2x^2 + x - 4 \qquad r(x) = 0$

63. $Q(x) = x^5 - 8x^4 + 64x^3 - 512x^2 + 4096x - 32{,}768$
$\quad r(x) = 262{,}080$

65. $Q(x) = x + 3 \qquad r(x) = -4x - 8$

67. $Q(x) = x^2 - 5x + 7 \qquad r(x) = -15$

69. length $= 3x^3 + 2x^2 - x + 4 \qquad r(x) = 0$

71. $f(-2) = -207$ **73.** $g(1) = 0$

75. no   **77.** yes

79. $P(x) = x(x + 2)(x - 4)^2$ **81.** $P(x) = x^2(x + 3)(x - 2)^2$

83.

POSITIVE (REAL)	NEGATIVE (REAL)
1	1

85.

POSITIVE (REAL)	NEGATIVE (REAL)
5	2
5	0
3	2
3	0
1	2
1	0

87. The possible rational zeros are: $\pm 1, \pm 2, \pm 3, \pm 6$.

89. The possible rational zeros are: $\pm\frac{1}{2}, \pm 1, \pm 2, \pm 4, \pm 8, \pm 16,$
$\pm 32, \pm 64$.

91. The possible rational zeros are $\pm\frac{1}{2}, \pm 1$. The rational zero is $\frac{1}{2}$.

93. The possible rational zeros are $\pm 1, \pm 2, \pm 4, \pm 8, \pm 16$. The
rational zeros are $1, 2, -2, 4$.

95. a.

POSITIVE (REAL)	NEGATIVE (REAL)
1	0

b. $\pm 1, \pm 5$
c. -1 is a lower bound, 5 is an upper bound.
d. There are no rational zeros.
e. not possible
f.

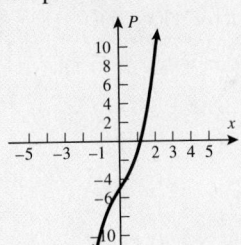

97. a.

POSITIVE (REAL)	NEGATIVE (REAL)
3	0
1	0

b. $\pm 1, \pm 2, \pm 3, \pm 4, \pm 6, \pm 12$
c. -1 lower bound; 12 upper bound
d. 1, 2, and 6 are zeros.
e. $P(x) = (x - 1)(x - 2)(x - 6)$
f.

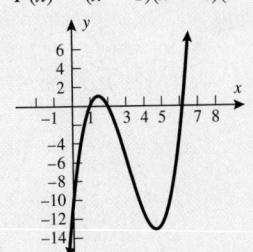

99. a.

Positive (Real)	Negative (Real)
2	2
0	2
2	0
0	0

b. $\pm 1, \pm 2, \pm 3, \pm 4, \pm 6, \pm 8 \pm 12, \pm 24$

c. -3 lower bound; 8 upper bound

d. $-2, -1, 2,$ and 6 are zeros.

e. $P(x) = (x + 1)(x + 2)(x - 2)(x - 6)$

f.

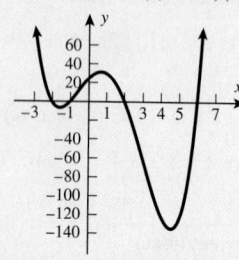

101. $P(x) = (x - 5i)(x + 5i)$

103. $P(x) = [x - (1 - 2i)][x - (1 + 2i)]$

105. $2i$ and $3 - i$

107. $-i$ and $2 + i$ (multiplicity 2)

109. $4, -1, -i$ $P(x) = (x - 4)(x + 1)(x - i)(x + i)$

111. $3i, 1 \pm i$ $P(x) = (x + 3i)(x - 3i)[x - (1 - i)][x - (1 + i)]$

113. $P(x) = (x - 3)(x + 3)(x - 3i)(x + 3i)$

115. $P(x) = (x - 1)(x - 2i)(x + 2i)$

117. VA: $x = -2$ and HA: $y = -1$

119. VA: $x = -1$, HA: none, slant: $y = 4x - 4$

121. VA: none, HA: $y = 2$

123.

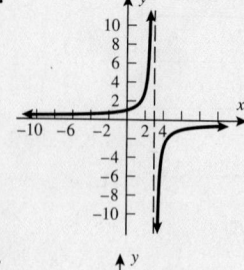

125.

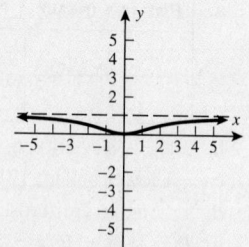

127.

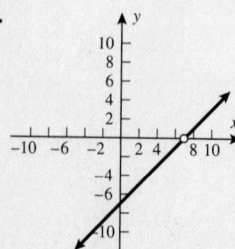

129. a. $(480, -1211)$
b. $(0, -59)$
c. $(-12.14, 0)$ and $(972.14, 0)$
d. $x = 480$

131. x-intercepts: $(-1, 0), (0.4, 0)(2.8, 0)$; zeros at $-1, 0.4, 2.8,$ each with multiplicity 1.

133. a linear function

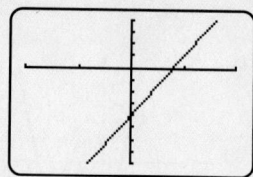

135. a. $-2, 3, 4$ **b.** $P(x) = (x + 2)^2(x - 3)(x - 4)$

137. $\frac{7}{2}, -2 \pm 3i$; $P(x) = (2x - 7)(x + 2 + 3i)(x + 2 - 3i)$

139. a. yes, it is one-to-one   **b.** $f^{-1}(x) = \dfrac{x + 3}{2 - x}$

c.

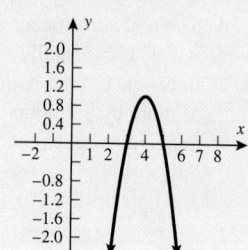

Practice Test

1.

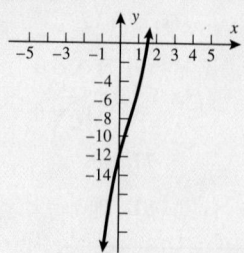

3. vertex $\left(3, \frac{1}{2}\right)$

5. $f(x) = x(x - 2)^3(x - 1)^2$

7. $Q(x) = -2x^2 - 2x - \frac{11}{2}$

$r(x) = -\frac{19}{2}x + \frac{7}{2}$

9. Yes, $x - 3$ is a factor of the polynomial.

11. $P(x) = (x - 7)(x + 2)(x - 1)$

13. Yes. Complex zeros.

15. $P(x) = 3x^4 - 7x^2 + 3x + 12$
Factors of $a_0 = 12$: $\pm 1, \pm 2, \pm 3, \pm 4, \pm 6, \pm 12$
Factors of $a_n = 3$: $\pm 1, \pm 3$
Possible rational zeros: $\pm 1, \pm 2, \pm 3, \pm 4, \pm 6, \pm 12, \pm \frac{1}{3}, \pm \frac{2}{3}, \pm \frac{4}{3}$

17. $\frac{3}{2}, \pm 2i$

19. Given the points (0, 300), (2, 285), (10, 315) (52, 300), you can have a polynomial of degree 3 because there are 2 turning points.

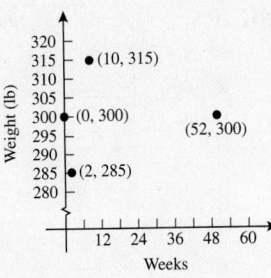

21. Given the points (1970, 0.08), (1988, 0.13), (2002, 0.04), (2005, 0.06), the lowest degree polynomial that can be represented is a third-degree polynomial.

23. a. (0, 0) is both x-intercept and y-intercept.
  b. VA: $x = \pm 2$
  c. HA: $y = 0$
  d. none
  e.

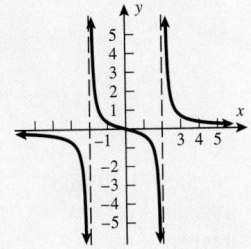

25. a. x-intercept: (3, 0); y-intercept: $\left(0, \frac{3}{8}\right)$
  b. VA: $x = -2; x = 4$
  c. HA: $y = 0$
  d. none
  e.

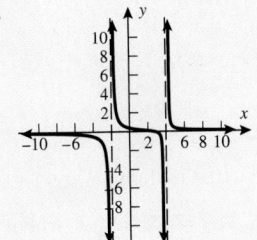

27. a. $x^2 - 3x - 7.99$
  b. $y = (x - 1.5)^2 - 10.24$
  c. $(-1.7, 0)$ and $(4.7, 0)$
  d. Yes, they agree.

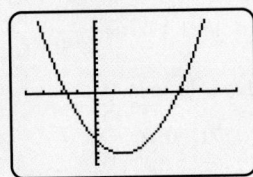

Cumulative Test

1. $f(2) = \frac{15}{2}, f(-1) = -5, f(1 + h) = 4 + 4h - \dfrac{1}{\sqrt{h + 3}},$

$f(-x) = -4x - \dfrac{1}{\sqrt{2 - x}}$

3. $f(-3) = \frac{7}{2}, f(0) = -\frac{5}{2}, f(1)$ undefined, $f(4) = -\frac{7}{18}$

5. $\dfrac{1}{\sqrt{x + h} + \sqrt{x}} + \dfrac{2x + h}{x^2(x + h)^2}$

7. domain: $(-\infty, 10) \cup (10, \infty)$
  range: $[0, \infty)$
  increasing: $(3, 8)$
  decreasing: $(-\infty, 3) \cup (10, \infty)$
  constant: $(8, 10)$

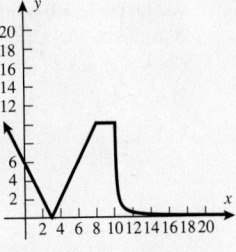

9. $-\frac{1}{28}$

11. neither

13. Right 1 unit; up 3 units

15. $g(f(-1)) = 0$

17. $f(x) = (x + 2)^2 + 3$

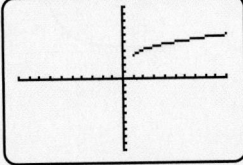

19. $Q(x) = 4x^2 + 4x + 1, r(x) = -8$

21. Factors of $a_0 = 6$: $\pm 1, \pm 2, \pm 3, \pm 6$
  Factors of $a_n = 12$: $\pm 1, \pm 2, \pm 3, \pm 4, \pm 6, \pm 12$
  Possible rational zeros: $\pm 1, \pm 2, \pm 3, \pm 6, \pm \frac{1}{2}, \pm \frac{3}{2}, \pm \frac{1}{3}, \pm \frac{2}{3},$
  $\pm \frac{1}{4}, \pm \frac{3}{4}, \pm \frac{1}{6}, \pm \frac{1}{12}$

  Testing the zeros: $P(-2) = 0, P\left(-\frac{3}{4}\right) = 0, P\left(\frac{1}{3}\right) = 0$

23. $P(x) = (x + 1)(x - 2)(x - 4)$

25. HA: $y = 0$; VA: $x = -2$ and $x = 2$

27. x-intercepts $(-1, 0)$; vertical asymptotes: $x = 0, x = \frac{3}{2}$; horizontal asymptote: $y = 0$; Yes

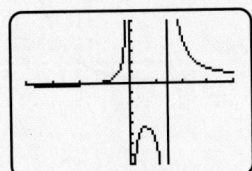

CHAPTER 3

Section 3.1

1. $\frac{1}{25}$ **3.** 4 **5.** 27 **7.** 9.7385 **9.** 7.3891

11. 0.0432 **13.** 27 **15.** 16 **17.** 4

19. 19.81 **21.** f **23.** e **25.** b

27. $(0, 1), \left(-1, \frac{1}{6}\right),$ and $(1, 6)$;
  domain: $(-\infty, \infty)$
  range: $(0, \infty)$
  HA: $y = 0$

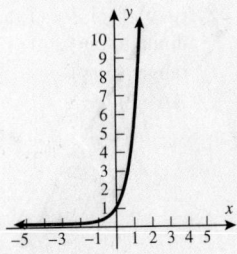

29. $(0, 1)$, $(-1, 10)$, and $(1, 0.1)$
  domain: $(-\infty, \infty)$
  range: $(0, \infty)$
  HA: $y = 0$

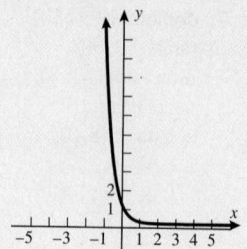

31. $(0, 1)$, $\left(-1, \dfrac{1}{e}\right)$, $(1, e)$
  domain: $(-\infty, \infty)$
  range: $(0, \infty)$
  HA: $y = 0$

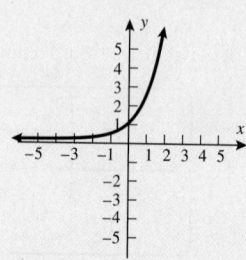

33. $(0, 1)$, $(-1, e)$, and $\left(1, \dfrac{1}{e}\right)$
  domain: $(-\infty, \infty)$
  range: $(0, \infty)$
  HA: $y = 0$

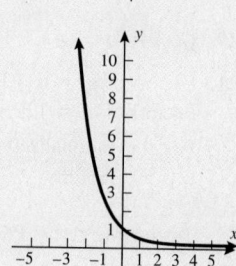

35. $(0, 0)$, $(1, 1)$, and $(2, 3)$
  domain: $(-\infty, \infty)$
  range: $(-1, \infty)$
  HA: $y = -1$

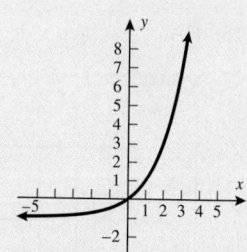

37. $(0, 1)$, $\left(-1, 2 - \dfrac{1}{e}\right)$, and
  $(1, 2 - e)$
  domain: $(-\infty, \infty)$
  range: $(-\infty, 2)$
  HA: $y = 2$

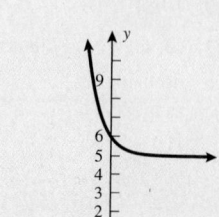

39. $(0, 6)$, $(-1, 9)$, $(1, 5.25)$
  domain: $(-\infty, \infty)$
  range: $(5, \infty)$
  HA: $y = 5$

41. $(0, e - 4)$, $(-1, -3)$, and
  $\left(1, e^2 - 4\right)$
  domain: $(-\infty, \infty)$
  range: $(-4, \infty)$
  HA: $y = -4$

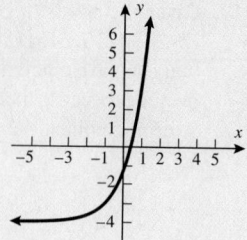

43. $(0, 3)$, $\left(-1, \dfrac{3}{\sqrt{e}}\right)$, and $\left(1, 3\sqrt{e}\right)$
  domain: $(-\infty, \infty)$
  range: $(0, \infty)$
  HA: $y = 0$

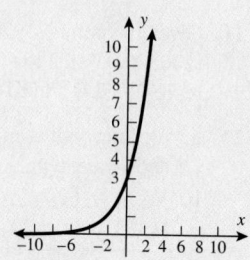

45. $(0, 5)$, $(-1, 9)$, and $(1, 3)$
  domain: $(-\infty, \infty)$
  range: $(1, \infty)$
  HA: $y = 1$

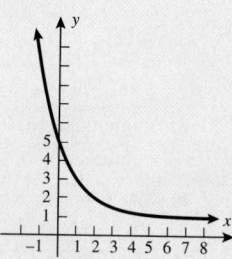

47. 10.4 million

49. $f(t) = 1500 \cdot 2^{t/5}$, where t is the number of years since it was purchased. Thirty years later: \$96,000/acre.

51. 168 milligrams   **53.** 2 mg

55. \$3,031.46   **57.** \$3,448.42

59. \$13,011.03   **61.** \$4,319.55

63. \$13,979.42   **65.** $4^{-1/2} = \dfrac{1}{4^{1/2}} = \dfrac{1}{2}$

67. 2.5% needs to be converted to a decimal, 0.025.

69. false   **71.** true

73.

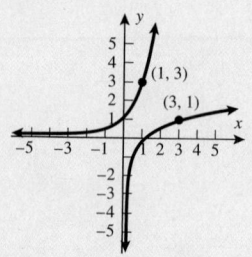

75.

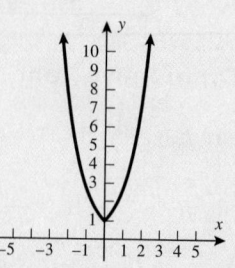

77. $(0, be - a)$ and $y = -a$

79. domain: $(-\infty, \infty)$

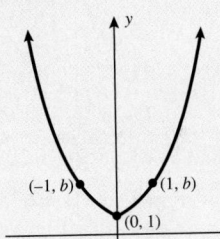

$(-1, b)$ $(1, b)$ $(0, 1)$

81.

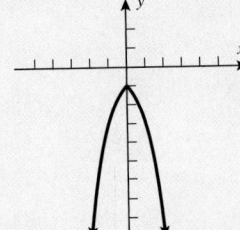

83. HA: $y = e$

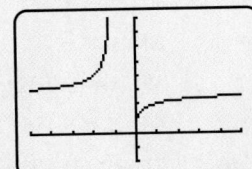

85. These graphs are very similar.

87. HA: $y = e$, $y = e^2$, $y = e^4$

89. odd function

91. $\cosh^2 x - \sinh^2 x = \left(\dfrac{e^x + e^{-x}}{2}\right)^2 - \left(\dfrac{e^x - e^{-x}}{2}\right)^2$

$\qquad = \dfrac{e^{2x} + 2 + e^{-2x}}{4} - \dfrac{e^{2x} - 2 + e^{-2x}}{4}$

$\qquad = \dfrac{4}{4} = 1$

Section 3.2

1. $81^{1/4} = 3$ **3.** $2^{-5} = \frac{1}{32}$ **5.** $10^{-2} = 0.01$

7. $10^4 = 10,000$ **9.** $\left(\frac{1}{4}\right)^{-3} = 64$ **11.** $e^{-1} = \frac{1}{e}$

13. $e^0 = 1$ **15.** $e^x = 5$ **17.** $x^z = y$

19. $y^x = x + y$ **21.** $\log 0.00001 = -5$ **23.** $\log_5 78,125 = 7$

25. $\log_{225} 15 = \frac{1}{2}$ **27.** $\log_{2/5}\left(\frac{8}{125}\right) = 3$ **29.** $\log_{1/27} 3 = -\frac{1}{3}$

31. $x = \ln 6$ **33.** $\log_y x = z, y > 0$ **35.** 0

37. 5   **39.** 7   **41.** -6

43. undefined   **45.** undefined   **47.** 1.46

49. 5.94   **51.** undefined   **53.** -8.11

55. $(-5, \infty)$ **57.** $\left(-\infty, \frac{5}{2}\right)$ **59.** $\left(-\infty, \frac{7}{2}\right)$

61. $(-\infty, 0) \cup (0, \infty)$ **63.** domain: $\mathbb{R}$

65. $(-2, 5)$ **67.** b

69. c   **71.** d

73. domain: $(1, \infty)$
range: $(-\infty, \infty)$

75. domain: $(-2, \infty)$
range: $(-\infty, \infty)$

77. domain: $(-2, \infty)$
range: $(-\infty, \infty)$

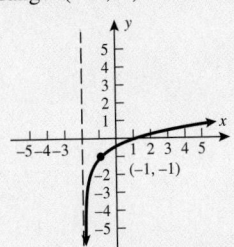

$(-1, -1)$

79. domain: $(0, \infty)$
range: $(-\infty, \infty)$

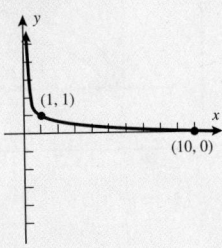

$(1, 1)$ $(10, 0)$

81. domain: $(-4, \infty)$
range: $(-\infty, \infty)$

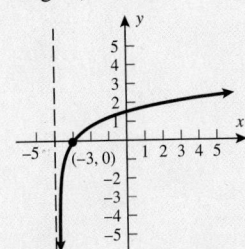

$(-3, 0)$

83. domain: $(0, \infty)$
range: $(-\infty, \infty)$

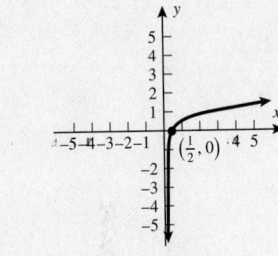

$\left(\frac{1}{2}, 0\right)$

85. 60 decibels   **87.** 117 decibels

89. 8.5 on the Richter scale   **91.** 6.6 on the Richter scale

93. 3.3 pH

95. normal rainwater: 5.6 pH; acid rain/tomato juice: 4.0 pH

97. 3.6 pH   **99.** 13,236 years old   **101.** 25 dB loss

103. $\log_2 4 = x$ is equivalent to $2^x = 4 \left(\text{not } x = 2^4\right)$.

105. The correction that needs to be made is

$x + 5 > 0$
$x > -5$
domain: $(-5, \infty)$

107. false   **109.** true

111. domain: (a, ∞)
range: $(-\infty, \infty)$
x-intercept: $(e^b + a, 0)$

113.

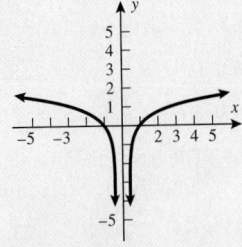

115. Symmetric with respect to $y = x$.

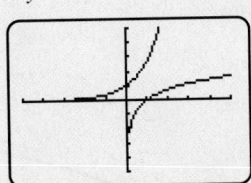

117. $(1, 0)$ and V.A. $x = 0$ are common.

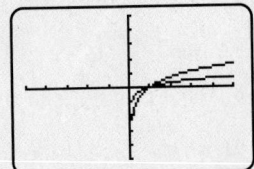

119. Graphs of $f(x)$ and $g(x)$ are the same with domain $(0, \infty)$.

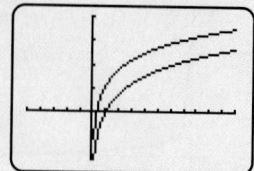

121. $f'(x) = e^x$

123. a. $f^{-1}(x) = \ln x$

   b. $(f^{-1})'(x) = \dfrac{1}{x}$

Section 3.3

1. 0

3. 1

5. 8

7. -3

9. $\dfrac{3}{2}$

11. 5

13. $x + 5$

15. 8

17. $\dfrac{1}{9}$

19. $\dfrac{7}{x^3}$

21. $3 \log_b x + 5 \log_b y$

23. $\frac{1}{2} \log_b x + \frac{1}{3} \log_b y$

25. $\frac{1}{3} \log_b r - \frac{1}{2} \log_b s$

27. $\log_b x - \log_b y - \log_b z$

29. $2 \log x + \frac{1}{2} \log(x + 5)$

31. $3 \ln x + 2 \ln(x - 2) - \frac{1}{2} \ln(x^2 + 5)$

33. $2 \log(x - 1) - \log(x - 3) - \log(x + 3)$

35. $\frac{1}{2} \ln(x + 5) - \frac{1}{2} \ln(x - 1)$

37. $\log_b(x^3 y^5)$

39. $\log_b \dfrac{u^5}{v^2}$

41. $\log_b x^{1/2} y^{2/3}$

43. $\log \dfrac{u^2}{v^3 z^2}$

45. $\ln \dfrac{x^2 - 1}{(x^2 + 3)^2}$

47. $\ln \dfrac{(x + 3)^{1/2}}{x(x + 2)^{1/3}}$

49. 1.2091

51. -2.3219

53. 1.6599

55. 2.0115

57. 3.7856

59. 110 decibels

61. 5.5 on the Richter scale $\left(\text{total energy: } 4.5 \times 10^{12} \text{ joules}\right)$

63. 0.0458

65. 16 times

67. $3 \log 5 - \log 5^2 = 3 \log 5 - 2 \log 5 = \log 5$

69. The bases are different, they must be the same in order to simplify the expression.

71. true

73. false

75. false

79. $6 \log_b x - 9 \log_b y + 15 \log_b z$

81. $\log_b \left(\dfrac{a^2}{b^3}\right)^{-3} = \log_b \dfrac{b^9}{a^6} = 9 \log_b b - 6 \log_b a = 9 - \dfrac{6}{\log_a b}$

83. yes

85. no

87. no

89. yes

91. $f'(x) = \dfrac{2}{x}$

93. $f'(x) = -\dfrac{2}{x}$

Section 3.4

1. $x = \pm 2$

3. $x = -4$

5. $x = -\dfrac{3}{2}$

7. $x = -1$

9. $x = 3$ or $x = 4$

11. $x = 0$ or $x = 6$

13. $x = 1$ or $x = 4$

15. $x \approx 1.918$

17. $x \approx 1.609$

19. $x \approx 13.863$

21. $x \approx 2.096$

23. $x \approx -0.303$

25. $x \approx 0.896$

27. $x \approx 0.223$

29. $x \approx \pm 2.282$

31. $x \approx -0.904$

33. $x = 0$

35. $x \approx 1.946$

37. $x = 0$

39. $x \approx 0.477$

41. $x = 40$

43. $x = \dfrac{9}{32}$

45. $x = \pm 3$

47. $x = 5$

49. $x = 6$

51. no solution

53. $x = \dfrac{25}{8}$

55. $x = -\dfrac{4}{5}$

57. $x = 47.5$

59. $x \approx \pm 7.321$

61. $x \approx -1.432$

63. $x \approx -1.25$

65. $x \approx 8.456$

67. $x = 0.3028$

69. $x \approx 3.646$

71. $t = 31.9$ years

73. $t \approx 19.74$ years

75. 3.16×10^{15} joules

77. 1 W/m^2

79. $t \approx 4.61$ hours

81. 15.89 years

83. 6.2

85. The correction that needs to be made is that the 4 should have been divided first, giving the following:

$$e^x = \frac{9}{4} \qquad x = \ln\left(\frac{9}{4}\right)$$

87. The correction that needs to be made is that $x = -5$ should be removed from the solutions. The domain of the logs cannot include a negative solution.

89. true

91. false

93. false

95. $x = \dfrac{1 + \sqrt{1 + 4b^2}}{2}$

97. $t = -5 \ln\left(\dfrac{3000 - y}{2y}\right)$

99. $f^{-1}(x) = \ln\left(x + \sqrt{x^2 - 1}\right) \qquad x \geq 1$

101.

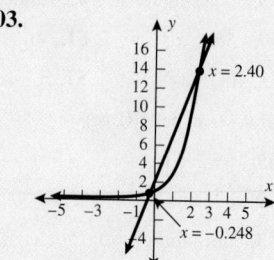

103.

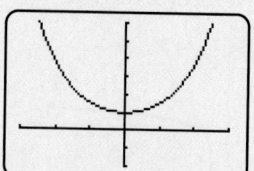

105. $(-\infty, \infty)$, symmetric with respect to the y-axis

107. $\ln\left(x + \sqrt{x^2 + 1}\right)$

109. $\ln y = x \ln 2$

Section 3.5

1. c; iv   **3.** a; iii   **5.** f; i

7. 94 million   **9.** 5.5 years (middle of 2008)

11. 799.6 million cell phone subscribers

13. $455,000   **15.** 332 million colonies of phytoplankton

17. 1.19 million   **19.** 13.53 ml   **21.** 7575 years

23. 131,158,556 years   **25.** 105 °F

27. 3.8 hours before 7 A.M.   **29.** $19,100

31. **a.** 84,520   **b.** 100,000   **c.** 100,000

33. 29,551 cases   **35.** 1.89 years   **37.** $r = 0$ (on-axis)

39. **a.**

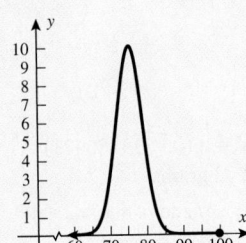

    b. 75

    c. 4

    d. 4

41. **a.** 18 years   **b.** 10 years

43. r is the decimal form of a percentage ($r = 0.07$).

45. true   **47.** false   **49.** more time

51. 10.9 days   **53.** $k_1 = k_2 + \ln\left(\dfrac{2 + c}{c}\right)$

55. **a.** For the same periodic payment, it will take Wing Shan fewer years to pay off the loan if she can afford to pay biweekly.

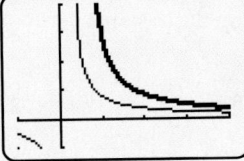

    b. 11.58 years

    c. 10.33 years

    d. 8.54 years, 7.69 years

57. $Pe^{kx} \cdot \dfrac{e^{kh} - 1}{h}$ **59.** $e^x + 1$

Review Exercises

1. 17,559.94   **3.** 5.52   **5.** 24.53   **7.** 5.89

9. 73.52   **11.** 6.25   **13.** b   **15.** c

17. The y-intercept is at $(0, -1)$. The horizontal asymptote is $y = 0$.

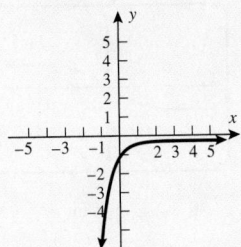

19. The y-intercept is at $(0, 2)$. The horizontal asymptote is $y = 1$.

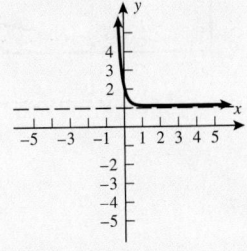

21. The y-intercept is at $(0, 1)$. The horizontal asymptote is $y = 0$.

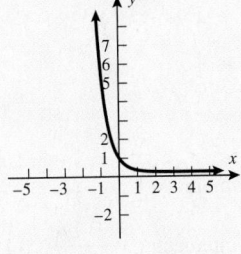

23. The y-intercept is at $(0, 3.2)$. The horizontal asymptote is $y = 0$.

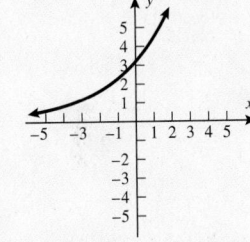

25. $6144.68   **27.** $23,080.29   **29.** $4^3 = 64$

31. $10^{-2} = \frac{1}{100}$ **33.** $\log_6 216 = 3$ **35.** $\log_{2/13} \frac{4}{169} = 2$

37. 0   **39.** -4 **41.** 1.51   **43.** -2.08

45. $(-2, \infty)$ **47.** $(-\infty, \infty)$ **49.** b   **51.** d

53.

55.

57. pH = 6.5   **59.** 50 dB   **61.** 1

63. 6   **65.** $a \log_c x + b \log_c y$

67. $\log_j r + \log_j s - 3 \log_j t$ **69.** $\frac{1}{2} \log a - \frac{3}{2} \log b - \frac{2}{5} \log c$

71. 0.5283   **73.** 0.2939   **75.** $x = -4$ **77.** $x = \frac{4}{3}$

79. $x = -6$ **81.** -0.218 **83.** no solution

85. $x = 0$ **87.** $x = \frac{100}{3}$ **89.** $x = 128\sqrt{2}$

91. $x \approx \pm 3.004$ **93.** $x \approx 0.449$ **95.** $28,536.88

97. $t = 16.6$ years   **99.** 3.23 million   **101.** 6250 bacteria

103. 56 years   **105.** $15.76 \approx 16$ fish

107. 343 mice   **109.** $e^{\sqrt{2}} \approx 4.11$ **111.** $(2.376, 2.071)$

113. $(1, \infty)$

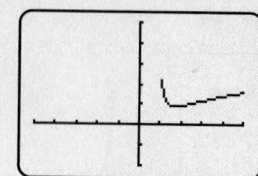

115. $(-\infty, \infty)$, symmetric with respect to the origin, HA at $y = -1$ and $y = 1$.

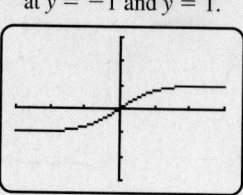

117. a. $y = 4e^{-0.038508\,t} = 4(0.9622)^t$ **b.** $y = 4(0.9622)^t$

    c. Yes, they are the same.

Practice Test

1. x^3 **3.** -4 **5.** $x \approx \pm 2.177$

7. $x \approx 7.04$ **9.** $x = 4 + e^2 \approx 11.389$

11. $x = e^{e^1} \approx 15.154$ **13.** $x = 9$

15. $x = \dfrac{-3 + \sqrt{9 + 4e}}{2} \approx 0.729$

17. $x \approx -0.693$ **19.** domain: $(-1, 0) \cup (1, \infty)$

21. y-intercept: $(0, 2)$ **23.** x-intercept: $\left(\dfrac{3 + 1/e}{2}, 0\right)$

    HA: $y = 1$ VA: $x = \dfrac{3}{2}$

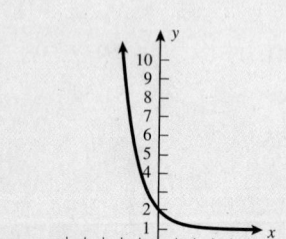
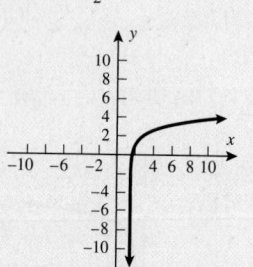

25. \$8,051.62 **27.** 90 dB

29. $7.9 \times 10^{11} < E < 2.5 \times 10^{13}$ joules

31. 7800 bacteria **33.** $2.75 \approx 3$ days

35. $(-\infty, \infty)$, symmetric with respect to the origin, no asymptotes

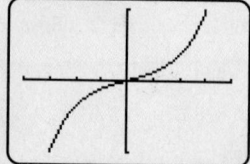

Cumulative Test

1. domain: $(-\infty, -3) \cup (3, \infty)$ **3.** $f(x) = \dfrac{1 - x^2}{1 + x^2}, g(x) = e^x$

    range: $(0, \infty)$

5. $f(x) = -\dfrac{4}{9}(x + 2)^2 + 3$ **7.** $x = \ln 9$

9. a. 1 **b.** 5 **c.** 1

    d. undefined **e.** domain: $(-2, \infty)$; range: $(0, \infty)$

    f. Increasing: $(4, \infty)$; decreasing: $(0, 4)$; constant: $(-2, 0)$

11. yes, one-to-one **13.** $(1, -1)$

15. $Q(x) = -x^3 + 3x - 5, r(x) = 0$

17. VA: $x = 3$; slant asymptote: $y = x + 3$

19.

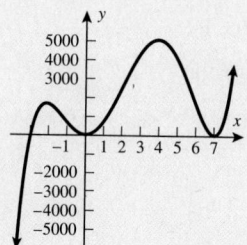

21. 5 **23.** $x = \dfrac{1}{2}$

25. 8.62 years **27. a.** $N = 6(0.9755486421)^t$

                        b. 2.72 grams

CHAPTER 4

Section 4.1

1. a. 72° **b.** 162° **3. a.** 48° **b.** 138° **5. a.** 1° **b.** 91°

7. $\dfrac{2}{11}$ or ≈ 0.18 **9.** $\dfrac{1}{50}$ or 0.02 **11.** $\dfrac{1}{8}$ or 0.125

13. $\dfrac{\pi}{6}$ **15.** $\dfrac{\pi}{4}$ **17.** $\dfrac{7\pi}{4}$

19. $\dfrac{5\pi}{12}$ **21.** $\dfrac{17\pi}{18}$ **23.** $\dfrac{13\pi}{3}$

25. $-\dfrac{7\pi}{6}$ **27.** -20π **29.** 30°

31. 135° **33.** 67.5° **35.** 75°

37. 1620° **39.** 171° **41.** $-84°$

43. 229.18° **45.** 48.70° **47.** $-160.37°$

49. 198.48° **51.** 0.820 **53.** 1.95

55. 0.986 **57.** QII **59.** negative y-axis

61. negative x-axis **63.** QI **65.** QIII

67. QII **69.** 52° **71.** 268°

73. 330° **75.** $\dfrac{5\pi}{3}$ **77.** $\dfrac{11\pi}{9}$

79. ≈ 1.42 **81.** $\dfrac{2\pi}{3}$ ft **83.** $\dfrac{5}{2}$ in.

85. $\dfrac{11\pi}{5}\ \mu$m **87.** $\dfrac{200\pi}{3}$ km **89.** 2.85 km²

91. 8.62 cm² 93. 0.0236 ft² 95. $\frac{2}{5}$ m/sec

97. 272 km/hr 99. 9.8 m 101. 1.5 mi

103. $\frac{5\pi}{2}\frac{\text{rad}}{\text{sec}}$ 105. $\frac{2\pi}{9}\frac{\text{rad}}{\text{sec}}$

107. 6π inches per second

109. $\frac{\pi}{4}$ mm per second 111. 26.2 cm

113. 653 in. (or 54.5 ft)

115. 1440° 117. 70 mph 119. 10 rad/sec

121. Angular speed needs to be in radians (not degrees) per second.

123. true 125. true 127. 110°

129. $\frac{15}{2}\pi$ square units 131. 68° 48′ 48″ 133. $\frac{\pi}{5}$

135. $\frac{\pi}{3}$

Section 4.2

1. $\frac{\sqrt{5}}{5}$ 3. $\sqrt{5}$ 5. 2

7. $\frac{2\sqrt{10}}{7}$ 9. $\frac{7}{3}$ 11. $\frac{3\sqrt{10}}{20}$

13. a 15. b 17. c

19. $\frac{\sqrt{3}}{3}$ 21. $\sqrt{3}$ 23. $\frac{2\sqrt{3}}{3}$

25. $\frac{2\sqrt{3}}{3}$ 27. $\frac{\sqrt{3}}{3}$ 29. $\sqrt{2}$

31. 0.6018 33. 0.1392 35. 0.2588

37. −0.8090 39. 1.3764 41. 0.4142

43. 1.0034 45. 0.7002 47. $a \approx 18$ ft

49. $a \approx 5.50$ mi 51. $c \approx 12$ km 53. $\alpha \approx 62°$

55. $\beta = 58°, a \approx 6.4$ ft, $b \approx 10$ ft

57. $\alpha = 18°, a \approx 3.0$ mm, $b \approx 9.2$ mm

59. $\beta = 35.8°, b \approx 80.1$ mi, $c \approx 137$ mi

61. $\alpha \approx 56.0°, \beta \approx 34.0°, c \approx 51.3$ ft

63. $\alpha \approx 55.480°, \beta \approx 34.520°, b \approx 24{,}235$ km

65. $c \approx 27.0$ in., $a \approx 24.4$ in., $\alpha = 64.6°$

67. $a = 88$ ft

69. 260 ft (262 rounded to two significant digits)

71. 11° (she is too low) 73. 80 feet

75. 170 meters 77. 0.000016° 79. 4414 feet

81. 136.7° 83. 121 feet 85. 24 feet

87. The opposite of angle y is 3 (not 4).

89. Secant is the reciprocal of cosine (not sine).

91. true 93. false 95. 0

97. 0 99. $\frac{2}{3}$

101. a. 2.92398 b. 2.92380; (b) is more accurate.

103. 0.70274, 0.70281; (b) is more accurate.

105. $\frac{6 - 2\sqrt{3}}{3}$ 107. $\sqrt{3} - 1$

Section 4.3

	$\sin\theta$	$\cos\theta$	$\tan\theta$	$\cot\theta$	$\sec\theta$	$\csc\theta$
1.	$\frac{2\sqrt{5}}{5}$	$\frac{\sqrt{5}}{5}$	2	$\frac{1}{2}$	$\sqrt{5}$	$\frac{\sqrt{5}}{2}$
3.	$\frac{4\sqrt{41}}{41}$	$\frac{5\sqrt{41}}{41}$	$\frac{4}{5}$	$\frac{5}{4}$	$\frac{\sqrt{41}}{5}$	$\frac{\sqrt{41}}{4}$
5.	$\frac{2\sqrt{5}}{5}$	$-\frac{\sqrt{5}}{5}$	-2	$-\frac{1}{2}$	$-\sqrt{5}$	$\frac{\sqrt{5}}{2}$
7.	$-\frac{7\sqrt{65}}{65}$	$-\frac{4\sqrt{65}}{65}$	$\frac{7}{4}$	$\frac{4}{7}$	$-\frac{\sqrt{65}}{4}$	$-\frac{\sqrt{65}}{7}$
9.	$\frac{\sqrt{15}}{5}$	$-\frac{\sqrt{10}}{5}$	$-\frac{\sqrt{6}}{2}$	$-\frac{\sqrt{6}}{3}$	$-\frac{\sqrt{10}}{2}$	$\frac{\sqrt{15}}{3}$
11.	$-\frac{\sqrt{6}}{4}$	$-\frac{\sqrt{10}}{4}$	$\frac{\sqrt{15}}{5}$	$\frac{\sqrt{15}}{3}$	$-\frac{2\sqrt{10}}{5}$	$\frac{2\sqrt{6}}{3}$
13.	$-\frac{2\sqrt{29}}{29}$	$-\frac{5\sqrt{29}}{29}$	$\frac{2}{5}$	$\frac{5}{2}$	$-\frac{\sqrt{29}}{5}$	$-\frac{\sqrt{29}}{2}$

15. QIV 17. QII 19. QI

21. QI 23. QIII 25. $\sin\theta = -\frac{4}{5}$

27. $\tan\theta = -\frac{60}{11}$ 29. $\sin\theta = -\frac{84}{85}$ 31. $\tan\theta = \sqrt{3}$

33. $-\frac{\sqrt{3}}{3}$ 35. $\frac{2\sqrt{3}}{3}$ 37. 1

39. −1 41. 0 43. 1

45. 1 47. possible 49. not possible

51. possible 53. possible 55. possible

57. $-\frac{1}{2}$ 59. $-\frac{\sqrt{3}}{2}$ 61. $\frac{\sqrt{3}}{3}$

63. 1 65. −2 67. 1

69. 30° and 330° 71. 210° and 330° 73. 90° and 270°

75. 270° 77. 110° 79. 143°

81. $322°$ **83.** $140°$ **85.** $340°$

87. $1°$ **89.** $335°$ **91.** 1.3

93. $12°$

95. The reference angle is made with the terminal side and the negative x-axis (not y-axis).

97. true   **99.** false   **101.** false

103. true   **105.** $-\frac{3}{5}$

107. $y = -(\tan\theta)(x - a)$ **109.** $-\dfrac{a}{\sqrt{a^2 + b^2}}$

111. $-\dfrac{\sqrt{a^2 - b^2}}{b}$ **113.** 0

115. 0; a calculator yields an error message because $\tan 270°$ is undefined.

117. 0

119. Does not exist, because we are dividing by zero.

121. $\dfrac{12 + 3\sqrt{2} + \sqrt{3}}{6}$ **123.** $\frac{8}{3}$

Section 4.4

1. SSA   **3.** SSS   **5.** ASA

7. $\gamma = 75°, b \approx 12\,\text{m}, c \approx 14\,\text{m}$

9. $\beta = 62°, a \approx 163\,\text{cm}, c \approx 215\,\text{cm}$

11. $\beta = 116.1°, a \approx 80\,\text{yd}, b \approx 257\,\text{yd}$

13. $\gamma = 120°, a \approx 6.9\,\text{m}, b \approx 6.9\,\text{m}$

15. $\alpha = 97°, a \approx 118\,\text{yd}, b \approx 52\,\text{yd}$

17. $\beta_1 \approx 20°, \gamma_1 \approx 144°, c_1 \approx 9; \beta_2 \approx 160°, \gamma_2 \approx 4°, c_2 \approx 1$

19. $\alpha \approx 40°, \beta \approx 100°, b \approx 18$

21. no triangle

23. $\beta = 90°, \gamma = 60°, c = 16$

25. $\beta \approx 23°, \gamma \approx 123°, c \approx 15$

27. $\beta_1 \approx 21.9°, \gamma_1 \approx 136.8°, c_1 \approx 11.36$
$\beta_2 \approx 158.1°, \gamma_2 \approx 0.6°, c_2 \approx 0.16$

29. $\beta \approx 62°, \gamma \approx 2°, c \approx 0.3$

31. $\beta_1 \approx 77°, \alpha_1 \approx 63°, a \approx 458$
$\beta_2 \approx 103°, \alpha_2 \approx 37°, a \approx 308$

33. $\alpha \approx 31°, \gamma \approx 43°, c \approx 1.28$

35. $1246\,\text{ft}$ **37.** $1.7\,\text{mi}$

39. $1.3\,\text{mi}$ **41.** $26\,\text{ft}$

43. $270\,\text{ft}$ **45.** $64°$

47. $17\,\text{m}$ **49.** $60\,\text{ft}$

51. There is no solution to $\sin\beta = 1.113$.

53. false   **55.** true   **57.** true

63. $X = 68°, B \approx 6.9, C \approx 10.3$

65. $Y = 48.3°, Z = 26.7°, C = 10.2$

67. $Y = 25.5°, Z = 89.5°, C = 28.4$

69. $27\,\text{in.}$ **71.** $22\,\text{m}$

Section 4.5

1. $b \approx 5, \alpha \approx 47°, \gamma \approx 33°$ **3.** $a \approx 5, \beta \approx 158°, \gamma \approx 6°$

5. $a \approx 2, \beta \approx 80°, \gamma \approx 80°$ **7.** $b \approx 5, \alpha \approx 43°, \gamma \approx 114°$

9. $b \approx 7, \alpha \approx 30°, \gamma \approx 90°$ **11.** $\alpha \approx 93°, \beta \approx 39°, \gamma \approx 48°$

13. $\alpha \approx 51.3°, \beta \approx 51.3°, \gamma \approx 77.4°$

15. $\alpha \approx 75°, \beta \approx 57°, \gamma \approx 48°$ **17.** no triangle

19. $\alpha \approx 67°, \beta \approx 23°, \gamma \approx 90°$ **21.** $\gamma \approx 105°, b \approx 5, c \approx 9$

23. $\beta \approx 12°, \gamma \approx 137°, c \approx 16$ **25.** $\alpha \approx 66°, \beta \approx 77°, \gamma \approx 37°$

27. $\gamma \approx 2°, \alpha \approx 168°, a \approx 13$

29. 55.4 sq units   **31.** 0.5 sq units   **33.** 23.6 sq units

35. 6.4 sq units   **37.** 4408.4 sq units **39.** 97.4 sq units

41. 25.0 sq units   **43.** 26.7 sq units   **45.** 111.64 sq units

47. $111{,}632{,}076$ sq units   **49.** no triangle

51. $2710\,\text{mi}$ **53.** $1280\,\text{mi}$ **55.** $16\,\text{ft}$

57. $21.67°$ **59.** about $83°$ **61.** $47{,}128\,\text{ft}^2$

63. $23.38\,\text{ft}^2$ **65.** 10.86: about 11 sq units

67. Should have used the smaller angle β in Step 2.

69. false   **71.** true   **73.** true

77. $\sqrt{\dfrac{1 - \cos\left[2\cos^{-1}\left(\frac{1}{4a}\right)\right]}{2}}$ or $\dfrac{\sqrt{15}}{4}$ **81.** 0.69 sq units

83. $A = 39.0, Y = 51.9°, Z = 85.1°$

85. $X = 33.0°, Y = 43.4°, Z = 103.6°$

87. $C = 4.3, X = 45.8°, Y = 71.4°$

89. $333\,\text{mi}$ **91.** $282\,\text{m}$

Review Exercises

1. a. $62°$ **b.** $152°$ **3. a.** $55°$ **b.** $145°$

5. a. $0.99°$ **b.** $90.99°$ **7.** $\dfrac{3\pi}{4}$

9. $\dfrac{11\pi}{6}$ **11.** $\dfrac{6\pi}{5}$

13. 9π **15.** $60°$ **17.** $225°$

19. $100°$ **21.** $1800°$ **23.** $150°$

25. 240π in./min ≈ 754 in./min

27. $\dfrac{2\sqrt{13}}{13}$ **29.** $\dfrac{\sqrt{13}}{2}$ **31.** $\dfrac{3}{2}$

33. b   **35.** b   **37.** c

39. 0.6691 **41.** 0.9548 **43.** 1.5399

45. 1.5477 **47.** 75 feet

	$\sin\theta$	$\cos\theta$	$\tan\theta$	$\cot\theta$	$\sec\theta$	$\csc\theta$
49.	$-\dfrac{4}{5}$	$\dfrac{3}{5}$	$-\dfrac{4}{3}$	$-\dfrac{3}{4}$	$\dfrac{5}{3}$	$-\dfrac{5}{4}$
51.	$\dfrac{\sqrt{10}}{10}$	$-\dfrac{3\sqrt{10}}{10}$	$-\dfrac{1}{3}$	-3	$-\dfrac{\sqrt{10}}{3}$	$\sqrt{10}$
53.	$\dfrac{1}{2}$	$\dfrac{\sqrt{3}}{2}$	$\dfrac{\sqrt{3}}{3}$	$\sqrt{3}$	$\dfrac{2\sqrt{3}}{3}$	2
55.	$-\dfrac{\sqrt{5}}{5}$	$\dfrac{2\sqrt{5}}{5}$	$-\dfrac{1}{2}$	-2	$\dfrac{\sqrt{5}}{2}$	$-\sqrt{5}$
57.	$-\dfrac{\sqrt{7.2}}{3}$	$-\dfrac{\sqrt{7.2}}{6}$	2	$\dfrac{1}{2}$	$-\dfrac{\sqrt{7.2}}{1.2}$	$-\dfrac{\sqrt{7.2}}{2.4}$

59. $-\dfrac{1}{2}$ **61.** $-\dfrac{\sqrt{3}}{3}$ **63.** $-\dfrac{2\sqrt{3}}{3}$

65. $-\dfrac{\sqrt{2}}{2}$ **67.** $\sqrt{3}$ **69.** $-\sqrt{2}$

71. $-\dfrac{2\sqrt{3}}{3}$ **73.** $\gamma = 150°, c \approx 12, b \approx 8$

75. $\gamma = 130°, a \approx 1, b \approx 9$

77. $\beta = 158°, a \approx 11, b \approx 22$

79. $\beta = 90°, a \approx \sqrt{2}, c \approx \sqrt{2}$

81. $\beta = 146°, b \approx 266, c \approx 178$

83. $\beta \approx 26°, \gamma \approx 134°, c \approx 15$ or $\beta \approx 154°, \gamma \approx 6°, c \approx 2$

85. $\beta \approx 127°, \gamma \approx 29°, b \approx 20$ or $\beta \approx 5°, \gamma \approx 151°, b \approx 2$

87. no triangle

89. $\beta \approx 15°, \gamma \approx 155°, c \approx 10$ or $\beta \approx 165°, \gamma \approx 5°, c \approx 2$

91. $\alpha \approx 42°, \beta \approx 88°, c \approx 46$

93. $\alpha \approx 51°, \beta \approx 54°, \gamma \approx 75°$

95. $\alpha \approx 42°, \beta \approx 48°, \gamma \approx 90°$

97. $\beta \approx 28°, \gamma \approx 138°, a \approx 4$

99. $\beta \approx 68°, \gamma \approx 22°, a \approx 11$

101. $\alpha \approx 51°, \beta \approx 59°, \gamma \approx 70°$

103. $\beta \approx 37°, \gamma \approx 43°, a \approx 26$

105. $\beta \approx 4°, \gamma \approx 166°, a \approx 28$

107. no triangle   **109.** $\beta \approx 10°, \gamma \approx 155°, c \approx 10.3$

111. 141.8 sq units   **113.** 51.5 sq units   **115.** 89.8 sq units

117. 41.7 sq units   **119.** 5.2 in.

Practice Test

1. 6000 feet

3. The first is an exact value of the cosine function; the second is an approximation.

5. QIV   **7.** $585°$ **9.** $\dfrac{15\pi}{4}$ sq in.

11. $a \approx 7.8, c \approx 14.6,$ and $\gamma = 110°$

13. $\alpha \approx 35.4°, \beta \approx 48.2°,$ and $\gamma \approx 96.4°$

15. no triangle

17. $b \approx 1.82, c \approx 4.08, \gamma = 50°$ **19.** 57 sq units

Cumulative Test

1. $-\dfrac{5}{8}$ **3.** $-2x - h$ **5.** 1

7. $f(x) = -2x^2 + 7$

9. VA: $x = 2$; slant asymptote: $y = x + 2$

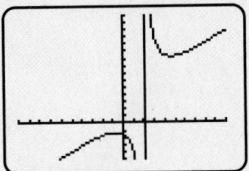

11. $\$37,250$ **13.** 0.435

15. $15\sqrt{2}$ ft ≈ 21.21 ft   **17.** $\dfrac{12\pi}{5}$

19. $\sqrt{3}$ **21.** 1.6616

23. $\alpha = 138°, b \approx 9$ cm, $c \approx 8$ cm

CHAPTER 5

Section 5.1

1. $-\dfrac{\sqrt{3}}{2}$ **3.** $-\dfrac{\sqrt{3}}{2}$ **5.** $\dfrac{\sqrt{2}}{2}$ **7.** -1

9. $-\sqrt{2}$ **11.** $\sqrt{3}$ **13.** 2 **15.** $-\dfrac{\sqrt{3}}{2}$

17. $-\dfrac{\sqrt{3}}{2}$ **19.** $-\dfrac{\sqrt{2}}{2}$ **21.** $-\dfrac{\sqrt{3}}{2}$ **23.** $\dfrac{\sqrt{2}}{2}$

25. 1   **27.** $\dfrac{\sqrt{2}}{2}$ **29.** 0   **31.** -2

33. $\dfrac{\sqrt{3}}{3}$ **35.** $\dfrac{\pi}{6}, \dfrac{11\pi}{6}$ **37.** $\dfrac{4\pi}{3}, \dfrac{5\pi}{3}$

39. $0, \pi, 2\pi, 3\pi, 4\pi$ **41.** $\pi, 3\pi$

43. $\dfrac{3\pi}{4}, \dfrac{7\pi}{4}$ **45.** $\dfrac{3\pi}{4}, \dfrac{5\pi}{4}$ **47.** $0, \pi, 2\pi$

49. $\dfrac{\pi}{2}, \dfrac{3\pi}{2}$ **51.** $\dfrac{7\pi}{6}, \dfrac{11\pi}{6}$ **53.** $\dfrac{\pi}{6}, \dfrac{11\pi}{6}$

55. 22.9 °F   **57.** 99.11 °F   **59.** 2.6 feet

61. 135 lb   **63.** 10,000 guests

65. Used the *x*-coordinate for sine and the *y*-coordinate for cosine; should have done the opposite.

67. true   **69.** false   **71.** true

73. odd   **75.** $\dfrac{\pi}{4}, \dfrac{5\pi}{4}$ **77.** $\dfrac{\pi}{4}, \dfrac{3\pi}{4}, \dfrac{5\pi}{4}, \dfrac{7\pi}{4}$

79. $\dfrac{\pi}{3} + n\pi, \dfrac{2\pi}{3} + n\pi$, where *n* is an integer

81. Yes, $\theta = \dfrac{\pi}{4}, \dfrac{3\pi}{4}, \dfrac{5\pi}{4}, \dfrac{7\pi}{4}$

83. $\sin 423° \approx 0.891$ and $\sin(-423°) \approx -0.891$

85. $6.314, -6.314$ **87.** 0.5   **89.** 0.8660

91. 2   **93.** $\dfrac{3 - \sqrt{3}}{3}$

Section 5.2

1. c   **3.** a   **5.** h   **7.** b

9. e   **11.** $A = \dfrac{3}{2}, p = \dfrac{2\pi}{3}$

13. $A = 1, p = \dfrac{2\pi}{5}$ **15.** $A = \dfrac{2}{3}, p = \dfrac{4\pi}{3}$

17. $A = 3, p = 2$ **19.** $A = 5, p = 6$

21. 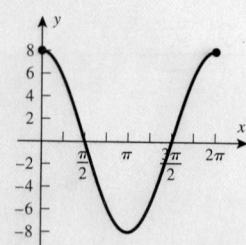   **23.**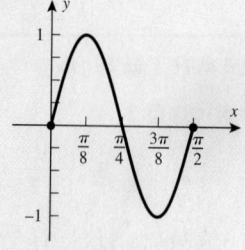

25. 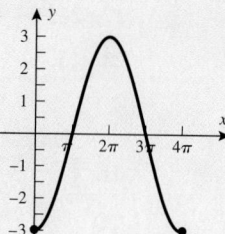   **27.**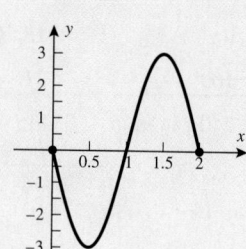

29. 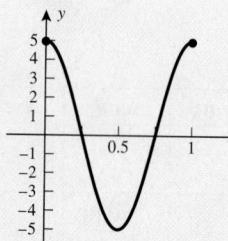   **31.**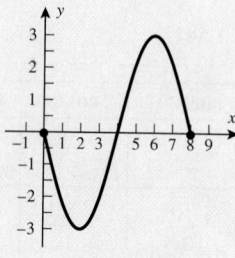

33. 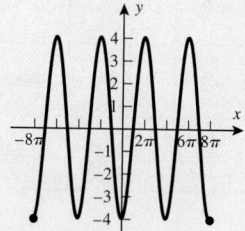   **35.**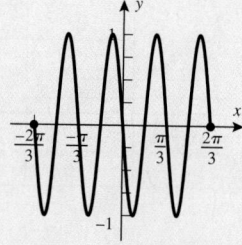

37. 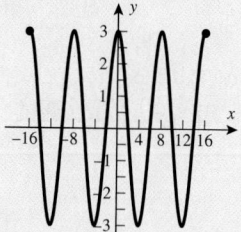   **39.**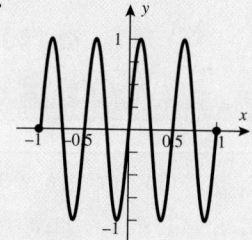

41. $y = -\sin(2x)$ **43.** $y = \cos(\pi x)$

45. $y = -2\sin\left(\dfrac{\pi}{2}x\right)$ **47.** $y = \sin(8\pi x)$

49. amplitude: 2; period: 2; phase shift: $\dfrac{1}{\pi}$ (right)

51. amplitude: 5; period: $\dfrac{2\pi}{3}$; phase shift: $-\dfrac{2}{3}$ (left)

53. amplitude: 6; period: 2; phase shift: -2 (left)

55. amplitude: 3; period: π; phase shift: $-\dfrac{\pi}{2}$ (left)

57. amplitude: $\dfrac{1}{4}$; period: 8π; phase shift: 2π (right)

59. amplitude: 2; period: 4; phase shift: 4 (right)

61.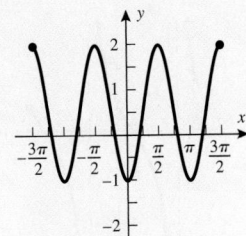

63.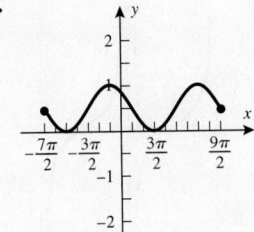

85.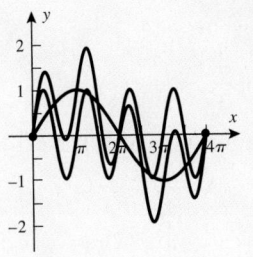

87.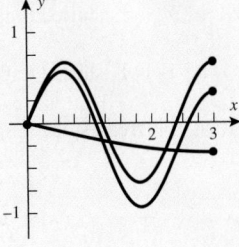

65.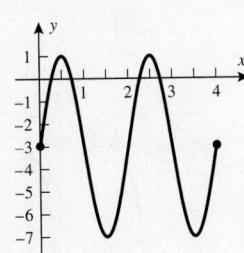

67.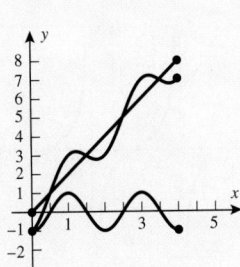

89.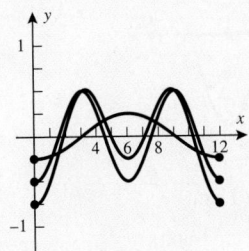

91.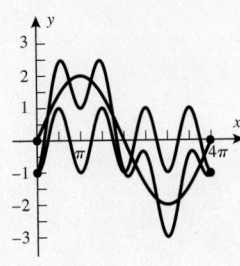

69.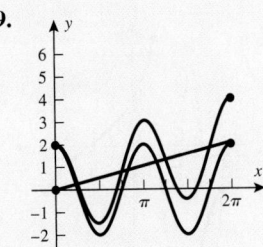

71.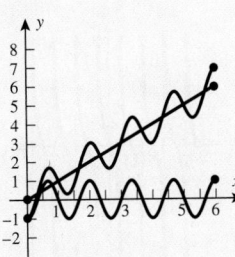

93.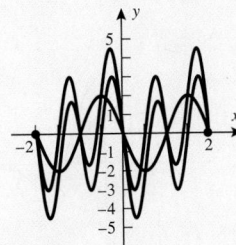

95. amplitude: 4 cm; mass: 4 g

97. $\frac{1}{4\pi}$ cycles per second

99. amplitude: 0.005 cm; frequency: 256 hertz

73.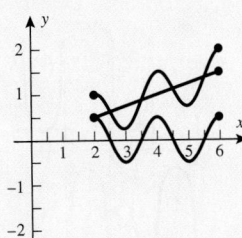

75.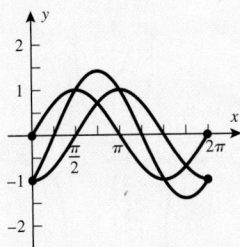

101. amplitude: 0.008 cm; frequency: 375 hertz

103. 660 m/sec   **105.** 660 m/sec

107. $y = 25 + 25 \sin\left[\frac{2\pi}{4}(t-1)\right]$ or $y = 25 - 25\cos\left(\frac{\pi t}{2}\right)$;

intensity is 0 candelas when $t = 4$ min.

109. Forgot to reflect about the x-axis.

77.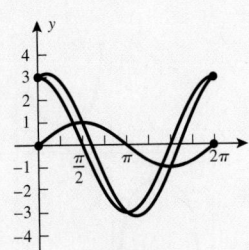

79.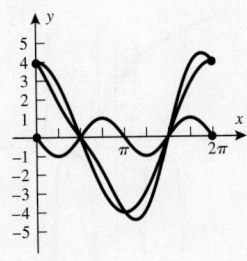

111. true   **113.** false   **115.** $(0, A)$

117. $x = \frac{n\pi}{B}$, where n is an integer   **119.** $\left(0, -\frac{A}{2}\right)$

121. $\left(\frac{\pi(3+4n)}{2B}, 0\right)$, where n an integer   **123.** $\left[-\frac{5A}{2}, \frac{3A}{2}\right]$

125. no   **127.** coincide

81.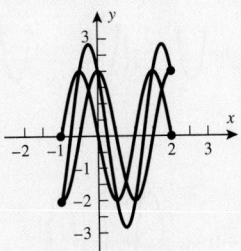

83.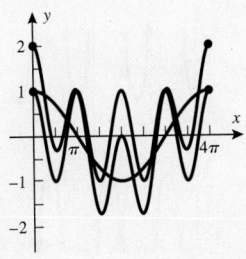

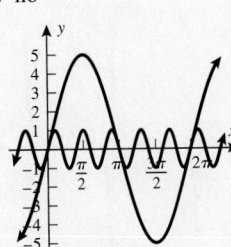

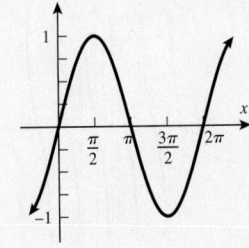

129. a. Y_2 is Y_1 shifted to the left $\dfrac{\pi}{3}$.

b. Y_2 is Y_1 shifted to the right $\dfrac{\pi}{3}$.

131. As t increases, the amplitude goes to zero for Y_3.

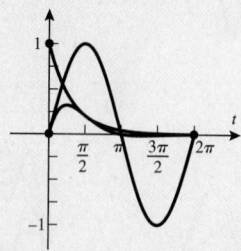

133. a. Y_2 is Y_1 shifted upward by 1 unit.

b. Y_2 is Y_2 shifted downward by 1 unit.

135. 5

137.

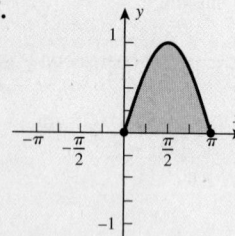

Area = 2

139.

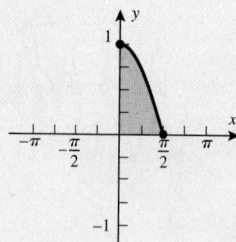

Area = 1

Section 5.3

1. b   **3.** h   **5.** c   **7.** d

9.

11.

13.

15.

17.

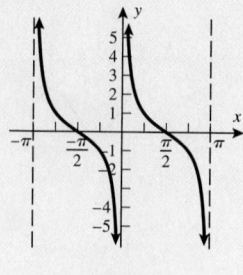

19.

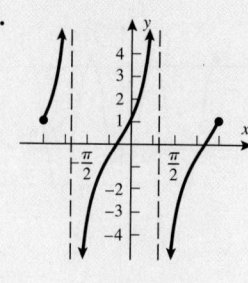

21.

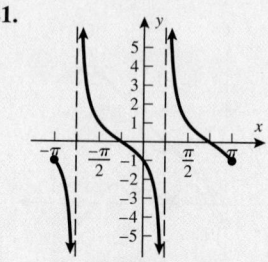

23.

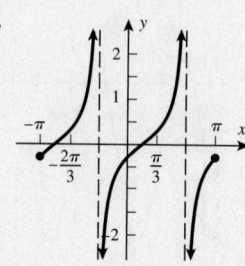

25.

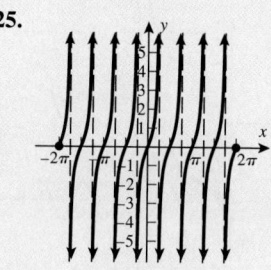

27.

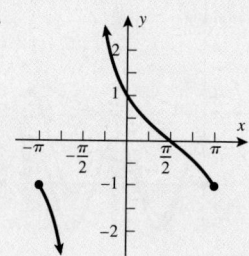

29.

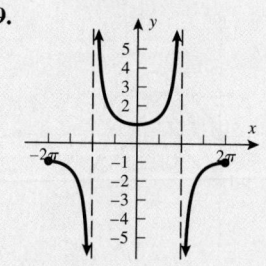

31.

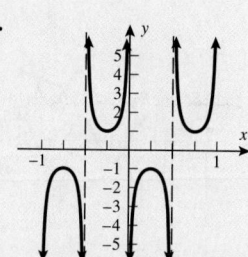

33.

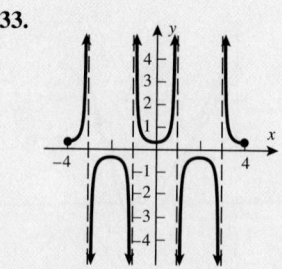

35.

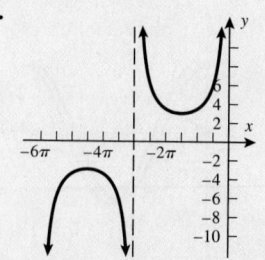

37.

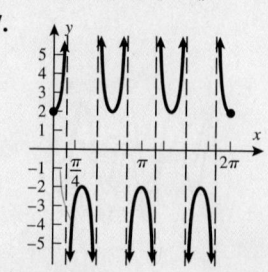

39.

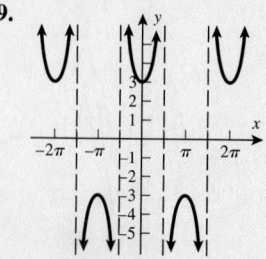

41.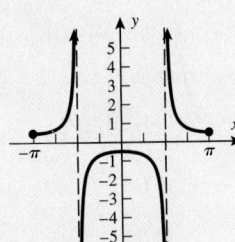

43.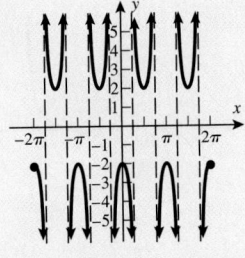

45.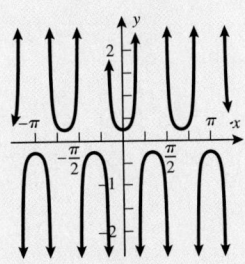

47.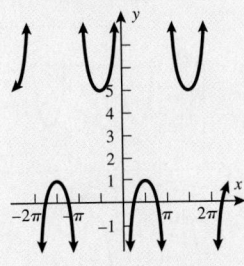

49.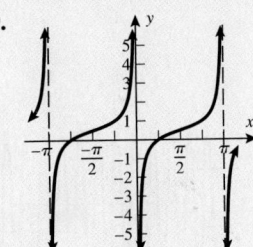

51.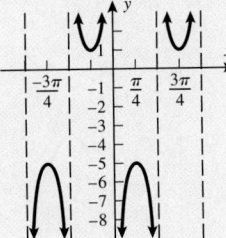

53.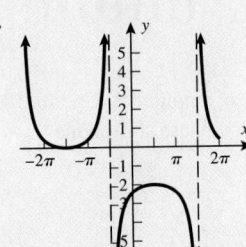

55.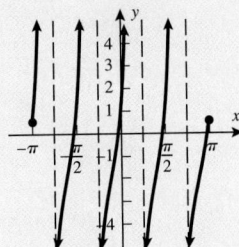

57. domain: all real numbers x, such that $x \neq n$, where n is an integer

range: all real numbers

59. domain: all real numbers x, such that $x \neq \dfrac{2n+1}{10}\pi$, where n is an integer

range: $(-\infty, -2] \cup [2, \infty)$

61. domain: all real numbers x, such that $x \neq 2n\pi$, where n is an integer

range: $(-\infty, 1] \cup [3, \infty)$

63. domain: $\{x : x \neq 4n + 6, n \text{ an integer}\}$

range: all real numbers

65. domain: $\{x : x \neq n, n \text{ an integer}\}$

range: $\left(-\infty, -\dfrac{5}{2}\right] \cup \left[-\dfrac{3}{2}, \infty\right)$

67. 48 m

69. a. −5.2 mi   **b.** −3 mi   **c.** 0 mi   **d.** 3 mi   **e.** 5.2 mi

71. Forgot the amplitude of 3.   **73.** true

75. $n =$ integer

77. $x = -\pi, -\dfrac{\pi}{2}, 0, \dfrac{\pi}{2}, \pi$

79. $\left(\dfrac{n\pi - C}{B}, 0\right)$, where n an integer

81. Infinitely many. The tangent function is an increasing, periodic function and $y = x$ is increasing.

83. $A = \sqrt{2}$

85. π

87.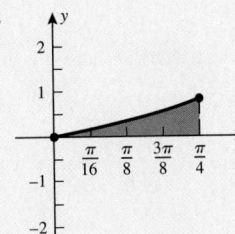

Area $= -\ln\left(\dfrac{\sqrt{2}}{2}\right)$

89.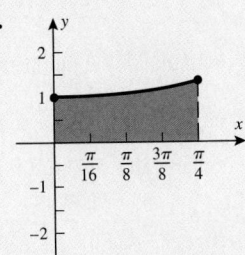

Area $= \ln(\sqrt{2} + 1)$

Review Exercises

1. $-\dfrac{\sqrt{3}}{3}$

3. $-\dfrac{1}{2}$

5. 1

7. −1

9. −1

11. $\dfrac{1}{2}$

13. $\dfrac{\sqrt{2}}{2}$

15. 1

17. $-\dfrac{\sqrt{3}}{2}$

19. $-\dfrac{\sqrt{3}}{3}$

21. 2π

23. $y = 4\cos x$

25. 5

27. amplitude: 2
period: 1

29. amplitude: $\dfrac{1}{5}$
period: $\dfrac{2\pi}{3}$

31.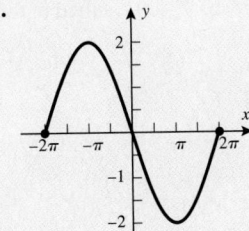

33.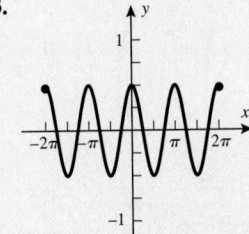

	AMPLITUDE	PERIOD	PHASE SHIFT	VERTICAL SHIFT
35.	3	2π	$\dfrac{\pi}{2}$ (right)	up 2 units
37.	4	$\dfrac{2\pi}{3}$	$-\dfrac{\pi}{4}$ (left)	down 2 units
39.	$\dfrac{1}{3}$	2	$\dfrac{1}{2\pi}$ (right)	down $\dfrac{1}{2}$ unit

41.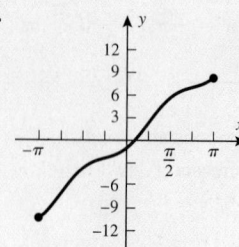

43.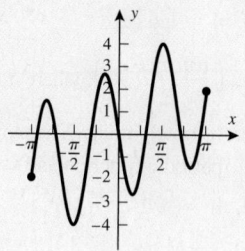

7.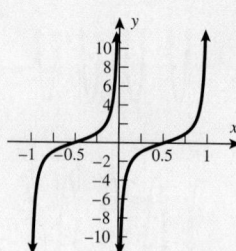

9. $x = \dfrac{n\pi}{2}$ or $\left(\dfrac{n\pi}{2}, 0\right)$, where n is an integer.

11. $(-\infty, -4] \cup [2, \infty)$

45. domain: all real numbers such that $x \neq n\pi$, where n is an integer
range: all real numbers

47. domain: all real numbers such that $x \neq \dfrac{2n+1}{4}\pi$, where n is an integer

range: $(-\infty, -3] \cup [3, \infty)$

49. domain: $\left\{ x : x \neq \dfrac{6n+7}{6}, n \text{ an integer} \right\}$

range: $\left(-\infty, -\dfrac{3}{4}\right] \cup \left[-\dfrac{1}{4}, \infty\right)$

13.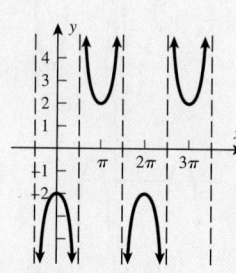

15. true

17. $y = 4\sin\left[2\left(x + \dfrac{3}{2}\right)\right] - \dfrac{1}{2}$

51.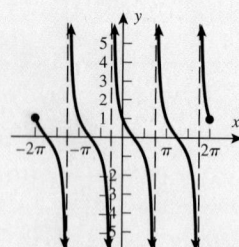

53.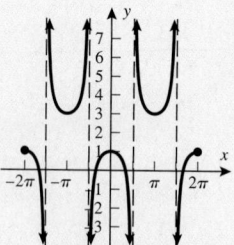

19.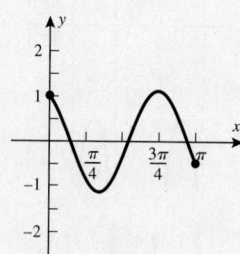

21. a.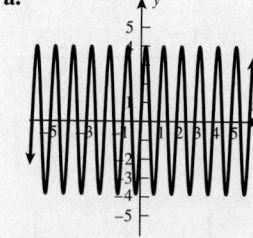

b. amplitude: 4, period: 1, phase shift: none

55.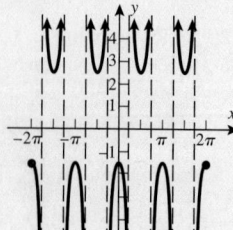

57. -0.9659

59. a. Y_2 is Y_1 shifted to the left by $\dfrac{\pi}{6}$ unit.

b. Y_2 is Y_1 shifted to the right by $\dfrac{\pi}{6}$ unit.

61. 5

23. a. $y = 5\cos\left[2\left(x - \dfrac{\pi}{2}\right)\right] + 6$

b. amplitude: 5, period: π, phase shift: $\dfrac{\pi}{2}$ right

c.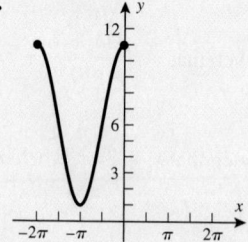

Practice Test

1. amplitude: 5, period: $\dfrac{2\pi}{3}$

3.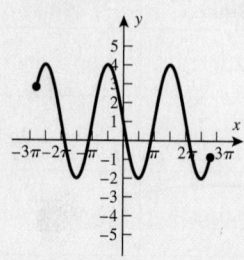

5.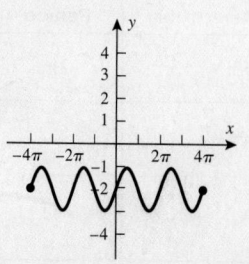

Cumulative Test

1. $(-5, \infty)$

3. $y = 2|x + 6| + 4$

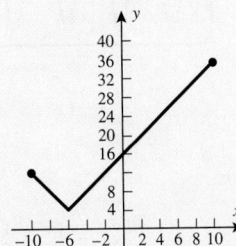

5. $f^{-1}(x) = \dfrac{2 + 5x}{1 - 3x}$

domain of f: $\left(-\infty, -\frac{5}{3}\right) \cup \left(-\frac{5}{3}, \infty\right)$,

range of f: $\left(-\infty, \frac{1}{3}\right) \cup \left(\frac{1}{3}, \infty\right)$

domain of f^{-1}: $\left(-\infty, \frac{1}{3}\right) \cup \left(\frac{1}{3}, \infty\right)$,

range of f^{-1}: $\left(-\infty, -\frac{5}{3}\right) \cup \left(-\frac{5}{3}, \infty\right)$

7. $Q(x) = 3x^2 - \frac{5}{2}x + \frac{9}{2}, r(x) = \frac{9}{2}x + \frac{1}{2}$

9. $P(x) = (x + i)(x - i)(x + \sqrt{5})(x - \sqrt{5})$

11. \$3,381.88

13. $3\ln a - 2\ln b - 5\ln c$

15. $x = 1$

17. $a = 12.47$ in., $b = 6.36$ in., $\alpha = 63°$

19. first solution: $\beta = 71.17°, \gamma = 40.83°, c = 16.92$ m
second solution: $\beta = 108.83°, \gamma = 3.17°, c = 1.43$ m

21. $\cos\theta = -\dfrac{\sqrt{3}}{2}, \tan\theta = -\dfrac{\sqrt{3}}{3}, \cot\theta = -\sqrt{3},$

$\sec\theta = -\dfrac{2\sqrt{3}}{3}, \csc\theta = 2$

23. $\dfrac{2}{\pi}$

CHAPTER 6

Section 6.1

1. $30°$

3. $90° - x$

5. $60°$

7. $\cos(90° - x - y)$

9. $\sin(70° - A)$

11. $\tan(45° + x)$

13. $\sec(30° + \theta)$

15. 1

17. $\csc x$

19. -1

21. $\sec^2 x$

23. 1

25. $\sin^2 x - \cos^2 x$

27. $\sec x$

29. 1

31. $\sin^2 x$

33. $\csc^2 x$

35. $-\cos x$

37. 1

65. conditional

67. identity

69. conditional

71. conditional

73. conditional

75. identity

77. conditional

81. $\sec\theta$

83. The $\cos x$ and $\sin x$ terms do not cancel. The numerators become $\cos^2 x$ and $\sin^2 x$, respectively.

85. This is a conditional equation. Just because the equation is true for $\dfrac{\pi}{4}$ does not mean it is true for all x.

87. false

89. QI, QIV

91. QIII and QIV

93. No, let $A = 30°$ and $B = 60°$.

95. No, take $A = \dfrac{\pi}{4}$.

97. $a^2 + b^2$

101. $\sec\theta$

103. $\cos(A + B) = \cos A \cos B - \sin A \sin B$

105. $\sin(A + B) = \sin A \cos B + \cos A \sin B$

107. $|a|\cos\theta$

109. $|a|\tan\theta$

Section 6.2

1. $\dfrac{\sqrt{6} - \sqrt{2}}{4}$

3. $\dfrac{\sqrt{6} - \sqrt{2}}{4}$

5. $-2 + \sqrt{3}$

7. $\dfrac{\sqrt{2} + \sqrt{6}}{4}$

9. $2 + \sqrt{3}$

11. $2 + \sqrt{3}$

13. $\sqrt{2} - \sqrt{6}$

15. $\dfrac{4}{\sqrt{2}(1 + \sqrt{3})} = \sqrt{6} - \sqrt{2}$

17. $\cos x$

19. $-\sin x$

21. 0

23. $-2\cos(A - B)$

25. $-2\sin(A + B)$

27. $\tan(26°)$

29. $\dfrac{1 + 2\sqrt{30}}{12}$

31. $\dfrac{-6\sqrt{6} + 4}{25}$

33. $\dfrac{3 - 4\sqrt{15}}{4 + 3\sqrt{15}} = \dfrac{192 - 25\sqrt{15}}{-119}$

35. identity

37. conditional

39. identity

41. identity

43. identity

45. conditional

47. identity

49. identity

51. conditional

53. $y = \sin\left(x + \dfrac{\pi}{3}\right)$

55. $y = \cos\left(x - \dfrac{\pi}{4}\right)$

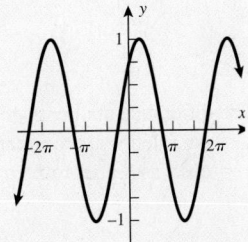

57. $y = -\sin(4x)$

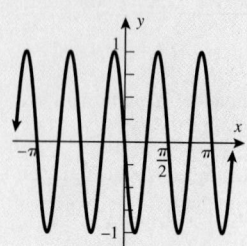

59. $y = \tan\left(x + \dfrac{\pi}{4}\right)$

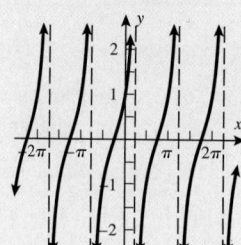

61. $y = \tan\left(\dfrac{\pi}{6} + x\right)$

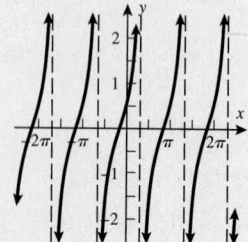

63. $\dfrac{\sqrt{2}}{2}\left(1 + x - \dfrac{x^2}{2!} - \dfrac{x^3}{3!} + \dfrac{x^4}{4!} + \dfrac{x^5}{5!} - \dfrac{x^6}{6!} - \dfrac{x^7}{7!} + \cdots\right)$

67. $\cos(kz - ct) = \cos(kz)\cos(ct) + \sin(kz)\sin(ct)$; when $\dfrac{z}{\lambda} = $ integer, then $kz = 2\pi n$ and the $\sin(kz)$ term goes to zero.

69. Tangent of a sum is not the sum of the tangents. Needed to use the tangent of a sum identity.

71. false   **73.** false

77. $A = n\pi$ and $B = m\pi$, where n and m are integers, or A and B, where $B = A \pm 2n\pi$.

79. a.

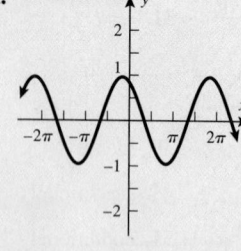

b.

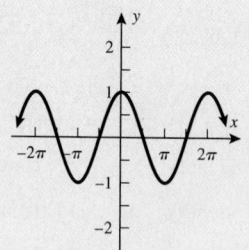

c.

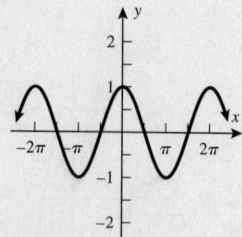

The difference quotients of $y = \sin x$ better approximate $y = \cos x$ as h goes to zero.

81. a.

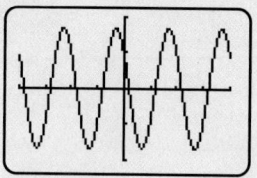

b.

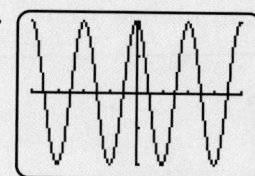

c.

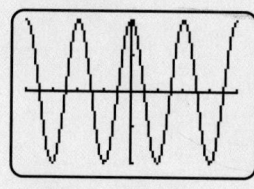

$2\cos(2x)$

83. $\tan x = -\tan y$ **85.** $\tan x = \dfrac{2 - \tan y}{1 + 2\tan y}$

Section 6.3

1. $-\dfrac{4}{5}$ **3.** $\dfrac{120}{119}$ **5.** $\dfrac{120}{169}$

7. $-\dfrac{4}{3}$ **9.** $\dfrac{\sqrt{19}}{10}$ **11.** $\dfrac{119}{120}$

13. $\tan 30° = \dfrac{\sqrt{3}}{3}$ **15.** $\dfrac{1}{2}\sin\left(\dfrac{\pi}{4}\right) = \dfrac{\sqrt{2}}{4}$ **17.** $\cos(4x)$

19. $-\dfrac{\sqrt{3}}{3}$ **21.** $-\dfrac{\sqrt{3}}{2}$ **23.** $-\dfrac{\sqrt{3}}{2}$

41. $y = \cot x$

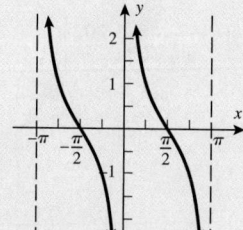

43. $y = \sec(2x)$

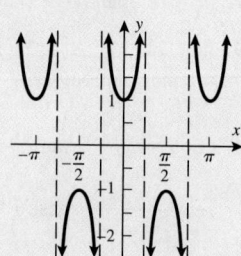

45. $y = \dfrac{1}{2}\sin(4x)$

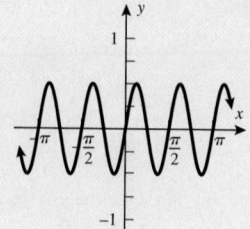

47. $y = 1 - \dfrac{1}{2}\sin(2x)$

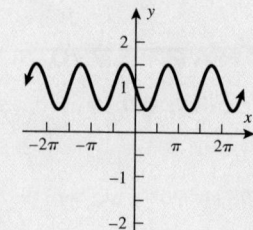

49. $y = 2\sin x - 3\cos(2x)$

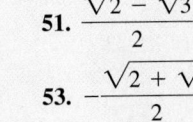

51. $\dfrac{\sqrt{2 - \sqrt{3}}}{2}$

53. $-\dfrac{\sqrt{2 + \sqrt{3}}}{2}$

55. $\dfrac{\sqrt{2 - \sqrt{3}}}{2}$

57. $\sqrt{3 + 2\sqrt{2}}$

59. $-\dfrac{2}{\sqrt{2 + \sqrt{2}}}$

61. $1 - \sqrt{2}$ or $\dfrac{-1}{\sqrt{3 + 2\sqrt{2}}}$

63. $-\dfrac{2}{\sqrt{2 - \sqrt{2}}}$

65. 1

67. $\dfrac{2\sqrt{13}}{13}$

69. $\dfrac{3\sqrt{13}}{13}$

71. $\dfrac{\sqrt{5} - 1}{2}$ or $\sqrt{\dfrac{3 - \sqrt{5}}{2}}$

73. $\sqrt{\dfrac{3 + 2\sqrt{2}}{6}}$ or $\dfrac{1 + \sqrt{2}}{\sqrt{6}}$

75. $-\dfrac{\sqrt{15}}{5}$

77. $\sqrt{\dfrac{1 - \dfrac{24}{\sqrt{601}}}{2}}$

79. $-\sqrt{\dfrac{1 - \sqrt{0.91}}{1 + \sqrt{0.91}}}$

81. $\sqrt{\dfrac{7}{3}}$ **83.** $\cos\left(\dfrac{5\pi}{12}\right)$ **85.** $\tan 75°$ **87.** $-\tan\left(\dfrac{5\pi}{8}\right)$

101. $y = 2 + 2\cos x$

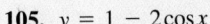

103. $y = \cos x$

105. $y = 1 - 2\cos x$

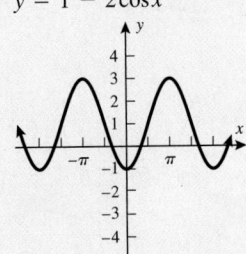

107. $y = |\tan x|$

109. 22,565,385 lb

113. $\sqrt{2}$ ft

115. Sine is negative (not positive).

117. $\sin\left(\dfrac{x}{2}\right)$ is positive, not negative, in this case.

119. false   **121.** false

123. false   **125.** false

129. no   **133.** $0 < x < \pi$

135. yes   **137.** yes

139. $\tan(2x)$ **141.** $\sin(3x)$

Section 6.4

1. $\frac{1}{2}[\sin(3x) + \sin x]$ **3.** $\frac{5}{2}[\cos(2x) - \cos(10x)]$

5. $2[\cos x + \cos(3x)]$ **7.** $\frac{1}{2}[\cos x - \cos(4x)]$

9. $\frac{1}{2}\left[\cos\left(\dfrac{2x}{3}\right) + \cos(2x)\right]$ **11.** $-\frac{3}{2}[\cos(1.9x) + \cos(1.1x)]$

13. $2\left[\sin(2\sqrt{3}x) - \sin(4\sqrt{3}x)\right]$

15. $2\cos(4x)\cos x$ **17.** $2\sin x\cos(2x)$

19. $-2\sin x\cos(\frac{3}{2}x)$ **21.** $2\cos(\frac{3}{2}x)\cos(\frac{5}{6}x)$

23. $2\sin(0.5x)\cos(0.1x)$ **25.** $-2\sin(\sqrt{5}x)\cos(2\sqrt{5}x)$

27. $2\cos\left(\dfrac{\pi}{24}x\right)\cos\left(\dfrac{5\pi}{24}x\right)$ **29.** $-\tan x$

31. $\tan(2x)$ **33.** $\cot\left(\dfrac{3x}{2}\right)$

43. average frequency: 443 Hz; beat frequency: 102 Hz

45. $2\sin\left[\dfrac{2\pi ct}{2}\left(\dfrac{1}{1.55} + \dfrac{1}{0.63}\right)10^6\right]\cos\left[\dfrac{2\pi ct}{2}\left(\dfrac{1}{1.55} - \dfrac{1}{0.63}\right)10^6\right]$

47. $2\sin\left[\dfrac{2\pi(1979)t}{2}\right]\cos\left[\dfrac{2\pi(439)t}{2}\right] = 2\sin(1979\pi t)\cos(439\pi t)$

49. $\dfrac{25(\sqrt{6} - \sqrt{3})}{3} \approx 5.98$ ft^2

51. $\cos A\cos B \neq \cos(AB)$ and $\sin A\sin B \neq \sin(AB)$. Should have used product-to-sum identity.

53. false   **55.** true

57. Answers will vary. Here is one approach:

$$\underbrace{\sin A\sin B}_{\frac{1}{2}[\cos(A-B) - \cos(A+B)]}\sin C = \frac{1}{2}\left[\underbrace{\cos(A - B)\sin C}_{\frac{1}{2}[\sin(A-B+C) + \sin(C-A+B)]} - \underbrace{\cos(A + B)\sin C}_{\frac{1}{2}[\sin(A-B+C) + \sin(C-A+B)]}\right]$$

$$\sin A\sin B\sin C = \frac{1}{4}[\sin(A - B + C) + \sin(C - A + B) - \sin(A + B + C) - \sin(C - A - B)]$$

61. $y = 1 - \frac{3}{2}\cos\left(\frac{7\pi}{6}x\right) + \frac{3}{2}\cos\left(\frac{5\pi}{6}x\right)$

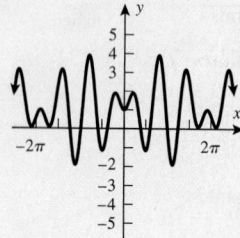

63. $y = -\frac{1}{2}\cos\left(\frac{3\pi}{2}x\right) - \frac{1}{2}\cos\left(\frac{\pi}{6}x\right)$

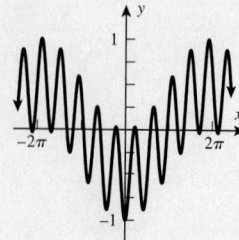

65. $y = \sin(4x)$

67. Y_1 and Y_3

69. $\cos x = \cos y - \frac{2}{5}$

71. $\sec x = 2\sin y$

Section 6.5

1. $\frac{\pi}{4}$

3. $-\frac{\pi}{3}$

5. $\frac{3\pi}{4}$

7. $\frac{\pi}{6}$

9. $\frac{\pi}{6}$

11. $-\frac{\pi}{3}$

13. 0

15. π

17. $60°$

19. $45°$

21. $120°$

23. $30°$

25. $-30°$

27. $135°$

29. $-90°$

31. $90°$

33. $57.10°$

35. $62.18°$

37. $48.10°$

39. $-15.30°$

41. $166.70°$

43. -0.63

45. 1.43

47. 0.92

49. 2.09

51. 0.31

53. $\frac{5\pi}{12}$

55. not possible

57. $\frac{\pi}{6}$

59. $\frac{2\pi}{3}$

61. $\sqrt{3}$

63. $\frac{\pi}{3}$

65. not possible

67. 0

69. $-\frac{\pi}{4}$

71. not possible

73. $\frac{2\pi}{3}$

75. $-\frac{\pi}{4}$

77. $\frac{\sqrt{7}}{4}$

79. $\frac{12}{13}$

81. $\frac{3}{4}$

83. $\frac{5\sqrt{23}}{23}$

85. $\frac{4\sqrt{15}}{15}$

87. $\frac{11}{60}$

89. $t \approx 0.026476$ sec or 26 ms

91. 173.4; June 22–23

93. $11.3 \approx 11$ years

95. $\tan\theta = \dfrac{\dfrac{7}{x} + \dfrac{1}{x}}{1 - \dfrac{7}{x}\cdot\dfrac{1}{x}} = \dfrac{8x}{x^2 - 7}$

97. 0.70 m; 0.24 m

99. $\theta = \pi - \tan^{-1}\left(\dfrac{150}{x}\right) - \tan^{-1}\left(\dfrac{300}{200 - x}\right)$

101. The wrong interval for the identity was used. The correct domain is $\left[-\dfrac{\pi}{2}, \dfrac{\pi}{2}\right]$.

103. $\cot^{-1}x \neq \dfrac{1}{\tan^{-1}x}$

105. false

107. false

109. $\frac{1}{2}$ is not in the domain of the inverse secant function.

111. $\dfrac{\sqrt{6} - \sqrt{2}}{4}$

113. 0

115. a. $\left(\dfrac{\pi}{4}, \dfrac{5\pi}{4}\right)$

b. $f^{-1}(x) = \cos^{-1}(x - 3) + \dfrac{\pi}{4}$
domain: [2, 4]

117. a. $\left(\dfrac{\pi}{12}, \dfrac{7\pi}{12}\right)$

b. $f^{-1}(x) = \dfrac{\pi}{12} + \dfrac{1}{2}\operatorname{arccot}(4x - 8)$
domain: all real numbers

119. The identity $\sin(\sin^{-1}x) = x$ only holds for $-1 \leq x \leq 1$.

121. $\left[-\dfrac{\pi}{2}, 0\right) \cup \left(0, \dfrac{\pi}{2}\right)$

123. a. $\dfrac{720}{1681}$ **b.** 0.42832 **c.** yes

125. $1 + x^2$

127. $x\sqrt{x^2 - 1}$

Section 6.6

1. $\dfrac{3\pi}{4}, \dfrac{5\pi}{4}$

3. $\dfrac{7\pi}{6}, \dfrac{11\pi}{6}, \dfrac{19\pi}{6}, \dfrac{23\pi}{6}$

5. $n\pi$, where n is any integer

7. $\dfrac{7\pi}{12}, \dfrac{11\pi}{12}, \dfrac{19\pi}{12}, \dfrac{23\pi}{12}$

9. $\dfrac{7\pi}{3} + 4n\pi$ or $\dfrac{11\pi}{3} + 4n\pi$, where n is any integer

11. $-\dfrac{4\pi}{3}, -\dfrac{5\pi}{6}, -\dfrac{11\pi}{6}, -\dfrac{\pi}{3}, \dfrac{\pi}{6}, \dfrac{2\pi}{3}, \dfrac{7\pi}{6}, \dfrac{5\pi}{3}$

13. $-\dfrac{4\pi}{3}, -\dfrac{2\pi}{3}$

15. $\dfrac{\pi}{6} + \dfrac{n\pi}{4}$, where n is any integer

17. $-\pi, -\dfrac{5\pi}{3}, -\dfrac{\pi}{3}$

19. $-\dfrac{\pi}{2}, -\dfrac{7\pi}{6}, -\dfrac{11\pi}{6}$

21. $\dfrac{\pi}{6}, \dfrac{\pi}{3}, \dfrac{7\pi}{6}, \dfrac{4\pi}{3}$

23. $\dfrac{\pi}{12}, \dfrac{7\pi}{12}, \dfrac{13\pi}{12}, \dfrac{19\pi}{12}$

25. $\dfrac{\pi}{3}, \dfrac{2\pi}{3}, \dfrac{4\pi}{3}, \dfrac{5\pi}{3}$

27. $\dfrac{\pi}{3}$

29. $\dfrac{\pi}{4}, \dfrac{3\pi}{4}, \dfrac{5\pi}{4}, \dfrac{7\pi}{4}$

31. $\dfrac{\pi}{2}, \dfrac{3\pi}{2}, \dfrac{\pi}{3}, \dfrac{5\pi}{3}$

33. $\dfrac{3\pi}{2}, \dfrac{7\pi}{6}, \dfrac{11\pi}{6}$

35. $\dfrac{\pi}{2}$

37. $0, \pi$

39. $\dfrac{\pi}{12}, \dfrac{5\pi}{12}, \dfrac{7\pi}{12}, \dfrac{11\pi}{12}, \dfrac{13\pi}{12}, \dfrac{17\pi}{12}, \dfrac{19\pi}{12}, \dfrac{23\pi}{12}$

41. $115.83°, 154.17°, 295.83°, 334.17°$

43. $333.63°$

45. $29.05°, 209.05°$

47. $200.70°, 339.30°$

49. $41.41°, 318.59°$

51. $56.31°, 126.87°, 236.31°, 306.87°$

53. $9.74°, 80.26°, 101.79°, 168.21°, 189.74°, 260.26°, 281.79°, 348.21°$

55. $80.12°, 279.88°$

57. $64.93°, 121.41°, 244.93°, 301.41°$

59. $15°, 45°, 75°, 105°, 135°, 165°, 195°, 225°, 255°, 285°, 315°, 345°$

61. $\dfrac{\pi}{4}, \dfrac{5\pi}{4}$ **63.** π **65.** $\dfrac{\pi}{6}$ **67.** $\dfrac{\pi}{3}$

69. $\dfrac{\pi}{4}, \dfrac{3\pi}{4}, \dfrac{5\pi}{4}, \dfrac{7\pi}{4}$

71. $\dfrac{\pi}{2}, \dfrac{3\pi}{2}$

73. $0, \dfrac{\pi}{4}, \pi, \dfrac{7\pi}{4}$

75. $\dfrac{\pi}{6}, \dfrac{5\pi}{6}, \dfrac{7\pi}{6}, \dfrac{11\pi}{6}, \dfrac{\pi}{2}, \dfrac{3\pi}{2}$

77. $\dfrac{\pi}{6}, \dfrac{\pi}{3}, \dfrac{7\pi}{6}, \dfrac{4\pi}{3}$

79. $\dfrac{\pi}{6}, \dfrac{5\pi}{6}, \dfrac{7\pi}{6}, \dfrac{11\pi}{6}$

81. $\dfrac{3\pi}{2}$

83. $\dfrac{2\pi}{3}, \dfrac{4\pi}{3}$

85. $\dfrac{\pi}{3}, \dfrac{5\pi}{3}, \pi$

87. $\dfrac{\pi}{24}, \dfrac{5\pi}{24}, \dfrac{13\pi}{24}, \dfrac{17\pi}{24}, \dfrac{25\pi}{24}, \dfrac{29\pi}{24}, \dfrac{37\pi}{24}, \dfrac{41\pi}{24}$

89. $57.47°, 122.53°, 323.62°, 216.38°$

91. $30°, 150°, 199.47°, 340.53°$

93. $14.48°, 165.52°, 270°$

95. $111.47°, 248.53°$

97. $\dfrac{\pi}{3}, \dfrac{5\pi}{3}$

99. March

101. $A = \dfrac{1}{2}h(b_1 + b_2)$
$= \dfrac{1}{2}(x \sin\theta)\{[x + (x\cos\theta + x + x\cos\theta)]\}$
$= x^2 \sin\theta(1 + \cos\theta)$

103. 2001 **105.** $24°$ **107.** $\dfrac{3}{4}$ sec

109. $(0, 1), \left(\dfrac{\pi}{3}, \dfrac{3}{2}\right), (\pi, -3), \left(\dfrac{5\pi}{3}, \dfrac{3}{2}\right)$

111. Extraneous solution. Forgot to check.

113. Can't divide by $\cos x$. Must factor.

115. false **117.** true **119.** $\dfrac{\pi}{6}, \dfrac{5\pi}{6}, \dfrac{7\pi}{6}, \dfrac{11\pi}{6}$

121. $\dfrac{\pi}{6}$ or $30°$ **123.** no solution

125. $5 + 2n$, where n is any integer

127. $x = \dfrac{\pi}{6} \approx 0.524$ and $x = \dfrac{5\pi}{6} \approx 2.618$

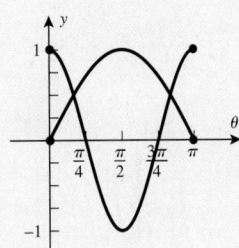

129. no solution

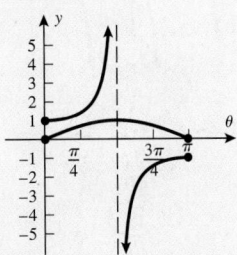

131. no solution

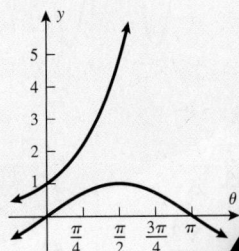

133. $x \approx 7.39$

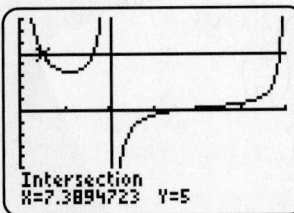

135. $x \approx 1.3$

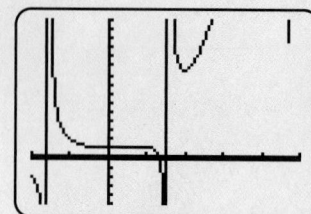

137. 2.21911   **139.** $0, 2\pi$ **141.** $-\dfrac{\pi}{3}, 0, \dfrac{\pi}{3}$

Review Exercises

1. $\cos 60°$ **3.** $\cot 45°$ **5.** $\csc 60°$

7. $\sec^2 x$ **9.** $\sec^2 x$ **11.** $\cos^2 x$

13. $-(4 + 2\csc x + \csc^2 x)$ **21.** identity

23. conditional   **25.** identity   **27.** $\dfrac{\sqrt{2} - \sqrt{6}}{4}$

29. $\dfrac{\sqrt{3} - 3}{3 + \sqrt{3}} = \sqrt{3} - 2$ **31.** $\sin x$

33. $\tan x$ **35.** $\dfrac{117}{44}$ **37.** $-\dfrac{897}{1025}$ **39.** identity

41. $y = \cos\left(x + \dfrac{\pi}{2}\right)$ **43.** $y = \tan\left(\dfrac{2x}{3}\right)$

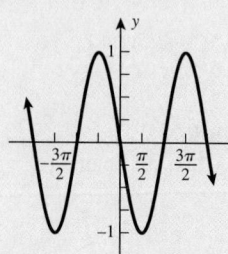

$x = -\dfrac{3\pi}{4}$ $x = \dfrac{3\pi}{4}$

47. $\dfrac{671}{1800}$ **49.** $\dfrac{336}{625}$

61. $-\dfrac{\sqrt{2 - \sqrt{2}}}{2}$

7. $\dfrac{7\sqrt{2}}{10}$

1125

77. $y = \left| \sin\left(\dfrac{\pi}{24}x\right) \right|$ **79.** $y = -\tan\left(\dfrac{x}{2}\right)$

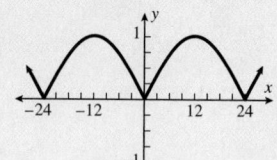   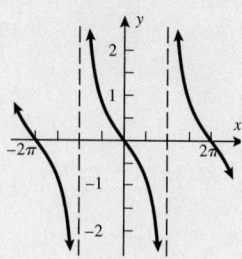

81. $3[\sin(7x) + \sin(3x)]$ **83.** $-2\sin(4x)\sin x$

85. $2\sin\left(\dfrac{x}{3}\right)\cos x$ **87.** $\cot(3x)$ **93.** $\dfrac{\pi}{4}$

95. $\dfrac{\pi}{2}$ **97.** $\dfrac{\pi}{6}$ **99.** $-90°$

101. $60°$ **103.** $-60°$ **105.** $-37.50°$

107. $22.50°$ **109.** 1.75 **111.** -0.10

113. $-\dfrac{\pi}{4}$ **115.** $-\sqrt{3}$ **117.** $\dfrac{\pi}{3}$

119. $\dfrac{60}{61}$ **121.** $\dfrac{7}{6}$ **123.** $\dfrac{6\sqrt{35}}{35}$

125. $\dfrac{2\pi}{3}, \dfrac{5\pi}{6}, \dfrac{5\pi}{3}, \dfrac{11\pi}{6}$

127. $-\dfrac{3\pi}{2}, -\dfrac{\pi}{2}$ **129.** $\dfrac{9\pi}{4}, \dfrac{21\pi}{4}$ **131.** $\dfrac{\pi}{3}, \dfrac{2\pi}{3}, \dfrac{4\pi}{3}, \dfrac{5\pi}{3}$

133. $\dfrac{3\pi}{8}, \dfrac{7\pi}{8}, \dfrac{11\pi}{8}, \dfrac{15\pi}{8}$ **135.** $0, \pi, \dfrac{3\pi}{4}, \dfrac{7\pi}{4}$

137. $80.46°, 170.46°,$ $260.46°, 350.46°$ **139.** $90°, 138.59°,$ $221.41°, 270°$

141. $17.62°, 162.38°$ **143.** $\dfrac{\pi}{4}, \dfrac{5\pi}{4}$

145. $\dfrac{\pi}{3}, \pi$ **147.** $0, \dfrac{\pi}{6}, \pi, \dfrac{11\pi}{6}$

149. $\dfrac{3\pi}{2}$ **151.** π **153.** $\dfrac{\pi}{2}, \dfrac{3\pi}{2}$

155. $90°, 135°, 270°, 315°$

157. $0°, 360°$ **159.** $60°, 90°, 270°, 300°$

161. (a) and (c)   **163.** $3\cos(3x)$

165. Y_1 and Y_3 **167.** Y_1 and Y_3 **169.** Y_1 and Y_3

171. a. $-\dfrac{3}{5}$ **b.** -0.6 **c.** yes   **173.** 0.5787

Practice Test

1. $x = \dfrac{(2n+1)}{2}\pi$, where n is an integer

3. $-\dfrac{\sqrt{2-\sqrt{2}}}{2}$ **5.** $\sqrt{\dfrac{3}{10}} = \dfrac{\sqrt{30}}{10}$

7. $\cos(10x)$ **9.** $\cos\left(\dfrac{a+b}{2}\right)$

11. $20\cos x\cos 3$ **13.** $\theta = \begin{cases} \dfrac{4\pi}{3} + 2n\pi \\ \dfrac{5\pi}{3} + 2n\pi \end{cases}$, where n is an integer

15. $14.48°, 90°, 165.52°, 270°$

17. conditional

19. $-\dfrac{\sqrt{26}}{26}$

21. $\cot\left(\dfrac{\pi}{6}x + \dfrac{\pi}{8}\right)$

23. One-to-one on the interval $\left[c - \dfrac{1}{2}, c\right) \cup \left(c, c + \dfrac{1}{2}\right]$.

$f^{-1}(x) = \dfrac{1}{\pi}\csc^{-1}\left(\dfrac{x-a}{b}\right) - \dfrac{c}{\pi}$.

25. $\dfrac{8}{3} + 8n, \dfrac{16}{3} + 8n$, where n is an integer

27. $\dfrac{\pi}{2} + 6n\pi, \dfrac{7\pi}{2} + 6n\pi$, where n is an integer

29. a. $\dfrac{3\sqrt{10}}{10}$ **b.** 0.94868 **c.** yes

Cumulative Test

1. a. $\dfrac{\sqrt{3}}{2}$ **b.** $\sqrt{3}$ **c.** -2

3. a. $\dfrac{2\pi}{3}$ **b.** $-\dfrac{\pi}{6}$ **c.** $\dfrac{5\pi}{6}$

5. even

7. $f \circ g = \dfrac{1}{x^3} - 1$; domain: $x \neq 0$

9. $\left(-\dfrac{6}{5}, -\dfrac{3}{5}\right)$

11. $Q(x) = 5x - 4, r(x) = -5x + 7$

13. HA: $y = 0.7$; VA: $x = -2, x = 3$

15. $(-3, \infty)$ **17.** 4 **19.** 0.4695

21. $-\dfrac{7\pi}{12}$ **23.** conditional   **25.** $\dfrac{5}{12}$

27. 1.3994

CHAPTER 7

Section 7.1

1. $\sqrt{13}$ **3.** $5\sqrt{2}$ **5.** 25

7. $\sqrt{73}; \theta = 69.4°$ **9.** $\sqrt{26}; \theta = 348.7°$ **11.** $\sqrt{17}; \theta = 166.0°$

13. $8; \theta = 180°$ **15.** $2\sqrt{3}; \theta = 60°$ **17.** $\langle -2, -2 \rangle$

19. $\langle -12, 9 \rangle$ **21.** $\langle 0, -14 \rangle$ **23.** $\langle -36, 48 \rangle$

25. $\langle 6.3, 3.0 \rangle$ **27.** $\langle -2.8, 15.8 \rangle$ **29.** $\langle 2.6, -3.1 \rangle$

31. $\langle 8.2, -3.8 \rangle$ **33.** $\langle -1, 1.7 \rangle$ **35.** $\left\langle -\dfrac{5}{13}, -\dfrac{12}{13} \right\rangle$

37. $\left\langle \dfrac{60}{61}, \dfrac{11}{61} \right\rangle$ **39.** $\left\langle \dfrac{24}{25}, -\dfrac{7}{25} \right\rangle$ **41.** $\left\langle -\dfrac{3}{5}, -\dfrac{4}{5} \right\rangle$

43. $\left\langle \dfrac{\sqrt{10}}{10}, \dfrac{3\sqrt{10}}{10} \right\rangle$ **45.** $7\mathbf{i} + 3\mathbf{j}$ **47.** $5\mathbf{i} - 3\mathbf{j}$

49. $-\mathbf{i} + 0\mathbf{j}$ **51.** $2\mathbf{i} + 0\mathbf{j}$ **53.** $-5\mathbf{i} + 5\mathbf{j}$

55. $7\mathbf{i} + 0\mathbf{j}$ **57.** vertical: 1100 ft/sec
horizontal: 1905 ft/sec

59. 2801 lb   **61.** 11.7 mph; 31° west of due north

63. 52.41° east of north; 303 mph   **65.** 250 lb

67. vertical: 51.4 ft/sec
horizontal: 61.3 ft/sec

69. 29.93 yd   **71.** 10.9°   **73.** 1156 lb

75. magnitude: 351.16; angle: 23.75° from 180° N force

77. tension: $5\sqrt{2}$; $|\mathbf{u}| = 5$

79. a. 0 **b.** $10 + 3\sqrt{2} + \sqrt{10}$ units

81. 8.97 Nm   **83.** 31.95 Nm

85. magnitude: 8.67; direction: 18.05° counterclockwise of S

87. Magnitude is never negative. Should not have factored out the negative but instead squared it in finding the magnitude.

89. false   **91.** true

93. vector   **95.** $\sqrt{a^2 + b^2}$

99. $\langle -6, 4 \rangle = \dfrac{1}{2}\langle -8, 4 \rangle - 2\langle 1, -1 \rangle$

103. directly

```
[[8][-5]]+3[[-7]
[11]]
        [[-13]
        [28 ]]
```

matrix

```
[[8][-5]]+3[[-7]
[11]]
        [[-13]
        [28 ]]
[A]+3[B]
        [[-13]
        [28 ]]
■
```

105. directly

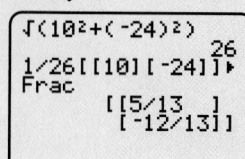

matrix

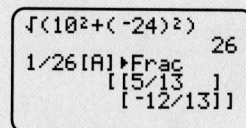

107. 183, 100.4°   **109.** $|\mathbf{v}(t)| = 1$ **111.** $\langle 1, 2t + h \rangle$

Section 7.2

1. 2   **3.** −3   **5.** 42

7. 11   **9.** −13a   **11.** −1.4

13. 98°   **15.** 109°   **17.** 3°

19. 30°   **21.** 105°   **23.** 180°

25. no   **27.** yes   **29.** no

31. yes   **33.** yes   **35.** yes

37. 400 ft-lb   **39.** 80,000 ft-lb   **41.** 1299 ft-lb

43. 148 ft-lb   **45.** 1607 lb   **47.** 694,593 ft-lb

49. $\mathbf{u} \cdot \mathbf{v} = 49,300$, and it represents the total cost of buying the prescribed number of 10-packs of both types of battery.

55. a. $v_1 = -14.72$ **b.** 64.57°

57. $\mathbf{n} = \langle r, 0 \rangle$; $\mathbf{u} \cdot \mathbf{n} = r|\mathbf{u}|$ **59.** −2

61. The dot product of two vectors is a scalar (not a vector). Should have summed the products of components.

63. false   **65.** true   **67.** 17

75. a. $-2\mathbf{u}$ **b.** $-c\mathbf{u}$

77. Any vector that is orthogonal (perpendicular) to $\mathbf{u}$.

79. −6 is minimum and 6 is maximum.

81. −1083   **83.** 31.43°   **85.** 47°

87. $\mathbf{u}(t) \cdot \mathbf{v}(t) = 2t^2 - 3t^3$ **89.** $\theta = 1.52$ rad

Section 7.3

1.

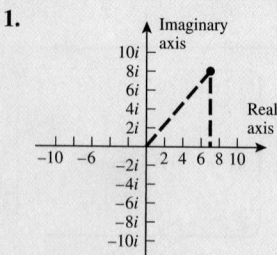

3.

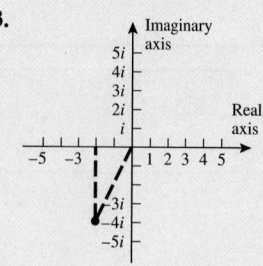

5.

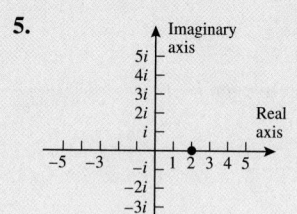

7.

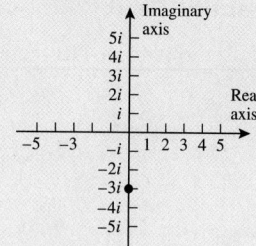

9. $\sqrt{2}\left[\cos\left(\dfrac{7\pi}{4}\right) + i\sin\left(\dfrac{7\pi}{4}\right)\right] = \sqrt{2}(\cos 315° + i\sin 315°)$

11. $2\left[\cos\left(\dfrac{\pi}{3}\right) + i\sin\left(\dfrac{\pi}{3}\right)\right] = 2(\cos 60° + i\sin 60°)$

13. $4\sqrt{2}\left[\cos\left(\dfrac{3\pi}{4}\right) + i\sin\left(\dfrac{3\pi}{4}\right)\right] = 4\sqrt{2}(\cos 135° + i\sin 135°)$

15. $2\sqrt{3}\left[\cos\left(\dfrac{5\pi}{3}\right) + i\sin\left(\dfrac{5\pi}{3}\right)\right] = 2\sqrt{3}(\cos 300° + i\sin 300°)$

17. $3(\cos 0 + i\sin 0) = 3(\cos 0° + i\sin 0°)$

19. $\dfrac{\sqrt{2}}{2}\left[\cos\left(\dfrac{5\pi}{4}\right) + i\sin\left(\dfrac{5\pi}{4}\right)\right] = \dfrac{\sqrt{2}}{2}(\cos 225° + i\sin 225°)$

21. $2\sqrt{3}\left[\cos\left(\dfrac{5\pi}{4}\right) + i\sin\left(\dfrac{5\pi}{4}\right)\right]$

23. $5\sqrt{2}\left[\cos\left(\dfrac{3\pi}{4}\right) + i\sin\left(\dfrac{3\pi}{4}\right)\right]$

25. $\sqrt{58}(\cos 293.2° + i\sin 293.2°)$

27. $\sqrt{61}(\cos 140.2° + i\sin 140.2°)$

29. $13(\cos 112.6° + i\sin 112.6°)$

31. $10(\cos 323.1° + i\sin 323.1°)$

33. $\dfrac{\sqrt{13}}{4}(\cos 123.7° + i\sin 123.7°)$

35. $5.59(\cos 24.27° + i\sin 24.27°)$

37. $\sqrt{17}(\cos 212.84° + i\sin 212.84°)$

39. $4.54(\cos 332.31° + i\sin 332.31°)$

41. −5   **43.** $\sqrt{2} - \sqrt{2}i$ **45.** $-2 - 2\sqrt{3}i$

47. $-\dfrac{3}{2} + \dfrac{\sqrt{3}}{2}i$ **49.** $1 + i$ **51.** $-3\sqrt{2} + 3\sqrt{2}i$

53. $2.1131 - 4.5315i$ **55.** $-0.5209 + 2.9544i$

57. $5.3623 - 4.4995i$ **59.** $-2.8978 + 0.7765i$

61. $0.6180 - 1.9021i$ **63.** $-0.8682 - 4.9240i$

65. a. 59.7 mi   **b.** $59.7(\cos 83.2° + i\sin 83.2°)$ **c.** 6.8°

67. a. $A\text{-}B: 0 - 2i$, $B\text{-}C: 3 - 3i$, $C\text{-}D: 0 - i$

   b. $A\text{-}D: 3\sqrt{5}(\cos 297° + i\sin 297°)$

69. $2\sqrt{13}\,(\cos 56.31° + i\sin 56.31°)$

71. The point is in QIII (not QI).

73. true   **75.** true   **77.** 0°

79. $|b|$ **81.** $a\sqrt{5}\,(\cos 296.6° + i\sin 296.6°)$

83. $\pi\left(\dfrac{1}{2} - \dfrac{\sqrt{3}}{2}i\right)$ **85.** $z = 8.79(\cos 28° + i\sin 28°)$

87. $-z = r[\cos(\theta + \pi) + i\sin(\theta + \pi)]$

89. $1.414, 45°, 1.414(\cos 45° + i\sin 45°)$

91. $2.236(\cos 26.6° + i\sin 26.6°)$ **93.** $35,323°$

95. $r = 5$ **97.** $r = 2\sin\theta$

Section 7.4

1. $-6 + 6\sqrt{3}i$ **3.** $-4\sqrt{2} - 4\sqrt{2}i$ **5.** $0 + 8i$

7. $\dfrac{9\sqrt{2}}{2} + \dfrac{9\sqrt{2}}{2}i$ **9.** $0 + 12i$ **11.** $\dfrac{3}{2} + \dfrac{3\sqrt{3}}{2}i$

13. $-\sqrt{2} + \sqrt{2}i$ **15.** $0 - 2i$ **17.** $\dfrac{3}{2} + \dfrac{3\sqrt{3}}{2}i$

19. $-\dfrac{5}{2} - \dfrac{5\sqrt{3}}{2}i$ **21.** $4 - 4i$ **23.** $-64 + 0i$

25. $-8 + 8\sqrt{3}i$ **27.** $1,048,576 + 0i$

29. $-1,048,576\sqrt{3} - 1,048,576i$

31. $2(\cos 150° + i\sin 150°)$ and $2(\cos 330° + i\sin 330°)$

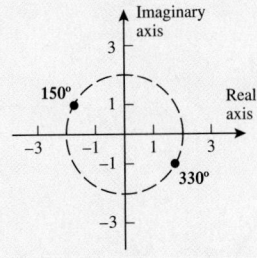

33. $\sqrt{6}\,(\cos 157.5° + i\sin 157.5°)$,
$\sqrt{6}\,(\cos 337.5° + i\sin 337.5°)$

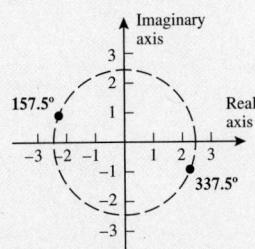

35. $2(\cos 20° + i\sin 20°), 2(\cos 140° + i\sin 140°),$
$2(\cos 260° + i\sin 260°)$

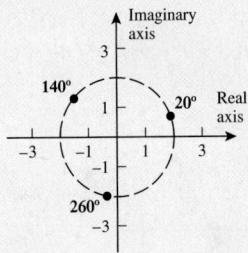

37. $\sqrt[3]{2}(\cos 110° + i\sin 110°), \sqrt[3]{2}(\cos 230° + i\sin 230°),$
$\sqrt[3]{2}(\cos 350° + i\sin 350°)$

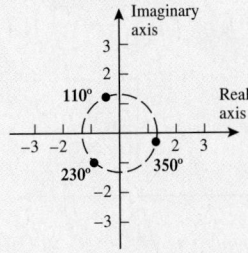

39. $2(\cos 78.75° + i\sin 78.75°), 2(\cos 168.75° + i\sin 168.75°),$
$2(\cos 258.75° + i\sin 258.75°), 2(\cos 348.75° + i\sin 348.75°)$

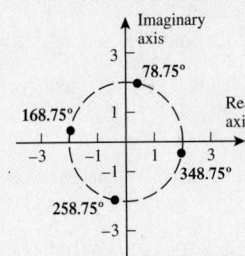

41. $x = \pm 2, x = \pm 2i$

43. $-2, 1 - \sqrt{3}i, 1 + \sqrt{3}i$

45. $\sqrt{2} - \sqrt{2}i, \sqrt{2} + \sqrt{2}i, -\sqrt{2} - \sqrt{2}i, -\sqrt{2} + \sqrt{2}i$

47. $1, -1, \dfrac{1}{2} + \dfrac{\sqrt{3}}{2}i, \dfrac{1}{2} - \dfrac{\sqrt{3}}{2}i, -\dfrac{1}{2} + \dfrac{\sqrt{3}}{2}i, -\dfrac{1}{2} - \dfrac{\sqrt{3}}{2}i$

49. $-\dfrac{\sqrt{2}}{2} + \dfrac{\sqrt{2}}{2}i, \dfrac{\sqrt{2}}{2} - \dfrac{\sqrt{2}}{2}i$

51. $\sqrt[4]{2}\left[\cos\left(\dfrac{\pi}{8} + \dfrac{\pi k}{2}\right) + i\sin\left(\dfrac{\pi}{8} + \dfrac{\pi k}{2}\right)\right], k = 0, 1, 2, 3$

53. $2\left[\cos\left(\dfrac{\pi}{5} + \dfrac{2\pi k}{5}\right) + i\sin\left(\dfrac{\pi}{5} + \dfrac{2\pi k}{5}\right)\right], k = 0, 1, 2, 3, 4$

55. $\pi^2\left[\cos\left(\dfrac{\pi}{14}+\dfrac{2\pi k}{7}\right)+i\sin\left(\dfrac{\pi}{14}+\dfrac{2\pi k}{7}\right)\right], k=0,1,2,3,4,5,6$

57. $(\cos 45°+i\sin 45°), (\cos 117°+i\sin 117°),$
$(\cos 189°+i\sin 189°), (\cos 261°+i\sin 261°),$
$(\cos 333°+i\sin 333°)$

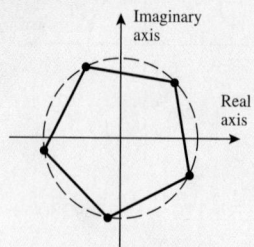

59. $\left[\cos\left(5\dfrac{\pi}{18}+\dfrac{\pi k}{3}\right)+i\sin\left(5\dfrac{\pi}{18}+\dfrac{\pi k}{3}\right)\right], k=0,1,2,3,4,5$

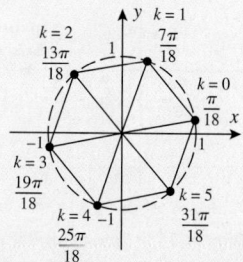

61. Reversed the order of angles being subtracted.

63. Use De Moivre's theorem. In general, $(a+b)^6 \neq a^6+b^6$.

65. true **67.** false **69.** true

71. true **81.** $2^{\frac{n+m}{2}}e^{\frac{\pi}{4}(m-n)i}$

83. $\cos 66°+i\sin 66°, \cos 138°+i\sin 138°, \cos 210°+i\sin 210°,$
$\cos 282°+i\sin 282°, \cos 354°+i\sin 354°$

85. $\cos 40°+i\sin 40°, \cos 100°+i\sin 100°, \cos 160°+i\sin 160°,$
$\cos 220°+i\sin 220°, \cos 280°+i\sin 280°,$
$\cos 340°+i\sin 340°$

87. $3(\cos 15°+\sin 15°), 3(\cos 135°+\sin 135°),$
$3(\cos 255°+\sin 255°)$

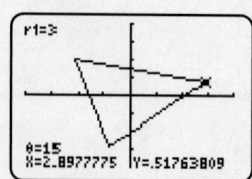

89. $\cos(2\theta)+i\sin(2\theta)=(\cos\theta+i\sin\theta)^2$
$=\cos^2\theta+2i\sin\theta\cos\theta+i^2\sin^2\theta$
$=(\cos^2\theta-\sin^2\theta)+i(2\sin\theta\cos\theta)$

$\cos(2\theta)=\cos^2\theta-\sin^2\theta$

91. $\cos(3\theta)+i\sin(3\theta)$
$=(\cos\theta+i\sin\theta)^3$
$=\cos^3\theta+3i\cos^2\theta\sin\theta+3i^2\cos\theta\sin^2\theta+i^3\sin^3\theta$
$=(\cos^3\theta-3\cos\theta\sin^2\theta)+i(3\cos^2\theta\sin\theta-\sin^3\theta)$
$=[\cos^3\theta-3\cos\theta(1-\cos^2\theta)]+i[3(1-\sin^2\theta)\sin\theta-\sin^3\theta]$
$=[\cos^3\theta-3\cos\theta+3\cos^3\theta]+i[3\sin\theta-3\sin^3\theta-\sin^3\theta]$
$=[4\cos^3\theta-3\cos\theta]+i[3\sin\theta-4\sin^3\theta]$
$\cos(3\theta)=4\cos^3\theta-3\cos\theta$

Section 7.5

The answers to Exercises 1, 3, 5, 7, and 9 are all plotted on the same graph below:

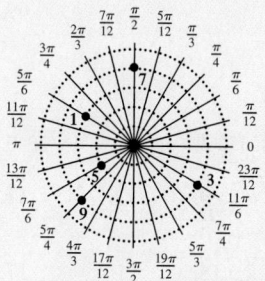

11. $\left(4,\dfrac{\pi}{3}\right)$ **13.** $\left(2,\dfrac{4\pi}{3}\right)$ **15.** $\left(4\sqrt{2},\dfrac{3\pi}{4}\right)$

17. $(3,0)$ **19.** $\left(2,\dfrac{7\pi}{6}\right)$ **21.** $(2,-2\sqrt{3})$

23. $\left(\dfrac{\sqrt{3}}{2},-\dfrac{1}{2}\right)$ **25.** $(0,0)$ **27.** $(-1,-\sqrt{3})$

29. $\left(\dfrac{\sqrt{2}}{2},-\dfrac{\sqrt{2}}{2}\right)$ **31.** d **33.** a

35. **37.**

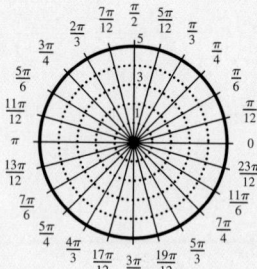 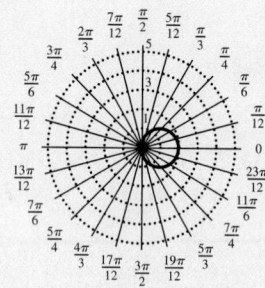

39.

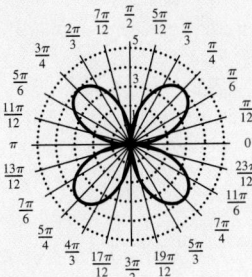

41.

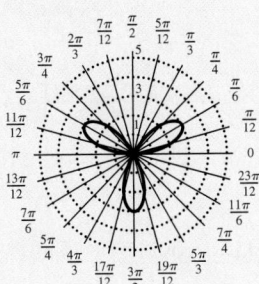

59.

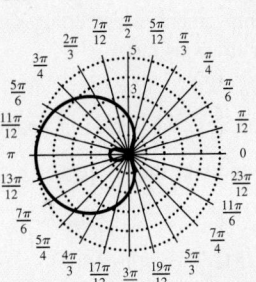

61.

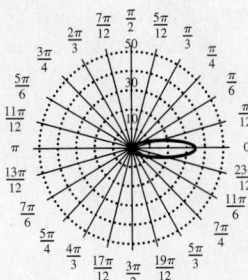

43.

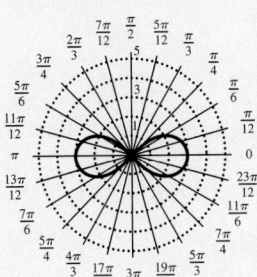

45.

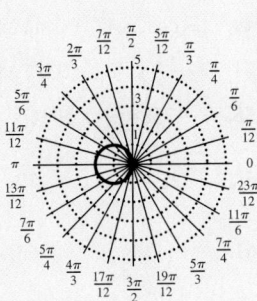

63.

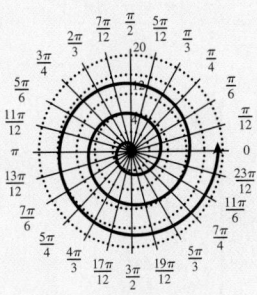

47.

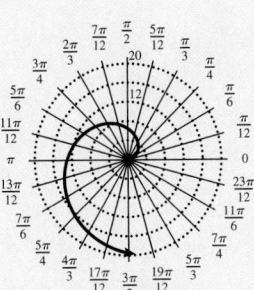

49.

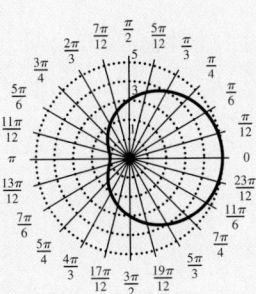

65.

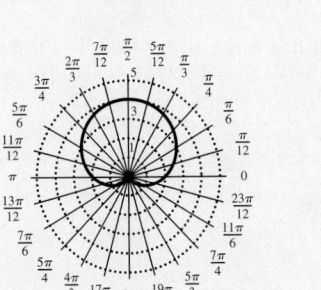

Multiplier 4

Multiplier $\frac{1}{4}$

51. Line: $y = -2x + 1$

53. Circle: $(x - 1)^2 + y^2 = 9$

67.

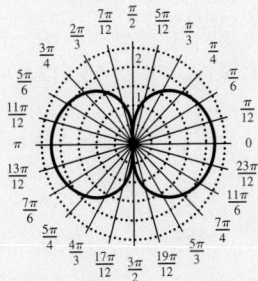

55.

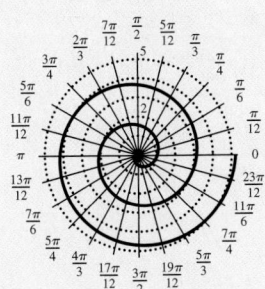

57.

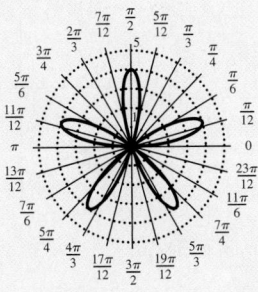

69. For (a)–(c) all three graphs generate the same set of points, as seen below:

1131

Note that all three graphs are figure eights. Extending the domain in (b) results in twice as fast movement, while doing so in (c) results in movement that is four times as fast.

71. 6 times
73. a. $r = 8\cos 3\theta$
    b. 50 times, because the period is 2π.

75. The point is in QIII; the angle found was the reference angle (needed to add π).

77. true
79. $r = \dfrac{a}{\cos\theta}$
81. $(-a, \theta \pm 180°)$

83. The graphs intersect when $\theta = \dfrac{\pi}{3}, \dfrac{2\pi}{3}, \dfrac{4\pi}{3}, \dfrac{5\pi}{3}$.

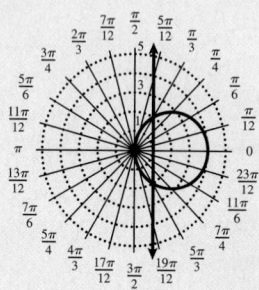

85. $x^3 - y^3 - 2axy = 0$

87. It is a circle of radius a and center $(a\cos b,\ a\sin b)$.

89. The inner loop is generated beginning with $\theta = \dfrac{\pi}{2}$ and ending with $\dfrac{3\pi}{2}$.

91. The very tip of the inner loop begins with $\theta = \dfrac{\pi}{2}$, then it crosses the origin (the first time) at $\theta = \cos^{-1}\left(-\frac{1}{3}\right)$, winds around, and eventually ends with $\theta = \dfrac{3\pi}{2}$.

93. 67.5°, 157.5°, 247.5°, 337.5°

95. $\left(2\sqrt{2}, \dfrac{\pi}{4}\right), \left(-2\sqrt{2}, \dfrac{5\pi}{4}\right)$

97. $\left(\dfrac{2 - \sqrt{2}}{2}, \dfrac{3\pi}{4}\right), \left(\dfrac{2 + \sqrt{2}}{2}, \dfrac{7\pi}{4}\right)$

Review Exercises

1. 13
3. 13
5. 26; 112.6°

7. 20; 323.1°
9. $\langle 2, 11 \rangle$
11. $\langle 38, -7 \rangle$
13. $\langle 2.6, 9.7 \rangle$

15. $\langle -3.1, 11.6 \rangle$
17. $\left\langle \dfrac{\sqrt{2}}{2}, -\dfrac{\sqrt{2}}{2} \right\rangle$
19. $(5\mathbf{i} + \mathbf{j})$

21. −6
23. −9
25. 16
27. 59°
29. 49°

31. 166°
33. no
35. yes
37. no
39. no

41.

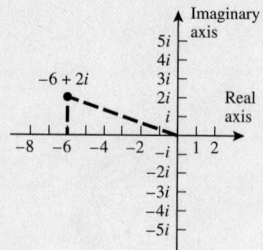

43. $2(\cos 315° + i\sin 315°)$

45. $8(\cos 270° + i\sin 270°)$

47. $61(\cos 169.6° + i\sin 169.6°)$

49. $17(\cos 28.1° + i\sin 28.1°)$

51. $3 - 3\sqrt{3}i$
53. $-1 + i$

55. $-3.7588 - 1.3681i$

57. $-12i$
59. $-\dfrac{21}{2} - \dfrac{21\sqrt{3}}{2}i$

61. $-\dfrac{\sqrt{3}}{2} + \dfrac{1}{2}i$
63. −6

65. −324
67. $16 - 16\sqrt{3}\,i$

69. $2(\cos 30° + i\sin 30°), 2(\cos 210° + i\sin 210°)$

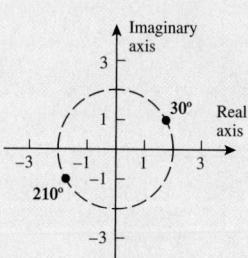

71. $4(\cos 45° + i\sin 45°), 4(\cos 135° + i\sin 135°),$
$4(\cos 225° + i\sin 225°), 4(\cos 315° + i\sin 315°)$

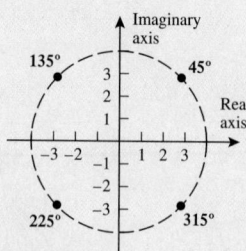

73. $-6, 3 + 3\sqrt{3}i, 3 - 3\sqrt{3}i$

75. $\dfrac{\sqrt{2}}{2} + \dfrac{\sqrt{2}}{2}i, \dfrac{\sqrt{2}}{2} - \dfrac{\sqrt{2}}{2}i, -\dfrac{\sqrt{2}}{2} + \dfrac{\sqrt{2}}{2}i, -\dfrac{\sqrt{2}}{2} - \dfrac{\sqrt{2}}{2}i$

All the points from Exercises 77–81 are graphed on the single graph below.

77. $\left(2\sqrt{2}, \dfrac{3\pi}{4}\right)$

79. $\left(10, \dfrac{7\pi}{6}\right)$

81. $\left(2, \dfrac{3\pi}{2}\right)$

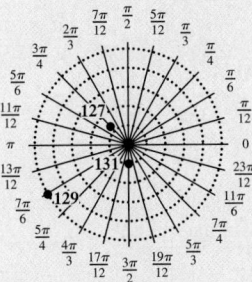

83. $\left(-\dfrac{3}{2}, \dfrac{3\sqrt{3}}{2}\right)$ **85.** $(1, \sqrt{3})$ **87.** $\left(-\dfrac{1}{2}, -\dfrac{\sqrt{3}}{2}\right)$

89.

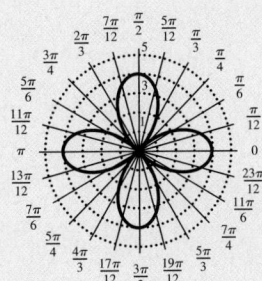

91.

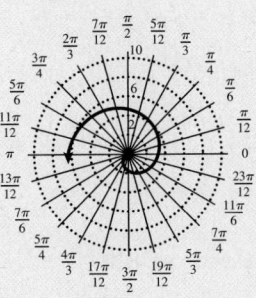

93. 65, 293° **95.** 140° **97.** 12, 246°

99. $2(\cos 30° + \sin 30°), 2(\cos 120° + \sin 120°),$
$2(\cos 210° + \sin 210°), 2(\cos 300° + \sin 300°)$

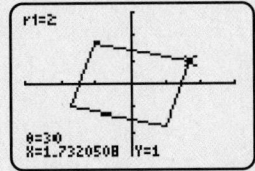

101. 10°, 50°, 130°, 170°, 250°, 290°

Practice Test

1. magnitude $= 13, \theta = 112.6°$

3. a. $\langle -14, 5\rangle$; **b.** -16

5. 9   **7.** $32{,}768\left(-1 + i\sqrt{3}\right)$

9. $\left(\dfrac{-3\sqrt{3}}{2}, -\dfrac{3}{2}\right)$ **11.** $\left(15\sqrt{5}, 333.4°\right)$

13.

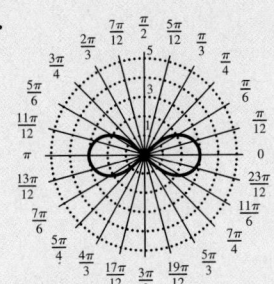

17. -20

19. $\sqrt{2}$

21. 50,769 lb at an angle of 12° from the tug pulling with a force of 27,000 lb

23. $4\left[\cos\left(\dfrac{3\pi}{8} + \dfrac{\pi k}{2}\right) + i\sin\left(\dfrac{3\pi}{8} + \dfrac{\pi k}{2}\right)\right], k = 0, 1, 2, 3$

25. 112°

Cumulative Test

1. $(f \circ g)(x) = \dfrac{-19 - 12x}{3x + 5}$
domain of f: $(-\infty, \infty)$
domain of g: $\left(-\dfrac{5}{3}, \infty\right)$
domain of $f \circ g$: $\left(-\dfrac{5}{3}, \infty\right)$

3. standard form: $f(x) = -\dfrac{1}{3}(x + 1)^2 + 2$
general form: $f(x) = -\dfrac{1}{3}x^2 - \dfrac{2}{3}x + \dfrac{5}{3}$

5. VA: $x = 1$ **7.** $\log_{625} 5 = \dfrac{1}{4}$
HA: none
SA: $y = x - 1$

9. $\sin\theta = \dfrac{3}{8}, \cos\theta = \dfrac{\sqrt{55}}{8}, \tan\theta = \dfrac{3\sqrt{55}}{55}$
$\cot\theta = \dfrac{\sqrt{55}}{3}, \sec\theta = \dfrac{8\sqrt{55}}{55}, \csc\theta = \dfrac{8}{3}$

11. $c = 13.1$ m, $\alpha = 56.1°, \beta = 73.9°$

13. amplitude: $\dfrac{1}{3}$, period: $\dfrac{\pi}{2}$, phase shift: $\dfrac{\pi}{4}$, vertical shift: 4

15. $\dfrac{25\sqrt{3} - 48}{11}$

CHAPTER 8

Section 8.1

1. $(8, -1)$ **3.** $(1, -1)$ **5.** $(1, 2)$

7. $\left(\frac{32}{17}, \frac{11}{17}\right)$ **9.** no solution

11. infinitely many solutions   **13.** infinitely many solutions

15. $(1, 3)$ **17.** $(6, 8)$ **19.** $(-6.24, -2.15)$

21. $(2, 5)$ **23.** $(-3, 4)$ **25.** $\left(1, -\frac{2}{7}\right)$

27. $\left(\frac{19}{7}, \frac{11}{35}\right)$ **29.** infinitely many solutions

31. $(4, 0)$ **33.** $(-2, 1)$ **35.** $(3, -2)$

37. $\left(\frac{75}{32}, \frac{7}{16}\right)$ **39.** $(4.2, -3.5)$ **41.** c

43. d

45. The solution is $(0, 0)$.   **47.** The solution is $(-1, -1)$.

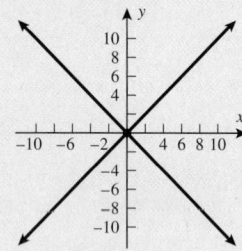   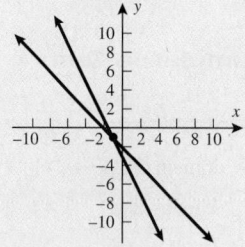

49. The solution is $(0, -6)$.   **51.** There are no solutions to this system of equations.

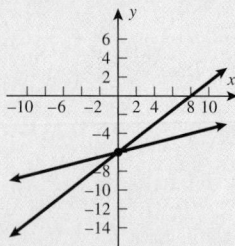   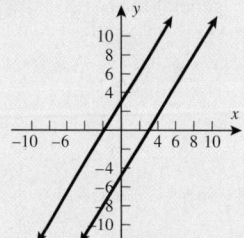

53. 15.86 ml of 8% HCl
21.14 ml of 15% HCl

55. $300,000 of sales

57. 169 miles highway, 180.5 miles city

59. Average plane speed is 450 mph; average wind is 50 mph.

61. $3,500 invested at 10% and $6,500 invested at 14%

63. The company must sell 8 CD players to break even.

65. $8 type I; $9 type II

67. 9,329,125 female; 8,999,215 male

69. Every term in Equation (1) is not multiplied correctly by -1.

71. The negative was not distributed through both terms:
$-(-3y - 4) = 3y + 4$.

73. false   **75.** false   **77.** $A = -4, B = 7$

79. 8 cups of 2% and 96 cups of 4%

81. $m = -1, b = 2$ **83.** $b = 2, c = 6$

85. $\left(-\sqrt{3}, -2\right), \left(-\sqrt{3}, 2\right), \left(\sqrt{3}, -2\right), \left(\sqrt{3}, 2\right)$

87. The point of intersection is approximately $(8.9, 6.4)$.   **89.** The lines coincide. Infinitely many solutions.

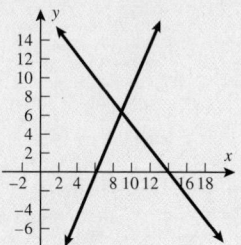   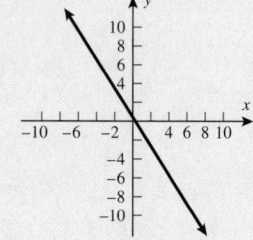

91. $(2.426, -0.059)$ **93.** $A = \frac{3}{2}, B = -\frac{1}{2}$

95. $A = \frac{4}{5}, B = \frac{1}{5}$

Section 8.2

1. $x = -\frac{3}{2}, y = -3, z = \frac{9}{2}$ **3.** $x = -2, y = \frac{9}{2}, z = \frac{1}{2}$

5. $x = 5, y = 3, z = -1$ **7.** $x = \frac{90}{31}, y = \frac{103}{31}, z = \frac{9}{31}$

9. $x = -\frac{13}{4}, y = \frac{1}{2}, z = -\frac{5}{2}$ **11.** $x = -2, y = -1, z = 0$

13. $x = 2, y = 5, z = -1$ **15.** no solution

17. no solution

19. $x = 1 - a, y = -\left(a + \frac{1}{2}\right), z = a$

21. $x = 41 + 4a, y = 31 + 3a, z = a$

23. $x_1 = -\frac{1}{2}, x_2 = \frac{7}{4}, x_3 = -\frac{3}{4}$

25. no solution

27. $x_1 = 1, x_2 = -1 + a, x_3 = a$

29. $x = \frac{2}{3}a + \frac{8}{3}, y = -\frac{1}{3}a - \frac{10}{3}, z = a$

31. $x = a, y = \frac{20}{3} - \frac{13}{3}a, z = 5 - 3a$

33. 6 Mediterranean chicken sandwiches
3 six-inch tuna sandwiches
5 six-inch roast beef sandwiches

35. $a = -32 \text{ ft/sec}^2, v_0 = 52 \text{ ft/sec}, h_0 = 0 \text{ ft}$

37. $y = -0.0625x^2 + 5.25x - 50$

39. $10,000 in the money market, $4,000 in the mutual fund, $6,000 in the stock

41. 33 pairs of regular model skis
72 pairs of trick skis
5 pairs of slalom skis

43. total points for first game: 885
total points for second game: 823
total points for third game: 883

45. Eagle: 1250, Birdie: 3750, Bogey: 5000

47. 10 seconds: 2, 20 seconds: 3, 40 seconds: 1

49. To eliminate a variable they must be opposite in sign ($\pm$). Equation (2) + Equation (3) yields a $2x$.

51. true   **53.** false   **55.** $a = 4, b = -2, c = -4$

57. $a = -\frac{55}{24}, b = -\frac{1}{4}, c = \frac{223}{24}, d = \frac{1}{4}, e = 44$

59. no solution   **61.** $x_1 = -2, x_2 = 1, x_3 = -4, x_4 = 5$

63. $A = 0, B = 1, C = 1, D = 0$

65. $x = 41 + 4a, y = 31 + 3a, z = a$

67. same result as Exercise 65   **69.** $\left(-\frac{80}{7}, -\frac{80}{7}, \frac{48}{7}\right)$

71. $A = 2, B = 3, C = -4$ **73.** $A = \frac{4}{3}, B = -1, C = -\frac{1}{3}$

Section 8.3

1. 2×3 **3.** 1×4 **5.** 1×1

7. $\begin{bmatrix} 3 & -2 & 7 \\ -4 & 6 & -3 \end{bmatrix}$ **9.** $\begin{bmatrix} 2 & -3 & 4 & -3 \\ -1 & 1 & 2 & 1 \\ 5 & -2 & -3 & 7 \end{bmatrix}$

11. $\begin{bmatrix} 1 & 1 & 0 & 3 \\ 1 & 0 & -1 & 2 \\ 0 & 1 & 1 & 5 \end{bmatrix}$ **13.** $\begin{bmatrix} -4 & 3 & 5 & 2 \\ 2 & -3 & -2 & -3 \\ -2 & 4 & 3 & 1 \end{bmatrix}$

15. $-3x + 7y = 2$
$\quad\ \ x + 5y = 8$
17. $-x = 4$
$\quad 7x + 9y + 3z = -3$
$\quad 4x + 6y - 5z = 8$
19. $x = a$
$\quad y = b$

21. The matrix is not in row–echelon form.

23. The matrix is in reduced row–echelon form.

25. The matrix is not in row–echelon form.

27. The matrix is in reduced row–echelon form.

29. The matrix is in row–echelon form.

31. $\begin{bmatrix} 1 & -2 & -3 \\ 0 & 7 & 5 \end{bmatrix}$ **33.** $\begin{bmatrix} 1 & -2 & -1 & 3 \\ 0 & 5 & -1 & 0 \\ 3 & -2 & 5 & 8 \end{bmatrix}$

35. $\begin{bmatrix} 1 & -2 & 5 & -1 & 2 \\ 0 & 1 & 1 & -3 & 3 \\ 0 & -2 & 1 & -2 & 5 \\ 0 & 0 & 1 & -1 & -6 \end{bmatrix}$ **37.** $\begin{bmatrix} 1 & 0 & 5 & -10 & -5 \\ 0 & 1 & 2 & -3 & -2 \\ 0 & 0 & -7 & 6 & 3 \\ 0 & 0 & 8 & -10 & -9 \end{bmatrix}$

39. $\begin{bmatrix} 1 & 0 & 4 & 0 & 27 \\ 0 & 1 & 2 & 0 & -11 \\ 0 & 0 & 1 & 0 & 21 \\ 0 & 0 & 0 & 1 & -3 \end{bmatrix}$ **41.** $\begin{bmatrix} 1 & 0 & -8 \\ 0 & 1 & 6 \end{bmatrix}$

43. $\begin{bmatrix} 1 & 0 & 0 & -2 \\ 0 & 1 & 0 & -1 \\ 0 & 0 & 1 & 0 \end{bmatrix}$ **45.** $\begin{bmatrix} 1 & 0 & 0 & 2 \\ 0 & 1 & 0 & 5 \\ 0 & 0 & 1 & -1 \end{bmatrix}$

47. $\begin{bmatrix} 1 & 0 & -2 & 1 \\ 0 & 1 & -2 & 2 \end{bmatrix}$ **49.** $\begin{bmatrix} 1 & 0 & 1 & 1 \\ 0 & 1 & 1 & -\frac{1}{2} \\ 0 & 0 & 0 & 0 \end{bmatrix}$

51. $x = -7, y = 5$

53. $x - 2y = -3$ or $x = 2a - 3, y = a$

55. no solution

57. $x = 4a + 41, y = 3a + 31, z = a$

59. $x_1 = -\frac{1}{2}, x_2 = \frac{7}{4}, x_3 = -\frac{3}{4}$

61. no solution

63. $x_1 = 1, x_2 = a - 1, x_3 = a$

65. $x = \frac{2}{3}a + \frac{8}{3}, y = -\frac{1}{3}a - \frac{10}{3}, z = a$

67. no solution

69. $x_1 = -2, x_2 = 1, x_3 = -4, x_4 = 5$

71. $(1, -2)$ **73.** no solution   **75.** $(-2, 1, 3)$

77. $(3, -2, 2)$ **79.** no solution

81. $x = \frac{a}{4} + 3, y = \frac{7a}{4} + 1, z = a$

83. $z = a, x = \frac{72 - 11a}{14}, y = \frac{13a + 4}{14}$

85. $x = 1, y = 2, z = -3, w = 1$

87. red dwarf: 960, yellow: 2,880,000, blue: 8

89. 2 chicken, 2 tuna, 8 roast beef, and 2 turkey–bacon

91. initial height = 0 ft, initial velocity = 50 ft/sec, acceleration = -32 ft/sec^2

93. $y = -0.053x^2 + 4.58x - 34.76$

95. $5,500 in the money market, $2,500 in the mutual fund, and $2,000 in the stock

97. $x = 25, y = 40, z = 6$

99. $x = 25, y = 30, z = 45$ (25 general tickets, 30 reserved tickets, and 45 end zone tickets)

101. $a = -\frac{22}{17}, b = -\frac{44}{17}, c = -\frac{280}{17}$

103. The correct matrix that is needed is
$$\left[\begin{array}{ccc|c} -1 & 1 & 1 & 2 \\ 1 & 1 & -2 & -3 \\ 1 & 1 & 1 & 6 \end{array}\right]$$

After reducing this matrix, the correct solution should be
$$\left[\begin{array}{ccc|c} 1 & 0 & 0 & 2 \\ 0 & 1 & 0 & 1 \\ 0 & 0 & 1 & 3 \end{array}\right]$$

105. Row 3 is not inconsistent. It implies $z = 0$.

107. false   **109.** true   **111.** false   **113.** false

115. $f(x) = -\frac{11}{6}x^4 + \frac{44}{3}x^3 - \frac{223}{6}x^2 + \frac{94}{3}x + 44$

117. 35 hours   **119.** $(-1, 2), (2, 5),$ and $(3, -1)$

121.
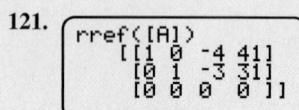
```
rref([A])
   [[1 0 -4 41]
    [0 1 -3 31]
    [0 0 0  0 ]]
```

123. a.

```
rref([A])
  [[1 0 0 -.23653...
   [0 1 0 .928846...
   [0 0 1 6.08846...
```

$y = -0.24x^2 + 0.93x + 6.09$

b.
```
QuadReg
 y=ax²+bx+c
 a=-.2365384615
 b=.9288461538
 c=6.088461538
■
```

$y = -0.24x^2 + 0.93x + 6.09$

125. $c_1 = \frac{3}{4}, c_2 = -\frac{3}{4}$ **127.** $c_1 = 2, c_2 = -3, c_3 = 1$

Section 8.4

1. 2×3 **3.** 2×2 **5.** 3×3

7. 4×4 **9.** $x = -5, y = 1$

11. $x = -3, y = -2, z = 3$ **13.** $x = 6, y = 3$

15. $\begin{bmatrix} -1 & 5 & 1 \\ 5 & 2 & 5 \end{bmatrix}$ **17.** $\begin{bmatrix} -2 & 4 \\ 2 & -2 \\ -1 & 3 \end{bmatrix}$

19. not defined   **21.** not defined

23. $\begin{bmatrix} -2 & 12 & 3 \\ 13 & 2 & 14 \end{bmatrix}$ **25.** $\begin{bmatrix} 8 & 3 \\ 11 & 5 \end{bmatrix}$

27. $\begin{bmatrix} -3 & 21 & 6 \\ -4 & 7 & 1 \\ 13 & 14 & 9 \end{bmatrix}$ **29.** $\begin{bmatrix} 3 & 6 \\ -2 & -2 \\ 17 & 24 \end{bmatrix}$

31. not defined   **33.** $[0 \quad 60]$

35. $[-6 \quad 1 \quad -9]$ **37.** $\begin{bmatrix} 7 & 10 & -8 \\ 0 & 15 & 5 \\ 23 & 0 & -7 \end{bmatrix}$

39. $\begin{bmatrix} 12 & 20 \\ 30 & 42 \end{bmatrix}$ **41.** $\begin{bmatrix} -4 \\ -4 \\ -16 \end{bmatrix}$

43. not defined

45. Yes, B is the multiplicative inverse of A.

47. Yes, B is the multiplicative inverse of A.

49. Yes, B is the multiplicative inverse of A.

51. $\begin{bmatrix} 0 & -1 \\ 1 & 2 \end{bmatrix}$ **53.** $\begin{bmatrix} -\frac{1}{13} & \frac{8}{39} \\ \frac{20}{39} & -\frac{4}{117} \end{bmatrix}$

55. $\begin{bmatrix} \frac{1}{2} & \frac{1}{2} & 0 \\ \frac{1}{2} & 0 & \frac{1}{2} \\ 0 & -\frac{1}{2} & -\frac{1}{2} \end{bmatrix}$ **57.** A^{-1} does not exist.

59. $\begin{bmatrix} -\frac{1}{2} & -\frac{1}{2} & \frac{5}{2} \\ \frac{1}{2} & \frac{1}{2} & -\frac{3}{2} \\ 0 & -1 & 1 \end{bmatrix}$ **61.** $\begin{bmatrix} \frac{1}{2} & \frac{1}{2} & 0 \\ \frac{3}{4} & \frac{1}{4} & -\frac{1}{2} \\ \frac{1}{4} & \frac{3}{4} & -\frac{1}{2} \end{bmatrix}$

63. $x = 2, y = -1$ **65.** $x = \frac{1}{2}, y = \frac{1}{3}$

67. $x = 0, y = 0, z = 1$

69. Cannot be solved because A is not invertible.

71. $x = -1, y = 1, z = -7$ **73.** $x = 3, y = 5, z = 4$

75. $A = \begin{bmatrix} 0.70 \\ 0.30 \end{bmatrix}$ $B = \begin{bmatrix} 0.89 \\ 0.84 \end{bmatrix}$

a. $46A = \begin{bmatrix} 32.2 \\ 13.8 \end{bmatrix}$ Out of 46 million people, about 32 million said yes they have tried to quit smoking and about 14 million people said they have not tried to quit smoking.

b. $46B = \begin{bmatrix} 40.94 \\ 38.64 \end{bmatrix}$ Out of 46 million people, about 41 million people believed smoking would increase their chances of developing lung cancer and about 39 million people believed smoking would shorten their lives.

77. $A = \begin{bmatrix} 0.589 & 0.628 \\ 0.414 & 0.430 \end{bmatrix}$ $B = \begin{bmatrix} 100M \\ 110M \end{bmatrix}$

$AB = \begin{bmatrix} 127.98M \\ 88.7M \end{bmatrix}$ tells us that there are 127.98M registered voters and of those 88.7M voted.

79. $A = [0.45 \quad 0.5 \quad 1], B = \begin{bmatrix} 7,523 \\ 2,700 \\ 15,200 \end{bmatrix}, AB = \$19,935.35$

81. $AB = \begin{bmatrix} \$0.228 \\ \$0.081 \\ \$0.015 \end{bmatrix}$ represents the cost per mile of a gas, hybrid, or electric car.

83. JAW   **85.** LEG   **87.** EYE

89. Matrix multiplication is not performed element by element. Instead, a row-by-column method should be used. The correct matrix multiplication would produce the answer:
$$\begin{bmatrix} 3 & 2 \\ 1 & 4 \end{bmatrix}\begin{bmatrix} -1 & 3 \\ -2 & 5 \end{bmatrix} = \begin{bmatrix} -7 & 19 \\ -9 & 23 \end{bmatrix}$$

91. A is not invertible because the identity matrix was not reached on the left side of the augmented matrix.

93. false   **95.** true   **97.** false

99. $A^2 = AA = \begin{bmatrix} a_{11} & a_{12} \\ a_{21} & a_{22} \end{bmatrix}\begin{bmatrix} a_{11} & a_{12} \\ a_{21} & a_{22} \end{bmatrix}$
$$= \begin{bmatrix} a_{11}^2 + a_{12}a_{21} & a_{11}a_{12} + a_{12}a_{22} \\ a_{21}a_{11} + a_{22}a_{21} & a_{21}a_{12} + a_{22}^2 \end{bmatrix}$$

101. $x = 9$

103. $A^2 = \begin{bmatrix} 2 & 2 \\ 2 & 2 \end{bmatrix}$ $A^3 = \begin{bmatrix} 4 & 4 \\ 4 & 4 \end{bmatrix}$
$A^4 = \begin{bmatrix} 8 & 8 \\ 8 & 8 \end{bmatrix}$ $A^n = \begin{bmatrix} 2^{n-1} & 2^{n-1} \\ 2^{n-1} & 2^{n-1} \end{bmatrix} = 2^{n-1} \cdot A$

105. must have $m = p$

109. $ad - bc = 0$

111. $\begin{bmatrix} 33 & 35 \\ -96 & -82 \\ 31 & 19 \\ 146 & 138 \end{bmatrix}$ **113.** not defined

115. $\begin{bmatrix} 5 & -4 & 4 \\ 2 & -15 & -3 \\ 26 & 4 & -8 \end{bmatrix}$

117. $\begin{bmatrix} -\frac{115}{6008} & \frac{431}{6008} & -\frac{1067}{6008} & \frac{103}{751} \\ \frac{411}{6008} & -\frac{391}{6008} & \frac{731}{6008} & -\frac{22}{751} \\ \frac{57}{751} & \frac{28}{751} & -\frac{85}{751} & \frac{3}{751} \\ -\frac{429}{6008} & \frac{145}{6008} & \frac{1035}{6008} & \frac{12}{751} \end{bmatrix}$

119. $\begin{bmatrix} \frac{1}{4x} & \frac{1}{4x} \\ \frac{1}{4y} & -\frac{1}{4y} \end{bmatrix}$ **121.** $\begin{bmatrix} \cos\theta & -\sin\theta \\ \sin\theta & \cos\theta \end{bmatrix}$

Section 8.5

1. -2 **3.** 31 **5.** -28

7. -0.6 **9.** 0 **11.** $x = 5, y = -6$

13. $x = -2, y = 1$ **15.** $x = -3, y = -4$

17. $x = -2, y = 5$ **19.** $x = 2, y = 2$

21. $D = 0$, inconsistent or dependent system

23. $D = 0$, inconsistent or dependent system

25. $x = \frac{1}{2}, y = -1$ **27.** $x = 1.5, y = 2.1$

29. $x = 0, y = 7$ **31.** 7

33. -25 **35.** -180

37. 0 **39.** 238

41. 0 **43.** $x = 2, y = 3, z = 5$

45. $x = -2, y = 3, z = 5$ **47.** $x = 2, y = -3, z = 1$

49. $D = 0$, inconsistent or dependent system

51. $D = 0$, inconsistent or dependent system

53. $x = -3, y = 1, z = 4$ **55.** $x = 2, y = -3, z = 5$

57. $x = -2, y = \frac{3}{2}, z = 3$ **59.** The points are collinear.

61. 6 units^2 **63.** 6 units^2

65. $y = 2x$ **67.** $I_1 = \frac{7}{2}, I_2 = \frac{5}{2}, I_3 = 1$

69. The mistake is forgetting the sign array (alternating signs).

71. D_x should have the first column replaced with the column of constants, and D_y should have the second column replaced with the column of constants.

73. true   **75.** false   **77.** abc

79. -419 **85.** -180 **87.** -1019

89. $x = -6.4, y = 1.5, z = 3.4$ **91.** r

93. $-\rho^2 \sin\phi$

Section 8.6

1. d   **3.** a   **5.** b

7. $\dfrac{A}{x-5} + \dfrac{B}{x+4}$ **9.** $\dfrac{A}{x-4} + \dfrac{B}{x} + \dfrac{C}{x^2}$

11. $2x - 6 + \dfrac{3x+33}{x^2+x+5}$ **13.** $\dfrac{Ax+B}{x^2+10} + \dfrac{Cx+D}{(x^2+10)^2}$

15. $\dfrac{1}{x} - \dfrac{1}{x+1}$ **17.** $\dfrac{1}{x-1}$

19. $\dfrac{2}{x-3} + \dfrac{7}{x+5}$ **21.** $\dfrac{3}{x-1} + \dfrac{4}{(x-1)^2}$

23. $\dfrac{4}{x+3} - \dfrac{15}{(x+3)^2}$ **25.** $\dfrac{3}{x+1} + \dfrac{1}{x-5} + \dfrac{2}{(x-5)^2}$

27. $\dfrac{-2}{x+4} + \dfrac{7x}{x^2+3}$

29. $\dfrac{-2}{x-7} + \dfrac{4x-3}{3x^2-7x+5}$

31. $\dfrac{x}{x^2+9} - \dfrac{9x}{\left(x^2+9\right)^2}$

33. $\dfrac{2x-3}{x^2+1} + \dfrac{5x+1}{\left(x^2+1\right)^2}$

35. $\dfrac{1}{x-1} + \dfrac{-3x-1}{2\left(x^2+1\right)} + \dfrac{1}{2(x+1)}$

37. $\dfrac{3}{x-1} + \dfrac{2x+5}{x^2+2x-1}$

39. $\dfrac{1}{x-1} + \dfrac{1-x}{x^2+x+1}$

41. $\dfrac{1}{d_o} + \dfrac{1}{d_i} = \dfrac{1}{f}$

43. $-\dfrac{11}{4}e^{2t} + \dfrac{7}{4}e^{-2t}$

45. The irreducible quadratic term $\left(x^2+1\right)$ should have $Bx + C$ in the numerator, not just B.

47. false   **49.** true   **51.** false

53. $\dfrac{1}{x-1} - \dfrac{1}{x+2} + \dfrac{1}{x-2}$

55. $\dfrac{1}{x} + \dfrac{1}{x+1} - \dfrac{1}{x^3}$

57. $\dfrac{x}{x^2+1} - \dfrac{2x}{\left(x^2+1\right)^2} + \dfrac{x+2}{\left(x^2+1\right)^3}$

59. yes   **61.** no

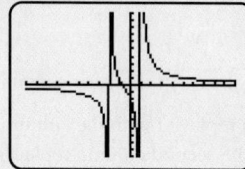 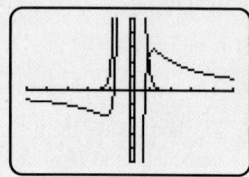

63. yes

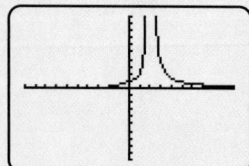

65. $\displaystyle\sum_{k=1}^{\infty}\left(\dfrac{3}{k} - \dfrac{3}{k+3}\right)$

67. $\displaystyle\sum_{k=1}^{\infty}\left[\dfrac{1}{k^2} - \dfrac{1}{(k+1)^2}\right]$

Section 8.7.

1. d   **3.** b

5.

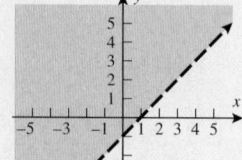

7.

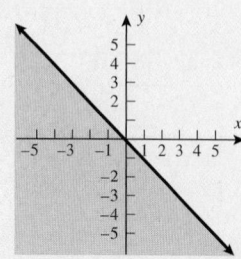

9.

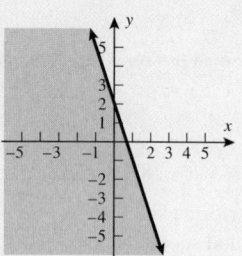

11.

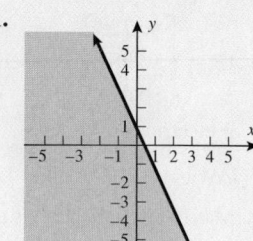

13.

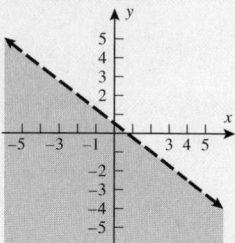

15.

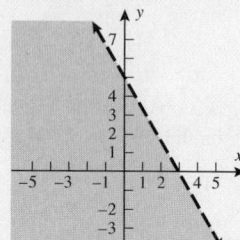

17.

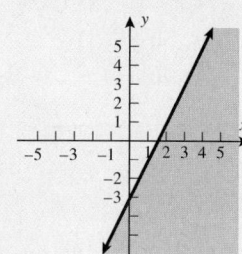

19.

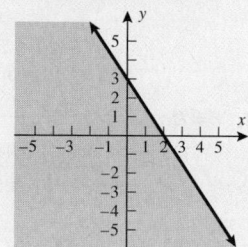

21.

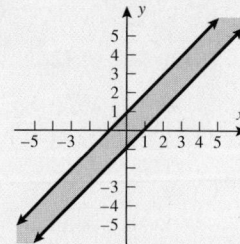

23. no solution

25.

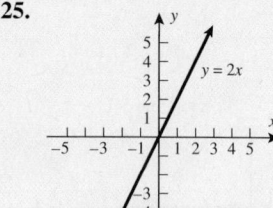

27.

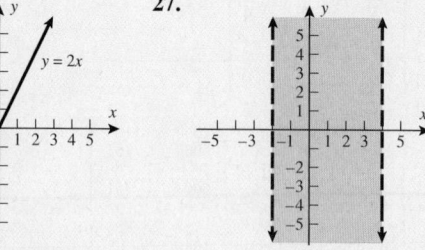

29.

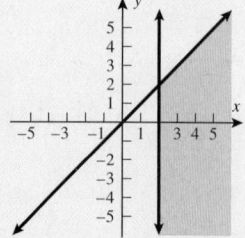

31.

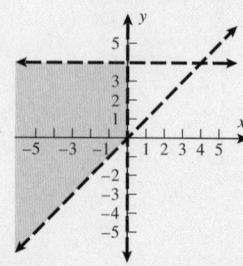

33.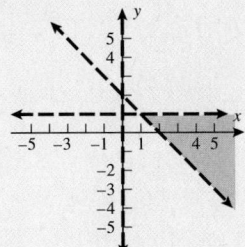

35.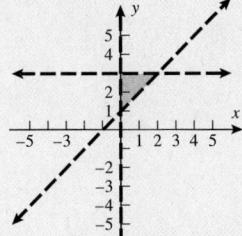

37.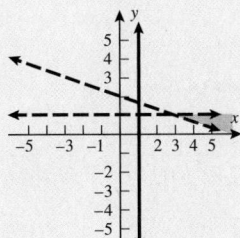

39.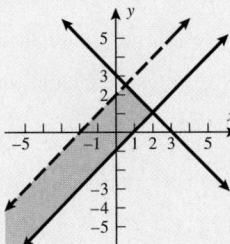

41.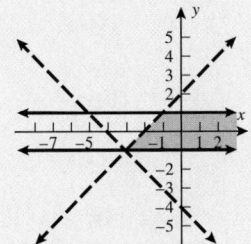

43.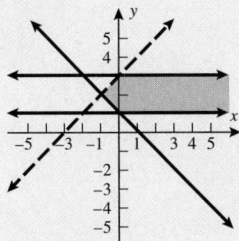

45. no solution

47.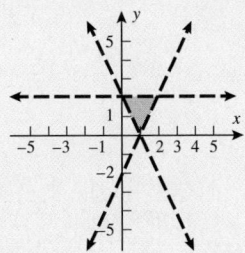

49.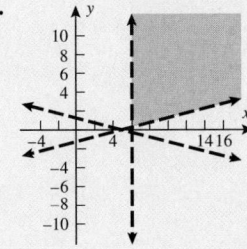

51. $f(x, y) = z = 2x + 3y$
$f(-1, 4) = 10$
$f(2, 4) = 16$ (MAX)
$f(-2, -1) = -7$ (MIN)
$f(1, -1) = -1$

53. $f(x, y) = z = 1.5x + 4.5y$
$f(-1, 4) = 16.5$
$f(2, 4) = 21$ (MAX)
$f(-2, -1) = -7.5$ (MIN)
$f(1, -1) = -3$

55. minimize at $f(0, 0) = 0$

57. no maximum

59. minimize at $f(0, 0) = 0$

61. maximize at $f(1, 6) = 2.65$

63. $P \leq 80 - 0.01x$
$P \geq 60$
$x \geq 0$

65. $20,000

67. $\begin{aligned} x + 20y &\leq 2400 \qquad x \geq 0 \\ 25x + 150y &\leq 6000 \qquad y \geq 0 \end{aligned}$

69. 50 Charley T-shirts, 130 Frances T-shirts (profit = $950)

71. 0 desktops, 25 laptops (profit = $7,500)

73. 3 first-class cars and 27 second-class cars

75. Let x represent the number of pairs of regular skis and y represent the number of pairs of slalom skis.
The profit function $P = 25x + 50y$.
The constraints are $x \geq 200$
$\qquad\qquad y \geq 80$
$\qquad\qquad x + y \leq 400$
$\qquad\qquad x \geq 0 \quad y \geq 0$

The feasible region is shown with the vertices indicated:

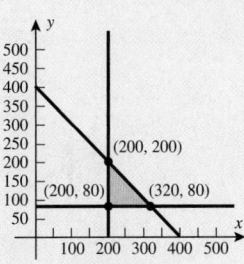

Evaluate the objective function at each vertex:
At (200, 80), $P = 25(200) + 50(80) = 9000$
At (200, 200), $P = (200) + 50(200) = 15,000$
At (320, 80), $P = 25(320) + 50(80) = 12,000$

Thus, the ski producer should produce 200 pairs of regular skis and 200 pairs of slalom skis to maximize profits.

77. Because points were not tested, the wrong half-plane was shaded.

79. true   **81.** false   **83.** false

85. The solution is a shaded rectangle.

87.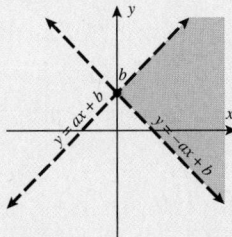

89. The maximum is a and occurs at $(0, a)$.

91. 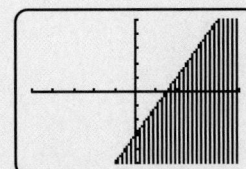   **93.**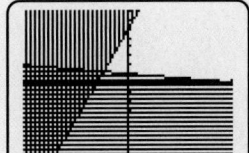

95.

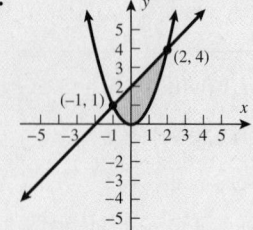

97.

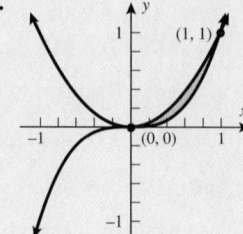

59. $\begin{bmatrix} 3 & 5 & 2 \\ 7 & 8 & 1 \end{bmatrix}$

61. $\begin{bmatrix} 9 & -4 \\ 9 & 9 \end{bmatrix}$

63. $\begin{bmatrix} 4 & 13 \\ 18 & 11 \end{bmatrix}$

65. $\begin{bmatrix} 0 & -19 \\ -18 & -9 \end{bmatrix}$

67. $\begin{bmatrix} -7 & -11 & -8 \\ 3 & 7 & 2 \end{bmatrix}$

69. $\begin{bmatrix} 10 & -13 \\ 18 & -20 \end{bmatrix}$

71. $\begin{bmatrix} 17 & -8 & 18 \\ 33 & 0 & 42 \end{bmatrix}$

73. $\begin{bmatrix} 10 & 9 & 20 \\ 22 & -4 & 2 \end{bmatrix}$

75. Yes, B is the multiplicative inverse of A.

Review Exercises

1. $(3, 0)$ **3.** $\left(\frac{13}{4}, 8\right)$ **5.** $(2, 1)$ **7.** $\left(\frac{19}{8}, \frac{13}{8}\right)$

77. Yes, B is the multiplicative inverse of A.

9. $(-2, 1)$ **11.** $\left(12, \frac{35}{6}\right)$

79. $\begin{bmatrix} \frac{2}{5} & -\frac{1}{5} \\ \frac{3}{10} & \frac{1}{10} \end{bmatrix}$

81. $\begin{bmatrix} 0 & -\frac{1}{2} \\ 1 & 0 \end{bmatrix}$

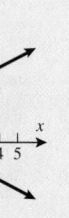

83. $\begin{bmatrix} -\frac{1}{6} & \frac{7}{12} & -\frac{1}{12} \\ \frac{1}{2} & -\frac{1}{4} & -\frac{1}{4} \\ \frac{1}{6} & -\frac{1}{12} & -\frac{5}{12} \end{bmatrix}$

85. $\begin{bmatrix} 0 & -\frac{2}{5} & \frac{1}{5} \\ 1 & -\frac{2}{5} & \frac{1}{5} \\ -\frac{1}{2} & \frac{3}{10} & \frac{1}{10} \end{bmatrix}$

13. $(3, -2)$ **15.** $(-1, 2)$

87. $x = 5, y = 4$ **89.** $x = 8, y = 12$

17. c   **19.** d

91. $x = 1, y = 2, z = 3$ **93.** -8

95. 5.4 **97.** $x = 3, y = 1$

21. 10.5 milliliters of 6%; 31.5 milliliters of 18%

99. $x = 6, y = 0$ **101.** $x = 90, y = 155$

23. $x = -1, y = -a + 2, z = a$

103. 11   **105.** $-abd$

25. no solution

107. $x = 1, y = 1, z = 2$ **109.** $x = -\frac{15}{7}, y = -\frac{25}{7}, z = \frac{19}{14}$

27. $y = -0.0050x^2 + 0.4486x - 3.8884$

111. $\dfrac{A}{x-1} + \dfrac{B}{(x-1)^2} + \dfrac{C}{x+3} + \dfrac{D}{x-5}$

29. $\begin{bmatrix} 5 & 7 & | & 2 \\ 3 & -4 & | & -2 \end{bmatrix}$

31. $\begin{bmatrix} 2 & 0 & -1 & | & 3 \\ 0 & 1 & -3 & | & -2 \\ 1 & 0 & 4 & | & -3 \end{bmatrix}$

113. $\dfrac{A}{x} + \dfrac{B}{4x+5} + \dfrac{C}{(2x+1)} + \dfrac{D}{(2x+1)^2}$

33. row–echelon   **35.** not row–echelon

115. $\dfrac{A}{x-3} + \dfrac{B}{x+4}$ **117.** $\dfrac{Ax+B}{x^2+17} + \dfrac{Cx+D}{(x^2+17)^2}$

37. $\begin{bmatrix} 1 & -2 & | & 1 \\ 0 & 1 & | & -1 \end{bmatrix}$

39. $\begin{bmatrix} 1 & -4 & 3 & | & -1 \\ 0 & -2 & 3 & | & -2 \\ 0 & 1 & -4 & | & 8 \end{bmatrix}$

119. $\dfrac{4}{x-1} + \dfrac{5}{x+7}$ **121.** $\dfrac{1}{2x} + \dfrac{15}{2(x-5)} - \dfrac{3}{2(x+5)}$

123. $\dfrac{-2}{x+1} + \dfrac{2}{x}$ **125.** $\dfrac{5}{x+2} - \dfrac{27}{(x+2)^2}$

41. $\begin{bmatrix} 1 & 0 & | & \frac{3}{5} \\ 0 & 1 & | & -\frac{1}{5} \end{bmatrix}$

43. $\begin{bmatrix} 1 & 0 & 0 & | & -4 \\ 0 & 1 & 0 & | & 8 \\ 0 & 0 & 1 & | & -4 \end{bmatrix}$

45. $x = \frac{5}{4}, y = \frac{7}{8}$ **47.** $x = 2, y = 1$

127.

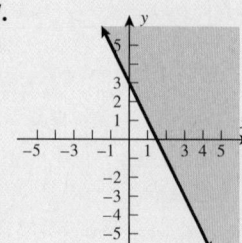

129.

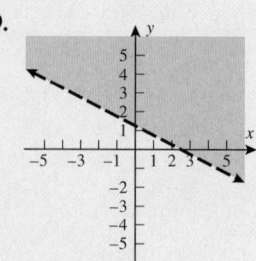

49. $x = -\frac{74}{21}, y = -\frac{73}{21}, z = -\frac{3}{7}$

51. $x = 1, y = 3, z = -5$

53. $x = -\frac{3}{7}a - 2, y = \frac{2}{7}a + 2, z = a$

55. $y = -0.005x^2 + 0.45x - 3.89$ **57.** not defined

131.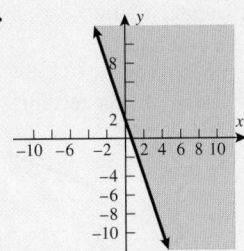

133.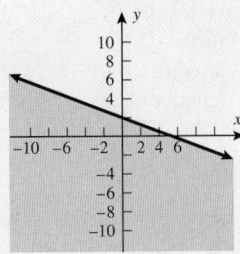

135. no solution

137.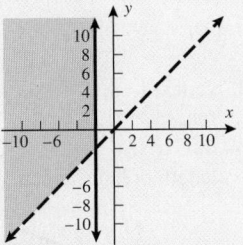

139.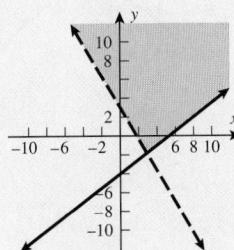

147. $(2, -3)$

149. $(3.6, 3, 0.8)$

151. a.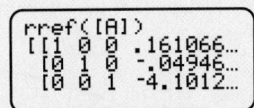

$y = 0.16x^2 - 0.05x - 4.10$

b.

$y = 0.16x^2 - 0.05x - 4.10$

153. $\begin{bmatrix} -238 & 206 & 50 \\ -113 & 159 & 135 \\ 40 & -30 & 0 \end{bmatrix}$

155. $x = 2.25, y = -4.35$

157. $x = -9.5, y = 3.4$

159. yes

161.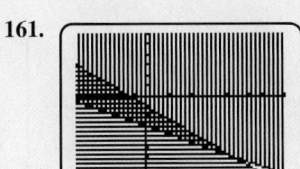

163. $z(1.8, 0.6) = 12.06$

Practice Test

1. $(7, 3)$

3. $x = a, y = a - 2$

5. $x = 1, y = -5, z = 3$

7. $\begin{bmatrix} 6 & 9 & 1 & | & 5 \\ 2 & -3 & 1 & | & 3 \\ 10 & 12 & 2 & | & 9 \end{bmatrix}$

9. $\begin{bmatrix} 1 & 3 & 5 \\ 0 & 1 & -11 \\ 0 & 7 & 15 \end{bmatrix}$

11. $x = -\frac{1}{3}a + \frac{7}{6}, y = \frac{1}{9}a - \frac{2}{9}, z = a$

13. $\begin{bmatrix} -11 & 19 \\ -6 & 8 \end{bmatrix}$

15. $\begin{bmatrix} \frac{1}{19} & \frac{3}{19} \\ \frac{5}{19} & -\frac{4}{19} \end{bmatrix}$

17. $x = -3, y = 1, z = 7$

19. -31

21. $x = 1, y = -1, z = 2$

23. $\dfrac{5}{x} - \dfrac{3}{x + 1}$

25. $\dfrac{1}{3x} + \dfrac{2}{3(x - 3)} - \dfrac{1}{x + 3}$

27.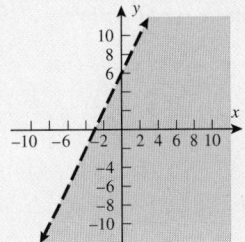

29.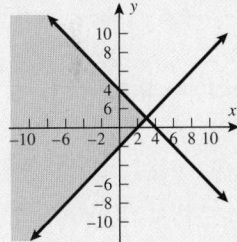

31. The minimum value occurs at $(0, 1)$, $f(0, 1) = 7$.

33. $(12.5, -6.4)$

Cumulative Test

1. 27

3. $2x + h - 3$

5. $(15, 6)$

7. VA: $x = 3$; HA: $y = -5$

9. -1

11. 2.585

13. $\dfrac{\sqrt{3}}{2}$

15. $\sin\theta = \dfrac{2\sqrt{29}}{29}, \cos\theta = -\dfrac{5\sqrt{29}}{29}, \tan\theta = -\dfrac{2}{5}, \cot\theta = -\dfrac{5}{2},$

$\sec\theta = -\dfrac{\sqrt{29}}{5}, \csc\theta = \dfrac{\sqrt{29}}{2}$

141. Minimum of z is 0 and occurs at $(0, 0)$.

143. Minimum of z is -30 and occurs at $(0, 6)$.

145. 10 watercolor, 30 geometric (profit = \$390)

17.

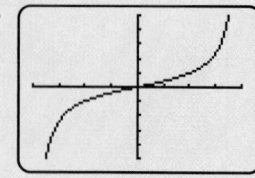

19. Domain: all real numbers such that $x \neq n\pi$, where n is an integer. Range: all real numbers.

21. $-\sqrt{3} + 2\sqrt{2}$ or $-\sqrt{2} - 1$

23. $\alpha = 36.3°; b = 123$ m; $c = 78.1$ m

25. $3, -\dfrac{3}{2} - \dfrac{3\sqrt{3}}{2}i, -\dfrac{3}{2} + \dfrac{3\sqrt{3}}{2}i$

27. $CB = \begin{bmatrix} 72 & -18 & 54 \\ 26 & -2 & 4 \end{bmatrix}$

29. $x = \dfrac{3}{11}, y = -\dfrac{2}{11}$

31.

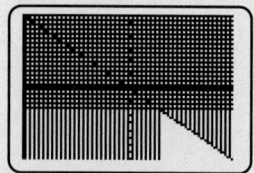

CHAPTER 9

Section 9.1

1. hyperbola   **3.** circle   **5.** hyperbola
7. ellipse   **9.** parabola   **11.** circle

Section 9.2

1. c   **3.** d   **5.** c
7. a   **9.** $x^2 = 12y$ **11.** $y^2 = -20x$

13. $(x - 3)^2 = 8(y - 5)$ **15.** $(y - 4)^2 = -8(x - 2)$

17. $(x - 2)^2 = 12(y - 1)$ **19.** $(y + 1)^2 = 4(x - 2)$

21. $(y - 2)^2 = 8(x + 1)$ **23.** $(x - 2)^2 = -8(y + 1)$

25. vertex: $(0, 0)$
focus: $(0, 2)$
directrix: $y = -2$
length of latus rectum: 8

27. vertex: $(0, 0)$
focus: $\left(-\dfrac{1}{2}, 0\right)$
directrix: $x = \dfrac{1}{2}$
length of latus rectum: 2

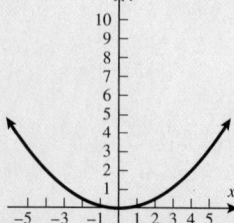

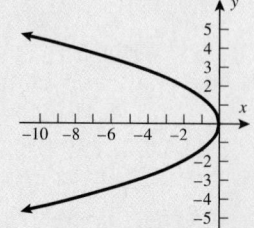

29. vertex: $(0, 0)$
focus: $(0, 4)$
directrix: $y = -4$
length of latus rectum: 16

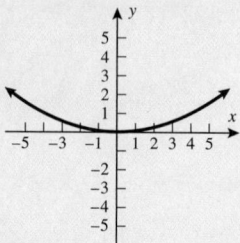

31. vertex: $(0, 0)$
focus: $(1, 0)$
directrix: $x = -1$
length of latus rectum: 4

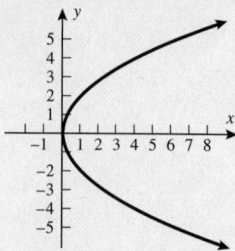

33. vertex: $(-3, 2)$
focus: $(-2, 2)$
directrix: $x = -4$
length of latus rectum: 4

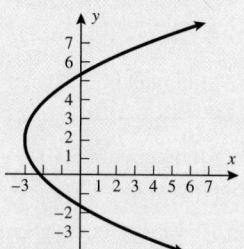

35. vertex: $(3, -1)$
focus: $(3, -3)$
directrix: $y = 1$
length of latus rectum: 8

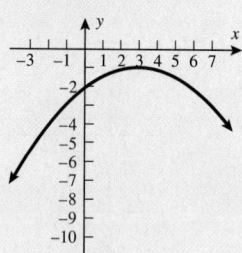

37. vertex: $(-5, 0)$
focus: $\left(-5, -\dfrac{1}{2}\right)$
directrix: $y = \dfrac{1}{2}$
length of latus rectum: 2

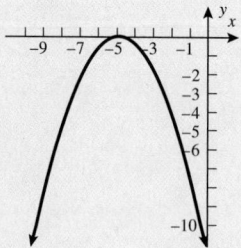

39. vertex: $(0, 2)$
focus: $\left(\dfrac{1}{2}, 2\right)$
directrix: $x = -\dfrac{1}{2}$
length of latus rectum: 2

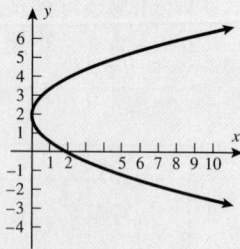

41. vertex: $(-3, -1)$
focus: $(-1, -1)$
directrix: $x = -5$
length of latus rectum: 8

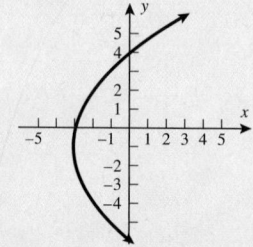

43. vertex: $\left(\dfrac{1}{2}, \dfrac{5}{4}\right)$
focus: $\left(\dfrac{1}{2}, 1\right)$
directrix: $y = \dfrac{3}{2}$
length of latus rectum: 1

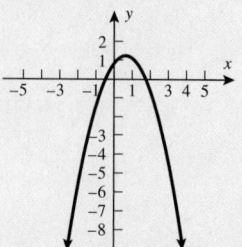

1142

45. The focus will be at (0, 2) so the receiver should be placed 2 feet from the vertex.

47. $x = \frac{1}{8}y^2$, $-2.5 \le y \le 2.5$

49. $x^2 = 160y$, $-50 \le x \le 50$

51. Yes. The opening height is 18.75 feet, and the mast is only 17 feet.

53. focal length $= 374.25$ ft
equation: $x^2 = 1497y$ (measurement in feet)

55. 55

57. The correction that needs to be made is that the formula $y^2 = 4px$ should be used.

59. true

61. false

63. $x = h - 1$

65. $\left(6, \frac{13}{2}\right)$

69. $(y - 6)^2 = 8(x - 2)$, $(y - 6)^2 = -8(x - 6)$

71. $(-2, 3)$, $(2, 3)$

73.

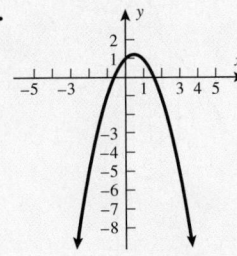

75. The vertex is located at $(2.5, -3.5)$. The parabola opens to the right.

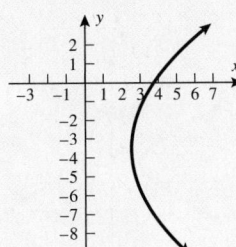

77. The vertex is located at $(1.8, 1.5)$. The parabola opens to the left.

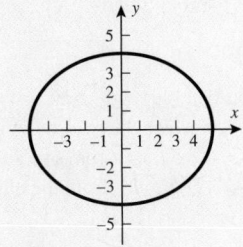

79. $(-2, 3)$, $(2, 3)$

81. $(1, 7)$, $(10, 10)$

Section 9.3

1. d

3. a

5.

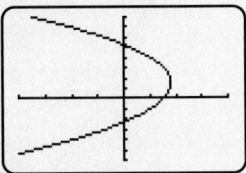

center: $(0, 0)$
vertices: $(\pm 5, 0)$

7.

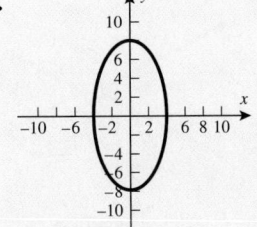

center: $(0, 0)$
vertices: $(0, \pm 8)$

9.

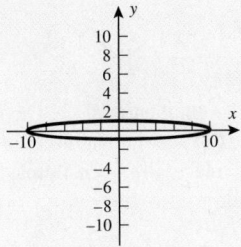

center: $(0, 0)$
vertices: $(\pm 10, 0)$

11.

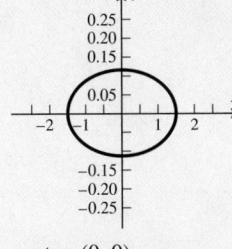

center: $(0, 0)$
vertices: $\left(\pm \frac{3}{2}, 0\right)$

13.

center: $(0, 0)$
vertices: $(0, \pm 4)$

15.

center: $(0, 0)$
vertices: $(\pm 2, 0)$

17. $\frac{x^2}{36} + \frac{y^2}{20} = 1$

19. $\frac{x^2}{7} + \frac{y^2}{16} = 1$

21. $\frac{x^2}{4} + \frac{y^2}{16} = 1$

23. $\frac{x^2}{9} + \frac{y^2}{49} = 1$

25. c

27. b

29.

31.

33.

35.

37.

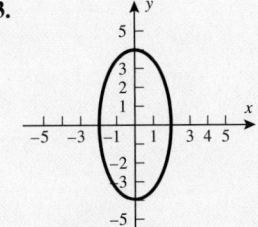

39. $\frac{(x - 2)^2}{25} + \frac{(y - 5)^2}{9} = 1$

41. $\frac{(x - 4)^2}{7} + \frac{(y + 4)^2}{16} = 1$

43. $\frac{(x - 3)^2}{4} + \frac{(y - 2)^2}{16} = 1$

45. $\dfrac{(x + 1)^2}{9} + \dfrac{(y + 4)^2}{25} = 1$ **47.** $\dfrac{x^2}{225} + \dfrac{y^2}{5625} = 1$

49. a. $\dfrac{x^2}{5625} + \dfrac{y^2}{400} = 1$ **b.** Let $x = 60$; then $y = 12$. The field extends 15 feet in that direction so the track will not encompass the field.

51. $\dfrac{x^2}{5,914,000,000^2} + \dfrac{y^2}{5,729,000,000^2} = 1$

53. $\dfrac{x^2}{150,000,000^2} + \dfrac{y^2}{146,000,000^2} = 1$

55. straight line **57.** 2 mg/cm^3

59. $a = 6$ and $b = 4$ is incorrect. In the formula, a and b are being squared; therefore, $a = \sqrt{6}$ and $b = 2$.

61. false **63.** true **65.** 3 **67.** 1

69. Pluto: $e \approx 0.25$; Earth: $e \approx 0.02$

71. $x^2 + 3y^2 = 28$

73. $8x^2 + 9y^2 - 32x + 54y + 41 = 0$

75. 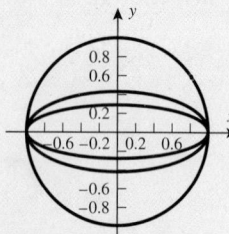 **77.**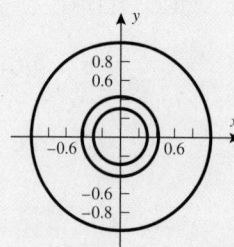

Ellipse becomes more elongated.

Circle gets smaller as c increases.

79. As c decreases, the major axis along the x-axis increases.

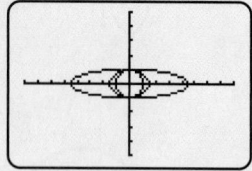

81. maximum value: -3; minimum value: -7

83. maximum value: 1; minimum value: -17

Section 9.4

1. b **3.** d

5. 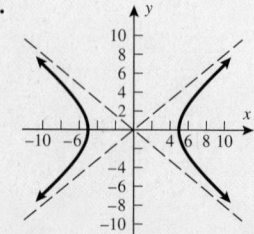 **7.**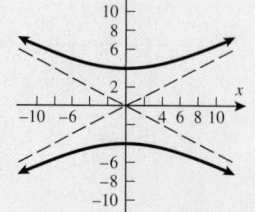

9. 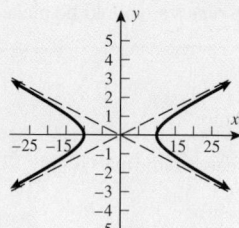 **11.**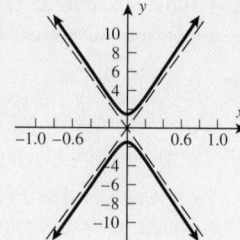

13. 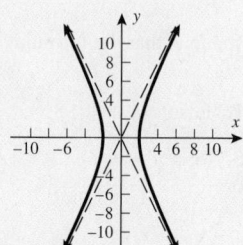 **15.**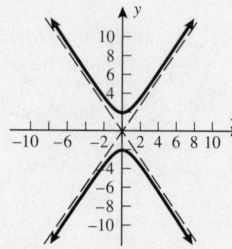

17. $\dfrac{x^2}{16} - \dfrac{y^2}{20} = 1$ **19.** $\dfrac{y^2}{9} - \dfrac{x^2}{7} = 1$

21. $x^2 - y^2 = a^2$ **23.** $\dfrac{y^2}{4} - x^2 = b^2$

25. c **27.** b

29. 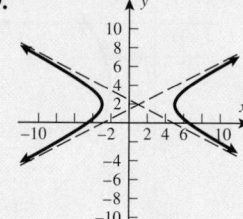 **31.**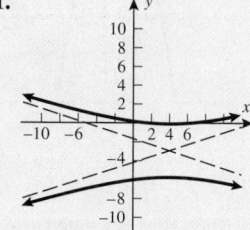

33. 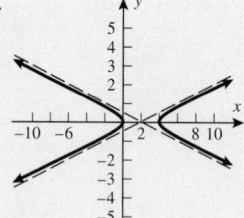 **35.**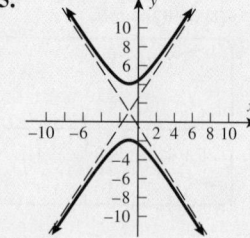

37.

39. $\dfrac{(x - 2)^2}{16} - \dfrac{(y - 5)^2}{9} = 1$

41. $\dfrac{(y + 4)^2}{9} - \dfrac{(x - 4)^2}{7} = 1$

43. The ship will come ashore between the two stations (28.5 miles from one and 121.5 miles from the other).

45. 0.000484 seconds

47. $y^2 - \dfrac{4}{5}x^2 = 1$

49. $(76, 50)$

51. $(109.4, 60)$

53. The transverse axis is vertical. The points are $(\pm 3, 0)$. The vertices are $(0, \pm 2)$.

55. false

57. true

59. $(-p, -q), (-p, q), (p, -q)$ **61.** $r > q$

63. $x^2 - y^2 = a^2$ or $y^2 - x^2 = a^2$; note that $a = b$.

65. $y = \frac{4}{3}x + \frac{10}{3}$ and $y = -\frac{4}{3}x + \frac{2}{3}$

67. $\left(-\sqrt{34}, 0\right); \left(\sqrt{34}, 0\right)$

69.

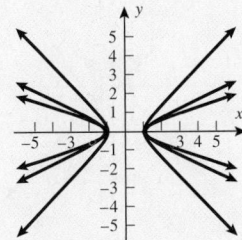

As c increases, the hyperbola narrows.

71. As c decreases, the vertices of the hyperbolas located at $\left(\pm\dfrac{1}{c}, 0\right)$ are moving away from the origin.

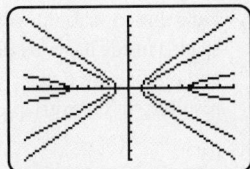

75. $-\dfrac{2x + h}{\sqrt{1 + x^2} + \sqrt{1 + (x + h)^2}}$

Section 9.5

1. $(2, 6), (-1, 3)$ **3.** $(1, 0)$

5. no solution   **7.** $(0, 1)$

9. $(-0.63, -1.61), (0.63, -1.61)$

11. no solution   **13.** $(1, 1)$

15. $\left(-\sqrt{2}, -2\sqrt{2}\right), \left(2\sqrt{2}, \sqrt{2}\right), \left(\sqrt{2}, 2\sqrt{2}\right), \left(-2\sqrt{2}, -\sqrt{2}\right)$

17. $(-6, 33), (2, 1)$ **19.** $(3, 4) (-2, -1)$

21. $(0, -3), \left(\frac{2}{5}, -\frac{11}{5}\right)$ **23.** $(-1, -1), \left(\frac{1}{4}, \frac{3}{2}\right)$

25. $(-1, -4), (4, 1)$ **27.** $(1, 3), (-1, -3)$

29. $(-4, -1), (4, 1)$

31. $(-2, -1), (2, 1), (2, -1), (-2, 1)$

33. $(2, 4)$ **35.** $\left(\frac{1}{2}, \frac{1}{3}\right), \left(\frac{1}{2}, -\frac{1}{3}\right)$

37. no solution   **39.** no solution

41.

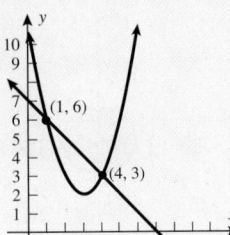

43. 3 and 7

45. 8 and 9; -8 and -9

47. 8 cm $\times$ 10 cm

49. 400 ft $\times$ 500 ft or $\frac{1000}{3}$ ft $\times$ 600 ft

51. Professor: 2 m/s
Jeremy: 10 m/s

53. 60 mph   **55.** 3762

57. Can't use elimination. Should have used substitution.

59. false   **61.** false

63. $2n$ **65.** $\left.\begin{array}{l} y = x^2 + 1 \\ y = 1 \end{array}\right\}$ other answers possible

67. $(-2, -1), (-2, 1), (2, 1), (2, -1), (1, 2), (1, -2),$ $(-1, -2), (-1, 2)$

69. no solution

71. no solution   **73.** $(-1.57, -1.64)$

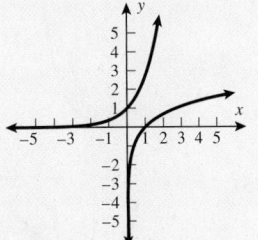   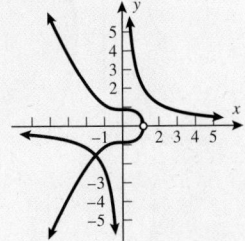

75. $(1.067, 4.119), (1.986, 0.638),$ $(-1.017, -4.757)$

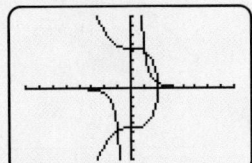

77. $y = -\dfrac{\sqrt{8 - x^2}}{2}$ **79.** $y = \dfrac{3}{x\sqrt{x}}$

Section 9.6

1. b   **3.** j   **5.** h   **7.** c   **9.** d   **11.** k

13.

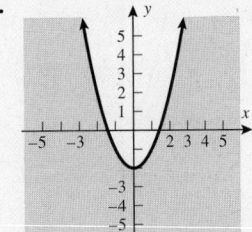

15.

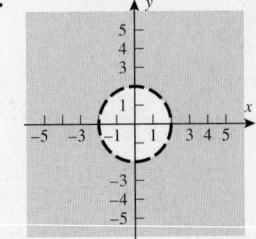

17.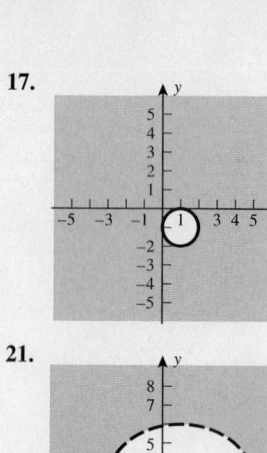

19.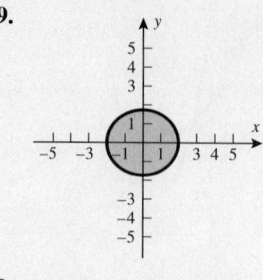

41.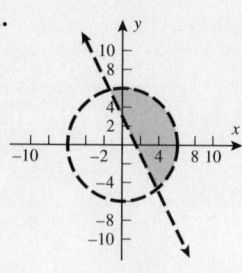

43.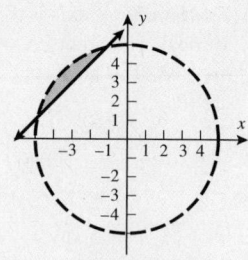

21.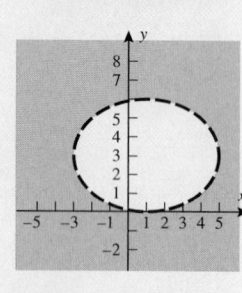

23.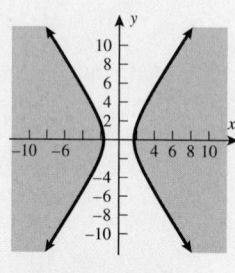

45.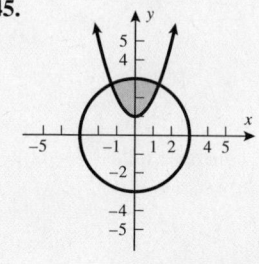

47.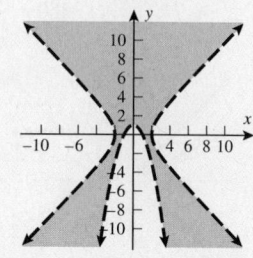

25.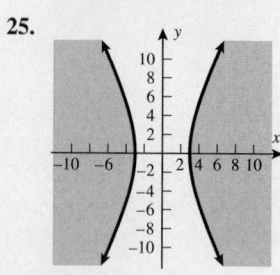

27.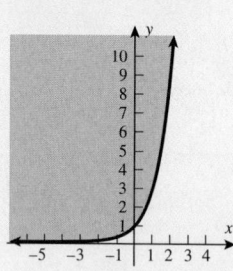

49.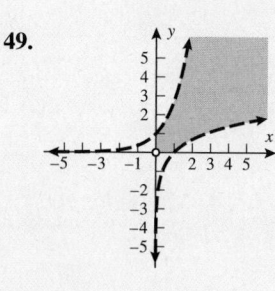

51. $\dfrac{9\pi}{2}$

53. 2π square units

55. 90 square units

57. No solution (should have shaded inside the inner circle and outside the outer circle, resulting in no overlap).

29.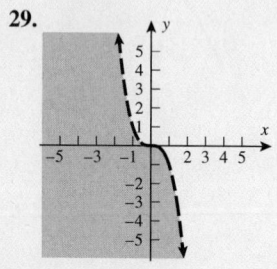

31.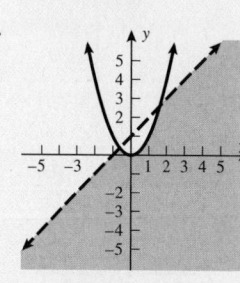

59. false

61. true

63. true

65. false

67. $b \geq a$

69. $a = 36$

71. $h = 0$

33.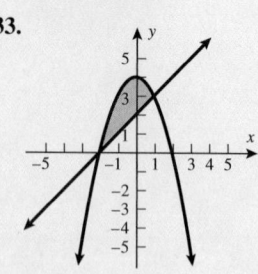

35.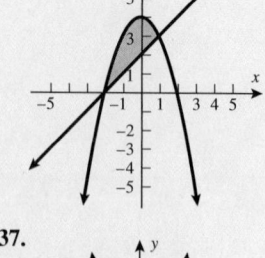

73.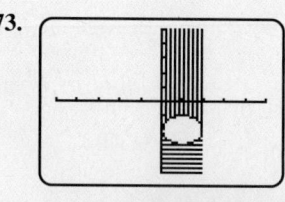

75.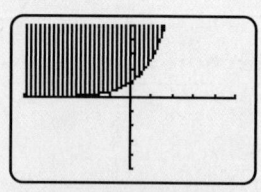

77.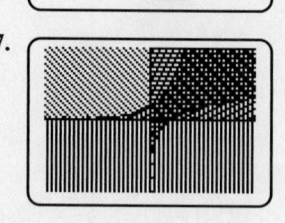

79.

37.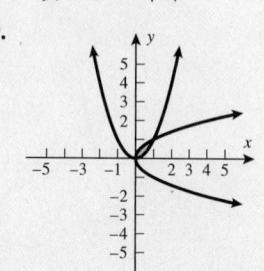

39.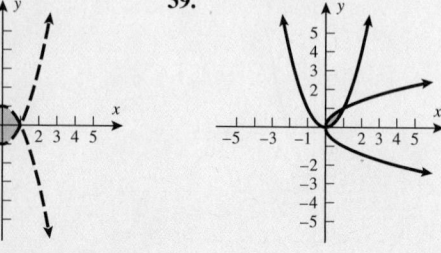

81.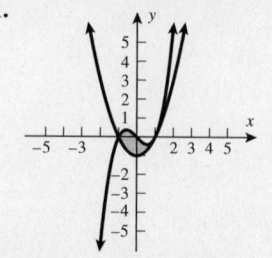

83.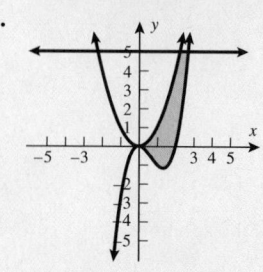

Section 9.7

1. $\left(3\sqrt{2}, \sqrt{2}\right)$

3. $\left(-\dfrac{3\sqrt{3}}{2} + 1, \dfrac{3}{2} + \sqrt{3}\right)$

5. $\left(-\dfrac{1}{2} - \dfrac{3\sqrt{3}}{2}, \dfrac{\sqrt{3}}{2} - \dfrac{3}{2}\right)$

7. $\left(\dfrac{3\sqrt{3}}{2}, \dfrac{3}{2}\right)$

9. hyperbola; $\dfrac{X^2}{2} - \dfrac{Y^2}{2} = 1$

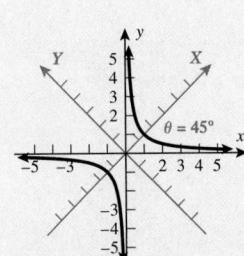

11. parabola; $2X^2 - 2Y - 1 = 0$

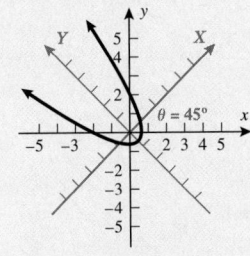

13. hyperbola; $\dfrac{X^2}{6} - \dfrac{Y^2}{2} = 1$

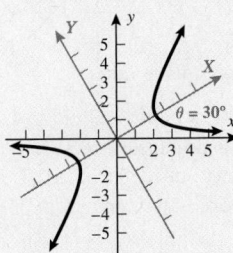

15. ellipse; $\dfrac{X^2}{2} + \dfrac{Y^2}{1} = 1$

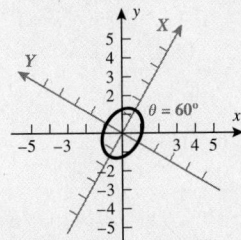

17. parabola; $2X^2 - 2Y - 1 = 0$

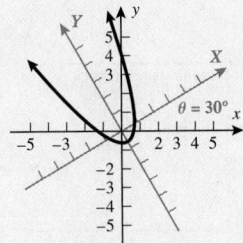

19. ellipse; $\dfrac{X^2}{1} + \dfrac{Y^2}{9} = 1$

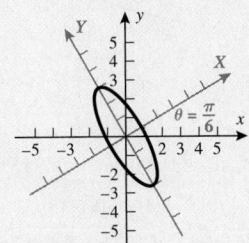
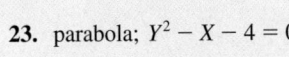

21. hyperbola; $\dfrac{X^2}{3} - \dfrac{Y^2}{2} = 1$

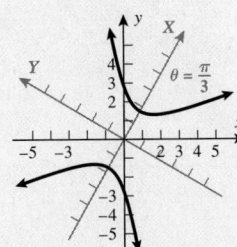

23. parabola; $Y^2 - X - 4 = 0$

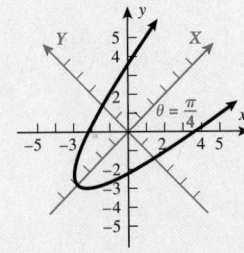

25. 45°

27. 60°

29. 30°

31. 45°

33. 15°

35. ≈40.3°

37. ≈50.7°

39. $\dfrac{X^2}{4} + \dfrac{Y^2}{9} = 1$; rotation of 60°

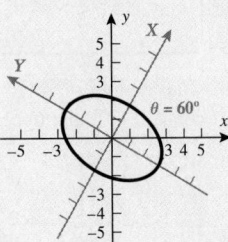

41. $\dfrac{X^2}{9} - \dfrac{Y^2}{1} = 1$; rotation of 45°

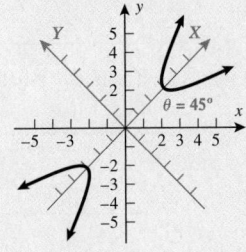

43. $X^2 - Y - 3 = 0$; rotation of 30°

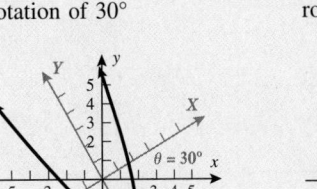

45. $\dfrac{X^2}{25} + \dfrac{Y^2}{4} = 1$; rotation of 30°

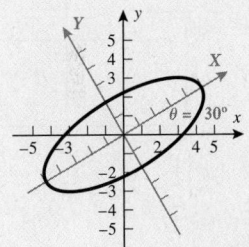

47. $X^2 + 4X - Y = 0$; rotation of 45°

49. true

51. true

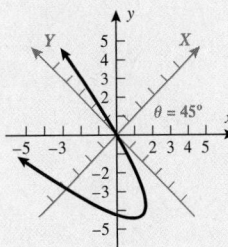

53. a. For 90°, the new equation is $\dfrac{x^2}{b^2} + \dfrac{y^2}{a^2} = 1$.

b. For 180°, the original equation results.

55. $a < 0$ hyperbola; $a = 0$ parabola; $a > 0, a \neq 1$ ellipse; $a = 1$ circle

57. Amount of rotation is 30°.

a.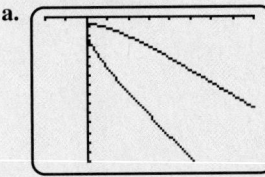

b.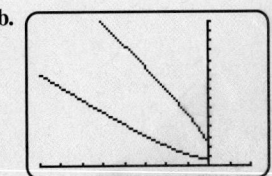

59. The amount of rotation is about 36°.

a. 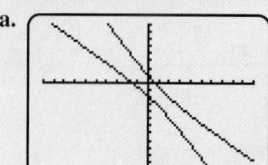 **b.**

61. a. The amount of rotation is about 63°. It is a parabola.   **b.** The amount of rotation is about 19°. It is a hyperbola.

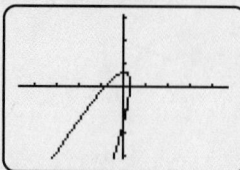   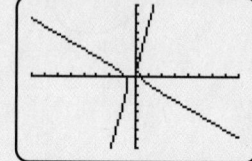

c. The amount of rotation is about 26.5°.

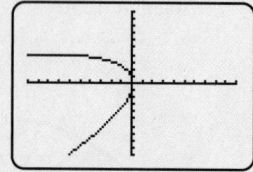

63. $\left(-\sqrt{2}, -2\sqrt{2}\right), \left(\sqrt{2}, 2\sqrt{2}\right)$

65. $(-1, -3), (1, 3), (-3, -1), (3, 1)$

Section 9.8

1. $r = \dfrac{5}{2 - \sin\theta}$ **3.** $r = \dfrac{8}{1 + 2\sin\theta}$ **5.** $r = \dfrac{1}{1 + \cos\theta}$

7. $r = \dfrac{6}{4 + 3\cos\theta}$ **9.** $r = \dfrac{12}{3 - 4\cos\theta}$ **11.** $r = \dfrac{3}{1 - \sin\theta}$

13. $r = \dfrac{18}{5 + 3\sin\theta}$ **15.** parabola   **17.** ellipse

19. hyperbola   **21.** ellipse   **23.** parabola

25. hyperbola

27. parabola; $e = 1$ **29.** hyperbola; $e = 2$

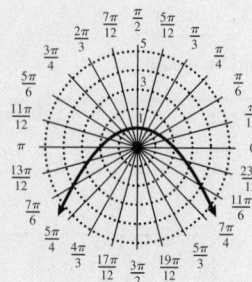   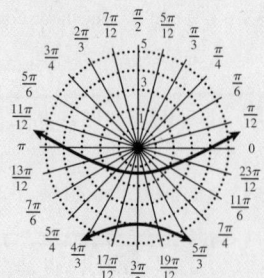

31. ellipse; $e = \frac{1}{2}$ **33.** parabola; $e = 1$

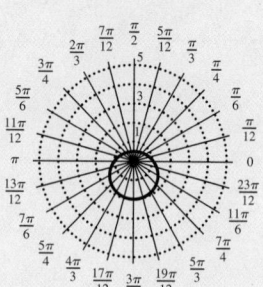   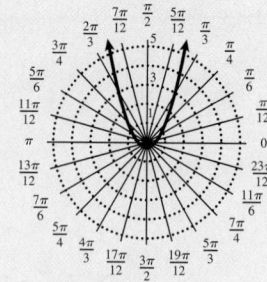

35. ellipse; $e = \frac{1}{3}$ **37.** hyperbola; $e = \frac{3}{2}$

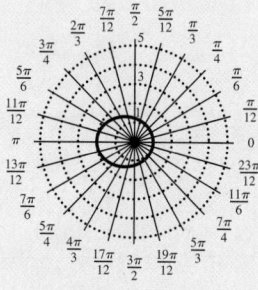   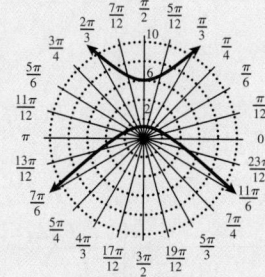

39. parabola; $e = 1$ **41.** hyperbola; $e = 3$

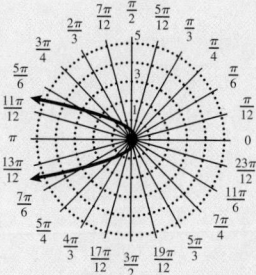   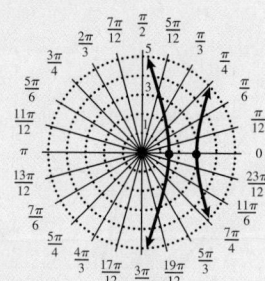

43. $e = 0.248$; $a = 5{,}913{,}500{,}000$ km;

$$r = \frac{5{,}913{,}500{,}000(1 - 0.248^2)}{1 - 0.248\cos\theta}$$

45. $r = \dfrac{75{,}000{,}000(1 - 0.223^2)}{1 - 0.223\cos\theta}$

47. a. $(-0.0167, 0)$ **b.** $(0.0167, \pi)$ **49.** $(0, -15.406)$

51. Conic becomes more elliptic as $e \rightarrow 1$ and more circular as $e \rightarrow 0$.

55. $\dfrac{2ep}{1 - e^2}$ **57.** $\left(\dfrac{\dfrac{ep}{1 - e} - \dfrac{ep}{1 + e}}{2}, \pi\right)$

67. a. With θ step $= \pi/3$, plot points $(2, 0)$, $(14.93, \pi/3)$, $(14.93, 2\pi/3)$, $(2, \pi)$, $(1.07, 4\pi/3)$, $(1.07, 5\pi/3)$, and $(2, 2\pi)$.

b. With θ step $= 0.8\pi$, plot points $(2, 0)$, $(4.85, 0.8\pi)$, $(1.03, 1.6\pi)$, $(40.86, 2.4\pi)$, $(1.26, 3.2\pi)$, and $(2, 4\pi)$.

69. $\dfrac{\pi}{4}, \dfrac{5\pi}{4}$

71. $\dfrac{\pi}{6}, \dfrac{5\pi}{6}$

21.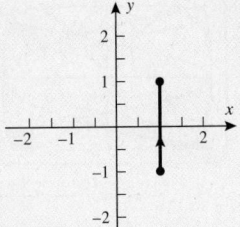

23. Arrow is in different directions, depending on t.
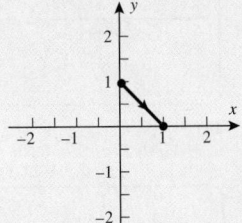

Section 9.9

1.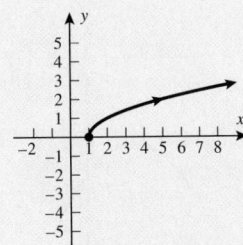

3.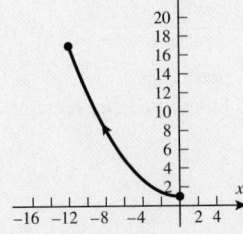

25.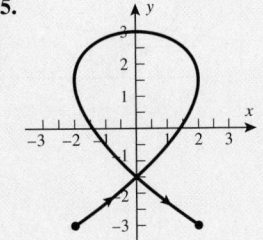

27.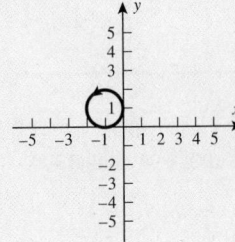

5.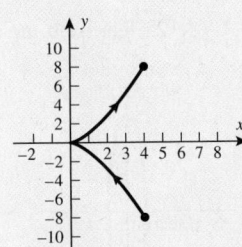

7.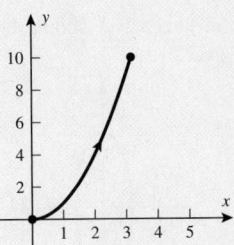

29.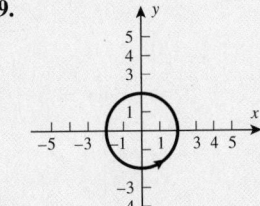

31. $y = \dfrac{1}{x^2}$

33. $y = x - 2$

35. $y = \sqrt{x^2 + 1}$

37. $x + y = 2$

9.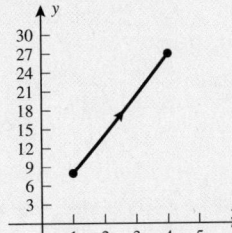

11.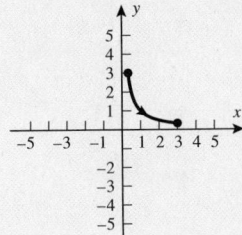

39. $x + 4y = 8$ **41.** 17.7 seconds

43. yes

45. height: 5742 ft; horizontal distance: 13,261 ft

47. 125 seconds

13.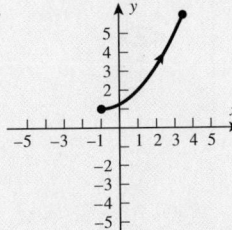

15.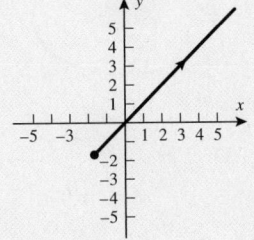

49.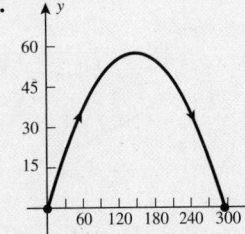

51.

t	x	y
0	$A + B$	0
$\dfrac{\pi}{2}$	0	$A + B$
π	$-A - B$	0
$\dfrac{3\pi}{2}$	0	$-A - B$
2π	$A + B$	0

53. The original domain must be $t \geq 0$; therefore, only the part of the parabola where $y \geq 0$ is part of the plane curve.

17.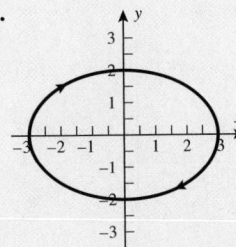

19.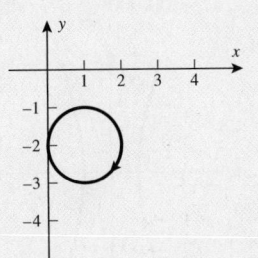

55. false   **57.** quarter circle in QI

61. $\dfrac{x + y}{2a} + \dfrac{y - x}{2b} = 1$ **63.** $y = b\sqrt[6]{x}$

65.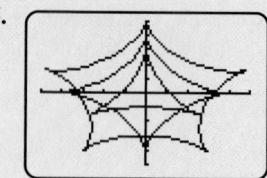

67.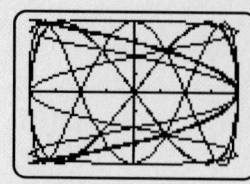

69. $a = 2, 0 \le t < 2\pi; a = 3, 0 \le t < 2\pi$

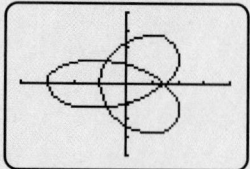

71. $(2, 3), (-3, 8)$

73. $(-100, -96)$ and $(1300, -1664)$

Review Exercises

1. false

3. true

5. $y^2 = 12x$

7. $y^2 = -20x$

9. $(x - 2)^2 = 8(y - 3)$

11. $(x - 1)^2 = -4(y - 6)$

13. focus: $(0, -3)$
directrix: $y = 3$
vertex: $(0, 0)$
length of latus rectum: 12

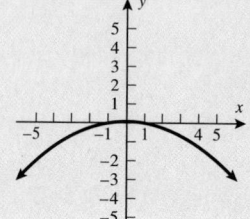

15. focus: $\left(\frac{1}{4}, 0\right)$
directrix: $x = -\frac{1}{4}$
vertex: $(0, 0)$
length of latus rectum: 1

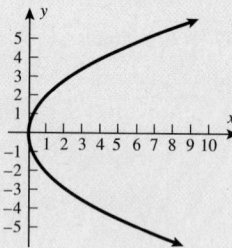

17. focus: $(3, -2)$
directrix: $x = 1$
vertex: $(2, -2)$
length of latus rectum: 4

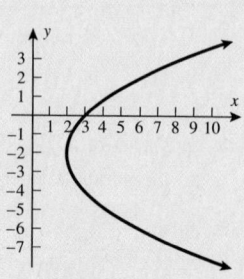

19. focus: $(-3, -1)$
directrix: $y = 3$
vertex: $(-3, 1)$
length of latus rectum: 8

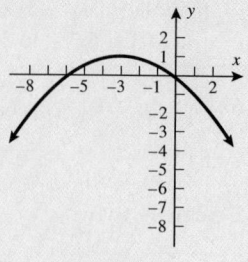

21. focus: $\left(-\frac{5}{2}, -\frac{79}{8}\right)$
directrix: $y = -\frac{71}{8}$
vertex: $\left(-\frac{5}{2}, -\frac{75}{8}\right)$
length of latus rectum: 2

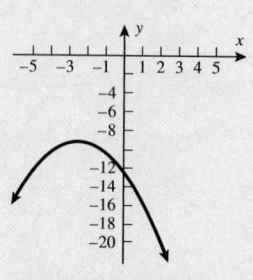

23. The receiver should be placed $\frac{25}{8} = 3.125$ feet from the vertex.

25.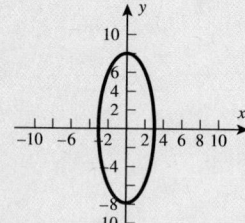

27.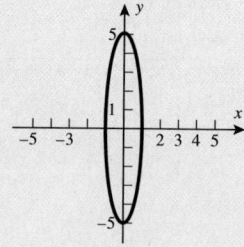

29. $\dfrac{x^2}{25} + \dfrac{y^2}{16} = 1$

31. $\dfrac{x^2}{9} + \dfrac{y^2}{64} = 1$

33.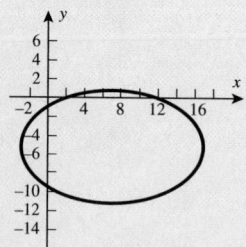

35.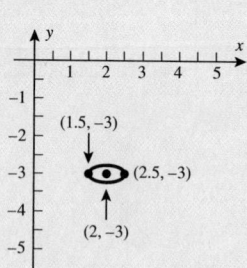

37. $\dfrac{(x - 3)^2}{25} + \dfrac{(y - 3)^2}{9} = 1$

39. $\dfrac{x^2}{778,300,000^2} + \dfrac{y^2}{777,400,000^2} = 1$

41.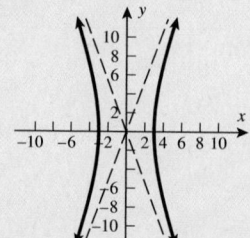

43.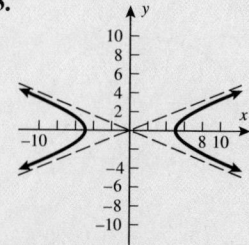

45. $\dfrac{x^2}{9} - \dfrac{y^2}{16} = 1$

47. $\dfrac{y^2}{9} - x^2 = 1$ (Other answers are possible.)

49.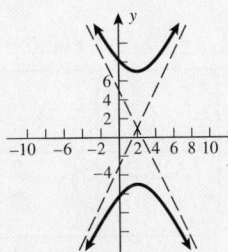

51.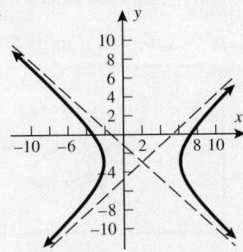

53. $\dfrac{(x-4)^2}{16} - \dfrac{(y-3)^2}{9} = 1$

55. The ship will get to shore between the two stations: 65.36 miles from one station and 154.64 miles from the other.

57. $(1, -4)$ and $(-2, -7)$ **59.** $(-1, 2)$ and $(1, 2)$

61. no solution **63.** no solution

65. $(-3, -2)$ and $(2, 3)$

67. $\left(-\dfrac{1}{2}, -\dfrac{1}{\sqrt{7}}\right), \left(-\dfrac{1}{2}, \dfrac{1}{\sqrt{7}}\right), \left(\dfrac{1}{2}, -\dfrac{1}{\sqrt{7}}\right), \left(\dfrac{1}{2}, \dfrac{1}{\sqrt{7}}\right)$

69.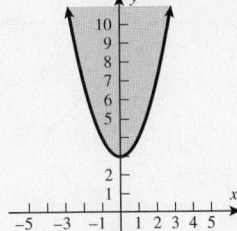

71.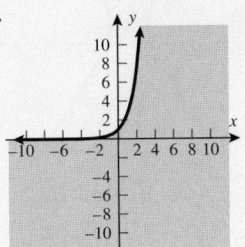

73.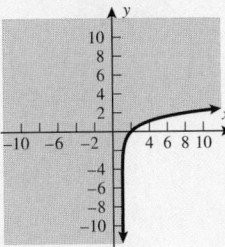

75.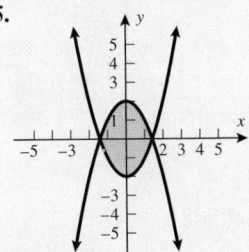

77.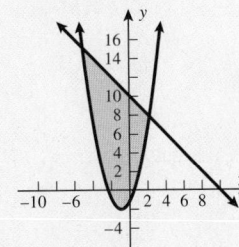

79.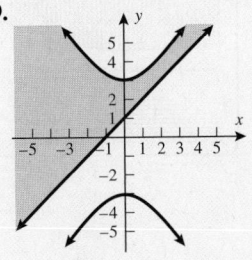

81. $\left(-\dfrac{3}{2} + \sqrt{3}, \dfrac{-3\sqrt{3}}{2} + 1\right)$

83. $\dfrac{X^2}{4} - \dfrac{Y^2}{4} = 1$

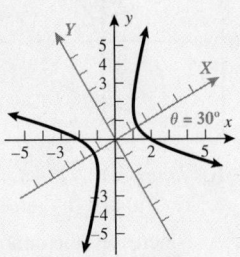

85. $60°$ **87.** $X^2 + 4 = Y$

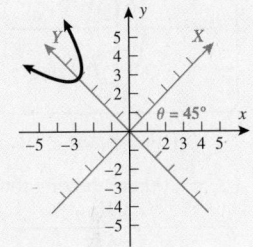

89. $r = \dfrac{21}{7 - 3\sin\theta}$ **91.** hyperbola

93. $e = \dfrac{1}{2}$; vertices $\left(\dfrac{4}{3}, 0\right), (4, \pi)$ or in rectangular form $\left(\dfrac{4}{3}, 0\right)$, $(-4, 0)$

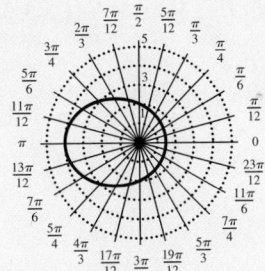

95.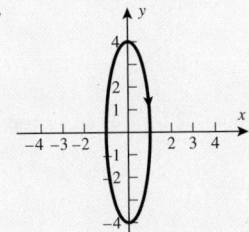

97.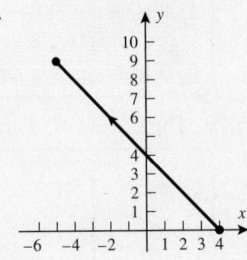

99. $x = 4 - y^2$ **101.** $y = 2x + 4$

103. The vertex is located at $(0.6, -1.2)$. The parabola opens down.

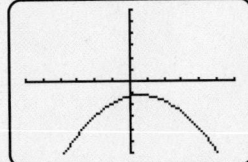

105. $y = -1.4 \pm \sqrt{8.81 - 3x}$

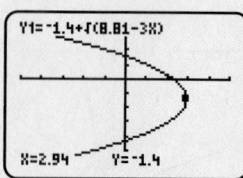

b. Vertex at $(2.94, -1.4)$ opens to the left.
c. Yes, (a) and (b) agree with each other.

107. As c increases, the minor axis along the x-axis decreases.

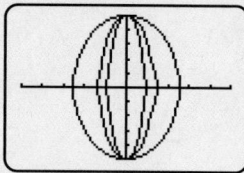

109. As c increases, the vertices of the hyperbolas located at $\left(\pm \dfrac{1}{2c}, 0 \right)$ are moving toward the origin.

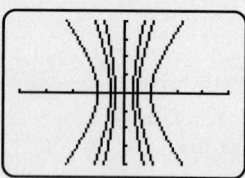

111. $(0.635, 2.480), (-0.635, 2.480), (-1.245, 0.645), (1.245, 0.645)$

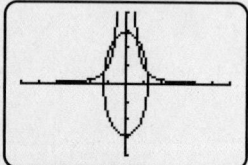

113.

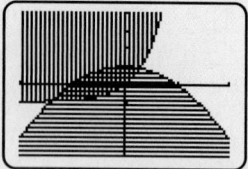

115. a. The amount of rotation is about $22.5°$. It is an ellipse.

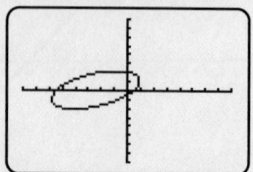

b. The amount of rotation is about $11.6°$. It is a hyperbola.

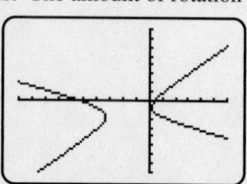

117. With θ step $= \dfrac{\pi}{4}$, points $(2, 0), \left(1.06, \dfrac{\pi}{4} \right), \left(0.89, \dfrac{\pi}{2} \right)$, $\left(1.06, \dfrac{3\pi}{4} \right), (2, \pi), \left(17.22, \dfrac{5\pi}{4} \right), \left(-8, \dfrac{3\pi}{2} \right), \left(17.22, \dfrac{7\pi}{4} \right)$, and $(2, 2\pi)$ are plotted.

119. $a = 2, b = 3, t$ in $[0, 2\pi]$ $a = 3, b = 2, t$ in $[0, 2\pi]$

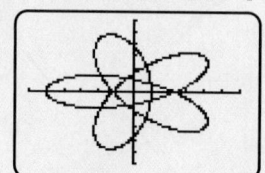 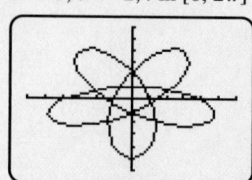

Practice Test

1. c   **3.** d   **5.** f

7. $y^2 = -16x$ **9.** $(x + 1)^2 = -12(y - 5)$

11. $\dfrac{x^2}{7} + \dfrac{y^2}{16} = 1$ **13.** $\dfrac{(x - 2)^2}{20} + \dfrac{y^2}{36} = 1$

15. $x^2 - \dfrac{y^2}{4} = 1$ **17.** $\dfrac{y^2}{16} - \dfrac{(x - 2)^2}{20} = 1$

19.   **21.**

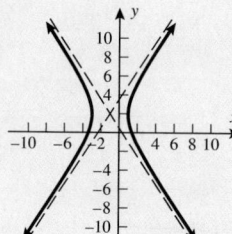

23. $x^2 = 6y; -2 \le x \le 2$

25.   **27.**

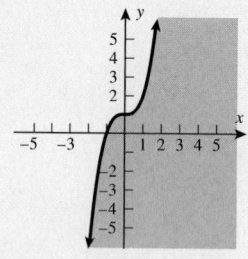

29. ellipse: $e = \frac{2}{3}$ **31.** 5.3 seconds, 450 ft

33.

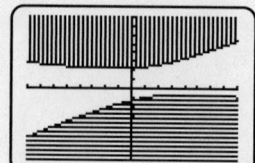

35. The vertex is located at $(-2.1, 1.2)$. The parabola opens upward.

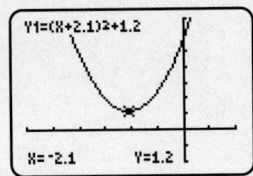

Cumulative Test

1. $-6, 3$ **3.** -7

5. $f(x) = \frac{1}{3}(x - 7)^2 + 7$ **7.** $-4, 4$

9. -1.9626 **11.** $\dfrac{1}{\cos\theta} + \sin\theta$

13. $\dfrac{\pi}{3}, \dfrac{2\pi}{3}, \dfrac{4\pi}{3}, \dfrac{5\pi}{3}$ **15.** $\langle -5.1, 14.1 \rangle$

17. soda: \$1.29; soft pretzel: \$1.45

19.

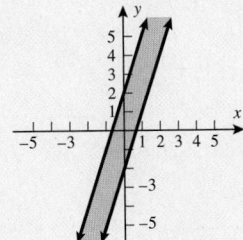

21. $\begin{bmatrix} 7 & -16 & 33 \\ 18 & -3 & -17 \end{bmatrix}$ **23.** $\dfrac{(x-6)^2}{9} + \dfrac{(y+2)^2}{25} = 1$

25. $(2, 4), (4, 2)$

27.

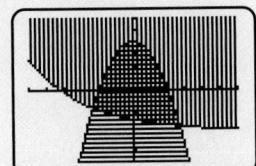

CHAPTER 10

Section 10.1

1. $a_1 = 1$ $a_2 = 2$ $a_3 = 3$ $a_4 = 4$

3. $a_1 = 1$ $a_2 = 3$ $a_3 = 5$ $a_4 = 7$

5. $a_1 = \frac{1}{2}$ $a_2 = \frac{2}{3}$ $a_3 = \frac{3}{4}$ $a_4 = \frac{4}{5}$

7. $a_1 = 2$ $a_2 = 2$ $a_3 = \frac{4}{3}$ $a_4 = \frac{2}{3}$

9. $a_1 = -x^2$ $a_2 = x^3$ $a_3 = -x^4$ $a_4 = x^5$

11. $a_1 = -\frac{1}{6}$ $a_2 = \frac{1}{12}$ $a_3 = -\frac{1}{20}$ $a_4 = \frac{1}{30}$

13. $a_9 = \frac{1}{512}$ **15.** $a_{19} = -\frac{1}{420}$ **17.** $a_{100} = 1.0201$

19. $a_{23} = 23$ **21.** $a_n = 2n$ **23.** $a_n = \dfrac{1}{n(n+1)}$

25. $a_n = \dfrac{(-1)^n 2^n}{3^n}$ **27.** $a_n = (-1)^{n+1}$ **29.** 72

31. 812 **33.** $\frac{1}{5852}$ **35.** $83, 156, 160$

37. $\dfrac{1}{n(n+1)}$ **39.** $(2n+3)(2n+2)$

41. $a_1 = 7$ $a_2 = 10$ $a_3 = 13$ $a_4 = 16$

43. $a_1 = 1$ $a_2 = 2$ $a_3 = 6$ $a_4 = 24$

45. $a_1 = 100$ $a_2 = 50$ $a_3 = \frac{25}{3}$ $a_4 = \frac{25}{72}$

47. $a_1 = 1$ $a_2 = 2$ $a_3 = 2$ $a_4 = 4$

49. $a_1 = 1$ $a_2 = -1$ $a_3 = -2$ $a_4 = 5$

51. 10 **53.** 30 **55.** 36 **57.** 5

59. $1 - x + x^2 - x^3$ **61.** $\frac{109}{15}$

63. $1 + x + \dfrac{x^2}{2} + \dfrac{x^3}{6} + \dfrac{x^4}{24}$ **65.** $\frac{20}{9}$

67. not possible   **69.** $\displaystyle\sum_{n=0}^{6} \dfrac{(-1)^n}{2^n}$

71. $\displaystyle\sum_{n=1}^{\infty} (-1)^{n-1} n$ **73.** $\displaystyle\sum_{n=1}^{5} \dfrac{(n+1)!}{(n-1)!} = \sum_{n=1}^{5} n(n+1)$

75. $\displaystyle\sum_{n=0}^{\infty} \dfrac{(-1)^n x^n}{n!}$ or $\displaystyle\sum_{n=1}^{\infty} \dfrac{(-1)^{n-1} \cdot x^{n-1}}{(n-1)!}$

77. $A_{72} \approx \$28{,}640.89$. A_{72} represents the total balance in 6 years (or 72 months).

79. $a_n = 20 + 2n$. The paralegal's salary with 20 years of experience will be \$60 per hour.

81. $a_n = 1.03a_{n-1}$ $n = 1, 2, \ldots$ and $a_0 = 30{,}000$

83. Approximately 10.7 years, $a_n = 1000 - 75n$ $n = 0, 1, 2, \ldots$

85. $A_1 = \$100$ $A_2 = \$200.10$ $A_3 = \$300.30$ $A_4 = \$400.60$ $A_{36} = \$3{,}663.72$

87. $7; 7.389$ **89.** $0.095306; 0.095310$

91. $6!$ is not equal to $(3!)(2!)$.

93. $(-1)^{n+1}$ is evaluated incorrectly. The sign is incorrect on each term.

95. true   **97.** false   **99.** false

101. $a_1 = C$ $a_2 = C + D$ $a_3 = C + 2D$ $a_4 = C + 3D$

103. 1 and 1

105. $a_{100} \approx 2.705$ $a_{1000} \approx 2.717$ $a_{10{,}000} \approx 2.718$

107. $\frac{109}{15} \approx 7.27$ **109.** monotonic nondecreasing

111. not monotonic

Section 10.2

1. arithmetic; 3 **3.** not arithmetic **5.** arithmetic; -0.03

7. arithmetic; $\frac{2}{3}$ **9.** not arithmetic

11. $a_1 = 3$ $a_2 = 1$ $a_3 = -1$ $a_4 = -3$ arithmetic; -2

13. $a_1 = 1$ $a_2 = 4$ $a_3 = 9$ $a_4 = 16$ not arithmetic

15. $a_1 = 2$ $a_2 = 7$ $a_3 = 12$ $a_4 = 17$ arithmetic; 5

17. $a_1 = 0$ $a_2 = 10$ $a_3 = 20$ $a_4 = 30$ arithmetic; 10

19. $a_1 = -1$ $a_2 = 2$ $a_3 = -3$ $a_4 = 4$ not arithmetic

21. $a_n = 6 + 5n$ **23.** $a_n = -6 + 2n$ **25.** $a_n = -\frac{2}{3} + \frac{2}{3}n$

27. $a_n = en - e$ **29.** 124 **31.** -684

33. $a_1 = 8$ $d = 9$ $a_n = 8 + 9(n-1)$ or $a_n = 9n - 1$

35. $a_1 = 23$ $d = -4$ $a_n = 23 - 4(n-1)$ or $a_n = 27 - 4n$

37. $a_1 = 1$ $d = \frac{2}{3}$ $a_n = 1 + \frac{2}{3}(n-1)$ or $a_n = \frac{1}{3} + \frac{2}{3}n$

39. 552 **41.** -780 **43.** 51

45. 416 **47.** 3875 **49.** -66.5 or $-\frac{133}{2}$

51. 630 **53.** $-\frac{817}{2} = -408.5$ **55.** 1,368

57. Colin $347,500, Camden $340,000

59. 850 seats

61. 1101 glasses in the bottom row. There are 20 fewer glasses in every row.

63. 1600 feet **65.** 210 oranges

67. 23 seats in the first row; 1125 total seats

69. The correct general term is $a_n = a_1 + d(n-1)$.

71. $n = 11$ (not 10) **73.** false **75.** true

77. $\dfrac{(n+1)(2a+nb)}{2}$ **79.** 27,420 or $\frac{1}{4}R$ **81.** 5050

83. 2500 **85.** 18,850 **87.** 1010 **89.** 1.204

Section 10.3

1. yes; $r = 3$ **3.** no

5. yes; $r = \frac{1}{2}$ **7.** yes; $r = 1.7$

9. $a_1 = 6$ $a_2 = 18$ $a_3 = 54$ $a_4 = 162$ $a_5 = 486$

11. $a_1 = 1$ $a_2 = -4$ $a_3 = 16$ $a_4 = -64$ $a_5 = 256$

13. $a_1 = 10{,}000$ $a_2 = 10{,}600$ $a_3 = 11{,}236$
$a_4 = 10{,}000(1.06)^3 = 11{,}910.16$
$a_5 = 10{,}000(1.06)^4 \approx 12{,}624.77$

15. $a_1 = \frac{2}{3}$ $a_2 = \frac{1}{3}$ $a_3 = \frac{1}{6}$ $a_4 = \frac{1}{12}$ $a_5 = \frac{1}{24}$

17. $a_n = 5(2)^{n-1}$ **19.** $a_n = (-3)^{n-1}$

21. $a_n = 1000(1.07)^{n-1}$ **23.** $a_n = \frac{16}{3}\left(-\frac{1}{4}\right)^{n-1}$

25. $a_7 = -128$ **27.** $a_{13} = \frac{4096}{3}$

29. $a_{15} = 1000\left(\frac{1}{20}\right)^{14} \approx 6.10 \times 10^{-16}$

31. $\frac{8191}{3}$ **33.** 59,048

35. $\frac{20}{9} = 2.\overline{2}$ **37.** 6560

39. 16,383 **41.** 2

43. $-\frac{1}{4}$ **45.** not possible, diverges

47. $-\frac{27}{2}$ **49.** $\dfrac{1{,}000{,}000}{95} \approx 10{,}526$

51. $\frac{2}{3}$ **53.** 100

55. $44,610.95

57. $2000(0.5)^n$; in 4 years when she graduates, it will be worth $125, and after graduate school it will be worth approximately $16.

59. On the fifth rebound the jumper will reach a height of approximately 17 feet.

61. about 58,640 students

63. about 66 days; $9,618 would be paid in January.

65. $3,877.64 (answers depend on rounding)

67. 26 weeks: $13,196.88 52 weeks: $26,811.75

69. $501,509

73. $r = -\frac{1}{3}$ not $\frac{1}{3}$ **75.** $a_1 = -12$ (not 4)

77. false **79.** true

81. If $|b| < 1$, then the sum is $a \cdot \dfrac{1}{1-b}$ or $\dfrac{a}{1-b}$.

83. $\frac{47}{99}$ **85.** $-375{,}299{,}968{,}947{,}541$

87. The sum is $\dfrac{1}{1-x}$ for $|x|$ less than or equal to $\frac{1}{2}$.

89. convergent, sum: 3

91. convergent, sum: $\frac{1}{2}$

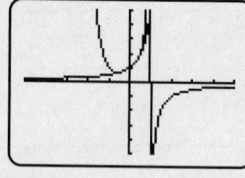

Section 10.4

25. 7 **27.** 31

31. false **39.** $\frac{255}{256}$, yes

Section 10.5

1. 35 **3.** 45 **5.** 1 **7.** 1 **9.** 17,296

11. $x^4 + 8x^3 + 24x^2 + 32x + 16$

13. $y^5 - 15y^4 + 90y^3 - 270y^2 + 405y - 243$

15. $x^5 + 5x^4y + 10x^3y^2 + 10x^2y^3 + 5xy^4 + y^5$

17. $x^3 + 9x^2y + 27xy^2 + 27y^3$

19. $125x^3 - 150x^2 + 60x - 8$

21. $\dfrac{1}{x^4} + \dfrac{20y}{x^3} + \dfrac{150y^2}{x^2} + \dfrac{500y^3}{x} + 625y^4$

23. $x^8 + 4x^6y^2 + 6x^4y^4 + 4x^2y^6 + y^8$

25. $a^5x^5 + 5a^4x^4by + 10a^3x^3b^2y^2 + 10a^2x^2b^3y^3 + 5axb^4y^4 + b^5y^5$

27. $x^3 + 12x^{5/2} + 60x^2 + 160x^{3/2} + 240x + 192x^{1/2} + 64$

29. $a^3 + 4a^{9/4}b^{1/4} + 6a^{3/2}b^{1/2} + 4a^{3/4}b^{3/4} + b$

31. $x + 8x^{3/4}y^{1/2} + 24x^{1/2}y + 32x^{1/4}y^{3/2} + 16y^2$

33. $r^4 - 4r^3s + 6r^2s^2 - 4rs^3 + s^4$

35. $a^6x^6 + 6a^5x^5by + 15a^4x^4b^2y^2 + 20a^3x^3b^3y^3 + 15a^2x^2b^4y^4 + 6axb^5y^5 + b^6y^6$

37. 3360   **39.** 5670   **41.** 22,680

43. 70   **45.** 3,838,380   **47.** 2,598,960

49. $\dbinom{7}{5} \neq \dfrac{7\,!}{5!}$ $\dbinom{7}{5} = \dfrac{7\,!}{5!2!}$

51. false   **53.** true   **55.** true

61. The binomial expansion of
$(1 - x)^3 = 1 - 3x + 3x^2 - x^3$

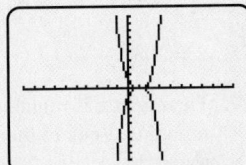

63. As each term is added, the series gets closer to $y = (1 - x)^3$. When $x > 1$, that's not true. The graphs for $-1 < x < 1$:

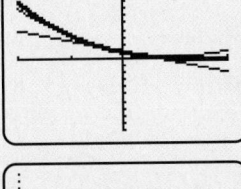

The graphs for $1 < x < 2$:

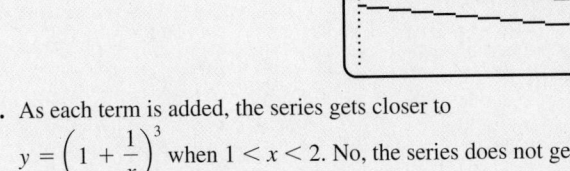

65. As each term is added, the series gets closer to
$y = \left(1 + \dfrac{1}{x}\right)^3$ when $1 < x < 2$. No, the series does not get

close to $y = \left(1 + \dfrac{1}{x}\right)^3$ for $0 < x < 1$.

67. $\displaystyle\sum_{k=1}^{n}\binom{n}{k}x^{n-k}h^{k-1}$ **69.** $f'(x) = \displaystyle\sum_{k=0}^{n-1}\dfrac{n!}{k!(n-1-k)!}x^{n-1-k}$

Review Exercises

1. $a_1 = 1$ $a_2 = 8$ $a_3 = 27$ $a_4 = 64$

3. $a_1 = 5$ $a_2 = 8$ $a_3 = 11$ $a_4 = 14$

5. $a_5 = \dfrac{32}{243} \approx 0.13$ **7.** $a_{15} = -\dfrac{1}{3600}$

9. $a_n = (-1)^{n+1}3n$ for $n = 1, 2, 3, \ldots$

11. $a_n = (-1)^n$ for $n = 1, 2, 3, \ldots$

13. 56   **15.** $\dfrac{1}{n + 1}$

17. $a_1 = 5$ $a_2 = 3$ $a_3 = 1$ $a_4 = -1$

19. $a_1 = 1$ $a_2 = 2$ $a_3 = 4$ $a_4 = 32$

21. 15   **23.** 69   **25.** $\displaystyle\sum_{n=1}^{7}\dfrac{(-1)^n}{2^{n-1}}$ **27.** $\displaystyle\sum_{n=0}^{\infty}\dfrac{x^n}{n!}$

29. $A_{60} \approx 36{,}629.90$, which is the amount in the account after 5 years.

31. Yes, the sequence is arithmetic. $d = -2$

33. Yes, the sequence is arithmetic. $d = \dfrac{1}{2}$

35. Yes, the sequence is arithmetic. $d = 1$

37. $a_n = -4 + 5(n - 1)$ or $a_n = 5n - 9$

39. $a_n = 1 - \dfrac{2}{3}(n - 1)$ or $a_n = -\dfrac{2}{3}n + \dfrac{5}{3}$

41. $a_1 = 5$ $d = 2$ $a_n = 5 + 2(n - 1)$ or $a_n = 3 + 2n$

43. $a_1 = 10$ $d = 6$ $a_n = 10 + 6(n - 1)$ or $a_n = 4 + 6n$

45. 630   **47.** 420

49. Bob: \$885,000   Tania: \$990,000

51. Yes, the sequence is geometric, $r = -2$.

53. Yes, the sequence is geometric, $r = \dfrac{1}{2}$.

55. $a_1 = 3$ $a_2 = 6$ $a_3 = 12$ $a_4 = 24$ $a_5 = 48$

57. $a_1 = 100$ $a_2 = -400$ $a_3 = 1600$ $a_4 = -6400$ $a_5 = 25{,}600$

59. $a_n = 7 \cdot 2^{n-1}$ **61.** $a_n = (-2)^{n-1}$

63. $a_{25} = 33{,}554{,}432$ **65.** $a_{12} = -2.048 \times 10^{-6} = -\dfrac{100}{5^{11}}$

67. 4,920.5   **69.** 16,400   **71.** 3

73. \$60,875.61   **79.** 165   **81.** 1

83. $x^4 - 20x^3 + 150x^2 - 500x + 625$

85. $8x^3 - 60x^2 + 150x - 125$

87. $x^{5/2} + 5x^2 + 10x^{3/2} + 10x + 5x^{1/2} + 1$

89. $r^5 - 5r^4s + 10r^3s^2 - 10r^2s^3 + 5rs^4 - s^5$

91. 112   **93.** 37,500   **95.** 22,957,480

97. $\dfrac{5369}{3600} \approx 1.49$ **99.** $\dfrac{34{,}875}{14} \approx 2491.07$

101. The series will sum to $\dfrac{1}{1 + 2x}$.

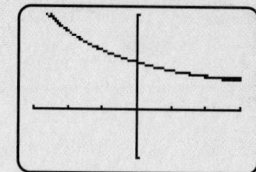

103. 99900, yes

105. As each term is added, the series gets closer to $y = (1 + 2x)^4$ when $-0.1 < x < 0.1$. No, the series does not get close to $y = (1 + 2x)^4$ for $0.1 < x < 1$.

Practice Test

1. $a_n = x^{n-1}$ $n = 1, 2, 3, \ldots$

3. $S_n = \dfrac{1 - x^n}{1 - x}$ **5.** $|x| < 1$ **7.** $1 - \left(\frac{1}{4}\right)^{10} \approx 1$

9. $\displaystyle\sum_{n=1}^{100}(5n - 3) = 24{,}950$

11. 2520   **13.** 455   **15.** 20,254

17. $x^{10} + 5x^7 + 10x^4 + 10x + \dfrac{5}{x^2} + \dfrac{1}{x^5}$

21. $\frac{5}{3}$ **23.** $\displaystyle\sum_{n=1}^{\infty}\dfrac{5}{2(n + 3)}$

25. $128 - 1344x + 6048x^2 - 15{,}120x^3 + 22{,}680x^4 - 20{,}412x^5 + 10{,}206x^6 - 2187x^7$

27. 184,756

Cumulative Test

1. $\dfrac{x^2 + xh + 2x + h}{(x + 1)(x + h + 1)}$

3. $f(x) = (x - 5)(x + 1)(x - i)(x + i)$

5. $Q(x) = -6x^3 + 21x + 2;\ R(x) = -63x - 13$

7. 2.678   **9.** $\dfrac{\sqrt{2}}{2}$ **11.** $0, \dfrac{2\pi}{3}, \dfrac{4\pi}{3}, 2\pi$

13. $2(\cos 315° + i \sin 315°)$

15. $x = 3a - 2, y = 7a - 5, z = a$ or $(3a - 2, 7a - 5, a)$

17. no solution   **19.** 0

21. $x^2 + y^2 < 4$ **23.** 3

25. 35

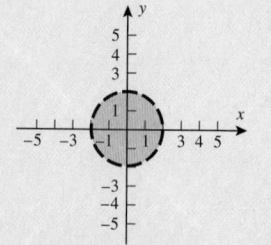

APPENDIX

Section A.1

1. rational   **3.** irrational   **5.** rational

7. irrational

9. a. 7.347 **b.** 7.347   **11. a.** 2.995 **b.** 2.994

13. a. 0.234 **b.** 0.234   **15. a.** 5.238 **b.** 5.238

17. 4   **19.** 26   **21.** -130

23. 17   **25.** 3   **27.** $x + y - z$

29. $-3x - y$ **31.** $\frac{3}{5}$ **33.** 68

35. $-9x - y$ **37.** -10 **39.** 5

41. $\frac{19}{12}$ **43.** $\frac{1}{2}$ **45.** $\frac{23}{12}$

47. $\frac{1}{27}$ **49.** $\dfrac{x}{3}$ **51.** $\dfrac{x}{21}$

53. $\dfrac{61y}{60}$ **55.** $\frac{11}{30}$ **57.** $\frac{4}{3}$

59. $\frac{3}{35}$ **61.** $\dfrac{12b}{a}$ **63.** $\frac{3}{4}$

65. $\dfrac{4xy}{3}$ **67.** $\dfrac{8x}{y}$ **69.** $\frac{2}{3}$

71. $\frac{3}{25}$ **73.** \$ 9,176,366,000,000

75. \$30,203

77. The mistake is rounding the number that is used in the rounding. Look to the right of the number; if it is less than 5, round down.

79. The -2 did not distribute to the second term.

81. false   **83.** true   **85.** $x \neq 0$

87. $-7x + 12y - 75$ **89.** irrational   **91.** rational

Section A.2

1. 256   **3.** -243 **5.** -25

7. -16 **9.** 1   **11.** $\frac{1}{10}$ or 0.1

13. $\frac{1}{64}$ **15.** -150 **17.** 5

19. -54 **21.** x^5 **23.** $\dfrac{1}{x}$

25. x^6 **27.** $64a^3$ **29.** $-8t^3$

31. $75x^5y^5$ **33.** $\dfrac{y^2}{x^2}$ **35.** $-\dfrac{1}{2xy}$

37. $\dfrac{16}{b^4}$ **39.** $\dfrac{a^4}{81b^6}$ **41.** $\dfrac{1}{a^6 b^2}$

43. $x^8 y^2$ **45.** $\dfrac{x^{14}}{4y}$ **47.** $\dfrac{y^{10}}{-32x^{22}}$

49. $-\dfrac{y^{24}}{a^{12} \cdot x^3}$ **51.** 2^{26} **53.** 2.76×10^7

55. 9.3×10^7 **57.** 5.67×10^{-8} **59.** 1.23×10^{-7}

61. 47,000,000   **63.** 23,000   **65.** 0.000041

67. 2.0×10^8 miles   **69.** 0.00000155 meters

71. In the power property, exponents are added (not multiplied).

73. In the first step, $\left(y^3\right)^2 = y^6$. Exponents should have been multiplied (not raised to power).

75. false   **77.** false

79. a^{mnk} **81.** -16

83. 156   **85.** 5.6 acres/person on Earth

87. 2.0×10^{-1} or 0.2   **89.** same

91. 5.11×10^{14}

Section A.3

1. $-7x^4 - 2x^3 + 5x^2 + 16$ degree 4

3. $-6x^3 + 4x + 3$ degree 3

5. 15   degree 0   **7.** $y - 2$ degree 1

9. $-x^2 + 5x + 5$ **11.** $2x^2 - x - 18$

13. $4x^4 - 10x^2 - 1$ **15.** $2z^2 + 2z - 3$

17. $-11y^3 - 16y^2 + 16y - 4$ **19.** $-4x + 12y$

21. $x^2 - x + 2$ **23.** $-4t^2 + 6t - 1$

25. $35x^2 y^3$ **27.** $2x^5 - 2x^4 + 2x^3$

29. $10x^4 - 2x^3 - 10x^2$ **31.** $2x^5 + 2x^4 - 4x^3$

33. $2a^3 b^2 + 4a^2 b^3 - 6ab^4$ **35.** $6x^2 - 5x - 4$

37. $x^2 - 4$ **39.** $4x^2 - 9$

41. $-4x^2 + 4x - 1$ **43.** $4x^4 - 9$

45. $2y^4 - 9y^3 + 5y^2 + 7y$ **47.** $x^3 - x^2 + x + 3$

49. $t^2 - 4t + 4$ **51.** $z^2 + 4z + 4$

53. $x^2 + 2xy + y^2 - 6x - 6y + 9$ **55.** $25x^2 - 20x + 4$

57. $6y^3 + 5y^2 - 4y$ **59.** $x^4 - 1$

61. $ab^2 + 2b^3 - 9a^3 - 18a^2 b$

63. $2x^2 - 3y^2 - xy + 3xz + 8yz - 5z^2$

65. $P = 11x - 100$ **67.** $P = -x^2 + 200x - 7500$

69. $V = (15 - 2x)(8 - 2x)x = 4x^3 - 46x^2 + 120x$

71. a. $P = 2\pi x + 4x + 10$ feet

   b. $A = \pi x^2 + 4x^2 + 10x$ square feet

73. $F = \dfrac{3k}{100}$

75. The negative was not distributed through the second polynomial.

77. true   **79.** false   **81.** $m + n$

83. $2401x^4 - 1568x^2 y^4 + 256y^8$ **85.** $x^3 - a^3$

87. $(2x + 3)(x - 4)$ and $2x^2 - 5x - 12$

Section A.4

1. $5(x + 5)$ **3.** $2\left(2t^2 - 1\right)$

5. $2x(x - 5)(x + 5)$ **7.** $3x\left(x^2 - 3x + 4\right)$

9. $x(x - 8)(x + 5)$ **11.** $2xy\left(2xy^2 + 3\right)$

13. $(x + 3)(x - 3)$ **15.** $(2x - 3)(2x + 3)$

17. $2(x - 7)(x + 7)$ **19.** $(15x - 13y)(15x + 13y)$

21. $(x + 4)^2$ **23.** $\left(x^2 - 2\right)^2$ **25.** $(2x + 3y)^2$

27. $(x - 3)^2$ **29.** $\left(x^2 + 1\right)^2$ **31.** $(p + q)^2$

33. $(t + 3)\left(t^2 - 3t + 9\right)$ **35.** $(y - 4)\left(y^2 + 4y + 16\right)$

37. $(2 - x)\left(4 + 2x + x^2\right)$ **39.** $(y + 5)\left(y^2 - 5y + 25\right)$

41. $(3 + x)\left(9 - 3x + x^2\right)$ **43.** $(x - 5)(x - 1)$

45. $(y - 3)(y + 1)$ **47.** $(2y + 1)(y - 3)$

49. $(3t + 1)(t + 2)$ **51.** $(-3t + 2)(2t + 1)$

53. $\left(x^2 + 2\right)(x - 3)$

55. $(a^3 - 8)(a + 2) = (a + 2)(a - 2)\left(a^2 + 2a + 4\right)$

57. $(3y - 5r)(x + 2s)$ **59.** $(5x + 2y)(4x - y)$

61. $(x - 2y)(x + 2y)$ **63.** $(3a + 7)(a - 2)$

65. prime   **67.** prime

69. $2(3x + 2)(x + 1)$ **71.** $(3x - y)(2x + 5y)$

73. $9(2s - t)(2s + t)$ **75.** $(ab - 5c)(ab + 5c)$

77. $(4x + 5)(x - 2)$ **79.** $x(3x + 1)(x - 2)$

81. $x(x - 3)(x + 3)$ **83.** $(y - 1)(x - 1)$

85. $\left(x^2 + 2\right)\left(x^2 + 3\right)$ **87.** $(x - 6)(x + 4)$

89. $x(x + 5)\left(x^2 - 5x + 25\right)$ **91.** $(x + 3)(x - 3)\left(x^2 + 9\right)$

93. $2(3x + 4)$ **95.** $(2x - 15)(x + 2)$

97. $-2(8t - 1)(t + 5)$

99. $\left(x^2 - 9\right) \neq (x - 3)^2$; instead, $x^2 - 9 = (x - 3)(x + 3)$

101. false   **103.** true

105. $\left(a^n - b^n\right)\left(a^n + b^n\right)$

107. $8x^3 + 1$ and $(2x + 1)\left(4x^2 - 2x + 1\right)$

Section A.5

1. $x \neq 0$ **3.** $x \neq 1$ **5.** $x \neq -1$

7. $p \neq \pm 1$ **9.** no restrictions

11. $\dfrac{x - 9}{2(x + 9)}$ $x \neq -9, -3$ **13.** $\dfrac{x - 3}{2}$ $x \neq -1$

15. $\dfrac{2(3y + 1)}{9y}$ $y \neq 0, \frac{1}{2}$ **17.** $\dfrac{y + 1}{5}$ $y \neq \frac{1}{5}$

19. $\dfrac{3x + 7}{4}$ $x \neq 4$ **21.** $x + 2$ $x \neq 2$

23. 1 $x \neq -7$ **25.** $\dfrac{x^2 + 9}{2x + 9}$ $x \neq -\frac{9}{2}$

27. $\dfrac{x + 3}{x - 5}$ $x \neq -2, 5$ **29.** $\dfrac{3x + 1}{x + 5}$ $x \neq -5, \frac{1}{2}$

31. $\dfrac{3x + 5}{x + 1}$ $x \neq -1, 2$ **33.** $\dfrac{2(5x + 6)}{5x - 6}$ $x \neq 0, \frac{6}{5}$

35. $\dfrac{2}{3}$ $x \neq 0, \pm 1$ **37.** $3(x + 2)$ $x \neq -5, 0, 2$

39. $\dfrac{t - 3}{3(t + 2)}$ $t \neq -2, 3$ **41.** $\dfrac{3t(t^2 + 4)}{(t - 3)(t + 2)}$ $t \neq -2, 3$

43. $\dfrac{3y(y - 2)}{y - 3}$ $y \neq -2, 3$

45. $\dfrac{(2x + 3)(x - 5)}{2x(x + 5)^2}$ $x \neq \pm 5, 0$

47. $\dfrac{(2x - 7)(2x + 7)}{(4x + 3)(3x - 5)}$ $x \neq -\frac{3}{4}, \pm\frac{5}{3}, \frac{7}{2}$

49. $\dfrac{x}{4}$ $x \neq 0$ **51.** $\dfrac{x + 2}{2}$ $x \neq \pm 2$

53. $\dfrac{x + 1}{5}$ $x \neq \pm 1$ **55.** $\dfrac{1}{2(1 - p)}$ $p \neq \pm 1, 2$

57. $\dfrac{6 - n}{n - 3}$ $n \neq -6, \pm 3$ **59.** $\dfrac{2t(t - 3)}{5}$ $t \neq -1, 2$

61. $\dfrac{5w^2}{w + 1}$ $w \neq 0, \pm 1$

63. $\dfrac{(x - 3)(x - 4)}{(x - 2)(x - 9)}$ $x \neq -7, -5, 2, 4, 9$

65. $\dfrac{x}{5x + 2}$ $x \neq -\frac{5}{3}, \pm\frac{2}{5}, -\frac{1}{4}, 0$

67. $\dfrac{13}{5x}$ $x \neq 0$

69. $\dfrac{5p^2 - 7p + 3}{(p + 1)(p - 2)}$ $p \neq -1, 2$

71. $\dfrac{4}{5x - 1}$ $x \neq \frac{1}{5}$

73. $\dfrac{3y^3 - 5y^2 - y + 1}{y^2 - 1}$ $y \neq \pm 1$

75. $\dfrac{x^2 + 4x - 6}{x^2 - 4}$ $x \neq \pm 2$

77. $\dfrac{x + 4}{x + 2}$ $x \neq \pm 2$ **79.** $\dfrac{12a + 7b}{a^2 - b^2}$ $a \neq \pm b$

81. $\dfrac{7x - 20}{x - 3}$ $x \neq 3$ **83.** $\dfrac{1 - x}{x - 2}$ $x \neq 0, 2$

85. $\dfrac{x}{3x - 1}$ $x \neq 0, \pm\frac{1}{3}$ **87.** $\dfrac{x + 1}{x - 1}$ $x \neq 0, \pm 1$

89. $\dfrac{x(x + 1)}{(x - 1)(x + 2)}$ $x \neq -2, \pm 1$

91. $A = (pi) \cdot \dfrac{(1 + i)^5}{(1 + i)^5 - 1}$ **93.** $R = \dfrac{R_1 R_2}{R_1 + R_2}$

95. $x + 1$, where $x \neq -1$ **97.** false

99. false

101. $\dfrac{(x + a)(x + d)}{(x + b)(x + c)}$ $x \neq -b, -c, -d$

103. No, there is a hole at $x = -7$.

105. a. $\dfrac{(x - 1)(x + 2)}{(x - 2)(x + 1)}$ **c.** $x \neq -1, x \neq -2, x \neq 2$

Section A.6

1. 10   **3.** -12 **5.** -6

7. 7   **9.** 1   **11.** 0

13. not a real number   **15.** -3 **17.** 4

19. -2 **21.** -1 **23.** 27

25. $-4\sqrt{2}$ **27.** $8\sqrt{5}$ **29.** $2\sqrt{6}$

31. $2\sqrt[3]{6}$ **33.** $\sqrt{21}$ **35.** $40\,|x|$

37. $2|x|\sqrt{y}$ **39.** $-3x^2y^2\sqrt[3]{3y^2}$ **41.** $\dfrac{\sqrt{3}}{3}$

43. $\dfrac{2\sqrt{11}}{33}$ **45.** $\dfrac{3+3\sqrt{5}}{-4}$ **47.** $-3-2\sqrt{2}$

49. $-3(\sqrt{2}+\sqrt{3})$ **51.** $\dfrac{2}{3}(3\sqrt{2}-2\sqrt{3})$

53. $\dfrac{5\sqrt{5}-2}{11}$ **55.** $\dfrac{(\sqrt{7}+3)(\sqrt{2}+\sqrt{5})}{-3}$

57. x^3y^4 **59.** $\dfrac{1}{y^2}$ **61.** $x^{7/6}y^2$

63. $\dfrac{x^{8/3}}{2}$ **65.** $x^{1/3}(x-2)(x+1)$

67. $7x^{3/7}\left(1-2x^{3/7}+3x\right)$ **69.** $4\sqrt{5}\approx 9$ seconds

71. $d\approx 9.54$ astronomical units.

73. The 4 should also have been squared.

75. false **77.** false **79.** a^{mnk} **81.** $\frac{1}{2}$

83. $\dfrac{\left(\sqrt{a}-\sqrt{b}\right)^2}{(a-b)^2}=\dfrac{a-2\sqrt{ab}+b}{(a-b)^2}$ **85.** 3.317

87. a. $10\sqrt{2}-8\sqrt{3}$ **b.** 0.2857291632 **c.** yes

Section A.7

1. $4i$ **3.** $2i\sqrt{5}$ **5.** -4

7. $8i$ **9.** $3-10i$ **11.** $-10-12i$

13. $2-9i$ **15.** $10-14i$ **17.** $2-2i$

19. $-1+2i$ **21.** $12-6i$ **23.** $96-60i$

25. $-48+27i$ **27.** $-102+30i$ **29.** $5-i$

31. $13+41i$ **33.** $87+33i$ **35.** $-6+48i$

37. $\frac{56}{9}-\frac{11}{18}i$ **39.** $37+49i$ **41.** $4-7i;\ 65$

43. $2+3i;\ 13$ **45.** $6-4i;\ 52$ **47.** $-2+6i;\ 40$

49. $-2i$ **51.** $\frac{3}{10}+\frac{1}{10}i$ **53.** $\frac{3}{13}-\frac{2}{13}i$

55. $\frac{14}{53}-\frac{4}{53}i$ **57.** $-i$ **59.** $-\frac{9}{34}+\frac{19}{34}i$

61. $\frac{18}{53}-\frac{43}{53}i$ **63.** $\frac{66}{85}+\frac{43}{85}i$ **65.** $-i$

67. 1 **69.** $21-20i$ **71.** $-5+12i$

73. $18+26i$ **75.** $-2-2i$ **77.** $8-2i$ ohms

79. multiplied by the denominator $4-i$ instead of the conjugate $4+i$

81. true **83.** true **85.** $(x-i)^2(x+i)^2$

87. $41-38i$ **89.** $\frac{2}{125}+\frac{11}{125}i$

Review Exercises

1. a. 5.22 **b.** 5.21 **3.** 4 **5.** -2

7. $-\dfrac{x}{12}$ **9.** 9 **11.** $-8z^3$

13. $\dfrac{9}{2x^2}$ **15.** 2.15×10^{-6} **17.** $14z^2+3z-2$

19. $45x^2-10x-15$ **21.** $15x^2y^2-20xy^3$

23. $x^2+2x-63$ **25.** $4x^2-12x+9$

27. x^4+2x^2+1 **29.** $2xy^2(7x-5y)$

31. $(x+5)(2x-1)$ **33.** $(4x-5)(4x+5)$

35. $(x+5)\left(x^2-5x+25\right)$ **37.** $2x(x-3)(x+5)$

39. $\left(x^2-2\right)(x+1)$ **41.** $x\neq \pm 3$

43. $x+2\quad x\neq 2$ **45.** $\dfrac{t+3}{t+1}\quad t\neq -1,2$

47. $\dfrac{(x+5)(x+2)}{(x+3)^2}\quad x\neq -3,1,2$

49. $\dfrac{2}{(x+1)(x+3)}\quad x\neq -1,-3$

51. $\dfrac{10x-25}{20x-59}\quad x\neq 3,\frac{59}{20}$ **53.** $2\sqrt{5}$

55. $-5xy\sqrt[3]{x^2y}$ **57.** $26\sqrt{5}$ **59.** $-3-\sqrt{5}$

61. $2+\sqrt{3}$ **63.** $\dfrac{9x^{2/3}}{16}$ **65.** $5^{1/6}$

67. $13i$ **69.** $-i$ **71.** 12

73. $-33+56i$ **75.** $\frac{2}{5}+\frac{1}{5}i$ **77.** $\frac{38}{41}-\frac{27}{41}i$

79. $-\frac{10}{3}i$ **81.** rational **83.** 1.945×10^{-6}

85. $(2x+3)^3$ and $8x^3+36x^2+54x+27$

87. None, because $x^2-3x+18$ is prime.

89. a. $\dfrac{x(x-4)}{(x+2)(x-2)}$

b.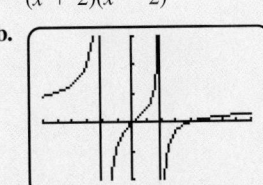

c. $x\neq -2,\ x\neq 0,\ x\neq 2$

91. a. $2\sqrt{5} + 2\sqrt{2}$

b. 7.30056308

c. yes

93. $2868 - 6100i$

95. 1.6×10^{14}

Practice Test

1. 4

3. -97

5. $2i|x|\sqrt{3}$

7. $\dfrac{y^5 z^{1/2}}{x^{7/2}}$

9. $-7\sqrt{2}$

11. $2y^2 - 12y + 20$

13. $(x + 4)(x - 4)$

15. $(2x + 3y)^2$

17. $(2x + 1)(x - 1)$

19. $t(2t - 3)(t + 1)$

21. $(x - 3y)(x + 4y)$

23. $3(3 + x)(x^2 - 3x + 9)$

25. $\dfrac{5x - 2}{x(x - 1)}$ $\quad x \neq 0, 1$

27. $\dfrac{1}{(x + 1)(x - 1)}$ $\quad x \neq \pm 1$

29. $-\dfrac{1}{x + 3}$ $\quad x \neq \frac{5}{2}, \pm 3$

31. $-8 - 26i$

33. $\dfrac{2 - 27\sqrt{3}}{59}$

35. $-\dfrac{1}{x(x + 1)}$ $\quad x \neq -1, 0, 1$

37. 2.330

APPLICATIONS INDEX

SUBJECT INDEX

Variation in sign, and Descartes's rule of signs, 241–242
Vectors
    addition of, 625, 627, 630–634
    algebraic interpretation of, 625–627
    angle between, 640–642
    components of, 629
    defined, 624
    dot product multiplication, 639–644
    equality of, 624–625, 627
    geometric interpretation of, 624–625
    operations on, 628
    properties of, 628
    resultant, 630–634
    unit, 629–630
Velocity
    apparent and actual, 630–632
    defined, 624
Velocity vectors, and resultant vectors, 630–632
Vertex/vertices
    of angles, 364
    of ellipses, 847, 849–852
    of hyperbolas, 860, 863
    of inequalities graphs, 805–807
    of parabolas, 197, 202, 833
Vertical asymptotes
    defined, 495–496
    in graphing, 267–273
    locating, 263–264
Vertical components, of vectors, 629
Vertical lines
    equation of, 73
    slope and, 75–76
Vertical line test, for functions, 109–110
Vertical shifts
    defined, 144
    graphing, 143–147, 149, 296–297, 312–313, 477–479, 508–510

W
Whole numbers, characteristics of, 993–994
Wiles, Andrew, proof of Fermat's last theorem, 970
Word problems, solving, 7
Work
    defined, 644
    modeling, 686
    problems, 14
    vectors in calculation, 643–644

X
x-axis
    defined, 59
    reflection about, 147–149, 312–313
    symmetry about, 64–67
x-coordinates
    defined, 59
    in distance formulas, 60–61
    x-intercepts, 62–63
x-intercepts. *See also* Intercepts; Symmetry; Zeros
    defined, 62
    of parabolas, 198

Y
y-axis
    defined, 59
    reflection about, 147–149, 312–313
    symmetry about, 64–67, 126–128, 527
y-coordinates
    defined, 59
    in distance formulas, 60–61
    y-intercepts, 62–63
y-intercepts. *See also* Intercepts; Symmetry
    defined, 62
    of parabolas, 198

Z
Zero
    in exponents, 1009
    properties of, 1001–1002
Zero-exponent property, defined, 1009
Zero matrix/matrices, 751–752
    defined, 751
Zero product property
    defined, 1002
    and factorable equations, 38–39
    in factoring quadratic equations, 19–20
    solving trigonometric equations with, 601–603
Zero row, in Pascal's triangle, 979
Zeros
    complex, 251–256
    irrational, 245–248
    multiplicity of, 217–222
    of polynomials, 48–51, 216–218
    and rational inequalities, 52–53
    real. *See* Real zeros
    remainder and factor theorems, 236–238
Zero vector, defined, 628

CONIC SECTIONS

Parabola

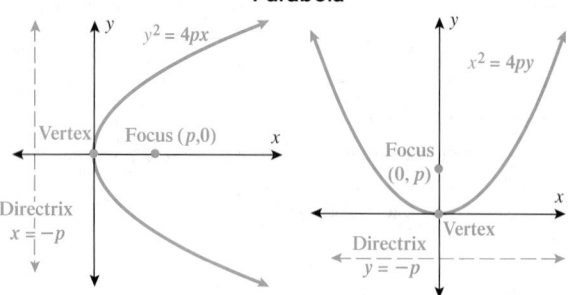

Ellipse

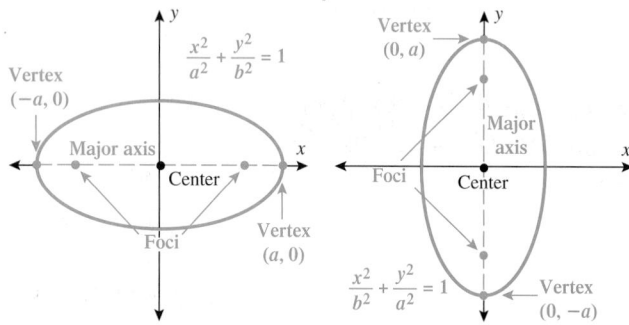

Hyperbola

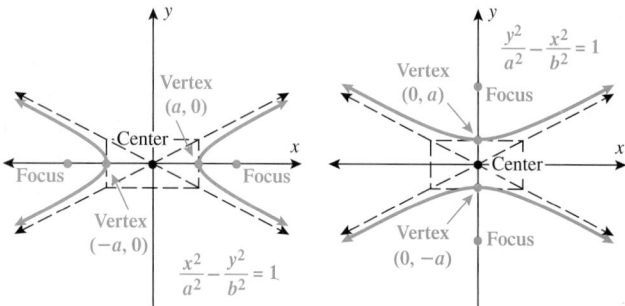

SEQUENCES

1. Infinite Sequence:

$$\{a_n\} = a_1, a_2, a_3, \ldots, a_n, \ldots$$

2. Summation Notation:

$$\sum_{i=1}^{n} a_i = a_1 + a_2 + a_3 + \cdots + a_n$$

3. nth Term of an Arithmetic Sequence:

$$a_n = a_1 + (n - 1)d$$

4. Sum of First n Terms of an Arithmetic Sequence:

$$S_n = \frac{n}{2}(a_1 + a_n)$$

5. nth Term of a Geometric Sequence:

$$a_n = a_1 r^{n-1}$$

6. Sum of First n Terms of a Geometric Sequence:

$$S_n = \frac{a_1(1 - r^n)}{1 - r} \quad (r \neq 1)$$

7. Sum of an Infinite Geometric Series with $|r| < 1$:

$$S = \frac{a_1}{1 - r}$$

THE BINOMIAL THEOREM

1. $n! = n(n - 1)(n - 2) \cdots 3 \cdot 2 \cdot 1;$
$1! = 1; \ 0! = 1$

2. $\displaystyle \binom{n}{r} = \frac{n!}{r!(n - r)!}$

3. Binomial theorem:

$$(a + b)^n = \binom{n}{0}a^n + \binom{n}{1}a^{n-1}b$$
$$+ \binom{n}{2}a^{n-2}b^2 + \cdots + \binom{n}{n}b^n$$

PERMUTATIONS, COMBINATIONS, AND PROBABILITY

1. $_nP_r$, the number of permutations of n elements taken r at a time, is given by

$$_nP_r = \frac{n!}{(n - r)!}.$$

2. $_nC_r$, the number of combinations of n elements taken r at a time, is given by

$$_nC_r = \frac{n!}{(n - r)!r!}.$$

3. *Probability of an Event:* $P(E) = \dfrac{n(E)}{n(S)}$, where

$n(E) =$ the number of outcomes in event E and
$n(S) =$ the number of outcomes in the sample space.

INVERSE TRIGONOMETRIC FUNCTIONS

$y = \sin^{-1}x$	$x = \sin y$	$-\dfrac{\pi}{2} \leq y \leq \dfrac{\pi}{2}$	$-1 \leq x \leq 1$
$y = \cos^{-1}x$	$x = \cos y$	$0 \leq y \leq \pi$	$-1 \leq x \leq 1$
$y = \tan^{-1}x$	$x = \tan y$	$-\dfrac{\pi}{2} < y < \dfrac{\pi}{2}$	x is any real number
$y = \cot^{-1}x$	$x = \cot y$	$0 < y < \pi$	x is any real number
$y = \sec^{-1}x$	$x = \sec y$	$0 \leq y \leq \pi, y \neq \dfrac{\pi}{2}$	$x \leq -1$ or $x \geq 1$
$y = \csc^{-1}x$	$x = \csc y$	$-\dfrac{\pi}{2} \leq y \leq \dfrac{\pi}{2}, y \neq 0$	$x \leq -1$ or $x \geq 1$

GRAPHS OF THE INVERSE TRIGONOMETRIC FUNCTIONS

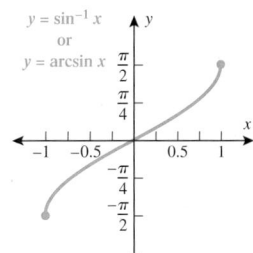

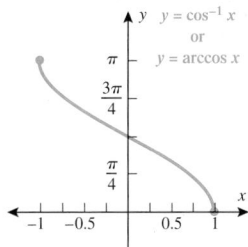

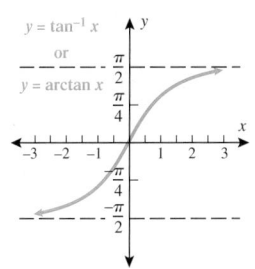

VECTORS

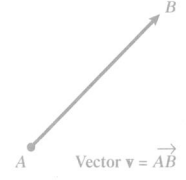

Vector Addition

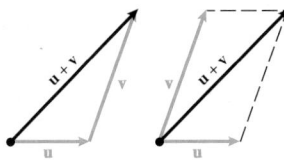

Scalar Multiplication

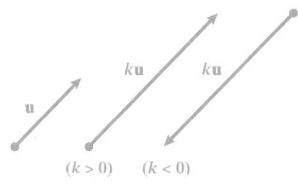

For vectors $\mathbf{u} = \langle a, b \rangle$ and $\mathbf{v} = \langle c, d \rangle$, and real number k,

$$\mathbf{u} = a\mathbf{i} + b\mathbf{j}$$
$$|\mathbf{u}| = \sqrt{a^2 + b^2}$$
$$\mathbf{u} + \mathbf{v} = \langle a + c, b + d \rangle$$
$$k\mathbf{u} = \langle ka, kb \rangle$$
$$\mathbf{u} \cdot \mathbf{v} = ac + bd$$
$$\cos\theta = \frac{\mathbf{u} \cdot \mathbf{v}}{|\mathbf{u}||\mathbf{v}|}$$

$$\text{Comp}_{\mathbf{v}}\,\mathbf{u} = |\mathbf{u}|\cos\theta = \frac{\mathbf{u} \cdot \mathbf{v}}{|\mathbf{u}|}$$

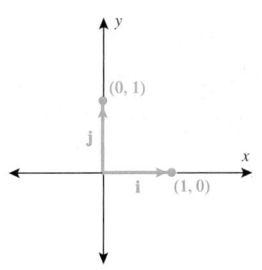

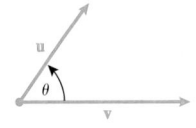

GRAPHS OF THE TRIGONOMETRIC FUNCTIONS

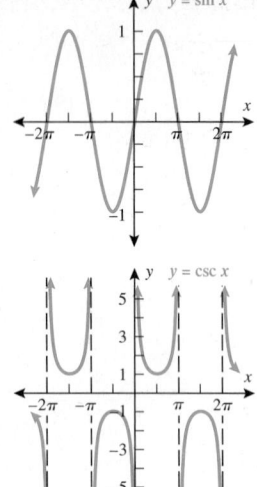

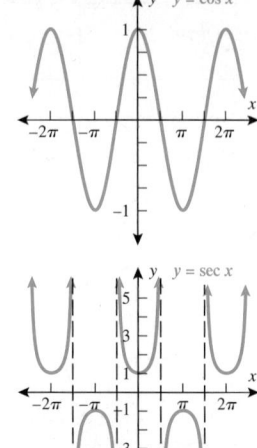

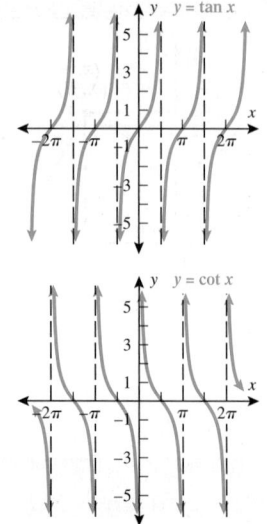

AMPLITUDE, PERIOD, AND PHASE SHIFT

$$y = A\sin(Bx + C) \qquad\qquad y = A\cos(Bx + C)$$

$$\text{Amplitude} = |A| \qquad \text{Period} = \frac{2\pi}{B}$$

$$\text{Phase shift} = \frac{C}{B} \begin{cases} \text{left} & \text{if } C/B > 0 \\ \text{right} & \text{if } C/B < 0 \end{cases}$$

$$y = A\tan(Bx + C) \qquad\qquad y = A\cot(Bx + C)$$

$$\text{Period} = \frac{\pi}{B}$$

$$\text{Phase shift} = \frac{C}{B} \begin{cases} \text{left} & \text{if } C/B > 0 \\ \text{right} & \text{if } C/B < 0 \end{cases}$$

POLAR COORDINATES

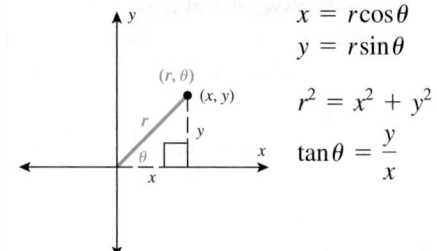

$$x = r\cos\theta$$
$$y = r\sin\theta$$

$$r^2 = x^2 + y^2$$

$$\tan\theta = \frac{y}{x}$$

COMPLEX NUMBERS

For the complex number $z = a + bi$
the **conjugate** is $\bar{z} = a - bi$
the **modulus** is $|z| = \sqrt{a^2 + b^2}$
the **argument** is θ, where $\tan\theta = b/a$

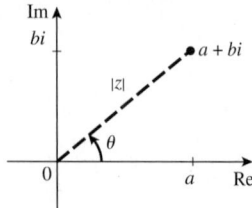

Polar (Trigonometric) form of a complex number

For $z = a + bi$, the **polar form** is

$$z = r(\cos\theta + i\sin\theta)$$

where $r = |z|$ is the modulus of z and θ is the argument of z

DeMoivre's Theorem

$$z^n = [r(\cos\theta + i\sin\theta)]^n = r^n(\cos n\theta + i\sin n\theta)$$

Nth Root Theorem

$$z^{1/n} = [r(\cos\theta + i\sin\theta)]^{1/n}$$
$$= r^{1/n}\left(\cos\frac{\theta + 2k\pi}{n} + i\sin\frac{\theta + 2k\pi}{n}\right)$$

where $k = 0, 1, 2, \ldots n - 1$

ROTATION OF AXES

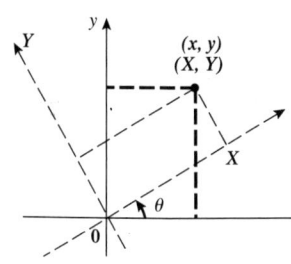

Rotation of axes formulas

$$x = X \cos\theta - Y \sin\theta$$
$$y = X \sin\theta + Y \cos\theta$$

Angle-of-rotation formula for conic sections

$$\cot(2\theta) = \frac{A - C}{B} \quad \text{or}$$

$$\tan(2\theta) = \frac{B}{A - C}$$

SUMS OF POWERS OF INTEGERS

$$\sum_{k=1}^{n} 1 = n \qquad \sum_{k=1}^{n} k = \frac{n(n + 1)}{2}$$

$$\sum_{k=1}^{n} k^2 = \frac{n(n + 1)(2n + 1)}{6} \qquad \sum_{k=1}^{n} k^3 = \frac{n^2(n + 1)^2}{4}$$

THE DERIVATIVE OF A FUNCTION

The **average rate of change** of f between a and b is

$$\frac{f(b) - f(a)}{b - a}$$

The **derivative** of f at a is

$$f'(a) = \lim_{x \to a} \frac{f(x) - f(a)}{x - a}$$

$$f'(a) = \lim_{h \to 0} \frac{f(a + h) - f(a)}{h}$$

THE AREA UNDER THE GRAPH A CURVE

The **area under the graph of** f on the interval $[a, b]$ is the limit of the sum of the areas of approximating rectangles

$$A = \lim_{n \to \infty} \sum_{k=1}^{n} \overbrace{\underbrace{f(x_k)}_{\text{height}} \underbrace{\Delta x}_{\text{width}}}^{\text{Area of rectangle, } R_k}$$

where

$$\Delta x = \frac{b - a}{n}$$

$$x_k = a + k \Delta x$$

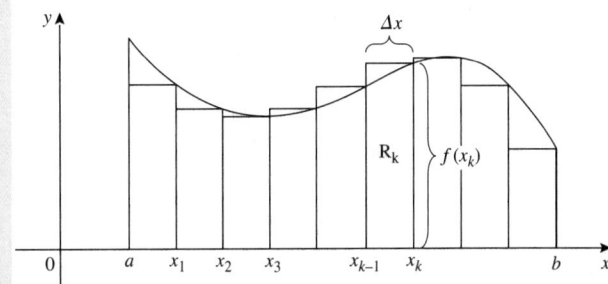

RIGHT TRIANGLE TRIGONOMETRY

$$\sin\theta = \frac{\text{opp}}{\text{hyp}} \qquad \csc\theta = \frac{\text{hyp}}{\text{opp}}$$

$$\cos\theta = \frac{\text{adj}}{\text{hyp}} \qquad \sec\theta = \frac{\text{hyp}}{\text{adj}}$$

$$\tan\theta = \frac{\text{opp}}{\text{adj}} \qquad \cot\theta = \frac{\text{adj}}{\text{opp}}$$

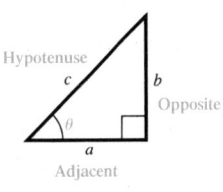

TRIGONOMETRIC FUNCTIONS IN THE CARTESIAN PLANE

$$\sin\theta = \frac{y}{r} \qquad \csc\theta = \frac{r}{y}$$

$$\cos\theta = \frac{x}{r} \qquad \sec\theta = \frac{r}{x}$$

$$\tan\theta = \frac{y}{x} \qquad \cot\theta = \frac{x}{y}$$

EXACT VALUES OF TRIGONOMETRIC FUNCTIONS

x degrees	x radians	$\sin x$	$\cos x$	$\tan x$
$0°$	0	0	1	0
$30°$	$\dfrac{\pi}{6}$	$\dfrac{1}{2}$	$\dfrac{\sqrt{3}}{2}$	$\dfrac{\sqrt{3}}{3}$
$45°$	$\dfrac{\pi}{4}$	$\dfrac{\sqrt{2}}{2}$	$\dfrac{\sqrt{2}}{2}$	1
$60°$	$\dfrac{\pi}{3}$	$\dfrac{\sqrt{3}}{2}$	$\dfrac{1}{2}$	$\sqrt{3}$
$90°$	$\dfrac{\pi}{2}$	1	0	—

ANGLE MEASUREMENT

π radians $= 180°$

$s = r\theta \quad A = \frac{1}{2}r^2\theta \quad$ (θ in radians)

To convert from degrees to radians, multiply by $\dfrac{\pi}{180°}$.

To convert from radians to degrees, multiply by $\dfrac{180°}{\pi}$.

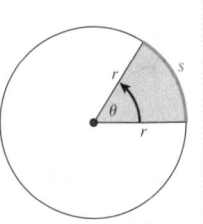

OBLIQUE TRIANGLES

Law of Sines

In any triangle,
$$\frac{\sin\alpha}{a} = \frac{\sin\beta}{b} = \frac{\sin\gamma}{c}.$$

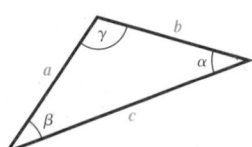

Law of Cosines

$a^2 = b^2 + c^2 - 2bc\cos\alpha$
$b^2 = a^2 + c^2 - 2ac\cos\beta$
$c^2 = a^2 + b^2 - 2ab\cos\gamma$

CIRCULAR FUNCTIONS ($\cos\theta$, $\sin\theta$)

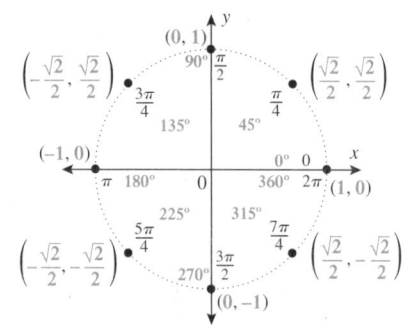